PROCEEDINGS 1998

DESIGN AUTOMATION CONFERENCE®

DESIGN AUTOMATION CONFERENCE®

35th DAC

ANNIVERSARY

35 Years of Excellence

Moscone Center, San Francisco, CA
June 15 - 19, 1998

Sponsored by

EDA CONSORTIUM

sigda

PROCEEDING OF THE 35th DESIGN AUTOMATION CONFERENCE

The Association for Computing Machinery, Inc.
1515 Broadway
New York, NY 10036

ACM Order Number 477980
ISBN 0-89791-964-5

Library of Congress Number 85-644924

IEEE Catalog Number 98CH36175
ISBN 0-7803-4409-X (Casebound Edition)
ISBN 0-7803-4410-3 (Microfiche Edition)

ISSN 0738-100X

Additional copies of 1998 or prior Proceedings may be ordered prepaid from

ACM Order Department
P.O. Box 11414
Church Street Station
New York, NY 10286

ACM European Service Center
108 Cowley Rd.
Oxford, OX41JF, U.K.
Phone: 44-1865-382338
Fax: 44-1865-381338
E-mail: ACM_Europe@acm.org

Phone: 1-800-342-6626
(U.S.A. and Canada)
+1-212-626-0500
(All other countries)
Fax: +1-212-944-1318
E-mail: acmpubs@acm.org

Additional copies of this publication are available from

IEEE Service Center
P.O. Box 1331
445 Hoes Lane
Piscataway, NJ 08855-1331

1-800-678-IEEE
1-908-981-1393
1-908-981-9667 (Fax)
833-233 (Telex)

Printed in the U.S.A.

General Chair's Welcome

Welcome to the 35th Design Automation Conference!

This marks a significant milestone in the history of DAC. It started as a small special interest group at the IBM SHARE meeting back in 1963, blossomed as the Design Automation Workshop in 1964, and finally ripened as the Design Automation Conference in 1975. The vendor exhibits, as we know them today, originated in 1983.

Ever since its beginning, DAC has maintained excellence in its technical program and has become the premier conference for the presentation of research and development work in design automation of integrated circuits and electronic systems. DAC has also become the premier forum for the EDA industry to exhibit the leading edge products and make new product announcements.

For the past several years, the DAC Executive Committee has been focusing on expanding the technical program to meet the needs of the design engineers. This year, we received a record number of technical papers on design methodology. It gives me great pleasure to say that about 40% of the technical program now concentrates on design methodology. To complement the expansion of the technical program, we have also expanded the exhibit area by introducing a special area on the floor, called Silicon Village, which is dedicated to Silicon vendors to address the growing interdependence between Silicon and EDA vendors. This will provide an opportunity for the attendees to interact with both EDA vendors and Silicon vendors at the same venue.

Another new attraction at DAC this year is the University Design Contest, a competition where university researchers submitted system designs to be judged on the basis of innovation in their design flows and their use of EDA tools. The top contestants will present their designs in a special session - a great place for the practicing design engineers to interact with the university researchers.

These Proceedings, representing an outstanding expanded technical program were assembled under the leadership of this year's Co-Technical Program Chairs Randy Bryant and Jan Rabaey. Three hundred ninety papers from around the world were submitted to the design tools and design methods tracks, and were reviewed by over 700 professionals along with detailed examination by the Technical Program Committee, yielding 142 papers presented at the conference. The papers are complemented by nine panel sessions and 8 embedded tutorials. Special this year is a session to commemorate the 35th Design Automation Conference which provides both a retrospective of our past accomplishments and a view to the future.

I want to thank all the people who contributed to DAC maintaining the premier conference status that it enjoys: the Executive Committee, the Technical Program Committee, the EDA Industry Committee, DAC's sponsors, MP Associates, and especially the exhibitors, authors, speakers, session organizers and session chairs. DAC is sponsored by ACM/SIGDA, IEEE Circuits and Systems Society, and EDA Consortium. Their members represent the breadth of DAC's participants and we are thankful for their continued and active support.

Welcome to San Francisco and the 35th Design Automation Conference. We wish you a very productive and fun-filled week, and trust you will find these proceedings to be a valuable information reference for many years to come.

Basant R. Chawla
General Chair, 35th Design Automation Conference

EXECUTIVE COMMITTEE

GENERAL CHAIR
Basant R. Chawla
Lucent Technologies
283 King George Rd., E4D43
Warren, NJ 07059
(908) 559-5281
bchawla@lucent.com

VICE CHAIR
Mary Jane Irwin
Penn State Univ.
Dept. of CS and Engr.
220 Pond Lab.
University Park, PA 16802-6106
(814) 865-1802
mji@cse.psu.edu

TECHNICAL PROGRAM CO-CHAIR, DESIGN TOOLS
Randal E. Bryant
Carnegie Mellon Univ.
School of CS
Pittsburgh, PA 15213
(412) 268-8821
randy.bryant@cs.cmu.edu

TECHNICAL PROGRAM CO-CHAIR, DESIGN METHODS
Jan M. Rabaey
Univ. of California
Dept. of EECS, 511 Cory Hall
Berkeley, CA 94720
(510) 643-8206
jan@eecs.berkeley.edu

FINANCE CHAIR
Giovanni De Micheli
Stanford Univ.
Gates Computer Science Bldg.
Rm. 333
Stanford, CA 94305-9030
(650) 725-3632
nanni@galileo.stanford.edu

TUTORIAL CHAIR
Antun Domic
Synopsys, Inc.
700 E. Middlefield Rd.
Mountain View, CA 94043-4033
(650) 943-5088
domic@synopsys.com

ELECTRONIC SYSTEMS INDUSTRY CHAIR
Bryan D. Ackland
Lucent Technologies, Bell Labs.
101 Crawfords Corner Rd.
Rm. 4E-508
Holmdel, NJ 07733-1900
(732) 949-7248
bda@lucent.com

EDA INDUSTRY CHAIR
Thomas P. Pennino
Lucent Technologies, Bell Labs.
101 Crawfords Corner Rd.
Rm. 1M-415
Holmdel, NJ 07733
(732) 949-7340
tpennino@lucent.com

ELECTRONIC MEDIA CHAIR
Michael Lorenzetti
Mentor Graphics Corp.
8005 SW Boeckman Rd.
Wilsonville, OR 97070-7777
(503) 685-0740
mike_lorenzetti@mentorg.com

PUBLICITY CHAIR
Abbie Kendall
OrCAD
9300 SW Nimbus Ave.
Beaverton, OR 97008
(503) 671-9500
abbie@orcad.com

EXECUTIVE COMMITTEE (cont.)

EUROPE/MIDDLE EAST REPRESENTATIVE
Gerry Musgrave
Brunel Univ.
Dept. of EEE
Uxbridge, UB8 3PH, UK
(44) 1-895-203-251
gerry.musgrave@brunel.ac.uk

PAST CHAIR
Ellen J. Yoffa
IBM Corp.
T.J. Watson Research Ctr.
P.O. Box 218, Rm. 33-109
Yorktown Heights, NY 10598
(914) 945-3270
yoffa@watson.ibm.com

ACM REPRESENTATIVE
James Cohoon
Univ. of Virginia
Dept. of Computer Science
Olsson Hall
Charlottesville, VA 22903
(804) 982-2210
cohoon@virginia.edu

ASIA/INDIA/S. PACIFIC REPRESENTATIVE
Fumiyasu Hirose
Fujitsu Labs. Ltd.
CAD Lab.
4-1-1 Kamikodanaka, Nakahara-ku
Kawasaki 211, Japan
(81) 44-754-2663
hirose@flab.fujitsu.co.jp

IEEE-CAS REPRESENTATIVE
Michael Lightner
Univ. of Colorado
Dept. of ECE, Campus Box 425
Boulder, CO 80309-0425
(303) 492-5180
lightner@boulder.colorado.edu

EDAC REPRESENTATIVE
Lorie Bowlby
111 W. St. John St., Ste. 200
San Jose, CA 95123-1104
(408) 226-7240
lorieb@pacbell.net

EXHIBIT MANAGER
Marie R. Pistilli
MP Associates, Inc.
5305 Spine Rd., Ste. A
Boulder, CO 80301
(303) 530-4562
marie@dac.com

CONFERENCE MANAGER
P.O. Pistilli
MP Associates, Inc.
5305 Spine Rd., Ste. A
Boulder, CO 80301
(303) 530-4562
pat@dac.com

Technical Program Committee

Randal E. Bryant
Design Tools Co-Chair
Carnegie Mellon Univ.
School of CS
Pittsburgh, PA 15213
(412) 268-8821
randy.bryant@cs.cmu.edu

Bryan D. Ackland
Lucent Technologies, Bell Labs.
Rm. 4E-508
101 Crawfords Corner Rd.
Holmdel, NJ 07733-1900
(732) 949-7248
bda@lucent.com

David T. Blaauw
Motorola, Inc.
Advanced Design Tech.
Bridgepoint, Plaza 1
5918 W. Courtyard Dr., Ste. 330
Austin, TX 78730
(512) 794-4356
blaauw@adttx.sps.mot.com

Ivo Bolsens
IMEC
VSDM, Kapeldreef 75
Leuven, BE B-3001 Belgium
(32) 16-281-211
bolsens@imec.be

Anantha Chandrakasan
Massachusetts Inst. of Tech.
Dept. of EE, Rm. 38-107
50 Vassar St.
Cambridge, MA 02139
(617) 258-7619
anantha@mtl.mit.edu

Mojy C. Chian
Harris Semiconductor
P.O. Box 883
MS 62BO22
Melbourne, FL 32902
(407) 724-7782
mchian01@harris.com

Nanette Collins
Consultant
37 Symphony Rd., Unit A
Boston, MA 02115
(617) 437-1822
nanette@nvc.com

Jan M. Rabaey
Design Methods Co-Chair
Univ. of California
Dept. of EECS, 511 Cory Hall
Berkeley, CA 94720
(510) 643-8206
jan@eecs.berkeley.edu

Jason Cong
Univ. of California
4711 Boelter Hall
Dept. of CS
Los Angeles, CA 90095
(310) 206-2775
cong@cs.ucla.edu

Anders Forsen
Ericsson Radio Systems AB
RCUR-T/N, Kista
Stockholm, S-16480 Sweden
(46) 8-7572-541
anders.forsen@era.ericsson.se

Patrick Groeneveld
Magma Design Automation
1025A Terra Bella Ave.
Mountain View, CA 94043
(415) 938-6970
patrick@magma-da.com

Rajesh K. Gupta
Univ. of California
444 Computer Science, 208B IREF
Irvine, CA 92697
(714) 824-8052
gupta@uci.edu

Randolph E. Harr
Synopsys, Inc.
700 E. Middlefield Rd.
Mountain View, CA 94043-4033
(415) 694-1927
rharr@synopsys.com

TingTing Hwang
Tsing Hua Univ.
Dept. of Computer Science
Hsin-Chu, 30043 Taiwan ROC
(886) 35-715-131
tingting@cs.nthu.edu.tw

Technical Program Committee (cont.)

Takahide Inoue
Sony Corp.
530 Cottonwood Dr.
Milpitas, CA 95035
(408) 955-4279
inoue@itc.sca.sony.com

Andrew B. Kahng
Univ. of California
Dept. of CS, 3713 Boelter Hall
Los Angeles, CA 90024-1596
(310) 206-7073
abk@cs.ucla.edu

Timothy Y. Kam
Intel Corp.
Strategic CAD Labs., JFT-102
5200 NE Elam Young Pkwy.
Hillsboro, OR 97124-6497
(503) 264-7536
tkam@ichips.intel.com

David Ku
Escalade Corp.
2575 Augustine Dr.
Santa Clara, CA 95054
(408) 654-1617
ku@escalade.com

Andreas Kuehlmann
IBM Corp.
T.J. Watson Research Ctr.
P.O. Box 218
Yorktown Heights, NY 10598
(914) 945-3458
kuehl@watson.ibm.com

Luciano Lavagno
Cadence Design Systems, Inc.
2001 Addison St., 3rd Fl.
Berkeley, CA 94704-1103
(510) 647-2810
luciano@cadence.com

Sharad Malik
Princeton Univ.
Dept. of EE
Princeton, NJ 08544
(609) 258-4625
sharad@ee.princeton.edu

Alan Mantooth
Analogy, Inc.
9205 SW Gemini Dr.
Beaverton, OR 97075-1669
(503) 626-9700
alan@analogy.com

Teresa Meng
Stanford Univ.
Gates Computer Science
Bldg. 301
Stanford, CA 94028
(415) 725-3636
meng@mojave.stanford.edu

Mike Murray
Acuson
1220 Charleston Rd.
MS L-1, Box 7393
Mountain View, CA 94039
(415) 694-5876
mikem@acuson.com

Kunle Olukotun
Stanford Univ.
Gates Computer Science
Gates 3A, Rm. 302
Stanford, CA 94305-9030
(415) 725 3713
kunle@ogun.stanford.edu

Hidetoshi Onodera
Kyoto Univ.
Dept. of Electronics & Comm.
Sakyo-ku
Kyoto, 606-01 Japan
(81) 75-753-5314
onodera@kuee.kyoto-u.ac.jp

Janusz Rajski
Mentor Graphics Corp.
8005 SW Boeckman Rd.
Wilsonville, OR 97070-7777
(503) 685-4797
rajski@wv.mentorg.com

James A. Rowson
Alta Group of
Cadence Design Systems, Inc.
555 N. Matilda Ave.
Sunnyvale, CA 94086
(408) 523-4157
jimr@altagroup.com

Louis Scheffer
Cadence Design Systems, Inc.
555 River Oaks Pkwy.
Bldg. 2, MS 2B1
San Jose, CA 95134
(408) 944-7114
lou@cadence.com

Technical Program Committee (cont.)

Sunil D. Sherlekar
Silicon Automation Systems
3008, 12th B Main, 8th Cross
HAL 2nd Stage, Indiranagar
Bangalore, 560008 India
(91) 80-528-1461
sds@sasi.com

Richard Smith
Cadence Design Systems, Inc.
5215 N. O'Connor Rd., Ste. 1000
Irving, TX 75039
(972) 889-0033
dsmith@cadence.com

Vivek Tiwari
Intel Corp.
2200 Mission College Blvd.
MS-RN 5-09
Santa Clara, CA 95052-8119
(408) 765-0589
vivek_tiwari@ccm.sc.intel.com

Kazutoshi Wakabayashi
NEC Corp.
C&C Research Labs.
4-1-1 Miyazaki
Kawasaki, 216 Japan
(81) 44-856-2134
wakaba@sbl.cl.nec.co.jp

Neil Weste
Macquarie Univ.
Electronics Dept.
Sydney, 2109 Australia
(61) 2-850-9149
new@mpce.mq.edu.au

Andrew T. Yang
Univ. of Washington
Dept. of EE, FT-10
Seattle, WA 98195
(206) 543-2932
atyang@uwcad.ee.washington.edu

Kenji Yoshida
Toshiba Corp.
580-1 Horikawa-Cho Saiwai-ku
Kawasaki, 210 Japan
(81) 44 548 2400
kyoshida@eecvun.eec.toshiba.co.jp

Yervant Zorian
LogicVision, Inc.
101 Metro Dr., Third Fl.
San Jose, CA 95110
(408) 453-0146
zorian@lvision.com

Panel Sub-Committee

Nanette Collins
Consultant
37 Symphony Rd., Unit A
Boston, MA 02115
(617) 437-1822
nanette@nvc.com

Takahide Inoue
Sony Corp.
530 Cottonwood Dr.
Milpitas, CA 95035
(408) 955-4279
inoue@itc.sca.sony.com

Andrew B. Kahng
Univ. of California
Dept. of CS, 3713 Boelter Hall
Los Angeles, CA 90024-1596
(310) 206-7073
abk@cs.ucla.edu

Mike Murray
Acuson
1220 Charleston Rd.
MS L-1, Box 7393
Mountain View, CA 94039
(415) 694-5876
mikem@acuson.com

1998 Best Paper Award

This year, awards are made for the best papers in four categories. Winners are determined from detailed reviews of the accepted papers in the technical sessions. Each award is accompanied by a plaque and a cash award of $400. The awards are given by ACM/SIGDA (Special Interest Group on Design Automation), IEEE/CAS (Institute of Electrical and Electronics Engineers/Circuits and Systems Society) and EDA Consortium (Electronic Design Automation Companies).

PHYSICAL DESIGN AND LOGIC SYNTHESIS

Paper 17.1: "Generic Global Placement and Floorplanning"
Authors: Hans Eisenmann, Frank M. Johannes
Affiliation: Technical Univ. of Munich, Munich, Germany

HIGH-LEVEL SYNTHESIS, VERIFICATION AND CODESIGN

Paper 32.1: "A Decision Procedure for Bit-Vector Arithmetic"
Authors: Clark W. Barrett, David L. Dill, Jeremy R. Levitt
Affiliation: Stanford Univ., Stanford, CA

MODELING, SIMULATION AND ESTIMATION

Paper 13.1: "A Fast Hierarchical Algorithm For 3-D Capacitance Extraction"
Authors: Weiping Shi, Jianguo Liu, Naveen Kakani; Tiejun Yu
Affiliation: Univ. of North Texas, Denton, TX; Univ. of North Carolina, Charlotte, NC

DESIGN METHODOLOGY

Paper 9.1: "OCCOM: Efficient Computation of Observability-Based Code Coverage Metrics for Functional Verification"
Authors: Farzan Fallah, Srinivas Devadas; Kurt Keutzer
Affiliation: Massachusetts Inst. of Tech., Cambridge, MA; Synopsys, Inc., Mountain View, CA

Paper 39.1: "Functional Verification of a Multiple-Issue, Out-of-Order, Superscaler Alpha Processor—The Alpha 21264 CPU Chip"
Authors: Michael Quinn, Scott Taylor
Affiliation: Digital Equipment Corp., Hudson, MA

Advancement in Computer Science and Electrical Engineering Undergraduate Scholarships

The objective of the ACSEE Scholarship program is to increase the pool of professionals in Electrical Engineering and Computer Science from under-represented groups (Women, African American, Hispanic, Native American, and Physically Challenged). In 1989, ACM Special Interest Group on Design Automation (SIGDA) began providing the program. Beginning in 1993, the Design Automation Conference provides the funds for the scholarship and SIGDA continues to administer the program for DAC. DAC normally funds two $4000 scholarships renewable up to 5 years to graduating high school seniors. In 1997 three awards were made.

The 1998 winners will be announced at the Conference.

1997 DAC ACSEE Undergraduate Scholarships

DAC $4K: Bryan Neil Ramirez, Crook, CO - attending Colorado School of Mines
DAC $4K: Christine Joyce Mina, Minot, MA - attending Worcester Polytechnic Institute
DAC $4K: Diala William Abboud, Charlotte, NC - attending Univ. of North Carolina at Charlotte

For more information about the ACSEE Scholarship, please contact Dr. Cherrice Traver, EE/CS Department, Union College, Schenectady, NY 12308 email: traverc@doc.union.edu.

Design Automation Conference Graduate Scholarships

Each year the Design Automation Conference sponsors several $24,000 scholarships to support graduate research and study in Design Automation (DA), with emphasis in "design and test automation of electronic and computer systems". Each scholarship is awarded directly to a university for the Faculty Investigator to expend in direct support of one or more DA graduate students.

The criteria for granting such a scholarship expanded in 1996 to include financial need. The criteria are: the academic credentials of the student(s); the quality and applicability of the proposed research; the impact of the award on the DA program at the institution; and financial need. Preference is given to institutions that are trying to establish new DA research programs.

Information on next year's DAC scholarship award program will be available on the DAC World Wide Web page at: http://www.dac.com/scholarship.html.

Design Automation Conference Graduate Scholarship Awards

- Prof. Rajesh K. Gupta of the University of California, Irvine, CA, for Ali Dasdan. Their project is entitled, "Framework for Timing Constraint Analysis and Debugging for Embedded Systems".

- Prof. John Lillis of the University of Illinois, Chicago, IL, for Sung-Woo Hur and Prashanthi Malireddy. Their project is entitled, "New Techniques for Timing-Driven Placement".

- Prof. Frank Vahid of the University of California, Riverside, CA, for Tony Givargis. Their project is entitled, "Interface Exploration for Core-Based Systems".

The IEEE CAS supported DAC Graduate Scholarship is awarded to:

- Prof. Soha Hassoun of the Tufts University, Medford, MA, for Alexndros Margomenos, Terry Orfanos and Edmund Sullivan. Their project is entitled, "Microarchitectural Optimizations and Synthesis".

Design Automation Conference Graduate Scholarship Committee

The 1998 DAC Scholarship Committee was comprised of the following people:

James P. Cohoon, University of Virginia (Chair)
Michael Lightner, University of Colorado
Jeffrey S. Salowe, Cadence Design Systems, Inc.

1998 SIGDA Meritorious Service Awards

For contributions in producing SIGDA CD ROM's-
Archiving the knowledge of the Design Automation Community

Jason Cong, University of California, Los Angeles, CA
Bryan Preas, Xerox PARC, Palo Alto, CA
Kathy Preas, KP Publications on CD-ROM, Palo Alto, CA
Chong-Chian Koh, University of California, Los Angeles, CA
Cheng-Kok Koh, University of California, Los Angeles, CA

1998 IEEE Fellows

William Joyner - IBM Corp., T.J. Watson Research Ctr., Yorktown Heights, NY
Michael Lightner - Univ. of Colorado, Boulder, CO
Michael Nakhala - Carleton Univ., Ottawa, ON, Canada
Rob A. Rutenbar - Carnegie Mellon Univ., Pittsburgh, PA
Karem Sakallah - Univ. of Michigan, Ann Arbor, MI

CALL FOR PAPERS

36th DESIGN AUTOMATION CONFERENCE®

ERNEST N. MORIAL CONVENTION CENTER • NEW ORLEANS, LA • JUNE 21 - 25, 1999

DAC is the premier conference devoted solely to the field of Design Automation. All aspects of the use of computers as aids to the design process are welcome, from conceptual design to manufacturing. Five types of submissions are invited: regular papers, special topic sessions, panels, tutorials, and design contest entries. All types of submissions should be sent to the Program Chair, **postmarked NO later than October 9, 1998**.

TOPICS OF INTEREST

Authors are invited to submit original technical papers describing recent and novel research or engineering developments in all areas of design automation. The **DESIGN TOOLS TRACK (T)** is devoted to contributions to the research and development of design tools and the supporting algorithms. The **DESIGN METHODS TRACK (M)** deals with contributions to the research and development of design methodologies and applications of design automation tools to designs. Topics of interest include, but are not limited to:

DESIGN TOOLS TRACK:

T0.1 Fundamental CAD Algorithms, e.g., BDDs, graph coloring, partitioning
T1.1 Electrical-level circuit and timing simulation
T1.2 Discrete simulation
T1.3 Critical path analysis and timing verification
T1.4 Power estimation
T2.1 Testing, fault modeling and simulation, TPG, test validation and DFT
T2.2 Design and implementation verification (excluding layout verification)
T3.1 Floorplanning and placement
T3.2 Global and detailed routing
T3.3 Module generation and compaction, transistor sizing and cell library optimization, layout verification
T4.1 Technology independent, combinational logic synthesis
T4.2 Technology dependent logic synthesis, library mapping, interactions between logic design and layout
T4.3 Sequential and asynchronous logic synthesis and optimization
T4.4 High-level synthesis
T5.1 Hardware Description Languages
T5.2 Hardware/Software co-design, partitioning, system-level specification and design aids
T5.3 Software synthesis and retargetable compilation
T5.4 Hardware/Software co-simulation
T6.1 Interconnect and packaging modeling and extraction
T6.2 Signal integrity and reliability analysis
T6.3 Analog and mixed-signal design tools and RF
T6.4 Microsensor and microactuator design tools
T6.5 Statistical design and yield maximization
T7.1 Frameworks, intertool communication, WWW-based tools and databases

DESIGN METHODS TRACK:

The Design Methods track (M) deals with innovative methodologies for the design of electronic circuits and systems, as well as creative experiences with design automation in state-of-the-art designs. Submissions for this track will be judged on how effectively they teach new art in the development and application of new tools and techniques to real-world design problems.

M1 Design methodologies and case studies for specific design tasks
M1.1 Design entry and specification
M1.2 Simulation, analysis, modeling and estimation
M1.3 Verification, test and debugging
M1.4 Physical design, module generation, design for manufacturing
M1.5 Logic and high-level synthesis and optimization
M1.6 System-level design, embedded-system design and co-design
M1.7 Other
M2 Design methodologies and case studies for specific application domains and platforms
M2.1 Configurable computing, FPGA and rapid prototyping
M2.2 Systems on a chip
M2.3 Microprocessor and multiprocessor
M2.4 DSP, data-paths, multimedia and communication
M2.5 Wireless and data networking
M2.6 Other (MCM, optical, consumer)
M3 Performance and technology driven design techniques
M3.1 Deep sub-micron: signal integrity, interconnect and modeling
M3.2 High-performance design: timing, clocking and power distribution
M3.3 Low power design
M3.4 Mixed-signal, analog, and RF
M3.5 Process technology development, extraction, modeling
M3.6 Other (MEMS, sensors, new devices)
M4 Integration and management of DA systems
M4.1 Management of DA systems, design interfaces, standards
M4.2 Distributed, networked, and collaborative design
M4.3 Intellectual property, design re-use and design libraries

Watch the WWW for updates! (http://www.dac.com)

REQUIREMENTS FOR SUBMISSION OF PAPERS

Previously published papers, including workshop proceedings, will not be considered. Each submission should include one cover page and eleven (11) stapled copies of the complete manuscript.
The one cover page should include:

- Name, affiliation, and complete address for each author
- A designated contact person including his/her telephone number, fax number, and email address
- A designated presenter, should the paper be accepted
- A list of topic numbers preceded by the letter T (Tools Track) or M (Methods Track), ordered by relevancy, most clearly matching the content of the paper
- The following signed statement: "All appropriate organizational approvals for the publication of this paper have been obtained. If accepted, the author(s) will prepare the final manuscript in time for inclusion in the Conference Proceedings and will present the paper at the Conference".

To permit a blind review, do not include name(s) or affiliation(s) of the author(s) on the manuscript. Include:

- Title of paper
- 60-word abstract indicating significance of contribution
- The complete text of the paper in English, including all illustrations and references, not exceeding 4000 words. The papers will be reviewed as finished papers. Preliminary submissions will be at a disadvantage.

Notice of acceptance will be mailed to the contact person by February 19, 1999. Authors of accepted papers must sign a copyright release form.

PANELS, TUTORIALS, SPECIAL TOPIC SESSIONS

Proposals should not exceed two pages in length and should describe the topic and intended audience. They must include a list of all participants, including the moderator for panels. **For proposal instructions, send a one-line email message to proposals@dac.com with "proposal" in the subject field, or www.dac.com.**

Special Topic Sessions may be either independent papers with a common theme or a set of closely related papers describing an overall system. In both cases, independent reviews of each paper and evaluation of the session as a whole will be used to select sessions. Proposals for Special Topic Sessions should be submitted along with the list of papers to be included in the session and should describe the session's theme. **These proposals and paper submissions must be postmarked no later than October 9, 1998.**

UNIVERSITY DESIGN CONTEST

Submissions of original electronic designs (circuit or system), developed at universities and research organizations after June 1997 and resulting in operational implementations are invited. Submissions should contain the title of the project, a 60-word abstract and a complete description of the design, not exceeding 4000 words in text. The submission should clarify the originality, distinguishing features, and the measured performance metrics of the design. Proof-of-implementation in the form of die or board photographs and measurement data is a must. Submitted designs should not have received awards in other contests. Submissions will be reviewed by a special committee of experts. Selected designs will be presented and exhibited at the conference.

PROGRAM CHAIR

MP Associates, Inc.
ATTN: Technical Program Co-Chairs
Randal E. Bryant/Bryan Ackland
5305 Spine Rd., Suite A
Boulder, CO 80301
For information call: (303) 530-4333

Sponsored by:

REVIEWERS

A total of 390 manuscripts were submitted to the 1998 DAC. The Conference Executive and Technical Program Committees wish to acknowledge the time and effort spent by the following people who reviewed these manuscripts and returned the review forms completed. Our thanks to all of those who participated and contributed to the success of the Conference.

Magdy S. Abadir
Miron Abramovici
Khosrow Adibsamii
Gaurav Aggarwal
David G. Agnew
Vishwani D. Agrawal
Takashi Aikyo
Robert Aitken
Kahlid M. Al-Ruwaihi
Akhtar Ali
Salahuddin Almajdoub
Charles J. Alpert
Joachim Altmeyer
Tod Amon
Kurt Antreich
Rafael Aquino
Guido Araujo
Pranav Ashar
William W. Au
Michael S. Austwick
Rosa Maria Badia
Franklin M. Baez
Brian Bailey
Smita Bakshi
Felice Balarin
Erich Barke
John K. Bartholomew
John G. Bate
Robert K. Beachler
Derek L. Beatty
James A. Beausang
James E. Beck
Bernd Becker
Peter Beerel
Dirk Behrens
Bill Bell
Nikolaos Bellias
Tarek Ben Ismail
Carina E. Ben-Zvi
Luca Benini
Reinaldo Bergamaschi
Michel Berkelaar
E. Berrebi
Narasimha Bhat
Sandeep Bhatia
Subhrajit Bhattacharya
Sudipta Bhawmik
Siddharth Bhingarde
Peter Bingley
Stephen Blythe
Alessandro Bogliolo
Pascal Bolcato
Cristiana Bolchini
Massimo Bombana
Carsten Borchers
Gaetano Borriello
Bhaskar Bose
Eric Bracken
Ansgar Bredenfeld
Forrest D. Brewer
Jay B. Brockman
Thomas Buechner
Giacomo Buonanno
Jerry R. Burch
Timothy M. Burks
Gianpiero Cabodi
Paolo E. Camurati
Ariel Cao
Wanlin Cao
Robert Carragher
Don Carter
Francky Catthoor
Patrizia Cavalloro
Mustafa Celik
Krishnendu Chakrabarty
Abhijeet Chakraborty
Tapan J. Chakraborty
Tharini Chakraborty
Srimat T. Chakradhar
Sreejit Chakravarty
Rabih N. Chamoun
V. Chandramouli
K.C. Chang
Shih-Chieh Chang
Yao-Wen Chang
Karam S. Chatha
Chien-In Henry Chen
Chih-Tung Chen
Dahe Chen
Guohua Chen
Howard H. Chen
Sao-Jie Chen
Wenfeng Chen
Xinghao Chen
Yirng-An Chen
David I. Cheng
Shu-Mei Cheng
Wu-Tung Cheng
Brian V. Chess
Chun-Ping George Chi
Massino Chiappone
Anton V. Chichkov
Venkat Chiluvuri
Massimiliano Chiodo
Eli Chiprout
Silvia Chiusano
Hoon Choi
Kiyoung Choi
Pai Chou
Tan-Li Chou
Amit Chowdhary
Salim Chowdhury
Malgorzata Chrzanowska-Jeske
Johan Cockx

REVIEWERS (cont.)

Paul Coene
John M. Cohn
Timothy W. Colleran
Alan J. Coppola
Fulvio Corno
Jose Luis Correia Neves
Jordi Cortadella
Donald Cottrell
Olivier R. Coudert
Bernard Courtois
Vinay P. Dabholkar
Ajay J. Daga
Joseph P. Damore
Hiroshi Date
Bharat P. Dave
Richard A. Davies
Kaushik De
Gjalt De Jong
Linda Debrunner
Anirudh Devgan
Allen M. Dewey
Sujit Dey
Olivier Deygas
Nagu R. Dhanwada
Abhijit Dharchoudhury
Ajit T. Dingankar
Rainer Doemer
Antun Domic
Jack E. Donovan
Bernard J. Doray
Rolf Drechsler
Anthony D. Drumm
Mario Dufresne
John Duh
Nikil D. Dutt
Klaus Eckl
Stephen A. Edwards
Hans Eisenmann
Cindy Eisner
Ibrahim M. Elfadel
Norman J. Elias
Gary Ellis
Frank Elof
Bernhard Eschermann
Nong Fan
Rong Fan
Amir H. Farrahi
Erin P. Fassio
Gary K. Fedder
Peter Feldmann
Joel Ferguson
Fabrizio Ferrandi
Josef Fleischmann
Paulo Flores
Marie-Lise Flottes
Paul Franzon
Mark S. Fredrickson
Norbert Froehlich
Robert C. Frye
Hiroshige Fujii
Takashi Fujii
Masahiro Fujita
Franco Fummi
George Gadelkarim
Anthony J. Gadient
Daniel D. Gajski
Neeta Ganguly
Shantanu Ganguly
Joseph Ganley
Andreas Ganz
David S.-W. Gao
Christophe Gauthron
Catherine H. Gebotys
Daniel Geist
Vassilios Gerousis
Andreas Gerstlauer
Indradeep Ghosh
Ekaterini E. Gikas
Dimitris Gizopoulos
James A. Goda
Nicole Goeckel
Manish Goel
Rich Goldman
Nanda Gopal
Nobuyuki Goto
Sriram Govindarajan
Ravender Goyal
F. Gail Gray
Gary S. Greenstein
Paul Grojean
J.P. Grossman
John S. Grout
Lon Grover
Xinli Gu
Carlo Guardiani
Wolfgang Guenther
Michaela Guiney
Kiran Gullapalli
Pei-Ning Guo
Rohini Gupta
Subodh Gupta
Mohan Guruswamy
Mitch R. Gusat
Paul Gutwin
Ian A. Guyler
Dong S. Ha
Ibrahim N. Hajj
Reinhard V. Hanxleden
Ikuo Harada
Klaus Harbich
Justin Harlow
Soha Hassoun
Gagan Hasteer
David J. Hathaway
Lei He
Jim Heaton
Lars Hedrich
Carl Hein
Sybille Hellebrand
Shankar Hemmady
Stefan Hendricx
Manfred Henftling
Harry Hengster

REVIEWERS (cont.)

Joerg Henkel
Keerthi Heragu
Hiroyuki Higuchi
Dwight D. Hill
Ken Hines
Lynwood Hines
Kanji Hirabayashi
Mokhtar Hirech
Kei Hirose
Pei-Hsin Ho
Shervin Hojat
Ulrich Holtmann
Inki Hong
Seongsoo Hong
Wei Hong
You-Pyo Hong
Harry Hsieh
Hong-Yean Hsieh
Pao-Ann Hsiung
Frank F. Hsu
Yaun-Chung Hsu
Alan J. Hu
Xiaobo Hu
Ching-Chao Huang
Pujiang Huang
Ying-Min I. Huang
Jose L. Huertas Diaz
Christophe P. Hui-Bon-Hoa
Jos A. Huisken
Brad Hutchings
Michael Hutton
Yeanyow Hwang
Takaki Ichimori
Paolo Ienne
Akihiko Inoue
Atsuki Inoue
C. Norris Ip
Mary Jane Irwin
Shinya Ishihara
Tohru Ishihara
Balakrishnan Iyer
Mahesh A. Iyer
Neil G. Jacobson
Margarida Jacome
Geert Jahssen
Jae-Young Jang
Alvin Jee
Niraj K. Jha
Pradip Jha
Yi-Min Jiang
Frank M. Johannes
Eric N. Johnson
Eric W. Johnson
Vess L. Johnson
K.D. Jones
Luli Josephson
Raju Joshi
Jing-Yang Jou
Attila Jurecska
Knut Just
Hilary J. Kahn
Asawaree Kalavade
Sudhakar Kale
Priyank Kalla
Robert L. Kanzelman
William H. Kao
Ashish Kapoor
Sharad Kapur
Jean-Michel Karam
Arvind K. Karandikar
Osamu Karatsu
Maddumage Karunaratne
Mark A. Kassab
Srinivas Katkoori
Meenakshi Kaul
Masamichi Kawarabayashi
Robert J. Kaye
Wuudiann Ke
Holger Keding
Martin Keim
Pratibha Kelapure
Christoph Kern
Kevin J. Kerns
John E. Kerro
Manpreet Khaira
Sanjay A. Khan
Sunil Khatri
Kei-Yong Khoo
Bruce Kim
Choon Kim
Jaewon Kim
Shinji Kimura
Darko Kirovski
Polen Kission
Robert H. Klenke
Alfred Koelbl
Cheng-Kok Koh
Hisao Kolzumi
Venk Kommu
Alex Kondratyev
Susanto M. Kong Woei
Andrzej Krasniewski
Byron Krauter
Harish Kriplani
Shankar Krishnamoorthy
Stanley J. Krolikoski
Kayhan Kucukcakar
Prabhakar Kudva
Yuji Kukimoto
Balakrishna Kumthekar
David S. Kung
Wolfgang Kunz
Ganesh Lakshminarayana
Rudy Lauwereins
Jens Leenstra
Christian Legl
Daksh Lehther
Guang-Tsai Lei
William C. Lester Jr.
Rainer Leupers
Regis Leveugle
Steven Levitan
Samuel Levitin

REVIEWERS (cont.)

Jeremy R. Levitt
Yanbing Li
Stan Liao
Luigi Licciardi
John Lillis
David D. Ling
Antonio Lioy
Paul E. Lippens
Fang-Jou Liu
Yu Liu
Arun N. Lokanathan
Michael Lorenzetti
Aiguo Lu
Gabriele Luculli
Yufeng Luo
Fadi Maamari
Alberto Macii
Enrico Macii
J. Christophe Madre
Serge Maginot
Naresh Maheshwari
Enrico Malavasi
Yossi Malka
Nazanin Mansouri
Henri J. Maramis
P.J. Marchand
Gilberto Marchioro
Diana Marculescu
Radu Marculescu
Igor L. Markov
Grant E. Martin
Michael Martin
Tom Martin
Peter Marwedel
Alida Mascitell
Anmol Mathur
Masataka Matsui
Yusuke Matsunaga
Peter C. Maxwell
Robert L. Maziasz
Kevin W. McCauley
Patrick McGuinness
Grace McNally
Uwe B. Meding
Mahesh Mehendale
Renu Mehra
Amit Mehrotra
Sharad Mehrotra
Gaurav Mehta
Stephen Meier
Christoph Meinel
Kevin L. Melocco
Noel Menezes
Prem R. Menon
Scott F. Midkiff
Paolo Miliozzi
Fumihiro Minami
Shin-ichi Minato
Salvador Mir
Manmohan Mittal
Hiroshi Miyashita
Toshiaki Miyazaki
Peter Moceyunas
Paul Molitor
Robert Molyneaux
Delfin Y. Montuno
Daniel G. Moritz
Vasily G. Moshnyaga
Dinos Moundanos
Klaus D. Mueller-Glaser
Ashutosh S. Mujumdar
Pradipto Mukherjee
Rajarshi Mukherjee
Tamal Mukherjee
Fidel Muradali
Masami Murakata
Hiroshi Murata
Takahiro Murooka
Gerry Musgrave
Lode Nachtersaele
Benoit Nadeau-Dostie
Sudip K. Nag
Surendra Nahar
Vinod T. Nair
Yuichi Nakamura
J. Narasimhan
Naren Narasimhan
Unni Narayanan
Danial Neebel
Mahadevamurty Nemani
Horacio Neto
Tuyen V. Nguyen
Lisa M. Noack
Adrian Nunez-Aldana
Emil S. Ochotta
Peter Odryna
Kyung S. Oh
Seong Yong Ohm
Mitsuyasu Ohta
Makiko Okumura
Arlindo Oliveira
Yukihito Oowaki
Neven Orhanovic
Ross B. Ortega
Iyad E. Ouaiss
Andras Pahi
Parimal Pal Chaudhuri
Sarala Paliwal
Wenwei Pan
Rajendran V. Panda
Shipra Panda
Manish Pandey
Maurizio Paolini
Christos A. Papachristou
Abelardo Pardo
Rubin A. Parekhji
Keshab K. Parhi
Nish P. Parikh
In-Cheol Park
Enric Pastor
Janak H. Patel
Rakesh J. Patel
Lalit M. Patnaik

REVIEWERS (cont.)

Michael Payer
Stefan Pees
Marco A. Pena
Marek A. Perkowski
Wim Philipsen
Joel R. Phillips
Marc Picquendar
Karen L. Pieper
Lawrence T. Pileggi
Alain Pirson
Carl P. Pixley
Luis A. Plana
Paul Ploeger
Francois Pogodalla
Irith Pomeranz
Massimo Poncino
Andras Poppe
Guido Post
Miodrag M. Potkonjak
Dhiraj K. Pradhan
Mukul R. Prasad
Pierre Pribetich
Paolo Prinetto
Iksoo Pyo
Isa S. Qamber
Giu Qinru
Stefano Quer
Ivan P. Radivojevic
Vijay Raghavendra
Anand Raghunathan
Richard Raimi
Rajesh Raina
Salil Raje
Suresh Rajgopal
Shankar Raman
Srilata Raman
Jerome Rampone
Rajeev K. Ranjan
Shishpal S. Rawat
Alain Raynaud
Philippe P. Raynaud
Bill Read
Maurizio Rebaudengo
Mark A. Rempel
Marta Rencz
Karla Reynolds
Michael Riepe
Bernhard M. Riess
Ann M. Rincon
Ramine Roane
Oriol Roig
Andreas Ropers
Wolfgang Rosenstiel
Charles Rosenthal
Mark Rossman
Kaushik Roy
Rabindra K. Roy
Elizabeth Rudnick
Adoracion Rueda
Charlie Rupp
Harald H. Sack
P. Sadasivan
Omer G. Sahhan
Hans Sahm
Alexander Saldanha
Bill Salefski
Jeffrey S. Salowe
Raul San Martin
Fermin Sanchez Carraledo
Jagesh Sanghavi
Ambar Sarkar
Koichi Sato
Janardhan H. Satyanarayana
Gabriele Saucier
Eric Sax
Vikram Saxena
Riccardo Scarsi
Patrick R. Schaumont
Ulf Schlichtmann
Stefan Schmerler
Christoph Scholl
Endric Schubert
Bernd Schuermann
Donatella Sciuto
Andrew Seawright
Carl Sechen
Reza Sedaghat-Naman
Joel Seidman
Masatoshi Sekine
Ellen M. Sentovich
Jonjen Sern
Dorothy E. Setliff
Pradip Shah
William D. Sharon
Wen-Zen Shen
Narendra V Shenoy
Der-Y Sheu
Hyongkyoon Shin
Thomas R. Shiple
Allan Silburt
Julio Silva
L. Miguel Silveira
Kanwar J. Singh
Vigyan Singhal
Mukund Sivaraman
Joseph P. Skudlarek
Eric Skuldt
Anna Slobodova
Ross Smith
Mani Soma
LeiLei Song
Mohamed Soufi
Larry P. Soule
Lambert Spaanenburg
Jim Sproch
Giovanni A. Squillero
Rajagopala Srinivasan
Vinoo N. Srinivasan
Mysore Sriram
Mani Srivastava
Brian Stacey
Jason W. Staiert
George Stamoulis

REVIEWERS (cont.)

Christian Stangier
Balsha R. Stanisic
Don Stark
Guenter Stenz
Jack A. Stinson
Joern Stohmann
Leon Stok
Neal Stollon
Noel Strader
Lixin Su
Peter R. Suaris
Ravi Subramanian
Ashok Sudarsanam
Stephen Sugiyama
Wei-Kai Sun
Peter Sutton
Gitanjali M. Swamy
Joseph M. Swenton
V. Szekely
Frank Szorc
Thomas G. Szymanski
Masayoshi Tachibana
Paul Tafertshofer
Haruyuki Tago
Yutaka Tamiya
Hiroshi Tanimoto
Alexander Taubin
Ramesh C. Tekumalla
Chin-Chi Teng
Shashidhar Thakur
Michael Theobald
Thorsten Theobald
Michael W. Tian
Adwin H. Timmer
Yosef Gavriel Tirat-Gefen
Steve Tjiang
Horia Toma
Hiroyuki Tomiyama
Masahiko Toyonaga
Quan Tran
Tuan Anh Tran
Stephen Trimberger
Yatin Trivedi
Anne-Marie Trullemans-Anckaert
Huan-Chih Tsai
Chi-Ying Tsui
Raymond Y. Tsui
Emre Tuncer
Bogdan Tutuianu
Jon G. Udell
Sanjay Upreti
Shmuel Ur
Frank Vahid
Radha Vaidyanathan
Carlos Valderrama
David Van Campenhout
N.P. Van Der Meijs
Arjan Van Genderen
Jagadeesh Vasudevamurthy
Ranga Vemuri
Raja Venkatachalam
Ingrid Verbauwhede
Wim Verhaegh
Diederik Verkest
Serge Vernalde
Idalina J. Videira
Roberto Vietti
Ashok Vittal
Jeroen P. Voeten
Sarma Vrudhula
Duncan M. Walker
Robert A. Walker
David Wallace
Jeffrey Walrath
Piet Wambacq
Chuan-Yu Wang
Duen-Jeng Wang
Jimmy S. Wang
Seongmoon Wang
Yosinori Watanabe
Joseph C. Watkins
Ivan L. Wemple
Yen-Cheng Wen
Manfred Wiesel
Joerg Wilberg
Bruce Wile
Markus Willems
John Willis
Steven J.E. Wilton
Bruce Winter
Alain D. Wittmann
Anthony S. Wojcik
Wayne Wolf
Yaron Wolfsthal
Allen C.-H. Wu
Chang Wu
Qing Wu
Min Xu
Songjie Xu
Alexandre Yakovlev
Masaaki Yamada
Ryuichi Yamaguchi
Koichi Yamashita
Lawrence R Yang
Hiroto Yasuura
Yibin Ye
Jenny Yee
Ti-Yen Yen
Dah-Cherng Yuan
Frank Y. Yuan
Jun Yuan
Li-Pen Yuan
Joel T. Yuen
Roberto Zafalon
Amir M. Zarkesh
Peter W. Zepter
Min Zhao
Hai Zhou
Lixin Zhou
Jiabi J. Zhu
Avi Ziv
Vojin Zivojnovic
Kristian M. Zoerhoff
Xiao Zuonan

OPENING KEYNOTE ADDRESS

William J. Spencer
Chairman of the Board
SEMATECH, Inc.
Austin, TX

DESIGN AUTOMATION CAN HELP THE SEMICONDUCTOR INDUSTRY ADDRESS ITS MANY CHALLENGES

Silicon technology has been the economic driver for much of the Information Age, by providing a 25-30% year-over-year improvement in cost per unit of performance. This amazing performance has been fueled by constantly improving semiconductor technology, including major gains in design automation. For the industry to remain on the historic productivity curve, break-throughs will be required in several aspects of semiconductor manufacturing technology often linked strongly to design.

Bill Spencer is currently Chairman of SEMATECH, a research and development consortium consisting of sixteen international corporations involved in semiconductor manufacturing. From 1990-1997, he served as President and Chief Executive Officer of SEMATECH. Prior to 1990, he was Group Vice President and Senior Technical Officer at Xerox Corporation in Stanford, Connecticut, as well as Vice President and Manager of the Xerox Palo Alto Research Center (PARC). He was Director of Systems Development and also Director of Microelectronics at Sandia National Laboratories from 1973 to 1981, prior to joining Xerox. He began his career at Bell Telephone Laboratories in 1959. He received his Ph.D. and M.S. from Kansas State University, and an A.B. from William Jewell College in Missouri.

Spencer is also a Research Professor of Medicine at the University of New Mexico, where the first implantable electronic drug delivery systems were developed jointly with Sandia National Labs. For this work, he received the Regents Meritorious Service Medal and later a doctor of science degree from William Jewell College. He is currently a Director of Adobe Systems, Investment Corporation of America, SRI International and the Austin Symphony. He is also a member of the Board of Trustees of the Computer Museum and William Jewell College.

He has served on several National Research Council studies in the areas of technology, trade, corporation and competition. In 1998, he will co-chair, with Dick Thornburgh, an NRC workshop on "Harnessing Technology for America's Future Economic Growth". He will also serve as a Regents Professor at the University of California at Berkeley.

THURSDAY KEYNOTE ADDRESS

George H. Heilmeier
Chairman Emeritus
Bellcore
Morristown, NJ

FROM POTS TO PANS: TRANSITION IN THE WORLD OF TELECOMMUNICATIONS FOR THE LATE 90s AND BEYOND

The telecommunications industry is undergoing transitions at an unprecented rate from a provider of plain old telephone service (POTS) to a provider of "pretty awesome new services" (PANS). This presentation will discuss the forcing factors and the technical/business transitions that are occuring at the industry, network, service and operations systems level. Key technical issues facing the industry as well as some agenda setting themes will also be discussed along with some of the implications for the microelectronics industry.

Dr. Heilmeier, world renowned for his pioneering work in the development of liquid crystal displays (LCDs) and the recipient of numerous awards including the National Medal of Science and the IEEE Medal of Honor, was drafted out of Texas Instruments in 1991 to fill the top job at Bellcore. Over the past six years, Heilmeier has piloted the company's successful transition from the nation's largest resource consortium to a competitive, profitable, growing, commercial enterprise with an expanded client base of more than 800 customers in the U.S. and abroad.

Prior to joining Bellcore in March 1991, as the company's President and Chief Executive Officer, Heilmeier was Senior Vice President and Chief Technical Officer of Texas Instruments, Inc. Dr. Heilmeier, a native of Philadelphia, holds a B.S. in electrical engineering from the University of Pennsylvania and M.A., M.S.E., and Ph.D. degrees in solid-state electronics from Princeton University. He has also been awarded honorary doctorates from Stevens Institute and Technion, the Israel Institute of Technology.

Table of Contents

Session 3
University Design Contest

Chair: Mary Jane Irwin
Organizer: Jan M. Rabaey

Session 4
Embedded System Design and Exploration

Chair: Ivo Bolsens
Organizers: James A. Rowson, Anders Forsen

Session 5
Taming Noise in Deep-Submicron Digital Designs

Chair: Nagaraj NS
Organizers: Kenneth L. Shepard, Nagaraj NS

Panel: Taming Noise in Deep Submicron Digital Integrated Circuits

Chair: Nagaraj NS
Organizers: Kenneth Shepard, Takahide Inoue

Session 6
Control and Data Driven High Level Synthesis

Chair: Kayhan Kucukcakar

Organizers: David Ku, Timothy Kam

Session 7
Synthesis Flow in Deep Submicron Technologies

Chair: Ralph H. J. M. Otten

Organizers: Sharad Malik, Randal E. Bryant

Session 8
Environment for Collaborative Design

Chair: Takahide Inoue

Organizers: Richard Smith, Takahide Inoue

Session 9
New Methods in Functional Verification

Chair: Rajesh K. Gupta

Organizers: Vivek Tiwari, Kenji Yoshida

Session 10
Panel: Hardware/Software Co-Design: The Next Embedded System Design Challenge

Session 11
System-Level Power Optimization

Session 12
Boolean Methods

Session 13
Extraction and Modeling for Interconnect

Session 14
Processor Design and Simulation

Session 15
Panel: How Much Analog Does a Designer Need To Know for Successful Mixed-Signal Design?

Session 16
Performance Modeling and Characterization for Embedded Systems

Session 17
Advances in Placement and Partitioning

Session 18
Parasitic Device Extraction and Interconnect Modeling

Session 19
Design Optimization for DSP

Chair: James A. Rowson
Organizers: Anders Forsen, Ivo Bolsens

Session 20
Panel: User Experience with High Level Formal Verification

Chair: Gerry Musgrave
Organizers: Randal E. Bryant, Gerry Musgrave

Session 21
Bridging the Gap Between Simulation and Formal Verification

Chair: Andreas Kuehlmann
Organizers: Randal E. Bryant, Sharad Malik

Session 22
Logic Optimization

Chair: Albert Wang
Organizers: Jason Cong, TingTing Hwang

Session 23
Routing for Performance and Crosstalk

Chair: Sachin S. Sapatnekar
Organizers: Patrick Groenveld, Andrew B. Kahng

Session 24
Practical Optimization Methodologies for High Performance Design

Chair: Vivek Tiwari
Organizers: Vivek Tiwari, Kenji Yoshida

Session 25
RF IC Design Methodology

Chair: Mojy C. Chian
Organizers: Bryan D. Ackland, Mojy C. Chian

Session 26
Theory and Practice in High Level Synthesis

Chair: Steve Tjiang
Organizers: David Ku, Timothy Kam

Session 27
BDD Approximation Techniques

Chair: Andreas Kuehlmann
Organizers: Andreas Kuehlmann, Kunle Olukotun

Session 28
Interconnect Modeling and Timing Simulation

Chair: Andrew T. Yang
Organizers: Andrew T. Yang, Hidetoshi Onodera

Session 29
Low Power Design Using Multiple Thresholds and Supplies

Chair: Bryan D. Ackland
Organizers: Anantha Chandrakasan, Bryan D. Ackland

Session 30
Panel: Technical Challenges of IP and System-on-Chip: The ASIC Vendor Perspective

Chair: A. Richard Newton
Organizers: Andrew Graham, Andrew B. Kahng

Session 31
Software Synthesis and Retargetable Compilation

Chair: Kurt Keutzer

Organizers: Luciano Lavagno, Sharad Malik

Session 32
Formal Methods in Functional Verification

Chair: Kunle Olukotun

Organizers: Kunle Olukotun, Andreas Kuehlmann

Session 33
Core Test and BIST

Chair: Janusz Rajski

Organizers: Yervant Zorian, Janusz Rajski

Session 34
Interconnect Analysis and Reliability in Deep Sub-Micron

Chair: Kenji Yoshida

Organizers: Kenji Yoshida, David T. Blaauw

Session 35
Panel: Design Productivity: How to Measure It, How to Improve It

Session 36
Timing Analysis

Session 37
New Techniques in State Space Explorations

Session 38
Advanced ATPG Techniques

Session 43
Technology Mapping for Programmable Logic

Chair: Jonathan Rose

Organizers: Jason Cong, TingTing Hwang

Session 44
Power Dissipation and Distribution in High Performance Processors

Chair: David T. Blaauw

Organizers: Anatha Chandrakasan, Jan M. Rabaey

Session 45
Challenge in the Test on System-On-A-Chip Era

Chair: Prab Varma

Organizers: Prab Varma, Takahide Inoue

Session 46
Controller Decomposition for Power and Area Minimization

Chair: Fabio Somenzi

Organizers: Timothy Kam, Luciano Lavagno

Session 47
IP Protection Technologies

Chair: Tom VandenBerge

Organizers: Richard Smith, Takahide Inoue

Session 48
Case Studies of New Design Methods

Chair: Anders Forsen

Organizers: Ivo Bolsens, James A. Rowson

Panel: Customers, Vendors, and Universities: Determining the Future of EDA Together

Chair: Thomas Pennino - Lucent Technologies, Holmdel, NJ.
Organizer: Mike Murray - Acuson Corporation, Mountain View, Ca.

Panelists:

Aart de Geus - Synopsys, Inc., Mountain View, Ca.

Jack Harding - Cadence Design Systems, Inc., San Jose, Ca.

Walden Rhines - Mentor Graphics Corporation, Wilsonville, Or.

Prof. Robert Brodersen - University of California, Berkeley, Ca.

Johan Danneels - Alcatel Microelectronics, Zaventem, Belgium

Gadi Singer - Intel Corporation, Santa Clara, Ca.

Abstract

Progress in the EDA industry depends on the cooperative efforts of customers, vendors, and universities. DAC's opening panel brings together leaders from these communities to discuss how EDA can continue to support rapid innovation in the electronics industry. What are the design challenges ahead? What new EDA technologies and working relationships are required for customers and universities to meet these challenges? What do customers want in a vendor? What do vendors want in a customer? How can universities contribute, and what can customers and vendors do in return? Come join us for a lively and thought-provoking session on the issues which will shape the future of EDA.

Asynchronous Interface Specification, Analysis and Synthesis*

Michael Kishinevsky
Intel Corporation
Hillsboro, OR, USA

Jordi Cortadella
Technical University of Catalonia
Barcelona, Spain

Alex Kondratyev
The University of Aizu
Aizu-Wakamatsu, Japan

Abstract

Interfaces, by nature, are often asynchronous since they serve for connecting multiple distributed modules/agents without common clock. However, the most recent developments in the theory of asynchronous design in the areas of specifications, models, analysis, verification, synthesis, technology mapping, timing optimization and performance analysis are not widely known and rarely accepted by industry.

The goal of this tutorial is to fill this gap and to present an overview of one popular systematic design methodology for design of asynchronous interface controllers. This methodology is based on using Petri nets (PN) *a formal model that, from the engineering standpoint, is a formalization of timing diagrams (waveforms) and from the system designer standpoint is a* concurrent *state machine, in which local components can perform independent or interdependent concurrent actions, changing their local states asynchronously. We will introduce this model informally based on a simple example: a VME-bus controller serving reads from a device to a bus and writes from the bus into the device.*

1 Specification with Petri Nets

1.1 From timing diagrams to Petri Nets

Figure 1 depicts the interface of a device with a VME bus. The behavior of the controller is as follows: a request to read from or write into the device is received by one of the signals DSr or DSw respectively. In a read cycle, a request to read is done through signal LDS. When the device has the data ready ($LDTACK$), the controller must open the transceiver to transfer data to the bus (signal D). In the write cycle, data is first transferred to the device. Next, a request to write is done (LDS). Once the device acknowledges the reception of the data ($LDTACK$) the transceiver must be closed to isolate the device from the bus. Each transaction must be completed by a return-to-zero of all interface signals, seeking for a maximum parallelism between the bus and the device operations. Figure 2 shows a timing diagram of the read cycle and Figure 3 the corresponding Petri Net. All events in this Petri Net are interpreted as signal transitions: rising and falling signal transitions are labeled with "+" and "−" respectively. Petri Nets with such signal interpretations are called *Signal Transition Graphs (or* STG*s)* [17].

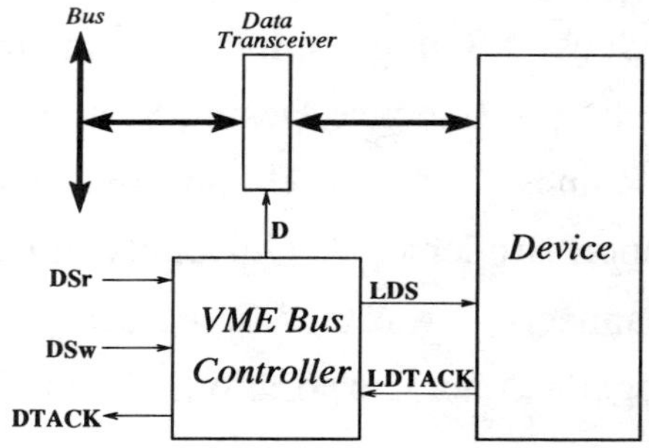

Figure 1: VME bus controller

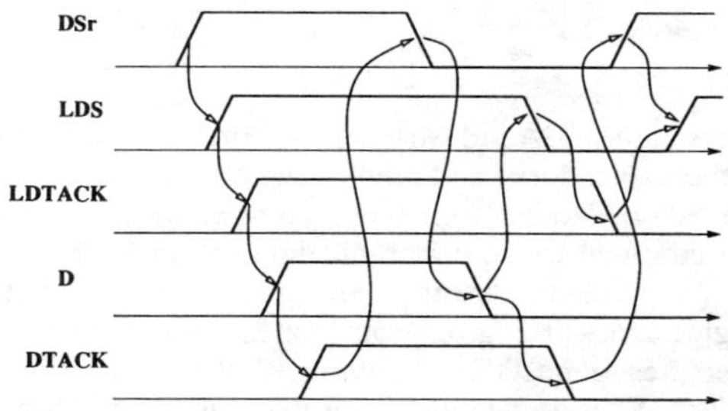

Figure 2: Waveforms for the READ cycle

A PN has two types of vertices: places (denoted by circles) and transitions (denoted by boxes), and arcs from places to transitions and from transitions to places. Places correspond to local states of the system and are used for keeping information about system resources and conditions for execution of transitions. Places can keep tokens (denoted by black dots). A token in a place indicates that a resource is available or a condition satisfied. In general more than one token can be kept in a place, but we will consider only the simplest case: a place cannot contain more than one token. A set of all places currently marked with a token corresponds to a current *global* state of the net. Such global states are called *markings.* The initial marking of the PN in Figure 3 is $\{p_0, p_1\}$.

1.2 Token game and concurrency

Transitions correspond to system events (signal transitions in the example). A transition is *enabled* if all input places contain a token. In the initial marking of the PN in Figure 3

*Work partially supported by ACiD-WG (Esprit 21949) and CICYT TIC 95-0419

DAC 98, San Francisco, California

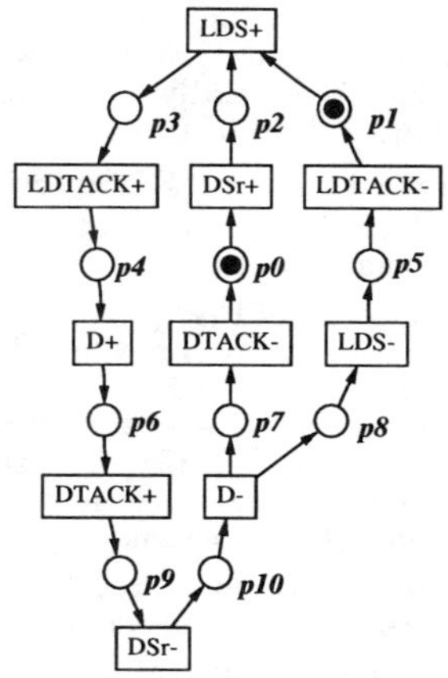

Figure 3: STG for the READ cycle

only one transition, $DSr+$, is enabled; another one, $LDS+$, is not: only place p_1 among two of its input places, p_1 and p_2, contains a token. Every enabled transition can fire. Firing removes one token from every input place of the transition and puts one token to each of its output places. Firing of a transition is an atomic instantaneous operation, while some unspecified time can pass between enabling and firing of the transition. After the firing of transition $DSr+$ the net moves to a new marking $\{p_1, p_2\}$ and then $LDS+$ becomes enabled.

This process of moving tokens around (a.k.a. token game) in a few steps will fire transition $D-$. This leads the net into the marking $\{p_7, p_8\}$. In this marking *two* transitions $DTACK-$ and $LDS-$ become enabled. Since their input places are different they do not conflict for tokens and cannot disable each other. This represents *concurrency* between $DTACK-$ and $LDS-$. In total, there are four pairs of concurrent transitions: $(DTACK-, LDS-)$, $(DTACK-, LDTACK-)$, $(DSr+, LDS-)$, and $(DSr+, LDTACK-)$, where concurrency is a potential to fire at the same time.

1.3 State graphs

Playing the token game one can generate a *Transition System (TS)* – an abstract state graph in which each arc between a pair of states is labeled with the corresponding fired transition. Figure 4 depicts a TS for the READ cycle. Each state in the TS generated from a PN corresponds to a marking, which is shown at the left from the corresponding state. A TS with states labeled with markings is called a *reachability graph* of a PN. For Signal Transition Graphs each state of the corresponding TS also can be associated with a binary code of signal values, which are shown at the right from the states[1]. A TS with states labeled with binary codes of signals is called a *state graph* of an STG.

1.4 Choice and arbitration

The environment of the device has a choice to request the read or the write operation. Similarly, if an arbitration within the device is involved, then the device itself can internally make a non-deterministic choice between two requests. Choice is expressed in PNs by choice places as shown in Figure 5. Here places p_0 and p_3 are choice places, places p_1 and p_2 merge alternative branches of the behavior and all other places are removed from the figure, since they have only one input and one output arc (they are called implicit places

[1]for the sake of readability we separate with dots left handshake signals, right handshake signals, and data transceiver control signal; enabled signals are marked with an asterisk.

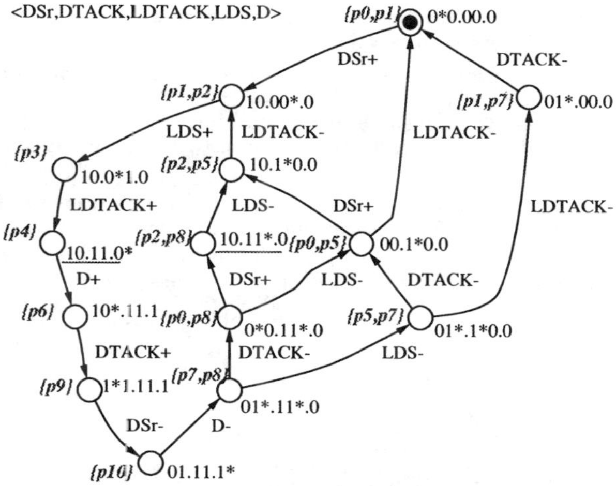

Figure 4: RG and SG for the READ cycle

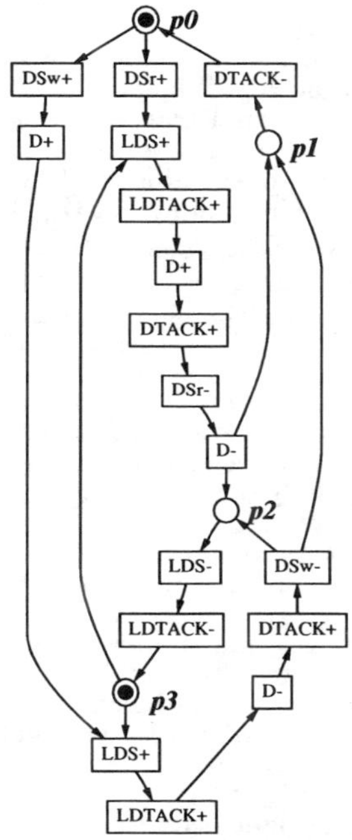

Figure 5: STG for READ and WRITE cycles

and are represented by arcs between two transitions). In the initial marking $\{p_0, p_3\}$ two input transitions are enabled, $DSw+$ and $DSr+$, but as soon as one of them fires another becomes disabled, since the token disappears from place p_0.

2 Analysis and verification

2.1 Properties

Analysis and verification are used at different stages of design.

- *Property verification.* After specifying the design it is required to check implementability properties to answer the following question: "Can the specification be implemented with an asynchronous circuit?" [14, 16]. Other properties of the specification can be of interest as well, e.g., absence of deadlocks, fairness in serving requests, etc. General purpose verification techniques can be employed for this analysis [19].

- *Implementation verification.* After design is done fully automatically or (especially) with some manual intervention it is often desirable to check that the implementation is correct with respect to the given specification [10, 24].
- *Performance analysis and separation between events* is required (a) for determining latency and throughput of the device and (b) for logic optimization based on timing information [12, 22] (see also Section 5).

Properties required for implementability include:

- *boundedness* of the PN to guarantee that the specified state space is finite;
- *consistency* of an STG to ensure that rising and falling transitions alternate for each signal;
- *completeness of state encoding* to check that there are no conflicts in definition of Boolean functions for each non-input (i.e. output and internal) signals;
- *persistency* of the STG to verify that (a) no *non-input* signal transition can be disabled by another signal transition and (b) no *input* signal transition can be disabled by a non-input signal transition. The former ensures that no short glitches, known as *hazards*, can appear at the gate outputs, while the latter ensures that no hazards can occur at inputs of the device.

If all the above properties are satisfied, then the STG specification can be implemented as a, so-called, *speed-independent* circuit [20] [2]. Speed-independence means no hazards under any variations of gate delays if variations of some critical wire delays after forks (so-called isochronic forks) stay within reasonable bounds (e.g., within one gate delay).

Let us illustrate two of the above properties with an example. Two states in the TS in Figure 4 are underlined. They correspond to the different markings, $\{p_4\}$ and $\{p2, p8\}$, but their binary codes are equal, 10110. Moreover, enabling conditions in these two states for output signals LDS, and D are *different*. Therefore, the implied value of the next state Boolean function for signal LDS for vector 10110 should be 1 (for the first state) and 0 (for the second state). This is a conflict in the definition of the function. To resolve this conflict two methods can be employed: (a) inserting an additional state signal whose value should distinguish two conflict states or (b) concurrency reduction. In the first case one feasible solution is to insert rising transition of the additional state signal right before $LDS+$ and its falling transition right before $D-$. So conflicting states will be associated with different values of the new state signal. In the second case, a possible solution is to remove the conflicting state $\{p2, p8\}$ from the specification. The environment should usually stay untouched for the compositional reasons, therefore delaying input signals is not allowed. Hence, signal transition $DTACK-$ can be delayed until $LDS-$ fires. The automatic techniques for solving the state encoding problem are presented, e.g., in [6, 27].

To illustrate the persistency property let us consider transitions $DSw+$ and $DSr+$ in Figure 5 assuming for a moment that they are *output signals* to be implemented. Both are simultaneously enabled and disable each other after firing. Such behavior cannot be implemented without hazards unless special mutual exclusion elements (arbiters) are used.

2.2 Techniques

There are several techniques for fighting with the "state explosion problem" in analysis of Petri Net-like specifications.

- Symbolic Binary Decision Diagram-based (BDD) [3] traversal of a reachability graph allows its implicit representation which is generally much more compact than an explicit enumeration of states [24].
- Partial order reductions ([11], stubborn sets [26], identification method [14]) ignores many (or even most) of the states for analysis of certain properties.
- Structural properties of PNs (e.g., place invariants) can provide fast upper approximation of the reachability space [21, 9] and also can be used for dense variable encoding of states in the reachability graph. Structural reductions are useful as a preprocessing step in order to simplify the structure of the net before traversal or analysis, keeping all important properties.
- Unfoldings [19, 16] are finite *acyclic* prefixes of the PN behavior, representing all reachable markings. They are often more compact than the reachability graph and due to the acyclic property are well-suited for extracting ordering relations between places and transitions (concurrency, conflict and preceding). Different types of unfoldings are also used for performance analysis [12].

More details on the applicability of these techniques can be found in [13].

3 Logic Synthesis

The goal of logic synthesis is to derive a gate netlist that implements the behavior defined by the specification. For simplicity, we will illustrate this step by synthesizing a speed-independent circuit for the read cycle of the VME bus (see Figure 3).

The main steps in logic synthesis are the following:

- Encode the SG in such a way that the complete state coding property holds. This may require the addition of internal signals.
- Derive the *next-state* functions for each output and internal signal of the circuit.
- Map the functions onto a netlist of gates.

3.1 Complete State Coding

As mentioned in Section 2.1, the SG of Figure 4 has state conflicts. A possible method to solve this problem is to insert new state signals that disambiguate the encoding conflicts. Figure 6 depicts a new SG in which a new signal, *csc0*, has been inserted. Now, the next-state functions for signals LDS and D can be uniquely defined. The insertion of new signals must be done in such a way that the resulting SG preserves the properties for implementability.

3.2 Next-State Functions

When an SG fulfills all the implementability properties, a next-state function can be derived for each non-input signal.

Given a signal z, we can classify the states of the SG into four sets: positive and negative *excitation regions* ($\mathsf{ER}(z+)$ and $\mathsf{ER}(z-)$) and *quiescent regions* ($\mathsf{QR}(z+)$ and $\mathsf{QR}(z-)$).

A state belongs to $\mathsf{ER}(z+)$ if $z = 0$ and $z+$ is enabled in that state. In this situation, the value of the signal is denoted by 0^* in the SG. A state belongs to $\mathsf{QR}(z+)$ if s in stable 1 state. These definitions are analogous for $\mathsf{ER}(z-)$ and $\mathsf{QR}(z-)$.

The next-state function for a signal z is defined as follows:

$$f_z(s) = \begin{cases} 1 & \text{if } s \in \mathsf{ER}(z+) \cup \mathsf{QR}(z+) \\ 0 & \text{if } s \in \mathsf{ER}(z-) \cup \mathsf{QR}(z-) \\ - & \text{otherwise} \end{cases}$$

[2] Also called *quasi-delay-insensitive* in the literature [18, 2]

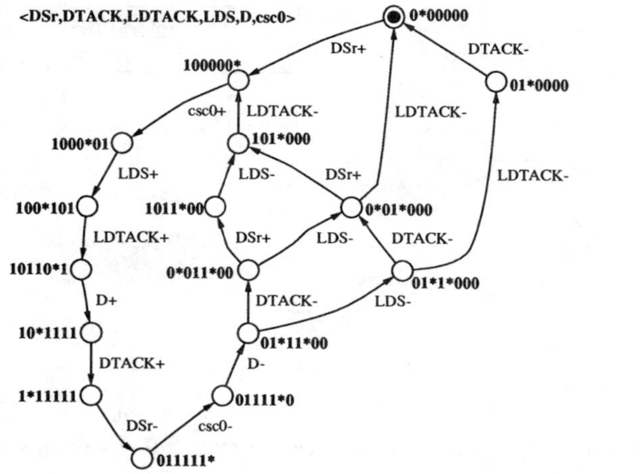

Figure 6: SG for the READ cycle with complete state coding.

where s denotes the binary code of a state. The fact that $f_z(s) = -$ indicates that there is no state with such code in the SG and, thus, s can be considered as a *don't care* condition for boolean minimization.

Once the next-state function has been derived, boolean minimization can be performed to obtain a logic equation that implements the behavior of the signal. In this step it is crucial to make an efficient use of the don't care conditions derived from those binary codes not corresponding to any state of the SG. For the example of Figure 6, the following equations can be obtained:

$$D = LDTACK \cdot csc0; \quad LDS = D + csc0$$
$$DTACK = D; \quad csc0 = DSr \cdot (csc0 + \overline{LDTACK})$$

A well known result in the theory of asynchronous circuits is that any circuit implementing the next-state function of each signal with only one atomic complex gate is speed independent. By atomic gate we mean a gate without internal hazardous behavior [14, 17]. Two possible hazard-free gate mappings for the next-state function of the READ cycle example are shown in Figure 7,a and b.

However, there could be two obstacles in the actual implementation of the next state functions: (a) a logic function can be too complex to be mapped into one gate available in the library; (b) the solution requires the use of gates which are not typically present in standard synchronous libraries. The second is the case with solution Figure 7,a. A gate pictured as a circle with "C" is a so-called C-element [20]: a popular asynchronous latch with the next state function $c = ab + c(a + b)$. Its output, c, goes high (low) if both inputs, a and b, go high (low); otherwise, it keeps the previous value.

3.3 Hazards

A crucial problem which makes solution of logic decomposition problem for asynchronous design difficult is a problem of *hazards* [25, 23]. Recent development in [23] shows that if the so-called *Fundamental mode* is acceptable (input cannot change until all internal circuit activity stabilizes), then most of the known methods of logic minimization can be gracefully extended to asynchronous hazard-free minimization. These results can further be extended to FSMs [29].

Unfortunately, the Fundamental mode is often too restrictive and in particular is not satisfied for logic implementing signal functions in synthesis using STGs.

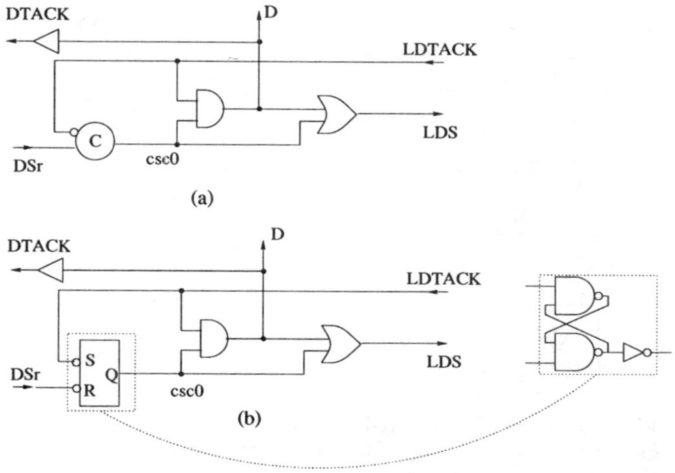

Figure 7: Implementations with latches

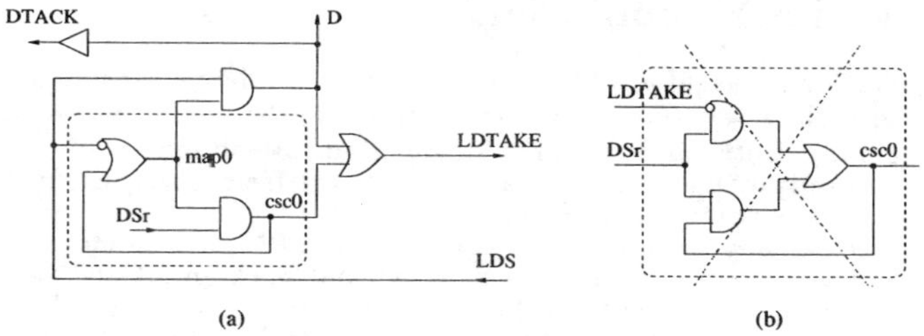

Figure 8: Implementation with two-input gates

3.4 Decomposition and Technology Mapping

One of the partial solutions to the logic decomposition for non-fundamental mode, called the monotonous cover requirement [1, 15], allows one to decompose any function into two-level combinational logic and a latch. This does not solve however a problem of breaking gates if the fan-in or fan-out is too large. The latest results [4, 5] allow one to obtain a hazard-free decomposition (and then map the decomposed solution into the available library) without [4] or with [5] gate sharing into gates with restricted fan-in.

Applying method from [5] two other correct solutions can be found for mapping the control for READ cycle into two inputs gate library: solution in Figure 7,b uses a standard reset dominant RS-latch instead of the C-element; solution in Figure 8,a uses *only* combinational gates. This solution seems to be a standard synchronous decomposition for the function of signal $csc0 = DSr \cdot (csc0 + \overline{LDTACK})$:

$$map0 = csc0 + \overline{LDTACK}; \quad csc0 = DSr \cdot map0$$

Note, however, that signal $map0$ is also fed to gate $D = LDS \cdot map0$. It is only because of this multiple acknowledgment of $map0$ by two different gates, that this solution for the READ cycle control is hazard-free: every rising transition at $map0$ is acknowledged by signal D, while every falling transition – by signal $csc0$. Another synchronous decomposition for $csc0$ presented in Figure 8,b is hazardous and cannot be accepted.

The technique for decomposition and technology mapping from [5] is based on using candidates for decomposition extracted by algebraic factorization and boolean relations and inserting hazard-free signals with multiple acknowledgment.

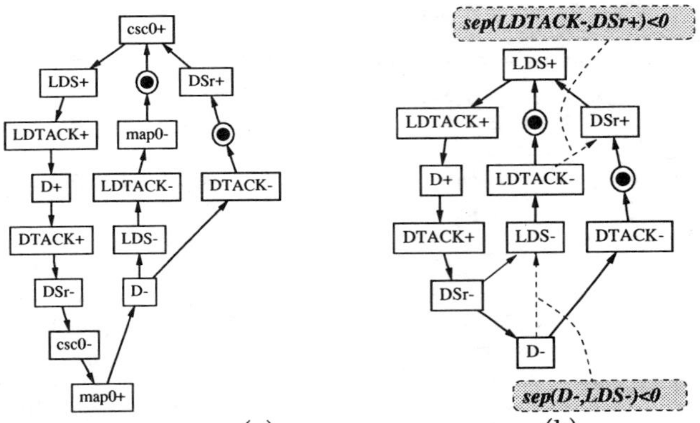

(a) (b)

Figure 9: (a) STG extracted for the two-input combinational gate circuit, (b) timing STG with separation constraints for the optimized circuit

4 Back annotation

State regions [8] are sets of states such that they correspond to a place (*regions*) or a transition of the PN (*excitation regions*). Entry and exit arcs for a region correspond to input and output transitions of a place. Apart from being useful for state exploration regions provide another important feature: at any step of the design process a PN corresponding to the current TS can be extracted and back-annotated to the designer. This is useful both for interactions with the design process and for the performance and timing analysis of the circuit. An example of a PN extraction is shown in Figure 9,a.

5 Timing Optimization

The power of optimization based on timing information is two-fold.

- Timing constraints always *reduce* the set of reachable states and hence increase the number of don't care states [22]. Moreover this concurrency reduction does not introduce new dependencies between signals since it is fully based on timing not on logic ordering.
- Using timing requirements it is possible to *extend* the set of states in which signal is enabled *without* changing the set of reachable states: signal transition enabling does not cause signal firing if other enabling signals are known to be (or can be made) faster.

Let us illustrate how timing information can increase the flexibility in logic optimization by example of the READ cycle. Assume first that, as a part of the initial specification, it is given that the reset at the right side handshake is always faster than the next read request at the left side handshake, formally: maximal separation[12] between transitions $LDTACK-$ and $DSr+$ is negative, *Sep*($LDTACK-, DSr+$) < 0. Then there is no need in the additional state encoding signal and the circuit is simplified to Figure 10,a.

Assume next that the physical design level tools achieve control over the delay information using gate and transistor sizing, placement and routing, and constraining interconnect delays. Then the logic-level synthesis tools can perform logic optimization at the same time generating separation constraints that must be implemented by the physical level tools. For example, it is possible to start enabling of $LDS-$ right after $DSr-$ instead of $D-$ given that the requirement *Sep*($D-, LDS-$) < 0 will be satisfied. This requirement is satisfied if the maximal delay of $D-$ is smaller than the minimal possible delay of $LDS-$ that can be implemented, e.g., by transistor sizing or delay padding. The resulting circuit corresponding to both timing requirements is shown in Figure 10,b. Back-annotation to an extended PN with relational timing constraints (so-called *lazy* PNs) can be done for the circuits optimized based on timing information (see Figure 9,b).

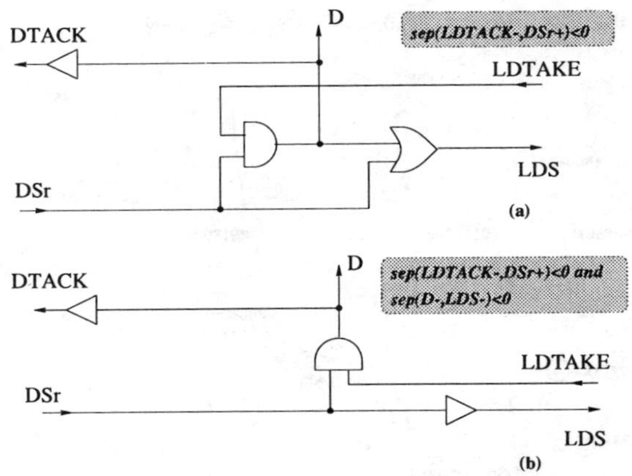

Figure 10: Circuits for the READ cycle after timing optimization

6 Other Design Techniques

This paper has presented a design methodology based on Petri net specifications of the behavior of a circuit. However, other models have been proposed in the literature. Among them, we can point up the methods based of burst-mode machines [29] and on syntax-directed [2] or transformation-based [18] translation from process algebras.

Burst-mode machines work under the so-called *fundamental mode* assumption, i.e. after each burst of inputs events accepted by the system, the environment allows the circuit to stabilize before reacting to the output events. This assumption is realistic for many applications and enables the utilization of combinational logic minimization methods for synchronous circuits with ad-hoc extensions to prevent hazardous behavior [23].

Translation from process algebras has been proposed for formalisms derived from CSP. Syntax-directed translation derives a netlist of components that implement the behavior of each of the constructs of the language (parallel/sequential composition, choice, communication, synchronization, etc.). The size of the resulting circuit is linearly dependent on the size of the input description. This fact enables designers and tools to predict the circuit's performance and complexity parameters at the earliest steps of the design process.

Other efforts have been devoted to map asynchronous specifications into standard HDLs aiming at the simulation and validation with commercial tools [28].

7 Summary

In the last few years, the techniques for asynchronous designed have matured. Among the applications for asynchronous design we can point up asynchronous interfaces, high-performance computing, low-power and low-emission design, etc. There are also applications at the system level, e.g. hardware-software co-design.

Recently there has been an increasing interest of few but large-scale industries (e.g. Intel, Philips, Sharp, ARM, Cogency, SUN, HP) in asynchronous design targeting at different goals: low power, high performance, etc.

Asynchrony introduces a new paradigm in logic design. Asynchronous circuits are much more difficult to design and, for this reason, it is crucial to provide CAD tools to handle the most difficult tasks automatically. Most of the steps of the design process presented in this tutorial are supported by the tool `petrify` available at `http://www.lsi.upc.es/~jordic/petrify`.

For a more complete tutorial in PN-based design of asynchronous control circuits we refer to [7]. For further information on asynchronous design, the reader can look at the Asynchronous Logic Home Page (`http://www.cs.man.ac.uk/amulet/async/index.html`) and the proceedings of the ASYNC Symposiums.

An extended version of this paper can be found in [13].

References

[1] P. A. Beerel and T. H-Y. Meng. Automatic gate-level synthesis of speed-independent circuits. In *Proceedings of the International Conference on Computer-Aided Design*, November 1992.

[2] Kees van Berkel. *Handshake Circuits: an Asynchronous Architecture for VLSI Programming*, volume 5 of *International Series on Parallel Computation*. Cambridge University Press, 1993.

[3] Randal Bryant. Symbolic boolean manipulation with ordered binary-decision diagrams. *ACM Computing Surveys*, 24(3):293–318, September 1992.

[4] S. Burns. General conditions for the decomposition of state holding elements. In *International Symposium on Advanced Research in Asynchronous Circuits and Systems, Aizu, Japan*, March 1996.

[5] J. Cortadella, M. Kishinevsky, A. Kondratyev, L. Lavagno, E. Pastor, and A. Yakovlev. Decomposition and technology mapping of speed-independent circuits using boolean relations. In *Proceedings of the International Conference on Computer-Aided Design*, pages 220–227, November 1997.

[6] J. Cortadella, M. Kishinevsky, A. Kondratyev, L. Lavagno, and A. Yakovlev. A region-based theory for state assignment in speed-independent circuits. *IEEE Transactions on Computer-Aided Design*, 16(8):793–812, August 1997.

[7] J. Cortadella, M. Kishinevsky, A. Kondratyev, L. Lavagno, and A. Yakovlev. Synthesis of control circuits from STG specifications. In handouts of the Summer School on Asynchronous Circuit Design, August 1997. `http://www.lsi.upc.es/~jordic/petrify/refs/summer97.ps.gz`.

[8] J. Cortadella, M. Kishinevsky, L. Lavagno, and A. Yakovlev. Synthesizing Petri nets from state-based models. In *Proceedings of the International Conference on Computer-Aided Design*, pages 164–171, November 1995.

[9] J. Desel and J. Esparza. *Free-choice Petri Nets*, volume 40 of *Cambridge Tracts in Theoretical Computer Science*. Cambridge University Press, 1995.

[10] David L. Dill. *Trace Theory for Automatic Hierarchical Verification of Speed-Independent Circuits*. ACM Distinguished Dissertations. MIT Press, 1989.

[11] P. Godefroid. Using partial orders to improve automatic verification methods. In E.M Clarke and R.P. Kurshan, editors, *Proc. International Workshop on Computer Aided Verification*, 1990. DIMACS Series in Discrete Mathematica and Theoretical Computer Science, 1991, pages 321-340.

[12] H. Hulgaard, S. M. Burns, T. Amon, and G. Borriello. An algorithm for exact bounds on the time separation of events in concurrenct systems. *IEEE Transactions on Computers*, 44(11):1306–1317, November 1995.

[13] M. Kishinevsky, J. Cortadella, A. Kondratyev, and L. Lavagno. Asynchronous interface specification, analysis and synthesis. Technical Report LSI-98-14-R, Technical University of Catalonia, March 1998. `http://www.lsi.upc.es/dept/techreps/ps/R98-14.ps.gz`.

[14] M. A. Kishinevsky, A. Y. Kondratyev, A. R. Taubin, and V. I. Varshavsky. *Concurrent Hardware. The Theory and Practice of Self-Timed Design*. John Wiley and Sons Ltd., 1994.

[15] A. Kondratyev, M. Kishinevsky, B. Lin, P. Vanbekbergen, and A. Yakovlev. Basic gate implementation of speed-independent circuits. In *Proceedings of the Design Automation Conference*, pages 56–62, June 1994.

[16] A. Kondratyev, M. Kishinevsky, A. Taubin, and S. Ten. Analysis of Petri nets by ordering relations in reduced unfoldings. *Formal Methods in System Design*, 12(1):5–38, 1997.

[17] L. Lavagno and A. Sangiovanni-Vincentelli. *Algorithms for synthesis and testing of asynchronous circuits*. Kluwer Academic Publishers, 1993.

[18] A. Martin. Programming in VLSI: From communicating processes to delay-insensitive circuits. In C. A. R. Hoare, editor, *Developments in Concurrency and Communications*, The UT Year of Programming Series. Addison-Wesley, 1990.

[19] K. McMillan. *Symbolic Model Checking*. Kluwer Academic Publishers, 1993.

[20] David E. Muller and W. S. Bartky. A theory of asynchronous circuits. In *Proceedings of an International Symposium on the Theory of Switching*, pages 204–243. Harvard University Press, April 1959.

[21] T. Murata. Petri Nets: Properties, analysis and applications. *Proceedings of the IEEE*, pages 541–580, April 1989.

[22] Chris J. Myers and Teresa H.-Y. Meng. Synthesis of timed asynchronous circuits. *IEEE Transactions on VLSI Systems*, 1(2):106–119, June 1993.

[23] Steven M. Nowick and David L. Dill. Exact two-level minimization of hazard-free logic with multiple-input changes. *IEEE Transactions on Computer-Aided Design*, 14(8):986–997, August 1995.

[24] Oriol Roig, Jordi Cortadella, and Enric Pastor. Verification of asynchronous circuits by BDD-based model checking of Petri nets. In *16th International Conference on the Application and Theory of Petri Nets*, volume 815 of *Lecture Notes in Computer Science*, pages 374–391, 1995.

[25] S. H. Unger. *Asynchronous Sequential Switching Circuits*. Wiley-Interscience, John Wiley & Sons, Inc., New York, 1969.

[26] Antti Valmari. Stubborn sets for reduced state space generation. *Lecture Notes in Computer Science; Advances in Petri Nets 1990*, 483:491–515, 1991.

[27] P. Vanbekbergen, B. Lin, G. Goossens, and H. De Man. A generalized state assignment theory for transformations on Signal Transition Graphs. *Journal of VLSI Signal Processing*, 7(1-2):101–116, 1994.

[28] Peter Vanbekbergen, Albert Wand, and Kurt Keutzer. A design and validation system for asynchronous circuits. In *Proc. ACM/IEEE Design Automation Conference*, June 1995.

[29] K. Y. Yun and D. L. Dill. Automatic synthesis of 3D asynchronous state machines. In *Proceedings of the International Conference on Computer-Aided Design*, November 1992.

Automatic Synthesis of Interfaces between Incompatible Protocols

Roberto Passerone
Department of EECS
University of California at Berkeley
roby@eecs.berkeley.edu

James A. Rowson
Alta Group of Cadence
Sunnyvale, California
jimr@altagroup.com

Alberto Sangiovanni-Vincentelli
Department of EECS
University of California at Berkeley
alberto@eecs.berkeley.edu

Abstract

At the system level, reusable Intellectual Property (or IP) blocks can be represented abstractly as blocks that exchange messages. The concrete implementations of these IP blocks must exchange the messages through complex signaling protocols. Interfacing between IP that use different signaling protocols is a tedious and error prone design task. We propose using regular expression based protocol descriptions to show how to map the message onto a signaling protocol. Given two protocols, an algorithm is proposed to build an interface machine. We have implemented our algorithm in a program named PIG that synthesizes a Verilog implementation based on a regular expression protocol description.

1 Introduction

As time to market pressures and product complexities climb, the pressure to reuse complex building blocks (also known as Intellectual Property, or IP) also increases. Today, most IP is available only at the RTL level. This is problematic because of verification speeds and the variety of signaling conventions used to for interfacing.

This paper addresses the problem of synthesizing interfaces between communicating IPs that use different signaling conventions. For this problem, a description of the entire behavior of the IP is not only cumbersome, but it introduces unnecessary details that may even hamper the design process. In [11] Interface-Based design was proposed as a methodology that attempts to separate the communication from the behavior for IP. To separate the communication, the blocks must be abstracted to a transaction or messaging level. With abstracted communication, improvement in simulation performance was shown with a simulator named Cheetah. However, the abstraction level that is appropriate for fast simulation is not efficient for implementation. By separating communication and behavior we can have a single abstract model that represents many concrete IPs that have different signaling conventions on their interfaces. For the purposes of this work, we are assuming that the concrete IPs have fixed signaling conventions and that in order to get two IPs to communicate, as specified at the abstract level, we will have to synthesize a machine to convert from one set of conventions (or protocol) to another.

Our problem formulation could then be expressed as: *Given two communicating design actors exchanging data, and a description of the two protocols that each one of them uses to transfer the data, determine an interface so that data transfers are consistent with both protocols.*

A protocol may in general be given regardless of its physical implementation. However, we take a few simplifying assumptions on the implementation in order to limit the size of the design space. Besides, only the first two are really essential and they are often verified in practice.

1. The communication is *point-to-point*, that is the communication media (the interface we want to synthesize) is not shared with other modules in the system. The algorithm could however be used as a building block to a more general approach.

2. The two sides exchange data in the form of the same *data type*; more specifically, the data type is translated into a string of bits, therefore we require that the number of bits that are transmitted be the same as the number of bits that must be received.

3. During the transfer, part of the data will be temporarily stored in the interface: to simplify the solution of the problem we have considered only a very simple implementation of the storage, namely an internal register long enough to store the entire data type.

4. We assume that the modules willing to exchange the data *are driven by the same clock* and are therefore fully synchronous. However, we believe that the algorithm is applicable to asynchronous and event based systems with minor modifications.

As a final remark, it must be pointed out that in this first version of the algorithm we are only considering a single transaction; the extension to a stream of data requires the analysis of the interface storage allocation. Given the assumptions, the output of our algorithm is a *finite state machine* and a *data path* consisting of the internal register. Reading from and writing to the register can be done in a single clock cycle.

DAC 98, San Francisco, California
©1998 ACM 0-89791-964-5/98/06..$5.00

As shown in Figure 1, the communication between an abstract producer and receiver must be implemented in the concrete world using an interface. We will synthesize this interface by using a declarative protocol description that relates the abstract communication to the signaling conventions on our concrete IP.

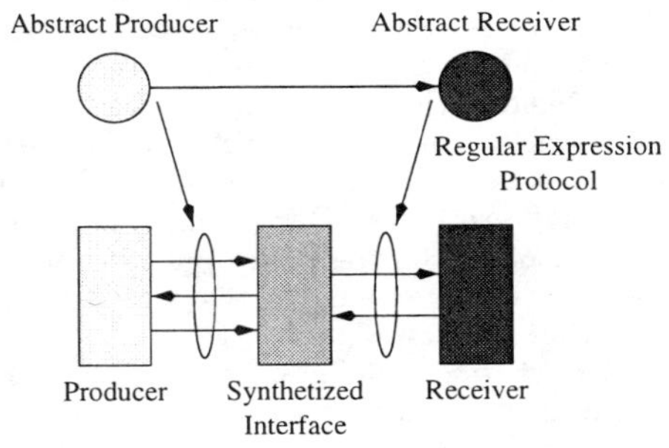

Figure 1: Problem formulation

This paper is organized as follows. First Section 2 gives a brief description of previous and related work; Section 3 describes the input language of our interface synthesis tool, while Section 4 presents the algorithm; some experiments and results are reported in Section 5 and finally Section 6 concludes the paper and presents some line of future work.

2 Related work

The problem of interface synthesis has been addressed in the literature. One of the first work is that of Borriello [3] [4], who introduces the "event graph" to establish the correct synchronization and data sequencing. The limitation of this approach is that the two protocols should be made compatible by manually assigning labels to the data on both sides, since the specification of the protocols is given at a very low level of abstraction using waveforms. Jane Sun [13] extends the approach by providing a library of components that frees the user from considering lower level details, but doesn't solve the problem we mentioned.

A different approach is that taken by Gajski et al. [8]: first, the protocol specification is reduced to the combination of five basic operations (data read/write, control read/write, time delay); the protocol description is then broken into blocks (called *relations*) whose execution is guarded by a condition on one of the control wires or by a time delay; finally the relations of the two protocols are matched into sets that transfer the same amount of data. Although this algorithm is able to account for data width mismatch between the two modules, the procedural specification of the protocols makes it difficult to adapt different data sequencing, so that only the synchronization problem is solved.

A third approach, similar to the one used in this work, is that of Akella and McMillan [2]: the protocols are described as 2 FSMs, while a third FSM represents the valid transfer of data. The product machine is taken and pruned of the invalid/useless states. The limitation here is that no data width mismatch can be handled and that the designer must manually enter the intended behaviour of the interface in the form of the third FSM (called the C-machine).

In our approach some of the mentioned limitations are overcome: the correspondence between pieces of data on the two sides is automatically resolved, the interface can translate between different sequencing of the data and no intended behavior must be introduced to describe the interface process. However, we assume that the two communicating parties are driven by the same clock, a limitation that is not present in [3] and in [8].

As we will see later, the two protocols are described using regular expressions. The use of regular expression for hardware description in general and for protocol description in particular is not new and has been already presented in the literature. The grammar-based specification employed in this work has been inspired mainly by the work of Seawright and Brewer [12] who demonstrated how effective regular expression could be for protocol and control intensive specifications. Hemani et al. also report in [9] the use of grammar-based specification for the synthesis of hardware for data communication protocols, although the specific problem of interface synthesis is not addressed.

The use of derivatives of regular expressions was introduces by Brzozowski [5]. The particular way in which the derivatives are computed in our work resembles that introduced by Coelho et al. in their work on *control-flow expressions* [6].

3 Protocol specification

As stated in the general problem formulation, the input to the algorithm is a description of the protocol used by the two modules. In our case, we assume that each module has a set of ports (data and control) over which the transfer occurs. We define a *protocol* as the *legal sequences of values that may appear on the ports from the onset to the end of the data transfer*.

If we order the ports in some arbitrary way, we can define a *symbol* in the protocol as a tuple composed of the values on the ports listed in their order. For simplicity of specification, ports that represent buses can be bundled together and assigned a single numerical value. Under this assumption, a protocol is simply a set of strings of symbols, or, in other words, a language in the alphabet of all the values that a symbol may assume. Our choice is to describe such a language with regular expressions: this way we can only use regular protocols, but a one to one correspondence can be easily established with finite state machines.

As mentioned before, symbols take values over the set of all possible tuples of values of the ports. If we included in the alphabet all the values that the data can take, then even very simple protocols would be expressed with exponentially growing regular expressions. For example, if a block has a connection to its environment with 8 wires transmitting a byte, then the protocol would be the set of all possible values over 8 wires, namely 256 different strings of 1 symbol. Since the interface is not concerned with the value that the data takes, but only with the control flow of the protocol, we introduce in the regular expression's alphabet a new symbol meaning *any value*. This choice is also unsatisfactory: in case of protocols where data is sequenced over a certain set of wires (e.g. a serial line), the interface must know what part of the data type is currently being transfered. Therefore we introduce a symbol with the meaning: *any value the data type takes on a certain subset of its string of bits*. For simplicity in parsing, we only allow the specification of such a *reference* over intervals (possibly a single bit) on the string of bits representing the data type. For example, if a data type `D` is composed of 20 bits, the syntax `D[10:5]` is a reference to the value of the data type from bit 5 to bit 10. The occurence of a reference in the protocol tells the interface either to store the value in the internal register or to output the value previously stored in the internal register at the specified location.

A very simple programming language describes this information as a text file. To ease the protocol specification, the token can be thought of as a typical composite datatype, either an array or a record structure composing other types. Usually there is a rich set of primitive types available, although in the case of our work we limited the primitives to only one type representing a single bit.

The ports are simply listed with their direction and their data type (a bit or a set of bits). Since a protocol is independent of the direction of the transfer of data, the direction of the port is specified as either `master` or `slave`, the actual direction being resolved when the synthesis of the interface is requested. This way an input device and an output device that use the same protocol share the same description.

Regular expressions can be expressed hierarchically using a regular grammar [1]. In addition to its name, each rule of the grammar has a list of parameters whose value can be specified by the parent expression. In each rule, the references to values (as in `D[10:5]`) can only be expressed in terms of the rules parameters which act as local variables (very much like a function call). For the top level rule of the grammar, the only parameter is the token to be transfered so that, ultimately, all references are made with respect to it.

A symbol in the alphabet of the regular expression is a comma-separated list of values and references enclosed in braces, as in { `0, 4, D[10:5]` } for a 3 port specification. The number of values and their type must match that of the port list for all rules, regardless of their level in the hierarchy of the specification.

Symbols and regular expressions can be composed using any combination of the standard operators, including $*$ for Kleene closure (0 or more of the referenced symbol), $+$ for semi-closure (1 or more), | for choice and a comma for sequential expressions. Recursion is not allowed, except in the form of tail recursion when the Kleene closure operator is used.

The following two examples show how the token `yow` can be mapped to two different protocols:

```
type byte bit[7:0];
type yow { byte a; byte b; }

protocol serial of type yow {

 master bit start, byte bus;

 term null() { 0, - }
 term one(byte b) { 1, b }
 term two(byte b) { 0, b }

 serial(yow y) { null()*,one(y.a),two(y.b) }
}

protocol handshk of type yow {

 master bit trigger, byte bus

 term wait(bit t) { t, - }
 term get(bit t, byte b) { t, b }

 handshk(yow y) { wait(0)*,get(1,y.a)+,get(0,y.b)+ }
}
```

This first snippet of input code defines a protocol for type `yow` named `serial`. The interface being defined has two ports that can be used within the protocol, a pin named `start` and byte-wide bus named `bus`. Both of these ports are driven by the master side of the interface. There are three sub-rules defined in the grammar: `null`, `one`, and `two`. Each of these patterns must provide a value for all the ports, with don't care being represented by a dash. Parameters are listed with their data type after the name of the rule. The top level rule is a regular expression composition of calls to the terminal rules. In this case, `null` is expected 0 or more times (a Kleene closure), followed sequentially by a `one` with the `a` field of the `yow` token, followed by a `two` with the `b` field. As mentioned earlier, rules can be nested hierarchically so that `serial` could be used as a building block for more complex protocols. Our example `serial` protocol waits for a value of one on the `start` pin with an associated byte on `bus`, followed immediately by a zero on the `start` pin and the other byte on `bus`.

The second input code defines protocol `handshk` for type `yow`. Here the protocol starts with `trigger` with value 0. Byte `a` is transfered when `trigger` goes to a 1 and byte `b` is transferred when `trigger` goes back down to a 0. Notice the difference with the previous protocol: in this case the time spent on the first byte is not known, whereas in the previous case it was specified as a single clock cycle. In the first example the pin `start` only identifies the start of the transfer, in the second example the change of value in `trigger` identifies the transfer of the first *and* of the second byte.

4 Synthesis Algorithm

The goal of the synthesis algorithm is to obtain a finite state machine that when placed between the two modules implementing the specified protocols would make the communication possible. The problem is that of recognizing a given regular language on the producer side and to generate a proper string contained in the other regular language on the other side; the difficulty of the problem is maintaining the same "meaning" (i.e. preserving the data contents of the message) while trying to optimize some parameters (the transfer latency, the size of the storage in the interface process, etc.).

The approach that we will undertake is the following: first the regular expressions representing the two protocols are translated into two automata that recognize the corresponding regular language. These two automata form the bulk of the interface. Then the product of the two automata is taken so that only the legal sequence of operations is retained, the signals are translated into inputs and outputs, and the non-determinism that arises is resolved following one or more of the following rules:

1. never output a piece of data that has not yet been received

2. transfer the data

3. minimize (optimize) the latency

Any remaining non determinism is broken usin arbitrary choices based on the order the states are generated.

A simple example will help illustrate the construction of the product machine. Consider the two protocols described in section 3: their corresponding finite automata are shown in Figure 2. Suppose we wish to transfer data from the `handshk` to the `serial` protocol. The product machine construction starts from the state corresponding to the pair of initial states. Following that, all possible transitions are explored to see which ones should be included in the product.

Figure 3 shows the process as it evolves in the product machine exploration.

From the initial states (Figure 3.a) the set of all possible transitions includes a loop to the initial state itself or a transition to a pair of states where either one of the two automata has advanced one position. The pair of states shown on the right (Figure 3.b) is an instance of a forbidden transition, because we are trying to output the first byte that has not been received yet. That transition is therefore not included in the product machine. The transition on the left (Figure 3.c) can be safely taken since the first byte is received and nothing is produced.

From this state (and also from the initial state) we could move to the pair of states shown in the right transition (Figure 3.d): although this pair of states is legal, the serial protocol requires that the product machine move on the second shaded pair of states even when the second byte is not received, which is not legal. Since we have no control over this transition, then also the legal state should be considered illegal and the transition should not be taken. This condition cannot be detected immediately, and therefore it implies a backtrack in the algorithm. The alternative is therefore either to loop if the second byte is not received or to move forward with the left transition when it arrives. In the second case (Figure 3.e), the product machine allows a non deterministic choice: we can either transition to the first shaded pair or directly to the second shaded pair. Since our objective is to minimize the latency, the final product machine shown in Figure 4 moves directly to the second pair before reaching the final state that concludes the transaction.

The following paragraphs describe each step in detail.

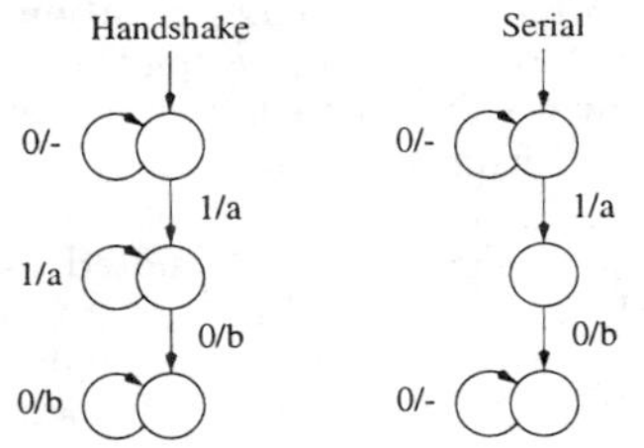

Figure 2: Finite automata

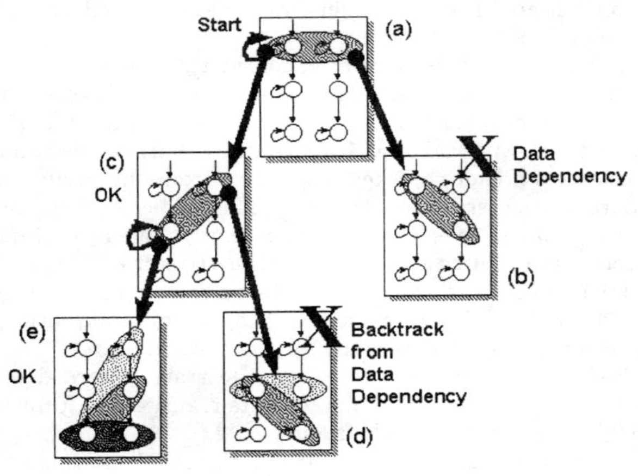

Figure 3: Product machine computation

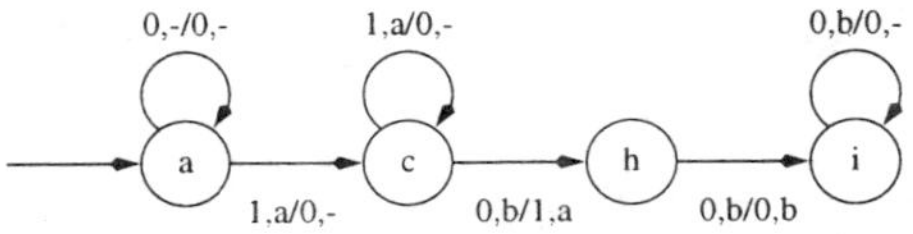

Figure 4: Final result

4.1 Automata generation

To translate the regular expression into a finite automaton we follow an approach based on the derivatives of regular expressions ([5]). A different algorithm would first build a non-deterministic automaton and then use the subsets construction to obtain the deterministic version (as described in [7]). The advantage of the derivative approach is that at each step we know how far we are in the protocol transaction, since we start constructing the automaton from the initial state rather than from some unspecified internal state. Exploiting this feature, we are able to characterize each state with the amount of data that has been transferred along the ports, an information that will be essential during the subsequent product machine construction.

As pointed out in [5], the challenge in computing the derivatives is recognizing whether two regular expressions (two derivatives) represent the same language, even though expressed in different form. In our algorithm, we represent the derivative of the regular expression with respect to a certain string as the set of symbols in the regular expression itself that may follow the given string (obtained by traversing a tree representing the regular expression using the operators as guides). Clearly, two identical sets of symbols represent the same derivative, thus the check for equivalence is reduced to a check of equality between sets; unfortunately the converse is not true, so that the worst case complexity of the algorithm is exponential in the number of symbols in the regular expression (the same as the subsets construction), and the finite automaton that is obtained may not be minimum. However, in all the practical cases that we have run through the algorithm, the state explosion was not observed: in fact this algorithm can easily factor any common term found at the beginning of the different branches of a choice operator, or detect the reconvergence after the choice was taken, thus following the style that designers naturally employ when specifying a protocol. More details can be found in [10].

4.2 Product Computation Algorithm

At the end of the previous step we have two deterministic finite automata that recognize a transaction on either side of the interface. The algorithm then proceeds with the construction of a subset of the product machine that performs the correct transfer of data. Moreover, the input/output nature of the signals is taken into account, so that a finite state machine rather than a finite automaton is eventually created.

An important issue that should be carefully examined is non determinism. The algorithm starts with two deterministic finite automata, so that the product is still a deterministic finite automaton. However, when signals are partitioned into inputs and outputs and a finite state machine is built, only the input set contributes to the condition on each arc, and therefore non determinism may arise (these machines are known as pseudo non deterministic, in that their corresponding finite automaton is deterministic). An example will help clarify this concept: if we consider a

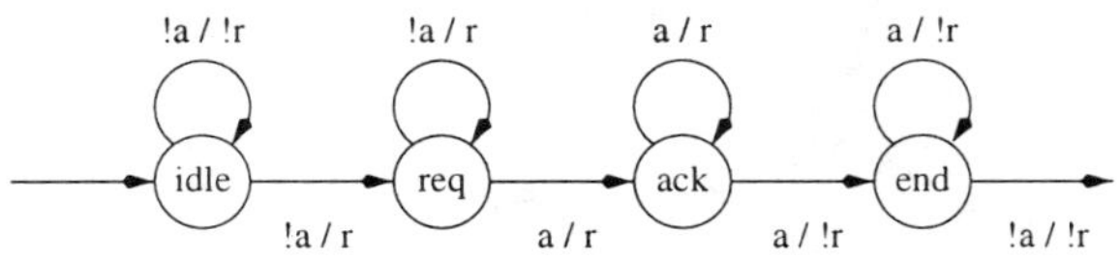

Figure 5: Non deterministic finite state machine

request-acknowledge protocol, the initial finite automaton would contain both the request and the acknowledge signals as input. The specification usually doesn't define the number of clock cycles intervening between the edges of these signals; when an FSM is built using this model, for example on the receiver side of the interface, the request signal is considered as an output and therefore the transition between an idle state and a request state, or between an acknowledge state and an end state (see Figure 5), becomes non deterministic. Obviously the final implementation must be deterministic, so that any choice must be taken at "compile" time. The nature and quality of these choices directly affect the quality of the interface in terms of correctness, latency, storage requirements and therefore performance. It is in fact the presence of non determinism that gives us the degrees of freedom necessary in the optimization process.

There can be basically two approaches to construct the required subset of the product machine: start from nothing and add states, or start from the entire product machine and remove the unwanted states. As shown in the simple example earlier in this section, in this implementation of the algorithm we follow the first approach (although a version using the reverse approach is under consideration). The algorithm systematically explores all paths going from the initial state to the final states, removing those that are not permissible. Considering the way non determinism is resolved, the following statement can be made:

Proposition 1 The Product Computation Algorithm always finds the correct interface between two protocols, if one exists. In addition, it always returns the minimum latency interface.

A more formal proof of this statement can be found in [10].

The algorithm that we use is a depth first recursive search implemented by a procedure called `explore`, outline in Figure 6. The process is started by the creation of the initial pair of states and a call to its `explore` function. Three data structures are used to support the computation: a *stack* records all the states that have been visited along the path to the current state, used to detect loops in the product machine; a *pool* records all the states that were found to be illegal either because of data inconsistency or because they lead to data inconsistency (this data structure acts like a cache that prevents the algorithm from re-exploring paths that are already known to be inconsistent); an *FSM* is used to collect all the states that will eventually be part of the product machine.

For each pair of states, i.e. a state in the product machine, a pair of bitsets is used to record the amount of data that has been received and that has been sent, or has to be sent. These values are updated each time a transition is taken between two pairs of states, and is used to check the data consistency.

The `explore` function returns an object to the caller which may assume different symbolic values: **Success** means that the state will certainly lead to a succesful transfer of the data and a companion number indicates the minimum

```
explore( ) {
  If on stack return ImmediateLoop
  If inconsistent return Fail
  If already in the FSM return previous result
  If previously explored and failed return Fail
  Push state on the stack
  For all pairs of outgoing transitions {
    Compute new state
    explore( new state )
  }
  Resolve non-determinism
  Compute return result
  Update data structures
  return
}
```

Figure 6: Function `explore`

number of clock cycles that it takes to get to the end of the transaction; **Fail** means that the state is either illegal or leads to an illegal state; **ImmediateLoop** indicates that the state has already been explored whereas **FutureLoop** that the state leads to a loop in one of its future transitions; **LoopSuccess** is returned when the state may either lead to a loop or to a succesful transfer of the data (depending on the behavior of the outside environment). In order to compute the return value, `explore` starts by checking if the state was already explored and returns immediately to avoid multiple computations. Otherwise the state is pushed on the stack and for each pair of outgoing transitions the new state is created and explored (after updating the data consistency bitsets). The return value of each exploration is stored and the state is popped from the stack.

Then the set of outgoing transitions is partitioned into equivalence classes: each class is denoted by the same input label (i.e., the non deterministic transitions are grouped together). For each equivalence class, only one transition is chosen to be part of the final implementation. Here is where choices can be taken to resolve non determinism and to optimize performance. In particular, a transition whose exploration returned a **Success** or **LoopSuccess** is always chosen against a **Loop** or a **Fail**. Ties between successful transitions are broken considering the number of clock cycles to the end of the transaction.

The final result is computed considering the exploration result from all the transitions that survived the determinization. Since now all transitions are deterministic, a **Fail** in any of them will make `explore` return a **Fail** to the calling function even though other transitions from the same state lead to a successful transaction. This is because we now have no control over the transition that is taken, therefore correctness requires us never to reach the state. If there is no **Fail**, then either **Success** or **LoopSuccess** is returned with a number of clock cycle equal to the minimum over all transitions plus 1 (to account for the present state).

Before returning, if successful, the state is recorded in the FSM data structure along with its transitions. If unsuccessful, it is inserted in the failing pool.

When the last `explore` returns the FSM contains the product machine. Since states are added to the FSM starting from the last state (because of the recursive call), we may add states that are unreachable from the start state. A final clean up procedure takes care of removing all dangling states.

From - To	States	Rec.	Full Rec.
handshake - serial	4	14	5
serial - handshake	3	13	5
atm ser. - atm req-ack	1438	2983	1490
atm req-ack - atm ser.	111	3416	1680
atm ser. rev - atm req-ack	116	235	116

Table 1: Experimental results

5 Experimental results

The algorithm has been implemented in about 4500 lines of Java code in a program called PIG. PIG is able to load the textual description of the protocols, parse them, create the finite automata with the derivatives approach and produce the subset of the product machine as previously outlined. In addition to that, the Verilog code in the form of a finite state machine can be generated to be simulated with the corresponding models of the producer and the consumer.

A few examples have been run through the entire process. Table 1 shows a summary of some of the performance data. In particular we are interested in the number of states that are visited during the product machine construction. The columns *Recursion* and *Full Rec.* show the number of times that the `explore` function was called during a computation, and the number of times that the explored state was not found in any of the auxiliary data structures. Although the number of explorations can be more than the total number of states in the product machine, full exploration is necessary only for far less states.

Example `atm` is a large case exapmple where a serial synchronous protocol was interfaced to a 4 phase request-acknowledge protocol and vice versa for a packet consisting of 56 bytes. The serial side transfers data 8 bits at a time, while the request-acknoledge uses a 32 bit bus. Both protocols result in a finite automaton with 57 states each. In all cases PIG produces the correct result and the latency is minimized. The first case contains many states because the output of the data is concurrent with the input of the data in the interface. On the other hand, in the second case the interface needs to wait until all the data from the request-acknoledge protocol has been received (because it must be output serially without interruption), and only then can the output start: in this case less choice means less states, but also a higher latency. As an experiment, we've tried reversing the order of the data of the serial protocol in a serial to req-ack interface: in this case, too, the interface has to wait till the end of the input phase before starting the output.

Notice that although the number of states in the FSM is very different for the first two atm examples, the number of full explorations is about the same. It's also very intersting to notice in the last example how the huge number of illegal states and the use of the failing pool requires us to visit only a very limited number of states in the product machine, thus obtaining the result really fast.

6 Conclusions and future work

The experimental results show that the proposed algorithm is a viable tool to synthesize an FSM to be used as an interface between two protocols. Still, there are a lot of limitations to be overcome and a lot of interesting extensions. Some of the assumption could be relaxed: in particular we are looking into ways to extend the methodology to asynchronous protocols, and to let many parties get involved in the communication. Moreover, an analysis of the product machine should let us easily factor it into two distinct machines (since that is where we started from) and trade-off the number of states with the number of *conditions* that must be setup between *anchor points*. The problem of a stream of data versus a single transaction should also be addressed.

An interesting extension would be the use of this methodology to create an interface with modules that are implemented in software in embedded systems. In that case, the references to values could be replaced by functions that know how to fetch the corresponding data.

7 Acknowledgments

The authors would like to thank Ellen Sentovich, Luciano Lavagno, Gaetano Borriello and Luca Carloni for many suggestions and discussions.

References

[1] A. V. Aho, R. Sethi, and J. D. Ullman. *Compilers Principles, Techniques and Tools.* Addison-Wesley, Reading, MA, 1988.

[2] J. Akella and K. McMillan. Synthesizing converters between finite state protocols. In *Proceedings of the International Conference on Computer Design*, pages 410–413, Cambridge, MA, October 14 - 15 1991.

[3] G. Borriello. *A New Interface Specification Methodology and its Applications to Transducer Synthesis.* PhD thesis, University of California at Berkeley, Berkeley CA, 1988.

[4] G. Borriello and R. H. Katz. Synthesis and optimization of interface transducer logic. In *Proceedings of the International Conference on Conputer Aided Design*, November 1987.

[5] J. A. Brzozowski. Derivatives of regular expressions. *Journal of the Association for Computing Machinery*, 11(4):481–494, October 1964.

[6] C. N. Coelho and G. D. Micheli. Analysis and synthesis of concurrent digital circuits using control-flow expressions. *IEEE Transactions on Computer-Aided Design of Integrated Circuits and Systems*, 15(8):854–876, August 1996.

[7] J. E. Hopcroft and J. D. Ullman. *Introduction to Automata Theory, Languages, and Computation.* Addison-Wesley, Reading, MA, 1986.

[8] S. Narayan and D. D. Gajski. Interfacing incompatible protocols using interface process generation. In *Proceedings of the 32nd Design Automation Conference*, pages 468–473, San Francisco, CA, June 12 - 16 1995.

[9] J. Oberg, A. Kumar, and A. Hemani. Grammar-based hardware synthesis of data communication protocols. In *Proceedings of the 9th International Symposium on System Synthesis*, pages 14–19, La Jolla, CA, November 6 - 8 1996.

[10] R. Passerone. *Automatic Synthesis of Interfaces between Incompatible Protocols.* M.S. Thesis, University of California at Berkeley, 1997.

[11] J. A. Rowson and A. L. Sangiovanni-Vincentelli. Interface-based design. In *Proceedings of the 34th Design Automation Conference*, pages 178–183, Anaheim, CA, June 9 - 13 1997.

[12] A. Seawright and F. Brewer. Clairvoyant: A synthesis system for production-based specification. *IEEE Transactions on VLSI Systems*, 2:172–185, June 1994.

[13] J. S. Sun and R. W. Brodersen. Design of system interface modules. In *Proceedings of International Conference on Computer Aided Design*, pages 478–481, 1992.

Automated Composition of Hardware Components

James Smith
Stanford University
Gates Building, Room 326
Stanford, California 94305–9030
redskins@aglaia.stanford.edu

Giovanni De Micheli
Stanford University
Gates Building, Room 333
Stanford, California 94305–9030
nanni@galileo.stanford.edu

Abstract

In order to automate design reuse, methods for composing system components must be developed. The goal of this research is to automate the process of generating interfaces between hardware subsystems. The algorithms presented here can be used to generate a cycle–accurate, synchronous interface between two hardware subsystems given an HDL model of each subsystem. These algorithms have been implemented in the POLARIS *hardware composition tool and have been used to generate an interface between a MIPS microprocessor and the SRAM that comprises its secondary cache. Interface generation for the MIPS R4000 is described.*

1. Introduction

The increasing complexity of electronic systems is forcing designers to consider, if not implement, design reuse and intellectual property sharing. As this methodology matures, a new breed of tools will be required to automate component selection, subsystem scheduling, and system composition. This paper presents a mechanism for composing hardware blocks that communicate with different protocols, given an HDL description of these protocols, while providing hooks for implementing arbitration algorithms.

Interface synthesis has focused on optimizing high level communication between subsystems given a set of communication constraints ([ErHeBe93], [JeElOb94], [KaLe94]). Interface modeling languages such as that developed by [ObKuHe96] allow a designer to explore a interface design space and generate a synthesizable description of the interface. [GuRo94] describes a methodology for modeling an interface with a behavioral description suitable for high level synthesis. Other interface synthesis research, such as that performed in [GaGl96], has investigated specifying and scheduling communication between hardware and software subsystems. The research presented here focuses on generating a low–level, synthesizable description of synchronous interfaces between hardware components. Similarly, [ChOrBo95] describes a mechanism for creating the glue logic between two hardware components, but requires a functional description of component ports. In this paper, we present the POLARIS tool, which converts a subsystem's communication protocol into a standard scheme given an HDL description of the subsystem. An HDL model of the resulting interface is generated.

The basic purpose of an interface is to facilitate the movement of data. Data could be addresses, commands, values destined for a memory location, or some combination of these descriptions. In order to allow hardware subsystems that follow different protocols for moving data to communicate with one another, the tool presented here maps these protocols into a standard communication scheme. This scheme is then implemented in an interface architecture that is general enough to accomodate the requirements of any target interface. The terminology *client* is used in this paper to indicate a component that is sending data and *resource* indicates a component that is receiving data.

2. Overview

The algorithms presented here are used to generate synchronous component interfaces. The components may operate at different frequencies and may employ unidirectional or bidirectional busses. Bidirectional busses, such as those employed by PCI or VME, are handled by treating the bus as two unidirectional busses and combining the resulting interface controllers. Multi–way interfaces that allow multiple clients to interact with multiple resources are synthesized by dynamically establishing a point to point link between the client and the resource (Fig. 1).

The interface architecture is described in Section 3. From the component model described in Section 4.1, a state machine is synthesized to map the component's communication protocol into a standard protocol that other interfaces can understand (Section 4.2). The data formats that a client component employs are translated into formats that the resource component can understand (Section 4.3).

The POLARIS hardware composition tool requires the user to supply an HDL description of the component being interfaced to, the name(s) of the ports across which data is tranferred, and the names of ports that the interface does not have access to (*uncontrollable* ports). The names of the uncontrollable ports (e.g. *reset*) must be supplied so that POLARIS does not manipulate these ports when creating a component interface.

3. The Interface Architecture

Unlike the hardware interface synthesis research performed by [MaHa95], this research links hardware components through a standard architecture rather than by attempting to map one component interface into another. This allows interfaces to be synthesized for a broad range of components. In addition, it allows multiple components to be linked via the same interface.

3.1 Architectural Blocks

The interface architecture includes a state machine for protocol conversion, a send and receive buffer for transaction information (which must be saved while a resource is unavailable), and an arbiter to govern access to a resource (Fig. 1). Although hooks are provided to allow the implementation of an optimized arbitration algorithm, the details of the arbitration scheme are not required for interface synthesis.

When the state machine for protocol conversion detects that a component is sending data, the data is placed in the receive buffer for that interface. This data will be passed to the send buffer for the resource interface. Correspondingly, when the state machine detects that there is data in its send buffer, it executes the necessary signal assignments to transfer the data to the resource component. Sizing of these queues is currently a manual task, but could be automated in the

DAC 98, San Francisco, California
©1998 ACM 0-89791-964-5/98/06..$5.00

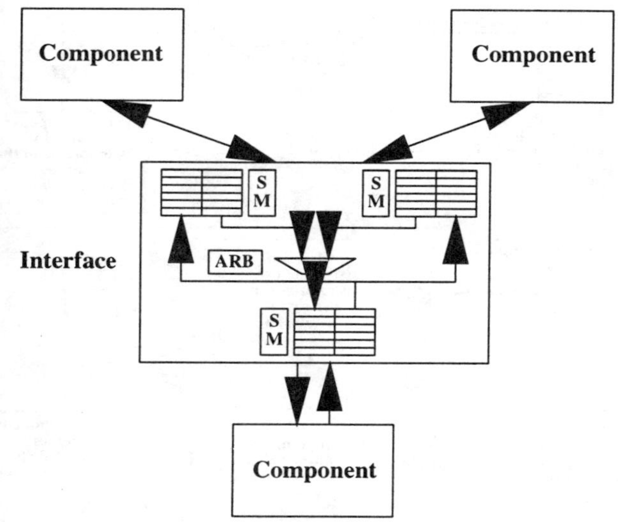

Fig. 1: High level view of a three component implementation of the interface architecture.

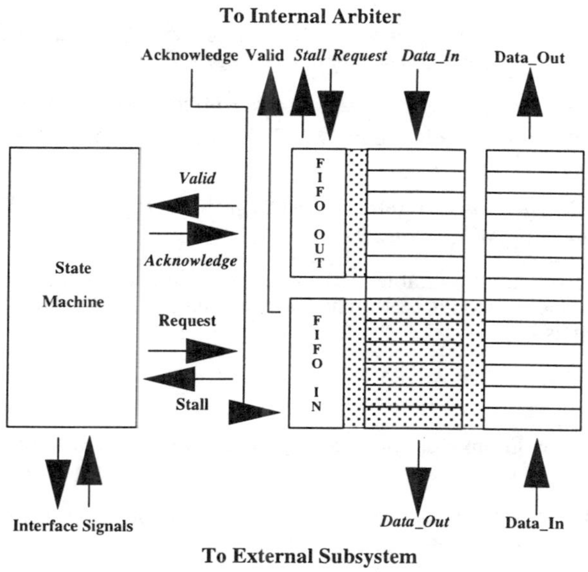

Fig. 2: Interface communication scheme.

manner of [AmBo91]. The default arbiter implements a round robin arbitration scheme to select a client receive buffer that will transfer its data into the appropriate resource send buffer.

3.2 Communication Scheme

The standard protocol employs four control signals each for the receive buffer and the send buffer (Fig. 2). For the each buffer, the signals are implemented as follows: (1) *Request* – input to the buffer controller that indicates data is being sent to the interface; (2) *Stall* – output from the buffer controller that indicates that the buffer is full (the state machine must prevent other components or interfaces from sending any more data); (3) *Valid* – output from the buffer controller that indicates that valid data is in the buffer and ready to be transferred; (4) *Acknowledge* – input to the buffer controller that indicates that data has been read from the buffer (the buffer controller will increment the read address). Resources are scheduled in the arbiter by selectively *Acknowledge*ing the client *Valid* signal. Data is transferred through four unidirectional busses, two for sending data to and receiving data from the arbiter, and two for sending data to and receiving data from the external component.

4. State Machine Generation

Given a component model that describes bus functionality (or a superset of bus functionality), conditions for transferring data to or from that component are determined. A sequence(s) of assignments to component ports is determined that will cause these conditions to become true. After the required assignments to component ports have been determined, they are executed on the component model, so as to resolve the values of control ports that are inputs to the synthesized interface. Once the values of all necessary input and output ports have been resolved, a state machine is generated that executes the required assignments and monitors the necessary control ports. This state machine provides a mapping from the communication protocol for a system component to the standard protocol that allows inter–component communication. The collection of operations necessary to perform a data transfer is referred to as a *function.*

4.1 Component Model

The component is abstracted as a list of assignments to variables and conditions for each assignment to be executed.

Definition 1. A *component* is described by the list of tuples $\{\mathcal{C}_i := <v_i, X_i, A_i, \sigma_i>\}$ where $X_i = \{\chi_{ij}\}$ are the conditions under which the values $A_i = \{\alpha_{ij}\}$ are assigned to the variable v_i. For the assignment to v_i, we assume there are n_i possible values $\{\alpha_{ij}, j= 1,2, \ldots, n_i\}$ each selected by one and only one condition $\chi_{ij} \in \{0, 1\}$, and σ_i indicates whether the assignment is combinational or synchronous. Thus,

$$v_i = \Sigma\, \chi_{ij} * \alpha_{ij}.$$

Variables that are not component ports are referred to as *internal* variables.

4.2 Protocol Conversion Algorithm

The state machine for protocol conversion is represented, analogous to the Moore model, as a collection of state assignments and conditions for state transitions.

Definition 2. The *state machine for protocol conversion* is described by the tuple $\mathcal{SM} := <\mathcal{S}, \mathcal{T}>$ where

$\mathcal{S} := \{S_i\}$ is a list of states,

$\mathcal{T} := \{(\tau_{ij} = 1) \Rightarrow S_i \rightarrow S_j\}$ is a list of conditions governing state transitions.

The algorithm for generating the state machine for protocol conversion is completed in five steps: (1) generate a sequence $\mathcal{S}_f$ of *functional states* that cause a function to be executed, (2) generate a sequence $\mathcal{S}_\chi$ of *exit states* that cause an executing function to be halted ($\mathcal{S} = \mathcal{S}_f \cup \mathcal{S}_\chi$), (3) generate the conditions $\{\tau_{ij}\}$ that govern the state transitions, (4) combine state sequences for multiple functions, and (5) reduce the number of states. The name of the component's data bus, referred to as the *target* variable, is supplied to initiate generation of the interface state machine. The algorithm is outlined in Fig. 3.

If the target variable is an input or bidirectional port, the component model is searched for a use of this variable such as:

if signalA = valueA then
signal <= target

if target = value and
signalB <= valueB then ...

The conditions under which the target variable (*target*) is used are the *zeroth order input conditions.*

Definition 3. The set of *zeroth order input conditions* for the target variable v_x is $I_0 := \{\chi_{ij}\}$ where χ_{ij} or α_{ij} is dependent on v_x.

In the previous example, v_x = {*target*} and I_0 = {*signalA* = *valueA*, *signalB* = *valueB*}

If the target variable is an output or bidirectional port, the component model is searched for an assignment to this port such as:

if signalA = valueA then
target <= signal

The conditions for these assignments are the *zeroth order output conditions.*

Definition 4. The set of *zeroth order output conditions* for the target variable v_x is $O_0 := \{\chi_{xj}\}$.

In the previous example, v_x = {*target*} and O_0 = {*signalA* = *valueA*}.

4.2.1 Executing a Function

Assignments that satisfy zeroth order conditions must be executed as part of the functional state sequence. That is, if I_0 = {*signalA* = *valueA*}, then state S_0 must contain the assignments {*signalA* <= *valueA*}. If *signalA* is a component port, then a single assignment can be made to allow that function to be completed. However, if *signalA* is an internal variable of the component, then conditions must be determined that will cause that variable assignment. For example, if a component description contained the sequential statements:

if externalSignalA = valueA and
internalSignal = value then
signal <= target

if externalSignalB = valueB then
internalSignal <= value

then the following state sequence is generated:

State 1: *extSignalB* <= *valueB*
/* causes *intSignal* <= *value* in State 0 */

State 0: *extSignalA* <= *valueA*

These conditions that must be satisfied to cause the internal variable assignment are the *nth order conditions.*

Definition 5. The set of *nth order input conditions* for the target variable v_x is $I_n := \{\chi_{kl}\}$ where, given the (n−1)th order input conditions for v_x, i.e. I_{n-1},
there exists an $\chi_{ij} \in I_{n-1}$
such that $(\chi_{kl} = 1) \Rightarrow (\chi_{ij} = 1)$.

In the previous example, v_x = {*target*}, I_0 = {*externalSignalA* = *valueA and internalSignal* = *value*} and I_1 = {*externalSignalB* = *valueB*}. An analogous definition applies to *nth order output conditions.* If conditions of all orders can be satisfied by a sequence of assignments to component ports, the interface state machine can deterministically drive a component into functional states.

If an internal variable assignment requires an nth order condition that in turn requires the same assignment, the variable assignment is *uncontrollable* and can not be deterministically driven to that value. For example, if a component description contained the sequential statements:

if reset = TRUE or intSignalA = valueA then
target <= value
intSignalB <= valueB,

if intSignalB = valueB then
intSignalA <= valueA,

then *intSignalA* <= *valueA* is uncontrollable if *reset* is uncontrollable.

Definition 6. An internal variable assignment ($v_k = \alpha_{kl}$) is *uncontrollable* if $\chi_{kl} \in I_n$ and
(1) $\chi_{kl} \in I_m$ where m < n, or
(2) χ_{kl} is dependent on another uncontrollable assignment.

If an uncontrollable internal variable assignment is encountered, no nth order conditions for that assignment are generated. The thread of states that required the assignment is thus aborted (Fig. 3). This prevents the algorithm from looping indefinitely on a component description such as the one described in the previous example.

A component can not be deterministically driven to a functional state when an uncontrollable signal is encountered. However, the functional states can be detected by examining component control signals (Section 4.2.3).

4.2.2 Exiting a Function

In the same way that a sequence of states is determined to execute a function, another sequence of states is determined to end that function. That is, when the data transfer is completed, the interface must exit its corresponding data transfer state.

A component exits a data transfer state when an assignment is made that contradicts a zeroth order condition. This can be

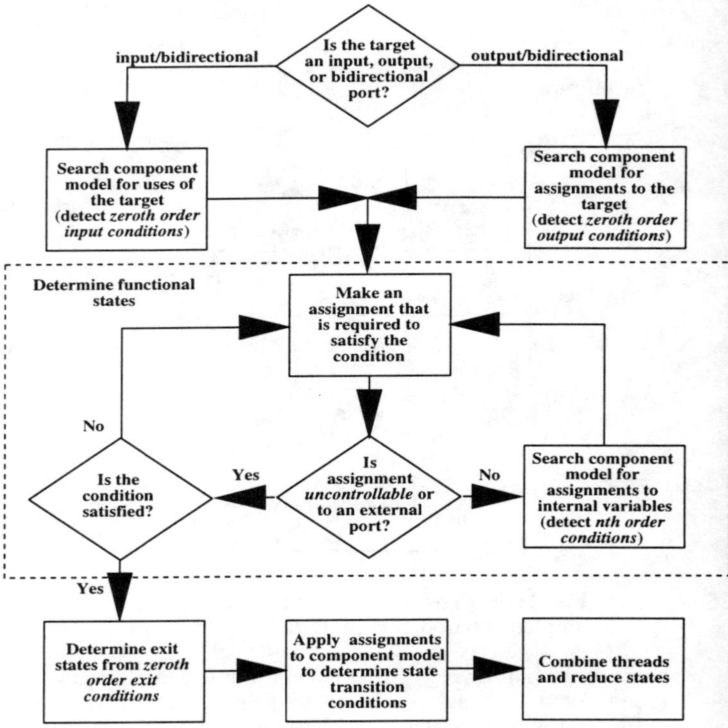

Fig. 3: The algorithm for state machine generation

achieved with assignments to component ports or internal variables. For example, if a component description contained the statement:

if externalSignalA = valueA and
internalSignal = value then
signal <= target

then satisfying the condition *not (externalSignalA = valueA and internalSignal = value)* must cause the data transfer state in the interface to be exited. Such conditions are the *zeroth order exit conditions.*

Definition 7. The set of *zeroth order exit conditions* for the target variable v_x is $\mathcal{E}_0 := \{\chi_{kl}\}$ where, given the zeroth order input (or output) conditions for v_x, i.e. I_0,

there exists an $\chi_{xj} \in I_0$,
such that $(\chi_{kl} = 1) \Rightarrow (\chi_{xj} = 0)$.

Nth order exit conditions are generated and satisfied in the same manner that nth order input and output conditions are generated and satisfied.

The exit state sequence is combined with the data transfer state sequence to obtain a state sequence that can execute a function and be reset.

4.2.3 Generating State Machine Conditions

After the conditions for executing and exiting a function have been completely satisfied or found to be uncontrollable, the required assignments are executed on the component model. If the component drives a port to a valid value in a cycle, then a conditional statement governing the state transition must be added to the state corresponding to that cycle. For example, if the component description contained the sequential statements:

if externalSignalA = valueA and
internalSignal = value then
signal <= target

if uncontrollableSignal = valueX and
externalSignalB = valueB then
internalSignal <= value
externalSignalC <= valueC

then the condition $\tau_{10} = \{externalSignalC = valueC\}$ must be satisfied to allow the state transition from state $S_1 = \{externalSignalB <= valueB\}$ to state $S_0 = \{externalSignalA <= valueA\}$.

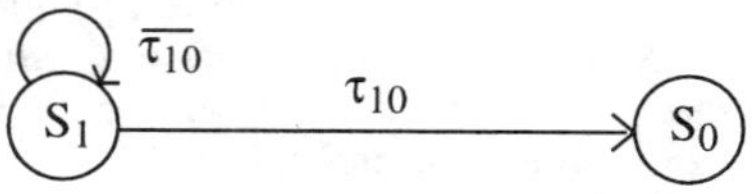

Fig. 4: Conditions governing state transitions.

Thus, if the interface state machine can not deterministically drive a component into a particular state (in the example above, the component state corresponding to S_1), it will be able to determine when the component has reached that state by evaluating its status signals.

4.2.4 Combining Multiple Threads of Execution

A function's execution may require or allow more than one sequence to be executed in parallel. For example, the component description,

if internalSignalA = valueA and
internalSignalB = valueB then
signal <= target

requires multiple variables (*internalSignalA, internalSignalB*) to be set in tandem. This results in multiple threads of execution being generated, which must be combined into a single state machine.

ANDing two state sequences is a straightforward combination of state assignments and state transition conditions. The final states in a sequence and their predecessors are ANDed. An example is shown in Fig. 5.

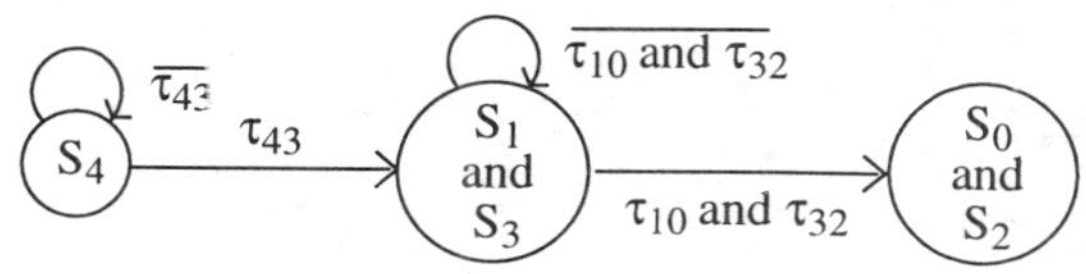

Fig. 5: The results of ANDing the two state sequences (S_0, S_1) and (S_2, S_3, S_4).

If any two ANDed states contain contradictory assignments the threads are discarded.

In ORing two threads, only the head states are combined. Previous states are not ORed to prevent state sequences that contain a portion of each thread from being executed. An example is shown in Fig 6.

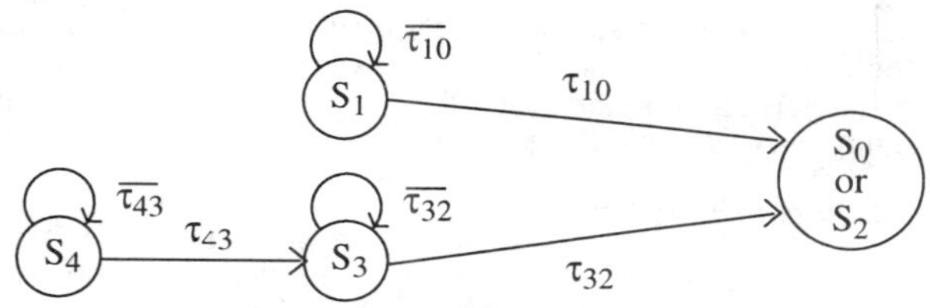

Fig. 6: The results of ORing the the two state sequences (S_0, S_1) and (S_2, S_3, S_4).

Duplicated states are removed as shown in the next section.

4.2.5 State Reduction

A component will frequently share states between execution threads. For example, a component may contain a state in which it polls its subsystems for writes, and executes a different sequence for each write:

if state = POLL_STATE then
case (subsystem)
subA_write:
state <= subA_WRITE_STATE
subB_write:
state <= subB_WRITE_STATE

The number of states is reduced by joining threads of execution at points where their respective states are *congruent.*

There are two sets of requirements that states can satisfy to be considered congruent. First, two states are congruent if they stem from congruent previous states and the conditions for entry into both states are the same. Second, two states are considered to be congruent if they contain the same variable assignments and one of them is an exit state. The second requirement allows a state machine to be reset once it has executed a functional sequence.

Definition 8. States S_i and S_j are *congruent* if either

(1) for all x, there exists a y
such that $\tau_{xi} = \tau_{yj}$ and
and S_x and S_y are congruent, or
(2) assignments of S_i = assignments of S_j
and S_i satisfies $\mathcal{E}_0$

```
module r4600_interface (SysADi, SysADCi, SysCmdi, SysCmdPi,
        SysADo, SysADCo, SysCmdo, SysCmdPo,
        RdRdy, WrRdy, ExtRqst, Release, ValidIn, ValidOut,
        Reset, Clock, SysOe,
        CPUwr, CPUrd, CPUrsp, CPUdataI, CPUdataO, CPUack, CPUvalid, CPUaddr);

 input [63:0] SysADi;
 input [7:0] SysADCi;
 input [8:0] SysCmdi;
 input SysCmdPi;
 output [63:0] SysADo;
 output [7:0] SysADCo;
 output [8:0] SysCmdo;
 output SysCmdPo;
 input RdRdy, WrRdy;
 input ExtRqst;
 output Release;
 input ValidIn;
 output ValidOut;
 input Reset;
 input Clock;
 output SysOe;
 input CPUwr;
 input CPUrd;
 input CPUrsp;
 input [63:0] CPUdataI;
 output [63:0] CPUdataO;
 output CPUack;
 output CPUvalid;
 input [63:0] CPUaddr;
 reg [63:0] SysADo;
 reg [7:0] SysADCo;
 reg [8:0] SysCmdo;
 reg SysCmdPo;
 reg Release;
 reg ValidOut;
 reg SysOe;
 reg [63:0] CPUdataO;
 reg CPUack;
 reg CPUvalid;
 reg [3:0] bus_state;

 parameter S_SEND_IDLE = 4'b0000, S_SEND_ADDR = 4'b0001, S_SEND_DATA = 4'b0010,
       S_SEND_SPIN = 4'b0011, S_RECV = 4'b0100, S_RECV_SPIN = 4'b0111;

 parameter H_READ = 2'b00, H_WRITE = 2'b01, H_NULL = 2'b10;

always @ (posedge Clock) begin
  ValidOut = 0;
  Release = 0;
  SysOet = 1;
  CPUack = 0;
  if (Reset) begin
    bus_state = S_SEND_IDLE;
  end else begin
    case (bus_state)
      S_SEND_IDLE: begin
        if ((CPUwr && WrRdy) || (CPUrd && RdRdy)) begin
          bus_state = S_SEND_ADDR;
        end else if (ExtRqst) begin
          Release = 1;
          bus_state  = S_RECV_SPIN;
          SysOe = 0;
        end
      end
      S_SEND_ADDR: begin
        if (CPUwr) begin
          SysCmdo = 9'b001000000;
          bus_state = S_SEND_DATA;
        end else if (CPUrd) begin
          SysCmdo = 9'b000000000;
          bus_state = S_SEND_IDLE;
        end
      end
      S_SEND_DATA: begin
        SysADo = CPUdataI;
        if (!CPUwr) begin
          SysCmdo = 9'b100000000;
          bus_state = S_SEND_IDLE;
        end else begin
          SysCmdo = 9'b110000000;
        end
        ValidOut = 1;
        CPUack = 1;
      end
      S_SEND_SPIN: begin
        SysOe = 0;
        bus_state = S_SEND_IDLE;
      end
      S_RECV: begin
        SysOe = 0;
        if (ValidIn) begin
          if (SysCmdi[8]) begin
            CPUdataO = SysADi;
          end else case (SysCmdi[6:5])
            H_NULL:
              bus_state = S_SEND_SPIN;
          endcase
        end
      end
      S_RECV_SPIN: begin
        SysOe = 0;
        bus_state = S_RECV;
      end
    endcase
  end
end
endmodule
```

Fig. 7: Verilog description of a simple MIPS SysAD interface (bidirectional pads for SysAD not shown).

Congruent states are combined by creating a state in which the assignments of both states are executed. If one of the states is an exit state, then the conditions for entrance into the newly created state are ORed.

4.3 Datapath Translation

The state machine for protocol conversion allows two components to communicate by translating their control signals into the standard interface. However, components often employ different datapaths that must be reconciled if they are to communicate with one another. Translating datapaths between interfaces imposes two requirements: (1) datapath widths must be reconciled, and (2) addresses must be extracted from a transaction so that client data can be directed to the appropriate resource.

Datapath widths are determined from the component description. When data is read from a client receive buffer, it is read into a register that is the width of the resource datapath. If the resource datapath is wider than the client datapath, an *Acknowledge* is returned to the client interface for every word that is popped off the client receive buffer. If the resource datapath is thinner than the client datapath, an *Acknowledge* is returned to the client interface when the resource register is filled.

Address extraction is currently a manual task. It can be performed using a simplified version of the structures that ([ChOrBo95], [MaHa95]) used to achieve interface synthesis. These two employed structures called "signal sequences" (SEQs) and "protocol flow graphs" (PFGs), respectively, to model the bit patterns that are required to interact with a component. In the case of POLARIS, a sequence of higher level descriptions can be provided to allow the interface to detect which cycles or bits are transmitting an address.

5. Example – Simple MIPS SysAD Interface

This example demonstrates how POLARIS generates a state machine that converts the communication protocol of a MIPS processor with a simplified SysAD interface to the standard protocol. This generated interface can communicate with a similarly synthesized interface for another component such as RAM, a DSP, etc. (not described here). The HDL model of the SysAD interface (not including the bidirectional buffer for SysAD) is given in Fig. 7. In addition to the HDL model, the bus for data transferral (*SysAD*) and a list of uncontrollable ports (Reset, *CPUwr, CPUrd, etc.*) is supplied to the tool.

Upon determining that the target variable (*SysAD*) is a bidirectional port, POLARIS first creates the state machine for input through the target variable. The component model is searched for the for the assignment * <= *SysAD*. The set of zeroth order input conditions I_0 = {*Reset = 0 and bus_state = S_RECV and Valid_In = 1 and SysCmdi[8] = 1*} is returned. State S_0 = {*Reset = 0, bus_state = S_RECV, Valid_In = 1, SysCmdi[8] = 1*} is created. Since *bus_state* is an internal variable and not uncontrollable, it is made the new target bus and the component model is searched for the assignment *bus_state <= S_RECV*. The set of first order input conditions I_1 = {*Reset = 0 and bus_state = S_RECV_SPIN*} is returned. State S_1 = {*Reset = 0, bus_state = S_RECV_SPIN*} is created, and the component model is searched for the assignment *bus_state <= S_RECV_SPIN*. The set of second order input conditions I_2 = {*Reset = 0 and bus_state = S_SEND_IDLE and ((CPUwr and WrRdy) or (CPUrd and RdRdy)) = 0 and ExtRqst = 1*} is returned and states S_2 = {*Reset = 0, bus_state = S_SEND_IDLE, WrRdy = 0, ExtRqst = 1*} and S_3 = {*Reset = 0, bus_state = S_SEND_IDLE, RdRdy = 0, and ExtRqst = 1*} are created (note that CPUwr and CPUrd are uncontrollable and states corresponding theirassignment are not generated). This process continues until nth order conditions can be completely controlled from external ports or require uncontrollable assignments (e.g. *Reset = 0, bus_state = S_SEND_IDLE*).

The zeroth order exit conditions are determined by negating the zeroth order input conditions. The set of zeroth order exit conditions is $\mathcal{E}_0$ = {*ValidIn = 0 or SysCmd[8] = 0 or bus_state != S_RECV or Reset = 1*}. This exit condition generates three exit states, two of which satisfy the first congruency criteria ({*ValidIn=0*} and {*SysCmd[8] = 0*}).

Now that all possible complete sequences of states have been determined, the state assignments are executed to determine the valid values on external ports during each state. Since *Release* and *ValidOut* are driven to valid values during state S_1, condition τ_{10} = (*Release = 0 and ValidOut = 0*) must

```
if (Reset) begin
  state = S_0;
end else begin
  case (state)

    S_3: begin
      if (valid)
        state = S_2;
    end;

    S_2: begin
      WrRdy = 0;
      RdRdy = 0;
      ExtRqst = 1;
      if (Release && !ValidOut)
        state = S_1;
    end;

    S_1: begin
      if (!Release && !ValidOut)
        state = S_0;
    end;

    S_0: begin
      if (valid) begin
        ValidIn = 1;
        SysCmd[8] = 1;
        SysAD = Data_Out;
        acknowledge = 1;
      end else  begin
        if (...) begin
          ValidIn = 1;
          SysCmd[8] = 1;
          SysCmd[6:5] = H_NULL;
          state = S_3;
        end else begin
          ValidIn = 0;
          SysCmd[8] = 0;
        end;
      end;
    end;

  endcase;
end;
```

Fig. 8: Resulting state machine that allows writes to the MIPS core through the SysAD bus.

be satisfied to complete the transition $S_1 \rightarrow S_0$. Similarly, $\tau_{21} = \tau_{31} = (Release = 1 \text{ and } ValidOut = 0)$ must be satisfied to complete the transitions $S_2 \rightarrow S_1$ and $S_3 \rightarrow S_1$.

Given the assignments for each state $\mathcal{S}$ and the conditions $\mathcal{T}$ for transitioning between states, the state sequences are joined at congruent states. The three sequences generated by the exit states all contain congruent states leading up to the functional state. Thus, the threads are joined at each of these states. However, the exit states that perform {*ValidIn = 0*} and {*ValidIn = 1, SysCmd[8] = 0, SysCmd[6:5] = H_NULL*} are mutually exclusive, thus there are multiple branches exiting the functional state. According to the second congruency criteria, the exit state from {*ValidIn = 0*} is congruent to state S_0 and the exit state from {*ValidIn = 1, SysCmd[8] = 0, SysCmd[6:5] = H_NULL*} is congruent to state S_3. The state machine for transferring data to the MIPS processor is given in Fig. 8.

The process above is repeated for data transferred from the MIPS processor to the interface. The only difference being that, initially, the functional states are determined by searching the expression list for *SysAD* <= *. The output state machine and input state machine are joined at states S_2 and S_3 in state reduction. The physical characteristics of the synthesized interface, not including the send and receive buffers, are shown in Fig. 9. The speed, area and power consumption of the bus control logic synthesized above is not optimized in this research because this logic is rarely a significant factor in the overall speed, area and power consumption of the interface.

6. Conclusion

Reusing existing high level blocks is becoming a necessity for designing complex systems. Composing blocks that are developed by different design groups with different communication protocols is an imperative in automating design reuse and IP sharing. The composition architecture presented here provides a means to compose synchronous blocks while providing hooks for optimizing system performance by prioritizing component communication. The example provided illustrates its ability to generate an interface between hardware blocks.

The techniques described above allow a designer to automatically generate an HDL model of an interface between two or more blocks given an HDL description of the corresponding blocks. In implementing these techniques, parsing of the input HDL has been completed in a front end module that can be adapted to different coding styles or represenatations. While overspecified input HDL descriptions can be used to geneate an interface, incompletely specified input blocks will yield interfaces with ambiguous states. However, these states are detectable. Interfaces between blocks with multiple busses can be generated when

Process	LSI LCB007
Gate Count	1464
Max. Operating Frequency	166 MHz
Est. Area	.92 mm^2
Est. Power Consumption	348mW

Fig. 9: Physical Charactersistics of the Automatically Generated Interface to a MIPS R4000 (Does not include 256B receive and send buffers)

the control of these busses is separate. Extensions for blocks that contain multiple busses with shred control is a future topic of research.

Future research will seek to expand the techniques presented here to generate interfaces between software and hardware components in a system that implements memory mapped I/O. The communication protocol for software interfaces is restricted by the instruction set architecture of the microprocessor on which the software is running. Instead of communicating with a hardware subsystem by assigning values to and reading values from ports, the software driver communicates by writing and reading hardware registers. Thus, the hardware/software composition requires the interface state machine generator described above and another layer, similar to [BoDeLi96], that encapsulates port assignments in microprocessor operations.

Acknowledgements

This research is sponsored by ARPA under grant DATB63–95–C0049.

References

[ErHeBe93] R. Ernst, J. Henckel, and T. Benner, "Hardware/Software Cosynthesis for Microcontrollers", *IEEE Design and Test of Computers*, p. 64–75, December 1993.

[JaElOb94] A. Jantsch, P. Ellervee, J. Oberg, A. Hemani, and H. Tenhunen, "Hardware/Software Partitioning and Minimizing Memory Interface Traffic", *Proceedings of EuroDac*, p. 226–231, 1994.

[KaLe94] A. Kalavade and E.A. Lee, "A Global Criticality/Local Phase Driven Algorithm for the Constrained Hardware/Software Partitioning Problem", *Proceedings of the 3rd International Workshop on Hardware/Software Codesign*, p. 42–48, 1994.

[ObKuHe96] Johnny Oberg, Anshul Kumar, Ahmed Hemani, "Grammar-based Hardware Synthesis of Data Communication Protocols", *9th International Symposium on System Synthesis*, p. 14–19, 1996.

[GuRo94] P. Gutberlet, W. Rosenstiel, "Specification of Interface Components for Synchronous Data Paths", *7th International Symposium on System Synthesis*, p. 134–139, 1994.

[GaGl96] Michael Gasteier, Manfred Glesner, "Bus-Based Communication Synthesis on System-Level", *9th International Symposium on System Synthesis*, p. 65–70, 1996.

[ChOrBo95] Pai Chou, Ross B. Ortega, Gaetano Borriello, "Interface Co-Synthesis Techniques for Embedded Systems", *Proceedings of the IEEE/ACM International Conference on Computer-Aided Design*, pp.280–287, 1995.

[MaHa95] Jan Madsen and Bjarne Hald, "An Approach to Interface Synthesis", *8th International Symposium on System Synthesis*, p. 16–21, 1995.

[AmBo91] T. Amon and G. Borriello, "Sizing Synchronization Queues: A Case Study in Higher Level Synthesis", *Proceedings of the 28th Design Automation Conference*, Jun 1991.

[BoDeLi96] I. Bolsens, H. DeMan, B. Lin, K. Van Rompaey, S. Vercauteren, D. Verkest, "Hardware–Software Codesign of Digital Telelcommunication Systems", Proceedings of the IEEE, p 391–418, March 1997.

Multilevel Integral Equation Methods for the Extraction of Substrate Coupling Parameters in Mixed-Signal IC's

Mike Chou Jacob White
Department of EECS, Massachusetts Institute of Technology
Cambridge, MA. 02139

Abstract—
The extraction of substrate coupling resistances can be formulated as a first-kind integral equation, which requires only discretization of the two-dimensional contacts. However, the result is a dense matrix problem which is too expensive to store or to factor directly. Instead, we present a novel, multigrid iterative method which converges more rapidly than previously applied Krylov-subspace methods. At each level in the multigrid hierarchy, we avoid dense matrix-vector multiplication by using moment-matching approximations and a sparsification algorithm based on eigendecomposition. Results on realistic examples demonstrate that the combined approach is up to an order of magnitude faster than a Krylov-subspace method with sparsification, and orders of magnitude faster than not using sparsification at all.

I. Introduction

The design of single chip mixed-signal systems is now an active area of research, driven by the relentless quest for high-level integration and cost reduction. A major challenge for mixed-signal design tools is the accurate modeling of the parasitic noise coupling through the common substrate between the high-speed digital and high-precision analog components [22], [11], [15]. Fast switching logic components inject current into the substrate, causing voltage fluctuations which can affect the operation of sensitive analog circuitry through the body effect. Since the bulk substrate behaves resistively up to a frequency of a few gigahertz [6], [23], it is sufficient to solve Laplace's equation inside the substrate with proper boundary and interface conditions. Examples of this approach [11], [22], [13], [20], [25] includes Finite Element (FEM) and Finite Difference (FD) methods. Although the resulting linear systems are *sparse*, such methods are impractical for complex layouts because the number of unknowns resulting from three-dimensional volume-meshing of the entire substrate is too large.

Integral equation based techniques such as the Boundary Element Method (BEM) have been applied with some success to the modeling of substrate coupling [19], [5], [24]. Since only the two-dimensional substrate contacts are discretized, BEM methods dramatically reduce the size of the linear system to be solved. However, they produce *dense* matrices, which are too expensive to store or factor directly. To address this difficulty, heuristic partitioning schemes [7], [19] were proposed to sparsify the matrix *inverse* by setting direct admittances to contacts outside a user-defined region to zero. This approach requires too much user intervention and, more importantly, results in errors that are difficult to control and quantify.

Iterative schemes, combined with a sparsification algorithm to compute dense matrix-vector products efficiently, can be effective for solving large BEM systems. In [24], GMRES [17], a Krylov-subspace based iterative method, was combined with a fast multipole [8] algorithm for substrate resistance extraction. However, accuracy is compromised since the multipole algorithm cannot handle substrate edge effects. A more serious difficulty is that Krylov-subspace iterative methods converge slowly when applied to large BEM systems, which tend to be ill-conditioned [18], [21], [12]. Hundreds of iterations may be required per solution for large problems.

Multigrid methods, or more generally, multilevel methods, are known to be the most efficient iterative techniques in the solution of elliptic partial differential equations (PDE's) [1], [9], [2] due to their fast convergence. However, multilevel methods are not well-developed for first-kind integral equations [12] defined over complicated geometries, as is our case here. In this paper, we address this void by developing a multigrid iterative solver, and then integrating it with sparsification algorithms specially tuned to accurately account for substrate edge effects. Section II summarizes the BEM formulation for substrate coupling resistance extraction. Our new multilevel algorithm is presented in Section III. Computational results are given in Section IV, where comparisons to Krylov-subpace iterative methods are also made. Finally we give our conclusions and acknowledgements in Section V.

II. Background and Previous Work

In the electrostatic approximation [6], [24], the substrate is modeled as a stratified medium composed of several homogeneous layers characterized by their conductivities shown in Figure 1. Three contacts are shown in gray. For this work, the substrate backplane is assumed to be grounded electrically.

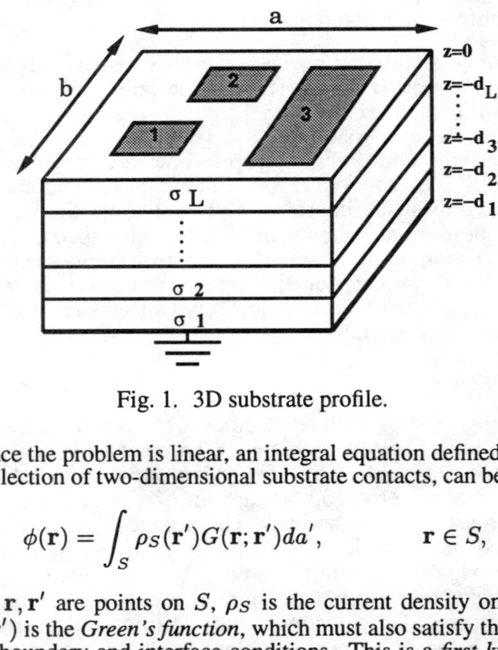

Fig. 1. 3D substrate profile.

Since the problem is linear, an integral equation defined over S, the collection of two-dimensional substrate contacts, can be written

$$\phi(\mathbf{r}) = \int_S \rho_S(\mathbf{r}')G(\mathbf{r};\mathbf{r}')da', \qquad \mathbf{r} \in S, \tag{1}$$

where $\mathbf{r}, \mathbf{r}'$ are points on S, ρ_S is the current density on S, and $G(\mathbf{r};\mathbf{r}')$ is the *Green's function*, which must also satisfy the appropriate boundary and interface conditions. This is a *first-kind* integral equation [12] which forms the basis for the boundary-element method (BEM) used in [19], [7], [24].

To numerically solve (1), the domain S discretized into N disjoint, rectangular *panels* $\{p_i\}$ such that $S = \bigcup_{i=1}^{N} p_i$. An example of panel discretization for a three-contact layout is given in Figure 2. In the Galerkin scheme [10], the current density on each panel p_i is assumed to be uniform. Then N linear equations are constructed by evaluating the *average* potential $\phi(\mathbf{r})$ over each panel p_i. The result is a discretized version of (1)

$$v = Pq, \tag{2}$$

DAC 98, San Francisco, California

where q and v are length-N vectors with q_i denoting the total current on panel i and v_j denoting the average potential on panel j. P is an $N \times N$ matrix given by

$$P_{ij} = \frac{1}{a_i a_j} \int_{p_i} \int_{p_j} G(\mathbf{r}; \mathbf{r}') da da' \tag{3}$$

where a_i and a_j are the areas of panels i and j respectively. P is often called the *coefficient-of-potential* matrix. We note here that P is *dense* since current injected into any panel i produces a non-zero potential at every other panel j.

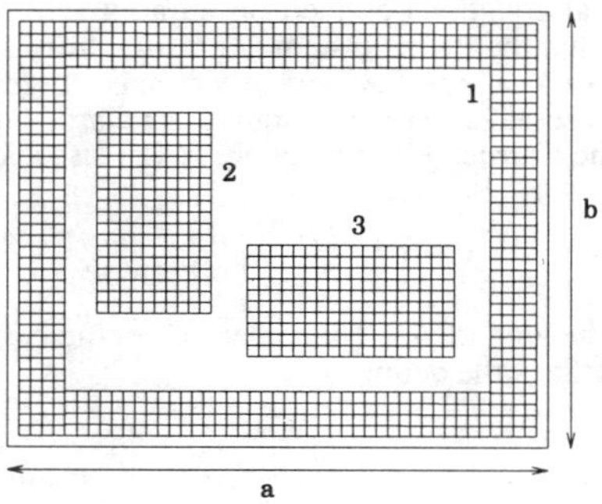

Fig. 2. Example of contact discretization.

For the extraction of substrate coupling resistances, the voltages at the contacts, each corresponding to a circuit node, are specified. We are required to solve for the detailed current distribution, which can be summed over each contact to yield the current flow into each node. This is analogous to the capacitance extraction problem [14]. The linear system (2) is to be solved for q given v.

The Green's function $G(\mathbf{r}; \mathbf{r}')$ for the bounded substrate with grounded backplane was shown in [6] to be a double infinite series of cosines in x and y. By truncating $G(\mathbf{r}; \mathbf{r}')$ to a finite $M \times M$ series and substituting it into (3), it was also shown in [6] that each entry P_{ij} can be constructed from linear combinations of appropriate terms from a two-dimensional $M \times M$ array $\{F_{lm}\}$, which is computed once and for all with a Type-1 Discrete Cosine Transform (DCT). Although this allows individual entries of P to be computed, direct *solution* of (2) still requires $\mathcal{O}(N^3)$ CPU time and $\mathcal{O}(N^2)$ memory since P is dense. This limits the size of the problem to a few hundred panels. In the next section, we develop an efficient *iterative* solver based on multiresolution analysis.

III. Matrix-Free, Multilevel Solution of First-Kind Integral Equations

The efficiency of multigrid iterative methods is a direct result of the fact that convergence rate is independent of discretization, and hence problem size. This is to be contrasted with Krylov-subspace iterative methods, whose convergence rates deteriorate with increasing mesh refinement, or equivalently, worse matrix conditioning. In this section, we develop a multigrid method for the discretized first-kind integral equation (2), as well as sparsification techniques necessary to avoid dense matrix-vector multiplication at each level in the multilevel algorithm.

To best present the method, we first describe the simpler case of a uniformly discretized contact that covers the entire substrate. We then describe the modifications needed for many irregularly shaped contacts.

A. *Basic Multigrid Algorithm for Uniform Grids*

In this section, we assume that the integral equation is defined over the *entire* substrate $\Omega = [0, a] \times [0, b]$

$$\phi(\mathbf{r}) = \int_{\Omega} \rho_S(\mathbf{r}') G(\mathbf{r}; \mathbf{r}') da', \qquad \mathbf{r} \in \Omega, \tag{4}$$

and that Ω is discretized into a uniform array of $M \times M$ panels. We assume further that M is a power of two, *i.e.* $M = 2^l$ for integer l. We refer to this discrete BEM system as a level l, or *fine-grid*, representation of (4) $P_{\{l\}} \cdot q_{\{l\}} = v_{\{l\}}$. The number of panel unknowns, and hence the size of the linear system, is then $N_l = M^2$. Suppose we also discretize (4) using a coarser, uniform $(M/2) \times (M/2)$ array of panels, yielding a discrete linear system of size $N_{l-1} = N_l/4$. This results in a level $(l-1)$, or *coarse-grid*, representation $P_{\{l-1\}} \cdot q_{\{l-1\}} = v_{\{l-1\}}$. See Figure 3 for the two discretizations.

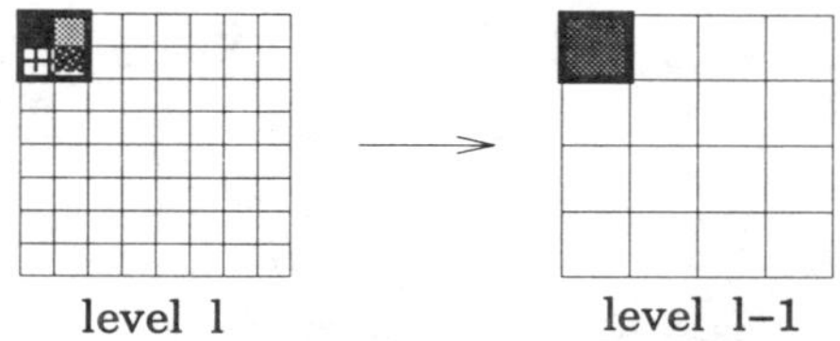

Fig. 3. Two-level Representation and Restriction for Uniform Grid Problem.

Algorithm 1 (Two-Grid Method (TGM))

Set $k \equiv 1, q_{\{l\}}^{(1)} \equiv 0$.
Repeat {
  Fine-Grid Smoothing:
    Solve $D_{\{l\}} \cdot \Delta q_{\{l\}}^* = -P_{\{l\}} \cdot q_{\{l\}}^{(k)} + v_{\{l\}}$
      for $\Delta q_{\{l\}}^*$.
    Compute intermediate guess $q_{\{l\}}^* = q_{\{l\}}^{(k)} + \Delta q_{\{l\}}^*$.
  Compute residual $u_{\{l\}} = P_{\{l\}} \cdot q_{\{l\}}^* - v_{\{l\}}$.
  Project to coarse grid $u_{\{l-1\}} = r u_{\{l\}}$.
  Coarse-Grid Correction:
    Solve for $\Delta q_{\{l-1\}}$ in $P_{\{l-1\}} \cdot \Delta q_{\{l-1\}} = u_{\{l-1\}}$.
    Project to fine grid $\Delta q_{\{l\}} = p(\Delta q_{\{l-1\}})$.
    Update intermediate guess $q_{\{l\}}^{(k+1)} = q_{\{l\}}^* - \Delta q_{\{l\}}$.
  Set $k = k + 1$.
} **Until** residual norm $\|u_{\{l\}}\| < \varepsilon$.

Solving the fine-grid problem by direct matrix factorization is impractical for large N_l since $P_{\{l\}}$ is dense. However, it may be possible to factor the smaller matrix $P_{\{l-1\}}$ corresponding to the coarse-grid problem, since $N_{l-1} = N_l/4$. This motivates our development of a *two-grid method* (TGM), in which the problem is solved *iteratively* at level l with the help of direct solution at level $(l-1)$. The two principal algorithmic components, analogous to TGM for PDE's [1], [9], [2], are the *smoothing* operator and the *intergrid transfer*, or restriction-prolongation, operators. In our TGM iteration for solving $P_{\{l\}}\, q_{\{l\}} = v_{\{l\}}$, the *error* in the k-th iterate, $q_{\{l\}}^{(k)}$, is *smoothed* by carefully solving a series of local problems. This first stage is typically called "fine-grid smoothing", and results in an intermediate guess $q_{\{l\}}^*$. Next, we compute the *residual* $u_{\{l\}} = P_{\{l\}} \cdot q_{\{l\}}^* - v_{\{l\}}$ and project it onto the coarse grid via $u_{\{l-1\}} = r u_{\{l\}}$, where r is a *restriction* operator. Then we solve explicitly the coarse-grid problem $P_{\{l-1\}} \cdot (\Delta q_{\{l-1\}}) = u_{\{l-1\}}$ for $\Delta q_{\{l-1\}}$, and project the result onto the fine grid via $\Delta q_{\{l\}} = p(\Delta q_{\{l-1\}})$, where p is a *prolongation* operator. Finally, the intermediate guess on the fine grid is updated to yield the $(k+1)$-st iterate $q_{\{l\}}^{(k+1)} = q_{\{l\}}^* - \Delta q_{\{l\}}$. This second stage is termed "coarse-grid correction" and is responsible for "long-range" interactions. The fine-grid smoothing/coarse-grid correction cycle is repeated until the norm of the residual $u_{\{l\}}$ is below some tolerance. TGM is summarized in Algorithm 1, where the matrix D is described below.

To derive the smoothing operator, we first make the important observation that the BEM matrix P is derived from a Green's function $G(\mathbf{r}; \mathbf{r}')$ that is sharply peaked as $\mathbf{r} \to \mathbf{r}'$, but is *smooth* otherwise, *i.e.* $|\mathbf{r} - \mathbf{r}'| > d$ for some distance d. We seek an *operator splitting* at level l

$$P_{\{l\}} = D_{\{l\}} + S_{\{l\}}, \tag{5}$$

such that $D_{\{l\}}$ captures the short-range, sharply-peaked portion of $P_{\{l\}}$, and $S_{\{l\}}$ captures the long-range, smooth portion of $P_{\{l\}}$. See Figure 4 for a rough depiction.

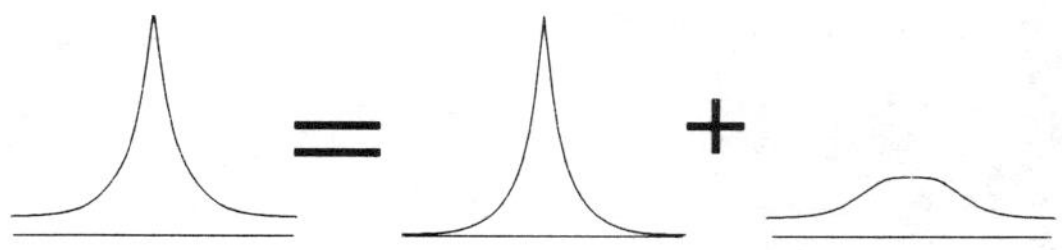

Fig. 4. Operator Splitting $P = D + S$.

Given (5), we define the smoothing operator as the result of solving

$$D_{\{l\}} \cdot q^*_{\{l\}} = -S_{\{l\}} \cdot q^{(k)}_{\{l\}} + v_{\{l\}} \tag{6}$$

for the vector $q^*_{\{l\}}$. Equation (6) defines a *fixed-point* iteration [21], since the condition $q^{(k)}_{\{l\}} = q_{\{l\}}$, where $q_{\{l\}}$ is the exact solution of the fine-grid problem, would lead to $q^*_{\{l\}} = q_{\{l\}}$. Since it is necessary that the above smoothing step be done cheaply, we require that $D_{\{l\}}$ be easy to invert, or that $D^{-1}_{\{l\}}$ has a *sparse* representation, since

$$q^*_{\{l\}} = D^{-1}_{\{l\}} \cdot \left(-S_{\{l\}} \cdot q^{(k)}_{\{l\}} + v_{\{l\}}\right). \tag{7}$$

Here, we construct directly a sparse matrix $D^{-1}_{\{l\}}$ based on the *preconditioner* developed in Fastcap [14], a BEM-based capacitance extraction program. Our particular implementation is outlined as follows. For each panel i, a local coefficient-of-potential matrix P_{loc} involving only panel i and its immediate neighbors is constructed. For the uniform grid problem, the size of P_{loc} is at most 9×9. This small matrix is easily inverted to yield P^{-1}_{loc}, whose elements from the *row* corresponding to panel i are then extracted and *stamped* into corresponding locations in the i-th *row* of $D^{-1}_{\{l\}}$. Hence, the matrix $D^{-1}_{\{l\}}$ contains at most 9 non-zero entries per row and is *sparse*. We recall that the panel-to-panel interaction coefficient can be computed inexpensively from a $M \times M$ DCT array described in Section II.

Because we do not construct $D_{\{l\}}$ directly, it may seem at first glance that the matrix $S_{\{l\}} \equiv P_{\{l\}} - D_{\{l\}}$ is difficult to obtain. But by subtracting $(D_{\{l\}} \cdot q^{(k)}_{\{l\}})$ from (6) and then multiplying through by $D^{-1}_{\{l\}}$, the result

$$\begin{aligned} \Delta q^*_{\{l\}} &= D^{-1}_{\{l\}} \cdot \left(-P_{\{l\}} \cdot q^{(k)}_{\{l\}} + v_{\{l\}}\right) \\ q^*_{\{l\}} &= q^{(k)}_{\{l\}} + \Delta q^*_{\{l\}}. \end{aligned} \tag{8}$$

can be used to compute $q^*_{\{l\}}$. This requires only operators $D^{-1}_{\{l\}}$ and $P_{\{l\}}$ which are readily available.

In addition to the smoothing operator, we require transfer operators r and p between the two grids. They are trivial in the case of uniform grids, where a coarse-grid panel, called a *parent*, is composed of four fine-grid panels, called *kids*. See Figure 3. Recall that in the Galerkin formulation, q_i is the net current on panel i, and v_j the average potential on panel j. For the restriction operator r mapping from level l to level $(l-1)$, the net current on a parent is simply the *sum* of the currents on the four kids, and the average potential on a parent is the average of the four kid potentials. The prolongation operator p mapping from level $(l-1)$ to level l is defined as the *adjoint*, or transpose, of the restriction r [2].

The *multigrid method* (MGM) is the generalization of the two-grid method to an arbitrary number of levels, and this is done exactly as in the standard multigrid literature. Instead of solving the coarse-grid problem explicitly at level $(l-1)$, which may still be too expensive, we apply a similar smoothing-correction cycle at level $(l-1)$. In the same manner, the correction cycle at level $(l-1)$ becomes a smoothing-correction cycle at level $(l-2)$, and so on. The integral equation (4) is now discretized at all levels $\{l_{min}, \ldots, l_{max}\}$. Only at the coarsest level $(l = l_{min})$ is the system $P_{\{l\}}\ q_{\{l\}} = v_{\{l\}}$ solved explicitly. Hence, each multigrid iteration is best described as a *recursive* function call.

B. Hierarchical Basis Functions for Complicated Domains

For integral equations defined on an arbitrarily shaped region S, we recall that the linear system (2) results from a Galerkin discretization based on *constant strength* panels. Figure 2 shows an example of discretization using square panels of equal size. We define here the *characteristic function*, $\mathcal{X}_i(\mathbf{r})$,associated with each panel p_i

$$\mathcal{X}_i(\mathbf{r}) = \begin{cases} 1/a_i & \text{if } \mathbf{r} \in p_i \\ 0 & \text{otherwise} \end{cases} \tag{9}$$

where a_i is the area of p_i. The Galerkin coefficients P_{ij} given in (3) is equivalent to the definition

$$P_{ij} = \int_S \int_S G(\mathbf{r};\mathbf{r}') \cdot \mathcal{X}_i(\mathbf{r}) \cdot \mathcal{X}_j(\mathbf{r}')\,da\,da' \tag{10}$$

where the integrations are now over the entire surface S.

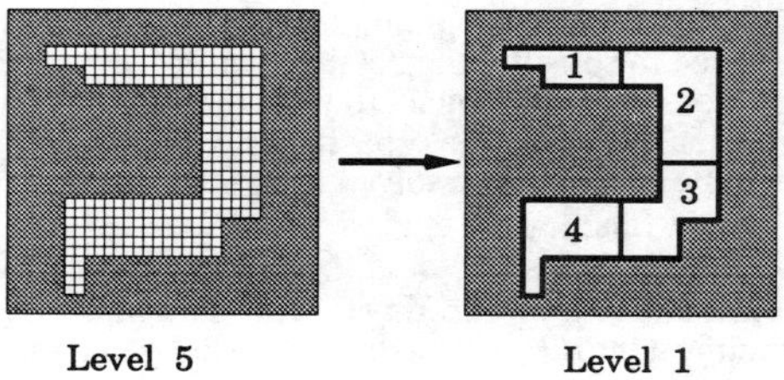

Fig. 5. Hierarchical basis functions.

The explicit use of characteristic functions allows us to construct coarser-level representations of the integral equation defined over complicated geometries. For simplicity, let us assume that at the finest level l_{max}, each panel coincides with a *cell* on the regular $M \times M$ grid, where $M = 2^{l_{max}}$. Of course, not all of the M^2 cells are occupied by panels. A panel at a *coarser* level $l = \{(l_{max} - 1), \ldots, 0\}$ is *defined* in the following manner. At each level l, the substrate is represented by a regular $2^l \times 2^l$ array of level-l cells. For each non-empty level-l cell, a level-l panel is defined as the *union* of all finest-level panels within that cell. The k-th panel at level l is denoted by $p^{\{l\}}_k$, and is associated with a characteristic function $\mathcal{X}^{\{l\}}_k$ defined similarly as in (9). Given the set of hierarchical basis functions $\{\mathcal{X}^{\{l\}}_k\}$, we can now easily define discrete representations of the integral equation $P^{\{l\}} \cdot q^{\{l\}} = v^{\{l\}}$ at each level l, where the Galerkin matrix elements $P^{\{l\}}_{ij}$ are computed as

$$P^{\{l\}}_{ij} = \int_S \int_S G(\mathbf{r};\mathbf{r}') \cdot \mathcal{X}^{\{l\}}_i(\mathbf{r}) \cdot \mathcal{X}^{\{l\}}_j(\mathbf{r}')\,da\,da' \tag{11}$$

Figure 5 shows a finest-grid discretization at level 5 being mapped into a coarse representation with only four panels at level 1. An efficient, hierarchical algorithm for the actual computation of the integrals in (11) is described in the next section.

C. Sparsification via Eigendecomposition and Moment-Matching

In this section, we first describe how to compute the matrix-vector product $P^{\{l_{max}\}} \cdot q^{\{l_{max}\}}$ efficiently at the finest level via an *eigendecomposition* technique. Then we propose a moment-matching algorithm which is used in conjuction with eigendecomposition to compute $P^{\{l\}} \cdot q^{\{l\}}$ efficiently at at all coarser levels $l = \{(l_{max} - 1), \ldots, 0\}$.

We first define a *surface eigenfunction* $\varphi(\mathbf{r})$ as an eigenfunction of the integral operator defined over the entire substrate surface $\Omega = [0, a] \times [0, b]$

$$\lambda \cdot \varphi(\mathbf{r}) = \int_{\Omega} \varphi(\mathbf{r}')G(\mathbf{r}; \mathbf{r}')da', \qquad \mathbf{r} \in \Omega. \tag{12}$$

It has been shown [4] that the surface eigenfunctions are

$$\varphi_{ij}(x, y) = \cos\left(\frac{i\pi x}{a}\right) \cos\left(\frac{j\pi y}{b}\right), \tag{13}$$

where the *eigenvalues* λ_{ij} are given in [7], [6]. Although the eigenfunctions in (13) also appear in [7], [6], they were used only to construct the panel-to-panel interaction coefficients P_{ij}. The dense matrix-vector multiplication $P \cdot q$ still requires $\mathcal{O}(N_p^2)$ operations in [7], where N_p is the number of panels. In contrast, we use the eigenfunctions to *expand* the global current density $J(x, y)$ and show that the $P \cdot q$ product can be computed in $\mathcal{O}(2 \cdot M^2 \cdot \log_2(M))$ operations using the DCT. If we assume that all panels are minimum sized cells on the $M \times M$ substrate grid, and that 10% of the cells are occupied by actual panels (*i.e.* $N_p = (0.1)M^2$), then the cost of computing the dense $P \cdot q$ product directly is $\mathcal{O}(0.01 \times M^4)$. At $M = 128$, eigendecomposition is already an order of magnitude faster than direct multiplication.

We define a prototype characteristic function centered about the origin in one dimension

$$\Theta_a(x) = \begin{cases} M/a & \text{if } |x| \leq a/2M \\ 0 & \text{otherwise} \end{cases}, \tag{14}$$

It is then clear that the panel characteristic functions $\mathcal{X}_i(x, y)$ can be obtained by combining and shifting $\Theta_a(x)$ and $\Theta_b(y)$. The global current density resulting from a Galerkin panel discretization is then

$$J(x, y) = \sum_{i=0}^{M-1} \sum_{j=0}^{M-1} q_{ij} \cdot \Theta_a\left(x - \frac{(i + 1/2)\,a}{M}\right) \cdot \Theta_b\left(y - \frac{(j + 1/2)\,b}{M}\right). \tag{15}$$

If we can expand $J(x, y)$ in (15) in terms of the eigenfunctions

$$J(x, y) = \sum_{i=0}^{\infty} \sum_{j=0}^{\infty} a_{ij}\, \varphi_{ij}(x, y), \tag{16}$$

i.e. if the coefficients $\{a_{ij}\}$ can be quickly computed, then (12) immediately leads to

$$\Phi(x, y) = \sum_{i=0}^{\infty} \sum_{j=0}^{\infty} \lambda_{ij}\, a_{ij}\, \varphi_{ij}(x, y). \tag{17}$$

Given the potential everywhere (17), the *average* panel potential $\overline{\Phi}_{pq}$ at the (p, q) position in the cell array is

$$\overline{\Phi}_{pq} = \sum_{i=0}^{\infty} \sum_{j=0}^{\infty} C_{ij} \lambda_{ij} a_{ij} \cos\left(\frac{(p + 1/2)\pi i}{M}\right) \cdot \cos\left(\frac{(q + 1/2)\pi j}{N}\right), \tag{18}$$

where C_{ij} are normalization coefficients. It can be shown [4] that if (16) and (18) are truncated to a finite series, the $M \times M$ coefficients $\{a_{ij}\}$ can be computed via a forward Type-2 DCT, and then the $M \times M$ array $\{\overline{\Phi}_{pq}\}$ via an inverse Type-2 DCT.

To calculate a single Galerkin coefficient $P_{ij}^{\{l\}}$ between two *coarse-grid* panels defined by (11), it is sufficient to perform a double summation of the panel-to-panel coefficients $P_{ij}^{\{l_{max}\}}$ at the finest level. However, this leads to an $\mathcal{O}(N_p^2)$ algorithm. Instead, we make the observation that when two coarse panels $p_i^{\{l\}}$ and $p_j^{\{l\}}$ are "well-separated", their interaction coefficient $P_{ij}^{\{l\}}$ can be computed approximately by leaving out much of the detail in the characteristic functions $\mathcal{X}_i^{\{l\}}$ and $\mathcal{X}_j^{\{l\}}$. Similar ideas have been used extensively in multipole-accelerated algorithms[16], [14], [24]. For the multipole approximation used in [24], it was necessary to assume a substrate Green's function which has translational invariance and which can be fitted to a sum of polynomials in $(1/r)$, where $r = |\mathbf{r} - \mathbf{r}'|$. In contrast, we develop here a fast moment-matching method which can be used in combination with the DCT to accelerate the coarse-grid computations *and* account properly for all the substrate boundary effects.

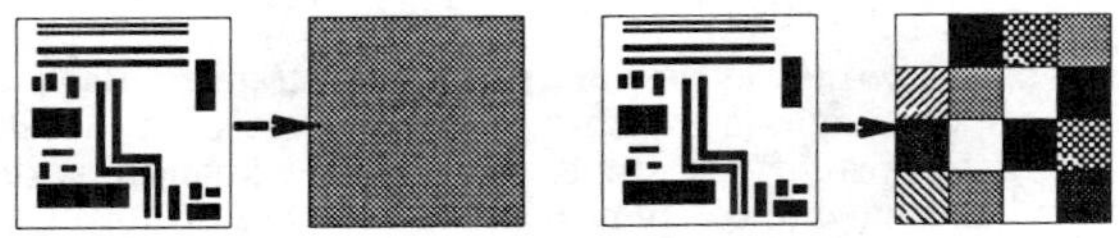

Fig. 6. Moment-matched representation of coarse panel.

If we are interested in the potential at a panel $p_j^{\{l\}}$ far from $p_i^{\{l\}}$, it may suffice to approximate $\mathcal{X}_i^{\{l\}}$ by distributing the current uniformly over the entire level-l *cell* as shown in Figure (6) on the left. This idea can be refined by matching higher-order moments of the characteristic function $\mathcal{X}_i^{\{l\}}$ with a regular $2^k \times 2^k$ array of characteristic functions associated with the cells at level $(l + k)$. For the choice $k = 2$, each coarse panel is associated with a set of $4 \times 4 = 16$ coefficients, as shown in Figure (6) on the right. If we let $\overline{\mathcal{X}}_i^{\{l\}}$ be the approximate characteristic funtion generated from the 4×4 cell array, the moment matching conditions require that the moments $Q_{\alpha\beta}^{(\gamma)}$ and $\overline{Q}_{\alpha\beta}^{(\gamma)}$ defined by

$$Q_{\alpha\beta}^{(\gamma)} = \int \mathcal{X}_i^{\{l\}} x^\alpha y^\beta dx dy, \quad \overline{Q}_{\alpha\beta}^{(\gamma)} = \int \overline{\mathcal{X}}_i^{\{l\}} x^\alpha y^\beta dx dy, \tag{19}$$

match exactly up to a certain order γ. The Cartesian moments defined in (19) for each order γ requires that $\alpha + \beta = \gamma$ and that α takes on all values $\{\ 0, \ldots, \gamma\ \}$. It can be shown that the if the Cartesian moments match up to order γ, then the *difference* between the the potentials generated by $\overline{\mathcal{X}}_i^{\{l\}}$ and $\mathcal{X}_i^{\{l\}}$ is of order $(1/r^{\gamma+1})$, but the proof is too long to state here.

Since each coarse-grid panel $p_i^{\{l\}}$ is now associated with 16 *geometric coefficients*, and since these coefficients correspond to cell currents at level $(l + k)$, it is possible to compute the $P^{\{l\}} q^{\{l\}}$ product in a matrix-free manner by using the DCT on a $2^{l+k} \times 2^{l+k}$ grid, similar to the finest level computation. However, since the nearest-neighbor panel interactions are not well approximated by moment matching, they are computed directly.

IV. Computational Results

We present numerical experiments comparing two iterative methods for solving (2): our new multigrid (MG) algorithm and the standard Generalized Minimal RESidual algorithm (GMRES [17]) without preconditioning. Since (2) results from a first-kind integral operator (1), the smallest eigenvalues of the matrix $P_{\{l\}}$ approach zero with increasing mesh refinement [12] and $P_{\{l\}}$ becomes more *ill-conditioned*. It is well-known that Krylov-subspace based iterative methods such as GMRES or CG (Conjugate Gradient) converge slowly for ill-conditioned linear systems [21]. Although it is possible to appply *preconditioning* to accelerate GMRES convergence as was done in [14], an increasing number of iterations is still required for finer discretizations. We demonstrate that the multigrid algorithm resolves this difficulty by retaining a *constant* convergence rate per iteration, independent of mesh refinement, and hence problem size. Thus, for a given relative error tolerance, the number of multigrid iterations required is *fixed.*

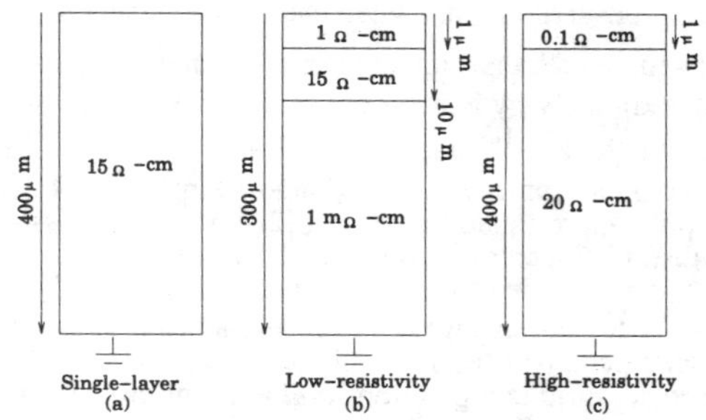

Fig. 7. Example substrate profiles.

In Figure 7, we display the three vertical substrate profiles used in this section: the single-layer substrate, the low-resistivity substrate, and the high-resistivity substrate. The lateral dimensions of the substrate is assumed to be 1mm × 1mm (or 1000 μm × 1000 μm).

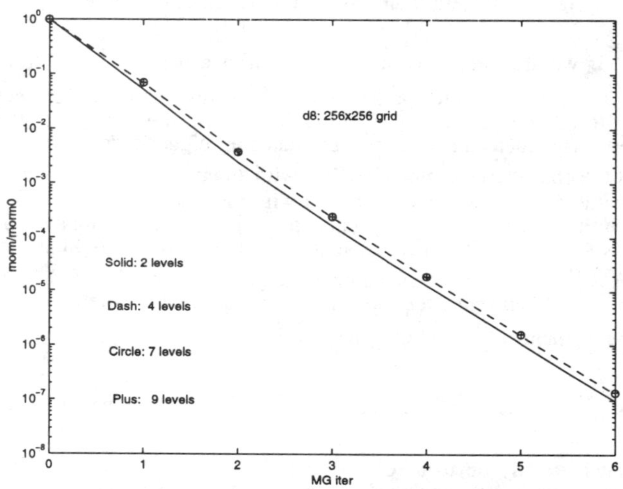

Fig. 8. Two-grid vs. Multigrid convergence.

The efficiency of multigrid algorithms in general arises from the fact that the *smoothing operator* at each level reduces the corresponding error components by the *same* numeric ratio [2]. This is demonstrated for the single-layer substrate example discretized with a 256 × 256 grid, and with a cell occupancy factor of one-half. The normalized residual, $||P_{\{l\}}q_{\{l\}}^{(k)} - v_{\{l\}}||/||v_{\{l\}}||$, is plotted versus the MG iteration count in Figure 8 for the two-grid method TGM (2 levels) and the multigrid method MGM (9 levels). The same convergence rate of about an order of magnitude per iteration is observed for multigrid methods of varying *depths*. We note that MGM requires only the *application* of operators P_l at various levels (and the solution of a scalar equation at the coarsest level $l = 0$), whereas TGM requires *solution* of the system at level $(l-1)$. Since the size of the linear system decreases *geometrically* with the level index, the cost of an MGM iteration is a constant multiple of an operator application at the finest level, or equivalently, a single GMRES iteration. Using the algorithms developed in Section III, we have observed this factor to be three to four in our implementation.

The crucial feature of multigrid schemes is that the convergence rate is independent of discretization, and hence problem size. We perform our next experiment with the single-layer substrate on five

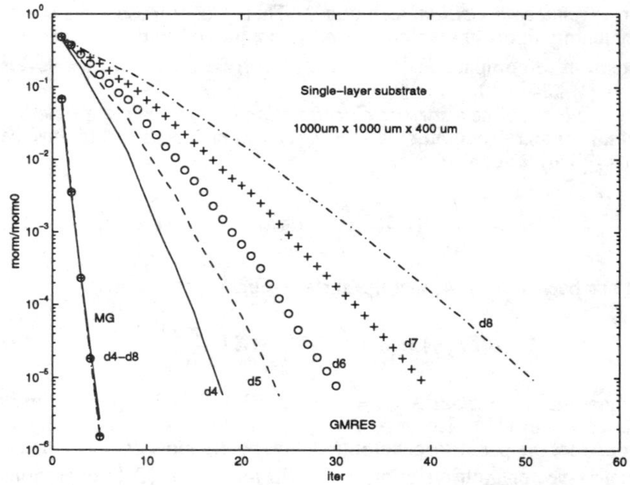

Fig. 9. Effect of mesh refinement on convergence.

test layouts of increasing mesh refinement, labeled $d = 4$ through $d = 8$. The $d = 4$ layout is discretized on a 16 × 16 grid, and the $d = 8$ layout is discretized on a 256 × 256 grid. The multigrid method with maximum depth ($l_{min} = 0, l_{max} = d$) is applied to solve each problem. The observed MG convergence rate is indeed independent of mesh size, as shown by the residual versus iteration plot in Figure 9. Also displayed are the GMRES convergence rates, which deteriorate with increasing mesh refinement as expected. Since the cost of a single MG iteration is a constant multiple of that of a GMRES iteration (three to four in our case), it is clear that MG is superior to GMRES, especially for large problems requiring fine discretization.

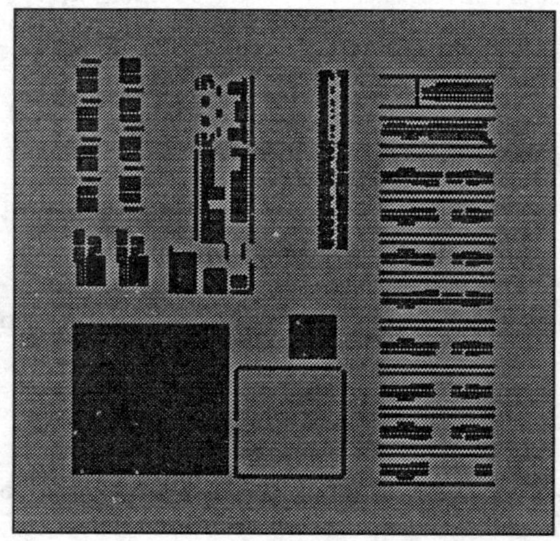

Fig. 10. PLL active area layout.

To show that the multigrid approach can be applied to realistic problems, we perform substrate parameter extraction on a a Phase Lock Loop (PLL) frequency synthesizer circuit [3] on a 1mm × 1mm chip. There are 478 substrate contacts defined by the active layer mask shown in Figure 10. Discretized with the help of a 1024 × 1024 grid, the total number of panels, or minimum-size cells, is $N_l = 183905$. This corresponds to roughly 20% of the chip area. The resolution thus achieved is about 1 micron.

The GMRES algorithm with sparsification via eigendecomposition [4] is used as a timing benchmark. We demonstrate the efficiency of MG versus GMRES by comparing the CPU times required to extract the 478 × 478 substrate conductance matrix. The substrate profile is assumed to be (a) single-layer, (b) low-resistivity, or (c) high-resistivity shown in Figure 7. We require convergence to a tolerance of 1e−3 in the relative residual norm for each of the 478 solves. The timing results are summarized in Table I. It is seen from the total extraction time that MG is faster than GMRES by almost an order of magnitude for the low-resistivity and high-resistivity substrates. More significant gains will be seen for even larger problems requiring finer meshes.

	T_{setup}	T_{iter}	N_{iters}	T_{solve}	T_{total}
MG (a)	317s	28.5s	3	85.5s	11.4h
GMRES (a)	11.2s	7.32s	50	366s	48.6h
MG (b)	343s	28.4s	4	114s	15.2h
GMRES (b)	29.9s	8.1s	95	771s	102h
MG (c)	333s	28.0s	6	168s	22.3h
GMRES (c)	23.0s	8.4s	180	1512s	201h

TABLE I
COMPUTATIONAL COST FOR PLL SUBSTRATE EXTRACTION.

A possible limitation of the multigrid algorithm is that it resolves only ill-conditioning caused by mesh refinement. It is less effective in dealing with ill-conditioning caused by the apparent loss of the groundplane, as seen in a slowdown of MG convergence for the high-resistivity case (c). To maintain the optimal MG convergence rate, it is necessary to *explicitly solve* the problem at some level $l_{min} > 0$.

V. CONCLUSIONS AND ACKNOWLEDGEMENTS

In this paper we described a multigrid method for solving the first-kind integral formulation of the substrate extraction problem. We believe that the ideas proposed here can be generalized to solving other problems arising from first-kind integral equations defined over complicated surfaces, such as BEM capacitance extraction [14].
The authors wish to thank Ranjit Gharpurey of Texas Instruments and Edoardo Charbon of Cadence Design Systems for helpful discussions and for providing the PLL layout example.

ACKNOWLEDGEMENTS

This work was supported by ARPA, MURI, SRC, and grants from Harris Semiconductor and HP.

REFERENCES

[1] A. Brandt. Multi-level adaptive solutions to boundary-value problems. *Mathematics of Computation*, 31(138):333–390, April 1977.

[2] W. L. Briggs. *A Multigrid Tutorial.* Society for Industrial and Applied Mathematics, Philadelphia, 1987.

[3] E. Charbon, R. Gharpurey, R.G. Mayer, and A. Sangiovanni Vincentelli. Semi-analytical techniques for substrate characterization in the design of mixed-signal ics. In *International Conference on Computer Aided-Design*, pages 455–462, San Jose, CA, November 1996.

[4] J. P. Costa, M. Chou, and L. M. Silveira. Efficient techniques for accurate modeling and simulation of substrate coupling in mixed-signal ic's. In *DATE'98 - Design, Automation and Test in Europe, Exhibition and Conference*, pages 892–898, Paris, France, February 1998.

[5] R. Gharpurey and R.G. Meyer. Modeling and analysis of substrate coupling in integrated circuits. In *IEEE 1995 Custom Integrated Circuits Conference*, pages 125–128, 1995.

[6] Ranjit Gharpurey. *Modeling and Analysis of Substrate Coupling in Integrated Circuits.* PhD thesis, Department of Electrical Engineering and Computer Science, University of California at Berkeley, Berkeley, CA, June 1995.

[7] Ranjit Gharpurey and Robert G. Meyer. Modeling and analysis of substrate coupling in integrated circuits. *IEEE Journal Solid-State Circuits*, 31(3):344–353, March 1996.

[8] L. Greengard. *The Rapid Evaluation of Potential Fields in Particle Systems.* M.I.T. Press, Cambridge, Massachusetts, 1988.

[9] W. Hackbusch. *Multi-Grid Methods and Applications.* Springer-Verlag, Berlin Heidelberg New York Tokyo, 1985.

[10] R. F. Harrington. *Field Computation by Moment Methods.* Macmillan, New York, 1968.

[11] T. A. Johnson, R.W. Knepper, V. Marcellu, and W. Wang. Chip substrate resistance modeling technique for integrated circuit design. *IEEE Transactions on Computer-Aided Design of Integrated Circuits*, CAD-3(2):126–134, 1984.

[12] R. Kress. *Linear Integral Equations.* Springer-Verlag, 1989.

[13] Sujoy Mitra, R. A. Rutenbar, L. R. Carley, and D. J. Allstot. A methodology for rapid estimation of substrate-coupled switching noise. In *IEEE 1995 Custom Integrated Circuits Conference*, pages 129–132, 1995.

[14] K. Nabors and J. White. Fastcap: A Multipole Accelerated 3-D Capacitance Extraction Program. *IEEE Transactions on Computer-Aided Design*, pages 1447–1459, November 1991.

[15] Bram Nauta and Gian Hoogzaad. How to deal with substrate noise in analog cmos circuits. In *European Conference on Circuit Theory and Design*, pages Late 12:1–6, Budapest, Hungary, September 1997.

[16] V. Rohklin. Rapid solution of integral equation of classical potential theory. *J. Comput. Phys.*, 60:187–207, 1985.

[17] Y. Saad and M. H. Schultz. Gmres: A generalized minimum residual algorithm for solving nonsymmetric linear systems. *SIAM J. Stat. Comp.*, 7:856–869, 1986.

[18] A. H. Schatz, V. Thomée, and W. L. Wendland. *Mathematical Theory of Finite and Boundary Element Methods.* Birkhäuser Verlag, Basel/Boston/Berlin, 1990.

[19] T. Smedes, N. P. van der Meijs, and A. J. van Genderen. Extraction of circuit models for substrate cross-talk. In *International Conference on Computer Aided-Design*, pages 199–206, San Jose, CA, November 1995.

[20] B.R. Stanisic, N.K. Verghese, R.A. Rutenbar, L.R. Carley, and D.J. Allstot. Addressing substrate coupling in mixed-mode ic's: Simulation and power distribution synthesis. *IEEEJSSC*, 29(3):226–238, March 1994.

[21] J. Stoer and R. Bulirsch. *Introduction to Numerical Analysis.* Springer-Verlag, second edition, 1993.

[22] D.K. Su, M.J. Loinaz, S. Masui, and B.A. Wooley. Experimental results and modeling techniques for substrate noise in mixed-signal integrated circuits. *IEEEJSSC*, 28(4):420–430, April 1993.

[23] Nishath Verghese. *Extraction and Simulation Techniques for Substrate-Coupled Noise in Mixed-Signal Integrated Circuits.* PhD thesis, Department of Electrical and Computer Engineering, Carnegie Mellon University, Pittsburgh, PA, August 1995.

[24] Nishath K. Verghese, David J. Allstot, and Mark A. Wolfe. Verification techniques for substrate coupling and their application to mixed-signal ic design. *IEEE Journal Solid-State Circuits*, 31(3):354–365, March 1996.

[25] Ivan L. Wemple and Andrew T. Yang. Mixed-signal switching noise analysis using voronoi-tesselation substrate macromodels. In 32nd *ACM/IEEE Design Automation Conference*, pages 439–444, San Francisco, CA, June 1995.

2.2

Phase Noise in Oscillators: A Unifying Theory and Numerical Methods for Characterisation

Alper Demir Amit Mehrotra* Jaijeet Roychowdhury
{alpdemir, mehrotra, jaijeet} @research.bell-labs.com

Bell Laboratories
Murray Hill, New Jersey, USA

Abstract

Phase noise is a topic of theoretical and practical interest in electronic circuits, as well as in other fields such as optics. Although progress has been made in understanding the phenomenon, there still remain significant gaps, both in its fundamental theory and in numerical techniques for its characterisation. In this paper, we develop a solid foundation for phase noise that is valid for any oscillator, regardless of operating mechanism. We establish novel results about the dynamics of stable nonlinear oscillators in the presence of perturbations, both deterministic and random. We obtain an exact, nonlinear equation for phase error, which we solve without approximations for random perturbations. This leads us to a precise characterisation of timing jitter and spectral dispersion, for computing which we develop efficient numerical methods. We demonstrate our techniques on practical electrical oscillators, and obtain good matches with measurements even at frequencies close to the carrier, where previous techniques break down.

1 Introduction

Oscillators are ubiquitous in physical systems, especially electronic and optical ones. For example, in radio frequency (RF) communication systems, they are used for frequency translation of information signals and for channel selection. Oscillators are also present in digital electronic systems which require a time reference, i.e., a clock signal, in order to synchronise operations.

Noise is of major concern in oscillators, because introducing even small noise into an oscillator leads to dramatic changes in its frequency spectrum and timing properties. This phenomenon, peculiar to oscillators, is known as *phase noise* or *timing jitter*. A perfect oscillator would have localized tones at discrete frequencies (i.e., harmonics), but any corrupting noise spreads these perfect tones, resulting in high power levels at neigbouring frequencies. This effect is the major contributor to undesired phenomena such as interchannel interference, leading to increased bit-error-rates (BER) in RF communication systems. Another manifestation of the same phenomenon, jitter, is important in clocked and sampled-data systems: uncertainties in switching instants caused by noise lead to synchronisation problems. Characterising how noise affects oscillators is therefore crucial for practical applications. The problem is challenging, since oscillators constitute a special class among noisy physical systems: their *autonomous* nature makes them unique in their response to perturbations.

Considerable effort has been expended over the years in understanding phase noise and in developing analytical, computational and experimental techniques for its characterisation (see Section 3 for a brief review). Despite the importance of the problem and the large number of publications on the subject, a consistent and general treatment, and computational techniques based on a sound theory, appear to be still lacking. In this work, we provide a novel, rigorous theory for phase noise and derive efficient numerical methods for its characterisation. Our techniques and results are general; they are applicable to *any* oscillatory system, electrical (resonant, ring, relaxation, etc.) or otherwise (gravitational, optical, mechanical, biological, etc.). The main ideas behind our approach, and our contributions, are outlined in Section 2. We apply our numerical techniques to a variety of practical oscillator designs and obtain good matches against measurements.

The paper is organised as follows. In Section 2, we present some preliminaries and an overview of the main results of the paper, and in Section 3, we give a brief review of the previous work. In Section 4, we consider the traditional approach (linearisation) to analysing perturbed nonlinear systems, and show how this procedure is not consistent for autonomous oscillators. In Section 5, we derive a nonlinear equation that exactly captures how perturbations result in phase noise. In Section 6, we solve this equation with random perturbations and arrive at a stochastic description of phase deviation, from which we derive timing jitter. Next, in Section 7, we use this stochastic characterisation to calculate the correct shape of the oscillator's spectrum with phase noise. In Section 8, we derive several quantities commonly used in oscillator design to quantify jitter and spectral properties. In Section 9, we address the problem of computing these quantities efficiently and develop numerical methods that can easily be implemented in existing simulators. Finally, in Section 10, we apply our methods to practical electrical oscillators. All proofs and discussion of mathematical background are omitted due to space limitations.

2 Preliminaries and overview

The dynamics of any autonomous system without undesired perturbations can be described by a system of differential equations:[1]

$$\dot{x} = f(x) \tag{1}$$

where $x \in \mathbb{R}^n$ and $f(\cdot) : \mathbb{R}^n \rightarrow \mathbb{R}^n$. We assume that $f(\cdot)$ satisfies the conditions of the Picard-Lindelőf existence and uniqueness theorem for initial value problems [2]. We consider systems that have an asymptotically orbitally stable[2] periodic solution $x_s(t)$ (with period T) to (1), i.e., a stable limit cycle in the n-dimensional solution space. We are interested in the response of such systems to a small state-dependent perturbation of the form $B(x)b(t)$ where $B(\cdot) : \mathbb{R}^n \rightarrow \mathbb{R}^{n\times p}$ and $b(\cdot) : \mathbb{R} \rightarrow \mathbb{R}^p$. Hence the perturbed system is described by

$$\dot{x} = f(x) + B(x)b(t) \tag{2}$$

Let the exact solution of the perturbed system in (2) be $z(t)$.

Although our eventual intent is to understand the response of the oscillator when $b(t)$ is random noise, it is useful to consider first the case when $b(t)$ is a known deterministic signal. We carry out a rigorous analysis of this case in Section 5 and obtain the following results:

1. the unperturbed oscillator's periodic response $x_s(t)$ is modified to $x_s(t + \alpha(t)) + y(t)$ by the perturbation, where:
   (a) $\alpha(t)$ is a changing time shift, or *phase deviation*, in the periodic output of the unperturbed oscillator.
   (b) $y(t)$ is an additive component, which we term the *orbital deviation*, to the phase-shifted oscillator waveform.
2. $\alpha(t)$ and $y(t)$ can always be chosen such that:
   (a) $\alpha(t)$ will, in general, keep increasing with time even if the perturbation $b(t)$ is always small.
   (b) the orbital deviation $y(t)$, on the other hand, will always remain small.

*visiting from the University of California, Berkeley.

DAC 98, San Francisco, California

[1] For notational simplicity, we use the ODE formulation throughout the paper to describe the dynamics of an autonomous system. The results and the numerical methods we present can be extended [1] for the MNA (Modified Nodal Analysis) formulation (i.e., DAE formulation) given by $d/dt\, q(x) + f(x) = 0$.

[2] After any small disturbance that does not persist, the system asymptotically settles back to the original limit cycle. See [2] for a precise definition of this stability notion.

These results concretise existing intuition amongst designers about oscillator operation. Our proof of these facts is mathematically rigorous; further, we derive equations for $\alpha(t)$ and $y(t)$ which lead to *qualitatively different results* about phase noise compared to previous attempts. This is because our results are based on a new nonlinear perturbation analysis that is valid for oscillators, in contrast to previous approaches that rely on linearisation. We show in Section 4 that analysis based on linearisation is not consistent for oscillators and results in non-physical predictions.

Next, we consider the case where the perturbation $b(t)$ is random noise – this situation is important for determining practical figures of merit like zero-crossing jitter and spectral purity (i.e., spreading of the power spectrum)[3]. Jitter and spectral spreading are in fact closely related, and both are determined by the manner in which $\alpha(t)$, now also a random process, spreads with time. We consider random perturbations in detail in Sections 6 and 7, and establish that:

1. the average spread of the jitter (mean-square jitter) increases precisely *linearly* with time.
2. the power spectrum of the perturbed oscillator is a *Lorentzian*[4] about each harmonic.
3. a *single scalar constant c* is sufficient to describe jitter and spectral spreading in a noisy oscillator.
4. the oscillator's output is a *stationary* stochastic process.

These results have important implications. The Lorentzian shape of the spectrum implies that the power spectral density at the carrier frequency and its harmonics has a finite value, and that the total carrier power is preserved despite spectral spreading due to noise. Previous analyses based on linear time-invariant (LTI) or linear time-varying (LTV) concepts erroneously predict infinite noise power density at the carrier, as well as infinite total integrated power. That the oscillator output is stationary is surprising at first sight, since oscillators are nonlinear systems with periodic swings, hence it might be expected that output noise power would change periodically as in forced systems. However, it must be remembered that while forced systems are supplied with an external time reference (through the forcing), oscillators are not. Cyclostationarity in the oscillator's output would, by definition, imply a time reference. Hence the stationarity result reflects the fundamental fact that noisy autonomous systems cannot provide a perfect time reference.

3 Previous work

A great deal of literature is available on the phase noise problem. Here we mention only some selected works. Most investigations of electronic oscillators aim to provide insight into frequency-domain properties of phase noise, in order to develop rules for designing practical oscillators; well-known references include [4, 5, 6, 7, 8]. Usually, these approaches apply linear time-invariant (LTI) analysis to high-Q or quartz-crystal type oscillators designed using standard feedback topologies. Arguments based on deterministic perturbations are used to show that the spectrum of the oscillator response varies as $1/f^2$ times the spectrum of the perturbation. While often of great practical importance, such analyses often require large simplifications of the problem, and skirt fundamental issues such as why noisy oscillators exhibit spectral dispersion whereas forced systems do not.

Attempts to improve on LTI analysis have borrowed from linear time-varying (LTV) analysis methods for forced (nonoscillatory) systems (e.g., [9, 10, 11, 12]). LTV analyses can predict spectra more accurately than LTI ones in some frequency ranges; however, LTV techniques for forced systems retain nonphysical artifacts of LTI analysis (such as infinite output power) and provide no real insight into the basic mechanism generating phase noise.

Possibly the most general and rigorous treatment of phase noise to date has been that of Kärtner [13]. In this work, the oscillator response is decomposed into phase and magnitude components, and a differential equation is obtained for phase error. By solving a linear, small-time approximation to this equation with stochastic inputs, Kärtner obtains the correct Lorentzian spectrum for the power spectral density due to phase noise. Despite these advances, certain gaps remain, particularly with respect to the derivation and solution of the differential equation for phase error.

[3]The deterministic perturbation case is also of interest, for, e.g., phenomena such as mode locking in forced oscillators. We consider this case elsewhere [3].

[4]A Lorentzian is the shape of the squared magnitude of a one-pole lowpass filter transfer function.

Recently, Hajimiri [14] has proposed a phase noise analysis based on a conjecture for decomposing perturbations into two (orthogonal) components, generating purely phase and amplitude deviations respectively. While this intuition is similar to Kärtner's approach [13], other aspects of Hajimiri's treatment (e.g., stochastic characterisation for phase deviation and the spectrum calculation) are essentially equivalent to LTV analysis. Unfortunately, the conjecture for orthogonally decomposing the perturbation into components that generate phase and amplitude deviations, while intuitively appealing, can be shown to be invalid [15]. Design intuition resulting from the conjecture about noise source contributions can also be misleading.

In summary, the available literature often identifies basic and useful facets of phase noise separately, but lacks a rigorous unifying theory clarifying its fundamental mechanism. Furthermore, existing numerical methods for phase noise are based on forced-system concepts which are inappropriate for oscillators and can generate incorrect predictions.

4 Perturbation analysis using linearisation

The traditional approach to analysing perturbed nonlinear systems is to linearise about the unperturbed solution, under the assumption that the resultant deviation[5] will be small. Let this deviation be $w(t)$, i.e., $z(t) = x_s(t) + w(t)$. Substituting this expression for $z(t)$ in (2), replacing $f(x_s(t)+w(t))$ by its first order Taylor series expansion, and approximating $B(x)$ with $B(x_s)$ (assuming $w(t)$ "small"), we obtain

$$\dot{w} \approx \left.\frac{\partial f(x)}{\partial x}\right|_{x_s(t)} w(t) + B(x_s(t))b(t) = A(t)w(t) + B(x_s(t))b(t) \quad (3)$$

where the Jacobian $A(t) = \left.\frac{\partial f(x)}{\partial x}\right|_{x_s(t)}$ is T-periodic. Here, we used the fact that $x_s(t)$ satisfies (1). Now, we would like to solve for $w(t)$ in (3) to see if our assumption that it is small is indeed justified. For this, we use results from Floquet theory [2, 16] as follows[6].

The state transition matrix for the homogeneous part of (3) is given by

$$\Phi(t,s) = U(t)\exp(D(t-s))V(s) = \sum_{i=1}^{n} u_i(t)\exp(\mu_i(t-s))v_i^T(s) \quad (4)$$

where $U(t)$ is a T-periodic nonsingular matrix, $V(t) = U^{-1}(t)$ and $D = diag[\mu_1, \ldots, \mu_n]$, where μ_i are the *Floquet (characteristic) exponents.* $\exp(\mu_i T)$ are called the *characteristic multipliers.* $u_i(t)$ are the columns of $U(t)$ and $v_i^T(t)$ are the rows of $V(t) = U^{-1}(t)$.

Remark 4.1 *$\{u_1(t), u_2(t), \ldots, u_n(t)\}$ and $\{v_1(t), v_2(t), \ldots, v_n(t)\}$ both span $\mathbb{R}^n$ and satisfy the* biorthogonality conditions *$v_i^T(t)u_j(t) = \delta_{ij}$ for every t. Note that, in general, $U(t)$ itself is not an orthogonal matrix.*

Let us first consider the homogeneous part of (3), the solution of which is given by

$$w_H(t) = \sum_{i=1}^{n} u_i(t)\exp(\mu_i t)v_i^T(0)w(0) \quad (5)$$

where $w(0)$ is the initial condition. Next, we will show that one of the terms in the summation in (5) does not decay with t.

Lemma 4.1

- *The unperturbed oscillator (1) has a non-trivial T-periodic solution $x_s(t)$ if and only if the number 1 is a characteristic multiplier of the homogeneous part of (3), or equivalently, one of the Floquet exponents satisfies* $\exp(\mu_i T) = 1$.
- *The time-derivative of the periodic solution $x_s(t)$ of (1), i.e., $\dot{x}_s(t)$, is a solution of the homogeneous part of (3).*

[5]By deviation we refer to the difference between the solutions of the perturbed and unperturbed systems.

[6]The reader who is unfamiliar with Floquet theory is encouraged to review it before continuing.

Remark 4.2 *One can show that if 1 is a characteristic multiplier, and the remaining $n-1$ Floquet exponents satisfy $|\exp(\mu_i T)| < 1$, $i = 2,\dots,n$, then the periodic solution $x_s(t)$ of (1) is asymptotically orbitally stable, and it has the asymptotic phase property [2].*[7] *Moreover, if any of the Floquet exponents satisfy $|\exp(\mu_i T)| > 1$, then the solution $x_s(t)$ is orbitally* unstable.

Without loss of generality, we choose $\mu_1 = 0$ and $u_1(t) = \dot{x}_s(t)$.

Remark 4.3 *With $u_1(t) = \dot{x}_s(t)$, we have $v_1^T(t)\dot{x}_s(t) = 1$ and $v_1^T(t)u_j(t) = 0$, $j = 2,\dots,n$. $v_1(t)$ will play an important role in the rest of our treatment.*

Next, we obtain the particular solution of (3), given by

$$w_P(t) = \sum_{i=1}^{n} u_i(t) \int_0^t \exp(\mu_i(t-r)) v_i^T(r) B(x_s(r)) b(r) dr \tag{6}$$

The first term in the above summation is given by $u_1(t)\int_0^t v_1^T(r)B(x_s(r))b(r)dr$, since $\mu_1 = 0$. If the integrand has a nonzero average value, then the deviation $w(t)$ in (3) will grow unbounded. Hence, the assumption that $w(t)$ is small becomes invalid and the linearised perturbation analysis is inconsistent.

When the perturbation $b(t)$ is a vector of uncorrelated white noise sources, one can show that the variances of the entries of $w(t)$ can grow unbounded. Thus, the assumption that the deviation $w(t)$ stays small[8] is also invalid for the stochastic perturbation case.

5 Nonlinear perturbation analysis for phase deviation

As seen in the previous section, traditional perturbation techniques do not suffice for analysing oscillators. In this section, a novel nonlinear perturbation analysis suitable for oscillators is presented.

The new analysis proceeds along the following lines:

1. Rewrite (2) with the (small) perturbation $B(x)b(t)$ split into two small parts $b_1(x,t)$ and $\tilde{b}(x,t)$:
$$\dot{x} = f(x) + b_1(x,t) + \tilde{b}(x,t) \tag{7}$$
2. Choose the first perturbation term $b_1(x,t)$ in such a way that its effect is to create only *phase errors* to the unperturbed solution. In other words, show that the equation
$$\dot{x} = f(x) + b_1(x,t) \tag{8}$$
is solved by $x_p(t) = x_s(t+\alpha(t))$ for a certain function $\alpha(t)$, called the *phase deviation*. It will be seen that $\alpha(t)$ can grow unboundedly large with time even though the perturbation $b_1(x,t)$ remains small.
3. Now treat the remaining term $\tilde{b}(x,t)$ as a small perturbation to (8), and perform a consistent traditional perturbation analysis in which the resultant deviations from $x_p(t)$ remain small. I.e., show that $z(t) = x_s(t+\alpha(t)) + y(t)$ solves (7) for a certain $y(t)$ that *remains small* for all t. $y(t)$ will be called the *orbital deviation*.

We start by defining $\alpha(t)$ concretely through a differential equation.

Definition 5.1 *Define $\alpha(t)$ by*

$$\frac{d\alpha(t)}{dt} = v_1^T(t+\alpha(t))B(x_s(t+\alpha(t)))b(t), \quad \alpha(0) = 0 \tag{9}$$

Remark 5.1 *$\alpha(t)$ can grow unbounded even if $b(t)$ remains small. For example, consider the case where $b(t)$ is a small positive constant $\varepsilon \ll 1$, $B \equiv 1$, and $v_1(t)$ is a constant k. Then $\alpha(t) = k\varepsilon t$.*

Having defined $\alpha(t)$, we are in a position to split $B(x)b(t)$ into $b_1(x,t)$ and $\tilde{b}(x,t)$:

Definition 5.2 *Let*

$$b_1(x,t) = c_1(x,t)u_1(t+\alpha(t)), \quad \textit{and} \tag{10}$$

$$\tilde{b}(x,t) = B(x)b(t) - b_1(x,t) = \sum_{i=2}^{n} c_i(x,t)u_i(t+\alpha(t)), \tag{11}$$

where the scalars $c_i(x,t) = v_i^T(t+\alpha(t))B(x)b(t)$

Note that $b_1(x,t)$ is obtained by projecting the original perturbation along the time-varying direction $u_1(t+\alpha(t))$. u_i, v_i are the Floquet vectors in Remark 4.1.

Lemma 5.1 *$x_p(t) = x_s(t+\alpha(t))$ solves (8).*

Lemma 5.1 states that the $b_1(x,t)$ component causes deviations only along the limit cycle, i.e., phase deviations. Next, we show that the remaining perturbation component $\tilde{b}(x,t)$ perturbs $x_p(t)$ only by a small amount $y(t)$, provided $b(t)$ is small.

Lemma 5.2 *For $b(t)$ sufficiently small, the mapping $t \mapsto t+\alpha(t)$ is invertible.*

Definition 5.3 *Let $b(t)$ be small enough that $\hat{t}(t) = t+\alpha(t)$ is invertible. Then define $\hat{b}(\cdot)$ by $\hat{b}(\hat{t}) = b(t)$, and $y(t)$ by*

$$y(t) = \sum_{i=2}^{n} u_i(\hat{t}) \int_0^{\hat{t}} \exp(\mu_i(\hat{t}-r)) v_i^T(r) B(x_s(r)) \hat{b}(r) dr \tag{12}$$

where $\hat{t} = t+\alpha(t)$.

Remark 5.2 *Note that the index of the summation in (12) starts from 2. Since $|\exp(\mu_i T)| < 1, i \geq 2$ (due to asymptotic orbital stability), this implies that $y(t)$ is within a constant factor of $b(t)$, hence small.*

Theorem 5.1 *If $b(t)$ is small (implying that $y(t)$ in Definition 5.3 is also small), then $z(t) = x_p(t) + y(t)$ solves (7) to first order in $y(t)$.*

6 Stochastic characterisation of the phase deviation α

We now find the probabilistic characterisation of the phase deviation α (Definition 5.1) as a stochastic process when the perturbation $b(t)$ is a vector of uncorrelated[9] Gaussian white noise sources. We will treat (9) as a stochastic differential equation [17, 18].

We will follow the below procedure to find an adequate probabilistic characterisation of the phase deviation α for our purposes:

1. We first calculate the time-varying *probability density function* (PDF) $p_\alpha(\eta,t)$ of α defined as
$$p_\alpha(\eta,t) = \frac{\partial \mathrm{P}(\alpha(t) \leq \eta)}{\partial \eta} \qquad t \geq 0$$
where $\mathrm{P}(.)$ denotes the *probability measure*, and show that it becomes the PDF of a Gaussian random variable asymptotically with t. A Gaussian PDF is completely characterised by the mean and the variance of the random variable. We show that $\alpha(t)$ becomes, asymptotically with time, a Gaussian random variable with a constant (as a function of t) mean and a variance that is linearly increasing with time.[10]
2. The time-varying PDF $p_\alpha(\eta,t)$ does not provide any correlation information between $\alpha(t)$ and $\alpha(t+\tau)$ that is needed for the evaluation of its spectral characteristics. We then calculate this correlation to be
$$\mathrm{E}[\alpha(t)\alpha(t+\tau)] = m^2 + c\min(t,t+\tau)$$
where m and c are scalar constants.
3. We then show that $\alpha(t_1)$ and $\alpha(t_2)$ become *jointly* Gaussian asymptotically with time, which does not follow immediately from the fact that they are individually Gaussian.

[7] Note that this is a sufficient condition for asymptotic orbital stability, not a necessary one. We assume that this sufficient condition is satisfied by the system and the periodic solution $x_s(t)$.

[8] The notion of "staying small" is quite different for a stochastic process than the one for a deterministic function. For instance, a Gaussian random variable can take arbitrarily large values with nonzero probability even when its variance is "small". We say that a stochastic process is "bounded" when its variance is bounded, even though some of its sample paths (representing a nonzero probability) can grow unbounded.

[9] The extension to correlated noise sources is trivial. We consider uncorrelated noise sources for notational simplicity. Moreover, various noise sources in electronic devices usually have independent physical origin, and hence they are modeled as uncorrelated stochastic processes.

[10] The fact that $\alpha(t)$ is a Gaussian random variable for every t does not imply that α is a Gaussian stochastic process. Individually Gaussian random variables are not necessarily jointly Gaussian.

Starting with the stochastic differential equation (9) for α, one can derive a partial differential equation, known as the *Fokker-Planck equation* [18, 19], for the time-varying PDF $p_\alpha(\eta,t)$. The Fokker-Planck equation for $\alpha(t)$ takes the form

$$\frac{\partial p_\alpha(\eta,t)}{\partial t} = -\frac{\partial}{\partial\eta}\left(\lambda p_\alpha(\eta,t)\frac{\partial v^T(t+\eta)}{\partial\eta}v(t+\eta)\right) + \frac{1}{2}\frac{\partial^2}{\partial\eta^2}\left(v^T(t+\eta)v(t+\eta)p_\alpha(\eta,t)\right) \tag{13}$$

where $v^T(t) = v_1^T(t)B(x_s(t))$, and $0 \le \lambda \le 1$ depends on the definition of the stochastic integral [18] used to interpret the stochastic differential equation in (9). We would like to solve (13) for $p_\alpha(\eta,t)$. It turns out that $p_\alpha(\eta,t)$ becomes a *Gaussian* PDF asymptotically with *linearly* increasing variance. We show this by first solving for the *characteristic function* $F(\omega,t)$ of $\alpha(t)$, which is defined by

$$F(\omega,t) = \mathrm{E}\left[\exp\left(j\omega\alpha(t)\right)\right] = \int_{-\infty}^{\infty}\exp\left(j\omega\eta\right)p_\alpha(\eta,t)d\eta$$

Since both $v_1^T(.)$ and $B(x_s(.))$ are T-periodic in their arguments, $v^T(.)$ is also periodic in its argument with period T. Hence we can expand $v^T(t)$ into its Fourier series: $v^T(t) = \sum_{i=-\infty}^{\infty} V_i^T\exp(ji\omega_0 t)$ where $\omega_0 = 2\pi/T$.

Lemma 6.1 *The characteristic function of $\alpha(t)$, $F(\omega,t)$, satisfies*

$$\frac{\partial F(\omega,t)}{\partial t} = \sum_{i=-\infty}^{\infty}\sum_{k=-\infty}^{\infty} V_i^T V_k^* \exp(j\omega_0(i-k)t) \left(-\lambda\omega_0 i\omega - \frac{1}{2}\omega^2\right)F(\omega_0(i-k)+\omega,t) \tag{14}$$

where $*$ denotes complex conjugation.

Theorem 6.1 *(14) has a solution that becomes the characteristic function of a Gaussian random variable asymptotically with time:*

$$\lim_{t\to\infty} F(\omega,t) = \exp\left(j\omega\mu(t) - \frac{\omega^2\sigma^2(t)}{2}\right) \tag{15}$$

solves (14), where $\mu(t) = m$ is a constant, and $\sigma^2(t) = ct$ where

$$c = \frac{1}{T}\int_0^T v^T(t)v(t)dt. \tag{16}$$

The variance of this Gaussian random variable increases linearly with time, exactly as in a Wiener process.

Remark 6.1 *$\alpha(t)$ becomes, asymptotically with t, a Gaussian random variable with mean $\mu(t) = m$ and variance $\sigma^2(t) = ct$.*

Lemma 6.2

$$\mathrm{E}\left[\alpha(t)\alpha(t+\tau)\right] = \begin{cases} \mathrm{E}\left[\alpha^2(t)\right] & \text{if } \tau \ge 0 \\ \mathrm{E}\left[\alpha^2(t+\tau)\right] & \text{if } \tau < 0 \end{cases}$$

Corollary 6.1 *Asymptotically with t*

$$\mathrm{E}\left[\alpha(t)\alpha(t+\tau)\right] = m^2 + c\min(t,t+\tau)$$

Definition 6.1 *Two real valued random variables Ψ_1 and Ψ_2 are called* jointly *Gaussian if for all $a_1, a_2 \in \mathrm{IR}$, the real random variable $a_1\Psi_1 + a_2\Psi_2$ is Gaussian.*

Theorem 6.2 *Asymptotically with time, $\alpha(t_1)$ and $\alpha(t_2)$ become* jointly *Gaussian.*

The stochastic characterisation of the phase deviation α we obtained in this section can be summarized by Remark 6.1, Lemma 6.2, Corollary 6.1 and Theorem 6.2. These provide adequate information for a practical characterisation of the effect of phase deviation α on the signal generated by an autonomous oscillator, e.g., its spectral properties, as we will see in Section 7 and Section 8.

7 Spectrum of an oscillator with phase noise

Having obtained the asymptotic stochastic characterisation of α, we now compute the power spectral density (PSD) of $x_s(t+\alpha(t))$. We first obtain an expression for the non-stationary autocorrelation function $R(t,\tau)$ of $x_s(t+\alpha(t))$. Next, we demonstrate that the autocorrelation becomes independent of t asymptotically. This implies our main result, that the autocorrelation of the oscillator output with phase noise contains no non-trivial cyclostationary components, confirming the intuitive expectation that a noisy autonomous system cannot have periodic cyclostationary variations because it has no perfect time reference. Finally, we show that the PSD of the stationary component is a summation of Lorentzian spectra, and that a single scalar constant, namely c in (16), is sufficient to characterize it.

We start by calculating the autocorrelation function of $x_s(t+\alpha(t))$, given by

$$R(t,\tau) = \mathrm{E}\left[x_s(t+\alpha(t))x_s^*(t+\tau+\alpha(t+\tau))\right] \tag{17}$$

Definition 7.1 *Define X_i to be the Fourier coefficients of $x_s(t)$: $x_s(t) = \sum_{i=-\infty}^{\infty} X_i\exp\left(ji\omega_0 t\right)$.*

Lemma 7.1

$$R(t,\tau) = \sum_{i=-\infty}^{\infty}\sum_{k=-\infty}^{\infty} X_i X_k^* \exp\left(j(i-k)\omega_0 t\right)\exp\left(-jk\omega_0\tau\right) \mathrm{E}\left[\exp\left(j\omega_0\beta_{ik}(t,\tau)\right)\right] \tag{18}$$

where $\beta_{ik}(t,\tau) = i\alpha(t) - k\alpha(t+\tau)$.

To evaluate the expectation in the above Lemma, it is useful to consider first the statistics of $\beta_{ik}(t,\tau)$.

Lemma 7.2

$$\lim_{t\to\infty}\mathrm{E}\left[\beta_{ik}(t,\tau)\right] = (i-k)m \tag{19}$$

$$\lim_{t\to\infty}\mathrm{E}\left[(\beta_{ik}(t,\tau))^2\right] - \left(\mathrm{E}\left[\beta_{ik}(t,\tau)\right]\right)^2 = \begin{matrix}(i-k)^2ct + k^2c\tau \\ -2ikc\min(0,\tau)\end{matrix} \tag{20}$$

where m and c are defined in Theorem 6.1. Also, $\beta_{ik}(t,\tau)$ becomes Gaussian asymptotically with t.

Using the asymptotically Gaussian nature of $\beta_{ik}(t,\tau)$, we are now able to obtain a form for the expectation in (18).

Lemma 7.3 *If $c > 0$, the characteristic function of $\beta_{ik}(t,\tau)$ is asymptotically independent of t and has the following form:*

$$\lim_{t\to\infty}\mathrm{E}\left[\exp\left(j\omega_0\beta_{ik}(t,\tau)\right)\right] = \begin{cases} 0 & \text{if } i \ne k \\ \exp(-\frac{1}{2}\omega_0^2 k^2 c|\tau|) & \text{if } i = k \end{cases} \tag{21}$$

Lemma 7.4

$$\lim_{t\to\infty} R(t,\tau) = \sum_{i=-\infty}^{\infty} X_i X_i^* \exp\left(-ji\omega_0\tau\right)\exp\left(-\frac{1}{2}\omega_0^2 i^2 c|\tau|\right) \tag{22}$$

The spectrum of $x_s(t+\alpha(t))$ can now be determined as follows:

Lemma 7.5 *The spectrum of $x_s(t+\alpha(t))$ is determined by the asymptotic behaviour of $R(t,\tau)$ as $t\to\infty$. All non-trivial cyclostationary components are zero, while the stationary component of the spectrum is given by:*

$$S(\omega) = \sum_{i=-\infty}^{\infty} X_i X_i^* \frac{\omega_0^2 i^2 c}{\frac{1}{4}\omega_0^4 i^4 c^2 + (\omega + i\omega_0)^2} \tag{23}$$

There is also a term $X_0 X_0^*\,\delta(\omega)$ due to the DC part of $x_s(t)$, which is omitted in (23).

8 Phase noise/timing jitter characterisation

Single-sided spectral density and total power

The PSD $S(\omega)$ in (23) (defined for $-\infty < \omega < \infty$, hence called a double-sided density) is a real and even function of ω, because the periodic steady-state $x_s(t)$ is real hence its Fourier series coefficients X_i in Definition 7.1 satisfy $X_i = X^*_{-i}$. The *single-sided* spectral density (defined for $0 \le f < \infty$) is given by

$$S_{ss}(f) = 2S(2\pi f) = 2\sum_{i=-\infty}^{\infty} X_i X_i^* \frac{f_0^2 i^2 c}{\pi^2 f_0^4 i^4 c^2 + (f + i f_0)^2} \tag{24}$$

where we substituted $\omega = 2\pi f$ and $\omega_0 = 2\pi f_0$. The *total power* (i.e. the integral of the PSD over the range of the frequencies it is defined for) in $S_{ss}(f)$ is the same as in $S(2\pi f)$, which is

$$P_{tot} = \text{Total power in } S_{ss}(f) = \int_0^\infty S_{ss}(f)\mathrm{d}f = \sum_{i=1}^{\infty} 2\,|X_i|^2 \tag{25}$$

Remark 8.1 *The phase deviation $\alpha(t)$ does* not *change the total power in the periodic signal $x_s(t)$, but it alters the power density in frequency, i.e., the power spectral density. For the perfect periodic signal $x_s(t)$, the power spectral density has δ functions located at discrete frequencies (i.e., the harmonics). The phase deviation $\alpha(t)$ spreads the power in these δ functions in the form given in (24), which can be experimentally observed with a spectrum analyzer.*

Single-sideband phase noise spectrum in dBc/Hz

In practice, we are usually interested in the PSD around the first harmonic, i.e., $S_{ss}(f)$ for f around f_0. The *single-sideband* phase noise $\mathcal{L}(f_m)$ (in dBc/Hz) that is very widely used in practice is defined as

$$\mathcal{L}(f_m) = 10\log_{10}\left(\frac{S_{ss}(f_o + f_m)}{2\,|X_1|^2}\right) \tag{26}$$

For "small" values of c, and for $0 \le f_m \ll f_0$, (26) can be approximated as

$$\mathcal{L}(f_m) \approx 10\log_{10}\left(\frac{f_0^2 c}{\pi^2 f_0^4 c^2 + f_m^2}\right) \tag{27}$$

Furthermore, for $\pi f_0^2 c \ll f_m \ll f_0$, $\mathcal{L}(f_m)$ can be approximated by

$$\mathcal{L}(f_m) \approx 10\log_{10}\left(\left(\frac{f_0}{f_m}\right)^2 c\right) \tag{28}$$

Notice that the approximation of $\mathcal{L}(f_m)$ in (28) blows up as $f_m \to 0$. For $0 \le f_m < \pi f_0^2 c$, (28) is not accurate, in which case the approximation in (27) should be used.

Timing jitter

In some applications, such as clock generation and recovery, one is interested in a characterisation of the phase/time deviation $\alpha(t)$ itself rather than the spectrum of $x_s(t + \alpha(t))$ that was calculated in Section 7. In these applications, an oscillator generates a square-wave like waveform to be used as a clock. The effect of the phase deviation $\alpha(t)$ on such a waveform is to create *jitter* in the *zero-crossing* or *transition* times. In Section 6, we found out that $\alpha(t)$ (for an autonomous oscillator) becomes a Gaussian random variable with a linearly increasing variance $\sigma^2(t) = ct$. Let us take one of the transitions (i.e., edges) of a clock signal as a reference (i.e., trigger) transition and synchronize it with $t = 0$. If the clock signal is perfectly periodic, then one will see transitions exactly at $t_k = kT$, $k = 1, 2, \ldots$ where T is the period. For a clock signal with a phase deviation $\alpha(t)$ that has a linearly increasing variance as above, the timing of the kth transition t_k will have a variance (i.e., mean-square error) $\mathrm{E}\left[(t_k - kT)^2\right] = ckT$. The spectral dispersion caused by $\alpha(t)$ in an oscillation signal can be observed with a spectrum analyzer. Similarly, one can observe the timing jitter caused by $\alpha(t)$ using a sampling oscilloscope. McNeill in [20] experimentally observed the linearly increasing variance for the timing of the transitions of a clock signal generated by an autonomous oscillator, as predicted by our theory.

Noise source contributions

The scalar constant c appears in all of the characterisations we discussed above. It is given by

$$c = \frac{1}{T}\int_0^T v_1^T(\tau)B(x_s(\tau))B^T(x_s(\tau))v_1(\tau)\mathrm{d}\tau \tag{29}$$

where $B(.) : \mathbb{R}^n \to \mathbb{R}^{n\times p}$ represents the *modulation* of the intensities of the noise sources with the large-signal state. (29) can be rewritten as

$$c = \sum_{i=1}^{p} \frac{1}{T}\int_0^T [v_1^T(\tau)B_i(\tau)]^2\mathrm{d}\tau = \sum_{i=1}^{p} c_i \tag{30}$$

where p is the number of the noise sources, i.e., the column dimension of $B(x_s(.))$, and $B_i(.)$ is the ith column of $B(x_s(.))$ which maps the ith noise source to the equations of the system. Hence, c_i represents the contribution of the ith noise source to c. Thus, the ratio

$$\frac{c_i}{c = \sum_{i=1}^{p} c_i} \tag{31}$$

can be used as a *figure of merit* representing the contribution of the ith noise source to phase noise/timing jitter.

Phase noise sensitivity

One can define

$$c_s^{(k)} = \frac{1}{T}\int_0^T [v_1^T(\tau)e_k]^2\mathrm{d}\tau \tag{32}$$

(where $1 \le k \le n$ and e_k is the kth unit vector) as the *phase noise/timing jitter sensitivity* of the kth equation (i.e., node), because e_k represents a unit intensity noise source added to the kth equation (i.e., connected to the kth node) in (1).

9 Numerical methods

From Section 6, Section 7 and Section 8, for various phase noise characterisations of an oscillator, one needs to calculate the steady-state periodic solution $x_s(t)$, and the periodic vector $v_1(t)$ in (29). Without providing details, we will present the outline of a time-domain method for computing the periodic vector $v_1(t)$[11]. The procedure for calculating $v_1(t)$ in the time domain is as follows:

1. Compute the large-signal periodic steady-state solution $x_s(t)$ for $0 \le t \le T$ by numerically integrating (1), possibly using a technique such as the shooting method [21].
2. Compute the state-transition matrix $\Phi(T,0)$ by numerically integrating $\dot{Y} = A(t)Y,\;\; Y(0) = I_n$ from 0 to T, where the Jacobian $A(t)$ is defined in (3). Note that $\Phi(T,0) = Y(T)$.
3. Compute $u_1(0)$ using $u_1(0) = \dot{x}_s(0)$. Note that $u_1(0)$ is an eigenvector of $\Phi(T,0)$ corresponding to the eigenvalue 1.
4. $v_1(0)$ is an eigenvector of $\Phi^T(T,0)$ corresponding to the eigenvalue 1. To compute $v_1(0)$, first compute an eigenvector of $\Phi^T(T,0)$ corresponding to the eigenvalue 1, then scale this eigenvector so that $v_1(0)^T u_1(0) = 1$ is satisfied.
5. Compute the periodic vector $v_1(t)$ for $0 \le t \le T$ by numerically solving the adjoint system

   $$\dot{y} = -A^T(t)y \tag{33}$$

   using $v_1(0) = v_1(T)$ as the initial condition. Note that $v_1(t)$ is a periodic steady-state solution of (33) corresponding to the Floquet exponent that is equal to 0, i.e., $\mu_1 = 0$. It is not possible to calculate $v_1(t)$ by numerically integrating (33) *forward* in time, because the numerical errors in computing the solution and the numerical errors in the initial condition $v_1(0)$ will excite the *modes* of the solution of (33) that grow without bound. However, one can integrate (33) *backwards* in time with the "initial" condition $v_1(T) = v_1(0)$ to calculate $v_1(t)$ for $0 \le t \le T$ in a numerically stable way.
6. Then, c is calculated using (29).

[11] We also developed a frequency domain numerical method based on an harmonic balance formulation.

10 Examples

Oscillator with a bandpass filter and a nonlinearity [22]

This oscillator (Figure 1) consists of a Tow-Thomas second-order bandpass filter and a comparator [22]. If the OpAmps are considered to be ideal, it can be shown that this oscillator is equivalent (in the sense of the differential equations that describe it) to a parallel RLC circuit in parallel with a nonlinear voltage-controlled current source (or equivalently a series RLC circuit in series with a nonlinear current-controlled voltage source). In [22], authors breadboarded this circuit with an external white noise source (intensity of which was chosen such that its effect is much larger than the other internal noise sources), and measured the PSD of the output with a spectrum analyzer. For $Q = 1$ and $f_o = 6.66$ kHz, we performed a phase noise characterisation of this oscillator using our numerical methods, and computed the periodic oscillation waveform $x_s(t)$ for the output and $c = 7.56 \times 10^{-8}$ sec^2.Hz. Figure 2(a) shows the PSD of the oscillator output computed using (24), and Figure 2(b) shows the spectrum analyzer measurement[12]. The single-sideband phase noise spectrum using both (27) and (28) is in Figure 3 . Note that (28) can not predict the PSD accurately below the cut-off frequency $f_c = \pi f_0^2 c = 10.56$ Hz (marked with a $*$ in Figure 3) of the Lorentzian.

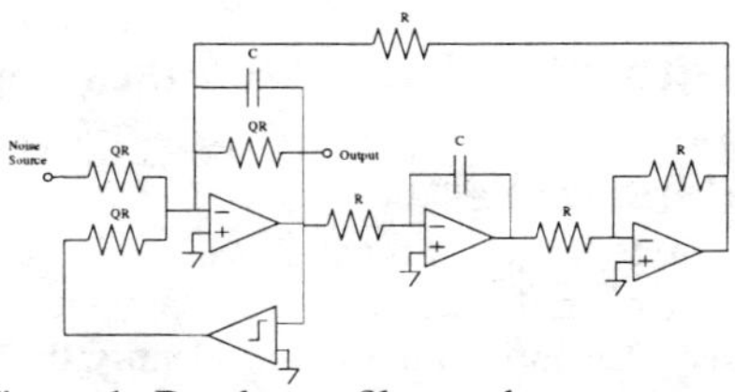

Figure 1: Band-pass filter and a comparator

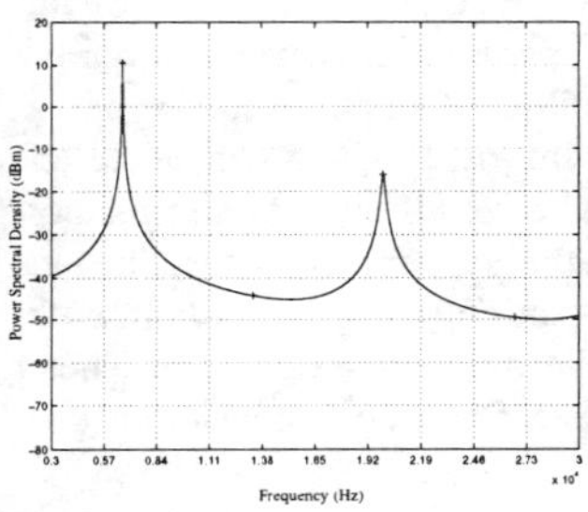

(a) Computed PSD (4 harmonics)

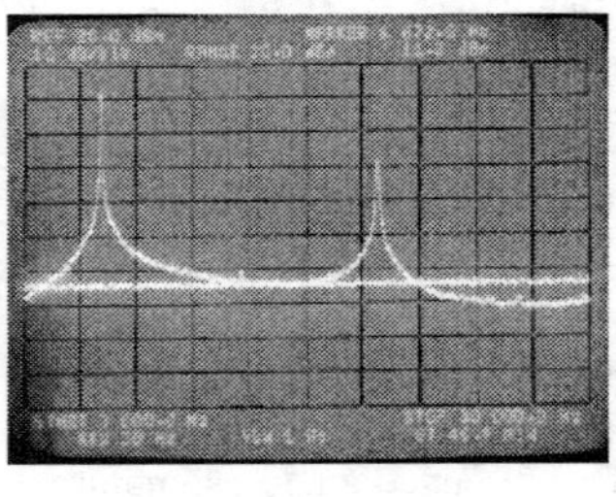

(b) Measured PSD [22]

Figure 2: Computed and measured PSD

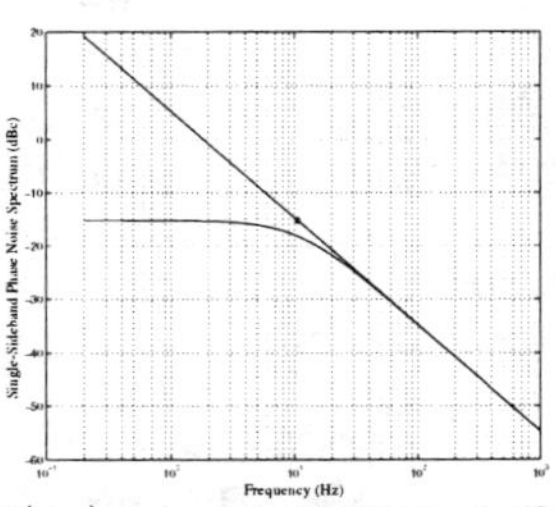

Figure 3: $\mathcal{L}(f_m)$ computed with both (27) and (28)

Ring oscillator

The ring oscillator circuit is a three stage oscillator with fully differential ECL buffer delay cells (differential pairs followed by emitter followers). This circuit is from [20]. [20] and [23] use analytical techniques to characterize the timing jitter/phase noise performance of ring-oscillators with ECL type delay cells. Since they use analytical techniques, they use a simplified model of the circuit and make several approximations in their analysis. [20] and [23] use time-domain

R_c (Ω)	r_b (Ω)	I_{EE} (μA)	f_o (MHz)	c (sec^2.Hz $\times 10^{-15}$)
500	58	331	167.7	0.269
2000	58	331	74	0.149
500	1650	331	94.6	0.686
500	58	450	169.5	0.182
500	58	600	169.7	0.151
500	58	715	167.7	0.142

(a) Phase noise characterisation

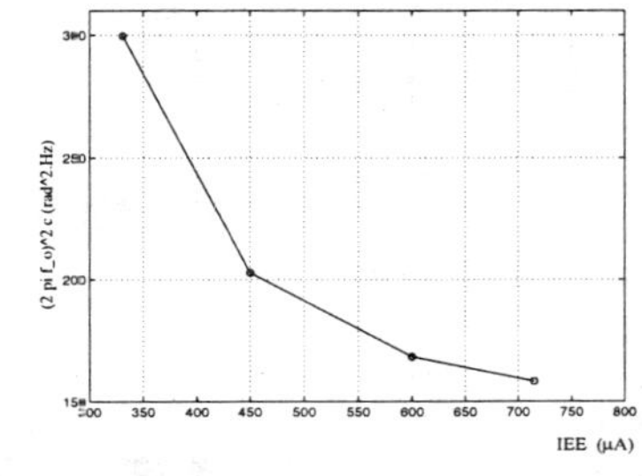

(b) Phase noise performance versus I_{EE}

Figure 4: Ring-oscillator

Monte Carlo noise simulations to verify the results of their analytical results. They obtain qualitative and some quantitative results, and offer guidelines for the design of low phase noise ring-oscillators with ECL type delay cells. However, their results are only valid for their specific oscillator circuits. We will compare their results with the results we will obtain for the above ring-oscillator using the general phase noise characterisation methodology we have proposed which makes it possible to analyze a complicated oscillator circuit without simplifications. We performed several phase noise characterisations of the bipolar ring-oscillator. The results are shown in Figure 4(a), where R_c is the collector load resistance for the differential pair (DP) in the delay cell, r_b is the zero bias base resistance for the BJTs in the DP, I_{EE} is the tail bias current for the DP, and f_o is the oscillation frequency for the three stage ring-oscillator. Note that the changes in R_c and r_b affect the oscillation frequency, unlike the changes in I_{EE}. Figure 4(b) shows a plot of $(2\pi f_o)^2\, c$ versus I_{EE} using the data from Figure 4(a). This prediction of the dependence of phase noise/timing jitter performance on the tail bias current is in agreement with the analysis and experimental results presented in [20] and [23] for ring-oscillators with ECL type delay cells. Note that larger values for $(2\pi f_o)^2\, c$ indicate *worse* phase noise performance.

References

[1] A. Demir. Floquet theory and phase noise in oscillators with differential-algebraic equations. Technical memorandum, Bell Laboratories, Murray Hill, January 1998.

[2] M. Farkas. *Periodic Motions*. Springer-Verlag, 1994.

[3] J. Roychowdhury. Multi-time analysis of mode-locking and quasi-periodicity in forced oscillators. Technical memorandum, Bell Laboratories, Murray Hill, 1997.

[4] D.B. Leeson. A simple model of feedback oscillator noise spectrum. *Proceedings of the IEEE*, 54(2):329, February 1966.

[5] W.P. Robins. *Phase Noise in Signal Sources*. Peter Peregrinus, 1991.

[6] U.L. Rohde. *Digital PLL Frequency Synthesizers: Theory and Design*. Prentice-Hall, 1983.

[7] J.R. Vig. *Quartz Crystal Resonators and Oscillators for Frequency Control and Timing Applications*. Army Research Laboratory, 1994.

[8] B. Razavi. Analysis, modeling and simulation of phase noise in monothilic voltage-controlled oscillators. In *Proc. IEEE Custom Integrated Circuits Conference*, May 1995.

[9] E. Hafner. The effects of noise in oscillators. *Proceedings of the IEEE*, 54(2):179, February 1966.

[10] K. Kurokawa. Noise in synchronized oscillators. *IEEE Transactions on Microwave Theory and Techniques*, MTT-16:234–240, 1968.

[11] V. Rizzoli, A. Costanzo, F. Mastri, and C. Cecchetti. Harmonic-balance optimization of microwave oscillators for electrical performance, steady-state stability, and near-carrier phase noise. In *IEEE MTT-S International Microwave Symposium Digest*, May 1994.

[12] M. Okumura and H. Tanimoto. A time-domain method for numerical noise analysis of oscillators. In *Proceedings of the ASP-DAC*, January 1997.

[13] F. Kärtner. Analysis of white and $f^{-\alpha}$ noise in oscillators. *International Journal of Circuit Theory and Applications*, 18:485–519, 1990.

[14] A. Hajimiri and T.H. Lee. A state-space approach to phase noise in oscillators. Technical memorandum, Lucent Technologies, July 1997.

[15] A. Demir and J. Roychowdhury. On the validity of orthogonally decomposed perturbations in phase noise analysis. Technical memorandum, Bell Laboratories, Murray Hill, 1997.

[16] R. Grimshaw. *Nonlinear Ordinary Differential Equations*. Blackwell Scientific, 1990.

[17] L. Arnold. *Stochastic Differential Equations: Theory and Applications*. John Wiley & Sons, 1974.

[18] C.W. Gardiner. *Handbook of Stochastic Methods for Physics, Chemistry and the Natural Sciences*. Springer-Verlag, 1983.

[19] H. Risken. *The Fokker-Planck Equation*. Springer-Verlag, 1989.

[20] J.A. McNeill. *Jitter in Ring Oscillators*. PhD thesis, Boston University, 1994.

[21] K.S. Kundert, J.K. White, and A. Sangiovanni-Vincentelli. *Steady-State Methods for Simulating Analog and Microwave Circuits*. Kluwer Academic Publishers, 1990.

[22] A. Dec, L. Toth, and K. Suyama. Noise analysis of a class of oscillators. To be published in IEEE Transactions on Circuits and Systems, 1997.

[23] T.C. Weigandt, B. Kim, and P.R. Gray. Analysis of timing jitter in cmos ring-oscillators. In *Proc. IEEE ISCAS*, June 1994.

[12]The PSDs are plotted in units of dBm.

Efficient Analog Test Methodology Based on Adaptive Algorithms

Luigi Carro, Marcelo Negreiros
Universidade Federal do Rio Grande do Sul - Departamento de Engenharia Elétrica
{carro, negreiro}@iee.ufrgs.br

Abstract

This papers describes a new, fast and economical methodology to test linear analog circuits based on adaptive algorithms. To the authors knowledge, this is the first time such technique is used to test analog circuits, allowing complete fault coverage. The paper presents experimental results showing easy detection of soft, large-deviation and hard faults, with low cost instrumentation. Components variations from 5% to 1% have been detected, as the comparison parameter (output error power) varied from 300% to 20%.

1 Introduction and motivation

Faults in analog circuits can cause different symptoms at the circuit outputs, from short and opening of components to slow deviations of the operation point caused by a degradation in passive or active component characteristics. Presently, there are some reported works that generate a minimum excitation set in order to detect these type of faults [1], [2]. The general idea is to test the response of the circuit to a given frequency. Deviations on circuit parameters caused by any soft or hard fault will affect the output response, either in its module or phase. The test engineer must supply a minimum number of excitation frequencies in order to detect various components faults.

In our methodology, the idea of exciting the circuit with different signals is maintained, by observing the output of the analog circuit and comparing it with an expected output. However, instead of exciting the circuit with a minimum set of frequencies in order to cover hard and soft faults (like in [1]), we excite the circuit with infinite frequencies at the same time (white noise). This way, we not only guarantee complete coverage of faults, but we are also able to detect in an easy way fine grain details like phase or amplitude variations at the output. Deviations of 1% from nominal value of the circuit components have been detected. Any analog function that can be represented in the s-plane by a pole-zero polynomial fraction N(s)/D(s) can be tested.

DAC 98, San Francisco, California
©1998 ACM 0-89791-964-5/98/06..$5.00

This paper is organized as follows: section 2 presents the overall idea of our approach, with mathematical explanation of its principles. Section 3 analyses a set of examples, were we show the success of the approach by detecting faults which are not easily detected. In section 4 we present some actual results of the methodology, followed by some limitations (section 5). Finally, section 6 presents our conclusions and future work.

2 Testing with adaptive procedures

2.1 The Adaptive Tester

In [3] and [4] DSP (Digital Signal Processing) techniques had been used to detect misbehaviors of AD converters. Basically, the Fast Fourier Transform was used to identify the presence of harmonics showing signal degradation. Recently, Spina and Upadhyaya ([5]) have used white noise to train an Artificial Neural Network to detect fault patterns in analog circuits. Although we use a similar technique (adaptive filters being related to neural networks as sum of products sequence of operations), we do not try to characterize all circuit patterns, nor we require any knowledge about the system under test. In our approach we first recognize the circuit under test as a plant, with all its pole-zero characteristics, like an analog signature. Then, we apply specific algorithms to detect any deviations from this first obtained analog signature.

The theory regarding linear plant recognition is quite settled [6]. Recently, with the advent of fast DSP microprocessors and boosted by digital communication problems, adaptive filtering took place as a good mathematical framework to solve not only line equalization problems, but plant recognition as well. The idea of an analog plant recognition by the use of an adaptive algorithm is shown in figure 1. A plant is submitted to an excitation, the same being done with the adaptive filter. Their output is compared, and the amount of error helps the adaptive filter change its coefficients in order to track plant performance.

After some time, all filter coefficients have stabilized, and the plant has a twin plant implemented in the filter. Now, one has only to change the output of the error signal in order to transform this adapted plant to a tester, as shown in figure 2. Once we substitute the plant by a circuit to be tested, the previously adapted filter will output an error signal showing how much the new plant is equal to the previous (fault free) one. It is interesting to notice that this explains the robustness of the methodology. The basic idea is an analogy to the detection of variations in a bridge circuit. Any fault in the analog circuit, being it soft, hard or catastrophic, will change the plant under test. In other words, any change in the plant signature (its pole-zero characteristics) will increase the power of the error signal. The tester will detect this change and compare it to a previously defined threshold, according to the tolerated variation of the circuit.

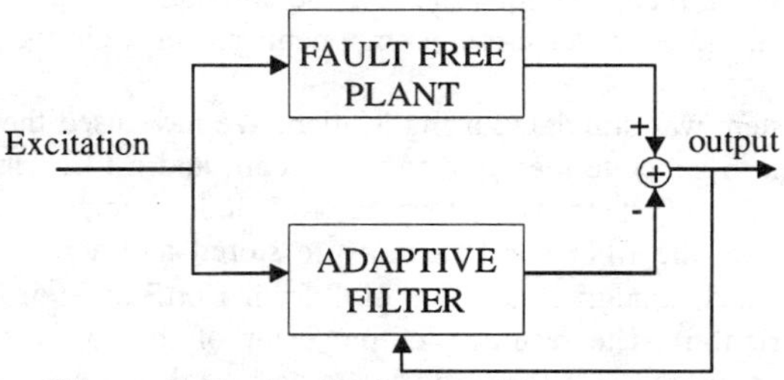

Figure 1. Basic plant recognition structure

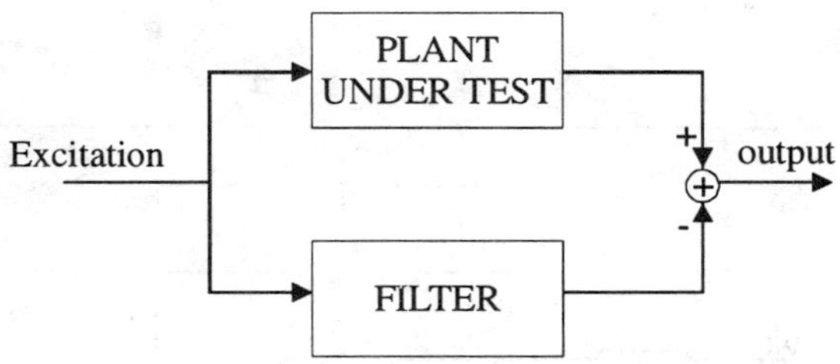

Figure 2. Plant being tested with filter

2.2 The adaptive algorithm and filter topology

A finite impulse response (FIR) digital filter ([7]) can be represented by a polynomial on its input signals delayed by a fixed amount of sampling times, like

$$Y(n) = a0*X(n) + a1*X(n-1) + ... + am-1*X(n-m+1) \quad (1).$$

Y(n) is the output at discrete time **n**, while **m** is the number of filter taps. In an adaptive digital filter all coefficients (a0 ... am-1) are determined at runtime. The questions to be answered are how to discover the correct set of coefficients, so that the output of the filter performs the desired functions and, also, the depth **m** of the filter. In case of an adaptive filter, the output should exactly mimic the output of the plant. There are available different algorithms ([8-10]), and their basic trade-off regards the amount of computations that must be performed against the speed of coefficient convergence.

The filter adapts to any plant, provided it has enough taps (pair coefficient-sample) to represent all poles and zeros of the original plant. Actually, the adaptive filter matches the impulse response of the plant. To decide which kind of signal to be used in order to excite the couple filter-plant, one must note that this signal must last long enough for the filter to converge (the error signal must go below a certain threshold), and must be rich regarding its frequency components, in order to provide excitation of the plant at all frequencies of interest.

In our case, we choose one of the simplest adaptive algorithms, the Least Mean Square or LMS, showed in figure 3.

```
begin
ai(0)=0, i = 0...m-1
loop
    Y(n)=SUM(ai(n)*X(n-i)) , i = 0...m-1          /* filter */
    error(n) = d(n) - Y(n)                        /* error */
    ai(n+1) = ai(n) + μ*error(n)*X(n-i)           /* coefficients */
end
```

Figure 3. LMS algorithm

Once one knows the plant to be mirrored, the sampling frequency should be determined by an evaluation of the plant fastest response to be tracked by the test. In other words, the frequency response of the plant must be under the Nyquist limit, fs/2. The maximum sampling frequency is a function of the data acquisition system.

The designer can now define the number of taps to be used by a set of relations shown in figure 4. To define the number of taps, one should know the sampling frequency (fs) and have an estimate for the duration of the plant impulse response (Tir). In figure 4 we show an example impulse response (after convergence), approximated by a filter with 40 taps.

The μ parameter can be chosen inside its upper and lower bounds ([8,9]), which are related to the power of the input signal applied to the system. The μ parameter can be a trade-off between speed of convergence (large μ) or precision in the error signal (smaller μ).

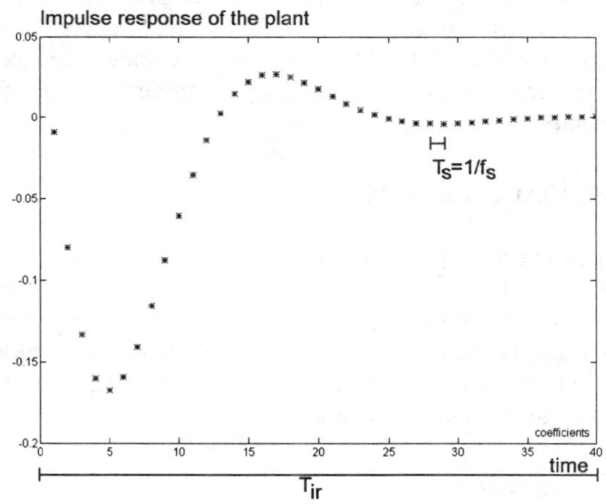

Ntaps = Tir/Ts
Ts=1/fs

Figure 4. Relations to define the minimum number of taps

2.3 Excitation Signal and the Tester

In order to excite the plant conveniently to detect the largest number of faults, the input signal should be such that all poles and zeros of the plant are properly excited. Any variation in this set of poles and zeros caused by any kind of fault will change the basic plant, being detected. The best signal would be one having all frequencies represented in its spectrum.

Our first approach was to use the impulse response of the system, for in an impulse signal all frequencies are present. The impulse response of a circuit applied to circuit diagnosis was partially used in [11]. The approach of using the impulse response of the circuit, however, is quite dangerous. An impulse can move some linear circuits out of their linear region of operation. Also, defining the correct finite amplitude of a theoretically infinite impulse signal while maintaining linearity could be quite tricky. A better input signal was needed.

White noise has, by definition, an equally distributed power in its frequency spectrum. White noise can be easily generated using a random number generator and a DA converter. Moreover, one can be quite sure about its energy frequency spectrum definition, since we can sample it as it is being fed to the circuit under test and measure its frequency distribution. This way, white noise was the second natural choice.

An AD converter, a PC and a DA converter can tackle the test to be carried constituting a low cost tester. White noise is generated in the PC, and the plant signal is sampled and processed in the same PC. The test engineer can take two tracks. The first one is to define a good plant at an abstract level based on his/her design, compute filter coefficients by any mathematical tool (Matlab, Matcad, etc) and to use this as the reference plant. Another strategy is to develop of a fault free prototype, and then use the system to adapt the filter and save the discovered coefficients to be used in the test procedure. It is important to notice that the choice of white noise guarantees that all frequencies are conveniently excited. This causes that all frequency dependent faults are detected, without the need of complex circuit analysis to define the minimum set of test frequencies.

3 Worked examples

In order to better explain the proposed test methodology, two test circuits will be used. The first one is a simple integrator, followed by a biquad filter, an example a little more complex, also chosen because published data was available in order to make comparisons. This circuit was built and tested for different variations in some of its components.

3.1 Integrator

The integrator used is shown in figure 5. The transfer function of this circuit can be found from elementary circuit analysis to be:

$$H(s) = \frac{-1}{R_1 C} \cdot \frac{1}{S + \frac{1}{R_2 C}}$$

For the simulation of this plant in the digital domain we take its z-transform ([6,7,9]), in series with a zero-order hold:

$$H(z) = \frac{-R_2}{R_1} \cdot \frac{1-e^{-kT}}{Z+e^{-kT}}$$

$$k = \frac{1}{R_2 C}$$

where T is the sampling period. From this equation we have obtained the recursive equations for the simulation of the plant, using the values for the circuit components presented in figure 5.

The system was simulated using Matlab. We have used the rand function to generate the input to the circuit, and a FIR adaptive filter with 60 taps was adapted for 3000 samples. After conversion, the filter coefficients were stored and a new plant was defined, changing the value of C from 1.0µF to 0.95µF (a -5% variation). The resulting output error of the two cases is shown in figure 6 (for the identified system) and in figure 7 (-5% variation). The estimated output error power ratio has been found to be 13.3475 for this case. Both figures are at the same scale.

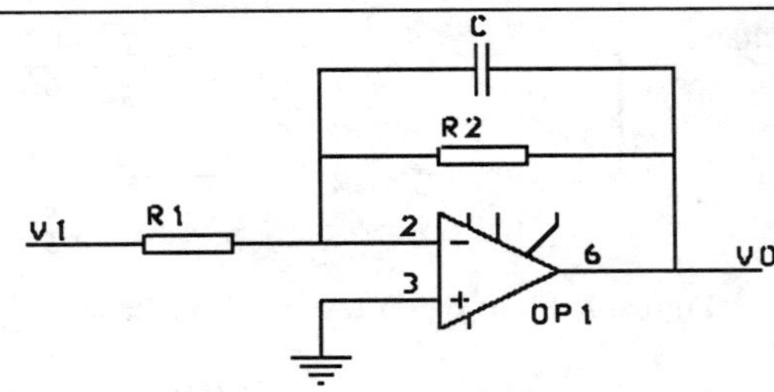

R1=1e3; R2=1e3; C=1e-6; T=0.1e-3;

Figure 5. Simple Integrator

3.2 Biquad filter

The biquad filter was used to validate the methodology and a prototype circuit was build from discrete components. The circuit is shown in figure 8, with the nominal values that were used. We have used a PC with a DSP board and stereo audio analogue interface as the test system. The sampling frequency used was 16KHz. The DSP system generated the input signal for the system, and a 2-channels AD sampled the input and the output of the plant.

We have used 40 taps for the adaptive filter, and adapted the system during 3000 samples. All data processing after acquisition was made in the Matlab environment. Table 1 presents the output error power ratio obtained for variations in components R2 and C2. The error power ratio shows how easily detected is a fault. A ration of 1 means an exact match between the plant and the adaptive filter. We cover soft, hard and catastrophic faults like a short or open circuit. All results were generated from real acquired data using variations introduced in the original system. For the case with no variations, the same data file was used, but at different time intervals.

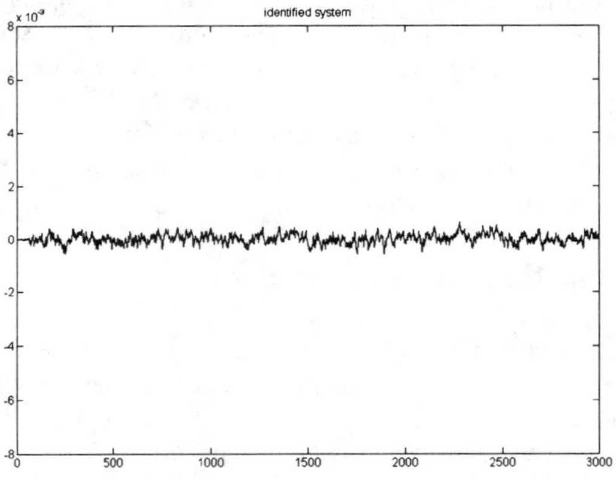

Figure 6. Error output of a good plant (identified system)

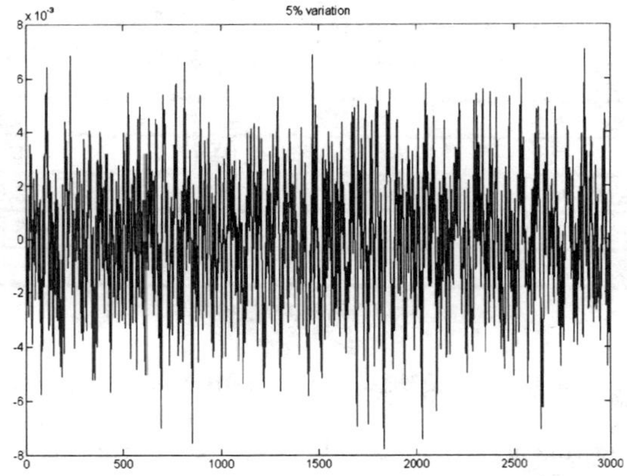

Figure 7. Error output of the integrator with a variation of 5% in C

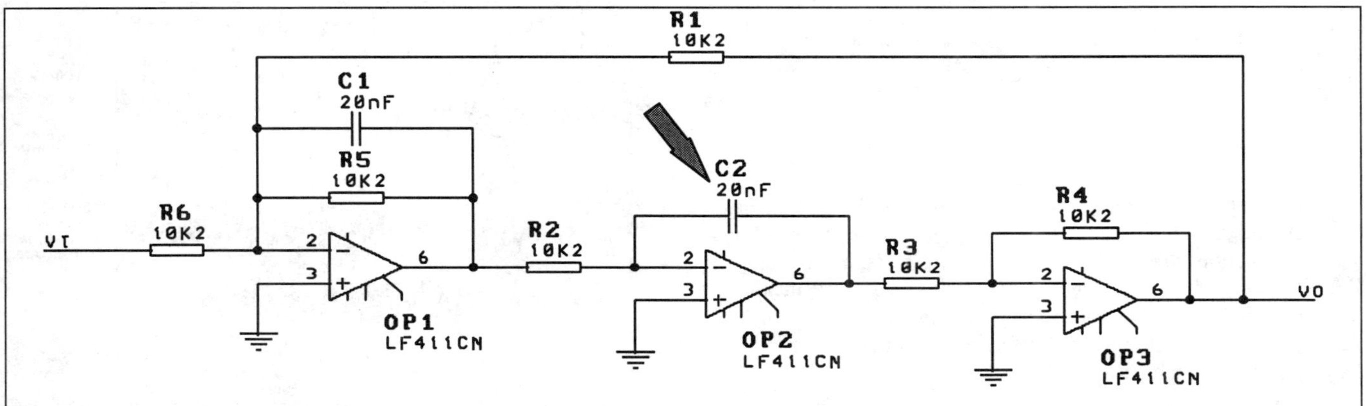

R1,...,R6=10K2 C1,C2=20nF

Figure 8. Biquad schematic

All data presented used 30000 samples to evaluate the output error power ratio. It should be notice that a small 1% variation on R2 caused a relatively large (19%) variation in the test variable, that is, the output error power ratio. This shows why faulty plants are easily recognized.

3.3 Comparison with other approaches

In case of the Biquad all tests of the proposed methodology detected smaller component deviations when compared to the same example using other approaches. According to the results presented in [2], only variations greater than 28.9% in R2 or in C2 could be detected, forcing the comparison parameters out of its 5% tolerance border. In the proposed method, we were able to detect 1% variations in both components, with comparison parameter variations of 20%, well beyond the 5% tolerance border.

One should also mention that the test is inherently simple and fast: there is not the need to choose the correct set of input frequencies, neither to verify more than one variable like the gain or phase. Acquisition time was 6s, and processing time was 30 seconds inside the Matlab environment, in a 66MHz 486. With a dedicated program and a faster machine test time can be further reduced.

4 Real-life examples

In this section we present the robustness of the method by developing some real-life examples, including on-line testing of systems, the sensitivity of circuits to open-loop gain and bandwidth as well as fault detection at the presence of complex signals like voice.

4.1 Testing of a 4th-order filter

We applied the proposed testing method to switched capacitors circuits, using the MF10 chip as a test vehicle [12]. This chip is a programmable filter with a state-variable approach. Filter programming is done by the use of 4 external resistors. In this test, a 5% variation in one of the resistor caused a 80% variation in the error power ratio. A variation in the clock input frequency was also detected. When we moved the clock from 200KHz to 50KHz the output error ratio was 50.89 times higher than the normal case (4989% variation).

	R2	C2
-50%	73.33	74.08
+5%	4.47	4.11
+1%	1.19	1.21
short	126.00	72.06
open	72.06	125.40
no variation	0.9905	0.9905

Table 1. Output error power ratio

4.2 Testing a Biquad with non-ideal opamp

Although most works regarding the testing of analog circuits focus on the testing of small deviations of components, the opamp itself is not generally addressed. Let us take as an example a basic integrator composed of a resistor R and a capacitor C. If the differential gain (Ad) of the opamp is considered to be finite, than the transfer equation is

$$H(s) = \frac{1}{R \cdot C \cdot s + \dfrac{1 + R \cdot C \cdot s}{Ad}}$$

The above equation clearly shows that, in case the opamp has some malfunction that diminishes its open-loop gain, then the transfer function will reflect this property. In our methodology this is easily detected, for the gain shortage will mean a difference in feedback impedance, changing the plant transfer function. Table 2 shows the detection of opamp non ideal behavior, like limited gain or limited bandwidth.

opamp characteristic	value	error variation
open loop gain	1000v/v	35%
open loop gain	50v/v	818%
gain-bandwidth product	1MHz	50%
gain-bandwidth product	0.01MHz	159%

Table 2. Detection of opamp non ideal behavior

4.3 On-line testing

The presented test methodology can be also used on-line, that is, while the circuit is operating. In this case, a general DSP processor or a simple dedicated adaptive FIR integrated circuit must be used. The adaptive phase can be developed at power-on, or alternatively at any time during the operating life of the equipment, to take into account slow varying parameters like temperature.

Training of the digital filter is done in much the same way. Although the training is done with a frequency rich signal like white noise, one might wonder what will ever happen in case the signal of interest has itself many frequency components. That is, a signal with frequencies spread all over the spectrum could make the adaptive filter loose its selectivity against what is a fault or not. However, this is not the case. The linear adapter has shown itself robust, as shown in figure 9. The scope screen presents the operation of a filter trained with white noise. The first wave is a voice signal, the second the error signal of the training phase with white noise. The third wave presents the error signal at the presence of the voice signal. As the plant is working without faults, the relative power of the error is very close to 1. In figure 10 the same filter has a 50% variation in one of the capacitors. As it can be seen, the RMS error signal with some voice input is much larger.

5 Limitations of the approach

Although conceptually simple and easy to use, the methodology is based on some assumptions that must be fulfilled. The first is that an AD converter with enough resolution and speed is available. This, however, regarding specific niches, is readily achieved. For example, for audio applications, a simple acquisition board at extremely low costs connected to almost any personal computer can be used.

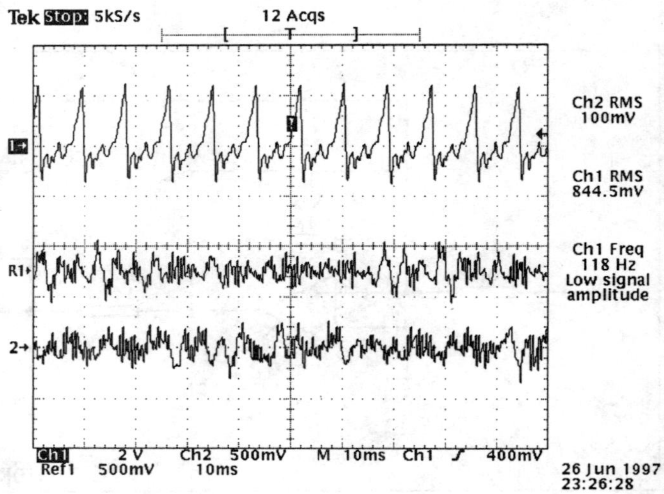

Figure 9. Good plant trained with noise

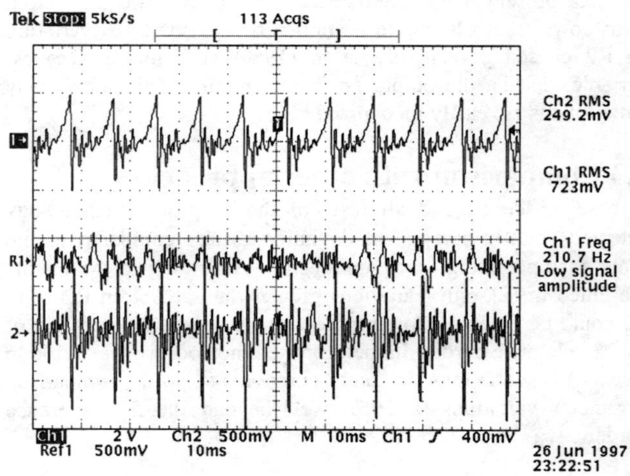

Figure 10. Error signal with voice input

The performance requests of the DA are somewhat relaxed. This is because since we need white noise, a large number of bits is not required, but the linear pass band must be higher than the one of the plant under test.

Testing time is quite fast, since although the filter might have a large number of coefficients to be settled, this is done only once every time a circuit is to be tested. Testing during production can be done with a number of excitation states which is proportional to the number of filter taps, guaranteeing full frequency sweep. For example, testing the Biquad took only 36 seconds in a 66MHz 486 PC inside the Matlab environment, and used part of a 6 seconds long data acquisition file.

Since adaptive algorithms are mathematically based on the Z transform, they can have a DC component. In our case, however, we still can not detect gain faults separately, since they could be mixed with some low frequency pole or zero variations.

This test is not able to identify the faulty component. This does not limit its range of use, since production testing is not intended to determine faulty components. The testing methodology can also serve as a high precision mechanism to evaluate the degree of similarity between mathematical models and prototype during design.

Finally, it should be mentioned that for circuits working at very small frequencies one would have to have a huge number of taps. To avoid this, the test engineer must use a smaller sampling frequency. Also, at high frequencies, one might have troubles with the acquisition system. In this case, the acquisition AD and the white noise DA converter must have enough bandwidth to excite all high frequency poles. If processing is not developed at real time, it can be done at any PC without problems. If on-line testing is required, power consumption of the adaptive filter must be taken into account.

6 Conclusions and future work

This paper has shown a new test strategy able to precisely detect soft, hard or catastrophic faults in an analog circuit. The method has low cost and high precision, and is straightforward to implement in any personal computer with a data acquisition board, or in a DSP board for on-line operation.

In comparison with published work, the method is quite robust, and is able to detect minor components variations without problems, up to less than 1%. Also, the error signal is 1 to 2 orders of magnitude greater than the traditional approach, being easier to detect. Since the methodology does not assume any set of particular frequencies, it can be integrated to any analog circuit described as a transfer function.

In our future work we intend to expand the methodology to include AD and DA converters, as well as DC signal components. Finally, when the original plant is known by the test engineer, the possibility of detecting the specific faulty component by a modification in the convergence will be investigated.

7 References

[1] SLAMANI, M.; KAMINSKA, B. Multifrequency Analysis of Faults in Analog Circuits. **IEEE Design & Test of Computers**, Summer 95, p.70-80.

[2] AYARI, B.; HAMIDA, B.H.; KAMINSKA, B. Automatic Test Vector Generation for Mixed-Signal Circuits. In: European Design & Test Conference, **proceedings**... Paris, France, 1995. p.458-463.

[3] BEN-HAMIDA, N.; AYARi, B.; KAMINSKA, B. Testing of Embedded A/D converters in Mixed-Signal Circuit. In: IEEE International Conference on Computer Design, **proceedings**... Austin, Texas, 1996. p.135-136.

[4] MIELKE, J. Frequency Domain testing of ADCs. **IEEE Design & Test of Computers**, Spring 96, p.64-69.

[5] SPINA, R.; UPADHYAYA, S. Linear Circuit Fault Diagnosis Using Neuromorphic Analyzers. IEEE Transactions on Cirucits and Systems - II: Analog and Digital Signal Processing, v44, n.3, March 1997. p. 188-196.

[6] OGATA, K. **Modern Control Engineering**. Prentice Hall, 1970. 929p.

[7] PROAKIS, J.G.; MANOLAKIS, D. G. **Introduction to Digital Signal Processing**. Macmillan Publishing Company, New York, 1988. 944 p.

[8] HAYKIN, S. **Adaptive filter theory**. 2.ed. Englewood Cliffs, N.J.: Prentice-Hall, 1991. 854p.

[9] WIDROW, B.; STEARNS, S.D. **Adaptive signal processing**. Englewood Cliffs, N.J.: Prentice-Hall, 1985. 474p.

[10] MULGREW, B.; COWAN, C. F. **Adaptive Filters and Equalisers**. Kluwer Academic Press, 1988.

[11] SU, C.; CHIANG, S.; JOU, S.-J. Impulse Response fault model and Fault Extraction for Functional Level Analog Circuit Diagnosis. In: IEEE International Conference on CAD, **proceedings**...1995, p. 631-636.

[12] NATIONAL SEMICONDUCTORS. Data Acquisition Databook. National Semiconductor Corporation, 1993. p.7-197.

General AC Constraint Transformation for Analog ICs

B. G. Arsintescu†, E. Charbon, E. Malavasi, U. Choudhury, W. H. Kao
Cadence Design Systems, San Jose, CA 95134 U.S.A.
†Delft University of Technology, Mekelweg 4,2628CD Delft, The Netherlands

Abstract

The problem of designing complex analog circuits is attacked using a hierarchical top-down, constraint-driven design methodology. In this methodology, constraints are propagated automatically from high-level specifications to physical design through a sequence of gradual transformations. Constraint transformation is a critical step in the methodology, since it determines in large part the degree to which specifications are met. In this paper we describe how constraint transformations can be efficiently carried out using hierarchical parameter modeling and constrained optimization techniques. The process supports complex high-level specification handling and accounts for second-order effects, such as interconnect parasitics and mismatches. The suitability of the approach is demonstrated through an 4th order active filter test case.

1. Introduction

The design of analog circuits is a difficult task due to the complex performance specifications and the impact of second-order effects on performance. The task is even more strenuous, when analog circuits need be implemented in a mixed analog/digital environment, due to the interactions between the various system components. The extreme criticality of analog circuits and the difficulty to meet the specifications is generally the source of long design iterations and multiple fabrication runs.

Recently, a number of attempts have been made to cope with this problem. The common denominator of the proposed approaches is the use of a design methodology to ensure that high-level specifications be translated hierarchically onto module parameters and finally onto a suitable layout. The top-down design methodology used is fully described by (a) the design paradigm, (b) the mechanisms supporting the paradigm and (c) the tool-set for all design tasks. We consider a specific design paradigm often used in top-down methodologies, known as constraint-based design paradigm. The main mechanism supporting such paradigm is responsible for the translation of high-level specifications onto lower levels of hierarchy and is known as *constraint transformation*. At each level of hierarchy all pertinent constraints are then enforced by a *constraint-driven tool-set* [1].

In this paper we formalize the constraint transformation mechanism for a class of linear, continuous time circuits. Figure 1 shows how constraint transformation operates at each hierarchical level of the design by mapping a set of constraints from a higher level onto a lower level [2]. Eventually, after the transformation and control of each hierarchical level is complete, a set of constraints for the lowest hierarchical level, namely physical design, is found. Constraint transformation consists of two phases:

1. *model generation* consists of creating a model relating performance measures at a given level of hierarchy to lower-level design parameters,
2. *constraint generation* consists of associating every design parameter with an actual bound.

A number of techniques have been proposed in the literature to address the problem of constraint generation. In [3] each high-level performance is associated with the root of a *design tree*, whose nodes represent either lower-level performance variables or independent design parameters. A path representing the parameters to adjust for performance improvement is computed by an iterative decision process. The process, based on qualitative reasoning [4] and circuit sensitivities, is aimed at satisfying all high-level specifications leaving the selection of all design parameters entirely to designer expertise. Recently, [5] proposed to solve a Constraint Satisfaction Problem (CSP) on a pre-defined design tree to obtain intervals for each parameter of the design. An alternative, proposed by [6, 7], consists of obtaining bounds on the performance variables of a given level of hierarchy as a byproduct of design opti-

DAC 98, San Francisco, California

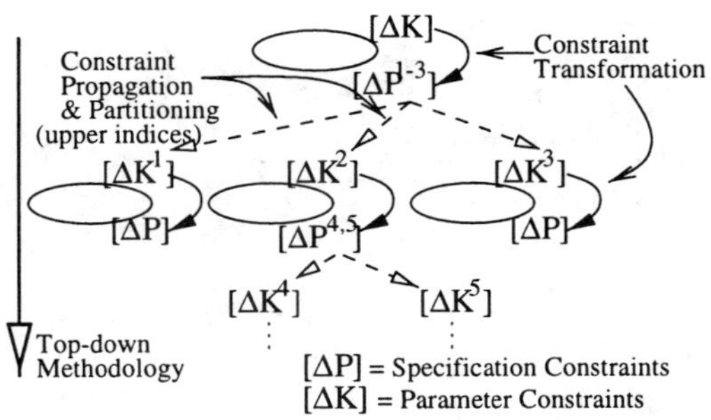

Figure 1. Top-down, constraint-based design methodology

mization. Constrained optimization alone [8, 9] allows for the removal of all non-critical design parameters and the calculation of parasitic constraints when the bottom level is reached. This methodology is powerful, since it does not assume an initial dependency tree for a given design. However at each level of hierarchy one needs a behavioral or device-level simulator, which in turn determines the performance variables that can be handled. Common to all mentioned approaches is that they rely on hand-crafted constraint models or on a model database.

In this paper we propose a semi-analytical approach to the problem of constraint transformation for a class of linear continuous-time circuits. Assuming that the circuit be fully designed *a priori*, model generation (phase 1) is performed as following. Sensitivities with respect to design parameters and parasitic effects are computed directly, using the analytic expression of the transfer function. At each level of hierarchy, the high-level design specifications are automatically transformed into constraints for the subsequent levels. Constraint generation (phase 2) is performed using conventional variable allocation methods for both deterministic and non-deterministic cases [8, 10, 11]. A 4th order active filter example is presented at the end of this paper to illustrate the advantages of our approach. In a top-down design methodology, continuous control at every step of the design is enabled using our constraint transformation method.

The paper is organized as follows. In Section 2 the constraint transformation problem (phase 1) is defined. Section 3 discusses the techniques used for the problem solution, while test examples are presented in Section 4.

2. General description

Let us define $[\mathbf{K}]$, the vector of all n performance measures considered in the design. Moreover, let $[\mathbf{K_0}]$ be the nominal performance vector and $[\Delta\mathbf{K}] = [\mathbf{K}] - [\mathbf{K_0}]$ the performance degradation of the circuit with respect to its nominal. Similarly, let us define $[\mathbf{P}]$ as the vector of all m design parameters, $[\mathbf{P_0}]$ the nominal parameter vector and $[\Delta\mathbf{P}] = [\mathbf{P}] - [\mathbf{P_0}]$ the vector of all parameter variations. The problem of finding vector $[\Delta\mathbf{P}]$ and a mathematical relation $[\mathbf{S}]$ between $[\Delta\mathbf{K}]$ and $[\Delta\mathbf{P}]$ corresponds to the model generation problem.

Problem 1 Model Generation Problem *For a given circuit with performance vector* $\mathbf{K}$*, find parameter vector* $\mathbf{P}$ *and* $n \times m$ *sensitivity matrix* $\mathbf{S}$ *s.t.:*

$$[\Delta\mathbf{K}] = [\mathbf{S}] \cdot [\Delta\mathbf{P}]. \tag{1}$$

Alternatively, the problem of deriving a set of bounds on the components of $[\Delta\mathbf{P}]$, given $[\overline{\Delta\mathbf{K}}]$, the constraints on performances, corresponds to the constraint generation problem.

Problem 2 Constraint Generation Problem *For a given circuit with performance vector* $\mathbf{K}$*, find a set of bounds* $\Delta\mathbf{P}^{bound}$ *on the components of* $\Delta\mathbf{P}$*:*

$$\mathbf{0} \leq [\Delta\mathbf{P}] \leq [\Delta\mathbf{P}]^{bound} : \quad \mathbf{0} \leq [\Delta\mathbf{K}] \leq [\overline{\Delta\mathbf{K}}]. \tag{2}$$

The combination of the above problems (phases 1 and 2 in Section 1) formally defines the *constraint transformation problem.* A number of approaches to solve problem 2 can be found in the literature [8, 10, 11]. In this paper we will focus on Problem 1.

The linear approximation in Equation (1) is valid for small variations in $[\Delta\mathbf{P}]$. A general method for finding matrix $[\mathbf{S}]$ for linear continuous time circuits is described hereafter. The method can handle hierarchical constraint transformation, related to AC performance functions only. AC performance functions are specified with respect to one or more transfer functions. The analytic expression of frequency transfer function $H(j\omega)$ of a given circuit is:

$$H(\omega) = \frac{\prod(j\omega - z_i)}{\prod(j\omega - p_i)}, \tag{3}$$

where p_i, z_i are the poles and zeroes of the transfer function respectively. The constraint transformation method will find the sensitivity matrix for any linear continuous time circuit using the analytic expression of the transfer function, based on the following assumptions:

(i) There is a finite number of poles and zeroes in the frequency domain of interest.

(ii) Simulation techniques to compute the values and multiplicity of the poles and zeroes exist.

(iii) A linear approximation of the transfer function around the nominal working point can be done.

The AC constraints are bounds given to the transfer function as shown in Figure 2. Two important classes of AC constraints are the following:

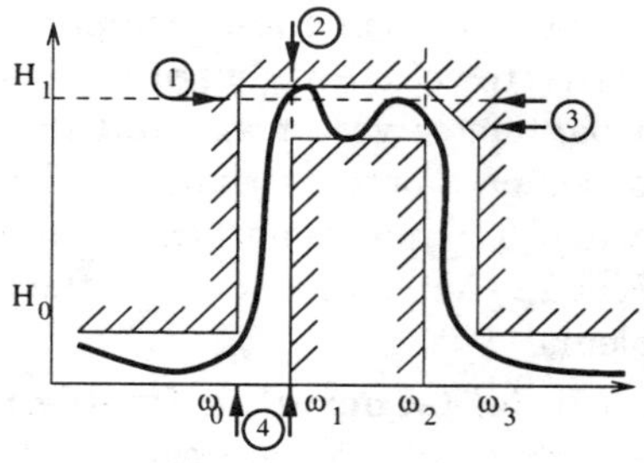

Figure 2. Transfer function constraints

1. **Transfer function constraints:** $\mathcal{C}(H(\omega))$ @ $\omega = \omega_1$;
2. **Frequency constraints:** $\mathcal{C}(\omega)$ @ $H(\omega) = H_1$.

Since these constraints refer to the behavior of $H(\omega)$ for a given $\omega = \omega_0$ or at a given value H_0, they are called punctual constraints.

The following classes of AC constraints are defined over a range of frequencies or transfer function values:

3. **Transfer function constraints:** $\mathcal{C}(H(\omega))$ @ $\omega \in (\omega_1, \omega_2)$;
4. **Frequency constraints:** $\mathcal{C}(\omega)$ @ $H(\omega) \in (H_0, H_1)$

Hence, they are called range constraints. For example, the low frequency gain, defined as $\mathcal{C}(H(\omega)) = |H(\omega)| \geq H_0 @\omega \to 0$, is a punctual constraint on the transfer function. Unity-Gain Bandwidth, defined as $\omega \geq \omega_0 @|H(\omega)| = 1$, is a punctual constraint on frequency. The maximum ripple constraint, defined as $\max_{(\omega_0,\omega_1)} H(\omega) - \min_{(\omega_0,\omega_1)} H(\omega) \leq H_\Delta @\omega \in (\omega_1, \omega_2)$ is a range constraint on the transfer function.

3. Constraint transformation techniques

The method consists of finding an analytic expression of the linear approximation around the nominal working point for the constraint function $\mathcal{C}$. The linear approximation is done by derivating the constraint function $\mathcal{C}(\cdot)$ with respect to the circuit parameters $[\mathbf{P}]$:

$$K = \mathcal{C}(\cdot) \Rightarrow \Delta K = \frac{\partial \mathcal{C}}{\partial P} \cdot \Delta P. \qquad (4)$$

3.1. Punctual Constraints

Let us consider a transfer function constraint $\mathcal{C}(H(\omega))$ @ $\omega = \omega_0$. The function $\mathcal{C}(\cdot)$ should be continuous and differentiable and the transfer function is continuous and differentiable in ω_0. Hence, a linear approximation around ω_0 is obtained as follows:

$$\Delta K_i = \left(\frac{\partial \mathcal{C}}{\partial H}(H(\omega_0))\right) \cdot S_P^H\Big|_{\omega=\omega_0} \cdot \Delta P. \qquad (5)$$

where $\frac{\partial \mathcal{C}}{\partial H}$ is the analytic expression of the derivative of function $\mathcal{C}$ and $\frac{\partial H(\omega)}{\partial P} = S_P^H$ is the sensitivity of $H(\omega)$ with respect to a circuit parameter P. This sensitivity can be calculated using a circuit simulator (e.g. SPICE) or derived analytically when the poles ans zeroes sensitivities $S_P^{p_i}$ and $S_P^{z_i}$ are available.

From Equation(5), since $\frac{\partial \mathcal{C}}{\partial H}(H(\omega_0))$ and $S_P^H|_{\omega=\omega_0}$ are known, we can conclude that the constraint transformation problem is solved for all the transfer function constraint problems at a fixed frequency.

For frequency constraints, the frequency ω_0 s.t. $H(\omega_0) = H_0$ is first calculated. When the solution is not unique, the problem is transformed into a system of simultaneous problems for all solutions. The composed derivative of the transfer function is:

$$S_P^H\Big|_{\omega=\omega_0} = \frac{\partial H}{\partial P}_{\omega=\omega_0} = \left(\frac{\partial H}{\partial \omega} \cdot \frac{\partial \omega}{\partial P}\right)_{\omega=\omega_0}. \qquad (6)$$

The analytic expression of the slope of the transfer function can be easily obtained by derivation and the frequency constraint problem is formulated as:

$$\Delta K_i = \frac{\partial \mathcal{C}}{\partial \omega}(\omega_0) \frac{1}{\frac{\partial H}{\partial \omega}(\omega_0)} \cdot S_P^H\Big|_{\omega=\omega_0} \cdot \Delta P. \qquad (7)$$

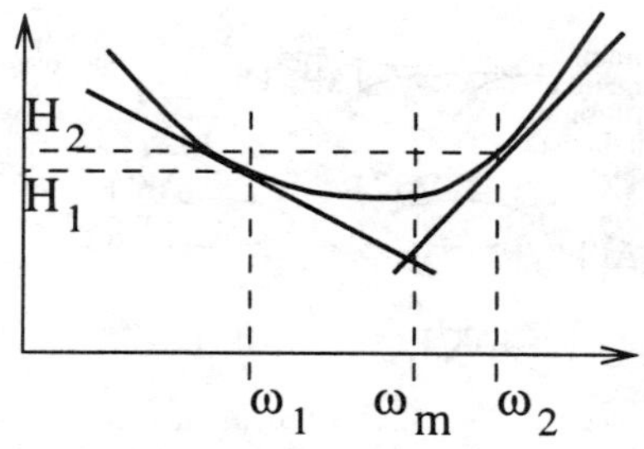

Figure 3. Transfer function extreme points

The sensitivity of K_i with respect to $P, S_P^{K_i} = \frac{\partial C(\omega)}{\partial P}$ in Equation (7) cannot be calculated for the extremes of the transfer function (i.e. maximum and minimum points), where $\frac{\partial H}{\partial \omega} = 0$. For the cases when a constraint on frequency is given at those extremes, the following approximation is considered: the neighbor poles and/or zeroes around the frequency of the maximum are selected, then the frequency of the maximum is approximated piecewise linearly as shown in Figure 3. The position of the extreme ω_m given in Equation (8) depends not only on the position of the poles and/or zeroes around the extreme point, but also depends on the slopes of the transfer function in the selected points:

$$\omega_m = \frac{(H_2 + \frac{\partial H}{\partial \omega}(\omega_2) \cdot \omega_2) - (H_1 + \frac{\partial H}{\partial \omega}(\omega_1) \cdot \omega_1)}{\frac{\partial H}{\partial \omega}(\omega_1) - \frac{\partial H}{\partial \omega}(\omega_2)}. \qquad (8)$$

Given a constraint $K_i = \mathcal{C}(\omega_m)$, the transformation problem is completely solved, and yields from Equation (8):

$$\Delta K_i = \frac{\partial \mathcal{C}}{\partial \omega}(\omega_m) \cdot \frac{S_P^H|_{\omega=\omega_2} - S_P^H|_{\omega=\omega_1}}{\frac{\partial H}{\partial \omega}(\omega_1) - \frac{\partial H}{\partial \omega}(\omega_2)} \cdot \Delta P. \qquad (9)$$

Equations (7, 9) provide a complete solution for all frequency constraint problems. The problem is completely solved given the analytic functions for the slope of the transfer function and for the constraint function and the required circuit sensitivities.

3.2. Range Constraints

The constraints transformation problem is easily extended to the range constraint problems under the hypotheses (i), (ii) and (iii) of Problem 1, when $\mathcal{C}(\cdot)$ is a continuous and differentiable function. Equations (5, 7, 9) hold over an interval when the constraint function $\mathcal{C}(\cdot)$ is monotonic. Otherwise, the analysis should be done for each monotonic sub-domain of $\mathcal{C}$, resulting in a constraint system.

The range constraints are expressed as a norm rather than a function. The constraint transformation method will be extended to deal with two norm operators.

1. Unity norm: $||H(\omega)||_1 \triangleq \int_{\omega_1}^{\omega_2}(H(\omega)\delta\omega)$. E.g., the total nonlinearity is calculated as an integral of the transfer function over a given interval.
2. Infinite norm:$||H(\omega)||_\infty \triangleq \max_{\omega_1 \le \omega \le \omega_2} H(\omega)$. E.g., the ripple constraint is the difference between the transfer function maximum and minimum within the interval.

For the unity norm $||\cdot||_1$, let us consider the transfer function is also dependent on circuit parameters $P : H(\omega, P)$. The two linear operators $\partial(), \int()$ can be swapped when ω and [**P**] are independent:

$$\frac{\partial}{\partial P}\left(\int_{\omega_1}^{\omega_2} H(\omega)\delta\omega\right) = \int_{\omega_1}^{\omega_2} \frac{\partial H}{\partial P}\delta\omega \Rightarrow$$

$$\frac{\partial}{\partial P}(||H(\omega)||_1)\Delta P = ||S_P^H||_1 \cdot \Delta P. \qquad (10)$$

If ω and[**P**] are not independent, the operators cannot be swapped. If the operators are not linear, the transformation is dependent on the sign and monotonicity of both operators, since the triangle formula should be used for linear approximation: $||a + b|| \le ||a|| + ||b||$.

For the infinite norm $\mathcal{C} = ||\cdot||_\infty$, the general problem is transformed in a system of simultaneous punctual problems. The solution of the problem is the maximum among the *numerical values* of the punctual problems. The expression for the linear approximation around the nominal point is:

$$\frac{\partial}{\partial P}(\max H(\omega)) = \left\{\max_{\omega_i}(\frac{\partial H}{\partial P}|_{\omega=\omega_i})\right\}, \qquad (11)$$

where ω_i are all the frequencies at which $H(\omega)$ has a local maximum. Approximate values for the frequencies ω_i can be calculated using Equation (8). The constraint transformation is completely solved calculating sensitivities in the extreme points ω_i:

$$\Delta K_i = \max_{\omega_i}\left\{S_P^H|_{\omega_i} \cdot \Delta P\right\}. \qquad (12)$$

When minimum values are required, the equation $\min(f) = -\max(-f)$ is used. The ripple constraint can be transformed into two systems of maximum constraints as in Equation (12). The general expression of the ripple constraint is:

$$\mathcal{R} = c_1 \max_{1 \le i \le k}(H(\omega_i)) - c_2 \min_{1 \le j \le l}(H(\omega_j)), \qquad (13)$$

where c_1, c_2 are constants, at least one different than zero and i, j are indices over frequencies of maximum and minimum of the transfer function, respectively.

Similar equations are obtained for the frequency range constraints. Hence, the domain of definition for the constraint function $\mathcal{C}$ is extended from continuous differentiable functions to linear norms independent of the parameters and infinity norm. No further extension was pursued because the definition domain for $\mathcal{C}$ already covers the need for AC constraint transformations.

3.3. Hierarchical transformations

The constraint transformation method is easily extended to hierarchical circuits. For simplicity but without loss of generality, let us consider a circuit with two levels of hierarchy. At the highest level, the circuit consists of a parametric model of the circuit for the performance function K_i represented as $K_i = \mathcal{K}_i([\mathbf{P^1}], \ldots, [\mathbf{P^k}])$, where $[\mathbf{P^j}]$ are the parameter vectors associated with the modules $\mathcal{M}_j, 1 \le j \le k$. A linear approximation of degradation ΔK_i can be derived with respect to the parameter vectors $[\mathbf{P^j}]$:

$$\Delta K_i = \frac{\partial}{\partial P^j}(\mathcal{K}_i([\mathbf{P^1}], \ldots, [\mathbf{P^m}])) \cdot [\Delta\mathbf{P^j}]. \qquad (14)$$

Once a constraint $[\Delta\mathbf{P^j}]$ is found for this hierarchical level, it becomes the specification vector $[\Delta\mathbf{K^j}]$ for the child block. The hierarchical transformation is repeated until the leaves of the hierarchy are reached.

A pole-zero analysis is required to obtain an analytic expression for the transfer function. The high-level constraints are transformed to a set of punctual transfer function and/or frequency constraints. For hierarchical constraints, the parametric cell constraints are transformed hierarchically down to device parameters. Sensitivity analysis is performed by the simulator for device parameter transformations. Constraint generation can be performed top-down at any depth in the hierarchy or can be propagated bottom-up for verification.

4. Examples

Three examples of filter design [12] are presented to verify the effectiveness of the constraint transformation

$$\begin{bmatrix} \Delta\omega_0 \\ \Delta BW \\ \Delta Q \\ \Delta H(\omega_0) \end{bmatrix} = \left(\begin{bmatrix} \frac{2\sqrt{p_1p_2}(p_1+p_2)^2}{p_2-p_1} \\ \frac{4p_1(p_1+p_2)^2}{p_2-p_1} \\ \frac{4p_1(1-\frac{1}{2}\sqrt{\frac{p_2}{p_1}})}{p_2-p_1} \\ 0 \end{bmatrix} [\mathbf{S}]^T_{p_1} + \begin{bmatrix} \frac{2\sqrt{p_1p_2}(p_1+p_2)^2}{p_1-p_2} \\ \frac{4p_2(p_1+p_2)^2}{p_1-p_2} \\ \frac{4p_2(1-\frac{1}{2}\sqrt{\frac{p_1}{p_2}})}{p_1-p_2} \\ 0 \end{bmatrix} [\mathbf{S}]^T_{p_2} + \begin{bmatrix} 0 \\ 0 \\ 0 \\ 1 \end{bmatrix} [\mathbf{S}]^T_{\omega_0} \right) [\Delta\mathbf{P}]$$

Figure 4. Second order filter transformation

technique. The first example deals with a transformation to device parameters and the following ones with hierarchical transformations. Constraint allocation for one of the hierarchical examples shows the usefulness of the method for design reuse and mismatch characterization.

Let us consider a black-box circuit with the following transfer function resulting from pole-zero analysis:

$$H(\omega) = \frac{H_0 j\omega}{(j\omega + p_1)(j\omega + p_2)}, \tag{15}$$

corresponding to a second order bandpass filter. In a first approximation, the analytical expressions for the high-level constraints vector of this filter are:

$$\left[\omega_0^2, BW, Q, H\right]^T = \left[p_1 \cdot p_2, p_1 + p_2, \tfrac{\omega_0}{BW}, H(\omega_0)\right]^T. \tag{16}$$

Given a device parameters vector $[\mathbf{P}] = [P_1, \ldots P_n]^T$, the complete transformation from biquad constraints into parameters is shown in Figure 4. Note that only three sensitivity calculations are necessary to complete the transformations, for $j\omega = p_1$, $j\omega = p_2$ and $\omega = \omega_0$. When $p_1 = p_2$, the transfer function reaches its maximum for $j\omega = p_1$, hence its derivative is zero in that point. The pole-zero analysis does not usually result in multiple poles, but rather in very close ones. However, when multiple poles occur, they are represented as $p_1 = p - \Delta p, p_2 = p + \Delta p$, such that the analysis can be done in the vicinity of the maximum.

A behavioral model of the biquad filter will improve the quality of the transformation. A more detailed high-level constraint vector $[\mathbf{K}]$ is shown in Figure 5. Since $OTA2, OTA3$ are identical, the matching parameters P_{o2}, P_{o3} are transformed into mean ($P_o = \frac{P_2+P_3}{2}$) and deviation parameters ($P_{o,\Delta} = \frac{P_2-P_3}{2}$). The resulting vector of low-level constraints is a collection of OTA AC parameters, resistor and capacitors parameters and mismatches between identical parameters:

$$\begin{aligned}[\mathbf{P_m}] = [&g_{m1}, g_{01}, A_{V1}, g_m, g_0, A_V, \omega_z, \omega_p, \\ &g_{m,\Delta}, A_{V,\Delta}, C, C_\Delta, g_2, \ldots]^T,\end{aligned} \tag{17}$$

where ω_z, ω_p are the dominant zero and pole respectively of the OTA. Note that OTA1 is a different design than OTA2 and OTA3. Such behavioral model yields the analytic transformation in Figure 6, where $A \ldots G$ are constants depending on the parameters of the modules. For the allowed high-level variations $\overline{[\Delta\mathbf{K}]}$, constraint generation was performed using PARCAR[8]. Feasible bounds $\overline{[\Delta\mathbf{P}]}$ were obtained for all modules.

Let us now consider the 4th order filter in Figure 7.a. Its architecture is based on two biquadratic sections (as depicted in Figure 5) connected in cascade. The pole frequencies are controlled by OTA2 of each biquad and the associated capacitors. The quality factor is controlled (ideally) by the resistor and the OTA1 of each biquad. The transfer function H_4 of the 4th order filter is the product of the transfer function of the biquads, H_1, H_2 respectively:

$$H_4(\omega) = H_1(\omega) \cdot H_2(\omega). \tag{18}$$

In order to have a similar behavior for both biquads their AC gain was designed to be unity $|H(\omega_0)| = 1$. In Figure 7.b a list of high-level constraint transformations to biquad parameters is given. The values of the bandwidth BW and quality factor Q for the 4_{th} order filter were taken from filter design tables. The parameter vector $[\mathbf{P}]_\mathbf{b}$ created by the biquad modules parameters yields:

$$[\mathbf{P_b}] = [\omega_{10}, \omega_{20}, BW_1, BW_2, Q_1, Q_2, |H_1|, |H_2|]^T. \tag{19}$$

Based on Figure 7.b relations, a sensitivity matrix $[\mathbf{S_{4ord}}]$ is created to transform the constraint vector $[\mathbf{K_{4ord}}]$ into biquad parameter constraints. The hierarchical constraint transformation for the 4th order filter through biquad modules constraints $[\Delta\mathbf{P_m}]$ (OTAs, Cs', Rs' constraints in Equation(17)) down to device parameters $[\mathbf{P_d}]$ yields:

$$\begin{aligned} &[\Delta\mathbf{K_{4ord}}] = [\mathbf{S_{4ord}}] \cdot [\Delta\mathbf{P_b}]; \\ [\Delta\mathbf{P_b}] = [\Delta\mathbf{K_b}] &\Rightarrow [\Delta\mathbf{K_b}] = [\mathbf{S_b}] \cdot [\Delta\mathbf{P_m}]; \\ [\Delta\mathbf{P_m}] = [\Delta\mathbf{K_m}] &\Rightarrow [\Delta\mathbf{K_m}] = [\mathbf{S_m}] \cdot [\Delta\mathbf{P_d}]. \end{aligned} \tag{20}$$

For each module, the parameter vector becomes the new constraint vector, transformed later into device

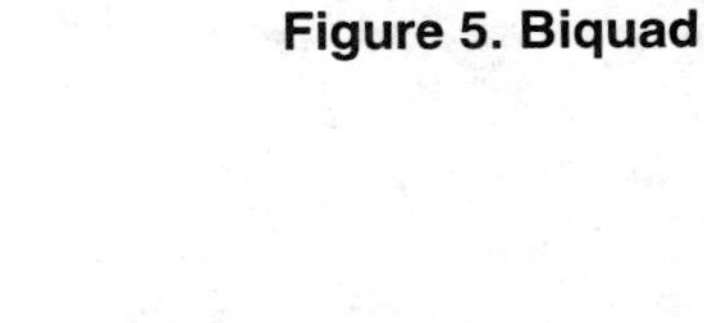

$$[\mathbf{K}]^\mathbf{T} = \begin{bmatrix} \sqrt{\frac{g_{m2}g_{m3}}{C_1C_2}} \\ \frac{g_2}{C} \\ \frac{\sqrt{g_{m2}g_{m3}}}{g_2}\sqrt{\frac{C_2}{C_1}} \\ \frac{g_{m1}}{g_2}\sqrt{\frac{g_{m2}}{g_{m3}}\frac{C_2}{C_1}} \end{bmatrix}$$

Figure 5. Biquad Constraints

$$
\begin{bmatrix} \Delta\omega_0 \\ \Delta BW \\ \Delta Q \\ \Delta|H(\omega_0)| \end{bmatrix} = \begin{bmatrix} A\frac{2g_m+g_{m,\Delta}}{g_m} & A\frac{g_m}{g_{m,\Delta}} & -B\frac{2C+C_\Delta}{C} & -B\frac{C}{C_\Delta} & 0 & 0 \\ 0 & 0 & -\frac{g_2}{(C+C_\Delta)^2} & -\frac{g_2}{(C+C_\Delta)^2} & 0 & \frac{1}{(C+C_\Delta)} \\ D\frac{2g_m+g_{m,\Delta}}{g_m} & D\frac{g_m}{g_{m,\Delta}} & -E\frac{C+C_\Delta}{C} & -E\frac{C+C_\Delta}{C_\Delta} & 0 & -\frac{Q}{g_2^2} \\ F\frac{g_{m,\Delta}}{g_m^{3/2}} & -F\frac{\sqrt{g_m}}{g_{m,\Delta}} & G\frac{C_\Delta}{C^{3/2}} & -G\frac{\sqrt{C}}{C_\Delta} & \frac{H_0}{g_{m1}^2} & -\frac{H_0}{g_2^2} \end{bmatrix} \begin{bmatrix} \Delta g_m \\ \Delta g_{m,\Delta} \\ \Delta C \\ \Delta C_\Delta \\ \Delta g_{m1} \\ \Delta g_2 \end{bmatrix}, \quad \overline{[\Delta\mathbf{K}]} = \begin{bmatrix} 100KHz \\ 50KHz \\ 4 \\ 2dB \end{bmatrix}
$$

$$
[\mathbf{P_0}] = [\ 300\mu A/V,\ 0\%,\ 4pF,\ 0\%,\ 15\mu A/V,\ 15\mu A/V\ ]^T
$$

$$
\overline{[\Delta\mathbf{P}]} = [\ [-3.68,7.4]\mu A/V,\ [-8.7,5.88]\mu A/V,\ [-0.94,0.94]pF,\ [-52,64]fF,\ [-1.97,1.5]\mu A/V,\ [-1.74,1.74]\mu A/V\ ]^T
$$

Figure 6. Transformation and allocation matrix

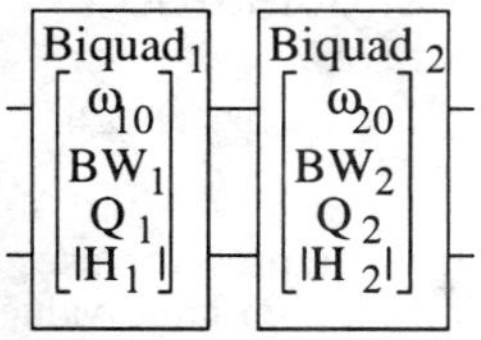

(a)

K_i	$\mathcal{K}(\cdot)$
ω_0	$\sqrt{\omega_{10}\omega_{20}}$
BW	$f(BW_1, BW_2)$
Q	$f(Q_1, Q_2)$
$\|H\|@\omega_0$	$\|H_1\| \cdot \|H_2\|$

(b)

Figure 7. Fourth order bandpass filter

constraint vector. When the modules are identical, the set of constraints passed from high-level to low-level decreases in size. In our example, the full design of the 4th order filter reduces to specifications on two OTA structures, one resistor and one capacitor specifications. Moreover, for physical design, matching constraints are already transformed from high-level to device parameters.

Such transformation method can be easily automated. When behavioral models are not available, high-level constraint are transformed directly into device parameters. Moreover, the method finds the sensitivity calculations necessary and sufficient to model the transformation.

5. Conclusions

An automated constraint transformation method has been used for hierarchical design of analog circuits using a top-down design methodology. Performance specification variations are mapped onto constraint bounds for the physical realization of the circuit through the generation and transformation of intermediate constraints at every design hierarchy level. The transformation of constraints within the design exploits hierarchical parameter modeling, the analytic expression of the transfer function and constrained optimization techniques. The objective is the efficient handling of complex performance specifications, while accounting for second-order effects. A realistic 4th order filter has been used to illustrate the technique and to show the effectiveness of the proposed constraint transformation techniques.

References

[1] E. Malavasi, E. Charbon, E. Felt, and A. L. Sangiovanni-Vincentelli. Automation of IC layout with analog constraints. *IEEE Trans. on Computer Aided Design*, CAD-15(8):923–942, August 1996.

[2] H. Chang, E. Charbon, U. Choudhury, A. Demir, E. Felt, E. Liu, E. Malavasi, A. Sangiovanni-Vincentelli, and I. Vassiliou. *A Top-down, Constraint-Driven Design Methodology for Analog Integrated Circuits.* Kluwer Academic Publ., Boston, MA, 1996.

[3] C. A. Makris and C. Toumazou. Analog IC Design: Part II - Automated Circuit Correction by Qualitative Reasoning. *IEEE Trans. on Computer Aided Design*, CAD-14(2):239–254, February 1995.

[4] J. de. Kleer. How circuits work. In *Qualitative Reasoning About Physical Systems.* D. G. Bobrow Ed., Elsevier Science Pub. B.V. (North-Holland), 1985.

[5] N. R. Dhanwada and R. Vemuri. Constraint allocation in analog system synthesis. In *VLSI '98, Chennai, India*, to appear, January 1998.

[6] I. Vassiliou, H. Chang, A. Demir, E. Charbon, P. Miliozzi, and A. L. Sangiovanni-Vincentelli. A video driver system designed using a top-down, constraint-driven methodology. In *Proc. IEEE International Conference on Computer Aided Design*, pages 463–468, November 1996.

[7] H. Chang, E. Felt, and A. L. Sangiovanni-Vincentelli. Top-down, constraint-driven design methodology based generation of a second order Σ-Δ A/D converter. In *Proc. IEEE Custom Integrated Circuit Conference*, pages 533–536, May 1995.

[8] U. Choudhury and A. L. Sangiovanni-Vincentelli. Automatic generation of parasitic constraints for performance-constrained physical design of analog circuits. *IEEE Trans. on Computer Aided Design*, CAD-12(2):208–224, February 1993.

[9] G. Gad-El-Karim and R. S. Gyurcsik. Use of performance sensitivities in analog cell layout. In *Proc. IEEE International Symposium on Circuits and Systems*, pages 2008–2011, June 1991.

[10] B. G. Arsintescu and R. Otten. Constraints space management for the layout of analog ICs. In *Proc. Design Automation & Test in Europe*, 1998.

[11] E. Charbon, P. Miliozzi, E. Malavasi, and A. L. Sangiovanni-Vincentelli. Generalized constraint generation in the presence of non-deterministic parasitics. In *Proc. IEEE International Conference on Computer Aided Design*, pages 187–192, November 1996.

[12] J. Silva-Martinez, M. Steyaert, and W. Sansen. *High-Performance CMOS Continuous-Time Filters.* Kluwer Academic Publ., Boston, MA, 1993.

Design Methodology used in a Single-Chip CMOS 900 MHz Spread-Spectrum Wireless Transceiver

Jacob Rael
UCLA Electrical Engineering
Box 951594
Los Angeles, CA 90095-1594
310-206-2797
jjrael@icsl.ucla.edu

Ahmadreza Rofougaran
UCLA Electrical Engineering
Box 951594
Los Angeles, CA 90095-1594
310-206-2797
rofo@icsl.ucla.edu

Asad Abidi[1]
UCLA Electrical Engineering
Box 951594
Los Angeles, CA 90095-1594
310-825-9490
abidi@icsl.ucla.edu

1. ABSTRACT

This paper describes the simulation and layout techniques used and developed in the design of a single-chip CMOS 900 MHz spread-spectrum wireless transceiver.

1.1 Keywords

Digital radio, transceiver, spread spectrum communication, RF CMOS, ISM frequency band.

2. INTRODUCTION

Most CAD tools cater to digital IC design and enable an entire system to be simulated at a behavioral level. When the designer is satisfied with a particular architecture, a netlist can be automatically synthesized and laid out. This is in sharp contrast to tools for analog IC design; no single platform exists for the automated design of an RF transceiver.

This paper describes the design flow and CAD tools used in the design, simulation, and layout of a single-chip 900 MHz CMOS wireless transceiver. Large sections of this chip were simulated and laid-out using tools meant for digital circuit design. These tools did not have features to assist in analog layout and simulation, but allowed us to take advantage of the extensive silicon generators developed by our digital design group. We were forced to improvise when the existing tools could not meet our needs, or when no tool existed for the task. With this design flow, we produced a chip with measured results on first silicon very close to simulation.

Working in a university environment places a non-traditional set of values on the design flow. First, the entire chip must work. Second, students who only wished to participate for a short time should be allowed to design an individual block of the system, fabricate it, and test it separately. Finally, individual students can work a particular cell from design to schematic entry to layout, which contrasts with industry where circuit designers and CAD personnel are often distinct.

Our design flow allowed students to work in parallel during design and especially during final layout. All these requirements were met by improvising with a disjoint set of tools.

DAC 98, San Francisco, California
©1998 ACM 0-89791-964-5/98/06..$5.00

3. Project Description

This 900 MHz all-CMOS spread-spectrum transceiver uses a minimum of off-chip components, several on-chip large-valued inductors, and a high level of integration. The transmitter on the chip sends digital data at 160 kb/s in the 902-928 MHz ISM band. The receiver downconverts a prescribed channel in the ISM band, de-hops it, and provides a digital baseband output. Details are extensively documented [1,5,6] (Figure 1).

In the receive mode, the transceiver downconverts and dehops the entire ISM band detected at the antenna. This received signal is amplified, filtered, and passed to a limiter which acts as a single-bit analog-to-digital converter (ADC). A separate chip demodulates and synchronizes the signal.

In transmit mode, a digital word specifying the hopping channel and the frequency-shift keyed (FSK) data offset is applied to the direct digital frequency synthesizer (DDFS). The resulting digital sinewave is converted into an analog signal by a pair of digital-to-analog converters (DAC) and translated in a single-sideband modulator to the carrier frequency, amplified and transmitted by the power amplifier.

4. Design Methodology

4.1 Specifications

The system was designed in ALTA VISTA'S SPW. Rough specifications, such as desired hopping rate, channel capacity, BER, and data rate, were assigned to the RF blocks. The boundary between the analog and the digital sections was adjusted to find an optimum balance between estimated power dissipation and dynamic range.

Using noise figure, gain, and linearity for individual blocks, the overall system performance was calculated using MATLAB formulas for the cascaded noise figure, cascaded IIP3, and cascaded gain. For blocks not meeting the initial specifications (e.g. the noise figure of the switched capacitor filter), the properties of other blocks were adjusted (e.g. gain after the mixer) so the overall system met the design requirements.

MATLAB formulas were written to calculate the cascaded noise figure and cascaded IIP3, as standard microwave software tools assume an input and output impedance of 50 Ω and underestimate noise contributions of downstream blocks operating at high impedance levels by as much as 6 dB (Figure 2).

Individually designed building blocks, when connected together

[1] Other authors are G. Chang, J. Y-C. Chang, M. Rofougaran, M. Djafari, and P. J. Chang.

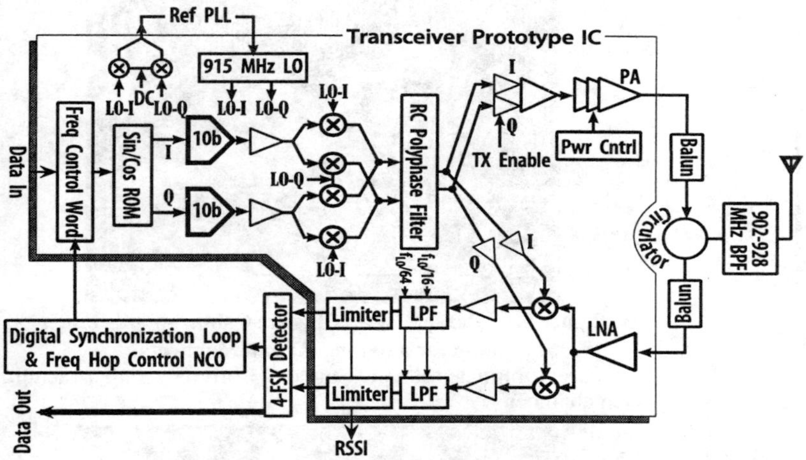

Figure 1. Transceiver block diagram

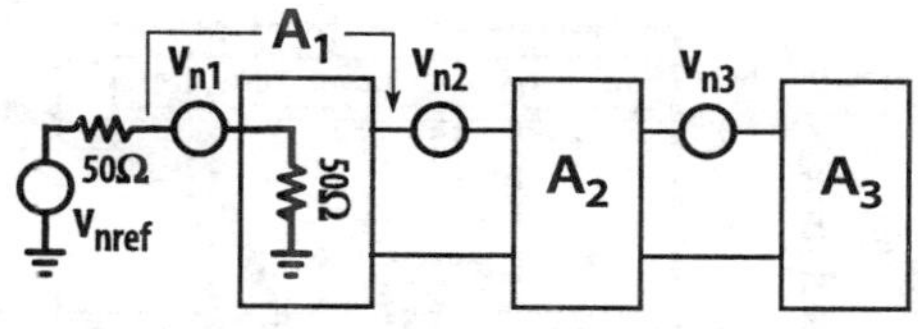

Figure 2. Cascade of integrated receiver blocks

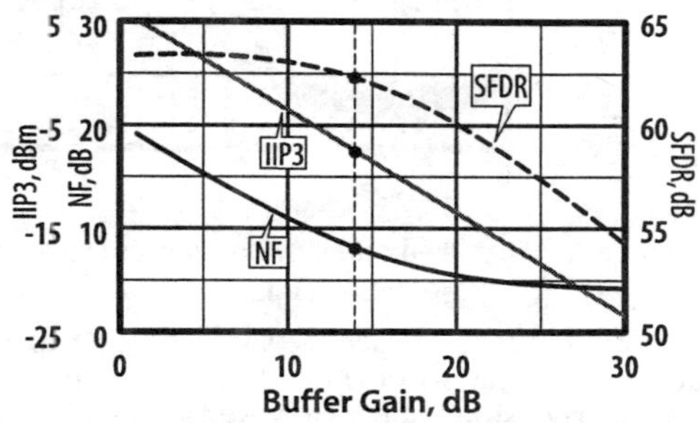

Figure 3. Receiver NF, IP3, and SFDR versus mixer buffer gain. Chosen value is highlighted

in an integrated receiver, seldom give the required total performance. Usually the overall dynamic range is not satisfactory. When this happens, the blocks must be redesigned, or intermediate gain stages must be strategically placed to improve the cascade dynamic range. As an example, take the SC filter in this receiver. Preceded only by the 20 dB voltage conversion gain of the LNA and mixer, its input noise leads to an unacceptably high 17 dB receiver NF. The filter's noise can be lowered only at a disproportionate (quadratic) expense of current drain and chip area. A better solution is to insert a baseband amplifier after the downconversion mixer, to lower the contribution of filter noise at the receiver input.

The resulting trade-offs can be difficult to visualize. Behavioral models in MATLAB, were used to simulate the results and are illustrated in Figure 3. The receiver noise figure, input intercept point, and SFDR are plotted versus buffer gain. With too much gain, the cascade intercept point falls proportionally, while the receiver noise figure flattens out to a constant value set by the LNA, and the overall dynamic range suffers. The buffer gain was selected to obtain the smallest noise figure at the point where the SFDR is close to maximum.

4.2 Cell Design

Individual cells were simulated in HSPICE to achieve low power dissipation, wide dynamic range, and low noise. The circuit was documented using the schematic capture tool in POWERVIEW. The individual cell was then laid out in MENTOR GRAPHICS' GDT and a layout-versus-schematic check was performed using TANNER LVS.

Some common problems with amateur layouts include inadvertent shorts between VDD and GND, flipped lines in differential circuits, and cells which don't connect. Unique to a wireless transceiver, 2 pairs of lines representing differential quadrature signals typically connect separate blocks. If any of these signals are accidentally switched, the entire chip fails. LVS was used to identify these potential problems. A uniform labeling convention was adopted: I, Ibar, Q, and Qbar. A top-level SPICE simulation verified that the labels were correctly assigned. A cell would not pass LVS unless the quadrature channels were properly labeled and checked. After the chip was laid-out and passed LVS, the extracted netlist was simulated to verify that the differential quadrature signals were connected correctly. The on-chip line lengths of these four signals were well-matched within each cell and between cells.

Each cell was placed in a pad ring, and fabricated for individual testing. Layout and testing of individual cells is frowned upon in industry because of the cost and the time involved, but it was necessary here because RF-CMOS was a new field without benefit of existing publications on circuits, layout techniques, and testing issues.

4.3 Circuit Extraction

The digital tool GDT extracts the SPICE netlist from the layout in a format incompatible with our simulation tools and LVS. The tool could extract a SPICE net list in two different modes. In the first mode, the interdigitated transistors were lumped as single devices. This results in the wrong junction capacitance. In the other mode, all hierarchal information is lost, making it virtually impossible to resolve LVS errors.

There were two additional problems. First, the spiral inductors were extracted as if they were interconnect. To manually fix this problem, the SPICE netlist must be edited, some nodes renamed, and an inductor inserted between these nodes. This method is time consuming and prone to error. Second, the extraction tool could not differentiate between parasitic capacitors and desired capacitors. Typically, the capacitor extraction is turned off for an LVS check, however this leads to errors in circuits that contain capacitors as passive components.

Our solution to both of these problems was to define new layers in the GDT layout tool. These layers were placed on top of inductors or capacitors which force the underlying devices to be extracted whether the parasitic capacitance extraction is enabled or not. Further, dummy inductors were placed in the schematic. Manual editing of the extracted file was still required, but now only to enter the inductance value.

4.4 Switched Capacitor Filter

It is difficult to simulate the noise of a switched capacitor filter. This is because the noise analysis in HSPICE is performed with an AC simulation on a linear, time-invariant model at the operating point of the circuit. In a switched capacitor filter, switches turn on and off periodically. This is a time-varying, sampled-data circuit.

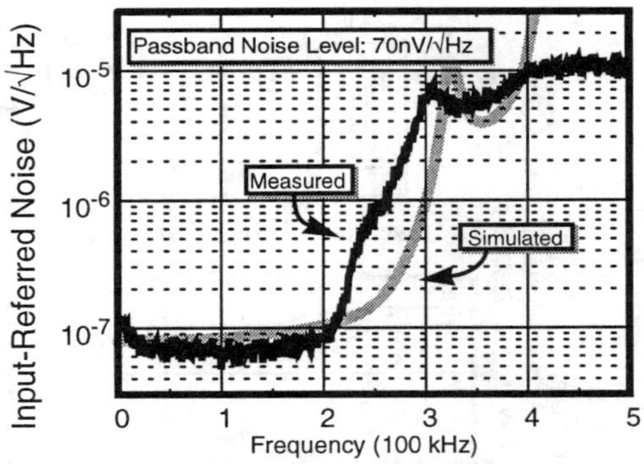

Figure 4. Measured and simulated noise spectrum of the switched capacitor filter

Our solution was to replace the switches and capacitors with an equivalent continuous-time resistor. At first, this procedure may seem overly simplistic. However, according to switched capacitor theory, at frequencies much lower than the sample rate, a switched capacitor resistor is thermo-dynamically equivalent to a real resistor. The resulting filter comprising resistors, capacitors, and op amps was analyzed by an AC simulation to verify the frequency response, and a noise analysis was also done. In this way, a problem was identified in an early version of the filter. The input noise of the filter peaked at the location of one of the data tones used in our 4-FSK modulation. The gain distribution among filter stages was modified to flatten the noise spectrum in the range of 20 to 160 kHz. A graph comparing the simulated and the measured noise spectrum is shown in Figure 4. The slight discrepancy in the frequency of the passband edge arises from incomplete op amp settling, which in turn is traceable to optimistic MOS model parameters used in the design (see 4.10).

4.5 On-chip Inductors and Low Loss Routing

On-chip inductors of 30 or 50 nH were used extensively in the transceiver to reduce power dissipation, tune out parasitic capacitances, and meet the phase noise specification in the VCO (Figure 6). The inductors are suspended in an oxide layer above the substrate to reduce the parasitic capacitance of the inductor, and to extend the self-resonance frequency to a few gigahertz. The technique used to suspend the inductors was also used to isolate RF signals on long interconnects from the lossy substrate. A close-in view of the inductors and the suspended high frequency routing lines is shown in Figure 5. First-order models were used to simulate circuits with the on-chip inductors (Figure 7). A MATLAB program calculates the values of the resistance, capacitance, and inductance, the last based on equations developed by Greenhouse [3].

4.6 Post-process Etching Treatment

The fully fabricated die are etched before packaging, requiring no prior modifications to the CMOS process. The technology file of the layout software is modified to create a new layer consisting of vias stacked one upon another. After chip fabrication, this layer creates an opening in the passivation down to the substrate. This opening becomes an entry point for the etchant (Figure 8). The passivation everywhere else acts as a mask for the etchant.

The etchant must not corrode aluminum, otherwise the pads will be removed. XeF_2 is used as the etchant here [2]. This gas-phase procedure works much better than liquid-phase etching. The XeF_2, a rocksalt-sized crystal at room temperature and

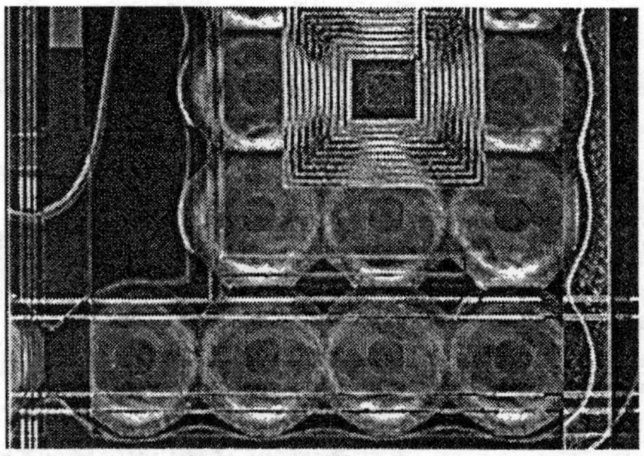

Figure 5. Enlarged view of Figure 6 showing pits under inductor and long RF interconnects

pressure, sublimates at a pressure of 4 Torr, requiring a vacuum etch chamber.

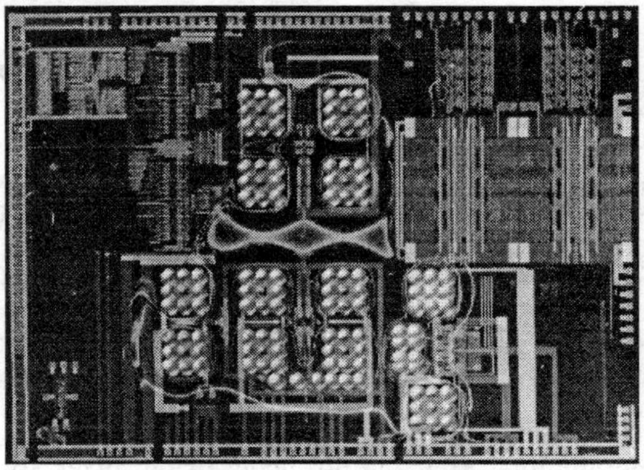

Figure 6. Photo of etched Chip

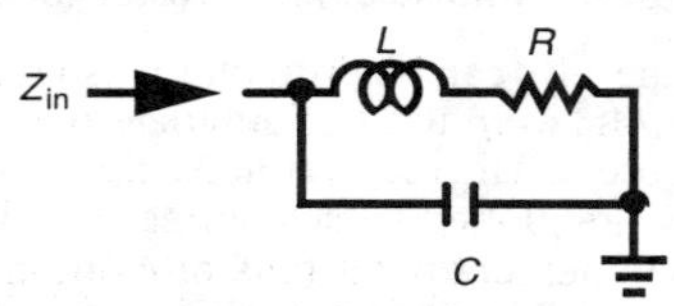

Figure 7. SPICE model of inductor

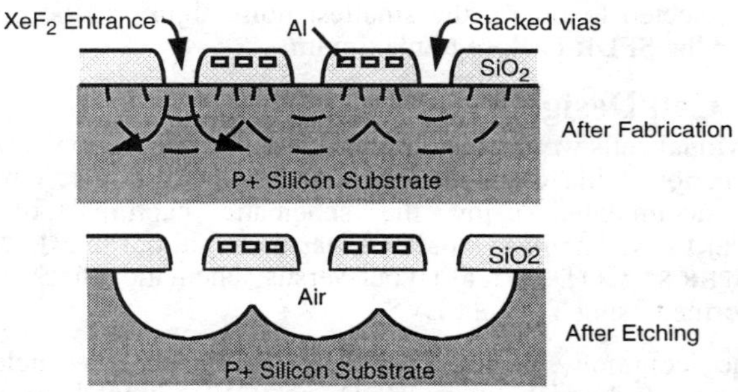

Figure 8. XeF_2 isotropic silicon etchant

4.7 Mixers

The intercept point (IP3) of the mixers was simulated exactly as it would be measured in the lab. Two tones were applied at the mixer input, and downconverted. The third order non-linearities in the mixer produce a set of sidebands around the downconverted tones. An FFT was performed on the transient output of HSPICE, and the amplitude of the down-converted tones was compared to the amplitude of the sidebands.

The input signal level must be chosen carefully for fast and accurate simulation. If the input is too small, the sidebands will not appear above numerical noise. If too large, the amplitude of the output tones may compress. This can be seen in the plot of the measured IIP3 performance of the LNA, Mixer, and buffer combination (Figure 9). The upper curve is the measured output power of the down-converted tones. At higher input power levels, the output is compressed due to the non-linearities. The lower line is the measured power of the sidebands and rises at a rate of three times the output tones. The input power where the extrapolated lines cross is the IIP3. Experience shows that at a well chosen signal level, a single simulation accurately predicts the IP3. To calculate the IIP3 from a single data point, the following equation was used:

$$IIP3 = P_{in} + \frac{\Delta}{2}$$

where Δ is the defined in Figure 9 and P_{in} is the input power. If the mixer was simulated with a P_{in} of -30 dBm and the measured Δ was 46 dB, the IIP3 would be -7 dBm.

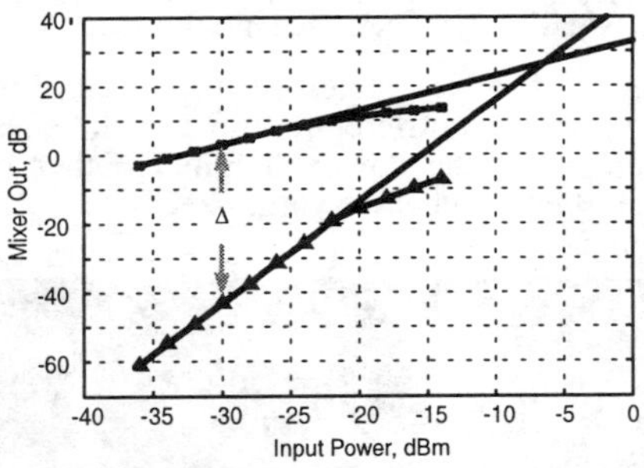

Figure 9. Front-end IIP3 measurement

This simulation can be very long. A small step size is required for the high frequency components and a long simulation time to resolve the small differences in frequency. For example, two tones at 915 MHz and 915.160 MHz require more than 250 k simulation points for a few cycles of 160 kHz, the difference frequency, to appear between the two input tones. As the frequency response of the mixer is flat past 10 MHz, a wider frequency difference can be used to lower the stiffness of this simulation: with 915 MHz and 920 MHz inputs, the total number of points drops to 8,000, a factor of 30 in simulation time.

4.8 Cell Generators Used In the Layout

The project made extensive use of silicon generators written in the MENTOR GRAPHICS LX language. Additionally, LX code was used during final assembly of the complete system.

The DDFS, a 100 MHz static CMOS logic block, was generated with a silicon compiler in LX. The cell was designed by a different group of students but since it was a stand-alone part, it could be imported and routed just like any other cell.

Of particular interest is the layout of the 10-bit digital-to-analog converter (DAC) (Figure 10) and the switched capacitor filter (Figure 11). Both were generated using Lx. When changes were made in the design, the Lx input file was edited to reflect the changes, and the layout was regenerated. This proved especially useful in the switched capacitor filter when different gain distributions and pole locations were being iteratively evaluated in the final transceiver.

4.9 Top-level Layout

Once all of the individual cells were verified, final layout

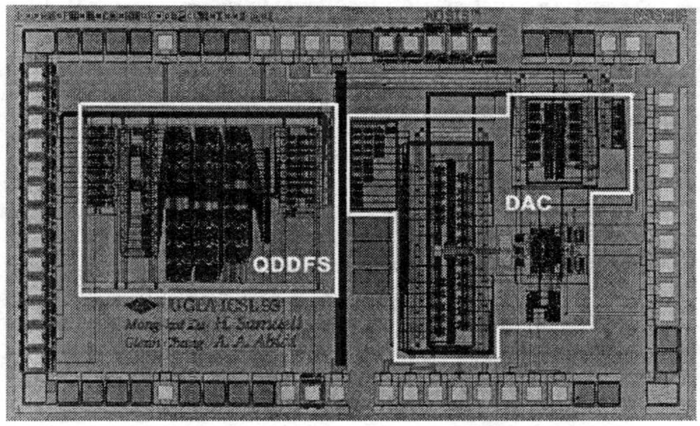

Figure 10. Layout of DDFS and DAC

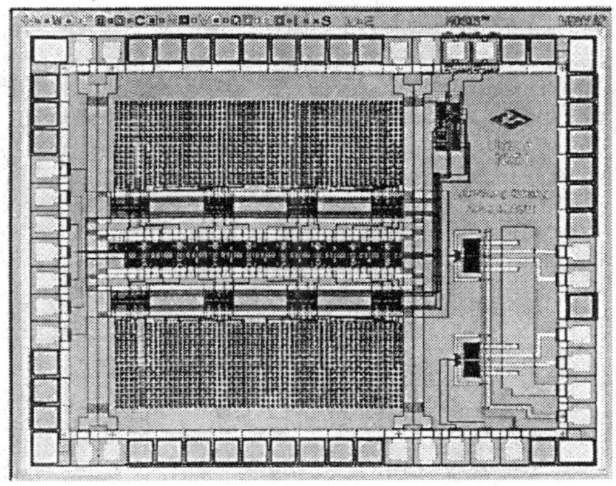

Figure 11. Layout of switched-capacitor filter

began. However, the earlier standalone prototypes were still evolving into final form. Information about the incomplete cells, such as size, terminal names, and terminal location were used for the final assembly. LX-code placed the tentative cell and connected it to the neighboring cell and to the pad ring. Later, the tentative layout was replaced with the final layout and the LX-code was modified to reflect any changes in the location of the terminals.

For example, this technique was very helpful during a last-minute change to the VCO. A buffer added to the VCO shifted its center frequency. To compensate, the layout of the inductor used in the VCO was modified. During the VCO modification, the terminals and the outer dimensions were not altered, so no change was necessary in the LX-code controlling the top-level layout. The only top-level change consisted of the old VCO file with the new layout file. Because, the top-level was not altered, LVS did not have to be re-checked.

4.10 Unforeseen Problems

There were other important problems in the design flow. Standard (Level 13) MOS model parameters did not predict the measured results very well. Instead, we used a more accurate model (Level 28), but its extracted parameters were five years old. We found through trial-and-error that this model's worst-case parameters represented the current process. The final transceiver was simulated using these parameters.

Another problem not anticipated was the RF loss in the substrate at input and output pads. This was discovered when the measured performance of a particular standalone cell did not match simulation. The pad structure was simulated in

PISCES, verifying that at high frequencies there is a substantial resistive component in the pad impedance. A lumped model was formed from the PISCES simulation and included in the HSPICE simulation. A new RF pad was designed that shielded the signal from the substrate (Figure 12).

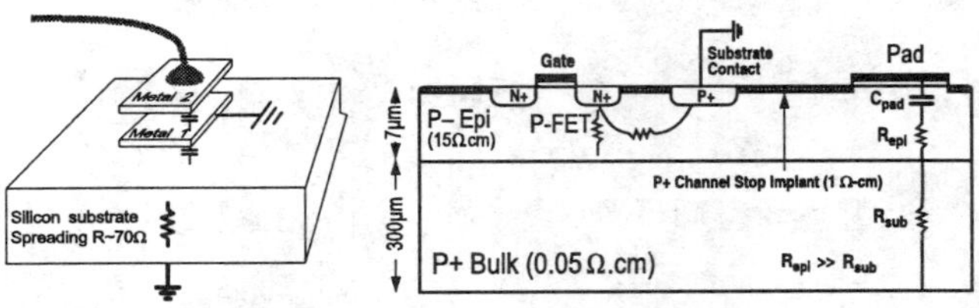

Figure 12. Substrate resistance under RF pads is shielded by Metal 1

Gate resistance of the input FET can dominate the noise figure of an LNA. CMOS models typically do not model the gate resistance. As CMOS devices can be laid out in a variety of different ways, the gate resistance must be hand calculated. An estimated value of the gate resistance was calculated and inserted into the HSPICE simulations as a series resistor (Figure 13).

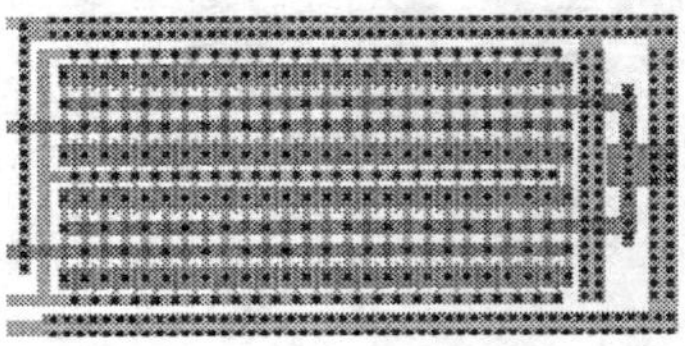

Figure 13. Transistor layout to minimize gate resistance

4.11 Packaging Issues

High frequency circuits use a multi-point ground system. In this transceiver, all connections made to the ground plane should be kept as short as possible [4]. The metallized cavity of the package was connected to the ground plane of the test board. The die was mounted using conductive epoxy to the metallized base of a ceramic package. A photo of the die and the package is shown in Figure 14.

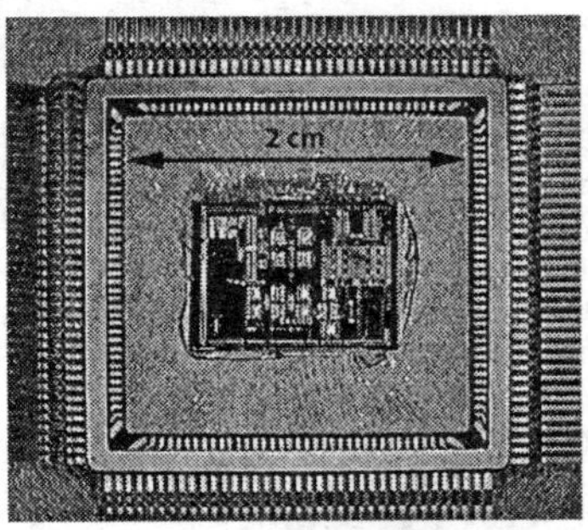

Figure 14. Photo of die and chip package

Qualitative agreement between measured and simulated input reflection coefficient over 0.4 to 1.4 GHz was found using a simple model for the package (obtained from the manufacturer), bondwire, and pad (Figure 15). From simulations, it was found that a simple matching network gives a good wideband match to 50 Ω.

4.12 Test Board Design

ORCAD was used to lay out the four-layer board. The top and bottom layers were used for signals, the middle two layers were

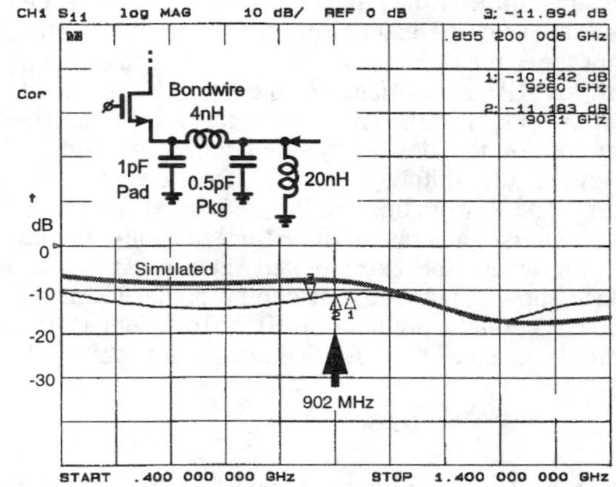

Figure 15. Model of pad, bondwire, and package

used for supply and ground. Separating the signal layers increases the isolation between the signal traces.

Only two pairs of signal traces required carefully matched delays. These carry the differential input to the LNA, and the differential output of the power amplifier. Other high speed lines used in the frequency synthesizer were also laid-out carefully using 50 Ω traces.

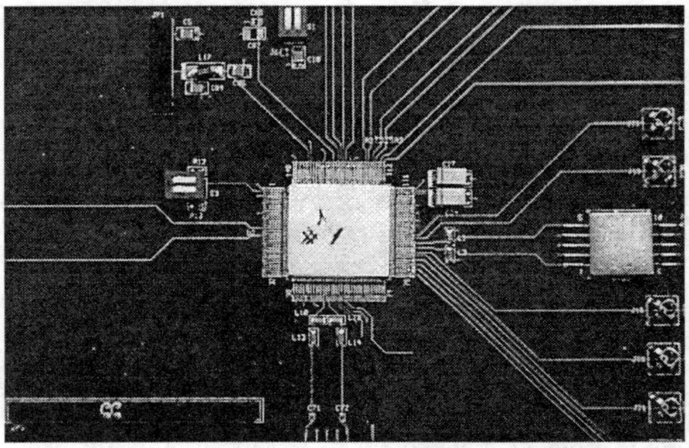

Figure 16. Photo of top of test board

The top of the board is relatively free of off-chip components. The only visible elements are two baluns, 6 chip inductors, and chip capacitors, Figure 16. The matching components were placed as close to the chip as possible. Decoupling capacitors were placed close to the chip on all supply lines and are clearly visible in the photo of the bottom of the test board.

This PCB is simple when compared to other RF PCBs with their shielding and large number of support components. No shielding is necessary here because of the high level of integration.

5. Measured Results

The transceiver worked at first silicon. An annotated photograph of the final chip is shown in Figure 17. The transceiver was fully characterized and the results are documented elsewhere [5-6]. Measured results for the transmitter are given in Table 1, and for the receiver in Table 2. The cascade effect of the various blocks is seen in a signal level diagram (Figure 18). For example, the mixer is the limiting component for the IIP3 and the filter is the limiting component

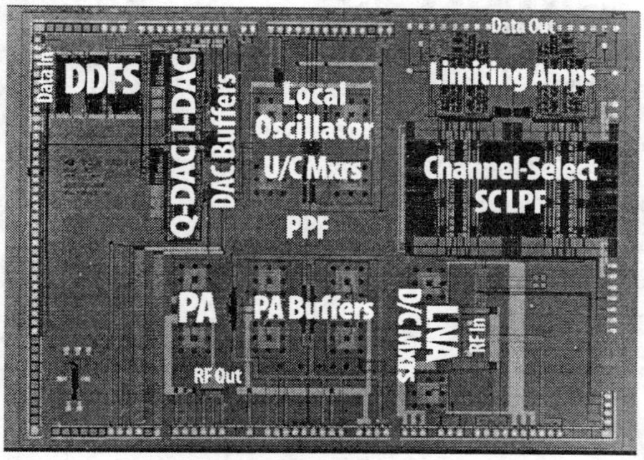

Figure 17. Chip Photo

Operating Frequency Band	902-928 MHz
Multiple Access	Fast Frequency Hopping
Frequency Hop Rate	20 kHz to 150 kHz
Data Rate	160 kHz with 4-FSK
PA Output Level	-17 to +13 dBm
Spurious Levels	-55 dBc at +3 dBm output
Output reflection coefficient	-15 dB
Oscillator Pulling	< 200 kHz over PA output range
Current Drain when active	100 mA from 3 V

Table 1. Measured Transmitter Characteristics

for the noise floor. From the plot it is clear that none of the components were over-designed for the specifications. The spurious free dynamic range (SFDR) is also given for each component and the entire receiver. It is noteworthy that this level diagram, constructed from measurements or simulations of individual prototypes, very accurately predicts the measured performance of the total receiver (Table 2). The measured transmitter output spectrum is shown in Figure 19.

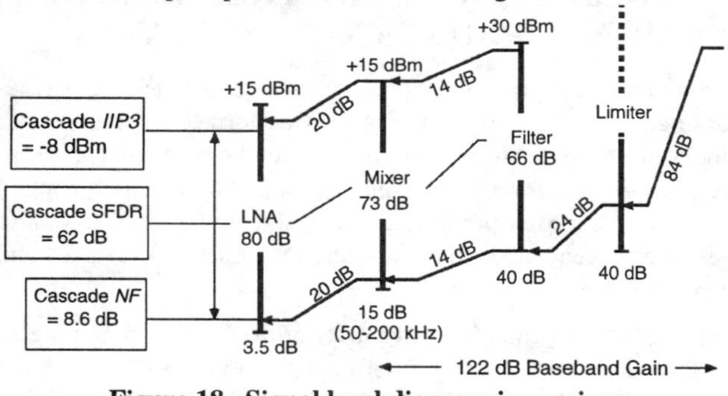

Figure 18. Signal level diagram in receiver

6. Conclusions

Using the above design flow, the first all-CMOS 900 MHz transceiver was designed, simulated, and laid out. In the absence of an integrated design environment, design flow was pieced together with existing tools. The transceiver worked at first silicon and yields a performance competitive with the state-of-the-art, after allowing for the modest f_T of the technology of implementation.

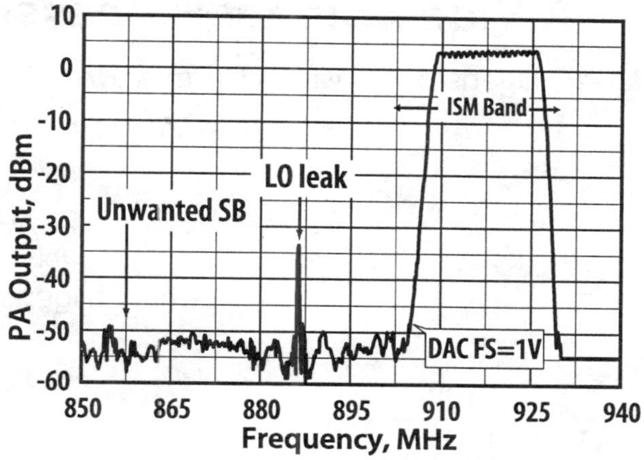

Figure 19. Measured transmitter output spectrum showing frequency hopping in ISM band

Total Noise Figure	8.5 dB
Minimum Detectable Signal	1.5 µV without coding
Total IIP3	-8.5 dBm
Total IIP2	+22 dBm
Input return loss	10 dB
Current Drain when active	120 mA from 3V
IC Technology	1-µm CMOS, 2M1P, n-well

Table 2. Measured Receiver Characteristics

7. References

[1] Abidi, A., Rofougaran, A., Chang, G., Rael, J., Chang, J., Rofougaran, M., Chang, P., Djafari, M. The Future of CMOS Wireless Transceivers, from the experience of realizing A Single-Chip 900 MHz Spread-Spectrum Wireless Transceiver in 1µm CMOS in *Int'l Solid State Circuits Conf.* (San Francisco CA, 1997), IEEE, 118-119.

[2] Chu, P.B., Chen, J. T., Yeh, R., Lin, G., Huang, J., Warneke, B. A., Pister, K. S. J. Controlled pulse-etching with xenon difluoride in *IEEE Proc. Nineth Int. Conf. on Solid-State Sensors and Actuators*, - TRANSDUCERS '97 (Chicago IL., June 1997), 665-668.

[3] Greenhouse, H. M. Design of planar rectangular microelectronic inductors. *IEEE Transactions on Parts, Hybrids and Packaging* (vol. PHP-10, no. 2, June 1974), 101-10.

[4] Ott, H. W. *Noise reduction techniques in electronic systems.* Wiley, New York NY, 1988.

[5] Rofougaran, A., Chang, G., Rael, J., and others. A Single-Chip 900 MHz Spread-Spectrum Wireless Transceiver in 1-µm CMOS - Part 1: Architecture and Transmitter Design. *IEEE Journal of Solid State Circuits* (to be published April 1998).

[6] Rofougaran, A., Chang, G., Rael, J., and others. A Single-Chip 900 MHz Spread-Spectrum Wireless Transceiver in 1-µm CMOS - Part 2: Receiver Design. *IEEE Journal of Solid State Circuits* (to be published April 1998).

A Video Signal Processor for MIMD Multiprocessing

Jörg Hilgenstock Klaus Herrmann Jan Otterstedt* Dirk Niggemeyer Peter Pirsch

Laboratorium für Informationstechnologie
Universität Hannover
Schneiderberg 32
30167 Hannover, Germany
Tel: +49 511 762 5058
E-mail: hilgenst@mst.uni-hannover.de

ABSTRACT

The video signal processor AxPe1280V has been developed for implementation of different video coding applications according to standards like ITU-T H.261/H.263, and ISO MPEG-1/2. It consists of a RISC processor supplemented by a coprocessor for convolution-like low-level tasks. RISC and coprocessor have been implemented in a standard cell design combined with full-custom modules. The processor was fabricated in a 0.5 μm CMOS technology and has a die size of 82 mm^2. It provides a peak performance of more than 1 giga arithmetic operations per second (GOPS) at 66 MHz. For processing of very computation-intensive algorithms or high data rates, several processors can be bus-connected to form a MIMD multiprocessor system.

1. INTRODUCTION

For real-time coding and transmission of video data, several international standards have been developed. These include ITU-T H.261/H.263 [1, 2] for video telephone, ISO MPEG-1 [3] for multimedia, and ISO MPEG-2 [4] for digital TV. All these standards utilize hybrid coding techniques [5]. They combine computation-intensive low-level tasks, like motion estimation or discrete cosine transform, with data-dependent medium-level tasks, like variable length coding, variable threshold or quantization.

The performance requirements of these standards are determined on one hand by the complexity of the employed algorithms. On the other hand, they are mainly based on image size and frame rate. Altogether, the performance requirements for real-time coding range from a few hundred million operations per second (MOPS) for a video telephone decoder, up to several giga arithmetic operations per second (GOPS) for a digital TV encoder. For these applications, a cost-effective and small system implementation is required. Therefore a VLSI implementation, which can provide the needed processing power, is essential.

*now with Siemens Semiconductors, Munich, Germany.

Two implementation strategies can be followed: First, an implementation of the codec's architecture dedicated to the envisaged application field can be used to reduce the required implementation area [6]. The drawback of this approach is, that the implementation of a modified coding algorithm often necessitates a redesign of the involved circuits.

Alternatively, programmable processors can be used. They allow flexible adaption to different coding applications and standards. A disadvantage of programmable solutions is the increased silicon area. Especially the use of general purpose digital signal processors for the above mentioned video coding applications often leads to an unacceptable system size. Therefore, the programmable video signal processor architecture AxPe has been developed, which has been optimized for hybrid video coding algorithms and can be used in a scalable multiprocessor system.

In the following the video signal processor AxPe1280V is presented. Section 2 introduces the architecture of the processor. The implemented test concept is presented in Section 3. Section 4 and 5 describe the design process and implementation details of the chip. In Section 6, a AxPe-based MIMD multiprocessor system is presented.

2. ARCHITECTURE OF THE VIDEO SIGNAL PROCESSOR AXPE1280V

The compression and decompression of digital video signals according to hybrid coding schemes is performed on rectangular blocks of 8*8 or 16*16 pels. In the encoder the entire image is segmented, and low-level and medium-level coding tasks are applied on block level. After processing, a data stream for transmission to a decoder is generated. In the decoder this process is reversed and the images are reconstructed.

The most computation-intensive tasks of the video coding process are low-level tasks. Over 70% of the total computation power are consumed by this group of tasks. Because they have a convolution-like structure and a deterministic program flow, the data accesses and number of execution cycles are predetermined and so the controlling is relatively simple. Less computation-intensive are medium-level tasks. They have a less deterministic program flow with many data dependent branches, which makes parallelization on data level more difficult.

An efficient implementation of programmable processors for video coding systems can be achieved by adaptation of the processor architecture to the coding algorithms. One possible approach is the coprocessor concept. Computation-intensive operations, which are common to all coding schemes, are mapped onto a specialized coprocessor. A programmable processor controls the coprocessor and executes all other parts of the algorithms.

Based on this approach, the architecture of the video signal processor AxPe1280V (Figure 1) has been developed. The processor consists of two main modules: A RISC processor and a low-level coprocessor. Both modules are adapted to a subclass of coding tasks. The RISC processor is used for computation of medium level tasks, like quantization or Huffman coding, and performs all control tasks. For fast processing of computation-intensive, convolution-type low-level coding tasks, like discrete cosine transformation or motion estimation, the microprogrammable coprocessor is used. An I/O control unit performs all communication with external devices. It features a bidirectional 32 bit data bus as well as a 16 bit data output bus. This additional 16 bit output bus was implemented to increase the achievable communication bandwidth of the AxPe1280V and to simplify arbitration in a multiprocessor system consisting of several processors.

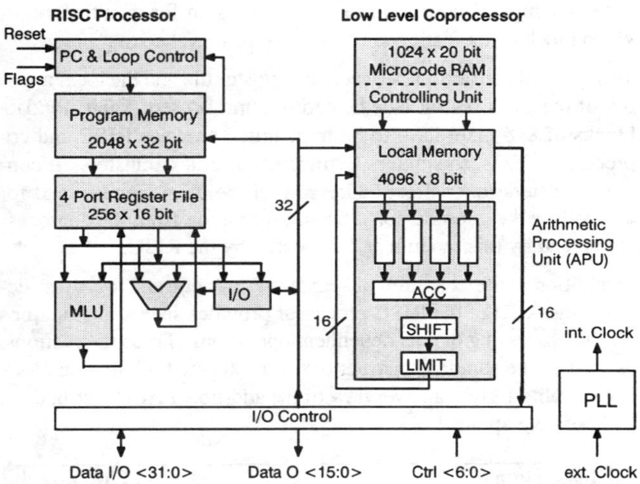

Figure 1: Architecture of the video signal processor AxPe1280V

2.1 Coprocessor

The main modules of the coprocessor are a local memory of 4096 bytes, a fourfold parallel arithmetic processing unit and a microprogrammable control unit. The local memory has been integrated to reduce transfers with external memories and to provide the arithmetic processing unit with enough bandwidth for parallel processing. The microprogrammable control unit masters the operation of the complete coprocessor.

2.1.1 Arithmetic Processing Unit

The architecture of the arithmetic processing unit is dedicated to the processing of convolution-type low-level tasks. To achieve a high data throughput, the inherent parallelism of these tasks must be exploited. Low-level tasks, like 2D discrete cosine transform or motion estimation access blocks of 8*8 or 16*16 pixels respectively. These tasks can be separated into 1D operations accessing neighbouring pixels. The presented arithmetic unit makes use of this characteristic by processing four pixels in four identical data paths. Each data path consists of a pipeline of subtracter, complementer, multiplier, and mode shifter. The output of the data paths is fed into a common multioperand accumulator (Figure 2).

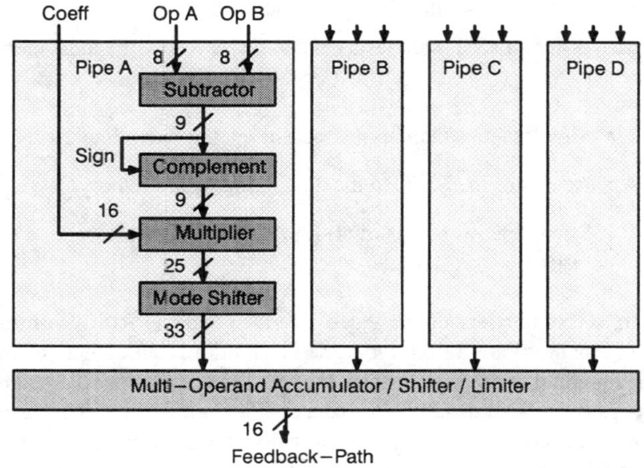

Figure 2: Architecture of the arithmetic processing unit

A generic description of all operations of the arithmetic processing unit is

$$\frac{1}{2^N} \sum \pm(A - B) * Coeff * 2^s \qquad s \in (0, 8)$$

The operands A and B of all data paths have an 8 bit representation. For most video coding tasks like motion estimation or FIR filtering, 8 bit arithmetic is sufficient, and therefore the data paths have been primarily adapted to these needs. Nevertheless, the mode shifters between multiplier output of each data path and inputs of the multioperand accumulator support 16 bit operations in two consecutive clock cycles. During the first clock cycle, the lower bytes of all operands are processed. The mode shifters are switched transparent. Processing of the upper bytes of the operands is performed during the following clock cycle. To adjust the results of the multipliers before accumulation, the mode shifters perform an 8 bit arithmetic left shift during this cycle.

Only two different modes - transparent mode and 8 bit arithmetic left shift - have to be provided by the mode shifters. Therefore, an implementation of the mode shifters increases the gate count of the arithmetic processing unit only marginally. More significant with respect to silicon area is, that 9x16 bit multipliers are required for the realization of 16 bit arithmetic.

Due to the different operation modes, the presented arithmetic processing unit combines fast fourfold parallel 8 bit arithmetic with high accuracy 16 bit arithmetic. Operating in 8 bit mode with a clock rate of 66 MHz, a peak performance of 1056 MOPS is achieved by the arithmetic processing unit. Because all 16 bit operations require two clock cycles, the effective parallelism in the 16 bit mode is reduced to two. Therefore, a single arithmetic processing unit provides a peak performance of 528 MOPS in 16 bit mode.

2.1.2 *Local Memory*

The local memory supports on-chip storage of input data and intermediate results of the arithmetic processing unit. Furthermore, the local memory serves as a 32 bit memory mapped interface between the coprocessor and the RISC processor core.

Three different access requests have to be served by the local memory:

- Simultaneous data in- and output via the video interface.
- Access of the RISC to the data in the local memory.
- Simultaneous access of all four data paths and storage of the results.

While the transfers of the video interface and the RISC with the local memory are 32 bit wide, the bandwidth requirements of the arithmetic unit are significantly higher. Each of the four data paths needs two operands (2*8 bit) and a coefficient of 16 bit. This leads to a read access of 128 bit. Additionally, the result of the accumulator/shifter/limiter (16 bit) has to be stored in the local memory.

The local memory was realized as dual-port RAM. To serve all memory requests, it can operate in the above mentioned three access modes. With integration of this 4096 byte local memory, parts of the processed video image can be kept locally. This is especially advantageous for motion estimation in hybrid coding applications, as it allows to carry out block matching within a search range of up to +/- 22 pixels horizontally and vertically without reloading, which leads to a significant decrease in overall processing time.

2.1.3 *Microprogrammable Control Unit*

The microprogrammable control unit features a 1024x20 bit microcode RAM and masters the operation of the local memory and the arithmetic processing unit. Each clock cycle it generates two addresses for the ports of the local memory and a control word for the arithmetic processing unit. Loop-based processing is supported by this unit, so that operations like 2-D discrete cosine transform or half-pel motion estimation can be executed on the coprocessor autonomously.

2.2 RISC Processor Core

The RISC processor core of the video signal processor is used for medium-level and control tasks. These tasks are characterized by data-dependent operations, i.e., the control flow of these tasks shows a high degree of data-dependent branches.

Several RISC cores are available for embedded applications and can be implemented as hard or soft macros. Because they are usually designed for control applications, they do not fulfill the requirements of video processing applications very well. However, controlling is a main task the RISC in the AxPe has to perform.

For the AxPe investigations were carried out to evaluate the necessary performance for medium-level processing [7]. Because the coprocessor can execute low-level tasks autonomously in parallel to the computation of medium-level tasks in the RISC, a good balance between the computation times of both units can reduce the number of stall cycles. The investigations have shown that on one hand commercially available cores offer more features, than are required. For example data caching speeds up access to large external memories, but this is not essential for hybrid video coding applications as data access patterns are known in advance.

On the other hand the performance for processing of medium-level tasks can be improved significantly by speeding up kernel routines. Especially the performance of quantization and Huffman coding can be improved by adapted execution units. Therefore, an adapted RISC core (Figure 1, left part) for the AxPe has been designed, which complements the low-level coprocessor.

The developed 32 bit RISC core is based on a harvard architecture with a 16 bit data path. The RISC processes one scalar instruction per clock cycle. Instructions are fetched from a 2048 * 32-bit on-chip program memory and executed in a four stage instruction pipeline, which consists of the stages `fetch, decode&read, execute, write`. In the `decode&read` stage operands are read from a 256 * 16 bit register file for being executed either by the ALU or by the medium-level unit (MLU). The register file can be addressed either directly or indirectly. For indirect addressing three indirect address registers are used, two for sources and one for destination. Additionally, a irregular indirect addressing scheme is supported by an implemented RAM-based address table. Because video coding schemes require a meander scan for quantization of video blocks, this feature speeds up the run level coding process.

To support data transfer between the register file and the local memory of the coprocessor, a dedicated I/O unit is used. Typically data blocks of 8x8 pixel have to be transmitted between RISC and coprocessor. Thus, this unit is optimized for block transfers. It contains registers for indirect addressing of the local memory and for controlling the coprocessor. Status information from the coprocessor is read by this module and evaluated by the RISC.

In addition to the execution of standard arithmetic and logic instructions, the ALU of the RISC processor provides special instructions for the processing of data dependent operations. These instructions, as well as the standard instructions, are executed within one clock cycle. Table 1 gives an overview of the additional ALU instructions and possible applications.

Instruction		Required for Task
$A-\|B\|$		Variable Length
$A+\|B\|$		Coding
$A-B$	$if\ (A<0)$	Quantization
$A+B$	$if\ (A>0)$	
A	$if\ (A=0)$	
$A-B$	$if\ (A<0)\ \&\ (A\ even)$	
$A+B$	$if\ (A>0)\ \&\ (A\ even)$	
A	$else$	

Table 1: Additional ALU instructions

The second execution unit in the RISC processor, the medium-level unit (MLU) has been implemented to speed up multiplications, which frequently occur in quantization and inverse quantization tasks. These tasks represent the most computation-intensive part of medium-level processing. A 17*17 multiplier for signed fixed-point multiplications and a shifter were implemented in this unit. A postponed limiter supports range control and rounding of the result and detects an overflow or underflow (Figure 3). The MLU is highly configurable and can not only be used for this special purpose.

The instruction memory of the RISC is addressed by the PC&Loop-

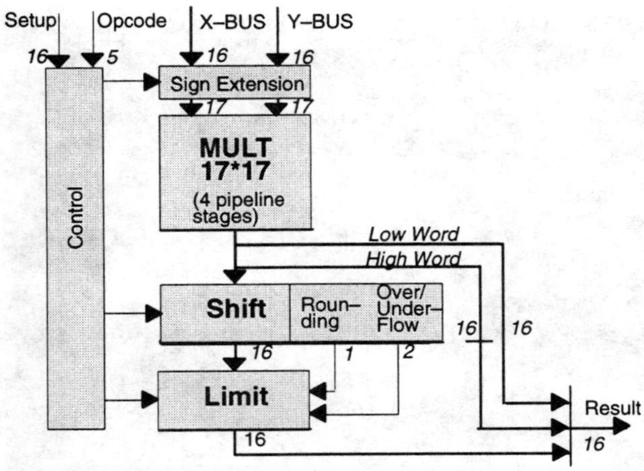

Figure 3: Architecture of the medium-level unit

Control unit, which supports hardware controlled instruction loops. To execute loops, a single register has to be initialized with the loop count by a single instruction. After invoking a hardware controlled loop, the unit performs a cyclic execution of a specific code segment without any overhead for branches. This feature reduces the amount of control instructions in medium-level tasks, which are typically based on loops of a fixed loop count.

3. INTEGRATION OF TEST STRUCTURES

The AxPe1280V contains test structures, that allow to carry out several tests on the integrated logic, memory and communication network. It features a scan path for bus tests (ScBT), which surrounds the core similar to a boundary scan path and is utilized for bus tests.

Furthermore, the AxPe1280V is equipped with a built-in self-test (BIST). The BIST is performed and controlled locally, because access to the modules from the outside is very limited and is also subject to defects. Therefore, a programmable fault-tolerant BIST controller has been integrated within the video signal processor. A scan path plus four signals secured by a parity bit are sufficient for programming and controlling the BIST controller. If the BIST controller detects an error in itself or in its control signals, or if the test of the processor fails, it moves into an isolation state. In this state the processor is prevented from accessing the output bus, the scan paths of this module are bypassed and the local clock is frozen, preventing dynamic power dissipation within the disabled processor. The isolation state can only be left by applying a reset instruction to the BIST controller. Together these measures ensure that only a sufficiently small number of potential defects remains within the chip, which could affect a multiprocessor system based on the AxPe1280V.

All logic parts of the video signal processor have been analyzed with respect to testability, which can be achieved by pseudorandom test patterns generated by BILBOs (Built-In Logic Block Observers) [8]. Selected system registers have been replaced by modified BILBOs, which — beside their normal function as registers — can perform tasks as a TPG (Test Pattern Generator), TAE (Test Answer Evaluator), or as a stage of a scan chain. BILBOs running in the TPG mode generate pseudorandom test patterns. BILBOs acting as TAEs are employed to collect and compress the test answers of the tested logic.

The control signals of all BILBOs in the RISC processor are generated by the BIST controller. All control signals of the BILBOs in the coprocessor (arithmetic processing unit, control unit, and local memory) as well as some additional coprocessor test control signals are set by the RISC processor.

The video signal processor including BIST properties has an area overhead of about 3.5% on top of a processor containing a scan path as its only test feature.

4. DESIGN PROCESS

In the beginning of this project several hybrid coding algorithms were analyzed according to the employed operations and their specific performance requirements. This led to the presented coprocessor concept. Parts of the architecture were modeled in the programming language C and extensive simulations with several filtering and coding algorithms were carried out. Bottlenecks of the initial concept were identified and eliminated.

After the detailed architecture specification was set up, the design of the video signal processor AxPe1280V was carried out by a core team of four engineers supported by several graduate students doing their project work or thesis in this project. This approach required to segment the chip design into many relatively small functional units which could be designed and simulated independently. For this reason a detailed functional and interface specification of the modules before implementation was crucial.

Most modules of the RISC processor and parts of the control logic of the coprocessor were specified in Verilog on register transfer level (RTL). For implementation of the arithmetic units like ALU, multiplier, multioperand accumulator, and barrel shifter two alternatives were considered: A standard-cell design in Verilog and synthesis or the full-custom implementation of these modules. On one hand a standard-cell implementation of these units would have reduced design time, but on the other hand would have led to a significant increase in the required silicon area for these modules. Experiments showed that a standard-cell implementation of the arithmetic would also decrease the overall performance of the video signal processor drastically compared to a full-custom implementation. For these reasons the arithmetic modules were designed especially for the AxPe1280V in full custom logic with Cadence tools. Other full-custom modules like memories and an analogue PLL were from a library developed by Philips semiconductors.

Each implemented module of the processor was simulated on functional level. After this initial verification, several modules were connected to larger units and simulated again. To integrate the instantiated full-custom modules in this simulation, behavioral Verilog models were created for these modules. More and more Verilog modules were connected together and a simulation of the complete RISC processor and coprocessor was carried out. Several test programs were developed in assembler and executed on the description of the processor. Their evaluation unveiled the remaining errors in the Verilog implementation and also helped to optimize parts of the architecture.

The verified Verilog modules on register transfer level were synthesized with Synopsys' design compiler. To achieve good synthesis results and to reduce the wiring overhead in the standard-cell layout, small modules containing not more than 2000 gates were synthesized independently. As the processor was already divided in relatively small functional blocks, no additional segmentation of these modules was necessary. The synthesized netlist of each module was again simulated with the Verilog simulator and timing was optimized.

For the final implementation, the synthesized and full-custom modules were integrated with the Cadence Design Framework II schematic tool and the test logic (see Section 3) was added. The layout of the AxPe1280V was also carried out with Cadence Design Framework II. Standard-cell logic was placed in the middle of the chip and the full-custom modules at the border. This approach was chosen, because it was not recommended to route signals over full-custom memories in this 3 metal layer process. Furthermore, a distribution of standard-cell logic over the whole chip would have caused wiring delay and additional silicon area for routing.

5. PROCESSOR IMPLEMENTATION

The processor has been fabricated in a 0.5 μm 3 metal layer CMOS technology by Philips semiconductors. The circuit consists of about 1.3 M transistors and has a die size of 82 mm^2. It is available in a QFP160 package. The power supply is 3.3 V for the core and the pad drivers. The power consumption is typically less than 0.7 W at 66 MHz. Because the complete processor is realized with static registers and memory cells it can be switched in stand-by mode. In stand-by mode the core's power consumption falls below 17 mW. This feature allows to integrate the video signal processor into mobile communication systems, e.g., notebook-based video telephones. A PLL which multiplies the external clock by a programmable factor of 1, 1.5, 2, 3, and 6 has been integrated in the AxPe1280V to allow a flexible system integration of the processor.

Extensive tests have been carried out with the AxPe1280V and have unveiled that the achievable clock rate is mainly limited by the on-chip memory in the RISC processor. With enhancements in these memories, their controlling, and by fabricating the chip in a new technology, the maximum clock rate can increase significantly.

Figure 4 shows a die microphotograph of the developed processor. The main characteristics of the chip are given in Table 2.

Technology	0.5 μm CMOS, 3 metal layer
Die Size	82 mm^2
Full-Custom Modules	RAM, Arithmetic
Standard-Cell Logic	Controlling
Transistors	1.3 M
On-Chip Memory	123 kbit
Maximum Clock Rate	66 MHz
Power Consumption	0.7 W @ 66 MHz
Computational Power	1056 MOPS
Max. Input Data Rate	132 Mbyte/s
Max. Output Data Rate	66 Mbyte/s

Table 2: Main characteristics of the AxPe1280V

6. MIMD MULTIPROCESSOR SYSTEM

For several video coding applications the performance of one AxPe1280V is sufficient. When connecting several processors to a MIMD (multiple instruction, multiple data) multiprocessor system (Figure 5), real-time processing of more complex applications

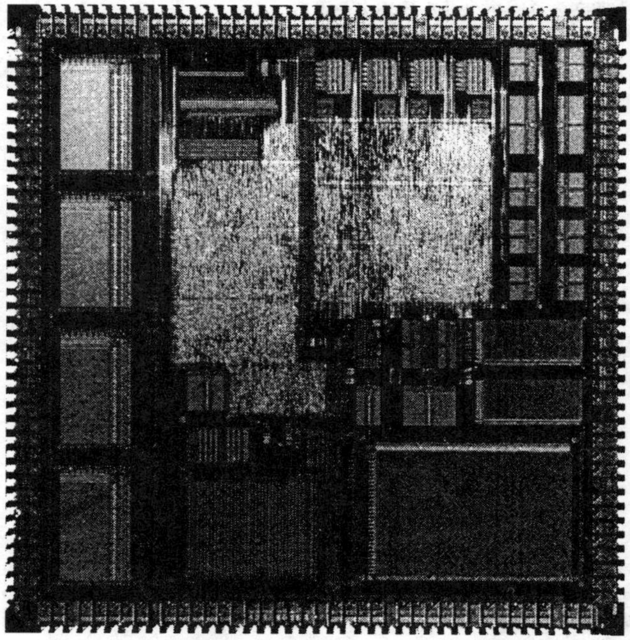

Figure 4: Die microphotograph of the video signal processor

is supported. The image data is distributed sequentially to the processors in the multiprocessor system and all processors perform the hybrid coding scheme in parallel on macro block level. Because each processor features on-chip program and data memories, the data is processed asynchronously and the overall structure of the multiprocessor system is MIMD based.

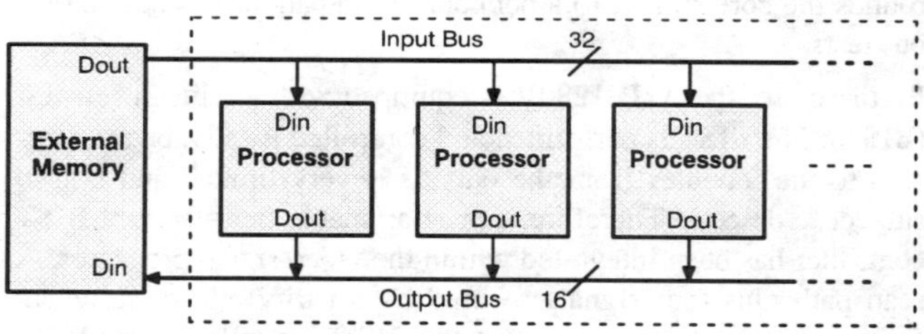

Figure 5: MIMD-based multiprocessor system

When using a SPMD scheme (single program, multiple data) a simple technique for bus arbitration can be applied (Figure 6). One processor loads and stores data from or to the external data memory at a specific period of time, while all other processors compute on locally stored data. This allows a high utilization of the buses and stall cycles can be eliminated. For block-based video coding applications like ISO MPEG-2, where the communication requirements between the processors can be neglected, the performance of the multiprocessor system is almost direct proportional to the number of processors. This speedup is only limited by the bandwidth of the input and output bus. Table 3 gives an overview on the number of processors required for different hybrid coding applications.

A multiprocessor system based on up to 6 AxPe1280V has been implemented as a PCI board for personal computers. Sample applications, that have been programmed on this system, are video coder/decoder according to ITU-T H.261 and H.263 and several proprietary video processing applications.

Application	Standard	Source rate	Number of AxPe		
			codec	encoder	decoder
Video telephone	H.261 / H.263	QCIF, 10 Hz 0.38 Mbyte/s	1	—	—
		CIF, 10 Hz 1.52 Mbyte/s	2	—	—
		CIF, 30 Hz 4.56 Mbyte/s	4	—	—
Video CD	MPEG 1	SIF, 25 Hz 3.80 Mbyte/s	—	3	2
Digital TV	MPEG 2	CCIR601, 25 Hz 20,74 Mbyte/s	—	16	6
Image coding	JPEG	CIF, 10 Hz 1.52 Mbyte/s	—	1	1

Table 3: Number of AxPe processors required for hybrid coding applications

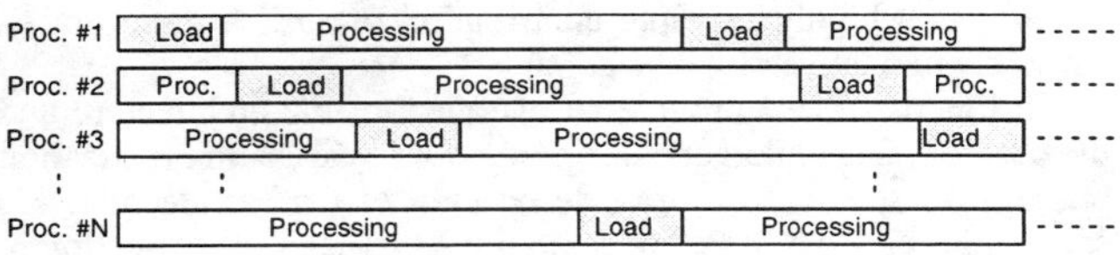

Figure 6: Loading scheme for MIMD-based multiprocessor system

The AxPe1280V has been designed to be used as building block for a monolithic multiprocessor system. Such a system with 9 AxPe640V, a previous version of the presented AxPe1280V, has already been implemented [9]. The monolithic integration is supported by the separate I/O buses of the processor and the sophisticated self-test capabilities of each processor, as discussed in Section 3.

7. CONCLUSION

The presented video signal processor AxPe1280V is optimized for block-based video coding standards like ITU-T H.261/H.263, and ISO MPEG-1/2. It consists of a RISC processor core for medium-level video processing and control tasks, and a coprocessor for computation-intensive convolution-like tasks.

The processor was designed in standard-cell logic combined with full-custom modules for arithmetic and memories, and was realized in a 0.5 μm 3 metal layer CMOS technology. It provides an arithmetic performance of more than 1 GOPS at a clock rate of 66 MHz.

For applications requiring higher computational power, the processor can be integrated into a scalable MIMD-based multiprocessor system, consisting of several AxPe1280V. For block-based video coding algorithms, almost a linear speedup can be achieved by parallelization on video data level. Such a multiprocessor system has been implemented on a PCI board for personal computers.

Current activities include the optimization of software tools for AxPe program development and the implementation of new applications on the AxPe multiprocessor system.

8. ACKNOWLEDGMENTS

The authors would like to thank Klaus Gaedke and Hartwig Jeschke for their contributions to this project. Furthermore we thank Axel Werner and Jens Castagne from Philips Semiconductors ASIC Support Centre, Hamburg, for providing full custom modules, the layout, and the fabrication of the processor. This work is supported by FhG under contract no. T/F41B/T0183/P1307.

9. REFERENCES

[1] ITU-T Recommendation H.261, "Video Codec for audiovisual Services at p x 64 kbits", International Telecommunication Union, March 1993

[2] Draft ITU-T Recommendation H.263, "Video Coding for Low Bit Rate Communication", International Telecommunication Union, December 1995

[3] ISO/IEC JTC1/SC2/WG11 MPEG-90/176 Rev. 2, 1990.

[4] ISO/IEC JTC1/SC29/WG11 CD 13818-2, 1994.

[5] H.G. Musmann, P. Pirsch, H.-J. Grallert, "Advances in Picture Coding", Proc. IEEE, Vol. 73, No. 4, pp. 523-548, 1985.

[6] Y. Okada et. al. "An 80 mm^2 MPEG2 Audio/Video Decode LSI", IEEE International Solid-State Circuits Conference, Vol. 40, pp.264-265, Feb. 1997.

[7] K. Herrmann, M. Seifert, K. Gaedke, H. Jeschke, P. Pirsch,"A RISC Core for a Monolithic Video Signal Processor", in Rabaey, J. et al., Eds., Workshop for VLSI Signal Processing, Oct. 1994, pp. 368-377, IEEE, New York, NY, 1994.

[8] B. Könemann, J. Mucha, G. Zwiehoff, "Built-In Logic Block Observation Techniques", Proc. 1979 Intern. Test Conf., Cherry Hill, NJ, USA, October 1979, pp. 37-41.

[9] J. Otterstedt, K. Gaedke, K. Herrmann, M. Kuboschek, H.U. Schroeder, A. Werner, " A 16 cm^2 Monolithic Multiprocessor System Integrating 9 Video Signal-Processing Elements", IEEE International Solid-State Circuits Conference, Vol. 39, pp. 306-307, Feb. 1996.

Realization of a Programmable Parallel DSP for High Performance Image Processing Applications

Jens Peter Wittenburg Willm Hinrichs Johannes Kneip Martin Ohmacht
Mladen Bereković Hanno Lieske Helge Kloos Peter Pirsch

Laboratorium für Informationstechnologie, Universität Hannover
Schneiderberg 32
30167 Hannover, Germany
Tel: +49 511 762 5018
E-mail: wittenburg@mst.uni-hannover.de

ABSTRACT

Architecture and design of the HiPAR-DSP, a SIMD controlled signal processor with parallel data paths, VLIW and novel memory design. The processor architecture is derived from an analysis of the target algorithms and specified in VHDL on register transfer level. A team of more than 20 graduate students covered the whole design process, including the synthesizable VHDL description, synthesis, routing and backannotation as the development of a complete software development environment. The 175mm^2, $0.5\mu\text{m}$ 3LM CMOS design with 1.2 million transistors operates at 80 MHz and achieves a sustained performance of more than 600 million arithmetic operations.

1. INTRODUCTION

The ongoing trend towards growing complexity in communication and image processing systems calls for the development of flexible signal processors. Monolithic architectures capable to process a wide range of algorithms and a performance sufficient to meet real time conditions will form the core of future multimedia platforms. An ideal processor platform has to meet a number of sometimes contradicting requirements: For a maximum of efficiency, the design has to be adapted to the target algorithms, should be generic to allow the porting of a high level language compiler - mandatory for the minimization of implementation cost of complex applications. Parallelization potential of the algorithms should be exploited as far as possible to boost performance, but keeping hardware cost as low as possible. Some fundamental problems, as the high bandwidth parallel access to shared memory segments, are not yet satisfactorily solved in currently available DSP architectures.

The target of the design project described in this paper is twofold: First, this design is an innovative processor architecture, proving that a parallel, homogenous DSP architecture can achieve a processing power hitherto addressed only by dedicated designs. This was achieved by an appropriate combination of measures known from standard processors with a novel controlling and memory concept. Second, it served as an ambitious tutorial project for graduate students, which worked on the project for 6-12 months during their final thesis in microelectronics or information technology. During the design process, all aspects of VLSI design from architectural specification, HDL description, synthesis, verification, routing and backannotation were actively performed by the students, making the project a highly valuable contribution towards a more practically oriented university education.

In this paper, we provide an overview of the architecture and the design it has undergone during the four year lasting realization phase. We start with a brief description of the architecture, it's specialities and the basic considerations that lead to it. Section 3. describes the HDL design on RTL and the synthesis on an 0.5μm 3 layer metal CMOS standard cell technology. Section 4. covers verification aspects while section 5. describes the backend design process and the problems students had to cope with during routing, backannotation and final verification of a complex sub-μ design. Section 6. gives an overview of the HiPAR-DSP's software development environment, also developed within the project. Finally, we present the layout and performance data for a selected number of algorithms for the processor, which is currently manufactured.

2. ARCHITECTURE

Fig. 1 shows a block diagram of the HiPAR-DSP architecture. The homogenous core consists of four (in a future version 16) identical 16 bit data paths, each containing a local register file, an ALU, a shift/round unit and a 36 bit multiply/accumulate unit. The data paths are centrally controlled by a RISC control unit, operating on a 96 bit very long instruction word (VLIW), which is read from an instruction cache assigned to the control unit. A two stage memory hierarchy was implemented. First, data paths have concurrent access to shared data in shape of a matrix (Fig. 2), which are stored in an on-chip memory, the matrix memory. Second, each data path may address private data autonomously, which is stored in a cache memory assigned to each data path. An autonomously operating DMA control unit serves all cache misses and may perform prefetch data transfers to the matrix memory. A test and debug interface is used for the *in system* software debugging by accessing all internal registers via the fabric scan chains and test access port, while the processor is halted.

The HiPAR-DSP architecture meets the special requirements of image processing algorithms implemented in parallel. By combining instruction word level parallelism (VLIW), data parallelism (multiple data paths, split ALU capability) and transfer parallelism (autonomous DMA unit), a high degree of concurrent processing can be achieved for various algorithms [1][2]. The SIMD controlling principle leads to low hardware implementation cost compared with MIMD or multithreaded approaches. The required flexibility for more complex algorithms is achieved by the implementation of a three stage autonomous controlling scheme for the data paths, providing as addressing and execution autonomy as the concurrent

DAC 98, San Francisco, California
©1998 ACM 0-89791-964-5/98/06..$5.00

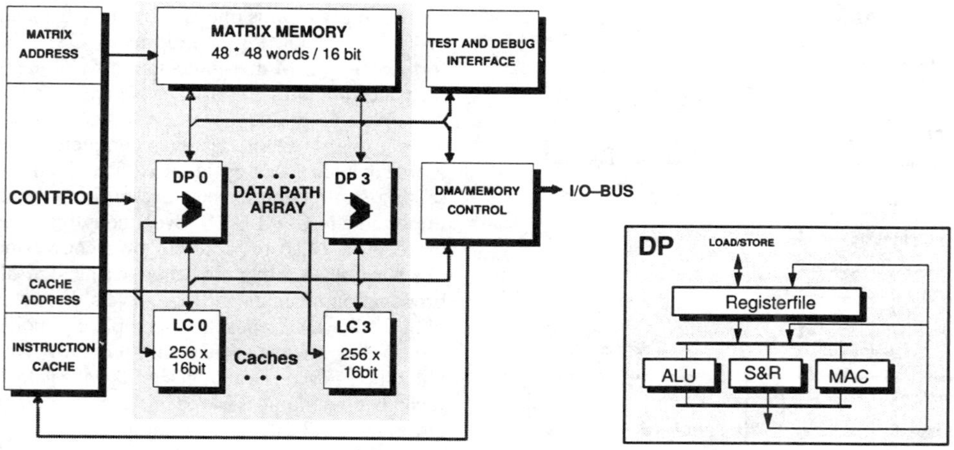

Figure 1: HiPAR-DSP architecture and structure of a data path (S&R: Shift-and-Round unit)

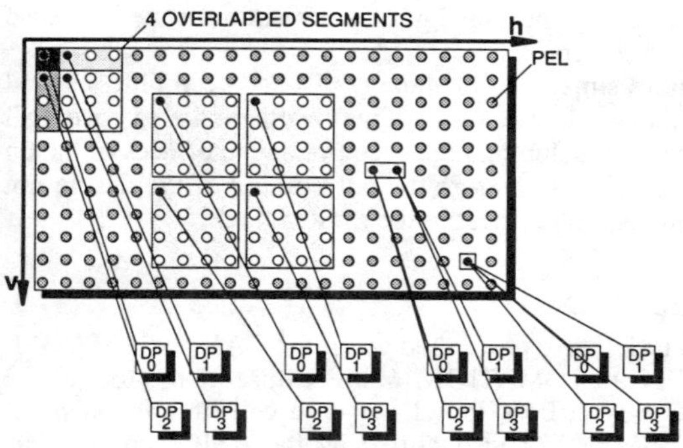

Figure 2: Concurrent access of data paths to the matrix memory. The matrix memory has a virtual 2D address space and provides a conflict free access to the data stored in it.

evaluation of local conditions.

The requirement of a high bandwidth access to shared data is solved by the matrix memory architecture. Internally, the data are distributed to nine single ported memory banks by use of a distribution function that provides a conflict free access to data from any address of it's virtual 2D address space [3]. The generic structure of the processor and the implementation of a stack mechanism made the porting of the GNU C++ compiler feasible. The architecture was derived from a thorough analysis of the targeted image processing algorithms. Available high-level design tools focus on mainly dedicated realizations, but were not of much help in the design of a generic programmable architecture. So, our experience in the parallel implementation of image processing application and in VLSI design were the driving forces leading to the final structure of the HiPAR-DSP.

3. FRONTEND

VHDL was chosen as hardware description language for design entry and verification at RT-Level. On one hand, technically the preference of VHDL to Verilog turned out to be slightly disadvantageous, because of nonexisting VHDL (VITAL) gate level models of the standard cells, which forced us to employ different simulators for functional and gate-level sign-off simulations. On the other hand, in Europe the number of commercial VHDL installation exceeds the number of Verilog installations by far, which justifies the preference of VHDL at European universities.

For designs with a complexity of more than several hundred thousand gates complexity a hierarchical design approach becomes absolutely mandatory. The SYNOPSYS DESIGN COMPILER used for RT-Level synthesis and timing optimization in this project is hardly able to optimize more than 5 to 10 thousand gates at a time, depending on regularity. On physical level, the routing tools enable efficient routing of blocks up to about 30 thousand gates without losses in gate density for the chosen 0.5 μm 3 LM standard cell library. Thus, logic- and physical hierarchy are not necessarily identical. Nevertheless, it is reasonable to take the physical level into account when defining the logic hierarchy, e.g., VHDL-modules to be designed independently. Conversely, logic hierarchies usually provide a certain degree of locality for interconnections, which lead to better routing results and lower delays caused by wire capacitances on the physical level.

Partitioning started from an initial logic hierarchy derived from the processor's block diagram and a black box floorplan based on course estimations of the silicon effort. After first synthesis steps with much more accurate area calculation, we were able to derive a logic partitioning of the processor. All major parts of the architecture were designed independently, while none of the resulting blocks in the floorplan exceeded the size of 30.000 gates. The subunits: data path, controller, data cache controller, instruction cache controller and DMA were VHDL-designed, synthesized and verified as independent design projects by graduate students. A project management of two to three research assistants was responsible for the supervision of design constraints (clock speed and area) and the definition and specification of the interfaces.

Due to the pipelined nature of the architecture, we were able to achieve a clock frequency of about 100 MHz from initial synthesis, with wiring capacities calculated from a heuristic wire load model delivered with the standard cell library. The arithmetic units, especially 32 bit adders and the 17x17 bit multiply & accumulate (MAC) unit required special treatment, since simple synthesized versions of the adders (riple carry) and multipliers (brown array) did not fulfill the timing requirements. Following two measures to increase arithmetic performance were investigated first:

- Instantiation of advanced hierarchical architectures (fast carry lookahead in case of the adders, booth coded wallace tree for the multipliers) from the SYNOPSYS DESIGN WARE libraries.
- Generation of Arithmetic Macro Cells using the COMPASS data path compiler, which allows to generate layout optimized cells from bit slice descriptions.

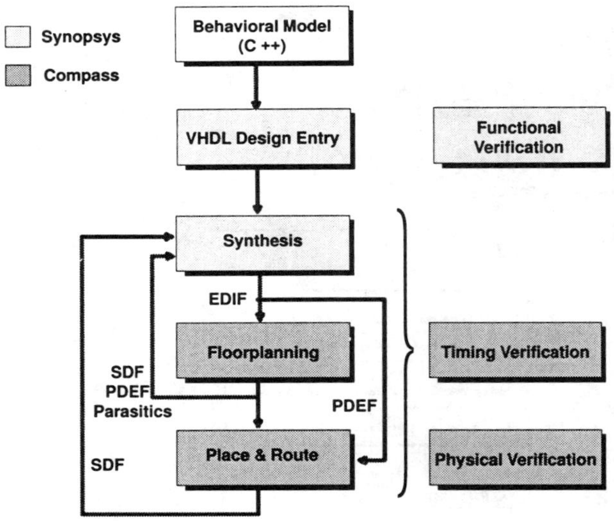

Figure 3: Simplified design flow and interfaces. SYNOPSYS tools (VSS, DESIGN COMPILER) were used for all frontend tasks, while COMPASS tools (PATHFINDER, CHIP TIME, QSIM) were employed for all backend activities [5][6].

Both methods lack the flexibility to insert pipeline stages at user defined places. This leads to inefficient realizations, where a complete second adder has to be introduced in case of the multiply/accumulator To solve these problems we designed own advanced arithmetic architectures by structural VHDL description. Instead of elements from the standard cell library we used generic cells from the SYNOPSYS GTECH library. Then we used DESIGN COMPILER to map and optimize the modules, which allowed us to keep the designflow consistent. This approach not only kept the required flexibility but surprisingly also by far outperformed all DESIGN WARE instantiations for multiplier, 32 bit adders and barrel shifters in terms of effciency (A * T product).

4. VERIFICATION

The processor design started with a high level behavioral model written in C ++. This model served as cycle precise simulator with the possibility to simulate complete algorithms. In addition it enabled access to all logical registers and memories. C ++ has been chosen as modeling language due to the high performance and portability. Because the abstraction level of this model was much too high for direct breakdown to RTL level by refinement, HDL-modeling would not have been of advantage anyway. The behavioral model has been extensively verified against the specification. Graphical user interface and efficient interfaces to object code enabled simulation of several image- and signal processing algorithms (e.g., complex FFT of variable length, grey level histogramming, hough transform) with realistic input data. Additionally synthetic command sequences were generated to verify rarely used special cases. Thus the verification coverage of the behavioral model could be judged as high enough to classify the model as "known good device" for all subsequent verification steps.

Functional verification of the RT-level VHDL-code took place using SYNOPSY's VSS VHDL simulator. Verification strategy followed the typical bottom up approach, starting in the lowest levels of the hierarchy and ending at top level with simulation of the complete architecture using the same stimuli already employed for verification of the behavioral model. The existence of this known good reference model significantly simplified verification on all levels of hierarchy.

As already mentioned, sign-off verification of the employed library using VHDL is impossible due to the lack of suitable standard cell models. COMPASS' proprietary QSIM is the alternative. Unfortunately QSIM does not support flexible testbenches and therefore does not allow text-I/O as we used for functional verification. To solve this problem we inserted certain VHDL tasks dumping logic levels at I/O interfaces (of complete design or submodule) for each executed clock cycle into a file. Such interface dumps representing stimuli and expected results can be converted to a format importable to COMPASS. Even though this seems to be a standard problem, we had to write our own conversion utilities (C++) for this purpose. Another challenge during sign-off simulation was the propagation of unknown logic levels due to uninitialized registers with feedback. Because of incompletely modeled standard cells (a multiplexer always propagating unknown when its select input is undefined - even if all inputs are equal - is the most common case) it can become impossible to initialize such registers. Controlled by attributes synthesis of a working synchronous reset can be forced for problematic registers. However, uninitializable registers can (and usually do) occur after each resynthesis, since implementing resets to all registers would cause non neglectable overhead. The major amount of work during sign-off verification related to those uninitializable registers.

After sign off simulation the final netlist had been fully verified against the behavioral description. This verification also covers all timing information including the asynchronous interfaces of the on chip memories, which represent the only timing of the design not fully synchronous with respect to the master clock.

5. BACKEND

The backend tasks were performed using the CAD-tools of COMPASS DESIGN AUTOMATION. Main feature of the tool is the tight interaction with PASSPORT standard cell libraries also developed by COMPASS and qualified by the ASIC vendor. The netlists generated by the synthesis tool SYNOPSYS were transferred via an EDIF interface to the COMPASS tools. All on-chip RAMs were generated by COMPASS RAM compilers. The resulting macro cells as well as boundary scan and pad cells were added in COMPASS schematic editor to complete the netlist for the HiPAR-DSP. Scan logic for test purposes was automatically inserted by the COMPASS tools following a two step approach. First all D-Flipflops were exchanged by scanable counterparts. The cell order in the scan paths had not defined at that point. This task was performed later after cell placement but before routing. This two step approach minimizes the routing overhead of scan logic significantly.

The final layout of the HiPAR-DSP is shown in Fig. 4. On-chip RAM cells for the matrix memory, data cache and instruction cache are placed close to their corresponding controllers. Each memory is served by the DMA/memory controller. Main silicon area is occupied by the global RISC controller including a base register file and the four data paths with local register files. Due to the wide internal data and address busses of the load/store architecture between data paths, matrix memory and data caches, a large vertical routing channel runs through the chip. The power pads are evenly distributed in the pad ring to handle simultaneously switching outputs. 20 power pad pairs are provided for core supply and six AC/DC pads for padring power.

Placement & route of the HiPAR-DSP were carried out using the COMPASS PATHFINDER tool. Starting from the final floorplan, placement was controlled by high-level timing constraints (e.g., targeted clock frequency). Therefore the placement tool has to employ its own static timing analysis engine. Clock tree compilation takes place after the placement has been completed. Since the position of the registers is not finally fixed at an earlier level, this ensures the best possible delay estimation to achieve a prop-

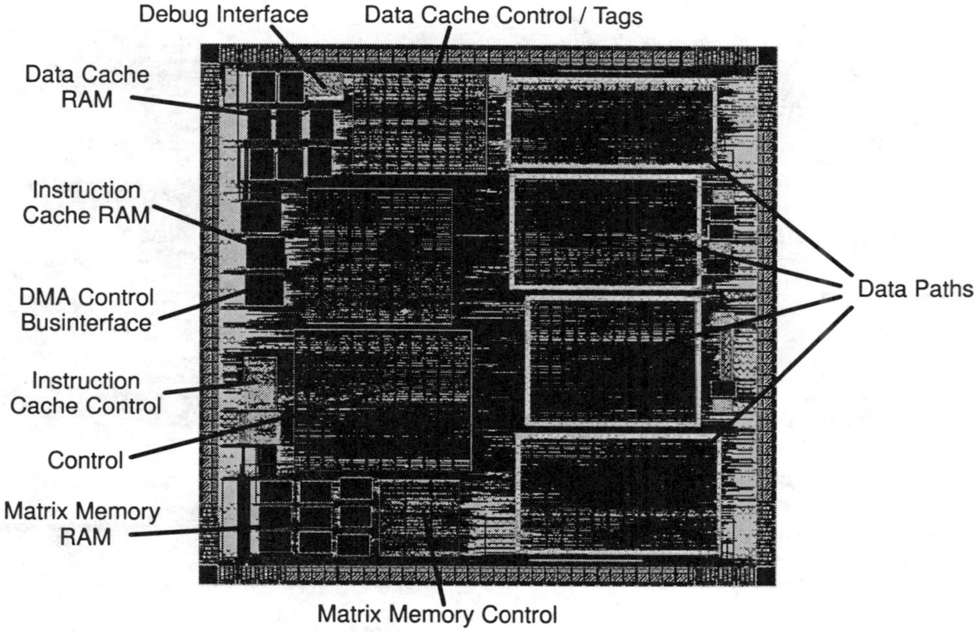

Figure 4: Layout of the HiPAR-DSP

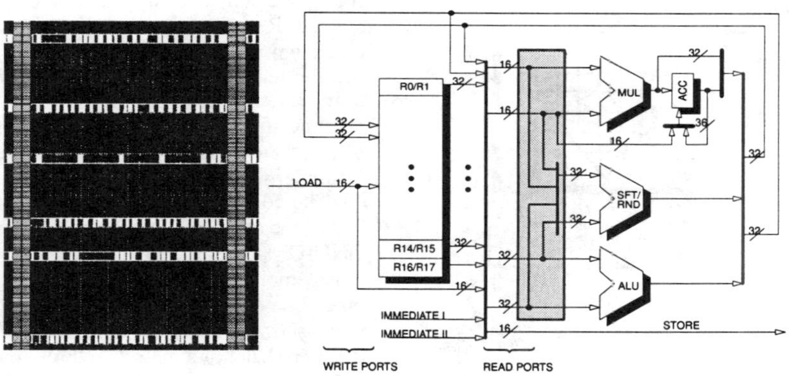

Figure 5: **a)** Typical segment of registerfile layout. Only the narrow horizontal strips represent active cell areas

b) Schematic overview of data paths using less register file ports. The number of register file read ports has been limited to two 32bit- and three 16bit read ports.

erly balanced clock tree. After clock tree compilation and 2nd step of scan path insertion, the bin based PATHFINDER router is invoked to connect the placed cells. A completely flat routing approach showed best results in terms of gate density, since global optimization saved block wiring overhead. Routing of the DSP required about one week CPU-effort on a 200 MHz Ultra SPARC Workstation equipped with 1 GByte main memory.

The RISC-typical registerfiles were one of the most critical elements concerning die size and clock frequency [4]. Especially the number of the register file ports necessary to distribute data according to the VLIW concept of the data paths, caused large hardware costs. Full custom optimized blocks of any kind were not available. The enormous amount of wiring between the standard cell rows of the data path layout is shown in Fig. 5a. In a straight forward implementation register file ports according to in- and output of arithmetic/logic- and I/O-units would result in five 32bit and three 16bit read ports respectively two 32bit and one 16bit write ports. Achieving acceptable register file size and speed while not drastically decreasing performance required careful evaluation of each possible measure. A proper balance had to be found between software requirements, performance and silicon area. The result is shown in Fig. 5b. The general purpose character has been kept with the exception that the two16bit operands of the MAC must not be both even or odd. Some further combinations of concurrent operations are no longer possible (e.g., ALU operation and 32bit Shift) but are rarely found in the targeted applications anyway. VLIW restrictions can be easily handled by the compiler. The improvements in size and speed for the implemented register file are shown in Table 1.

Even though floorplanning and thus estimation of wiring delays has been carried out from the beginning of the frontend design, achievement of timing convergence turned out to be the most challanging part of backend design. Several steps of resynthesis had to be carried out in addition to the timing driven cell placement:

Table 1: Size and speed of port unlimited and implemented register file based on $0.5\mu m$ 3LM CMOS. The average gate density for typical standard cell design is about 4000 gates / mm^2

number of ports	area per data path	gate density	max clock speed
2 32bit + 1 16bit WRITE 5 32bit + 3 16bit READ	$35mm^2$	910 $\frac{gates}{mm^2}$	55 MHz
2 32bit + 1 16bit WRITE 2 32bit + 3 16bit READ	$17mm^2$	1600 $\frac{gates}{mm^2}$	80 MHz

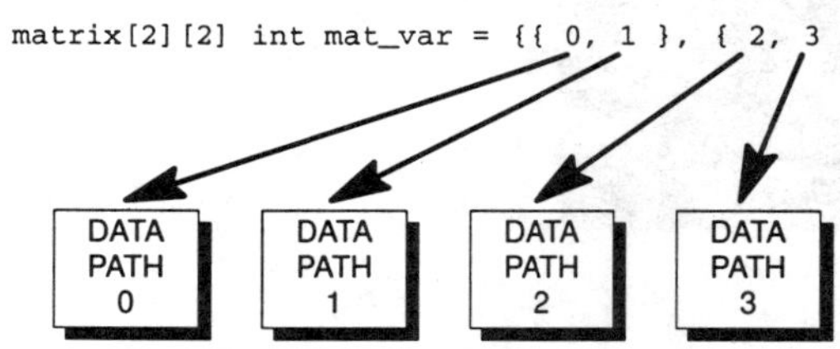

Figure 6: Matrix type variable definition and distribution to data paths

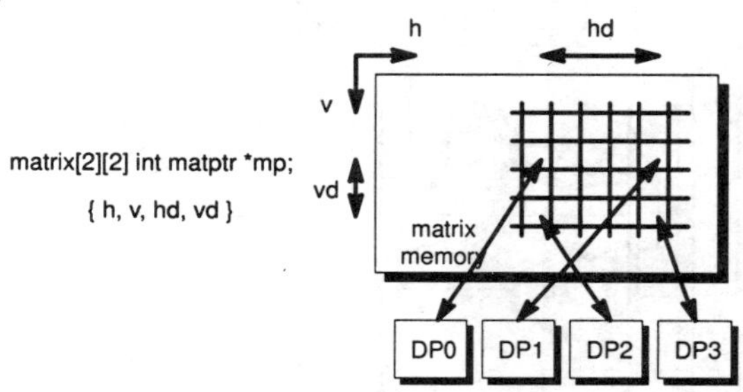

Figure 7: Access to matrix memory

- Resynthesis using custom wire load models generated from the estimated floorplan
- Reoptimization (rebuffering) of the design using delay information backannotated from an initial placement. Therefore clustering information (PDEF), annotated timing (SDF) and parasitics (RC-Delays) have to be backannotated into the SYNOPSYS DESIGN COMPILER. The synthesis tools then start from the annotated timing and uses clustering and parasitic information, where nets have to be changed. Due to the different hierarchy in the SYNOPSYS domain (e.g. missing memories, pads, scan paths) and differing naming conventions between the tools, backannotation across the interface between COMPASS and SYNOPSYS is no trivial task.
- In place optimization again using backannotated delay data. For final placement, positions of inserted buffer trees were forward annotated to the placement tool, which is able to place such buffers without having to discard the existing placement.

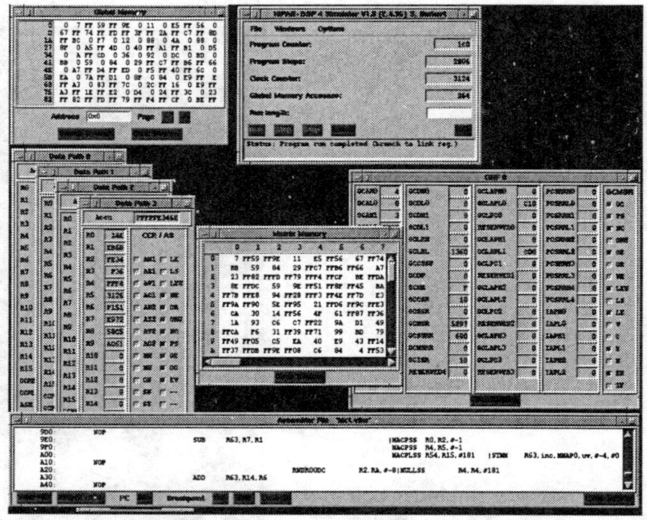

Figure 8: Screen shot of the interactive simulation environment

6. SOFTWARE DEVELOPMENT ENVIRONMENT

Starting with the first specification phase of the HiPAR-DSP, concepts for a high level language development environment have been created. Resulting from this tight cooperation, the HiPAR-DSP architecture provides some features, e.g., efficient operations on a parallel, cache buffered stack, supporting high level programmability for imperative languages like C++ or PASCAL. During further specification phases and the beginning of the implementation phase, we developed an optimizing assembler and adapted a C++ Compiler for the HiPAR-DSP.

The compilation process consists of three phases. The front-end of the development systems consists of an extended port of the GNU-C++-Compiler, which supports parallel operations and matrix memory access using language extensions. The scalar variable types like int, long, etc. have been extended to compound, matrix types shown in Fig. 67, consisting of a set of scalar variables ordered in shape of a matrix. Expressions using variables of this type result in parallel operation of all data paths on different components of this type. This allows existing C or C++ code to be easily ported to the HiPAR-DSP by simply adapting variable types in time critical code sections to the new features of the system.

During the second phase of the compilation process, an optimizing assembler automatically generates VLIW assembler instructions from the sequential code exploiting the VLIW parallelism and the micropipelines, using advanced inter basic block scheduling and software pipelining. In the last phase, machine readable binary code is generated.

The correct behavior of programs developed for the HiPAR-DSP can be verified using an interactive simulator, which is shown in Fig 8. The simulator operates clock cycle based and allows the user to inspect and modify all programmable features of the HiPAR-DSP architecture. Besides standalone debugging, in-circuit debugging using the JTAG interface is supported, allowing rapid application development and testing in the target environment.

The compiler, assembler and debugger, combined with the so far developed application library form a powerful development platform that enables the user to create applications efficiently in a way he is used from industrial design.

Table 2: Performance data for selected algorithms. Data from cycle true simulation of assembler coded programs, cycle time from backannotated layout data.

ALGORITHM	DESCRIPTION	PROCESSING TIME ($f_{clock} = 80MHz$)
Normalized Correlation	512 x 512 input image size 32 x 32 mask size	2.25 s
Gray level Histogram	512 x 512 image size, 256 gray level	2.75 ms
complex FFT	256 x 2 samples (r,i)	35 μs
3x3 FIR-Filter	non-separable, 512x512 pixels	21.4 ms
Edge Detection	512 x 512 pixels, vertical and horizontal sobel filters and non maximum suppression	11.5 ms
Connected Component Labeling	512 x 512 pixels	25.7 ms
Skeletonization	512 x 512 pixels	21.2 ms
Parallel Search	1024 elements	625 ns

Table 3: Implementation data of the HiPAR-DSP

Parallel Data Paths	4 parallel data paths (13/32 bit integer)
On-chip memories	9*0.5 kbyte matrix-memory, 4*0.5kbyte data cache, 3kbyte instruction cache (Σ:9.5kbyte)
Technology	Atmel 0.5 μm CMOS 3LM standard cell with full custom memories
Transistors	1.2 million
Area	175 mm^2
Power consumption	$\sim$ 5 watt (estimated)
Clock frequency	80 MHz (typically), 50 MHz (worst case)

7. DESIGN RESULTS AND PERFORMANCE DATA

The design process started in 1993 with first architectural considerations. The project was intended to last 3 years until silicon. With specification completed end of 1993, the HDL design lasted until end of 1994. A lot of time was lost due to two unexpected changes in manufacturer and technology, partly requiring overworking of large sections of the HDL code (which is not so manufacturer-independent as one might believe, if efficient synthesis has to be achieved). A high effort of time went into the setup of a working back end design flow, which was flawed with tool bugs and incompatibilities. The result was a project delay of more than 18 months. There were 22 student thesis written within the project. About half of the students worked on the VLSI design, the other half on the software environment and algorithm implementations. Together with an average of 3 scientists working on the project, the total effort was about 32 men year, not too much for a design of that size and the huge software package developed for it. Table 3. gives an overview of the final chip data.

We have started practical programming of image processing algorithm cores on the C++ processor model already at a very early stage of the design to evaluate potential for the further optimization of instruction set and controlling scheme. The resulting code also formed an important part of the later verification on RTL and gate level. Table 2. shows an overview of the HiPAR-DSP's performance, gained from these optimized assembler written implementations. The performance of C++ written programs varies with the experience of the programmer and the use of the extended processor specific data types. In average, about 1.5 times slower execution is achieved with C++ implemented software.

8. CONCLUSION

A novel programmable DSP architecture has been proposed and implemented. The innovative parallel architecture reaches high performance while keeping a flexibility comparable to other commercially available DSPs. All design phases - beginning from the programming of a first behavioral processor model followed by VHDL implementation, synthesis, placement, route and backannotation and finally ending with the generation of GDS-II mask tapes - have been carried out at the university. The complexity of the architecture required state of the art ASIC technology, design flow and EDA-Tools. Various difficulties in the establishment of the design flow have been mastered. Most recent features especially of the backend tools turned out to be utterly necessary to fulfil the demands of deep submicron design. Currently the HiPAR-DSP is in production phase at the Atmel Fab in Rousset, France. First samples will be available by time of DAC'98. The project impressively demonstrates the possibility to design programmable digital signal processors under university conditions. Complexity, performance and tool support is at least comparable to those of commercially available counterparts.

References

[1] K. Rönner, J. Kneip: "Architecture and Applications of the HiPAR Video Signal Processor", Transactions on Circuits and Systems on Video Technology, 2/1996

[2] J. Kneip, M. Ohmacht, K. Rönner, P. Pirsch: "Architecture and C++-programming environment of a highly parallel image signal processor", Microprocessing and Microprogramming, Vol. 41 (1995), pp. 391-408.

[3] J. Kneip, K. Rönner, P. Pirsch: "A Data Path Array with Shared Memory as Core of a High Performance DSP", Proc. Int. Conf. Application Specific Array Processors (ASAP)'94, pp. 271-282, Aug. 1994.

[4] J. P. Wittenburg, M. Ohmacht, W. Hinrichs, J. Kneip, P. Pirsch: "HiPAR-DSP: A Parallel VLIW RISC Processor for Real Time Image Processing Applications", Proc. Int. Conf. on Algorithms And Architectures for Parallel Processing (ICA3PP)'97, pp. 155-162, Melbourne Dec. 1997

[5] Synopsys Online Documentation, Version 1997.01, Synopsys Inc.

[6] Compass Online Documentation, Version V9R2, Avant! Inc.

A Multiprocessor DSP System Using PADDI-2

Roy A. Sutton[1], Vason P. Srini[2], Jan M. Rabaey[1]

University of California, EECS Dept., Berkeley
Data Flux Systems Inc.

ABSTRACT

We have integrated an image processing system built around PADDI-2, a custom 48 node MIMD parallel DSP. The system includes image processing algorithms, a graphical SFG tool, a simulator, routing tools, compilers, hardware configuration and debugging tools, application development libraries, and software implementations for hardware verification. The system board, connected to a SPARCstation via a custom Sbus controller, contains 384 processors in 8 VLSI chips. The software environment supports a multiprocessor system under development (VGI-1). The software tools and libraries are modular, with implementation dependencies isolated in layered encapsulations.

CR Descriptors: B.7.1 [Integrated Circuits]:Types and Design Styles - VLSI; C.1.2 [Processor Architectures]:Multiple Data Stream Architectures - Parallel processors; D.1.3 [Programming Techniques]:Parallel Programming; I.3.1 [Computer Graphics]:Hardware Architecture; I.4.3 [Image Processing]:Enhancement.

1 INTRODUCTION

We have constructed a system built around a custom parallel DSP chip, PADDI-2[1] (Programmable Arithmetic Devices for high speed DIgital signal processing[2]). The system consists of a multiprocessor DSP board, a host interface controller, and a complete software infrastructure for mapping, configuring, and accessing the board. The host interface controller is in-circuit reconfigurable and is synthesized from VHDL.

An algorithm is mapped to the multiprocessor hardware and is also coded completely in software for verification. The hardware and software implementations operate on a single data set and the result is compared.

University of California, EECS Dept., Berkeley, CA 94720[1]
Data Flux Systems Inc., 1678 Shattuck Ave., MS 292, Berkeley, CA 94709[2]

This project is partially funded by USAF, Eglin AFB, FL under the contracts F08630-94-C-0074 and F08630-95-C-0101. Additional support comes from a Lucent Technologies/Sandia National Laboratories CRFP Fellowship.

DAC 98, San Francisco, California
©1998 ACM 0-89791-964-5/98/06..$5.00

Numerous software tools have been developed to assist in mapping DSP algorithms to the hardware and for verifying the result. The process of mapping an algorithm to the hardware begins with manual partitioning using a graphical SFG entry tool. The partitioned algorithm is then simulated and routed using another graphical tool. The result is then compiled to a configuration bitstream, downloaded to the multiprocessor board, and run using configuration and I/O tools. The board software is built upon an application programming interface and driver library. The software has been designed in a modular manner to facilitate compilation under other computer operating systems and to reduce the dependence on PADDI-2. Implementation details have been encapsulated in replaceable software libraries.

There are three custom hardware component types: a parallel DSP processor (PADDI-2), the multiprocessor system board, and a computer interface controller.

2 HARDWARE

PADDI-2

A goal of PADDI-2 is to achieve high performance (billions of operations per second) by interconnecting numerous simple processors in an MIMD fashion. It is intended for signal processing applications that contain abundant parallelism[3][4][5]. It's architecture draws on ideas from earlier architectures[4][6][7][8][9].

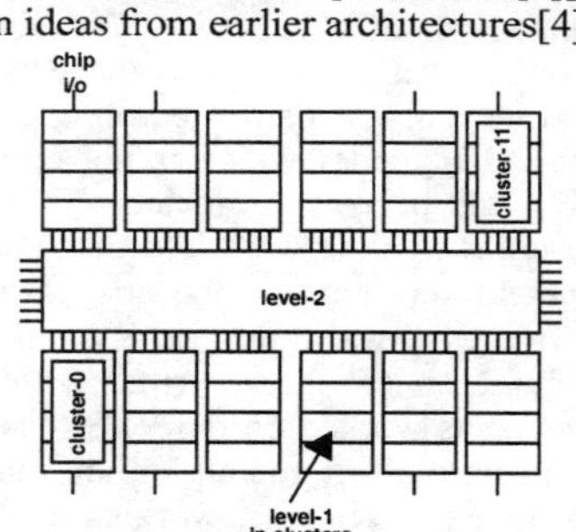

As shown above, it is a 48 node multiprocessor chip, composed of 12 clusters, each with 4 simple "nanoprocessors." Two levels of interconnect are present. There are six level-1 data buses and four one-bit control buses within each cluster. These buses are used for local communication within a cluster and for communication from within a cluster to outside a cluster. They extend into a second level crossbar. The level-2 interconnect has 16 data buses and eight one-bit control buses and are used to provide connectivity between clusters. Each nanoprocessors has 17 programmable switches available to make connections to level-1 buses. Clusters have 50 switches within each level-1 to level-2 crossbar. The eight outer most clusters have dual purpose nanoprocessors which provide off-chip I/O.

Six general purpose registers, three scratch pad registers, an eight element instruction memory, and a simple control unit is present within each nanoprocessor. All instructions have a single cycle latency with branches delayed by one cycle. A two stage pipeline is used. The processor implements 12 instructions including: *add, add with carry-in, subtract, subtract with carry-in, logical or, logical and, logical xor, logical invert, arithmetic up shift, arithmetic down shift, booth multiply, and select.* Multiplication can be done in eight cycles on a single processor or eight processors can be pipelined to deliver a result every cycle. The add and subtract with carry-in instructions allows data precision in multiples of 16. Each processor fetches instructions from its eight-word instruction memory. Data may come from internal registers or external sources via two-word input queues (the registers of the processor can be configured as upto three queues). The execution of all the processors can be halted via global control.

Data communication between all processors, irrespective of position, takes one cycle and uses a hardware synchronization handshake protocol. With each clock cycle, handshake lines are precharged. If a sender or receiver is not ready, it is pulled low (wired or). This will stall all participants. The line will remain charged if, and only if, all participants are ready to pass a data token. This mechanism eliminates the need for global synchronization and retiming during the mapping process.

PADDI-2 has a 7464-bit scan chain that traverses the die, providing static configuration for each programmable resource, including the processors and interconnect. In addition to configuration loading, the scan chain is also used for debugging. After a global halt, the status of each processor can be retrieved and examined under global control. PADDI-2 has been fabricated using a 0.8 micron CMOS process with two levels of metal. The die size is 11.5 mm X 11.5 mm and is housed in a 208 pin ceramic package. Additional details can be found in Yeung's Ph. D. thesis [1].

Multi-processor DSP board

Multiple PADDI-2 chips have been integrated within a test-board. This board has eight independent chips, each with a 128KB static RAM. The interconnection of the I/O between chips is hardware configurable via jumpers. Each chip can be configured to connect one or more ports to its four logical neighbors in a 4x2 torus fashion. Multiple boards can be daisy chained into a single logical device and connected to one host adapter. Since synchronization is done locally in hardware, this method of scaling works well. Alternatively, additional interface controllers, with one or more multiprocessor boards can be added to increase available computation. System software has been implemented to address any number or configuration of boards. Port connectors are provided to facilitate "plug-in" daughter-card and peripherals such as sensors, displays, and other devices.

Due to the wide variation in communication requirements for the algorithms of interest, no single optimal interconnect topology exists. Our approach is to provide board-level options for interconnect configuration over a common range. In the configuration shown above, the chips are connected in serial fashion with each connected to an external memory. Two ports are used to send and receive data to and from the host interface controller via a private 16-bit data bus.

Address generation and read write control for the memories is performed under nanoprocessor control. This provides flexible memory access. One common access method is the delay line mode. Another example, which demonstrates this flexibility, is a multiple delay line implemented using one physical memory. The address generator requires 6 instruction lines using the nanoprocessor instruction set.

A board design goal was to demonstrate and evaluate the efficiency and flexibility of PADDI-2. At 50 MHz, each board can provide 20 GOPS of computation. It is being used to explore PADDI-2's architecture, its strengths and weaknesses, so as to provide guidance in our next generation architecture.

Host Computer Interface Controller

The multiprocessor board is connected to a workstation using a custom interface controller. It has two main components: a field programmable gate array and a host bus-interface chip. We have designed an adapter for the Sun Microsystem SBus. The low level SBus protocol is implemented using an LSI L64853A interface chip. Global control is implemented using a field programmable gate array (Xilinx 4010) and is in-host reconfigurable. Its purpose is three fold. It realizes the protocol translation between the SBus interface chip and the private bus. It performs arbitration for the private bus. And, it provides global control for the processors. The FPGA controller logic is specified using VHLD and is synthesized. This enables us to explore alternative protocols, arbitration schemes, and global control techniques without re-fabricating hardware. Mature controller realizations (configuration bitstreams) are cataloged for reuse.

3 SOFTWARE

The software components of the system can be classified into eight general types. These type are identified in the table below and each is discussed in the following sections.

Software Libraries in Java	(High-level Validation)
SFG Graphical Interface	(Visual Entry and Manipulation)
Routing Tools	(Interactive/Automatic PR)
Simulation Tools	(Functional Verification)
Compilation Tools	(Instruction Stream Configuration)
Board Access Tools	(Communication with Hardware)
Board Debugging Tools	(Interrogation of Hardware Status)
Development Libraries	(Tools Development Libraries)

Development Libraries

A set of application development libraries has been implemented. Six libraries have been developed to provide standard interfaces upon which other tools are built. A low level I/O, a link level I/O, a scan chain bit manipulation, a controller access, and a general purpose parsing package have been developed. These packages provide non-overlapping function and interact in a layered fashion. If any hardware component changes, only those libraries directly affected by the change need to be modified. For example, if we were to replace PADDI-2 with VGI-1 in the existing multiprocessor board, only the scan chain manipulation library would need modification.

Board Debugging Tools

Several utilities have been written to provide parameterized stimulation of all hardware components. For example, to verify that the chip scan chains are operational, a utility has been written that generates test vectors. It can provide numerous, in type and quantity, test patterns for a variable number of boards and chips. As a second example; one of the interface controller configurations supports parameterizeable timing. When this version is used to configure the interface controller, a test tool is available to dynamically modify various controller timing parameters. This was used to optimize controller efficiency and function. Numerous other debugging tools exist, but will not be described.

Board Access Tools

Four major board access tools have been developed. One is used to download the configuration bitstreams to the FPGA in the interface controller. A second is used for configuration of the PADDI-2 processor array. The third is used to provide I/O streams to algorithms running on the array. A fourth exists which helps debug algorithms that use external data memory.

The tools all have a "self documenting" command line which provides a usage synopsis. It provides command description, revision, argument keywords, argument keyword data types, argument descriptions, and argument defaults, when run with no options. All tools can be readily invoked within scripts and other programs to perform complex sequences.

Compilation Tools

To compile an algorithm, one writes assembly code for each function in the signal flow graph. These instruction streams are assigned to specific nanoprocessors within a chip using an index file. This is also where the level-1 and level-2 interconnect is configured. All switches are, by definition, *off* and are explicitly turned *on* as needed. A separate index is created for each chip used in an algorithm. The compilation tools generate configuration bitstreams for a chip. Each assembly file is compiled by the assembler (*asm*) to an intermediate object file. A netlist is generated from the index (*gen_vsm*). This netlist and intermediate object files are combined (*pgen*) to generate the chip configuration bitstream. This process is ordered by a *makefile* which itself can be automatically generated (*mkmf_set*). This is required for each chip on a board. Chip configuration bitstreams are assigned to physical chips using a meta-index file. This file simply is an index which relates chip configurations to chip numbers. A *makefile* is also used hierarchically to ensure that each chip configuration is current. This *makefile* can also be automatically generated (*mkmf_mset*). The entire source for an algorithm can be built using a single call to *make* on the meta-index *makefile*.

Simulation Tools

A functional simulator has been developed to assist in application development. The simulator is written in *Java* for portability. It processes compiled assembly code to initialize the simulated chip state. A nanoprocessor simulator is instantiated for each assembled object used in the index file. The index file is also used to establish the communication interconnections between instances. The simulator, like the hardware, can be run, halted, and stepped. At any time the processor state can be viewed. All register, queue, output port, condition code, flag register, current pc, previous pc, and next pc, and processor stall information is available. The state of each nanoprocessors is summarized in separate state windows and nanoprocessor communication can be observed and recorded. Chip I/O is also simulated. With this, the same data sets processed by the hardware can also be processed by the simulator.

VGI-1 has on-chip data memory and independent I/O processors. Extensions to support these and the other additional function are under development. The implementation organization is general enough to support similar designs which use hardware synchronization to communicate between processors. The network primitives support arbitrary hierarchical networks topologies with programmable routing.

Routing Tools

One of the more difficult tasks in programming multiprocessor systems (of this class) is establishing the required inter-processor communication. To assist, a graphical routing assistant tool has been developed which performs the allocation as a user requests and assigns routing resources. The interconnect network is constructed from a specification data set which describes the resource available on the chip.

The algorithm index file is read to identify nanoprocessor instruction sequences. Nanoprocessor assignment and interconnect resources are interactively allocated graphically. The assistant keeps track of all available and allocated resources. Routes are allocated from a displayed list of available routes computed based on a user selected processor pair. Previously allocated resources can be released and reallocated in any order. We have a slow tool for automated PR which is guaranteed to find a mapping, if one exists, by using exhaustive methods.

SFG Graphical Interface

A graphical signal flow graph manipulation tool has been developed to help in algorithm mapping and visualization. SFG's are represented using nodes, edges, and graph objects. All objects accept arbitrary property tagging (selectably visible or invisible) and support hierarchical composition. For example, a node can contain other objects, such as another graph, or collection of graphs. Nodes may be moved around at will, grouped into clusters (subgraphs) and collapsed into a single node. This tool aids in interactive SFG partitioning, manual placement, and routing. It works in conjunction with the other mapping tools by sharing a common I/O data representation.

A SFG can be interactively partitioned, placed, and routed with this tool. One selects groupings which can be mapped to clusters, assigning properties to the nodes and edges to allocate level-1 resources. Next groupings are assigned to clusters by adding properties to the higher-level node which contains the grouped subgraphs. Finally, properties are added to the edges which interconnect groupings to allocate the required level-2 routing resources.

Alternatively, a partial mapping can be obtained from other tools, such as the slow autorouter, and completed manually by visually identifying the objects which have not been assigned (those missing an assignment property).

Software Libraries in Java for Image Processing

Our algorithm mapping flow has two paths which branch from the algorithm and re-converge with an algorithm input data stream. One which targets hardware realizations and another which targets software realizations. This algorithm-in-software path provides system level verification of algorithm-in-hardware mappings. The output of both implementations is compared and differences are readily observed. A subtraction of one image from the other is one such useful comparison.

Several image processing algorithms have been implemented. Enhancement algorithms include median filters, local averaging, max/min sharpening, and blurring filters. For edge detection, we have the discrete differencing, the Robert's edge detector, the Prewitt edge detector, and the Sobel edge detector. Others include the fast fourier transform and the inverse transform. Others, such as the discrete cosine transform and wavelet transform, are under development.

4 SYSTEM USAGE

Currently the task of going from algorithm function to nanoprocessor assembly code is manual, and uses macro libraries, however, techniques are being investigated to support more automated mapping flows. Given the simplicity of a nanoprocessor, code generation is not difficult. The processor assembly code, node-to-processor assignment, and the interconnect routing assignment are processed to produce a configuration bitstream which is downloaded to configure the array. A separate assembly source file is required for each unique node of the SFG. An index is used to map assembly code to nanoprocessors.

A Mapping Technique

There are numerous techniques for mapping. The efficiency of the technique chosen is highly dependent on the algorithm description. We assume the algorithm to be described by an SFG and that each node does not exceed the context or port capacity of a nanoprocessors. PADDI-2's two level interconnect architecture encourages a two phase mapping process; intra-cluster routing, followed by inter-cluster routing. The discussion below is an example of one mapping scheme.

The task of intra-cluster routing involves first partitioning then placing and routing subgraphs of the algorithms SFG. Node count, port count, and interconnect topology limits the selection of subgraphs. Once all nodes of the algorithms' signal flow graph have been assigned to a subgraph, this phase is complete.

The routing resource within a cluster is shared to provide internal and external I/O. For example, a level-1 bus can be used to connect two nodes within the cluster or it can be used to connect a node within a cluster to one outside the cluster. Because of this, there is a dependency between intra- and inter-cluster routing. In some cases, intra-cluster routing obstructs inter-cluster routing and in these cases, one might identify alternative intra-cluster routing to avoid such obstructions. An obstruction is not revealed until the second routing phase.

In the second phase, groupings are assigned to clusters and interconnected as required by the partitioned SFG. This involves assigning subgraphs to clusters and allocating level-2 interconnect resources for the inter-cluster communication. Once all subgraphs have been assigned and interconnected, then mapping is complete.

5 SUMMARY

We have integrated a complete DSP system built around PADDI-2. This hardware/software design consists of a DSP, system boards, driver libraries, access tools, configuration tools, algorithm development tools, and verification software. We have produced a flexible functional simulator which works for both PADDI-2 and VGI-1 and provides routing and placement assistance in a Java based graphical tool. Our software development libraries have been designed for portability. New board configurations and hosts environments can be supported with minimal porting effort. Our multiprocessor DSP array has been designed to be modular and scalable and can directly incorporate sensor daughter cards. Several innovations exist in the PADDI-2 chip architecture. A flexible interconnect switch has been designed which requires approximately 17% the die area. Additionally, a robust hardware local synchronization technique has been implemented which reduces programming difficulties present in many other multiprocessor systems. Consistency was used in the software development methodology and command utility interfaces.

The PADDI-2 chip was designed and verified using the Lager tool set and other VLSI design tools developed at the University of California, Berkeley. The custom boards were designed using Lager in conjunction Racal-Redac's to describe, place, and route board components. VHDL synthesis is performed using View-Logic and mapped using XACT. Java programs were developed using JDK. Web formatted code documentation[10] were automatically generated using EXT.

Acknowledgments

We would like to acknowledge contributions of other research group members. Arthur Abnous implemented the PADDI-2 assembler. Nelson Chow worked on various parts of the board design and fabrication. Alfred Yeung delivered PADDI-2 an accompanying compiler. Matt Armstrong implemented the Java simulator and routing assistant. Erek Borowski coded numerous software implemented algorithms in Java and the SFG graphical interface. Stas Frumkin assisted with board layout and routing.

The help provided by Brian Richards, Fred Burghardt, and Sue Mellers is appreciated. The encouragement and support provided by Bob Brodersen is greatly appreciated. STTR USAF funding is monitored by Robb Brown. The suggestions provided by Pat Coffield are appreciated.

REFERENCES

[1] Yeung, A. K. W., "PADDI-2 Architecture and Implementation," Ph. D. Thesis, University of California, Berkeley, CA, June, 1995.

[2] Chen, D. C., "Programmable Arithmetic Devices for High Speed Digital Signal Processing," Electronic Research Laboratory, Memorandum No. UCB/ERL M92/49, University of California, Berkeley, CA 94720, May 14, 1992.

[3] Lee, S. Y. and J. K. Aggarwal, "Parallel 2-D Convolution on a Mesh Connected Array Processor," IEEE Transactions on Pattern Analysis and Machine Intelligence, vol. PAMI-9, no. 4, pp. 590 -594, July, 1987.

[4] Sunwoo, M. H. and J. K. Aggarwal, "Flexibly Coupled Multiprocessors for Image Processing," 1988 Intl. Conf. on Parallel Processing, St. Charles, IL, Aug. 1988.

[5] Webb, J. A., L. G. Hamey, and I. C. Wu, "An Architecture Independent Programming Language for Low-level Vision," Computer Vision, Graphics, and Image Processing, vol. 37, no. 11, pp. 1751 - 1762, Pittsburgh, PA, Nov., 1989.

[6] Stellakis, H. M. and E. S. Manolakos, "A Tri-Array for the Realtime Computation of Higher Order Moment Estimates," VLSI Signal Processing V, Edited by K. Yao, R. Jain, W. Przytula, J. Rabaey, IEEE New York, pp. 510 - 519, New York, 1992.

[7] Robert, M., M. Paindavoine, and P. Gorria, "Architectures for Integration of Real Time Image Processing Systems," VLSI Signal Processing V, Edited by K. Yao, R. Jain, W. Przytula, J. Rabaey, IEEE New York, pp. 267 - 276, New York, 1992.

[8] Intel, "iWARP Forum'91," iWARP Users Meeting, Crystal City, VA, Sept., 1991.

[9] Borkar, S. and et al, "Supporting Systolic and Memory Communication in iWARP," Proceedings of the 17th International Symposium on Computer Architecture, pp. 70 - 81, Seattle, WA, May, 1990.

[10] http://infopad.EECS.Berkeley.EDU/spartan/

Design and Implementation of the NUMAchine Multiprocessor*

A. Grbic, S. Brown, S. Caranci, R. Grindley, M. Gusat, G. Lemieux,
K. Loveless, N. Manjikian†, S. Srbljic‡, M. Stumm, Z. Vranesic and Z. Zilic§

Dept. of Electrical and Computer Engineering, University of Toronto, Canada

Abstract

This paper describes the design and implementation of the *NUMAchine* multiprocessor. As the market for CC-NUMA multiprocessors expands, this research project provides a timely architectural design and cost-effective prototype. The key to the successful implementation of our 48-processor prototype is the use of off-the-shelf components and programmable logic devices. Since this machine will serve as a research vehicle for parallel software development, a number of hardware features to enhance experimentation have been included in the design.

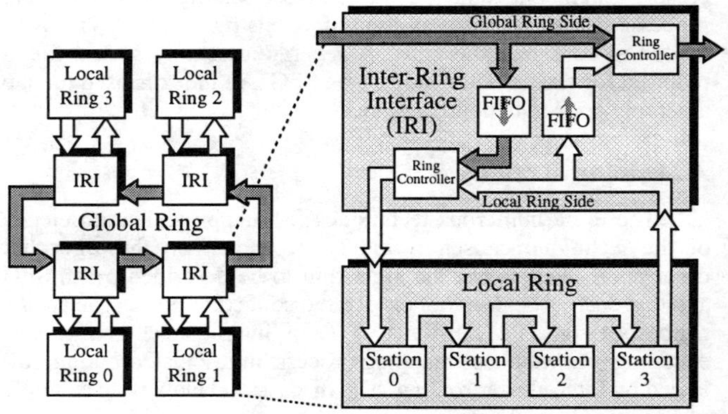

Figure 1: NUMAchine Architecture

1 Introduction

This paper describes the hardware implementation of a multiprocessor called NUMAchine [7]. In NUMAchine, processors, caches, and memory are physically distributed throughout the system. The memory is shared by all processors, but the access latency depends on location. Hardware automatically maintains coherent copies of data throughout the system. This type of multiprocessor is known as a CC-NUMA (Cache-Coherent Non-Uniform Memory Access) distributed shared-memory multiprocessor. CC-NUMA multiprocessors are quickly gaining commercial acceptance for use as powerful compute, file, network and web servers. The shared-memory model eases the transition from uniprocessor-based software. In addition, multiprocessors that support sequential consistency provide ease-of-programming by extending the already familiar uniprocessor programming model.

There has been considerable research interest into how to design CC-NUMA multiprocessors to achieve maximum performance. The Stanford DASH [5] multiprocessor is a mesh-based system that uses separate request and response communication networks. The follow-on project, FLASH [4], is centered around the MAGIC chip, a customized RISC engine acting as coherence, network, and memory controller. Both DASH and FLASH use MIPS-based computing nodes. In contrast, the MIT Alewife project [1] uses the custom-designed Sparcle processor that tolerates memory latency by quickly switching among multiple threads.

In addition to high performance, we are also concerned with cost, ease-of-programming and providing features for parallel software research. The NUMAchine architecture is hierarchical and modular so that small systems can be affordably extended into larger ones, up to a few hundred processors. A directory-based hardware cache coherence protocol scales efficiently with system size by exploiting the hierarchical architecture. The architecture is also exploited to efficiently implement a sequentially-consistent memory model. As well, NUMAchine includes novel hardware features that enhance experimentation. First, non-coherent operations and directory manipulation features permit direct comparison between hardware and software cache coherence. Second, block data transfers and coherence operations reduce cache pollution, bandwidth use, and overhead. Third, a hardware barrier register supports efficient parallel synchronization. Fourth, flexible and non-intrusive performance monitoring hardware improves the observability and accuracy of software-only measurement schemes.

The 48-processor prototype system is based on 150 MHz MIPS R4400 processors. It has a peak performance of 1.7 GFLOPS and a peak bandwidth of 400 MB/s at any point in the interconnection network. The prototype is constructed entirely with off-the-shelf components and programmable logic devices (PLDs) without compromising our goal of a 50 MHz system clock. The use of PLDs allows for a low cost implementation and rapid resolution of problems. PLDs also provide the flexibility to implement new ideas by modifying some of the novel hardware features described above.

2 NUMAchine Architecture

This section gives an overview of the NUMAchine multiprocessor. Computing nodes are clustered into bus-based units called *stations*.

---

*This work was supported by the Strategic Grant #STR0149404 from the Natural Sciences and Engineering Research Council of Canada

†now with Queen's University at Kingston, Canada

‡with University of Zagreb, Croatia

§now with Lucent Technologies, Allentown, PA, USA

A NUMAchine station uses a bus to connect four processor modules, up to two memory modules, a network interface module, and up to two I/O modules. Depending on the address, processors transparently access either local or remote memory. Remote memory requests are handled by the network interface, which routes them to the correct remote station. The network interface also includes a third-level *network cache* for holding data from remote memories.

Stations are interconnected by a hierarchical arrangement of unidirectional bit-parallel rings, as shown in Figure 1. This architecture provides modularity and flexibility because stations may be added to rings as necessary, and the depth of the hierarchy can be extended for more rings. Hierarchical rings, as an alternative to meshes, have been shown to perform well for systems with up to 128 processors and workloads with medium to high memory access locality [6].

A slotted-ring protocol transfers data packets across the network and then reassembles these packets at the remote station. Each ring packet is a portion of a bus transaction augmented with additional routing information. The routing information is contained in a set of *routing masks*, one for each level of the hierarchy [3]. The bits in a routing mask correspond to siblings in that level of the hierarchy. If a particular bit is set, a copy of the packet descends to the children in that group. Packets only travel as high as necessary in the network to reach all the targets. This routing mask mechanism provides a concise way to multicast and broadcast data, and is used by the cache coherence protocol to maintain the directories.

The choice of a ring topology for the interconnection network provides three important benefits. First, the unique path between any two points in the network maintains the relative ordering of packets between the same source and destination. Second, the ring topology provides an efficient method for multicasting and broadcasting. Third, the point-to-point ring connections simplify wiring and allow for a simple routing scheme. The first two benefits allow for an efficient cache coherence implementation, while the third allows fast data transmission rates.

The NUMAchine write-back/invalidate cache coherence protocol is hierarchical, ownership-based, and uses directories to maintain information on cache lines. The directories maintain cache line state information and locations, represented by routing masks. The routing mask scheme permits the directory to scale logarithmically as the system size increases [3]. The protocol supports sequential consistency by exploiting the order-preserving properties of the ring hierarchy.

3 Design Methodology

Our design methodology relies on the use of commodity components to satisfy cost constraints and to reduce design time. These components include two types of PLDs: FPGAs and CPLDs. PLDs permit hardware-software co-design and allow the hardware to be later modified to support new features.

Our design flow consisted of multiprocessor simulations to define the architecture, design partitioning into multiple printed-circuit boards (PCBs), and bottom-up design of each PCB. The simulations guided the selection of architectural parameters, such as the number of ring levels, stations per ring, and processors per station. These parameters then guided the partitioning into multiple PCBs. The bottom-up design methodology for each PCB consisted of the following steps: (1) logic partitioning, (2) assignment of logic to commodity components or PLDs, (3) design entry for PLDs, (4) high-level schematic design entry, (5) board- and system-level simulation, and (6) PCB layout.

The CAD tools used for hardware design were Cadence Logic Workbench (LWB), Cadence Allegro Layout and Altera MAX+plus II. The two Cadence tools provide a state-of-the-art system for the implementation of digital systems. We used LWB for board-level schematic capture and simulation, and Allegro for PCB layout. MAX+plus II supports logic implementation in Altera PLDs using graphical and text-based design entry methods, as well as full timing simulation. The bottom-up PCB design approach resulted from difficulties encountered in integrating LWB and MAX+plus II, as documented in an earlier paper [2].

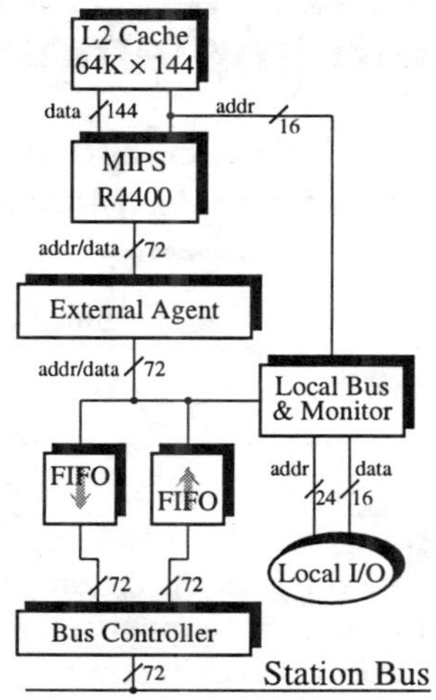

Figure 2: Processor Board

4 Hardware Implementation

Our overriding design goal is high performance within our cost constraints. All boards, busses, and rings operate at 50 MHz, and processors run internally at 150 MHz. Achieving such speeds with current PLDs is a major challenge: clever design methods and careful hand-tuning of circuits is often necessary. Experience guides how and when to apply these manual optimizations [8].

The use of field-programmable devices enabled the design of custom controllers without the higher cost of full-custom chip design. Reprogrammability allowed rapid resolution of problems in the design. Furthermore, the rapid pace of technological improvement in programmable devices enabled, as well as encouraged, prototyping and revisions to integrate multichip designs into a single chip for equivalent cost. The end result is an operational system in less time and at a lower cost than full-custom design.

4.1 Processor Board

The NUMAchine processor board uses a MIPS R4400 64-bit processor running at 150 MHz, with 1 MB of second-level (L2) cache as shown in Figure 2. The 128-bit L2 cache interface runs at 75 MHz and is controlled entirely by the processor, while the 64-bit[1] external (system) interface runs at 50 MHz. To communicate with external memory, the R4400 requires an *External Agent* (EA) circuit. The EA accepts requests from the processor, and forwards external requests for retrieving or invalidating L2 cache lines to the processor. The EA also serves as a bridge to an on-board local bus. The FIFO buffers between the EA and the station bus hold bursts of incoming requests/responses and outgoing write-back data from the processor. A *mailbox* register feature in the FIFOs allows a read request to bypass write-backs waiting in the queue. Read requests are given higher priority and issued to the memory as soon as possible. The bus controller manages data transfers between the FIFOs and the station bus.

[1] 72-bit connections are shown in the figures. These include data and parity/ECC bits

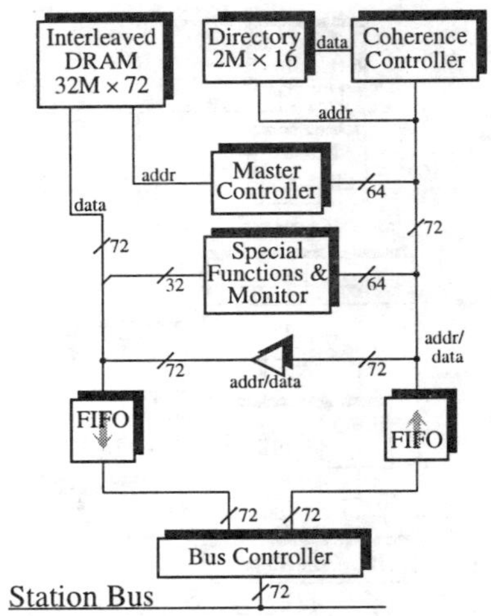

Figure 3: Memory Board

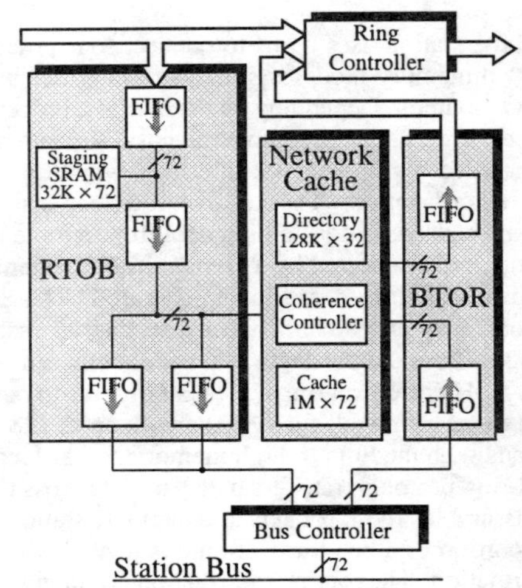

Figure 4: Network Interface Board

The on-board local bus provides a serial port, boot PROM, local hardware monitoring, and a connection to a diagnostic board for bootstrap debugging. The local bus also contains interrupt and barrier registers. The interrupt register provides an efficient mechanism for externally-generated interrupts, while the barrier register supports efficient global synchronization for parallel applications.

4.2 Memory Board

A block diagram of the Memory board is shown in Figure 3. FIFOs receive and send packets to the bus through the *Bus Controller*. The *Master Controller* provides control signals for the FIFOs and coordinates the other functional blocks. A *DRAM controller* manages up to 256 MB of local DRAM. The *Coherence Controller* maintains a directory in SRAM and implements all of the coherence actions and state transitions for the cache coherence protocol. For 128-byte cache lines, 4 MB of directory SRAM are needed for 256 MB of DRAM. The *Special Functions and Monitoring* unit manages special hardware functions on address ranges, described in Section 4.7, and generates interrupts.

For high performance, the DRAM memory is 4-way interleaved. Furthermore, data access is overlapped with cache coherence operations, and successive memory requests are pipelined. For each memory request, the master controller starts the Cache Coherence and the DRAM controllers simultaneously. While the DRAM is accessed, the directory entry is updated and the header packet for the response is placed in the outgoing FIFO. Data packets from DRAM are then placed in the outgoing FIFO (64 bits every clock cycle). There is sufficient buffering to permit processing of the next request as soon as the directory entry has been updated for the previous request. A single memory board can nearly saturate the bus with response data. When two memory boards are used together on a bus, they are interleaved at the cache-line level to increase parallelism and reduce total latency.

4.3 Network Interface Board

The main components of the Network Interface board are the *BTOR* (Bus TO Ring) interface, the *RTOB* (Ring TO Bus) interface, and the *Network Cache*, as shown in Figure 4. The Network Cache consists of the Coherence Controller, 8 MB of SDRAM to cache data from remote memories, and a directory (512 KB of SRAM). The BTOR controller moves data from the incoming bus FIFO to the outgoing ring FIFO or into the network cache input buffers. It also moves data from the network cache output buffers to the outgoing ring FIFO. The *Ring Controller* sends this data as individual 64-bit ring packets mixed in with existing ring traffic. At the destination, packets are transferred from the ring into the incoming ring FIFO. These packets are reassembled into cache lines by the RTOB controller in the *Staging SRAM* and then sent to the network cache input buffers or to one of the outgoing bus FIFOs. The network cache alternates packet processing between the two sides. On each request, the Coherence Controller updates the directory information according to the NUMAchine protocol and may generate responses to the BTOR and/or RTOB sides.

The datapaths on the Network Interface board were initially implemented using discrete buffers. The large number of components required for this implementation complicated the layout of the PCB. In the following revision of the board, we were able to implement this datapath with FPGAs and realize a net savings in cost. This redesign also yielded improved performance by adding more parallelism to the datapaths.

4.4 I/O Board

The I/O board contains a 100 MHz MIPS R4650 processor, up to 32 MB of DRAM storage, 4 DMA channels, and a 33 MHz PCI bus. The PCI bus connects up to 4 SCSI controllers and one PCI expansion slot. The R4650 manages all data transfers and can also execute protocols for parallel I/O and networking. The I/O board is designed to facilitate parallel I/O research through a flexible bus interface and efficient polling/interrupts.

4.5 Station Bus

We chose the Futurebus+ physical backplane for NUMAchine stations because it provides a wide bus with up to 14 card slots. We did not, however, use the Futurebus+ protocol because it is overly complex for our purposes. In addition, the Futurebus+ protocol chips available during our early design work did not meet our 50 MHz speed requirement. We designed a custom, synchronous, split-transaction protocol with a centralized, pipelined arbiter. The core datapath width is 64 bits; a 128-bit bus was considered, but architectural simulations indicated that the marginal improvement in performance did not justify the cost and complexity.

The synchronous protocols for NUMAchine necessitate low-skew clock distribution. For each station, we generate a central clock signal which is then replicated and distributed to all boards using differential ECL for low-skew fanout and noise immunity. Additionally, phase-locked loops are employed to reduce the clock skew.

4.6 Rings

A local ring in NUMAchine is assembled by connecting together the Network Interface boards of up to four stations with high-density, controlled-impedance ribbon cables. The Ring Controller on the Network Interface board adds only a single cycle delay to traffic traveling to the next station. When a packet reaches its final destination, it is removed from the ring and a new packet can be injected by the Ring Controller. In addition, it can simultaneously source and sink data at the maximum link transmission rate of 400 MB/s.

The Ring Controller is implemented as a single CPLD and a number of discrete tri-state registers. The ring clock rate of 50 MHz is limited by the speed at which our design can multiplex the data between the ring traffic and new traffic from the station. A full-custom implementation would allow for a much faster ring clock rate.

To assemble a larger NUMAchine system, the ring hierarchy must be expanded to another level. Rather than constructing four Inter-Ring Interface boards as shown in Figure 1, which would be limited to the same speed as the local rings, we decided to construct it in a more centralized fashion. A Global Ring (GR) backplane was designed with two types of daughter cards. A Local Ring Interface (LRI) daughtercard provides the local ring controller. Signals from four LRIs are redistributed on the GR backplane to six DataPath (DP) daughtercards and a central controller. Each DP implements an 18-bit slice of the global ring using FIFOs and an FPGA. This organization clusters the global ring logic so that a clock rate of 80 MHz can be achieved for a maximum data rate of 640 MB/s.

4.7 Novel Hardware Features

Dedicated monitoring circuitry is included in all parts of NUMAchine, including the processor, memory, and interconnect. Applications can reconfigure the FPGA-based monitoring hardware for data collection without resetting the machine. Most of the monitoring circuits comprise a large FPGA, fast 64K×32 SRAMs, and supporting CPLDs. A novel concept that is included throughout the monitor is that of a *phaseID*, which is derived from a writable 4-bit register on each processor board. The phaseID, which is attached to *all* transactions leaving the processor, can be used by an application to separate performance data into fine-grained phases. The monitoring hardware on each processor board can separately count monitoring events based on transaction type, phaseID, or address range. The monitoring hardware on the memory board can use these filters as well as the originating processor and cache line state.

Additional specialized hardware also provides uncached operations, non-coherent operations, and special functions. Uncached operations bypass the secondary and network caches. Non-coherent operations still use caches, but bypass the cache coherence protocol when transferring cache lines. Special functions provide low-level control including directory read/write operations, returning modified cache lines to memory, invalidating copies of cache lines system-wide, multicasting of cache lines, block data transfers, prefetching into network caches, and forced write-backs to home memory. In addition, the prototype supports 64-byte or 128-byte cache lines.

4.8 Performance

Table 1 gives the measured latency of read requests on the prototype hardware in 150 MHz processor cycles and 50 MHz system cycles. Note that for remote requests, 1 hop across the ring was used for the data in the table because the numbers were measured on a two-station system. Although not measured, the additional latency to traverse the global ring is estimated to be 24 system cycles.

Level of hierarchy	Procr. cycles (6.67ns)	Sys. cycles (20ns)
L1 cache	1	
L2 cache	6	
Local memory	135	45
Local network cache	165	55
Other L2 cache	255	85
Rem. mem. (same ring)	594	198

Table 1: Measured memory access latencies

5 Current Status

Currently, a 4-station (16-processor) NUMAchine system is operational. Further information, including photographs, can be found on the WWW.[2] A custom parallel operating system with a UNIX-like interface, called Tornado, has been developed. The operating system boots, allows logins, and runs various applications. All of the boards for the NUMAchine prototype have been fabricated. The hardware components are being integrated together to form the complete system.

References

[1] A. Agarwal et al. The MIT Alewife Machine: Architecture and Performance. In *Proc. of the 22nd Annual ISCA*, pages 2–13, June 1995.

[2] S. Brown et al. Experience in Designing a Large-Scale Multiprocessor usign Field-Programmable Devces and Advanced CAD Tools. In *Proc. of the 33rd DAC*, pages 427–432, Las Vegas, NV, June 1996.

[3] K. Farkas, Z. Vranesic, and M. Stumm. Scalable cache consistency for hierarchically-structured multiprocessors. *J. of Supercomputing*, pages 345–368, 1995.

[4] J. Kuskin et al. The Stanford FLASH Multiprocessor. In *Proc. of the 21st Annual ISCA*, pages 302–313, Chicago, IL, May 1994.

[5] D. Lenoski et al. The DASH prototype: Implementation and performance. In *Proc. of the 19th Annual ISCA*, pages 92–103, Gold Coast, Australia, May 1992.

[6] G. Ravindran and M. Stumm. A Performance Comparison of Hierarchical Ring- and Mesh-Connected Multiprocessor Networks. In *Proc. of the Third International Symposium on HPCA*, pages 58–69, San Antonio, Texas, February 1997.

[7] Z. Vranesic et al. The NUMAchine Multiprocessor. Technical Report CSRI-324, Computer Systems Research Institute, University of Toronto, 1995.

[8] Z. Zilic et al. Designing for High Speed-Performance in CPLDs and FPGAs. In *The 3rd Canadian Workshop on Field-Programmable Devices (FPD'95)*, pages 108–113, May 1995.

[2] `http://www.eecg.toronto.edu/parallel`

Design and Specification of Embedded Systems in Java Using Successive, Formal Refinement

James Shin Young, Josh MacDonald, Michael Shilman, Abdallah Tabbara, Paul Hilfinger, and A. Richard Newton

Department of Electrical Engineering and Computer Sciences
University of California, Berkeley

Abstract

Successive, formal refinement is a new approach for specification of embedded systems using a general-purpose programming language. Systems are formally modeled as Abstractable Synchronous Reactive systems, and Java is used as the design input language. A policy of use is applied to Java, in the form of language usage restrictions and class-library extensions, to ensure consistency with the formal model. A process of incremental, user-guided program transformation is used to refine a Java program until it is consistent with the policy of use. The final product is a system specification possessing the properties of the formal model, including deterministic behavior, bounded memory usage, and bounded execution time. This approach allows systems design to begin with the flexibility of a general-purpose language, followed by gradual refinement into a more restricted form necessary for specification.

1 Introduction

A variety of languages are used in the design and specification of hardware and software for embedded electronic systems[5]. General purpose languages such as C and C++ are usually used for the design of software, while hardware description languages, such as Verilog and VHDL, are used in the design of hardware. Embedded systems design requires integration of hardware and software components, often a difficult task if they are described using dissimilar languages. Instead of using different languages for design of hardware and software, a single language for system specification is desired. We propose a methodology that allows a single-language platform to be used for early design, as well as for guiding subsequent refinement, partitioning, and synthesis steps to obtain a final implementation.

In particular, we describe an approach for using Java as a design and specification language for embedded systems. Java is a pragmatic choice for several reasons. Since it is a member of the C "family" of languages, it is familiar to designers. Unlike C and C++, it provides standard language and library support for concurrency. Its treatment of arrays permits better static and dynamic error checking than is conveniently feasible in C and C++. Finally, while Java's expressive power is comparable to C++, it is a much simpler language, which reduces the difficulty of program analysis, optimization, and transformation. Using Java also has many practical benefits– its widespread adoption by the science and engineering community promises a large base of support, in the form of compilers, debuggers, development environments, and class libraries.

Embedded systems function as sub-components of larger systems, and their design requisites are often quite different from general-purpose computing systems, the domain for which Java is typically used. Many embedded systems are *reactive*[1], operating at speeds dictated by their external environments, and must be reliable and predictable. Therefore, embedded systems should behave deterministically, and operate within bounded resources, including time and memory. However, Java programs in general guarantee neither determinacy nor bounded resource usage.

To address this fundamental incompatibility, our approach anchors programs on formal models of computation. The Abstractable Synchronous Reactive (ASR) model is one we have developed which has properties suitable for embedded systems specification. Restrictions on the usage of Java are applied in order to ensure the programs are consistent with the ASR model. This principle of language restriction is one area where our approach differs most from other proposals to use Java for system specification, such as that of Helaihel and Olukotun[10].

However, restricting the use of Java limits its flexibility and expressiveness. Designers may be discouraged if the resulting language is overly restrictive. We have addressed this concern by developing a methodology for specification development called *successive, formal refinement* (SFR). SFR supports the transformation of a program written in a general-purpose programming language into one that can be embedded within a more restricted model of computation. It consists of a series of static analyses of programs, coupled with an iterative process of incremental, semi-automated program transformation. SFR enables a system to be designed initially in an unrestricted manner using a general-purpose programming language, and then be gradually refined into a restricted form for use as a system specification.

[1] This term was first introduced by Pnueli[9].

DAC 98, San Francisco, California

In this paper, we present the SFR methodology and describe how it is applied to Java programs to obtain a specification suitable for embedded systems, using the ASR model as a formal basis. We have developed a set of tools, collectively known as *JavaTime*, to aid analysis and transformation of Java source code, and used it to demonstrate application of SFR to a number of examples.

2 Successive, Formal Refinement

Successive, formal refinement rests on two foundations: a *policy of use* on the source language, and a procedure of *incremental transformation*. In general, programs expressed in a general-purpose language cannot be expressed as equivalent systems in a specific model of computation. In SFR, a policy of use is imposed on the source language, S, to make it consistent with the target model, T. A policy of use consists of *restrictions* and *extensions*. The restrictions removes portions of S incompatible with T, while the extensions introduce semantics present in T that have no equivalent in S. The result is a new language S', as shown in Figure 1, whose programs are expressible in T.

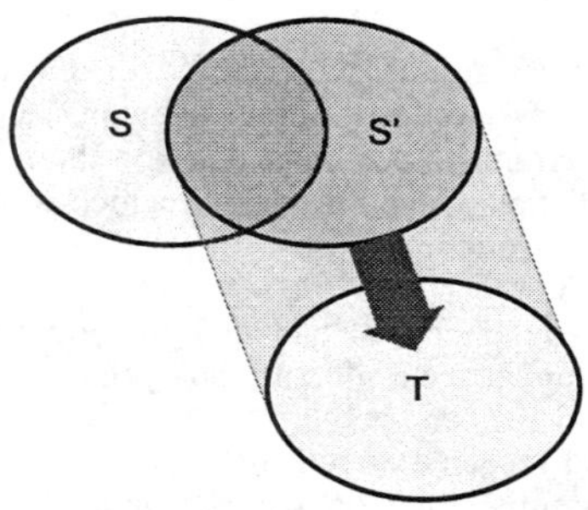

Figure 1. A policy of use applied to language S yields S', compatible with model T.

In this methodology, a user's initial program P is contained within language S (Figure 2a), and cannot be mapped directly onto T. A process of incremental program transformation is used to refine it into a program in language S', to make it valid with respect to the policy of use.

The program is analyzed to verify that the rules in the policy of use are satisfied. If a violation is found, the user is presented with information regarding the nature of the error, and a list of suggested solutions for fixing the problem, including automated program transformations when possible. The user can then modify the program manually or allow the tools to alter it automatically.

This process of analysis and modification is repeated until the program complies with all rules in the policy of use. At this point, the altered version of the program, P', is contained within S' (Figure 2b). Because S' is constructed to be compatible with T, P' corresponds to a system in T.

The SFR process may be applied with respect to a variety of target models, each with its own policy of use. This enables one to use a single general-purpose language for designing systems in an implementation-independent manner, and choose from different models according to the desired implementation style. In addition to the ASR model described in this paper, others might include

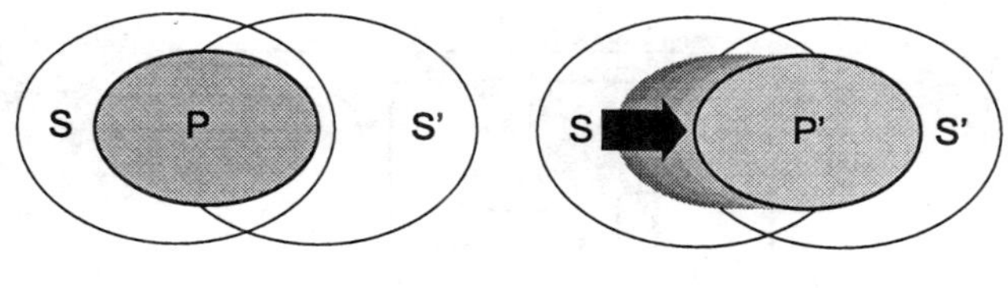

Figure 2. (a) Original program is in language S. (b) Program is embedded in S' through successive, formal refinement.

dataflow, discrete-event, and hierarchical finite state machine models, or models defined by the user. This provides a design language and environment that is highly expressive, and yet can be customized to suit the needs of a particular implementation medium.

Systems for program construction through transformations exist for the development of software[1][13], but ours is the first effort we know of to apply such techniques for the purpose of formal modeling of mixed hardware/software systems. Transformation systems for software typically are used to improve program performance while preserving semantics, while in the SFR methodology transformations are used to restrict and alter a program's semantics.

3 The Abstractable Synchronous Reactive Model

We use the *Abstractable Synchronous Reactive* (ASR) model to specify reactive embedded systems. In the ASR model, time is divided into hierarchically nested *instants*. Systems are represented as collections of *functional blocks*, *channels*, and *delay elements*. Blocks serve to generate output values based on their inputs; channels carry set-valued data between blocks; and delay elements carry data between successive instants in time. ASR systems are *reactive*; that is, they interact with their environments by reacting to the inputs presented to it and producing outputs. The operation of these systems is driven by their environments; if no inputs are provided to a system, it would simply sit idle. Figure 3 shows a graphical representation of a system in the ASR model. The white rectangles are functional blocks, while the shaded rectangle represents a delay element. Channels are drawn as directed edges.

The ASR model is based on the *synchronous* approach to reactive systems, in which systems are said to compute instantaneously and time is divided into instants[2]. Synchronous languages, such as Esterel[3], Lustre[7], and Statecharts[8], also use this approach as the foundation for their formal models of computation. The ASR model is distinguished from other synchronous models in that time may be hierarchically abstracted, as well as function.

Functional blocks calculate output values based on the input values. Inputs and outputs are members of ordered sets, and blocks are restricted to compute only continuous functions between these domains. These idealized blocks produce their outputs synchronously with their inputs, meaning that inputs and their resulting outputs appear in the same instant. Likewise, propagation of values through channels also occurs within a single instant. Channels

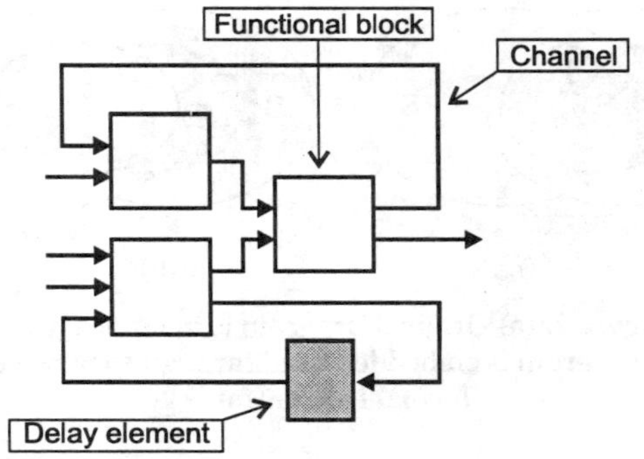

Figure 3. An ASR system.

connect the outputs and inputs of blocks and delay elements together.

Blocks and channels cannot hold values across instants. Instead, values are carried through time by delay elements. At each instant, a delay element's output is equal to the value of its input at the previous instant. From a block's point of view, inputs originating from delay elements appear indistinguishable from external inputs.

It is possible for the input to a block to depend on its output, by introducing cycles in the system graph with no delay elements in the path. To resolve such cyclic dependencies, ASR uses a fixed-point semantics fashioned after the scheme presented by Edwards[4].

Time is viewed as a partially ordered set of instants. The instants may be nested, yielding a hierarchical structure of time, as shown in Figure 4. Thus, although blocks and channels are said to execute instantaneously, their operation can be composed of a set of "sub-instants". Such abstraction of time can be useful in hiding details of implementation[12]. For example, communication of a message between two processors may be viewed as a single instant, rather than as a multitude of instants representing the detailed protocol activities. The amount of physical time associated with an instant is not predetermined. It is associated with the design later, as the solution to a set of temporal constraints imposed by components used to implement the design or via constraints imposed by the system's environment. In that way, the nested refinement described above is symbolic and need not involve "nanoseconds" until later in the design process.

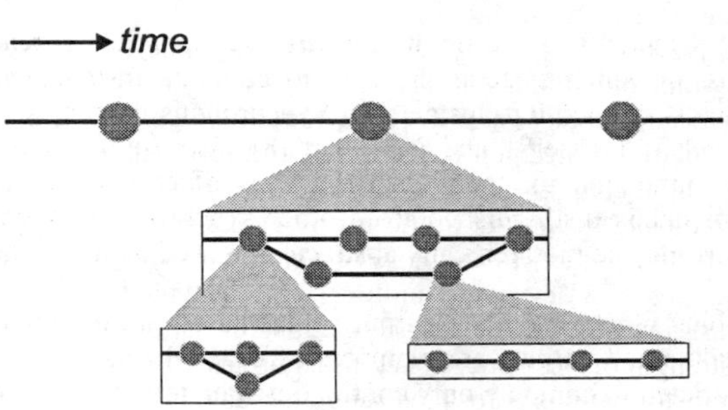

Figure 4. Abstraction of instants in time.

The ASR model has properties that make it useful for specifying embedded systems. ASR systems are deterministic; that is, any particular input can produce only one possible output. Also, they are compositional– an aggregation of blocks is functionally equivalent to a single block, while a collection of blocks and delay elements is equivalent to a system containing one block and one delay element (Figure 5). ASR systems may be represented as a hierarchical composition of ASR subsystems.

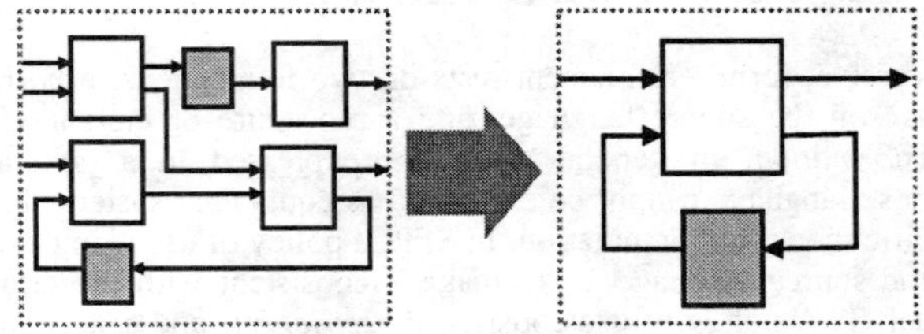

Figure 5. Abstraction of ASR systems in space.

4 Specifying Embedded Systems Using Java

Our approach for design and specification of reactive embedded systems uses Java as the design input language, and SFR is applied to refine programs to be consistent with the ASR model. To begin, we must first clarify the manner in which we interpret Java programs for specification purposes.

Java's ability to dynamically load and link objects prevents a standard Java compiler from globally analyzing and optimizing a program, as any of the classes the program uses may be individually modified and replaced at run time. To enable more detailed analysis of a system's behavior, we assume all the code used to describe it is known at compile time. A class used in describing a particular ASR block cannot be modified without analyzing the other classes to determine the overall effect on the block's behavior.

The Java language specification[6] divides the execution of a Java program into loading, linking, initialization, and finalization. In our approach, the loading, linking, and initialization steps are considered to describe the structure of the system, not the behavior; the program's actions following loading, linking, and initialization are considered to be the specification of the system's behavior. Finalization is disallowed, as it may be considered as representing the termination or destruction of the system.

Java allows concurrency to be specified in the form of user-managed threads that communicate with one another by modifying and reading shared variables. Thus, in general we consider Java programs to describe partial orders of events in a system, as shown in Figure 6.

4.1 Determining the policy of use

In the SFR methodology, a policy of use, consisting of restrictions and extensions, is imposed on the source language to reconcile it with the target model. The user interacts with tools to verify a program's compliance with the policy, and rectify violations through an interactive process of incremental program transformation. In this application, the source is the Java language and the target is the ASR model of computation.

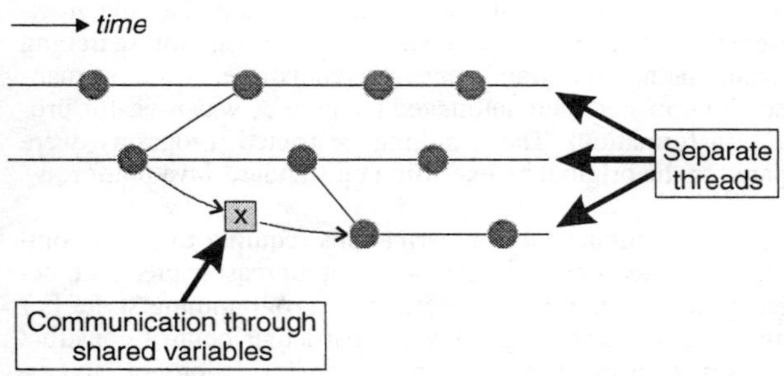

Figure 6. Java programs specify a partial order.

Compliance with restrictions is verified through static analyses of source code. Program transformations may be applied automatically to the abstract syntax tree of the Java program, or manually if the tools are not able to provide a satisfactory transformation. Extensions are made through model-specific class libraries, so that programs undergoing refinement remain valid Java programs throughout, and may remain usable by standard Java development tools. Others have utilized class libraries to model systems[11], but the user-written code is not restricted and verified to ensure the libraries are used correctly.

We may determine the necessary restrictions and extensions by comparing the capabilities and limitations of Java and the ASR model:

- Java programs may be nondeterministic, while ASR systems must be deterministic.
- Java programs may require infinite amounts of memory to execute. In comparison, the memory available to ASR systems is finite.
- The ASR model separates time into hierarchically nested instants, while Java has no notion of time.
- ASR systems operate by reacting to their environments through well-specified inputs and outputs. In Java, no clear distinction between the system and its environment is made.

Therefore, restrictions must be placed on Java programs to guarantee they are determinate and use bounded memory; extensions are needed to distinguish inputs and outputs; and we must define the boundaries of instants.

4.2 Designing an ASR system

A class called `ASR` is provided to the user for describing ASR systems using Java. It must be used as the base class of a specification, and serves to model the functionality of an ASR system in Java. From the environment's point of view, an object subclassed from it looks like a "black box," operated by providing it with inputs, which causes the system to produce outputs. The `ASR` class includes output and input ports, used to convey signals to and from the environment. To specify the functionality of the block, the user writes code to fill in the `run` method Figure 7 shows a conceptual sketch of the resulting Java object.

The lifecycle of an object subclassed from the `ASR` class is considered to be divided into two parts: *initialization* and *behavior.* Initialization includes the invocation of its constructor and execution of static initialization routines. It represents the creation and initialization of the object; if implemented as a hardware/software system, this might correspond to the fabrication of the system and power-on reset behavior.

The object's operational behavior is defined with respect to its `run` method. The object sits idle until its `run` method is invoked

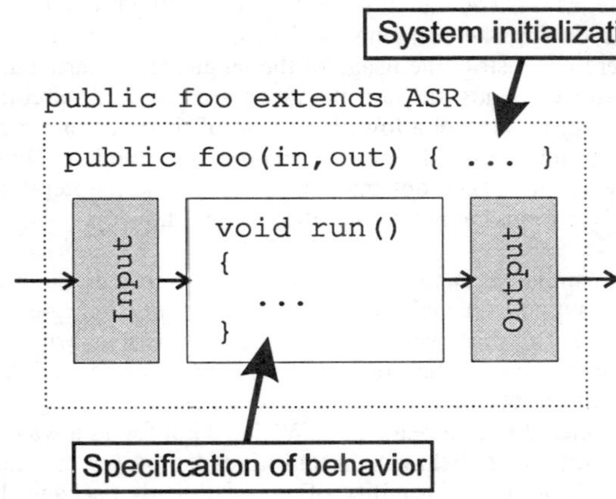

Figure 7. Encapsulation of Java design in ASR class.

by the environment, at which time it produces a set of output values and returns to its idle state. It is the responsibility of the environment to provide the appropriate inputs. With respect to time, an instant is initiated by the invocation of the component's `run` method, and ended when the method completes. The sequence of actions performed in the `run` method can be considered as instants nested within an instant, and are not visible to the object's environment. That is, the execution of the `run` method appears atomic to external observers.

4.3 Restricting the initial design

A specification of an ASR block consists of a single class that extends `ASR`, and all the classes that this primary class may require, either directly or indirectly. The code invoked by the `run` method of the main class must be analyzed and restricted with respect to the policy of use of the ASR model. Unlike standard Java's dynamic loading model, classes that constitute a specification are considered in our analyses to be bound at compile time, and cannot be restricted and verified independently of one another. This is because verification of a block's behavior requires that all dependencies between classes and methods be statically identified.

The amount of memory in an ASR system is fixed, but Java programs may be written such that they require an unbounded amount of memory, for example if dynamic data structures are used. Therefore, one important restriction is that objects may be instantiated only during initialization. This is accomplished by scanning the code and identifying possible allocation statements. One problem that may be encountered is the use of linked structures, and these types of complex structures should be checked for and eliminated in favor of statically allocated data structures.

We must also take care that an ASR object's internal state may not be externally accessible by requiring the object's variables to be `private`. This prevents external modification or observation of the object's state, which undermines data encapsulation and abstraction, and may result in unpredictable behavior.

Addition of the notion of instants motivates an additional restriction: Computation of the output must be bounded in time; otherwise, the system's execution would never advance to the next instant. For arbitrary Java programs, verifying bounded execution is impossible, as it is equivalent to solving the halting problem. At the risk of reducing the computational capabilities of the block, we must therefore restrict the usage of the language. In particular, calculable upper bounds on loop iterations are required, circular method invocations are not allowed, and use of methods that may halt or indefinitely suspend thread execution is forbidden. Thus, `while` and `do while` loops may not be used, and the iteration variable in `for` loops cannot be modified within the loop.

Nondeterminism is easily to implement in a Java program, due to its integrated support for multithreading. For example, in Figure 8, threads *A*, *B*, and *C* share variable *X*. *A* and *B* write to *x*, while *C* reads the value from *x*. The order in which the three threads access *x* may differ between different executions of the program, and may produce different behaviors. We have not found a way to guarantee deterministic behavior in multithreaded programs without severely limiting the usability of Java's threads package. In the current policy of use, direct use of Java threads is prohibited, and concurrency is obtained through specification of separate functional blocks.

The restrictions given in the policy of use are conservative– they are sufficient to ensure that a Java program is consistent with the ASR model, but may not be necessary. That is, there are programs that violate our restrictions, but are expressible as ASR systems.

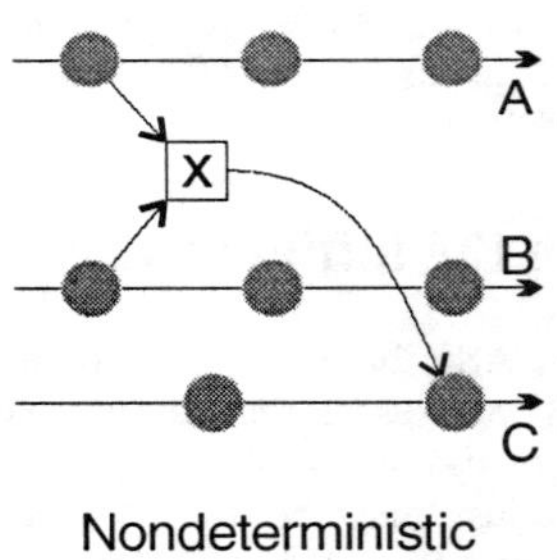

Figure 8. Nondeterministic thread interaction.

5 Experiments

We have applied our proposed methodology to a number of design examples, and found that restriction of Java programs is feasible and beneficial to performance. *JavaTime*, a set of tools to support the SFR methodology has been developed. *JavaTime* takes as input the program source code, and allows analyses, restrictions, transformations, and interpretations to be associated with a design.

A number of publicly available Java programs were taken as examples of unrefined designs to be implemented in an embedded system. The examples were incrementally transformed until they conformed to the restrictions set forth in the ASR model's policy of use. Identification of policy violations were aided by the *JavaTime* tools, which provide a flexible environment for searching and manipulating a program's abstract syntax tree. A mix of manual, semi-automated, and automated techniques was used for program transformation. The resulting restricted programs were compared to the original by execution on standard Java platforms.

In general, the number of transformations required to attain compliance with the policy of use was not unreasonable, and depended greatly on the original designer's programming style. For example, one designer might favor a particular control construct or data structure more than another. Transformations of `while` loops, which are prohibited in the policy of use, to restricted `for` loops was possible in the majority of cases, and limitation of new object instantiations to the initialization phase was found to be feasible for the examples as well.

		Unrestricted program	Restricted version	Restricted/ unrestricted
Sun jdk 1.0.2	Initialization time (s)	1.4	3.5	2.50
	Reaction time (s)	11.8	8	0.68
Café JIT	Initialization (s)	0.5	2.1	4.20
	Reaction time (s)	3.6	1.2	0.33
	Program size (k)	19.7	21.2	1.08

Table 1. Comparison of unrestricted vs. restricted versions of JPEG design example.

Table 1 shows results obtained from transforming our largest design example, a JPEG compression/decompression program. The JPEG algorithm was encapsulated using the `ASR` base class, and images were input as arrays of integers representing its red, green, and blue components. The *initialization time* reflects the amount of time spent in the object's constructor, while the *reaction time* is the average time required to process the 130x135 pixel test image. If multiple images are to be processed, the initialization time is incurred only once, while the reaction time is incurred for each image processed. For typical embedded systems, the reaction time is of primary concern, as initialization occurs relatively infrequently. These results were measured for the programs running on both a conventional Java Virtual Machine, and a Just-In-Time compiler. The program size given is the size of the Java `.class` files. The machine used is a 150MHz Pentium processor with 32 MB RAM and Windows 95.

Under both execution environments, the restricted version takes longer to initialize, but once initialization is complete, processes images faster than the unrestricted version. The performance differential reflects the original program's extensive use of dynamic data structures. The ASR policy of use disallows dynamic memory allocation, thus the restricted version we created uses only static data structures created during initialization. Hence, the restricted program is much slower to initialize, but performs no additional memory allocation once activated. While a significant performance improvement is obtained, the restricted program is roughly equal in size to the original.

These experiments show that a program can be modified to make it consistent with the ASR policy of use. Furthermore, execution of the modified example in standard Java environments resulted in increased system initialization times but shorter response times. Such a characteristic is consistent with the needs of reactive embedded systems, which maintain an ongoing dialogue with their environments. Hence, higher startup overhead is tolerable in exchange for faster response times. This suggests that the SFR methodology might also be used as a directed optimization strategy for programs running on standard Java Virtual Machines.

6 Summary

We have presented successive formal refinement, a new methodology for design and specification of embedded systems This methodology allows systems to be designed initially using a general-purpose programming language and refined to a form consistent with a formal model of computation, within a single-language platform.

Refinement of a design with respect to the ASR model yields a program that is deterministic, uses bounded memory, executes in bounded time, and is compositional with other systems in the model. A program with these properties is suitable for use as a specification for embedded, reactive systems.

A system, called *JavaTime*, was created to aid in the program analyses and transformations required for SFR. The methodology was applied to a number of design examples, and found to be applicable and successful in improving performance on standard Java execution platforms.

Future development will be carried out along several lines. Development and implementation of more sophisticated Java program analysis techniques will enable a wider variety of transformations. Work on advanced user interface and system visualization tools will also better the quality of interaction a user has with the system in refining a design.

To support a variety of application domains, policies of use will be developed for additional models of computation, and the interaction of components specified with respect to different models explored. To provide a complete system design solution, the *JavaTime* system should be coupled with verification and synthesis tools.

Acknowledgments

This research was supported by DARPA, under contract DABT63-95-C-0074-NEWTON-06/96, NASA, under contract NCC 2-999, Synopsys, and Intel. Their support is gratefully acknowledged.

References

[1] F. Bauer, B. Moller, H. Partsch, and P. Pepper, "Formal Program Construction by Transformations– Computer-Aided, Intuition-Guided Programming," *IEEE Trans. Software Engineering*, vol. 15, no. 2, pp. 165-180, Feb. 1989.

[2] A. Benveniste and G. Berry, "The Synchronous Approach to Reactive and Real-Time Systems," *Proc. IEEE*, 79(9):1270-1282, September 1991.

[3] G. Berry and G. Gonthier, "The ESTEREL synchronous programming language: design, semantics, implementation," *Science of Computer Programming*, vol. 19, pp. 87-152, 1992.

[4] Stephen A. Edwards, *The Specification and Execution of Heterogeneous Synchronous Reactive Systems*. Ph.D. Thesis, University of California, Berkeley, 1997. Available as UCB/ERL M97/31. `http://ptolemy.eecs.berkeley.edu/papers/97/sedwardsThesis/`

[5] D. D. Gajski, F. Vahid, S. Narayan, and J. Gong, *Specification and Design of Embedded Systems*. Englewood Cliffs, NJ: Prentice Hall, 1994.

[6] J. Gosling, B. Joy, and G. Steele, *The Java Language Specification*. Reading, MA: Addison-Wesley, 1996.

[7] CN. Halbwachs, *et. al.* The synchronous data flow programming language LUSTRE. *Proc. of the IEEE*, 79(9):1305-1320, 1991.

[8] D. Harel, "Statecharts: a visual approach to complex systems," *Science of Computer Programming*, 8:231-274, 1987.

[9] D. Harel and A. Pnueli, "On the Development of Reactive Systems," vol. 13 of *NATO ASI Series. Series F, Computer and Systems Sciences*, pp. 477-498. Springer-Verlag, 1985.

[10] R. Helaihel and K. Olukotun, "Java as a Specification Language for Hardware-Software Systems," *Proc. ICCAD '97*, pp. 690-697, 1997.

[11] S. Liao, S. Tjiang, and R. Gupta, "An Efficient Implementation of Reactivity for Modeling Hardware in the Scenic Design Environment," *Proc. 34th DAC*, pp. 70-75, 1997.

[12] J. Rowson and A. Sangiovanni-Vincentelli, "Interface-Based Design," *Proc. 34th DAC*, pp. 178-183, 1997.

[13] D. Smith. "KIDS: A Semiautomatic Program Development System," *IEEE Trans. Software Engineering*, vol. 16, no. 9, pp. 1024-1043, Sept. 1990.

4.2

Efficient System Exploration and Synthesis of Applications with Dynamic Data Storage and Intensive Data Transfer

Julio Leao da Silva Jr., Chantal Ykman-Couvreur, Miguel Miranda, Kris Croes, Sven Wuytack, Gjalt de Jong*, Francky Catthoor, Diederik Verkest, Paul Six, Hugo De Man

IMEC, Kapeldreef 75, B-3001 Leuven, Belgium
* Alcatel, Francis Wellesplein 1, B-2018 Antwerp, Belgium

Abstract

Matisse is a design flow intended for developing embedded systems characterized by tight interaction between control and data-flow behavior, intensive data storage and transfer, dynamic creation of data, and stringent real-time requirements. Matisse bridges the gap from a system specification, using a concurrent object-oriented language, to an optimized embedded single-chip HW/SW implementation. Matisse supports stepwise system-level exploration and refinement, memory architecture exploration, and gradual incorporation of timing constraints before going to traditional tools for HW synthesis, SW compilation, and HW/SW interprocessor communication synthesis. Application of Matisse on telecom protocol processing systems shows significant improvements in area usage and power consumption.

1 Introduction

Due to tight interaction between control and data flow parts, protocol processing applications are best specified, analyzed, and synthesized simultaneously. Also due to area, power, performance and flexibility reasons, protocol processing applications are usually (partly) realized as embedded systems incorporating both hardware and embedded software on a single chip.

Protocol processing applications require manipulation of large amounts of dynamically created and destroyed data. Hence, in the design of protocol processing applications, a large part of the area is due to memory units, and the power is dominated by data storage and transfer [16]. Therefore the design decisions related to the data types are crucial in the entire design trajectory. In Matisse, a system is specified in an object-oriented approach that supports exploration of abstract data types into data structures, abstracting from detailed memory implementations. This allows to heavily reduce design time. Despite the higher level of abstraction of our input with respect to e.g., high-level synthesis (HLS), we still achieve more efficient implementations due to the incorporation of a dynamic memory management refinement and physical memory management step (Section 2) on top of HLS, as demonstrated by our experiments (Section 6).

The dynamic data types are opposed to the (multi-dimensional) array signal streams present in DSP applications that can be largely analyzed at compile time. So the memory-oriented synthesis techniques proposed for DSP applications are not directly suited. Also DSP models tend to be data-flow oriented and less-suited for mixed control/data-flow behaviors found in protocol processing applications.

On the other hand, approaches intended for reactive control systems lack support for complex types and their data (storage and transfers). Work on system synthesis of protocol applications has been dealing with very particular aspects, e.g., solving timing constraints [9] and memory allocation for minimizing memory interface traffic during HW/SW partitioning [8]. No global approach exists.

In the SW community, distributed programming languages are used for programming general-purpose multiprocessor systems or distributed networks of workstations [1, 10, 14]. They rely on elaborate run-time environments and are intended for SW implementation on large computing environments. While their underlying models are strongly related to our model, the implementation target for protocol processing applications, i.e., an optimized embedded single-chip HW/SW realization, requires a different approach.

Since none of the existing approaches fulfills the previously described requirements, we have developed a system model [2] and proposed a global design flow [3], specific to data storage and transfer intensive applications with dynamically created data types. In this paper, we describe the prototype tool support that allows to automate the most critical parts of our design flow. First, an overview of the system model and design flow is presented. Next, each step in the design flow is described. The outputs of such a design flow can be used as inputs to traditional HW synthesis tools, SW compilers and interface synthesis tools. Finally, our design flow is applied to industrial applications from the ATM network components domain, in cooperation with Alcatel.

2 Matisse Methodology

We first introduce the Matisse language and model, and then discuss the Matisse design flow.

2.1 Language and model

Protocol processing applications are conceptually seen as sets of concurrent tasks that access data (defined as sets of records potentially with access keys). These applications with their environment are specified using the Matisse language [2]. It is a *concurrent object-oriented* specification language, extended from the widely used programming language C++. We summarize here the main features of the underlying Matisse model.

DAC 98, San Francisco, California

Tasks and concurrency - It is possible to specify tasks, called *active objects*, and data, called *passive objects*. Tasks have their own local virtual memory space and default thread of control. They are created using *activenew*. Concurrency is specified at the task level, by the concurrent execution of the default thread of control of all created tasks. Data may be created and destroyed in the local virtual memory space of the tasks, either at compile time or at run time.
Communication - Within one task, communication is specified using C++ pointers. Between tasks, communication is specified using *global pointers*. Except for their potentially higher cost of implementation, global pointers are used just like C++ pointers.
Synchronization - To avoid simultaneous accesses to data *atomic* functions should be used. Whenever several threads are calling an atomic function, it is executed the required number of times in a sequential order. Also the execution of an atomic function never interleaves with the execution of another atomic function within one task.

2.2 Design flow

The Matisse design flow is depicted in Figure 1. Below we present the goals of each step in the design flow. Two new major contributions to this design flow and prototype tool support are detailed in Sections 3 and 4.

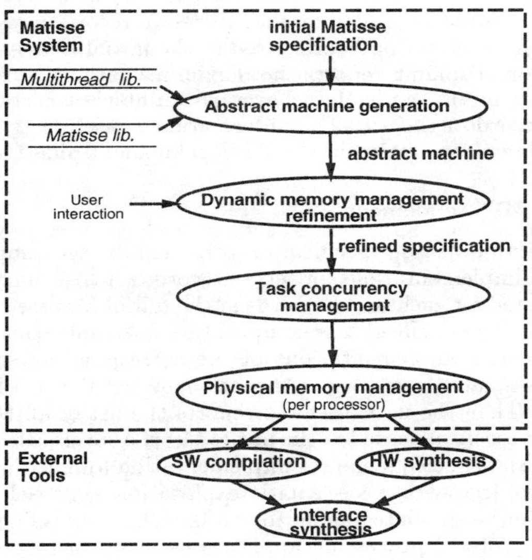

Figure 1: Matisse design Flow

Abstract machine generation (Section 3) - The main goal is to obtain a library-based executable specification, called Abstract Machine (AM), suitable for simulation and usable as input for exploration, and refinement of the system specification. The AM also allows profiling for further use in the design flow. The AM consists of a set of communicating concurrent tasks, an ultra-light OS to manage the execution of these tasks and a user interface to allow interactive refinement of the specification.
Dynamic memory management (DMM) refinement (Section 4) - In Matisse, the designer define sets of records as Abstract Data Types (ADTs), without low-level specification details. Protocol processing applications in the OSI layers 2, 3 and 4 are usually characterized by algorithms that operate on large Data Structures (DSs), which are dynamically allocated. When implementing these applications on a chip, efficient organization and implementation of sets of records is crucial and dynamic memory allocation must be handled efficiently in terms of time and number of memory accesses (power). Hence both must be refined before synthesis.
Task concurrency management - The goal of task concurrency management is to meet the overall real-time requirements imposed to the application being designed, by making decisions at the task abstraction level without incorporating low-level operation details already. Currently performed manually, the automation of this step is under investigation.
Physical memory management (PMM) for each HW and SW processor (Section 5) - Protocol processing applications require large storage capacities and very high I/O bandwidth to achieve the real-time requirements. Distributed memory architectures allow to exploit parallelism, thus alleviating memory access bottlenecks. This step aims to synthesize area and power efficient distributed memory architectures and memory management units, meeting the real-time requirements. We have developed techniques mainly suited for HW processors. Further investigation is ongoing to tune these techniques to SW processors.
Finally, **software compilation** proceeds using traditional software compilers, **hardware synthesis** proceeds using High-Level Synthesis (HLS) tools and **interface synthesis** generates SW device drivers for each SW processor and VHDL specifications of the necessary HW blocks allowing communication between HW and SW processors. The interface synthesis is performed using the system integration tool-box *CoWare* [6, 7] that also allows cosimulation of the synthesized system.

3 Abstract Machine Generation

The AM executable model is generated according to the flow depicted in Figure 2. The input to the abstract machine generation is the initial Matisse specification of the protocol processing application together with its environment. Any set of records is defined as a Matisse ADT, available in the Matisse library. Any dynamically allocated data structure is defined as a passive class, whose accesses to data elements are provided by `read` and `write` access member functions. The abstract machine generation produces two main outputs: an Abstract Parse Tree (APT) and C++ executable code. The APT is used as input to the task concurrency management step. The C++ executable code is used as input to the DMM refinement step. The abstract machine generation consists of three phases: an input phase, an expansion phase, and an output phase described in the following subsections.

3.1 Input phase

The **parser** reads the initial Matisse specification and builds an APT.
The **concurrency analyzer** traverses the APT and collects information about concurrency, communication, and synchronization. This information is obtained by identifying the nodes derived from the Matisse keywords `active`, `activenew`, `global`, and `atomic`. For the moment, only concurrency at the task level is identified.
The **data structure analyzer** traverses the APT and collects information about ADTs and other dynamically allo-

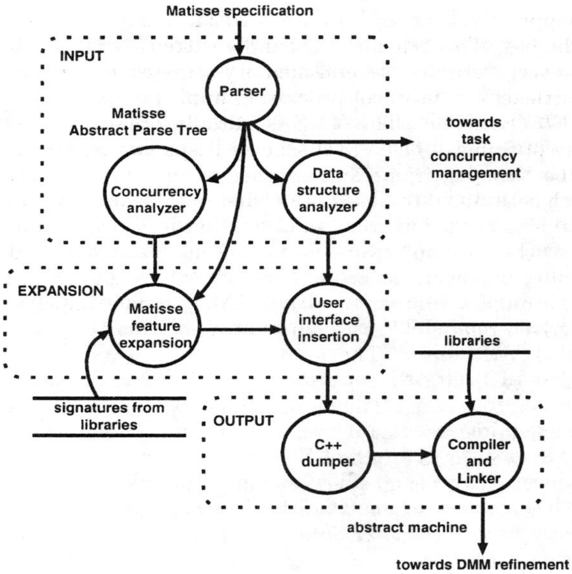

Figure 2: Abstract Machine Generation

cated DSs. This information is obtained by identifying the nodes derived from the *new* keyword and by determining in which active object they are created.

3.2 Expansion phase

The **Matisse feature expansion** expands the Matisse keywords into plain C++. Its inputs are the APT generated by the parser, the information collected by the concurrency analyzer, and signatures from the non-preemptive multi-thread library and from the Matisse library. The multi-thread library provides functions to manipulate (create, lock, and unlock) semaphores and to manipulate (create, start, stop, and yield) tasks. The Matisse library provides an abstract class from which all active classes will be derived. The output is a C++ APT, generated using the above mentioned libraries and expanding the following Matisse keywords: *active*, *atomic*, *activenew*, and *global*. The insertion of yields and semaphores to avoid deadlocks is automatically done.
The **User interface insertion** adds function calls to the C++ APT to allow designer interactivity in the DMM refinement step.

3.3 Output phase

In this phase, the C++ APT derived from the expansion phase is traversed. The corresponding C++ source code is dumped, compiled and linked with the multi-thread and Matisse libraries. The executable code is used as input to the DMM refinement, described in the next section.

4 Dynamic Memory Management Refinement

The system to be designed is initially specified using the Matisse language, abstracting low-level specification details. Any set of records is declared as a Matisse ADT. All other dynamically allocated DSs are declared as passive classes. They are dynamically allocated by the C++ operation `new`, and deallocated by the C++ operation `delete`. The DMM refinement step is completed interactively by the designer with a systematic methodology in three phases: initialization, ADT refinement, and virtual memory management (VMM) refinement, as depicted in Figure 3.

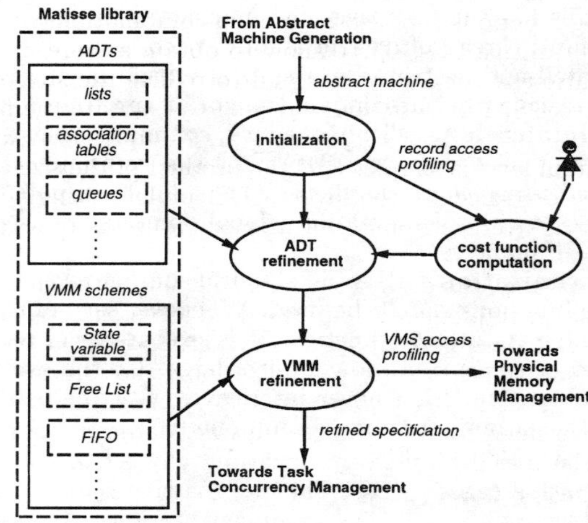

Figure 3: DMM refinement

4.1 Initialization

The AM is simulated to validate the functionality of the initial Matisse specification and to derive record access profiling for each set of records used in the initial Matisse specification. Profiling reports the maximum number of records present in the set and the average number of accesses to any record of the set. This information is used for the cost function computation in the ADT refinement phase.

4.2 ADT refinement

The initial Matisse specification is refined, by selecting optimized implementations for sets of records. A cost function is evaluated for each set of records in the initial Matisse specification. The result of this computation is a table containing all possible implementations of the corresponding set and the associated cost (e.g. estimation of the power dissipation). Then the user interface reports the list of all Matisse ADT objects allocated in the initial Matisse specification and the designer selects one set implementation for each Matisse ADT object with a systematic exploration approach. The AM can be simulated again to validate this first refinement.

In protocol processing applications, a set of records accessed by one or more keys can be best organized as a hash table or an association table. An association table can be implemented in different ways, based on different primitive data structures. These implementations have different characteristics in terms of memory occupation, number of memory accesses to locate a record in the set, and corresponding number of addresses to be generated, and power dissipation. A set of records accessed without using a key can be best organized as a general container, a list or a queue.

We represent possible implementations of sets accessed without using a key in a generic model, implementing lists or queues. Sets accessed by one or more keys are represented in another generic model, implementing association tables possibly with hashing functions. These generic models are based on primitive DSs, such as linked lists, arrays, binary trees, and pointer arrays (which are arrays of pointers to dynamically allocated records). They can be combined to create more complex layered DSs, in several ways: keys can be split into subkeys, several keys can be combined into one super key, the order in which the keys are used to access the different primitive DSs can vary, the number of layers also is not fixed and several keys can be associated with any layer. Hashing functions can also be used in combination with any of the primitive DSs. They allow to obtain a more uniform distribution of the key values and to reduce the size of the primitive DS. The cost function used to estimate the power dissipation of a given association table implementation at the system level is simple. It is the product of the size of the memory occupied by the layered DS implementing the set and of the average number of memory accesses to a record in the set.

In this way, the designer can make a motivated choice for the best set implementation, based on a cost function, during the DMM refinement step. However, finding the best set implementation for a given application is not trivial if the solution space is large. Therefore an optimization methodology extended from [18] is used in the Matisse design flow.

In the initial Matisse specification, any set of records is defined as a Matisse ADT. They are passive classes defined in the Matisse library of ADT implementations. Access synchronization inside the set is ensured through atomic functions. Therefore, an active object that accesses a set of records does not have to worry about synchronization whenever records are inserted in the set or removed from the set.

4.3 VMM refinement

In [4], the memory management task in hardware design of telecommunication network systems is addressed. It bridges the gap between the algorithmic system specification, written in a programming language like C++, and commercially available behavioral synthesis tools. In the Matisse design flow, the virtual memory refinement phase is extended from [4], to take into account concurrent allocation into the virtual memories, as explained below.

In the virtual memory refinement phase, DSs, dynamically allocated in the system specification, are mapped into Virtual Memory Segments (VMSs). Each VMS is a collection of frames. Each frame is a memory chunk large enough to store an object of the specified DS. Since a given DS may be used multiple times in different circumstances in the system to be designed, several VMSs of different sizes and with different VMM schemes may be used for each DS.

For each VMS a VMM scheme is needed. This VMM scheme is responsible for allocating frames to store newly created objects, deallocating frames whenever the stored objects are destroyed, keeping track of the used and free frames in the VMS, and detecting memory overflow allocation failures. VMM schemes are defined as passive classes in the Matisse library. A VMM scheme is responsible for the protection of its VMS. Synchronization of concurrent allocations inside the VMS is ensured through the use of atomic functions.

Currently three VMM schemes are available in the Matisse library:

1. State variable scheme. For each frame of the VMS, a state bit keeps track of its state (either free or used). This scheme is very fast: allocating a frame does not require separate memory accesses and can usually be implemented in a single clock cycle. However, the hardware behavior can become complex for large VMSs.

2. Free-list scheme. The VMS is initially organized into a linked list of free frames. Whenever a frame is needed to store an object in the VMS, the first free frame of the list is popped. Whenever a frame is freed, it is pushed on top of the list. This scheme can handle extremely large VMSs. However, it requires several memory accesses, which may be a bottleneck in protocol processing applications.

3. FIFO scheme. The VMS is initially organized into an array of frames. The VMM scheme keeps two pointers, one to the "last in" frame and another to the "first in" frame, which is assumed to be also the "first out" frame. Whenever a frame is needed to store an object in the VMS, the one on top of the "last in" frame is used. This scheme is used for memory management inside FIFO queues.

Through a simulation of the AM, profiling information is derived related to the virtual memory behaviour and the needed VMS sizes. The user interface reports the list of all DSs and the designer selects the VMS size and a VMM scheme for each DS. The operators `new` and `delete` become redefined operators in the refined Matisse specification. Finally, VMS access profiling is derived by simulating the AM, taken into account all specification refinements.

Other VMM schemes are being investigated and a systematic exploration has been proposed.

5 Physical Memory Management

Usually, for data-intensive algorithms the cycle budget available is insufficient to perform all the memory accesses sequentially. Hence a number of accesses have to be done in parallel. The signals accessed in parallel have to be assigned to different memories or they have to be accessed through different ports of a multiport memory. Many different orders of the memory accesses are possible for the given cycle budget. Manually exploring all different ordering possibilities and memory configurations for area and power efficiency is a very tedious task. Therefore, a partially automated methodology based on a three-step approach has been developed. First, exploration of the different VMS layout possibilities is performed since it heavily influences the optimal memory architecture as it affects the number of memory accesses and the word width. This step is still under investigation. Afterwards a partial order of the memory transfers that leads to final schedules with minimal number of access conflicts is generated [17]. Finally the most efficient memory configuration for area and power, including the data to memory assignment decisions, is determined [13].

Address manipulation is a crucial component of any architecture which deals with data transfer intensive algorithms. The efficient access to memories within real-time constraints requires an optimized mapping of address expressions in the algorithm onto address arithmetic optimized for both area and power within a given throughput. High-level address optimization is a very error-prone and tedious task if done manually. Therefore, a methodology [11] has been developed to reduce the cost overhead for address generation for both custom and instruction-set processors. This methodology includes address expression splitting/clustering, induction variable analysis, target architecture selection, and global scope algebraic optimizations. The most time consuming steps are supported with prototype tools. In addition, high-level controller synthesis and optimal partitioning of the arithmetic unit are under development for the synthesis of custom memory management units.

6 Results for Industrial ATM Applications

Two real-life protocol processing applications have been used as test drivers for the Matisse system. One is the Segment

Protocol Processor (SPP) [16] which is an ATM component implementing the AAL3/4 of a connection transport server. The other is the STORM application, which stands for **S**TM1 level to **T**ransputer netw**O**rk **R**elayer with **M**ultiple protocol handling. It deals with protocol conversion between an ATM transport layer and a transputer network.

The main functionality of the SPP can be viewed as a set of concurrent processes manipulating data which are allocated at run-time. The data contain information about packets flowing through the ATM network. Packets are streams of ATM cells, and different ATM cells are interleaved with each other. The data consist of a FIFO queue of ATM cells, an association table of IPI records, accessed by two keys, link (LID) and multiplexing (MID) identifiers, a list of active IPI records, a list of non-active IPI records, and lists of routing records. This is depicted in Figure 4.

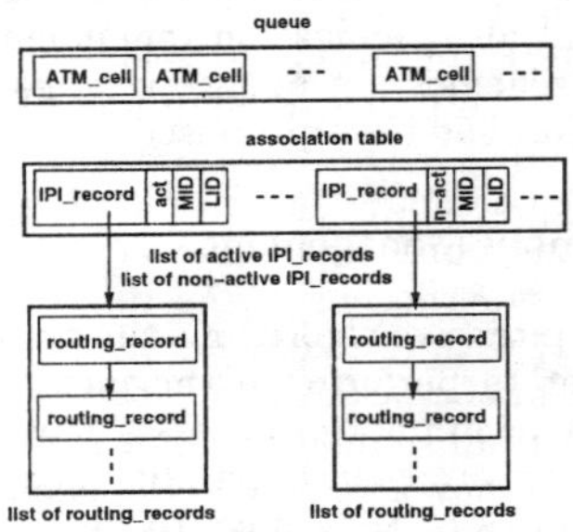

(a) Initial specification

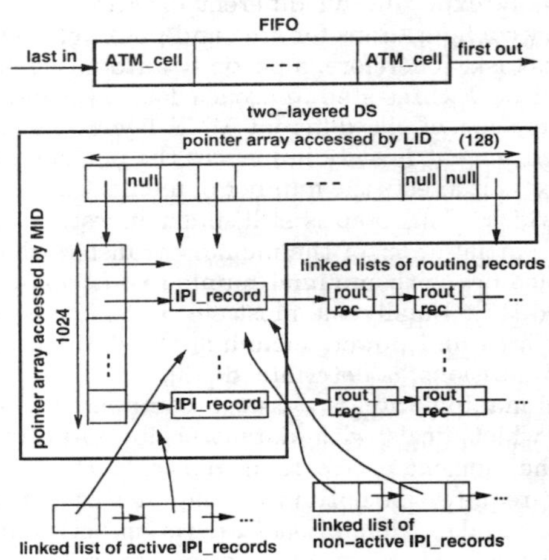

(b) Refined specification

Figure 4: Organization of the SPP data

The first step is the ADT refinement for these sets of records. From simulation of the underlying AM, efficient implementations for these sets of records are selected. For a 17 bit address space and a memory width of 64 bits, and for $0 \leq LID \leq 127$ and $0 \leq MID \leq 1023$, record access profiling indicates that at most half of the LID values are used at the same time, and that IPI_records are accessed twice on average. Based on these profiling results, the cost function computation generates a table containing 32 possible implementations of the association table, using the method mentioned in Section 4.2. The best implementation is a two-layered data structure. The first layer is a pointer array, accessed by the LID key, and pointing to the pointer arrays of the second layer, accessed by the MID key. From the SPP behavior we derive that the queue of ATM cells is best implemented as a FIFO, the list of active IPI records is best implemented as a linked list where new elements are pushed to the head, whereas the other lists are best implemented as linked lists where new elements are pushed to the tail.

The second step is the VMM refinement. The following VMSs are needed: a VMS for storing the IPI records, a VMS for storing the pointer arrays of the association table, a VMS for storing the list elements, and a VMS for storing the ATM cells in the FIFO. From simulation of the AM, a size and a VMM scheme is derived for all these required VMSs. The free-list scheme is selected by the designer for all schemes except for the VMS of ATM cells, that follows the FIFO scheme.

In contrast to HLS tools, the Matisse design flow uses PMM as a step prior to scheduling and allocation to minimize the required memory bandwidth and to support the memory allocation/assignment decisions. Afterwards, the detailed scheduling can be done, constrained to the memory architecture obtained.

Since our task concurrency management is currently done manually and due to limitations of our prototype tool for generating the internal format for the PMM step, the test drivers have been simplified for the PMM step to contain only one conditional trace of the complete control-flow.

To show the advantage of the PMM step on top of HLS, we have constrained the SPP driver to perform all 38 accesses to the stored data in only 17 cycles. After the PMM step, the optimal memory architecture resulted in three single port memories [15]. Afterwards, the detailed scheduling was done using Synopsys' Behavioral Compiler (Synopsys-BC) [5]. The compiler was able to meet the cycle budget, given our memory architecture and data-to-memory assignment decisions as constraints. Doing the same experiment without first performing our PMM step and using the conventional design flow, six single port memories were required to meet the cycle budget. Hence, for this example, the required memory bandwidth using our preprocessing step is only half that of the conventional one. This obviously leads to a significant reduction of the area cost related to the data storage.

Conventional HLS also heavily benefits from preprocessing it with our high-level address optimization methodology [12]. To show the impact, a functional description of the MMU's arithmetic unit of the simplified test driver has been obtained and synthesized. Component count and area figures for the addressing and the rest of data processing (Data path column) are shown in Table 1.

Block	Addressing				Data-path	
	Synopsys-BC		Preprocessed		Synopsys-BC	
	units $\breve{o}e/oe$	Area $\breve{o}e/oe$	units	Area	units	Area
Operators	7/5	0.44/0.23	2	0.49	2	0.19
Mux.logic	—	0.27/0.37	—	0.28	—	0.13
Registers	14/13	0.82/0.75	0	0.00	1	0.12
Controller	1/1	0.49/0.41	1	0.06	1	0.11
Total	—	2.02/1.76	—	0.83	—	0.44

Table 1: Results for SPP: Synopsys-BC and preprocessed.

Using our preprocessing step on top of Synopsys-BC (Preprocessed column) we achieve a factor of 2.4 (0.83:2.02) area saving with respect to the architecture generated by making direct use of Synopsys-BC (column $\overline{oe}$). On the other hand, by performing operation expansion (oe) before HLS and then making direct use of Synopsys-BC (column oe), the resulting area figure improve only 15% (2.02:1.76). The reason is that conventional HLS allocates registers inside the data-path block to store intermediate data resulting from operation-level time-multiplexing optimizations. These registers are not present when the multiplexing is done at the cluster level since the associated expression is computed at once. The (potential) overhead in the arithmetic is reduced by identifying system-level similarity across clusters to exploit global scope common-subexpression elimination and to limit the overhead in the multiplexing logic.

The controller also benefits from the much lower overhead of registers (Synopsys-BC - 14/13, Preprocessed - 0) in the data-path and of the larger granularity of the resulting arithmetic components (Synopsys-BC - 7/5, Preprocessed - 2), thus requiring much less decoding.

Note that the data processing (Data-path column) is a factor of 4.5 (0.44:2.02) smaller in total area cost with respect to the overhead of the addressing when making direct use of the Synopsys-BC compiler.

To show the exploration freedom available for VMS layout possibilities, the test driver based on the STORM application was selected. Different VMS layouts result from different choices for field packing possibilities into a common VMS word. For the STORM driver, three different VMS layouts have been used for the exploration. Two of them represent the extreme cases of the layout possibilities: each field in a separated VMS word or, all fields packed into a single word. In between these two extreme cases, many other VMS layouts are possible, e.g, by grouping the fields of a data structure such that they fit as well as possible into 32 bit words. This case corresponds to the row labeled *word/32_bit* in Table 2.

	1 mem	2 mem	3 mem	4 mem
word/field	Not Possible	Not Possible	Not Possible	A: 1.00 P: 1.00
word/32_bit	Not Possible	A: 0.87 P: 0.96	A: 0.90 P: 0.60	A: 0.91 P: 0.50
word/all_fields	A: 0.84 P: 0.92	A: 0.81 P: 0.64	A: 0.83 P: 0.39	A: 0.84 P: 0.34

Table 2: Optimal memory architecture for different VMS and storage cycle budget of 6 cycles.

The entries in Table 2 contain relative area and power numbers for the total memory architecture resulting from the PMM step. The area numbers are relative to the maximum area number of the table, whereas the power numbers are relative to the maximum power number.

From this table we can conclude that, for this example, the area and power decrease when the fields are packed into fewer words. Also fewer memories need to be allocated in case the fields are packed into fewer words. Note also that the power decreases when more memories are allocated, while the area exhibits a minimum between the two extremes.

7 Conclusions

In this paper, we have addressed the support for system design exploration for protocol processing applications used in telecom networks. Using the Matisse language the designer is able to write a system specification, which abstracts low-level details (enabling design time reduction), and is easily retargetable to different embedded hardware/software realizations (enabling system level exploration and reuse). The Matisse design flow assists the designer to: functionally validate the specification, explore the design space at system level for different ADT implementations and VMM schemes, and explore different memory architectures for mixed hardware/software realizations. We have demonstrated (Section 6) that despite the higher level of abstraction of our input with respect to e.g., high-level synthesis (HLS), we achieve more efficient implementations. Currently, we are investigating a systematic methodology for task concurrency management and how to specify timing constraints in the Matisse language itself.

Acknowledgments We would like to thank Bill Lin (UC San Diego) and Mark Genoe (Alcatel) for many insightful discussions. This research has been partly funded by the Flemish IWT and Alcatel in the HASTEC project, by the Esprit project No.21929 (MEDIA), and by a Brazilian Government Fellowship (CAPES).

References

[1] K. M. Chandy and C. Kesselman. *CC++: A declarative concurrent object-oriented programming notation.* MIT Press, 1993.

[2] J. Leao da Silva Jr., Ch. Ykman-Couvreur, and G. de Jong. Matisse: A concurrent and object-oriented system specification language. In *Int. Conf. on VLSI*, 1997.

[3] J. Leao da Silva Jr., Ch. Ykman-Couvreur, G. de Jong, B. Lin, and H. De Man. A system design methodology for telecommunication network applications. In *Seventh Great Lakes Symposium on VLSI*, 1997.

[4] G. de Jong, B. Lin, C. Verdonck, S. Wuytack, and F. Catthoor. Background memory management for dynamic data structure intensive processing systems. In *Proc. of the Intnl. Conf. on Computer-Aided Design*, pages 515–520, November 1995.

[5] D.Knapp, T.Ly, D.MacMillen, and R.Miller. Behavioral synthesis methodology for hdl-based specification and validation. In *Proc. of the Design Automation Conference*, June 1995.

[6] I. Bolsens et al. Hardware-software codesign of digital telecommunication systems. *IEEE Proceedings*, April 1997.

[7] CoWare Inc. http://www.coware.com/.

[8] A. Jantsch et al. Hardware/software partitioning and minimizing memory interface traffic. In *Proc. of the EuroDAC*, 1994.

[9] Kuchcinski et al. Embedded system synthesis by timing constraint solving. In *Proc. of the Intnl. Symposium on System Synthesis*. Antwerp, Belgium, 1997.

[10] B. Meyer. Systematic concurrent object-oriented programming. *Communications of the ACM*, September 1993.

[11] M. Miranda, F. Catthoor, M. Janssen, and H. De Man. Adopt: Efficient hardware address generation in distributed memory architectures. In *IEEE/ACM Proc. of the Intnl. Symposium on System Level Synthesis*, 1996.

[12] M.Miranda, F.Catthoor, M.Janssen, and H.De Man. High-level address optimisation and synthesis techniques for data-transfer intensive applications. *IEEE Transactions on VLSI Systems*, submitted 1997.

[13] P.Slock, et al. Fast and extensive system-level memory exploration for atm applications. In *Proc. of the Intnl. Sympopsium on System Synthesis*, September 1997.

[14] B. Selic, G. Gullekson, and P. T. Ward. *Real-Time Object-Oriented Modeling.* Wiley, 1994.

[15] S.Wuytack, F.Catthoor, G.De Jong, and H.De Man. Minimizing the required memory bandwidth in vlsi system realizations. *IEEE Transactions on VLSI Systems*, submitted 1997.

[16] Y. Therasse, et al. VLSI architecture of a SDMS/ATM router. *Annales des Telecommunications*, 48(3-4), 1993.

[17] S. Wuytack, F. Catthoor, G. de Jong, B. Lin, and H. De Man. Flow graph balancing for minimizing the required memory bandwidth. In *Proc. of the Intnl. Sympopsium on System Synthesis*, pages 127–132, November 1996.

[18] S. Wuytack, F. Catthoor, and H. De Man. Transforming set data types to power optimal data structures. *IEEE Transactions on Computer-Aided Design*, 15(6):619–628, June 1996.

Design Space Exploration Algorithm For Heterogeneous Multi-processor Embedded System Design

Ireneusz Karkowski and **Henk Corporaal**
Delft University of Technology
Information Technology and Systems
Mekelweg 4, 2628 CD Delft, The Netherlands

Abstract

Single-chip multi-processor embedded system becomes nowadays a feasible and very interesting option. What is needed however is an environment that supports the designer in transforming an algorithmic specification into a suitable parallel implementation. In this paper we present and demonstrate an important component of such an environment - an efficient design space exploration algorithm. The algorithm can be used to semi-automatically find the best parallelization of a given embedded application. It employs functional pipelining [13] and data set partitioning [16] simultaneously with source-to-source program transformations to obtain the most advantageous hierarchical parallelizations.

1 Introduction

As the application area of the embedded processors widens, the demands on their performance are constantly growing. Until now, instruction level parallelism has been successfully exploited to satisfy these high performance requirements. Practice shows however that increasing the number of concurrently operating functional units of typical ILP (instruction level parallel) architectures above a certain level does not necessarily lead to significant performance gains [9]. Instead, high hardware costs and inefficient use of this hardware occurs. The advent of sub-micron processing, allowing integration of millions of transistors on a single carrier, has brought new opportunities in the embedded system design. A multi-processor embedded system becomes nowadays a very interesting alternative. This both in terms of the hardware cost and performance. Especially, if the system consists of several (different) ASIPs (application specific instruction set processor), each with functionality optimized for the subtasks which they have to perform. Code partitioning among the processors leads then to exploitation of the course-grain parallelism (task parallelism and parallelism in loops [4]), while the fine-grain (instruction level) parallelism [9] is exploited locally by each of the processors.

In the past several environments for the embedded system design have been realized (a lot of references can be found in [7, 8, 12]). Also a number of papers specifically about multi-processor system design have been published [3, 2, 15]. None of them however addresses the problem of the automatic extraction of the parallelism from the system specification.

In this paper we propose a new approach to mapping of an embedded application written in ANSI C onto a cost-efficient heterogeneous multi-processor. Its uniqueness lies in a combination of the state of the art automatic ASIP synthesis software with a coarse- and fine-grain parallelism exploitation methodology.

The paper is organized as follows. Section 2 states the problem. Section 3 is devoted to the introduction of different parallelization methods. The system design space exploitation algorithm is presented in section 4. The performance of the algorithm is demonstrated on the frequency tracking system in section 5. Section 6 concludes the paper.

2 Problem statement

Multi-processor system design involves finding a mapping Γ of a program graph $G_P(V, E)$ onto an architecture template graph $G_A(P, C)$, such that the resulting hardware-software solution satisfies the price-performance specification for the design. In these graphs, V is the set of program statements, E the set of data and control dependencies between the statements, P the set of processing elements and C the interconnection network between them. An example multi-processor architecture can include a set of ASIPs, possibly with local memories, communicating via a combination of fast interprocessor links and/or shared memory. In our case the ASIPs are designed using the MOVE framework [5]. The architectural variables defining the design space are shown in table 1.

Our goal is to find a design trade-offs curve Π, connecting a set of Pareto points in the 2-dimensional cost-performance space [11] for a given application (see example in figure 10; as the cost usually the chip area, power dissipation or number of packaging pins is used). Having it, we can select a cost-efficient and feasible (satisfying the constraints) solution.

Symbol	Description
N	Number of processors
Σ	Set of parallelization methods used
Θ	Set of parallelized program parts (i.e. loops)
Ξ	Mapping of partitions to processors
Φ	Mapping of data transfers to inter-processor communication hardware
Ψ	Hardware configuration of ASIPs

Table 1: Variables in the design space of the heterogeneous multi-processors.

3 Parallelization methods

Basically, we can distinguish two *parallelization modes*. They can be defined as follows:

Definition 1 *In* **operation-parallel** *mode different operations of a single threaded program are executed in parallel. The* **data-parallel** *mode involves applying one or more operations to many data items in parallel.*

Figure 1 demonstrates these modes when applied to a single FOR loop with 9 iterations containing three operations A, B, C. Notation A_i denotes an instance of the operation A executed in iteration i. Parallelization can be obtained by either partitioning the execution of this loop in horizontal or vertical direction. In the first case iterations $i = 1..3$ would be executed on processor 1, iterations $i = 4..6$ on the second one, etc. (*data-parallel mode*). Alternatively, the vertical partitioning may result in the *operation-parallel mode* in which operations A, B, C may be executed in parallel (A on the first processor, B on the second one, etc.). Note that not

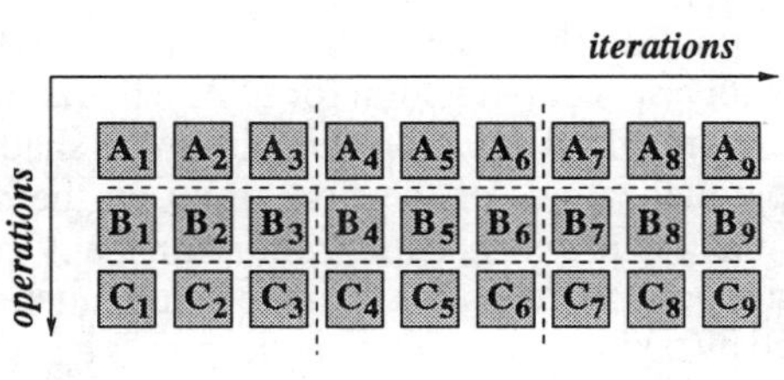

Figure 1: Operation-parallel and data-parallel parallelization modes - vertical and horizontal partitioning.

only the loops of a program suit well for parallelization. Also straight-line code sections can often be parallelized (coarse- and fine-grain).

Often direct parallelization is not advantageous. A series of code transformations [16] are necessary to enable efficient parallelization. For example, to allow a better *index set partitioning* of a loop nest, a combination of loop interchange and tilling can be used. If we take these code transformations into account then parallelization methods can be defined as follows:

Definition 2 *A* **parallelization method** *is an element of the following set:*

$$\{P\} = \{T\}^* \otimes \{M\},$$

where T is the set of the code transformations and M the set of the parallelization modes.

4 Design Space Exploration

Our design space exploration algorithm takes as input the system specification in ANSI C, accompanied by several parameters, as for example the maximal number of processors available and a set of parallelization methods to be attempted (as defined section 3). Pushing the design in a certain direction is possible as well, for example by specifying that some program fragments have to be parallelized in a certain way. Such extra directives prune the search space. In addition, the algorithm requires providing profiling information. It can be obtained by code profiling on an oversized ASIP architecture and with software compilation options for maximal ILP exploitation (the performance metrics obtained in this way represent optimistic bounds).

To explain the algorithm, we define the following:

Definition 3 *A* context graph *$C(V, E)$ of a program is a directed graph with tree kinds of vertices (procedure, loop and block). Block vertices represent code segments between loops and procedure calls. There is an edge between two vertices if they are nested within each other, or if one calls another. The* root *vertex of $C(V, E)$ is the procedure $main()$. With every vertex a coefficient $l(v)$ is associated. It represents the total number of clock cycles (latency) required to execute its code (including successor vertices). Also every edge carries one coefficient - $f(e)$, which takes a value different than 1 in case it targets a procedure vertex for a procedure called from more than one place in the program. In such case this value represents a fraction of the vertex latency which is due to that calling site.*

Definition 4 *A* speedup function *$SF_v(N)$ of vertex v gives the local speedup at this vertex as a function of the number of processors used at v (example speedup functions can be seen in figure 8).*

4.1 System design space exploration algorithm

The algorithm for system design exploration is shown in figure 2. It has been optimized to avoid repeatedly performing the most time consuming tasks in every step. Therefore the general code transformations[1] and data dependency analysis are done at the very beginning. For the same reasons the final parallelization and code generation is performed at the end only.

The algorithm performs the following four major steps:

1. After some initialization code, calculate the speedup function for the *main* procedure: for each number of available processors N, find a set of applied parallelization method(s) Σ and a set of parallelized program fragments Θ which optimize the overall speedup. As side effect the speedup functions for all other vertices in $C(V,E)$ are calculated.

2. Code partitions for each point in the speedup function, which are the result of the parallelization, are mapped onto available processors, data transfers onto inter-processor communication links (Ξ, Φ). The code partitions are mapped onto the processors using the assumption that the serial parts of the program are always executed on processor 1. At barriers (moments when parallel execution is started or terminated) the code partitions are assigned to available processors.

3. For each point in the speedup function $SF_{main}(N)$, a design trade-offs curve Π_N between the cost and the performance is calculated (by exploring the design spaces of all processors, see figure 9).

4. The Π_N curves are combined to form a single design trade-offs curve Π (see figure 10).

```
Apply a set of general code transformations;
Do data dependency analysis;
Build the context graph C(V,E);
for ( all v ∈ V and N = 1 ... N_max ) SF_v(N) = 1;
SF_main=SPEEDUP(main);
Map code partitions to processors, data transfers
  to communication links for points in SF_main;
Do design space exploration of the processors
     for points in SF_main;
Combine results into a design trade-offs curve Π;
```

Figure 2: System design space exploration algorithm

```
SpeedupFunction SPEEDUP(v){
   for ( all { w | w = succ(v) ∧ l(w)/l(main) ≥ α} )
      SF_w=SPEEDUP(w);
   PARALLELIZE(v,SF_v);
   Return SF_v;
}
```

Figure 3: The SPEEDUP function

The function called SPEEDUP is presented in figure 3. As can be seen, the speedup functions for all vertices in the program are calculated recursively. Only vertices with $p(v) = \frac{l(v)}{l(main)} \geq \alpha$ are considered. The value of the coefficient α should exclude vertices which form a marginal part of the total computation time. Accelerating vertices with low $p(v)$ is very unlikely to deliver any significant speedup even if many processors are used.

4.2 Parallelization

After the speedup functions of the successors are calculated (first loop inside the SPEEDUP procedure), the parallelizations at the vertex v are attempted. Three different situations are possible (see figure 4; notation $Part_v(R,N)$ denotes a solution obtained for N processors and parallelization p with R partitions at the vertex v, $R \leq N$):

1. *Vertex-only parallelization*: only the vertex v itself is parallelized (top-left); i.e. $R = N$.

2. *Successors-only parallelization*: only the successor vertices of v are parallelized (top-right); i.e. $R = 1$.

3. *Hierarchical parallelization*: the vertex v itself is parallelized, in combination with successor vertices (bottom-left and bottom-right); i.e. $1 < R < N$.

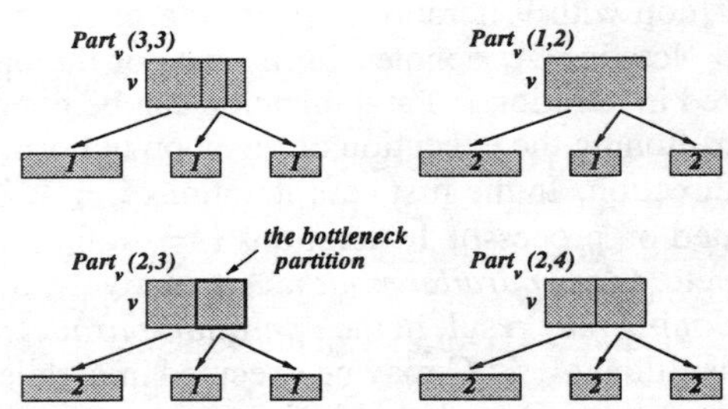

Figure 4: Parallelizations at v

Figure 5 presents the space of the possible parallelizations, together with a speedup function for v. The hierarchical parallelizations (the third situation) lie within the shaded triangle, the parallelizations of the first group on the diagonal, while the *successor-only parallelizations* on the N-axis. The dashed lines point to the points in the speedup function which are defined by taking maximum over the points lying on these lines in the parallelizations space.

At v we consider only hierarchical parallelizations in the *operation-parallel mode*[2]. This is not an essential limitation; it is expected that in most cases hierarchical *data-parallel mode* parallelizations do not deliver better results than the single level ones.

[1] A transformation is general if it can be applied to most loops, independent of their structure.

[2] A parallelization of a loop nest with n loops is not considered a hierarchical parallelization and is attempted.

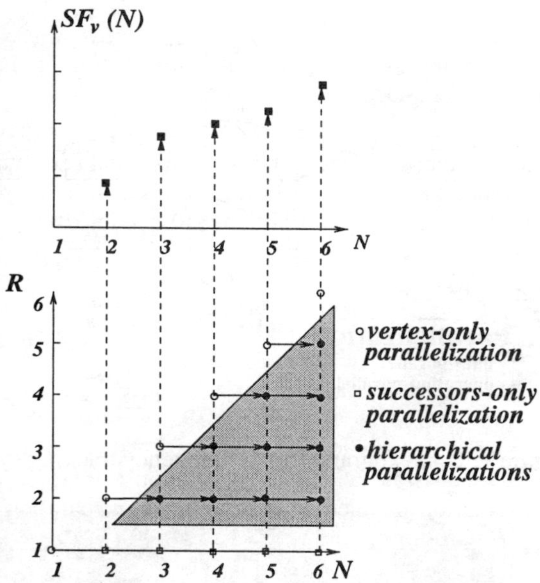

Figure 5: The space of possible parallelizations at v.

The number of processors used at vertex v, $N_{proc}(v)$ can be calculated using the following formulas:

$$N_{proc}(v) = \sum_{b \in B(v)} N_{proc}(b) \quad (1)$$

$$N_{proc}(b) = \max_{\{w | w = succ(v) \wedge w \subset b\}} N_{proc}(w) \quad (2)$$

where $B(v)$ is the set of partitions at the vertex v and $N_{proc}(b)$ the number of processors used inside successor nodes of the partition b.

To obtain exact results for the hierarchical parallelizations a substantial number of alternatives (combinations of the points in the speedup functions of the successor vertices) would have to be considered. It can be shown that the number of alternatives K can be calculated using the following formula:

$$K = \sum_{N=2}^{N_{max}} [2 + \sum_{R=2}^{N-1} (N - R + 1)^z] \quad (3)$$

where R is the number of partitions at v and z the number of successor vertices of v. For example if $z = 6$ and $N_{max} = 6$ then, using this formula, we obtain $K = 26270$ combinations. In practice this number will be slightly smaller since some alternatives cannot result in a legal parallelization, but still we would have to run our parallelization algorithms for the majority of them. This can very easily result in unacceptably long run times.

The following example shows a more practical method of obtaining a legal parallelization:

Example 1 *Suppose that we are at a point to find optimal parallelization of a vertex v onto 4 processors, with two partitions at v (i.e.* $Part_v(2, 4)$*). We can start with the previously obtained parallelization* $Part_v(2, 3)$ *(see the bottom-left part of figure 4). Since the parallelization method p is* operation-parallel *both partitions at v will be executed in parallel. One of these partitions will probably have larger latency (in the figure it appeared to be the right one). Only by accelerating this "bottleneck" partition further speedup can be obtained. Therefore we allocate one extra processor to one of the successor vertices of that partition (bottom-right part of figure 4).*

To avoid the high computational complexity we decided to use an approximation algorithm to calculate the best parallelizations at v (the PARALLELIZE procedure in figure 6). In its body, the *apply(v,p,R)* procedure applies a parallelization method p at v with R partitions at v and returns the obtained parallelization X. The obtained speedup depends on the parallelization method used and is calculated using the *sp(p,X)* procedure.

First, appropriate points from the speedup functions of the successor vertices are selected and the speedup S calculated (first FOR loop). Subsequently parallelizations $p \in P$ (recall def. 2) on $N = 2..N_{max}$ processors are tried. In the first inner loop, a parallelization of the vertex v itself is attempted. Then, if the parallelization method is *operation-parallel* the hierarchical ones are tried. We follow the methodology from the example 1. The optimal previous solution $Part_v(R, N-1)$ is used as the starting point (following arrows in figure 5). We identify the bottleneck partition at v and attempt using an extra processor on its successor vertices to further speedup v.

```
PARALLELIZE(v,SF_v) {
  /***** successors-only  parallelization *****/
  for (N = 2; N ≤ N_max; N = N + 1) {
    S = l(v)/ Σ_{w|w=succ(v)} l(w)/SF_w(N)
    SF_v(N)=max (SF_v(N), S);
  }
  for ( all p ∈ P ) {
   for (R = 2; R ≤ N_max; R = R + 1) {
  /***** vertex-only parallelization **********/
      for ( all {w | w = succ(v)} ) Select SF_w(1);
      X=apply(v,p,R);
      SF_v(R)=max (SF_v(R), sp(p,X));
      if ( p is operation-parallel ) {
  /***** hierarchical parallelization  ********/
        for (N = R + 1; N ≤ N_max; N = N + 1) {
          Identify bottleneck part. b at v in X;
          for ( all {w | w = succ(v) ∧ w ⊂ b} )
             Select next point in SF_w;
          X=apply(v,p,R);
          SF_v(N)=max (SF_v(N), sp(p,X));
  }}}}
}
```

Figure 6: Parallelization of the vertex v

4.3 Computational complexity

The maximal number of the parallelization try-outs necessary per vertex and per parallelization method is a small constant, which depends only on the number of points in the speedup function (equal to the number of points in the area $R \geq 2$ in figure 5):

$$Y = \sum_{n=2}^{N_{max}} (n-1) = \frac{N_{max} \cdot (N_{max}-1)}{2} \quad (4)$$

For example for $N_{max} = 6$ we obtain $Y = 15$ tries, which is much less than the number of combinations possible (recall eq. 3). This together with the fact that each vertex is visited only once has a positive effect on the total computational complexity of the algorithm. This complexity is $O(N_{max}^2 \cdot |V| \cdot |P|)$.

5 Case study

In this section we present an example application of our algorithm - the frequency tracking embedded system from [6]. The system specification contains about 2k lines of ANSI C source code. In its main loop, the program reads a stream of samples (complex numbers) and uses LMS (adaptive signal enhancement) to determine instantaneous frequency estimates. A 1024-point FFT (fast Fourier transform) is then used to determine the frequency response of the adaptive filter every 100 input samples.

In the experiments we used a combination of tools belonging to the MOVE automatic processor generation framework [5], and the SUIF parallelizing compiler [1]. For functional pipelining a set of new tools operating on SUIF (Stanford University Intermediate Format) has been implemented [13].

First, we compiled the frequency tracking program with the **gcc-move** compiler, then scheduled it with the MOVE scheduler [9]. The options for maximal ILP exploitation (including software pipelining) and oversized architecture (large number of move busses and FUs) were used. The generated code was simulated using the move simulator to obtain detailed profiling information. A number of general code transformations has been applied to this code. For example the *lms()* and *fft()* functions were inlined inside the FOR loop bodies *F20* and *F15*. The context graph including computation distribution information is presented in figure 7. For the sake of readability, only the loop and procedure vertices are shown. Vertices which are marked with '*' are suitable for *data-parallel mode*, while the ones with '#' for *operation-parallel mode* parallelization. The detailed data dependency vectors were generated using the combined static & dynamic methodology described in [14]. Communication overhead was estimated assuming the availability of fast bidirectional interprocessor links only.

Subsequently the SPEEDUP procedure was called to generate the speedup functions. The following parameters were used: $N_{max} = 4$, $\alpha = 0.01$. Several vertices with $p(v) < \alpha$ were skipped (F57, F91, F121, F72). Total of 297 parallelizations were attempted. The generated speedup functions for the most important vertices in the context graph are presented in figure 8. As can be seen from the speedup function for the whole program (SF_{main}), an overall speedup varying from

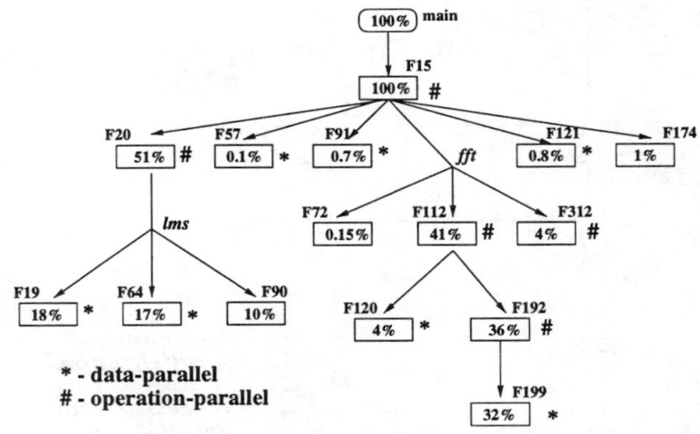

Figure 7: Context graph for the frequency tracking system

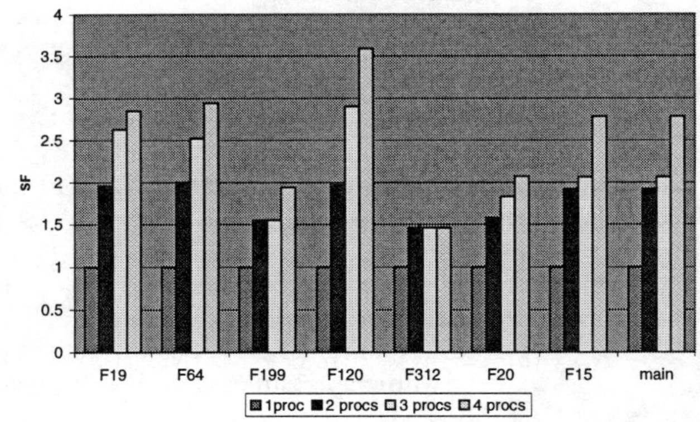

Figure 8: Speedup functions for selected vertices from figure 7

1.92 to 2.78 could be obtained. The point for $N_{proc} = 2$ involves *operation-parallel* mode parallelization of the F15 loop, while the speedup of 2.78 on 4 processors can be obtained by applying in addition parallelizations to the loops F19 and F64 in *lms*, and to the loops F120, F199 and F312 in *fft*.

Next the design trade-offs curves Π_N for $N = 1..4$ were generated using the *Explore* tool of the MOVE framework. Figure 9 presents these curves. To combine them we have to select Pareto points from each curve. In our case this results in the design trade-offs curve presented in figure 10. Note that points for implementations with larger number of processors lie to the right of the implementations with smaller N.

After obtaining the combined speedup function one of the points had to be selected. The specified timing constraint was 150 ms (required speedup of 1.48). The point for the configuration with only 2 processors turned out to be sufficient (marked in figure 10). Note, that none of the single processor solutions, even with many FUs, meets the timing constraints. The obtained multi-processor is presented in figure 11. Besides standard components, processor 1 included 2 ALUs, 1 FPU, 2 Load-Store units and 8 move busses. The

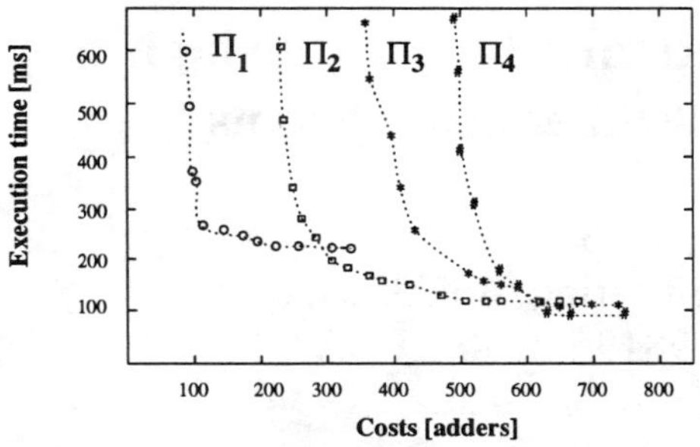

Figure 9: The design trade-offs curves for different numbers of processors used.

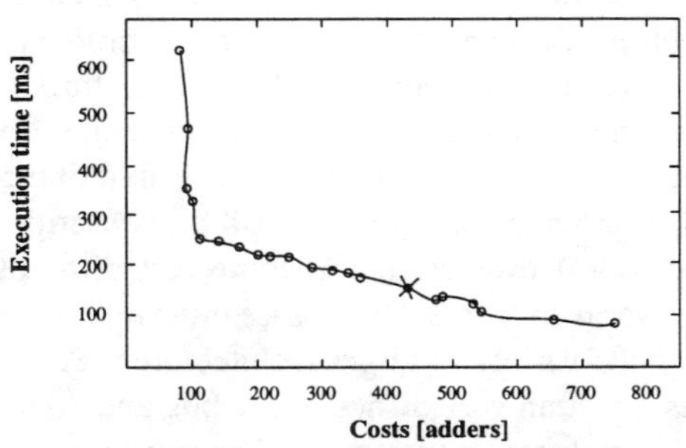

Figure 10: The combined design trade-offs curve.

second processor turned-out to be smaller - included only 1 ALU, 1 FPU, 1 Load-Store unit and 5 move busses. Both processors were equipped with small instruction caches and local memories. Also a fast bidirectional interprocessor communication link has been included; each *COMM* unit contains a FIFO supporting asynchronous communication. Note also that the multi-ported (integer) register file can easily be split into four dual ported register files (see [10]).

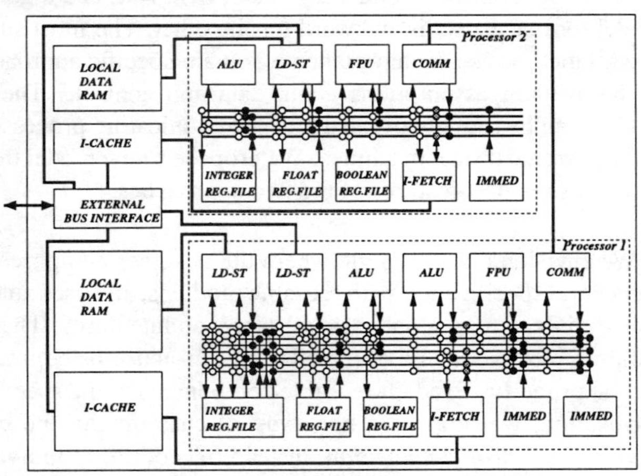

Figure 11: The instant frequency tracking multi-processor.

6 Conclusions

In this paper we proposed a new design space exploration algorithm for semi-automatic mapping of the embedded application onto a cost-efficient heterogeneous multi-processors. Its uniqueness lies in a combination of the state of the art automatic ASIP synthesis software with a coarse- and fine-grain parallelism exploitation methodology. The computational complexity of the algorithm is linear in the number of loops in the program and in the number of applied parallelization methods. Its applicability was demonstrated on a case study of the frequency tracking system.

The presented approach can be easily extended to handle real-time reactive embedded systems with many subtasks. We can consider each subtask separately when calculating the speedup functions. After that, perform the mapping of the task partitions onto the processors and do the processors design space exploration. Once the trade-offs functions are calculated we should check them for possible subtask constraints violations and then combine them and select points which satisfy the global constraints.

References

[1] Saman P. Amarasinghe, Jennifer M. Anderson, Christopher S. Wilson, Shin-Wei Liao, Brian R. Murphy, Robert S. French, Monica S. Lam, and Mary W. Hall. Multiprocessors From a Software Perspective. *IEEE micro*, pages 52–61, June 1996.

[2] J. Axelsson. Architecture synthesis and partitioning of real-time systems: A comparison of three heuristic search. In *International Workshop on Hardware-Software Codesign*, Braunschweig, Germany, March 1997.

[3] A. Bender. Design of an optimal loosely coupled hetergeneous multiprocessor system. In *Proceedings of The European Design & Test Conference*, 1996.

[4] Henk Corporaal. *Microprocessor Architectures; from VLIW to TTA*. John Wiley, 1997. ISBN 0-471-97157-X.

[5] Henk Corporaal and Hans Mulder. MOVE: A framework for high-performance processor design. In *Supercomputing-91*, pages 692–701, Albuquerque, November 1991.

[6] P.M. Embree. *C Language Algorithms for Real-Time DSP*. Prentice-Hall, 1995.

[7] Daniel D. Gajski, Frank Vahid, and Sanjiv Narayan. *Specification and design of embedded systems*. PTR Prentice-Hall, Englewood Cliffs, 1994.

[8] R.K. Gupta. *Co-synthesis of hardware and software for digital embedded systems*. Kluwer Academic, Boston, 1995.

[9] Jan Hoogerbrugge. *Code generation for Transport Triggered Architectures*. PhD thesis, Delft Univ. of Technology, February 1996.

[10] Johan Janssen and Henk Corporaal. Partitioned Register Files for TTAs. In *MICRO-28*, Michigan, November 1995.

[11] Ireneusz Karkowski. *Performance Driven Synthesis of Digital Systems*. PhD thesis, Delft University of Technology, December 1995.

[12] Ireneusz Karkowski. Computer aided embedded systems design. In *Proceedings of the third annual conf. of ASCI*, Heijen, The Netherlands, 2-4 June 1997.

[13] Ireneusz Karkowski and Henk Corporaal. Fp-map - an approach to the functional pipelining of embedded programs. In *Proceedings of 4th International Conference on High Performance Computing*, Bangalore, India, 18-21 December 1997.

[14] Ireneusz Karkowski and Henk Corporaal. Overcoming the limitations of the traditional loop parallelization. In *Proceedings of the HPCN'97*, April 1997.

[15] J. Teich, T. Blickle, and L. Thiele. An evolutionary approach to system-level synthesis. In *International Workshop on Hardware-Software Codesign*, Braunschweig, Germany, March 1997.

[16] Michael Wolfe. *High Performance Compilers for Parallel Computing*. Addison-Wesley Publishing Company, 1996.

Modal Processes: Towards Enhanced Retargetability through Control Composition of Distributed Embedded Systems *

Pai Chou and Gaetano Borriello
Department of Computer Science and Engineering, Box 352350
University of Washington, Seattle, WA 98195-2350 USA
{chou,gaetano}@cs.washington.edu

Abstract

To explore different points in the design space of an embedded system, it is important to be able to compose a design from reusable design components, and then map the resulting system description onto several possible target architectures with different partitionings of functionality. Today's specification models support composition styles that work well for data communication but not for control communication between concurrent processes to be mapped onto a distributed architecture. We propose a new retargetable system specification model that combines the best properties of process-based and hierarchical-FSM-based methods for modular composition of data and control. The model lends itself to automated synthesis of the run-time system for coordinating tasks on different processors in the system. The model and synthesis method are illustrated with several examples of embedded systems.

1 Introduction

Embedded systems are increasingly being implemented as distributed systems with heterogeneous processors. Distributed architectures are motivated by applications that must interact with multiple elements of the environment concurrently. For example, typical automobiles are now controlled by a distributed system that coordinates everything from the braking system and engine to the dashboard and climate control. These systems also tend to be heterogeneous, so that designers have more flexibility in optimizing the design to their specific objectives of cost, size, power, and performance. It is often imperative that designers explore many points in the design space, but at the same time, they are given less time to complete their designs. To meet these conflicting goals, successful designers must maximize design reuse and work at the highest possible level of abstraction.

*This work was supported by DARPA DAAH04-94-G-0272

DAC 98, San Francisco, California
©1998 ACM 0-89791-964-5/98/06..$5.00

Today, most designs are written in *implementation languages* such as C, Ada-95 (for avionic or military applications), or Java. These languages allow reuse through a *procedural interface*, and object-oriented languages further enable specialization and extension through inheritance. They work well for a wide variety of algorithmic descriptions on a single processor. However, they become very difficult to manage when several concurrent processes must interact with each other on a wide variety of target architectures. Programs that use threads are often very difficult to debug and do not behave consistently on different platforms. The run-time system also consumes high overhead that can be prohibitive on low-cost embedded processors. Furthermore, today's methodology relies heavily on legacy code and components that severely limit the optimizations a designer can explore.

What makes the development of a retargetable specification method difficult is the underlying dichotomy between data and control flow. Virtually all practical embedded systems contain both control and data aspects of behavior, and a good model for one aspect is awkward for the other. The most successful methods so far have taken a domain specific approach by fixing their assumptions about data and control. These models can be roughly divided into communicating processes and hierarchical state machines. We propose a new model that combines the best features of the two approaches.

We envision a design style where the designer composes a behavioral specification with reusable modules, and uses automated tools to map it onto several target architectures. These building blocks may be designed either in-house or be intellectual property. To adapt reusable components to the specific application, we propose a new way of customizing the behavior by *control composition*, instead of modifying individual components. By deriving a complete implementation of the system on the target architecture, designers would be able to closely evaluate not only the mix of components and their communication topology and protocols, but also the allocation of system tasks to different processors. Such tools would support systems with arbitrary topologies that use a rich set of components and interfaces.

2 Motivating examples

Embedded systems are very diverse. We have selected two examples with different features that will allow us to illustrate our specification model.

2.1 Tessellator robot

The first example is a simplified version of the space shuttle's heat-shield tile tessellator robot. It can be controlled either manually with a joystick or automatically to inspect and replace tiles on the hull of the spacecraft. The mode of operation is selected by a toggle switch, and defaults to manual mode. For safety reasons, the robot moves only when a human operator is present and holding the deadman switch on the joystick. The robot must always halt whenever the switch is released. The joystick controls the heading and acceleration of the robot. In autonomous mode, the robot uses its sonar and bumper sensor for maneuvering. It moves forward normally, but if the sonar detects an obstacle, then the robot turns 45 degrees and continues forward. If its bumper is hit then the robot goes in reverse until two seconds after the bumper has been released continuously, and then turns 45 degrees and moves forward in the same manner as the sonar's reaction. The bumper overrides the sonar, switching from auto to manual mode overrides both, and releasing the deadman button stops all activities.

2.2 controller for a human-powered airplane

The RAVEN human-powered airplane [11] includes an embedded system to control its rudder and elevators. It runs one of several control algorithms based on user commands and regularly samples sensor values. The flight data can be logged in a flash memory module and transfered over a serial line for analysis after landing. The logger has two timing resolution options. The LCD-based user interface, which has seven command screens, is the most complex part of the design because it interacts with all other components in the system. Some commands are local to the user interface (*e.g.* changing screens, editing parameters) while others affect the system's modes of operation.

3 Related work

Existing models offer ways of composing modules to form a complete design. The ways existing models organize behavior can be divided into two styles: *functional encapsulation* and *temporal (state) encapsulation*.

3.1 functional encapsulation

Most systems are composed in a style we call *functional encapsulation*, as exemplified by process based models. Each process or module is an encapsulation of logically related functionality. Processes are a natural way of organizing design components, and they can be composed by message passing or signaling. Examples include synchronous dataflow (SDF) and dynamic dataflow (DDF) [6], CSP (strictly message passing) [9], and many concurrent FSM variants like CFSM [1] and SDL [12]. For design space exploration, processes define the granularity for partitioning. A designer can experiment with different assignments of processes to processors, and tools can help with system integration by synthesizing interprocessor communication [10]. Process models promote a modular design style – as long as the composition is limited to *data*. The modularity breaks down for control composition.

Data is concrete, while control is more abstract and can have many implementations: implicitly with the program counter or explicitly encoded and manipulated as data. Currently, control must be *encoded* as a command or a named signal in order to be communicated, and the receiver must be ready to interpret the command and change its control flow accordingly. This has been done successfully for specific domains, such as the subsumption architecture [4], where processes in a chain either send their own commands to the lower level or pass commands from above when overridden. Outside the specific domain, though, the problems with this approach are that data primitives are too low level for control and control is handled in an *ad hoc* manner by user code.

Control composition inherently requires processes to make use of a common set of states. To be modular, each process must maintain its own copy of the shared states. Each process is burdened with the responsibility of ensuring the coherence of their replicated states. Existing functional encapsulation models lack high-level primitives for expressing the state coherence requirements on the processes. Instead, state-coherence is handled imperatively in terms of transmitting state changes, interpreting the commands, and actually making those changes. Such low level code is embedded and scattered throughout each process and can be a major source of error.

3.2 temporal encapsulation

An alternative to functional encapsulation is to focus on the composition of control based on hierarchical state machines, as exemplified by hierarchical state machine models like StateCharts [7] and scoped watchdogs like Esterel [2]. Instead of embedding states in a process (in terms of the program counter or in state variables), explicit states and transitions are used to structure the behavior. Explicit states describe the behavior of the entire system and can be related in several ways: as parent/child, mutually exclusive, or parallel. Control composition by temporal encapsulation can be more modular than functional encapsulation because neither the preempting nor preempted state needs to know about each other, and different compositions do not require modifications to the behavior by explicitly anticipating the transition command and interpreting it. Also, there is no state coherence problem because there is only one set of states.

Temporal encapsulation can be difficult to partition for mapping onto a distributed architecture. It breaks process modularity because logically related functionality must be scattered in different states. Also, these models rely on the *perfect synchrony hypothesis* for control composition, but it is impractical for many distributed systems. As a result, languages that support temporal encapsulation also support process-like, parallel composition style using *choice* and signaling statements that use asynchronous composition at the system level, such as CRP [3] and ModuleCharts [8]. However, these are effectively processes without control modularity.

4 The modal process model

We propose a model that provides primitives for composing control independently of the communication semantics. The unit of composition is a *modal process*. It encapsulates functionality and defines the granularity of partitioning. Its behavior is organized into *modes*. Modes can be viewed as states of an incompletely specified state machine and serve as the interface for modular control composition. A distinguishing feature is that modes are managed by the run-time system code that we synthesize, according to the coherence requirements on the modes specified explicitly, rather than being embedded in user code and tied to the message passing semantics. By separating the specific synchronization requirements of the application from reusable behavior, this approach not only helps eliminating a whole class of bookkeeping errors related to state coherence, but also enhance module reusability in different applications and over different architectures.

4.1 modal processes

In our terminology, a **process** is a container of event handler routines. It may be derived from another model that assumes shared memory, message passing, event-driven, time-driven, dataflow, *etc.* In general, events may include I/O, interprocess communication, timing, dataflow, and mode transitions. We assume nonblocking communication, but the *synchrony* can be parameterized.

A **modal process** is a process with several modes, where a **mode** is, informally, a way of handling events. Each mode defines a collection of handlers to use while the mode is active. Different modes may pick different handlers to handle an event or ignore it. Modes allow a modal process to have several possible behaviors.

Although modes are similar to states, there are several differences. A traditional FSM has a transition function δ that maps a state $q_i \in Q$ to the next state q_{i+1}, or $\delta : Q \times I \rightarrow Q$. The state space of a modal process is a subset of the powerset of its modes, that is, $Q \subseteq 2^M$. When a handler finishes execution, it may return its transition request (A, D), where $A \subseteq M$ is a set of modes to activate and $D \subseteq M$ is a set to deactivate, and $A \cap D = \emptyset$. The requests from different handlers are merged.

For syntactic convenience, designer do not need to actually specify all of the modes to activate and deactivate on every transition, but they can rely on *mode constraints* to automatically activate or deactivate a set of modes. In fact, hierarchical state machines are a special case where two states can be mutually exclusive or related as parent and child. Mutual exclusion is an automatic deactivation of the current mode on

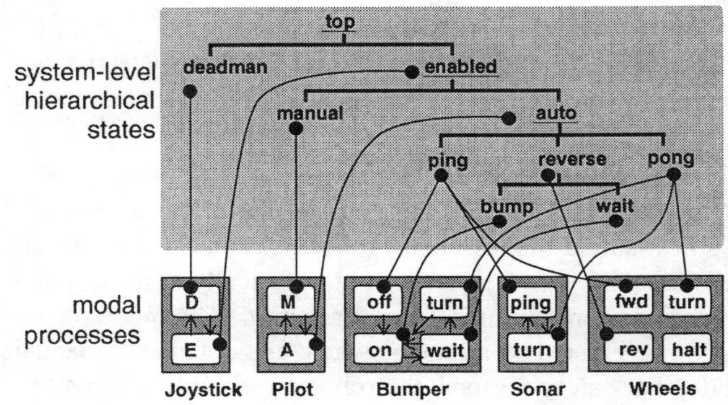

Figure 1: Control composition of the robot specification.

activating the new mode. A parent is automatically activated on activating a child, and a child is automatically deactivated on deactivating its parent. Unlike a state machine, a modal process is *incompletely specified*: it defines the maximum allowed state space and the minimum transition space. In fact, it may define no transitions at all. Control composition will restrict the state space and expand the transition space, as described in the next subsection.

Robot example

The robot has five processes: Joystick, Pilot, Bumper, Sonar, and Wheels (Fig. 1). Joystick has two modes, D for deadman and E for enabled. The Pilot process has two modes: M for manual and A for auto pilot. The Sonar process has two modes: Ping to pulse the sonar every two seconds and Turn for turning after obstacle detection. The Bumper process has four modes: off, on when the bumper is hit, wait when waiting for two more seconds, and turn like the Sonar. The Wheels process has modes for going forward, reverse, turning, and halt, but it does not define any transitions of its own.

RAVEN example

This system has five processes. The logger is capable of logging fast LF, slow LS, uploading data UL, or being off. The altimeter can measure air altitude or perform calibration. The control algorithm process has two modes for manual (M) or autopilot (A). The user interface process has two sets of mutually exclusive modes. One set is for setting the logging modes, and the other set for plane operations: manual, autopilot, upload logger data, calibrate, analog diagnostics, and digital diagnostics. The ground sensor detects whether the plane is in the air or on the ground.

4.2 system-level states

Control composition is done by *binding* modes to common states that can be organized hierarchically. A *system-level control state*, or simply **state** for short, is a boolean variable. When one process requests to activate a mode, it also causes

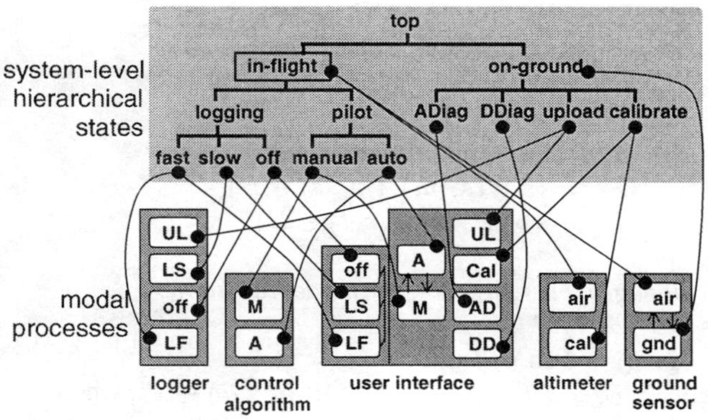

†Transition edges not completely shown between modes of the user interface process. Also, processes that perform analog/digital diagnostics are not shown. A framed box around a state, such as in-flight, indicates a fork (parallel children states). Data composition is not shown.

Figure 2: Control composition of the RAVEN human powered airplane controller.

other processes to activate their modes that are bound to the same state. A transition consists of an exit phase followed by an entry phase. A process may choose to be notified of the transition with a special internal mode-exit event to give it a chance to clean up, or a mode-entry event to set up.

Robot example

(Fig. 1) The system's state hierarchy has two states at the top level for deadman and enabled. In enabled, we have two nested states manual and auto. The auto state has three nested states: forward, turning, and reverse. The reverse state has two modes for bumped and wait. Note that Bumper and Wheels processes do not fully specify the transitions between their own modes; instead, they acquire some transitions through binding. By binding Bumper's turn mode to the same state as Sonar's turn mode, Bumper effectively reuses Sonar's behavior, which will take it back to off mode when Sonar returns to ping. The Wheels process has four behaviors but does not decide when to transition between them.

The binding between states and modes is many-to-many and may be asymmetric. A state activates *all* modes that are bound to it. On the other hand, a mode can be bound to several states, and *any* of them can activate the mode.

4.3 synchrony

Synchrony is the characterization of their relative progress or correlation to the events in the system. Different semantics can be obtained by changing the synchronization behavior.

Transition-synchronous semantics means that all parallel state machines make progress in lock-step, such as Esterel. **Event-synchronous** semantics, used by discrete event systems, means that all parallel state machines see the same set of events simultaneously, but it makes no restrictions on how many transitions each state machine can make within a time step. It is possible to combine the two. Synchronous composition can yield deterministic and predictable implementations, though they may be impractical for distributed architectures.

In **asynchronous** composition, each process can make arbitrary amount of progress asynchronously to other processes. Mode transitions are asynchronous to the communication. A special case is **communication synchronous**, where processes make progress asynchronously until they need to communicate. Asynchronous semantics is more realistic for distributed systems, and in fact synchronous models have come to rely on asynchronous compositions at the system level. For example, CRP [3] is essentially a set of locally-synchronous Esterel components that are composed asynchronously as CSP processes at the system level. StateMate [8] offers similar composition: locally-synchronous StateCharts components are connected together asynchronously in the Module-Charts.

Mode synchronous semantics requires different processes to synchronize on a mode change if their modes are affected, so that the composed system behaves in a coherent manner. This is particularly useful for distributed systems because most of the time the systems make independent but coherent progress by running in the same mode context without synchronization. The occasional synchronizations ensure that the entire system (or subsystem) is in a coherent new context before it is allowed to make further progress. In **data-synchronous** semantics, mode changes happen synchronously to dataflow. It allows the system to operate in a logically coherent manner by pipelining mode changes along dataflow, thereby relaxing the instantaneous state-coherence requirement and eliminating the need for additional synchronization.

Examples

Although both the robot and RAVEN are control oriented and have similar auto-pilot and manual modes, they have different synchronization requirements. The robot can be implemented correctly with asynchronous composition, while the RAVEN needs to be at least data-synchronous. On the robot, the sonar and bumper processes are not synchronized with each other or with the joystick, but the only requirement is respose time. The RAVEN, however, operates on streams of sensor/actuator values from the user interface to the control algorithm and the logger, similar to SDF. Theoretically the mode can change every iteration, but an implementation only needs to ensure that each process operates on the data in consistent modes. A rendezvous type of synchronization would be unnecessarily costly and even preclude pipelined implementations.

4.4 timing constraints

Another use of modes is to scope timing constraints. A system may have different sets of timing requirements depending on what mode it is in. Event detection may impose polling constraints in one mode, but in other modes, the event may be ignored and therefore need not be dispatched at all. Event sensitivity information can be extracted from the specification at compile time and is used by the real-time scheduler to determine the work load for each mode configuration. Handlers may be constrained by rate or related by precedence, with minimum or maximum separation requirements between the times of two observable events. Our modal process model is not tied to any particular scheduling model, and most types of timing constraints can be specified, as long as the available schedulers support them. These include both priority-driven schedulers, such as EDF and rate monotonic, and static scheduling techniques that can meet more complex intramodal and intermodal constraints [5].

5 Synthesis

We implement the modal process abstraction with a mix of compile-time transformations and synthesized run-time support. The first step partitions the hierarchical states of the system onto the processors and determines interprocessor communication needed for mode transitions. We also synthesize *mode managers* as run-time support for maintaining state coherence among modal processes that can reside on different processors.

5.1 representation

The system states are represented as a graph $H = (S, E, r)$. S is the set of vertices that represent states. $E \subset C \times S \times S$ is the set of edges that constrain the activation. C is the set of activation relations {AA, AD, DA, DD}: (AA, a, b) requires that when a is activated (denoted $a\uparrow$), b must be activated, too (*i.e.*, $a \uparrow \Rightarrow b \uparrow$), while (AD, a, b) requires b to be deactivated instead, namely $a \uparrow \Rightarrow b \downarrow$. Similarly, (DA, a, b) means $a \downarrow \Rightarrow b \uparrow$ and (DD, a, b) is $a \downarrow \Rightarrow b \downarrow$. Finally, $r \in S$ is the root (initial) state.

A modal process $\pi_i \in \Pi$ is (M_i, T_i), where M_i is the set of modes and $T_i \subseteq M_i \times M_i$ is the set of transitions between modes. We define $M = M_1 \cup M_2 \cup \ldots M_{|\Pi|}$. The binding between the modes and the states is a set of relations $B \subseteq S \times M \cup M \times S$.

A *partitioning* is a function that maps a process to its processor number $[1, n]$. Because a mode is contained by a process, we overload the function P to map a mode to its processor ID, without ambiguities, namely $P : M \rightarrow [1, n]$.

5.2 state partitioning

State partitioning involves projecting the system states onto the individual processors. Some replication may be required, although it is not always necessary to duplicate those states that have no binding to any of the modes on the given processor except those required structurally. In addition to projecting the states, this procedure also returns E_C, the interprocessor communication edges for mode transitions. The algorithm is shown in Fig. 3.

We make a copy of a state if one of the modes on the processor has a binding to that state or its descendent. The resulting

PartitionState($H = (S, E, r), P : M \rightarrow [1, n],$
  $B \subseteq S \times M \cup M \times S$) {
  // copy states and edges for processor i
  for ($i = 1$ **to** n) {
    // project states onto processor i
    $S^i := \{s^i | \forall(s, m_i), (m_i, s) \in B\}$
    $E^i := \{(c, s^i, t^i) | (c, s, t) \in E\}$
  }
  // edges for interprocessor mode transitions
  $E_C := \{\}$
  for ($i = 1$ **to** n, $j = 1$ **to** n, $j \neq i$) {
    $E_C := E_C \cup \{(c', s^i, v^j) | (c, s^i, t^i) \in T^i$, where
      $s \rightarrow u$ and $v \rightarrow t$ are the shortest paths
      and $c = c'' \cdot c', (c'', s, u)$
  }
  return $H^1 \ldots H^n, E_C$
}

Figure 3: State partitioning algorithm.

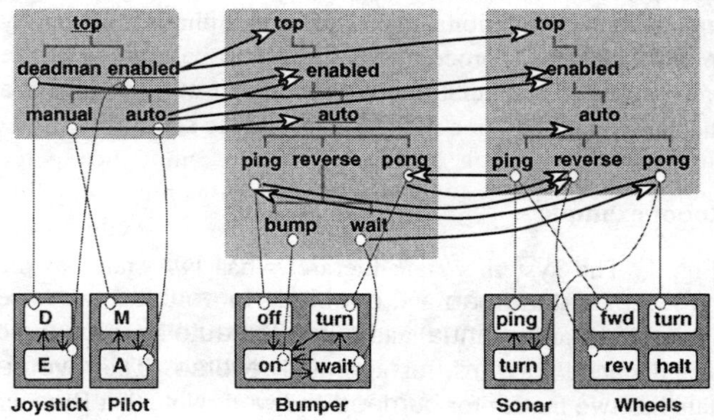

Figure 4: Control communication of the robot as partitioned onto three processors

graph is necessarily connected as a tree. The edges E^i are exactly those needed to connect the replicated states locally. The second part of the procedure builds the set of edges E_C for interprocessor mode transitions. It projects the local *transition path* (c, s^i, t^i) to a remote host j to obtain (c', u^j, v^j). If the projected path is empty then the mode change has no effect on that processor, and therefore no communication is necessary. The projection can be obtained by taking the transitive closures of the paths. The activation operators can compose as follows:

	AA	AD	DA	DD
AA	AA	AD	∅	∅
AD	∅	∅	AA	AD
DA	DA	DD	∅	∅
DD	∅	∅	DA	DD

Robot example

Consider partitioning the Robot onto three processors. To project the (manual, enabled, auto) path on the bumper's processor, we add the edges (manual^1, auto^2) and (manual^1, auto^3) since they are the terminal vertices. On the other hand, the path (auto, enabled, manual) terminates at enabled on processors 2 and 3, because they do not have the replicated manual state. Paths like (wait, reverse, bump) yield empty projections on processor 1 and therefore no edges are added from processor 2 to 1.

5.3 mode manager

The mode manager is synthesized as part of the run-time support for modal processes. It maintains the state hierarchy on its own processor and services the mode change requests. When an event is dispatched, the mode manager calls those handlers that are mapped by the local processes' modes. The order in which these handlers are called can be defined statically by the user or by the scheduling algorithm. It is possible that one or more handlers can request a mode transition, and they are serviced and resolved after all pending handlers are called. Local transitions can be implemented with either static path enumeration or dynamic path generation. The tradeoffs are determinism vs. code size.

By default, interprocess mode transitions assume mode-synchronous semantics, which is implemented with a three-phase synchronization. After the sender communicates the mode change, all receivers must acknowledge the completion of the exit phase. A receiver may also have a pending transition, and may need to either override the sender's request with a NACK or nullify its own transition and ACK. Transition conflicts can be resolved in a number of ways, and we currently support static priorities. For example, for a control-synchronous version of the robot, suppose the sonar tries to transition from pong to ping at the same time the bumper wants to go to bumped. This can be statically resolved in favor of the bumper.

If a sender receives a NACK or decides to nullify its pending transition, it continues collecting ACKs and sends a consolidated ACK to the receiver. If a sender receives all of the ACKs (statically determined count) then it sends out a "go-ahead" event that tells all receivers to proceed with the entry phase of the transition. This may be very strict but it can be used to implement the semantics of Esterel and StateCharts.

6 Conclusions

We propose the modal process model for capturing control and data behavior with real-time constraints by composing *modal processes* and automatically synthesizing run-time support software. Modes provide not only an interface for control composition between processes, but also a way of systematically scoping timing constraints. By enforcing a separation between system states and process modes, this model combines the data modularity of processes with the control modularity of hierarchical state machines. The decoupling between control and data enhances reuse and allows us to automate the most error prone aspects of distributed systems implementation, namely, synchronization and real-time. We believe that this approach will lead to a higher level of retargetability for distributed embedded systems by enabling the designer to better explore the design space. We are currently developing analysis tools to help the designer determine the best choice of synchronization mechanism and integrate hardware components into the modal process model.

References

[1] F. Balarin, H. Hsieh, A. Jurecska, L. Lavagno, and A. Sangiovanni-Vincentelli. Formal verification of embedded systems based on CFSM networks. In *Proc. 33rd Design Automation Conference*, pages 568–571, June 1996.

[2] A. Benveniste and G. Berry. The synchronous approach to reactive and real-time systems. *Proceedings of the IEEE*, 79(9):1270–1282, September 1991.

[3] G. Berry, S. Ramesh, and R. K. Shyamasundar. Communicating reactive processes. In *Conference Record of the Twentieth Annual ACM SIGPLAN-SIGACT Symposium on Principles of Programming Languages*, pages 85–98, Jan. 1993.

[4] R. A. Brooks and J. H. Connell. Asynchronous distributed control system for a mobile robot. In *Proceedings of the SPIE - The International Society for Optical Engineering*, volume 727, pages 77–84, 1987.

[5] P. Chou, E. A. Walkup, and G. Borriello. Scheduling issues in reactive real-time systems. *IEEE Micro*, pages 37–47, August 1994.

[6] G.C.Sih and E. A. Lee. A compiled-time scheduling heuristic for interconnection-constrained heterogeneous processor architectures. *IEEE Trans. on Parallel and Distributed Systems*, 4(2), February 1993.

[7] D. Harel. StateCharts: a visual formalism for complex systems. *Science of Programming*, 8(3):231–274, June 1987.

[8] D. Harel, H. Lachover, A. Naamad, A. Pnueli, M. Politi, R. Sherman, A. Shtull-Trauring, and M. Trakhtenbrot. STATEMATE: a working environment for the development of complex reactive systems. *IEEE Trans. on Software Engineering*, 16(4):403–414, April 1990.

[9] C. A. R. Hoare. *Communicating Sequential Processes*. Prentice-Hall, 1985.

[10] R. Ortega and G. Borriello. Communication synthesis for embedded systems with global considerations. In *Proc. Codes/CACHE*, pages 69–73, 1997.

[11] *The RAVEN Project*. http://ihpva.org/Raven/, September 1997.

[12] The SPECS Consortium and J. Bruijning. Evaluation and integration of specification languages. *Computer Networks and ISDN Systems*, 13:75–89, 1987.

Design Methodologies for Noise in Digital Integrated Circuits

Kenneth L. Shepard
Department of Electrical Engineering
Columbia University, New York, NY 10027

Abstract

In this paper, we describe the growing problems of noise in digital integrated circuits and the design tools and techniques used to ensure the noise immunity of digital designs.

1 Introduction

Noise has become a metric in the design of digital integrated circuits of comparable importance to area, timing, and power for four principle reasons: increasing interconnect densities, faster clock rates, more aggressive use of high-performance circuit families, and scaling threshold voltages. Increasing interconnect densities imply a significant increase in coupling capacitance as a fraction of self-capacitance. Faster clock rates imply faster on-chip slew times. These two effects combine to make capacitive coupling a growing source of noise on-chip. Many high-performance circuit styles try to speed up one transition (usually falling) at the expense of the other and assign logical evaluates to the faster edge. Any circuit that utilizes these techniques we refer to as a *skewed-evaluate* circuit (e. g. domino). Skewed-evaluate circuits have noise sensitivities directly related to the threshold voltages of the transistors responsible for the evaluate transitions (usually n-FETs). Threshold voltages are, however, scaling lower to maintain drive in the presence of scaling supply voltages. These effects combine to produce more sources of on-chip noise due to switching circuits as well as less immunity to this noise. More details of these technology trends can be found in Reference [1].

Noise has two deleterious effects on digital design. When noise acts against a normally static signal, it can transiently destroy the logical information carried by the static node in the circuit. If this ultimately results in incorrect machine state stored in a latch, functional failure will result. When noise acts simultaneously with a *switching* node, this is manifest as a change in the timing (delay and slew) of the transition (a noise-on-delay effect[2]). We are concerned with the former effect in this paper.

Successful design methodologies to ensure noise immunity incorporate a three-level strategy. The first line of defense is a set of noise avoidance rules to guide circuit and interconnect design. These rules should prevent most noise problems but not introduce too many constraints on area or timing. This is followed by detailed static noise analysis on the design to find all possible noise failures. Because of the potential of false paths in static noise analysis, some failures flagged by static noise analysis can be allowed after careful circuit simulation.

DAC 98, San Francisco, California

2 Noise fundamentals

In digital circuits, an *evaluation node* is any node of the circuit that carries logical information by means of an analog voltage level. Evaluation nodes are generally chosen as the inputs and outputs of the logic gates of the circuit, which are generally abstractions for sets of channel-connected components (CCCs); that is, transistors connected through their sources and drains. Noise in the context of CMOS digital integrated circuits refers to any deviation from the nominal supply or ground voltages at nodes which should otherwise represent stable logic 1 or 0. Although noise causes these analog voltages to vary, the system still functions as long as the voltages fall into a valid range. If this is not the case, then the circuits' correct functioning cannot be certain. The complexity of noise analysis comes from the fact that the voltage ranges that represent valid logic levels depend on the precise time-domain characteristics of the noise appearing on the evaluation nodes as well as the sensitivity of receiving circuits to this noise.

The *noise stability* metric is used to determine if the noise appearing on a given evaluation node is of sufficient magnitude to risk functional failure. Noise stability requires that each restoring logic gate, when acted upon by a noise stimulus, must have a dc-noise time-domain sensitivity that is always less than one[1, 3]. This is a generalization of dc noise margins[4] to take into account the fact that restoring logic gates are much more effect at rejecting ac noise than dc noise.

It is convenient to classify noise according to the voltages' relationship to the rails: V_H noise reduces an evaluation node voltage below the supply; V_H^* noise increases an evaluation node voltage above the supply level; V_L noise increases an evaluation node voltage above the ground level; and V_L^* noise decreases an evaluation node voltage below the ground level. These noise classifications are useful because circuits generally propagate noise types in well-defined way. For example, a CMOS inverter is sensitive to V_L and V_H noise on its input, propagating it as V_H and V_L noise, respectively, to its output. We refer to V_H^* and V_L^* noise as *bootstrap* noise.

To develop a comprehensive strategy for noise analysis, we must consider all the possible sources of noise on chip. For this purpose, noise can be classified as power-supply noise, circuit noise, or interconnect noise. Power-supply noise refers to noise appearing on the on-chip power and ground distribution network produced by the current demands of the switching circuits. Circuit noise includes noise propagated onto an evaluation node from the driving gate or charge-sharing effects onto the output of the driving gate[1]. For silicon-on-insulator (SOI) circuits, circuit noise might also include the effects of the parasitic bipolar in floating-body MOS devices[5], but we will not consider SOI devices further here. Interconnect noise refers to noise appearing as a result of capacitive or inductive coupling in the interconnect or as a result of poor impedance matching between the devices and interconnect.

3 Noise-reducing design techniques

Designers can use a number of techniques and practices to control noise in digital circuits. Controlling power supply noise involves careful design of the power and ground distribution network and adequate decoupling capacitance. To achieve effective supply decoupling, the ratio of on-chip decoupling capacitance to total effective switching capacitance is generally greater than 10:1[6]. To control circuit and interconnect noise, circuit design rules are commonly employed. Some examples: disallow single n-FET or p-FET pass gates because of the V_T voltage drop they cause; disallow pass gates at the ends of long wires; disallow long wire runs feeding domino gate inputs; and disallow high beta static circuits feeding low beta static circuits, or vice versa.

In addition, design rules can require "baby-sitting" devices added to internal nodes in n-FET pull-down stacks of domino gates to mitigate charge-sharing noise, and half-latches added to dynamic nodes. Rules of this sort are sometimes accompanied by a circuit-checking tool. In addition to tuning gate widths, one can tune gate lengths to balance performance against leakage noise. Weaker devices at the top of the n-FET stack can also reduce charge-sharing noise in the presence of a half-latch device.

In addition to circuit techniques and optimization, there are also many ways to reduce interconnect noise in the design of the on-chip wiring network. For example, one can increase spacing between wires or route signal lines alternately with power or ground. In the Digital 21264 design[7], two metal layers are devoted exclusively for the distribution of power and ground. These ground planes more tightly control on-chip inductance and reduce coupling interactions.

4 Static noise analysis

Design rules can be used to prevent most noise problems in the design phase but are very dependent on design style and circuit topology. Making them rigid enough to preclude noise failures would be exordinarily difficult and result in a significant penalty to area and performance. Instead, design rules are augmented with a *static noise analysis* of the design. In this type of analysis, the noise stability of a ditial circuit is determined by performing small simulations on an individual CCC basis and combining these simulation results statically.

Several fundamental assumptions guide the technique:

1. Gate inputs can be replaced by grounded capacitors. This creates a clean partitioning between one CCC and the next and is a technique commonly employed in fast circuit simulation engines[8].

2. Worst-case *sensitization* conditions drive the CCC simulations used for calculating circuit noise and interconnect noise. By this, we mean how the transistor gates are biased and internal node voltages are preset before the noise stimulus or switching waveform is applied. We will describe the conditions that guide this sensitization in more detail below.

3. We assume that the superposition principle applies in adding (in the time domain) circuit noise and interconnect noise sources. For noise sources small enough to satisfy the noise stability requirement, active FET channels (that is, those attempting to hold nodes to their static level) are biased in the triode regions of their current-voltage characteristics, justifying this "linear" assumption. In particular, charge-sharing noise and propagated noise can be calculated on a "single-input" changing basis and superposed with the coupled noise calculations to find the total noise. The sensitization producing the producing the largest amplitude output noise is used. Noise sources can only be combined when the sensitization conditions are mutually satisfiable.

4. Worst-case temporal relationships are defined by superposing the peak responses of the circuit and interconnect noise for each allowable noise (V_L, V_H, V_L^*, V_H^*) type. When timing information is known, it can be used to reduce pessimism in coupled noise analysis by disallowing the simultaneous switching of signals. Hazard-free logic constraints can also be used to disallow simultaneous switching.

5. A noise stability check is performed across every restoring logic gate in the design. Noise stability violations are assumed to be a sufficient condition for finding the circuit to be nonfunctional.

6. Power-supply integrity analysis is performed independently and is generally characterized by dc bounds on the local power supply variation. In calculating propagated noise, collapsed rails are used, characterized by dc values V_{DD}^{min} and Gnd^{max}. For switching waveforms producing charge-sharing or coupled noise, expanded rails are used, characterized by dc values V_{DD}^{max} and Gnd^{min}. In doing a noise stability check, expanded rails are also used.

7. Drivers on switching perpetrator nets (which we refer to as *secondary nets*) are modelled as ideal voltage sources. This presumes that the noise-on-delay effect has been handled elsewhere.

8. In the case that the circuit contains feedback (as in a latch circuit), the feedback loop is broken at a restoring logic gate. The worst possible dc noise that can be propagated without producing a stability violation in the "broken" gate is assumed. To determine the magnitude of this dc *noise-limited* propagated noise, the subunity gain criterion is applied to a dc voltage transfer characteristic with supplies defined by V_{DD}^{min} and Gnd^{max}. The use of noise-limited propagation in one restoring gate of a logic loop breaks the cyclic dependence in determining the noise potentially circulating the loop.

9. Noise-limited propagation is also used for restoring logic gates which have a stability violation. This allows the noise analysis to continue, despite the violation, and places the burden on fixing the noise problem on the circuits driving the violating gate.

The key abstraction in static noise analysis is the noise graph, a directed graph containing all of the circuit's evaluation nodes connected by segments that move and transform noise. In many ways, this graph is analogous to the timing graph used in static timing analysis. There are three types of segments in a noise graph: restoring segments, propagate segments, and node injection segments. *Restoring segments* cross gates that at some dc bias point have a small-signal gain greater than one. Noise is propagated across restoring segments; in addition, a noise stability check must also be performed. *Propagate segments* connect nodes, between which there is subunity gain at all dc bias points. Noise-stability checking is not required across propagate segments. Each restoring and propagate segment in the noise graph is labelled by the type of noise propagated by the segment. For example $L \rightarrow H$ indicates that the segment propagates V_L noise and transforms it into V_H noise. The *node injection segment* can introduce noise directly onto an evaluation node, superposing with the propagated noise. Coupled interconnect noise and charge-sharing noise are both modelled as node-injection segments. Once the noise graph is constructed, the loops of the graph are broken and the graph is topologically sorted for traversal. Assumption 8 is used to propagate noise across each of the "loop snips." The graph is then searched in a breadth-first fashion to propagate noise through the network, and in the case of restoring segments, to perform the sensitivity tests required to ensure noise stability.

In general, transistor path-based functional extraction[9] guides three main types of sensitization required for Assumption 2: sensitization for coupled noise calculation on the output node of a CCC, sensitization for noise stability and propagated noise calculation from a given input, and sensitization for charge-sharing noise calculation from a switching waveform on a given input. In all three cases, allowable sensitizations are determined by the Boolean satisfiability of constraint relations determined by this functional extraction. Logic conditions between the input variables (denoted as f_{input}), when they exist, must be included in these constraint relations. The constraint equations differ slightly for the case of precharge logic to account for the state established by the reset condition.

We define a path function $f_{P_{i,j}}$ as the logical condition for the path from i to j to conduct. For example, to sensitize for noise appearing on the output, O, of a CCC due to capacitive coupling to a given node, D, in the CCC, there are different logic constraint functions depending on whether we are interested in V_H or V_L noise at the output. In particular, let us consider the sensitizations that allow V_L noise to appear on O due to capacitive coupling to D. In this case, the V_L noise at O is produced by a net switching from ground to V_{DD}, capacitively coupled to node D. The constraint relation in this case is: $f_{P_{O,Gnd}} \cdot f_{P_{O,D}} \cdot f_{input}$. All input sensitizations must satisfy this constaint to inject V_L noise onto node O due to coupling onto node D from a net switching from ground to V_{DD}. Similar sensitization relations exist for charge-sharing noise and propagated noise.

In general, to find the noise appearing at the output of a given CCC, according to Assumption 3, we must find the sensitization producing the largest amplitude output noise for each noise type (V_L or V_H). Stability violations that occur due to noise appearing on any input are reported independently of the sensitization conditions required for the maximum output noise.

We seek to justify and clarify the concepts of static noise analysis by means of a simple, though comprehensive, example. Consider the circuit of Figure 1, in which a domino gate drives a latch. The dynamic node (E) of the domino gate is capacitively coupled to another switching net. In Figure 2, we show circuit simulation results of a functional fail that results from noise. As shown in Figure 2(a), A switches high, while B is still low, resulting in charge-sharing noise on node E. Node I switches, adding coupled noise onto node E. At the same time, noise appears on nodes C and D as shown in Figure 2(b). Power-supply noise is causing the rails to expand during the evaluate phase of the domino gate as can be expected to typically happen since the voltage rails often collapse transiently during precharge. These noise sources *together* are enough switch the output of the dynamic gate, F, and change the state of the latch (node G), as shown in Figure 2(c). Since the latch should have a logic 1 as its output, but instead has a logic 0, functional failure of this hardware will result.

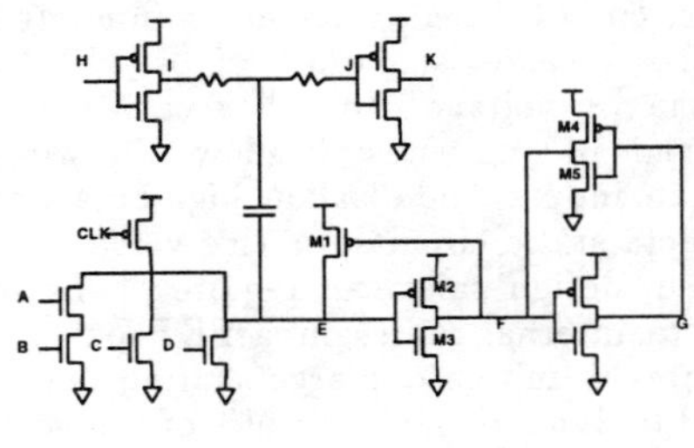

Figure 1: Example circuit for noise analysis consisting by a domino gate driving a latch.

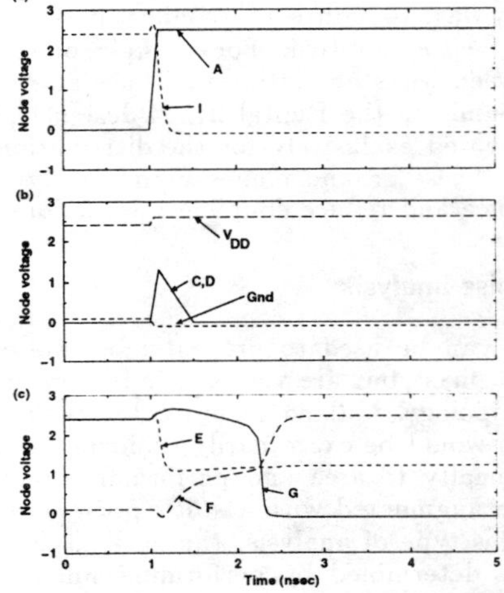

Figure 2: Circuit simulation of a functional failure due to noise. (a) Driver output I is switching as is node A. (b) Coupling noise appears on inputs C and D along with power-supply noise on the voltage rails. (c) The dynamic node (E) falls, switching the output inverter of the domino gate (F) and the latch output (G).

The noise failure demonstrated in Figure 2 is critically dependent on the contributions of all of the noise sources at work: power-supply noise, charge-sharing noise, coupling noise, and propagated noise. Figure 3 shows how this noise fail would not occur in the absence of any of these noise sources. Figure 3(a) shows the node voltages E, F, and G in the absence of input noise on either node C or node D. In Figure 3(b), we show the voltages in the case that there is no coupling noise; that is, node I does not switch. In Figure

3(c), we show the voltages in the case that node A does not switch. We show the voltages in the case that there is no power supply noise in Figure 3(d).

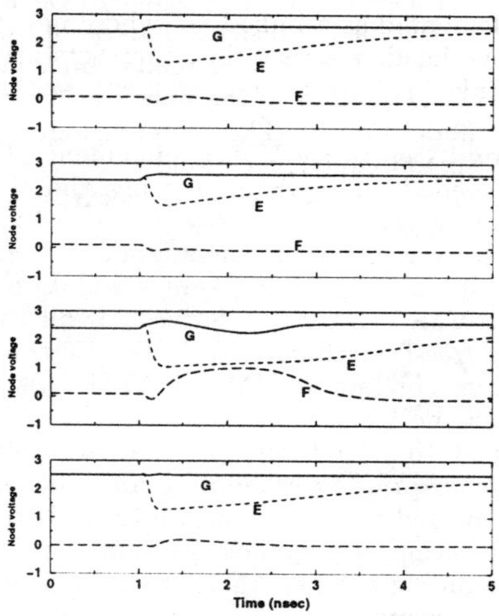

Figure 3: Noise failure will not occur in the absence of (a) injected noise on either input C or input D, (b) coupling noise, (c) charge-sharing noise, or (d) power-supply noise.

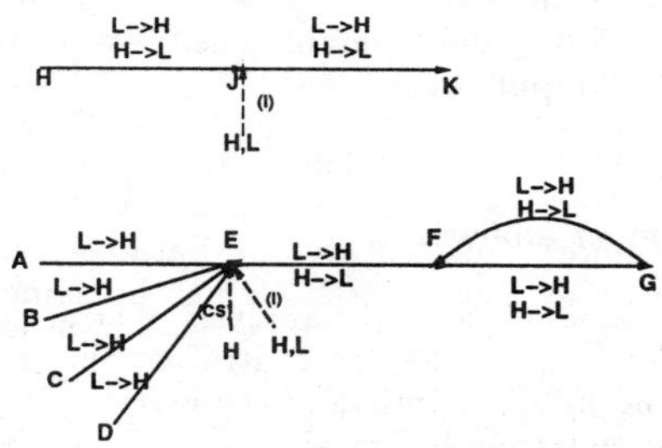

Figure 4: Noise graph for the example circuit.

We now show how a static noise analysis of this network precisely predicts this fail using the assumptions outlined above. The noise graph for this circuit is shown in Figure 4. Restoring segments are denoted by solid lines, while dashed lines denote node-injection segments. We must first calculate the worst-possible noise which can appear on node E according to Assumption 4. This will be, according to Assumption 3, the superposition of the charge-sharing noise injected by the switching of node A, the noise propagated from node C, the noise propagated from node D, and the interconnect coupling noise injected by the switching of node I.

To calculate each of these noise components, we establish the network shown in Figure 5. By Assumption 1, gate inputs are treated as linear capacitors tied to ground. The driver at I is replaced by a independent voltage source by Assumption 7. We now calculate each noise source independently. The sensitization producing the worst total V_H noise response at node E has A, C, D, and F all set to 0 and B set to 1. CLK is 1 for the evaluate phase of this dynamic gate. In Figure 6(a), we show the coupling noise appearing on node E. In this case, node I is switched from

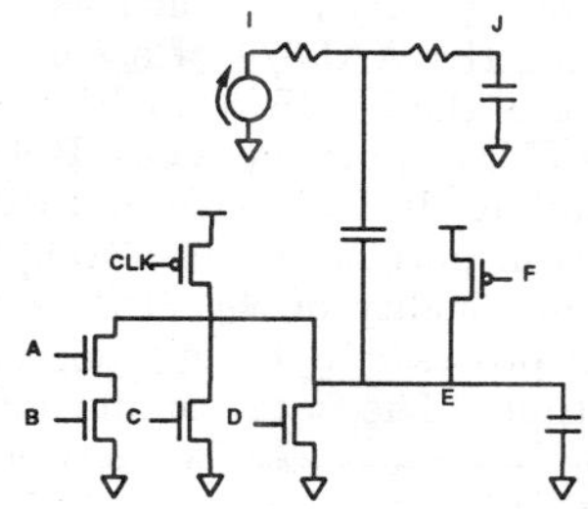

Figure 5: Network for the simulations to compute the noise at node E.

V_{DD}^{max} to Gnd^{min}. V_{DD}^{min} is used for supply and Gnd^{max} is used for ground for the CCC itself, following Assumption 6. In Figure 6(b), we show the charge-sharing noise calculated on node E due to the switching of node A. Figure 6(c) shows the noise propagated to node E due to V_L noise injected onto node C. The exact same curve results for V_L noise propagated to node E from node D under comparable sensitization conditions. In Figure 6(d), we show the noise that results by superposing in the time-domain the results of graphs (a)-(c) with (c) contributing twice. Following Assumption 3, all of the peak noises are aligned. We compare this result with a full simulation, shown as the dashed curve in Figure 6(d). They are almost the same, demonstrating the accuracy of the superposition aspect of Assumption 3.

Having calculated the total noise appear on node E, following the noise graph, we must now propagate this noise across the inverter and perform a noise stability check, following Assumption 5. Figure 7 shows the network for this analysis. Node G is set to 1. The capacitor C_1 represents the gate capacitance of transistor M1, while C_2 represents the gate capacitance of transistors M2 and M3. In Figure 8, the noise propagated to node F is shown as well as the time-domain sensitivity of this noise to dc variations on node E. The sensitivity at the peak position is almost exactly -1, indicating that the inverter is biased to the verge of a noise instability. Static noise analysis would then flag this as a possible noise failure.

Given that a noise instability occurs in propagating to node F, a noise-limited value of this propagated noise should be used at node F, following Assumptions 8 and 9. The dc voltage transfer characteristic of the inverter between nodes E and F is used for the calculation of the noise-limited propagated noise at node F, in this case, 0.23 V. In handling the latch, the restoring segment from G to F is "snipped". Noise limited propagation is assumed out of the inverter defined by transistors M4 and M5. This noise is superposed with the noise propagated from E to F, and the sum propagated across the restoring segment from F to G with the associated noise stability check. To complete the analysis, a noise stability check is also performed at the M4-M5 inverter with the noise propagated to node G.

There are three main sources of pessimism in this analysis. The first is the conservatism of the metric; that is, a circuit can still function even in the presence of a noise instability. For example, in the circuit of Figure 1, the noise instability detected across the M3-M4 inverter would not result in an erroneous value stored in the latch in the absence of the half-latch M1 device. Despite this fact, the noise instability violation is indicative of a serious design weakness which should be corrected. The second source of pessimism comes from the worst-case temporal correlation of Assump-

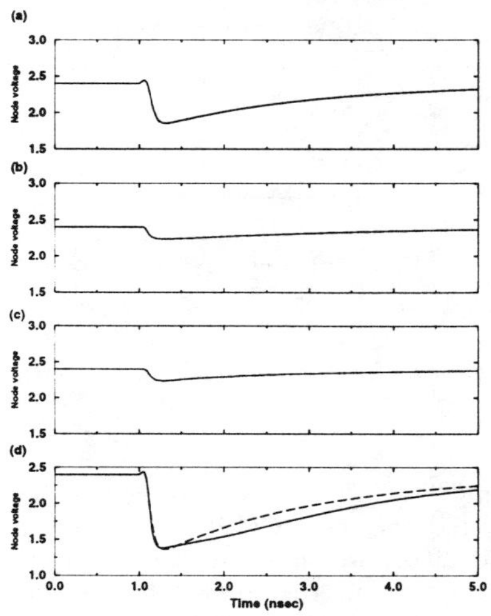

Figure 6: Noise calculation at node E: (a) coupled noise due to switching driver at I; (b) noise due to charge-sharing from switching input A; (c) noise propagated from node C or D; and (d) superposed noise. The solid curve in (d) comes from a strict time-domain sum. The dashed curve shows the exact result from circuit simulation.

tion 3. Not only are we assuming that all possible sources of noise are acting, but that they are acting together in time in the worst possible way. False paths can exist in static noise analysis which can be disallowed by both timing constraints and logic constraints as described in more detail in Section 7. The last major source of pessimism comes from Assumption 8. Using noise-limited propagation at a loop snip can introduce more noise into the circuit than would be introduced if the actual noise appearing on the input of the loop segment were propagated to the output. Because of these sources of pessimism, some fails identified by static noise analysis may be allowed after careful circuit simulation.

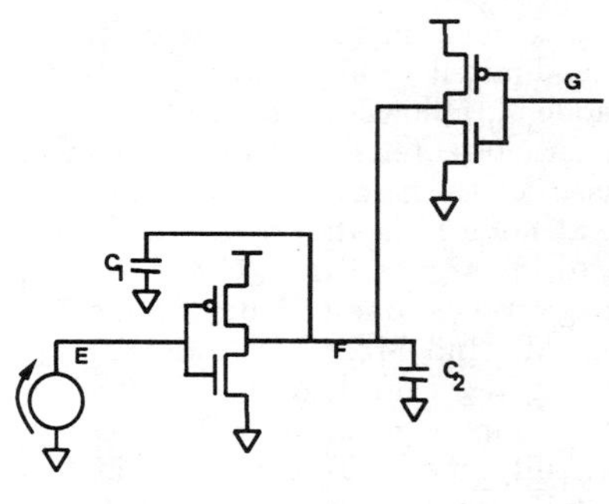

Figure 7: Network to perform the stability check and propagate the noise from node E to node F.

5 Power supply integrity analysis

Power and ground integrity analysis is also an important part of noise verification. There are two components to power-supply noise. The first is the variations in the dc power supply and ground levels due to the resistance of the power and ground distribution network. To calculate these IR drops, we can apply a separate analysis of the chip's spatial current demands against a supply- and ground-rail resistance extraction[10]. A second type of power supply noise is delta-I noise, produced by the simultaneous switching of off-chip drivers and internal circuits, usually synchronized with clock activitiy. Detailed transient analysis of the power grid involves applying models of the circuits' current demands to a detailed RLC extraction of the power grid combined with the package model[11, 10]. The current models usually take the form of Norton equivalent circuits at designated points in the power or ground distribution hierarchy, usually on a designated via layer.

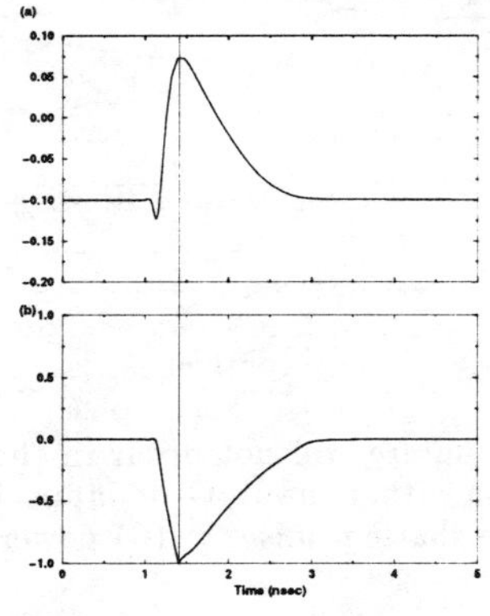

Figure 8: Noise calculation at node F: (a) propagated noise at node F, and (b) time-domain sensitivity of this noise to dc noise at the input.

6 Interconnect analysis

Central to noise analysis is the analysis of large coupled RLC networks. Two key ideas are central to this analysis: net complexes and hybrid admittance-transfer multiport macromodels. The net complex as shown in Figure 9 allows one to create a local environment to analysis the coupling for a given net in the design[12]. The primary net of the complex is the one for which we are trying to calculate the noise. The complex also includes secondary nets with significant coupling to the primary net. Coupling between the significant secondary nets and nets other than those already in the net complex are grounded and disconnected so that no current flows through them. This means that mutual inductances to these nets are discarded and coupling capacitors are treated as capacitors tied to ground. When ports of the resulting net complex terminate on MOS gates, we replace these loads with linear capacitors, which allows us to reduce the number of driving ports. At these "tap points", we instead calculate a transfer function from each of the ports. Krylov-subspace reduced-order modelling techniques, such as Arnoldi[13] and Lanczos[14], can be used to reduce these net complexes with ports and taps into hybrid admittance-transfer multiport macromodels.

Because of the performance limitations of interconnect RC delays, much practical effort has been focussed on reducing them through technology and design. Next generation technologies can be expected to make increasing use of low-resistivity metals (copper) and low-dielectric-constant insulators. Designers increasingly use wide, thick upper level metal for power and ground distribution and for long signal

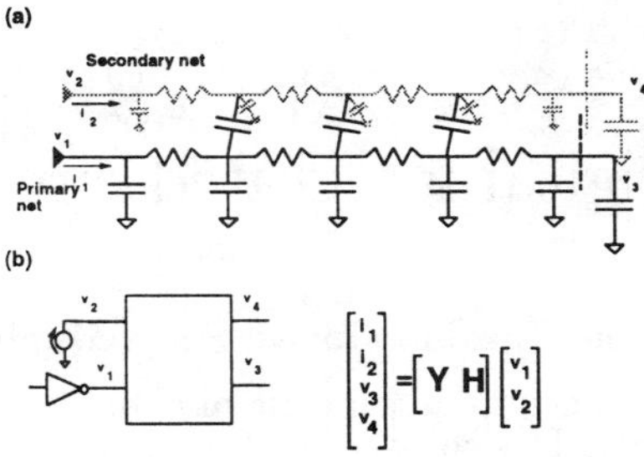

Figure 9: Modelling of interconnect: (a) a typical net complex consisting of a primary net coupled (in this case) to a single secondary net, (b) the hybrid admittance-transfer multiport model for this net complex

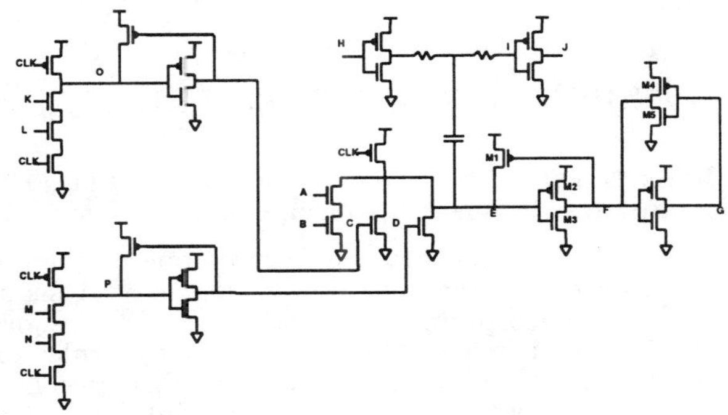

Figure 10: Figure 1 with the driving circuits included.

runs. These efforts, however, have resulted in the growing importance of inductance and inductive coupling in the timing and noise and noise analysis of signal lines for three important reasons. First, inductance must be incluced to accurately predict rise and fall times and delays in timing analysis. Secondly, if an inductive net is overdriven, an underdamped ringing response can be observed. This ringing is noise to subsequent receiving circuits and can potentially result in functional failure of the design. Lastly, inductive coupling along with capacitive coupling can be a significant source of noise on quiet nets due to the switching of nearby perpetrators. In the case of inductive interconnect, when a ringing response is possible, the temporal relationship that guides superposition in static noise analysis is an open question.

7 False paths in static noise analysis

False paths in static noise analysis correspond to cases in which signals cannot switch simultaneously because of non-overlapping arrival time windows or because of hazard-free logic constraints. Within the context of static noise analysis, these cases are easy to detect within the analysis associated with a single CCC. More complex situations, however, span multiple CCCs. Consider again the example of Figure 1 but include the circuits driving nodes C and D as shown in Figure 10. Now assume that the source of the noise appearing at C and D is charge-sharing noise due to the switching of M and K from 0 to 1 with L and N set to 0. Timing orthogonality[12] or hazard-free logic constraint information for switching signals H, A, M, and K could now be used to limit the pessimism associated with the noise calculation at node E. These false paths are usually eliminated with designer involvement.

8 Conclusions

Noise immunity has become of comparable importance to area, timing, and power in the design of digital integrated circuits. The noise stability criterion can be used to recognize possible failures and static noise analysis provides a technique for checking this criterion on a chip-wide basis. A successful design methodology for noise combines static noise analysis with design rules and power-supply integrity analysis. Hybrid multiport admittance-transfer macromodels are used to model the coupled RLC on-chip interconnect networks.

References

[1] K. L. Shepard and V. Narayanan. Conquering noise in deep submicron digital design. *IEEE Design and Test of Computers*, pages 51 – 62, January/March 1998.

[2] F. Dartu and L. T. Pileggi. Calculating worst-case gate delays due to dominant capacitance coupling. In *Proceedings of the IEEE/ACM Design Automation Conference*, pages 46 – 51, June 1997.

[3] K. L. Shepard and V. Narayanan. Noise in deep submicron digital design. In *Proceedings of the IEEE/ACM International Conference on Computer-Aided Design*, pages 524–531, San Jose, CA, November 1996.

[4] L. A. Glasser and D. W. Dobberpuhl. *The Design and Analysis of VLSI Circuits*. Addison-Wesley, 1985.

[5] Pong-Fei Lu et al. Floating-body effects in partially depleted SOI CMOS cicuits. *IEEE Journal of Solid-State Circuits*, 32(8):1241–1253, August 1997.

[6] D. Dobberpuhl et al. A 200 MHz 64b Dual-Issue CMOS Microprocessor. *IEEE Journal of Solid-State Circuits*, 27(11):1555 – 1567, 1992.

[7] Bruce A. Gieseke et al. A 600 MHz superscalar RISC microprocessor with out-of-order execution. In *Proceeding of the International Conference on Solid-State Circuits*, pages 176 – 177, 1997.

[8] Anirudh Devgan and Ronald A. Rohrer. Adaptively controlled explicit simulation. *IEEE Trans. CAD*, 13(6):746 – 761, 1994.

[9] A. Kuehlmann, A. Srinivasan, and D. P. Lapotin. Verity – a formal verification program for custom CMOS circuits. *IBM Journal of Research and Development*, 39(1/2):149 – 165, 1995.

[10] K. L. Shepard, S. Carey, E. Cho, B. Curran, R. Hatch, D. Hoffman, S. McCabe, G. Northrop, and R. Seigler. Design Methodology for the G4 S/390 Microprocessors. *IBM Journal of Research and Development*, 21(4/5):515 – 548, 1997.

[11] H. H. Chen. Minimizing chip-level simultaneous switching noise for high-performance microprocessor design. In *Proceedings of the IEEE International Symposium on Circuits and Systems*, volume 4, pages 544 – 547, 1996.

[12] K. L. Shepard, V. Narayanan, P. C. Elmendorf, and Gutuan Zheng. Global Harmony: Coupled Noise Analysis for Full-Chip RC Interconnect Networks. In *Proceedings of the IEEE International Conference on Computer-Aided Design*, 1997.

[13] L. Miguel Silveira, Mattan Kamon, Ibrahim Elfadel, and Jacob White. A Coordinate-Transformed Arnoldi Algorithm for Generating Guaranteed Stable Reduced-Order Models of RLC Circuits. In *IEEE/ACM International Conference on Computer-Aided Design*, pages 288 – 294, San Jose, CA, November 1996.

[14] Peter Feldmann and Roland W. Freund. Reduced-order modeling of large linear subcircuits via a block lanczos algorithm. In 32nd *ACM/IEEE Design Automation Conference*, pages 474–479, San Francisco, California, June 1995.

Panel: Taming noise in deep submicron digital integrated circuits

Chair: Nagaraj NS, Texas Instruments, Dallas, TX
Organizers: Kenneth Shepard, Columbia University, New York, NY and Takahide Inoue, Sony, Milpitas, CA

As technology scales into the deep submicron regime, noise immunity is becoming a metric of comparable importance to area, timing, and power for the analysis and design of digital VLSI chips. Are functional failures due to noise really a problem in a static CMOS design? Are design rules in the circuits and interconnect sufficient to protect against noise failures? Do design rules targetted to ensure noise immunity result in excessive penalty to performance and area due to their inherent conservatism? Is inductance in the interconnect really a problem? How much do we really need to account for capacitive coupling, inductance, and inductive coupling in delay analysis?

Chris Houghton, Digital Equiqment Corporation, Hudson, MA For years the digital IC design community has analyzed on-chip interconnect for capacitive and resistive effects. This has been done for both timing (capacitive loading on local interconnect and RC analysis of longer signals) and to assure integrity (capacitive coupling analysis). This is direct recognition that functional failures due to noise really are a problem.

Approaching the problem of on-chip noise with an armada of design rules does not assure success. Design rules are certainly necessary as the basis for any good design but they are not sufficient to protect against failures due to noise. An example of a "design rule" that no longer assures success is that of using only static CMOS circuit topologies. Although static circuits can have significantly more noise immunity than dynamic structures, in today's technologies (where coupled loading is a large portion of total loading) static structures can also fail. Even if static circuits mitigated all worry of noise, exclusive use of static design is often not an option in high performance circuits.

No rule set will, on its own, assure the integrity of a design. Even if it could, there certainly would be a penalty in circuit size, performance or power. Detailed circuit analysis *is* required to isolate timing and "signal integrity" problems. Given the large number of circuits on today's chips, the bulk of this analysis must be done with the help of EDA tools.

With the operating frequencies of ICs reaching the point where multi-Gigahertz on-chip interconnect bandwidths are required to accurately propagate signals, the question of what, beyond the standard RC analysis of yesterday, is needed today (and certainly tomorrow!) is raised. A popular question concerning on- chip signaling these days is: "is inductance in the interconnect really a problem?". The answer is "yes", but "maybe" is also valid. Recent improvements in the conductivity of on-chip metalization do not necessarily lessen the problems with inductance. In fact, in some situations higher conductivity interconnect can amplify the inductive effects. If the IC's "physical architecture" does not properly accommodate signal current return paths and does not control mutual inductance effects, then the answer is "yes, inductance is a problem" for even the 500 MHz designs of yesterday. The answer is a softer "maybe" for some "local" on-chip interconnect if the designer assures at least minimal accommodation of signal returns and minimization of coupled loading (w.r.t. self loading). This is true even in designs over 1 GHz designs.

Technology has come to our rescue in the past and hopefully will continue to in the future. Finer pitch interconnect and extra metalization layers have allowed for direct inclusion of reference structures (signal return paths). Recent and/or planned CMOS technology features will further ease high frequency design constraints, (e.g. higher conductivity on-chip metalization, low-κ dielectric layers and SOI). Flip-chip packaging has reduced the maximum length of global on-chip interconnects, reduced on-chip power supply noise and on-chip global clock skew. However, before we rely solely on technology as the panacea, consider the improvements that present design and analysis techniques yield within any given set of technology constraints.

Barbara Chappell, Intel Corporation, Hillsboro, OR With the shrinking voltages and increasing currents in deep-submicron IC's, performance for many important applications is too severely compromised by the usual worst-casing methods for assuring noise margins. However, modeling noise effects with sufficient accuracy to allow relieving of worst-case costs is difficult to do in any general way. What is the correct approximation model and analysis method is dependent on many design-specific elements - such as the circuit topologies to be supported, the performance targets, and the design optimization methods. Consequently, CAD developers need to work closely with designers in making their design tools noise-aware.

Xiaonan Zhang, Metaflow Technologies, Inc., La Jolla, CA As CMOS technology advances, intrinsic gate delay decreases and interconnect wire delay increases with respect to overall circuit delay. The coupling capacitance between minimum pitch wires on a $0.25\mu m$ CMOS IC can account for over 80% of the total capacitance of a wire. This makes interconnect crosstalk noise one of the biggest challenges in VLSI design today. I will discuss this noise problem from a circuit designer's point of view.

Crosstalk due to capacitive coupling is a complex phenomenon. One type of coupling-capacitance-induced crosstalk noise can be pervasive in today's interconnect intensive VLSI chips. It can occur in the following manner. When the voltage of an aggressor signal changes as the voltage of a victim signal is in transition through the high gain region of a receiver circuit (This is near $V_{DD}/2$ in a typical static CMOS design), if the switching directions are opposite of each other, a ripple or a small glitch is formed in the victim signal. This "high-gain crosstalk" can affect a circuit in many different ways:

- In a static CMOS design, this glitch increases the wire delay considerably (the signal goes across the receiving threshold back and forth). It also increases the receiving circuit delay since it alters the effective input rise/fall time. This causes speed-related logic errors.
- Sometimes, the receiving circuit amplifies the glitch to a larger glitch. This causes failures in edge-sensitive logic. The glitch can propagate through many gates.
- If the signal feeds into a dynamic logic gate, it can discharge the storage charge during the evaluation phase and cause a logic error.

Capacitive-coupling-induced interconnect crosstalk noise is a major problem in deep sub-micron VLSI design. The worst noise problem is high gain crosstalk. Circuit design techniques may provide viable solutions to the problem. However, this requires mostly custom design solutions. It is important to realize, crosstalk due to coupling capacitance between different wires on a chip is not simply a signal integrity problem, it can cause several other types of failures. We have observed the effects of crosstalk not only in dynamic circuits and analog circuits but also in static digital CMOS circuits. Minimizing crosstalk noise has become a major challenge to today's CAD tool developers and circuit designers. Copper wires will alleviate the problem to a certain extent. However, they do not eliminate the problem. High gain crosstalk will remain a challenge to engineers and scientists working on deep sub-micron and future generation VLSI designs.

John MacDonald, Sun Microsystems, Palo Alto, CA With metal trace dimensions a small fraction of a micron and rise and fall times a small fraction of a nanosecond, both capacitive and, to some degree, inductive coupling between signals become a significant problem for digital CMOS designs.

While most designers are aware of the damage that noise can do to the functionality of dynamic or precharged circuits, coupling noise can also cause functional failures in static CMOS. The effect that coupling has on propagation delay is fairly well understood but remains poorly modeled. As capacitive coupling is becoming the dominate component of interconnect loading, the standard practice of modeling it as a grounded capacitor with a "Miller" multiplier is becoming a leading source of inaccuracy in modeling delay. While we are not yet seeing significant problems on chip with inductive effects such as ringing or reflection, inductive coupling between signals is sometimes significant and should also be considered.

With diligence, design rules can be established and verification tools utilized to keep designs out of trouble. Nevertheless, the cost in area and performance is high, particularly considering the conservativism that design rules and tools rely upon.

John McBride, Hewlett-Packard, Fort Collins, CO Controlling "noise" has long been a concern of digital circuit designers. Whether that noise be radio-frequency-interference (RFI) noise, voltage supply noise, or signal integrity, noise has always been a concern. It is nothing new. PC board designers have been dealing with noise issues for a long time. Now signal speeds have increased and spacing has decreased to the point where digital ICs are highly susceptible to noise, too. The first coupling to show up is capacitive coupling, and the extraction and analysis of capacitive couple is fairly established at this point. Because designers have a knowledge of capacitive coupling, they know not to route long lines on inputs to dynamic gates, diffusion inputs to latches, and other sensitive circuits. Even static logic is becoming more susceptible to all types of noise as the supply voltage drops to 2.5V and beyond. Designers can take methodological steps to prevent capacitive coupling, and then use analysis to see if they were right, and if there are any hidden problems. There can always be improvements in the way designers analyze coupling, but compared to other problems in developing digital ICs, capacitive coupling is well understood and can be thoroughly analyzed.

Inductive coupling in digitial ICs is another story. The extraction of inductance is painfully slow. Extracting the inductance of all the signals on a large IC is completely infeasible at this time. Yet, inductance has been shown to be at least a contributing factor in the electrical problems of some current high-speed ICs. As signal edge rates increase, inductive coupling will get worse. Even though copper interconnect make inducantance easier to eliminate, copper will not be available to most designers for quite a while. Even then, inductance will still be a concern. Like capacitive coupling, many of the instances of inductive coupling can be stopped by methodology used in the design. Unlike capacitive coupling, the methodology is virtually the only way, today, to to eliminate inductive coupling problems. With every process, the design team needs to analyze the process for inductive coupling, and the team needs to carefully select the right guidelines for wire spacing, wire width, metal fill, wire length, ground returns, etc. These guidelines should be applied by all circuit designers, and checked for adherence to the guidelines. All "long" nets that drive sensitive circuitry should be extracted and analyzed. ("long" is defined by the design team). Attacking the problem primarily through the methodology has inherent downsides, namely that performance and signal density will suffer unnecessarily on some signals, unnecessary work will be performed on some signals, and there will always be some signals that are missed. However, until the extraction and analysis of inductive coupling catch up to where capacitive coupling are today, methodology will continue to be the primary defense against inductive coupling.

Bob Masleid, IBM, Austin, TX No advance position statement.

FACT: A Framework for the Application of Throughput and Power Optimizing Transformations to Control-flow Intensive Behavioral Descriptions *

Ganesh Lakshminarayana and Niraj K. Jha

Department of Electrical Engineering, Princeton, NJ 08544

Abstract

In this paper, we present an algorithm for the application of a general class of transformations to control-flow intensive behavioral descriptions. Our algorithm is based on the observation that incorporation of scheduling information can help guide the selection and application of candidate transformations, and significantly enhance the quality of the synthesized solution. The efficacy of the selected throughput and power optimizing transformations is enhanced by the ability of our algorithm to transcend basic blocks in the behavioral description. This ability is imparted to our algorithm by a general technique we have devised. Our system currently supports associativity, commutativity, distributivity, constant propagation, code motion, and loop unrolling. It is integrated with a scheduler which performs implicit loop unrolling and functional pipelining, and has the ability to parallelize the execution of independent iterative constructs whose bodies can share resources. Other transformations can easily be incorporated within the framework. We demonstrate the efficacy of our algorithm by applying it to several commonly available benchmarks. Upon synthesis, behaviors transformed by the application of our algorithm showed up to 6-fold improvement in throughput over an existing transformation algorithm, and up to 4.5-fold improvement in power over designs produced without the benefit of our algorithm.

1 Introduction

The explosive growth in the embedded systems market has fuelled interest in the automated synthesis of digital systems from functional specifications. These systems often need to be compact, and satisfy throughput and power constraints. Transformations which can improve these design metrics have recently gained prominence. Transformational techniques replace a behavioral description by a structurally dissimilar, but functionally equivalent, description. The new description usually results in designs which are better than those synthesized from the original description, with respect to specific design metrics.

Research in compiling techniques has identified a large number of program transformations which enhance performance and reduce size of programs on general-purpose computers. A comprehensive survey of these techniques is provided in [1]. These include algebraic transformations such as constant propagation, commutativity, associativity, distributivity, strength reduction, and common sub-expression elimination [2], loop-optimizing transformations such as loop un-switching, loop distribution, loop fusion, loop coalescing, loop peeling, and loop unrolling, software pipelining, speculative execution, code motion, redundant code elimination, and interprocedural optimizations such as inlining, cloning, *etc.* The idea of using program execution profiles to guide performance-optimizing transformations has been proposed in [3] and [4]. In most of these approaches, scheduling succeeds transformation application. We demonstrate that, in high-level synthesis, scheduling information can guide the application of transformations and enhance their effectiveness. In addition, we support power-optimizing transformations, unlike compiler-related optimization techniques for which speed is the primary objective.

In high-level synthesis, transformations for throughput and power optimization for data-flow intensive (DFI) behavioral descriptions [5], [6], as well as control-flow intensive (CFI) behaviors [7], [8] have been presented. None of these transformation techniques for CFI behaviors targets power optimization. With the exception of [8], these systems do not incorporate scheduling information into the transformation application process.

In this paper, we present a framework for transforming CFI behavioral descriptions to synthesize throughput- or power-optimized designs. At the heart of our approach is a procedure we have devised for recognizing candidates for transformation application across the scope of basic blocks, and applying the transformation while preserving functionality. This procedure drives the transformation process, which is closely integrated with scheduling. Our algorithm begins with a scheduled behavioral description, specified in the form of a state transition graph (STG). Profiling information is used to partition the STG into blocks, which are separately optimized by transformation application. This enables our algorithm to direct its focus on the critical sections of the behavior. Since the identified blocks can have arbitrary control flows within them, our algorithm has a complete picture of the trade-offs involved, unlike many other algorithms that only target simple structures such as straight-line code, or single-entry loops. Each block is transformed by the repeated application of data- and control-flow transformations, to optimize for the desired objective (throughput or power). Within a block, scheduling is interleaved with transformation application. The scheduling information, available at each step, allows effective identification of bottlenecks and the selection of appropriate transformations to eliminate them. We currently support the application of commutativity, constant propagation, associativity, distributivity, code motion, and loop unrolling. The scheduling algorithm we use performs implicit loop unrolling, and functional pipelining, and parallelizes the execution of independent loops. The framework we have developed can, however, easily be customized by the addition of user-specified transformations.

2 Preliminaries

In this section, we discuss control-data flow graphs (CDFGs), which describe the input behavior, and STGs, which describe the schedule. We then outline a high-level power estimation technique which obtains a fast estimate of the power consumption of a design, using an STG, user-specified input traces, and a set of library mod-

*This work was supported in part by NSF under Grant No. 9319269 and in part by Alternative System Concepts, Inc. under an SBIR contract from Air Force Rome Laboratories.

```
TEST1 ( int c1, int c2){
  int i = a = 0;
  while (c2 > i){ // > 1
    if (i < c1) { // < 1
      t1 = a + 7; // +1
      a = 13 * t1; // *1
    }
    else {
      a = a + 17; // +2
    }
    i = i + 1; //++1
    x[i] = a; // S
}}
```

(a)

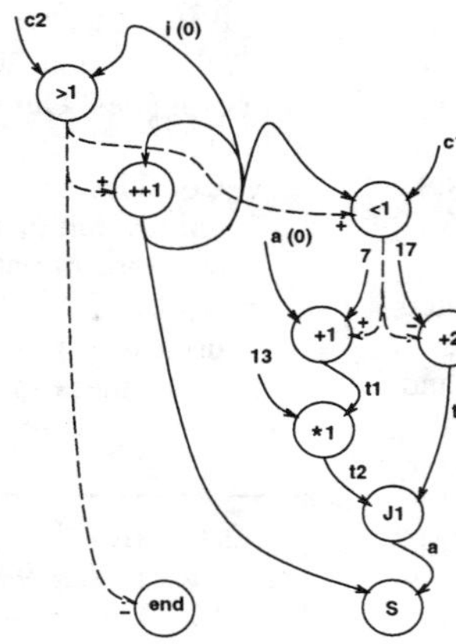

(b)

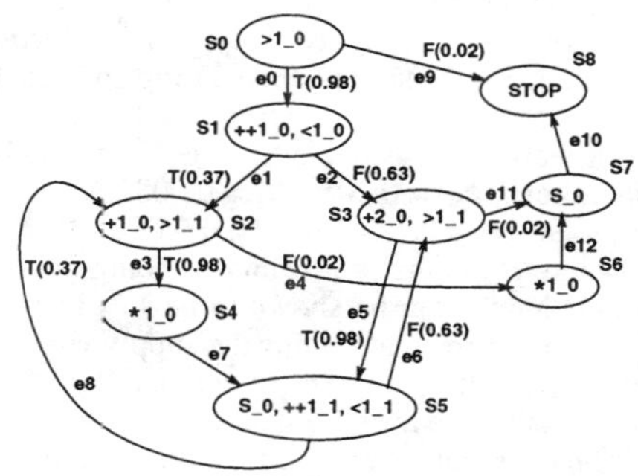

(c)

Figure 1: (a) A fragment of code called TEST1 written in a high-level language, (b) its CDFG representation, and (c) an STG representation of its schedule

ules characterized for delay and power.

2.1 CDFG and STG models

A CDFG is a directed graph, whose nodes represent operations, and edges represent dependencies of two types: data and control. An edge represents a data dependency if the source node of the edge produces data that the sink consumes. Existence of a control dependency between nodes implies that the execution of the sink node depends on the outcome of the execution of the source node. Our CDFG model uses a token passing semantic [9]. An operation consumes the tokens available at its inputs and generates a token at its output. The execution of an operation is contingent upon the arrival of tokens at its inputs.

Figures 1(a) and 1(b) show a behavioral description written in a high-level language and the CDFG corresponding to the description, respectively. Statements in the behavioral description are shown annotated with the operation(s) in the CDFG they represent. Data dependencies in the CDFG are indicated by continuous arcs, and control dependencies, by broken arcs. An operation, whose execution depends on a condition evaluating to *true* (*false*), is marked with a + (−). A *join* operation assigns to its output, the value at either of its input edges. Such an operation executes when a token is available at either input. Also included in our CDFG representation is a *select* operation with three inputs, l, r, and s, and an output o, which assigns to o, the value at its l (r) input ports, if the value at s is *true* (*false*).

On completion, the scheduler outputs the STG describing the schedule. Figure 1(c) represents the STG obtained by scheduling the CDFG shown in Figure 1(b). The STG is a directed graph whose nodes represent states and edges represent transitions between states. Nodes in the STG have information about the operations executed in the corresponding state, and edges capture the conditions under which a state transition takes place. Many schedulers support *loop unrolling*, a performance optimization technique, which allows operations from different iterations of a loop to be performed concurrently. In such cases, operations in a state are annotated with the iteration of the loop they represent. For example, the *zero*th iteration of operation $x[i] = a$ (**S_0**), and the first iteration of operations $i = i + 1$ (++1_1) and $i < c1$ (<1_1) are executed in state **S5**. The number in parentheses placed adjacent to an edge represents the probability that the corresponding edge is taken.

2.2 High-level power estimation

Application of transformations targeted towards reducing power consumption requires a high-level power consumption model whose predictions are correlated with the final power consumption of the design, and which is fast enough to be applied repeatedly in the inner loop of the transformation algorithm. In this section, we illustrate our power estimation technique with an example. Note that though this procedure is used to estimate the power consumption of designs during the transformation process, the final power measurements, used for our experimental results, are obtained from actual layouts.

Table 1: Functional unit selection, allocation, and component information

FU type	*Ops.*	*#*	$\frac{E}{V_{dd}^2}$	*Delay*	*Area*
comp1	>1, <1	2	1.1	12	1.3
cla1	+1, +2	2	1.3	10	1.5
incr1	++1	1	0.7	13	1.1
w_mult1	* 1	1	2.3	23	3.9
reg1		n/a	0.3	3	1.0
mem1	S_0	1	1.9	15	8.1

Problem statement: Given (a) a scheduled behavioral description in the form of an STG, (b) functional unit selection information (information about the type of functional unit which performs an operation), (c) a library of functional units and storage elements (registers, memories) characterized for area, delay, and power, (d) the clock period of the design, and (e) a set of typical input traces, estimate the power consumption of the final design.

We extend the method proposed in [5] to handle CFI designs. The average power consumption of the circuit is computed as the average energy consumption divided by the *average schedule length*. The average schedule length is defined as the average time, in cycles, to complete one execution of the behavior, and is computed from the STG and the input traces, using the method presented in [10]. The total energy consumed by the circuit is computed as the sum of the energies consumed by the functional units, registers, interconnect, and the controller. The energy, E, consumed by a functional unit or register is expressed as $E = C_{type} \times V_{dd}^2 \times N_{ops}$, where C_{type} is a constant which depends on the functional unit or register type, V_{dd} is the supply voltage, and N_{ops}, for a functional unit (register) is the number of operations executed (number of read/write accesses). Further details about the power model can be found in [5]. The following example illustrates our power estimation procedure.

Example 1: Consider the problem of estimating the power consumption of a design based on the STG shown in Figure 1(c). The library elements available, the energy dissipation, delay, and area associated with each element, functional unit selection, and allocation information (numbers and types of functional units allowed for synthesis) are summarized in Table 1. The clock period constraint is $25ns$. Simulation of the behavioral description with the input traces yields the following branch probabilities: the while loop "**while** $(c2 > i)$" closes with a probability of 0.98, and the branch "**if** $(i < c1)$" is taken with a probability of 0.37. Note that once the branch probabilities are determined, they can be *repeatedly used*. Therefore, simulation is done only once during an execution of the algorithm.

The state probabilities, evaluated using the method presented in [10], are as follows $P_{S0} = 0.008$, $P_{S1} = 0.008$, $P_{S2} = 0.153$, $P_{S3} = 0.259$, $P_{S4} = 0.149$, $P_{S5} = 0.404$, and $P_{S6} = 0.003$, $P_{S7} = 0.008$, and $P_{S8} = 0.008$, where P_{Si} is the probability of being in state Si. The average schedule length can be shown to be 119.11 cycles. Note that scheduling is performed assuming the supply voltage to be $5V$. If the supply voltage were different, functional unit delays would change, thus impacting the average schedule length.

Using the above probabilities, one can derive the average energy consumption per iteration of the design. To do this, we need to find the average number of operations executed by each type of functional unit, and the average number of variable accesses, per iteration of the design. These numbers are derived as the sums of per-state numbers of operations executed by different functional unit types, and variables accessed, weighted by the state probabilities. For example, the number of operations executed by functional units of type *incr1* is given by $119.11 \times (P_{S1} \times 1 + P_{S5} \times 1) = 48.95$. Since the energy consumed by the incrementer while performing one operation is $0.7 \times V_{dd}^2$ units (see Table 1), the average energy consumption, per iteration, by the incrementer is $34.27 \times V_{dd}^2$ units. The energy consumption due to the comparators, adders, multipliers, registers, and memory, and can be found to be $108.75 \times V_{dd}^2$, $63.64 \times V_{dd}^2$, $41.70 \times V_{dd}^2$, $99.38 \times V_{dd}^2$ and $93.10 \times V_{dd}^2$ units, respectively, using the same technique. The total energy consumption is $665.58V_{dd}^2$ units, after accounting for the contribution due to the interconnect and controller. We first compute the average power consumption by assuming the supply voltage to be $5V$. This is done by dividing the average energy consumption by the average time taken (119.11 cycles at $5V$), and evaluates to $\frac{665.58 \times V_{dd}^2}{119.11 \times cycle\_time}$ *units*.

The quadratic dependence of power consumption on supply voltage implies that significant reductions in power consumption can result if the supply voltage is scaled down. We estimate the scaled supply voltage in the following manner. As mentioned in Section 1, our algorithm starts with a scheduled behavior. Transformations and scheduling are then applied in an interleaved fashion. The initial schedule given to us is regarded as the "base" case, and all comparisons are made with respect to it. When we optimize for power, we scale the supply voltage until the average schedule length drops to that of the untransformed design. For example, if the STG shown in Figure 1(c) is derived from the transformed behavior, and the schedule derived from the initial (untransformed) behavior has an average schedule length of 151.30 cycles, the supply voltage is scaled according to the equation [1]

$$\frac{V_{dd}\_initial/(V_{dd}\_initial - V_t)^2}{V_{dd}\_new/(V_{dd}\_new - V_t)^2} = 119.11/151.30$$

If $V_{dd}\_initial$ is assumed to be $5V$ and the threshold voltage V_t is assumed to be $1V$, then V_{dd}_new evaluates to $4.29V$, which yields a power estimate of $665.58 \times 4.29^2/(151.30 \times cycle\_time) = 80.96/cycle\_time$ units for the design. Note that the term representing the average schedule length changes from 119.11 to 151.30 to reflect the change in the supply voltage. This is because an average schedule length of 119.11 *cycles* at $5V$ is equivalent to an average schedule length of 151.30 at $4.29V$. Thus, our aim is to keep the performance of the untransformed and transformed scheduled behaviors the same while reducing power. ∎

[1] $Delay = k \times V_{dd}/(V_{dd} - V_t)^2$ [11].

3 Motivational Examples

In this section, we motivate the key features of our algorithm with examples. Our algorithm is based on the following key observations:

- Considering scheduling information during transformation application can significantly enhance the quality of the chosen transformations. This is illustrated through Example 3.
- Applying transformations across basic block boundaries is critical in producing high-quality solutions, and maintaining correctness in the process can be complex and non-trivial. This is illustrated through Example 3.

```
TEST2 (int k){
L1   for i = 0 to 99{ //< 1
       t1 = y[i] + k; // +1
       k = k + t1; // +2 }
L2   for j = 0 to 99{ // < 2
       t2 = x[j] - k; // -2
       k = k - t2; // -3}
L3   for m = 0 to 399{ // < 3
       t3 = y1[m] + y2[m]; // +3
       t4 = y3[m] + y4[m]; // +4
       t5 = t3 - t4; // -1
       x1[m] = t5; }
     return;}
```

(a)

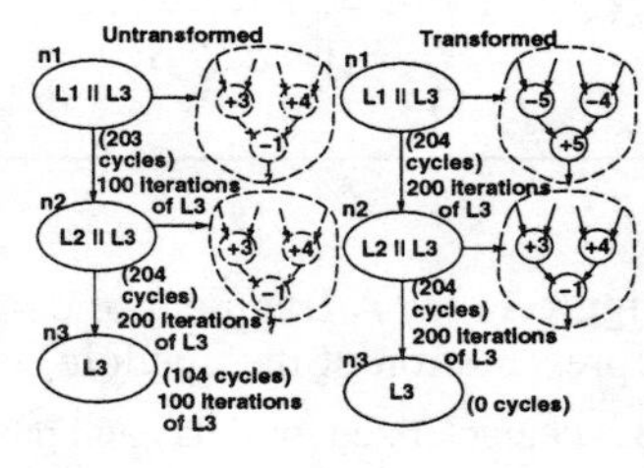

(b) (c)

Figure 2: (a) A behavior, TEST2, (b) a schedule for TEST2, and (c) a schedule derived after transforming TEST2

Example 2: Consider the problem of synthesizing the behavior shown in Figure 2(a) to optimize for throughput. Suppose the resources available to implement this behavior are two adders, two subtracters, two comparators, and two incrementers, all of which execute in one cycle. Arrays x, $x1$, y, $y1$, $y2$, $y3$, and $y4$ are assumed to be mapped to separate memories, and can be simultaneously accessed. Figures 2(b) and 2(c) describe schedules for the untransformed and transformed behavior, respectively. Since the STGs for the actual schedules are rather involved, the given figures provide the basic features of the schedule, without giving a state-by-state description. Nodes in Figure 2(b) correspond to loops in the behavior. A node is annotated with its name (placed outside the ellipse representing it), and the loops it corresponds to (placed inside the ellipse). In this schedule *L1* and *L2* first execute in parallel, then *L2* and *L3*, and finally *L3* executes alone.

Nodes *n1* and *n2* are also annotated with the number of iterations of *L3* they represent. While *L3* is able to execute 200 iterations in node *n2*, it can only execute 100 iterations in node *n1*. An inspection of Figure 3(a) reveals the reason for this bottleneck. This figure shows concurrent execution of loops *L1* and *L3* over two iterations. Since contention for the adders and subtracters is the reason for the bottleneck, we only show the schedules for the addition and subtraction operations which compete for resources.

Since there are no inherent data dependencies that slow down the execution of *L3*, we can explore methods to transform its body in such a manner that its resource requirements are tailored to the environment in which it is embedded. One way to achieve this would be to rewrite $(y1[m] + y2[m]) - (y3[m] + y4[m])$ in the form $(y1[m] - y3[m]) + (y2[m] - y4[m])$, as shown in Figure 2(c) (see the CDFG enclosed in broken boundaries placed adjacent to node *n1*). In this figure, operation -5 represents $y1[m] - y3[m]$, operation -4 represents $y2[m] - y4[m]$, and operation +5 represents their addition. In this case, the body of loop *L3* would require two subtraction operations and one addition operation. Since *L1* consumes one adder per cycle, one adder and two subtracters are available. In this

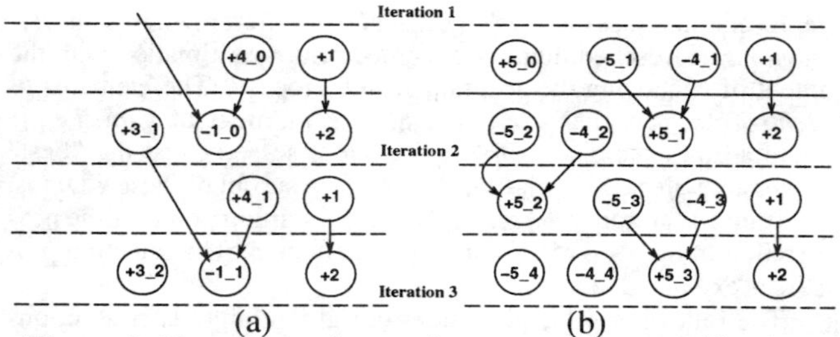

Figure 3: Transformations to improve resource utilization

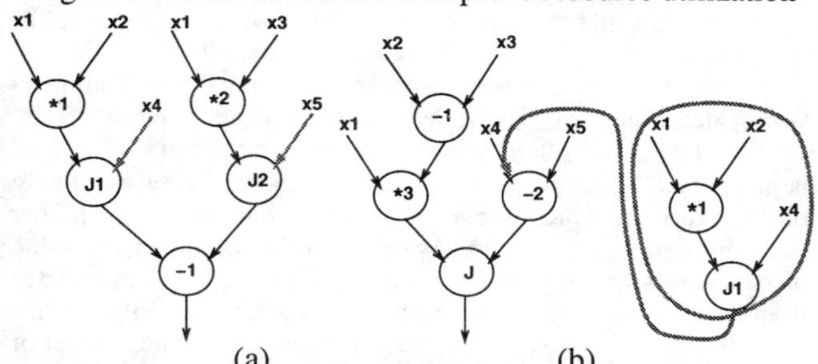

Figure 4: Transformations across basic blocks

scenario, one iteration of *L3* can begin every cycle, as illustrated in Figure 3(b). Note that this transformation is applied only to node *n1*, and node *n2* remains unaltered.

The schedule shown in Figure 2(c) has an expected length of 408 cycles, a 1.25-fold improvement over the schedule presented in Figure 2(b). This improvement is a direct consequence of considering scheduling information while applying transformations. Note that if this behavior were optimized for power, the improvement in throughput could be traded off for a 25% reduction in power consumption through V_{dd}-scaling. ∎

Example 3: Consider the problem of transforming the CDFG shown in Figure 4(a) to optimize for throughput under the following allocation constraints: one multiplier and two subtracters. All functional units are assumed to take one cycle to execute. The join operations embedded in the CDFG indicate that different threads of execution are possible, depending upon which of the *join* operation's inputs propagates to its output. In this case, since there are two distinct *join* operations, four threads of execution are possible. Out of the four possible threads, not all would occur in practice, because the occurrence of a token on one input might preclude the occurrence of a token on another, *i.e.*, some input pairs might be mutually exclusive. In this example, the pairs $\{x2, x5\}$ and $\{x3, x4\}$ are assumed to be mutually exclusive.

Transformational techniques typically identify specific structures in the CDFG which are good candidates for transformation application, and alter these structures in the manner dictated by the selected transformation. In this example, let us suppose we choose to utilize the distributivity of multiplication over subtraction to replace a structure, *Source*, of the form $a \times b - a \times c$ by a structure, *Target*, of the form $a \times (b - c)$. An inspection of the CDFG reveals that, if the *join* operations in the CDFG were configured to select their left inputs, *i.e.*, if *J1* selected the result of operation $*1$ and *J2* selected the result of operation $*2$, the data and control flows within the resultant CDFG would be identical to *Source*. This implies that under some conditions, the CDFG can be made to appear isomorphic to *Source*. Let *C* denote the condition under which this occurs. If *Source* were replaced by *Target* when *C* were to evaluate to *true*, then the resultant behavior would take only two cycles to execute as there is one subtraction, ($t1 = x2 - x3$), and one multiplication, $x1 \times t1$, as opposed to the three cycles it would otherwise need (two multiplications, occurring in sequence on one multiplier, with the results feeding a subtraction). If *C* were to occur with a high probability, the ability to perform this substitution would, clearly, be desirable.

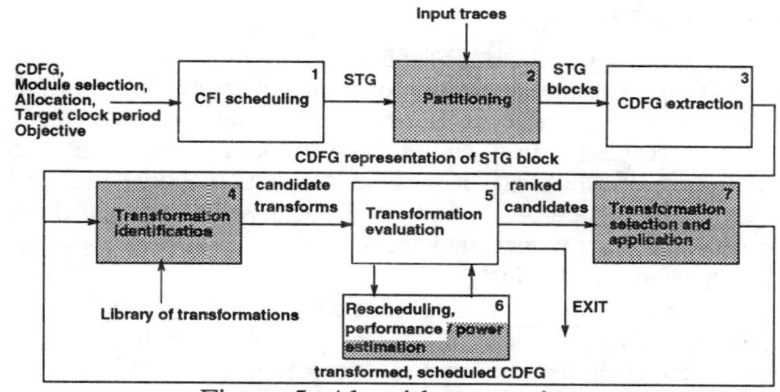

Figure 5: Algorithm overview

In performing this substitution, the following two issues need to be considered.

- The transformed CDFG should be functionally equivalent to the original CDFG for every thread of execution encountered.
- The transformed CDFG should be as compact as possible, *i.e.*, no redundant operations should be performed.

Figure 4(b) shows the CDFG obtained upon transforming the CDFG shown in Figure 4(a) (for now, the part of the CDFG shown in the grey enclosure, and the grey edge it sources can be neglected). It can be verified that, for every thread of execution encountered, the output of the transformed CDFG is functionally equivalent to the output of the original CDFG. Note that, if *x2* and *x5* were not mutually exclusive, the transformed behavior would not be functionally equivalent to the original one, as the original CDFG would compute $x1 \times x2 - x5$ upon the arrival of tokens at inputs *x1*, *x2*, and *x5*, and the transformed CDFG would not produce an output. To ensure correctness, in the case when *x2* and *x5* are not mutually exclusive, the left input to operation -2 should come from the grey edge. ∎

4 The Algorithm

In this section, we outline the working of our algorithm. Details can be found in [12]. Figure 5 illustrates the basics of our algorithm. In this figure, the steps shown shaded have been enhanced by our algorithm. Other steps correspond to well-researched problems, and are not described. The inputs consist of a behavioral description, specified as a CDFG, functional unit selection and allocation information, target clock period, typical input traces, and synthesis objective (performance or power). The behavior is first scheduled, using an existing CFI scheduler (step **1** in Figure 5). The output of the scheduling process is an STG. The input traces provided are used to analyze the STG to determine the probabilities with which individual states are encountered. The STG is then partitioned into blocks (step **2**). Individual blocks (groups of states) in the partition are then transformed and rescheduled in an iterative fashion using the following technique. First, the functionality of the block is captured in the form of a CDFG (step **3**), *i.e.*, we identify the portion of the CDFG which corresponds to the STG block. We then determine multiple ways in which the CDFG can be transformed, using a library of transformations (step **4**). The key strength of this process lies in its ability to recognize and apply transformations across the scope of basic blocks. The efficacy of each considered transformation is then assessed (steps **5** and **6**). This is done by applying the considered transformation, scheduling the CDFG, and using the schedule information to estimate the throughput or power (see Section 2.2). A subset of candidates is then chosen for further exploration (step **7**).

4.1 Partitioning

In this section, we describe the procedure followed to partition the STG prior to the application of transformations. The starting point is an STG and a set of user-specified input traces. The output of the process is a set of *disjoint* "STG blocks". An STG block is a collection of states that satisfies the following properties: (a) any state, S, can form a block, and (b) if a state, S', is an immediate predecessor or an immediate successor of any state in a block, $\mathcal{B}$, then $\mathcal{B} \cup S'$ is a valid block.

The first step in partitioning is the derivation of transition probabilities on the edges of the STG, *i.e.*, the probability that an edge, e, is taken, given that the current state is *Source*(e)[2]. This is done by simulating the CDFG representing the input behavior with the input traces provided. The simulation yields the number of times each branch in the CDFG is encountered, from which the probability of a branch can be computed. Once the probability of each CDFG branch is known, the transition probabilities on the edges of the STG can be computed as the product of the probabilities of the CDFG branches which constitute the STG edge. These probabilities are used to evaluate the probability of being in a particular state, using the technique described in [10].

The derived state probabilities are used to rank transitions in decreasing order of *relative frequency*, as follows: the relative frequency of a transition along edge e in the STG is computed as the product of the probability of being in *Source*(e) and the probability of e being taken, given that the current state is *Source*(e). Edges, whose relative frequency exceeds a threshold are chosen for further processing.

From the chosen set of edges, STG blocks are formed in the following manner: the edge set is traversed in decreasing order of frequency. The source and sink nodes of the first edge are grouped into a block, $\mathcal{B}$. If the source (n_{source}) and sink (n_{sink}) nodes of any subsequent edge, e_{next}, do not belong to any existing block, a new block with contents n_{source} and n_{sink} is created. If the source (sink) node of e_{next} belongs to an existing block, $\mathcal{B}'$, but the sink (source) node, n_{sink} (n_{source}), does not, $\mathcal{B}'$ is augmented with the addition of n_{sink} (n_{source}). If e_{next} has its source node in one block, and its sink node in another, the two blocks are fused into one. The created blocks are optimized by the application of transformations, as described in the following section.

4.2 Controlling the application of transformations

As mentioned in the previous section, the partitioned STG is modified by the application of transformations chosen from a library. In this section, we describe the procedure used to control the application of transformations to an STG block.

Figure 6 describes the part of our algorithm which corresponds to steps **3**, **4**, **5**, **6** and **7** in Figure 5. The inputs are the STG block, B, to be modified, functional unit selection information, allocation constraints, clock period, the transformation library, and the objective function (throughput or power). The approach we use combines some aspects of both simulated annealing and iterative improvement. At each stage of the algorithm, we maintain a set, *In_set*, of solutions which serve as starting points for future exploration. These solutions are CDFGs which are the result of the application of a sequence of transformations to the CDFG derived from the input STG block. The neighborhood of each element of *In_set* is explored to identify candidate transformations. The loop in statement **5** of the pseudocode in Figure 6 represents iteration through *In_set*, and statement **6** represents exploration of the neighborhood of an element, g, of *In_set* by identifying candidate transformations which can be applied to g. As shown in statement **7**, CDFGs obtained by the applications of the candidate transformations are grouped into a single set, *Behavior_set*. Statements **8-10** represent the assessment of the efficacy of different CDFGs in *Behavior_set*. Rescheduling is performed prior to estimation with the intent of enhancing the accuracy of the process. The elements of *Behavior_set* are then ranked in decreasing order of gain, *i.e.*, if there are N candidate CDFGs, the one assessed to be the "best" ("worst") will rank 0 ($N-1$). A fixed-size subset of these CDFGs, selected and stored in *In_set*, serves as the starting point for the next iteration of the loop represented by statement **3**. The selection process is explained next.

The following principle guides our algorithm: better solutions are selected with a higher probability. In the initial phase of the algorithm, bad solutions get selected with relatively high probabilities, and as the algorithm proceeds, their probabilities of selection drop. We, therefore, generate a stream of unique random numbers whose size equals the size of the required subset. The ratio of the probabilities of selection of two solutions, which rank $n1$ and $n2$ is given by e^{-kn1}/e^{-kn2}. The parameter k is low at the beginning of the algorithm, which implies that poorer solutions have a higher probability of being selected. As the algorithm proceeds, the value of k increases, favoring better solutions. The algorithm terminates if an iteration of the loop represented by statement **3** does not improve the quality of solution. k is a linear function of the number of executions of the loop represented by statement **3**.

The rationale behind our algorithm is as follows: we try to simultaneously explore multiple regions of the solution space. If at any point, we have multiple possible solutions, simulated annealing or iterative improvement would choose one, based on a heuristic, while our algorithm would choose more solutions, thus increasing the probability of generating a better solution.

```
Apply_transforms (STG_BLOCK B, ALLOCATION_CONSTRAINT C,
 TRANSFORM_LIBRARY T_lib, CLOCK_PERIOD clk,
 FUNCTIONAL UNIT SELECTION fu_select, OBJECTIVE obj){
0  CDFG G = extract_CDFG(B);
1  SET <CDFG> In_set = {G};
2  while (stopping criterion not satisfied){
3    for(i = 0; i < MAX_MOVES; i++){
4      SET <CDFG> Behavior_set = Φ;
5      foreach element (g, In_set){
6        SET <CDFG> beh_set = Identify_and_apply
           _candidate_transformations(g, T_lib) ;
7        Behavior_set = Behavior_set ∪ beh_set;
       }
8      foreach element (b, Behavior_set) {
9        Reschedule CDFG b using fu_select;
10       Estimate objective function;//power or throughput
11       Store best solution seen so far;
       }
12     LIST <CDFG> Behavior_list = Sort (Behavior_set);
       //Sort candidate CDFGs in decreasing order of objective function
13     Select a fixed-size subset, s, of Behavior_set
       //based on a random variable distributed as c × k × e^(-kx)
       //where c is a constant, k is a monotonically increasing
       //function of i, and x is a uniform random variable.
14     In_set = s;
}}}
```

Figure 6: Controlling the application of transformations

5 Experimental Results

The techniques described in this paper were implemented in a program called *FACT*, written in C++. We evaluated this program by using it to transform several commonly available benchmarks, to optimize for throughput and power. We compare our method, *FACT*, with the technique presented in [7] (*Flamel*), and another technique, which will be called *M1* for the remainder of this discussion. *Flamel* applies the same transformation suite as our method does, and also has the ability to transcend basic blocks in its optimization, which makes it an ideal candidate for comparison. All

[2] *Source*(e) represents the state that sources e.

Table 2: Throughput and power results

Circuit	Clk	T-opt.			P-opt	
		M1	Fl	FACT	M1	FACT
	ns	T	T	T	P	P
GCD	25	6.3	10.1	16.9	2.8	0.9
FIR	25	167	167	1000	7.6	1.7
Test2	25	2.0	2.0	2.5	11.3	8.4
SINTRAN	25	1.3	1.7	2.5	11.4	4.0
IGF	25	0.2	0.3	0.3	9.1	7.0
PPS	25	125	333	333	9.9	3.6

Table 3: Allocation constraints for the examples in Table 2

Circuit	a1	sb1	mt1	cp1	e1	i1	n1	s1
GCD	-	2	-	1	1	-	-	-
FIR	1	4	1	-	-	-	4	-
Test2	2	2	-	2	-	2	-	-
SINTRAN	4	4	5	1	-	1	2	-
IGF	1	1	2	1	-	1	-	1
PPS	5	-	-	-	-	-	-	-

methods have access to our in-house scheduling algorithm [13], which can perform the following transformations in an integrated fashion: loop unrolling, functional pipelining (even across **if** constructs), and *concurrent loop optimization* (the ability to parallelize the execution of independent loops, possibly with unknown bounds). Method *M1* just takes the input CDFG through behavioral synthesis, giving it access to only those transformations supported by our scheduling algorithm. Note that recent work on power-optimizing transformations, *e.g.* [5], has focussed on DFI descriptions, making them inapplicable to many of our benchmarks.

Comparisons were performed with respect to throughput and power in the following manner. The CDFGs were synthesized, placed, and routed, using our in-house synthesis tool. Throughput was estimated as the inverse of the expected execution time between consecutive iterations of the CDFG, using techniques presented in [10]. Power estimation was performed using a transistor-level netlist, extracted from the layout, using a switch-level simulator, IRSIM-CAP. The inputs to the power estimator were derived using a zero-mean Gaussian sequence, which was subsequently passed through an autoregressive filter to introduce the desired level of temporal correlation.

Table 2 summarizes the results obtained. *Clk* represents the clock period constraint. The columns labeled *T-opt* and *P-opt* represent, respectively, the CDFGs produced by the application of transformations aimed at throughput and power optimization. Minor column *M1* represents method *M1*, and minor columns *Fl* and *FACT* represent CDFGs which have been transformed by the application of techniques presented in [7] and this paper, respectively. Columns *T* and *P* represent the throughput (measured in $cycles^{-1} \times 1000$) and power (measured in *mW*), respectively, of the circuits synthesized from the corresponding CDFG, measured in the manner outlined before (*e.g.*, for *FIR*, *FACT* produces a throughput of 1 $cycle^{-1}$ in the throughput optimization mode). All CDFGs were synthesized under the same allocation constraints, to meet the same target clock period. We used a library of functional units which consists of (a) an adder, *a1*, with a delay of 10*ns*, (b) a subtracter, *sb1*, with a delay of 10*ns*, (c) a multiplier, *mt1*, with a delay of 23*ns*, (d) a less-than comparator, *cp1*, with a delay of 10*ns*, (e) an equality comparator, *e1*, with a delay of 5*ns*, (f) an incrementer, *i1*, with a delay of 5*ns*, (g) a multi-bit inverter, *n1*, with a delay of 2*ns*, and (h) a shifter, *s1*, with a delay of 10*ns*. The allocation constraints for an example can be found by looking up the entry corresponding to the example in Table 3. For example, the allocation constraint for *GCD* is two *sb1*, one *cp1*, and one *e1*.

Of our examples, *GCD* computes the greatest common divisor of two numbers, *FIR* is a finite impulse response filter, *SINTRAN* evaluates the sine transform, *IGF* evaluates the incomplete gamma function, *PPS* is a parallel prefix sum, and *Test2* is the example shown in Figure 2(a).

The results obtained indicate that the designs derived from *FACT*, in the throughput optimization mode, produced on an average a 2.7-fold improvement in throughput over designs produced using *M1* and 2.1-fold improvement over designs derived from *Flamel*. The power-optimized designs produced by *FACT* consumed on an average 62.1% less power than circuits synthesized from *M1*, for the same throughput (the throughput was made the same as *M1*'s throughput in the throughput-optimized case). We do not compare the results of *FACT* in the power optimization mode with *Flamel* since the latter does not have the ability to apply power-optimizing transformations.

6 Conclusions

In this paper, we presented a framework, *FACT*, for applying throughput and power-optimizing transformations to CFI behavioral descriptions. Our algorithm interleaves scheduling with transformation application, which allows accurate identification of performance or power bottlenecks, and the selection of transformations to eliminate them. The applied transformations can span several basic blocks, thus improving their optimizing capabilities. This ability is the result of a general technique we have devised for recognizing and applying a general class of transformations across the scope of basic blocks. The CDFGs produced by *FACT*, upon synthesis, have achieved upto 4.5-fold improvement in power and upto 6-fold improvement in throughput over designs synthesized without the benefit of our technique.

References

[1] D. F. Bacon, S. L. Graham, and O. J. Sharp, "Compiler transformations for high-performance computing," *ACM Computing Surveys*, vol. 26, pp. 346–420, Dec. 1994.

[2] A. V. Aho and J. D. Ullman, *Principles of Compiler Design*. Narosa Publishing House, New Delhi, 1985.

[3] P. P. Chang, S. A. Mahlke, and W. W. Hwu, "Using profile information to assist classic code optimizations," *Software Practice & Experience*, vol. 21, pp. 1301–1321, Dec. 1991.

[4] K. Pettis and R. C. Hansen, "Profile guided code positioning," in *Proc. SIGPLAN Conf. Programming Language Design & Implementation*, pp. 16–27, June 1990.

[5] A. P. Chandrakasan, M. Potkonjak, R. Mehra, J. Rabaey, and R. W. Brodersen, "Optimizing power using transformations," *IEEE Trans. Computer-Aided Design*, vol. 14, pp. 12–51, Jan. 1995.

[6] M. B. Srivastava and M. Potkonjak, "Power optimization in programmable processors and ASIC implementations of linear systems: Transformation-based approach ," in *Proc. Design Automation Conf.*, pp. 343–348, June 1996.

[7] H. Trickey, "Flamel: A high-level hardware compiler," *IEEE Trans. Computer-Aided Design*, vol. CAD-6, pp. 259–269, Mar. 1987.

[8] A. Nicolau and R. Potasman, "Incremental tree height reduction for high level synthesis," in *Proc. Design Automation Conf.*, pp. 770–774, June 1991.

[9] S. Amellal and B. Kaminska, "Functional synthesis of digital systems with TASS," *IEEE Trans. Computer-Aided Design*, vol. 13, pp. 537–552, May 1994.

[10] S. Bhattacharya, S. Dey, and F. Brglez, "Performance analysis and optimization of schedules for conditional and loop-intensive specifications," in *Proc. Design Automation Conf.*, pp. 491–496, June 1994.

[11] N. Weste and K. Eshraghian, *Principles of CMOS VLSI Design*. Addison-Wesley, Reading, MA, 1988.

[12] G. Lakshminarayana and N. K. Jha, "Throughput and power optimizing transformations for control-flow intensive behavioral descriptions," Tech. Rep. CE-J98-004, Princeton University, 1998.

[13] G. Lakshminarayana, K. S. Khouri, and N. K. Jha, "*Wavesched:* A novel scheduling technique for control-flow intensive behavioral descriptions," in *Proc. Int. Conf. Computer-Aided Design*, pp. 244–251, Nov. 1997.

Incorporating Speculative Execution into Scheduling of Control-flow Intensive Behavioral Descriptions *

Ganesh Lakshminarayana †, Anand Raghunathan ‡, and Niraj K. Jha †

† Dept. of Electrical Engineering
Princeton University
Princeton, NJ 08544

‡ C&C Research Laboratories
NEC USA
Princeton, NJ 08540

Abstract

Speculative execution refers to the execution of parts of a computation before the execution of the conditional operations that decide whether it needs to be executed. It has been shown to be a promising technique for eliminating performance bottlenecks imposed by control flow in hardware and software implementations alike. In this paper, we present techniques to incorporate speculative execution in a fine-grained manner into scheduling of control-flow intensive behavioral descriptions. We demonstrate that failing to take into account information such as resource constraints and branch probabilities can lead to significantly sub-optimal performance. We also demonstrate that it may be necessary to speculate simultaneously along multiple paths, subject to resource constraints, in order to minimize the delay overheads incurred when prediction errors occur. Experimental results on several benchmarks show that our speculative scheduling algorithm can result in significant (upto seven-fold) improvements in performance (measured in terms of the average number of clock cycles) as compared to scheduling without speculative execution. Also, the best and worst case execution times for the speculatively performed schedules are the same as or better than the corresponding values for the schedules obtained without speculative execution.

1 Introduction

Speculative execution refers to the execution of a part of a computation before it is known if the control path to which it belongs will be executed (for example, execution of the code after a branch statement before the branch condition itself is evaluated). It has been used to overcome, to some extent, the scheduling bottlenecks imposed by control-flow. There has been previous work on speculative execution in the areas of high-level synthesis [1, 2, 3] as well as high-performance compilation [4, 5].

Previous work [1, 2, 3] in high-level synthesis has attempted to locate single or multiple paths for speculation *prior* to scheduling. This paper presents techniques to integrate speculative execution into scheduling during high-level synthesis of control-flow intensive designs. In that context, we demonstrate that not using information such as resource constraints and branch probabilities while deciding when to speculate can lead to significantly sub-optimal performance. We also demonstrate that it is necessary to perform speculative execution along multiple paths at a fine-grain level during the course of scheduling, in order to obtain maximal benefits. In addition, we present techniques to automatically manage the additional speculative results that are generated by speculatively executed operations. We show how to incorporate speculative execution into a generic scheduling methodology, and in particular present the results of its integration into an efficient scheduler *Wavesched* [6]. Experimental results for various benchmarks and examples are presented that indicate upto seven-fold improvement in performance (average number of clock cycles required to perform the computation).

*This work was supported in part by NSF under Grant No. 9319269 and in part by Alternative System Concepts, Inc. under an SBIR contract from Air Force Rome Laboratories.

DAC 98, San Francisco, California

2 Background and Motivation

Scheduling tools typically work using one or more intermediate representations of the behavioral description, such as a *data flow graph* (DFG), *control flow graph* (CFG), or *control-data flow graph* (CDFG). In this paper, we use the CDFG as the intermediate representation of a behavioral description, and *state transition graphs* (STGs) to represent the scheduled behavioral description, as explained in later sections. In addition to the behavioral description, our scheduler also accepts the following information:

- A constraint on the number of resources of each type available (*resource allocation* constraints).
- The target clock period for the implementation, or constraints that limit the extent of data and control chaining allowed.
- Profiling information that indicates the branch probabilities for the various conditional constructs present in the behavioral description.

We now present some motivational examples to illustrate the use of speculative execution during scheduling.

Example 1: Consider a part of a behavioral description and the corresponding CDFG fragment shown in Figure 1, that contains a *while* loop. The CDFG contains vertices corresponding to operations of the behavioral description, where solid lines indicate data dependencies, and dotted lines indicate control dependencies. Control edges in the CDFG are annotated with a variable that represents the result of the conditional operation that generates them. For example, the control edges fed by operation > 1 are marked c in Figure 1. The initial values of variables i and $t4$ used in the loop body are indicated in parentheses beside the corresponding CDFG data edges.

Let us now consider the task of scheduling the CDFG shown in Figure 1. Suppose we have the following constraints to be used during scheduling.

```
...
i := 0;
t4 := 0;
while (k > t4) {
  i := i + 1;
  t1 := M1[i];
  t2 := t1 * C1;
  t3 := t2 * C2;
  t4 := t3 + C3
  M2[i] := t4;
}
...
```

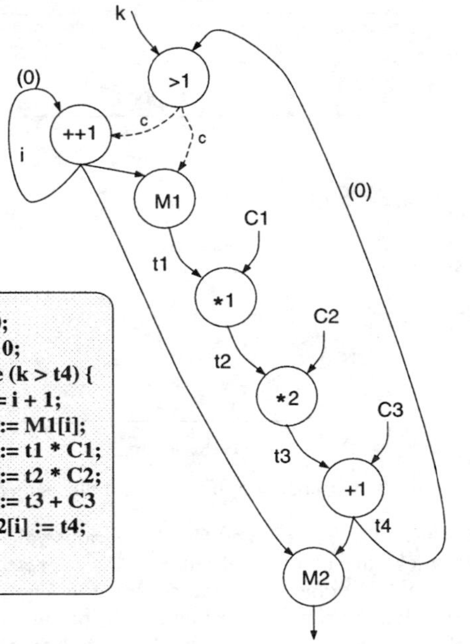

Figure 1: A CDFG to illustrate speculative execution

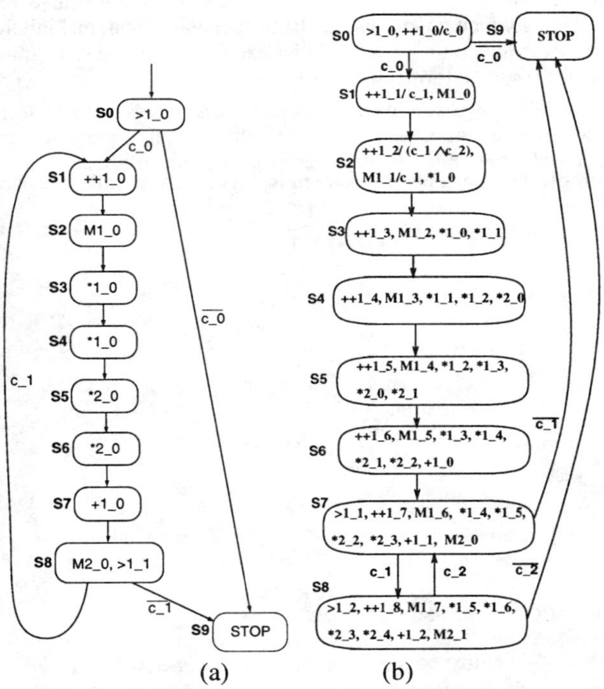

Figure 2: (a) Non-speculative schedule for the CDFG of Figure 1, and (b) schedule incorporating speculative execution

- The target clock period allows the execution of +, ++, >, and memory access operations in one clock cycle, while the $*$ operation requires two clock cycles. In addition, we assume that the $*$ operation will be implemented using a 2-stage pipelined multiplier.
- No operation chaining is allowed, since it leads to a violation of the target clock period constraint (in general, however, our algorithm can handle chaining).
- The aim is to optimize the performance of the design as much as possible. Hence, no resource constraints are specified for the purposes of illustration for this example. This is not a limitation of our scheduling algorithm, which does handle resource constraints as described in later sections.

A schedule for the CDFG that does not incorporate speculative execution is shown in Figure 2(a). This schedule can be obtained by applying either the loop-directed scheduling [7] technique or the *Wavesched* [6] technique to the CDFG. Vertices in the STG represent schedule states, that directly correspond to states in the controller of the RTL implementation. Each state is annotated with the names of the CDFG operations that are performed in that state, including a suffix that represents a symbolic iteration index of the CDFG loop that the operation belongs to. For example, consider operation >1 of the CDFG. When >1 is encountered the first time during scheduling, it is assigned a subscript 0, resulting in operation $>1\_0$ in the STG of Figure 2(a). In general, multiple copies of an operation may be generated during scheduling, corresponding to different conditional paths, or different iterations of a loop. For example, operation $>1\_1$ in the STG of Figure 2(a) corresponds to the execution of the first unrolled instance of CDFG operation >1. An edge in the STG represents a controller state transition, and is annotated with the conditions that activate the transition.

Each iteration of the loop in the scheduled CDFG requires *eight* clock cycles. For this example, the data dependencies among the operations within the loop require them to be performed serially. In addition, the control dependencies between the comparison operation >1 and operations $++1$ and $M1$, together with the inter-iteration data dependency [1] from +1 to >1, prevent the parallel computation of multiple loop iterations, even when loop unrolling is employed.

A schedule for the CDFG of Figure 1 that incorporates speculative execution is shown in the STG of Figure 2(b). This schedule was derived by techniques we present in later sections. Speculatively executed operations are annotated with the conditional operations whose results they depend upon, using the following notation. *op/cond* represents an operation *op* that is executed assuming that the *speculation condition cond* will evaluate to *true*. The speculation condition *cond* could, in general, be an expression that is a conjunction of the results of various conditional operations in the STG. For example, consider operation $++1\_1/c\_1$ in state $S1$ of Figure 2(b). This is a speculatively executed operation, that corresponds to the second instance of CDFG operation $++1$ in the schedule, and assumes that the result of conditional operation $>1\_1$, which is executed only in state $S7$, is going to be *true*. States $S7$ and $S8$ represent the steady state of the schedule. Note that, when in the steady state, a new iteration is initiated every cycle, as opposed to once in eight cycles. ∎

The following example illustrates the impact of branch probabilities and resource constraints on the performance of speculatively derived schedules and makes a case for the integration of speculation into the scheduling process.

Example 2: Consider the example CDFG shown in Figure 3. The select operation *Sel1* selects the data operand at its l (r) port if the value at its s port is 1 (0). Figure 4 shows three different schedules that use speculative execution, that were generated using different resource constraints and branch probabilities. The STG of Figure 4(a) was generated assuming the following information. Available resources consist of one incrementer (++), one adder(+),

[1] An intra-iteration data or control dependency is between operations that correspond to the same iteration of a loop, while an inter-iteration dependency is between operations in different (*e.g.*, consecutive) iterations. We refer to intra-iteration data and control dependencies simply as data and control dependencies.

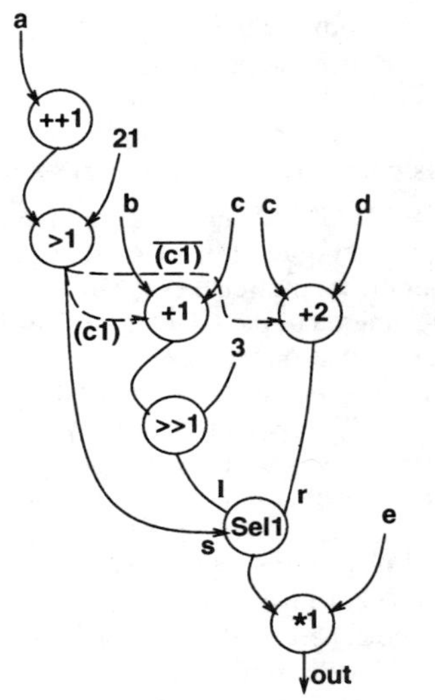

Figure 3: CDFG demonstrating the effect of resource constraints and branch probabilities on speculative execution

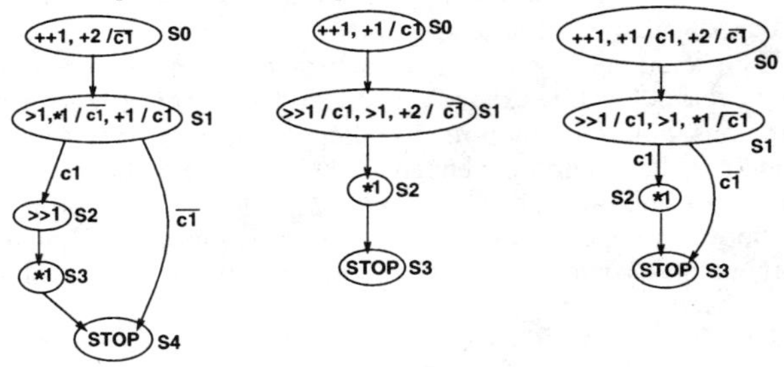

(a) (b) (c)

Figure 4: Three speculative schedules derived using different resource constraints or branch probabilities

one comparator (>), one shifter (>>), and one multiplier (*), all of which require one cycle. Also, the probability of comparison > 1 evaluating to *false* is higher than it evaluating to *true*. Since the result of > 1 evaluates to *false* more often, the schedule of Figure 4(a) gives preference to executing operations from the corresponding control path (*e.g.*, +2). As a result, +2 is scheduled to be performed on the sole adder in state $S0$, as opposed to +1, even though the data operands for both operations are available. The average number of clock cycles, CC_a, required for the STG in Figure 4(a) can be calculated as follows.

$$CC_a = 4.P(c1) + 2.(1 - P(c1)) = 2.P(c1) + 2 \quad (1)$$

In the above equation, $P(c1)$ represents the probability that the result of comparison > 1 evaluates to *true*.

The STG of Figure 4(b) was derived with the same information as above, except that it was assumed that comparison > 1 evaluates to *true* more often than it evaluates to *false*. Hence, operation +1 is given preference over operation +2 and is scheduled in $S0$. The average number of clock cycles, CC_b, required for the STG in Figure 4(b) is given by the following expression

$$CC_b = 3.P(c1) + 3.(1 - P(c1)) = 3$$

Suppose the resource constraints were relaxed to allow two adders. The speculative schedule that results is shown in Figure 4(c). The average number of clock cycles, CC_c, required for the STG in Figure 4(c) is given by the following expression.

$$CC_c = 3.P(c1) + 2.(1 - P(c1)) = P(c1) + 2 \quad (2)$$

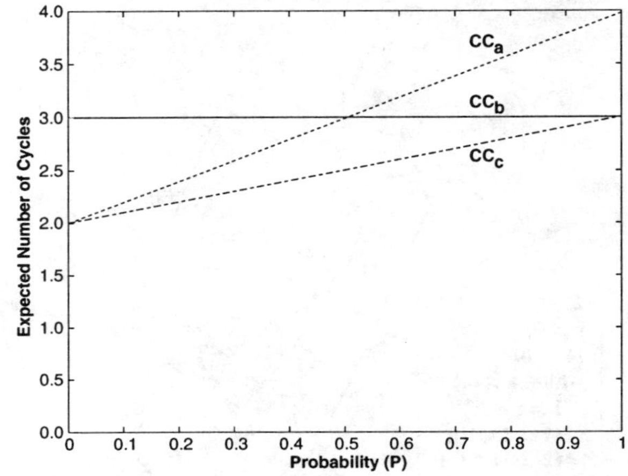

Figure 5: Comparison of the speculative schedules

The values of CC_a, CC_b, and CC_c for various values of P ranging from 0 to 1 are plotted in Figure 5. As expected, the schedule of Figure 4(a) outperforms the schedule of Figure 4(b) when $P(c1) < 0.5$, and the schedule of Figure 4(b) performs better when $P(c1) > 0.5$. Moreover, the schedule of Figure 4(c), which was derived using one extra adder, outperforms the other two schedules for all values of $P(c1)$. Thus, we can conclude that branch probabilities and resource constraints do influence the trade-offs involved in deciding which conditional paths to speculate upon, making the case for the integration of speculative execution into the scheduling step where such information is available. ■

The following example illustrates that it is necessary to perform speculative execution along multiple paths, in a fine-grained manner, in order to obtain maximal performance improvements.

Example 3: The schedules shown in Figure 4 were all generated by speculatively executing operations from both the conditional paths of the CDFG in a fine-grained manner, as allowed by the resource constraints. For the purpose of comparison, we scheduled the CDFG shown in Figure 3, assuming the same scheduling information that was assumed to derive the schedule of Figure 4(b). However, in this case, we restricted the scheduler to allow speculative execution along only one path. The resulting schedule is shown in Figure 6. The average number of clock cycles, CC_d, required for the STG in Figure 6 is given by the following expression.

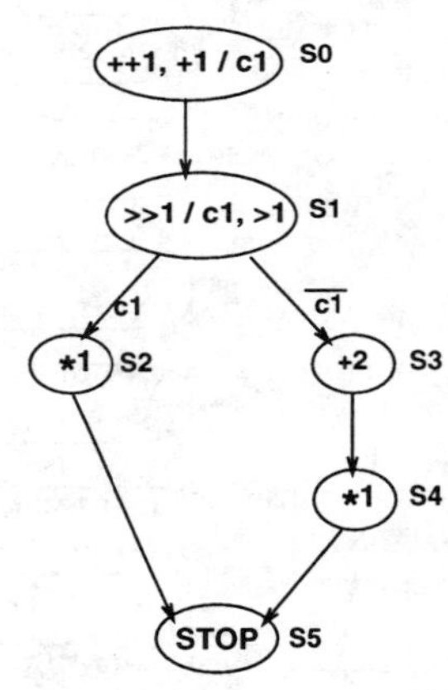

Figure 6: Speculation along a single path

$$CC_d = 3.P(c1) + 4.(1 - P(c1)) = 4 - P(c1) \quad (3)$$

Comparing the expression for CC_d to the expression for CC_b from the previous example indicates that $CC_d \geq CC_b$ for all feasible values of $P(c1)$. Thus, in this example, simultaneously speculating along multiple paths according to resource availability results in a schedule that is provably better than one derived by speculating along only the most probable path. Our scheduling algorithm automatically decides the best paths to speculate upon for the given resource constraints and branch probabilities. ∎

3 The Algorithm

In this section, we present the changes that need to be made to a generic scheduling algorithm to support speculative execution.

3.1 A generic scheduling algorithm

Figure 7 shows the pseudocode for a generic scheduling algorithm. The inputs to the scheduler are a CDFG, *G*, to be scheduled, the target clock period of the design, allocation constraints, which specify the numbers and types of functional units available, and module selection information, which gives the type of functional unit an operation is mapped to. The output of the scheduler is an STG which describes the schedule. At any point, a generic scheduler maintains (a) the set of unscheduled operations whose data and control dependencies have been satisfied, and can therefore be scheduled (*Schedulable_operations*), and (b) the set of operations which are unscheduled (*Unscheduled_operations*). The scheduling process proceeds as follows: an operation from *Schedulable_operations* is selected for scheduling in a given state (statement **2**). The selection should honor allocation and clock cycle constraints. The manner in which the selection is done varies from one scheduling algorithm to another. The selected operation, *op*, is scheduled in the state. Since *op* no longer belongs to either *Schedulable_operations* or *Unscheduled_operations*, it is removed from these sets (statements **4** and **5**). Also, the scheduling of *op* might render some of the operations in its fanout schedulable. The routine *Compute_schedulable_successors* (statement **6**) identifies such operations, and these operations are subsequently included in the set *Schedulable_operations* (statement **7**).

```
Generic_scheduler(CDFG G, ALLOCATION_CONSTRAINT K,
      MODULE_SELECTION_INFO M_inf, CLOCK_PERIOD clk) {
   SET<OPERATION> Unscheduled_operations;
   SET<OPERATION> Schedulable_operations;
1  while (|Unscheduled_operations| > 0) {
2     op = Select_schedulable_operation
                 (Schedulable_operations, K, M_inf, clk);
      //Select an operation for scheduling. The selected
      // operation must honor allocation and clock cycle constraints
3     Schedule(op);
4     Unscheduled_operations.remove_operation(op);
5     Schedulable_operations.remove_operation(op);
6     SET<OPERATION> schedulable_successors = Compute-
        _schedulable_successors(op);//Find the set of operations
      // in op's fanout which become schedulable when op is scheduled
7     Schedulable_operations.append(schedulable_successors);
      //Augment Schedulable_operations by addition of
      //operations in schedulable_successors
}}
```

Figure 7: Pseudocode for a generic scheduling algorithm

3.2 Incorporating speculative execution into a generic scheduler: An overview

We now provide an overview of the changes that need to be made to incorporate speculative execution into the framework of the generic scheduler shown in Figure 7.

To support speculative execution, the generic scheduler shown in Figure 7 needs to be modified as follows (the details of these steps are provided in Section 3.3).

1. When an operation is scheduled, one needs to recognize *all* its schedulable successors, including the ones which can be speculatively scheduled. In addition, speculatively executed operations and their successors need to be specially marked. Clearly, procedure *Compute_schedulable_successors* needs to be augmented to consider such cases. Note that, at any stage, *every* speculatively schedulable operation is added to the list of schedulable operations. However, few of them are actually scheduled. Operations which are not worth being speculated on are ignored, and eventually removed from the list of schedulable operations, using procedures described later in this section.

   Example 4: Consider the CDFG fragment shown in Figure 8. We assume that operation *op0* is scheduled, operation *op2* has just been scheduled, and operations *op1*, *op3*, *Sel1*, and *op4* are unscheduled. The output of the routine *Compute_schedulable_successors(op2)* must include operation *op4*, which can now be speculatively executed, *i.e.*, its operands can be assumed to be the results of operations *op2* and *op0*. ∎

2. When operations are scheduled, control and data dependencies of speculatively executed operations are resolved. This would potentially validate or invalidate speculatively performed operations. Operations which are validated should be considered "normal", *i.e.*, they need not be specially marked any longer. Operations in *Unscheduled_operations* and *Schedulable_operations* which are invalidated need no longer be considered for scheduling. They can, therefore, be removed from these sets. In general, the resolution of the control or data dependencies of a speculatively performed operation creates two separate threads of execution, which correspond to the success and failure of the speculation.

   Example 5: Consider again, the CDFG fragment shown in Figure 8. Suppose operations *op0*, *op2* and *op4* have been scheduled, and operation *op3* is unscheduled. Operation *op4* uses as its operands, the results of operations *op2* and *op0*. Assume that operation *op1* has just been scheduled. If *op1* evaluates to *true*, then the execution of *op4* can be considered fruitful, because the operands chosen for its computation are correct. Therefore, *op4*, and its scheduled and schedulable successors need not be considered conditional on the result of *op1* anymore, and the data structures can be modified to reflect this fact. If, however, *op1* evaluates to *false*, then *op4* should use as its operands, the results of operations *op3* and *op0*, thus invalidating the result of our speculation. Therefore, schedulable operations, whose computations are influenced by the result computed by *op4* are invalid, and can be removed from the set *Schedulable_operations*. ∎

3. The set, *Schedulable_operations*, from which an operation is selected for scheduling, contains operations whose execution is speculative, *i.e.*, whose results are not always useful. The selection procedure, represented by the routine *Select_schedulable_operation()* (statement **2**), needs to be modified to account for this fact. For example, operations, whose execution is extremely improbable, would make poor selection candidates, as the resources consumed by them might be

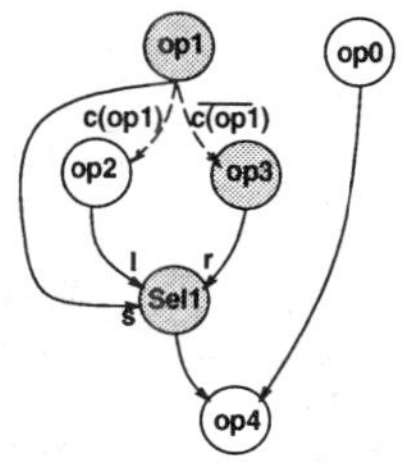

Figure 8: A CDFG fragment illustrating speculative execution

better utilized by operations whose execution is more probable. Also, operations, which fall on critical paths, would be better candidates for selection than those on off-critical paths.

3.3 Incorporating speculative execution into a generic scheduler: A closer look

In this section, we fill in the details of the changes outlined in Section 3.2. This is preceded by a formal treatment of concepts related to speculative execution.

A scheduler which supports speculative execution works with *conditioned operations* as its atomic schedulable units, just as a normal scheduler uses operations. Therefore, the fanin-fanout relationships between operations, captured by the CDFG, need to be defined for conditioned operations. Since all speculatively performed operations are conditioned on some event, the adjective "speculatively performed" when applied to an operation, implies that it is conditioned on some event or combination of events.

As mentioned in Section 3.2, when an operation is scheduled, its schedulable successors need to be computed.

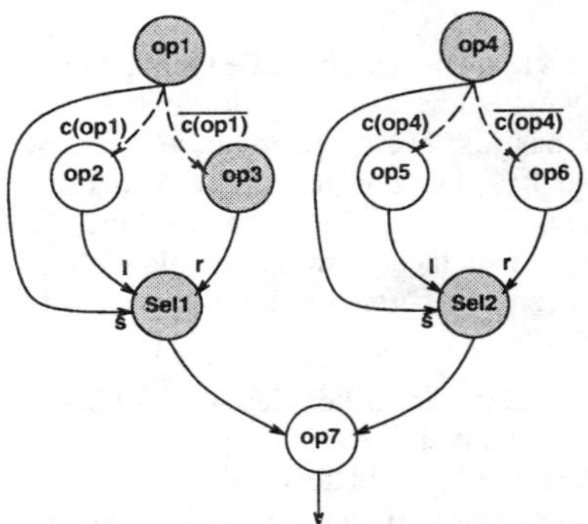

Figure 9: Illustrating the scheduling of successors of speculatively performed operations

Example 6: Consider the CDFG fragment shown in Figure 9. Assume that operations *op5* and *op6* have been scheduled, operations *op1*, *op3*, and *op4* are unscheduled, and *op2* has just been scheduled. It is now possible to schedule *two* versions of operation *op7*, with the first version, $op7'$, using *op2* and *op5* as its operands, and the second, $op7''$ using *op2* and *op6*. $op7'$ is conditioned on $c(op1) \wedge c(op4)$, and $op7''$ is conditioned on $c(op1) \wedge \overline{c(op4)}$. The following analysis presents a structured means of identifying such relationships. ∎

We now present a result which helps derive fanin-fanout relationships among speculatively performed operations.

Lemma 1: Consider an operation, *op*, whose fanins are *op1*, *op2*, ..., *opn*. If the fanins of *op* have been speculatively scheduled, so can *op*. In particular, if the *i*th fanin, *opi*, is conditioned on C_i, then *op* would be conditioned on $\bigwedge_{i=1}^{n} C_i$. ∎

We now present details of Steps 1, 2, and 3, outlined in Section 3.2.

Step 1: This step addresses the issue of deriving all schedulable successors of a scheduled operation, *op0*. The result of Lemma 1 is used for this procedure.

Observation 1 Every set, $S = \{op0, op1, \ldots, opi\}$ of scheduled operations, which satisfies the following condition sources a schedulable operation.

Condition: There exits an operation, *fanout*, in the CDFG, all of whose fanins are reachable from the outputs of the operations in *S* through paths which consist exclusively (if at all) of select operations. The path connecting the output of an operation *opj* in *S* to an input of *fanout* is denoted by *Pj*, and the operations on *Pj* are $Selj_1$, $Selj_2$, ..., $Selj_{aj}$. Note that *aj* can equal 0. C_j represents the condition that path *Pj* is selected, *i.e.*, the result of operation op_j is propagated through path *Pj* to the appropriate input of *fanout*. Operation *fanout* is conditioned on $\bigwedge_{k=0}^{i}(C(opk) \wedge C_k)$ where $C(opk)$ represents the expression *opk* is conditioned on. ∎

Observation 1 can be used to infer the schedulable successors of an operation. The procedure *Compute_schedulable_successors*, which is called in statement **6** of the pseudocode shown in Figure 7, is appropriately augmented.

So far, we have described the technique used to identify all schedulable successors of an operation. This was accomplished by tagging operations with the conditions under which their results would be valid. Note that our procedure allows us to speculate on *all* possible outcomes of a branch, and *arbitrarily deeply* into nested branches. If integrated with a scheduler which supports loop unrolling, the speculation could also cross loop boundaries. We now present the technique used to validate or invalidate speculatively performed operations whose dependencies have just been resolved.

Step 2: Suppose operation op_s, which resolves a condition *c*, has just been scheduled. The resolution of *c* results in the creation of two different threads of execution, where (i) *c* = *true*, and (ii) *c* = *false*. The following procedure is carried out for every operation, *op*, which belongs either to the set, *Schedulable_operations*, or the set of scheduled operations. Let *op* be conditioned on $C = \bigwedge_{j=1}^{i} c_j$. In the *true* (*false*) branch, *C* is evaluated assuming a value of 1 (0) for *c*, and the resultant expression is the new expression that *op* is conditioned on.

Step 3: We now describe the procedure employed by the scheduler to select an operation to schedule, from a pool of schedulable operations, *Schedulable_operations*. *Schedulable_operations* can contain operations which are conditioned on different sets of events, *i.e.*, we can choose different paths to speculate upon. We need to decide the "best" candidate to map to a given resource, where, by best, we mean the operation whose mapping on the given resource would minimize the expected number of cycles for the schedule. Formally, the problem can be stated as follows: given (i) a partial schedule, (ii) a functional unit, *fu*, (iii) a set of operations, *S* (some of which may be speculative), which can execute on the functional unit, and (iv) typical input traces, select the operation, which, if mapped to *fu*, would minimize the expected number of cycles.

The above problem has been proven to be NP-complete, *even for conditional- and loop-free behavioral descriptions* [8]. We, therefore, use the following heuristic, whose guiding principle has been successfully employed by several scheduling algorithms [9]. The heuristic is based on the following premise: operations in the CDFG which feed primary outputs through long paths are more critical than operations which feed primary outputs through short paths and, therefore, need to be scheduled earlier. The rationale behind this heuristic is that operations which belong to short paths are more mobile than those on long paths, *i.e.*, the total schedule length is less sensitive to variations in their schedules. The length of a path is measured as the sum of the delays of its constituent operations.

In data-dominated descriptions, with no loops and conditional operations, the longest path between any pair of operations is fixed. In control-flow intensive descriptions, some paths could be input-dependent. Therefore, the longest path between a pair of operations must be defined with respect to a given input. For example, for the CDFG shown in Figure 3, the longest path connecting primary input *c* with output *out* depends upon the value taken by operation > 1. Since our scheduling algorithm is geared towards minimizing the average execution time, we use the expected length of the longest path from an operation to a primary output as a metric to rank different operations. We use the notation $\lambda(op)$ to denote this quantity for operation *op*.

Speculation adds a new dimension to this problem: the result computed by an operation is not guaranteed to be useful. For an

Table 1: Expected number of cycles, number of states, best- and worst-case number of cycles results

Circuit	E.N.C.		#states		bc		wc	
	WS	SP	WS	SP	WS	SP	WS	SP
Barcode	450	227	8	3	4	1	502	251
GCD	95	49	6	6	3	3	515	260
Test1	580	80	10	10	1	1	4050	265
TLC	507	507	4	4	507	507	507	507
Findmin	522	265	4	3	2	2	472	237

Table 2: Allocation constraints for the examples in Table 2

Circuit	add1	sub1	mult1	comp1	eqc1	inc1
Barcode	2	-	-	3	3	3
GCD	-	2	-	1	2	-
Test1	2	-	4	1	-	1
TLC	-	-	-	1	1	1
Findmin	-	-	-	2	2	1

operation, *op*, we account for this effect by multiplying the probability that an operation's output is utilized with $\lambda(op)$ to derive a metric of an operation's *criticality*. This is expressed by means of the following equation:

$$criticality(op) = \lambda(op) \times \prod_{j=1}^{i} P(c_j) \qquad (4)$$

where *criticality(op)* measures the desirability of scheduling *op*, $\prod_{j=1}^{i} P(c_j)$ is the product of the probabilities of the events that *op* is conditioned on, and $\lambda(op)$ is as defined above.

4 Experimental Results

The techniques described in this paper were implemented in a program called *Wavesched-spec*, written in C++. We evaluated this program by using it to produce schedules for several commonly available benchmarks. These schedules were compared against those produced by the scheduling algorithm, *Wavesched* [6], without the use of speculative execution, with respect to the following metrics: (a) expected number of cycles, (b) number of states in the STG produced, (c) the smallest number of cycles taken to execute the behavioral description, and (d) the largest number of cycles taken to execute the behavioral description. In general, finding the largest number of cycles taken to execute a behavioral description is a hard problem. However, for the examples considered in this paper, static analysis of the description was sufficient to find the number.

Table 1 summarizes the results obtained. The columns labeled *E.N.C.*, *#states*, *bc*, and *wc* represent, respectively, the expected number of cycles, the number of states in the STG produced, smallest number of cycles taken to execute the STG, and the largest number of cycles taken to execute the STG. Minor columns *WS* and *SP* represent schedules produced by *Wavesched* and *Wavesched-spec*, respectively. We used a library of functional units which consisted of (a) an adder, *add1*, (b) a subtracter, *sub1*, (c) a multiplier, *mult1*, (d) a less-than comparator, *comp1*, (e) an equality comparator, *eqc1*, and (f) an incrementer, *inc1*. Unlimited numbers of single-input logic gates (*OR*, *AND*, and *NOT*) were assumed to be available. All functional units except *mult1*, which executes in two cycles, take one cycle to execute. The allocation constraints for an example can be found by looking up the entry corresponding to the example in Table 2. For example, the allocation constraints for *GCD* are two *sub1*, one *comp1*, and two *eqc1*.

The expected number of cycles for the final design was measured by simulating a VHDL description of the schedule using the Synopsys VSS simulator. The input traces used for simulation were obtained as zero-mean Gaussian sequences.

Of our examples, *Barcode*, *GCD*, *TLC*, and *Findmin* are borrowed from the literature. *Test1* is the example shown in Figure 1. *Barcode* represents a barcode reader, *GCD* computes the greatest common divisor of its inputs, *TLC* represents a traffic light controller, and *Findmin* returns the index of the minimum element in an array.

The results obtained indicate that *Wavesched-spec* produced an average expected schedule length speedup of 2.8 over schedules obtained using *Wavesched*. Note that *Wavesched* [6] was reported to have achieved an average speedup of 2 over schedules produced by existing scheduling algorithms, such as path-based scheduling [10], and loop-directed scheduling [7]. To get an idea of the area overhead of this technique, we obtained a 16-bit RTL implementation for the *GCD* example using an in-house high-level synthesis system, for the schedules produced by *Wavesched-spec* and *Wavesched*. These RTL circuits were technology-mapped using the MSU library, and the area of the gate-level circuits were obtained. The area overhead for the circuit produced from *Wavesched-spec* was found to be only 3.1%. We also note that for *Wavesched-spec*, the number of cycles in the shortest and longest paths is smaller than or equal to the corresponding number for *Wavesched*.

5 Conclusions

In this paper, we presented a technique for incorporating speculative execution into scheduling of control-flow intensive designs. We demonstrated that in order to fully exploit the power of speculative execution, one needs to integrate it with scheduling. We introduced a node-tagging scheme for the identification of operations which can be speculatively scheduled in a given state, and a heuristic to select the "best" operation to schedule. Our techniques were fully integrated into an existing scheduling algorithm which can support implicit unrolling of loops, functional pipelining of control-flow intensive behaviors, and can parallelize the execution of independent loops whose bodies share resources. Experimental results demonstrate that the presented techniques can improve the performance of the generated schedule significantly. Schedules produced using speculative execution were, on an average, 2.8 times faster than schedules produced without its benefit.

References

[1] U. Holtmann and R. Ernst, "Experiments with low-level speculative computation based on multiple branch prediction," *IEEE Trans. VLSI Systems*, vol. 1, pp. 262–267, Sept. 1993.

[2] K. Wakabayashi and H. Tanaka, "Global scheduling independent of control dependencies based on condition vectors," in *Proc. Design Automation Conf.*, pp. 112–115, June 1991.

[3] U. Holtmann and R. Ernst, "Combining MBP-speculative computation and loop pipelining in high-level synthesis," in *Proc. European Design & Test Conf.*, pp. 550–556, Mar. 1995.

[4] J. A. Fisher, "Trace scheduling: A technique for global microcode compaction," *IEEE Trans. Computers*, vol. 30, pp. 478–490, July 1981.

[5] S. A. Mahlke *et al.*, "Sentinel scheduling: A model for compiler-controlled speculative execution," *IEEE Trans. Computers*, vol. 11, pp. 376–408, Nov. 1993.

[6] G. Lakshminarayana, K. S. Khouri, and N. K. Jha, "Wavesched: A novel scheduling technique for control-flow intensive behavioral descriptions," in *Proc. Int. Conf. Computer-Aided Design*, pp. 244–250, Nov. 1997.

[7] S. Bhattacharya, S. Dey, and F. Brglez, "Performance analysis and optimization of schedules for conditional and loop-intensive specifications," in *Proc. Design Automation Conf.*, pp. 491–496, June 1994.

[8] M. Garey and D. Johnson, *Computers and Intractibility*. W.H. Freeman & Company, New York, 1979.

[9] R. Jain, A. Majumdar, A. Sharma, and H. Wang, "Empirical evaluation of some high-level synthesis scheduling heuristics," in *Proc. Design Automation Conf.*, pp. 686–689, June 1991.

[10] R. Camposano, "Path-based scheduling for synthesis," *IEEE Trans. Computer-Aided Design*, vol. 10, pp. 85–93, Jan. 1991.

The DT-Model: High-Level Synthesis Using Data Transfers

Shantanu Tarafdar and Miriam Leeser
Department of Electrical and Computer Engineering
Northeastern University
Boston, Massachusetts 02115

Email: shantanu@ece.neu.edu
mel@ece.neu.edu

Abstract

We present a new model for formulating the classic HLS subproblems: scheduling, allocation, and binding. The model is unique in its use of data-transfers as the basic entity in synthesis. A data transfer represents the movement of one instance of data and contains the operation sourcing the data and all the operations using it. Our model compels the storage architecture of the design to be optimized concurrently with the execution unit. We have built a high-level synthesis system, Midas, based on our data transfer model. Midas generates designs with smaller storage and data transfer requirements than other HLS systems.

1 Introduction

The high memory and data-transfer bandwidth required by modern day applications, like those in multimedia, is available only through application specific integrated circuit implementations. Advances in fabrication technology have raised the allowable complexity of an ASIC to the point where high level synthesis (HLS) tools, which synthesize ASIC architectures given the required behavior, are essential for designing efficient architectures.

Some HLS systems synthesize the execution unit of the architecture first, and then the storage and data transfer subsystem as a secondary and independent step [6]. More recent HLS systems use simple models for storage and interconnect to account for effects on them during operation scheduling and binding [4]. Most of these systems have a post-processing step to convert the simple model into a set of register files and interconnect [1]. Relegating this to a post-processing stage precludes certain optimizations like data duplication, regeneration and preemptive routing [2].

We propose a new model for formulating the classic HLS subproblems of scheduling, allocation, and binding [5]. The model uses data transfers to formulate the subproblems as opposed to the traditional approach of using operations. This raises the status of data-related issues in HLS by forcing storage and data-transfer architectures to be considered during scheduling and binding.

DAC 98, San Francisco, California
©1998 ACM 0-89791-964-5/98/06..$5.00

Our experimental results show that architectures synthesized using the DT-model have lower storage and data transfer requirements than those of a competitor HLS system that uses the more traditional operation-based formulation of the HLS subproblems.

2 Motivating the DT-model

The design of the DT-model for HLS is motivated by the characteristics exhibited by the custom architectures for applications with high storage and data-transfer demands that are becoming more common today. The goal of the DT-model is to provide a framework for an HLS system tuned to synthesizing architectures with the following characteristics:

1. Data transfers are many, but short.
2. Data is broadcast where possible.
3. Data is stored close to where it is produced and used.

Architectures synthesized using the DT-model will show lower data-transfer, storage bandwidth and storage size requirements as a result of better modeling data broadcasts and lifetime. In the presence of floorplanning, the DT-model provides a mechanism for synthesizing architectures that exhibit spatial locality in data processing, storage and motion and, thereby, lower ASIC areas.

The importance of including data transfer optimizations in the scheduling and binding phases of HLS has been stressed by other researchers [2, 3]. Our model is unique in the way it formulates HLS around data-transfers.

3 The data-transfer model

Data transfers (DTs) are extracted from the dataflow graph (DFG) that describes the behavior being synthesized. A DT corresponds to the motion of a single piece of data and contains the operation sourcing the data and all the operations using it. Figure 1 illustrates a DT and its relation to the DFG that contains it. The DTs extracted from a DFG are *primary* DTs. Each DFG node with outgoing edges is the source operation of a distinct primary DT and each node may belong to multiple DTs.

3.1 DT-cost: The physical significance of a data transfer

The cost of transferring data has a spatial and a temporal component. The spatial component is the area of interconnect needed to transfer the data. The temporal component

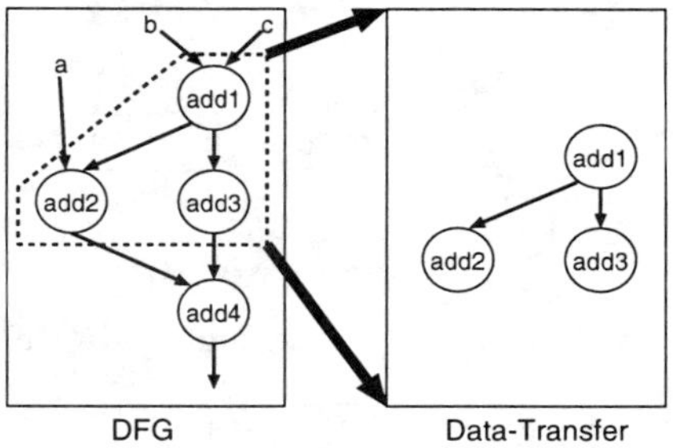

Figure 1: Definition of a data-transfer (DT)

is the length of time for which the interconnect must be reserved. The product of the two components is a measure of how "expensive" a DT is in terms of its utilization of ASIC area. This is the DT-cost.

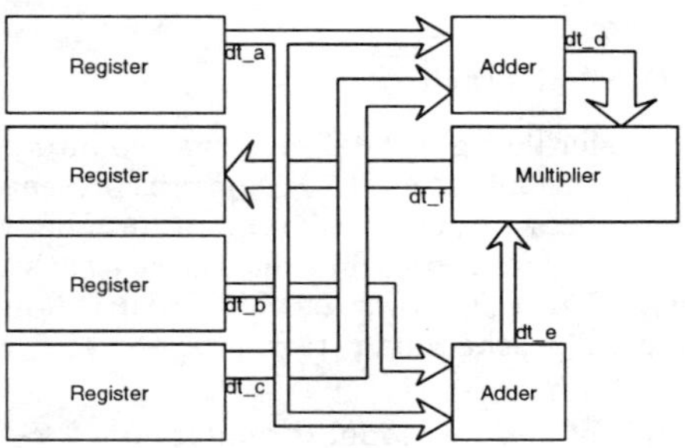

Figure 2: Cost of DTs at a cstep in an architecture

Figure 2 is a snapshot at a control step in the operation of an ASIC. DTs are indicated on the floorplan of the ASIC by arrows. Areas of the arrows reflect DT costs

3.2 Scheduling, binding, and placing a DT

The prime scheduling directive in the DT-model is that all the operations of a DT be scheduled in the same cstep. Doing so eliminates the need to store the data. In case all the operations *cannot* be scheduled in the same cstep, the primary DT must be partitioned into a set of secondary DTs which implies data must be stored and the storage accessed. Binding a DT involves binding its functional operations, like multiplications and additions, to corresponding functional units (FUs) and its storage access operations to storage elements like register files. In addition, the data must be routed by binding it to terminals of FUs and ports of storage elements to which the DT is bound, as well as to a bus in the architecture. Placing a DT means placing the functional units, storage elements and bus to which the DT is bound within a floorplan for the ASIC architecture.

3.3 Partitioning a DT

DT partitioning is central to the DT-model. A DT needs to be partitioned when it is impossible to schedule all its operations in the same cstep. The operations of the primary DT are partitioned among a set of secondary DTs. The one containing the source operation of the primary DT is called the source, secondary DT and the rest are destination, secondary DTs. The source DT has a storage write operation and the destination DTs have storage read operations as their sources. Figure 3 illustrates DT partitioning.

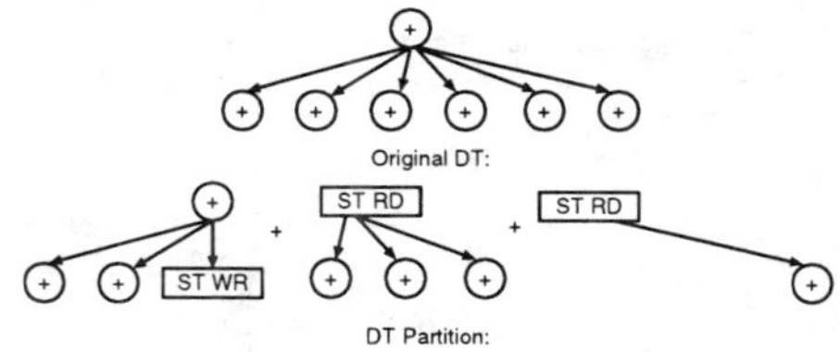

Figure 3: DT partitioning

3.4 Effects of partitioning

Partitioning has three implications: (1) a variable needs to be held in storage, (2) a variable needs to be written to storage, and (3) a variable needs to be read from storage and transfered each time.

The first implication impacts the size of storage, since the variable represented by the original DT now needs a location in a storage unit where it will be stored. The second and third affect the bandwidth of the storage unit, since they call for the availability of read and write ports on the storage unit, as well as the number of buses in the architecture. Careful partitioning can result in good storage and data-transfer architectures.

One way to perform partitioning is to minimize the number of data transfers partitioned. By partitioning a DT only when it is otherwise unscheduleable, we minimize the number of variables being sent to storage. When we *do* perform partitioning, minimizing the size of the partition, minimizes the number of times the variable is accessed in storage. The net effect is a reduction in storage size and bandwidth.

3.5 When is partitioning needed

Partitioning is required when scheduling all the operations of a data transfer in the same control step would result in an inconsistency. Three conditions leading to inconsistency are described below.

C-Step period violation In Figure 4, the path through the source addition and destination multiplication is 11ns. This is longer than the 10ns cstep period and so the DT must be partitioned into two as shown.

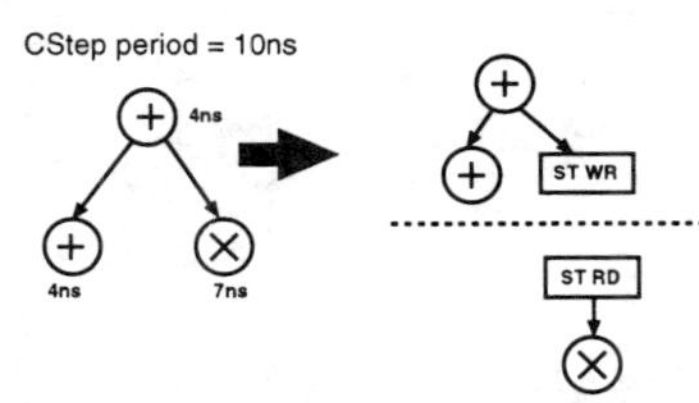

Figure 4: Inconsistency: C-step period violations

Resource allocation violation In Figure 5, the DT would require four adders if all its operations were to be scheduled in the same control step. But the given resource allocation for adders is two. The DT must be partitioned as shown for scheduling to be viable.

Local temporal inconsistency In Figure 6, we have a chain of three data transfers. Let DT1 have been scheduled at cstep 1 and DT3 at cstep 10. Since n2 is part of DT1 and

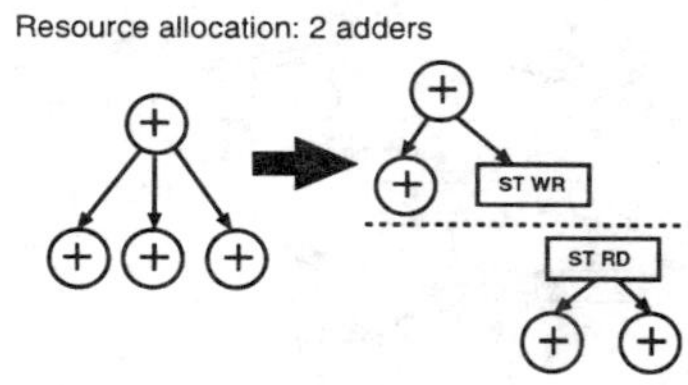

Figure 5: Inconsistency: Resource allocation violations

n3 is part of DT3, n2 is scheduled at cstep 1 and n3 is scheduled at cstep 10. These two nodes are also operations of DT2, but are not scheduled in the same cstep. Therefore DT2 is now inconsistent and must be partitioned into DT2a and DT2b for the system to regain scheduling consistency.

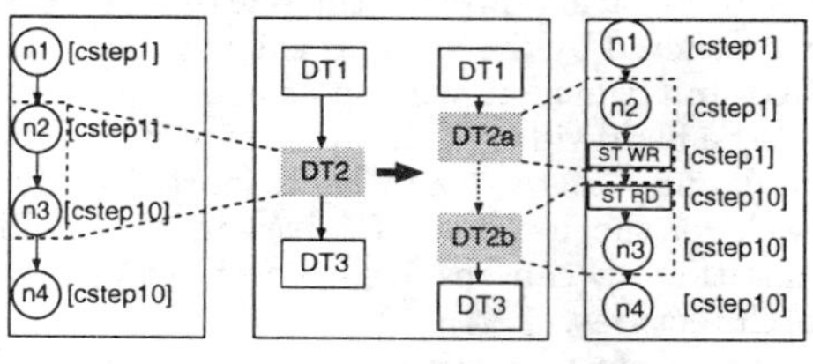

Figure 6: Local temporal inconsistency

3.6 Scheduling constraints

As in the operation-based model for HLS, there are scheduling constraints between DTs which dictate temporal relationships between pairs of DTs in the schedule of our synthesized architecture.

Precedence constraints In the box to the left in Figure 7, the destination of DT1 is the source operation of DT2. Therefore, DT1 must be scheduled before DT2. In this form of precedence constraint both DTs must still be scheduled in the same cstep. Within the cstep, DT1 is scheduled before DT2.

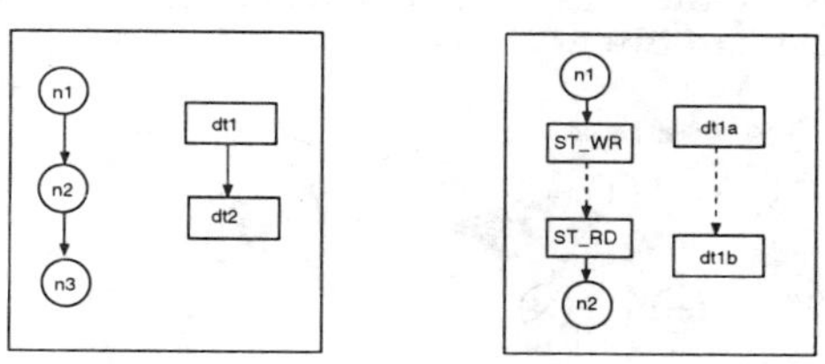

Figure 7: Precedence constraints between DTs

There is a second form of precedence constraint which occurs between DTs of a DT partition. The source DT must be scheduled at least one cstep prior to any of the destination DTs. This is shown in the box to the right in Figure 7.

Concurrency constraints In Figure 8, DT1 and DT2 share a destination operation. If all the operations in both DTs must be scheduled in the same cstep, DT1 and DT2 must be scheduled at the same cstep.

Figure 8: Concurrency constraints between DTs

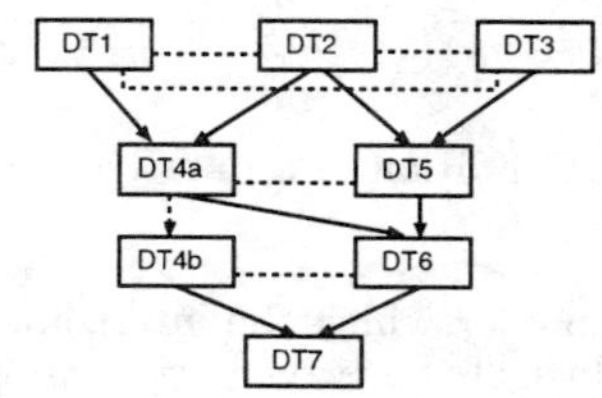

Figure 9: A DT flow graph

3.7 The DT flow graph

The data transfer flow graph (DTFG) is a counterpart of the DFG from operation-based HLS. Nodes represent DTs, directed edges represent precedence constraints between DTs, and undirected edges represent concurrency constraints between them. Figure 9 shows a DTFG. All DT-model synthesis actions, like scheduling, take place on the DTFG.

3.8 Benefits of the dt-model formulation

The DT-model raises the status of the data-related subproblems of HLS. Since DTs need to be consistent before they can be scheduled and bound, since DT partitioning directly impacts the storage unit and since DT scheduling and binding force data and operation scheduling and binding to be performed simultaneously, the data issues of high level synthesis can no longer be ignored or treated separately. An HLS system based on the DT-model is forced to consider the impact of scheduling and binding on the storage and data-transfer architectures. In addition, the DT-model allows data broadcasts and their spatial costs to be modeled more accurately.

4 Results

To test the efficacy of the DT-model, we constructed two time-constrained HLS systems that used identical synthesis algorithms. The only difference was that one, Midas, used the DT-model to formulate HLS and the second did not. We based the DT-model's performance on existing criteria for judging storage and data transfer subsystems: the numbers of registers, storage read ports, storage write ports and buses in the synthesized architecture.

We synthesized architectures using both systems for five input behaviors, including the 5th order elliptic filter and a motion estimator. For each input, we varied the allowed schedule length from 30 control steps to the minimum possible. For each input and schedule length pair, the basic optimization goal of synthesis was to minimize the number of a single resource type. The resource types we considered were each type of functional unit, storage size, storage read ports, storage write ports, and buses. For each synthesized design, we extracted the functional unit utilization, the storage size, the storage bandwidth, the number of buses and the number of data transfers in the architecture.

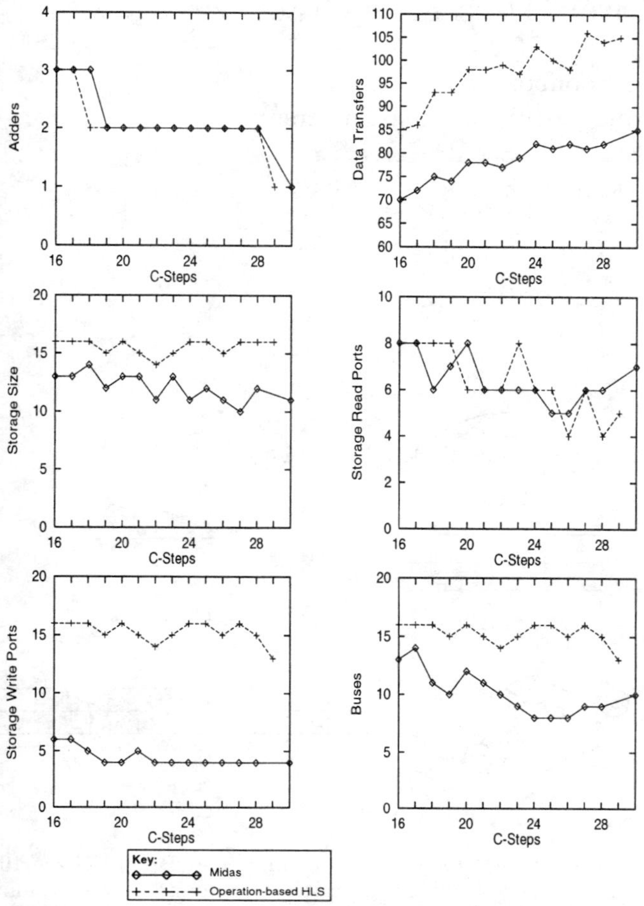

Figure 10: Elliptic filter (adders minimized)

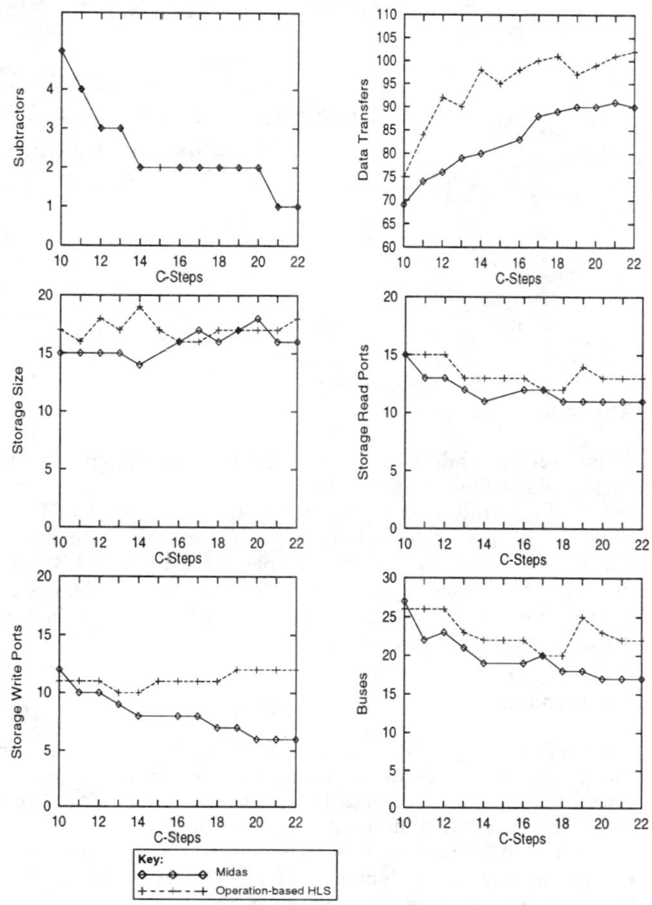

Figure 11: Motion estimator (subtractors minimized)

In Figure 10, we see results of runs on the elliptic filter benchmark. As can be seen, Midas consistently generates architectures with lower numbers of registers, storage read and write ports, and buses. The numbers of data transfers in the resulting designs is smaller too.

In Figure 11, we see the results of runs on the motion estimation example. Midas synthesizes architectures with nearly as few instances of the resource type being minimized, in this case subtractors. In addition, it synthesizes smaller storage and data-transfer architectures as a welcome side-effect. We observed this trend in all of our test cases.

5 Conclusions

We have presented a data-transfer model for formulating the classic subproblems of high-level synthesis. The model introduces an additional subproblem, DT-partitioning, and compels the core high-level synthesis algorithms – scheduling, allocation, and binding – to consider their effects on storage size, storage bandwidth, and bus size in the synthesized design. Our experiments have demonstrated that Midas, an HLS system based on the data-transfer model, generates smaller storage and data transfer subsystems than an HLS system that does not use the data-transfer model but is otherwise identical.

Acknowledgments

We would like to thank Sun Microsystems and the Rational Software Corporation, whose generous donations made this work possible. The work was funded in part by NSF grant CCR-9696196.

References

[1] T. Kim and C. L. Liu. A new approach to the multiport memory allocation problem in data path synthesis. *Integration, The VLSI Journal*, 19(3):133–60, November 1995.

[2] D. Laneer, M. Cornero, G. Goosens, and H. De Man. Data Routing: A Paradigm for Efficient Data-Path Synthesis and Code Generation. *Proceedings of the 7th International Symposium on High Level Synthesis*, pages 17–22, 1994.

[3] H. D. Lee and S.-Y. Hwang. A scheduling algorithm for multiport memory minimization in datapath synthesis. *Proceedings of the ASP-DAC'95/CHDL'95/VLSI'95*, pages 93–100, 1995.

[4] J. L. V. Meerbergen, P. E. R. Lippens, W. F. J. Verhaegh, and A. Van Der Werf. Phideo: High Level Synthesis for High Throughput Applications. *Journal of VLSI Signal Processing*, 9:89–104, May 1995.

[5] S. Tarafdar. *A Data-Transfer Model For High Level Synthesis And Its Application In Storage And Interconnect Optimization.* PhD thesis, Cornell University, 1998.

[6] C.-Y. Wang and K. K. Parhi. High-Level DSP Synthesis Using Concurrent Transformations, Scheduling, and Allocation. *IEEE Transactions on Computer-Aided Design of Integrated Circuits and Systems*, 14(3):274–295, March 1995.

Rate Optimal VLSI Design from Data Flow Graph*

Moonwook Oh and Soonhoi Ha
Department of Computer Engineering, Seoul National University
Kwanak-ku Shinlim-dong San 56-1, Seoul, Korea
{mwoh, sha}@iris.snu.ac.kr

Abstract

This paper considers the rate optimal VLSI design of a recursive data flow graph (DFG). Previous research on rate optimal scheduling is not directly applicable to VLSI design. We propose a technique that inserts buffer registers to allow overlapped rate optimal implementation of VLSI. We illustrate that nonoverlapped schedules can be implemented by a simpler control path but with a larger unfolding factor, if exists, than overlapped schedules.

1 Introduction

A data flow graph (DFG) is a useful specification method in many application fields including digital signal processing, parallel programming, high level synthesis and so on. Now that DFG plays a significant role as an algorithm specification method in high level system design, it becomes more important how to implement VLSI from a given DFG. Currently many existing tools support the synthesis of VLSI from DFG in various ways [1][2][3].

High level DFG transformation prior to VLSI implementation can play an important role in increasing the speed or in reducing the area or power consumption of a synthesized circuit [4]. Among these transformations are pipelining, retiming, parallelizing, unfolding, and so on.

If a DFG has any feed-back loop with delay(s), we call it a recursive DFG. In a recursive DFG, the current output depends on the result of previous iterations. By this fundamental property, any recursive DFG has an inherent lower bound on the achievable iteration period referred to as the *iteration bound* [5]. This iteration bound T_∞ is given by

$$T_\infty = max[\frac{T_l}{D_l}]$$

where the maximum is taken over all loops in the DFG, and T_l is the sum of the computation times of all nodes in loop l, and D_l is the number of delay elements in loop l [6]. We call a DFG schedule *rate optimal* if the iteration period

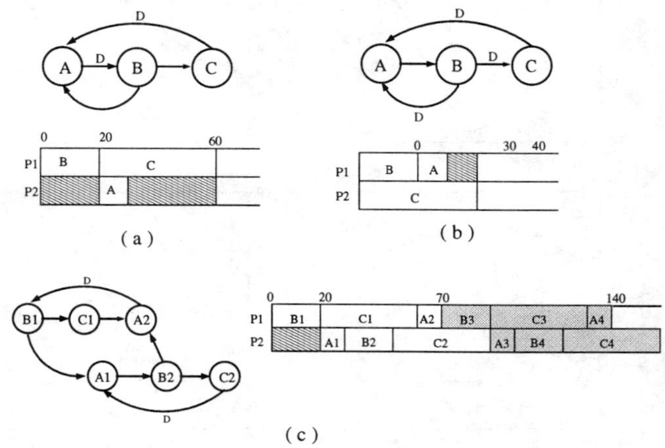

Figure 1: **(a)** Node execution time of A, B, and C are 10, 20, and 40 respectively. The Iteration bound is 35 time unit and the critical path time (or the iteration period) is 60 time unit. **(b)** A retimed DFG of (a). The critical path time is reduced to 40 time unit. **(c)** 2-unfolded DFG. The iteration period is 70 time unit which is 35 × 2. It has an overlapped rate optimal schedule.

equals the iteration bound. The DFG in Fig.1(a) has an iteration period of 60 time units and it is much larger than the iteration bound of 35 time units. The retimed DFG in depicted in Fig.1(b) and it still does not have a rate optimal schedule.

Unfolding is suggested to produce a rate optimal scheduling of a recursive DFG [6] . If we unfold any DFG by *optimum unfolding factor*, $\alpha_{prg} = lcm[D_l]$[1] which is the least common multiple of the number of loop delays in the original DFG, we can always obtain a rate optimal schedule. After unfolding was introduced, much research was studied to find algorithms that produce unfolding factors smaller than α_{prg} [7][8][9][10]. It is important to have a smaller unfolding factor because a large unfolding factor gives a large number of nodes in the unfolded DFG and as a consequence it increases the complexity of scheduling polynomially. One of the smallest unfolding factors was proposed by [8] as α_{gprg}[2]

*This research was supported in parts by Ministry of Education through ISRC under contract number 96-E-2103

[1] *prg* means *perfect rate graph* which is introduced in [6]. Perfect rate graph has only 1 delay element for each loop.

[2] *gprg* means *generalized perfect rate graph.* A DFG is gprg iff its iteration period is greater than or equal to maximum node computation time.

DAC 98, San Francisco, California
©1998 ACM 0-89791-964-5/98/06..$5.00

and it is given by

$$\alpha_{gprg} = \lceil \frac{max[n_i]}{T_{ip}} \rceil$$

where n_i is the execution time of node i and T_{ip} is the required iteration period. Fig.1(c) shows $\alpha_{prg} = \alpha_{gprg} = 2$ unfolded DFG and its schedule. Now the iteration period is 35 and satisfies the rate optimal schedule.

While unfolding guarantees a rate optimal schedule of a recursive DFG, most of the previous works [6][7][8][9][10] assumed multiprocessor systems as their implementation platform and relatively little attention has been paid to the implementation of rate optimal VLSI. In this paper, we are concerned with rate optimal VLSI design from unfolded recursive DFGs. In section 2, we give a demonstration that VLSI has an additional constraint for the execution of a delay-free path, therefore overlapped schedules of previous works are not directly applicable in VLSI design. Section 3 explains the proposed technique which inserts buffer registers to an unfolded recursive DFG and makes it feasible to implement overlapped schedules in VLSI. In section 4 we review the possibility of a nonoverlapped rate optimal schedule in VLSI design. We show the pros and cons of nonoverlapped schedules. In section 5 we provide an example to show the effect of buffering registers and to compare between overlapped and nonoverlapped rate optimal schedules in VLSI. The conclusion is stated in section 6.

2 Scheduling Problem in VLSI Design

Schedules of DFG are said to be overlapped if any task of the $(n+1)$st iteration is scheduled before the completion of nth iteration. The multiprocessor schedule in Fig.1(c) corresponds to an overlapped schedule. Overlapped schedules imply that the critical path of DFG is longer than the iteration period. Most unfolding algorithms are likely to give overlapped schedules because they do not take into consideration the critical path of DFG while they seek for unfolding factors [6][7][8][9][10] .

Remark 1 *Most unfolding algorithms are likely to give overlapped schedules.*

While an overlapped schedule can be easily implemented in multiprocessor systems, this is not the case in VLSI design. Let us assume that we want to design a VLSI which executes the schedule in Fig.1(c). We assume that all nodes are combinational logic and that only delay elements correspond to registers. Then we can easily notice that B3 can not start its execution at time unit 70 because the value of B1 should be maintained until the execution of C2 completes (at time unit 90). This constraint is added from the characteristics of VLSI but not from the semantics of DFG, which allows B3 to start at time unit 70 after the executions of B1 and A2 are completed.

Remark 2 *In VLSI design, all nodes in a delay-free path are combinational logic and they must not change their output values until the last node finishes its operation.*

Now we know that previous research results on rate optimal multiprocessor schedules are not directly applicable in VLSI design. Then, is it impossible to implement an overlapped schedule in VLSI? The answer is 'no'. We propose a technique with multiple clock timings and the insertion of buffer registers to make an overlapped rate optimal schedule in VLSI design. Though research of [11] considers VLSI implementation, it is different from ours since it allows each node has internal registers to make it sequential.

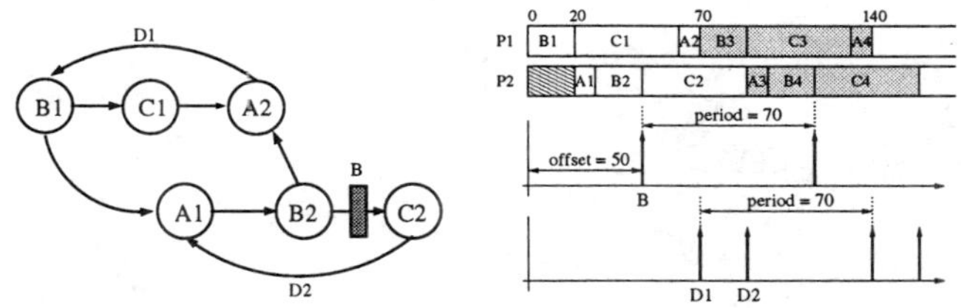

Figure 2: The DFG has a buffer register on arc (B2,C2). Because the buffer register preserves the computation result of ancestor nodes, node B1 can start the next iteration before node C2 finishes. The control signal diagram below the schedule shows activation timing of buffer register **B**.

3 Overlapped Rate Optimal Schedule in VLSI Design.

If we insert buffer registers to the delay-free arcs they preserve the value of the source node and set all the ancestor nodes free from the constraint in Remarks 2. In general, insertion of registers to a recursive DFG might corrupt the original functionality. This is the reason why we can not pipeline a recursive DFG. But we can insert buffer registers to DFG without corrupting the original semantics by carefully scheduling their activations.

Theorem 1 *If a DFG has an iteration period of T_{ip} and node A produces its valid output at time offset of k ($k \geq 0$) each iteration interval, a buffering register in $A \rightarrow B$ may have a periodical activation schedule, $T_{ip} \cdot i + k (i = 0, 1, 2, \ldots)$ without corrupting the original functionality.*

Fig.2 depicts the modified DFG of Fig.1(c) after inserting a buffer register and the activation schedule of the buffer register. We can verify that an overlapped rate optimal schedule is feasible now in VLSI implementations.

Note also that delay registers on arc (A2,B1) and on arc (C2,A1) are clocked the first time at 70 time units and 90 time units respectively, although clocking periods (70 time units) are same.

Below is suggested algorithm which decides where to locate buffering registers and their activation schedules. The input is a DFG which has an overlapped rate optimal schedule and the outputs are a new DFG with buffer registers and the schedules of them.

Algorithm 1 BufferRegistering
(DFG_{org} : input, DFG_{new}, schedule : output)

Step 1: *Let schedule be an empty set.*
Let T_∞ = iteration bound of DFG_{org}.
Step 2: *If there is any path of execution time $> T_\infty$*
then let the path $P_{critical}$.
else return.
end if
Step 3: *Let $T_{elapsed} = t = 0$.*
Step 4: *While traversing $P_{critical}$ from start to end,*
$T_{elapsed} = T_{elapsed} + N_{curr}$
(N_{curr} is the execution time of the current node)
$t = t + N_{curr}$
If $t > T_\infty$ then
place a buffer register before the current node.
put $T_\infty \cdot i + (T_{elapsed} - N_{curr})$ to schedule.
$t = 0$.
end if
Step 5: *goto Step 2.*

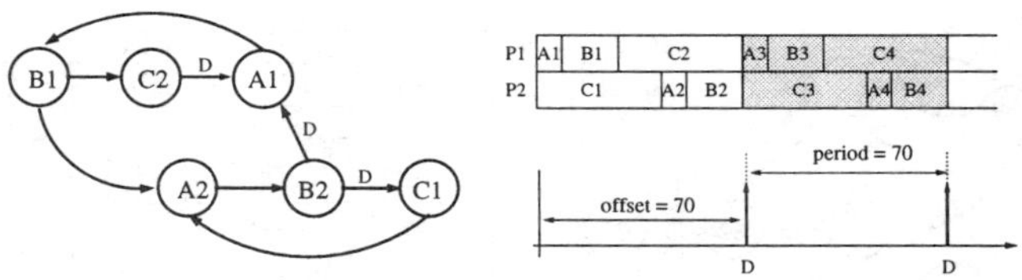

Figure 3: The DFG is obtained by 2-unfolding of the DFG in Fig.1(b). The schedule shows nonoverlapped rate optimal schedule of the DFG. Compared to the DFG in Fig.1(c), the critical path time is reduced from 90 to 70 time units

	Rate opt.		Non-rate opt.
	Overlapped	Nonoverlapped	
Unfolding factor	$\lceil \frac{max[n_i]}{T_\infty} \rceil$ small	$\lceil \frac{CPT}{T_\infty} \rceil$ large	N/A
Buffer register	essential	unnecessary	unnecessary
Schedulability	guranteed	no gurantee	guranteed
Data path	simple	complex	N/A
Control path	complex	simple	simple

Table 1: A Comparison between Implementation Choices for VLSI Design.

Algorithm 1 implies that the number of buffer registers decreases when the original DFG is transformed so that it may have less delay-free paths whose execution times are longer than the iteration bound. Therefore we apply the retiming technique before unfolding to reduce the number of buffer registers. Minimizing the number of buffer registers is another problem to be solved as a future work.

We suggest an overall process from a DFG to a VLSI implementation which has an overlapped rate optimal schedule.

Algorithm 2 OverlappedRateOptimalVLSI
(DFG_{org} : input, VLSI : output)

Step 1: Retime and unfold the DFG_{org} and call it DFG_{tmp}
Step 2: Schedule DFG_{tmp} rate optimally.
Step 3: **BufferRegistering***(DFG_{tmp} , DFG_{new}, schedule)*
Step 4: Synthesize data path for DFG_{new}.
Step 5: Synthesize control path using schedules in Step 2 and Step 3.
Step 6: Return.

There are many novel works which deal with the unfolding of DFG and give quite small unfolding factors that promise good quality of final implementations. Now thanks to Algorithm 2, we can enjoy all the benefits of previous works for DFG unfolding in multiprocessing area while pursuing rate optimal VLSI design.

4 Nonoverlapped Rate Optimal Schedule in VLSI Design

In this section we review the nonoverlapped rate optimal schedule in VLSI design. We show the pros and cons of nonoverlapped schedules and clarify that unfolding cannot guarantee a nonoverlapped rate optimal schedule.

A schedule of DFG is said to be nonoverlapped if all tasks of the nth iteration complete before the $(n+1)$st iteration starts. Fig.3 depicts a DFG and its nonoverlapped rate optimal schedule. Compared with Fig.1(c) it has nice features for VLSI design. **(1)** The schedule in Fig.3 has a latency equal to the iteration period while the schedule in Fig.1(c) has a longer latency. **(2)** All the delay registers in Fig.3 have the same activation schedules but we must feed different control signals in case of Fig.1(c). **(3)** Finally, no buffer registers are needed.

In a nonoverlapped schedule, minimum iteration period that can be achieved is the computation time associated with the critical path of DFG [6]. So, to achieve a nonoverlapped rate optimal schedule, we should reduce the critical path time of a DFG below the iteration bound. In the case of a nonrecursive DFG, pipeline transformation is widely used to reduce the critical path time. It can achieve a critical path time as short as the maximum node computation time. But for recursive DFGs, as we mentioned in section 1, pipelining is not applicable and retiming also fails to guarantee a critical path time smaller than the iteration bound.

In research of [12] unfolding with retiming is examined to achieve nonoverlapped rate optimal scheduling of recursive DFG. They applied α_{cpt}[3] -unfolding iteratively, until the unfolded DFG has a critical path time shorter or equal to the iteration bound.

$$\alpha_{cpt} = \lceil \frac{CPT}{T_\infty} \rceil$$

Remark 3 *If we try to build a nonoverlapped rate optimal schedule, it always requires a larger or equal unfolding factor compared to an overlapped schedule. (It is obvious because the critical path time is always greater than or equal to maximum node computation time)*

Unfortunately, unfolding does not always provide a nonoverlapped rate optimal schedule [12]. We proved that unfolding over α_{prg} is useless for nonoverlapped rate optimal scheduling. But in this paper, the formal proof is not provided.

Now we know that, an nonoverlapped schedule has many fancy features for VLSI design but it is not always achievable by unfolding. Table 1 summarizes some key points discussed in this section and in the previous section. Overlapped and nonoverlapped rate optimal schedule have their own pros and cons when they are implemented in VLSI.

5 Example

In this section we will clarify important points of this paper by showing an example.

Let us assume we want to implement a time critical algorithm, depicted in Fig.4(a) [6], in a VLSI so that it has the execution speed of 16 time units per iteration. This execution speed can be obtained by no other means except rate optimal implementation. (One should notice that the DFG in Fig.4(a) has the iteration bound of $(N_A + N_D + N_E)/2 = 32/2 = 16$ time units.) Formerly, when we had no support of buffer register insertion, an overlapped multiprocessor schedule was not implementable in VLSI and we had to find a nonoverlapped rate optimal schedule. For this example, there exists an unfolding factor which gives an nonoverlapped schedule. If we run the algorithm in [12], we can see it is $\alpha_{cpt} = 6$. The 6-unfolded DFG is illustrated in Fig.4(b).

Although we obtained a rate optimal VLSI by 6-unfolding, it requires much area cost. If we insert a buffer register according to Algorithm 1, the existing algorithms for overlapped multiprocessor schedules can be applied to VLSI design. When we run one of them [8], a rate optimal schedule

[3] cpt means critical path time.

is obtained by only $\alpha_{gprg} = 2$ unfolding. It means that the area cost is reduced by almost 1/3, compared to nonoverlapped implementation. Fig.4(c) shows the 2-unfolded DFG which is much simpler than that of Fig.4(b).

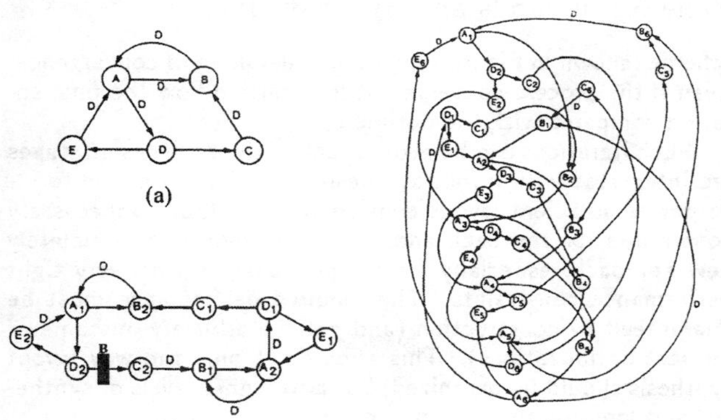

Figure 4: **(a)** Original DFG. Node A, B, C, D, E require 20, 5, 10, 10, 2 time unit respectively. The iteration bound is 16, the maximum node execution time is 20 and the critical path time is 20 time units. **(b)** A 6-unfolded DFG which allows nonoverlapped rate optimal schedule. A critical path is (D1,E1,A2,D3,E3,A4,D5,E5,A6) and requires 96 time units, which corresponds to 6 × 16. **(c)** A 2-unfolded DFG which has an overlapped rate optimal schedule. It is much simpler than the DFG in (b).

6 Conclusion

We reviewed the previous research on unfolding and rate optimal scheduling in the perspective of VLSI design. For an overlapped rate optimal schedule, we showed that the results of multiprocessor schedulings are not directly applicable to VLSI implementations and suggested an algorithm which inserts buffer registers so that an overlapped schedule may be feasible.

We also clarified that nonoverlapped schedules can be implemented by a simpler control path, although they are likely to require larger unfolding factors than overlapped schedules.

In this paper, we did not address the area overhead of the inserted buffer registers. Minimizing the number of buffer registers is to be studied as a future work.

References

[1] Synopsys Inc., 700 E. Middlefield Rd., Mountain View, CA 94043, USA. *COSSAP User's Manual : VHDL Code Generation.*

[2] M. C. Williamson and E. A. Lee. Synthesis of parallel hardware implementations from synchronous dataflow graph specifications. In *30th Asilomar Conference on Signals, Systems, and Computers*, Pacific Grove, California, USA, November 1996.

[3] C. Y. Wang and K. K. Parhi. The mars high-level dsp synthesis system. In *VLSI Design Methodologies for Digital Signal Processing Architectures.* Kluwer Academic Press, 1994.

[4] K. K. Parhi. High-level algorithm and architecture transformations for dsp synthesis. *Journal of VLSI Signal Processing*, 9:pp 121–143, 1995.

[5] R. Reiter. Scheduling parallel computations. *Journal of ACM*, 15(4):pp. 590–599, October 1968.

[6] K. K. Parhi. Static rate-optimal scheduling of iterative data-flow program via optimum unfoliding. *IEEE Trans. on Computers*, 40(2), February 1991.

[7] L. G. Jeng and L. G. Chen. Rate-optimal dsp synthesis by pipeline and minimum unfolding. *IEEE Trans. on VLSI Systems*, 2(1), March 1994.

[8] D. J. Wang and Y. H. Hu. Fully static multiprocessor array realizability criteria for real-time recurrent dsp applications. *IEEE Trans. on Signal Processing*, 42(5), May 1994.

[9] L. G. Jeng and L. G. Chen. A globally static rate optimal scheduling from recursive dsp algorithms. In *Proc. ICASSP*, pages pp. 1005–1008, May 1991.

[10] L. F. Chao and E. H. M. Sha. Unfolding and retiming data-flow dsp programs for risc multiprocessor scheduling. In *Proc. ICASSP*, volume 5, pages pp. 565–568, 1992.

[11] C. Y. Wang and K. K. Parhi. High-level dsp synthesis using concurrent transformations, scheduling and allocation. *IEEE Trans. on CAD of IC and System*, 14(3), March 1995.

[12] L. E. Lucke and K. K. Parhi. Data-flow transformations for critical path time reduction in high-level dsp synthesis. *IEEE Trans. on CAD of IC and Systems*, 12(7), July 1993.

7.1

PLANNING FOR PERFORMANCE

Ralph H.J.M. Otten(*) and Robert K. Brayton
University of California at Berkeley, California
(*)Also at Delft University of Technology, The Netherlands and Synopsys Inc.

Abstract

A shift is proposed in the design of VLSI circuits. In conventional design, higher levels of synthesis produce a netlist, from which layout synthesis builds a mask specification for manufacturing. Timing analysis is built into a feedback loop to detect timing violations which are then used to update specifications to synthesis. Such iteration is undesirable, and for very high performance designs, infeasible. The problem is likely to become much worse with future generations of technology. To achieve a non-iterative design flow, we propose that early synthesis stages should use "wireplanning" to distribute delays over the functional elements and interconnect, and layout synthesis should use its degrees of freedom to realize those delays. In this paper we attempt to quantify this problem for future technologies and propose some solutions for a "constant delay" methodology.

1 Introduction

In the past three decades, layout synthesis has relied on wire length and area minimization under the constraints of a technology file (design rule set) to generate masks for chips that showed acceptable functionality, yield and performance. Interconnect served merely as the realization of the net list and its influence on performance was negligible. As technology moves deeper into sub-micron feature sizes, and more components are integrated on a single chip, interconnect effects become more problematic.

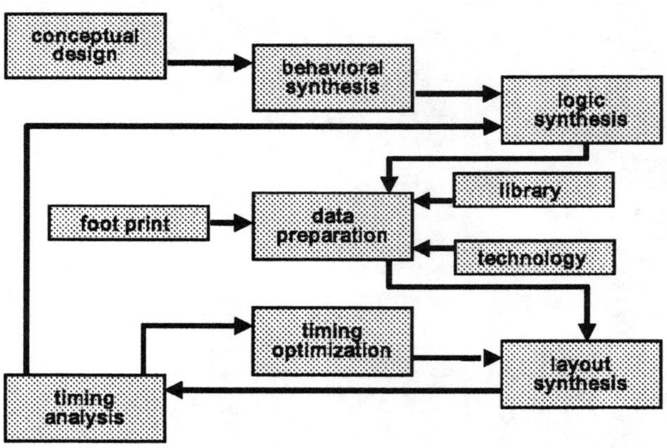

Figure 1: Synthesis with timing constraints

Up to now the effect of wiring on delay was determined by timing analysis tools that detect timing violations and produce either input for timing optimization procedures (such as transistor sizing, buffer insertion and fanout trees) or an updated specification file for higher level synthesis, expecting an improved gate and netlist for layout synthesis. Essentially the back-end of the design process has become a slow iterative scheme (shown in Figure 1), with no guarantee of convergence. Even if the process converges, it is uncertain how the final solution compares with the optimum.

Such iterations can be avoided only if the early design stages are integrated with layout synthesis, or at least are able to incorporate sufficient layout considerations without unnecessary constraints for the back end. This will require a completely new approach, especially for complex designs with very tight performance constraints. The required performance must be guaranteed by construction (and not the arbitrary outcome of indirect optimizations). This affects not only the way layout synthesis should be organized, but also higher levels of synthesis, and logic synthesis in particular.

2 Global wires

In recent years many sophisticated models for interconnect delay have been developed[10]. The complexity of these models and/or the size of the look-up tables used inhibits their use during synthesis, when the geometry of the interconnect is unknown, and when only estimates of length and topology are available. In these early stages only simple models such as Elmore's first moment matching can be used effectively. This model is the basis for almost all methods for reducing delay in point-to-point interconnection with unidirectional signal flow. The most common method is to split the wire into segments buffered by inverters[1]. This method, like any other, has its fundamental limitations which are often reached well before required performance has been achieved. In the next section we will study the limits of repeater insertion.

2.1 Critical lengths and critical delay

We use a first order model for a generic restoring buffer driving a capacitive load through a homogeneous line of length l given in Figure 2. [1].

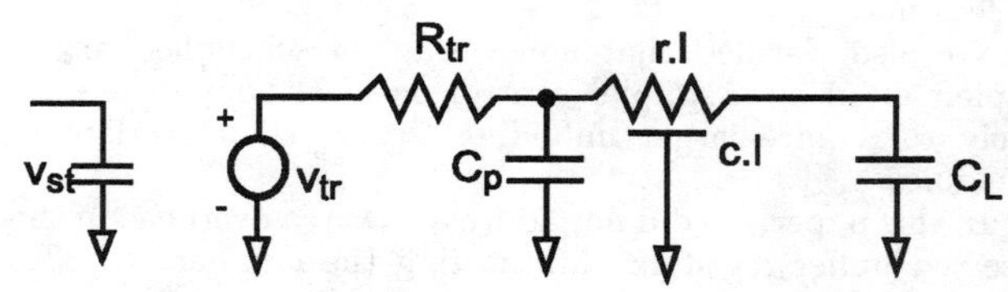

Figure 2: Generic restoring buffer model

The repeater is represented as a voltage source controlled by the voltage v_{st} at the input capacitance. This voltage source switches instantaneously when the fraction denoted by x, $0 \leq x \leq 1$, of the total swing has been reached. The switching at the voltage source is a perfect voltage step. The parasitic capacitance C_p is mainly composed of the drain capacitance of the transistors. The line is assumed uniform. If there is fanout, no resistance is assumed after branching.

Starting from this simple model, a general formula for the delay between the switching of the buffer and completing the

[1] Delay can also be reduced by "tapering" the wire, causing some difficult layout problems. Swing reduction with regenerative reaction to smaller voltage changes at the end of the line is another method that can speed up communication at the cost of increased noise sensitivity.

DAC 98, San Francisco, California

x fraction of the swing at the end of the line can be derived, similar to [1]:

$$\tau = b(x)R_{tr}(C_L + C_p) + b(x)(cR_{tr} + rC_L)l + a(x)rcl^2.$$

R_{tr} is the equivalent transistor resistance. The constants a and b depend on the switching model of the repeater, that is on x.[2] This should be chosen so that when we divide the line in n equal parts by inverters the delays of the sections can be added. The size of the inverters is s (in multiples of the minimum size inverter), which makes $R_{tr} = r_o/s$, $C_L = s.c_o$ and $C_p = sc_p$. The initial driver of the line is assumed to have the same size, possibly after cascading up from smaller initial drivers for optimum speed. The total delay for n such sections of length l/n is

$$T = n.\tau = n\left[br_o(c_o + c_p) + b(c\frac{r_o}{s} + rc_o s)\frac{l}{n} + arc\frac{l^2}{n^2}\right] =$$

$$= br_o(c_o + c_p)n + b(c\frac{r_o}{s} + rc_o s)l + arc\frac{l^2}{n} \qquad (1)$$

Now we can ask for the values of s and n which give the minimum delay. For n we obtain

$$\frac{\partial T}{\partial n} = br_o(c_o + c_p) - arc\frac{l^2}{n^2} = 0$$

or the optimum length of each section is

$$l_{crit} = \frac{l}{n_{opt}} = \sqrt{\frac{\frac{b}{a}r_o c_o\left(1 + \frac{c_p}{c_o}\right)}{rc}} = \frac{P}{\sqrt{rc}}. \qquad (2)$$

Accepting that $r_o c_o$ and $r_o c_p$ are process constants makes the optimum distance between inverters only dependent on the rc product per unit length of the wire.

THEOREM 1 *The length of a section in an optimally segmented[3] line is inversely proportional to* $\sqrt{rc}$.

P depends on the process and the delay model (x) only. Since r and c differ from layer to layer, these distances also differ from layer to layer.

Substituting n_{opt} in (1) yields

$$T(n_{opt}) = \left(2\sqrt{abrcr_o c_o\left(1 + \frac{c_p}{c_o}\right)} + \frac{bcr_o}{s} + brc_o s\right)l$$

which shows

THEOREM 2 *The delay of a line that is optimally segmented is linear in its length.*

The optimum repeater size is obtained as

$$\frac{\partial T}{\partial s} = b(rc_o - \frac{r_o c}{s^2})l = 0 \Rightarrow s_{opt} = \sqrt{\frac{r_o\, c}{c_o\, r}}$$

which is independent of n, the number of inverters used. By substituting the optimum repeater size and the optimum number of sections into (1) we find the delay of the line is

$$T(l) = 2l\sqrt{rcr_o c_o}\left(b + \sqrt{ab\left(1 + \frac{c_p}{c_o}\right)}\right)$$

which of course is linear in l. More surprisingly, substituting the critical length shows that the delay of a section of critical length does not depend on the line resistance and capacitance:

$$\tau_{crit} = 2br_o c_o\left(1 + \sqrt{\frac{b}{a}\left(1 + \frac{c_p}{c_o}\right)}\right) = 2br_o c_o\left(1 + \frac{P}{\sqrt{r_o c_o}}\right) \qquad (3)$$

and therefore only depends on the process (and the model), but not on the wiring layer.

[2] If $x = 0.9$ (90%-swing) then $a = 1.0$, $b = 2.3$. Mostly, we use ($x = 0.5$) yielding $a = 0.4$, $b = 0.7$.

[3] We call it an optimally segmented line rather than an optimally buffered line because this length is independent of the buffer size s.

THEOREM 3 *The delay of a section in an optimally buffered line is the same for all layers.*

Note that all derivations were made on a chain of inverters driving an uniform wire. Using this in more general networks, with different fanouts and branch-off geometries is therefore only an approximation, which can be made more accurate if isolation techniques are used to offset fanout effects.

2.2 Model justification

Since we use the results only for point-to-point connections (that is without branching) between restoring circuits, first moment matching is accurate enough. However, some remarks concerning the model parameters are in order.

The via resistance shows up as a resistor in series with the R_{tr} in Figure 2. It is reasonable that this scales with the size of the inverter[4], and hence can be absorbed in r_o and the formulas do not change. Although $r_o c_o$ is no longer a process constant and a layer dependence is introduced in the critical delay, experiments show that the via resistance, even to the top layer, is negligible and has hardly any effect on the wave forms.

The line was assumed to have constant capacitance per unit length. For advanced technologies, this is dominated by capacitance to sections of interconnection, especially neighboring sections in the same layer. Since these may undergo voltage changes, the value is not even constant. The latter effect may cause variations in the effective capacitance by up to a factor of 3. To make use of the derivations in the previous section, before the geometry of the wiring is known, requires the enforcement of a routing style which produces a time-invariant homogeneous line[5]. In addition, its resistance and capacitance should be known a priori.

The remaining problem with the model is the determination of the effective transistor resistance. It is reasonable that such a resistance exists if we only consider one waveform and a fixed x. The most practical way to obtain useful parameters is to simulate a ring of an odd number of buffer sections with large transistors (100 times minimum size) after extracting c very accurately. Then we optimize speed by varying the value of l for each section to obtain l_{crit}. This will give P of equation 2 since r is known quite accurately. With the length of each section fixed at l_{crit} the ring is optimized next for speed once more, now by varying s. This will yield τ_{crit} and by equation 3 therefore $r_o c_o$. Since we can accurately calculate c_o from the transistor geometry[6], we get

$$r_o = \frac{1}{4c_o}\left(\sqrt{P^2 + \frac{2\tau_{crit}}{b}} - P\right)^2.$$

2.3 Numerical data

To quantify what this all means we performed some calculations using a fictitious, but well reviewed technology file based on [11], and an extraction program [2][7] for solving exact 3-dimensional field problems. The critical lengths and the critical delay are given in Table 1[8]. Each layer has its own critical

[4] The contact resistance (the largest part) will scale if the contact area grows with buffer size, and also the cross section of the via is likely to scale then.

[5] One (possibly drastic) way of achieving this is to shield each signal line with neighboring lines tied to ground.In addition to reliable characterization, this style eliminates most cross-coupling noise problems.

[6] Theoretically, we don't have to do this since we obtained s_{opt} in the second optimization and $c_o = r_o c/(rs^2_{opt})$. However, since τ_{crit} is likely to be insensitive to s at the optimum, the value of s_{opt} is probably not very accurate, even though τ_{crit} is accurate.

[7] For more information, consult http://cas.et.tudelft.nl/research/space.html

[8] Table 1 is by Amit Mehrotra of UC Berkeley with independent corroboration by studies by Lixin Su, Sunil Khatri, and Dennis

length, and the values are pairwise close. Such a pair is called a *tier*. With a bit of process tuning the critical lengths within a tier can be made almost the same where the difference is mainly in the "between layers" capacitance. The higher tiers having longer critical lengths than the lower ones.

critical parameter	feature size 0.25 μ	feature size 0.10 μ
$l_{crit}(m1)$	10440	6757
$l_{crit}(m2)$	10600	7162
$l_{crit}(m3)$	36000	43446
$l_{crit}(m4)$	38400	45135
$l_{crit}(m5)$	63200	64932
$l_{crit}(m6)$	62000	56892
$l_{crit}(m7)$		97581
$l_{crit}(m8)$		93378
$T(l_{crit})$	205ps	80ps

Table 1: Critical wire lengths measured in feature size units.

Note that the critical length, measured in the feature size, changes much less than proportional with the feature size! This may come as a surprise, but is mainly due to velocity saturation effects, and therefore represents a trend that will affect smaller feature sizes even more. Other recent studies also indicate that even with scaling down in the logic blocks, the gate delay will continue to dominate the performance [4].

Today's synthesis is capable of handling blocks with up to 10000 gates. A square with side lengths of $l_{crit}(m1 - m2)$ can contain on the order of a hundred of these blocks in a 0.1μ technology. So, even with careful extrapolation, this means that fairly complex blocks can still be designed while mainly controlling gate delay.

3 Wire planning

The term *wire planning* was coined more than ten years ago "to describe an approach that first focuses on determining an optimal plan for global wiring" [3]. In its original context its task was mainly to identify groups of nets each connecting to (almost) the same set of modules. The most common example are *buses*: when identified routing complexity can be reduced considerably by handling bus wires as groups inside *data path generators*.

Another, new task for wire planning is indicated by the computations and derivations of Section 2. Although complex modules can still be designed using present day logic synthesis methodologies, controlling mainly the gate delays to achieve performance, at the chip level, future technologies will involve many (hundreds to thousands) of such complex modules, and at this level wire delay begins to dominate. Wire planning should produce a location for these modules with the main concern that timing constraints on input/output paths are met. This requires knowledge of the global interconnection structure and the performance implications of the functionality of the modules. In the early "conceptual" stages of a design there is not much more than an awareness of size-speed trade-offs. This accuracy depends completely on the design experience in the team. In general, all that can be said is that these interpolated delay-area relations appear convex. Delay in synchronous systems is often measured in terms of the number of clock cycles to complete the computation of the module. Obviously, a module that takes more cycles to do its computation never requires more area than one that takes fewer cycles. Although defining speed of modules in general is not possible, the reasoning will always be similar, whether speed in a given technology is obtained by sizing, parallelism, or other means.

Sylvester. The associated technology files can be found at http://www-cad.eecs.berkeley.edu/Respep/Research/nexsis/strawman

Knowing module functionality and interconnection structures implies a decomposition. Initially, such decompositions emerge solely on the basis of functional considerations, with little regard for their impact on both the performance of the product as well as the efficiency of synthesis steps later on. Therefore, while the design evolves into a hierarchical description acceptable for behavioral synthesis, wire planning tools should aid in quick analyses and proposals for function duplication, absorption and decomposition as well as module (re)locations and (partial) pad planning.

Examples of wire planning tasks are establishing the existence of a module placement in which no path from input to output has to make detours, assigning time budgets to modules such that area is minimized, establishing the existence of a valid retiming and producing a valid minimal-area retiming, assigning wire sections to layers so that feasible time budgets are preserved, and encouraging floorplans that lead to efficient optimizations in later stages of the design.

The final result of conceptual design aided by wire planning is a composition of a network of blocks and interconnections along with well established time budgets and delays. Considering the data concerning critical lengths, blocks will be small compared to these critical "units". They can be treated without internal distributed delays, and their wiring is mostly realized in the lower levels of metal. The tools can aid in creating subsets of regular grids with blocks at grid points and predefined wire segments on the grid lines. The latter enables good characterization of these segments, and routing consists of "using the available segment" rather than "placing segments". By the time that synthesis begins to create the gate and net lists, the delay on the "global" wires is quite well established and therefore also the timing budget that remains for the blocks.

3.1 Monotonic wire plans

Consider a high level description of a design described as a *functional network* modeled by as a directed acyclic graphs with primary inputs as sources, primary outputs as sinks and "functions" on the other nodes. There is an arc from one node to another if the "result" of the former is used as an argument in the latter. If a primary output depends on a primary input, there must be a path connecting them, possibly passing through other blocks, and possibly sharing some with other paths. Total delay is the sum of the delay in the blocks and the delay in the wires. If wires are composed entirely out of sections with critical delay, the total wire delay on a path is a multiple of the critical delay, and is invariant with respect to how the functional units are distributed over the restoring sites (*end points of critical sections*). If a functional block is placed at each "grid point" along a path then no repeaters are necessary. A *wire plan* in this context is a position for all the nodes in the functional network and a pin assignment for all primary inputs and outputs. Such a wire plan is called *monotonic* if all interconnections can be made so that the $\mathcal{L}_1$-length ("Manhattan length") of each input/output path is equal to the $\mathcal{L}_1$-distance ("Manhattan distance") between the two associated i/o pins. Under the model this is the fastest possible wire plan for a functional network with that pin assignment[9] having its wires in a given tier.

For a given pin assignment a monotonic wire plan may not exist. This existence question has been answered in [5] as

[9] Under a model where interconnections have capacitance but negligible resistance, a monotonic wireplan has the minimum total wire capacitance. This can be useful when power is a major concern and may be relevant for logic synthesis when a pin assignment is given [5].

follows. The *support* of a node is the set of primary inputs connected to that node by a directed path. The *range* of a node is the set of primary outputs connected to it by a directed path The *inbox* of a node is the smallest iso-rectangle containing its support, and the *outbox* is the smallest iso-rectangle containing its range. A *bridge* of a node is a minimum $\mathcal{L}_2$-length line connecting its inbox with its outbox. Using these ideas and and working out a few special cases leads to Theorem 4:

THEOREM 4 *Every node in a monotonic wire plan must be placed within the smallest iso-rectangle containing its bridge.*

A simple proof by induction then yields:

THEOREM 5 *A functional network has a monotonic wire plan with respect to a given pin assignment if and only if every node has a unique bridge.*

This makes it very easy to find out whether such a wire plan exists: we only have to check on a node by node basis whether each node in the network has a unique bridge. Such a check is extremely simple since

THEOREM 6 *A node has a unique bridge if*

1. *the support or the range contains a single pin, or*
2. *the range is contained in an iso-line while the support is on a single line perpendicular to that, or*
3. *the output box is in the "projection" of the input box, that is the two boxes have disjoint support in both axes, except for at most one point.*

Note that a placement conformant to Theorem 4 is not necessarily a monotonic wire plan. A valid placement, but possibly having nodes at the same position, is assigning each node the point which the output box has in common with the bridge[10].

Of course, certain deviations from strict monotonicity may be necessary or desirable, because of availability of space or for sharing functionality with other paths. However, deviation from monotonicity can only be allowed if the timing requirements are not violated. Note that monotonicity can always be obtained by duplicating functionality, synthesizing faster blocks, and absorbing functions in their fanout. In the extreme, a monotonic wire plan always exists if each output is produced by a single node.

Once the wireplan for a functional network has been determined, which means that the delay on the arcs of this network is known, the remaining time budgets have to be distributed over the function nodes. If the same graph is a suitable model for this task, and the sources and sinks have arrival times and required times assigned to them, a simple (quasi-)convex optimization problem can be used to answer questions such as "what is the smallest network that does not violate any timing constraints?" Size is in this case the sum of the areas assigned to each node according to its area-delay trade-off.

3.2 Valid retiming

At this level, our goal is to assign delays (measured in units of clock periods) to wires and to get an appropriate retiming of the chip. We assume that the dependency graph is acyclic. At each input and output, a number of latches are put in a consistent[11] way to reflect the number of cycles allowed between the input and the computed output.

Given a wire plan, and therefore the distance k_{ij} between two (point) modules, measured in number of critical lengths at the highest tier, we want to see if it can be retimed so that all connections can be made in a single cycle. For example, if a wire needs 2 cycles even if assigned to the highest level of metal, then a solution is to split the wire into two parts separated by a latch. The following problem is formulated. Given an integer assignment on each wire, k_{ij} does there exist a retiming of the circuit such that whenever there is wire between module i and j the number w_{ij} of latches assigned to that wire, satisfies

$$k_{ij} \le w_{ij}?$$

If the answer is yes, the placement is *valid* and has a *valid retiming*.

A second problem is, given a placement, to find a valid retiming, if it exists, which minimizes the total area. Here the area-delay tradeoff curves with cycles versus area for each module is used. At the start, each module is modeled with two nodes separated by a edge. Now a retiming will move some of peripheral latches to internal edges. After a retiming, the number of latches in a module edge may be changed. This affects the area of the module through its area-delay trade-off curve. We call this the minimum-area valid retiming problem.

Finally, we look for the placement which gives the least area of all the minimum area valid retimings[12].

3.3 Layer assignment

Here we assume that an RTL description of the design is available. Our aim is to assign a delay number δ_{ij} to each wire, $i \rightarrow j$ in such a way that all delays from chip input or latch output to chip output or latch input can be made in the assigned number of clock periods for that connection. We assume that if a path is allowed a delay longer than a single clock period it is divided into an appropriate number of pseudo-nodes. Thus every combinational path must be within a single clock period. Let π be such a path. We assign half the clock period to this path, reserving the other half for the gates on the path. So

$$\sum_{ij\in\pi} \delta_{ij} = \frac{1}{2}$$

where δ_{ij} will be the delay assigned to connection $i \rightarrow j$. We solve the *zero-slack distribution problem* where all paths satisfy the above equations.

Next, from the different levels of metal, we create a number of *wire types*[13]. Each wire type has a critical length l_{crit}. These are sorted in ascending order and denoted as,

$$l_1 < l_2 <, \ldots, < l_m$$

Now given a placement, let the Manhattan length of wire $i \rightarrow j$ be denoted by d_{ij}. Wire $i \rightarrow j$ can be seen to be in a rectangle bounded by (x_i, y_i) and (x_j, y_j). This rectangle is assigned a cost C^k_{ij} which depends on the wire type k assigned to the wire. A wire is more costly if its type is on a higher level or if its rectangle is nearer the center of the chip. We minimize

$$\sum C^k_{ij}$$

Here $i \rightarrow j$ is assigned to the kth wire type if and only if

$$\frac{d_{ij}}{l_k} \le \delta_{ij} \le \frac{d_{ij}}{l_{k+1}}$$

[10] Also the points that input boxes have in common with the bridge is a feasible set.

[11] A similar notion was used with the so-called "peripheral" retiming.

[12] What is missing in this formulation is how the assignment of wires to layers plays a role. In the above formulation, only the top level wire type is considered. The fact that a wire can be placed on a lower level of wire and still meet its timing obligation is not considered. A possible answer is to modify the total area cost function, to penalize wires that are put on higher layers.

[13] For example, each level of metal k may be divided into a number of *wire types* (k, i) where i denotes the number of wires tied together on level k to construct a "thicker" wire. This gives us a finer grained resolution for having wires that reach different distances in the same time. Another way to create a wire type is to assign a wire to multiple layers.

Thus wire $i \rightarrow j$ will be assigned to a wire type where the delay on the wire, $\frac{d_{ij}}{l_k} \leq \delta_{ij}$. Therefore for any path π we have

$$\sum_{ij \in \pi} \frac{d_{ij}}{l_k} \leq \frac{1}{2}$$

If a solution exists, then all wire segments can be assigned a fixed wire delay, δ_{ij}[14]. A solution may not exist if there is no placement where all the delays can be met. In that case, we may have to return to the higher level wireplanning problem and even possibly alter the chip latency.

We also note in the derivation of the critical wire length for each kind of wire, that the optimum sized driver was used. For a wire of less than critical length, we have a choice in the size of the driver which gives more flexibility to adjust the drivers to meet the assigned delays.

4 Layout synthesis

To complement an approach based on wire planning, layout synthesis should realize the functional blocks in such a way that the delays in the blocks do not exceed their timing budgets, or rather keep them right on target. Since logic synthesis knows the budgets after wire planning and the range of available gates, it should deliver a gate list with an assigned specified delay for every gate. Layout synthesis should produce a network in which each gate causes exactly that delay. This is called *constant delay synthesis* [6]. Given a fixed delay for a gate, its size becomes a function of the output capacitance.

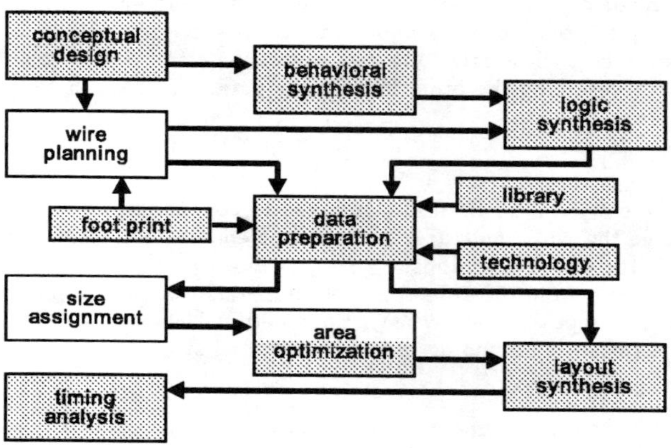

Figure 3: Constant delay flow

This of course is not without consequences for the back-end tools. Sizes now have to be assigned according to the results of logic synthesis, and scale only with the imposed capacitances on the outside. Timing optimization is out of the question: buffer insertion can only serve as an area reduction trick. New cell libraries have to be developed to adequately reflect the demands of delay-based requirements. And finally, the layout generation must be capable of handling a variety of cell sizes, and absorb changes in sizes efficiently.

[14] We have only indirectly accounted for the number of available wires of wire type k by assigning a wire to the least level which gives the required delay. This assumes that wires at the lower levels are the more plentiful. The cost of a wiring rectangle is weighted by its relative overlap with the center of the chip, but the number of wires in the wiring rectangle is not accounted for in this formulation.

4.1 Fixed delays

Again starting from the model of Figure 2, but not including the wire (Figure 4) leads to a delay formula[12] which is the sum of two terms, the *effort delay* and the *parasitic delay*:

$$\tau = bR_{tr}C_L + bR_{tr}C_p = br_oc_o\frac{C_L}{C_{in}} + br_oc_p = \frac{g}{f} + p.$$

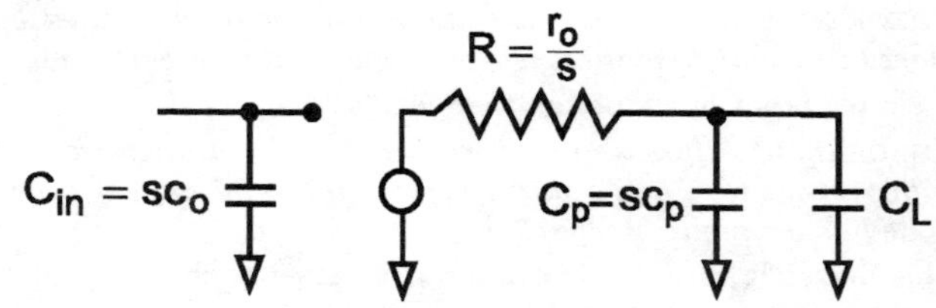

Figure 4: Gate model for obtaining a size independent delay expression

The parasitic delay $p = br_oc_p$ is independent of size. The effort delay g/f is a product of *computing effort* $g = br_oc_o$, and *restoring effort*

$$\frac{1}{f} = \frac{C_L}{C_{in}}$$

The computing effort is also size independent, but in general depends on the function, topology and relative transistor dimensioning of the gate. The important observation is that τ can be kept constant by fixing $f = C_{in}/C_L$. This leads to a new paradigm in synthesis [6, 9]: any delay imposed by synthesis can be realized, provided that the sizes of the gates can be continuously adjusted.In [6] the authors show that the size of a gate varies linearly with the load under constant delay. This enables size assignment after logic synthesis has fixed the *scaling factor* f for all gates and of course the net list.

4.2 Cell generation and shape assignment

Wire plans perform their analyses on point placements or sequences, most likely under the presence of larger blocks that may be pre-placed. This position information along with a possibly partial pin assignment must be preserved during the layout synthesis when the results of size assignment and area optimization become available. This requires efficient and robust floorplan optimization. These qualities heavily depend on the floorplan to be optimized. This can only be achieved by maintaining sliceability throughout the design, from the early wire planning stage down to the determination of the final dissection. This presumed "restriction" is amply offset by the guaranteed optimum in its class.

Cell generation is most likely the big challenge in a constant delay approach. The set of functions can be quite small, but extensive research is necessary to determine which sizes should be made available. Ultimately, a library of cell layout generators seems to be the way to go. In addition, yield is also an issue here.

5 Conclusions

In synthesizing high performance chips using present day design practices, the meeting of timing constraints necessitates an iteration which is not guaranteed to converge. In future technologies, unless these global delays are planned up front, convergence will be even more of a problem and even if convergence is achieved, the answer is likely to be far from optimum. This suggests a shift in design methodology where a global wire plan is put in place beginning at the conceptual stage of the design. We propose an approach in which a wire plan is created before the functionality of the blocks in that plan has been fixed. This allows for better control over the performance of the total design. Inherent to such an approach is that wire delay is accurately known wherever it has impact. This means that "global wires" should be well characterized a priori, which requires a strict layout styles. We have chosen to use a minimum width optimally buffered interconnections with a fairly stable electrically environment. Adherence to this style provides delays linear with distance, and thus invariance over equal length paths. Sharing functional blocks is likely to cause detours in one or more paths, causing additional delay. Creating a wire plan may distribute units all over the chip, thus abandoning the principle of easily recognizable and recoverable blocks, in exchange for exact knowledge of delay on connections and control over delay in the blocks. Enforcing delays in the blocks means that sizes become uncertain, and with uncertainties in size also distances become uncertain. If a block cannot be synthesized with the required delay in the available space, then the wire plan cannot be realized. Thus, reliable predictors in the early stages must be developed to obtain a non-iterative design flow. Of course, existence of solutions can never be guaranteed under too strict timing requirements, but we postulate that this new methodology can find solutions for a broader range of specifications than the current methods.

Acknowledgment: The vision that synthesis should be conducted with guaranteed performance was developed by Lukas van Ginneken. The authors are grateful to the wire planning group, Philip Chong, Wilsin Gosti, Hiroshi Murata,and Mukul Prasad, at Berkeley for stimulating discussions, to the Nexsis group at Berkeley, in particular Amit Mehrotra, Sunil Khatri, Subarna Sinha, and Philip Chong, who made the studies that quantified the critical lengths and to the SRC under grant 324 and the California Micro program, 96-091, with industrial support from Synopsys, Cadence and Fujitsu.

References

[1] H.B. Bakoglu, *Circuits, interconnections, and packaging for vlsi*, Addison-Wesley Pub Co, 1990

[2] F.Beeftink, A.J. van Genderen, N.P. van der Meijs *Accurate and efficient layout-to-circuit extraction for high speed mos and bipolar/bicmos Integrated circuits* ICCD, Oct. 1995

[3] R.K. Brayton, C.-L. Chen, J.A.G. Jess, R.H.J.M. Otten, L.P.P.P. van Ginneken, *Wire planning for stackable designs*, Proceedings 1987 International Symposium on VLSI Technology, Systems and Applications, Taipeh, Taiwan, pp 269-273, May 1987

[4] P.D. Fisher, *Clock cycle estimation for future microprocessor generations*, 1998

[5] W.Gosti, *Wire planning in logic synthesis*, 1998

[6] J. Grodstein, E. Lehman, H. Harkness, B. Grundmann, Y. Watanabe, *A delay model for logic synthesis of continuously-sized networks*, ICCAD, Nov. 1995

[7] Y. Kukimoto, R.K. Brayton, P. Sawkar, *Delay-optimal technology mapping by dag covering*, DAC, June 1998

[8] R.H.J.M. Otten, *Layout compilation*, in *Design systems for vlsi circuits*, edited by G. DeMicheli, A. Sangiovanni-Vincentelli and P.Antognetti, pp.439-472, Martinus Nijhoff Publishers, 1987

[9] R.H.J.M. Otten, L.P.P.P. van Ginneken, N.V. Shenoy, *Speed: new paradigms in design for performance*, ICCAD, Nov. 1996

[10] L. Pileggi, *Delay metrics*, ISPD98

[11] Semiconductor Industry Association, *The national technology roadmap for semiconductors: technology needs*, California, U.S.A., 1997

[12] I. Sutherland, R. Sproull, *The theory of logical effort: designing for speed on the back of an envelope*, in *Advanced Research in VLSI*, UC Santa Cruz, 1991

A DSM Design Flow: Putting Floorplanning, Technology-Mapping, and Gate-Placement Together *

Amir H. Salek, Jinan Lou, Massoud Pedram
Department of Electrical Engineering - Systems
University of Southern California
Los Angeles, CA 90089
{amir, jlou, massoud}@zugros.usc.edu

ABSTRACT - This paper presents an integrated design flow which combines floorplanning, technology mapping, and placement using a dynamic programming algorithm. The proposed design flow consists of five steps: maximum tree sub-structure formation, levelized cluster tree construction, minimum area implementation using 2-D shape functions, critical path identification, and repeated application of simultaneous floorplanning, technology mapping and gate placement along the timing critical paths. Experimental results obtained from an extensive set of benchmarks demonstrate the effectiveness of the proposed flow.

1. INTRODUCTION

The strong desire for ever-increasing levels of integration and higher performance often transcends the capabilities of traditional design tools and flows in the IC design arena. In particular, new deep-sub-micron (DSM) design considerations, such as the dominance of interconnect delays and signal integrity issues, have forced IC designers to re-evaluate the existing design methodologies and techniques. To address the DSM design challenges, one can increase the lookahead capability of high-level tools or develop new algorithms for optimally solving larger portions of the overall design problem simultaneously. This latter unification-based approach is, in our view, more promising. Indeed, the current state of CAD tools and algorithms has evolved to a point where it is both possible and necessary to combine certain steps of the logic synthesis and physical design processes.

This paper introduces an integrated approach for simultaneous floorplanning, technology-mapping, and detailed gate placement, of row-based standard-cell layout styles. The unique contributions of the paper are:

•A new design flow, *SiMPA, which simultaneously performs floorplanning, technology mapping, and placement.

•A new data structure, k-way levelized cluster tree, which represents the hierarchy of the circuit for *SiMPA.

* This work was supported in part by SRC under contract No. 98-DJ-606, and NSF under contract No. MIP-9628999.

•A new global area optimizer, *SiMPA-E, which optimizes chip area via simultaneous floorplanning, technology mapping, and gate placement.

•A new critical path optimizer, *SiMPA-R, which for a given number of critical paths effectively trades area for delay while simultaneously considering all floorplanning, technology mapping, and gate placement solutions.

The rest of this paper is organized as follows. In section 2, background and motivation behind this work are given. Section 3 describes our methodology for unifying floorplanning, technology mapping, and gate placement. In sections 4 and 5, our experimental results and concluding remarks are presented.

2. Background

Technology Mapping - It is well-known that the general technology mapping (TM) problem is NP-hard [HS96]. Keutzer in [Ke87] approximated the general TM problem by a set of tree-covering sub-problems and optimally mapped each, with respect to gate area, by a polynomial dynamic programming-based (DP) algorithm (KA). Later, Rudell in [Ru89] extended this idea to the minimum delay mapping problem. Subsequently, Touati et al. provided a mapper capable of minimizing area under delay constraints [TMBW90]. Chaudhary and Pedram presented a method for finding a set of all non-inferior mappings for a tree with different area-arrival time trade-offs [CP92]. All of the above mentioned methods assume the dominance of gate area and delay over interconnects.

Linear Placement - Linear placement (LP) is defined as the assignment of the vertices of a graph to the open slots located on a line so as to minimize a desired cost function. MINCUT is the LP problem where the cost function is the maximum cutwidth. It has been proven that this problem is NP-complete for general graphs [GJ79] but that it is polynomial in cases where the subject graph is a tree. Lengauer, in [Le82], developed an approximation algorithm (LA) for tree MINCUT problem whose cutwidth is within a factor of two of the optimal value. Later, Yannakakis introduced a polynomial optimal algorithm (YA) for tree MINCUT problem using the DP strategy [Ya85].

SiMPA - To address the high performance requirements of DSM designs, SiMPA (Simultaneous Technology Mapping and Linear Placement Algorithm) integrates technology mapping and linear placement by combining DP-based YA/LA with KA. This algorithm allows TM to access accurate physical information and LP to guide logic optimization. SiMPA-E ('E' stands for *exact total area*) is a combination of YA and KA which optimally finds the minimum total (wiring+gate) area implementation of a given tree circuit [LSP97]. SiMPA-D ('D' stands for *disjoint combination based*) combines LA and KA and optimizes total area and total delay by generating three dimensional trade-off curves for tree circuits [LSP98]. FPD-SiMPA (FPD stands for *floorplan driven*) is a design flow which incorporates SiMPA-D and SiMPA-E (SiMPA-D/E) for building two dimensional implementations for DAG-structure circuits [LSP98]. The outline of this design flow is as follows: 1. Partition the initial DAG into a set of tree clusters. 2. Map and place each cluster. 3. Floorplan the clusters on a two dimensional plane. 4. Perform global routing followed by timing analysis. 5. Trade off area for delay using SiMPA-D/E for each timing or area critical cluster.

Floorplanning - It refers to the placement and sizing of flexible blocks and is a common feature in many hierarchical design environments. In general, floorplanning algorithms are based on one of the following methods: mathematical programming, rectangular dualization, and combinatorial based methods [Sh93]. Recently, two novel floorplanning techniques based on sequence pair [MFNK96] and bounded slicing grid structures have been proposed [NFMK96]. In this paper, a combinatorial based floorplanning method, like that in [PK92], is combined with SiMPA through the use of an appropriately defined hierarchy.

Acyclic Partitioning - Acyclic partitioning prevents nets from creating directed cycles among the parts. A two-way acyclic partitioning algorithm was introduced in [IPFC93], then extended to multi-way acyclic partitioning in [CLB94]. The latter is used in this work when delay optimization is added to the proposed algorithm (*SiMPA-R). The acyclic property is necessary because it allows simple handling of the timing dependencies among the parts by following a topological ordering.

3. Simultaneous Floorplanning, Technology Mapping, and Gate Placement (*SiMPA)

SiMPA-D and SiMPA-E are only capable of manipulating tree-structure circuits. However, such a structure is atypical of most circuits and therefore in FPD-SiMPA, a DAG circuit is clustered into tree clusters and each cluster is manipulated by SiMPA-D/E separately. In other words, the design flow divides the problem into a set of smaller sub-problems and then combines the sub-solutions to build the final solution. This approach can be classified as a member of the divide-and-conquer family of algorithms and consequently inherits the intrinsic shortcomings associated with this family. Generally, FPD-SiMPA restricts SiMPA to work within the individual tree clusters only, and employs the floorplanner for taking advantage of the possible inter-cluster optimizations opportunities. The floorplanner, however, is subject to the optimization decisions already made within the clusters which are not necessarily appropriate for the overall circuit optimization.

*SiMPA, the proposed algorithm in this work, addresses this deficiency by combining SiMPA with the floorplanning design step. Hence, the flooplanner controls all the intra-cluster and inter-cluster optimization opportunities and so the decisions made inside and outside the clusters are the best in the global sense. Another important property of *SiMPA is that it eliminates delay budgeting (slack distribution) in the design flow. *SiMPA, due to its integrated nature, has no need for this step. It provides designers with a conceptually simpler and more precise design tool which outputs the global trade-off picture of the design using fewer heuristics.

The way FP is merged into SiMPA is different from the way TM and LP were merged in SiMPA. TM and LP both work at the same hierarchy level (inside the tree clusters) and so it was possible to implement their combination via interleaved calls from one to another at every step of the DP-based algorithm. FP on the other hand, works at a higher level of hierarchy (among the tree clusters), and therefore, its integration into SiMPA must be achieved by other means. The floorplanner uses SiMPA to capture the shape functions of all the tree clusters. These are subsequently used by the bottom-up floorplanner to calculate the global shape function for the whole circuit.

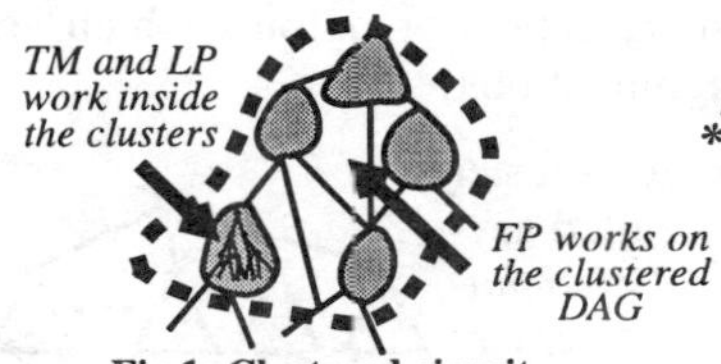

Fig.1: Clustered circuit

Fig.2: *SiMPA

*SiMPA comes in two flavors:

- *SiMPA-E which targets the total minimum area by generating a set of non-inferior solutions (w.r.t. shape) for the whole circuit.
- *SiMPA-R which removes the timing violations for a given implementation and a design hierarchy by simultaneously replacing and remapping the clusters on timing critical paths. This algorithm is capable of computing accurate shape/delay trade-off curves for the whole chip while speeding up one critical path of the circuit at a time.

In our flow, *SiMPA-E is first called to generate a minimum area implementation of the subject circuit, then *SiMPA-R is called to remove the timing violations one path at a time.

In *SiMPA, a special data structure is used for the representation of the design hierarchy. The first sub-section below describes and analyzes this representation. Next, *SiMPA-E and *SiMPA-R are presented and discussed.

3.1 K-LCT: A K-Way Levelized Cluster Tree

Any bottom-up technique requires, at least implicitly, a hierarchy to work with. For the case of BearFP (a DP-based floorplanner), the hierarchy is represented by a *cluster tree* which shows a grouping of macro-cells into first-level clusters, first-level clusters into second-level super clusters, and so on, until the root of the cluster tree is reached [PK92]. *SiMPA, which includes a floorplanner similar to

BearFP, likewise requires a data structure for the extraction and representation of the hierarchical organization of the tree clusters. In contrast to BearFP which uses a simple k-way tree by including the most connected blocks as sibling nodes under the same parent node, *SiMPA requires a more elaborate hierarchy and a corresponding suitable data structure to represent it. This novel data structure is called *k-way levelized cluster tree (k-LCT)* and is introduced and discussed below.

Definition1: For any given rooted tree, $T(V, E)$, $\forall v \in V$, the following terms are defined:

- *parent(v)*: immediate parent of *v*.
- *children(v)*: set of the immediate children of *v*.
- *descendents(v)*: the set of all nodes in the subtree rooted by *v*, including *v* itself.
- $leaves(v) = \{ u \mid u \in descendent(v) \wedge u$ is a leaf node $\}$.

Definition2: Given a directed acyclic graph $G(V', E')$, a rooted tree $C(V, E)$ with root *r* is a cluster tree for *G* if there exists a one-to-one function Γ: $leaves(r) \rightarrow V'$. Members of *E* and *E′* are called *cluster-edges* and *DAG-edges*, respectively. Note that for a given circuit, the corresponding DAG does not include the primary inputs or outputs which are later handled in a pad assignment phase.

Definition3: $C(V, E)$, a rooted cluster tree of a given $G(V', E')$, is called acyclic, if for any $v \in V$ and $1 < i < |children(v)|$ the following statement holds: $\forall v_1, \ldots, v_i \in children(v)$, and $\forall (u_1, \ldots, u_i) \in leaves(v_1) \times \ldots \times leaves(v_i)$ then $\{(\Gamma(u_1), \Gamma(u_2)), \ldots, (\Gamma(u_{i-1}), \Gamma(u_i)), (\Gamma(u_i), \Gamma(u_1))\} \not\subset E'$.

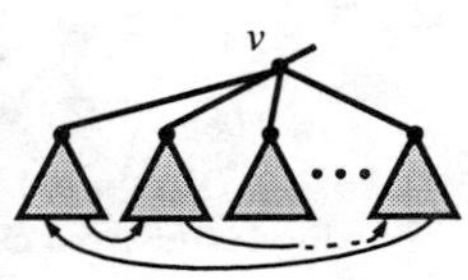

Fig.3: A cycle

Definition4: $C(V, E)$, a rooted cluster tree of a given $G(V', E')$, is a *k-LCT* if: **1)** $\forall v \in V$, $|children(v)| \le k$, **2)** $\forall (v \in V \wedge v \notin leaves(r)), \forall (u \in leaves(r) \wedge u \notin descendents(v)), \forall w \in leaves(v), (\Gamma(u), \Gamma(w)) \notin E'$.

In the following, $C(V, E)$ (with root *r*) is assumed to be a k-LCT of a given circuit $G(V', E')$ unless otherwise is stated. Property two of Definition4 states that in a k-LCT, all the DAG-edges connect nodes to those in the same or higher level, hence guaranteeing a k-LCT to be an acyclic structure. This property empowers *SiMPA to calculate all needed wire lengths signal arrival times while synthesizing the circuit; details will be discussed later on in this paper. Fig.4. which demonstrates a k-LCT for a simple circuit can be used to verify the acyclicness property of k-LCT's.

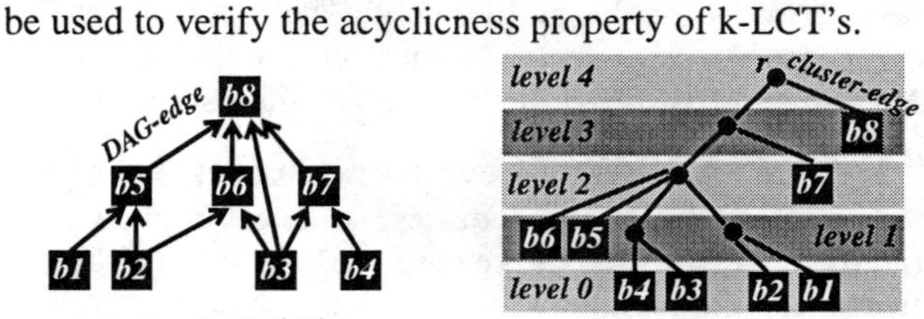

A circuit G(V′,E′) A 4-LCT for G, C(V,E)

Fig.4

Lemma1: For any edge in $G(V', E')$, say (u,v) (let $\upsilon = \Gamma^{-1}(u)$ and $\omega = \Gamma^{-1}(v)$), we have $\upsilon \in descendents(parent(\omega))$.

Theorem1: $C(V, E)$, a k-LCT of a given $G(V', E')$ is an acyclic cluster tree.

The following lemmas introduce transformations on k-LCT's for the generation of a new equivalent k-LCT with different properties; such as a more balanced k-LCT which generally generates more efficient floorplans.

Lemma2: In *C*, any leaf node can be raised as high as the level where its lowest-level fanout resides.

Lemma3: In *C*, any leaf node can be lowered as low as the level where its highest-level fanin resides.

Lemma4: In *C*, any two nodes which have the same parent node and the total number of their immediate children is less than *k* can be merged together.

The 4-LCT in Fig.5. is equivalent to that shown in Fig.4, generated using a sequence of transformations as mentioned above. This tree is more desirable than that of Fig.4, since DP-based floorplanner tends to produce better results when the number of the internal nodes is $log_k(n)$. Interested readers are referred to [SLP98] for the proof of the theorems.

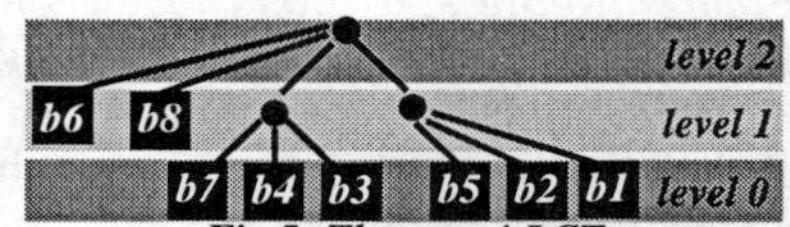

Fig.5: The new 4-LCT

3.2 *SiMPA-E for Global Area Optimization

*SiMPA-E is a combination of SiMPA and floorplanning which targets total area optimization. If a bottom-up floorplanning scheme with shape propagation ability is used, *SiMPA-E is capable of finding a global height versus width trade-off curve for the whole network. This curve is then used in finding the solution satisfying the user-specified aspect ratio requirement.

Total area for a one-dimensional standard-cell layout is calculated by $A = W \,.\, (h + \beta \,.\, c)$, where *W* is the sum of the width of all the cells in a row, *h* is the cell height, a constant determined by the ASIC library being used, β is the minimum distance between the two adjacent wires' centers, and *c* is the maximum cutwidth (a.k.a. cut-density) of the design.

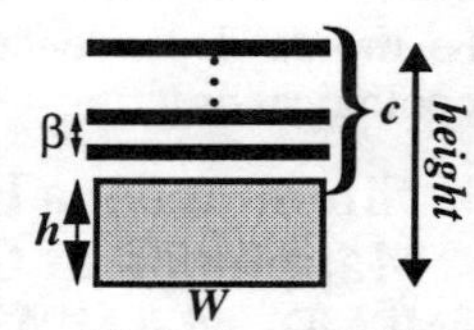

Fig.6: Area model

Lemma5: Cutwidth versus gate-area (*c* versus *g*) curves are transformed into height versus width curves, according to the following equations: $height = h + c \times \beta$ and $W = g/h$.

As shown in Fig.7, *SiMPA-E first partitions the given decomposed circuit into a set of maximal tree sub-networks (clusters). Then, SiMPA-E produces the cutwidth/gate-area trade-off curves, which are subsequently translated to height/width curve (shape function) using Lemma1, for each cluster. During the

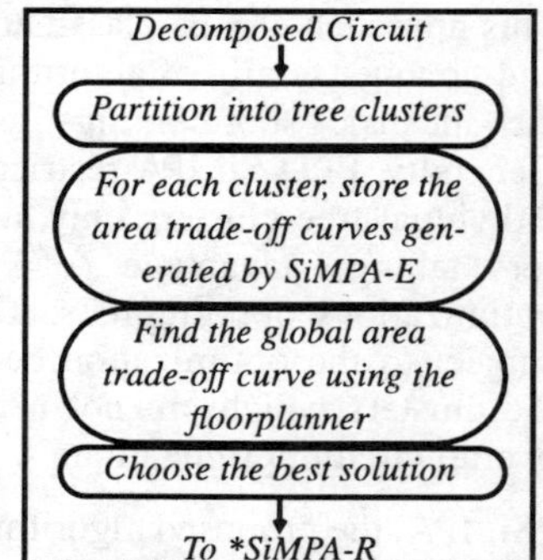

Fig.7: *SiMPA-E

third step, the set of all the shape functions is passed to a bottom-up floorplanner employed by the algorithm. The floorplanner takes these shape functions and generates the global 2-D trade-off curve. Finally, the best floorplan is picked from this curve. Note that, solutions for all the clusters are generated simultaneously from the best solution for the root of the cluster tree. Once the best global solution is found, the cluster sub-solutions which were used to construct the solution are traced back and assigned as the internal implementation of each cluster. Few observations relating to the operation of *SiMPA-E are:

Observation1: In the shape function for a cluster, the detailed physical and logical implementations of every solution are known prior to executing the floorplanner. This information is then used throughout the floorplanning process for the exact area and delay calculations of the whole or part of the design.

Observation2: *SiMPA-E achieves global area optimization by taking into account detailed design information for all the design choices and comparing solution qualities based on exact physical and logical information. Further, *SiMPA-E does not need to use any area estimators which are usually based on heuristic techniques.

Observation3: Any floorplanner which can properly use the set of calculated shape functions can be employed in this flow. However, the global height versus width trade-off curve is built most effectively using a bottom-up floorplanner with shape propagation capability. The use of this type of floorplanner, as mentioned earlier, needs a hierarchy on the clusters required for satisfying the *acyclicness* property if critical paths are intended to be resynthesized using *SiMPA-R. In this work, we employ k-LCT's (with k=4) to satisfy all these requirements.

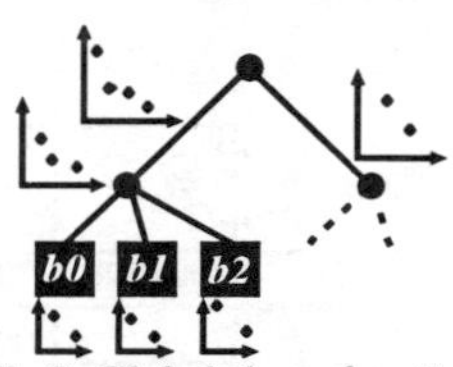

Fig.8: *Global shape function generation*

Observation4: SiMPA-E calculates the shape function (including gate and wire areas) for every tree cluster and therefore all optimum non-inferior points are included in the shape function. However, the area taken by inter-cluster wiring is an entirely different issue. Bottom-up floorplanning can be made to estimate (probabilistically or constructively) the inter-cluster (inter-block) connection lengths. These estimates are not exact and hence claims of optimality for *SiMPA-E are subject to inaccuracies in estimating the inter-block connection lengths and area requirements. The choice of a k-LCT hierarchy in *SiMPA-E leads not only to advantages in *SiMPA-R (discussed later), but also restricts inter-cluster interconnects which in turn reduces the impact of the inaccuracies suffered due to the use of a bottom-up approach.

Observation5: For every leaf and internal node of the cluster tree (hierarchy), *SiMPA-E stores the calculated shape functions for future reference by *SiMPA-R. This information is needed by the resizing process of the clusters on the timing critical path. The process may open up space for the timing critical clusters by optimally resizing the non-critical ones in order to allow trading space for better timing behavior inside the critical clusters.

Theorem2: *SiMPA-E is a polynomial algorithm.

3.3 *SiMPA-R for Eliminating Timing Violations

Although the global shape function generated by *SiMPA-E captures the best implementation in terms of area, it is likely for the resulting implementation(s) to violate user specified timing requirements. In the *SiMPA flow, it is thus essential to employ a technique to eliminate the timing critical paths by resynthesizing and trading area for delay. After using *SiMPA-E and choosing an area optimized implementation, *SiMPA-R is employed for the purpose of fixing the timing violations.

3.3.1 Critical Clusters Initial Manipulation

The output of *SiMPA-E is a fully mapped, two-dimensionally placed implementation of the given circuit. *SiMPA-R first identifies all critical paths and selects the one with the most negative slack and marks all the tree clusters on this critical path as critical. For every critical leaf node in the k-LCT, the internal nodes on the unique path to the root of the k-LCT are also marked so that floorplanning, technology mapping, and placement can be redone. We refer to the resulting representation as a marked k-LCT. Fig.10 demonstrates a simple example in which the critical path is shown by thicker arrows in the clustered circuit. In the 4-LCT the critical leaves and nodes are marked by white boxes and black circles, respectively.

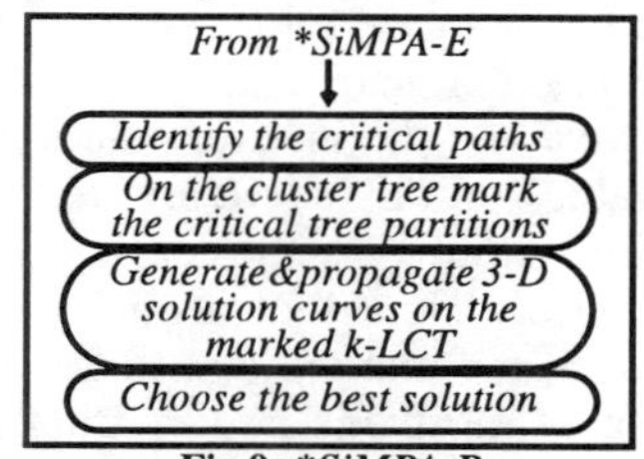

Fig.9: ***SiMPA-R**

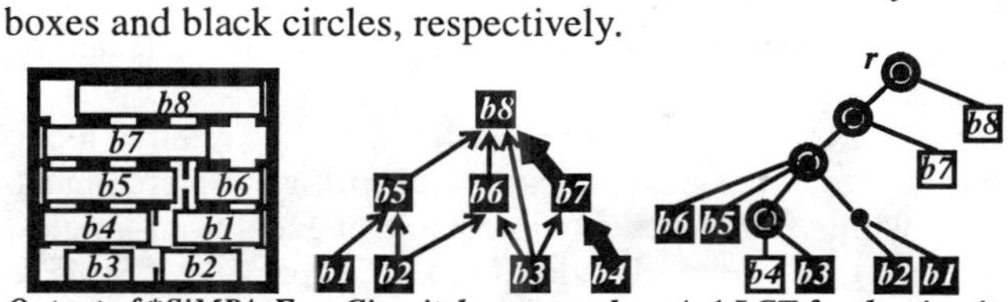

*Output of *SiMPA-E*   *Circuit decomposed into tree clusters*   *A 4-LCT for the circuit*

Fig.10

Lemma6: In a marked k-LCT, there is at most one critical internal node and one critical leaf node among the children of every internal node.

3.3.2 3-D Curve Generation and Propagation

Having marked the k-LCT, *SiMPA-R begins the generation and propagation of the 3-D (height, width, and the critical signal ready time at the root of a subtree) solution curves from the lowest level marked internal node up towards the root. At each step, the algorithm detects the immediate children of the current internal node, *v*, and retrieves or generates their corresponding 2/3-D solution curves in order to examine the solution possibilities for all the floorplan templates. An immediate child of *v*, say *u*, is either a non-critical internal/leaf node, a critical internal node, or a critical leaf node. In the first case (a non-critical internal/leaf node), there is no marked node among the *descendents(v),* indicating the absence of a critical signal. Therefore, the corresponding solution curve is the 2-D curve already calculated and stored by *SiMPA-E. For the case of a critical internal node, a 3-D solution curve is already calculated by the previous calls to the same procedure. The situation is a bit different for a critical leaf node; in this case one of the incoming signals is the critical signal which may be provided by its marked sibling's descendents. Consequently, the arrival time of the signal depends on the implementation chosen for the sibling node. However, all the physical and logical implementation information up to that point are available and therefore for every floorplan template and block-to-room assignment, the possible critical signal arrival times can be precisely calculated.

Using these arrival times, SiMPA-D generates a 3-D solution curve for that critical cluster. These 3-D curves (for every template and assignment) are assembled into one curve and the inferior solutions are pruned out. The timing values on this curve are the critical signal's ready times at the output of v for each possible implementation of the subtree. Note that according to Lemma6, there exists at most one marked leaf node and therefore SiMPA-D is run for at most one node during processing of the immediate children of every v. If v has no immediate marked leaf node, the timing values of the critical internal node are the critical signal's ready times. As for the geometrical calculations, the height and width of the solutions of v are calculated from the solutions to the sub-problems in ways similar to *SiMPA-E's method.

Now, few remarks relating to *SiMPA-R:
•Every solution generated by *SiMPA keeps the pointers to its constituent sub-solutions in order to retrieve all the corresponding design information for the best solution selected in the final stage.
•For every template and room assignment, the relative locations of the blocks are known. Accordingly, interconnect routes within v are known (subject to using a bottom-up global routing algorithm. In addition, k-LCT provides *SiMPA with a proper structure which helps avoid signal cycles among *children(v)*, and therefore all timing information can be calculated according to the topological order of *children(v)*.
•This procedure proceeds recursively until it reaches the root in the k-LCT. The trade-off curve of the root reveals the global design options from which the best is selected and traced down using the stored pointers.

Theorem3: *SiMPA-R is a polynomial time algorithm.

4. Experimental Results

The sequence of steps consisting of technology independent optimization, technology mapping, placement, global and detailed routing is the conventional design flow. To verify the effectiveness of the simultaneous approach, results are presented in Table1 for two setups: I) Conventional flow. II) Technology mapping, maximal tree clustering, floorplanning (BearFP), linear placement of each cluster using YA, followed by TimberWolf and YACR. The presented results show only small differences between the two runs which points to the fact that *SiMPA's efficiency is due to its simultaneous approach to the design process rather than just improved mapping and placement algorithms.

Our next set of experiments compares the performance of the conventional design flow with our proposed DSM methodology on a number of benchmarks using a CASCADE standard cell library (0.5u HP CMOS process). Gate and wire delays are calculated using a 4-parameter delay equation (similar to that in [LSP97]) and the Elmore delay model, respectively. These experiments were run in the SIS environment [SSLM92] on an Ultra-2 Sun Sparc workstation with 256MB memory. The area and delay reported here are total chip area and delay after detailed routing. Table2 compares the results of *SiMPA with the conventional flow showing an average of 34% performance improvement with small area penalty. In these experiments, the floorplanning was performed by our preliminary k-LCT floorplanner prototype, TimberWolf was used for global routing, and YACR was used for detailed routing. The runtimes for *SiMPA remained comparable to the ones for the conventional flow, as shown in Table3 for a number of benchmark circuits.

	I.		*II.*		*II over I*	
Circuit	***Area***	***Delay***	***Area***	***Delay***	***Area Ratio***	***Delay Ratio***
alu2	*2459*	*13.13*	*2402*	*13.10*	*0.98*	*1.00*
apex7	*1254*	*6.45*	*1260*	*6.61*	*1.00*	*1.02*
cm150a	*226*	*3.22*	*221*	*3.18*	*0.98*	*0.99*
cm151a	*168*	*2.97*	*172*	*2.91*	*1.03*	*0.98*
cm162a	*229*	*2.50*	*234*	*2.57*	*1.02*	*1.03*
duke2	*3283*	*10.16*	*3241*	*9.75*	*0.99*	*0.96*
k2a	*7555*	*15.96*	*7570*	*19.51*	*1.00*	*1.22*
rot	*4748*	*8.91*	*4651*	*12.12*	*0.98*	*1.36*
table3	*5014*	*55.17*	*5030*	*66.34*	*1.00*	*1.20*
C1908	*4288*	*17.46*	*4477*	*16.67*	*1.04*	*0.95*
C880	*2673*	*9.53*	*2496*	*10.61*	*0.93*	*1.11*
C1355	*2452*	*8.84*	*2517*	*9.69*	*1.03*	*1.10*
C3540	*10101*	*27.63*	*9674*	*36.09*	*0.96*	*1.31*
			Average Ratios:		***1.00***	***1.09***

Table 1: Verifying the effectiveness of the simultaneous approach

5. Conclusion

This paper presents a novel design flow *SiMPA, which performs simultaneous floorplanning, technology mapping and linear placement. *SiMPA is a dynamic-programming based algorithm which first computes the shape function of every tree cluster using SiMPA-E and then utilizes a bottom-up floorplanner to generate higher level shape based on a k-way levelized cluster tree. After the minimum area solution is chosen and the critical paths are identified, the clusters on each critical path are re-floorplanned, re-mapped and re-placed simultaneously to achieve improved timing. This new design flow yields very high quality circuits in terms of post layout chip area and delay.

	Conventional Flow		*SiMPA		Ratios	
Circuit	Area	Delay	Area	Delay	Area	Delay
alu2	2459	13.13	2017	8.60	0.82	0.65
alu4	4258	22.89	3725	11.89	0.87	0.52
apex6	4554	12.01	5106	6.82	1.12	0.57
apex7	1254	6.45	1317	4.31	1.05	0.67
b9	620	3.62	598	2.57	0.96	0.71
cm150a	226	3.22	215	1.42	0.95	0.44
cm151a	168	2.97	94	1.46	0.56	0.49
comp	641	2.84	544	2.22	0.85	0.78
con1	83	1.55	93	0.76	1.12	0.49
cordic	376	2.74	266	1.90	0.71	0.69
dalu	6356	19.45	5959	14.75	0.94	0.76
duke2	3283	10.16	2708	8.64	0.82	0.85
f51m	300	5.76	370	4.42	1.23	0.77
k2a	7555	15.96	9421	12.37	1.25	0.78
lal	560	4.09	537	2.60	0.96	0.64
misex3	3374	12.89	3559	10.73	1.05	0.83
mux	221	3.03	234	1.46	1.06	0.48
pcle	471	3.66	300	2.41	0.64	0.66
pcler8	602	3.90	570	3.00	0.95	0.77
ritex	337	2.16	299	1.84	0.89	0.85
ritex1	90	1.09	94	0.77	1.04	0.71
rot	4748	8.91	4864	8.20	1.02	0.92
term1	711	3.92	778	2.72	1.09	0.69
z4ml	267	3.45	182	1.66	0.68	0.48
5xp1	641	7.66	546	4.23	0.85	0.55
9sym	811	7.71	1019	5.62	1.26	0.73
Z5xp1	398	7.08	459	5.79	1.15	0.82
Z9sym	876	7.77	1068	5.14	1.22	0.66
b12	393	3.81	355	1.97	0.90	0.52
bw	1017	9.33	845	5.77	0.83	0.62
clip	689	7.85	768	3.61	1.11	0.46
e64	1602	8.29	1648	9.13	1.03	1.10
inc	587	7.78	577	3.45	0.98	0.44
misex1	293	5.90	313	2.61	1.07	0.44
misex2	517	4.03	514	2.48	0.99	0.62
misex3c	2629	19.50	2765	8.54	1.05	0.44
rd53	226	3.16	155	2.00	0.69	0.63
rd73	335	3.78	346	2.73	1.03	0.72
rd84	857	7.01	767	4.61	0.89	0.66
sao2	815	6.20	778	3.49	0.95	0.56
squar5	256	3.65	221	2.36	0.87	0.65
table3	5014	55.17	5356	53.51	1.07	0.97
vg2	543	3.66	462	2.67	0.85	0.73
xor5	96	1.56	92	1.10	0.96	0.71
C17	40	0.78	41	0.37	1.02	0.47
C432	2810	11.91	1329	12.28	0.47	1.03
C1908	4288	17.46	2538	14.17	0.59	0.81
C880	2673	9.53	2377	10.74	0.89	1.13
C1355	2452	8.84	2684	7.84	1.09	0.89
C499	2304	8.64	2702	6.55	1.17	0.76
C2670	4678	8.40	4597	7.17	0.98	0.85
C3540	10101	27.63	6985	22.20	0.69	0.80
C5315	11232	16.03	10201	16.22	0.91	1.01
C7552	12738	24.53	14158	11.77	1.11	0.48
				Average Ratios:	0.96	0.66

Table 2: *SiMPA versus the conventional flow

Circuit	Runtime (sec) Conventional	Runtime (sec) *SiMPA	Ratio
alu2	146	178	1.22
apex7	63	67	1.06
cm150a	15	22	1.47
cm151a	11	15	1.36
cm162a	16	17	1.06
duke2	146	152	1.04
k2a	519	753	1.45
rot	215	312	1.45
table3	243	265	1.09
C1908	255	291	1.14
C880	155	199	1.28
C1355	191	243	1.27
C3540	663	711	1.07

Table 3: Runtimes

6. References

[CP92]K. Chaudhary, and M. Pedram "A near-optimal algorithm for technology mapping minimizing area under delay constraints," In *Proceeding 29th Design Automation Conference,* pages 492-498, June 1992.

[CLB94]J. Cong, Z. Li, and R. Bagrodia, "Acyclic multi-way partitioning of boolean networks," In *Proceedings of Design Automation Conference,* pages 670-675, June 1994.

[GJ79]M. R. Garey and D. S. Johnson, *Computers and Intractability,* Bell Telephone Lab Inc., 1979.

[HS96] G. D. Hachtel, and F. Somenzi, *Logic synthesis and verification algorithms.*Kluwer Academic Publishers, Norwell, MA, 1996.

[IPFC93]S. Iman, M. Pedram, C. Fabian, and J. Cong, "Finding uni-directional cuts based on physical partitioning and logic restructing," *4th ACM/SIGDA Physical Design Workshop,* pages 187-198, April 1993.

[Ke87]K. Keutzer "DAGON: Technology mapping and local optimization," In *Proceedings of Design Automation Conference,* pages 341-347, June 1987.

[Le82]T. Lengauer, "Upper and lower bounds on the complexity of the min-cut linear arrangement problem on trees," In *SIAM J. Alg. Disc. Meth.,* vol. 3, no. 1, pages 99-113, March 1982.

[LSP97]J. Lou, A. H. Salek, and M. Pedram,"An exact solution to simultaneous technology mapping and linear placement problem," In *Proceedings International Conference on Computer-Aided Design,* pages 671-675, November 1997.

[LSP98]J. Lou, A. H. Salek, and M. Pedram,"An integrated flow for technology remapping and placement of sub-half-micron circuits," In *Proceedings of Asia and South Pacific Design Automation Conference,* pages 295-300, February 1998.

[MFNK96]H. Murata, K. Fujiyoshi, S. Nakatake, and Y. Kajitani,"VLSI module placement based on rectangle packing by the sequence-pair," *IEEE Transactions on Computer-Aided Design of Integrated Circuits and Systems,* 15(12), pages 1518-1524, December 1996.

[NFMK96]S. Nakatake, K. Fujiyoshi, H. Murata, and Y. Kajitani,"Module placement on BSG-structure and IC layout applications," In *Proceedings of International Conference on Computer-Aided Design,* pages 484-491, November 1996.

[PK92]M. Pedram and E. Kuh, "Bear-FP: A Robust Framework For Floorplanning," In *International Journal of High Speed Electronics,* Vol. 3, No. 1, pages 137-170, 1992.

[Ru89]R. Rudell "Logic synthesis for VLSI design," Memorandum UCB/ERL M89/49, Ph.D. Dissertation, University of California at Berkeley, April 1989.

[Sh93]N. Sherwani, *Algorithms for VLSI Physical Design Automation,* Kulwer Academic Publishers, 1993.

[SLP98]A. H. Salek, J. Lou, and M. Pedram, "Simultaneous floorplanning, technology mapping, and gate-placement for deep submicron designs," *University of Southern California Technical Report,* 1997.

[SSLM92] E. M. Sentovich, K. J. Singh, L. Lavagno, C. Moon, R. Murgai, A. Saldanha, H. Savoj, P. R. Stephan, R. K. Brayton, and A. Sangiovanni-Vincentelli," SIS: A system for sequential circuit synthesis", *Memorandum No. UCB/ERL M92/41,* Electronics Research Laboratory, College of Engineering, University of California, Berkeley, CA 94720, May 1992.

[St83] L. Stockmeyer, "Optimal orientation of cells in slicing floorplan designs," *Information and Control,* volume 57, pages 91-101, 1983.

[TMBW90]H. J. Touati, C. W. Moon, R. K. Brayton, and A. Wang "Performance oriented technology mapping," In *Proc. Sixth MIT Conf. Advanced Res. VLSI,* pages 79-97, April 1990.

[Ya85]M. Yannakakis, "A polynomial algorithm for the min-cut linear arrangement of trees," In *Journal of the Association for Computing Machinery,* vol. 32, no. 4, pages 950-988, October 1985.

Framework Encapsulations: A New Approach to CAD Tool Interoperability

Peter R. Sutton*
Cooperative Research Centre for Satellite Systems
Queensland University of Technology
2 George St. Brisbane QLD 4001 Australia
Phone: +61-7-3864-2458
p.r.sutton@ieee.org

Stephen W. Director*
College of Engineering, University of Michigan
1221 Beal Avenue
Ann Arbor MI 48109
Phone: +1-734-647-7010
director@umich.edu

ABSTRACT

Today's complex leading-edge design processes require the use of multiple CAD tools that operate in multiple frameworks making management of the complete design process difficult. This paper introduces the concept of framework encapsulations: software wrappers around complete CAD frameworks that allow the design data and flow management services of a framework to be utilized by a common design process management tool. This concept has been applied to the Minerva II Design Process Manager, enabling Minerva II to manage the design process across multiple CAD frameworks, and potentially multiple design disciplines.

1. INTRODUCTION

In today's complex design processes, designers require multiple CAD point tools in order to perform the various tasks that make up a design process. In "leading-edge" design processes, where "best-of-class" point tools are required, CAD tools may be obtained from several different sources: commercial CAD tool vendors, in-house development groups and academic research groups. Due to the different sources, these tools often operate within disparate CAD framework environments, making it difficult to manage the entire design process. This problem is exacerbated by the trend towards multidisciplinary concurrent design, where different parts of a system (e.g., hardware and software, or product and manufacturing processes) are designed in parallel by different groups of designers using different sets of CAD tools and environments. The aim of this work is to realize *unified design process management*, that is, simplified management of the design process across multiple CAD frameworks and potentially across multiple design disciplines.

CAD frameworks can be thought of as design environments which provide *design management services*. Early frameworks provided *design data management* and *design tool management* services - managing data format translations, tool command lines, and then on to design database services including the management of configurations, libraries and versions. More recent CAD framework systems, both commercial and research (e.g., [14],[16] and [17]), have provided further design management services which can be called *design flow management* or *design methodology management* [10]. These services provide for the automation of sequences of tool executions and for the specification of, and enforcement of, repeatable design steps. These sequences are usually represented by directed graphs called *flows* to show the movement of data between tools. CAD frameworks which provide services at this level, will be referred to in this paper as *framework executive systems*, due to their ability to manage the execution of sequences of tasks.

*Both authors were previously with Carnegie Mellon University. This work was supported in part by the SRC under contract 98-DC-068.

DAC 98, San Francisco, California
©1998 ACM 0-89791-964-5/98/06..$5.00

Design process management [7] is a level of design management support above that provided by today's CAD frameworks. Design process management involves support for, and control of, all aspects of the solution of design problems - including conceptual (or exploratory) design, problem decomposition, backtracking, constraint propagation and management, and design history management. Design process management can be considered a new level of abstraction of design management services [1]. Just as the increasing complexity of VLSI design is driving design tools to operate at higher levels of abstraction (e.g., at the behavioral level instead of at the RTL or gate levels), the increasing complexity of design management means design management tools must evolve to operate at a higher level of abstraction: abstracting away some of the details of design flow and data management.

A design process manager (i.e., a design process management tool) providing such services must be both *discipline and framework independent*. Such a design process manager is not intended to replace existing CAD frameworks - there is no need to reimplement the lower level design management services. A design process manager should instead *manage* and *supplement* those lower level services. These requirements have led to the development of the concept of a *framework encapsulation*: a software wrapper around a complete CAD framework that enables the framework's design management services to be used and managed by a design process management tool. Just as tool encapsulations allow CAD point tools to be used within a CAD framework, framework encapsulations allow complete frameworks to be used from within a design process management tool.

Framework encapsulations have been implemented in a prototype tool called the *Minerva II Design Process Manager*. Minerva II is based on the Minerva Design Process Planner and Manager, originally developed by Jacome [7],[8]. Minerva II is a discipline-independent tool which implements some of the design process management features described above and, through framework encapsulations, can use the design flow and data management services provided by various CAD frameworks.

The remainder of this paper is organized as follows. Section 2 describes existing tool interoperability approaches. Section 3 describes the concept of framework encapsulations and issues regarding their implementation. Section 4 describes the use of framework encapsulations within Minerva II. Finally, Section 5 summarizes the paper and draws some conclusions.

2. EXISTING TOOL INTEROPERABILITY APPROACHES

The most closely related previous work in this area is that concerning *tool encapsulations* or, more generally, *tool interoperability* - the ability of CAD point tools to work together.

There are two important aspects to tool interoperability: the ability of tools to *share* design data and the ability to *be controlled* by a framework executive system. This section considers existing approaches to these aspects of tool interoperability in more detail.

2.1 Design Data Sharing

The tasks that CAD point tools perform are not independent, meaning that design data must be able to be shared between the tools. There are three common methods for sharing data: common file formats, procedural interfaces and messaging interfaces.

Common data formats, such as CIF for VLSI layouts, allowed tools to interoperate prior to the existence of CAD frameworks. More recent standards, such as EDIF 3 0 0 [6] continue to allow tools to work together, now usually within or between CAD frameworks.

CAD tools may also share data via a procedural or programming interface (PI) to an information provider (or server) process. An example is the CAD Framework Initiative Design Representation (CFI-DR) standard [2]. A PI standard allows design tools to interact with any information provider that implements the standard interface routines. Procedural interfaces are also used whenever design data is stored in a proprietary database, e.g., that of a commercial CAD framework. The interface may be provided in more than one language: typically a C interface is provided, along with an extension language interface for scripted access.

A similar approach to the PI is a messaging interface. In this case, a tool communicates with a whole suite of tools via a standard messaging architecture (such as ToolTalk from SunSoft [12]) using a predefined message dictionary (such as the CFI message set [3]). Messages allow separate processes to work together, but instead of a tool working with a static information provider (as in the PI case), the tool works with a dynamic suite of tools: any tool may answer requests, and new tools may be added to the system without any of the other tools being modified.

Standards (common file formats, standard procedural interfaces, and message dictionaries) may be defined to allow tools to work together. On the "leading edge", however, standards are not necessarily helpful: technology and tool capabilities are advancing more rapidly than the standardization process; tool vendors perceive no commercial advantage in fully supporting interoperability; and some "standards", such as Verilog and VHDL, are not fully portable due to incompatible implementations.

2.2 Design Tool Control

In order for a CAD framework executive system to control tool invocation, a standard interface to the tool must be provided. There are two methods by which this interface can be provided: *tool integration* and *tool encapsulation*. These approaches can together be referred to as *tool incorporation* [5]. Tool incorporation allows CAD frameworks to manage the execution of the tools, abstracting away the tool invocation details (e.g., filenames and command line arguments) and simplifying usability by the designer, possibly including the tool in a design flow. Data sharing, as discussed in Section 2.1 above, is one aspect of controlling this invocation - an executive system must be able to manage tool input and output. The two tool incorporation approaches are discussed below.

2.2.1 Tool Integration

Tool integration describes the process whereby a CAD point tool is tightly connected to a CAD framework. The source code of the point tool must be modified so as to implement communication with the CAD framework. The tool can directly obtain data to operate on, indicate completed results and, if desired, signal any "interesting" events during tool execution. Tool integration is often not possible due to source code unavailability and, even if possible, is likely to be an expensive process, requiring knowledge of the tool implementation. Tool integration provides great flexibility, however, in that integrated tools and frameworks can communicate with each other at will (via procedural or messaging interfaces). This can lead to improved functionality over encapsulated tools, which may only be able to communicate with a framework at the start and finish of execution. Tool integration, however, usually restricts the tool to working within a particular framework.

2.2.2 Tool Encapsulation

Tool encapsulation describes the process whereby a CAD point tool is loosely interfaced with a CAD framework. A layer (or wrapper) of software (also called a tool encapsulation) between the framework and the tool is developed, perhaps automatically [5]. The software wrapper treats the tool as a black box - working only through the tool's defined interfaces. These interfaces will include the tool input (e.g., data files, command line arguments, environment variables and standard input), the tool output (e.g., data files, return codes and standard output and error), and sometimes intermediate information. On the framework side, the software wrapper communicates with the framework using a predefined procedural or messaging interface. At a minimum, this interface will provide hooks so that the tool encapsulation may obtain the data to operate on and return the result to the framework. The framework may also restrict the language(s) in which the software wrapper is written by only providing a communication interface for a limited set of languages.

2.3 Design Process Management Implications

The ideas of the tool interoperability approaches discussed above can be applied to the problem of making a design process management tool work with several, independent CAD frameworks. Whole frameworks could be incorporated into a design process management tool via integration or encapsulation. Most CAD frameworks are commercial products without readily available source code, meaning that the only practical approach is encapsulation, i.e., *framework encapsulation*. Just as tool encapsulations allow a CAD framework to communicate with a tool and utilize the tool's functionality, framework encapsulations allow a design process management tool to communicate with a CAD framework and utilize its functionality. The concept of framework encapsulations is described in the following section.

3. FRAMEWORK ENCAPSULATIONS

> *Definition 1:* A *framework encapsulation* is a software wrapper around a CAD framework executive system that provides a common interface to the design management services provided by the CAD framework for use by a design process manager.

Framework encapsulations allow design process management software to be independent of CAD frameworks, yet utilize the design data and flow management services of the frameworks. Fig. 1 shows possible interaction between a designer, design process manager, framework encapsulations, framework executive systems and CAD point tools.

Section 3.1 describes the specific CAD framework design management services which need to be provided (through a framework encapsulation) to a design process manager. Section 3.2 describes some issues regarding the implementation of framework encapsulations and Section 3.3 describes the framework encapsulations which have been implemented.

3.1 Required Framework Services

A design process manager has specific requirements of a CAD framework executive system, the most important of which is the ability to perform tasks (i.e., execute tools and flows) in order to solve problems. These tasks may be formalized in a design plan [7]:

Definition 2: A *design plan* is a partially ordered sequence of tasks which can be executed to solve a design problem.

Definition 3: A *partially ordered sequence of tasks* is a set of tasks in which conditions *may* be imposed on the order in which (some of) the tasks must be carried out.

The following sections enumerate the specific services which a framework encapsulation should make available to a design process manager. The services are considered in two categories, corresponding to the lower level framework services. The specifics of how these requirements are met are discussed in Section 3.2.

3.1.1 Design Flow Management

In order to solve design problems, a design process management tool requires knowledge of what problems a framework executive system is capable of solving and how (in terms of flows or tool executions) they will be solved. This is resolved on a per-problem basis through the process of design plan generation:

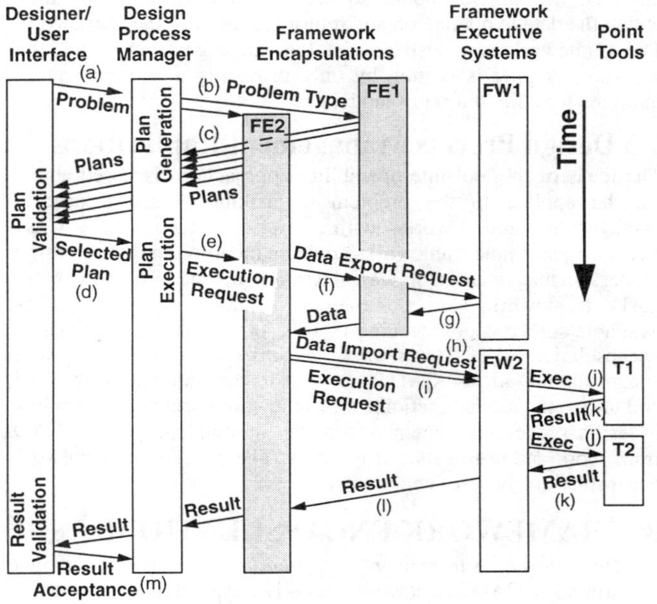

Figure 1. Possible communications over time between a designer, design process manager (DPM), framework encapsulations (FE1 and FE2), framework executive systems (FW1 and FW2) and CAD point tools (T1 and T2). (a) The designer selects a problem to solve. (b) The problem type is passed to the framework encapsulations. (c) FE1 and FE2 return possible plans to the DPM which passes them back to the user. (d) The user selects a plan. (e) The DPM sends an execution request to the relevant FE (FE2). (f) Assuming this execution requires data stored in another framework (FW1), FE2 generates a data export request of FW1 via FE1. (g) FW1 returns the data to FE2 via FE1. (h) FE2 imports the data into FW2. (i) FE2 can then pass an execution request to FW2. (j) FW2 executes the appropriate tools which make up the design plan. (k) FW2 records the results of their execution. (l) The overall plan result is passed back to FE2, the DPM and the designer. (m) The designer may accept or reject the plan result.

Definition 4: Design plan generation is the determination of a design plan to solve a design problem within a particular framework executive system, i.e., which tasks can be performed by which tools in what order so as to solve the design problem.

Along with the ability to generate possible plans within a framework executive system, a framework encapsulation must also be able to perform design plan execution, using appropriate data values as arguments. When a framework executive system completes execution of a plan, this must be detected by the framework encapsulation, and the execution result returned to the design process manager.

3.1.2 Design Data Management

Whereas a framework executive system can manage design flows independently of the data *values* being operated on[1], a design process manager may require data values to help control the design process, for example, initiating iteration if a constraint is not met. For this reason, framework encapsulations need to be able to extract data values from the encapsulated framework for use by the design process manager.

Because tools from multiple CAD frameworks often need to be used in the solution of a design problem, data transfer between CAD frameworks must be supported. This necessitates a data import/export mechanism.

3.2 Implementation Issues

There are a number of issues regarding the implementation of the above services within framework encapsulations. These issues are discussed below.

3.2.1 Design Plan Generation

For a framework encapsulation to determine whether a framework executive system can solve a certain problem it must determine whether it can generate (or has stored) a design plan that solves that problem. To do this, a framework encapsulation needs specific information about the problem to be solved, namely, the *objective* of the problem (e.g., placement at the layout level of abstraction), the *type of artifact* being operated on (e.g., operational amplifier), the *input data types* (e.g., transistor level description, process technology) and the *required output data types* (e.g., layout). In the parlance of Minerva II, this information is considered a "problem type" and is described in detail in [13].

There are several approaches a framework encapsulation could take to design plan generation. One would be for the framework encapsulation to contain complete knowledge of all of the plans that a framework executive system is capable of performing (or those that are supported) along with which tools can perform which tasks in each plan. Such a framework encapsulation would require maintenance to update possible plans every time a change is made to the CAD framework, e.g., a new tool is added.

Another design plan generation approach would be to dynamically build plans in response to information about the problem to be solved. Design plan generation in this context bears some relationship to planning in the artificial intelligence sense [9] - deriving a sequence of actions to change the state of a system. AI planning technology may serve as an implementation technology for plan generation within a framework encapsulation, although in most cases, simple algorithmic approaches would suffice.

[1] Executive systems must know meta-data about the data, e.g., the type and location of the data, and the name of the tool which created the data. It is the data (values) themselves that are not needed by the executive system.

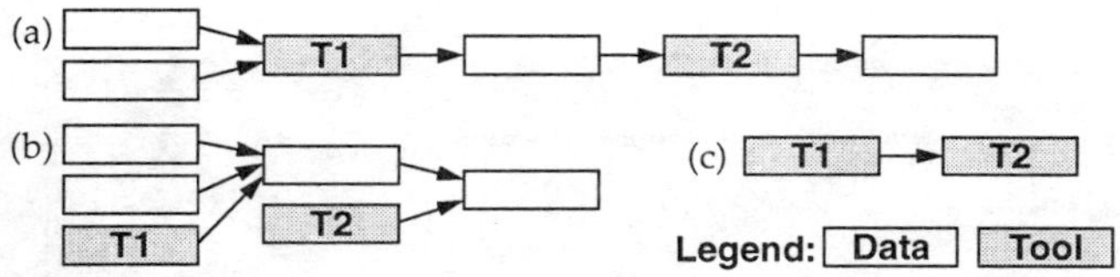

Figure 2. CAD framework flow representations. (a) Bipartite flow graph, showing data "flowing into" tools which produce data. (b) Task graph [14], showing tools and data as dependencies of data to be created. (c) Tool flow, where data is implicit, as used by some commercial frameworks.

Other approaches to design plan generation could fall between these two extremes. For example, a framework encapsulation may store a set of plan-templates, which are filled in by querying the framework for the tools that are available to perform the tasks that make up each plan. It has the advantage over the first approach that maintenance is only needed when the tasks that the framework can perform are changed and not when new tools are added.

The output from plan generation is a series of plans, each of which should be capable of solving the problem. (Multiple framework encapsulations may return plans, and each framework encapsulation may return more than one plan if desired.) Each plan is a sequence of tasks that can be represented as a flow or flows to capture the ordering constraints within the plan. CAD frameworks use many different flow representations - some examples are shown in Fig. 2. Framework encapsulations are responsible for mapping between the representation used by the framework and that used by the design process manager (which is a bipartite flow graph (Fig. 2(a)) in the case of Minerva II).

Each plan may also have choices associated with elements in the flow, for example, multiple tools may be available to perform some task, or multiple data instances may be available for some input (e.g., tool configuration options). Before plan execution, specific instances of such items would need to be chosen.

3.2.2 Design Plan Execution

Once design plans are generated by framework encapsulations, selected plans must then be executed. The design process manager will pass the selected design plans (and associated tool and data choices) to the relevant framework encapsulation for execution. Each unique combination of a design plan with selected tool and data items is considered by the design process manager to be a separate solution alternative; however, the underlying framework encapsulation need not execute them that way. Plans which have tasks in common may share the result of a single task execution. A framework encapsulation is also free to execute all tasks in all design plans, even if some tasks are shared between plans.

3.2.3 Execution Result Return

Upon the completion of plan execution, the result of the execution must be returned from the framework encapsulation to the design process manager. There are two aspects to the execution result: the completion status and the output data.

The completion status specifies whether the plan was executed successfully. "Success", in this case, does not necessarily mean that the output data is correct or useful; it means that CAD tools performing the tasks in the plan did not return errors[1]. In the case of an editing task, for example, it means the editor quit successfully - it does not mean that the edited data meets particular requirements, or is even syntactically correct[2]. Plan failure may occur for several reasons, e.g., tool failure or data incompatibility. Minerva II returns notification of a failure but may not be able to provide the reason - this depends on the capabilities of the executive system. In some cases, a designer may have to investigate the failure at a lower level in the design management hierarchy, e.g., examining tool execution log files.

Output data from a successful design plan execution is typically stored in a file or a CAD framework database. The framework encapsulation is responsible for returning this data to the design process manager. In some cases the actual data may be returned (e.g., simple strings and numerical values), whereas in other cases, a pointer to the data (e.g., filename or database object-id) is returned.

3.2.4 Data Value Extraction

Data value extraction refers to a framework encapsulation obtaining data from a CAD framework for use by the design process manager in *controlling* the design process. As outlined above, execution result return is one form of data value extraction. Our approach will not consider any other form of data value extraction. This is based on the assumption that data values within a CAD framework are created as the result of some task, and that if this task (or a data value it created) was of importance to the design process it would be managed by the design process manager.

Data value extraction should not be confused with data import and export, which refer to data transfer for the purpose of making data available for use by other CAD frameworks. Data import and export are discussed in the following section.

3.2.5 Data Import and Export

If a tool from one framework executive system is to access the data contained within a different framework executive system, then the owning framework must be able to export data (via its framework encapsulation) to the other framework - which must be able to import it.

The framework executive systems could be running on separate hosts, necessitating network communication. The simplest data import/export mechanism (and the one used in Minerva II) is to use files in a shared file-system such as AFS or NFS. If the design data of interest is not already available as a shared file (e.g., it is stored within a CAD framework database), then the framework encapsulation is responsible for converting the database-stored data into a file with a common format that may be imported by the other framework encapsulation.

3.3 Framework Encapsulation Implementations

Several CAD frameworks have been encapsulated into the Minerva II Design Process Manager using the method described above. These include:

- the Hercules Task Manager [14], configured to perform tasks in the area of analog integrated circuit design;
- the Hercules Task Manager, configured to perform software development tasks; and
- the Cadence Design Framework II.

Although these framework encapsulations are limited in terms of the number of tasks encapsulated[3], they illustrate that the concept applies to the frameworks of different types.

[1] On UNIX systems, this can be interpreted as a tool exit status of 0.

[2] Of course, a check for syntactical, or other, correctness could be a task within a design plan, or could be part of a tool encapsulation itself.

[3] Description of a *task* being encapsulated refers to a task which is able to be returned to a design process manager in a generated plan. A framework encapsulation need not allow all tasks which a framework can perform to be available for plan generation.

The major cost in constructing a framework encapsulation is in implementing the communication mechanism between the framework encapsulation and the CAD framework. For the frameworks above, this effort was on the order of one week. Once this mechanism is built, and one task within a framework has been encapsulated, it is relatively simple to encapsulate further tasks. Similarly, once one data import or export function has been encapsulated, it is relatively simple to encapsulate others. Once this is completed, it may be possible to automate the framework encapsulation process using a simple configuration file as input.

The frameworks are programmed, or interfaced to, using different extension languages: Hercules can be controlled using Tcl [11], and the Cadence Design Framework II with SKILL [4]. The framework encapsulation for each of the frameworks is written in Tcl along with the extension language for that framework. For example, the Cadence Design Framework II framework encapsulation consists of Tcl routines which, when called, generate SKILL code to be sent to a Cadence UNIX process. Framework encapsulations are implemented as Tcl *packages* which can be dynamically loaded into a Tcl interpreter, such as one of the Tcl interpreters within the Minerva II design process manager.

4. MINERVA II

The Minerva II Design Process Manager is a discipline and framework independent software tool providing design process management capabilities serving multiple designers working with multiple CAD frameworks, possibly in multiple design disciplines. Minerva II is configured to manage the solution of problems in a particular discipline by a description written in DDDL (**D**esign **D**iscipline **D**escription **L**anguage [15]). In this section we briefly describe how Minerva II interacts with framework encapsulations and designers[1]. The main events in this interaction are plan generation, plan validation, plan execution, and result validation. These are shown in the communication sequence of Fig. 1 and are described in the following sections, followed by a discussion of the advantages and limitations of Minerva II.

4.1 Plan Generation

Plan generation occurs within all of the framework encapsulations loaded into the Minerva II server. Each is passed information about the type of problem that is being addressed so that any plans which address that problem type may be returned, usually within a fraction of a second. These plans are then subject to validation/ selection by the designer before execution.

4.2 Plan Validation

Plan validation (or selection) within Minerva II is performed by the designer in a window like that shown in Fig. 3. Each plan is labelled with the name of the framework (encapsulation) which generated it and the textual description of the plan. Also shown is a graphical representation of the flow to be executed by the framework if this plan is selected. This flow may have associated choices to be made, e.g., tools to be placed in particular nodes of the flow. These choices are presented in the listboxes to the right of each flow. Plans are selected by checking the button on the left of the plan. Any number of plans may be selected for execution. Where multiple options are given for a plan and it is desired to execute the plan more than once with different options, the plan may be duplicated (using the "Duplicate Plan" button shown in Fig. 3) and different options chosen. After suitable plans (and options) have been selected, these are returned to the respective framework encapsulations for execution.

[1] Further details may be found in [13] and [15].

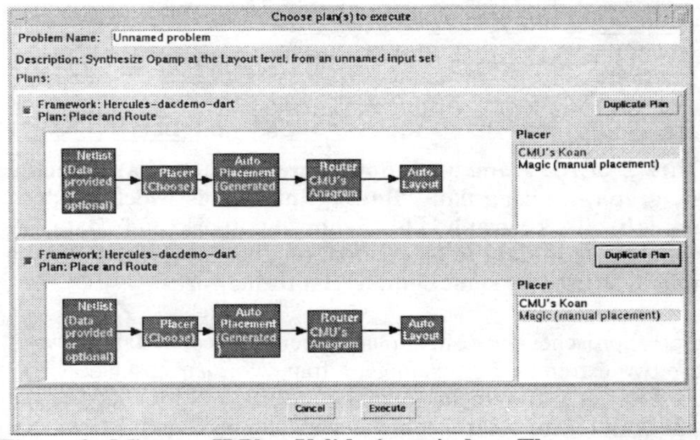

Figure 3. Minerva II Plan Validation window. The one returned plan has been duplicated using the "Duplicate Plan" button so that both tool choices may be made (i.e., two plans executed).

As mentioned earlier, the flow representation used by Minerva II is a traditional bipartite flow representation. The displayed flow is a mapping from the representation used by the encapsulated framework. For example, the plan in Fig. 3 (there really is only one, which has been duplicated) was generated by the framework encapsulation for the Hercules Task Manager. The Hercules task graph representation of that same flow is shown in Fig. 4.

4.3 Plan Execution

Plan execution takes place within the encapsulated CAD framework executive systems. The tools executed as part of the plan may or may not be interactive. If user interaction is required, it is the framework encapsulation's responsibility to ensure that the tool display is correct, as it is possible for the tool to be executing on a remote host.

4.4 Execution Result Validation

A plan execution can have two possible outcomes: *failure*, meaning that the tool execution failed, e.g., the tool returned an error status; and *success*, meaning that a new object version is to be created with new properties generated as part of the execution. The results of a plan execution through a framework encapsulation are presented to a designer as a new object version that has been created (Fig. 5). The designer may then accept or reject the result.

4.5 Advantages and Limitations

Framework encapsulations allow for unified design process management by a design process manager such as Minerva II. Unified design process management has several advantages over traditional approaches. Designers are able to focus upon the design

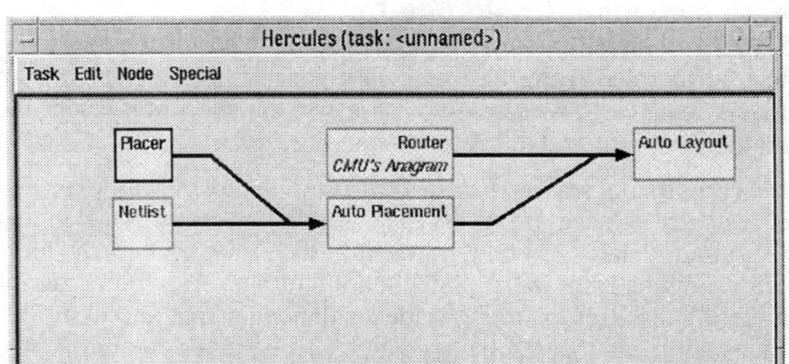

Figure 4. Hercules Task Manager representation of the design flow shown in Fig. 3.

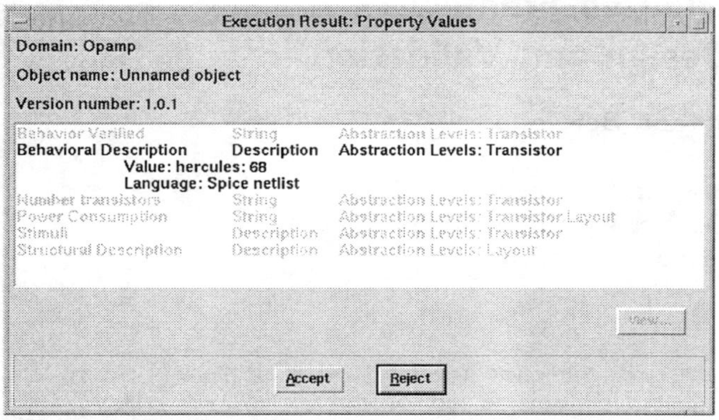

Figure 5. Plan Execution Result Validation in Minerva II.

problems being solved, rather than on flow, tool or data manipulations. Interactions between designers working in different disciplines (e.g., hardware and software design) and different frameworks can be managed easily. Consistency constraints can be monitored so that inconsistencies can be detected earlier and backtracking commenced sooner. Both backtracking and problem decomposition may be managed. All of these features can lead to shorter design cycles.

There are some disadvantages to our approach. Writing framework encapsulations and the DDDL description of a design discipline takes time. However, as was noted earlier, the major cost in encapsulating a framework is implementing the communication mechanism between the framework encapsulation and the framework executive system. Once that is operational, encapsulation of additional capabilities is a relatively simple process.

Just as tool encapsulations may not allow access to the complete functionality of a tool, framework encapsulations may not allow access to the complete functionality of a CAD framework. In order to perform additional operations it may be necessary to use a CAD framework independently from the design process manager. Similarly, just as an encapsulated point tool may not provide a suitable interface to completely automatically control tool function or invocation, a CAD framework executive system may not implement all the necessary capabilities (or at least, may not provide an interface to allow all operations to be performed programmatically). In this case, a *partial encapsulation* may be provided with manual intervention required to perform some steps.

5. CONCLUSIONS

Existing tool interoperability approaches do not suffice for today's leading-edge design processes where multiply-sourced CAD tools and CAD frameworks are required. Attempts to standardize design data interfaces (procedural and messaging interfaces and file formats) have, in general, not been accepted and are unlikely to be accepted due to rapid technological change and vendor-perceived competitive advantage reasons.

The approach we have taken in this paper leaves existing tools operating within existing environments but introduces a new level of design management: design process management. A design process management tool communicates with existing CAD framework executive systems through framework encapsulations. Framework encapsulations are closely related to, but a step above, tool encapsulations used in CAD frameworks. Framework encapsulations encapsulate several aspects of lower level design management services for use by a design process management tool. These aspects include design plan generation, design plan execution with result return, and data import and export.

To illustrate the applicability of this approach, framework encapsulations have been applied to a design process management tool called Minerva II. Framework encapsulations of the Hercules Task Manager and the Cadence Design Framework II allow Minerva II to control the design process at a high level - managing tool interoperability across multiple CAD frameworks and simplifying design management.

6. REFERENCES

[1] J. B. Brockman, T. F. Cobourn, M. F. Jacome, and S. W. Director, "The Odyssey CAD Framework," *IEEE DATC Newsletter on Design Automation*, Spring 1992.

[2] CAD Framework Initiative, Inc., *Design Representation Programming Interface: Electrical Connectivity*, Version 1.4.0, Oct. 1994.

[3] CAD Framework Initiative, Inc., *CFI Message Dictionary*, Version 1.0 (Initial Draft), Jan. 1994.

[4] Cadence Design Systems, Inc., *SKILL Quick Reference*, Version 4.3, Mar. 1994.

[5] J. Câmara and H. Sarmento, "AutoCap: An Automatic Tool Encapsulator," in *Proceedings of IFIP WG 10.5 - 4th International Working Conference on Electronic Design Automation Frameworks*, pp. 31-40, Nov. 1994.

[6] Electronic Industries Association. *Electronic Design Interchange Format Version 3 0 0*. Doc. EIA-618 (4 volumes), Dec. 1993.

[7] M. F. Jacome, *Design Process Planning and Management for CAD Frameworks*. Ph.D. Thesis, Carnegie Mellon University, Department of Electrical and Computer Engineering (Report CMUCAD-93-65), Nov. 1993.

[8] M. G. Jacome and S. W. Director, "A Formal Basis for Design Process Planning and Management," *IEEE Transactions on Computer-Aided Design of Integrated Circuits and Systems*, vol. 15, no. 10, pp. 1197-1211, Oct. 1996.

[9] N. A. Kartam and D. E. Wilkins, "Towards a Foundation for Evaluating AI Planners," *Artificial Intelligence for Engineering Design, Analysis and Manufacturing* (AI EDAM), vol. 4, no. 1, pp. 1-13, 1990.

[10] S. Kleinfeldt, M. Guiney, J. K. Miller, and M. Barnes, "Design Methodology Management," *Proceedings of the IEEE*, vol. 82, no. 2, pp. 231-250, Feb. 1994.

[11] J. K. Ousterhout, *Tcl and the Tk Toolkit*. Reading, Massachusetts: Addison-Wesley Publishing Company, 1994.

[12] Sun Microsystems, *The ToolTalk Service - a SunSoft White Paper*, Sunsoft Inc., Jun. 1991.

[13] P. R. Sutton, *A Framework and Discipline Independent Approach to Design Process Management*, Ph.D. Thesis, Carnegie Mellon University, Department of Electrical and Computer Engineering (Report CMUCAD-97-18), May 1997.

[14] P. R. Sutton, J. B. Brockman, and S. W. Director, "Design Management Using Dynamically Defined Flows," in *Proceedings of the ACM/IEEE Design Automation Conference*, pp. 648-653, June 1993.

[15] P. R. Sutton and S. W. Director, "A Description Language for Design Process Management," in *Proceedings of the 33rd Design Automation Conference*, ACM/IEEE, pp. 175-180, June 1996.

[16] K. O. ten Bosch, P. Bingley, and P. van der Wolf, "Design Flow Management in the NELSIS CAD Framework," in *Proceedings of 28th ACM/IEEE Design Automation Conference*, pp. 711-716, June 1991.

[17] P. van den Hamer and M. A. Treffers, "A Data Flow Based Architecture for CAD Frameworks," in *Proceedings of 1990 IEEE International Conference on Computer-Aided Design* (ICCAD), pp. 482-485, Nov. 1990.

A Geographically Distributed Framework for Embedded System Design and Validation

Ken Hines and Gaetano Borriello

Department of Computer Science & Engineering, Box 352350
University of Washington, Seattle, WA 98195-2350
{hineskj,gaetano}@cs.washington.edu

Abstract

The difficulty of embedded system co-design is increasing rapidly due to the increasing complexity of individual parts, the variety of parts available and pressure to use multiple processors to meet performance criteria. Validation tools should contain several features in order to keep up with this trend, including the ability to dynamically change detail levels, built in protection for intellectual property, and support for gradual migration of functionality from a simulation environment to the real hardware. In this paper, we present our approach to the problem which includes a geographically distributed co-simulation framework. This framework is a system of nodes such that each can include either portions of the simulator or real hardware. In support of this, the framework includes a mechanism for maintaining consistent versions of virtual time.

1 Introduction

Embedded system co-design is becoming increasingly difficult due to a number of factors:

- Performance criteria are harder to meet with single processor designs, suggesting that multi-processor designs should become more common. Unfortunately there are almost no tools available that address the specific issues of these sorts of systems.
- Individual components (both hardware and software) in an embedded systems design are themselves becoming more complex, and problems that occur because of subtle interactions may not show up until the entire system is assembled.
- Many off-the-shelf solutions that could be used in an embedded system design are becoming available. However, it is difficult for a designer to evaluate these in the context of a specific system.

To address these issues, we identify several desirable features for an embedded system validation framework. First, it should allow the designer to view all parts of the system, including hardware and software, at several levels of detail - both in behavior and in communication *and* to dynamically switch between these detail levels. Second, it should facilitate the inclusion of intellectual property (IP), such as algorithms, new processors, special purpose ICs, etc. without compromising the internals of the IP. Third, it should support geographically distributed design groups and as well as tools which may also be geographically distributed. Fourth, it should allow the system functionality to be gradually migrated to physical hardware while still allowing the entire system to be modeled with the newly included hardware. Finally, it should include debugging support for the parts of the system that are in hardware, the parts in software, the parts that are in simulation, as well as the system as a whole.

Most of these have been individually addressed, for example, Mentor Graphics' Seamless CVE [9], Viewlogic's EagleI [5], and the previous version of Pia all allow dynamic changes in the detail levels. Viper technology allows the use of IP in simulation, through use of encrypted, unsynthesizable models. There is also research into including actual hardware in simulation, for example, the UWTester [10], CATFISH [11] and EagleI.

In this paper, we present the approach taken in Pia, the co-simulator of the Chinook project [3, 12]. Pia provides a distributed hardware-software co-simulator and tools for schematic capture as well as a means of connecting these to synthesis tools and actual hardware. To improve the speed of simulation and to reduce network bandwidth, Pia allows for multiple levels of detail and provides a mechanism for dynamically switching between them. There are other projects involving frameworks for geographically distributed electronic design, such as WELD [1], but Pia differs from these in that its primary focus is in facilitating hardware/software co-design through geographically distributed co-simulation and in integrating remotely located hardware into a co-simulation environment (rather than in facilitating distributed design of arbitrary hardware)

Parts vendors have already begun to use the Internet to provide users with access to parts for evaluation. Intel, for example now has a remote evaluation facility [8] that permits designers to evaluate various i960 processors over the web. Users can enter. compile and run programs on the desired processor and observe the results - all over the web.

The Pia framework pushes this concept a little further and allows the user to patch web based components into a simulated circuit for more extensive evaluation.

This work was supported by ARPA contract DAAH04-94-G-0272 and a Mentor Graphics graduate fellowship

DAC 98, San Francisco, California

2 The Pia system

The Pia simulation system is a set of *Pia nodes* that can be interconnected through a network. Each node contains a number of sockets and each socket can facilitate a connection to a design tool such as a simulator or a compiler, or a device such as a processor, an ASIC or an FPGA. Design tools can have built in support for Pia sockets (as do all the Chinook tools), but if not, the tools can be connected through a customized wrapper. The Pia system maintains consistent versions of virtual time for all simulator and hardware components, regardless of where they are physically located. Fig. 1 shows an example system with several items connected together.

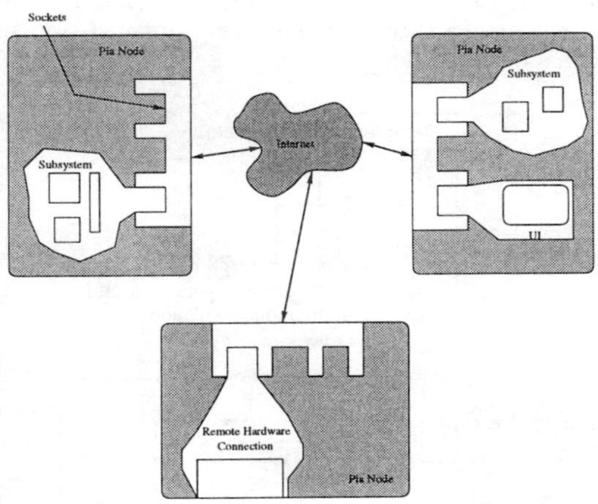

Fig. 1: Several Pia nodes connected through the Internet. This shows how simulator subsystems, user interfaces etc. are connected together

A distinguishing feature of Pia is its ability to dynamically alter the level of detail represented by the simulator *at runtime*. In order for this to work, the simulator must have instructions (either stored in a library, or provided by the user) to guide it through each of the basic communication actions at several levels of detail. These basic actions can include anything from putting a byte of data on a bus to sending a complete packet across a network. This can be a big advantage when the communication occurs between components that are not located on the same host since it allows the designer to reduce the communication bandwidth at times when detail isn't required. Given several communication methods for each action, Pia can dynamically switch between them at *safe points* in the execution, that is points in the method where the state of the interface is stable and consistent.

We are in the process of building a library of standard communication protocols, each with several built-in detail levels. In the cases where the user must provide additional instructions for levels of detail not currently in any library, we allow these to be entered as a set of assertions which describe the activating conditions, and results of any action [7].

2.1 Pia on a single host

This section gives a brief overview of Pia's operation on a single host. It is intended to provide background for the distributed version of Pia, but for the complete details see [6]. From the designer's point a view, a system simulated through Pia consists of *components, interfaces, ports* and *nets*. Components are containers for some basic functionality, interfaces connect components to ports, and ports are interconnected through nets. A component would typically be used to represent such things as embedded processors running programs, ASICs, FPGAs, and so forth. Currently in Pia, processors running software are represented by a component which has as its behavior the actual software (in Java) that would run on the embedded software. Specific processors are characterized by their timing characteristics (in the form of a basic block timing estimator) and by their external interfaces. Currently, the basic block timing estimation is performed by hand, but we hope to eventually take advantage of other research in this area. Basically, the timing estimates are embedded in the source code, and when the simulator encounters one of these, it updates a version of virtual time. There is no reason that the component can't be an instruction set simulator of a particular processor, but we have not yet devoted any effort to either implementing such components or adapting an existing ISS to Pia.

The single host version of Pia uses a two level hierarchal view of virtual time which includes a *system time* as well as *local times* for each component. When a component is activated, it is allowed to continue until it is ready to receive a value from another component. When this happens, the component must pause until until system time equals its own local time. System time is always required to be less than or equal to all local times, so that when a component is restarted, it is certain that its view of the world is up to date.

2.1.1 Synchronization between components

This technique works without any problems when the receiving component has distinct modes for data receipt and for computation. Many components fit this model, for example, reactive components, components that poll for new data, and so forth. This model can also work fairly well for components where the phases are less distinct (for example, where the data might be received during the computational phase through an interrupt).

If we can statically determine which addresses in a processor's local memory are either written or read by interrupt handlers, we can statically mark those locations as *synchronous*. This means that the component will have to ensure that its local time matches system time when it reads or writes to any of these locations (this is the same requirement we apply to all receives).

If we cannot statically determine such addresses, the simulator can make the optimistic assumption and treat all memory as *safe*. When the system detects a violation of this assumption it can dynamically mark the relevant addresses as synchronous, then rewind using Pia's *checkpoint* and *restore* facilities.

2.1.2 Checkpoint and restore facilities

The idea behind Pia's checkpoint and restore facilities is that components occasionally store images of their state at particular points in the execution. On encountering consistency problems, the simulator can restore previous images and reexecute more conservatively, as we demonstrated in the discussions on interrupts above. Although Pia's current checkpoint facility saves complete component images, we plan to look into incremental checkpoints at some point in the future. A checkpoint request *does not* require all components to save images with the same local time, instead components save at the earliest local time possible after the request.

These semantics introduce the danger of a *domino effect* hazard [13]. This effect occurs if it's possible for a state restoral to require any component to load more than one checkpoint to obtain a causally consistent state. In the worst case, this can force all processes to roll back to their initial state. We avoid this by requiring each component to save a checkpoint *before* receiving any messages after a checkpoint request from the scheduler. This prevents any messages from the future of the checkpoint on one component from influencing the state before the checkpoint on another.

2.1.3 Switching detail levels

Changes in detail levels or *runlevels* are triggered by any one of three things. First - the user may have directly altered a runlevel (usually through a detail level slider) second, there may be a *switchpoint* defined in the simulation run control file, and third, the designer may have included imperative runlevel switch statements in the source code. A switchpoint is an expression of the form `@<condition>:<action>` that tells the simulator when and how to change runlevels. An example of a switchpoint follows:

```
@I2CComponent.localTime>=67.0:
    I2CComponent->hardwareLevel,
        VidCamComponent->byteLevel
```

This tells the simulator that as soon as it finds that "I2CComponent" shows a localtime of 67 or later, it should change the runlevel of I2CComponent to "hardwareLevel" and the runlevel of VidCamComponent to "byteLevel". This particular switchpoint seems unusual since it is based only on I2CComponent's version of virtual time. There is no way of knowing what time VidCamComponent will think it is when the switch occurs. The condition can include conjuncts and disjuncts of conditions across multiple components.

2.2 Pia nodes and subsystems

As we described it earlier, a Pia system is a collection of distributed Pia nodes, with interconnections being determined at runtime. Each Pia node contains one or more *subsystems* and each subsystem contains some fragment of the embedded system design under test. Associated with each subsystem is a scheduler object, which is primarily responsible for enforcing the local timing semantics. By itself, a Pia node with a single subsystem behaves very much like the single host version of Pia described earlier. Its primary duty is to schedule components and to ensure that the *subsystem time* is always less than or equal the *local times* of all components in the subsystem.

Currently, components, interfaces and ports are all atomic. In other words, components, interfaces and ports will all be contained in a single subsystem.

2.2.1 Interconnection of subsystems

Pia nodes are interconnected through Java's RMI interface. Each node serves as both a client and a server, and handles all inter-node communication so that it is hidden from the user. Since nets are the only user object that can be split across subsystems, they are the only objects that require any special handling. A net that connects components on two different subsystems is split into two nets, one for each subsystem, and each net includes an extra (hidden) port that connects bus events to the subsystem upon which it resides.

Between each pair of communicating subsystems is a *channel*, across which all communication occurs. Each channel is associated with a pair of dummy components (one on each subsystem). Each of the hidden ports is the property of one of these channel components. In essence, each change in net's value is registered with the corresponding channel component, and the component is responsible for performing any required actions.

Channel components are not self contained, rather, they are proxies for the subsystems on the opposite side of the channel. As such, they may be responsible for coordinating run levels between the components, as well as insuring the consistency of time across channels. Channel components do not have a thread of their own, but instead use the subsystem's own thread.

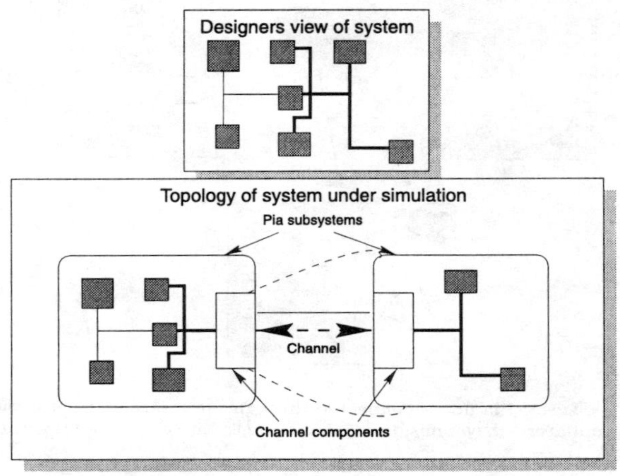

Fig. 2: A pair of Pia subsystems. This figure shows a channel and a pair of channel components which perform interfacing between nodes. The dark net is split between the subsystems

Fig. 2 shows how this works. The dark net in the top picture is split across two subsystems, so hidden ports and channel components are introduced into the split. When moving a set of components from one subsystem to another, the split in the relevant nets can be determined by a cut of the component graph. Essentially, a boundary is drawn around all components that are moved, and any net that crosses this boundary is split. If performed repeatedly and locally, this could force some nets to pass through subsystems which contain no components relevant to the net, so a global view of the system must be consulted when performing each split.

2.2.2 Managing virtual time between different nodes

In general, it isn't possible to both maximize parallelism and guarantee continuous system consistency. The reason for this is that in order to guarantee system consistency, no subsystem is allowed to have a virtual time that is later than the time-stamp of a message that has not yet been delivered to the appropriate component. Since messages arrive asynchronously, we don't know in advance when this may happen, so to maintain consistency, we must be conservative and allow a subsystem to advance only when we are certain that no messages will arrive with an earlier time-stamp.

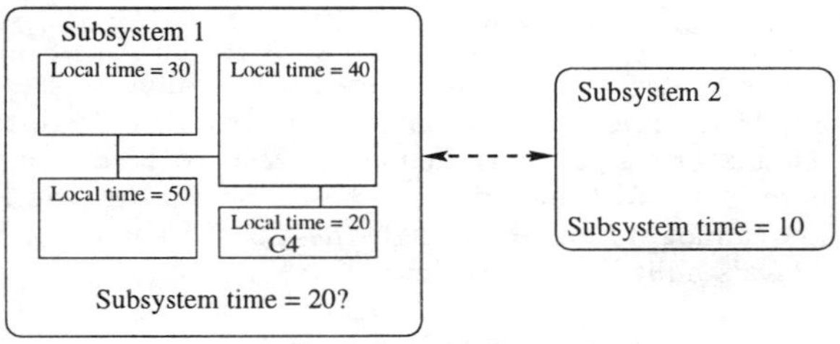

Fig. 3: $Subsystem_1$ must stall to maintain continuous consistency.

On the other hand, if we want to maximize parallelism, no subsystem may stall *unless* it *will* receive a message with a time-stamp that is *equal to or later than* subsystem time. In the general case, this would require clairvoyance.

This is true even if all components in the subsystem have distinct modes for computation and reception of data. Fig. 3 shows a system in which all components in $Subsystem_1$ are blocked until subsystem time catches up. If this were a single host simulator, the subsystem would be able to advance its own virtual time to 20, and activate component C_4. In this case if it does that, there may be a consistency violation because $Subsystem_2$ may send a value to C_4 with a time-stamp of, say, 15. If our goal is to maintain consistency, $Subsystem_1$ can do no work until we are certain $Subsystem_2$ will not send any messages with a time-stamp less than 20, even though all components in $Subsystem_1$ have a distinct receive phase, it still needs to wait for $Subsystem_2$.

If there isn't much communication expected between subsystems, it is often reasonable for a subsystem to continue as if there were no asynchronous messages, but to save state occasionally. If any such messages occur, then the affected subsystem must restore a previously saved copy of system state, and continue from there.

Pia allows for both possibilities through *conservative* and *optimistic* channels. The rule is that a processor cannot proceed past the time on a conservative channel unless it is certain that it will not receive any messages on this channel with time-stamps earlier than its local time.

2.2.3 Conservative Channels

Before a subsystem can advance its version of virtual time, it must first make sure that no conservative channels will send it any messages with an earlier time-stamp. To ensure this, each subsystem can request a *safe time* from the subsystem on the far end of the channel. The safe time of component C_i indicates the latest time to which the opposite subsystem is allowed to advance to without consulting C_i again.

This is illustrated in Fig. 4. If SS_1 is ready to advance its own subsystem time it must first get safe times from both SS_2 and SS_3. Once it has these, it must compare these to the time value of the next event it has scheduled. If the event time is less than all the reported safe times, SS_1 is allowed to advance its own system time to the value of the event, and deliver it.

The time a subsystem reports is essentially its own subsystem time with all restrictions from the opposite processor removed. If this were not the case, there would be deadlock, since no subsystem would be allowed to advance its subsystem time at all. A set of interconnected subsystems must make a directed graph with only simple cycles. A simple cycle is simply a bidirectional edge. The reason for this is

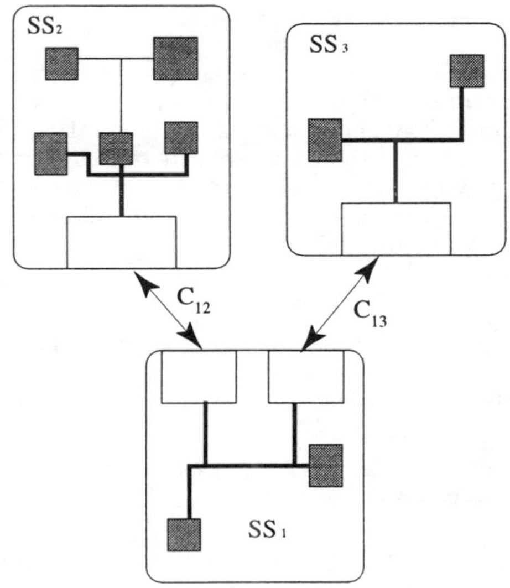

Fig. 4: Three subsystems are shown

that it is computationally hard to eliminate self-restriction on the fly for general graphs.

2.2.4 Optimistic Channels

Subsystems linked by optimistic channels are not restricted from updating their virtual time beyond the safe time of the subsystem on the opposite side of the channel. For example, if channel C_{12} were an optimistic channel, SS_1 would only need to consult SS_3 before advancing local time. This requires each subsystem to occasionally save state so that it can fully recover if a consistency error occurs. This is usually covered by the interrupt checkpoint and restore mechanism, so the only impact could be more expensive restores if optimistic channels are poorly placed.

2.2.5 Managing checkpoints in distributed Pia

As we mentioned above, a checkpoint and restore system that does not require all components to save state simultaneously risks a domino effect. Although this is straight forward for single host Pia, we need to address the issue again when we distribute the simulator. This is because there may be some delay between the time when the scheduler requests a checkpoint, and all components receive the request.

Since all channels between subsystems are FIFO channels, we can solve this problem with the Chandy-Lamport algorithm [2]. After a subsystem receives (or generates) a checkpoint request, it performs a local checkpoint and transmits a *mark* on all of its outgoing channels. Upon receipt of a mark, a subsystem immediately performs a local checkpoint, before receiving anything else on that same channel. Following this, before transmitting anything else on a particular outgoing channel, the subsystem sends a mark. To ensure that each subsystem performs the local checkpoint only once per request, each mark contains an identifier, and that identifier is also sent with each mark generated in response, such that a subsystem can ignore marks with that have the same identifier as checkpoints already performed.

2.3 Connecting Pia to real hardware

Adding real hardware to a simulation requires a hardware/software stub to be attached to the hardware through some means. One possibility is to use a DEC Pamette board [4] to provide the hardware side of this, and the software side could be written using the Pamette control library. This hardware/software stub serves to match semantics between the hardware and the simulator, and must provide certain functionality:

- It must be able to set and read time on the hardware. The setting of time could be implemented as software translation of actual time given by the hardware.
- It must be able to either stall or otherwise idle the hardware.
- It must be able to buffer interrupts generated by hardware and pass these up to the simulator.

All of these functions could be implemented in software if the hardware allows the user to start the clock and stop it after a measured number of ticks, but this is likely to be the case only if the hardware was designed with connection to Pia in mind.

Connecting Pia to an actual embedded processor can be greatly facilitated by small server which resides on the embedded system. If the system already includes a Java virtual machine, then this is fairly straightforward and the stub can allow migration of objects from a simulated component the actual component.

3 Implementation in Java

There are a number of reasons that caused us to choose Java as the implementation language for Pia including Java's built in support for concurrent threads, distribution, and flexible loading of modules.

3.1 Imposing the Pia scheduling semantics on the Java VM

Since Java is a threaded language, it would seem natural to use the Java VM's thread package to provide for concurrence between components. In general, however, this would mean adopting the scheduling semantics of the VM. Since the scheduler offers no scheduling guarantees, (if there are two runnable threads with different priorities, the scheduler will occasionally run the thread with the lower priority) they are inappropriate for use in Pia.

An alternative to this is to essentially define a scheduler class that chooses which thread to run, and then tricks the VM scheduler into running that thread. There are a couple of ways we can do this, both based on ensuring that that the VM scheduler finds only one runnable thread at a time. First of all, we can have the scheduler *suspend* all threads, and *resume* only the one that should be running. The other way is to have all the threads queue up on mutexes and have the scheduler signal the one it wants to run. The latter method is the one used in Pia although the former should be equivalent.

3.2 The Pia class loader

The class loader used in Pia is designed to allow a user to recompile and reload a component without having to restart the simulator. Pia's class loader is able to load components on demand from arbitrary URLs on the Internet. If a class cannot be found through the custom channels, Pia uses Java's built in class loader.

4 An example embedded system

We now introduce an example system which we will use to further illustrate the concepts of this paper, and to obtain some performance metrics. The example we will use is the "WubbleU" application, a suggested benchmark for embedded system design tools [14]. WubbleU is essentially a hand held Web Browser, or, more accurately, a Web Browser that consists of a hand held unit and a wireless connection to a dedicated server. The specification allows for a certain amount of flexibility in assigning tasks to the server or handheld unit. In fact, that is something a designer may want play with, given available parts and software modules.

An implementation of WubbleU can contain several forms of Intellectual Property. For example there may be special integrated circuits (GSM chips, JPEG chips), software (Java VMs, Handwriting recognition software), as well as various compression and communication algorithms. Also, it is likely that some parts of the hardware will be ready before others and possible that a designer would like to try various synthesis tools.

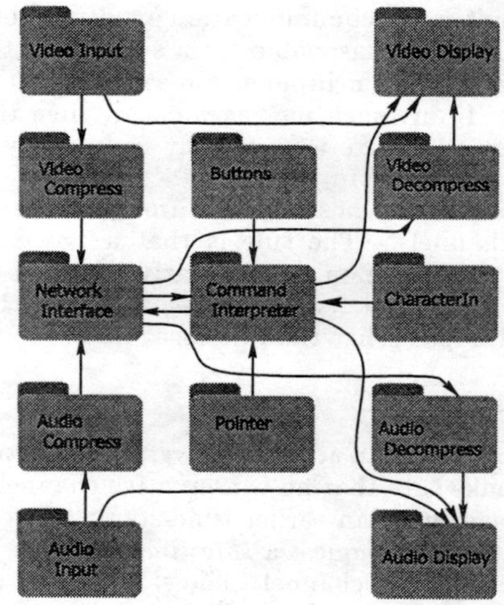

Fig. 5: A communication flow diagram for the WubbleU handheld web browser

Fig. 5 shows a high level communication graph of the modules in WubbleU. The nodes in this graph may be implemented in either hardware or software. We will focus on an a particular implementation that includes a simple cellular connection to a server which connects to the Internet, and most of the functionality is on the handheld unit. This is not necessarily the best implementation, but it works for the purposes in this paper.

The cellular connection is controlled by an ASIC which transfers packets to the system through DMA. This chip is our candidate for remote operation. Fig. 6 shows a block diagram of this circuit, as well as the topology of this circuit on

Location	Detail level	simulation time
N/A	HotJava	.54 seconds
local	word passage	130.2 seconds
local	packet passage	43.1 seconds
remote	word passage	604 seconds
remote	packet passage	80.3 seconds

Tab. 1: Time and simulation overhead on several configurations of the WubbleU example

the simulator. In this architecture, all processes are mapped to the processor, with the exception of whe network interface which was mapped to the cellular communication chip. In this case, both Pia nodes were running on Linux/Pentium Pro 200MHz work stations, both on the same subnet. Ideally, we would like to perform this experiment using the real hardware, but we are do not yet have any functional hardware servers. Besides, the main point of this experiment is to show that even with reduction in simulation performance introduced by the Internet we can still obtain reasonable overall performance by changing abstraction levels for this link.

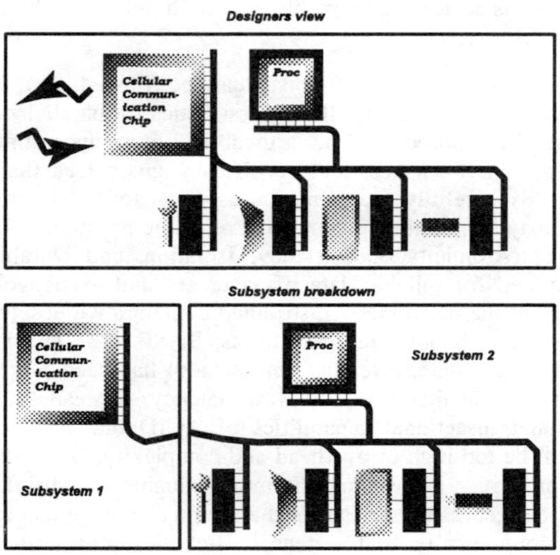

Fig. 6: A possible architecture for the WubbleU system, and it's simulation topology

The test performed is the loading of the Pia homepage (http://www.cs.washington.edu/homes/hineskj/Pia.html) which contains approximately 66KB of data, including graphics. For comparison, we also timed loading this page with Sun's HotJava browser which we use as a rough reference for estimating simulation overhead in each case. The transfer modes that we used for each of the simulation tests were *word passage* where individual four byte words were passed across the network, and *packet passage* where the data was sent across the channel in 1KB packets. These same modes were used for local testing, where all parts of the simulation are in a single subsystem, and the results are all shown in Table 1.

We notice that there is a large performance hit in all the simulation, but that is to be expected. The loading time under the remote packet passing abstraction level is actually quite reasonable, and is actually fast enough to allow the designer to play with the simulated hardware. The reason that the performance improvement is much less dramatic when all components are simulated locally is because communication between the other components is still rendered with a high level of detail. It's possible, given specification style and detail level, that this local performance could approach that of HotJava.

5 Conclusions and Future work

We believe that a geographically distributed environment for coordination design effort and validation of embedded systems is an important addition to the designers repertoire of tools. Although this solution could conceivably cause problems with performance, we showed how the principles of selective focus introduced in [6] can be used to offset this.

Current work is in the extension of Pia to include a debugger and changing the checkpoint mechanism to use incremental rather than total checkpoints. Additional current and future work involves setting up Pia socket versions of hardware servers, and building additional examples.

References

[1] Chan, F. L., Spiller, M. D., and Newton, A. R. Weld - an environment for web-based electronic design. In *Proceedings of the 35th Annual Design Automation Conference* (1998).

[2] Chandy, K., and L., L. Distributed snapshots: Determining global states in distributed systems. *ACM Transactions on Computer Systems 3*, 1 (1985), 63–75.

[3] Chou, P., and Borriello, G. Software architecture synthesis for retargetable real-time embedded systems. In *Codes/CASHE '97* (1997).

[4] DEC Pamette Board http://www.research.digital.com/SRC/pamette/.

[5] http://www.viewlogic.com/products/eagletools.html.

[6] Hines, K., and Borriello, G. Dynamic communication models in embedded system co-simulation. In *Proceedings of the 34th Design Automation Conference* (June 1997).

[7] Hines, K., and Borriello, G. Optimizing communication in hardware-software co-simulation. In *Codes/CASHE '97* (1997), IEEE, ACM.

[8] Intel Remote Evaluation Facility, http://developer.intel.com/design/i960/testcntr/ref/INDEX.HTM.

[9] Klein, R. Miami: a hardware software co-simulation environment. In *Proceedings. Seventh IEEE International Workshop on Rapid System Prototyping. Shortening the path from specification to prototyping* (June 1996).

[10] McKenzie, N. R., Ebeling, C., McMurchie, L., and Borriello, G. Experiences with the mactester in computer science and engineering education. *IEEE Transactions on Education.* (February 1997), 12–21.

[11] Mueller, A., Groetker, T., Post, G., and Meyr, H. Catfish - a configurable atm testbench for interfacing simulation and hardware. *DATE'98* (1998).

[12] Ortega, R., and Borriello, G. Communication synthesis for embedded systems with global considerations. In *Codes/CASHE '97* (1997).

[13] Russell, D. L. State restoration in systems of communicating processes. *IEEE Transactions of Software Engineering SE-6*, 2 (March 1980), 183–194.

[14] WubbleU hand held PDA benchmark for co-design, http://www it.dtu.dk/jan/WubbleU.

WELD – An Environment for Web-Based Electronic Design

Francis L. Chan
fchan@eecs.berkeley.edu

Mark D. Spiller
mds@eecs.berkeley.edu

A. Richard Newton
rnewton@eecs.berkeley.edu

Department of Electrical Engineering and Computer Sciences
University of California at Berkeley

Abstract

Increasing size and geographical separation of design data and teams has created a need for a network-based electronic design environment that is scaleable, adaptable, secure, highly available, and cost effective. In the WELD project we are evaluating aspects of the network integration and communication infrastructure needed to enable such a distributed design environment. The architecture of WELD and the components developed to implement the system, together with performance results, are described and evaluated.

1 Introduction

Advances in areas such as software methodology, operating systems, storage systems, and programming languages have often had an enormous impact on EDA. The explosive growth in the development of wide-area network infrastructure over the past few years indicates an opportunity for the industry and the field of computer science in general to make a leap to a new generation of capabilities. Specifically, we envision the entire EDA community organized as an integrated distributed environment [39] that offers users the ability to create an evolvable, customizable and adaptable "virtual" design system that can couple tools, libraries, design, and validation services. Beyond that, the system could also provide manufacturing, consulting, component acquisition, and product distribution, encompassing the developments of companies, universities, and individuals throughout the world.

In this paper we present an approach to the network infrastructure developed for such a distributed design system. The motivation and description of the WELD project [47] is given in Section 2. Comparisons to related work are documented in Section 3. A high level view of the WELD architecture and its key components is introduced in Section 4 and results and experiences from our efforts are described in Section 5.

DAC 98, San Francisco, California

2 The WELD Project

The goal of the WELD project is to pull together emerging technologies, such as network communications, visualization [45], alternative interfaces, and new algorithmic approaches to design, to provide the basis for a next-generation EDA system. Working sets of modern VLSI designs, which approach two gigabytes of data today and will continue to grow, are forcing design teams to become larger and more distributed [35], indicating a great need for a collaborative, distributed design system that is scaleable, adaptable, secure, highly available, and cost effective.

The World Wide Web [49] today can be viewed as a "bad multicomputer", which has all the possibilities for parallelization of a normal multiprocessor, but a great deal more uncertainty in the latency, performance, and consistency guarantees that are offered. By carefully analyzing the usage of today's users and tools, it may be possible to somewhat relax the traditional ACID properties (**A**tomicity, **C**onsistency, **I**solation, and **D**urability) [15] required for reliable data management and exchange and make use of the network for distributed collaborative design. A new set of semantics referred to as BASE [13] (**B**asically **A**vailable, **S**oft state, **E**ventual consistency) has been proposed that stresses availability and low latency, instead of the traditional transactional capabilities of ACID that have often proven to be too high in overhead and complexity. In BASE, a fast, approximate answer may be more valuable initially than a late-but-correct response. Part of the design challenge facing the WELD group was to find a general, flexible architecture that features both properties and addresses as wide a range of EDA requirements as possible.

To be successful, the WELD system must be built upon an open design environment that is *scaleable and adaptive*, where system service and performance are not sacrificed with the addition of tools, users and different technologies, and the use of components guarantees flexibility and extensibility. In the current implementation, users can easily extend features or incorporate into the system existing tools and technologies, regardless of programming language, operating system base, or development environment. A *wide range of computing resources,* from high-end workstations to mobile PDAs, are supported in WELD through the use of servers and proxies located both locally and in the network.

Although the Web potentially offers substantial performance improvements for suitably partitioned or iterative designs [4], additional infrastructure will be required to assure *robustness*, especially in the area of data management. Intermediate state of long processes must be stored so that design states can be recovered efficiently and incrementally after any hardware, software, or network failure. *Security* measures[1] must also be available so that data security is never comprised.

Applications and services in a distributed environment should manifest *high availability,* providing users with consistent on-demand access regardless of time or location. The Web offers several built-in advantages to a distributed design system. The wide distribution of Java-enabled browsers enables the system to be readily accessible to a *large user base*, making it fertile ground for innovation and wide-area collaboration. In addition, Java clients allow for *consistent and intuitive user interfaces* and *platform independence,* thus reducing start-up and training time.

3 Related Work

Our architecture and infrastructure for a distributed design environment can be compared to CAD frameworks [17, 48]. Comparable features include:

- A design database – distributed data manager and client object management package
- A design data manager – versioning, security and meta-data handling capabilities
- A design process manager – server wrapper and distributed workflow system.

However, our approach differs from others' work[2] in a number of fundamental ways. Most of the past efforts in the area of CAD frameworks are involved in introducing new systems, techniques or extensions in specific areas of design data management [33, 46], design meta-data management [14, 24, 34], and flow, process and tool management [12, 16, 18, 25, 43]. WELD, however, concentrates on providing reliable, scaleable connection and communication mechanisms for distributed users, tools and services. It also extends the traditional client/server model by treating both tools and data as potentially mobile components. This component-based approach facilitates both the fragmentation of large data sets into manageable and transferable segments as well as the separation of sophisticated tools into logical components that can be run on either clients, servers, or proxies. While we are also involved in the development of EDA applications [47], the goal of the work presented in this paper is to provide the enabling and enhancing mechanisms that leverage existing systems and toolkits. The infrastructure was deliberately engineered such that no restrictions or assumptions are placed on data representation, design methodologies or data and tool usage, instead allowing policy choices to be built on top of the infrastructure. Finally, the WELD project avoids the tight coupling found between many frameworks and their operating environments (i.e. UNIX and NFS) by using platform-independent standards such as Java, sockets, proxies, and generic string-based communication protocols.

[1] Security features can be layered upon the base WELD architecture.

[2] A number of external efforts that are relevant to our research include the REUBEN system [26], CREW [8], the Design Agent system [3], and Active Documents [38].

4 System Architecture

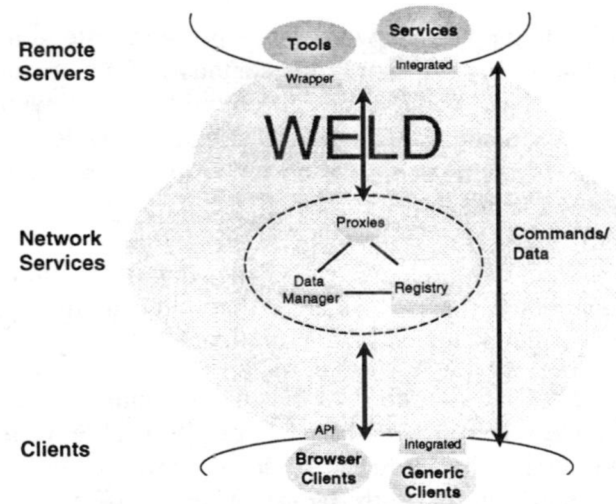

Figure 1: WELD System Architecture.

The three-tier WELD architecture[3] (Figure 1) consists of:

- *Remote Servers* – network-accessible tools or services. These can be either command-line tools encapsulated by server wrappers or tools with built-in support for socket connections and WELD communication protocols.
- *Network services* – infrastructure components found in the network that are available on demand to applications. Examples include the *Distributed Data Manager*, *Proxies* and *Registry Service.*
- *Clients* – applications that use the WELD infrastructure to access network resources. These can be either Java *Browser Clients*, which incorporate the WELD Java client packages and are run via Java-enabled browsers, or *Generic Clients*, which are developed in socket-enabled languages such as Java, C, C++, or PERL and use the client protocol(s).

The *Client-Server Communication Protocol* and *Client-Database Communication Protocol* are the mechanisms that enable these network entities to communicate with each other.

4.1 Server Infrastructure

A *Server Wrapper* package [41] was implemented to provide an interface between normal UNIX applications and the WELD system, abstracting away the system internals from application integrators. Its network code allows programs to switch between the role of a client, server, and proxy in the WELD system. Also included is the capability to send heartbeats and periodic updates to remote systems, providing status and failure information.

4.2 Network Services

The *Distributed Data Manager* provides data services to remote applications via generic network socket connections. It utilizes the WELD client-database communication protocol to serialize and mirror Java object hierarchies to an object-oriented database server in the network. The data manager facilitates

[3] The WELD architecture is covered in greater detail in [5, 40].

5.3 Experiences and Measurements

5.3.1 Data Rates

Most of the WELD components were developed and tested on a high-speed local area network more representative of a company Intranet than a widely distributed design environment. Several measurements were made to gauge the effects of wide area networks on the system using a design process requiring interaction between a machine at Berkeley and a machine at North Carolina State University (NCSU). Although these measurements are not comprehensive, they provide useful order-of-magnitude estimates for performance that indicate that the length of the link is not a current system bottleneck.

Figure 4 shows a set of data rate comparisons for both local and distant data transfers in C++. While the results (Table 1) for local server with large buffer transfer, local server with byte parsing, and remote server with large buffer transfer, scale linearly, the fact that byte parsing, even in C++, delayed the system more than the wide-area network was surprising.

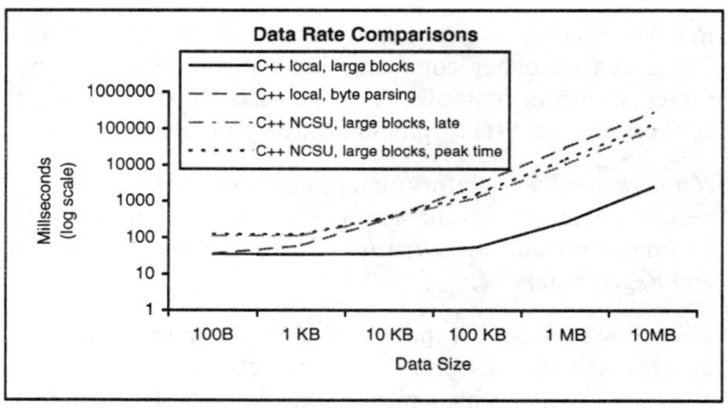

Figure 4: Data Rate Comparisons

Data Rate Comparisons												
Data Size	Local, large buffer			Local, byte parsing (Time in ms)			NCSU, large buffer			Ratios		
	Avg.	Min.	Max	Avg.	Min.	Max.	Avg.	Min.	Max.	Local	Parse	NCSU
100 B	36	24	51	37	27	49	114	101	130	1	1	3
1 KB	35	24	46	61	50	71	114	101	130	1	2	3
10 KB	36	26	48	332	323	345	359	347	374	1	9	10
100 KB	56	44	66	2917	2895	2951	1243	1209	1286	1	52	22
1 MB	270	238	359	30510	29736	31789	10482	9959	11718	1	113	39
10 MB	2446	2303	2717	284207	282755	286085	97506	97332	97745	1	116	40

Table 1: Data Rate Comparisons

Figures 5 and 6 compare the network performance of C++ and various Java implementations. These results indicate that, for the same hardware platform, the basic network capabilities for Java applications (Sun Java Development Kit (JDK) 1.02 [21]) and Java applets (Netscape 3.01 on a Sun Sparc 20) are comparable to that of C++.

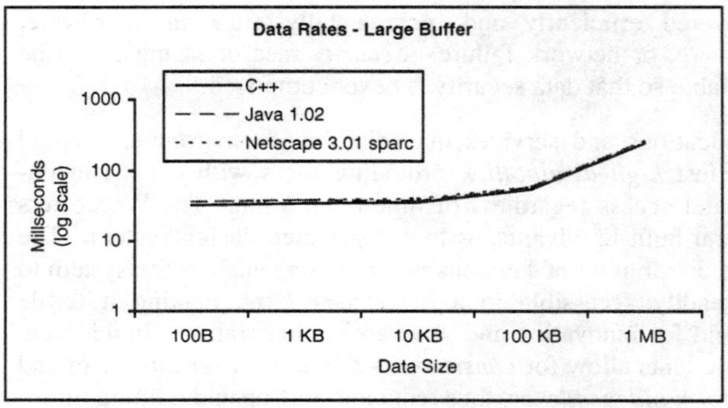

Figure 5: Data Rates – Large Buffer

5.3.2 Java

Prototyping Java applications provided a number of insights with regards to using Java and a distributed server system for computing-intensive (EDA) applications. Java and Java-based OCT were well-received within a group of users (UC Berkeley CAD Group) currently developing applications primarily in C. This was due to Java's object-oriented nature, ease of programming for both general programming logic and user interfaces, good abstraction of I/O and networking, useful documentation, library functions, and reasonable performance.

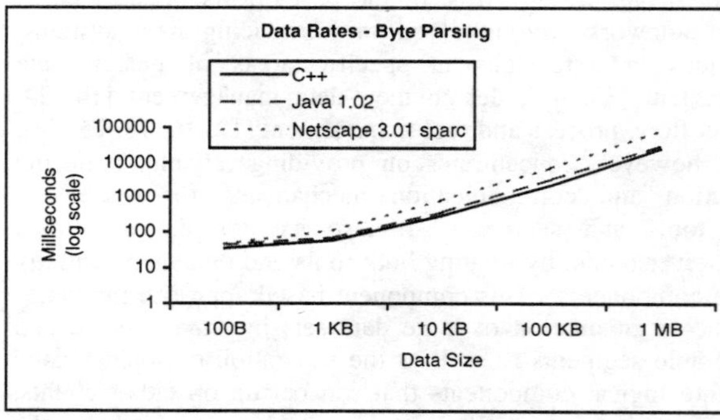

Figure 6: Data Rates – Byte Parsing

Most of WELD group's Java development was implemented using the JDK, which offered favorable performance and promised a portable Java front-end. Java performance was initially anticipated to be a system bottleneck, but extensive application usage indicated that it was reasonable, and will only improve as just-in-time compiler [23] technology advances. In addition, the focus on a proxy-based architecture allows the off-loading of work onto proxies, and can thus reduce many (Java) clients to lightweight communications and user interface tools. Fine object granularity, and the resulting large number of objects transmitted, turned out to be more of an issue than Java performance.

In order to reduce the resulting network and processing overhead, *DataObjects*, generic persistent objects specializing in string-based data storage, were implemented. Despite this work, customizability in the addition of new message types as well as the transfer and translation of data between different tools. It is also accessible via multiple interfaces, including the WELD protocols and HTTP.

In the WELD architecture, *Proxies* are software programs that reside in the network and act as transparent, "intelligent" agents between clients, network servers, and other proxies. Although originally used as a means to circumvent the Netscape Java security model[4], proxies offer a wide range of other possibilities, which include translating between data formats, filtering and redirecting of data flow, tracking and monitoring network tool usage, caching, and security.

The *Registry Service* server dynamically maintains information on the availability of network resources and their status, including uptime, history, failure rate, etc.

4.3 Client Infrastructure

The *Java Client Persistent Object Management Package* [6] works in tandem with the data manager. It enables Java (client) objects to be managed and manipulated by a network data server. In addition to object storage and retrieval[5], it extends remote data backend capabilities, such as querying and versioning, to client users, through a set of extensible APIs.

A *Java Network Client Package* [6] was also developed to allow Java applets and applications to communicate with WELD network resources.

5 Results and Experiences

5.1 Java-Based OCT - A CAD Data Manager

The first application of the data server and persistent object management package was a Java-based version [6] of the Berkeley OCT Tools. OCT serves as the data manager for a large number of VLSI/CAD applications following the Berkeley CAD framework [30]. The Java persistent object package provides a layer of abstraction for the data server, presenting capabilities including object saving and loading, as well as object traversal with an interface similar to that of the OCT generator. A data schema was created that follows the OCT convention, in which objects can be arbitrarily attached to each other, providing a general mechanism for data organization upon which developers can impose specific policies. Several applications that involved the construction, display, and partitioning of netlists were built upon this infrastructure.[6]

5.2 The Distributed Workflow System

The *Distributed Workflow System* [41] pulls together many of the WELD components developed, including the server wrapper, client packages, proxies, registry server, and data manager. The result is a cohesive, distributed design system that uses and tests the limits of the network infrastructure, as well as sheds light on many of the criteria discussed in Section 2, such as ACID/BASE properties, flexibility, and ease of use. Components (see Figure 2) include a Java front-end (Workflow Monitor, see Figure 3) in which workflows are created, displayed, and tracked, (2) tool servers, which actually perform operations upon data, (3) a registry that tracks distributed resources, (4) a database server that stores the final and intermediate results, and (5) a central control proxy (workflow server) that receives the workflow graph from the client and then handles workflow execution. Tools in the system in the 1997 DAC demonstration [7] included (1) simple programs, such as rwho, finger, email [27], (2) UC Berkeley CAD tools, including the Nova state optimization tool [30], VIS [44], SIS [36], POLIS [32], and (3) a commercial CAD tool, the Synopsys Design Compiler [10].

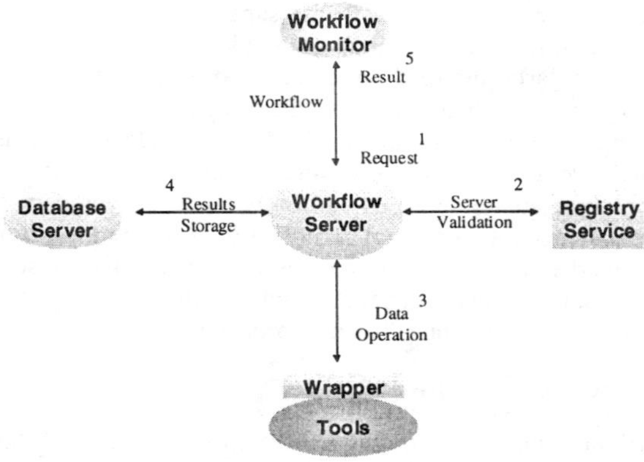

Figure 2: Architecture of the Distributed Workflow System. The numbers indicate the flow of events during a workflow execution.

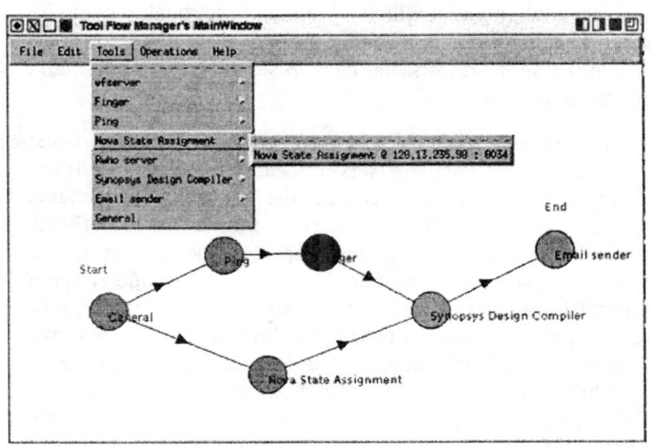

Figure 3: Workflow Monitor of the Distributed Workflow System.

---

[4] Java applets can only make socket connections to the server from which they were downloaded.

[5] Other efforts in the area of Java object and database connectivity include the Java Database Connectivity API [22] and various commercial persistent Java object systems [20, 28, 29].

[6] EE 244, a graduate-level course on CAD in UC Berkeley Fall 1996, used the Java-based OCT package for several class assignments [11].

bottlenecks[7] in the manipulation of large amounts of information still proved to be significant. We are exploring the tradeoffs in the use of current object communication mechanisms, including CORBA [2] and DCOM [31], and new techniques in dynamic creation and distribution of components, such as Java applets and ActiveX [9] controls, to address this issue.

5.3.3 Network Overhead

Network overhead (setting up and closing of sockets, input streams, and output streams), was another anticipated source of poor performance, but proved to be largely insignificant (around 10ms) when compared to the actual transmission of data in a LAN environment. However, these costs are higher in a WAN environment (in the order of 100ms), thus creating an incentive to reduce network transmission overhead by bundling messages together and using a connection-oriented protocol. Policies regarding object granularity during saving and loading objects can also be implemented to reduce the number of network transactions and/or length of messages. For example, unmodified child or parent objects of modified objects need not be saved. Similarly, the load operation need only load a certain number of levels of hierarchy, leaving further retrieval to the user, as opposed to exhaustively loading all related objects. While we provide the infrastructure necessary for different implementation strategies, it is currently up to developers to strike a balance among program complexity, execution time, and network overhead. We are implementing an adaptive system which manages this tradeoff and optimizes expected performance using a learning-based approach [1].

5.3.4 Server Technology

The distributed nature of tools and placement of the workflow and registry servers as proxies made it very easy to organize, manage, and select resources, as well as track tool information (such as function, failure rates, help information, and sample designs). The proxy workflow server simplified the task of both the clients and the actual tool servers by offering a convenient, scaleable middle-point at which additional functionality could be added. For instance, it offers, transparent to the user, fault tolerance (using heartbeats and backup servers), replication, and progress-monitoring.

While the object-oriented nature and built-in capabilities made the object-oriented database server ideal for storing persistent Java objects, limitations included the difficulty of schema evolution and significant object translation overhead. Alternatively, a data service built upon a file system would provide an extremely portable solution, but would require additional infrastructure to handle many necessary capabilities, such as versioning, transactions, object linking, ids, and multiple hierarchy. A relational database would, in turn, offer many built-in capabilities, but would require a great deal of additional modifications to represent object hierarchies. Since the beginning of the WELD project, object-oriented and object-relational databases [19] have advanced and may offer improved tradeoffs. However, we believe a system with generic, simple data back-ends and sets of proxies that dynamically manipulate application data, will provide the best all-around solution to the network-data management problem, and is being explored.

6 Conclusion

Computing infrastructure will continue to play a large role in determining overall EDA tool integration strategies and collaboration. Developments in the different network layers, especially server and middleware, have the potential to enable distributed collaborative EDA design systems that are scaleable, adaptable, and secure. The development efforts of the WELD project towards this goal to date include a data manager, a server wrapper package, Java client packages, proxy services and a distributed workflow system, all built with scalability, flexibility, and extensibility in mind. Prototyping efforts indicate that such a system, based upon Java front-ends in conjunction with proxies and back-end servers, is indeed feasible, although there are still issues to be addressed

Acknowledgements

This work was supported in part by DARPA under contract DABT63-95-C-0074, and by Hewlett Packard, Hughes Research, and the State of California MICRO program. Their support is gratefully acknowledged.

References

[1] Special Issue on Bayesian Networks: *Communications of the ACM*, vol. 38, no. 3, March 1995.

[2] R. Ben-Natan., "CORBA: A Guide to Common Object Request Broker Architecture," McGraw-Hill, New York, 1995.

[3] O. Bentz, "An Information-centric Design Exploration and Implementation Server", Ph.D. dissertation, UC Berkeley, 1997.

[4] A. Casotto, A. R. Newton, and A. Sangiovanni-Vincentelli, "Design Management Based on Design Traces," *Proceedings of the 27th ACM/IEEE Design Automation Conference*, pp. 136-141, 1990.

[5] F. Chan, "Architecture and Infrastructure for a Distributed Design Environment – A Client Perspective," MS Report, UC Berkeley, May 1997.

[6] Francis Chan's research page, http://www-cad.eecs.berkeley.edu/~fchan/research.

[7] P. Clark and R. Goering, "Web-based design hoists new sail," Electronic Engineering Times, June 16, 1997, issue 958. Demonstration materials at http://www-cad.eecs. berkeley.edu/weld/dac97.

[8] CREW, The Collaboratory for Research on Electronic Work, http://www.crew.umich.edu.

[9] A. Denning, *ActiveX Controls Inside Out*, Microsoft Press, Redmond, 1997.

[10] The Design Compiler Family Datasheet, http://www. synopsys.com/products/logic/design_comp_ds.html.

[11] EECS 244 Fall 96 home page, http://www-cad/~newton/ courses/.

[7] Current implementation requires input streams to be parsed byte-by-byte, and cast by Java from byte streams to strings. Both operations are very expensive.

[12] N. Filer, M. Brown and Z. Moosa, "Integrating CAD tools into a framework environment using a flexible and adaptable procedural interface," *European Design Automation Conference with EURO-VHDL*, pp. 200-205, September 1994.

[13] A. Fox, et al., "Extensible Cluster-Based Scalable Network Services," *Proceedings of the 16th ACM Symposium on Operating Systems Principles (SOSP-16)*, St. Malo, France, October 1997.

[14] S. T. Frezza, S. Levitan and P. Chrysanthis, "Requirements-based design evaluation," *Proceedings of the 32nd ACM/IEEE Design Automation Conference*, pp. 76-81, June 1995.

[15] J. Gray, "The Transaction Concept: Virtues and Limitations," *Proceedings of VLDB*, pp.144-154, Cannes, France, September 1981.

[16] J. W. Hagerman and S. W. Director, "Improved tool and data selection in task management," *Proceedings of the 33rd ACM/IEEE Design Automation Conference*, pp. 181-184, June 1996.

[17] D. S. Harrison, et al., "Electronic CAD Frameworks," *Proceedings of the IEEE*, vol. 78, no. 2, pp. 393-417, February 1990.

[18] A. Hoeven, et al., "A flexible access control mechanism for CAD frameworks," *European Design Automation Conference with EURO-VHDL*, pp. 188-193, September 1994.

[19] The Informix's Universal Web Architecture, http://www.informix.com/informix/bussol/uwa/ uwa.htm.

[20] Infoscape, http://www.infoscape.com.

[21] Java Development Kit, http://java.sun.com/products/jdk/1.1/index.html

[22] The JDBC page, http://java.sun.com/products/jdbc/index.html.

[23] Symantec Just In Time Compiler Performance Analysis, http://www.symantec.com/jit/jit_pa.html.

[24] E. W. Johnson and J. B. Brockman. "Incorporating design schedule management into a flow management system," *Proceedings of the 32nd ACM/IEEE Design Automation Conference*, pp. 88-93, June 1995.

[25] S. Kleinfeldt, et al., "Design methodology management," *Proceedings of the IEEE*, vol. 82, no. 2, pp. 231-250, February 1994.

[26] H. Lavana, et al., "Executable Workflows, A Paradigm for Collaborative Design on the Internet," *Proceedings of the 34th ACM/IEEE Design Automation Conference*, June 1997.

[27] Descriptions of finger, mail, and rwho are available from section one of the online UNIX Programmer's Manual.

[28] NetDynamics, http://www.netdynamics.com.

[29] "Objectivity for Java," *The Insider: News and Views for Objectivity Users*, July 1997.

[30] OCTTOOLS-5.2 Reference Manual, UC Berkeley, 1993.

[31] D. Platt, *The Essence of COM and ActiveX,* Prentice Hall Computer Books, Jan. 1998.

[32] The Polis Project, http://www-cad.eecs.berkeley.edu/Respep/Research/hsc.

[33] W. Schettler and S. Heymann, "Towards support for design description languages in EDA frameworks," *Proceedings of the IEEE International Conference on Computer-Aided Design*, pp. 762-767, November 1994.

[34] J. Schubert, A Kunzmann and W. Rosentiel, "Reduced design time by load distribution with CAD framework and methodology information," *European Design Automation Conference with EURO-VHDL*, pp. 314-319, September 1995.

[35] SEMATECH, http://www.sematech.org/member/division/dsgn/ecad/home.htm.

[36] E. M. Sentovich et al., "SIS: A System for Sequential Circuit Synthesis," Technical Report UCB/ERL M92/41, May 1992.

[37] The Silicon Integration Initiative (formerly the CAD Framework Initiative, CFI), http://www.si2.org.

[38] M. J. Silva, "Active Documentation for VLSI design," Ph.D. dissertation, available as Technical Report UCB/CSD-94-843, UC Berkeley, December 1994.

[39] M. D. Spiller and A. R. Newton, "EDA and the Network", *Proceedings of the IEEE International Conference on Computer-Aided Design*, pp. 470-476, November 1997.

[40] M. D. Spiller, "Architecture and Infrastructure for a Distributed Design Environment – A Server Perspective," MS Report, UC Berkeley, November 1997.

[41] Mark D. Spiller's research page, http://www-cad.eecs.berkeley.edu/~mds/research.

[42] The Vela Project, http://www.cbl.ncsu.edu/vela/.

[43] I. Videira, P. Verissimo and H. Sarmento, "Efficient communication in a design environment," *Proceedings of the 33rd ACM/IEEE Design Automation Conference*, pp. 169-174, June 1996.

[44] The VIS Group, "VIS: A system for verification and synthesis," *Proceedings of the International Conference on Computer Aided Verifications*, pp. 428-432, November 1996.

[45] The DARPA Intelligent Collaboration & Visualization program (IC&V), http://www.ito.darpa.mil/ResearchAreas96/IntellCollabVisual.html.

[46] F. R. Wagner, L. Golendziner, and M. R. Fornari, "A tightly coupled approach to design and data management," *European Design Automation Conference with EURO-VHDL*, pp. 194-199, September 1994.

[47] The WELD Project, at http://www-cad.eecs.berkeley. edu/weld.

[48] P. van der Wolf, *CAD Frameworks Principles and Arcitecture*, Kluwer Publishers, Boston, 1994.

[49] The WWW consortium, http://www.w3.org/WWW/.

OCCOM: Efficient Computation of Observability-Based Code Coverage Metrics for Functional Verification

Farzan Fallah, Srinivas Devadas
Department of EECS
MIT, Cambridge

Kurt Keutzer
Department of EECS
University of California, Berkeley

Abstract—
Functional simulation is still the primary workhorse for verifying the functional correctness of hardware designs. Functional verification is necessarily incomplete because it is not computationally feasible to exhaustively simulate designs. It is important therefore to quantitatively measure the degree of verification coverage of the design. Coverage metrics proposed for measuring the extent of design verification provided by a set of functional simulation vectors should compute statement execution counts (controllability information), and check to see whether effects of possible errors activated by program stimuli can be observed at the circuit outputs (observability information). Unfortunately, the metrics proposed thus far, either do not compute both types of information, or are inefficient, i.e., the overhead of computing the metric is very large.

In this paper, we provide the details of an efficient method to compute an Observability-based Code COverage Metric (OCCOM) that can be used while simulating complex HDL designs. This method offers a more accurate assessment of design verification coverage than line coverage, and is significantly more computationally efficient than prior efforts to assess observability information because it breaks up the computation into two phases: Functional simulation of a modified HDL model, followed by analysis of a flowgraph extracted from the HDL model. Commercial HDL simulators can be directly used for the time-consuming first phase, and the second phase can be performed efficiently using concurrent evaluation techniques.

I. INTRODUCTION

Verification is increasingly perceived as the major bottleneck in integrated circuit design. On many design teams the staff involved in verification is larger than the staff involved in the implementation of the design. For most designs the hardest verification problem is verifying the correctness of the initial register-transfer level RTL description coded into a hardware description language HDL. Formal verification techniques promise to greatly ease the problem of verifying the equivalence of a design across many steps of implementation but this still leaves unresolved the problem of getting the initial HDL description correct. Formal techniques for language containment and property checking are making some progress on this problem; however, there is no indication that these techniques will be able to offer comprehensive verification across a wide variety of designs. Thus, it appears that simulation will continue to be the workhorse for design verification for some time to come.

Improving the speed of simulation and allowing for parallel and distributed approaches to simulation are two well-motivated approaches to improving verification. Unfortunately these do not help a designer assess how comprehensively the simulation vector set verifies the design. As a result there is a well-defined need for tools that enable designers to assess the comprehensiveness of verification coverage offered by a simulation vector set.

The analogy that we use to guide our work is that of fault simulation. A fault simulator first enables you to assess that your fault coverage is high enough. This is critical to ensuring that your design is adequately tested. A fault simulator can then enable you to perform vector-set compaction. By assessing the faults detected by each vector, a minimal-length vector set can be chosen which covers all faults. By analogy, we are interested in providing coverage metrics that allow designers to assess the comprehensiveness of their simulation vector set. Next we aim to help them diagnose what part of their design may be inadequately verified by the current vector set. This will help guide the writing of additional vectors. Finally, we aim to help them compact their vector sets.

Currently, the state-of-the-art in coverage metrics is found in commercial tools which principally rely on either line-coverage or path-coverage metrics inherited from software testing. In software testing, given a set of program stimuli, coverage metrics such as line coverage, branch coverage and path coverage are used for software quality assurance. Most coverage metrics in software testing [2] are based on the activation of statements, branches or sequences of statements and do not address observability requirements; the fact that a statement with a bug has been activated by input stimuli does not mean that the observed outputs of the program will be incorrect. Exceptions are the sensitivity analysis methods of Voas [16], and the impact analysis methods of Goradia [7] (cf. Section II).

Similar approaches can be taken in hardware; however, hardware offers much less observability of *meaningful* data [1] through primary outputs than software does through inspection of memory contents. Therefore, coverage metrics proposed for measuring the extent of hardware design verification provided by a set of functional simulation vectors should compute statement execution counts (controllability information), and check to see whether effects of possible errors activated by program stimuli can be observed at the circuit outputs (observability information). Unfortunately, the coverage metrics for hardware proposed thus far, either do not compute both types of information, or are inefficient, i.e., the overhead of computing the metric is very large. In Figure 1, we indicate the levels of performance currently achievable using various metrics, normalized to HDL simulation time. Computing simple controllability metrics adds some overhead [2]; somewhat higher overhead is incurred when computing observability information at the HDL level using a specialized tag simulator [5], or when using gate-level fault simulators.

A simple example of why observability information is important to the designer is shown in Figure 2. Assume that model A and model B are both exercised thoroughly using a functional vector set. Controllability metrics will report 100% statement coverage for both models. However, it may be that statements in model A are only exercised with vectors for which $c = 0$, implying that the variables assigned in these statements *never* affect the observed output F for any simulated vector.

In this paper, we provide the details of an efficient method to compute an Observability-based Code COverage Metric (OCCOM) that can be used while simulating complex HDL designs. The computed coverage information serves as a diagnostic aid to the designer; it helps to debug and design, and/or create better functional tests.

We circumvent the drawbacks of previously-proposed coverage metrics in our new approach of two-phase OCCOM computation. The two phases are:

1. We automatically modify the given HDL model, in particular adding new variables and moving statements out of conditionals, and *simulate the given vectors using commercial* HDL *simulators*. Thus, current (future) simulation technology is

DAC 98, San Francisco, California

[1] By meaningful, we mean that the designer should be able to easily differentiate incorrect values from correct values for a given input stimulus.

[2] This data on controllability metrics is consistent with our information on commercial tools such as Simulation Technology/Summit's Vericov or Advanced Technology Corporation's Covermeter.

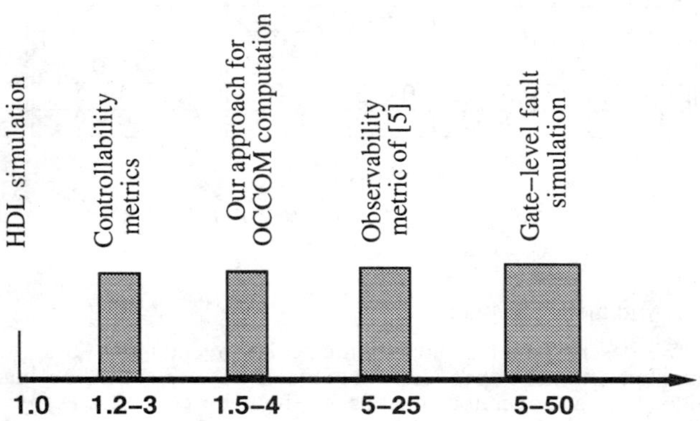

Fig. 1. Coverage Metric Computation Overheads

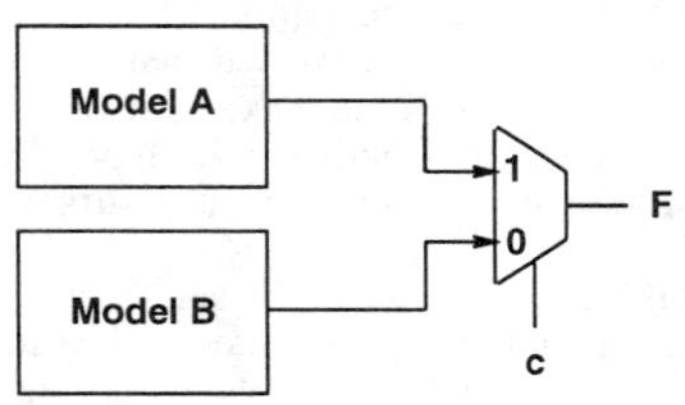

Fig. 2. Observability Versus Controllability Coverage

(can be) directly exploited. There is a loss of simulation efficiency due to the addition of new variables, but this is not large (cf. Section VI).

2. We create a flowgraph from the modified HDL model, and use the results of simulation to determine coverage under OCCOM. Using concurrent evaluation techniques, this phase can be executed efficiently.

We describe related work in Section II. In Section III, we describe our overall strategy, and present the simulation calculus. The phase of modifying the given HDL model to handle conditionals is described in Section IV. In Section V, we describe the method for OCCOM computation given the results of simulation. Experiments on a set of benchmarks are described in Section VI, that show the new method has significantly less overhead than previous methods. Ongoing work is briefly described in Section VII.

II. Related Work

We describe representative coverage analysis methods from the software test literature, and representative HDL coverage analysis work.

A. Software Testing

A *control flowgraph* is a graphical representation of a program's control structure [2, Chapter 3]. A control flowgraph is comprised of *processes*, *decisions*, and *junctions*. A process is a sequence of program statements uninterrupted by either decisions or junctions. A process has one entry and one exit. A decision is a program point at which the control flow can diverge. A junction is a point in the program where the control flow can merge.

A path in the control flowgraph is a sequence of processes that starts at an entry, junction, or decision and ends at another, or possibly the same, junction, decision or exit.

Given a set of program stimuli, one can determine the statements activated by the stimuli by applying the stimuli to the control flowgraph. The *line coverage* metric measures the number of times every process (and therefore constituent statements) is exercised by the program stimuli. In the case of *branch coverage*, we measure the number of times each branch is taken, or not taken, under the set of program stimuli. *Path coverage* measures the number of times every path in the control flowgraph is exercised by the set of program stimuli. A potential goal of software testing is to have 100% path coverage, which implies branch and line coverage. However, 100% path coverage is a very stringent requirement and the number of paths in a program may be exponentially related to program size.

There are other coverage metrics as well, for example, multi-condition coverage, loop coverage and relational operator coverage that are variants of branch coverage. The Generic Coverage Tool GCT [13] can automatically determine the coverage of an arbitrary set of program stimuli.

These coverage metrics require activation but say nothing about the observability conditions required to see the effect of possible errors in the activated statements. The path coverage metric will satisfy observability requirements if paths from program inputs to program outputs are exercised and the values of variables are such that the erroneous value is not masked (this is analogous to side inputs having non-controlling value in fault propagation). However, the path coverage metric does not explicitly evaluate whether the effect of an error is observable at an output.

Sensitivity and impact analysis take into account observability requirements. Voas [16] estimates probabilities that a fault on a location will result in an output error under a specific input distribution. The program is executed for a large number of inputs to estimate the propagation probability for a *single* location. Impact analysis [7], estimates "impact strengths" of all entity instances in an execution, in a time proportional to the execution time. If x is an input to y, then the impact strength of x depends on whether a change in the value of x from $V(x)$ to $V(x')$ propagates to y or not. However, the impact strength does not appear to depend on the values of other inputs, say z, that affect y.

B. HDL Coverage Analysis

Coverage analysis techniques proposed for general HDL models include – guaranteed coverage of every statement in an HDL model [4], evaluation of transition coverage of a test set [10], and abstraction of models and semantic control over transition coverage [6]. These metrics do not directly address observability requirements.

Observability requirements are addressed in the metric proposed in [5]. In order to compute observability information, variables are tagged during simulation, and a simulation calculus is used to calculate the coverage provided by an arbitrary set of functional vectors. There are several drawbacks with the tag simulation algorithm and calculus presented in [5]. Firstly, the calculus for only a subset of VERILOG [15] was developed. In particular, nested if statements cannot be handled, neither can looping constructs used in popular HDL's. Secondly, the efficiency of the simulation algorithm leaves much to be desired because the calculus dictated the use of an augmented simulator whose speed is much slower than the compiled-code speed of commercial simulators. The speed can be improved by incorporating all the optimizations currently present in commercial HDL simulators, but this is a huge undertaking.

Ho *et al.* [9] use transition tours on the implementation control FSM to automate test generation of corner cases, for validation of an embedded dual-issue pipelined processor. The issue of error coverage is not addressed. A formal metric for coverage computation is presented in [8]. This method has several attractive features, including a guarantee of design error coverage, however, it has not been automated and has only been applied to a processor example.

Design-specific coverage analysis and test generation approaches have been proposed for processors (e.g., [1], [11], [12]).

III. Computing Coverage: Basic Strategy

A. Context

Given an HDL model and a set of functional vectors, computing controllability metrics during functional simulation is relatively easy. Assuming an event-driven simulator, counters can keep track of how many times, if any, each statement is executed, or action taken. We, however, wish to compute observability metrics as well.

A **tag** at a location represents the possibility that an incorrect value was computed at that location. These tags on variables are not tied to particular design errors; they serve as a mechanism for extending standard coverage metrics to include observability requirements [5]. Each location corresponds to an assigned variable in some statement in the HDL model. Our goal, given a set of functional vectors and an HDL model, is to determine if a tag **injected** in any particular location is **propagated** to the circuit output. That is, we want to see if incorrectly computed values are propagated to circuit outputs, or not. [3] This is dependent on the data values at other locations in the circuit; other data values may *block* the tag from reaching any circuit output. The quality of a vector set is determined by how many injected tags are propagated to the output. The percentage of propagated tags is what we call **code coverage** under the metric.

Note that we will use the **single tag** model, where the effects of exactly one injected tag are computed, for many different injected tags. Each injected tag can be thought of creating a distinct "faulty" HDL model, and several such faulty models are processed in parallel by the simulation algorithm. (This is akin to concurrent fault simulation [3].)

B. Our Two-Phase Approach

Computing an observability metric for a collection of logic gates is similar to fault simulation using the D-calculus [14]. An error at the input of an AND gate is propagated to the output only if the other inputs are at a 1, or have errors of the same polarity. (Similarly for an OR gate.) Generalizing this notion to HDL's requires us to handle three main issues:

- Arithmetic operations,
- Conditional and looping constructs, and
- Concurrency constructs.

We can define a calculus to handle HDL models, for example, augmenting the calculus of [5], but the calculus should be easily computable. Designers wish to simulate many thousands of vectors on complex HDL models. Commercial HDL simulators incorporate many optimizations to improve logic simulation speed. A new calculus implies that the simulation engine has to be modified, and this can result in a dramatic loss in efficiency, unless a major effort to incorporate optimizations is made. In order to avoid this problem, we propose a two-phase approach to computing our coverage metric OCCOM.

The two-phase approach is:

1. We modify the given HDL model. The modifications are necessary because of conditionals in the HDL model. In particular, we will add new variables and extract statements out of conditional clauses. We then simulate the modified HDL model using a standard HDL simulator. Note that tags do not play a part in this phase.
   Simulating the modified HDL model will give us more information than simulating the original model. This extra information is used in the second phase to perform tag propagation.
2. In the second phase, we will actually perform tag injection and propagation. We extract a flowgraph from the HDL model. Tag injection simply corresponds to introducing a tag on an edge in the graph and tag propagation corresponds to selectively traversing paths from the edge to the output nodes.

We will describe Phase 1 in Section IV, and Phase 2 is described in Section V. We now describe the tag simulation calculus. Note that this calculus is not used in Phase 1, but only in Phase 2. Nevertheless, in order to understand why we are modifying the HDL description, we need to first describe the tag simulation calculus.

C. Tag Simulation Calculus

A tag is represented by the symbol Δ which signifies a possible change in the value of the variable due to an error. We will consider both positive and negative tags, $+\Delta$ written simply as Δ, and $-\Delta$. Later we will also consider tags of unknown sign (cf. Section V).

C.1 Logic Gates

We will first define the calculus for Boolean logic gates. This is identical to the D-calculus [14]. The calculus for a two-input AND gate and an INVERTER are shown in Table I. The four possible values at each input are $\{0, 1, 0+\Delta, 1-\Delta\}$. (Note that $0-\Delta \equiv 0$ and $1+\Delta \equiv 1$.) The entries are self-explanatory.

Using the above calculus any collection of Boolean gates comprising a combinational logic module can be tag simulated.

[3] While there is full observability of internal locations in the HDL model during simulation, the particular data values at internal locations may be incomprehensible to the designer. For example, the designer may be able to verify that the output of a Wallace tree multiplier is an incorrect 6, for inputs of 4 and 4, but will not be able to determine if an arbitrary internal wire is at a faulty 1 or 0.

AND	0	1	$0+\Delta$	$1-\Delta$
0	0	0	0	0
1	0	1	$0+\Delta$	$1-\Delta$
$0+\Delta$	0	$0+\Delta$	$0+\Delta$	0
$1-\Delta$	0	$1-\Delta$	0	$1-\Delta$

INVERT
1
0
$1-\Delta$
$0+\Delta$

TABLE I
Δ-CALCULUS FOR AND GATE AND INVERTER

C.2 Arithmetic Operators

We describe the tag simulation calculus procedurally for some common arithmetic operators. All modules are assumed to be n bits wide. For each operator op, after the simulator computes $v(f) = v(a)\ \langle op \rangle\ v(b)$, we tag $v(f)$ with a positive or negative Δ and we write it as $v(f)+\Delta$ or $v(f)-\Delta$.

Adder : If all tags on the adder inputs are positive, and if the value $v(f) < MAXINT$, we will assign the adder output $v(f)+\Delta$. $MAXINT$ is the maximum value possible for f. Similarly, if all tags are negative. If both positive and negative tags exist at adder inputs, the output is assumed to be unknown sign tag.

Multiplier : All tags have to be of the same sign for propagation. A positive Δ on input a is propagated to the output f provided $v(b) \neq 0$ or if b has a positive Δ. We will assign the multiplier output $v(f)+\Delta$. Similarly for negative Δ.

$>$ *Comparator* : If tags exist on inputs a and b, they have to be of opposite sign, else the tag is not propagated to the output. Assume a positive tag on a alone, or a positive tag on a and a negative tag on b. If $v(a) \leq v(b)$ then the tag(s) is (are) propagated to the output, else the tag(s) is (are) not. We will assign the comparator output $0+\Delta$. Similarly for other tags and other kinds of comparators.

C.3 Conditionals

When a conditional is encountered by the tag simulator, there are two cases:

- There is no tag in the control condition. In this case simulation proceeds normally. In the appropriate processes, statements are executed and tags (if any) are propagated/injected.
- If there is a tag on the control condition $expr$ it means that the tag will result in the incorrect branch being taken. Under the tag condition, some assignments will be missed, and others incorrectly made. Unfortunately, we do not have information regarding what happens when the incorrect branch is taken; we only have information regarding the simulation of the correct branch. [4]

As an example of the second case, consider,

```
if ( y )
        x = exp1 ;
else z = exp2 ;
```

If `y = 1` and there is a negative tag on `y`, we have to consider the case where `x` is not assigned, and `z` is assigned. While we know the values of `x` prior to and after the assignment, we only know the old value of `z`, since the simulator does not compute `exp2`.

Nested conditionals and concurrency constructs exacerbate the above situation. We describe how we handle various cases in the next section by modifying the HDL model prior to simulation.

IV. Phase 1: Modifying the HDL Model

We consider various commonly occurring cases of conditionals and events in VERILOG [15], and show how to automatically modify the HDL model such that the HDL simulation produces enough information to compute coverage in Phase 2. The modification is illustrated for several commonly-occurring cases in the following sections. The modifications all require the addition of new variables.

[4] The approach of [5], somewhat arbitrarily, tags all the variables in the assignments that would be missed under the tag condition.

A. Simple Conditional

Consider the code on the left-hand-side below. The transformed code is shown to the right.

```
if (cexp)
      x = expr1 ;
else  x = expr2 ;
```

```
x1 = expr1 ;
x2 = expr2 ;
if (cexp)
      x = x1 ;
else  x = x2 ;
```

Consider the case of a tag on cexp. During the simulation of the modified code, we will compute the values of both expr1 and expr2, and store them in the new variables x1 and x2. We know the old value of x, and we will know the new values of x corresponding to the execution of both the *then* and *else* clauses, regardless of the value of cexp. This will allow us to correctly propagate positive or negative tags on cexp in Phase 2 (cf. Section V).

B. Nested Conditionals

The case of nested conditionals is more complicated. Further, we have to consider the situation where variables such as x are assigned values that depend on the old values (e.g., increment operation). As before, we show the original code to the left, and the transformed code to the right. The transformed code will compute the necessary information to perform propagation of tags on cexp1 or cexp2.

```
if (cexp1) begin
  if (cexp2)
    x = expr1 ;
  if (cexp3)
    x = x + expr2 ;
end
```

```
x2 = x ;
x1 = expr1 ;
if (cexp2)
  x2 = x1 ;
x3 = x2 + expr2 ;
if (cexp3)
  x2 = x3 ;
if (cexp1)
  x = x2 ;
```

C. Simple Loop

Consider the simple loop shown to the left-hand-side below. The interesting case is where there is a tag on the variable N. The transformed code is shown to the right.

```
for(i=0; i<N; i++)
  x = x + expr ;
```

```
x1 = x ;
for(i=0; i<N+1; i++)
  begin
    x1 = x1 + expr ;
    if (i < N)
      x = x1 ;
  end
```

If there is a negative tag on N, then that corresponds to the situation that the loop is iterated fewer than N times. If there is a positive tag on N, then that corresponds to the situation where the loop is iterated more than N times. To minimize simulation overhead, we have chosen to perform exactly one more iteration of the loop. Given the extra conditional in the modified loop, the two pieces of code result in the same value for x. However, x1 will contain the value for the case where the loop is iterated N + 1 times.

Similar transformations can be made for while and repeat loops.

We do not inject tags on the loop counter variable i. We assume that errors on i are reflected by changes in the number of loop iterations, i.e., N. Propagating tags on variables such as x does not require additional information.

D. Conditional with Event

Events in VERILOG correspond to setting variables that affect other processes. Consider the code on the left-hand-side below.

```
if (cexp) begin
  x = expr1 ;
  --> ec1 ;
  y = expr2 ;
end
else  x = expr3 ;
```

```
x1 = expr1 ;
y1 = expr2 ;
x2 = expr3 ;
if (cexp) begin
  x = x1 ;
  --> ec1 ;
  y = y1 ;
end
else  x = x2 ;
```

This is similar to the simple conditional case, where we are concerned with a tag on cexp. The additional complication is the event ec1. We should not move the event outside the *then* clause, because we have no way of undoing the effect of the event in the simulator, if the clause is not going to be executed.

E. Loop and Conditional

A loop inside a conditional is handled like any other statement.

A conditional inside a loop requires a more complicated transformation.

```
for(i=0; i<N; i++)
  if (cexp)
      x = x + expr1 ;
  else  x = expr2 ;
```

```
x3 = x ;
for(i=0; i<N+1; i++)
  begin
    x1 = x3 + expr1 ;
    x2 = expr2 ;
    if (cexp)
          x3 = x1 ;
    else  x3 = x2 ;
    if (i < N)
      x = x3 ;
  end
```

We wish to propagate tags on both iteration count variable N, and the conditional expression cexp in Phase 2. We need to transform the code such that there is enough information to propagate either tag. The right-hand-side code corresponds to the modified version.

Consider a tag on N, and no tag on cexp. This is similar to the simple loop case, except that we are computing values in both clauses of the conditional. The case of no tag on N and a tag on cexp is the same as the simple conditional case.

While we are considering the single tag model, it may be that a tag injected on a variable y is propagated to both N and cexp. The propagation of tags to variable x (Phase 2, Section V) is dependent on the signs of the tags on N and cexp, and also on the values (computed during simulation) of the new xi variables.

The propagation of tags is the subject of the next section.

V. Phase 2: Graph Analysis

We describe, given the information regarding variable values for the simulated functional vectors, how to perform concurrent tag propagation. We work with a flowgraph data structure that enables efficient tag propagation.

A. Graph Structure

We create a graph $G(V, E, L(E))$ from the given HDL model. Each vertex $v \in V$ corresponds to a variable in the HDL model. Each edge $e(v, w) \in E$ is a directed edge from source node v to destination node w. The edge implies a data dependence between the nodes. Node w depends on v.

The label of an edge $l(e) \in L$ contains the information below:

- Line number: The edge exists because of a data dependence. A particular line of the *modified* HDL file containing the description of the model is associated with the edge. For example, given a = b + c in Line 154, the edges from b to a and c to a will both have Line 154 as a label.
- Conditional expression: We have a conditional expression which has to be true, in order for the data dependency to exist. For example, given

  ```
  if (x > y)
      a = b + c ;
  ```

  the conditional expression for the edge from b to a would be x > y. Since we have removed nested conditionals in the modified HDL model, we do not need to conjunct multiple expressions. The expression will correspond to the conditional clause of a single if statement in the original HDL model.
- Indices for array variables: Arrays are common in HDL models. There will be a separate tag associated with each array element, resulting in an array of tags. Given

  ```
  a[i] = b[i+4] ;
  ```

... a single node for the array b and the array a ... edge from b to a. We will associate information corresponding to the array indices with the edge from b to a. In general, we require two indices, one for b and one for a. We will need to compute these indices dynamically during tag propagation.

- Type of dependence: Dependence can be, normal, conditional, through task or function call, and through module instantiation. For example, given

```
if (x > y)
    a = b + c ;
```

there is an edge from x to a and y to a. These edges are conditional edges. The edges from b and c to a are normal edges.
A task or function call dependency is a dependency between inputs and outputs of a task enable or function call. If there is a tag on one of inputs or the tag-injected line is inside the task or function, there may be tags propagated to the output of the task or function.

- Propagation multiplier: We will have a multiplier associated with each edge that will be used to multiply tags propagating down the edge. These multipliers can be expressions. For example, given a = b - c, the edge between b and a has a $+1$ multiplier, but the edge from c to a has a -1 multiplier. (This is because a positive tag on c should result in a negative tag on a.)
There is an additional complication for edges whose dependence type is conditional. Consider

```
if (x > y)
        a = a1 ;
else    a = a2 ;
```

In this case, the conditional dependence edge from x to a will have the expression a1 - a2 as the propagation multiplier. We are only interested in the sign of a1 - a2, not its magnitude.

We will create a single observable vertex O in the graph that does not correspond to any variable. The variables in each "display" or "monitor" statement in the HDL model will be connected to O via edges. Each edge will have a Line number corresponding to the display statement. The conditional expression will be always true, and the propagation multiplier will be $+1$ or -1. If array variables are displayed, the index of the variable will be attached as a label for the edge.

B. Concurrent Tag Propagation

Tag propagation is performed on the graph structure, and uses information obtained from the simulation trace of the modified HDL model. For each vector, we determine all the tags that can be propagated to the observable vertex O using concurrent tag propagation.
The steps are:

1. Each edge in the graph is first determined to be active or inactive for the given vector. An active edge is an edge that can propagate a tag from a predecessor node to the successor node. Whether an edge is active or not depends on the tag simulation calculus (cf. Section III-C). For example, given the statement $C = A \times B$, if B is 0, then the edge from A to C will be inactive. A complication is that edges corresponding to the control statements may be active or inactive depending on the sign of the tag which is propagated, as well as the value of the control predicate under the given vector. For example, in

```
if ( y )
        x = exp1 ;
else    x = exp2 ;
```

there is an edge with conditional dependence from y to x. This edge is inactive if exp1 - exp2 = 0 for the given vector, and is marked active otherwise. The propagation multiplier of the edge is $+1$ if expr1 - expr2 > 0, and -1 if expr1 - expr2 < 0. If y = 1, then a positive tag on y cannot be propagated to x from y, but a negative tag can.
2. We delete all inactive edges in the graph.
3. We inject a positive and negative tag on the observable vertex O.
4. Starting from the observable vertex O, and assuming positive (negative) tags on the vertex, we traverse edges in the graph backwards, in the reverse order of simulation trace, determining all nodes $n \in V$ that can reach the observable vertex. The vertex O is reachable from node n, if before traversing any of n's fanin edges, all traversed paths from n to O have propagation multipliers of the same sign. (The propagation multiplier of a path is simply the product of the propagation multipliers of the constituent edges.) Note that paths with multipliers of different signs, might result in tag cancellation, and we conservatively assume that the tag is not propagated.
5. Line numbers of edges encountered during the backward traverse, determine the tags which are observable in O.

VI. EXPERIMENTS

A. Metric Comparison

The examples used in Table II correspond to various algorithms and processors implemented in VERILOG. Example **dcnew** is a train system relay, **eight queens** is an algorithm to solve the eight queens problem in chess, **crd** is a traffic controller, **amp** is a general-purpose processor, and **coherence** is a cache coherence protocol.

The results in Table II serve to illustrate the usefulness of OCCOM in relation to vanilla controllability measures. We gave a simple example of how controllability information can be misleading in Figure 2. In Table II, we give statistics comparing line coverage and OCCOM coverage for designer-generated functional tests (labeled Directed), and randomly-generated tests (labeled Random).

It is clear that the extra observability information in OCCOM is required to distinguish between the "good" design-generated test and the random tests. In general, the designer generates tests by exercising chosen paths through the HDL model. If a path from inputs to outputs is exercised by an input vector sequence, the tags injected on the assignment statements on the path will almost always[5] be propagated to the output. (Using a path coverage metric has the disadvantage that an exponential number of paths exist in the design, and one cannot possibly exercise all paths. Further, many of these paths may be false, i.e., may not be exercisable.)

B. Performance Comparison

We compare the performance of our OCCOM computation methodology against functional simulation, and the method of [5]. The results are summarized in Table III. Example **arbiter** is a bus arbiter, **counter** is a 3-bit counter, **crd** is a traffic controller, **ctlp3** is the logic for a Dining philosophers problem, and **8251A** is an USART. The time required to simulate the original HDL model (prior to modification) for 10,000 vectors using a commercial VERILOG simulator is given as a baseline for comparison purposes. The same simulator and the same vectors were used for OCCOM computation. OCCOM computation is broken up into two parts; time required to simulate the modified HDL model, and the time required for tag propagation. All the times correspond to seconds on an UltraSparc-2.

We could not run the simulator of [5] for some of the examples, since the simulator does not handle certain VERILOG constructs such as nested if statements. Our method is faster than the specialized tag simulation because it allows for the use of highly-optimized commercial HDL simulator and concurrent tag propagation.

Our method is significantly faster than fault simulation. Note that in order to use the fault simulation strategy, the HDL model has to be converted to gate-level form, and this in itself is a time-consuming operation.

C. Diagnosis: Feedback to Designer

OCCOM coverage obtained for a given HDL model can be used to debug the model, or create better functional tests.

Assume that a certain coverage has been obtained for an HDL model. The tags that are not propagated to the outputs are examined, one by one. For each such tag, we have the information regarding whether the line on which the tag was initially injected was executed (controllability information). We determine all the vectors (if any)

[5] One exception is when tag propagation is blocked because of arithmetic overflow or underflow.

Ex	#Lines	#Tags	Directed			Random		
			#Vec	#OCCOM Coverage%	Line Coverage%	#Vec	OCCOM Coverage%	Line Coverage%
eight queens	220	146	179	87%	92%	2000	67%	90%
dcnew	353	256	338	86%	100%	3000	77%	100%
crd	120	88	156	94%	100%	2000	67%	90%
amp	288	198	168	68%	92%	3000	68%	92%
coherence	530	346	442	85%	94%	5000	54%	90%

TABLE II
COMPARING OCCOM AND A LINE COVERAGE METRIC

Ex	HDL Simulation	OCCOM			Specialized Simulator [5]
		HDL	Graph	Total	
arbiter	9 s	13 s	20 s	33 s	150 s
counter	7.5 s	10 s	1.5 s	11.5 s	25 s
crd	19 s	21 s	19 s	40 s	100 s
ctlp3	11.5 s	13 s	10 s	23 s	NA
8251A	15 s	22 s	20 s	42 s	NA

TABLE III
COMPARING TIMES REQUIRED TO COMPUTE METRICS

for which the line was executed. For each vector, the tag propagation path is examined to determine the **blocking** statement(s). The values of the variables in the blocking statements need to be changed in order to propagate the tag further. The designer determines a legal change in variable values and creates a new functional test, which is tag-simulated, and the process repeated. If the designer is unable to come up with such a test, then it may be that there is a redundancy in the HDL model, or that there is a design error.

VII. ONGOING WORK

There are three general directions of work to be explored in improving verification-coverage metrics for hardware designs: exploring new error models, computationally improving coverage analysis and providing diagnostics.

Associating an error with every possible assignment captures most of the errors that can at all be discovered by structural analysis, but probably produces more errors than are necessary to be analyzed. It would be useful to find an error model that produces fewer errors to be analyzed. Any changes in the error model will require a rethinking of the calculus used for propagating error values. We are exploring other error models and associated simulation calculi.

ACKNOWLEDGEMENTS

F. Fallah and S. Devadas were supported in part by a grant from Schlumberger Foundation. We thank Abhijit Ghosh for his contributions to this project.

REFERENCES

[1] A. Aharon, D. Goodman, M. Levinger, Y. Lichtenstein, Y. Malka, C. Metzger, M. Molcho, and G. Shurek. Test Program Generation for Functional Verification of PowerPC Processors in IBM. In *Proceedings of the* 32nd *Design Automation Conference*, June 1995.

[2] B. Beizer. *Software Testing Techniques*. Van Nostrand Rheinhold, New York, second edition, 1990.

[3] M. A. Breuer and A. D. Friedman. *Diagnosis and Reliable Design of Digital Systems*. Computer Science Press, 1976.

[4] K-T. Cheng and A. S. Krishnakumar. Automatic Functional Test Generation Using the Extended Finite State Machine Model. In *Proceedings of the* 30th *Design Automation Conference*, pages 86–91, June 1993.

[5] S. Devadas, A. Ghosh, and K. Keutzer. An Observability-Based Code Coverage Metric for Functional Simulation. In *Proceedings of the International Conference on Computer-Aided Design*, pages 418–425, November 1996.

[6] D. Geist, M. Farkas, A. Landver, Y. Lichtenstein, S. Ur, and Y. Wolsfthal. Coverage-Directed Test Generation Using Symbolic Techniques. In *Proceedings of the Int'l Conference on Formal Methods in CAD*, November 1996.

[7] T. Goradia. Dynamic Impact Analysis: A Cost Effective Technique to Enforce Error Propagation. In *Proceedings of Int'l Symposium on Software Testing and Applications*, March 1993.

[8] A. Gupta, S. Malik, and P. Ashar. Toward Formalizing a Validation Methodology Using Simulation Coverage. In *Proceedings of the* 34th *Design Automation Conference*, pages 740–745, June 1997.

[9] R. C. Ho, C. H. Yang, M. A. Horowitz, and D. L. Dill. Architecture Validation for Processors. In *Proceedings of the* 22nd *Annual Symposium on Computer Architecture*, June 1995.

[10] Y. Hoskote, D. Moundanos, and J. A. Abraham. Automatic Extraction of the control flow machine and application to evaluating coverage of verification vectors. In *Proceedings of the Int'l Conference on Computer Design*, pages 532–537, October 1995.

[11] K. D. Jones and J. P. Privitera. The Automatic Generation of Functional Test Vectors for Rambus Designs. In *Proceedings of the* 33rd *Design Automation Conference*, pages 415–420, June 1996.

[12] M. Kantrowitz and L. M. Noack. I'm Done Simulating; Now What? Verification Coverage Analysis and Correctness Checking of the DECchip 21164 ALPHA microprocessor. In *Proceedings of the* 33rd *Design Automation Conference*, pages 325–330, June 1996.

[13] B. Marick. *The Craft of Software Testing*. Prentice-Hall, Englewood Cliffs, N. J., 1995.

[14] J. P. Roth. Diagnosis of Automata Failures: a Calculus and a Method. *IBM journal of Research and Development*, 10:278–291, July 1966.

[15] D. E. Thomas and P. R. Moorby. *The Verilog Hardware Description Language*. Kluwer Academic Publishers, Boston, MA, second edition, 1994.

[16] J. M. Voas. PIE: A Dynamic Failure-Based Technique. *IEEE Transactions on Software Engineering*, 18(8):717–727, August 1992.

User Defined Coverage - A Tool Supported Methodology for Design Verification

Raanan Grinwald, Eran Harel, Michael Orgad, Shmuel Ur, Avi Ziv
IBM Research Lab in Haifa
MATAM
Haifa 31905, Israel
email: {grinwald, harel, orgad, sur, aziv@vnet.ibm.com}

Abstract

This paper describes a new coverage methodology developed at IBM's Haifa Research Lab. The main idea behind the methodology is a separation of the coverage model definition from the coverage analysis tool. This enables the user to define the coverage models that best fit the points of significance in the design, and still have the benefits of a coverage tool. To support this methodology, we developed a new coverage measurement tool called Comet. The tool is currently used in many domains, such as system verification and micro-architecture verification, and in many types of designs ranging from systems, to microprocessors, and ASICs.

1 Introduction

Functional verification comprises a large portion of the effort in designing a processor [1]. The investment in expert time and computer resources is huge, and so is the cost of delivering faulty products [4]. In current industrial practice, most of the verification is done by generating a massive amount of tests by random test generators [1, 2, 9]. The use of advanced random test generators can increase the quality of generated tests, but it cannot detect cases in which some areas of the design are not tested, while other parts are tested repeatedly.

The main technique for checking and showing that the testing has been thorough, is called test coverage analysis [10]. Simply stated, the idea is to create, in a systematic fashion, a large and comprehensive list of tasks and check that each task was covered in the testing phase. Coverage can help in monitoring the quality of testing, and assist in directing the test generators to create tests that cover areas that have not been tested before.

Many coverage tools exist in the market. Most coverage tools are program-based and check code coverage of execution of programs. Examples include code coverage in C [8] or VHDL [12], and covering states and transitions of state machines [13]. The main disadvantage of code coverage is that it does not "understand" the application domain. It is, therefore, very hard to tune the tools to areas which the user thinks are of significance. To overcome this problem, some domain dependent tools have been developed which measure specific functional events [3]. In all the tools we are familiar with, the coverage models are hard-coded. Therefore, the user is forced to use a tool that does not fit his needs, or put a lot of effort into developing a new tool for his coverage needs.

In this paper, we present a new methodology for coverage that was developed at IBM's Haifa Research Lab, and a coverage measurement tool called Comet (COverage MEasurement Tool) that was developed to support this methodology. The main idea behind the new methodology is separation of the coverage model definition from the tool. This separation enables the user to use a single tool for most of his coverage needs. The user can define coverage models that fit the design in the best way, while enjoying all the benefits of a coverage tool, such as data collection and processing, creation of coverage reports, and generation of regression suites with high coverage. Moreover, the user can change the scope or depth of coverage during the verification process, starting from simple coverage models in the early stages of verification and, based on those models, define deeper and more complex models later on.

Our coverage methodology and Comet are used by several sites in IBM in designs ranging from systems, to microprocessors, and ASICs. The tool is currently used in many different domains, such as architectural verification of microprocessors, micro-architecture verification of units in a processor, system and unit verification of a communication ASIC, and verification of coherency and bus protocols.

The rest of the paper is organized as follows: In Section 2, we describe the functional coverage process and provide some terminology used in this paper. In Section 3, we describe our new coverage methodology in detail. In Section 4, we describe Comet, our coverage tool, and the coverage process as is done in Comet. Finally, Section 5 concludes the paper.

2 Functional Coverage Process

Coverage, in general, can be divided into two types: program-based and functional. Program-based coverage concentrates on measuring syntactic properties in the execution, for example, that each statement was executed, or each transition in a state machine taken. This makes syntactic coverage a generic method which is usually easy to measure. Functional coverage focuses on the functionality of the program,

DAC 98, San Francisco, California

and it is used to check that every aspect of the functionality is tested. Therefore, functional coverage is design and implementation specific, and is much harder to measure. Currently, functional coverage is done mostly manually.

The manual process through which one gains confidence that testing has been thorough, is composed of the following steps: first, one prepares a list of testing requirements in a test plan which contains the events that have to happen during testing. Then, one executes the tests and checks that for every requirement there is a test in which the particular event happened and that this test has been executed successfully. In general, this is a very labor intensive effort, as the tests are crafted in order to fulfill specific requirements. If a test fulfilled a requirement "by chance", it will usually not be noticed.

Automation can be added to this process in the following way: a *coverage task*, a binary function on a test or a trace which specifies whether an event occurs or not, is created for each test requirement. Examples of coverage tasks are: "statement 534 was executed", and "the instructions in the first two stages of a pipe were add and divide writing to the same target". A cohesive group of coverage tasks is called a *coverage model.*

The second example of a coverage task can be generalized into coverage models over attributes. Each task is an instantiation of these attributes and the coverage model is the cross product of the possible values. For example, (first instruction, second instruction) is a coverage model over two attributes, and (ADD, DIV) is a task in that model. Comet uses coverage models of this type.

In order to find out if a coverage task happens in a test, we create a trace called *event trace.* Rows in the event trace usually contain the values of the attributes at the time the row was produced. Coverage is measured by activating the coverage models on the event trace.

We divide coverage models into two types: *snapshot* and *temporal.* Snapshot models are coverage models whose attributes are taken from a single row in the event trace, for example, the combination of two instructions that are at different stages of a pipe at the same time. Temporal coverage models are coverage models whose attributes are taken from different rows in the event trace. A coverage task in these models is a specific scenario that occur during the test, for example, the combination of two instructions that were fetched by the processor with a distance of less than 5 cycles between them.

Often, some of the tasks in the coverage model, defined by the cross product of the attributes, are illegal tasks, that is, coverage tasks that should not occur. For example, two instructions that write to the same resource should never be in the write-back stages of pipes in the same cycle. Comet reports to the user on any illegal tasks found in the event trace. It also reports coverage statistics on which tasks out of the task coverage list (TCL), a list of all the (legal) tasks in the coverage model, have been covered.

3 User Defined Coverage

Coverage measurement and the use of coverage as an indicator of the quality of testing and reliability of the design are growing rapidly. More tools for coverage measurement are becoming available, both for software and hardware testing. These tools provide the user with many features that are necessary for an efficient coverage measurement, such as data gathering, updating of the coverage task list, and reports on the coverage status.

A common property of most of these tools is that the coverage models which the tool is designed to handle, are hard-coded into the tool. Therefore, in order to make the tools applicable to many users, the models that are implemented in these tools are generic. This leaves a user that wants to cover models that are specific to his design with two options: to measure coverage manually or to build his own coverage tool for these models. Both options require a large effort and are error prone. Therefore, specific models are rarely used.

We propose a different methodology for coverage. This methodology calls for separation of the coverage model from the coverage measurement tool. The idea behind the methodology is that most of the functionality provided by existing coverage tools, such as data gathering and coverage reports, is similar in all tools. The main difference between tools is the models they implement. Therefore, a single general purpose tool, that will be oblivious to the coverage model and provide all the functionality of existing tools, can be used to provide all the coverage needs of a user, both for generic and specific models.

To provide all the needed functionality of a coverage tool to a specific model, the tool has to be aware of the exact specification of the model. Therefore, one of the inputs to the tool is the definition of the model. The definition of a model has to be done in a language which is simple enough for a user to use, yet rich enough to describe all the models that the user wants to cover. Our experience shows that a language that contains the predicates used in first order temporal logic (*and, not, exist, for all, before, after,* etc.) combined with the use of simple arithmetic (natural numbers, arithmetic and relational operators), is sufficient.

The advantages of such a general purpose coverage tool are enormous. First, it allows its users to define their own models, according to their specific needs, and still enjoy all the functionality of a dedicated coverage tool without the need to develop such a tool. It, therefore, provides users with the means to measure coverage on models which were not available to them before. The tool also allows organizations to use a single tool for most of their coverage needs, and not be forced to buy and maintain a set of tools, one for each coverage model. Another advantage for explicit and external model definition is that it enables sharing of coverage models between projects. Finally, the tool enables its users to adapt their coverage models to their testing resources, and refine these models during testing. For example, users that can afford only quick and dirty testing can define coarse grain models with a small amount of tasks, while users that want to do comprehensive testing can define finer grain models [11].

4 Comet (COverage MEasurement Tool)

Based on the methodology described above, we developed a coverage measurement tool called Comet. Comet enables users to define their own coverage models, gather and process traces to measure the coverage of these models, and generate coverage reports on the results of the measurement. Comet depends on a relational database in order to supply a comprehensive and stable environment needed for the coverage measurement process itself and the analysis of the coverage results.

The relational database provides to Comet all the data management it needs. It stores the traces on which coverage is measured and the TCLs, and provides access and processing of the data using standard SQL [5]. It also enables the

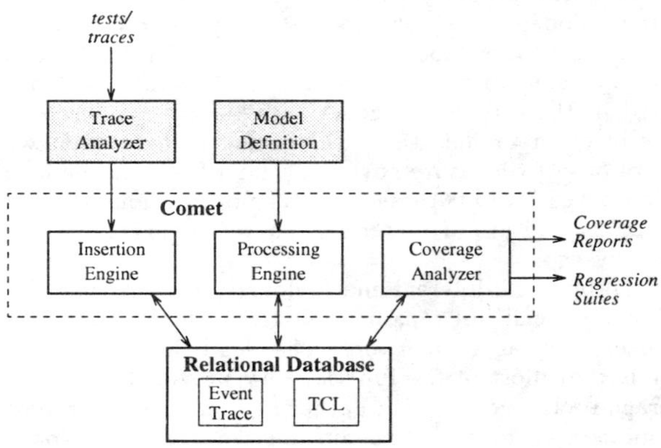

Figure 1: Comet Structure

user to produce non-standard reports, that are not provided by Comet, by directly querying the database itself. Another advantage of the relational database is that it enables us to implement our requirements for model specification using standard SQL [5], without the need to invent a new specification language and write an interpreter for it.

Comet consists of three major parts, as shown in Figure 1: the *Insertion Engine,* whose task is to insert events from input traces into the database, the *Processing Engine,* whose task is to process the traces in the database in order to detect coverage tasks according to the model definition, and the *Coverage Analyzer,* which analyzes the measurement results and prepares coverage analysis reports according to the user definition.

Since Comet is oblivious to the coverage model and it can receive many types of traces, it requires two user-provided additions that do not exist in other tools. The first addition is definition of the coverage model which is done using Comet's GUI. The second addition is a trace analyzer that converts the format of the traces to a standard format that Comet can handle.

Below, we explain in more detail the coverage measurement process, as it is done with Comet. We use two examples of models to illustrate the main features of the tool, one that was used to build a set of architectural tests for PowerPC processors, called *IIwI* (Interleaved Instructions with Interrupts), and the other which was used during the testing of a branch prediction unit in a processor, called *branch unit pipe.* The two examples also help to illustrate the versatility of Comet as a coverage measurement tool.

4.1 Model Definition

The first and most important step in the coverage process is deciding what to cover or, more precisely, on what coverage models to measure coverage. In order to make coverage successful and use it to influence the verification process, it is important to choose the correct types of coverage models. First, it is important to choose coverage models for areas which the user thinks are of significance. Next, the size of the model should be chosen according to the testing resources. The model size (number of tasks) should not be too large, such that it is impossible to cover the model, given test time and resource constraints. From our experience, we found that the best way to create effective models, is to start with small models and later on to refine or combine them to create bigger and more complex models.

The definition of a coverage model in Comet is done in three steps: defining the attributes of the model, specifying how to extract the attributes from the event trace, and separating between the legal and illegal tasks for the model. These steps are explained in detail below.

4.1.1 IIwI model

The IIwI coverage model was designed as a part of an Architecture Verification Suite (AVS) for PowerPC processors. Since incomplete or wrong cleanup after interrupts is a source of many bugs in processors, we decided that it is important to test all possible pairs of instructions with all types of interrupts between them. The IIwI model is an example of a temporal model. The model is defined by a scenario of three events: the first instruction, the second instruction, and an interrupt between the two instructions, with additional constraints on the timing of the events.

The first step in the definition of the IIwI model is selection of the attributes of the model. The natural choice for attributes for the model are the instruction name for the first and second instructions, and the interrupt type for the interrupt. The problem with that choice is the size of the model. There are more than 400 PowerPC instructions and 30 types of interrupts, and therefore more than 5 million tasks (400x400x30). We, therefore, decided to group the instructions into 34 groups, each group containing instructions of similar type. This is done using the grouping option of Comet. The attributes of the IIwI model can be seen in the top part of Figure 2. The attributes are `First` and `Second`, which are obtained from the `InstName` attribute in the event trace using the `IIwIGroup` grouping function, and `Interrupt`, which is the `InterruptBasic` attribute, taken directly from the event trace.

The second part of model definition is to provide Comet with a method to extract tasks from the event trace. This is done by specifying a `WHERE` clause of an SQL query, that defines the conditions on the events that create the tasks. Since the IIwI is a temporal model, with tasks consisting of events occurring at three different times, the SQL query involves three views of the event trace, denoted by `e1` for the first instruction, `e3` for the second one and `e2` for the interrupt. The conditions for the IIwI model appear at the bottom of Figure 2. The conditions are divided into 3 sub-conditions. The first states the instructions that are of interest to us (they do not belong to the `DontCare` group). The second sub-condition deals with the timing of the instructions (the first instruction occurs before the second one with a maximal distance of 6 instructions, and the interrupt is between them). The last sub-condition states that the first and second instructions are not interrupted, while the middle event is an interrupt.

After defining the attributes of the model and its conditions, we need to create the TCL containing all the legal tasks of the model. This step is explained below, using the branch unit pipe model.

4.1.2 Branch unit pipe model

The branch unit pipe model is part of a set of coverage models we designed to measure coverage on the branch unit of a processor. This model was designed to check the amount and locations of branch instructions in a nine stage pipe inside the branch unit. The specification of the unit is for a maximum of seven simultaneous branches in the pipe.

The branch unit pipe model is a snapshot model, and its attributes are taken directly from the event trace. The first

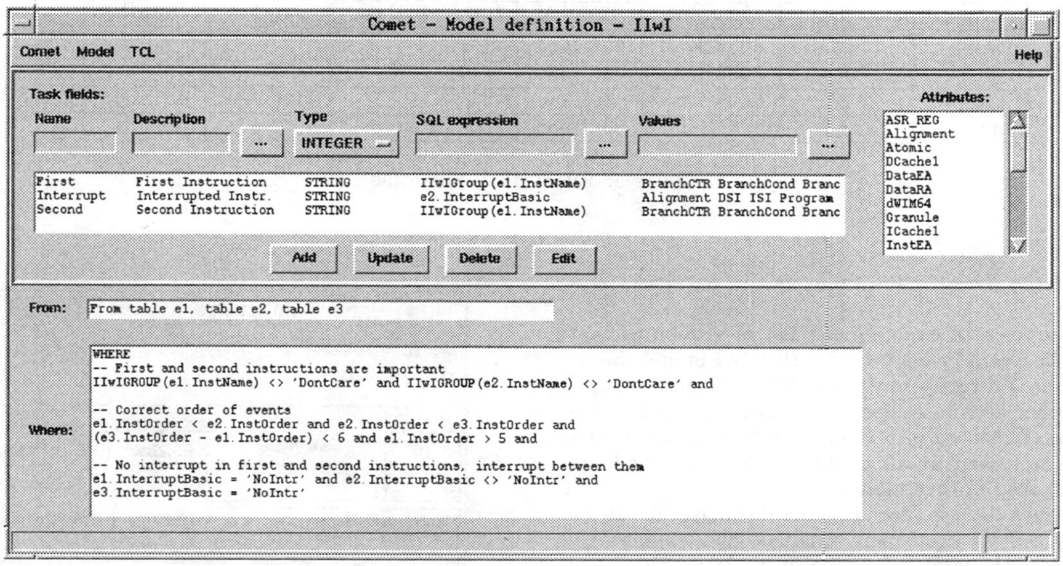

Figure 2: IIwI model definition

Attribute	Description	Values
a1	Stage of 1st branch	0–9
a2	Prediction of 1st branch	0–1
b1	Stage of 2nd branch	0–8
b2	Prediction of 2nd branch	0–1
c1	Stage of 3rd branch	0–7
c2	Prediction of 3rd branch	0–1
d1	Stage of 4th branch	0–5
d2	Prediction of 4th branch	0–1
e1	Stage of 5th branch	0–4
f1	Stage of 6th branch	0–4
g1	Stage of 7th branch	0–2
field13	Reset type	0–2

Figure 3: Branch unit pipe model attribute definition

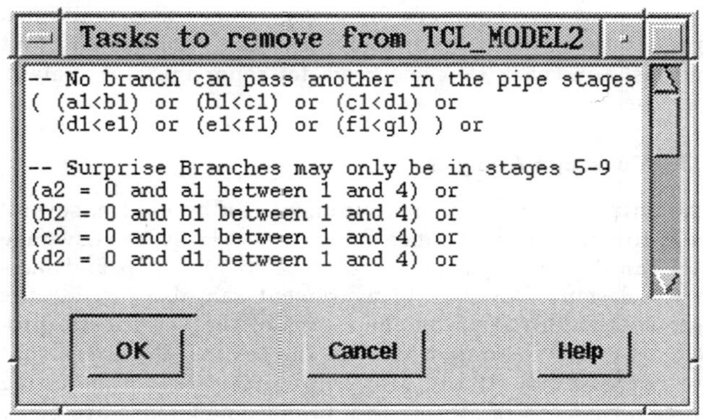

Figure 4: Restrictions for branch unit pipe model

two steps in the model definition are similar to the steps for the IIwI model, and thus are omitted. Figure 3 gives the list of attributes for the model. The attributes describe, for each possible branch in the pipe, the stage it is in and whether it is a predicted or a surprise branch. Another attribute for the model is a reset signal, that can have three values.

Specification of the unit and its environment puts many restrictions on combinations of possible locations and types of branches in the pipe. The restrictions affect which tasks in the model are legal and which are not. Separation between the legal and illegal tasks is done while building the TCL of the model, which is a list of all legal tasks in the model. Figure 4 shows the window for restriction specification with some restrictions for the branch unit pipe model.

4.2 Data Collection

After defining the coverage models, the next step is to collect event data from tests and measure the coverage. The events can come directly from the test programs or from trace files that are created by tracing the behavior of the design during the test. For example, the input for the IIwI model are the instructions and interrupts in the test, which are provided directly by the test generator, while the input for the branch unit pipe model are the stages of instructions in the branch pipe, which are provided by monitoring the pipe during execution of a test.

The decision on which attributes to monitor is taken in parallel with the model definition or even before it, and is affected by the requirements of the model and the ability to monitor the attributes. The exact format of the trace also depends on the environment and mechanisms that are used to generate the test or the trace.

To measure coverage, the traces are inserted into a table called **EventTrace** in the database. Since Comet needs to handle many types of traces, the insertion of data is done in two steps. The first step is to convert the data to a single format using the Trace Analyzer, and the second step is to insert the data into the database using the Insertion Engine (See Figure 1). The Trace Analyzer is specific to each type of trace and should be provided by the user, usually in the form of a simple script. The Insertion Engine, that writes

InstName	InstOrder	InstEA	InstRA	InterruptBasic	DataEA	DataRA	...
mtmsr	0	8085B660	03C1664	-	0	0	...
isync	1	8085B664	03C1660	-	0	0	...
ld	2	8085B668	03C166C	-	21466048	0263D48	...
fnmsub.	3	8085B66C	03C1668	FPUI	0	0	...

Figure 5: Event Trace example

the trace into the `EventTrace` table in the database, is a part of Comet.

Figure 5 provides an example of a list of attributes that are stored in the `EventTrace` table for the IIwI model. Some attributes are used directly by the model, such as InterruptBasic, while other attributes are used indirectly, such as InstName that is translated into the instruction group. Other attributes are not used at all by the IIwI model, and are stored for the sake of other models.

After the trace data is stored in the database, it can be processed by Comet's Processing Engine. This engine executes the query of the model on the trace to detect the list of tasks that occurred in the test. For each of these tasks, if it appears in the TCL, Comet increases its coverage count. If the task does not appear in the TCL, Comet reports it as an illegal task, as explained in the next section. Besides updating the coverage counters for tasks, Comet also maintains other data for each trace and each task, such as the first and last time each task occurred, for reporting and analysis purposes.

4.3 Coverage Analysis

The output of coverage measurement can be used in several ways to improve the verification process. First, the coverage tool can detect illegal events that occur and help find bugs in the design. Coverage measurement can also provide the user with information on the status of the verification process. It can help to find holes in the testing, i.e. areas that are not covered. It can also help detect when the testing process runs out of steam, and no more new tasks are being covered. Coverage measurement can also assist the verification process more directly, by directing the test generation program to create tests that check uncovered areas, and by creating good regression suites.

4.3.1 Detection of Illegal Events

The most direct way in which coverage can assist the verification process, is by detecting illegal events. Detection of illegal events can be difficult, since they may consist of a series of sub-events, each occurring in different parts of the design or at different times.

Comet detects illegal events during processing of the event trace. If it discovers a coverage task that is not part of the TCL for the model it is currently processing, it marks this task as an illegal task. For example, if in a given cycle the branch unit pipe contains $a(1) = 4$ and $a(2) = 0$, Comet will detect it as an illegal task, since all the tasks with these values were removed from the TCL due to the 2nd restriction in Figure 4.

When such an illegal task is detected, Comet stores it in a special table, and notifies the user of detection of the task. Along with the illegal task, Comet also stores other information that can help the user to identify the task easily. This information includes the test in which the task was detected, the cycle in which it occurred, and the restrictions that the task violates.

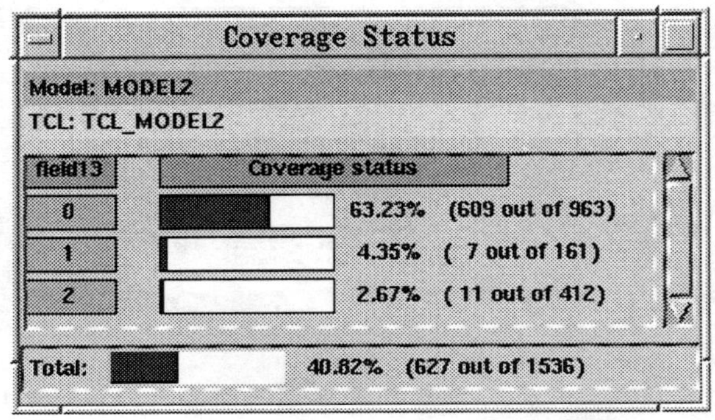

Figure 6: Simple coverage report for branch unit pipe

4.3.2 Coverage Reports

Analysis of coverage data can provide the user with vital information on the status of the verification process. For example, by looking at a map of covered and uncovered tasks, the user can detect areas that have not been tested. By examining the progress of coverage over time, one can find out when the testing is not producing any new uncovered tasks. We can also see the effects of changes in the testing strategy on the progress of coverage.

Comet provides several analysis and reporting mechanisms to the user. The most basic one is examination of the coverage map in a specific cross section, for example, the coverage of a single attribute. Figure 6 shows the coverage of the `field13` attribute of the branch unit pipe model. The figure shows that the coverage when `field13` is 0 is high (more than 60% of the legal tasks with this value are covered), while the coverage for the other two values is very low.

This type of coverage analysis can be used to improve the quality of testing by directing the test generator to generate tests that will reach tasks that are not currently covered. It can also assist the designer in finding new restrictions, which were overlooked previously, and thus, get better understanding of the design and the environment.

Another type of report that Comet can generate is the progress in coverage as a function of time (measured either by testing days, testing cycles, or the number of tests). This type of report can help detect the affects of changes in the testing strategy on the progress of coverage and the time when testing stops being effective since it is not reaching new tasks. For example, Figure 7 shows the progress in coverage of the IIwI model as a function of the number of generated tests. The figure shows that, after about 30,000 tests, the rate of new tasks became low. This caused us to

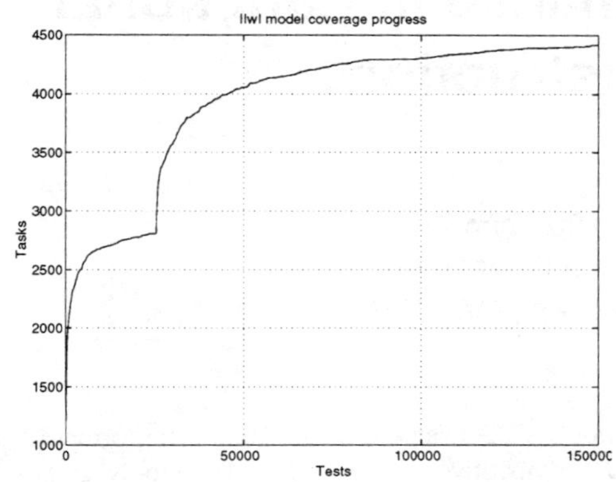

Figure 7: Coverage progress for IIwI model

change the test specification for the test generator which, in return, increased the rate of new tasks dramatically.

Comet can generate other types of reports, such as a list of uncovered tasks, tasks that were covered more than a minimal number of times, and the number and list of new tasks that were covered between certain dates.

4.3.3 Creation of Regression Suites

Coverage can also be used to find good regression suites. A regression suite [10] is a set of tests that is run periodically or after major changes to the design, to check that no new bugs were introduced to the design. A regression-suite must, on one hand, be economic so that it can be executed many times. On the other hand, a regression suite must be comprehensive, so that it will be likely to discover new bugs. A good example for a compact and comprehensive regression suite is a minimal set of tests that covers all the tasks in a coverage model (or a set of models). We have developed a new algorithm for creating small regression suites incrementally on the fly [6]. The algorithm creates small regression suites that cover all the tasks that were covered in the past. For example, we created a regression suite of 742 tests for 5 models of microprocessor pipelines containing 9810 tasks out of 500,000 tests used. We incorporated the algorithm into Comet.

5 Conclusions

In this paper, we presented a new methodology for coverage, that separates the coverage model from the coverage tool. The advantage of this methodology is that it enables us to define coverage models according to the specific needs of each unit, while keeping the benefits of a coverage tool, such as data collection and processing, and coverage analysis.

Based on this methodology, we developed a general purpose coverage measurement tool called Comet. Comet enables us to do coverage work on many types of designs, ranging from ASICs to systems, and in many verification domains, such as system verification and micro-architecture verification. Our experience from using Comet as our coverage tool, shows that it can be used for a large majority of our coverage needs, and that its users can benefit from the advantages of a coverage tool without losing the flexibility of defining their own models.

We are currently working to enhance the capabilities of Comet in many areas, such as automatic feedback to the test generator, enhanced reports capabilities, and capabilities to process very large amounts of data (insertion and processing of traces from hundreds of simultaneous sources). Another area which we are working on, is using data mining techniques [7] for analysis of coverage status.

References

[1] A. Aharon, D. Goodman, M. Levinger, Y Lichtenstein, Y. Malka, C. Metzger, M. Molcho, and G. Shurek. Test program generation for functional verification of PowerPC processors in IBM. In *Proceedings of the 32nd Design Automation Conference*, pages 279–285, June 1995.

[2] A.M. Ahi, G.D. Burroughs, A.B. Gore, S.W. LaMar, C.R. Linand, and A.L. Wieman. Design verification of the HP9000 series 700 PA-RISC workstations. *Hewlett-Packard Journal*, 14(8), August 1992.

[3] J. Baumgartner and R. Raghavan. Method to compute test coverage in complex computer system simulation. *IBM Technical Disclosure Bulletin*, 40(3):1–4, March 1997.

[4] B. Beizer. The Pentium bug, an industry watershed. *Testing Techniques Newsletter, On-Line Edition*, September 1995.

[5] J.S. Bowman, S.L. Emerson, and M. Drnovsky. *The practical SQL handbook: using structured query language.* Addison-Wesley, 1996.

[6] E. Buchnik and S. Ur. Compacting regression-suites on-the-fly. In *Proceedings of the 4th Asia Pacific Software Engineering Conference*, December 1997.

[7] P. Cabena, P. Hadjinian, R. Stadler, J. Verhees, and A. Zanasi. *Discovering Data Mining, from Concept to Implementation.* Prentice Hall, 1997.

[8] J.R. Horgan, S. London, and M.R. Lyu. Achieving software quality with testing coverage measures. *Computer*, 27(9):60–69, September 1994.

[9] A. Hosseini, D. Mavroidis, and P. Konas. Code generation and analysis for the functional verification of microporcessors. In *Proceedings of the 33nd Design Automation Conference*, pages 305–310, June 1996.

[10] B. Marick. *The Craft of Software Testing, Subsystem testing Including Object-Based and Object-Oriented Testing.* Prentice-Hall, 1985.

[11] S. Ur and A. Ziv. Off-the-shelf vs. custom made coverage models, which is the one for you? In *proceedings of STAR98: the 7th international conference on software testing analysis and review*, May 1998.

[12] Vhdlcover: How thoroughly have you simulated your vhdl code? http://www.veda.co.uk/vhdlcov.html.

[13] Visual StateScore. http://www.summit-design.com/products/vcd/index.html.

Enhanced Visibility and Performance in Functional Verification by Reconstruction

Joshua Marantz

Ikos Systems, Inc.
josh@ikos.com
1-781-370-1714

230 Third Ave.
4th Floor North
Waltham, MA 02154

1. ABSTRACT

Cycle simulators, in-circuit emulators, and hardware accelerators have made it possible to rapidly model the functionality of large digital designs. But these techniques provide limited visibility of internal design nodes, making debugging hard. Simulators run slowly when all nodes are traced. Emulators provide full visibility only with limited depth, or with greatly reduced speed. This paper discusses software techniques for increasing design visibility while reducing tracing overhead in simulation, and achieving 100% visibility in emulation without reducing speed or compromising depth.

Keywords

Functional simulation, emulation, reconstruction, visibility.

2. INTRODUCTION

It is common to debug large designs with fast simulators "post-mortem." During simulation, a very large trace file is written, containing value-changes for every node in the design, or a subsystem within the design. Upon completion, the designer has rapid access to the entire simulation history, using a commercial waveform viewer.

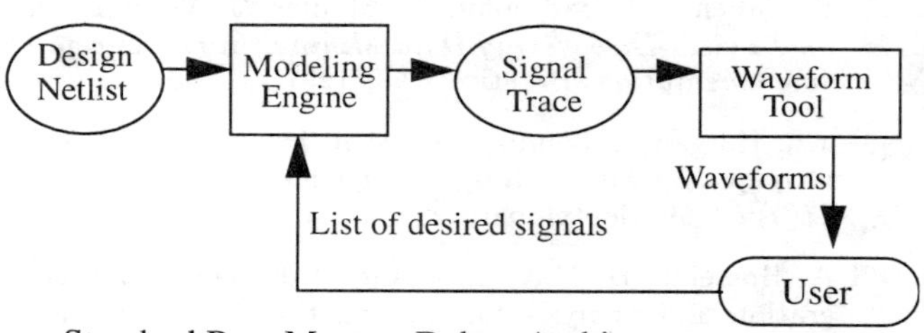

Standard Post-Mortem Debug Architecture

There are several reasons why this architecture is popular:

- Decoupling simulation from analysis speeds analysis.
- Long simulations can be run in the background.
- Expensive simulation licenses can be better managed by a group of designers, while relatively cheap waveform-viewer licenses can be purchased for everyone.
- Verilog's VCD (Value Change Dump) file format [2] has created a competitive market for waveform tools, greatly improving their quality. (Though each waveform vendor also has a proprietary optimized file format which is generated using PLI routines [3,10,12,13])
- The same waveform tool can be used with different simulators (event-driven, cycle-based, gate-optimized, RTL-optimized, analog, ...) and with emulation.

Performance Issues with Post-Mortem Debugging

The drawback of post-mortem debugging is the performance overhead of tracing a large number of nodes during simulation. The resultant VCD files (or vendor-specific optimized format) are very large; they take a long time to write during simulation, and a long time to read when debugging.

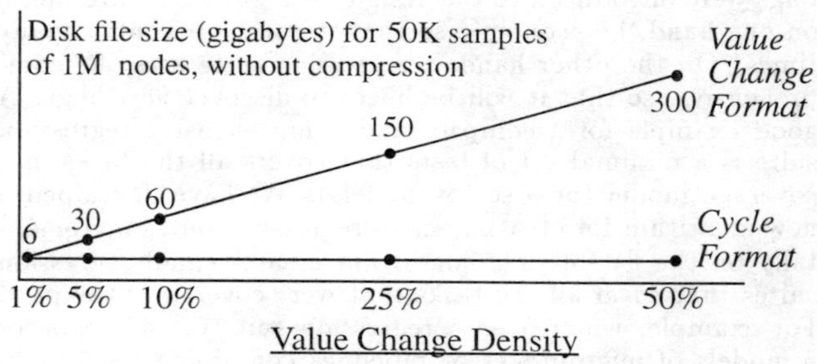

Depending on the value change density (defined here as the average number of events per signal per clock edge), a 50,000 cycle simulation of a million gate design can require tens, perhaps hundreds, of gigabytes. This assumes 64-bit (8 byte) time-stamps, 4-byte signal values, and no vendor-specific compression. Allowing for an order of magnitude or so reduction in file size for compression and storage of aggregate signals in one value-change, the file size is still going to be huge -- hundreds of megabytes to tens of gigabytes. This I/O bottleneck reduces the benefit of a fast simulator for the debug phase.

The storage requirement is reduced if two-state data is stored at all cycle boundaries (samples), rather than using timestamped value-changes. In this *Cycle Format,* about 6 gigabytes is required regardless of value-change density. This is still an awkward file to maintain, back up, write, and read. In addition, commercial waveform tools developed for timing simulations don't typically accept this format, and a cumbersome data-expansion and file-translation process would have to occur in order to debug.

Post Mortem Debugging in Emulation

An in-circuit emulator is a programmable hardware device that models a user design in the context of a *target system.* The target system may be the exact board for which the chip is intended, or it may contain modifications to facilitate emulation. The emulator is usually cabled to the target system as a surrogate for a chip that may not yet be fabricated.

The target system supplies clock inputs and other stimuli to the emulator. The emulator must respond to this stimuli in *real time.* If a configured emulator is determined to run at 1Mhz, then it can't be allowed to lag behind, or it will miss its inputs and fail.

Commercial emulators can also be used in a mode where the clock is controlled by a host computer. For the purposes of this discussion, we shall call this *hardware-accelerated simulation.* For the moment, we are concerned with real-time, target-system driven emulation.

Streaming samples of a million signals at 1Mhz into a disk drive is not practical, so RAM must be used, either in a commercial logic analyzer or on-board. In either case, 6 gigabytes of RAM is expensive, and difficult to route to all design nodes in real time. Consequently, emulation vendors are forced to compromise on either sample depth, visibility, or speed. If speed is compromised, then in-circuit emulation may be impossible due to target-system slowdown limitations, arising from dynamic logic, timed bus protocols, etc. If visibility or depth is compromised, then design debugging is hard.

Visibility Obscured by Optimization

Simulators and emulators may transform the design to one that is functionally equivalent, but whose intermediate nodes may be altered or optimized away. The user can either disable these optimizations or lose visibility. Again, performance is pitted against debugging.

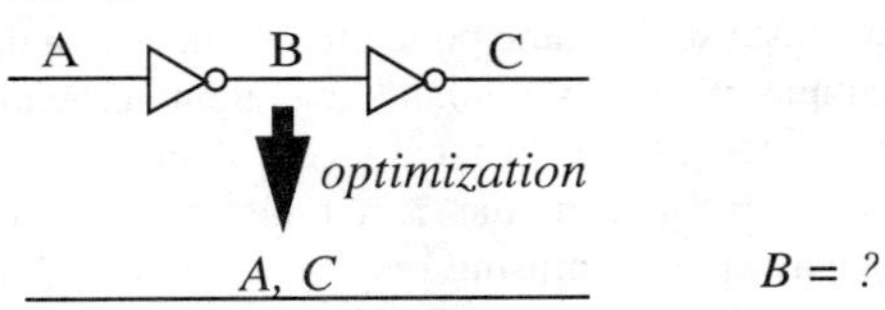

In this paper, we discuss the selection of a subset of nodes for tracing that allows all design nodes to be transparently *reconstructed on demand* in a commercial waveform tool. Reconstruction allows modeling engines to optimize designs without compromising visibility. It reduces real-time storage requirements enough to enable 100% visibility in a real-time emulation of a one million gate design for a time-window of 50,000 samples.

In section 3 we describe the software reconstruction techniques. In section 4 we discuss reconstruction in the context of functional simulation. In section 5 we discuss its application to an in-circuit emulator. In section 6 we discuss further work: event-based storage, reconstruction techniques with regard to timing simulation, and RTL-based modeling.

3. On-Demand Reconstruction

The premise of reconstruction is that when debugging, designers may need access to *any* signal in the design, but will not need access to *all* of them. Storing all the signals on disk or in memory is a waste of time and space. It is better to save a minimal subset of signals and interpolate the stored results only when the designer indicates interest in them. If this can be done quickly, the tools provide an illusion of having traced all nodes without incurring the overhead.

Using the waveform tool, the designer will find that any signal he chooses will be available for post-mortem analysis of its history. However, no designer will want to look at more than a small fraction of the signals in a design. No human can make sense of a million waveforms, and no waveform tool can display that quantity of data. So in all practicality, the enormous storage cost of full visibility will never be paid in full.

Choosing the Subset

In a digital design, it is simple to choose the subset of signals which must be traced during simulation or emulation:

- State Outputs (flip-flops, latches, memory)
- Primary Inputs
- Asynchronous Feedback Loops -- one signal in the feedback path must be traced.

This set of nodes comprises a *basis set* from which all others may be combinatorially derived. We call this set of nodes the *basis nets.* Armed with a trace of these nets, the *basis trace*, the reconstruction engine can deliver on the promise of full visibility without storing a full trace.

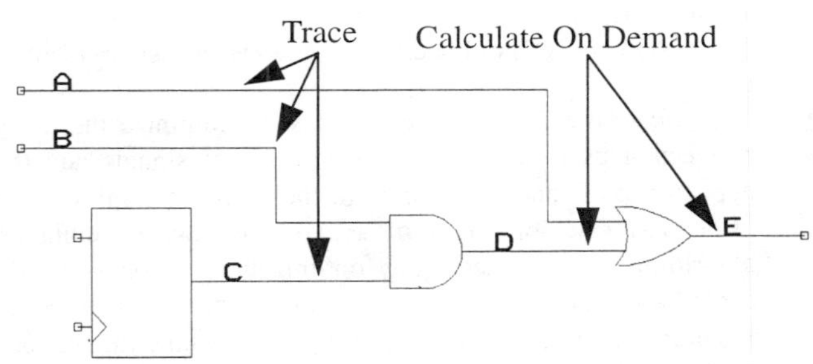

In large chip designs, we have observed that approximately 10% of all nodes are state outputs; 100,000 for a million net design. There are typically on the order of a thousand or fewer primary inputs, due to pin limitations on most chip packages. Most designs that are well-suited for fast functional simulation have a relatively small number of asynchronous feedback loops. To a first approximation, 10% of the design nodes need to be traced. The remaining 90% are combinational logic outputs that can be reconstructed on demand, and never stored to disk.

Reconstructing Combinational Logic

To provide an environment in which the combinational logic outputs can be reconstructed on demand from an interactive waveform

... debug architecture is modified:

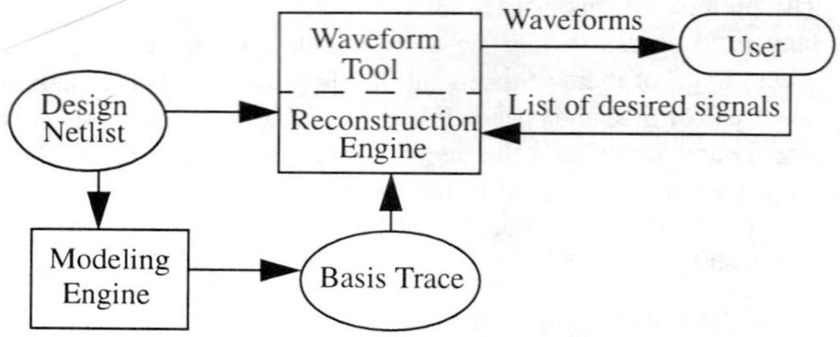

Modified Post-Mortem Debug Architecture

In this architecture, the waveform tool is linked to a *reconstruction engine* with access to the design netlist. When the designer selects a set of signals for the waveform viewer, a cone of logic is traced back to the basis nets. A set of equations is constructed from this cone of logic, and is applied to the stored logic and input traces. For the schematic above, the equations are:

- D = B & C
- E = D | A

The equation output is sent to the waveform tool for graphical display. There is no need to save this data on disk. Depending on the complexity of the logic cone, it might be reconstructed from an in-memory basis trace faster than it could be retrieved from disk.

Several issues must be addressed to ensure sufficient performance to maintain the illusion of a full trace.

- The design netlist must be kept in memory to facilitate rapid construction of logic cones for desired combinational outputs. The in-memory design representation must conserve memory usage to avoid running out of virtual memory on 32-bit machines when a large design is loaded.
- The design netlist should be loaded quickly to encourage interactive use.
- The equation evaluation must be fast to minimize the delay between a user's request for a new set of signals and the appearance of their waveforms in the graphics display. The evaluation resembles half of a two-phase cycle simulation algorithm [10]. Hence many optimizations associated with cycle-based simulators can be applied to reconstruction, including the use of two-state logic and machine instructions for logic operations.

If performance is poor, users will avoid using reconstruction, and will suffer the debugging pain of huge trace files, incomplete visibility, or a reduced number of samples.

Partial Reconstruction

In some cases, the designer may find that recording the entire basis trace is burdensome, and may wish to focus on one or more design subsystems. In this case, reconstruction is still very interesting.

By tracing the inputs to a subsystem, as well as the state outputs and asynchronous feedback points within the subsystem, full visibility within the subsystem can be achieved at a cost that's much lower than that of tracing all nodes. In a chip design, subsystem inputs are not bounded in number like the primary inputs are. However, the total number of them, in most cases, should still be manageable.

Waveform Tool Integration

The reconstruction engine needs to be very tightly integrated with the waveform tool. This avoids a potential bottleneck in conveying the large amount of waveform data for combinational outputs to the graphics engine. It also avoids a user interface glitch where a user might have to manually transfer the set of desired signals from the waveform tool to the reconstruction engine, followed by a disk write and file reload. If this flow is not transparent to the user, it shatters the illusion of having traced all design nodes and makes debugging cumbersome.

Ideally, the waveform tool selected should be one the designer is already accustomed to using. Otherwise, established debugging habits must be broken, making the solution less effective. Unfortunately, this requirement is at odds with the tight integration. There is no standard API for waveform tools.

At the expense of speed, VCD files can transmit waveforms from the reconstruction engine to most waveform viewers. If the waveform tool supports a scripting language such as TCL, the interaction between the waveform tool and the reconstruction engine can potentially be hidden, avoiding the UI glitch.

Risks

Aside from performance, the biggest risk in using a reconstruction engine is that an inconsistency with the modeling engine (simulator or emulator) could result in incorrect data being displayed. Potential pitfalls include:

- Behavior of undriven nets
- Behavior of multiply driven nets
- Behavior of temporally undriven tri-state busses
- Behavior of unstable asynchronous feedback loops

When two-state logic is used, the modeling and reconstruction engines must use the same policy for dealing with these conditions. For example, if a two-state modeling engine grounds undriven nets and lets tri-state busses float to 1 when all drivers are off, the reconstruction engine must do the same thing. Otherwise, the waveform display may show confusing results, where the Q output of a flop does not match its D input following a clock edge.

4. RECONSTRUCTION IN FUNCTIONAL SIMULATION

The main benefit of reconstruction for functional simulation is that it creates a more favorable speed-visibility curve.

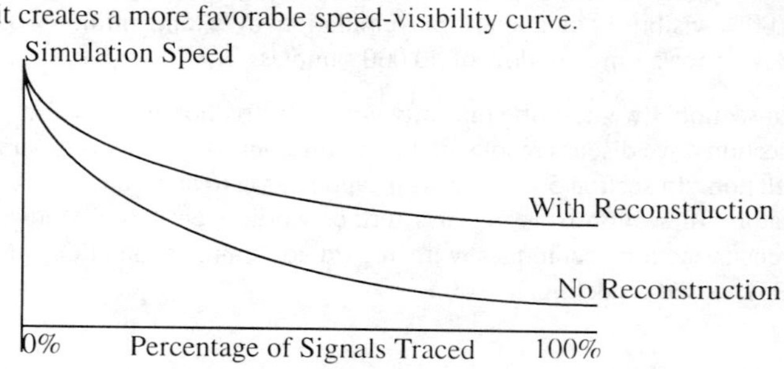

Sim Speed vs. Visibility, with & without Reconstruction

In simulators where combinational outputs are optimized away, or converted to Binary Decision Diagrams [6,9], reconstruction is *required* to achieve full visibility, even for a small number of samples. Reconstruction is thus ideal for cycle simulators. It is particularly compelling when the simulator transforms the design into a functionally equivalent one, so that "lost" signals can be recovered during debug.

However, reconstruction can be easily used for event simulations without timing. Of course, if the simulator provides multi-valued logic (e.g. 0, 1, X, Z), then the reconstruction engine will have to perform equation evaluation using table lookups rather than machine instructions.

Beyond the reconstruction engine and simulator, a trace-selection mechanism is needed. It takes a list of interesting subsystems from the designer (possibly the entire design), finds the appropriate basis nets, and enables tracing for those nets in the simulator. In Verilog, this could be implemented via a PLI call and controlled by the user from the test bench.

Simulation Results

We integrated a reconstruction engine with Cadence Verilog-XL 2.3.3. We tested it using three gate level designs synthesized to a library of timing-free primitives. The designs have multiple asynchronous clock domains, asynchronous feedback loops, wide multi-ported RAMs, including asynchronous FIFOs, and include both latches and flops. We simulated each design four times to determine how tracing impacts performance:

- No Trace - no signals are traced, no file is saved
- VCD Trace - write standard Value Change Dump file [2]
- VPD Trace - write Virsim's optimized binary format [11]
- Basis Trace - write cycle-based, compressed binary file with values for basis nets only.

We used an unloaded Sun Ultra 2/200 with 1 gig of memory, spooling trace files to a file-server over NFS. The simulations fit comfortably in physical memory.

Table 1: Simulation Time (mm:ss) / File Size

Design	25k vectors 200k nets	650 vectors 320k nets	5k vectors 12k nets
File Size			
VCD Trace	2100 Meg	167 Meg	35 Meg
VPD Trace	579 Meg	65 Meg	19 Meg
Basis Trace	12 Meg	10 Meg	1 Meg
Simulation Time			
No Trace	16:00	4:30	0:27
VCD Trace	54:00	6:58	1:08
VPD Trace	61:00	14:17	1:40
Basis Trace	30:00	11:01	0:54
Number Basis Nets	22,000	39,500	950

The times listed are measured in wall-clock time via the Unix "time" command. The number of nets given includes the internals of library elements, which may not be needed for debugging. The simulation time includes netlist compilation, which is about a minute. The file sizes given each include space required for design hierarchy. A network disk was used for all software, design, and output files.

From this data, we conclude that using the basis trace approach has a profound impact on trace file size. But simulation speed is very design and data dependent.

File Size Conclusions

The basis trace files were consistently dramatically smaller than the VCD and VPD files. This arises from storing fewer signals, and the use of a two-state cycle-based format rather than value changes. X and Z values are stored as exceptions.

Simulation Speed Conclusions

In two of the three designs tested, using the reconstruction engine improved simulation performance. In the 200k net design, the basis trace simulation ran twice as fast as VCD or VPD. In the 12k net design, it simulated 30% faster than VCD and 60% faster than VPD. Some of the speed improvement may be due to the reduced file system overhead for the smaller files, particularly using network file servers. Another source of speed improvement is the need to examine the values of about 90% fewer nets. The ratio of 1 basis net for every 10 nets in the design appears to hold consistently.

In the 320k net design, the basis trace simulation was slower than the VCD trace. This is a large design with a fairly small number of vectors. Much of the design is not covered by these vectors, and remains dormant. This favors an event-based storage approach, where we are employing a cycle-based mechanism. Thus, while the basis trace file is smaller than the VPD or VCD files, the ratio is 6:1, rather than the 50:1 or 20:1 that we see in the other designs.

This data dependency is an artifact of the current implementation, rather than a fundamental limitation of reconstruction. An event-based basis-trace storage mechanism optimized for four-state logic would likely yield consistently superior results to VPD and VCD, both in file size and simulation time.

Waveform Results

We used Summit Design's Virsim waveform browser, with its optimized VPD trace file, to characterize debug performance. Virsim was also used with the reconstruction engine and the basis trace file to compare performance, using a tight API-based integration

When using the basis trace to analyze waveforms, a Netlist database file must be read. The size of the database for the three designs is given in Table 2. The time to read this file is included in the basis trace invoke time. The netlist connectivity information is required by the reconstruction engine to recover values for untraced nodes on demand.

Data is not included for Virsim displaying data from VCD files because the it does not read VCD files directly; a translation tool is used to convert to the vendor's format.

Several operations were timed, in order to convey a feel for system performance in typical debug scenarios. The clock nets, busses, and modules chosen mostly had a medium to high value change density, although this was not measured.

Table 2: Waveform Analysis Speed

Design	25k vectors 200k nets	650k vectors 320k nets	5k vectors 12k nets
Invoke Time			
VPD Trace	13 sec	22 sec	10 sec
Basis Trace	30 sec	40 sec	7 sec
Netlist DB file size	25 Meg	30 Meg	2 Meg
Add 3-4 clock nets			
VPD Trace	12 sec	1 sec	1 sec
Basis Trace	3 sec	3 sec	1 sec
Add 32-bit bus			
VPD Trace	1 sec	1 sec	1 sec
Basis Trace	1 sec	1 sec	1 sec
Add 2-3k nets			
VPD Trace	27 sec	13 sec	
Basis Trace	22 sec	20 sec	
Add 5-10k nets			
VPD Trace	28 sec	65 sec	22 sec
Basis Trace	47 sec	11 sec	15 sec
Memory Footprint			
VPD Trace	118 Meg	120 Meg	47 Meg
Basis Trace	115 Meg	114 Meg	52 Meg

From this we conclude that there is comparable performance between a debug environment where all signals are traced in a large VPD file, compared to reading a smaller basis trace file and reconstructing the combinational outputs. Which method is faster is dependent on the number of equations that must be evaluated in order to reconstruct the desired nets, locality of reference within the large trace file, etc.

For large designs, basis-trace invocation time is penalized because of the netlist database read step. The cost of this is justified when the number of vectors is large, or when multiple simulation runs are debugged in the context of a single netlist revision.

The memory footprint data was noted after all the operations in Table 2. This verifies that holding the design netlist in memory is not problematic.

The main benefits of reconstruction in simulation are:

- It can increase simulation performance without reducing visibility (or increase visibility with minimal speed loss.)
- It creates dramatically smaller trace files.

5. RECONSTRUCTION IN EMULATION

Commercial emulators today offer the user several choices for design-node visibility:

- Most emulators allow a probe-set of 1k-16k nodes for 10k-100k samples using external or internal logic analyzers [4, 7,8]. The probe-set is hard to choose prior to going "in circuit". When a problem is detected, it may be hard to debug if you didn't choose wisely. Re-running the test may prove difficult if the erroneous behavior was due to a combination of internal state and real world stimulus that is not easily repeatable.
- Some emulators provide a mechanism for retrieving the value of all nodes for one sample by using shadow registers and a "readback" feature available on some FPGAs [8,14]. It is often difficult to debug problems from a single snapshot. You can answer "What?" but not "Why?"
- Record all nodes for all time by slowing the emulator to reduce bandwidth requirements to bulk storage. It is unlikely that such a configuration would run fast enough to respond to a target system, so we label this *simulation acceleration*.
- Some emulators allow rapid reconfiguration of the probe-set, without taking the emulator out of circuit. This is convenient, but you still may have the wrong set of probes selected when real-world stimulus causes erroneous behavior, and you may not get a second chance. You cannot rewind the real world.

Why is 100% Visibility in Emulation Hard?

The fundamental limitation of visibility in emulation is *bandwidth*. Consider an emulation of a million gate design, running at 1Mhz. Assuming that only 2-state values are recorded on clock edges, brute force 100% visibility of a 50ms (50,000 samples) snapshot would require substantial resources:

$$Storage = 10^6 nodes \times 50ms \times 1Mhz = 5 \times 10^{10} bits$$

$$Bandwidth = 10^6 nodes \times 1Mhz = 10^{12} bits/\sec$$

The high storage requirement (6.25 gigabytes) makes it impractical to satisfy with fast RAM. The high bandwidth requirement makes it difficult to transmit to a disk in real time. Recall that the emulator cannot be allowed to pause at any time for saving data because it must respond in real time to an external clock.

Using the reconstruction engine cuts down the nets we need to trace by 90%, reducing the trace memory requirement to 625 megabytes. This is a more manageable quantity of memory to distribute in an emulator box.

The bandwidth requirement is also cut down by 90% to 100 gigabits/sec. This can be spread across the emulation FPGAs or processors,

There are ways to further reduce the fast RAM requirements without compromising visibility, depth, or emulation speed. They are beyond the scope of this paper.

Limitations

Thanks to reconstruction, visibility of all nodes in a real-time emulation is possible. But the sample depth available is still limited by the depth of the trace memory in the logic analyzer.

6. FUTURE WORK

Experiment with Event-based Storage Mechanism

The results obtained using the two-state optimized cycle-storage mechanism indicated uniformly superior file-size, but simulation speed has proved more data dependent. An event-based storage mechanism should be incorporated into the basis trace and reconstruction engine to see the impact on simulation speed and file size over a wide range of value change densities and a predominance of 4-state logic. We believe that this will result in a consistent speed and file-size advantage over a full VCD or VPD trace.

Reconstruction with RTL Modeling Engines

There are several species of RTL modeling engines. The first is based on direct compilation or interpretation, where the RTL code is compiled to C, native code, p-code, or emulation processor instructions [7]. The second is compile-to-gates, where a fast synthesis engine is used to translate the RTL description to a gate-level description with a translation table between RTL names and gate names.

In compile-to-gates, the reconstruction engine can be used with little modification. A translation table should be used in the waveform viewer so that the user is presented with familiar design names.

To support direct compilation, the reconstruction engine could use a Control Data Flow Graph in place of the logic cone to generate the equation list for the desired nodes. This graph can be used prior to simulation to enumerate the basis nets. Alternatively, with the extra semantic knowledge available from an RTL flow analysis, a smaller basis set can potentially be found, offering further reductions in trace overhead [5].

Timing Simulation

Adding timing capabilities to the reconstruction engine is not prohibitively complex, although it's not clear whether the result would be fast enough to be usable interactively.

In a timing simulation that includes back-annotated net delays from layout, state elements will all change when their clock edge arrives, and each state element may see its clock edge at a slightly different time. A timing accurate waveform display will need to reflect this.

Consequently, the basis trace would have to be value-change based, rather than cycle based. As we have noted above, this approach may prove superior in any case.

The equation-list would need to be augmented with a delay map from the basis nets to the user's desired combinational outputs. This would allow reconstruction to report the correct transition times for each signal to the waveform viewer.

If the user requests waveforms for a large number of signals simultaneously, it might be prohibitively slow to evaluate all equations every time any state element transitions. We may then want to partition the logic cone in some fashion, triggering the evaluation of each a given partition when any of its state inputs change. This *selective trace* methodology [1] is another step along the spectrum from pure cycle-based simulation to pure event-based simulation.

The question of whether the end result of these changes will be fast enough may be superseded by the question of whether designers wish to debug functionality in a timing simulation environment.

7. CONCLUSION

The reconstruction engine presented here provides an opportunity to dramatically improve the debugging environment for both fast functional simulation and emulation. Simulation users can see more of their design with reduced tracing overhead. Emulation users can achieve 100% visibility with significant depth in a real-time emulation. This represents an opportunity for unprecedented debugging power of hard-to-reproduce system-level problems.

8. ACKNOWLEDGEMENTS

Thanks to Jason Campbell, Pat Julik, and crew at Summit Design for providing the tight integration path into Virsim and for extensive support and cooperation. Thanks to Charley Selvidge, Mark Seneski, and others at Ikos for ideas, infrastructure, test cases, and draft review.

9. REFERENCES

[1] M. A. d'Abreu, "Gate Level Simulation", IEEE Design and Test, pp 63-71, December, 1985A.L. Sangiovanni-Vincentelli et al., "Verification of Electronic Systems", 33rd DAC, pp 106-111, 1996

[2] Cadence Design Systems Inc., "Verilog-XL Reference Manual", Version 2.3, 1995

[3] Design Acceleration Inc., Signalscan Database, http://www.designacc.com/Products/sst2db/sst2db.html

[4] Ikos Systems Inc, VirtuaLogic Data Sheet, http://www.ikos.com/products/vsli/index.html

[5] D. Kirovski, M. Potkonjak, "A Quantitative Approach to Functional Debugging", ICCAD 1997, pp 170-173

[6] P.C. McGeer, et al, "Fast Discrete Function Evaluation using Decision Diagrams", ICCAD 1995, pp 402-407

[7] Quickturn Design Systems, Cobalt Data Sheet, http://www.quickturn.com/products/cobalt.htm

[8] Quickturn System Realizer Data Sheet, http://www.quickturn.com/products/systemrealizer.htm

[9] A.L. Sangiovanni-Vincentelli et al., "Verification of Electronic Systems", 33rd DAC, pp 106-111, 1996

[10] Speedsim Inc, Cycle simulation technology, http://www.speedsim com/technology/cbs.htm

[11] Summit Design Inc, Virsim Data Sheet, http://www.sd.com/products/verification/virsim.html

[12] Systems Science Inc, Magellan Data Sheet, http://www.systems.com/

[13] Veritools, Undertow Data Sheet, http://www.veritools-web.com/towv.htm

[14] XILINX, Inc., "The XC4000 Data Book", Aug. 1992

Virtual Chip: Making Functional Models Work on Real Target Systems

Namseung Kim, Hoon Choi, Seungjong Lee, Seungwang Lee, In-Cheol Park, Chong-Min Kyung

Department of Electrical Engineering

KAIST, Taejon, 305-701, Korea

Abstract

As design complexity increases, functional verification becomes a crucial issue to ensure design correctness at an early design stage. Traditional methods for verifying functional designs are based on the HDL simulation, which is becoming the bottleneck of the design cycle because of the increasing design complexity. The accurate verification ability at the architectural level through a large set of the test programs and real world applications is a foundation for the next design step. In this paper, we describe how to verify a functional model on a real target system. The proposed methodology called *virtual chip* makes it possible not only to check the functional correctness on real systems, but also to explore design space by measuring the performance effectiveness of various architecture parameters under real applications. Experimental results show that functional models can be verified on real systems using complicated application programs. The proposed functional verification method is faster than HDL simulation and even comparable to emulation.

1. Introduction

According to advances in design methodology and semiconductor technology, the complexity of a design is so rapidly increasing that more than one million transistors can be integrated on a single chip. This makes it difficult to verify such a complex design. In addition to the design complexity, lifetime of a design becomes shorter than ever before. It is very important to reduce the design time as minimal as possible in order to fit the design in a narrow time-to-market window. Thus, an efficient verification methodology is essential to make the first silicon chip work on its target system. This is the major reason why verification receives significant attentions from academia and industries.

It is well-known fact the cost associated with debugging increases exponentially if the bugs are found at the later design stage. This means that a trivial design bug at the functional level design stage leads to critical defects in the market. Therefore, we have to test the functionality of the design at an early design stage in order to avoid critical problems that may be amplified at the later design stage. In addition, we need to guarantee the functionality on the target system where the designed chip will work. To reduce the total development time, the chip and its target system should be concurrently developed [1]. To verify the functionality on the target system is an important problem to be solved, but it is quite difficult because only a functional model not adequate for the board-level testing is available at an early design stage.

Previous functional verification methodologies rely on the test vector simulations and the consistency check between design stages [3,4]. Even with a great deal of effort for verification, some design bugs are often detected while running applications on the fabricated chip [3]. This suggests the necessity of executing the real world applications. Since there are many obstacles in simulating real world applications, it is practically difficult to simulate the functional design for such a complex application, if not impossible. To overcome the difficulty in simulating real world applications, emulation has received spotlights in state-of-art chip designs [2,3,8,9]. However a gate-level design of the chip needs to be completed before starting the emulation. Thus this method can be employed only at the final verification stage.

To run real world applications with simulation, we need to construct a HDL simulation environment including system models and system model interfaces such as various kinds of peripherals. An accurate simulation model is inevitable to execute operating systems and application programs perfectly. Building the accurate models of simulation environment is as difficult and time-consuming as the chip design. Moreover it is really difficult to guarantee the correctness of these modeling with a conviction. Some system models have been applied only to limited area such as checking the protocol of a multiprocessor [1]. Thus, we need a verification method that can test the functional design on target systems with directly using all hardware peripherals instead of software peripheral models.

There has been significant effort to reduce the development time [4,6], but previous works only focused on the reduction of simulation time and on the automatic synthesis of a gate-level design. There are no reported works that have attempted to simulate a functional model with a real target system in order to convince the functionality of the design. Needless to say, significant reduction of the development time can be achieved from the correct design specification of the desired function that is conformed on the target systems. In this paper, we propose a new functional verification methodology called *Virtual Chip* that enables the verification of functional models under real target systems and real application programs at an early design stage. The proposed methodology is also applicable to measuring the effectiveness of design alternatives to be explored.

The reminder of this paper is organized into four sections. In the following section, we introduce the concept of *Virtual Chip* and describe the design partition problem between hardware and software components of the virtual chip framework. In section 3, we present a virtual Pentium chip implementation as an example. We compare the performance of the proposed concept with other verification frameworks in section 4.

DAC 98, San Francisco, California

2. Virtual Chip

In this section, we introduce a new concept called *Virtual Chip* to make a functional model work on a real target system. A virtual chip is a functional model of a chip to be designed, whose external signals are supported by hardware. As shown in Fig. 1, a virtual chip consists of two modules; the first module is the Pin Signal Generator (PSG) that generates and receives external pin signals of the chip to be designed, and the second one executes the functional model of the chip on a host computer. The PSG implemented in hardware using a rapid hardware prototyping method such as FPGA (Field Programmable Gate Array) can be regarded as the real chip on the target system, but its objective is only to imitate the pin signals of the real chip. Actual functional processing of the chip is performed in the second module that is implemented in high level languages such as C.

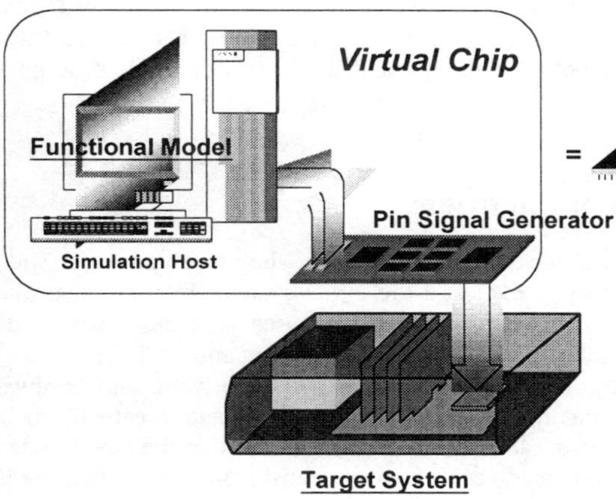

Figure 1: Block Diagram of Virtual Chip

The second module running on the host computer is composed of Functional Interfacing Module (FIM) and Functional Processing Module (FPM), as shown in Fig. 2. FPM contains core functional model of the chip to be designed. The role of FIM is to interface FPM and PSG. How they work together is described below. Whenever FPM needs a resource access to the target system, it sends a request to PSG, which emulates pin signals of the real chip, through FIM. When PSG receives a request from FPM, it translates the request into real chip signals according to the desired chip specification. Also PSG informs the target system's requests to FPM through FIM when it receives a request for the processing results of FPM from the target system. Since the external I/O bandwidth of the simulation host is usually smaller than the PSG-to-target system bandwidth, FPM cannot send/receive all the information to/from PSG at a time. Therefore, to manage the I/O bandwidth difference, FIM is required for information transfer between FPM and PSG.

Although a virtual chip might be slower than a real chip, we can make the functional model work on the target system with this method. The only difference between a real chip and a virtual chip is that processing speed of the former is much slower than the latter. But this does not mean that the virtual chip is analogous to the real chip working at slower clock frequency, because the external pin signals generated by the PSG can be made as fast as those of the real chip working in normal speed. Since external pin signals generated by hardware keep similar speed to those of the real chip for each memory access, the virtual chip can play the same role as the real chip if the target system does not depend on processing speed.

Henceforth we focus on how to design and implement the PSG and the FIM because how to model the functionality of a design has been well explored and researched.

2.1 Pin Signal Generator

The PSG is logically consisted of three units: Internal Interface Unit (IIU), External Interface Unit (EIU) and Operational Buffer Unit (OBU). IIU plays a role of interfacing with the FIM, EIU interfaces with the target system, and OBU handles the difference of operating frequency and I/O interface channel bandwidth between the FIM and the target system. Figure 2 shows a block diagram of the PSG.

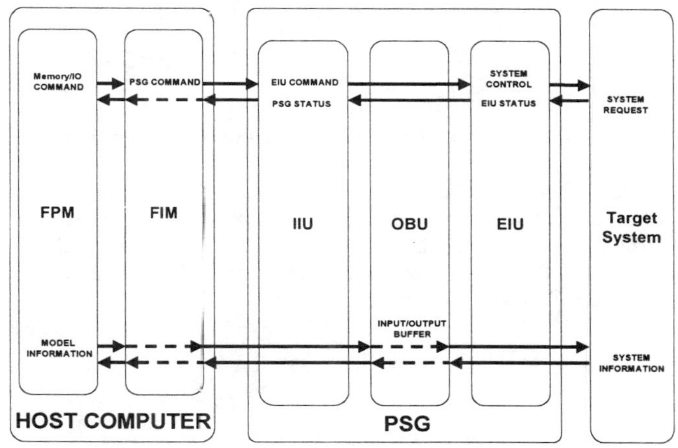

Figure 2. Block Diagram of Pin Signal Generator

2.1.1 Internal Interface Unit

Figure 3 illustrates the operation of Internal Interface Unit (IIU) in detail. IIU receives PSG control commands and information needed to generate one bus cycle from FIM. When IIU receives a command containing what bus cycle type is to be performed, it keeps the command in a temporary register. After that, FIM sends information needed for the bus cycle. In this case, IIU controls Operational Buffer Unit (OBU) to receive the information. Since the IIU knows the current bus cycle type, it can decide how many information transfers are needed. If the information received is sufficient, IIU invokes EIU to execute the bus cycle by passing the stored command. Also the IIU receives system requests and bus information through EIU and OBU, then sends these to FPM through FIM.

Since IIU is closely working with FIM, IIU is synchronized with the host computer that executes the FIM. However, EIU should work at the clock frequency of the target system to generate proper pin signals. As these two clocks are usually very different, IIU communicates with EIU using an asynchronous handshaking protocol as mentioned in section 2.2 in detail.

2.1.2 External Interface Unit

Figure 4 represents External Interface Unit (EIU). Once EIU receives an operational command from IIU, it generates control signals and bus signals that should occur in a proper sequence. In addition, it monitors the status of the target system and operational status of the EIU itself. EIU transfers bus information of the target system to IIU through OBU, and sends directly operational status of the target system and EIU to IIU via a status communication channel.

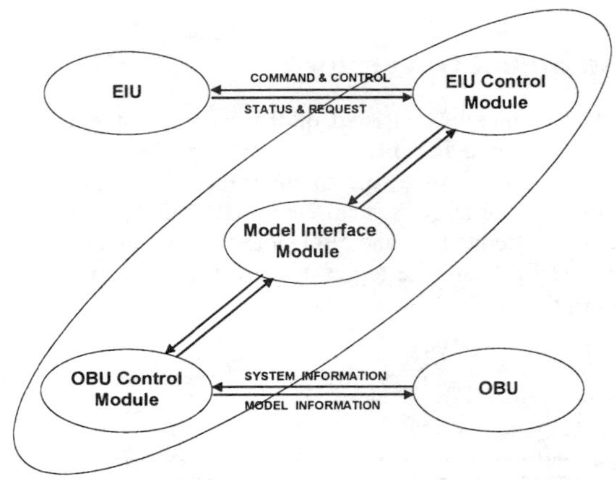

Figure 3. Block Diagram of Internal Interface Unit

One of the most important purposes of the EIU is to emulate external I/O signals of the chip to be designed in order to make the virtual chip be recognized as an real chip. Therefore EIU is implemented in hardware whose functionality and external I/O interfaces are identical to the planned chip.

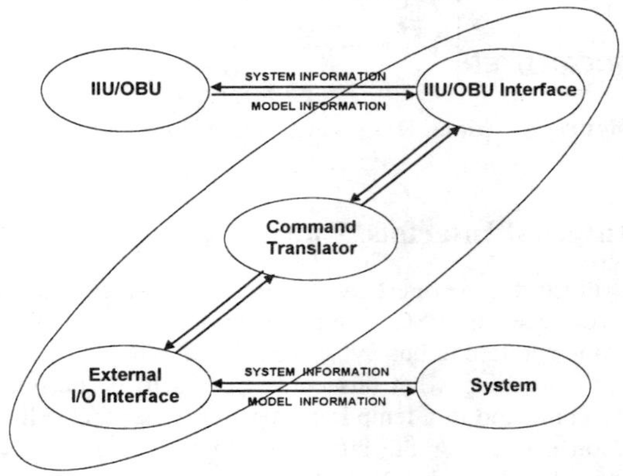

Figure 4. Block Diagram of External Interface Unit

OBU plays a considerable role in managing the inconsistency of operation frequency between FIM and the target system. Furthermore it buffers communication channel bandwidth difference among PSG, FIM and the target system.

2.2 Functional Interface Module

Because of the operating frequency and I/O bandwidth difference between FPM-to-PSG and PSG-to-target-system, we need interface software that schedules and controls communication between FPM and PSG. FPM sends behavioral information and receives target system information through the virtual chip port access routines that are part of the FIM.

There are two communication mechanisms we can use: polling and interrupt. We can apply the polling to obtain the status of the PSG and the target system using communication registers. With a definite time interval, PSG monitors and stores the status of the target system and the PSG itself. When required, the functional model reads the communication registers in PSG through FIM. When there is an urgent system request such as system interrupt and reset, we can use an interrupt mechanism to inform the functional model. If PSG receives a system interrupt from the system board, it interrupts FIM and notifies to FPM that a system interrupt occurred. We can replace the interrupt with the polling, but there are some performance degradation and system consistency problems if the FIM has to poll for unexpected system requests.

We can make the functional model relatively independent to the external systems because FPM can communicate external system through FIM and it hides physical I/O systems such as the target system and PSG. With this feature, the concept of virtual chip can be used as a general verification method that can be applied to various design stages as well as the initial functional verification stage. However this is closely related to the hardware/software partitioning between the virtual chip and its design description model.

2.3 Design Partition

The more complex a design becomes, the more transfer traffic is required between FIM and the target system. In addition to that, the design complexity of PSG increases. Therefore we need a proper partitioning between the PSG design and the FIM.

We can consider a trade-off between software and hardware module to obtain efficiency in PSG design and eventually in the verification using the virtual chip. In a case that the design size is large and system is very complex, transfer traffics, between FIM and PSG, required to control the PSG and the target system increase significantly. However I/O bandwidth between FIM and PSG is usually fixed if the FIM is running on a commercial computer, and is usually less than that between PSG and the target system in most cases. As a result, increasing the number of I/O operations between FIM and PSG degrades performance of a virtual chip. To enhance virtual chip performance, we have to reduce I/O traffic between FIM and PSG as few as possible.

One feasible method to reduce the traffic is to make the PSG undertake considerable portions of the FPM's role at the expense of increasing the PSG design complexity such that it can generate a bus cycle automatically with minimum guidance from the FPM.

3 Case Implementation : Virtual Pentium Processor

We have applied the proposed virtual chip concept to implementing a virtual Pentium processor. The FPM and FIM of Pentium processor were described in C language, and the corresponding PSG was designed in Verilog HDL. Finally, we implemented the PSG using 4 ALTERA FPGAs.

Since Pentium processor has 29 address pins, 64 data pins and tens of control I/O pins, PSG and FIM require around one hundred bits of information to request one bus cycle to the target system (Pentium Motherboard) [8]. Accordingly a number of I/O communications between the FIM and the PSG is required for one memory access, because our design was based on 16-bit ISA bus.

For communications between the FIM and the PSG, we used both the polling and interrupt mechanism. To obtain information on system status and PSG status, the polling mechanism is employed, and the interrupt is used to notify an urgent system request to the FPM. The IIU is designed to work with minimal ISA operations so that I/O traffic and its design complexity are significantly reduced. We entrust the FIM for the rest of the IIU control function.

The EIU was designed to support Pentium processor's basic bus cycles and external I/O interfaces. The bus cycles supported are non-pipelined read/write cycle, lock cycle, and interrupt acknowledge cycle. With these bus cycles, we could perform the functional verification of our Pentium processor C model.

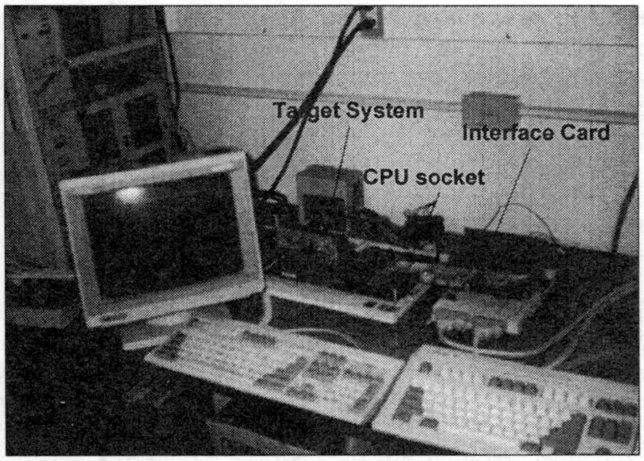

Figure 5: Close up of Virtual Pentium Processor

The operating frequency of Pentium processor functional model is in a range of one hundred kHz, so that, to interface between the PSG and the target system, we reduced system clock rate to 500kHz for the functional model to process Pentium system's requests properly. There are two major problems in testing the PSG, the FPM and the FIM under the normal operating frequency (i.e., at 133MHz or 166MHz). If the frequency of a system board is more than 10MHz, the timer interrupt request occurs so frequently that FPM consumes all of its simulation capability to process the timer services. The other problem is that EIU implemented using FPGAs could operate up to 10MHz. Therefore the operating frequency of system clock is limited to 10MHz. For the host computer, we used a Pentium-Pro 200MHz system

4. Result and Discussion

We could verify the functional model of Pentium processor using the virtual Pentium processor. Several applications including DOS are successfully executed on the virtual Pentium processor, as shown in Fig. 5. The IIU is designed to operate about 8MHz which is the same frequency of ISA-bus clock, and the counterpart EIU at 500kHz-system-clock frequency.

Sim/Emulator	Run Platform	Run Time	# of Inst./sec
VCS (RT-Level)	UltraSparc-II	24 hrs	12
Quickturn	1M3000 Box	25 min	13K
Virtual Chip	Pentium-Pro	140 min	2.4K

Table 1: Performance Comparison

To count the number of instructions executed per second for each verification system, we executed about one million instructions for VCS simulation and twenty million instructions for Quickturn and Virtual Chip. Table 1 shows that the virtual processor is more than 100 times faster than HDL simulation. These results suggest that the virtual chip can be successfully applied to functional level verification system at early design stage, and can be regarded as an emulator in functional level. This is a very different concept to the traditional emulators working with a gate-level design.

5. Conclusion

In this paper, we have described the importance of functional verification at an early design stage, and proposed a new concept called virtual chip in order to verify functional models on the target systems. The virtual chip is composed of hardware and software components: the hardware unit called PSG is responsible for generating pin signals of the chip to be developed, while the software unit simulates the behavior of the chip. Besides verifying functional correctness at the early design stage, we can evaluate the performance of the target design for various design parameters and determine the future design strategy and configurations.

The PSG interfaces the target system and the functional model that are belonging to completely different test domains, and presents fast and convenient simulation and verification environments. As an application of virtual chip, we have implemented a virtual Pentium processor and could successfully verify our behavior C model on real system boards.

6. References

[1] Lawrence Yang, David Gao, Jamshid Mostoufi, "System Design Methodology of UltraSPARC-I", Proc. 32th Design Automation Conference, Jun. 1995.

[2] Val Popescu, Bill McNamara, "Innovative Verfication Strategy Reduces Design Cycle Time for High-End SPARC Processor", Proc. 33th Design Automation, Jun 1996.

[3] Micahel Kantrowitz, Lisa M. Noack, "I'm Done Simulating; Now What? Verification Coverage Analysis and Correctness Checking of the DEC chip 21164 Alpha microprocessor", Proc. 33th Design Automation Conference, Jun. 1996

[4] James Monaco, David Holloway, Rajesh Raina, "Functional Verification Methodology for the PowerPC 604 Microprocessor", Proc. 33th Design Automation Conference, Jun. 1996

[5] Gopi Ganapathy et al., "Hardware Emulation for Functional Verification of K5", Proc. 33th Design Automation Conference, Jun. 1996.

[6] Joon-Seo Yim, Yoon-Ho Hwang et al., "A C - based RTL Design Verification Methodology for Complex Microprocessor", Proc. 34th Design Automation Conference, Jun. 1997.

[7] "Pentium Processor User's Manual", Intel, 1993.

[8] James Gateley et al. "UltraSPARC-I Emulation", Proc. 32th Design Automation Conference, Jun. 1995.

[9] Francoise Casaubieilh et al. "Functional Verification Methodology of Chameleon Processor", Proc. 33th Design Automation Conference, Jun. 1996.

PANEL: HARDWARE/SOFTWARE CO-DESIGN: THE NEXT EMBEDDED SYSTEM DESIGN CHALLENGE

Chair: Pete Heller, Consultant - Collett International, Inc., Santa Clara, Calif.
Organizers: Diane Orr and Kristin Hehir - Tsantes & Associates, Inc., Campbell, Calif.

With the proliferation of consumer electronics, the number of embedded systems is growing dramatically, according to Collett International Inc.'s research. At the same time, embedded systems are growing in size and complexity. Another major trend is to increasingly implement functionality in software. Design teams are using software to differentiate products, increase flexibility, respond to changing standards, enable inexpensive upgradability, and get products to market sooner. This confluence of forces is causing design teams to confront a host of new challenges and to even re-evaluate their fundamental design practices. The challenges of designing hardware and software in concert are now coming to the forefront. Some of the topics to be discussed include: Short of producing synthesizable output, can hardware/software co-design provide meaningful value? Can automatic hardware and software partitioning tools replace engineering judgement? Who is responsible for hardware/software co-design? System architects? Hardware designers? Software designers? Can today's tools truly optimize system functionality and performance tradeoffs between hardware and software implementation? Are commercial real-time operating systems impacting hardware and software tradeoff decisions?

James A. Rowson, Alta Group of Cadence Design Systems, Sunnyvale, CA

The consumer-driven, low-cost, fast time-to-market mentality that pervades electronics design these days has begun to force many companies to rethink their methodology for product development. A new consensus on design methodology is beginning to emerge where the functionality of the product is described separately from the architecture of the product. Many alternative mappings of function onto architecture must be explored quickly in order to get a cost-effective implementation. This function-architecture co-design will utilize reuse to sharply improve design time. Synthesis will be used to glue large IP together. The central task in function-architecture co-design will actually be the design exploration, performance analysis phase where the feature, cost, performance, and power will all be defined.

Guido Arnout, CoWare, Inc., Santa Clara, CA

Whether you call it an embedded system, a system-on-a-chip, a tightly integrated consumer device or simply a system, one thing is becoming clear to chip designers: systems, large or small, are implemented as a combination of software and hardware. We all know what it takes to develop software and hardware independently of each other. So it appears that hardware/software co-verification is essential in those cases where hardware and software must be designed closely together. But, what exactly are we co-verifying? Co-verification implies that we first have fully implemented the part of the system that will be executed as software running on a processor and what will be implemented as dedicated hardware. But, how do we get to the right partitioning? Don't we first have to define what the system is supposed to be doing before we dabble in hardware/software partitioning, and its performance tradeoffs, before we try to synthesize anything, before we co-verify the result of all this work? The key to co-design is to first describe and fully debug the complete system functionality as an architecture-independent executable specification that can be re-used throughout the implementation process. Only then do partitioning, synthesis and detailed co-verification really come into play.

Fred Rose, Honeywell, Inc., Minneapolis, MN

Much of the tool development work for embedded hardware/software co-design focuses on reasonably small, high-volume systems (such as consumer applications). In larger embedded systems (such as avionics), many factors affect the choice of partitioning functions across a hardware and software architecture. These factors include fault tolerance, redundancy, reliability, life cycle costs, legacy hardware and software issues, and new capabilities such as configurable computing and multiple domains such as optical and MEMS devices. System designers for these systems need flexible co-design tools which can interface with domain-specific notations and provide robust analytic capability. The co-design tools must support a spiral model of development, which means multiple views of the design and multiple layers of abstraction. Co-design tools are needed for design space exploration as well as for validation and verification.

Vess L. Johnson, Omniview Design, Inc., Pittsburgh, PA

As system and chip designs become more complex, the ability to evaluate different architectures to optimize performance has suffered. What is required is a way to evaluate hardware-software tradeoffs at the architectural stage of the design process. The impact of multiple processors, types of processors, memory sizing, bus structure, etc. as well as hardware/software partitioning must be examined very early in the design process. Hardware and software subsystems need to be modeled separately and the allocation of functions should be performed between the hardware and software components in an optimal manner.

Such performance optimization at the early architectural stage of the design process will result in major reductions of expensive design iterations later in the design cycle, yielding dramatic savings in cost and time to market.

Mark Medovich, Sun Microsystems, Inc., Palo Alto, CA

Embedded system design has been a testimonial to the existence of engineering disorders. Engineers often complain about a host of ailments, among which are "reinvention of the wheel" syndrome, and its cousin, "not invented here" syndrome. There are almost no standard design methodologies in place which provide an easy migration from one generation of an embedded system design to its next incarnation. The embedded system designer and design team manager have long since realized that embedded system design is too low level for complex designs which need to be deployed and maintained over a long term product life cycle. Design and development environments for individual components within systems, be they hardware or software, are often non-homogenous. That is, if multiple processors, ASICs, and FPGAs are implemented on a system board which is networked or connected in some way with different system boards, the associated development environments are radically fractured. In design development we use tools for synthesis, simulation, analysis, emulation, test, design capture and layout, assemblers, debuggers, linkers, loaders, compilers, executives, kernels, embedded operating systems, full blown operating systems, and so on. This plethora of design environments represents a very intense management challenge. There is hope. Embedded tools of the future will provide a homogenous, top down approach to system design and cure embedded development disorders.

John Fogelin, Wind River Systems, Alameda, CA

In consideration of the ever-expanding market of embedded software, the geometric increase in software content, and the stringent market windows that must be satisfied, the time has never been better to focus on a fundamental means to compress the design and implementation cycle of an embedded application. Proponents of hardware/software co-design contend that de-serializing the hardware and software design phases offers the greatest opportunity to streamline application development. While progress in the past decade has enabled full simulation of OS/application interaction well in advance of stable hardware, and also even enabled simulation environments that allow software simulators to effectively integrate hardware simulators for full simulation of software I/O with ASIC designs, a principal challenge remains. Without the emergence of a uniform embedded application development methodology, the hardware CAD systems and software IDEs still largely speak an entirely different language. With regard to up-coming innovation, it will be in those disciplines of application design for which a uniform language exists, with little semantic variance, that the full benefits of hardware/software co-design will be realized. With the mathematical underpinnings of most DSP algorithms, those applications characterized by a high degree of signal filtering, compound filtering, signal transforms, etc. will likely be the earliest benefactors of this emerging design technique.

Power Optimization of Variable Voltage Core-Based Systems

Inki Hong†, Darko Kirovski†, Gang Qu†, Miodrag Potkonjak†, and Mani B. Srivastava‡
†Computer Science Department, University of California, Los Angeles, CA 90095-1596 USA
‡Electrical Engineering Department, University of California, Los Angeles, CA 90095-1596 USA

Abstract

The growing class of portable systems, such as personal computing and communication devices, has resulted in a new set of system design requirements, mainly characterized by dominant importance of power minimization and design reuse. We develop the design methodology for the low power core-based real-time system-on-chip based on dynamically variable voltage hardware. The key challenge is to develop effective scheduling techniques that treat voltage as a variable to be determined, in addition to the conventional task scheduling and allocation. Our synthesis technique also addresses the selection of the processor core and the determination of the instruction and data cache size and configuration so as to fully exploit dynamically variable voltage hardware, which result in significantly lower power consumption for a set of target applications than existing techniques. The highlight of the proposed approach is the non-preemptive scheduling heuristic which results in solutions very close to optimal ones for many test cases. The effectiveness of the approach is demonstrated on a variety of modern industrial-strength multimedia and communication applications.

1 Introduction

1.1 Motivation

The growing class of portable systems, such as personal computing and communication devices, demands data- and computation-intensive functionalities with low power consumption. For such systems, power consumption is the primary design goal since the battery life is a primary constraint on power, due to the fact that the battery technology has not followed the progress pace of the semiconductor industry. Recent advances in power supply technology along with custom and commercial CMOS chips that are capable of operating reliably over a range of supply voltages make it possible to create processor cores with supply voltage that can be varied at run time according to application timing constraints. The variable voltage processor core can be made to operate at different optimal points along its power vs. speed curve in order to achieve much higher energy efficiency than existing techniques for a wider class of applications. Such systems also require design flexibility which result in the need for implementation on programmable processor platform. In fact, embedded software running on RISC and DSP processor cores has emerged as a leading implementation methodology for such applications as speech coding, modem functionality, video compression and communication protocol processing [15]. Current semiconductor technology allows the integration of programmable processors and memory structures on a single die, which enables the implementation of a system on a single chip. Similarly, the exponential growth of both applications and implementation technology has outpaced the design productivity of the traditional synthesis process. The shrinked time-to-market window has exacerbated the situation. There is a wide consensus that only through reuse of highly optimized cores the demands of the pending applications and the potential ultra large scale integration may be matched. Therefore, low power core-based system-on-chip, consisting of a variable voltage programmable processor core and a memory hierarchy, attracted much attention of virtually all silicon vendors.

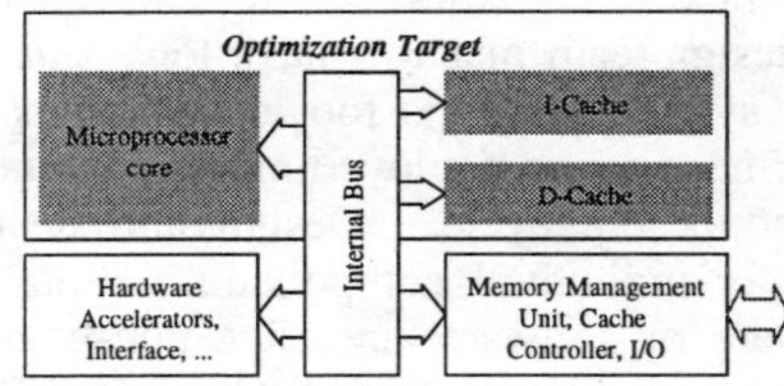

Figure 1: A typical core-based application-specific system-on-chip.

Our synthesis technique targets typical modern application specific system-on-chip, consisting of a variable voltage processor core, instruction and data cache, and a number of optional hardware accelerators and control blocks as depicted in Figure 1. The distribution of power dissipation by the components of application-specific system-on-chip depends on the actual applications running on the system. However, extensive studies indicate that the power consumption of the processor and cache cores accounts for significant portion of the overall power consumption of the described system-on-chip [6]. Therefore, in this paper we focus on the power optimization of the processor and cache cores.

The most effective way to reduce power consumption of a processor core in CMOS technology is to lower the supply voltage level, which exploits the quadratic dependence of power on voltage [2]. Reducing the supply voltage however increases circuit delay and decreases clock speed. The resulting processor core consumes lower average power while meeting the deadlines. This technique is ineffective when tight deadlines are present in systems. Another power optimization technique for processor cores is the system shutdown [15]. The system shutdown technique, though usable even in the presence of tight deadlines, is inferior to the supply voltage reduction technique for the cases when both techniques can be applied. The limitations of the techniques arise due to the fact that systems are designed with a fixed supply voltage. The supply voltage reduction technique attempts to find a single optimal voltage level for the entire processor operation, while the system shutdown technique makes a binary runtime decision whether to turn on or off the power supply.

The goal of the research presented in the paper is to develop the design methodology for the low power core-based real-time system-on-chip based on dynamically variable voltage hardware. The key challenge is to develop effective scheduling techniques that treat voltage as a variable to be determined, in addition to the conventional task scheduling and allocation. Our synthesis technique also addresses the selection of the processor core and the determination of the instruction and data cache size and configuration so as to fully exploit dynamically variable voltage hardware, which will

DAC 98, San Francisco, California

result in significantly lower power consumption for a set of target applications than existing techniques.

1.2 Motivational Example

To illustrate the key point of the proposed dynamically variable voltage approach, we consider a set of tasks as a motivational example, shown in Table 1. Two independent computations $Task_A$ and $Task_B$ need to be executed on an embedded processor core in the time interval [0, 20]. Each task can be executed immediately after its arrival and is required to be finished by its deadline time. Preemption is not allowed due to the high context-switching cost.

task	arrival	deadline	execution time at 3.3 V
A	0	6	5
B	3	20	5

Table 1: The characteristics of the 2 tasks used to illustrate the motivation for dynamically variable voltage approach.

Assume the maximum supply voltage to be $(V_{dd})_{ref} = 3.3$ volts. Power is normalized to its value at the reference point, i.e., $P(3.3\ volts) = 1$ Watt. Reducing supply voltage results in increased circuit delay and to a good accuracy, the circuit delay is given by $k \times \frac{V_{dd}}{(V_{dd}-V_t)^2}$, where V_t is the threshold voltage, and k is a constant [2]. We assume a typical value of 0.8 $volts$ for the threshold voltage. The power consumption is given by $P = \alpha C_L V_{dd}^2 f$, where f is the system clock frequency, V_{dd} is the supply voltage, C_L is the load capacitance and α is the switching activity [2]. We now consider the application of shutdown, supply voltage reduction, both shutdown and supply voltage reduction, and dynamically variable voltage approach, respectively.

With the shutdown technique, the system will operate at $V_{dd} =$ 3.3 volts. The $Task_A$ is executed in the interval [0, 5]. The $Task_B$ is executed in the interval [5,10]. The processor can be shut down for the interval [10, 20] and then be resumed for the next task. The duty cycle of the processor is 50 %, so the average power consumption is 0.5 Watts.

With the supply voltage reduction technique, the system will operate at a lower but fixed supply voltage. The tight deadline on $Task_A$ means that supply voltage can not be lowered less than $V_{dd} = 2.97$ volts. Thus, the system operates at $V_{dd} = 2.97$ volts with $P(2.97\ volts) = 0.67$ Watts. $Task_A$ is executed in the interval [0, 6]. $Task_B$ is executed in the interval [6,12]. The average power consumption is 0.67 Watts. Since the system can be shut down during the interval [12, 20], the average power consumption can be lowered to $0.67 \times \frac{12}{20} = 0.40$ Watts, which results in 20 % power reduction, compared to the shutdown technique.

With the variable voltage hardware, one can schedule the two tasks such that the $Task_A$ is executed in the interval [0, 5] at 2.97 volts with $P(2.97\ volts) = 0.67$ Watts and the $Task_B$ is executed in the interval [6,20] at 1.95 volts with $P(1.95\ volts) = 0.11$ Watts. The average power consumption is $\frac{0.67\times6+0.11\times14}{20} = 0.28$ Watts, which is 44 % lower than the shutdown technique and 30 % lower than the supply voltage reduction in combination with the shutdown technique.

The preceding example illustrates that it is always better to reduce voltage than to shutdown, and that dynamically variable supply voltage helps unleash this potential.

1.3 What is New?

In this paper, we develop the first approach for power minimization of variable voltage core-based application specific systems. The power minimization is addressed at several synthesis tasks including task scheduling, instruction and data cache size determination, and processor core selection. We explore static scheduling algorithms that treat voltage as an optimization degree of freedom for the applications with real-time constraints. By selecting the most efficient voltage profile in the presence of multiple timing constraints, our algorithms result in much larger savings in energy than other techniques such as the system shutdown and the supply voltage reduction.

1.4 Paper Organization

The rest of the paper is organized in the following way. Section 2 presents the related work. Section 3 explains the necessary background material on variable voltage systems. The design flow of the novel synthesis approach is presented in Section 4. In Section 5 the optimization problems are recognized and their computational complexities are established. In the same section the synthesis algorithms are proposed and explained in detail. Section 6 presents experimental data and discussion to evaluate the effectiveness of the approach. The paper is concluded in Section 7.

2 Related Work

We review the research results relevant to low power systems based on dynamically variable voltage hardware.

Supply voltage reduction coupled with architecture level parallelism and pipelining to compensate for lower clock rate due to voltage reduction [2] works well for applications such as signal processing where throughput is the sole metric of speed. Several researchers have addressed the issue of power in event-driven systems, and proposed various techniques for shutting down the system or parts of the system [8, 15].

At the technology level, efficient DC-DC converters that allow the output voltage to be rapidly changed under external control have recently been developed [13]. At the hardware design level, research work has been performed on chips with dynamically variable supply voltage that can be adjusted based on (i) process and temperature variations, and (ii) processing load as measured by the number of data samples queued in the input (or output) buffer [1].

There has been research on task-level scheduling strategies for adjusting CPU speed so as to reduce power consumption. Most of the existing work is in the context of non-real-time workstation-like environment [7, 17]. Yao et al. [19] described an minimum energy schedule for scheduling with **preemption** for independent processes with deadlines.

A number of research groups have addressed the use of multiple (in their software implementation restricted to two or three) different voltages [3, 9, 12, 14]. Interestingly, although they used the term "variable voltage", they actually addressed scheduling when a fixed number of simultaneously available voltages are used. Therefore, there is no similarity between the proposed research and these efforts beyond accidental syntax similarity.

3 Preliminaries

In this Section we describe the task model and hardware and power models for processor cores, caches and variable voltage hardware.

3.1 Computational and Timing Models

A set J of independent tasks is to be executed on a system-on-chip. Each task $j \in J$ is associated with the following parameters: a_j its arrival time, d_j its deadline, p_j its period, and r_j its execution time at the nominal(maximum possible) voltage level V_{ref}. Without loss of generality we assume that all tasks have identical periods. Since context switching overhead is usually high for modern systems, we assume no task preemption.

3.2 Power Model

It is well known that there are three principal components of power consumption in CMOS integrated circuits: switching power, short-circuit power, and leakage power. In CMOS technology, switching power dominates power consumption. It is also known that reduced voltage operation comes at the cost of reduced throughput [2]. The maximum rate at which a circuit is clocked monotonically decreases as the voltage is reduced. From these equations together with the observation that the speed is proportional to f

and inversely proportional to the gate delay, the power vs. speed curve can be derived. By varying V_{dd}, the system can be made to operate at different points along this curve. Due to the convexity of the power vs. speed function, it is always better to run the computation at a constant speed if there is a single computation that needs to be executed by some deadline. In the proposed scheduling heuristic, we use this observation so that each task is executed at a single voltage.

Processor core	Clock (MHz)	MIPS	Technology (μm)	Area (mm^2)	Power diss. (mW)
ARM7 LPower	27	24	0.6	3.8	45 (3.3V)
LSI TR4101	81	30	0.35	2	81 (3.3V)
LSI CW4001	60	53	0.5	3.5	120 (3.3V)
LSI CW4011	80	120	0.5	7	280 (3.3V)
Motorola 68000	33	16	0.5	4.4	35 (3.3V)
PowerPC 403	33	41	0.5	7.5	40 (3.3V)

Table 2: The performance, area, and power data for a subset of processor cores

3.3 Architecture and Hardware Model

Several factors combine to influence system performance: instruction and data cache miss rates and penalty, processor performance, and system clock speed. Power dissipation of the system is estimated using processor power dissipation per instruction and the number of executed instructions per task, supply voltage, energy required for cache read, write, and off-chip data access as well as the profiling information about the number of cache and off-chip accesses. The approach that we use to realistically address the factors leverages on existing cache and processor models.

Data on microprocessor cores have been extracted from manufacturer's datasheets as well as from the CPU Center Info web site [4]. A sample of the collected data is presented in Table 2.

We use CACTI [18] as a cache delay estimation tool with respect to the main cache design choices: size, associativity, and line size. The energy model for cache is adopted from [16]. We use a recently developed compiler strategy which efficiently minimizes the number of cache conflicts that considers direct-mapped caches [10]. We consider only direct-mapped caches based on the results of our experimentation that 2-way set associative caches do not dominate comparable direct-mapped caches in a single case. We estimated the cache miss penalty based on the operating frequency of the system and external bus width and clock for each system investigated. This penalty ranged between 4 and 20 system clock cycles. Write-back is used instead of write through due to its superior performance and power savings. We consider caches with single access port. The nominal energy consumption per single off-chip memory access, $98nJ$, is assumed [11]. A subset of the cache model data is given in 3.

	No optimizations			Block buffering, sub-banking and Gray code addressing [16]		
Cache size	8B	16B	32B	8B	16B	32B
512B	0.3301	0.3777	0.4683	0.2913	0.3219	0.4012
1KB	0.3561	0.3943	0.4626	0.2953	0.2993	0.3452
2KB	0.4223	0.4442	0.4894	0.3171	0.2706	0.2710
4KB	0.6513	0.6662	0.6942	0.4557	0.3336	0.2729
8KB	1.146	1.15.6	1.175	0.7686	0.5044	0.3474
16KB	2.158	2.164	2.174	1.412	0.8693	0.5298
32KB	4.198	4.202	4.209	2.702	1.608	0.9223

Table 3: A subset of the cache power consumption model: power consumption (nJ) estimation for various direct-mapped caches with variable line sizes at 5V

3.4 Variable Voltage Hardware

The variable voltage is generated by the DC-DC switching regulators in the power supply which are essentially a feedback control system. The time for the voltage to reach a steady state at the new voltage is a strong function of the feedback loop parameters and the LC output filter in the switching regulator. [13] reported efficient DC-DC switching regulators with fast transition times. DC-DC converters with faster transition times may be obtained with the cost of increased power dissipation within the converter. Dual to the supply voltage variation is the accompanying variation of the clock frequency. The clock frequency also takes time to stabilize at the new value. The time and power overhead associated with voltage switching is negligible [13] and not considered in the analysis.

3.5 Summary of Theoretical Results

The *non-preemptive* scheduling of a set of independent tasks with arbitrary arrival times and deadlines on a *fixed voltage* processor is an NP-complete task [5]. Therefore, the *non-preemptive* scheduling of a set of independent tasks with arbitrary arrival times and deadlines on a *variable voltage* processor is also an NP-complete task. Yao, Demers and Shenker [19] have provided the optimal *preemptive* static scheduling algorithm for a set of independent tasks with arbitrary arrival times and deadlines. This solution serves as a lower bound for *non-preemptive* scheduling solutions. This lower bound plays a crucial role in speeding up our branch and bound algorithm for the *non-preemptive* scheduling problem.

4 Design Methodology for Variable Voltage Systems: Global Design Flow

In this Section we describe the global flow of the proposed synthesis system and explain the function of each subtask and how these subtasks are combined into a synthesis system.

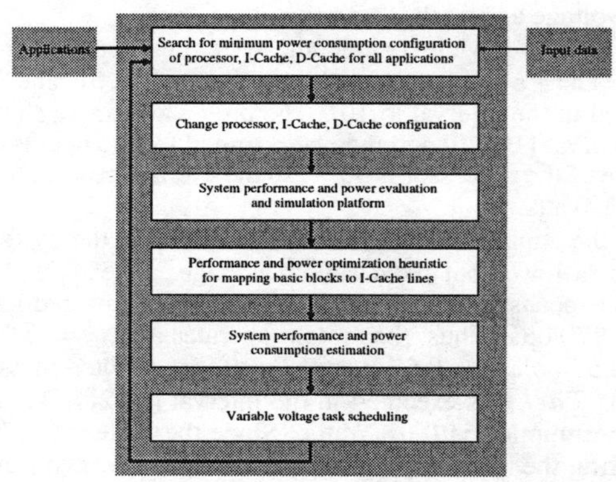

Figure 2: The global flow of the synthesis approach.

Sort processor cores in a list L in an increasing order of $\frac{EnergyPerInstruction}{SystemClockFrequency}$ at the nominal voltage 5V.
Delete the dominated processor cores from L
For each processor core in L in the order of appearance
  For I-cache = 512B..32KB and CacheLineSize = 8B..512B
    For D-cache = 512B..32KB and CacheLineSize = 8B..512B
      Check bounds; if exceeded **break**;
      If (current I- and D-cache configuration has never been evaluated)
        Evaluate performance and power consumption of the cache structure
      Using the cache system analysis, evaluate the system power consumption using variable voltage task scheduling
      Memorize Configuration C if power consumption is minimal

Figure 3: Pseudo code for the resource allocation procedure.

Figure 2 illustrates the synthesis system. The goal is to choose the configuration of processor, I-cache, and D-cache and the variable voltage task schedule with minimum power consumption which satisfy the requirements of multiple non-preemptive tasks. To accurately predict the system performance and power consumption for target applications, we employ the approach which integrates the optimization, simulation, modeling, and profiling tools. The synthesis technique considers each non-dominated microprocessor core and competitive cache configuration, and selects the hardware setup which requires minimal power consumption and satisfies the individual performance requirements of all target applications. The application-driven search for a low-power core and cache system requires usage of trace-driven cache simulation for each promising point considered in the design space. We attack this problem by carefully scanning the design space using search algorithms with sharp bounds and by providing powerful algorithmic performance and power estimation techniques. We use the system performance and power evaluation and simulation platform based on SHADE, DINEROIII and a custom analyser [10, 11].

5 Synthesis Algorithms

The optimization problems encountered in the synthesis of a low power variable voltage system on a chip and competitive optimization algorithms are described in this Section.

5.1 Resource Allocation

In this phase of the synthesis approach, a search is conducted to find an energy-efficient system configuration. The search algorithm is described using the pseudo-code shown in Figure 3. Since performance and power evaluation of a single processor, I- and D-cache configuration requires a trace-driven simulation, the goal of our search technique is to reduce the number of evaluated cache systems using sharp bounds for cache system performance and power estimations. However, a particular cache system is evaluated using trace-driven simulation only once since the data retrieved from such simulation can be used for overall system power consumption estimation for different embedded processor cores with minor additional computational expenses.

The algorithm excludes from further consideration processor cores dominated by other processor cores. One processor type dominates another if it consumes less power at higher frequency and results in higher MIPS performance at the same nominal power supply. The competitive processors are then sorted in ascending order with respect to their power consumption per instruction and frequency ratio. Microprocessors which seem to be more power-efficient are, therefore, given priority in the search process. Later on this step provides sharper bounds for search termination.

The search for the most efficient cache configuration is bounded with sharp bounds. A bound is determined by measuring the number of conflict misses and comparing the energy required to fetch the data from off-chip memory due to measured conflict misses and the power that would have been consumed by twice larger cache for the same number of cache accesses assuming zero cache conflicts. We terminate further increase of the cache structure when the power consumption due to larger cache would be larger than the energy consumed by the current best solution. Similarly, another bound is defined at the point when the energy required to fetch the data from off-chip memory due to conflict cache misses for twice smaller cache with the assumption of zero-energy consumption per cache access, is larger than the energy required for both fetching data from cache and off-chip memory in the case of the current cache structure. We abort further decrease of the cache structure if the amount of energy required to bring the data due to additional cache misses from off-chip memory is larger than the energy consumed by the current best solution.

When evaluating competitive hardware configurations, the target applications are scheduled with the variable voltage task scheduler using the predicted system performance and power on the configuration considered. Before the non-preemptive task scheduling is performed, more efficient optimal preemptive scheduler is applied to get a lower bound. If the result is worse than the current best solution, the configuration is worse than the current best one.

5.2 Task Scheduling

As described in Section 3, the non-preemptive scheduling of a set of independent tasks with arbitrary arrival times and deadlines on a *variable voltage* processor is an NP-complete task. We have developed an efficient and effective heuristic for the general variable voltage task scheduling problem, which leverages least-constraining most-constrained heuristic paradigm in order to obtain competitive solutions. The algorithm is described using the pseudo-code shown in Figure 4.

The algorithm consists of two phases. In the first phase all tasks are scheduled at the nominal voltage. If a feasible schedule is found in the first phase, in the second phase the supply voltages are adjusted for low-power. These two phases are repeated $LOOPS$ times and the best solution after the iterations is chosen as the final solution. Each iteration of the two phases generates different solutions since the objective functions used in the first phase to guide the search strategy are randomized by a random offset.

In the first phase, the task scheduling at the nominal voltage is done in the following way. Initially, the time period in which the tasks appear is divided into time regions such that each time region has its start or end time equal to a start or end time of some task and there does not exist a task which has either an arrival or deadline time within the time region. Then, for each time region, an objective function $OBJ(TimeRegion)$ is computed as:

$$OBJ(TimeRegiontr) = maxVoltage(tr) \cdot aveVoltage(tr)$$

where the functions $maxVoltage(tr)$ and $aveVoltage(tr)$ assume that all tasks are scheduled from their arrival until their deadline regardless of time overlaps and return the maximum and average voltage for all tasks alive at time region tr, respectively. The objective functions (or constraints) of time regions are used to compute the constraints of tasks. For each task an objective function $OBJ(Task)$ is computed as

$$OBJ(Taskt) = \sum_{tr \in [a_t, d_t]} OBJ(tr) \cdot \frac{r_t^2}{(d_t - a_t)^2}.$$

The most-constrained task t, which does not include in its life-time another tasks' life-time, is then scheduled in the interval equal to its run-time at the nominal voltage which spans over a subset of time regions with the lowest sum of objective functions among any feasible subset of time regions. When comparing the subset of time regions, we only consider the cases that the start time of the task is only at the start time of each time region in the task's life-time or the end time of the task is only at the end time of each time region. Upon scheduling t, other tasks' deadlines and arrival times are updated in the following way. If the scheduled task's life-time is included in the life-time of some other task s and the task s can be scheduled at both partitions of its life-time upon deleting the time regions that contains t, then both partitions of the task's life-time are evaluated for scheduling. Objective functions are computed for all the time-frames with its length equal to the task's nominal voltage run-time such that the start time of the time frame is only at the start time of each time region or the end time of the time frame is only at the end time of each time region, shown in the lower part of Figure 5. The partition which includes the time frame with the minimal sum of objective functions of containing time regions is selected. Otherwise, each remaining task's arrival time and deadline is updated if its lifetime intersects with the lifetime of the task t. This process is repeated until all tasks are scheduled at the system nominal voltage.

We explain the process using the example shown in Figure 5. Six tasks T[0], T[1], T[2], T[3], T[4] and T[5] are shown with their arrival, deadline, and nominal voltage run-times. The set of tasks

Given: a set of tasks T, where each task $t \in T$ is characterized by its arrival time a_t, its deadline d_t, and its execution time r_t at 5V

Repeat $LOOPS$ times

Scheduling tasks at the nominal voltage.

Repeat
Create(TR) >> Create a set TR of time regions $tr_i[start_i, end_i]$ where $start_i$ and end_i correspond to an arrival or deadline time of any two not necessarily distinct tasks and there does not exist another task t_o which has $start_i < a_{t_o}, d_{t_o} < end_i$.
For each $tr_i \in TR$
Compute its objective function $OBJ(TimeRegion\ tr_i)$
For each task $t_i \in T$
Compute $OBJ(Task\ t_i)$
Select the task $t \in T$ such that it has the maximum $OBJ(t)$ and there does not exist another task s such that $a_s > t_a$ and $d_s < d_t$
Find the sequence of SEQ time regions $[tr_1, tr_{SEQ}]$ such that $a_t \leq start_{tr_1}$ and $d_t \geq end_{tr_{SEQ}}$ and $[tr_1, tr_{SEQ}]$ has the smallest $\sum_{i=1}^{SEQ} OBJ(tr_i)$
If $end_{tr_{SEQ}} - start_{tr_1} = r_t$
Schedule t from $start_{tr_1}$ to $end_{tr_{SEQ}}$
Else If $OBJ(tr_1) > OBJ(tr_{SEQ})$
Schedule t from $OBJ(tr_{SEQ}) - r_t$ to $OBJ(tr_{SEQ})$
Else Schedule t from $OBJ(tr_1)$ to $OBJ(tr_1) + r_t$
Delete t from list of tasks T
Create(TR)
For each $tr_i \in TR$
Compute its objective function $OBJ(TimeRegion\ tr_i)$
For each task $s \in T$
If scheduling of task t splits the life-time of task s into two time portions p_1 and p_2 both larger than the nominal voltage run-time
For each portion p_i find the sequence of time regions $[tr_1, tr_{SEQ_i}]$ which comprises time period greater or equal to r_s and has minimal $S = \sum_{i=1}^{SEQ_i} OBJ(tr_i)$
Assign the task to the portion which has greater S
Else update the task arrival and deadline times
until all tasks are scheduled.

Voltage adjusting of scheduled tasks for low-power

For each task t
Compute the earliest start st_t and latest finish fi_t times according to the existing scheduling order.
Compute the minimal voltage v_t at which the task can run if expanded from st_t to fi_t
Repeat
For each task t
If there exists task $s \in T$ such that $fi_s = st_t$ and $v_s < v_t$ and $d_s > a_t$
Schedule s from st_s to $max(st_s + r_s(v_t) + \frac{st_t - st_s - r_s(v_t)}{2}, a_t)$
Schedule t from $max(st_s + r_s(v_t) + \frac{st_t - st_s - r_s(v_t)}{2}, a_t)$ to d_t
If there exists task $s \in T$ such that $fi_t = st_s$ and $v_s < v_t$ and $a_s < d_t$
Schedule t from st_t to $min(d_s - r_s(v_t) - \frac{d_s - r_s(v_t) - fi_t}{2}, d_t)$
Schedule s from $min(d_s - r_s(v_t) - \frac{d_s - r_s(v_t) - fi_t}{2}, d_t)$ to d_s
until no further energy consumption improvements
Compute system energy E. **If** $E < BestE$, $BestE = E$

Return (BestE)

Figure 4: Pseudo code for the variable voltage task scheduling.

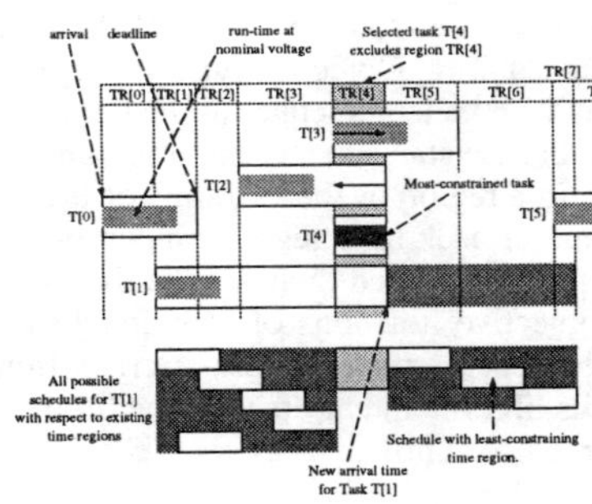

Figure 5: Scheduling tasks at nominal voltage.

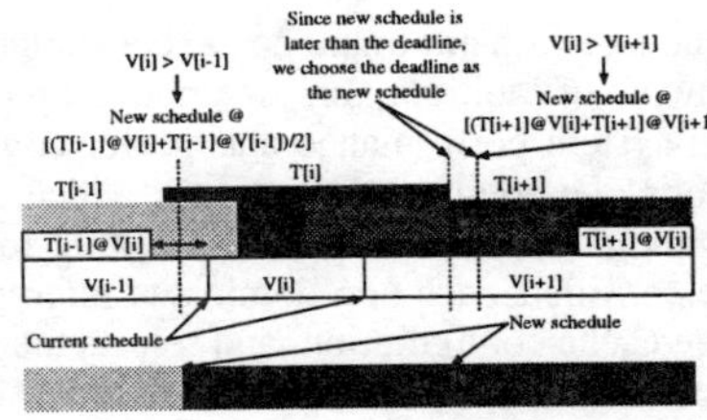

Figure 6: Adjusting the voltage of scheduled tasks for low-power.

creates a set of nine time regions TR[0] through TR[8]. Based on objective functions computed, task T[4] is selected as the most-constrained. Once this task is scheduled the arrival times and deadlines of tasks T[2] and T[3] are updated as shown (see arrows in Figure 5). However, task [0] can be scheduled at both TR[1, 2, 3] and TR[5, 6, 7]. There are four possible schedules with respect to the existing time regions at TR[1, 2, 3] and three (actually five, but only three distinct) at TR[5, 6, 7] as shown in the bottom part of Figure 5. The least-constraining schedule (marked in Figure 5) is in the TR[5, 6, 7] partition since during this time region there are no alive tasks. Therefore, the life-time of task T[1] is reduced to TR[5, 6, 7].

Once tasks are scheduled, the procedure which tunes the supply voltage of each task is invoked. It iteratively traverses the set of tasks and tries to lower the voltage of each task until no energy dissipation improvement occurs. Whenever a task t has a neighboring task s with a life-time that intersects the life time of t and supply voltage v_s lower than v_t, the border between the tasks is moved according to Figure 6. For example, in Figure 6, task T[i] has been assigned a supply voltage higher than task T[i-1]. The new arrival time of T[i] and deadline time of T[i-1] is moved to the arithmetic mean of the current arrival/deadline time and the arrival/deadline time as if T[i-1] is supplied with T[i]'s voltage V[i]. For a fixed task scheduling this voltage adjustment algorithm results in solutions arbitrarily close to the optimal solution.

6 Experimental Results

We used nine public domain applications to demonstrate the effectiveness of the approach. JPEG software from the Independent JPEG Group implements JPEG baseline, extended-sequential, and progressive compression processes. GSM software, obtained from Technische Universität at Berlin, is an implementation of the European GSM 06.10 provisional standard for full-rate speech transcoding, prI-ETS 300 036, which uses RPE/LTP (residual pulse excitation/long term prediction) coding at 13 kbit/s. EPIC (Efficient Pyramid Image Coder), obtained from University of Pennsylvania, implements lossy image compression and decompression utilities. The remaining benchmarks (Mipmap, Osdemo, and Texgen), obtained from University of Wisconsin, use a 3-D graphic library called Mesa. Mipmap is a simple mipmap texture mapping example. Osdemo is a demo program that draws a set of simple polygons using the Mesa 3-D rendering pipe. Finally, Texgen renders a texture mapped version of the Utah teapot.

In order to evaluate the efficiency of our algorithmic approach for variable voltage system synthesis, we conducted a set of experiments. We experimented our variable voltage scheduling algorithm by scheduling sets of applications with their time constraints on a number of configurations and comparing the obtained results with a lower bound obtained using Yao's preemptive scheduling algorithm with assumed zero preemption cost. The results are presented in Table 4. The low-power processor-cache application-mix champion configurations are described in the first five columns of Table 4. The configurations are assumed to have operational dynamically variable [0.8, 3.3] V supply voltage. The next four pairs of columns present the power consumption lower bound and the result obtained

Core	I-Cache		D-cache		Energy (J) 10 Tasks		15 Tasks	
	Cache	Block	Cache	Block	LB	New	LB	New
M68000	1KB	128B	2KB	128B	2.46	2.50	2.61	2.66
ARM7LP	1KB	128B	2KB	128B	2.42	2.47	2.50	**2.54**
TR4101	512B	64B	4KB	32B	2.43	2.50	2.52	2.58
CW4001	512B	64B	1KB	128B	2.39	**2.44**	2.49	**2.54**
PC403	1KB	32B	2KB	32B	2.48	2.55	2.75	2.78
CW4011	1KB	64B	2KB	64B	2.50	2.59	2.72	2.77
Average					*2.44*	*2.51*	*2.58*	*2.65*
Core	I-cache		D-cache		20 Tasks		25 Tasks	
	Cache	Block	Cache	Block	LB	New	LB	New
M68000	1KB	128B	2KB	128B	11.1	11.9	27.3	**27.1**
ARM7LP	1KB	128B	2KB	128B	12.3	13.1	29.9	31.8
TR4101	512B	64B	4KB	32B	10.6	**11.0**	27.3	27.9
CW4001	512B	64B	1KB	128B	13.8	15.0	35.0	37.4
PC403	1KB	32B	2KB	32B	14.1	14.4	37.2	38.7
CW4011	1KB	64B	2KB	64B	14.8	15.4	39.6	40.7
Average					*12.8*	*13.4*	*32.7*	*33.9*
Core	I-cache		D-cache		35 Tasks		50 Tasks	
	Cache	Block	Cache	Block	LB	New	LB	New
M68000	1KB	128B	2KB	128B	59.9	64.1	141.7	142.0
ARM7LP	1KB	128B	2KB	128B	64.4	64.5	139.9	140.2
TR4101	512B	64B	4KB	32B	63.7	63.9	127.3	127.9
CW4001	512B	64B	1KB	128B	69.4	73.7	135	136.3
PC403	1KB	32B	2KB	32B	53.0	**53.4**	109.7	110.7
CW4011	1KB	64B	2KB	64B	55.2	55.4	100.2	**100.6**
Average					*61.0*	*62.5*	*125.6*	*126.3*

Table 4: Results on a set of architectures and application mixes: LB - Lower Bound, New - Our Approach.

using our algorithm for four different sets of applications extracted from the nine applications.

Our non-preeptive scheduling algorithm resulted in solutions which were only **1.47%** in average higher than the lower bound determined using Yao's preemptive scheduling algorithm over series of more than 10^4 schedulings. An average standard deviation of voltages per schedule is **0.61V** over the same set of scheduling actions. The average maximal voltage per scheduling was $3.12V$, while the average minimal voltage was $1.08V$. The results also indicate the importance of effective allocation for more complex systems such that while for the 10-task mix the best configuration is only 3% better than the average, our allocation technique finds for the case of 50-task mix a configuration with 25% power savings.

7 Conclusion

We developed the design methodology for the low power core-based real-time system-on-chip based on dynamically variable voltage hardware. The key contribution was to develop effective scheduling techniques that treat voltage as a variable to be determined, in addition to the conventional task scheduling and allocation. Our synthesis technique also addressed the selection of the processor core and the determination of the instruction and data cache size and configuration so as to fully exploit dynamically variable voltage hardware, which resulted in significantly lower power consumption for a set of target applications than existing techniques. The highlight of the proposed approach was the non-preemptive scheduling heuristic which resulted in solutions very close to optimal ones for many test cases. The effectiveness of the approach was demonstrated on a variety of modern industrial-strength multimedia and communication applications, such as MPEG and JPEG encoders and decoders.

REFERENCES

[1] A. Chandrakasan, V. Gutnik, and T. Xanthopoulos. Data driven signal processing: an approach for energy efficient computing. In *International Symposium on Low Power Electronics and Design*, pages 344–352, 1996.

[2] A. P. Chandrakasan, S. Sheng, and R.W. Broderson. Low-power CMOS digital design. *IEEE Journal of Solid-State Circuits*, 27(4):473–484, 1992.

[3] J.-M. Chang and M. Pedram. Energy minimization using multiple supply voltages. In *International Symposium on Low Power Electronics and Design*, pages 157–162, 1996.

[4] http://infopad.eecs.berkeley.edu/CIC/.

[5] M.R. Garey and D.S. Johnson. *Computer and Intractability: A Guide to the theory of NP-Completeness*. W. H. Freeman & Co., New York, NY, 1979.

[6] R. Gonzalez and M. Horowitz. Energy dissipation in general purpose microprocessors. *IEEE Journal of Solid-State Circuits*, 31(9):1277–1284, 1996.

[7] K. Govil, E. Chan, and H. Wasserman. Comparing algorithms for dynamic speed-setting of a low-power CPU. In *ACM International Conference on Mobile Computing and Networking*, pages 13–25, 1995.

[8] C. Hwang and A.C.-H. Wu. A predictive system shutdown method for energy saving of event-driven computation. In *IEEE/ACM International Conference on Computer-Aided Design*, pages 28–32, 1997.

[9] M. C. Johnson and K. Roy. Datapath scheduling with multiple supply voltages and level converters. *ACM Transactions on Design Automation of Electronic Systems*, 2(3), 1997.

[10] D. Kirovski, C. Lee, W. Mangione-Smith, and M. Potkonjak. Application-driven synthesis of core-based systems. In *IEEE/ACM International Conference on Computer-Aided Design*, pages 104–107, 1997.

[11] D. Kirovski, C. Lee, W. Mangione-Smith, and M. Potkonjak. Synthesis of power efficient systems-on-silicon. In *Asia and South Pacific Design Automation Conference*, 1998.

[12] Y.-R. Lin, C.-T. Hwang, and A.C.-H Wu. Scheduling techniques for variable voltage low power designs. *ACM Transactions on design Automation of Electronic Systems*, 2(2):81–97, 1997.

[13] W. Namgoong, M. Yu, and T. Meng. A high-efficiency variable-voltage CMOS dynamic DC-DC switching regulator. In *IEEE International Solid-State Circuits Conference*, pages 380–381, 1997.

[14] S. Raje and M. Sarrafzadeh. Variable voltage scheduling. In *International Symposium on Low Power Design*, pages 9–14, 1995.

[15] M. Srivastava, A. P. Chandrakasan, and R. W. Brodersen. Predictive system shutdown and other architectural techniques for energy efficient programmable computation. *IEEE Transactions on VLSI Systems*, 4(1):42–55, 1996.

[16] C.-L. Su and A.M. Despain. Cache design trade-offs for power and performance optimization: a case study. In *International Symposium on Low Power Design*, pages 63–68, 1995.

[17] M. Weiser, B. Welch, A. Demers, and S. Shenker. Scheduling for reduced CPU energy. In *USENIX Symposium on Operating Systems Design and Implementation*, pages 13–23, 1994.

[18] S.J.E. Wilton and N.P. Jouppi. CACTI: an enhanced cache access and cycle time model. *IEEE Journal of Solid-State Circuits*, 31(5):677–688, 1996.

[19] F. Yao, A. Demers, and S. Shenker. A scheduling model for reduced CPU energy. In *IEEE Annual Foundations of Computer Science*, pages 374–382, 1995.

Policy Optimization for Dynamic Power Management

G. A. Paleologo[#] L. Benini [†] A. Bogliolo [*] G. De Micheli [†]

[#] Stanford University
Dept. of Engineering-Economic Systems and
Operations Research, Stanford, CA 94305-4023

[*] Università di Bologna
Dip. Informatica, Elettronica,
Sistemistica
Bologna, ITALY 30165

[†] Stanford University
Computer Systems Laboratory
Stanford, CA 94305-4070

Abstract

Dynamic power management schemes (also called policies) can be used to control the power consumption levels of electronic systems, by setting their components in different states, each characterized by a performance level and a power consumption. In this paper, we describe power-managed systems using a finite-state, stochastic model. Furthermore, we show that the fundamental problem of finding an optimal policy which maximizes the average performance level of a system, subject to a constraint on the power consumption, can be formulated as a stochastic optimization problem called policy optimization. Policy optimization can be solved exactly in polynomial time (in the number of states of the model). We implemented a policy optimization tool and tested the quality of the optimal policies on a realistic case study.

1 Introduction

Battery-operated portable appliances impose tight constraints on the power dissipation of their components. Designers struggle to meet increasingly tight power budgets as complexity and performance requirements are pushed forward by user demand. Numerous computer-aided design techniques [11] for low power have been proposed to help designers reduce time to market and improve the quality of their products. The vast majority of CAD techniques for low power targets digital VLSI circuits, i.e. chip-level designs. Unfortunately, almost every portable electronic appliance is far more complex than a single chip. Portable devices such as cellular telephones and laptop computers contain tens or even hundreds of components. In most electronic products, digital components are responsible for only a fraction of the power consumed. Analog, electromechanical and optical components are a significant fraction of the total, and are often responsible for the largest contributions to the power budget. For example, the power breakdown for a well-known laptop computer [14] shows that, on average 36% of the total power is consumed by the display, 18% by the hard drive, 18% by the wireless LAN interface, 7% by non-critical components (keyboard, mouse etc.), and only 21% by digital VLSI circuitry (mainly memory and CPU). Reducing the power in the digital logic portion of this laptop by *10X* would reduce the overall power consumption by less than 19%. Laptop computers are not an isolated case. Almost all electronic appliances are complex and heterogeneous systems containing a wide variety of devices that do not fall within the scope of the available computer-aided power optimization techniques. Nevertheless, designers have reacted promptly to the new challenges posed by low-cost, low-power portable appliances. Battery lifetime (or time between recharges) is steadily increasing and the physical dimensions of portable devices are progressively shrinking. These surprising results are achieved thanks to a well-balanced mix of technological innovation, architectural design and optimization. One of the most successful techniques employed by designers at the system level is *dynamic power management* [2]. This technique reduces power dissipation by selectively turning off (or reducing the performance of) system components when they are idle (or partially unexploited). Building a complex system that supports dynamic power management is a difficult and error-prone process. Long trial-and-error iterations cannot be tolerated when fast time to market is the main factor deciding the success of a product. Unfortunately, system-level computer-aided design environments are still in their infancy, and EDA vendors are lagging far behind the needs of this segment of the electronic industry. To compensate for this lack of support, several system developers and vendors [10, 9] are aggressively pursuing a long-term, wide-scope strategy to greatly simplify the task of designing large and complex power-managed systems. The strategy is based on a standardization initiative known as the *Advanced Configuration and Power Interface* (ACPI). ACPI specifies an abstract and flexible interface between power-manageable hardware components (VLSI chips, disk drivers, display drivers, etc.) and the *power manager* (the system component that controls when and how to turn on and off functional resources). The ACPI interface specification enormously simplifies the task of controlling the operating conditions of the system resources but it does not provide any insight on how and when to power manage them. We call *power management policy* (*policy* for brevity) an algorithm that takes decisions upon the state of operation of system components. The most aggressive policy (that we call *eager* policy) turns off every system component as soon as it becomes idle. Whenever the functionality of a component is required to carry out a system task, the component must be turned on and restored to its fully functional state. The transition between the inactive and the functional state requires time and power. As a result, the eager policy is often unacceptable because it greatly degrades performance and may not decrease power dissipation. The choice of the policy that minimizes power under performance constraints (or maximizes performance under power constraint) is a new kind

DAC 98, San Francisco, California

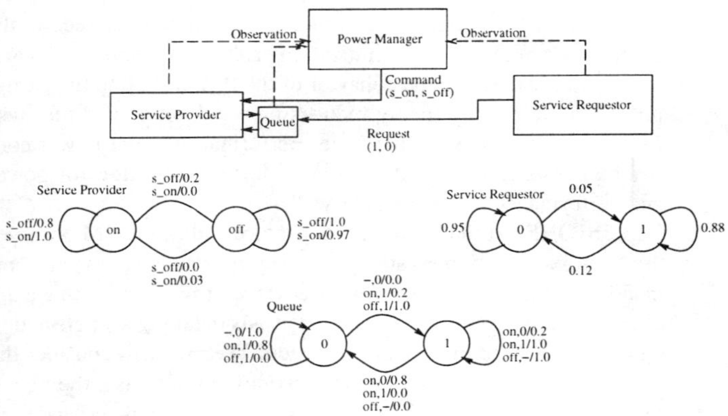

Figure 1: The abstract system model. Components are modeled as Markov chains whose state transition graphs are represented for a simple example situation. The service requestor (SR) has only two states, 0 and 1, representing the number of requests per time period sent to the provider. The queue of the service provider (SQ) has two states, 0 and 1, representing the number of requests to be serviced. The service provider (SP) has two states, on and off, representing its functional state. When on, it serves up to a request per time period taken from the queue. When off it does not serve any request. SR evolves independently, while the transition probabilities of SP depend on the command issued by the power manager (PM) and those of SQ depend on the state of both SP and SR.

of constrained optimization problem which is of great relevance for low-power electronic systems. We call this problem *policy optimization* (PO). Several heuristic power management policies have been investigated in the past [5, 8, 12] but no strong optimality result has been proved. In this paper we propose a stochastic model for the rigorous formulation of policy optimization and we describe a procedure for its *exact* solution. The procedure can be employed to explore the power vs. performance tradeoff curve. The global optimum solution of PO is computed in polynomial time by solving a linear optimization problem. This strong optimality result critically depends on the modeling assumptions. We assess the soundness of our assumptions by constructing the stochastic model for a real-life device (a disk drive) under a realistic workload. We then apply our optimization algorithm and compute optimal policies. The performance and power dissipation of the policies are then validated against simulation. Moreover, the optimal policies are compared with heuristic solutions.

The paper is organized as follows: in Section 2 we outline our stochastic model, starting from a qualitative description, then moving to a more rigorous mathematical formulation. The policy optimization problem is then formulated and a procedure for its solution is described. In Section 3 we carry out modeling and policy optimization for a realistic case study, namely a commercial disk drive with advanced power management support under workload conditions obtained from actual usage traces. In Section 4 we summarize our findings and outline future direction of research.

2 Stochastic model

We introduce a modeling approach that is aimed at providing support for system-level optimization of power-managed systems. The key features of our model are:

• **Generality**. Any power-managed electronic appliance (not only computer systems) can be modeled. Cellular phones, pagers, digital cameras are a few examples of portable appliances for which power is a major concern.

• **High level of abstraction**. The choice of abstraction level is probably the most important decision in system-level power modeling. In order to manage complexity, irrelevant (or marginally relevant) information should be abstracted. In our formulation, resources are described by abstract power and performance models.

• **Non-determinism**. The uncertainty on estimation caused by the abstraction process (and the consequent information loss) is mapped to the inherent uncertainty in the estimation of the expected value of random variables.

In the following we will consider a discrete-time (i.e. slotted time) setting. Thus, time is described by an infinite series of discrete values $t_n = Tn$, where T is the time resolution (or *period*), $n \in \mathbb{N}_+$. We model a system with a single user (or service requestor) whose requests are enqueued in a single queue and serviced by a single service provider. A power manager controls over time the behavior of the service provider (Figure 1). In more detail, the components are described in the following subsections.

A **SERVICE REQUESTOR** (SR). This unit sends requests to the Service Provider. The SR is modeled as a Markov Chain with transition matrix P_{SR}, where the observed variable is the number of requests s_r (with $s_r \in \{0, 1, \ldots, S_r - 1\}$) sent to the SR during time interval t_n. We assume that the process and all its relevant parameters are known.

As a simple example, consider the SR with a maximum of one request per period (i.e., two states), as shown in the right hand side of Figure 1.

$$P_{SR} = \begin{array}{c} \\ 0 \\ 1 \end{array} \begin{array}{c} \begin{array}{cc} 0 & 1 \end{array} \\ \begin{pmatrix} 0.95 & 0.05 \\ 0.12 & 0.88 \end{pmatrix} \end{array}$$

The example matrix models a "bursty" workload. There is a high probability (0.88) of receiving a request at time t_{n+1} if a request was received at time n, and the mean duration of a stream of requests is equal to $1/0.12 = 8.33$ periods.

A **SERVICE PROVIDER** (SP). The SP is a device which serves incoming requests from a workload source. In each time interval, it can be in only one *state*. Each state $s_p \in \{1, 2, \ldots, S_p\}$ is characterized by a performance level and by a power consumption level (to be defined later). In the simplest example, we could have two states ($S_p = 2$): *on* and *off*. At each period, transitions between power states are controlled by a *power manager* (PM) through *commands* $a \in A = \{1, 2, \ldots, N_a\}$. For example, we can define two simple commands: switch on (*s_on*) and switch off (*s_off*). When a specific command is issued, the SP will move to a new state in the next period with a fixed probability dependent only on the command a itself, and on the departure and arrival states. In other terms, after being given a transition command by the power manager, the SP can remain in its current state during the next period with a non-zero probability. This aspect of the model takes into account the uncertainty in the transition time between states caused by the abstraction of functional information. Our probabilistic model is equivalent to the assumption that the evolution in time of power states is modeled by a Markov process that depends on the commands issued by the PM. Thus, we define one transition matrix for each command a. Back to our example with two states and two commands, we could have the following transition matrices:

$$P_{SP}(s\_on) = \begin{array}{c} \\ on \\ off \end{array} \begin{array}{c} \begin{array}{cc} on & off \end{array} \\ \begin{pmatrix} 1 & 0 \\ 0.03 & 0.97 \end{pmatrix} \end{array} \quad P_{SP}(s\_off) = \begin{array}{c} \\ on \\ off \end{array} \begin{array}{c} \begin{array}{cc} on & off \end{array} \\ \begin{pmatrix} 0.8 & 0.2 \\ 0 & 1 \end{pmatrix} \end{array}$$

The Markov chains corresponding to the matrices are pictorially represented in Figure 1. Note that the transition time from *off* to *on* when the *s_on* command has been issued is a geometric random variable with average equal to $1/0.03 = 33$ periods.

Each power state has a specific *power consumption rate* c. It is a function both of the state and the command performed on the state: $c(s_p, a)$. Dependency on the state is obvious: power dissipation depends on the operational state of the SP (for our example $c(on, a) \geq c(off, a)$). Dependence on the command issued is also essential: referring to our example, $c(off, s\_on) \geq c(off, s\_off)$, because the activation process of the device requires additional energy consumption. Data on the power dissipation in various operating conditions is usually provided in the data-sheets of the SPs, or it can be directly measured.

A **QUEUE**. When service requests arrive during one period, they are buffered in a queue of length $(S_q - 1)$. In our treatment we will consider a FIFO discipline, although other disciplines can be modeled. The request is processed and serviced within the same period with a probability dependent on the power state of the SP. In this way we model the non-deterministic service time of a request as a geometric random variable, similarly to the exponential service time for the $G/M/1$ class in queueing theory [6]. It follows that also the queue length (denoted by s_q, with $0 \leq s_q \leq S_q$) is a Markov process with transition matrix $P_{SQ}(s_p, s_r)$. We refer again to our example: if $q(on) = 0.8$ and $q(off) = 0$, and the buffer contains at most one request, we have (on the bottom of Figure 1):

$$P_{SQ}(on, 0) = \begin{matrix} & 0 & 1 \\ 0 & 1.0 & 0.0 \\ 1 & 0.8 & 0.2 \end{matrix} \quad P_{SQ}(on, 1) = \begin{matrix} & 0 & 1 \\ 0 & 0.8 & 0.2 \\ 1 & 0.8 & 0.2 \end{matrix}$$

$$P_{SQ}(off, 0) = \begin{matrix} & 0 & 1 \\ 0 & 1.0 & 0.0 \\ 1 & 0.0 & 1.0 \end{matrix} \quad P_{SQ}(off, 1) = \begin{matrix} & 0 & 1 \\ 0 & 0.0 & 1.0 \\ 1 & 0.0 & 1.0 \end{matrix}.$$

A **POWER MANAGER** (PM). This component communicates with the service provider and attempts to set its state at the beginning of each period, by issuing commands chosen among a finite set A. In our example, the commands can be *s_on*, and *s_off*. It contains all proper specifications and collects all relevant information (by observing SR and PM) needed for implementing a power management policy. The consumption of the power manager is assumed to be much smaller than the consumption of the subsystems it controls and it is not a concern here. The state of the system composed of the SP, the SR and the queue is then a triple $s = (s_r, s_p, s_q)$. Being the composition of three Markov chains, s will be a Markov chain (with $S = S_r \times S_p \times S_q$ states), whose transition matrix $P(a)$ depends on the command a issued to the SP by the PM. Hence, the system is fully described by a set of N_a transition matrices, one for each command.

In the above description no mention is made of the energy source (i.e., the battery). In this paper our goal is to minimize (or bound) the average power consumption of the SP, and not to maximize the expected battery life. This choice has several advantages: it allows us not to consider the details of the power source (path-dependent discharge characteristics, possible recharge "on the run"), while still retaining the primary feature of minimizing (or constraining) the consumption level. A second desirable feature of the model is that in this it can be used to independently maximize policies in the case of multiple SPs, and consider the average aggregate consumption rate as the only relevant quantity for dimensioning the power source.

Given our stochastic system model, we can now proceed to the formulation of the policy optimization problem. Our objectives are: i) to formally describe the behavior of the PM; ii) to find the particular behavior that optimally reduces power dissipation under performance constraints (or optimizes performance under power constraints). Hence, we need to provide formal definition for power and performance cost metrics as well.

DECISIONS. At the beginning of time period n, the PM observes the "history" H_n of the system (i.e., the sequence of states and commands up to $n-1$) and controls the Service Provider by taking a *decision* δ_n. A *deterministic decision* consists in taking a single action on the basis of the history of the system. Yet, we will consider the much broader set of *randomized decisions*: in this case, the Power Manager assign probabilities to every available command and then chooses the command to issue, according to this probability distribution. In this way, even if the same decision is taken in different periods, the actual commands issued could be different. Analytically, the decision $\delta(H_n)$ can be represented by a set of real numbers, $\{p_a(H_n) | 0 \leq p_a(H_n) \leq 1, \sum_a p_a(H_n) = 1\}$, where each p_a represents the probability of issuing command a given that the history of the system is H_n. A deterministic decision is the special case with $p_{a'} = 1$ for some command a'. The definition of $P(a)$ extends naturally to the case of a randomized decision δ, and we denote the transition matrix P_δ. We stress again the generality of such a class of decisions (randomized and history-dependent). Consider the example of the previous section. Suppose that the PM observes the following history: $s_1 = (0, on, 0)$, $s_2 = (1, off, 0)$ (states in period 0, 1), and $a(0) = s\_off$ (action taken at time 0). Then, a possible decision at time 1, when state s_2 is observed, could consist in assigning probabilities of performing commands *s_on* and *s_off* equal to $p(s\_on) = 0.44$, $p(s\_off) = 0.56$ respectively.

POLICIES. Over an finite time horizon (up to time interval n), the decisions taken by the PM are a finite discrete sequence $\delta_1, \delta_2, \ldots, \delta_n$. We call this sequence a *policy* π. The policy π is the independent variable of our optimization problem. If a policy $\pi = (\delta_1, \delta_2, \ldots, \delta_n)$ is adopted, we define $P_\pi^n := P_{\delta_1} P_{\delta_2} \cdots P_{\delta_n}$; this is simply the transition matrix from period 0 to period n under policy π. Among all policies the *stationary policies* play an important role: in this case δ does not depend on the entire history H_n. For stationary policies, decisions are a function of the current state of the system $s = (s_r, s_p, s_q)$ and the function does not change over time (otherwise, stationarity would be lost). A generic randomized stationary policy can be represented as a $S \times N_a$ matrix M_π. An element $m_{s,a}$ of M_π is the probability of issuing command a given that the state of the system is s.

COST METRICS. It is now possible to define the metrics of relevance in the PO problem. In their most general form, they are function both of the state s and of δ_s, i.e., the decision we take when we are in state s. An early example of cost metric is the *expected power consumption level* $c(s_p, \delta_s)$ per unit time, introduced earlier in this section, which represents the expected consumption per period when the SP is in state s_p and decision δ_s is taken. A second parameter of interest is the *performance penalty* per unit time $d(s_q)$ which depends on the queue length (number of jobs in the queue). A natural way to define the performance penalty is the queue length: $d(s) = s_q$. Finally, we consider the *request loss* per period $b(s_r, s_q)$. It takes value 1 when a request arrives ($s_r = 1$) and is forced to balk because the queue buffer is full ($s_q = S_q$); otherwise it is 0. For notational convenience, define the consumption, performance penalty and request loss vectors

$$c_\delta := \begin{pmatrix} c(1,\delta_1) \\ \vdots \\ c(S,\delta_S) \end{pmatrix} \quad d_\delta := \begin{pmatrix} d(1) \\ \vdots \\ d(S) \end{pmatrix} \quad b_\delta := \begin{pmatrix} b(1) \\ \vdots \\ b(S) \end{pmatrix}.$$

For example the sth element of each vector is the expected value of the consumption rate (or performance penalty, or request loss) when the system is in state s and decision δ is taken. Analogous considerations apply to the other vectors. Using the above vectors, we can express the expected value of a cost metric in period n, when decision δ_n is taken, given that the system is in state s in period 0: it is given by the sth element of the vector $P_\pi^n d_{\delta_n}$.

When performing policy optimization, we want to maximize the average performance level over a long time period while keeping the average consumption and request loss below some levels specified by the user. This can be formally expressed as follows:

$$\text{PO}: \min_\pi \lim_{N\to\infty} \frac{1}{N} \sum_{n=1}^{N} P_\pi^n d_{\delta_n}$$

$$\text{s.t.} \quad \lim_{N\to\infty} \frac{1}{N} \sum_{n=1}^{N} P_\pi^n c_{\delta_n} \le C_M$$

$$\lim_{N\to\infty} \frac{1}{N} \sum_{n=1}^{N} P_\pi^n b_{\delta_n} \le B_M$$

the objective function is an S-dimensional vector. Its sth element is the average (over N time periods) of the expected performance level per period, given that the system starts in state s in period 0, and C_M and B_M are the maximum expected consumption per period and the maximum expected request loss per period respectively (as specified by the user). Strictly speaking, we should solve S optimization problems [4], one for each initial state. Moreover, the optimization is carried the set of *all* possible policies. Hence, solving PO appears to be a formidable task. Fortunately, the this is not the case, thanks to the following result of general validity.

Theorem *[4, 7] When the constraints in PO are active and the Markov process is unichain (i.e., for each policy there is a single communicating class), there exists a unique optimal policy for all the optimization problems of PO, which is stationary, Markovian (and possibly randomized). Furthermore, if constraints are inactive, the policy is deterministic.*

The above theorem has three implications. First, and most importantly, under our modeling assumptions the optimal policies are stationary. This result rules out policies based on statistics taken in periods before the period during which the decision is taken. Second, a stationary policy can be described and implemented in a compact fashion, because no time dependence must be taken into account, and decisions at any time n can be taken only by observing the current state of the system. Third, the optimal policy can be efficiently computed. Several algorithms are available to solve PO when no constraint is active. For constrained problems, it is possible to solve PO by solving a linear programming problem, which is solvable in polynomial time in the number of states. Hordijk and Kallenberg [7], and references therein, describe a solution procedure. The solution procedure is not outlined here because of space limitation. It is however important to mention that the matrix M_π describing the optimum policy that solves PO can be extracted from the solution of the optimization problem in a straightforward fashion. If M_π is available, the decision procedure implemented by the PM is the following: i) observe the state s_n of the system (SP, SR and queue) at time n, and ii) select the command to be issued with probabilities given by the $s(n)^{th}$ row of matrix M_π. For instance, the optimum policy for the system in Figure 1 is represented by a 8×2 matrix M_π. Assume that the row of M_π corresponding to state $s = (on, 0, 0)$ is $m_{s,\cdot} = (0.12, 0.88)$. If the system is in state s, the PM will issue the *s_on* command with probability 0.12 or the *s_off* command with probability 0.88. Concluding the section, we want to stress the fundamental assumptions that allows us to compute optimal policies in polynomial time. First, both SR and SP must be modeled by Markov chains with known parameters; also the cost metrics must be known. Second, the time horizon (i.e., the number of time intervals over which the policy is valid) must be long enough to be a good approximation of an infinite horizon. These assumptions will be checked in the next section for a realistic case study.

State	ΔT	$Power$
active	NA	$2.5W$
idle	$1ms$	$1W$
idleLP	$40ms$	$0.8W$
standby	$2.2sec$	$0.3W$
sleep	$6sec$	$0.1W$

Table 1: State, transition time to active and power dissipation for a hard disk driver

3 Case Study

We tested the validity of our approach by calculating optimal policies for a commercially-available hard disk drive [13] and contrasting them against heuristic policies. The hard disk drive can be operated in 5 different energy states, as shown in Table 1. In four of the five states, the device cannot perform data reads or writes, hence they are all *inactive* states. The inactive states span the tradeoff between power consumption and time required to return to the fully operational *active* state.

More in detail, in the *idle* state the disk is spinning, but some of the electronic components of the drive are turned off. The transition from *idle* to *active* is extremely fast, but only marginal power is saved in the *idle* state. The *low-power idle* state is similar to the *idle* state, but it has decreased power dissipation (and increased transition time to the *active* state). In the *standby* and *sleep* state, the disk is spun down, hence the transition to the *active* state is not only slow, but it causes additional power consumption (the additional current absorbed by the motor to accelerate the disk). It is important to mention that the transition times of Table 1 are explicitly declared as *typical* in the data sheets. In other works, they can be interpreted as expected values of random variables. The simplified transition graph of the service provider that models the disk drive is shown in Figure 2 (a). Several transitions between inactive states (and a few transient states) have been omitted for the sake of readability. The figure shows only the transitions from and to the *active* state, which have a major impact on power and performance. Assume that the disk is in one of the inactive states and a file read or write request is received. The request must be enqueued (in our model, the queue has length two) because the SP is inactive and cannot serve it right away. The PM observes the state of the system

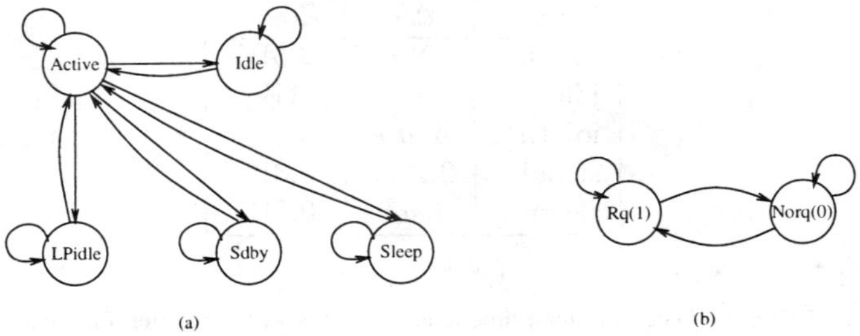

Figure 2: (a) Simplified transition graph for the disk drive. (b) Transition graph for the workload model

and issues a command for transitioning the resource to the active state.

The exit from the inactive states upon assertion of a command from the PM is not instantaneous. Our Markov model has been set up so that the expected exit time from an inactive state when a wakeup command has been issued is equal to the (experimental) average exit time from that inactive state (Table 1). Hence, the self-loops of the idle states have widely varying conditional proabilities. For example the conditional probability of the self-loop from *idle* when a wakeup command has been issued is much smaller than the conditional probability of the self-loop from *sleep* when the same command has been issued. The probabilities on the self-loops of the inactive states model the "inertia" of the system in waking up. Similarly, the self-loop on the active state models the inertia of the system in going to the inactive states. Its conditional probability depends on the command issued by the PM.

The model of the workload source can be automatically extracted from traces of disk accesses (we used the traces provided in [1]). The Markov model of the SR is extracted from usage traces with a modeling procedure that will be only briefly described because of space limitations. Given a time resolution T (proportional to the shortest wakeup time of the SP), the arrival times of the disk access requests are discretized (i.e., the trace is converted into a binary stream that has value 1 in position k if a request is received between time kT and time $(k+1)T$, zero otherwise). For the time resolution and the usage traces of this case, a Markov chain with two states as shown in Figure 2 (b) was sufficient to characterize the system. stream. At any time, the PM can issue a command to the SP. In our case study, the manager can chose among 5 commands: GO_ACTIVE, GO_IDLE, GO_IDLELP, GO_STBY, GO_SLEEP. The final product of the policy optimization is a matrix with five columns (one for each command) and as many rows as the number of states of the SP, SR and queue system. The complete model of the system (which is slightly more complex than the simplified version described here) has a total of 48 states. The power cost metric is obtained from the data sheets of the disk drive (summarized in Table 1). The performance metric is defined by assigning a cost to each state of the system where the SP is inactive and the queue is full. Roughly speaking, our metric represents a performance *penalty* which is negative and worsens as the average waiting time for a request increases. Finally the request loss metric is defined by assigning a cost to the state of the system where the queue is full and a new request arrives. Notice that request loss in our abstract models does not imply that a read or write request is actually lost (i.e., the system malfunctions). It simply represents an undesirable operating condition where the operating system controlling the drive is required to do extra work to guarantee correct service because the drive's response is very slow. The tool for policy optimization is built around an advanced LP solver based on an interior point algorithm [3]. We implemented an optimization shell that can compute trade-off curves between performance and power. Each point on the curve is a solution of a PO problem with different constraints (i.e., a *Pareto-efficient point*). Additionally, we implemented a simulation tool for validation of the PO results. The simulator can compute the power consumption, performance degradation and request loss of the system by simulating actual usage traces as they are processed by the SP controlled by the policy computed by the policy optimization tool. The cost metrics obtained by simulation should match the values computed by the optimizer. It is important to remember that the policies computed by the optimizer are global optimum ones. In other words, the only sources of error in our procedure come from the modeling process. Simulation is important to assess the quality of the modeling assumptions, and not the quality of the optimization. We tested the two fundamental modeling assumptions, namely the infinite-horizon assumption and the Markov assumption for the SR. Moreover, we compared the optimal policies with heuristic ones. The results of our experiments are shown in Figure 3(a).

The continuous line is the trade-off curve spanned by the optimal policies computed by the optimizer. Its computation took less than 1 minute on a SUN UltraSPARC workstation. The curve is obtained by maximizing performance with varying power constraint and fixed request loss (smaller than 50%). Notice how performance is traded off for power up to a minimum point where power cannot be reduced further because the constraint on request loss is violated. The circles in Figure 3(a) represent the results of simulation of the policies computed by the optimizer with the actual trace and for a finite time (10^6 time slots). The distance of the circles from the curve is a measure of the inaccuracies of the modeling process. It is visually obvious that the inaccuracies are very small, and that the simulated points lie almost perfectly on the tradeoff curve.

The triangles in Figure 3(a) represent heuristic solutions to the PO problem. The downwards triangles represent deterministic policies based on timeouts. Timeout-based policies are widely used for disk power management [5]. They are based on a simple heuristic that shuts down the disk when it has been inactive for a time longer than the timeout T_{to}. The choice of T_{to} is based on simulations and on designer's experience. The upwards triangles are randomized policies where the timeout value and the inactive state are chosen randomly with a given probability distribution. The randomized policies are the heuristic version of the optimal policies computed by our tool.

Although we cannot claim that our heuristic policies are the best that any experienced designer can formulate, some of our policies perform nearly as well as the optimum ones. The important point to notice is that it is much more difficult to control the power and performance produced by the heuristic policies. For example, we could not generate valid policies in the leftmost part of the power vs. performance plane, because the constraint on request loss was always violated. Moreover, it is possible to produce heuristic policies that produce "reasonable" results (such as those around power dissipation of $2.0W$), but there is no way for the designer to estimate if the results can be improved.

In Figure 3(b) we study the dependency of the policies by the workload statistics. The tradeoff curve computed with the Markov model of our usage trace is shown in continuous line. A new tradeoff curve computed with another SR model (i.e., a Markov chain with differ-

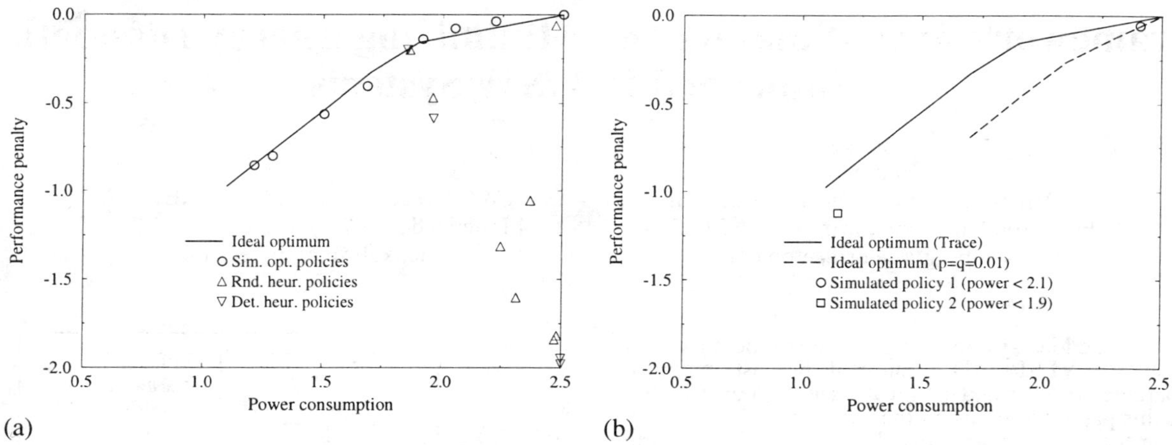

Figure 3: (a) Power consumption versus performance penalty for optimal and heuristic policies. (b) Dependency of optimal policies from SR model.

ent transition probabilities) is shown in dashed line. Clearly, the optimal policies are sensitive to the SR model. To further stress this point, we simulated two optimal policies computed with the second SR model by applying the original trace. The figure shows that both policies perform suboptimally (they are both below the continuous line). Moreover, they provide average power consumptions completely different from those used as optimization constraints. This experiment points out the need of an accurate SR model for obtaining high-quality policies.

4 Conclusion and future work

The identification of optimal power management policies for low-power system is a critical issue that has been addressed using common sense and heuristic solutions. In this work we provided a mathematical framework for the formulation and solution of the *policy optimization* problem. Our approach is based on a stochastic model of power-managed devices and workloads. The constrained policy optimization problem can be solved exactly in our modeling framework. Policy optimization can be cast into a linear programming problem and solved in polynomial time by efficient interior point algorithms. Moreover, tradeoff curves of power versus performance can be computed. The soundness of our modeling assumptions (and consequently the practicality of our power management policies) has been tested on a realistic case study. Our experimental results show that our stochastic model is robust and the optimal policies are flexible and power-efficient. Several extensions to this work are under way. First, we are planning to verify the quality of the results by implementing our power management policies in a portable PC and measuring their power and performance impact. Second, we are extending our model to deal with systems consisting of multiple interacting resources. Finally, we are studying adaptive algorithms that can compute optimal policies in systems where workloads are highly non-stationary and the service provider model changes over time.

Acknowledgements

This work was supported by NSF under contract MIP-9421129 and Toshiba Corp.

References

[1] Auspex File System Traces, available at *http://now.cs.berkeley.edu/Xfs/AuspexTraces/auspex.html* (1993).

[2] L. Benini and G. De Micheli, *Dynamic Power Management: Design Techniques and CAD Tools*, Kluwer (1997).

[3] J. Czyzyk, S. Mehrotra, and S. Wright, "PCx User Guide", *Technical Report OTC 96/01*, Optimization Technology Center, May, 1996.

[4] C. Derman, *Finite State Markov Decision Chains*, Academic Press, (1970).

[5] R. Golding, P. Bosh et al, "Idleness is not sloth", in *Proceedings of Winter USENIX Technical Conference*, pp.201-212 (1995).

[6] D. Gross and C. M. Harris, *Fundamentals of Queueing Theory*, Wiley (1985).

[7] A. Hordijk, and L. C. M. Kallenberg, "Constrained Undiscounted Stochastic Dynamic Programming", *Mathematics of Operations Research*, Vol. 2, pp. 276-289 (1984).

[8] C.-H. Hwang and A. C.-H. Wu, "A Predictive System Shutdown Method for Energy Saving of Event-Driven Computation", in *Proceedings of the ICCAD*, pp. 28-32 (1997).

[9] Intel, Microsoft and Toshiba, "Advanced Configuration and Power Interface specification", available at *http://www.intel.com/ial/powermgm/specs.html* (1996).

[10] Microsoft, "OnNow: the evolution of the PC platform", available at *http://www.microsoft.com/hwdev/pcfuture/ONNOW.HTM* (1997).

[11] W. Nebel and J. Mermet (Eds.), *Low power design in deep submicron electronics*, Kluwer (1997).

[12] M. Srivastava, A. Chandrakasan and R. Brodersen, "Predictive system shutdown and other architectural techniques or energy efficient programmable computation", *IEEE Transactions on Very Large Scale Integration Systems* vol. 4, no. 1, pp. 42-55, March 1996.

[13] Technical specifications of hard drive IBM Travelstar VP 2.5-inch, available at *http://www.storage.ibm.com/storage/oem/data/travvp.htm* (1996).

[14] S. Udani and J. Smith, "The power broker: intelligent power management for mobile computing", *Technical report MS-CIS-96-12, Dept. of Computer Information Science, University of Pennsylvania* (1996).

A Framework for Estimating and Minimizing Energy Dissipation of Embedded HW/SW Systems

Yanbing Li
Department of Electrical Engineering
Princeton University, Princeton, NJ 08544.
yanbing@ee.princeton.edu

Jörg Henkel
C&C Research Laboratories, NEC USA
4 Independence Way, Princeton, NJ 08540
henkel@ccrl.nj.nec.com

Abstract Embedded system design is one of the most challenging tasks in VLSI CAD because of the vast amount of system parameters to fix and the great variety of constraints to meet. In this paper we focus on the constraint of low energy dissipation, an indispensable peculiarity of embedded mobile computing systems. We present the first comprehensive framework that *simultaneously* evaluates the tradeoffs of energy dissipations of software and hardware such as caches and main memory. Unlike previous work in low power research which focused only on software or hardware, our framework optimizes system parameters to minimize energy dissipation of the overall system. The trade-off between system performance and energy dissipation is also explored. Experimental results show that our *Avalanche* framework can drastically reduce system energy dissipation.

1 Introduction

The design of embedded systems is a challenging task for today's VLSI CAD environments. As opposed to a general purpose computing system, an embedded system performs just *one* particular application that is known a priori. Therefore, the system can be designed with respect to the particular application to have lower cost, higher performance, or be more energy-efficient. Energy efficiency is a hot topic in embedded system design. As mobile computing systems (e.g. cellular phones, laptop computers, video cams, etc.) become more popular, how to length the battery life of these systems becomes a critical issue.

From the design process point of view, many of the embedded systems can be integrated on just one chip (*systems on a chip*) using core based design techniques. Previous work in core-based system design has mainly focused on performance and cost constraints. Some recent work has been presented in co-synthesis for low power [1, 2]. However, the *trade-off* in energy dissipation among software [1], memory and hardware has not yet been explored. This is a challenging and indispensable task for the design of low power embedded systems. Consider for example, that the use of a bigger cache can reduce the number of cache misses and speed up the software execution, which may cause less energy dissipation on the processor. On the other hand, a larger cache size also causes bigger switching capacitance for cache accesses and therefore increases the cache energy dissipation per access.

In this paper we present our framework *Avalanche*, the first framework that explores the design space of hardware/software systems in terms of overall system energy dissipation. Since embedded system design usually has multiple constraints such as performance and power, our framework

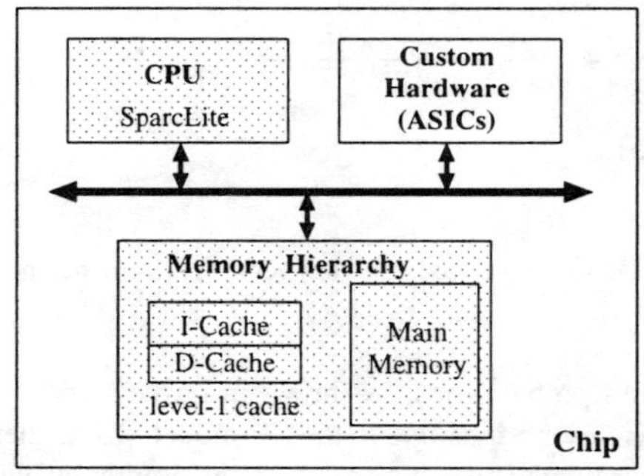

Figure 1: Target architecture of an embedded system

evaluates performance as well and optimizes for the best energy-performance trade-off.

This paper is structured as follows: Sec.2 reviews some of the related work in energy estimation and optimization for embedded systems. Sec.3 describes our model for embedded system energy dissipation. In Sec.4 we present our approach for energy dissipation optimization under timing constraints, and energy and performance trade-off optimization. Experimental results are presented in Sec.5.

2 Related Research

Energy estimation and optimization has been studied for both software and hardware. Tiwari and Malik [3] investigated the energy dissipation during the execution of programs running on different processor cores. Ong and Ynn [4] showed that the energy dissipation may drastically vary depending on the algorithms running on a dedicated hardware. A power and performance simulation tool for a RISC design has been developed by Sato et al. [5]. Their tool can be used to conduct architecture-level optimizations.

Further work deals with energy dissipation from a hardware point of view. Gonzales and Horowitz [6] explored the energy dissipation of different processor architectures (pipelined, un-pipelined, super-scalar). Kamble and Ghose [7] analyzed cache energy consumption. Itoh et al. studied SRAM and DRAM energy dissipation and low power RAM design techniques [14]. Panda et al [8] presented a strategy for exploring on-chip memory architecture in embedded systems with respect to performance only. Optimizing energy dissipation by means of high-level transformations has been addressed by Potkonjak et al. [9].

While estimating or optimizing power, these previous work only focuses on one component of the system at a time. A comprehensive approach that takes into consideration the mutual impacts of software *and* hardware in terms of energy dissipation — as it is actually the case in an embedded hardware/software system — has not been addressed so far.

3 System Model and Design Flow

In this section we present our *energy estimation model* of an embedded *system–on–a–chip*. It is based on an architecture template shown in Fig.1, which comprises a processor

[1] We use the term *software energy dissipation* for the energy that is dissipated within a processor core.

DAC 98, San Francisco, California

core, an instruction cache, a data cache, a main memory, and a custom hardware part (ASICs). We assume that hardware/software partitioning has already been performed and the custom hardware is fixed, therefore it adds a *constant* amount of energy to our model. During the design space exploration, we change the software and the cache/memory part, by performing high-level transformations on software and changing the cache and/or main memory parameters. When either of these components changes, the energy dissipation of other components is influenced and so is the overall system energy.

3.1 Analytical Cache Memory Model

We deploy a cache energy model based on transistor-level analysis. The model consists of an input decoder, a tag array and a data array. Attached to the tag array are column multiplexers whereas data output drivers are attached to the data array. A SRAM cell in data and tag array comprises six CMOS transistors. The switching capacitances in the equations derived below, are obtained by the tool *cacti* [10].

Only the energy portions in the bit lines for read and write ($E_{bit,rd}$ and $E_{bit,wr}$), in the word lines ($E_{word,rd/wr}$), in the decoder (E_{dec}) and in the output drivers (E_{od}) contribute essentially to the total energy. The according effective capacitances are:

$$\begin{aligned} C_{bit,rd} &= N_{bitl} \cdot N_{rows} \cdot (C_{SRAM,pr} + C_{SRAM,rd}) \\ &\quad + N_{cols} \cdot C_{pr\_logic} \end{aligned} \tag{1}$$

where $C_{SRAM,pr}$, $C_{SRAM,rd}$ and C_{pr_logic} are the capacitances of the SRAM cell affected by precharging and discharging and the capacitance of the precharge logic itself, respectively. N_{rows} is the number of rows (number of sets) in the cache. The number of bit lines is given by N_{bitl}:

$$\begin{aligned} N_{bitl} &= (T \cdot m + St + 8 \cdot L \cdot m) \cdot 2 \\ N_{cols} &= m \cdot (8 \cdot L + T + St) \end{aligned}$$

where m means a m–way set associative cache, L is the line size in bytes, T is the number of tag bits and St is the number of status bits in a block frame. $C_{bit,wr}$ is defined in a similar manner as $C_{bit,rd}$.

The effective wordline capacitance is given by:

$$C_{word} = N_{cols} \cdot C_{word,gate} \tag{2}$$

where $C_{word,gate}$ is the sum of the two gate capacitances of the transmission gates in the 6–transistor SRAM cell. For simplification, we do not include the equations for C_{dec} and C_{od} here. Apparently, the switched capacitance is directly related to the cache parameters (Eq. 2).

Finally, the total energy dissipated within the cache (i-cache or d-cache) during the execution of a software program is related to the number of total cache accesses N_{acc}, as well as the number of hits and misses for cache reads and writes:

$$\begin{aligned} E_c &= 0.5 \cdot V_{DD}^2 (N_{acc} \cdot C_{bit,rd} + N_{acc} \cdot C_{word} \\ &\quad + a \cdot C_{bit,wr} + b \cdot C_{dec} + c \cdot C_{od}) \end{aligned} \tag{3}$$

where a, b and c are complex expressions that depend on read/write accesses and, in parts on statistical assumptions. $a \cdot C_{bit,write}$, $b \cdot C_{dec}$ and $c \cdot C_{od}$ [2] are the effective capacitances to switch when writing one bit, during decoding of an access and during output, respectively.

The implemented cache model has a very high accuracy (compared to the real hardware) since every switching transistor within the cache has been taken into consideration (even if this is not transparent through our equations because of the simplification). All the capacitances are obtained by running *cacti* [10] and are derived for a $0.8\mu m$ CMOS technology. The calculation of the capacitances within *cacti* has been proofed against a *Spice* simulation.

3.2 Main Memory Energy Model

For energy analysis of the main memory, we use the model for DRAM described by Itoh, *et al.* [14]. The energy source for DRAM mainly includes: the RAM array, the column decoder, the row decoder and peripherals.

$$I_a = m \cdot i_{act} + m(n-1) \cdot i_{hld} + m \cdot i_{dec} + n \cdot i_{dec} + I_{peri} \tag{4}$$

Eq.4 shows the current drawn during each memory access. Note that during each access, m cells are selected. $m \cdot i_{act}$ is the active current of the m selected cells. $m(n-1) \cdot i_{hld}$ is the data retention current of the $m \cdot (n-1)$ cells that are not selected. $m \cdot i_{dec}$ and $n \cdot i_{dec}$ are the currents drawn on column and row decoder, respectively. I_{peri} represents the current on peripheral circuits. The equations show that energy dissipation of each memory access is directly related to the size of the memory. For the total energy dissipation, i_{active} is the dominating component. At high clock frequencies, i_{hld} is negligible [14].

3.3 Software Energy and Performance Model

For software energy estimation we deploy a behavioral simulator ([16]) that we enhanced by values of the current drawn during the execution of an instruction. Those current values are obtained from [12]. The total SW program energy is:

$$\begin{aligned} E_{prg} &= T_{w\_c} \cdot V_{DD} \cdot \sum_{i=0}^{N-1} (I_{instr,i} \cdot N_{cyc,i}) + \\ &\quad T_{cyc} \cdot V_{DD} \cdot (\underbrace{N_{miss,rd} \cdot N_{cyc,rd\_pen} \cdot I_{instr,nop}}_{\text{data write miss penalty}} + \\ &\quad \underbrace{N_{miss,wr} \cdot N_{cyc,wr\_pen} \cdot I_{instr,nop}}_{\text{data write miss penalty}} + \\ &\quad \underbrace{N_{miss,fetch} \cdot N_{cyc,fet\_pen} \cdot I_{instr,nop}}_{\text{instruction fetch miss penalty}}) \end{aligned} \tag{5}$$

where V_{DD} is the voltage supply, I_{instr} is the current that is drawn during the execution of instruction i at the processor pins, $N_{cyc,i}$ is the number of cycles the instruction needs for execution and N is the total number of instructions of the program. T_{w_c} is the execution time of the application assumed that there is a cache as specified.

The three additional portions within the brackets refer to the energy dissipated in the penalty cycles when occurs a data cache write miss, a data read miss and an instruction fetch miss, respectively. We assume that the energy dissipated within processor is negligible after the program has been executed (through gated clock).

Let T_{w/o_c} be the execution time of a program running on the processor core (simulated by a behavior compiler) without cache, the corrected execution time (i.e. including cache behavior) is estimated by:

$$\begin{aligned} T_{w\_c} &= T_{w/o\_c} + T_{cyc} \cdot (N_{miss,rd} \cdot N_{cyc,rd\_pen} + N_{miss,wr} \cdot \\ &\quad N_{cyc,wr\_pen} + N_{miss,fet} \cdot N_{cyc,fet\_pen}) \end{aligned} \tag{6}$$

[2] The capacitances of the output drivers are derived for an on-chip cache implementation i.e. we assume that all resources like processor, cache and main memory are implemented on just one chip.

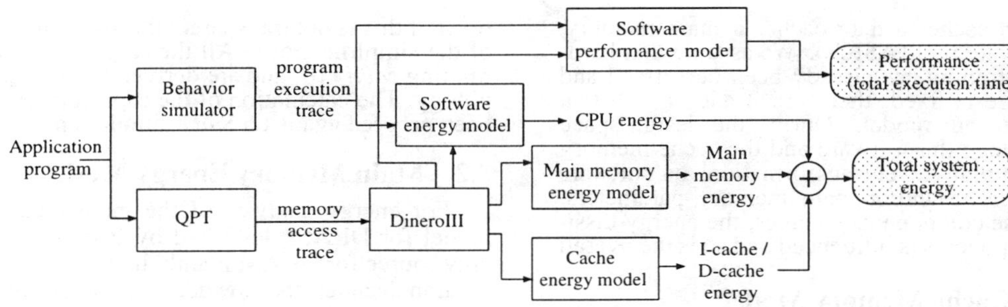

Figure 2: Design flow of the estimation part of our *Avalanche* framework

3.4 Design Flow of Our Framework

Using the above energy models and timing models, the estimation design flow (the energy optimization part is not shown) of our framework is shown in Fig.2. The input is an application program. It is fed into a behavioral model of the target processor that simulates the program and delivers a program trace to the *software energy model* and the *software performance model*. At the mean time, the input program is also fed into the memory trace profiler *QPT* [13] which generates the memory access trace to be used by *Dinero*[13]. *Dinero* provides the number of demand fetches and demand misses (for data and instructions). These numbers are then used: by the *software performance model* to get the total execution time with cache miss penalty considered (Eq.6); by the *software energy model* to adjust the software energy with the stalls caused by cache misses (Eq.5); and by the *cache* and *main memory energy models* (Eq.3 and Eq.4) to calculate the energy dissipation by the memory components based on the actual number of instruction/data cache accesses and main memory accesses.

4 System-level Energy Optimization

To optimize the system energy, we explore the design space in the dimensions of software and cache/memory. As mentioned in Sec.3, our framework assumes that the hardware (ASIC) is fixed. It changes the software by performing various high-level transformations. It changes the cache/main memory by modifying their parameters such as size, associativity, etc. When one component (software, cache or memory) changes, it not only affects the energy consumption of itself, but also that of other components in the system; it not only affects the power, but also the performance. The interesting aspect is that the change of overall system energy and performance can not be easily predicted unless comprehensive system analysis is performed. We now discuss some scenarios of software and cache/memory changes and their possible impacts on energy and performance:

- **Software transformation:** suppose a transformation can be performed on the software to lower the *software energy*. However, this transformation may change the cache/main memory access pattern and result in ambiguous changes of the caches or main memory energy and the performance. In some cases, software transformations may increase the code size so that a larger main memory is required to accommodate the new program; therefore, the energy of each memory access increases.
- **Cache:** when a larger instruction and/or data cache is used, in general, there are less cache misses and the *system performance* is improved. The *software energy* decreases because less cache misses imply less main memory access penalties. The energy of the main memory is decreasing because of less accesses. However, the energy dissipated by the caches increases due to its increased size, and the system energy change is ambiguous.

```
test1(...)                                          test2()
{ ... ...                                           { ... ... }
  test2();    /*call 3*/
  ... ... }
main()
{ int i, j;
  ... ...
  for (i=0; i<100; i++) {          /* loop 1 */
    test1(...);                    /* A */
    for(j=0; i<100; j++)           /* loop 2 */
       test1(...);                 /* B */
  }
  ... ... }
```

Figure 3: Program example for software transformations.

- **Main memory:** When a bigger main memory is used, the energy dissipation of the main memory increases because of its larger size (Eq. 4), but the energy of other parts is usually not affected.

4.1 Software Transformation and Energy

Many source-level transformations have been proposed for the purpose of improving performance. However, they may have some side-effects other than performance improvement, such as a bigger main memory requirement due to increased code size. This will lead to larger energy dissipation due to larger capacitances to switch for each access. Here we have a brief look at some of the commonly used transformation techniques and analyze their impacts on energy and performance.

Procedure calls are costly in most architectures. *Procedure in-lining* can help improve performance and save software energy by eliminating the overhead associated with calls and returns. For example, suppose we have a SPARC architecture that features up to 8 register windows. For each new procedure call a new window is required and released after the return from the procedure. However, if the depth of procedure calls (i.e. a consecutive number of calls without returns) exceeds the available number of register windows, an interrupt is released for the operating system to process the spilling of register contents to the main memory. This is time consuming. A side effect of in-lining is the increased code size, especially when the procedure is called from different points within the program.

Loop unrolling is another transformation technique. It can help to increase the instruction level parallelism and eliminate control overhead. Similar to procedure in-lining, it also results in code size increase. Another possible impact is that an unrolled loop may no longer fit in the instruction cache so that it possibly will be slowed down. Other techniques include *software pipelining, recursion elimination, loop optimization,* etc. [15], whose impacts on both the software and cache/memory accesses may make it hard to judge the change of the overall system energy dissipation.

4.2 Software Transformation Selection Algorithm

When a designer is concerned about both performance and power, a sophisticated approach is mandatory to choose which transformations to perform, and in what order. In order to find the combination and sequences of transformations that yield the most energy savings under memory size constraints, we designed a *transformation-selection* algorithm. Given a

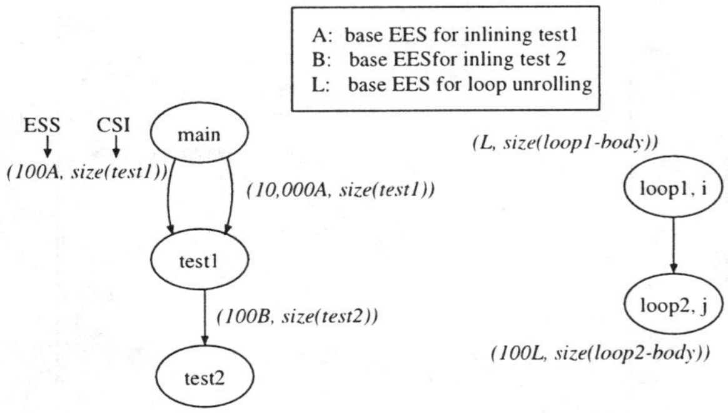

(a) Procedure call graph (b) Loop nesting graph

Figure 4: Procedure calling graph and loop graph of the program example.

set of available transformations techniques, the algorithm needs to:

1. identify *which* transformations can be applied and *where*, and evaluate these choices of transformations.
2. choose the combination and the order of the transformations that obtain the best energy improvement without violating a memory size limit.

Currently, we have implemented procedure in-lining and loop unrolling using SUIF [11]. However, our *transformation-selection* algorithm is applicable to general types of transformations as long as their particular characteristics are defined (see later). The algorithm is independent of the transformations themselves.

In the first step of the algorithm, to evaluate the impacts of the transformations, we developed some heuristic measures to characterize the estimated-energy-saving (EES) and code size increase (CSI) incurred by these transformations. EES is the estimated energy improvement while performing a certain transformation. Fig.3 shows a code segment that contains two loops and three procedure calls. The EES for inlining the procedure *test1* at location *A* is 100 times the base EES of inlining *test1*. At location *B*, EES is 10,000 times the base EES. Similar considerations apply to loop unrolling.

To identify the calling relationships, a *procedure calling graph* is constructed (Fig.4) for each program. In the calling graph, a node represents a procedure and a directed edge represents a procedure call. Multiple edges between nodes may exist, reflecting that a procedure can be called from different locations. The edges have been assigned the attributes *EES* and *CSI*. Since our algorithm does not support recursion, the procedure calling graph is *acyclic*. After inlining has been applied, the edge corresponding to the call is removed. For loop unrolling, a similar graph is created, in which a node represents a loop, an edge indicates one loop is nested in another. However, unlike the procedure calling graph, the nodes are labeled instead of the edges because the nodes are where the transformations are applied to.

Note that the possible transformations are not independent of each other. For the example in Fig.3: if *loop 1* is unrolled, 100 new instances of *test1* calls will be generated and new calling edges in the procedure calling graph need to be added. It is important for the algorithm to not only choose the best combination of the transformations, but also the right order.

In the second step, the algorithm, 1) prioritizes all possible transformations according to a heuristic measure — the EES/CSI ratio; 2) a probability is assigned to each transformation according to its priority value; 3) in each transformation step, randomly select a transformation based on the probabilities, perform the transformation, and update the procedure and loop graphs; 4) repeat 3) until the memory limit is reached. This algorithm is called repeatedly by the system-level energy optimization algorithm (Sec.4.3).

```
Inputs: source_software, design_goal;
Variable: solution_pool, current_sw, new_sw,
     current_design, new_design, tmp_design;
1.  Static analysis of program:
2.    identify all possible transformations;
3.    construct procedure graphs / loop graphs;
4.    generate possible cache / memory sizes;
5.  Energy optimization:
6.    for each memory size m_size {
7.      current_sw = source_sw;
8.      current_design = best design with
              current_sw from solution_pool;
9.      do{
10.       new_sw = transformation_select
                  (current_sw, m_size);
11.       new_design = (new_sw,0,0,m_size);
12.       A=set of i_cache/d_cache sizes for new_sw;
13.       for each (dcache, icache) in A {
14.          tmp_design = (new_sw,d_cache,
                     i_cache,m_size);
15.          new_design = choose one by design_goal
                 (tmp_design, new_design);
16.       }
17.       if new_design better than current_design{
18.          save new_design in solution_pool;
19.          current_sw = new_sw;
             current_design= new_design; }
20.       else continue;
21.     } while( !stop_condition)
22.   }
23.   output: designs from solution_pool
              that satisfies design_goal.
```

Figure 5: System-level energy optimization algorithm.

4.3 System-Level Energy Optimization Algorithm

We now formally define the problem of our system-level energy optimization algorithm. We assume that:
1. Hardware/software partitioning has already been done and application specific hardware is synthesized and therefore fixed.
2. A processor has been chosen.
3. We are given an initial version of the software.
4. The user specifies one *optimization goal*. The algorithm is designed for minimizing energy. However, as power is usually not the sole concern in the design process, we allow three different optimization goals:

- *Goal I*: minimized power.
- *Goal II*: minimized power under performance constraints.
- *Goal III*: multiple objective optimization

Goal III is to find a set of solutions within performance *and* energy constraints. This will provide important trade-off information to the designer. The designer can review different design options and choose the most suitable one. For example, there are two designs A and B, with design A being 20% faster than design B, but Design B consuming just 1% less energy. They both meet the performance constraints. *Goal I and II* will discard A, although it might be a better choice for the designer.

The algorithm returns the optimized new system configuration of the target system architecture (Fig.1): the transformed program, the data cache and instruction cache sizes and other parameters, and the main memory size. The energy dissipation of each component and the performance is also delivered. For *Goal-III*, a set of designs is returned, with percentage data indicating energy and performance difference between two designs adjacent in terms of energy dissipation.

Fig.5 shows the pseudo-code for the optimization algorithm. It consists of two main steps:

1. **Static analysis of the application program** (lines 1-4), includes
   - generating the procedure calling graph and loop graph, as described in Sec.4.2, and
   - generating the set of feasible cache and memory sizes and configurations based on the current version of the program.

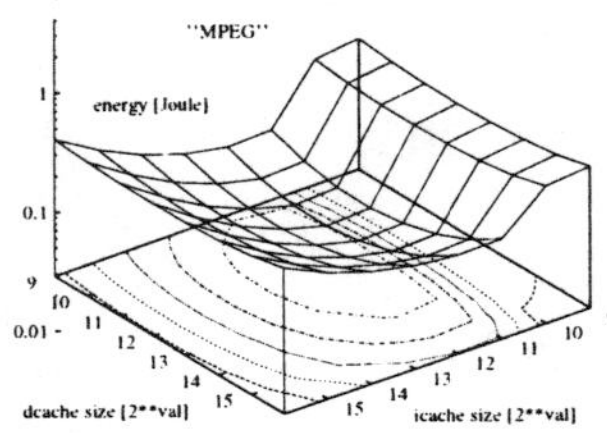

(a) MPEG: energy

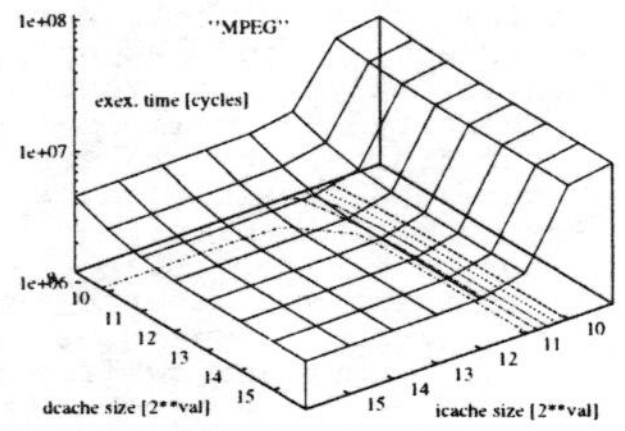

(b) MPEG: exec time

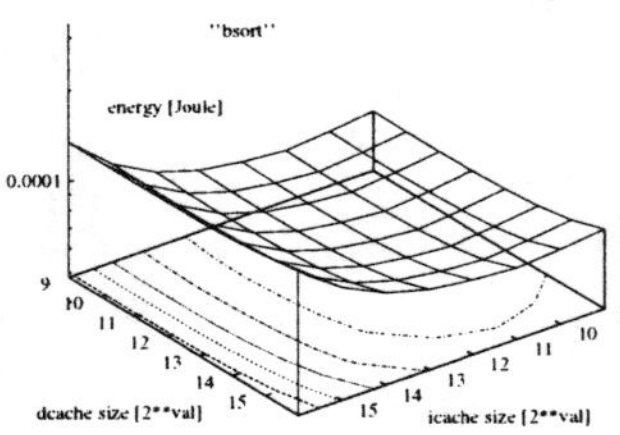

(c) bsort: energy

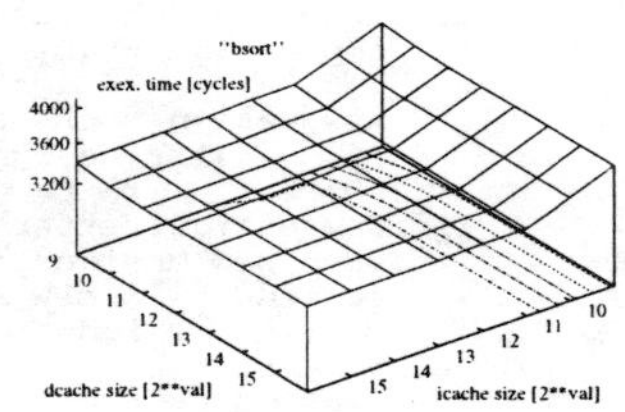

(d) bsort: exec time

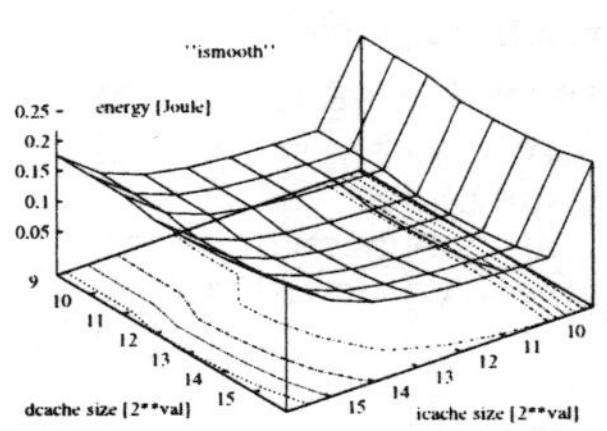

(e) i-smooth: energy

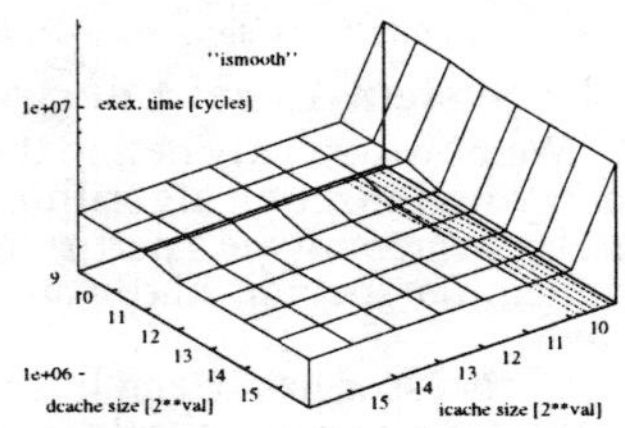

(f) i-smooth: exec time

Figure 6: Energy and execution time vs. instruction/data cache size, for applications *MPEG*, *bsort* and *ismooth*.

2. **Optimization step** (lines 6-23): choose the design, i.e. set of transformations and the cache and memory parameters, to meet the constraints and optimization goal.

In the algorithm, we limit the maximum memory size to four times of the original memory size (of the original, not transformed software program) because as shown by our experiments, the energy overhead of a very large memory usually out-weighs the energy saving provided by software transformations. We generate a set of designs for each possible memory size (lines 6-22), and select design(s) that meet the design goal (*I, II or III*) in line 23. A design is represented as a quadruple of software, instruction cache, data cache and main memory.

To construct designs for a certain memory size, we perform software transformations using the algorithm described in Sec. 4.2 (line 10), and then decide the subset of feasible instruction/data caches for the transformed software (lines 11-12). The best instruction/data caches are chosen based on the designer's goal (line 13-16). The transformed software, the best suited cache sizes and parameters and the new memory size makes up a new design.

If the new design has a better quality than the previous one, then it is saved in a solution pool (line 17-19) and will be used in the next iteration. Otherwise, the transformation is discarded (line 20) and a new transformation is performed on the previous version of the software. The process is repeated until a stop criteria is met (line 21): there is no improvement in a given number of consecutive iterations, or the total number of iterations reaches a preset limit.

An important issue in the algorithm is evaluating the quality of two designs. The evaluation depends on the design

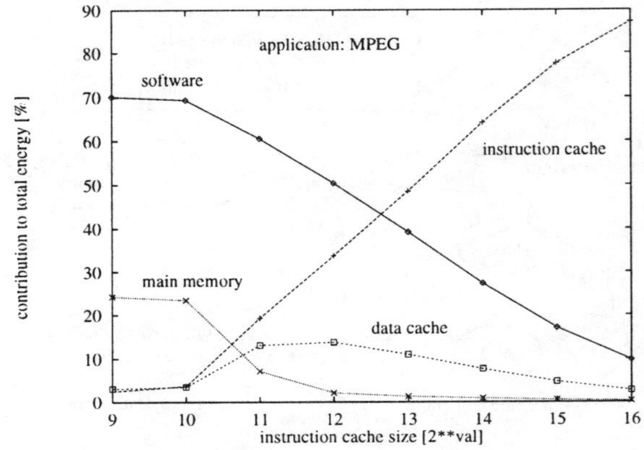

Figure 7: Contribution of software, d-cache, i-cache and main memory to total energy dissipation, at fixed data cache size (4k).

goal: for *minimizing energy* (*Goal I*), energy is the sole standard; for minimized energy under a performance constraints (*Goal II*), the performance constraint is to be met first then energy is considered; for *Goal III*, for designs falling within energy and performance constraints, we use the *Pareto optimality* measure to discard solutions that are both higher in energy and slower in performance. Note the energy and performance data are obtained using the models described in Sec.3.

5 Experiments and Results

As will be shown in this section, our framework can be deployed to support a system designer through a design space exploration or for automated optimization according to the designers goals (energy, performance). We used data-dominated applications (including an MPEG encoder consisting of about 200kB of C source code), in which caches and memory play a key role in energy and performance.

5.1 Design space explorations

Since the variety of system parameters to choose is very large, we fixed all parameters except for the sizes of data cache and instruction cache. Each application in Fig.6 has been dedicated two figures showing the total energy dissipation of the whole system (left) and the number of clock cycles to perform that application on our target architecture (right). The varied parameters in each figure are the data cache size (left axis) and the instruction cache size right axis). A number of 12 for example means a cache size of 4K ($= 2^{12}$) byte. Figures a) and b) show the results of the MPEG encoder. Here, both large and small cache sizes lead to a high system energy dissipation. In case of large caches, the caches' energy dominates the system energy; while in case of smaller cache sizes, the software energy is dominating — from Eq.5, large number of cache misses (due to small caches) result in poor performance and large energy consumed in the miss penalty cycles. The right figure shows in fact that the program execution time raises drastically for small instruction cache sizes (< 1024 byte). Remarkably, the least energy consuming configuration (data cache size and instruction cache size 4k each) is also one of those with the highest performance (i.e. small number of clock cycles). This is a behavior that would possibly not be expected (and is not the case for the other applications). In addition, Fig.7 reveals the contribution (in percent) of each component to whole system energy dissipation (i.e. software program, caches, main memory). In that figure, the data cache size has been fixed whereas the instruction cache size varies.

The experiments conducted with the *bsort* (c) and d)) application (bubble sort) show mainly that there is almost no dependency on data cache size in terms of system energy dissipation and system performance. More dependencies can be observed by changing the instruction cache size. Obviously, a large instruction cache size leads to a large system energy

appl.	objective	orig. architecture		optim arch. Goal I
		w/o cache	fixed cache	
bsort	energy [J]	0.33	0.30	23.21E-3
	energy improv. [%]	n/a	-9.09	-93.21
	time [# cyc $\times 10^6$]	19.4	17.6	1.4
	exec. improv. [%]	n/a	-9.52	-92.98
eg2	energy [J]	0.31	0.28	19.41E-3
	energy improv. [%]	n/a	-9.68	-93.73
	time [# cyc $\times 10^6$]	17.9	16.5	1,186.8
	exec. improv. [%]	n/a	-8.34	-93.39
ismooth	energy [J]	1.03	0.96	67.411E-3
	energy improv. [%]	n/a	-6.80	-93.45
	time [# cyc $\times 10^6$]	60.1	56.2	3.9
	exec. improv. [%]	n/a	-6.60	-93.46
itimp	energy [J]	2.97	2.78	186.01E-3
	energy improv. [%]	n/a	-6.40	-93.73
	time [# cyc $\times 10^6$]	173.8	162.5	8.8
	exec. improv. [%]	n/a	-6.50	-94.91

Table 1: Optimization of architecture through *GOAL I* compared to original architecture (no cache) and an achitecture with non–optimized cache size

dissipation also. But as opposed to the *MPEG* encoder, a small instruction cache size does not lead to a larger system energy dissipation as a consequence of a larger program execution time. Rather than that, the performance decreases (more cycles due to the mid-right figure on page 6).

An additional different behavior is shown by the application *ismooth*, (e) f)) an image smoothing application. As could be observed, the behavior of a system in terms of energy and performance is hard to predict and therefore needs powerful tools for analyzing and optimizing.

5.2 Optimizing system-level energy dissipation

In Sec.4.3 we have presented our algorithms for three different design goals. For all following experiments we have chosen the clock cycle time to 30ns. The behavior simulation tool for evaluating software performance is a SPARC simulator [16]. The same conventions were used for the experiments in Sec.5.1. All other system parameters are subject to change through the optimization process.

Table.1 shows the results for *Goal I* compared to two original architectures called *w/o cache* (no cache) and the architecture called *fixed cache* (a small standard cache; same size for i-cache and d-cache). For all applications, system energy dissipation (Joule) and execution time (number of clock cycles) is given. Additionally, the relative improvement $(value - value_{ref})/value_{ref} * 100$ for both objectives is given. Apparently, a negative percentage number means an improvement.

As shown in Table.1, *Goal I* yields remarkable improvements in energy and performance. The "fixed cache" architecture is somewhere between "w/o cache" and the optimized architecture. As investigations have shown, in parts, the drastic improvements in performance are due to the architecture: we assume a cache miss penalty of 20 clock cycles, which is a quite typical value. The contribution of software transformation techniques solely (as described in section 4.1) to energy and performance improvements varies in most of the shown cases between 5% and 10%.

Table.2 shows the results yielded with our algorithms for improving energy dissipation under performance constraints (*Goal II* and a multiple objective optimization *Goal III*). In the latter case, the designer is provided with a set of different solutions where he can choose from since design constraints are not completely defined every time. The algorithm for *Goal II* is searching the design space around a given performance constraints i.e. it searches for design configurations

appl.	objective	w/o cache	optimized architecture			
			GOAL II	GOAL III		
bsort	energy [J]	0.33E-3	22.4E-3	0.27	0.24	0.21
	e-improv. [%]	n/a	-93.2	-17.6	-26.8	-36.1
	time [# cyc $\times 10^6$]	19.4	1.4	15.9	14.1	12.3
	t-improv. [%]	n/a	-93.0	-18.4	-27.5	-36.8
eg2	energy [J]	0.31	19.4E-3	0.25	0.23	-
	e-improv. [%]	n/a	-93.7	-18.2	-24.8	-
	time [# cyc $\times 10^6$]	18.0	1.2	14.8	13.6	-
	t-improv. [%]	n/a	-93.4	-17.6	-24.3	-
ismooth	energy [J]	1.03	67.4E-3	0.10	0.72	0.48
	e-improv. [%]	n/a	-93.4	-90.1	-29.8	-53.5
	time [# cyc $\times 10^6$]	60.1	3.9	3.6	42.1	28.0
	t-improv. [%]	n/a	-93.5	-94.0	-30.0	-53.5
itimp	energy [J]	2.97	0.19	2.4	2.2	-
	e-improv. [%]	n/a	-93.7	-19.9	-25.3	-
	time [# cyc $\times 10^6$]	173.8	8.8	139.1	129.5	-
	t-improv. [%]	n/a	-94.9	-20.0	-25.5	-

Table 2: Optimization of architecture through *GOAL II* and *GOAL III* compared to original architecture that has no cache

with minimum energy dissipation while not exceeding the budget of clock cycles to execute. Here also, large improvements could be yielded as shown in Table.2.

The computation time for determining *one* design point (fixed system parameters) is in the range of 3-5 minutes. A whole optimization run is between 2 and 10 hours on an Ultra Sparc.

6 Conclusions

We have presented our *Avalanche* framework for estimating and optimizing the energy dissipation of embedded systems. It is the first approach that trades off the energy dissipation of software against the energy dissipation of system resources like caches and main memory. Through various experiments we have shown that it is not straightforward to judge the change of the total system energy when various system parameters are varied and software transformation are performed. Our *Avalanche* framework provides a powerful tool for low power design at system level. Experimental results have shown significant improvements (up to $\approx 95\%$ energy cut) in energy dissipation.

References

[1] D. Kirovski, M. Potkonjak, *System-Level Synthesis of Low-Power Hard Real-Time Systems*, Proc. DAC'97, pp.697-702, 1997.

[2] B.P. Dave, G. Lakshminarayana, N.K. Jha, *COSYN: Hardware-Software Co-Synthesis of Embedded Systems'* Proc. DAC'97, pp.703-708, 1997.

[3] V. Tiwari, S. Malik, A. Wolfe, *Instruction Level Power Analysis and Optimization of Software*, Kluwer Academic Publishers, Journal of VLSI Signal Processing, pp. 1–18, 1996.

[4] P.-W. Ong, R.-H. Ynn, *Power-Conscious Software Design – a framework for modeling software on hardware*, IEEE Proc. of Symp. on Low Power Electronics, pp. 36–37, 1994.

[5] T. Sato, M. Nagamatsu, H. Tago, *Power and Performance Simulator: ESP and its Application for 100 MIPS/W Class RISC Design*, IEEE Proc. of Symp. on Low Power Electronics, pp. 46–47, 1994.

[6] R. Gonzales, M. Horowitz, *Energy Dissipation in General Purpose Processors*, IEEE Proc. of Symp on Low Power Electronics, pp. 12–13, 1995.

[7] M.B. Kamble, K. Ghose, *Analytical Energy Dissipation Models For Low Power Caches*, IEEE Proc. of Symposium on Low Power Electronics and Design, pp. 143–148, 1997.

[8] P.R. Panda, N. D. Dutt, A. Nicolau, *Architectural Exploration and Optimization of Local Memory in Embedded Systems*, Proc. of IEEE International Symposium on System Synthesis, pp. 90–97, 1997.

[9] I. Hong, D. Kirovski, M. Potkonjak, *Potential-Driven Statistical Ordering of Transformations*, Proc. DAC'97, pp.347-352, 1997.

[10] S.J.E Wilton, N.P. Jouppi, *An Enhanced Access and Cycle Time Model for On-Chip Caches*, DEC, WRL Research Rep. 93/5, 1994.

[11] Stanford Compiler Group, *The SUIF Library: A set of core routines for manipulating SUIF data structures*, Stanford University, 1994.

[12] V. Tiwari, *Logic and system design for low power consumption*, PhD thesis, Princeton University, Nov. 1996.

[13] M. D. Hill, J. R. Laurus, A. R. Lebeck et al., *WARTS: Wisconsin Architectural Research Tool Set*, Computer Science Department University of Wisconsin.

[14] K. Itoh, K. Sasaki and Y. Nakagome, *Trends in Low-Power RAM Circuit Technologies*, Pro. of the IEEE, VOL. 83, No. 4, 1995.

[15] H. Mehta, R. Owens, M.J. Irwin, R. Chen and D. Ghosh, *Techniques for Low Power Software*, IEEE Proc. of Symposium on Low Power Electronics and Design, pp. 72–75, 1997.

[16] W. Ye, R. Ernst, Th. Benner, J. Henkel, *Fast Timing Analysis for Hardware-Software Co-Synthesis*, Proc. ICCD, pp.452–457, 1993.

Using Reconfigurable Computing Techniques to Accelerate Problems in the CAD Domain: A Case Study with Boolean Satisfiability

Peixin Zhong, Pranav Ashar, Sharad Malik and Margaret Martonosi
Princeton University and NEC CCRL
pzhong, sharad, mrm@ee.princeton.edu, ashar@ccrl.nj.nec.com

Abstract

The Boolean satisfiability problem lies at the core of several CAD applications, including automatic test pattern generation and logic synthesis. This paper describes and evaluates an approach for accelerating Boolean satisfiability using configurable hardware. Our approach harnesses the increasing speed and capacity of field-programmable gate arrays by tailoring the SAT-solver circuit to the particular formula being solved. This input-specific technique gets high performance due both to (i) a direct mapping of Boolean operations to logic gates, and (ii) large amounts of fine-grain parallelism in the implication processing. Overall, these strategies yields impressive speedups (>200X in many cases) compared to current software approaches, and they require only modest amounts of hardware. In a broader sense, this paper alerts the hardware design community to the increasing importance of input-specific designs, and documents their promise via a quantitative study of input-specific SAT solving.

1 Introduction

While hardware accelerators for CAD are not new, the increasing capacity and speed of field-programmable gate arrays (FPGAs) offers flexibility and computing power not available before. The hardware can be reused for different applications and can be easily updated to accommodate new developments in algorithms. This paper introduces an novel approach for accelerating the Boolean satisfiability (SAT) problem using configurable computing hardware. In our implementation of the SAT problem, we have harnessed this flexibility by implementing the SAT solver circuit on an "input-specific" basis. That is, a circuit is generated especially for each SAT formula to be solved. Since Boolean satisfiability approaches are logical-operation-intensive, the hardware approach can show large advantages by mapping the SAT expressions directly to logic gates, and by harnessing large amounts of parallelism in the evaluation of the logic.

While the bulk of the configurable computing applications currently implemented have been fairly simple data-centric systolic designs, our design shows the potential of custom computing on difficult problems with complex control structures. We have implemented a tree search algorithm, as opposed to the generally-simpler signal processing pipelines common to configurable computing thus far. In addition, we have also implemented a hardware version of nonchronological backtracking similar to that seen in software SAT solvers [11]. This requires complex control structures to manage the backtracking steps. These implementations demonstrate configurable computing's potential for impact on a much broader set of applications, particularly in CAD domain, than may initially have been considered.

DAC 98, San Francisco, California

2 The SAT Problem

The Boolean satisfiability (SAT) problem is a well-known constraint satisfaction problem with many applications in computer-aided design of integrated circuits, such as test generation, logic verification and timing analysis. Given a Boolean formula, the goal is either to find an assignment of 0-1 values to the variables so that the formula evaluates to 1, or to establish that no such assignment exists.

The Boolean formula is typically expressed in conjunctive normal form (CNF), also called product-of-sums form. Each sum term (clause) in the CNF is a sum of single literals, where a literal is a variable or its negation. In order for the entire formula to evaluate to 1, each clause must be satisfied, *i.e.*, at least one of its literals should be 1.

An assignment of 0-1 values to a subset of variables (called a partial assignment) might satisfy some clauses and leave the others undetermined. If an undetermined clause has only one unassigned literal in it, that literal must evaluate to 1 in order to satisfy the clause. In such a case, the corresponding variable is said to be *implied* to that value. A variable is considered free if neither assigned nor implied. A conflict or contradiction arises if the same variable is implied to be different values. This means that the corresponding partial assignment cannot be a part of any valid solution.

Most current SAT solvers are based on the Davis-Putnam algorithm [4]. This is a backtrack search algorithm. The basic algorithm begins from an empty assignment. It proceeds by assigning a 0 or 1 value to one free variable at a time. After each assignment, the algorithm determines the direct and transitive implications of that assignment on other variables. If no contradiction is detected after the implication procedure, the algorithm picks the next free variable, and repeats the procedure (forward search). Otherwise, the algorithm attempts a new partial assignment by complementing the most-recently assigned variable (backtrack). If this also leads to contradiction, this variable is reset to the free value and the next most-recently assigned variable is complemented. The algorithm terminates when: (i) no free variables are available and no contradictions have been encountered (a solution has been found), or (ii) it wants to backtrack beyond the first variable, which means all possible assignments have been exhausted and there is no solution to the problem.

Determining implications is crucial to pruning the search space since it allows the algorithm to skip entire regions of the search space corresponding to contradictory partial assignments. Every implied variable corresponds to one less free variable to branch on. Unfortunately, detecting implications in software is very slow. Each clause containing the newly assigned or implied variable is scanned and updated sequentially, with the process repeated until no new implications are detected.

Our intuition for hardware speedup potential in the SAT algorithm stems from recognizing that the implication procedure central to the algorithm is both highly parallelizable and easily mapped to basic logic gates. Our entire hardware architecture is designed to take advantage of this parallelism. Section 4 details our mapping of the algorithm onto reconfigurable hardware in a formula-specific manner.

Recent software implementations of the SAT algorithm have enhanced it in several ways while maintaining the same basic flow [8, 3, 12, 11]. The contribution of the GRASP work [11] is notable since it applies nonchronological backtracking and dynamic clause addition to prune the search space further. Very significant improvements in run time are reported. Recognizing nonchronological backtracking as an important feature, Section 5 shows how it can be mapped to reconfigurable hardware with significant improvements in run time.

3 The Promise of Configurable Computing

Special-purpose hardware accelerators for CAD problems are not new. This approach, however, usually suffers from long development time and high cost. The growth of field programmable logic devices allows application-specific hardware without the costs of fabrication. The SRAM-based programmable logic devices (such as Xilinx FPGAs) allow the same chip to be reprogrammed to different circuits an unlimited number of times. One successful application of such devices is logic emulation. A logic design is mapped to an FPGA array and simulated before it is actually fabricated. An emulator is many orders of magnitude faster than simulation on a general purpose computer.

The development of FPGAs has also led to the advent of field-programmable custom computing machines. The abundant programmable logic components and routing resources are used to form special-purpose computers exploiting fine-grain parallelism. Such custom computing machines have achieved very high performance in signal processing, genetic analysis, and cryptography applications. Previous work mostly concentrated on data-centric problems with relatively simple logic control.

Increasing density in programmable logic opens up exciting opportunities for hardware acceleration for complex computations. As feature sizes shrink even further, 1M-gate FPGAs will be feasible by roughly the year 2001 [10]. With programmability and high integration densities, we can create a machine specialized not only to the application but also to the input data associated with a specific problem.

Our choice of Boolean satisfiability (SAT) as a case study here was guided by two main factors:

- It is a search-intensive application, as opposed to data-intensive applications with simple control found in many previous studies of configurable computing. It is also a basic NP-complete problem that is core to many important problems. Gaining experience with this problem will help us accelerate a wide range of applications.

- It contains a significant amount of bit-level logic operations. This makes it very amenable for mapping to configurable hardware.

At this time, we know of three interesting proposals for solving SAT using reconfigurable hardware [13, 1, 9].

Suyama *et al.* [13] have proposed their own SAT algorithm - instead of the Davis-Putnam approach - for implementation in reconfigurable hardware. The search uses full assignment to explore the space. In their paper, they acknowledge that this is not as efficient as a backtrack search algorithm.

The proposal by Abramovici and Saab [1] is also not an implementation of the Davis-Putnam algorithm. Their proposal basically amounts to an implementation of a PODEM-based [6] algorithm in reconfigurable hardware. Therefore it is difficult to handle a generic CNF formula directly. On the other hand, our approach is more general and can exploit heuristics not available to PODEM.

Rashid *et al.* [9] implement small instances of the PODEM algorithm on a Xilinx XC6216. While mapping and compilation issues are presented in detail, there are currently no performance results for their approach.

4 FPGA Mapping: Basic Backtrack Search

4.1 Hardware Organization

Our hardware implementation has two parts: (i) the implication circuit (ii) a state machine to manage the backtrack search. Given a SAT formula, a VHDL description of the hardware is generated automatically.

The speedup over software arises from our implementation of the implication circuit. It finds all direct implications of newly-assigned or newly-implied variables in a single clock cycle. Consequently, all transitive implications of a new variable assignment can be determined in a few cycles. Since all clauses are examined in one cycle, parallelism is quite considerable in practice. For the aim-100-6_0-yes1-1 example in the DIMACS benchmark suite, this approach yields an average of 41.9 effective clause evaluations per cycle.

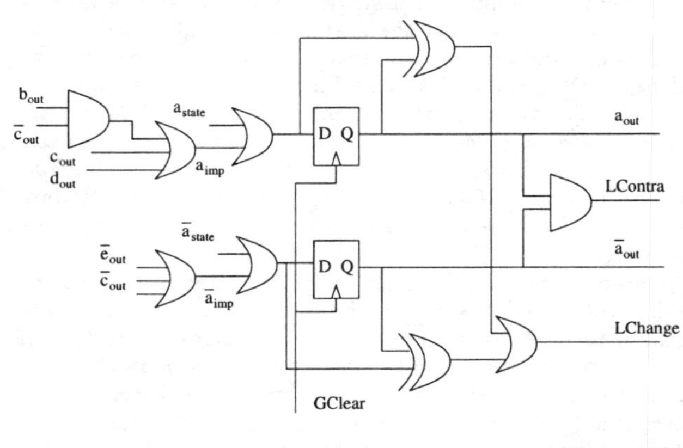

Figure 1: Circuit for implication and conflict detection.

Figure 1 shows the details of the implication circuit. The state of each variable in the circuit is encoded into two bits, $(v, \bar{v})$. A free literal is $(0,0)$. The 1 value of the variable is represented by $(1,0)$ and the 0 value by $(0,1)$.

Implications are easy to determine. For a clause with n literals, if n-1 literals are 0, the last literal is implied to 1. In our encoding, given the clauses $(\bar{c}+a)(\bar{d}+a)(\bar{b}+c+a)(e+\bar{a})(c+\bar{a})$, a is implied by $c+d+b\bar{c}$ and $\bar{a}$ is implied by $\bar{e}+\bar{c}$, as shown in Figure 1. If both v and $\bar{v}$ are implied or assigned to be true, there is a contradiction. No solution exists with this partial assignment.

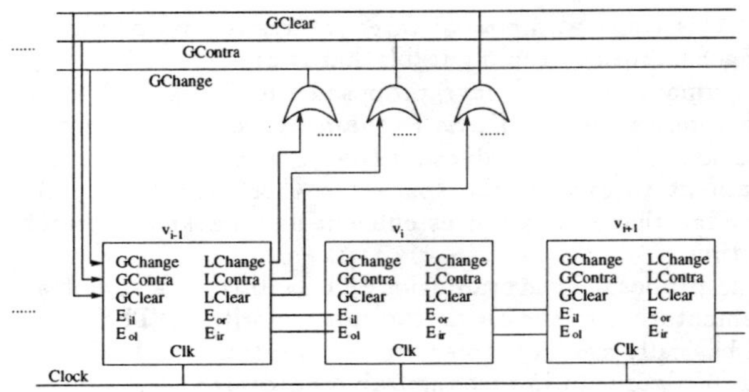

Figure 2: Global Circuit Topology

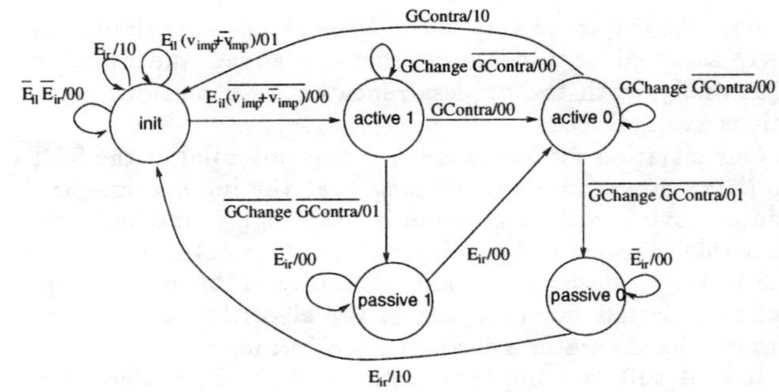

Figure 3: State Diagram for Backtracking Machine

The transitive implication may include cyclic loops. Since the implication is monotonic, changing from 0 to 1 due to our 2-bit encoding, it will never cause oscillation. A D flip-flop is used with each literal, so new implications will propagate one level in one cycle. The LChange signal checks new value changes. Implications finish when no new changes are found.

In order to avoid heavy communication with a host computer, search control is also implemented in hardware. A state machine is implemented for each variable and they are connected according to the order of search, as in Figure 2. Each box in the figure contains the implication circuit and the state machine for each variable. This distributed topology keeps global signals to a minimum and reduces hardware costs. At any instant, only one state machine is in control. Once that state machine has finished processing, it asserts E_{or} to transfer control to the state machine on the right (if progressing forward in the computation) or it asserts E_{ol} to pass control to the left (if backtracking). Each state machine is aware of whether its variable has been assigned, implied or is free.

The state machine for a single variable is shown in Figure 3. The five states in the state machine are encoded by three bits. Two bits correspond to the values of the positive and negative literals of its variable. The third bit indicates whether this particular state machine is active. The inputs to each state machine are the Enable signals from its left and right neighbors, and the global contradiction (GContra) and change (GChange) signals. State machine outputs are the enable signals, E_{ol} and E_{or} that pass control to the left or right.

System operation is fairly simple. After initialization, all the state machines are in the `init` state. When a state machine receives an enable signal, it changes its states. If control is transferred from the left, it can only be in the `init` state. If this variable's value is not implied, it asserts value 1 (`active 1`) and determines all the transitive implications of that assignment. If a contradiction is detected, it will try the value 0 instead (`active 0`) and repeat the implication step. If the implication settles without a contradiction, it transfers control to the state machine on its right and it transitions to the `passive 1` state. From `active 0`, if a contradiction is detected, it will backtrack. Its state is reset to `init` and the control is passed to the left. On the other hand, if implications settle without a contradiction, then it moves from `active 0` to `passive 0` and passes the control to the right.

If, as above, control is transferred to a state machine from the left, but its variable's value has already been implied by a previous assignment, then it merely passes control to the state machine on its right on the next clock cycle and remains in the `init` state. Its output value will be maintained by the implication.

If control is transferred to a state machine from the right, the system is backtracking. If the variable has been implied by some previous assignment, the state machine is in the `init` state now. It simply passes control to the variable on its left. If the variable currently has the assigned value 1 (`passive 1`), then it is changed to 0 (`active 0`). The implications are computed. If the variable already has the assigned value 0, we have already tried both possible values for this variable. In this case, the state machine resets to `init` and backtracks further by transferring control to the left.

Finding a Solution A solution has been found when the rightmost state machine further attempts to pass control to the right. No solution exists if the leftmost state machine attempts to backtrack by further passing control to the left.

4.2 Performance Results

In this section we compare the performance of our hardware implementation of the basic Davis-Putnam SAT algorithm to its implementation in GRASP [11]. In order to estimate the hardware performance, we use simulation to count the exact number of hardware clock cycles needed to solve each problem and divide by clock rate to compute the run time. We have used a VHDL model to verify design correctness on moderate examples. However, since VHDL simulation can be slow, we have generated a C-language hardware model that is significantly faster then general VHDL simulation. We use this for performance estimation of long-running problems. Like a VHDL simulation, the C-language simulation is also capable of counting the number of hardware clock cycles required exactly.

The number of hardware clock cycles is translated into run time based on a clock rate of 1.33 MHz. This is a clock rate achieved on a large example on an IKOS FPGA board [7] as detailed in the section on hardware implementation issues. The GRASP runs were timed on a Sun 5 workstation with 64 MB of RAM and a 110 MHz processor.

As discussed earlier, our hardware implementation currently requires variables be ordered statically prior to the hardware implementation. Our ordering strategy places earlier the variables with more appearances in the formula.

We compared the performance for all the examples in the DIMACS SAT benchmark suite [5]. There are 240 problems in the DIMACS suite. Some of them take a very long time to solve and neither GRASP nor our C-model simulation gets a result in reasonable amount of time (several hours). There are 134 instances in which at least one program finished.

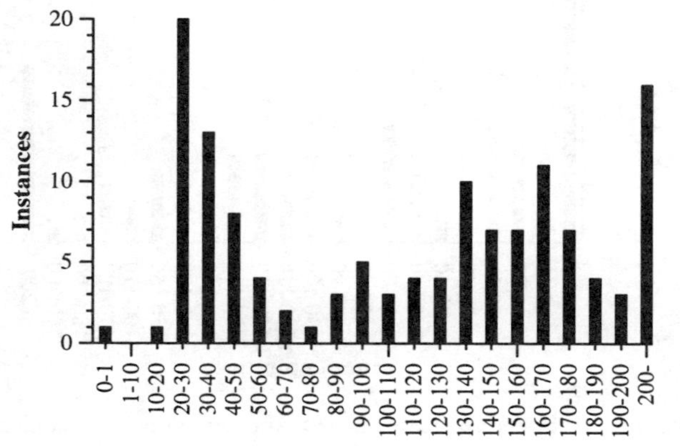

Figure 4: Speedup ratio histogram comparing our FPGA approach to a GRASP run, both with basic backtracking.

In all examples for which the implementation of the basic Davis-Putnam algorithm completes in GRASP, our hardware implementation also completes — usually in much shorter time. Figure 4 shows a histogram in which each bar corresponds to a speedup range. Speedup is defined as the ratio of the GRASP run time to the hardware run time. (That is, a 10X speedup means that our approach is ten times faster than GRASP.) The height of the bar corresponds to the number of DIMACS SAT examples for which the corresponding speedup is obtained.

The histogram indicates that *more than 90% of the examples have speedup greater than 20X* and *more than 45% of the examples have speedup greater than 100X*. The results clearly demonstrate that utilizing parallelism by direct logic mapping in the SAT algorithm achieves significant speedup even when the hardware clock is much slower than that of the general purpose computer.

5 FPGA Mapping: Nonchronological Backtrack

5.1 Algorithm

When the Davis-Putnam algorithm backtracks to the most recently assigned variable, its backtracking is said to be *chronological*. On the other hand, when an algorithm jumps over several previously-assigned variables to a variable more than one level above the current variable, the backtracking is said to be *nonchronological*. In order to jump directly to a previous level, the algorithm must first determine that no combination of values on the skipped variables will result in a satisfying assignment. GRASP is a recent SAT implementation with nonchronological backtracking. The GRASP work demonstrated that nonchronological backtracking can lead to significant reductions in run time. We describe an algorithm to achieve a similar goal in hardware, but with less control complexity than GRASP.

GRASP maintains a data structure called the implication graph, from which one derives the transitive implications leading to the contradiction. From the contradiction, it traces back to a set of assignments contributing to the contradiction. It jumps to the most recently assigned variable in the set. It is difficult, however, to traverse the implication

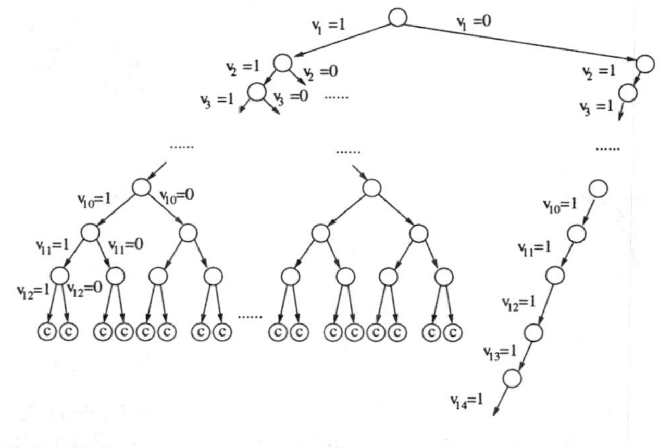

Figure 5: Search example for basic backtracking.

circuit in hardware. As a result, we devised the following alternative.

Rather than analyzing the implications, we take advantage of the fact that determining implications is very fast in our hardware. When a contradiction occurs for both assignments of the variable, it should backtrack. If we reset this variable to be free, the observed contradiction will disappear, but may recur when the same variable is assigned a value in the future. We want to change the value of the variable that really contributes to the observed contradiction.

If a contradiction was detected for both assignments to v_i, with i being its level, our procedure works its way back up the levels, one at a time. At each level j ($< i$), the procedure calls the implication routine twice, once for each v_i value, while v_j has a flipped value. The variables v_k between i and j ($j < k < i$) are left unassigned when the implication routine is called, i.e. both literals v_k and $\bar{v_k}$ are set to 0. The algorithm must backtrack to v_j if a no-contradiction case was found for one of the two assignments. If both cause contradiction, the variable v_j can be skipped, and the procedure repeats this step for variable v_{j-1}.

This procedure requires only $2n$ calls to the implication routine if the algorithm reverts back by n levels. Since the procedure does not call the implication routine an exponential number of times, and since the implication hardware is very fast, this analysis for nonchronological backtracking can be expected to be very fast also. In comparison, GRASP has the overhead of maintaining the implication graph data structure and analyzing it when a contradiction is found, but it does not do the $2n$ implications.

The following simple example shows how our approach works. Suppose we have the following formula: $(v_1' + v_{12}' + v_{13})(v_1' + v_{12}' + v_{13}')(v_1' + v_{12} + v_{14})(v_1' + v_{12} + v_{14}')\cdots$. Assume that no assignment for variables $v_2 - v_{11}$ leads to a contradiction. The basic search tree for this formula is shown in Figure 5. In the figure, each solid arrow represents assigning a new value and determining its implications. It begins by assigning $v_1 = 1$ and goes on until $v_{11} = 1$. Finally, $v_{12} = 1$ is tried. This leads to a contradiction since v_{13} is required to be both 0 and 1 simultaneously. Similarly, $v_{12} = 0$ also leads to a contradiction. The normal backtrack procedure would reset v_{12} to unknown and it would backtrack to v_{11}. The conflict disappears temporarily but will appear again whenever v_{12} is set.

Figure 6(a) shows our version of nonchronological backtracking. Figure 6(b) shows how GRASP directly backtracks to v_1. The dotted arrows indicate skipping and do

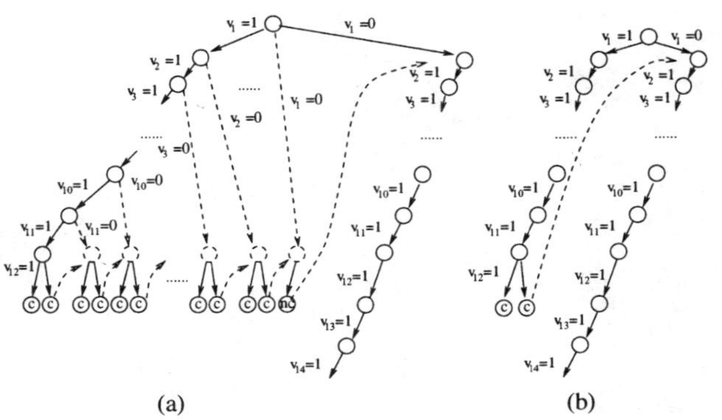

Figure 6: Search examples for nonchronological backtracking. (a) Algorithm in our hardware algorithm. (b) Software algorithm used in GRASP.

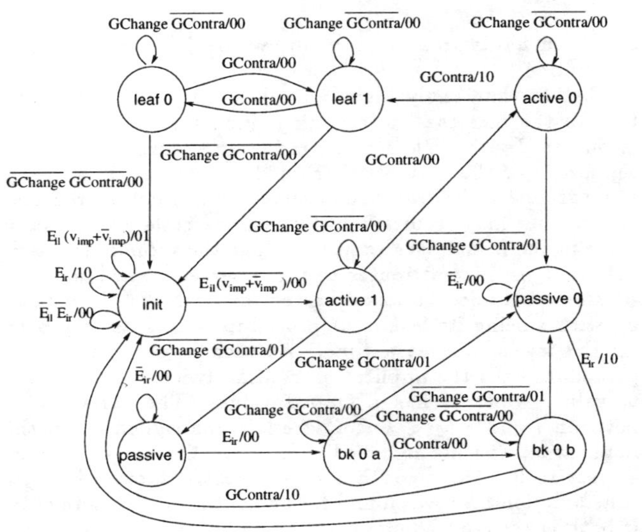

Figure 7: State diagram for nonchronological backtracking machine

not incur any computation; they simply show what values are assumed before an implication calculation. For both methods, the search is the same before reaching v_{12}. Backtracking differs, however. When backtracking at each level, our implication procedure is called twice. That is, we backtrack to v_1 after effectively calling the implication procedure only 22 times, rather than 4082 attempts with the basic backtracking approach. While this simple example is mainly pedagogic, dramatic speedup potential also exists in real benchmarks. For example, in the aim-100-1_6-yes1-1 benchmark from the SAT suite, chronological backtracking visits 606,578 partial assignments; the nonchronological backtracking algorithm only visits 1384 partial assignments.

5.2 Hardware Organization

The only modifications required to implement our nonchronological backtracking algorithm are to the state machine. No new global signals are added and the implication circuit is not modified.

The state machine to perform the nonchronological backtracking is shown in Figure 7. It has more states and transitions to reflect the complex backtracking procedure outlined

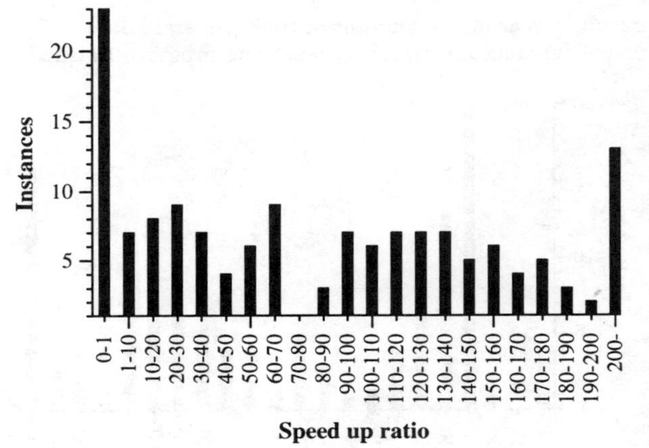

Figure 8: Speedup ratio histogram for nonchronological backtracking

previously. Although the state machine has become bigger, the extra states require only one more encoding bit. Since most of the circuit is the same and since the state machine is not instrumental in determining the clock cycle time, the cycle time from the basic algorithm remains valid here.

5.3 Performance Results

We compare our runtime with nonchronological backtracking to a GRASP implementation with nonchronological backtracking. Similar to Figure 4, the results are graphed in Figure 8. Due to the improved performance, there are 149 problems that finished with either GRASP of our approach.

While GRASP's more complicated nonchronological backtracking allows it to garner a bigger improvement from this technique than our hardware does, we still obtain a median speedup over GRASP of 63.5X. From the full DIMACS suite, we offer 100X or greater speedups on 63 of the examples.

6 Implementation Issues

6.1 Envisioned Usage

Depending on the problem characteristics, the configurable computing implementation can often offer significant speedups over a software SAT solver. In order for the acceleration to be useful, however, it must offer performance advantages even after hardware compile time and configuration time are considered. For this reason, we envision that the configurable hardware techniques will mainly be used on SAT problems with very long GRASP run-times (hours or days) or in cases where GRASP aborts. In such cases, the hardware synthesis times required will be acceptable. Furthermore, current commercial CAD tools for FPGAs are quite general and therefore quite slow in this usage; it is not uncommon for synthesis and place-and-route times to exceed one hour. Since our designs are quite regular and easy-to-route, however, compiler tools specialized for our template design could be much faster, and could thus broaden the range of problems for which hardware acceleration is useful. Overall, we envision a system where easy SAT problems are still solved in software. Very large problems, ones that often timeout today, will invoke the hardware compiler and configure an FPGA board to assist in solving them.

The envisioned hardware platform is a high-performance workstation with an FPGA board attached via an external connection or the I/O bus. Easy problems are handled by software running on the workstation itself, while longer-running problems are handled by the FPGA board. Requirements for the FPGA board itself are partly a function of the hardware size of the problems being solved; this is discussed in the following subsection.

6.2 Hardware Resources

Hardware requirements for our approach are a function of the size and complexity of the formula being solved. The hardware usage is estimated in terms of CLBs of Xilinx XC4000 series FPGAs.[14]. The SAT solver's CLB requirements varies widely with the formula to be solved. Across the DIMACS benchmarks, the median CLB requirements are 3655. Overall, relatively few (less than 30%) of these problems will fit within the about 2000 CLB limit of current FPGA chips. However, 202 of 240 problems require fewer than 20000 CLBs and should fit on fewer than ten FPGAs.

Partitioning the design across FPGAs is straightforward because of the regular topology used. In our experiments, we have used the partitioning and pin multiplexing techniques in the Ikos SLI logic emulation system originally developed as part of the MIT Virtual Wires effort [2, 7]. This partitioning software can multiplex several inter-chip signals to use the same physical pins, thereby circumventing the pin limitations that can often limit the CLB utilization of FPGA designs. Furthermore, the software performs this multiplexing with minimal impact on hardware cycle times. We have compiled many of the DIMACS problems for the IKOS system. Their clock rate ranges from 700KHz to 2 MHz. We use a 1.33 MHz clock rate as a typical value for performance comparison. If SAT is implemented on a directly-connected array, the clock rate is expected to be 10 to 20 MHz.

7 Future Work

As evidenced by the organization of our paper, we began our work by implementing the basic Davis-Putnam procedure in reconfigurable hardware. Recent software implementations of the Davis-Putnam procedure have grown to be very sophisticated however [11]. To beat them consistently, our implementation should match that sophistication. We zeroed in on nonchronological backtracking as a key requirement and have been able to implement it with a small increase in hardware. The GRASP implementation still contains two additional features that our implementation does not:

1. The ability to add clauses on the fly plays a significant role in GRASP's ability to solve hard problems. Variable relationships derived in the search help prune search tree in the future. It is hard to change the circuit and reroute the FPGA in the middle of the algorithm. We are working on low cost ways of implementing this feature based on multiple configuration contexts.

2. The ability to choose the next decision variable on the fly also contributes to the efficiency of the software implementations. While this is very simple to do in software, it can be very hardware intensive in reconfigurable hardware, as evidenced by the Abramovici and Saab proposal [1]. In practice, we have found that a good static ordering of variables works quite well. If dynamic ordering can provide significant improvement, we will implement it as well.

8 Conclusions

Overall, the contributions of this paper are two-fold. First, we provide a system design for formula-specific Boolean satisfiability solutions based on a configurable hardware implementation. Our design's hardware requirements are quite modest compared to other recent proposals. Moreover, our hardware performance results indicate that the configurable hardware approach offers dramatic improvements over even the best current software-based techniques. We have observed speedups of well over 200X compared to software SAT solutions on several of the DIMACS SAT benchmarks.

The second contribution of this paper is much broader. We present our SAT-solver as a case study of a class of *input-specific* configurable hardware applications. These applications aggressively harness the increasing integration levels and reconfigurability of current FPGAs by performing template-driven hardware designs for each problem/data set being solved. Their amenability to parameterized, automated design makes it easy and fast to compile configurations for them. Overall, this paper alerts the hardware design community to an increasingly-important application style, and documents its significant promise via a detailed case study on formula-specific SAT-solving.

References

[1] M. Abramovici and D. Saab. Satisfiability on Reconfigurable Hardware. In *Seventh Intl. Workshop on Field Programmable Logic and Applications*, Sept. 1997.

[2] J. Babb, R. Tessier, and A. Agarwal. Virtual Wires: Overcoming pin limitations in FPGA-based logic emulators. In *Proc. IEEE Workshop on FPGA-based Custom Computing Machines*, pages 142–151, Apr. 1993.

[3] S. Chakradhar, V. Agrawal, and S. Rothweiler. A transitive closure algorithm for test generation. *IEEE Trans. on Computer-Aided Design of Integrated Circuits and Systems*, 12(7):1015–1028, July 1993.

[4] M. Davis and H. Putnam. A Computing Procedure for Quantification Theory. *Journal of the ACM*, 7:201–215, 1960.

[5] DIMACS. DIMACS Challenge benchmarks and UCSC benchmarks. Available at ftp://Dimacs.Rutgers.EDU/pub/challenge/sat/benchmarks/cnf.

[6] P. Goel. An Implicit Enumeration Algorithm to Generate Tests for Combinational Logic Circuits. *IEEE Tran. on Computers*, C30(3):215–222, March 1981.

[7] IKOS Systems. Virtual Logic SLI Documentation. Version 1.6.

[8] T. Larrabee. Test Pattern Generation Using Boolean Satisfiability. In *IEEE Trans. on Computer-Aided Design*, volume 11, pages 4–15, January 1992.

[9] A. Rashid, J. Leonard, and W. H. Mangione-Smith. Dynamic Circuit Generation for Solving Specific Problem Instances of Boolean Satisfiability. In *Proceedings IEEE Workshop on FPGA-based Custom Computing Machines*, Apr. 1998.

[10] J. Rose and D. Hill. Architectural and Physical Design Challenges for One Million Gate FPGAs and Beyond. In *Proc. 1997 ACM/SIGDA Fifth Intl. Symp. on Field-Programmable Gate Arrays*, Feb. 1997.

[11] J. Silva and K. Sakallah. GRASP-A New Search Algorithm for Satisfiability. In *IEEE ACM Intl. Conf. on CAD-96*, pages 220–227, Nov. 1996.

[12] P. Stephan, R. Brayton, and A. Sangiovanni-Vincentelli. *Combinational Test Generation Using Satisfiability.* Department of Electrical Engineering and Computer Sciences, University of California at Berkeley, 1992. UCB/ERL Memo M92/112.

[13] T. Suyama, M. Yokoo, and H. Sawada. Solving Satisfiability Problems on FPGAs. In *6th Int'l Workshop on Field-Programmable Logic and Applications*, Sept. 1996.

[14] Xilinx Corp. The Programmable Logic Data Book. Xilinx Corp. San Jose, CA, 1996.

Fast Exact Minimization of BDDs

Rolf Drechsler Nicole Drechsler Wolfgang Günther

Institute of Computer Science
Albert-Ludwigs-University
79110 Freiburg im Breisgau, Germany
drechsle@informatik.uni-freiburg.de

Abstract

We present a new exact algorithm for finding the optimal variable ordering for reduced ordered Binary Decision Diagrams (BDDs). The algorithm makes use of a lower bound technique known from VLSI design. Up to now this technique has been used only for theoretical considerations and it is adapted here for our purpose. Furthermore, the algorithm supports symmetry aspects and makes use of a hashing based data structure. Experimental results are given to demonstrate the efficiency of our approach. We succeeded in minimizing adder functions with up to 64 variables, while all other previously presented approaches fail.

1 Introduction

Recently, several design methods have been proposed that are based on *ordered Binary Decision Diagrams* (BDDs) [7]. The resulting circuits have very nice properties, like e.g. testability [2, 1] and low power [17]. For synthesis approaches based on *Pass Transistor Logic* (PTL) BDDs seem to be a good starting point. First promising results on how to transform a decision diagram to a circuit based on PTL have been reported in [24, 9, 3].

One drawback of BDDs is that they are very sensitive to the variable ordering, i.e. the size of the representation may vary from linear to exponential. Therefore in the last few years several methods have been presented to determine good orderings. However, finding the optimal variable ordering starting from a given BDD representation is known to be NP-hard [5].

Existing methods for the determination of good variable orderings can be classified into three categories. The first are initial heuristics starting from a circuit [13], the second are gradual improvement heuristics based on the exchange of variables in the BDD [15, 14, 21, 19, 20], and the third are exact methods to find an optimal ordering [12, 15, 16]. Obviously, it is desirable to determine the exact result. In applications like the ones discussed above it is even more important to find the *best* variable ordering since it turned out by experiments that the best greedy approaches are up to a factor of two worse than the optimal result. But exact methods were so far only applicable to functions with less than 20 variables, due to their exponential runtime behaviour. Since for larger functions no exact method can be applied, in the past few years several methods have been proposed that are time consuming, but get high quality results, e.g. based on simulated annealing [18, 4] and evolutionary algorithms [10]. The drawbacks of these methods were that they had very bad runtime behaviour or/and were only applicable to small functions. Additionally, these probabilistic approaches are sensitive to several chosen parameters and for this are not very robust.

In this paper we present an approach to exact BDD minimization based on the classical algorithm from [12] with branch&bound technique [15, 16]. The key part of our algorithm is a lower bound technique used in [8] for (theoretical) lower bound proofs. This technique has been improved. It is used for the first time in a practical application, i.e. a theoretical method has been implemented as an automated tool. Moreover, we tuned the underlying data structures with respect to execution time and memory requirement. Our algorithm works well for multi-output functions and makes use of symmetries (see also [16, 22]). By this, we were able to find the optimal variable ordering for all benchmarks from LGSynth93 with less than 24 variables. Furthermore, for partially symmetric functions, like adders, we were able to determine the exact result up to 64 variables. For those functions all previously presented approaches fail.

The paper is structured as follows: First, we introduce our notations and definitions. We then review earlier approaches on exact minimization of BDDs in Section 3 and explain the differences to our work. The lower bound that is used in our algorithm is given in Section 4 where also a sketch of the algorithm and some technical details are illustrated. In Section 5 we show our experiments. Finally, the work is summarized.

2 Preliminaries

Boolean variables can assume values from $\mathbf{B} := \{0, 1\}$ and are denoted by Latin letters. Indices from a given index set $\{1, .., n\}$ can be used for the variables, e.g. $x_1, .., x_n$. In the following we consider Boolean functions $f : \mathbf{B}^n \to \mathbf{B}^m$ over the variable set $X_n = \{x_1, \ldots, x_n\}$. A subset of variables is denoted by I ($I \subseteq X_n$).

As well-known, each Boolean function $f : \mathbf{B}^n \to \mathbf{B}$ can be represented by a *Binary Decision Diagram* (BDD) [7],

DAC 98, San Francisco, California
©1998 ACM 0-89791-964-5/98/06..$5.00

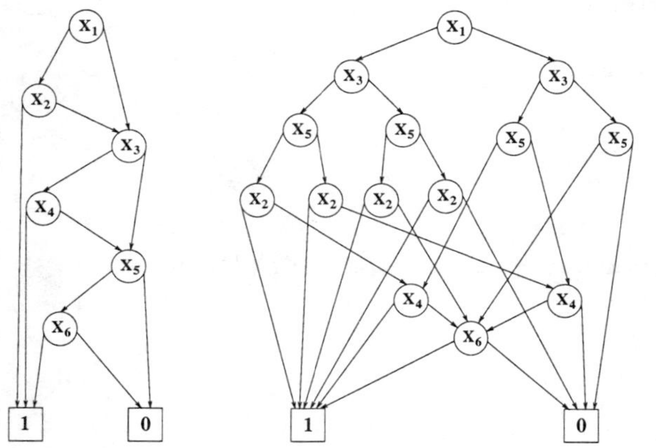

Figure 1: BDDs of the function $f = x_1x_2 + x_3x_4 + x_5x_6$

i.e. a directed acyclic graph where a Shannon decomposition

$$f = \overline{x}_i f_{x_i=0} + x_i f_{x_i=1} \quad (1 \leq i \leq n)$$

is carried out in each node.

A BDD is called *ordered* if each variable is encountered at most once on each path from the root to a terminal node and if the variables are encountered in the same order on all such paths. A BDD is called *reduced* if it does not contain vertices either with isomorphic sub-graphs or with both edges pointing to the same node. Reduced and ordered BDDs are canonical, i.e. for each Boolean function the BDD can be uniquely determined.

BDDs are defined analogously for multi-output functions $f: \mathbf{B}^n \to \mathbf{B}^m$ as for the case of single-output functions: A BDD G_j for each component function f_j $(1 \leq j \leq m)$ is used for the *shared* BDD representation G for f. The variable ordering is fixed for all G_js.

For functions represented by reduced, ordered BDDs efficient manipulations are possible [7]. In the following only reduced, ordered BDDs are considered and for briefness these graphs are called BDDs.

It has been shown in [5] that improving the variable ordering of BDDs is NP-complete. However, the BDD's size largely depends on the variable ordering, i.e. it may vary from linear to exponential.

Example 1 Let $f = x_1x_2 + \ldots + x_{2n-1}x_{2n}$. If the variable ordering is given by $(x_1, x_2, \ldots, x_{2n})$ the size of the resulting BDD is $2n$. On the other hand if the variable ordering is chosen as $(x_1, x_{n+1}, x_2, x_{n+2}, \ldots, x_{2n})$ the size of the BDD is $\Theta(2^{n-1})$. Thus, the number of nodes in the graph varies from linear to exponential depending on the variable ordering. In Figure 1 the BDDs of the function $f = x_1x_2 + x_3x_4 + x_5x_6$ with variable orderings $(x_1, x_2, x_3, x_4, x_5, x_6)$ and $(x_1, x_3, x_5, x_2, x_4, x_6)$ are illustrated. The left (right) outgoing edge of each node marked by x_i denotes $f_{x_i=1}$ $(f_{x_i=0})$. As can be seen the choice of the variable ordering largely influences the size of the BDDs.

In the following we make use of *Complement Edges* (CEs) [6] without mentioning it further. (Note that all results directly transfer to BDDs without CEs.)

In this paper we denote a permutation of variables with π. If $x_i = \pi(k)$ for a variable x_i, then x_i is the kth element of the variable ordering π, i.e. x_i is in the kth *level* of the BDD. For a set of variables $I \subseteq X_n$, let $\Pi(I)$ be the set of permutations π on X_n whose first $|I|$ members constitute I.

The number of nodes of a BDD for function f and permutation π is denoted by $\#nodes(f, \pi)$. The number of nodes in level k marked with variable x_i under permutation π is given by $\#nodes_{x_i}(f, \pi)$.

We now consider the following problem:

> *How can we determine an optimal variable ordering for a BDD representing a given Boolean function f, such that the BDD's size is a global minimum?*

3 Previous Work

To make the paper self-contained we give an overview on existing exact minimization algorithms for BDDs. Since improving the variable ordering is NP-complete, all exact minimization algorithms presented have exponential runtime behaviour.

The trivial idea is to construct the BDDs for all $n!$ variable orderings. The optimal variable ordering is the one leading to the smallest BDD. This approach is only applicable to very "small" functions with up to 10 variables.

In [12] an exact algorithm has been presented where the number of variable orderings which have to be considered could be reduced significantly. The main idea behind that algorithm is the following lemma:

Lemma 1 Let $f: \mathbf{B}^n \to \mathbf{B}^m$, $I \subseteq X_n$, $k = |I|$, and $x_i \in I$. Then there exists a constant c such that $\#nodes_{x_i}(f, \pi) = c$ for each $\pi \in \Pi(I)$ with $\pi(k) = x_i$.

More informally the lemma states that the number of nodes in a level is constant, if the corresponding variable is fixed in the ordering and no variables from the lower and upper part are exchanged. Notice, that this holds independently of the ordering of the variables in the upper and lower part of the BDD.

The lemma can be used as follows: for $I \subseteq X_n$, let

$$min\_cost_I = \min_{\pi \in \Pi(I)} \sum_{i=1}^{|I|} \#nodes_{x_i}(f, \pi),$$

and let π_I denote some permutation leading to that minimum. Assume for a fixed $I \subseteq X_n$ with $|I| = k$ that we know $min\_cost_{I'}$ for all $I' \subseteq I$ with $|I'| = k - 1$. Then we get

$$min\_cost_I = \min_{x_i \in I}(min\_cost_{I \setminus \{x_i\}} + \#nodes_{x_i}),$$

with $\#nodes_{x_i}$ being the number of nodes in level $n - k + 1$ under some permutation π with $\pi(I \setminus \{x_i\}) = I \setminus \{x_i\}$ and $\pi(n - k + 1) = x_i$. So the optimal ordering can be computed iteratively by computing for increasing k's min_cost_I for each k-element subset I, until $k = n$. The representation in [12] is done by extended truth tables.

Remark 1 For increasing k the considered level $n - k + 1$ moves up. By this the BDD is build bottom up. In fact, the BDD can also be built top down. As $\Pi(I) = \Pi(X_n \setminus I)$, Lemma 1 can be used the same way in that case. This is an essential aspect of our algorithm, because of the calculation of the lower bound (see next section). Notice that all exact algorithms presented so far use a bottom up construction.

```
1   compute_optimal_ordering (f(x_1, ···, x_n)) {
2     π_∅ = an initial ordering;
3     insert I = ∅, min_cost_∅=0, π_∅ into table;
4     for(k = 1; k ≤ n; k++) {
5       k: level under consideration;
6       next_table = ∅;
7       for(each I ∈ table) {
8         set_permutation (π_I);
9         for(each x_i in X_n \ I) {
10          if (x_i is not top of ((symmetry group of x_i) ∩ (X_n \ I)) )
11            goto 9 and consider next x_i;
12          shift x_i to level k;
13          π: current ordering;
14          I' = I ∪ {x_i};
15          cost = min_cost_I + #nodes_{x_i}(f, π);
16          if (I' ∉ next_table or cost < min_cost_I') {
17            store I', cost, π into next_table;
18            upper_bound = update_upper_bound();
19            lower_bounds_I' = compute_lower_bound(x_i);
20          }
21          if (#nodes(f, π) > 1.5 · (size before shifting x_i))
22            undo shifting of x_i;
23        }
24      }
25      clear all elements in next_table with lower_bounds_I' ≥ upper_bound;
26      table = next_table;
27    }
28    set_permutation(ordering of upper_bound);
29  }
```

Figure 2: Sketch of the algorithm

An extension of [12] has been presented in [15]. This approach makes use of BDDs instead of truth tables. Additionally, the search space is reduced by a branch&bound strategy, where parts of the search tree are pruned.

The most efficient algorithm presented so far is an extension of [15] described in [16]. There the search space is reduced if the function has symmetric variables and a better lower bound is used. Additionally, the exchange of variables is performed efficiently and the algorithm works well for multi-output functions. The algorithm also uses hash tables as the underlying data structure. For non-symmetric functions, execution times are speeded up by a factor of two in comparison to [15], for symmetric functions even more. In the following only comparisons to [16] are given, since this is the fastest and most memory effective algorithm presented so far.

4 Exact Minimization

In this section we describe the flow of our exact algorithm. In principle, the branch&bound mechanism from [16] is used. The gain of our algorithm results from the following:

- The lower bound which is used to prune the search space is derived from a theoretical technique in circuit complexity (see [8]). Our method obtains much better lower bounds and by this speeds up the algorithm tremendously. As a consequence of our lower bound calculation, the BDD *has to* be build top down instead of bottom up (see Remark 1). Details are given in Section 4.2.
- The BDD's size may become exponential while processing one subset $I \subseteq X_n$. This growth is reduced by undoing the sifting of some specified variables.
- The size of the hash table is dynamically adapted. Thus, we are able to reduce the memory used during computation.

4.1 Algorithm

A sketch of the exact algorithm is given in Figure 2. First, the BDD for a given Boolean function is constructed using sifting operations during symbolic simulation. The present BDD size is used as an initial upper bound for the exact minimization procedure[1].

For each subset $I \subseteq X_n$ the best variable ordering π_I and the corresponding BDD size min_cost_I are stored in hash table *table*. Lines 2 and 3 of Figure 2 show the initialization phase of *table*. In line 4 the iteration over all levels in the BDD starts. Then each $I \subseteq X_n$ which has an entry in *table* is considered (line 7). (I has no entry in *table*, if the lower bound was larger or equal than the upper bound in previous steps.) The remaining variables marked by x_i which are not in I are then shifted to level k. Notice, that each symmetry group is considered only once (see also lines 10/11). The cost of the current ordering is computed in line 15. Lines 16 to 20 describe the computation of *min_cost* for the current subset I' and then these subsets are stored in *next_table*. Obviously, the new upper bound is given by the current

[1] The algorithm can often be further speeded up, if the upper bound is determined by a preprocessing based on evolutionary algorithm principles [11].

minimal BDD size. The lower bounds for all I's (line 19) are determined as given in Section 4.2. To save time and memory during computation, in line 21 and 22 the shifting of variable x_i is reset if the size of the BDD grew too much by the last shifting. If the lower bound for subset I' is larger or equal than the current upper bound, I' need not to be considered anymore, i.e. it is deleted from *table* (line 25).

In the last iteration ($k = n$), $min\_cost_{X_n}$ is computed, which is the minimal BDD size over all variable orderings.

4.2 Lower Bound Technique

In this section we describe our lower bound technique that is the key part of our algorithm. In [8] lower bounds for the size of BDDs have been proven using lower bound techniques from VLSI design. First, it is shown how to adapt this theoretical technique.

The argument of [8] is shortly reviewed giving a lower bound on the BDD's size for $f : \mathbf{B}^n \to \mathbf{B}$. (Notice that the same argumentation holds for multi-output functions.) Let (L, R) be a partition of X_n. Then, any input assignment $x : X_n \to \mathbf{B}$ can be split into a *left input assignment* $l : L \to \mathbf{B}$ and a *right input assignment* $r : R \to \mathbf{B}$. We use the notation $x = lr$. A set $\mathbf{F}_{(L,R)}$ of left input assignments is called a *fooling set* for the partition (L, R) iff for each two distinct $l, l' \in L$ there is a right input assignment r with $f(lr) \neq f(l'r)$. Fixing a real parameter $\omega \in (0, 1)$ and a subset $Y \subseteq X_n$ a partition (L, R) is called a *balanced partition* iff

$$\lfloor \omega \cdot |Y| \rfloor \leq |Y \cap L| \leq \lceil \omega \cdot |Y| \rceil.$$

The main tool in [8] for proving lower bounds on the size of BDDs is:

Lemma 2 If, for each balanced partition (L, R) (with respect to some ω and Y), f has a fooling set $\mathbf{F}_{(L,R)}$ of size c, then the BDD for f has size at least c.

Following [8] the proof of this lemma is based on

Lemma 3 If, for one partition (L, R), f has a fooling set $\mathbf{F}_{(L,R)}$ of size c, then for any variable ordering π with

$$L = \{\pi(i) : 1 \leq i \leq |L|\},$$

$$R = \{\pi(i) : |L| < i \leq n\},$$

the BDD with variable ordering π has at least c nodes in levels $|L| + 1, \ldots, n$.

For our application this can be used as follows: For some $I \subseteq X_n$, the size c of the fooling set $\mathbf{F}_{(I, X_n \setminus I)}$ equals the number of nodes in levels $|I| + 1, \ldots, n$ referenced directly from the nodes in levels $1, \ldots, |I|$.

When computing the lower bound for a multi-output function $f : \mathbf{B}^n \to \mathbf{B}^m$, we already know the minimum number min_cost_I of nodes in levels $1, \ldots, |I|$. So the lower bound can be computed as

$$lower\_bound = min\_cost_I + \max\{c + m_R, n - |I|\} + 1,$$

where m_R is the number of output nodes in levels $|I| + 1, \ldots, n$ and $n - |I|$ is the number of variables in $X_n \setminus I$, since there will be at least one node for each variable in $X_n \setminus I$. The constant node is always needed.

The computation of $c + m_R$ is done by traversing the levels $|I| + 1$ to n of the BDD, counting for each node of these levels the number of references from a visited node.

```
compute_lower_bound(level: k) {
   for(i = k+1; i ≤ n; i++)
      for(any node v in level i)
         copy reference count of v;
   lb1 = 0;
   for(i = k+1; i ≤ n; i++)
      for(any node v in level i) {
         if (references(v) ≥ 1)
            lb1++;
         decrement references of successors;
      }
   lb2 = n − k;
   return (min_cost_I + max(lb1, lb2) + 1);
}
```

Figure 3: Computation of lower bound

The number $c + m_R$ then is the number of nodes for which the counted number of references and the actual reference count differ[2].

A sketch of the algorithm is given in Figure 3.

4.3 Data Structure

The use of an efficient data structure largely influences the performance of our algorithm. Therefore the most important details are outlined.

For each set $I \subseteq X_n$ the best variable ordering π_I and the corresponding BDD size min_cost_I are stored in a hash table. I is used as a key and table size s is always chosen prime. We use double hashing, with first hash function $h_1(k) = (k \bmod s)$ and second hash function $h_2(k) = 1 + (k \bmod (s - 2))$. Every time the number of elements in the hash table exceeds $0.7 \cdot s$, the table size is increased to the next prime $\geq 2 \cdot s$.

The data structure also considers symmetry aspects of the function to be minimized (see also [16, 22, 23]). Each symmetry group is stored as a circular list where each variable points to the next variable in the same symmetry group. Symmetry groups are determined by symmetric sifting [19, 20]. This also helps to get a better upper bound by an improved initial ordering. To check whether a given variable x_i is the topmost of its symmetry group intersected with a set $S \subseteq X_n$ of variables, we check whether x_i is the one with the smallest index among those in S belonging to x_i's symmetry group. Therefore we only have to follow the circular list from x_i. If a variable x_j ($j \neq i$) occurs with $x_j \in S$ and index less than x_i, then x_i is not the topmost one. If $j = i$, then we have visited all variables of the symmetry group, and x_i was the topmost variable.

5 Experimental Results

All experimental results have been carried out on a *SUN Ultra 1-140* using an upper memory limit of 300 MByte and a runtime limit of 120.000 CPU seconds. Our algorithm has been integrated in the CUDD package [23]. We compare our algorithm to the original exact algorithm in the CUDD package, called JUNON [16]. By this we guarantee that both algorithms are run in the same system environment. Our algorithm has been implemented as the program *FīzZ*.

[2] In commonly used BDD packages, the actual reference count is stored for each node to detect unused nodes.

name	in	out	opt	S	JUNON time	JUNON space	F$\widehat{\imath z}$Z time	F$\widehat{\imath z}$Z space
adder8	16	8	36	8	7.3s	2M	1.1s	<1M
adder12	24	12	56	12	779s	289M	24.4s	<1M
adder16	32	16	76	16	—	—	190s	2M
adder20	40	20	96	20	—	—	882s	4M
adder24	48	24	116	24	—	—	5498s	15M
adder28	56	28	136	28	—	—	26489s	34M
adder32	64	32	156	32	—	—	120104s	77M
mult5	10	10	388	10	5.1s	<1M	7.3s	<1M
mult6	12	12	1098	12	45.9s	<1M	121s	<1M
mult7	14	14	3082	14	493s	<1M	2119s	<1M
mult8	16	16	8658	16	4992s	2M	59219s	3M

Table 1: Arithmetical circuits

In a first series of experiments we consider arithmetic functions, i.e. adders and multipliers (see Table 1). In the first column the name of the function is given. *in* (*out*) denotes the number of inputs and outputs of a function. Column *opt* shows the number of BDD nodes that are used for the minimum representation. In column S the number of symmetry sets is given. E.g., a 1 in this column denotes that the function is totally symmetric, and if the number is equal to the number of inputs then the function has no symmetric variables. In columns time and space the runtime in CPU seconds and the space requirement in MByte for JUNON [16] and our approach are given, respectively. For the multipliers JUNON obtains smaller runtimes, since in this case lower bound techniques do not work well, due to exponential size of the multiplier under all orderings [8], and furthermore the computation of the lower bound is more time consuming in our algorithm. On the other hand for the adder function our algorithm clearly outperforms the approach from [16]. The argument of the lower bound works well for adder functions, since many orderings have exponential size. Thus the construction of BDDs with "worse" orderings can be cancelled in an early construction phase. We can minimize the adder function with **32** inputs in less than 200 CPU seconds, while JUNON fails, since it needs too much memory. Due to the efficient data structure described above, F$\widehat{\imath z}$Z can handle this example within 2 MByte of main memory. The limiting resource for F$\widehat{\imath z}$Z with respect to adder functions is the execution time as can be seen in example *adder32* whose runtime is almost in the given range. Nevertheless, F$\widehat{\imath z}$Z is able to compute minimum BDDs for adder functions with up to **64** inputs.

In a second series of experiments we applied our algorithm to a larger set of benchmark circuits from LGSynth93. The results are given in Table 2. As can easily be seen our algorithm is much faster in most cases (see e.g. *pcle*) and also uses much less memory (see e.g. *s400*). It clearly turns out that our lower bound technique is more powerful. Even though the lower bound computation is more expensive in one step, the overall performance is much better since large areas of the search space do not have to be considered. Especially for larger functions with more than 20 variables often JUNON cannot determine the exact result within the given time bound. In contrast our algorithm can handle these difficult problem instances due to its lower bound technique.

Based on our algorithm for several circuits the exact BDD sizes could now be determined, while only heuristical minimization results have been known before.

6 Conclusions

We presented a new exact algorithm for determining the optimal variable ordering for BDDs. It is based on a lower bound technique known from VLSI design. Furthermore, we described an efficient data structure. Experimental results have been reported that clearly demonstrate the efficiency of the approach. A comparison to the best algorithm known so far shows that runtime can often be speed up by a factor of 400 and additionally, the memory requirements can be reduced. Furthermore, we were able to compute exact results for problem instances where all previously published methods fail.

References

[1] P. Ashar, S. Devadas, and K. Keutzer. Path-delay-fault testability properties of multiplexor-based networks. *Integration the VLSI Jour.*, 15(1):1–23, 1993.

[2] B. Becker. *Synthesis for Testability: Binary Decision Diagrams*, volume 577 of *LNCS*. Symp. on Theoretical Aspects of Comp. Science, 1992.

[3] V. Bertacco, S. Minato, P. Verplaetse, L. Benini, and G. De Micheli. Decision diagrams and pass transistor logic synthesis. In *Int'l Workshop on Logic Synth.*, 1997.

[4] B. Bollig, M. Löbbing, and I. Wegener. Simulated annealing to improve variable orderings for OBDDs. In *Int'l Workshop on Logic Synth.*, pages 5b:5.1–5.10, 1995.

[5] B. Bollig and I. Wegener. Improving the variable ordering of OBDDs is NP-complete. *IEEE Trans. on Comp.*, 45(9):993–1002, Sept. 1996.

[6] K.S. Brace, R.L. Rudell, and R.E. Bryant. Efficient implementation of a BDD package. In *Design Automation Conf.*, pages 40–45, 1990.

[7] R.E. Bryant. Graph - based algorithms for Boolean function manipulation. *IEEE Trans. on Comp.*, 35(8):677–691, 1986.

[8] R.E. Bryant. On the complexity of VLSI implementations and graph representations of Boolean functions with application to integer multiplication. *IEEE Trans. on Comp.*, 40:205–213, 1991.

name	in	out	opt	S	JUNON		F$\widehat{\text{iz}}$Z	
					time	space	time	space
cc	21	20	46	21	—	34M	689s	32M
cm150a	21	1	33	21	315s	34M	3607s	32M
cm163a	16	5	26	13	59s	<1M	7.3s	<1M
cmb	16	4	28	3	0.2s	<1M	0.2s	<1M
comp	32	3	95	16	—	—	48631s	99M
cordic	23	2	42	11	152s	140M	23.8s	<1M
cps	24	102	971	21	—	289M	45809s	48M
i1	25	16	36	16	—	—	186s	9M
lal	26	19	67	21	—	—	4415s	64M
mux	21	1	33	21	318s	34M	3619s	32M
parity	16	1	17	1	<0.1s	<1M	<0.1s	<1M
pcle	19	9	42	18	23275s	9M	57.1s	2M
pm1	16	13	40	10	14.3s	<1M	3.5s	<1M
s208.1	18	9	41	18	6822s	5M	56.5s	<1M
s298	17	20	74	17	3427s	2M	88.5s	2M
s344	24	26	104	24	—	289M	10820s	99M
s349	24	26	104	24	—	289M	10746s	99M
s382	24	27	119	24	—	289M	6554s	67M
s400	24	27	119	24	—	289M	6568s	67M
s444	24	27	119	24	—	289M	6709s	74M
s526	24	27	113	24	—	289M	8443s	86M
s820	23	24	220	21	—	140M	17320s	53M
s832	23	24	220	21	—	140M	17854s	53M
sct	19	15	48	18	26210s	9M	58.9s	2M
t481	16	1	21	8	2.4s	<1M	1.2s	<1M
tcon	17	16	25	17	2159s	2M	3.3s	<1M
ttt2	24	21	107	24	—	289M	6727s	74M
vda	17	39	478	17	4177s	2M	633s	2M

Table 2: Comparison of exact algorithms

[9] P. Buch, A. Narayan, A.R. Newton, and A.L. Sangiovanni-Vincentelli. On synthesizing pass transistor networks. In *Int'l Workshop on Logic Synth.*, 1997.

[10] R. Drechsler, B. Becker, and N. Göckel. A genetic algorithm for variable ordering of OBDDs. *IEE Proceedings*, 143(6):364–368, 1996.

[11] R. Drechsler and N. Göckel. Minimization of BDDs by evolutionary algorithms. In *Int'l Workshop on Logic Synth.*, 1997.

[12] S.J. Friedman and K.J. Supowit. Finding the optimal variable ordering for binary decision diagrams. In *Design Automation Conf.*, pages 348–356, 1987.

[13] H. Fujii, G. Ootomo, and C. Hori. Interleaving based variable ordering methods for ordered binary decision diagrams. In *Int'l Conf. on CAD*, pages 38–41, 1993.

[14] M. Fujita, Y. Matsunaga, and T. Kakuda. On variable ordering of binary decision diagrams for the application of multi-level synthesis. In *European Conf. on Design Automation*, pages 50–54, 1991.

[15] N. Ishiura, H. Sawada, and S. Yajima. Minimization of binary decision diagrams based on exchange of variables. In *Int'l Conf. on CAD*, pages 472–475, 1991.

[16] S.-W. Jeong, T.-S. Kim, and F. Somenzi. An efficient method for optimal BDD ordering computation. In *International Conference on VLSI and CAD*, 1993.

[17] L. Lavagno, P. McGeer, A. Saldanha, and A.L. Sangiovanni-Vincentelli. Timed shannon circuits: A power-efficient design style and synthesis tool. In *Design Automation Conf.*, pages 254–260, 1995.

[18] M.R. Mercer, R. Kapur, and D.E. Ross. Functional approaches to generating orderings for efficient symbolic representations. In *Design Automation Conf.*, pages 624–627, 1992.

[19] D. Möller, P. Molitor, and R. Drechsler. Symmetry based variable ordering for ROBDDs. *IFIP Workshop on Logic and Architecture Synthesis, Grenoble*, pages 47–53, 1994.

[20] S. Panda, F. Somenzi, and B.F. Plessier. Symmetry detection and dynamic variable ordering of decision diagrams. In *Int'l Conf. on CAD*, pages 628–631, 1994.

[21] R. Rudell. Dynamic variable ordering for ordered binary decision diagrams. In *Int'l Conf. on CAD*, pages 42–47, 1993.

[22] D. Sieling. Variable orderings and the size of OBDDs for partially symmetric Boolean functions. In *SASIMI*, pages 189–196, 1996.

[23] F. Somenzi. *CUDD: CU Decision Diagram Package Release 2.1.2.* University of Colorado at Boulder, 1997.

[24] K. Yano, Y. Sasaki, K. Rikino, and K. Seki. Top-down pass-transistor logic design. *IEEE Jour. of Solid-State Circ.*, 31(6), June 1996.

Boolean matching for large libraries*

Uwe Hinsberger
HighTec EDV-Systeme GmbH†
Feldmannstr. 98
66119 Saarbrücken, Germany

Reiner Kolla
Lehrstuhl für Informatik V‡
Universität Würzburg
Zwinger 34
97070 Würzburg, Germany

Abstract

Boolean matching tackles the problem whether a subcircuit of a boolean network can be substituted by a cell from a cell library. In previous approaches [7, 10, 8] each pair of a subcircuit and a cell is tested for NPN equivalence. This becomes very expensive if the cell library is large. In our approach the time complexity for matching a subcircuit against a library L is almost independent of the size of L. CPU time also remains small for matching a subcircuit against the huge set of functions obtained by bridging and fixing cell inputs; but the use of these functions in technology mapping is very profitable. Our method is based on a canonical representative for each NPN equivalence class. We show how this representative can be computed efficiently and how it can be used for matching a boolean function against a set of library functions.

1 Introduction

Translating an abstract network consisting of primitive boolean operators (AND, OR, NOT) into an optimized circuit making the best use of the powerful cells available in modern target technologies is a complex problem. Therefore Mailhot and De Micheli [7] proposed to solve it by means of two different passes: *boolean matching* and *covering*. In [5] we presented an efficient algorithm for optimal duplication free DAG-covering and heuristics for the duplication of logic. At first, we tested these methods for minimizing area and depth of l-LUT-based FPGAs because in this case the computation of cell candidates is a very easy task: each connected subcircuit with up to l source nodes and one sink is a candidate. But our approach is more general; it is applicable for arbitrary technologies with single output cells and for a wide class of cost functions. The TEMPLATE system (technology mapping platform) represents a generic implementation of this approach [4]: new cost functions can be incorporated by deriving them from an abstract base class; for supporting new target technologies only the module for computing cell candidates has to be adapted. Thus, for supporting standard cell libraries we had to filter the subcircuits which can be substituted by a single cell, i.e. we needed a function for boolean matching.

This paper describes the TEMPLATE approach to boolean matching and its advantages over existing approaches. We start with a formal description of the problem and a short summary of previous work.

2 The problem

Let f be a subcircuit's function and $L=\{g_1,\ldots,g_n\}$ the set of the cell functions available in our target technology. The goal is to know which cell functions $g_{i_1},\ldots,g_{i_k}$ are suited for the realization of the subcircuit. A function g_{i_j} is suited if it can be transformed into f by an appropriate permutation of its inputs and by an appropriate phase assignment to inputs and output.

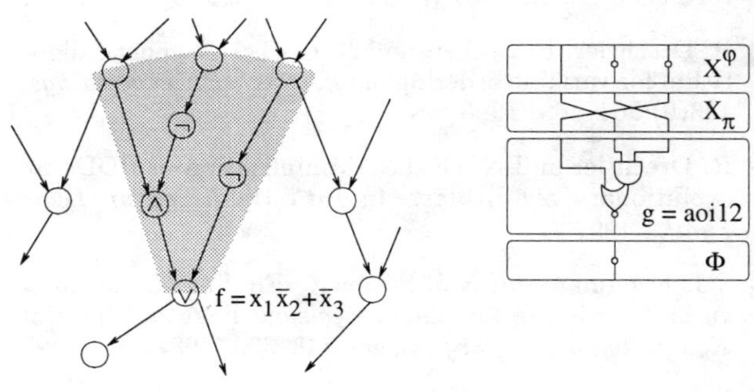

Figure 1: Matching a subcircuit against a cell

In [7] the notion *NPN equivalence* (input negation, input permutation, output negation) was introduced for this property:

$$f \equiv g \Longleftrightarrow f = \left(g \circ X_\pi \circ X^\varphi\right)^\Phi$$

for an appropriate permutation $\pi \in S_n$ and an appropriate phase assignment $\varphi \in \mathcal{B}^n$ ($\Phi \in \mathcal{B}$) to the inputs (output). Herein

$$X_\pi(x_1,\ldots,x_n) := (x_{\pi(1)},\ldots,x_{\pi(n)})$$
$$X^\varphi(x_1,\ldots,x_n) := (x_1^{\varphi_1},\ldots,x_n^{\varphi_n})$$

*This work was supported by Deutsche Forschungsgemeinschaft under grant Ko 1362/1-2

†http://www.hightec-rt.com

‡http://sirion.informatik.uni-wuerzburg.de

DAC 98, San Francisco, California

Previous approaches treat the problem, how $f \stackrel{?}{\equiv} g$ can be decided without enumerating all triples (φ, π, Φ): [7] use symmetry and unateness for reducing the search space. Don't cares can be considered by means of the so-called matching compatibility graph. [10] besides symmetry exploit the satisfy count of the cofactors. Cells belonging to the same NPN equivalence class are stored together; thus a subcircuit's function has to be matched only against one element of each class. [8] use the satisfy count of additional subfunctions and refine these signatures for further reduction of the search space.

We will start with the question whether f can be matched against a set of functions $L = \{g_1, \ldots, g_n\}$ without matching f against each g_i. This is important if there are many cells and/or if L contains all functions that can be obtained by configuring cells (i.e. by fixing some cell inputs to constant values and/or by adding bridges between some inputs). The next section explains the basic idea of an algorithm for matching f against L with time complexity almost independent of $|L|$. Section 4 presents an efficient implementation of this idea for the simpler problem of permutation equivalence; in section 5 we show how symmetry can be used to speed up this implementation. Section 6 generalizes to NPN equivalence and section 7 describes how we deal with configurations of cells. We conclude in section 8 with experimental results.

3 The basic idea

Our approach is based on the definition of exactly one canonical representative $R[f]$ for each NPN equivalence class $[f]$. Using this definition, matching a subcircuit's function f against a library L can be done as follows: first we compute the canonical representative $R[f]$ of $[f]$; then we use $R[f]$ as a key for searching in a hash table or a trie containing all canonical representatives of cells of L. Both passes are almost independent of the size of L.

Before we describe how $R[f]$ is defined and how it can be computed efficiently let us first discuss which data structure should be used for boolean functions. In our application the dimensions of the functions are rather small ($\leq 6..10$); therefore we can work with truth tables, i.e. we represent a boolean function $f : \mathcal{B}^n \longrightarrow \mathcal{B}$ as a bit string

$$T(f) := T_0(f) \ldots T_{2^n-1}(f)$$

with $T_i(f) := f(i_{n-1}, \ldots, i_0)$; herein $(i_{n-1}, \ldots, i_0)$ denotes the binary representation of i. Bit strings have the advantage, that they are very compact for small numbers of inputs, and that can be used easily as a key for hashing. For other applications our results can be generalized to BDDs: all operations which we need are available on BDDs; also the $\leq$ relation introduced later on for truth tables can be evaluated efficiently on an implicit representation of truth tables.

We exploit the fact that there is a total order on bit strings, namely the lexicographic order $\leq_{lex}$. We use $\leq_{lex}$ for defining a total order on boolean functions: $f \leq g$ if and only if $T(f) \leq_{lex} T(g)$. As $\leq$ is total there is exactly one maximal element in each subset of the boolean functions, esp. in each NPN equivalence class $[f]$. We call the maximal element the representative $R[f]$ of $[f]$:

$$R[f] := \max_{g \in [f]} g = \max_{(\varphi, \pi, \Phi)} \left(f \circ X_\pi \circ X^\varphi \right)^\Phi$$

Of course the naive approach for computing $R[f]$, i.e. trying out all triples (φ, π, Φ), can be used only for small examples. The main part of the paper explains, how $R[f]$ can be computed efficiently.

4 Permutation equivalence

We start in this section with the simpler problem of permutation equivalence. Two boolean functions are permutation equivalent

$$f \equiv_P g \iff f = g \circ X_\pi$$

The corresponding equivalence class of f is denoted by $[f]_P$ and the representative of this class is defined as

$$R[f]_P := \max_{g \in [f]_P} g = \max_{\pi \in S_n} \left(f \circ X_\pi \right)$$

Just as well we can maximize $f \circ X_{\pi^{-1}}$ and we will prefer this alternative for reasons explained later on. The tree in figure 2 illustrates the naive approach for computing $R[f]_P$: the nodes of the k-th level ($k = 0, \ldots, n$) represent the injections

$$\gamma : \{1, \ldots, k\} \longrightarrow \{1, \ldots, n\}$$

$S_{k,n}$ denotes the set of these injections. Thus the leaves correspond to the permutations $S_{n,n} = S_n$; each leaf π is labeled with the truth table of the function $f \circ X_{\pi^{-1}}$. The occurrences of the maximal truth table are underlined.

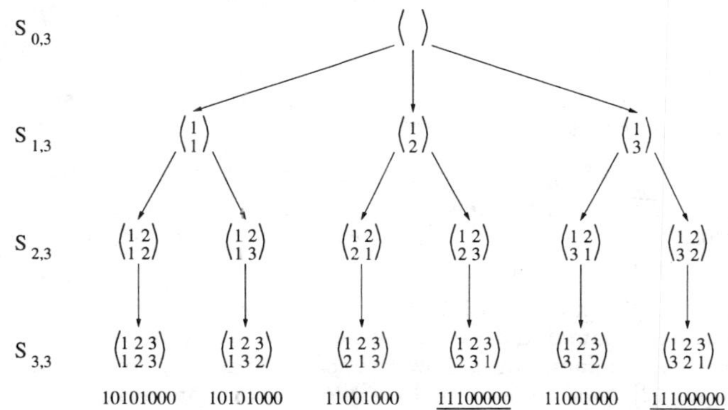

Figure 2: Naive approach for computing $R[aoi12]_P$

Now we examine the question how this search space can be explored in an implicit manner; i.e. how can we find the maximal leaf label without traversing the entire tree. We first look for a common characteristic of all leaves of a subtree with root $\gamma \in S_{k,n}$.

Lemma 1 *Let $0 \leq k \leq n$ and $\pi, \sigma \in S_n$ two permutations with $\pi(i) = \sigma(i)$ for all $i \in \{1, \ldots, k\}$. Then*

$$(f \circ X_{\pi^{-1}})_{\bar{x}_{k+1} \ldots \bar{x}_n} = (f \circ X_{\sigma^{-1}})_{\bar{x}_{k+1} \ldots \bar{x}_n}$$

Lemma 1 states that the truth tables of all leaves in the subtree rooted at $\gamma \in S_{k,n}$ are identical in the first 2^k bits. A formal proof of this lemma can be found in [3]. For an intuitive argument look at the truth table of $f \circ X_{\pi^{-1}}$:

x_n ... x_{k+1}	x_k ... x_1	$f\left(x_{\pi^{-1}(1)}, ..., x_{\pi^{-1}(n)}\right)$
0	0 ... 0	T_0
	⋮ ... ⋮	⋮
	1 ... 1	T_{2^k-1}
0 ... 1	0 ... 0	T_{2^k}
⋮ ⋮	⋮ ⋮	⋮
1 ... 1	1 ... 1	T_{2^n-1}

Substituting π by a permutation σ with the above property does not alter $T_0 \ldots T_{2^k-1}$ since in the first 2^k rows of the table $x_{\pi^{-1}(i)} = x_{\sigma^{-1}(i)}$ holds for all i:

$$\text{if } \pi^{-1}(i) \leq k \text{ then } \pi^{-1}(i) = \sigma^{-1}(i)$$
$$\text{if } \pi^{-1}(i) > k \text{ then } x_{\pi^{-1}(i)} = 0 = x_{\sigma^{-1}(i)}$$

We use this property to assign a label T^γ to each node $\gamma \in S_{k,n}$. This label provides relevant information about all leaf labels in the subtree rooted at γ:

$$T^\gamma := T\left(\left(f \circ X_{\pi^{-1}}\right)_{\overline{x_{k+1}} \ldots \overline{x_n}}\right)$$

where π is an arbitrary permutation with $\pi(i) = \gamma(i)$ for all $i \in \{1, ..., k\}$. Lemma 1 guarantees that T^γ is independent of the choice of π and thus well-defined. Let

$$S_{k,n}^{max} := \left\{\gamma \in S_{k,n} \mid T^\gamma = \max_{\delta \in S_{k,n}} T^\delta\right\}$$

We will show that it is not necessary to explore the subtree rooted at γ if $\gamma \notin S_{k,n}^{max}$:

If $\gamma \notin S_{k,n}^{max}$ there exists an injection $\delta \in S_{k,n}$ with $T^\delta > T^\gamma$. Using the definition of T^δ and T^γ we can even say that $T_0^\delta ... T_{2^k-1}^\delta > T_0^\gamma ... T_{2^k-1}^\gamma$. All leaf labels in the subtree rooted at δ (γ) start with $T_0^\delta ... T_{2^k-1}^\delta$ ($T_0^\gamma ... T_{2^k-1}^\gamma$ respectively). Thus the leaf labels in the subtree rooted at γ cannot be maximal and we do not need to explore this subtree. Figure 3 demonstrates how the search space for computing $R[aoi12]_p$ can be reduced by exploiting this property.

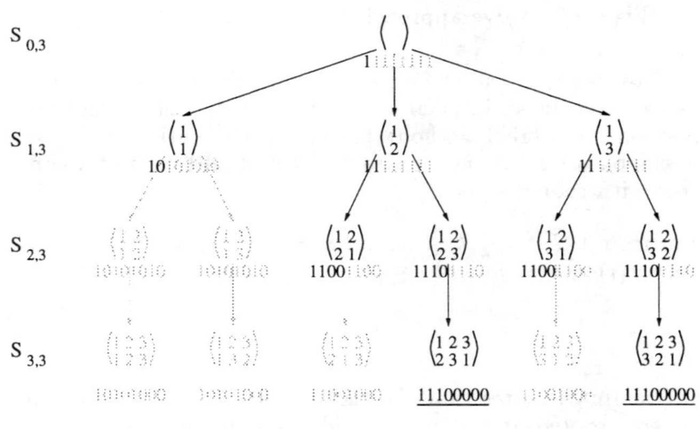

Figure 3: Reducing search space using $S_{k,n}^{max}$

In this example there remain two of the eight permutations: $\binom{123}{231}$ and $\binom{123}{321}$ both maximize $f \circ X_{\pi^{-1}}$; this is not astonishing because the function $aoi12$ is symmetric in x_2 and x_3. For totally symmetric functions the situation is even worse; in this case we gain nothing by using only $S_{k,n}^{max}$. That is why the next section demonstrates how our approach can be extended for exploiting symmetry.

5 Using variable symmetry

The method presented in the previous section computes *all* leaves with maximal label. But in many applications we need just one such leaf. The key to a more efficient implementation is the observation that $f \circ X_\sigma$ is not changed by permuting variables of the same symmetry class. Therefore we may assume that σ orders the variables of the same symmetry class in a special way:

Lemma 2 *There is a permutation $\pi \in S_n$ with*

- *π maximizes $f \circ X_{\pi^{-1}}$*
- *if x_i and x_j belong to the same symmetry class of f and $i < j$, then $\pi^{-1}(i) < \pi^{-1}(j)$*

As an immediate consequence we need to consider only subsets of $S_{k,n}$: let $\tilde{S}_{k,n}$ the set of injections $\gamma \in S_{k,n}$ such that for all $i < j$ with x_i and x_j belonging to the same symmetry class the following holds: if $\gamma^{-1}(j)$ is defined then $\gamma^{-1}(i)$ is defined too and $\gamma^{-1}(i) < \gamma^{-1}(j)$. Similar as in section 4 we define

$$\tilde{S}_{k,n}^{max} := \left\{\gamma \in \tilde{S}_{k,n} \mid T^\gamma = \max_{\delta \in \tilde{S}_{k,n}} T^\delta\right\}$$

Lemma 2 guarantees that $S_{n,n}^{max} \cap \tilde{S}_{n,n}^{max}$ is not empty. For $k \geq 1$, the nodes corresponding to $\tilde{S}_{k,n}^{max}$ are children of nodes in $\tilde{S}_{k-1,n}^{max}$. Thus the search tree can be further reduced as shown in figure 4.

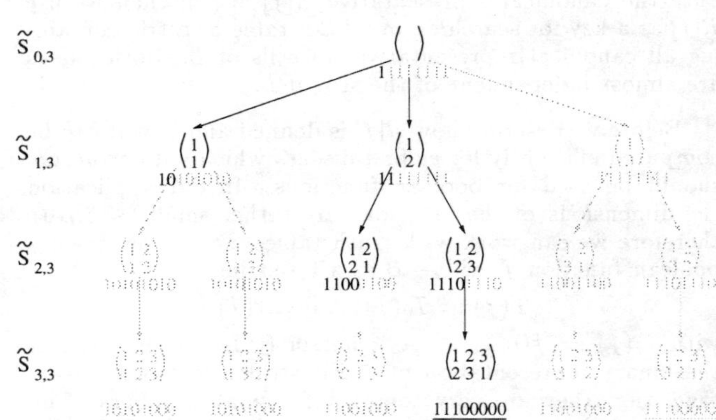

Figure 4: Reducing search space using $\tilde{S}_{k,n}^{max}$

In this example the approach enumerates just one permutation. In general $\tilde{S}_{k,n}^{max}$ may contain several elements because our current implementation does not exploit group symmetry. So for $aoi222$ there still remain 3! elements. However a generalization of our approach for considering group symmetry is straightforward.

6 Generalization to NPN

In order to compute the representative $R[f]$ of a NPN equivalence class we have to consider a tree of (partial) assignments $A_{k,n} := \mathcal{B}^k \times S_{k,n} \times \mathcal{B}$ instead of injections $S_{k,n}$. The leaves (φ, π, Φ) are labeled with $T((f \circ X_{\pi^{-1}} \circ X^\varphi)^\Phi)$. In order to find a characteristic for a subtree let us have a look at the truth table of a leaf (φ, π, Φ):

$x_n \;\ldots\; x_{k+1}$	$x_k \;\ldots\; x_1$	$f^{\Phi}\left(x_{\pi^{-1}(1)}^{\varphi_{\pi^{-1}(1)}}, \ldots, x_{\pi^{-1}(n)}^{\varphi_{\pi^{-1}(n)}}\right)$
0	$0 \;\ldots\; 0$	T_0
	$\vdots \quad \vdots$	$\vdots$
	$1 \;\ldots\; 1$	T_{2^k-1}
$\vdots$	$\vdots \quad \vdots$	$\vdots$
1	$0 \;\ldots\; 0$	$T_{2^n-2^k}$
	$\vdots \quad \vdots$	$\vdots$
	$1 \;\ldots\; 1$	T_{2^n-1}

It is easy to see that now modifying $(\varphi_n, \ldots, \varphi_{k+1})$ and/or $\pi(n), \ldots, \pi(k+1)$ may alter the first 2^k entries of this truth table; e.g. choosing

$$\varphi_i' := \begin{cases} \varphi_i & \text{for } 1 \le i \le k \\ \overline{\varphi}_i & \text{for } k < i \le n \end{cases}$$

$$\pi'(i) := \pi(i) \text{ for all } 1 \le i \le n$$

swaps $T_0 \ldots T_{2^k-1}$ and $T_{2^n-2^k} \ldots T_{2^n-1}$ (as well as other "*blocks*" of length 2^k). An important observation is that modifications of $(\varphi_n, \ldots, \varphi_{k+1})$ and/or $\pi(n), \ldots, \pi(k+1)$ always rearrange complete blocks of length 2^k; there are no rearrangements within the blocks. That is why the leaves of a subtree rooted at the same partial assignment $a \in A_{k,n}$ are labeled with tables consisting of the same blocks:

Lemma 3

Let $(\varphi, \pi, \Phi), (\psi, \sigma, \Psi) \in A_{n,n}$ two assignments with

- $\varphi_i = \psi_i$ *for all* $i \in \{1, \ldots, k\}$
- $\pi(i) = \sigma(i)$ *for all* $i \in \{1, \ldots, k\}$
- $\Phi = \Psi$

Let $T = T[1] \ldots T[2^{n-k}]$ and $S = S[1] \ldots S[2^{n-k}]$ be the truth tables associated with (φ, π, Φ) and (ψ, σ, Ψ) respectively; herein $T[i]$ and $S[i]$ denote the blocks of length 2^k. Then the following holds:

$$\{T[1], \ldots, T[2^{n-k}]\} = \{S[1], \ldots, S[2^{n-k}]\}$$

Lemma 3 states that the tables T and S at two arbitrary leaves of a common subtree consist of the same blocks; thus, they may differ only in the arrangement of these blocks. Furthermore each block can be moved to the beginning of the table by choosing an appropriate leaf of the subtree:

Lemma 4 *Let $(\varphi, \pi, \Phi) \in A_{n,n}$ an assignment and $T = T[1] \ldots T[2^{n-k}]$ the corresponding truth table consisting of blocks $T[i]$ with $len(T[i]) = 2^k$. Then for each block $T[j]$, $j \in \{1, \ldots, 2^{n-k}\}$, there is an assignment $(\psi, \sigma, \Psi) \in A_{n,n}$ such that*

- $\psi_i = \varphi_i$ *for all* $i = 1, \ldots, k$
- $\sigma(i) = \pi(i)$ *for all* $i = 1, \ldots, k$
- $\Phi = \Psi$

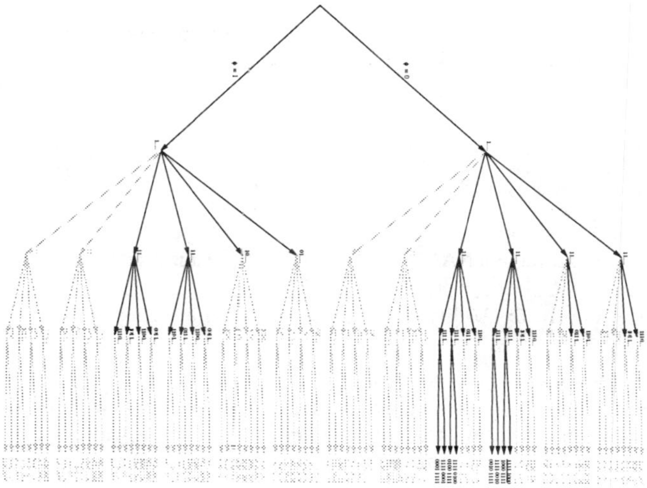

Figure 5: Computing $R[aoi12]$

- $T((f \circ X_{\sigma^{-1}} \circ X^{\psi})^{\Psi}) = T[j]S[2] \ldots S[2^{n-k}]$

Among all blocks of a subtree the maximal one is the most interesting: it represents a 2^k bit prefix of the maximal label we can get for a leaf of that subtree. Therefore we can use it as a criterion for determining whether a subtree can be excluded from search or not. For a partial assignment $(\varphi', \pi', \Phi') \in A_{k,n}$ let

$$T^{(\varphi', \pi', \Phi')} := \max_{(\epsilon_{k+1}, \ldots, \epsilon_n)} T\left(\left(f \circ X_{\pi^{-1}} \circ X^{\varphi}\right)^{\Phi}_{x_{k+1}^{\epsilon_{k+1}} \ldots x_n^{\epsilon_n}}\right)$$

where $(\varphi, \pi, \Phi) \in A_{n,n}$ denotes an arbitrary assignment with $\Phi = \Phi'$, $\varphi_i = \varphi_i'$ and $\pi(i) = \pi'(i)$ for all $i \in \{1, \ldots, k\}$. We define

$$A_{k,n}^{max} := \left\{a \in A_{k,n} \mid T^a = \max_{a' \in A_{k,n}} T^{a'}\right\}$$

It is easy to see that for $k \ge 1$ all nodes of $A_{k,n}^{max}$ are children of elements of $A_{k-1,n}^{max}$. Analogously to section 4 $A_{n,n}^{max}$ represents the set of all leaves with maximal label.

If we just need one leaf with maximal label we can do the same as in section 5: we may confine ourselves to

$$\tilde{A}_{k,n} := \mathcal{B}^k \times \tilde{S}_{k,n} \times \mathcal{B} \text{ and}$$

$$\tilde{A}_{k,n}^{max} := \left\{a \in \tilde{A}_{k,n} \mid T^a = \max_{a' \in \tilde{A}_{k,n}} T^{a'}\right\}$$

Figure 5 demonstrates the substantial reduction of the search space resulting from this approach.

7 Working with configurations

By using the representative as a key in a hash table we can match a boolean function against a cell library with time complexity almost independent of the library size. This permits us to work efficiently with very large libraries and to take into account the huge function set originating from the configuration of library cells: a cell can be configured by fixing some inputs to constant values and/or bridging some inputs (i.e. feeding them with the same signal, either in positive or negative phase).

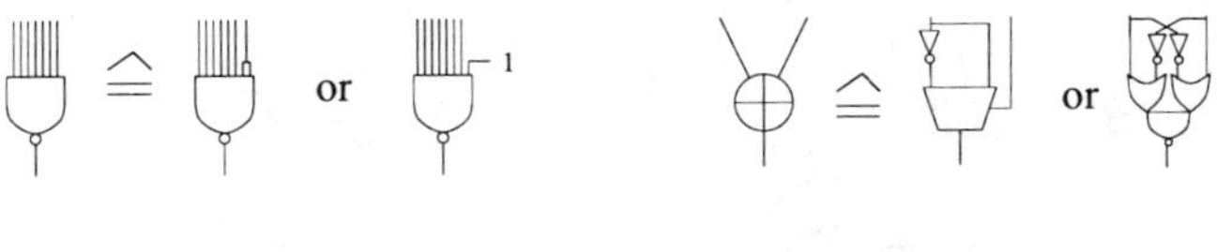

(a) realization of a NAND7 (b) realization of an XOR

Figure 6: Additional functions by configuring cells

Up to now such methods have been applied only to special technologies as multiplexer-based FPGAs. But we found out that extending the set of functions in such a way is also very effective in technology mapping onto standard cell libraries. The small library *lib2* of the MCNC benchmark set [11] consists of 27 cells and can realize only 13 equivalence classes; by using configurations of cells, the number of equivalence classes increases to 30; especially all two-dimensional and 70% of the three-dimensional classes become feasible. The effect on the mapping results will be presented in the next section.

Within our approach the library, i.e. the set of feasible functions, is organized as a directed acyclic graph. Each node represents a feasible NPN equivalence class. It is accessible by hashing with the corresponding representative as a key. It is annotated with a list of the cells that belong to the class; this list is empty if the class can be realized only by configuring cells with more inputs. Configuration possibilities are given by the edges of the graph: there is an edge from a n-dimensional class $[f]$ to a $(n+1)$-dimensional class $[g]$ if $R[f]$ can be realized by means of one or more configurations of $R[g]$. Thus, finding realizations of a subcircuit consists of computing the representative $R[f]$ of the subcircuit's function f, finding the corresponding node v and exploring (one, some or all) paths starting in v. Figure 7 shows this graph for the library *lib2*.

8 Experimental results

The methods presented in this paper have been implemented in C++ and integrated in the TEMPLATE system. The tables 1 and 2 summarize the results obtained with this implementation. Our experiments used the library *lib2* and the NOR-networks distributed with the 1989 MCNC benchmark

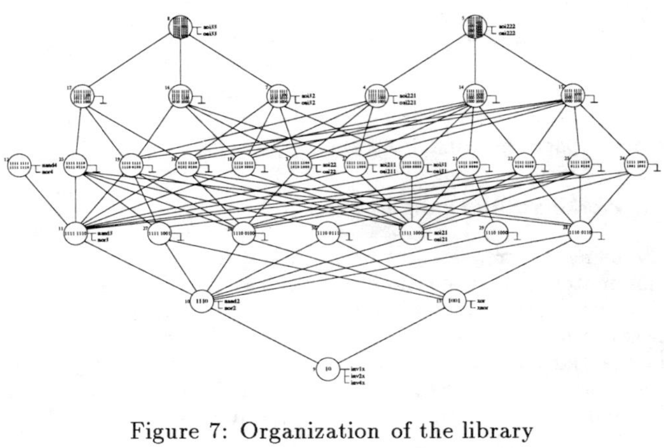

Figure 7: Organization of the library

network	NORs	cuts	compar	result	CPU	$\frac{\text{compar}}{\text{sec}}$
C1355	619	2198	$163 \cdot 10^6$	9320	123	$1.33 \cdot 10^6$
C1908	718	2550	$179 \cdot 10^6$	20324	145	$1.23 \cdot 10^6$
C2670	1100	4210	$369 \cdot 10^6$	30778	416	$0.89 \cdot 10^6$
C3540	1161	5350	$417 \cdot 10^6$	54820	522	$0.80 \cdot 10^6$
C499	579	1806	$132 \cdot 10^6$	8488	113	$1.17 \cdot 10^6$
C5315	1938	10637	$984 \cdot 10^6$	85086	1169	$0.84 \cdot 10^6$
C6288	2400	3744	$5 \cdot 10^6$	14976	74	$0.07 \cdot 10^6$
C7552	3078	12442	$1081 \cdot 10^6$	86054	932	$1.16 \cdot 10^6$
alu4	720	3206	$261 \cdot 10^6$	37820	421	$0.62 \cdot 10^6$
apex6	831	2222	$136 \cdot 10^6$	26930	241	$0.56 \cdot 10^6$
des	4679	23301	$1733 \cdot 10^6$	329124	2163	$0.80 \cdot 10^6$
frg2	1137	6212	$475 \cdot 10^6$	79278	547	$0.87 \cdot 10^6$
k2	1971	12313	$1019 \cdot 10^6$	188256	345	$2.95 \cdot 10^6$
pair	1674	6408	$461 \cdot 10^6$	68122	633	$0.73 \cdot 10^6$
rot	689	2170	$132 \cdot 10^6$	24988	334	$0.40 \cdot 10^6$
too_large	748	4598	$389 \cdot 10^6$	70748	372	$1.05 \cdot 10^6$
vda	926	5711	$461 \cdot 10^6$	85684	160	$2.88 \cdot 10^6$
x3	763	4110	$308 \cdot 10^6$	42388	431	$0.71 \cdot 10^6$

Table 1: Performance of our matching algorithm

set; we just list the larger networks ($\geq$ 500 NOR gates) for lack of space.

Table 1 shows the results obtained for the matching phase alone: beside the name of the circuit and the number of NOR gates in the initial network we list the number of cuts used as an input for our matching algorithm; these cuts have been computed by means of a maxflow algorithm as proposed in [9]. The fourth column represents the size of the search space, i.e. the number of comparisons which a naive implementation would need for matching each cut against each class of the library. Then we list the size of the result, i.e. the number of cell candidates computed by our program. The last two columns inform about the efficiency of the approach, giving first the CPU time on a HP 735/125 in seconds and then the number of comparisons per second. The central observation is that our implementation implicitly manages about 10^6 comparisons per second.

Table 2 demonstrates the effectiveness of exploiting cell configurations and of using boolean matching instead of graph or tree pattern matching [6, 2]. In the sum over all benchmarks we need 8% less area than MIS/SIS [1, 2]; for the circuit C1355 area is even reduced by 50%. Notice that

network	TEMPLATE		SIS 1.3	
	area	CPU	area	CPU
C1355	870	270	1748	48
C1908	1330	322	1640	62
C2670	1965	687	2145	116
C3540	2808	942	3020	141
C499	870	228	1172	39
C5315	4420	1864	4546	379
C6288	6264	186	7910	286
C7552	5631	1697	6917	573
alu4	1710	743	1753	80
apex6	1637	466	1623	67
des	11909	4388	12142	2833
frg2	3529	1144	3235	236
k2	5902	1590	6247	534
pair	3601	1090	3891	198
rot	1594	449	1540	71
too_large	2250	904	2172	183
vda	2818	691	3123	154
x3	2376	751	2100	214
sum	61484	18412	66924	6214

Table 2: Results for area optimization

all area values are scaled down by 464 (the greatest common divisor of these values) for the sake of clearness, and that we mapped the NOR networks without any preprocessing. Both, SIS and TEMPLATE, could profit from technology independent optimization performed before. We observed that our results can be improved further by about 2% by using duplication of logic; but then the CPU times become huge because there are too many cuts. Even without duplication of logic we need three times more CPU than SIS. However within our approach matching time would not increase (significantly) if we used a much larger library than *lib2*.

References

[1] R.K. Brayton, R. Rudell, A. Sangiovanni-Vincentelli, and A.R. Wang. MIS: A Multiple-Level Logic Optimization System. *IEEE Transactions on Computer-Aided Design*, CAD-6(6):1062–1081, November 1987.

[2] E. Detjens, G. Ganot, A. Sangiovanni-Vincentelli, and A. Wang. Technology mapping in MIS. In *Proceedings of the International Conference of Computer Aided Design*, pages 116–119, 1987.

[3] U. Hinsberger and R. Kolla. Matching a Boolean Function against a Set of Functions. Technical Report No. 185, Preprint-Reihe, Institut für Informatik, Universität Würzburg, 1997.

[4] U. Hinsberger and R. Kolla. TEMPLATE: a generic TEchnology Mapping PLATform. Technical Report No. 186, Preprint-Reihe, Institut für Informatik, Universität Würzburg, 1997.

[5] U. Hinsberger and R. Kolla. Optimal Technology Mapping for Single Output Cells. In *Proceedings of the 5th Great Lakes Symposium on VLSI*, pages 14–19, March 1995.

[6] K. Keutzer. DAGON: Technology Binding and Local Optimization by DAG Matching. In *Proceedings of the 24th Design Automation Conference (DAC87)*, pages 341–347, June 1987.

[7] F. Mailhot and G. de Micheli. Technology Mapping Using Boolean Matching and Don't Care Sets. In *Proceedings of the European Design Automation Conference*, pages 212–216, 1990.

[8] J. Mohnke and S. Malik. Permutation and Phase Independent Boolean Comparison. In *Proceedings of the European Design Automation Conference (EDAC93)*, pages 86–92, 1993.

[9] R. Murgai, Y. Nishizaki, N. Shenoy, R.K. Brayton, and A. Sangiovanni-Vincentelli. Logic Synthesis for Programmable Gate Arrays. *27th ACM/IEEE Design Automation Conference*, pages 620–625, 1990.

[10] H. Savoj, M.J. Silva, R.K. Brayton, and A. Sangiovanni-Vincentelli. Boolean Matching in Logic Synthesis. In *Proceedings of the European Design Automation Conference (EURODAC)*, pages 168–174, 1992.

[11] S. Yang. Logic Synthesis and Optimization Benchmarks User Guide (Version 3.0). Technical report, Microelectronics Center of North Carolina, P.O. Box 12889, Research Triangle Park, NC 27709, January 1991.

A Fast Hierarchical Algorithm for 3-D Capacitance Extraction

Weiping Shi, Jianguo Liu†, Naveen Kakani and Tiejun Yu‡
Dept. of Computer Science, Univ. of North Texas, Denton, TX 76203
† Dept. of Mathematics, Univ. of North Texas, Denton, TX 76203
‡ Dept. of Mathematics, Univ. of North Carolina, Charlotte, NC 28223

Abstract

We present a new algorithm for computing the capacitance of three-dimensional perfect electrical conductors of complex structures. The new algorithm is significantly faster and uses much less memory than previous best algorithms, and is kernel independent.

The new algorithm is based on a hierarchical algorithm for the n-body problem, and is an acceleration of the boundary-element method for solving the integral equation associated with the capacitance extraction problem. The algorithm first adaptively subdivides the conductor surfaces into panels according to an estimation of the potential coefficients and a user-supplied error bound. The algorithm stores the potential coefficient matrix in a hierarchical data structure of size $O(n)$, although the matrix is size n^2 if expanded explicitly, where n is the number of panels. The hierarchical data structure allows us to multiply the coefficient matrix with any vector in $O(n)$ time. Finally, we use a generalized minimal residual algorithm to solve m linear systems each of size $n \times n$ in $O(mn)$ time, where m is the number of conductors.

The new algorithm is implemented and the performance is compared with previous best algorithms. For the $k \times k$ bus example, our algorithm is 100 to 40 times faster than FastCap, and uses 1/100 to 1/60 of the memory used by FastCap. The results computed by the new algorithm are within 2.7% from that computed by FastCap.

1 Introduction

In this paper, we study the capacitance extraction problem of three-dimensional perfect electrical conductors of complex structures. Fast and accurate capacitance estimation is important in the design of high-performance integrated circuits [8, 9, 10, 11, 12, 13, 15]. Two examples of three-dimensional complex structures for which capacitance strongly affects performance are DRAM cells and chip carriers used in high density packaging [10].

DAC 98, San Francisco, California

1.1 Integral Equation Approach

We assume the conductors are embedded in a homogeneous dielectric medium, though our techniques can be extended to multiple dielectrics by using multi-layered Green functions. Since our algorithm is kernel independent, there is no need to introduce dielectric-dielectric interface conditions and additional variable, as required by kernel-dependent algorithms such as FastCap [11].

The capacitance of an m-conductor geometry can be summarized by an $m \times m$ capacitance matrix $\mathbf{C}$. The diagonal entries C_{ii} of $\mathbf{C}$ are positive, representing the self-capacitance of conductor i, and the non-diagonal entries C_{ij} are negative, representing the coupling capacitance between conductors i and j. To determine the j-th column of the capacitance matrix, we need only solve for the surface charges on each conductor produced by raising conductor j to unit potential while grounding rest of the conductors. Then C_{ij} is numerically equal to the charge on conductor i. This procedure is repeated m times to compute all columns of $\mathbf{C}$.

Each of the m potential problems can be solved using an equivalent free-space formulation where the conductor-dielectric interfaces are replaced by a charge layer of density σ. Assuming a homogeneous dielectric, the charge layer in the free-space problem will be the induced charge in the original problem if σ satisfies the integral equation

$$\psi(x) = \int_{surfaces} \sigma(x') \frac{1}{4\pi\epsilon_0 \|x - x'\|} da', \tag{1}$$

where $\psi(x)$ is the known conductor surface potential, da' is the incremental conductor surface area, $x, x' \in \Re^3$, and $\|x - x'\|$ is the Euclidean distance between x and x'. The kernel of the integral equation is $1/\|x - x'\|$. For simplicity, we will temporarily omit the scale factor $1/4\pi\epsilon_0$ and reintroduce it later in Section 4.

We use Galerkin scheme to numerically solve (1) for σ. In this approach, the conductor surfaces are divided into n small panels, and it is assumed that on each panel A_i, a charge q_i is uniformly distributed. Then for each panel A_i, an equation is written which relates the known potential on A_i, denoted by v_i, to the sum of the contribution of potential from charges on all n panels $A_1, \ldots, A_n$. The result is a dense linear system

$$\mathbf{Pq} = \mathbf{v}, \tag{2}$$

where $\mathbf{q} \in \Re^n$ is the vector of panel charges, $\mathbf{v} \in \Re^n$ is the vector of known panel potentials, and $\mathbf{P} \in \Re^{n \times n}$ is the potential coefficient matrix where each entry

$$P_{ij} = \frac{1}{area(A_i)} \int_{x_i \in A_i} \frac{1}{area(A_j)} \int_{x_j \in A_j} \frac{1}{\|x_i - x_j\|} da_j da_i, \tag{3}$$

for panels A_i and A_j. Matrix $\mathbf{P}$ is known to be positive, symmetric and positive definite.

The linear system of (2) has to be solved to compute panel charges, and the capacitances are derived by summing the panel charges. Direct methods based on triangularization of matrix $\mathbf{P}$, such as Gaussian elimination and Cholesky factorization, require $O(n^3)$ operations. Iterative algorithms normally require $O(n^2)$ operations per iteration. These approaches are inefficient if the number of panels is more than several thousands. In fact, any algorithm that uses the explicit representation of matrix P must use at least $\Omega(n^2)$ time and memory since the matrix is size n^2.

Several algorithms have been proposed to solve (2) in sub-quadratic time. The FastCap algorithm of Nabors and White [10, 12] uses $O(n)$ time, and is based on the $O(n)$ time multipole algorithm for the n-body problem [6]. In this paper we propose a different $O(n)$ time algorithm, which is based on a different $O(n)$ time hierarchical algorithm for the n-body problem by Appel [1]. Efficient algorithms that are not based on the n-body problem include an FFT algorithm of Phillips and White [13], and the singular value decomposition (SVD) algorithm of Kapur and Zhao [9, 8]. The FFT algorithm and the SVD algorithm are about twice as fast as FastCap for test examples [13, 8], though the running time of both algorithms are $O(n \log n)$.

1.2 The n-Body Problem

In the n-body problem, each of the n particles exerts a force on all the other $n-1$ particles, implying $\binom{n}{2}$ pairwise interactions [1, 6]. The capacitance extraction problem shares many similarities with the n-body problem. Both the gravitational force and the electro-magnetic field decreases as the distance increases, and the principle of superposition applies to both problems. The principle of superposition states that the potential due to a cluster of particles is the sum of the potential due to each individual particle.

There are two types of algorithms for the n-body problem: the hierarchical algorithm of Appel [1] (1985) and the multipole algorithm of Greenberg and Rohklin [5] (1987). When Appel first published his paper, he thought the time complexity was $O(n \log n)$. Later, a careful analysis shows the time complexity is $O(n)$ [3]. At about the same time, Greengard and Rohklin [5] proposed the multipole algorithm, and proved its time complexity is $O(n)$. The capacitance extraction algorithm FastCap is related to the multipole algorithm, while our algorithm is related to Appel's algorithm [1], and a radiosity algorithm [7]. Hanrahan, Salzman and Aupperle [7] used the ideas in Appel's algorithm to compute the reflection of light among a set of objects in computer graphics. Their experience shows Appel's algorithm is not only fast, but also very easy to implement.

The hierarchical algorithm of Appel [1] uses the following two key ideas to compute all the forces on a particle: 1) For practical considerations, the forces acting on a particle need only be calculated to within the given precision. 2) The force due to a cluster of particles at some distance can be approximated, within the given precision, with a single term. However, there are several differences between the capacitance extraction problem and the n-body problem, which prevent us from blindly adopting the n-body solution. One major difference is that in the n-body problem the objects are particles, while in the capacitance extraction problem, the objects are continuous conductor surfaces. Therefore, the hierarchical data structures are formed differently. The n-body algorithm begins with n particles and clusters them into larger and larger groups. Our algorithm begins with a set of panels and subdivides the panels into smaller and smaller panels. (We can also start with a set of given small panels and build a hierarchical data structure using the same idea.) Another difference is that the self-capacitance is very important in our problem while there is no such concept in the n-body problem.

1.3 Our Contribution

In Section 2, we show how to build a hierarchical data structure for the potential coefficient matrix by adaptively subdividing the conductor surfaces into subpanels according to the estimation of the coefficient and a user-supplied error bound. This is different from the multipole algorithm where the conductor surfaces are divided in a pre-processing stage according to the geometry of each conductor individually. Our new algorithm subdivides the surfaces into panels of variable sizes, and it is guaranteed that all coefficients are calculated to the same precision. More importantly, the coefficient matrix contains $O(n)$ block entries, where n is the number of panels.

In Section 3, we show how to multiply the potential coefficient matrix $\mathbf{P}$ with any vector in $O(n)$ time, using the hierarchical data structure. The hierarchical data structure is of size $O(n)$, but it implicitly stores the whole $n \times n$ potential coefficient matrix. As a result, the multiplication can be done in $O(n)$ time through three traversals of the hierarchical data structure. Then we use the Generalized Minimum Residual (GMRES) method to solve the linear systems.

In Section 4, we present the experimental results that show the new algorithm is significantly faster than the multipole algorithm FastCap and uses significantly less memory, with a small difference in the result compared with FastCap.

2 Potential Matrix Approximation

This section describes the recursive refinement procedure which decomposes a large panel into a hierarchy of small panels, and builds a hierarchical representation of the potential matrix. We first describe the procedure, then estimate the number of interactions that need to be considered, and finally analyze the error in the resulting potential matrix.

```
Refine(Panel Ai, Panel Aj)
{
  Pij = PotentialEstimate(Ai, Aj);
  Ri = longest side of Ai;
  Rj = longest side of Aj;

  if ((Pij*Ri < Peps) AND (Pij*Rj < Peps))
    RecordInteraction(Ai, Aj);
  else if (Ri > Rj) {
    Subdivide(Ai);
    Refine(Ai.left, Aj);
    Refine(Ai.right, Aj);
  } else {
    Subdivide(Aj);
    Refine(Ai, Aj.left);
    Refine(Ai, Aj.right);
  }
}
```

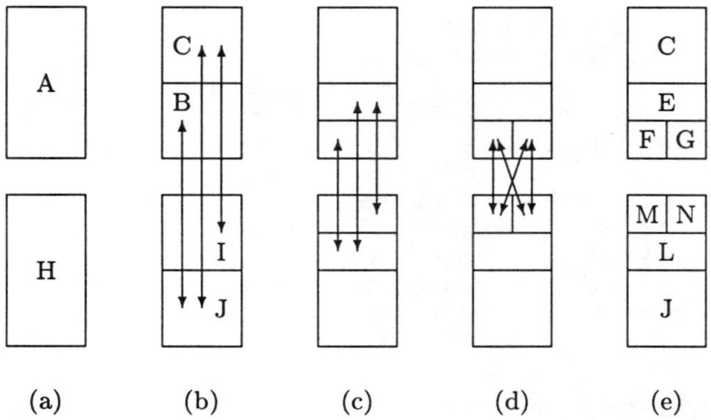

Figure 1: Partition conductor surfaces into panels.

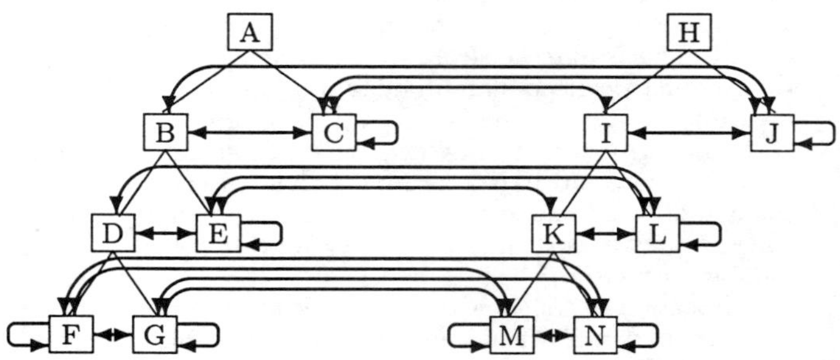

Figure 2: Potential coefficient matrix stored as links.

Procedure **PotentialEstimate** returns an estimate of the potential factor from panel **Ai** to panel **Aj** defined by (3). We use the closed form expressions derived by Wilton, et al [16] and numerical integration to compute the potential factor. If the estimated potential factor is less than the user provided error bound **Peps** (P_ϵ), then the panels are allowed to interact at this level of detail. The recursion is terminated and the interaction is recorded between the two panels. However, if the potential factor estimate is too large, then the estimate may not be accurate. In this case, the panel with the larger area, say **Ai**, is to be subdivided into **Ai.left** and **Ai.right**. The procedure **Refine** is called recursively. Procedure **Subdivide** subdivides a panel into two new panels. The subdivision hierarchy is stored in a binary tree with each node having two pointers **left** and **right** pointing to the two subpanels. Since a panel may be refined against many panels, the actual subdivision of a panel may have occurred previously. When this happens, **Subdivide** does not perform any task.

Figure 1 shows the refine process of two conductor surfaces. We start with two conductor surfaces A and H in (a). Assume the coefficient estimates between A and H are greater than the user provided error bound P_ϵ, we subdivide A into $B \cup C$, and then subdivide H into $I \cup J$ in (b). Now assume the estimates BJ, CI and CJ are less than P_ϵ, but estimate BI is greater than P_ϵ. Therefore, we record interactions BJ, CI and CJ at this level, while further subdivide panels B and I. The final panels are shown in (e). We also compute the self-capacitance at this time. Note that if we use uniform grid partition, then there would be a total of 16 panels. Also note that we consider mainly the interaction with other panels, instead of the geometry.

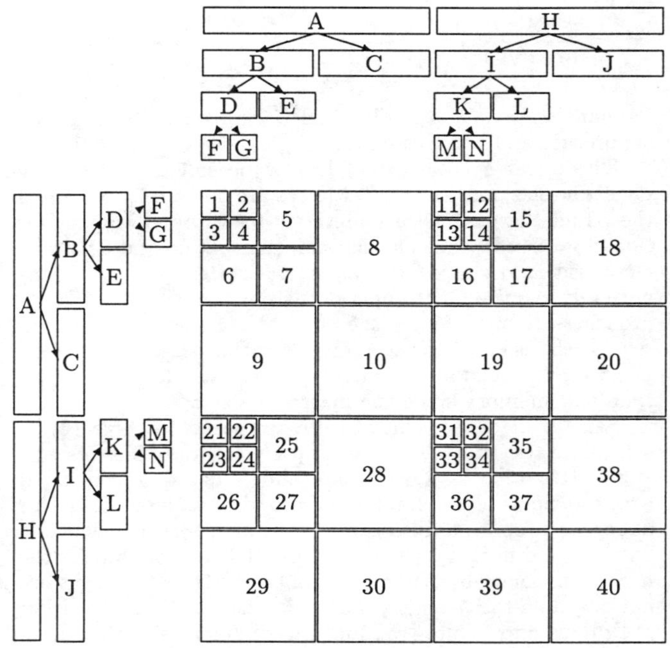

Figure 3: Potential coefficient matrix with block entries.

Figure 2 shows the hierarchical data structure produced by **Refine**, and associated potential coefficients stored as links between the nodes in the hierarchy. Each binary tree represents one conductor surface, and each leaf node represents one final panel that is not further divided in the discretization. The combination of all the leaf nodes completely cover all surfaces of the conductors. Each horizontal link represents one pair of potential coefficients defined in (3). Each self-link represents one self-potential coefficient.

Figure 3 shows the block matrix represented by the links of Figure 2. Each block entry represents one interaction between panels. Note that there are total 8 panels for the two conductors, so a traditional coefficient matrix would require 64 entries. However, the block matrix has only 40 block entries, each stored as one floating-point number.

We now show the block matrix always contains $O(n)$ block entries, while the traditional matrix that explicitly lists all pairwise interactions must contain n^2 entries. For simplicity, consider n panels of about the same size on the surface, and a binary tree constructed above the panels by merging adjacent panels recursively. Therefore, panels on the same level of the tree are of the same size. The error criterion in **Refine** says that two panels can interact directly only if `Pij*R < Peps` where `R` is the length of the longest side of the two panels. Since P_{ij} is asymptotically $1/r$ where r is the distance between the two panels, the two panels can interact only if $R/r < P_\epsilon$, or $r > R/P_\epsilon$. For any fixed P_ϵ, this criterion is equivalent to the criterion that two panels at the same level in the tree can interact only if there are k other panels between them on that level, for some fixed constant k. On the other hand, if panels A_i and A_j are too far, the ancestor of A_i would have interacted with the ancestor of A_j. Therefore, A_i and A_j cannot interact either. This argument applies to any level of the tree. Therefore, each node in the tree interacts with a constant number of other nodes, and the total number of block entries is $O(n)$. Esselink did a detailed analysis on the number of interactions by Appel's algorithm for the general case, and compared it with the multipole algorithm [3].

In Figure 2, it shows that each panel undergoes a constant number of interactions with other panels, regardless of their level in the tree. Large panels that are far apart interact directly, in the same way that small panels near each other interact directly.

Finally, we analyze the relationship between the termination criteria and the accuracy of the computed potential factors. The termination criteria causes the potential coefficient corresponding to each interaction to have approximately the same magnitude. This is because if an estimated potential factor were larger, the panels would be subdivided. More importantly, the termination criteria also places an upper bound on the error associated with the potential factor integral between any two interacting panels.

Consider a panel A which contains two sub-panels A_1 and A_2 of similar shape and size. Let the radius of the smallest sphere that contains both A_1 and A_2 be R. Consider a point x' of distance r from the center of the sphere, for some $r > R$. The potential at x' due to the charge on panels A_1 and A_2, with uniform charge densities σ_1 and σ_2 respectively, is

$$\int_{x_1 \in A_1} \frac{\sigma_1}{||x' - x_1||} da_1 + \int_{x_2 \in A_2} \frac{\sigma_2}{||x' - x_2||} da_2. \tag{4}$$

If we treat A_1 and A_2 as a single panel A with uniform charge density $(\sigma_1 + \sigma_2)/2$, then the potential at x' will be:

$$\int_{x \in A} \frac{\sigma_1 + \sigma_2}{2} \frac{1}{||x' - x||} da. \tag{5}$$

Assume without loss of generality $\sigma_2 \geq \sigma_1$, then the difference between (4) and (5) is

$$\begin{aligned} & \frac{\sigma_2 - \sigma_1}{2} \left(\int_{A_1} \frac{1}{||x' - x_1||} da_1 - \int_{A_2} \frac{1}{||x' - x_2||} da_2 \right) \\ \leq \; & \frac{\sigma_2 - \sigma_1}{2} \int_{A_1} \left(\frac{1}{||x' - x_1||} - \frac{1}{||x' - x_1|| + R} \right) da_1 \\ \leq \; & \frac{\sigma_2 - \sigma_1}{2} \frac{R}{r} \int_{A_1} \frac{1}{||x' - x_1||} da_1. \end{aligned}$$

Therefore the relative error is at most a constant times R/r. Since P_{ij} is in proportion with $1/r$, the condition `Pij*R < Peps` in procedure `Refine` implies that the error of the approximation for every entry is bounded by a constant times P_ϵ. Therefore, as P_ϵ goes to 0, the error goes to 0. Also, all entries in the coefficient matrix are approximated to the same precision, which is a constant times P_ϵ.

In summary, our algorithm computes the potential factor matrix to within a fixed error tolerance. In the process it organizes the potential factor matrix into blocks. The estimated potential factor associated with each block has the same error as other blocks. Furthermore, the total number of blocks grows linearly in the number of elements.

3 Solve Linear Systems

3.1 Fast Matrix-Vector Multiplication

To solve the linear system $\mathbf{Pq} = \mathbf{v}$, an iterative algorithm requires the multiplication of the coefficient matrix $\mathbf{P}$ with a vector, which normally takes $O(n^2)$ time. However, because our $\mathbf{P}$ is represented by $O(n)$ blocks, each matrix-vector multiplication can be done in $O(n)$ time. The multiplication proceeds in three steps. To help understand the algorithm, please keep in mind $\mathbf{P}$ is the coefficient matrix, $\mathbf{q}$ is the given charge vector, and the product $\mathbf{Pq}$ is the potential vector which we want to compute, though the algorithm works for any matrix and vector.

In the following pseudo-codes, each panel `A` has two pointers `left` and `right`, pointing to the two sub-panels, and two fields `charge` and `potential`. For readers not familiar with recursive tree traversal, please see [2].

The first step computes the charge for all interior nodes in the tree. The charge of an interior node is the sum of charges of its children panels. The charge of a leaf node `Ai` is given by `Qi` from $\mathbf{q}$. This calculation can be done in a single depth-first traversal of the tree propagating the charge upward. To compute the charge for each node, the charges of its children are computed first, and then the charge of the node equals the sum of the charge of its children's. The time for computing the charge for all nodes is linear in terms of the number of nodes in the tree.

```
AddCharge(Panel Ai)
{
  if (Ai is leaf)
    Ai.charge = Qi;
  else {
    AddCharge(Ai.left);
    AddCharge(Ai.right);
    Ai.charge =
      Ai.left->charge + Ai.right->charge;
  }
}
```

The second step computes for each panel A_i the potential due to its interacting panels. This can be computed by adding up the product of potential coefficient P_{ij} and charge at A_j, for all A_j that has interaction with A_i. The time for computing the charge for all nodes is linear in terms of the number of blocks in the coefficient matrix.

```
CollectPotential(Panel Ai)
{
  for all Aj such that AiAj has interaction {
    Ai.potential = Ai.potential + Aj.charge*Pij;
    if (Ai is not leaf) {
      CollectPotential(Ai.left);
      CollectPotential(Ai.right);
    }
  }
}
```

The third step distributes the potential from the interior nodes to the leaves. This is done by another depth first traversal of the tree which propagates potential down to the leaf nodes. Each interior node adds its accumulated potential to its children's potential, recursively. The time of this step is linear in terms of the number of nodes in the tree.

```
DistributePotential(Panel Ai)
{
  if (Ai is not leaf) {
    Ai.left->potential =
      Ai.left->potential + Ai.potential;
    Ai.right->potential =
```

```
        Ai.right->potential + Ai.potential;
      DistributePotential(Ai.left);
      DistributePotential(Ai.right);
    }
  }
```

The total time for the matrix-vector product is linear in terms of the number of nodes and links. It is well known that for any binary tree with n leaves, there are exactly $n-1$ interior nodes. Therefore, the time is $O(n)$, where n is the number of panels that are not further subdivided.

3.2 Generalized Minimum Residual Method

We use the Generalized Minimum Residual (GMRES) method with restart [14] to solve the system of equations. The basic idea behind the GMRES method for solving an $n \times n$ linear system is to project the problem onto a Krylov subspace $\mathcal{K}_k$ of dimension $k < n$ using the orthonormal basis constructed by a scheme due to Arnoldi [4], solve the k-dimensional subproblem using a standard approach, and then recover the solution of the original problem from the solution of the projected problem. The Arnoldi scheme involves the coefficient matrix only multiplicatively. The dimension of the Krylov subspace is usually small. Since we can multiply $\mathbf{P}$ with any vector in $O(n)$ time, each iteration can be computed in $O(n)$ time.

4 Experimental Results

The new algorithm is implemented and simulation results are reported in Tables 1 to 3. Both the new algorithm and the FastCap algorithm (for expansion order 0 and 2) are compiled and executed on a SUN Ultra SPARC Enterprise 4000. FastCap(0) is the fastest in the FastCap package, and is about twice as fast as FastCap(2). However FastCap(0) has 5% to 10% relative error with respect to FastCap(2).

The test examples are $k \times k$ bus crossing conductors for $k = 2$ to 6, taken from the FastCap paper [10]. Each bus in $k \times k$ example is scaled to $1m \times 1m \times (2k+1)m$. The distance between buses on the same layer is $1m$, and the distance between layers is $1m$. The constant $4\pi\epsilon_0$ is $111.27pF/m$, according to [10]. Our GMRES reduces the two-norm of the residual to 1% of the initial residual, the same condition used by the conjugate residual algorithm in FastCap.

The difference or error of capacitance matrices is defined as follows. Let the capacitance matrix computed by FastCap (2) be $\mathbf{C}$ and the capacitance matrix computed by another program be $\mathbf{C}'$. Then the difference is estimated in the Frobenius norm [4]: $||\mathbf{C} - \mathbf{C}'||/||\mathbf{C}||$. This is the standard way to measure the difference between two matrices.

Table 1 compares the new algorithm with FastCap, and Table 2 and 3 show the first row of the capacitance matrix computed by the new algorithm and by FastCap. Here is a summary of the comparison:

1) $P_\epsilon = 0.01$. Compared with FastCap(2), the new algorithm is 100 to 40 times faster, and uses 1/100 to 1/60 of the memory. The error is less than 2.7% with respect to FastCap(2). Compared with FastCap(0), our algorithm is 40 to 14 times faster, and is three times more accurate. (The memory usage for FastCap(0) appears to be incorrect.)

2) $P_\epsilon = 0.003$ and compared with FastCap(2). the new algorithm is 5 to 3 times faster, and uses 1/6 to 1/8 of the memory. The error is less than 0.8% with respect to FastCap(2).

Table 1: Comparison for the bus problem. Time is seconds, memory is MB, and error is with respect to FastCap(2).

	Test Problems				
	2x2	3x3	4x4	5x5	6x6
			FastCap (0)		
Time	4.3	11.0	41.3	39.0	84.4
Memory	12	16	86	24	97
Panel	792	1620	2736	4140	5832
Error	8.6%	5.0%	5.2%	8.5%	10.0 %
			FastCap (2)		
Time	10.1	22.9	51.1	161.8	231.9
Memory	5.6	16	26	46	62
Panel	792	1620	2736	4140	5832
			New Algorithm ($P_\epsilon = 0.01$)		
Time	0.1	0.4	0.9	2.2	5.8
Memory	0.06	0.2	0.3	0.5	0.9
Panel	160	408	576	720	1584
Error	2.1%	1.7%	1.8%	1.7%	2.7%
			New Algorithm ($P_\epsilon = 0.003$)		
Time	2.0	7.0	19.2	45.2	83.6
Memory	0.9	2.0	3.9	6.4	9.4
Panel	1888	3504	6720	10960	13152
Error	0.4%	0.2%	0.8%	0.6%	0.4%

We do not have access to the precorrected FFT algorithm [13], the MEI algorithm [15], and the SVD algorithm [8]. However, based on their relative performance to FastCap for the $k \times k$ bus problem, our algorithm is much faster than these algorithms as well.

5 Conclusion

This paper presents a hierarchical algorithm that is significantly faster than previous best algorithms, and uses much less memory. The new algorithm is kernel independent, and therefore, is even more efficient when applied to multilayered dielectrics. The new algorithm does not require preprocessing to partition the conductor surface into panels, instead, it automatically partitions the panels according to a user supplied error bound. The new algorithm provides continuous tradeoff of time with precision by changing the error bound P_ϵ. The new algorithm is very simple, and provides many rooms for further improvements.

There are several major differences between our hierarchical algorithm and the multipole algorithm FastCap. 1) We partition the conductor surfaces according to an estimation of the coefficients for the actual problem, while FastCap considers the geometry of each conductor separately. As a result, we use fewer panels than FastCap does, yet our precision is higher for the same expansion order. 2) To achieve high precision, we divide the conductor surfaces into smaller panels, while FastCap uses high order expansion terms. Experimental results show that it is less expensive to use more panels than to use more expansion terms. 3) We organize

Table 2: Comparison for 4x4 bus problem. Error is with respect to FastCap(2).

	First Row of Capacitance Matrix (pF)								Error	Time (sec)	Memory (MB)
	C_{11}	C_{12}	C_{13}	C_{14}	C_{15}	C_{16}	C_{17}	C_{18}			
FastCap											
$l=0$	394.5	-124.0	-0.175	-2.471	-52.15	-43.39	-43.08	-52.92	5.2%	41.3	86
$l=2$	405.2	-137.8	-11.91	-8.079	-48.36	-40.09	-40.01	-48.45		51.1	26
New Algo											
$P_\epsilon=0.01$	401.6	-139.6	-9.180	-6.480	-48.46	-37.80	-37.53	-48.46	1.8%	0.9	0.3
$P_\epsilon=0.003$	407.9	-135.7	-13.08	-8.720	-48.79	-41.20	-41.21	-48.72	0.8%	19.2	3.9

Table 3: Comparison for 5x5 bus problem. Error is with respect to FastCap(2).

	First Row of Capacitance Matrix (pF)										Error	Time (sec)	Mem (MB)
	C_{11}	C_{12}	C_{13}	C_{14}	C_{15}	C_{16}	C_{17}	C_{18}	C_{19}	$C_{1,10}$			
FastCap													
$l=0$	505.5	-142.8	-10.48	-20.31	2.21	-54.51	-50.13	-46.04	-50.32	-55.01	8.5%	39.0	24
$l=2$	484.5	-166.1	-13.62	-6.17	-6.54	-48.84	-40.12	-40.12	-40.21	-48.90		161.8	46
New Algo													
$P_\epsilon=0.01$	476.5	-168.2	-12.97	-1.91	-4.48	-47.37	-38.52	-37.36	-38.41	-47.69	1.7%	2.2	0.5
$P_\epsilon=0.003$	486.3	-165.5	-15.05	-6.77	-6.01	-48.78	-40.49	-40.52	-40.52	-48.83	0.6%	45.2	6.4

the coefficient matrix more efficiently using the hierarchical data structure. Some feathers, such as the adaptive evaluation defined in FastCap [12], is free under our data structure.

Acknowledgment The authors wish to thank Weishi Sun, Steve Tate and Dian Zhou for discussions and references, and anonymous referees for improving the presentation.

References

[1] A. A. Appel, "An efficient program for many-body simulation," *SIAM J. Scientific and Statistical Computing*, Vol. 6, No. 1, 1985, 85–103.

[2] T. H. Corman, C. E. Leiserson, and R. L. Rivest, *Introduction to Algorithms*, MIT Press, 1990.

[3] K. Esselink, "The order of Appel's algorithm," *Information Processing Letters*, Vol. 41, 1992, 141–147.

[4] G. H. Golub and C. F. Van Loan, *Matrix Computations*, 2nd ed., Johns Hopkins University Press, 1989.

[5] L. Greengard and V. Rohklin, "A fast algorithm for particle simulations," *J. Comp. Phys.*, Vol. 73, 1987, 325–348.

[6] L. Greengard, *The Rapid Evaluation of Potential Fields in Particle Systems*, MIT Press, Cambridge, Massachusetts, 1988.

[7] P. Hanrahan, D. Salzman, L. Aupperle, "A rapid hierarchical radiosity algorithm," *Computer Graphics (Proc. SIGGRAPH'91)*, Vol. 25, No. 4, July 1991, 197–206.

[8] S. Kapur and D. E. Long, "IES^3: A fast integral equation solver for efficient 3-dimensional extraction," *Proc. 1997 ICCAD*, 448–455.

[9] S. Kapur and J. Zhao, "A fast method of moments solver for efficient parameter extraction of MCMs," *Proc. 34th DAC*, June 1997, 141–146.

[10] K. Nabors and J. White, "FastCap: A multipole accelerated 3-D capacitance extraction program," *IEEE Trans. CAD*, Vol. 10, No. 11, Nov. 1991, 1447–1459.

[11] K. Nabors and J. White, "Multipole-accelerated 3-D capacitance extraction algorithms for structures with conformal dielectrics," *Proc. 29th DAC*, 1992, 710–715.

[12] K. Nabors, et al., "Preconditioned, adaptive, multipole-accelerated iterative methods for three-dimensional first-kind integral equations of potential theory," *SIAM J. Sci. Comput.*, Vol. 15, No. 3, May 1994, 713–735.

[13] J. R. Phillips and J. White, "A precorrected FFT method for capacitance extraction of complicated 3-D structures," *Proc. 1994 ICCAD*, 268–271.

[14] Y. Saad and M. H. Schultz, "GMRES: A generalized minimal residual algorithm for solving nonsymmetric linear systems," *SIAM J. Sci. Stat. Comput.*, Vol. 7, No. 3, July 1986, 856–869.

[15] W. Sun, W. W.-M. Dai, and W. Hong, "Fast parameters extraction of general 3-D interconnects using geometry independent measured equation of invariantce," *Proc. 33rd DAC*, 1996.

[16] D. R. Wilton, et al., "Potential integration for uniform and linear source distributions on polygonal and polyhedral domains," *IEEE Trans. Antennas and Propagation*, Vol. AP-32, No. 3, March 1984, 276–281.

Boundary Element Method Macromodels for 2-D Hierarchical Capacitance Extraction†

E. Aykut Dengi
Motorola Inc.
3501 Ed Bluestein Blvd., Austin, TX 78721

Ronald A. Rohrer
Intersouth Partners
1000 Park Forty Plaza, RTP, NC 27709

Abstract

A 2-D hierarchical field solution method was recently introduced for capacitance extraction for VLSI interconnect modeling. In this paper, we present several extensions to the method including a Boundary Element Method (BEM) formulation for creating macromodels, which provides a better trade-off between accuracy and efficiency, as well as parameterized elements, which allow the analysis of gridless designs with reasonable accuracy and a small library size.

1. Introduction

Capacitance extraction is a critical step in submicron integrated circuit interconnect modeling. In [4] a hierarchical 2-D capacitance extraction method for VLSI circuits was introduced. The method uses a geometrical hierarchical partitioning on the 2-D vertical cross-section of a VLSI circuit. These partitions are characterized at their interfaces by macromodel capacitance matrices which can be combined to yield a global capacitance matrix for a given set of conductors or the coupling capacitance values for a given conductor (Fig. 1). A library of such macromodels, which is sufficient to describe any vertical cross section for a given technology, is built as a pre-processing step to improve runtime extraction efficiency. Irregular conductor geometries as well as conformal dielectrics are handled at the preprocessing stage, enhancing runtime efficiency without affecting accuracy.

To aid understanding, [4] introduced the method in a Finite Difference Method formulation, but the method is in fact best implemented using the Boundary Element Method (BEM). The efficiency of the runtime global solution is proportional to the number of ports on each element. In the finite difference formulation, the number of ports is intimately related to the discretization used inside the element. Hence, when we discretize the conductor surfaces finely, as we should, we end up with a large number of ports. In this paper, we present the BEM formulation, which allows us to effectively decouple the discretization on the conductors and the discretization at the boundary (hence the number of ports), leading to a better trade-off between accuracy and efficiency.

The rest of the paper is organized as follows: In Section 2, the basic BEM formulation for macromodels is presented; Section 3 describes how these macromodels can be combined to yield capacitance values for a given configuration of conductors; Section 4 discusses how the choice of different interpolation functions for the BEM affects accuracy and efficiency; Section 5 presents how a versatile library can be built as a pre-processing step for maximum efficiency and accuracy; Finally in Section 6, we offer some conclusions.

†. This work was supported by the National Science Foundation under Grant MIP-9216942, and by the Semiconductor Research Corporation under Contract DC-068.

2. BEM Formulation

With BEM we attain the most accuracy for a given number of nodes, because no error is incurred inside the domain being analyzed and all the error is due to the approximation of flux density and potential at the boundaries of the domain. The inside of the domain need not be discretized using the BEM [2]. In the traditional use of the method, there is a performance penalty due to two factors: BEM gives rise to a full matrix, and it is expensive to calculate the entries of this matrix and to invert it. In our methodology, however, some of these expenses are all shifted to the pre-processing library building phase where they can be afforded, as we shall see below.

The BEM relates the potential at the conductor boundaries to the charge distribution at those boundaries. However, in order to repeat what we did with Finite Differences, using the BEM we must artificially partition the problem domain into cells to be characterized separately. In FD, we chose points that already existed in the grid for the FD solution and created a boundary. In BEM, however, there are no nodes in the dielectric between conductors, so we must create them artificially. In this case, we have not only the charge distribution on conductors, but also the flux density at the artificial interfaces created.

The creation of such artificial boundaries may seem redundant, or even detrimental to accuracy at first. However, we have reasons to believe that we may gain by using our method as opposed to a straight forward BEM implementation:

1. The charge density in the regular BEM varies rapidly near conductor corners, necessitating more nodes, whereas potential and flux density vary smoothly at the artificial boundaries we have chosen so that fewer nodes suffice;
2. In regular BEM, it is not possible to form the potential matrix [9] as a pre-processing step, and it is very costly at run-time, therefore macromodeling is not possible;
3. Regular BEM generates dense potential matrices to be inverted,

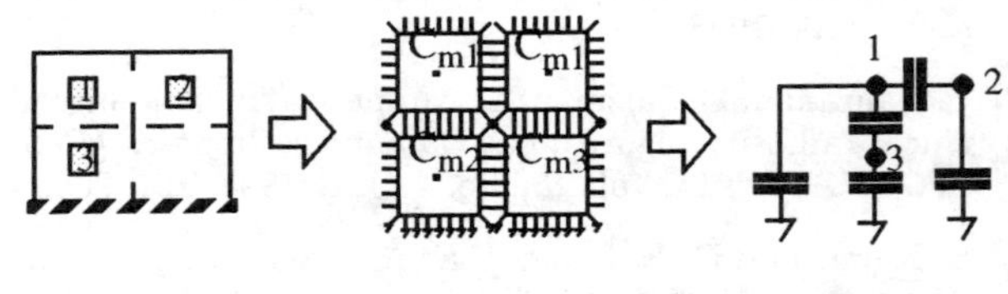

Figure 1. Hierarchical field solution combines the individual macromodels eliminating nodes at their common edges.

whereas our method generates block-sparse matrices;

4. BEM can handle only absorbing boundary conditions, which may not be appropriate to the problem domain.

Such partitioning of the problem domain using the BEM, referred to as multi-zone BEM, has been used in mechanics [7] and the capacitance extraction of SRAM cell vertical cross-sections [5]. [7] uses the multi-zone BEM method on elasticity problems. [5] uses the multi-zone BEM to model multiple dielectric layers (assigning them to different zones) and to break up zones with large aspect ratios, which tend to cause numerical problems, into sections with smaller aspect ratios. Both references compute the entries of the global capacitance (using Green's functions) matrix at runtime and do not consider macromodeling. We give a brief summary of the DBEM as used by our method below. Detailed background can be found in, e.g.,[2].

For any point x at the boundary G (Fig. 2), the following equations hold [2]:

$$c(\xi)u_0(\xi) + \int_\Gamma u_0(x)f^*(\xi, x)d\Gamma(x) = \int_\Gamma f_0(x)u^*(\xi, x)d\Gamma(x) \quad (1)$$

$$u^*(\xi, x) = \frac{1}{2\pi}ln\left(\frac{1}{r(\xi, x)}\right) \quad (2)$$

$$f^*(\xi, x) = \frac{\partial}{\partial n}u^*(\xi, x)\,. \quad (3)$$

If the boundary is smooth, $c(\xi) = 1/2$. We pick N collocation points ξ on Γ and use the subscript i to mark the quantities at ξ_i, as in Fig. 2. We use the same set of interpolation functions ϕ to approximate u_0 and f_0:

$$u_0 \cong u = \sum_{k=1}^{N}\phi_k u_k\,,\; f_0 \cong f = \sum_{k=1}^{N}\phi_k f_k \quad (4)$$

We may further pick interpolation functions such that

$$\phi_k(x) = 0 \text{ if } x \notin \Gamma_k\,,\; k \in \{1, 2 \ldots N\} \quad (5)$$

where Γ_k are segments such that

$$\bigcup_k \Gamma_k = \Gamma \text{ and } \xi_k \in \Gamma_k \quad (6)$$

Then, we can write Eq. (1) in matrix form:

$$Hu = Gf\,,\; H_{ij} = c_i\delta_{ij} + \int_{\Gamma_j}\phi_j(x)f^*(\xi_i, x)d\Gamma(x)\,,$$

$$G_{ij} = \int_{\Gamma_j}\phi_j(x)u^*(\xi_i, x)d\Gamma(x) \quad (7)$$

where $c_i = 1/2$ for smooth boundaries. And finally,

$$C = G^{-1}H \quad (8)$$

This C matrix contains all the individual nodes on the boundary and the conductors. We condense the conductor nodes into one

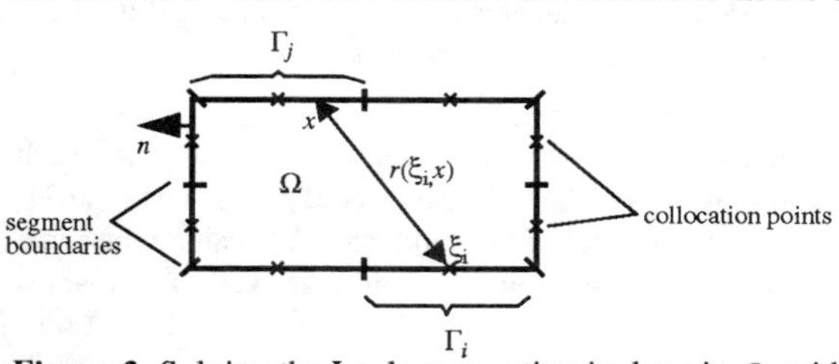

Figure 2. Solving the Laplace equation in domain Ω with boundary Γ using the BEM. For constant elements, the collocation point is in the middle of each segment.

without loss of accuracy by integrating the flux f on each conductor to calculate q, the total charge on that conductor:

$$C_m = RCP^T\,,\; R = \begin{bmatrix} I_n & 0 \\ 0 & R_c \end{bmatrix}\,,\; r_i = \int_\Gamma \phi_i(x)d\Gamma(x)\,, \quad (9)$$

$R_{c(k \times (N-n))} =$

$$\begin{bmatrix} r_{n+1} & r_{n+2} & \cdots & r_{n+l_1} & 0 & \ldots & 0 & 0 & \ldots & 0 & 0 \\ 0 & 0 & 0 & 0 & r_{n+l_1+1} & \cdots & r_{n+l_1+l_2} & 0 & \ldots & 0 & 0 \\ 0 & \ldots & \ldots & \ldots & \ldots & \ldots & \ldots & \ldots & \ldots & 0 & 0 \\ 0 & \ldots & \ldots & \ldots & \ldots & \ldots & \ldots & \ldots & \ldots & \ldots & r_N \end{bmatrix} \quad (10)$$

where n is the number of terminals on the boundary, k is the number of conductors in the macro l_i is the number of nodes on the ith conductor and r_i is the total charge on the segment i. P is defined as:

$$P = \begin{bmatrix} I & 0 \\ 0 & P_c \end{bmatrix}\,,\; P_c = p_{ij}$$

$$p_{ij} = \begin{cases} 1 & \text{if node j is on conductor i} \\ 0 & \text{otherwise} \end{cases} \quad (11)$$

We can now partition C_m:

$$C_m v = \begin{bmatrix} C_{bb} & C_{bc} \\ C_{cb} & C_{cc} \end{bmatrix}\begin{bmatrix} v_b \\ v_c \end{bmatrix} = \begin{bmatrix} f_b \\ q_c \end{bmatrix} \quad (12)$$

where subscript b denotes nodes on the artificial boundary, subscript c denotes nodes corresponding to each conductor inside the macromodel, v denotes potentials at nodes/conductors, f denotes flux density at nodes and q denotes total charge on conductors

3. Global Solution

We obtain a global solution using BEM macromodels in much the same as with FD macromodels in [4]. We stencil the macromodel of each element into a global capacitance matrix and eliminate the nodes at the interfaces between elements through a partial LU-decomposition:

$$C = C_{22} - C_{21}C_{11}^{-1}C_{12}\,, \quad (13)$$

where the subscript 1 indicates the nodes that are eliminated and the subscript 2 the nodes that remain. Using Eqn. (13), we can obtain a capacitance matrix for a set of conductors; alternatively, we can obtain a macromodel for the given configuration by preserving the nodes at the outer boundary of the configuration.

Because the individual library elements are built *a priori*, the conductors and dielectric interfaces inside an element macromodel can be discretized as finely as needed without increasing the size of C_m. Thus, more time can be spent during pre-processing for condensing the conductor nodes and eliminating the inner dielectric interface nodes. This improves the accuracy of the macromodel by arbitrarily reducing the error due to the discretization of the conductors without having a negative impact on subsequent extraction runtime.

4. Order of Interpolation Functions

In Eq. (4), we specify a set of interpolating functions, but we so far have not discussed how the choice of such functions affects our method. The most common interpolation function is zeroth-order piecewise constant. It has been used extensively and with success in conjunction with the Indirect (or regular) BEM (IBEM). One reason for its common use is the ease of implementation. The integrals used in the construction of matrices H and G become more complicated as the order of the interpolation function increases. However, it is also well-known that higher-order interpolation functions lead to a higher rate of asymptotic convergence for Direct BEM (DBEM) methods [10]†.

Comparisons of different interpolation functions have been carried out in literature for the DBEM in general (see, for instance, [10]), and specifically for our method in [3]. Here, we will illustrate the convergence peoperties of different interpolation functions on an example. The configuration in Fig. 3 is solved with constant, linear partially discontinuous, linear discontinuous and quadratic elements. The convergence of each is graphed in Fig. 4. The number of nodes per element is determined by the discretization at the boundary of the element (Fig. 3b) and is a good measure of the efficiency of the solution because it determines the size of the global capacitance matrix. The dashed lines in the logarithmic plots are inserted for reference. The absolute value of the slope of an error curve is equal to the order of the asymptotic convergence for the corresponding method. The discretization dm for the conductors inside the elements (see Fig. 3b) affects the accuracy but does not have an effect on the efficiency of the global solution. For the results in Fig. 4, $dm = 0.05$.

We see that in terms of asymptotic convergence, quadratic elements have the best properties. However, in IC applications, we are not really after errors on the order of 10^{-4}, because we know that our geometrical parameters and the process variations present much greater variations. Therefore asymptotic convergence is of limited use in a practical application. We are more interested in which method yields a given accuracy most economically. We can attain reasonable accuracy (~1%) for the lowest discretization for both diagonal and off-diagonal entries of the capacitance matrix. We also see that quadratic elements intrinsically have a larger minimum number of nodes per element. Experience suggests that in fact, the discontinuous linear elements offer good compromises among efficiency, accuracy and ease of implementation. As our method matures, higher orders of interpolation functions may prove in the future to provide better trade-offs.

The number of nodes on the conductor is not the only determinant of accuracy we can improve without a run-time performance

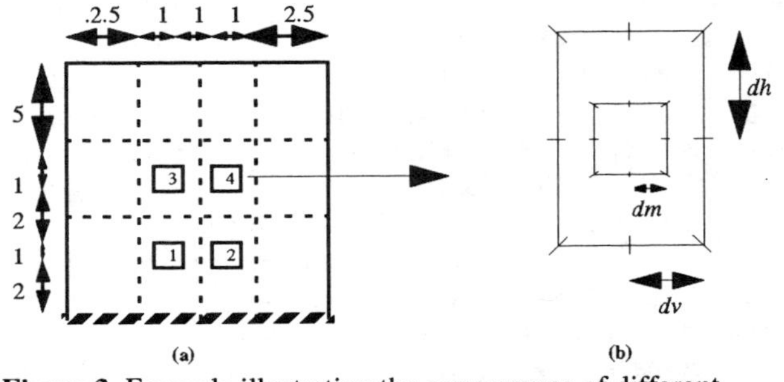

Figure 3. Example illustrating the convergence of different interpolating functions. (a) shows the dimensions of the configuration and (b) defines the variables for a single element

†. The difference between IBEM and DBEM is that only charge (or flux density) is discretized in the former and both flux density and potential in the latter.

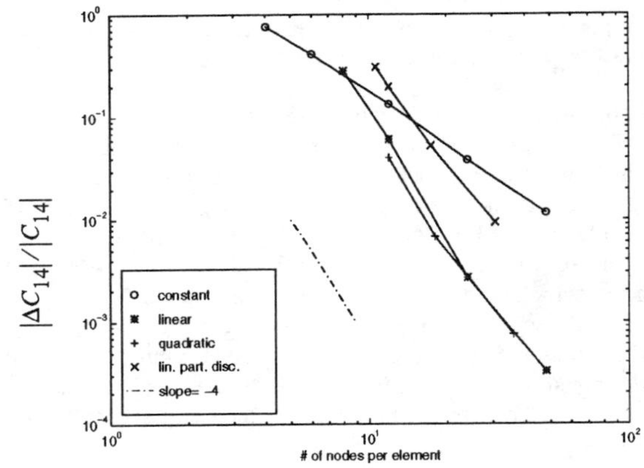

Figure 4. The relative error in one component of the capacitance matrix (C_{14}) for the configuration in Fig. 3is plotted for constant, discontinuous linear, partially discontinuous linear and quadratic elements for varying discretization.

penalty. In [5], the accuracy of integration (in the computation of the capacitance matrix elements) is identified as a major determinant of accuracy. Note that a high accuracy integration with many grid points on the conductor would be very costly if executed at run-time as in regular BEM [5]. Our method shifts this computational cost to pre-processing. The advantage of our method is more pronounced for more complicated conductor geometries.

In order to demonstrate this, we examine the simple problem in Fig. 5. An imaginary process with two distinct dielectric layers is modeled: layer 1 with a dielectric constant $\varepsilon = 1$ and layer 2 with $\varepsilon = 1.5$. The top metal layer has an irregular cross-section. Conductor 1 is slightly shifted to the right to render the problem non-symmetric. The relative error is expressed in percentages as $100(|C_x^{\text{approx}} - C_x^{\text{exact}}|/C_{11}^{\text{exact}})$ in Fig. 5(b). We see that even for a very coarse discretization (with a total of 30 nodes in the global solution) the error is below 1.5%.

5. Library Building

One of the advantages of the hierarchical field solution approach is that a significant part of the computational cost can be shifted to the pre-processing stage by building a library of macromodels which can be combined to describe any vertical cross-section allowed for a given technology [4]. Only three basic elements (Fig. 6) are sufficient to span each unique interconnect layer [4], but more complex elements may be added to the library for improved efficiency and accuracy. A detailed discussion on the trade-offs between library size, efficiency and accuracy can be found in [3]. Here, we shall present two aspects of library building that prove to be crucial in practice: parameterized macromodels and interfacing macormodels with different discretizations.

Parameterized Macromodels

Instead of coming up with several unique elements for each conductor width, we can make the macromodel parameters of an element functions of the width of the conductor. We can do the same with respect to spacing between conductors as well. The macromodel capacitance matrices C_m in Fig. 6 may be general functions. However, a piecewise-linear function is often enough for accuracy because the range of parameters is limited. Moreover, the accuracy

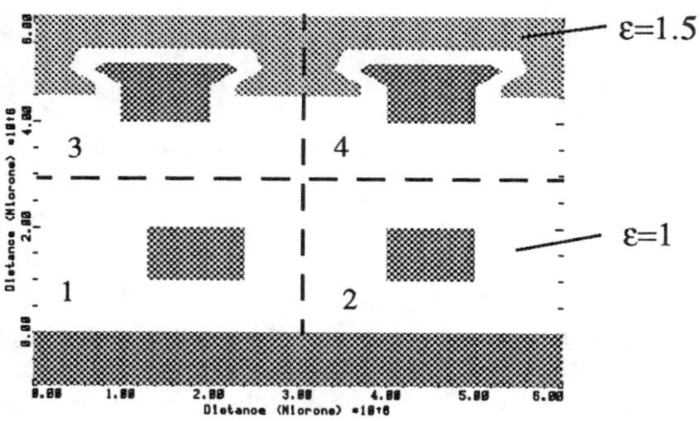

(a)

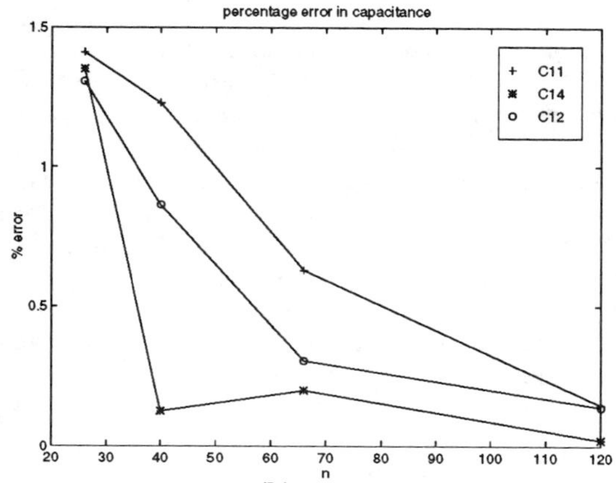

(b)

Figure 5. Convergence with improving discretization (a) The vertical cross-section for an imaginary process. The dielectric constant is taken as 1. The top metal layer has an irregular cross-section and there is a conformal dielectric with a dielectric constant 1.5. Dashed lines indicate the element boundaries used. (b) shows the convergence in absolute terms of C11, the total capacitance of conductor 1, and C14, the coupling capacitance between conductors 1 and 4. n is the number of variables in the global capacitance matrix formed.

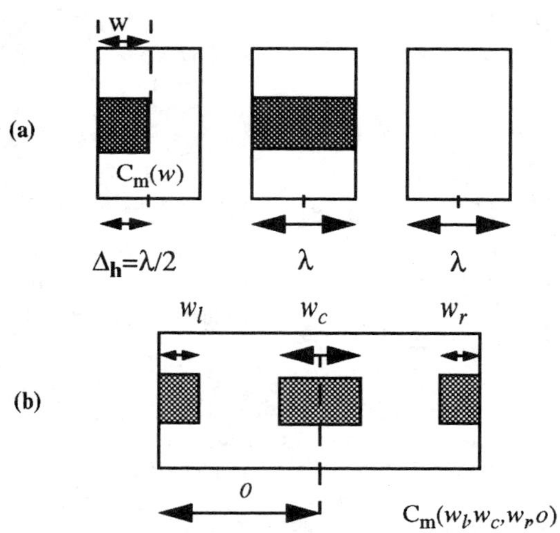

Figure 6. **(a)** The minimum set of three basic elements that span a layer of a given technology; **(b)** a complex element.

can be arbitrarily improved by increasing the number of segments and/or increasing the order of the interpolating function. In the case of one linear segment:

$$\bar{C}_m(w) = \bar{C}_{m0} + (w - w_0)\Delta C_w,\ w_{min} < w < w_{max}. \qquad (14)$$

$$|\bar{C}_m(w) - C_m(w)| < \varepsilon \qquad (15)$$

It is possible to use many numerical techniques to compute C_{m0} and ΔC_w, such as a least-squares fit, interpolation or analytic differentiation. It is always possible to pick w_{min}, w_{max} such that Eq. (15) holds. For large parameter variations and rapidly changing functions, this would mean that we would need more segments for a given error bound. Macromodel matrix functions of more than one variable (Fig. 6b) can be similarly handled.

By running many experiments [3], we have found that the parameterization error is insignificant (1-3%) and is dominated by the error due to the discretization of the boundaries for practical purposes. Complex elements (e.g., Fig. 6b) tend to decrease the error due to discretization and reduce the number of unknowns in the global solution matrix, resulting in high efficiency and accuracy. Therefore, there is a good reason to build a library of parameterized and possibly complex elements to minimize the number of interfaces between conductors while maintaining a reasonable library size.

Interfacing Elements with different discretizations

In defining library elements, we have imposed limitations so that all elements that can be neighbors have matching ports [4]. This is a significant limitation when we require a higher level of accuracy on the capacitance of some nets than others, e.g., in a net-by-net extraction methodology. We need a finer discretization closer to the net of interest (NOI) than farther away from it. However, the limitation that all elements have the same discretization does not allow us to use a coarser grid as we move away from the NOI. This causes inefficiency, especially for large problems. There are two solutions to this problem:

1. We can have elements with edges that have different discretizations. For instance, the left vertical edge of an element could have twice as many ports as the right vertical edge. We could use such an element with the left edge facing the NOI (Fig. 7). This solution results in a larger library because we now need to store *versions* of the same element, which have different discretizations.
2. We can try to interface elements with non-matching ports through interpolation. In classic BEM literature, this is not considered good practice [2], because BEM minimizes the error in potential and flux density at nodes. Therefore, the non-node parts of an element, of which interpolation would make use, may have larger errors.

We will describe the second approach because of its positive impact on performance. The essence of the problem is demonstrated in Fig. 8. As discussed above, given potential at ports, capaci-

Figure 7. Elements with different discretization on different edges: **(a)** shows an element for which the left edge has a higher discretization; **(b)** shows this element used with the left edge facing the NOI, which has high discretization on all edges .

tance macromodel matrices yield flux density. We see that four ports of element one (E1) correspond to only two ports of element two (E2). We are trying to satisfy the compatibility conditions at the interface. Therefore, the potential (flux density) on E1 must be equal to the potential (flux density) on E2. We use a *least-squares* fit to obtain the potential on E2, v_2. Given v_2, C_2 yields f_2, which is supported by the two ports of E2. Linear interpolation is used to obtain f_1. By doing this, we constrain the degrees of freedom of f_1: The two segments are constrained to act as one. This is equivalent to a matrix transformation:

$$T: C_2^{(2\times 2)} \rightarrow P^{(4\times 2)} C_2^{(2\times 2)} Q^{(2\times 4)} = \tilde{C}_2^{(4\times 4)} \quad (16)$$

which transforms C_2 into a 4x4 matrix, $\tilde{C}_2$, which is compatible with C_1 (Fig. 9(a)).

We shall examine an example that is in line with the net-by-net extraction scheme. We have a net of interest for which we would like to extract the capacitance (Fig.10). We have already computed the coupling between the NOI and the left lateral neighbor, and we have discovered that the neighbor on the right will make a small but not negligible difference in the capacitance of the NOI. We try using the technique outlined above and use a transformed low-accu-

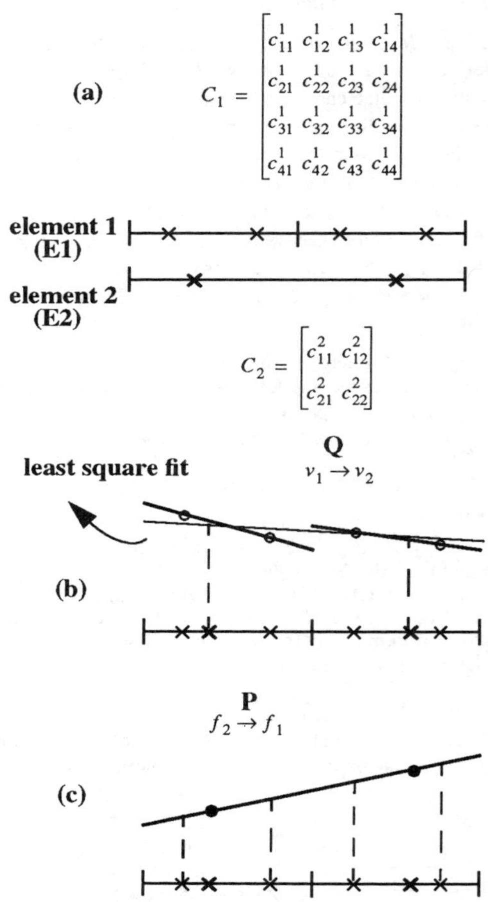

Figure 8. Matching elements with different discretization. (a) The elements are represented by their capacitance macromodel matrices, which have different sizes. (b) The potential on element 1 is mapped onto the potential on element 2. (c) The flux density on element 2 is mapped onto the flux density on element 1.

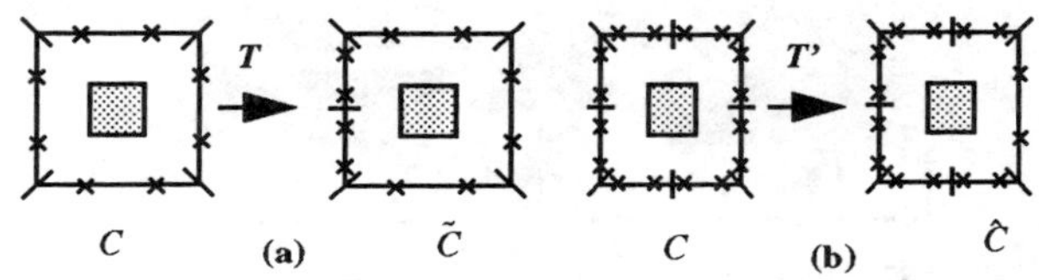

Figure 9. Transforming elements: (a) The matrix transformation *T* creates a new *version* of an element by increasing the discretization on one (or more) of the edges; (b) The matrix transformation *T'* creates a new *version* of an element by decreasing the discretization on one (or more) of the edges.

racy model for the right neighbor (Fig.10d). We compare the results with those of higher accuracy. To emphasize the necessity of the new technique, we also do the extraction using low accuracy models for all three elements (Fig.10e). We see that the proposed technique is a good compromise between computational cost and accuracy because it provides accuracy only where needed. Note that the percentage error for the mixed extraction (Fig.10d) for the coupling between the NOI and the right lateral neighbor is roughly equal to the error of the same for the extraction with coarse elements (Fig.10e). This is to be expected because the transformation does not increase the accuracy of the coarse model used. If we use the relative error measure:

$$E(C_{ij}^{nl}) = \frac{C_{ij}^{nl} - C_{ij}^{0}}{C_{ii}^{0}} \quad (17)$$

which is more meaningful in the net-by-net extraction context, we see that indeed we have good overall accuracy for the column corresponding to the NOI (2nd column) compared to the coarse extrac-

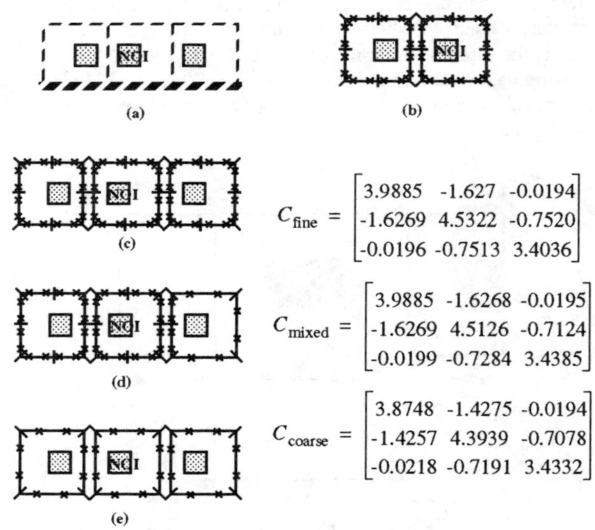

Figure 10. Example to illustrate the accuracy of mixing elements of different discretizations: (a) shows the configuration and the decomposition into elements – in this case, three instances of one unique library element; (b) shows the initial extraction, taking into account only the NOI and the left lateral neighbor; (c) shows the resulting capacitance matrix when only elements with high discretization are used; (d) shows the use of mixed elements; (e) shows the solution using only elements with coarse discretization.

tion:

$$\Delta C^{\mathrm{rel}}_{\mathrm{mixed}} = \begin{bmatrix} -1\times10^{-7} & 3\times10^{-5} & 2\times10^{-5} \\ 7\times10^{-7} & -0.0043 & 0.0116 \\ -9\times10^{-5} & 0.0051 & 0.0102 \end{bmatrix},$$

$$\Delta C^{\mathrm{rel}}_{\mathrm{coarse}} = \begin{bmatrix} 0.0285 & 0.0440 & 3\times10^{-5} \\ -0.0505 & 0.0262 & -0.0014 \\ 0.0005 & 0.0020 & 0.0016 \end{bmatrix}. \qquad (18)$$

There also exists the dual transformation, which we shall call T'. We define the new transformation:

$$T' : C_2^{(4\times4)} \to Q^{(2\times4)} C_2^{(4\times4)} P^{(4\times2)} = \hat{C}_2, \qquad (19)$$

where this time P interpolates the *potential* and Q performs a least-squares fit to match the *flux density* (Fig. 9(b)). We can use the dual transformations together to our advantage. We can use T when we need an element with only one accurate edge, and T' when we need one with only one coarse edge.

6. Conclusion

The Boundary Element Method implementation of the 2-D hierarchical field solution technique introduced in [3] is more efficient than the Finite Difference implementation. Among many options available for the BEM implementation, the choice of linear discontinuous elements seems to offer a good compromise between efficiency and accuracy. Parameterized elements allow the analysis of gridless designs with reasonable accuracy and a small library size. A pair of linear transformations allow us to use elements with different discretizations together. Using these transformations, elements with lower accuracy can be used where high accuracy is not required, resulting in further gains in efficiency.

7. Bibliography

[1] Anand, M.B. *et al.* "Fully Integrated Back End of the Line Interconnect Process," *1994 VLSI Multilevel Interconnect Conference,* June 7-8, 1994, IEEE 1994 pp. 15-21

[2] Brebbia, C.A; Telles, J.C.F; Wrobel, L.C. *Boundary Element Techniques*, Springer-Verlag, Berlin, Heidelberg, 1984

[3] Dengi, E.A.; "A Parasitic Capacitance Extraction Method for VLSI Interconnect Modeling", PhD Thesis, Carnegie Mellon University, March 1997

[4] Dengi, E.A.; Rohrer, R.A.; "Hierarchical 2-D Field Solution for Capacitance Extraction for VLSI Interconnect Modeling", *to appear at 34th Design Automation Conference,* Anaheim, CA, June 9-13, 1997

[5] Fukuda, S; Shigyo, N; Kato, K; Nakamura, S. "A ULSI 2-D Capacitance Simulator for Complex Structures Based on Actual Processes," *IEEE Trans. Computer-Aided Design,* TCAD-9 no.1, pp.39-47, Jan. 1990

[6] Gray, L.J; Lutz, E. "On the treatment of corners in the boundary element method," *J. Comput. Appl. Math.* Vol. 32, pp. 369-386, 1990

[7] Kane, J.H; Kumar, B.L.K; Saigal, S. "An Arbitrary Condensing, Noncondensing Solution Strategy for Large Scale, Multizone Boundary Element Analysis," *Computer Meth. Appl. Mech. Eng.* 79(1990) pp. 219-244

[8] Patterson, C; Sheikh, M.A. "Interelement continuity in the boundary element method," in *Topics in Boundary Element Research, Vol. 1,* C.A. Brebbia (ed.), Springer-Verlag, Berlin, 1984

[9] Ruehli, A.E; Brennan, P.A. "Efficient Capacitance Calculations for Three Dimensional Multiconductor Systems," *IEEE Trans. Microwave Theory Tech.* MTT-21, pp. 76-82, 1973

[10] Wendland, W.L; Yu, D. "Adaptive boundary element methods for strongly elliptic integral equations," *Numer. Math.,* Vol. 53, pp. 539-558, 1988

[11] Yan, G; Lin, F. "Treatment of corner node problems and its singularity," *Engineering Analysis with Boundary Elements* Vol.13, pp.75-81, 1994

Efficient Three-Dimensional Extraction Based on Static and Full-Wave Layered Green's Functions

Jinsong Zhao Wayne W.M. Dai
Department of Computer Engineering
UC Santa Cruz
Santa Cruz, CA 95064

Sharad Kapur David E. Long
Bell Laboratories, Lucent Technologies
600 Mountain Avenue
Murray Hill, New Jersey 07974

Abstract

Integral equation approaches based on layered media Green's functions are often used to extract models of integrated circuit structures. The primary advantage of these approaches over equivalent-source based schemes is the dramatic reduction in problem size. When combined with an SVD-accelerated scheme for the solution of the associated dense linear system, this leads to a substantial speedup.

In this paper we derive and solve for these multilayered 3D Green's functions using a transmission line circuit analog. A generalized image method for an arbitrary number of layers is presented. This method is rapidly convergent for near-field interactions. For the far field, a Chebyshev interpolation approach is adopted, where a database is precomputed (using a Fast Hankel Transform) and stored. The combination of these two approaches leads to an extremely efficient scheme for the generation of Green's functions.

We combine the SVD-accelerated integral equation solver IES^3 with the multilayered Green's function approach, apply it to the extraction of IC parasitics and passive components, and we demonstrate its speed, accuracy and versatility via a number of examples.

1 Introduction

Extraction and modeling in multilayered circuit environments play a significant role in the design of modern ICs, packages and printed circuit boards. Shrinking circuits and increasing operating frequencies have made it critical that three dimensional effects be modeled accurately. General purpose field solvers based on finite difference or finite element schemes (e.g., Raphael, Ansoft) can be used. However, they require volume discretization, and it is difficult to enforce radiating boundary conditions for open regions. This results in large computation times and memory use. Integral equation schemes (such as FastCap [10]) based on equivalent sources can also be used. They work by introducing additional equations to enforce boundary conditions at region interfaces; this can result in a prohibitive increase in problem size. In this paper, we present a general-purpose method based on layered Green's functions which exploits the radial invariance of the media. The principal advantage of this scheme is that the unknowns are restricted to only the source region, i.e., no additional equations need to be introduced to enforce boundary conditions. This leads to a substantial reduction in problem size compared to equivalent-source based schemes. The layered Green's functions are combined with the SVD-accelerated integral equation solver IES^3 [6, 7] (which is not restricted to any specific kernel), giving a dramatic speedup over all competing approaches.

Layered Green's functions decouple the process structure from the circuit geometry and can be precomputed, compressed and stored. Once the process variables are fixed, the Green's functions remain unchanged for each new extraction. Layered Green's functions have traditionally been used in a 2.5D simulation context where the sources are confined to infinitely thin sheets. This approach has been popular in the microwave and antenna communities. For these communities, 2.5D modeling of the structures is adequate because conductor thickness is much smaller than the width. However, in IC and packaging contexts this assumption is no longer valid. The thickness of the conductors is often on the same order as the widths.

In the spectral domain it can be shown that the full wave Green's function can be decoupled into three sets of one-dimensional wave equations [8]. Poisson's equation for the static Green's function leads to a one-dimensional wave equation [3]. These one-dimensional wave equations have a transmission line circuit analog. The boundary conditions to be enforced are obtained by solving the equivalent circuit. In this paper, we derive and solve for these multilayered 3D Green's functions. A generalized image method for an arbitrary number of layers is presented. The implementation is based on a heap data structure which dynamically sorts the images for the rapid computation of the near field. While the image method is always numerically stable, it is efficient only in the near field; in the far field slow convergence makes it impractical. For the far field, a database is efficiently constructed with a Fast Hankel Transform and compressed with an adaptive Chebyshev interpolation procedure. The singularity in the Green's function in the near field obviates the use of any 3D interpolation procedure. The combination of the two approaches, i.e., an image scheme for the near-field and a lookup scheme for the far field, leads to an extremely efficient implementation. The typical cost of building the database for a given process takes less than one hour for 3D problems and a few minutes for 2.5D. The cost of evaluating a single interaction between a source and observation pair is on the same order of that of the free-space Green's func-

DAC 98, San Francisco, California

tion evaluation. When combined with the integral equation solver IES[3], this leads to a substantial speed up over competing approaches (such as those implemented in FastCap and Raphael).

2 Green's Functions for Layered Media

The following discussion on layered Green's functions is presented in the context of static capacitance extraction. The full-wave extension of the techniques developed in this paper is straightforward with only a few changes.

2.1 Wave Equation and Transmission Line Equivalence

It is well known that the basic equations in electromagnetics are the wave propagation equations. A complete solution in 3D is computationally infeasible. However, in layered media, by taking advantage of radial invariance, Maxwell's equations can be reduced to three independent one-dimensional wave equations via TE and TM decomposition [8]. Matching boundary conditions for a one-dimensional problem can be simplified by solving an analogous transmission line circuit.

A uniform transmission line can be characterized by three parameters: the propagation constant γ, the characteristic impedance Z_c, and the line length l. Voltage and current propagate along the line based on the Telegrapher's equations:

$$\frac{dv}{dz} = -\gamma Z_c i \tag{1}$$

$$\frac{di}{dz} = -\gamma Y_c v + i_s \tag{2}$$

where i_s is the current source density, and $Y_c = 1/Z_c$ is the characteristic admittance. The boundary conditions for the transmission line circuit at the source-free interfaces are that voltage and current be continuous, i.e., $v_+ = v_-$ and $i_+ = i_-$. These conditions are automatically satisfied by enforcing KCL and KVL.

In layered media, the potential ϕ satisfies the propagation equation

$$\left(\frac{\partial^2}{\partial z^2} - \gamma^2\right)\phi = -\alpha\phi_E \tag{3}$$

where ϕ_E is the source of excitation. ϕ satisfies the following boundary conditions at the media interface:

$$\phi_+ = \phi_- \tag{4}$$

$$\beta_+\dot{\phi}_+ = \beta_-\dot{\phi}_- \tag{5}$$

where the dot denotes the derivative with respect to z. This is equivalent to solving a transmission line with the same propagation constant. We reduce the problem of solving for ϕ to the problem of solving for the voltage on a line with the following transformation.

$$v \sim \phi \tag{6}$$

$$i \sim -\beta\dot{\phi} \tag{7}$$

$$Y_c \sim \beta\gamma \tag{8}$$

$$i_s \sim \alpha\beta\phi_E \tag{9}$$

The equivalent transmission line circuit is constructed as follows:

1. For terminal layers such as ground, open layers, etc:
   (a) For electric or magnetic walls use a short or open circuit;
   (b) For an open layer, use a terminating resistor with resistance given by the equivalent characteristic impedance;
   (c) For an imperfect layer, use a resistor with resistance given by equivalent surface impedance.
2. For the intermediate layers, use a transmission line with the equivalent characteristic impedance, propagation constant, and length.

2.2 Green's Functions

A Green's function is the response of a physical system due to a point exciting source. More specifically, the Green's function is the underlying solution of the associated partial differential equation. For capacitance extraction problems in layered media the associated equation to be solved is Poisson's equation, i.e., the Green's function G satisfies

$$\nabla^2 G(z,\rho) = -\frac{1}{2\pi\epsilon\rho}\delta(z')\delta(\rho') \tag{10}$$

where (z,ρ) and (z',ρ') are the observation and source points in cylindrical coordinates. Fourier transforming in the radial direction yields the spectral domain Green's function $\tilde{G}$

$$\left(\frac{\partial^2}{\partial z^2} - \gamma^2\right)\tilde{G}(z) = -\frac{1}{2\pi\epsilon}\delta(z'). \tag{11}$$

For a fixed spectral number γ, this equation can be mapped to an equivalent transmission line equation as described in the previous subsection.

One set of transmission line equations is sufficient to describe the static Green's functions because of the one dimensional source *charge*. On the other hand, for full-wave problems, the three-dimensional source *current* results in three different sets of transmission line equations, each with its own characteristic impedance. Furthermore, because of retardation, the singularities and discontinuities in the spectral domain Green's functions involve additional complications that will not be discussed in detail here but that have been successfully tackled [8].

3 Computations of 3D Green's functions

The Green's functions can be easily derived and solved in the spectral domain. However, since the integral equation solution is usually obtained in the space domain, a transformation from the spectral domain to the space domain is required. This involves the use of a two-dimensional Fourier transform, which, due to radial symmetry, reduces to a one dimensional Hankel Transform.

More specifically, for a fixed z and z', if the impedance transfer function of the equivalent transmission line from the source to the observation point is Z, then the space domain Green's function is:

$$G(\rho) = \frac{1}{2\pi\epsilon_0}\int_0^\infty Z(\gamma)J_0(\rho\gamma)d\gamma. \tag{12}$$

This transform is computationally expensive. Although efficient codes such as the Fast Hankel Transform alleviate this problem, it is still prohibitive to perform these computations on-line.

Fortunately, the Green's functions in the far field are smoothly varying as a function of distance and can be precomputed and compressed (via interpolation) off-line. An adaptive Chebyshev interpolation procedure is used to exploit the particular nature of the $1/r$-like dependence in the Green's functions. We can tabulate and store the Chebyshev coefficients in a relatively small database.

The $1/r$ singularity in the Green's function in the near field obviates the use of any 3D interpolation procedure (due to excessive storage costs). Instead, it is more efficient to use a method based on images and to perform this computation on-line. The method of images is based on an infinite series expansion which is equal to the Hankel Transform integral. The image method [14] is generalized for an arbitrary number of layers and is briefly described in the following subsection.

3.1 Generalized image method

The image method is one of the most popular approaches for capacitance extraction in multilayered media [14]. But its use is usually limited to media with only a small number of layers for the following reasons:

1. For more than two layers, the infinite image series expression is relatively hard to derive and implement;
2. There is no robust convergence testing procedure for the image series. Moreover, in the far field an enormous number of images are required to converge to a reasonable tolerance.

For multilayered media in modern printed or lithographied circuits, the traditional image method is usually abandoned because of the above difficulties [15, 10]. Instead, an equivalent-source formulation is employed where the dielectric interfaces are discretized.

Wave-tracing is a generalized image method to obtain all the possible images by inspecting the physical meaning of the spectral domain Green's function [16]. Recall the mechanism in the transient response of transmission line circuits in time domain. At the initial time, the current source generates starting voltage waves (up and down) at the excitation point. The two waves can independently propagate. If the wave meets an impedance discontinuity, it splits into a reflected wave and a transmitted wave according to the boundary condition. The reflected and transmitted waves then independently propagate. Each wave is attenuated by the splitting process and the distance it travels. Waves passing the observation point are called *images.*

We developed a Prioritized Wave-Tracing method using a priority queue to store and retrieve the wave information. Rather than a naive implementation which recursively traces waves, each wave is instead assigned a key value which is calculated from the wave's amplitude and distance. Waves are dynamically sorted in a Priority Queue (PQ) with the strongest ones at the head. The pseudo-code is as follows:

```
Prioritized Wave-Tracing
insert initial waves in the PQ;
while the last image contribution is not small enough {
  Extract the wave at the head of the PQ;
  if the wave passes the observation point then
        accumulate Green's function;
        record image information;
  split the wave into reflected and transmitted;
  assign the key to the two new waves;
  insert them into the PQ;
}
```

The key advantage of this implementation is that only the dominant waves are traced, and that a good stopping criterion is available.

The image method is not feasible for the far field since the number of images grows exponentially. For a typical MCM structure at distance 10μm the number of images required to achieve accuracies of 10^{-4} and 10^{-3} is shown in Figure 1.

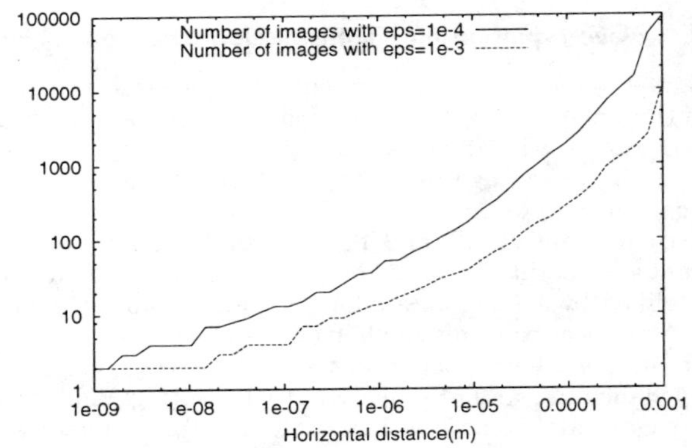

Figure 1: The number of images grows exponentially in the far field. This example is for a typical MCM-D technology with 5 dielectric layers and a ground plane.

3.2 The Fast Hankel Transform

In this subsection we discuss the method used to compute the space domain Green's function for the far field. Traditionally one of the following three techniques has been used.

- Discrete complex image [2, 16]. This method approximates the smooth part of spectral domain Green's function by a sum of complex exponentials. However, serious accuracy and stability problems make it impractical for more than two layers.
- FFT schemes [5, 4]. These methods enforce Dirichlet or Neumann boundary conditions on a bounding box, and use 2D FFTs to perform the transformation. They require extremely fine sampling to cover sufficient spectral content to achieve reasonable accuracy.
- Numerical integration [9]. As mentioned, the two dimensional Fourier transform of the Green's function in the spectral domain is reduced to a one-dimensional Hankel transform. Numeric evaluation of this transform, although robust, is relatively time consuming. However, the use of the Fast Hankel Transform (as discussed below) reduces the cost significantly.

The spectral domain Green's function can be split into two parts: A region near the origin which may contain poles and branch points, and a smooth tail. For the first part, adaptive Gaussian integration along a contour excluding the poles is used. The tail is smoothly decaying and oscillating when convolved with the Bessel function kernel. The Fast Hankel Transform [1] provides an efficient procedure for computing the Hankel Transform from 0 to ∞ of a smooth function f. In order to apply this technique, we extend the tail back to the origin by matching the function and its first few derivatives at the start of the tail with a polynomial. We then correct for the contributions of the polynomial in the initial region.

The Fast Hankel Transform is a digital filter for evaluating the following integral:

$$f(\rho) = \int_0^\infty F(k_\rho) J_k(k_\rho \rho) dk_\rho. \qquad (13)$$

By a change of variables, $\rho = e^x, k_\rho = e^{-y}$, the integral reduces to a convolution, which is then computed using

$$g(x) = \sum_{i=1}^{N} w_i G(x - a_i) \qquad (14)$$

where $g(x) = e^x f(e^x)$, $G(x) = F(e^{-x})$, and where w_i and a_i are sets of weights and abscissa respectively.

3.3 3D Green's function database

A space domain Green's function with radial invariance has three variables: z, z', and ρ. For a 3D problem every required value cannot be stored because of memory constraints. Instead we use an interpolation strategy that takes advantage of the fact that the majority of source and observation points lay on a fixed number of sheets rather than on the side walls of the conductor. The database stores the interaction between every pair of sheets. We use the database with a mixed-dimension interpolation scheme. Adaptive Chebyshev interpolation is applied in the radial direction and Lagrange interpolation is used for the other two. The accuracy of the interpolation is controlled. More importantly, the cost of evaluating a Green's function is on the same order as that of the free-space Green's function.

Figure 2 shows the various interpolation schemes.

1. For sheet to sheet interactions such as between points A and D, we use 1D interpolation on ρ.
2. When only one point is on a sheet (e.g., points A and C) we use 2D interpolation.
3. When neither point is on a sheet (e.g., points B and C) we use 3D interpolation.

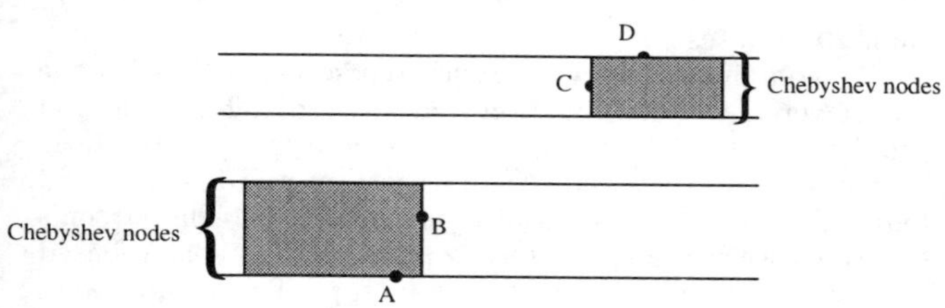

Figure 2: Various interpolation schemes for multilayer structure.

4 Experimental Results

In this section, we present experimental results for IES^3 using the layered static and full-wave Green's functions. All tests for IES^3 and FastCap were run on an SGI machine (200 MHz R10000 CPU).

4.1 Comparison to other tools

We begin by comparing IES^3 to the FMM-based FastCap program (version 2) [10] for two examples: a 1x1 bus crossing and a 3x3 bus crossing in a layered medium. A comparison to Raphael [13] is also made for the 1x1 bus structure.

The criteria for the discretization of the conductors (and for FastCap, the ground plane and dielectric interfaces) was to ensure convergence to within 1% accuracy. For FastCap it was sufficient to make the ground plane twice the extent of the conductors. Both examples were run with the dielectric layer information shown in Figure 4. The sizes of the layers are typical for a CMOS process but the dielectric constants were arbitrarily chosen. The ratios are small to ensure that FastCap does not exhibit numerical ill conditioning [12]. FastCap was optimized at the same level of our program and was run in its default mode. The relative tolerance for IES^3 was set to give the same accuracy as FastCap's default two-term multipole expansions. Table 1 summarizes the results of the experiments.

For each structure we show the additional panels required by FastCap for the discretization of the ground plane and dielectric interfaces. It was interesting to see that the two approaches for modeling the dielectric interfaces give good agreement.

Since the number of unknowns for running a problem in FastCap is dependent on the size and discretization of the ground and dielectric interfaces, it is difficult to make an accurate timing comparison. Typically, our algorithm is three to six times faster. More importantly, once the layer structure is fixed the accuracy of the solution is only dependent on discretization of the conductors surfaces.

In addition, the 1x1 bus crossing example was run in Raphael, which is based on finite differences. This tool enforces the Neumann boundary condition on a bounding box. In this case the box was made to be four times the size of the structure. Raphael converged to within 2% of IES^{3}and FastCap answers.

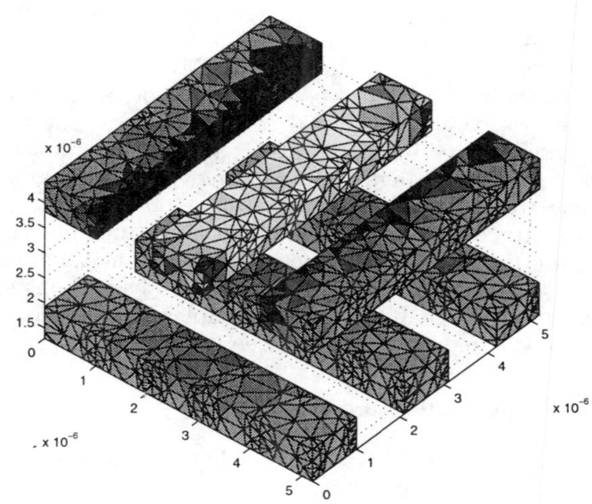

Figure 3: Discretization of and charge distribution on the 3x3 bus. All wires are $1\mu m \times 5.2\mu m \times$ the layer thickness and are separated by $1.1\mu m$.

4.2 Full-wave simulation of MCM inductors

In this subsection we present results of a 2.5D full-wave simulator [7] using the full-wave layered Green's functions and

Example	Conductor panels	IES^3 Time	FastCap Ground panels	Dielec. panels	Total panels	Time	Difference
1x1 bus crossing	1172	31	637	2548	4357	191	0.19%
3x3 bus crossing	4384	177	637	2548	7569	511	0.38%

Table 1: Comparison of IES^3 to FastCap (time in seconds)

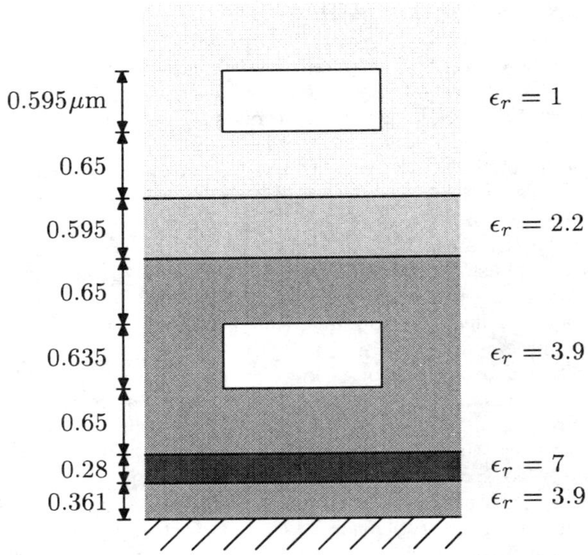

Figure 4: The layer structure for the 1x1 and 3x3 bus crossing examples

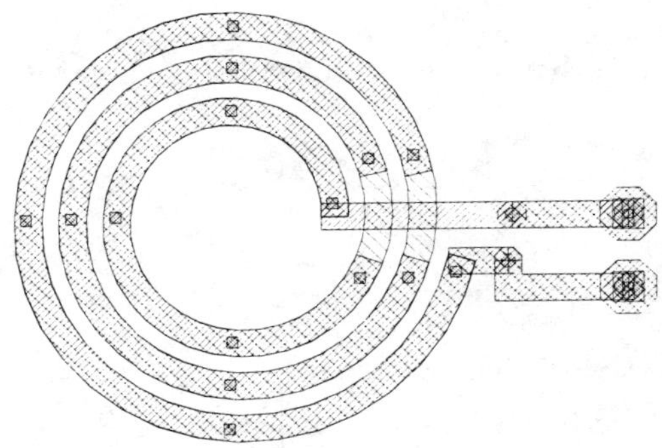

Figure 5: Layout of the two metal layer MCM inductor.

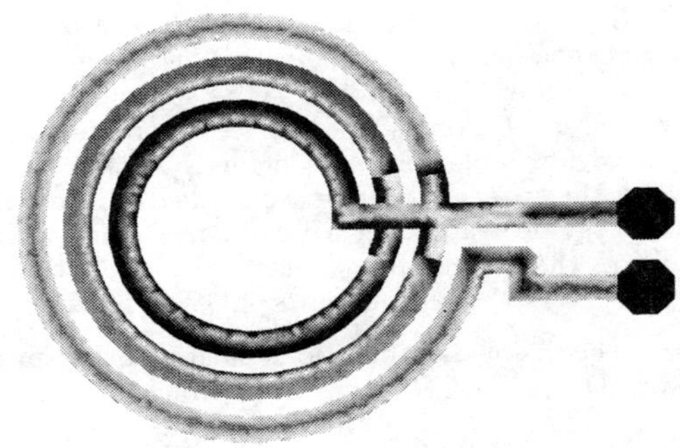

Figure 6: Density of current flow in the inductor near resonance.

IES^3. The simulator has been run on a number of examples, including capacitors, inductors, and filters on CMOS and MCM substrates. In this section, we report simulation results for a inductor on a seven layer MCM substrate. The layout is shown in Figure 5. The inductor is constructed with the top two metal layers of the process. The inductor was discretized into 1962 panels and the impedance was calculated over a range of frequencies. The simulation of a single frequency point takes about 100 seconds. Current density over the structure near resonance are shown in Figure 6. Comparisons between the simulation results and measurement are shown in Figure 7. The graphs show the magnitude and phase of the complex impedance. The self resonant frequency is accurately captured.

5 Conclusion and future work

In this paper we used a transmission line analog to derive the spectral domain Green's function for multilayered media. To efficiently evaluate the corresponding three-dimensional space domain Green's function, we proposed a combination of two techniques. For near-field interactions, we use a generalized image method. In the far field, we construct a database using a Fast Hankel Transform and adaptive Gaussian quadrature. The database is efficiently represented with adaptive Chebyshev interpolation. The time to evaluate the layered Green's function is usually about twice the time required to evaluate the free-space Green's function. The cost of precomputing the database for a typical process is less than 20 minutes.

We combined the SVD-accelerated solver IES^3 with the multilayered Green's function and applied it to extraction problems on CMOS, MCM and GaAs substrates. For static extraction, we compared IES^3 to FastCap and Raphael and found IES^3 to be substantially more efficient. The Green's functions were also used in a 2.5D full-wave solver based on IES^3 to simulate an MCM inductor. The results were compared to measurement.

Future work can be summarized as follows:

1. Combining the layered Green's functions with an equivalent-source formulation to handle conformal dielectrics and other irregular boundaries.

2. Sensitivity analysis of process variations via the layered Green's functions.

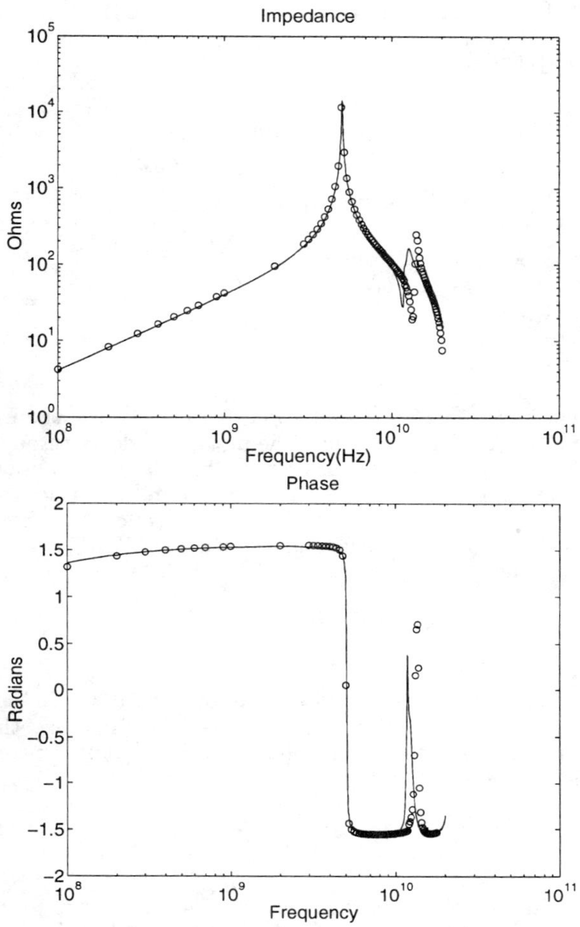

Figure 7: Comparison of simulation (circles) to measurement (lines)

References

[1] W. Anderson. Fast Hankel transforms using related and lagged convolutions. *ACM Transactions on Mathematical Software*, 8(4):344–70, Dec. 1982.

[2] Y. Chow, J. Yang, and G. Howard. Complex images for electrostatic field computation in multilayered media. *IEEE Transactions on Microwave Theory and Techniques*, 39(7):1120–5, July 1991.

[3] R. Crampagne, M. Ahmadpanah, and J.-L. Guiraud. A simple method for determining the Green's function for a large class of MIC lines having multilayered dielectric structures. *IEEE Transaction on Microwave Theory and Techniques*, 26(2):82–87, Feb. 1978.

[4] R. Gharpurey and R. Meyer. Modeling and analysis of substrate coupling in integrated circuits. *IEEE Journal of Solid-State Circuits*, 31(3):344–53, March 1996.

[5] A. Hill and V. Tripathi. An efficient algorithm for the three-dimensional analysis of passive microstrip components and discontinuities for microwave and millimeter-wave integrated circuits. *IEEE Transactions on Microwave Theory and Techniques*, 39:83–91, Jan 1991.

[6] S. Kapur and D. Long. IES^3: A fast integral equation solver for efficient 3-dimensional extractions. In *37th International Conference on Computer Aided Design*, Nov 1997.

[7] S. Kapur, D. E. Long, and J. Zhao. Efficient full-wave simulation in layered, lossy media. In *Proceedings of the IEEE 1997 Custom Integrated Circuits Conference*, May 1998.

[8] K. Michalski and J. Mosig. Multilayered media Green's functions in integral equation formulation. *IEEE Transaction on Antenna and Propagation*, 45(3):508–519, March 1997.

[9] J. Mosig and F. Gardiol. Analytic and numerical techniques in the green's function treatment of microstrip antennas and scatterers. In *Inst. Elect. Eng. Proc.*, volume 130, pages 175–182, March 1983. Pt. H.

[10] K. Nabors and J. White. Multipole-accelerated capacitance extraction for 3-D structures with multiple dielectrics. *IEEE Transactions on Circuits and Systems*, 39(11):946–954, Nov. 1992.

[11] J. R. Shewchuk. Triangle: Engineering a 2D quality mesh generator and Delaunay triangulator. In *First Workshop on Applied Computational Geometry*, pages 124–133. Association for Computing Machinery, May 1996.

[12] J. Tausch and J. White. Multipole accelerated capacitance calculation for structures with multiple dielectrics with high permittivity ratios. In *Proceedings of 33rd Design Automation Conference*, pages 367–70, Las Vegas, NV, June 1996.

[13] Technology Modeling Associates, Inc, Sunnyvale, California. *Raphael Reference Manual*, May 1996.

[14] W. T. Weeks. Calculation of coefficients of capacitance of multiconductor transmission lines in the presence of a dielectric interface. *IEEE Transactions on Microwave Theory and Techniques*, 18(1):35–43, Jan. 1970.

[15] C. Wei, R. Harrington, J. Mautz, and T. Sarkar. Multiconductor transmission lines in a multilayered dielectric media. *IEEE Transactions on Microwave Theory and Techniques*, MTT-32(4):439–50, April 1984.

[16] J. Zhao, W. Dai, R. C. Frye, and K. L. Tai. Green function via moment matching for rapid and accurate substrate parasitics evaluation. In *Proceedings of the IEEE 1997 Custom Integrated Circuits Conference*, Santa Clara, CA, USA, May 1997. New York, NY, USA: IEEE.

Robust Elmore Delay Models Suitable for Full Chip Timing Verification of a 600MHz CMOS Microprocessor

Nevine Nassif
Digital Equipment Corporation, Hudson MA
nevine.nassif@digital.com

Madhav P. Desai
Indian Institute of Technology, Mumbai, India
madhav@ee.iitb.ernet.in

Dale H. Hall
Digital Equipment Corporation, Hudson MA
dale.hall@digital.com

Abstract

In this paper we introduce a method for computing the Elmore delay of MOS circuits which relies on a model of the capacitance of MOS devices and a model of the Elmore delay of individual MOS devices. The resistance of a device is not explicitly modelled. The Elmore models are used to compute the Elmore delay and the 50% point delay of CMOS circuits in a static timing verifier. Elmore delays computed with these models fall within 10% of SPICE and can be computed thousands of times faster than if computed using SPICE. These models were used to verify critical paths during the design of a 600MHz microprocessor.

1 Introduction

The static timing verifier developed and used to verify the 21264 Alpha microprocessor [3] uses a family of delay models to compute the delay through MOS circuits. Simple complementary structures and some latches are modeled with precharacterized delays. The Elmore delay is one of the simplest delay models used in the static timing verifier [4] and we use it to compute the delays across complex complementary structures and arbitrary non-complementary structures. We need to use the Elmore delay because many of the structures used in the Alpha microprocessor design are arbitrary complementary and noncomplementary structures and it is impractical to characterize the delay for each one of these. The static timing verifier uses exhaustive path tracing techniques to verify every path in the microprocessor design and to identify critical paths (slow paths which violate setup constraints) and race paths (fast paths which violate hold constraints). Static timing verification is an important and necessary part of the design of the 21264 Alpha CPU. Because of the size (more than 15M transistors) and speed (600MHz) of this CPU, it is important for us to be able to filter out as many non-violating paths as possible before having to verify them or analyze them with more time consuming algorithms [1]. It was, therefore, necessary for us to radically improve the accuracy of our simple delay models before resorting to more complicated models or simulation techniques. In addition, because we use the Elmore delay model to compute the maximum delay through a MOS circuit, we had to ensure that the delays were always pessimistic. In other words, the delays computed using the Elmore delay model have to always be greater than those computed using a circuit simulator such as SPICE[7].

DAC 98, San Francisco, California
©1998 ACM 0-89791-964-5/98/06..$5.00

The Elmore delay which was defined [2] as

$$\tau = \int_0^\infty t\frac{dv}{dt}dt \tag{1}$$

has been used extensively as a measure of delay for RC interconnect and for MOS circuits. Given an RC tree, the Elmore delay at an output node i can be computed using the following equation [9]:

$$T_{D_i} = \sum_k R_{ki}C_k \tag{2}$$

where R_{ki} is defined as the resistance which the two paths from the input to node i and the input to node k have in common. Equation 2 is also used to compute the Elmore delay of MOS circuits [8]. The effective resistance of each device and the capacitive load seen by each device are used to approximate a MOS circuit by an RC circuit and the Elmore delay is computed. More efficient and general algorithms for computing the Elmore delay have also been developed [5][6].

The resistance and capacitance models which are used to approximate the MOS circuit are a great source of error in computing the Elmore delay. We decided to focus on these models in order to improve the accuracy of the Elmore delay in our timing verifier. Our goal was to use the Elmore delay model to compute 50% point delays which are always greater than SPICE and within 20%.

We began our modelling effort by attempting to define an effective resistance. Our objective was to write the resistance as a function of device size, the slope of the input signal on the gate, and the capacitive load. We wanted to model the resistance using a simple equation in order to make the computation of the resistance as fast as possible. In other words, we wanted to stay away from table lookup algorithms or complicated models.

In our quest for a simple model for resistance we moved away from Equation 2 and instead focused on the voltage waveform whose time constant is the Elmore delay. We realized that we should not even bother modelling the effective resistance and instead model the Elmore delay of a device as a function of device size, slope of the input signal on the gate, and capacitive load. Since we were interested in using the Elmore delay only as an estimate for the maximum delay of our circuit, we also wanted to ensure that the Elmore delay calculated using the Elmore models was always greater than or equal to the Elmore delay computed using SPICE. The equations we derive are in fact very simple and accurate to within 5% of SPICE. As a result, we can compute the Elmore delay to within 10% of SPICE for most of our circuits.

In Section 2 we review the Elmore delay, describe the characterization of the Elmore delay and introduce the resulting model. In Section 3 we describe how we use the model for the Elmore delay within our timing verifier and how we use it to compute maximum 50% point delays in MOS circuits. In Section 4 we present results comparing the Elmore delay computed with the help of precharacterized delays and the Elmore delay computed using SPICE. We also compare the 50% point delay computed with Elmore delay model to the 50% point delay computed using SPICE.

2 A Model for the Elmore Delay

2.1 The Elmore Delay

The Elmore delay was introduced half a century ago [2] as a measure of delay. The Elmore delay can be used to model the waveforms in an RC network. If the node voltage at a node i in the RC circuit is $v_i(t), 0 \le t < \infty$, then the Elmore delay model approximates $v_i(t)$ as:

$$v_i^{elmore}(t) = v_i(\infty) + (v_i(0) - v_i(\infty))\, e^{-t/\tau}. \tag{3}$$

The time constant τ, which is also the Elmore delay, is chosen so that the area under the curve $v_i(\infty) - v_i^{elmore}(t)$ is the same as the area under the curve $v_i(\infty) - v_i(t)$. In other words:

$$\tau = \frac{\int_0^\infty (v_i(\infty) - v_i(t))dt}{v_i(\infty) - v_i(0)} \tag{4}$$

See Shi [10] for details.

2.2 Characterizing the Elmore Delay

For a falling waveform at the output of a MOS circuit where $v_i(\infty) = 0$ and $v_i(0) = V_{DD}$, the Elmore delay reduces to:

$$\tau = \frac{\int_0^\infty v_i(t)dt}{V_{DD}} \tag{5}$$

Similarly, for a rising waveform where $v_i(0) = 0$ and $v_i(\infty) = V_{DD}$, the Elmore delay reduces to:

$$\tau = \int_0^\infty dt - \frac{\int_0^\infty v_i(t)dt}{V_{DD}} \tag{6}$$

From both Equations 5 and 6 we can see that the quantity that must be characterized in order to model the Elmore delay is

$$\int_0^\infty v_i(t)dt \tag{7}$$

Let us consider the discharge path shown in Figure 1. Let $v_{DS}(t)$ represent the source-drain voltage across an NMOS device for $t > 0$. The Elmore delay at the output node y is

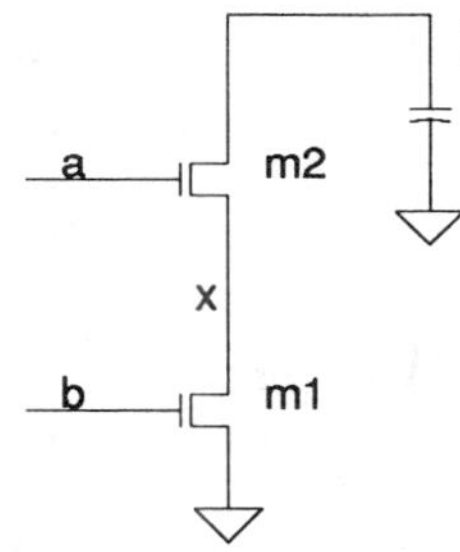

Figure 1: Discharge path through $m1$ and $m2$

computed by "summing up" the Elmore delay from V_{SS} to node x and the Elmore delay from node x to node y:

$$\begin{aligned} \tau &= \frac{1}{V_{DD}} \int_0^\infty v_y(t)dt \\ &= \frac{1}{V_{DD}} \left[\int_0^\infty v_{DS_{m1}}(t)dt + \int_0^\infty v_{DS_{m2}}(t)dt \right] \\ &= \tau_{m1} + \tau_{m2} \end{aligned} \tag{8}$$

Therefore if we can model the Elmore delay for a single MOS device, we can use the model to approximate the Elmore delay for a discharge path in a MOS circuit.

In order to get accurate models we used SPICE to simulate several configurations of PMOS and NMOS devices. These configurations which are shown in Figure 2, include 1-high, 2-high and 3-high stacks, where one device is turning on (i.e the gate voltage is transitioning) and the remaining devices are already on. We found it necessary to model each case separately and to differentiate between:

1. NMOS (PMOS) devices whose source was tied to V_{SS} (V_{DD}) and NMOS (PMOS) devices whose source was tied to the drain of another NMOS (PMOS) device;

2. NMOS (PMOS) devices whose drain is tied to the drain of a PMOS (NMOS) device and NMOS (PMOS) devices whose drain is tied to the source of an NMOS (PMOS) device;

3. NMOS (PMOS) devices whose source and drain are tied to NMOS (PMOS) devices.

4. NMOS (PMOS) devices for which the gate voltage is high (low) and NMOS (PMOS) devices for which the gate voltage is rising (falling).

For each simulation we varied the load seen at the output node and the sizes of the devices over the ranges allowed by the design rules imposed on the circuit designers. For each simulation of a MOS device we computed the following two quantities from the current and voltage waveforms:

$$\tau_m V_{DD} = \int_0^\infty v_{DS}(t)dt \tag{9}$$

and

$$q = \int_0^\infty i_D(t)dt \tag{10}$$

where $i_D(t)$ represents the drain current for $t > 0$ and q is the charge discharged by the device.

In Figure 3 we show a plot of $\tau_m V_{DD}$ as a function of q for various device widths. The linear relationship between

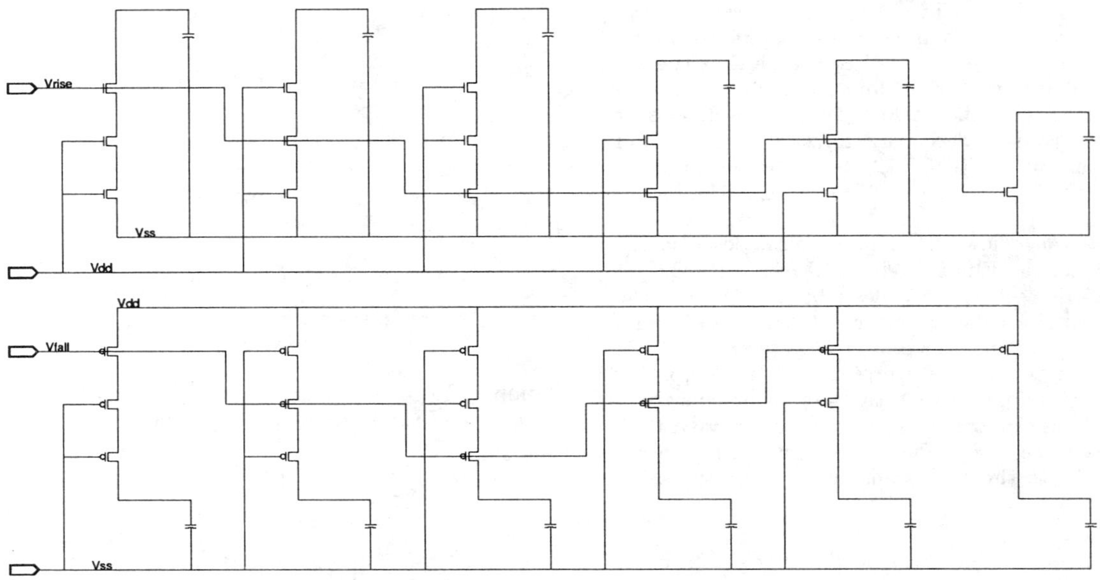

Figure 2: Configurations for which τ_m is modeled.

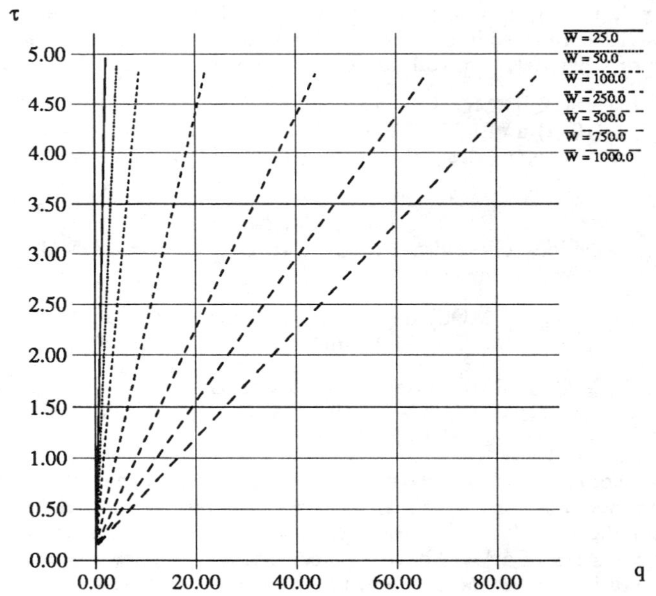

Figure 3: τ_m vs q

$\tau_m V_{DD}$ and q helps to make the modelling of the Elmore delay for MOS devices very simple.

We wanted to write $\tau_m V_{DD}$ as a function of q and the device width W. For each of the configurations shown in Figure 2 we were able to fit our simulation data to equations of the following form:

$$\tau_m V_{DD} = k_2 \frac{q}{W} + k_1 \frac{1}{W} + k_0 \qquad (11)$$

In addition the coefficients k_0, k_1, and k_2 are chosen so that $\tau_m V_{DD}$ computed using Equation 11 is always greater than $\tau_m V_{DD}$ computed from SPICE waveforms. The maximum error in $\tau_m V_{DD}$ when computed using equations of the form of Equation 11 compared to the data simulated and measured using SPICE was less than 5% for more than two thirds of the configurations shown in Figure 2. The maximum error across the samples taken in our simulations for the remaining configurations was less than 10%. It was necessary to model each configuration with a separate equation in order to obtain this accuracy.

3 Computing the Elmore Delay

In order to compute the Elmore delay for a given structure, the timing verifier identifies the discharge path of that particular structure, computes the charge which must be discharged through each device due to its load capacitance, computes the Elmore delay for each device along the path, and sums up the total delay. For symmetric waveforms the Elmore delay is the 50% point delay. Unfortunately, our voltage waveforms, examples of which are shown in Figures 4 and 5 are rarely symmetric. We therefore use a waveform model to map the Elmore delay into a 50% point delay.

Consider the circuit shown in Figure 6. If input b is rising, the discharge path will be from the output node y through node x to V_{SS}. Therefore the Elmore delay τ_y at the output node y will be the sum of the Elmore delay through devices $m1$ and $m2$ as shown by Equation 8. τ_{m1} is a function of W_{m1} and q_{m1}. W_{m1} is the width of $m1$ and q_{m1} is

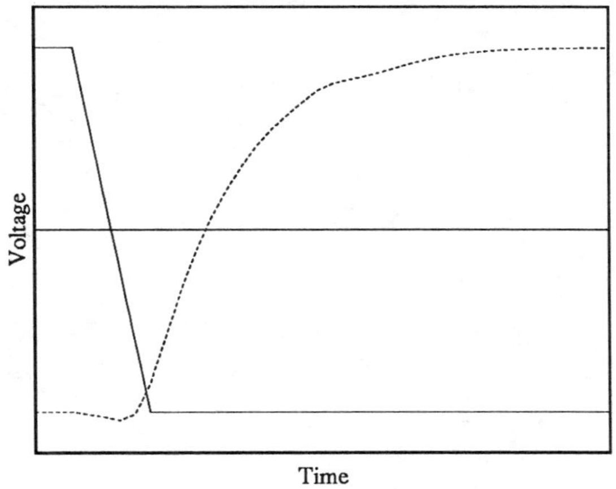

Figure 4: Example of a rising waveform

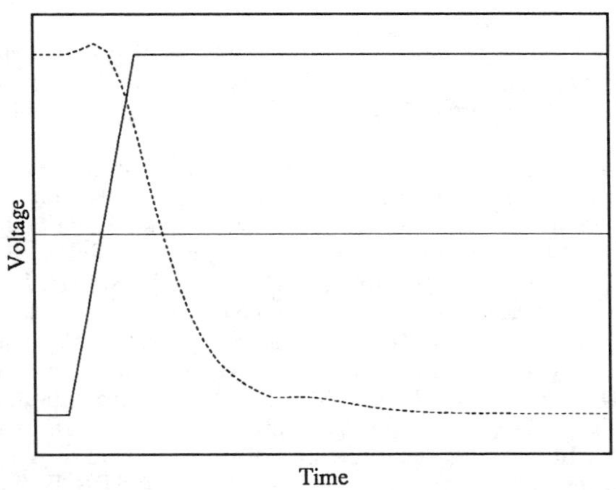

Figure 5: Example of a falling waveform

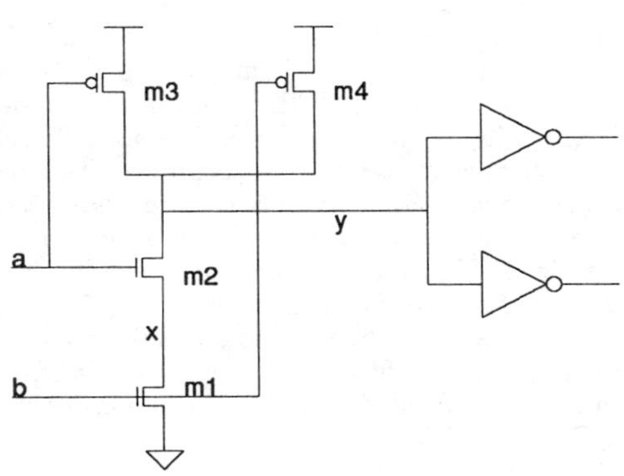

Figure 6: Example

Input	Rising	Falling
a1	6.1 %	3.9 %
b2	5.1 %	3.0 %
d3	3.4 %	5.6 %
c4	1.5 %	2.7 %
a5	4.0 %	3.0 %
d6	2.0 %	3.0 %

Table 1: % Difference in the Elmore delay beween model and SPICE

the charge that must be discharged through $m1$. Similarly, τ_{m2} is a function of W_{m2} and q_{m2}.

In order to use the models derived for the Elmore delay effectively we also had to model the charge seen by the drain of each device as accurately as possible. Here again we chose models for the charge such that the models yield values that are greater than the data obtained from SPICE simulations. Once we have computed the charge, we use the Elmore model for an NMOS device which is turning on and whose scurce is connected to V_{SS} in order to compute τ_{m1} and the Elmore model for an NMOS which is on and whose drain is connected to the output node to compute τ_{m2}.

After computing the Elmore delay we use a set of precharacterized waveform equations to map the Elmore delay into the 50% point delay. These equations are also based on a set of pre-simulated configurations and are a function of the Elmore delay and the slope of the input waveform.

On an AlphaServer 8400 Model 5/300 (300 MHz CPU) we can compute more than 20,000 delays/sec for circuits of the complexity shown in Figure 7. This includes identifying the discharge path, computing the charge at every node, computing the Elmore delay, and computing the 50% point delay. On the other hand we can only compute about 4 delays/sec with SPICE for these circuits after the input stimulus has been manually set up.

4 Results

For the example circuits shown in Figure 7, we show in Table 1 the % difference between the Elmore delays computed using the models of the Elmore delay that we described in Section 2 and the Elmore delay computed using SPICE. For all the examples shown the difference between SPICE and the precharacterized Elmore model is less than 10%. In 75% of our test cases the Elmore delay computed using the precharacterized model was greater than the Elmore delay computed using SPICE by less than 6%. For the remaining cases the difference was less than 10%.

In Tables 2, 3, and 4 we show the % difference between the 50% point delays computed for the same circuits using the Elmore model and SPICE. In Table 2 the input waveform has a typical transition time, in Table 3, the input waveform has a fast transition time, and in Table 4, the input waveform has a slow transition time. In 45% of our test cases the 50% point delay computed using our delay model is within 5% of SPICE, in 25% of our test cases it is between 5% and 10% of SPICE, and in 23% of our test cases it is between 10% and 20% of SPICE. For the remaining 7% of our test cases the 50% point delay was between 20% and 30% of SPICE.

One weakness still remains in this method in going from the Elmore delay to the 50% point delay. Our goal was to use the Elmore delay model to compute 50% point delays which are always greater than SPICE and within 20%. We

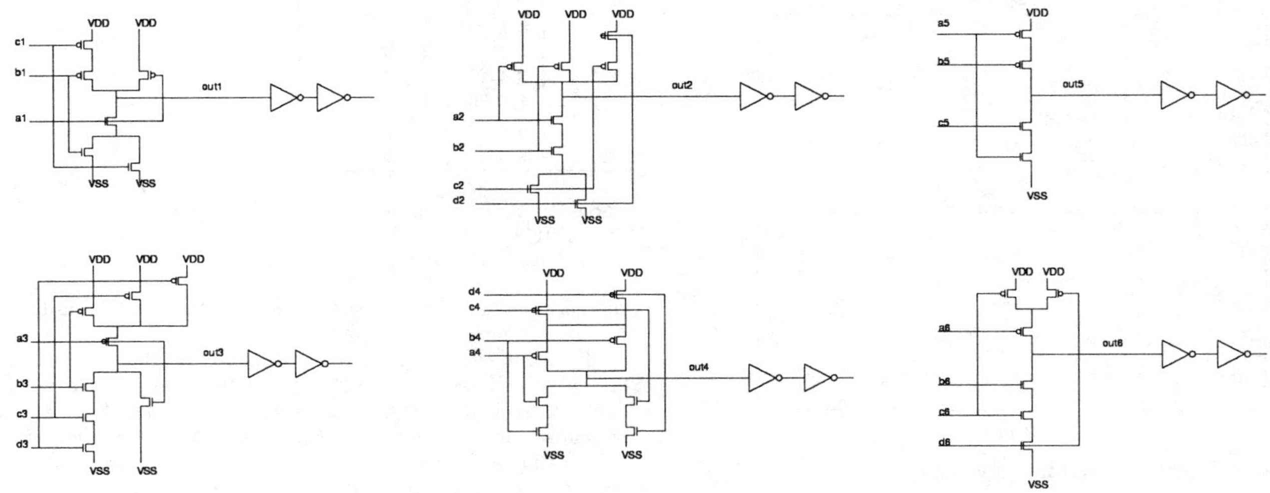

Figure 7: Sample circuits

Input	Rising	Falling
a1	3.9 %	1.2 %
b2	1.3 %	11.2 %
d3	0.8 %	23.1 %
c4	0.4 %	6.2 %
a5	7.9 %	1.0 %
d6	2.4 %	18.6 %

Table 2: % Difference in 50 % point delay for typical input slope

Input	Rising	Falling
a1	5.5 %	4.1 %
b2	0.7 %	13.1 %
d3	3.6 %	22.7 %
c4	3.8 %	4.5 %
a5	9.0 %	11.6 %
d6	2.7 %	7.0 %

Table 3: % Difference in 50% point delay for fast input slope

Input	Rising	Falling
a1	0.3 %	1.0 %
b2	0.3 %	11.8 %
d3	5.6 %	23.6 %
c4	4.4 %	2.4 %
a5	6.3 %	0.9 %
d6	0.0 %	30.0 %

Table 4: % Difference in 50% point delay for slow input slope

have fallen short of this goal for about 7% of our circuits where the error is greater than 20%. However in all cases we compute a delay which is greater than the delay computed with SPICE.

In Figure 8 a section from a critical path which was detected by the timing verifier is shown. The path contains circuits whose delays are precharacterized and circuits for which the Elmore model is used to compute the delay. For the section of the path through the latch between CLK and I_H and I_L and through the simple complementary logic between Y_L and OUT the delay is computed in the timing verfier using precharacterized delays. For the section of the path between I_H/I_L and Y_H/Y_L the Elmore model is used. The 50% point delay from CLK to OUT rising calculated using the various delay models in the timing verifier is 19% greater than the delay calculated using SPICE. The 50% point delay from CLK to OUT falling computed in the timing verifier is 11% greater than the delay computed using SPICE.

5 Conclusions

We have succeeded in modelling the Elmore delay of a device accurately and ensuring that the Elmore delay is always pessimistic. For almost all of our circuits the Elmore delay computed using the model described in this paper is greater than the Elmore delay computed using SPICE by less than 10%. We have successfully incorporated the model into our static timing verifier and used it to compute the 50% point delays of all circuits whose delays could not easily be precharacterized. The speed and accuracy of this model makes full chip timing verification of 600MHz, 15 million transistor microprocessor design feasible since the majority of non-critical paths can be filtered out by the timing verifier and do not need to be verified using time consuming circuit simulations.

6 Acknowledgements

The authors would like to thank Jim Farrell for his help.

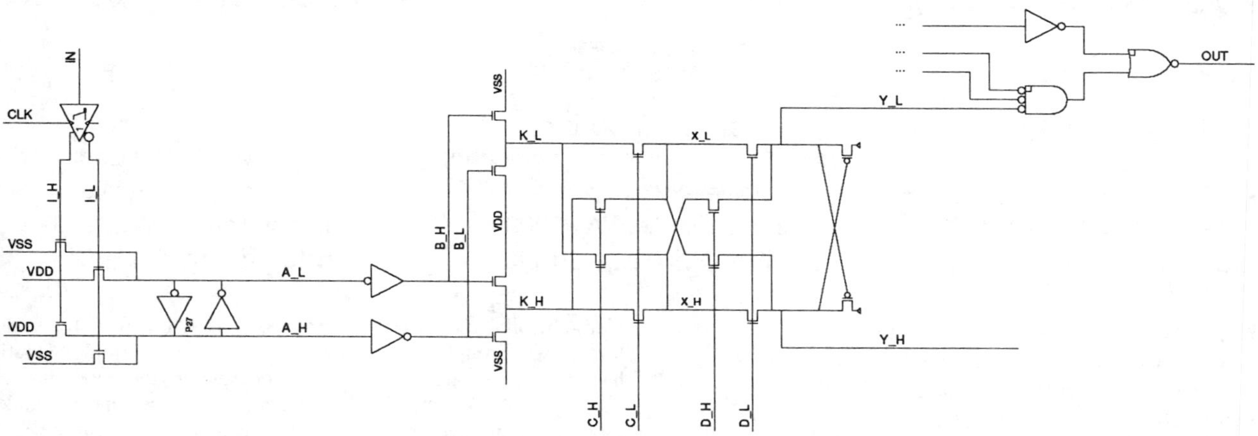

Figure 8: Critical path in 21264 design

References

[1] M.P. Desai, Y.T. Yen, A Systematic Technique For Verifying Critical Path Delays in a 300MHz Alpha CPU Design Using Circuit Simulation, *Proceedings of the 33rd Design Automation Conference*, pp. 125–130, June 1996.

[2] W.C. Elmore, The Transient Response of Damped Linear Networks with Particular Regard to Wideband Amplifiers, *Journal of Applied Physics*, Vol. 19, pp. 55–63, January 1948.

[3] B. Gieseke, et. al., A 600MHz Superscalar RISC Microprocessor with Out-of-Order Execution, *ISSCC Digest of Technical Papers*, pp. 176–177, February 1997.

[4] W.J. Grundmann, D. Dobberpuhl, R.L. Allmon, N.L. Rethman, Designing High Performance CMOS Microprocessors Using Full Custom Techniques, *Proceedings of the 34th Design Automation Conference*, pp. 722–727, June 1997.

[5] T.M. Lin, C.A. Mead, Signal Delay in General RC Networks, *IEEE Transactions on Computer-Aided Design*, Vol. CAD-3, No. 4, pp.331–349, October 1984.

[6] D. Martin, N.C. Rumin, Delay Prediction from Resistance-Capacitance Models of General MOS Circuits, *IEEE Transactions on Computer-Aided Design of Integrated Circuits and Systems*, Vol. 12, No. 7, pp. 997–1003, July 1993.

[7] L.W. Nagel, SPICE2: A Computer Program to Simulate Semiconductor Circuits, *Rep. No. ERL-M520, Electronics Research Laboratory, University of California, Berkeley*, May 1975.

[8] J.K. Ousterhout, A Switch-Level Timing Verifier for Digital MOS VLSI, *IEEE Transactions on Computer-Aided Design*, Vol. CAD-4, No. 3, pp. 336–349, July 1985.

[9] J. Rubinstein, P. Penfield, Jr., M.A. Horowitz, Signal Delay in RC Tree Networks, *IEEE Transactions on Computer-Aided Design*, Vol. CAD-2, No. 3, pp. 202–211, July 1983.

[10] C.-J. Shi, Analysis, Sensitivity, and Macromodelling of the Elmore Delay in Linear Networks for Performance-Driven VLSI design, *Int. J. Electronics*, Vol. 75, No. 3, pp. 467–484, 1993.

A Top-down Design Environment for Developing Pipelined Datapaths

Robert McGraw
RAM Laboratories
119 N. El Camino Real, Suite 175
Encinitas,CA 92024
rmcgraw@adnc.com

James H. Aylor
Department of Electrical Engineering
University of Virginia
Charlottesville, VA 22903
jha@virginia.edu

Robert H. Klenke
Department of Electrical Engineering
University of Virginia
Charlottesville, VA 22903
rhk2j@virginia.edu

1. Abstract

This paper presents a design environment for cycle-based systems, such as microprocessors, that permits modeling of these systems at various levels, from the abstract system level, through the detailed RTL level, to an actual implementation. The environment allows the models to be refined to lower levels in a step-wise manner. The environment provides the ability to obtain meaningful metrics from abstract models of a processor's architecture. This capability allows design alternatives to be evaluated earlier in the design cycle, thus eliminating costly redesign and reducing the processor time to market.

2. Introduction

Currently within the design community there is an increasing interest in the development of methodologies which reduce the time to market for a given system under development. One area of particular concern deals with the development of application specific processors [1]. With integrated circuits projected to reach the size of over 100 million transistors per die by the turn of the century [2], this increasing complexity must be handled properly so as not to adversely affect processor design time. One way to address this problem of complexity management is through the use of a top-down design methodology.

Top-down design methodologies have been used to design digital hardware design since the early 1970's [3]. A top-down design methodology follows a design from the top level, usually the specification level, of detail down to a detailed implementation. Model refinement in these methodologies works by having each level of detail serve as the design specification for the level of detail immediately below. It is acknowledged that if this hierarchical chain can be verified from one level of detail to the next, the resulting behavioral implementation will be "*right the first time*" [4]. Being able to develop systems that work on the first pass in a timely manner helps address the time to market problem. Unfortunately, there exists a lack of modeling environments which promote complete top-down design and refinement of processors from the system level.

This paper presents a timed cycle-based design environment which is geared toward the development of pipelined datapaths for processors and other synchronous systems. This cycle-based environment permits the processor designer to model and hierarchically refine pipelined processor datapaths from the system level down through the RTL level until a behavioral implementation has been developed. This paper focuses on the modeling and development of pipelined datapaths because most modern processor architectures contain considerable pipelining. The remainder of this paper is organized as follows; Section 2 presents a background of existing processor design environments. Section 3 presents an overview of the new design environment proposed herein. Section 4 describes the intermediate level modeling capability of the environment that provides a link between the abstract system level of modeling and the detailed functional level model. Finally, Section 5 presents an example of modeling a MIPS R4000 processor using the environment and Section 6 presents some conclusions.

DAC 98, San Francisco, California

3. Existing environments and methods

For a processor design environment to completely support top-down design and refinement, the environment must have some means of developing a system level processor model, some means of refining the system level model to the RTL level, and some means of providing abstract control to the datapath in order to obtain meaningful results from the model. At each level of design detail, different architectural analyses can be performed as detailed in Figure 1. For instance, at the system level, datapath control is often provided through the use of random distributions to exercise all model paths. Resulting analyses which can be performed include determination of cycle time and critical paths. At the RTL level, the design is very detailed and control is provided by a explicit control unit. At the RTL level, all functional and detailed performance metrics can be obtained. The need for a methodology and environment which supports the modeling and refinement of both system level and RTL level datapath models has been expressed in the literature [5,6].

Existing commercial design methodologies use a variety of tools to analyze designs at varying levels of detail. For example, Sun Microsystems uses architecture-specific simulators such as the UltraSPARC Performance Simulator (UPS) [7] to examine architectural trade-offs at a functional level. The UPS is a trace-driven simulator designed to simulate the Ultra-SPARC microarchitecture at a functional, RTL level of modeling. IBM uses several modeling tools to satisfy different parts of its design methodology during the development of its PowerPC line of processors. IBM examines architectural trade-offs at the functional level of detail by using the Basic RISC Architecture Timer (BRAT) [8]. The BRAT tool is an architecture-specific simulator. IBM also developed processor models using Verilog and their propriety Design Structure Language (DSL) which were used to analyze architectural trade-offs at the both the system and functional levels.The DEC design methodology for the 100 MHz CISC NVAX processor and the 200 MHz RISC Alpha AXP 21064 processor [9,10] included the analysis of the processor architecture starting at the RTL level using Digital's in-house hardware description

Design Flow

Modeling Level	Datapath Control	Model Input	Type of Analyses
System Level	Distribution-based	Distribution-based	Cycle, setup and hold times, Critical Path Analysis
Intermediate System/RTL	Instruction-based Control provided through reservation tables or RTL level descriptions	Instruction Trace or Instruction Mix Driven	Cycle-time, Critical Path Analyses, Latency, Throughput, Concurrency, Register Setup and Hold Times, Determination of MIPS bounds
Register Transfer Level (functional)	Based on Modeled Control Unit or Datapath Information	Instruction Trace or Instruction Mix Driven	All Performance and Functional Analyses

Figure 1. Design flow for methodologies

language (DECSIM) based simulator. AMD's Am29000 processor was modeled at the functional level, using a specifically designed, C-based architectural language, and at the gate level to guide logic design [11]. Additional processor design methodologies currently being used deal with design starting at the functional level [12,13]. These methodologies are similar to bottom-up design strategies in that they are often based on existing architectures.

Using existing methods, the complete design of a processor cannot be performed in the same modeling and simulation environment. The current methodologies require the construction of multiple, disjoint, processor models at the system level and RTL level [14]. Typically these methodologies create system level models on a processor-by-processor basis using some type of modeling language or hardware description language. When a more detailed model of the processor is required, a new model is developed at the RTL level. There are several reasons for the creation of multiple models. First, processor modeling tools above the RTL level are relatively non-existent. Second, along with the need to address design refinement issues, existing environments do not have suitable methods for controlling an abstract datapath to produce meaningful results.

Several approaches have attempted to address processor and datapath modeling above the RTL level. Zhang and Grunbacher [5] have developed a Petri Net based design approach for pipelined processors. This approach allows for a design to be modeled at the system level through the use of Petri Nets. In addition, Razouk [6] has developed a timed Petri Net approach for detailed modeling of a processor design at the system level. Unfortunately, these methods lack the ability to link system level modeling environments to the RTL level of development.

This paper presents a timed-cycle-based design environment which allows for the modeling of processor datapaths above the RTL level. In particular, this methodology and environment provides datapath development constructs and control methods which link system level and RTL level models together through the use of an intermediate system/RTL level modeling domain. This intermediate system/RTL level domain, detailed in the shaded row of Figure 1, consists of a model of execution and datapath control methods which allow for the analysis of pipelined datapaths.

4. Environment and Model of Execution

A timed cycle-based processor design environment which specifically addresses the development of pipelined datapaths has been constructed. This environment supports system level processor modeling using abstract datapath constructs and mechanisms to control the datapaths. This environment addresses model refinement issues by providing modeling constructs and abstract control methods which bridge the system/RTL level modeling gap.

This design environment is based on the ADEPT performance modeling environment [15]. ADEPT is based on the VHSIC Hardware Description Language (VHDL) and provides a modeling environment where high-level models can be refined down to an implementation in an integrated manner. In the ADEPT environment, a system model is constructed by interconnecting a collection of *ADEPT modules*. The modules model the information flow, both data and control, through a system. Each ADEPT module is implemented in VHDL and communicates with other modules by exchanging *tokens* which represent data being transmitted in the system. The ADEPT modeling modules communicate via a four-state token passing protocol (present, acknowledged, released, removed). This protocol provides fully interlocked handshaking between elements. This type of asynchronous handshaking protocol is needed because the communications between the existing ADEPT modules is inherently asynchronous in nature. The VHDL code generated by ADEPT can be simulated using any IEEE 1076-87 compliant VHDL simulator. Facilities and programs to collect and analyze the simulation results are provided as part of the ADEPT system.

4.1 Design Flow and Datapath Control

The timed cycle-based environment augments the existing ADEPT environment to allow for the modeling of cycle-based systems, while still including the concept of asynchronous delay for combinational elements. The design flow using this timed cycle-based environment takes an instruction set architecture and refines it using modeling constructs of increasing detail through the RTL level down to a behavioral implementation. The cycle-based modeling constructs support datapath modeling at, or above, the RTL level. In addition, the existing capabilities for mixed-level modeling in ADEPT [15] allow RTL level models to be refined to an actual implementation in a step-wise manner. The modeling levels which are supported by the cycle-based modeling constructs include an abstract system level, an intermediate system/RTL level, and an RTL level. These modeling levels are unique in that they are exercised using different means of datapath control as more detail is entered into the datapath model. Figure 1 denotes the modeling levels along with methods for controlling those levels, means of exercising models developed at those levels, and the types of functional and performance analyses which can be performed at each level.

The system level modeling domain supports a very high-level of abstraction (almost all data and control have been abstracted away). This particular level of design can be equated to a "block-diagram" level of design detail. At this particular level, all of the clocked elements (registers, memories) are required to be present in the design. These elements receive or source the information tokens on every cycle. In addition, the combinational elements, between the clocked elements, are modeled simply as delay elements. The system level modeling constructs currently included in the environment include clocked register constructs and various routing elements which mainly deal with value-less (uninterpreted) tokens. An example of a system level model of a four-stage pipeline is shown in Figure 2(a). Datapath routing at this level is accomplished by using various stochastic methods. At this level, datapath control is provided by using stochastic distributions to make routing decisions as tokens arrive *at the routing elements*. The main goal of modeling at such a level is to ensure that the information flow between clocked elements meets the cycle time requirements.

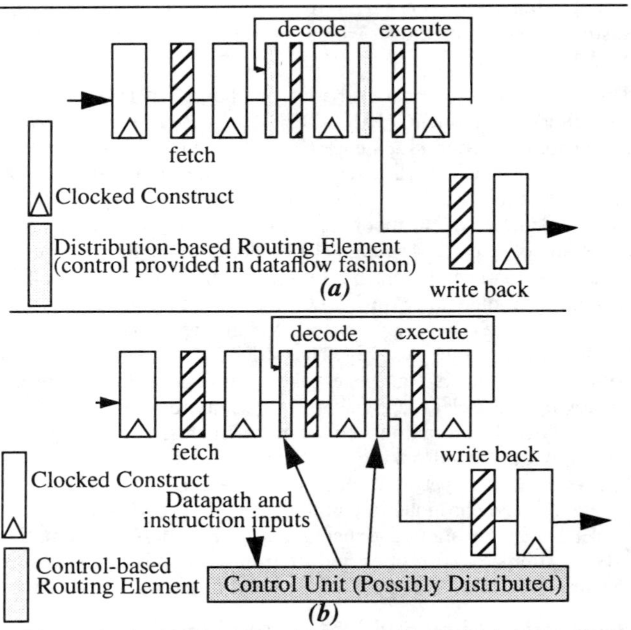

Figure 2. 4-stage pipeline (a) system level, (b) intermediate

The RTL level of modeling is much more detailed than the system modeling level. The RTL level modeling constructs include various clocked memory and register elements along with processor routing elements. An example of a RTL level model is shown in Figure 2(b). These RTL level modeling constructs model existing hardware elements such as multiplexers, demultiplexers and combinational logic using a one to one mapping of hardware signals to tokens or token values. The RTL level constructs are value-based. The constructs at the RTL modeling level route tokens based primarily on token values. The control for the RTL level datapaths is typically provided through some modeled control unit. In addition to having the responsibility of routing tokens from register to register, the RTL modeling level unclocked constructs also have the capability to operate on data (found on the token color fields).

The intermediate system/RTL level modeling constructs are the key to the environment in that they provide a link, through refinement, between the system level of modeling and RTL level of modeling. This intermediate system/RTL level modeling constructs and control methods are discussed in Section 4. By providing constructs which gradually incorporate more detail, the cycle-based design environment facilitates step-wise refinement from the system level to the RTL level.

4.2 Models of Execution

In order to communicate between various cycle-based modeling constructs, each construct must have a consistent model of execution. The model of execution refers to the way in which the modeling constructs of this environment communicate with each other. Because the modeling constructs must actually represent real systems or elements in a synchronous environment, models of execution for two types of elements are needed: clocked constructs (for synchronous elements) and unclocked constructs (for combinational elements).

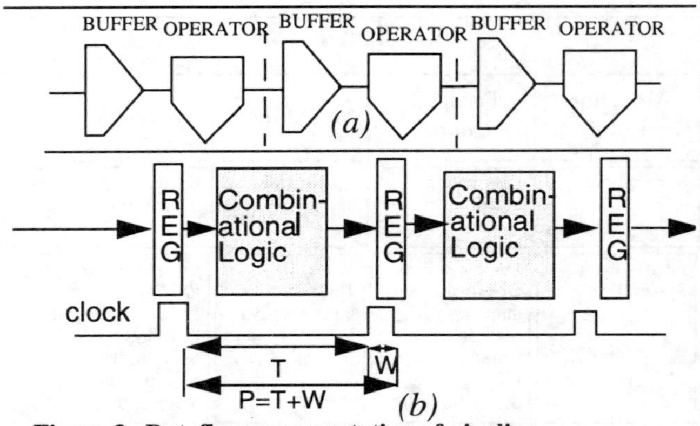

Figure 3. Dataflow representation of pipeline

Typically, existing processor datapaths can be represented using pipelined stages in a manner similar to Figure 3 [16]. Such a pipelined architecture is often implemented through stages of clocked elements (registers) followed by unclocked elements (combinational elements) as shown in Figure 3(b). Existing cycle-based environments typically map this pipeline architecture into the representation of Figure 3(a). Figure 3(a) shows each pipeline stage as being comprised of buffer elements followed by some type of operator element. The concept of buffering of information between modules is important because it is this buffering which separates the pipeline stages. In terms of the ADEPT four-way handshaking protocol, the buffer is the element that acknowledges the receipt of the token at the next stage of the pipeline. The operator element is viewed as an element which simply "operates" on arriving information before passing it on to subsequent pipeline stages. The operator modules do not buffer or acknowledge the receipt of information. The operator elements have an asynchronous delay representing combinational blocks and are known as the unclocked elements. In addition, the buffer elements are only allowed to acknowledge receipt of information on cycle boundaries. These are known as the clocked elements.

The model of execution for the unclocked elements is fairly straightforward. The unclocked constructs operate via the four-way interlocking handshake for asynchronous elements. These constructs map their inputs to their outputs using some type of control mechanism. This control mechanism may require inputs to be joined, synchronized, or forked in order to map them to the outputs. These constructs are also unbuffered in that they do not generate an acknowledgment upon the receipt of information. These constructs simply operate on arriving information and pass the information to the next construct.

The model of execution for clocked constructs is more complicated. The clocked elements are synchronized by some clock signal (to identify the cycle boundaries), yet these constructs must maintain a four-way interlocking handshake so they can communicate with the unclocked elements. In addition, these elements must contain buffering in order to acknowledge receipt of information at the cycle boundaries for each pipeline stage. For this reason, the model of execution for the clocked elements handles the four-way handshake, the buffering and acknowledgment of information, and the synchronizing of the inputs and outputs with respect to some type of clock signal.

The model of execution for both the unclocked and clocked

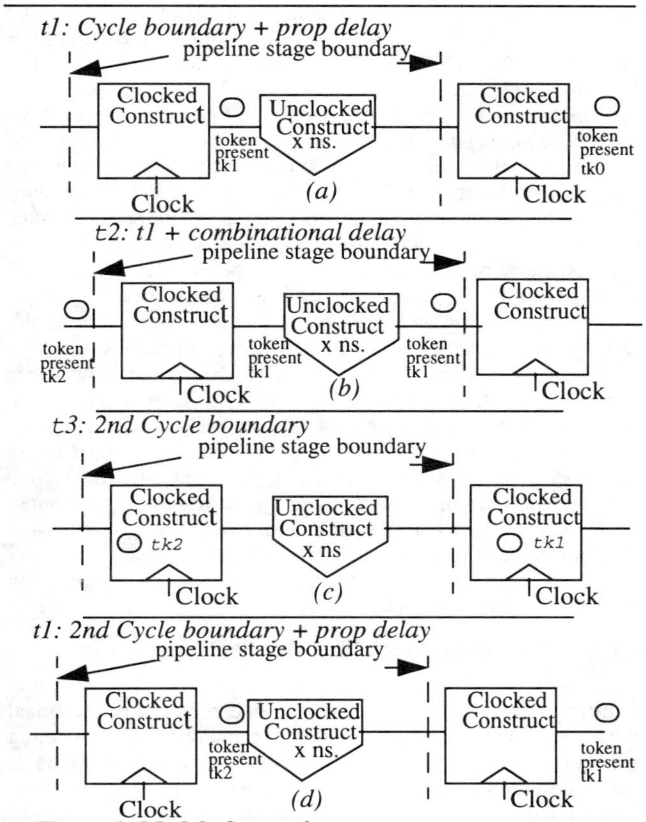

Figure 4. Model of execution

constructs for the cycle-based design environment is demonstrated in Figure 4 using a single pipeline stage. Figure 4(a) shows the clocked constructs outputting a token (*tk0,tk1*) at the cycle boundary after the propagation delay of the clocked constructs. This is represented by the token being present at the outputs of the clocked constructs. Figure 4(b) shows the token (*tk1*) propagating through the unclocked constructs after a delay of ***X*** (equal to the unclocked combinational delay). This results in the token, *tk1*, being present at the output of the unclocked construct. Because the unclocked constructs are unbuffered elements and do not generate their own acknowledgment, the input of the unclocked constructs still has token, *tk1*, present. The tokens remain in these "present" states until the cycle boundary is reached.

At the cycle boundary (determined when the clock signal is enabled), the clocked constructs copy the values on their input tokens to an internal token and finish the four-state handshake on their inputs. The clocked constructs then place their internal tokens on their outputs after accounting for the propagation delay *only* if those outputs are clear. This is the normal operating model of execution for the clocked constructs.

The models of execution for the clocked and unclocked constructs were verified using several basic architecture configurations. These basic configurations included linear pipelines, linear pipelines with single feedback loops, and linear pipelines with multiple feedback loops. In addition, each model of execution's ability to handle a stalled pipeline (due to resource contention issues or multiple cycle delay stages) has also been examined and verified.

5. Intermediate System/RTL Level Modeling

This new environment is set apart from the existing environments in that it provides an intermediate system/RTL level of modeling constructs which bridges the system level to RTL level modeling gap for abstract processor datapaths.

5.1 Intermediate System/RTL Level Modeling Constructs

Datapaths developed using the intermediate system/RTL level modeling constructs can provide the designer with a more detailed datapath analysis than can be found using only a system level model. While continuing to allow the designer to perform cycle-time and critical path analyses, datapaths which are developed using the more detailed intermediate system/RTL level modeling constructs also allow the designer to examine concurrency issues and perform latency and throughput analyses. Also, the system/RTL modeling level can permit the designer to obtain an estimated value for *instructions per second* before a detailed design or a complete compiler for the processor are developed.

The intermediate system/RTL modeling level constructs route the datapath information based on the desired datapath routes needed to satisfy a particular instruction. Typically these datapaths will be exercised using a statistical instruction mix, although an instruction trace can also be used. Each element of the system/RTL level datapath receives the active instruction, or instructions, for the current cycle. Because a modeled control unit is typically absent at early stages of the design cycle, the datapath control must be solely based on this instruction and its associated instruction fields. The current instruction is provided through the use of a colored information token. The control for the system/RTL level datapaths is dependent upon this current instruction and provided in two ways, depending upon the type of datapath and analyses required by the designer. Control for un-pipelined datapaths is provided based on the register transfer description for each instruction. Control for pipelined datapaths is provided using the reservation tables which describe the stage to stage information flow for each instruction. The reservation table-based control methods and modeling constructs are described in Section 5.2.

5.2 Reservation Table Control Methods

The goals of analyzing such a pipelined datapath would be to obtain latency and throughput information as well as a bounds for *instructions per second* for a pipelined execution unit under a given instruction mix or workload. One way of providing the control information for pipelined units is to make use of design methods concerning the design of pipelined execution units. In order to analyze the operation of pipelined execution units (such as integer pipelines and floating point units) system designers often use reservation tables [17,18]. Reservation tables are used to specify the use of given resources used by a instruction as it proceeds through the pipeline. Reservation tables can be used to determine instruction latency, or how long an instruction has to wait at the "head" of the pipeline before entering without causing resource contentions. These reservation tables can be used to give the designer a rough idea of attainable throughput and latency metrics concerning any pipelined unit.

The system/RTL level modeling constructs allow the designer to encode these reservation tables in a file. Figure 5 shows the reservation table for an integer instruction for the four-stage DLX pipeline of Figure 6[19]. The intermediate system/RTL level model of the four-stage pipeline is shown in Figure 6. The coded reservation tables are accessed by the intermediate system/RTL level routing elements and used to control the pipelined datapaths on a cycle-by-cycle basis.

The reservation tables are employed at the pipeline's clocked constructs to control instruction initiation within the pipeline. The pipehead_cyc element, shown in Figure 7, is the clocked element which has been developed to govern instruction initiations. The pipehead_cyc modeling construct requires four generic properties: *i_tag*, *trig_tag*, *delay*, and *filename1*. The *i_tag* property specifies the token color tag on which the instruction information is contained. The *delay* property specifies the propagation delay tokens encounter while passing through the pipehead_cyc element. The *filename1* property specifies the file which contains the names of the coded reservation table data files and reference numbers for all pipeline reservation tables which are used in the pipeline model.

The pipehead_cyc element is placed at the head of the top-level pipeline. When instructions arrive at the pipehead_cyc construct, they are checked for resource conflicts with all resources in the pipeline. First, the pipeline status reservation table is accessed. This pipeline status reservation table contains the status information (stage and cycle markings) for the pipeline referenced by the pipehead_cyc construct. This pipeline status reservation table is intersected with the reservation table of the incoming instruction to determine if a resource contention will occur if that instruction is initiated. If a resource contention will occur if the instruction is initiated, then no initiation is made for that cycle and the instruction is left on the pipehead_cyc element's input. This allows the same instruction to be presented on the subsequent cycle.

The reservation tables are also utilized at the pipeline's unclocked routing elements to control the stage-to-stage routing for each instruction within the pipeline. The unclocked routing units have their outputs bound (using defined properties and net interconnections) to different stages of the modeled pipeline. Tokens arriving at the unclocked elements are routed by accessing their reservation tables, and identifying the stage(s) to which they should be routed for that cycle. An example of such a routing element is the piperoute2 element, which is shown in Figure 8. The piperoute2 is used to route tokens internal to the pipeline execution units using reservation tables. This element requires several

cycle / stage	1	2	3	4	5
Fetch	X				
Decode		X		X	
Ex			X		
WB					X

Coded Reservation Table

i1
1 0 0 0 0
0 1 0 1 0
0 0 1 0 0
0 0 0 0 1

Figure 5. Reservation Table and Coded File for Figure 6

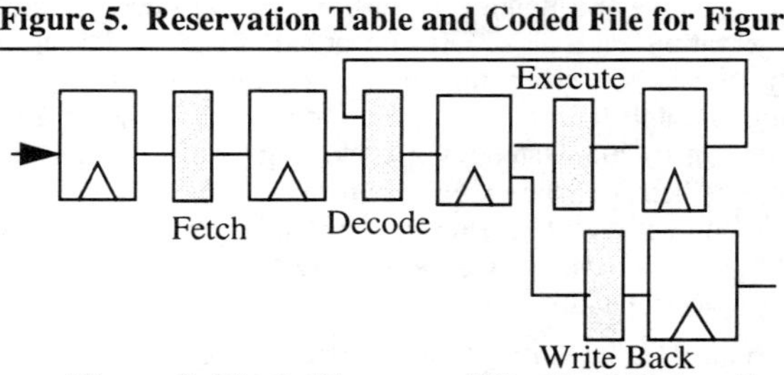

Figure 6. Block Diagram of Pipelined Datapath

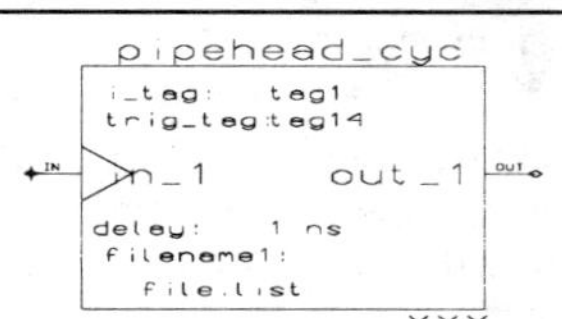

Figure 7. Pipehead_cyc Modeling Construct

generics in order use reservation tables to assist in routing tokens. The *i_tag* and *filename1* properties are the same as those found in the pipehead_cyc element. In addition to these properties, the piperoute2 element also has output binding properties *Outbindings1* and *Outbindings2*, a *max_inst* property, a *maxclkcyc* property, a *pipelength* property and a *cyc_no_tag* property. The *max_inst* property specifies the maximum number of instructions to be handled by the pipeline containing this particular element. The *maxclkcyc* property specifies the maximum number of cycles required to complete any instruction. The *pipelength* property specifies the number of stages in the pipeline. The *cyc_no_tag* property specifies the token color tag field which contains the cycle count for each instruction. The *Outbindings* properties specify the stage connectivity for each output. These properties are arrays which list the stages which connect to the current stage. For example, the four-stage DLX pipeline of Figure 6 contains a piperoute2 construct after its decode stage (stage 2). This routing element is required because information needs to be routed to the write back (stage 4) or execution (stage 3) stages after the decode stage. For this reason, the *Outbindings* properties of the piperoute2 construct are assigned to stages 3 and 4 respectively. When the token arrives at the piperoute2 element, its reservation table is accessed and the token is routed to the output which is referenced in the reservation table for that cycle.

5.3 Hierarchical Modeling Using Reservation Tables

In order to facilitate top-down design and refinement, the timed cycle-based environment has the capability of modeling hierarchical pipelines using the reservation table control methods. The control methods developed allow for the insertion of a "low-level" pipeline into a top-level pipelined datapath.

The pipelined cycle-based constructs were used to model a five-stage DLX pipeline with multiple execution units. Figure 9 shows the five-stage DLX pipeline (fetch, decode, execute, memory access, write-back) where execution unit 2 (Ex-2) represents a multi-function floating point unit that has its own underlying reservation table. Ideally, once this pipelined stage has been designed, it would be desired to hierarchically replace the single Ex-2 stage in the top-level pipeline with the 3-stage multifunction pipeline. In addition, it is also necessary to perform all routing at this "lower-level" by routing this 3-stage pipeline locally using its own reservation table. The original top-level reservation table is then altered to reflect the extra cycles spent in the multi-function

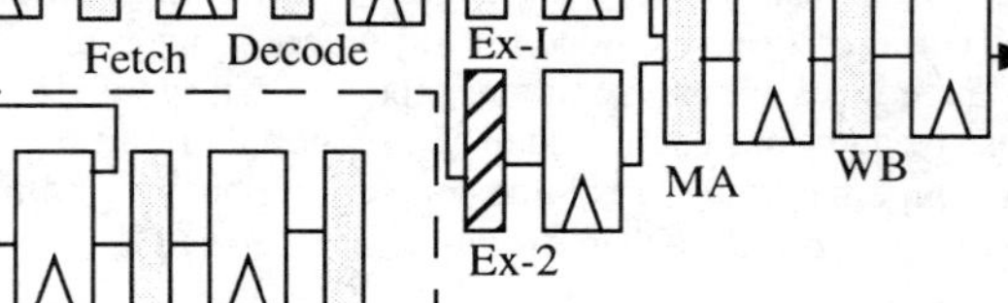

Figure 8. Piperoute2

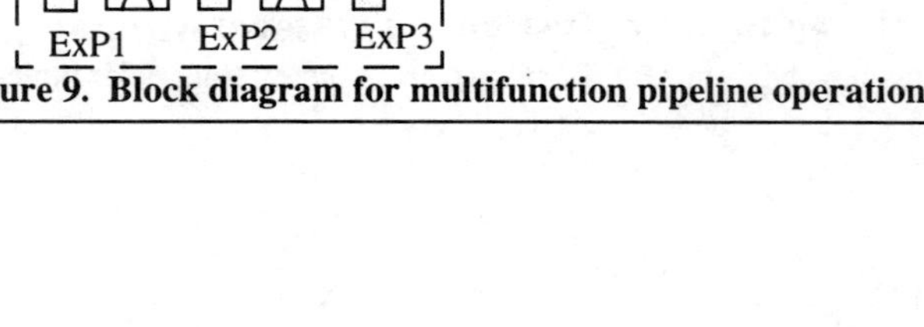

Figure 9. Block diagram for multifunction pipeline operation

Ex-2 unit.

The pipehead_cyc modeling constructs have the capability to examine all levels of pipeline hierarchy to decide when instruction initiations can be performed at the head of the top-level pipeline. By allowing the pipehead_cyc modeling construct at the head of the top-level pipeline to examine the reservation tables for all levels of pipeline hierarchy, refined lower-level pipelines can be simply "plugged in" to the top-level datapath model.

6. Example --- MIPS R4000 Processor

The timed cycle-based modeling environment's intermediate system/RTL modeling level has been verified through the use of several examples. One such example involves the modeling of a MIPS R-4000 processor [20]. The MIPS R-4000 is a 9-stage pipelined processor. These stages include: instruction fetch 1, instruction fetch 2, register fetch, execution, data fetch 1, data fetch 2, tag check and, write back. The execution units used for the execution pipeline stage include both an single stage integer execution unit and a floating point unit consisting of 8 pipelined stages. The MIPS R-4000 was modeled in hierarchical fashion with separate reservation tables developed for the integer and floating point execution units.

The reservation tables for both the top-level MIPS DLX pipeline and the internal pipeline execution unit were developed by hand for each instruction. During simulation, each modeling element accesses these reservation tables to determine if an instruction can be initiated and where information needs to be routed to. The MIPS R-4000 model was exercised using the SPEC95 benchmarks [21]. The benchmark instruction traces were obtained by compiling the SPEC95 source code on a Silicon Graphics MIPS R4000 machine and outputting the symbolic assembly instruction trace. This assembly language instruction trace was then mapped to instruction tokens entering the DLX pipeline. The MIPS R-4000 DLX model was simulated using a Mentor Graphics QuickVHDL simulator on a Sun Sparc-10 workstation. The simulation showed that the R-4000 processor executing the SPEC benchmark tomcatv.f, had a millions of instructions per second (MIPS) value of 75.76 using a 100 MHz clock. The published performance rating for the 100 MHz SGI R4000 was 76.5 indicating the abstract model was useful in obtaining a ballpark performance metric for millions of instructions per second. The model simulates at 16.4 cycles per minute of CPU time. It should be noted that this model did not take into account such issues as cache hits and misses and interrupts. Because this model was constructed at the intermediate system/RTL modeling level, statistical probabilities were used to help predict instruction branching. Exact branching values were not used because this model is an abstract model and does not contain the detail required to obtain those values. By using distributions to predict when branches could by taken, the model was able to use the SPEC benchmark traces in sequence to obtain a representative workload.

7. Summary and Conclusions

This paper presented a timed cycle-based design environment which provides a means for modeling and simulating processor datapaths at high levels of design abstraction. This environment was made possible by developing modeling constructs and abstract control methods which facilitate the modeling and control of processor datapaths above the RTL level. The methods for controlling the abstract processor datapath models are rooted in existing processor design methods and have been extended to assist in exercising meaningful processor models at early stages of the design. By obtaining meaningful metrics from abstract models of the processor's architecture, design decisions can be evaluated earlier in the design cycle, thus eliminating costly redesign and reducing the processor time to market.

8. References

[1] Proceedings, First Annual RASSP Conference, August 1994.

[2] Heaton, J., "Simulation - A Key to Smart Design,". Proceedings, Institution of Electrical Engineers, 1995.

[3] Rose, Charles. "The What and How of Top-Down System Design" TD Technologies, 1993.

[4] Transcend Promotional Document, TD Technologies, Inc.

[5] Zhang, Q.and H. Grunbacher. "Petri Nets Modeling in Pipelined Microprocessor Design," Applications of Theory of Petri Nets, pp. 582-591. 1993.

[6] Razouk, Rami R. "The Use of Petri Nets for Modeling Pipelined Processors," 25th ACM/IEEE Design Automation Conference. pp. 548-553.

[7] Tremblay, Maturana, Inoue, and Kohn. "A Fast and Flexible Simulator for Micro-architecture trade-off analysis on Ultra-Sparc-I," 32nd Design Automation Conference, 1995. pp. 2-6.

[8] Poursepanj, et al."The PowerPC 603 Microprocessor: Performance Analysis and Design Trade-offs," IEEE Spring COMPCON 1994 pp. 316-323.

[9] Dutton, Todd A. "The Design of the DEC 3000 Model 500 AXP Workstation," IEEE Digest of Papers, COMPCON, Spring 1993. pp 449-454.

[10] Peng, Donchin, Yen. "Design Methodology and CAD Tools for the NVAX Microprocessor," IEEE International Conference on Computer Design: VLSI in Computers and Processors, 1992 pp. 310-313.

[11] "Simulation in the design of the Am29000 microprocessor," Electronic Engineering. November 1987. pp. 44-52.

[12] Taylor, Rekow, Radke, and Thompson. "A 100 MHz Floating Point/ Integer Processor," IEEE 1990 Custom Integrated Circuits Conference. pp. 24.5.1-24.5.4.

[13] Narita, Arakawa, Uchiyama, and Kawasaki. "Design Methodology for GMicro/500 TRON Microprocessor," IEEE International Conference on Computer Design: VLSI in Computers and Processors, 1993 pp. 253-257.

[14] Franke, D., Purvis, M., "Hardware/Software Codesign: A Perspective", 13th International Conference on Software Engineering, May 1991, pp. 344-352.

[15] Klenke, R. H., M. Meyassed, J. H. Aylor, B. W. Johnson, R. Rao, A. Ghosh, "An Integrated Design Environment for Performance and Dependability Analysis," Proceedings of the ACM Design Automation Conference, June 1997 pp. 184-189.

[16] Kogge, Peter M. The Architecture of Pipelined Computers. Hemisphere Publishing Corporation, 1981.

[17] Stone, Harold S. High-Performance Computer Architecture. Addison-Wesley Publishing. 1993.

[18] Hayes, John P. Digital System Design and Microprocessors. McGraw-Hill, Inc. 1984.

[19] Hennessey, John L. and David A. Patterson. Computer Architecture: A Quantitative Approach. Morgan Kaufmann, Publishers, San Francisco, Ca. 1996.

[20] MIPS R-4000 User's Manual, Silicon Graphics, Inc.

[21] Standard Performance Evaluation Corporation. 1995 Benchmarks.

Validation of an Architectural Level Power Analysis Technique

Rita Yu Chen   Robert M. Owens   Mary Jane Irwin
Raminder S. Bajwa†
Department of Computer Science and Engineering
The Pennsylvania State University
†Semiconductor Research Laboratory
Hitachi America Ltd.

Abstract

This paper presents a technique used to do power analysis of a real processor at the architectural level. The target processor integrates a 16-bit DSP and a 32-bit RISC on a single chip. Our power estimator provides power consumption data of the architecture based on the instruction/data flow stream. We demonstrate the accuracy of the estimator by comparing the power values it produces against measurements made by a gate level power simulator for the same benchmark set. Our estimation approach has been shown to provide very efficient, accurate power analysis at the architectural level.

1 Introduction

As power dissipation has become a critical issue in many VLSI systems, power analysis at the architectural level has become more important because of its efficiency. It also allows the designer to experiment with design tradeoffs to lower power consumption. Most of the research in this area falls in the category of *empirical methods* which "measure" the power consumption of existing implementations and produce models based on those measurements. This *macromodeling* technique can be subdivided into three subcategories.

The first approach introduced in [7] is a *fixed-activity* macromodeling strategy called the Power Factor Approximation (PFA) method. The energy models are parameterized in terms of complexity parameters and a PFA proportionality constant. Thus, the intrinsic internal activity is captured through this PFA constant. This approach implicitly assumes that the inputs do not affect the switching activity of the hardware block.

To remedy this weakness of the fixed-activity approach, *activity-sensitive* empirical energy models have been developed. They are based on predictable input signal statistics, such as used in the SPA method [3, 4, 5]. Although the individual models built in this way are relatively accurate (the error rate is $10\% - 15\%$), overall accuracy may be sacrificed for the reasons of unavailable correct input statistics or an inability to model the interactions correctly.

DAC 98, San Francisco, California
©1998 ACM 0-89791-964-5/98/06..$5.00

The third empirical approach, *transition-sensitive* energy models, is based on input transitions rather than input statistics. The method presented in [2] assumes an energy model is provided for each functional unit - a table containing the power consumed for each input transition. The authors give a scheme for collapsing closely related input transition vectors and energy patterns into clusters, thereby reducing the size of the table for each functional unit. A significant reduction in the number of clusters, and also the size of the tables and the effort necessary to generate them, can be obtained while keeping the maximum error within 30% and the root mean square error within $10\% - 15\%$. After the energy models are built, it is not necessary to use any knowledge of the unit's functionality or to have prior knowledge of any input statistics during the analysis. However, this approach has never been used to analyze real commercial processors. Thus, it remains to be validated as a viable technique. Our work follows the third approach. The accuracy of our power estimator will validate the correctness of this methodology.

Our simulator imitates the behavior of the processor in each clock cycle as it executes a set of benchmark programs. It also collects power consumption data on the functional units (ALU, MAC, etc.) exercised by the instruction and its data. The results of our power estimator are compared with the power consumption data provided by [6]. This comparison confirms the accuracy of our power estimation technique.

The rest of this paper consists of four sections. Section 2 overviews the processor architecture and describes the simulation strategy. Section 3 presents the power estimation approach. Section 4 covers the power benchmarks and discusses the validation results. Finally, section 5 draws the conclusion and suggests the related future work.

2 Architectural Simulation of the Processor

2.1 Processor Overview

The processor considered in this research integrates a 32-bit RISC processor and a 16-bit DSP on a single chip. Figure 1 shows its main block diagram. The processor core includes the X-memory, the Y-memory, the buses, the CPU engine and DSP engine. The CPU engine includes the instruction fetch/decode unit, the 32-bit ALU, the 16-bit Pointer Arithmetic Unit (PAU), the 32-bit Addition Unit (AU) for PC increment, and 16 general purpose 32-bit registers. As the ALU calculates the X-memory address, the PAU can calculate the Y-memory address. Additional registers in the

CPU support hardware looping and modulo addressing. The DSP engine contains the MAC, the ALU, the shifter, and 8 registers (six 32 bits wide and two 40 bits wide).

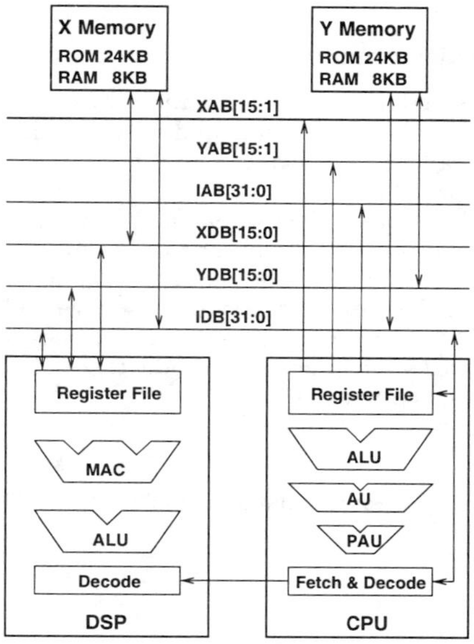

Figure 1: Block diagram of the processor

2.2 Architectural Simulation

An overview of the power estimator is shown in Figure 2. The inputs of the estimator are a benchmark program and energy models for each functional unit. The next section will introduce the details of energy model production. This section will only focus on the structural simulation of the architecture. The architectural simulator consists of several parts: the assembler, the controller, and the datapath implemented as many functional units (the small blocks labeled as m).

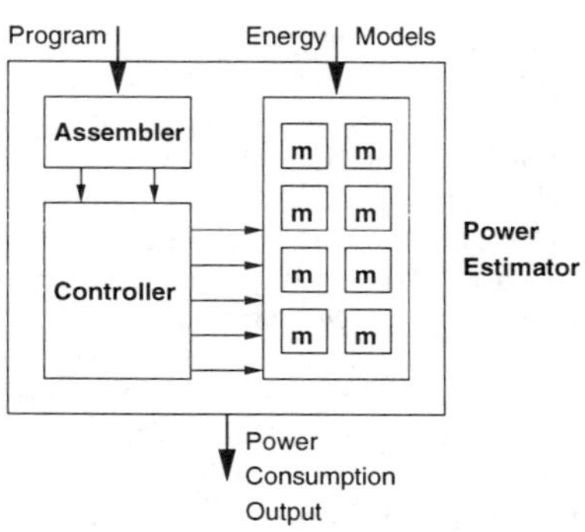

Figure 2: An overview of the power estimator

The first part of the simulation effort was to build an assembler which generates machine codes from benchmark programs written in the assembly language of the processor. The second part is the controller generating control signals for each functional unit when instructions are running on the simulator. These signals will be used to control the behavior of the functional units. For example, when a load instruction puts data from the memory to the IDB bus, the "MEMtoIDB" control bit will be set. The third part of the simulator specifies each functional unit in the datapath. As the control signals are set, the functional units are activated. For each unit, there is a function routine which gathers all event activities. Figure 3 shows an example of these specification functions. In this example, when a control signal is set, the IDB will receive data from one of its sources. This input data to the functional unit will be used to determine its power consumption.

```
toIDB()
{
  if ctr_bit[MEMtoIDB] is set
     read data from memory;
     put the data on IDB;
  if ctr_bit[MWBtoIDB] is set
     MWB buffer puts data on IDB;
  if ctr_bit[BIFtoIDB] is set
     BIF (Bus Interface) puts data on IDB;
}
```

Figure 3: Event activities in the IDB

Finally, the simulator will provide the precise behavior and functional unit state information. The behavior information describes which instruction is executed in each pipeline stage. The functional unit state information shows the content of each register or buffer, the data transferred on each active bus, and the function to be performed by each arithmetic unit (ALU, PAU, AU, MAC, etc.).

During the design of the simulator, the implementation of the pipeline was complicated because of the processor's unusual architecture. And it is necessary to use two program counters and special PC increment units, because the instruction stream for the processor is a mixture of 16-bit and 32-bit instructions. The simulator also implemented instruction loop buffers in the CPU engine.

3 Power Estimation of the Processor

Because dynamic power consumption is strongly dependent on the input switching of the functional units, using only the average power definitely causes a loss of accuracy. Our analysis technique overcomes this loss of accuracy by taking data and instruction streams into account.

The processor is separated into many functional units. The energy consumed by each functional unit per access can be computed as follows [1]

$$E_m = \frac{1}{2} C_m Vdd^2 \qquad (1)$$

where C_m is the switch capacitance per access of the functional unit and Vdd is the supply voltage. For each active functional unit, C_m is calculated from the energy model of the unit based on the previous and present input vectors. The functional units can be grouped into two classes: *bit-dependent functional units* and *bit-independent functional units*. They have different energy characterization methods. Here, the terms energy and power are used interchangeably.

In the bit-dependent functional units, the switching of one bit affects other bit slice's operations. Typical bit-dependent functional units are adders, multipliers, decoders, multiplexers, etc. Their energy characterization is based on a lookup table consisting of a full energy transition matrix where the row address is the previous input vector, the column address is the present input vector, and the matrix value is the switch capacitance. All combinations of previous and present input vectors are contained in one table. A major problem is that the size of this table grows exponentially in the size of the inputs. A clustering algorithm solves this problem and the sub-problems associated with it by compressing similar energy patterns. The details of this algorithm can be found in [2]. The capacitance data in the energy characterization table is obtained from a switch level simulation of the functional unit. The accuracy of switch level analysis is not as good as circuit level simulation tools, but is much faster.

In the bit-independent functional units, the switching of one bit does not affect other bit slice's operations; for example, registers, logic operations in the ALU, memories, etc. The total energy consumption of the functional unit can be calculated by summing the energy dissipation of the individual bits. To model the energy dissipation of a bus, load capacitance of each bit is used as an independent bit characterization. Furthermore, each bit is assumed to have the same load capacitance. The product of the load capacitance and the number of bits is the C_m in equation (1).

Some energy models are not built from the complete functional unit but from smaller subcells of these units. For example, because a 4:1 multiplexer can be made from three identical 2:1 multiplexers, its energy dissipation can be calculated by using the energy model of a 2:1 multiplexer. This reduces $2^{6^2} = 4096$ table entries to $2^{3^2} = 64$ entries. After building the subcell energy model, energy routines will implement those operations needed to calculate the functional unit power dissipation from subcells.

Because we did not have access to the design of the control unit of the processor, average data is used to characterize its power consumption. Although the accuracy is not as good as if we were able to build the transition dependent energy model, this average data reduces simulation time since there is no table lookup involved.

4 Validation Results

4.1 The Power Benchmarks

A set of simple synthetic programs listed in Table 1 are used as the benchmarks to validate our power analyzer. These are the same benchmarks used to test the processor in [6]. Power consumption of the instruction loop buffer (ILB), memories, buses, ALU, the multiplier and other functional units are collected as the benchmarks are run. Data values used were those which maximized the switching in the datapath.

Program	*Use ILB*	*Main Loop Operation*
pow032	No	padd pmuls movx+ movy+
pow033	Yes	padd pmuls movx+ movy+
pow035a	Yes	padd pmuls movx movy
pow034	No	padd pmuls
pow035	Yes	padd pmuls
pow035b	Yes	padd
pow035d	No	nop
pow035c	Yes	nop

Table 1: The Power Benchmark

In Table 1, "padd pmuls movx+ movy+" contains four separate operations executed simultaneously. They are an addition, a multiplication, two loads and two address increments. "Padd pmuls movx movy" is the same as the previous instruction except it contains no address increments.

4.2 Validation Results

The power simulation results generated by our architectural level power analyzer are compared with the data gathered by [6]. Overall, the average error rate of power analysis by our simulator is 8.98%.

Figure 4 compares the estimated (by our simulator) and reported (by [6]) power dissipation of the processor core including the CPU engine and the DSP engine. The power consumption data of each program is normalized to that of the *pow034* benchmark. As you can see, for most of the benchmarks our simulator produces results very close to those reported data in [6].

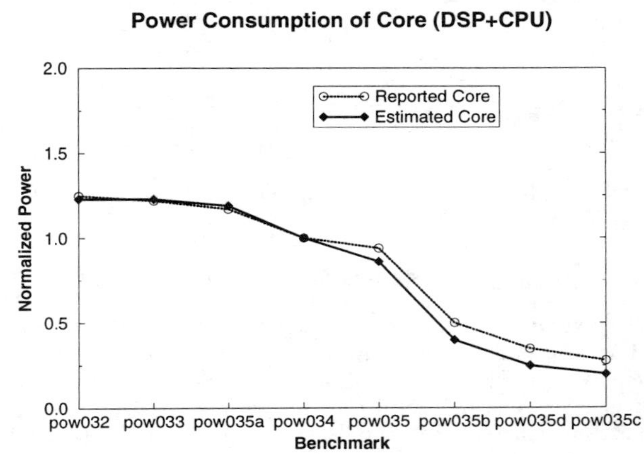

Figure 4: Core (DSP+CPU) estimated and reported power

In both the case of the *pow035d* benchmark and the *pow035c* benchmark, the power consumption is underestimated by our simulator. Possible reasons are that our power estimator has not considered clock power and that the power consumed by the control unit has not been accurately estimated. These two benchmarks use the least power overall, so that clock and control unit power account for a higher percentage of the total power consumption.

In order to verify the above suspicion, the simulation accuracy of each major functional unit (the DSP engine, the CPU engine, the memory and the top-level buses) was also analyzed and is shown in Figure 5. The estimated power distribution rate is compared to that reported by [6]. The distance from a symbol to the 1:1 line indicates the accuracy. The power simulation of the top-level buses is the least accurate of the four units - the average error rate is 10.2%. One of the reason for this is that the "top level interconnect" in [6] consists not only of the buses we considered but also the control and clock lines. The random control logic is also a large part of the total power consumption, but we were not able to build the energy model for the control unit since we did not have access to the design. Thus, the accuracy of the CPU engine power estimation is not as good as that of the DSP engine and memory, though the average error rate is only 5.7%.

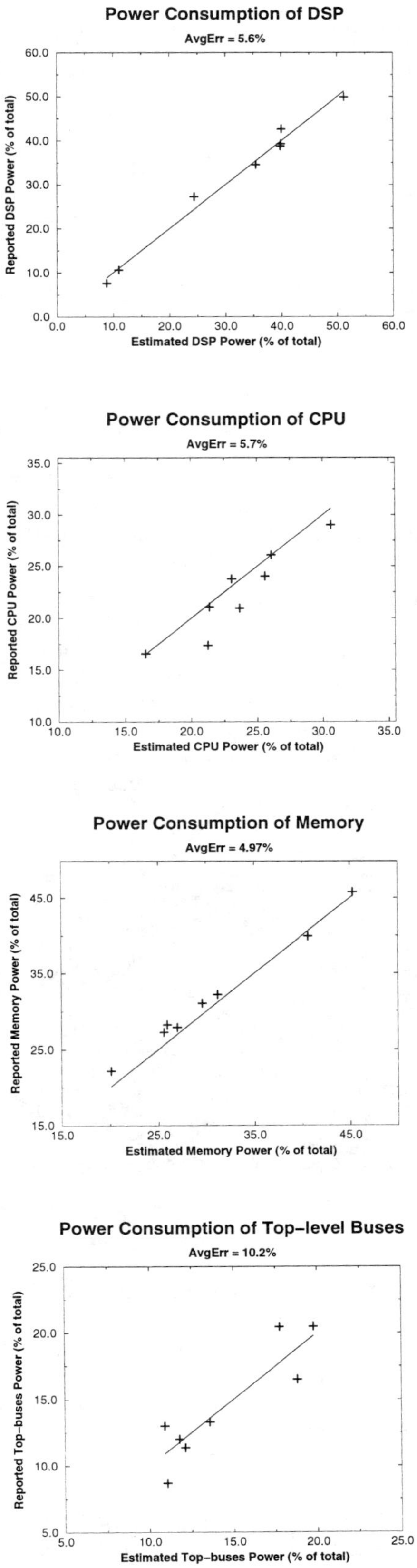

Figure 5: Power distribution comparison

5 Conclusion and Future Work

In this work, we implemented an accurate, high level power estimator for a real processor. Our estimation technique needs to do energy characterization for each functional unit (ALU, MAC, etc.) only once. When the operations of the functional units are specified by the instruction stream, the power consumed in each functional unit is calculated from its energy model. The simulation results clearly verified the correctness of our power analysis methodology. Without loss of accuracy, the running time of power estimation for each benchmark program was less than 90 CPU seconds. This time saving feature of our power estimator is beneficial for reducing the design cycle time. The structural simulator is also applicable for power optimization research at the architecture, system software, and application software levels.

The following are some of the factors which have not been taken into account during the power estimation:

- Clock power
- Transition dependent control logic power
- I/O pad power
- Varying signal arrival times
- The effects of functional unit placement and routing

Finding a better interconnect energy model also is a valuable research goal. Our future work to improve our power estimator will include all of the above. We are also investigating, running and testing more complicated benchmarks.

References

[1] H. Mehta, R.M. Owens, and M.J. Irwin. Instruction level power profiling. In *International Conference on Acoustics, Speech and Signal Processing* (1996).

[2] H. Mehta, R.M. Owens, and M.J. Irwin. Module energy characterization using clustering. In *Proceedings of 33rd Design Automation Conference* (June 1996).

[3] P. Landman, and J. Rabaey. Power estimation for high level synthesis. In *EDAC-EUROASIC* (1993), pp. 361–366.

[4] P. Landman, and J. Rabaey. Black-box capacitance models for architectural power analysis. In *1994 International Symposium on Low Power Electronics and Design* (April 1994), pp. 165–170.

[5] P. Landman, and J. Rabaey. Activity-sensitive architectural power analysis for the control path. In *1995 International Symposium on Low Power Electronics and Design* (April 1995), pp. 93–98.

[6] R. Bajwa, N. Schumann, and H. Kojima. Power analysis of a 32-bit RISC microcontroller integrated with a 16-bit DSP. In *1997 International Symposium on Low Power Electronics and Design* (1997), pp. 137–142.

[7] S.R. Powell, and P.M. Chau. Estimating power dissipation of VLSI signal processing chips: The PFA technique. In *VLSI Signal Processing IV* (1990), pp. 250–259.

Design Methodology of a 200MHz superscalar microprocessor: SH-4

Toshihiro Hattori, Yusuke Nitta, Mitsuho Seki, Susumu Narita,
Kunio Uchiyama, Tsuyoshi Takahashi, Ryuichi Satomura

Hitachi, Ltd.
1-280 Higashi-Koigakubo, Kokubunji,
Tokyo 185-8601, JAPAN
+81-423-23-1111

hattori@crl.hitachi.co.jp

1. ABSTRACT

A new design methodology focusing on high speed operation and short design time is described for the SH-4 200MHz superscalar microprocessor. Random test generation, logic emulation, and formal verification are applied to logic verification for shortening design time. Delay budgeting, forward / back annotation, and clock design are key features for timing driven design.

1.1 Keywords

Microprocessor, design methodology, verification, timing

2. INTRODUCTION

Recently, LSI design is becoming more difficult and complicated, not only because of the increase in the number of transistors in LSI's and new design constraints required by deep submicron processes, but also because the designer must take into account many factors. Some targets of LSI design are listed below.

(1) High speed operation

(2) Minimizing chip fabrication cost (minimizing area)

(3) Low power dissipation (for portable use)

(4) Minimizing the design time (to shorten time to market)

(5) High reliability

Priority of these targets is different for each LSI design. It depends on the application and market ,etc..

To satisfy these difficult targets, the design methodology becomes more important for an LSI design project. The design methodology of each LSI design must be focused on the targets of the LSI. For example, in case of designing LSI's for portable equipment, a design methodology focusing on low power dissipation must be applied.

Hitachi has been releasing its original RISC architecture SuperH series microprocessors. And we have been developing the new SuperH microprocessor SH-4 which focuses on multi-media equipment for home use. The significant targets of SH-4 are (1) high speed operation and (4) short design time. Of course, other targets are also important for SH-4. This paper will present the design methodology of SH-4, especially focusing on 200MHz operation and short design time.

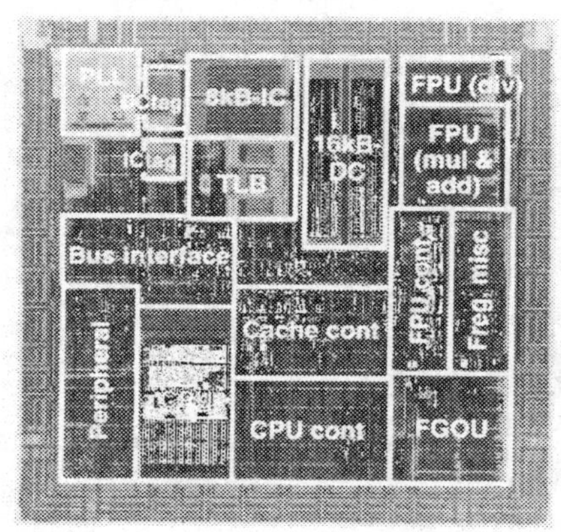

Fig. 1 Microphotograph of SH-4

3. Overview of the SH-4 microprocessor

SH-4 is a 200MHz CMOS 2-issue superscalar microprocessor. Fig. 1 shows a microphotograph of the chip. In this chip 3.2 M transistors are implemented in a die area of 7.6 mm x 7.6 mm. This chip is designed for low-cost graphic and embedded applications. Fig. 2 shows the specifications of SH-4, which runs at 360MIPS(Dhrystone) and 1.4GFLOPS(peak) performance.

SH-4 implements new inner product instructions and it makes possible excellent 3 dimensional computer graphics performance at a consumer price. According to a simple benchmark, SH-4's 3D CG geometry performance is 5.0M polygon/sec.

The process used is 0.25 μm CMOS, with Leff= 0.20 μm, and 5 metal layers. The metal pitch is 0.98 μm for layers 1, 2, and 3, and 1.96 mm for layers 4, and 5. The fourth and fifth layers are used for long (> 4 mm) wires and Vcc, GND because of their low resistivity.

Technology	0.25 μm CMOS, 5 Layer Metals
Voltage	1.8 V (I/O: 3.3 V)
Frequency	200 MHz (internal) / 100,66, etc. MHz (I/O)
Performance	360MIPS (Dhrystone), 1.4 GFLOPS (peak)
Cache	8/16 KB (Inst./Data) Direct-mapped
TLB	4/64-entry (Inst./Unified) Fully-associative
Interfaces	SRAM, DRAM, SDRAM, burst ROM, PCMCIA
Peripherals	DMAC, SCI, RTC, TMU,INTC,EMU

Fig. 2 SH4 Specifications

4. Overview of SH-4 Design Methodology

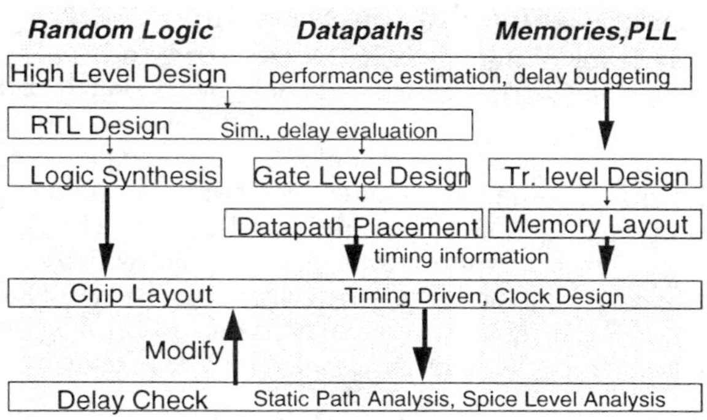

Fig. 3 Design Flow of SH-4

Fig. 3 shows the rough design flow of SH-4. SH-4 architecture simulator SH-4fun was implemented in the C-language. This simulator is used for performance estimation and generating expected behavior for test programs. SH-4 is mainly described in register transfer level Verilog language. First and main logic verification is done at this RTL description. A logic synthesis tool generates the gate level Verilog description. Timing verification is mainly based on the static path analysis of every path between FF's. Some analog circuits, for example memory array and PLL, are verified by a circuit simulator. Memory megacells are designed using a circuit simulator and manual layout tool as same as conventional LSI's. In the SH-4 design, only datapaths in FPU and CPU are performed as structured elements. Other datapaths are layouted as random logic, because timing constraints for these datapaths are not so severe and to layout together with other random logic can reduce the chip area. But, because the datapaths in FPU and CPU are critical for chip operation speed, these datapaths are semi-manually layouted. Their gate level logic is not synthesized from RTL description, but designed manually and modified by generating a new cell library. Semi-manual placement is applied for the purpose of data flow regularity. All random logic were placed at once preserving logic structure, and interblock and intra-block signals are routed at once, because of shortening design time and performing timing driven layout throughout the chip.

5. Logic Verification Methodology

Fig. 4 shows SH-4 logic verification flow. Because increasing transistor number and complicated architectures such as superscalar make logic verification difficult, a logic verification methodology is very important for shortening design time.

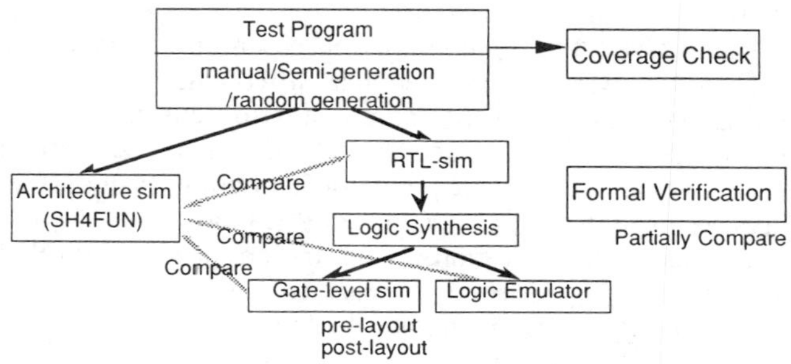

Fig. 4 Logic Verification Flow

5.1 Pseudo-board simulation model

Because preparing input 0/1 stimulus and expected output is very difficult for a large scale microprocessor, we built a pseudo-board in the simulation environment. The pseudo-board contains SH-4 and external memory, some of which data is programmed, and external circuits such as serial interface, and interrupt control description. In this environment, designers only make test programs in assembly language and the architecture simulator SH-4fun can generate expected results easily.

5.2 Random Test Generation

In logic simulation, quality of logic verification deeply depends on quality of the test programs. In our project, a test program design team separate from the logic design team is organized to make test programs in order to check logic functions. However, for the complicated microprocessor, it is impossible to make perfect test programs which can verify all functions of the LSI. Making many good test programs is an expensive and time consuming job. In addition to designer's well-considered test programs, we adapted semi-automatically generated test programs and random generated test programs. Semi-automatically generated test programs are generated by a script which varies the data value of test programs using a template program. The random test generator varies many factors of chip situation and the instructions themselves are randomly generated. But, by simple random test programs, it is hard to find logic bugs. We made many random test generators for specific purposes.

5.3 Logic Emulation

In order to verify a complicated microprocessor, the number of test programs and steps of each test program must be increased. Especially, because random generated test programs are easily made, the number of test sequences increases rapidly. However, to simulate the large scale microprocessor requires much CPU time. Fig. 5 shows the typical simulation speed using a compiled-type high-speed Verilog simulator on a high performance workstation. To overcome this situation, logic emulation is used. We use SimExpress[1] as the logic emulator. Conventional logic emulators have a problem in that the building the netlist inside the logic emulator is very difficult. Not only does netlist compilation and implementation into FPGA take 2days using 10 workstations, but also to fixing timing / skew trouble inside the logic emulator requires a long time. In some cases, the logic emulation environment using the final netlist was set up after first silicon had come out. SimExpress can build the logic emulation environment in 3~5 hours using one workstation from the final fixed netlist. Thus, we can use the logic emulator for not only system level final verification, but also for logic verification before tape out. In case of SH-4 verification, the emulation speed was limited to 100K cycle / second because we used two systems connecting by wires to emulate the large scale whole SH-4. But, practically emulation speed is limited by data transfer time from / to the emulator. Both a test program which requires 3 hours and another test program which requires 1 hour in logic simulation take only 20 seconds in logic emulation.

System	Compile Time	Execution Speed
Gate-level simulation	5Hr.	0.9 cycle/sec
RTL simulation	3Hr.	15 cycle/sec
Gate-level emulation	5Hr.	85K cycle/sec

Fig. 5 Logic Simulation/Emulation Performance

5.4 Formal Verification

Logic verification by simulation / emulation cannot be perfect verification because it is performed under limited test programs. If we must verify the equivalence between two logic descriptions, the formal verification technique is useful. In our project, formal verification is used in the case below.

(i) Verification of equivalence between RTL datapath description and gate-level datapath description which is designed manually.

(ii) Verification of equivalence between before and after logic modification, for example, clock tree generation by layout tool.

(iii) Verification of equivalence before and after minor logic change for performance such as buffer insertion to avoid minimum delay violation.

[1] SimExpress is a product of Mentor Graphics Cooperation.

6. Timing Design

The most significant target of SH-4 design was the 200MHz high-speed operation. To achieve this target, we have taken care of the timing at every design phase. Fig. 6 shows the timing design flow of SH-4. We will describe some features below.

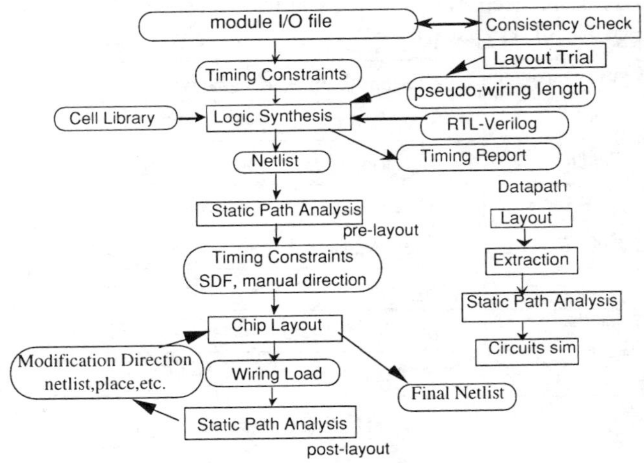

Fig. 6 Timing Design Flow

6.1 Delay Budget for each module

The key factor of timing design is the timing driven logic synthesis. But, logic synthesis can only be performed at the module level. In these cases, the most difficult is definition of timing constraints for inter-module delay. To perform good logic synthesis, accurate module i/o timing constraints are very important. In our project, the so-called module I/O file describes signal arrival time, target delay, and wiring load, which are checked for their consistency by the in house tool. This master file describes the delay budget of each module and makes chip level path delay satisfy the requirements.

6.2 Chip Level Static Path Analysis

In previous microprocessors, datapaths were layouted as structured elements. But this may allow an escape of a real critical path through a datapath from static path analysis. In our project, datapaths are layouted as white boxes for chip level static path analysis; all paths including CPU/FPU datapath can be listed up in the timing report.

6.3 Back Annotation & Forward Annotation

The timing-driven logic synthesis system requires estimated wiring load for the synthesizing net and wire load for the terminals of the synthesizing module. These constraints affect the synthesized netlist. We performed layout trials using preliminary netlists and evaluated back annotated wiring lengths for inter-module nets and estimate average wiring length for internal nets of the module. Results of the chip level static path analysis were forward annotated to the layout tool as timing constraints. Also the final results of net capacitance was back annotated to static path analysis and some modifications were added to the layout.

7. Clock Distribution Design

SH-4 has three clocks, internal (processor) clock (ICLK), bus-clock (BCLK), and peripheral-clock (PCLK). The frequency ratio of ICLK/BCLK and BCLK/PCLK are programmable; for example ICLK/BCLK can be 2,3,4,6, and 8. The reason why we use this three clock scheme is that the reduced bus- and peripheral-clock frequencies saves the maximum clock distribution power by 23% compared with a 200MHz single clock design. In order to perform at high speed, we must also reduce the clock skews between these different frequency clocks.

In order to realize this complicated clock distribution logic, logic design and physical design of clocks are performed at the layout design phase. At the logic design netlist, the clock net is expressed with only one gated clock buffer, which has imaginary large drivability, and drives hundreds of FF's in a logic module.

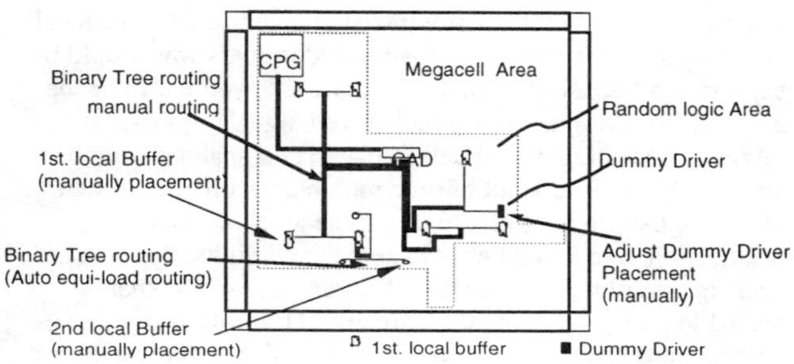

Fig. 8 Chip Level Clock Distribution

As we mentioned previously, minimizing clock skew and minimizing path delay must be pursued in order to design high speed LSI's. However, sometimes these two targets require different solutions. For example, in order to minimize clock skews, a model in which FF's are placed uniformly in the chip is adequate. But in order to minimize path delay, FF's which are at edges of critical paths must be placed near by. To minimize clock skew and minimize path delay simply, we separate the design phase and at each design phase only one target is focused upon.

(1)Perform top level clock distribution logic carefully, and pre-place chip level buffers.

(2)Place random cells only considering timing driven.

(3) Synthesize the clock tree and place clock buffers according to the placement of FF's.

(4)Manually modify the placement of clock buffers because clock tree synthesis and an automatic clock buffer placement create poor results near megacells.

(5)Route top level clock tree(from CPG to 1st local buffers) nets carefully by hand before signal routing because clock nets must avoid detours. Use Metal 4 and Metal 5 for some top level clock nets which drive large loads because of their low resistivity.

(6)Apply automatic zero skew routing for the clock nets from 1st local buffer to 2nd local buffers.

(7) Execute signal routing after clock routing. Direct signal nets not to run near some critical clock nets.

(8) Perform delay/skew analysis with wiring load because a skew value is very sensitive to adjacent wiring capacitance.

(9)Do final modification by adding extra segments of the net to increase wiring load and so on.

Using these careful design, we achieve less than 200pS clock skew for multi-phase FF's including FF's inside megacells according to the static path calculation. However, taking account of process scattering and power supply drop inside the chip ,etc., the worst case skew estimation for the racing problem is larger than 200pS. Measured clock skew from the sampling points in a fabricated chip is less than 100pS.

8. Conclusion

A design methodology for SH-4, which is a high performance superscalar microprocessor for home multi-media equipment, has been described. Pseudo-board simulation model, random test generation, logic emulation, and formal verification are applied in order to shorten design time. Combination of random test generation and logic emulation can verify the netlist very quickly. Formal verification for partial circuits can eliminate re-simulation. RTL delay estimation and delay budget for each synthesizing module help high quality timing driven logic synthesis. To achieve timing driven design, static timing analysis for whole chip and circuit simulation for detailed circuits are used to evaluate the timing. Back annotation from preliminary layout and forward annotation from logic level static path analysis are key features for reducing machine cycle of the chip. Clock distribution logic and layout is designed carefully by automatic tool and manual modification. Clock skews for multi-phase clock is designed to be less than 200pS and measured clock skew from sampling points in a fabricated chip is less than 100pS.

9. References:

[1] Arakawa, F., et al., "SH4 RISC Microprocessor for Multimedia," Hot CHIPS IX, Aug., 1997.

[2] Ishibashi, K., et al., "The Design of 300MIPS Microprocessor with a Full Associative TLB for Hand-held PC OS," Symp. VLSI Circ. Digest of Technical Papers, pp. 9-10, June, 1997.

[3] Nishii, O. et al., "A 200MHz 1.2W 1.4GFLOPS Microprocessor with Graphic Operation Unit, " Int. Solid-State Circ. Conf., Feb.,1998

[4] Osada, K., et al., "A 2ns Access, 285MHz, Two-Port Cache Macro using Double Global Bit-Line Pairs," ISSCC Digest of Technical Papers, pp. 402-403, 494, Feb., 1997.

PANEL: HOW MUCH ANALOG DOES A DESIGNER NEED TO KNOW FOR SUCCESSFUL MIXED-SIGNAL DESIGN?

Organizer: Georgia Marszalek – ValleyPR, Foster City, CA

Chair: Stephan Ohr - Electronic Engineering Times, Westfield, NJ

Inevitably, what designers call "systems-on-a-chip" technology will need to include basic analog functions. These include analog signal conditioning circuits (amplifiers and filters), go/no-go trigger circuits (analog comparators) and power management devices. There are a number of techniques for implementing these circuits in a largely-CMOS digital system: The designer can use special processes (like BiCMOS or BCD) to implement precision analog circuits on a CMOS substrate. Alternately, the designer can dramatically re-cast the analog function so that it can be implemented with digital CMOS processes. But how can a digital designer make choices between these when he knows so little about the basic functions these circuits perform? This panel addresses the system integration issue from a "what-do-you-need-to-know?" perspective. It includes analog industry veterans who decry the dwindling of analog circuit design skills, and EDA tools experts who envision digital designers rendering analog designers with high-level language statements and behavioral models. The panel also includes outspoken mixed-signal designers, conversant in both analog design skills and EDA tools, who are not hesitant to point out the ideological underpinnings of both schools of thought.

Felicia James, Texas Instruments, Dallas, TX
With the continuing trend to higher levels of integration, the issues with interactions on one piece of silicon are essentially analog problems. At smaller geometries more traditional analog functions can move to digital implementations but many analog challenges remain. Smaller geometries and lower supply voltages force new analog solutions. A successful mixed-signal designer not only must have an understanding of digital design but clearly needs a strong foundation in analog. The same skills, which have historically identified an experienced analog design engineer, continue to be critical for success in today's mixed-signal designs.

Ken Kundert, Cadence Design Systems, San Jose, CA
There is a continuum of mixed-signal design — from analog circuits with a touch of digital to huge digital chips with a bit of analog stuck in the corner. Interesting, the most difficult challenges tend to occur in the middle where there is a large amount of both analog and digital circuitry that interacts in a complex manner. A key to improving the effectiveness and productivity of designers of all types of mixed-signal circuits is a MS-HDL such as Verilog-AMS. When designing largely analog circuits, the MS-HDL allows the digital circuitry to be included easily and simulated efficiently. When designing complex mixed-signal circuits, a MS-HDL supports a top-down design style and allows verification of large systems using mixed-level simulation. Finally, a MS-HDL allows predominantly digital systems to verified by including accurate and efficient models of the analog portion into the digital simulation.

Lavi Lev, Silicon Graphics, Mountain View, CA
Here is scary thought: If digital chip designers had to worry about analog, they'd need to actually know how transistors work. From a CPU and System-on-a-Chip point of view, analog parts will be implemented as cell modules, to be designed by the few that can still read a non linear equation and know what it actually mean. On the reliability side, analog circuits are more sensitive and therefore a cell-like approach will probably be favored by design managers who want to cap their level of anxiety to where it is today. Given this design methodology, a design team will have few analog experts that will design those cells, and all the rest will have to know how to interface with those engineers (and the cells they design).

Maq Mannan, IEEE 1364 chairman, DSM Technologies, San Jose, CA
In order to design high quality mixed-signal products, the ideal world would have true "mixed-signal" designers who would be experts in all nuances of analog designs and would also be top notch digital designers. But true "mixed-signal" designers are rare, and educating digital design experts on analog would be very costly in the amount of time required. In order to design mixed-signal products effectively some level of cross-fertilization is required between analog and digital designers. The big question is how much? With the technology that is available today, can this dependence and cross-fertilization be reduced in time but expanded in content? That would be the key to producing high quality mixed-signal products.

Rob Rutenbar, Carnegie Mellon University, Pittsburgh, PA
Judging only by the ferocity with which semiconductor companies today recruit circuit-savvy students, anybody who has good circuit design skills is immensely valuable. (And, don't believe for a minute that "digital" and "analog" circuits are vastly different at GHz speeds.) The real problem is that large digital systems are increasingly specified in ways vastly different than analog. They use executable high-level specifications, rely on synthesis, and struggle to reuse prior designs (e.g., datapaths) or generate them on the fly Synthesis, cell libraries, semicustom layout, etc., all insulate the digital side from some (not all) low-level circuit details. This is not true in analog, where most transistors are still lovingly handcrafted. The realistic solution is not to turn all digital system designers into analog circuit experts. The solution is analog specification, synthesis and layout tools that make analog design more abstract-able, less like writing assembly language.

Bob Dobkin, Linear Technology, Milpitas, CA
Mixed mode simulation techniques have made great strides in the ability to simulate a full system. The combination of behavioral models, macro models combined with low level simulation yield excellent results in many cases. But simulation techniques still fail to provide detailed results in acceptable times when there is a combination of many transistors and a long loop response. Data converters, for example, take too long to simulator at the transistor level. Switching regulators (or chopped-stabilized amplifier) with a large difference in the time constant between the overall loop and the switching frequency are extremely difficult to simulate — and would certainly tax a digital designers skills.

Hierarchical Algorithms for Assessing Probabilistic Constraints on System Performance *

G. de Veciana, M. Jacome and J.-H. Guo
Department of Electrical and Computer Engineering
University of Texas at Austin, Austin, Texas 78712
Tel: (512) 471-2051 Fax: (512) 471-5532
E-mail: {gustavo,jacome,jianhuei}@ece.utexas.edu

Abstract

We propose an algorithm for assessing probabilistic performance constraints for systems including components with uncertain delays. We make a case for designing systems based on a probabilistic relaxation of performance constraints, as this has the potential for resulting in lower silicon area and/or power consumption. We consider a concrete example, an MPEG decoder, for which we discuss modeling and assessment of probabilistic throughput constraints.

1 Introduction

This paper discusses models and algorithms to support statistical relaxations of worst case constraints on system performance. Consider, for example, a system which is designed to meet a delay constraint d and suppose the critical path's delay D^p is in fact random. A design based on a *worst case analysis* would ensure that $\mathsf{P}(D^p > d) = 0$. In our view, for a number of application domains, such designs may be unnecessarily conservative. For example, suppose the design constraint d can be relaxed in the sense that it can be violated but only rarely, say $\mathsf{P}(D^p \geq d) \leq 10^{-6}$. Such a relaxation of design constraints will in turn allow for a larger set of acceptable design solutions with hopefully less demanding performance requirements and/or power consumption. Note that even when performance constraints are truly worst case, in the sense that the system malfunctions if they are not met, it is reasonable, and possibly beneficial, to still relax the performance constraints – say, to the same level of certainty as the probability of failure of the system's components. The examples in §4 suggest that one might expect to benefit significantly from a probabilistic relaxation of worst case constraints for systems comprising a large number of non-deterministic components.

Uncertainty in the performance of a system's components may have a variety of origins. For example, for high-level system representations, such as those used in system level and hardware/software codesign, uncertainty may be due to a looping index whose exact initial value is unknown, *e.g.*, data dependent [1, 3, 6, 11]. A number of current system-level models use hierarchy and aggregation as a means of controlling complexity [3, 11]. If such approaches are to succeed, and one is to reason efficiently about the required performance of such systems during design space exploration, it will become increasingly critical to capture the performance variability of aggregated system elements.

We propose to use probability distributions to model uncertainty. The distributions may either be derived from statistical models for the underlying source of variability, estimated based on experimental data, or gathered through simulation/profiling. Realizing that characterizing distributions is in itself a challenging and expensive task, in §3.1 we propose a crude model based on knowing the mean and upper and lower bounds on delay. This simple characterization of non-determinism is shown to be conservative for assessing system performance and eliminates (in some cases) the need for obtaining detailed statistical information on component delays.

In §2 we formulate a *probabilistic critical path* problem, and propose an approximate algorithm for assessing probabilistic constraints on systems represented by directed acyclic graphs (DAGs) with random edge weights. Therein we discuss related work. In §3 we discuss modeling distributions as well as a set of transformations (reductions) of typical high-level system models to obtain DAGs which are in turn amenable to analysis with the proposed algorithm. Synthetic examples exhibiting significant differences between worst case and probabilistic requirements are presented in §4. In §5 we discuss a concrete example, showing how our approach might be applied to modeling and assessing throughput constraints for an MPEG decoder. We conclude with a discussion of research/implementation directions we are currently pursuing.

2 Algorithm for Assessing Probabilistic Constraints

Consider a weighted directed acyclic graph $G(\mathcal{V}, \mathcal{E})$. Suppose a source node $s \in \mathcal{V}$ is selected and let $\mathcal{P}_s$ denote the set of paths starting at s. Naturally, a path p is an ordered set of adjacent edges in the graph, *i.e.*, $p \subset \mathcal{E}$. The standard *critical path* problem assumes that edges have fixed weights, and determines the longest path, in terms of cumulative weight, in $\mathcal{P}_s$. Algorithms to solve this problem are well known and have a runtime complexity of $\Theta(|\mathcal{V}| + |\mathcal{E}|)$. Suppose random weights (delays) D_e, with arbitrary distributions, are associated with edges $e \in E$ of the graph. We pose an analogous *probabilistic critical path* problem as follows: given a delay

*This work was supported by NSF Career Grants NCR-9624230 and MIP-9624321 and a Texas Higher Education Coordinating Board Grant ATP-088.

constraint d, identify which path is *most likely* to violate the constraint and what is its probability of failure.

Assuming edge delays are mutually independent, a path's delay, denoted by $D^p = \sum_{e \in p} D_e$, has a distribution given by the convolution of its edge's delay distributions. The probability that a path p fails to meet the delay constraint d is denoted by

$$\pi_d(p) = \mathsf{P}(D^p \geq d).$$

We state the probabilistic critical path problem as follows:

Problem 1 *Find $p^* \in \mathcal{P}_s$, not necessarily unique, such that $\pi_d(p^*) \geq \pi_d(p)$ for all $p \in \mathcal{P}_s$, and determine the probability $\pi_d(p^*)$ that the constraint will be violated.*

In general this problem is difficult to solve, primarily because the edge weights of a path are not additive (due to the convolution) and thus cannot be decomposed along the path as is usual when using dynamic programming approaches. In fact, by adapting a Lemma 2 in [5], one can show that Problem 1 is NP-hard, suggesting that one should seek good heuristics.

A word of warning is in order. There are two reasons why the probabilistic critical path problem should not be interpreted as the equivalent of the standard critical path problem when the weights are random. First, the problem is predicated on specifying a constraint d with respect to which a probabilistic critical path is identified. Second, and more subtly, we compare the performance of *individual* paths with each other, rather than assessing the maximum of the delays across all paths. Whereas in the case with deterministic weights these two problems are equivalent, in a graph with random weights they certainly are not. The discussions on modeling in §3 and examples in §4 further elucidate this point.

2.1 Previous Work

To our knowledge there is no previous work addressing the above problem. Our work was inspired from recent work in network routing considering the uncertainty in the delay or bandwidth availability at remote links [5]. The flavor of the approach, in particular our use of the Chernoff bound to perform constraint analysis for a simple example, can be found in [6][page 112]. However, both our formulation of the probabilistic critical path problem as well as the proposed systematic algorithm are new. We also note that there are efficient algorithms for determining the most reliable path through a network with unreliable links, a problem which arises in voice recognition and Viterbi decoding applications. However such problems are significantly easier (can be reduced to a the traditional shortest paths problems) than the problem we address here.

2.2 Approximate Algorithm

We propose an algorithm to solve the probabilistic critical path problem based on an approximate formulation as a convex optimization problem. The edge weight distributions D_e on the DAG are represented via a parametric weight $\Lambda_e(\theta) = \log \mathsf{E} \exp[\theta D_e]$ for $\theta \geq 0$, *i.e.*, the moment generating function of the delay distribution on the edge. This results in a collection of DAGs with deterministic weights $\Lambda_e(\theta)$ parameterized by θ. Our algorithm solves an optimization problem over this set of parameterized DAGs by using the standard critical path algorithm. The derivation of the algorithm can be found in [2]

Initialization: check that the problem is "well posed" by verifying that:

1. the constraint d exceeds the critical path delay for the graph where the weights are given by the mean edge delays;
2. and, the constraint d is bounded by the critical path delay for the graph with weights given by the maximum (possibly infinite) delay on each edge.

Optimization: determine the maximum $f^*(d)$ and the optimizers $\hat{\theta}$ and $\hat{p}$ for the following optimization problem

$$f^*(d) = \sup_{\theta \geq 0}(\theta d - \max_{p \in \mathcal{P}_s} \sum_{e \in p} \Lambda_e(\theta)) = \hat{\theta} d - \sum_{e \in \hat{p}} \Lambda_e(\hat{\theta}). \quad (1)$$

Note that evaluating $\max_{p \in \mathcal{P}_s} \sum_{e \in p} \Lambda_e(\theta)$ for some θ requires determining the critical path in a graph with edge weights $\Lambda_e(\theta)$.

Result: a *guaranteed* upper bound on the probability of failure, $\pi_d(p^*) \leq \exp[-f^*(d)]$, and a candidate path $\hat{p}$ most likely to violate the constraint.

Standard line search methods, see *e.g.*, [9], can be used to determine the supremum, which will be achieved in the cases of interest, in (1). However some care should taken in selecting an efficient algorithm since each evaluation of $f(\theta) = \max_{p \in P_s} \Lambda^p(\theta)$ requires solving a standard critical path problem incurring a runtime cost $\Theta(|\mathcal{V}| + |\mathcal{E}|)$. Note that, whatever optimization algorithm is selected, any stopping criterion will yield an *upper* bound on the failure probability. Given a solution one can use refined asymptotics to improve the estimate and explore the *sensitivity* of the probability of failure to the constraint d, see [2].

3 Modeling Issues

3.1 Modeling edge delay distributions

In practice it may be difficult to characterize the edge delay distributions, or equivalently the corresponding functions $\Lambda_e(\theta) = \log \mathsf{E} \exp[\theta D_e]$. Fact 3.1 below shows that one can find a simple uniform bound on $\Lambda_e(\theta) \leq \bar{\Lambda}_e(\theta)$, given a *minimal amount* of information on the distribution of D_e. This in turn allows us to replace the weights on such edges by an upper bound, resulting in a conservative estimate for the probability of failure.

Fact 3.1 (See *e.g.*, **[10])** *Suppose that bounds, $l_e \leq D_e \leq u_e$, are known for the edge (or path) delay, as well as an upper bound m_e on the average delay $\mathsf{E}D_e \leq m_e \leq u_e$. Let $\bar{D}_e$ be a Bernoulli random variable on $\{l_e, u_e\}$ with mean m_e, i.e., $\mathsf{P}(\bar{D}_e = u_e) = \frac{m_e - l_e}{u_e - le}$, and $\mathsf{P}(\bar{D}_e = l_e) = 1 - \frac{m_e - l_e}{u_e - l_e}$, then $\forall \theta$ we have that $\Lambda_e(\theta) \leq \bar{\Lambda}_e(\theta) = \log \mathsf{E} \exp[\theta \bar{D}_e]$.*

3.2 Hierarchical representations and reductions to directed acyclic graphs

In general a hierarchical high-level system models are respresented by control/data flow graphs (CDFGs) [3]. Such graphs allow the represention of flow of control constructs, including branching, looping, and possible synchronization requirements, see *e.g.*, [3, 1, 11]. Rather than formally defining such a modeling framework we will exhibit some cases that arise and the manner in which they are reduced to a corresponding DAG. Below we show how the delay weights

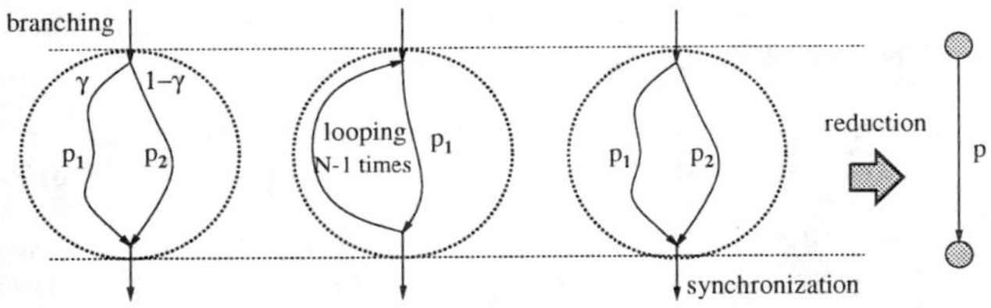

Figure 1: Probabilistic branching, looping, and synchronization reductions.

associated with traversing the nodes in Fig. 1 can be reduced to *equivalent path weights*. Generalizations of these cases to more than two non-intersecting (independent) sub-paths should be clear from the discussions below.

3.2.1 Reducing nodes with probabilistic branches.

Consider the left node in Fig. 1. Within the node, a branch is modeled probabilistically, in the sense that one of the two sub-paths, p_1 or p_2, is selected at random. Suppose the branching probability is γ then the weight for the sub-path p through this node is given by $\Lambda^p(\theta) = \gamma\Lambda^{p_1}(\theta) + (1-\gamma)\Lambda^{p_2}(\theta)$. Note that if a branch is not modeled probabilistically (due to lack of information) then both paths would be kept in the eventual DAG.

3.2.2 Reducing nodes with iterations or feedback loops.

Consider the middle node in Fig. 1. It represents a node in which there is an uncertain number of iterations through path p_1, which can be modeled via a loop index random variable, say $N > 0$. Let $D(n)$, denote the delay for the nth loop execution. Suppose these delays have a common distribution, $D(n) \sim D^{p_1}$, and are mutually independent and independent of the loop index N.

The delay D^p for a sub-path through this node is $D^p = \sum_{n=1}^{N} D(n)$, *i.e.*, a random sum of random variables. The weight for the node can be computed by conditioning on N to obtain $\Lambda^p(\theta) = \log \mathsf{E}[\exp \theta D^p] = \log M_N(\mathsf{E}[\exp \theta D^{p_1}])$, where $M_N(z) = \mathsf{E}[z^N]$ is the probability generating function of the loop index's probability mass function.

For example, suppose the loop index, N, is modeled by a geometric distribution with parameter γ, *i.e.*, after completion of any iteration the probability of looping back is $1-\gamma$. In this case $M_N(z) = \frac{\gamma z}{1-z(1-\gamma)}$. Of course if the loop index is deterministic, *i.e.*, $\mathsf{P}(N = n) = 1$, then the corresponding DAG would unravel the loop, *i.e.*, the weight for a path through this node would be $\Lambda^p(\theta) = n \times \Lambda^{p_1}(\theta)$.

3.2.3 Reducing nodes with synchronization constraints.

In general synchronization is the most difficult abstraction to handle, particularly in a setup with random delays. Consider the rightmost example in Fig. 1, where two paths p_1 and p_2 must synchronize prior to leaving the node. The delay incurred in this node, D^p, is given by $D^p = \max[D^{p_1}, D^{p_2}]$, the maximum of the delay along the two paths. The weight for this node would be $\Lambda^p(\theta) = \log \mathsf{E} \exp[\theta \max[D^{p_1}, D^{p_2}]]$. Unfortunately there is no general way to compute this metric, without explicitly computing the distributions for the maximum of the delays for the two paths.

Notice that whereas for graphs with deterministic delays we need only consider the worst case path to deal with synchronization, in the case of random path delays, both paths contribute to the characteristics of the synchronization time

	m	$n = 10$	$n = 20$	$n = 80$
Exact	1/2	1.07×10^{-2}	2.01×10^{-4}	2.69×10^{-14}
Chernoff	1/2	2.52×10^{-2}	6.35×10^{-4}	1.63×10^{-13}
Exact	1/4	2.96×10^{-5}	1.61×10^{-9}	1.35×10^{-34}
Chernoff	1/4	7.38×10^{-5}	5.45×10^{-9}	8.81×10^{-34}

Table 1: Constraint violation for synthetic example.

– it is this coupling that makes such nodes difficult to address. In special cases of interest ad hoc approximations can deal with a limited amount of synchronization, see [2] – some of these are discussed below.

Paths with delays on bounded intervals and known means. Suppose the delays on p_1 and p_2 have upper and lower bounds and known means, *i.e.*, $l^{p_i} \le D^{p_i} \le u^{p_i}$ with $\mathsf{E}D^{p_i} = d^{p_i}$ for $i = 1, 2$. In this case the synchronization time satisfies the following inequalities:

$$l^p = \max[l^{p_1}, l^{p_2}] \le D^p \le \max[u^{p_1}, u^{p_2}] = u^p,$$

and $\mathsf{E}D^p = d^p \le d^{p_1} + d^{p_2}$. Based on Fact 3.1 a conservative approximation for the weight of D^p is that of a Bernoulli random variable $\bar{D}^p$ with $\mathsf{P}(\bar{D}^p = u^p) = \frac{\min[d^{p_1}+d^{p_2}, u^p]}{u^p} = \alpha$, and $\mathsf{P}(\bar{D}^p = l^p) = 1-\alpha$. More explicitly we have that $\Lambda^p(\theta) \le \bar{\Lambda}^p(\theta) = \log \mathsf{E} \exp[\theta \bar{D}^p]$ where

$$\bar{\Lambda}^p(\theta) = \log \{\exp[\theta l^p](1 + \alpha(\exp[\theta(u^p - l^p)] - 1))\}.$$

By using the delay metric $\bar{\Lambda}^p(\theta)$ for this node we can proceed safely knowing we will still obtain an upper bound on performance. Note that if $d^{p_1} + d^{p_2} \ge u^p$ then this reduces to using the worst case upper bound on synchronization. However when $d^{p_1} + d^{p_2} < u^p$ we can still glean some information on the probabilistic behavior of that node.

Last resort conservative bound. If the paths in the node are "short" relative to the critical path of the graph then a simple upper bound can be devised by noting that, $D^p \le D^{p_1} + D^{p_2}$, so it follows that $\Lambda^p(\theta) \le \Lambda^{p_1}(\theta) + \Lambda^{p_2}(\theta)$. This is likely to be conservative in the probabilistic sense, yet may still be reasonable when compared to the results obtained using the worst case edge delay.

4 Synthetic Examples: Why use probabilistic v.s. worst case critical paths?

For simplicity let us assume that all edge delays are independent and identically distributed Bernoulli random variables with mean m, *i.e.*, $\mathsf{P}(D_e = 1) = m$ and $\mathsf{P}(D_e = 0) = 1-m$. Suppose there is a single path through the DAG representing a system and it has n edges so $D^p = \sum_{i=1}^{n} D_i$. Clearly the worst case critical path would have a length of n. A probabilistic analysis might consider the likelihood that the delay exceeds 90 % of the worst case delay, *i.e.*, $\mathsf{P}(D^p \ge 0.9n)$. Table 1 exhibits some results for this setup, where both the length n of the path and the mean m of the edge delays are varied.

When $m = 1/2$ and the path is relatively long, say $n = 20, 80$ the probability of failure are $O(10^{-4})$ and $O(10^{-14})$ respectively, possibly small enough to be neglected. Thus a delay constraint which is 90 % of the worst case, is very likely to be met. For $m = 1/4$ even a path with a moderate number of elements $n = 10$ has a small probability of failure

$O(10^{-5})$ again showing that a probabilistic relaxation of the constraint is likely to be advantageous.

Based on this simple example it should be clear that as we consider increasingly large systems with many uncertain elements, the gains of a probabilistic relaxation of constraints will accrue. Moreover if the delay distributions are such that the average performance is significantly smaller than the worst case bounds, *e.g.*, 75 % smaller when $m = 1/4$, then probabilistic constraints are likely to allow a significant relaxation over the worst case critical path. The failure probabilities in Table 1 were computed exactly based on Bernoulli distributions and via the Chernoff bound used by our algorithm. Clearly the results compare favorably, and as expected the Chernoff bound gives an upper bound on the failure probability. In summary these examples show that if indeed there is sufficient uncertainty in the performance of elements on a reasonably large graph, the proposed method is likely to pay off handsomely if one can allow for a probabilistic relaxation of constraints.

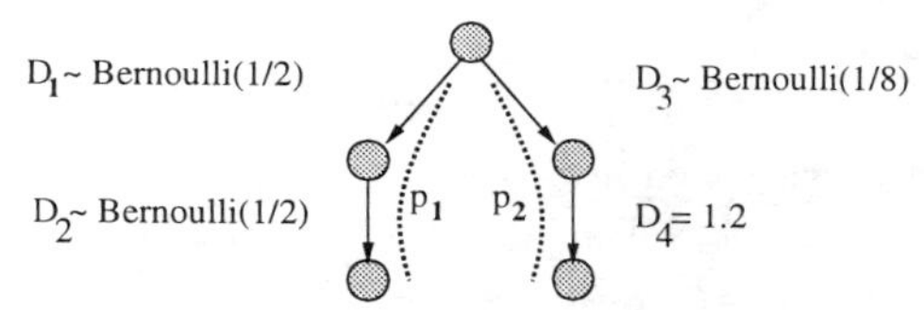

Figure 2: Probabilistic versus worst case critical paths.

In general the probabilistic and conventional critical paths need not coincide. Indeed, consider the graph in Fig. 2, where three edges have independent Bernoulli distributions on $\{0, 1\}$ with means $1/2, 1/2, 1/8$ and the fourth is deterministic with mean 1.2. The worst case critical path is obviously p_2 with a delay of 2.2. Now, given a delay constraint $d = 1.5$ one can easily show that the probability of violation is largest on p_1, *i.e.*, $\mathsf{P}(D^{p_1} \geq 1.5) = 1/4 > 1/8 = \mathsf{P}(D^{p_2} \geq 1.5)$. This suggests that a designer optimizing a system based on worst case information *may* be addressing the wrong path, at least when a probabilistic relaxation of system constraints is possible.

5 Probabilistic Constraints and MPEG Video Decoders

In this section we illustrate the practical interest of the proposed algorithm for probabilistic constraint analysis by considering MPEG video decoders [7, 4, 8]. The MPEG decoder was chosen due to the presence of non-deterministic (data dependent) delays in some of the key decoding sub-tasks. This example illustrates how the inherently variable nature of these tasks makes it interesting to assess probabilistic throughput constraints.

5.1 Background on MPEG-2

A video stream consists of a sequence of pictures or frames sampled at a given rate. Three basic types of pictures are defined: *intra-coded* pictures, which are coded without reference to other pictures; *forward/backward predictively coded* pictures, which can use motion prediction from a past/future picture; and *bidirectionally-predictively coded* pictures, which can use motion prediction from both past and future pictures. These are referred to as I, P and B pictures respectively.

Pictures are in turn subdivided into a number of *macroblocks* - a 16 by 16 pixel region. Depending on the picture type, a good match might be sought between its macroblocks and other pictures in the sequence, based on computing *motion vectors.* Thus a macroblock can be:

- *causal (forward coded):* defined from a previous picture, – allowed for macroblocks within P and B-pictures;
- *non-causal (backward coded):* defined from a future picture – allowed for macroblocks within B-pictures only;
- *interpolative (bidirectionally coded):* defined from a past and a future picture – allowed for macroblocks within B-pictures only.

Non-motion compensated macroblocks, are allowed for all types of pictures, and are said to be *intra-coded.*

As the MPEG-2 decoder reads the bitstream, it identifies the start and type of a coded picture, and then decodes each macroblock in the picture, as shown in Fig. 3.

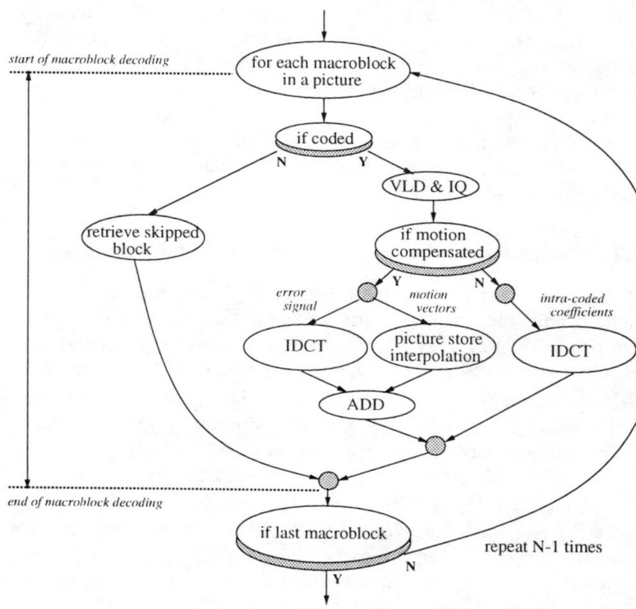

Figure 3: Macroblock decoding in an MPEG decoder.

frame		type of macroblock coding			
type	fraction	skip	intra	forw/back	bidir
I	1/15	0	1	0	0
P	4/15	0.0173	0.0658	0.9169	0
B	10/15	0.0848	0.0050	0.2226	0.6876

Table 2: Estimates of branching probabilities for MPEG macroblock decoding.

In Fig. 3, N represents the number of macroblocks in a picture – for the streams being considered a picture is comprised of 330 macroblocks. The shaded ellipses in Fig. 3 represent basic flow of control decisions taken during the decoding of each macroblock within a picture. Table 2 shows estimates of branching probabilities for these decision points. These estimates were generated by running a software decoder on a collection of MPEG video traces. The first two columns in Table 2 identify the type of picture (I, P, or B) and the percentage of occurrence of that particular type of picture in the fixed sequence of pictures considered for our

MPEG-2 decoder. The third column in the table gives the branching probability for the first decision point in Fig. 3, i.e., the probability that a macroblock will be skipped within a P or a B picture (note that all macroblocks within an I pictures are intra-coded). The three last columns in Table 2 give the probability that a given macroblock will be intra-coded, forward/backward coded, or bidirectionally coded, for I, P and B pictures.

The performance of an MPEG-2 decoder is determined by the individual performance of five key modules: Variable Length Decoding (VLD), Inverse Quantization (IQ), Inverse DCT (IDCT), Pixel Interpolation (PI), and Pixel Add (PA) [7]. However not all the modules are executed for every macroblock. In particular, as shown in Fig. 3, none of the modules is executed for non-coded (or skipped) macroblocks, and the PI and PA Modules are not executed for intra-coded macroblocks. Moreover, the processing done by the PI and PA Modules for bidirectionally coded macroblocks is twice of that required by forward or backward coded macroblocks, since one additional reference macroblock needs to be considered in the first case. (This extra-processing is the reason for the separation of bidirectionally coded macroblocks from the two other types of motion compensated macroblocks in the control flow graph shown in Fig. 3.)

The algorithm-level descriptions of the MPEG-2 modules referred to above have been the focus of extensive studies on optimizations/transformations geared towards performance enhancement [7, 4]. In our example, we have adopted the set of highly optimized algorithmic descriptions discussed in [7]. In these behavioral descriptions, the VLC and IQ modules are merged in order to save on write/read cycles to memory.

5.2 Using Probabilistic Constraint Analysis to Guide the Design of an MPEG2 video Decoder

The objective in this example is to define/specify the RTL architecture (functional units and registers/memory) for the key MPEG-2 decoder modules referred to above, so as to derive a decoder supporting a throughput of 30 frames/sec (which translates into a 33.3 ms decoding time per picture).

Design Option 1 The modules' RTL descriptions of our initial design, referred to as Option 1, were directly derived from the modules' algorithmic descriptions given in [7]. The scheduling of operations within each module was strictly performed based on data dependencies, *i.e.*, the performance of such modules is never compromised by resource sharing. Memory blocks were assumed to be implemented by RAMs with a single read port (with two cycle read operations) and a single write port.

Table 3 shows the resulting execution delays (in # cycles) for the various MPEG-2 decoding modules. As mentioned previously, the execution delays of the PI and PA modules are given separately for bidirectionally coded and for forward/backward coded macroblocks.

A crucial observation needs to be made with respect to the numbers shown in Table 3 for the VLD+IQ Module. The execution delay of that module for each macroblock depends on the number of non-zero DCTs per macroblock, and is thus data dependent. In [7], the average size of VLCs in typical MPEG-2 bitstream was reported to be about 4.5 bits which in turn translates to an average of 30 non-zero DCTs per macroblock, *i.e.*, an average of 484 cycles per macroblock for Option 1. We have used a crude model for the delay of the VLD+IQ module given by a Gaussian distribution with this mean (see Table 3) and a standard deviation of 20 % of the mean, to account for the variability in the stream.

	block type	**Option 1**	**Option 2**
VLD+IQ (avg)	any	484	436
IDCT	any	2,304	1,152
Pixel Interpolation	forw/back	320	160
	bidir	640	320
Picture Add	forw/back	512	256
	bidir	1024	512

Table 3: Time estimates (# cycles per macroblock) for MPEG modules.

In the upper part of Fig. 4, we show the decoding time distributions for I, P, and B pictures for design Option 1, derived using the execution delays per macroblock (in # cycles) given in Table 3, the branching probabilities given in Table 2, and the previously mentioned model for the VLC+IQ block. Table 4 shows the corresponding average and worst case decoding times (in # cycles) for the three types of pictures, and also the worst case and average decoding time considering all picture types (given on the last row of the table).

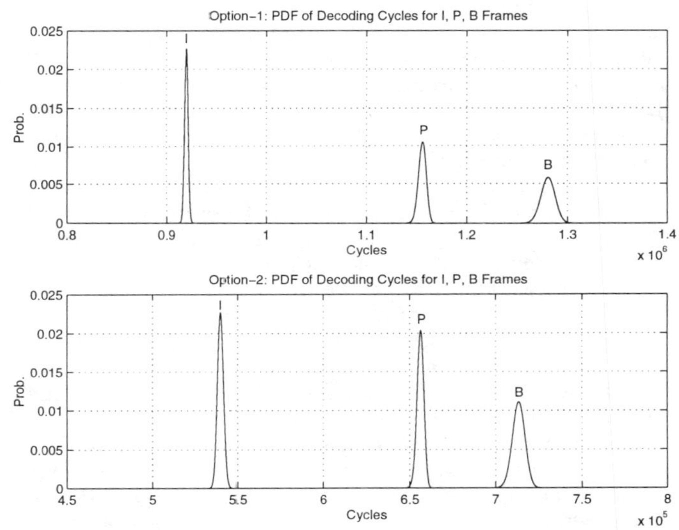

Figure 4: Decoding time distributions, for I,P and B Frames for Options 1 and 2

	type	Option 1	Option 2
Average / Maximum ($\times 10^5$)	I	9.200 / 10.159	5.399 / 6.357
	P	11.559 / 12.500	6.564 / 7.506
	B	12.807 / 13.684	7.134 / 8.011
	all	12.234 / 13.684	6.866 / 8.011

Table 4: Average and worst case decoding times, in # cycles for I,P and B frames.

The maximum combinatorial delay for our module's RTL descriptions was determined to be 43 ns (for a 0.7 μm standard-cell library). So, for a 43 ns clock, our Option 1 design led to an average delay per picture of 52.6 ms (i.e., the decoder would only sustain a throughput 19 pictures per second), which is below the target of 33.3 ms per picture (i.e., the desired throughput of 30 pictures per second). Moreover, the resulting design exhibited a worst case picture decoding delay of 58.8 ms. Option 1 was thus clearly insufficient in terms of performance, and was dropped.

Design Option 2 A second implementation, which we will call Option 2, was then developed, taking advantage of the fact that the computations performed by the IDCT, the PI, and the PA Modules can be easily parallelized. A new design was developed that: (1) has two parallel IDCT units (i.e., can compute simultaneously two 8x8 2-D IDCTs, each of which is done as a loop whose body computes an 8-point IDCT); (2) has two parallel pixel interpolation and pixel add units; and, (3) uses RAMs with two parallel read ports (still with a two cycle read operation).

Table 3 shows the resulting execution delays (in # cycles) for the various decoder modules for Option 2. The bottom part of Fig. 4 shows the decoding time distributions for I, P, and B pictures for the new design. Table 4 shows the resulting average and worst case picture decoding delays for the three types of pictures, and also the worst case and average decoding delays for all picture types.

The maximum combinatorial delay was determined to be 46 ns, using the same standard-cell library, thus leading to an average decoding delay per picture of 31.5 ms, now below the target delay of 33.3 ms per picture, and to a worst case delay of 36.5 ms. Note, however, that the (relative) gap between the average and the worst case delays for Option 2 has increased significantly with respect to that for Option 1 (see last row of Table 4). Indeed, in the Option 2 design we have increased the decoder performance by introducing some parallelism in the IDCT, PI, and PA Modules. As a result, the relative percentage of time spent on the heavily data dependent VLD+IQ Module, with respect to the total decoding time, has increased significantly, leading to more significant delay variations across pictures.

It is in cases such as the above that the interest of the systematic algorithm for assessing probabilistic constraints proposed in this paper becomes obvious. Indeed, in order to adequately evaluate the suitability of the decoder design under discussion, a key piece of information (to be given to the designer) is the probability that the target delay of 33.3 ms will be exceeded by the particular decoder design. Note that, based on such a probability, and depending on the specific timing requirements of the application for which the MPEG-2 decoder is being developed, the outcome of the evaluation might be radically different. Specifically, the Option 2 design could be considered an adequate solution, could be an unnecessarily expensive solution (in terms of area and/or average power consumption), or could still require further performance improvements. Table 5 shows the probability of violating the decoding time constraint and the Chernoff bound, for I, P and B frames and overall obtained by our algorithm. The exact numbers are exhibited to show the quality of the approximations (upper bound) provided by the algorithm. Based on these numbers, the designer would proceed, either by performing yet another iteration at the RTL level, or by starting the physical design of the decoder.

6 Conclusions

In this paper we have formulated a probabilistic critical path problem on a DAG with random weights and proposed a novel approximate algorithm for determining the likelihood that a constraint is satisfied. Through a discussion using synthetic and real examples, we have made a case for the importance/relevance of assessing probabilistic constraints on system performance, whenever the application domain is amenable to some level of constraint relaxation. Specifically, the ability to analyze the system model so as to derive less aggressive performance requirements on its various components has the potential to reduce the final cost and power consumption of the system. Our algorithm is currently being implemented in an environment for assisting algorithm and architecture-level design space exploration during system-level design [11].

frame type	**Exact**	**Chernoff Bound**
I	0	N/A
P	8.364×10^{-12}	1.162×10^{-10}
B	5.095×10^{-4}	4.103×10^{-3}
all	3.397×10^{-4}	2.735×10^{-3}

Table 5: Probabilities of violating decoding time requirement of 33.3 ms for Option 2, given a 46 ns clock.

References

[1] In G. De Micheli and M. Sami, editors, *Hardware/Software Codesign*. Kluwer Academic, 1996.

[2] G. de Veciana, M. Jacome, and J.-H. Guo. Hierarchical algorithms for assessing probabilistic constraints on system performance. *Tech. Rep.*, Dec. 1997.

[3] D. Gajski et. al. *Specification and Design Of Embedded Systems*. PTR Prentice Hall, 1994.

[4] V. Bhaskaran *et al.* Algorithmic and architectural enhancements for real-time MPEG-1 decoding on a general purpose risc workstation. *IEEE Trans. Circ. & Syst. Video Tech.*, 5(5):380–86, Oct. 1995.

[5] R. Guérin and A. Orda. QoS-based routing in networks with inaccurate information: theory and algorithms. *IBM Research Report 20515*, 1996.

[6] R. Gupta. *Co-synthsis of Hardware and Software for Digital Embedded Systems*. Kluwer Academic, 1995.

[7] W. Lee and Y. Kim. MPEG-2 video decoding on programmable processors: computational and architectural requirements. In *Proc. SPIE*, pages 265–87, 1995.

[8] N. Liu. MPEG decoder architecture for embedded applications. *IEEE Trans. Consumer Elect.*, 42(4):1021–28, Nov. 1996.

[9] D. G. Luenberger. *Linear and Nonlinear Programming*. Addison-Wesley, 1989.

[10] D. Mitra and J. A. Morrison. Multiple time scale regulation and worst case processes for ATM network control. *Proc. 34th CDC*, pages 353–357, 1995.

[11] H. Peixoto and M. Jacome. Algorithm and architecture level design space exploration using hierarchical data flows. In *Proc. 11th Intern. Conf. on Application-specific Systems, Architectures and Processors*, pages 271–82, July 1997.

A Tool for Performance Estimation of Networked Embedded End-Systems

Asawaree Kalavade
DSP and VLSI Systems Research Dept.,
Bell Labs, Holmdel, NJ 07733.
kalavade@bell-labs.com

Pratyush Moghé
Network and Service Management Research Dept.
Bell Labs, Holmdel, NJ 07733.
pmoghe@bell-labs.com

Abstract

Networked embedded systems are expected to support adaptive streaming audio/video applications with soft real-time constraints. These systems can be designed in a cost efficient manner only if their architecture exploits the "leads" suggested by clever compile-time performance estimators. However, performance estimation of networked embedded systems is a non-trivial problem. The computational requirements of such systems show statistical variations that stem from several interacting factors. At the slowest time scale, applications can adapt to network bandwidth by configuring the processing functionality of their tasks (e.g. compression parameters). Also, there could be significant execution time variations within a task. Thus, it is tricky to compute the net processing demand of several such applications on a system architecture, especially if the system schedules these applications using prioritized run-time schedulers.

In this paper, we describe an analytical tool called AsaP for fast performance estimation of such embedded systems. AsaP builds approximate models of these systems and characterizes the processing load on the system as a stochastic process. The output of AsaP is an exact distribution of the processing delay of each application. This is a powerful result that can be leveraged for efficient design of multimedia networked systems requiring soft real-time guarantees. It is also the first known framework that quantifies the effect of run-time schedulers (FCFS, RM, EDF) on the performance of such systems.

1. Introduction

Examples of networked embedded end-systems include slim hosts like personal digital assistants and network computers [1]. In their next generation, these systems are expected to support multiple concurrent continuous media (CM) applications like audio, video, and graphics. These applications are computation-intensive, periodic, and becoming network-aware. Figure 1 shows two networked embedded systems. The transmitter captures data, encodes it, packetizes it using the particular transport protocol, and sends it over the network. One candidate protocol is RTP (Real-time Transport Protocol [2]), where packets are tagged with timing information. A complementary protocol, RTCP (Real-time Transport Control Protocol), is used by the receiving end to provide feedback regarding packet loss, delay, jitter, etc. This feedback is used by the sender system to *adapt* audio and video applications. If the network is congested, a video application manager adapts the video application to send at a lower frame rate, trading off bandwidth for perceptual quality [3]. When network conditions improve, a higher frame rate may be used. Different frame rates are referred to as *adaptation levels*. The audio application can also have adaptive behavior; one of PCM, ADM, LPC, and GSM compression algorithms can be selected at run time depending on network feedback [4]. These algorithms trade off output bit rates with computation load. For example, the relative processing loads of PCM and LPC is 1:100. This adaptive behavior is one characteristic of "smart networked" embedded systems.

In addition to adaptive behavior, the applications running on these end-systems also tend to have varying performing constraints ranging from relatively strict real-time (e.g. audio) to tolerating looser constraints (e.g video). The heterogeneous demands of these applications are typically managed using sophisticated run-time schedulers on the end-systems.

Our focus in this paper is on the system-level design of such systems. We believe that estimating the performance of a set of adaptive applications on a particular architecture is one of the key problems in system-level design of networked end-systems. Further, the estimation has to account for the particular run-time scheduling policy that may be used to prioritize applications. Performance estimation is difficult since:

1. Each adaptation level offers a different processing load.
2. Data- and architecture-dependent variations (cache misses, bus contention, communication overhead, ...) lead to variable execution times of tasks *within* each application.

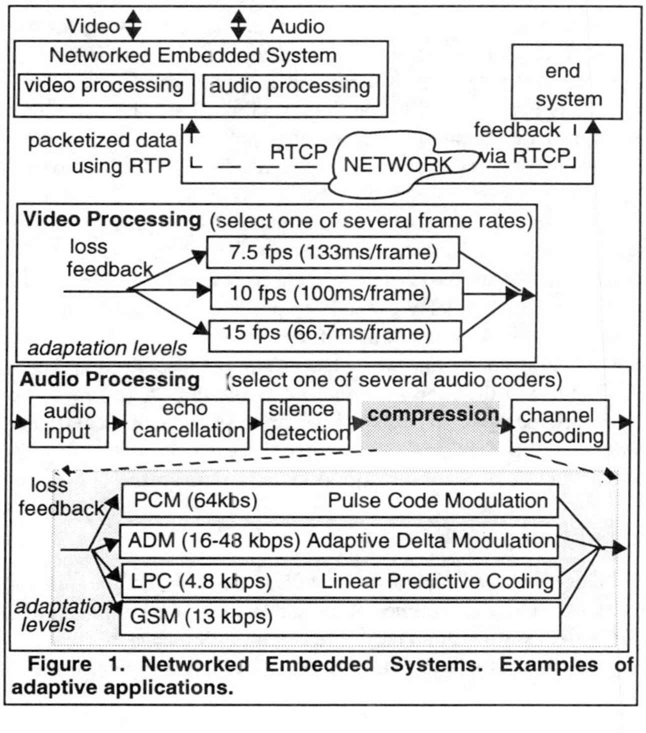

Figure 1. Networked Embedded Systems. Examples of adaptive applications.

3. The scheduling policy used by the *run-time scheduler* and application priorities impact the processing delay.

The contribution of this paper is to propose an analytical framework to estimate the performance of a set of concurrent adaptive applications on a given architecture and run-time scheduler, wherein the dynamics at any of the above three categories can be captured systematically. The output of performance estimation is several metrics including the **processing delay** (time taken to process one iteration of the application), probability of soft deadline misses, and resource utilization. Our approach is as follows. The first step consists of modeling the adaptation of each application according to network feedback by a time-varying dynamic process called the **adaptation process**. The transitions of this process indicate the probability of jumping from one adaptation level to another and approximate the behavior of the network state and the adaptation algorithm being employed. The second step is to model each adaptive application as a task graph with several alternative execution paths, each path corresponding to a particular adaptation level. Only one of several paths in the task graph is activated at run time, depending on the current adaptation level. This provides the link between the adaptation levels and their processing requirements. Unfortunately, this dependence makes the analysis of the computation complicated. We assume that task-level changes (order of milliseconds) are much faster than changes in the adaptation levels (order of minutes). This assumption makes it much simpler to characterize the state of computation on the end-system between two consecutive changes in the adaptation process. In the third step we characterize the state of computation between changes in the adaptation levels by another process called the **computation process**. The computation process captures dynamics due to the tasks and the run-time scheduler. To characterize this process, we need to capture the task-level resource requirements within each application. We assume that each task in an application has an execution time with an arbitrary known distribution. This distribution captures data-dependent variations as well as other architecture-dependent variations such as cache misses. Communication costs are modeled by adding dummy nodes between tasks. The run-time environment on the end-system is assumed to activate tasks to be run according to a specified scheduling policy. Policies such as earliest deadline first, first-come first served, and rate-monotonic scheduling can be modeled. We show that the computation process is a semi-Markov stochastic process. The exact processing delay distribution of the applications can be derived in terms of the stationary distribution of the jumps made by the computation and adaptation processes. The processing delay distribution can be used to compute the *excess delay probability*, the probability that the delay exceeds a specified threshold. This is a powerful metric for computing soft deadline misses. Other metrics such as average delay and worst-case delay can also be computed.

The rest of the paper is organized as follows. In Section 2, we discuss related work. In Section 3, we propose our analytical framework (*AsaP*). In Section 4, we describe three experiments that demonstrate the use of AsaP for exploring design tradeoffs.

2. Related Work

There has been considerable work reported in the literature on estimation of execution times of individual tasks and entire applications in both software and hardware communities. Li *et al.* [5] study the problem of estimating the worst-case execution time of tasks in cached systems using integer linear programming. Gupta *et al.* [6] use stochastic techniques to analyze execution times of single tasks, especially to estimate the execution times of loops with non-deterministic behavior. Henkel *et al.* [7] propose fast simulation-based techniques, for use in hardware/software partitioning, that estimate the execution times for tasks mapped to hardware and software.

Kim *et al* [8] and Li *et al* [9] compute the distribution of processing delay of an application by convolving density functions of individual tasks. Our approach differs in several aspects: First, we support a more general model for tasks; execution times of tasks are not constrained to have an identical distribution [8] but can have an arbitrary distribution. Second, their methods assume a single application; we support concurrent applications under different run-time scheduling policies. Finally, our approach is to model the system-wide behavior through a single stochastic process. The statistics fall out of the characterization; intensive convolutions are not needed.

The work by Yen *et al.* [10] focuses on computing the worst case processing delay of concurrent applications, where each application is made up of tasks and the execution time of a task is specified as a tuple, representing the lower and upper bound. Note that our formulation, in contrast, models the execution time of a task as a *distribution,* which is more powerful in expressing variations in execution time; worst case execution times can be one or two orders of magnitude larger than the actual execution times [8]. We compute the distribution of the processing delay, in addition to the worst-case execution time. We comment more on this in the results section.

In a related area of real-time operating systems (RTOS's), static analysis of RTOS's using deterministic models has been studied extensively [11][12] The RTOS community has also focussed on designing schedulers to give guarantees to multimedia tasks [13][14]. The problem of automated generation of RTOS's has been addressed in the context of hardware/software codesign [15][16][17]. To the best of our knowledge, our paper is the first that analytically explores the combined impact of different scheduling policies, task variations, and network feedback on the effective processing time of concurrent applications using a probabilistic framework.

3. Modeling and analysis methodology

3.1. Application specification and architecture

We assume that each application is specified by a task graph consisting of nodes representing task-level computations and arcs specifying precedences between nodes (Figure 2). When multiple arcs merge into a node, the node is set to run only when it receives

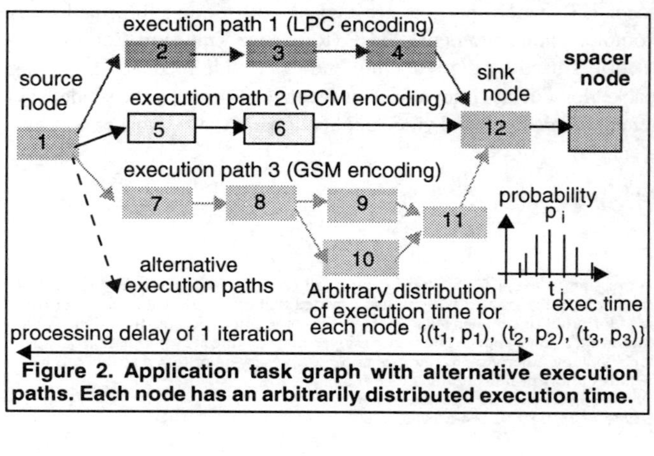

Figure 2. Application task graph with alternative execution paths. Each node has an arbitrarily distributed execution time.

data from all its incident arcs. Without loss of generality, we assume that the execution time of a node is a discrete random variable that takes one of the values $t_1, t_2, .. t_k$ with probabilities $p_1, p_2, .. p_k$ respectively, where these probabilities sum to 1. Such a distribution can approximately model execution time variations due to data and architectural dependencies. Note that the framework can model dependencies in execution time across different nodes; we will not go into details due to lack of space. A final note on distributions: Although they are dependent on the nature of input data, we believe that reasonable distributions can be obtained for specific classes of applications such as video conferencing and action-dominated movies. Analysis using such traffic models has been quite successful for network design.

Due to the repetitive behavior of CM applications, we assume that data samples arrive at the source node at a fixed period dictated by the input rate (e.g. 33ms for 30fps video). We assume non-pipelined execution, i.e., the next iteration of an application starts only after the current iteration finishes. The deadline constraint on each application is assumed to be equal to the period. The actual time when new samples are taken in by the source node for processing depends on when the earlier samples finish processing. This behavior is modeled through the artifice of a dummy **spacer node** that is assumed to run in each iteration after the sink node. The execution time of the spacer node changes based on the elapsed time in each iteration. If the processing delay of an iteration is less than the inter-arrival period, the spacer node models the idle wait of the application until new samples arrive. If the processing delay is more than the period, new samples are made to wait and the execution time of the spacer node is 0.

An application consists of several alternative execution paths corresponding to adaptation levels. To model concurrent applications, task graphs of individual applications are combined into a single *aggregated graph*. Note that the measurement interval for the aggregated graph is *not* the LCM of the periods, which has been traditionally used in literature [12]. Since the execution times of tasks are variable, the LCM does not account for all possible scenarios (also observed by Wolf *et al.* [10]). Instead, we define a *regeneration state*, which is the time when all applications finish execution at the same time. Section 3.3 explains how to compute this. The processing delay is computed over the regeneration interval.

Figure 3 shows the assumed system architecture consisting of programmable processors and hardware accelerators communicating over a user-selected communication fabric. In this work we assume that the number and types of processors and hardware accelerators and the mapping of nodes is known. Each processor runs a RTOS that selects the next task to run. The framework can model different scheduler policies (e.g. rate monotonic, earliest deadline first, ...). Tasks are assumed to be non-preemptive.

3.2. Adaptation process

The adaptation process approximates the changes in the adaptation level of applications by an adaptation process. We approximate the possibility of different feedback reports in an adaptation

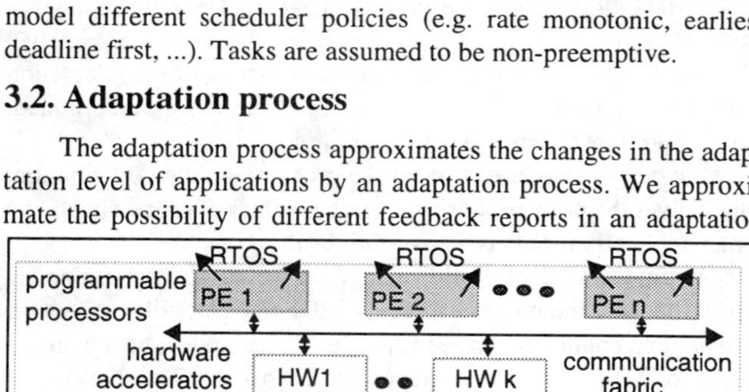

Figure 3. System architecture.

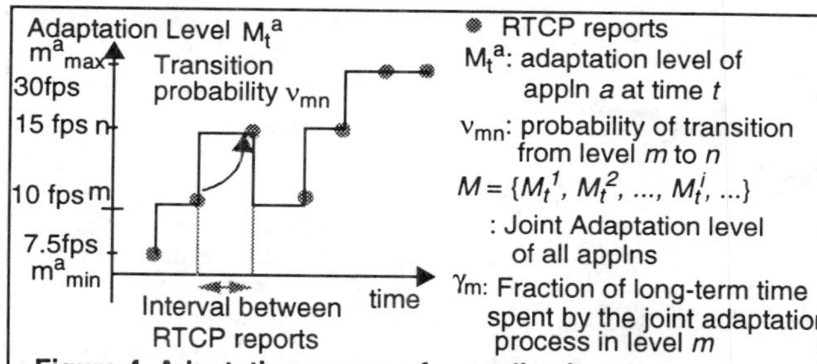

Figure 4. Adaptation process for application a.

level by different probabilities (Figure 4). Transitions between adaptation levels m and n are assumed to occur with a probability v_{mn}. This probability is determined by profiling the adaptation of applications [3]. Note that these transition probabilities are time-homogeneous; this may not be a valid approximation in practice, but we believe it is reasonable as a starting point. The interval between jumps corresponds to the interval between consecutive RTCP reports [2] and is modeled as a uniformly distributed random variable. If M_t^a represents the adaptation level of application a at instant t, the joint process $M = \{M_t^1, M_t^2, .. M_t^k\}$ represents the joint adaptation level of all the applications currently running on the end-system. Once the transition probabilities v are computed, the stationary distribution Λ_m is computed by solving the linear equations: $\Sigma \Lambda_m v_{mn} = \Lambda_m$, and $\Sigma \Lambda_m = 1$. Due to constant intervals between jumps, the steady-state distribution γ_m is equal to Λ_m. Intuitively, the steady state distribution γ_m is the fraction of long-term time spent by the joint adaptation process M in level m.

3.3. Computation process

We model the state of computation in the end-system for a particular combination of adaptation levels as a **computation process.** This process is shown to be a *semi-Markov stochastic process*. Such a process has a one-step memory and the stationary probability distribution of its jumps to a particular state can be exactly computed. This in turn enables us to compute the processing delay of an application. Now for the details.

Let us define a few variables that characterize the state of computation as a function of time. Let the transition sequence $A = \{A_n\}$, n = 0,1,.. be the time-instants when the state of the computation changes. Define a vector sequence $Y = \{Y_n\}$, n = 0,1,.. where $Y_n = (I, w_I)$ = (set of nodes ready and waiting at transition time A_n^-, the delay incurred by these nodes thus far), where A_n^- denotes the time just prior to the nth transition time. Define the sequence $Z = \{Z_n\}$, n = 0,1,.. where $Z_n = (J, r_J)$ = (set of nodes running at A_n^-, remaining execution time of these nodes). Sequences Z and Y capture information about running and waiting nodes. Define a sequence $U = \{U_n\}$ n = 0,1,.. where $U_n = \{(App, m^{App}, t^{App})$ for all applications$\}$ = (application, adaptation level of application at A_n^-, elapsed time of application at A_n^-). t^{App} is set to zero at the start of each application iteration, and is incremented at transitions when nodes in that iteration finish running or finish waiting. (Y, Z, U) models the state of computation in the system and is called the computation process.

Figure 5 illustrates the computation process for a simple example with two concurrent applications. For simplicity of illustration, suppose that the distribution of execution time for node 4 is $\{(t_{4a}, pa), (t_{4b}, 1\text{-}pa)\}$. All other nodes i are assumed to have deterministic execution times t_i. Initially, nodes 1 and 2 start running on processors $P1$ and $P3$. The state at A_0 is: $\{(), ((1,t_1), (2,t_2)), (1, m_1, 0), (2,$

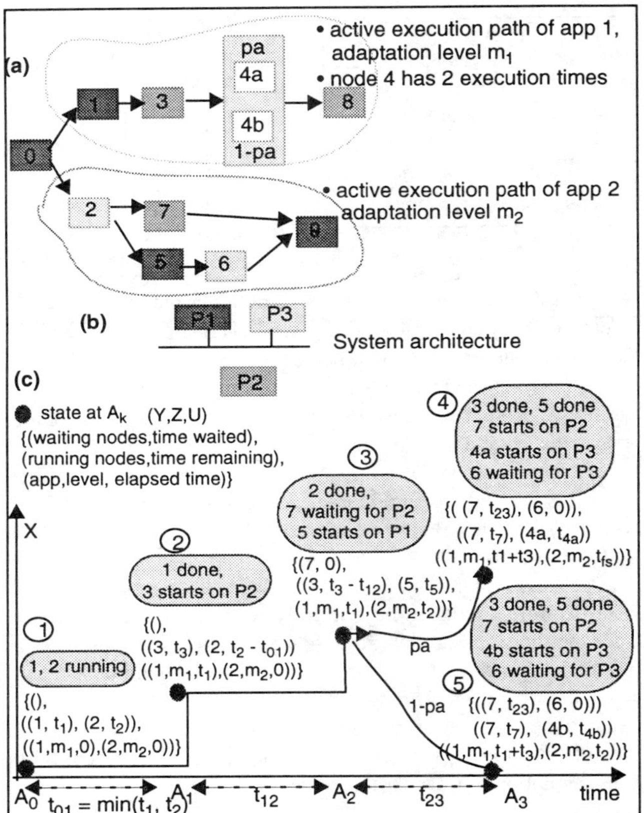

Figure 5. a) Selected execution paths for two apps b) architecture, mapping c) Computation process.

m_2, 0)} which is elaborated as: {(no waiting nodes), (1 running with remaining time t_1, 2 running with remaining time t_2), (application 1 in level m_1 and elapsed time 0, application 2 in level m_2 and elapsed time 0). Suppose that $t_1 < t_2$. At t_1, 1 finishes and there is a state change in the system. Only one transition is possible at this time: 3 starts running on *P2* with remaining time t_3. 2 is still running with remaining time t_2-t_1. At time t_2, 2 finishes. 5 and 7 are "ready". Since 7 is mapped to *P2* which is currently running 3 (tasks are non-preemptive), 7 goes on the waiting list, while 5 starts running. Suppose that 3 and 5 finish at t_1+t_3, 7, 6, and 4 are ready to run. Since 4 and 6 are mapped to the same processor, the run-time scheduler selects one based on priorities. Suppose that a static priority scheme is used, where application 1 has a higher priority. In this case, node 4 is set to run and node 6 goes on the waiting list. (Other scheduling policies can similarly be modeled. Note that the selection of the next node to run is typically made using information carried in the current state only; thus the computation process retains the Markovian property.) There are two possible execution times for 4, leading to two possible states at this point, with probabilities *pa* and 1-*pa*. This process is continued until all the applications end at the same time instant. This is called the regeneration point, after which the computation process repeats its evolution. Note that the dependencies in execution times across different nodes can be modeled in this framework by introducing additional paths. Although this increases the size of the state space, the method is quite fast and has been used to analyze realistic systems quite efficiently, as demonstrated in the results section.

We now state the following theorem that characterizes the computation process. We omit the proof due to lack of space.

Theorem: *The joint sequence (Y, Z, U) represents the state of computation at each transition and is a Markov chain. By joining the state of computation between transitions by straight lines, we obtain the computation process X. X is a continuous-time stochastic process and since its underlying jump process (Y, Z, U) is Markovian, X is a semi-Markov process* [18][19].

The implications of this theorem are that the performance statistics can be computed on this process by analyzing the behavior of a single traversal through the state space of X using the Markov renewal theory. This is much faster than simulating the system, since several simulation runs are needed to get reliable statistics of the processing delay, due to variable execution times. Our approach is much faster since a single exploration of the state space can be done efficiently even for a large state space. Further, this approach is much more efficient than the convolution approach [8][9], which is especially unwieldy for arbitrary distributions.

Also, for a given set of task-level distributions, the processing delay distribution computed by the method is exact. We are currently evaluating how effectively these distributions capture real applications. The simulations done so far show a close match with computed results.

A second key observation is that the computation process is Markovian under several scheduling policies. An exact proof is omitted for lack of space. Consequently, the impact of different scheduling policies can be analyzed within this framework.

3.4. Performance metrics

We first compute the stationary distribution of the computation process. Roughly speaking, the stationary distribution specifies the probability of a transition into each state, given that a jump occurs.
Stationary distribution of *(Y, Z, U)* We have stated earlier that, for a particular adaptation value *M*, the joint process (*Y*, *Z*, *U*) is a discrete-time Markov chain. This means that when (*Y*, *Z*, *U*) jumps from state (*i*, *j*, *a*) to (*k*, *l*, *c*) its future evolution is independent of the past, given state (*i*, *j*, *a*). This chain is therefore completely specified in terms of its transition probability function defined by R_{ija}^{klc}, which is the probability that (*Y*, *Z*, *U*) moves from a state (*i*, *j*, *a*) to (*k*, *l*, *c*) in a single jump. This one-step probability can be determined for a particular set of applications from the computation process. The stationary distribution π_{ija}^{M} is the probability that (*Y*, *Z*, *U*) jumps to state (*i*, *j*, *a*), given that (*Y*, *Z*, *U*) changes states and can be computed from *R*, since it satisfies the equations: $\Sigma_{(i,j,a) \text{ in } S}\, \pi_{ija}^{M} R_{ija}^{klc} = \pi_{klc}^{M}$ for all (*k*,*l*,*c*) in state space *S*, and $\Sigma_{(k,l,c) \text{ in } S}\, \pi_{klc}^{M} = 1$.

To obtain the stationary distribution unconditioned of adaptation level *M* we assume that the adaptation process converges to a steady-state distribution γ, as described earlier. Here, γ_M is roughly the long-term fraction of time the adaptation process spends in level *M*. γ can be computed from the transition matrix of the adaptation process and the interval of RTCP reports. Assuming γ is determined, the unconditioned stationary distribution $\pi_{ija} = \Sigma_{all\ M}\, \pi_{ija}^{M} \gamma_M$.
Next we derive an expression for the processing delay distribution, which gives the probabilities for different values of processing delay for an application. This is a powerful result.
Processing Delay distribution : Suppose we wish to determine the probability that the processing delay (PD) of a particular execution path *m* of an application exceeds a value *T*. Let node *b* be the predecessor to the spacer node on the execution path *m*. Then, Pr{PD of path m > *T*} = *max*[Pr{PD of path m > *T*, for all M}]. Note that the

right hand side is a *max* operator over all adaptation vectors of which path m is an element. The expression inside the *max* operator is the ratio of the number of times the processing delay of the path m exceeds T to the number of times path m is activated. Recall that $\pi_{ija}{}^{M}$ represents the probability of a jump of the underlying process (Y, Z, U) to the state (i, j, a) conditioned on a jump. If we focus only on the jumps to state where node b begins running, it is clear that:

$$\Pr\{\text{PD of path m} > \text{T}\} = \frac{\text{number of times PD} > \text{T}}{\text{number of times path m activated}}$$

$$= \frac{\sum_{\forall ija} \pi_{ija}{}^{M} \Big|_{b \in J,\, r_b = t_b,\, t^{App} + r_b > T}}{\sum_{\forall ija} \pi_{ija}{}^{M} \Big|_{b \in J,\, r_b = t_b}}$$

Having computed the processing delay of path m in the application, the processing delay of the entire application is computed by choosing the worst-case processing delay among all paths.

Excess delay probability Pr_{excess}: The probability of processing delay exceeding the deadline D, is simply computed by setting T equal to D in the above expression.

The above analytical framework has been implemented as a software prototype called *AsaP*. *AsaP* is about 10k lines of C code and has a Tcl/Tk GUI. Applications can be specified either textually or by using the Ptolemy [21] front-end. The state space for a given set of applications is generated using a recursive procedure. The transition probability matrix R is computed from the generated state space. The Jacobi iterative power series method [20] is used to solve for the stationary distribution π. Several other performance metrics such as nodal wait and resource utilization can also be derived.

4. Results

We illustrate the use of *AsaP* for system-level design with three examples. In the first example, the processing delay for an adaptive video application is computed for different adaptation levels. This is used to compute the maximum sustainable adaptation level. In the second example, the variation in the processing delay due to multiple concurrent applications is quantified. This shows how to trade off quality and computation between audio and video applications. In the third example, the impact of different run-time scheduling policies on the processing delay is studied. In all three cases, we demonstrate results for a single processor system, although the analytical framework can also handle multiprocessor systems.

4.1. Example 1: Impact of adaptation levels on processing delay of MPEG

We use a MPEG video encoder as an example of an adaptive application, where adaptation levels correspond to frame rates of 7.5fps, 10 fps, and 15 fps. Figure 6 shows the application and the execution time distributions. The execution times for each node were obtained by profiling the Berkeley MPEG-1 video encoder [22] on

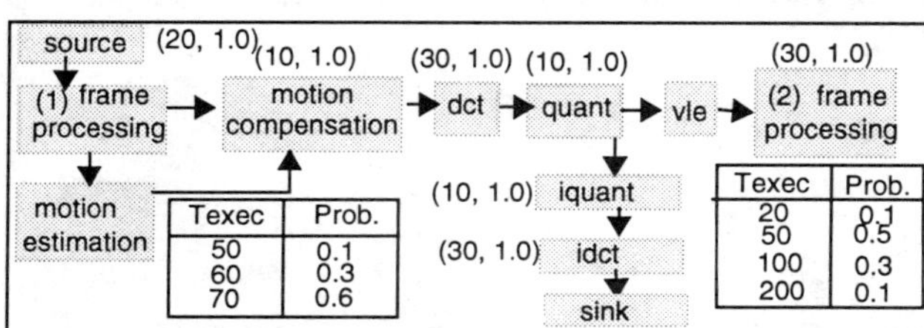

Figure 6. MPEG video encode. Exec. time dist. in 10^4 cycles.

a table-tennis sequence. We are currently exploring systematic methods to compute task-level distributions.

The adaptation level changes were profiled based on experimental results reported in [3], where a video conferencing application VIC [23] is modified to adapt its bit rate based on network feedback. The steady state distribution for each adaptation level was calculated by profiling the adaptation level changes over a period of 300 seconds (Table 1). Although such a profile should actually be computed over a much larger time window, this is an illustrative example.

The output of *AsaP* is shown in the last two columns of Table 1. For example, in level 3, the excess delay probability is 0.1, i.e. 10% of the samples miss their deadlines. The last column in Table 1 represents the effective miss probability for each level, by factoring in the probability of the level itself. While the deadlines are missed quite often in level 4, level 4 itself is much less frequent. Figure 7

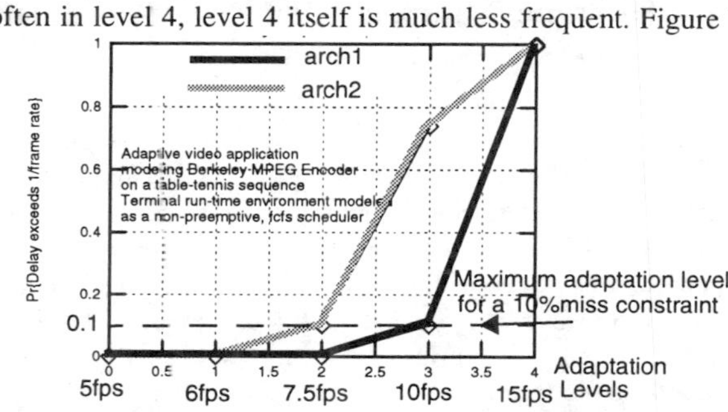

Figure 7. Excess delay probability vs. adaptation levels for two architectures. *arch1* 1.3 times as fast as *arch2*.

plots the probability of deadline misses for different adaptation levels on two architectures, *arch1* and *arch2*. The processing delay indicates sustainable adaptation levels. Thus, for a 10% acceptable miss rate, *arch1* supports adaptation levels 1 through 3, while *arch2* supports levels 1 and 2 only.

Adaptation Level	steady state distribution γ	Pr_{excess}	γ x Pr_{excess}
0 (5 fps)	0.0727	0.0	0.0
1 (6 fps)	0.0909	0.0	0.0
2 (7.5 fps)	0.2181	0.0	0.0
3 (10 fps)	0.5454	0.1	0.05454
4 (15 fps)	0.0727	0.994	0.0722

Table 1: (Ex. 1) Adaptation levels (frame rates), steady-state distribution of adaptation process, excess delay probability.

4.2. Example 2: Impact of adaptation levels on processing delay of concurrent applications

In this experiment, we consider two concurrent applications: video (adapting between 7.5 fps and 15 fps) and audio (adapting between PCM and LPC, with a constant period of 20ms). Table 2 summarizes the processing delay for different combinations of audio algorithms and video frame rates. This data is used to determine feasible combinations of adaptation levels. Consider the combination (LPC, 7.5fps), with audio missing 24% of its deadlines. If a higher frame rate is desired, switching to (LPC, 15fps) leads to an unacceptable 46% miss rate for audio. The application manager should instead step down to (PCM, 15fps). Such statically computed information may be used to design "smart" application managers.

(audio, video)	(PCM, 7.5fps)	(PCM,15fps)	(LPC,7.5fps)	(LPC,15fps)
(Deadlines 10^4 cycles)	(200, 1330)	(200, 660)	(200, 1330)	(200, 660)
Pr_{excess}	(0.00046, 0)	(0.01, 0)	(0.241, 0)	(0.462, 0.007)
PD_{ave} in 10^4 cycles	(16.2, 344.1	(32.1, 344.1)	(146.2, 509.7)	(198, 500.93)
PD_{max} in 10^4 cycles	(240, 486)	(240, 486)	(440, 675)	(450, 680)
Number of states	19856	16070	6457	20767
State space time	1393.29 s	856.96 s	122.16 s	1446.27 s
Analysis time	1948.53 s	862.37 s	494.03 s	352.11 s

Table 2: (Example 2) Performance metrics for audio (PCM, LPC) and video (7.5fps, 15fps).

4.3. Example 3: Impact of run-time schedulers

Table 3 summarizes the processing delay distribution when video is at 15fps and audio uses LPC (LPC 15fps) under three different run-time scheduling policies. In FCFS, nodes are set to run in the order in which they become "ready". RMS and EDF are prioritized schemes, where nodes with higher priority are selected to run first. In RMS, priorities of the nodes are set according to rates. In this particular example, audio has a higher priority over video. In EDF, the priorities are computed at run-time; the node whose "due date" is earliest gets higher priority. Note how shifting from FCFS to RMS improves the mean and worst-case delays and probability of deadline misses for audio, while degrading the performance of video. Using EDF seems to be a good compromise for both applications. To reiterate the importance of a distribution of the processing delay against just the worst case value, consider the (LPC, 15fps) case for RMS scheduler in Table 2. The worst case processing delay is 290. However, the distribution tells us that the mean is 118, the standard deviation is 29, and only 2% of all samples miss deadlines. This is much more valuable than knowing that the worst case execution time is 290. Typically, CM applications have constraints such as "no more than 10% samples should miss deadlines". Using worst-case leads to very conservative and hence inefficient architecture design.

Perf. metrics (audio, video)	FCFS	RMS	EDF
Pr_{excess}	(0.462, 0.007)	(0.0235, 0.52)	(0.175, 0.227)
PD_{ave}	(198.0, 501.0)	(118.4, 685.9)	(169.3, 615.7)
σ	(77.5, 54.3)	(29.34, 94.72)	(61.59, 58.07)
PD_{max}	(450,680)	(290,980)	(380,840)

Table 3: Example 3: Impact of run-time scheduling policy on the performance delay of LPC audio and video at 15fps

5. Conclusions

We have identified system-level design issues for next-generation networked embedded systems in the context of adaptive applications and run-time schedulers. Applications adapt in order to reduce the output bit rate. However, adaptations impose different processing demands on the end terminal. This variation, in addition to task-level processing variations, manifests by making the processing delay of each application variable and unpredictable. Run-time scheduling policies also impact the performance of the end-system. We propose an analytical framework to estimate the performance of a set of concurrent adaptive applications for a particular architecture and run-time scheduler. The framework can efficiently compute the exact distribution of the processing delay of applications. This result can be used to synthesize cost-efficient solutions for systems requiring soft real-time guarantees. We demonstrate the use of this framework to explore the system-level tradeoffs in selecting adaptation levels for applications, designing smart application managers, and in selecting an appropriate run-time environment.

6. References

[1] PDA Buyer's Guide, Pen Computing Magazine, vol. 3, no. 11, July/August 1996, pp. 84.

[2] H. Schulzrinne, *et al.* "RTP: A Transport Protocol for Real-Time Applications", RFC 1889, Audio/Video Transport WG, Jan. 1996.

[3] I. Busse, *et al,* "Dynamic QoS Control of Multimedia Applications based on RTP", *Computer Communications* 19:1, Jan. 1996, pp. 49-58.

[4] J.-C. Bolot, A. Vega-Garcia, "Control Mechanisms for Packet Audio in the Internet", *Proc. IEEE Infocom '96*, CA, April 1996, pp. 232-9.

[5] Y. S. Li, S. Malik, A. Wolfe, "Performance Estimation of Embedded Software with Instruction Cache Modeling", *ICCAD 95.*

[6] R. K. Gupta, *Co-synthesis of Hardware and Software for Digital Embedded Systems*, Kluwer Academic, 1995.

[7] J. Henkel *et al,* "The Interplay of Run-time Estimation and Granularity in HW/SW Partitioning", *Proc. Codes96*, USA, March 1996, pp. 52-8.

[8] J. Kim, K. G. Shin, "Execution Time Analysis of Communicating Tasks in Distributed Systems", *IEEE Trans. on Computers*, vol. 45, no. 5, May 1996, pp. 572-9.

[9] Y. Li, J. Antonio, "Estimating the Execution Time Distribution for a Task Graph in a Heterogeneous Computing System", *Proc. Sixth Heterogeneous Computing Workshop* (HCW '97), Switzerland, pp. 172-84.

[10] Ti Yen, W. Wolf, "Performance Estimation for Real-time Distributed Embedded Systems", *Proc. of ICCD 95*, pp. 64-69.

[11] C. L. Liu, J. W. Layland, "Scheduling Algorithms for Multiprogramming in a Hard Real Time Environment", *JACM*, v20, 1973, pp. 46-61.

[12] K. Ramamritham *et al.*, "Scheduling Algorithms and Operating Systems Support for Real-Time Systems", *Proc. of the IEEE*, vol. 82, no. 1, Jan. 1994, pp. 55-66.

[13] D. K.Y. Yau, *et al.* "Adaptive Rate-Controlled Scheduling for Multimedia Applications", *Proc. ACM Multimedia'96*, Boston, MA, Nov. 1996.

[14] R. Yavatkar, K. Lakshman, "A CPU Scheduling Algorithm for Continuous Media Applications", *Proc. of NOSSDAV*, April 1995, pp.223-6.

[15] V. Mooney *et al.*, "Run-Time Scheduler Synthesis for Hardware-Software Systems and Application to Robot Control Design", *CODES 97.*

[16] F. Balarin *et al.*, "Automatic Generation of a Real-Time Operating System for Embedded Systems", *Proc. of CODES 97.*

[17] P. Chou *et al.*, "Software Scheduling in the Co-synthesis of Reactive Real-time Systems", *Proc. of DAC*, June 1994, pp 1-4.

[18] E. Cinlar, *Introduction to Stochastic Processes*, Prentice-Hall, 1975.

[19] S. Ross, *Introduction to Probability Models*, Academic Press., 1985.

[20] V. L. Wallace, R. S. Rosenberg, "Markovian Models and Numerical Analysis of Computer System Behavior", AFIPS Spring Joint Computer Conf. Proceedings, pages 141-148, 1966.

[21] Ptolemy at *http://ptolemy.eecs.berkeley.edu*

[22] *http://bmrc.berkeley.edu/projects/mpeg/mpeg_encode.html*

[23] S. McCanne, V. Jacobson, "*vic*: a flexible framework for packet video", *Proc. of ACM Multimedia* '95, CA, Nov 1995, pp. 511-522.

Rate Derivation and Its Applications to Reactive, Real-time Embedded Systems

Ali Dasdan
Dept. of Comp. Sci.
University of Illinois
Urbana, IL 61801
dasdan@cs.uiuc.edu

Dinesh Ramanathan
Advanced Technology Group
Synopsys, Inc.
Mountain View, CA 94043
dinesh@synopsys.com

Rajesh K. Gupta
Dept. of Info. & Comp. Sci.
University of California
Irvine, CA 92697
rgupta@ics.uci.edu

Abstract

An embedded system (the system) continuously interacts with its environment under strict timing constraints, called the external constraints, and it is important to know how these external constraints translate to time budgets, called the internal constraints, on the tasks of the system. Knowing these time budgets reduces the complexity of the system's design and validation problem and helps the designers have a simultaneous control on the system's functional as well as temporal correctness from the beginning of the design flow. The translation is carried out by first deriving the rate of each task in the system, hence the term "rate derivation", using the system's task structure and the rates of the input stimuli coming into the system from its environment. The derived task rates are later used to derive and validate the rest of the internal as well as external constraints. This paper proposes a general task graph model to represent the system's task structure, techniques for deriving and validating the system's timing constraints, and a hardware/software codesign methodology that puts everything together.[1]

1 Introduction

An embedded system is typically reactive and real-time in nature because it continuously has to react to the stimuli coming from its environment and it also has to do this interaction under strict timing constraints. As these timing constraints define the timing characteristics of the system's interaction with its environment, i.e., those of the system's external behavior, they are called *external (timing) constraints.*

A typical embedded design flow starts with the requirements specification phase, which describes what the system's external behavior must be without describing how the system works internally. The latter is for the first time expressed at a high level in the architectural design phase by means of a task structure diagram, which describes a decomposition of the system into manageable components or *tasks*. Since the system has many activities occurring in parallel, these tasks are usually concurrent tasks.

The problem of designing a functionally and temporally correct system is a difficult one. The current practice for this problem is based on trial and error guided by engineering experience [7]. Despite the importance of the temporal correctness, most design methods pay significantly more attention to structural and behavioral aspects of embedded systems than to timing aspects [9]. The main concern in a typical design flow is first to design a functionally correct system and then to focus on its temporal correctness. For example, a typical design flow proceeds in the following four stages: (i) The system is structured into its tasks. (The issues related to task structuring are explained in [9].) (ii) Each task is designed one by one. (iii) The tasks are integrated together to make up the system. (iv) The system is first checked for functional correctness. If it is functionally correct, its temporal correctness is checked. In case of any timing violations, the designers will go back to fine tune or modify some tasks and/or the task structure of the system until all the violations are eliminated or a compromise between the designers and the customer is reached in case some violations are too difficult to eliminate.

However, the above design flow has many problems such as: (i) The designers design the tasks without knowing their time budgets. This seriously hinders the exploration of the design space for better task implementations in terms of performance, cost, and power. (ii) The integration stage occurs very late in the design flow; Any problems detected at this stage are very expensive to correct in terms of design time and cost.

A solution to the above problems should provide the time budgets on the system's tasks very early in the design flow and help the designers keep the system's correctness under control throughout the design flow. This paper offers such a solution. The main idea is to derive the time budgets on the system's tasks, called the *internal (timing) constraints*, from the system's task structure and its external constraints, and to use them to validate the design choices at every stage of the design flow. This reduces the complexity of ensuring temporal correctness from system level to task level and enables the designers to have a simultaneous control on the system's functional and temporal correctness from very early stages in the design flow onwards.

Figure 1 illustrates the above discussion. In this figure, (a) shows the system and its environment. The system gets the stimuli from the environment through its sensors and responds back to the environment through its actuators. This interaction is constrained by the external constraints labeled

[1]This work is supported by an NSF Career Award MIP 95-01615.

DAC 98, San Francisco, California
©1998 ACM 0-89791-964-5/98/06..$5.00

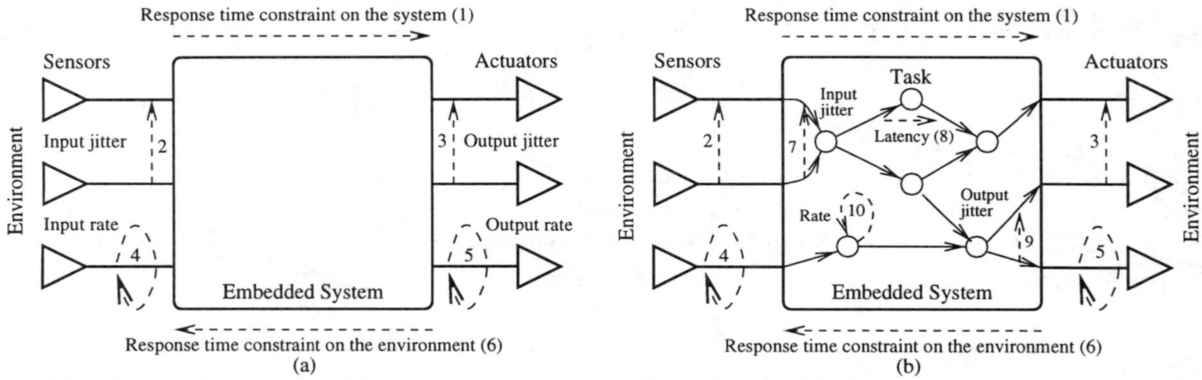

Figure 1: (a) An embedded system with its external constraints (labeled 1-6). (b) The same system with its task structure and internal constraints (labeled 7-10).

1-6, which are the (worst-case) *response time constraints* on the system (1) and on the environment (6), the *input* (2) and *output* (3) *jitter constraints*, and the *input* (4) and *output* (5) *rate constraints*. The jitter constraints define the time separation between two different stimuli or two different responses, whereas rate constraints define the time separation between two successive arrivals of the same stimulus or two successive departure of the same response. Note that the constraints 2, 4, and 6 are imposed on the environment, which we assume are already satisfied, and the constraints 1, 3, and 5 are imposed on the system, which we will validate.

In Figure 1, (b) shows the same system but this time with its task structure and internal constraints. The task structure is described using a task graph of tasks (circles) and their interactions (arcs). The system's internal constraints, called *bounds* for easy reference, are labeled 7-10, which are the *latency bound* (8), the *input* (7) and *output* (9) *jitter bounds*, and *rate bound* (10). Note that these bounds correspond to the time budgets on the tasks and are external to the tasks.

In this paper, we derive the system's internal constraints from its task structure, which is assumed to be given, and its external constraints. However, we do not use all of the external constraints for derivation; we use only the input rate constraints. This means that we assume only that the environment satisfies its timing constraints. We use the input rate constraints to first derive the rate bound of each task, hence the term "rate derivation", and then use them to derive and/or validate the rest of the internal and external constraints.

We perform the derivation under very general task interaction semantics, those that can be found in real-life embedded systems. For this, we propose a task graph model, called the *generalized task graph* model. This model is causality-based and can handle all of the existing causality relations as well as cyclic task dependencies.

The complexity of the internal constraint derivation problem is very high because (i) the external constraints are usually a few in number, (ii) they are defined on sets of tasks rather than for individual tasks, and (iii) tasks can interact in a rather complex manner. To reduce this complexity, this paper assumes that we are given *all* of the system's input rate constraints. Once we know all of them, we do not need the rest of the system's external constraints for derivation; instead, we will validate them. However, if we do not know all of the input rate constraints, we can use some of the system's external constraints as if they are already satisfied.

We perform rate derivation in a hierarchical manner. We first partition the system's task graph into its cyclic components and represent each component using a "super task" in the resulting component graph. We then perform rate derivation in this acyclic component graph using a method called RADHA (RAte Derivation and High-level Analysis). After that, we perform rate derivation in each cyclic component using a method called RATAN (RATe ANalysis). Finally, the results of these two are combined. RADHA and RATAN are also the names of the corresponding software tools [5].

We will present the above methods in a hardware/software codesign methodology. It starts with the system's task structure expressed in a generalized task graph and its external constraints, and ends with its hardware, software, and interface synthesis. This methodology is obtained by incorporating RADHA and RATAN into a standard hardware/software codesign methodology, and gives an overall view of our contributions and also of the organization of this paper. In the rest of this paper, we will first present this methodology (Section 2) and introduce our task graph model (Section 3). After that, we explain RADHA (Section 4.1) and RATAN (Section 4.2) and finally give their applications (Section 5).

2 Proposed Hardware/Software Codesign Methodology

The proposed codesign methodology is given using the flow diagram in Figure 2. The steps in the methodology are labeled by 1-5. The steps 2, 4, and 5 also exist in many other codesign methodologies, e.g., POLIS [2]. The steps 1 and 4 correspond to RADHA and RATAN. RATAN can also be used in step 1 [6]. As RADHA and RATAN will be explained later, we will briefly discuss the other steps now. Step 2 has three substeps: (i) The tasks are written in a hardware description language (HDL) or a programming language at behavioral level. This is actually the first step in the existing codesign methodologies. (ii) These tasks are then partitioned into hardware and software tasks to determine their implementation platforms, e.g., see [11]. (iii) The latency of each task and the task communication delays are estimated [16, 14]. Step 4 synthesizes all the tasks on their respective platforms together with the interface between these platforms [2, 4, 14]. Step 5 performs modifications in the design, e.g., in case the final design does not satisfy its functional and/or temporal correctness requirements. This steps shows that the methodology is an iterative one as it should be.

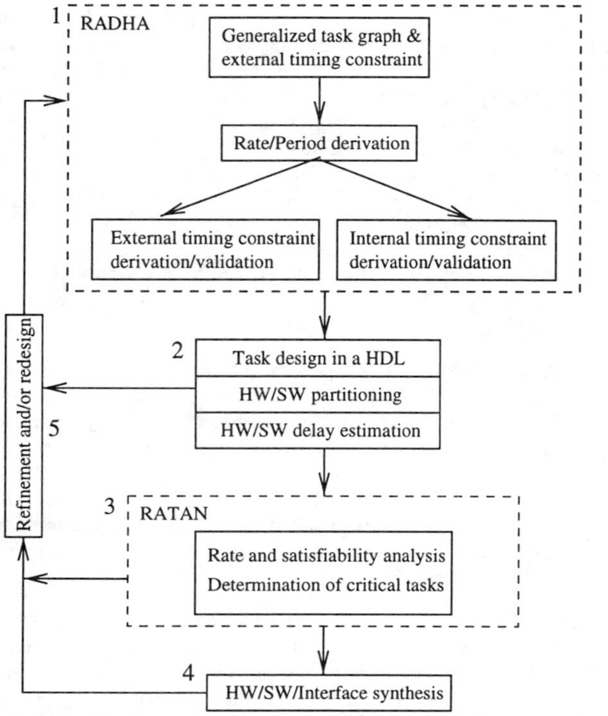

Figure 2: Hardware/software codesign methodology using RADHA and RATAN.

3 Generalized Task Graph Model

We model an embedded system using a hierarchical directed graph called a *generalized task graph* (the task graph for short). This model builds upon the models in [3, 8, 10, 12, 13, 15, 17, 18, 20, 21], as discussed in [6]. Each node corresponds to a task, and each arc corresponds to an asynchronous, unidirectional communication channel between its *producer* and *consumer*. This model actually represents the system's data/control flow diagram where each node is either a data or a control transformation, and each arc is either a data or a control flow, as explained in [9]. The sensors and actuators of the system are also included in its task graph, in which they are referred to as *input* and *output tasks*, respectively.

Each arc prescribes a dependence relation between its end nodes. The hierarchy in the task graph has two levels, the *top* and *bottom levels*, as depicted in Fig. 3. The bottom level consists of strongly connected components (SCCs) of the actual system graph (or the initial task graph) that have at least one arc. The top level is the component graph of the initial task graph such that every node with a self-loop, called a *super task*, corresponds to a SCC at the bottom level, and every node without a self-loop corresponds to the same node in the initial task graph. The hierarchy is mandated by the fact that the rate derivation techniques for acyclic and cyclic components are different. In the sequel, we will refer to each graph or SCC at any level of the hierarchy as "the task graph" within its context. Note that any directed graph can be represented using the above hierarchy.

We assume that each node represents a periodic or sporadic task, each of which has a rate interval bounded from above and below. Recall that a task can have a latency bound, a rate bound, and jitter bounds. The rate of a task gives the number of its executions per unit time. Its reciprocal is equal to the task's *period*. We assume integer values for periods.

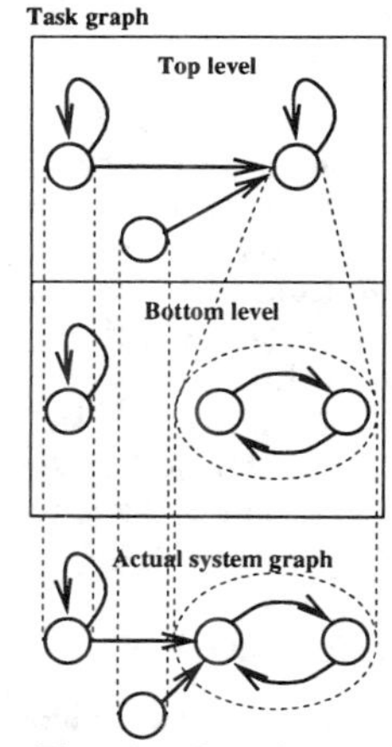

Figure 3: Two-level task graph model.

Both data and control flows through a channel is modeled using token flows. The granularity of every token is *channel-specific*, and once the designers determine the tokens in each channel, they are indistinguishable within and across channels for timing analysis purposes. The task behavior is causality-based such that it uses either AND-, OR-, or AND/OR-causality. Note that each causality relation also implies a timing relation. We say that a task is *enabled* when its predecessors have sent all the tokens that are needed to make the task ready to run. We now classify the tasks by asking three questions:

(i) How many predecessors are needed to enable a task?

1. an AND task needs all of its predecessors,
2. an OR task needs one of its predecessors, and
3. an AND/OR task needs some (more than one) of its predecessors.

(ii) Does a task allow any loss of tokens coming from its predecessors?

1. an AND/unskipped or OR/unskipped task does not allow any loss of tokens.
2. an AND/skipped or OR/skipped task may allow an intentional loss of tokens.

(iii) What happens to the tokens coming from the rest of the predecessors of an OR task?

1. an OR/unskipped task uses all the tokens before it completes.
2. an OR/skipped/joint task ignores the rest completely for its current execution.
3. an OR/skipped/disjoint task uses every token to start a new execution.

The above task types define the input behavior of the tasks. Their output behavior can be different based on what is required of them by their consumers. Moreover, no order is imposed for reading the input channels although a particular order can be imposed using task partitioning. If inputs arrive simultaneously for an OR task, only one of them enables the task.

4 Rate Derivation

Rate derivation refers to deriving the task rates or rate bounds (in terms of rate intervals) at the top and bottom levels. We next use the task rates to derive the task periods because the task periods are actually the ones used for further derivation and validation. As the rate of a task is equal to the reciprocal of its period or vice versa, it is easy to derive one from the other, and so we will use the terms "rate derivation" and "period derivation" interchangeably. We now discuss the rate derivation at the top level followed by the one at the bottom level. We assume that every task in the task graph is concurrent.

4.1 Rate Derivation for Acyclic Graphs

We now present the rate derivation algorithm to derive the rate of each task at the top level. This algorithm is part of RADHA. The algorithm assumes that a task graph and the rate of every input task are given. The algorithm proceeds in topological order of the nodes from the input tasks towards the output tasks. The derivation is done using a rate equation for each task type. This equation is then used to derive period intervals. The running time of the algorithm is polynomial in the size of the task graph; however, the algorithm may use expensive operations such as computing the least common multiple of two integer intervals.

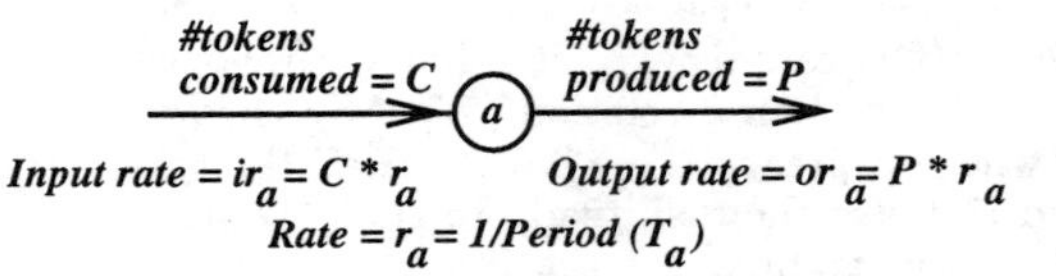

Figure 4: Basic rate relationships.

Figure 4 depicts the definitions of the *input* and *output rates* of a task in relation to its rate as well as the number of tokens produced and consumed. If the period is in seconds, the rate is in executions/seconds, and the input and output rates are in tokens/seconds. We will use these in the algorithm without further definition. The derivations are always done to find the smallest rate and integer period for the task in question and to ensure bounded channel capacities. In the following algorithm, knowing the number of tokens produced/consumed for each channel improves the accuracy of the derivation. Since the granularity of a token is channel-specific, these numbers are relatively easy to obtain from the data/control flow diagram of the system. In case these numbers cannot be obtained, using value 1 for each of them is a safe choice. More examples on how to specify and use these numbers are given in [6].

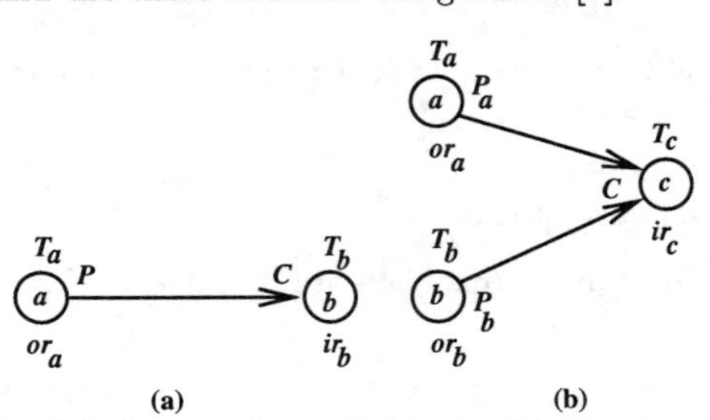

Figure 5: (a) One predecessor case. (b) Two predecessor case.

The rate derivation algorithms is as follows. The rate equations below directly follow from the definition of the corresponding task.

One predecessor: Consider the task graph in Figure 5(a). We want to derive the rate of task b from that of task a. The rate equation is $or_a = ir_b$, which implies that $P/[T_a(l), T_a(u)] = C/[T_b(l), T_b(u)]$, where l and u correspond to the lower and upper bound, respectively.

Two predecessors: Consider the task graph in Figure 5(b). We want to derive the rate of task c from those of tasks a and b. The derivation is different for each task type, as explained below:

1. **AND/skipped.** The rate equation is $r_c = \min\{r_a, r_b\}$, which leads to $T_c(l) = \max\{T_a(l), T_b(l)\}$ and $T_c(u) = \max\{T_a(u), T_b(u)\}$.

2. **AND/ and OR/skipped.** The rate equation is $ir_c = or_a + or_b$, which leads to $T_c(l) = \min\{LCM(i,j)\}$ and $T_c(u) = \max\{LCM(i,j)\}$, where $LCM(i,j)$ denotes the least common multiple of i and j such that $i \in [T_a(l), T_a(u)]$ and $j \in [T_b(l), T_b(u)]$. Here, $LCM(i,j)$ is computed by first enumerating the period intervals for tasks a and b and then using the Euclid's algorithm for the greatest common divisor (GCD), as $GCD(i,j) * LCM(i,j) = i * j$.

3. **OR/skipped/joint.** At least one predecessor of an OR/skipped/joint task generates tokens for it but it is not required that every predecessor of the task will do so. Thus, the rate equation is $\min\{r_a, r_b\} \leq r_c \leq \max\{r_a, r_b\}$, which leads to $T_c(l) = \min\{T_a(l), T_b(l)\}$ and $T_c(u) = \max\{T_a(u), T_b(u)\}$.

4. **OR/skipped/disjoint.** It is difficult to write a rate equation for this case. Instead, we will find the smallest and the largest time separations between token arrivals. A conservative period interval for task c is then $[1, \min\{T_a, T_b\}]$.

Three or more predecessors:

1. **AND or OR tasks.** This is a straightforward generalization of the above derivations.

2. **AND/OR tasks.** This is similar to the AND/skipped case where the "AND part" and the "OR part" behave like tasks a and b. Hence, we derive the period equation directly, which is $T_c = \max\{T_{ANDpart}, T_{ORpart}\}$, where AND and OR parts should be replaced by the actual predecessors.

Note that as the rate equations imply, the above derivations still hold for all the task types except for the case (iv) even if we assume different initial start times for tasks a and b. For that case, the time separation between the initial start times should also be taken into account. Also note that the above derivations also hold when periods are real numbers except for the AND/unskipped case, for which $LCM(i,j)$ should be replaced by $i * j$.

Related Work. Most of the previous works assume that the internal constraints of the system are known in advance [19], and there are surprisingly few works on their derivation [1, 7, 19]. Amongst them, [7] is the most comprehensive one. This work introduces the problem for real-time systems; however, its task semantics is restrictive. Furthermore, due to its assumptions, it is not useful at the level where RADHA is used; [7] can instead be used before scheduling analysis within software synthesis (in step 4) of the codesign methodology in Figure 2.

4.2 Rate Derivation for Cyclic Graphs

We now briefly describe the rate derivation algorithm to derive the rate of each task at the bottom level. This algorithm is part of RATAN, and is explained in detail in [17]. This algorithm assumes that the tasks in a given each SCC at the bottom level are either all AND tasks or all OR/skipped/joint tasks; the case with a combination of tasks is still an open problem [10]. It also assumes that each task starts its very first execution independently. Consider a SCC with only AND tasks. The algorithm uses the fact that the rate of each task in the SCC is the same and is equal to reciprocal of the *maximum cycle mean* of the SCC [17]. The maximum cycle mean of a graph is equal to the maximum of the *cycle means* over all the cycles. The mean of a cycle is defined as the average weight of any arc in the cycle. The weight of an arc is in turn equal to the sum of the latency of its producer and its channel delay. Arc weights are intervals, so are the task rates and periods. A cycle that determines the maximum cycle mean of the graph is called a *critical cycle*. Similarly, the tasks on this cycle are *critical tasks*. Since there are efficient algorithms to find the maximum cycle mean of a SCC [17], we can derive the rate of every task in the SCC. The above definitions are analogous for the OR/skipped/joint case.

5 Applications of Rate Derivation

We now consider the use of the task periods to derive latency bounds and validate response time constraints at the top level, and validate rate constraints at both top and bottom levels. More about the derivation and validation of the other internal and external constraints can be found in [6].

5.1 Deriving Latency Bounds

As a task executes *once* in its period, the lower bound of its period imposes an upper bound on its latency. Since the period is now derived, the designers know an upper bound on the time budget of the task. We need to derive a lower bound for OR/unskipped tasks because such tasks have to wait until they read all the tokens generated for their current execution. Consider Figure 5(b), if task c is an OR/unskipped task, its latency should be greater than $(T_c(u) - \min\{T_a(l), T_b(l)\})$. Note that a self-loop on a task can be used to ensure that the task waits on itself.

5.2 Validating Response Time Constraints

The worst-case response time of the system is at least as large as the longest path delay through the system. Here, the longest path delay is denoted by D, and is obtained using the derived task periods. If the response time constraint is specified a priori, it must be greater than D. This comparison amounts to validating the specified response time constraint.

The algorithm to find L is a modification of the standard longest path algorithm. The modification arises because of the following reason: The delay of a path of tasks in the task graph is not always equal to the sum of the upper bounds of their periods. For example, consider task c in Figure 5(b). Suppose $T_a = 2$ and $T_b = 3$, and that task c is an AND/unskipped task. Then, $T_c = LCM(2, 3) = 6$. The delay of the path with tasks a and c seems to be equal to $T_a + T_c = 2 + 6 = 8$; however, task a has to execute three times for one execution of task c. making the path delay equal to $3 * T_a + T_c = 3 * 2 + 6 = 12$. This reasoning also holds for the other task types.

5.3 Validating Rate Constraints

Recall that the rate derivation algorithms at the top and bottom levels derive the rate of each task (including output tasks). The validation of a rate constraint that is specified a priori on a task is performed by comparing it with the derived rate of the task. If the specified constraint subsumes the derived one, the specified one is said to be satisfied. If not, it is said to be violated. Violations can be eliminated by speeding up or slowing down the tasks involved. Note that for the bottom level, violations can only occur for the critical tasks.

So far, we have not mentioned the relationship between the results of the rate derivation at the top level and those at the bottom level. Recall that the top level has a super task for each SCC at the bottom level. As explained in Section 4.2, a SCC at the bottom level behaves like a single task, justifying the use of super tasks. Consider a super task and its corresponding SCC. The rate derivation algorithm for the top level derives a period interval for the super task. At the same time, the derivation algorithm for the bottom level derives a period interval for the SCC. Now, these two intervals should be the same; otherwise there is a violation. Note that the rate derivation algorithm for the bottom level assumes that the arc weights are known, which is possible after step 2 in the codesign methodology. If they are not known, we cannot derive a period interval for the SCC. However, we can still use the relationship between the super task and its SCC as follows: By the definition of the maximum (minimum) cycle mean of a SCC with all AND (OR/skipped/joint) tasks, the upper (lower) bound of the period interval of the super task defines an upper (lower) bound on the mean of *every* cycle in the SCC. This result can be used to derive additional latency bounds on the tasks at the bottom level.

6 Experimental Results

We have applied the rate derivation algorithms to two examples: the task graph in [7] and the dashboard controller in [2]. The application to the task graph in [7] is discussed in [6]. We now discuss the application to the dashboard controller. The derivation for the dashboard controller will be done using real numbers.

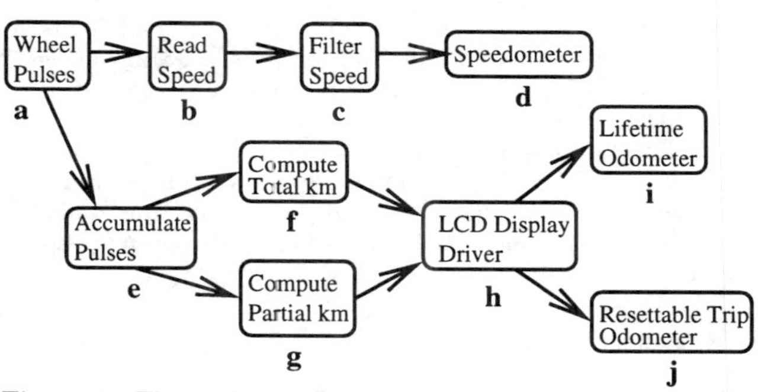

Figure 6: The task graph for the dashboard controller. It consists of the tasks for the speedometer and two odometers.

Figure 6 gives the task graph for the dashboard controller. The speedometer (task d) registers vehicle speed in the range of 0-260 km/h where any speed value less than 5 km/h is regarded as zero. The odometers (tasks i and j) register distance traveled at increments of 0.1 km starting from 0 km. The dashboard controller gets four pulses from task a for every rotation of the tire. The tire travels 0.66 m per rotation.

In the above task graph, task h is an OR/skipped/disjoint task, and the other tasks are all AND/unskipped tasks. As we need the rate of the input coming into the system, we will first derive the period interval for the input task, task a. Task a generates one pulse or one token per one quarter rotation of the tire. The vehicle travels this distance in (0/66/4 m)/v seconds when it travels at a speed of v km/h. As $v \in [5, 260]$ km/h, task a will have the following period interval $T_a = [(0.66m/4)/(260km/h), (0.66m/4)/(5km/h)] = [2.28, 118.80]$ ms. Task b measures the instantaneous speed by counting the tokens coming from task a. To compute a speed value, task b needs at least two tokens. We know this fact because of the functionality of task b although we do not know how this task will be implemented. Hence, the period interval for task b is $T_b = 2 * T_a = 2 * [2.28, 118.80] = [4.56, 237.60]$ ms. Tasks c and d take one speed value from their respective predecessors, process it, and produce another speed value. This means that they have the same period interval as T_b.

The derivation for the odometers proceeds as follows: From the functionality of task e, we know that it needs to accumulate enough tokens to compute an increment of 0.1 km. Now, let us derive the number of tokens that task e needs. As task a produces one token for every 1/4 rotation (0.66 m/4), task e needs to collect $\lceil 0.1\,km/(0.66\,m/4) \rceil = 607$ tokens. This amounts to a period interval of $T_e = 607 * T_a = 607 * [2.28, 118.80]$ ms $= [1.38, 72.11]$ s. Tasks f and g have the same period as task e because they need to use one token per execution, which we know from their functionalities. Task h passes every token it gets from task f to task i and every token from task g to task j. Also, it does this every time it gets a token, regardless of the identity of its predecessors. Hence, it is an OR/skipped/disjoint task. We do not know the exact time separation between token arrivals from tasks f and g, but from their period intervals, we know that the smallest time separation can be some $\epsilon > 0$, and that the largest time separation can be 72.11 s, which is the upper bound of the period interval for tasks f and g; thus, $T_h = [\epsilon, 72.11]$ s. Finally, we know that tasks i and j consumes one token per execution, so they have the same period interval as task h.

We can validate the above derivations by the implementation parameters given in [2]. One parameter is the period value selected for task b. This value is chosen to be 250 ms in [2]. As the derived period interval for T_b is $[4.56, 237.60]$ ms, we validate this choice. Another parameter is the response time of the odometer. This value is measured to be at most 363 s. The response time that we derive is $5 * 72.11 = 360.55$ s, which is satisfied by the implementation. The advantage of the rate derivation here is that it derives the time budgets far before the implementation, aiding in the design exploration.

7 Conclusions

This paper addresses the problem of deriving the system's internal constraints from its task structure and external constraints so that the designers can keep the system's functional as well as temporal correctness under control from very early stages in the design flow onwards. This paper proposes a generalized task graph model suitable for modeling real-life embedded systems, timing constraint derivation and validation techniques, and a timing-driven hardware/software codesign methodology. This methodology provides all the time budgets before the implementation of the system.

References

[1] Audsley, N. C., and et al. Data consistency in hard real-time systems. Tech. Rep. YCS 203, Univ. of York, England, June 1993.

[2] Balarin, F., and et al. *Hardware-Software Co-Design of Embedded Systems: The POLIS Approach.* Kluwer Academic Publ., Boston, MA, USA, 1997.

[3] Buck, J. T., and Lee, E. A. The token flow model. In *Advanced Topics in Dataflow Computing and Multithreading,* (1993), L. Bic, G. Gao, and J. Gaudiot, Eds., IEEE.

[4] Chou, P., Walkup, E. A., and Borriello, G. Scheduling for reactive real-time systems. *IEEE Micro* (Aug. 1994), 37–47.

[5] Dasdan, A., Mathur, A., and Gupta, R. K. RATAN: A tool for rate analysis and rate constraint debugging for embedded systems. In *Proc. Euro. Design and Test Conf.* (1997), IEEE, pp. 2–6.

[6] Dasdan, A., Ramanathan, D., and Gupta, R. K. A methodology for interactive design and validation of embedded real-time systems. Submitted to the acm trans. on design automation of electronic systems., 1998.

[7] Gerber, R., Hong, S., and Saksena, M. Guaranteeing real-time requirements with resource-based calibration of periodic processes. *IEEE Trans. Software Eng. 21,* 7 (July 1995), 579–92.

[8] Gillies, D. W., and Liu, J. W.-S. Scheduling tasks with AND/OR precedence constraints. *SIAM J. Comput. 24,* 4 (Aug. 1995), 797–810.

[9] Gomaa, H. *Software Design Methods for Concurrent and Real-Time Systems.* Addison-Wesley, Reading, MA, USA, 1993.

[10] Gunawardena, J. Periodic behavior in timed systems with AND,OR causality. Tech. Rep. STAN-CS-93-1462, Stanford Univ., 1993.

[11] Gupta, R. K. Special issue on partitioning methods for embedded systems. *Design Automation for Embedded Systems 2,* 2 (Mar. 1997), 123–261.

[12] Jeffay, K. The real-time producer/consumer paradigm: A paradigm for the construction of efficient, predictable real-time systems. In *Proc. ACM/SIGAPP Symp. on Applied Computing* (Feb. 1993), pp. 796–804.

[13] Karp, R. M., and Miller, R. E. Properties of a model for parallel computations: Determinacy, termination, and queueing. *SIAM J. Appl. Math. 14,* 6 (Nov. 1966), 1390–1411.

[14] Ku, D., and Micheli, G. D. *High Level Synthesis of ASICs Under Timing and Synchronization Constraints.* Kluwer Academic Publ., Boston, MA, USA, 1992.

[15] Lee, E. A., and Messerschmitt, D. G. Synchronous data flow. *Proc. IEEE 75,* 9 (Sept. 1987), 1235–45.

[16] Malik, S., Martonosi, M., and Li, Y.-T. S. Static timing analysis of embedded softare. In *Proc. 34st Design Automation Conf.* (1997), ACM/IEEE, pp. 147–52.

[17] Mathur, A., Dasdan, A., and Gupta, R. K. Rate analysis of embedded systems. *ACM Trans. on Design Automation of Electronic Systems 4,* 2 (Apr. 1999).

[18] Puchol, C., and Mok, A. K. The integration of control and dataflow structures in distributed hard real-time systems. In *Proc. 2nd Wrkshp on Parallel and Distributed Real-Time Syst.* (Apr. 1994), IEEE, pp. 104–7.

[19] Seto, D., and et al. On task schedulability in real-time computer-controlled systems. In *Proc. 17th IEEE Real-Time Systems Symp.* (Dec. 1996), pp. 13–21.

[20] Yakovlev, A., and et al. On the models for asynchronous circuit behavior with OR causality. *Formal methods in System Design 9,* 3 (Nov. 1996), 189–233.

[21] Yen, T.-Y. Hardware-software co-synthesis of distributed embedded systems. PhD thesis, Princeton Univ., 1996.

Generic Global Placement and Floorplanning

Hans Eisenmann and Frank M. Johannes

http://www.regent.e-technik.tu-muenchen.de

Institute of Electronic Design Automation
Technical University Munich
80290 Munich
Germany

Abstract

We present a new force directed method for global placement. Besides the well-known wire length dependent forces we use additional forces to reduce cell overlaps and to consider the placement area. Compared to existing approaches, the main advantage is that the algorithm provides increased flexibility and enables a variety of demanding applications. Our algorithm is capable of addressing the problems of global placement, floorplanning, timing minimization and interaction to logic synthesis. Among the considered objective functions are area, timing, congestion and heat distribution. The iterative nature of the algorithm assures that timing requirements are precisely met. While showing similar CPU time requirements it outperforms Gordian by an average of 6 percent and TimberWolf by an average of 8 percent in wire length and yields significantly better timing results.

1 Introduction

Automated cell placement for VLSI circuits has always been a key factor for achieving designs with optimized area usage and timing behavior. Beyond this, the deep-submicron era is posing new challenges onto a placement tool: Meeting timing specifications is becoming more difficult, floorplanning requires that larger designs are placed in less time and a variety of additional physical and geometrical constraints must be fulfilled simultaneously.

A common formulation of the placement objective is to minimize wire length under the constraint that cells don't overlap each other. Early formulations of the placement problem use forces for reducing the overlaps between cells [1]. Complexity and convergence problems quickly led to the development of more powerful methods. The current state-of-the-art placement tools for handling large designs can be classified into two categories based on how they make sure that the placement is free of overlaps. The first class consists of algorithms which keep the placement free of overlaps during the whole placement process. Among them, the simulated annealing method showed excellent results [2]. The second class of algorithms is based on a hierarchical subdivision of the placement area with a corresponding partitioning of the set of cells. Within this approach, a min-cut objective [3, 4] has been used successfully and the combination with a quadratic objective function showed best results [5, 6, 7]. Partitioning based methods typically yield a placement with small overlaps between cells, a so-called global placement, which has to be made free of overlaps by a final placement step.

Decreasing feature sizes require that timing issues are considered during layout synthesis. Among various techniques to reduce circuit delays, placement algorithms can be extended to be timing driven. Timing driven placement aims at minimizing the wire length of nets along critical paths. Net based approaches transform path constraints to net constraints or net weights [8] whereas path based approaches can consider the length of the longest path directly [9].

The mixed-block/cell placement occurs in floorplanning applications. It is typically converted into a block placement problem by assigning cells to flexible blocks. Flexible block placement can be solved efficiently [10]. Alternatively, the locations of the blocks are set constant before or during the placement process.

We present a force directed method which uses a new approach of dealing with cell overlaps. We use the well-known force directed formulation and apply additional forces to reduce cell overlaps and to distribute cells over the placement area. Our iterative approach has the advantage that no hard constraints are used during the placement procedure. This property makes our approach much more powerful than partitioning based methods which make irreversible decisions at early stages based on incomplete or inaccurate information. Moreover, avoiding hard constraints gives the algorithm the flexibility to address a variety of applications and objective functions. For example, our algorithm is the first one which is able to handle large mixed block/cell placement problems without treating blocks and cells differently. Contrary to existing state-of-the-art methods, our approach can assure that timing requirements are precisely met.

The rest of the paper is organized as follows: In section 2, we formulate the placement task using the well-known wire length dependent forces and add an additional force to each cell. This formulation transforms the placement problem into the problem of finding the additional forces. Section 3 addresses the problem of determining the additional forces to reduce cell overlaps and to adapt the placement to the placement area. The results from section 2 and 3 are used in section 4 to formulate an iterative algorithm. Section 5 illustrates applications for generic cell placement, floorplanning, timing minimization and interaction with logic synthesis. In section 6, comparisons with state-of-the-art algorithms demonstrate the superiority of our approach with regard to wire length and timing optimization.

DAC 98, San Francisco, California
©1998 ACM 0-89791-964-5/98/06..$5.00

2 Problem Formulation

In this section, we begin with the standard formulation of the placement problem where a quadratic objective function is used to model wire length. We introduce additional forces working on each cell. This formulation transforms the problem of finding cell coordinates into the problem of finding forces. The benefit of our formulation is that it inherently incorporates wire length minimization. The problem of determining additional forces is addressed in section 3.

2.1 Quadratic Objective Function and Additional Forces

Let n be the number of movable cells in the circuit, x_i the x-coordinate of the center of cell i and y_i the y-coordinate of the center of cell i. A placement of the circuit can be described by the $2n$-dimensional vector $\vec{p} = (x_1, \ldots, x_i, \ldots, x_n, y_1, \ldots, y_i, \ldots, y_n)^T$.

As usual, we are modeling the circuit connectivity as a graph. Cells are modeled as vertices and nets are modeled as edges. A net connecting k cells yields a clique in the graph. The clique consists of $k(k-1)/2$ edges with weight $1/k$ connecting each vertex with all other vertices.

We formulate the cost of an edge as the squared Euclidean distance between its adjacent vertices multiplied with the weight of the edge. The squared Euclidean distance between two movable cells i and j is $(x_i - x_j)^2 + (y_i - y_j)^2$. If cell i is connected to a fixed cell with coordinates (x_f, y_f), the distance is $(x_i - x_f)^2 + (y_i - y_f)^2$. Consequently, the objective function sums up the cost of all edges and can be written in matrix notation [11]

$$\frac{1}{2}\vec{p}^T C \vec{p} + \vec{d}^T \vec{p} + const \tag{1}$$

by using the $2n \times 2n$ symmetric matrix C and the $2n$-dimensional vector $\vec{d}$. For example, the x-part of the connection between two movable cells i and j is $(x_i - x_j)^2 = x_i^2 - 2 \cdot x_i \cdot x_j + x_j^2$. The first term contributes to the diagonal of C at row i, the second term causes negative entries at row i, column j and at row j, column i. The third term is a contribution to the diagonal of C at row j. In case of a fixed connection, $(x_i - x_f)^2$ evaluates to $x_i^2 - 2 \cdot x_i \cdot x_f + x_f^2$. The first term is a contribution to the diagonal of C, the second term gives a negative entry at $\vec{d}$ in row i and the third term affects the constant part of (1). The objective function (1) is minimized by solving the linear equation system

$$C\vec{p} + \vec{d} = 0 \tag{2}$$

This formulation is equivalent to modeling nets as springs and calculating the state of equilibrium. In detail, row i (row $i+n$) of equation system (2) states that the force working on cell i is zero in x direction (y direction).

2.2 Additional Forces

In the following, we introduce additional forces working on each cell. We extend equation (2) with the force vector $\vec{e}$ to model constant additional forces:

$$C\vec{p} + \vec{d} + \vec{e} = 0 \tag{3}$$

The force vector $\vec{e}$ contains the additional forces working on the cells in the x and y direction. Equation (3) transforms the problem of finding a placement into the problem of finding additional forces $\vec{e}$ and calculating the placement $\vec{p}$ according to (3). It is easy to show that the introduction of forces does not restrict the solution space, i.e. any given placement can fulfill equation (3) if the additional forces $\vec{e}$ are chosen appropriately.

3 Determination of the Additional Forces

In this section, we present a specific choice of how additional forces can be used. We take advantage from the fact that equation (2) already inherently minimizes wire length and formulate a specific choice for the additional forces from which we derive a unique set of forces.

3.1 Specific Choice of Additional Forces

Additional forces shall be used to distribute the cells evenly over the placement area.

Motivation: Solving equation (2) gives the global optimum with regard to squared wire length. However, equation (2) neither considers the overlaps of the cells nor the placement area. Therefore, the resulting placement is overlapping and not well distributed over the placement area in general. We use the additional forces in equation (3) to remove cell overlaps and to adapt the placement to the placement area.

3.2 Requirements

In the following, we derive four requirements for the additional forces.

1. For a given placement, the additional force working on a cell depends only on the coordinates of the cell.

   Motivation: If the task of the forces is to reduce the overlaps of the cells, it is only natural that a cell at the same place as another cell gets the same force.

2. Regions with higher density are the sources of the forces. Regions with lower density are the sinks.

   Motivation: Forces should move cells away from high density regions and lead them to low density regions.

3. The forces do not form circles.

   Motivation: The purpose of the forces is to improve the distribution. A circular force does not improve the distribution and is of no use.

4. In infinity, the force should be zero.

   Motivation: This requirement prevents the existence of a constant force.

3.3 Mathematical Formulation and Solution

We now show that these four requirements are sufficient to uniquely determine the forces. The basic background of how the requirements can be put into mathematical formulations can be found in [12].

The first requirement allows us to write the additional force $\vec{f_i}$ at cell i as a function of x and y: $\vec{f_i} = \vec{f}(x,y)|_{x=x_i, y=y_i}$

For the second requirement we define a supply-and-demand model for describing the density at a given point. We first define the rectangle function $R(z)$:

$$R(z) = \begin{cases} 1 & \text{if } -1/2 < z < 1/2 \\ 0 & \text{otherwise} \end{cases}$$

The width of cell i is denoted by w_i, the height of cell i is h_i. The function $a_i(x,y)$ describes the area of cell i and is defined as $a_i(x,y) = R(\frac{x-x_i}{w_i}) \cdot R(\frac{y-y_i}{h_i})$. The value of $a_i(x,y)$ is one for a point (x,y) which is covered by cell i and zero otherwise.

Using W as the width of the placement area and H as its height, the area function $A(x,y)$ of the placement area can be defined similarly as $A(x,y) = R(\frac{x}{W}) \cdot R(\frac{y}{H})$. The value of $A(x,y)$ is one for a point (x,y) within the placement area and zero otherwise.

With the definition of s as the quotient of total cell area and placement area we define a function for the density $D(x,y)$:

$$D(x,y) = \sum_i a_i(x,y) - s \cdot A(x,y) \text{ with } s = \frac{\sum_i w_i \cdot h_i}{W \cdot H} \quad (4)$$

In other words: The density at a certain point within the placement area is the number of cells which cover the point, minus s.

The first term of equation (4) can be seen as the demand of area by the cells. The second term is the supply of area from the placement area scaled by factor s. The integral of $D(x,y)$ over the whole area is zero. This is achieved by scaling the supply with s. Let us note that $D(x,y) > 0$ at locations with higher density than desired and $D(x,y) < 0$ at locations with lower density than desired.

Using a proportional constant k, requirement 2 gives [12]

$$\vec{\nabla}\vec{f}(x,y) = k \cdot D(x,y) \quad (5)$$

Requirement 3 means that $\vec{f}(x,y)$ is conservative, i.e. there exists a scalar function $\Phi(x,y)$ with

$$\vec{\nabla}\Phi(x,y) = \vec{f}(x,y) \quad (6)$$

Combining (5) and (6) results in Poisson's equation

$$\Delta\Phi(x,y) = k \cdot D(x,y) \quad (7)$$

with boundary conditions from requirement 4

$$\lim_{|\vec{r}| \to \infty} |\vec{\nabla}\Phi(x,y)| = 0 \text{ with } \vec{r} = (x,y)^T \quad (8)$$

This standard problem has a unique solution for $\vec{f}(x,y)$ [13] :

$$\vec{f}(x,y) = \frac{k}{2\pi} \int_{-\infty}^{\infty} \int_{-\infty}^{\infty} D(x',y') \frac{\vec{r} - \vec{r}'}{|\vec{r} - \vec{r}'|^2} dx'\, dy' \quad (9)$$

3.4 Interpretation

In the case of modeling cells as points and subdividing the placement area into places with unit area, the above integral becomes a discrete sum and equation (9) can be interpreted informally to make its meaning and its nature more clear:

1. The force working on a cell is the superposition of the forces originating from other cells and places.
2. The force exerted on a cell by another cell (place) is repelling (attracting) with a strength proportional to the inverse of their Euclidean distance.
3. The direction of the force is parallel to the straight line between the cells (between the cell and the place, resp.).

4 Basic Algorithm

We now combine the results from section 2 and 3. The forces in section 2 have been assumed to be constant whereas section 3 states that forces depend on the location of the corresponding cell. We resolve this conflict by introducing an iterative algorithm which determines the forces according to the current placement. These forces are set constant and used to calculate a new placement. The new placement then is the base for the next iteration step and so on.

We first describe a single iteration step called 'placement transformation'. Then, we show how a global placement can be achieved by successively applying placement transformations.

4.1 Placement Transformation

We call one step of the algorithm a placement transformation. The input placement can be arbitrary. The transformation step can be applied to fully overlapping placements as well as nearly legal placements. We do not consider information from the past like the number of iterations or a certain cooling schedule.

We determine the forces $\vec{f}_i$ of the current placement according to equation (9). We choose the proportional constant k so that the maximum strength of all forces $\vec{f}_i$ is equivalent to the force of a net with length $K \cdot (W + H)$ with a constant parameter K as user parameter. The parameter K determines how strongly the additional forces influence the placement and therefore determines the speed of convergence and the quality of the results. Next, we add the determined additional forces $\vec{f}_i$ to the force vector $\vec{e}$: $(\Delta e_i, \Delta e_{n+i})^T = \vec{f}_i$. Matrix C is set up as described in section 2.1. We apply a linearization scheme for adjusting netweights [14] and solve equation (3) by using a conjugate gradient approach with preconditioning. During the solving step of equation (3) we assume that the additional forces $\vec{e}$ remain constant.

4.2 The Iterative Algorithm

The complete algorithm consists of three key elements:

1. Initialization: In the initial routine, all cells are placed at the center of the placement area and the forces $\vec{e}$ are set to zero.
2. Iteration loop: We iteratively apply placement transformations. We use K to control the desired speed of the algorithm. We choose $K = 0.2$ for standard behavior while $K = 1.0$ is used for fast operation. Each iteration makes the distribution of the cells more even and adapts the placement more to the placement area.
3. Stopping criterion: We stop the iterations when there exists no empty square within the placement area which is larger than four times the average area of a cell. Our experiments showed that this criterion is sufficient for a desired even distribution of cells.

5 Application to Placement Tasks

In this section we describe what placement tasks can be addressed by the algorithm. We also describe which extensions have to be made for the different applications.

Timing Optimization We apply an iterative net weighting approach in order to optimize the timing behavior. We use the Elmore delay model based on the half perimeter of the enclosing rectangle as net delay. At iteration step m, each net j has a criticality $c_j^{(m)}$. We initialize $c_j^{(0)}$ with zero and update it as follows.

Before each placement transformation, we carry out a longest path search for timing analysis. This gives us the maximum delay and the minimum slack for each net. Please note that our approach does not rely on this special model as it works well with any kind of timing analysis.

We define the criticality of net j at step m as

$$c_j^{(m)} = \begin{cases} (c_j^{(m-1)} + 1)/2 & \text{if net } j \text{ is among the 3 percent most critical nets} \\ c_j^{(m-1)}/2 & \text{otherwise} \end{cases}$$

The criticality describes how critical a net tends to be in general and is used later for our weight updating scheme. A net which is critical at step m contributes 50 % to its criticality, a net which is critical at step $m-1$ contributes 25 % and so on. For example, if a net was never critical its criticality is zero whereas an always critical net has a criticality of one. This scheme effectively reduces oscillations of netweights.

Each net j has an associated weight $w_j^{(m)}$. The initial weights $w_j^{(0)}$ are set to one. We multiply the weight $w_j^{(m-1)}$ of net j with $1+c_j^{(m)}$ and get the new weight $w_j^{(m)}$. For example, a net which has never been critical has $c_j^{(m)} = 0$ and keeps its weight $w_j^{(m)} = w_j^{(m-1)}$. The weight of a net which has always been critical is multiplied by a factor of 2: $w_j^{(m)} = 2 \cdot w_j^{(m-1)}$.

Our timing optimization especially benefits from the fact that even in late stages the placement has the ability to change globally because there are no hard constraints. This property makes our approach much more powerful than other methods where decisions at early stages have to be made based on incomplete or inaccurate information.

Meeting Timing Requirements The task of meeting a given timing requirement is different from optimizing the longest path delay. A timing requirement is a constraint under which other criteria should be optimized.

The typical approach with existing methods is iterative on a coarse level. The timing behavior of the resulting placement is compared to the specifications. Then, the input data is changed to make the method more aggressive or less aggressive. Additional placement runs are carried out until the desired timing behavior is obtained.

In order to avoid these multiple placement runs, we propose an extension to the described algorithm which directly meets timing requirements. First, we run the basic non-timing driven algorithm until it converges. We now have an area optimized placement. Then, we carry out a net weight adaption before each placement transformation. This net weight adaption is the same as described for timing optimization. In order to get a timing/area tradeoff curve we record the wire length and timing data for each step. We stop after the timing requirements are met. Since we used the resulting placement for timing analysis we can assure that the placement meets precisely the timing requirements.

Compared to existing methods, our approach has two major advantages: It guarantees that the timing requirements are precisely met if it is possible at all. No further reruns of the placement algorithm are necessary. Secondly, our two-phase approach provides a tradeoff curve showing which timing can be achieved at which area cost.

Mixed Block Placement and Floorplanning Using placement programs for floorplanning can yield problem sizes of several hundreds of thousands of cells. Floorplanning is a highly interactive task and the placement step is carried out multiple times resulting in the need for fast placement. Moreover, floorplanning poses the problem of placing big blocks and small cells simultaneously. Existing approaches are either not able to cope with the large problem size or disregard the dimension of the big blocks (at least at some step of the algorithm). Simulated annealing methods suffer from the fact that moving big cells is computationally intensive and partitioning based methods disregard the dimension of the big cells.

Contrary to that, our algorithm is the first one which is able to handle large mixed block/cell placement problems without treating blocks and cells differently.

Congestion and Heat Driven Placement Since the modeling of the force sources is done by a supply and demand model, it is straightforward to extend the procedure to incorporate congestion information. Before each placement transformation a routing estimation is executed. Then, a congestion map is determined which is used in combination with the density $D(x,y)$ to calculate additional forces. With this approach, the placement and the congestion map converge simultaneously and yield a placement which directly considers the resulting congestion. Additionally, by replacing the congestion map with a heat map we can use the same approach to avoid hot spots in the layout.

ECO and Interaction with Logic Synthesis Our approach is well suited for netlist changes during placement and for considering netlist changes after placement. This feature is typically needed by the application of ECO, gate resizing techniques and fully automated interaction with logic synthesis. The main requirement is that the existing placement is disturbed as little as possible.

Our method starts from the given placement and introduces additional forces according to the density deviations arising from netlist changes. Any changes in the netlist result in additional forces which move the surroundings slightly in order to adapt to the changed situation. The placement of cells relative to each other is preserved. The deviations in density are typically small which leads to small additional forces resulting in small changes for the placement. Compared to other methods, our approach has the advantage that an incrementally changed netlist results in small changes in the placement.

6 Experimental Results

We used the benchmark set from [15] for comparison. Table 1 shows the parameters of the circuits. Wire length is measured by summing up the half perimeter of the enclosing rectangle for each net. CPU times are measured in seconds for an Alphastation 250/4-266. CPU times of other approaches are scaled according to [16].

6.1 Standard Cell Benchmarks

We compare our results to the state-of-the-art methods Gordian/Domino [14, 17] and TimberWolf [2]. As final placer for the proposed method we used Domino [17]. The reported CPU times include the time used for final placement. The Gordian/Domino values are taken from [2]. We take the TimberWolf values from [18, 19] which are summarized in [2].

Since results for circuit *struct* have not been published, we ran TimberWolf 1.3.0 and Gordian/Domino 9.4 in their default configuration to obtain the values for circuit *struct*.

Table 1 lists the absolute values. The columns 'wire length' show the wire length in meters and 'CPU' the CPU time in seconds. We used the standard mode ($K = 0.2$) of our approach.

For better comparison, we also list the results relative to other methods in table 2. The columns 'improvement' give the wire length improvement in percent. Positive values mean that our approach is better. Columns 'rel. CPU' give the CPU time of our approach divided by the (scaled) time of the compared approach. Therefore, values smaller than 1.0 mean that our approach is faster.

One can see that our results are comparable with TimberWolf while using one third of the runtime. A comparison under similar runtime conditions (approximately 40 percent slower) outperforms TimberWolf by 7.9 percent and Gordian/Domino by 6.6 percent in wire length on the average.

We further investigated the quality/runtime tradeoff of our approach. Our motivation is that a fast placement is useful to achieve a placement estimation during the floorplanning phase. We compared the fast ($K = 1.0$) and the standard ($K = 0.2$) mode of our

circuit	# cells	#nets	# rows	T.-Wolf [19] wire length [m]	T.-Wolf [19] CPU [s]	T.-Wolf [18] wire length [m]	T.-Wolf [18] CPU [s]	Go./Do. [17] wire length [m]	Go./Do. [17] CPU [s]	Our Approach wire length [m]	Our Approach CPU [s]
fract	125	147	6								
primary1	752	904	16	-	-	0.99	22	0.88	17	0.87	37
struct	1888	1920	21	0.364	272	-	-	0.377	30	0.338	40
primary2	2907	3029	28	-	-	3.72	125	3.68	92	3.72	152
biomed	6417	5742	46	1.62	885	1.88	216	1.95	264	1.78	284
industry2	12142	13419	72	13.53	3285	14.34	925	15.80	958	14.6	1283
industry3	15059	21940	54	42.84	4504	44.67	877	44.97	1034	45.1	1605
avq.small	21854	22124	80	5.41	4587	6.13	1302	5.68	1744	4.91	1741
avq.large	25114	25384	86	5.86	5501	6.81	1560	6.21	2108	5.38	2031

Table 1: Benchmarks: Wire Length and CPU Time

circuit	T.-Wolf [19] % improvement	T.-Wolf [19] rel. CPU	T.-Wolf [18] % improvement	T.-Wolf [18] rel. CPU	Go./Do. [17] % improvement	Go./Do. [17] rel. CPU
primary1	-	-	+12.1	1.68	+1.1	2.17
struct	+7.1	0.14	-	-	+11.5	1.33
primary2	-	-	0.0	1.21	-1.1	1.65
biomed	-9.8	0.32	+5.3	1.31	+8.7	1.07
industry2	-7.9	0.39	-1.8	1.38	+7.5	1.33
industry3	-5.2	0.35	-0.9	1.83	-0.2	1.55
avq.small	+9.2	0.37	+19.9	1.33	+13.5	0.99
avq.large	+8.1	0.36	+20.9	1.30	+13.3	0.96
average	+0.2	0.32	+7.9	1.43	+6.6	1.38

Table 2: Comparisons to Other Approaches: Wire Length Improvement and Relative CPU Times

approach. Using the fast mode, we can calculate a placement in approximately one third of the time compared to the standard mode. The average wire length increase is 6 percent. Within 10 minutes, the fast approach is capable of obtaining a legal placement for 25 000 cells which could not be obtained by any other approach.

6.2 Timing

For better comparison to existing methods, we introduce a lower bound for the length of the longest path by setting all wire lengths to zero and performing a timing analysis. This lower bound can only be reached if all nets of the longest path have length zero which means that all cells would be interconnected by abutment. Then, we calculate the longest path of a placement without timing optimization and subtract the lower bound. This gives the optimization potential of the placement. Next, we calculate the reduction of the longest path by using timing optimization. Dividing this value by the optimization potential tells us how much the method could exploit the optimization potential, i.e. how good it is. Measuring the performance in the proposed way helps us to reduce the influence of differences in net models, timing models and timing parameters.

We compare our method to [20] and [21] which has been shown to be superior to [22]. Since having big nets in the longest path is not realistic we disregard nets with more than 60 pins for timing analysis in our approach. Therefore, the lower bound used for our approach is different from the lower bound used for the other approaches in the case of 'avq.small' and 'avq.large'. Using the same lower bound would give unrealistic good results.

As in [21], we assume a capacitance per length of 242 pF/m and a resistance per length of 25.5 kΩ/m. Table 3 shows the results. Columns 'without timing' list the results without timing optimization, 'with timing' the results for timing optimization. All timing values are in nanoseconds.

Table 4 demonstrates the capability of exploiting the optimization potential. The columns 'lower bound' list the lower bound in nanoseconds which we used for calculating the optimization potential. The columns 'exploitation' show the exploitation of the optimization potential. Higher values mean better timing optimization. Columns 'relative CPU' contain the CPU time requirement relative to our approach. Values larger than one mean that the compared method is slower.

On the average, the compared methods can exploit the optimization potential by up to 42 percent whereas our method utilizes 53 percent of the optimization potential and requires less CPU time.

7 Conclusions

The upcoming placement tasks pose new challenges to placement tools. They must be capable of placing larger netlist in less time, of considering more geometrical and physical constraints, and of dealing with more complicated timing requirements. Moreover, they must provide a flexible interface for manual changes as well as for changes introduced by logic synthesis tools.

To address these tasks, we proposed an iterative force directed method. A new approach has been presented which uses additional forces to remove cell overlaps and to adapt the placement to the placement area. The approach has the advantage that no hard constraints are used during the placement procedure. This property makes our approach much more powerful than existing methods which make irreversible decisions at early stages that are based on incomplete or inaccurate information. The avoidance of hard constraints gives the algorithm the flexibility to address a variety of applications and objective functions. For example, our algorithm is the first one which is able to handle large mixed block/cell placement problems without treating blocks and cells differently. Contrary to state-of-the-art approaches, our approach can assure that timing requirements are precisely met. While using comparable or less CPU time, our approach outperforms Gordian and TimberWolf by an average of 6 and 8 percent in wire length and yields significantly better timing results.

8 Acknowledgments

The authors thank the reviewers for their valuable suggestions. The first author thanks Kurt Antreich for his continuous support.

circuit	TimberWolf [20] without timing [ns]	with timing [ns]	CPU [s]	Speed [21] without timing [ns]	with timing [ns]	CPU [s]	Our Approach without timing [ns]	with timing [ns]	CPU [s]
fract	208	131	500	-	-	-	21.2	19.8	3
struct	907	449	1450	108	101	32	92.5	90.1	50
biomed	-	-	-	126	65	218	48.6	35.7	397
avq.small	1106	798	31993	827	580	3792	102	80	2791
avq.large	-	-	-	836	605	7185	113	94	3642

Table 3: Timing Results: Longest Path and CPU Time

circuit	TimberWolf [20] lower bound [ns]	exploi-tation	rel. CPU	Speed [21] lower bound [ns]	exploi-tation	rel. CPU	Our Approach lower bound [ns]	exploi-tation	rel. CPU
fract	18.5	40%	164	-	-	-	18.5	51%	1
struct	84.0	55%	28	84.0	29%	0.64	84.0	28%	1
biomed	-	-	-	27.0	61%	0.55	27.0	59%	1
avq.small	142	31%	11	142	36%	1.35	69.9	68%	1
avq.large	-	-	-	142	33%	1.97	79.9	57%	1
average		42%	67		40%	1.13		53%	1

Table 4: Relative Timing Results: Exploitation of Optimization Potential and relative CPU requirements

References

[1] N. Quinn and M. Breuer, "A force directed component placement procedure for printed circuit boards," *IEEE Trans. CAS*, vol. CAS-26, pp. 377–388, June 1979.

[2] W.-J. Sun and C. Sechen, "Efficient and effective placement for very large circuits," *IEEE Trans. CAD*, vol. 14, no. 3, pp. 349–359, 1995.

[3] A. Dunlop and B. Kernighan, "A procedure for placement of standard-cell VLSI circuits," *IEEE Trans. CAD*, vol. CAD-4, pp. 92–98, Jan. 1985.

[4] D. J.-H. Huang and A. B. Kahng, "Partitioning based standard cell global placement with an exact objective," in *ISPD*, pp. 18–25, 1997.

[5] C.-K. Cheng and E. S. Kuh, "Module placement based on resistive network optimization," *IEEE Trans. CAD*, vol. CAD-3, pp. 218–225, July 1984.

[6] R.-S. Tsay, E. S. Kuh, and C.-P. Hsu, "PROUD: A fast sea-of-gates placement algorithm," in *ACM/IEEE DAC*, vol. 25, pp. 318–323, 1988.

[7] J. M. Kleinhans, G. Sigl, F. M. Johannes, and K. J. Antreich, "GORDIAN: VLSI placement by quadratic programming and slicing optimization," *IEEE Trans. CAD*, vol. CAD-10, pp. 356–365, Mar. 1991.

[8] A. E. Dunlop, V. D. Agrawal, D. N. Deutsch, M. F. Jukl, P. Kozak, and M. Wiesel, "Chip layout optimization using critical path weighting," in *ACM/IEEE DAC*, vol. 21, pp. 133–136, 1984.

[9] M. A. B. Jackson and E. S. Kuh, "Performance-driven placement of cell based IC's," in *ACM/IEEE DAC*, vol. 26, pp. 370–375, 1989.

[10] R. Otten, "Efficient floorplan optimization," in *IEEE ICCD*, pp. 499–501, Oct. 1983.

[11] K. M. Hall, "An r-dimensional quadratic placement algorithm," *Management Science*, vol. 17, pp. 219–229, Nov. 1970.

[12] R. Ellis and D. Gulick, *Calculus*. Harcourt Brace Jovanovich, 1991.

[13] W. E. Williams, *Partial Differential Equations*. Oxford University Press, 1980.

[14] G. Sigl, K. Doll, and F. M. Johannes, "Analytical placement: A linear or a quadratic objective function?," in *ACM/IEEE DAC*, 1991.

[15] "www.cbl.ncsu.edu/benchmarks/layoutsynth92/"

[16] "performance.netlib.org/performance/html/PDStop.html"

[17] K. Doll, F. M. Johannes, and K. J. Antreich, "Iterative placement improvement by network flow methods," *IEEE Trans. CAD*, vol. 13, pp. 1190–1200, Oct. 1994.

[18] W.-J. Sun and C. Sechen, "Efficient and effective placement for very large circuits," in *IEEE/ACMICCAD*, pp. 170–177, 1993.

[19] W.-J. Sun and C. Sechen, "A loosely coupled parallel algorithm for standard cell placement," in *IEEE/ACMICCAD*, pp. 137–144, 1994.

[20] W. Swartz and C. Sechen, "Timing driven placement for large standard cell circuits," in *ACM/IEEE DAC*, 1995.

[21] B. M. Riess and G. G. Ettelt, "Speed: Fast and efficient timing driven placement," in *IEEE ISCAS*, 1995.

[22] A. Srinivasan, K. Chaudhary, and E. S. Kuh, "RITUAL: A performance driven placement algorithm," *IEEE Trans. CAS*, vol. CAS-39, pp. 825–840, Nov. 1992.

Congestion Driven Quadratic Placement

Phiroze N. Parakh, Richard B. Brown, Karem A. Sakallah
University Of Michigan
1301 Beal Ave, EECS - ACAL
Ann Arbor, MI. 48109-2122
1-(734)-647-0301
{phiroze, brown, karem}@umich.edu

1. ABSTRACT

This paper introduces and demonstrates an extension to quadratic placement that accounts for wiring congestion. The algorithm uses an A* router and line-probe heuristics on region-based routing graphs to compute routing cost. The interplay between routing analysis and quadratic placement using a growth matrix permits global treatment of congestion. Further reduction in congestion is obtained by the relaxation of pin constraints. Experiments show improvements in wireability.

1.1 Keywords

Congestion, Quadratic placement, Relaxed pins, Global routing, Routing models, Supply-demand.

2. INTRODUCTION[1]

The drive toward higher levels of integration has invalidated classical assumptions in VLSI CAD. Previous abstractions have proven to be inadequate, thus changing the intent of CAD from optimization of gates to the management of routing resources. A design that effectively uses its resources requires smaller area because inefficient utilization leads to more wire crossings, increased wire length and poor channeling.

With the advent of overcell routing, the goal of every place and route methodology has been to utilize all available active area to prevent spilling of routes into channels. It is this overflow of routes that accounts for an increase in area. "Congestion aware" global routers avoid this by constructing a set of routing trees for all nets in the design to match the global supply. Further heuristics are then applied to manage local congestion to enhance and improve the final route quality. The concept of supply and demand to drive global routers is favorably demonstrated in [13]. The objective of this global router is defined in terms of route uniformity and route density. However, this scheme and others like it are unlikely to achieve optimality because the quality of a routing solution is largely determined by the input placement, with congestion occurring where demand exceeds supply. Thus the need rises to model congestion within placement.

[1]This work was supported by ARPA/ARO. Contract: DAAH04-94-G-0327

Typical placement objectives involve reducing net-cut costs or minimizing wire length. Due to their constructive nature, min-cut based strategies minimize the number of net crossings but fail to uniformly distribute them [14]. Quad-partitioning schemes, first demonstrated in [8] did account for some global routing resources, but did not address supply nor account for internal routing. Congestion-driven placement based on multi-partitioning was proposed in [10], but was limited in the number of partitions due to the use of pre-computed Steiner trees. To our knowledge, congestion driven placement has not used congestion at a global level to update global placement. Alternatives to min-cut based schemes attempt to minimize wire length. The use of minimal wire length as a metric to guide placement has been successful in achieving good placement. However, it only *indirectly* models congestion and the behavior of the router. Reducing the global wire length helps reduce the *wiring demand*, but does not affect the *wiring supply*. It is entirely feasible for a minimum wire length solution to require more tracks for routing through a region, than are available. Therefore, schemes such as quadratic placement which are based solely on wire length minimization cannot adequately account for congestion. A simulated annealer in [1] used wire distribution functions and net bounding boxes for routing-resource demand modeling. However, the complexity of its wire model precluded a global view and forced the analysis to be performed in smaller blocks.

The quadratic fomulation for squared wirelength can be solved for a global minimum. However the solution contains much cell overlap which must be resolved by partitioning with appropriate constraints. This methodology is incapable of producing *sufficient non-overlap* to account for routing resources. This paper develops a framework that drives quadratic placement to relieve congestion while simultaneously solving for minimal wire length. Supply-demand weights are computed through appropriate modeling of routing resources on a region-based routing graph. These weights influence the quadratic placer into growing (or shrinking) regions based on resource demand. Repeated partitioning during quadratic placement allows the analysis to converge upon a dense routing grid. The dynamic behavior of this graph supports increasingly accurate supply-demand modeling at each iteration of the algorithm. The current implementation of the supply-demand algorithm creates global routes between regions using an A* search [2] algorithm and computes internal costs using a line-probe [3] heuristic. However, this component of the algorithm can be replaced with any appropriate route model. Further reductions in congestion, wire-length and area are obtained by the inclusion of relaxed pin constraints into the formulation.

Section 3 of the paper presents the key concepts in quadratic placement. Section 4 describes region-updating during quadratic placement along with the necessary framework for congestion analysis on a region-based grid. Results of this new methodology are presented in Section 5, followed by conclusions in Section 6.

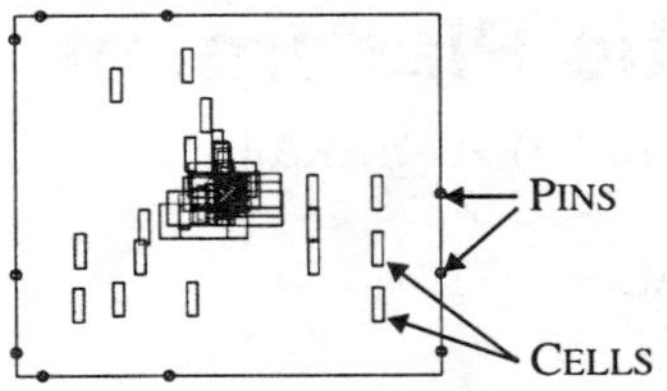

Figure 1. First iteration of Quadratic placement

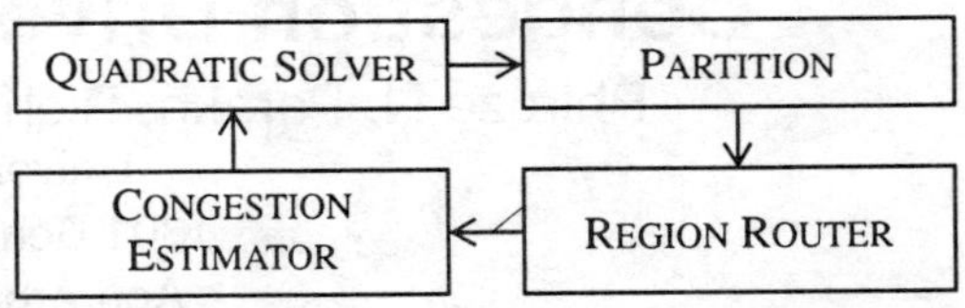

Figure 2. Congestion driven Placement methodology

3. QUADRATIC PLACEMENT

Quadratic placement engines [7, 6, 9] solve an unconstrained minimization problem, the objective function $\Phi_q(\boldsymbol{x}, \boldsymbol{y})$ of which is the squared wirelength

$$\Phi_q(x, y) = \sum_{i,j} a_{ij}(x_i - x_j)^2 + \sum_{i,j} a_{ij}(y_i - y_j)^2 \qquad (1)$$

where $\boldsymbol{x}$ and $\boldsymbol{y}$ are coordinate vectors for the n cells and $a_{ij} \in \{0, 1\}$ represents connectivity between cells i, j. This can be represented in two systems of linear equations, separable in x and y. By expansion and proper assimilation of common terms, (1) can be rewritten in the more familiar form[1] of

$$\Phi_q(\boldsymbol{x}) = \frac{1}{2}\boldsymbol{x}^T Q\boldsymbol{x} + \boldsymbol{d}_x^T\boldsymbol{x} \qquad (2)$$

and optimally minimized by solving $\nabla\Phi_q(\boldsymbol{x}, \boldsymbol{y}) = 0$, where $\boldsymbol{Q}$ is the $n \times n$ Laplacian matrix for the netlist. The vector $\boldsymbol{d}_x$ originates from the locations of the pins and can be adjusted to include port locations on the cells. The solution for Φ_q tends to group around the origin [Figure 1]. In order to disperse the cell locations and resolve overlaps, the solutions are partitioned and iterated upon.

The quadratic placer used in this work makes use of the GORDIAN algorithm by Klienhans et. al [7]. Initially, Φ_q is constructed and minimized to obtain the $\boldsymbol{x}$ and $\boldsymbol{y}$ coordinates. These vectors are used to aid the partitioner in generating new regions. New center of gravity constraints are then imposed upon each region. The process is repeated, alternating in horizontal and vertical directions until each region represents an individual cell. Generating center-of-gravity constraints for each region has the effect of distributing the cells by forcing the area-weighted mean location of cells within a region to correspond to the region's center of gravity. These constraints are embedded into (2) through $\boldsymbol{Z}$, yielding the unconstrained linear system

$$\psi(\boldsymbol{x}_I) = \frac{1}{2}\boldsymbol{x}_I^T\boldsymbol{Z}^T\boldsymbol{C}\boldsymbol{Z}\boldsymbol{x}_I + \boldsymbol{c}^T\boldsymbol{x}_I \qquad (3)$$

on the $\boldsymbol{x}_I$ independent cells. Which can be minimized by setting $\nabla\psi(\boldsymbol{x}_I) = 0$, and solving

$$\boldsymbol{Z}^T\boldsymbol{C}\boldsymbol{Z}\boldsymbol{x}_I = -\boldsymbol{c} \qquad (4)$$

Quadratic placement algorithms require pin locations, else they return a trivial solution with all cells at the origin. Graphically, the pins stretch the network of cells to induce a valid placement solution, and therefore, they have a strong impact on the solution. While it may be prudent to assign pin locations based on floorplanning, or through other top-down pin-placement approaches, the location (or order) of these pins may aggravate the placement. Consider the fact that incorrect pin ordering could twist the design and thus create congestion. In order to reduce congestion, the placement should be able to influence the pin locations.

[1] Shown for the x-dimension only.

4. CONGESTION DRIVEN PLACEMENT

Figure 2 illustrates the congestion-driven placement methodology. Fundamental to this technique is the need to model congestion and to influence the quadratic placer with a congestion metric. As in the GORDIAN algorithm, placement produced by minimizing the wire length metric is iteratively partitioned into regions, and placed with new center of gravity constraints. Before each successive placement, internal route estimation and a region-based global route are performed on each region to estimate supply-demand ratios. These ratios are used to influence the quadratic placer into growing (or shrinking) regions based on resource demand. Furthermore, congestion induced by incorrect pin ordering is relieved by relaxing the pin constraints to a single dimension. Experimental observations show that computing the internal and external routes takes much less time than a single minimization of Φ_q.

4.1 Growing and Shrinking Regions

For q regions, we define $\boldsymbol{G}$ as an $\langle n - q \times n - q\rangle$ diagonal *growth* matrix with entry g_{ii} equal to the region weight for the independent cell $\boldsymbol{x}_i$. If we let $\boldsymbol{x}_i' = \boldsymbol{G}\boldsymbol{x}_i$ and substitute for $\boldsymbol{x}_i$ in (4), we can obtain a global minimum of Φ_q by solving:

$$(\boldsymbol{G}^{-1}\boldsymbol{Z})^T\boldsymbol{C}(\boldsymbol{Z}\boldsymbol{G}^{-1})\boldsymbol{x}_i = -c' \qquad (5)$$

with $c'^T = (\boldsymbol{C}\boldsymbol{x}_0 + \boldsymbol{d}_x)^T\boldsymbol{G}^{-1}\boldsymbol{Z}$. We know that $\boldsymbol{Z}$ forms a basis for the positive definite $\boldsymbol{C}$ [7], and since $\boldsymbol{G}$ is a diagonal matrix with positive diagonal entries, $\boldsymbol{G}^{-1}\boldsymbol{Z}$ will have the same form as $\boldsymbol{Z}$ and thus (5) is a positive definite system. Standard SOR or Krylov subspace solvers [12, 5] can be used to solve the linear system.

The use of $\boldsymbol{G}$, causes cell positions to reflect supply-demand ratios. Resource limited regions are expanded (or reduced) to account for the wiring demand imposed upon them. The horizontal ratio is used to stretch the region in $\boldsymbol{y}$, while the vertical ratio stretches the region in $\boldsymbol{x}$. The growth factors disperse modules within a region and influence other cell positions while permitting the solver to minimize Φ_q. Reduction in congestion occurs due to the ability of the model to transform horizontal routes into vertical and vice-versa. In Figure 3, if a region is deemed vertically congested, then a vertical expansion is performed. At this point, internal horizontal nets could become vertical, thus relieving congestion. Alternatively, if the region is shrunk horizontally, during the next iteration more nets from the previous iteration could become vertical, further relieving congestion.

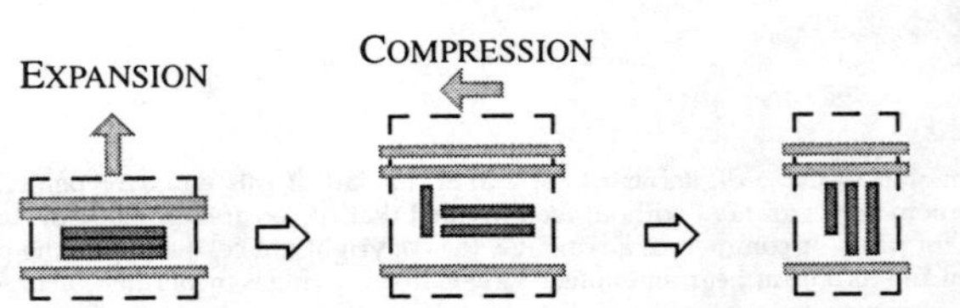

Figure 3. Example of region growth relieving congestion

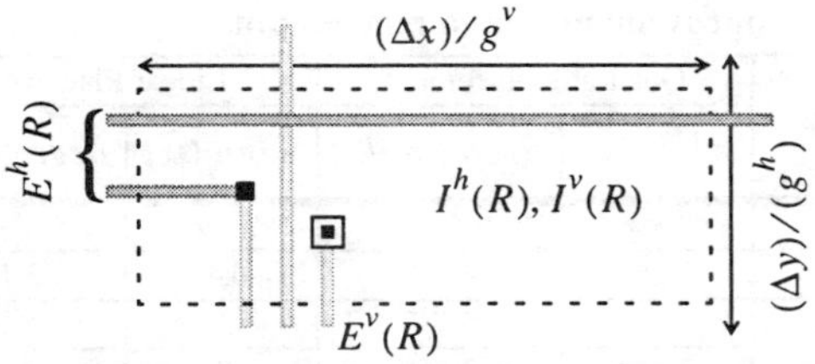

Figure 4. Region supply and demand components

4.2 Computing the Growth Matrix

After each minimization of ψ, the current regions are partitioned to generate a new set of regions. For each congestion analysis, a new routing graph is constructed using these regions. An important characteristic of this routing graph is its *dynamic* behavior. With each successive placement and partitioning, the granularity of the routing graph increases, thus increasing the quality of the analysis, until a detailed routing graph is achieved. Furthermore, since the regions reflect well-connected cells, they form a more appropriate routing graph than one produced for an arbitrary pitch. Together, these features permit the appropriate amount of supply and demand analysis at each iteration.

The supply created by each region R (Figure 4) is separated into horizontal and vertical components: R_s^h and R_s^v, and cannot exceed the capacity for the region, which is determined by the horizontal and vertical metal pitch (g^v, g^h). Internal blockages, reflecting pre-allocated routing channels and port locations on cells diminish the supply. Each cell $\mu \in R$ contributes μ_f to the total number of vertical and horizontal feed-thrus for the region. This amount is normalized to the width R_w (or height) of the region, thus producing the incident supply of a region

$$R_s^v = \frac{\sum_{\mu \in R} \mu_f^v}{(\sum_{\mu \in R} \mu_w)/R_w} \qquad (6)$$

where μ_w is the cell width. The horizontal component is expressed identically. Demand on each region is comprised of external, $E(R)$, and internal, $I(R)$, components

$$R_d^v = E^v(R) + I^v(R) \qquad (7)$$

The external cost originates from nets that span multiple regions and is computed by the region router. It is the sum of all region-routes that intersect it and includes routes that terminate within it. In order to compute the internal-route cost, a heuristic that mimics a line-probe router is used. Figure 5 illustrates the line-probe internal [3] cost for horizontal and vertical tracks. Once the supply and demand values for a region are determined, their difference determines the number of extra routing tracks required and therefore, the amount of region growth required to satisfy demand.

$$w_R^v = 1 - \frac{(R_s^v - R_d^v) \cdot g^v}{\Delta x_R} \qquad (8)$$

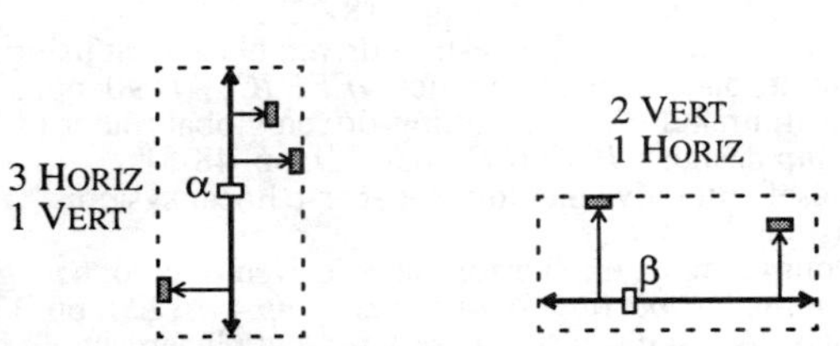

Figure 5. Internal routing model

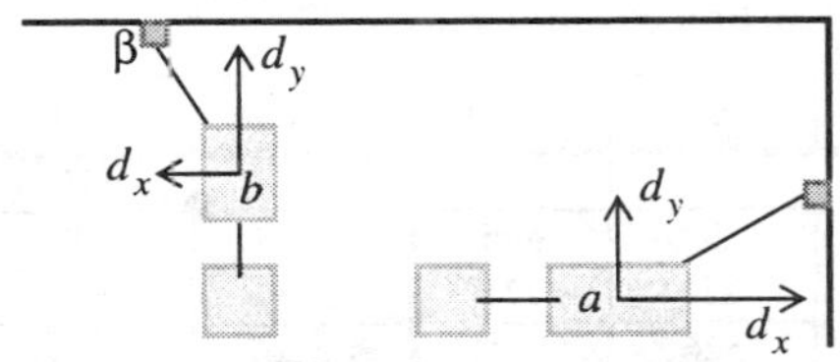

Figure 6. Effect of fixed pin locations

4.3 Relaxed Pins

The task of optimally assigning pin locations to a large block of standard cells is NP-complete. Consider the fact that the optimal pin location permits an optimal placement which is a known NP-complete problem. While classical top-down floorplanning does aid in establishing the locations of pins on a block, it does not consider the resultant effect on the block. A poor ordering of pins could, therefore, unduly "torque" the block, creating congested regions. This effect can be very pronounced in quadratic placers, where the placement solution, ultimately, is derived from the locations of the pins. From (2), we know that $\Phi(x)$ and $\Phi(y)$, are functions of d_x and d_y respectively, with each fixed pin imparting a d_x and d_y component to the layout. Thus pin α in Figure 6 pulls module a up as well to the right. It is obvious that module a will be unduly moved to minimize the vertical component on it. This effect can be nullified by setting the d_y component due to α, and likewise the d_x component for β, to zero. This has the effect of allowing α to track a vertically and β for b horizontally. In our experiments this approach produced wirelength reductions of as much as 7% for quadratic placement and 10% for linearized quadratic placement. Similar reductions in area were achieved, thus showing a reduction in congestion.

5. IMPLEMENTATION AND RESULTS

5.1 The Region Router

Routing algorithms are numerous. The algorithm for this application should accurately estimate supply and demand while maintaining computational efficiency. It is critical that the region router used in this methodology match the characteristics of the detailed router. Discrepancies between the characteristics of the two could lead to incorrect assumptions, causing the placer to create inappropriate placement for the final route. In this preliminary implementation, routing on the region-based graph was performed by an A^* algorithm [2]. The algorithm initially sorts the sinks in order of criticality (manhattan distance to source). The most critical sink is selected and routed to the nearest edge on the current tree. The tree is then updated with the new edges and the criticality of the remaining nodes are redetermined. The process terminates when all nodes are connected. In our methodology, it is not essential for the router to be congestion aware, as the placer will yield solutions that permit direct routes.

Two implementations of the sink lists can be considered: region-

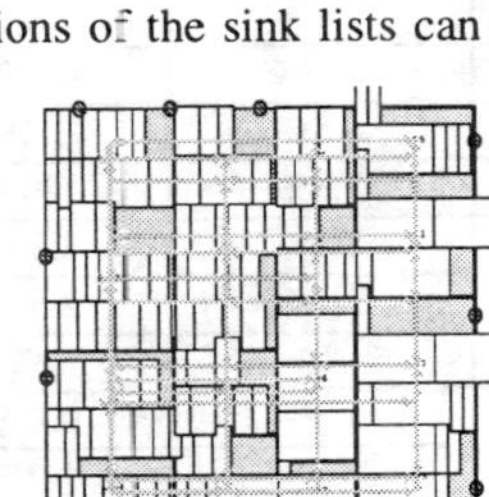

Figure 7. Global router on s298 in region center mode

Table 1: Benchmark Information

Module	Cells	Edges	Module	Cells	Edges
s298	133	404	s1494	653	2057
s444	202	595	s5378	2958	7527
s641	398	974	s9234	5825	14251
s1238	526	1602	s13207	8620	21122
s1423	731	2042	s15850	10369	25207
s1488	659	2057	s35932	17793	49517

based (Figure 7) and cell based. In the former, region centers are the sink locations for inter-region nets, while, in the latter, cell locations are used. An advantage of using region centers is the potential reduction in the number of sinks. However, this is accompanied by a loss of detail, as certain regions may not get intersected. This effect gets reduced at finer levels of partitioning until the region centers correspond very closely to the cell locations. In the current implementation routes are initially performed to cell locations. At later stages, a switch to region mode occurs, thus maintaining accuracy and efficiency.

5.2 Results

The congestion-driven algorithm, implemented in C, was incorporated into an existing quadratic placer provided by Cascade Design Automation. The numerical routines for sparse matrix operations [12] were obtained through use of the Meshach library [4]. A subset of netlists from the ISCAS-89 benchmark (Table 1), were placed and subsequently analyzed.

We express global flow uniformity as the average supply-demand excess over all edges, and local flow uniformity using standard deviation. When comparing two placements, a lower average reflects lower global congestion and a tighter standard deviation is representative of greater uniformity. The goal of a routing resource driven placement algorithm should be to minimize the global, local and maximum flow excess. The final analysis is performed on a uniform grid. The demand-supply excess, Δ_{DS}, is computed for each grid node and is processed to determine the global maximum, $max\{\Delta_{DS}\}$, average, $\overline{\Delta_{DS}}$ and standard deviation, $\sigma(\Delta_{DS})$. Table 2 lists the improvements in congestion, as measured by $max\{\Delta_{DS}\}$, $\overline{\Delta_{DS}}$ and $\sigma(\Delta_{DS})$ over the benchmark set. With larger designs such as s15850 and s35932, the local reductions in congestion tend to have less effect on the global average and uniformity. $wlen$ is the total route length as determined by the A* global router which also creates the internal routes during the final analysis.

Table 2: Improvement over quadratic placement

Benchmark	$wlen$	$max\{\Delta_{DS}\}$	$\overline{\Delta_{DS}}$	$\sigma(\Delta_{DS})$
s298	0%	0.5%	2.2%	3.7%
s444	4%	14.9%	26.5%	17%
s641	6%	5.0%	21.8%	15.6%
s1238	2%	9.1%	18.7%	15.6%
s1423	2%	15.7%	6.6%	12.4%
s1488	2%	8.5%	11.1%	-1.9%
s1494	2%	7.8%	-6.0%	6.8%
s5378	5%	8.1%	8.3%	3.0%
s9234	2%	2.4%	2.5%	4.1%
s13207	4%	-8%	7%	3.0%
s15850	-1%	5.8%	0.4%	-0.8%
s35932	0%	12.8%	-0.1%	4.4%

Table 3: Improvement due to relaxed pins

Module	Quadratic placement		Linear Placement	
	$wlen_q(\mu)$	$area_q(mils)$	$wlen_l(\mu)$	$area_l(mils)$
s298	0.8%	-8.4%	9.3%	6.5%
s641	2.2%	-0.2%	10.4%	3.1%
s1238	7.1%	5.7%	-3.7%	-8.8%
s1488	3.6%	2.5%	1.9%	2.3%
s1494	2.1%	2.8%	1.9%	2.4%
s5378	7.2%	4.2%	3.2%	-1.1%
s9238	7.0%	4.5%	3.9%	-2.4%

Table 3 shows the effectiveness of relaxed pins in wire length reduction for a sample set of small to medium sized netlists. The effect of the enhancement on both quadratic and linearized quadratic placement is shown. The wirelength and area reductions were obtained after detailed routing using a router provided by Cascade Design Automation.

6. CONCLUSION

This paper presents a methodology for integrating congestion into quadratic placement. The inclusion of a growth matrix into the quadratic placement formulation permits the use of a routing model to resolve congestion while concurrently minimizing wirelength. Further reduction in congestion, reflected in wirelength and area reduction, is obtained by the relaxation of pin constraints. This implementation with an A* search router and line-probe heuristics showed up to 20% reductions in average demand-supply excess. Ongoing experiments have shown it capable of improving linearized quadratic placement, but a more accurate model of routing resources is required. The methodology holds promise for greater improvements in congestion, with the possibility of including crosstalk into the routing cost.

7. REFERENCES

[1] C. L. E. Cheng, "RISA: accurate and efficient placement routability modeling". *1994 IEEE/ACM ICCAD*, pp. 690-5.

[2] D. K. Boese et. al, "High-performance routing trees with identified critical sinks". *30th DAC, 1993*, pp 182-7.

[3] D. W. Hightower, "A solution to Line Routing Problems on the Continuous Plane", *6th Design Automation Workshop* 1969, pp. 172-174.

[4] E. S. David et. al, "Meshach: Matrix Computations in C". *Proceedings of the Center for Mathematics and its Applications*. Australian National University, vol. 32, 1994.

[5] G. H. Golub et. al, "Matrix computations", The Johns Hopkins University Press, 2nd ed.

[6] G. Sigl et. al, "Analytical placement: a linear or a quadratic objective function?". *28th ACM/IEEE DAC*, pp. 427-432.

[7] J. M. Kleinhans et. al, "GORDIAN: VLSI placement by quadratic programming and slicing optimization". *1991 IEEE TCADICS*, vol.10, no.3, pp. 356-5.

[8] P. R. Suaris et. al, "A quadrisection-based combined place and route scheme for standard cells". *1989 IEEE TCADICS*, vol.8, no.3, pp. 234-244.

[9] R. S. Tsay et. al, "PROUD: A Fast sea-of-gates placement algorithm". *25th DAC 1988*, pp 318-323.

[10] S. Mayrhofer et. al, "Congestion-driven placement using a new multi-partitioning heuristic". *IEEE ICCAD* 90, pp. 332-5.

[11] S. Prasitjutrakul et. al, "A timing-driven global router for custom chip design". *1990 IEEE ICCAD*, pp. 48-51.

[12] S. Yousef, "Iterative methods for sparse linear systems". PWS Publishing Company.

[13] W. Dongsheng et. al, "Performance-driven interconnect global routing". *Proc. 6th Great Lakes Symp. on VLSI*, pp. 132-6.

[14] Y. Saab, "A fast clustering-based min-cut placement algorithm with simulated-annealing performance". *1996 VLSI Design*, vol.5, no.1, pp. 37-48.

Potential_NRG: Placement with Incomplete Data

Maogang Wang, Prithviraj Banerjee and Majid Sarrafzadeh
Department of Electrical and Computer Engineering
Northwestern University, Evanston, IL 60208

Abstract

Traditional placement problems are studied under a fully specified cell library and a complete netlist. However, in the first, e.g., 2 years of a 2-3 year microprocessor design cycle, the detailed netlist is unavailable. For area and performance estimation, layout must nevertheless be done with incomplete information. Another source of incompleteness comes from reuse of instances from earlier design generations; these instances and their parameters will change as the project evolves. The problem of *placement with incomplete data* (PID) can be abstracted as having to place a circuit when p_c% of the cells and p_n% of the nets are missing. The key challenge in PID is how to add missing cells and nets.

In this paper, two "patching-methods" for adding missing nets and cells are proposed. The methods are called *abstraction* and *fusion*.

Experimental results are very interesting and illurstrative. First, they show that PID is a difficult problem and an arbitrary (and perhaps intuitively sound) method may not produce high-quality results. Experiments verify that the abstraction method is a very good predictor and that fusion is not because circuits produced by abstraction attain much of the properties of the origianl circuits. Summary Table 3 in Section 4 shows that when a circuit has 10% incompleteness, abstraction can predict the final total wirelength with an error of 5.8%, while fusion has a 67.8% error in predicting the wirelength in the same circuit.

1 Introduction

The placement problem has been an interesting and challenging problem for many years. Recently, performance-driven, issues related to deep submicron, reliability and many more factors make the placement problem even more challenging.

DAC 98, San Francisco, California

Numerous approaches to the placement problem have been proposed, such as force directed [1, 6], simulated annealing [8], and partitioning-based placement [3. 2. 5]. A constructive placement method that employs resistive networks as a working domain was proposed in [4]. In [9], there is a comparison between linear and quadratic cost functions. The linear cost function used in the GordianL placement tool achieves results with up to 20% less area than the quadratic cost function of the original Gordian procedure. Therefore, in this paper, we will use the half-perimeter cost function for area minimization.

Here we study the placement problem with incomplete data (PID). This problem is of fundamental importance since VLSI design is getting more and more complicated. A typical VLSI design will take 2 to 3 years to complete. In the first 2 years, a detailed netlist is not available. However, for area estimation and performance estimation, several placement runs are done with such incomplete information. Traditional placement problems are based on a fully specified cell library and a complete netlist. Obviously, this cannot handle this problem.

Another source of incompleteness will come from reuse of instances from earlier generations. The hardware instances are modified to some extent and more information may be added as the project moves forward. So even if we do have a complete cell library and netlist, this cell library and netlist will not be the same library and netlist as in the final version of the design. The designer will surely want to know how much he or she can trust the placement result. PID is the problem of trying to place the final design while only a part of the netlist and a fraction of the cell library is known.

In our experiments, we use MCNC benchmarks. We randomly delete p_n% of the nets and p_c% of the cells. We then try to re-construct the netlist using different "patching-methods". We compare the wire-length of the original circuit against the wire-length of the "patched" circuit as a basis to judge the quality of the proposed patching-methods. Experiments on different values of p_c and p_n and different benchmarks have been performed.

This paper is organized as follows: In Section 2 we formally described the *placement with incompleteness* problem followed by a detailed summary of approaches and implementations in Section 3. In Section 4, experimental results will be shown and discussed, followed by conclusions in Section 5.

2 Problem Formulation

The best way is to study PID problem is starting from a complete circuit. This circuit, including the complete netlist and the cell library, will be no doubt the only fair metric to measure how good any kind of PID algorithm is. A good algorithm will give a good prediction on area or performance using only part of the original circuit information.

Starting from a complete circuit $\mathcal{C}$, we can build an incomplete circuit $\mathcal{C}_i$ simply by deleting or modifying cells and nets from the original circuit $\mathcal{C}$. There are two numbers to quantitively measure the incompleteness, p_c and p_n. We denote by p_c the percentage of cells we deleted or modified from $\mathcal{C}$ to get $\mathcal{C}_i$. Similarly, p_n is the percentage of nets we deleted or modified from $\mathcal{C}$ to get $\mathcal{C}_i$. Now we are ready to describe the *placement with incompleteness* problem formally:

Given a complete circuit $\mathcal{C}(\mathcal{M},\mathcal{N})$, an incomplete circuit $\mathcal{C}_i(\mathcal{M}_i,\mathcal{N}_i)$ is formed by deleting a number of cells and nets from $\mathcal{C}(\mathcal{M},\mathcal{N})$. Two real numbers p_c and p_n satisfy the following conditions:

1. $0 < p_c < 1$, $0 < p_n < 1$.
2. $|\mathcal{M} \cap \mathcal{M}_i| = (1 - p_c) \cdot |\mathcal{M}|$
3. $|\mathcal{N} \cap \mathcal{N}_i| = (1 - p_n) \cdot |\mathcal{N}|$

We are going to make predictions on issues like final chip area, power consumption, etc., of the circuit $\mathcal{C}(\mathcal{M},\mathcal{N})$ based only on the circuit $\mathcal{C}_i(\mathcal{M}_i,\mathcal{N}_i)$ and parameters p_c, p_n. These predictions will be compared to the actual value of the circuit $\mathcal{C}(\mathcal{M},\mathcal{N})$.

This above description concludes the formulation of the placement problem with incomplete data. There will be numerous ways to obtain such predictions for circuit $\mathcal{C}$. In this paper, we are actually trying to "reconstruct" the circuit $\mathcal{C}$ from the incomplete circuit $\mathcal{C}_i$. Specifically, we are going to find a set of modules $\mathcal{M}_a$ and a set of nets $\mathcal{N}_a$. We use $\mathcal{M}_a$ and $\mathcal{N}_a$, together with $\mathcal{M}_i$ and $\mathcal{N}_i$, to form a new circuit $\mathcal{C}'((\mathcal{M}_a \cup \mathcal{M}_i), (\mathcal{N}_a \cup \mathcal{N}_i))$.

Figure 1 shows an example of a complete circuit $\mathcal{C}$ with 8 cells and 10 nets. The incomplete circuit $\mathcal{C}_i$ can be obtained by deleting cells $M4$, $M6$ and all the nets attached to them. Figure 2 shows the result of $\mathcal{C}_i$. In this case, $p_c = 25\%$ and $p_n = 20\%$ since only 2 nets are gone. We can add two new dummy cells, $N1$ and $N2$ and some dummy nets to get circuit $\mathcal{C}'$. Figure 3 shows a possible way to get $\mathcal{C}'$. Of course, most probably, $\mathcal{C}'$ will not be exactly the same as the original circuit $\mathcal{C}$. Our hope is that by using a "clever"way to get $\mathcal{C}'$, the placement result of $\mathcal{C}'$ will be very close to the placement result of $\mathcal{C}$.

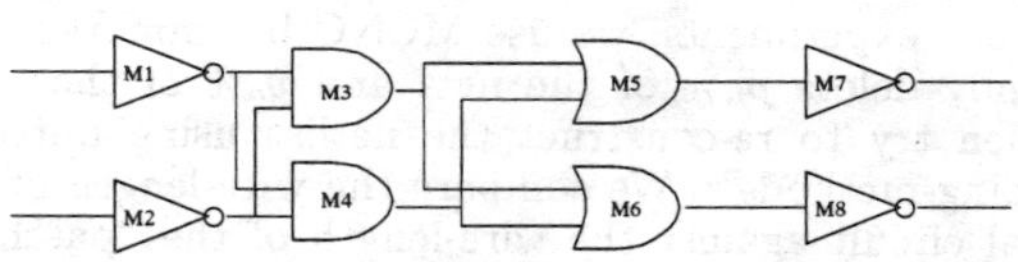

Figure 1: Original circuit $\mathcal{C}$.

In the next section, we will show our approaches to find $\mathcal{M}_a$ and $\mathcal{N}_a$.

Note that in this paper, we only try to make a prediction on the total net lengths, which is consistant with the final chip area. "Half perimeter bounding box" is what we will use to measure the length of each net.

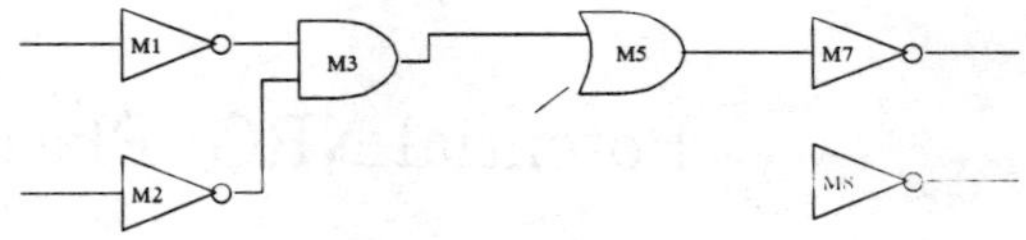

Figure 2: Incomplete circuit $\mathcal{C}_i$.

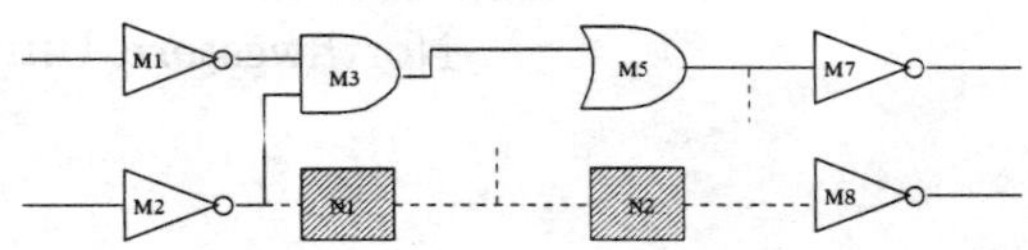

Figure 3: Reconstructed circuit $\mathcal{C}'$.

3 Our Approach and Implementation

Following the discussion in the previous section, sets $\mathcal{M}_a$ and $\mathcal{N}_a$ should be "similar" to sets $\mathcal{M}_i$ and $\mathcal{N}_i$ in circuit $\mathcal{C}_i$. The question is, what does "similar" mean here. Our interpretation is that they have similar statistics. For example, two module sets should have similar size distribution curves (to be discussed later).

There are a lot of possible statistics which we can use. Among these, we chose three of them to be the statistics for the module set. They are:

1. The size distribution of cells.
2. The number of terminals distribution on each cell.
3. The I/O type distribution of all terminals.

There is no question that the size of cells is the most important information in area estimation. Besides this, the number of terminals on each cell is also important since it determines the connectivity of each cell. The larger this number is, the larger the total final area will be. An appropriate I/O type distribution of terminals will assure a netlist similar to the original circuit. Without this I/O type distribution, any pair of terminals can be hooked up, leaving too much freedom to deal with.

The above is a brief description of these statistics and reasons why they are important to us. Based on these three statistics, we can write the psudo-code for finding missing cells, set $\mathcal{M}_a$ as following:

```
1. AddCell(circuit C_i, p_c) {
2.     Get these three statistics discussed above.
3.     n_c = p_c · |M_i|
4.     for i = 1 to n_c
5.         Create a cell data structure.
6.         Determine the size of this cell.
7.         Determine number of terminals on this cell
8.         Determine each terminal's type.
9. }
```

In Step 6, 7 and 8, we make use of those three statistics created in step 2.

After this procedure is done, we have added a set of cells $\mathcal{M}_a$ to the circuit $\mathcal{C}'$. These newly added cells are isolated since no nets are connected to them. The next step is to add some nets to the incomplete circuit $\mathcal{C}_i$. We will refer to all the existing nets already in $\mathcal{C}_i$ as "old nets" and all the nets we will add later as "new nets". Similarly, all the cells previously existing in $\mathcal{C}_i$ are called "old cells" and all the cells added by AddCell() are called "new cells".

Things are not so easy when finding the set of new nets, $\mathcal{N}_a$. The reason for this is because the set of nets is more like a graph, Similarity for graphs is very hard to define.

In the actual implementation, we use only one statistic for the set of nets, the number of terminals in each net. It is not clear what other statistics are also good, so we stick to this one for now. Even after deciding to use only one statistic, it is still difficult to get the whole missing netlist, the set $\mathcal{N}_a$.

The first thing we need to know is the number of nets to be added. This can be found by using the input parameter p_n.

After that, a natural scheme will be similar to what we did on adding new cells: Look at all the cells in the circuit, including those "old" cells in C_i and those "new" cells added by AddCell(), and create a list $\mathcal{L}$ of all the terminals unattached to any nets. Initially, this list $\mathcal{L}$ will include all terminals from all new cells and those unattached terminals in the old cells.

As the example shown in Figure 3, this list $\mathcal{L}$ will contain two pins in $N1$, two pins in $N2$ and the input pin in $M8$.

After the list $\mathcal{L}$ is created, for each to be added new net, first determine the number of terminals in it using the distribution function. Then assign terminals from the list $\mathcal{L}$ to this new net and delete those assigned terminals from the list $\mathcal{L}$. Repeat this procedure on the next to be added new net.

One of these following three cases will happen during this procedure.

1. List $\mathcal{L}$ is empty after all nets have been added. Then we have finished our reconstruction procedure on the circuit C'.

2. There are no terminals left in list $\mathcal{L}$ when adding a new net. There is nothing we can do in this situation unless we add some new cells. In this case, we will just stop adding new nets.

3. There are still some terminals left in list $\mathcal{L}$ after all nets have been added.

Actually, case 3 will happen a lot in real life. This is because of the way we get the circuit C_i. For example, if in the original circuit C, a 4-pin-net is connected to cell A, B, C and D, and later we only delete cell D from the circuit C to form the circuit C_i while retaining cells A, B and C, then this net will still exist in the incomplete circuit C_i, except that now it becomes a 3-pin-net instead of a 4-pin-net.

This fact suggests that some terminals in newly added cells should be assigned to an old net as well. That is why the number of terminals in list $\mathcal{L}$ is almost always larger than the total number of terminals in all new nets.

Based on this analysis, if case 3 happens, for each terminal left in the list $\mathcal{L}$, we will randomly pick either an old net or a new net and assign this terminal to it.

We will refer to this patch method as *fusion*.

However, *fusion* is not the only way to accomplish such a task. We have an alternative patching method, *abstraction*, described as the following: if case 3 happens, instead of assigning the remaining terminals to an existing net as in *fusion*, we will keep adding new nets and assign these terminals to these new nets until the list $\mathcal{L}$ becomes empty.

Since most terminals initially in list $\mathcal{L}$ are just terminals in new cells, in *abstraction*, all new cells are interconnected and rarely is there a connection between new cells and old cells.

We briefly summarize the difference between *fusion* and *abstraction* as in the following: In *fusion*, new cells are attached to both new nets and old nets, so they are connected to both old cells and new cells; while in *abstraction*, new cells are only attached to new nets, new cells form kind of a cluster since most of their nets are internal nets.

Another difference is that the circuit C' created by *abstraction* will have more nets than its counterpart created by *fusion*.

Intuitively, *fusion* makes much more sense than *abstraction*, since it appears that *fusion* will generate a circuit more like the original one, circuit C, while *abstraction* will almost never look like C.

Contrary to people's intuition, experimental results show that *abstraction* offers much more accurate prediction than *fusion* does. In the next section, we will present this interesting experimental result, followed by an explanation.

4 Experimental Results and Explanation

Fusion and *abstraction* were implemented using C++. All experiments were run on a Sun Sparc-20 workstation. Three MCNC benchmark standard cell circuits were used for experiments. Table 1 shows some basic statistics for these three benchmark circuits. We delete all the I/O pads from all three of these circuits, allowing us to concentrate on the effects of incomplete data on core cells. As shown in Table 1, circuit Nprimary1 is the MCNC benchmark circuit primary1 without I/O pads, and Nprimary2 and Nstruct are benchmark primary2 and struct without I/O pads, respectively.

We feed the reconstructed circuit to the high quality placement tool called NRG [7]. Since this is a statistical approach, we needed to run a good number of trials to get the data. So we run NRG in fast mode.

Circuit	# Cells	# Pins	# Nets	WL
Nprimary1	750	3062	1156	642853
Nprimary2	2907	11681	3671	3215512
Nstruct	1888	5407	1920	374940

Table 1: Statistics for tested benchmark circuits.

For each benchmark circuit $C(\mathcal{M},\mathcal{N})$, for a specified degree of incompleteness p_c. We first delete $p_c \cdot |\mathcal{M}|$ cells and the nets attached to them to form C_i. Then we run our two reconstruction procedures, *fusion* and *abstraction*, to get a "reconstructed" circuit C' (*fusion* and *abstraction* were discussed in detail in Section 3). After that, we use the NRG placement tool to get the total wirelength of C' and compare this value with the value of the original circuit C.

Table 2 shows the total wirelength results achieved from the reconstructed circuit $Nprimary1'$. For each degree of incompleteness, column "WL" shows the average total wirelength over all trials; column "dev." shows the standard deviation of all trials, and "%err" shows the percent error compare to the real value from the original complete circuit Nprimary1.

It's clear that *fusion* has a much bigger error than *abstraction* for any percentage of incompleteness.

Since this is a non-deterministic procedure, standard deviation is another important factor to look at. However, there is not a clear pattern in this figure. All the data points are spread in the range between 0.7% and 1.7%. Standard deviation does not increase as the percentage of incompleteness increases. This is determined by the nature of *fusion* and *abstraction*. Since *fusion* and *abstraction* rely heavily on the statistics from the incomplete circuit C_i, the increase of incompleteness really does not affect the degree of randomness in reconstructing circuit C'. That is why the standard deviation does not increase along with the percentage of incompleteness.

Similar results for Nprimary2 and Nstruct were obtained from the experiments. Table 3 is a summary table of these three circuits. All the percent errors in Table 3 are the average percent errors among these three circuits.

From experimental results shown above, we can conclude that *abstraction*'s prediction is much more accurate than *fusion*'s. That is very surprising and counter-intuitive.

The reason why *abstraction* is better than *fusion* is because *fusion* mantains the original properties of the circuit.

A big conclusion from this is that the netlist topology is very important. It will have a huge effect on the total wirelength. In this particular *placement with incomplete data* problem, we should not try to change the circuit topology of existing netlist.

% in	Fusion			Abstraction		
	WL	dev.	%err	WL	dev.	%err
0%	642853	6782	0%	642853	6782	0%
5%	740452	11697	15%	609263	9470	-5%
10%	874490	14284	36%	605948	6398	-6%
15%	941024	12486	43%	593153	7697	-8%
20%	1080158	16305	68%	597267	4250	-7%
25%	1202426	10309	87%	599750	8384	-7%
30%	1231592	11368	92%	560518	9623	-13%

Table 2: Wirelength results for Nprimary1 after reconstructing the circuit.

% of incomp.	%err for fusion	%err for abstraction
0%	0%	0%
5%	43.9%	4.8%
10%	67.8%	5.8%
15%	91.8%	8.5%
20%	118%	9.1%
25%	141%	9.1%
30%	155%	16.0%

Table 3: Average wirelength prediction results for all three testing circuits.

5 Conclusions and Future Work

We formally proposed a new problem in placement area, placement with incomplete data. This problem arises along with the increasing complexity of current VLSI designs. It will become more and more important in the future.

We tried two approaches: *fusion* and *abstraction* as described in detail in Section 2. We want to see how good they can predict the final wirelength/ chip area using only an incomplete netlist and library.

Experimental results in Section 4 show that *abstraction* is a much better patching method than *fusion*. With 10% incompleteness in the circuit, *abstraction* can still predict the wirelength with an error of 5.8%; while *fusion* can have an error more than 50% even with a 5% incompleteness in the circuit.

Comparisons between *fusion* and *abstraction* shows that the netlist topology is essential to the final wirelength. In order to get a good prediction, we should try to preserve the original topology as much as possible.

References

[1] H. Anway, G. Farnham, and R. Reid. "Plint Layout System for VLSI Chip Design". In *Design Automation Conference*, pages 449–452. IEEE/ACM, 1985.

[2] M. A. Breuer. "A Class of Min-cut Placement Algorithms". In *Design Automation Conference*, pages 284–290. IEEE/ACM, 1977.

[3] M. A. Breuer. "Min-cut Placement". *J. Design Automation and Fault-Tolerant Computing*, 1(4):343–382, 1977.

[4] C. K. Cheng and E. S. Kuh. "Module Placement Based on Resistive Network Optimization". *IEEE Transactions on Computer Aided Design*, 3(3):218 225, 1984.

[5] A. E. Dunlop and B. W. Kernighan. "A Procedure for Placement of Standard Cell VLSI Circuits". *IEEE Transactions on Computer Aided Design*, 4(1):92–98, January 1985.

[6] C. P. Hsu et al. "APLS2: A Standard Cell Layout System for Double-layer Metal Technology". In *Design Automation Conference*, pages 443–448. IEEE/ACM, 1985.

[7] M. Sarrafzadeh and M. Wang. "NRG: Global and Detailed Placement". In *International Conference on Computer-Aided Design*. IEEE/ACM, November 1997.

[8] C. Sechen. *VLSI Placement and Global Routing Using Simulated Annealing*. Kluwer, B. V., Deventer, The Netherlands, 1988.

[9] G. Sigl, K. Doll, and F. M. Johannes. "Analytical Placement: A Linear or a Quadratic Objective Function". In *Design Automation Conference*, pages 427–431. IEEE/ACM, 1991.

Performance-Driven Multi-FPGA Partitioning Using Functional Clustering and Replication

Wen-Jong Fang and Allen C.-H. Wu
Department of Computer Science, Tsing Hua University
Hsinchu, Taiwan, 300, Republic of China
{Email: chunghaw@cs.nthu.edu.tw}

Abstract

This paper presents a new performance-driven partitioning method for multi-FPGA designs. The proposed method consists of three steps: (1) functional-cluster formation, (2) slack computation, and (3) set-covering-based partitioning with functional replication. The proposed method performs multi-FPGA partitioning by taking into account path delays and design structural information. We introduce a functional replication technique which performs circuit replications at the functional-cluster level instead of the cell level for delay and interconnect minimization. Experimental results on a number of benchmarks and industrial designs demonstrate that the proposed method achieves high-performance and high-density multi-FPGA partitions.

1 Introduction

In general, the traditional HDL-based design flow from initial specification to multiple-FPGA systems consists of two independent design phases. In the first phase, a synthesizer is used to transform an HDL design specification into a CLB-based netlist. In the second phase, a circuit-level partitioner is used to partition the CLB netlist into FPGA chips. The FPGA-based partitioning problem is quite different from the classical ASIC partitioning problem. Typically, mapping a design onto a set of FPGAs is predominately constrained by pin limitations. This often results in FPGA partitions with very low logic utilization.

In the past several years, many partitioning approaches and algorithms have been proposed to solve the FPGA partitioning problem. Most of the proposed algorithms [1, 2, 3, 4] focus mainly on cost minimization. There are only few algorithms targeted to delay minimization [5, 6]. Roy and Sechen [5] proposed a two-phased natural and adaptive clustering method, followed by a simulated-annealing-based N-way partitioning process for timing optimization of multi-chip designs. Sawkar and Thomas [6] devised a set-covering-based partitioning algorithm to minimize the number of chip crossings on each circuit path. All of the above approaches perform circuit-level partitioning on flattened gate-level or CLB-level netlists. They do not take into account design-hierarchy information.

In this paper, we present a new performance-driven partitioning method for multi-FPGA designs. The proposed method performs multi-FPGA partitioning by taking into account path delays and design structural information. Experimental results on a number of benchmarks and industrial designs are reported to demonstrate the effectiveness of the proposed method.

†This work was supported by the National Science Council of R.O.C. under Grant NSC 87-2215-E-007-015.

DAC 98, San Francisco, California

2 Problem Description

In our previous study [7], an integrated HDL synthesis and partitioning method was proposed for multi-FPGA partitioning. This method produces good partitions with high logic and low IO-pin utilizations. However, timing issues are not considered in this method. The research described in this paper extends this approach by focusing on timing optimization.

In order to achieve timing improvement, conventional methods usually impose higher weights on the critical paths during the partitioning process so that the delays on these critical paths can be improved by making sure components on these paths end up on the same chip. However, there are two potential problems when using this approach. First, when critical paths are improved, some noncritical paths may become critical due to excessive inter-chip routing delays. To guarantee a timing-violation-free multi-FPGA design, path delays have to be taken into account throughout the entire partitioning process. Second, due to limited I/O pins of FPGA chips, partitioning a design onto a set of FPGAs is usually constrained by the pin limitation. Forcing critical paths to be assigned to the same chip may increase the number of cut lines between chips. This often results in a partition using more FPGA chips. In order to reduce the cost, IO-pin limitations must be considered throughout the entire partitioning process.

Recent research results [8, 9] have shown that the cell replication approach is very effective in reducing of the number of cut edges in a partitioned circuit. The cell replication approach has been particularly useful for improving logic utilization of pin-limited devices such as multi-FPGAs [2, 6]. In addition, cell replication has also

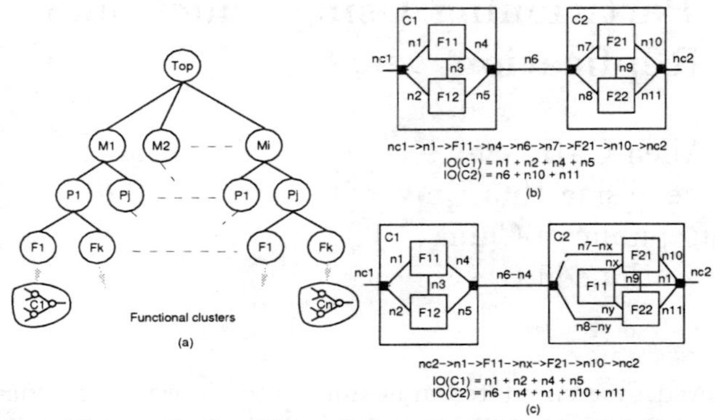

Figure 1: Overview of the proposed approach.

proved effective to partitioning for delay minimization [10].

By taken into account the above considerations, we propose a functional clustering and replication approach for performance-driven multi-FPGA partitioning.

3 The Proposed Method

3.1 Overview

The design flow consists of three steps: (1) functional-cluster formation, (2) slack computation, and (3) set-covering-based partitioning with functional replication. In the first step, a synthesizer takes a Verilog design specification as an input and generates an HDL-based structural tree with functional clusters for the design, as shown in Figure 1(a). In the second step, the system performs path-delay analysis and computes slack values for functional clusters. In the third step, a performance-driven partitioner uses a set-covering-based algorithm with functional replication to perform multi-FPGA partitioning. Finally, the system generates a multi-FPGA design in XNF and Verilog formats.

The partitioning algorithm takes into account both path delay and logic utilization during the covering process. In addition, the algorithm incorporates a functional replication technique to improve timing as well as logic utilization of multi-FPGA designs. The main advantage of the functional replication method is that it can fully exploit the design structural property during the cloning process. We use the example in Figures 1(b) and (c) to explain the functional replication procedure. After initial partitioning, functional clusters $\{F11, F12\}$ and $\{F21, F22\}$ are assigned into two chips $C1$ and $C2$, respectively. The critical path from $nc1$ to $nc2$ includes one cross-chip connection $n6$. By replicating $F11$ in $C2$, the cross-chip connection can be eliminated which results in fast timing, as shown in Figure 1(c). Using functional replication for delay minimization may reduce or increase the number of IO-pins used on a partitioned chip. For example, before replicating $F11$ in $C2$ (Figure 1(b)), the IO-pins used in $C1$ and $C2$ are $n1+n2+n4+n5$ and $n6+n10+n11$, respectively. Figure 1(c) shows the IO-pin usage after replicating $F11$ in $C2$, in which the IO-pin used in $C1$ stays the same but that in $C2$ changes to $n6-n4+n1+n10+n11$, where $n4$ is the number of connections between $F11$ and all the functional clusters in $C2$ (e.g., $F21$ and $F22$), and $n1$ the number of external connections between $F11$ and all the other nodes, excluding $C2$. If the number of $F11$'s external connections is less than or equal to it's connections to $C2$, the number of IO-pins used in $C2$ will be reduced or stay the same (e.g., $n4 \leq n1$). This is an ideal case because both the path delay and the number of IO-pin used can be reduced. However, in some cases (e.g., $n4 > n1$), we have to pay an IO-pin usage penalty in order to reduce path delay.

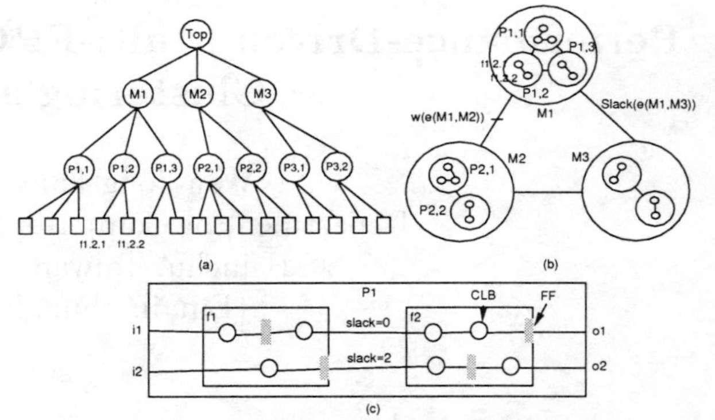

Figure 2: A connected hypergraph:(a) the structural tree, (b) the corresponding graph, (c) slack computation.

3.2 Functional-cluster formation

We use the functional-based synthesis method [7] to synthesize the design in a fine-grained way so that a set of CLB-based clusters can be formed according to the structural nature of the design. We use a structural tree to represent the structural hierarchy of the HDL description of a design. In a structural tree, the root node represents the top-level design, each intermediate node a higher-level language construct such as *modules*, *processes*, and *tasks*, and each leaf node a circuit cluster, as shown in Figure 1(a). The detailed functional-cluster formation procedure can be found in [7].

3.3 Slack Computation

After constructing a structural tree and forming the clusters, we convert the structural tree into a connected hypergraph. Figures 2(a) and (b) show a structural tree and its corresponding connected hypergraph. We first use a breadth-first search method to enumerate signal paths in a design and then compute the slacks for all paths by traversing the graph at the leaf-node level. In the slack computation, we use a unit delay model to approximate the path delays. Subsequently, the slacks between nodes at higher-level are computed using the formula: $Slack(e(M_k, M_l)) = MIN\{Slack(e(n_i, n_j)), n_i \in M_k \text{ and } n_j \in M_l\}$.

The slack between two hypernodes is defined as the minimum slack between any two nodes embedded within the hypernodes. For example, in Figure 2(c), the slack values between functional clusters $f1$ and $f2$ are 0 and 2. Hence, the slack value between functional nodes $f1$ and $f2$ is 0 ($Slack(e(f1, f2)) = 0$).

3.4 Set-Covering with Functional Replication

The performance-driven FPGA partitioning problem is defined as: Given a connected hypergraph G and the

CLB/IO-pin constraint of the FPGA chips, find a multi-FPGA partition such that the timing is optimized and the number of FPGAs is minimized.

We formulate the performance-driven FPGA partitioning problem into a set-covering problem. Let $G = (V, E)$ be a connected hypergraph. $V = V_m \cup V_p \cup V_f$, where V_m is a set of module nodes, V_p a set of process nodes, and V_f a set of functional nodes. C_{cover} denotes a set of FPGA chips. For a node v_i, $clb(v_i)$ and $io(v_i)$ represent the number of CLBs and IO pins of v_i. CLB and IO denote the CLB and IO-pin constraints of the FPGA chips. The pseudo code of the set-covering with functional replication algorithm ($SCFR$) is listed below.

The inputs to the algorithm include a connected hypergraph G and the CLB and IO-pin constraints (CLB and IO) of the FPGAs used. The output is the number of required FPGA chips. The algorithm first computes the score after assigning the nodes at the top level into a chip. If the number of CLBs and IO pins by assigning a node v_i into a chip c_k satisfies the CLB and IO-pin constraints of the chip, then the score of the pair (c_k, v_i) is calculated as a function of three values: the interconnect-closeness ($Conn(\{c_k, v_i\})$) between the node and the nodes already assigned to the chip, the ratio between the number of CLBs and IO pins, and the criticality ($Crit(\{c_k, v_i\})$) between node v_i and the nodes already assigned to the chip c_k. Otherwise, the score of (c_k, v_i) is zero. The criticality is inversely proportion to the slack between the node v_i and the nodes already assigned to the chip c_k. α, β, and γ are three parameters set by the user to express a preference of one term over the other.

```
Algorithm SCFR(G, CLB, IO) {
  C_cover ← c_new;
  V_cluster = V;
  top_level = V_m;
  while (V_cluster ≠ φ) {
    for (all v_i ∈ V_cluster(top_level))
      for (all c_k ∈ C_cover)
        if (clb({c_k ← v_i}) ≤ CLB and
        io({c_k ← v_i}) ≤ IO) then
          score(c_k, v_i) = α Conn(c_k, v_i) +
          β clb({c_k, v_i})/io({c_k, v_i}) + γ Crit({c_k, v_i});
        else
          score(c_k, v_i) = 0;
    if (all score(c_k, v_i) ≠ 0) then {
      Select the pair of {c_k, v_i} with the highest score;
      V_cluster = V_cluster - v_i;
      c_k = c_k ∪ v_i; }
    else {
      if (top_level ≠ V_f) then
        top_level = Next_Level(top_level);
      else {
        Function_Duplication(c_k, V_f);
        C_cover = C_cover ∪ c_new;
        top_level = V_m; }}}}
```

After computing the scores for all nodes, if some of the nodes can be assigned to any chips, then the one with the highest score will be selected and assigned to a chip. Otherwise, none of the current nodes can be assigned to any chips due to the CLB and IO-pin constraints of chips. In this case, there are two possible solutions. First, if all the current nodes are not leaf nodes, then procedure $Next\_Level()$ is invoked to expand the covering process to the nodes in the next lower level so that the covering process can be performed on nodes with a finer granularity. Second, if none of the leaf nodes (i.e., the functional nodes) can be covered by any of the allocated FPGA chips, then procedure $Functional\_Replication$ is invoked to improve path delay and relax IO-pin usage of chips. Subsequently, a new chip c_{new} is allocated and the covering procedure is continued, starting from the nodes at the top level (i.e., module nodes).

Table 1: Characteristics of the benchmarks.

Designs	#Lines of code	#M	#P	#F
Deaign1	136	7	5	548
Design2	167	9	5	595
Design3	248	11	10	1,023
Design4	404	1	7	57
Design5	4,084	20	30	1,095
Design6	6,298	36	167	5,148

In our approach, a simple heuristic is used for functional replication which is described below. Given a chip c_k and the set of functional nodes V_f, the following steps are repeated until no more nodes can be replicated.

1. Delay and IO-pin gain computations. If a replication solution is feasible (i.e., **if** $(clb(\{c_k \leftarrow v_i\}) \leq CLB$ **and** $io(\{c_k \leftarrow v_i\}) \leq IO)$, then the delay and IO gains of replicating a node $v_i \in V_f$ to c_k is computed. Let $DG(v_i, c_k)$ and $IOR(v_i, c_k)$ denote the delay gain and IO-pin reduction of replicating v_i to c_k. The overall gain by replicating a node is defined as $Gain(v_i, c_k) = \alpha_1 DG(v_i, c_k) + \alpha_2 IOR(v_i, c_k)$. α_1 and α_2 are two parameters set by the user to express a preference of one term over the other.
2. Functional replication. The algorithm will replicate the functional node with the highest gain into c_k.

The algorithm stops when all the nodes are covered by FPGA chips. The time complexity of the $SCFR$ algorithm is described in the following manner. Let n be the number of functional nodes and m the number of chips used. It takes $O(n * m)$ time to compute the scores. Hence, the $SCFR$ algorithm takes $O(n^2 * m)$ time. The functional replication procedure takes $O(n^2)$ time.

4 Experiments

We have implemented the performance-driven multi-FPGA partitioning algorithm in the C programming language running on SUN and HP workstations.

We have tested our methodology on six designs, as shown in Table 1 in which $\#M$, $\#P$, and $\#F$ denote teh number of module, process, and functional nodes. The six designs are described in Verilog. The first design ($Design1$) is an ALU design, and the second design ($Design2$) is the fifth-order elliptic filter benchmark. The bit-width of the two designs is 32. $Design3 - 6$ are four industrial designs: a DSP core, a 32-bit floating-point multiplier, a controller, and a communication chip.

We have conducted experiments on the six designs using three methods, namely, the module-based synthesis and modified RFM partitioning method (MRFM), the functional-based synthesis and hierarchical set-covering method (FHSC) [7], and the functional-based synthesis and hierarchical set-covering with functional replication method (FHSCFR). In the worst path-delay calculation,

Table 2: Delay, required-chip, and run-time comparisons.

MRFM/FHSC/FHSCFR							
Design	Chip Type	#CLBs	IOU	#Chips	Delays	RT(Syn)	RT(Par)
	XC4013	761/761/763	.83/.37/.37	2/2/2	910/730/700	121/254/254	6/31/35
1	XC4010	761/761/763	.89/.44/.44	2/2/2	1,150/730/700	121/254/254	7/36/43
	XC4005	761/779/781	.95/.86/.86	9/6/6	1,180/865/835	121/254/254	15/52/59
	XC4013	1,680/1,741/1746	.92/.78/.78	10/5/5	1,030/940/940	207/381/381	84/40/46
2	XC4010	1,680/1,799/1804	.96/.81/.81	12/7/7	1,240/970/970	207/381/381	99/59/65
	XC4005	1,680/1,876/1881	.97/.88/.88	25/13/13	1,660/1,030/1,030	207/381/381	101/122/130
	XC4013	822/829/1,044	.97/.45/.63	3/2/2	1,195/985/955	238/472/472	11/33/41
3	XC4010	822/833/1,125	.93/.58/.74	4/3/3	1,345/1,135/955	238/472/472	13/39/56
	XC4005	822/921/1,352	.94/.78/.83	11/8/8	1,555/1,225/1,145	238/472/472	24/61/89
	XC4013	1,153/1,196/1,413	.93/.87/.75	6/5/5	1,465/1,225/1,225	674/989/989	33/58/64
4	XC4010	1,153/1,228/1,522	.93/.82/.76	10/6/6	1,555/1,285/1,285	674/989/989	39/60/67
	XC4005	1,153/1,377/1,784	.98/.91/.85	18/13/13	1,735/1,475/1,475	674/989/989	45/67/77
	XC4013	798/803/803	.90/.71/.71	9/3/3	390/220/220	563/1,757/1,757	21/73/75
5	XC4010	798/829/934	.98/.79/.77	12/4/5	390/340/310	563/1,757/1,757	25/87/95
	XC4005	798/978/1,140	.96/.88/.84	21/9/11	420/415/355	563/1,757/1,757	32/143/162
	XC4013	2,964/3,050/3,342	.96/.74/.77	16/6/6	535/490/430	661/2,453/2,453	151/207/224
6	XC4010	2,964/3,247/3,691	.98/.89/.78	23/13/15	625/540/490	661/2,453/2,453	135/472/493
	XC4005	2,964/3,678/4,198	.99/.93/.83	41/27/33	685/685/625	661/2,453/2,453	154/833/895

we assume that delays for each CLB, wire inside the same chip, and cross-chip wire are 5, 10, and 20 delay-unit, respectively.

Table 2 shows the results in which *Type*, *#CLBs*, *#IOU*, *#Chips*, *Delay*, *RT(Syn)*, and *RT(Par)* represent chip types, the number of CLBs, I/O utilization, the number of chips used, the worst path delay, synthesis runtime (in seconds), and partitioning runtime (in seconds), respectively. We have used three types of FPGA chips: (1) the XC4013 with 192 I/O pins and 576 CLBs, (2) the XC4010 with 160 I/O pins and 400 CLBs, and (3) the XC4005 with 112 I/O pins and 196 CLBs.

From the results in Table 2, the following observations can be made. First, the FHSC method produced designs using in average 10% more CLBs than that produced by the MRFM method. This overhead is due to the inability of the former to optimize logic across boundaries of finer HDL blocks. Nevertheless, the FHSC method produced designs using less chips, less IO-pins, and fast timing compared to that produced by the MRFM method. The FHSC method produced designs in average 25% faster than the MRFM method. This improvement is due to the former method performs partitioning procedure on the functional-cluster-level and functional clusters has a property of preserving closely-connected components in the same cluster. Second, the FHSCFR method produced designs using 1% to 40% more CLBs than that produced by the FHSC method. This overhead is due to the cloning process. For *Design*1 − 4, both the FHSC and FHSCFR methods used the same number of chips. But, the FHSCFR method achieved the same or better IO-pin utilization and timing compared to the FHSC method. For *Design*5 − 6, the FHSCFR method achieved better IO-pin utilization and timing with cost penalties ranged from 1 to 6 chips compared to the FHSC method. Table 2 also shows the run-time comparison between the three methods. The run time was measured on a SPARC20 workstation. The results show that the average run-time of the MRFM method is shorter than the FHSC and FHSCFR methods.

5 Conclusions

In this paper, we have presented a new performance-driven partitioning method for multi-FPGA designs. We have introduced a functional replication technique which performs circuit replications at the functional-cluster level instead of the cell level for delay and interconnect minimization. The proposed method performs multi-FPGA partitioning by taking into account path delays and design structural information. Experimental results on a number of benchmarks and industrial designs demonstrate that the proposed method achieves high-performance and high-density multi-FPGA partitions.

References

[1] N.-S. Woo and J. Kim, "An Efficient Method of Partitioning Circuits for Multiple-FPGA Implementation," *Proceedings of the 30th DAC*, pp. 202-207, 1993.

[2] R. Kuznar, F. Brglez, and B. Zajc, "Multi-way Netlist Partitioning into Heterogeneous FPGAs and Minimization of Total Device Cost and Interconnect" *Proceedings of the 31th DAC*, pp. 238-243, 1994.

[3] N.-C. Chou, L.-T. Liu, C.-K. Cheng, W.-J. Dai, and R. Lindelof, "Circuit Partitioning for Huge Logic Emulation Systems," *Proceedings of the 31st DAC*, pp. 244-249, 1994.

[4] D. J.-H. Huang and A. B. Kahng, "Multi-Way System Partitioning into a Single Type or Multiple Types of FPGAs," *Proceedings of 3rd International Symposium on FPGAs*, pp. 140-145, 1995.

[5] K. Roy and C. Sechen, "A Timing Driven N-Way Chip and Multi-Chip Partitioner," *Proceedings of ICCAD93*, pp. 240-247, 1993.

[6] P. Sawkar and D. Thomas, "Multi-way Partitioning for Minimum Delay for Look-up Table Based FPGAs," *Proceedings of the 32st DAC*, pp. 201-205, 1995.

[7] W.-J. Fang and A. C.-H. Wu, "Multi-Way FPGA Partitioning by Fully Exploiting Design Hierarchy," *Proceedings of 34th DAC*, pp. 518-521, 1997.

[8] C. Kring and A. R. Newton, "A Cell-Replicating Approach to Mincut-Based Circuit Partitioning," *Proceedings of ICCAD91*, pp. 2-5, 1991.

[9] J. Hwang and A. El Gamal, "Min-Cut Partitioning in Partitioned Networks," *IEEE Trans. on CAD*, vol. 14, no. 1, pp. 96-106, Jan. 1995.

[10] L. T. Liu, M. Shih, N.-C. Chou, C.-K. Cheng, ang W. Ku, "Performance-Driven Partitioning Using Retimiming and Replication," *Proceedings of ICCAD93*, pp. 296-299, 1993.

Multi-pad Power/Ground Network Design for Uniform Distribution of Ground Bounce*

Jaewon Oh
Sun Microsystems, Inc.
901 San Antonio Road, MS: USUN03-202
Palo Alto, CA 94303-4900
(408) 774-8657
joh@eng.sun.com

Massoud Pedram
University of Southern California
Dept. of Electrical Engineering - Systems
Los Angeles, CA 90089-2562
(213) 740-4458
massoud@zugros.usc.edu

ABSTRACT

This paper presents a method for power and ground (p/g) network routing for high speed CMOS chips with multiple p/g pads. Our objective is not to reduce the total amount of the ground bounce, but to distribute it more evenly among the pads while the routing area is kept to a minimum. We first show that proper p/g terminal to pad assignment is necessary to reduce the maximum ground bounce and then present a heuristic for performing simultaneous assignment and p/g net routing. Experimental results demonstrate the effectiveness of our method.

1. INTRODUCTION

Ground bounce refers to inductively induced voltage fluctuation in the internal ground lines of a chip (V_{ss}). Similarly, "power bounce" and "signal bounce" are voltage fluctuations in the power (V_{dd}) and signal (e.g data bus) lines, respectively. Since their characteristics are similar, we will focus on the ground bounce.

The major component of inductance comes from the chip's package pin and the lead wire that connects the pin to the pad on the die. In current technology, inductance on power/ground (p/g) nets can be ignored. However, this inductance may be comparable to the package inductance in the future as the minimum feature size decreases. For now, we assume that package inductance is dominant. We also ignore mutual inductance between the lead traces of the package.

P/g routing is done in two phases: (a) construction of the net topologies, and (b) determination of width of the wire segments. The topologies connect the p/g terminals of each cell to the p/g pads on the chip. Our work deals with the first task (see also [3], [4]). The second task (see also [5], [6]) determines the width of the wire segments so that reliability constraints such as voltage drop and electromigration constraints are met.

The topologies can be general graph, multi-pad tree or single-pad tree as shown in Figure 1. In [1] and [2], it was shown that general graphs and multi-pad trees can always be reduced to single pad trees which have lower area while satisfying the same reliability constraints. Therefore we restrict our topologies to single-pad trees. The topology generators in [3] and [4] focused on planarity of the p/g net. However, the planarity can be achieved under special terminal and pad configurations (locations and number of pads and terminals). As our target is high performance chip design, we assume that there are enough metal layers and planarity is of no concern.

The remainder of this paper is organized as follows. In section 2, we give the background for ground bounce. We define the p/g network design problem and present a heuristic technique for solving it in sections 3 and 4. Sections 5 and 6 give our experimental results and concluding remarks.

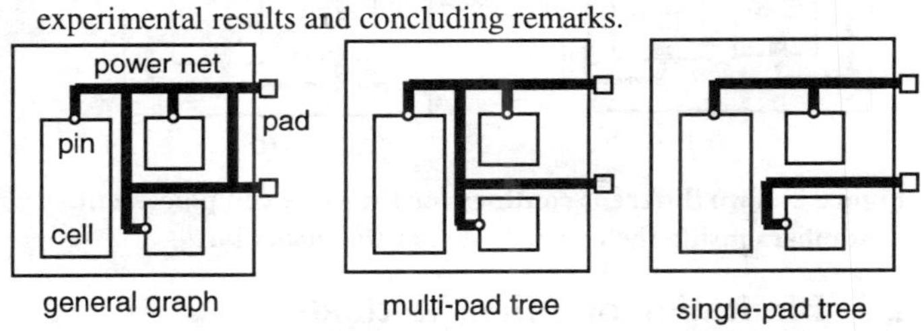

Figure 1: Various power/ground net topologies (taken from [2])

2. BACKGROUND

2.1 Delta-I noise

It is not uncommon to see ASIC chips that are designed correctly both in functionality and timing, but do not work when they are plugged into the sockets. It is likely that the designer did not fully appreciate the bounce noises. The bounce noise, also called simultaneous switching noise or delta-I noise, arises when multiple transistors switch simultaneously. The spontaneous current demand (di/dt) appears as a voltage drop (L di/dt) across the inductor which is present between the system board's ground and the ground pad of the chip (similarly with power pads). Together with the capacitance in the current path, the voltage on the ground experiences LC oscillation. For on-chip drivers, the power pad simultaneously experiences the same noise with opposite polarity to that of the ground bounce noise. This is manifested in a variety of transient and permanent circuit malfunctions, including the appearance of undesirable glitches,

* This work was funded in part by DARPA under contract no. F33615-95-C1627 and SRC under contract no. 98-DJ-606.

DAC 98, San Francisco, California

the flipping of state bits in registers and memory, and additional signal delay due to reduced voltage across the power and ground. For more detailed explanation and analytical delta-I noise estimations, see [7], [8] and [9].

2.2 Distribution of delta-I noise

Our problem domain is p/g routing in sea-of-gates or full-custom designs. P/g routing refers to the process of building connections between the power/ground terminals of the cells and the pads. Pads are usually placed around the periphery of the chip die. Associated with the p/g terminals of every cell are the amount of maximum current changes (di/dt). We assume that these numbers are given. They can be obtained from simulations or analytical estimations. When two terminals are connected, we add up the di/dt of the two terminals. This is not always correct when the two terminals are correlated such that the maximums of di/dt do not occur at the same time. For simplicity, we assume that peak di/dt is simply additive.

The next step is to assign each terminal to a pad. Since we have chosen single pad tree topology, two terminals assigned to the same pad belong to the same tree. If the peak di/dt of each part is severely unbalanced, then some pads may experience too much noise while other pads experience little noise. For example in Figure 2, the partition in (b) is balanced while (a) is not. In (a), the pad A has di/dt almost twice that of pad B. In (b), both pads have nearly the same di/dt.

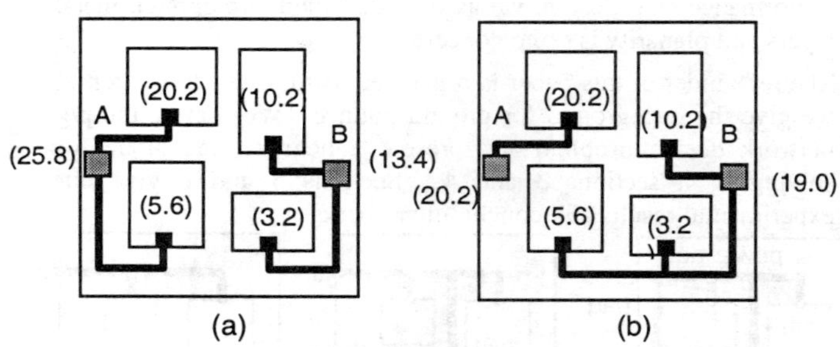

Figure 2: Two different routings for the same cell placement. Numbers inside the parentheses are the peak di/dt.

2.3 Limitation of Other Methods

On-chip decoupling capacitance (decap) helps reduce the ground bounce. The decap is placed across the V_{dd} and V_{ss} close to the cell as shown in Figure 3. The spontaneous current demand by the on-chip drivers is partially supplied by the decap rather than all the current being supplied from the outside of the chip.

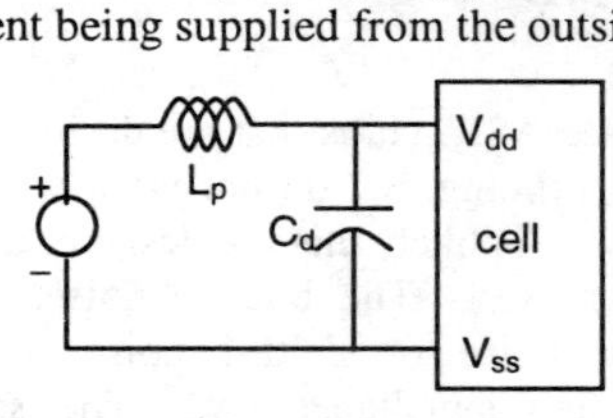

Figure 3: Decoupling capacitance C_d and pin inductance L_p

The decap is usually implemented using unused transistors in the vicinity of the cell. For large amount of di/dt, bigger capacitor should be used. Since the decap occupies silicon area, it may not always be possible to find such a big area near the cell. Besides, large decap may lead to unreliable operation of the chip as described next. Suppose a pad has exceptionally large amount of di/dt. Then in the routing tree that includes the pad, there will be a lot of on-chip decaps attached to tree to reduce the current transient of the cells. As a result, the capacitance of the pad becomes very large. In Figure 3, the voltage source is as good as short for the AC current. Therefore the decap C_d and the inductance L_p form a LC tank. The tank has a resonance frequency of

$$\omega = \frac{1}{\sqrt{L_p C_d}}$$

If C_d is sufficiently large, ϖ can be close to the operating frequency of chip. In that case, the resonance occurs at the power lines over many cycles, i.e. a large voltage fluctuation builds up at the power lines and causes the circuit to fail. Therefore we cannot indefinitely increase the size of the decaps to reduce the ground bounce.

Wire sizing (increasing width) of the p/g net to meet the IR voltage drop and the electromigration constraints worsens the ground bounce problem. Since wire sizing reduces the resistance of the net, the amount of damping on the LC oscillation is reduced, therefore increasing the ground bounce. Hence wire sizing is not a solution to the ground bounce problem.

Another popular method for ground bounce reduction is to connect all the ground pads together to create a ring. This has the effect of sharing currents between pads. When a pad has excessive current flow, the current can be absorbed by the nearby pads. However, due to finite resistance of wires between pads, the excessively loaded pad still gets more ground bounce than its neighbors. Besides, if all the nearby pads also get excessive current, having connections among them is not very helpful.

3. Problem Definition

In the previous section, we have seen that uneven current distribution among pads cannot be fixed easily. So we need a method of balanced partitioning of pins based on their current requirements. At the same time, the partitioning should be done such that the routing area in each part is small. We define the following notations:

M: number of pads

Y: number of terminals

N: number of pins (= M + Y)

$P = \{p_1, p_2, \ldots, p_N\}$: all the pins

$P = P_1 \cup P_2 \cup \ldots \cup P_M$: disjoint partition of pins

where each P_i contains exactly one pad

We call this problem the "Minimum Cost Balanced Partitioning (MCBP)" problem.

Definition. Minimum Cost Balanced Partitioning (MCBP): Given a number M of pads and a number Y of cell terminals with their geometrical locations and a set of current requirements (di/dt) for the terminals, partition the set of pins into M parts where each part has exactly one pad such that the total routing cost is minimized and di/dt of each part is no more than a bound B.

Since there is no net in the input of our problem (in fact, determining the nets is our objective), it does not fall on the category of traditional net-based partitioning problems. It is instead similar to the bin-packing problem. It can be shown that MCBP is NP-complete [12].

The parts should be formed such that they are balanced and the routing cost for each part is minimized. To achieve these two objective simultaneously, we propose a "Min-Forest Heuristic" as follows.

4. Min-Forest Heuristic

In our Min-Forest heuristic, the routing area for each part is approximated as the cost (sum of the edge lengths) of Minimal Spanning Tree (MST). Spanning tree approximation of the routing area is certainly an overestimation of the real Steiner tree routing, but it can give at least a comparative measure of routing areas of different parts. Starting from the pads, we grow an MST for each pad using the well-known Prim's algorithm [11]. At any moment during the growth of trees, there are exactly M MSTs (hence the name Min-Forest). When all the pins are included, these MSTs naturally constitute the partitioning solution. During the growth of trees, each MST competes to acquire a next available terminal that is not included in any MST. Some MSTs may compete to acquire the same terminal while others may compete to acquire different terminals. The winner of this competition is the one which makes the cost function smallest. These are the basics of our Min-Forest heuristic.

In the beginning, the parts $P_1, P_2, \ldots, P_M$ contain one pad each. Let $T_1, T_2, \ldots, T_M$ be the MSTs for each part. Also let $|T_i|$ be the cost of MST and $|P_i|$ be the size of the part P_i. The size of a part is the sum of di/dt of all the terminals in that part. To achieve both balanced partitioning and minimum routing cost, we propose the following cost function.

$$\frac{\sum_{i=1}^{M} |T_i|}{\sum_{i \neq j}^{M} |P_i||P_j|} \qquad (4.1)$$

The numerator favors low routing cost while the denominator favors balanced partitioning. This is similar to the cost function of ratio-cut partitioning problem [10] except that the numerator is the sum of the routing costs of each part rather than the cut size. Although this cost function helps obtain a balanced partitioning, it alone cannot return a balanced partitioning to the degree we want. So in our heuristic, we add a constraint whereby the growth of an MST stops if its size is more than $(1+\varepsilon)$ Q/M where Q is the sum of all the di/dt on the chip and ε is a user defined parameter. For small ε, we get highly balanced partitioning. For large ε, the balance is simply determined by the cost function, and the degree of the balance may not be what we want.

In Prim's algorithm, a single tree grows starting from the root node by continuously adding *light edges*. A light edge (u, v) is an edge whose edge length is minimum where u is a node already included in the tree and v is a node not included yet. Let v be light node of the tree. During the growth of the forest of MSTs, each MST has a light node that is not included in any MST. For each MST, we tentatively include its light node and evaluate the cost function of (4.1). Then we choose an MST which makes the cost function the lowest and let the MST acquire its light node. This procedure is iterated until all the nodes are included. Next is the summary of the Min-Forest heuristic.

Min-Forest(P: pins, ε: number)
Initialize each T_i to one pad tree;
Repeat
  Select T_i and its light node u_i such that $|P_i \cup \{u_i\}| \leq (1+\varepsilon)Q/M$
    and the resulting cost function (4.1) is minimum;
  $T_i \cup T_i + \{u_i\}$;
Until all the nodes are included;

5. Results

We implemented our algorithm in C++ on an Ultra Sparc 2. The benchmarks are generated randomly and tested. In the benchmarks, pads are located evenly around the periphery, and the locations of the terminals and their di/dt values are randomly and independently generated. For benchmark size 120 or lower, our program runs in sub-seconds. Below is the characteristics of our benchmarks.

Benchmark	# pins	# pads
B20	20	5
B30	30	4
B80	80	8
B120	120	10

Table 1: Benchmark set

5.1 Comparison with the conventional methods

We first consider a case where $|P_i|$ = # of pins in P_i. Traditional p/g routing involves the partitioning of the entire chip area such that all the parts have approximately equal number of terminals and then assigns the closest pad from each part. Therefore, our Min-Forest heuristic with the above the size definition is a close approximation of the traditional method since each MST tries to acquire terminals in the vicinity of the pad where the MST started with. However, since the traditional method ignores the current, we can expect unbalanced di/dt distribution. Next we consider the case $|P_i|$ = sum of di/dt. We tested our heuristic for both partition size definitions on B30 and B120 with $\varepsilon = 0.2$.

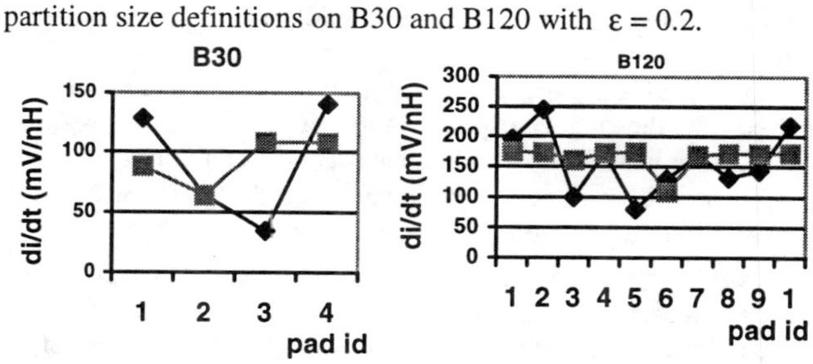

Figure 4: Current distribution comparison between the conventional method and our proposed method. Conventional method (diamond dots) vs. our method (square dots).

As can be seen in Figure 4, our method consistently returns more balanced current distributions compared with the conventional method. However due to enforcement of balanced current distribution in our method, the routing area is increased. This is shown in Figure 5.

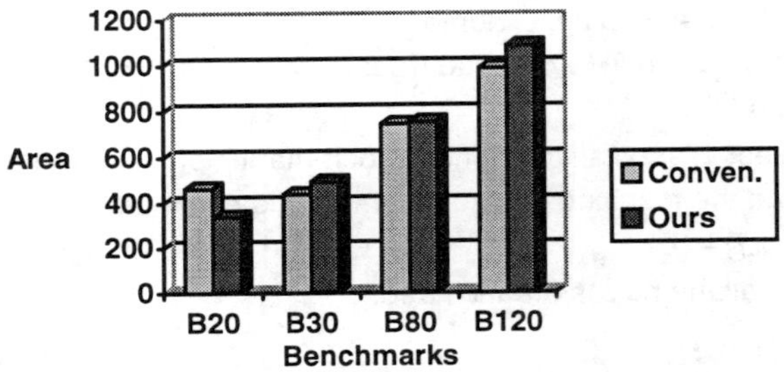

Figure 5: Comparison of the routing area between the conventional method and our proposed method.

5.2 Effects of ε

The effect of ε on partition balance is shown in Figure 6. We computed standard deviations of the di/dt for various values of ε. As can be seen from the figure, ε should be low to get balanced partitioning. However, low ε tends to result in large tree costs as can be seen in Figure 7.

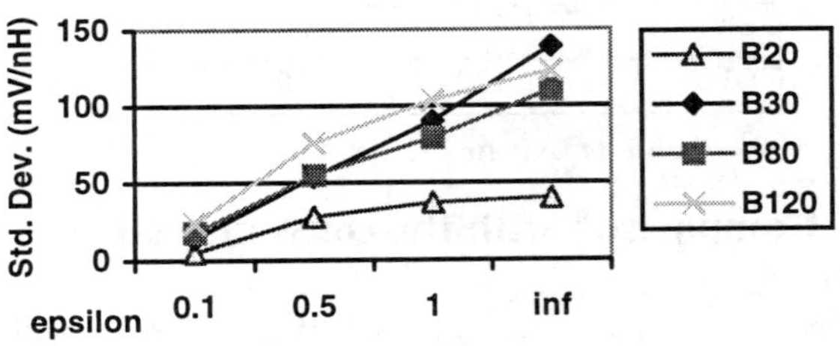

Figure 6: Standard deviations for different value of ε

6. CONCLUSION

We proposed a Min-Forest heuristic for the p/g net routing and partitioning for uniform distribution of ground bounce across the pads. We have shown that proper partitioning is essential in reducing the ground bounce as other methods have limitations in doing the job.

For a complete p/g routing, our algorithm can be separately applied on power and ground. If different p/g nets are used for on-chip and off-chip drivers, the algorithm can also be separately applied for these nets. Our algorithm can be used without modification when the pin inductances are different or when noise margins are different across the pins. In such cases, we can just enforce different partition sizes for each tree.

Our work is not a complete substitute for other ground bounce reduction techniques. Use of decaps or other techniques are still needed to reduce the total amount of ground bounce. However, if our work is used together with those techniques, it will give better distribution of noise and less chance of circuit failure.

Future work will include the extension of our method under the presence of activity correlation between cells (di/dt is not simply additive) and using other topologies such as multi-pad trees.

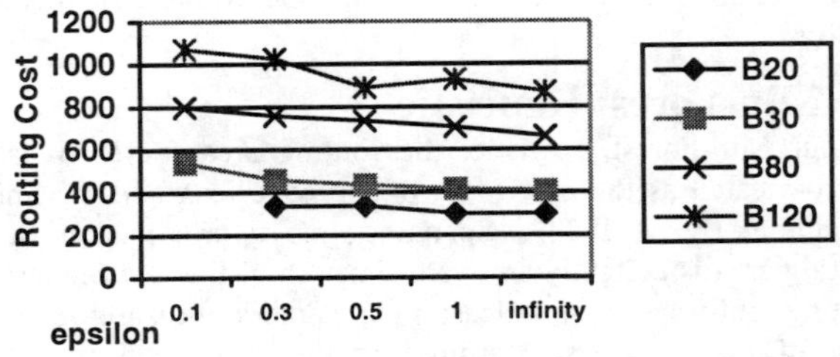

Figure 7: Routing cost vs. epsilon tradeoff. Missing data means no feasible solution at that epsilon

7. References:

[1] K.-H. Erhard and F.M. Johannes, "Power/ground networks in VLSI: are general graphs better than trees?," *Integration, the VLSI journal*, vol. 14, pp. 91-109, 1992.

[2] K.-H. Erhard, F.M. Johannes and R. Dachauer, "Topology Optimization Techniques for Power/Ground Networks in VLSI," *Proc. European Design Automation Conference*, pp. 362-367, 1992.

[3] Zahir A. Syed and Abbas El Gamal, "Single Layer Routing of Power and Ground Networks in Integrated Circuits," *Journal of Digital Systems*, vol. 6, no. 1, 1982.

[4] H. Cai, "Multi-pads, single layer power net routing in VLSI circuits," *Proc. ACM/IEEE Design Automation Conference*, pp. 183-188, 1988.

[5] S. Chowdhury and Melvin A. Breuer, "Optimum Design of IC Power/Ground Nets Subject to Reliability Constraints," *IEEE Transactions on Computer-Aided Design*, vol. 7, no. 6, pp. 787-796, July, 1988.

[6] Robi Dutta, Malgorzata Marek-Sadowska, "Automatic Sizing of Power/Ground Networks in VLSI," *Proc. IEEE/ACM Design Automation Conference*, pp. 783-786, 1989.

[7] A. Vaidyanath, B. Thoroddsen, J.L. Prince, "Effect of CMOS Driver Loading Conditions on Simultaneous Switching Noise," *IEEE Trans. on Components, Packaging, and Manufacturing Technology – Part B*, vol. 17, no. 4, pp.480-485, Nov. 1994.

[8] Y. Yang, J.R. Brews, "Design Trade-Offs for the Last Stage of an Unregulated, Long-Channel CMOS Off-chip Driver with Simultaneous Switching Noise and Switching Time Consideration," *IEEE Trans. on Components, Packaging and Manufacturing Technology – Part B*, vol. 19, no. 3, pp. 481- 486, Aug. 1996.

[9] P. Larsson, C. Svensson, "Noise in Digital Dynamic CMOS Circuits," *IEEE J. of Solid-State Circuits*, vol. 12, no. 6, pp. 655-662, June 1994.

[10] C. K. Cheng and Y.-C. A. Wei, "An Improved Two-way Partitioning Algorithm with Stable Performance," *IEEE Trans. on CAD*, vol. 10, no 12, pp. 1502-1511, Dec. 1991.

[11] T. Cormen, C. Leiserson and R. Rivest, "Introduction to Algorithms," *The MIT Press*, pp. 337-343, 1990.

[12] See our web site at http://atrak.usc.edu/~joh

Layout Extraction and Verification Methodology for CMOS I/O Circuits *

Tong Li
Dept. of Computer Science
Coordinated Science Laboratory
Univ. of Illinois at Urbana-Champaign

Sung-Mo (Steve) Kang
Dept. of Electrical and Computer Engineering
Coordinated Science Laboratory
Univ. of Illinois at Urbana-Champaign

ABSTRACT

This paper presents a layout extraction and verification methodology which targets reliability-driven I/O design for CMOS VLSI chip, specifically to guard against electrostatic discharge (ESD) stress and latchup. We propose a new device extraction approach to identify devices commonly used in CMOS I/O circuits including MOS transistors, field transistors, diffusion and well resistors, diodes and silicon controlled rectifiers (SCRs), etc. Unlike other extractors, our extractor identifies circuit-level netlist based on the specified ESD stress condition. In addition, novel techniques are proposed for the identification of parasitic bipolar junction transistors (BJTs).

I. INTRODUCTION

I/O circuits provide protection against external hazards such as electrostatic discharge (ESD). ESD is a high current (in excess of 1 A) event [1]. On-chip I/O protection circuits often consist of parasitic devices and the layout of these devices does not conform to normal design rules. Layout extraction of I/O circuits is critically important for design validation since it provides a link between layout and circuit analysis. The hitherto emphasis of computer-aided design has been on the chip core circuitry, whereas the design and layout of I/O cells have heavily relied on design expertise and guidelines. Commercial layout extraction programs are only capable of performing limited extraction and verification of the I/O layout. The limitations of commercial extraction tools stem from the following:

- They perform device extraction based on *boolean* operations of layout masks. The tools can not extract parasitic devices such as field devices, SCRs or the parasitic BJTs which are often unintended side effects of the layout and not recognized by designers. But they may play an important role during ESD stress events.
- They only extract circuit schematics under normal operating condition, i.e. when the chip is powered up. However, the circuit is typically not powered during ESD events. An ESD event can generate stress current in excess of 1 A. The normal device models are not applicable because devices are operating in high current regimes or are biased differently from normal operating modes. Therefore, the schematic for circuit simulation must be determined according to the ESD stress condition.

The proposed extraction and verification methodology for CMOS I/O is outlined in Fig. 1. The program takes layout input in CIF or GDSII format and generates the input decks for the circuit simulator iETSIM [2]. The extraction program consists of three major parts:

- **Device Extraction**: A generic device extraction approach is employed to extract all devices used in I/O circuits.
- **Stress Annotation**: Given a specified ESD stress condition, the device bias conditions are identified and the circuit schematic is extracted. The ESD stress condition is defined by specifying an I/O pad zapped positively or negatively with respect to another pad (V_{dd}, V_{ss}, or any other I/O pad).
- **BJT Extraction**: The program extracts parasitic BJTs and detects *critical* ones. A critical BJT is defined as a BJT which is unintended byproduct of the layout. A critical BJT's turning-on can induce circuit failures.

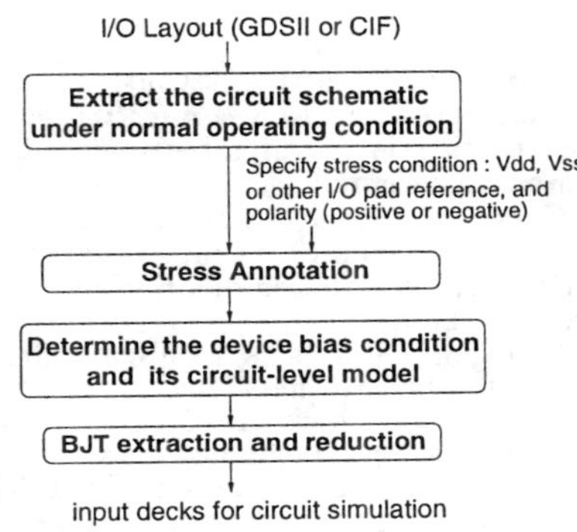

Figure 1: The extraction flow diagram.

The presentation of this paper is organized as follows. Section 2 describes the intermediate representation for the layout. Section 3 presents the device extraction and layout verification approach. Section 4 describes the circuit stress annotation and the extraction of the circuit schematic for a specified ESD stress condition. Section 5 describes the extraction of parasitic BJTs. Section 6 demonstrates the effectiveness of the methodology using industrial layout examples and presents circuit simulation results. Section 7 describes the extraction strategy for the full chip I/O frame. Finally, we present summary.

II. INTERMEDIATE REPRESENTATION

The typical representation of a VLSI layout is the geometrical description of masks, such as CIF format. Additional abstractions are needed for extracting complex device structures. The

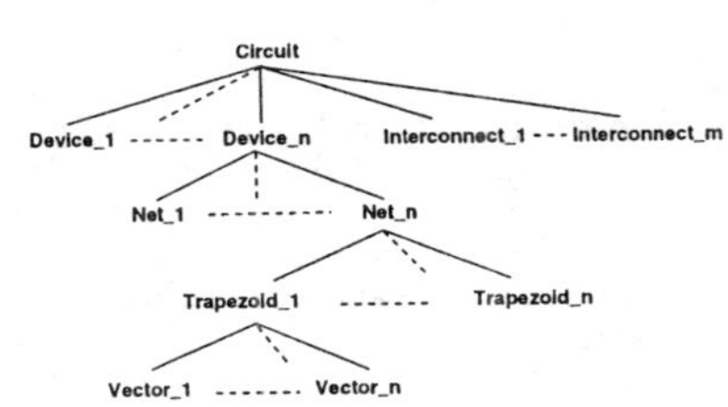

Figure 2: The hierarchical representation for a circuit.

general structure of a circuit is mapped into the hierarchical form shown in Fig. 2. A circuit contains devices and interconnect nets. A device is made up of a set of nets. A net is composed of connected trapezoids, and vectors make up trapezoids.

*This research was supported in part by Semiconductor Research Corp. (SRC97-DP-109), JSEP (N00014-97-J1270) and Rome Laboratory (F30602-97-1-0006).

DAC 98, San Francisco, California

The hierarchy represents the layout at different abstraction levels. Vector and trapezoid belong to geometrical domain. Net is the structural representation. Device and circuit are functional abstractions. Such a hierarchical representation is geometry-conserving. In other words, the geometrical information for any level of abstraction is retained through the hierarchical structure.

For device extraction, the mask layer data must be transformed into new layer representations. We use an edge-based scan line algorithm to perform *boolean* operations on the masks

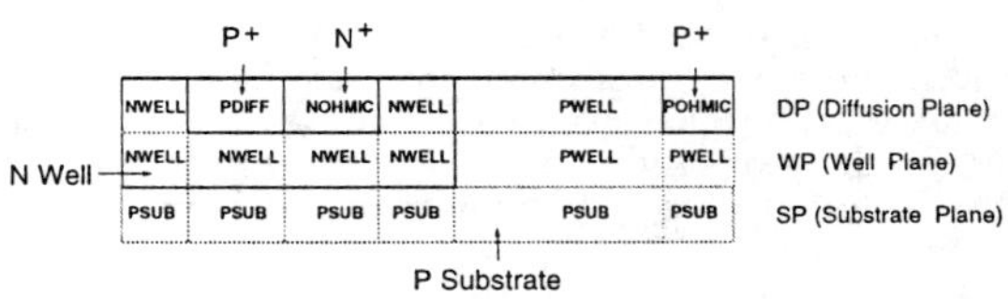

Figure 3: Substrate represented by non-overlapping trapezoids.

[3][4]. As the trapezoid representation is needed for net abstraction, vectors for a polygon are decomposed into trapezoids [4]. The silicon substrate is divided into non-overlapping trapezoids as shown in Fig. 3.

Next, connected trapezoids are grouped into nets to provide a structural abstraction for geometrical layout. The scan line algorithm extracts nets according to a *connectivity matrix* which specifies the connectivity relationships between different layers. The device is composed of nets. There are three kinds of nets, namely *substrate net, interconnect net,* and *contact net.* An interconnect net is a set of connected interconnect layer trapezoids. The substrate net is composed of adjacent substrate layer trapezoids of the same kind.

The extraction of contact nets is handled differently from interconnect nets and substrate nets, because contacts do not abut each other. Contacts which are connected to the same substrate and the same interconnect net make up one contact net. In Fig. 4, there are two substrate nets Sub1(NDIFF) and Sub2(POHMIC) and three interconnect nets Inter1, Inter2 and

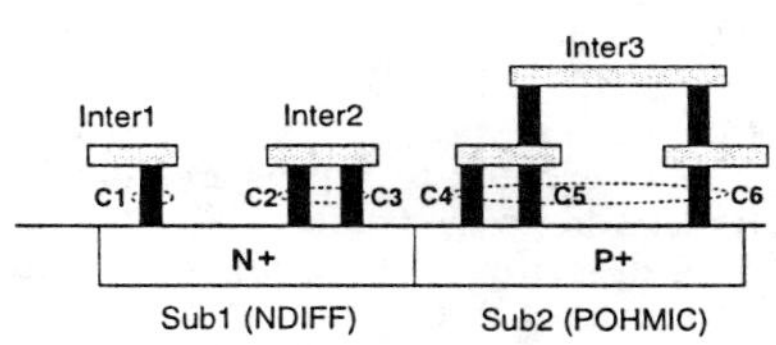

Figure 4: Extraction of contact net.

Inter3. Three contact nets, which are denoted by dashed circles, are extracted, namely CN1, CN2 and CN3. CN1 consists of C1. CN2 is composed of C2 and C3. CN3 consists of C4, C5, and C6.

III. Device Extraction and Layout Verification

A. Device Extraction

To help explain the device extraction approach, we introduce the concept of *device graph* (DG). Each device graph is used to describe one type of device. The basic structure of the device graph is defined as $DG = (N, R, s)$, where N is the set of nets in the device, R represents the set of relationships between the nets, and s is the *seed* net of the graph. The relationships in R can be categoried as below

- R1: Adjacency
- R2: Electrical connectivity
- R3: Geometrical position

The device graph proposed in this work is more general than the one introduced in [5] for analog bipolar verification. Fig. 5 depicts device graphs for PMOS transistor (the extraction of well contacts is necessary for the PMOS circuit model covering the snapback regime) and n-diffusion resistor. In graphical notation, a net is represented by a filled circle, and multiple nets by double circles. A solid line or dashed lines indicate the adjacency relationship between two nets. A contact net is connected with dashed lines when it is considered optional. The seed net is marked by an arrow. It is the location from which the DG can be constructed by the breadth-first neighbor searching.

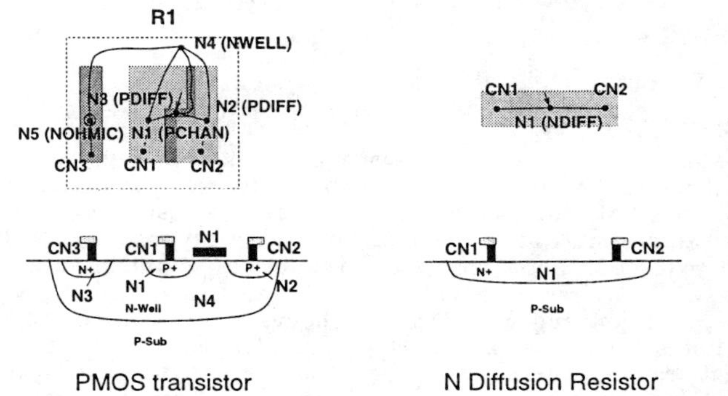

Figure 5: Device graphs for the PMOS transistor and the n-diffusion resistor.

MOS transistors and diffusion resistors can be completely specified with the adjacency relationship R1. However, specifications for R2 and R3 are necessary for identification of other types of devices. The device graph for the thick field device (TFD) is illustrated in Fig. 6. R2 specifies that nets N2 and N3 can not be at the same electrical potential, and they can not belong to one MOS transistor if they are associated with MOS transistors. In addition, either of the two interconnect nets connected to N2 or N3 must be exposed to ESD stress. R3 specifies that the distance between net N2 and N3 must be less than or equal to 2 μm. The distance is process technology dependent. The device graph for an LVSCR (low-voltage trig-

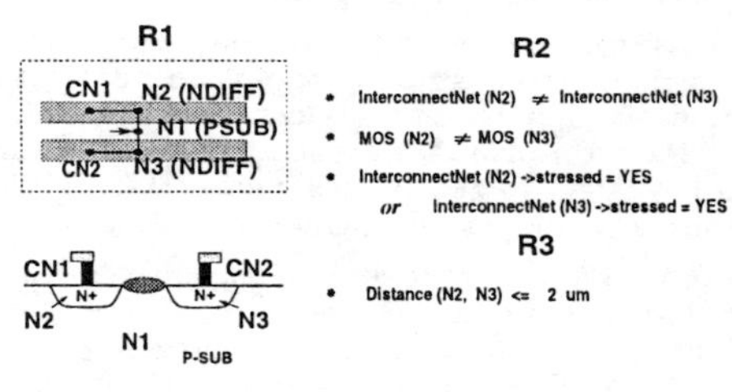

Figure 6: Device graph for the thick field device.

gered SCR) is given in Fig. 7. The LVSCR is a commonly used protection device in deep submicron CMOS technologies [6].

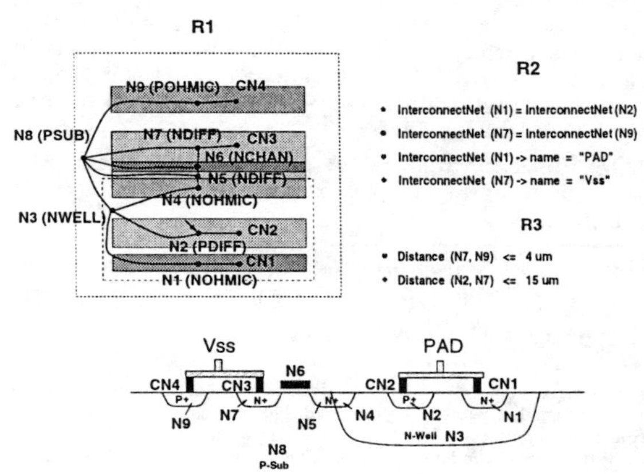

Figure 7: Device graph for the LVSCR.

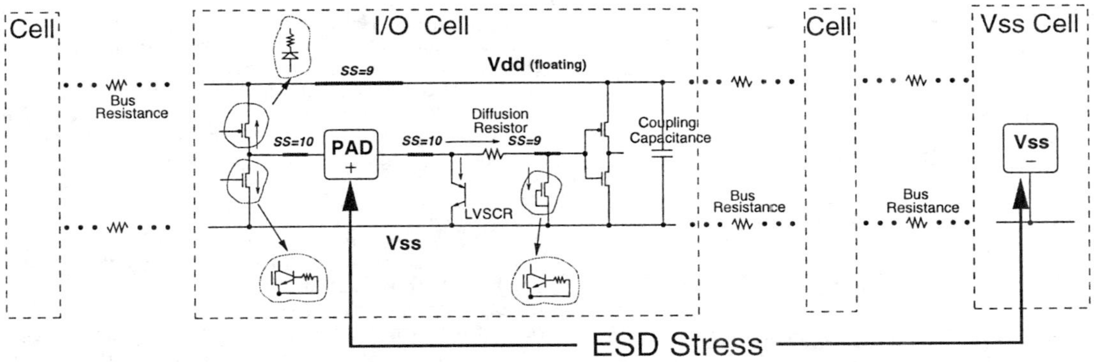

Figure 8: Stress annotation to identify the device's bias condition, and determine its circuit model.

B. Layout Verification

Normal design rules can often be described by geometrical relationships between mask layers. Since I/O layout does not conform to normal design rules, a different set of rules is often developed for I/O circuits [7]. The descriptions of I/O rules often involve specification of layout geometries associated with layout structures and electrical circuit properties. I/O layout verification using commercial tools is limited and often resort to extensive use of user-defined marking layers. This requires that circuit designers and layout engineers have in-depth understanding of protection circuits. Still, the majority of I/O design rules can only be checked by expert's visual inspection. I/O rules may not be fully verified in industrial products because of the lack of CAD tools.

With device graph, I/O design rule checking can be easily performed by searching for *error devices*. A design rule can be described by an error device graph. For instance, the error device graph for a latchup path can be defined similarly as the DG for an LVSCR. The detection of an error device will signal a design or layout error. Layout verification using device graphs can check the design rules described by both electrical and geometrical properties. Typical I/O rule examples are guard ring rules and driver contact rules (the contact-to-gate spacings for protection devices are often more than the minimum spacing) [7]. Most of the I/O rules can be checked by detecting error devices, one exception being the parasitic BJT detection. The detection of critical parasitic BJTs is very important for I/O verification. However, it requires knowledge of ESD stress conditions. Thus a different treatment is needed and will be described in section 5.

C. Cluster Extraction

Extracted I/O circuit schematics can often be simplified by clustering devices. For example, merging of multifinger MOS transistors into a single MOS transistor results in a simplified circuit topology and thus faster electrical simulation. Similarly, clustering can also be done at the circuit level. Subcircuits after clustering form a subcircuit cluster. In Fig. 9, circuit schematic of a typical output protection circuit is simplified by performing device clustering and subcircuit clustering. Clustering not only

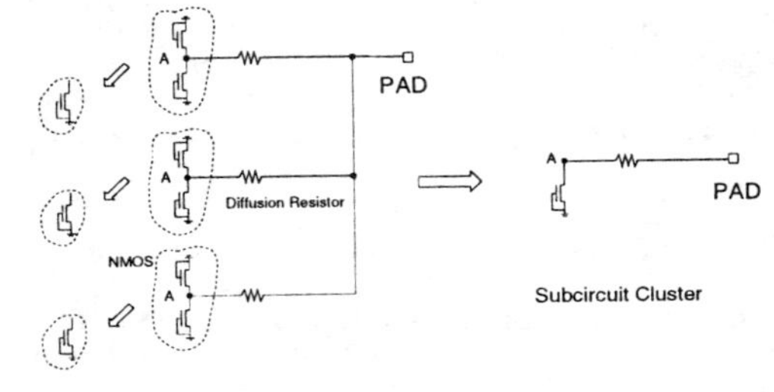

Figure 9: An example for device and subcircuit clustering. All nodes labeled A are considered at the same electrical potential.

can facilitate device reduction but also identify critical devices among peers, such as the lateral BJT reduction which will be discussed in section 5.

IV. Stress Annotation

Under ESD testing the chip is not powered up. According to the Mil Standard 883C method 3015.7 [8], a Human Body Model (HBM) test should zap all possible pin combinations of a chip for both positive and negative stresses, and the chip must pass 2KV HBM-ESD level for all stresses. When the ESD zapping is performed between two pads, all other pads are kept floating. Each I/O cell must pass the required protection level with respect to V_{dd}, V_{ss} and any other pad for both polarities.

Since the behavior of a device is bias-dependent, its circuit model should be determined according to its operating condition. We propose a static analysis technique, called *stress annotation*, to determine each device's applicable circuit model. Fig. 8 illustrates the stress annotation performed on an output circuit for a specified stress condition, i.e. positive stress on the pad with respect to V_{ss}. First, the circuit schematic for the layout at the normal operating condition is extracted. Then, the stress annotation is conducted using a breadth-first search to propagate the stress current from the stressed pad. The stress current passes through forward-biased pn junctions and semiconductor resistors. The search stops when a reverse-biased pn junction is reached. Each interconnect net in the current path is annotated with its *stress strength* (SS). The relative voltage levels between two interconnect nets can be compared by checking their stress strengths. In addition, stress annotation will help identify parasitic BJT devices, which will be discussed in the next section. Starting with an initial value, such as 10 used in Fig. 8, the stress strength is reduced by one whenever the stress current passes through a resistive device.

After the bias condition is determined from the stress annotation, each device's circuit model can be determined. When a device is under ESD current stress, a high current model must be used, such as an NMOS model which covers the snapback regime [9] and a resistor model which considers velocity saturation effect [10]. When a pn junction in a transistor is forward-biased under ESD stress, it operates as a diode. When the junction formed between the drain diffusion of the driver PMOS transistor and the n-well is forward biased, the n-well is charged up and the high current is propagated to the V_{dd} power line via the well contact. For this case, we substitute the PMOS transistor with the serial combination of a forward-biased diode and a well resistor. This is the applicable circuit model for the PMOS transistor under this specific bias condition.

V. BJT Extraction

In CMOS technologies, all devices are laid out in a single substrate and affected by substrate currents. During an ESD event, the currents injected into the silicon substrate can propagate far from the source and be collected. The substrate coupling effects can be modeled by parasitic BJTs and substrate resistance networks.

A. Lateral BJT Extraction

Ideally, the ESD current should be conducted through intentionally designed protection devices such as TFDs, LVSCRs or NMOS transistors. However, parasitic BJTs often come into being as the side effects of the layout or design. The turn-on of lateral BJTs has been a common cause of chip ESD failures [1][12][15]. The extraction of parasitic BJTs requires knowledge of electrical relationships. Without electrical relationships, any two diffusions or wells can potentially form a BJT. The total possibilities for BJTs to occur can be $2\binom{n}{2}$, where n is total number of diffusions or wells within a single cell, $\binom{n}{2}$ denotes the possible number of pairs and the constant 2 accounts for both polarities (emitter and collector). However, with the known device bias conditions and interconnect potentials from the stress annotation, the number of possible BJTs can be much less. An example of parasitic BJT identification is shown in Fig. 10. Once the stress annotation has been performed, the forward-biased pn junctions are identified as possible emitters of BJTs. Those diffusions or wells which are stressed with high voltages are identified as possible collectors (checking the lower voltages for the extraction of lateral PNPs). Note that different BJTs will be extracted for different stress conditions. If there are p emitters and q collectors, the number of possible BJTs is pq. However, not all of them will conduct significant currents. Further reduction is needed to detect active BJTs among them. The gain, β, of the parasitic BJT is the most important parameter in the BJT models, such as the Gummel-Poon model. Hence, it is used as a criterion for reduction. β is inversely proportional to the square of the base width W. The set of reduction rules is listed below while the referenced configurations are illustrated in Fig. 11.

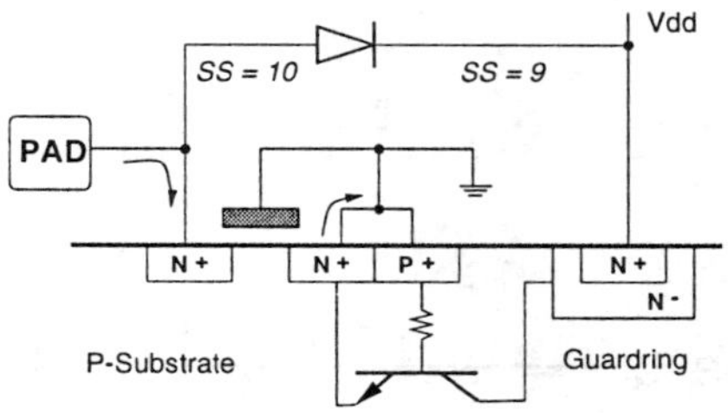

Figure 10: Extraction of lateral parasitic BJTs. The source of NMOS transistor (emitter) and guard ring (collector) form a parasitic BJT.

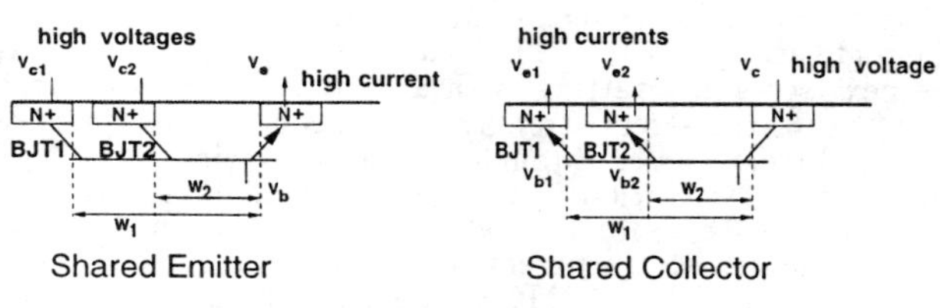

Figure 11: Configurations of parasitic BJTs for reduction.

1. Shared emitter rule:
   (a) If $V_{c1} \leq V_{c2}$, and $\beta_1 < \beta_2$, remove BJT1.
   (b) If $V_{c1} < V_{c2}$, and $\beta_1 \leq \beta_2$, remove BJT1.
   (c) Otherwise, no reduction.
2. Shared collector rule:
   (a) If $V_{e1} \geq V_{e2}$, and $\beta_1 < \beta_2$, remove BJT1.
   (b) If $V_{e1} > V_{e2}$, and $\beta_1 \leq \beta_2$, remove BJT1.
   (c) Otherwise, no reduction.
3. Minimum β rule (technology dependent):
   (a) If $\beta < \beta_{threshold}$ ($W > W_{threshold}$), no BJT.

Note that the reduction rules consider V_{cb}. To the first order, the collector current is independent of the value V_c. In the shared emitter configuration, there exist two current paths, from C2 to C1 then to the shared emitter, or from C2 to the emitter. $V_{c1} < V_{c2}$ and $\beta_1 \leq \beta_2$ indicates that the path from C2 to the emitter is less resistive, so that the current is shunted through C_2. The minimum β rule is technology dependent.

B. Vertical BJT Extraction

Vertical BJTs can also impact the ESD circuit performance. Current injected by vertical BJTs into the substrate will raise the substrate potential and may change the substrate bias conditions of all other devices. When a diffusion in a well is forward-biased, a vertical BJT will be extracted under either of the two situations, i.e. the diffusion is at high potential while the substrate is grounded, or the diffusion is grounded while the substrate is under the negative stress. The diode in Fig. 8 is replaced by the vertical BJT Q1 as shown in Fig. 12. The lateral BJT Q2 formed by the drain and source of the PMOS transistor is extracted according to the lateral BJT extraction method. There are two current paths from PMOS drain diffusion to V_{dd}, i.e. through Q1 and Q2. When the well resistance is high, Q2 conducts a significant amount of current [11]. If the size of PMOS transistor is small and can not sustain the high current, circuit failure will occur.

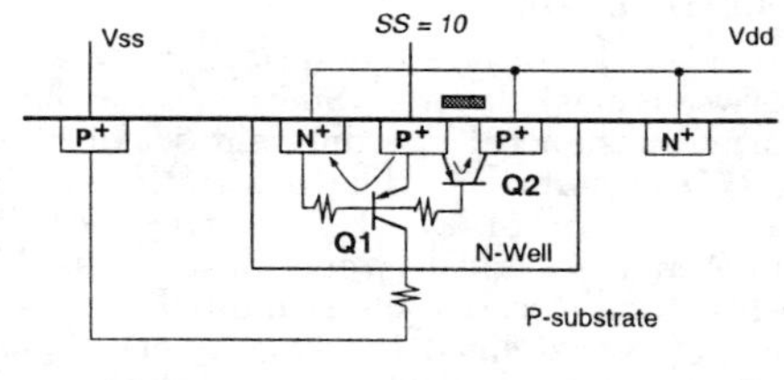

Figure 12: BJTs formed by the PMOS transistor in Fig. 8.

VI. Walk-Through Examples

In this section, three industrial layout examples from the literature are used to illustrate the proposed extraction methodology.

A. An Input Protection Circuit

A typical input protection circuit consists of two stages, namely, the primary protection and the secondary protection. The secondary protection is provided by gate-grounded NMOS (GGNMOS) transistors to protect the gate oxide of the input transistor. The circuit uses a TFD as the primary protection and a diffusion resistor as the isolation resistor. Two different layouts for the input protection circuit are given in Fig. 13 and 14. Although both layouts have the same circuit schematic under the normal operating conditions, it has been found experimentally that there exists a lateral parasitic BJT in layout 2 under the positive stress condition from the pad w.r.t. V_{ss} [12][13]. The nodes in the device graphs are numbered in the layout. A lateral BJT is denoted by the node numbers of its collector

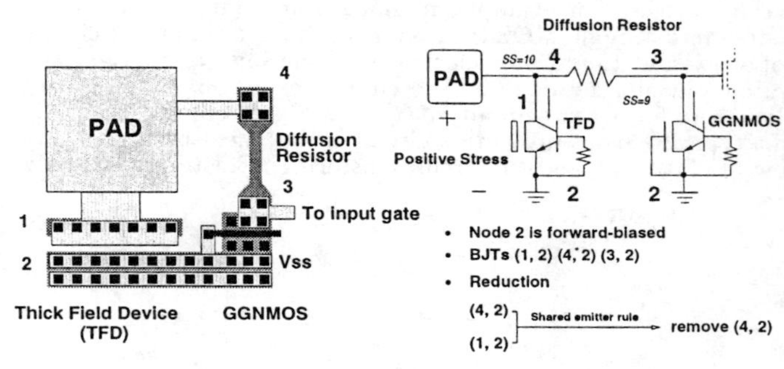

Figure 13: Analysis of layout 1.

and emitter. For example, (1, 2) represents the BJT formed by node 1 as its collector and node 2 as its emitter.

First, the stress annotation is conducted on the circuit schematic. Since the drains of TFD and GGNMOS are under positive stress, the high current models including the bipolar action must be used. Next, the BJTs are extracted for the two layouts as explained below.

- Layout 1: As shown in Fig. 13, diffusion 2 is forward-biased after the TFD or GGNMOS is triggered, while diffusions 1, 3 and 4 are under high voltage stress. Therefore, three lateral BJTs may be formed. They are denoted as (1, 2) (3, 2) and (4, 2). To perform the BJT reduction, BJTs (4, 2) and (1, 2) are clustered to form the shared emitter configuration. BJT (4, 2) is eliminated according to the shared emitter rule. Since BJTs (3, 2) and (1, 2) have already been included in the TFD and NMOS high current models, **no extra parasitic BJT exists in layout 1.**

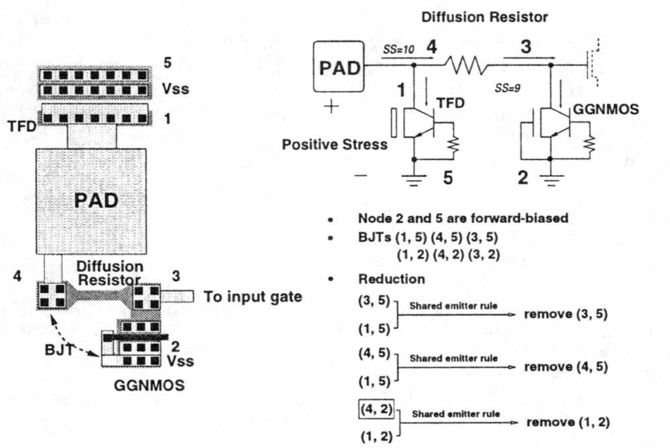

Figure 14: Analysis of layout 2. Lateral BJT (4, 2) is detected under the positive stress to pad w.r.t. V_{ss}.

- Layout 2: As illustrated in Fig. 14, diffusions 2 and 5 are forward-biased. They become the sources for minority carrier injection into the substrate. Since diffusions 1, 3 and 4 are stressed with high voltages, there are six possible lateral BJTs, namely (1, 5) (4, 5), (3, 5), (1, 2), (4, 2) and (3, 2). The BJTs are reduced similar to layout 1. Because the TFD and NMOS high current models include BJTs (1, 5) and (3, 2), **parasitic BJT (4, 2) is detected in layout 2.**

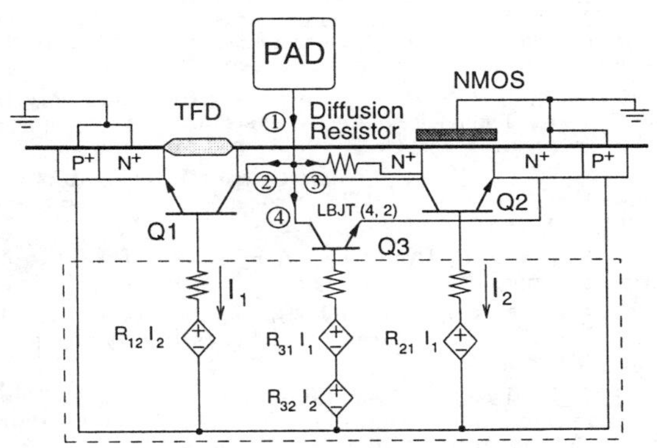

Figure 15: The circuit schematic with the substrate resistance network under positive stress.

The circuit schematic for layout 2 is drawn in Fig. 15 along with the cross-section of silicon substrate. The substrate resistance network model has been developed in [14]. Circuit simulation results in Fig. 16 indicate that there is indeed a significant amount of stress current conducting through the lateral

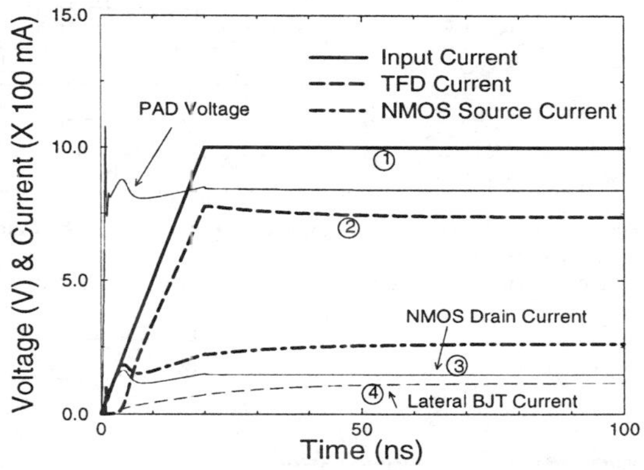

Figure 16: Simulation results under a ramped current input. Most of the stress current flows through the TFD. Part of the source current is collected at the junction connected to pad through the lateral BJT (4, 2).

parasitic BJT. This confirms the experimental and failure analysis results reported in [12][13].

B. An Input Protection Circuit – II

The circuit layout and details of extraction are shown in Fig. 17. It uses an LVSCR instead of a TFD as the primary protection device. Although the circuit schematic is same as the one in the previous example, the circuit failure mechanism is different. The failure site is circled in the layout when the pad is stressed negatively with respect to V_{dd} [15]. It indicates that a parasitic BJT (5, 2) turns on and collector failure occurs.

The negative current is propagated from the pad to perform the stress annotation. The GGNMOS is now stressed negatively so that the drain becomes the emitter of its associated BJT. Under negative stress conditions, the diffusion resistor to substrate junction is forward-biased and injects minority carriers into the substrate. We divide the diffusion resistor into three segments. The n-well denoted by node 5 is at high potential. Therefore, there may exist six BJTs (5, 4) (5, 3) (5, 2) (5, 1) (7, 4) (6, 1). After the reduction process, BJT (5, 2) is finally detected. The circuit simulation can be performed similarly as for example 1.

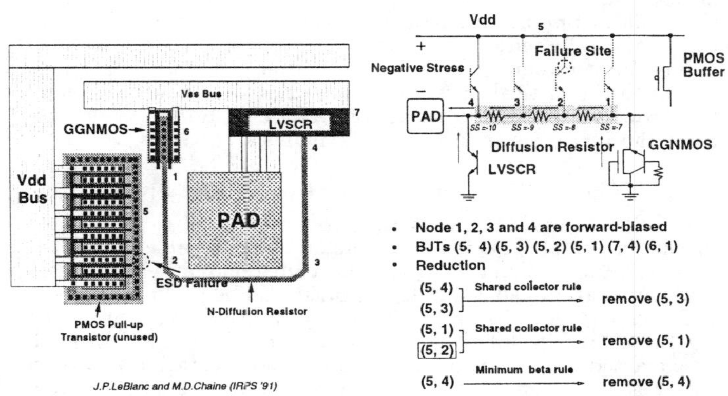

Figure 17: Analysis of the input protection circuit – II.

C. An I/O Protection Circuit

The last example illustrates the extraction of vertical BJTs and its role for accurate simulation. The circuit is composed of a gate-coupled NMOS device (consists of NMOS transistors, an NMOS capacitor and a well resistor) and a lateral diode

Circuit Layout	Stress [1] Condition	No. of Lateral BJTs				No. of Vertical BJTs		
		(-, -, -)	(c, -, -)	(c, a, -)	(c, a, r) [2]	(-, -, -)	(c, -, -)	(c, a, -)
Ex1 (layout 1)	V_{ss} (+)	12	12	3	**0**	0	0	**0**
Ex1 (layout 2)	V_{ss} (+)	20	20	6	**1**	0	0	**0**
Ex2	V_{dd} (-)	42	30	6	**1**	6	1	**0**
Ex3	V_{ss} (+)	210	42	3	**0**	4	1	**1**

(c, a, r) stands for (clustering, annotation, reduction). A dash denotes that the corresponding operation is not performed.
[1] The stress is from pad to the reference node.
[2] The number of critical parasitic BJTs. The intentionally designed BJTs are excluded.

Table 1: Summary of the BJT extraction results.

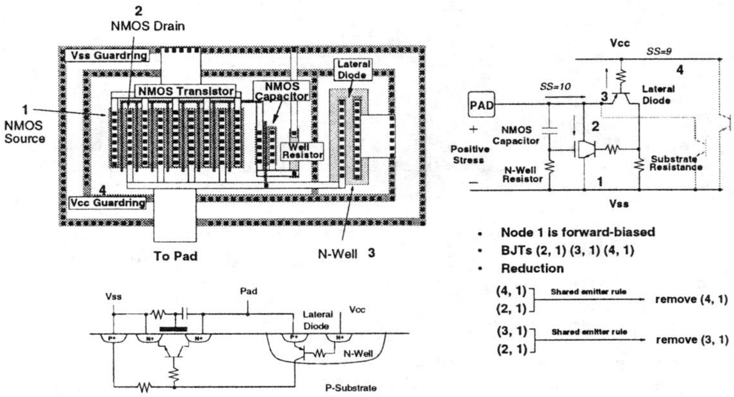

Figure 18: Analysis of the I/O protection circuit. Lateral BJT (2, 1) is included in NMOS snapback model. Therefore, no parasitic lateral BJT is detected.

connected to V_{cc} [16]. The circuit schematic is extracted for the positive stress from pad with respect to V_{ss}. After we have performed the stress annotation, we proceed to extract lateral BJTs and perform reduction as shown in Fig. 18. Finally, we extract the vertical BJTs. As the pn junction in the lateral diode is forward-biased and resides in the n-well, the diode is replaced by a vertical BJT to consider the substrate coupling effect. The work on substrate coupled simulation using iETSIM was published in [16].

The extraction results for the above three examples are summarized in Table 1. The BJT extraction procedure involves three major operations, namely device clustering, stress annotation and BJT reduction. As we can see, each step can greatly reduce the number of BJTs, and eventually the procedure pinpoints to the critical ones which are printed in bold. By using the proposed systematic procedure, we can automatically extract the I/O circuits for full ESD analysis.

VII. Full Chip Frame Extraction

The ESD standard requires the protection level to be verified for the full chip. This is due to the fact that reliable individual I/O cells do not guarantee the full chip reliability [1]. Therefore, verification using simulation must also be performed at the I/O frame level. The program employs a hierarchical approach for the extraction of the chip I/O frame. First, each I/O cell is extracted once and its intermediate representation shown in Fig. 2 is stored in an object-oriented database. The extraction of the full chip I/O frame only utilizes the interconnect relationships among I/O cells. It identifies the bus architecture (single bus or separate internal and external buses, the power bus protection structure) and extracts bus resistances and power bus coupling capacitances shown in Fig. 8. Then, stress annotation is performed on the extracted I/O frame circuit. The extraction of possible parasitic BJT devices must be done for every two adjacent cells.

VIII. Summary

We have presented a novel extraction and verification methodology for CMOS I/O cell and chip frame. To our knowledge, the layout extractor presented in this paper is the first comprehensive layout extractor developed for reliability-driven CMOS I/O design. With short turn-around time, extensive ESD protection circuit simulations can be performed by using our extractor and simulator. Thus the design and layout flaws such as the existence of a critical parasitic BJT can be detected before the design and layout are committed to silicon.

References

[1] C. Duvvury and A. Amerasekera, "ESD: A Pervasive Reliability Concern for IC Technologies," *Proc. of the IEEE,* vol. 81, no. 5, pp. 690–702, May 1993.

[2] C. Diaz, S. M. Kang and C. Duvvury, "Circuit-level Electrothermal Simulation of Electrical Overstress Failures in Advanced MOS I/O Protection Devices," *IEEE Trans. on CAD,* vol. 13, no. 4, pp. 482–493, April 1994.

[3] U. Lauther, "An $O(NLogN)$ Algorithm For Boolean Mask Operations," *Proc. ACM/IEEE Design Automation Conf.*, pp. 555–562, 1981.

[4] K. Chang, S. Nahar and C. Lo, "Time-Efficient VLSI Artwork Algorithms in GOALIE2, " *IEEE Trans. on CAD,* vol. 8, no. 6, June 1989.

[5] E. Barke, " A Layout Verification System For Analog Bipolar Integrated Circuits," *Proc. ACM/IEEE Design Automation Conf.*, pp. 353–359, 1983.

[6] A. Chatterjee and T. Polgreen, "A Low-Voltage Triggering SCR for On-Chip ESD Protection at Output and Input Pads," *IEEE Electron Device Letters,* vol. 12, no. 1, pp. 21–22, 1991.

[7] A. Kapoor, D. Suh and P. Bendix, "ESD/Latchup Design Rules," Techinical Report, LSI Logic, 1996.

[8] MIL-STD-883C, Electrostatic Discharge Sensitivity Classification, Technical Report, Notice 8, DoD, March 1989.

[9] A. Amerasekera, S. Ramaswamy, M. Chang and C. Duvvury, "Modeling MOS Snapback and Parasitic Bipolar Action for Circuit-Level ESD and High Current Simulations," *International Reliability Physics Symp.*, pp. 318–326, 1996.

[10] G. Krieger and P. Niles, " Diffused Resistors Characteristics at High Current Density Levels – Analysis and Applications," *IEEE Trans. on Electron Devices*, Vol. 36, pp. 416–423, 1989.

[11] C. Duvvury and R. Rountree, "Output ESD Protection Techniques for Advanced CMOS Processes," *EOS/ESD Symp.*, pp. 206–211, 1988.

[12] C. Duvvury and R. Rountree, "A Synthesis of ESD Input Protection Scheme," *EOS/ESD Symp.*, pp. 88–97, 1991.

[13] Y. Fong and C. Hu, "Internal ESD Transients in Input Protection Circuits," *IEEE International Reliability Symp.*, pp. 77–81, 1989.

[14] T. Li, C. H. Tsai, E. Rosenbaum and S. M. Kang, "Modeling, Extraction and Simulation of CMOS I/O Circuits under ESD Stress," to appear in *IEEE International Symp. on Circuits and Systems*, 1998.

[15] J. P. LeBlanc and M. D. Chaine, "Proximity Effects of Unused Output Buffers on ESD Performance," *IEEE International Reliability Symp.*, pp. 327–330, 1991.

[16] S. Ramaswamy, C. Duvvury, A. Amerasekera, V. Reddy and S.-M. Kang, "EOS/ESD analysis of high-density logic chips," *EOS/ESD Symp.*, pp. 285-290, 1996.

A Mixed Nodal-Mesh Formulation for Efficient Extraction and Passive Reduced-Order Modeling of 3D Interconnects

Nuno Marques+ Mattan Kamon* Jacob White* L. Miguel Silveira+

+ INESC/Cadence European Laboratories
Dept. of Elect. and Comp. Eng.
Instituto Superior Técnico
1000 Lisboa, Portugal

* Research Laboratory of Electronics
Dept. of Elect. Eng. and Comp. Science
Massachusetts Institute of Technology
Cambridge, MA 02139

Abstract

As VLSI circuit speeds have increased, reliable chip and system design can no longer be performed without accurate three-dimensional interconnect models. In this paper, we describe an integral equation approach to modeling the impedance of interconnect structures accounting for both the charge accumulation on the surface of conductors and the current traveling in their interior. Our formulation, based on a combination of nodal and mesh analysis, has the required properties to be combined with Model Order Reduction techniques to generate accurate and guaranteed passive low order interconnect models for efficient inclusion in standard circuit simulators. Furthermore, the formulation is shown to be more flexible and efficient than previously reported methods.

1 Introduction

As VLSI circuit speeds have increased, reliable and accurate chip and system design can no longer be performed without accurate three-dimensional interconnect models. In the past, designers had to cope with tools where 2D models, used for modeling long and uniform interconnect structures, were glued together with 3D models that were necessary for handling discontinuities such as vias through planes, chip-to-board and board-to-board connectors. Not only was this mixing of models cumbersome but, as circuit density continues to increase, such discontinuities become more prevalent. In recent years much effort has been devoted to the study and development of fast and accurate algorithms for computing full 3-D interconnect models directly from Maxwell's equations.

Many of these structures are small compared to a wavelength, and much work has been directed at rapidly solving for the inductance and capacitance of these structures. However inductance and capacitance are not necessarily decoupled quantities, and for higher frequencies a distributed model is necessary. In this paper, we describe an integral equation approach to modeling the impedance of interconnect structures accounting for both the charge accumulation on the surface of conductors and the current traveling along conductors. When high accuracy is desired, the models generated can become excessively large and difficult to solve for a continuous range of frequencies. The need for reduced-size models leads us to consider Model Order Reduction (MOR) techniques, which have been developed in the field of parameter extraction [1, 2, 3, 4, 5, 6, 7, 8, 9].

Our approach, which is based on a combination of nodal analysis formulation with a mesh analysis formulation, has significant advantages over previously reported methods, both in extraction speed and model size, making it possible to generate guaranteed passive low order models for efficient inclusion in a circuit simulator such as SPICE or SPECTRE. Combining our formulation with acceleration techniques such as the Fast Multipole Method [10, 11] or the Precorrected-FFT [12], allows the accurate analysis of larger, more complex three-dimensional geometries than previously possible.

In Section 2 we discuss the integral formulation and discretization from which we derive the circuit equations that describe the interconnect effects. In Section 3 we describe a nodal analysis-based formulation that allows the application of recent model order reduction techniques in order to directly generate passive reduced-order models. We then present an extension to this formulation based on a combination of nodal and mesh analysis that can be used to accelerate model construction. In Section 4 we present results of using the new extraction tool and analyze the performance of the generated reduced-order models. Finally, conclusions are drawn in Section 5.

2 Mathematical Formulation

Parasitic extraction for a set of conductors involves determining the relation between the terminal (or port) currents and the terminal voltages. Given a structure of k terminal pairs, the admittance matrix which relates the terminal currents and the terminal voltages is defined as

$$\boldsymbol{Y}_t(\omega)\boldsymbol{V}_t(\omega) = \boldsymbol{I}_t(\omega), \tag{1}$$

where sinusoidal steady-state at frequency ω is assumed. $\boldsymbol{Y}_t(\omega) \in \mathbb{C}^{k\times k}$, and $\boldsymbol{I}_t, \boldsymbol{V}_t \in \mathbb{C}^k$ are the terminal current and voltage vectors, respectively [13]. If it is possible to compute the currents given the voltages at the terminals then, by adding voltage sources to all terminals in the circuit, $\boldsymbol{Y}_t(\omega)$ can be computed one column at a time. To do this, we set entry i of $\boldsymbol{V}_t$ to one, the others to zero, and solve for $\boldsymbol{I}_t$, which will be the ith column of $\boldsymbol{Y}_t(\omega)$.

To derive a relation between voltages and currents, we resort to an integral equation approach derived directly from Maxwell's equations and similar to the Partial Element Equivalent Circuit (PEEC) method [14]. At each point inside the conductors, we have [15]

$$\boldsymbol{E} = -\nabla\varphi - \frac{\partial \boldsymbol{A}}{\partial t} \tag{2}$$

where

$$\varphi(\boldsymbol{r}, t) = \frac{1}{4\pi\epsilon}\int_{V'} \frac{\rho(\boldsymbol{r}')}{\|\boldsymbol{r}-\boldsymbol{r}'\|}dv' \tag{3}$$

$$\boldsymbol{A}(\boldsymbol{r}, t) = \frac{\mu}{4\pi}\int_{V'} \frac{\boldsymbol{J}(\boldsymbol{r}')}{\|\boldsymbol{r}-\boldsymbol{r}'\|}dv'. \tag{4}$$

In deriving equations, the Electromagneto-Quasistatic assumption was considered. This approximation corresponds to assuming that the influence of the charge and currents at any point in the structure is instantaneously felt everywhere. That is to say, we consider the speed of light as infinite. So, our model will be valid only for structures small compared to a wavelength. (In practice, however, one can analyze structures on the order of a wavelength, because the interaction terms where retardation time is significant correspond to weak coupling).

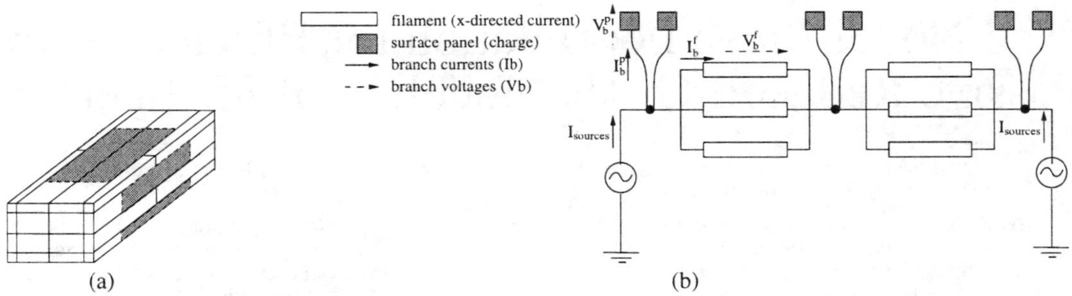

Fig. 1. a) Results of discretization operation applied to a conductor showing volume filaments (with its cross section decreasing toward the surfaces to properly capture skin and proximity effects) and charge panels (only some panels are shown); b) A simplified version of the correspondent electrical circuit – nodal analysis quantities are described.

To compute a model from this formulation, a discretization operation is performed on the conductors. Following the PEEC approach, the interior of each conductor is divided into a grid of filaments where each filament is assumed to have a constant current density with the direction of its length, and the surfaces of the conductors are covered with panels, where each panel is assumed to have a constant charge density. For conductors that are very long and thin we can assume that the current running along its length is much larger than the current running along the other two orthogonal directions. In those cases, a filament discretization along a single direction is performed. For planar conductors, the filament discretization must be performed along two coordinate directions and for volume-like conductors, filaments are set along the three-coordinate directions. This discretization operation allows us to generate an equivalent electrical "circuit" made up of filaments and panels. Figure 1-a) shows the discretization operation applied to a long and thin conductor (filaments along one direction only) and Figure 1-b) shows the electrical circuit associated with it.

To generate the constitutive relations for these elements, we apply the Galerkin method to (2), after discretizing the equation (see [14] for details). A constitutive relation for the filaments is obtained in the form of

$$Z_L I_b^f \equiv (R + j\omega L) I_b^f = V_b^f \tag{5}$$

where $I_b^f \in \mathbb{C}^f$ is the vector of f filament currents,

$$R_{ii} = \frac{l_i}{\sigma a_i} \tag{6}$$

is the $f \times f$ diagonal matrix of filament DC resistances, and

$$L_{ij} = \frac{\mu}{4\pi a_i a_j} \int_{V_i} \int_{V'_j} \frac{\mathbf{l_i} \cdot \mathbf{l_j}}{\|\boldsymbol{r} - \boldsymbol{r}'\|} dV' \, dV \tag{7}$$

is the $f \times f$ dense, symmetric positive semidefinite matrix of partial inductances. $V_b^f = \Phi_{n_1} - \Phi_{n_2}$ is the vector of voltages given as the difference between the node potentials, Φ_n, at the two ends of the filament.

In our formulation, and unlike the PEEC method, the coefficients of the potential matrix P, which relates node potentials to panel charges, are computed such that charge conservation is verified on every node. In our case, for p panels, $P \in \mathbb{R}^{p \times p}$ will be computed as:

$$P_{ij} = \frac{1}{A_j 4\pi\epsilon_0} \int_{p'_j} \frac{1}{\|\boldsymbol{r}_{p_i} - \boldsymbol{r}'\|} dV', \tag{8}$$

that is in the form of partial coefficients of potential - where p'_j is the surface of panel j, A_j its area and $\boldsymbol{r}_{p_i}$ is the center of each panel p_i. Note that we allow multiple panels on each node - as is the case for nodes on edges or apices of conductors. Furthermore, we shall see that this allows us to establish a set of equations amenable to generate passive reduced-order models.

Since the current flowing onto the panels is given by $I_b^p = \frac{d}{dt} q_p$, $q_p \in \mathbb{C}^p$ being the charge on each of the p panels, for the sinusoidal steady state we can write $q_p = I_b^p/(j\omega)$. Additionally, since the panel node voltages are voltages relative to infinity, we can view the panel branches as connecting the panel node to the zero potential node at infinity. Then the panel branch voltages are given by $V_b^p = \Phi_p - 0 = \Phi_p$, where Φ_p are the panel potentials. Combining, we get a relation between the panel currents and their voltages,

$$V_b^p = \frac{1}{j\omega} P I_b^p \tag{9}$$

With (5) and (9) we can write the constitutive relations for the elements as a single matrix in $\mathbb{C}^{b \times b}$, $b = f + p$,

$$V_b = \begin{bmatrix} V_b^f \\ V_b^p \end{bmatrix} = \begin{bmatrix} Z_L & 0 \\ 0 & P/(j\omega) \end{bmatrix} \begin{bmatrix} I_b^f \\ I_b^p \end{bmatrix} = Z I_b. \tag{10}$$

where $I_b \in \mathbb{R}^b$ is the vector of branch (filaments and panels) currents and V_b is the vector of branch (filaments or panels) voltages. See Figure 1-b) for an illustration of these quantities.

Applying voltage sources to the circuit, we can now solve it, and extract the desired terminal currents, thus producing the required admittance model. To that end we will use a Modified Nodal Analysis technique. Note that only voltage sources can be used because one needs to connect all the sources to the zero potential node at infinity and that node is only connected to capacitors. This is necessary in order to be able to obtain DC results, which are desirable if we want to use the model in time-domain simulations.

Kirchoff's Current Law, which implies that the sum of the branch currents leaving each node in the network must be zero, can be written as

$$\begin{bmatrix} A & -N \end{bmatrix} \begin{bmatrix} I_b \\ I_{src} \end{bmatrix} = \begin{bmatrix} 0 \end{bmatrix} \tag{11}$$

where $A \in \mathbb{R}^{n \times b}$ is the sparse nodal incidence matrix summing the filament and panel currents in each node, $N \in \mathbb{R}^{n \times n_{src}}$ is the sparse matrix summing the currents through the voltage sources, n is the number of nodes (excluding the one for the point at infinity), b is the number of branches (filaments plus panels), and n_{src} is the number of voltage sources in the circuit. $I_{src} \in \mathbb{R}^{n_{src}}$ is the vector of currents leaving each voltage source (always connected to ground) and entering the terminal nodes. Note the 0 in the right hand side due to the absence of current sources in the circuit.

Applying Kirchoff's Voltage Law to the circuit, we obtain

$$\begin{bmatrix} A^T \\ N^T \end{bmatrix} \begin{bmatrix} V_n \end{bmatrix} = \begin{bmatrix} V_b \\ V_{src} \end{bmatrix} \tag{12}$$

where V_n is the vector of voltages at each node in the network, V_{src} are the known source voltages. Note that V_{src} is exactly the terminal voltage vector V_t, from (1). Combining (12) with (11) and

(10) yields the system of equations

$$\begin{bmatrix} Z & -A^T & 0 \\ A & 0 & -N \\ 0 & N^T & 0 \end{bmatrix} \begin{bmatrix} I_b \\ V_n \\ I_{src} \end{bmatrix} = \begin{bmatrix} 0 \\ 0 \\ V_{src} \end{bmatrix} \tag{13}$$

One approach to coupling the above models with circuits is to directly include an admittance model, for instance some version of (13), into a circuit simulator. This approach has the drawback that the size of the system in (13) can easily be very large if high-accuracy is desired. Even computing the admittance matrix directly, as described at the beginning of this section is inefficient, as it requires t solutions of a system with a very large matrix. The use of iterative methods could reduce the computational cost of solving this system but the resulting model would be valid only at a single frequency. However to design with this admittance model it is often necessary to perform coupled simulation with nonlinear devices, such as CMOS drivers and receivers. Nonlinear devices require time domain simulation, and thus the admittance information is necessary from DC up to the highest frequency of interest in the circuit. Thus, it is essential to have models valid for a continuous range of frequencies.

3 Guaranteed Passive Model Order Reduction and Interconnect Simulation

Recently, Model Order Reduction (MOR) algorithms [1, 2, 5, 9, 6] have been presented to solve this problem. The basic idea of MOR techniques is to reduce the size of the system described by the circuit equations, usually written in a convenient state-space form, to a much smaller one that still captures the dominant behavior of the original system. This approximation is then used to generate a model amenable to be inserted in a circuit simulator such as SPICE or SPECTRE. The field of MOR has matured significantly in the past few years. Recently, MOR techniques, such as the PRIMA algorithm [6], have been presented, that produce guaranteed stable and passive reduced-order models.

To apply this algorithm, the circuit equations are written in the state-space form

$$\begin{aligned} s\mathcal{L}\,x &= -\mathcal{R}\,x + B\,u \\ y &= B^T x. \end{aligned} \tag{14}$$

where $s = j\omega$. To generate passive reduced order models both $\left(\mathcal{R} + \mathcal{R}^T\right)$ and $\left(\mathcal{L} + \mathcal{L}^T\right)$ must be positive semidefinite. It was shown in [16] that it is possible to derive a state-space form with these properties when Mesh analysis is applied to the circuit. We will now describe how a state-space form with these properties can also be derived using Modified Nodal Analysis.

3-1 *Nodal Formulation*

To derive a state space form of (13), the powers of the Laplace variable $s = j\omega$ must all be to the first powers only. However, the constitutive relation Z in (10) contains terms with both s and $1/s$. To separate the $1/s$ power, note that as seen in Figure 1-b), the branch currents can be separated into two types, $I_b = [I_b^f; I_b^p]$ where I_b^f represents the currents in filaments and I_b^p represents the currents onto panels. Also, A can be split into $[A_e; A_i]$ where A_e corresponds to the n_e external nodes (an external node is a node that has at least one panel connected to it) and A_i corresponds to the n_i internal nodes (an internal node is a node to which no panels are connected). Based on this categorization, (11) becomes

$$\begin{bmatrix} A_e & B_e & -N_e \\ A_i & 0 & 0 \end{bmatrix} \begin{bmatrix} I_b^f \\ I_b^p \\ I_{src} \end{bmatrix} = \begin{bmatrix} 0 \\ 0 \end{bmatrix}. \tag{15}$$

The zero-blocks in (15) correspond to B_i and N_i, which are always null since there are no panels or voltage sources connected to internal nodes. We can also rewrite (12) separating the branch voltages in $V_b = [V_b^f; V_b^p]$. V_n^e and V_n^i correspond to, respectively, the voltage in external and internal nodes:

$$\begin{bmatrix} A_e^T & A_i^T \\ B_e^T & 0 \\ N_e^T & 0 \end{bmatrix} \begin{bmatrix} V_n^e \\ V_n^i \end{bmatrix} = \begin{bmatrix} V_b^f \\ V_b^p \\ V_{src} \end{bmatrix} \tag{16}$$

The desired equation in state-space form can then be obtained using (15) and (16) and the constitutive relations expressed in (10). Using the first equation in (15) and the constitutive relation for the panels, we can write

$$A_e I_b^f + sB_e P^{-1} V_b^p - N_e I_{src} = 0 \tag{17}$$

Furthermore, using the second relation in (16), we have

$$A_e I_b^f + sB_e P^{-1} B_e^T V_n^e - N_e I_{src} = 0 \tag{18}$$

Now using the constitutive relation for filaments with the first relation in (16), yields

$$(R + sL)I_b^f - A_e^T V_n^e - A_i^T V_n^i = 0 \tag{19}$$

Finally, using (18) and (19) plus the relations not yet used in (15) and (16), we obtain the desired state-space form:

$$s \begin{bmatrix} L & 0 & 0 & 0 \\ 0 & B_e P^{-1} B_e^T & 0 & 0 \\ 0 & 0 & 0 & 0 \\ 0 & 0 & 0 & 0 \end{bmatrix} \begin{bmatrix} I_b^f \\ V_n^e \\ V_n^i \\ I_{src} \end{bmatrix} = \tag{20}$$

$$- \begin{bmatrix} R & -A_e^T & -A_i^T & 0 \\ A_e & 0 & 0 & -N_e \\ A_i & 0 & 0 & 0 \\ 0 & N_e^T & 0 & 0 \end{bmatrix} \begin{bmatrix} I_b^f \\ V_n^e \\ V_n^i \\ I_{src} \end{bmatrix} + \begin{bmatrix} 0 \\ 0 \\ 0 \\ V_{src} \end{bmatrix}.$$

which can readily be written as (14).

It was shown in [17] that this system has the properties mentioned above as necessary to generate guaranteed passive reduced order models. Model Order Reduction techniques can thus be applied directly to either this formulation or the one developed in [16] using Mesh Analysis. Considering an expansion point at $s = 0$, for which the matrix $\mathcal{R}$ needs to be inverted, the sizes of the resulting reduced-order models produced by those formulations is similar for similar accuracy. However, because the number of states in the nodal formulation is smaller, it is, for most cases, faster to obtain a reduced-order model using this formulation. Indeed, the number of states in the nodal case is

$$f + n + n_{src} \tag{21}$$

where f is the number of filaments, n the number of nodes in the network, and n_{src} is the number of sources; the number of states in the mesh formulation is

$$m + p = f + p - n + 1 \simeq f + 2p - n \tag{22}$$

where m is the total number of meshes and p is the total number of panels. So the nodal formulation is smaller if $p > n$, which is always the case for a discretization along one or two directions.

As described in [18], and as we shall see in the results section, interconnect models derived using the Electromagneto-Quasistatic assumption, which captures skin effect, have in general a large number of real poles that have a weak effect on the system behavior. Many of these poles are near the origin and moment-matching around $s = 0$ tends to match these non-dominant poles first, and only then consider more important outward poles. Thus it might be useful to consider using other expansion points. An obvious choice would be $s = \infty$. For certain structures, it turns out that the size of the reduced models obtained with expansions around $s = \infty$ can be significantly smaller, for similar accuracy, than the sizes of those obtained with expansions around $s = 0$. Therefore, the ability to perform expansions around any point in the complex plane, including $s = \infty$, is an important property of the nodal formulation. For expansions around $s = \infty$, the matrix $\mathcal{L}$ from (14) will require inversion. Unlike mesh analysis, in the nodal formulation $\mathcal{L}$ is non-singular if there are no internal nodes, a common situation since for most structures discretizing conductors along one direction is sufficient. Note that in order to derive a model based on an expansion around $s = \infty$, one needs to consider current sources, instead of voltage sources, and so the $\boldsymbol{I}_{src}$ unknowns do not appear in the system.

3-2 *Mixed Nodal-Mesh Formulation*

Computing a reduced-order model from (14) with an algorithm such as PRIMA, using any expansion point, corresponds to solving a linear system with a matrix that involves $\mathcal{R}$ and $\mathcal{L}$ in some form. If $s = 0$ is considered then the system matrix is $\mathcal{R}^{-1}\mathcal{L}$. In the case $s = \infty$ presented above, $\mathcal{L}$ must be inverted directly to form the system matrix (note that $\mathcal{L}$ is never inverted explicitly: if $\mathcal{L}^{-1}\boldsymbol{b} = \boldsymbol{x}$ is needed, $\mathcal{L}\boldsymbol{x} = \boldsymbol{b}$ is solved for $\boldsymbol{x}$ using an iterative algorithm such as GMRES [19]). In any case, the smaller the condition number of $\mathcal{L}$, the faster the convergence will be.

The form of the $\mathcal{L}$ matrix reveals one of the differences between the Nodal and Mesh formulations. In the latter the block matrices describing the filament constitute relations appear together with the mesh matrix $\boldsymbol{M}$, thus the block $\boldsymbol{M}\boldsymbol{L}\boldsymbol{M}^T$ in $\mathcal{L}$, while in the former $\boldsymbol{L}$ is used directly, as shown in (20).

It is known that under an appropriate choice of preconditioner, the condition number of a system obtained from the matrix $\boldsymbol{M}\boldsymbol{L}\boldsymbol{M}^T$ can be *significantly smaller* than that obtained from the partial inductance matrix $\boldsymbol{L}$ (see [20] for a proof). Thus faster system solution using $\mathcal{L}$ can be accomplished in the nodal formulation if the block $\boldsymbol{L}$ were replaced by a term like $\boldsymbol{M}\boldsymbol{L}\boldsymbol{M}^T$ in $\mathcal{L}$, as in the mesh formulation. Such a substitution implies the combination of a mixed nodal and mesh analysis method taking advantage of the best characteristics of each system. Although this mixed nodal-mesh formulation was developed to be combined with PEEC-style discretizations, the authors have been made aware that the resulting approach is quite similar to a specialized basis function methods presented in [21].

Using Mesh Analysis, we know that

$$\boldsymbol{M}\boldsymbol{V}_b = \boldsymbol{V}_s \tag{23}$$

$$\boldsymbol{M}^T\boldsymbol{I}_m = \boldsymbol{I}_b. \tag{24}$$

Eqn. (23) is based on Kirchoff's Voltage Law, which implies that the sum of branch voltages around each mesh in the network must be zero. $\boldsymbol{V}_b$ is the vector of branch voltages, $\boldsymbol{V}_s$ is the mostly zero vector of source voltages and $\boldsymbol{M}$ is the mesh incidence matrix. Eqn. (24), on the other hand, is based on Kirchoff's Current Law, and establishes a relation between the mesh currents $\boldsymbol{I}_m$ and the branch currents $\boldsymbol{I}_b$. Consider now separating the branch terms in (23) and (24) for filaments ($\boldsymbol{V}_b^f, \boldsymbol{I}_b^f$) and panels ($\boldsymbol{V}_b^p, \boldsymbol{I}_b^p$). Three different types of meshes in the circuit are then considered: those that contain only filaments - $\boldsymbol{M}_f$ (with mesh currents $\boldsymbol{I}_m^b$), those that contain only panels - $\boldsymbol{M}_p$ (with mesh currents $\boldsymbol{I}_m^p$), and those that contain both filaments and panels - $\boldsymbol{M}_{fs}, \boldsymbol{M}_{ps}$ (with mesh currents $\boldsymbol{I}_m^s$). Rewriting Eqns. (23) and (24) in terms of these variables leads to

$$\begin{bmatrix} \boldsymbol{M}_f & \\ & \boldsymbol{M}_p \\ \boldsymbol{M}_{fs} & \boldsymbol{M}_{ps} \end{bmatrix} \begin{bmatrix} \boldsymbol{V}_b^f \\ \boldsymbol{V}_b^p \end{bmatrix} = \begin{bmatrix} 0 \\ \boldsymbol{V}_s^p \\ 0 \end{bmatrix} \tag{25}$$

$$\begin{bmatrix} \boldsymbol{M}_f^T & & \boldsymbol{M}_{fs}^T \\ & \boldsymbol{M}_p^T & \boldsymbol{M}_{ps}^T \end{bmatrix} \begin{bmatrix} \boldsymbol{I}_m^f \\ \boldsymbol{I}_m^p \\ \boldsymbol{I}_m^s \end{bmatrix} = \begin{bmatrix} \boldsymbol{I}_b^f \\ \boldsymbol{I}_b^p \end{bmatrix} \tag{26}$$

Using the first equation in (26), multiplying the first line of (20) $\boldsymbol{M}^f$, and using $\boldsymbol{I}_m^{f+s}$, where

$$\boldsymbol{M}^f = \begin{bmatrix} \boldsymbol{M}_f \\ \boldsymbol{M}_{fs} \end{bmatrix} \qquad \boldsymbol{I}_m^{f+s} = \begin{bmatrix} \boldsymbol{I}_m^f \\ \boldsymbol{I}_m^s \end{bmatrix}$$

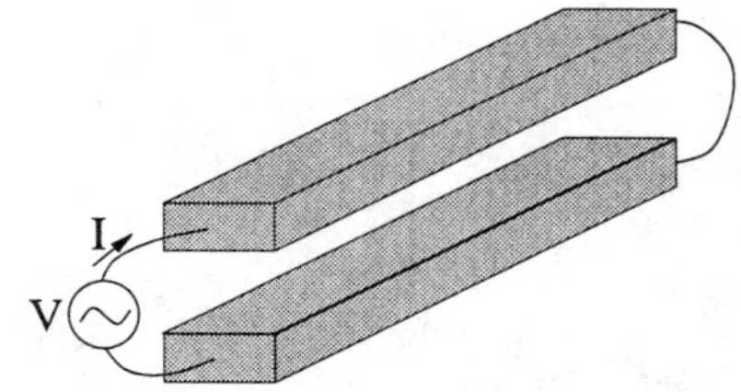

Fig. 3. Example geometry to compare models obtained at various expansion points.

the final state space-form is, then,

$$s\begin{bmatrix} \boldsymbol{M}^f\boldsymbol{L}\boldsymbol{M}^{fT} & 0 & 0 & 0 \\ 0 & \boldsymbol{B}_e\boldsymbol{P}^{-1}\boldsymbol{B}_e^T & 0 & 0 \\ 0 & 0 & 0 & 0 \\ 0 & 0 & 0 & 0 \end{bmatrix} \begin{bmatrix} \boldsymbol{I}_m^{f+s} \\ \boldsymbol{V}_n^e \\ \boldsymbol{V}_n^i \\ \boldsymbol{I}_{src} \end{bmatrix} =$$
$$-\begin{bmatrix} \boldsymbol{M}^f\boldsymbol{R}\boldsymbol{M}^{fT} & -\boldsymbol{M}^f\boldsymbol{A}_e^T & -\boldsymbol{M}^f\boldsymbol{A}_i^T & 0 \\ \boldsymbol{A}_e\boldsymbol{M}^{fT} & 0 & 0 & -\boldsymbol{N}_e \\ \boldsymbol{A}_i\boldsymbol{M}^{fT} & 0 & 0 & 0 \\ 0 & \boldsymbol{N}_e^T & 0 & 0 \end{bmatrix} \begin{bmatrix} \boldsymbol{I}_m^{f+s} \\ \boldsymbol{V}_n^e \\ \boldsymbol{V}_n^i \\ \boldsymbol{I}_{src} \end{bmatrix}$$
$$+\begin{bmatrix} 0 \\ 0 \\ 0 \\ \boldsymbol{V}_{src} \end{bmatrix} \tag{27}$$

or, equivalently, adding the output equations

$$\begin{aligned} s\mathcal{L}\,\boldsymbol{x} &= -\mathcal{R}\,\boldsymbol{x} + \boldsymbol{B}\,\boldsymbol{V}_t \\ \boldsymbol{I}_t &= \boldsymbol{B}^T\boldsymbol{x}. \end{aligned} \tag{28}$$

In this mixed formulation, again both the $\mathcal{R}$ and $\mathcal{L}$ matrices satisfy the conditions required for passive MOR. The size of the system is the same as that of the nodal formulation. Also as for nodal, $\mathcal{R}$ and $\mathcal{L}$ (subjected to the same restrictions) are non-singular, meaning that any expansion point can be used in the model order reduction.

4 Results

In this section examples will be considered that allow the study of the efficiency of MOR techniques when applied to the extraction of 3D interconnects using the formulation proposed in this paper. First, models of a simple structure generated using expansion points at $s = 0$ and $s = \infty$ will be explored and compared. Then a model of a connector example, will be analyzed in order to verify the formulation. To attest to the accuracy of the formulation and the reduction algorithms, results of a time-domain analysis of the connector model will be compared with experimental measurements.

4-1 *A simple two-conductor example*

Consider the simple geometry made up of two long and thin conductors, as shown in Figure 3, whose dominant behavior is similar to a transmission line. Both conductors are $1cm$ long, $37\mu m$ wide, $13\mu m$ in height and the distance between conductors is $17\mu m$. For this example a discretization is chosen such that 416 panels are used to cover all surfaces. Each conductor is divided into 10 segments along its length and each segment is divided into 15 filaments to properly capture skin effect. Figure 2 shows the efficiency of reduced models, compared with the full model, with expansion points at $s = 0$ and $s = \infty$. The full system has 323 states for an expansion point $s = 0$ (voltage sources are used in this case; for $s = \infty$ current sources are used). For this example a 62th-order model was generated. Figure 2-a) and c) show the positioning of the system and reduced-order model poles, for expansions around $s = 0$ and $s = \infty$ respectively, while Figure 2-b) and d) show the magnitude of the frequency dependent impedance of both the full system and the reduced models for each situation. Note that, because we have to use different sources for $s = 0$ and $s = \infty$, the poles in the figure relative to $s = 0$ are the admittance poles, while in that relative to $s = \infty$, they are impedance poles.

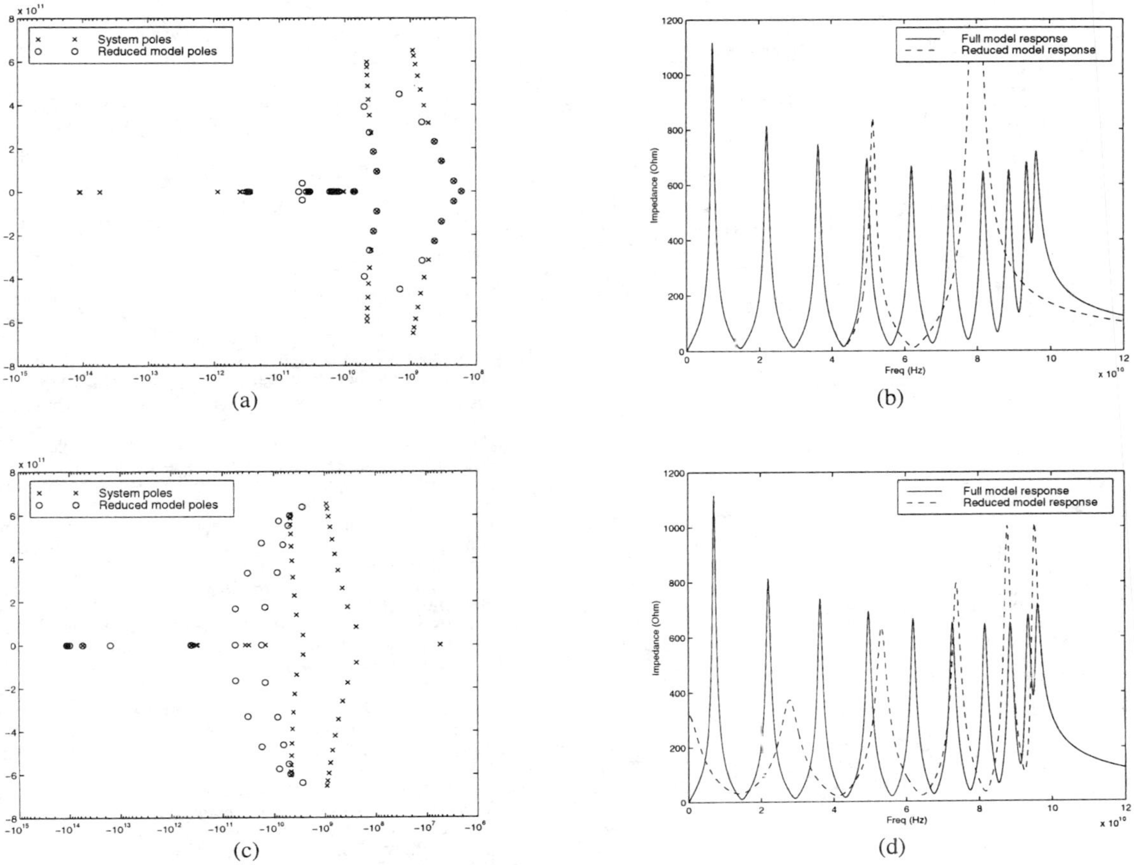

Fig. 2. Results obtained for the example geometry using the nodal-mesh formulation using an expansion point of $s = 0$ ((a) and (b)) and an expansion point of $s = \infty$ ((c) and (d)). In (a) the admittance poles are shown, while in (c) the impedance poles are plotted.

It is important to note the large number of poles of the full system along the real axis. Almost all of these poles have a weak effect on the final response (they have very small residues). For this example we found that only around 15% of the poles are relevant. These poles appear due to the large number of filaments connected in parallel (capacitances in panels are summed in each node, in the formulation, and so do not contribute with weak poles). These non-dominant poles are responsible for the large orders required for the reduced models shown since, for expansions around $s = 0$ or $s = \infty$, there is always a large cluster of such poles "nearby" and the MOR algorithm will capture them. However only about 22 of these poles are necessary to maintain the accuracy of the model as shown in the figures.

This simple geometry demonstrates some of the differences from expanding at different points. For $s = 0$, when the order of the reduced models is increased, we obtain a model that is accurate for an increasingly larger frequency range. However, for $s = \infty$, all the reduced models are inaccurate until full accuracy is suddenly achieved for a specific order, smaller than the original system size. This is because the algorithm, in the $s = \infty$ case, tries to first match all of the weak poles near -10^{14} before matching any of the dominant ones while in the $s = 0$ case the distance between the dominant and the weak poles near zero is smaller.

4-2 *Connector example*

Figure 4 presents an eighteen-pin portion of a real backplane connector structure. The connector is composed of 18 pins with a ground shield around and between the conductors. A discretization operation is performed using 582 filaments and 864 panels. Figure 5 shows the results obtained for the self-impedance of a single pin using the mixed nodal-mesh formulation with different order reduced models computed using an expansion around $s = 0$. The full system has 658 states. In the figure, models with orders 182 and 420 are compared to the exact response. Similarly to what was found for the first structure presented, only a small number of these poles are relevant, in this case 78 and 110 respectively.

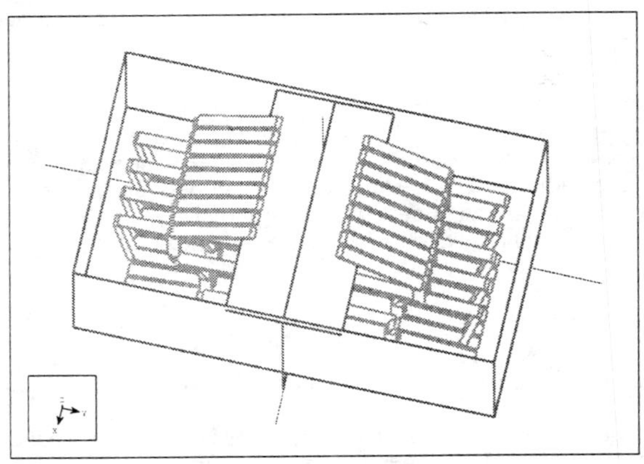

Fig. 4. A 3D connector.

Finally, in Figure 6 we present a comparison between results of a time-domain simulation obtained with a reduced model generated with our formulation and experimental measurements. For this ex-

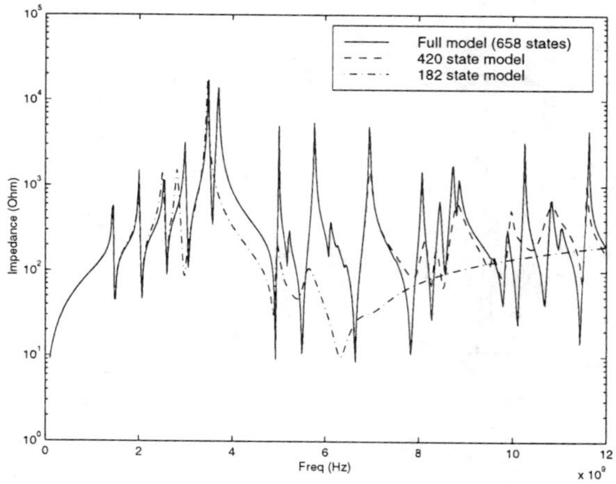

Fig. 5. Various reduced order models for the connector

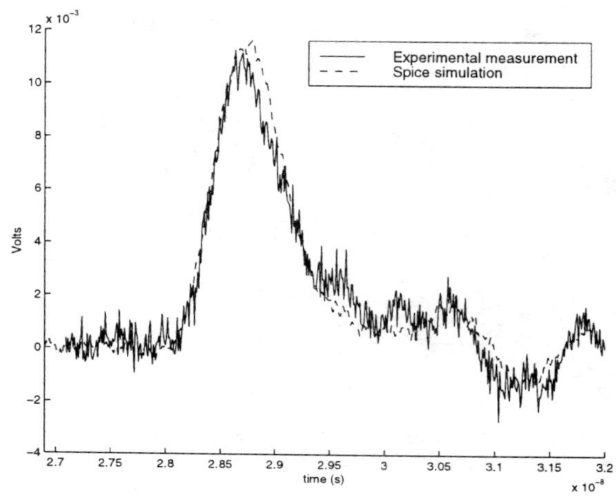

Fig. 6. Comparison between measured time-domain waveforms and simulated results obtained using a reduced-order model of the connector example.

periment all pins are connected to ground through resistors. Then a noisy input is connected in series with one of these resistors and a step with a $500ps$ rise-time is imposed on it. The voltage waveform at an adjacent pin is collected. As can be seen from the plot the waveforms are qualitatively similar and acceptable accuracy is obtained. If higher accuracy is required, a finer discretization can be used.

5 Conclusions

The main contribution of this paper is the presentation of a modeling approach based on a mixed nodal-mesh formulation. This formulation, when combined with recent Model Order Reduction algorithms, allows for the generation of guaranteed passive reduced order models of three-dimensional interconnect structures. The mixed nodal-mesh method takes advantage of the flexibility introduced by the nodal analysis, namely by maintaining relatively small model sizes and the ability to use expansion points at any frequency, and adds to it the good conditioning properties of the mesh analysis matrices. Important properties of the new formulation, such as the ability to use any expansion point in the complex plane, were illustrated with an example. Also, a comparison between measured data and the time-domain response of a generated model of a real 3-D connector structure, was used to validate the accuracy of the formulation. Further research is still required in order to improve the model generation process such that only the set of dominant poles is captured.

Acknowledgments

The authors would like to thank Mike Tsuk of Digital Equipment Corp. for the practical example. This work was partially supported by the National Science Foundation, the Portuguese JNICT programs PRAXIS XXI and FEDER under contracts 2/2.1/T.I.T/1661/95 and 2/2.1/T.I.T/1639/95 and grant BM-8592/96.

References

[1] Lawrence T. Pillage and Ronald A. Rohrer. Asymptotic Waveform Evaluation for Timing Analysis. *IEEE Trans. CAD*, 9(4):352–366, April 1990.

[2] P. Feldmann and R. W. Freund. Efficient linear circuit analysis by Padé approximation via the Lanczos process. *IEEE Trans. on Computer-Aided Design of Integrated Circuits and Systems*, 14:639–649, 1995.

[3] L. Miguel Silveira, Mattan Kamon, and Jacob K. White. Efficient reduced-order modeling of frequency-dependent coupling inductances associated with 3-d interconnect structures. In *Proc. 32nd Design Automation Conf.*, pages 376–380, San Francisco, CA, June 1995.

[4] K. J. Kerns, I. L. Wemple, and A. T. Yang. Stable and efficient reduction of substrate model networks using congruence transforms. In *Int. Conf. on Computer Aided Design*, pages 207 – 214, San Jose, CA, November 1995.

[5] L. Miguel Silveira, Mattan Kamon, Ibrahim Elfadel, and Jacob K. White. A coordinate-transformed Arnoldi algorithm for generating guaranteed stable reduced-order models of rlc circuits. In *Int. Conf. on Computer Aided-Design*, pages 288–294, San Jose, California, November 1996.

[6] Altan Odabasioglu, Mustafa Celik, and Lawrence Pileggi. Prima: Passive reduced-order interconnect macromodeling algorithm. In *Int. Conf. on Computer Aided-Design*, pages 58–65, San Jose, California, November 1997.

[7] I. M. Elfadel and David D. Ling. Zeros and passivity of Arnoldi-reduced-order models for interconnect networks. In 34nd *Design Automation Conf.*, pages 28–33, Anaheim, California, June 1997.

[8] I. M. Elfadel and David. L. Ling. A block rational arnoldi algorithm for multipoint passive model-order reduction of multiport rlc networks. In *Int. Conf. on Computer Aided-Design*, pages 66–71, San Jose, California, November 1997.

[9] Eric Grimme. *Krylov Projection Methods for Model Reduction*. PhD thesis, Coordinated-Science Laboratory, University of Illinois at Urbana-Champaign, Urbana-Champaign, IL, 1997.

[10] L. Greengard and V. Rokhlin. A fast algorithm for particle simulations. *Journal of Comp. Physics*, 73(2):325–348, December 1987.

[11] K. Nabors and J. White. Fast capacitance extraction of general three-dimensional structures. *IEEE Trans. on Microwave Theory and Techniques*, June 1992.

[12] J. R. Phillips. Error and complexity analysis for a collocation-grid-projection plus precorrected-FFT algorithm for solving potential integral equations with Laplace or Helmholtz kernels. In *Proc. 1995 Copper Mountain Conf. on Multigrid Methods*, April 1995.

[13] C. Desoer and E. Kuh. *Basic Circuit Theory*. McGraw-Hill, New York, 1969.

[14] Albert E. Ruehli. Equivalent circuit models for three-dimensional multiconductor systems. *IEEE Trans. on Microwave Theory and Techniques*, MTT-22(3):216–221, March 1974.

[15] E. C. Jordan and K. G. Balmain. *Electro-Magnetic Waves and Radiating Systems*. Prentice-Hall, Englewood Cliffs, N.J., 1968.

[16] M. Kamon, N. Marques, and J. White. FastPep: A Fast Parasitic Extraction Program for Complex Three-Dimensional Geometries. In *Int. Conf. on Computer Aided-Design*, San Jose, California, November 1997.

[17] Nuno Marques, Mattan Kamon, Jacob White, and L. Miguel Silveira. An efficient algorithm for fast parasitic extraction and passive order reduction of 3d interconnect models. In *DATE'98 - Design, Automation and Test in Europe, Exhibition and Conference*, pages 538–548, Paris, France, February 1998.

[18] Mattan Kamon, Nuno Marques, L. Miguel Silveira, and Jacob K. White. Generating reduced order models via peec for capturing skin and proximity effects. In *Proc. IEEE 6th Topical Meeting on Electrical Performance of Electronic Packaging, Monterey, CA*, San Jose, California, October 1997.

[19] Y. Saad and M. H. Schultz. GMRES: A generalized minimal residual algorithm for solving nonsymmetric linear systems. *SIAM Journal on Scientific and Statistical Computing*, 7:856–869, July 1986.

[20] Mattan Kamon. Efficient techniques for inductance extraction of complex 3-d geometries. Master's thesis, Massachusetts Institute of Technology, February 1994.

[21] S. M. Rao, D. R. Wilton, and A. W. Glisson. Electromagnetic scattering by surfaces of arbitrary shape. *IEEE Trans. Antennas Propagat.*, AP-30(3):409–418, May 1997.

Layout Based Frequency Dependent Inductance and Resistance Extraction for On-Chip Interconnect Timing Analysis

Byron Krauter
IBM Corporation
Austin, TX 78758-3493
krauter@austin.ibm.com

Sharad Mehrotra
IBM Corporation
Austin, TX 78758-3493
mehrotra@austin.ibm.com

Abstract—It is well understood that frequency independent lumped-element circuits can be used to accurately model proximity and skin effects in transmission lines [7]. Furthermore, it is also understood that these circuits can be synthesized knowing only the high and the low frequency resistances and inductances [4]. Existing VLSI extraction tools however, are not efficient enough to solve for the frequency dependent resistances and inductances on large VLSI layouts, nor do they synthesize circuits suitable for timing analysis. We propose a rules-based method that efficiently and accurately captures the high and low frequency characteristics directly from layout shapes, and subsequently synthesizes a simple frequency independent ladder circuit suitable for timing analysis. We compare our results to other simulation results.

1 Introduction

Increasing clock speeds, die sizes, and power dissipations have driven VLSI manufacturers to abandon the simple scaling approach of interconnect wiring. Instead, they employ a hierarchy of metal wiring levels. Thinner wiring levels are used at the circuit level where density is required, and thicker layers at the top or global levels in order to route low-skew clock trees, low-loss power distribution busses, and the fastest signal interconnects. This trend, coupled with the recent introduction of copper wiring (because its resistivity is approximately half that of aluminum wiring), has made on-chip inductance modeling necessary for clocks and the fastest signal interconnects. Lumped RC circuits are no longer adequate to model these interconnects because $j\omega L$, the inductive impedance of the line is now comparable to R, the resistance of the line at frequencies near the clock frequency of the chip. Clocks lines, because there are so few, can generally be manually routed and analyzed, and their arrival times subsequently stamped into timing analysis. Fast signal interconnects, however because they may number in the thousands, must be automatically routed and analyzed by the static timing analysis tools.

The standard model in timing analysis for an interconnect wire is the symmetric RC T-model shown in Fig. 1. (For increased frequency response the single T-model can be replaced with a network of cascaded T-models each representing a shorter segment of the wire.) Although the resistance R actually represents the loop resistance of the wire plus its ground return path, R is typically set to the dc resistance of the wire only as the ground return path resistance is generally much smaller than that of the wire itself. Interconnect networks, which typically comprise of multiple interconnect wires, multiple interconnect vias, and at least one sink and source point, are analyzed using either the Elmore delay metric or AWE based methods.

To first order, inductive effects can be included in timing analysis by replacing the RC T-model in Fig. 1 with the RLC T-model shown in Fig. 2. The resistance R again represents the loop resistance of the wire plus its ground return path, while the inductance L represents the loop inductance of the wire plus its ground return path. This first order model however, is often insufficient because R and L are frequency dependent due to the manner in which most VLSI chips distribute power.

Wiring tracks are scarce and valuable in most VLSI designs. Hence, most VLSI chips use gridded co-planar power distribution structures where power distribution takes as little as 20% of the wiring tracks. Because current loops always seek the path of least impedance, this sparsity makes the loops formed by interconnect currents and their equal and opposite ground return currents wide at low frequencies (when $R >> j\omega L$) and narrow at high frequencies (when $R \approx j\omega L$). This variation in width is proximity effect and when the series impedance of the interconnect wire plus its ground return path is expressed in terms of a simple series RL circuit, yields frequency dependent resistance and inductance terms.

The changes in resistance and inductance can be substantial. Consider the interconnect wire and ground return cross-section depicted in Fig. 12. The loop resistance and inductance of this system were analyzed using FastHenry [2] and the results of this analysis are shown in Fig. 3. Because of proximity effect (and later skin effect), the loop inductance L decreases from about 9.7 nH/cm at low frequencies to about 5.5 nH/cm at high frequencies. The loop resistance R on the other hand, continually increases from its dc value of 54.5 ohms/cm and never saturates. In fact at very high frequencies when skin effect dominates, R is proportional to the square root of the frequency.

In addition, these changes occur over the frequency range of interest. Using (2), the approximate bandwidth of an interconnect

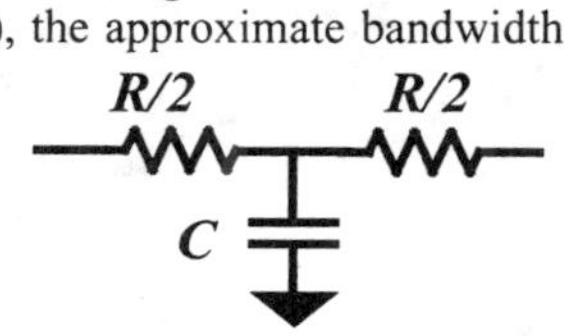

FIGURE 1: RC T-model for interconnect wiring.

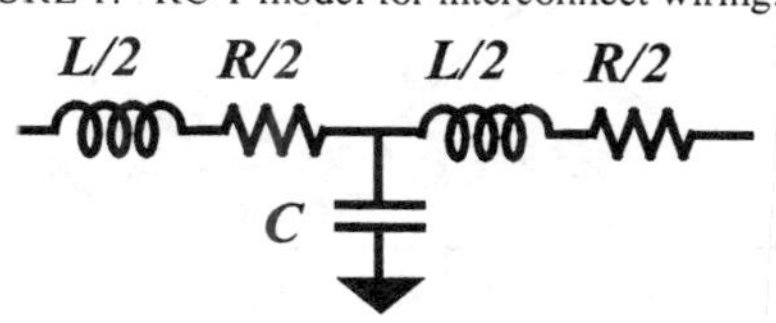

FIGURE 2: RLC T-model for interconnect wiring.

signal with a 100 psec edge rate is 3 GHz. In Fig. 3, as frequency ranges from dc to about 3 GHz, R increases by 50% while L decreases by 32%. Finally, these changes are due primarily to proximity effect because the skin depth at 3 GHz (with $\rho = 2e\text{-}6\ \Omega$-cm) is still greater than the thickness of the interconnect wiring in Fig. 12 (1.3 μm using (3) versus 1.0 μm).

Before static timing analysis can move beyond RC interconnects and into the realm of RLC interconnects, three challenges with respect to RLC interconnects must be addressed:

- When are Transmission Line Effects Important
- Modeling Proximity Effects Using Lumped Circuit Models
- Layout Extraction to Capture Proximity Effects

Substantial progress has been made on the first two challenges. First, Deutsch et al [1] have given guidelines for RLC modeling of on-chip interconnections. Consider a point to point net with the following characteristics: length l, characteristic impedance Z_0, resistance per unit length R, total capacitive load C_T, and source impedance Z_{drive}. If the following conditions are satisfied:

$$\frac{Rl}{2Z_0} < 1 \qquad Z_{drive} < Z_0 \qquad Z_{drive}C_T < \frac{RCl}{2} \tag{1}$$

then inductance needs to be included in the delay simulation.

Second, the frequency dependence of R(f) and L(f) due to proximity and skin effects can be modeled with frequency independent lumped-element circuit models up to an arbitrary maximum frequency using simple ladder networks. (A commonly used network topology, employed by Wheeler in [7], is shown in Fig. 4.) Furthermore it has been demonstrated that accurate RL ladder models can be synthesized knowing just the high and low frequency behavior [4].

1.1 Layout Extraction to Capture Proximity Effects

Because of proximity effects in the return path, layout based extraction must consider not only the nearby ground return conductors, but all ground return conductors within a predetermined radius. While it is possible using shapes processing, to isolate the ground return conductors in the vicinity of a signal or clock line, no simple methods exist for computing the frequency dependent resistance and inductance of these multi-conductor systems.

The most commonly used procedure today, is a brute-force three step process. First individual conductor cross-sections are sub-divided into sections smaller than the skin depth at the maximum frequency of interest. With this discretization uniform current density can be assumed in each section. For all practical purposes, the maximum frequency of interest is given by

$$f_{max} = \frac{1}{\pi t_r} \tag{2}$$

where t_r is the minimum rise time, and the skin depth is

$$\delta = \sqrt{\frac{1}{\pi\mu\sigma f_{max}}} \tag{3}$$

where μ is the magnetic permeability constant and σ is the conductivity of the conductor.

Next, resistance and inductance values are computed for each conductor cross-section assuming uniform current density in each cross-section. This allows inductance formulas developed at the turn of the century [5] to be employed for computing the L matrix:

$$L_{self} = 2l\left(ln\left(\frac{2l}{w+t}\right) + 0.5 + 0.22\frac{(w+t)}{l}\right)$$
$$L_{mutual} = 2l\left(ln\left(\frac{2l}{d}\right) - 1 + \frac{d}{l}\right) \tag{4}$$

Finally, the coupled RL circuit depicted in Fig. 5 is analyzed at several points in the frequency domain. This technique was first demonstrated by Weeks et al in [6] and is the foundation of frequency dependent extraction tools such as FastHenry [2] today.

To briefly recap this procedure, the frequency-dependent branch circuit equations for the conductors depicted in Fig. 5 are given as

$$Z(\omega)I_b(\omega) = [R + j\omega L]\, I_b(\omega) = V_b(\omega) \tag{5}$$

These branch equations are combined with the circuit topology constraints and the resultant circuit equations solved at multiple frequencies. For example if mesh analysis is employed, the final circuit equations are given as

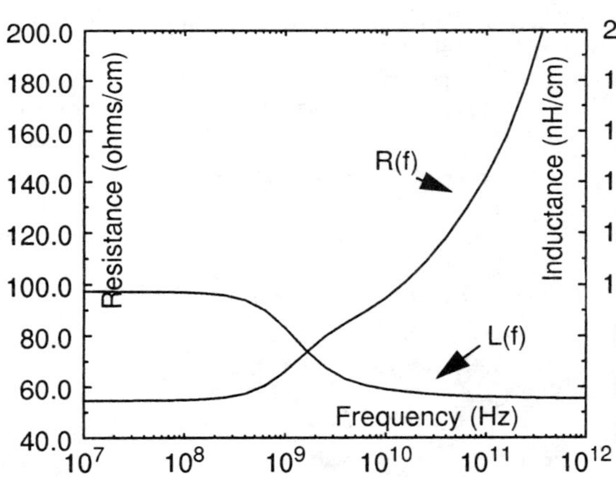

FIGURE 3: L(f) and R(f) for cross-section depicted in Fig. 12.

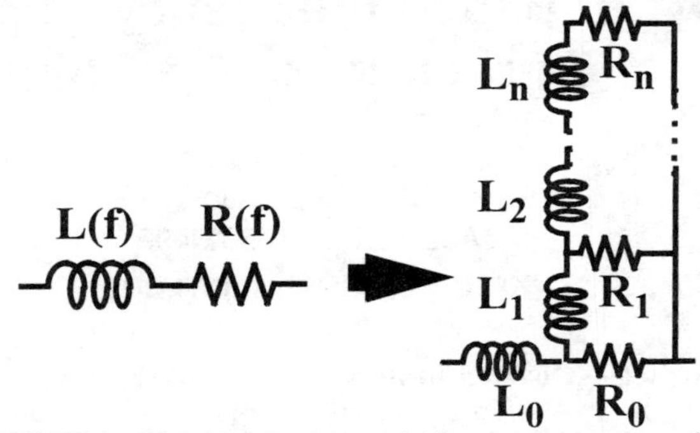

FIGURE 4: RL ladder circuit to model frequency dependence due to proximity and skin effects.

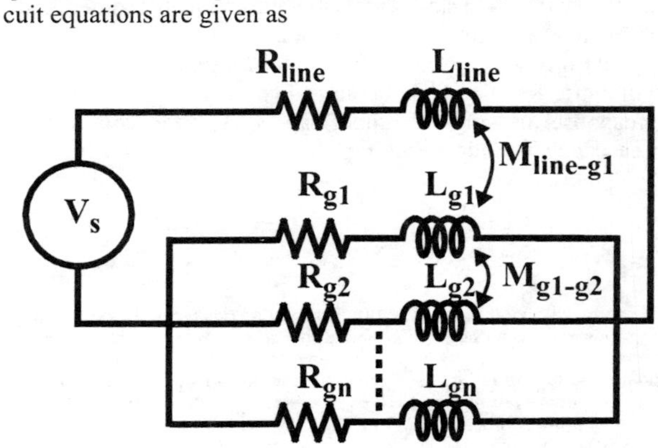

FIGURE 5: Coupled RL circuit model for wire and ground returns.

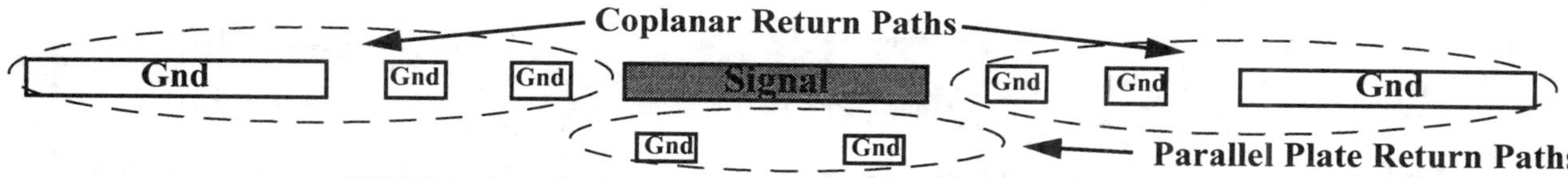

FIGURE 6: Specialized structure of on-chip signals requiring R(f) and L(f) models.

$$MZM^TI_m = V_s \qquad (6)$$

where I_m and V_s are the mesh currents and applied source voltages.

Any direct solution of this problem necessarily involves complex arithmetic and requires $O(n^3)$ operations to factor (or invert) the fully-dense impedance matrix. Although iterative methods can be employed to reduce the complexity to $O(n^2)$ on large matrices, todays extraction tools are still slow because solutions are required at multiple frequencies and complex arithmetic is used at every frequency point. Layout-based extraction of frequency dependent inductance and resistance models can only be made practical if the frequency dependent behavior can be inferred from fewer solutions of (5) and (6), these solutions can be obtained without inverting large inductance matrices, and complex arithmetic can be avoided.

In the next two sections, we give justification for, and demonstrate simple methods for computing frequency dependent inductance and resistance directly from layout geometry. These methods require fewer solutions of (5) and (6), do not require any matrix inversion or complex arithmetic, and have $O(n^2)$ complexity (where n is the number of conductors in the return path).

2 Approach

The two central problems for frequency dependent inductance and resistance extraction are 1) finding a simple method for computing the frequency dependent resistance R(f) and inductance L(f) of a signal conductor with its multiple ground return conductors and 2) synthesizing an accurate ladder circuit that captures this behavior over the desired frequency range. While recent work [3,4] has simplified the task considerably by synthesizing accurate ladder networks from the high and low frequency characteristics only (R_{dc}, L_{dc}, $L(\omega_{max})$, and $R(\omega_{max})$), finding $R(\omega_{max})$ has remained costly and the synthesis procedure is iterative.

Fortunately, on-chip signal conductors requiring R(f) and L(f) modeling are few (only the long-wide wires at the top wiring levels where $R \sim j\omega L$ at the maximum frequency of interest), and these few signals form specialized structures with their ground return paths. Consider the signal and ground return conductors depicted in Fig. 6. To a first approximation, this signal and its ground return conductors form a coplanar transmission line. The presence of thinner ground return conductors directly below the signal line can be treated as a second order effect. (These conductors will be referred to as parallel plate return conductors in the subsequent paragraphs).

At low frequencies (that is, frequencies at or below where proximity effects occur), return currents seek the path of least resistance, and hence are carried primarily on co-planar return paths because of their greater combined cross-section, and hence greater combined conductivity. At high frequencies, currents redistribute to minimize the loop inductance. Parallel plate return conductors and the closest coplanar return conductors become important at high frequencies when their close proximity to the signal conductor becomes an important factor in minimizing total loop inductance.

Our extraction and synthesis approach is tailored to this coarse behavior. Consider the hybrid ladder circuit shown in Fig. 7. The co-planar system, because of its importance at both low and high frequencies, is modeled as a ladder circuit (R_0, R_1, L_0 and L_1 in Fig. 7). Although we chose a simple ladder so that we could synthesize it from easily-derived high and low frequency characteristics, it still captures the proximity effects of the coplanar system, and hence captures frequency dependent behavior well into the GHz range. The parallel plate system because of its high frequency impact only, is modeled as a simple shunt impedance ($R_{pp} + j\omega L_{pp}$) across the high frequency loop inductance L_0 of the coplanar transmission line system.

2.1 Co-planar Transmission Line Extraction

Our approach further simplifies the task of capturing the frequency dependent inductance (and resistance) of the co-planar transmission line system by breaking the task into high and low frequency problems and using different methods (and approximations) to solve for the current distributions at these extremes. By breaking up the problem in this way we can avoid complex arithmetic. The following paragraphs detail our overall approach to the modeling of a co-planar transmission line, beginning with our approach at low frequencies, continuing to our approach at high frequencies, and concluding with our approach to ladder network synthesis.

2.1.1 Low Frequency Inductance (L_{dc}) and Resistance (R_{dc})

At low frequencies the impedance of the line and its coplanar return path is dominated by resistance. The ground return currents will distribute so as to minimize the total resistance of the signal conductor and the ground return path. While this would imply that the current could return anywhere on the chip, and even in the package, we choose a radius for the ground return such that the resistance of the return conductors is a small percentage of the line resistance. This keeps the computational cost of modeling low, without sacrificing accuracy. If we assume a radius for this ground return path, the equivalent circuit for the signal line and ground return path is shown in Fig. 8-a where each parallel segment in the ground system represents a distinct return conductor.

Choosing the source current to be unity, the branch current in ground return conductor j is

$$i_j = \frac{R_{GND}}{R_{gj}} \qquad (7)$$

where R_{gj} is the resistance of the jth ground conductor and R_{GND} is the parallel resistance of the ground return network in Fig. 8-a. To compute the equivalent loop inductance of this system we equate the total energy stored in the magnetic field for the two systems in

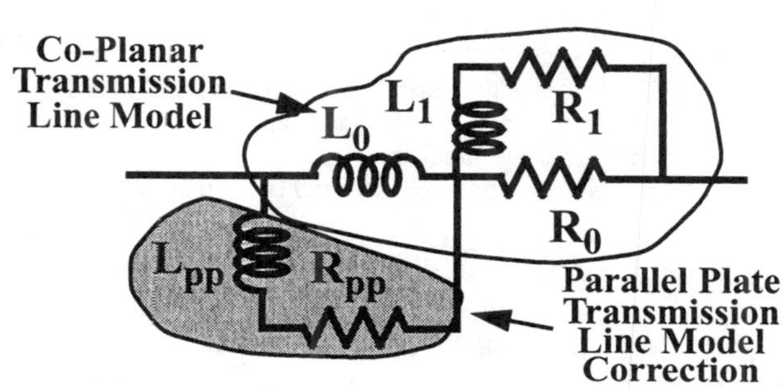

FIGURE 7: Hybrid ladder circuit for coplanar transmission line with minor parallel plate return paths.

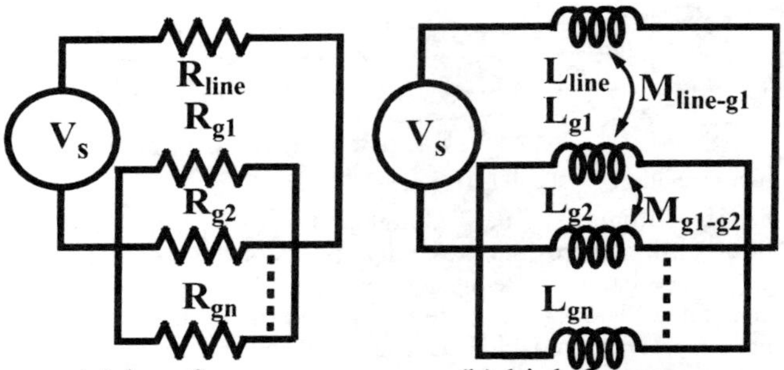

(a) low frequency (b) high frequency

FIGURE 8: Approximate circuit models to determine low and high frequency current distributions.

Fig. 5 and Fig. 9:

$$i\, L_{loop}\, i = [i_b]^T [L]\, [i_b] \quad (8)$$

Choosing the source current to be unity yields the following equations for R_{loop} and L_{loop}.

$$R_{loop} = R_{line} + R_{GND}$$
$$R_{GND} = (R_{g1}^{-1} + \ldots R_{gn}^{-1})^{-1}$$
$$L_{loop} = \sum_{i=0}^{n} \alpha_i \sum_{j=0}^{n} \alpha_i L_{ij} \quad (9)$$
$$\alpha_0 = 1 \quad \text{and} \quad \alpha_i = \frac{-R_{GND}}{R_{gi}} \quad \text{for } i > 0$$

Interestingly, if we assume $R >> j\omega L$ and solve the nodal equations for the system in Fig. 5, we also arrive at the above result for the effective (or loop) inductance and resistance. (The proof has been omitted due to space considerations.)

2.1.2 High Frequency Inductance (L_{hf})

At high frequencies the impedance of the line and the return path is dominated by inductance. Although inductance decreases monotonically with frequency, due to the opposing nature of skin and proximity effects, L(f) quickly saturates near its minimum value well before classical skin effect has occurred. Therefore, a practical limit for high frequency inductance can be obtained with moderate discretizations (i.e. subdivisions greater than the skin depth at the highest frequency of interest).

The high-frequency ground return current distribution (and subsequently the high-frequency loop inductance) can be determined by solving for the branch currents for the circuit in Fig. 8-b (which contains inductances only). If we formulate our circuit equations using mesh analysis, select as the n mesh loops the n ground return conductor plus signal line loops, and choose the source voltage to be unity, this problem can be expressed as

$$MLM^T I_m = 1 \quad (10)$$

where $L \in R^{(n+1)(n+1)}$ is the partial inductance matrix, $M \in R^{n(n+1)}$ is the mesh matrix, and $I_m \in R^n$ is given by the ground return branch currents I_b. If we label the signal line as conductor 0 and ground return lines as conductors 1 to n, (10) can be written as

$$\mathcal{L} I_b = 1 \quad (11)$$

where

$$\mathcal{L}_{ij} = L_{ij} + L_{00} - L_{i0} - L_{j0} \quad (12)$$

To compute the equivalent loop inductance of this system we normalize the ground return branch currents obtained from (11) and

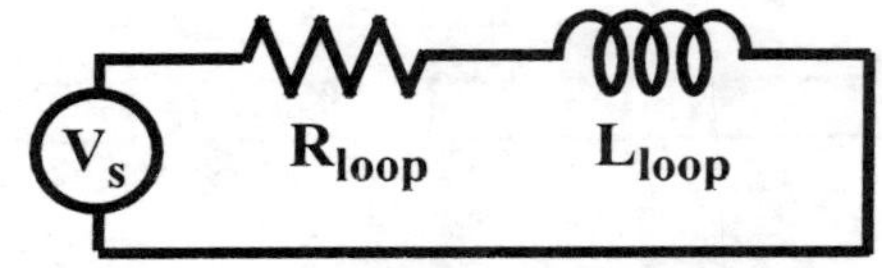

FIGURE 9: Simplified circuit model to find L_{loop} by equating magnetic energy stored with $I^T[L]\,I$.

again equate the total energy stored in the magnetic field for the two systems in Fig. 5 and Fig. 9 (as in (8)).

It should be noted that the preferred solution method for (11) is a function of the size of the matrix $\mathcal{L}$. When $\mathcal{L}$ is small, direct methods are preferred. When L is large, an iterative method should be used to avoid $O(n^3)$ complexity.

2.1.3 High Frequency Resistance

Unlike inductance, resistance does not saturate at high frequencies. It continues to increase with frequency and it eventually approaches a $f^{1/2}$ dependence as skin effects begin to dominate. While there are no high frequency approximations one can employ to capture resistance R(f) at a particular frequency, we have observed that the cross-over frequency ω_c (where $R(\omega_c) = \omega_c L(\omega_c)$) corresponds roughly to R_{dc}/L_{hf}. (This observation suggests the increase in resistance from R_{dc} matches the increase in inductance from L_{hf} at the cross-over frequency). Moreover, slight perturbations with the addition of high resistance parallel plate return paths do not appreciably alter this crossover frequency.

2.1.4 Ladder Synthesis for Coplanar Transmission Lines

To model the frequency dependent change in return path resistance and inductance, we choose to fit the simple ladder model shown in Fig. 10. For this network, the frequency dependent R and L are given as:

$$R(\omega) = \frac{R_0 R_1 (R_0 + R_1) + \omega^2 R_0 L_1^2}{(R_0 + R_1)^2 + \omega^2 L_1^2} \quad (13)$$

$$L(\omega) = L_0 + \frac{L_1 R_0^2}{(R_0 + R_1)^2 + \omega^2 L_1^2} \quad (14)$$

To determine R_0, R_1, L_0 and L_1, the resistance and inductance at ω=0 are equated to R_{dc} and L_{dc} while the inductance at $\omega=\infty$ is equated to L_{hf}. Again, a high frequency resistance is not computed. Instead, we observe that the cross-over frequency where $R(\omega) = \omega L(\omega)$ corresponds roughly to R_{dc}/L_{hf}. By forcing this equality for the ladder network, the elements of the network are as follows:

$$R_0 = \frac{R_{dc} L_{dc}}{L_{hf}} \quad (15)$$

$$L_0 = L_{hf} \quad (16)$$

$$R_1 = \frac{R_{dc} L_{dc}}{(L_{dc} - L_{hf})} \quad (17)$$

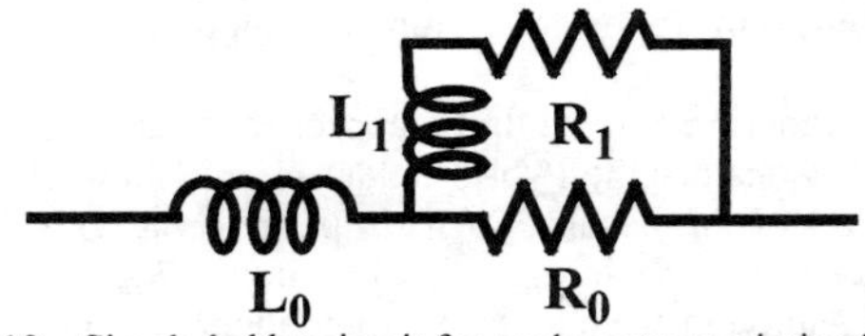

FIGURE 10: Simple ladder circuit for coplanar transmission line.

FIGURE 11: Signal line with narrow coplanar ground returns.

$$L_1 = \frac{L_{dc}^2}{(L_{dc} - L_{hf})} \qquad (18)$$

2.2 Parallel Plate Transmission Line Corrections

Because of their relatively small cross-sectional area, and hence low conductivity, the parallel plate return paths modify the total loop impedance at high frequencies only. For this reason, their effect can be modeled as a simple shunt impedance ($R_{pp} + j\omega L_{pp}$) across the high frequency loop inductance of the larger coplanar transmission line system. Since the coplanar ladder network already includes the line resistance, R_{pp} is the simply the parallel resistance of the parallel plate return conductors only, while L_{pp} is the total loop inductance of the signal conductor plus its parallel plate return conductors.

In summary our approach can be described as follows:

- Compute R_{dc} and L_{dc} using (9) with a wide return radius. Include all return conductors within the wide radius.
- Compute high frequency current distribution using (10) for the line and its coplanar return conductors only. Normalize the line current and compute L_{hf} using (8), (i.e. $L_{hf} = L_{loop}$).
- Compute R_0, R_1, L_0 and L_1 using (15) - (18).
- Compute L_{pp} for the signal line and parallel plate return conductors (if any exist) using (10) and (8). If the signal is much wider than the parallel plate return conductors, assume the signal cross-section is the same as (a mirror image of) the parallel plate return conductor.
- Compute R_{pp}, the parallel resistance of the parallel plate return conductors.
- Assemble hybrid ladder network as shown in Fig. 7.

3 Results

In this section, we present results for several typical on-chip signal and clock nets with coplanar (and sometimes parallel plate) power distribution to demonstrate the accuracy of the low and high frequency inductance approximations and the efficacy of the hybrid ladder network in capturing the frequency dependent resistance and inductance.

3.1 Signal Line with Narrow Co-planar Ground Returns

The first test case (depicted in Fig. 11) considers a signal conductor with multiple narrow coplanar ground return paths only. Although this example is not representative of real layouts (where ground return paths have various widths and are present on multiple wiring levels), it demonstrates how our simple extraction methods can capture the four points needed to synthesize the ladder circuit for coplanar transmission lines (namely R_{dc}, L_{dc}, L_{hf}, and the cross-over frequency ω_c where $R(\omega) = \omega L(\omega)$).

R_{dc}, L_{dc}, L_{hf}, and the cross-over frequency ω_c were computed for various combinations of signal width, numbers of ground return paths, and return path conductors width, using (9) for R_{dc} and L_{dc}, (10) and (8) for L_{hf}, and R_{dc}/L_{hf} for the cross-over frequency ω_c. Individual conductors were not further subdivided, and the high frequency inductance L_{hf} was computed iteratively and converged within 2-5 iterations. These same four points were also computed using FastHenry. R_{dc} and L_{dc} were taken at the lowest frequency simulated, 10 MHz, while L_{hf} was taken at the highest frequency simulated, 100 GHz. Individual conductors were subdivided into 2x2 cross-sections.

Table 1 compares the FastHenry simulation results to the simple computations using the coarser discretization. In all cases, the two methods match to within a few percent for R_{dc}, L_{dc}, L_{hf}, and the cross-over frequency ω_c.

3.2 Signal Line with Narrow & Wide Coplanar Ground Returns

The second test case (depicted in Fig. 12) considers a wide signal line with both narrow and distant wide co-planar return paths. Such a configuration is common for long, critical signal lines running on the highest level metal layer, parallel to the wide power distribution lines. This test case demonstrates how the simple ladder circuit depicted in Fig. 10 can capture the frequency dependent behavior of a coplanar transmission line into the GHz range.

The simple ladder model was synthesized from R_{dc}, L_{dc}, and L_{hf} using (15)-(18). R_{dc} and L_{dc} were computed using (9) and the high frequency inductance L_{hf} was computed using (10) and (8) without subdividing any of the conductors. Table 2 compares the computed values for R_{dc}, L_{dc}, L_{hf}, and ω_c with those obtained from Fas-

TABLE 1. R_{dc}, L_{dc}, L_{hf}, ω_c for 5.0 mm Coplanar Transmission Line in Fig. 11. (Signal width & thickness = 2.0 & 1.0 μm.)

Power bus width, thickness, & pitch = 1.0, 1.0 & 5.0 μm								
# of power busses	FastHenry				Computed			
	L_{dc} nH	R_{dc} Ω	L_{hf} nH	f_c GHz	L_{dc} nH	R_{dc} Ω	L_{hf} nH	f_c GHz
2	2.77	75.0	2.64	4.45	2.77	75	2.64	4.51
4	2.97	62.5	2.58	3.81	2.97	62.5	2.58	3.86
6	3.17	58.3	2.56	3.62	3.18	58.3	2.56	3.62
8	3.36	56.3	2.56	3.54	3.36	56.3	2.56	3.50
Power bus width, thickness, & pitch = 2.0, 1.0 &10.0 μm								
# of power busses	FastHenry				Computed			
	L_{dc} nH	R_{dc} Ω	L_{hf} nH	f_c GHz	L_{dc} nH	R_{dc} Ω	L_{hf} nH	f_c GHz
2	3.54	62.5	3.42	2.89	3.54	62.5	3.43	2.90
4	3.71	56.3	3.34	2.70	3.70	56.3	3.34	2.68
6	3.89	54.2	3.32	2.65	3.89	54.2	3.32	2.60
8	4.06	53.1	3.31	2.63	4.06	53.1	3.31	2.55

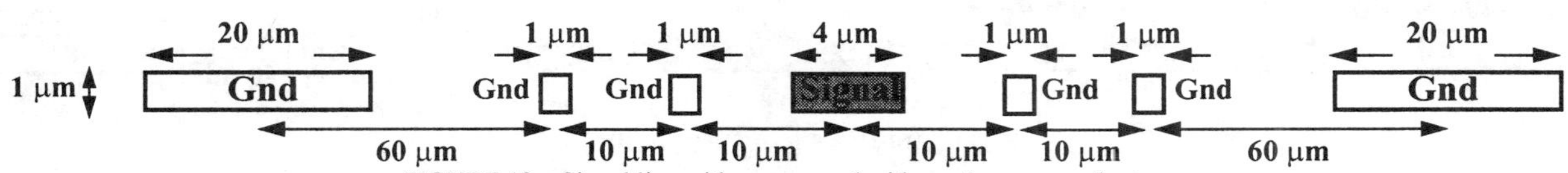

FIGURE 12: Signal line with narrow and wide coplanar ground returns.

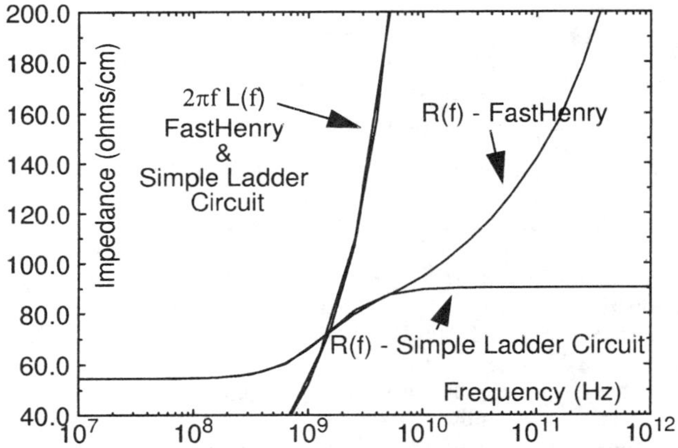

FIGURE 13: Impedance vs. frequency for cross-section in Fig. 12

tHenry simulations. Fig. 13 shows the frequency dependent impedance predicted for this configuration by both the simple ladder circuit and FastHenry simulations. (In order to capture frequency dependence well beyond the GHz range, the FastHenry simulations used a much finer discretization: 40x2 for the wide ground returns, 4x2 for the narrow ground returns, and 8x2 for the signal conductor.) Note, the simple ladder model not only predicted the cross-over frequency well, it also accurately predicted $R(\omega)$ and $j\omega L(\omega)$ up to approximately 6.0 GHz.

TABLE 2. R_{dc}, L_{dc}, L_{hf}, ω_c for 1.0 cm Line in Fig. 12.

FastHenry				Computed			
L_{dc} nH	R_{dc} Ω	L_{hf} nH	f_c GHz	L_{dc} nH	R_{dc} Ω	L_{hf} nH	f_c GHz
9.72	54.5	5.53	1.53	9.70	54.5	5.84	1.49

3.3 Wide Signal Line with Co-planar and Parallel Plate Ground Returns

The last test case (depicted in Fig. 14) considers a very wide signal line with narrow local-coplanar ground returns, wide distant-coplanar ground returns, and a local, narrow ground return, parallel to and running underneath the signal line. Such a configuration is typical of clock distribution spines running on the highest metal layer. This last example demonstrates how the hybrid ladder model depicted in Fig. 7 can capture the frequency dependent behavior of more complex structures into the GHz range.

The hybrid ladder model was synthesized using the procedure outlined in Section 2.2. R_{dc} and L_{dc} (6.95 Ω/cm and 5.75 nH/cm respectively) were computed considering all the ground return conductors, while L_{hf} (4.55 nH/cm) was computed considering only the coplanar ground return conductors. R_{pp} was set to the resistance of the parallel plate return conductor (80 Ω/cm), and L_{pp} to the loop inductance of the conductor and a mirror image of it located at the height of the signal conductor above (4.94 nH/cm).

Fig. 15 shows the frequency dependent behavior predicted by the hybrid ladder circuit and FastHenry (using a finer discretization). Again the ladder model predicted the cross-over frequency well and it also accurately predicted $R(\omega)$ and $j\omega L(\omega)$ up to 2.0 GHz, approximately one decade above the cross-over frequency.

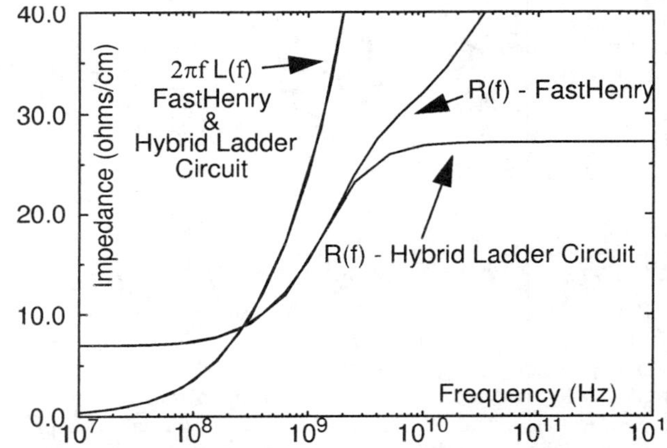

FIGURE 15: Impedance vs. frequency for cross-section in Fig. 14

4 Conclusions

In summary, we've demonstrated an effective technique for modeling frequency dependent inductance and resistance for on-chip interconnects. Frequency dependence is captured with lumped circuit elements (thereby creating a model suitable for timing analysis). Furthermore, this model is synthesized via closed form equations, using easily-obtained low and high-frequency characteristics. Both complex arithmetic and matrix inversion are avoided.

References

[1] A.Deutsch, G.V.Kopcsay, P.Restle, et al, "When are Transmission-Line Effects Important for On-Chip Interconnections?" *IEEE Transactions on Microwave Theory and Techniques*, vol 45, No. 10, Oct. 1997.

[2] M.Kamon, M.J.Tsuk, and J.K.White, "FASTHENRY: A Multipole-Accelerated 3-D Inductance Extraction Program," *IEEE Transactions on Microwave Theory and Techniques*, 42, No. 9, Sept. 1994.

[3] S.Kim and D.P.Neikirk, "Compact Equivalent Circuit Model for the Skin Effect," *1996 IEEE-MTT-S International Microwave Symposium,* San Francisco, June17-21, 1996.

[4] D.P.Neikirk, Private communication.

[5] E.B.Rosa, "The Self and Mutual Inductance of Linear Conductors," Bulletin of the National Bureau of Standards, vol. 4, pp. 301-344, 1908.

[6] W.T.Weeks, L.L.Wu, M.F.McAllister, and A.Singh, "Resistive and Inductive Skin Effect in Rectangular Conductors," *IBM Journal of Research and Development,* **23**, No.6, 652-660 (November 1979)

[7] H.A.Wheeler, "Formulas for the Skin-Effect," *Proceedings of the Institute of Radio Engineers*, vol. 30, pp. 412-424, Sept. 1942

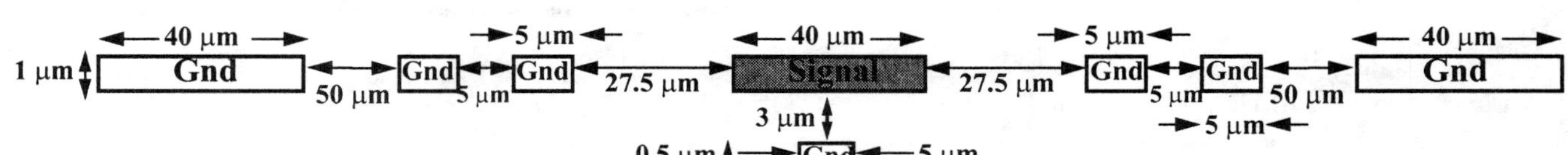

FIGURE 14: Wide signal line with coplanar and parallel plate ground returns.

A Methodology for Guided Behavioral-Level Optimization

Lisa Guerra
Advanced VLSI Architecture Group
Rockwell Semiconductor Systems
Newport Beach, CA

Miodrag Potkonjak
Computer Science Department
University of California
Los Angeles, CA

Jan Rabaey
EECS Department
University of California
Berkeley, CA

Abstract — Optimization at the early stages of design are crucial. However, due to an overwhelming number of design and optimization options, design exploration is often conducted in a qualitative, ad-hoc manner. This paper presents a methodology and interactive environment for guiding the exploration process. A prototype targeting behavioral-level optimization for datapath-intensive ASIC implementations has been developed. The key to the approach is encapsulated knowledge about the various optimizations and a set of techniques to automatically extract the "essence" of a design description. At each stage in the exploration process, the system suggests and ranks potential optimizations, both in terms of immediate and longer-term impact. It also provides evaluations of the design and of the likely affects each optimization will have on metrics like power and performance. In the new approach, the designer is responsible for making the actual optimization selections. However, using the provided guidance, designers can make decisions in a more informed manner, and therefore can explore the design solution space more effectively. The effectiveness of the approach is demonstrated on a number of designs.

1.0 Introduction

Traditionally, concentration has been placed in the development of point tools and methodologies which address a single task within the optimization process. For example, in behavioral-level optimization, in addition to the approaches proposed for partitioning, template matching, and clock selection, there exist more than a hundred transformations [Bac94, Par95]. Some optimizations are restricted to optimization of a specific class of computation (e.g. only linear computations). Some focus on power optimization while others target area or throughput. Furthermore, some concentrate on reduction of just a specific sub-metric; for example, different power optimization techniques have been developed for reducing capacitance, for reducing activity, and for enabling voltage scaling.

Effective *global* optimization, or optimization of a global objective function on current-day complex and heterogeneous applications, clearly requires the coordinated application of a *collection* of techniques. While many of the problems being addressed by the individual techniques are in themselves computationally intractable, an added layer of complexity clearly exists in deriving an entire optimization trajectory. This added complexity is due to the strong interdependency that exists between techniques. Exploring different trajectories is crucial, as a large variability in attainable improvements exists. Exploring all possible sequences of actions inevitably results in combinatorial explosion, however, and thus is not a feasible option. Furthermore, static scripts are not sufficient since their effectiveness is strongly dependent on the particular design.

DAC 98, San Francisco, California

This paper proposes a methodology for enabling systematic and effective global optimization. The approach involves interactively and quantitatively guiding the designer's exploration process. At each stage in the exploration process, the system suggests and ranks potential optimizations, taking into consideration not only immediate benefits, but benefits which may be gained by successive optimizations. It provides evaluations of the design and of the likely affects each optimization will have on metrics like power, cost, and performance. Using the guidance environment, designers maintain a more global view of the exploration space, can make decisions in a more informed manner, and thus can more easily and quickly discover effective trajectories of optimizations.

2.0 Preliminaries

Based on the proposed methodology, an environment targeting behavioral-level design optimization for semi-custom ASIC implementations has been developed. This section presents the domain that we have focused on.

Applications are described in either the Silage high-level language or using restricted C++, then automatically parsed into a flowgraph representation. The flowgraph consists of nodes representing data operators or sub-graphs, and edges representing the data, control, and timing precedences. The flowgraph computation model is homogeneous synchronous data flow [Lee87]. Hyper behavioral synthesis [Rab91] tools are used to map the flowgraph to an architectural netlist implementation. The ASIC architecture model consists of an unlimited number of parallel, time-multiplexed execution units. Registers are clustered into register files, which are tied to the input ports of the execution units.

Before mapping, optimizations are applied to improve the design's area, power, or throughput. They can be transformations on the flowgraph itself, manual re-writes of the input specification, or library and parameter changes. Optimization examples include time and for-loop unfolding, algebraic optimizations, pipelining and retiming, voltage scaling, clock selection, operator chaining, and hardware module selection. A good overview of the behavioral synthesis area can be found in [McF90].

3.0 Related Work

Before proceeding, this section outlines related efforts in design guidance, design planning, and in performing optimization globally across the boundaries of individual techniques. A design guidance system, the Clio Design Advice System, was proposed by Lopez, Jacome, and Director in 1992 [Lop92]. This system provides design advice consisting of *static* pieces of *qualitative* information about different design options. The advice database operates as the collective memory of the entire community of designers utilizing the framework. There have also been several related works in the area of "design planning." The ADAM Design Planning System [Kna91], for example, is a knowledge-based planner that constructs a sequence of design synthesis tasks in response to a given set of specifications. The ADAM system concentrates on the design flow of tools (e.g. module selection, allocation) to synthesize an RTL design. Other related work on includes the work of [Goo94] on global design methodologies for heterogeneous ICs and work on flow management [Kle94].

In the area of optimization ordering, there have been a number of approaches in the CAD and compilers areas. These, however, only address fixed sets of techniques. The techniques include peephole optimization [McK65], pre-defined script-based optimization [Bra84], theory-based ordering [Wol91], generic probabilistic techniques [Pot94, Cha95], bottleneck identification and elimination approaches, and approaches using enabling and disabling transformations, [Whi90, Pot92, Hua93].

The key novelties of this work are 1) a combination of *quantitative* and qualitative design advice is given, 2) design advice that is specific to the design at hand is given, 3) the designer remains an integral part of the design process, 4) the approach is extendible to include new optimizations. While the work of [Gue94] presents the use of properties for guidance through estimation, in this work, we present a complete environment for their use in guided optimization.

4.0 Proposed Methodology and System Overview

4.1 Methodology Overview

Fig. 1 depicts an overall view of the interactive guided optimization process. The designer enters a design specification, constraints, and goals; for example, a C++ description of a DCT algorithm, the required sample rate, and the goal of power optimization could be specified. The design guidance system analyzes the specified design and generates design feedback. The user interprets this information and directs the exploration by selecting a specific most promising direction from among the suggestions or by proposing another alternative. Exploration continues, with control alternating between the system and user. Note that the system keeps a global view of the design space, adapting rankings as new information is gained and thus suggesting backtracking to previous design versions as soon as the current path loses promise.

During the exploration process, the user is presented with a visual trace of the exploration (e.g. Fig. 4). This display is essentially a rooted tree representation of all the optimization sequences that have been applied so far. The root of the tree is the original design, the edges represent specific optimizations, and the nodes are the intermediate design versions. The display also shows all the optimizations suggested for further exploration and their ranking, and has links to detailed information for each.

Key advantages of the methodology are that designer expertise can be integrated into the optimization process, and the designer can learn from the system. The experienced designer is given the

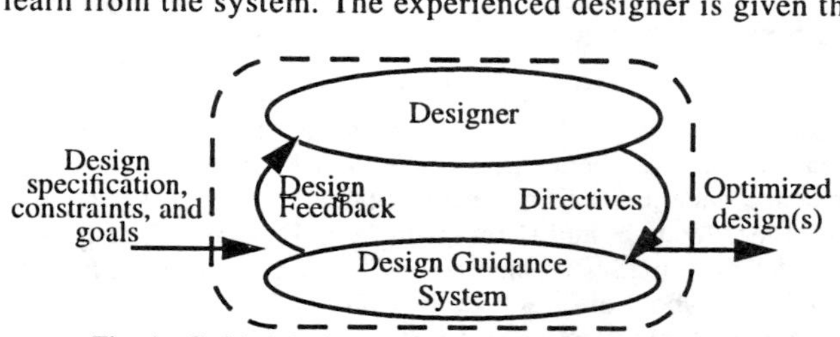

Fig. 1. Guided design space exploration and optimization

freedom to synergistically combine the information provided by the system with his/her intuition, specific design ideas, and common sense. Instead of following suggestions, (s)he may modify parameters with which an optimization is invoked, may select poorly ranked suggestions, or may provide an altogether new alternative. Alternatively, (s)he can choose to either blindly follow suggestions or use the system as resource of encapsulated optimization knowledge.

Furthermore, the approach allows the use of not only the available automated techniques, but also non-automated ones. The system can propose a technique which the user can apply manually through code re-write. Using the accompanying predictions, the designer evaluates the time investment to potential reward ratio. The use of non-automated techniques greatly increases optimization power since it expands the library of techniques the designer is likely to apply from just the available ones to include all techniques that have been characterized.

Fig. 2 shows the main components of the design guidance system. The input to the system is a directive specifying a particular optimization and design version. The output is the updated design trace. The system has automated optimizations that it can apply for the designer. It also has estimators [Rab91, Meh94, Gue94] for determining the throughput, area, and power of any of the design versions. Most importantly, it has components for generating design guidance. The components contributing to this are shown in Fig. 2 in bold, and will be the primary topic of Section 4.2. These are the keys to creating guidance: a mechanism for characterizing or extracting the "essence" of a design, a pre-characterized library of optimizations, and a mechanism for evaluating and ranking options.

Before delving into the details of these blocks, we summarize the key features of the environment:

- provides estimates of the design's power, area, and throughput
- suggests relevant optimizations
- provides reasonings for their application and predictions of their effects
- applies automated optimizations
- enables the use of non-automated techniques
- maintains a global view of the exploration space, backtracking when necessary
- allows designers to integrate their expertise in the exploration
- provides optimization knowledge that designers can learn from

4.2 Guidance System

The backbone of the new methodology is guidance generation.

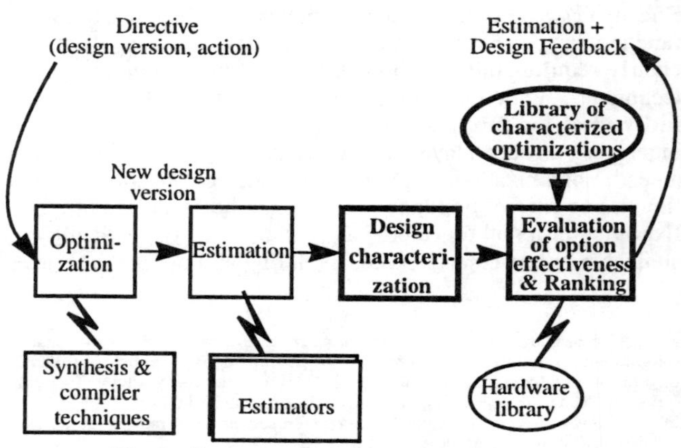

Fig. 2. Components of the design guidance system

The basis of the system is a library of optimization techniques. These techniques are characterized into models of performance. These models produce cost analysis based on properties that can be extracted from the design. This section will explain the characterization of both designs and optimizations. It will also explain how this information is used to produce ranked suggestions.

4.2.1 Design Characterization

A key requisite for providing useful design feedback is a mechanism for analyzing the design being optimized. Previous research has shown the use of design characteristics, or design properties, for predicting a variety of design metrics of the final implementation [Jam87, Gue94]. Effectiveness of various optimizations can also be linked to a design's properties. The properties *identify* situations in which certain optimization techniques will work well and can be used in predicting their effectiveness. Classes of properties used in this work are outlined below. Examples of their use in optimization characterization are shown in the next section.

- Size measures includes quantities such as the number of functional operations, the bitwidth, the number of I/O and delay operations, and the number of memory accesses.
- Timing measures include the critical path, the ratio of sample rate to the critical path, the number and types of nodes on the ε-critical network, the delays of various operators, and statistics of the scheduling slack.
- Concurrency measures estimate the number of operation and interconnect accesses that can be executed concurrently.
- Uniformity metrics capture the degree to which resource accesses are evenly distributed in time.
- Temporal properties capture information about the lifetimes of variables in the computation.
- Regularity metrics quantify the degree to which common patterns appear in the control data flow graph.
- Cyclic properties include the iteration bound, the number of strongly connected components and the critical paths of each, and the percentage of operations in cycles.
- Other metrics include characterizing whether the computation is linear or non-linear and recursive or non-recursive.

When not already available, algorithms were developed for quantifying the various design properties. In the developed environment, properties are automatically extracted and are available for the interested designer to view (Fig. 3a). This raw feedback (without the guidance layer) can in itself be a great design aid that relieves the designer of mundane time-consuming calculations (e.g. of the critical path, operation counts, concurrency, etc.) that (s)he would have otherwise computed by hand. The precise definition of the assumed properties are beyond the scope of this paper, but can be found in [Gue94, Gue96].

4.2.2 Optimization Characterization

Another key step in the methodology involves identifying and characterizing relevant optimizations. An optimization is quantified by an encapsulated command-line call for its execution, by a set of trigger conditions, and by information regarding its immediate and longer term optimization potential. This information, or design *knowledge,* is encapsulated in a library of optimization characterizations. To date we have characterized over 25 optimizations, with focus on their impact on throughput, power, and area. Sample optimization characterizations can be found in [Gue96]. Characterized optimization actions can be either an atomic synthesis task (e.g. clock selection, module selection, retiming) or a static script of tasks.

(a)

Timing View

Critical Path 7

Loop Critical Path N/A
Iteration Bound 7.00
Number of Strongly Connected Components 1
Ratio of Iteration Bound to Critical Path 1.00

Epsilon (E) Critical Network Size (%)

	=	*	+	input	Overall
E=0.10	0.00	0.00	100.00	0.00	43.86
E=0.25	0.00	33.33	100.00	0.00	61.40

Overall Epsilon (E) Critical Network Size (%)

E	0.00	0.05	0.10	0.15	0.20	0.25	0.50	0.75	1.00
	43.9	43.9	43.9	61.4	61.4	61.4	80.7	98.2	100.0

Computational Requirements

	=	*	+	input	Total
Number	1	30	25	1	57
Percent	1.8	52.6	43.9	1.8	100

(b)

Area (i) = (1.18 * Max Concurrency (i) + 1.0) * AreaPerUnit (i)

RESOURCE	MAX CONCURRENCY	UNIT AREA (sq-mm)	AREA (sq-mm)	PLOTS
-	0.50	0.0778	0.1235	
*	1.50	2.134	5.896	
+	1.00	0.07713	0.1678	
input	2.50	0.0	0.0	

Accuracy

Model accuracy:
R–squared = .87; Average value = 1.91; RMS error = .67
For accuracy of the concurrency metric, see

Close | Feedback

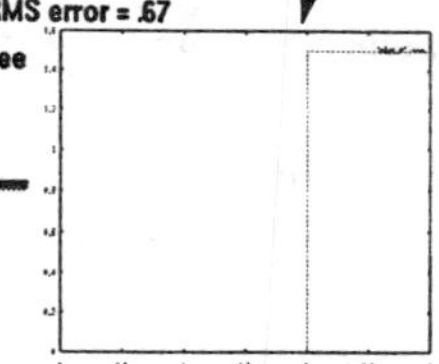

Fig. 3. a) Property and b) property-based estimation windows

The encapsulated command line calls are included so that an optimization can be automatically applied (with the appropriate parameters) when the designer selects it. The trigger conditions aid in quick pruning of options. If an option's triggers are not satisfied, the option is eliminated from consideration, and time need not be wasted generating feedback regarding its effects. Design characteristics play a major role in trigger definition. Triggers can be any boolean function of the design's properties. Example triggers include "Is the computation linear?" (for optimizations that only work on linear computations), "Does the computation have delays?" (for the delay retiming optimization), and "Does the computation have multiplications by constants?" (for constant multiplication expansion into adds and shifts).

The feedback to aid in selecting the next optimization to apply are encapsulated in the form of text and parameterized rules/equations (as a function of the design's properties). There are 3 main types of feedback that are characterized. The first type of feedback involves predicting the immediate effect that the action has on the design metrics. The second type of feedback is information on the enabling effect that the action has to enhance subsequent optimizations. We characterize enabling potential for only those optimizations identified as typical enablers/enhancers of subsequent optimizations. For these enablers, predictions are made regarding their 1 or 2-step look-ahead predictions when applied in combination with a few common transformations. The third type of feedback includes qualitative hints about the order in which to apply the optimizations (e.g. "this optimization is often effective after applying optimization x"). All information used in the characterization is assumed to be obtained through the literature and through design experience.

The feedback is both presented textually to the designer and used in option ranking. While the ranking uses only the quantitative feedback pertaining to immediate and potential affects on the cost functions, the textual feedback can additionally contain more qualitative information (for example, information about typical optimizations with which an optimization works well).

4.2.3 Ranking

In addition to suggesting and predicting the effect of feasible options, another key component of design guidance involves their ranking. Options are ranked in order of their overall predicted effectiveness in helping to meet the specified constraint and optimization objectives. Ranking is done between all proposed optimization actions along all partially-explored optimization paths. Note that if a quantitative prediction of an optimization's effectiveness is not available, then it is not given a ranking. The designer uses the rankings in conjunction with the textual feedback to aid in selecting an optimization and specific design version on which to apply it.

There are three basic factors used to evaluate an action a_i's effectiveness on design version d_j — the resulting performance as compared to existing solutions; its potential to enhance subsequent optimization; and the resulting feasibility in meeting constraints. As an example, for the goal of power minimization under a throughput constraint T_c, measures of these three factors are defined below:

ImmediateImprovement$_{ij}$ = (lowest power so far - power$_{ij}$) / (lowest power so far)

if (a_i is not an enabler) {

EnablingPotential$_{ij}$ = 0

} else {

EnablingPotential$_{ij}$ = 1/k, where k is a_i's rank among actions sorted in order of enabling potential on d_j }

if constraints are met (predicted throughput > T_c) {

Infeasibility$_{ij}$ = 0

} else {

Infeasibility$_{ij}$ = (predicted critical path / available time) }

The immediate improvement function is a measure of how the resulting power after application of a_i to d_j will compare to the lowest power attained so far among all solutions. The one with the lowest predicted power will have the highest immediate improvement function. The enabling potential is 0 if a_i is not an enabler. Otherwise it is a function of the action's predicted enabling potential as compared to other actions. The infeasibility function is 0 if constraints are met, otherwise it is a measure of how far the current achievable throughput is from the minimum throughput constraint.

The basic effectiveness R_{ij} of action a_i on design version d_j can now be formed as a linear combination of these factors. The actions are ranked in decreasing order of R_{ij}:

$$R_{ij} = \alpha_1 t_{ij} ImmedImprovement_{ij} + \frac{1}{t_{ij}} EnablingPotential_{ij} - (\alpha_2 t_{ij})^{\alpha_3 t_{ij}} Infeasibility_{ij}$$

The α_1, α_2, and α_3 are constants empirically derived to be 4, 2, and 5, respectively. The value t_{ij} is an indicator of how far the application of action a_i is into the particular optimization trajectory:

$$t_{ij} = (1 + length(d_j)) / \text{total number of applicable actions} \qquad 0 < t_{ij} \leq 1$$

It is defined as the ratio of the number of unique actions that have been applied along a given path to the total number of actions. The functions $1/t_{ij}$, $\alpha_1 t_{ij}$, and $(\alpha_2 t_{ij})^{\alpha_3 t_{ij}}$ are used to weight the sub-components of the ranking function as follows. In the beginning of an exploration trajectory, the actions with high enabling potential are favored. As the exploration progresses, the guidance favors more immediate improvements. Towards the end, ensuring feasibility becomes vitally important.

4.2.4 Interface and Implementation

An environment has been developed to demonstrate the interactive guided exploration approach. The environment's graphical user interface is written using Tcl and Tk. Embedded functions which perform the design analysis, optimization analysis, and ranking are implemented in C. The parameters passed to these functions include the design parameters and constraints, and pointers to the design (in flowgraph form), hardware library, and optimization characterization library. All invocations of existing tools are made through the framework. The Tcl-Tk interface handles all the software system glue logic and management of exploration history.

From the main window of the guidance environment, design management, estimation, or guided exploration is invoked. Design management tasks include loading the design, editing the algorithm input description, setting parameters and constraints, or exiting the tool. From the design management menu, the designer can also view the design's property metrics (Fig. 3a). Under estimations, Hyper estimation for area, speed, and power is encapsulated [Rab91, Meh94], as well as property-based area estimation. A sample property-based estimation window is shown in Fig. 3b. The plot sub-window shows a concurrency graph of subtraction operations over time. Guided design space exploration is invoked by depressing the guided exploration button, and selecting a desired performance objective. A sample screen shot for guidance will be shown in the next section.

5.0 Putting It All Together

5.1 Example

To illustrate the key ideas and the effectiveness of the new methodology, this section presents the highlights of a sample "session" of an ASIC design of a General Electric (GE), state-space, 5-state, 32-bit linear controller [Cha93]. The goal is to provide a feel for the system, the types of feedback and guidance that it provides, and the underlying design methodology. Assuming a required sample rate of 1.2 MHz, direct synthesis using the Hyper behavioral level synthesis tools and Berkeley Low-Power library, gives a design with estimated power [Meh94] of 441 mW at 5 volts using a 1.2 μm technology.

Upon commencement of the guided optimization exploration session, the system characterizes the design then prunes a number of non-applicable optimizations. For example, library selection is not triggered since the low-power library has more effective modules with respect to power than the alternate dpp library [Bro92]. Clock selection is not suggested since it has already been set efficiently by the user. The user-specified description does not have common sub-expressions or for-loops so CSE and for-loop unfolding are not applicable. Finally, direct voltage scaling is not triggered since there is no difference between the critical path and the available time.

The system proposes and ranks a number of actions that may result in immediate or subsequent power reduction. In this case, suggested actions are pipelining, constant multiplication replacement with additions and shifts, time loop unfolding, and the "maximally fast" script [Pot92]. For each action it provides feedback to guide the designer's selection. Information regarding pipelining, for example, includes a bound on critical path improvement attainable by pipelining; in this case, it can reduce the critical path from 12 clock cycles to the iteration bound of 6 clock cycles. A power prediction is also provided based on an empirical voltage-delay curve [Cha95]; in this case, the critical path reduction has enabled a voltage reduction to 2.8 volts, giving an estimated power reduction of 3.2. Constant multiplication expansion into additions and shifts is another potential optimization. There are many constant multiplications: 36 out of 67 total operations. Furthermore, their bitwidth of 32 results in a large difference between multiplier and adder power dissipation, delay, and area. The maximally fast script is also an attractive alternative since it can reduce the critical path to at least 4. The maximally fast static transformation script combines several algebraic and redundancy manipulation transformations for critical path and operation reduction.

While each of the mentioned tasks will result in immediate power reduction, the system proposes time loop unfolding as the most attractive alternative. This technique reduces the critical path to the iteration bound of the design, but is favored not for it's immediate benefit (which is actually very similar to pipelining's), but rather for the long-term improvement potential that it provides for greater subsequent optimization. In particular, as indicated by the tool's analysis, the computation is linear (formal degree is 1) and therefore unfolding enables effective application of the "maximally fast" technique simultaneously on several iterations of the computation. The specific unfolding factor is selected by the designer. The system provides a table (partial table shown in Table 1) of suggested voltage and number of operations for various unfolding factors (closed form equations exists for computing the estimated critical path and number of operations). The suggested voltage is based on maintaining some slack between the resulting critical path after voltage scaling and the available time (a slack of 10% is used). Based on this table and qualitative feedback that control overhead increases with greater unfolding, the designer selects an unfolding factor of six to balance achieving both speedup for voltage scaling and reduction in capacitance.

Application of time loop unfolding followed then by the suggested maximally fast script reduces both the effective critical path by a factor of 21 and more importantly reduces the effective number of operations (the number of operations per iteration) by a factor of 2. This in turn enables application of the MCM technique [Pot94] for conversion of constant multiplications to series of shifts and

Unfolding factor	New critical path	Speedup	Suggested voltage	Number of additions and multiplications
0	4	3	2.2	66
5	.67	18	1.2	33.5
6	.57	21	1.2	33.43

Table 1: GE controller example — the impact on attainable voltage and operation counts for various unfolding factors.

additions, reducing the effective capacitance per operation. The final steps are suggested by the environment so that immediate improvements are maximized. They involve lowering the voltage from 5 to 1 volts and performing clock selection for efficient utilization of the clock period. The optimized design has a power of 2.9 mW. For this example, not only is power reduced by a factor of 151, but area was also reduced by a factor of 2.

Fig. 4 shows a snapshot of the exploration session. The left-most justified actions are those that were initially suggested (unfolding,

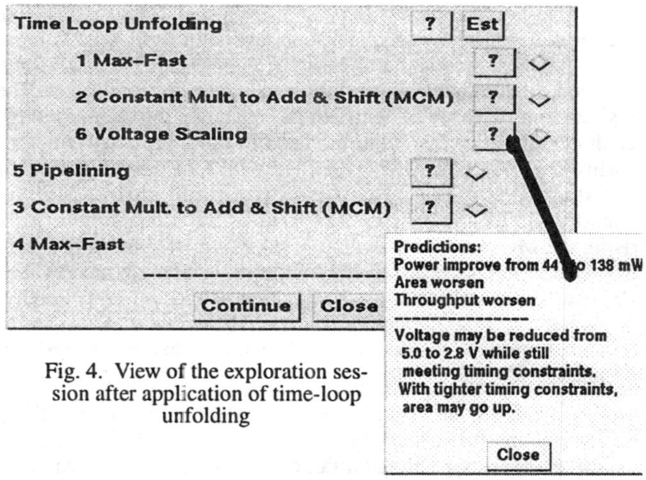

Fig. 4. View of the exploration session after application of time-loop unfolding

pipelining, MCM, and max-fast). Application of time-loop unfolding has already been done through selection of the appropriate radio button and the "Continue" button. Several optimizations are now suggested for application to the new unfolded design version (e.g. max-fast, MCM, voltage scaling). Note that ranking is done on both optimizations suggested for application to the new unfolded design (ranks 1, 2, and 6) as well as to the original design (ranks 3, 4, 5). Specific feedback regarding a suggested action is provided by selecting the corresponding information button marked with a "?" (see sub-window in Fig. 4). The information for voltage scaling is displayed.

5.2 Experimental Results

Table 2 presents the obtained improvements in energy and area. The designs include a 6th order IIR filter, a bandpass filter, cordic, 5 and 7 point convolutions, an 8-point fft, an image convolver, a

Design	Initial cap. (C)	Final C	C reduction	Energy improvement	Area improvement	Optimization steps
6 IIR	9.08	0.38	23.9	260.2	5.67	U4, MF, MCM + CS, AIB, MP, V, CS
bandpass	12.26	1.54	7.96	86.7	2.20	CM + CS, U7, V, CS
conv5	11.20	0.04	280	3049.0	32.57	CM + CS, U5, V, CS
conv7	3.34	0.04	83.5	909.3	26.9	CM + CS, U6, V, CS
cordic	49.96	32.76	1.53	16.6	1.08	U8, MP, V, CS
fft8	5.37	1.02	5.26	57.3	2.49	U5, CM, V, CS
imagec	38.82	36.08	1.08	11.7	0.95	U2,MP, V, CS
modem	38.67	0.80	48.3	526.4	16.19	CM + CS, U4, AIB, P2, V, CS
steam	137.58	85.50	1.61	17.5	0.49	CM + CS, U3, AIB, P2, V, CS
volterra 2	7.42	1.03	7.20	78.5	4.36	U4, CM + CS, AIB+ MP, V, CS
volterra 3	59.90	13.64	4.39	47.8	2.77	U3, CM + CS, AIB+ MP, V, CS
wangd	15.90	2.41	6.60	71.85	2.86	U4, CM, V, CS

Table 2: Experimental Results: minimizing power under a fixed throughput constraint

modem example, a large controller for a steam plant, 2nd and 3rd order non-linear Volterra filters, and the Wang 8-point DCT. Note that most designs have feedback, and several including the Volterra filters, image convolver, and cordic are nonlinear. The latter 2 are also control dominated with a high percentage of conditionals. The last column of Table 2 identified the optimization steps used for each design. The abbreviations used in the table are defined as follows: UN - unfolding by a factor N; CM - substitution of constant multiplications by addition and shifts; MCM - substitution of constant multiplications by addition and shifts using the MCM transformation; MF - maximally fast algorithm; MP - maximal pipelining; PN - pipelining using N stages; R - retiming; AIB - associativity for iteration bound; A - associativity; V - voltage scaling; CS - clock selection. Note that a "+" is used to indicate optimizations that are suggested together as a script (e.g. CM+CS).

For all examples, we assume that clock selection has been initially performed and that a starting voltage of 3.3 Volts is used. For all examples, the final voltage attained was 1 Volt. The maximum, minimum, average and median capacitance improvements were by a factor of 280, 1.08, 7.68, and 33.6 respectively. The maximum, minimum, average and median energy improvements were by factors of 3049, 11.7, 126.8 and 83.6, respectively. Simultaneously for area, the maximum, minimum, average and median improvements were 32.57, 0.49, 7.88 and 2.86 times, respectively. Note that although the resulting sequences of optimization steps are in many cases similar, rarely are they identical. Some of the key issues the design guidance feedback helped in included determining the unfolding factor, determining how much to pipeline (if at all), freeing the designer from many manual calculations, and eliminating optimizations. While in these examples we lowered voltage always to 1 V, in practice the designer will likely target a variety of voltages.

Obtained improvements when targeting exclusively area reduction for the same set of designs are also available, but were omitted here due to space.

6.0 Summary

A novel methodology for guided design space exploration has been presented. It has been applied for behavioral-level optimization of datapath-intensive semi-custom ASIC implementations. The approach encourages systematic and quantitative as opposed to ad-hoc global optimization across point optimizations. The intelligence of both the system and designer can be leveraged upon. The methodology and environment facilitates these goals using a library which characterizes design optimizations, mechanisms for ranking design alternatives, and techniques for analysis of designs.

One of the key unanswered issues concerns the portability of the presented ideas to new application areas (e.g. control-intensive applications) and architecture models. A change in either would require a new design characterization where some metrics are unchanged, some are modified, and some are altogether new. It would also require establishing the relations between these metrics and performance. The hope is that with each new domain addressed, the number of new metrics needed will decrease, and at some point, a core superset of metrics will be established. Other areas of future work include continued growth and improvement of the optimization "knowledge database," continued research on ranking and partitioning, and the use of automated searches to complement the user-driven exploration.

7.0 References

[Bac94] Bacon, D.F., Graham, S.L., Sharp, O.J. Compiler transformations for high performance computing. *ACM Computing Surveys*, Vol. 26, No. 4, 345-420.

[Bra84] Brayton, R.K., et al. *Logic Minimization Algorithms for VLSI Synthesis.* Kluwer, Boston, MA, 1984.

[Cha93] Chatterjee, A., Roy, R.K., d'Abreu, M.A. Greedy hardware optimization for linear digital circuits using number splitting and refactorization. *IEEE Trans. on VLSI Systems*, Vol.1, No.4, 423-431, 1993.

[Cha95] Chandrakasan, A., et al. Optimizing power using transformations. *IEEE Trans. on CAD*, Vol. 14, No. 1, 1995.

[Goo94] Goossens, G., Bolsens, I., Lin, B., Catthoor, F. Design of heterogeneous ICs for mobile and personal communication systems. *ICCAD*, 524-531, 1994.

[Gue94] Guerra, L., Potkonjak, M., Rabaey, J. System-level design guidance using algorithm properties. *VLSI Signal Processing Workshop*, 73-62, 1994.

[Gue96] Guerra, L. Behavioral-level design guidance for ASIC implementations. Ph.D. dissertation, UC Berkeley Dept. of EECS, 1996.

[Hua93] Huang, S., Rabaey, J. Maximizing the throughput of high performance DSP applications using behavioral transformations. *European Design & Test Conf.*, 25-30, 1994.

[Jam87] Jamieson, L. Characterizing parallel algorithms, in *The Characteristics of parallel algorithms*, L. Jamieson, D. Gannon, R. Douglass (eds.), MIT Press, Cambridge, Mass., 1987.

[Kle94] Kleinfeldt, S., et al. Design methodology management. *Proc. of the IEEE*, Vol.82, No.2, 231-50, 1994.

[Kna91] Knapp, D. W., Parker, A. C. The ADAM design planning engine. *IEEE Trans. on CAD*, Vol.10, No.7, 829-46, 1991.

[Lee87] Lee, E.A. and Messerschmitt, D.G. Static scheduling of synchronous data flow programs for digital signal processing. *IEEE Trans. on Computers*, Vol. 36, No. 1, 24-35, 1987.

[Lei83] Leiserson, C. and Saxe, J. Optimizing synchronous systems. *Journal of VLSI and Computer Systems*, Vol.1, No.1, 1983.

[McF90] McFarland, M.C., Parker, A.C., Camposano, R. The high-level synthesis of digital systems. *Proc. of the IEEE*, Vol. 78, No. 2, 301-317, 1990.

[McK65] McKeeman, W.M. Peephole optimization. *Comm. of the ACM*, Vol. 8, No. 7, 443-444, 1965.

[Meh94] Mehra, R. and Rabaey, J. Behavioral level power estimation and exploration. *Int'l Workshop on Low-Power Design*, 197-202, 1994.

[Par95] Parhi, K.K. High-level algorithm and architecture transformations for DSP synthesis. *Journal of VLSI Signal Processing*, Vol. 9, No. 1-2, 121-143, 1995.

[Pot92] Potkonjak, M. and Rabaey, J. Maximally fast and arbitrarily fast implementation of linear computations. *ICCAD*, 304-308, 1992.

[Pot94] Potkonjak, M., Srivastava, M., and Chandrakasan, A. Efficient substitution of multiple constant multiplications by shifts and additions using iterative pairwise matching. *Design Automation Conference*, 189-194, 1994.

[Rab91] Rabaey, J., et al. Fast prototyping of data path intensive architectures. *IEEE Design and Test*, Vol. 8, No. 2, 40-51, 1991.

[Whi90] Whitfield, D. and Soffa, M.L. An approach to ordering optimizing transformations. *ACM Symposium on Principles and Practice of Parallel Programming*, 137-147, 1990.

[Wol91] Wolf, M. and Lam, M. A loop transformation theory and an algorithm to maximize parallelism. *IEEE Trans. on Parallel and Distributed Systems*, Vol. 2, No. 4, 452-471, 1991.

A Programming Environment for the Design of Complex High Speed ASICs

Patrick Schaumont Serge Vernalde Luc Rijnders Marc Engels Ivo Bolsens
IMEC vzw, Kapeldreef 75, B-3001 Leuven

Abstract

A C++ based programming environment for the design of complex high speed ASICs is presented. The design of a 75 Kgate DECT transceiver is used as a driver example. Compact descriptions, combined with efficient simulation and synthesis strategies are essential for the design of such a complex system. It is shown how a C++ programming approach outperforms traditional HDL-based methods.

1 Introduction

In this contribution, we present a programming environment based on C++ that supports simulation, verification and synthesis of complex high speed ASICs for digital telecommunications. It is part of a larger environment that targets an automated synthesis path from the Matlab algorithm level to the VHDL architecture level [8].

In order to introduce the requirements put on to such an environment, a recent design experience will be documented. The design consists of a digital radiolink transceiver ASIC, residing in a DECT base station (figure 1). The chip processes DECT burst signals, received through a radio frequency front-end `RF`. The signals are equalized to remove the multi-path distortions introduced in the radio link. Next, they are passed to a wire-link driver `DR`, that establishes communication with the base station controller `BSC`. The system is also controlled locally by means of a control component `CTL`.

The specifications that come with the design of the digital transceiver ASIC in this system are as follows:

- The equalization involves complex signal processing, and is described and verified inside a high level design environment such as Matlab.
- The interfacing towards the control component `CTL` and the wire-link driver `DR` on the other hand is described as a detailed clock-cycle true protocol.
- The allowed processing latency is, due to the real time operation requirements, very low: a delay of only 29 DECT symbols (25.2 μsecs) is allowed. The complexity of the equalization algorithm, on the other hand, requires up to 152 data multiplies per DECT symbol to be performed. This implies the use of parallel data processing, and introduces a severe control problem.
- The scheduled design time to arrive from the heterogeneous set of specifications to the verified gate level netlist, is 18 person-weeks.

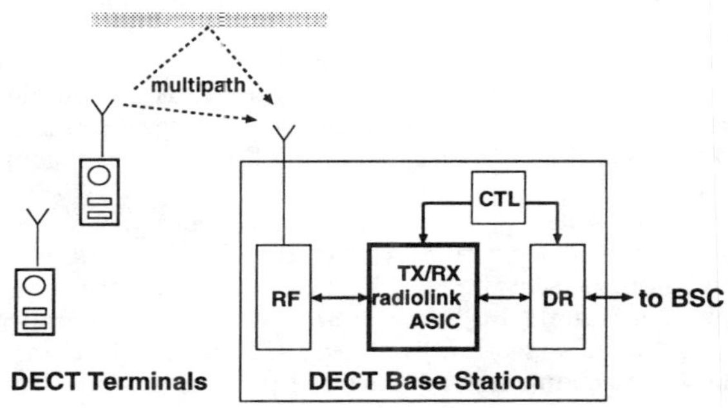

Figure 1: DECT Base Station Configuration

The most important degree of freedom in this design process is the target architecture, which must be chosen such that the requirements are met. Due to the critical design time, a maximum of control over the design process is required. To achieve this, we use a programming approach to implementation, in which the system is modeled in C++. The object oriented features of this language allows it to mix high-level descriptions of undesigned components with detailed clock-cycle true, bit-true descriptions. In addition, appropriate object modeling allows the detailed descriptions to be translated to synthesizable HDL automatically. Finally, verification test-benches can be generated automatically in correspondence with the C++ simulation.

The result of this design effort is a 75 Kgate chip with a VLIW architecture, including 22 datapaths, each decoding between 2 and 57 instructions, and including 7 RAM cells. The chip has a 194 mm^2 die area in 0.7 μ CMOS technology.

The C++ programming environment allows it to obtain results faster than existing approaches. Related to register transfer design environments such as [10], it will be shown that C++ enables more compact, and consequently less error prone descriptions of hardware. High level synthesis environments [9] could solve this problem but have to fix the target architecture on beforehand. As will be described in the case of the DECT transceiver design, sudden changes in target architecture can occur due to hard initial requirements, that can be verified only at system implementation.

On the other hand, recent works in the research community propose to use high level languages as the specification for hardware synthesis [4, 1] and modeling [12, 13]. This makes us believe that a programming approach goes beyond the capabilities of traditional HDL.

In this paper, we will present our design environment as follows. First, the system machine model is introduced (section 2). This model includes two types of description: high-level *untimed* ones and detailed *timed* blocks (section 3). Using such a model, a simulation mechanism is constructed (section 4). It will be shown that the proposed approach outperforms current environments in code size and simulation speed. Following this, HDL code generation issues (section 5) and hardware synthesis strategies (section 6) are described. A summary of the contributions is given in 7.

2 System Machine Model

Due to the high data processing parallelism, the DECT transceiver is best described with a set of concurrent processes. Each process translates to one component in the final system implementation.

At the system level, processes execute using dataflow simulation semantics. That is, a process is described as an iterative behavior, where inputs are read in at the start of an iteration, and outputs are produced at the end. Process execution can start as soon as the required input values are available.

Inside of each process, two types of description are possible. The first one is a high level description, and can be expressed using procedural C++ constructs. A firing rule is also added to allow data-flow simulation [7, 6].

The second flavor of processes is described at register transfer level. These processes operate synchronously to the system clock. One iteration of such a process corresponds to one clock cycle of processing.

For system simulation, two schedulers are available. A data-flow scheduler is used to simulate a system that contains only untimed blocks. This scheduler repeatedly checks process firing rules, selecting processes for execution as their inputs are available.

When the system also contains timed blocks, a cycle scheduler is used instead. The cycle scheduler manages to interleave execution of multi-cycle descriptions

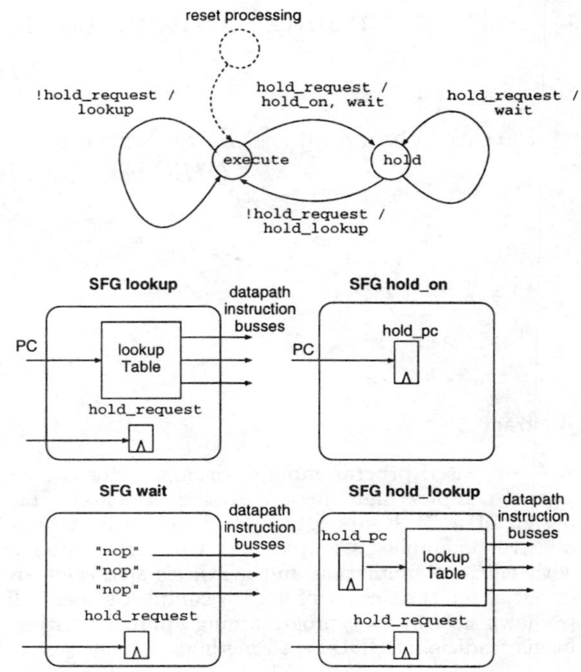

Figure 2: Cycle True Model Example

but can incorporate untimed blocks as well. The ability to mix both high level descriptions and detailed clock cycle true descriptions is essential in maintaining an executable system specification at all times.A detailed description of the cycle scheduler operation is given in section 4.

In the following section, we will focus on the modeling of detailed, timed descriptions, and motivate why the proposed model is at the right level for a telecommunications ASIC.

3 Cycle-true Descriptions

Detailed process descriptions reflect the hardware behavior of a component at the same level of the implementation. To gain simulation performance and coding effort, several abstractions are made.

- Finite Wordlength effects are simulated with a C++ fixed point library. It has been shown that the simulation of these effects is easy in C++ [5, 11]. Also, the simulation of the quantization rather than the bit-vector representation allows significant simulation speedups.
- The behavior is modeled with a mixed control/data processing description, under the form of a finite state machine coupled to a datapath. This model is common in the synthesis community [2]. In high throughput telecommunications circuits such as the ones in the DECT transceiver ASIC, it most often occurs that the desired component architecture is known before the hardware description is made. The FSMD model works well for these type of components.

Sig Class

```
class sig {
  Value value;
  char  *name;
public:
  sig(Value v);
  sig operator +(sig v);
  virtual Value simulate();
  virtual void gen_code(ostream &os);
};

sig sig::operator +(sig v) {
  sigadd add;
  add.left  = &v;
  add.right = this;
  return add;
}

Value sig::simulate() {
  return value;
}

sig::gen_code(ostream &os) {
  os << name;
}
```

Derived Operator Class

```
class sigadd : public sig {
  sig *left;
  sig *right;
public:
  Value simulate();
   void gen_code(ostream &os);
};

Value sigadd::simulate() {
  return left->simulate() +
         right->simulate();
}

sigadd::gen_code(ostream &os) {
  os << left->cg()
     << " + "
     << right->cg();
}
```

Figure 3: C++ SFG construction class

The two aspects, wordlength modeling and cycle true modeling, are available in the programming environment as separate class hierarchies. Therefore, fixed point modeling can be applied equally well to high level descriptions.

As an illustration of cycle true modeling, a part of the central VLIW controller description for the DECT transceiver ASIC is shown in figure 2. The top shows a Mealy type finite state machine. As actions, the signal flow graph descriptions below it are executed. The two states `execute` and `hold` correspond to operational and idle states of the DECT system respectively. The conditions are stored in registers inside the signal flow graphs. In this case, the condition `hold_request` is related to an external pin.

In state `execute`, instructions are distributed to the datapaths. Instructions are retrieved out of a lookup table, addressed by a program counter. When `hold_request` is asserted, the current instruction is delayed for execution, and the program counter `PC` is stored in an internal register. During a hold, a `nop` instruction is distributed to the datapaths to freeze the datapath state. As soon as `hold_request` is removed, the stored program counter `hold_pc` addresses the lookup table, and the interrupted instruction is issued to the datapaths for execution.

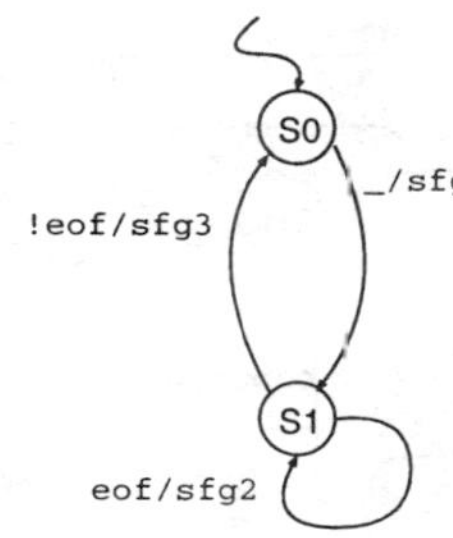


```
fsm f;
initial s0;
state s1;

s0 << always     << sfg1 << s1;
s1 << cnd(eof)   << sfg2 << s1;
s1 << !cnd(eof)  << sfg3 << s0;
```

Figure 4: C++ FSM description

We next discuss the C++ objects that allow us to capture clock cycle true behavior.

3.1 Signals and Signal Flow Graphs

Signals are the information carriers used in construction of a timed description. Signals are simulated using C++ `sig` objects. These are either plain signals or else registered signals. In the latter case the signals have a current value and next value, which is accessed at signal reference and assignment respectively. Registered signals are related to a clock object `clk` that controls signal update. Both types of signals can be either floating point values or else simulated fixed point values.

Using operations, signals are assembled to expressions. By using the overloading mechanism as shown in figure 3, the parser of the C++ compiler is reused to construct the signal flow graph data structure.

A set of `sig` expressions can be assembled in a signal flow graph (SFG). In addition, the desired inputs and outputs of the signal flowgraph have to be indicated. This allows to do semantical checks such as dangling input and dead code detection, which warn the user of code inconsistency.

An SFG has well defined simulation semantics and represents one clock cycle of data processing.

3.2 Finite State Machines

After all instructions are described as SFG objects, the control behavior of the component has to be described. We use a Mealy-type FSM model to do this.

Again, the use of C++ objects allow to obtain very compact and efficient descriptions. Figure 4 shows a graphical and C++-textual description of the same FSM. The correspondence is obvious. To describe an equivalent FSM in an event driven HDL, one usually has to follow the HDL simulator semantics, and for example use multi-process modeling. By using C++ on the other hand, the semantics can be adapted depending on the type of object processed, all within the same piece of source code.

3.3 DECT Transceiver Architecture

The FSMD model described above was used to construct an ASIC architecture shown in figure 5. It consists of a central (VLIW) controller, a program counter controller and 22 datapath blocks. Each of these are modeled with the combined control/data processing

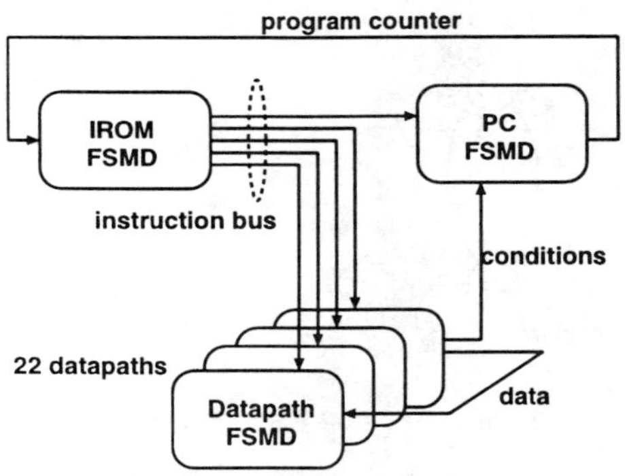

Figure 5: DECT Transceiver System Architecture

shown above. They exchange data signals that depending on the particular block, are interpreted as instructions, conditions or signal values. By means of these interconnected FSMD machines, a more complex machine was constructed.

It is now motivated why this centrally controlled architecture was chosen. For the DECT transceiver, there is a severe latency requirement. Originally, a data-flow target architecture was chosen which is common for these types of telecommunications signal processing. In such an architecture, the individual components are controlled locally and data driven. However, the extreme latency requirement (29 DECT symbols) required the introduction of global exceptions. In a data driven architecture however, such global exceptions are very difficult to implement. This is far more easy in a central control architecture, where they take the form of a jump in the instruction ROM. Because of these difficulties, the target architecture was changed from data driven to central control.

The FSMD machine model allowed to reuse the datapath descriptions and only required the control descriptions to be reworked. This architectural change was done *during* the 18-week design cycle, which shows the flexibility of using C++ as modeling mechanism.

4 The Cycle Scheduler

Whenever a timed description is to be simulated, a cycle scheduler is used instead of a data-flow scheduler. The cycle scheduler creates the illusion of concurrency between components on a clock cycle basis.

The operation of the cycle scheduler is best illustrated with an example. In figure 6, the simulation of one cycle in a system with three components is shown. The first two, components 1 and 2, are timed descriptions constructed using `fsm` and `sfg` objects. Component 3 on the other hand is described at high level using a firing rule and a behavior. In the DECT transceiver, such a loop of detailed (timed) and high level (untimed) components occurs for instance in the RAM cells that are attached to the datapaths. In that case, the RAM cells are described at high level while the datapaths are described at clock cycle true level.

The simulation of one clock cycle is done in three phases. Traditional RT simulation uses only two; the first being an evaluation phase, and the second being

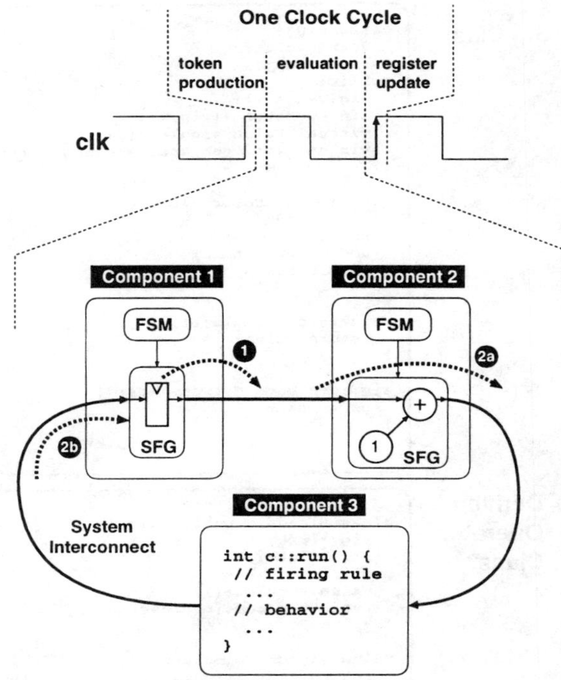

Figure 6: Cycle Scheduler Operation

a register update phase.

The three phases used by the cycle scheduler are a token production phase, an evaluation phase and a register update phase. The three-phase simulation mechanism is needed to avoid apparent deadlocks that might exist at the system level. Indeed, in the example there is a circular dependency in between components 1, 2, and 3, and a data-flow scheduler can no longer select which of the three components should be executed first. In data-flow simulation, this is solved by introducing *initial tokens* on the data dependencies. Doing so would however require us to devise a buffer implementation for the system interconnect, and introduce an extra code generator in the system.

The cycle scheduler avoids this by creating the required initial tokens in the token production phase. Each of the phases operates as follows.

0) At the start of each clock cycle, the `sfg` descriptions to be executed in the current clock cycle are selected. In each fsm description, a transition is selected, and the `sfg` related to this transition are marked for execution.

1) **Token production** phase. For each marked `sfg`, look into the signal flow graph, and identify the outputs that solely depend on registered signals and/or constant signals. Evaluate these outputs and put the obtained tokens onto the system interconnect.

2a) **Evaluation** phase (case a). In the second phase, schedule marked `sfg` and untimed blocks for execution until all marked sfg have fired. Output tokens are produced if they are directly dependent on input tokens for timed `sfg` descriptions, or else if they are outputs of untimed blocks.

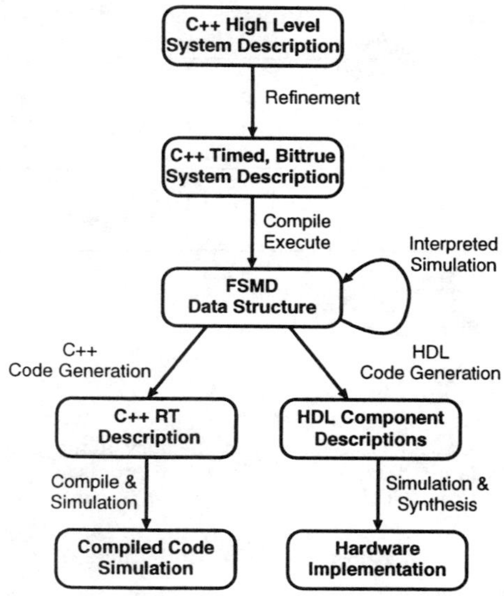

Figure 7: Code Generation and Simulation

2b) **Evaluation** phase (case b). Outputs that are however only dependent on registered signals or constants will not be produced in the evaluation phase.

3) **Register update** phase. For all registered signals in marked `sfg`, copy the next-value to the current-value.

The evaluation phase of the three-phase simulation is an iterative process. If a preset amount of iterations have passed, and there are still unfired components, then the system is declared to be deadlocked. This way, the cycle scheduler identifies combinatorial loops in the system.

5 Code Generation and Simulation Strategy

The clock-cycle true, bit-true description of system components serves a dual purpose. First, the descriptions have to be simulated in order to validate them. Next, the descriptions have also to be translated to an equivalent, synthesizable HDL description.

In view of these requirements, the C++ description itself can be treated in two ways in the programming environment. In case of a *compiled code* approach, the C++ description is translated to directly executable code. In case of an *interpreted* approach, the C++ description is preprocessed by the design system and stored as a data structure in memory.

Both approaches have different advantages and uses. For simulation, execution speed is of primary importance. Therefore, compiled code simulation is needed. On the other hand, HDL code generation requires the C++ description to be available as a data structure that can be processed by a code generator. Therefore, a code generator requires an interpreted approach.

We solve this dual goal by using a strategy as shown in figure 7. The clock-cycle true and bit-true description of the system is compiled and executed. The description uses C++ objects such as signals and finite state machine descriptions which translate themselves to a control/data flow data structure.

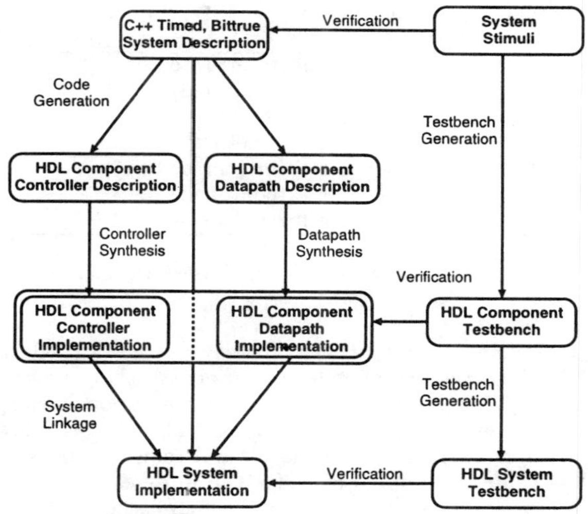

Figure 8: Hardware Synthesis Strategy

This data structure can next be interpreted by a simulator for quick verification purposes. The same data structure is also processed by a code generator to yield two different descriptions.

- A C++ description can be regenerated to yield an application-specific and optimized compiled code simulator. This simulator is used for extensive verification of the design because of the efficient simulation runtimes.
- A synthesizable HDL description can also be generated to arrive at a gate-level implementation.

The simulation performance difference between these three formats (interpreted C++ objects, compiled C++, and HDL) is illustrated in table 1. Simulation results are shown for the DECT header correlator processor, and also the complete DECT transceiver ASIC.

The C++ modeling gains a factor of 5 in code size (for the interpreted-object approach) over RT-VHDL modeling. This is an important advantage given the short design cycle for the system. Compiled code C++ on the other hand provides faster simulation and smaller process size then RT-VHDL.

For reference, results of netlist-level VHDL and Verilog simulations are given also.

6 Synthesis Strategy

Finally, we document the synthesis approach that was used for the DECT transceiver. As shown in figure 8, the clock-cycle true, bit-true C++ description can be translated from within the programming environment into equivalent HDL.

For each component, a controller description and a datapath description is generated, in correspondence with the C++ description. This is done because we rely on separate synthesis tools for both parts, each one optimized towards controller or else datapath synthesis tasks.

Design	Size (Gates)	Type	Source Code (# lines)	Simulation Speed (cycles/sec)	Process Size (Mbyte)
HCOR	6K	C++ (interpreted obj)	230	69	3.8
		C++ (compiled)	1.7K	819	2.7
		VHDL (RT)	1.6K	251	11.9
		VHDL (netlist)	77K	2.7	81.5
DECT	75K	C++ (interpreted obj)	8K	2.9	20
		C++ (compiled)	26K	60	5.1
		Verilog (netlist)	59K	18.3	100

Table 1: Performances of interpreted and compiled approaches

For datapath synthesis, we rely on the Cathedral-3 back-end datapath synthesis tools [3], that allow to obtain a bit-parallel hardware implementation starting from a set of signal flow graphs. These tools allow operator sharing at word level, and require run times less than 15 minutes even for the most complex, 57-instruction datapath of the DECT transceiver.

Controller synthesis on the other hand is done by logic synthesis such as Synopsys DC. For pure logic synthesis such as FSM synthesis, this tool produces efficient results. The combined netlists of datapath and controller are also post-optimized by Synopsys DC to perform gate-level netlist optimizations. This divide and conquer strategy towards synthesis allows each tool to be applied at the right place.

During system simulation, the system stimuli are also translated into test-benches that allow to verify the synthesis result of each component. After interconnecting all synthesized components into the system netlist, the final implementation can also be verified using a generated system test-bench.

7 Conclusions

In this paper, we illustrated the use of C++ for system synthesis by the design of a DECT transceiver. As systems grow more complex, designers tend to rely on higher levels of modeling to increase design cycle speed. The presented programming environment allows to use this higher level of modeling for both simulation and synthesis. The description style for synthesis is a behavioral register transfer model. The writing of HDL is avoided through code generation from C++. The design environment is also lightweight, and uses only a C++ compiler and library to do the system capture and simulation. Finally, as the C++ library is generic, it is currently being reused for several demonstrator designs done, inluding an upstream cable modem, an image compressor and a wireless LAN modem.

Acknowledgements

The authors wish to thank Francky Catthoor and Hugo De Man from IMEC for the constructive discussions and review by the writing of this paper. Also, the design efforts of Radim Cmar from IMEC were invaluable in the construction of the DECT transceiver. Marc Engels is a senior research assistant of the Belgium National Fund for Scientific Research. This work was carried out under the Flemish Government Impulse Program for Information Technology (IT-ISIS).

References

[1] G. Dejong, B. Lin, C. Verdonck, S. Wuytack, and F. Catthoor. Background memory management for dynamic data structure intensive processing systems. In *Proc ICCAD 95*, pages 515 – 520, November 1995.

[2] D. Gajski, N. Dutt, A. Wu, and S. Lin. *High Level Synthesis*. Kluwer Academic Publishers, 1992.

[3] W. Geurts, F. Catthoor, S. Vernalde, and H. De Man. *Accelerator Data-Path Synthesis for High-Throughput Signal Processing Applications*. Kluwer Academic Publishers, 1996.

[4] R. K. Gupta and S. Y. Liao. Using a programming language for digital system design. *IEEE Design and Test of Computers*, pages 72 – 80, April-June 1997.

[5] S. Kim, K. Kum, and W. Sung. Fixed-point optimization utility for c and c++ based digital signal processing programs. In *Workshop on VLSI Signal Processing '95*, pages 197–206. Osaka, November 1995.

[6] R. Lauwereins, M. Engels, M. Ade, and J.A. Perperstraete. Grape-2: A tool for the rapid prototyping of multi-rate asynchronous dsp applications on heterogeneous multiprocessors. *IEEE Computer*, 28(2):35 – 43, February 1995.

[7] E. A. Lee and D. G. Messerschmidt. Static scheduling of synchronous data flow programs for digital signal processing. *Transactions on Computers*, C-36(1):24 –35, January 1987.

[8] P. Schaumont, S. Vernalde, M. Engels, and I. Bolsens. Synthesis of multi-rate and variable rate digital circuits for high throughput telecom applications. In *Proc. EDTC 1997*.

[9] Synopsys Inc., 700 E. Middlefield Rd, Mountain View, CA 94043. *BC User's Manual*.

[10] Synopsys Inc., 700 E. Middlefield Rd, Mountain View, CA 94043. *DC User's Manual*.

[11] M. Willems, V. Bursgens, H. Keding, T. Grotker, and H. Meyr. System level fixed-point design based on an interpolative approach. In *Proc. 34th Design Automation Conference*. Anaheim, CA, 1997.

[12] J-S Yim, Y-H Hwang, C-J Park, H Choi, W-S Yang, H-S Oh, I-C Park, and C-M Kyung. A c-based rtl design verification methodology for complex microprocessor. In *Proc. 34th Design Automation Conference*. Anaheim, CA, 1997.

[13] V. Zivojnovic, S. Pees, and H. Meyr. Lisa - machine description language and generic machine model for hw/sw co-design. In W. Burleson, K. Konstantinides, and T. Meng, editors, *VLSI Signal Processing IX*, 1996.

Media Architecture: General Purpose vs. Multiple Application-Specific Programmable Processor

Chunho Lee†, Johnson Kin‡, Miodrag Potkonjak† and William H. Mangione-Smith‡
†Computer Science Department, UCLA, ‡Electrical Engineering Department, UCLA

Abstract

In this paper we report a framework that makes it possible for a designer to rapidly explore the application-specific programmable processor design space under area constraints. The framework uses a production-quality compiler and simulation tools to synthesize a high performance machine for an application. Using the framework we evaluate the validity of the fundamental assumption behind the development of application-specific programmable processors. Application-specific processors are based on the idea that applications differ from each other in key architectural parameters, such as the available instruction-level parallelism, demand on various hardware components (e.g. cache memory units, register files) and the need for different number of functional units. We found that the framework introduced in this paper can be valuable in making early design decisions such as area and architectural trade-off, cache and instruction issue width trade-off under area constraint, and the number of branch units and issue width.

1 Introduction

Since 1992, microprocessors account for 23% of total semiconductor sales. By 1998, these chips will account for 27% of total sales, estimated at $242 billion by an industry expert [14]. The increasing share of microprocessors in semiconductor market is due to a new phase of silicon integration, thanks mainly to deep submicron fabrication technology. For example, SA-1100 from Digital Semiconductor [31] incorporates many system functions such as a memory controller, color LCD driver, PCMCIA interface, IrDA and USB communication channels, and extensive power management into a single chip. One major implication of this advanced technology is that almost all semiconductor manufacturers are moving into the microprocessor business. This shift will force manufacturers to focus on microprocessors, specifically media processors, that are cheaper and more aggressively optimized for specific applications. A challenge to microprocessor designers will be to design a microprocessor that executes a targeted application very well yet can achieve large economies of scale.

Recent advances in compiler technology and microprocessor architecture for instruction-level parallelism (ILP) have significantly increased the ability of microprocessors to exploit the opportunities for parallel execution that exist in many programs. Key ILP compiler technologies, such as trace scheduling [8], superblock scheduling [24], tree-scheduling [3], hyperblock scheduling [25] and software pipelining [15] are in the process of migrating from research labs to product groups. Commercial microprocessor manufacturers have introduced a number of products based on new architectures that present hardware structures that are well matched to ILP compilers. Architectural enhancements found in commercial products include predicated instruction execution, VLIW, split register files and multi-gauge arithmetic (or variable-width SIMD) [5, 12, 17, 32]. Most of the multimedia extensions of general-purpose processors also adopt similar architectural enhancements [20, 28].

The arrival of production quality ILP compilers and commercial DSPs with VLIW architecture stimulated the idea of so-called custom-fit processors [9]. Researchers have argued that applications differ from each other in, for example, the available ILP, demand on various hardware components such as cache memory units, register files, and the number of functional units. The presumption is that a microprocessor can be designed by adding hardware components tailored to a specific application so that it can execute the single application well.

Previous discussions and research efforts on multiple application mediaprocessors and single application processors have lacked a well-defined and agreed-upon set of benchmarks. No systematic study on programmable processor design space exploration using a comprehensive benchmark suite has been reported. We have developed a framework to rapidly explore the design space by focusing on the trade-off between cost and generality of possible designs under area constraints. We note that the advances in compiler and architecture technologies should be incorporated in a programmable processor design space exploration framework.

This paper is organized as follows. The next section briefly surveys related works and summarizes the contributions of this work. Machine model, benchmarks, tools and example set of results obtained using the tools are described in Section 3. Our approach in this work is explained in Section 4 in detail. Section 5 formulates the search problem defined in the previous section in precise terms. The search strategy and algorithm is described in Section 6. Experimental results of the tools and algorithms for the system-level synthesis are reported in Section 7. Finally, Section 8 draws conclusions.

2 Related Works and Our Contributions

The works on synthesis and evaluation of application-specific programmable processors has been conducted independently in two research communities, computer-aided design and architecture. There is, however, a strong converging trend of the two areas due to recent technological advances and application trends. In this section we survey the related works in these two fields.

Since the early 90's, there have been a number of efforts related to the design of application-specific programmable processors and application-specific instruction sets. Comprehensive surveys of the works on computer-aided design of application-specific programmable processors can be found in [11, 26, 27]. In particular, a great deal of effort has been made in combining retargetable compilation technologies and design of instruction sets [1, 21, 22, 23, 30]. Several research groups have published results on the topic of selecting and designing instruction set and processor architecture for

DAC 98, San Francisco, California
©1998 ACM 0-89791-964-5/98/06..$5.00

particular application domains [18, 33].

Early works in the area of processor architecture synthesis tended to employ *ad hoc* methods on small code kernels, in large part due to the lack of good retargetable compiler technology. Conte and Mangione-Smith [6] presented one of the first efforts that focused on large application codes written in high-level languages (SPEC). While they had a similar goal to ours, i.e. evaluating performance efficiency by including hardware cost, their evaluation approach was substantially different. Fisher, Faraboschi and Desoli [9] studied the variability of applications-specific VLIW processors using a highly advanced and retargetable compiler. However, their study considered small program kernels rather than complete applications. They also focused on finding the best possible architecture for a specific application or workload, rather than understanding the difference between best architectures across a set of applications.

Unlike most previous works, we use a set of complete applications written in a high-level language as benchmarks. We incorporate the role of cache memory units in machine performance into the machine model which, is essential for realistic design.

The main contribution of the research presented in this paper is the framework that makes it possible for designers to rapidly explore programmable processor design space for a given set of applications under area constraints. It enables a designer to evaluate the trade-off between area and generality of possible designs.

3 Experiment Platform

To estimate the cost of a machine configuration, we adopt a simple model described in [2]. Given the area of the issue unit, the cost of any machine configuration is a linear function of the number of function units for branch, memory, and ALU operations. For a superscalar machine, the issue unit area cannot be estimated using a simple linear model since it requires more complex logic for the on-the-fly code scheduling. We assume that the issue unit area will take $O(n^2)$ space since the complexity of dependency checking algorithm is $O(n^2)$. When a VLIW machine is considered, the issue unit area is of linear or sub-linear complexity.

The architecture chosen for the analysis, the PowerPC 604 [29], is a four-issue processor. The 604 has two simple integer ALUs and one complex integer ALU, one floating-point unit, one branch unit, and one memory unit. We assume that machine configurations with an issue width smaller than the baseline machine have at least one complex integer ALU. The area of the complex integer unit is assumed to be half of the baseline integer unit (two simple integer units and one complex integer unit). The area of issue unit is scaled down based on the area complexity ($O(n^2)$). We did not include floating-point units in any machine configurations as the benchmarks have mostly integer operations. Finally, we scaled the area to 0.35μ technology rather than the original 0.5μ technology. A set of example machine configurations and their respective area estimates are shown in Table 1.

The set of benchmarks used in this experimentation is composed of complete applications which are publically available and coded in a high-level language. The collection, known as MediaBench is composed of 21 applications culled from available image processing, communications, cryptography and DSP applications. A brief summary of benchmarks used is shown in Table 2. More detailed descriptions of the benchmarks can be found in [19].

We use the IMPACT tool suit [4] to collect performance measurements of benchmarks on various machine configurations. The IMPACT C compiler is a retargetable compiler with code optimization components especially developed for multiple-instruction-issue processors. The target machine for the IMPACT C can be described using the high-level machine description language (HMDES). A HMDES file supplied by a user is compiled by the IMPACT machine description language compiler. IMPACT provides cycle-level simulation of both the processor architecture and implementation. Figure 1 shows the flow of simulation using IMPACT tools.

Configuration	Issue	IALU	Branch	Mem	Cache	Total
(1, 1, 1, .5, .5)	0.74	3.67	6.13	5.52	1.53	31.56
(2, 2, 1, 2, 1, 1)	2.94	7.33	6.13	11.03	2.54	43.95
(4, 4, 1, 4, 2, 2)	11.76	14.66	6.13	22.07	4.55	73.15
(8, 8, 1, 8, 4, 4)	47.04	29.32	6.13	44.14	8.56	149.17
(4, 4, 2, 4, 8, 8)	11.76	14.66	12.26	22.07	16.55	91.28
(8, 8, 2, 8, 4, 4)	47.04	29.32	12.26	44.14	8.56	155.30
(8, 8, 4, 8, 8, 8)	47.04	29.32	24.52	44.14	16.55	175.55

Table 1: Machine configuration examples and their area estimates (mm^2): a machine configuration consists of (issue width, number of ALUs, number of branch units, number of memory units, size of instruction cache(KB), size of data cache(KB)). The total area includes the overhead area of 13.98 mm^2.

Benchmark	Instr.	Source	Description
JPEG encoder	13.9	Independent	JPEG image
JPEG decoder	3.8	JPEG Group	encoding/decoding
MPEG encoder	1,121.3	MPEG Simulation	MPEG-2 movie
MPEG decoder	175.5	Group	encoding/decoding
GSM encoder	184.2	Technische	European wireless
GSM decoder	73.0	Universität, Berlin	voice coding standard
G.721 encoder	274.1	Sun Microsystems,	CCITT voice
G.721 decoder	511.7	Inc.	coding standard
PGP encryption	169.9	Massachusetts Institute	encryption/
PGP decryption	155.3	of Technology	decryption
Pegwit encryption	34.0	George Barwood	encryption/
Pegwit decryption	18.5		decryption
Mipmap	47.6	University of	3-D rendering
OS-demo	9.0	Wisconsin	examples using Mesa
Texgen	83.9		graphics library
Rasta	24.4	ICSI at UC Berkeley	Voice recognition
EPIC encoder	50.3	University of	Wavelet image
EPIC decoder	7.2	Pennsylvania	encoding/decoding
ADPCM encoder	6.8	Jack Jansen	speech compression
ADPCM decoder	5.9		and decompression

Table 2: Applications used in the experiment. Dynamic instruction counts were measured on a SPARC-5, and are reported in terms of millions of instructions

4 Approach

We collected run-times (expressed as a number of cycles) of the benchmarks on 175 different machine configurations (25 cache configurations for 7 processor configurations). First we build executables of the benchmarks on seven different architectures. They are machines with a single branch unit and one of the one-, two-, four-, and eight-issue units, machines with two branch units and one of the four- and eight-issue units, and machines with four branch units

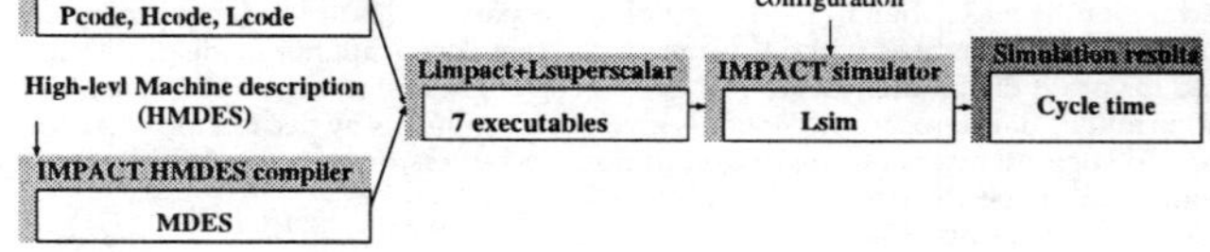

Figure 1: Performance measurement flow using IMPACT tools

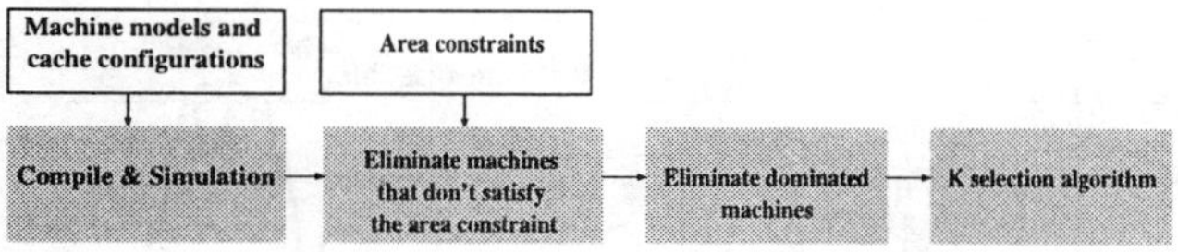

Figure 2: Global design flow

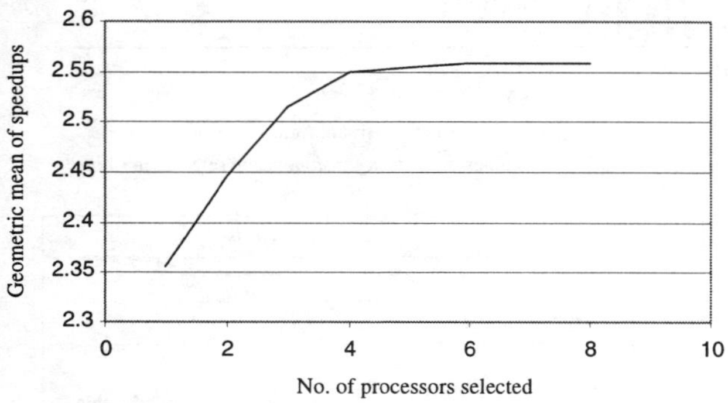

Figure 3: Performance vs. number of selected processors: Area constraint $84mm^2$

and a eight-issue unit. The IMPACT compiler generates aggressively optimized code to increase ILP for each architecture. The optimized code is consumed by the Lsim simulator. We simulate the benchmarks for a number of different cache configurations. For each executable of a benchmark, we simulate 25 combinations of instruction and data caches ranging from 512 bytes to 8 KB.

Measured run-times of benchmarks through simulations are normalized with respect to a baseline machine with one branch unit, one-issue unit, 512 bytes of instruction cache, and 512 bytes of data cache.

Simulation results feed the search engine for each area constraint, we eliminate all machines that do not satisfy the area requirement. From the machines that satisfy the area requirement, we eliminate all the dominated machines, i.e. a machine that runs slower than or equal to the speed of another machine for *all* benchmarks. Finally, we perform a K-selection algorithm (see Section 5) to select a set of machine configurations that run the benchmark set best. Figure 2 shows the global flow of design process.

The search space is relatively small due to the area constraints and the number of dominated implementations. Nevertheless, there is a possibility that the search space explodes for alternative hardware models. We use an efficient branch-and-bound based algorithm to avoid this problem.

Figure 3 shows an example measurement on performance vs. number of selected processors. Empirical results suggests that there are break points of diminishing returns for adding architectures. This will be elaborated in more detail in the next section.

5 Selection Problem Formulation

Given an area constraint and the performance of benchmarks on machines that fit into the given area, we select a subset of the machines in such a way that the geometric mean of speed-ups across all the benchmark is maximized and the subset size is kept small.

We normalize the run time with respect to a baseline since we are not interested in the sum of run-times [16]. We use geometric mean to summarize the selected machines since we normalize the measurements [13].

We define the problem using more formal Garey-Johnson format [10].

Selection problem

Instance: Given a set of n benchmarks, a_i, $i = 1, 2, ..., n$, k machine configurations, m_j, $j = 1, 2, ..., k$, the speed-up factors E_{ij} of the benchmarks a_i, $i = 1, 2, ..., n$ on the machines m_j, $j = 1, 2, ..., k$ with respect to a baseline machine and constants K and C,

Question: Is there a set M of K machine configurations, c_p, $p = 1, 2, ..., K$, such that $\sqrt[n]{\prod_{i=1}^{n} \max_{j \in M} E_{ij}} \geq C$?

To determine the constant K we divide the problem into two sub-problems, namely, w-selection problem and K-selection problem. Starting from $w = 1$ we iteratively increase w until the benefit of increasing w is less than a given threshold ρ. Formally the sub-problems are stated as follows.

w-selection problem: Given a set of n benchmarks, a_i, $i = 1, 2, ...,$ n, k machine configurations, m_j, $j = 1, 2, ..., k$, the speed-up factors E_{ij} of the benchmarks a_i, $i = 1, 2, ..., n$ on the machines m_j, $j = 1, 2, ..., k$ with respect to a baseline machine and constants w,

Maximize :

$$D_w = \sqrt[n]{\prod_{i=1}^{n} \max_{j \in P} E_{ij}}, \qquad (1)$$

where P is the selected machine set of size w.

The size of the machine set is determined by an iterative test of comparing D_w and D_{w+1}. Since the D is monotonic, we continue to evaluate D and compare them using Equation 2 until we reach a point where the benefit of the set size increase diminishes to a certain degree.

K-selection problem:

$$\min\{w | \rho \leq \frac{D_{w+1} - D_w}{D_w}, w = 1, 2, ...k - 1\}, \qquad (2)$$

where D_w is given by Equation 1 and ρ is a cut-off ratio.

6 Solution Space Exploration: Strategy and Algorithms

Due to space limitation we present only the top level algorithm for system-level synthesis in Figure 4. Considering that the run-time of simulations for 20 benchmarks on 175 machine configurations is about a week, we can tolerate a longer search time to find an optimal selection result. Generally, the size of the search problem can be dramatically reduced by eliminating machine configurations that do not satisfy a given area constraint and those that are dominated by at least one other machine. Consequently, a smaller number of machines needs to be considered.

The search for optimum solution is organized using an implicit enumeration method. In particular, we adopt a branch-and-bound algorithm shown in Figure 4 to speed up the selection.

The algorithm consists of branching and evaluation. The branching step takes the current state of search and generates a number of new nodes by adding an available machine (i.e., one that has not been considered in particular search path of the search tree). It determines if adding a machine to the current state of selection can

```
for a set of machine configurations {
  for a set of benchmarks {
    generate an executable of a benchmark for the target machine;
    measure speed-up factors with respect to the baseline machine;
  }
}
eliminate all machines that don't satisfy area constraint;
eliminate all machines that are dominated by at least one other machine;
R = 1;
i = 1;
D_{i-1} = 1;
while ( (R >= ρ) && (i <= k) ) {
  D_i = branch-and-bound( D_{i-1}, i );
  R = (D_i - D_{i-1}) / D_i;
  D_{i-1} = D_i;
  i++;
}
```

Figure 4: System-level synthesis of application-specific programmable machines

result in a better solution than the current best solution found. Initially, the current best solution is set to the previous best solution. The previous best solution is the best solution found for the machine set size less than the current search size by one. The bounding function compares the current node and a candidate processor with the best node of the same size found. The node size is the number of processors. If the current node and the candidate are dominated by the best node then we bound the search. We compute the lower bound of the geometric mean of the maximum speed-up factors of each benchmark. The lower bound is obtained by using a steepest descent algorithm. The steepest descent algorithm selects machines in the order that the biggest improvement can be achieved. If the estimate is greater or equal to the current best solution, we have an opportunity to find a better solution than the current best solution by exploring the search path. Otherwise, there is less chance of obtaining a better solution. We sort the search order based on the lower bound so as to increase the bounding rate.

7 Experimental Results

We evaluate the tools and algorithms by running experiments from $30mm^2$ to $200mm^2$. The implementation technology is assumed to be 0.35μ. For each area constraint, we obtain the optimum set of machine configurations for cut-off values (ρ's as defined in Equation 2) 0.1, 0.05, 0.01, and 0.005.

Figure 5(a) shows experimental results using the cut-off value of 0.05. The thicker line shows the number of machines that are left after eliminating dominated machines. The thinner line indicates the number of machines required to cover all the benchmarks under area constraints. We see that more machine configurations are needed when less area is given. In other words, with less area, each processor configuration is more specialized. On the other hand, the more area we have, the more general the processor we can design. The results suggest that when more than $100mm^2$ of area is available, there is little advantage in having more than one architecture. Moreover, for the given compiler technology and benchmarks, there is no need to have more than $100mm^2$ of area since the speed-up increase achieved by machines greater than $100mm^2$ are minimal.

The overall performance comparison between all configurations and selected configurations are shown in 5(b). There are three distinctive points where the speed-up increase rate changes. Up to the area $57mm^2$, we see rapid performance increase, which is mainly due to increased amount of cache memories. From $57mm^2$ to $101mm^2$, the measurement shows modest increase of performance due to increased issue width. For the processors larger than

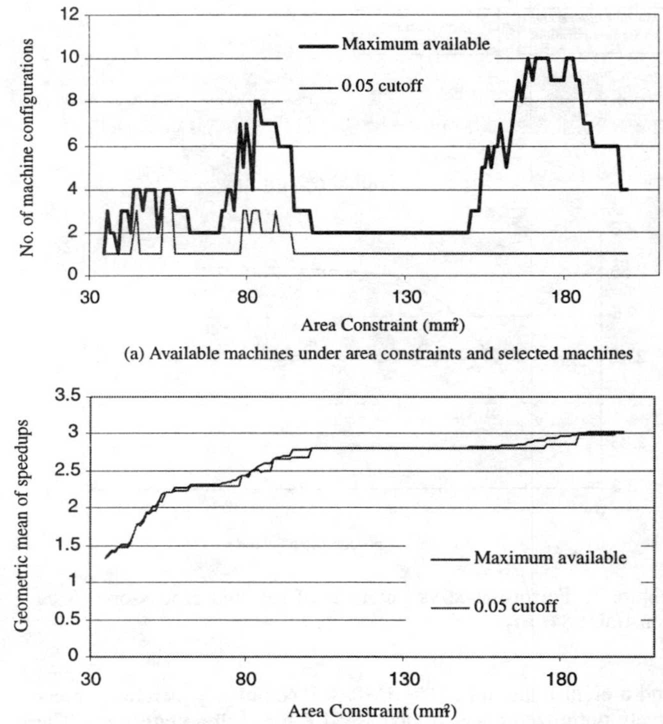

(a) Available machines under area constraints and selected machines

(b) Performance of *all* available machines and selected machines

Figure 5: Quadratic complexity issue unit area model: Selected processor configurations and their performance (cut-off value: 0.05)

$101mm^2$, the performance increase is minimal. One of the underlying reasons that causes the phenomenon is that the ILP found by the compiler and hardware scheduler is fully exploited by having a certain amount of hardware, thereby performance increase possibility is exhausted. The limitation of performance increase in the face of increased area illustrates either the limitation of the current compiler technology or the inherent lack of ILP in the benchmarks, which we are not able to answer at this point. Note, however, that the measurement is not for a single processor. Smaller area cases tend to have more than one architectures which are more application specific.

Experimental results for the cut-off values 0.1, 0.05, 0.01, and 0.005 are given in Figure 6. Smaller cut-off values result in machine configuration sets that are more tuned to each application. In general, however, smaller cut-off values do not result in significant performance increase. This conclusion is valid when the available area is bigger than $100mm^2$. In most cases, the cut-off value of 0.05 gives good trade-off between the number of machine configurations and performance.

A sample set of speed-up factors of each benchmark are shown in Table 3. They are snapshots of experimental results summarized by the line graphs in Figure 6. The table contains maximum speed-up factors for three cut-off values (0.05, 0.01, and 0.005) and three area constraints ($85mm^2$, $100mm^2$, and $169mm^2$). Note that the area constraints are not actual areas. We find machines under the given area constraints. Table 4 gives the number of machine configurations selected for the snapshot cases shown in Table 3. The actual areas of the selected machines are given in column 4 of the table. The combination of components for the selected machines are shown in column 3.

Figure 7 shows the results when the liner complexity issue unit

Benchmark	cut-off:0.05			cut-off:0.01			cut-off:0.005		
	84	100	169	84	100	169	84	100	169
JPEG enc	3.02	3.51	3.87	3.02	3.65	3.87	3.02	3.65	3.87
JPEG dec	3.56	3.91	4.27	3.56	4.23	4.27	3.56	4.23	4.27
MPEG enc	2.45	2.56	2.57	2.45	2.56	2.57	2.45	2.42	2.57
MPEG dec	3.05	3.55	3.68	3.05	3.55	3.95	3.05	2.56	3.95
GSM enc	2.22	2.30	2.31	2.22	2.30	2.32	2.24	2.94	2.32
GSM dec	2.89	2.99	3.00	2.89	2.99	3.19	2.91	2.99	3.19
G.721 enc	2.68	3.38	3.38	2.88	3.38	3.38	2.88	3.38	3.38
G.721 dec	2.71	3.44	3.44	3.04	3.44	3.44	3.04	3.44	3.44
PGP enc	2.78	3.08	3.21	2.83	3.15	3.21	2.83	2.26	3.21
PGP dec	2.88	3.22	3.38	3.02	3.28	3.38	3.02	2.15	3.46
Pegwit enc	2.64	2.66	2.82	2.64	2.79	2.82	2.64	3.28	2.82
Pegwit dec	2.63	2.68	2.84	2.63	2.82	2.84	2.62	2.79	2.84
Mipmap	2.81	2.69	2.79	2.81	2.94	3.10	2.81	3.56	3.10
Osdemo	2.22	2.40	2.47	2.25	2.42	2.47	2.25	2.82	2.47
Texgen	2.11	2.31	2.40	2.11	2.40	2.40	2.11	2.42	2.40
Rasta	2.13	2.22	2.27	2.13	2.26	2.31	2.13	2.31	2.31
EPIC enc	1.93	1.78	2.17	1.93	2.06	2.17	2.04	2.17	2.30
EPIC dec	1.70	1.76	1.77	1.70	1.76	1.83	1.70	1.76	1.83
ADPCM enc	2.30	2.31	2.31	2.30	2.31	2.43	2.30	2.42	2.44
ADPCM dec	2.33	2.34	2.42	2.33	2.34	2.55	2.34	2.40	2.65

Table 3: Snapshots of speed-ups of benchmarks at various cut-off values for the given area constraint ($84mm^2$, $100mm^2$, $169mm^2$) (NOTE: Actual area of designs are given in Table 4)

Cut-off Value	Max area	Configuration	Actual area
0.05	84	(2, 2, 1, 2, 8, 8)	61.11
		(4, 4, 1, 4, 4, 4)	81.89
		(4, 4, 2, 4, 1, 4)	83.24
	100	(4, 4, 2, 4, 8, 4)	94.36
	169	(4, 4, 2, 4, 8, 8)	100.71
0.01	84	(2, 2, 1, 2, 8, 8)	61.11
		(4, 4, 1, 4, 4, 4)	81.89
		(4, 4, 1, 4, 8, 1)	83.46
		(4, 4, 2, 4, 1, 4)	83.24
	100	(4, 4, 1, 4, 8, 8)	94.58
		(4, 4, 2, 4, 8, 4)	94.36
	169	(4, 4, 2, 4, 8, 8)	100.71
		(8, 8, 2, 8, 4, 4)	166.33
0.005	84	(2, 2, 1, 2, 8, 8)	61.11
		(4, 4, 1, 4, 1, 8)	83.46
		(4, 4, 1, 4, 4, 4)	81.89
		(4, 4, 1, 4, 8, 1)	83.46
		(4, 4, 2, 4, 1, 4)	83.24
		(4, 4, 2, 4, 4, 1)	83.24
	100	(4, 4, 1, 4, 8, 8)	94.58
		(4, 4, 2, 4, 4, 8)	94.36
		(4, 4, 2, 4, 8, 4)	94.36
	169	(4, 4, 2, 4, 8, 8)	100.71
		(8, 8, 2, 8, 1, 8)	168.68
		(8, 8, 2, 8, 4, 4)	166.33
		(8, 8, 4, 8, 1, 2)	166.70

Table 4: Machine configuration and actual area of designs for the snapshot cases in Table 3: Column 3 shows machine configurations in the form of (issue width, number of ALUs, number of branch units, number of memory units, size of instruction cache(KB), size of data cache(KB)). Actual areas are given in the 4th column.

area model is assumed. The results suggest that the machine configuration selection problem has no strong dependence to an issue area model used. Although we observe that there is shift to smaller areas, the results are essentially equivalent to the results based on the quadratic complexity issue unit area model.

In summary, we found that under the machine models and machine configuration choices described in this paper, when more than $100mm^2$ of area is available, there is no advantage in having more than one architecture to be able to build application-specific systems for all the benchmarks. Moreover, for the given compiler technology and benchmarks, there is very little need to have more than $100mm^2$ of area since the speed-up increase achieved by machines greater than $100mm^2$ are minimal one. Notable exception is that for highly cost sensitive designs we observe a need for a small number of specialized architectures which occupy smaller areas.

8 Conclusion

The arrival of production quality ILP compilers and commercial DSPs with VLIW architecture stimulated the idea of programmable processors that are aggressively tuned to specific applications. The assumption behind the idea is that there are ways of designing programmable processors that can exploit the run-time characteristics of specific applications. The run-time characteristics include the available ILP, demand on various hardware components such as cache memory units, register files, and the number of functional units. Under the assumption, a microprocessor can be designed by adding hardware components tailored to a specific application so that it can execute the single application extremely well.

We ran extensive experiments on a framework based on the key paradigms of CAD and architecture communities. This combination enabled us to gain valuable insights about design and use of application specific programmable processors for modern applications. We evaluated 175 machine configurations on 20 benchmarks under the area constraint ranging from $30mm^2$ to $200mm^2$. For each area constraint, we obtain the optimum set of machine configurations for a number of cut-off values. The run time of the entire synthesis process was about a week on a single HP workstation.

We found that the framework introduced in this paper can be

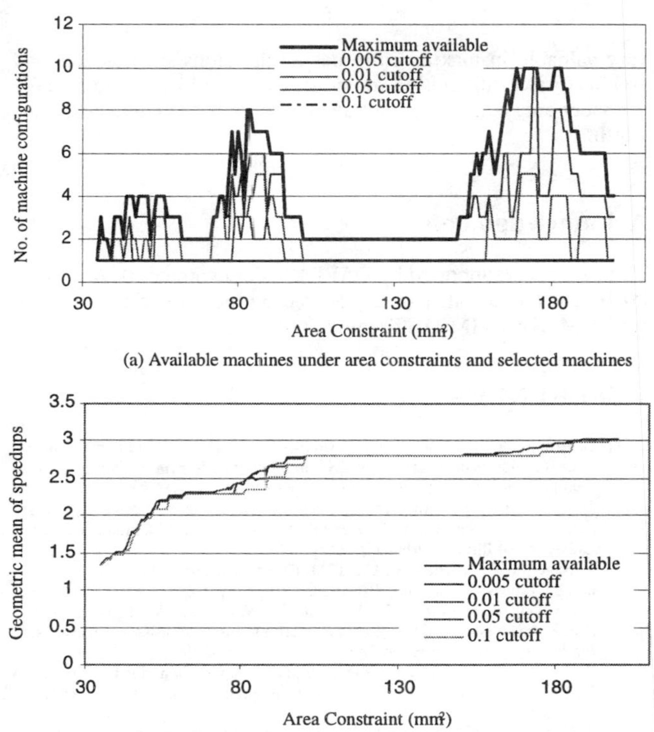

(a) Available machines under area constraints and selected machines

(b) Performance of *all* available machines and selected machines

Figure 6: Quadratic complexity issue unit area model: selected processor configurations and their performance for 4 cut-off values

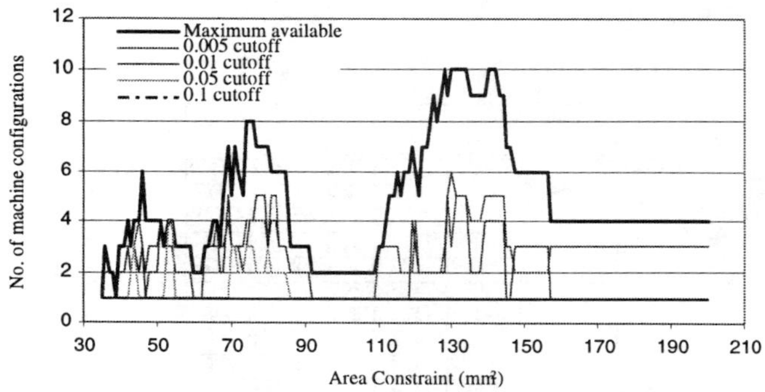

(a) Available machines under area constraints and selected machines

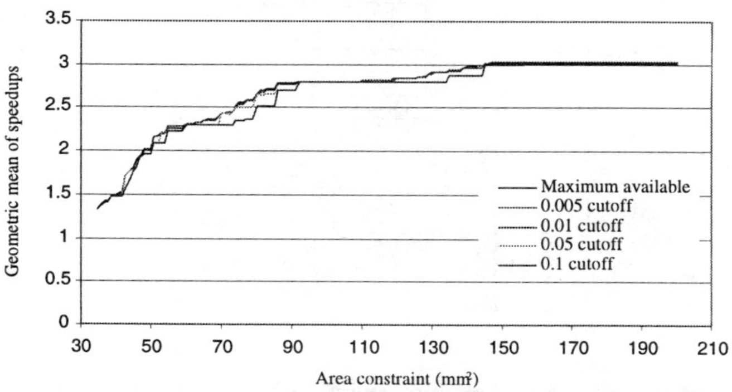

(b) Performance of *all* available machines and selected machines

Figure 7: Linear complexity issue unit area model: selected processor configurations and their performance for 4 cut-off values

very valuable in making early design decisions such as area and architectural configuration trade-off, cache and issue width trade-off under area constraint, and the number of branch units and issue width.

Acknowledgments

This work was supported by DARPA under contract F04701-97-C-0010. The authors also greatly appreciate the assistance of Professor W. M. Hwu's IMPACT team at UIUC.

REFERENCES

[1] G. Araujo, A. Sudarsanam, and S. Malik. Instruction set design and optimizations for address computation in DSP architectures. In *Proceedings of International Symposium on System Synthesis*, pages 102–107, 1996.

[2] D. C. Argyres. Performance and cost analysis of the execution stage of superscalar microprocessors. Master's thesis, Department of Computer Science, University of Illinois, Urbana IL, May 1995.

[3] S. Banerjia, W. A. Havanki, and T. M. Conte. Treegion scheduling for highly parallel processors. In *Euro-Par*, Passau, Germany, 1997.

[4] P. P. Chang, S. A. Mahlke, W. Y. Chen, N. J. Warter, and W. m. W. Hwu. IMPACT: An architectural framework for multiple-instruction-issue processors. In *International Symposium on Computer Architecture*, 1991.

[5] R. P. Colwell, R. P. Nix, J. J. O'Donnell, D. B. Papworth, and P. K. Rodman. A VLIW architecture for a trace scheduling compiler. In *Proceedings of ASPLOS-II*, pages 180–192, 1982.

[6] T. Conte and W. Mangione-Smith. Determining cost-effective multiple issue processor designs. In *International Conference on Computer Design*, 1993.

[7] T. M. Conte, K. N. P. Menezes, and S. W. Sathaye. A technique to determine power-efficient, high-performance superscalar processors. In *Proceedings of the Twenty-Eighth Hawaii International Conference on System Sciences*, volume 1, pages 324–333, 1995.

[8] J. A. Fisher. Trace scheduling: A technique for global microcode compaction. *IEEE Transactions on Computing*, C-30:478–490, 1981.

[9] J. A. Fisher, P. Faraboschi, and G. Desoli. Custom-fit processors: Letting applications define architectures. In *International Symposium on Microarchitectures*, Paris, France, 1996.

[10] M. R. Garey and D. S. Johnson. *Computers and Intractability: A Guide to the Theory of NP-Completeness*. W. H. Freeman and Company, New York, NY, 1979.

[11] G. Goossens, J. Van Praet, D. Lanneer, W. Geurts, , et al. Embedded software in real-time signal processing systems: design technologies. *Proceedings of the IEEE*, 85(3):436–454, March 1997.

[12] C. Hansen. MicroUnity's MediaProcessor architecture. *IEEE Micro*, 17:34–41, 1997.

[13] J. L. Hennessy and D. A. Patterson. *Computer Architecture: A Quantitative Approach*. Morgan Kaufman, San Francisco, CA, 1993.

[14] R. D. Hof and O. Port. Silicon dreams. *Business Week International Edition*, May 13 1996.

[15] P. Y. Hsu. Highly concurrent scalar processing. Technical Report CSG-49, Coordinated Science Laboratory, University of Illinois at Urbana-Champaign, 1986.

[16] R. Jain. *The Art of Computer Systems Performance Analysis*. Wiley, 1991.

[17] P. Kalapathy. Hardware-software interactions on MPACT. *IEEE Micro*, 17:20–26, 1997.

[18] K. Kim, R. Karri, and M. Potkonjak. Heterogeneous built-in resiliency of application specific programmable processors. In *International Conference on Computer-Aided Design*, pages 406–411, 1996.

[19] C. Lee, M. Potkonjak, and W. H. Mangione-Smith. Mediabench: A tool for evaluating and synthesizing multimedia and communications systems. In *International Symposium on Microarchitectures*, 1997.

[20] R.B. Lee and M.D. Smith. Media processing: A new design target. *IEEE Micro*, 17:6–9, 1997.

[21] R. Leupers and P. Marwedel. Retargetable generation of code selectors from HDL processor models. In *European Design and Test Conference*, pages 140–144, 1997.

[22] S. Liao, S. Devadas, K. Keutzer, and S. Tjiang. Instruction selection using binate covering for code size optimization. In *International Conference on Computer-Aided Design*, pages 393–399, 1995.

[23] C. Liem, T. May, and P. Paulin. Instruction-set matching and selection for dsp and asip code generation. In *The European Design and Test Conference*, pages 31–37, 1994.

[24] W. m. W. Hwu, S. A. Mahlke, W. Y. Chen, P. P. Chang, N. J. Warter, R. A. Bringmann, R. G. Ouellette, R. E. Hank, T. Kiyohara, G. E. Haab, J. G. Holm, and D. M. Lavery. The superblock: An effective technique for VLIW and superscalar compilation. *Journal of Supercomputing*, 1993.

[25] S. A. Mahlke, D. C. Lin, W. Y. Chen, R. E. Hank, and R. A. Bringmann. Effective compiler support for predicated execution using the Hyperblock. In *International Symposium on Microarchitecture*, 1992.

[26] P. Marwedel. Processor-core based design and test. In *Asia and South Pacific Design Automation Conference*, pages 499–502, 1997.

[27] P.G. Paulin, C. Liem, M. Cornero, F. Nacabal, , et al. Embedded software in real-time signal processing systems: application and architecture trends. *Proceedings of the IEEE*, 85(3):419–435, March 1997.

[28] A. Peleg and U. Weiser. MMX technology extension to the Intel architecture. *IEEE Micro*, 16(4):42–50, August 1996.

[29] S. P. Song and M. Denman. The PowerPC 604 RISC microprocessor. Technical report, Motorola Inc. and International Business Machines, 1994.

[30] A. Sudarsanam and S. Malik. Memory bank and register allocation in software synthesis for ASIPs. In *International Conference on Computer-Aided Design*, pages 388–392, 1995.

[31] J. Turley. SA-1100 puts PDA on a chip. *Microprocessor Report*, 11(12), September 1997.

[32] J. Turley and H. Hakkarainen. TI's new 'C6x DSP screams at 1,600 MIPS. *The Microprocessor Report*, 11:14–17, 1997.

[33] W. Zhao and C.A. Papachristou. An evolution programming approach on multiple behaviors for the design of application specific programmable processors. In *European Design and Test Conference ED&TC 96*, pages 144–150, 1996.

Panel: User Experience with High Level Formal Verification

Convenors: Randy E Bryant - CMU, USA and Gerry Musgrave - Brunel University, UK

Formal Verification is a "hot topic" for the user and vendor community. It has moved from the research community to the industrial domain in a very short time. Everyone wants to know more about how effective the techniques are. This experienced user panel will attempt to address your concerns in an open and frank way. They will give their personal opinions and not the commercial hype that so often heralds a new era.

Formal verification methods is a generic term which currently covers equivalence checking, both structural and sequential, model checking and formal proof reasoning. To date only the first category has begun to be accepted as an adjunct to the verification process particularly at the gate level. Yet the higher levels of abstraction have the potential for greater fertility from use of the other classifications of formal tools.

Leading edge industries are beginning to explore the value of the more sophisticated techniques. Often their learning curve has been via the structural equivalence checking route, which when operational in an ASIC division has valuable lessons for the user and CAD communities.

Higher level techniques would be those which ensure consistency of constraints such as, reset requirements, safety requirements, determinacy and other parameter factors across all levels of implementation. It moves this form of validation from the post design phase to the proactive design aid. This is a much bigger change in design philosophy compared to equivalence checking yet it is a long way from the purist view of formal verification. This latter approach requires a formal proof engine to validate all design decisions against a reasoned specification. Hence the road map for formal verification has just started and there is a long way to go. But the confidence expressed by the many vendor announcements indicates that this is a highway we will be travelling along.

The panel will address such questions as: -

- How effective are these techniques in solving the simulation bottleneck?
- How easy is it to embed them in the existing design flow or do we have to change the design flow to gain maximum improvement in time-to-market?
- Is there a re-education requirement for the design community in order to benefit from these tools?
- How do we transform from using formal verification tools as a post design checking to become a proactive design aid in achieving quality designs in a shorter time?
- Are VHDL and VERILOG specifications adequate input to these design decision aids?
- Are these the tools that will give IP real credibility?

These and many more questions, mainly from the audience, will be put to our international panellists who have had experience in the field from a number of industry sectors.

The Panel:

Fumiyasu Hirose, Fujitsu, Japan

Michael Payer, Siemens Semiconductors, Germany

Pierre Aulagnier, Cisco, USA

Alan Silbert, Nortel, Canada

John Van Tassel, TI, USA

Chair G. Musgrave
Uxbridge, UK

What's Between Simulation and Formal Verification? (Extended Abstract)

David L. Dill
Stanford University

Abstract

This embedded tutorial surveys some possibilities for verification techniques that combine conventional simulation and ideas, techniques, and algorithms from formal verification, to obtain better functional test coverage of large designs.

1 Introduction

There is a desperate need for an immediate practical solution to the problem of verifying large designs. Simulation-based verification has not been able to keep up with increasing design complexity. In spite of increasing simulation and emulation speed, the number of problem cases that can be covered by manually generated functional test vectors or pseudo-random functional testing is declining. The amount of human and computational effort devoted to verification is increasing rapidly, time to market is being delayed, and bad bugs are showing up after tape-out and in products.

Many designers hope that formal verification will solve the problem. Formal verification will help (indeed, it will be indispensible) in niches where it is particularly effective. However, it will not provide a general solution to the verification problem in the near future. Except for equivalence checking tools, which are targetted at low-level design errors (e.g. in hand-optimized net-lists), formal verification tools do not currently scale well to large designs. Effective use of these techniques requires reducing the core computational problem by a variety of methods, none of which is supported well by current design practices. For example, model checking a small part of a design needs a detailed specification of the interface behavior of the block being verified, so that only legal inputs are checked. Real designs almost never have up-to-date, detailed specifications of the interfaces between low-level blocks, so verification often requires painful reverse engineering of the interface specification. Another approach to complexity reduction is to abstract the circuit behavior. But abstracting parts of a complex design is a frustrating process, because of false error reports resulting from unanticipated dependencies on design details which are changed by the abstraction.

DAC 98, San Francisco, California
©1998 ACM 0-89791-964-5/98/06..$5.00

I believe that design practices and formal verification techniques will evolve together to resolve these and other problems, at least in part. However, there is every reason to believe that it will be a long, difficult journey to the point where formal verification tools displace traditional simulation-based verification.

It is likely that the most practically effective verification methods in the medium-term future will be *"semi-formal"* methods, which combine ideas from conventional and formal verification to achieve much better test coverage of designs than conventional verification, while avoiding the scaling and methodology problems inherent in formal verification methods. Although work with along these lines has been going on for several years, the collection of ideas has not yet evolved into a "field of study." Even a modest attempt to compare techniques reveals that many obvious ideas have not yet been tried. We should expect research in this area to bloom in the next few years, accompanied by rapid commercialization and use of the best ideas.

2 The spectrum

If we imagine a spectrum, as in Figure 1, ranging from conventional simulation to full formal verification. Near the simulation end of the spectrum is conventional simulation coupled with *coverage analysis*, which provides data on how thoroughly a design has been exercised. Coverage metrics can be used to guide the manual definition of test vectors, as well as evaluating the effectiveness of pseudo-random strategies.

Quite a variety of metrics have been proposed. HDL-based measures include line coverage, statement coverage, branch and conditional coverage, and path coverage. There are coverage metrics that record all the values that selected variables have taken, whether each bit has changed in each direction (toggle coverage). Other metrics are based on state machine coverage, including state and transition coverage of individual state machines or of combined state machines. Commercial tools are available to report on many of these coverage metrics. Coverage can also be based on an abstraction of the design, such as a state machine provided by the designer; this would allow a designer to indicate exactly the cases that he or she feels need to be exercised.

Another class of metrics comes from the viewpoint of manufacturing test: try to characterize a class of "design faults," analogous to fault models from test, and measure the coverage of these "faults" achieved by a test case. Sometimes, fault models from manufacturing test are used, unmodified, as a rough measure of coverage for functional veri-

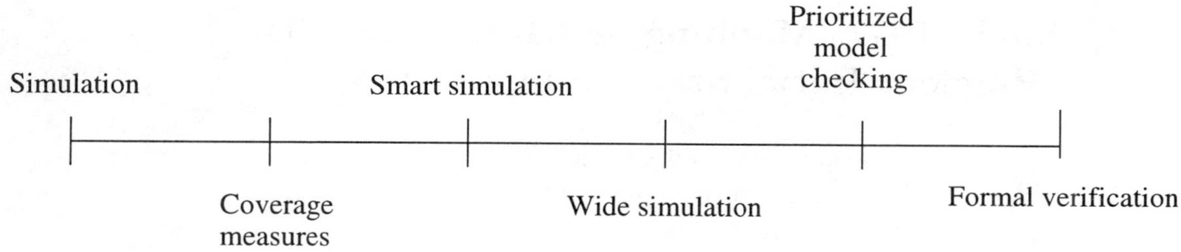

Figure 1: A spectrum of verification techniques.

fication (this is easy because of the ready availability of fault simulators).

A bit further to the right on our spectrum are techniques that I will dub *"smart simulation,"* which *generate* functional tests based on a coverage metric. Test generation can be done off-line, storing a set of test vectors in a file, or on-line, generating new inputs to the design on-the-fly. Most existing research seems to be driven by the desire to cover an abstract state machine, provided by the designer or as part of a requirements specification, or selected "corner cases" that are part of the specification of the design. It seems that a set of test vectors could be generated to maximize any given coverage model, although most such ideas have not been explored.

Even further towards the formal end of the spectrum are techniques that do *"wide simulation"* by symbolically representing large sets of states in relatively few simulations. A variety of symbolic representations could be used. Current systems use Boolean decision diagrams (BDDs) or logical expressions which have operators on higher-level data types than bits or bit vectors, such as integers, or uninterpreted functions to abstract away datapath elements.

Finally, relatively close to the formal verification end of the spectrum are model-checking techniques that explore the state space in parts, possibly using heuristics that try to find design errors early in search. As with the other approaches, only the initial possibilities have been explored.

3 Concluding remarks

It will be a major challenge to produce meaningful research results in this area, especially for academic researchers, because of the difficulty of quantifying the quality of results produced by a particular idea. The ultimate measure of quality is bug-finding ability on real designs, but measuring this reliably will require significant resources. Dealing with large designs requires lots of engineer or researcher time, especially when using prototype tools. Even getting access to realistic designs is difficult for researchers in academia. Moreover, conclusions drawn from a small number of designs will be suspect, because all of these techniques are likely to be very sensitive to the particular characteristics of a design and its designers.

It seems clear to me from conversations with my colleagues that semi-formal verification techniques will soon receive a great deal of attention from researchers. There will be commercial interest and major practical impact in the not-too-distant future as well.

What does this portend for formal verification? It will grow in importance, but not provide a universal solution. When applicable, formal verification provides value that other techniques cannot match. It can achieve vastly greater behavioral coverage, and (subject to assumptions about models and requirements) prove the absences of some classes of errors. However, hybrid verification methods may clear a long-term path to greater acceptance of formal verification by encouraging changes in design methodology that make formal verification easier.

A full version of this paper, with references, will appear on my website: `http://verify.stanford.edu`.

Optimal FPGA Mapping and Retiming with Efficient Initial State Computation

Jason Cong and Chang Wu
Department of Computer Science
University of California, Los Angeles, CA 90095

Abstract

For sequential circuits with given initial states, new equivalent initial states must be computed for retiming, which unfortunately is NP-hard. In this paper we propose a novel *polynomial time algorithm for optimal FPGA mapping with forward retiming* to minimize the clock period with guaranteed initial state computation. It enables a new methodology of separating forward retiming from backward retiming to avoid time-consuming iterations between retiming and initial state computation. Our algorithm compares very favorably with both of the conventional approaches of separate mapping followed by retiming [1, 8] and the recent approaches of combined mapping with retiming [12, 2]. It is also applicable to circuits with partial initial state assignment.

1 Introduction

Retiming is a well known technique to reduce the clock period by repositioning flipflops (FFs) in sequential circuits originally proposed by Leiserson and Saxe [8]. Many studies have been done on combining retiming with logic optimization, circuit partitioning and technology mapping, e.g., [11, 16, 10]. For lookup-table (LUT) based FPGA designs, a significant advancement was made by Pan and Liu [12] on simultaneous optimal technology mapping with retiming for clock period minimization. For designs with initial states, however, equivalent initial states need to be computed for retiming, which unfortunately is NP-hard [14].

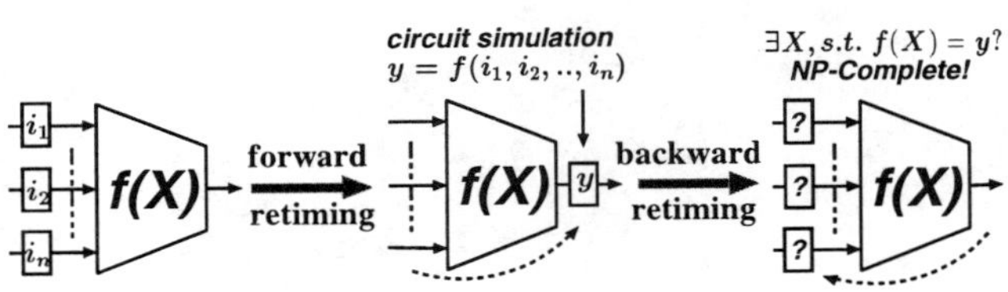

Figure 1: Retiming and initial state computation. Each small rectangle represents a FF.

A general retiming procedure usually involves two kinds of FF movements. *Forward retiming* moves FFs from the inputs of some gates to their outputs, while *backward retiming* moves FFs in the opposite direction. As shown in Figure 1, the equivalent initial state of forward retiming can be computed easily using circuit simulation. The initial state for backward retiming, however, needs to solve a satisfiability problem which is in general NP-complete.

Several heuristics for initial state computation based on ATPG (automatic test pattern generation) techniques [17, 4, 9] or STG (state transition graph) traversal [15] have been proposed in recent years. However, none of them guarantees to find an equivalent initial state for a given retiming. For example, L. Stok, et. al. [4] proposed to try to exclude backward retiming by minimizing the maximum retiming value. If backward retiming is still needed for a target clock period, ATPG based justification was proposed to compute equivalent initial states with possible iterations of retiming and initial state computation. However, ATPG itself is also NP-complete [5, 6]. We also observe that minimizing the maximum retiming value as formulated in [4] does not always lead to a simpler initial state computation problem. Our experience shows that unless a complete forward retiming solution can be found, the effort of initial state computation depends on the number of nodes with backward retiming, instead of the maximum retiming value. Alternatively, Touati and Brayton [15] proposed an initial state computation algorithm based on STG traversal. Their algorithm can compute equivalent initial states under the condition that there exists an input sequence for the original initial state going back to itself in the STG of the original circuit [15]. However, unless provided externally, to determine the existence and compute such an input sequence needs to search the STG which may have an exponential number of nodes. Although representing the reset logic explicitly can avoid the STG traversal [15], the extra cost of the reset logic separated from FFs after retiming can be high. Furthermore, such kind of reset logic may restrict possible retiming of the original circuit, thus reduce the potential of retiming optimization.

In order to avoid the high complexity of initial state computation for *general* retiming, we propose to perform LUT-based FPGA mapping with forward retiming. Our main contribution is the development of the first polynomial time optimal algorithm for FPGA mapping with forward retiming, which has immediate benefit of guaranteed equivalent initial state computation in linear time. With this algorithm, we can try to push FFs back to PIs as much as possible to enlarge the solution space of mapping with forward retiming, thus achieve better results. One important advantage of our approach is that during the preprocessing of backward retiming we only need to focus on the initial state computation without considering the effect on the clock period. As a result we can still achieve effective retiming for circuit optimization without time-consuming iterations between retiming and initial state computation, avoiding the slow STG traversal procedure or any extra reset logic in existing approaches [9, 17, 14, 15]. Our algorithm compares

DAC 98, San Francisco, California
©1998 ACM 0-89791-964-5/98/06..$5.00

very favorably with both of the conventional approaches of separate mapping with retiming [1, 8] and the recent approaches of combined mapping with retiming [12, 2]. It is also applicable to circuits with partial initial state assignment.

The rest of the paper is organized as follows. Section 2 is the problem formulation and definitions and an overview of our algorithm. Our algorithm is presented in Section 3. The experimental results are presented in Section 4 and the conclusions and our future work in Section 5. Due to the page limit, the details of the algorithm and the proofs of the theorems are left out. They are available in [3].

2 Problem Formulation and Definitions

Given a *sequential* circuit, the technology mapping problem for K-LUT based FPGAs is to construct an *equivalent* circuit consisting of K-LUTs and flipflops (FFs), such that both of the circuits generate the same output sequence for any input sequence, starting from their individual initial states, respectively. We propose to study the following problem in this paper.

Problem 1 *Given a sequential circuit with initial state s_0, find an equivalent LUT circuit with initial state s_0^r and its clock period is minimum under forward retiming.*

As in [12, 2], instead of solving the optimization Problem 1 directly, we shall solve its decision version and use binary search to get the minimum clock period.

Problem 2 *Given a sequential circuit with initial state s_0 and target clock period Φ, find an equivalent LUT circuit with initial state s_0^r and the clock period of no more than Φ under forward retiming.*

In this paper, sequential circuits are represented as retiming graphs. The retiming graph $G(V, E, W)$ of a sequential circuit is a directed graph, where V is the set of nodes, E is the set of edges and W is the set of edge weights [8]. Each node in V represents a gate, a primary input (PI) or a primary output (PO) in the original circuit. Each edge $e(u, v)$ in E represents a directed connection from node u to node v. The *weight* $w(e)$ of an edge e is the number of FFs on the connection represented by the edge. The *path weight* $w(p)$ of a path p is the sum of all edge weights on the path. Under the unit-delay mode[1], *delay* $d(v) = 1$ for every internal node v, or 0 for every PI or PO. *Path delay* $d(p)$ of a path p is the sum of all node delays on the path.

For a target clock period Φ, we define the edge length $length(e)$ to be $d(v) - \Phi \cdot w(e)$ for an edge $e(u, v)$. The path length $length(p)$ is $\sum_{e\in p} length(e)$. In an LUT network M, a node's **l-value** is defined to be $l_M(v) = max\{length(p)\}$ for all paths p from PIs to v. It can be proved that,

Theorem 1 *Given an LUT network M and a target clock period Φ, the clock period of M under forward retiming is no more than Φ if and only if the l-values $l_M(v) \leq \Phi$, for every LUT $v \in M$.*

Let M be one solution of mapping with forward retiming, thereafter called *FRT mapping solution*. In M, the output of an LUT rooted at v is generally $r(v)(\geq 0)$ cycles ahead

[1] We use the unit-delay mode and assume synchronous sequential circuits with single-phase clock and edge-triggered flipflops in this work.

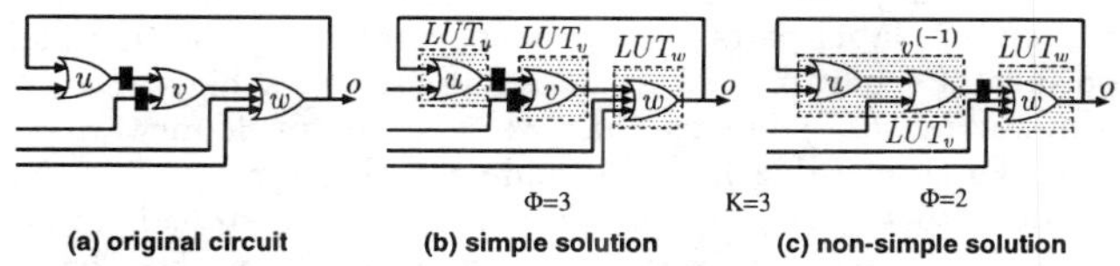

Figure 2: Simple vs. non-simple mapping solutions. Each dotted block represents a 3-LUT.

the output of v in the original circuit due to forward retiming. A mapping solution with $r(v) \leq 0$ for every LUT root v is called a *simple mapping solution* [12]. One important property of simple mapping solution is that the path weight of any path from PIs to a node is the same as that in the original circuit. Thus, we only need to focus on reducing the path delay for path length and clock period minimization. It was shown in [12] that, there always exists a *simple* mapping solution whose clock period will be the minimum after a second step of *general* retiming. However, there may not exist any simple FRT mapping solutions with the minimum clock period after optimal *forward* retiming. For the example shown in Figure 2(a) with $K = 3$, it is not difficult to verify that there does not exist a simple FRT mapping solution with $\Phi = 2$ under forward retiming, while there does exist a non-simple FRT mapping solution with $\Phi = 2$ as shown in Figure 2(c). As a result, the minimum l-values among all possible FRT mapping solutions are much more difficult to compute, because the path weight from PIs to a node will also be changed which makes the monotone property of l-values shown in [2] no longer holds. To overcome these difficulties we separate the l-value into two values such that the first one has the monotone property defined in [2] and the second one can be computed according to the first one.

Definition 1 *Given a mapping solution M and a target clock period Φ, the number of FFs moved forward across LUT v, denoted $r_M(v)$, is called the* **forward retiming value** *of v in M.*[2] *The* **s-value** *$l_M^s(v) = l_M(v) - \Phi \cdot r_M(v)$ represents the l-value in the corresponding simple mapping solution of M after pushing back the $r_M(v)$ FFs to the inputs of v, where $l_M(v)$ is the l-value of v in M.*

According to Theorem 1 we have that,

Corollary 1 *A mapping solution M has a clock period of no more than a given Φ under forward retiming, if and only if $l_M^s(v) + \Phi \cdot r_M(v) \leq \Phi$ for all the LUT roots v in M.*

Definition 2 *Given a circuit and a target clock period Φ, if there exists a FRT mapping solution with clock period of no more than Φ under forward retiming (called a* feasible solution*), the* **s-label** *$L^s(v)$ is defined to be the minimum $l_M^s(v)$ among* all feasible FRT mapping solutions *and, the* **r-label** *$R(v)$ is defined to be the minimum $r_M(v)$* among all feasible FRT mapping solutions *with $l_M^s(v) = L^s(v)$. The* **node label pair** *is defined to be a 2-tuple $(L^s(v), R(v))$.*

We shall show that $L^s(v)$ has the important monotone property in § 3.2. According to Corollary 1, we can solve the decision Problem 2 by computing the node label pairs and checking if $L^s(v) + \Phi \cdot R(v) \leq \Phi$ holds for every node.

[2] In [8], the retiming value $\Re(v) \leq 0$ corresponds to forward retiming at node v. We define $r_M(v) \geq 0$ for FRT mapping.

After getting the minimum clock period Φ_{min} and the corresponding label pairs, we construct a mapping solution and perform a separate forward retiming step to achieve Φ_{min}.[3] Minimizing $R(v)$ means that we want to push forward as few FFs as possible for each node v to leave the maximum freedom to the subsequent forward retiming step, because those FFs pushed forward can no longer be pushed back. On the other side, $R(v)$ cannot be *too* small because we also want to construct larger LUTs to minimize $L^s(v)$ by pushing forward some FFs. Our algorithm can simultaneously minimize both $L^s(v)$ and $R(v)$ to compute optimal solutions.

3 TurboMap-frt Algorithm

Our algorithm, named TurboMap-frt, computes optimal mapping with forward retiming solutions for synchronous sequential circuits with given initial states to minimize the clock period. It performs binary search on the clock period from 1 to $|V|$.[4] For a given clock period, a procedure named FRTcheck is used to decide whether there exists a feasible solution through label computation. An overview of the label computation procedure can be described as follows. First, we assign an initial lower-bound on the value of each node label pair. Then we iteratively update those lower-bounds until all of them converge to the values of node label pairs if there exists a feasible solution, or stop if we conclude there is no feasible solution. Expanded circuits are used to represent all possible LUTs under node duplication and forward retiming for label pair update.

Before presenting the details of our algorithm, we review the definition of K-cuts here. In a directed graph with one sink and one source, a *cut* $(X, \overline{X})$ is a partition of the nodes in the graph such that the sink is in $\overline{X}$ and the source is in X. The *node cut-set* $V(X, \overline{X})$ is the set of nodes in X that are connected directly to nodes in $\overline{X}$. If $|V(X, \overline{X})| \leq K$, $(X, \overline{X})$ is called a *K-feasible cut*, or *K-cut* in short, and $\overline{X}$ is called a K-feasible cone. A cut is a *min-cut*, if $|V(X, \overline{X})|$ is minimum among all the cuts in the graph.

3.1 LUT Formation

To compute an optimal solution we need to be able to search all possible LUTs for every node under node duplication and forward retiming. Pan and Liu [12] proposed to search all possible LUTs rooted at a node on expanded circuits of the node. An *expanded circuit* $\mathcal{E}_v$ for a node v is a directed acyclic graph (DAG) rooted at v^0 and constructed from the original circuit with node duplication as shown in Figures 3b – 3e. One important property of $\mathcal{E}_v$ is that for any node, denoted u^w, every path from u^w to the root v^0 passes exactly w FFs. Pan and Liu [12] showed that *to examine all K-LUTs for a node v, it sufficed to examine all the K-LUTs that can be derived from the K-cuts in a large enough expanded circuit* $\mathcal{E}_v^{Kn}$ *of v* due to the one-to-one correspondence between K-cuts and K-LUTs.[5]

However, the one-to-one correspondence between K-cuts on $\mathcal{E}_v^{Kn}$ and K-LUTs rooted at v under forward retiming no longer holds. As shown in Figure 3(d), to cluster the nodes in the dotted box to form an LUT we have to move the FF backward from the fanout of b^1 to the fanins of b^1, because the path $i_1^0 \rightarrow a^0 \rightarrow c^0$ does not have any FF to push forward. Thus, such a cluster is not a valid LUT formation for FRT mapping. For FRT mapping, we construct a smaller expanded circuit for every node based on its *maximum forward retiming value* so that there is a one-to-one correspondence between K-LUTs to K-cuts on the expanded circuits.

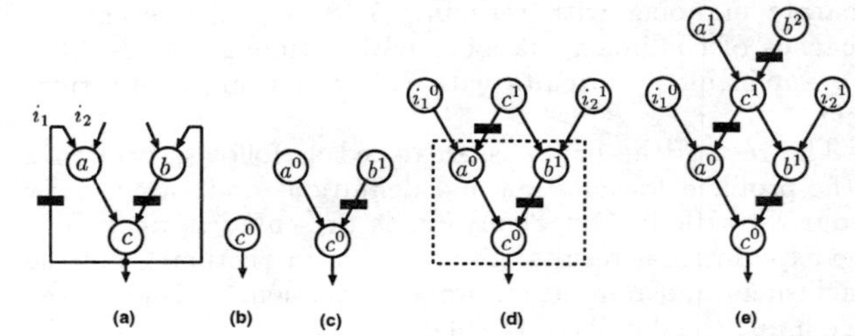

Figure 3: Expanded circuits for LUT representation.

Definition 3 *The* **maximum forward retiming value** *$frt(v)$ of a node v in a retiming graph is the maximum number of FFs which can be moved forward from the (transitive) fanins of v to the output of v.*

Lemma 1 $frt(v) = min\{w(p) \mid \forall\ p : PI \rightsquigarrow v\}$.

Since the edge weight $w(e) \geq 0$, the set $\{frt(v)\}$ of all the nodes in a circuit can be computed in $O(|V|^2)$ time by Dijkstra's shortest path algorithm.

Now we define a set of expanded circuits of a node v for FRT mapping. The expanded circuit $\mathcal{F}_v^i$ for a given upper-bound i of forward retiming value of node v is a sub-DAG of $\mathcal{E}_v^{Kn}$ (defined in [12]) with root v^0 such that, u^w is an internal node of $\mathcal{F}_v^i$ if and only if u^w is an internal node of $\mathcal{E}_v^{Kn}$ and $w \leq i$, and u^w is a leaf of $\mathcal{F}_v^i$ if and only if u^w is either a leaf of $\mathcal{E}_v^{Kn}$ or a fanin of an internal node of $\mathcal{F}_v^i$ and $w > i$. Notice that if two leaves ${u_1}^{w_1}$ and ${u_2}^{w_2}$ have an edge in $\mathcal{E}_v^{Kn}$, the edge will not appear in $\mathcal{F}_v^i$ because a leaf cannot have any inputs. For example, the shaded area of Figure 4(b) represents $\mathcal{F}_c^1$ for the circuit in Figure 4(a), where the heavy shaded nodes are leaves of $\mathcal{F}_c^1$.

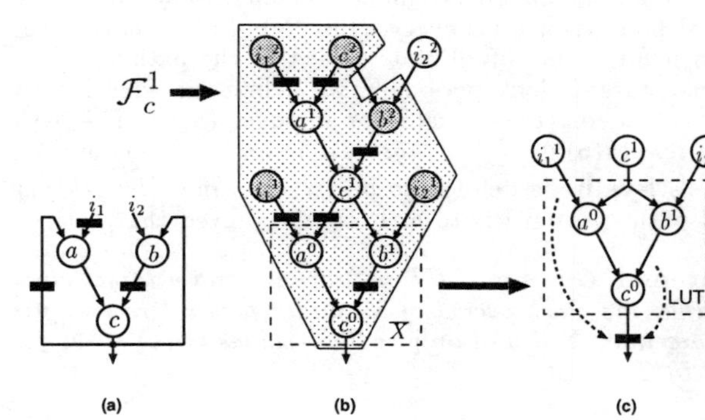

Figure 4: LUT formation for FRT mapping based on the expanded circuits.

When i is set to $frt(v)$, we have that:

[3] Notice that an optimal forward retiming after mapping can be computed easily based on [8] by adding linear constraints $\Re(v) \leq 0$ for every node v. In fact, we can compute a forward retiming directly from the node label pairs as shown in § 3.3.

[4] Much smaller upper-bound can be computed by performing depth-optimal mapping (FlowMap [1]) on combinational subcircuits following by forward retiming.

[5] It was shown that the expanded circuit had no more than $O(Kn^2)$ nodes and $O(K^2n^2)$ edges for a circuit with n gates [12].

Theorem 2 *Every K-feasible cut $(X, \overline{X})$ on $\mathcal{F}_v^{frt(v)}$ corresponds to a K-LUT rooted at v under node duplication and forward retiming, where all the leaves are in X and the root is in $\overline{X}$. On the other hand, any K-LUT in a FRT mapping solution corresponds to a K-feasible cut on $\mathcal{F}_v^{frt(v)}$.*

To further demonstrate our approach, let us look at the circuit shown in Figure 4(a). Notice that its only difference with the circuit in Figure 3(a) is one extra FF on edge (i_1, a) which makes $frt(c) = 1$. Now the 3-cut shown in Figure 4(b) can form a 3-LUT as shown in Figure 4(c).

3.2 Iterative Label Computation

For a target clock period Φ, we compute all node label pairs of a circuit to decide the existence of a feasible FRT mapping solution. First, we assign an initial lower-bound $(0,0)$ on the value of label pair for every PI and $(-\infty, 0)$ for the other nodes. Then we iteratively update the lower-bounds until they converge to the values of node label pairs. If, however, the lower-bounds cannot converge to finite values after $|V|^2$ iterations, we conclude that there is no feasible FRT mapping solution for the given Φ. The pseudo-code of the FRTcheck algorithm is shown in Figure 5.

```
FRTcheck(G(V,E,W), Φ)
1    assign initial lower-bound (0,0) for each PI and
     (−∞,0) for the other nodes
2    converge ← FALSE, i ← 0
3    while (not converge and i ≤ |V|²)
4        converge ← TRUE
5        for each node v ∈ V
6            (l^s_new(v), r_new(v)) ←LabelUpdate(v, Φ)
7            if (l^s_new(v) > l^s(v))
8                l^s(v) ← l^s_new(v)
9                converge ← FALSE
10       i ← i + 1
11   return (converge? TRUE : FALSE)
```

Figure 5: Label computation for a target clock period Φ.

The most important part of our algorithm is the update of the lower-bound pair $(l^s(v), r(v))$ for every node v (procedure LabelUpdate, line 6 in Figure 5). For each node v, we update $l^s(v)$ at first and then $r(v)$ through the computation of min-height-min-weight K-cuts on $\mathcal{F}_v^{frt(v)}$.

Definition 4 *For a K-cut $(X, \overline{X})$ in an expanded circuit of node v, the* **cut-weight** *is $w(X, \overline{X}) = max\{w \mid \forall u^w \in \overline{X}\}$.*

Definition 5 *Given a set of lower-bounds $(l^s(v), r(v))$, where $l^s(v) \leq L^s(v)$ and, $r(v) \leq R(v)$ when $l^s(v) = L^s(v)$, the* **cut-height** *of a K-cut $(X, \overline{X})$ in the expanded circuit $\mathcal{F}_v^{frt(v)}$ of node v is defined to be*

$$h(X, \overline{X}) = \max\{l^s(u) - \Phi \cdot w + 1 \mid \forall u^w \in V(X, \overline{X})\}.$$

To compute a tighter lower-bound $l^s_{new}(v)$ for node v, we compute $\mathcal{L}^s(v) = max\{l^s(u) - \Phi \cdot w(e) \mid \forall e(u,v) \in G\}$ on the original circuit G. Then we decide whether there exists a K-cut on $\mathcal{F}_v^{frt(v)}$ with height of no more than $\mathcal{L}^s(v)$ based on the max-flow K-cut computation on a partial flow network presented in [2].[6] If there does not exist such a K-cut, we set $l^s_{new}(v) = \mathcal{L}^s(v) + 1$ and $r_{new}(v) = 0$, because it can then be proved that $L^s(v) > \mathcal{L}^s(v)$.

[6] We refer readers to [2] for the detail.

If, however, there does exist such a K-cut with height of no more than $\mathcal{L}^s(v)$, we then compute a K-cut with the minimum cut-weight among those with height of no more than $\mathcal{L}^s(v)$ by binary search on the cut-weight from 0 to $frt(v)$ as follows. (Recall that $R(v)$ is the minimum $r_M(v)$ among all the feasible FRT mapping solutions M with $l^s_M(v) = L^s(v)$. To minimize $w(X, \overline{X})$ is to guarantee that $r_{new}(v) = w(X, \overline{X}) \leq R(v)$ when $l^s_{new}(v) = L^s(v)$.)

For a given $w \in [0, frt(v)]$, to decide whether there is a K-cut with height of no more than $\mathcal{L}^s(v)$ and weight of no more than w, we construct an expanded circuit $\mathcal{F}_v^w$ and decide whether there exists a K-cut on $\mathcal{F}_v^w$ with height of no more than $\mathcal{L}^s(v)$. If there does exist such a K-cut, clearly it is a K-cut with height of no more than $\mathcal{L}^s(v)$ and cut-weight of no more than w.

Let $(X, \overline{X})$ be such a K-cut with height of $\mathcal{L}^s(v)$ and weight of w_{min} computed above. If $\mathcal{L}^s(v) + \Phi \cdot w_{min} \leq \Phi$, we update the new lower-bound to be $(\mathcal{L}^s(v), w_{min})$. Otherwise, we update the new lower-bound to be $(\mathcal{L}^s(v) + 1, 0)$.

It can be proved that $l^s_{new}(v) \leq L^s(v)$ and, $r_{new}(v) \leq R(v)$ when $l^s_{new}(v) = L^s(v)$. In other words, the updated pair is still a lower-bound on the value of the label pair of v. This result is based on the monotone property of $L^s(v)$. Given a sequential circuit G which has a feasible FRT mapping solution for a target clock period Φ, a set of $L^s(v)$ is *monotone* if, $L^s(u) - \Phi \cdot w(e) \leq L^s(v)$ for any edge $e(u,v) \in G$.

Theorem 3 (Monotone Property) *Given a sequential circuit G which has a feasible FRT mapping solution for target clock period Φ, the set $\{L^s(v) \mid v \in G\}$ is monotone.*

Let one *iteration* denote the computation process where the lower-bound $(l^s(v), r(v))$ is updated once for every node v (lines 5 ~ 9 in Figure 5). Based on the monotone property we can prove that:

Lemma 2 *Given a sequential circuit which has a feasible FRT mapping solution for a target clock period Φ, the inequality $l^s(v) \leq l^s_{new}(v) \leq L^s(v)$ holds all the time after any number of iterations.*

Lemma 3 *Given a sequential circuit G which has a feasible FRT mapping solution for a target clock period Φ, $r_{new}(v) \leq R(v)$ when $l^s_{new}(v) = L^s(v)$.*

As a result, we have the following theorem:

Theorem 4 *For a sequential circuit which has a feasible FRT mapping solution for a target clock period Φ, starting from the initial lower-bounds $(0,0)$ for PIs and $(-\infty, 0)$ for the other nodes, the inequalities 1. $l^s(v) \leq L^s(v)$ and, 2. $r(v) \leq R(v)$ when $l^s(v) = L^s(v)$ hold all the time after any number of iterations of label update.*

Based on the above label pair update procedure, we can prove the optimality of FRTcheck algorithm as follows.

Lemma 4 *If there is a feasible FRT mapping solution for a target clock period Φ, the 2-tuples $(l^s(v), r(v))$ computed by FRTcheck will converge to the node label pairs $(L^s(v), R(v))$ for all nodes after no more than $|V|^2$ iterations.*

Lemma 5 *For a target clock period Φ, if FRTcheck returns TRUE, there must exist a feasible FRT mapping solution.*

Theorem 5 *For a sequential circuit with n gates, the optimal FRT mapping solution with the minimum clock period can be computed in the worst-case time complexity of* $O(K^3n^5\log^2 n)$ *and worst-case space requirement of* $O(K^2n^2)$.

Notice that the above result is based on the worst-case scenario that the expanded circuits have $O(K^2n^2)$ edges [12] and we need go through n^2 iterations (lines 5~9 in Figure 5). In practice, the expanded circuits have much fewer than Kn edges using the efficient max-flow computation on partial flow networks [2] and the number of iterations for each Φ is around 5 ~ 15 based on a computation order proposed in [2]. Practically, the runtime of our algorithm is in the order of $K^2n^2\log^2 n$, the space requirement is in the order of Kn. Our experiments show that the optimal solution for s38417 with 9800 gates and 1500 FFs can be computed in 20 minutes on a Sun Ultra2 with 256MB memory.

3.3 Mapping Generation

After computing Φ_{min} based on binary search and obtaining the label pairs $(L^s(v), R(v))$ of all nodes v under Φ_{min}, the last step is to generate the mapping solution based on the K-cuts computed during the label computation and perform forward retiming with initial state computation.

First, we get all the LUT roots in the final mapping solution based on the K-cuts computed during the label computation. In the final mapping solution, all the POs of the original circuit are LUT roots and, if v is an LUT root, all the nodes in the node-cut set of the K-cut of v are LUT roots. Those LUT roots can be computed as follows: starting with a first-in-first-out (FIFO) queue with all the POs, we repeatedly extract nodes from the head of the queue until the queue is empty. For each node extracted from the queue, we mark it as an LUT root and put all the nodes in its node-cut set to the end of the queue.

Second, we create a new equivalent network by connecting the K-feasible cones $\overline{X}_v$ of the K-cuts $(X_v, \overline{X}_v)$ of the LUT roots v. Suppose each LUT has $O(C)$ nodes for a constant $C \ll n$, where n is the number of nodes in the original network, the constructed network has $O(Cn)$ nodes. We compute a forward retiming $\{\Re(v)\}$ on the new network, where $\Re(v) = \lceil \frac{L^s(v)}{\Phi_{min}} \rceil - 1 \leq 0$ for LUT roots v and $\Re(u) = \Re(v) + w$ for $u^w \in \overline{X}_v$ or $\Re(u) = 0$ for $u^w \in V(X_v, \overline{X}_v)$ in each LUT rooted at v. After retiming, all the FFs within each $\overline{X}_v$ will be moved forward outside the cone, because $w^r(u^w \rightsquigarrow v^0) = w + \Re(v) - \Re(u) = 0$ for all $u^w \in \overline{X}$. So we can collapse each $\overline{X}_v$ into a K-LUT and the final circuit has a clock period of no more than Φ_{min} as follows.

Theorem 6 $\{\Re(v)\}$ *is a legal forward retiming to achieve clock period* Φ_{min}.

Since the retiming is a forward retiming, an equivalent initial state can be computed in linear time with circuit simulation based on the approach in [15]. For a K-bounded circuit with n gates and $O(Kn)$ edges, the first step of getting LUT roots can be done in $O(Kn)$ time. The forward retiming on the constructed circuit can be done in $O(CK^2n^2)$ time since $|\Re(v)| \leq O(Kn)$ for each v and each K-input node can be simulated in $O(K)$ time [15].[7] So the mapping generation with forward retiming and initial state computation can be finished in $O(CK^2n^2)$ time.

[7] To be precise, a node can be simulated in $O(l)$ time, where l is the number of variables to represent the function of the node, for example, the number of literals on cover table representation or the number of variables in the BDD representation.

4 Experimental Results

The TurboMap-frt algorithm has been implemented in C language and incorporated into the SIS package [13]. Our test set consists of 14 MCNC FSMs and 4 ISCAS'89 benchmarks. The initial circuits are shown in Column Original of Table 1, in which N means the number of nodes and F means the number of FFs, respectively, in the circuits.

Our experiments were performed on a Sun Ultra2 with 256MB memory. K was set to be 5. All the mapping results of TurboMap-frt were computed and verified by *verify_fsm* of SIS [13], except the 4 largest ones which were verified by simulations with input sequences of 3008 random vectors due to huge amount of memory (>1GB) needed by *verify_fsm* for those cases. We compared TurboMap-frt with FlowMap-frt and TurboMap [2]. FlowMap-frt is based on the FlowMap algorithm [1]. It first map each combinational subcircuit bounded by FFs independently with the FlowMap algorithm. Then it merges the mapped LUT subcircuits with the original FFs and performs a postprocessing of forward retiming for clock period minimization. TurboMap [2] computes *optimal* mapping with *general* retiming solutions for synchronous sequential circuits. The initial states of TurboMap solutions were computed with SIS [13] based on the algorithm in [15]. The experimental results are shown in Table 1, where Columns LUT and FF list the numbers of LUTs and FFs, respectively, in the final mapping solutions by each approach. Columns Φ list the clock periods of the results. Columns CPU list the CPU time. Those marked with $\star$ are examples for which SIS failed to compute equivalent initial states for TurboMap solutions. Column Best lists the best valid solutions (with computed equivalent initial states) by TurboMap and FlowMap-frt.

The results show that comparing with TurboMap-frt, FlowMap-frt computed results with 20.2% larger clock period. Though TurboMap [2] can compute results with 2.8% smaller clock period, there are 10 out of 18 solutions by TurboMap SIS concludes no equivalent initial states exist or cannot find them due to huge memory requirement (>300MB) and long runtime (>2hours). Even if the best valid solutions by TurboMap or FlowMap-frt are used for comparison, TurboMap-frt still outperforms the two algorithms by 8.6% on the clock periods. On area part, TurboMap-frt used 8.4% and 1.9% less LUTs as compared with FlowMap-frt or TurboMap, respectively. TurboMap-frt also reduced 6% FFs vs. TurboMap. But TurboMap-frt used 33.8% more FFs as compared with FlowMap-frt. In general we think simultaneous mapping with retiming leads to smaller clock period but tends to use more FFs.

5 Conclusions and Future Work

For sequential circuits with initial states, we presented a new algorithm TurboMap-frt for FPGA mapping with forward retiming and initial state computation to minimize the clock period. Unlike previous retiming algorithms which compute retiming at first and then try to compute equivalent initial states, we compute optimal mapping with forward retiming solutions with equivalent initial states in one step. Our algorithm enables a new methodology of performing backward retiming separately before forward retiming. Since we can compute the optimal mapping with forward retiming solutions, backward retiming can be used to push back all the FFs to PIs as much as possible as long as the equivalent initial states can be computed without taking into consideration of the impact on the final clock period. This approach is also applicable to sequential circuits with partial initial state assignments by allowing FFs with unassigned initial state to be moved backward or forward freely with other FFs.

The experimental results show that TurboMap-frt is very efficient and effective comparing with conventional design flow of mapping each combinational subcircuits separately.

circuit	Original	FlowMap-frt				TurboMap				Best	TurboMap-frt			
	N/F	Φ	LUT	FF	CPU	Φ	LUT	FF	CPU	Φ	Φ	LUT	FF	CPU
bbara	28/10	4	13	10	0.2	3	12	7	0.4	3	3	12	12	0.2
bbtas	15/5	2	7	5	0.1	1	6	4	0.2	1	1	6	4	0.1
dk16	162/5	14	101	5	0.9	14	103	14	3.8	14	14	103	9	1.7
dk17	42/5	2	10	5	0.2	1	6	3	0.4	1	1	6	3	0.2
ex1	140/5	8	83	5	0.7	8	92	21	1.9	8	8	92	20	1.3
ex2	16/7	2	9	7	0.2	⋆1	4	3	0.2	2	1	4	3	0.1
keyb	134/5	10	75	5	0.6	10	79	5	1.6	10	10	81	5	1.0
kirkman	106/5	6	48	5	0.7	⋆5	57	24	1.2	6	5	57	14	0.8
planet1	348/6	19	213	6	2.0	⋆19	201	18	12.5	19	19	199	37	5.0
s1	107/5	7	58	5	0.5	7	63	11	1.2	7	7	56	6	0.7
sand	327/17	16	176	17	1.8	⋆15	178	30	10.6	16	15	176	12	4.3
scf	516/7	14	325	7	2.8	⋆13	304	20	19.8	14	13	301	27	8.8
sse	74/4	7	42	4	0.4	6	45	10	0.9	6	6	44	8	0.5
styr	281/5	17	163	5	1.6	⋆16	168	8	5.2	17	17	168	12	3.2
s5378	1503/164	4	421	204	7.9	⋆4	444	301	51.5	4	4	427	261	40.3
s9234.1	1299/135	6	462	161	8.5	⋆4	498	217	>7200	6	5	441	203	58.8
s15850.1	3801/515	10	1240	504	30.3	⋆8	1161	732	>7200	10	10	1166	621	205.6
s38417	9817/1464	8	3526	1464	561.5	⋆6	3420	2264	1201.8	8	6	3301	2573	1210.6
geomean		7.0	100	15	1.4	5.6	94	24	7.4	6.3	5.8	92	23	2.8
%		+20.2	+8.4	-33.8	-47.6	-2.8	+1.9	+6.0	+167.0	+8.6	1	1	1	1

Table 1: Comparison of TurboMap-frt with FlowMap-frt and TurboMap for 5-LUT. "geomean" lists the geometric means of the results by each approach. The runtime was recorded on a Sun Ultra2 with 256MB memory. Those marked with ⋆ are circuits that SIS failed to compute initial states for TurboMap solutions. Column Original lists the original circuit size, where N and F represent the numbers of nodes and FFs, respectively, in the original circuits.

It can also produce results very close to the optimal mapping with general retiming solutions in terms of both area and clock period.

In the future we plan to extend our work for library based technology mapping with (forward) retiming for high performance gate array and standard cell designs. A general framework on retiming with multiple clock designs was proposed recently by Legl, et. al. [7]. We plan to accommodate our approach into this framework as well.

6 Acknowledgements

This work is partially supported by National Science Foundation Young Investigator Award MIP9357582 and grants from Xilinx, Lucent Technologies and Quickturn Design Systems under the California MICRO program.

References

[1] J. Cong and Y. Ding. FlowMap: An Optimal Technology Mapping Algorithm for Delay Optimization in Lookup-Table Based FPGA Designs. *IEEE Trans. on Computer-Aided Design of Integrated Circuits And Systems*, 13(1):1–12, 1994.

[2] J. Cong and C. Wu. An Improved Algorithm for Performance Optimal Technology Mapping with Retiming in LUT-Based FPGA Design. In *IEEE International Conference on Computer Design*, pages 572–578, 1996.

[3] J. Cong and C. Wu. *Optimal FPGA Mapping and Retiming with Efficient Initial State Computation.* UCLA-CSD 980016, Technique Report, March 1998.

[4] G. Even, I. Y. Spillinger, and L. Stok. Retiming Revisited and Reversed. *IEEE Trans. on Computer-Aided Design of Integrated Circuits And Systems*, 15(3):348–357, 1996.

[5] H. Fujiwara and S. Toida. The Complexity of Fault Detection: An Approach to Design for Testability. In *FTCS-12*, pages 101–108, 1982.

[6] S. Kundu, L. M. Huisman, I. Nair, and V. Iyengar. A Small Test Generator for Large Designs. In *International Test Conference*, pages 30–40, 1992.

[7] C. Legl, P. Vanbekbergen, and A. Wang. Retiming of Edge-Triggered Circuits with Multiple Clocks and Load Enables. In *International Workshop on Logic Synthesis*, 1997.

[8] C. E. Leiserson and J. B. Saxe. Retiming Synchronous Circuitry. *Algorithmica*, 6:5–35, 1991.

[9] N. Maheshwari and S. S. Sapatnekar. Minimum Area Retiming with Equivalent Initial States. In *IEEE International Conference on CAD*, pages 216–219, 1997.

[10] S. Malik, K. J. Singh, R. K. Brayton, and A. Sangiovanni-Vincentelli. Performance Optimization of Pipelined Logic Circuits Using Peripheral Retiming and Resynthesis. *IEEE Trans. on Computer-Aided Design of Integrated Circuits And Systems*, 12(5):568–578, 1993.

[11] G. D. Micheli. Synchronous Logic Synthesis: Algorithms for Cycle-Time Minimization. *IEEE Trans. on Computer-Aided Design of Integrated Circuits And Systems*, 10(1):63–73, 1991.

[12] P. Pan and C. L. Liu. Optimal Clock Period FPGA Technology Mapping for Sequential Circuits. *ACM Transactions on Design Automation of Electronic Systems*, 4(1), 1999. http://www.acm.org/todaes/V4N1/L166/paper.ps.gz.

[13] E. Sentovich, K. Singh, L. Lavagno, C. Moon, R. Murgai, A. Saldanha, H. Savoj, P. Stephan, R. Brayton, and A. Sangiovanni-Vincentelli. *SIS: A System for Sequential Circuit Synthesis.* Electronics Research Laboratory, Memorandum No. UCB/ERL M92/41, 1992.

[14] L. Stok, I. Spillinger, and G. Even. Improving Initialization through Reversed Retiming. In *Proc. Euro. Design & Test Conference*, pages 150–154, 1995.

[15] H. Touati and R. K. Brayton. Computing the Initial States of Retimed Circuits. *IEEE Trans. on Computer-Aided Design of Integrated Circuits And Systems*, 12(1):157–162, 1993.

[16] H. Touati, N. Shenoy, and A. Sangiovanni-Vincentelli. Retiming for Table-Lookup Field-Programmable Gate Arrays. In *FPGA'92*, pages 89–94, 1992.

[17] H. Yotsuyanagi, S. Kajihara, and K. Kinoshita. Retiming for Sequential Circuits with a Specified Initial State and Its Application to Testability Enhancement. *IEICE Trans. INF. & SYST.*, E78-D(7):861–867, July 1995.

M32: A Constructive Multilevel Logic Synthesis System*

Victor N. Kravets Karem A. Sakallah
Department of Electrical Engineering and Computer Science
University of Michigan, Ann Arbor, MI 48109
{vkravets, karem}@eecs.umich.edu

Abstract

We describe a new constructive multilevel logic synthesis system that integrates the traditionally separate technology-independent and technology-dependent stages of modern synthesis tools. Dubbed M32, this system is capable of generating circuits incrementally based on both functional as well as structural considerations. This is achieved by maintaining a dynamic structural representation of the evolving implementation and by refining it through progressive introduction of gates from a target technology library. Circuit construction proceeds from the primary inputs towards the primary outputs. Preliminary experimental results show that circuits generated using this approach are generally superior to those produced by multi-stage synthesis.

I. Introduction and Motivation

In this paper we describe a new multilevel logic synthesis system, M32, that departs in several important respects from current practice in logic synthesis technology. The development of M32 was motivated by the oft-cited refrain that wires are starting to dominate active logic in determining the area and speed of deep submicron ICs, and that current synthesis flows are biased primarily towards optimizing gates. M32 was designed to address this bias by intertwining the traditionally separate stages of technology-independent Boolean optimization and technology-dependent mapping in a constructive synthesis strategy that is cognizant of the structural implications of optimization decisions. Such generation of the technology specific circuit from the initial steps makes the task of assessing final design quality more accurate from the initial steps of synthesis, and allows exploration of degrees of freedom which are lost in the split between technology-independent and technology-dependent stages of synthesis.

The remainder of this paper is organized as follows. Section II gives a brief description of prior work. A synthesis algorithm based on the constructive synthesis methodology is then presented in Section III. It overcomes the limitations of previous constructive approaches and benefits from recent advances in synthesis technology, allowing it to handle much larger circuits. Its prototype implementation in the M32 synthesis system has been demonstrated for combinational circuits containing several thousand gates. An example illustrating execution of the algorithm is given in Section IV. Experimental results are discussed in Section V.

II. Prior Work

Combinational multilevel logic synthesis is the process of implementing a set of logic expressions using cells from a technology library, each with a prescribed function and physical characteristics [26]. Most of the current logic synthesis systems divide the logic synthesis process into technology-independent [4] and technology-dependent stages [16, 10]. The technology-independent stage focuses on partitioning the logic, whereas the technology-dependent stage chooses appropriate gates from the library to implement the partitioned logic. Such multi-stage approaches to complex optimization problems are common in electronic design automation (e.g. placement followed by routing) and are usually necessitated by the difficulty of solving these problems simultaneously.

This "serialization" of the optimization process implies that decisions made in earlier stages must necessarily be based on loose estimates of what later stages can accomplish. At the same time, the solutions produced by early stages place limitations on the degrees of freedom to improve final implementation of a design. For two-stage logic synthesis, decisions made during the technology-independent stage [5, 7, 6] significantly determine the structure of a circuit. They are made with no regard for the downstream technology. When the technology characteristics become available in the mapping stage it is often too late to augment the effects of these decisions to improve circuit quality.

The back-annotated approaches, which perform resynthesis after technology specific information is extracted from the mapped circuit, compensate partially for this problem. This is typically done by re-synthesizing critical sections identified during the *sign-off* process [14, 31, 2, 20, 13, 27]. While this yields improvement in circuit quality, technology-independent and technology-dependent transformations still remain disconnected. In [19], the authors address this problem by dynamically modifying the set of AND2/INV decompositions while deleting others based on the actual cost function used in technology mapping. It allows technology-independent transformations to be part of the technology mapping. However, these transformations do not exhibit global knowledge about circuit structure and functionality. Recently, De Micheli [21] discussed concurrent technology mapping and optimization by utilizing the don't-care sets of the bound portion of the network.

The potentially detrimental effects of serializing logic synthesis become more pronounced in fine-feature CMOS ICs where interconnections tend to dominate both area and delay [17]. For example, changes in the average topological wire length of a circuit, can significantly contribute to the post-layout delay and routing area [18]. It is therefore critical for a synthesis tool to have accurate estimation of wiring from the initial stages of synthesis. In [29], the authors pursued this problem by relying on the traditional synthesis methods, accounting for routing area during logic decomposition using heuristics for minimizing the fanout ranges of a decomposed network. Another method of minimizing routing cost by deriving and maintaining an order among the primary inputs of the circuit was proposed in [24, 1]. These approaches operate during the early phase of synthesis, and remain removed from the final implementation circuit. In [23] an attempt is made to account for the interconnections during the technology mapping phase of synthesis. The idea is to generate a "companion" placement solution for the circuit before it is mapped. This placement is then dynamically updated as the mapping process tries to evaluate the cost of a matching gate. The algorithm estimates the interconnection contribution to the area and delay by referring to the dynamically updated placement of the network. Using this technique, the authors were able to generate circuits with shorter wire lengths and

*This research was supported by DARPA under contract DAAH04-94-G-0327

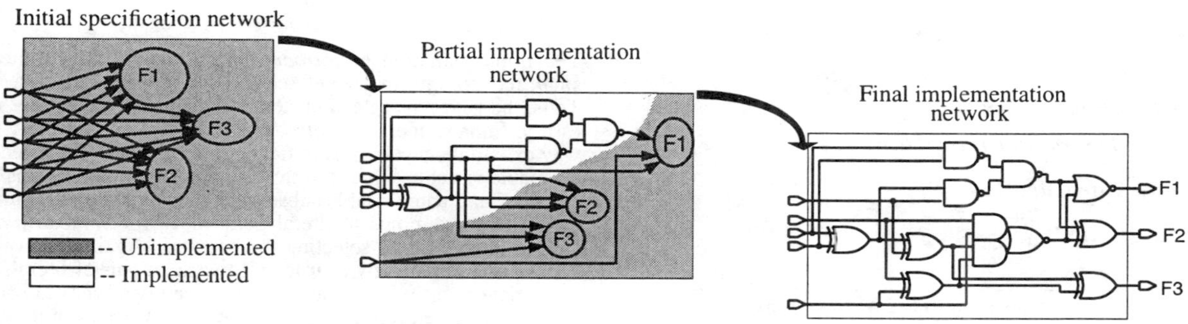

Fig. 1. Illustration of the constructive technology-dependent synthesis

smaller area.

Several attempts to synthesize networks incrementally are reported in the literature. Davidson presented a branch-and-bound algorithm for NAND network synthesis [9]. The algorithm constructs a network realization incrementally starting from the primary outputs. In each iteration, the algorithm extends a partial solution by introducing a new NAND gate together with its fanin connections. Another incremental synthesis procedure for arbitrary gates was presented by Schneider and Dietmeyer [25]. In each step, their algorithm finds the circuit package (i.e. gate type in a technology library) that satisfies some goal, such as area and delay, while meeting fanin, loading and power constraints. Such approaches, however, did not yield practical synthesis tools: their exponential run time complexity rendered them useless except for very small circuits.

III. The M32 Logic Synthesis System

M32 is a multilevel technology-dependent constructive logic synthesis system. It relies on extended algebraic decomposition techniques which define a feasible space of transformations which can be applied to sum-of-products expressions. The M32 algorithm performs circuit transformations that account for the topological characteristics of the evolving network structure in the context of a specific gate library. The system is currently geared towards performance-oriented synthesis which also minimizes average topological wire length. In this section we first give an overview of the system, and then describe its main features in detail.

A. System Overview

The overall synthesis process in M32 is depicted in Fig. 1. The state of the synthesis process is modeled by a *Boolean network* η [3] that captures an evolving structural representation of the functions being synthesized. Each vertex v in such a network has an associated Boolean function f_v that computes the logic value at the vertex's output in terms of the logic values on its inputs. A vertex is considered to be *implemented* if its function is equivalent to one of the functions in a given gate library L, and *unimplemented* otherwise. To insure an *onto* mapping from L to the vertices in η, the library must, at a minimum, have inverter and pass-through wire gates as well as any 2-input gate that makes it functionally complete.

Reading its functional specification F as a set of multi-output cubes, M32 constructs an initial *specification* network consisting of implemented vertices corresponding to the primary inputs and unimplemented vertices corresponding to the specified output functions $f_1, f_2, \ldots, f_n$. As the synthesis process evolves, the functions of unimplemented vertices are successively decomposed in terms of those of already implemented vertices, resulting in a series of *partial implementation* networks. The decomposition is closely tied with the given gate library, and leads to the creation of new implemented vertices, i.e. vertices that correspond to library gates. Each implemented vertex v introduces a new variable y_v which can now be used to simplify the functions of unimplemented vertices. The effect of these successive decompositions is an expansion of η from the PIs towards the POs as more implemented vertices are created and as the functional complexity of unimplemented vertices is reduced. This constructive creation process makes it possible to control the structural complexity of the evolving implementation. The synthesis process terminates when all vertices become implemented yielding a *final implementation* network.

The main loop in the M32 synthesis algorithm is shown in the pseudo-code of Fig. 2. The algorithm treats F and η as globally accessible structures. In each iteration, the functions of unimplemented vertices are examined and one of them, f_k, is selected for a decomposition step. An *atomic divisor* P is extracted from f_k by an appropriate division procedure. The divisor is subsequently implemented by a small set of vertices corresponding to gates from L leading to the expansion of the implemented part of η. The decomposition step is completed by substituting the variables y_v of the newly created vertices into the functions of the unimplemented vertices. An unimplemented vertex becomes implemented when its function reduces to a single literal. Thus, the iteration stops, signalling completion of the synthesis process, when all the functions in F have been reduced to single literals.

This algorithm has several features that distinguish it from commonly-used synthesis methods:

- It interleaves functional decomposition and technology mapping throughout the synthesis process
- It considers the structural implications of candidate decompositions
- It selectively applies Boolean transformations to improve synthesis quality without adversely affecting run time efficiency

The remainder of this section is devoted to detailed descriptions of the four main routines of the algorithm.

B. SelectFunction

The order in which the functions of F are decomposed clearly affects the final synthesized implementation. In each iteration of the algorithm, the set of vertices that are candidates for decomposition is the subset of unimplemented vertices whose fanins are already implemented. The *SelectFunction* routine identifies the next function to decompose by greedily choosing the candidate vertex whose function has the largest number of literals in its cube representation. This choice is motivated by the expectation that larger, richer, functions yield better divisors. In this context, divisor P_1 is considered better than divisor P_2 if the vertices in its implementation subcircuit represent better opportunities for sharing among the unimplemented vertices. More sophisticated selection strategies can be easily envisioned, especially when the initial specification is a multi-level network.

C. GenerateDivisor

Like most modern synthesis algorithms, M32 relies on an efficient division procedure to decompose a function f into the form $pq + r$. The most commonly used division procedure is weak division [7] which is applied to an SOP expression for f and is equivalent to excluding all Boolean transformations except for the distributive law. While primarily motivated by the need for a fast division operation, weak division has been empirically shown to yield acceptable decompositions in practice. Still, the quality of the divisors, as measured by the total number of literals in the resulting factored form, can be improved by the judicious application of

```
while exists f_i ∈ F such that f_i
        is not a single literal do {
         k ← SelectFunction( ) ;
         P ← GeneratateDivisor(f_k) ;
         ẏ_v ← IntroduceGates(P) ;
         Substitute(P, ẏ_v) ;
}
```

Fig. 2. Synthesis loop in the M32 system

additional Boolean transformations. The division operation in M32 augments the distributive law with the annihilation ($a'a = \mathbf{0}$) and idempotency ($aa = a$) properties to generate better decompositions. This additional flexibility comes at a modest computational cost.

It is interesting to note that limiting allowable transformations of Boolean expressions to the above three properties (distributivity, annihilation, and idempotency) guarantees that any generated factored form will reproduce the original SOP cover under *flattening* [15]. This, in turn, implies that different initial SOP representations of a function can lead to different decompositions. Removal of this bias requires the deployment of the entire arsenal of Boolean transformations, i.e. operating in the unrestricted functional domain. In general, the attendant improvement in the quality of such unrestricted decomposition comes at a steep computational cost to be practical. M32 partially compensates for this bias by intertwining decomposition with mapping to a specific gate library while managing the structural attributes of the evolving implementation.

Divisor selection in M32 is accomplished through successive factorization, using the distributive law, of repeated literals from an SOP expression f. The annihilation and idempotency transformations are subsequently invoked to modify the resulting quotient and reduce the literal cost of the decomposition. This process is iterated until each literal appears only once in the factored form. To account for the structural implications of particular decompositions, literals are chosen based on a structural cost metric that is computed according to:

$$\text{cost}(\dot{x}) = \sum_{\dot{y}_v \in \text{support}(f/\dot{x})} \frac{[\text{depth}(f/\dot{x}) - \text{depth}(\dot{y}_v)]}{\text{occurrence}(\dot{y}_v, f/\dot{x})} / \text{size}(f/\dot{x}) \qquad (1)$$

where $\dot{x}$ is a candidate literal in expression f and $f/\dot{x}$ is the quotient resulting from algebraic division of f by $\dot{x}$. The term $\text{support}(f/\dot{x})$ in the formula denotes the set of literals appearing in $f/\dot{x}$. The depth of a literal, $\text{depth}(\dot{y}_v)$, is the topological depth of its corresponding vertex v in η; the depth of a primary input is defined to be 0. Similarly, the depth of expression $f/\dot{x}$, or a set of literals, is the maximum depth of any of its literals plus 1. Given $f/\dot{x}$, $\text{occurrence}(\cdot)$ denotes the number of times a literal $\dot{y}_v$ occurs in the expression, and $\text{size}(\cdot)$ is the total number of literal occurrences in it. *GenerateDivisor* creates a divisor by successively selecting the next candidate literal with least cost according to (1).

D. IntroduceGates

Unlike conventional technology mappers that operate on an intermediate "optimized" Boolean network obtained in a prior technology-independent phase, M32 closely ties its creation of gates with functional decomposition. As soon as a divisor P is found by *GenerateDivisor, IntroduceGates* proceeds to map it to the given gate library. The mapping process is also different from those used in conventional synthesis tools: no intermediate *subject graph* is constructed. Similar to *GenerateDivisor*, this procedure is aware of the structural implications of its choices, and involves iterating the following steps until P is fully implemented by library gates:

1. A gate from the technology library with its suitable variable support in P is selected for instantiation as vertex v
2. The vertex v, along with possible fanin inverters, is instantiated and added to η; variable y_v is associated with the new vertex v
3. The divisor P is re-expressed in terms of y_v

The implementation of *IntroduceGates* is currently limited to a small technology library defined by $L = \{wire, INV, NAND2\}$. Thus, the gate type selection step in the above procedure is unnecessary. Enhancements to *IntroduceGates* that are currently underway extend its capabilities to richer gate libraries. The selection of a suitable variable support in step 1 is also simplified by this library choice. Candidate variable subsets are determined by examining all associative groupings of literal pairs in P for *boolean matching* [22] to NAND2, and selecting the best pair. Again, quality in this context is estimated by a structural metric: a pair of literals which instantiate a vertex v of least depth is greedily selected. (The depth of v in η is determined from the depth of vertices in its support, while accounting for the possible presence of an inverter on v's fanin lines.) To break ties between the candidate supports we have used estimated signal arrival times based on the SIS-1.2 *library model*. In our experiments we have used nominal delay and capacitance was taken from the MCNC library.

After a gate support and its phases are determined, it is introduced into the circuit by establishing its fanin connections and placing an inverter (either a new one or reusing a previously introduced one) on each connection which has negative input phase. No inverter is placed on the gate output regardless of the output phase.

Finally, divisor P is re-expressed in terms of this newly introduced gate, leading to a reduction in its size. The process continues until P is reduced to a single literal. This literal is then returned by the *IntroduceGates* routine, and is later used to substitute it for P in F.

E. Substitute

At each iteration of the algorithm routine *Substitute* re-expresses functions $f_1, \ldots, f_n$ of the unimplemented vertices in network η in terms of the newly introduced vertices. The substitutions applied to the functions are based on the division operation. If $pq + r$ is the result of dividing f by p then substitution re-expresses f as $\dot{y}_v q + r$, where $\dot{y}_v$ is a new literal. As in the *GenerateDivisor* step, the division operation utilizes the distributive, idempotent and annihilation laws. In addition, cube reduction using the sharp product $A\#B \equiv AB'$ [11] is used to maximize the sharing of potential divisors.

Substitute re-expresses $f_1, \ldots, f_n$ in two steps: (1) substituting literal $\dot{y}_v$ ($\dot{y}'_v$) for the P (P') function; and (2) substituting variables of newly introduced vertices for their gate functions, modulo the phase assignment on its inputs determined in *IntroduceGates*. These steps are applied to each of the functions $f_1, \ldots, f_n$ individually. The routine performs substitution selectively to minimize topological cost. In the first step *Substitute* uses the literal returned by the *IntroduceGates* routine, which corresponds to the output vertex of a subcircuit implementing P. The second step of the routine implements substitutions of finer granularity. Substitution of gate functions in this step is performed for each of the vertices in their topological order.

Note that substitution using the distributive law only does not require step (1), since it is subsumed by step (2). We should also point out that the division under the used properties may not be unique. This is illustrated in the example below, where annihilation gives rise to two distinct decompositions:

Example 3.1 Suppose $f = abc'e' + a'bcd + ac'de' + a'cde$, and let $P = ac'e' + a'cd$ be a divisor of f. We then can have two different quotients $q_1 = b + c'd + ce$ and $q_2 = b + ad + a'e$. Thus either substitution $f = y_v(b + c'd + ce)$ or $f = y_v(b + ad + a'e)$ is feasible. ■

IV. An M32 Synthesis Example

Synthesis of a full adder will be used as an example of the M32 system. M32 first reads the functional specification

$$z = a \oplus b \oplus c = a'b'c + a'bc' + ab'c' + abc \qquad (2)$$

$$c_{out} = ab + ac + bc \qquad (3)$$

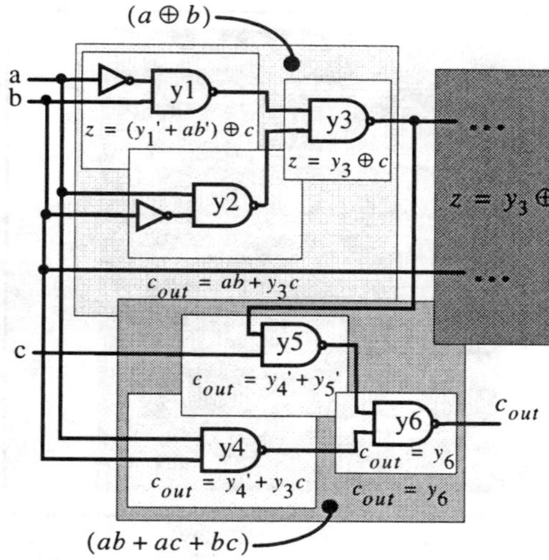

Fig. 3. M32 synthesis steps of the full adder

of the circuit to be synthesized. Thus, $F = \{z, c_{out}\}$ is the set of unimplemented vertices to be synthesized. The algorithm then selects function z since it has more literals then c_{out}. A subexpression P selected from z by the *GenerateDivisor* routine is $a'b + ab'$. *GenerateDivisor* selects this divisor since it has the least cost, with the assumption that signal c arrives later than the two other input signals a and b. This expression is then implemented using *IntroduceGates* by introducing gates y_1, y_2, and y_3 through the following sequence of transformations of P:

$$P \equiv a'b + ab' \rightarrow y_1' + ab' \rightarrow y_1' + y_2' \rightarrow y_3 . \quad (4)$$

The *Substitute* function would then substitute y_3 for $ab' + a'b$ (complement substitution is also tried) in both z and c_{out}, giving:

$$z = cy_3' + c'y_3 \quad (5)$$

$$c_{out} = ab + cy_3 . \quad (6)$$

Note that $ab' + a'b$ is not an algebraic divisor of $ab + ac + bc$. Therefore substitution based on weak division alone would not bring any changes to c_{out}. The *Substitute* routine makes substitution in c_{out} possible due to the cube reduction based on the sharp product operation. The M32 system detects that the division becomes possible if cubes ac and bc are reduced to $a'bc$ and abc respectively. This is a valid transformation since the result of $ac\#ab'c$ and $bc\#a'bc$, which is in both cases cube abc, is covered by the cube ab. It is now easy to see that $a'b + ab'$ divides expression $ab + ab'c + a'bc$, representing the same carry-out function c_{out}: $ab + (ab' + a'b)c$.

The next step in the *Substitute* routine is to see if any of the local functions of vertices y_1, y_2 or y_3 can be used to re-express either z or c_{out}. No further substitutions are possible in this case. This completes the first iteration of the algorithm in Fig. 2. On the next iteration of the loop either the unimplemented vertex for c_{out} or z can be selected, since both of their functions have same number of literals. Fig. 3 depicts execution of the algorithm with the assumption that c_{out} is selected. c_{out} is completely implemented on this iteration of the algorithm through the following sequence of transformations:

$$c_{out} \equiv ab + y_3c \rightarrow y_4' + y_3c \rightarrow y_4' + y_5' \rightarrow y_6 . \quad (7)$$

The final implementation of the circuit has 9 NAND2 gates and 4 inverters.

V. Experimental results

We evaluated the performance of M32 by synthesizing a set of circuits selected from the MCNC benchmarks [30] and comparing the results against SIS-1.2 [12]. The benchmarks were first minimized using ESPRESSO [4] prior to multilevel synthesis in M32 or SIS-1.2. We used the `delay` script in SIS-1.2 which is based on the *clustering script* proposed in [28] and is targeted towards technology-independent minimization of circuit delay. Both systems used the minimal gate library $L = \{wire, INV, NAND2\}$, and delay of the implementations obtained by both systems was estimated using the *library* delay model of SIS-1.2 and parameters from the `mcnc.genlib` library.

TABLE I: Pre-layout synthesis results

Circuit Name				SIS-1.2				M32				Norm.
	Inp.	Outp.	Cubes	Gates	Lev.	Compl.	Delay	Gates	Lev.	Compl.	Delay	Delay
z4ml	7	4	59	75	12	1.96	17.6	44	8	1.55	11.6	0.65
vda	17	39	793	1573	29	2.48	62.3	1115	16	1.56	34.3	0.55
inc	7	9	42	152	20	2.20	30.2	133	10	1.55	17.2	0.56
count	35	16	184	311	17	2.55	24.1	201	13	2.19	21.1	0.87
ldd	9	19	70	139	13	1.96	18.9	110	10	1.84	15.4	0.81
b9	41	21	141	176	12	1.69	17.8	177	10	1.60	14.8	0.83
ex4	128	28	620	690	20	1.73	28.0	665	15	1.43	19.3	0.68
cordic	23	2	1180	182	18	1.65	23.4	133	11	1.39	14.6	0.62
cps	24	109	654	2162	35	2.52	61.9	1625	17	1.93	41.3	0.66
duke2	22	29	120	743	20	2.17	35.0	534	13	1.68	24.8	0.70
vg2	25	6	110	239	16	1.88	21.6	152	11	1.36	15.5	0.71
apex2	39	3	438	564	33	2.92	44.6	484	18	1.83	25.4	0.56
sqrt8	8	4	88	79	14	1.84	18.3	76	12	1.72	17.6	0.96
bw	5	28	110	232	16	1.91	28.9	206	9	1.58	18.3	0.63
clip	9	5	167	240	26	2.55	31.3	309	14	1.75	17.5	0.55

TABLE II: Post-layout synthesis results

Circuit Name	SIS-1.2			M32			Improvement M32 / SIS-1.2		
	Total area (mil^2)	Routing length (μm)	Delay (ns)	Total area (mil^2)	Routing length (μm)	Delay (ns)	Total area	Routing length	Delay
z4ml	33.5	6592	2.66	21.5	4212	1.81	0.64	0.63	0.68
vda	1159.2	326400	7.45	715.6	189921	5.87	0.61	0.58	0.78
inc	77.6	17041	4.21	62.2	14039	2.52	0.80	0.82	0.59
count	137.7	28675	3.63	81.5	16941	3.62	0.59	0.59	0.99
ldd	60.7	12220	2.96	48.0	9123	2.18	0.80	0.74	0.73
b9	74.9	15136	2.88	76.9	15519	2.10	1.02	1.02	0.72
ex4	311.7	66272	4.16	294.9	60746	3.15	0.94	0.91	0.75
cordic	70.8	13996	3.55	54.5	10751	1.98	0.77	0.76	0.55
cps	1414.4	393693	8.19	1032.2	271116	5.74	0.72	0.68	0.70
duke2	414.3	106240	4.87	273.1	65001	3.80	0.65	0.61	0.78
vg2	100.5	20847	3.20	63.1	12526	2.15	0.63	0.60	0.67
apex2	283.0	68251	6.89	237.5	54155	4.67	0.90	0.79	0.67
sqrt8	34.4	7245.9	2.74	32.1	6640	2.41	0.94	0.91	0.87
bw	111.3	24798	3.71	99.4	22616	2.46	0.89	0.91	0.66
clip	116.8	25814	5.87	156.5	34502	3.71	1.34	1.33	0.63
						Average Improvement:	0.81	0.78	0.71

Table I compares the generated circuits in terms of the number of gates, levels of logic, topological complexity (i.e. average topological wire length) [18] and estimated pre-layout delay. The results in this table suggest the following observations:

- Even though minimization of gate count is not a primary objective in the M32 system, it generates implementations with fewer gates in all but two cases. In some cases the reduction in gate count is almost 50%.
- M32-generated circuits have consistently fewer logic levels, in several cases being almost half as deep as SIS-generated circuits.
- The topological complexity of M32-generated circuits is consistently lower than that of SIS-generated circuits. The metric is similar to the fanout range suggested by Vaishnav and Pedram [29] for controlling routing complexity during technology-independent logic synthesis. The topological complexity however, should not be interpreted as a lone judge of circuit quality. This becomes apparent from the post-layout results in Table II.
- The pre-layout circuit delays of M32-generated circuits are consistently lower than those of SIS-generated circuits, the average improvement in delay being about 30%.

The run times of M32 were comparable to or better than those of SIS-1.2 for all benchmarks suggesting that the use of more powerful decompositions and substitutions is computationally feasible.

To get a better indication of synthesis quality, the netlists produced

TABLE III: Results from mapping to a richer technology library

Circuit	SIS				M32			
	NAND2/INV		mcnc.genlib*		NAND2/INV		mcnc.genlib*	
Name	Area	Delay	Area	Delay	Area	Delay	Area	Delay
z4ml	142	17.6	131	12.0	84	11.6	85	11.2
vda	2731	62.3	2188	21.6	1870	34.3	1558	20.1
inc	268	30.2	223	13.8	236	17.2	189	10.8
count	521	24.1	394	14.4	340	21.1	277	17.2
ldd	237	18.9	198	12.7	194	15.4	167	12.1
b9	280	17.8	224	12.6	297	14.8	261	10.8
ex4	1149	28.0	888	18.5	1078	19.3	843	15.6
cordic	301	23.4	244	17.2	216	14.6	163	11.1
cps	3641	61.9	2862	29.5	2753	41.3	2368	20.0
duke2	1263	35.0	1039	18.6	898	24.8	779	15.6
vg2	394	21.6	297	12.9	239	15.5	174	14.1
apex2	963	44.6	797	28.9	797	25.4	649	21.5
sqrt8	133	18.3	111	12.4	130	17.6	110	11.5
bw	395	28.9	333	14.3	349	18.3	328	11.3
clip	413	31.3	338	23.9	521	17.5	437	15.7

*Technology mapping was done using the SIS-1.2 command `map -s -n 1 -AFG -p`

from M32 and SIS-1.2 were laid out using the Epoch standard cell place and route tools [8] from Cascade Design Automation. The layouts were generated using cells in a 0.5μm CMOS process with two layers of metal, and allowing over-cell routing. I/O pins were distributed around the perimeter of the standard cell block. Delays were computed using the Epoch static timing analyzer TACTIC. These results, shown in Table II, indicate a 23% average improvement in total area, routing length and post-layout delay for M32-generated circuits. The layouts generated for a representative circuit, the `cordic` benchmark, are shown in Fig. 4.

It is important to note that for some of the M32-generated circuits, such as `b9` and `clip`, smaller average topological wire length does not imply reduced wiring when compared to SIS-1.2. We conjecture that this is due to the larger degree of reconvergent fanout in these circuits, since gates in the M32 algorithm become available across the entire logic.

The `cordic` benchmark is also used in Fig. 5 to highlight the constructive nature of M32's synthesis algorithm and to illustrate its ability to dynamically adjust the implementation topology. The two variants shown in the figure were forced to diverge after the 28th iteration of the synthesis loop (1/3 of total iterations for the Figure 5 (a) solution). The gates marked with ● in both variants correspond to the common portion of the implemented schematics. The implementation in part (a) of the figure corresponds to the results shown in Table I and Table II, and reflects the incorporation of topological complexity constraints. The implementation in part (b) was generated by relaxing these constraints after the 28th iteration. This incremental synthesis capability can prove invaluable when the generated netlists marginally fail to meet specifications and must be fine tuned in the neighborhood of a given solution.

Table III shows a final experiment designed to assess the effect of using a richer gate library on the quality of the generated circuits. Since the only library that is currently supported by our prototype implementation of M32 is the simple NAND2/INV library, we had to resort to a less-than-ideal work-around to generate circuits based on other libraries. Using the SIS-1.2 `map` command, the NAND2/INV circuits produced by M32 were technology-mapped to the `mcnc.genlib` library. The same mapping process was also applied to the NAND2/INV circuits generated by SIS-1.2. The columns labeled "Area" and "Delay" record, respectively, the active areas and circuit delays of each implementation as reported by the mapper.

An examination of these results indicates that, overall, the circuits produced from M32 are still faster and smaller than those produced from SIS-1.2. However, the improvement is not as pronounced as it was for the NAND2/INV library. This outcome is hardly surprising since the above mapping process is decidedly antithetical to the constructive synthesis philosophy of M32 and undoes many of its gains. Specifically, mapping by tree covering is ill-suited to a

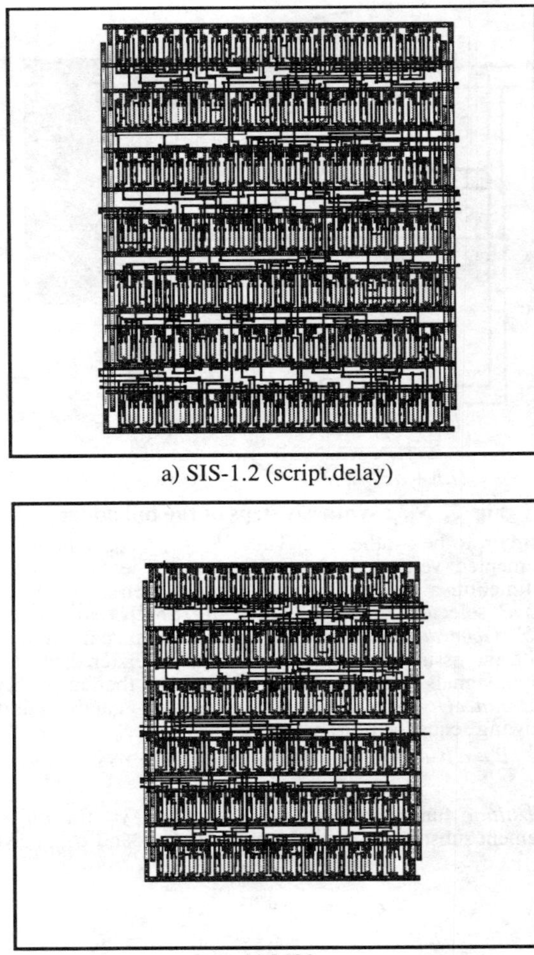

a) SIS-1.2 (script.delay)

b) M32

Fig. 4. `cordic` relative layout areas (shown to the same scale)

highly optimized DAG and we conjecture that exact DAG covering may have produced better results. It must be noted, however, that DAG covering is notoriously difficult and no published algorithms that can effectively handle large circuits have been demonstrated. We believe that a more natural approach for solving this problem is the extension of the M32 algorithm to handle arbitrary gate libraries directly. This effort is currently underway.

VI. Conclusions and Future Work

The M32 synthesis approach outlined in this paper is a promising alternative to conventional multi-stage logic synthesis algorithms. Initial results from a prototype implementation are encouraging and suggest that further exploration of this method is worthwhile. We are currently examining a number of extensions and variants including:

- Experimentation with other structural complexity metrics
- Support of arbitrary gate libraries
- Synthesis of partially-specified functions
- Exploration of other functional representations, such as BDDs, to enable more powerful Boolean transformations

Ultimately, we would like to integrate physical optimization (placement and routing) with logic synthesis for better management of interconnect effects on both area and delay.

References

[1] P. Abouzeid, K. Sakouti, G. Saucier, and F. Poirot. Multilevel synthesis minimizing the routing factor. In *Proc. 27th DAC*, pp. 365–368, June 1990.

[2] K. Bartlett, W. Cohenand A. de Geus, and G. Hachtel. Synthesis and optimization of multilevel logic under timing

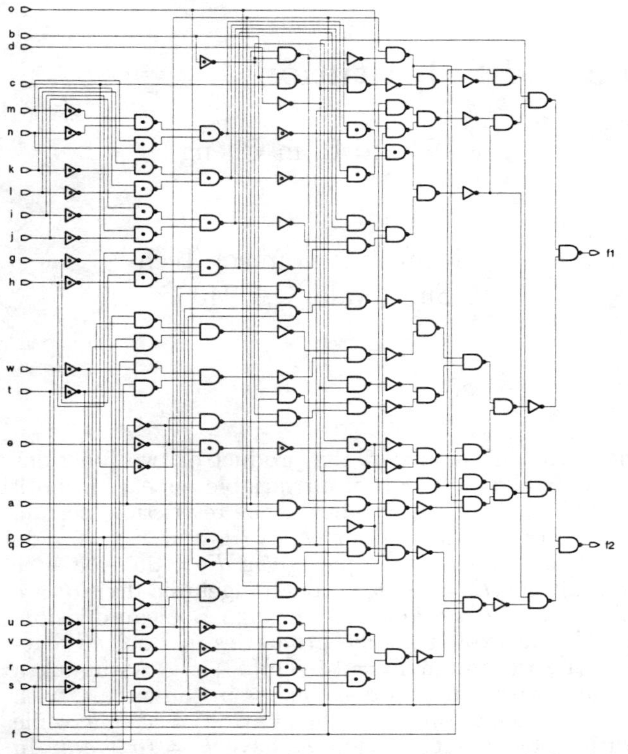

a) Synthesis with topological constraints: 133 gates, 11 levels

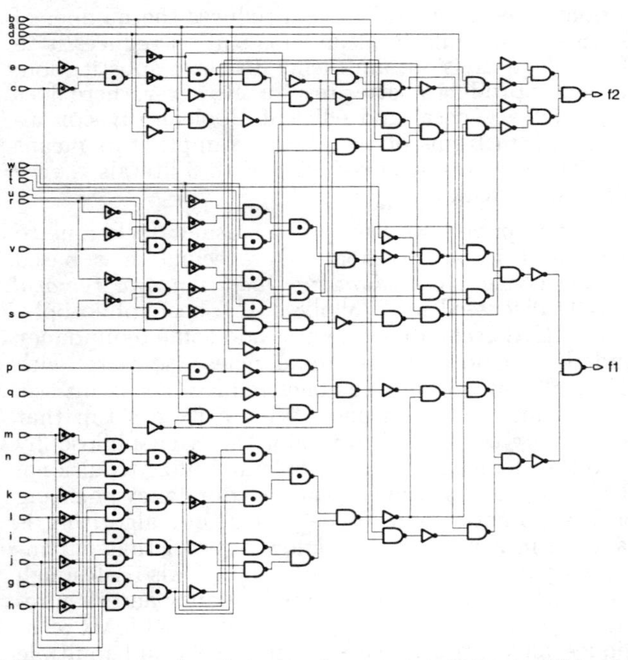

b) Synthesis with relaxed structural constraints: 116 gates, 13 levels

Fig. 5. Two incrementally-different implementations of the `cordic` circuit

constraints. *IEEE Trans. CAD IC*, CAD-5, 1986.

[3] R. Brayton, E. Detjens, S. Krishna et al. Multiple-level optimization system. In *Proc. ICCAD*, Nov. 1986.

[4] R. K. Brayton, G. D. Hachtel, L. A. Hemachandra et al. A comparison of logic minimization strategies using ESPRESSO. In *Proc. IEEE Intern Symp. Circ. Systems*, pp. 42–48, May 1982.

[5] R. K. Brayton and C. McMullen. Synthesis and optimization of multistage logic. In *Proc. ICCD*, pp. 23–28, 1984.

[6] R. K. Brayton, C. McMullen, G. D. Hachtel, and A. Sangiovanni-Vincentelli. *Logic Minimization Algorithms for VLSI Synthesis*. Kluwer Academic Publishers for VLSI Synthesis, 1984.

[7] R. K. Brayton, R. Rudell, A. Sangiovanni-Vincentelli, and A. Wang. MIS: A multiple-level logic optimization system. *IEEE TCAD IC*, 6:1062–1081, Nov. 1987.

[8] Cascade Design Automation, Bellevue, WA 98006. *EPOCH User's Manual, ver. 3.2*, 1995.

[9] E. Davidson. An algorithm for NAND decomposition under network constraints. *IEEE TC*, C-18(12):1098–1109, Dec. 1969.

[10] E. Detjens, G. Gannot, R. Rudell, et al. Technology mapping in MIS. In *Proc. ICCAD*, pp. 116–119, Nov. 1987.

[11] D. L. Dietmeyer. *Logic design of digital Systems*. Allyn and Bacon, Boston, MA, 1978.

[12] E. M. Sentovich et. al. SIS: A system for sequential circuit synthesis. Technical Report UCB/ERL M92/41, UC Berkeley, Electronics Research Laboratory, May 1992.

[13] J. P. Fishburn. LATTIS: An iterative speedup heuristic for mapped logic. In *Proc. 29th DAC*, pp. 488–491, June 1992.

[14] D. Gregory, K. Bartlett, A. de Geus, and G. Hachtel. Socrates: A system for automatically synthesizing and optimizing combinational logic. In *Proc. 23rd DAC*, pp. 79–85, 1986.

[15] G. Hachtel, R. M. Jacoby, K. Keutzer et al. On properties of algebraic transformations and the synthesis of multifault-irredundant circuits. *IEEE TCAD IC*, 11(3):313–321, March 1992.

[16] K. Keutzer. DAGON: technology binding and local optimization by DAG matching. In *Proc. 24th DAC*, pp. 341–347, June 1987.

[17] K. Keutzer, A. R. Newton, and N. Shenoy. The future of logic synthesis and physical design in deep-submicron process geometries. In *Proc. ISPD*, pp. 218–224, 1997.

[18] V. N. Kravets and K. A. Sakallah. Constructive multilevel logic synthesis under properties of boolean algebra. Technical Report CSE-TR-359-98, March 1998.

[19] E. Lehman, Y. Watanabe, J. Grodstein and H. Harkness. Logic decomposition during technology mapping. In *Proc. ICCD*, pp. 263–271, 1995.

[20] P. McGeer, R. Brayton, and A. Sangiovanni-Vincentelli. Performance enhancement through the generalized bypass transform. In *Proc. ICCAD*, pp. 184–187, Nov. 1991.

[21] G. De Micheli. *Synthesis and Optimization of Digital Circuits*. McGraw-Hill, Inc., 1994.

[22] L. Benini G. De Micheli. A survey of boolean matching techniques for library binding. *ACM TODAES*, 2(3), 1997.

[23] M. Pedram and N. Bhat. Layout driven technology mapping. In *Proc. 28th DAC*, pp. 99–105, June 1991.

[24] G. Saucier, J. Fron, and P. Abouzied. Lexicographical expressions of boolean functions with applications to multilevel synthesis. *IEEE TCAD IC*, 12:1642–1654, Nov. 1993.

[25] P. R. Schneider and D. L. Dietmeyer. An algorithm for synthesis of multiple-output combinational logic. *IEEE TC*, C-17(2):117–128, Feb. 1968.

[26] K. Scott and K. Keutzer. Improving cell libraries for synthesis. In *Proc. IEEE Custom Integrated Circuits Conf.*, pp. 721–724, 1994.

[27] K. J. Singh and A. Sangiovanni-Vincentelli. A heuristic algorithm for the fanout problem. In *Proc. 27th DAC*, pp. 357-360, June1990.

[28] H. J. Touati, H. Savoj, and R. K. Brayton. Delay optimization of combinational logic circuits and partial collapsing. In *Proc. 28th DAC.*, pp. 188–191, June 1991.

[29] H. Vaishnav and M. Pedram. Minimizing the routing cost during logic extraction. In *Proc. 32nd DAC*, pp. 70-75, June 1995.

[30] S. Yang. *Logic synthesis and optimization benchmarks user guide – version 3.0*. MCNC, Research Triangle Park, NC, January 1991.

[31] K. Yoshikawa, H. Ichiryu, H. Tanishita et al. A depth-decreasing heuristic for combinational logic; or how to convert a ripple-carry adder into a carry-lookahead adder or anything in-between. In *Proc. 28th DAC*, pp. 112–117, June 1991.

Efficient Boolean Division and Substitution

Shih-Chieh Chang
Dept. of CS and IE
National Chung Cheng University
Chia-Yi, TAIWAN, ROC

David Ihsin Cheng
Ultima Interconnect Tech.
Sunnyvale, CA 94086

Abstract

Boolean division, and hence Boolean substitution, produces better result than algebraic division and substitution. However, due to the lack of an efficient Boolean division algorithm, Boolean substitution has rarely been used. We present an efficient Boolean division and substitution algorithm. Our technique is based on the philosophy of redundancy addition and removal. By adding multiple wires/gates in a specialized way, we tailor the philosophy onto the Boolean division and substitution problem. From the viewpoint of traditional division/substitution, our algorithm can perform substitution not only in sum-of-product form for but also in product-of-sum form. Our algorithm can also naturally take all types of don't cares into consideration. As far as substitution is concerned, we also discuss the case where we are allowed to decompose not only the dividend but also the divisor. Experiments are presented and the result is promising.

1 Introduction

In multi-level logic synthesis, an important step in minimizing the area of a circuit is *substitution* [4] (or *resubstitution* [10]). Substitution refers to the step where a function is simplified in complexity by using an additional input that was not previously in the function's immediate fanins. Substitution can reduce the complexity of a function because part of the function is replaced by the additional input that represents some existing function in the circuit. The expression of the existing function is therefore shared and reused. To perform substitution, the concept of *division* plays a major role. Given two Boolean functions F and D, if we can express F in the form $F = Q \cdot D + R$, where $\cdot$ and $+$ respectively represent the Boolean AND and Boolean OR operators, then we say that F can be *divided* by D and that functions Q and R are, respectively, the *quotient* and the *remainder*.

Substitution can be algebraic or Boolean, depending on if the underlining division is algebraic or Boolean. In algebraic division [4], logic expressions are treated as algebraic polynomials, with some restrictions placed on the manipulations of the polynomials. In particular, the product of two functions $F \cdot G$ is algebraic only if no variable appears in both F and G. As a consequence of the restriction, certain Boolean identities such as $x \cdot \overline{x} = 0$ and $x \cdot x = x$ do not exist. As an example, given $F = \overline{a}b + bc + ac$ and divisor $D = a + b$, through algebraic division we obtain $F = c(a + b) + \overline{a}b$. Through Boolean division, which can exploit all the properties in Boolean algebra [1], we obtain $F = (\overline{a} + c)(a + b)$. Assuming a node d with function $a + b$ exists in the circuit, with algebraic substitution we then have $F = cd + \overline{a}b$, while with Boolean substitution we have $F = (\overline{a} + c)d$. In this example, function F has 5 literals[1] before substitution. Algebraic substitution reduces the number of literals to 4, while Boolean substitution reduces it to 3. Boolean division, and hence Boolean substitution, in theory produces better results. However, there does not exist a general and efficient Boolean division algorithm. In terms of the above example, this means that the best result of reducing F to 3 literals is very difficult to achieve.

In this paper, we first present a new technique to perform Boolean division. Our technique is based on the concept of *redundancy addition and removal* (RAR) discussed in [2][3][5][8]. The basic philosophy of the RAR technique is to first add some redundancy and then remove other redundancies elsewhere, with the goal that the removed ones reduce the circuit size more than the added one. With a fixed setup that is specially configured, we tailor the RAR philosophy onto the Boolean division problem. Unlike traditional RAR techniques, which require redundancy checking on the potential wire to be added, our algorithm is tailored in a way that we know a priori that our interested potential wire is redundant. Also, although quite effective on adding one redundancy and then removing other redundancies, the traditional RAR techniques have little success on trying to add multiple wires/gates. In our algorithm, the traditional RAR philosophy is tailored to add multiple wires/gates in a specific way particularly for the Boolean division problem.

As far as substitution is concerned, knowing how to perform division is only the first step. The second step is to choose potential divisors. Traditionally, substitution on a function F is done by going through the existing nodes in the circuit and treating each of them

[1] In *factored form* [4], as opposed to sum-of-product form.

DAC 98, San Francisco, California

as a potential divisor of F. Division is tried on each potential divisor and substitution is carried out when the trial is favorable. Since it is up to the underlining division algorithm to conclude if a divisor is good or not, the algorithm may miss some "good" divisors. In the example mentioned earlier, let us say the node with function $a+b$ does not exist and, instead, a node with function $D = a+b+x$ exists. Since function F does not depend on variable x, a traditional division algorithm would quickly conclude that the quotient of function F divided by D is 0, and therefore no substitution would occur. However, if we slightly change the circuit structure by decomposing $a+b+x$ to two nodes $n_1 = a+b$ and $n_2 = n_1+x$, function F can then be substituted with node n_1. We will use the term *basic division* to refer to the scenario where the given divisor is not allowed to be decomposed, and the term *extended division* for the scenario where the divisor is allowed to be decomposed, certainly with some purpose in mind. In the above example where function $F = \bar{a}b+bc+ac$ is divided by $D = a+b+x$, we would say that under basic division the quotient is 0. For the same F and D we would also say that under extended division the subexpression $a+b$ can be extracted out as a new divisor, and with the new divisor $a+b$ the quotient is $\bar{a}+c$. From this viewpoint, all the traditional division algorithms perform only basic division, while our algorithm presented in this paper performs extended division.

Traditional substitution approaches operate on each node's internal sum-of-product data structure, and hence can only perform substitution/division in the sum-of-product form. In contrast, our algorithm operates on circuit structure directly. Given an initial circuit, the first step of our algorithm is to decompose each node's internal sum-of-product form into a two-level AND and OR gates. The circuit then, in general, has a level of AND gates, followed by a level of OR gates, and so on. As a result, in addition to the traditional sum-of-product type of substitution, our algorithm can also perform substitution in the flavor of product-of-sum form. In other words, in two-level form, whether the dividend/divisor are a bunch of ANDs followed by an OR, or a bunch of ORs followed by an AND are completely symmetric to us. For example, let $F = (\bar{a}+b)(b+c)(a+c)$ and $D = (a)(b)$ be existing nodes. With our algorithm we can quickly substitute D into F and obtain $F = D+\bar{a}c$, i.e., $F = ab+\bar{a}c$. Performing substitution in such a manner is completely not possible in the traditional approaches because of the strong attachment to the underlining sum-of-product expression, while in our technique performing substitution through sum-of-product form or product-of-sum form are basically the same.

2 Redundancy addition and removal

The most related work to our Boolean substitution algorithm is the technique of *redundancy addition and removal* (RAR). Here we provide a detailed review. In [2][3][5][8], the technique of RAR is proposed and applied to general multi-level logic optimization. The basic philosophy in RAR is to add some redundancy first and then try to remove other redundancies elsewhere, with the goal that the removed ones reduce the circuit size more than the added one. We review the technique with an example circuit. Fig. 1(a), without the dotted wire, shows an irredundant circuit. The dotted wire $g5 \rightarrow g9$ is a redundant wire, i.e., adding the wire does not change the circuit's functionality. However, once this wire is added, the two thick wires, $g1 \rightarrow g4$ and $g6 \rightarrow g7$, become redundant. In this case, we can remove these two redundant wires without changing the circuit's functionality. After removing these two wires, we then have the circuit shown in Fig. 1(b), which is smaller in size.

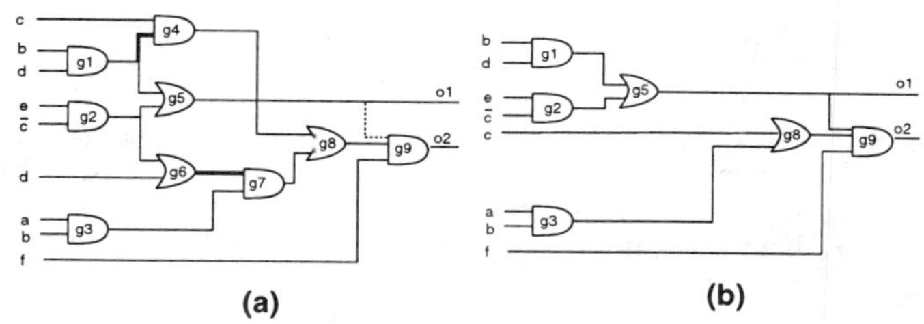

Figure 1: The RAR technique

In general, the RAR technique first decides, based on some cost function, some existing irredundant wire that is the target to be removed. Then the technique searches for some non-existing wire, sometimes called a candidate connection, that once added can remove the target wire. Finally the technique checks if the candidate connection is redundant, i.e., if adding the non-existing wire preserves the circuit's functionality. Only when the candidate connection is verified as redundant, we can then add the connection and further remove the target wire. Note that most of the RAR techniques only try to incrementally add one wire at a time. Due to a high search space, efforts that try to add multiple wires/gates and remove even more wires/gates have only little success (e.g.: [2]).

3 Basic division

Given a function F and a divisor D, we use the term *basic division* to refer to the scenario where the divisor is not allowed to be decomposed, and use the term *extended division* to refer to the scenario where the divisor can be freely decomposed, with some optimization goal in mind. In this section we focus on basic division.

3.1 SOS and POS of a function

We first need some definitions. A *product term*, or *cube*, is a set of literals AND'ed together. A *sum term* is a set of literals OR'ed together. A function f_1 *contains* a function f_2 if the on-set of f_1 contains the on-set of f_2. As an example, function (cube) a contains function (cube) ab; function (sum term) $a+b$ contains function (sum term) a. Furthermore, we define SOS and POS of a function as follows:

- **SOS:** Given a function F in two-level sum-of-product form, we say a function G, also in sum-of-product form, is a *sum-of-subproduct*, or SOS, of F if every cube in F is contained by at least one cube in G.

- **POS:** Given a function F in two-level product-of-sum form, we say a function G, also in product-of-sum form, is a *product-of-subsum*, or POS, of F if every sum term in F contains at least one sum term in G.

For example, $D = a+b$ is a SOS of $F = \overline{a}b+bc+ac$ because every cube in F is contained by either cube a or cube b in D. For another example, $D' = a+b+x$ is also a SOS of the above F, since adding more cubes to D does not change the original containment relationship in F. On the other hand, function $E = ab+c$ is not a SOS of F, since cube $\overline{a}b$ is not contained in any cube in function E.

On the POS side, for example, $D = (a)(b)$ is a POS of $F = (\overline{a}+b)(b+c)(a+c)$ because every sum term in F contains either sum term a or sum term b in D. For another example, function $D' = (a)(b)(x+y)$ is also a POS of the above F, since adding more sum terms to D does not change the original containment relationship in F. On the other hand, function $E = (a)(b+c)$ is not a POS of F, since sum term $\overline{a}+b$ does not contain any sum term in function E.

The concepts of SOS and POS play a central role in our algorithm, and we now look at some of their simple properties.

Lemma 1 *Let function G be a SOS of function F. Then $F = F \cdot G$.*

Lemma 2 *Let function G be a POS of function F. Then $F = F + G$.*

These two lemmas establish the ground where we can tailor the technique of redundancy addition and removal (RAR) onto our substitution problem. To illustrate the concept, we take the example of $F = \overline{a}b + bc + ac$ and $D = a + b$ from Section 1. Since D is a SOS of F, by Lemma 1, the new function $F_{new} = (a+b)(\overline{a}b+bc+ac)$ must be equivalent to the original function F. From the RAR viewpoint, we have successfully "added" a redundancy into the circuit. Focusing on the original F part inside F_{new}, we then try to remove as many redundancies as possible, and can quickly arrive $F = (a+b)(\overline{a}+c)$. Symmetric to the SOS case, we can perform similar operations on POS. Let $F = (\overline{a}+b)(b+c)(a+c)$ and $D = (a)(b)$. Since D is a POS of F, by Lemma 2, the new function $F_{new} = (a)(b) + (\overline{a}+b)(b+c)(a+c)$ must be equivalent to the original function F. Focusing on removing redundancies from the original F part inside F_{new}, we then quickly have $F = ab + \overline{a}c$.

3.2 Performing basic division

Given a function F and a divisor D, in this section we present an algorithm that performs basic Boolean division, i.e., $F = Q \cdot D + R$. The best way to explain our algorithm is to discuss it with an example. Fig. 2(a) shows two nodes, which correspond to $F = \overline{a}b+bc+ac+y$ and $D = a+b$. Since our central idea is based on the SOS concept, the first step to

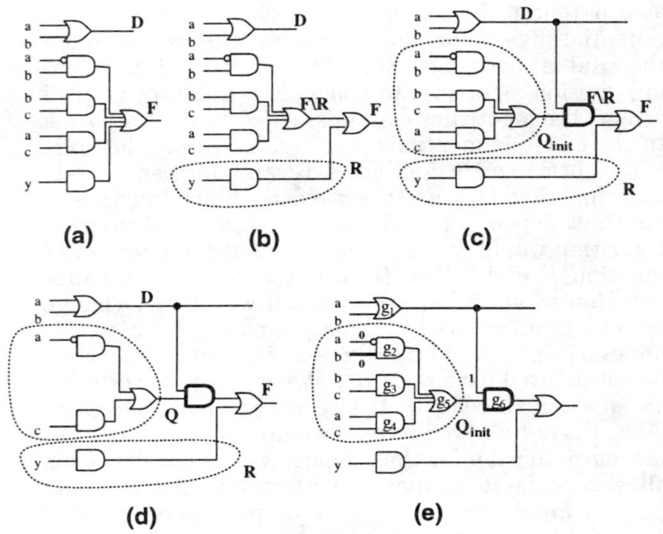

Figure 2: Basic division

perform F divided by D is to take out from F all the cubes that are not contained by any cube in D, and such cubes will be our final remainder term R. Among the four cubes in F, y is the only such cube since $y \not\subseteq a$ and $y \not\subseteq b$. Fig. 2(b) shows the circuit structure after we form the remainder y, where we use dotted circle R to indicate the remainder region and $F \backslash R$ to denote the resulting function with cube y taken out from F. Since every cube in $F \backslash R$ is now contained by at least one cube in D, D is a SOS of $F \backslash R$. By Lemma 1, $F \backslash R$ would stay unchanged if AND'ed with D. This fact is shown in Fig. 2(c) with an extra bold AND gate and the shift of $F \backslash R$ from before this AND gate to after this AND gate. From the viewpoint of the RAR technique, we have successfully added a redundancy and the circuit still has the same functionality. Now the region marked by the circle Q_{init} is highly redundant. The final step is to perform redundancy removal on the Q_{init} region and we reach the final result shown in Fig. 2(d), which is of the form $F = Q \cdot D + R$. To show how redundancy removal is done, we duplicate the circuit snapshot shown in Fig. 2(c) to Fig. 2(e) and remark some nodes. Let us illustrate how wire $b \rightarrow g_2$, the thick wire in Fig. 2(e), is detected as a redundant wire. For wire $b \rightarrow g_2$ stuck-at-1 fault to be testable, b must be 0 to activate the fault. For the fault effect to propagate through gate g_2, a must be 0. For the fault effect to propagate through gate g_6, g_1 must be 1. Since $a = 0$ and $b = 0$ implies $a + b = 0$, gate g_1 must be 0, which is a conflict. A conflict during the implication process means the fault $b \rightarrow g_2$ stuck-at-1 is untestable, and therefore wire $b \rightarrow g_2$ can be replaced by a constant 1. Our basic division algorithm works as illustrated by the above example. In summary, our algorithm consists of three steps. The first step of our algorithm is to decompose the dividend F so that the cubes that make the divisor D not a SOS of F form the remainder R. The second step is to AND D with $F \backslash R$, which does not change the functionality of $F \backslash R$ by Lemma 1. The third step is to remove all the redundancies inside the $F \backslash R$ region.

Note that it is the redundancy addition and removal steps that make our technique Boolean. Comparing to the traditional RAR techniques, however, a major difference lies on the fact that we know *a priori* that the added wires/gates are redundant because of the SOS property in Lemma 1. In other words, unlike the traditional RAR techniques, we do not need to check if the added wire/gate are redundant or not. Furthermore, as mentioned in Section 2, there has been little success in works trying to generalize the RAR technique to adding multiple wires/gates. What our algorithm does is essentially a tailored version of the RAR philosophy onto the substitution problem, with a fixed configuration of multiple wires/gates addition. Also note that since the added wires/gates are known to be redundant a priori, the most time-consuming step in our algorithm is only on the redundancy removal step. With different implication methods (e.g.:[6][7][9]) we can actually adjust the tradeoff between the run time and the quality of result. For example, we can limit our implication process only inside a small region, the $F \backslash R$ region plus the D region. As far as substitution is concerned, most of the reconvergences and implication conflicts would occur in this small region. Limiting the implication process inside this small region would greatly reduce the time required as opposed to a traditional redundancy removal process. On the other hand, we can certainly spend more time to perform implications to gates outside this small region, and thereby naturally incorporating any external or internal don't cares into consideration. In the extreme case, we can even adopt some quite exhaustive implication technique such as recursive learning [7] to incorporate a large amount of internal don't cares. We do not discuss the details here but simply point out the existence of such a flexibility on various implication algorithms. Finally, as can be seen from the above example, our algorithm operates on circuit structure directly, rather than manipulating expressions like traditional approaches. As mentioned earlier, we therefore are not limited to doing substitution only in terms of the traditional sum-of-product viewpoint. With the POS concept, we can also perform substitution on two functions when they are both in the product-of-sum form. Instead of using the SOS concept and Lemma 1, we can use the POS concept and Lemma 2, and the same philosophy as illustrated above would apply directly. As a simple example, imagine a circuit that is identical to the one shown in Fig. 2 with all the AND gates changed to OR gates and vice versa. With our algorithm it is as easy as was illustrated in this section, while in a traditional substitution technique all the sum-of-product expressions form a complete new problem whose result is difficult to predict. Since conceptually SOS and POS are symmetric, throughout the remaining of this paper we do not go into the details of the case for POS.

4 Extended division

The previous section presented our algorithm that performs basic division, where a divisor is not allowed to be decomposed. Given a function F and a divisor D, under basic division we seek to reexpress F as $F = Q \cdot D + R$. This means we are only allowed to decompose F but not on D. In this section we present an algorithm that performs what we call extended division. Given a function F and a divisor D, under extended division we are allowed to decompose not only F but also D, with the purpose of minimizing the number of literals in substitution. In essence, we first want to separate the cubes in D into two groups, the *core divisor* D_C and the *remaining divisor* D_R. Once this separation is determined, we decompose the original divisor D into two nodes such that $D = D_C + D_R$. Decomposing into a new node for the core divisor D_C means that D_C, a subexpression that was originally embedded in the given divisor D, is now exposed and can be used for substitution. We then apply our basic division algorithm in the previous section on function F and core divisor D_C to obtain the result. For example, given function $F = \overline{a}b + bc + ac + y$ and divisor $D = a + b + x$, we decompose the divisor D into the core divisor $D_C = a + b$ and the remaining divisor $D_R = x$. Applying our basic division algorithm on F and D_C, we then obtain the same result as illustrated in the previous section. It should be clear that the most important thing here is to intelligently determine the core divisor D_C, since once D_C is determined an extended division reduces to a basic division.

Recall that during our basic division algorithm, it is the step of redundancy removal that really performs the minimization process. Looking back in Fig. 2(c), whenever we remove a wire from the cubes in the Q_{init} region, we effectively reduce a literal in the final quotient. What we would like to have is a core divisor that is able to remove the most wires. To determine the core divisor D_C with a given function F and a given divisor D, our basic idea is to have each wire in the cubes of F "vote" for a candidate core divisor. For each wire w in the cubes of F, we perform implications to see which cubes in divisor D are able to remove wire w. For example, let function $F = abd + \overline{a}c + cd + ae$ and divisor $D = ab + c + f + bd + e$, whose circuit structure is shown in Fig. 3(a). In Fig. 3(a), we name divisor D's five cubes y_1, y_2, y_3, y_4 and y_5; we also name function F's four cubes x_1, x_2, x_3 and x_4, which are respectively driven by gates g_1, g_2, g_3 and g_4. Consider wire $a \to g_1$ stuck-at-1 fault. We have the following implications.

$a = 0$ (activate fault)	$\Longrightarrow$	$y_1 = 0$
$b = 1$ & $d = 1$ (allow fault thru g_1)	$\Longrightarrow$	$y_4 = 1$
$x_2 = 0$ (allow fault thru g_5) & $a = 0$	$\Longrightarrow$	$c = 0$
$c = 0$	$\Longrightarrow$	$y_2 = 0$

Assume for now that we somehow have determined a core divisor D_C. This core divisor, in our specialized configuration for basic division, feeds into a gate similar to the bold AND gate g_6 in Fig. 2(e) of the previous section. This means that if we want any fault effect in the Q_{init} region to propagate through the bold AND gate, this core divisor D_C must have a value 1 during the fault's implication process. In the case of extended division, if the core divisor that we eventually deter-

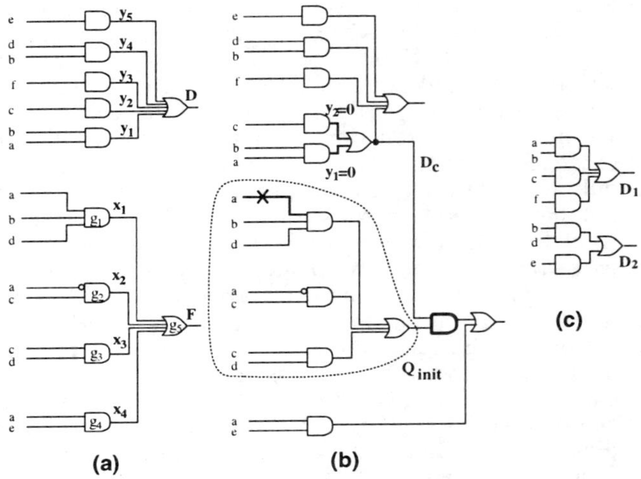

Figure 3: Extended division

mine has implication value 0 for a particular fault, the fault must be untestable because a conflict will occur with the required assignment of 1 mentioned above. We illustrate this point by continuing the example for wire $a \rightarrow g_1$ stuck-at-1 fault. We focus on the results that appear on the y_i's side whose implication values are 0. In this case, we have $y_1 = 0$ and $y_2 = 0$. Assuming we eventually choose $y_1 + y_2$ as our final core divisor, i.e., $D_C = y_1 + y_2 = ab + c$, then our basic division algorithm in the previous section would change the circuit structure to the one shown in Fig. 3(b), where $D_C = ab+c$ is connected to the bold AND gate. Following the basic division algorithm in the previous section, we would try to remove as many wires as possible in the Q_{init} region. When we again perform implications for the fault $a \rightarrow g_1$ stuck-at-1 fault, shown with a cross in Fig. 3(b), we know that y_1 and y_2, and hence D_C, all have implication value 0. This creates a conflict because, as stated earlier, for the fault effect of $a \rightarrow g_1$ stuck-at-1 to propagate through the bold AND gate, D_C must be assigned 1. In other words, if we do choose $y_1 + y_2$ as our core divisor, we expect wire $a \rightarrow g_1$ to be removed in the subsequent basic division. Now, in determining the core divisor, different wires have different implication values on the y_i's side in Fig. 3(a). In some sense, this means that each wire "votes" for a candidate core divisor. In the above example, wire $a \rightarrow g_1$ votes for candidate core divisor $y_1 + y_2$. This should become clear if we look at the complete situation after each wire performs implications on the example circuit shown in Fig. 3(a). Table 1(a) lists all the y_i's that have implication value 0 for each wire.

We explain the interpretation of Table 1(a) by examples. The meaning of the second row is that we expect wire $b \rightarrow g_1$ to be removed if we choose $y_1 + y_2 + y_4 + y_5$ as the core divisor. For simplicity, we also say that wire $b \rightarrow g_1$ *votes* for candidate core divisor $y_1 + y_2 + y_4 + y_5$. Similarly, the meaning of the fourth row is that we do not expect wire $a \rightarrow g_2$ to be removed, regardless of whatever core divisor we

wire	$y_i = 0$
$a \rightarrow g_1$	y_1, y_2
$b \rightarrow g_1$	y_1, y_2, y_4, y_5
$d \rightarrow g_1$	y_4, y_5
$a \rightarrow g_2$	
$c \rightarrow g_2$	y_1, y_2
$c \rightarrow g_3$	y_1, y_2
$d \rightarrow g_3$	y_4, y_5
$a \rightarrow g_4$	y_1
$e \rightarrow g_4$	y_4, y_5

(a) initial

wire	$y_i = 0$
$a \rightarrow g_1$	y_1, y_2
$b \rightarrow g_1$	y_1, y_2, y_4, y_5
$d \rightarrow g_1$	y_4, y_5
$a \rightarrow g_2$	
$c \rightarrow g_2$	y_1, y_2
$c \rightarrow g_3$	y_1, y_2
$d \rightarrow g_3$	
$a \rightarrow g_4$	
$e \rightarrow g_4$	y_4, y_5

(b) final

Table 1: Vote table

choose, and hence has no candidate core divisor. The remaining entries of Table 1(a) can be interpreted in a similar way.

The above voting scheme demonstrates our criteria for choosing a good core divisor. From the RAR technique's viewpoint, however, one more thing we need to make sure is that a candidate core divisor is indeed a redundant wire which we can eventually "add" to the circuit. This is done by checking if the candidate core divisor voted by a wire w is a SOS of the cube that is connected to wire w. For example, from the first entry in Table 1(a), the candidate core divisor of wire $a \rightarrow g_1$ is $y_1 + y_2 = ab + c$. The cube that is connected to wire $a \rightarrow g_1$ is $x_1 = abd$. Since the candidate core divisor $ab + c$ is a SOS of cube abd, we know eventually if we add core divisor $ab + c$ into the circuit, the added wire will be a redundant wire and therefore the circuit functionality would not change. In Table 1(a), the only candidate core divisors that do not hold for this condition are wires $d \rightarrow g_3$ and $a \rightarrow g_4$. The candidate core divisor for wire $d \rightarrow g_3$ is $y_4 + y_5 = bd + e$, which is not a SOS of the corresponding cube $x_3 = cd$. On the case of wire $a \rightarrow g_4$, candidate core divisor $y_1 = ab$ is not a SOS of the corresponding cube $x_4 = ae$. We therefore need to delete these two entries in Table 1(a) and we have our final *vote table*, shown in Table 1(b). To finalize the choice of the core divisor, various heuristics can be used. We reduce the above choice problem to a maximal clique problem in graph theory. Due to space limit, we omit the details here.

Applying our extended division algorithm to the substitution problem, we want to point out that we can actually do more than what the above discussion shows. In the above formulation, we focus only on one existing node D. In the case of substitution, we actually have freedom to select our core divisor from among many circuit nodes. As an example of how this generalization works, imagine the given divisor in the above example, $D = ab + c + f + bd + e$, does not exist in our circuit and instead, two nodes $D_1 = ab + c + f$ and $D_2 = bd + e$ exist, as shown in Fig. 3(c). When function $F = abd + \bar{a}c + cd + ae$ is given and we want to search for a good divisor between D_1 and D_2 with extended division, we can temporarily pretend that all the five cubes are from the same node, and therefore the flow is identical to the example shown in this section. Each wire in the cubes of D_1 and D_2 votes for a candidate core divisor and we have an identical vote

bench	init	sis		basic		ext.		ext.+GDC	
	lit.	lit.	cpu	lit.	cpu	lit.	cpu	lit.	cpu
C1355	560	560	1.35	560	1.96	558	1.98	558	7.59
C1908	679	632	2.61	598	1.65	598	1.70	596	3.32
C2670	939	849	3.65	840	2.63	831	2.80	828	10.03
C3540	1635	1618	7.86	1300	17.39	1345	16.82	1332	315.26
C499	560	560	0.82	560	1.94	558	1.96	558	6.50
C5315	1961	1927	8.02	1870	5.94	1864	6.44	1856	24.11
C6288	4212	4212	37.19	3747	20.64	3747	21.10	3747	21.71
C7552	2627	2534	15.94	2382	13.13	2390	14.38	2367	113.62
alu4	864	857	81.63	821	16.57	833	15.70	829	328.24
apex6	817	817	1.46	803	3.57	802	3.79	800	24.81
dalu	1809	1765	17.07	1671	37.09	1609	38.07	1587	895.06
des	4920	4918	80.63	4728	110.30	4734	115.31	4733	3325.34
frg2	1177	1135	6.82	1066	14.63	1071	15.04	1058	253.36
i10	2772	2757	13.23	2567	44.18	2558	46.17	2518	608.61
i8	1491	1491	14.04	1211	34.82	1221	39.21	1216	850.84
i9	749	749	3.07	730	10.77	730	14.16	687	413.90
rot	758	763	1.73	710	1.95	706	1.90	703	5.00
t481	1395	1313	58.80	1114	17.37	1111	15.66	1108	195.26
term1	357	349	2.93	319	1.40	323	1.31	309	5.45
x3	975	966	3.36	932	4.77	946	5.01	930	27.62
x4	522	492	1.56	488	2.21	469	2.62	457	16.81
s1196	626	624	3.10	577	5.80	573	5.89	569	83.57
s13207	3569	3552	28.84	2894	37.57	2836	36.54	2828	931.91
s1423	652	652	2.58	621	1.87	624	1.88	621	6.55
s1488	762	758	4.97	722	18.48	714	17.25	710	467.52
s1494	756	754	5.05	724	16.97	713	16.09	709	453.01
s15850	4306	4269	29.81	3789	34.16	3760	35.71	3751	605.06
total	47134	46547	503	42642	511	42496	526	42200	10347
%		1.2%		9.5%		9.8%		10.5%	

Table 2: Experimental results

table as shown in Table 1(b). The only slight modification we need is in the final maximal clique formulation, where we need to model the fact that some cubes in the second column of Table 1(b) originally come from a different node. Note that, as is also the case for basic division, we can perform extended division in terms of sum-of-product form as well as product-of-sum form. Instead of focusing on the cubes that have implication value 0, we would then focus on the sum terms that have implication value 1. The rest of the algorithm applies similarly.

5 Experimental results

We have implemented our algorithm and applied it to the substitution problem. Our implementation has three configurations: basic division, extended division without considering global internal don't cares, and extended division with global internal don't cares taken into consideration. We performed experiments on MCNC and ISCAS benchmarks within SIS [10] environment. We first run the script

eliminate 0; simplify; gcx;

on each benchmark to obtain the initial circuit. We then compare our algorithm with the algebraic resubstitution "resub -d" in SIS. Table 2 shows the comparison between SIS and our results. The first column shows the name of the circuit. The second column shows the initial literal count. The columns labeled "sis" is the result of running "resub -d" command, with subcolumns "lit." and "cpu" reporting the number of literals and CPU time, respectively. All literal counts are in factor form. Take the circuit *C2670* as an example, after running the above script, initially the circuit had 939 literals, shown in the second column. After running "resub -d" the literal count was reduced to 849 literals. With our basic division it is reduced to 840 (Column "basic"). The extended division reduced it to 831 (Column "ext."). The extended division with global don't cares taken into account brought it down to 828 (Column "ext.+GDC"). The last two rows show the summation of each column and the percentage of improvement compared to the initial literal count. As the table indicates, our division algorithms outperformed the traditional division and substitution. In terms of run time, "resub -d", the basic division, and the extended division without global don't cares spent similar CPU times. The extended division with global don't cares spent the most time.

6 Conclusion

In this paper we first presented an efficient algorithm, based on the philosophy of redundancy addition and removal (RAR), for performing Boolean division. With the concept of SOS and POS, we tailored the RAR philosophy to the Boolean division problem. The tailoring enables us to add multiple wires/gates in a specialized configuration and remove more wires/gates. Applying our Boolean division algorithm, our algorithm can perform substitution not only in the traditional sum-of-product form, but also in product-of-sum form. We then generalize our basic division to what we call extended division. Extended division allows us to decompose not only on the dividend but also on the divisor. Furthermore, our technique is able to naturally incorporate all types of don't cares into consideration. We also presented some experimental results to verify the effectiveness of our algorithm.

References

[1] R.K. Brayton and C. McMullen, "The Decomposition and Factorization of Boolean Expressions," *Proc. ISCAS,* pp. 49-54, 1982.

[2] S.C.Chang and M.Marek-Sadowska, "Perturb and simplify: multi-level Boolean network optimizer," *Proc. ICCAD,* pp. 2-6, Nov. 1994.

[3] S.C. Chang, L. VanGinneken, and M. Marek-Sadowska, "Fast Boolean Optimization by Rewiring," *Proc. ICCAD,* pp. 262-259, 1996.

[4] G. De Micheli, "Synthesis and Optimization of Digital Circuits," *Mc Graw Hill Text,* 1994.

[5] L.A.Entrena and K.T.Cheng, "Combination and sequential logic optimization by redundancy addition and removal," *IEEE TCAD,* Vol. 14, no. 7, pp. 909-916, Jul. 1995.

[6] T. Kirkand and M.R. Mercer, "A Topological Search Algorithm For ATPG," *Proc. DAC,* pp. 502-508, 1987.

[7] W. Kunz and D.K. Pradhan, "Recursive Learning: An Attractive Alternative to the Decision Tree for Test Generation in Digital Circuits," *Proc. ITC,* pp. 816-825, 1992.

[8] W. Kunz and D.K. Pradhan, "Multi-Level Logic Optimization by Implication Analysis," *Proc. ICCAD,* pp. 6-13, 1994.

[9] M. Schulz and E. Auth, "Advanced Automatic Test Pattern Generation and Redundancy Identification Techniques," *Proc. FTCS,* pp. 30-34, 1988.

[10] E.Sentovich et. el. "SIS: A System for Sequential Circuit Synthesis" *Memorandum UCB M92/41, UC, Berkeley.*

Delay-Optimal Technology Mapping by DAG Covering*

Yuji Kukimoto Robert K. Brayton
Department of Electrical Engineering and Computer Sciences
University of California, Berkeley, CA 94720
{kukimoto,brayton}@eecs.berkeley.edu

Prashant Sawkar
Strategic CAD Laboratories
Intel Corporation, Hillsboro, OR 97124
psawkar@ichips.intel.com

Abstract

We propose an algorithm for minimal-delay technology mapping for library-based designs. We show that subject graphs need not be decomposed into trees for delay minimization; they can be mapped directly as DAGs. Experimental results demonstrate that significant delay improvement is possible by this new approach.

1 Introduction

In 1987 Keutzer [7] proposed an algorithmic approach to the technology mapping problem, in which he observed similarity between this problem and the code optimization problem for programming languages and adapted an existing tree-covering technique for the latter problem to technology mapping. This approach soon dominated rule-based approaches and became the de facto standard.

In Keutzer's formulation a technology-independent circuit and each gate in a given library are decomposed into NAND2-INV circuits. The decomposed circuit is called a *subject graph* while each decomposed gate is called a *pattern graph*. The technology mapping problem can then be formulated as covering the subject graph by using pattern graphs. A subject graph is a directed acyclic graph (DAG) in general since it is derived from a given network. Keutzer showed that if a subject graph is a DAG, graph covering for *minimum area mapping* is NP-hard [8]. He further proposed a linear-time dynamic programming algorithm which guarantees optimal results for the case where a subject graph and pattern graphs are trees. These observations led naturally to the following three-step approach: (1) Decompose a subject DAG into a disjoint set of trees (2) Solve the technology mapping problem optimally for each tree (3) Glue the results together. This separation of the problem again has become a standard approach due to the theoretical justification about the complexity of DAG covering.

Inspired by Keutzer's result technology mapping has been studied extensively to optimize different criteria. Rudell [12] worked on *minimum-delay* technology mapping and showed that if loading effects are completely ignored, the minimum-delay mapping problem for subject *trees* can be solved optimally by dynamic programming in linear time. He also considered the minimum-delay mapping problem for trees under loading effects and showed that by maintaining the best mapping for each possible load at each node the same dynamic programming approach can guarantee optimal results. Touati [13] further refined this idea later by combining the optimal tree mapping with sophisticated buffer tree construction. An interesting fact is that they directly started looking at *tree covering* without studying the complexity of *DAG covering* for minimum delay.

In parallel to these works on library-based technology mapping the emergence of FPGAs posed a new technology mapping problem in early 90's. LUT-based FPGAs can implement any function of k inputs by a single LUT, where k is a fixed constant depending on a given technology. It is not practical to follow the same approach as library-based technology mapping since one needs to generate pattern graphs for all 2^{2^k} k-input functions. Based on this observation many ideas have been proposed for the FPGA mapping problem again under different cost criteria [4]. As for minimum area mapping Levin *et al.* [10] and Farrahi *et al.* [6] proved that the problem is NP-hard for $k = 4$ and $k \geq 5$ respectively. Minimum-delay mapping, on the other hand, was shown for LUT-based FPGAs to be solvable in polynomial time by Cong *et al.* in [1, 2]. Here the given circuit is directly mapped without decomposing its DAG structure to trees unlike conventional library-based mapping.

In this paper we consider the minimum-delay technology mapping problem for library-based designs where a subject graph is a DAG. Careful analysis of [2] shows that the basic dynamic programming approach in [2] is not specific to FPGA mapping and can be easily adapted to library-based mapping. This leads to a linear time algorithm for minimum-delay DAG covering under load-independent delay models. As far as we know, this is the first result that shows that the minimum-delay technology mapping problem for DAGs can be solved optimally in polynomial time. This implies that tree decomposition in performance-oriented mapping is not necessary; a given subject DAG can be directly mapped optimally. We experimentally confirmed that the additional solution space explored by this direct approach finds significantly faster mappings especially under a rich library.

2 Delay-Optimal Technology Mapping for FPGAs

We will have a close look at the FlowMap algorithm presented in [2] since this gives the basis of our proposed algorithm.

Assume that a network is decomposed into a k-bounded network [2], which is a Boolean network where the number of fanins of each node is less than or equal to k. If a given network is not k-bounded, simple decomposition can yield an equivalent k-bounded network. In the following we assume that an LUT has a unit delay and that wiring delay is negligible.

The key idea of the FlowMap algorithm is in the labeling procedure that labels each node of the DAG its optimal depth achievable. The algorithm visits each node in the network in a topological order. All primary inputs are labeled 0 assuming that they are available at $t = 0$. At each intermediate node the goal is to investigate all cuts of size less than or equal to k in the fanin cone of the node and to find the best delay realizable at the node. Each such cut represents a mapping of the node. More specifically the node can be implemented by a single LUT whose inputs are the nodes forming the cut. The constraint on the size of cuts comes from the fact that an LUT can implement any function of up to k-inputs. Since nodes are visited in a topological order, by the time the current node is examined the optimal depths of all the nodes in its transitive fanin are available. Therefore the optimal depth of the current node x can be computed as follows by dynamic programming.

$$optimal\_depth(x) = \min_{\text{cut } X:|X|\leq k} \max_{x_i \in X} (optimal\_depth(x_i) + 1)$$

Notice that this cost criterion meets *the principle of optimality* of dynamic programming. The cut X that realizes the optimal depth is stored at the node along with the depth. Although explicit enumeration of all valid cuts is possible by a brute-force approach, the complexity is pseudo polynomial $O(n^k)$, where n is the number of nodes in a given network. [2] showed that this optimal depth com-

*This work is partially supported by SRC-98-DC-324.

DAC 98, San Francisco, California

putation at each node can be formulated as a network flow problem whose runtime is strongly polynomial with respect to k.

Once all the nodes have been labeled by their optimal depths, the network is traversed backward from primary outputs to primary inputs. At each primary output an LUT is created whose fanins are the same as the best cut stored at the node. The LUT creation is repeated for each of those fanins. This process is continued until either a primary input or a node whose output is already available in the mapping is reached. An important fact is that intermediate nodes are automatically duplicated in an optimal way to guarantee optimal depths while in tree mapping no duplication is allowed.

The complexity of the entire algorithm is $O(kmn)$, where m is the number of edges in the network.

3 Delay-Optimal Technology Mapping for Library-Based Designs

Although the FlowMap algorithm was originally developed for FPGAs, the basic principle of the labeling procedure is not necessarily specific to those architectures[1]. In this section we will show how the FlowMap algorithm can be easily adapted to the standard library-based technology mapping under a load-independent delay model, where each gate has an intrinsic delay and loading has no effect in delays. This extension leads to a linear time algorithm for delay-optimal technology mapping of DAG networks. We assume that a given network is decomposed into a subject DAG as usual. Therefore, the optimality of delay is claimed with respect to this subject DAG.

3.1 Computation of Optimal Delay at Internal Nodes

The only difference between FPGAs and library-based designs is how an internal node is mapped. In FPGAs all the local mappings that cover an internal node and part of its fanin cone are examined by enumerating all k-cuts of the fanin cone, which gives the best possible delay realized at the node. This step needs to be modified for library-based designs so that all successful matches for a given set of pattern graphs are systematically examined. However this can be easily done by mimicking the standard pattern matching step used in conventional technology mapping. More precisely once the fanin cone of the node is extracted, the standard matching procedure against pattern graphs can be applied to exhaustively check all the successful matches. This way the best delay achievable at each intermediate node can be computed in a similar way to [2]. The only difference is that actual pin-to-pin delays of gates specified in a given library need to be used in our case instead of unit delay in [2]. As with FPGA mapping, the principle of optimality is still valid here.

Notice that as long as delay is optimized, any DAG pattern graph can be used directly without losing the optimality, i.e. it is not necessary to restrict the library to pattern graphs of trees and leaf DAGs. General DAG patterns are problematic only in the context of area optimization.

3.2 Pattern Matching

We now examine how pattern matching is done between a subject graph and a pattern graph.

Pattern matching between a subject graph and a pattern graph in the context of technology mapping was studied extensively by Keutzer [7] and Rudell [12]. A match between a subject graph $G_s = (V_s, E_s)$ and a pattern graph $G_p = (V_p, E_p)$ is defined as follows [12].

Definition 1 *A (standard) match of a pattern graph $G_p = (V_p, E_p)$ on a subject graph $G_s = (V_s, E_s)$ is a* one-to-one *mapping of the pattern graph nodes into the subject graph nodes $I : V_p \mapsto V_s$ such that:*

1. $\forall e = (e_1, e_2) \in E_p, (I(e_1), I(e_2)) \in E_s$.
2. $\forall v \in V_p, |i(v)| \neq 0 \Rightarrow |i(v)| = |i(I(v))|$, *where* $i(v) = \{w \mid (w, v) \in E\}$ *for* $G = (V, E)$.

The first condition requires that the edge relationship in the pattern graph is completely preserved in the subject graph. The second condition constrains the in-degree of a non-primary-input node in the pattern graph to be the same as that of the matching node in the subject graph. Notice that it is allowed for a subject-graph node covered by an intermediate pattern-graph node to have fanout to a node which is not covered by the pattern graph. However, in the conventional tree-covering based approach such a match is invalid, i.e. all fanouts of a subject-graph node matched with an intermediate pattern-graph node need to be covered by the same pattern. A match satisfying this additional constraint is called an *exact* match [12] and defined as follows.

Definition 2 *An exact match of a pattern graph $G_p = (V_p, E_p)$ on a subject graph $G_s = (V_s, E_s)$ is a one-to-one mapping of the pattern graph nodes into the subject graph nodes $I : V_p \mapsto V_s$ such that:*

1. $\forall e = (e_1, e_2) \in E_p, (I(e_1), I(e_2)) \in E_s$.
2. $\forall v \in V_p, |i(v)| \neq 0 \Rightarrow |i(v)| = |i(I(v))|$
3. $\forall v \in V_p, |i(v)| \neq 0$ *and* $|o(v)| \neq 0 \Rightarrow |o(v)| = |o(I(v))|$, *where* $o(v) = \{w \mid (v, w) \in E\}$ *for* $G = (V, E)$.

Rudell proposed an algorithm called `graph_match` [12] for the general case where both a subject graph and a pattern graph are DAGs. We can simply use this matching algorithm to enumerate all successful standard matches instead of exact matches.

Although a constraint that a mapping is *one-to-one* is posed in the above two definitions by Rudell, this can be safely dropped as follows, which leads to the definition of a larger class of matches.

Definition 3 *An* extended *match of a pattern graph $G_p = (V_p, E_p)$ on a subject graph $G_s = (V_s, E_s)$ is a mapping of the pattern graph nodes into the subject graph nodes $I : V_p \mapsto V_s$ such that:*

1. $\forall e = (e_1, e_2) \in E_p, (I(e_1), I(e_2)) \in E_s$.
2. $\forall v \in V_p, |i(v)| \neq 0 \Rightarrow |i(v)| = |i(I(v))|$

The only difference between extended matches and standard matches is that in extended matches the requirement of a mapping from the pattern graph nodes into the subject graph nodes being one-to-one is dropped. Therefore extended matches subsume standard matches. This relaxation of the requirement allows duplication of subject graph nodes while searching for a match by unfolding a DAG structure. Figure 1 shows an example where a pattern graph is matched successfully as an extended match but not as a standard match. Assume that a two-input node is a NAND2 gate and a single-input node is an inverter. Consider pattern matching at the top node of the subject graph shown on the left against the pattern graph on the right. An extended match exists by mapping both m and m' to n while a standard match does not since such a mapping violates the one-to-one mapping property. A simple modification to the `graph_match` algorithm makes the algorithm search all extended matches instead of all standard matches without changing its asymptotic complexity.

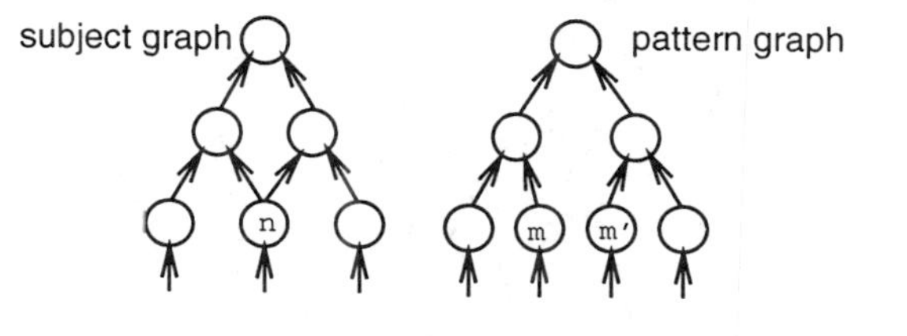

Figure 1: Standard Match vs. Extended Match

[1] It is interesting to note that [2] has a comment as follows. "Our result makes a sharp contrast with the fact that the conventional technology mapping problem in library-based designs is NP-hard for general Boolean networks."(page 2 [2])

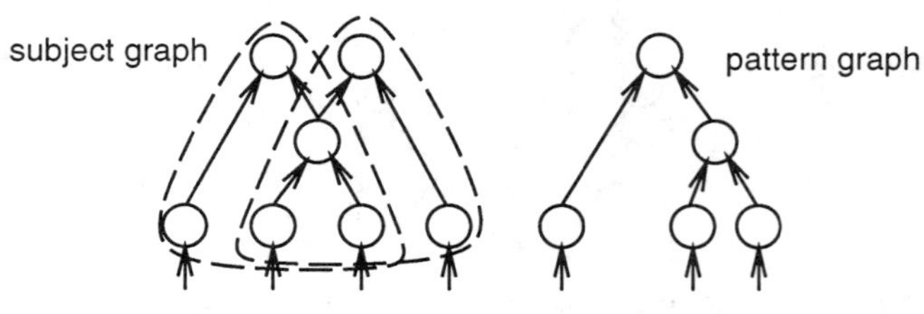

Figure 2: Duplication of Subject-Graph Nodes in DAG Mapping

3.3 Constructing an Optimum Mapping

Once a (best delay, best gate)-pair is computed at each node, a delay-optimal network can be constructed in exactly the same way as in [2]. We maintain a queue which contains nodes to be created in the final mapping. This queue is initialized to the set of all primary outputs. A node is picked up from the queue and the best gate at the node is created in the mapping. Each fanin node of the gate is then inserted to the queue if the fanin is not a primary input and does not yet have a corresponding gate in the mapping. Once the queue becomes empty, the mapping is complete.

3.4 Complexity of DAG Mapping for Delay Minimization

An application of `graph_match` to enumerate all successful matches at a node is $O(p)$ [12], where p is the number of nodes in the entire *unique* pattern graphs [2]. Since this procedure is called once at each node in a subject graph, the complexity of the labeling step is $O(sp)$, where s is the number of nodes in the subject graph. The final step of constructing a delay-optimal mapping only costs $O(s)$. Therefore the complexity of DAG mapping is $O(sp)$. Since p is a constant defined by a given technology, the procedure is linear in the size of a subject graph.

3.5 Comparison between DAG Mapping and Tree Mapping

In the past, performance-oriented technology mapping has been done by a combination of tree covering and buffer tree construction [13]. A fundamental limitation of this conventional tree-covering approach is that the search space is highly limited by the structure of a given subject graph since multiple-fanout points in the subject graph are completely preserved in the final results. On the other hand, since DAG mapping does not respect initial multiple-fanout points at all, it can explore a strictly larger search space. In other words multiple-fanout points are created as the result of delay optimization as we see later in Figure 2. Buffering techniques proposed in the literature can be directly used in conjunction with DAG covering to speed up such multiple-fanout points.

Another major difference is how subject-graph nodes are duplicated during technology mapping. DAG mapping can duplicate subject-graph nodes while creating final mappings whereas in tree mapping no duplication is allowed since each subject-graph node is covered only once by a single pattern. In some sense, subject-graph node duplication is limited to the buffer tree construction phase in the tree-mapping-based approach.

Figure 2 illustrates how duplication of subject-graph nodes helps reduce the delay of a mapping. Consider a subject graph shown on the left. Suppose that a pattern graph on the right is available in a given library. If tree mapping is invoked on this subject graph, the pattern graph is of no use since there is no exact match between the subject graph and the pattern graph. If, on the other hand, DAG mapping is employed, the two output nodes in the subject graph are implemented as in Figure 2. The cone rooted at the middle node in the subject graph is duplicated in this mapping, which makes effective use of the pattern graph possible.

This example also illustrates how multiple-fanout points are created in DAG mapping. Since the middle node of the subject graph with multiple fanouts is an internal node of each of the matchings in Figure 2, the mapped circuit does not inherit the multiple fanout point. On the other hand, the two primary inputs of the subject graph in the middle have multiple fanouts in the mapped circuit while each of the inputs has a single fanout in the subject graph.

4 Extensions

We will examine how the approach of Section 3 can be generalized to sequential circuits so that optimal cycle time is guaranteed in conjunction with retiming. We only consider sequential circuits with edge-triggered latches all of which are controlled by a single clock.

This problem was studied for LUT-based FPGAs by Pan and Liu [11]. Given a k-bounded network consider the following three-step transformation: (1) Retime an initial circuit, (2) Perform technology mapping of the combinational portion of the circuit, (3) Retime the resulting mapped circuit. [11] proposed a polynomial-time algorithm for computing the minimum cycle-time mapping among all the mapped circuits obtained by the above transformation, which was later improved in [5]. The key ingredient is a polynomial-time decision procedure which determines whether there exists a mapping whose cycle time is less than or equal to a given value. This procedure is used repeatedly to guide a binary search to determine the minimum cycle time achievable by retiming and optimal technology mapping. The core of this decision procedure is again a labeling scheme quite similar to the one used in FlowMap [2]. All k-cuts at each intermediate node are explored by considering retiming possibility. This is again done implicitly by converting the original problem to a flow network problem. This step of examining all k cuts can be replaced by pattern matching as was done for combinational mapping. All the other theories hold without any modification. Details are omitted here due to page limitation.

So far optimality is guaranteed in terms of a subject graph constructed arbitrarily from a given circuit by decomposition. Since a single subject graph is chosen among a huge number of different decompositions without knowing an actual library to be used, it is likely that many potentially good mappings are simply not explored due to this initial choice. Lehman *et al.* [9] have recently resolved this issue by encoding various decompositions into a single extended subject graph called *mapping graph* and performing technology mapping on it. Since this technique is orthogonal to our technique, the two can be combined to produce even better results. In fact, we have recently become aware that the actual implementation of [9] performed DAG covering similar to ours although [9] discusses their algorithm for subject trees [14]. It is interesting to know how much of delay improvement in [9] is due to DAG covering.

5 Experimental Results

To show the effectiveness of this approach the technology mapper of SIS was extended so that delay-optimal mapping is obtained for combinational circuits by DAG covering [3]. As discussed in the previous sections the delay model used in this experiment is the intrinsic delay model where a fixed, load-independent delay is given between each input and the output of a gate. This is in fact the delay model used in [9]. Although loading effects are certainly an important factor in delays, there are several justifications. In design scenarios where continuous sizing of any gate is permissible,

[2]Note that p is not equal to the number of nodes in the entire pattern graphs since during matching a single pattern graph is tried for all possible permutations of its inputs. p is thus the number of nodes in the expanded pattern graphs. See [12] for details.

[3]In this experiment we used `graph_match` for only finding standard matches instead of extended matches. Therefore the optimality of the results is claimed with respect to standard matches. So far we have not been able to see any major difference in mapping quality between the use of standard matches and extended matches.

circuit	Delay		Area		CPU time	
	tree	DAG	tree	DAG	tree	DAG
C2670	11.54	9.43	1552	2008	2.3	2.6
C3540	17.20	14.00	2075	2926	3.1	3.7
C5315	16.55	13.04	3687	4275	5.4	6.0
C6288	56.99	41.95	4107	9291	4.9	5.9
C7552	14.23	11.06	4983	6452	6.8	8.4

Table 1: Tree mapping vs. DAG mapping for `lib2.genlib`

circuit	Delay		Area		CPU time	
	tree	DAG	tree	DAG	tree	DAG
C2670	27	18	2998	4568	2.0	2.0
C3540	42	30	4007	6640	2.7	2.8
C5315	46	33	6817	8352	4.6	4.8
C6288	125	120	7782	7121	4.3	4.4
C7552	39	28	9552	11149	6.0	6.3

Table 2: Tree mapping vs. DAG mapping for `44-1.genlib`

circuit	Delay		Area		CPU time	
	tree	DAG	tree	DAG	tree	DAG
C2670	22	10	2314	3943	92.2	159.7
C3540	28	13	2983	6148	128.2	255.6
C5315	31	15	5115	6685	220.4	341.5
C6288	125	42	7694	14775	155.1	229.5
C7552	27	13	7062	13267	248.7	491.0

Table 3: Tree mapping vs. DAG mapping for `44-3.genlib`

one way to capture this flexibility in technology mapping is to approximate this flexibility by having many discrete size gates. This approach, however, is known to be very expensive. The approach taken in [9] is to pick a single delay for each gate and perform technology mapping by ignoring loads. Each gate in the final mapping is then continuously sized by considering actual loads so that the delay matches the one associated with the gate. Even without the capability of continuous sizing, the buffer tree construction methods of [13] can be used later at multiple fanout points to reduce load dependency of delays. Therefore the use of this delay model is at least justified as an approximation to the minimum-delay mapping problem under realistic delay models.

Table 1 shows the comparison of the quality of final circuits between the DAG mapping approach proposed in this paper and the standard tree mapping approach. In this experiment each benchmark circuit was first decomposed into a subject graph. We then applied the DAG mapping algorithm and the tree mapping algorithm on this same subject graph using MCNC gate library `lib2.genlib`[4]. The effectiveness of the DAG mapping algorithm is clear. We could obtain significantly faster circuits. CPU time was obtained on DEC AlphaServer 8400 5/300 and is reported in seconds. The increase of CPU time from tree mapping to DAG mapping is reasonable.

The same experiment was repeated using different libraries to see how the DAG mapping algorithm performs on rich libraries. MCNC libraries `44-1.genlib` and `44-3.genlib` were used in this comparison. The former library only contains 7 gates while the latter library has 625 gates, many of which are complex gates with many inputs [5]. `44-3.genlib` is a strict superset of `44-1.genlib`. Table 2 and Table 3 summarize the results of `44-1` and `44-3` respectively. We can see that the difference in mapping quality between the DAG and tree mapping approaches is further pronounced with the use of richer libraries. Complex gates are used more effectively in DAG covering than in tree covering since no tree decomposition is performed.

6 Conclusions

We have shown that the delay-optimal technology mapping problem for combinational circuits under load-independent delay models can be solved optimally in polynomial time without decomposing a subject DAG into trees. We experimentally showed that the proposed approach gives significant improvement in mapping quality compared against conventional tree mapping. Extensions of this technique to sequential circuits have also been discussed.

In this paper we focus on delay minimization without any area consideration. Therefore at each intermediate node the fastest mapping is simply created no matter how critical the node is. By constructing slower but smaller mapping for non-critical subnetworks we can have a better control over area increase. Cong *et al.* [3] already have results on area-delay tradeoff for FPGA mapping based on the FlowMap algorithm. An extension of this idea to library-based mapping is currently being investigated.

[4]Each gate has a non-zero load-dependent delay specified in `lib2.genlib`. In the experiment this was simply assumed to be zero.

[5]The largest gate has 16 inputs.

References

[1] J. Cong and Y. Ding. An optimal technology mapping algorithm for delay optimization in lookup-table based FPGA designs. In *Proceedings of IEEE International Conference on Computer-Aided Design*, pages 48–53, November 1992.

[2] J. Cong and Y. Ding. FlowMap: An optimal technology mapping algorithm for delay optimization in lookup-table based FPGA designs. *IEEE Transactions on Computer-Aided Design*, 13(1):1–12, January 1994.

[3] J. Cong and Y. Ding. On area/depth trade-off in LUT-based FPGA technology mapping. *IEEE Transactions on VLSI Systems*, 2(2):137–148, June 1994.

[4] J. Cong and Y. Ding. Combinational logic synthesis for LUT based field programmable gate arrays. *ACM Transactions on Design Automation of Electronic Systems*, 1(2):145–204, April 1996.

[5] J. Cong and C. Wu. An improved algorithm for performance optimal technology mapping with retiming in LUT-based FPGA design. In *Proceedings of IEEE International Conference on Computer Design*, pages 572–578, October 1996.

[6] A. Farrahi and M. Sarrafzadeh. Complexity of the lookup-table minimization problem for FPGA technology mapping. *IEEE Transactions on Computer-Aided Design*, 13(11):1319–1332, November 1994.

[7] K. Keutzer. DAGON: Technology binding and local optimization by DAG matching. In *Proceedings of 24th ACM/IEEE Design Automation Conference*, pages 617–623, June 1987.

[8] K. Keutzer and D. Richards. Computational complexity of logic synthesis and optimization. In *Proceedings of International Workshop on Logic Synthesis*, May 1989.

[9] E. Lehman, Y. Watanabe, J. Grodstein, and H. Harkness. Logic decomposition during technology mapping. In *Proceedings of IEEE/ACM International Conference on Computer-Aided Design*, pages 264–271, November 1995.

[10] I. Levin and R. Y. Pinter. Realizing expression graphs using table-lookup FPGAs. In *Proceedings of the European Design Automation Conference*, pages 306–311, September 1993.

[11] P. Pan and C. L. Liu. Optimal clock period FPGA technology mapping for sequential circuits. In *Proceedings of 33rd ACM/IEEE Design Automation Conference*, pages 720–725, June 1996.

[12] R. Rudell. Logic synthesis for VLSI design. Technical Report UCB/ERL M89/49, University of California, Berkeley, April 1989.

[13] H. J. Touati. Performance-oriented technology mapping. Technical Report UCB/ERL M90/109, University of California, Berkeley, November 1990.

[14] Y. Watanabe. Private communication, October 1997.

A Fast Fanout Optimization Algorithm for Near-Continuous Buffer Libraries

David S. Kung
IBM T. J. Watson Research Center
33-115, P. O. Box 218
Yorktown Heights, NY 10598 USA
914 945 3183
kung@watson.ibm.com

1. ABSTRACT

This paper presents a gain-based fanout optimization algorithm for near-continuous buffer libraries. A near-continuous buffer library contains many buffers in a wide range of discrete sizes and each buffer of a specific type satisfies a size-independent delay equation. The new fanout algorithm is derived from an optimal algorithm to a special fanout optimization problem for continuous libraries. The gain-based technique constructs fanout trees which have better timing at similar area cost. Since no combinatorial search over buffer sizes or fanout tree topologies is used, our execution time is up to 1000 times faster when compared to conventional fanout algorithms.

1.1 Keywords

Logic synthesis, gate-sizing, fanout optimization.

2. INTRODUCTION

Fanout optimization is the construction of a buffer tree to drive the large fanouts so that timing and capacitance violations are reduced to a minimum. Fanout optimization is a very effective logic transform but is typically the most time-consuming transform in technology-dependent timing optimization. It could be responsible for 70% of the slack improvement after an initial rough gate-sizing but might incur up to 50% of the time spent in timing optimization. Therefore finding a fast and effective fanout optimization algorithm deserves a high priority.

Unfortunately, the fanout optimization problem for libraries with discrete sizes has been proven to be NP-complete[1]. The majority of the previous work on this subject[2][3] relies on heuristics which assume some template for the buffer tree and then use some intelligent combinatorial search to select a good buffer size for each tree node and a good sink arrangement. These algorithms are polynomial in the number of sinks, the number of stages in the buffer tree and the number of buffers available in the library. As a result the runtime can be quite prohibitive when the number of sinks and the number of available buffers are large.

Reference [3] made the important observation that using a buffer library with almost continuous sizes greatly simplifies the fanout optimization problem. Armed with this insight, the strategy of this paper is as follows. Fanout optimization is first analyzed using a continuous buffer library. Since non-inverting buffers are usually constructed out of inverters, only inverters are assumed to be in the buffer library. *From now on, buffers and inverters are synonymous.* A continuous buffer library contains buffers which are available in continuous sizes and each buffer of a given type obeys a size-independent linear delay equation[5]

$$delay = p + l \cdot g \qquad (1)$$

where g (referred to as gain) is defined as $load/C_{in}$. p and l are independent of the size of the inverter and C_{in} is the input pin capacitance of the inverter. *In this paper, size is used interchangeably with input pin capacitance.* Using the concept of gain, a class of fanout optimization problem with a specific optimization objective is shown to be exactly solvable. A practical fanout optimization algorithm referred to as the gain-based algorithm is then derived from this optimal algorithm. The gain-based algorithm is applicable to near-continuous buffer libraries in addition to continuous buffer libraries. A near- continuous buffer library contains many buffers in a wide range of discrete sizes and each buffer of a given type obeys delay equation (1). Near-continuous buffer libraries can be realistically designed [6] and are gaining popularity as the buffer libraries of choice for high performance designs.

3. THE FANOUT PROBLEM

The inputs to the fanout problem are as follows:

- a fanout region which consists of a source and a set of N sinks, P sinks with positive polarity and Q sinks with negative polarity. Each sink, s_i, has a given polarity and drives a load, L_i. A timing constraint is imposed on each sink so that the delay from the source to sink, s_i, is less than or equal to D_i.
- a set of inverters with a continuous range of sizes.

The objective is to construct a fanout tree which presents the minimum load to the source such that the timing constraint at each sink is observed. This is in contrast to the standard optimization criterion which minimizes the area of the fanout tree so a discussion of the relevance of this alternative optimization criterion is in order. In high performance designs where delay optimization is critical, there is a tendency towards adopting a gain-based synthesis methodology[6]. We would like to argue that the minimum source load criterion is a reasonable objective within the gain-based framework.

In gain-based synthesis, library cells and gates in the network are parametrized by gain instead of size. The delay of a gate is a function of its gain and is independent of the load that it drives. Therefore it is relatively easy to find a gain assignment to satisfy the timing constraints. However, primary inputs have maximum load limits and each gate has a maximum size limit. These size constraints are difficult to satisfy. The size computation of a gate requires a propagation of the primary output loads through its fanout cone. The size of each gate in the fanin cone of the source gate (the gate that drives the source of the fanout region) is directly proportional to the source load of the fanout region. Minimizing the source load increases the chance of satisfying these size constraints.

This paper focuses on the minimum source load criterion and "optimal" in this paper carries the narrow sense of being optimal with respect to this criterion.

4. FANOUT TREES

To set the stage for theoretical analysis of fanout trees, several pertinent concepts will be defined.

Given two fanout trees T_1 and T_2, a sink s_i of T_1 is equivalent to a sink s_j of T_2 iff s_i and s_j have the same polarity, $L_i = L_j$, and the delay from the source of T_1 to s_i is equal to the delay from the source of T_2 to s_j.

T_1 is equivalent to T_2 iff

- $P_1 = P_2$ and $Q_1 = Q_2$,
- there exists a 1-1 correspondence between the sinks of T_1 and the sinks of T_2 under the equivalence relation,
- the load at the source of T_1 is equal to the load at the source of T_2.

The following lemma is an immediate consequence

Lemma 1

If T_1 is equivalent to T_2 and T_1 is an optimal solution to a fanout optimization problem, then T_2 is also an optimal solution to the fanout optimization problem.

Two important transformations on fanout trees, merge and split, are defined as illustrated by Fig. 1 where all the inverters have the same gain and the sinks of net5 is the union of the sinks of net2 and net3.

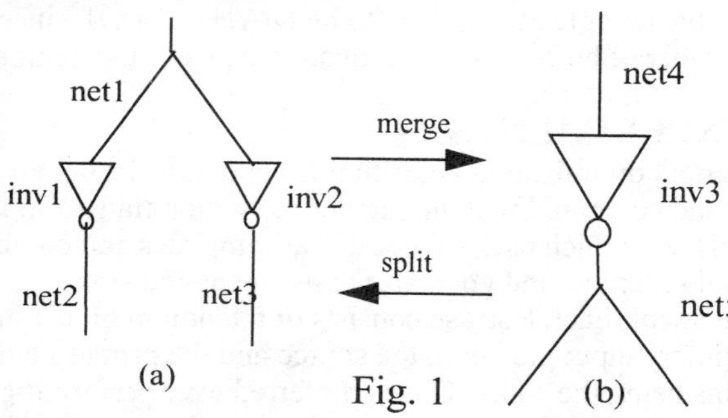

Fig. 1

Lemma 2

If T_2 is derived from T_1 by either the merge or the split transform, T_1 and T_2 are equivalent.

By repeated applications of the split transform, any fanout tree can be transformed into a tree for which each inverter has only one fanout. Fig. 2(a) is a fanout tree with 4 sinks of positive polarity and 1 sink of negative polarity and Fig. 2(b) is the fanout-free equivalent. Such a tree will be referred to as a *fanout-free tree* and is important for the following reason.

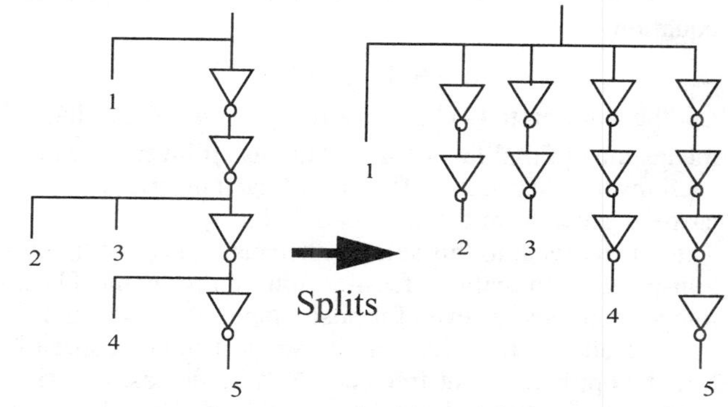

Fig. 2

Theorem 1

There exists an optimal fanout-free tree for a given fanout optimization problem.

Theorem 1 states that there is no loss of optimality by restricting the search to fanout-free trees. The next section focuses on finding the optimal fanout-free tree for a given fanout optimization problem.

5. FANOUT-FREE CHAINS

The fanout-free tree with N sinks is simply N fanout-free chains connected at the source. To find the optimal solution, it is sufficient to find the optimal solution to the following problem for each fanout-free chain:

Given a source, a sink with load L and a given polarity, and a source to sink delay constraint D, find the number of inverters, m, and their input pin capacitances, c_i, such that the input pin capacitance of the inverter connected to the source is minimized.

Let g_i be the gain of the ith inverter on the chain with the first inverter being the one connected to the source, the problem is to minimize the capacitance

$$c_1 = L/\prod_1^m g_i \tag{2}$$

under the constraint

$$\sum_1^m (p + lg_i) \le D \tag{3}$$

At first glance, there seems to be no simple solution to this problem since there are $m+1$ variables (including m itself) to solve for. However using the inequality of the arithmetic and geometric means the following theorem can be derived.

Theorem 2

c_1 is minimized when $g_1 = g_2 = \ldots = g_m = g$

Armed with theorem 2, the optimal fanout-free chain problem is reduced to finding the g and m which minimize

$$c_1 = Lg^{-m} \tag{4}$$

with the equality constraint

$$m = D/(p + lg) \tag{5}$$

Substituting (5) into (4), c_1 becomes

$$Lg^{D/(p + lg)} \tag{6}$$

The minimum of c_l occurs when g satisfies the following equation

$$\ln g = 1 + p/(lg) \quad (7)$$

Let the solution to (7) be γ and $\mu = D/(p + l\gamma)$. That means the optimal fanout-free chain has μ inverters and each inverter has gain γ. The size of each inverter can be easily derived from the sink load L and γ.

There is a caveat to this solution because μ is most likely non-integral. In reality a feasible solution for m must be an integer and must be even for positive polarity sinks and must be odd for negative polarity sinks. The procedure to find the optimal fanout-free chain is modified as follows:

- calculate the optimal γ and μ (could be fractional) as above,
- round off μ to the two nearest feasible integers accounting for polarity
- if $\mu < 1$ and sink polarity is negative the only feasible integer is 1, in which case the optimal fanout-free chain contains 1 inverter with gain $(D-p)/l$, otherwise
- let the 2 integers be μ_1 and μ_2, so the respective gains are $\gamma_1 = (D-\mu_1 p)/l$ and $\gamma_2 = (D-\mu_2 p)/l$
- assume $m = \mu_1$ and $g = \gamma_1$ is the solution with a smaller c_l
- the optimal fanout-free chain has μ_1 inverters each with gain γ_1
- the size of each inverter is computed from L and γ_1.

This fanout-free chain is optimal because c_l as a function of m, $L\left(\frac{D-mp}{lm}\right)^{-m}$, is convex for $0 \le m \le D/p$. The optimal fanout-free tree is obtained by finding the optimal solution for each of its fanout-free chains. The optimal gain value for each chain could be different. The optimal fanout-free tree is unacceptable as a buffer tree in practice because the first inverter of each chain might have a unrealistically small input pin capacitance. The next section discusses how to derive a practical fanout tree from this optimal solution.

6. THE FANOUT ALGORITHM

In this section we utilize the theoretical results from the previous sections to design a practical fanout optimization algorithm. In practice, inverter libraries do not offer continuous sizes and there are lower and upper bounds to the input pin capacitances. The fanout tree topology sought for is a tree with internal fanouts so that the sizes of the inverters fall into a realistic range. We learn from section 4 that if the gains assigned to each chain of the fanout-free tree are the same then the merge operation can be applied to create a desirable topology. Moreover, the optimal gain for each chain in the fanout-free tree is most likely not very different from γ because the absolute magnitude of the round-off in m is less than 2 and most of the times less than 1. Therefore a common gain close to γ assigned to each chain is likely to result in a near-optimal fanout-free tree. This observation leads to the concept of a *uniform gain fanout-free tree*.

The uniform gain fanout-free tree with respect to a fanout problem input and a uniform gain g is constructed as follows:

- for each sink in the fanout problem
  - compute μ by substituting g into (5)
  - round off μ to the biggest integer m less than μ accounting for polarity and such that the delay constraint for that chain is observed.
  - construct an inverter chain with m inverters each with gain g

The fanout optimization algorithm proceeds as follows:

- compute the solution, γ, to (7) given p and l from the inverter library,
- let G be a set of g's in the neighborhood of γ, e.g. if $\gamma = 3.4$ choose G = {3.0, 3.2, 3.4, 3.6, 3.8}
- for each g in G, construct a fanout tree using the following 5 steps
  - construct a uniform gain fanout-free tree with respect to the problem input and the uniform gain g
  - perform the merge operation iteratively on the uniform gain fanout-free tree until no merging is possible
  - from g and the sink loads compute the size of each inverter in the tree
  - locate the inverter in the library whose size is closest to the computed size
  - replicate inverters if the maximum inverter size available is exceeded
- choose the best fanout tree among the ones constructed

The gain-based algorithm is distinctly different from prior arts in that once a gain is selected, no combinatorial search is required in calculating the buffer sizes and the sink arrangement. The complexity of the algorithm is analyzed as follows. For each sink (5) is solved, from which the number of buffers for the tree is determined. Both operations are simple arithmetic operations and performing them on all the sinks requires complexity $O(N)$. Then each buffer in the fanout tree is selected by matching the closest size, incurring another $O(N*ln(K))$ (K is the number of inverters in the near-continuous inverter library) since in the worst case the number of buffers is proportional to the number of sinks. In reality the complexity is likely to be $O(N)+O(ln(K))$, since the number of buffers is approximately equal to the number of levels of buffering.

7. EXPERIMENTS

The fanout optimization algorithm is applied to 19 fanout cases and compared with the fanout correction transform in BooleDozer which uses LT trees[2] as templates and combinatorial search to find good size assignment and sink arrangement. Each testcase consists of a fanout region with the primary input pin being the source and the primary output pins being the sinks. This is preferred over performing fanout optimization on logic networks, in order to isolate the effect of fanout optimization. The fanout testcases are specified in a similar fashion as the fanout problem inputs discussed in section 3. Testcase (Q, P) refers to a fanout with P positive sinks and Q negative sinks. The first 10 testcases are randomly generated examples designed to stretch the limits of fanout optimization algorithms. The arrival time at the source is set to 0.0 while the sink required time is a random distribution over a prescribed range. The sink loads are also a random distribution over the range of input pin capacitances of the cells available in the library. The remaining 9 testcases are fanout regions extracted from industrial high performance designs. The library is discrete and has 30 inverters with varying sizes. For a fixed input slew, the linear delay equation (1) is a very good approximation for each inverter in the library.

The results are listed in Table 1. "slack" is the slack at the

	combinatorial			gain-based		
model	slack	area	cpu	slack	area	cpu
(5,5)	-0.01	36	13.8	0.01	32	0.12
(12,16)	0.05	60	45.8	0.05	48	0.14
(32,8)	-0.04	48	33.0	-0.01	86	0.20
(16,64)	-0.08	194	547	-0.08	172	0.40
(8,128)	0.0	266	1931	0.04	282	0.63
(28,34)	-0.06	120	189	-0.06	124	0.32
(56,44)	-0.35	210	280	-0.17	216	0.53
(12,5)	-0.04	32	5.53	-0.03	46	0.16
(7,31)	0.04	84	113	0.04	74	0.21
(18,47)	-0.03	140	272	-0.01	124	0.29
(12,20)	-0.24	28	22.6	-0.23	28	0.31
(15,23)	-0.31	34	24.1	-0.26	30	0.39
(16,17)	-0.16	28	23.2	-0.14	28	0.43
(16,18)	-0.11	28	23.4	-0.09	30	0.47
(20,17)	-0.19	26	20.0	-0.16	30	0.29
(21,12)	0.09	22	11.0	0.12	32	0.41
(25,21)	-0.37	30	32.6	-0.32	38	0.52
(6,28)	-0.52	48	27.2	-0.50	24	0.38
(8,24)	-0.45	36	22.3	-0.42	24	0.31

Table 1

source of the fanout tree, "area' is the area of the entire fanout tree and "cpu" is the run time on a 200MHz PowerPc workstation. The new fanout optimization algorithm is better in timing across the board with only a 1% increase in overall area. However, the biggest advantage is the runtime performance. Conventional combinatorial methods consume enormous computation time when the number of buffers becomes large (> 10) and when the number of sinks becomes large. The gain-based algorithm is more than 1000 times faster in such cases.

8. CONCLUSION

This paper has presented a very fast and effective fanout tree construction algorithm for a near-continuous buffer library. The algorithm is applicable to buffer libraries with discrete sizes as long as there is a wide range of buffer sizes and the size independent delay equation (1) is obeyed. Since fanout optimization is arguably the most important transform in timing optimization, the quality of logic synthesis can be substantially enhanced by this new algorithm. This leads to the issue of how a buffer library should be designed. Many buffer libraries are designed with only several (less than 5) buffers for each buffer type and buffers are constructed out of multi-stage inverters to increase the drive capability. These buffer designs make sense when fanout optimization algorithms in logic synthesis consist of combinatorial heuristics whose run time are unacceptable if the number of buffers is large. With the new algorithm, it is beneficial to design buffer libraries in the near-continuous style and allow synthesis the extra degrees of freedom to construct good fanout trees.

9. ACKNOWLEDGMENT

The author would like to thank L. Trevillyan, D. Brand, L. Stok and R. Puri for reading the manuscript.

10. REFERENCES

[1] C. L. Berman, J. L. Carter and K. F. Day, "The Fanout Problem: From Theory to Practice", Advanced Research in VLSI: Proceedings of the 1989 Decennial Caltech Conference, C. L. Seitz, editor, MIT Press, 1989, pp. 69-99.

[2] K. J. Singh and A. Sangiovanni-Vincentelli, "A Heuristic Algorithm for the Fanout Problem", Proceedings of the 27th DAC, 1990, pp. 357-360.

[3] K. Kodandapani, J. Grodstein, A. Domic and H. Touati, "A Simple Algorithm for Fanout Optimization using High-Performance Buffer Libraries", Proceedings of ICCAD, 1993, pp. 466-471.

[4] L. Stok et. al. "BooleDozer: Logic synthesis for ASICs", IBM Journal of Research and Development, vol. 40, no. 4, July 1996, pp. 407-430.

[5] I. Sutherland and R. Sproul, "The Theory of Logical Effort: Designing for Speed on the Back of an Envelope", Advanced Research in VLSI, Univ. of Calif. Santa Cruz, 1991.

[6] K. Shepard et. al. "Design Methodology for the S/390 Parallel Enterprise Server G4 microprocessors", IBM J. Res. Development, vol. 41 (1997), pp. 515-547.

Performance Driven Multi-Layer General Area Routing for PCB/MCM Designs

Jason Cong and Patrick H. Madden
UCLA Computer Science Department
Los Angeles, California, 90095

Abstract

In this paper we present a new global router appropriate for Multichip Module (MCM) and dense Printed Circuit Board (PCB) design, which utilizes a hybrid of the classical rip-up and reroute approach, and the more recent iterative deletion[9] method. The global router addresses performance issues by utilizing recent results in high performance interconnect design, while still effectively minimizing global congestion.

With experiments on the maze-routing component of our global router, we show that the choice of routing cost functions can have a significant impact on final solution quality. The results of a number of previously proposed routers may be improved dramatically by adopting the cost functions we suggest here. We also find little evidence of the "net ordering problem" when our cost functions and routing model are applied. The iterative deletion method is shown to improve global solution quality, particularly when high performance interconnect is required. We evaluate the performance of our global router by comparing the congestion of routes produced by our global router to those of a well known MCM router, V4R [14].

Our global router, MINOTAUR, supports arbitrary numbers of routing layers, differing capacities for each layer, preexisting congestion and obstacles, and high performance interconnect structures (including those which require variable width interconnect).

1 Introduction

Advances in VLSI fabrication technology have resulted in an increasing interest in interconnection and packaging technologies. For many systems, the delay due to circuit interconnections can dominate the total system delay [1, 5]. Interconnect optimization must be considered to achieve high performance at the system level.

One approach that provides many benefits to packaging level interconnection is the Multichip Module (MCM) technology. This technology increases practical packing densities, and removes a level of interconnection. In MCM designs, a number of bare chips are placed directly onto a substrate, with the substrate itself providing the circuit interconnections. Routing problems for MCM designs have many similarities to routing for dense Printed Circuit Boards (PCBs).

Routing within MCM substrates or PCBs, however, can be an extremely difficult problem with a potentially large number of interconnect layers. A Fujitsu supercomputer design, for example, has over 50 interconnect layers, each with a large routing capacity [10]. The MCM/PCB routing problem in general may be considered to be an extremely large general area routing problem, and cannot be decomposed into small channel or switchbox routing problems easily. Additionally, high performance designs must consider signal crosstalk (due to coupling capacitance between long parallel lines), and variable width wiring (to address impedance mismatches and wiring termination requirements).

The problem of MCM/PCB routing has been considered by a number of authors. One set of approaches perform global and detail routing in a single step.

An early MCM router was presented in [10], and performs a combination of 2D and 3D routing obtain a solution. Computation times for this method were relatively high.

In [13] the SLICE router was presented; it constructs detailed routes through a mixture of planar routing and two-layer maze routing. While this router was superior to a 3D maze router, it requires relatively high computation times. The V4R router [14] was subsequently presented, and efficiently routes layer-pairs one at a time using routes with no more than four vias each.

The M^2R[4] router addresses performance concerns by routing critical nets on a single layer. The remaining nets are routed using $x - y$ layer pair routing.

In [17], a linear programming approach was presented. Randomized rounding is used to obtain integer solutions. A small number of possible topologies are considered, and routes are restricted to a channel intersection graph.

The MCG [2] router considers a small number of possible routes for each net, and attempts to maximize the routing density on any layer-pair by selecting a set of compatible routes from among the unrouted nets. The results of this router are quite good, requiring fewer routing layers than many of the other MCM routers. Performance concerns are addressed by removing nets which exceed signal crosstalk thresholds, and routing them on other layers.

In [20], the MLR router was presented; it first performs layer assignment of nets, followed by Steiner optimal area routing using a hierarchical (α, β) approach (to find low-cost paths between pins).

The SEGRA [3] MCM global router has been described, and also approaches the routing problem by completing pairs of layers in each pass. A simple greedy heuristic is used to select which nets are to be extended or completed during each pass.

While the methods mentioned above consider global and detail routing in a single step, there have also been approaches which consider the problem in two steps.

In [21], the MCM substrate is decomposed into a set of "towers," rectangular regions containing portions of the routing surface through all layers. They first distribute routing density across a three dimensional surface, using hierarchical decomposition, then determine locations of nets on the boundaries of the towers, and finally obtain a routing for each tower using a detail router.

We approach the problem in a manner similar to [21], and consider the global routing problem in this paper. In contrast to these previous approachs, our MINOTAUR global router offers the following features.

- We utilize current high-performance variable-width interconnect optimization results, obtaining structures which address delay and signal integrity requirements [6]. We can in fact consider a *set* of high performance interconnect structures for each signal net, and perform congestion optimization from a global perspective.
- We optimize global congestion by not restricting the layers a route may use; some approaches route a single layer-pair at a time, and cannot obtain routing solutions which span multiple or non-paired layers.
- We combine the freedom and flexibility of maze routing solutions, with the global optimization abilities of the iterative deletion method.

While we consider MCM and PCB routing in this paper, the basic techniques used in our router can be easily applied to multilayer VLSI IC routing problems as well.

The remainder of this paper is organized as follows. In Section 2, we present our problem formulation. We also describe our approach to congestion estimation in this section. In Section 3, we describe the MINOTAUR global router. We detail our methods for applying rip-up and reroute in Section 3.1, the iterative deletion approach in Section 3.2, our support for high-performance interconnect structures in Section 3.3, and the integration of algorithms in Section 3.4. Experimental results are presented in Section 4, and we conclude our paper with Section 5.

2 Problem Formulation

Our routing model is similar to that of [21]; we divide the 3-dimensional routing surface into a set of "tiles," with each tile consisting of a number of layers. We use "tile-layer" to indicate a single layer within a tile. The routing substrate is modeled as a graph, with tile-layers as nodes, and both vias and the "borders" between tile-layers as edges. Preferred-direction routing is obtained by including (or excluding) border edges. Figure 1 shows an example of our routing model.

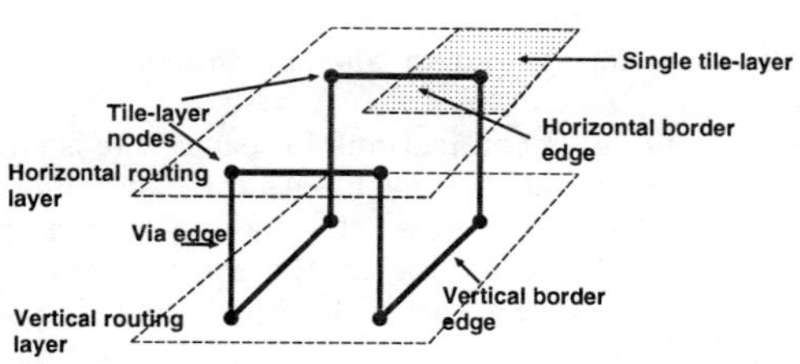

Figure 1: A two-layer four-tile routing problem.

The routing problem consists of determining the interconnect structure for a set of nets N. In addition to basic feasibility concerns (based on physical routing capacity constraints), there are a number of performance issues as well; many of these were detailed in [4]. In this paper, we are interested in meeting delay, signal integrity, crosstalk, and routing capacity constraints, while minimizing total wire length, maximum congestion, and routing resource costs.

We assume that a detail router will be used to complete the routings of the individual tiles, and determine solution feasibility through the maximum *congestion* of any tile-layer node, border edge, or via edge. We estimate solution congestion as the number of topology edges using a particular routing resource. Note that the border congestion and tile-layer congestion metrics are in general not comparable.

3 The MINOTAUR Global Router

Our approach is to first decompose each non-critical signal net into a set of edges. In most of our experiments, we use a simple minimum spanning tree algorithm, although we can also utilize Steiner tree or high performance interconnect algorithms. Steiner topologies had little impact on solution quality in our experiments. Nets which require high-performance interconnect structures are not decomposed into 2-pin edges, and are handled as described in Section 3.3.

For each edge of the non-critical nets, we wish to find a *path* through the modeled MCM substrate; we model a path as a set of tile-layer nodes, border edges, and via edges.

Our global router contains two basic algorithmic approaches: a traditional rip-up and reroute scheme, and a method based on iterative deletion. We first discuss the details of each, and then show how they are integrated.

3.1 Rip-Up and Reroute

Rip-up and reroute is one of the oldest approaches to the global routing problem; it generally requires long computation times and provides few guarantees as to solution quality. Despite these drawbacks, rip-up and reroute continues to be used in many routing applications. The rip-up and reroute portion of our global router is similar to that of [19], in that we iteratively change our congestion bounds to encourage convergence to a low-congestion solution.

Given the complexity of MCM and PCB designs, we believe that rip-up and reroute may be necessary for aggressive design. We implement rip-up and reroute in our router as follows. Each edge is first routed using a maze router based on Lee's classic algorithm [15]. We then iteratively rip-up and reroute each edge, continuing for a user-specified number of iterations, or until solution quality stabilizes. Details of our maze routing cost functions, and edge orderings used, are given below.

If high-performance interconnect structures are required, we select a single high-performance interconnect structure with the minimum area for the performance critical net, and do not modify it. The rip-up and reroute process operates only on the edges of non-critical nets, and attempts to find routes "around" the congestion of the critical nets. Congestion optimization of the high-performance nets by the iterative deletion method is more robust, and is detailed below.

3.1.1 Maze Routing Cost Functions

As is done in a number of other routers, we assign a cost to each tile-layer node, border edge, and via edge. Our objective is to assign costs to each resource such that we find a relatively short path (in terms of wiring distance), while avoiding heavily congested areas. We are therefore interested in knowing what cost functions result in the best performance, a question that has received relatively little attention in the literature.

In [19], the cost of routing in the most congested region is infinite, while the remaining regions have low cost. In [12], the cost of a routing path includes terms for the number of other routes that a path "touches" or "crosses," as well as two terms to capture the route length in preferred and non-preferred routing directions. In [18], steps in the preferred routing direction were assigned a cost of 1, while non-preferred steps were assigned a cost of 30. In [16], costs were piecewise-linear functions depending on a measure of the utilization and the capacity of each edge. Clearly, there are many possible ways to calculate the cost of using a particular resource.

In general, we expect resources which have high usage or large numbers of obstacles to have relatively high cost. We have explored three cost functions in depth, and describe them here.

We first consider a *step* function, similar to those of [19], [18], and [12], which provides a low cost if congestion falls below a specific threshold value. A high cost results if congestion is at or above the threshold value. Second, we considered a piece-wise linear *slope* function, similar to that of [16], which provides low cost when congestion is below a threshold, high cost if congestion is above *cap*, and a linearly interpolated cost when congestion is between the threshold and *cap* levels. For each iteration of rip-up and reroute, we determine the maximum and minimum congestion for any resource. We modify the *step* and *slope* functions to reflect current congestion levels by adjusting the *threshold* and *cap* parameters. We refer to this as *Adaptive Congestion Estimation (ACE)*, and perform experiments using *ACE-step* and *ACE-slope* cost functions. A third function, which we refer to as a "two-tier" function, is similar to the *slope* function when routing congestion is below a fixed physical capacity. If congestion is above physical capacity, routing costs are increased substantially. These cost functions are illustrated in Figure 2.

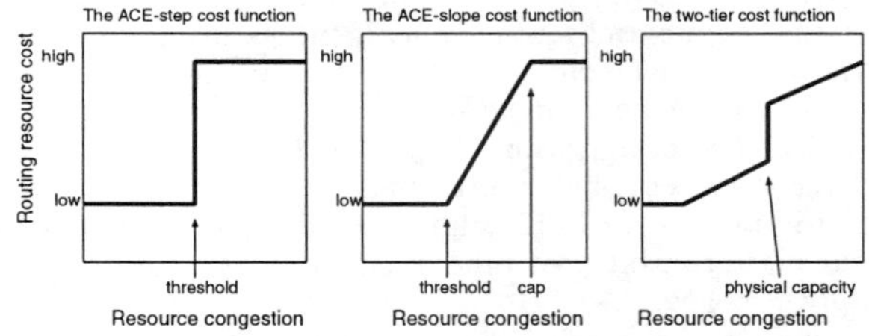

Figure 2: *ACE-step, ACE-slope*, and "two-tier" cost functions. These functions are used to determine resource cost during maze routing.

We find that the *ACE-slope* cost function is consistently superior to the *ACE-step* function. Detailed results are given in Section 4, Table 2. Many routing applications which utilize step-like functions ([19], [12],[18], for example) may obtain substantial improvements in results by adopting a slope-like cost function.

Our global router supports *separate* cost functions for each type of resource (tile-layer nodes, border edges, and via edges), and can optimize these independently.

3.1.2 The Net Ordering Problem

A perennial question in global routing problems has been *which net do we route (or reroute) first?* [9, 2]. To evaluate the importance of net ordering in our implementation, we compared five different edge orderings within the rip-up and reroute framework. Our orderings were low length first, low cost first, the reverse of these, and the default order of our design database.

Surprisingly, there was negligible difference in the results by the various orderings. While examples can be constructed in which certain orderings are clearly superior to others, we found little evidence that this occurs in our experiments. In our model, we have no hard upper limit on routing capacity, and perform rip-up and reroute until solution quality converges. Our study concludes that solution quality measured in our routing model is far more sensitive to the choice of routing cost functions than to the edge ordering.

3.2 The Iterative Deletion Method

Rip-up and reroute is computationally expensive. Through *iterative deletion*, we can obtain high quality results while using rip-up and reroute sparingly. In this section, we describe iterative deletion; Section 3.4 describes integration with rip-up and reroute. The use of iterative deletion was first proposed in [9], and was recently improved in [7].

In many iterative improvement methods, we repeatedly modify a solution, changing a small portion at each step. Rather than representing the problem with a single instance, and changing one aspect of it at each step, the iterative deletion method represents a *set* of possible states. Consider the example in Figure 3, which is a relatively simple planar routing problem with three nets.

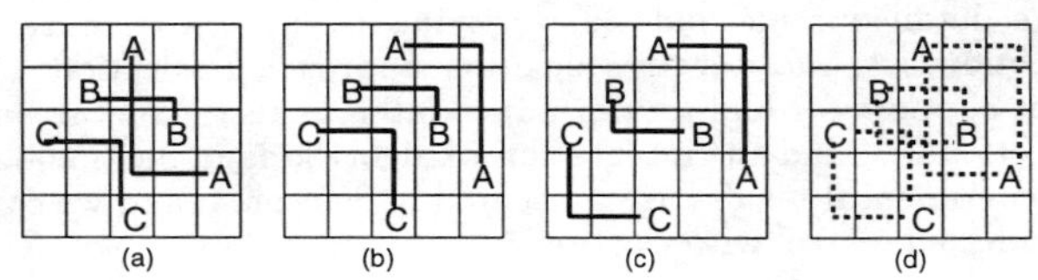

Figure 3: A simple planar routing problem (showing solutions with high congestion, high crosstalk, low crosstalk, and a set of redundant candidate routes).

Figure 3a has high congestion, while Figure 3b has high coupling capacitance (due to parallel wire segments). Figure 3c has both low congestion and low coupling capacitance, and is a *subset* of Figure 3d. The iterative deletion process begins with a redundant set of paths (such as Figure 3d), and iteratively removes a single high-cost redundant path while attempting to minimize congestion and crosstalk among the remaining connections. Interactions between paths (such as coupling capacitance) can also be modeled within this framework.

The set of possible connections for each edge is called the *path set*. In Figure 3d, the path sets for each edge consist of the two single-bend routes. By having several possible routes for each edge available at the start of the iterative deletion process, we can easily determine which paths are high quality. In the example, the lower-left routing for net C does not interfere with any other path, allowing the alternative to be removed without degrading solution quality. This approach has a number of significant features. First, we obtain a *global* estimate of congestion, and can determine not only where congestion is likely, but also where it is unavoidable. Second, crosstalk between paths can be modeled and considered during optimization. By selecting a set of routes that are compatible at a *global* level, we can obtain high quality solutions quickly.

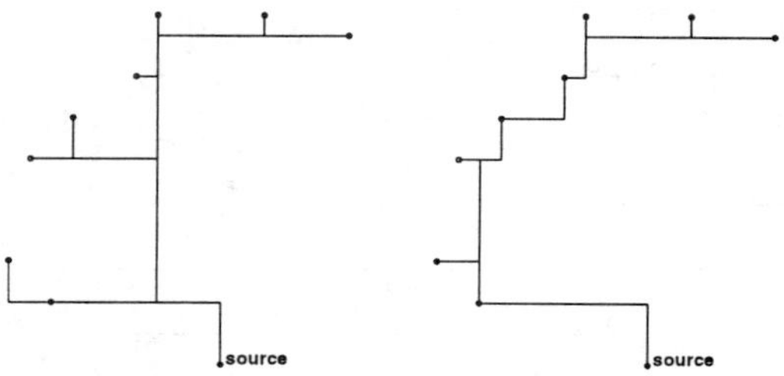

(a) *Tree 1* (b) *Tree 2*

Figure 4: An example of the topologies generated by the *RATS-tree* algorithm for a 9-pin net.

Iterative deletion was shown to be an effective method for selecting a set of channel segments in standard cell IC routing problems[7], and motivates our use of the method here. Our congestion estimation here is similar to the channel density estimation of [7]. We recalculate congestion levels after the removal of each redundant path.

3.3 Performance Optimization

Many approaches for interconnect optimization in high performance design have been proposed. [11] and [5] provide an overview of current interconnect optimization approaches. Recently, *Required-Arrival-Time Steiner trees (RATS-trees)* have been proposed [6] to provide a set of solutions allowing trade-off among signal delay, wave-form, and routing area. This algorithm uses a bottom-up dynamic programming approach, and can synthesize a *set* of high performance structures; Figure 4 shows two possible trees for a 9-pin net.

While the our basic iterative deletion approach considers alternate routings for each topology edge, we consider alternate *trees* for each critical net. Rather than removing a single path between a pair of pins during deletion, we instead remove one of the candidate interconnect trees generated by the *RATS-tree* algorithm. The *RATS-tree* algorithm and our global router support variable-width routing.

3.4 The Hybrid Approach

We have three approaches that form a hybrid between the rip-up and reroute approach, and the iterative deletion method. The first applies iterated rip-up and reroute, and then adds the single bend paths to the path set for each edge. We then perform iterative deletion on the path sets.

Our second approach begins by generating single-bend paths for each edge, and then applies iterative deletion to these path sets. After iterative deletion, we use the maze router to find a new candidate path for each edge, and add it to the path set for the edge. Multiple iterations of this process can produce high quality results quickly.

A third approach applies iterative deletion as above, followed by rip-up and reroute for non-critical nets.

4 Experimental Results

Experiments were performed with the MCC multichip module benchmark circuits *mcc1* and *mcc2*, and also on the three test cases from [14], summarized in Table 1. We report maximum and average congestion levels (C_{max} and C_{avg}) for either the tile-layer metric or the border metric (these two are

Benchmark	No. of Chips	No. of Nets	No. of Pins	Net cardinality 2	3	4	5+
test1	4	500	1000	500	0	0	0
test2	9	956	1912	956	0	0	0
test3	9	1254	2508	1254	0	0	0
mcc1	6	802	2495	608	87	50	57
mcc2	37	7118	14659	6699	415	4	0

Table 1: Benchmark circuit parameters. *test1*, *test2*, and *test3* are constructed examples, containing only two-pin nets.

not in general comparable). Average congestion should correlate reasonably well with wire lengths available after completion of detail routing. A complete set of results is available in [8]. We omit experimental results on edge ordering (in which no ordering outperformed another), an exhaustive set of results comparing cost function approaches, and experiments comparing the use of spanning tree and Steiner tree topologies (which also show negligible differences).

Maze routing (which is unoptimized) dominates the run time. MINOTAUR routes *mcc2* in one hour on a Sun Ultra-1, using a single layer 16 by 16 routing graph. Increasing graph resolution increases run times.

4.1 Comparison of Cost Functions

To compare cost function performance, we route each example as a single layer problem, using a 16 by 16 routing grid. We use a low routing cost of 1, and tested combinations with high routing costs of 2, 4, 8, and 16, *ACE* cost function threshold points at 0.2, 0.4, 0.6, and 0.8 of the range between the minimum and maximum congestion level, and also a threshold point at the average congestion level. A summary of these results is shown in Table 2, in which we show the best, worst, and "typical" performance of the rip-up and reroute method for either the step or slope cost functions, and the result of a single pass of iterative deletion considering the rip-up and reroute paths and the single-bend candidate paths. In the "typical" cost function configuration, we use a high routing cost of 16, and the cost function threshold point at the average congestion level.

While the congestion values during the rip-up and reroute process converged for all experiments, they converged to vastly different values depending on the cost function, and the *high*, *low*, and *threshold* values used to configure the cost function; maximum congestion levels differed by as much as a factor of 2.29. We find that the results using the slope cost function are superior to the step based cost function. When the *high*, *low*, and *threshold* values for tile costs are not well suited to the routing problem, a single pass of iterative deletion can improve the results considerably.

4.2 Comparison with V4R

In Table 3, we compare maximum tile-layer congestion values of our approach with the results of the MCM global router V4R [14]. Congestion estimation favors the V4R router, as the MINOTAUR result considers distinct *tree edges*, while the V4R result reported considers distinct nets. For the single layer comparisons, we merge layer assignments of the V4R routing into a single layer. Our global router obtains congestion reductions ranging from 35% to 48% for the single layer model, and reductions of 14% to 38% for the multi-layer model.

	Best RR		Best RR + ID		Worst RR		Worst RR + ID		Typical RR		Typical RR + ID	
Benchmark	C_{max}	C_{avg}	C_{max}	C_{avg}	C_{max}	C_{avg}	C_{max}	C_{avg}	C_{max}	C_{avg}	C_{max}	C_{avg}
test1 *slope*	27	23.7	27	23.7	28	23.7	28	23.7	27	24.3	28	24.0
step	33	24.7	30	24.1	58	23.7	33	23.7	33	24.8	30	23.1
test2 *slope*	49	46.3	51	46.1	51	46.5	51	46.2	50	47.0	52	46.5
step	57	49.1	54	47.7	112	45.9	59	45.9	61	48.4	53	47.3
test3 *slope*	62	59.0	64	59.0	63	59.9	65	59.5	63	60.3	65	59.7
step	67	60.9	67	60.2	111	59.0	73	59.0	76	60.6	67	59.9
mcc1 *slope*	59	49.3	60	49.1	67	48.5	63	48.3	60	50.1	61	49.6
step	83	51.3	69	49.6	123	48.5	81	48.5	89	51.5	68	49.9
mcc2 *slope*	295	216.0	291	211.7	363	186.3	344	186.0	303	215.6	293	210.9
step	350	209.1	312	203.8	524	192.1	376	188.0	416	212.3	356	205.2

Table 2: Experiments on MCM benchmarks considering a variety of cost functions. We report tile-layer maximum and average congestion C_{max} and C_{avg}, and show the best and worst performance using a classic rip-up and reroute approach, and the results of applying a single pass of iterative deletion to the rip-up and reroute result combined with the two single-bend candidate paths. We also show results using a "typical" cost function configuration.

	Single Layer		Multi Layer		
Benchmark	Minotaur	V4R	L	Minotaur	V4R
test1	27	52	4	16	26
test2	50	78	4	29	34
test3	63	102	4	33	45
mcc1	60	94	4	48	74
mcc2	303	496	6	196	228

Table 3: Comparison of maximum tile-layer congestion of the MINOTAUR global routing result with that of V4R. For the single-layer comparison, the V4R layers were merged into a single layer.

4.3 Experiments with Physical Capacity Constraints

Table 4 shows experimental results and routings using a "two-tier" cost function and differing physical routing capacity constraints. Assuming a wire pitch of 75μ, we have a physical capacity per border of 36 for *mcc1* and 125 for *mcc2*. We also consider cases with increased wire pitch, resulting in physical capacities of 24, 20, 16, and 10 for *mcc1*, and 100, 80, and 60 for *mcc2*, respectively. We apply either an *ACE-slope* cost function, the "two-tier" function, or an *ACE-slope* approach followed by a post-processing step using the "two-tier" cost function. When physical constraints are relatively loose, using a two-tier cost function or *ACE-slope* followed by post-processing can further improve the averge congestion (and expected wire length), as shown in the case with a routing capacity of 125 for *mcc2*. When an unrealistically tight capacity constrnt is given (as is the case for the capacity of 60 for *mcc2*) the two-tier cost function may give a solution that is inferior to the one obtained using *ACE-slope*.

4.4 Performance Optimization

To evaluate the impact of performance optimization on global solution quality, we applied the *RATS-tree* algorithm to a number of nets from the benchmarks *mcc1* and *mcc2*. The number of nets optimized, minimum and maximum number of *RATS-trees*, and average number of *RATS-trees* per net are shown in Table 5. For these experiments, we constructed Steiner trees for each net, and evaluated the delay for each using *SPICE*. $T\%$ of the pins with highest delay were selected, and we attempt to obtain an $X\%$ reduction in signal delay for these pins (no delay constraints are placed on the other nets or pins). Router performance is shown in Table

mcc1	C_{max}	C_{avg}	Mono.	Non-Mono.
ACE-slope	17	11.4	1597	97
two-tier (35)	27	11.2	1678	16
two-tier (24)	21	11.2	1652	42
two-tier (20)	17	11.3	1622	72
two-tier (16)	16	11.7	1515	179
two-tier (10)	22	11.6	1535	159
post-process (35)	26	11.2	1668	26
post-process (24)	21	11.2	1645	49
post-process (20)	17	11.3	1630	64
post-process (16)	16	11.3	1625	69
post-process (10)	17	11.4	1597	97
mcc2	C_{max}	C_{avg}	Mono.	Non-Mono.
ACE-slope	84	48.7	6874	667
two-tier (125)	103	43.8	7412	129
two-tier (100)	84	44.5	6957	584
two-tier (80)	81	48.8	6191	1350
two-tier (60)	98	54.6	5406	2135
post-process (125)	103	43.7	7474	127
post-process (100)	83	44.5	7251	290
post-process (80)	80	48.8	6759	782
post-process (60)	84	48.7	6874	667

Table 4: Four-layer routing experiments on *mcc1* and *mcc2* with the border congestion metric, using physical routing capacity constraints (shown in parenthesis). We use either the *ACE-slope* cost function, a "two-tier" function, or the *ACE-slope* cost function followed by improvement with a "two-tier" function as a post-processing step. The number of monotonic and non-monotonic paths are also shown.

Bench		Total	Min/max/avg. *RATS-trees*		
mark	T	Nets	$X = 5\%$	$X = 10\%$	$X = 15\%$
mcc1	5%	38	4/12/8.16	4/12/8.13	4/12/8.13
mcc1	10%	51	3/17/8.71	3/14/8.67	3/14/8.55
mcc2	5%	614	1/6/1.02	1/4/1.02	1/4/1.02
mcc2	10%	1278	1/6/1.01	1/4/1.01	1/4/1.01

Table 5: Number of nets with performance constraints, and the minimum, maximum, and average number of *RATS-trees* for each net.

Benchmark	T	Rip-up and reroute						Iterative deletion with RR postprocessing					
		X = 5%		X = 10%		X = 15%		X = 5%		X = 10%		X = 15%	
		C_{max}	C_{avg}	C_{max}	C_{avg}	C_{max}	C_{avg}	C_{max}	C_{avg}	C_{max}	C_{avg}	C_{max}	C_{avg}
mcc1	5%	34	22.6	34	22.6	34	22.6	32	22.6	33	22.7	33	22.7
mcc1	10%	41	22.3	41	22.3	41	22.3	33	22.8	33	22.8	33	22.8
mcc2	5%	184	93.6	184	93.6	184	93.6	184	93.6	184	93.6	185	93.6
mcc2	10%	223	93.3	223	93.3	223	93.3	222	93.3	222	93.3	222	93.3

Table 6: Performance optimization comparison of rip-up and reroute (using a single topology for critical nets) with iterative deletion followed by rip-up and reroute. In these experiments, we select *T%* of the high-delay pins, and attempt to obtain at least an *X%* delay reduction for each. In experiments with no performance optimization, rip-up and reroute obtains maximum and average congestion levels of 33 and 23.0 for mcc1, while congestion levels are 168 and 93.4 for mcc2. These results assume the single-layer routing model.

6.

We consider two routing approaches for this problem. The first approach selects a single minimum-area candidate *RATS-tree* for each constrained net, and then applies rip-up and reroute the the remaining nets. The second approach is a hybrid, and applies iterative deletion to select a single *RATS-tree* for each constrained net, and then applies rip-up and reroute. When a number of candidate topologies are available, the hybrid approach obtains a 19.5% reduction in maximum congestion. In this set of experiments, the minimum-area *RATS-trees* were quite similar under the different performance constraints, resulting in similar performance for the rip-up and reroute approach.

5 Conclusions

We have presented an MCM/PCB global router which integrates the classic rip-up and reroute approach, with the modern iterative deletion method.

We draw a number of conclusions from our experiments. We find that slope-based cost functions are superior to step-based cost functions, rip-up and reroute performance is heavily dependent on cost function parameters, and net ordering has little impact for our routing model. Iterative deletion is effective for congestion minimization when we have signal delay and integrity performance constraints, or when rip-up and reroute cost function parameters are not well chosen.

We are currently integrating our MCM global router with a performance driven variable width multi-layer detail router, and will report results of these experiments when they are complete. We are also considering the use of a small set of candidate routes such as those suggested in [2], allowing us to eliminate maze routing entirely. Our methods may also extend to handle general area routing for performance driven multi-layer VLSI IC design problems.

6 Acknowledgements

This work is partially supported by DARPA/ETO under Contract DAAL01-96-K-3600 managed by the US Army Research Laboratory. The authors would like to thank Mr. Cheng-Kok Koh for his assistance and for supplying performance constraints and the *RATS-tree* algorithm.

References

[1] H. B. Bakoglu. *Circuits, Interconnections, and Packaging for VLSI.* Addison-Wesley, 1990.

[2] J. D. Carothers, D. Li, and T. Hameenanttila. MCG: A correct-by-design multichip module router with crosstalk avoidance. In *Int. Workshop on Rapid System Prototyping*, pages 183–188, 1996.

[3] Y.-J. Cha, C. S. Rim, and K. Nakajima. A simple and effective greedy multilayer router for MCMs. In *Proc. Int. Symp. on Physical Design*, pages 67–72, 1997.

[4] J. D. Cho, K.-F. Liao, S. Raje, and M. Sarrafzadeh. M^2R: Multilayer routing algorithm for high-performance MCMs. *IEEE Trans. on Circuits and Systems*, 41(4):253–265, April 1994.

[5] J. Cong, L. He, C.-K. Koh, and P. H. Madden. Performance optimization of VLSI interconnect layout. *Integration, the VLSI Journal*, 21:1–94, 1996.

[6] J. Cong and C.-K. Koh. Interconnect layout optimization under higher-order RLC model. In *Proc. Int. Conf. on Computer Aided Design*, 1997.

[7] J. Cong and P. H. Madden. Performance driven routing for standard cell design. In *Proc. Int. Symp. on Physical Design*, pages 73–80, 1997.

[8] J. Cong and P. H. Madden. Performance driven multi-layer general area routing for pcb/mcm designs,. Technical Report 980013, UCLA CS Dept, March 1998.

[9] J. Cong and B. Preas. A new algorithm for standard cell global routing. In *Proc. Int. Conf. on Computer Aided Design*, pages 176–179, November 1988.

[10] A. Hanafusa, Y. Yamashita, and M. Yasuda. Three-dimensional routing for multilayer ceramic printed circuit boards. In *Proc. Int. Conf. on Computer Aided Design*, pages 386–389, 1990.

[11] A. B. Kahng and G. Robins. *On Optimal Interconnections for VLSI.* Kluwer Academic Publishers, 1994.

[12] K. Kawamura, T. Shindo, T. Shibuya, H. Miwatari, and Y. Ohki. Touch and cross router. In *Proc. Int. Conf. on Computer Aided Design*, pages 56–59, 1990.

[13] K.-Y. Khoo and J. Cong. A fast multilayer general area router for MCM designs. *IEEE Trans. on Circuits and Systems*, pages 841–851, 1992.

[14] K.-Y. Khoo and J. Cong. An efficient multilayer MCM router based on four-via routing. *IEEE Trans. on Computer-Aided Design of Integrated Circuits and Systems*, pages 1277–1290, 1994.

[15] C. Y. Lee. An algorithm for path connection and its applications. *IRE Trans. on Electronic Computers*, EC-10(3):346–365, 1961.

[16] R. Linsker. An iterative-improvement penalty-function-driven wire routing system. *IBM J. Research and Development*, 28(5):613–624, September 1984.

[17] S. Mehrotra, P. Franzon, and M. Steer. Performance driven global routing and wiring rule generation for high speed PCBs and MCMs. In *Proc. Design Automation Conf*, pages 381–387, 1995.

[18] H. Shin and A. Sangiovanni-Vincentelli. A detailed router based on incremental routing modifications: Mighty. *IEEE Trans. on Computer-Aided Design of Integrated Circuits and Systems*, CAD-6(6):942–955, November 1987.

[19] E. Shragowitz and S. Keel. A global router based on a multi-commodity flow model. *Integration, the VLSI Journal*, 5:3–16, 1987.

[20] D. Wang and E. S. Kuh. A new timing-driven multilayer MCM/IC routing algorithm. In *Proc. IEEE Multi-Chip Module Conf.*, pages 89–94, 1997.

[21] Q. Yu, S. Badida, and N. Sherwani. Algorithmic aspects of three dimensional MCM routing. In *Proc. Design Automation Conf*, pages 397–401, 1994.

Buffer Insertion for Noise and Delay Optimization

Charles J. Alpert
IBM Austin Research Laboratory
Austin, TX 78660
alpert@austin.ibm.com

Anirudh Devgan
IBM Austin Research Laboratory
Austin, TX 78660
devgan@austin.ibm.com

Stephen T. Quay
IBM Microelectronics Division
Austin, TX 78660
quayst@austin.ibm.com

Abstract

Buffer insertion has successfully been applied to reduce delay in global interconnect paths; however, existing techniques only optimize delay and timing slack. With the increasing ratio of coupling to total capacitance and the use of aggressive dynamic logic circuit families, noise is becoming a major design bottleneck. We present comprehensive buffer insertion techniques for noise and delay optimization. Our experiments on a microprocessor design show that our approach fixes all noise violations that were identified by a detailed, simulation-based noise analysis tool. Further, we show that the performance penalty induced by optimizing both delay and noise as opposed to only delay is 2%.

1. Introduction

Performance optimization has always been a critical step in the design of integrated circuits. Process technology scaling has made interconnect performance more dominant than transistor and logic performance. With the continued scaling of process technology, the interconnect resistance per unit length continues to increase, the capacitance per unit length remains roughly constant and logic delay continues to decrease. These trends have caused interconnect delay to become more dominant than logic delay. Process technology options, such copper wires, can only provide temporary relief. The trend of increasing interconnect dominance is expected to continue.

Interconnect-driven timing optimization techniques, such as wire sizing, buffer insertion and gate sizing have gained widespread acceptance in deep submicron design [7]. In particular, buffer insertion techniques have been successful in reducing interconnect delay. To the first order, interconnect delay is proportional to the square of the length of the wire. Inserting buffers effectively divides the wire into smaller segments, which makes the interconnect delay almost linear in terms length (plus the buffer delays). Additional advantages of buffer insertion will make this optimization even more pervasive as the ratio of device to interconnect delay continues to decrease.

Several works study delay-driven buffer insertion. Closed formed solutions for 2-pin nets are proposed in [1] [4] [5] and [9]. In [16], Van Ginneken develops a dynamic programming algorithm which finds the *optimal* buffer placement under the Elmore delay model [10]. In [12], Lillis et al. extends this algorithm to simultaneously perform wire sizing, while also minimizing the total number of buffers. Finally, Alpert and Devgan [1] propose a wire segmenting pre-processing algorithm to handle the one buffer per wire limitation of Van Ginneken's algorithm, which results in a smooth trade-off between solution quality and run time.

Although timing optimization has always been critical in the design process, present day design techniques and process technologies are making noise analysis and avoidance as important. The shrinking of the minimum distance between adjacent wires has caused an increase in their coupled capacitance. Furthermore, as the ratio of wire thickness to width continues to increase, so will the ratio of coupling to total capacitance. Coupling capacitance can cause a switching net to induce noise onto a neighboring net, resulting in an incorrect functional response. Further, the widespread use of dynamic logic circuits has made noise avoidance even more critical since these logic families are more susceptible to noise failure. It is no longer sufficient or even acceptable to optimize only for delay. Noise avoidance techniques must become an integral part of the performance optimization environment. Buffer insertion provides a suitable platform for optimizing both timing and noise.

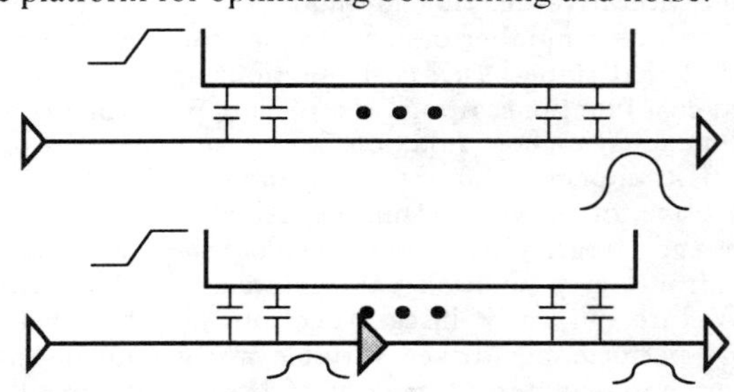

Figure 1 Noise on a victim net (a) without and (b) with a buffer.

Figure 1(a) shows the noise effect that an aggressor net (top) can have on a victim net (bottom). The coupling capacitance may cause an input signal on the aggressor net to induce a noise pulse on the victim net. If the resulting noise is greater than the tolerable noise margin of the sink, then an electrical fault results. Figure 1(b) shows how inserting a buffer can distribute the capacitive coupling between the two newly created wires, resulting in smaller noise pulses on the input of the inserted buffer and on the sink. If the amplitude of these noise pulses are less than he noise margins for the sink and the buffer, then the circuit will function correctly.

Noise analysis is typically performed through detailed circuit simulation or through reduced order interconnect analysis (e.g., AWE[13] and RICE[15]). Although the latter is more efficient, it is still too slow to be used within an optimization tool. Instead, we adopt the noise metric of [8].

The rest of the paper is as follows. Section 2 presents notation and definitions. In Section 3, we derive a formula for the maximum wire length such that no noise violation is induced and also present two optimal algorithms for noise

DAC 98, San Francisco, California
©1998 ACM 0-89791-964-5/98/06..$5.00

avoidance. Section 4 presents a third algorithm for minimizing delay such that all noise constraints are satisfied. Finally, Section 5 presents experimental results.

2. Preliminaries

A routing *tree* $T = (V, E)$ contains a set of $n-1$ *wires* E and a set of n *nodes* $V = \{\{so\} \cup SI \cup IN\}$ where so is the unique *source* node, SI is the set of *sink* nodes, and IN is the set of *internal* nodes. A wire $(e = (u, v)) \in E$ with length l_e is an ordered pair of nodes in which the signal propagates from u to v. Each node $v \in SI \cup IN$ has a unique *parent wire* $(u, v) \in E$. The tree is assumed to be *binary*, i.e., each node can have at most two children.[1] Let the left and right children of v be denoted by $T.left(v)$ and $T.right(v)$ respectively. Assume that if v has only one child, then it is $T.left(v)$. The path from node u to v, denoted by $path(u, v)$, is the set of wires that connect u to v. We are also given a buffer library $B = \{b_1, b_2, \ldots, b_m\}$.

A buffer insertion solution is a mapping $M: IN \rightarrow B \cup \{\bar{b}\}$ which either assigns a buffer or no buffer, denoted by $\bar{b}$, to each internal node of T.[2] Let $|M| = |\{v \in IN: M(v) \in B\}|$ denote the number of internal nodes with inserted buffers. Wires are segmented as in [1] to create as many internal nodes as necessary to form a reasonable set of buffer locations. Assigning k buffers to T induces $k+1$ nets, and hence $k+1$ subtrees, each with no internally placed buffers. For each $v \in V$, let $T(v) = (\{v\} \cup SI_{T(v)} \cup IN_{T(v)}, E_{T(v)})$, the *subtree rooted at* v, be the maximal subtree of T such that v is the source and $T(v)$ contains no internal buffers. Observe that if $v \in SI$, $T(v)$ contains only one node.

2.1 Delay Optimization

As in [1][12][16], we adopt the Elmore delay model [10] for interconnect delays and a linear model for gate delays. For each gate v, let C_v denote the input capacitance, R_v the resistance and K_v the intrinsic delay of v. Let C_e and R_e respectively denote the lumped capacitance and resistance for each wire $e \in E$. The capacitive *load* seen at node v is the total lumped capacitance $C_{T(v)}$ of $T(v)$, i.e.,

$$C_{T(v)} = \sum_{w \in SI_{T(v)}} C_w + \sum_{e \in E_{T(v)}} C_e . \quad (1)$$

The Elmore delay for a wire $e = (u, v)$ is given by $Delay(e) = R_e(C_e/2 + C_{T(v)})$. The delay through a gate $v \in \{so\} \cup B$ is given by $Delay(v) = K_v + R_v C_{T(v)}$. If $v \in IN, M(v) = \bar{b}$, then $Delay(v) = 0$. The total delay from $v \in V$ to $si \in SI$ is given by

$$Delay(v - si) = \sum_{e = (u, w) \in path(v, si)} Delay(e) + Delay(u) . \quad (2)$$

Each sink si has a given required arrival time $RAT(si)$, and assume that the input signal arrives at the source node at time zero. The condition $\forall si \in SI, Delay(so - si) \le RAT(si)$ must hold for the circuit to meet timing requirements. For every $v \in V$, let $q(v) = min_{si \in ds(v)} RAT(si) - Delay(v - si)$ be the *slack* at v where $ds(v)$ is the set of all sinks that are downstream from v. Observe that the circuit meets its timing if and only if $q(so) \ge 0$.

[1] A non-binary tree can be converted into a binary tree by inserting wires with zero resistance and capacitance where appropriate.

[2] A buffer placed on an internal node with degree d is interpreted as having one input, one output, and $d-1$ fanouts.

2.2 Noise Avoidance

In [8], Devgan proposes a coupled noise estimation metric which is an upper bound for RC and overdamped RLC circuits. The metric depends on the resistance of the victim net, the resistance of the gate driving the victim net, coupling capacitances to the aggressor nets, and the rise times and the slopes of the signals on the aggressor nets. For example, consider the three aggressor nets and the single 2-pin victim net in Figure 2. The wire in the victim net is segmented into seven new wires such that each new wire is completely coupled to either 0, 1 or 2 of the aggressor nets.

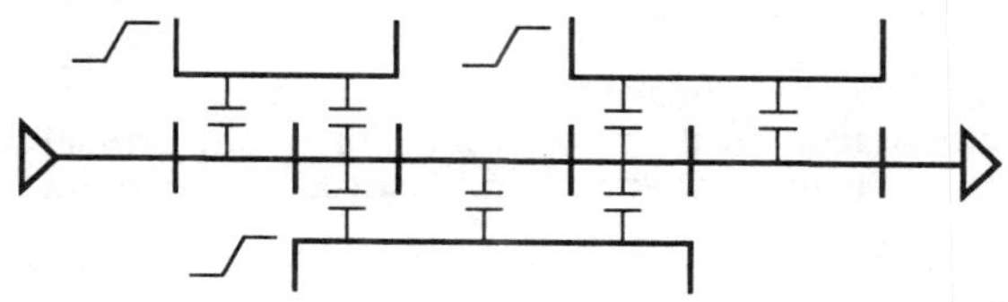

Figure 2 Wire segmenting scheme for multiple aggressor nets.

The coupling capacitance from an aggressor net can be modeled as some fraction of the wire capacitance of the victim net. Given t simultaneously switching aggressor nets $1, \ldots, t$ near wire e, let $\lambda_1, \ldots, \lambda_t$ be the ratios of coupling to wire capacitance from the aggressor nets to e, and let $\mu_1, \ldots, \mu_t$ be the *slopes* (i.e., power supply voltage over input rise time) of the aggressor net signals. The total *current* I_e induced by the aggressor nets on e is

$$I_e = C_e \sum_{j=1}^{t} (\lambda_j \cdot \mu_j) \quad (3)$$

Often, information about neighboring aggressor nets is incomplete, especially when buffer insertion is performed before routing. When performing buffer insertion in *estimation mode*, one might assume that (i) each wire is coupled to exactly one aggressor net, (ii) the slope of all aggressor nets is μ, and (iii) some fixed ratio λ of the total capacitance of each wire is due to coupling capacitance. Under these assumptions $I_e = \lambda\mu C_e$ for each wire e.

Let $I_{T(v)}$ be the total downstream current seen at v, i.e.,

$$I_{T(v)} = \sum_{e \in E_{T(v)}} I_e . \quad (4)$$

Each wire adds to the noise induced on the victim net. The amount of additional *noise* induced from a wire $e = (u, v)$ is given by $Noise(e) = R_e(I_e/2 + I_{T(v)})$. The total additional noise seen at a sink si starting at some upstream node v is given by

$$Noise(v - si) = I_{T(v)} R_v + \sum_{e \in path(v - si)} Noise(e) \quad (5)$$

where $R_v = 0$ if there is no gate at v. The path from v to si has no intermediate buffers. If an intermediate buffer is present, the noise computation begins from the output of the buffer, since the buffer is a restoring stage. Each node $v \in SI \cup B$ has a predetermined *noise margin* $NM(v)$. The condition $\forall v \in \{so\} \cup \{IN | M(v) \in B\}$, $\forall si \in SI_{T(v)}$, $Noise(v - si) \le NM(si)$ must hold for there to be no electrical faults. We define the noise slack for every $v \in V$ as $NS(v) = min_{si \in SI_{T(v)}} NM(si) - Noise(v - si)$. Noise slack serves equivalently as a noise margin for internal nodes. Noise constraints for downstream sinks in $T(v)$ are satisfied if and only if $NS(v)$ is greater than the noise seen at v. Observe that $NS(v) \ge 0$ for each $v \in \{so\} \cup \{IN | M(v) \in B\}$ if there are no noise violations.

2.3 Problem Formulations

We study two different buffer insertion problems. The first problem seeks to fix all noise violations with the fewest possible buffers. Delay is not considered

Problem 1: Given a tree $T = (\{so\} \cup SI \cup IN, E)$, a buffer library B, and noise margins $NM(v)$ for each $v \in SI \cup B$, find a solution $M\colon IN \to (B \cup \{\bar{b}\})$ which minimizes $|M|$, such that $NS(v) \ge 0$ for each $v \in \{so\} \cup \{IN | M(v) \in B\}$.

This problem may be useful for non-critical nets, for which delay optimization is unnecessary. For timing-critical nets, we must consider both noise and delay at the same time.

Problem 2: Given a tree $T = (\{so\} \cup SI \cup IN, E)$, a buffer library B, and noise margins $NM(v)$ for each $v \in SI \cup B$, find a solution $M\colon IN \to (B \cup \{\bar{b}\})$ which minimizes $q(so)$ such that $NS(v) \ge 0$ for each $v \in \{so\} \cup \{IN | M(v) \in B\}$.

A third formulation can seek to minimize the total number of buffers inserted by while satisfying both noise and timing constraints. Algorithm 3, which is used to solve Problem 2, can also be applied to address this third formulation using an extension of Lillis et al, [12] to Van Ginneken [16].

3. Noise Constrained Buffer Insertion

We begin with the simplest case of a wire with uniform width and neighboring coupling capacitance, as shown in Figure 3. For each wire e, let $R = R_e/l_e$ and $I = I_e/l_e$ respectively be the wire resistance per unit length and the current per unit length. Since current is a constant times wire capacitance, we can use a π-model to represent its distribution.

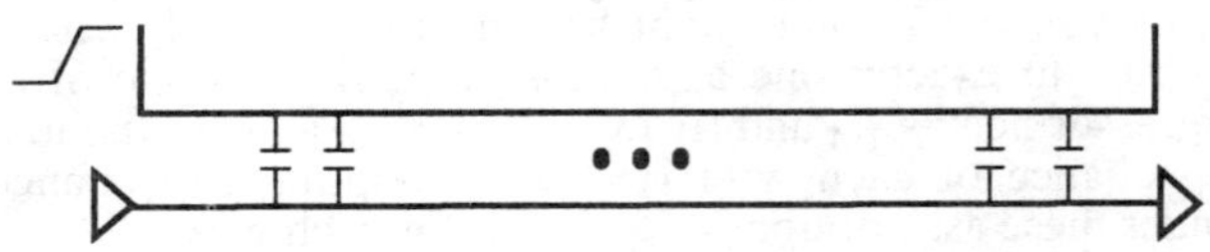

Figure 3 Uniform coupling capacitance for a single wire.

Theorem 1 For a given wire $e = (u, v)$ in a routing tree T, a buffer b needs to be inserted on e to satisfy noise constraints if and only if

$$l_e \le -\frac{R_b}{R} - \frac{I_{T(v)}}{I} + \sqrt{\left(\frac{R_b}{R}\right)^2 + \left(\frac{I_{T(v)}}{I}\right)^2 + \frac{2NS(v)}{IR}}. \tag{6}$$

Proof: For noise constraints to be satisfied, we must have $R_b(Il_e + I_{T(v)}) + Rl_e((Il_e/2) + I_{T(v)}) \le NS(v)$ which is a quadratic in l_e. Solving for l_e yields the theorem. □

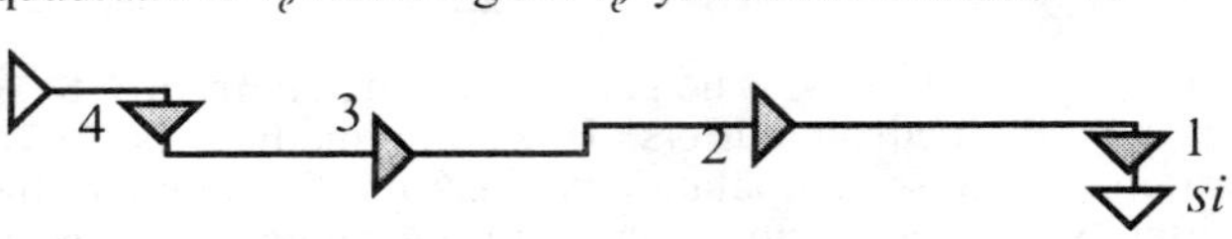

Figure 4 Iterative application of Theorem 1.

Theorem 2 A net that has been optimally buffered to minimize *delay alone* may be susceptible to noise violations.

Proof: Consider a wire $e = (u, v)$ in which u and v are gates in the buffered net (a similar analysis holds for a path from u to v). Let μ be the slope of an aggressor net, and let λ be the ratio of coupling to total capacitance for e. If there is a noise violation at v, then the following must hold.

$$l_e > -\frac{R_b}{R} + \sqrt{\left(\frac{R_b}{R}\right)^2 + \frac{2NM(v)}{RC\lambda\mu}} \tag{7}$$

Solving for $NM(v)$ yields

$$NM(v) < \frac{RC\lambda\mu}{2}\left(l_e^2 + \frac{2l_eR_b}{R}\right). \tag{8}$$

For any fixed values for the parameters on the right side of the inequality, a noise violation will occur if the noise margin is small enough. Even if $NM(v)$ is reasonably large, an aggressor net can have a very large slope μ and a high coupling ratio λ which would also cause a violation. □

3.1 Noise Avoidance for Single-Sink Trees

Theorem 1 suggests how to insert buffers for single-sink trees. Begin at the sink and work up the tree, updating the total downstream current and noise slacks of visited internal nodes. At each internal node, use Theorem 1 to decide if a buffer should be inserted. The algorithm terminates when the source node is reached.

Input: $T = (\{so\} \cup \{si\} \cup IN, E)$ ≡Routing tree
b ≡ buffer type
Output: $M\colon IN \to B$ ≡ Buffer Insertion Solution

1. Set $I_{T(si)} = 0$, $NS(si) = NM(si)$, $v = si$.
2. while $(v \ne so)$ do
   Let $e = (u, v)$ be the parent wire for v.
3. if $R_b(I_{T(v)} + I_e) + R_e(I_{T(v)} + I_e/2) < NS(v)$
   Set $I_{T(u)} = I_{T(v)} + I_e$,
   Set $NS(u) = NS(v) - R_e(I_{T(v)} + I_e/2)$. Set $v = u$.
4. else
   Let $I = I_e/l_e$, $R = R_e/l_e$.
   Set $l = -\frac{R_b}{R} - \frac{I_{T(v)}}{I} + \sqrt{\left(\frac{R_b}{R}\right)^2 + \left(\frac{I_{T(v)}}{I}\right)^2 + \frac{2NS(v)}{IR}}$.
   Create internal node w (with parent u and child v) on e at distance l from v. Set $M(w) = b$.
   Set $I_{T(w)} = 0$, $NS(w) = NM(b)$, and $v = w$.
5. $NS(so) = NS(so) - R_{so}I_{T(so)}$
   if $NS(so) < 0$, create internal node w at so. Set $M(w) = b$.

Figure 5 Algorithm 1, Noise Avoidance for Single-Sink Trees.

Algorithm 1, Noise Avoidance for Single-Sink Trees, is presented in Figure 5. The algorithm accepts a routing tree, and a single buffer type b. Step 1 initializes the current and noise slack of the sink node, then Steps 2-4 climb up the tree visiting each node in turn. Step 3 examines whether or not a buffer needs to be inserted on the current wire $e = (u, v)$, by computing the noise from placing a buffer at u. If this noise is less than the noise slack, no buffer needs to be inserted, so the algorithm computes the downstream current and noise slack for node u, then moves to the next wire. But, if the noise is larger then the noise slack, then a buffer must be inserted. Step 4 computes the maximum length that this buffer may be inserted from v, and inserts it there at a new internal node w. Finally, Step 5 computes the noise slack at the driver and inserts a buffer right after the driver if there is a noise violation (which can only occur if $R_{so} > R_b$). Optimality follows from the fact that buffers are always inserted their maximal distance up the tree, according to Theorem 1.

3.2 Noise Avoidance for Multi-Sink Trees

Some difficulty arises in extending Algorithm 1 to multiple sinks. Let $e_l(e_r)$, $I_l(I_r)$, and $NS_l(NS_r)$ respectively be the wire, current and noise slack for the left (right) branch of an internal node v with two children. It is possible that $R_b I_l \leq NS_l$, $R_b I_r \leq NS_r$ and $R_b(I_l + I_r) > min(NS_l, NS_r)$, i.e., the noise constraints for the left and right branches are satisfied but merging the left and right branches will induce a noise violation. Thus, a buffer must be placed on either the left or right branch immediately after v. One cannot immediately deduce which branch to choose since one needs to know the characteristics and location of the gate driving v. Since the algorithm is bottom-up, the location of this gate has not yet been determined.

Input: $v \equiv$ Current node to be processed
Output: $S \equiv$ List of candidate solutions for node v
Globals: $T = (\{so\} \cup SI \cup IN, E) \equiv$Routing tree
$B \equiv$ Buffer library

1. if $v \in SI$, then set $S = \{(0, NM(v), \bar{b})\}$
2. else if v has only one child then
   $S = Algorithm\ 2(T.left(v))$
3. else if v has two children
   $S_l = Algorithm\ 2(T.left(v))$
   $S_r = Algorithm\ 2(T.right(v))$
   Set $i = 1$ and $j = 1$.
4. while $j \leq |S_l|$ and $k \leq |S_r|$ do
   Let $\alpha_l = (I_l, NS_l, M_l)$ be the ith candidate in list S_l.
   Let $\alpha_r = (I_r, NS_r, M_r)$ be the jth candidate in list S_r.
5. if $R_b(I_l + I_r) > min(NS_l, NS_r)$ then
   Insert nodes w_l, w_r just after v on the left and right branches.
   Set $M_l^b = M_l$, $M_r^b = M_r$.
   Assign $M_l^b(w_l) = b$, $M_r^b(w_r) = b$
6. $S = S \cup \{(I_r, min(NM(b), NS_r), M_l^b \cup M_r)\}$
   $S = S \cup \{(I_l, min(NM_l, NS(b)), M_l \cup M_r^b)\}$
   else
7. $S = S \cup \{(I_l + I_r, min(NS_l, NS_r), M_l \cup M_r)\}$
8. if v is the source then
   for each $\alpha = (I_{T(v)}, NS(v), M) \in S$ do
   $NS(so) = NS(so) - R_{so} I_{T(so)}$
   if $NS(so) < 0$ create internal node w at so.
   Set $M(w) = b$.
   return S
9. Let $e = (u, v)$ be the parent wire of v
   for each $\alpha = (I_{T(v)}, NS(v), M) \in S$ do
   if $R_b(I_{T(v)} + I_e) + R_e(I_{T(v)} + I_e/2) < NS(v)$ then
   Set $\alpha = (I_{T(v)} + I_e, NS(v) - R_e(I_{T(v)} - I_e/2), M)$
   else
10. Let $I = I_e/l_e$, $R = R_e/l_e$.
   Set $l = -\frac{R_b}{R} - \frac{I_{T(v)}}{I} + \sqrt{\left(\frac{R_b}{R}\right)^2 + \left(\frac{I_{T(v)}}{I}\right)^2 + \frac{2NS(v)}{IR}}$
   Create internal node w (with parent u and child v) on wire e at distance l from v.
   $M(w) = b$, $\alpha = (I(l_e - l), NM(b) - IR_e(l_e - l), M)$
11. Prune S of inferior solutions and return S

Figure 6 Algorithm 2, Noise Avoidance for Multi-Sink Trees.

We propose to generate a set of *candidate* solutions for each node and propagate these candidate solutions up the tree, in the same spirit as in Van Ginneken's algorithm [16]. A *candidate* α is a 3-tuple $(I_{T(v)}, NS_{T(v)}, M)$ where $I_{T(v)}$ is the downstream current seen at v, $NS_{T(v)}$ is the noise slack for v, and M is the current solution for the subtree $T(v)$. Let $M = M_l \cup M_r$ denote the resulting solution from merging the left solution M_l and the right solution M_r for an internal node v with two children. For each w in $T(v)$, we assign $M(w) = b$ if either $M_l(w) = b$ or $M_r(w) = b$ and $M(w) = \bar{b}$ otherwise. Whenever a node with two children is encountered and a buffer needs to be inserted, both a left and right candidate is generated. The candidates are sorted in non-decreasing order by downstream current so that inferior solutions can be pruned [16]. Given two candidates $\alpha_1 = (I_1, NS_1, M_1)$ and $\alpha_2 = (I_2, NS_2, M_2)$, α_1 is *inferior* to α_2 if and only if $I_1 > I_2$ and $NS_1 \leq NS_2$.

Algorithm 2 is similar to Algorithm 1, except for Steps 4-7 which handle nodes with two children. Step 4 iterates through each candidate for the left branch and each candidate in the right branch using the Van Ginneken's linear merging technique. Step 5 tests whether merging the two candidates results in a noise violation, and if not, Step 7 merges the two sets of candidates without inserting a buffer. If there is a violation, then two new solutions, one with a buffer on the left branch and one a new buffer on the right branch, are generated and inserted into the current list of candidates. When the algorithm terminates, the solution(s) in S with the fewest number of buffers is chosen.

The algorithm returns an optimal solution to Problem 1 in time quadratic in $|V|$. As for Algorithm 1, one can obtain an optimal solution for a buffer library with multiple buffer types by selecting the buffer type with least resistance.

4. Optimizing Noise and Delay

To address Problem 2, we modify the approach of Van Ginneken [16] to include noise avoidance. Whenever Van Ginneken's algorithm considers inserting a buffer, we check the noise constraints; if they have been violated, the buffer is not inserted. Hence, our algorithm generates fewer solutions than Van Ginneken's algorithm since it *prunes* solutions which have noise violations.

Input: $T = (\{so\} \cup SI \cup IN, E) \equiv$Routing tree
$B \equiv$ Buffer library
Output: $\alpha \equiv$ Best candidate solution for source so

1. $T = Segment\_Wires(T, B)$ [1].
2. $S = Find\_Cands(so)$.
3. for each $\alpha = (C_{T(so)}, q(so), \boldsymbol{I_{T(so)}}, \boldsymbol{NS(v)}, M) \in S$ do
   Set $q(v) = q(v) - K_{so} - R_{so} C_{T(v)}$
   Set $\boldsymbol{NS(so) = NS(so) - I_{T(so)} R_{so}}$.
4. return $(C_{T(so)}, q(so), I_{T(so)}, NS(so), M) \in S$ with maximum $q(so)$ **such that** $\boldsymbol{NS(so) \geq 0}$

Figure 7 Algorithm 3: Optimizing Noise and Delay.

As before, a list of candidates is computed for each node, except that a *candidate* $\alpha = (C_{T(v)}, q(v), I_{T(v)}, NS(v), M)$ is now a 5-tuple where $C_{T(v)}$ is the load seen at v, $q(v)$ is the slack at v, $I_{T(v)}$ is the downstream current seen at v, $NS(v)$ is the noise slack at v, and M is the current solution. Figure 7 illustrates Algorithm 3, which is the same as Van Ginneken's[3] except for the modifications in boldface. Step 1 of Figure 7 segments the wires to generate sufficient possible buffer locations. Step 2 calls Find_Cands which

[3] The algorithm can be modified to handle inverting buffers [12].

returns a list of candidate solutions. Step 3 adds the driver delay and computes the noise slack, then the candidate with the best timing slack, such that noise constraints are satisfied, is returned in Step 4.

The Find_Cands procedure shown in Figure 8. It takes the node v as input, recursively computes the lists of possible candidates for all the nodes in $T(v)$, and then returns the candidates for v. Find_Cands consists of four main parts:

- Steps 1-4 constructs candidates for the children of v and merges them to form S, the set of candidates for v.
- Step 5 inserts considers each buffer type in the library and adds the buffer which yields the largest slack *such that noise constraints are satisfied*. A buffer will not be inserted if there is a noise violation. This step is the fundamental difference between Algorithm 3 and Van Ginneken's algorithm [16].
- Step 6 computes the new load, slack, current and noise slack for each candidate induced by the parent wire of v.
- Finally, Step 7 prunes inferior candidates from S, using the pruning schemes of [12][16].

Modifications for noise avoidance do not increase the $O(|V|^2|B|^2)$ time complexity of Van Ginneken's algorithm since the modifications are mostly constant time operations for bookkeeping and pruning.

Input: $v \equiv$ Current node to be processed
Output: $S \equiv$ List of candidate solutions for node v
Globals: $T = (\{so\} \cup SI \cup IN, E) \equiv$ Routing tree
$B \equiv$ Buffer library

$S = S_b = \varnothing$
1. if $v \in SI$ then $S = \{(C_v, RAT(v), \mathbf{0}, \boldsymbol{NM(v)}, M)\}$
2. else if v has only one child then
   for each $\alpha \in Find\_Cands(T.left(v)))$ do
      $S = S \cup \{\alpha\}$
3. else if v has two children
   $S_l = Find\_Cands(T.left(v))$
   $S_r = Find\_Cands(T.right(v))$
   Set $i = 1$ and $j = 1$
4. while $i \le |S_l|$ and $j \le |S_r|$ do
      Let $\alpha_l = (C_l, q_l, \boldsymbol{I_l}, \boldsymbol{NS_l}, M_l)$ be j[th] candidate in S_l.
      Let $\alpha_r = (C_r, q_r, \boldsymbol{I_r}, \boldsymbol{NS_r}, M_r)$ be k[th] candidate in S_r.
      Let $q = min(q_l, q_r)$, $\boldsymbol{NS = min(NS_l, NS_r)}$
      $S = S \cup \{(C_l + C_r, q, \boldsymbol{I_l + I_r}, \boldsymbol{NS}, M_l \cup M_r)\}$
      if $q_l \le q_r$ then $i = i + 1$. if $q_r \le q_l$ then $j = j + 1$.
5. if v is a feasible buffer location then
   for each buffer $b \in B$ do
      Find $(C_{T(v)}, q(v), \boldsymbol{I_{T(v)}}, \boldsymbol{NS(v)}, M) \in S$ that maximizes
         $q(v) - K_b - R_b C_b$ **such that** $\boldsymbol{NS(v) - I_{T(v)} R_b \ge 0}$
      if such a candidate exists then
         Set $M(v) = b$ and
         $S_b = S_b \cup (C_b, q(v) - K_b - R_b C_b, \mathbf{0}, \boldsymbol{NM(b)}, M)$
   $S = S \cup S_b$
6. Let $e = (u, v)$ be the parent wire for v.
   for each $\alpha = (C_{T(v)}, q(v), \boldsymbol{I_{T(v)}}, \boldsymbol{NS(v)}, M) \in S$ do
      Set $q = q(v) - R_e(C_e/2 + C_{T(v)})$
      Set $\boldsymbol{NS = NS(v) - R_e(I_e/2 + I_{T(v)})}$
      $S = S \cup \{(C_{T(v)} + C_e, q, \boldsymbol{I_{T(v)} + I_e}, \boldsymbol{NS}, M)\} - \alpha$
7. Prune S of inferior solutions and return S.

Figure 8 Find_Cands Procedure(v).

Theorem 3 If $|B| = \{b\}$ and $NM(b) \ge NM(si), \forall si \in SI$, then Algorithm 3 returns an optimal solution to Problem 2.

Proof: Refer to [2] for a proof. With multiple buffers in the buffer library, the optimality of Algorithm 3 is no longer guaranteed. However, in practice, Algorithm 3 generates solutions that are very close to optima; our experimental results in the next section strongly support this claim.

5. Experimental Results

The algorithm we use in our experiments is an extended version of Algorithm 3, called BuffOpt, which can trade-off between delay reduction, noise avoidance, and the total number of buffers [12]. We refer to the algorithm which optimizes only delay [1][12][16] as DOpt. DOpt is the same as Algorithm 3 but without the boldface text. For our experiments, we selected a set of 500 critical nets from a modern Power PC microprocessor design. Table 1 shows the distribution of the sizes of these nets. To verify BuffOpt, we also ran a detailed, simulation-based noise analysis tool, called 3dnoise [14]. 3dnoise was run both before and after BuffOpt and DOpt.To perform noise analysis, 3dnoise uses accurate moment-matching based techniques that are similar to RICE [15].

#Sinks	1	2	3	4	5-7	8-10	11-20	21-40	41-57
#Nets	165	33	33	38	55	69	63	39	12

Table 1: Sink distribution of the 500 test nets.

We ran BuffOpt, DOpt and 3dnoise all in estimation mode (see Subsection 2.2) assuming a 0.7 coupling to total capacitance ratio from a single aggressor net with rise time 0.25 nanoseconds and a power supply voltage of 1.8V. The tolerable noise margin for every gate was set to 0.8V. The buffer library contained 5 inverting and 6 non-inverting buffers of varying power levels. Our experiments show that

- BuffOpt eliminated all noise problems in the design,
- DOpt could not fix all noise problems, and
- the average delay penalty from using BuffOpt instead of DOpt was less than 2%.

5.1 BuffOpt Successfully Avoids Noise

We ran BuffOpt on the 500 nets, and observed that BuffOpt identified 423 noise violations and successfully inserted buffers to fix all of them. To verify BuffOpt's identification of noise violations, we ran 3dnoise on the nets before running BuffOpt. The accurate analysis of 3dnoise identified 386 nets with noise violations, all of which were also identified by BuffOpt. BuffOpt identified 423 - 386 = 37 more nets with violations, which shows that the noise metric [8] is slightly conservative. We also ran 3dnoise on the nets after running BuffOpt, and 3dnoise identified no noise violations. This data is summarized in Table 2.

	# Nets	# Optimized	# Noise Violations
Before BuffOpt	500	0	386
After BuffOpt	500	423	0

Table 2: Number of noise violations reported by 3dnoise before and after running BuffOpt.

5.2 Optimizing Delay Alone is Insufficient

We now compare BuffOpt to DOpt (optimal delay-driven buffer insertion) in terms of noise avoidance. Since BuffOpt never inserted more than four buffers on any net, we ran DOpt four times in which no solution was allowed to have more than k buffers where k ranged from 1 to 4. We denote one such run of DOpt by DOpt(k). Table 3 compares DOpt(k) with BuffOpt. TBI stands for "total buffers inserted" (i.e., $\sum d$ times the number of nets with d buffers)[4], and #NVs stands for "number of noise violations".

Method	# Nets with d Buffers					TBI	# NVs	Total CPU
	d=0	d=1	d=2	d=3	d=4			
None	500					0	386	N/A
DOpt(1)	7	493				493	150	216
DOpt(2)	5	19	476			971	37	613
DOpt(3)	5	16	77	402		1376	21	936
DOpt(4)	4	16	29	35	416	1843	13	1135
BuffOpt	77	161	232	28	2	717	0	980

Table 3: BuffOpt versus DOpt(k) for noise.

Observe that running DOpt(4) causes the addition of 1126 more buffers than BuffOpt, yet there are still 13 noise violations. Reducing the maximum number of buffers that can be inserted by DOpt only increases the number of noise violations and still causes more buffers to be inserted for $k \geq 2$. Thus, as shown by Theorem 2, noise avoidance *must* be integrated into the algorithm to guarantee no noise violations. Finally, observe that BuffOpt uses less CPU than DOpt(k) for $k \geq 3$. This occurs because BuffOpt prunes candidates with noise violation, which gives BuffOpt fewer total candidates to analyze.

5.3 The Delay Penalty is Small

# Buffers	0	1	2	3	4	Avg DR
# Times	77	161	232	28	2	423
BuffOpt	0	124.8	336.0	1043.4	45.5	301.1
DOpt	0	135.8	338.2	1053.1	50.3	307.1
Delay Penalty	0	8.77	0.64	0.93	10.50	1.99

Table 4: Average delay reduction from buffer insertion.

Finally, we compare DOpt to BuffOpt in terms of total delay. We first ran BuffOpt on the 500 nets and recorded the buffers inserted for each net. We then ran DOpt for the same number of buffers as BuffOpt inserted in order to make an apples to apples comparison. We computed the reduction in total delay for each net and averaged the results by the number of buffers inserted. The cumulative results are presented in Table 4. For example, there were 232 nets for which two buffers were inserted, and on average BuffOpt reduced delay by 336.0 picoseconds while DOpt reduced delay by 338.2 picoseconds. The overall delay penalty is $100(338.2 - 336.0)/338.2 = 0.64$ percent. The average delay reduction over all 423 nets is computed by taking the total delay reduction for all nets and dividing by 423.[5]

Observe from the last column that the overall average delay penalty is only 6 ps, or equivalently 1.99%, from avoiding noise. Thus, Buffopt is able to integrate noise into a delay-driven algorithm with virtually no loss in total delay.

Recall that Algorithm 3 is optimal when the buffer library contains a single buffer type, but we could not guarantee optimality for a larger buffer library. The DOpt results in Table 4 form an upper bound on the optimal solution to Problem 2 since DOpt is optimal for delay alone. Observe that even with a buffer library of size 11, BuffOpt solutions are virtually optimal since they are on average within 2% of an upper bound.

Acknowledgments

The authors would like to thank Joe Rahmeh for his help with 3dnoise and the PowerPC design data.

References

[1] C. J. Alpert and A. Devgan, "Wire Segmenting for Improved Buffer Insertion", *DAC*, 1997, pp. 588-593.

[2] C. J. Alpert, S. T. Quay and A. Devgan, "Comprehensive Buffer Insertion Techniques", submitted to *IEEE Trans. CAD*, April, 1998.

[3] C. J. Alpert, S. T. Quay and A. Devgan, "Comprehensive Buffer Insertion Techniques for Noise Avoidance", U.S. patent filed, 1997.

[4] H. B. Bakoglu, *Circuits, Interconnections, and Packaging for VLSI*, Addison-Wesley, 1990.

[5] C. C. N. Chu and D. F. Wong, "Closed Form Solution to Simultaneous Buffer Insertion/Sizing and Wire Sizing", *International Symposium on Physical Design*, 1997, pp. 192-197.

[6] C. C. N. Chu and D. F. Wong, "A New Approach to Simultaneous Buffer Insertion and Wire Sizing", *ICCAD*, 1997, pp. 614-621

[7] J. Cong, L. He, C.-K. Koh, and P. H. Madden, "Performance Optimization of VLSI Interconnect Layout", *Integration: the VLSI Journal*, 21, 1996, pp. 1-94.

[8] A. Devgan, "Efficient Coupled Noise Estimation for On-Chip Interconnects", *ICCAD*, 1997, pp. 147-151.

[9] S. Dhar and M. A. Franklin, "Optimum Buffer Circuits for Driving Long Uniform Lines", *IEEE Journal of Solid-State Circuits*, 26(1), 1991, pp. 32-40.

[10] W. C. Elmore, "The Transient Response of Damped Linear Network with Particular Regard to Wideband Amplifiers", *J. Applied Physics*, 19, 1948, pp. 55-63.

[11] M. Z.-W. Kang, W. W.-M. Dai, T. Dillinger, and D. P. LaPotin, "Delay Bounded Buffered Tree Construction for Timing Driven Floorplanning", *ICCAD*, 1997, pp. 707-712.

[12] J. Lillis, C.-K. Cheng and T.-T. Y. Lin, "Optimal Wire Sizing and Buffer Insertion for Low Power and a Generalized Delay Model", *IEEE Journal of Solid-State Circuits*, 31(3), 1996, pp. 437-447.

[13] L. T. Pillage and R. A. Rohrer. Asymptotic Waveform Evaluation for Timing Analysis. *IEEE Trans. CAD*. 9(4):352-366, April, 1990.

[14] J. Rahmeh, "The 3d-Noise User Guide", *IBM Internal Report*, 1997.

[15] C. Ratzlaff and L. T. Pillage, "RICE: Rapid Interconnect circuit Circuit Evaluator using Asymptotic Waveform Evaluation", *IEEE Trans. CAD*, pp. 763-776, June 1994.

[16] L. P. P. P. van Ginneken, "Buffer Placement in Distributed RC-tree Networks for Minimal Elmore Delay", *ISCAS*, 1990, pp. 865-868.

---

[4] For example, BuffOpt inserted $(0 \cdot 77) + (1 \cdot 161) + (2 \cdot 232) + (3 \cdot 84) + (4 \cdot 2) = 717$ buffers.

[5] E.g., for BuffOpt, the weighted average of 301.1 is given by $((161 \cdot 124.8) + (336.0 \cdot 232) + (1043.4 \cdot 28) + (45.5 \cdot 2))/423$

Table-Lookup Methods for Improved Performance-Driven Routing*

John Lillis
Dept. of EECS, U.I. Chicago
Chicago, IL 60607-7053
jlillis@eecs.uic.edu

Premal Buch
Magma Design Automation
Palo Alto, CA 94303
premal@magma-da.com

Abstract

The inaccuracy of Elmore delay [3] for interconnect delay estimation is well-documented. However it remains a popular delay measure to drive performance optimization procedures such as wire-sizing and topology construction. This paper studies the merits of incorporating "better-than-Elmore" delay measures into the optimization process. The proposed delay metrics use a table-lookup method to incorporate better load modeling and approximate the effect of signal slew. We demonstrate that the proposed metrics exhibit a much narrower error distribution than Elmore delay, eliminating Elmore's frequent gross delay over-estimation. Finally we show the improvement in solution quality which can be had by incorporating the new metrics into a timing driven topology construction algorithm.

1 Introduction

The advent of deep submicron (DSM) technology has produced a great deal of interest in CAD algorithms for interconnect delay optimization. Past work includes algorithms for buffer insertion, wire sizing, high-performance topology construction and various hybrids of these techniques. Most of these algorithms are based on Elmore delay or even simpler delay models. This paper explores the amount of improvement in end solution quality that may be had by incorporating a more accurate delay model into an optimization procedure.

For any particular optimization technique such as topology construction, an algorithm designer must make several key decisions:

(1) What *solution space* should be considered?

(2) What should be the precise *problem formulation*?

(3) What *delay measure* should be incorporated?

A reasonable taxonomy reflecting these decisions appears in Figure 1. The figure gives three partial orders, one for each decision above. The partial orders proceed from less-expansive solution spaces, simpler problem formulations, and simpler delay models on the left to progressively more ideal spaces, formulations and delay models on the right. An algorithm may either be heuristic or optimal for a particular triple of choices. For instance the P-Tree algorithm [5] is optimal for choices (1B, 2F, 3C). On the other hand, it can only be considered a heuristic solution to, for example, the ideal of (1D, 2F, 3F). Similarly, a shortest paths tree (physical path length) is optimal for (1A, 2A, 3A), but heuristic for more realistic combinations.

In principle, the merit of an optimization algorithm (in addition to such criteria as run-time) is measured by the quality of the solutions it produces versus those produced by a hypothetical "ideal" algorithm — e.g., one which finds optimal solutions over the entire space of Steiner trees (1D), under the min-cost timing feasible formulation (2F) and with Spice providing delay measurements.[1] In [2], the authors argued that the "fidelity" of Elmore is high – i.e., that an 'optimal' Elmore solution was near an 'optimal' Spice solution. This was done experimentally by finding spice optimal solutions through enumerative techniques for the the combination (1A, 2C, 3C) and (1A, 2C, 3F) – i.e., it was assumed that only the delay to an arbitrarily chosen sink was of importance and the remaining delays could be neglected. It was concluded that optimal Elmore solutions were very close in performance to the optimal Spice solutions. The variance of the Elmore/Spice and a 2-pole/Spice ratios were also statistically studied and shown to be nearly identical.

While the results in [2] are interesting and provide some degree of assurance that following Elmore can produce high quality solutions, we cannot conclude that, for instance, Elmore produces high quality solutions for the min-cost timing feasible problem formulation.

This paper explores this question by incorporating a "better-than-Elmore" delay model into an established timing-driven Steiner routing algorithm [5, 4]. It is our goal to give an idea of how much improvement can be had by incorporating more accurate models (or more precisely to give an empirical lower-bound on this possible improvement). In the context of Figure 1, this is achieved by comparing the results of two algorithms: one which solves the combination (1B, 2F, 3C) to one which attacks the combination (1B, 2F,

*This research was funded in part by grants from NSF project number MIP-9625910 and SRC task 324.012.

[1] Of course, the definition of the "ideal" algorithm can become arbitrarily complex when one considers such things as coupling capacitance, or false-paths at the logic level, etc.

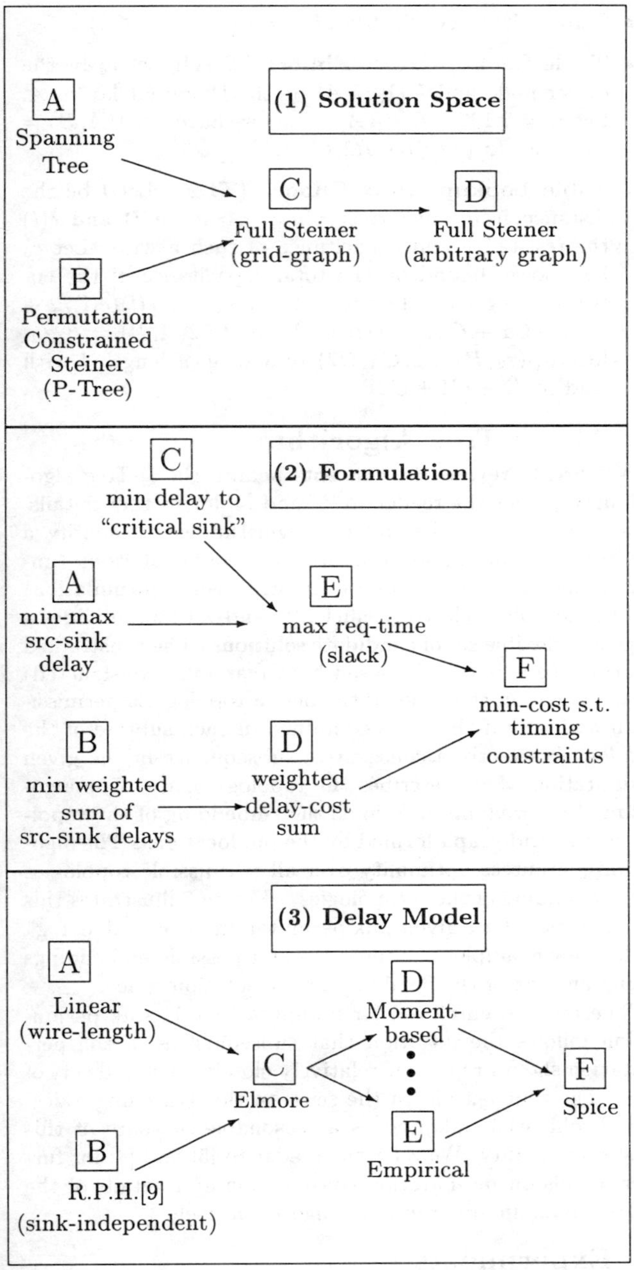

Figure 1: *Taxonomy of Timing-Driven Steiner Routing*

3E). Thus, we have a fixed solutions space and formulation while varying the delay model dimension.

We selected the permutation constrained solution space (1B) used by the P-Tree algorithm for several reasons. First, it has produced the best published results for formulation 2F under Elmore delay. Second, it provides a framework in which new delay models can be incorporated relatively easily (provided the models can be evaluated "bottom-up"). Additional background on the P-Tree algorithm is presented in Section 4.

2 The Nature of Elmore's Error

In RC trees two main factors contribute to Elmore's error [7]:

- its inability to account for resistive shielding (an increasingly important phenomenon in deep submicron)
- its inability to account for the effect of signal slew rate τ.

As a result, Elmore can grossly overestimate the delay to nearby sinks for which the resistive shielding effect is significant and the signal waveform is relatively sharp. Conversely, as the signal slew increases (i.e., as we travel to more distant sinks), Elmore becomes increasingly accurate.[2] The danger in an optimization algorithm is that this error will cause misidentification of critical paths and lead the algorithm to make incorrect decisions by optimizing delay to a false critical sink at the expense of the true critical sink.

3 Proposed Delay Models

Clearly it is desirable to improve upon the deficiencies of Elmore described in the previous section. In addition, for our purposes a proposed model should be relatively easily incorporated into an optimization procedure. With these two goals in mind, we have devised the table-lookup method summarized below.

Interpolation in a fairly sparse 5D table is used to estimate the delay of wire segments by interpolation as we proceed bottom-up. Each segment is restricted to be short enough so that it can be approximated by a lumped RC model. A wire link in the topology then corresponds to one or more segments. The delay of a segment AB depends on the resistance and capacitance of the segment AB, the downstream RC subtree rooted at B and the slew of the driving signal at A. All the relevant resistances and capacitances are captured with a pi-model approximation of the load at B (see Figure 2). We then estimate the delay from A to B (50%-to-50% delay) with the 5 parameters $R1, R2, C1, C2$ and τ. The lookup table is constructed for a range of $R1, R2, C1, C2$ and τ via multiple Spice runs. The values of $R1, R2, C1, C2$ and τ for a given segment AB are computed as follows:

- At each stage we estimate the signal slew τ at A by a multiple of the Elmore delay from the root to A (or an approximation thereof). The coefficient for the Elmore delay based slew estimate is determined as follows: the signal slew at an internal node A in an RC tree depends on the downstream RC subtree seen by node A. If the Elmore delay to this subtree is

[2]Indeed, for ramp inputs Elmore approaches the true 50% delay as τ tends toward infinity [7].

$ED(v_d, A)$, the voltage waveform at node A is given by $v_A(t) = V_{DD}(1 - e^{-t/ED(v_d,A)})$ under the assumption that a single dominant time-constant exists for the RC subtree rooted at node A. The signal slew τ_A at node A is then given by $\tau_A = t(v_A(t) = 0.9V_{DD}) - t(v_A(t) = 0.1V_{DD}) = 2.197ED(v_d, A)$. Note that the single dominant time-constant assumption is also made in deriving the the Elmore delay model, and the above formulation of the signal slew is thus exact under Elmore delay.

- For the pi-model, we accumulate the first three coefficients of the Taylor series expansion of the driving point admittance at node B, in a bottom-up fashion as in [6]. From this a pi-model approximation is derived for an RC tree (the subtree rooted at B in this case) using the algorithm of [6]. This pi-model of the RC sub-tree is combined with the resistance and capacitance of the segment AB to obtain the model in Figure 2. In this case, R1, the "driver" resistance, corresponds to the resistance of the segment AB, C2 is the sum of half the capacitance of the segment AB (i.e., AB itself is approximated by a pi-model) and the upstream branch capacitance of the pi-model of the RC subtree rooted at B, and R2 and C1 are the other two branch elements of this subtree's pi-model.

Note that when computing the delay across a segment driven by the root node, R1 only serves as the 'driver' resistance. By applying the above lookup procedure while travelling bottom-up, we can compute the delays along all paths from sinks to the root.

As an aside, we emphasize the importance of the often overlooked process of finding an appropriate scaling coefficient β for any delay model – i.e., for a model such as Elmore, we select β so as to "center" the error distribution (0.69 being the typical coefficient for Elmore). While a scaling coefficient will not alter the solutions produced by an algorithm designed for minimizing max source-to-sink delay, this is not the case for required-arrival-time (or just "required-time") formulations; in the first case, $d_1 < d_2 \Leftrightarrow \beta d_1 < \beta d_2$ while in the latter case each sink u introduces it's own offset in the form of its required-arrival-time q_u and we cannot claim such a property (i.e., the choice of β can alter the critical path).

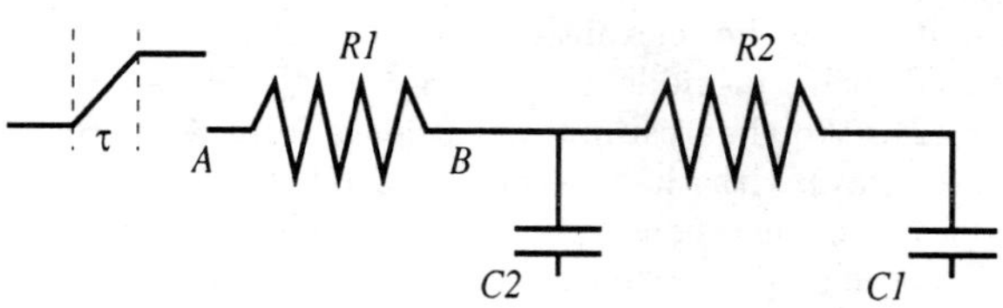

Figure 2: *Load modeling by a pi-model circuit*

To further simplify the incorporation of our new models into the P-Tree Steiner routing algorithm [5], we also introduced another model which uses an estimate of the Elmore delay from the driver to an arbitrary point in the tree. This is because in the bottom-up dynamic programming algorithm, we do not yet know the structure of the unexplored portion of the tree.

The three models for the delay from node A to node B (refer to Figure 2) are summarized below (note the only difference between TL1 and TL2 is the method used for approximating τ; the coefficient 2.197 follows the rationale presented in the previous section).

- **Scaled Elmore:** $ED_s(A, B) = \beta_E R1(C1 + C2)$
- **Table Lookup Exact Elmore (TL1):** Let v_d be the driver node and $ED(v_d, A)$ be the Elmore delay to A. Let $\tau_1 = 2.197 \cdot ED(v_d, A)$. Then we have, $TL1(A, B) = \beta_{TL1} \cdot$ **lookup**$(\tau_1, R1, R2, C1, C2)$.
- **Table Lookup Appx Elmore (TL2):** Let l be the distance from the driver pin to A and $r(l)$ and $c(l)$ the resistance and capacitance of such a wire. Let c_0 be a lower-bound on the total capacitance of the entire routing tree. Let $\tau_2 = 2.197(r_d c_0 + r(l)(c(l)/2 + c_0/2 + C1 + C2)$. Then we have, $TL2(A, B) = \beta_{TL2} \cdot$ **lookup**$(\tau_2, R1, R2, C1, C2)$ of a wire of length l with load $(c_0/2 + C1 + C2)$.

4 The P-Tree Algorithm

We briefly review the concepts behind the P-Tree algorithm. We refer the reader to [5] and [4] for further details. The main principle behind the algorithm is to identify a constrained solution space in which optimal solutions (under Elmore and the min-cost timing feasible formulation) can be identified relatively efficiently and yet still provides a large and flexible set of candidate solutions. The constrained solution space is that imposed by a (carefully constructed) permutation on the sinks of the net: a topology is permissible if and only if the sinks contained in each subtree of the topology is exactly a consecutive subsequence in the given permutation. This describes the topology space. The algorithm also *simultaneously* finds the embedding of the topology in the grid-graph formed by the pin locations. The algorithm guarantees optimality over all permissible topologies and embedding of those topologies.[3] Figure 3 illustrates this solution space for a given sink permutation "a, b, c, d, e, f, g" with a few examples of topologies and possible embeddings giving an idea of the flexibility of the solution space.

The method employed for finding a "good" sink permutation follows the intuition that subsequences of the permutation should represent relatively closely clustered sets of pins. The tour length (in the sense of the Travelling Salesman Problem) is adopted as a reasonable measure of this clustering quality. We refer the reader to [5] and [4] for further details on permutation construction and details of the P-Tree dynamic programming algorithm itself.

5 Experiments

After constructing the 5D lookup table (with 10 entries in each dimension or 100k total entries) from spice data we performed two sets of experiments.

Our first set of experiments studies the error distribution of the three models themselves. We first generated a large set of routing topologies of 4, 6 and 8 pin nets. These were drawn from routing regions of size 1cm x 1cm, 0.5cm x 0.5cm and 0.25cm x 0.25cm. For each net cardinality and

[3]The algorithm actually produces a family of topologies giving a performance vs. cost (total wire-length) tradeoff.

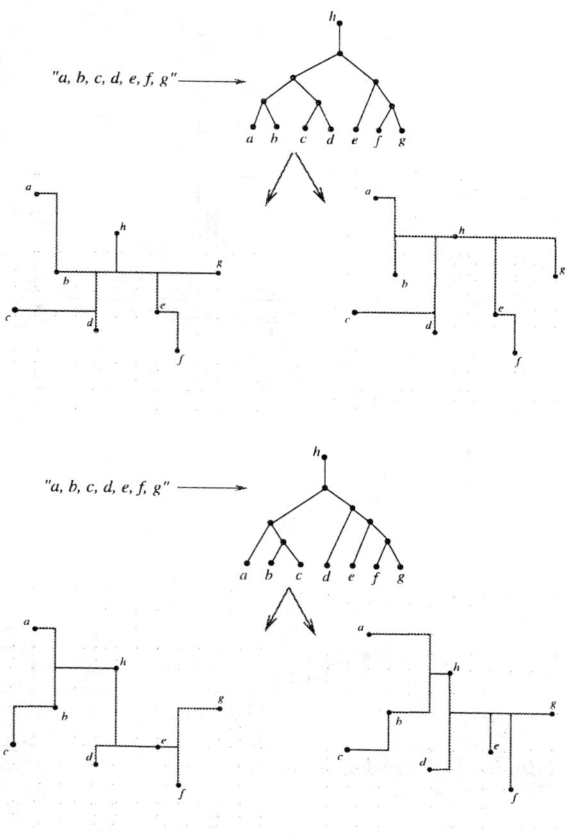

Figure 3: *The P-Tree Solution Space*

routing region size pair we generated 50 point sets and corresponding topologies using the P-Tree algorithm (not to produce high performance topologies, merely reasonable topologies for our testbed). Technology parameters (corresponding closely with $0.5\mu m$ technologies in the literature) were as follows:

driver resistance: 300Ω

wire resistance: $0.112\Omega/\mu m$

wire capacitance: $0.15fF/\mu m$

sink cap: $0.05fF/\mu m$

For each topology, the delay to each sink was computed for each model in addition to Spice. After this data was collected, scaling factors were selected so as to minimize the mean-squared error of each model versus Spice. Figure 4 shows histograms for each of the delay models; for each model there is one histogram showing error in pico-seconds vs. spice and another showing relative error as the ratio of the predicted delay to the spice delay. As can be seen, all three models produce reasonable delay predictions most of the time; however, as can be seen in the case of Elmore, there is a long tail of large over-estimations while the table-lookup models are more well-behaved. This trend is even more dramatic when only the topologies from the 1cm x 1cm routing region were considered as shown in Figure 5.

The standard deviation of the delay models to spice ratios were as follows:

Scaled Elmore:	0.321
TL1:	0.069
TL2:	0.073

As may be expected, the table-lookup methods produced markedly more stable delay predictions. It is also interesting that, despite its crude estimation of signal slew, TL2 behaves nearly as well as TL1.

The various scaling coefficients were as follows. For Elmore, the scaling coefficient β_E was 0.69 (matching the theoretically correct ln 2.) The scaling coefficients for TL1 and TL2 were $\beta_{TL1} = 1.13$ and $\beta_{TL2} = 1.15$.

Our second set of experiments give preliminary results on the amount of improvement in final solution quality we observed after we incorporated TL2 into the P-Tree algorithm (P-Tree$_{TL}$) versus those produced by Elmore-based P-Tree. We generated 100 6-pin Steiner routing instances with the technology parameters from the previous section. Required-times for each sink were drawn at random between 1ns and 3ns. Both algorithms were run on each instance and the resulting topologies were compared in terms of wire length and actual spice-computed required-times (i.e., slack at the root). This was done in two scenarios: first, the fastest topologies (as predicted by the respective algorithms) were compared; second, the fastest topology from each algorithm whose wire-length was not more than 25% greater than that of the min wire length solution (recall that the P-Tree algorithms produce a family of solutions with a wire-length vs. required-time tradeoff).

The results of these experiments are shown graphically in Figures 6 and 7. Each plot shows the difference in spice-computed required-time (in pico-seconds) versus the table-lookup/elmore wire-length ratios. Thus, points in the lower-right quadrant indicate that the table-lookup topology had both better performance *and* shorter wire length than the corresponding Elmore-based topology. With the exception of two outliers in Figure 6, the trend clearly favors the table-lookup method (also, the topologies in Figure 7 are more likely to be of practical interest because of their more modest wire-length overhead). Finally, we note that in a number of cases both algorithms produced identical topologies: where the fastest topologies were selected, 67/100 were identical, where only 25% overhead was allowed, 68/100 were identical.

6 Discussion

In this paper we have presented evidence that indeed some improvement in solution quality may be had by incorporating better-than-Elmore delay models into the performance-driven routing process. The proposed models themselves were shown to have significantly better error distribution than Elmore. While the routing solutions produced by each model agree a good portion of the time, a significant fraction favor the improved models substantially.

Future work along these lines may include a similar method for improved buffer modeling in the buffer insertion process (and to better model the driver in case presented in this paper where we modeled it as a resistor for simplicity). Performing similar experiments for $0.25\mu m$ technologies is also of interest since Elmore's error distribution presumably

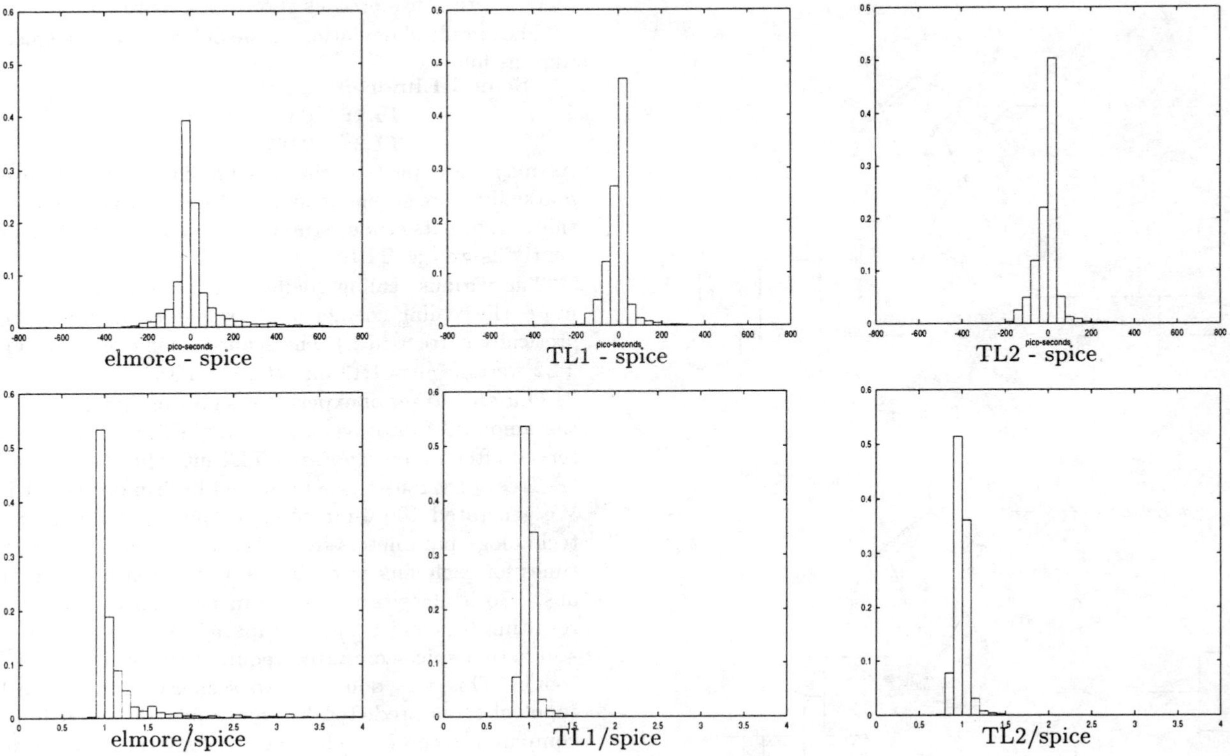

Figure 4: *Error Distributions Over All Topologies (all models scaled)*

will become even worse.

Finally we note that P-Tree with the table-lookup models does not ensure optimality under the given model (as it does for Elmore). This is a result of a basic pruning premise in P-Tree and many similar algorithms: if a candidate solution for a subtree is faster and has less capacitance than a second solution, the second is considered sub-optimal and discarded. This is correct for Elmore, but it is conceivable that the second solution's *effective capacitance* [8] is less than that of the first and therefore could contribute to an optimal global solution. Exploring this phenomenon is another possible direction for future work. The notion of effective capacitance may also provide a viable alternative to the table-lookup methods presented here.

Acknowledgement

The authors thank Professor Ernest Kuh of U.C. Berkeley for his encouragement and several helpful discussions.

References

[1] K. D. Boese, A. B. Kahng, B. A. McCoy, G. Robins, "Fidelity and Near-Optimality of Elmore-Based Routing Constructions," *Proc. Proc. IEEE Intl. Conf. Computer-Aided Design*, 1993.

[2] Boese, K.D.; Kahng, A.B.; McCoy, B.A.; Robins, G. "Near-optimal critical sink routing tree constructions." *IEEE Transactions on Computer-Aided Design of Integrated Circuits and Systems*, Dec. 1995, vol.14, (no.12):1417-36.

[3] W.C. Elmore, "The Transient Response of Damped Linear Network with particular Regard to Wideband Amplifiers," *J. Applied Physics* 19 (1948), pp 55-63.

[4] J. Lillis "Algorithms for Performance Driven Design of Integrated Circuits," Technical Report #CS96-492, CSE Dept., U.C. San Diego, Aug. 1996

[5] J. Lillis, C.-K. Cheng, T.-T. Lin, C.-Y. Ho, "New Techniques for Performance Driven Routing with Explicit Area/Delay Tradeoff and Simultaneous Wire Sizing," *Proc. 33rd ACM/IEEE Design Automation Conference*, Las Vegas, Jun. 1996, pp. 395-400.

[6] P. R. O'Brien and T. L. Savarino, "Modeling the Driving-Point Characteristic of Resistive Interconnect for Accurate Delay Estimation," *Proc. IEEE Intl. Conf. Computer-Aided Design*, Nov., 1989, pp. 512-515.

[7] R. Gupta, B. Tutuianu, L.T. Pileggi, "The Elmore delay as a bound for RC trees with generalized input signals." *IEEE Transactions on Computer-Aided Design of Integrated Circuits and Systems*, Jan. 1997, vol.16, (no.1):95-104.

[8] Qian, J.; Pullela, S.; Pillage, L. "Modeling the "Effective capacitance" for the RC interconnect of CMOS gates". *IEEE Transactions on Computer-Aided Design of Integrated Circuits and Systems*, Dec. 1994, vol.13, (no.12):1526-35.

[9] J. Rubinstein, P. Penfield, and M.A. Horowitz, "Signal Delay in RC Tree Networks," *IEEE Trans. on CAD* 2(3) (1983), pp 202-211.

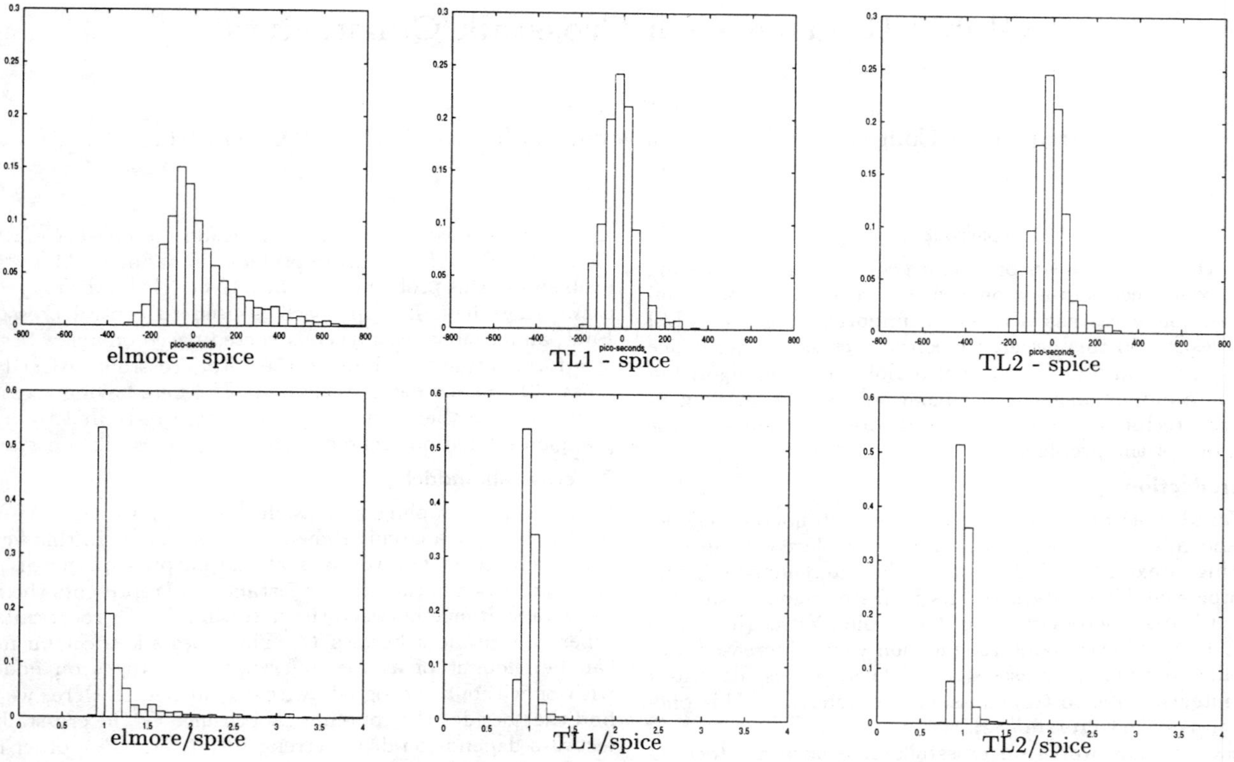

Figure 5: *Error Distributions Over Topologies Drawn From a 1cm x 1cm Routing Region (all models scaled)*

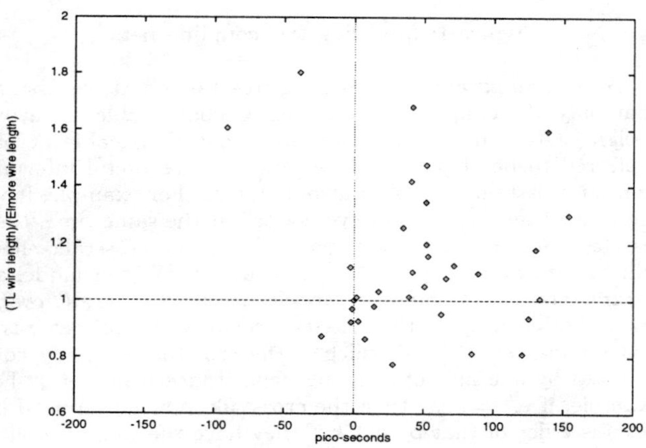

Figure 6: *Results for 'fastest' topologies produced by Elmore-based and table-lookup based P-Tree. X-axis is the difference between the table-lookup required time and the Elmore required-time (as computed by Spice).*

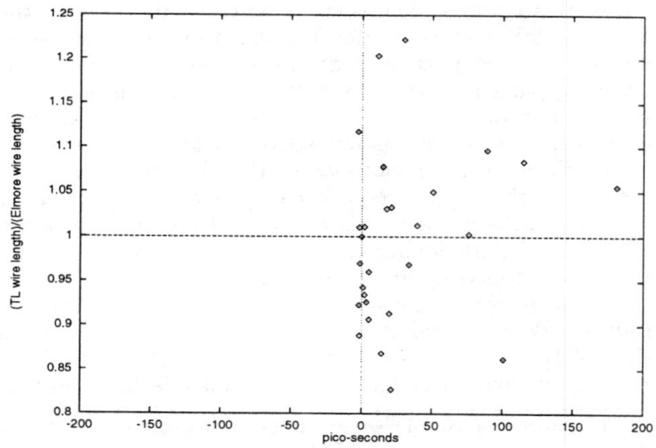

Figure 7: *Results for 'fastest' topologies with wire-length no more than 25% greater than that of the min wire-length solution.*

Global Routing with Crosstalk Constraints*

Hai Zhou and D.F. Wong
Department of Computer Sciences, University of Texas, Austin, TX 78712-1188

Abstract

Due to the scaling down of device geometry and increasing frequency in deep sub-micron designs, crosstalk between interconnection wires has become an important issue in VLSI layout design. In this paper, we consider crosstalk avoidance during global routing. We present a global routing algorithm based on a new Steiner tree formulation and the Lagrangian relaxation technique. We also give theoretical results on the complexity of the problem.

1 Introduction

With VLSI fabrication entering the deep sub-micron era, devices and interconnection wires are being placed at an ever increasing proximity. Reduction in the interconnection and transistor switching delays results in faster signal transition times. All these factors increase the coupling effect (inductive and capacitive) between interconnection wires. Increased coupling effect not only increases signal delays, but also decreases signal integrity due to transmission line behavior. This phenomenon is called *crosstalk* [2].

Most previous works on crosstalk avoidance are focused on detailed routing [3, 5, 6, 9, 11, 14], where the estimation of crosstalk is accurate but the flexibility to avoid it is restricted. Therefore, it is often not possible to find a crosstalk-feasible solution in detailed routing if the global routing is crosstalk-blinded.

In order to make a global routing solution crosstalk risk-free, Xue et al. [13] proposed a post global routing adjustment procedure to modify a given routing solution. It used a gridded model and assumed that the crosstalk only exists between adjacent wires. That is, if there is a spare track between two wires, then there is no crosstalk between them. The crosstalk risk estimation is divided into two coupled procedures: bound partitioning and net ordering. Bound partitioning partitions the crosstalk bound of each net into the regions it routes through. Net ordering orders the nets in each region to require as few spare tracks as possible. Their approach is totally region-based, that is, crosstalk risks are defined by the regions requiring extra tracks and the number of extra tracks needed. If a region still requires extra tracks after bound partitioning and net ordering, some nets in the region are ripped up and re-routed using other regions.

In this paper, we consider crosstalk avoidance during global routing. We use a crosstalk model which is more general than that of [13]. In order to have an accurate crosstalk estimation, we extend global routing to include layer and track assignments. Based on this extension, a novel *Crosstalk-Constrained Global Routing* problem is defined. Many subproblems of this problem are shown to be NP-hard. Therefore, a two-stage heuristic approach is used to control crosstalk. First, an initial solution is constructed based on a new Steiner tree formulation to minimize the total crosstalk. After that, crosstalk on each net is estimated. The nets having crosstalk violations are then ripped up. They are re-routed by a Lagrangian relaxation technique to satisfy crosstalk constraints.

*This work was partially supported by the Texas Advanced Research Program under Grant No. 003658288.

2 crosstalk model

Consider two coupling nets as shown in Figure 1(a). We can model them by a circuit shown in Figure 1(b). In this figure, v_1, v_2 represents the voltages at the outputs of the drivers, R_1, R_2 represents the driver resistances, C represents the coupling capacitance between the nets, and C_1, C_2 represents all other capacitances besides C. The crosstalk effect on net 2 can be thought of as the difference of voltages on node O with or without the signal switching in net 1. Here, we can find that besides the coupling capacitance C, the crosstalk effect also depends on driver strengths (R_1 and R_2), other load capacitances (C_1 and C_2), and the voltages (v_1 and v_2).

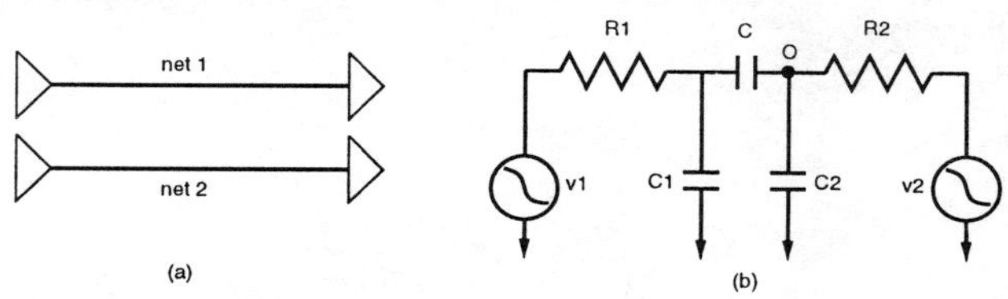

Figure 1: Modeling two coupling nets

Among all parameters affecting crosstalk effect, we assume that only the coupling capacitance is controllable in layout design. Other parameters are either not changeable or not preferred to be changed. However, they are useful information for crosstalk control in layout design. For example, if we know net 1 and net 2 will never switch at the same time, then we can safely route them together since the crosstalk effect will be zero even they are heavily coupled. We can model all parameters except coupling capacitance by a *crosstalk coefficient*. For each net i, the *crosstalk coefficient* from net j is a real number $e_{ij} \in [0, 1]$ which is the crosstalk on net i contributed by one unit of coupling capacitance from net j. For example, if $e_{ij} = 2e_{ik}$, then the crosstalk induced on net i by net j is twice of that by net k if they have the same coupling capacitance.

For a net, the crosstalk effects from other nets may not happen at the same time. But characterizing all cases requires exhaustive timing analysis, which in turn depends on crosstalk. Therefore, in the worse case, we can use the summation of all the effects from other nets as the total crosstalk on one net. That is,

$$X_i = \sum_{j \neq i} e_{ij} C_{ij}, \tag{1}$$

where C_{ij} is the coupling capacitance between net i and net j.

Generally speaking, each element in a chip is coupled with every other element. But coupling capacitance decreases rapidly when an element is out of the neighborhood of the other element. For interconnection, coupling capacitance between perpendicular wires is also very small. Therefore, we assume coupling capacitance only exists between neighboring parallel wires and is given by the following formula:

$$C = \alpha \frac{length}{distance^{\beta}}, \tag{2}$$

where β is a constant which was estimated to be about 2 in [10].

3 Problem formulation

In traditional design flow, routing is divided into global and detailed routing. In global routing, the set of regions a net goes through is decided. In detailed routing, the actual route in each region is then computed. From the previous section, we know coupling is critically dependent on wire adjacency. However, in traditional global routing, for each net, only routing regions it goes through are decided. The adjacency information in each region is not available. Therefore, it is not possible to get accurate crosstalk estimation in that situation.

In order to estimate crosstalk during global routing, some "detailed" information must be known. Of course, doing the whole task of global and detailed routing together would be very complicated. Therefore, the idea is to do a "simplified" detailed routing "on the fly" in the process of global routing. This detailed routing must be simple enough to be merged into the global routing process, and at the same time, must be detailed enough for crosstalk estimation.

In fact, each region in the global routing stage can be viewed as a multi-layer rectangular area. We also assume that each layer is composed of tracks with equal spacing and each track can be used only by one net. During global routing, besides deciding which regions a net goes through, layer and track assignments in each region are also decided. Formally, we are given an undirected graph $G = (V, E)$, where each edge e represents a region $r(e)$ which has a length $l(e)$ and may have k layers each with a capacity $c_i(e)$, for $1 \leq i \leq k$. Also specified are n nets each is a subset of V. An *extended global routing solution* is an edge labeling S of the form $S(e) = (s_1, s_2, \ldots, s_k)$, where each s_i for $1 \leq i \leq k$ is a sequence of nets. For $1 \leq i \leq n$, the edges including i must form a Steiner tree for net i. If $s_i = (\ldots, 2, 1, 3, \ldots)$, this means that nets 1, 2, and 3 are routed through region $r(e)$ and placed in layer i in the order $2, 1, 3$. For example, an extended global routing solution is shown in Figure 2, where each region has only one layer.

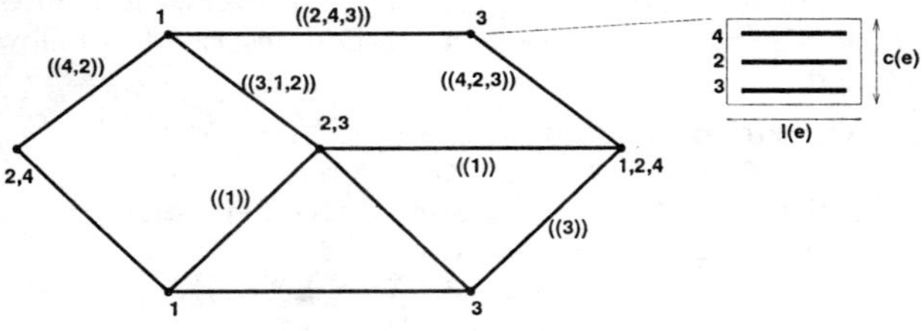

Figure 2: An extended global routing solution

Given an extended global routing solution, the adjacency information is known in each edge. Suppose nets i and j are adjacent on layer l of edge e, the coupling capacitance between them can be computed as $C = \alpha \frac{l(e)}{c_l(e)/|N_l(e)|}$, where $N_l(e)$ represents the set of nets on layer l of edge e. Therefore, crosstalk on each net can be computed. Using $X_i(S)$ to represent crosstalk on net i under an extended global routing solution S, the problem we need to solve can be defined as follows.

Problem 1 (Crosstalk-Constrained Global Routing (CCGR)) *Given a routing graph $G = (V, E)$, n nets, and the crosstalk constraints $C_1, \ldots, C_n$, find an extended global routing solution S such that $X_i(S) \leq C_i$ for all $1 \leq i \leq n$.*

4 A global router with crosstalk control

As will be shown in the next section, many sub-problems of the CCGR problem are NP-hard. We use a heuristic to deal with this problem. It is a sequential approach, that is, routing is done net by net. The approach consists of two stages. The first stage constructs a routing solution to minimizes the summation of crosstalk on all nets. Each time, a Steiner tree is computed for a net which minimizes the total crosstalk and its topology is then fixed for the remaining nets. In the second stage, crosstalk on each net in the initial solution is estimated. The nets having constraint violations are then ripped up. During re-routing, a method based on Lagrangian relaxation is used to balance crosstalk according to each net's constraint. Because there is no forbidden region in the process, a new route can have the same topology as the old one but with different layer/track assignment. This means layer/track re-assignment is also included in this procedure.

4.1 Minimum crosstalk Steiner tree

Since we use a sequential approach in our routing, the kernel procedure is how to construct a Steiner tree and decide the layer/track assignment in each region such that the total crosstalk is minimize. Formally, this can be defined as the following problem.

Problem 2 (Minimum Crosstalk Steiner Tree) *Given a routing graph $G = (V, E)$ and an extended global routing solution S for nets $1, \ldots, m-1$ which minimized the total crosstalk, find an extended global routing solution S' for nets $1, \ldots, m$ such that routing topologies for nets $1, \ldots, m-1$ are kept and $X(S') = \sum_{i=1}^{m} X_i(S')$ is minimized.*

Let $X^e(S)$ denote the summation of crosstalk on all nets in edge e under solution S. The objective function in the above problem can be written as follows.

$$X(S') = \sum_{e \in E} X^e(S') = \sum_{e \in T_m} X^e(S') + \sum_{e \notin T_m} X^e(S')$$
$$= \sum_{e \in T_m} (X^e(S') - X^e(S)) + \sum_{e \in E} X^e(S) \tag{3}$$

where T_m represents the Steiner tree for net m. For each edge e, $X^e(S') - X^e(S)$ is the difference of the minimum total crosstalks in edge e before or after inserting net m in the edge. Given a set of nets going through an edge, the minimum crosstalk in the edge can be computed by an optimal layer/track assignment. Therefore, we can compute $X^e(S') - X^e(S)$ and use it as the cost of edge e. Since $\sum_{e \in E} X^e(S)$ in (3) is a constant, the minimum crosstalk Steiner tree problem becomes a minimum Steiner tree problem.

Unfortunately, according to Theorem 1, layer/track assignment for crosstalk minimization in one region is NP-hard. Therefore, a heuristic is designed to do the job. It works as follows. First, we decide an ordering on all nets. Then, each net is added sequentially to the region according to its order. When adding one net, the existing nets keep their relative ordering, and the new net is inserted to a position which has the minimum increase of crosstalk.

If we use the same net ordering for sequential routing in the above heuristic, the layer/track assignments can be merged into the routing process. In this case, when computing edge costs, there is no need to do layer/track re-assignments for the existing nets. Suppose S is the given solution for nets $1, \ldots, m-1$, the cost of routing net m on edge e is the minimum increase of crosstalk by inserting net m in one sequence of $S(e)$. If e is chosen by a minimum Steiner tree algorithm, such insertion will give $S'(e)$. An example of the procedure is given in Figure 3.

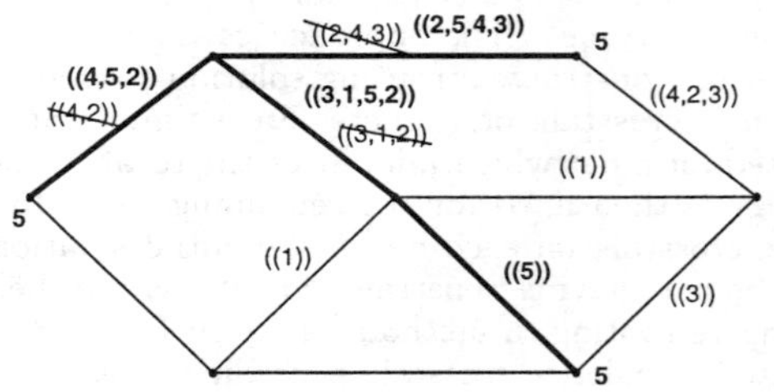

Figure 3: Routing net 5

In practice, crosstalk is not the only objective we want to minimize. We also need to consider wire length and congestion. In such case, we can use the following edge cost function.

$$cost(e) = \alpha \cdot length(e) + \beta \cdot overflow(e)^2 + \gamma \cdot xtalk(e)$$

where $overflow(e)$ is the extra space needed when compared with the capacity of the edge, and $xtalk(e)$ is the increase of crosstalk on the edge. Based on this cost function, any Steiner tree heuristic on general weighted graph can be used to construct a Steiner tree which simultaneously optimizes the wire length, congestion, and crosstalk.

The Steiner tree heuristic we use in our global router is based on the shortest path and the minimum spanning tree. According to [8], these heuristics are guaranteed to have solutions within $2(1-1/l)$ of the optimum, where l is the number of leaves in the optimal solution. We use Dijkstra's algorithm [4] for the shortest path and Prim's algorithm [4] for the minimum spanning tree. The reason we use Prim's algorithm instead of Kruskal's is that Dijkstra's algorithm and Prim's algorithm are both working in a tree expansion fashion, hence can be easily combined.

4.2 Rip-up and re-route based on Lagrangian relaxation

Given an initial solution constructed by the minimum crosstalk Steiner tree algorithm, nets with crosstalk violations are ripped up. We re-route them one net at a time in the order of decreasing violations. When routing a net, besides crosstalk on the current net, extra crosstalk can be induced on previously routed nets. Therefore, we need to solve the following problem.

Problem 3 (Re-Routing for Crosstalk Control) *Given a routing graph $G = (V, E)$ and an extended global routing solution S for nets $1, \ldots, m-1$ such that $X_i(S) < C_i$ for $1 \le i \le m-1$, find an extended global routing solution S' for nets $1, \ldots, m$ such that routing topologies of nets $1, \ldots, m-1$ are kept, the wire length is minimized, and $X_i(S') < C_i$ for all $1 \le i \le m$.*

The above problem requires to balance crosstalk on each net according to its constraint. Similar as in [15], we can use Lagrangian relexation to satisfy all constraints simultaneously. By relaxing constraints into objective function, we have the following problem.

Problem 4 (Lagrangian Sub-Problem (LSP)) *Given a routing graph $G = (V, E)$, an extended global routing solution S for nets $1, \ldots, m-1$, and constants $\mu_1, \ldots, \mu_m$, find an extended global routing solution S' for nets $1, \ldots, m$ such that the routing topology for nets $1, \ldots, m-1$ is kept, and $length(S') + \sum_{i=1}^{m} \mu_i (X_i(S') - C_i)$ is minimized. Where $length(S')$ is the wire length in solution S'.*

Let $length_i(S')$ be the wire length of net i under solution S'. The objective function of above problem can be written as follows.

$$length(S') + \sum_{i=1}^{m} \mu_i (X_i(S') - C_i)$$

$$= \sum_{i=1}^{m} (length_i(S') + \mu_i X_i(S')) - \sum_{i=1}^{m} \mu_i C_i$$

Since $\sum_{i=1}^{m} \mu_i C_i$ is a constant, by treating $length_i(S') + \mu_i X_i(S')$ as crosstalk on net i, the Lagrangian sub-problem becomes the minimum crosstalk Steiner tree problem, hence can be solved by the same approach.

Let $L(\mu_1, \ldots, \mu_m)$ denote the optimal solution of the Lagrangian sub-problem. For any $\mu_i \ge 0, 1 \le i \le m$, we have

$$L(\mu_1, \ldots, \mu_m) = \min_{S'} \{length(S') + \sum_{i=1}^{m} \mu_i (X_i(S') - C_i)\}$$

$$\le \min_{S'} \{length(S') + \sum_{i=1}^{m} \mu_i (X_i(S') - C_i) : X_i(S') \le C_i, 1 \le i \le m\}$$

$$\le \min_{S'} \{length(S') : X_i(S') \le C_i, 1 \le i \le m\}.$$

Therefore, if we have $\mu_1^*, \ldots, \mu_m^*$ such that

$$L(\mu_1^*, \ldots, \mu_m^*) = \min_{S'} \{length(S') : X_i(S') \le C_i, 1 \le i \le m\},$$

we know that S' is an optimal solution. Hence we need to solve the following problem, which is actually a convex program.

Problem 5 (Lagrangian Multiplier Problem (LMP))

$$\begin{array}{ll} \textit{Maximize} & L(\mu_1, \ldots, \mu_m) \\ \textit{subject to} & \mu_i \ge 0, \forall 1 \le i \le m \end{array}$$

Since $L(\mu_1, \ldots, \mu_m)$ as a function is not everywhere differentiable, we will use sub-gradient method described as follows to solve it.

Algorithm Sub-gradient Method for LMP
1. $k = 0; \mu_i = 0, \forall 1 \le i \le m;$
2. Solve the LSP by the Steiner tree heuristic;
3. If stop criteria is satisfied, halt;
4. Else $\mu_i = \max\{0, \mu_i + \theta_k (X_i(S') - C_i)\}$, $\forall 1 \le i \le m; k = k + 1;$ goto 2.

The general theory for sub-gradient method states that if $\theta_k \to 0$ and $\sum_{j=1}^{k} \theta_j \to \infty$ as $k \to \infty$, it will converge to an optimal solution [1]. But we must notice that an optimal solution to the LMP does not always give an optimal solution to the re-routing for crosstalk control problem.

Table 1: Experimental results

test problems	Traditional GR		post-GR layer/track assignment		our approach			
					first stage		second stage	
	#vio.	max. vio	#vio.	Max. vio	#vio.	Max. vio	#vio.	Max. vio
ami49.1.a	329	285.29	301	191.91	121	81.46	1	0.84
ami49.1.b	252	86.27	143	35.46	46	15.23	1	0.14
ami49.1.c	303	114.50	265	46.78	146	25.93	3	0.90
ami49.2	325	100.46	283	46.16	156	16.92	2	0.19
ami33.1	70	82.15	65	31.90	35	8.97	5	0.39
ami33.2	64	83.07	55	31.87	24	8.89	1	0.19
apte.a	19	21.93	11	7.56	3	4.54	0	0.00
apte.b	17	5.15	11	2.10	5	1.74	0	0.00
xerox.1	49	5.16	9	4.34	3	3.83	2	0.08
xerox.2	57	17.17	18	5.87	2	0.32	0	0.00
xerox.3	70	49.70	34	31.23	22	15.98	0	0.00

5 Complexity results

In the CCGR problem, both routing topology and adjacency relations in regions can be changed. Therefore, the following methods can be use to reduce crosstalk: *shielding* (inserting a spare wire between coupling nets), *track assignment* (changing neighboring relations of nets), *layer assignment* (distributing coupling wires to different layers), But all of them are difficult problems. We summarize them into the following theorem.

Theorem 1 *All of the following problems are NP-hard.*

Problem 6 (Shielding for Crosstalk Control) *Suppose the routing topology, layer/track assignments in all regions are fixed, and only one shielding wire can be inserted in each region. Find a shielding position in each region such that each net satisfies its crosstalk constraint.*

Problem 7 (Track Assignment for Crosstalk Minimization) *Suppose the routing topology is fixed, and so are track assignments in all regions except one. Find a track assignment in that region such that the total crosstalk on the nets is minimized.*

Problem 8 (Track Assignment for Crosstalk Control) *Suppose the routing topology is fixed, and so are track assignments in all regions except one. Find a track assignment in that region such that each net satisfies its constraint.*

Problem 9 (Layer Assignment for Crosstalk Control) *Suppose there are only two layers in each region, and the routing topology and net orderings in all regions are fixed. Find a layer assignment in each region such that each net satisfies its crosstalk constraint.*

6 Experimental Results

We implement the global router described in Section 4 on a Solaris/PC platform. Five circuits from the CBL/NCSU building block benchmarks: ami33, ami49, apte, hp, and xerox are used. Since these benchmarks do not come with placement information, we use a floorplaner based on simulated annealing [12] to give placements. Using different parameters, we obtain two different placements for both ami33 and ami49: ami33.1, ami33.2, ami49.1, ami49.2, and three different placements for xerox: xerox.1, xerox.2, xerox.3. Using different minimum track spacing and crosstalk constraints, different test problems are produced for the same placement, which are indicated by the suffixes: a, b, etc.

In order to measure the effectiveness of our global router, we compare it with other routing approaches. "Traditional GR" only considers wire length and overflow, and nets are ordered sequentially in each region. "Post-GR layer/track assignment" uses the same topologies as in traditional GR but use the greedy heuristic given in Section 4.1 to do layer/track assignment in each region. For each approach, the numbers of nets having crosstalk violations and the maximum violations are reported in Table 1. For our global router, both results after the first stage (i.e., minimizing total crosstalk) and the second stage (i.e., rip-up & re-route by Lagrangian relaxation) are reported. As we can see, after rip-up and re-route, almost all violations are solved. For those few nets which still have violations, the violations are so small that it could be easily solved in detailed routing.

Acknowledgements

We would like to thank Dr. Y.-Z. Liao and Dr. Yang Cai for their stimulating discussions.

References

[1] R.K. Ahuja, T.L. Magnanti and J. B. Orlin, *Network Flows: Theory, Algorithms, and Applications*. Prentice Hall, 1993.

[2] H.B. Bakoglu, *Circuits, interconnections and packaging for VLSI*. Addison-Wesley, 1990.

[3] K. Chaudhary, A. Onozawa and E.S. Kuh, A Spacing Algorithm for Performance Enhancement and Cross-talk Reduction. *ICCAD,* 1993.

[4] T.H. Cormen, C.E. Leiserson, and R.L. Rivest, *Introduction to Algorithms*. The MIT Press, 1990.

[5] T. Gao and C.L. Liu, Minimum Crosstalk Channel Routing. *ICCAD,* 1993.

[6] T. Gao and C.L. Liu, Minimum Crosstalk Switchbox Routing. *ICCAD,* 1994.

[7] M.S. Garey and D.S. Johnson, *Computers and Intractability: A Guide to the Theory of NP-Completeness*. W. H. Freeman, New York, 1979.

[8] F.K. Hwang, D.S. Richards, and P. Winter, *The Steiner Tree Problem*. North-Holland, 1992.

[9] D.A. Kirkpatrick and A.L. Sangiovanni-Vincentelli, Techniques for Crosstalk Avoidance in the Physical Design of High-Performance Digital Systems. *ICCAD,* 1994.

[10] T. Sakurai and K. Tamaru, Simple formulas for two and three dimensional capacitance. *IEEE Trans. Electron Devices,* 1993.

[11] S. Thakur, K.-Y. Chao and D.F. Wong, An Optimal Layer Assignment Algorithm for Minimizing Crosstalk for Three Layer VHV Channel Routing. *ISCAS,* 1995.

[12] D.F. Wong and C.L. Liu, A New Algorithm for Floorplan Design. *DAC,* 1986.

[13] T. Xue, E.S. Kuh, and D. Wang, Post Global Routing Crosstalk Risk Estimation and Reduction. *ICCAD,* 1996.

[14] H. Zhou and D.F Wong, An Optimal Algorithm for River Routing with Crosstalk Constraints. *ICCAD,* 1996.

[15] H. Zhou and D.F. Wong, Crosstalk Constrained Maze Routing Based on Lagrangian Relaxation. *ICCD,* 1997.

Timing and Crosstalk Driven Area Routing

Hsiao-Ping Tseng
University of Washington
Dept. of Electrical Engineering
Seattle, WA 98195
206-685-8678
hptseng@twolf.ee.washington.edu

Louis Scheffer
Cadence Design Systems, Inc.
555 River Oaks Parkway, Bldg. 2, MS2B2
San Jose, CA 95134
408-944-7114
lou@cadence.com

Carl Sechen
University of Washington
Dept. of Electrical Engineering
Seattle, WA 98195
206-685-8756
sechen@twolf.ee.washington.edu

ABSTRACT

We present a timing and crosstalk driven router for the chip assembly task that is applied between global and detailed routing. Our new approach aims to process the crosstalk and timing constraints by ordering nets and tuning wire spacing in a quantitative way. Our graph-based optimizer preroutes wires on the global routing grids incrementally in two stages - net order assignment and space relaxation. The timing delay of each critical path is calculated taking into account interconnect coupling capacitance. The objective is to reduce the delays of critical nets with negative timing slack values, by tuning net ordering and adding extra wire spacing. It shows a remarkable 8.4-25% delay reduction for MCNC benchmarks for wire geometric ratio=2.0, against a 33% delay reduction if interconnect interference disappear.

1 INTRODUCTION

One of the emerging deep submicron process issues is the increasing effect of coupling capacitance between interconnect on the same layers. Our capacitance extractor shows that for typical process parameters the intra-layer coupling capacitance contributes 25% of metal2 interconnect capacitance in a 0.25um process if the wire geometric ratio (height/width) is 1.0, and 33% if the wire geometric ratio is 2.0. Wire resistance and intra-layer coupling capacitance therefore increase as feature sizes shrink. Most of the existing routing frameworks only consider wire length as the constraint for path delay and ignore the neighboring switching correlation. Previous work in the crosstalk-aware routing problem is mostly in the gridded domain.

Crosstalk reduction efforts have been made at a detailed routing level for the river routing problem[1], the switchbox problem[3] and the channel routing problem[2]. However these algorithms are all in the net ordering domain with fixed spacing in which segments are rearranged during the postprocessing stage such that accumulated coupling capacitance of critical nets is reduced. The segment rearrangement heuristics of the above approaches only apply to nets in a local scope, such as in a switchbox or in a channel. On the other hand, Xue et al. [4] develop a post global routing crosstalk risk analyzer which estimates the possible coupling between sensitive nets and tries to reroute nets away from crosstalk risky zones. Chaudhary, et al. develop a general post-routing spacing algorithm [5] which measures the crosstalk effect by superposition of unweighted voltage glitches from driver to sink. We shall describe the difference in detail in Section 5.

The interconnect coupling capacitance causes signal integrity problems in two respects - timing faults and logic faults. There are many heuristics and techniques, such as wire permutation, wire shielding, buffer insertion and buffer sizing, which can decrease coupling capacitance or improve timing and potentially reduce logic fault hazards and timing fault hazards. In this paper, we mainly concentrate on improving the worst case timing.

The remainder of the paper describes our new timing and crosstalk driven area prerouter as follows. An overview of the problem and the new approach is given in Section 2. Section 3 briefly describes the RC extraction model and interconnect delay calculation. The net ordering algorithm and space relaxation algorithm are presented in Sections 4 and 5, respectively. We shall show the experimental results in Section 6.

2 OVERVIEW

2.1 Chip Assembly Area Routing Problem

The chip assembly routing problem is too complex to solve in one piece and is often solved by divide-and-conquer approaches - global routing, then detailed routing. The layout floorplan is first broken into a 3D global routing plane of global cells (*Gcells*), which have a typical size range of 10x10 - 20x20 tracks. A global router searches for the rough path of each signal on the global cell plane and determines which global cells each signal should traverse into. Based on the global routing information, a detailed router is able to efficiently complete the routing of one Gcell at a time. On the interface of two neighboring Gcells, there are net *crosspoints*. Gcells on the same row (column) can be glued together and seen as a *horizontal* (*vertical*) *routing channel*. However, the channel height of Gcell channels is fixed. From the practice of industrial chip assembly circuits, we observe that most signals traverse through multiple Gcells, usually by straight segments. The ordering of nets and spacing between them can be determined by the crosspoints on Gcell boundaries if all wires are straightened. Chip assembly is clearly a case in which crosspoint assignment (CPA) can determine the layout almost as well as the final routing. Kao et al. [6] develop a crosspoint assignment algorithm to improve the routability but it does not consider timing and crosstalk. We develop a two-stage crosspoint assignment algorithm considering timing and crosstalk constraints.

2.2 Timing Driven Crosspoint Assignment

The crosspoint assignment problem for a single Gcell is solved by a graph-based framework. The crosspoint of a signal determines the position of two segments connected to the crosspoint. Two consecutive crosspoints of the same signal with different positions on boundaries imply a jog is needed to connect them. A straightened segment of crosspoints at the same boundary location is represented as a *node* in the graph. The physical design rules of a

DAC 98, San Francisco, California
©1998 ACM 0-89791-964-5/98/06..$5.00

segment are realized by vertical constraints in the graph [10]. A *vertical constraint* (VC) is introduced from layout component c_i to c_j when (1) c_i and c_j have vertical overlapping on the same layer and (2) c_i is above c_j. A *directed* edge $d(c_i, c_j)$ is added from node u to node v if $c_i \in u$ and $c_j \in v$. The weight of directed edge $d(c_i, c_j)$ is equal to the minimum center-to-center distance of c_i and c_j. For example, if crosspoint c1 of node n1 is assigned above crosspoint c2 of node n2, a directed edge d(c1, c2) is added into the constraint graph between n1 and n2.

In practice, redundant jogs waste routing resources and thus increase the routing difficulty. To maximize the detailed routability for each global cell, the number of jogs should be minimized. The main effort in our work to enhance the routability is to straighten segments as much as possible. The crosspoint assignment problem that we aim to solve includes several optimization objectives - routability, crosstalk effect and path delay. In our delay calculation, the signal switching activities are taken into account for the interconnect coupling capacitance. Therefore the main optimization objectives fall into two groups - routability and path delay. We call the problem of assigning crosspoints on Gcell boundaries to optimize routability and path delay to be the Timing Driven Crosspoint Assignment (TDCA) problem.

2.3 Program Flow

Our approach is to adjust net ordering and relax wire spacing in a global scope and on a path delay basis. As shown in Figure 1, it has two stages - a net ordering algorithm to determine the crosspoint order on each boundary of the global routing cell, and a space relaxation algorithm to augment the wire spacing. The Elmore delay is calculated for each interesting signal in the worst case. The processing priority of a net at each stage is equal to its current worst delay minus the user specified maximum delay (timing constraint). In the first stage, crosspoints on each boundary are assigned to boundaries and their corresponding geometric constraints are inserted into a constraint graph while the coupling capacitance of the critical nets is minimized. After all crosspoints are assigned to boundaries, a valid routing solution is achieved. Based on this initial solution, the second-stage algorithm seeks to add space around critical nets and therefore further reduce delay.

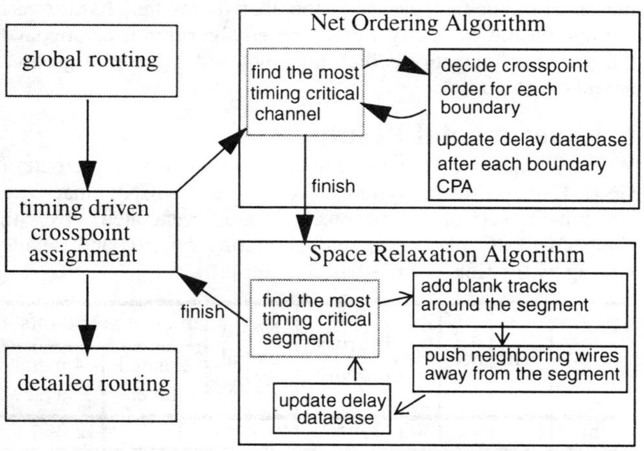

Figure 1: Flow of the algorithms

3 RC EXTRACTION AND WIRE DELAY

3.1 RC Extraction Model

From in-house capacitance extraction simulation, we observe that if a layer above or below is only 50% covered with metal, the vertical capacitance (also commonly called "substrate capacitance") to a 50% covered metal layer is close to that from a fully covered metal layer due to the fringing effect. A lookup table is constructed with respect to the wire width and distance to the neighboring wire. The wire geometry aspect ratio (height/width) was tested at 1.0, 1.5 and 2.0 for the lookup table. The wire capacitance for various wire geometry aspect ratios (AR1.0, AR1.5, AR2.0) versus the spacing to neighboring wires is drawn in Figure 2. As the picture indicates, the reduction in coupling capacitance due to extra spacing is very close to the minimum at 3c (one blank track). Therefore, our heuristics at most add one extra track of spacing around sensitive critical wires.

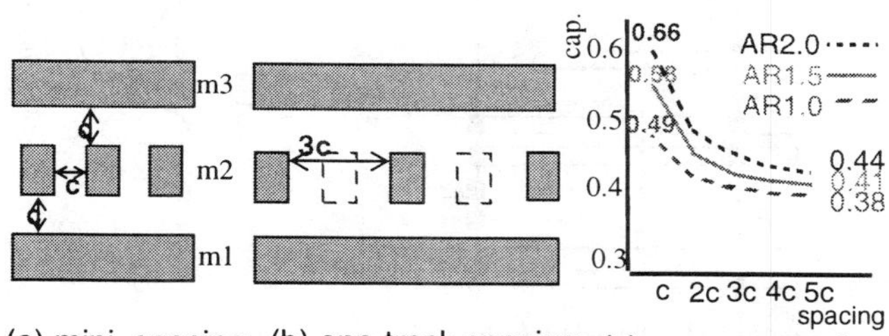

Figure 2: Coupling capacitance vs. wire spacing (c: minimum wire width)

A switching factor (SF) with regard to the timing (transition) windows of two signals is used in delay calculation. If the transition windows of two signals overlap but switch in the same direction, then for the best case the switching factor SF=0 and the effective coupling capacitance C_{eff} = 0C (for the worst case, SF=1 and C_{eff}=C). If one signal switches and the other is silent, then for both the best and worst case the switching factor SF=1 and C_{eff} = 1C. Conversely, if the transition windows of two signals overlap but switch in different directions, the switching factor is 2 and C_{eff} = 2C. Previous work [7] on predicting circuit switching activities has proved that it is very difficult and inaccurate to guess switching activities from circuit functionality. In this paper, we assume the switching factors of signals are provided by synthesis tools or the user. The effective capacitance of wire w is the summation of the contributions from the upper layer, the lower layer, and the intra-layer neighboring wires ($C_{j,w}*SF_{j,w}$), where $SF_{j,w}$ is the switching factor between two coupled signals through wires j and w.

3.2 Interconnect Delay

The Elmore delay calculation is used to estimate path delays. A Π model is used for each wire segment for delay calculation.

4 Net Ordering Algorithm

Path delays of interesting nets are calculated on the fly during the optimization process. Initially we assume each node of the interesting nets has the worst possible neighboring wires and then RC extraction is applied. A net having a negative timing slack value is called a *critical net* if its delay exceeds the user-specified timing requirement. Nodes (non-fragmented segments) on a critical net are called *critical nodes* and crosspoints of a critical node are called *critical crosspoints*. The net ordering algorithm always processes the most critical channel which has the longest segment of the most critical node. The assignment process starts from the most congested boundary in that channel and continues to proceed outward.

A crosspoint is *processed* if it is inserted into the graph. A node is processed if any one of its crosspoints is inserted into a graph. The crosspoints on a selected boundary are inserted into the graph in the following order:

- the crosspoints whose nodes are not movable
- the crosspoints whose nodes have been processed, starting from bottommost one
- the rest of the crosspoints

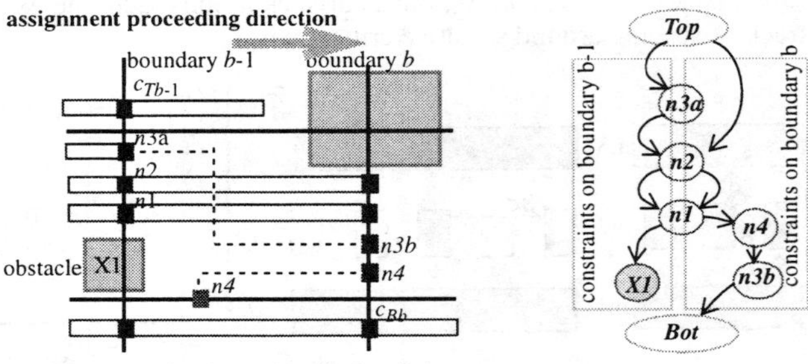

Figure 3: Crosspoint assignment

For each new crosspoint insertion, all the geometric intervals between the existing assigned crosspoints and obstacles are examined to satisfy design rules. The interval with the minimum cost is selected and the crosspoint is inserted into the graph. The cost function to assign crosspoint c_j in the interval (c_i, c_{i+1}) is equal to the coupling capacitance increase among interesting nets.

If none of these assignments is valid and there are intervals large enough to accommodate c_I, we break n_I between the previous boundary b-1 and the current boundary b. Then the new node of c_I is inserted into boundary b. If there is no interval large enough for c_I, we rip up all crosspoints on boundary b and break all nodes between boundary b and b-1. The new nodes of ripped crosspoints are assigned back to boundary b from the bottom in the ascending order of their old y positions because we want to preserve the net ordering inherited from boundary b-1 for better routability.

In Figure 3, boundary b-1 is first processed and then it proceeds to boundary b. The crosspoint for net *n3* on boundary b is tested and cannot be extended; therefore we break it into two nodes, *n3a* and *n3b*. Node *n3b* is then inserted into boundary b at the minimum cost. Node *n4* has a crosspoint at the bottom of Gcell and one at boundary b. Since the crosspoints of node *n4* are not processed, the crosspoint of *n4* on boundary b is inserted after *n3b* is inserted.

5 Space Relaxation Algorithm

In a graph, there could be multiple critical nodes to which we need to allocate routing resources. However, there may be multiple directed paths connecting two critical nodes in a channel graph, the space allocation for one node may degrade the possible space allocation of other critical nodes on the same directed path. To alleviate the ordering problem in space allocation, we allocate only partial space resources around a node at each iteration. At each iteration, the critical node n_c (the longest segment) of the most critical net N_c ($timing\_slack_{N_c} = min(timing\_slack)$) is selected for space relaxation. The available spaces around the selected node are allocated by two stages of relaxation optimizations - *self relaxation* and *neighbor relaxation*. The former has higher priority than the latter. Node n_c is said to be *self-relaxable* if it's movable in the graph. In the self relaxation stage, one of the upper and lower sides of n_c is selected, based on the least cost, and all neighbors on that side are moved outward by the a predetermined incremental spacing (e.g. 0.5 track). Its cost function is equal to the increase in the total capacitances of interesting nets due to this operation. In the neighbor relaxation stage, the neighboring node is selected, based on the least cost, to add space to n_c. The cost function is the same as that of self relaxation.

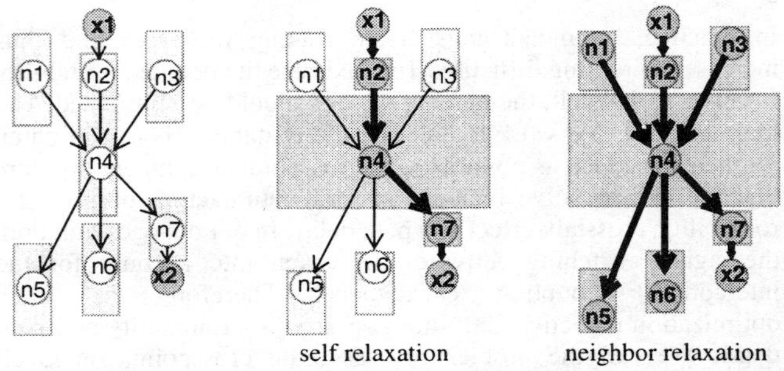

Figure 4: Self relaxation of node 4 and its neighbor relaxation

An example of self relaxation and neighbor relaxation on node 4 is shown in Figure 4. The light shaded region of each node in Figure 4(a) stands for the movable region, and the dark nodes in (b) are non-relaxable because extra spaces (dark shaded areas) are added. The relaxable neighbors *n1, n3, n5, n6* are further relaxed outward by adding extra spacing until none of them is movable.

Comparison with Previous Work - Chaudhary *et. al.* [6] proposed a graph-based spacing algorithm in 1993. The framework spaces out wires after detailed routing has been finished, contrary to the prerouting strategy of our framework prior to detailed routing. Our framework is different from his work in two aspects - the measure on crosstalk effect and the graph-based algorithm. Chaudhary's measure of the crosstalk effect is based on voltage glitches, instead of actual delay in our framework. In his work, the crosstalk effect from driver to sink is the unweighted superposition of all glitches of wires along the path. The cost function is therefore more for the measure of logic fault hazards rather than timing fault hazards as in our approach. Chaudhary's graph-based algorithm is analogous to the neighbor relaxation which does not add space to a group of neighbors at a time.

6 Experimental Results

We have implemented the timing and crosstalk driven prerouter in GNU C++ on Unix systems. To test the MCNC macro-cell benchmark circuits, placements are generated by the TimberWolfMC placer [8], global routing by [9], and detailed routing by the Kokanee tile-based router [10].

	cells	nets	pins	cross points	No. of global cells	mini. area (E6 mfs^2)	
						2 metal layer	4 metal layer
hp	11	83	309	1375	19 X 25	13	12.3
ami33	37	124	513	2611	35 X 42	2.44	1.89
apte	9	96	283	938	18 X 20	54.3	49.6
xerox	10	203	696	2132	19 X 23	26.8	24.6
ami49	49	408	953	13360	88 X 96	48.2	46.4

Table 1: Characteristics of benchmark circuits

The characteristics of the tested MCNC benchmark circuits are listed in Table 1. Area and length in tables are in the units of minimum feature size (*mfs*) of its fabrication process. In all of our tested circuits, we use *msf* = 0.25um. In Table 2, six groups of two-pin nets are selected as critical nets. The second row is the total wire length of the selected nets over the chip wire length. 'N+S' stands for turning on both Net ordering and Space relaxation algorithms, and 'N' stands for only turning on Net ordering. The average delay reduction of selected critical nets is shown by percentage. The net ordering algorithm accounts for 60-70% of the total improvement. The delay reduction increases when the wire

group of 2pin nets	A (7nets)	B (15)	C (14)	D (10)	E (17)	F (24 nets)	
wire length	1.7%	6.2%	5.8%	6.5%	3.4%	10.1%	
						N+S	N
AR1.0	-5	-7	-13	-6	-7	-10	-7
AR1.5	-6.8	-8.8	-15.4	-7.7	-9.9	-11	-8
AR2.0	-8.4	-10.7	-18.5	-9.5	-11.8	-13.9	-10.5

Table 2: delay reduction vs. num. of critical nets for ami33

aspect ratio increases. The combinations of six net groups are again tested for routability in Table 3. Notice that only 0.7-0.8% of Gcells (GC) have detailed routing failures. The failed switchboxes are merged with neighboring switchboxes and rerouted successfully. Notations 0.8N, 0.9N, 0.95N in Table 3 and Table 4 stand for 20%, 10%, 5% of total wire length being randomly selected for deletion. It shows that lower wire density gives more improvement from space relaxation. Five MCNC benchmark

	A (1.7%)		E (3.4%)		B&C (12%)			D&F (16.6%)		
	delay red.	failed GCs	delay	failed GCs	delay N+S*	delay N	failed GCs*	delay N+S*	delay N	failed GCs*
1.00N	-5	0.73	-7	0.83	-10	-7	0.77	-8.2	-6	0.77
0.95N	-6.3	0.8	-7.1	0.8	-11.8	-7	0.77	-8.9	-7.1	0.83

Table 3: routability on test cases of Table 2 at wire aspect ratio=1 for ami33

circuits are tested using three wire aspect ratio (1.0, 1.5, 2.0) for two metal layers in Table 4. The CPU time is collected on an Intel PentiumPro 200MHz using AR2.0 and wire density 100%. The

	AR1.0			AR1.5			AR2.0			CPU (secs)
	1.0N	0.9N	0.8N	1.0N	0.9N	0.8N	1.0N	0.9N	0.8N	
ami33	-9.5	-11	-13	-11	-11.5	-12.3	-13.9	-13.2	-14	7.4
ami49	-9.2	-10	-10.8	-13.6	-14.5	-15.3	-15.9	-18.8	-18.7	34
hp	-7.3	-8.7	-8.9	-10.8	-12.7	-14.2	-12.5	-15.2	-17	3.1
xerox	-11.4	-12.1	-13.8	-13.9	-16	-18.5	-15.8	-18.3	-22	7.2
apte	-9.7	-12.6	-12.8	-13.7	-11.8	-12.5	-16.4	-14.2	-14.8	2.7

Table 4: path delay reduction for MCNC benchmark circuits

results demonstrate the improvement increases when wire aspect ratio increases and when the wire density decreases.

7 Conclusion

We have presented a timing and crosstalk driven router for the chip assembly task that is applied between global and detailed routing. Our new approach aims to process the crosstalk and timing constraints by ordering nets and tuning wire spacing in a quantitative way. The new approach fits between global routing and detailed routing along the physical design flow. It is the first to address the timing and crosstalk driven area routing problem in the pre-detailed routing stage, in contrast to the previous approaches which are applied in the post-detailed routing stage. Our new approach enjoys a larger optimization solution space than the previous approaches whose solution space is highly limited by routed geometric constraints.

8 ACKNOWLEDGEMENTS

We would like to thank Professor Andrew Kahng of UCLA for his helpful opinions on interconnection delay analysis.

9 REFERENCES

[1] H. Zhou and D.F. Wong, "An Optimal Algorithm for River Routing with Crosstalk Constraints," 1996 International Conference on Computer-Aided Design, pp. 310-15, 1996.

[2] K, Jhang, S. Ha, and C. S. Jhon, "COP: A Crosstalk Optimizer for Gridded Channel Routing," *IEEE Transactions on Computer-Aided Design of Integrated Circuits and Systems*, pp. 424-9, Vol. 15, N. 5, April 1996.

[3] Tong Gao and C. L. Liu, "Minimum Crosstalk Switchbox Routing," 1994 International Conference on Computer-Aided Design, pp. 610-15, 1994.

[4] T. Xue, E. S. Kuh and D. Wang, "Post Global Routing Crosstalk Risk Estimation and Reduction," 1996 International Conference on Computer-Aided Design, pp. 302-9, 1996.

[5] K. Chaudhary, A. Onozawa, and E. S. Kuh, "A Spacing Algorithm for Performance Enhancement and Cross-talk Reduction," 1993 International Conference on Computer-Aided Design, pp. 697-702, 1993.

[6] W.C. Kao and T.M. Parng, "Cross Point Assignment with Global Rerouting for General-Architecture Designs," *IEEE Transactions on Computer-Aided Design of Integrated Circuits and Systems*, pp. 337-348, Vol. 14, No. 3, March 1995.

[7] D. A. Kirkpatrick and A. L. Sangiovanni-Vincentelli, "Digital Sensitivity: Predicting Signal Interaction using Functional Analysis," 1996 International Conference on Computer-Aided Design, pp. 536-541, 1996.

[8] W. Swartz and C. Sechen, "New Alogrithms for the Placement and Routing of Macro Cells," Digest of Technical Papers of 1990 IEEE/ACM International Conference on Computer-Aided Design, pp. 336-9, 1990

[9] L. C. E. Liu and C. Sechen, "Multi-layer Chip-level Global Routing Using an Efficient Graph-based Steiner Tree Heuristic," Proceedings. European Design and Test Conference, pp. 311-18, ED & TC 97, 1997

[10] H. P. Tseng, "Detailed Routing Algorithms for VLSI Circuits," Ph.D. Thesis, University of Washington, Seattle, 1997.

Process Multi-Circuit Optimization

Arun Lokanathan
Computer Science and Engineering
University of Notre Dame
Notre Dame, IN 46556
alokanat@cse.nd.edu

Jay Brockman
Computer Science and Engineering
University of Notre Dame
Notre Dame, IN 46556
jbb@cse.nd.edu

Abstract: This paper describes the implementation of a concurrent methodology for integrated circuit optimization that spans the fabrication process design and circuit design disciplines. Results from this methodology show substantial performance gains as compared to a separate process/circuit optimization and a substantial time savings over an "all-at-once" combined optimization.

1. Introduction

For a given semiconductor technology (CMOS, bipolar, BiCMOS, GaAs) and a given circuit topology (a particular datapath or a memory architecture, for example), the performances of an integrated circuit are generally determined by two distinct "camps" of designers: the device and fabrication process designers, and the circuit designers. Device and process designers control the electrical characteristics of the devices that they fabricate, while circuit designers seek to develop optimal circuits with those devices. Circuit optimization and process optimization are logically treated as separate problems, where circuit designers assume a fixed set of device model parameters (Spice device models) and adjust circuit geometries to enhance circuit performances, and process engineers strive to fabricate devices that conform to the circuit designers' assumptions. Caught in the middle, however, is the question of how to target the device characteristics. This question is becoming increasingly important as we move to systems-on-a-chip, such as mixed logic and memory (e.g. [13]), where different parts of a chip may "prefer" higher or lower threshold voltages, oxide thicknesses, etc. Advanced or exotic processing techniques may sometimes be able to produce different devices for different circuits, but in general this is quite expensive and should be avoided.

The problem of optimizing devices for multiple circuits on a single die is challenging for several reasons. First of all, individual device model parameters cannot be treated as independent design variables as they are correlated through the fabrication process. Secondly, device parameters cannot be properly chosen if the circuit designs themselves are mistargetted; in other words, device characteristics should ideally be optimized for the best circuits that could be designed with those devices. This demands that circuit, device, and process design be considered concurrently.

In theory, concurrent process, device, and circuit optimization could be represented as a "flat" problem, with the fabrication process controls and circuit geometries as the design variables and circuit performances as the objective, that is solved "all-at-once" using process, device, and circuit simulators to estimate state values. This, however, is practically infeasible for several reasons: obviously, because of the size of the problem and complexity of the tools involved, and more subtly, because it ignores the natural and logical organizational partitioning between the circuit and process design domains.

This paper describes the implementation and application of Process Multi-Circuit Optimization (PMCO), a strategy that enables process designers and circuit designers to work concurrently and coordinatedly toward the common goal of optimizing system-level IC performance. Results of benchmarking the concurrent methodology against separately optimized circuits and devices, as well as against an "all-at-once" joint optimization approach are presented. As examples of different circuits with different device needs, an output buffer is optimized for speed and power and an SRAM cell is optimized for access time and read and write reliability. Process and device simulations for a twin-tub CMOS process are done with pdFab ([11]) and circuit simulations are done with a commercial version of Spice. To facilitate the optimization, an extensible object-oriented environment has been developed that integrates the various simulation and optimization tools. Results show substantial aggregate circuit performance gains over separate optimizations, and substantial time savings over the all-at-once optimization.

2. Background and Related Work

Consider the problem of optimizing Φ, the aggregate performance of an integrated circuit composed of subcircuits with performances $f_1, f_2 \ldots f_m$, as shown in Figure 1. The subcircuit performance vectors $f_1, f_2 \ldots f_m$ depend upon the respective subcircuit geometries $G_1, G_2 \ldots G_m$ and the set of device model parameters D. The set of fab process inputs P, determines the values of the device model parameters, D. The goal of process-circuit co-optimization is to determine values for the process inputs P and circuit geometries $G_1, G_2 \ldots G_m$, that optimize the system performance, Φ. The

DAC 98, San Francisco, California
©1998 ACM 0-89791-964-5/98/06..$5.00

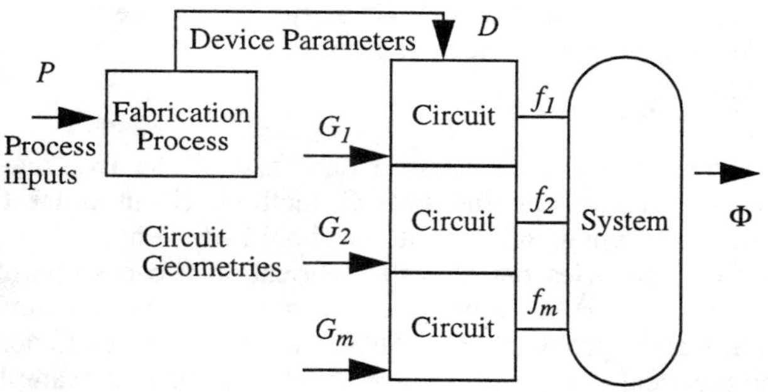

Figure 1. Variables in Integrated Circuit Optimization

system objective function must reflect the multiple performance requirements of the circuits that the IC system is composed of—common choices for Φ include linear combination of performances, sum of squares, or sum of exponentiated terms ([1]). We cast the IC optimization problem as a multi-objective optimization problem using a technique called goal programming ([6]).

In the past, work in the area of integrated circuit optimization has focussed predominantly on techniques for local design improvement in either the circuit design discipline or the process design discipline. In fact, circuit optimization is a relatively mature area; a survey of relevant literature can be found in [1]. Statistical circuit optimization that further accounts for process variations has also been studied extensively, and a number of formulations and techniques, based on traditional optimization approaches, have been proposed in the past to solve the problem ([2], [3], [4]). The examples in the literature almost exclusively demonstrate single-domain optimization. Circuit optimization examples adjust circuit geometries with a view to achieving acceptable circuit performances within some range of variation of device model parameters and/or operating conditions. Process optimization examples adjust process inputs to achieve acceptable device model parameter values. Previous work reported in [5] that does address optimization across the process and circuit design disciplines, uses a stochastic algorithm to optimize a process and single-circuit system in a sequential all-at-once manner. In contrast, the PMCO method is a concurrent deterministic method that may be used to optimize a fab process and a set of circuits in an independent yet coordinated manner.

Process Multi-Circuit Optimization (PMCO), the methodology of this paper, draws on ideas from the general area of Multidisciplinary Design Optimization (MDO) [7], and is based specifically on the Concurrent Subspace Optimization (CSSO) algorithm [8]. MDO algorithms start with a complex coupled system-level optimization problem, decompose the problem into smaller approximate subproblems that are solved independently (by domain-specific subspace design teams), and coordinate the solutions of the subproblems to obtain a system-level optimal solution.

3. Process Multi-Circuit Optimization

The PMCO method optimizes the overall system of Figure 1 by formulating a common system-level optimization problem, and then solving it in a concurrent manner. The IC optimization problem is cast using the goal programming formulation which attempts to simultaneously minimize (or maximize) multiple circuit performance functions below (or above) user-specifiable targets. The minimization of several functions $f_1, f_2 \ldots f_m$ is converted to the following scalar constrained nonlinear optimization problem:

$$
\begin{array}{ll}
\text{Minimize} & \gamma \\
\text{w.r.t.} & \gamma, P, G_1, G_2 \ldots G_m \\
\text{subject to} & f_1 - \gamma\alpha_1 \le \delta_1 \\
 & f_2 - \gamma\alpha_2 \le \delta_2 \\
 & \ldots \\
 & f_m - \gamma\alpha_m \le \delta_m
\end{array}
$$

The values $\delta_1, \delta_2 \ldots \delta_m$ are targets or goals that are set for the performances $f_1, f_2 \ldots f_m$ respectively. $\alpha_1, \alpha_2 \ldots \alpha_m$ are weights used to adjust the relative rigidity with which the targets need to be met. γ, a scalar, is introduced as an independent design variable as well as the objective function of the transformed optimization problem. The multiple objective functions of the original problem are then transformed into the constraints of the goal programming formulation. Assuming positive weights α_i, as γ is decreased, each of the performance functions $f_i(D(P), G_i)$, $i = 1 \ldots m$ need to be decreased in order to satisfy their respective constraints in the goal programming problem. This leads to all the functions f_i being simultaneously minimized, if such a point exists.

The overall flow of the PMCO method, first proposed in [9], is shown in Figure 2. The algorithm starts with the current set of values for the design variables $X^k = [P^k, G_1^k \ldots G_m^k]$ at the kth outer iteration. It first performs what is termed as a "system analysis" which evaluates the following: (i) values of device parameters, D^k, and device parameter sensitivities with respect to process inputs, $\nabla_P D^k$, via process simulation, (ii) values of the performances of each circuit, f_i^k and their partial sensitivities with respect to geometries and device model parameters, namely $\nabla_{G_i} f_i^k$ and $\nabla_D f_i^k$. In other words, the system analysis computes process and circuit function values and gradients at the current design point.

The algorithm then launches several independent subspace optimization tasks, one in the process domain and one corresponding to each circuit. Within the process optimization, the process designers calculate device model parameters as a function of process inputs via process

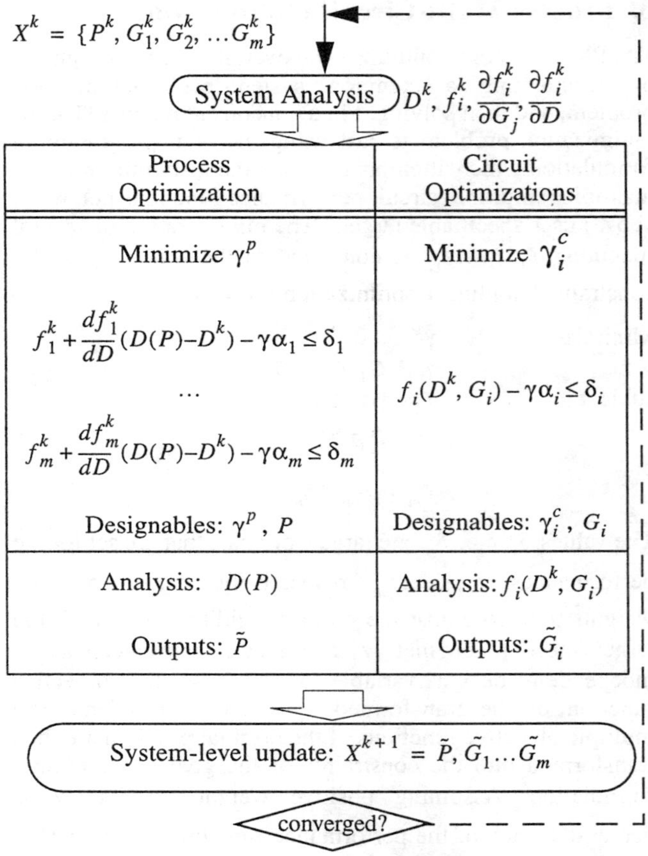

Figure 2. The Concurrent Process-Circuit Codesign Method

simulation. In order to approximate circuit performances, they use a linear mapping of device model parameters to circuit performances using the currently available gradient information, namely, the matrix $\nabla_D f_i^k$. Due to the linear $D \to f$ assumption, move limits must be imposed on the design variables of the process subspace optimization, namely the process inputs P. A user-selectable constant move limit (defaulting to 25% of the global range) is used in this work. In the circuit optimizations, circuit designers use the currently available device model parameter set, and adjust circuit geometries to optimize individual circuit performances. The process-optimal and circuit-optimal design vectors are concatenated to obtain the system design vector for the next iteration.

The key advantages of the PMCO method are (i) the subspace optimization tasks are of smaller dimensionality than the whole system-level problem, leading to robust convergence characteristics (in general, it is reasonable to expect optimization algorithms to converge better on smaller-dimensional problems) (ii) the process and circuit subspace optimization tasks may proceed concurrently, thereby decreasing design time, (iii) there is a certain degree of design freedom within the process and circuit optimization tasks—one may use different move limits, tolerance settings or even different optimization strategies in the different tasks. We will illustrate these benefits via a series of examples in the next section.

4. Results

A series of examples is used demonstrate the two main points that motivate the PMCO method: (i) in order to optimize an integrated circuit, we should adjust not only the circuit geometries but also the fabrication process control inputs, and (ii) in solving the combined process-circuit optimization problem, the PMCO method is more efficient and practical than an all-at-once approach. The first example optimizes the a CMOS buffer while a second optimizes an SRAM cell. A third example then optimizes a system composed of both the buffer and SRAM circuits.

The examples use a twin-tub CMOS process modeled by 75 processing steps in pdFab ([11]), which maps 120 process inputs to 26 Spice level 3 MOS transistor model parameters. Because only a few process variables generally impact device/circuit performances significantly, a simple sensitivity-based variable screening step was performed initially to identify the process inputs that had the most impact on the performances of the buffer and SRAM circuits ([14]).

In each example, the optimization problem is first solved by adjusting only circuit design variables and then solved by adjusting process inputs as well as circuit geometries. This brings out the benefit of combined process-circuit optimization over circuit-only optimization. To demonstrate the efficiency of the concurrent PMCO approach, the combined process-circuit optimization problem is solved using two methods: (i) an all-at-once method that treats the entire process-circuit system as a monolithic whole and "synchronously" optimizes the system over all the variables of the system (process and circuit variables), and (ii) the concurrent PMCO method.

The results indicate that while the choice of circuit geometries is responsible for large performance improvements, the performances of optimal-geometry circuits can be further improved significantly via an optimal choice of device characteristics (and hence process inputs). The third example also demonstrates the application of the concurrent methodology to multiple circuits as well as its capability to negotiate process trade-offs. Furthermore, the concurrent PMCO algorithm is found to be more efficient than the all-at-once approach in terms of elapsed time.

4.1 Optimization of CMOS Buffer

The buffer circuit is a 4-stage static CMOS inverter chain that may be used for driving off-chip loads or large on-chip capacitive loads such as clock networks—the designable geometries are the widths of the PMOS and NMOS transistors of inverter chain. The performances are delay and power dissipation. Both quantities are measured via a transient simulation in SmartSpice [12]. The delay is measured across the 50% Vdd points and the power is measured via a power meter along the lines of [10].

Consider the case for which the delay of the buffer needs to be minimized subject to a power constraint, as is the typical requirement for high-speed (as opposed to low-power) systems.

Minimize	$\text{delay}(P, G_1)$
subject to	$\text{power}(P, G_1) \leq 2.5\text{m}$
wrt.	P, G_1

where P is the set of 5 significant process variables determined via variable screening and G_1 is the set of 8 transistor widths.

Variable	Initial	Circuit Only	Process Circuit
Ntubbox/Temp(C)	1100.0	1100.0	1113.0
gateox2/Temp (C)	900.0	900.0	760.4
gateox3/Temp (C)	880.0	880.0	740.4
padox4/Temp (C)	1030.0	1030.0	1050.0
passiv/Temp (C)	840.0	840.0	834.0
buffer/wn5 (μm)	3.0	22.5	20.3
buffer/wn6 (μm)	3.0	51.7	44.5
buffer/wn7 (μm)	3.0	139.8	114.8
buffer/wn8 (μm)	3.0	297.0	263.1
buffer/wp5 (μm)	4.0	33.2	31.8
buffer/wp6 (μm)	4.0	84.9	80.1
buffer/wp7 (μm)	4.0	189.7	153.4
buffer/wp8 (μm)	4.0	445.5	434.9
buffer/power (mW)	1.32	1.94	2.30
buffer/delay (ns)	18.0	0.97	0.66

Table 1: CMOS Buffer Design Points

Table 1 lists the values of process inputs, circuit geometries, and circuit performances at the key design points. The initial design point (initial process, 3μ NMOS, 4μ PMOS transistors) resulted in a delay of 18.0 ns and a power dissipation of 1.32 mW. The performances were first optimized by adjusting circuit geometries alone, keeping the process inputs constant, as is done in conventional circuit optimization. As shown in the "circuit-only" column of the table, circuit optimization reduced buffer delay from 18.0 ns to 0.97 ns (a 95% reduction). The optimization problem was then solved by adjusting both the process inputs as well as the circuit geometries—the optimal design after process-circuit optimization produced a delay of 0.66 ns (an additional 32% reduction from the circuit-only optimal point). The 2.5 mW power constraint is satisfied by all the design points—the power dissipation is traded off, however, to decrease delay from the initial to the optimal points.

In order to compare the PMCO method with an all-at-once method, the above problem was solved using both methods. Table 2 lists the times spent by the all-at-once and concurrent methods. The all-at-once column is straightforward— the total time is the sum of time spent in function calls, time spent in gradient calls and time spent in line searches by the optimizer. Remember that every time the all-at-once method makes a function call, it involves a process simulation followed by a circuit simulation. For the PMCO column, we have to add the system analysis time with the larger of the process optimization time and circuit optimization time, over all iterations of the algorithm. To clarify this point, we illustrate the flow of the PMCO algorithm using the timeline diagram of Figure 3—the algorithm proceeds by iteratively performing a system analysis followed by concurrent process and circuit optimization tasks. The function calls, gradient calls and line search calls in the PMCO algorithm are made to the respective subspace approximations that involve only one simulation tool (either process or circuit, not both, as in the all-at-once case). The net result is that the all-at-once approach takes 10.79 hours whereas the concurrent optimization takes 7.97 hours to arrive at the optimal process-circuit point.

(Time in hours over all iterations)	All-at-once	PMCO	
		Process	Circuit
System Analyses	-	1.2	
Function calls	1.21	0.90	0.39
Gradient calls	4.90	3.10	1.97
Line searches	4.68	0.81	2.94
Total Time	10.79	7.97	

Table 2: Timing Analysis

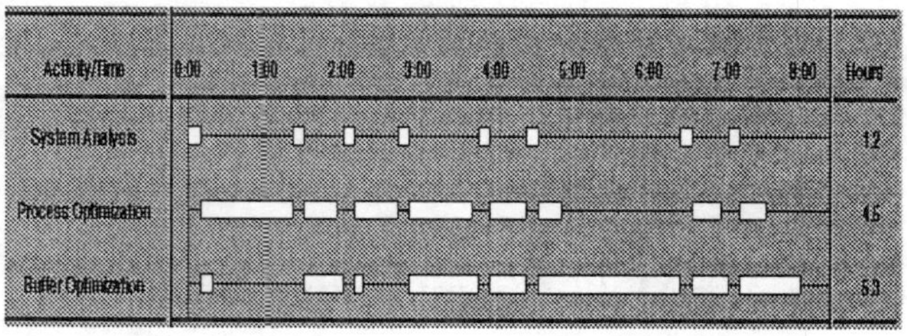

Figure 3. Time line for Concurrent Optimization

4.2 Optimization of SRAM Cell

This example optimizes the access time, read safety and write reliability of a standard 6-transistor SRAM cell (connected between two bit lines and a word line in the usual manner). Figure 4 shows the 2-dimensional space of bit line voltages divided into regions of safe read, write-0 and write-1. The quantity v_{sr0}, the minimum voltage required on a bit line to ensure a safe read-0 operation,

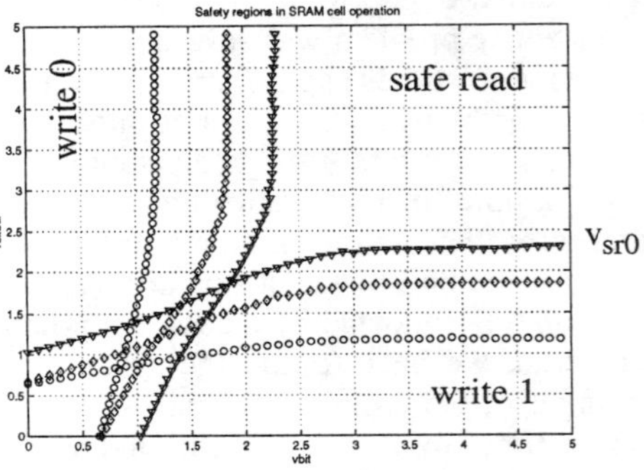

Figure 4. SRAM Cell Read Safety/Write Reliability

defines the boundary between the read and write regions. A higher v_{sr0} implies higher read safety but lower write reliability. We will require that v_{sr0} be made as low as possible but no lower than 1.5 V. This leads to the following optimization formulation:

Minimize	$\mathrm{accesstime}(P, G_2)$
and	$v_{sr0}(P, G_2)$
subject to	$v_{sr0}(P, G_2) \geq 1.5$
wrt.	P, G_2

The performances at the initial, circuit-optimal and combined process-circuit optimal design points are shown in Table 3 (the input variables have been left out for brevity). The initial access time of 2.66 ns was reduced to 0.97 ns via circuit optimization. Process optimization accounted for a further drop in access time from 0.97 ns to 0.79 ns. The quantity v_{sr0} was reduced in both cases from 2.27 V to 1.50 V, which maximizes read safety subject to the write reliability constraint. Combined process-circuit optimization took 8.02 hours via the all-at-once approach as opposed to 6.14 hours via the concurrent method.

	Initial	Circuit Only	Process Circuit
sram/accesstime (ns)	2.66	0.97	0.79
sram/vsr0 (V)	2.27	1.50	1.50

Table 3: SRAM Cell Optimization Design Points

4.3 Process Multi-Circuit Optimization

This example illustrates the applicability of the concurrent methodology to multiple circuits as well as its ability to address competing performance concerns via process trade-offs. It is required to find a set of process inputs and circuit geometries that minimize buffer delay as well as SRAM cell accesstime and read safety, while satisfying constraints on buffer power dissipation and SRAM cell write reliability. This leads to the following optimization formulation:

Minimize	$\mathrm{delay}(P, G_1)$
and	$\mathrm{accesstime}(P, G_2)$
and	$v_{sr0}(P, G_2)$
subject to	$v_{sr0}(P, G_2) \geq 1.5$
and	$\mathrm{power}(P, G_1) \leq 2.5\mathrm{m}$
wrt.	P, G_1, G_2

Variable	Initial	Circuit-only	Process Circuit
buffer/power	1.32 mW	1.94 mW	2.3 mW
buffer/delay	18.0 ns	0.97 ns	0.69 ns
sram/accesstime	4.93 ns	1.59 ns	0.82 ns
sram/vsr0	2.05 V	2.10 V	1.51 V

Table 4: Combined Optimization Design Points

Table 4 shows the circuit performances at the initial design point, the circuit-optimal design point and the process-circuit optimal design point. Circuit-only optimization improves the performances of both circuits—the buffer delay is reduced to 18.0 ns from 0.97 ns, the SRAM access time is reduced from 4.93 ns to 1.59 ns. v_{sr0}, however, is slightly increased to 2.10 V, and is undesirable for read safety—circuit-only optimization maximize read safety at the given initial process point. Combined process-circuit optimization, on the other hand, arrives at a combination of process inputs and circuit geometries that produces an SRAM v_{sr0} of 1.51V, thereby maximizing read safety as desired. Furthermore, it improves the buffer delay to 0.69 ns (a 28% improvement over the circuit-only optimal point) and improves the access time of the SRAM to 0.82 ns (a 48% improvement over the circuit-only optimal point). Process-circuit optimization via the concurrent PMCO method thus produces a significantly better design point than merely circuit optimization. Again, this example was solved using both the concurrent method and the all-at-once method—the concurrent method took 6.61 hours compared with 17.1 hours of the all-at-once approach. The reason for the time savings of the PMCO method over the all-at-once approach is that the PMCO method solves several smaller problems independently—a trend that may be expected to get better as the number of circuits being optimized increases.

Finally, Table 5 summarizes the key design points, namely, the initial point, the point after circuit optimization, and points after process optimization targeting the buffer alone, the SRAM cell alone and then both simultaneously. Observe that the process point that optimizes buffer alone is detrimental to the SRAM cell access time; the process point that optimizes the SRAM alone penalizes buffer delay. The PMCO method produces a process point that is beneficial to the system as a whole, in that while it sacrifices SRAM access time somewhat, it produces the fastest buffer and SRAM cell while satisfying constraints on power dissipation and SRAM read safety and write reliability.

Design Point	Buffer		SRAM cell		
	Delay	Power	Access time	vsr0	
Initial	18.0 ns	1.32 mW	4.93 ns	2.05 V	Slow buffer, slow SRAM, Unsafe read
After circuit optimization	0.97 ns	1.94 mW	1.59 ns	2.10 V	Room for improvement
Buffer-optimal process	0.66 ns	2.30 mW	1.56 ns	1.50 V	Fast buffer, safe SRAM read, slow SRAM
SRAM-optimal process	0.70 ns	2.48 mW	0.79 ns	1.50 V	Fast SRAM, safe SRAM read slow buffer
Joint optimal process	0.69 ns	2.30 mW	0.82 ns	1.50 V	Fast buffer, fast SRAM, safe SRAM read

Table 5: Process Trade-offs

5. Conclusion

This paper has demonstrated that, in order to design optimal integrated circuits, it is important not only for circuit designers to adjust circuit geometries but also for fabrication process engineers to adjust process controls for suitable device characteristics and better overall system performance. The results presented in this paper indicate that significant performance improvements can be achieved by performing combined process multi-circuit optimization. There are two distinct ways in which combined process-circuit optimization may be performed—either the entire system may be optimized all-at-once or the concurrent process multi-circuit optimization (PMCO) method may be used. The PMCO method produces similar design points as the all-at-once approach in significantly less time. The concurrent PMCO method is useful in optimizing a fab process targeting multiple circuits that may embody competing performance concerns.

6. Acknowledgments

This work was supported in part by grants from Motorola Semiconductor Products Sector, NSF and NASA.

7. References

[1] R. K. Brayton, and R. Spence, "*Sensitivity and optimization*", Elsevier Scientific, Amsterdam, 1980.

[2] J. C. Zhang and M. A. Styblinski, "Yield and Variability Optimization of Integrated Circuits", Kluwer Academic Publishers, 1995.

[3] K. J. Antreich, H. E. Graeb, and C. U. Weiser, "Circuit Analysis and Optimization Driven by Worst-Case Distances", *IEEE Transactions on CAD*, vol. 13, no. 1, January 1994.

[4] P. Feldmann and S.W. Director. Accurate and efficient evaluation of circuit yield and yield gradients. In *IEEE International Conference on Computer-Aided Design, ICCAD-90*, pages 120-123, November 1990.

[5] M. A. Styblinski and L. J. Opalski, "Algorithms and Software Tools for IC Yield Optimization Based on Fundamental Fabrication Parameters", *IEEE Transactions on CAD*, vol. CAD-5, no. 1, pp. 79-89, January 1986.

[6] R. K. Brayton, S. W. Director, G. D. Hachtel, L. Vidigal, "A New Algorithm for Statistical Circuit Design Based on Quasi-Newton Methods and Function Splitting", *IEEE Transactions on Circuits and Systems*, vol. CAS-26, pp. 784-794, 1979.

[7] R. J. Balling, and J. Sobieszczanski-Sobieski, "Optimization of Coupled Systems: A Critical Overview of Approaches", In *Proceedings of The AIAA/NASA/ USAF/ISSMO Symposium on Multidisciplinary Analysis and Optimization*, Panama City, FL, September 7-9, 1994.

[8] J. E. Renaud, "An Optimization Strategy for Multidisciplinary Systems Design", *Proceedings of the 9th International Conference on Engineering Design*, August 1993.

[9] Arun N. Lokanathan, Jay B. Brockman and John E. Renaud, "A Methodology for Concurrent Fabrication Process/Cell Library Optimization", In *Proceedings of the 33rd IEEE/ACM Design Automation Conference*, June 1996.

[10] Sung Mo Kang, "Accurate Simulation of Power Dissipation in VLSI Circuits", *IEEE Journal of Solid-State Circuits*, vol. SC-21, no. 5, October 1986.

[11] "*Pdfab User's Reference Manual*", PDF Solutions, San Jose, CA 1996.

[12] "SmartSpice/UTMOST III Users Manual", Silvaco International, Santa Clara, CA, 1995.

[13] P. M. Kogge, T. Sunaga, E. Retter, *et al*, "Combined DRAM and Logic Chip for Massively Parallel Applications," *16th IEEE Conference on Advanced Research in VLSI*, Raleigh, NC, IEEE Computer Society Press #PR07047, March 1995, pp. 4-16

[14] Arun N. Lokanathan, "Concurrent Process Multi-Circuit Optimization", Ph.D. thesis, University of Notre Dame, Notre Dame, IN, 1998.

Migration : A new technique to improve synthesized designs through incremental customization

Rajendran Panda, Abhijit Dharchoudhury, Tim Edwards, Joe Norton, and David Blaauw
Advanced Tools Group, Advanced System Technologies Lab., Motorola, Austin, TX 78730
email: panda@adttx.sps.mot.com

Abstract - A novel technique to explore the *performance vs design effort* trade-off is proposed. Starting from an optimally synthesized design, performance-critical cells are incrementally and optimally selected and custom-sized to generate this trade-off. Efficient algorithms for the optimal selection, and for improving the reuse of custom-sized cells in the design are given. Significant performance gains are shown in several real circuits through the addition of very few customized cells.

Keywords - migration, custom sizing, timing optimization

1. Introduction

Circuit designers today face two challenges of conflicting objectives - designing for optimum performance, and designing in a minimum time to reduce the 'time to market'. Though full-custom solutions are ideal for achieving optimum performance, they require the most time and manpower resources. On the other hand, automatically synthesized solutions are attractive in the required design effort, but often fall short of the performance goals. The common design practice of identifying *a priori* what portions of a design are to be custom-designed and what are to be synthesized can be very sub-optimal, in terms of both the achieved performance and the design resources spent. There is currently no systematic way of automatically, and optimally, exploring the trade-off between these two metrics. Addressing this need, we provide a technique to generate design solutions that lie on an optimal *performance vs design effort* trade-off, and let the designer select the solution that is appropriate for the situation.

Despite the advancements in automatic synthesis, full-custom solutions still show superior performance than their synthesized counterparts, as illustrated by a real design in Figure 1. The curve in the figure is the optimal *area vs performance* trade-off generated by a mature custom sizing tool using a TILOS[1]-like algorithm.

The x, y co-ordinates are performance (in terms of slack) and transistor area respectively, with points toward the left indicating better performance.

Also shown in the figure to the right of the curve is a standard-cell solution synthesized for optimum performance, using a commercial synthesis tool and an industrial library of 163 cells, and 2 - 4 drive strengths per cell type. Note the large performance difference between the two design methods. This is clearly due to the extreme flexibility in custom-sizing the devices, since the circuit structure is the same for these solutions. This is despite the fairly large size of the library used in synthesis. We have observed such performance differences consistently across numerous designs, and with different libraries. The library cells which are optimized for a wide range of loads may be suitable to instantiate a major chunk of logic which is not critical. However, while instantiating some logic on critical paths, the finite selection of drive strengths in the library, their fixed transistor sizes, and the fixed P/N ratio for a given drive strength become severe bottlenecks from the performance point of view.

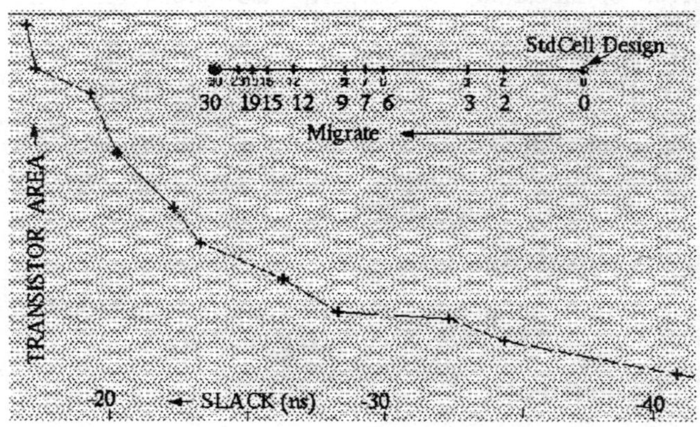

Figure 1. Performance gap between custom and standard cell designs.

The key idea in our proposal to optimize both the performance and the design effort is to stay close to the standard cell design paradigm, while also adding the flexibility of fine-tuning the device sizes of a selected few logic instances. This is very similar to what the designers have been doing, but with two important differences: (i) the custom–synthesis partition is now fine-grained, which is done at the cell (i.e. gate) level, and (ii) the partitioning decision is automatic and nearly optimal. The starting point is a synthesized design that has been optimized for performance and a given area. The performance is then improved in incremental steps, while maintaining the same area, by replacing a selected few critical cell instances with optimally sized custom cells. In Figure 1, this process is shown by the horizontal trade-off line which is migrating the design from the standard cell solution point to the left toward the full-custom trade-off curve. The number of custom cells needed at each step of migration is shown below the solution point. The main steps in the process are the optimal selection of the cell instances to be customized, and the optimal sizing of the selected instances. The latter step can be accomplished using sizing techniques proposed earlier, such as [1]. We address here the first part, viz. the optimal selection.

The remainder of the paper is organized as follows. In Section 2, we review the state-of-the-art in circuit optimization. Section 3 presents the proposed technique and the algorithms. Finally, Section 4 presents the results of several case studies and the conclusion.

2. Related Work

The topic of transistor and gate sizing for optimal performance/area/power has been very actively addressed. [1] proposed an optimization based on the posynomial relation between device sizes and circuit delays. [2] optimizes delay in a convex non-linear problem. Optimal gate sizing is addressed in [3] using linear programming techniques. [4] addresses the power-delay trade-off in gate sizing. [5] combines iterative and linear programming techniques for gate sizing. [6] proposes transistor sizing based on dynamic timing, and using non-linear solution techniques. A related problem of simultaneously optimizing the devices and wires has been studied in [7] and [8].

Earlier approaches assumed either custom or standard cell design paradigm and targeted at optimal *performance vs area/power* trade-off. In contrast, our work uniquely extends sizing techniques to explore the '*performance vs design effort*' trade-off.

3. Custom Cell Addition

We shall formulate the task as a constrained optimization problem, discuss its complexity, motivate our approach and then present effective heuristics. First, we introduce the relevant concepts and terminologies.

The overall objective is to improve the *delay graph* of a circuit, which is a DAG of timing arcs between nodes after all feedback arcs are broken. Every arc is marked with a value which is the propagation delay between its nodes, and there are distinct arcs for rising and falling transitions. A *path* is a sequence of edges in the delay graph. Associated with each arc is a *trigger* transistor, the device whose gate node is the source of the timing arc.

The sizing optimization is done iteratively on the transistor with the largest figure of merit, the merit being determined by a combination of *slack* and *sensitivity* metrics, given below :

Slack of a node is the difference between the required and the actual arrival times for the node, with large negative slack signifying large timing criticality.

Delay sensitivity, $\delta(m, a)$, of a transistor m w.r.t. an arc a is a measure of change in a's delay to a small change in the size of m[1], a large negative value indicating a favorable area/delay relation. Delay sensitivities are calculated directly from analytical timing relations.

The size of a device affects the delays of several arcs - all arcs through the device, and arcs from or to those nodes whose loading is affected by the size of that device. Therefore the merit of a transistor should consider its sensitivities w.r.t. every arc it can affect.

In our discussion, the term *cell* refers to a logic gate in the library, and *instance* refers to a logic block that can be mapped to a cell.

3.1. Formulation

We assume that the design effort (layout, extraction and characterization) needed to generate a new library cell is constant, and is independent of its logic functionality. Then the additional design effort to improve a given standard cell design can be measured by the number of new cells required. We can now formally pose the problem as a constrained optimization problem, as follows.

Given an initial standard-cell design, and a performance improvement target, find a minimum cardinality set of customized cells with which a subset of the cell instances of the design can be replaced to realize that performance improvement.

Let $\mathcal{G}$ be the delay graph under optimization and d be the delay of the longest (slowest) path in the circuit. To improve the performance of the circuit by a small amount, say, Δd, we need to improve a set P of the most critical paths whose delays are in the range $[d,\ d - \Delta d]$, such that their delays become equal to or less than $d - \Delta d$.

The problem is inherently complex for several reasons. (1) More than one arc in $\mathcal{G}$ has delay dependencies on the size of a transistor. So it becomes difficult to study the optimization of the identified critical paths in isolation. Changing the size of a transistor to improve a critical path in P can increase the delays in paths outside P. (2) The relative criticality of paths need to be considered in deciding which arcs of a path should be improved.

We consider a hypothetical, simplified, version of the above problem, purely to get a handle on the theory and to motivate our approach.

Consider a reduced delay graph $\mathcal{G}' \subset \mathcal{G}$ comprising only of timing arcs lying on the paths in P. Assume that (1) $\mathcal{G}'$ can be optimized without changing the delays in $\mathcal{G} - \mathcal{G}'$, (2) for every $a_i \in \mathcal{G}'$, there is a unique transistor m_i such that $\delta(m_i,\ a_j) = K$ for $i = j$ and $= 0$ otherwise in the region of improvement (Δd), and where K is a negative constant, and let M be the union of such transistors, corresponding to all the arcs in $\mathcal{G}'$, (3) for any $m \notin M$, and any arc $a \in \mathcal{G}'$, $\delta(m,\ a) = 0$ and (4) the transistors in M each belong to distinct cell instances in the design.

(1) above ignores the effect of sizing on the delays of arcs outside the selected critical paths. (2) and (3) together identify *one and only one* transistor for every arc on the critical paths that can be sized so that the arc's timing will

be improved. The equal delay sensitivity (K) eliminates the relative area merits of improving one arc over other. (4) establishes a simple cardinality relation between the instances sized and the transistors sized.

It is clear that, to improve $\mathcal{G}'$, every path in P must be covered (i.e. improved), by improving one or more edges in each path. Since improving an edge corresponds to sizing its corresponding transistor, this cover must be optimal to minimize the design effort. We refer to this problem as the *optimal selection problem.* We can easily show that the above simplified version of the problem is NP-complete, by reducing the *Hitting Set Problem*[9], known to be NP-complete, to this problem in polynomial time.

3.2. Optimal Selection

The foregoing formulation motivates an approach that improves the delays of a set of edges s.t. every path in P is covered (improved). This edge set is in fact the *hitting set* for P. The delay of each edge in the hitting set can be reduced by resizing the instance containing the trigger of that edge. Although resizing other instances may also improve an edge, the rationale for sizing the instance of the trigger is that the trigger and the transistors in its dc-component usually have larger sensitivities w.r.t. that edge than the transistors in other instances. Only in rare extreme cases when the trigger is already over-sized and is also driving no load or light load, this is not true.

To minimize the number of cells resized, the hitting set must be optimal. A good heuristic for the hitting set problem is the greedy covering of P using edges that lie on the most number of paths in P. We use such a greedy approach but with certain modifications so that the deviations from the above simplifying assumptions are properly accounted for. The simple greedy measure of an edge, which is the number of paths it covers, is modified by weights based on the slacks and sensitivities associated with the edges.

Without going into detail, we give below the merit function of a transistor that is used in determining the optimal set of instances to be resized. For an edge a, and a transistor m, let $S(a)$ be the slack of a, $\delta(m,a)$ be the delay-sensitivity of m w.r.t. a, $\gamma(m)$ be the input-capacitance-sensitivity of m, $R_{in}(m)$ be the drive resistance at the input (gate) of m, $S_{in}(m)$ be the slack at the input (gate) of m, and $r(m)$ be the total number of paths covered by the arcs in $\mathcal{G}'$ for which m is a trigger.

Then, the merit function $F(m)$ is taken as :

$$F(m) = r(m) \times w(m)$$

$$w(m) = \sum_{a \in \mathcal{G}'} S(a) \times \delta(m,a) + \gamma(m) \times R_{in}(m) \times S_{in}(m)$$

The $S(a)$ weights account for the relative criticality of the edges. Likewise, $\delta(m,a)$ considers relative gains of sizing, and $\gamma(m)$ considers the loading effects on the driver. The slacks are forced to be negative by suitably shifting their raw values so that the weighting function is not misinterpreted for an arc with positive slack and positive sensitivity.

3.3. Auto-grouping

Suppose multiple instances of a cell have identical drivers and loads, and the logic structure surrounding these instances are 'identical'. An example of such identical structures is a battery of inverters driving a bus, as shown in Figure 2. In such situations, if one instance is performance-critical, it is very likely that the other copies are also equally critical. Moreover, it is possible to resize all of them identically so that several signal paths in the design are improved with just one custom cell. This is the basic idea in improving reuse of customized cells. The designer can explicitly specify which are similar instances that can be grouped for sizing. When no explicit information is available, the auto-grouping algorithm detects such situations automatically.

As soon as a cell is selected for sizing, the auto-grouping mechanism looks at all other instances of this cell in the design, and verifies a set of conditions to detect if two instances are similar. The conditions verified are related to sizing. Specifically, we check to see if the instances have the same cell type, strength, loads and slacks at the output pins, and slopes at the input pins, within a specified tolerance. If so, the similar instances are grouped.

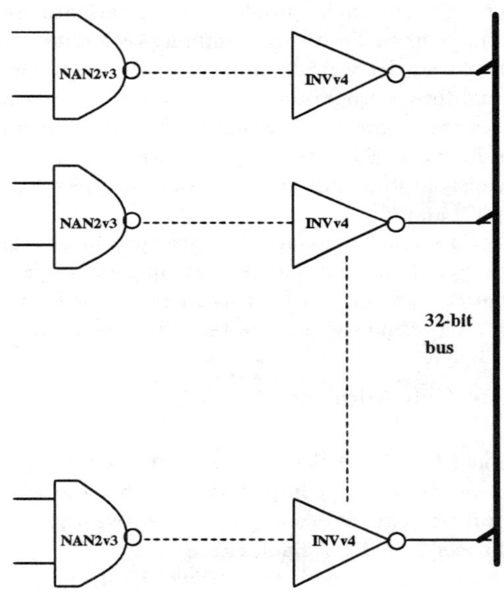

Figure 2. Multiple similar instantiation of cells

If the similar instances were not grouped, then the sizing algorithm would size them differently, though approximately equally, since the factors influencing the sizing algorithm, such as slacks, sensitivities, loads, etc. will change from one iteration to another, and are also dependent on the order in which the instances are sized. This will lead to more than one custom cell. The auto-grouping criteria are easily tested on sorted lists of instances. Thus the algorithm is very fast. □

The overall algorithm for migration is:

(1) Identify a set of top critical paths as the target set to improve. A minimum of 5 – 50 paths are targeted at each step of migration.
(2) Identify the triggers on the targeted critical paths.
(3) Merit the triggers using the cost function in Section 3.2.
(4) Greedily select the instances containing the triggers with the highest merit, until all the targeted paths are covered.
(5) For each selected instance, auto-group it with other identical instances elsewhere in the design.
(6) Resize the transistors of selected instances for optimal performance.

4. Results and Conclusion

The proposed technique was benchmarked using four real circuits drawn from different designs - two control blocks, one decoder logic, and one large microprocessor core. The initial synthesized circuits were obtained using a commercial synthesis tool, and running production quality scripts to optimize the delay under area constraints. Three different libraries were used in synthesizing the above four circuits. The proposed migration algorithm was then applied to these circuits, specifying a maximum of 20-50 cells in the design to be custom sized. The optimal selection and grouping of cells was done as described in Section 3, and the optimal sizing was done using a mature sizing optimization tool. For the migrated circuits, only the selected cells were allowed to be resized. The same sizing tool was used also to completely resize the circuits to ascertain the maximum performance gain achievable through sizing. The results are reported in Table 1. Due to the large size of the microprocessor core, a fully resized solution could not be obtained for this case.

In the table, column 2 shows the size of the circuit. Columns 3 and 4 show the size of the library used for synthesizing the initial circuit. Columns 5 – 7 show the delays respectively of the synthesized, fully resized, and migrated circuits. Column 10 shows the number of cells customized, and its percentage w.r.t. the total cell instances. Column 8 shows the percentage reduction in the delay of the migrated circuit w.r.t. the synthesized design. Column 9 compares the improvement obtained by migration to the best improvement possible by resizing the entire circuit. All delay comparisons are w.r.t. a fixed area, which is the area of the initial design.

Despite the high quality libraries used, the synthesized designs were 1.4 times to 2.1 times slower than the fully resized versions, indicating the sizing-related performance limitation of the standard cell design. But by adding very few (0.3% to 17%) custom cells, the migration obtained significant speed improvements (11.8% - 38.1%) over the synthesized design. The 11.8% delay reduction in the large microprocessor core is significant considering the small (0.3%) part of the design that required to be customized. This suggests that another application of migration can be in a design situation when timing problems creep up after placement and routing. In this situation, migration can supplement the in-place optimization efforts.

The effectiveness of the algorithm is evident in the fact that 45% to 71.4% of the performance gap between the full-custom and synthesis solutions could be reclaimed at the expense of as little as 0.3% - 17% custom design effort.

The above results indicate the usefulness of this approach in today's design situation, where minimizing design effort has become crucial to the success of a product. The technique adds another dimension to the trade-offs that a designer can exercise automatically.

REFERENCES

[1] Fishburn et. al. Tilos: A posynomial programming approach to transistor sizing. *ICCAD*, pp 326–328, 1985.

[2] Chuang, et. al. Timing and area optimization for standard-cell vlsi circuit design. *IEEE Trans. on CAD*, pp 308–320, 1995.

[3] Berkelaar et. al. Gate sizing in mos digital circuits with linear programming. *EDAC*, pp 217–221, 1990.

[4] Sapatnekar et. al. Power vs delay in gate sizing : Conflicting objectives? *ICCAD*, pp 463–466, 1995.

[5] Chen, et. al. An iterative gate sizing approach with accurate delay evaluation. *ICCAD*, pp 422–427, 1995.

[6] Conn, et. al. Optimization of custom mos circuits by transistor sizing. *ICCAD*, pp 174–180, 1996.

[7] Cong et. al. Simultaneous driver and wire sizing for performance and power optimization. *IEEE Trans. on VLSI Systems*, pp 408–425, 1996.

[8] Menezes, et. al. A sequential quadratic programming approach to concurrent gate and wire sizing. *ICCAD*, pp 144–151, 1995.

[9] Garey et. al. *Computers and Intractability*, 1979.

TABLE 1. Results of Migration

Circuit Name	No. of Trans.	Synthesized Design			Custom	Migrated Design			
		Library Size		Delay	Delay	Delay	Improvement		No. of
		Cells	Strengths	ns	ns	ns	% over synthesized	% of best gain	New Cells
Control 1	1059	163	2-4	37.3	17.4	23.1	38.1%	71.4%	30(17%)
Control 2	6239	163	2-4	41.0	28.5	34.5	15.8%	52.0%	20(2%)
Decoder	2926	430	3-4	12.0	7.1	9.8	18.3%	45.0%	30(5.5%)
Core	80542	175	2-4	22.8	n/a	20.1	11.8%	n/a	51(0.3%)

A Practical Repeater Insertion Method in High Speed VLSI Circuits

Julian Culetu
Sun Microsystems, Inc.
901 San Antonio Road
Palo Alto, CA 94303
iulian.culetu@Eng.Sun.com

Chaim Amir
Sun Microsystems, Inc.
901 San Antonio Road
Palo Alto, CA 94303
chaim.amir@Eng.Sun.com

John MacDonald
Sun Microsystems, Inc.
901 San Antonio Road
Palo Alto, CA 94303
john.macdonald@Eng.Sun.com

Abstract

In today's design of VLSI high speed circuits, frequency has a major impact on the number of repeaters that needs to be inserted. A microprocessor operating at less than 200Mhz might require several hundred repeaters, while one operating at greater than 500Mhz may require a number in the thousands. The following paper describes an efficient and simple way to automatically determine buffer placement based on maintaining equal transition time for all gate input signals across the net. A maximum allowable transition time is determined (limited by the frequency of the circuit), and correlated with the interconnect Elmore Delay. A Spice RC model having nodes with physical locations (X, Y coordinates) can be obtained by extraction tools providing standard parasitic format (SPF). This can then be used with the results of the algorithm for repeater placement to determine the exact physical location desired for each repeater.

1. Introduction

In a high performance VLSI design, signal propagation over on chip metal interconnect is significant due to RC delay. In general, the interconnect delay increases with the square of the length of the line. Delay can be reduced by inserting inverting or non-inverting buffers. A large scale integration design has certain objectives when it comes to inserting repeaters. These design objectives are (a) to minimize interconnect delay and (b) limit transition times while limiting impact on (c) area and (d) power.

Algorithms for post-layout buffer insertion in interconnect wiring were studied by van Ginneken [4] with further contributions by Kannan, Suaris and Fang [3] and by Lillis, Cheng and Lin [4, 6]. Given required arrival times at cell inputs and analyzing Elmore delay on a globally routed net, van Ginneken presented a dynamic programming algorithm that positioned and inserted buffers such that the latest possible arrival time at the interconnect driving point could be achieved. Buffers were then connected during a detail route stage. Work by Lillis, et. al. [1, 2] employed a similar approach in providing polynomial-time algorithms for simultaneous buffer insertion and wire sizing and, later, algorithms for power minimization and multiple-source nets.

In practical application there are additional considerations that must be applied in a large, high performance design. One such consideration is the necessity for signals propagated through wiring with RC degradation to achieve acceptable voltage levels for recognition of logic low and high. This translates into a limitation on signal transition or rise and fall times at gate inputs. In this paper we present a simple and efficient method using transition time limitation for post-layout buffer insertion and show how the solution achieved is near optimal, providing a practical trade-off among timing, power and area.

Our repeater insertion method starts with a Spice analysis to find the optimum delay and to limit the transition times while not losing track of area and power. The goal here is to find the wire sizing of each metal and buffer size in the design that best fit the above mentioned requirements (delay, area, and power). Once the buffer and the wire sizes have been selected, a correlation between transition time (measured in Spice) and the Elmore delay is performed (Section 2). The Elmore delay corresponding to the transition time (limited by the frequency of the design) can be determined and used as a limiting value in the repeater insertion algorithm. When inserting repeaters, equal transition times are maintained at the input of each gate in the net. The placement of repeaters can be easily obtained given that the extraction tools provide standard parasitic format (SPF) with X and Y coordinates for each node in the net (Section 3). The implementation of this method is described in Section 4 followed by the conclusions in Section 5.

DAC 98, San Francisco, California

2. Transition Time and Elmore Delay Correlation

2.1 Background

The Elmore delay, as described in [6], is an efficient way of calculating the delay for an RC tree. A first order approximation of RC delay at any node i on an RC tree is given by the Elmore time constant

$$T_i = \sum_{k=1}^{n} R_{ki} C_k \qquad (1)$$

where *Rki* is the resistance of the portion of the (unique) path between the input and node i, that is common with the (unique) path between the input and node *k*, and *Ck* is the capacitance at node *k*.

The first order approximation of the waveform at node *i* is

$$V_i(t) = V_{DD}\left(1 - e^{-t/T_i}\right) \qquad (2)$$

where *Ti* is the time constant Elmore delay given by (1). The time at which the voltage at node *i* reaches any value *Vx* is determined by

$$T_x = T_i \ln \frac{V_{DD}}{V_{DD} - V_x} \qquad (3)$$

A quick approximation of 20-80% (or 10-90%) transition time can be calculated by substituting *Vx* with voltages at 20 and 80% of

VDD.

$$T_i \cong 0.72 \cdot T_{20-80} \quad (4)$$

The ratio coefficient of the transition time (20-80%) based on first order approximation is about 1.39.

2.2 Transition Time vs Elmore Delay

The resulting ratio of the transition time vs the Elmore delay of the above equation (4) would not be accurate enough for this repeater insertion method. The reason being is that the optimum distance between repeaters, and the size of the repeater itself that will satisfy the above design requirements have been determined by running Spice simulations. Therefore using the above method for correlating Elmore delay and transition time is inaccurate. The idea is to determine a limiting value for the Elmore delay, that corresponds to the maximum transition time obtained by running Spice simulations.

Once the optimum wire lengths/widths, buffer sizes and maximum transition time allowed for a design, have been determined, a comparison of the Spice waveform vs the above first order waveform approximation (2) can be performed. This can be accomplished by finding the ratio coefficient between transition time (measured in Spice for optimum wire length/width and buffer size) and the Elmore delay.

To find the optimum delay in our design, that will not affect the area and power, a set of Spice simulations were run. This was done for different buffer and wire sizes. Since area and power for our processor were strict limitations, the buffer and wire sizes have been chosen accordingly. As a limit for metal 1-4 wire width/space, a double the minimum width (2w) and a triple the minimum space (3s) was enough to satisfy our area and power goals. A limitation to buffer sizes was also necessary due to the fact that a number of these were placed in the channels (32 x minimum size). Placing large buffers in the channels and making wires unnecessarily wide would present significant blockage to the routing. This would have a significant impact on the area, due to wire congestion. Figure 1 shows the result of running Spice on different metal 1-4 wire and buffer sizes.

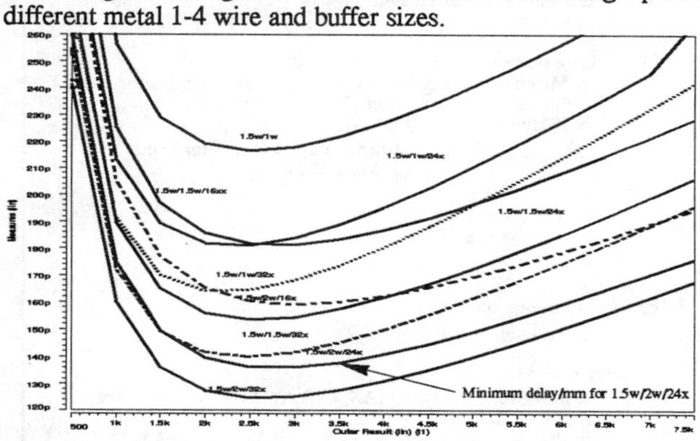

Figure 1. Delay per mm for 1.5w wire width, 1/1.5/2 w space, and 32/24/16 x minimum buffer size

As an observation, when the width or space between wires decreases the delay/mm increases. Different wire widths and spaces were chosen for our design based on needs. For instance, for critical nets that required optimum delays, wide widths and spaces were selected. In all the other nets, more conservative solutions were chosen such that area and power would not be affected. Table 1 shows some measurements of delay/mm and the optimum length between repeaters of 24 and 32 x minimum size.

As an example, it was found that, for a metal 1-4 wire with 1.5 w width, 2w space and 24 x minimum size buffer, the minimum delay per mm (including repeater delay) occurs at 2.5 - 3.0mm between the chosen sized repeaters (see figure 1). Looking at the curve (figure 1), the delay per mm does not change much between 2.5mm and 3.5mm. Since our main goal is to insert as less repeaters as possible while still be close to the optimum delay, 3.5 mm looks like a better option than 2.5 and 3.0mm.

Table 1: M1 - M4 Repeater data

Repeater Size			24x		32x	
width (um) w - min	space (um) w - min	pitch (um) w- min	l (mm)	d/l (ps/mm) rise/fall	l (mm)	d/l (ps/mm) rise/fall
1.0w	1.2w	1.1w	2.68	187/191	2.85	176/184
1.5w	1.0w	1.2w	2.75	183/183	3.02	168/172
	1.5w	1.5w	3.24	156/157	3.51	145/150
	2.0w	1.7w	3.65	139/141	3.90	130/136

Using this method, we have reduced the number of repeaters by ~29%. As a result area and power are reduced considerably.But there is a limit to how long a wire can be in between two repeaters at a certain frequency. For our example, as shown in figure 2, the waveform dies out due to RC degradation as the length is increased. Making the wire longer than 3.5 mm will considerably affect the transition time to a point where the waveform does not reach the voltage ramps (VDD and VSS). The length that we have chosen initially (3.5mm) looks safe.

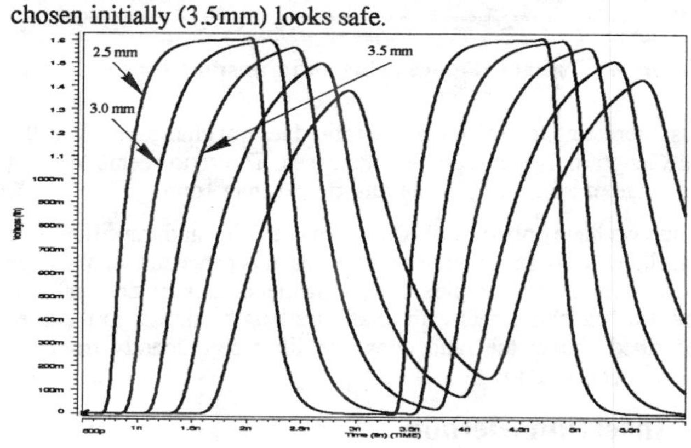

Figure 2. Transition time analysis

The voltage waveform corresponding to 3.5 mm is captured and shown in figure 3. Also the corresponding first order approximation (see equation (2)) of the same wire is overlapped on the same plot for comparison purposes. The Elmore delay constant Ti in (2) was calculated for the 3.5mm wire and used in the first order approximation. Thus Equation (2) becomes a simple exponential function of time. As can be observed, the waveforms almost match at the 50% of VDD point, but not at 20 and 80%

points.

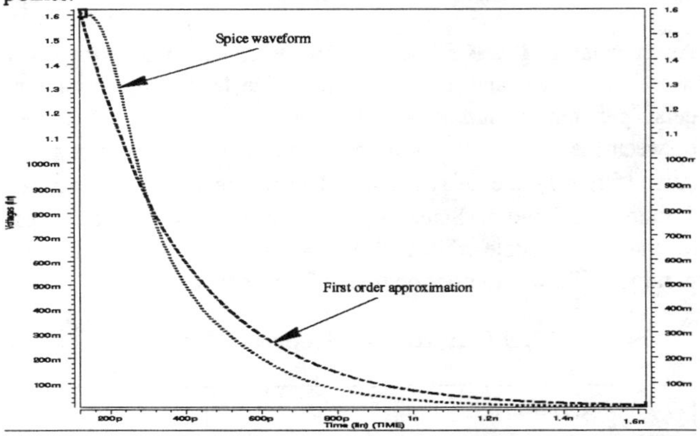

Figure 3. Spice waveform and first order approximation

To get the Elmore time constant that corresponds to the transition time at the load, a ratio of the Elmore delay and transition time was plotted with respect to the wire length (see figure 4). The ratio corresponding to 3.0 mm of wire is 0.98 and for 3.5mm is 1.01.

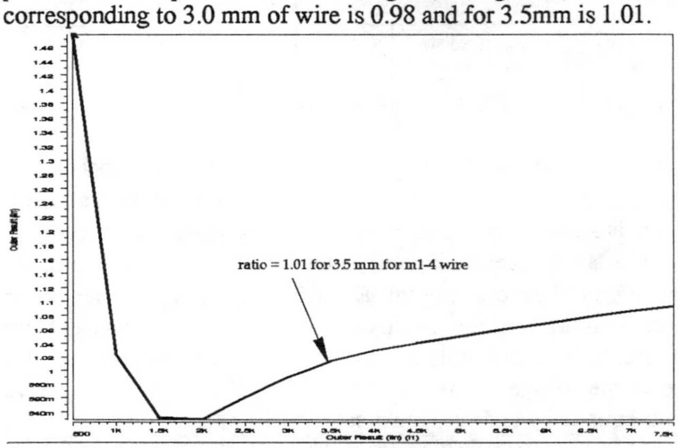

Figure 4. Ratio of Elmore delay and transition time vs length

It is interesting to notice that the ratio does not change much as the wire length between repeaters increases. The ratio seems to be in the range of ~0.9 - 1.1 for lengths greater than 1mm.

While we have only described the Elmore delay and transition time correlation only for point to point wires, this procedure can also be applied for more complex branching nets. Experiments showed that, for branching nets with total wire lengths similar to the ones described above, the ratio does not differ significantly from the above mentioned range (0.9 - 1.1).

3. Insertion Method

After the maximum transition time for a particular circuit performance has been determined, and the correlation with the interconnect Elmore delay has been performed, a maximum allowable RC delay is set as input to our algorithm. In addition an optimum repeater is determined by Spice simulations and a timing model is created. To obtain RC data from physical design, an RC extraction tool can be used, that can provide an SPF (Standard Parasitic Format) output to be used by the insertion tool. Standard Parasitic Format is a way to represent RC data and it is compatible with the Spice format. Block timing models are also needed to provide driver and load information.

In an SPF (Standard Parasitic Format) data, the RC nets are represented as lumped RC models (see figure 5). To be able to insert repeaters, there is a need for transforming the lumped RC model into a distributed model.

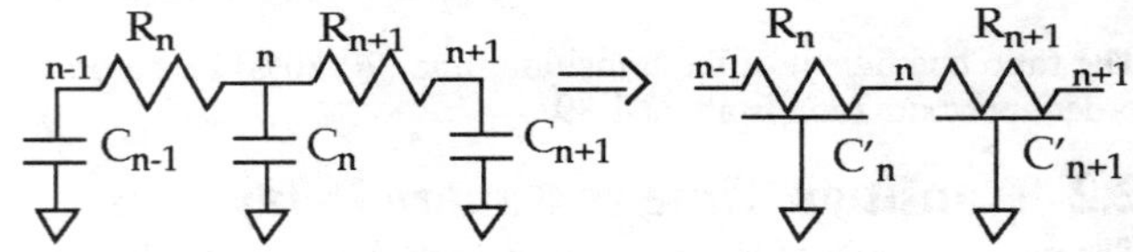

$$C'_n = C_{n-1} + \frac{R_n}{R_n + R_{n+1}} Cn; C'_{n+1} = C_{n+1} + \frac{R_{n+1}}{R_{n+1} + R_n} C_{n+1} \quad (5)$$

Figure 5. Lumped to distributed RC model transformation

Once the distributed capacitance has been calculated, the Elmore delay to each of the loads in the net from driver is calculated (figure 6). The repeaters are being inserted starting from the load with the highest Elmore delay towards the driver. While parsing through the RC net segment by segment, Elmore delay is calculated considering a repeater as driver of the segment. If the delay exceeds the maximum, a repeater is inserted in the segment, forming a new subnet. The process starts again from the highest Elmore delay load. In the case of branching, the assumed branching node is moved from an assumed repeater input to a repeater output. This continues until the maximum delay is exceeded, in which case a repeater is inserted with its output at the branching node (figure 7). The process continues until all the segments in the net have been explored.

```
INPUT: SPF RC net and MAX_DELAY
OUTPUT: SPF RC net with repeaters inserted

1: Read net
2: Convert lumped into distributed RC model
3: Compute Elmore delay to each load (in subnet)
4: Identify load with largest Elmore delay
5: Identify unprocessed segment from the load
6: Calculate Elmore delay assuming repeater as driver
7: IF delay > MAX_DELAY {
        a: Interpolate to determine point of repeater placement in the
        segment
        b: Insert repeater, creating subnet with driver, connected
        segments and loads, counting repeater as load
        c: Reset data structures
        d: Go back to (3)
   }
8: IF branching node present {
        a: Move an exiting branch from (assumed) repeater input to
        repeater output and find Elmore delay
        b: IF delay > MAX_DELAY {
                Move last branch back to repeater input
                Call (7b), (7c) and (7d)
        }
        c: IF more exiting segments {
                Go back to (8a)
        }
   }
9: IF more segments on path {
        a: Go back to (5)
   }
10: IF delay < f * MAX_DELAY {
        a: new_max_delay = MAX_DELAY + delay/(max # of
        repeaters in the path)
        b: IF new_max_delay > MAX_DELAY + p * MAX_DELAY {
                new_max_delay = Average Elmore delays
        }
        c: MAX_DELAY = new_max_delay
        d: Go back to (7c)
   }
11: Output repeater inserted nets
```

Figure 6. Repeater insertion algorithm

To optimize the total delay of the net with the repeaters inserted, there are some special cases where a repeater could be removed if it is too close to the driver. This is presented as step (10) in figure 6. The process of elimination is being done by resetting the new maximum delay to a new value lower than the initial one plus some set margin error.

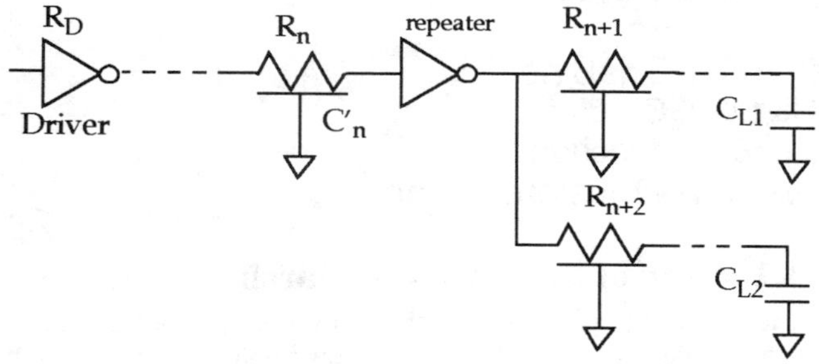

Figure 7. Net with repeater inserted

A Spice RC model having nodes with physical locations (X, Y coordinates) can be obtained by extraction tools providing standard parasitic format (SPF) data. This can then be used with the results of the above algorithm for repeater placement to determine the exact physical location desired for each repeater. This can be done by calculating the X, Y coordinates for the point at which the repeater is to be inserted along a segment extending between two nodes with coordinates given.

4. Implementation

The above method has been implemented in the design of Ultra Sparc III microprocessor at Sun Microsystems. A repeater map that shows the distribution of repeaters throughout the chip is shown in figure 9 (a similar map can be obtained which indicates the number of repeaters per square mm) as a result of running the repeater insertion program on SPF data with node coordinates. This is not to be considered as the final picture of where the repeaters are physically located in the chip, but rather the optimum locations from the timing point of view. Since placing the repeaters at those exact locations was not physically possible, a more realistic approach was to place blocks of repeaters in the very heavy areas, as shown in the map by the darker color indicating a large number of repeaters. The rest of the repeaters can be placed in the vicinity blocks.

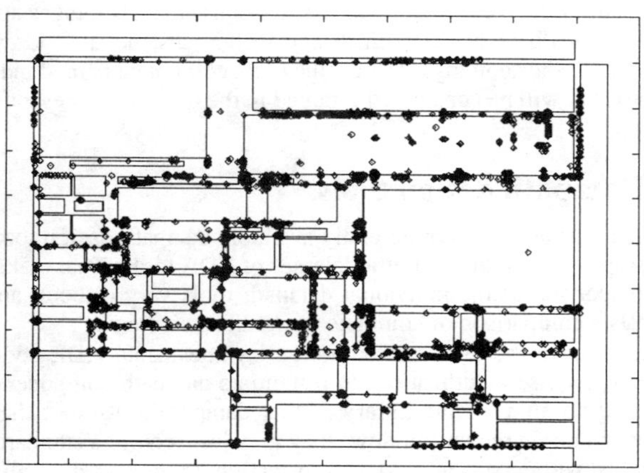

Figure 8. Repeater Map Distribution

After blocks of repeaters are placed in the floorplan in the designated locations (indicated by the dark color areas or by the number of repeaters per square mm) and the remaining ones in the logic blocks, nets are rerouted with a routing tool. Timing and distance constraints can be put on certain nets to enforce optimum routing, but even that will not give the needed results after the first shot. Several iterations are required through: Fullchip RTL, Routing, RC extraction, Fullchip Timing Analysis (mainly identifying the critical paths and long wires) and back, until a satisfying result is obtained in terms of area, power and timing.

5. Conclusion

This paper presents a practical repeater insertion method that satisfies the three main goals in designing a high speed microprocessor: delay, area and power. We have shown, step by step, how the method has been developed by first (a) identify the maximum wire length between repeaters and transition time by running spice simulations, (b) correlate the transition time with the Elmore delay based on the chosen wire lengths and repeaters resulting in a ratio coefficient, (c) use this coefficient to find the maximum Elmore delay to be used as an input to our repeater insertion program, (d) develop a simple and applicable algorithm that ensures equal transition time (limited by the frequency of the design), for all the gate input signals across the net and (e) a map of repeater distribution is to be obtained given a standard parasitic format (SPF) with node coordinates specified. The repeater insertion method has been implemented in the design of Ultra Sparc III microprocessor at Sun Microsystems.

6. Acknowledgments

As a closing statement, we would like to thank Fabian Klass, Farzad Chehrazi, and Tony Todesco from our physical design group for their help and support in the areas of circuit simulation and power analysis.

7. References

[1] J. Lillis, C.-K. Cheng, T.-T. Lin, "Optimal Wire Sizing and Buffer Insertion and Low Power and a generalized Delay Model, "*IEEE Journal of Solid State Circuits*, vol. 31, no. 3, March 1996, pp. 437-447.

[2] J. Lillis and C. -K. Cheng, "Timing Optimization for Multi-Source Nets:Characterization and Optimal Repeater Insertion," *Proc. 34th Design Automation Conf.*, 1997, pp. 214-219.

[3] L. N. Kannan, P.R. Suaris, H. -G Fang, "A Methodology and Algorithms for Post-Placement Delay Optimization," *Proc. ACM/IEEE Design Automation Conf.*, 1994, pp 327-332.

[4] L. P. P. van Ginneken, "Buffer Placement in Distributed RC-tree Networks for Minimal Elmore Delay," *Proc. International Symposium on Circuits an systems*, 1990, pp 865-868.

[5] M. A. Horowitz, "Timing Models for MOS circuits," Ph.D. thesis, Stanford Univ., Stanford, CA, Jan. 1984.

[6] P. Penfield and J. Rubinstein, "Signal Delay in RC tree networks", in *Proc. 18th Design Automation Conf.*, 1981, pp. 613-617.

[7] T. H. Cormen, C.E. Leiserson, R.A. Rivest, Introduction to Algorithms," 1990, MIT Press and McGraw-Hill Book Co., pp. 558-562.

[8] W. C. Elmore, "The transient analysis of damped linear networks with particular regard to wideband amplifiers," *J. Appl. Phys.*, vol. 19, no. 1, pp. 55-63, 1948.

Practical Experiences with Standard-Cell Based Datapath Design Tools

Do We Really Need Regular Layouts?

Paolo Ienne and Alexander Grießing
Siemens AG, HL LIB GC
Postfach 80 17 09, D–81617 München
{Paolo.Ienne, Alexander.Griessing}@hl.siemens.de

Abstract

Commercial tools for standard-cell based datapath design are here classed according to design flows, and the advantages of each class are discussed with the results of two test circuits. Algorithmic generation of netlists and of relative cell placement can help reducing area but, contrary to common belief, appears often detrimental to speed. Extraction of regularity from synthesized netlists is difficult and requires counterproductive simplifications to the synthesis process. Most promising are synthesis tools which can generate placement data; yet, no tool of this class appears ready today.

1 Introduction

Informally, *datapaths* are circuits where the same or similar logic is applied to several bits of a bus. These signals get processed in parallel by complex operators such as adders, multipliers, or simply by multiplexers, bitwise logic gates, etc. Typically the same piece of logic is repeated—almost unchanged—for all bits, or an operand gets implemented as a regular array structure (e.g., multipliers) [6].

Traditional *place & route* (P&R) tools are incapable to use this regularity to produce more compact and higher performance results. *Datapath design tools* are programs which in different ways help to exploit regularity: they may replace standard P&R tools—and to a certain extend preceding tools in the design flow—altogether and use some special library of datapath-cells, or they can simply produce placement constraints for traditional P&R tools.

Generally, datapath tools are supposed to help designers in achieving smaller and faster circuits. The obtained implementations are smaller because of the regular placement; they are also supposed to be faster because of several reasons. The first argument for the speed improvement is that with a regular arrangement of the basic cells, wiring will be reduced to a minimum. A second argument is the common belief that by carefully generating a netlist one can easily improve on synthesis results. A third and different argument says that by integrating in a datapath tool everything from synthesis down to layout generation, the more accurate timing information available at synthesis can be used to further improve the design. Finally, datapath tools could also improve controllability over the design process (e.g., clock skew, matching bus parasitics).

DAC 98, San Francisco, California
©1998 ACM 0-89791-964-5/98/06..$5.00

1.1 Aim of the present research

The aim of the present research is to assess whether any of the commercially available tools can tangibly improve the results of the standard synthesis-based design flow by producing regular layouts. Therefore research works such as [2, 3, 7] are outside the scope of this investigation.

Instead, the capabilities of seven datapath design tools have been tested by having synthesizable HDL descriptions of four datapath intensive designs and by using these descriptions (modified or not, as required) to generate the layouts with the datapath tools. The result of the last two test circuits are compared here to layouts obtained by synthesis in standard cells and normal P&R, that is, to layouts obtained with the standard synthesis-based flow.

Basic metrics for layout evaluation are the circuit area and speed. A rather qualitative but important element to asses is the additional complexity incurred by the designer to use a datapath tool for a design, compared to the standard synthesis flow. These three parameters—speed, area, and simplicity of flow—are in the main scope of the present research.

After a few preliminary tests, the present research concentrated on standard-cell based datapath designs: since the original interest was in introducing datapath tools in a language-based flow, typical standard-cell circuit techniques are the only viable ones for reliable design. On the other hand, tests on different layout styles indicated that mixing the standard cell layout with other styles usually leads to poor results: when the datapath area is not perfectly regular, large sections of the rectangle enclosing the datapath could be unused; the incompatible datapath layout style prevents control logic cells to mix into the empty spaces and these are therefore simply wasted. The small area advantage possible by redesigning the cells for abutment and in a "classic" datapath style is easily offset by such losses. Therefore, all the tests reported here are in our 0.35μm, 3.3V, 3 metal layer technology and use the same standard-cell library.

The purpose of the present paper is to discuss the advantages and problems of different datapath design techniques, not to address specific tool shortcomings. Hence, the tools will not be named individually, but will be considered grouped in the categories presented below.

2 Datapath design tools

The heterogeneity of commercially available datapath design tools well surpasses that of most other classes of EDA tools. That makes the comparison of the individual datapath tools very difficult and suggests a comparison of complete design flows instead.

The input to the tools can be normal synthesizable HDL code, or it may require modifications to instantiate datapath components belonging to an available library. Other datapath tools start their job much later in the design flow: they may just require a standard-cell netlist without any additional information, or they need a netlist accompanied by "suggested" relative placement data for all or some of the cells.

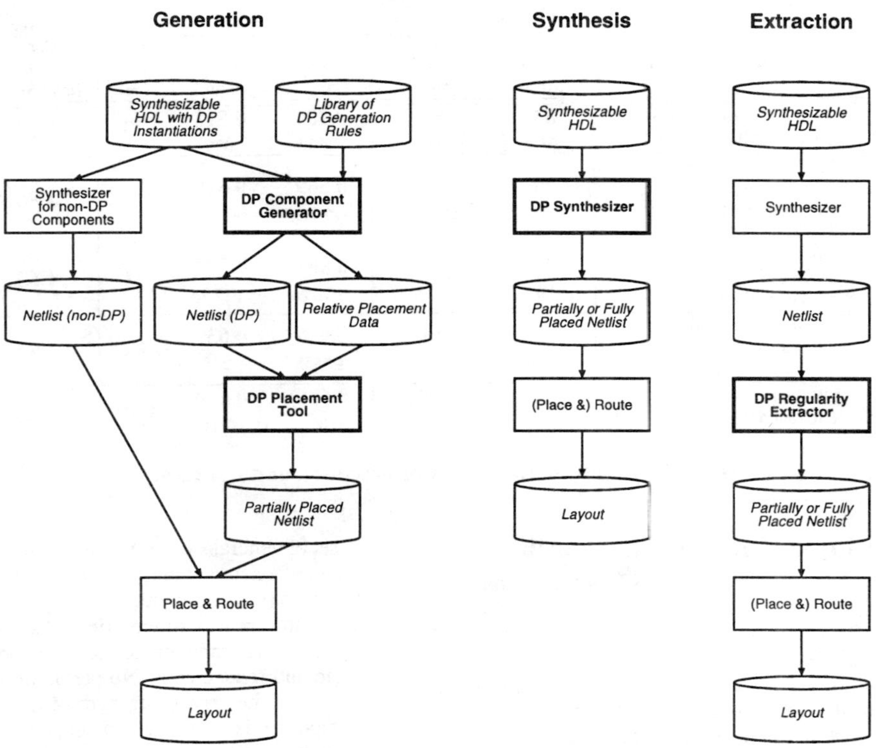

Figure 1. Basic design flow for the three main types of commercial datapath tools.

Similarly, the output of the various tools is also pretty different: some tools have a dedicated placement engine integrated (or are themselves part of P&R tools) and produce a final layout. Others just produce a placement suggestion for all or some of the cells, and this can be used in the P&R tool of user's choice (when supported) to complete the placement and route the design.

Another distinctive feature of the tools is the ability to handle mixed datapath and control logic. Some tools can separate among control logic and datapath components automatically, while others require as an input the datapath part completely independent from any control logic, leaving the partitioning of the design to the user.

2.1 Types of commercial datapath tools

The available datapath tools can be fundamentally classed in the three categories represented in Figure 1:

- **"Generation" tools.** The simplest tools are those more closely imitating a manual datapath design: from a list of datapath operators belonging to a library, the individual cell-level netlists are generated algorithmically. This *algorithmic generation* process is different from synthesis in that one wants to produce the mentioned netlist and a database with the desired relative placement information of the cells at the same time. In the simplest example, generation of an n-bit register requires instantiation of n registers (with trivial connections) and assignment of each register to the corresponding bit-slice of the datapath. A second part of the tool usually fixes the cell placement obeying the specified relative placement.
- **"Extraction" tools.** An approach which preserves as much as possible of the standard flow consists in extracting relative placement information from the results of normal synthesis. The ways this extraction from the final netlist is performed are usually not documented and essentially consist in a combination of techniques such as: (1) exploit levels of hierarchy in the netlist, (2) look for bus names, (3) look for telling instance name patterns, (4) look for identical cells sharing control signals, (5) obtain from the user hints on the "average" number of bits, and perhaps others. Clearly the possibility of using a normally synthesized netlist depends very much on the actual regularity of the design after synthesis; this implies that the synthesis process may have to be performed following some guidelines to maximize the amount of regularity available at the end.
- **"Synthesis" tools.** The most complex tools are special synthesizers dedicated to datapath designs, which produce from an HDL description the placement information together with the netlist. The tools of this category are potentially the most versatile since their scope spans over most of the design flow. These are the only ones which can easily profit from a more accurate modeling of the datapath components: during synthesis, an accurate wiring model can be chosen for each subdesign or even for each wire thanks to the available placement data. The potential weakness of these tools may be the inability to leverage the state-of-the-art synthesis capabilities or placement algorithms from other traditional tools.

The tools evaluated in the present research explore all the three different categories.

3 Evaluation results

In this section, the main test circuit is described and the relevant results reported. The results of the test case are supplemented by a discussion of the implementation of 16-bit multipliers with different techniques which throw additional insight into the difficulties of competing with synthesis techniques.

	Tools	Speech: Max f C only (MHz)	Speed: Max f Synth. STA (MHz)	Area: Total Area (mm^2)	Area: Cell Area (mm^2)	Area: Area Util. (%)	Power Cons. @66MHz (mW/MHz)
(A)	Traditional synthesis tool Traditional P&R	**83** 0%	**84** 0%	**0.708** 0%	**0.518** 0%	**73**	**2.12**
(B)	DP component generators Traditional synthesis tool DP placement tool Traditional P&R	**90** +8%	**91** +8%	**0.569** −20%	**0.520** +0%	**91**	**1.86**
(C)	DP component generators Traditional synthesis tool Traditional P&R	**99** +19%	**97** +15%	**0.808** +14%	**0.573** +11%	**71**	**1.63**
(D)	Traditional synthesis tool DP extraction tool Traditional P&R	**81** −2%	**85** +1%	**0.956** +35%	**0.639** +23%	**67**	**2.65**
(E)	DP synthesis tool (No routing)	—	**(104)** +24%	**(0.686)** −3%	**(0.640)** +24%	**(93)**	—

Table 1. Results of the processor execution unit test case.

3.1 Brief description of the main test circuit

The main test circuit is a simplified version of a DSP execution unit. It is mostly composed of: (1) registers (six 40-bit internal ones, and four 16-bit at the inputs); (2) several 40-bit multiplexers before and after the internal registers and four 16-bit to select sources from six input buses; (3) two simplified functional units, one composed of a simple ALU and another of a 16 × 16-bit 2's complement multiplier. The test circuit also contains some less regular datapath structures and some general control logic. The optimization purpose is to make the circuit run at the highest possible speed, of course with the smallest possible die area. After normal synthesis, the circuit contains approximately 4,500 cells corresponding to approximately 40k transistors.

3.2 Results on the main test circuit

Table 1 shows the results on the described test case. The first two columns report the maximum frequency for each implementation. They indicate respectively the post-layout measurements after parasitic capacitance extraction and the pre-layout estimation in the synthesis tool. The next three columns indicate respectively the total area of the implementation, the area of the cells alone, and the ratio between the latter and the former, i.e., the area utilization of the implementation. The last column contains an estimation of the power consumption obtained from post-layout data.

Implementation (A) is the reference, performed using a standard language-based flow. It is obtained by synthesizing a VHDL description with a traditional synthesizer. The ALU and the multiplier are standard implementations from the synthesis tool. The design is placed and routed using a traditional P&R tool.

Basic experiments with generation techniques. First attempts with only generation techniques proved disappointing: the standard solution is faster and in practice not larger. It was not larger because of the better optimized synthesized netlist, and faster because of at least two typical advantages of synthesis: (1) Automatic selection of the cells with the most appropriate drive strength. (2) Possibility of reducing the logic depth on the critical path (as it has been observed in some multiplexers for example, implemented with gates in the synthesized version and using symmetric multiplexer cells in the generated one).

Synergy of synthesis and generation tools. Implementation (B) (which is actually the identical result of two competitor tools for special datapath placement) has been achieved by trying to introduce some of the synthesis advantages into the datapath generation flow. These implementations have been produced as follows:

1. A set of general-purpose generators have been prepared. They include registers (multiple flip-flops) and two versions of multiplexers (a version with a privileged fast path and another one with more or less symmetric paths). Some of the larger multiplexers are implemented as selectors to share the decoding logic and reduce area. No particular attention is given to the choice of the driving strength of the cells, and each datapath component is available in a single strength only.
2. The above datapath components have been instantiated in the top netlist. All the other components—essentially the arithmetic unit and the multiplier—are synthesized.
3. After generation of the datapath netlists and of the relative placement data, the design is optimized with the synthesis tool. The optimization consists of:
    - tuning the driving strengths of the cells belonging to the datapath areas (i.e., registers, multiplexers);
    - full re-synthesis of the netlist sections not belonging to datapath components (i.e., ALU, multiplier, etc.);
    - re-tuning of the driving strengths of all cells in all areas.

In practice, the only additional operation which must be performed by the designer—compared to the standard synthesis flow—is the instantiation of the datapath components. The development of the datapath generators themselves can be thought as part of building a portable and general purpose library of datapath components.

Implementation (B) indicates that the relative placement information of simple components is sufficient to achieve a tangible improvement during P&R in area utilization and therefore in total area (as mentioned, the multiplier and the arithmetic unit are synthesized).

Generation alone does not help to speed up a design. Implementation (B) uses simple generation techniques for simple components to produce not only a netlist but also some relative placement data. Contrasting the speed of implementation (B) with (A) could suggest that these techniques are superior for timing-critical applications. One could believe that it is advantageous to replace synthesis in full with generation and achieve two goals at the same time: improve area (because of the regular placement not achievable through synthesis) and timing (because of some intrinsic property of the generation technique adopted and of the reduced wiring induced by the regular placement).

In fact, this appears to be incorrect: generation techniques can well improve the capabilities of synthesis, but used alone tend to have a rather negative impact on speed. This view is supported

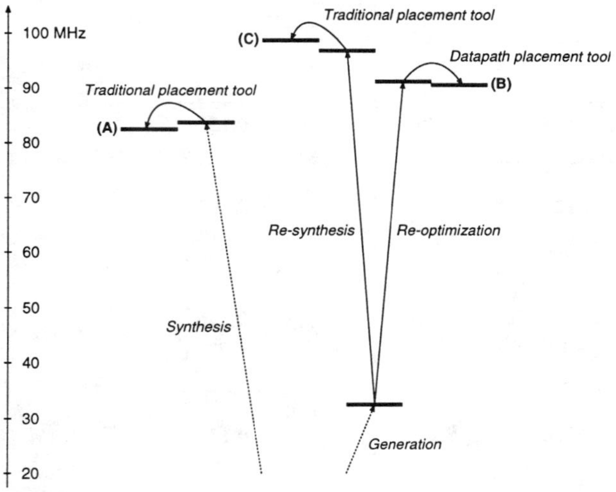

Figure 2. Evolution of speed after each design step of implementations (A), (B), and (C).

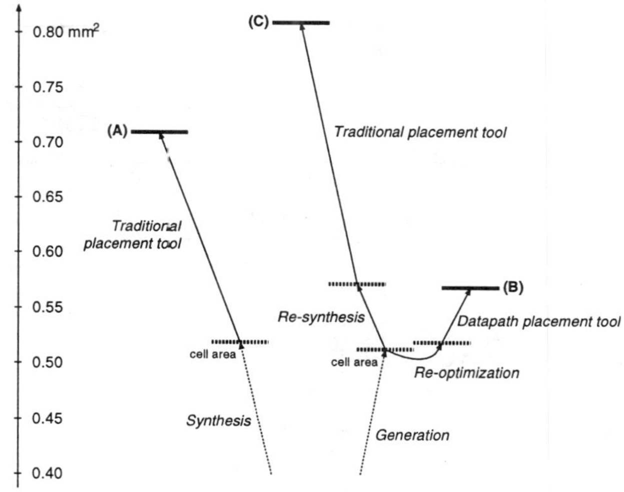

Figure 3. Evolution of area after each design step of implementations (A), (B), and (C).

by two experiences. The first concerns the generation of complex and timing critical blocks, such as multipliers. It appears that it is extremely hard to compete with the synthesis results simply by generating—however carefully—a netlist of standard cells: all the generated multipliers which were planned to replace the synthesized one in implementation (B) turned out to be slower and therefore not worth substituting. The important fact is that when these generated netlists have been re-synthesized in the synthesis tool, the timing has improved tangibly over the standard implementation of the synthesizer. This fact indicates that the intrinsic quality of the generated netlist is good, but generation alone cannot compete with synthesis. Since the regularity and the placement information get completely disrupted during re-synthesis, implementation (B) contains a synthesized multiplier traditionally placed by the P&R tool. Full details on the multiplier speed will be reported in Section 3.3.

The second experience indicating the importance for timing optimization of a synthesizer even in a generation-based flow is represented by implementation (C). This result has been produced by re-synthesizing the netlist used for implementation (B): the speed of the design improves further—about 15–20% better than the simple synthesis of implementation (A). However, as in the case of the mentioned multipliers, the regularity is lost and the placement is therefore performed by a traditional P&R tool (with the same area penalty as the reference implementation).

Synthesis, generation, or generation plus synthesis? The results of Table 1 and the results on the multipliers discussed in the next section seem to indicate that synthesis is today an almost essential step in a cell-based design flow to get the best performance from a given cell library.

Figure 2 highlights the contributions on speed of the different design steps for implementations (A), (B), and (C). It visually shows that the generation of simple components (multiplexers, registers, etc.) followed by a driver strength re-optimization leads to better timing results than direct synthesis (+8%). However, it also shows that a complete re-synthesis step after generation has a tangible impact, bringing the advantage over direct synthesis to +15%.

It is also interesting to notice that apparently the techniques used for implementation (B) do not have a visible impact on wiring delay. Implementations (A) and (C) are after placement only slightly slower or faster than the pre-placement calculation of the synthesis tool: implementation (B) has a behavior which is clearly similar. (One possible reason not to observe any intrinsic advantage of the regular layout is perhaps the fact that a large part of the design is still conventionally placed.)

Figure 3 presents a similar picture about the area contributions of the different implementation steps. Generation and synthesis look quite equivalent at the netlist stage, but the whole advantage of the regular placement of implementation (B) is now apparent: the routing overhead of this implementation is about one third of the one of implementations (A) and (C). It is also evident how the fastest implementation (C) is achieved at the expenses of a very large area (both because of large cell area and of irregular placement).

Extraction of the regularity from the synthesized netlist. With the good results achieved by synthesis, it is particularly interesting to see what extraction tools can do to leverage synthesis capabilities. Implementation (D) is the best result of a series of iterations with the only tested "extraction" tool. Clearly, extracting regularity from the final synthesized netlist is not an easy task. It is not easy because the best possible optimizations in the synthesis tool apparently destroy so much of the circuit regularity that the netlist gets useless for the tool. If, on the other hand, one tries to retain the regularity by stopping the optimization process slightly earlier, one can get better placement results (about 10–15% more density) but with a much larger cell area (up to 40% more) which basically offsets the advantage. In many cases the design is also slower. Implementation (D) represents the fastest design achieved. A tangible penalty on power consumption can be observed.

Datapath synthesis. With synthesis achieving very good results and post-synthesis extraction being limited in its possibilities, datapath synthesis appears as promising. Indeed, implementation (E) seems to indicate that direct synthesis in a datapath-aware tool can lead to good results. Unfortunately this implementation was never routed because the tool was in a preliminary version, and therefore results should be considered only indicative. Yet, some preliminary placements produced by the tool appeared qualitatively interesting; in a way, these indicative results support the view that merging the regularity extraction process with synthesis leads to the best solution. In the next section, the optimistic (E) timing results will be given some credibility.

	Design Technique	Speed				Area	
		Max *f* RC (MHz)	**Min *d* RC (ns)**	**Max *f* Synth. STA (MHz)**	**Min *d* Synth. STA (ns)**	**Total Area (mm^2)**	**Cell Area (mm^2)**
(A)	Synthesis tool	**118** 0%	**8.51** 0%	**128** 0%	**7.80** 0%	**0.160** 0%	**0.153** 0%
(B)	Custom Generators (Booth encoding, optimized adder) Synthesis tool (re-optimization)	—	—	**124** −3%	**8.08** +4%	—	**0.107** −30%
(Bbis)	Custom Generators (Booth encoding, optimized adder) Synthesis tool (full re-synthesis)	**143** +21%	**7.01** −18%	**152** +19%	**6.60** −15%	**0.164** +3%	**0.156** +2%
(C)	Custom Generators (Booth encoding, standard adder) Synthesis tool (re-optimization)	—	—	**106** −17%	**9.42** +21%	—	**0.110** −28%
(Cbis)	Custom Generators (Booth encoding, standard adder) Synthesis tool (full re-synthesis)	**136** +15%	**7.33** −14%	**143** +12%	**6.97** −11%	**0.151** −6%	**0.144** −6%
(D)	Custom Generators (no Booth enc., standard adder) Synthesis tool (re-optimization)	—	—	**106** −17%	**9.44** +21%	—	**0.127** −17%
(Dbis)	Custom Generators (no Booth enc., standard adder) Synthesis tool (full re-synthesis)	—	—	**157** +23%	**6.36** −18%	—	**0.192** +25%

Table 2. Comparison of different multipliers obtained by synthesis, by generation, and by a combination of both techniques.

3.3 Sixteen-bit multipliers

In the previous section, it has been mentioned that the best generated implementation (B) contains a synthesized multiplier. Indeed, several multiplier architectures have been explored with the purpose of including a generated multiplier made of regularly tiled standard cells. Table 2 shows the results of the most interesting multipliers achieved, all using the same standard cell library in the same technology. The first two pairs of columns report the maximum frequency and the critical path delay for each implementation. The two pairs contain the post-layout measurements after parasitic resistance and capacitance extraction and the pre-layout estimation in the synthesis tool, respectively. The remaining columns indicate the total area of the implementation and the area of the cells alone.

Power consumption has been measured for all the multipliers whose layout has been completed—multipliers (A), (Bbis), and (Cbis). The measured power consumption is—predictably—almost the same for all designs (the highest measured value is only about 3% more than the lowest value). This seems to reflect the fact that the final optimization technique for all the described multipliers is equivalent.

Satisfactory P&R results for the multiplier alone. Due to the globally poor generation results, no regular placement has been attempted, although for some of the multipliers tiling informations had been produced during generation. (The techniques originally developed to prepare relative placement information had been abandoned in the current of the experiments, when it has been clear that a regular placement could not lead to a tangible advantage.) The core area utilization after standard P&R is in all cases 95%.

Several generation techniques explored. Multiplier (A) is the standard implementation of the synthesis tool: it is the reference implementation and the one used in most designs of the previous section. The other multipliers (some not shown on the table) have been generated combining the following elements: (1) Partial product generation is built in the customary using either a Booth-2 encoding scheme or no encoding [5]. (2) Column compressors are generated with an algorithm similar to the one published in [4], and use the available timing information on the signals to add and on the standard cells to be used (several combinations have been tried). (3) Two types of final adders have been used, either with a fixed 8-bit group carry-select/carry-lookahead structure or with a generalized structure optimized with a timing-driven algorithm which should represent an improvement over [1].

Synthesis is very sensitive to a better architectural starting point. Two are the important results that can be evinced from Table 2 and emphasized in Figure 4: (1) Even the best attempts to generate a fast multiplier do not achieve the speed of the synthesizer's standard implementation (only multiplier (B) comes close to it with a tangible area advantage, but it should be remembered that all the generation techniques developed were aiming at delay reduction at any area cost). (2) Re-synthesizing the generated netlists systematically reduces the critical path delay of approximately 20–30%. Rather consistently, the better the starting re-optimized netlist, the better is the re-synthesized result. However, the speed improvement with re-synthesis usually reduces for the best starting netlists—for instance, −33% from (D) to (Dbis), but only −18% from (B) to (Bbis).

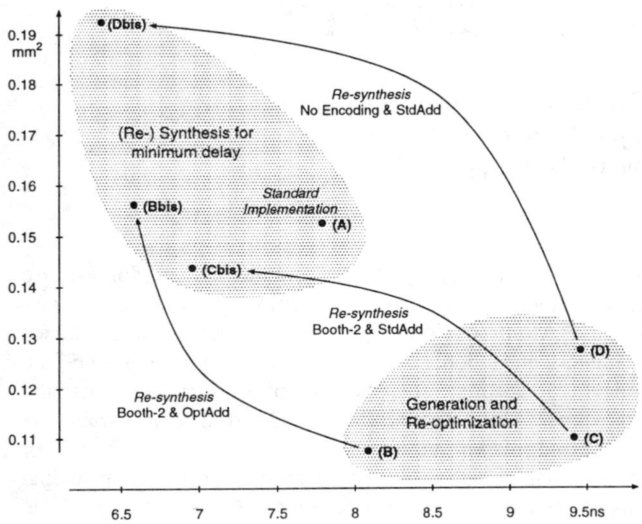

Figure 4. Cell area and maximum delay (as predicted by the sythesis tool) for the studied multipliers.

These results appear in agreement with those for the fastest implementations of the execution unit test case discussed in the previous section. The experiments performed suggest that the logic synthesis and optimization capabilities of state-of-the-art synthesizers are essentially unbeatable: no manually or algorithmically generated example described in the experiments performed for the present research could not get a significative improvement by re-synthesis. (One may notice that all generated netlists seem interesting area-wise compared to (A); yet, it should be noted again that (1) all the generated netlists have been produced disregarding area and trying to minimize delay, and (2) multiplier (A) has also been synthesized for highest speed and is therefore not in direct competition with the generated netlists.)

On the other hand, although manual or algorithmic generation alone hardly competes with synthesis—unless the design complexity is very low as in implementation (B) of Table 1—the starting logic mapping given to the synthesis tool appears to be highly determinant for the result. In fact, it seems that the initial mapping can be improved both for simple components (with no real "architectural" issues) and for complex ones (such as multipliers, where a first timing-driven mapping proves its superiority).

Impact of faster multipliers on the execution unit test case. It is useful to speculate on the impact of the present multipliers on the maximum speed of the execution unit test case. The fastest implementation (B) of Table 1 uses a 7.80ns multiplier in a 10.36ns design (97MHz). Multiplier (Bbis) of Table 2 has a maximum delay of 6.60ns. Simple calculations suggest that the maximum achievable speed for the execution unit test case is about 110MHz. Interestingly, this frequency is higher than the predicted frequency for the implementation with a datapath synthesizer (104MHz) and adds to the predicted value some credibility. Considering that the fast multiplier has essentially the same area as the slower one, the area of the design at 110MHz would be about 15–20% larger than the value predicted by the datapath synthesizer, thus emphasizing the potential interest of handling generation, synthesis, and directed placement in the same tool.

4 Conclusions

The present evaluation has shown that there is no panacea for the design of standard-cell based datapath structures. State-of-the-art synthesis tools and standard area-based P&R tools in technologies with more than 2 layers of metal now perform a job which strongly challenges manual and semi-manual techniques to preserve and make use of design regularity.

"Extraction" tools have big problems in performing a good job, at least without putting heavy constraints on the synthesis process. These constraints may compromise the final result by producing a negative effect which offsets the benefits of regular placement.

"Generation" tools are the only ones which really produced useful results in our experience. However, their beneficial effect can be seen exclusively in terms of reduced die area, and the indiscriminate use of algorithmic generation techniques for the netlist may have a very negative impact on the speed of the design. In fact, the present experiences show that when timing is the only metric of interest, pure synthesis of accurately generated netlist (without any placement data) achieves the best results. If, on the other hand, area is important, generation of netlist and relative placement data for very simple components may lead to a tangible advantage. The generated components must be simple enough so that they do not have a too negative impact on timing (or, equivalently, simple enough so that the timing improvement possible by re-synthesizing them is not too large).

From the above, it clearly appears that on timing-critical cell-based designs it is nowadays a severe penalty not to use a synthesizer. On the other side, regular placement is still beneficial (at least with 3 layers of metal) in situations of very high connectivity. It follows that adding the placement generation inside a datapath "synthesis" tool (or just having the synthesis tool preserve a given relative placement) might prove to be the best solution. Although some interesting preliminary results have been observed with a tool of this class, the technique is still immature and it represents an area where the best improvements are still to be expected.

References

[1] J. Fadavi-Ardekani. $M \times N$ Booth encoded multiplier generator using optimized Wallace trees. *IEEE Transactions on Very Large Scale Integration Systems*, VLSI-1(2):120–25, June 1993.

[2] M. Gansen, F. Richter, O. Weiss, and T. Noll. A datapath generator for full-custom macros of iterative logic arrays. In L. Tiehle et al., editors, *Proceedings of the International Conference on Application Specific Systems, Architectures and Processors*, Zurich, July 1997.

[3] L. L. Larmore, D. D. Gajski, and A. C.-H. Wu. Layout placement for sliced architecture. *IEEE Transactions on Computer-Aided Design*, 11(1):102–14, Jan. 1992.

[4] V. G. Oklobdzija, D. Villeger, and S. S. Liu. A method for speed optimized partial product reduction and generation of fast parallel multipliers using an algorithmic approach. *IEEE Transactions on Computers*, C-45(3):294–306, Mar. 1996.

[5] S. Vassiliadis, E. M. Schwarz, and B. M. Sung. Hard-wired multipliers with encoded partial products. *IEEE Transactions on Computers*, C-40(11):1181–97, Nov. 1991.

[6] N. H. E. Weste and K. Eshraghian. *Principles of CMOS VLSI Design*. VLSI System Series. Addison-Wesley, Reading, Mass., second edition, 1993.

[7] Q. Wu, C. Y. R. Chen, and B. S. Carlson. LILA: Layout generation for iterative logic arrays. *IEEE Transactions on Computer-Aided Design of Integrated Circuits and Systems*, 14(11):1359–69, Nov. 1995.

A STATISTICAL PERFORMANCE SIMULATION METHODOLOGY FOR VLSI CIRCUITS

Michael Orshansky, James C. Chen, and Chenming Hu
Department of Electrical Engineering and Computer Science,
University of California at Berkeley, Berkeley, CA, 94720

ABSTRACT

A statistical performance simulation (SPS) methodology for VLSI circuits is presented. Traditional methods of worst-case corner analysis lack accuracy and Monte-Carlo simulations cannot be applied to VLSI circuits because of their complexity. SPS methodology is accurate because no statistical information about the device parameter variation is lost. It achieves efficiency by analyzing the smaller circuit blocks and generating the performance distribution for the entire circuit. Circuit evaluation at any specified performance level is possible.

1 INTRODUCTION

Deep sub-micron technologies have made the problem of statistical modeling of the device and circuit behavior more critical. Shrinking dimensions make device characteristics more sensitive to stochastic process variation. Thus the relative spread of device and circuit behaviors is broadened. Combined with the continuing reduction of the design cycle, this creates an urgent need for methodologies that can accurately model and predict the statistical variations of circuit performances, such as speed and power consumption.

The industry's de facto standard in the field of statistical modeling is still limited to worst- and best-case analysis. The worst- and best-case SPICE models are produced by independently combining the outmost device parameter values. A circuit designer then uses this set of models to verify the design at the extremes of the process variations. What would be much more valuable, however, is the ability to analyze the circuit at any performance level, e.g. 25%, 75%, and 95%, and not only at the ill specified "best" and "worst" extremes. Moreover, this approach completely neglects parameter inter-correlations, and produces circuit performance predictions that are overly pessimistic (or optimistic).

A number of approaches have been taken toward solving this problem. The most straightforward of them are methods that identify the important device parameters and perform Monte-Carlo simulations for a given circuit [1].

DAC 98, San Francisco, California

However, the complexity of VLSI circuits makes Monte-Carlo based methods prohibitive. In addition to being very time-intensive, Monte-Carlo methods either assume that the device parameters are independent, or necessitate building elaborate models to account for the complex correlation of the parameters. Alternatively, the correlated device parameters can be expressed in terms of independent factors, or components, using statistical techniques such as principal component analysis [2]. While improving the simple Monte Carlo based methods, this approach is susceptible to the problems of statistical transformation techniques that are discussed below.

To overcome these limitations, a methodology based upon expressing the circuit characteristics in terms of the device model parameters has been proposed [3]. This approach reduces the number of device parameters under consideration using principal component analysis (PCA), and then employs response surface methodology to generate the statistical circuit performance characteristics. However, the use of PCA on device parameters from the deep sub-micron technologies is undesirable because it may fail to capture the complex correlations of the device parameters [5].

This is due to the fact that for the deep sub-micron CMOS technologies a combination of device physics, die location-dependence, optical proximity effect, microloading in etching and deposition, etc., may lead to heterogeneous and non-monotonic relationships among the device parameters that cannot be captured by the PCA-based device parameter characterization. For example, the same industrial technology can produce two opposite correlations between the threshold voltage and the channel length (Figure 1). One trend is purely stochastic: devices with shorter effective channel lengths have slightly higher Vt following the reverse short-channel effect dependence. The other trend is deterministic, i.e. die-location dependent: probably, due to variation of the lateral doping profile near wafer edge. PCA is a statistical technique that transforms correlated variables into uncorrelated variables, which graphically amounts to a rotation of the original axis. In the presence of variation arising from two or more mechanisms, a phantom averaging effect occurs leading to underestimation of the real level of parameter fluctuations. This can be clearly seen from Figure 2, where the larger number of stochastic data points 'outweighs' the systematic variation.

The proposed SPS methodology attempts to find a solution that avoids the drawbacks and problems mentioned above. It allows generation of statistical information at any specified

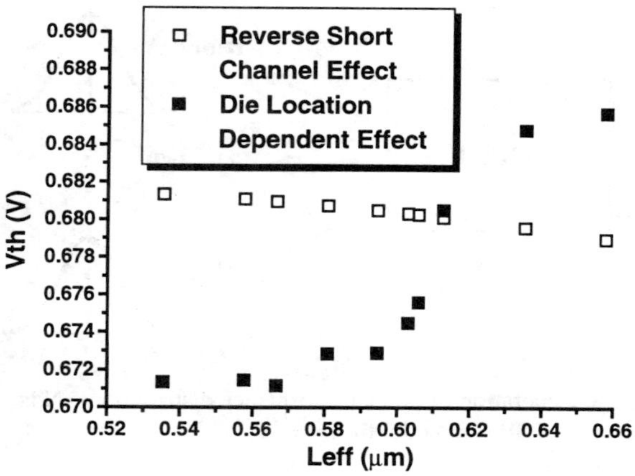

Figure 1. Different trends of Vth vs Leff dependence are found in deep sub-micron technologies.

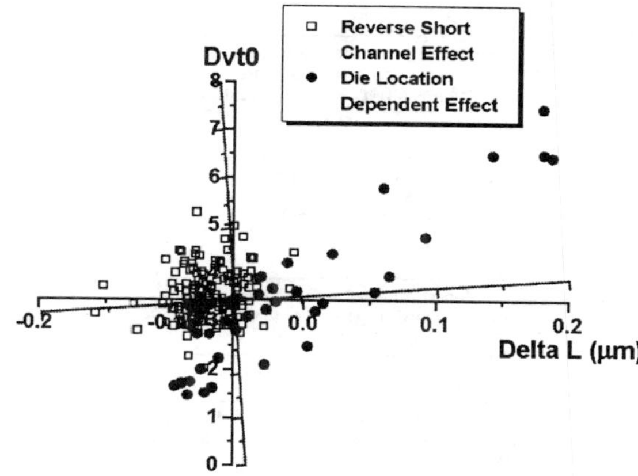

Figure 2. PCA-based methods lead to erroneous results when data coming from different sources of variation is present.

performance level. It does not employ any statistical techniques that may lead to the inadvertent loss of parameter correlations arising from the complex physical processing of the deep sub- micron technologies. This is accomplished with a direct-sampling method [5]. In order to avoid a large number of simulations of the whole circuit, and thus overcome the deficiency of Monte-Carlo based methods, it constructs a circuit model in terms of performance variables of the simpler blocks constituting the circuit. The computational effort is greatly reduced because circuits normally consist of only a limited number of distinctly behaving 'primitives', and only one representative of every such primitive needs to be analyzed. Also, because simulation time increases more than linearly with the number of nodes, this approach results in additional simulation time saving. A concise comparison of advantages and disadvantages of some of the reported methods and SPS methodology is summarized in Table 1.

2 METHODOLOGY

The proposed SPS methodology makes no assumptions about the degree of device parameter independence. This is achieved by using the direct sampling method (DSM). The main feature of DSM is the insistence that all sampled device characteristics from a test die be kept as a set, and be directly applied to all further analysis. No statistical operation is performed that can inadvertently lead to the effect of averaging out the device fluctuations arising from different mechanisms, which can happen if principal component analysis or factorial analysis are used [5]. In DSM a statistically significant number of dies is sampled and kept as a set that typically contains the device I-V parameters, and gate, overlap, junction, and interconnect capacitances. Because all the parameters from the same location are kept together, neither additional information nor analysis is required concerning their mutual correlation. The larger is N, the finer is the resolution of the performance distribution. Therefore, a large number of SPICE device parameters need to be efficiently generated. One possible approach is to extract SPICE parameters directly from Electrical-Test data, using an equation solver technique [6].

The flow chart of the sequence of steps in SPS methodology is given in Figure 3. The basic premise of SPS is the idea that a large circuit can be analyzed in terms of simpler blocks that constitute the circuit (Figure 4). The possibility of such

Method	Advantages	Disadvantages
Traditional Worst-Case Model File Generation	- Straightforward - Computationally efficient	- Not accurate - No statistical yield information
Monte-Carlo [1]	- Provides statistical performance information. - Accurate	- Very computationally expensive. Impractical for large circuits
RSM Model Construction In Terms Of Device Parameters [3]	- Provides statistical yield information - Computationally efficient	- Not suitable for complex correlation of deep-submicron device parameters
Optimization Approaches [4]	- Reasonably efficient - Evaluates circuit performance	- Provides no statistical yield information
SPS – Model Construction In Terms Of Circuit Blocks (Current Paper)	- Provides statistical yield information - Computationally more efficient than Monte-Carlo	- Must isolate the "building" blocks of the larger circuit - Approximate substitute for Monte-Carlo

Table 1. A comparison of different approaches to statistical circuit modeling.

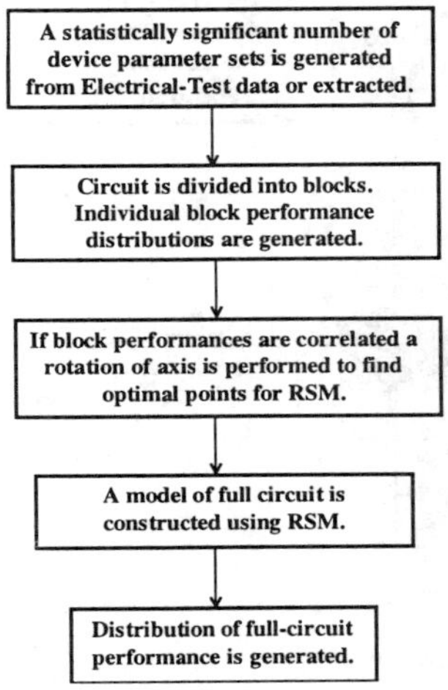

Figure 3. Steps involved in SPS methodology.

structural subdivision is important in two respects.

Digital circuits are usually well structured and can be broken down into the smaller building blocks such as inverters, NAND and NOR gates, clock generators, latches, etc... Thus, a large circuit normally is a combination of a limited number of statistically distinct building blocks. Statistically distinct blocks are those that produce different performance distributions for the identical set of input SPICE files. Two statistically distinct blocks, for example, would produce a performance (e.g. delay) value lying at a different level of statistical distribution, i.e. the same SPICE model file would lead to a simulation result at the 95th percentile and the 60th percentile of the performance distribution. It is reasonable to assume, as a first-order approximation, that the statistically identical blocks remain behaving similarly even when placed into the network of connections within the larger circuit. This assumption allows a great reduction in analysis effort, because now only a representative block needs to be analyzed, leaving aside a large number of its replications. Clearly, this advantage can be utilized only when a hierarchical block approach similar to SPS methodology is used.

In addition, because the simulation time of the SPICE-based circuit simulators is on the order of $O(M^{1.2})$ with M

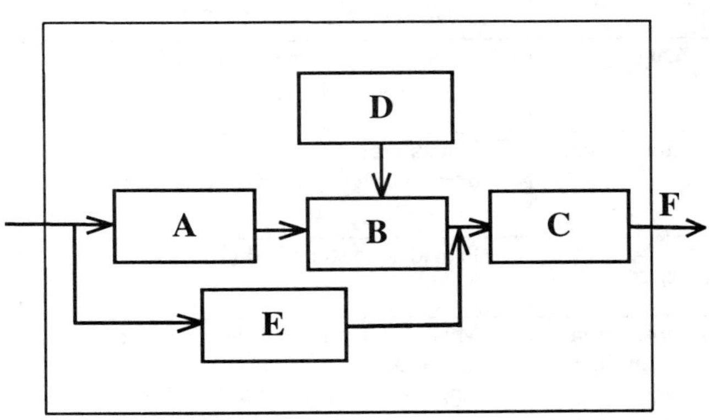

Figure 4. Large circuit is analyzed in terms of its building blocks.

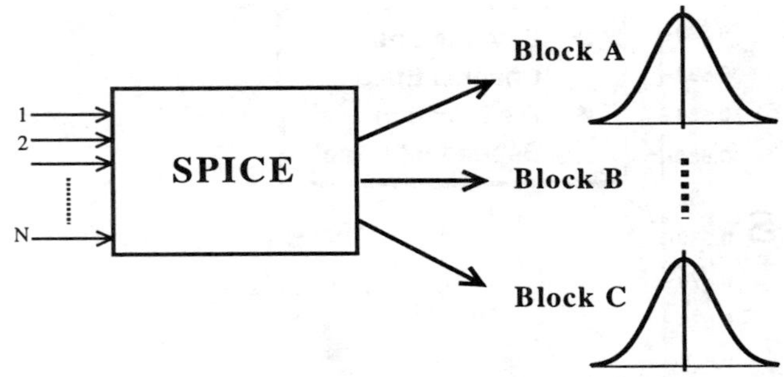

Figure 5. Generation of block performance distributions. N is the number of SPICE model files.

being the number of nodes, it is advantageous to break up a single set of nodes into a number of smaller sub-sets [7]. If $T(M)$ is the simulation time of M nodes, then:

$$T(M) > T_1(M_1)+T_2(M_2)+\ldots+T_n(M_n),$$

where $M_i<M$, and n is the number of blocks. With the increase of M, the advantage of breaking up the large circuit into blocks becomes greater.

The model is to be constructed using response surface methodology (RSM). Because the number of simulations of the whole circuit required to generate an accurate model using RSM is on the order of 2^n, it is beneficial to minimize the number of distinct blocks n. For example, several simple digital blocks, such as inverters, NAND and NOR gates, have been shown to be highly correlated with respect to speed [8]. (Even though they can behave differently with respect to another circuit performance characteristics, such as noise margins [9].) The blocks that are a priori known to be statistically similar n be excluded from the analysis; thus, reducing the required number of full circuit simulations.

Once the simpler blocks are identified, they are evaluated with respect to a particular circuit performance variable, e.g. speed, or power dissipation. Each block is simulated N times using SPICE, where N is a statistically significant number of samples from DSM. The distribution of random variables corresponding to each block is thus generated (Figure 5).

We next wish to find the distribution of the performance variable for the whole circuit based upon the generated block performance distributions. A model relating block performances A,B,C... and the whole circuit performance F is needed: F=f(A,B,C...). The analytic form of a function f (A,B,C...) is not available and has to be estimated by using response surface method. RSM involves running a limited number of full circuit simulations, and then fitting a function relating full circuit performance to block performances. The coefficients of the model are calculated using linear regression.

At this stage, the objective is to minimize the number of full circuit simulations without sacrificing the model's accuracy. The training points should be chosen to cover as much area as possible in the space of block performances, and a simple way to do it is to "sample" the space at the extremes of the distribution. If block distributions are correlated the choice of the optimal training point is not a trivial task. Therefore, the algorithm used to select such points for RSM requires that the block performances be uncorrelated. To

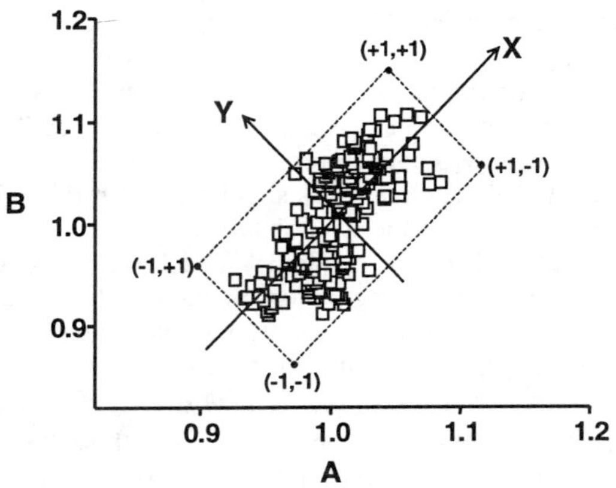

Figure 6. Rotation of axis may be necessary to determine the optimal training points for RSM.

are generated, an analysis is performed to see whether the blocks are correlated. If some of them are significantly correlated a rotation of axis is performed in such a way that the resultant variables are mutually independent (Figure 6). The training points are easily identified as corresponding to the extremes of the distribution relative to the new rotated axis.

Design of experiment theory helps to determine the optimal set of training points for the simulations in such a way that the best model is generated [10,11]. The most straightforward way is to perform the simulations with all possible combinations of the extremes of block performances. This approach is called full factorial analysis and requires a full set of 2^n simulations. It is possible to reduce this number. Fractional factorial designs require 2^{n-k} simulations (n- number of blocks, k- degree of fraction). We consider only two level factors, and denote the maximum and minimum values of each factor (block performance variable) as +1 or –1. Then, Table 2 lists the set of combinations of variables required by full factorial design and indicates which combinations should be used in half-factorial design (n=3, k=1).

It is important to see the limitations of fractional factorial design. Reducing the number of full-circuit simulations results in the loss of some accuracy. The main effects (coefficients of A, B, and C) and two-factor interaction effects (coefficients of $A \times B$, $B \times C$, and $A \times C$) can still be accurately represented. Only the three-factor interaction term is lost but since in most cases such a high-order interaction is negligible this loss is permissible [11].

The selection of optimal training points is achieved by considering the uncorrelated variables after the axis are rotated. To find the actual SPICE model file corresponding to a selected point no backward transformation is needed, however. Even though the coordinates were rotated, the correspondence between the points and the actual model file was kept. The rotation of axis was needed only to identify which of N models corresponds to the optimal training points for RSM. Figure 6 helps to illustrate this idea.

Simulation	A	B	C	A*B	B*C	A*C	A*B*C	Half-fraction
1	-1	-1	-1	+1	+1	+1	-1	
2	-1	-1	+1	+1	-1	-1	+1	*
3	-1	+1	-1	-1	+1	-1	+1	*
4	-1	+1	+1	-1	-1	+1	-1	
5	+1	-1	-1	-1	-1	+1	+1	*
6	+1	-1	+1	-1	+1	-1	-1	
7	+1	+1	-1	+1	-1	-1	-1	
8	+1	+1	+1	+1	+1	+1	+1	*

Table 2. Choice of training points for the half-factorial design.

The SPICE model files corresponding to the chosen training points are used to simulate the whole circuit, resulting in 2^{n-k} values of full-circuit delay. Using this set of data points together with the already simulated 2^{n-k} values for each of the building blocks, the coefficients of the model can be generated using linear regression. Once the model is generated, the distribution of the full-circuit performance can be easily calculated by the simple substitutions of N sets of block performance values of A,B,C... into the model. From the cumulative probability plot of full circuit distribution, the performance analysis can be straightforwardly performed. Because to each point on this plot there is a corresponding SPICE model, one can very easily use it to evaluate other performance characteristics of the whole circuit (such as power dissipation, noise margin, etc.) at a given level of speed distribution. In the cases of linear or quadratic models, the expected value and variance of the full-circuit distribution can also be directly calculated from the knowledge of the mean values and variance-covariance matrix of the block performances.

3 RESULTS AND DISCUSSION

To verify the proposed SPS methodology, a case study of a low-swing bus driver for low-power applications was performed [12]. This specific circuit was chosen because it combines blocks of digital as well as analog circuitry (Figure 7). We consider a part of the circuit dividing it into 3 blocks: the sense-amplifier (Sense), the latch (Latch), and the clock generator (Clock). The clock is generated through the use of ring-oscillators.

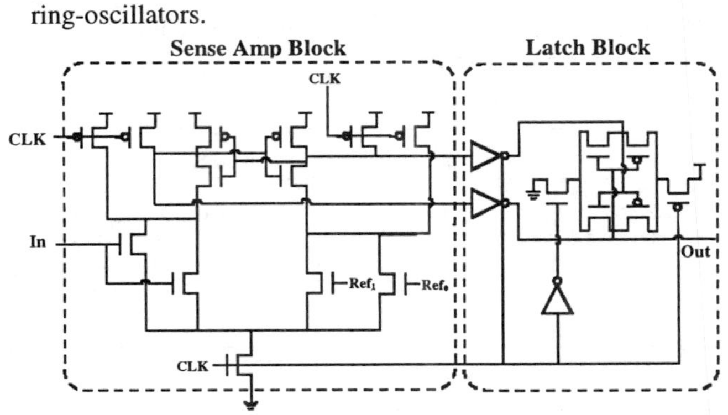

Figure 7. A bus driver circuit. Clock generator is not shown.

The individual block performance distributions were generated by considering the delay of each block, independently from other blocks. The total delay (Total) was then measured, being defined as time from arrival of the clock to the output signal. The simulations were performed using HSPICE [13].

The methodology was tested using a 0.5 μm CMOS production technology. BSIM3v3 IV device model parameters from two lots were efficiently generated from Electrical Test data using an equation solver [6]. In addition, device gate, overlap, and junction capacitances were extracted for each die, and kept as a set together with IV parameters. Altogether, N=214 sets of device parameters were prepared for use with direct sampling method.

Each block was then simulated N=214 times to generate the block performance distributions. Out of 214 simulated data points, 8 were filtered because of simulation failures. The scatter-plots of the 3 block performance distributions and their correlation matrix are shown in Figure 8. The characteristics of the individual distributions are presented in Table 3. The scatter-plots show that the sense amplifier and latch are correlated very weakly, while there is a stronger correlation between the clock generator and the sense-amp. In both cases the blocks are neither uncorrelated nor absolutely correlated. This finding justifies treating them as statistically distinct, and explicitly accounting for their contribution.

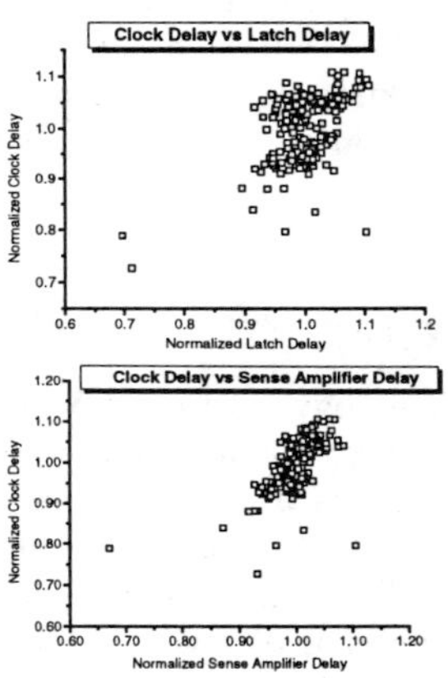

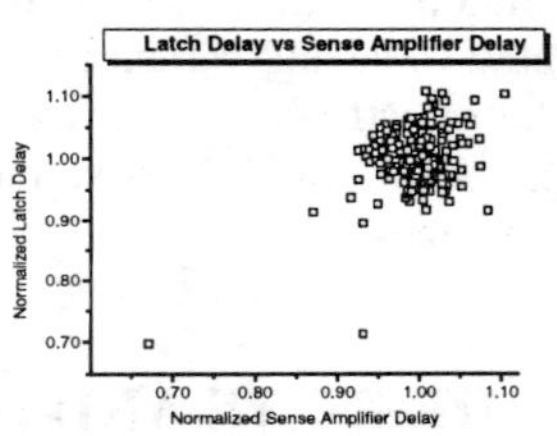

	Sense	Latch	Clock
Sense	1	0.41	0.62
Latch	0.41	1	0.51
Clock	0.62	0.51	1

Figure 8. Scatter plots of individual block distributions and their correlation matrix.

	Mean	Standard Deviation
Sense	$9.16 \cdot 10^{-10}$	$3.63 \cdot 10^{-11}$
Latch	$5.63 \cdot 10^{-10}$	$2.8 \cdot 10^{-11}$
Clock	$1.35 \cdot 10^{-8}$	$8.6 \cdot 10^{-10}$
Total	$1.07 \cdot 10^{-9}$	$4.4 \cdot 10^{-11}$

Table 3. Characteristics of block and full circuit distributions.

Because the block distributions are not independent, the rotation of axis is necessary in order to simplify identifying the optimal training points for RSM. Employing a half-factorial design, four training points, picked out of possible 8, were selected. These points correspond to the ones identified in Table 2. The whole circuit was simulated using the SPICE models corresponding to these 4 points.

A linear (rather than, say, quadratic) model was found to be sufficient. A simple linear regression function of the statistical package S-Plus was used to generate the coefficients of the linear model [14]. The fitted linear model is:

$$Total = 1.53 \cdot 10^{-10} + 0.034\ Clock + 0.157\ Sense + 0.58\ Latch$$

To verify the accuracy of the model, a full set of 208 simulations of the whole circuit was performed. Then the model was compared with the exact simulation result. The average error was found to be 1.33% with $\sigma = 1.22\%$. The scatter-plot of the model versus the exact simulation result is presented in Figure 9.

The model was also used to calculate the characteristics of the full-circuit performance distribution directly using the characteristics of the block performance distributions. For the particular case of a linear model, the expressions for mean and variance are of the form:

$$E[F] = C_0 + \sum_{i=1}^{3} C_i E[X_i]$$

$$D[F] = \sum_{i=1}^{3} C_i^2 D[X_i] + 2 \sum_{i<j} C_i C_j \Sigma_{ij} ,$$

where C_i are coefficients of the model, and Σ_{ij} are elements of the variance-covariance matrix, and X_i are random variables corresponding to blocks A,B, and C (i.e. Clock, Sense, and Latch). The actual and calculated characteristics are compared in Table 4. Good accuracy can be observed.

	Mean (Exact)	Std. Dev. (Exact)	Mean (Model)	Std. Dev. (Model)
Total	$1.08 \cdot 10^{-9}$	$4.51 \cdot 10^{-11}$	$1.07 \cdot 10^{-9}$	$4.4 \cdot 10^{-11}$

Table 4. Exact characteristics of full circuit distribution and characteristics calculated directly from the model.

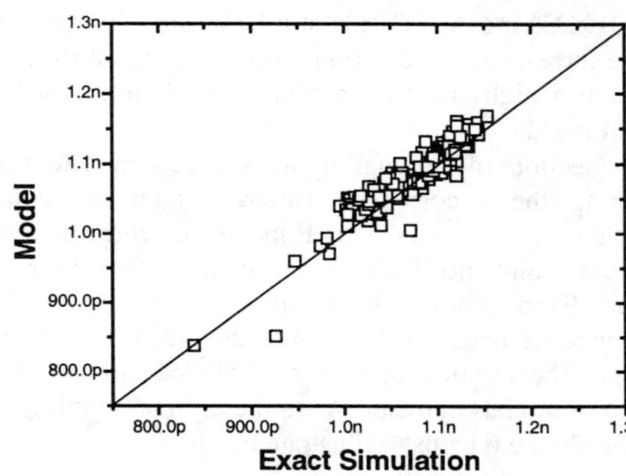

Figure 9. Good agreement between model and exact result.

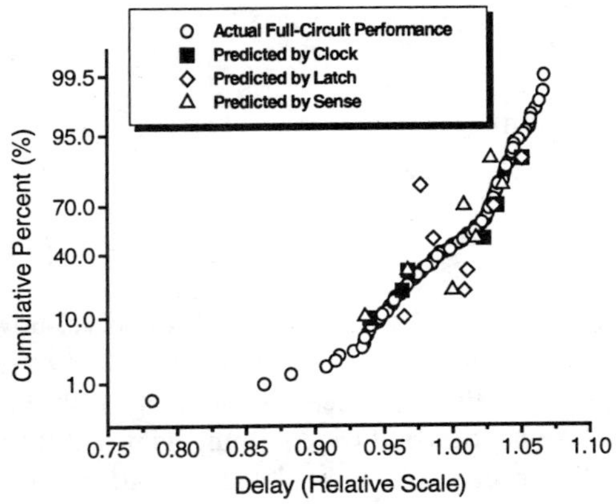

Figure 10. Blocks predict full circuit performance differently.

The model was used to generate the plot of cumulative probability function for the full-circuit delay, Figure 10. From this plot one can readily evaluate the performance of the full circuit at, for example, 2%, 10%, 50%, 90%, and 98% of speed distribution thus going further than the current approach of considering only the "worst", "best", and typical points. Because a particular SPICE model file corresponds to each data point on the plot, any other circuit performance characteristics (such as power dissipation, noise margin) can be evaluated for any specified speed performance level.

It is theoretically possible that the blocks of the circuit were well correlated, i.e. statistically identical. If it were so, statistical characterization of one of the blocks would be equivalent to the statistical characterization of the whole circuit. A SPICE file selected on the basis of the analysis of a single block at a given performance level (e.g. a latch) would correspond to the same performance level of the whole circuit. In the particular circuit considered, the blocks were only weakly correlated. Figure 10 illustrates the fact that the choice of the SPICE file at a given performance level is different for individual blocks and the whole circuit. Clearly, using a simplistic approach in this case would lead to notable prediction errors.

The simulation-time saving resulting from using SPS has been evaluated. The savings come from two distinct mechanisms. First, mere breaking down the whole circuit into a combination of smaller sub-circuits reduces simulation time of a SPICE-type simulator. Second, only one representative circuit has to be simulated for each statistically distinct block, while in the full circuit the number of its replications is likely to be significant. The simulation overhead required by the methodology has also to be taken into account.

To evaluate the time saving, we consider a 16 bit version of the bus driver architecture that was analyzed. Total computational time for the direct simulation of the whole circuit necessary to generate a statistical distribution of the same resolution:

$$T_{tot}^{direct} = N \cdot T(16 \times Full)$$

where N is the number of simulations, N=208, and $T(16 \times Full)$ is time to simulate 16 replications of the full driver circuit, corresponding to 16 bits.

If only time saving due to breaking down the circuit into smaller blocks is considered:

$$T_{tot}^{I} = N\left[T(16 \times Sense) + T(16 \times Latch) + T(16 \times Clock)\right] + 4T(16 \times Full)$$

Accounting, in addition, for replication-number reduction allowed by SPS, total time is:

$$T_{tot}^{II} = N\left[T(1 \times Sense) + T(1 \times Latch) + T(1 \times Clock)\right] + 4 \cdot T(16 \times Full)$$

The simulation results show that $T_{tot}^{I} \cong 0.89 \cdot T_{tot}^{direct}$, and $T_{tot}^{II} \cong 0.04 \cdot T_{tot}^{direct}$. In other words, 11% and 96% reductions in computational time are achieved. Clearly, the dramatic simulation time reduction comes from the replication-free approach of SPS.

4 CONCLUSION

A methodology for statistical modeling of VLSI circuits has been presented. It permits performance analysis at any specified yield level, and is suitable for use with deep sub-micron CMOS technology generations. SPS considerably reduces computer simulation time required to generate the statistical distribution of a large circuit by using a novel approach of hierarchical modeling.

5 REFERENCES

[1] R.Y.Rubinstein, *Simulation and the Monte Carlo Method*, John Wiley & Sons, 1981.

[2] J. Power, B. Donnellan, A. Mathewson, W. Lane, "Relating Statistical MOSFET Model Parameter Variabilities to IC Manufacturing Process Fluctuations Enabling Realistic Worst Case Design", *IEEE Trans. Semicon. Manuf.*, August 1994, pp. 306-318.

[3] E. Felt, S. Zanella, G. Guardiani, A. Sangiovanni-Vincentelli, "Hierarchical Statistical Characterization of Mixed-Signal Circuits Using Behavior Modeling," *1996 Proc. Of IC-CAD*, pp. 374-380.

[4] A. N. Lokanathan, J. B. Brockman, "Efficient Worst Case Analysis of Integrated Circuits," *1995 Proc. of CICC*, pp. 237-240.

[5] J. Chen, M. Orshansky, C. Hu, C.-P. Wan, "Statistical Circuit Characterization for Deep-Submicron CMOS Designs", *1998 Proc. ISSCC*, pp.90-91.

[6] J. Chen, C. Hu, C.-P. Wan, P. Bendix, A. Kapoor, "E-T Based Statistical Modeling and Compact Statistical Circuit Simulation Methodologies," *1996 Proc. of IEDM*, pp. 635-638.

[7] J. Vlach, K. Singhal, *Computer Methdos for Circuit Analysis and Design*, Van Nostrand Reinhold, 1994.

[8] J. Chen, C. Hu, Z. Liu, P. Ko, "Realistic Worst-Case SPICE File Extraction Using BSIM3," *1995 Proc. of CICC*, pp. 375-378.

[9] S. R. Nassif, "Statistical Worst-Case Analysis for Integrated Circuits", *Statistical Approach to VLSI*, pp. 233-253, 1994.

[10] G. E. P. Box, W. G. Hunter, J. S. Hunter, *Statistics for Experimenters: An Introduction to Design, Data Analysis, and Model Building*, John Wiley & Sons, 1978.

[11] S. M. Kang, Y. Leblebici, *CMOS Digital Integrated Circuits: Analysis and Design*, The McGraw-Hill Companies, Inc., 1996.

[12] M. Hiraki, H. Kojima, H. Misawa, T. Akazawa, Y. Hatano, "Data-Dependent Logic Swing Internal Bus Architecture for Ultralow-Power LSI's", IEEE Journal of Solid-State Circuits, Vol. 30, No.4, April 1995.

[13] Meta-Software, HSPICE User's Manual, 1996.

[14] MathSoft Inc, S-Plus, 1994.

RF IC Design Challenges

Behzad Razavi
Electrical Engineering Department
University of California, Los Angeles

Abstract

This paper describes the challenges in designing RF integrated circuits for wireless transceiver applications. Receiver architectures such as heterodyne, homodyne, and image-reject topologies are presented and two transmitter architectures, namely, one-step and two-step configurations are studied. The design of building blocks such as low-noise amplifiers, mixers, and oscillators is also considered.

I. Introduction

The number of cellular telephone users in the United States exceeded 50 million in June 1997, only 14 years after the introduction of this product. By contrast, the wired telephone took 77 years and the television 24 years to reach such market penetration. It is also believed that about 20% of the world's population has neven made a phone call . . .

The radio frequency (RF) and wireless market has suddenly expanded to unimaginable dimensions. Devices such as pagers, cellular and cordless phones, cable modems, and RF identification tags are rapidly penetrating all aspects of our lives, evolving from luxury items to indispensable tools. Semiconductor and system companies, small and large, analog and digital, have seen the statistics and are striving to capture their own market share by developing various RF products.

In addition to the market "push," the RF industry has also experienced a "pull" by the integrated circuits technology. Advances in silicon and GaAs devices continue to increase the level of integration and decrease the cost of RF circuits, allowing wireless products to compete with even their wired counterparts.

This paper describes the challenges posed to RF IC designers at the architecture and circuit levels. We first consider transceiver architectures such as heterodyne, homodyne, and image-reject receivers and one-step and two-step transmitters. Next, we study the design of various building blocks to understand the issues and trade-offs at the circuit level.

II. Transceiver Architectures

Complexity, cost, power dissipation, and the number of external components have been the primary criteria in selecting receiver architectures. As IC technologies evolve, however, the relative importance of each of these criteria changes, allowing approaches that once seemed impractical to return as plausible solutions. Since filter requirements prohibit channel selection at RF, receivers first translate the input spectrum to a much lower frequency.

A. Heterodyne Receivers

Heterodyne receivers downconvert the input to an "intermediate frequency" (IF), perform band-pass filtering and amplification, and translate the spectrum to a lower frequency again (Fig.1). In the case of phase or frequency modulation,

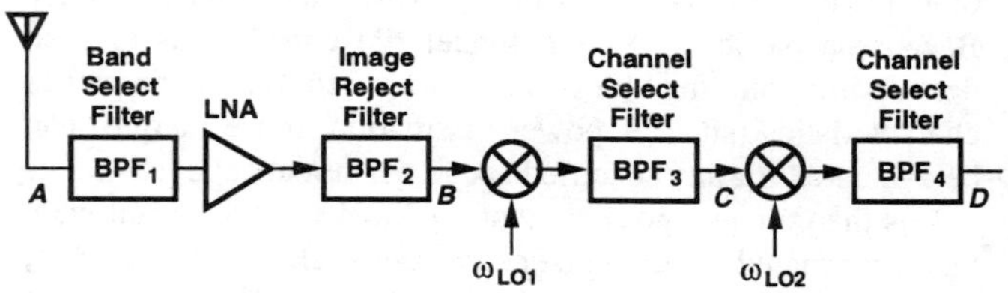

Fig. 1. Dual-conversion heterodyne receiver.

downconversion to the baseband requires both in-phase (I) and quadrature (Q) components of the signal. This is because the two sidebands of such signals carry different information and must be separated into quadrature phases in translation to zero frequency.

Perhaps the most important feature of the heterodyne receiver is its selectivity, i.e., the capability to process and select small signals in the presence of strong interferers. While selecting a 30-kHz channel at a center frequency of 900 MHz requires prohibitively large Q's, in Fig. 1 band-pass filtering is performed at progressively lower center frequencies.

Heterodyning nonetheless entails a number of drawbacks. The trade-off between image rejection and channel selection typically requires a relatively high IF, making it difficult to integrate the IF filter monolithically. Furthermore, the image-reject filter is a passive, bulky device that must be placed off-chip and driven as a 50-Ω load. This intensifies the trade-offs in the design of the low-noise amplifier.

In heterodyne architectures, the choice of the noise figure (NF), the third intercept point (IP_3), and the gain of each stage in the chain depends on those of the preceding and following stages, thereby demanding considerable iteration at the architecture and circuit levels to arrive at an acceptable distribution of gain in the receiver building blocks. Moreover, each mixer generates many spurious components whose frequencies are related to those of the RF and IF signals and the oscillators. Some of these components may fall in the desired channel, degrading the signal quality. Thus, the "frequency

DAC 98, San Francisco, California
©1998 ACM 0-89791-964-5/98/06..$5.00

planning" of the receiver directly impacts the performance of the receiver in the presence of interferers.

B. Direct-Conversion Receivers

Direct-conversion (homodyne) receivers translate the channel of interest directly to zero frequency (Fig. 2). For frequency- and phase-modulated signals, the downconversion must provide quadrature outputs so as to avoid loss of information.

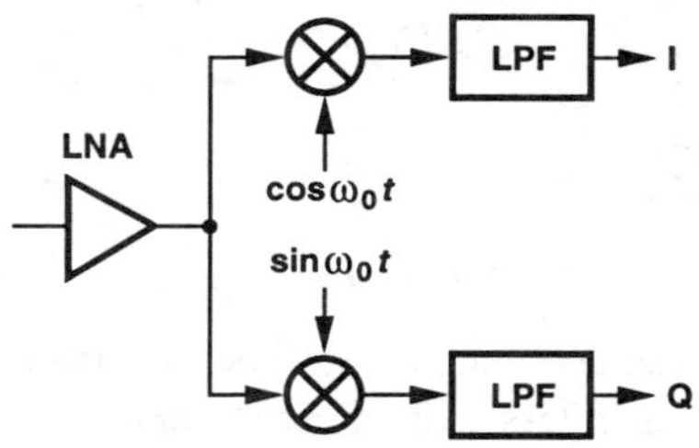

Fig. 2. Direct-conversion receiver.

Direct conversion offers two important advantages over a heterodyne counterpart. First, the problem of image is circumvented because $\omega_{IF} = 0$. As a result, no image filter is required and the LNA need not drive a 50-Ω load. Second, the IF filter and subsequent downconversion stages are replaced with low-pass filters and baseband amplifiers that are amenable to monolithic integration.

The simplicity of direct conversion nevertheless comes with a number of design issues. First, dc offsets due to mixing of the local oscillator (LO) leakage with itself corrupt the baseband signal and, more importantly, saturate the following gain stages [1, 2]. At the output of the mixer, such offsets can be as high as 10 mV whereas the signal may be as low as a few tens of microvolts. If the modulation scheme contains significant energy near dc, as is the case in most cellular and cordless phone standards, then ac coupling with practical values of capacitors and time constants severely degrades the signal. To remove dc offsets, periodic offset cancellation can be performed during idle times, but at the cost of kT/C noise [1].

Second, as depicted in Fig. 3, even-order distortion in the RF signal path generates low-frequency beats from large interferers. In the presence of mismatches and hence asymmetry in

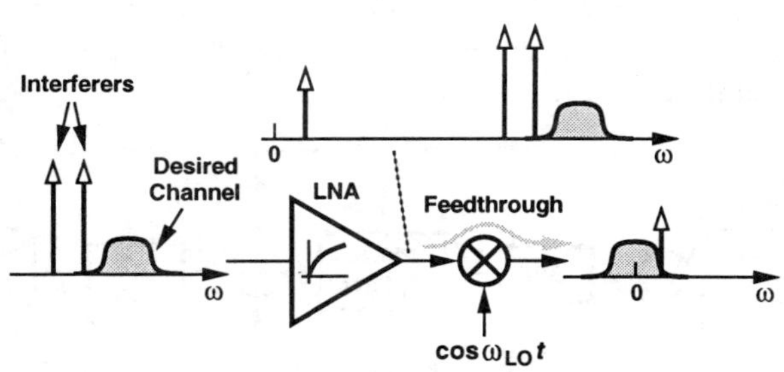

Fig. 3. Effect of second-order distortion in direct-conversion receivers.

the mixer, such components appear at the output, thus degrading the signal-to-noise ratio (SNR). This effect can be reduced by differential circuits or high-pass filtering the beats.

Owing to the limited gain provided by the LNA and the mixer, the downconverted signal is quite sensitive to noise. Especially problematic here is the flicker noise of baseband amplifiers and filters in CMOS technology, making it desirable to realize a *high* gain in the RF and IF sections.

Another issue in direct-conversion receivers is the phase and gain mismatch introduced by the mixers. Phase mismatch gives rise to *cross-talk* between demodulated quadrature waveforms, lowering the SNR because the I and Q data streams are usually uncorrelated.

In addition to the above issues, the leakage of the LO signal to the antenna creates interference in the band of other users and must therefore be sufficiently small (typically between −60 and −80 dBm).

While most of these difficulties can be resolved by means of circuits techniques, the dc offset and flicker noise problems continue to challenge designers.

C. Image-Reject Architectures

The issues related to the image-reject filter have motivated RF designers to seek other techniques of rejecting the image in a heterodyne receiver. One such technique originates from a single-sideband modulator introduced by Hartley [3]. Illustrated in Fig. 4, Hartley's circuit mixes the RF input with the

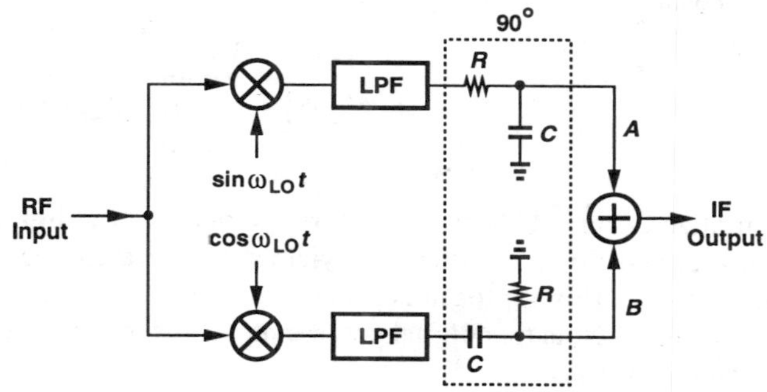

Fig. 4. Hartley image-reject receiver.

quadrature outputs of the local oscillator, low-pass filters and shifts the results by 90° before adding them together. It can be shown that the spectra at points A and B contain the desired band with the same polarity and the image with opposite polarity. The summed output is therefore free from the image.

The principal drawback of the Hartley architecture is its sensitivity to mismatches: with phase and gain imbalance, the image is only partially cancelled. Note that the effect of I/Q mismatch is much more severe here than in direct-conversion topologies. Also, the loss and noise of the shift-by-90° stage and the linearity of the adder are critical parameters. Furthermore, the variation of R and C introduces gain mismatch, limiting the image rejection ratio severely.

Shown in Fig. 5 is another image-reject architecture introduced by Weaver [4]. Replacing the 90° shift of Hartley's circuit with a second quadrature mixing operation, this technique provides an arbitrary translation of the signal band without image interference. It can be shown that the subtraction of the spectrum at point C from that at point D produces the signal while suppressing the image.

The Weaver architecture is also sensitive to mismatches, but it avoids the use of an RC-CR network, thereby achiev-

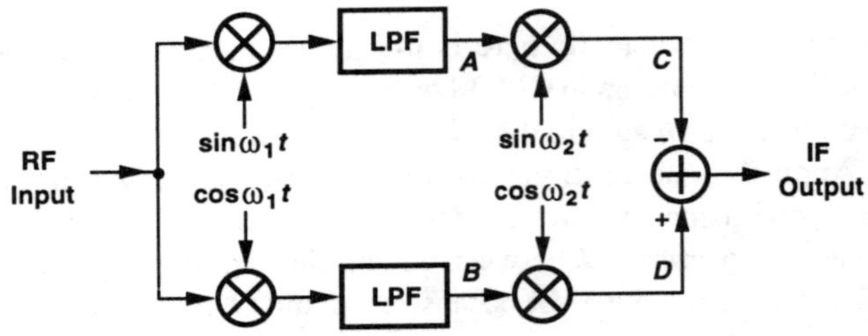

Fig. 5. Weaver image-reject receiver.

ing greater image rejection despite process and temperature variations.

D. Digital-IF Receivers

In the dual-conversion heterodyne architecture of Fig. 1, low-frequency operations such as the second set of mixing and filtering can be performed more efficiently in the digital domain. Shown in Fig. 6 is an example where the first IF signal

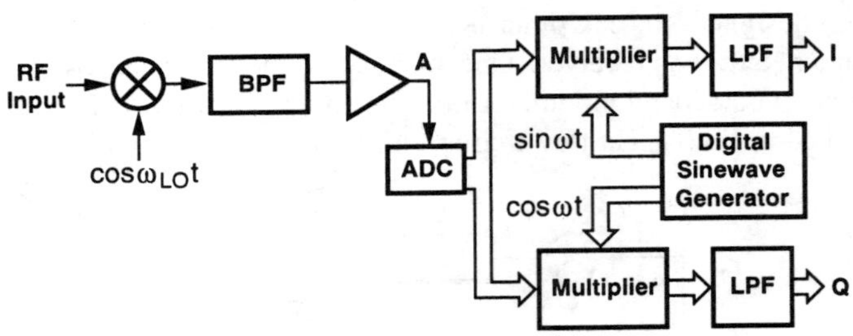

Fig. 6. Digital-IF receiver.

is digitized, "mixed" with the quadrature phases of a digital sinusoid, and low-pass filtered to yield the quadrature baseband signals. This approach is sometimes called a "digital-IF architecture." Note that digital processing avoids the problem of I and Q mismatch.

The principal issue in this approach is the performance required of the A/D converter. Since the signal level at point *A* in Fig. 6 is typically no higher than a few hundred microvolts, the quantization and thermal noise of the ADC must not exceed a few tens of microvolts. Furthermore, if the first IF bandpass filter cannot adequately suppress adjacent interferers, the nonlinearity of the ADC must be sufficiently small to minimize corruption of the signal by intermodulation. Additionally, the ADC must achieve an input bandwidth commensurate with the value of IF while consuming a reasonable amount of power.

The above requirements make it difficult to employ a Nyquist-rate ADC in the digital-IF architecture. Typical IF values of 50 to 200 MHz mandate sampling rates in the range of 100 to 400 MHz, and linearity, noise floor, and dynamic range requirements may necessitate resolutions greater than 14 bits. Such performance cannot be obtained in today's A/D converters even if cost and power dissipation are not critical. The idea is nonetheless the subject of active research [5, 6].

E. One-Step Transmitters

If the transmitted carrier frequency is equal to the local oscillator frequency, the architecture is called a "one-step" topology. In this case, modulation and upconversion occur in the same circuit. As shown in Fig. 7, the modulator is followed by a power amplifier and a matching network,[1] whose role is to provide maximum power transfer to the antenna and filter out-of-band components that result from the nonlinearities in the amplifier. Note that since the baseband signal is produced

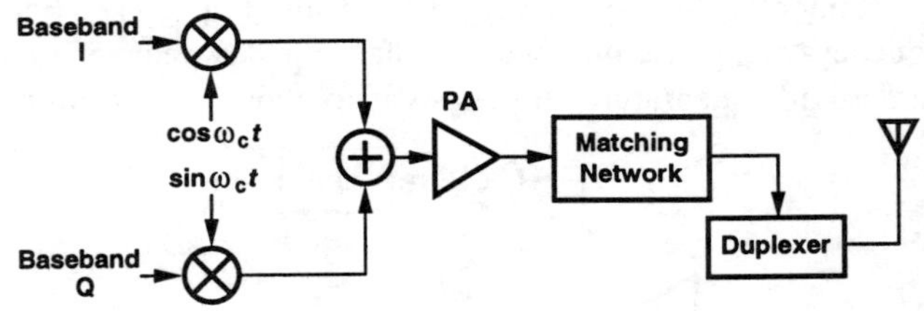

Fig. 7. One-step transmitter.

in the transmitter and hence is sufficiently strong, the noise of the mixers is much less critical here than in receivers.

The architecture of Fig. 7 suffers from an important drawback: disturbance of the transmit local oscillator by the power amplifier. Illustrated in Fig. 8, this issue arises because the PA output is a modulated waveform with high power and a

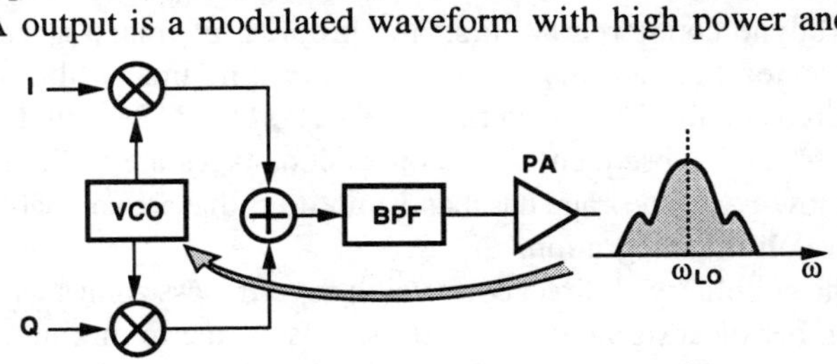

Fig. 8. Leakage of PA output to oscillator.

spectrum centered around the LO frequency. Despite various shielding techniques employed to isolate the VCO, the "noisy" output of the PA still corrupts the oscillator spectrum. This corruption occurs through a mechanism called "injection pulling" or "injection locking" [8]. The problem worsens if the PA is turned on and off periodically to save power.

The phenomenon of LO pulling is alleviated if the PA output spectrum is sufficiently higher or lower than the oscillator frequency. This can be accomplished by "offsetting" the LO frequency, i.e., by adding or subtracting the output frequency of another oscillator [9]. Fig. 9 shows an example where the output signals of VCO_1 and VCO_2 are mixed and the result is filtered such that the carrier frequency is equal to $\omega_1 + \omega_2$, far from either ω_1 or ω_2.

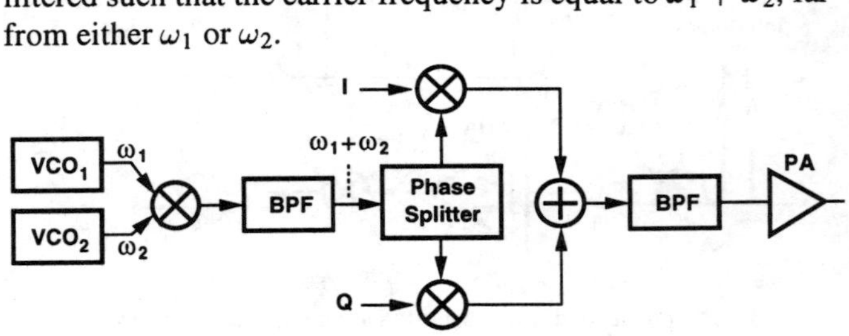

Fig. 9. One-step transmitter with offset LO.

[1] A band-pass filter may be interposed between the modulator and the PA to suppress the harmonics.

F. Two-Step Transmitters

Another approach to circumventing the problem of LO pulling in transmitters is to upconvert the baseband signal in two (or more) steps so that the PA output spectrum is far from the frequency of the VCOs. As an example, consider the circuit shown in Fig. 10. Here, the baseband I and Q

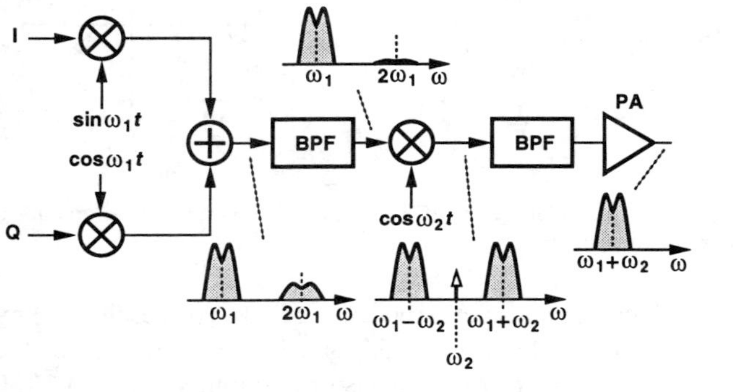

Fig. 10. Two-step transmitter.

channels undergo quadrature modulation at a lower frequency, ω_1, (called the intermediate frequency) and the result is upconverted to $\omega_1 + \omega_2$ by mixing and band-pass filtering. The first BPF suppresses the harmonics of the IF signal while the second removes the unwanted sideband centered around $\omega_1 - \omega_2$.

An advantage of two-step upconversion over the direct approach is that since quadrature modulation is performed at lower frequencies, I and Q matching is superior, leading to less cross-talk between the two bit streams. Also, a channel filter may be used at the first IF to limit the transmitted noise and spurs in adjacent channels.

The difficulty in two-step transmitters is that the bandpass filter following the second upconversion must reject the the unwanted sideband by a large factor, typically 50 to 60 dB. This is because the simple upconversion mixing operation produces both the wanted and the unwanted sidebands with *equal* magnitudes. Owing to the higher center frequency, this filter is typically a passive, relatively expensive off-chip device.

III. Transceiver Circuits

The design of an RF transceiver heavily depends on the performance of its constituent subcircuits. Low-noise amplifiers, mixers, oscillators, frequency synthesizers, modulators, and power amplifiers are the principal RF building blocks in a typical system, each exhibiting trade-offs between noise, linearity, gain, power dissipation, frequency of operation, and supply voltage.

In this section, we present recent work in the design of some of these building blocks.

A. Low-Noise Amplifiers and Mixers

As the first stages in the receive path, LNAs and mixers must process the signal with minimal noise and interference. In addition to the noise figure and third intercept point, the gain, port-to-port isolation, and power dissipation of these circuits impact the performance of a transceiver. The very low noise required of LNAs usually mandates the use of only one active device at the input without any (high-frequency) resistive feedback. In order to provide sufficient gain while driving 50 Ω, some LNAs employ more than one stage.

A bipolar LNA is shown in Fig. 11 [10], where the first stage utilizes a bond wire inductance $L_e = 1.5$ nH to degen-

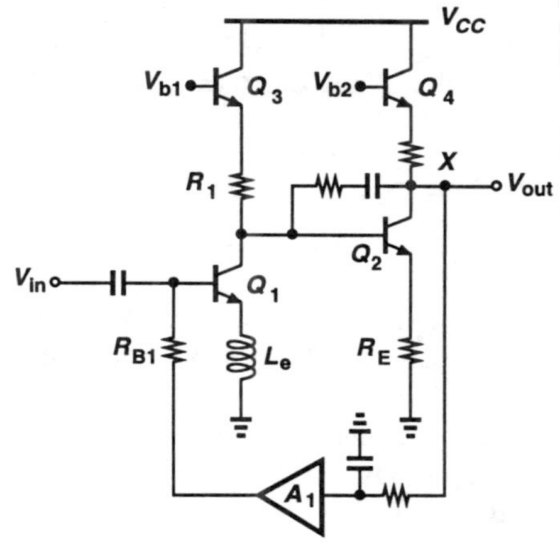

Fig. 11. LNA reported in [10].

erate the common-emitter amplifier, Q_1, without introducing additional noise. This technique both linearizes the LNA and makes it possible to achieve a 50-Ω input impedance. Bias voltages V_{b1} and V_{b2} and the low-frequency feedback amplifier A_1 are chosen so as to stabilize the gain against temperature and supply variations. The resistive feedback in the second stage improves the linearity and lowers the output impedance. The circuit exhibits a noise figure of 2.2 dB, an IP_3 of -10 dBm, and a gain of 16 dB at 900 MHz.

Another bipolar LNA designed to drive a 50-Ω load is depicted in Fig. 12 [11]. Employing negative feedback through

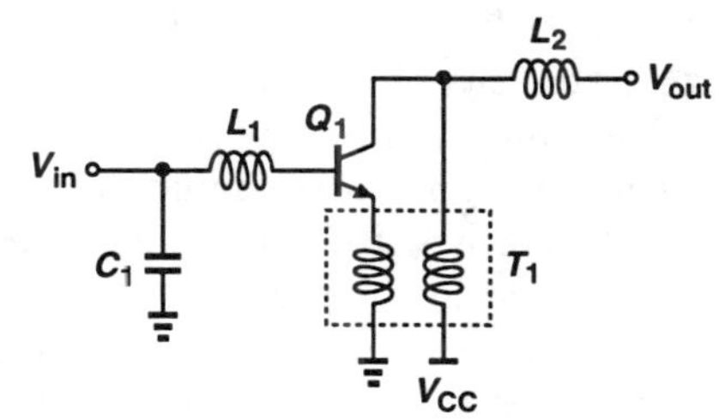

Fig. 12. LNA reported in [11].

a monolithic transformer to linearize the circuit, the LNA can operate with supply voltages as low as one V_{BE}. Interestingly, the transformer reduces the amplifier gain at both low and high frequencies, helping to stabilize the circuit. The external inductor L_1 and capacitor C_1 provide conjugate matching at the input.

Drawing 2 mA from a 1.9-V supply, the circuit of Fig. 12 achieves a noise figure of 2.8 dB and a gain of 9.6 dB at 1.9 GHz in an 11-GHz BiCMOS technology. The transformer feedback boosts the input IP_3 to -3 dBm.

A double-balanced bipolar mixer designed in conjunction with the above LNA is shown in Fig. 13 [11] . Here, an on-chip transformer both operates as a single-ended to differential converter and provides input matching. The bias current

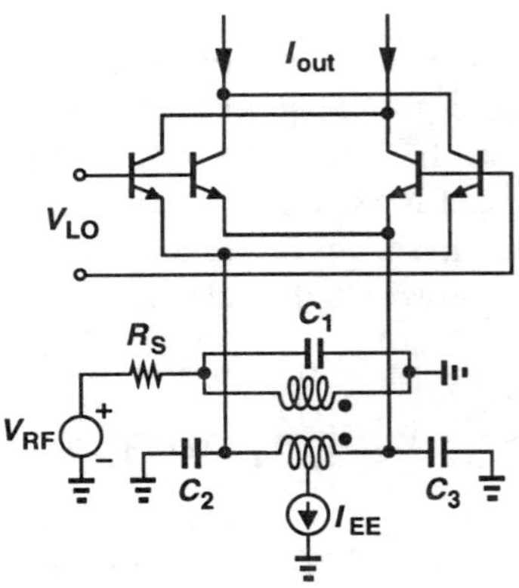

Fig. 13. Mixer reported in [11].

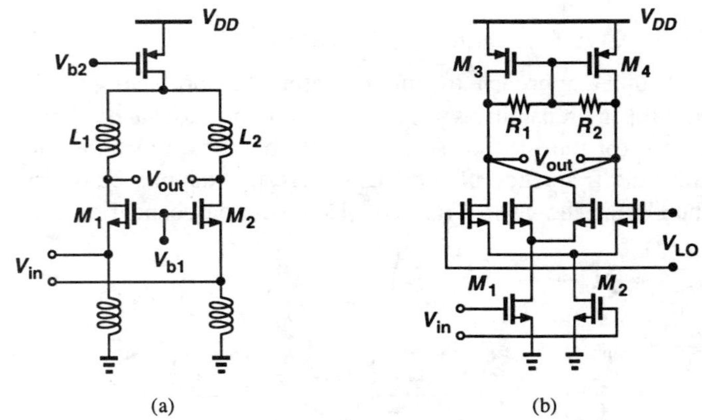

Fig. 15. (a) Common-gate CMOS LNA, (b) CMOS mixer with grounded-source input pair.

of the switching quad is established by I_{EE}, and capacitors C_1-C_3 effect resonance at the primary and secondary of the transformer.

A 1.9-GHz implementation of this configuration in an 11-GHz bipolar technology exhibits an NF of 10.9 dB with an IP_3 of +2.3 dBm while dissipating 5 mW from a 1.9-V supply [11].

Shown in Fig. 14 is a 1.5-GHz CMOS LNA employing on-chip and off-chip inductors [12]. In a manner similar to that

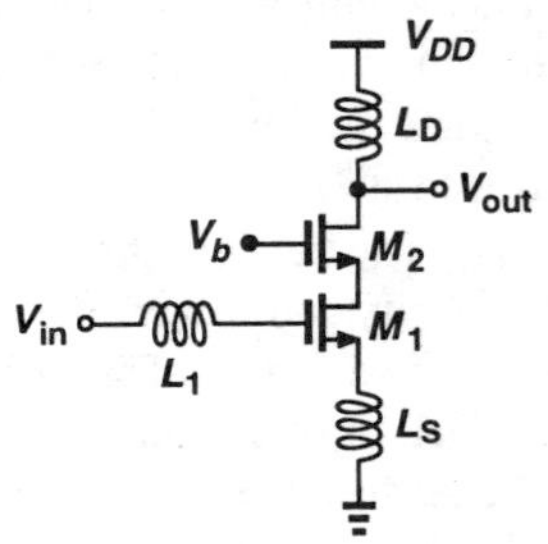

Fig. 14. LNA reported in [12].

described for the circuit of Fig. 11, this LNA incorporates L_S and L_1 to create conjugate matching at the input. At 1.5 GHz, the on-chip inductor L_D provides significant voltage gain even though its Q is less than 4. By contrast, a load resistor would require a large voltage drop to provide a comparable gain.

The common-gate transistor, M_2, plays two important roles by increasing the reverse isolation of the LNA: (a) it lowers the LO leakage produced by the following mixer; (b) it improves the stability of the circuit by minimizing the feedback from the output to the input. Note that the same circuit with no cascode device would be prone to oscillation.

The LNA of Fig. 14 is followed by another cascode stage so as to drive a 50-Ω load, with each stage drawing 10 mA. Fabricated in a 0.6-μm CMOS technology and operating from a 1.5-V supply, the circuit achieves a noise figure of 3 dB, a gain of 20 dB, and an input IP_3 of -10 dBm.

Fig. 15 [13] shows the simplified circuits of an LNA and a mixer designed for a 900-MHz direct-conversion receiver. Requiring no image-reject filtering by virtue of the zero IF, the receiver allows direct cascading of the two circuits, obviating the need for a 50-Ω interface between the LNA and the mixer. The LNA is configured as a differential common-gate topology, exhibiting an input impedance of 50 Ω (on each side) by proper sizing and biasing of M_1 and M_2. However, since the transconductance of each transistor is roughly equal to 1/(50 Ω), the noise figure cannot drop below a certain bound (2.2 dB with long-channel approximations).

The mixer resembles a Gilbert cell except that the sources of M_1 and M_2 are grounded. For square-law devices, this configuration does not produce third-order distortion whereas a differential pair biased at a constant tail current does. Thus, "grounded-source" MOS pairs potentially achieve a higher IP_3 than do regular differential pairs. The load of the mixer consists of self-biased current sources M_3 and M_4 and gain-setting resistors R_1 and R_2.

The LNA/mixer combination has been fabricated in a 1-μm CMOS technology. Consuming 27 mW from a 3-V supply, the circuit exhibits an overall noise figure of 3.2 dB and an IP_3 of +8 dBm.

B. Oscillators

The local oscillators used to drive downconversion and upconversion mixers are embedded in a synthesizer loop to achieve a precise frequency definition. Phase noise, sidebands (spurs), tuning range, and settling behavior of synthesizers are critical parameters in RF applications, creating severe trade-offs as the number of external components is reduced.

Most integrated RF oscillators are configured as a negative-G_m stage with inductive load. The idea is that an active circuit provides a negative resistance that cancels the finite loss in the inductors (and capacitors), thereby sustaining oscillation. While on-chip spiral inductors are attractive for higher levels of integration, various loss mechanisms limit the quality factor (Q) to approximately 4 in typical CMOS technologies. As depicted in Fig. 16, wire resistance and electric and magnetic coupling to the substrate contribute loss. Another important issue, particularly in CMOS circuits, is the upconversion of $1/f$ noise to the vicinity of the carrier frequency [14].

In bipolar technologies, spiral inductors exhibit slightly higher Q's because the substrate is lightly-doped, i.e., electric and magnetic coupling of the inductor to the substrate is less pronounced. Fig. 17 shows a bipolar implementation

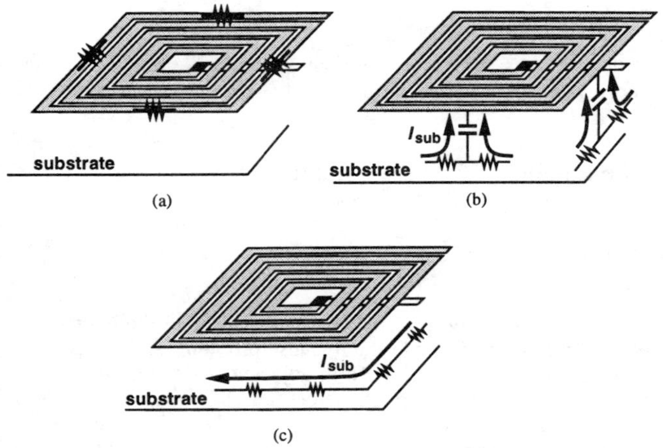

Fig. 16. Loss mechanisms in monolithic inductors.

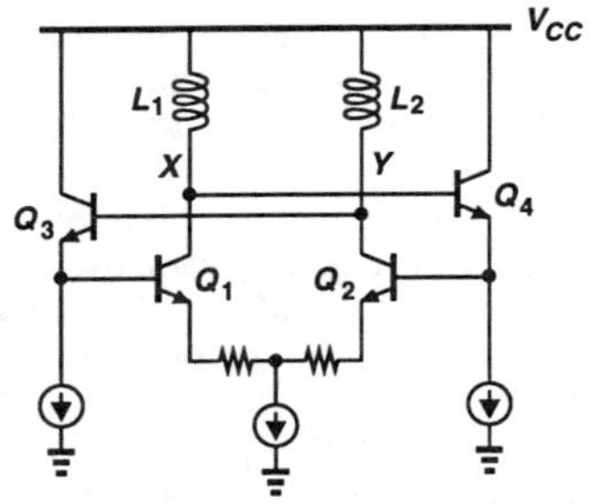

Fig. 17. Bipolar LC oscillator.

incorporating monolithic inductors with a Q of approximately 9 [15]. Using emitter followers in the loop to allow larger voltage swings at X and Y, the oscillator exhibits a phase noise of -105 dBc/Hz at 100 kHz offset.

Fig. 18 shows a CMOS VCO topology designed for 900 MHz and 1.8 GHz [16]. Here, the transconductance ampli-

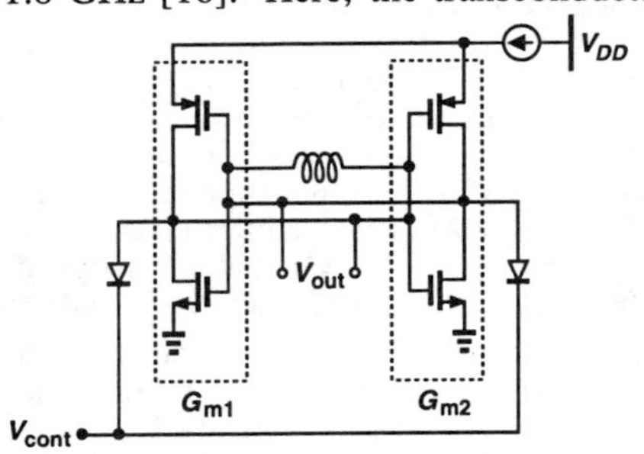

Fig. 18. CMOS negative-G_m VCO.

fier incorporates both NMOS and PMOS devices to achieve a higher transconductance for a given bias current. However, the additional capacitance contributed by the PMOS transistors limits the tuning range further. Drawing approximately 10 mW from a 3-V supply, the 900-MHz version of the oscillator exhibits a phase noise of -108 dBc/Hz at 100 kHz offset and a tuning range of 190 MHz [16].

An important issue in fully-monolithic LC oscillators is the trade-off between the phase noise and the tuning range [17]. For a given power dissipation, the relative phase noise decreases as the value of the tank inductance increases, but at the cost of making the capacitance of the transistors and the inductor a significant part of the tank. As a result, the variable component of the tank capacitance drops. Also, at low supply voltages, the variation obtained from a varactor diode becomes more limited.

At present, oscillators used in demanding applications such as cellular telephones still incorporate external resonators (inductors, microstrip lines, or filters) to achieve an acceptably low phase noise and an adequate tuning range. Nonetheless, the properties of inductors built on silicon substrates are under vigorous study.

REFERENCES

[1] B. Razavi, "Design Considerations for Direct-Conversion Receivers," *IEEE Trans. Circuits and Systems, Part II,* vol. 44, pp. 428-435, June 1997.

[2] A. A. Abidi, "Direct-Conversion Radio Transceivers for Digital Communications," *IEEE Journal of Solid-State Circuits,* vol. 30, pp. 1399-1410, Dec. 1995.

[3] R. Hartley, "Modulation System," US Patent 1,666,206, April 1928.

[4] D. K. Weaver, "A Third Method of Generation and Detection of Single-Sideband Signals," *Proc. IRE,* vol. 44, pp. 1703-1705, Dec. 1956.

[5] A. Hairapetian, "An 81 MHz IF Receiver in CMOS," *ISSCC Dig. Tech. Papers,* pp. 56-57, Feb. 1996.

[6] S. Jantzi, K. Martin, and A. Sedra, "A Quadrature Bandpass $\Sigma\Delta$ Modulator for Digital Radio," *ISSCC Dig. Tech. Papers,* pp. 126-127, Feb. 1997.

[7] B. Razavi, *RF Microelectronics,* Upper Saddle River, NJ: Prentice Hall, 1998.

[8] K. Kurokawa, "Injection Locking of Microwave Solid-State Oscillators," *Proc. IEEE,* vol. 61, pp. 1386-1410, Oct. 1973.

[9] T. D. Stetzler et al, "A 2.7-4.5 V Single Chip GSM Transceiver RF Integrated Circuit," *IEEE Journal of Solid-State Circuits,* vol. 30, pp. 1421-1429, Dec. 1995.

[10] R. G. Meyer and W. D. Mack, "A 1-GHz BiCMOS RF front-end integrated circuit," *IEEE Journal of Solid-State Circuits,* vol. 29, pp. 350-355, March 1994.

[11] J. R. Long and M. A. Copeland, "A 1.9GHz Low-Voltage Silicon Bipolar Receiver Front-End for Wireless Personal Communication Systems," *IEEE Journal of Solid-State Circuits,* vol. 30, pp. 1438-1448, Dec. 1995.

[12] D. K. Shaeffer and T. H. Lee, "A 1.5 V 1.5 GHz CMOS Low Noise Amplifier," *VLSI Circuits Symp. Dig. Tech. Papers,* pp. 32-33, June 1996.

[13] A. Rofougaran et al, "A 1-GHz CMOS Front End IC for a Direct-Conversion Wireless Receiver," *IEEE Journal of Solid-State Circuits,* vol. 31, pp. 880-89, July 1996.

[14] B. Razavi, "A Study of Phase Noise in CMOS Oscillators," *IEEE Journal of Solid-State Circuits,* vol. 31, pp. 331-343, March 1996.

[15] B. Jansen, K. Negus, and D. Lee, "Silicon Bipolar VCOs for 1.1 to 2.2 GHz with Fully-Integrated Tank and Tuning Circuits," *ISSCC Dig. of Tech. Papers,* pp. pp. 392-393, Feb. 1997.

[16] J. Craninckx, M. Steyaert, and H. Miyakawa, "A Fully Integrated Spiral LC CMOS VCO Set with Prescaler for GSM and DCS-1800 Systems," *Proc. CICC,* pp. 403-406, May 1997.

[17] B. Razavi, "A 1.8 GHz CMOS Voltage-Controlled Oscillator," *ISSCC Dig. Tech. Papers,* pp. 388-389, Feb. 1997.

Tools and Methodology for RF IC Design

Al Dunlop Alper Demir Peter Feldmann Sharad Kapur
David Long Robert Melville Jaijeet Roychowdhury

Bell Laboratories, Murray Hill, NJ

Abstract

We describe powerful new techniques for the analysis of RF circuits. Next-generation CAD tools based on such techniques should enable RF designers to obtain a more accurate picture of how their circuits will operate. These new simulation capabilities will be essential in order to reduce the number of design iterations needed to produce complex RF ICs.

1 Introduction

Design methodology and superior computer-aided design tools are key to success in the integrated circuit (IC) business. They are particularly important in the case of radio-frequency (RF) IC applications, where the digital IC divide-and-conquer design style, based on partitioning by functional blocks and abstraction levels, does not apply.

The goal of an RF designer is to get a manufacturable design that meets the specifications with minimum cost, under severe time-to-market constraints. Unlike traditional discrete-component RF design, prototyping is practically impossible, and the validation of a design can only be done by extensive simulation. Undetected mistakes in the design result in costly and time-consuming iterations that involve silicon processing.

Design methodology is the approach or philosophy used in creating a design. This includes the steps associated with the synthesis of the circuit that implements the desired functionality. This constructive aspect needs to be complemented by a verification methodology required to assure, with a high degree of confidence, the validity of the design before it is committed to silicon. The design and verification methodologies are implemented with the help of design tools. These tools aid in carrying out design and verification steps, i.e., producing mask layout, model extraction, and the various performance verification steps.

Typical specifications which must be met by RF ICs and, therefore, supported by the design and verification methodologies, include sensitivity, linearity, adjacent channel interference, and power level. These specifications depend on other performance measures such as noise figure, intercept point, and 1dB compression point. Verification tools need to be able to analyze the design at its various stages and predict the performance measures as accurately as possible.

A number of factors characterize RF-IC applications and render the verification process particularly difficult. One is the extreme range of frequencies, or time-scales, over which the circuit is operating. Typical applications have carrier frequencies in the 1-2 GHz range with modulating signals in the KHz range, a span of 6 orders of magnitude. The signals have a tremendous dynamic range. A radio will often have 80 dB of gain in the receiver path. Signals that must be received may be 60 dB weaker than the ones in adjacent frequency bands. Attenuations of more than 100dB for unwanted interferences are not unheard-of. At GHz frequencies, the passive components of an RF circuit are very significant and must be carefully modeled. Coupling through the interconnect and the substrate are prone to cause problems such as undesired interference (cross-talk) from adjacent structures. At these high frequencies, capacitors, inductors, on-chip wires, the silicon substrate, package, sockets, etc., must be analyzed at the electromagnetic level in order to capture and model all the significant effects. Finally, device noise determines the fundamental bounds of circuit performance and plays a significant role in RF circuit design. Device noise (flicker noise, shot noise, thermal noise, etc.) is a stochastic phenomenon resulting from the physics of the device and should be distinguished from deterministic phenomena such as crosstalk.

The verification methodology must rely on tools that are appropriate for the problems enumerated above. These tools must be efficient enough to permit a sufficiently extensive verification of the design. In some cases, the capability to perform more simulation results in the uncovering and fixing of more design flaws, thus avoiding costly processing iterations.

The wide range of operating frequencies imposes an insurmountable burden on the SPICE-type, time-domain simulators familiar to IC designers. It would take the simulation of millions of carrier cycles to analyze only one period of the modulating signal. Therefore novel, multi-scale simulation techniques must be employed. One such technique, the method of harmonic balance, is widely used and very familiar to microwave RF designers. Harmonic balance analysis is suitable for the analysis of circuits operating at widely separated frequencies and provides the required dynamic range. Unfortunately, the existing commercial implementations of harmonic balance simulators rely on algorithms that are not capable of handling the large circuits, consisting mainly of nonlinear elements, which characterize typical RF-IC applications. Moreover, harmonic balance cannot address all RF-IC verification problems of interest. Section 2 surveys an entire family of multi-scale simulation techniques (including harmonic balance) capable of evaluating all RF-IC circuit performance measures of interest. A novel implementation of harmonic balance suitable for the simulation of a full RF chip is described.

Device noise being among the most important design considerations, any verification methodology must rely on adequate tools that simulate the effects of noise in RF circuits. Noise sources and signals in RF circuits are modulated by time-varying signals and can only be modeled by cyclo-stationary and non-stationary stochastic processes. Analyzing the effect of noise in oscillators (phase noise/timing jitter) is particularly difficult. Analyzing the effects of phase noise in autonomous oscillators is described in Section 3.

The accurate analysis of passive linear structures at GHz frequencies must rely on the solution of the full set of Maxwell's equations, in all three dimensions, on domains having varied and complicated geometries. Section 4 surveys the techniques available for this problem and describes in detail a novel, very efficient method-of-moments approach. The models resulting from the analysis of the linear structures must be combined with the active components into a comprehensive simulation of the RF circuit.

DAC 98, San Francisco, California
©1998 ACM 0-89791-964-5/98/06..$5.00

Unfortunately some of the linear analysis tools can only produce models in the frequency domain. Of all the analysis methods, only harmonic balance can naturally handle these models. For other types of analysis, the linear model must be translated into a form suitable for time-domain simulations. The methodology to compute reduced-order models that are compatible with both time-domain and frequency-domain simulations is described in Section 5.

2 Circuit Simulation for RF ICs

Simulating RF ICs places new demands on transistor-level simulation tools: (1) The need to find steady-state and transient/envelope response with stimuli at widely separated frequencies or time-scales. Mixer simulation and two-tone intermodulation studies are classic examples of this scenario. (2) The need for large dynamic range in numerical results; Radio designs commonly deal with intentional signals separated in amplitude by 60-80 dB. Accurate prediction of spurious signals and feed-through requires a dynamic range in excess of 100 dB. Ideally, the numerical methods used should not be the limiting factor. (3) The simulation should be accomplished with reasonable computer resources.

The method of *harmonic balance* (HB) [16, 24] is already familiar to microwave RF designers and is becoming more well known among RF IC designers. Section 2.1 contains a discussion of the application of HB to a complete RF transmitter chip.

A new formulation, introduced even more recently, provides a general mathematical framework for the aforementioned issues. The *multi-rate partial differential equation (MPDE)* technique [4, 39, 40] represents signals as functions of more than one time variable. HB is a special case of the MDPE, which also allows multi-tone analysis of strongly nonlinear circuits (in fact, non-RF circuits such as power converters and switched-capacitor filters can also be treated effectively with the MPDE). Section 2.2 contains a discussion of the MPDE for strongly nonlinear RF circuits.

2.1 Harmonic Balance

The method of Harmonic Balance represents all circuit waveforms in the frequency domain. The method is particularly natural in the case of incommensurate multi-tone drive [45]. Early implementations focused on microwave circuitry [29], which often has a relatively small number of nonlinear components embedded in a large collection of linear elements. Unfortunately, RF integrated circuits do not really fit this model, since sophisticated semiconductor device equations require nonlinear modeling of the majority of components. Recent work by various authors [3, 10, 28, 31, 37] has demonstrated that Harmonic Balance can handle integrated designs containing many more nonlinear components than traditional implementations of the technique. Specifically, iterative linear algebra techniques [12] have been used to solve the large Jacobian matrix which results from linearization of the nonlinear equations.

Of course a large circuit can always be simulated in pieces. However, there are several disadvantages to this approach: (1) Circuit partitioning is inconvenient and error-prone, especially for extracted networks. (2) It is often difficult to preserve source and load impedances across partition boundaries. Nonlinear effects like load pull can be missed entirely. (3) Leakage and feed-through are a crucial concern in integrated RF designs and, by definition, cross module boundaries.

Figure 1 shows a typical output spectrum from a Harmonic Balance simulation. The simulated circuit was a large dual-conversion quadrature modulator chip designed for cellular applications. The left-most spectral component is a weak, but significant, spurious response from the local oscillator which might

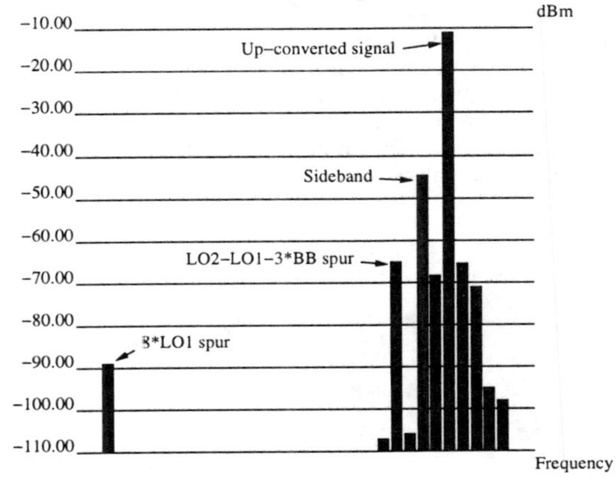

Figure 1: Modulator in-band spectrum

or might not be within the specifications of the chip. The spectrum display also shows a sideband component at a level of -35 dBc which was out of specification for this design, and was traced back to a layout imbalance. This effect was missed during conventional transient analysis. This was a big job—about 27 hours on a fast scientific workstation with approximately 500MB of memory—but still significantly faster and more useful than a transient analysis.

To summarize the significance of Harmonic Balance simulation for this example:

- The large range in driving frequencies [80 KHz and 1.62 GHz] would require a conventional transient analysis to run for several hundred *thousand* cycles. A conventional transient run *was* performed on this design, but with the base-band frequency set to 1 MHz (rather than 80 KHz) and required approximately the same amount of time as the Harmonic Balance run using the appropriate base-band frequency.
- The numerical dynamic range of the transient simulation was insufficient to pick up a weak spurious response at -78 dBc.

On the other hand:

- The memory and time required for Harmonic Balance simulation increase rapidly as more "tones" are added, i.e., driving sources at frequencies which are not in a simple harmonic relationship to one another. For example, predicting the intermodulation distortion of the entire modulator chain would require two different fundamental frequencies at base-band for a total for *four* tones; such a simulation would probably exceed available memory for this design, even with the memory compression techniques which were used in this example.
- Within the bounds of available numerical dynamic range, the time and memory requirements of transient simulation are not sensitive to the number of fundamental frequencies applied to the circuit.

2.2 Multi-Time Methods

The key to the MPDE formulation is the use of multivariate functions (functions of several time variables) to represent signals with separated time scales efficiently. To understand the concept, consider the product of a 1 Hz sine wave and a 1Ghz pulse train, given by:

$$y(t) = \sin(2\pi t)\ \text{pulse}\left(\frac{t}{10^9}\right) \qquad (1)$$

Figure 2 depicts $y(t)$, with the pulse period of 10^9 changed to 50 for viewing convenience. This quasi-periodic signal is expensive to represent in the time domain because 10^9 pulses of different shapes need to be sampled before the waveform repeats. It is this problem that makes traditional time-domain techniques like SPICE's transient analysis inefficient for such signals. Representation in the frequency domain as a two-tone signal is also inefficient because the pulses require many Fourier components for accuracy.

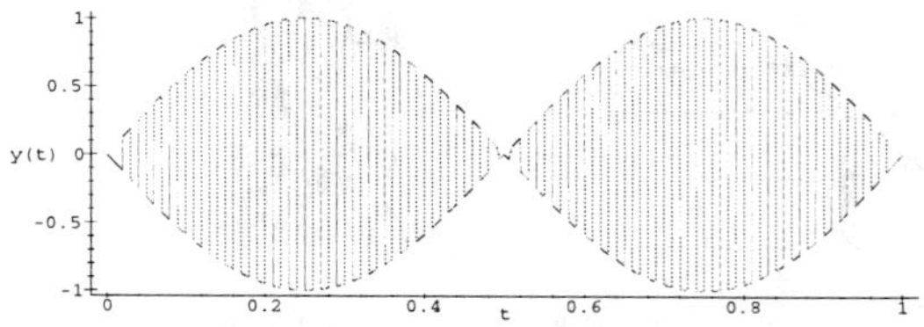

Figure 2: $y(t)$

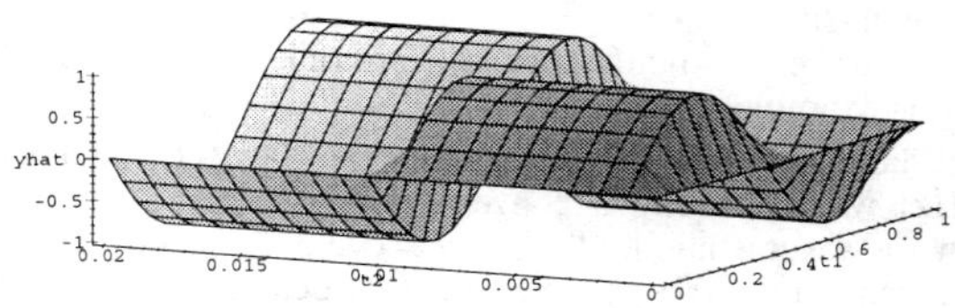

Figure 3: $\hat{y}(t_1,t_2)$

Consider, however, the function of two variables obtained by replacing the 'slow' time component by t_1 and the 'fast' time component by t_2:

$$\hat{y}(t_1,t_2) = \sin(2\pi t_1)\ \text{pulse}\left(\frac{t_2}{10^9}\right) \tag{2}$$

$\hat{y}(t_1,t_2)$, a *bi-variate form* of $y(t)$, is shown in Figure 3. Notice that it is easy to represent $\hat{y}$ accurately using relatively few numerical samples, in contrast to $y(t)$ in Figure 2. The number of samples does not depend on the separation of the two time scales, which merely determines the scaling of the axes. Moreover, $y(t)$ can be easily obtained by interpolation from samples of $\hat{y}(t_1,t_2)$, using the fact that $y(t) = \hat{y}(t,t)$ and that $\hat{y}(t_1,t_2)$ is periodic in each argument.

This observation is the basis of the MPDE formulation, in which all the waveforms in a circuit are represented in their bivariate forms (or multivariate forms if there are more than two time scales). The key to efficiency is to solve for these waveforms directly, without involving the numerically inefficient one-dimensional forms at any point. To do this, it is necessary to first describe the circuit's equations using the multivariate functions. The traditional form of a circuit's equations, used in all simulators, is the Differential-Algebraic Equation (DAE):

$$\dot{q}(x) + f(x) = b(t) \tag{3}$$

$x(t)$ is the vector of circuit unknowns (node voltages and branch currents); q denotes the charge/flux terms and f the resistive terms; $b(t)$ is the vector of excitations to the circuit (typically from independent voltage/current sources). It can be shown [40] that if $\hat{x}(t_1,t_2)$ and $\hat{b}(t_1,t_2)$ denote the bi-variate forms of the circuit unknowns and excitations, then the following MPDE is the correct generalization of (3) to the bi-variate case:

$$\boxed{\frac{\partial q(\hat{x})}{\partial t_1} + \frac{\partial q(\hat{x})}{\partial t_2} + f(\hat{x}) = \hat{b}(t_1,t_2)} \tag{4}$$

More precisely, if $\hat{b}$ is chosen to satisfy $b(t) = \hat{b}(t,t)$, and $\hat{x}$ satisfies (4), then it can be shown that $x(t) = \hat{x}(t,t)$ satisfies (3). Also, if (3) has a quasi-periodic solution, then (4) can be shown to have a corresponding bi-variate solution.

By solving the MPDE numerically in the time domain, strong nonlinearities can be handled efficiently. The following new methods have been developed for solving (4):

1. **Quasi-periodic time-domain methods (MFDTD and HS):** Quasi-periodic solutions are found by enforcing bi-periodic boundary conditions on the MPDE. In the Multivariate Finite Difference Time Domain (MFDTD), (4) is discretized on a grid in the t_1-t_2 plane by approximating the differentiation operators with a numerical differentiation formula. The resultant system of nonlinear equations, together with the bi-periodic boundary conditions, is solved using a nonlinear solution method. The grid is refined adaptively so that the solution is captured efficiently. Another purely time-domain method, Hierarchical Shooting (HS), is a generalization of the traditional shooting method to multiple time scales. Both MFDTD and HS are appropriate for circuits with no sinusoidal waveform components, such as power converters.
2. **Quasi-periodic mixed frequency/time method (MMFT):** In some circuits, the slow-scale signal path is often almost linear, while the fast-scale action is highly nonlinear. The linearity of the signal path can be exploited by expressing the slow scale components in a short Fourier series, and solving the mixed frequency/time system of equations. This Multivariate Mixed Frequency Time (MMFT) method is often more efficient for switched-capacitor filters and switching mixers.
3. **Time domain envelope methods (TD-ENV):** Envelope-type solutions can be generated from the MPDE by applying mixed initial/periodic boundary conditions. Novel time-domain methods based on FDTD or shooting along the fast time scale, and transient integration along the slow time scale, have been devised [40]. These techniques are capable of handling circuits with nonlinearities on a fast time scale, e.g., power converters, switched-capacitor filters, switching mixers, etc..

The above numerical techniques generate sparse matrices with near diagonal or block-diagonal structure, which makes it convenient to use iterative linear solution methods (e.g., [10, 31, 41])to solve large circuits efficiently.

The application of MMFT to a double-balanced switching mixer and filter circuit is described below. The RF input to the mixer was a 100kHz sinusoid with amplitude 100mV; this sent it into a mildly nonlinear regime. The LO input was a square wave of large amplitude (1V), which switched the mixer on and off at a fast rate (900Mhz).

The circuit was also simulated by univariate shooting for comparison. For MMFT, 3 harmonics were taken in the RF tone $f_1 = 100\text{kHz}$ (corresponding to the t_1 variable). The LO tone at $f_2 = 900\text{MHz}$ was handled by shooting in the t_1 variable. The output of the algorithm is a set of time-varying harmonics that are periodic with period $T_2 = \frac{1}{f_2}$. The first harmonic is shown in Figure 4(a). This plot contains information about all mix components of the form $f_1 + if_2$, i.e., the frequencies 900.1 Mhz, 1800.1 Mhz, etc. The main mix component of interest, 900.1 Mhz, is found by taking the fundamental component of the waveform in Figure 4(a). This has an amplitude of 60mV.

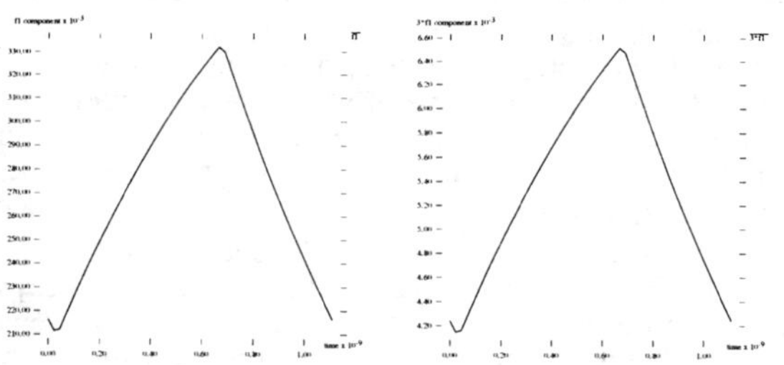

(a) First harmonic (b) Third harmonic

Figure 4: Switching Mixer: MMFT output

The third harmonic is shown in Figure 4(b). It contains information about the mixes $3f_1 + if_2$, i.e., the frequencies 900.3 Mhz, 1800.3 Mhz, etc.. The amplitude of the 900.3 Mhz component is about 1.1mV; hence the distortion introduced by the mixer is about 35dB below the desired signal.

The output produced by univariate shooting is shown in Figure 5. This run, using 50 steps per fast period, took almost 300 times as long as the new algorithm.

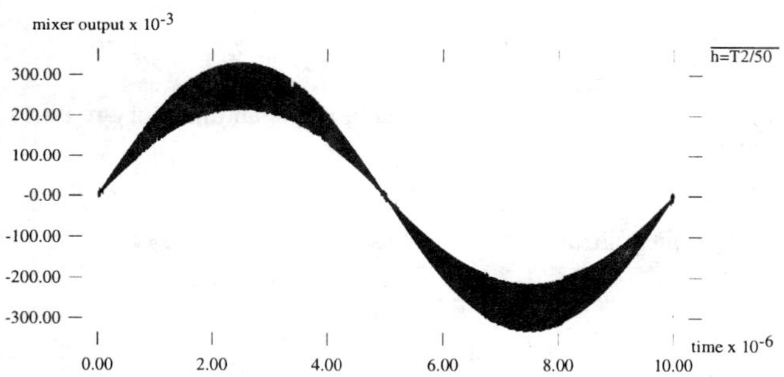

Figure 5: Mixer output from univariate shooting

3 Phase Noise in Oscillators

Oscillators are ubiquitous in physical systems, especially electronic and optical ones. In RF communication systems, they are used for frequency translation of information signals and for channel selection.

Noise is of major concern in oscillators, because introducing even small noise into an oscillator leads to dramatic changes in its frequency spectrum and timing properties. This phenomenon, peculiar to oscillators, is known as phase noise or timing jitter. A perfect oscillator would have localized tones at discrete frequencies (i.e., harmonics), but any corrupting noise spreads these perfect tones, resulting in high power levels at neighboring frequencies. This effect is the major contributor to undesired phenomena such as interchannel interference, leading to increased bit-error-rates (BER) in RF communication systems. Characterizing how noise affects oscillators is therefore crucial for practical applications. The problem is challenging, since oscillators constitute a special class among noisy physical systems: their autonomous nature makes them unique in their response to perturbations.

Considerable effort has been expended over the years [1, 18, 23, 25, 27, 30, 38, 46] in understanding phase noise and in developing analytical, computational and experimental techniques for its characterization. Despite the importance of the problem and the large number of publications on the subject, a consistent and general treatment, and computational techniques based on a sound theory, appear to be still lacking. We developed a novel, rigorous theory for phase noise which leads to efficient numerical methods for its characterization [5]. Our techniques and results are general; they are applicable to any oscillatory system, electrical or otherwise.

Understanding how perturbations affect stable oscillators is a crucial step in the analysis of phase noise. A nonlinear perturbation analysis that is valid for oscillators is required, in contrast to traditional linear perturbation analysis, which is not consistent for oscillators and results in non-physical predictions. With a rigorous nonlinear perturbation analysis, one can show that the effect of perturbations on an oscillator's response can be represented by a changing time shift, or phase deviation, in the periodic output of the unperturbed oscillator, and an additive component, called the orbital deviation, to the phase-shifted oscillator waveform. Moreover, the phase deviation and the orbital deviation can always be chosen such that the phase deviation will, in general, keep increasing with time even if the perturbation is always small, but the orbital deviation will always remain small. These results formalize existing intuition among designers about oscillator operation.

Considering random noise perturbations (e.g. thermal noise, $1/f$ noise), one sees that jitter and spectral spreading are closely related, and both are determined by the manner in which the phase deviation, a random process, spreads with time. The average spread of the jitter (mean-square jitter) increases without bound (precisely linearly for shot and thermal noise) with time. The power spectrum of the perturbed oscillator has a finite value at the carrier frequency and its harmonics, and the total carrier power is preserved despite spectral spreading due to noise. Previous analyses based on linear time-invariant (LTI) or linear time-varying (LTV) concepts erroneously predict infinite noise power density at the carrier, as well as infinite total integrated power. Furthermore, one can show that the oscillator's output is stationary. This might be surprising at first sight, since oscillators are nonlinear systems with periodic swings. Hence it might be expected that output noise power would change periodically as in forced systems. However, it must be remembered that while forced systems are supplied with an external time reference (through the forcing), oscillators are not. Cyclostationarity in the oscillator's output would, by definition, imply a time reference. Hence the stationarity result reflects the fundamental fact that noisy autonomous systems cannot provide a perfect time reference.

We obtain not only the above qualitative characterizations from a rigorous theory of phase noise in oscillators, but also correct computational techniques that are efficient for practical circuits. New numerical methods (in the time and frequency domains) for jitter/spectral dispersion, require only a knowledge of the steady state of the unperturbed oscillator and the values of the noise generators. Large circuits are handled efficiently. The separate contributions of noise sources, and the sensitivity of phase noise to individual circuit devices and nodes, can be obtained easily. We used the theory and numerical methods to analyze several oscillators, and compared the results against measurements. We obtained good matches even at frequencies close to the carrier, unlike most previous analyses.

4 Accurate Extraction

Extracting compact, accurate linear models for packages, interconnect, and components plays a significant role in modern RF designs. Models can be extracted in a variety of ways, but for the high accuracy that critical sections of RF designs demand, only direct numeric simulation suffices. At lower frequencies (far from resonance), capacitive or inductive coupling can be extracted with electrostatic or magnetostatic simulations. These simulations are generally simpler than full electromagnetic simulations, but at high frequencies (near resonances), electromagnetic simulation must be used. In either case, the problem is ultimately reduced to that of solving a linear system of equations $Ax = b$. This is accomplished by discretizing the physical structure into small elements. Interactions between elements give rise to the matrix A. Stimuli such as excitation voltages are reflected in the right-hand side b. Output from the simulator is typically an S parameter matrix, which can be used directly in a frequency-domain simulation. Alternatively, a circuit model can be constructed, using either parameter fitting or the model reduction techniques described in Section 5.

Simulation methods can be divided into two classes based on the type of matrix A that is involved. Methods in the first class use differential equation formulations. Finite-element (FE) [11], finite-difference (FD) [43], and finite-difference time-domain (FDTD) [44] approaches all fall into this class. In all of these, the matrix A is sparse. Methods from the second class use integral equations. The method-of-moments (MoM) approach [19] is based on integral equations. In this case, A is a dense matrix. However, an integral equation formulation allows us to apply Green's theorem, reducing volume integrals to surface integrals. This can reduce the matrix dimension significantly since the discretization only involves surfaces such as the boundary of a conductor or the interface between two dielectrics. The characteristics of the two classes are summarized in Table 1. Commercial tools such as Raphael and Ansoft are based on differential equations, while others such as Sonnet and Momentum use integral equations. For simulations that involve difficult-to-describe material variations (e.g., the doping profile of a MOSFET), the differential approach is clearly superior. For typical IC, board, or MCM simulations, where the material variation is simpler, the integral approach becomes attractive due to the use of surface discretizations. In these cases, the integral formulation often reduces the problem size by orders of magnitude. This reduction, combined with the advanced numerical methods described below, can yield MoM solvers that are ten or more times faster than finite-difference and finite-element tools. Thus, we believe that in the future integral equation methods will be the approach of choice for these problems.

	Differential	Integral
Matrix type	sparse	dense
Discretization	volume	surface
Matrix conditioning	poor	good

Table 1: Characteristics of classes of simulation methods

In recent years, numeric methods have been developed where the dense matrix A arising from an integral equation formulation can be represented implicitly and concisely. With these methods, the size of the representation for A is only $O(n)$ or $O(n \log n)$, where n is the dimension of the matrix. This is a significant reduction compared to the $O(n^2)$ storage that would be required to store A directly. Further, multiplication by A can be accomplished in time proportional to the size of the representation. Because integral equations usually give a relatively well-conditioned matrix A, Krylov-subspace iterative methods [41] can be used to quickly solve the linear system.

FastCap [33] and FastHenry [20] were the first electrostatic extraction tools based on this methodology. Multiplication by the matrix A in these tools is accomplished using the Fast Multipole Method [17]. The main weakness of these tools is that the interaction between discretization elements must have a $1/|\mathbf{r}-\mathbf{r}'|$ dependence, where $\mathbf{r}$ and $\mathbf{r}'$ are the positions of the elements in space. This "kernel" dependence means that all material boundaries must be discretized.

IES^3 [21] is a more recent kernel-independent scheme for compressing the matrix A. With IES^3, the matrix is recursively decomposed and compressed using the singular value decomposition. The interaction between well-separated groups of discretization elements is represented using a low-rank outer product. The interaction need not have a $1/|\mathbf{r}-\mathbf{r}'|$ dependence. In an extraction tool using IES^3, the effect of layered materials can be captured using a specialized Green's function [32]. IES^3 has been used in both electrostatic [21] and electromagnetic [22] simulators. Figure 6 shows how time and memory requirements scale only slightly faster than linearly with increasing problem size in an IES^3-based electromagnetic simulator. Comparisons of electromagnetic simulations to measurements for an integrated CMOS inductor on a lossy substrate are shown in Figure 7. We believe that in the near future, these techniques will make it possible to simulate critical multi-component assemblies such as the resonator shown in Figure 8.

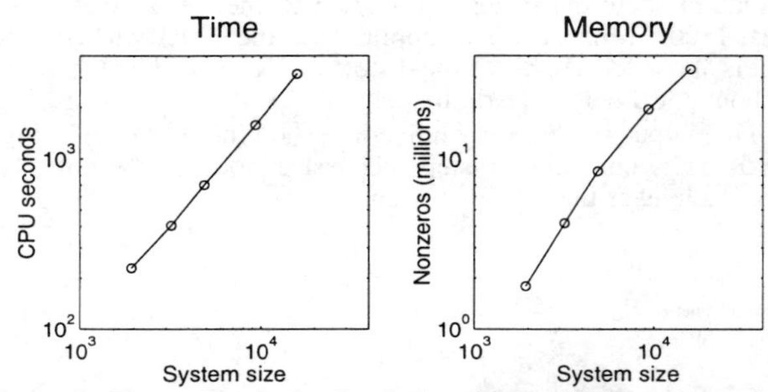

Figure 6: Electromagnetic simulation time and memory requirements

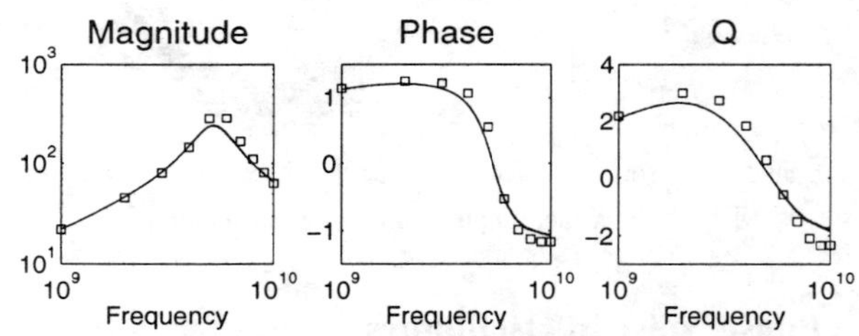

Figure 7: Comparison of inductor simulations and measurements

5 Reduced-Order Modeling

Integrated RF circuits often contain large linear sub-blocks. They can represent passive components such as capacitors, on-chip inductors, tapered RC lines, etc. They can also represent the parasitic effects of interconnect, substrate, power distribution lines, package, MCM, or board.

These linear components are sometimes modeled by equivalent lumped-element circuits composed of resistors, capacitors, inductors, and mutual couplings. Such equivalent models are generated by IC layout extraction tools. The lumped-element circuit model is, however, only adequate when signal wavelengths, (3cm for 10GHz) are large in comparison to circuit feature sizes.

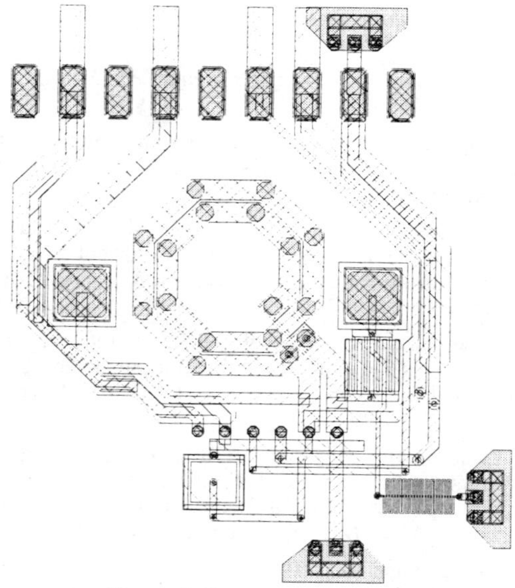

Figure 8: Resonator assembly

At current operating frequencies, 1-2GHz, such methods are adequate only at the chip level. As integrated circuits penetrate higher frequency applications, or even at current frequencies, at the package or printed circuit board levels, lumped-element circuit models become inaccurate. For these cases designers rely on frequency-domain models. These models are obtained either from measurements or, as described in the previous section, from field solvers.

The analysis of RF circuits containing such linear sub-blocks raises a number of problems and difficulties. First, the large size of the linear sub-blocks may render the total size of the problem infeasible for general RF circuit analysis of the type described in Section 2. For example, lumped element networks produced by layout extraction tools can reach sizes of millions of elements. Second, the linear sub-block may be modeled as a frequency-domain transfer function matrix while the rest of the nonlinear RF circuit is modeled as a set of nonlinear, time-domain, differential-algebraic equations. Of all the general RF circuit analysis methods, only the method of Harmonic Balance can efficiently handle a mixture of time-domain and frequency-domain methods.

A solution to both the size and the mixed domain problems is reduced-order modeling. The reduced-order model should capture with sufficient accuracy the interesting behavior of the original linear sub-block in the desired domain of interest, and, simultaneously, be much less expensive to evaluate. In addition, the reduced-order model should have efficient representations in both the time and frequency domains.

A methodology for reduced-order modeling that satisfies the above requirements is based on computing Padé-type approximations [15] to the frequency domain transfer function of the linear sub-block. The Padé approximation is a general method which attempts to approximate a function, which may not have an analytical expression and/or may be very difficult and expensive to compute, by a much simpler rational function. The coefficients of the approximating rational function are chosen such that the original function and the approximant match in the maximum number of the leading Taylor expansion coefficients.

The Padé approximation is particularly attractive for frequency-domain transfer functions because it captures very well the dominant poles and zeros of the linear system. In addition, the matched Taylor coefficients also have physical significance: they represent the moments of the time-domain circuit response. These properties, allow the substitution of the Padé approximant in the place of the original circuit with little significant loss of accuracy, but substantial efficiency gains.

The direct computation of Padè approximations [35, 36] is numerically unstable. Therefore, the preferred methods for the computation of Padé approximations of linear systems are Krylov-subspace techniques, in particular, variants of the Lanczos [26] and Arnoldi algorithms [2] which are efficient and numerically stable. Lanczos-type algorithms, as implemented in the Padé via Lanczos (PVL) algorithm [8, 9, 14, 13], produce the most efficient approximations. For the same order of approximation and computational effort they match twice as many moments as the Arnoldi algorithm [6, 34, 42]. However, in certain cases, Lanczos-based methods may produce non-passive reduced-order models of passive linear systems. In these cases post-processing is required to enforce the desired properties.

The Krylov subspace based reduced-order modeling algorithms produce one or two small matrices. These matrices can be used either to efficiently evaluate the linear sub-block's transfer functions at any desired frequency, or to formulate a small system of linear differential equations which model its time-domain behavior, and which can be solved in conjunction with the entire RF circuit.

Recently, reduced-order modeling techniques were also applied to the noise analysis problem [7]. The benefit is a significantly more efficient evaluation of noise power over a wide range of frequencies. Moreover, the entire noise behavior of a circuit block is captured in a compact form and can be used hierarchically in system-level simulations.

Conclusion

We have described powerful new techniques for the analysis of RF circuits. Next-generation CAD tools based on such techniques should enable RF designers to obtain a more accurate picture of how their circuits will operate. These new simulation capabilities will be essential in order to reduce the number of design iterations needed to produce complex RF ICs.

References

[1] A.A. Abidi and R.G. Meyer. Noise in relaxation oscillators. *IEEE Journal of Solid-State Circuits*, December 1983.

[2] W. E. Arnoldi. The principle of minimized iterations in the solution of the matrix eigenvalue problem. *Quart. Appl. Math*, 9:17–29, 1951.

[3] H. Brachtendorf, G. Welsch, and R. Laur. A simulation tool for the analysis and verification of the steady state of circuit designs. *Int. J. of Circuit Theory and App.*, 23:311–32, 1995.

[4] H.G. Brachtendorf, G. Welsch, R. Laur, and A. Bunse-Gerstner. Numerical steady state analysis of electronic circuits driven by multi-tone signals. *Electrical Engineering (Springer-Verlag)*, 79:103–112, 1996.

[5] A. Demir, A. Mehrotra, and J. Roychowdhury. Phase noise in oscillators: A unifying theory and numerical methods for characterisation. In *ACM/IEEE Design Automation Conference*, 1998.

[6] I.M. Elfadel and D.D. Ling. Zeros and passivity of arnoldi-reduced-order models for interconnect networks. In *Proc. 34nd ACM/IEEE Design Automation Conf*, 1997.

[7] P. Feldmann and R. W. Freund. Circuit noise evaluation by pade approximation based model reduction techniques. In *Tech. Dig. 1997 IEEE/ACM International Conference on Computer-Aided Design*, November 1997.

[8] P. Feldmann and R.W. Freund. Efficient linear circuit analysis by padé approximation via the lanczos process. *IEEE Trans. Computer-Aided Design*, 14:639–649, May 1995.

[9] P. Feldmann and R.W. Freund. Reduced-order modeling of large linear subcircuits via a block lanczos algorithm. In *Proc. 32nd ACM/IEEE Design Automation Conf*, pages 474–479, 1995.

[10] P. Feldmann, R. Melville, and D. Long. Efficient frequency domain analysis of large nonlinear analog circuits. In *Proc. of CICC*, 1996.

[11] R. L. Ferrari and P. P. Silvester. *Finite Elements for Electrical Engineers*. Cambridge Univ. Pr., 1996.

[12] R. Freund, G. Golub, and N. Nachtigal. Iterative solution of linear systems. *Acta Numerica*, pages 57–100, 1991.

[13] R. W. Freund and P. Feldmann. Reduced-order modeling of large linear passive multi-terminal circuits using matrix-pade approximation. In *DATE-98*, February 1998.

[14] R.W. Freund and P. Feldmann. Reduced-order modeling of large passive linear circuits by means of the sypvl algorithm. In *Tech. Dig. 1996 IEEE/ACM International Conference on Computer-Aided Design*, pages 280–287, November 1996.

[15] Jr. G.A. Baker and P. Graves-Morris. *Padé Approximants*. Cambridge Univ. Pr., New York, NY, 2nd edition, 1996.

[16] R. Gilmore and M. Steer. Nonlinear circuit analysis using the method of harmonic balance—a review of the art. *Int. J. on Microwave and Millimeter Wave Comp. Aided Eng.*, 1, Jan. 1991.

[17] L. Greengard and V. Rokhlin. A fast algorithm for particle simulations. *J. of Computational Physics*, 73(2):325–348, December 1987.

[18] A. Hajimiri and T.H. Lee. A state-space approach to phase noise in oscillators. Technical memorandum, Lucent Technologies, July 1997.

[19] R. F. Harrington. *Field Computation by Moment Methods*. IEEE Press, New York, 1991.

[20] M. Kamon, M. J. Tsuk, and J. White. FASTHENRY: A multipole-accelerated 3-d inductance extraction program. *IEEE Trans. on Microwave Theory and Techniques*, 42(9):1750–1758, September 1994.

[21] S. Kapur and D. E. Long. IES^3: A fast integral equation solver for efficient 3-dimensional extraction. In *Proc. of the Int. Conf. on Computer-Aided Design*, November 1997.

[22] S. Kapur, D. E. Long, and J. Zhao. Efficient full-wave simulation in layered, lossy media. In *Proc. of the 1998 Custom Integrated Circuits Conf.*, May 1998.

[23] F. Kärtner. Analysis of white and $f^{-\alpha}$ noise in oscillators. *International Journal of Circuit Theory and Applications*, 18:485–519, 1990.

[24] K. Kundert, J. White, and A. Sangiovanni-Vincentelli. *Steady-State Methods for Simulating Analog and Microwave Circuits*. Kluwer, Boston, MA, 1990.

[25] K. Kurokawa. Noise in synchronized oscillators. *IEEE Transactions on Microwave Theory and Techniques*, MTT-16:234–240, 1968.

[26] C. Lanczos. An iteration method for the solution of the eigenvalue problem of linear differential and integral operators. *J. Res. Nat. Bur. Standards*, 45:255–282, 1950.

[27] D.B. Leeson. A simple model of feedback oscillator noise spectrum. *Proceedings of the IEEE*, 54(2):329, February 1966.

[28] M. Rösch and K.J. Antreich. Schnell stationäre Simulation nichtlinearer Schaltungen im Frequenzbereich. *AEÜ*, 46(3):168–176, 1992.

[29] S. Maas. *Nonlinear Microwave Circuits*. Artech House, 1988.

[30] J.A. McNeill. *Jitter in Ring Oscillators*. PhD thesis, Boston University, 1994.

[31] R. Melville, P. Feldmann, and J. Roychowdhury. Efficient multi-tone distortion analysis of analog integrated circuits. In *Proc. of CICC*, 1995.

[32] K.A. Michalski and J.R. Mosig. Multilayered media Green's functions in integral equation formulations. *IEEE Transaction on Antenna and Propagation*, 45(3):508–519, March 1997.

[33] K. Nabors and J. White. Multipole-accelerated capacitance extraction for 3-d structures with multiple dielectrics. *IEEE Trans. on Circuits and Systems*, 39(11):946–954, November 1992.

[34] A. Odabasioglu, M. Celik, and L.T. Pileggi. Prima: Passive reduced-order interconnect macromodeling algorithm. In *Tech. Dig. 1997 IEEE/ACM International Conference on Computer-Aided Design*, November 1997.

[35] L.T. Pillage and R.A. Rohrer. Asymptotic waveform evaluation for timing analysis. *IEEE Trans. Computer-Aided Design*, 9:352–366, April 1990.

[36] V. Raghavan, R.A. Rohrer, L.T. Pillage, J.Y. Lee, J.E. Bracken, and M.M. Alaybeyi. Awe–inspired. In *Proc. IEEE Custom Integrated Circuits Conference*, May 1993.

[37] V. Rizzoli, A. Costanzo, P. Ghigi, F. Mastri, D. Masotti, and C. Cecchetti. Recent advances in harmonic-balance techniques for nonlinear microwave circuit simulation. *AEÜ*, 46(4):286–296, 1992.

[38] U.L. Rohde. *Digital PLL Frequency Synthesizers: Theory and Design*. Prentice-Hall, 1983.

[39] J. Roychowdhury. Efficient Methods for Simulating Highly Nonlinear Multi-Rate Circuits. In *Proc. IEEE Design Automation Conference*, 1997.

[40] J. Roychowdhury. Analysing Circuits with Widely-Separated Time Scales using Numerical PDE Methods. *IEEE Trans. Ckts. Syst. – I: Fund. Th. Appl.*, 1998. To appear.

[41] Y. Saad. *Iterative methods for sparse linear systems*. PWS, Boston, 1996.

[42] L.M. Silveira, M. Kamon, I. Elfadel, and J. White. A coordinate-transformed arnoldi algorithm for generating guaranteed stable reduced-order models of rlc circuits. In *Tech. Dig. 1996 IEEE/ACM International Conference on Computer-Aided Design*, pages 288–294, November 1996.

[43] G. D. Smith. *Numerical Solution of Partial Differential Equations: Finite Difference Methods*. Oxford Univ. Pr., 1986.

[44] A. Taflove. *Computational Electrodynamics: The Finite-Difference Time-Domain Method*. Artech House, 1995.

[45] A. Ushida and L. Chua. Frequency domain analysis of nonlinear circuits driven by multi-tone signals. *IEEE Trans. on Circuits and Sys.*, CAS-31:766–779, Sep. 1984.

[46] T.C. Weigandt, B. Kim, and P.R. Gray. Analysis of timing jitter in cmos ring-oscillators. In *Proc. IEEE ISCAS*, June 1994.

Electromagnetic Modeling and Signal Integrity Simulation of Power/Ground Networks in High Speed Digital Packages and Printed Circuit Boards

Frank Y. Yuan
Viewlogic Systems Group, Inc.
1385 Del Norte Road
Camarillo, California 93010, USA
fyuan@viewlogic.com

1. ABSTRACT

The electromagnetic modeling and parameter extraction of digital packages and PCB boards for system signal integrity applications are presented. A systematic approach to analyze complex power/ground structures and simulate their effects on digital systems is developed. First, an integral equation boundary element algorithm is applied to the electromagnetic modeling of the PCB structures. Then, equivalent circuits of the power/ground networks are extracted from the EM solution. In an integrated simulation scheme, the equivalent circuits are combined with signal nets, package models, device circuits, and other external circuitry for system level signal integrity analysis and simulation. This methodology has been implemented as software tools and applied to practical design problems. Effects related to power/ground networks, such as simultaneous switching noises, crosstalk, and ground discontinuity are analyzed for realistic designs.

2. INTRODUCTION

With the ever increasing speed and density of digital integrated circuit systems, effects due to field interaction among IC chips, packages and PCB boards have become more and more the limiting factors in high speed system design [1-3]. Signal integrity effects such as propagation delay, crosstalk, and simultaneous switching noises (SSN) require more accurate and efficient electromagnetic analysis and modeling beyond the traditional static method with a few lumped circuit elements. Particularly, the power/ground supply networks in digital circuits involve complex, large structures such as combinations of plane/partial planes and routed traces in multi-layer environment, and are often tightly coupled with other parts of the circuit systems, such as signal networks and chip packages. Characterization of true distributed effect and frequency dependence is crucial. On the other hand, due to the overwhelming complexity of digital circuit board, full-wave rigorous electromagnetic field solutions, such as those used in microwave analysis [4], become impractical for digital application because of the extremely high computational requirement, and the difficulty of integrating with time-domain circuit type simulators and design environments used by digital designers. Therefore, EM modeling and simulation methodologies that can handle large complex digital designs with reasonable accuracy and efficiency, and can provide good macro models over extended frequency bands, are very much desired and thus present a challenge to us [5-8].

This paper presents a systematic approach which includes an efficient method for the electromagnetic modeling of packages and power/ground structures, and an integrated process for the analysis and simulation of system signal integrity and SSN effect in practical digital designs. The focus of this work is to develop a fast, flexible and accurate electromagnetic model with a clear circuit interpretation for the power/ground network, and to integrate it with active device models, transmission line networks, and chip package parasitics in a time domain co-simulation scheme. This methodology has been implemented and applied to practical design problems. It should be able to handle the complexity of real IC/MCM/PCB designs within the practical computational constraints of an engineering workstation environment [7,8].

The paper is organized as follows: In Section 3, the electromagnetic modeling and analysis is presented. The general interconnects and power/ground structures are modeled as a multi-layered dielectric embedded with conductors. A mixed potential integral equation with layered Green's function is applied to the electromagnetic modeling of the structure, and is solved numerically using a boundary element approach. Section 4 describes the extraction of distributed equivalent circuits. A quasi-static approximation is introduced to simplify the complex frequency dependence of the power distribution system, while retaining the important AC characteristics. From the electromagnetic field solutions, equivalent circuits are constructed with frequency independent R, L, C circuit elements. This greatly reduces the required computational time, compared to a full wave solution, and makes a time-domain digital circuit simulation practical for real design applications. The distributed equivalent circuit models give accurate high frequency characteristics of the system up to a certain frequency limit that is well above most digital signal bandwidth, and can be used as macromodels for the power/ground supply networks. Section 5 describes the integrated simulation approach for system level signal integrity and SSN analysis. First, an efficient circuit solver is specifically developed for the equivalent circuit. A completely designed digital PCB system is partitioned into four major subsystems: chip devices such as drivers/receivers; chip packages; signal nets in the form of transmission lines; and power/ground supply nets in the form of equivalent circuits. Appropriate methods have been applied to model the different subsystems, and all the subsystems are linked together dynamically in a time-domain co-simulation. In Section 6, various examples are given to verify the approach against measurements and published results. Then, simultaneous switching noises and the effects of decoupling capacitors are analyzed in practical designs.

3. ELECTROMAGNETIC MODELING AND ANALYSIS

3.1 Electromagnetic Formulation

Consider the general structure of integrated circuits and printed circuit boards as multi-layer dielectric planar substrate embedded with signal traces and power/ground planes as arbitrarily shaped thin conductors. The planes and signal traces are connected to each other and to external power supplies through vias or ground pins. Currents injected by active devices and/or power voltages applied to the conductors produce induced charges and currents on the conductor surfaces, which in turn generate induced electromagnetic fields. The electric field $\mathbf{E}(\mathbf{r})$ can be expressed in terms of a vector potential $\mathbf{A}(\mathbf{r})$ and a scalar potential $\Phi(\mathbf{r})$ as

$$E(r) = -j\omega A(r) - \nabla\Phi(r) \tag{1}$$

where $\mathbf{A}(\mathbf{r})$ and $\Phi(\mathbf{r})$ are related to the charge and current densities $q(\mathbf{r})$ and $\mathbf{J}(\mathbf{r})$ by the following relation:

$$A(r) = \mu \int_V dr' G_A(r, r') \bullet J(r') \tag{2}$$

$$\Phi(r) = \frac{1}{\varepsilon} \int_V dr' G_\phi(r, r') q(r') \tag{3}$$

In the above equations, $\mathbf{G}_A$ and G_ϕ are the dyadic and scalar Green's functions for the vector and scalar potentials in the layered structure, respectively. The Green's functions have been derived analytically in forms of spectral integrals in general, and as infinite series for some special cases.

Characterizing the lossy conductor by its surface impedance Z_s, then the total electric field and surface current density $\mathbf{J}_s(\mathbf{r})$ satisfy the following impedance boundary condition over the conductor surfaces:

$$\mathbf{n} \times E_s(r) = Z_s \mathbf{n} \times J_s(r) \tag{4}$$

Also, the surface charge $q_s(\mathbf{r})$, surface current $\mathbf{J}_s(\mathbf{r})$, and the external current flowing into the ports of the conductors, $\mathbf{J}^i$, satisfy the general continuity relation:

$$\nabla \bullet J_s(r) + j\omega q_s(r) = \nabla \bullet J^i(r) \tag{5}$$

Combining the boundary condition with continuity equation, we have the following set of mixed potential integral equations:

$$Z_s \mathbf{n} \times J_s(r) + \mathbf{n} \times j\omega\mu \int_S dr' G_A(r, r') \bullet J(r') + \mathbf{n} \times \nabla\Phi(r) = 0 \tag{6}$$

$$\nabla \bullet J_s(r) + j\omega q_s(r) - \nabla \bullet J^i(r) = 0 \tag{7}$$

$$\Phi(r) - \frac{1}{\varepsilon} \int_S dr' G_\phi(r, r') q_s(r') = 0 \tag{8}$$

where S is the conductor surfaces and $r \in S$. This set of integral equations completely describes the potential, current, and charge distributions on the power and ground conductor planes. By solving these equations, the voltage and current distributions over the conducting planes can be obtained in relation to the injected currents or fixed port potentials for given excitations. Notice that the scalar potential is retained as an explicit variable in the above MPIE's. If we consider the scalar potential as a close approximation to voltage defined in a static sense, the above formulation enables us to obtain an explicit relation between port voltage and injected current, and thus a multi-port network representation or an equivalent circuit for the substrate system.

3.2 Boundary Element Method

A boundary element method is applied to solve the integral equations. The total conductor surface, domain S, is divided into N elementary sub-domains (elements) with quadrilateral shapes. As an example, Figure 1 shows the discretisation of a split MCM power plane where the 3.3 volt power net and the 5 volt power net complement each other. The current, charge and potential over S are expanded in terms of subsectional basis functions defined over each element, as

$$J_s(r) = \sum_i^N I_i T_i \quad q_s(r) = \sum_i^N Q_i \varphi_i \quad \Phi_s(r) = \sum_i^N V_i \phi_i \tag{9}$$

where I_i, Q_i and V_i are the unknown expansion coefficients, and $\mathbf{T}_i$, φ_i and ϕ_i are basis functions for current, charge and potential, respectively. Note that current is a vector quantity, thus $\mathbf{T}_i$ is a vector basis function with both x and y components. There are many possible choices for basis functions. The simplest is pulse function which will be used in the expansions of charge and potential distributions. To ensure the continuity of current, however, a bilinear expansion will be utilized for current expansion. After substitution of the above expansions into the integral equations, an inner product of a test function at each integral equation is taken over each subdomain (testing procedure or sampling). The choice of test functions is also a crucial matter. The simplest choice is a delta function, corresponding to the so-called point matching method. It is computationally fast and simple, but exhibits accuracy and stability problems. A more sophisticated choice utilizes the same basis functions as the test functions, called Galerkin's method, which leads to improved accuracy and stability at the expense of computational requirement. Both methods have been implemented in our simulation program and used when as appropriate.

Following the above boundary element procedure, the integral equations become a set of linear equations (in matrix notation) as

$$(Z_s + j\omega L)I - PV = 0 \tag{10}$$

$$P^T I + j\omega C V = J^i \tag{11}$$

where: $\mathbf{Z}_s$, $\mathbf{L}$, $\mathbf{P}$, and $\mathbf{C}$ are matrices; vector $\mathbf{V}$ represents the potential (voltage) while vector $\mathbf{I}$ represents the surface current on all the conductor surface elements; the element of $\mathbf{J}^i$ vector is the external current injected into each surface element, which except on terminal elements (ports) is zero. The elements of $\mathbf{Z}_s$, $\mathbf{L}$, $\mathbf{P}$, and $\mathbf{C}$ matrices in general involve two dimensional double surface integrals of Green's functions and test functions over the subdomains. Special techniques such as closed form formulas have been applied in the evaluation of those integrals. Depending on the specified boundary (terminal) condition, we can solve for either the unknown port voltages with specified port currents, or the unknown port currents with specified port voltages, along with the charge, surface current and potential all over the conductor planes.

4. QUASI-STATIC APPROXIMATION AND EQUIVALENT CIRCUIT EXTRACTION

4.1 Quasi-Static Approximation

The above integral equation/boundary element approach is a rigorous full-wave analysis as it has been presented up to this point. This formulation of the system, as represented by the matrix equations, has complex frequency dependence. All the matrix elements and operations are frequency dependent and thus need to be recomputed and updated for every single frequency component for wideband or transient applications.

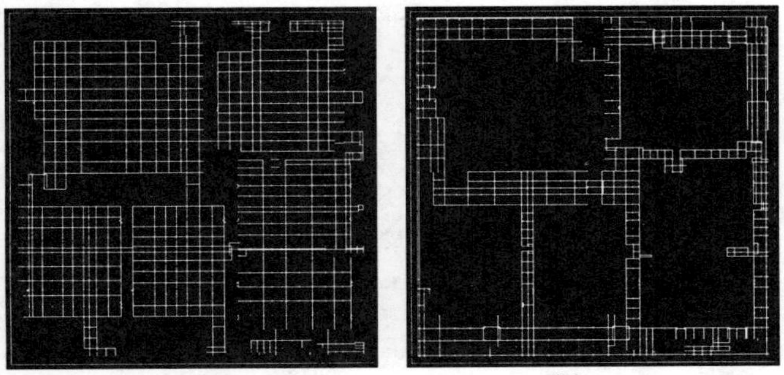

(a) (b)

Figure 1. Split MCM power planes and their discretisation. (a) 3.3 volt VCC0 net, (b) 5.0 volt VCC net. They share a common ground plane with a dielectric layer of 0.15 mm thick.

For a time-domain digital circuit simulation on the scale of MCM/PCB substrates, the above full-wave solution requires a tremendous amount of computation. A simplification in the frequency relation is therefore very much desired. For typical MCM/PCB substrate environments, the following relation provides a guide to the applicability of quasi-static methods:

$$\frac{h}{\lambda} = hf\sqrt{\varepsilon\mu} \ll 1 \qquad (12)$$

where h is the characteristic dimension of the sub-domain structure and f is the operating frequency. Thus for systems satisfying (12) it is reasonable to make the following quasi-static approximation. First, the higher order terms in the exponential retardation factor are neglected in both the vector potential Green's function and the scalar potential Green's function. For example, the free space Green's function $G(r) = exp(-jkr)/4\pi\varepsilon r$ is replaced by the static one as $G(r) = 1/4\pi\varepsilon r$. Under this approximation, the matrix elements in **L** and **C** become real and frequency independent. The only frequency dependence is now the $j\omega$ factor in the matrix equations which will be further simplified. Secondly for good conductors, we have

$$\mathbf{Z}_s + j\omega \mathbf{L} \approx R_s \qquad (\omega \to 0) \qquad (13)$$

$$\mathbf{Z}_s + j\omega \mathbf{L} \approx j\omega \mathbf{L} \qquad (\omega \gg 0) \qquad (14)$$

With this assumption, the surface resistance $\mathbf{Z}_s$ can be considered as the first order low frequency approximation for the total resistance, and the system becomes a pure resistive network at DC. Solving equations (10) and (11) under the above quasi-static approximation, we obtain the system solution in admittance form as $\mathbf{J}^i = \mathbf{Y}\mathbf{V}^i$, where $\mathbf{J}^i$ and $\mathbf{V}^i$ are the port currents and voltages, respectively, and **Y** is the system admittance matrix given by

$$Y = \left[j\omega C + P^T(Z_s + j\omega L)^{-1} P\right] \qquad (15)$$

Let $\mathbf{L}_T = (P^T L^{-1} P)$, and neglect surface loss, then the **Y** matrix has the following form as

$$Y = \frac{1}{j\omega}(L_T)^{-1} + j\omega C \qquad (16)$$

where $(L_T)^{-1}$ denotes a matrix with elements equal to the inverse of the corresponding elements of $\mathbf{L}_T$, not the inverse matrix of $\mathbf{L}_T$.

Similarly, an impedance matrix formula can be obtained under the same quasi-static approximation. Again, from equations (10) and (11) we have the system solution as $\mathbf{V}^i = \mathbf{Z}\mathbf{J}^i$, where

$$Z = \left[j\omega C + P^T(Z_s + j\omega L)^{-1} P\right]^{-1} \qquad (17)$$

Taking a Taylor expansion of (17) in frequency and keeping only the first and second order terms, we obtain the following

$$Z = j\omega(P^T L^{-1} P)^{-1} - (j\omega)^3 (P^T L^{-1} P)^{-1} C (P^T L^{-1} P)^{-1} \qquad (18)$$

Let $\mathbf{L}_R = (P^T L^{-1} P)^{-1}$, we finally have the simplified impedance matrix as

$$Z = j\omega L_R - (j\omega)^3 L_R C L_R \qquad (19)$$

In the above formula, all major matrix operations are frequency independent. More sophisticated expansion and manipulation in the above procedure could include the surface resistance term, and lead to similar results with simple frequency dependence.

4.2 Equivalent Circuit Extraction

From the impedance matrix or the admittance matrix, a distributed equivalent circuit representing the system can be constructed. Since a nodal formulation based on Kirchhoff's current law (KCL) is preferred over a mesh formulation, we consider the extraction of equivalent circuit from the admittance matrix **Y** as in Equation (16). In fact, the admittance matrix is the same **Y** matrix in a nodal formulation of the circuit in frequency domain, and can be used directly for frequency domain solutions [9]. To see more clearly the equivalent circuit in terms of its R, L and C elements, and to use it for more general circuit purposes, note the congruence between the admittance representation in Equation (16) and the frequency domain admittance expression for a parallel LC circuit branch, as given below in Equation (20). The equivalent circuit thus consists of branches between every pair of ports. Each branch has an inductance $\tilde{L}$ in series with a resistance $\tilde{R}$, and then in parallel with a capacitance $\tilde{C}$. A four node equivalent circuit with common ground is shown in Figure 2. The branch admittance between nodes m and n is given by

$$\tilde{Y}_{mn} = \frac{1}{\tilde{R}_{mn} + j\omega\tilde{L}_{mn}} + j\omega\tilde{C}_{mn} \qquad (20)$$

For circuit analysis, write KCL at each node. This leads to the same circuit equation as obtained by rewriting the admittance relation in terms of differential voltages between nodes

$$J_m = \tilde{Y}_{mm}V_m + \tilde{Y}_{1m}(V_m - V_1) + \ldots = \sum_{n=1}^{M} \tilde{Y}_{mn}V_{mn} \qquad (21)$$

The branch admittances $\tilde{Y}_{mn}$ are then determined by comparison with matrix **Y**, as following

$$\tilde{Y}_{mn} = -Y_{mn} \qquad (m \neq n) \qquad (22)$$

$$\tilde{Y}_{mm} = \sum_{n=1}^{M} Y_{nm} \qquad (23)$$

Correspondingly, the $\tilde{L}$ and $\tilde{C}$ values are obtained from the matrix elements in $\mathbf{L}_T$ and **C** as

$$\tilde{L}_{mn} = -L_{mn} \qquad (m \neq n) \tag{24}$$

$$\tilde{C}_{mn} = -C_{mn} \qquad (m \neq n) \tag{25}$$

while

$$1/\tilde{L}_{mm} = 0 \tag{26}$$

$$\tilde{C}_{mm} = \sum_{n=1}^{M} C_{nm} \tag{27}$$

Equation (26) follows from Kirchhoff's current law for the reference node. To a first order approximation, the resistance $\tilde{R}$ takes the corresponding values of the DC resistance matrix obtained through (15) as $(\boldsymbol{P}^T \boldsymbol{Z}_s^{-1} \boldsymbol{P})^{-1}$. For a real design where every external connection, such as power/ground pin, is selected as a circuit node, then the equivalent circuit is a N node network where N is the number of power/ground connections.

Since the equivalent circuit captures the wideband frequency behavior and distributed characteristics, it should correctly predict the wave effects such as propagation delay, resonance, and signal reflection on the power/ground plane structure. Numerical examples show that it indeed captures those effects correctly. However, it is fundamentally limited by the retardation effect, and does not include losses due to radiation.

5. CIRCUITS AND SYSTEM LEVEL SIMULATIONS

5.1 Circuit Solution

The equivalent circuits can be utilized to simulate the response of complex interconnect and power/ground structures, or be used as subcircuits or macromodels in integrated system level simulation. An efficient circuit simulator specifically for the equivalent circuit is developed in our simulation program which provides both time-domain and frequency domain capability. The time-domain circuit solution algorithm is developed based on modified nodal formulation [9]. The linear equivalent circuit system is characterized by

$$\boldsymbol{G}x + \boldsymbol{C}x' = w \tag{28}$$

In the above equation **G** consists of resistive and conductive elements while **C** consists of capacitive and inductive elements. Special formulation of the system equations eliminates the unnecessary internal inductance nodes and therefore greatly reduces the rank of the linear system. Both first order and second order integration methods are used in solving the linear circuit equation, providing good stability and accuracy with speed. Combined with uniform time step for the linear circuit portion, this approach gives us very efficient simulation time and thus enables us to analyze the large scale networks from realistic circuit boards and designs. General purpose circuit simulators such as SPICE can also be used for the simulation, but may not be very efficient.

Frequency domain simulations are useful for gaining insight of high frequency characteristics of the complex systems. In practice, experimental measurements and characterization of high-speed/high-frequency systems are mostly made in frequency domain in terms of S-parameters. It is thus used for verification of the simulation code's accuracy and limitation in comparison with experimental measurements and other benchmark results.

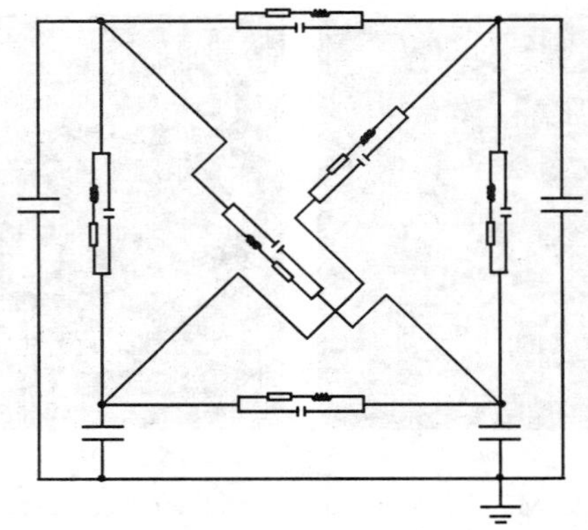

Figure 2. A four node equivalent circuit with common ground.

5.2 Integrated System Simulation

In our integrated simulation scheme, given a completely designed digital PCB board, the system can be partitioned into four major subsystems: chip devices such as drivers/receivers, chip packages, signal nets in form of transmission lines, and power supply nets in the form of planes/partial planes and routed traces. Proprietary behavioral device models, as well as IBIS or SPICE models can be used for IC chip devices or discrete components with various accuracy. Accurate device models are critical to correct simulation and many IC vendors provide models or data for their products in one or other format. Chip package modeling involves mostly parasitic extraction for parameters such as pin inductance and capacitance, and the package is modeled as a few circuit elements or a subcircuit connecting devices with signal traces and power/ground nets. The signal nets are modeled as multiconductor transmission lines (microstrip or strip lines). Fast 2-D field solver is used to extract the transmission line parameters, and an accurate and efficient modal analysis is applied to the time-domain simulation of signal propagation which includes crosstalk between multiple lines. The subsystems are very different in their operation principles, their characterizations and behaviors, and the ways they are analyzed or simulated. Various and appropriate methods have been applied to different subsystems along with focus on various aspects of interests such as signal crosstalk or package extractions. All the four subsystems interact dynamically with each other in the system operation, and therefore are combined together for system level simulation in our analysis, as shown in Figure 3 below.

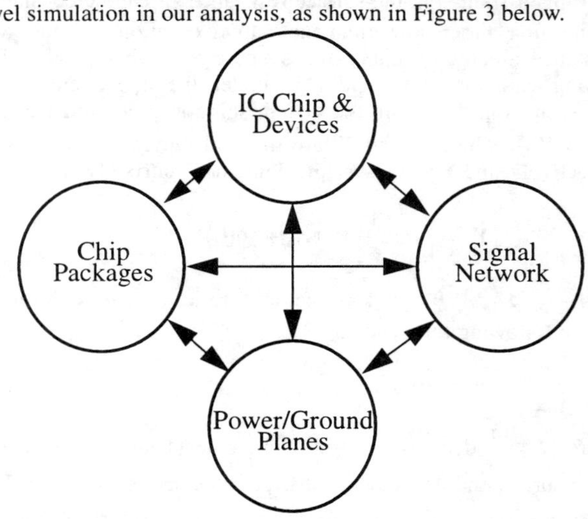

Figure 3. Partition and interaction of the four subsystems of a PCB circuit.

The details of the modeling and simulation methods in the other subsystems except power/ground networks are omitted here due to limited space, but will be presented separately in future publications.

In the case of simultaneous switching noise, when one or more drivers on a single or multiple chips switches, large transient currents are drawn through package pins/connections from the power distribution network (planes) and thus create noises or voltage fluctuations in the power supply, which in turn affect the operation of the switching devices as well as other devices. The switching currents therefore act as the excitation sources to the distributed power/ground planes and the transient noises propagate and resonate in the plane structures. In our co-simulation, since every power/ground pin is a node in the equivalent circuit, at every time step the driver Vcc and Gnd currents are imposed upon the power/ground net as source to calculate the ground noise responses, and these noises are fed back to the device and signal simulation, allowing dynamic interaction among all the subsystems.

6. VERIFICATION AND APPLICATIONS

The following examples first verify the approach against measurements and known results in both time and frequency domain analyses. Then applications on practical designs will be shown for the effect of simultaneous switching noises.

6.1 Comparison and Verification

As the first example, we simulated the case of L-shaped microstrip patch from [4]. A three port equivalent circuit gives good low frequency results comparable with [4]. To better capture the high frequency characteristics, a 14-node equivalent circuit is obtained and simulated to calculate the resonant response for the input impedance at node A. The first two resonant modes are found at f_0 = 1.02 GHz and f_1 = 1.65 GHz, respectively. Comparing with the full wave results f_0 = 0.998 GHz and f_1 = 1.56 GHz given by [4], it demonstrates that the equivalent circuit gives good frequency response up to the first several resonant modes, which are well above the bandwidth of most PCB digital circuits.

As the second example, the coupled microstrip line structure shown in Figure 4 was simulated. A 16-node equivalent circuit was extracted and used in the transient simulation of a pulse signal propagating along the lines. A voltage source with a 50 ohm output resistance was used as driver for the active line, while the other ends were matched with 50 ohm loads. A 5 volt pulse signal of 0.3 ns rise/fall time and 1.0 ns duration was applied at the near end input. Figure 5(a) shows near and far end responses on the active line, while Figure 5(b) shows the near and far end crosstalk on the passive line. Those results were compared with commercially available transmission line simulator and good agreements were obtained. It needs to be pointed out that for well known structure like microstrip line, more efficient and natural approaches exist and should be readily applied to similar problems. This example, however, demonstrates the correct prediction of propagation delay characteristics and coupling effects from the distributed equivalent circuit model extracted with a quasi-static approximation.

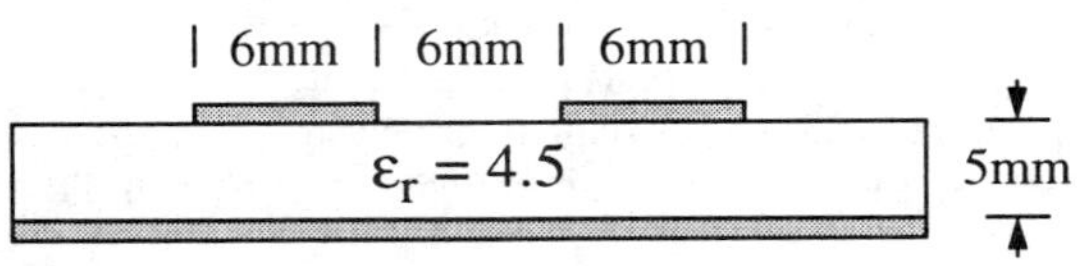

Figure 4. Cross section of the coupled microstrip line in example 2.

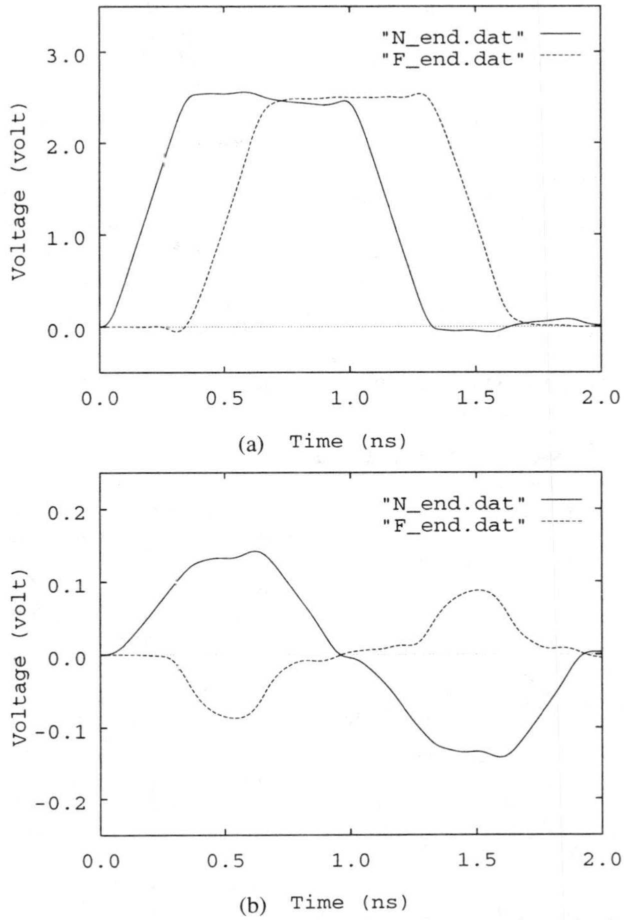

Figure 5. (a) Near and far end voltage waveforms on the active line, (b) near and far end crosstalk waveforms on the passive line.

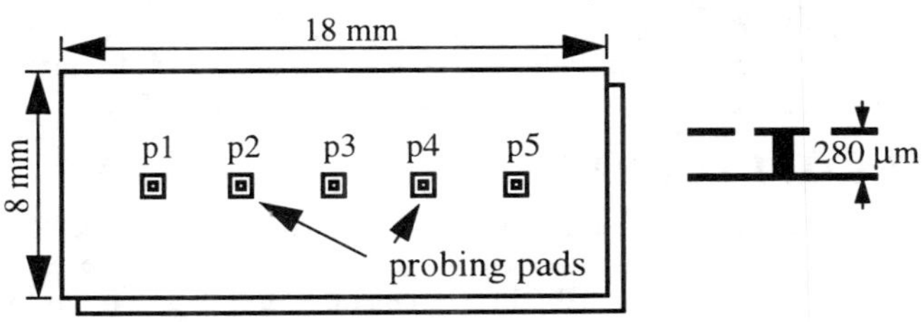

Figure 6. Test structure by HP Lab. The planes are made with 16 mΩ/sq. tungsten. The dielectric is alumina with ε of 9.6.

In the third example, we compare our simulation results with the measurement data for the test plane structure given by K. Lee et al in [3], as shown in Figure 6. In our simulation, a 42-node equivalent circuit was extracted from the field solution, and used in frequency domain calculation of S-parameters. Figure 7 compares the simulated and measured results, where the measured S21 was given in [3]. It is evident that the agreement is quite good up to about 10 GHz, and the simulation captured the essential characteristics of the system. Towards higher frequency, however, the simulated result shifted away from the measurement in a systematic fashion. This behavior is within our expectation due to the limit of quasi-static approximation. But for most current digital applications, as we pointed out before, 10 GHz is a sufficiently high estimate of the signal source bandwidth while the computational requirement is practical. A full-wave simulation, though possible under the current framework, would be too costly and unnecessary.

To further study the transient characteristic of the system, and to see the effect of high frequency limit on high speed digital signal, time domain simulations using both the equivalent RLC circuit and 2-D FDTD are carried out on this test structure. A pulse signal of 5 volt magnitude, 0.2 ns rise/fall time, and 1.0 ns duration is applied at Port 1. All five ports are terminated with 50 ohm load. Transient waveforms are calculated. In the FDTD simulation, a grid size of 1 mm by 1 mm and a time step of 10 ps are used. Figure 8 shows the voltage waveforms at Port 2, where solid lines are the results obtained by equivalent RLC circuit while dashed lines are FDTD results. Good agreement again is evident and demonstrates the validity of our approaches.

6.2 System Level Simulation of SSN

Power and ground noises due to the inductive effects in packages and PCB boards have become a great concern for practical digital system design. A most common practice in PCB design is the utilization of decoupling capacitors to reduce switching noises. However, due to the lack of analysis tools, the de-caps are used in a way of "play it safe" and "put as much as you could if there is space" in practical design. A major application for this work is to simulate the effect of de-caps and thus optimize the decoupling strategy which includes the placement, number, and value of de-caps necessary for noise reduction against design margin.

As a practical example of pre-layout evaluation, we simulated the SSN and decoupling effects of a 7x10 inch, six layer FR4 board with power and ground planes separated by 30 mil. A chip with sixteen CMOS drivers is placed on the board. The ground noises were simulated with different combination of drivers switching, and the effectiveness of decoupling capacitance were observed. As another practical application, a real customer design was used for post-layout system signal integrity simulation and evaluation of power/ground network. This four-layer board with twenty-six chips has two power/ground planes separated by 10 mil and has 55 Vcc pins and 80 Gnd pins. Limited by space, details are omitted here for these two examples. Refer to the presentation and [8] for more complete description of the designs and the simulation results.

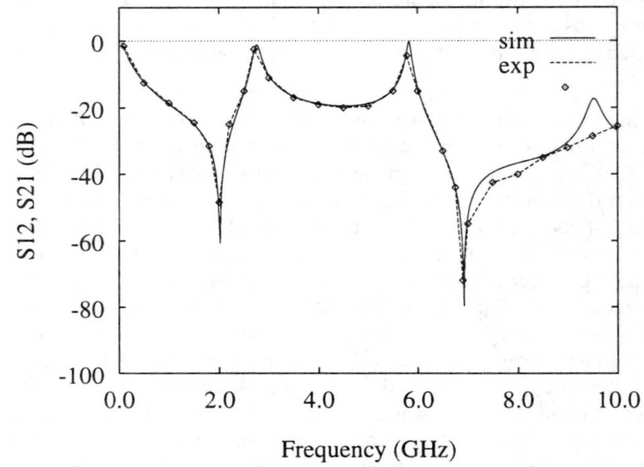

Figure 7. Comparison of simulated and measured S-parameters on the test plane structure.

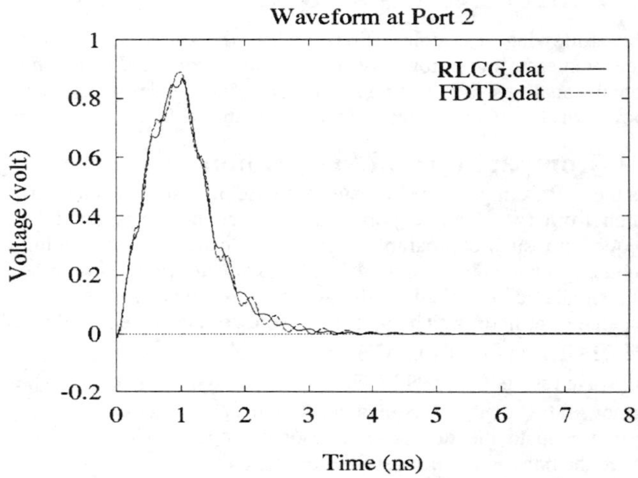

Figure 8. Comparison of transient waveforms at Port 2 on the test plane, obtained by FDTD and equivalent circuit simulations.

7. CONCLUSION

The modeling of power/ground networks and simulation method for system level signal integrity in high speed digital packages and MCM/PCB's are presented. This systematic approach enables us to analyze effects such as simultaneous switching noises, crosstalk, and the effectiveness of decoupling capacitors. The simulation results compare well with measurements and published results.

8. REFERENCES

[1] W. Becker *et al*, "Power Distribution Modeling of High Performance First Level Computer Packages," *IEEE 2nd Topical Meeting on Electrical Performance of Electronic Packaging*, pp. 203-205, Monterey, CA, Oct. 1993.

[2] K. Lee and A. Barker, "A Comparison of Power Supply Planes in Thick and Thin MCM's," *IEEE 3rd Topical Meeting on Electrical Performance of Electronic Packaging*, pp. 3-6, Monterey, CA, Nov. 1994.

[3] R. Senthinathan *et al*, "Reference Plane Parasitics Modeling and Their Contribution to Power and Ground Path "Effective" Inductance as Seen by the Output Drivers," *IEEE Trans. Microwave Theory Tech.*, vol. 42, pp. 1765-1773, Sept. 1994.

[4] J. R. Mosig, "Arbitrarily Shaped Microstrip Structures and Their Analysis with a Mixed Potential Integral Equation," *IEEE Trans. Microwave Theory Tech.*, vol. 36, pp. 314-323, Feb. 1988.

[5] A. E. Rueli, "Circuit Models for Three-dimensional Geometries Including Dielectrics", *IEEE Trans. Microwave Theory Tech.*, vol. 44, pp. 263-267, Feb. 1992.

[6] G. Coen *et al*, "Automatic Derivation of Equivalent Circuits for General Microstrip Interconnection Discontinuities", *IEEE Trans. Microwave Theory Tech.*, vol. 44, pp. 1010-1016, July. 1996.

[7] F. Yuan *et al*, "Analysis and Modeling of Power Distribution Networks and Plane Structures in Multichip Modules and PCB's," *IEEE Electromagnetic Compatibility Symposium Proceedings*, pp. 447-452, Atlanta, GA, Aug. 1995.

[8] F. Yuan, "Analysis of Power/Ground Noises and Decoupling Capacitors in Printed Circuit Board Systems" *IEEE Electromagnetic Compatibility Symposium Proceedings*, pp. 425-430, Austin, TX, Aug. 1997.

[9] J. Vlach and K. Singhal, *Computer Methods for Circuit Analysis and Design*, Van Nostrand Reinhold, New York, 1994.

Efficient Coloring of a Large Spectrum of Graphs

Darko Kirovski and Miodrag Potkonjak
Computer Science Department, University of California, Los Angeles, CA 90095-1596

ABSTRACT

We have developed a new algorithm and software for graph coloring by systematically combining several algorithm and software development ideas that had crucial impact on the algorithm's performance. The algorithm explores the divide-and-conquer paradigm, global search for constrained independent sets using a computationally inexpensive objective function, assignment of most-constrained vertices to least-constraining colors, reuse and locality exploration of intermediate solutions, search time management, post-processing lottery-scheduling iterative improvement, and statistical parameter determination and validation. The algorithm was tested on a set of real-life examples. We found that hard-to-color real-life examples are common especially in domains where problem modeling results in denser graphs. Systematic experimentations demonstrated that for numerous instances the algorithm outperformed all other implementations reported in literature in solution quality and run-time.

1 Introduction

The ever-increasing amount of resources encountered in contemporary designs with exponentially ascending complexities impacts optimization intensive development, compilation, and fabrication strategies. Since many problems related to resource sharing can be modeled as graph coloring, its importance is self evident. Possible applications include register assignment, cache line coloring, logic minimization, circuit testing, operations scheduling, etc [Cou97].

Graph coloring is an NP-complete problem. Previous work shows that so far algorithm development was focused on establishing a single heuristic search methodology expected to provide exceptional results on all instances. Such algorithms usually perform well only on small subset of graph model classes or real-life examples. For example, Johnson et al. [Joh91] simultaneously employed three distinct simulated annealing graph coloring techniques, and one multivariant successive augmentation XRLF algorithm on a set of various graph models. All implementations resulted in relatively low-quality solutions on some instances.

Recently, Coudert proposed a set of exact algorithms for graph coloring based on the knowledge of the maximum clique in a graph [Cou97]. Statistically, such reduction of a hard problem onto another hard problem is not a good problem solving strategy especially when the reduction result is finding a maximum clique in a graph, a problem that has been suggested, due to its difficulty, to hide information in some cryptographic protocols [Jue96]. Coudert applied his algorithm on a large set of real-life examples. However, he did not provide experimental results for most graph classes which correspond to difficult optimization problems.

Complex problems, such as graph coloring, require extensive exploration of the problem solution space. Deployment of a single search or constraint strategy usually leads to a limited adaptability to particular problem topologies. Moreover, finding strategies that meet search quality requirements for a large number of problem topologies is an extremely hard task. We have developed a new algorithm for graph coloring which outperforms all algorithms reported in literature with respect to both solution quality and search run-time. In this paper, we describe the key algorithm and software development ideas that had crucial impact on the algorithm's performance. The key algorithm development techniques are: divide-and-conquer paradigm (coloring one color class, e.g. one independent set of vertices at a time), global probabilistic search for the best intermediate solution (current color class) and its computationally inexpensive objective function computation, least-constraining and most-constrained heuristics, reuse and locality exploration of intermediate solutions, search timing management, post-processing iterative improvement using lottery-scheduling probabilistic search, and once the software is written, optimization of execution bottlenecks and statistical determination and validation of all parameters in the program.

The algorithm implementation strategy can be generalized for a subset of NP-hard problems such as graph partitioning, and feedback arc set. The algorithm is well tested on a number of real-life examples and graph models which correspond to possible real-life optimization problems. We found that hard-to-color real-life examples exist especially in domains where problem modeling results in denser graphs.

2 Related Work

Algorithms for graph coloring can be classified into three main classes: exact [Bre79, Pee83, Cou97], constructive with [Hal93] and without [Bre79, Lei79] guaranteed coloring quality, and iterative improvement coloring algorithms [Bol85, Cha87, Her87, Mor86, Joh91, Mor94, Fle96]. The best results up to date were reported for algorithms that utilize knowledge about the solution space topology in order to establish exceptional colorings. Such algorithms were implemented Johnson et al. [Joh91], Morgenstern [Mor94], etc. An efficient hybrid genetic tabu search algorithm has been developed by Fleurent and Ferland [Fle96]. The standard test example, random $V = 1000$ vertex graph with $p = 0.5$ probability of edge existence, averaged 85.5 for Johnson et al. They required 136 hours on a Sequent Balance 21000 single processor multicomputer. The recent attacks by both Morgenstern, and Fleurent and Ferland resulted in 84-colorings of the $G(1000, 0.5)$ graph, and took between 9 and 36 hours on five Sparc-2s upgraded with 80MHz CPUs for Morgenstern's mix of distributed XRLF and simulated annealing based rejectionless Metropolis move selection, and 41.35 hours on a Sparc-10 workstation for the tabu search genetic algorithm implementation.

3 The New Approach

The problem of graph coloring can be defined using the standard Garey-Johnson format [Gar79]:

PROBLEM: GRAPH K-COLORABILITY

INSTANCE: *Graph $G(V, E)$, positive integer $K \leq |V|$.*

QUESTION: *Is G K-colorable. i.e., does there exist a function $f : V \rightarrow 1, 2, 3, .., K$ such that $f(u) \neq f(v)$ whenever $u, v \in E$?*

We introduce two novel distinct graph coloring algorithms and propose their hybrid as a champion algorithmic solution to the graph coloring problem. The first algorithm is a successive augmentation of a global probabilistic **l**east-constraining **m**ost-constrained e**X**tended RLF (lmXRLF). The second is **l**ottery-**s**cheduling-driven **I**terative **I**mprovement algorithm (lsII). Their hybrid (lmXRLF/lsII) uses initially the lmXRLF algorithm to reduce the solution search space to a statistically determined reduced graph size checkpoint. Then, the lsII algorithm iteratively improves the solution until the user stops the search for improvement.

The performance evaluation and validation of a graph coloring algorithm is strongly dependent on the benchmark graph topologies. We used several diverse graph models: random, geometrical, k-cooked, and uniquely k-cooked graph model. The details about the generation algorithms for each of the used graph models are given in [Kir98].

3.1 lmXRLF

Problem solution space reduction is essential in fast solving of problems with large domains such as graph coloring. The lmXRLF algorithm performs an efficient solution space reduction by assigning the most-constrained vertices for coloring to the class of least-constraining colors, i.e. colors assigned early in the successive coloring process. The remaining vertices in the graph after each color assignment are expected to establish a graph coloring subproblem quantitatively and qualitatively easier to solve than the one in the previous step. Successive augmentation of the described technique, that colors the most-constrained vertices with the least-constraining color, evolves into a complete lmXRLF algorithm. The objective function that quantifies the vertex constraints has crucial impact on the algorithm's performance. We have determined experimentally the least-constraining most-constrained objective of this search. The pseudo-code of the algorithm is presented in Figure 1. In the rest of this subsection we describe the lmXRLF algorithm in detail.

The outermost loop of the algorithm drives the search for the most objective-efficient independent set. Once the underlying search returns the subset, all vertices in the independent set are colored with the next available color and deleted from the graph. The described graph reduction is unchangeable, i.e., deleted vertices are not revisited. The objective function that decides the independent set selection was experimentally determined and validated on a set of examples [Kir98]. The objective function found to perform the best on a set of benchmarks is:

$$ObjF(S) = \sum_{V \in S}(NoN(V)^2 \cdot \sum_{U \in Neigh(V)} NoN(U)^3)$$

where S is the evaluated independent set, $Neigh(V)$ returns the set of vertices adjacent to V, and $NoN(V)$ returns the number of vertices adjacent to V. Thus, vertices found to have large number of incident edges as well as neighborhood with large number of neighbors are treated as constrained. Finding a set of vertices which has the largest accumulated sum of constraints corresponds to the INDEPENDENT SET problem [Gar79].

Important property of the objective function is its local computation and usage of simple search operations such as *remove node* and *add node* from/to the current independent set. Local cost computation means that the objective function is not computed over the entire graph, but only over the nodes in the independent set. The random numbers and their squares and cubes used in the computation are precomputed to avoid complex arithmetic operations. This approach enables fast independent set evaluation and boosts the number of independent sets considered in the search.

The search for the most objective-efficient independent set is performed in the following way. The algorithm first creates a random initial independent set S. Then it iteratively randomly deletes vertices from S until there exists at least one node in the graph not adjacent to the remaining vertices in the set. The uncovered vertices are then randomly selected and added to S until it becomes an independent subset. We experimented greedy strategies for vertex exclusion/inclusion but they did not yield any performance improvement over the fully probabilistic search. Besides introducing larger computation delays between generating two independent sets, controlling the convergence of the greedy probabilistic search strategies is time-consuming and highly sensitive to graph topologies [Joh91]. On the other hand, fully probabilistic search showed unexpectedly good results for all experimented instances.

```
Color = 0
Repeat
  If Color = 0 then GlobalIterations = F · GLOBAL
  Else GlobalIterations = GLOBAL · (1 − ColoredVerts^2 / Verts^2)
  Generate a random independent set S and add it to the list of
  solutions ListOfBestSolutions with largest objective costs
  Repeat GlobalIterations times
    Delete randomly vertices from S until at least one vertex does
    not have a colored neighbor (not counting the deleted ones)
    Add randomly vertices which do not have colored neighbors to S
    If ObjF(S) > min(ObjF(ListOfBestSolutions_i)), i = 1..F
      Add S to the list
      Discard solution S_x with min(ObjF(S_x) from the list
  End Reapeat
  For each independent set S in the ListOfBestSolutions
    Repeat
      S+ = S
      Repeat LOCAL times
        S* = S
        Delete randomly vertices from S* until at least one vertex
        does not have a colored neighbor (not counting the deleted ones)
        Add randomly vertices which do not have colored neighbors to S*
        If ObjF(S*) > ObjF(S+) then S+ = S*
      S = S+
    until there is at least one improvement
    End Reapeat
  End For
  For each V_i ∈ max(ObjF(ListOfBestSolutions_i), i = 1..F)
    COLOR V_i with Color
    Discard V_i from further color consideration
  End For
  Delete all solutions S ∈ ListOfBestSolutions with at least one
  common V_i with max(ObjF(ListOfBestSolutions_i), i = 1..F)
  Color = Color + 1
until there are uncolored vertices
End Repeat
```

Figure 1: Pseudo-code of the ImXRLF algorithm.

During the problem reduction search, at each step a number of independent sets is generated. The search is terminated when $GlobalIterations$ successive randomly generated independent sets do not yield any solution improvement. Due to the randomized search, the locality of the most objective efficient color class is searched extensively in order to find better solution. While searching locally, $LOCAL$ number of independent sets are generated starting from the solution returned by the global search.

While searching for the most objective efficient independent set, a set of color class candidates is generated. When an extensive search is performed, candidates may have disjoint color classes. In order not to discard the solutions which may be objective-competitive and most probably not reachable in the next search iteration, a list of F most objective-efficient solutions is maintained throughout the search. When the search is stopped and the best independent set is colored and deleted from the graph, all other independent sets in the list are checked for mutual vertices and if conjunctive, deleted from the list. The remaining independent sets are a good starting point for the next search iteration.

Since the first iteration in the color class search is handicapped due to lack of previous results, the number of search iterations is adjusted to the number of solutions available from previous searches and the cardinality of the remaining graph. For the first color class $F \cdot GLOBAL$ candidates are generated. For every succeeding color class, the number of generated candidates during the global search equals $GLOBAL \cdot (1 - \frac{ColoredVerts^2}{Verts^2})$. Therefore, the program search time is managed in such a way that difficult to find subsets in the beginning of the search are given longer timeouts. A visual example how the search is performed is presented in [Kir98].

3.2 IsII

There exist instances (e.g. small or extremely dense graphs), for which iterative solution improvement algorithms, that use probabilistic search to traverse the solution space, perform exceptionally well. While searching, the best colorings are memorized and successive trials are conducted in order to improve upon the best memorized solution. At any time during the search, the vertex selection and coloring is non-deterministically greedy, i.e. some probability of traversing towards worse solutions always exists. The main disadvantage of employing such algorithms is the fact that they do not have fixed run-times. Moreover, the run-times for obtaining a solution may differ substantially from one graph to another with the same cardinalities and topology. Finally, search parameters, such as annealing rate, are difficult to establish and greatly vary depending on the graph topology [Joh91].

We developed a novel iterative improvement algorithm for graph coloring. The IsII algorithm tries to color the graph with K colors. The graph is initially, usually incorrectly, colored randomly using K colors. The incorrectly colored vertices are considered for recoloring using the steepest descent rule and the lottery-scheduling paradigm for vertex and color selection probability distribution [Wal94]. The recoloring step is iterated until feasible coloring is found. The algorithm is described using pseudo-code in Figure 2.

Each recoloring step consists of two phases. In the first phase, for each incorrectly colored vertex an objective is assigned describing the vertex' constraint. An incorrectly colored vertex is selected using a lottery scheduling driven probabilistic selection such that vertices with higher constraints are given proportional selection priority. As an objective function to quantify the vertex constraint, we used the number of adjacent vertices colored with the same color as the incorrectly colored vertex. In the second phase, the colors are assigned objectives which correspond to their constraints. The constraint of a color is computed according to the following formula:

$$ObjF(V, C) = \frac{1}{1 + VertsAdjTo(V, C)^{\alpha}}$$

where α is fixed parameter and $VertsAdjTo^{*}(V, C)$ returns the number of vertices adjacent to V and colored with C. According to the individual constraint of each color lottery scheduling driven probabilistic selection is used to select the one in which the selected incorrectly colored vertex is to be colored. Priority is given to colors rarely used in the incorrectly colored vertex' neighborhood. The algorithm iteratively performs the recoloring phases until a viable coloring is found.

Randomly color all V vertices with K colors
Repeat
  Generate random number
    $R \in [0, \sum_{V \in ICVerts(G)} VertsAdjTo(V, V.Color)$
  Select vertex V_w such that
    $R \geq \sum_{j=V_0..V_{[w-1]} \in ICVerts(G)} VertsAdjTo^{*}(V_j, V_j.Color)$
  and w is minimum
  Generate random number
    $Q \in [0, \sum_{i=1..COLORS} \frac{1}{1+VertsAdjTo(V_w, i)^{\alpha}}$
  Select color C_x such that
    $Q \geq \sum_{j=C_0..C_{[x-1]}} VertsAdjTo(V_w, C_j)$ and x is minimum
  Color V_w with C_x
  If $\sum_{V \in ICVerts(G)} VertsAdjTo(V, V.Color)$
  of the current coloring G is minimum
    Memorize G as the best coloring G^{+}
  If there is no improvement of
    $\sum_{V \in ICVerts(G)} VertsAdjTo(V, V.Color)$
  after $SUCC$ successive colorings
    Color all vertices with colors of G^{+}
until graph colored properly

$ICVerts(G)$ returns the set of incorrectly colored vertices in the graph G.
$VertsAdjTo(V, V.Color)$ returns the number of vertices adjacent to V colored with color $V.Color$.

Figure 2: Pseudo-code of the lsII algorithm.

The probability distribution using lottery scheduling is accomplished in the following way. A binary tree is built such that each candidate is assigned to one leaf. The nodes of the tree are assigned the sum of objectives of all leaves contained in the subtree to which the current node is the root. Then, a random number R is selected uniformly in the range of the objective $ObjF(root)$ assigned to the root of the entire tree. According to this number the tree is depth-search traversed to the left child if $R < ObjF(currentNode)$ or right otherwise, until a particular leaf is reached. The reached leaf is the lottery winner and the selected candidate. Implementation details for efficient lottery scheduling are presented in [Wal94].

Parameter α has strong impact on the convergence speed of the algorithm. Large values leave no room for non-greedy recolorings. This results in fast but inflexible convergence and usually oscillations. Smaller values result in longer convergence, but explore the search space more extensively. In our experiments, we dynamically converged α from starting $\alpha = 10$ to asymptotic $\alpha = 4$, and $SUCC = 1000$. An example how lottery scheduling is used in order to achieve probabilistic color perturbation is presented in [Kir98].

The developed lsII iterative improvement algorithm performs exceptionally well on graphs with relatively small number of edges (see Table 1). Graphs with large number of edges induce non-tolerable run-times which make the algorithm impractical for such graphs. General characteristic of all improvement algorithms is that their run-time is highly unpredictable. Standard deviations of typical run-times reported in literature reach the order of magnitude of the average search run-time [Mor94]. Therefore, we employ this algorithm only to small graph instances where the effect of extremely variant expected run-times can be neglected.

3.3 The Hybrid lmXRLF/lsII Algorithm

In order to explore the benefits of both the constructive and iterative improvement algorithm, we coordinated their application to the graph coloring problem. The flow of the algorithm is depicted in Figure 3. The lmXRLF algorithm is used to reduce the search space by peeling sets of color classes until a graph of edge cardinality equal to or less than CAR remains. Then, the lsII algorithm is invoked. The starting chromatic number K for the lsII algorithm is obtained as a result of the application of the lmXRLF algorithm to the remaining subgraph. Every time the lsII returns successful coloring it, is reinvoked with decremented trial chromatic number K. The coloring process is either timed-out or terminated by the user.

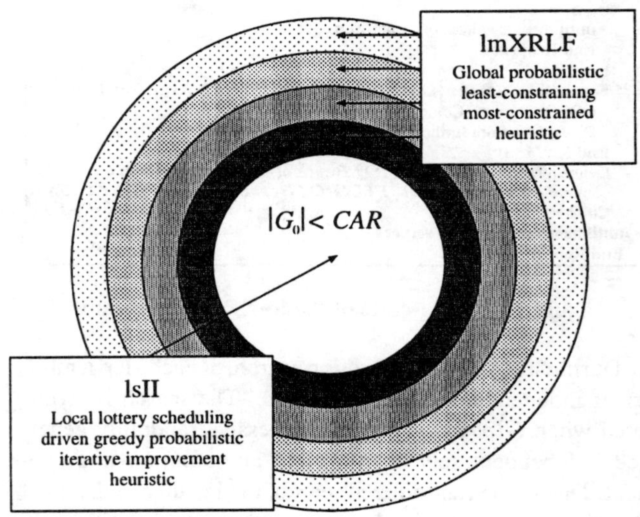

Figure 3: The global algorithm flow: coordinated lmXRLF/lsII deployment for graph coloring.

4 Experimental Results

In this section the conducted experiments are described and results are presented. Obtained colorings are presented for sets of random, k-cooked, uniquely colorable k-cooked, geometric [Kir98], real-life, and DIMACS benchmark graphs.

The independent set selection function was determined and validated by performing a statistic evaluation of performance measurements for a set of objective function candidates [Kir98]. Function $ObjF$ (see subsection 3.1) appeared to have statistically the best performance from the set of function candidates. Therefore, in the rest of the conducted experiments objective-efficient independent sets were evaluated using this function.

Table 1 shows the experimental results for coloring a set of random $R(V,p), p \in \{0.5, 0.1, 0.9\}$ graphs. First two columns describe the $G = GLOBAL$ and $L = LOCAL$ properties of the lmXRLF algorithm or signify that the lsII algorithm was used. The size of the window used to store the best independent sets was $F = 10$ for lmXRLF. Next eight columns present obtained average colorings and run-times in seconds for a set of 10 graph instances for each vertex count. The timings are shown with respect to the $GLOBAL$ and $LOCAL$ parameters. The simulations of the instances with vertex count equal to 125, 250, and 500 were executed on a 40MHz Sparc4, while the 1000-node graphs were colored on a 200MHz UltraII processor. Rows detoned with $\chi(G)$ specify probabilistically determined chromatic number of each graph.

Trials	$R(125, 0.5)$		$R(250, 0.5)$		$R(500, 0.5)$		$R(1000, 0.5)$	
$\chi(G)$	16		27		46		80	
G - L	Cols	Time	Cols	Time	Cols	Time	Cols	Time
1 - 1	21	0.08	35.6	0.33	61.8	2.16	113	3.28
10 - 10	18.9	0.80	31.1	3.27	53.9	18.3	-	-
100 - 100	18.3	5.18	30.7	21.9	51.8	117	-	-
10^3 - 100	18.3	6.41	30.3	25.3	51.1	133	90.6	150
10^4 - 100	18.2	9.67	30.2	35.3	50.9	174	-	-
10^5 - 100	18.2	46.2	29.9	172	50	625	85.9	2850
10^6 - 100	-	-	30	420	49.7	4612	85.1	18083
lsII	18	0.23	30	112	-	-	-	-
	17	117	29	-	-	-	-	-
Trials	$R(125, 0.1)$		$R(250, 0.1)$		$R(500, 0.1)$		$R(1000, 0.1)$	
$\chi(G)$	5		7		11		19	
1 - 1	7.7	0.01	11.5	0.03	18.2	0.16	29.3	0.25
10 - 10	6	0.17	9.1	0.79	14.9	5.53	-	-
100 - 300	6.1	2.28	9.2	9.47	14.4	64.2	-	-
10^3 - 300	6	2.54	9	10.1	14	65.8	23	30.4
10^4 - 300	6	3.31	9	12	14.1	72.3	-	-
10^5 - 300	6	13.9	9	38.7	13.9	172	22	1319
10^6 - 300	-	-	9	289	13.3	1045	21.8	9079
lsII	6	0.01	9	0.86	13	471	22	4516
	5	3.23	8	14.6	12	-	21	-
Trials	$R(125, 0.9)$		$R(250, 0.9)$		$R(500, 0.9)$		$R(1000, 0.9)$	
$\chi(G)$	40		70		122		217	
1 - 1	48.1	0.14	86.4	0.7	154	6.7	279	14.4
10 - 10	46.2	0.9	78	3.7	137	32	-	-
100 - 300	46	17	77	68	133	424	-	-
10^3 - 300	45.8	19	77.5	77	133	511	234	904
10^4 - 300	46	24.4	77.7	97	133	791	-	-
10^5 - 300	45.8	96.5	77.3	413	132	4481	231	10935
lsII	43	2.25	73	1117	-	-	-	-
	42	118	72	-	-	-	-	-

Table 1: Average random $p \in \{0.5, 0.9, 0.1\}$ graph colorings using lmXRLF and lsII.

During the simulations, for the 125-node graphs, we succeeded to obtain 17-colorings on two graphs for two different number of trials (10^5 and $3 \cdot 10^5$). For the set of 250-vertex graphs, we succeeded to obtain 29 coloring only for one graph, while for all others, we obtained 30-colorings. The 500-node graphs were all colored with 50 colors after less than 10 minutes, while more extensive searching resulted in three 49-colorings. We obtained 85-colorings after 30-minutes and reached 85.4 average after less than 1.9 hours on a sample of ten 1000-node graphs. We obtained one 84-coloring in 1.9 hours and two in 5+ hours. The lsII algorithm was applied only to the set of 125- and 250-vertex graphs. Although significantly better results were obtained on smaller graphs than with lmXRLF, competitive coloring of 500-vertex and larger graphs took more than 5+ hours. Moreover, finding 17-colorings on 125-graphs took from 5 to 300 seconds. In one out of twenty cases, the algorithm could not find a 17-coloring in less than 1 hour.

Using the hybrid lmXRLF/lsII algorithm, we were always able to obtain 84-colorings of our set of $R(1000, 0.5)$ graphs. The 84-colorings were obtained even for 10^5 trials of lmXRLF and $CAR = 8000$ for lsII. The obtaining of 84-colorings averaged 50-minutes.

We succeeded to obtain one 12- and two 21-colorings for the $R(500, 0.1)$ and $R(1000, 0.1)$ graphs. Johnson et al. reported for their XRLF implementation that it took 16.5 hours to get 13-coloring for $R(500, 0.1)$ and 35 hours for a 22-coloring of $R(1000, 0.1)$. The lmXRLF did not perform well for $p = 0.9$ type of graphs. lsII demonstrated more powerfull search capabilities on such graph samples.

p	$G = 1000;$ $L = 100; F = 6$				$G = 100000;$ $L = 100; F = 8$			
k	32	64	Single 64	Single 96	32	64	Single 64	Single 96
0.1	32	64	-	-	32	64	-	-
0.2	32	64	-	-	32	64	-	-
0.3	33.7	64	71.2	-	32.5	64	71.5	-
0.4	38.2	65	74.7	102	32	64	73	102
0.5	32.2	65.7	77.7	105.7	33	65	75.5	105.5
0.6	32	66.2	77.5	109.7	33.5	64.5	71	109.5
0.7	32	66	64.5	113	32	66	64.5	111
0.8	32	64.5	64	102	32	64	64	103
0.9	32	64	64	96.5	32	64	64	96.5

Table 2: Average k-cooked and uniquely colorable k-cooked graph colorings using lmXRLF.

The algorithm performance for coloring k-cooked and single solution k-cooked 500-node graphs is presented in Table 2. Table 2 presents obtained colorings for a set of 500-node with fixed k and variable p. Single solution graphs are created for two values of $k = 64, 96$, while regular cooked graphs are created for $k = 32, 64$. The table indicates the correlation between k, p, and the coloring difficulty when the solution is unique.

The lmXRLF algorithm was applied to the challenge graphs at the DIMACS site. Tables 3 and 4 contain a report on the obtained colorings using three different sets of search parameters. In both tables, the first column represents the file name. Next three columns represent graph's vertex and edge cardinalities, and the optimal coloring if known. Next three columns represent three different search trials with different parameter sets. Note that none of the coloring runs did not take more than 1.5 hours on a Sparc4. We report 84-coloring in 50 minutes of the DSJC1000.5 graph when lmXRLF/lsII algorithm was applied. Similarly to Coudert's experimental results [Cou97], we were able to solve ALL real-life examples at the DIMACS site optimally in less than a second (all parameters equal 1, except in few cases where $G = 100$). However, difficult-to-color real-life optimization problems exist, in particular, when the conceptual problem modeling results in higher vertex degrees. We present few examples extracted from the cache line coloring optimization challenge [Kir98].

$FILENAME$	V	E	k	A	B	C
DSJC1000.1	1000	99258	?	29	22	22
DSJC1000.5	1000	499652	?	116	89	85
DSJC125.1	125	1472	?	8	6	6
DSJC125.5	125	7782	?	22	19	18
DSJC125.9	125	13922	?	50	45	45
DSJC250.1	250	6436	?	11	9	9
DSJC250.5	250	31336	?	39	30	30
DSJC250.9	250	55794	?	85	77	77
DSJC500.1	500	24916	?	19	14	14
DSJC500.5	500	125248	?	64	51	50
DSJC500.9	500	224874	?	156	134	134
DSJR500.1	500	7110	?	15	13	13
DSJR500.1c	500	242550	?	97	96	96
DSJR500.5	500	117724	?	144	128	128
anna	138	986	11	11	11	11
david	87	812	11	11	11	11
flat1000_50_0	1000	245000	50	112	87	50
flat1000_60_0	1000	245830	60	112	89	61
flat1000_76_0	1000	246708	76	112	89	85
flat300_20_0	300	21375	20	39	20	20
flat300_26_0	300	21633	26	41	31	28
flat300_28_0	300	21695	28	41	33	32
fpsol2.i.1	496	11654	65	65	65	65
fpsol2.i.2	451	8691	30	30	30	30
fpsol2.i.3	425	8688	30	30	30	30
games120	120	1276	9	9	9	9
homer	561	3258	13	14	13	13
huck	74	602	11	11	11	11
inithx.i.1	864	18707	54	54	54	54
inithx.i.2	645	13979	31	31	31	31
inithx.i.3	621	13969	31	31	31	31
cacheline.1	822	22117	?	64	48	46
cacheline.3	585	9912	?	37	34	31
$A \rightarrow GLOBAL = 1; LOCAL = 1; F = 1$						
$B \rightarrow GLOBAL = 1000; LOCAL = 100; F = 6$						
$C \rightarrow GLOBAL = 100000; LOCAL = 100; F = 8$						

Table 3: lmXRLF-Algorithm performance for the DIMACS set of challenge graphs.

5 Conclusion

We have developed a novel algorithm design strategy which focuses on building an effective software implementation of a mix of various constraint quantification and problem search strategies rather than using a single heuristic. We demonstrated the efficacy of the strategy on the graph coloring problem. Extensive experimentations showed that the algorithm outperforms all other implementations reported in literature for order of magnitude in search run-time while maintaining the solution quality.

6 References

[Bol85] Bollobas, B., Thomason, A. Random Graphs of Small Order. Annals of Disc. Math., vol.28, pp. 47-57, 1985.

[Bre79] Brelaz, D. New methods to color the vertices of a graph. Comm. of the ACM, vol.22, (no.4), pp. 251-6, 1979.

[Cha87] Chams, M., et al. Some Experiments With Simulated Annealing for Coloring Graphs, European Journal of Operations Research, vol.32, pp.260-66, 1987.

[Cou97] Coudert, O. Exact coloring of real-life graphs is easy. 34th Design Automation Conference, pp. 121-126, 1997.

[Fle96] Fleurent, C., Ferland, J.A. Genetic and hybrid algorithms for graph coloring. Annals of Operations Research, vol.63, pp.437-461, 1996.

[Gar79] Garey, M.R., Johnson, D.S. Computers and intractability: a guide to the theory of NP-completeness. W. H. Freeman, San Francisco, CA, 1979.

[Hal93] Halldorsson, M.M. A still better performance guarantee for approximate graph coloring. Information Processing Letters, vol.45, (no.1), pp.19-23, 1993.

[Her87] Hertz, A., de Werra, D. Using Tabu Search Techniques for Graph Coloring, Computing, vol.39, pp.345-351, 1987.

[Joh91] Johnson, D.S., et al. Optimization by simulated annealing: an experimental evaluation. II. Graph coloring and number partitioning. Operations Research, vol.39, (no.3), pp.378-406, 1991.

[Jue96] Juels, A., Peinado M. Hidden cliques as cryptographic keys. Technical report, UC Berkeley, 1996.

[Kir98] Kirovski, D., Potkonjak, M. Exact Coloring of Many Real-Life Graphs is Difficult, but Heuristic Coloring is Almost Always Effective. Technical Report, CSD, UCLA, 1997.

[Lei79] Leighton, F.T. A Graph Coloring Algorithm for Large Scheduling Algorithms. Journal of Res. Natl. Bur. Standards, vol.84, pp.489-506, 1979.

[Mor86] Morgenstern, C., Shapiro, H. Chromatic Number Approximation Using Simulated Annealing. Unpublished, 1986.

[Mor94] Morgenstern, C. Distributed Coloration Neighborhood Search. DIMACS Series in Disc. Math., vol.0, 1994.

[Pee83] Peemoeller, J. A correction to Brelaz's modification of Brown's coloring algorithm. Communications of the ACM, vol.26, (no.8), pp.595-7, 1983.

[Wal94] Waldspurger, C.A.; Weihl, W.E. Lottery scheduling: flexible proportional-share resource management. The First USENIX Symposium on Operating Systems Design and Implementation, pp.1-11, 1994.

$FILENAME$	V	E	k	A	B	C
latin_square_10	900	307350	?	129	109	100
le450_15a	450	8168	15	21	17	17
le450_15b	450	8169	15	21	17	17
le450_15c	450	16680	15	29	22	21
le450_15d	450	16750	15	29	22	21
le450_25a	450	8260	25	26	25	25
le450_25b	450	8263	25	27	25	25
le450_25c	450	17343	25	33	28	28
le450_5a	450	5714	5	12	7	5
le450_5b	450	5734	5	13	7	5
le450_5c	450	9803	5	16	5	5
le450_5d	450	9757	5	16	5	5
miles1000	128	6432	42	45	43	42
miles1500	128	10396	73	75	73	73
miles250	128	774	8	9	8	8
miles500	128	2340	20	20	20	21
miles750	128	4226	31	33	32	32
mulsol.i.1	197	3925	49	49	49	49
mulsol.i.2	188	3885	31	31	31	31
mulsol.i.3	184	3916	31	31	31	31
mulsol.i.4	185	3946	31	31	31	31
mulsol.i.5	186	3973	31	31	31	31
myciel7	191	2360	8	8	8	8
queen15_15	225	10360	?	20	17	17
queen16_16	256	12640	?	23	18	18
school1	385	19095	14	39	16	14
school1_nsh	352	14612	14	35	16	14
zeroin.i.1	211	4100	49	49	49	49
zeroin.i.2	211	3541	30	31	30	30
zeroin.i.3	206	3540	30	30	30	30
cacheline.2	1121	287220	?	74	53	50
cacheline.4	742	15001	?	44	37	35
$A \rightarrow GLOBAL = 1; LOCAL = 1; F = 1$						
$B \rightarrow GLOBAL = 1000; LOCAL = 100; F = 6$						
$C \rightarrow GLOBAL = 100000; LOCAL = 100; F = 8$						

Table 4: lmXRLF-Algorithm performance for the DIMACS set of challenge graphs.

Arithmetic Optimization using Carry-Save-Adders

Taewhan Kim† William Jao‡ Steve Tjiang†

†Synopsys Inc.
700 E. Middlefield Rd.
Mountain View, CA 94043

‡Aimfast Corp.
846 Stewart Dr.
Sunnyvale, CA 94086

Abstract

Carry-save-adder(CSA) is the most often used type of operation in implementing a fast computation of arithmetics of register-transfer level design in industry. This paper establishes a relationship between the properties of arithmetic computations and several optimizing transformations using CSAs to derive consistently better qualities of results than those of manual implementations. In particular, we introduce two important concepts, *operation-duplication* and *operation-split*, which are the main driving techniques of our algorithm for achieving an extensive utilization of CSAs. Experimental results from a set of typical arithmetic computations found in industry designs indicate that automating CSA optimization with our algorithm produces designs with significantly faster timing and less area.

1 Introduction

Hardware designers have long applied many arithmetic optimization techniques for implementation of arithmetic functionality like addition, subtraction, and multiplication. Among them, carry-save-adder(CSA)[1] has proved a powerful mechanism to improve timing with little, if any, area penalty (even reduced area). Figure 1(a) shows the structure of an n-bit CSA.

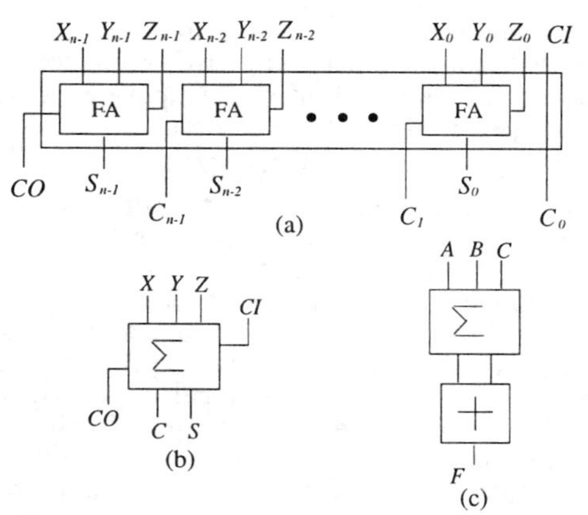

Figure 1: An n-bit CSA and an example of use

DAC 98, San Francisco, California

The n-bit CSA consists of n disjoint full adders(FAs). It consumes three n-bit input vectors and produces two outputs, i.e., n-bit sum vector S and n-bit carry vector C. We use the block symbol in Figure 1(b) to represent a CSA operation. Unlike the normal adders (e.g., ripple-carry adder(RCA) and carry-lookahead adder(CLA)), a CSA contains no carry propagation. Consequently, the CSA has the same propagation delay as only one FA delay (compared to RCA's n FA delay where n is the bit-width), and the delay is constant for any value of n. For sufficiently large n, the CSA implementation becomes much faster and also relatively smaller in size than the implementation of normal adders.[1]

There has been an extensive research work on the arithmetic optimizations in several areas[2, 3, 4]. They, however, focused on the transformation of operations using techniques such as simple algebraic manipulations and constant propagation; they did not address the problem of arithmetic optimization using CSAs. This paper introduces the concept of CSA transformation and presents an algorithm that effectively utilizes CSAs to derive consistently better quality of results than that of manual implementations.

We define a *CSA tree* to be a tree of CSA operators and one adder at the root of the tree. A CSA tree can be used to transform an arbitrary number of additions to produce two adding operands and the adder is used at the root of CSA tree to produce a final sum. In other words, an expression of N additions can be transformed to a CSA tree of depth $log_{1.5}(N)$ and the overall delay is $log_{1.5}(N)$ plus the delay of final adder (which is about $log_2(n)$ for a CLA). For example, expression $A+B+C$ can be transformed into one CSA and one adder as shown in Figure 1(c).

Based on several results of our experimentations, our transformation can be stated as follows: Given a non-cyclic dataflow graph of arithmetic operations, we wish to transform the computations using as many CSA operations as possible while preserving the functionality of the design.

[1]Note that a CSA performs the same functionality of the conventional adders in the sense that each reduces the number of adding operands by one, i.e., a CSA reduces from 3 to 2 and an adder from 2 to 1.

2 Applicability of Transformation

CSA transformation is not limited to addition only. We can transform other arithmetic operations like subtraction and multiplication into additions to produce longer chains of additions. We replace a subtraction by adding the negation[2] of the subtraction. That is, (x-y) in expression is changed into ($x + \bar{y} + 1$). For multiplication, we can use two possible options:

- **sum of products:**

  A multiplication is decomposed into a set of shift-and-add operations, which is definitely beneficial when it has a constant as input.[3] With full decomposition we can effectively extend the addition tree by merging the decomposed additions with another additions. However, decomposition can increase the number of CSA operations drastically as the bit-width increases.

- **partial multiplication and addition:**

  Two types of implementation for multiplication operation are typically found in designs[6]: (i) wallace tree model for fast timing and (ii) carry-save array model for small area. The wallace tree model consists of two blocks called *partial-mult*[4] and *final-add*. Partial-mult uses the two inputs of the multiplication as input and produces two outputs, in which by adding them the final result of multiplication is obtained.[5] Consequently, by decomposing into a partial-mult and a final-add we can merge the final-add with a descendent operation to transform into CSAs. This option increases only one operation, but is less flexible than the case of full decomposition.

3 An Algorithm

Our transformation algorithm consists of three major steps: (1) identification of operation tree to be transformed, (2) translation of the expression of the identified tree into an addition expression, and (3) conversion of the addition expression into a CSA tree. Given a non-cycle dataflow graph of arithmetic computations, our algorithm iteratively performs the three steps until there is no candidate expression of tree to be transformed. The following subsections describe the details of the three steps.

3.1 Candidate Identification

As explained in Sec. 2, the CSA transformation can be applied to any type of operations that can be converted to additions. A *candidate cluster* is basically a tree which contains operations like additions/subtractions/multiplications. Our algorithm is a bottom-up (from the output boundary of design toward the input boundary) approach. From the outputs of design it finds an operation which has not been transformed yet in the previous iterations. We refer the operation to *root*. The root is then expanded toward the input boundary of design to construct a tree of operations. Note that the type of root must be one of addition, subtraction, and multiplication operations.

Suppose that operation A is a leaf of the operation tree expanded so far and B is an operation in which its output is used as an input of operation A. We expand the current operation tree by including B if the following four conditions are satisfied:

- Condition 1: A must be one of addition and subtraction operations.
- Condition 2: B must be one of addition, subtraction, and multiplication operations.
- Condition 3: Output of B must be single fanout. That is, B drives only an input of A.
- Condition 4: There should be no "leak" of data values through the connections from B to A up-to the output bit-width of root of the tree.

Condition 1 ensures that only addition and subtraction are used as non-leaf operations of the tree, and Condition 2 ensures that only addition, subtraction and multiplication are used as leaf operations. Condition 3 ensures that we do not allow transforming non-tree structure of operations. (We solve the multiple fanout case in Sec. 4.1.) Finally, Condition 4 is required for preserving the functionality of design before and after transformation. The examples shown in Figure 2 clarify the concept of leakage of data values. Suppose operation A is also the root of the current tree. Figure 2(a) shows a case of upper-bit truncation between B and A. Because up-to 8 bits of data values must be preserved, the truncation does not allow merging operation B with A. Similarly, Figure 2(b) shows a case of lower-bit truncation between the two operations. It also prevents merging the two operations. However, Figure 2(c) shows a case of preserved data values up-to 8 bits from B through A. Consequently, the two operation can be merged and transformed into CSA operations without changing the functionality of design.

3.2 Conversion to Additions

With the identified cluster of operation tree obtained from Step 1, this step converts the expression of operation tree consisting of additions, subtractions, and multiplications into a tree of additions. What we are interested in this step is to extract all the operands from the converted addition expression. Table 1 summarizes all possible conversion rules for basic operations. Note that there exist three different rules to convert a multiplication. Which rule to apply depends on constraints such as number of operations to allow, word size of operation, and optimization goals. For example, rule 6 (full decomposition) decreases timing, but increases the number of CSAs drastically. On the other hands, rule 5 (partial decomposition) is not much effective to decrease timing than that of rule 6, but maintains the number of CSAs in a certain

[2] We used two's complement.

[3] We used the signed-digit(SD) encoding scheme[5] for the multiplication with a constant input to reduce the number of decomposed operations.

[4] It is a functionality of wallace tree compressor.

[5] We use *S(partial_mult(A,B))* and *C(partial_mult(A,B))* to denote the two outputs of the partial-mult of operation $A * B$.

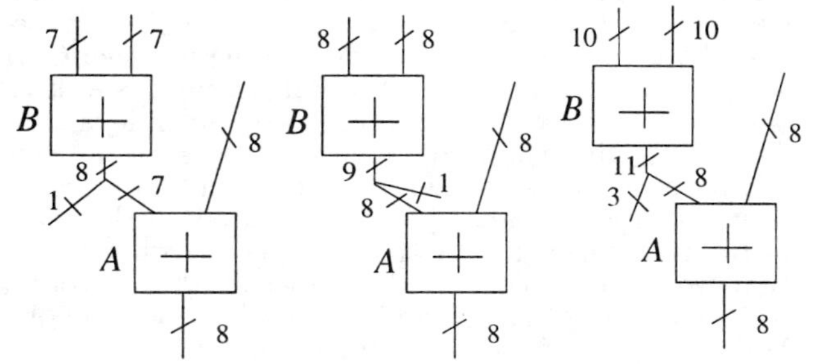

(a) upper-bit truncation (b) lower-bit trunction (c) preserved bit-flow upto 8 bits

Figure 2: Example of possible net-connections between two operations

	primitive exp.	new exp.
1	$x + y$	$x + y$
2	$x + y$ with c_{in}	$x + y + c_{in}$
3	x - y	$x + (-y)$
4	x - y with c_{in}	$x + (-y) + c_{in}$
5	$x * y$	$C(partial\_mult(x, y)) +$ $S(partial\_mult(x, y))$
6	$x * y$	$\sum_{i=0}^{n-1} AND(x, y_i) << i$
7	x * *const*	summing from SD encoding[5]

Table 1: Conversion rules to additions for primitive expressions

amount. In addition, rule 7 can be used as a preprocessing step which guarantees that the number of operations is always within half of bit-size of the constant.

We apply the rules in Table 1 to the original expression of tree to obtain a set of operands, some of which might need to negated (e.g. $-y$ in subtraction) or shifted (e.g. $AND(x, y_i) << i$ in multiplication). The operands may also include single-bit ones and constants. We then replace negated operands, e.g., $-x$ by $\bar{x}+1$ and all constants are folded.

3.3 Conversion to CSAs

All the operands obtained from Step 2 are to be summed up by CSA tree. In this step, we build up the tree. Our algorithm constructs CSAs iteratively one by one. Each iteration selects three operands as input of a new CSA and two new operands are created from the CSA. At the next iteration, the three operands used in the previous iteration are removed from the operand set and the new two operands are added to the set. Therefore, whenever one CSA is created the size of operand set is reduced by one. CSA construction proceeds until only two operands in the set remain. Finally, a normal adder adds the two remaining operands to produce a final output of the expression.

Because the primary objective of our transformation is to reduce timing, each iteration selects, from all the possible triples of operands in the set, the triple with the earlist arriving time. We use an efficient linear-time heuristic: Initially we sort operands in the set in a non-decreasing order according to the delay to the operands from input boundary. If two operands have the same delay, we give a higher priority to the one of smaller bit-width. At each iteration, our algorithm picks the first three operands from the sorted list and creates a CSA. The delays to two output operands of CSA are then computed and the operands are inserted to the correct positions of the sorted list. Our experimentation indicates that our operand selection approach is very efficient and yet, does not degrade the quality of results when compared with that of the exhaustive search of selections. In addition, the algorithm can easily tune to optimize area of the resultant CSA tree by sorting the operands according to the size of bit-width of operands. (If two operands have the same bit-width we resolve it by giving higher priority to the one of shorter delay.)

We handle one-bit operands and a constant in different ways to reduce the number of CSAs created. This can be achieved by utilizing as many carry inputs of CSAs as possible. Each CSA created can consume an one-bit operand as carry input. Therefore, it is desirable to assign as many one-bit operands to the carry inputs of CSAs as possible because otherwise, each requires one additional CSA. On the other hand, for the constant operand it can be used as a multi-bit operand or be decomposed into logic-1 values to be used as carry input of CSAs. Using constant as a multi-bit operand requires only one additional CSA. However, when there are sufficiently large number of CSAs for the multi-bit (non-constant) operands, the constant can be assigned to the carry inputs of CSAs, avoiding unnecessary creation of an additional CSA. Furthermore, when there is a mixture of operands of (non-constant) single-bit and logic-1 values, because logic-1 operand takes zero delay, it is desirable to assign the logic-1 operands to CSAs which are far from the root of final CSA tree, and to assign the non-constant operands to CSAs which are close to the root of tree to reduce the timing of the critical path of the final CSA tree.

Figure 3 illustrates the effects of different utilization of carry inputs of CSAs on the timing and area of final CSA tree. Suppose that A, B, C, D are mult-bit operands, E is single-bit, and the constant value is 2. We assume that the bit sizes of the multi-bit operands are the same. Also, we assume we have already computed delay to the operands as shown in Figure 3(a).[6] Figure 3 shows a set of possible transformations for the operands. Figure 3(a) shows the case that the single-bit operand E and constant 2 are not used as carry input of CSA. Consequently, four CSAs are created and timing of the final tree is worse than the other cases. On the other hand, Figures 3(b) and 3(c) show the cases that the single-bit operand and constant are used as carry input of CSA but not both, respectively. Each of them requires three CSAs and timing of the CSA trees is better than the previous one, but still there is a room to improve. Figures 3(d) and 3(e)

[6] We use notation $D(X)$ to denote the delay of operand X.

show the cases that the carry inputs of CSA tree are maximally utilized. However, according to the way of assigning single-bit operand and constant there is a big difference in timing. One guideline is that we need to assign the constant operands (i.e., logic-1) to earlier steps than the single-bit ones during the top down process of CSA tree construction. In fact, our algorithm produced the one in Figure 3(e) which is the best in terms of timing and area.

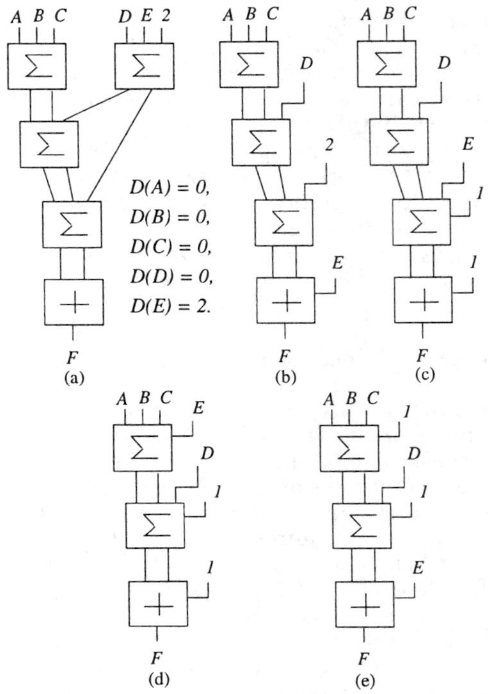

Figure 3: Effects of utilization of carry inputs on CSA tree

4 Extension to the Applicability of CSA Transformation

Many signal processings and data-intensive computations frequently contain an operation in which only the upper m bits among n bits of its output are fed input to a descendent operation. We call this partial use of an output the *lower-bit truncation* problem. Moreover, many designs often contain an operation whose output feeds several other operations. We call this the *multiple fanout* problem. In fact, the two problems correspond to the violations of Condition 4 (Figure 2(b)) and Condition 3 in Sec. 3.1, respectively. In the following, we provide solutions to the problems.

4.1 A Solution to the Multiple Fanout Problem

We begin with an example to demonstrate how our approach handles operations with multiple fanout. Suppose that Figure 4(a) is a partial structure of a design that we are going to transform into CSAs. Note that operation O_3 has multiple fanout. According to the algorithm described in Sec. 3, the first iteration will identify operation tree $tree_1$ and transform it into CSAs, and the second iteration will identify $tree_2$ and transform it into CSAs as show in Figure 4(b). Consequently, two CSAs and two adders (one for each CSA tree) are used. Suppose that a design contains a chain of operations in which n of them have multiple fanout. Then, our algorithm will identify n operation trees having each operation with multiple fanout as root. Consequently, n adders will be allocated on the chain of the transformed CSA trees. Here, our objective is that we want to reduce the number of adders from n to 1 and replace the remaining n-1 adders with CSAs. We accomplish this by introducing the concept of *CSA transformation without final adder.*

We use the algorithm in Sec. 3 for identifying expression tree. However, when an operation with multiple fanout is encountered (i.e., Condition 3), we mark the operation as a root of another operation tree and continue to expand the operation tree. Once we collect all the operation trees by crossing over operations with multiple fanout, we topologically sort the operation trees from input boundary of design to output boundary. We then transform the operation trees on the list one by one. For those trees with multiple fanout root, we do not allocate final adders in the resultant CSA trees. Instead, we generate final two outputs of each CSA tree. The two outputs are then used in two ways: (1) Both of them are used as input operand of the operations trees which depend on the operation tree corresponding to the transformed CSA tree without final adder and (2) A new adder is created and they are used as input of the adder. (we refer this process to as *operation-duplication.*) The output of adder is then used for the fanout of the root of operation tree corresponding to the CSA tree. Consequently, only one adder will be created at the end of the last operation tree of the sorted list of trees.

For example, in Figure 4(a) the first iteration will find two operation trees $tree_1$ and $tree_2$ in which there is a data flow from $tree_2$ to $tree_1$. $tree_2$ is then transformed into a CSA tree without final adder. It generates two outputs $T1$ and $T2$ and one new adder O_{dup} as shown in Figure 4(c). $tree_1$ is then transformed into a CSA tree with final adder by using $T1$, $T2$, D, and E as operands as shown in Figure 4(d). Note that the resultant transformation contains three CSAs and one adder in the two CSA trees and generates a faster timing than the one in Figure 4(b). However, the total area increases because one additional adder is created outside of the CSA trees.

4.2 A Solution to the Lower-bit Truncation Problem

We also begin with an example to demonstrate how our approach can solve the problem of lower-bit truncation. Figure 5(a) shows an example of lower-bit truncation. Note that only 8 upper-bits of the output of operation O_2 are used. Therefore, we cannot merge operation O_1 and O_2 into CSAs because there is a data leakage between the operations(i.e., Condition 4 in Sec. 3.1). We solve this problem by introducing the concept of *operation-split*: We split operation O_2 into two operations O_3 and O_4 as shown in Figure 5(b). This means that given the split operations our algorithm can identify an expression tree as

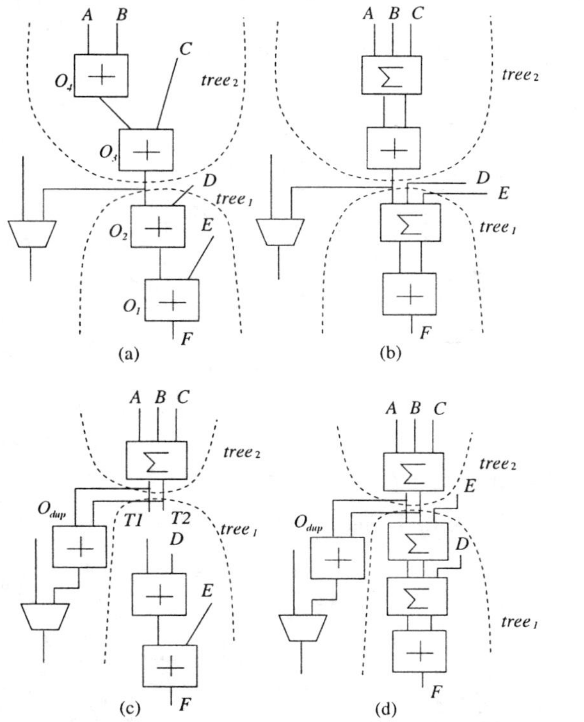

Figure 4: Examples of transformation of operation with multiple fanout

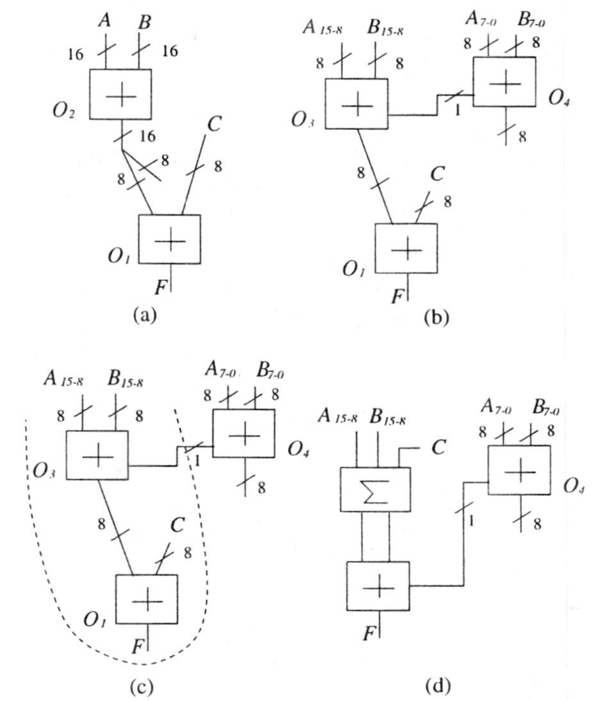

Figure 5: An example of transformation of operation with lower-bit truncation

shown in Figure 5(c) where C, upper 8-bits of A and B, and carry-out of operation O_4 are the operands to be added. Figure 5(d) shows the transformed CSA tree from the operation tree in Figure 5(c).

Unlike the case of solution of multiple fanout problem, the operation-split does not increase area. Furthermore, when the operation with lower-bit truncation has multiple fanout, the carry-out of lower one of two split operations can be extracted directly from the adder created from the solution of the multiple fanout problem.

5 Experimental Results

We tested our algorithm on a large number of arithmetic computations which are typically used in industrial designs. We used Design Compiler package from Synopsys Inc. to perform the implementation selection, tree-height minimization and logic optimizations for the designs transformed by our algorithm and the designs without using our algorithm[7], and compared their results.

- **Designs with additions, subtractions, and multiplications:**

  We tested our algorithm on designs with a mixture of additions, subtractions, and multiplications as shown in Table 2. We used 8-bit operands for non-constant inputs of multiplication and used 16-bit operands for the rest. We also assume that the arrival times of all input operands are 0. For multiplication operation, our algorithm fully decomposed it into additions. The results in Table 2 show a strong indication that our algorithm can extensively apply to the designs with additions, subtractions and multiplications.

- **Designs with multiple fanout:**

  We conducted our experimentation on three designs shown in Figure 6. (In all designs in the following, we assume that the arrival times of all input operands are 0.) Each design has operations with multiple fanout. We used two CSA transformation techniques: (i) without operation-duplication and (ii) with operation-duplication, and compared the results with that produced without CSA transformation. The results are summarized in Table 3. Note that operation-duplication can reduce timing of design much further, but it increases area. Consequently, the transformation with operation-duplication can be applied in a local way to those operations on the critical path of design to reduce timing with a minimal increase of area.

- **Designs with lower-bit truncation:**

  We also conducted our experimentation on the three designs shown in Figure 7. Note that the designs have operations with lower-bit truncation.

[7] The tree-height minimization was performed on non-CSA operations. We used *lcbg10pv* (0.35μ) technology[7] for all test cases.

We used two CSA transformation techniques: (i) without operation-split and (ii) with operation-split, and compared the results with that produced without CSA transformation. The results shown in Table 4 reflect that the operation-split is a very powerful technique to overcome the limitation of lower-bit truncation problem and extend the applicability of CSA transformation to produce much faster timing and less area.

Equations	RTL Design timing/area	CSA design timing/area	diff.
A * B + 1	6.18 ns 3430 units	5.88 ns 3410 units	-5% -1%
A * 3E3E (0011111000111110)	6.86 ns 2920 units	6.55 ns 2068 units	-5% -29%
A*B + C*D + E*F	7.21 ns 11225 units	6.84 ns 10141 units	-37% -10%
A*B - C + D	7.19 ns 5185 units	6.37 ns 4880 units	-10% -6%
A + B - C - D - E - F	7.94 ns 5328 units	6.03 ns 3697 units	-24% -31%

Table 2: Comparison of results for expressions with addition, subtraction and multiplication

Designs	RTL Design timing/area	CSA design w/o duplicate timing/area	CSA Design w/ duplicate timing/area
fanout_1	7.11 ns 4010 units	6.83 ns 3600 units	6.24 ns 4659 units
fanout_2	7.96 ns 6328 units	7.13 ns 6072 units	7.04 ns 7037 units
fanout_3	7.86 ns 9344 units	7.07 ns 8992 units	6.87 ns 9772 units

Table 3: Comparison of results for designs in Figure 6

Designs	RTL Design timing/area	CSA design w/o split timing/area	CSA Design w/ split timing/area
truncate_1	6.01 ns 2315 units	5.94 ns 2138 units	4.02 ns 1721 units
truncate_2	7.75 ns 4610 units	7.25 ns 4560 units	5.81 ns 4133 units
truncate_3	7.98 ns 8374 units	7.25 ns 8161 units	6.45 ns 7405 units

Table 4: Comparison of results for designs in Figure 7

6 Conclusions

We presented a new technique to optimize arithmetic circuits. The technique automatically expands circuits consisting of adders, subtractors, and multipliers into their carry-save-adder (CSA) representation which are then optimized. The representation obviates the need for implementation selection, the automatic selection of the best implementation among several adder/subtractor/multiplier implementations, a time-consuming part of arithmetic optimization.

Acknowledgment- We wish to thank Ron Miller, Hazem Almusa, Reiner Genevriere, Sean Huang, and Tai Ly for making this work successful.

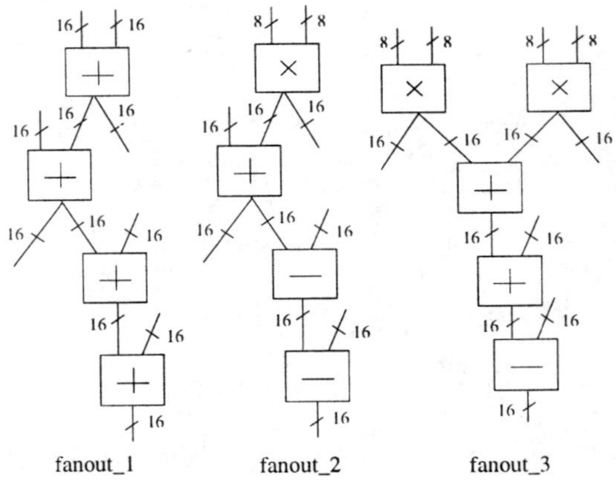

Figure 6: Designs with multiple fanout

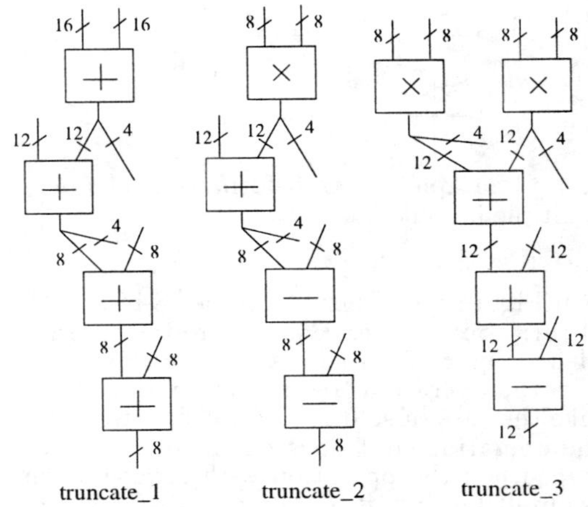

Figure 7: Designs with lower-bit truncation

References

[1] N. Weste and K. Eshraghian, *Principles of CMOS VLSI Design - A Systems Perspective*, Addition-Wesley Publishers, 1985.

[2] M. Potkonjak and J. Rabaey, "Optimizing Resource Utilization Using Transformations", *Proc. of ICCAD*, pp. 88-91, 1991.

[3] D. Lobo and B. Pangrle, "Redundant Operation Creation: A Scheduling Optimization Technique", *Proc. of DAC*, pp. 775-778, 1991.

[4] A. Aho and J. Ullman, *Principles of Compiler Design*, Addition-Wesley Publishers, 1977.

[5] K. Hwang, *Computer Arithmetic: Principles, architecture, and Design*, New York, 1979.

[6] Synopsys Inc., *DesignWare Components Databook*, 1996.

[7] LSI Logic Inc., *G10-p Cell-Based ASIC Products Databook*, 1996.

Synthesis of Power-Optimized and Area-Optimized Circuits from Hierarchical Behavioral Descriptions *

Ganesh Lakshminarayana and Niraj K. Jha

Dept. of Electrical Engineering, Princeton University, NJ 08544

Abstract

We present a technique for synthesizing power- as well as area-optimized circuits from hierarchical data flow graphs under throughput constraints. We allow for the use of complex RTL modules, such as FFTs and filters, as building blocks for the RTL circuit, in addition to simple RTL modules such as adders and multipliers. Unlike past techniques in the area, we also customize the complex RTL modules to match the environment in which they find themselves. We present a fast and efficient algorithm for mapping multiple behaviors onto the same RTL module during the course of synthesis, thus allowing our synthesis system to explore previously unexplored regions of the design space. These techniques are at the core of an iterative improvement based approach which can accept temporary degradation in solution quality in its quest for a globally optimal solution. The moves in our iterative improvement procedure explore optimizations along different dimensions such as functional unit selection, resource allocation, resource sharing, resource splitting, and selection and resynthesis of complex RTL modules. These inter-related optimizations are dynamically traded off with each other during the course of synthesis, thus exploiting the benefits that arise from their interaction. The synthesis framework also tackles other related high-level synthesis tasks such as scheduling, clock selection, and V_{dd} selection. Experimental results demonstrate that our algorithm produces circuits whose area and power consumption are comparable to or better than those produced using flattened synthesis, within much shorter CPU times. The efficacy of our algorithm in the power-optimization mode is illustrated by the fact that it produces circuits that consume upto 6.7 times less power than area-optimized circuits working at 5 *Volts* at area overheads not exceeding 50%.

1 Introduction

High-level synthesis is the process of deriving an optimized register-transfer level (RTL) architecture from a behavioral description, usually specified as a data flow graph (DFG) for data-dominated circuits. In this paper, we present a technique for synthesizing power- as well as area-optimized data-dominated circuits from hierarchical DFGs under throughput constraints. *Data-dominated* behaviors have a predominance of arithmetic operations, and an absence of control-flow constructs. They are commonly encountered in digital signal processing and image processing applications.

*This work was supported in part by NSF under Grant No. 9319269 and in part by Alternative System Concepts, Inc. under an SBIR contract from Air Force Rome Laboratories.

DAC 98, San Francisco, California
©1998 ACM 0-89791-964-5/98/06..$5.00

Hierarchical high-level synthesis can be divided into two subproblems, (i) deriving hierarchical information from a flattened behavioral description, and (ii) synthesizing a circuit using a hierarchical behavioral description. The latter problem, which is addressed in this paper, was also considered in [1, 2, 3, 4]. In [1] and [3], the problem of synthesizing area-optimized circuits from hierarchical behaviors was considered. The technique presented in [4], which was geared towards applying power-optimizing transformations, supports two levels of hierarchy. Past research in flattened high-level synthesis for low power has considered allocation and assignment [5, 6], module selection [7], scheduling [6], and transformations [4]. A technique which solved the problems of allocation, assignment, module selection, scheduling, clock selection, and V_{dd} selection simultaneously to produce power-optimized architectures was presented in [10]. We compare experimental results obtained by our method with those presented in [10].

The following general observation can be made about existing hierarchical high-level synthesis techniques: they handle large designs much faster than flattened high-level synthesis techniques, however, the solutions produced are generally of inferior quality. This is because fine-grained optimizations, which are within the scope of flattened synthesis techniques, are not visible while performing hierarchical synthesis. Our work aims at combining the global perspective afforded by hierarchical synthesis with the flexibility of flattened synthesis to produce compact, power-optimized architectures, within short periods of time. Our algorithm accepts as input a hierarchical DFG (arbitrarily deep hierarchies are allowed), and a library of modules, which can be *simple* (adders, multipliers) or *complex* (RTL modules implementing specific DFGs). Other inputs include typical input traces to aid power estimation, and a throughput constraint for synthesis. Our algorithm works iteratively, *i.e.*, it begins with an initial solution, which is iteratively refined. This can happen in four ways: (A) simple or complex modules in the initial or intermediate solution are replaced by new ones, chosen from the library, (B) complex modules are resynthesized by descending the hierarchy, (C) a pair of simple (complex) modules is combined into one simple (complex) module which executes the functionality of both its constituents, and (D) a simple (complex) module is split into multiple simple (complex) modules. Move *A* allows us to take advantage of user-specified simple and complex modules, and move *B* allows us to tailor the complex modules to their environments. Move *C* helps create compact circuits, and move *D* helps create new optimization opportunities. Note that in case of move *B*, modules, whose internal descriptions are not available or cannot be altered, are not resynthesized. Furthermore, the algorithm can support chained, multi-cycled, and pipelined functional units, and deal with multi-function arithmetic-logic units (ALU's). It enables escape from local minima by considering a sequence of moves at a time, where individual moves can have negative gain, *i.e.*, lead to temporarily degraded RTL architectures.

2 Background

In this section, we introduce the basic concepts used in our work. We first explain the notion of hierarchical behaviors and cir-

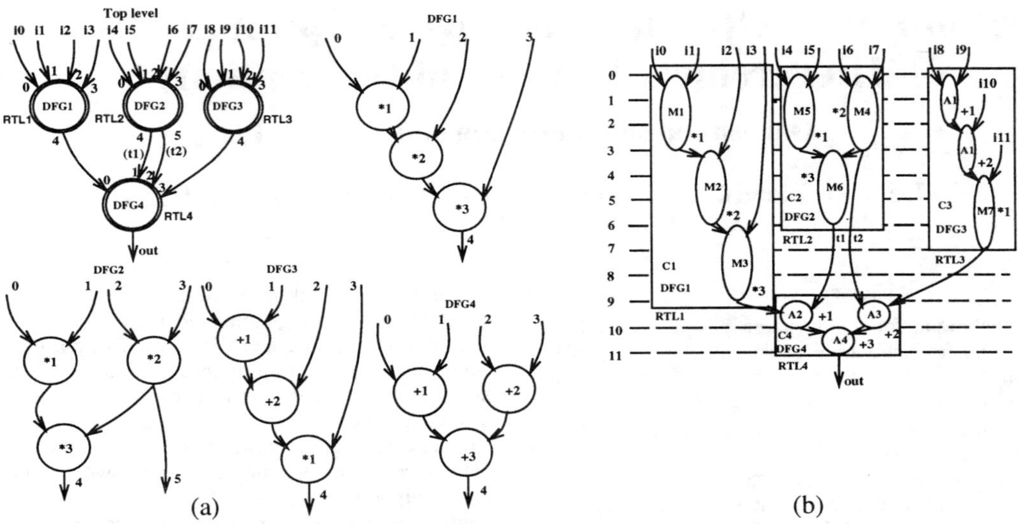

Figure 1: (a) A hierarchical DFG, and (b) a scheduled and assigned hierarchical DFG

cuits with an example.

Figure 1(a) shows a hierarchical DFG. The DFG labeled "Top level" is the *top-level DFG*. Nodes of the DFG can represent *operations*, like addition and multiplication, or *hierarchical nodes* like convolution and filtering. In general, if the input hierarchy is preserved, hierarchical nodes usually represent specific, well-characterized behaviors. However, if the hierarchy is the output of a partitioning process, hierarchical nodes could represent arbitrary behaviors. The edges in the top-level DFG entering and exiting hierarchical nodes are annotated with numbers which represent connectivity between the top-level edges and edges in the underlying DFG. For instance, the edge labeled 0 feeding node *DFG1* in the top-level DFG is connected to edge 0 in *DFG1*. The DFGs representing hierarchical nodes can in turn consist of simple and hierarchical nodes. Although the hierarchical DFG shown in Figure 1(a) is acyclic, the system is capable of handling loops at any hierarchical level.

A hierarchical node is implemented by an *RTL module*. Just as multiple operations can map to the same module library element, multiple hierarchical nodes can map to the same RTL module. An RTL module can itself be described hierarchically as an interconnection of RTL modules, functional units, multiplexers, and registers.

The *environment* of an RTL module for a hierarchical node mapped to it is an ordered set consisting of the arrival times of the inputs to the hierarchical node and the times when the outputs of the hierarchical node are *consumed*. The environment is defined only for RTL modules which are part of a scheduled, allocated, and assigned RTL circuit. The *profile* of an RTL module for a hierarchical node is defined as an ordered set consisting of the *expected* input arrival times and output arrival times. Unlike the environment, profile is defined for any module, irrespective of whether it is a part of an RTL circuit.

Example 1: Consider the hierarchical DFG shown in Figure 1(a) again. The hierarchical nodes are annotated with their names (placed inside the circle representing the node) and the names of instances of the RTL modules they are mapped to (placed outside the circle). For example, hierarchical node *DFG1* is mapped to instance *RTL1*. Figure 1(b) shows a schedule and assignment of the flowgraph in Figure 1(a). The environment of *RTL3* with respect to *DFG3* can be described by the set $Env(RTL3, DFG3) = \{0,\ 0,\ 0,\ 0,\ 9\}$ under the input/output ordering $\{0,\ 1,\ 2,\ 3,\ 4\}$. This environment assumes that the four inputs of *RTL3* arrive at time 0. Note that the output of *RTL3* is consumed by *RTL4* at time 9. The profile of *RTL3* with respect of *DFG3* can be represented as $Profile(RTL3, DFG3) = \{0,\ 0,\ 2,\ 4,\ 7\}$ under the same ordering. This is because the first two inputs of *RTL3* are expected at time 0, the third at time 2 and the fourth at time 4. When the inputs arrive in this fashion, the output is expected at time 7. Given the profile of a module and the input arrival times, the output arrival times can be computed as illustrated next. If the inputs at terminals 0, 1, 2, and 3 of *RTL3* arrive at times 2, 5, 3, and 7, respectively, then *RTL3* can begin operation at $max(2-0, 5-0, 3-2, 7-4) = 5$ cycles, which implies that the output is available after 12 cycles. In this example, each hierarchical node is mapped to a unique instance of an RTL module. In general, however, different hierarchical nodes could map to the same instance of an RTL module and functionally identical hierarchical nodes could map to different RTL modules or to different instances of the same RTL module. ■

At this stage, we formalize the notion of throughput constraint. For the class of applications we consider, the circuit should be able to process inputs at a constant rate, and it does not pay to be able to process inputs any faster. The sampling period is the number of time units allowed for completing one iteration of the design. The throughput is the inverse of the sampling period, *i.e.*, it is the number of samples processed per unit of time (clock cycles).

3 Motivational examples

In this section, we motivate key algorithmic features. The main constituents of our algorithm are the moves we use to refine the initial solution. As mentioned in Section 1, moves are of four types. The first two types of moves involve replacement or resynthesis of an existing module, and the next two types involve merging and splitting of modules. Note that modules can be simple or complex. Example 2 illustrates the benefits of the first two move types and Example 3 illustrates the benefits of the third.

Example 2: In this example, we illustrate the following: combining *coarse-grain* knowledge available at the higher levels of the hierarchy with *fine-grain* techniques traditionally associated with flattened high-level synthesis can significantly enhance solution quality. To this end, we perform moves of type *A* (which leverage off coarse-grain knowledge by replacing an instance of an RTL module by another one, better suited to the environment), and type *B* (which perform fine-grain resynthesis), on an example RTL circuit to demonstrate the power savings obtainable.

Consider the problem of synthesizing a power-optimized RTL circuit for the hierarchical DFG shown in Figure 1(a), under a sampling period constraint of 12 cycles. The simple modules present in the module library, their areas and delays (in number of clock cycles) are summarized in Table 1. These library modules also have

Table 1: Summary of functional unit and register properties

	add1	*add2*	*chained_add2*	*chained_add3*	*mult1*	*mult2*	*reg1*
Area	30	20	60	90	150	100	10
Delay	1	2	1	1	3	5	-

power models, which we omit for brevity [1]. The main fact to be noted is that, to perform the same sequence of operations, *mult2* consumes much less power than *mult1*. Figure 2 shows the complex modules available in the library. For each complex module, the DFG implemented, and the names of the instances of modules implementing individual operations are shown. For example, *RTL* module *C1* consists of three multipliers, *M1*, *M2*, and *M3*, which are of type *mult1*. Chains of functional units are shown enclosed within boundaries. For example, RTL module *C5* is a chain of three functional units of type *add1*. Note that a chain of adders can complete execution almost as fast as an individual adder, enabling a chain of three adders of type *add1* to complete in the same time, in cycles, as one adder of type *add1*. Also, we observe that the complex modules do not use all available resource sharing opportunities, *e.g.*, operations $+1$ and $+3$ in module *C4*, which can map to the same functional unit instance, are mapped to different instances. This is because these modules are power-optimized. Power optimization often (but not always) requires that operations be bound to different functional unit instances, even when the schedule permits the operations to share resources. A detailed discussion of the effect of resource sharing on power consumption can be found in [9]. The solution which is input to the procedure for the selection of moves of type *A* and *B* is shown in Figure 1(b). As we can see, DFGs *DFG1*, *DFG2*, *DFG3*, and *DFG4*, are mapped to complex modules of types *C1*, *C2*, *C3*, and *C4*, respectively.

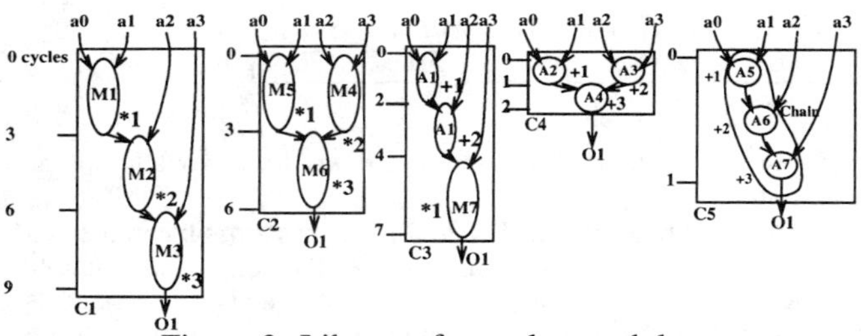

Figure 2: Library of complex modules

We now illustrate the application of moves of types *A* and *B* to the given solution. The first step is the identification of a set of modules to target. *RTL1*, *RTL2*, and *RTL3* are selected by this process. We then try to evaluate the *slack* available to the selected modules, *i.e.*, we ascertain the earliest input arrival times and the latest output arrival times whose satisfaction by the selected modules would ensure the schedulability of the implementation. For example, we can see that the outputs of *RTL2* can arrive nine cycles later than the inputs. In other words, *RTL2* can have a profile of $\{0,0,0,0,9,9\}$, though the profile for *RTL2* in the initial solution is $\{0,0,0,0,6,3\}$. We now apply a move of type *A*, followed by a move of type *B*. A move of type *A* would identify the "best" (lowest power solution if the objective is power optimization, and most compact solution if objective is area optimization) library module which can replace *RTL2*. In this example, the current implementation is indeed the best available in the library, therefore, no changes are made. We then apply a move of type *B*, *i.e.*, we resynthesize *RTL2* with the new, relaxed constraints, to see if we can obtain a better, lower power match for the required profile. The resynthesis procedure for *RTL2* would suggest the replacement of modules *M5* and *M4*, currently of type *mult1*, by *mult2*, which would significantly reduce power consumption. This illustrates the *utilization of high-level knowledge*, which results from a global view of the hierarchy, to *drive synthesis at lower hierarchical levels*, a key algorithmic feature.

The application of a move of type *A* can *change the DFG representing a hierarchical node*. For example, for the case of *RTL1*, module *C2* is recognized as a better candidate than module *C1* (note that *C1* and *C2* implement functionally equivalent behavior), and the substitution is performed. Recognizing such opportunities requires *knowledge provided by the user* regarding the functional equivalence of different DFGs. Many hierarchical DFGs are constructed out of several, commonly-used "building blocks" like *dot-product, butterfly, etc.*. These building blocks are well studied and have carefully analyzed behaviors, and a number of DFGs describing individual building blocks are available, each with its distinct advantages. A move of type *A* tries to select the "best" DFG which describes a hierarchical node, *i.e.*, the DFG which is best suited to the environmental constraints. ∎

A move of type *A* is followed by a move of type *B*, which attempts to improve the RTL module further. This is because the library element selected, though a good match, might not have been designed with the current environmental constraints in mind.

The example given ahead illustrates the benefits of moves of type *C*, which merge two modules into one. Candidate modules for merging can be either simple or complex.

Current hierarchical synthesis systems do not allow different (anisomorphic) DFGs to be implemented by the same RTL module. This often results in implementations with a large area. The main hurdles in performing this merging are as follows:

- It is difficult to determine the best candidate DFGs for merging prior to synthesis.
- Current research tackles the problem of designing an RTL circuit, which can support multiple behaviors, by just considering it as a multi-behavior high-level synthesis problem. Thus, a traditional technique would need to perform scheduling, allocation, and assignment afresh for every pair of modules which need to be merged. Clearly, the complexity of this approach would render it unusable in practice.

Our algorithm overcomes the first limitation because it is iterative, *i.e.*, during the course of the algorithm, different merging configurations are considered, and the best one is selected. Since we need to assess several sharing configurations during synthesis, our procedure to derive an RTL module which can execute multiple DFGs needs to be fast. We solve this problem by just *embedding the RTL modules, which implement the DFGs, into a new RTL module*. The schedule, assignment, *etc.*, for individual DFGs is unaltered, and the merged RTL module cannot execute the DFGs in parallel. The procedure for merging tries to find an embedding which honors clock cycle constraints, while minimizing the area of the merged RTL module.

Example 3: This example illustrates the area saving obtainable by implementing different DFGs on the same RTL module. In Figure 3, DFGs *DFG1* and *DFG2* are mapped to RTL modules *RTL1* and *RTL2*, respectively. Operations in the DFGs are shown annotated with the names of the modules they are mapped to, and variables are shown annotated with their names (unparenthesized) and the names of the registers they are mapped to (parenthesized). RTL module *NewRTL* can execute both the DFGs, and preserve the original schedules and assignments of the individual DFGs. All three RTL modules presented in this example were placed and routed using tools from the OCTTOOLS suite. Puppy was used

[1] We use the power models presented in [10]. Details of the power estimation methodology can be found in [8]

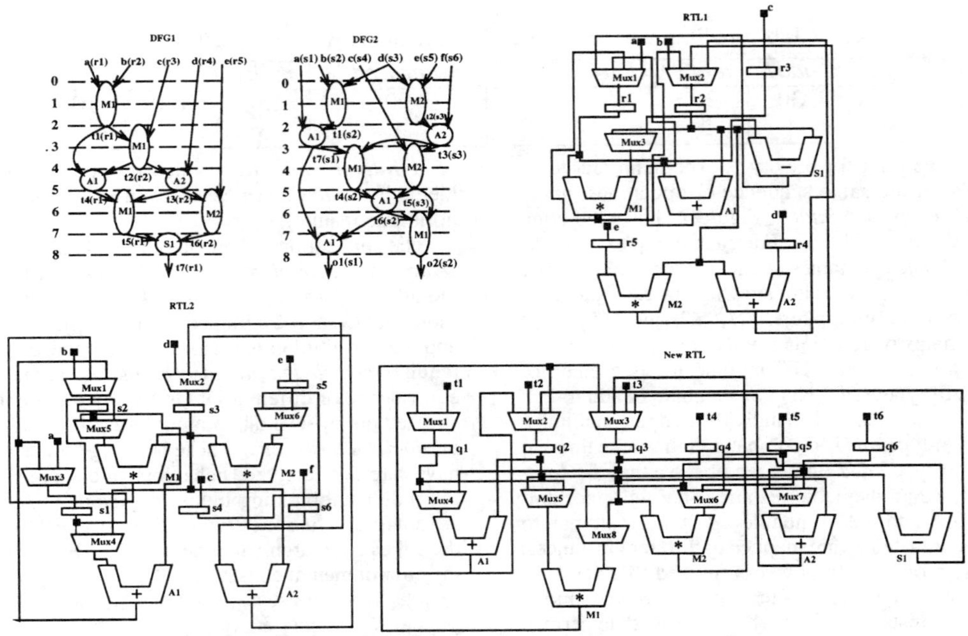

Figure 3: Mapping of two distinct DFGs to the same RTL module

Table 2: Labeling the new RTL module to implement DFG1 and DFG2

New RTL	*q1*	*q2*	*q3*	*q4*	*q5*	*q6*	*A1*	*A2*	*M1*	*M2*	*S1*
RTL1	-	*r1*	*r2*	*r3*	*r4*	*r5*	*A1*	*A2*	*M1*	*M2*	*S1*
RTL2	*s1*	*s2*	*s3*	*s4*	*s5*	*s6*	*A1*	*A2*	*M1*	*M2*	-
Library	*reg*	*reg*	*reg*	*reg*	*reg*	*reg*	*Add1*	*Add1*	*Mult1*	*Mult1*	*Sub1*
Area	*5*	*5*	*5*	*5*	*5*	*5*	*20*	*20*	*50*	*50*	*20*

for placement and Mosaico was used for routing. *RTL1* had an area of 57.94 units, *RTL2*, 53.89 units, and *NewRTL*, 61.67 units. This significant area saving is possible, without significant computational effort, due to the technique of *RTL embedding*, which we have devised. Our technique simply constructs a new RTL module, *NewRTL*, in which the original RTL modules can be embedded. The goal of our procedure is to find the minimum area embedding (including a measure of interconnect) which satisfies clock cycle constraints. The correspondence between the components of *NewRTL* and their counterparts in *RTL1* and *RTL2* is given in Table 2. The module library used and the area of the simple modules are also shown. ∎

4 The algorithm

In this section, we present a brief description of the basic components of our synthesis algorithm and their interactions. Details of individual components can be found in [8].

Figure 4 describes the top level of our synthesis algorithm. The algorithm accepts as input a DFG, a library of simple and complex RTL modules, an objective function to optimize (area or power), and a constraint that specifies when the primary inputs arrive and when the primary outputs are expected. The core iterative improvement procedure is enclosed in loops which iterate through the set of available V_{dd}'s and allowed clock periods. [2] The best solution for a given V_{dd} and clock period is obtained through variable-depth *iterative improvement*, a general optimization technique that starts with an initial solution and improves its quality through the application of a sequence of *moves* [11]. This algorithm derives its power from the ability to perform moves which worsen the quality of the solution, as an intermediate step, in striving for a globally optimal solution.

First, the library is searched to find the best possible implementation of the input behavior that satisfies the constraints imposed. If such an implementation cannot be found, one is synthesized using the routine INITIAL_SOLUTION. This routine maps each simple node in the DFG to the fastest implementation available in the library. DFGs which represent hierarchical nodes are handled in the same manner. Each operation is mapped to a separate functional unit, and each variable to a separate register, resulting in a completely parallel architecture. The initial solution is improved through the application of moves of types *A*, *B*, *C*, and *D*. When a move is performed, its validity is checked by scheduling to make sure that the throughput constraints are still met.

The scheduler is called whenever we need to check the validity of an assignment. Before scheduling, we derive an ordering for the operations that need to execute on the same functional unit or RTL module, and variables that need to be stored in the register. This ordering imposes extra dependencies in the DFG, which are modeled as edges. Under this scenario, scheduling of a node reduces to the problem of finding the longest path from a primary input to the node [10]. Scheduling of DFGs is a well-studied problem [12], and we do not discuss it further in this paper.

As mentioned in Example 2, moves of types *A* and *B* alter the implementation of a selected set of RTL modules and functional units with a view to minimizing the power consumption (or area) of the circuit. The application of these moves can be described by the algorithm shown in Figure 5. This algorithm corresponds to statement 7 of Figure 4. As is evident from the pseudocode shown in Figure 5, this process can be divided into three distinct parts: module group formation, constraint derivation, and resynthesis. Mod-

[2] The set of V_{dd}'s and clock periods can be drastically pruned by using a procedure from [10] to obtain the subset that needs to be considered.

```
SYNTHESIZE(DFG G, LIBRARY L, CONSTRAINT_SET C,
           OBJECTIVE Obj){
foreach supply voltage V_dd in the pruned supply voltage set {
  foreach clock period clk in the pruned clock period set {
    Cur_DP ← INITIAL_SOLUTION(G, L);
    //If the library cannot supply an implementation of
    //the behavior, generate an initial implementation
    Pass_gain ← ∞; Best_DP ← Cur_DP;
    while(Pass_Gain > 0){
      for(i = 0; i < MAX_MOVES; i++){
        MOVE m1 = GET_BEST_TYPE-
          _A_AND_B_MOVE(G, L, Cur_DP, C, Obj);
        MOVE m3 = GET_BEST_RESOURCE-
          _SHARING_MOVE(G, L, Cur_DP, C, Obj);
        if (Gain(m3, Obj) < 0)
          MOVE m3 = GET_BEST_RESOURCE_-
          SPLITTING_MOVE(G, L, Cur_DP, C, Obj);
        if (Gain(m1, Obj) > Gain( m3, Obj))
          IMPLEMENT_M1_AND_M2(Cur_DP, m1);
        else IMPLEMENT_M3(Cur_DP, m3);
      }
      Pass_Gain = Get sequence of moves
      with best cumulative gain;
      if (Pass_Gain ≤ 0)
        break;
      apply best move sequence to Cur_DP;
    }
    Best_DP = best solution seen so far;
}}}
```

Figure 4: Iterative improvement

ule group formation divides the modules in the current RTL implementation into groups for the steps that follow. This is followed by constraint derivation where each operation which is performed by a module belonging to the group under consideration is assigned a new constraint for synthesis. For instance, an addition operation which initially took two cycles to complete could be set a constraint of five cycles, or, a filter which accepted inputs in cycle 0 and produced an output in cycle 5 could be constrained to produce an output in cycle 7 due to this process. The new constraints must preserve schedulability of the implementation, *i.e.*, the new architecture, when rescheduled, should meet throughput constraints. Modules are then resynthesized to meet the new constraints. Note that the term "resynthesis" refers to alterations performed on RTL modules as well as choosing new functional units or RTL modules from the library. The relaxation in constraints might enable replacement of the library element performing the function by a slower, potentially lower power library element.

5 Experimental results

We have implemented the behavioral synthesis framework presented in the previous sections as the program *H-SYN* in 44,000 lines of C++. We have performed experiments to evaluate our techniques using several behavioral descriptions of digital signal and image processing applications. *H-SYN* reads in a textual description of the hierarchical DFG, and performs V_{dd} selection, clock selection, module selection, scheduling, allocation, and assignment to result in a highly power/area-efficient RTL circuit that consists of a datapath netlist, and a finite-state machine description of the controller. The controller and datapath netlists are merged and mapped to the MSU standard cell library using the SIS logic synthesis system. Placement and routing are performed using tools from the OCTTOOLS suite. A switch-level circuit extracted from the layout is simulated by a switch-level simulator, IRSIM, and the capacitance switched is recorded and used to compute the power.

We have synthesized flattened and hierarchical examples of several existing benchmarks to optimize for area and power.

```
GET_BEST_TYPE_A_AND_B_MOVE (DFG G, LIBRARY L,
 DATAPATH Cur_DP, CONSTRAINT_SET C,
OBJECTIVE Obj){
//Obj can be area or power
  Module_groups = Φ;
  Best_gain = -∞;
  Module_groups = FORM_MODULE_GROUPS(Cur_DP);
  //Module_groups is a set whose elements are sets of
  // modules which will be considered together for synthesis
  mgroup = Φ ;
  foreach(mgroup, Module_groups){
    Gain = 0;
    DERIVE_CONSTRAINTS(mgroup, Cur_DP);
    //derive constraints for resynthesis of each
    //node mapped to module group mgroup.
    foreach(M, mgroup){
      Resynthesize module M based on the constraints
      derived in the DERIVE_CONSTRAINTS routine
      Gain = Gain - COST(M, Obj);
    }
    if (Gain > Best_gain){
      store the current solution as the best solution;
      Best_gain = Gain;
}}}
```

Figure 5: Module selection moves

hier_paulin is a hierarchical DFG obtained by unrolling the well-known benchmark. *Paulin*. *dct* implements discrete cosine transform. *avenhaus_cascade*, *dct*, *iir*, and *lat* are hierarchical DFGs which represent filters and are a part of the HYPER package [4]. *test1* is the DFG in Figure 1(a). In all cases, the input hierarchy was preserved. To assess the efficacy of our technique, we compared our hierarchical algorithm to the flattened version of the same algorithm [10]. The flattened technique also takes advantage of resource sharing, module selection, V_{dd} selection, clock selection, and scheduling for low power, and is one of the most comprehensive techniques available today.

The results obtained are tabulated in Table 3. Column *L.F.* represents the *laxity factor* used for synthesis. Laxity factor is defined as the ratio of the given sampling period (inverse of specified throughput) to the minimum sampling period that can be attained for the input DFG using the library of modules provided. Columns *Flat* and *Hier* represent flattened and hierarchical versions of the same behavioral description. The circuits in column *A* were synthesized to optimize for area at a supply voltage of 5 *Volts* and subsequently voltage-scaled for low power operation. The circuits in column *P* were synthesized to optimize for power. Rows *A* and *P* represent the area and power consumption, respectively, of the synthesized circuits. The area (power) is normalized with respect to the area (power) of an area-optimized, non-V_{dd}-scaled architecture synthesized from a flattened input description at the same laxity factor. For example, if row *P* of column *P*, under major column *Hier* is x, then the power consumed by a circuit obtained from a hierarchical behavioral description, optimized for power, is x times the power consumed by a circuit obtained from a flattened behavioral description, optimized for area, and operating at 5 *Volts*. All entries in row *A* of column *A* under major column *Flat* are 1. This is because this column represents the area of an area-optimized circuit, synthesized from a flattened behavioral description, and V_{dd}-scaled for low power, normalized with respect to the area of an area-optimized circuit, synthesized from a flattened behavioral description, which operates at 5 *Volts*. Since V_{dd}-scaling makes no difference to area, these entries are 1.

The area-power-synthesis time trade-offs for different laxity factors for flattened (*Fl*) and hierarchical (*Hi*) behavioral descriptions are summarized in Table 4. In this table, the area and power ratios are with respect to flattened area-optimized architectures, and

Table 3: Area (normalized) and power (normalized) results

Circuit	A/P	L.F. = 1.2				L.F. = 2.2				L.F. = 3.2			
		Flat		Hier		Flat		Hier		Flat		Hier	
		A	P	A	P	A	P	A	P	A	P	A	P
avenhaus_cascade	A	1	1.26	1.08	1.34	1	1.39	1.13	1.47	1	1.50	1.08	1.54
	P	1	0.66	0.87	0.58	1	0.54	0.85	0.47	1	0.55	0.85	0.47
lat	A	1	1.36	0.91	1.34	1	1.27	1	1.39	1	1.07	0.96	1
	P	1	0.55	0.99	0.54	1	0.30	0.93	0.30	1	0.52	0.83	0.39
dct	A	1	1.33	0.83	1.0	1	1.18	0.84	0.91	1	1.04	0.74	0.96
	P	1	0.62	0.90	0.55	1	0.50	0.92	0.44	1	0.58	0.92	0.50
iir	A	1	1.21	1.07	1.28	1	1.13	1.01	1.23	1	1.13	0.99	1.23
	P	1	0.61	0.93	0.61	1	0.46	0.86	0.39	1	0.42	0.81	0.36
hier_paulin	A	1	1.20	1.05	1.32	1	1.23	1.13	1.34	1	1.17	1.16	1.33
	P	1	0.59	0.97	0.55	1	0.57	0.97	0.50	1	0.43	0.87	0.37
test1	A	1	1.34	1.23	1.85	1	1.37	1.38	1.70	1	1.52	1.42	1.65
	P	1	0.54	0.68	0.44	1	0.39	0.68	0.30	1	0.37	0.75	0.29

are averaged over all examples for the corresponding laxity factor. For example, if the entry in column *Hi*, major column *Area ratio* is x, then the average area of power-optimized circuits, synthesized from hierarchical behavioral descriptions is x times the average area of area-optimized circuits, synthesized from flattened behavioral descriptions, for that laxity factor. Under the major column *Power ratio*, comparisons of power-optimized architectures are made against area-optimized architectures at $5V$ (labeled $5V$) and those that have been V_{dd}-scaled to just meet the sampling period constraint (labeled V_{dd}-sc). Major column *Synth. time* gives the synthesis time, in seconds, for area- and power-optimized circuits, averaged over all examples for a given laxity factor. The experiments were conducted on an SGI Challenge workstation with 256 *MB* RAM.

The results obtained indicate that our hierarchical synthesis algorithm produces circuits whose area and power consumption are comparable to those of circuits produced from flattened descriptions. This is significant because most previous hierarchical high-level synthesis systems synthesize circuits which are significantly less compact than than their flattened equivalents [3]. The synthesis time, which is measured as the time to compile the behavioral description into an RTL implementation, is significantly less for hierarchical circuits.

For application-specific integrated circuits, which typically have a short design cycle (order of weeks), reduction in synthesis time can be extremely significant. For larger hierarchical behavioral descriptions, we expect the ratio of synthesis times for flattened and hierarchical synthesis to be even greater. Note that flattened synthesis can also take advantage of module selection, scheduling, resource sharing, clock and V_{dd} selection techniques available for hierarchical synthesis. The efficacy of our power optimization techniques is illustrated by the improvement in power obtained at a relatively low area overhead. Under hierarchical synthesis, our method results in upto 6.7-fold reduction in power at area overheads not exceeding 50% over area-optimized circuits working at 5 *Volts*. On an average, hierarchical power-optimized designs consumed 13.3% less power than flattened designs optimized for power, and hierarchical area-optimized designs had an area overhead of 5.6% over flattened, area-optimized designs.

6 Conclusions

In this paper, we presented a methodology for the synthesis of power- and area-optimized circuits from hierarchical behavioral descriptions. Our algorithm can uniformly handle arbitrarily deep hierarchies, and aims at using the high-level knowledge available at higher hierarchical levels to drive fine-grain optimization techniques at lower levels. Our approach allows for the use of libraries of specialized RTL modules, such as filters and FFT's, in an efficient manner. Furthermore, it allows for customization of these libraries if needed. Unlike previous hierarchical synthesis methods, our algorithm allows multiple DFGs to be executed on the same RTL module. This is made possible by the technique of RTL embedding which we have developed for the purpose. We implemented our ideas in a program named *H-SYN* written in C++. Experimental results performed on commonly available benchmarks, including some from the industry, demonstrated the efficacy of our techniques in synthesizing power-optimized circuits with relatively low area overheads within short run times.

Table 4: Summary of area (ratio), power(ratio) , and synthesis time (seconds) results

L.F	Area ratio		Power ratio				Synth. time (sec.)	
			5V		V_{dd}-sc			
	Fl	Hi	Fl	Hi	Fl	Hi	Fl	Hi
1.2	1.28	1.36	.51	.47	.60	.55	844	261
2.2	1.26	1.34	.31	.27	.46	.40	854	322
3.2	1.24	1.29	.19	.17	.48	.40	1029	357

References

[1] M. Potkonjak and J. Rabaey, "A scheduling and resource allocation algorithm for hierarchical signal flow graphs," in *Proc. Design Automation Conf.*, pp. 7–12, June 1989.

[2] D. S. Rao and F. J. Kurdahi, "Controller and datapath tradeoffs in hierarchical RT-level synthesis," in *Proc. Int. Symp. High Level Synthesis*, pp. 152–157, May 1994.

[3] D. S. Rao and F. J. Kurdahi, "Hierarchical design space exploration for a class of digital systems," *IEEE Trans. Computer-Aided Design*, vol. 1, pp. 282–294, Sept. 1993.

[4] A. P. Chandrakasan, M. Potkonjak, R. Mehra, J. Rabaey, and R. W. Brodersen, "Optimizing power using transformations," *IEEE Trans. Computer-Aided Design*, vol. 14, pp. 12–51, Jan. 1995.

[5] A. Raghunathan and N. K. Jha, "An ILP formulation for low power based on minimizing switched capacitance in datapath allocation," in *Proc. Int. Symp. Circuits & Systems*, pp. 1069–1073, May 1995.

[6] A. Dasgupta and R. Karri, "Simultaneous scheduling and binding for power minimization during microarchitecture synthesis," in *Proc. Int. Symp. Low Power Design*, pp. 69–74, Apr. 1995.

[7] L. Goodby, A. Orailoglu, and P. M. Chau, "Microarchitectural synthesis of performance constrained, low power VLSI designs," in *Proc. Int. Conf. Computer Design*, pp. 323–326, Oct. 1994.

[8] G. Lakshminarayana and N. K. Jha, "Hierarchical synthesis for low power," Tech. Rep., CE-J98-003, Princeton University, 1998.

[9] A. Raghunathan and N. K. Jha, "Behavioral synthesis for low power," in *Proc. Int. Conf. Computer Design*, pp. 255–270, Oct. 1994.

[10] A. Raghunathan and N. K. Jha, "An iterative improvement algorithm for low power datapath synthesis," in *Proc. Int. Conf. Computer-Aided Design*, pp. 597–602, Nov. 1995.

[11] C. H. Papadimitriou and K. Steiglitz, *Combinatorial Optimization: Algorithms and Complexity*. Prentice Hall, Englewood Cliffs, NJ, 1982.

[12] D. D. Gajski, N. Dutt, A. Wu, and S. Lin, *High-Level Synthesis: Introduction to Chip and System Design*. Kluwer Academic Publishers, Boston, 1992.

Approximation and Decomposition of Binary Decision Diagrams*

Kavita Ravi
University of Colorado

Kenneth L. McMillan
Cadence Design Systems

Thomas R. Shiple
Synopsys

Fabio Somenzi
University of Colorado

Abstract

Efficient techniques for the manipulation of Binary Decision Diagrams (BDDs) are key to the success of formal verification tools. Recent advances in reachability analysis and model checking algorithms have emphasized the need for efficient algorithms for the approximation and decomposition of BDDs. In this paper we present a new algorithm for approximation and analyze its performance in comparison with existing techniques. We also introduce a new decomposition algorithm that produces balanced partitions. The effectiveness of our contributions is demonstrated by improved results in reachability analysis for some hard problem instances.

1 Introduction

Symbolic state enumeration techniques based on Binary Decision Diagrams (BDDs [2]) have revolutionized formal verification [8, 4, 17, 1, 14]. They have two key features that make them suitable to the exploration of very large state graphs: They represent sets compactly, and they avoid explicit enumeration in image computation. Given the transition relation of a system, $R(x, y)$, and a set of states, $F(x)$, the set of states reachable in one step from states in F, $T(y)$, is computed by

$$T(y) = \exists_x [R(x, y) \cdot F(x)].$$

This simple formula is at the heart of efficient algorithms for reachability analysis, language containment, and model checking. However, numerous improvements must be applied to the basic idea in order to make it work for realistic problems. The common aim of these improvements is to control the size of the BDDs created and manipulated during state exploration. This has been achieved by keeping the transition relation in partitioned form [28, 3, 10, 22]; by controlling the BDD variable order [12, 24]; by abstracting the system to be verified [16, 13, 15, 7]; or by abandoning pure breadth-first search in favor of more flexible approaches [23, 5, 21, 19].

Abstractions and methods that mix breadth-first and depth-first search rely, sometimes in crucial ways, on operations that approximate and decompose BDDs. In this paper we present new algorithms for these problems, and compare new and existing techniques. Our contributions help high-density traversal [23] achieve significant acceleration over the conventional breadth-first approach.

Minimization and decomposition of boolean functions have been the subject of much research in the last half century. However, early work is mostly inapplicable to decision diagrams; hence, the first relevant results are the *constrain* [8] and *restrict* [9] algorithms. We discuss *restrict* briefly now for two reasons: First, to provide the background required in Section 3, and second, to illustrate an approach to manipulation of BDDs that is common to the algorithms introduced in this paper.

*This work was supported in part by NSF grant MIP-94-22268 and SRC contract 96-DJ-560.

DAC 98, San Francisco, California
©1998 ACM 0-89791-964-5/98/06..$5.00

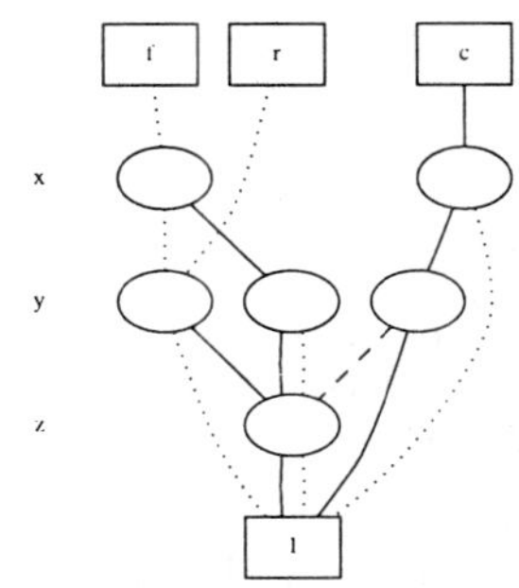

Figure 1: Remapping in *constrain* and *restrict*. Solid lines are *then* arcs. Dashed lines are regular *else* arcs, and dotted lines are complement arcs.

Throughout this paper we assume that a variable order is given—possibly derived dynamically. We use $|f|$ to denote the size of the BDD of function f under the given order; and $||f||$ to denote the number of minterms of function f. We only distinguish between a function and its BDD when it is necessary to avoid ambiguity.

The inputs to *restrict* are a BDD to be minimized, f, and the BDD of a *care set*, c. The output is a BDD, $f \Downarrow c$, that agrees with f wherever $c = 1$. The choice of $(f \Downarrow c)(x)$ when $c(x) = 0$ is heuristically made to reduce $|(f \Downarrow c)(x)|$. Consider Figure 1. One child of c is 0; hence the corresponding subgraph of $f = xf_t + x'f_e$ can be changed. If it is replaced by the subgraph of the other child—in the figure, f_e is replaced by f_t—not only the nodes that are only in the replaced child are eliminated, but the parent node becomes redundant as well. The process then recurs on f_t to yield r. (Which in this case is f'_e.) This *remapping* step, which *restrict* recursively applies while it traverses f and c, is its basic optimization technique; it is quite typical of BDD algorithms because it is a local transformation, whose purpose is to increase the sharing of nodes. We shall see that the remapping step plays a significant role in both our approximation and decomposition algorithms.

This paper is organized as follows: Section 2 discusses the problem of approximating BDDs and presents a new algorithm called *remapUnderApprox*. Section 3 is devoted to BDD decomposition. Section 4 presents experimental results obtained by applying the new approximation algorithm to reachability analysis, as well as a detailed comparison of various approximation techniques. Conclusions are in offered in Section 5.

2 Approximation

BDD approximation is the problem of deriving from a given BDD another BDD smaller in size, and whose function is at a low Hamming distance from the input BDD. (That is, differing from the input BDD in a small number of input assignments.) Let $\alpha(f)$ be the BDD produced by the application of approximation algorithm α to the BDD of f. Usually, the function of the approximating BDD is required to be either a subset or a superset of the input function, that is,

$$\alpha(f) \leq f \quad \text{or} \quad \alpha(f) \geq f.$$

For an *underapproximation* algorithm α (such that $\alpha(f) \leq f$), $\neg\alpha(\neg f) \geq f$. Hence, we only discuss underapproximation.

Underapproximation algorithms must trade off the size of the result for the distance from f. The two trivial solutions, 0 and f itself, are seldom useful. A natural way to rank different approximations is by their density, denoted by $\delta(\alpha(f))$ and defined in [23] by:

$$\delta(g) = \frac{||g||}{|g|}.$$

High density corresponds to a concise representation, and is therefore desirable. (For overapproximation, we want to maximize the density of the complement of the result.) However, in some circumstances, the number of nodes, or the number of minterms, and not just their ratio, are important.

Two algorithms for underapproximation were proposed in [23]; both are two-pass procedures, where the input is analyzed in the first pass and the result is produced in the second pass by replacing some arcs to internal nodes with arcs to the constant 0.

Heavy-branch subsetting (HB) determines how many minterms are in the function rooted at each internal node, and how many nodes would be eliminated by replacing arcs pointing to it. The second pass then proceeds from the root of the BDD and discards the "light branch" of each node, that is, the child with fewer minterms, until the size of the residual BDD crosses a given threshold. The result is a BDD with a string of nodes at the top, each with one child as the constant 0.

Short-path subsetting (SP) is based on the idea that short paths in a BDD correspond to large implicants of the function and use few nodes. Therefore in its first pass it determines the length of the shortest paths through each node, and then builds the result by discarding nodes with no short paths through them.

Both HB and SP run in time linear in $|f|$: This implies that only a limited study of the BDD is possible. Also, both are controlled by an upper bound on $|\alpha(f)|$, and though they both strive to increase density, they may occasionally decrease $\delta(\alpha(f))$ to reach the target $|\alpha(f)|$. Therefore, in analogy to the concept of *safe minimization* of [11], we introduce the following definition:

Definition 1 *An underapproximation algorithm α is* safe *if $\delta(f) \leq \delta(\alpha(f))$.*

It follows from Definition 1 that a safe overapproximation algorithm increases the density of the complement.

2.1 A Safe Underapproximation Algorithm

We now present a new algorithm for safe underapproximation whose main idea first appeared in [26, 25]. The algorithm is called *remapUnderApprox* (RUA) and differs from HB and SP in two main respects: 1) It computes for each node a lower bound on the increase in density that would follow from its replacement. 2) It can replace a node not just with a constant 0, but also with other subfunctions of f.

remapUnderApprox consists of three passes, as shown in Figure 2. The first pass, *analyze*, consists of a depth-first search of the BDD to be approximated. For each node, the number of minterms in the function rooted at the node, and the number of arcs pointing to it from other nodes in f (referred to as *functionRef* later) are computed. This information is collected in the data structure *info*, which is then updated in the second pass and used in the third pass to build the result; *info* also holds global information on the BDD

```
remapUnderApprox(f, threshold, quality) {
    info = analyze(f);
    info = markNodes(f, threshold, quality, info);
    return(buildResult(f, info));
}
```

Figure 2: Remapping underapproximation algorithm.

```
markNodes(f, threshold, quality, info) {
    queue = emptyPriorityQueue();
    enqueue(queue, f, level(f));
    while (queue not empty) {
        if (resultSize(info) ≤ threshold) return(info);
        node = dequeueFirst(queue);
        replacement = findReplacement(node,info);
        if (densityRatio(replacement, info) > quality)
            info = updateInfo(info, node, replacement);
        enqueueChildren(queue, node, replacement);
    }
    return(info);
}
```

Figure 3: Node marking procedure.

and its approximation, and other fields for each node that are used by *markNodes* and *buildResult*. In particular, for each node it holds its replacement status, which initially is *do-not-replace*.

2.1.1 Determining Node Replacement

The second step of *remapUnderApprox* is detailed in Figure 3. The BDD of f is traversed in top-down fashion using a priority queue, so that all nodes labeled by variable x are examined before any node labeled by variable y is examined, if x precedes y in the order. When a node is retrieved from the queue, it is tested for replacement. We have implemented three types of replacement. The first type, *remap*, replaces a node by one of its children as in *constrain* [8]: It can be applied if the function rooted at one child of the current node is contained in the function rooted at the other child. (That is the function rooted at the current node is unate in its top variable.) The current node is replaced by the smaller child: If $f = xf_t + x'f_e$ and $f_e \leq f_t$, for instance, then f is replaced by f_e. The second type of replacement, *replace-by-grandchild*, applies when the two children of the current node are labeled by the same variable, and they share either the "then" child or the "else" child. Under these conditions, the shared grandchild can be used to replace the current node: If $f = x(yf_{tt} + y'f_{te}) + x'(yf_{et} + y'f_{ee})$, and $f_{tt} = f_{et}$, then $yf_{tt} \leq f$ and yf_{tt} can replace f. The last and simplest replacement type is *replace-by-0*: it can always be applied.

Procedure *findReplacement* tries the replacements in the above order and selects the first that applies. The chosen replacement is then definitely accepted or rejected based on the impact that it has on the density of the result. The impact is computed by counting exactly the number of minterms that will be lost to the replacement, and by computing a lower bound to the number of nodes that will be saved. *findReplacement* computes these two quantities as detailed in the next section. It then returns the result in "replacement," which also holds the replacement type.

2.1.2 Estimating the Change in Density

The number of minterms lost if a node is removed from a BDD is computed as the product of two quantities: The minterm count of the function rooted at the node (considered as a function of all the variables in f) and the fraction of paths from the root of f to either constant that go through the current node. Procedure *markN-*

```
nodesSaved(g,info) {
        savings = 0;
        setLocalRef(info,g,functionRef(info,g));
        queue = emptyPriorityQueue();
        enqueue(queue, g, level(g));
        while (queue not empty) {
                v = dequeueFirst(queue);
                if (localRef(info,v) = functionRef(info,v)) {
                        savings++;
                        increaseLocalRefOfChildren(info,v);
                        enqueueChildren(queue, v);
                }
        }
        return(savings);
}
```

Figure 4: Node savings computation. Not shown is the initialization of *localRef* to 0 for all nodes.

odes computes this fraction during the top down traversal, so that for each node it may take into account the replacements for nodes higher in the BDD. The minterm count, on the other hand, is not affected by changes to the upper part of the BDD and therefore is computed during the first pass. The complexity of counting the minterms lost to replacements is linear in $|f|$ if this way of breaking down the computation is used.

The bound on the node savings, on the other hand, is simply computed by finding how many nodes are dominated[1] by the nodes that are eliminated. The eliminated nodes are the current node, for *replace-by-0*; the containing child of the current node for *remap*; either child of the current node for *replace-by-grandchild*. In addition, the current node is always eliminated. In all three cases the procedure of Figure 4 is used to find the dominated nodes: A top-down search of the BDD is started from each node g that is being eliminated. When a node v is retrieved from the queue all its predecessors reachable from g have been processed. Hence, *localRef* gives the number of predecessors of v reachable from g. On the other hand, *functionRef* gives the number of predecessors of v node reachable from f. If these two numbers are the same, then g dominates v. The total number of predecessors returned by *functionRef* is initialized by *analyze* and updated by *updateInfo*. The other important tasks of *updateInfo* are to record the type of replacement chosen, and to update the number of minterms and the estimate of the size of the result.

The nodes that are dominated by nodes that are going to be eliminated clearly contribute to the node savings of a replacement. However, further nodes may be saved due to sharing caused by the replacement in the part of the BDD above the current node. This is why we only have a lower bound on the reduction in size of the BDD.

Function *densityRatio* computes the ratio of the densities with and without the selected replacement and compares it to "quality." If parameter "quality" is greater than or equal to 1, then only replacements that increase the density are accepted. Values smaller or greater than 1 can be used when the result obtained for quality = 1 have too many or too few nodes, respectively.

Once all nodes have been marked with their replacement status, *buildResult* traverses f in depth-first manner and for each node returns the result of applying the selected replacement.

2.1.3 Discussion

The worst-case run time of *remapUnderApprox* is quadratic in $|f|$, due to the total cost of checking for each node the containment of each child in the other, and to the call to *nodesSaved*. In practice, however, the algorithm takes time linear in $|f|$ in most cases.

So far, we have ignored complement arcs for the sake of simplicity. The algorithm as outlined applies also in their presence, with a few modifications. First, the only replacement that can be applied to nodes reachable through paths of different complementation parity is *replace-by-0*. In addition, the estimate of the node savings is no longer guaranteed to be a lower bound if *replace-by-0* is applied to those nodes. Therefore, replacement is only applied to nodes reachable through paths of one complementation parity only. Minterm counts and fractions of paths are counted separately for the positive and negative phase, and the minterms lost to replacements are computed as a weighted sum.

The original *bddUnderApprox* algorithm of [26, 25] differs from *remapUnderApprox* in the following aspects:

- The cost function is a convex combination of the number of minterms and the number of nodes, instead of the ratio of the two numbers.
- Only *replace-by-0* is used. Because of this restriction, it is easy to replace nodes that are reachable through paths of different complementation parity (if complement arcs are used). The resulting algorithm is not safe, because replacement of one such node may cause splitting of a node higher in the BDD, but in most cases it gives denser results than the safe version.

2.2 Compound Algorithms

Given an approximation algorithm $\alpha(f)$, suppose a minimization algorithm $\mu(l,u)$ is also given such that $l \leq \mu(l,u) \leq u$. We say that $\mu(l,u)$ is *safe* [11] if $|\mu(l,u)| \leq |l|$ and $|\mu(l,u)| \leq |u|$. Then

$$\mu(\alpha(f), f)$$

is an approximation algorithm, which is safe if both α and μ are safe. Also, if α_1 and α_2 are approximation algorithms, then

$$\alpha_1(\alpha_2(f))$$

is an approximation algorithm, which is safe if both α_1 and α_2 are safe. These simple rules can be used to trade off CPU time for the increased density of the results. We call the resulting algorithms *compound* approximation algorithms. As an example one can repeatedly apply RUA starting with a quality factor greater than 1 and decreasing it at each iteration until it equals 1. This has the effect of mitigating the greediness of RUA. Other compound algorithms are discussed in Section 4. Methods that are not compound are *simple*.

3 Decomposition

Decomposition is another important approach to reducing the size of large BDDs. Decomposition of BDDs is closely related to finding efficient partitioned representations of a given boolean function. Partitioned representations [20, 3] may be derived in the process of building a BDD or by decomposing a given BDD. The former is easier to obtain when some structural information, such as the network, is provided. Auxiliary variables are introduced while constructing the BDD. They alleviate ordering constraints and reduce the size of the partitions. Alternately, a static analysis of the network may yield good decomposition points.

We concentrate on the problem of functional decomposition—decomposing a large BDD into a set of smaller ones, which can be combined to form the original BDD. Given a BDD, a variable order, and no prior knowledge of the nature of the function,

[1] A node v dominates a node w if all paths from the root of the BDD to w go through v.

decomposition techniques require an analysis of the given BDD. In the simplest case, these may be conjunctive or disjunctive decompositions. The goal is to reduce the shared size (to occupy less storage space) as well as the individual sizes (for easier manipulation) of the decomposed set as compared to the original BDD size.

Prior Work. To form conjunctive factors for a function f such that

$$f = x \cdot f_x + x' \cdot f_{x'}$$

we can create two factors g and h where

$$f = g \cdot h\ ;\ \ g = x + f_{x'}\ ;\ \ h = x' + f_x \tag{1}$$

Each of the 2 factors has a different cofactor set to 1. Disjunctive partitioning, in this approach, is completely symmetric to the conjunctive method and is obtained by setting the different cofactors to 0.

Cabodi *et al.* [6] and Narayan *et al.* [19] propose decompositions based on Equation 1. They pick a suitable set of variables and both produce disjunctive factors by computing cofactors of the function f with respect to all boolean assignments of the chosen variables. They use different cost functions for the choice of variables. The cost functions heuristically reduce the size of the decomposed BDDs. Cabodi *et al.* use a threshold size to derive the decomposed BDDs and Narayan *et al.* use a pre-specified number of partitions. These methods have been tested in reachability analysis where a balanced partition is important for efficient traversal. Cabodi *et al.* also observed that obtaining a disjunctive partition using dense subsets [23] does not yield balanced partitions.

McMillan [18] presents a different approach to decomposition. His method computes a canonical conjunctive decomposition based on exploiting conditional independence between variables. The algorithm performs successive existential abstraction and cofactoring to reduce the size of the decomposed representation. The number of factors produced are equal to the number of variables. The size of the decomposed representation is linear in the number of partitions and the size of the original BDD.

Our Approach. Our decomposition method is closely related to the first approach. In that method, variable x forms a cut in the BDD of f. Creating the factors amounts to setting the THEN children of nodes labeled x to 1 in g and the ELSE children to 1 in h. A more general strategy would be to set the opposite children of each node labeled x to 1 (instead of setting *all* positive or *all* negative cofactors to 1). A further generalization would be to relax the constraint of the choice of nodes—from those labeled x to an arbitrary set of nodes in f. We refer to these nodes as *decomposition points*.

The algorithm is illustrated in Figure 5. Given a set of *decomposition points*, our algorithm constructs the factors bottom up. At the decomposition points, factors are created using Equation 1. At every node above the decomposition point, factors are constructed by combining those of the children as shown below.

$$g = x \cdot g_T + x' \cdot g_E\ ;\ \ h = x \cdot h_T + x' \cdot h_E \quad \text{or}$$
$$g = x \cdot g_T + x' \cdot h_E\ ;\ \ h = x \cdot h_T + x' \cdot g_E$$

The algorithm maintains a cache for previously computed factors for each node. It also attempts to encourage node sharing by storing nodes created for each factor.

Decomposition Points. The choice of *decomposition points* is crucial to this method. We have experimented with two greedy approaches to pick these *decomposition points*.

Band: The first method is to pick nodes that are low enough in the BDD to achieve substantial reduction in the individual factor sizes. However, if the nodes are picked too low in the BDD, most of the recombination may be destroyed in building the factors. The required "middle band" is determined by using the distance of the node from the constant as a measure. This measure requires one pass of the BDD.

```
decomp (f) {
    if (f is 1 or 0) return (f, 1);
    if cacheLookup(f, g, h) return (g, h) ;
    v = topVar(f);
    if (f is decomposition point) return (v + f_e , v' + f_t);
    (g_t,h_t) = decomp (f_t);
    (g_e,h_e) = decomp (f_e);
    (g, h) = combine(g_t,h_t, g_e,h_e);
    return (g, h);
}
```

Figure 5: Decomposition algorithm.

Disjoint: Nodes with sufficiently disjoint children yield maximum reduction in individual sizes, yet leave the shared size small. Our second approach attempts to identify such nodes. For every node, the number of nodes shared between its children is measured. A node is chosen as a *decomposition point* when its children have little sharing and are balanced. Determining whether each node is a *decomposition point* requires one pass of the BDD per node. Hence the cost of this measure is quadratic in the number nodes of the BDD. However, in practice, only a fraction of the nodes are sampled for this measure.

We implemented a slightly different version of the algorithms of [6, 19] (referred to as *Cofactor*). Experiments and results of the comparison with our approach are presented in the next section.

4 Experimental Results

The algorithms described in Sections 2 and 3 have been implemented in the CUDD package [27] and tested in two sets of experiments, one using an extension of VIS-1.2 [1], and the other using CUDD's stand-alone test program. The first set of experiments regards the application of BDD approximation techniques to reachability analysis. The RUA algorithm described in Section 2 is applied to "high density" reachability analysis [23] in the same manner as the SP procedure. We experimented with a few circuits and present the results in Table 1.

Table 1 presents best-time results for high-density reachability analysis of four circuits. High density reachability analysis modifies breadth-first search to compute a subset of reachable states at each iteration of the traversal. It can be viewed as a mixed depth-first breadth-first exploration of the state space. At each iteration, image computation is provided with a dense subset that is extracted from the set of new states. This extraction is implemented with one of the approximation techniques. Additionally, the traversal computes a subset of the image of a given set by computing subsets of intermediate products of image computation. Subsetting is used when the size of the intermediate product exceeds some large threshold. Dynamic reordering is always turned on during these experiments. For all the experiments reported (except BFS on `am2910`), both BFS traversal and high density traversal completed with the exact set of reachable states.

Columns 1–3 of Table 1 provides the circuit name, the number of flip-flops and the reachable states of this circuit. The exact traversal times are reported in Column 4. `s1269`, although small in terms of the number of latches (37), took more than 256MB for exact traversal and hence, was run on a SUN Ultra-1 with 1GB of main memory. The main column labeled RUA indicates results where *remapUnderApprox* is used for extracting the subset of the new states as well as the subset of the intermediate products of image computation. Similarly, the column labeled SP refers to *short-path subsetting*. The sub-columns report run times for each method, the threshold used and (for RUA) the quality factor. The column labeled PImg reports two numbers: The first is a node limit that triggers approximation of partial products in image computation; the second is the threshold passed to the approximation procedure. The "NA"s in the table indicate that partial image computation was not required as the intermediate sizes always stayed small.

These experiments are the best runs for some parameter settings (threshold, quality, partial image factors). The relative merit of tuning the quality factor or adjusting the threshold cannot be studied in this table. The parameters were determined by trial and error. The speedup afforded by HD with RUA over breadth-first search is very large in two cases out of four. In our experiments, we noticed that when RUA had the fastest run times, the runs always corresponded to small BDD sizes of reachable states (under 50000 nodes). The intermediate sizes were significantly smaller compared to the original runs: RUA is very effective in reducing the sizes of large intermediate products of image computation. The fast run times and low memory occupation also support the conjecture that RUA extracts dense subsets, which is required for high-density reachability analysis to perform well.

In comparison to SP, there is no clear indication as to which method is better. With a 0 threshold, RUA is free to reduce the size of the given set arbitrarily. This works well for `s1269`, `s3330`, and `am2910`. In the case of `s5378opt`, SP is more effective for the given threshold. The differences among the various circuits seem due to the topology of their state graphs in ways that we only partly understand so far. However, we have observed that the method that performs better always has a higher average density of subset of new states. We conducted some experiments using SP for creating subsets of new states but RUA for partial image computation, and the run-times were faster than using SP for both. Using approximation procedures in reachability analysis has compounded effects on the statistics of the runs (reordering, growth of the reached set in number of states and nodes). If a procedure extracts the "wrong" subset of states for image computation, then traversal may progress very slowly or may not complete. Indeed, this is true for some parameter settings. Notice that "wrong" here does not necessarily mean "less dense." However, the fast run times in this table indicate that RUA is effective in spite of the cascaded effects and is practically applicable in efficient symbolic traversal techniques.

It is difficult to conduct a head-to-head comparison of different approximation techniques in the context of reachability analysis, due to the repercussions of any change in one step of the procedure on the successive steps. Therefore we present a second set of experiments, where we apply the approximation techniques to the outputs and next state functions of a collection of circuits. From a total of 7157 functions, we extracted the 336 that had 5000 nodes or more. In Tables 2 and 3 we report the geometric means of the numbers of nodes, the number of minterms, and the densities of several methods, as well as the number of cases in which each technique produced the densest result, either alone (wins) or together with other techniques (ties). Table 2 describes the simple methods and Table 3 describes compound methods. The results of UA are for the non-safe implementation, which on average outperformed the safe one, in spite of decreasing density in 3 of the 336 cases. Simple and compound methods are kept separate so that the relative strengths of SP and RUA can be better appreciated, because C1 never loses to RUA, and C2 never loses to SP.

In these experiments, the thresholds for UA and RUA were set to 0 and the quality factor was kept at 1. In general these were the most favorable values for these experiments. The sizes of the BDDs produced by RUA were used as thresholds for HB and SP. Though there is no guarantee that these thresholds produce the best results, experimentation showed that they did not put SP and HB at

Method	nodes	minterms	density	wins	ties
F	14449.4	1.06e+45	7.30e+40	0	2
HB	24.5	3.30e+42	1.35e+41	3	65
SP	41.9	2.48e+44	5.92e+42	6	7
UA	28.3	3.72e+44	1.32e+43	24	78
RUA	30.4	6.04e+44	1.99e+43	219	80

Table 2: Comparison of approximation methods I: Simple methods. F denotes the original function, HB heavy-branch subsetting, SP short-path subsetting, UA bddUnderApprox, and RUA remapUnderApprox.

Method	nodes	minterms	density	wins	ties
C1	30.3	6.14e+44	2.03e+43	125	87
C2	14.7	2.59e+44	1.76e+43	124	85

Table 3: Comparison of approximation methods II: Compound methods. C1 denotes RUA followed by minimization, and C2 denotes SP followed by RUA followed by minimization.

a disadvantage with respect to UA and RUA.

Regarding the compound methods, one should notice that C1 retained more minterms of F than RUA on average, and C2 retained more minterms than SP (in spite of halving the number of nodes). This increase is the effect of the minimization step.

We also ran experiments comparing the decomposition methods described in Section 3 for two-way decomposition. We implemented the approach of [6, 19] choosing as cofactoring variable the one that minimizes the size of the larger of the two cofactors. The cost of estimation of the cofactor sizes is linear in the product of the number of variables and the size of the function.

Table 4 shows results for the 3 methods—the above implementation, and our algorithm with the two different choices of decomposition points. The experimental setup is similar to Tables 2 and 3. There are 2 sets of results in this table—one for BDDs with at least 5000 nodes and the other for BDDs with at least 20000 nodes. Each method produced 2 factors, G and H, whose mean sizes are reported in the table. The column labeled "Shared" reports the mean shared size of G and H. The wins and ties of each method, reported in the last 2 columns, are based on the size of the larger of G and H.

Cofactor yields the best results for a large number of BDDs. *Band* performs a little better than *Disjoint* (additionally is cheaper to compute). One explanation for the results is that *Disjoint* and *Band*, although less restricted than *Cofactor* in the choice of *decomposition points*, perform a local search and make greedy choices. *Cofactor*, on the other hand, benefits from a global search for the smallest factors. For some of the larger BDDs, *Disjoint* does perform better than *Cofactor*, as shown in the lower half of the table. However, since the sample space is small, this result is not conclusive.

Min. Nodes = 5000 , $\|f\|$ = 11134.5, 279 BDDs					
Method	Shared	G	H	wins	ties
Cofactor	11352.1	7029.7	4906.3	192	4
Disjoint	11990.9	8637.5	8182.8	57	4
Band	13590.4	8244.0	7182.7	26	0
Min. Nodes = 20000, $\|f\|$ = 42872.9, 11 BDDs					
Cofactor	47952.0	26394.3	25813.0	2	1
Disjoint	42877.0	22575.4	22866.5	8	1
Band	52614.4	30728.8	26055.5	0	0

Table 4: Comparison of decomposition methods.

Ckt	FF	States	BFS time	RUA				SP		
				Th	Qual	PImg	Time	Th	PImg	Time
s3330	132	$7.2778e+17$	**3204**	0	1.0	50000/50000	**562**	7000	100000/100000	**1351**
s1269	37	$1.13134e+09$	**52691**	0	0.5	100000/50000	**290**	3000	100000/50000	**525**
s5378opt	121	$2.58061e+17$	**1454**	5000	1.4	NA	**1140**	4000	NA	**575**
am2910	99	$1.16057e+26$	$>$ **2 weeks**	0	1.0	NA	**217**	99	NA	**224**

Table 1: Reachability analysis results using BDD approximations.

5 Conclusions

We have presented new algorithms for BDD approximation and decomposition. Our new approximation algorithm produces significantly denser subsets than existing techniques thanks to a combination of versatile replacement techniques and efficient and accurate estimation of the impact of local transformations. We have also shown how approximation algorithms can be combined among themselves and with minimization procedures to yield even better results. Our new conjunctive decomposition algorithm efficiently balances the size of the conjuncts. In spite of being more general, however, it does not improve on the algorithm of [6, 20]: Further work is required on the selection of the decomposition points. Both the approximation and decomposition techniques have been incorporated in an experimental reachability analysis engine based on high-density traversal. The results are very promising, with substantial increases in speed over conventional breadth-first traversal. More work is required to increase the robustness and performance of the new traversal techniques, but the effectiveness of the algorithms discussed in this paper greatly increases our confidence in their eventual success.

Acknowledgments

We thank Dirk Grunwald for providing us with computing resources.

References

[1] R. K. Brayton et al. VIS: A system for verification and synthesis. In T. Henzinger and R. Alur, editors, *Eigth Conference on Computer Aided Verification (CAV'96)*, pages 428–432. Springer-Verlag, Rutgers University. 1996. LNCS 1102.

[2] R. E. Bryant. Graph-based algorithms for boolean function manipulation. *IEEE Transactions on Computers*, C-35(8):677–691, August 1986.

[3] J. R. Burch, E. M. Clarke, and D. E. Long. Representing circuits more efficiently in symbolic model checking. In *Proceedings of the Design Automation Conference*, pages 403–407, San Francisco, CA, June 1991.

[4] J. R. Burch, E. M. Clarke, K. L. McMillan, D. L. Dill, and L. J. Hwang. Symbolic model checking: 10^{20} states and beyond. In *Proceedings of the Fifth Annual Symposium on Logic in Computer Science*, June 1990.

[5] G. Cabodi, P. Camurati, L. Lavagno, and S. Quer. Disjunctive partitionining and partial iterative squaring: An effective approach for symbolic traversal of large circuits. In *Proceedings of the Design Automation Conference*, pages 728–733, Anaheim, CA, June 1997.

[6] G. Cabodi, P. Camurati, and S. Quer. Improved reachability analysis of large finite state machines. In *Proceedings of the International Conference on Computer-Aided Design*, pages 354–360, Santa Clara, CA, November 1996.

[7] H. Cho, G. D. Hachtel, E. Macii, B. Plessier, and F. Somenzi. Algorithms for approximate FSM traversal based on state space decomposition. *IEEE Transactions on Computer-Aided Design*, 15(12):1465–1478, December 1996.

[8] O. Coudert, C. Berthet, and J. C. Madre. Verification of sequential machines using boolean functional vectors. In L. Claesen, editor, *Proceedings IFIP International Workshop on Applied Formal Methods for Correct VLSI Design*, pages 111–128, Leuven, Belgium, November 1989.

[9] O. Coudert and J. C. Madre. A unified framework for the formal verification of sequential circuits. In *Proceedings of the IEEE International Conference on Computer Aided Design*, pages 126–129, November 1990.

[10] D. Geist and I. Beer. Efficient model checking by automated ordering of transition relation parititons. In D. L. Dill, editor, *Sixth Conference on Computer Aided Verification (CAV'94)*, pages 299–310, Berlin, 1994. Springer-Verlag. LNCS 818.

[11] Y. Hong, P. A. Beerel, J. R. Burch, and K. L. McMillan. Safe BDD minimization using don't cares. In *Proceedings of the Design Automation Conference*, pages 208–213, Anaheim, CA, June 1997.

[12] S.-W. Jeong, B. Plessier, G. D. Hachtel, and F. Somenzi. Variable ordering and selection for FSM traversal. In *Proceedings of the IEEE International Conference on Computer Aided Design*, pages 476–479, Santa Clara, CA, November 1991.

[13] R. P. Kurshan. *Computer-Aided Verification of Coordinating Processes*. Princeton University Press, Princeton, NJ, 1994.

[14] R. P. Kurshan. Formal verification in a commercial setting. In *Proceedings of the Design Automation Conference*, pages 258–262, Anaheim, CA, June 1997.

[15] W. Lee, A. Pardo, J. Jang, G. Hachtel, and F. Somenzi. Tearing based abstraction for CTL model checking. In *Proceedings of the International Conference on Computer-Aided Design*, pages 76–81, San Jose, CA, November 1996.

[16] D. E. Long. *Model Checking, Abstraction, and Compositional Verification*. PhD thesis, Carnegie-Mellon University, July 1993.

[17] K. L. McMillan. *Symbolic Model Checking*. Kluwer Academic Publishers, Boston, MA, 1994.

[18] K. L. McMillan. A conjunctively decomposed boolean representation for symbolic model checking. In R. Alur and T. A. Henzinger, editors, *8th Conference on Computer Aided Verification (CAV'96)*, pages 13–25. Springer-Verlag, Berlin, August 1996. LNCS 1102.

[19] A. Narayan, A. J. Isles, J. Jain, R. K. Brayton, and A. L. Sangiovanni-Vincentelli. Reachability analysis using partitioned ROBDDs. In *Proceedings of the International Conference on Computer-Aided Design*, pages 388–393, November 1997.

[20] A. Narayan, J. Jain, M. Fujita, and A. L. Sangiovanni-Vincentelli. Partition ROBDDs: A compact, canonical and efficiently manipulable representation for boolean functions. In *Proceedings of the International Conference on Computer-Aided Design*, pages 547–554, Santa Clara, CA, November 1996.

[21] A. Pardo and G. D. Hachtel. Automatic abstraction techniques for propositional μ-calculus model checking. In O. Grumberg, editor, *Ninth Conference on Computer Aided Verification (CAV'97)*. Springer-Verlag, Berlin, 1997. LNCS 1254.

[22] R. K. Ranjan, A. Aziz, R. K. Brayton, B. F. Plessier, and C. Pixley. Efficient BDD algorithms for FSM synthesis and verification. Presented at IWLS95, Lake Tahoe, CA., May 1995.

[23] K. Ravi and F. Somenzi. High-density reachability analysis. In *Proceedings of the International Conference on Computer-Aided Design*, pages 154–158, San Jose, CA, November 1995.

[24] R. Rudell. Dynamic variable ordering for ordered binary decision diagrams. In *Proceedings of the International Conference on Computer-Aided Design*, pages 42–47, Santa Clara, CA, November 1993.

[25] T. R. Shiple. *Formal Analysis of Synchronous Circuits*. PhD thesis, University of California at Berkeley, Electronics Research Laboratory, College of Engineering, University of California, Berkeley, CA 94720, October 1996. Memorandum No. UCB/ERL M96/76.

[26] T. R. Shiple, R. K. Ranjan, A. L. Sangiovanni-Vincentelli, and R. K. Brayton. Deciding state reachability for large FSMs. Technical Report UCB/ERL M97/73, University of California at Berkeley, August 1996.

[27] F. Somenzi. *CUDD: CU Decision Diagram Package*. ftp://vlsi.colorado.edu/pub/.

[28] H. Touati, H. Savoj, B. Lin, R. K. Brayton, and A. Sangiovanni-Vincentelli. Implicit enumeration of finite state machines using BDD's. In *Proceedings of the IEEE International Conference on Computer Aided Design*, pages 130–133, November 1990.

Approximate Reachability with BDDs using Overlapping Projections *

Shankar G. Govindaraju[1] , David L. Dill[1], Alan J. Hu[2], and Mark A. Horowitz[1]

[1] Computer Systems Laboratory
Stanford University
Stanford, CA 94305, USA

[2] Department of Computer Science
University of British Columbia
Vancouver, B.C, Canada V6T 1Z4

Abstract

Approximate reachability techniques trade off accuracy with the capacity to deal with bigger designs. Cho et al [3] proposed approximate FSM traversal algorithms over a *partition* of the set of state bits. In this paper we generalize it by allowing *projections* onto a collection of nondisjoint subsets of the state variables. We establish the advantage of having *overlapping projections* and present a new *multiple constrain* function for BDDs, to compute efficiently the approximate image during symbolic forward propagation using overlapping projections. We demonstrate the effectiveness of this new algorithm by applying it to several control modules from the I/O unit in the Stanford FLASH Multiprocessor. We also present our results on the larger ISCAS 89 benchmarks.

1 Introduction

Binary Decision Diagrams (BDDs) [1] have enabled formal verification to tackle larger hardware designs than before. However for many large design examples, even the most sophisticated BDD-based verification methods cannot produce exact results because of BDD-size blowup. One alternative is to trade accuracy for BDD size requirements, by using approximate verification algorithms.

Computing the set of reachable states from an initial set is a basic component of many verification algorithms and has other applications as well. An overapproximated reachable set can be viewed as an underapproximated *don't care* set, which has applications in the synthesis domain. It can be used to simplify symbolic model checking efforts, by preventing the model checking algorithms from exploring unreachable states.

*This work was supported by DARPA contracts DABT63-94-C-0054 and DABT63-96-C-0097. The content of this paper does not necessarily reflect the position of the policy of the Government and no official endorsement should be inferred.

DAC 98, San Francisco, California

1.1 Related Work

Cho et al [3] proposed approximate algorithms to do symbolic forward propagation. Their basic idea was to partition the set of state bits into *mutually disjoint* subsets, and then do a symbolic forward propagation on each individual subset. The individual subsets can be viewed as submachines which have in some ways been torn from other submachines. The original problem is thus reduced to doing the exact symbolic forward propagation over smaller submachines. This induces extra degrees of freedom for the submachines, and hence yields an overapproximation of the reachable state space.

They also propose different variants of the approximated symbolic forward propagation algorithm: MBM (Machine By Machine) and FBF (Frame By Frame) which basically differ in the way they model the interaction among the various submachines. FBF allows interactions among the submachines at each time frame of a least fixed point routine, and hence allows for tighter don't care *sequences* to constrain the other submachines. MBM on the other hand allows interaction only after a complete least fixed point has been computed for a submachine. As a result the sequencing information is lost when trying to constrain the other submachines. They further propose two variants of the FBF scheme, RFBF (Reached Frame By Frame) and TFBF (To Frame By Frame), which again differ in the constraint set posed to the various sub machines during the course of the least fixed point routine. Cho et al [3, 4] also propose heuristics on how to partition the set of state bits.

1.2 Contribution

In this paper, we improve on the approximate symbolic reachability analysis of Cho et al [3] by allowing for *overlapping projections*. We establish the need for overlapping projections, and propose a new *multiple constrain* function for BDDs, which allows us to compute efficiently the image of an *implicit* conjunction of BDDs with possibly overlapping support, using Boolean function vectors. We apply our algorithm on a real, large design and show its relative superiority over the FBF algorithms. Of course our scheme cannot be less accurate than the FBF method, since overlapping projections include disjoint partitions as a special case.

2 Background

We analyze synchronous hardware, given as a Mealy machine $M = \langle x, y, q_0, \mathbf{n} \rangle$, where $x = \{x_1, \ldots, x_k\}$ is the set of state variables, and y is the set of input signals. The set of states is given by $[x \to \mathcal{B}]$, where $\mathcal{B} = \{0,1\}$. The initial state $q_0 \in [x \to \mathcal{B}]$. The next state function is $\mathbf{n} : [x \to \mathcal{B}] \times [y \to \mathcal{B}] \to [x \to \mathcal{B}]$.

In our applications, sets can be viewed as predicates, since we can form the characteristic function corresponding to a set. BDDs can be used to represent predicates and manipulate them [2]. For example, let $R(x)$ be a predicate with support in x, we can compute the image of R under $\mathbf{n}$ as

$$Im(R(x), \mathbf{n}(x,y)) = \lambda x'.\exists x, y.(x' = \mathbf{n}(x,y)) \wedge R(x).$$

Im produces a predicate with support x', which is 1 *iff* x' is in the image of R under $\mathbf{n}$. The set of reachable states in M can be computed by a least fixpoint iteration [2]

$$Reach(M) = \text{ lfp } R.\lambda x.(q_0(x) \vee Im(R(x), \mathbf{n}(x,y))).$$

2.1 Approximation by Projections

Let $\mathbf{w} = (w_1, \ldots, w_p)$ be a collection of not necessarily disjoint subsets of x. Each subset will be referred to as a *block*. We define the operator $\alpha_j(R)$ which projects a predicate $R(x)$ onto the variables in w_j. Intuitively, $\alpha_j(R)$ represents a set of Boolean vectors that agree for the variables in w_j with some Boolean vector satisfying R. Let z consist of all of the Boolean variables in x that are *not* in w_j, then we can define α_j as

$$\alpha_j(R(x)) = \lambda x.\exists z.R(x).$$

From the explanation above, it should be clear that the set of Boolean vectors satisfying R is a subset of those satisfying $\alpha_j(R)$. This can be written using logical implication as $R \to \alpha_j(R)$. The approximation operator α projects a predicate $R(x)$ onto the various w_j's and returns the tuple,

$$\alpha(R(x)) = (\alpha_1(R), \ldots, \alpha_p(R)).$$

The concretization operator γ conjoins the collection of projections:

$$\gamma(R_1, \ldots, R_p) = \bigwedge_{j=1}^{p} R_j.$$

Lemma 1 *Given a collection of subsets* $(w_1, \ldots, w_p)$ *and a predicate* $R(x)$*,* $R \to \gamma(\alpha(R))$*.*

The proof for this lemma is simple since $R \to \alpha_j(R)$ for all j. Thus projecting a predicate R onto a collection of subsets, and then concretizing the projections by γ results in an overapproximation.

Let $\mathbf{R} = (R_1, \ldots, R_p)$ and $\mathbf{S} = (S_1, \ldots, S_p)$ be two equally sized tuples. We define the *join* operator between $\mathbf{R}$ and $\mathbf{S}$ as follows:

$$(R_1, \ldots, R_p) \sqcup (S_1, \ldots, S_p) = (R_1 \vee S_1, \ldots, R_p \vee S_p)$$

Note that $\gamma(\mathbf{R}) \cup \gamma(\mathbf{S}) \subseteq \gamma(\mathbf{R} \sqcup \mathbf{S})$. Hence the join operator is an approximation of set union.

The operator α allows us to represent a big BDD with support in x by a tuple of potentially smaller BDDs with limited support, at the cost of loss of accuracy. In contrast, γ can potentially result in a bigger BDD with bigger support, hence we would like to avoid computing $\gamma(R_1, \ldots, R_p)$ explicitly. We therefore need to compute the image of an implicit conjunction and return the result as an implicit conjunction of the elements of a tuple. Let Im_{ap} return the projected version of the image of an *implicit* conjunction of BDDs.

$$Im_{ap}(\mathbf{R}, \mathbf{n}) = \alpha(Im(\gamma(\mathbf{R}), \mathbf{n}(x,y)))$$

Using Im_{ap}, we can compute an overapproximation, $Reach_{ap}(M)$, of the reachable states for a machine M as follows:

$$Reach_{ap}(M) = \text{ lfp } \mathbf{R}.(\alpha(q_0) \sqcup Im_{ap}(\mathbf{R}, \mathbf{n}))$$

Note that the least fixpoint routine above starts with $\mathbf{R} = (0, \ldots, 0)$, and finally after reaching convergence, it returns a tuple $\mathbf{R}$ to $Reach_{ap}(M)$.The overapproximate reachable states set is the *implicit* conjunction $\gamma(Reach_{ap}(M))$.

Theorem 1 *For a given Mealy machine M,*

$$Reach(M) \to \gamma(Reach_{ap}(M)) \quad (1)$$

The proof relies on the observation that during computation of $Reach_{ap}(M)$, the image at *every* iteration of the least fixpoint routine is an overapproximation. The formal proof is omitted.

3 Overlapping Projections

3.1 Motivation for Overlaps

Overlapping projections can capture limited interactions among state machines while keeping the sizes of the BDDs under control. We discuss some common scenarios where this happens in this subsection. In contrast, disjoint partitions can only capture interactions among a set of state machines by including all of them in a single projection, which often leads to large variable subsets, that cause BDD size blowup.

Often, two rather big state machines have a small interface, which can be captured by adding extra blocks to our collection $\mathbf{w}$, that merely include the bits through which the two machines communicate. Design modules usually have a *master* FSM that communicates with a number of other *slave* FSMs. A collection of overlapping subsets allows us to capture the master-slave behavior by having blocks, where the master is paired with each of its slaves in *different* blocks. We can further capture the correlation between various FSMs by having small blocks with pairs of FSMs in them.

In each of the these cases, disjoint partitions require larger sized blocks to capture the same property. Thus overlapping subsets allow us to hit intermediate points in the *memory space* vs *strength of invariant* tradeoff curve, with disjoint partitions on one extreme and exact reachability on the other.

3.2 Multiple Constrained Image

The key step in symbolic forward propagation algorithms is image computation.

$$Im_{ap}(\mathbf{R}, \mathbf{n}) = (S_1, \ldots, S_p) = \alpha(Im(\gamma(\mathbf{R}), \mathbf{n}(x, y)))$$

We would like to be able to compute the S_j's separately, without computing $Im(\gamma(\mathbf{R}), \mathbf{n})$. Clearly S_j can only depend on the next state functions of the variables appearing in the jth block, w_j in $\mathbf{w}$. In our implementation, $\mathbf{n}(x, y)$ is represented as a set of predicates $\{n_i(x, y) \mid 1 \leq i \leq k\}$, where each predicate determines the value of a bit in the next state. Let $\alpha_j(\mathbf{n})$ be the subset of predicates determining the next state for the bits in w_j. Clearly, $S_j = Im(\gamma(\mathbf{R}), \alpha_j(\mathbf{n}))$

To avoid unnecessary BDD blowup, we want to avoid the explicit conjunction $\gamma(\mathbf{R})$. S_j can be computed, by forming the next state *relation* for block w_j and using early quantification [2, 8]. However this did not work when we tried it on our larger examples. Instead Coudert and Madre [5, 8] have shown how to compute the image of a Boolean function vector, using the *generalized cofactor* (also called *constrain*) operator ($\downarrow$). $(f \downarrow g)(x)$ has the same value as $f(x)$ when $g(x)$ holds, and usually results in a smaller BDD than f. Generalized cofactor allows us to cofactor a function, f, with respect to another function, g, and reduces to ordinary cofactor when g is a single variable. So $f \downarrow x_i$ is the cofactor of f when $x_i = 1$.

Coudert and Madre [5] show that $Im(\gamma(\mathbf{R}), \alpha_j(\mathbf{n}))$ $= Im(1, \alpha_j(\mathbf{n}) \downarrow \gamma(\mathbf{R})$. To avoid computing the large BDD for $\gamma(\mathbf{R})$, it is tempting to compute $\alpha_j(\mathbf{n}) \downarrow R_1 \downarrow R_2 \ldots \downarrow R_p$. This works well if the supports of R_i's are disjoint. (McMillan has shown [7] that **if** g and h have independent support, **then** $f \downarrow (g \wedge h) = (f \downarrow g) \downarrow h$). However since we have overlapping subsets, the naive method is incorrect. For example, consider $f = wxy, g = w$ and $h = \bar{w} \vee wxy$. $f \downarrow (g \wedge h) = 1$, and so $Im(g \wedge h, f) = \{1\}$. However $(f \downarrow g) \downarrow h = w \vee \bar{w}xy$, and $Im(1, (f \downarrow g) \downarrow h) = \{0, 1\}$ (we used the variable order $w < x < y$ for this example).

Instead, for overlapping projections, we propose the following method of *multiple constrain.* Let $(z_1, \ldots, z_p)$ be dummy state bits with corresponding next state functions $(R_1, \ldots, R_p)$. The multiple constrain method relies on the following key observation

$$Im(\gamma(R_1, \ldots, R_p), \alpha_j(\mathbf{n})) = \\ Im(1, [\alpha_j(\mathbf{n}), R_1, \ldots, R_p]) \downarrow z_1 \downarrow z_2 \ldots \downarrow z_p$$

In words, we first extend the Boolean function vector $\alpha_j(\mathbf{n})$ with $(R_1, \ldots, R_p)$, and compute the range of the extended vector to get the set of next states. Every point in the range of $[\alpha_j(\mathbf{n}), R_1, \ldots, R_p]$ will be "tagged" with the dummy variables z_i, which keep track of which of the R_i's were satisfied in the present state. The required image is the part of the range where all the dummy bits $(z_1, \ldots, z_p)$ are 1, where all the R_i's were satisfied by the present state. Selecting the cofactors where $z_1 = z_2 = \ldots = z_p = 1$ finds the BDD for the relevant part of the range while eliminating the dummy z_i variables. In our example, our *multiple constrain* would compute $(f \downarrow h) \downarrow (g \downarrow h) = 1$. Hence we get $Im(1, (f \downarrow h) \downarrow (g \downarrow h)) = \{1\}$ which matches $Im(f, g \wedge h)$.

We can optimize on the usual recursive co-domain partitioning algorithm [5], by avoiding computing the parts of the range that will be discarded. Hence, we start with $[\alpha_j(\mathbf{n}), R_1, \ldots, R_p]$, constrain the vector of predicates, $[\alpha_j(\mathbf{n}), R_1, \ldots, R_{p-1}]$, with R_p, then constrain the resulting $[\alpha_j(\mathbf{n}), R_1, \ldots, R_{p-2}]$, with the constrained R_{p-1} and so on till we constrain the resulting $\alpha_j(\mathbf{n})$ with the final R_1. Thereafter we can do the standard recursive image computation given by Coudert and Madre [5].

function $Im_{mc}((R_1, \ldots, R_p), (n_1, \ldots, n_m))$
$\mathbf{v} \leftarrow [n_1, \ldots, n_m, R_1, \ldots, R_p]$
for *j=p* **down to** *1* **by** *1* **do**
  $\mathbf{v} \leftarrow \mathbf{v} \downarrow v[m + j]$
endfor
return $Im(1, \{v[1], \ldots, v[m]\})$

Our least fixpoint routine starts with $\mathbf{R}$: $(0, \ldots, 0)$ and computes the tuple $Reach_{ap}$ as,

$$\text{lfp } \mathbf{R}.(\alpha(q_0) \sqcup (Im_{mc}(\mathbf{R}, \alpha_1(\mathbf{n})), \ldots, Im_{mc}(\mathbf{R}, \alpha_p(\mathbf{n}))$$

Our algorithm most closely resembles the RFBF algorithm proposed by Cho et al [3], but differs in that we allow for overlapping projections and compute the image for each block with our new Im_{mc} operator. It is also straightforward to do MBM, TFBF, TMBM [3] traversals using overlapping projections.

3.3 Choice of Collection of Subsets

Our scheme for choosing the collection of subsets is presently manual. First, we find the FSMs by inspecting the HDL source (we had access to the RTL description for our design examples). For each state bit x_i we compute a score by counting the number of predicates $n_j(x, y)$ it supports. To each machine, a score is assigned which is the sum of the scores of its state bits. The two machines (M_1, M_2) with the highest scores are identified as *master* FSMs. If the state bits of machines M_i and M_j support the bits of the *master* machine M_1 in their next state predicates, then M_i and M_j are *slaves* of M_1. The different slave machines for each of the master FSMs are identified. We then form blocks by pairing the master FSMs with their slaves. Thus, in this case, we would add the blocks (M_1, M_i) and (M_1, M_j) to the collection of subsets.

Often some FSMs are very small. The corresponding small blocks can then be aggregated with other blocks without running into intermediate image BDD size explosion. The converse problem is some FSM, M_i may have large state registers, resulting in big blocks. If so, we try to prune these blocks by exploiting the *small interface* phenomenon, described in Section 3.1. We also add a block with the master FSMs,to capture the *correlation* between the FSMs. We ensure that no block w_i in the collection $\mathbf{w}$ is a proper subset of another block $w_j \in \mathbf{w}$, since this would clearly be wasteful.

4 FLASH Design Example

The Stanford FLASH (Flexible Architecture for SHared memory) multiprocessor [6] efficiently integrates support for cache coherent shared memory and high performance message passing. Each node in FLASH contains a microprocessor (MIPS R10000), a portion of the machine's global memory, a port to the interconnection network, an I/O interface and a custom node controller called MAGIC (Memory And General Interface Connect). MAGIC handles all communication both within the node and among nodes, using a embedded superscalar, dual issue RISC processor core. We focus on the *control logic* in the I/O unit, since bugs more often than not reside in the control logic rather than the datapath. (The MAGIC chip design description has a rather clean division between the control and datapath). Table 1 gives a brief description of the various control modules in the I/O unit.

Table 1: Control Modules in I/O unit in FLASH

Module	State Bits	Input Bits
IOInboxQCtl	23	8
ReqDecode	37	27
ReqService	41	58
IOMiscBusCtl	44	18
PciInterface	88	55

5 Experiments

We built a LISP interface to David Long's BDD package. Our approximate algorithm returns a superset of the reachable states, which is also an invariant of the design. To quantify the size of the superset, we compute the satisfying fraction of the the superset. (Please refer to the appendix for the algorithm that was used to compute an upper bound on the satisfying fraction). Since projection induces an overapproximation, the smaller the satisfying fraction, the stronger the invariant.

We preset the maximum number of BDD nodes (BDD Node Limit) for each experiment. Given our partitioning heuristics, we try to get the strongest invariant using overlapping projections. We compare our results with the disjoint partition schemes. The same variable ordering was used for both the schemes. *Node Count* keeps track of the highest number of nodes that existed at a time during the experiment. The *Iter* column lists the number of iterations needed to reach the fixpoint. The last column under the heading *Relative* is the ratio between the satisfying fraction with disjoint partitions and the satisfying fraction with overlapping projections. Thus, higher the figures in the *Relative* column, the stronger is the invariant with overlapping projections.

Table 2a: *IOInboxQCtl Invariant Strengths:* Note that to improve upon the invariant with satisfying fraction 5.004883e-03, in the case of disjoint partitions, the BDD node count had to jump from 28,254 to 76,630. which is a 2.71 times increase in the BDD node count. The

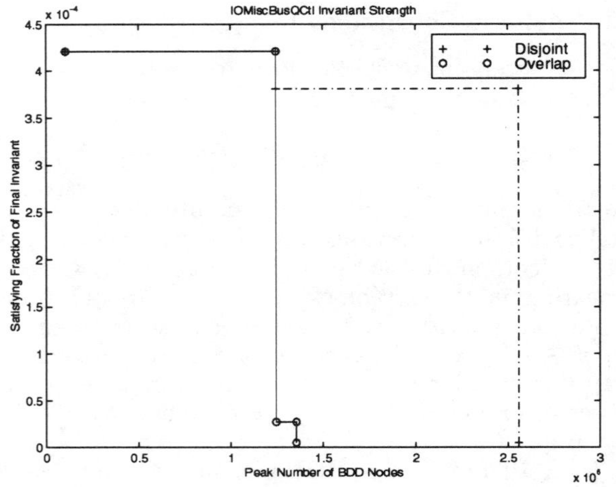

Figure 1: IOMiscBusCtl: Projections vs Partitions

last entry under disjoint partitions was computed with all the state variables in a single block, which clearly gives the strongest possible invariant. Overlapping projections produces the strongest invariant at much lower node count.

Table 2b: *ReqDecode Invariant Strengths:* For node limits of 1.0 million and 1.5 million, our algorithm with overlapping projections yields stronger invariants (by a factor of 1.252 and 1.562 respectively). Further, for the node limit of 1.0 million, our algorithm with overlapping projections uses lesser number of BDD nodes compared to the FBF runs with disjoint partitions, for a stronger invariant.

Table 2c: *ReqService Invariant Strengths:* Here, with disjoint partitions, the node count penalty goes up from 407,728 to 11,007,330 (a factor of 27) before we see any improvement in the strength of the invariant. The last entry under disjoint partitions was computed with all state variables in a single block, which extracts the strongest invariant. Note that the same invariant is extracted by our overlapping projections scheme at much lower node count penalty (1,995,304 nodes vs 11,007,330 nodes, which is lower by a factor of 5.517).

Table 2d: *IOMiscBusCtl Invariant Strengths:* Note that with node limits of 1.5 million and 2.0 million, our algorithm with overlapping projections yields significantly stronger invariants (by a factor of 13.973 and 71.329 respectively), at only incremental extra cost in terms of BDD node count. In Figure 1, we plot the satisfying fraction of the final invariant *vs* the peak number of BDD nodes. We see that the solid curve for overlapping projections is considerable below the other curve for disjoint partitions, indicating that overlapping projections give stronger invariants with lower BDD node counts compared to disjoint partitions.

Table 2e: *PciInterface Invariant Strengths:* The final invariant with overlapping partitions is much stronger (factor of 5.932) than that obtained with disjoint partitions. Note that even as the node limit is raised from 2 million to 25 million, there is no improvement in the disjoint partition case.

6 ISCAS Benchmarks

We have also tried our algorithm on the ISCAS 89 benchmark suite. We use the partitions used by Cho et al [3] to identify the FSMs in the design. We chose the variable subsets by adding small overlaps to some of their blocks. We are unable to report comparative figures for s35932 because we could not procure the partitions used by Cho et al for s35932. In the case of of s5378, our version of s5378 had 179 flip flops as opposed to the 164 flip flops in the one used by Cho et al. We report our results on the remaining benchmarks. The numbers under the *Disjoint Partitions* column correspond to the results we got by running TMBM algorithm [3] (s13207, s15850, s38584) and RFBF algorithm (s1238, s1423) on the partitions used by Cho et al [3] **Table 3**: *ISCAS-89 Invariant Strengths:* Note that we report orders of magnitude improvement in the strength of the invariant for s13207 and s38584. The numbers in Table 3 under overlapping projections are *upper bound estimates* of the satisfying fraction of the final invariant. Thus, the invariant with overlapping projections is stronger, *at least* by a factor equal to the figures under the *Relative* column. (TMBM algorithm starts off as TFBF and switches to MBM after a few iterations. Since we are using TMBM algorithm for some circuits, the *Iter* column in Table 3, lists the number of iterations of doing TFBF + the number of iterations in the outer greatest fixpoint of MBM).

7 Conclusions

In this paper we have proposed a new approximation scheme that enables us to do symbolic forward reachability analysis over an overlapping projection of the set of state bits. The approximation scheme results in tighter overapproximations compared to earlier schemes based on disjoint partitions. Our experiments show that a small amount of appropriately chosen overlaps in a given projection can substantially improve the quality of the overapproximation. Overlapping projections allow us to hit intermediate points in the *memory space* vs *strength of invariant* tradeoff curve, with disjoint partitions on one extreme and exact reachability on the other.

We have also proposed a new *multiple constrain* function for BDDs, that enables us to compute efficiently the image of an *implicit* conjunction of BDDs with possibly overlapping support, using Boolean function vectors.

We further need to look at automatic methods for choosing the collection of subsets, from gate level descriptions.

8 Acknowledgments

We thank David Long for his quick responses for BDD package support. We thank Enrico Macii for sending us the partitions that were used by Cho et al in their paper [3]. We further thank Jules Bergmann for helping us use his tool, *vex*, as a front end Verilog parser for the FLASH benchmarks.

9 Appendix

9.1 Approximating *Sat_Fr* of Superset

Given $\mathbf{S} : (S_1, \ldots, S_p)$, corresponding to the collection $\mathbf{w} : (w_1, \ldots, w_p)$, we want an upper bound on *sat_fr* of $\gamma(\mathbf{S})$. If elements of $\mathbf{S}$ have mutually disjoint support, we could compute *sat_fr* exactly as $\prod_{i=1}^{p} sat\_fr(S_i)$. However here, S_i's may have overlapping support. Our greedy algorithm computes *sat_fr* of a superset of $\gamma(\mathbf{S})$, by using the fact, $\exists x.(a \wedge b) \subseteq (\exists x.a) \wedge (\exists x.b)$.

A set Z is used to keep track of which variables to hide existentially before computing *sat_fr* of each block. At every step the BDD S_m, with the lowest *sat_fr* (after hiding existentially variables in Z from S_m), is picked. Its *sat_fr* is cumulatively multiplied to f, and variables in w_m are added to set Z.

$Z \leftarrow \emptyset; \quad f \leftarrow 1.0$
for *j=1* **up to** *p* **by** *1* **do**
  find m, s.t. $\forall i.(sat\_fr(\exists Z.S_m) \leq sat\_fr(\exists Z.S_i))$
  $f \leftarrow f \times sat\_fr(\exists Z.S_m)$
  $Z \leftarrow Z \cup w_m$
endfor
return f

An alternative approach, Monte Carlo simulation, appears to be ineffective because of the extreme sparseness of the state space covered by $\gamma(\mathbf{S})$.

References

[1] Bryant, R. E., "Graph-Based Algorithms for Boolean Function Manipulation," *IEEE Transactions on Computers*, Vol. C-35, No. 8, pp. 677-691, August 1986.

[2] Burch, J. R., Clarke, E. M., McMillan, K. L., Dill, D, L, and Hwang, L. J., "Symbolic Model Checking: 10^{20} States and Beyond," *Proceedings of the Conference in Logic in Computer Science*, pp. 428-439, 1990.

[3] Cho, H., Hachtel, G., Macii, E., Pleisser, B., and Somenzi, F., "Algorithms for Approximate FSM Traversal Based on State Space Decomposition," *IEEE Transactions on Computer-Aided Design of Integrated Circuits and Systems*, Vol. 15, No. 12, pp. 1465-1478, December 1996.

[4] Cho, H., Hachtel, G., Macii, E., Poncino, M., and Somenzi, F., "Automatic State Space Decomposition for Approximate FSM Traversal Based on Circuit Analysis," *IEEE Transactions on Computer-Aided Design of Integrated Circuits and Systems*, Vol. 15, No. 12, pp. 1451-1464, December 1996.

[5] Coudert, O., and Madre, J. C., "A Unified Framework for the Formal Verification of Sequential Circuits," *IEEE International Conference on Computer-Aided Design*, pp. 126-129, 1990.

[6] Kuskin, J., Ofelt, D., Heinrich, M., Heinlein, J., Simoni, R., Gharachorloo, K., Chapin, J., Nakahira, D., Baxter, J., Horowitz, M., Gupta, A., Rosenblum, M., and Hennessy, J., "The Stanford FLASH Multiprocessor," *Proceedings of the 21st International Symposium on Computer Architecture*, pp. 301-313, April 1994.

[7] McMillan, K. L., "A Conjunctively Decomposed Boolean Representation for Symbolic Model Checking," *In Proceedings of Computer-Aided Verification*, pp. 13-25, 1996.

[8] Touati, H. J., Savoj, H., Lin, B., Brayton, R. K., and Sangiovanni-Vincentelli, A., "Implicit State Enumeration of Finite State Machines using BDDs," *IEEE International Conference on Computer-Aided Design*, pp. 130-133, 1990.

Table 2: FLASH Design Example Results

Node Limit	Disjoint Partitions			Overlapping Projections			Relative
	Sat. Fr.	Iter	Node Count	Sat. Fr.	Iter	Node Count	
30000	5.004883e-03	20	28254	5.004883e-03	20	28254	1.000
60000	”	”	”	4.943967e-03	20	53740	1.012
70000	”	”	”	3.967404e-03	20	64462	1.262
80000	3.967404e-03	20	76630	3.967404e-03	20	64462	1.000

Node Limit	Disjoint Partitions			Overlapping Projections			Relative
	Sat. Fr.	Iter	Node Count	Sat. Fr.	Iter	Node Count	
50000	2.184883e-05	20	33408	2.184883e-05	20	33408	1.000
200000	2.107944e-05	20	134536	1.979293e-05	20	171448	1.065
1000000	1.274049e-05	33	980968	1.017604e-05	20	608726	1.252
1500000	”	”	”	8.156174e-06	20	1195109	1.562
2500000	3.168825e-06	25	2032890	3.168825e-06	25	2032890	1.000

Node Limit	Disjoint Partitions			Overlapping Projections			Relative
	Sat. Fr.	Iter	Node Count	Sat. Fr.	Iter	Node Count	
50000	1.658440e-02	34	23662	1.658440e-02	34	23662	1.000
500000	1.351655e-03	44	407728	1.053363e-03	37	470642	1.283
750000	”	”	”	1.039535e-03	40	537578	1.300
1800000	”	”	”	1.039460e-03	40	1776965	1.300
2000000	”	”	”	1.036219e-03	44	1995305	1.304
12000000	1.036219e-03	44	11007330	”	”	”	1.000

Node Limit	Disjoint Partitions			Overlapping Projections			Relative
	Sat. Fr.	Iter	Node Count	Sat. Fr.	Iter	Node Count	
1000000	4.210770e-04	4	104135	4.210770e-04	4	104135	1.000
1500000	3.810450e-04	4	1173863	2.726912e-05	4	1244294	13.973
2000000	”	”	”	5.342066e-06	4	1353024	71.329
3000000	5.342066e-06	4	2556733	”	”	”	1.000

Node Limit	Disjoint Partitions			Overlapping Projections			Relative
	Sat. Fr.	Iter	Node Count	Sat. Fr.	Iter	Node Count	
500000	3.574305e-03	21	283186	1.168328e-04	50	228390	30.593
1000000	1.041354e-05	55	598257	1.041354e-05	55	598257	1.000
2000000	9.463498e-07	71	1616055	6.311379e-07	71	1293062	1.499
25000000	”	”	”	1.595220e-07	71	20851528	5.932

Table 3: ISCAS 89 Benchmark Results

Circuit	Disjoint Partitions			Overlapping Projections			Relative
	Sat. Fr.	Iter	Node Count	Sat. Fr.	Iter	Node Count	
s1238	7.036972e-02	3	59360	6.320953e-03	4	73849	11.133
s1423	2.985005e-03	37	310461	2.193374e-03	248	1032286	1.361
s13207	3.421447e-106	10+6	161447	1.136200e-115	10+5	198779	3.3208e+08
s15850	5.840135e-102	10+5	271093	3.937940e-102	10+4	336048	1.483
s38584	6.494194e-41	10+2	646258	5.764063e-57	10+5	1853461	8.876e+15

Incremental CTL Model Checking Using BDD Subsetting*

Abelardo Pardo
Mentor Graphics Corporation
267 Boston Road, Suite 2
Billerica, MA, 01862, USA
abelardo_pardo@mentorg.com

Gary D. Hachtel
University of Colorado
ECEN Campus Box 425
Boulder, CO, 80309, USA
hachtel@colorado.edu

Abstract

An automatic abstraction/refinement algorithm for symbolic CTL model checking is presented. Conservative model checking is thus done for the full CTL language–no restriction is made to the universal or existential fragments. The algorithm begins with conservative verification of an initial abstraction. If the conclusion is negative, it derives a "goal set" of states which require further resolution. It then successively refines, with respect to this goal set, the approximations made in the sub-formulas, until the given formula is verified or computational resources are exhausted. This method applies uniformly to the abstractions based in over-approximation as well as under-approximations of the model. Both the refinement and the abstraction procedures are based in BDD-subsetting. Note that refinement procedures which are based on error traces, are limited to over-approximation on the universal fragment (or for language containment), whereas the goal set method is applicable to all consistent approximations, and for all CTL formulas.

1 Introduction

The success of formal verification in detecting incorrect designs has been proven over the last decade. However, limitations on the size of verifiable problems continue to be a serious drawback. Symbolic techniques based on BDDs (Reduced Ordered Binary Decision Diagrams [3]), such as symbolic model checking [4, 13] have significantly increased the size of the systems which can be verified. However, industrial strength examples, model checking based on automatic conservative approximation is required [10, 9, 14].

The main idea behind abstraction is to interpret the behavior of a system in a abstracted (and therefore simplified) system with a more compact representation. An abstraction can be seen as a relation between two systems. On one hand, the original system has the complete description of its behavior, whereas its abstraction preserves some of that behavior and abstracts the rest. The key issue is to know in advance what parts of a design are relevant, and how to automatically abstract the rest of it. In most cases, the problem in the abstract domain is solved conservatively, that is, only when the formula is proved true it can be concluded true in the original system.

In this paper we provide an algorithm to perform CTL symbolic model checking based on the generic abstraction paradigm of [14]. We explore the effect of two kinds of abstraction: (1) BDD subsetting in the pervasive pre-image computations, and (2) gross initial approximation of the overall partitioned transition relation, followed by restoration of the exact relation. We emphasize, that this excludes a vast panoply of alternative automatic abstraction methods. However, the experimental results show that even with this extremely limited set of abstractions, excellent results can be obtained for CTL model checking.

2 Automatic Abstraction

Abstraction has been seen as an effective technique to simplify the systems to be verified to the extent that they can be handled by conventional symbolic techniques. The main challenge for abstraction is the trade-off between the degree of automation and its effectiveness.

In [12], Long proposed a conservative abstraction paradigm that preserved the validity of the logic ∀CTL. This method was one-sided, in that it considered only upper bound approximations of the underlying Kripke structure. In [5], a procedure for approximate traversal of large systems was presented, based on automatic state space decomposition. This technique was similarly one-sided, and applied only to reachability analysis. Neither of these methods provided a procedure for automatically refining the approximation until verification was conclusive.

In [8], Kurshan described an abstraction paradigm called "localization reduction" in the context of ω-regular language containment based on reducing the parts of a system that are *irrelevant* with respect to the task being verified. A systematic procedure was sketched which refined the approximations based on error trace analysis. A related iterative approach to abstraction in language containment verification was presented in [1]. In contrast to Kurshan's method, this was BDD-based, and the refinement considered sets of error traces. However, the details of iterating their method to a definite conclusion were omitted.

In [10], a method was given for conservative CTL model checking. Although this approach used both upper and lower bounds, and included a complete procedure for refining the initial approximation, it was limited to ∀CTL. In [7] Kelb et al. proposed an abstraction mechanism for μ-calculus model checking based on two novel approximations, which were called universal and existential. The process re-

*This work was supported in part by NSF/DARPA grant MIP-9115432 and SRC contract 95-DJ-560.

quired the intervention of the user. Although this approach applied to the full μ-calculus, and used both upper and lower bounds, only one (very interesting) type of approximation was used, and no automatic refinement procedure was given.

Our work is mainly based on the automatic abstraction paradigm presented by Pardo et al. in [14], thus a brief summary of it is given. The paradigm provides the sufficient conditions for an approximation to provide a conservative abstraction.

The formulas are represented by its parse tree, where each vertex represents a sub-formula. We will refer indistinctly to a sub-formula or to the vertex representing it. The model checking procedure traverses this tree in postorder computing the set of states that satisfy each of the sub-formulas. In this paradigm, each node in the tree is labeled with a *polarity* which denotes the type of approximation (either an over or under-approximation) that can be produced at that level in the verification.

The paradigm follows a "lazy" strategy, in the sense that it creates an initial crude approximation and keeps refining it until the result is conclusive. This refinement process is guided by a derived "goal set" of states that need further resolution. The "goal set" refinement method bears a superficial resemblance to the error trace refinement methods of [8] and [1]. However, the method of [14] is uniformly applicable to over and under-approximations whereas error traces are available only for upper bounds applied to universal CTL formulas or language containment.

In the context of reachability analysis of sequential systems Ravi et al. presented in [15] a set of techniques to simplify the representation of a set as a BDD while creating either a superset or a subset.

The work presented in this paper combines these two mechanisms of goal-set based abstraction/refinement and consistent approximation by BDD subsetting into one single implementation. Our primary objective is to explore how BDD simplification techniques can be used in the context of automatic abstraction for CTL model checking.

3 CTL Symbolic Model Checking

CTL (Computational Tree Logic) is a modal logic proposed by Clarke et al. in [6]. Predicates in this logic enable reasoning about the behavior of a given model with respect to a set of atomic propositions over time. Examples of atomic proposition predicates are `reset = 1` or `busId = 0x24`. Let us assume a set of atomic propositions A. The set of CTL formulas ϕ are those generated by the following grammar:

$$\phi ::= p \in A | \neg\phi \mid \phi_1 \wedge \phi_2 | \mathbf{EX}(\phi_1) | \mathbf{EG}(\phi_1) | \mathbf{E}(\phi_1 \mathbf{U} \phi_2).$$

The semantics of CTL is defined with respect to infinite sequences of states in a Kripke structure.

Definition 3.1 *Let a Kripke structure be a 5-tuple $K = (S, Q, R, A, \lambda)$ in which S is a set of states, $Q \subseteq S$ is the set of initial states, $R \subseteq S \times S$ is a transition relation, A is a set of atomic propositions, and $\lambda : A \rightarrow 2^S$ is the labeling function that returns the set of states in which an atomic proposition is satisfied.*

Definition 3.2 *An infinite path π in a Kripke structure $K = (S, Q, R, A, \lambda)$ is a sequence of states $s_1, s_2, \ldots$ such that $s_1 \in Q$ and $\forall i \in \mathbb{N}^+, (s_i, s_{i+1}) \in R$.*

Given a Kripke structure $K = (S, Q, R, A, \lambda)$ and a set of atomic propositions A, we write $(K, s) \models \phi$ to denote that the state s of the structure K satisfies the CTL formula ϕ. Let $\pi = s_1, s_2, \ldots$ be an infinite path. The semantics of CTL formulas is defined as:

- $(K, s) \models p$, where $p \in A$ iff $s \in \lambda(p)$.
- $(K, s) \models \neg\phi$ iff s does not satisfy the formula ϕ in K.
- $(K, s) \models (\phi_1 \wedge \phi_2)$ iff s satisfies both ϕ_1 and ϕ_2.
- $(K, s) \models \mathbf{EX}(\phi_1)$ iff $\exists\pi$ such that $(K, s_2) \models \phi_1$.
- $(K, s) \models \mathbf{EG}(\phi_1)$ iff $\exists\pi$ such that all states satisfy ϕ_1.
- $(K, s) \models \mathbf{E}(\phi_1 \mathbf{U} \phi_2)$ iff $\exists\pi$ such that there exists a state s' which satisfies ϕ_2 and every state from s_1 until s' (excluding s') satisfies ϕ_1.

Different techniques have been studied to improve the efficiency of the model checking algorithm. Among them, the one that provided a significant breakthrough was the manipulation of states symbolically with BDDs [4]. In symbolic model checking, the size of the BDD representation of the sets of states manipulated has a big impact on the efficiency of the algorithm.

The algorithm presented in this paper uses 2-sided BDD reduction techniques like the 1-sided ones presented in [15] to gradually increase the accuracy of the verification. Conventional symbolic model checking algorithms provide either an exact or an approximate computation. Our algorithm explores the space of possible approximations while looking for a conclusive verification result.

4 The Abstraction/Refinement Algorithm

The algorithm consists of two stages. Let us assume that a certain CTL formula ϕ is to be verified. The first stage amounts to obtaining a *crude* approximation of the verification problem. This approximation will have the property of being conservative in the sense defined in Section 1. In the second stage, the verification process is recursively analyzed to obtain a measure of how the current result needs to be refined to achieve the conclusive verification of the formula. This measure will be a set of states which will be called the *goal set*. This refinement process involves the recomputation of certain parts of the verification but with an increased level of accuracy.

In order to apply a BDD based approximation at any point in the model checking algorithm, each temporal operator is expressed in terms of conjunction, negation, preimage computation (denoted by *PreImage*) and fix-point iteration (denoted by *lfp*). This translation is based on the equivalences $\mathbf{EG}(\phi_1) = \neg lfp(X, \neg(\phi_1 \wedge PreImage(\neg X)))$ and $\mathbf{E}(\phi_1 \mathbf{U} \phi_2) = lfp(X, \neg\phi_2 \wedge \neg(\phi_1 \wedge PreImage(X)))$, where $lfp(X, \phi)$ represents a least fix-point iteration of the formula ϕ over the variable X, and ϕ represents both a CTL formula and the set of states s such that $(K, s) \models \phi$. Figure 5 shows an example of the tree representation of a formula.

4.1 Initial Abstraction

In principle, the algorithm may approximate the evaluation of any type of sub-formula. Our objective is then to introduce the approximations gradually in the parts of the verification that are more time consuming. From empirical data, we have observed that most of the verification time is spent in the pre-image computation.

Let us assume that a relation R and a set of states S are given to compute the operation $PreImage(R, S)$. The algorithm uses two types of abstractions: (1) Apply BDD subsetting to the set S and reduce its size by a constant factor. This technique can be applied to produce either over-approximations or under-approximations. (2) Assuming that the relation R is given as a product of sub-relations, the algorithm chooses a subset of the sub-relations as an over-approximation of R. This type of abstraction is used only when an over-approximation is required.

At every evaluation step, the algorithm annotates if an approximation method has been used. In the second stage, these approximation will be refined in order to obtain a conclusive result. Due to the simplifications in this initial evaluation, the answer to the verification process is obtained usually in a negligible amount of CPU time.

4.2 Refining the Approximation

The refinement stage of the algorithm is applied only if the result of the initial approximation is inconclusive. In our implementation, the initial approximation is very significant, and consequently, in almost all the cases, the verification problem remains inconclusive.

In the second stage we have the choice of applying gradual refinements to the different approximations. That is, whenever an approximation needs to be refined, it can be replaced by a better approximation and the computation is repeated. In the implementation discussed here, the refinement of an operation is done directly to the exact computation.

The motivation for this decision is based on experimental data obtained by exploring the trade-off in time between approximate and exact computations. Experimental results showed that in the case of gradual refinements, the number of approximate evaluations was significantly higher, and more importantly, this number outweighed the fact that the approximation required much less computation time to be performed. As a consequence, the heuristic that has been implemented attempts to detect these situations.

The overall refinement algorithm has an operand named the *goal set* that *guides* the process towards the resolution of the verification problem (i.e a conclusive result.) The goal set of a sub-formula ϕ (denoted by f_ϕ contains the set of states by which the approximation has to be modified in order to progress towards a conclusive result. If the operation has been over-approximated, the set contains the states that need to be removed from the result. Conversely, if an operation has been under-approximated, the set contains the set of states that need to be included in the result. The rules to propagate the goal set through the different sub-formulas were given in [14].

In the remaining of the section, for the sake of clarity, it will be assumed that all trivial cases have already been explored. The different algorithms presented in this section represent the situation in which an operation needs to be refined and the refinement is non-trivial. Also, we will denote by $Sat(v)$ the set of states that satisfy the formula represented by v and by $Sat'(v)$ an approximation of this set.

4.2.1 Refinement of the Conjunction Operation

Let us assume a formula v of type conjunction with sub-formulas v_1 and v_2. The refinement procedure receives the Kripke structure K, v, and the goal set f_v. Figure 1 shows the pseudo-code for the refinement algorithm of the conjunction. The function $P(v)$ returns the polarity labeling of the formula v denoting if an over-approximation or an under-approximation is required. The algorithm returns the BDD representation of the remaining goal set after refinement.

```
funct RefineConjunction(K, v, f_v): set ≡
  if (The conjunction has been approximated) then
    return TestSatisfaction(v, f_v)
  endif
  f_v1 = SubFormulaRefine(K, v1, f_v)
  if (P(v) = +) then
    result = SubFormulaRefine(K, v2, f_v1)
  else
    f_v2 = SubFormulaRefine(K, v2, f_v)
    result = f_v1 ∨ f_v2
  endif
  Sat'(v) = Sat'(v1) ∧ Sat'(v2)
  return result
end
```

Figure 1: Refinement algorithm for the conjunction

If the conjunction has been approximated due to the size of the result then no further refinement attempt is made and the set of states that remain to be refined is returned by the function *TestSatisfaction*.

If the conjunction has not been approximated, the refinement needs to be propagated recursively to the sub-formulas. The sub-formula to be refined first is chosen trying to minimize the potential amount of refinement. The implemented heuristic checks first if any of the sub-formulas contains the variable of a fix-point. In that case the other sub-formula is chosen. If none of them contain the variable of a fix-point, the sub-formula with the smaller depth is chosen.

If the node must be over-approximated, the formula v_2 has to be refined with the goal set containing the states that could not be refined in v_1. The set denoted by f_{v_1} contains precisely these states. The set to be returned at the level of formula v is precisely the result of the refinement of v_2. If the node must be under-approximated, then both refinements of v_1 and v_2 have to be done with the goal set f_v.

After these refinements have been completed, the conjunction needs to be re-evaluated.

4.2.2 Refinement of the Pre-image Operation

The refinement procedure for the pre-image operation departs slightly from the common structure of the algorithm. It is assumed that a refinement fails, and the algorithm directly computes the propagation of the goal set f_v as $Image(R, f_v)$. The reason for this heuristic is totally based on experimental results. In all the examples in which this refinement was required, the image computation of the goal set was computed very efficiently. This design decision can be thought of as speculative execution. Figure 2 shows the pseudo-code for the algorithm just described.

```
funct RefinePreImage(K, v, f_v): set ≡
  f_v1 = Image(R, f_v)
  SubformulaRefine(K, v1, f_v1)
  if (There has been some refinement in v1) then
    ComputeExactPreImage(v)
  endif
  result = TestSatisfaction(v, f_v)
  return result
end
```

Figure 2: Refinement algorithm for the pre-image

The algorithm first propagates the goal set to the sub-formula v_1. No exact pre-image computation is performed to check that this propagation is required. If this propagation is indeed required, then the procedure saves the computation of an exact pre-image. If this propagation is not required, the image computation is performed but the refinement will succeed in the immediate sub-formula.

After the recursive call, if the refinement of the sub-formula succeeds (the function returns the empty goal set), the exact pre-image computation is then performed.

4.2.3 Refinement of the Fix-point Operation

The approximation of the fix-point computation has several important effects. There is no relation between the number of iterations in the exact computation and the number of iterations of an approximation. The refinement procedure may refine several or all of the iterations in the sequence. If the last iteration is refined, the convergence criteria may not hold, and the fix-point needs to be iterated further. After convergence is restored, the result of the fix-point might have changed, and therefore, the refinement process needs to be started again.

Because of these considerations, the refinement process of a fix-point operation is divided into two levels. At the internal stage, one of the individual iterations is refined. In the outer stage, a post-processing of the sequence is required to guarantee convergence. Figure 3 shows the pseudo code for the overall procedure.

```
funct RefineFixPoint(K, v, f_v): set ≡
   NewIterations = True; f'_v = f_v
   while (NewIterations ∧ f'_v ≠ ∅) do
      RefineIteration(K, v, f'_v, lastIteration)
      RestoreContainmentInIterations(v)
      PruneIterations(v)
      NewIterations = IterateFixPoint(v)
      f'_v = TestSatisfaction(v, f'_v)
   endwhile
   return f'_v
end
```

Figure 3: Refinement algorithm for a fix-point

The main loop of the procedure is executed if there has been new iterations produced or the evaluation of the fix-point can be further refined.

Let us denote by $(\mu_1, \ldots, \mu_n)$ the set of fix-point iterations computed so far, and $\mu' = (\mu'_1, \ldots, \mu'_n)$ the new set of iterations after the loop has been executed once.

The function *RefineIteration* computes the refinement for one specific iteration. If this refinement fails, it proceeds to refine the previous iteration. The main procedure shown starts by refining the last iteration of the sequence. The discussion of the function *RefineIteration* is deferred until later in this section.

The sequence of iterations satisfies the property that $\mu_{i-1} \subseteq \mu_i$. However, this property does not necessarily hold after some of the iterations have been refined. This situation happens when different approximations are produced every time the sub-formula is evaluated. After the refinement, the procedure *RestoreContainmentInIterations* propagates the effect of the local refinement to the entire sequence restoring the containment property. If the iterations have been over-approximated and iteration μ_i has been refined, the containment relation is restored by replacing every iteration $\mu_1, \ldots, \mu_{i-1}$ by its product with μ_i. Conversely, in the case of the refinement of an under-approximation, iterates $\mu_{i+1}, \ldots, \mu_n$ are replaced by its disjunction with μ_i.

The function *PruneIterations* detects early convergence cases in the new sequence of iterations. After the containment relation has been restored, it is possible that two arbitrary iterations in the sequence satisfy $\mu'_i = \mu'_{i+1}$, and they are not at the end of the sequence. In other words, the refinement of the iterations has provided a different fix-point value. In this case the iterations $\mu'_{i+2}, \ldots, \mu'_n$ can be disregarded and μ'_{i+1} is considered the new convergence point.

The procedure *IterateFixPoint* assures that the convergence of the fix-point is restored after refinement. This procedure simply re-evaluates the formula introducing approximations as in the first stage of the overall algorithm.

After the local refinement step has finished, the overall algorithm computes the new goal set. These steps are repeated until no further refinement can be applied, the refinement process has succeeded completely, or the refinement process did not produce any changes in the sequence of iterations. Figure 4 shows the pseudo-code for the procedure to refine one iteration.

```
funct RefineIteration(K, v, f_v, index) ≡
   x = ReadFixPointVariable(v)
   e(x) = ReadIteration(v, index - 1)
   SubFormulaEvaluate(K, v)
   result = SubFormulaRefine(K, v_1, f_v)
   if (No further refinement is appropriate) then
      StoreNewIteration(v_1, index)
      return result
   endif
   f'_v = ReadPropagatedGoalSet(v)
   RefineIteration(K, v, f'_v, e, index - 1)
   e(x) = ReadIteration(v, index - 1)
   SubFormulaEvaluateExact(K, v)
   StoreNewIteration(v_1, index)
   return TestSatisfaction(v, f_v)
end
```

Figure 4: Refinement algorithm for a fix-point iteration

The procedure receives as a parameter the index of the iteration to be processed. In the first part of the procedure, the sub-formula is re-evaluated with the value of the previous iteration in order to replicate in all the vertices of the sub-formula the values that were present at evaluation time. This re-evaluation is the main reason to use a table storing previous results in the sub-formulas.

Once the formula has been re-evaluated, the refinement is performed by calling the function *SubFormulaRefine*. After this function returns, there are several possible situations:

- The set *result* is empty, the refinement of the sub-formula has succeeded completely at refining the goal set f_v. In this case no further computation is required.

- The set *result* is non-empty, the formula can still be refined, but this refinement will not change the value of the verification. In this case a conclusive result has been reached with an approximated result.

- The set *result* is non-empty, but the formula has been evaluated with no approximation. In this case the result is conclusive.

If none of the three conditions is satisfied, the refinement is applied to the previous iteration.

5 An Example

In this section a simple example is presented to illustrate the techniques presented in the previous section. The system models a three element token ring network. Each element has a binary state variable which denotes if the token is present or not. Let us denote by *state*[*i*] the value of the *i*th state component. The formula that will be verified is expressed in CTL as

$$\mathbf{AG}(\mathit{state[0]} = \mathit{1} \Rightarrow \mathbf{AX}(\mathit{state[0]} = \mathit{0})), \qquad (1)$$

where $\mathbf{AG}(\phi) = \neg\mathbf{E}(\mathit{True}\mathbf{U}\neg\phi)$ and $\mathbf{AX}(\phi) = \neg\mathbf{EX}(\neg\phi)$. Intuitively, this formula verifies that whenever a state element has the token, it will not have it at the next time step.

Let us denote the possible states of the system by the decimal interpretation of the state elements being *state[0]* the least significant bit. Figure 5 shows the state transition graph of the system as well as the parse tree corresponding to the CTL formula.

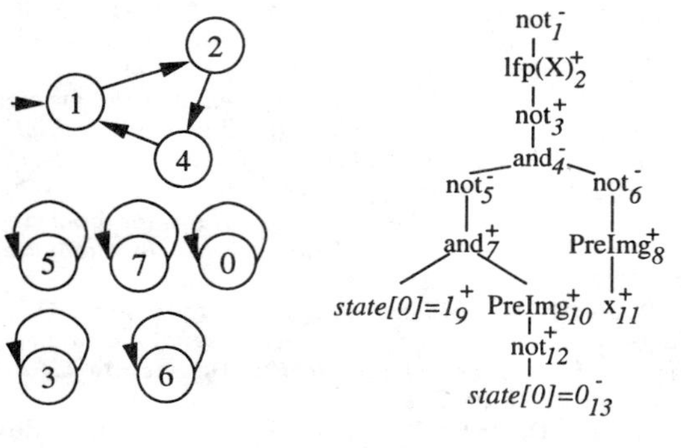

Figure 5: State graph and formula of the token ring example

Each vertex in the tree has been labeled with a unique number in order to refer to its verification result as well as the type of approximation provided (+ for over and − for under-approximation). We further assume that only the pre-image computations in vertices 8 and 10 are approximated.

Due to the size of the example, instead of considering the ordinary BDD subsetting as the approximation technique, we have approximated the computation by letting the behavior of the memory elements in *state*[1] and *state*[2] be free at all times.

After the first approximation has been obtained, the value of the function *Sat* for some of the vertices in the tree are

$$\begin{array}{ll} Sat(v_{13}) = \{0,2,4,6\} & Sat(v_5) = \{0,2,4,6\} \\ Sat(v_{12}) = \{1,3,5,7\} & Sat(v_2^1) = \emptyset \\ Sat(v_{10}) = S & Sat(v_2^2) = \{1,3,5,7\} \\ Sat(v_9) = \{1,3,5,7\} & Sat(v_2^3) = S \\ Sat(v_7) = \{1,3,5,7\} & Sat(v_2^4) = S, \end{array}$$

where S denotes the set of states $\{0,\ldots,7\}$.

The formula v_i in the table denotes the vertex labeled with the sub-index i in Figure 5. The values $v_2^1, v_2^2, v_2^3, v_2^4$ represent the sequence of iterations in the fix-point.

After the initial approximation, the result is the empty set, and the formula is declared false. Since some formulas have been approximated, this result is conservative. The refinement process is then started in v_1 with goal set $f_{v_1} = \{1\}$. The refinement process propagates until vertex v_4. The sub-formula chosen for refinement is v_5 because it does not contain any fix-point variable. For the same reason the refinement propagates from v_7 to v_9. Once in v_9 with $f_{v_9} = \{1\}$ the refinement fails because $f_{v_9} \neq \emptyset$ and the operation has been computed exactly. The program then proceeds to refine v_{10}.

At the level of v_{12}, $f_{v_{12}} = \{2\}$, and since v_{12} has been evaluated exactly, and $2 \notin Sat(v_{12})$ the refinement succeeds, and the pre-image in v_{10} is computed exactly. As a result, $Sat'(v_{10}) = \{2,3,4,5,6,7\}$, and since $f_{v_{10}} = \{1\}$, the refinement of v_{10} has succeeded. This new result propagates through vertices v_7 and v_5, such that $Sat'(v_7) = \{3,5,7\}$ and $Sat'(v_5) = \{0,1,2,4,6\}$.

The refinement then returns to v_4 and proceeds with v_6. In this case the refinement propagates until formula v_{11}, in which $f_{v_{11}} = \{2\}$ and the refinement propagates to the previous iteration.

The algorithm returns to v_2 and starts the refinement of iteration 3 with goal set $f_{v_2^3} = \{2\}$. Now the refinement is required only in v_6. This refinement succeeds and as a result, $v_2^3 = \{3,4,5,7\}$. Note that element 2 no longer belongs to the set. This refinement prompts the re-evaluation of v_2^4 and now $Sat'(v_2^4) = \{2,3,4,5,7\}$. As a consequence of these refinements the containment property among the iterations no longer holds. The containment is restored and the new sequence $(\emptyset, \{3,5,7\}, \{3,4,5,7\}, \{2,3,4,5,7\})$ is obtained.

This new sequence is no longer convergence so the algorithm keeps iterating until convergence is reached again. After the procedure *IterateFixPoint* is executed, two more iterations are added to the sequence, namely $Sat(v_2^5) = Sat(v_2^6) = S$.

Since the sequence of iterations has changed, the refinement process starts again. The process just described is repeated, including the restoration of the containment relation and the pruning of the iterations. As a result, the new obtained sequence is $(\emptyset, \{3,5,7\}, \{3,5,7\})$.

At this point $Sat(v_2) = \{3,5,7\}$ which satisfies the refinement of $f_{v_2} = \{1\}$, and the formula is proved true.

6 Experimental Results

The algorithm described has been implemented within the framework provided by VIS [2]. The results have been obtained in a Sun UltraSparc workstation with 170Mhz and 192Mb of RAM memory. For three examples we have compared the execution time using the incremental approach with that obtained with the model checker in VIS 1.2.

Production Cell: This system is a control circuit with 45 registers [11] for which 15 formulas have been verified. Reachability analysis has been used as a don't care condition in the verification process. Table 1 summarizes the results.

The left half of the table shows the time and number of pre-image operations taken by VIS. The right part shows the execution time with the new procedure as well as the number of the pre-image, approximate pre-image and exact image operations. The column labeled "Ratio" shows the ratio between the execution time with VIS and the incremental approach.

As it can be seen, there are certain instances in the specification in which the formula is proved in the first stage. This means that the initial crude approximation of our algorithm sufficed to prove the formula correct. The overall speedup factor is almost 2.

Ethernet model: The second example is a model of the Ethernet protocol to communicate a set of processors. It has 118 memory elements. The definition of this system

	VIS		Incremental		
Form.	*Pre*	Time	$Pre/\widetilde{Pre}/Img$	Time	Ratio
1	43	17.2	27/39/27	21.1	0.82
2	83	17.9	39/33/73	20.1	0.89
3	84	18.8	0/9/0	6.4	2.94
4	91	35.7	43/34/58	30.5	1.17
5	92	36.2	0/9/0	7.1	5.10
6	83	28.2	39/26/20	22.2	1.27
7	84	28.4	0/5/0	6.5	4.37
8	83	13.3	39/31/21	13.8	0.96
9	84	13.3	0/4/0	6.0	2.22
10	83	22.3	39/25/17	17.3	1.29
11	84	22.0	0/4/0	6.1	3.61
12	83	28.2	39/22/29	21.2	1.33
13	84	27.6	0/5/0	6.3	4.38
14	83	28.1	20/32/32	28.7	0.98
15	84	27.9	0/6/0	6.1	4.57
Total		414.8		219.4	1.89

Table 1: Verification of the production cell

includes several parameters that can be used to scale up the size of the design.

Table 2 shows the results for two version of the system with different internal parameters that have a significant impact in the execution time. In this case our algorithm did not verify any formulas with the initial approximation only, however, we see a consistent reduction in the number of pre-image computations. In some cases our method achieves speedups of up to a factor of 5.

	VIS		Incremental		
Form.	*Pre*	Time	$Pre/\widetilde{Pre}/Img$	Time	Ratio
13.1	45	416	24/18/22	345	1.20
13.2	45	344	23/16/14	89	3.85
13.3	61	486	27/45/21	632	0.77
13.4	61	479	27/28/21	339	1.41
13.5	47	346	21/16/20	384	0.90
13.6	47	314	22/17/21	152	2.07
Total		2414.6		1944.1	1.24
23.1	65	1056	31/26/25	1939	0.54
23.2	65	946	35/23/27	480	1.97
23.3	71	7911	55/133/76	1956	4.04
23.4	71	7884	24/33/23	1593	4.95
23.5	67	1794	35/23/27	1514	1.18
23.6	67	956	32/22/26	571	1.67
Total		20578.2		8056.2	2.55

Table 2: Verification of the Ethernet system

Landing Gear Control: The third example analyzed is a system controlling the landing gear of an aircraft. The system contains 231 memory elements. The only formula verified has been the resetability condition, which is defined as the possibility of driving the system from any state to its initial state. The execution results showed 22626 seconds for VIS compared with 11383 seconds for the incremental algorithm.

7 Conclusions

We have presented a novel incremental algorithm to perform CTL symbolic model checking. The idea is to initially create an approximation of the verification and gradually refine it until the formula is proved conclusively. The approximations are created by reducing the size of the BDDs while creating a superset or a subset of the original sets.

This paradigm makes no assumption about the structure of the system, so it can be applied automatically. The experimental results have shown that for three systems, there are instances in which the algorithm is significantly more efficient than conventional exact model checking. The presented procedures are straightforwardly extended to the past tense of the μ-calculus by simply adding Image to the logic.

As for future directions, we plan to extend the type of abstractions that can be used in this paradigm in order to obtain the improvements needed to handle designs with a memory element count around 10,000. We feel that this work constitutes a first step in that direction. Techniques likes the one presented in [10] are applicable in this paradigm. Furthermore, there has been significant advances in the area of BDD subsetting since [15], and not all them have been incorporated into our paradigm.

References

[1] F. Balarin and A. L. Sangiovanni-Vincentelli. An iterative approach to language containment. In C. Courcoubetis, editor, *Fifth Conference on Computer Aided Verification (CAV '93)*. Springer-Verlag, Berlin, 1993. LNCS 697.

[2] R. K. Brayton et al. VIS: A system for verification and synthesis. In T. Henzinger and R. Alur, editors, *Eigth Conference on Computer Aided Verification (CAV'96)*, pages 428–432. Springer-Verlag, Rutgers Univ., 1996. LNCS 1102.

[3] R. E. Bryant. Graph-based algorithms for boolean function manipulation. *IEEE Transactions on Computers*, C-35(8):677–691, August 1986.

[4] J. R. Burch, E. M. Clarke, K. L. McMillan, and D. L. Dill. Sequential circuit verification using symbolic model checking. In *Proceedings of the Design Automation Conference*, pages 46–51, June 1990.

[5] H. Cho, G. D. Hachtel, E. Macii, B. Plessier, and F. Somenzi. Algorithms for approximate FSM traversal based on state space decomposition. In *Proceedings of the Design Automation Conference*, pages 25–30, Dallas, TX, June 1993.

[6] E. M. Clarke and E. A. Emerson. Design and synthesis of synchronization skeletons using branching time temporal logic. In *Proceedings Workshop on Logics of Programs*, pages 52–71, Berlin, 1981. Springer-Verlag. LNCS 131.

[7] P. Kelb, D. Dams, and R. Gerth. Practical symbolic model checking of the full μ-calculus using compositional abstractions. Technical Report 95-31, Department of Computing Science, Eindhoven University of Technology, 1995.

[8] R. P. Kurshan. *Computer-Aided Verification of Coordinating Processes*. Princeton University Press, 1994.

[9] R. P. Kurshan. Formal verification in a commercial setting. In *Proceedings of the Design Automation Conference*, pages 258–262, Anaheim, CA, June 1997.

[10] W. Lee, A. Pardo, J. Jang, G. Hachtel, and F. Somenzi. Tearing based abstraction for CTL model checking. In *Proceedings of the IEEE International Conference on Computer Aided Design*, pages 76–81, San Jose, CA, November 1996.

[11] Thomas Lindner. *Case Study "Production Cell": A Comparative Study in Formal Software Development*, chapter 2, pages 9,21. FZI, 1994.

[12] D. E. Long. *Model Checking, Abstraction, and Compositional Verification*. PhD thesis, CMU, July 1993.

[13] K. L. McMillan. *Symbolic Model Checking*. Kluwer Academic Publishers, Boston, MA, 1994.

[14] Abelardo Pardo and Gary Hachtel. Automatic abstraction techniques for propositional μ-calculus model checking. In *9th Conference on Computer Aided Verification (CAV'97)*. Springer-Verlag, June 1997.

[15] K. Ravi and F. Somenzi. High-density reachability analysis. In *Proceedings of the IEEE International Conference on Computer Aided Design*, pages 154–158, San Jose, CA, November 1995.

PRIMO: Probability Interpretation of Moments for Delay Calculation†

Rony Kay and Lawrence Pileggi
Department of Electrical and Computer Engineering
Carnegie Mellon University
rkay@ece.cmu.edu, pileggi@ece.cmu.edu

Abstract

Moments of the impulse response are widely used for interconnect delay analysis, from the explicit Elmore delay (first moment of the impulse response) expression, to moment matching methods which create reduced order transimpedance and transfer function approximations. However, the Elmore delay is fast becoming ineffective for deep submicron technologies, and reduced order transfer function delays are impractical for use as early-phase design metrics or as design optimization cost functions. This paper describes an approach for fitting moments of the impulse response to probability density functions so that delays can be estimated from probability tables. For RC trees it is demonstrated that the incomplete gamma function provides a provably stable approximation. The step response delay is obtained from a one-dimensional table lookup.

1: Introduction

Due to CMOS technology trends, it is no longer possible to model all of the delays in terms of gates driving capacitive loads, since the RC charging delays for the interconnect can be the dominant delay component. Model order reduction methods using moments [10][15] or Krylov subspace methods [4][14][19] have effectively addressed the interconnect modeling problem for back-end design verification, however, all of these methods produce transfer function models for which a transcendental equation must be solved to obtain the delay. The implicit nature of these models makes them impractical for most front-end design applications or for use in the inner-loop during design optimization.

The Elmore delay [3], or first moment of the impulse response, provides a simple, explicit, delay approximation that is the *defacto*-standard metric for performance driven design applications. Elmore proposed in 1948 to approximate the median of the impulse response (50% delay of the step response) by the mean of the impulse response by noting the similarity between non-negative impulse responses and probability density functions. But the accuracy of this metric is sometimes unacceptable for RC interconnect problems encountered with today's CMOS technologies. The Elmore delay was shown to be an upper bound on the 50% step and ramp response delays for RC trees[6], but with relative accuracy that can be quite poor at times. Moreover, it can be shown that the first moment of the impulse response is sometimes incapable of even approximating delay sensitivities with respect to changing path resistances, which is becoming increasingly important due to resistive shielding effects. This lack of correlation between front-end delay metrics and back-end model order reduction methods can produce convergence problems for top-down design methodologies.

Some attempts have been made to formulate explicit solutions of 2nd order (two time constant) models, which would seem to be the obvious compromise between an Elmore approximation and a complete reduced-order model via moment matching. To apply such a two pole model, however, requires a moment matching formulation that characterizes the poles in a provably stable manner. Horowitz proposed the first stable two pole model in [1][7], but while it was stable, it can produce complex pole pairs for RC circuits. In addition, this two pole model requires nonlinear iterations to solve for the delay. Some two pole models were proposed in [8] and [21] which can be evaluated by closed-form approximations, however, these methods are lacking in terms of accuracy and generality.

This paper proposes an extension of Elmore's approximation to include matching of higher order moments of the probability density function. Specifically, using a time-shifted incomplete Gamma function approximation for the impulse responses of RC trees, the three parameters of this model are fitted by matching the first three central moments (mean, variance, skewness), which is equivalent to matching the first three moments of the circuit response (m1, m2, m3). Importantly, it is proven that such a gamma fit is guaranteed to be realizable and stable for the moments of an RC tree. Once the moments are fitted to characterize the Gamma function, the step response delay is obtained via a one-dimensional table lookup, thereby providing the same explicitness as the Elmore approximation.

2: Background

We begin with a review of the relationship between moments of a linear circuit response and moments of a random variable.

2.1: Moments of a Linear Circuit Response

Let $h(t)$ be a circuit impulse response in the time domain and let $H(s)$ be the corresponding transfer function. By definition, $H(s)$ is

† This work was supported in part by the Semiconductor Research Corporation under contract DC-068.

the Laplace transform of $h(t)$:

$$H(s) = \int_0^\infty h(t)e^{-st}dt \quad (1)$$

Applying a Taylor series expansion of e^{-st} about $s = 0$ yields:

$$H(s) = \int_0^\infty h(t)\left\{1 - st + \frac{1}{2!}s^2t^2 - \frac{1}{3!}s^3t^3 + \dots\right\}dt =$$

$$\sum_{i=0}^{\infty} \frac{(-1)^i}{(i)!} s^i \int_0^\infty t^i h(t)dt \quad (2)$$

Using the terminology in [15], the i-th circuit-response moment, $\tilde{m}_i$ is defined as:

$$\tilde{m}_i = \frac{(-1)^i}{(i)!}\int_0^\infty t^i h(t)dt \quad (3)$$

From (2) and (3), the transfer function $H(s)$ can be expressed as:

$$H(s) = \tilde{m}_0 + \tilde{m}_1 s + \tilde{m}_2 s^2 + \tilde{m}_3 s^3 + \tilde{m}_4 s^4 + \dots \quad (4)$$

Asymptotic Waveform Evaluation (AWE)[15] demonstrated how impulse response moments could be calculated recursively for a lumped, linear RLC circuit. Following the first moment calculation, all subsequent moments are calculated recursively from the same dc equivalent circuit. Path tracing algorithms, such as those used in RICE[17], can be used to calculate moments with extreme efficiency for interconnect circuits.

For RC trees, Penfield and Rubenstein proposed in [18] a path tracing algorithm to calculate the first moment, or Elmore delay. This was done by traversing the tree topology and summing up contributions of branches along a path P_i from the root to the node of interest. The contribution of branch j is the product of the branch resistance R_j and the total down stream capacitance Cds_j. Denoting $ds(j)$ the set of capacitances downstream of j, the Elmore delay ED_i of path i is

$$ED_i = \sum_{j \in P_i} R_j \cdot Cds_j = \sum_{j \in P_i}\left[R_j \cdot \sum_{k \in ds(j)} C_k\right] \quad (5)$$

In a similar manner, the nth moment of the impulse response along path *i*, m_i^n, is obtained by scaling the downstream capacitors at all nodes, *k*, by their *n-1*th moments, $C_k m_k^{n-1}$, yielding the following recursive expression:

$$\tilde{m}_i^n = -\sum_{j \in P_i}\left[R_j \cdot \sum_{k \in ds(j)} C_k \tilde{m}_k^{n-1}\right] \quad (6)$$

Any set of moments can be computed by two traversals of the RC tree in linear time.

It should be noted that the impulse response moments (series coefficients) in (3) are related to the probability theory moments by a $\frac{(-1)^i}{(i)!}$ term. It is this relation which forms the basis of PRIMO.

2.2: Moments of Probability Density Functions

A *probability function* is a real valued set function where the domain is a subset of the sample space, S, and the range is a real number in the interval $[0, 1]$. Generally, a function $Pr\{\circ\}$ should satisfy the three Kolmogorov axioms [11], or equivalent conditions, in order to be considered a probability function: (i) $Pr\{S\} = 1$; (ii) $Pr\{A\} \geq 0$ for all $A \subset S$; (iii) $Pr\{A \cup B\} = Pr\{A\} + Pr\{B\}$ if $A \cap B = \varnothing, A \subset S, B \subset S$.

The *distribution function* of a continuous random variable T - denoted $F_T(t)$ - provides the value of $Pr\{T \leq t\}$ for any real number $-\infty \leq t \leq \infty$. The associated *probability-density-function* (pdf) - denoted $f_T(t)$ - is the derivative of the distribution function with respect to t, thus

$$f_T(t) = \frac{dF_T(t)}{dt} \quad (7)$$

and

$$F_T(t) = \int_{-\infty}^{t} f_T(\tau)d\tau \quad (8)$$

The median, $t_{(0.5)}$, is defined by:

$$Pr\{T \leq t_{(0.5)}\} = F_T(t_{(0.5)}) = \int_{-\infty}^{t_{(0.5)}} f_T(t)\ dt = 0.5 \quad (9)$$

Whereas, the expected value or mean, $E[T]$, of a continuous random variable T with distribution $f_T(t)$ is

$$E[T] = \int_{-\infty}^{\infty} t f_T(t)\ dt \quad (10)$$

The mean is also the first moment of the distribution (or pdf). In general, the *i*-th moment m_i of the distribution is

$$m_i = E[T^i] = \int_{-\infty}^{\infty} t^i f_T(t)\ dt \quad (11)$$

Fig. 1 summarizes these distribution definitions. Note that we use '~' (tilda) to distinguish between the probability moments m_1, m_2, m_3 and circuit moments $\tilde{m}_1, \tilde{m}_2, \tilde{m}_3$.

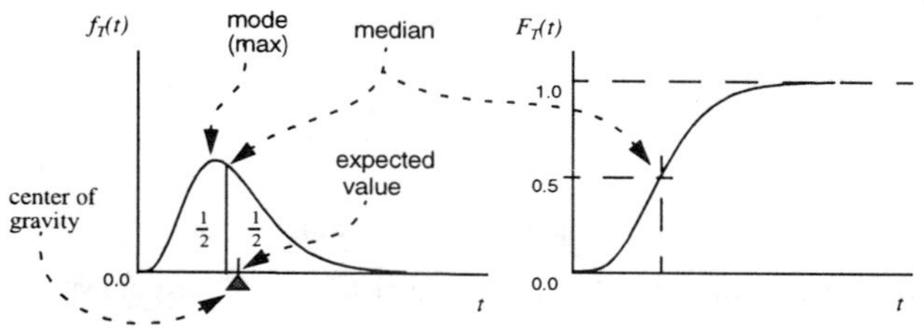

FIGURE 1. A probability density function (left) and corresponding distribution function (right).

2.3: The Central Moments

The moments of a pdf in (11) are more frequently viewed in terms of moments about the origin. The central moments of a distribution - denoted μ_i - are the moments around the expected value. More formally, central moments of a random variable T with probability density function $f_T(t)$ and expected value $E[T]$ are

$$\mu_i = \int_{-\infty}^{\infty} (t - E[T])^i f_T(t)\ dt = E[(T - E[T])^i] \qquad (12)$$

It is straight forward to show from (12) that the central moments as a function of moments about the origin are[6]:

$$\begin{aligned} \mu_0 &= 1 \\ \mu_1 &= 0 \\ \mu_2 &= m_2 - (m_1)^2 \\ \mu_3 &= m_3 - 3m_1m_2 + 2(m_1)^3 \\ \mu_4 &= m_4 - 4m_3m_1 + 6m_2(m_1)^2 - 3(m_1)^4 \end{aligned} \qquad (13)$$

From (3) and (11),

$$m_1 = -\tilde{m}_1\ ;\ m_2 = 2! \cdot \tilde{m}_2\ ;\ m_3 = -3! \cdot \tilde{m}_3 \qquad (14)$$

From which we can express the central moments - variance μ_2 and skewness μ_3 - in terms of the circuit response moments:

$$\begin{aligned} \mu_2 &= 2\tilde{m}_2 - (\tilde{m}_1)^2 \\ \mu_3 &= -6\tilde{m}_3 + 6\tilde{m}_1\tilde{m}_2 - 2(\tilde{m}_1)^3 \end{aligned} \qquad (15)$$

Unlike the moments of the impulse response, the central moments have geometrical interpretations:

- μ_0 is the area under the curve. It is generally unity, or else a simple scaling factor is applied.
- μ_2 is the *variance* of the distribution which measures the spread of the curve from the center. A larger variance reflects a larger spread of the curve.
- μ_3 is a measure of the *skewness* of the distribution; for a unimodal function its sign determines if the mode (global maximum) is to the left or to the right of the expected value (mean). Its magnitude is a measure of the distance between the mode and the mean.
- μ_4 is a measure of the *kurtosis* or *peakedness* of the distribution; for a unimodal function its relation to μ_2 reflects the portion of the total area under the curve that is attributable to the tails of the curve.

2.4: Connection Between Probability Density Functions and Circuit Responses

Any function $f(t)$ can be treated as a probability density function if it is defined in the range $[a, b]$ and satisfies

$$\begin{cases} f(t) \geq 0 \quad \forall t \\ \int_a^b f(t)\ dt = 1 \end{cases} \qquad (16)$$

If $f(t)$ is equal to zero outside of the range $[a,b]$, we can replace the integration limits in (16) with $-\infty$ and ∞. Elmore[3] was the first to apply moments for delay approximation of a limited class of circuit responses by observing that the impulse response of a circuit can be treated as a probability density function. He used this observation to justify the approximation of the 50% point of a monotonic step response (the median point of the impulse response) by the first moment (mean of the impulse response).

In [6] it was shown that the impulse response corresponding to an RC tree is unimodal with positive skew. From this it follows that the mode is less then the median which is less then the mean and vice versa [9] [13]:

$$(skew > 0)\ \ \textit{if and only if}\ \ (mode < median < mean)$$

This proved that the Elmore delay is an upper bound for the 50% step response delay[6], and was shown to hold for finite input signal rise time.

An important observation from [6] is that because of the variation in impulse response shapes along an interconnect path (*e.g.* from driver to load), the relative accuracy of the Elmore delay bound can be quite poor. Especially for the interconnects associated with deep submicron technologies, more than one moment is needed to capture the waveform shape-characteristics.

3: PRIMO

The approach proposed here is related to the work of Pearson [2] for approximating a histogram of discrete experimental data with a continuous function. The most important step in this process is finding the representative distribution family. The Gamma distribution, depicted in Fig. 2, is reasonably representative of the impulse responses in RC trees since it provides good "coverage" of bell shaped curves which are bounded on the left and exponentially decaying to the right.

The probability density function, $g_{\lambda,n}(t)$, of Gamma distribution is a function of one variable t and two parameters λ and n (real numbers):

$$g_{\lambda,n}(t) = \frac{\lambda^n t^{n-1} e^{-\lambda t}}{\Gamma(n)} \quad , \quad \Gamma(x) = \int_0^{\infty} y^{x-1} e^{-y}\, dy \quad (17)$$

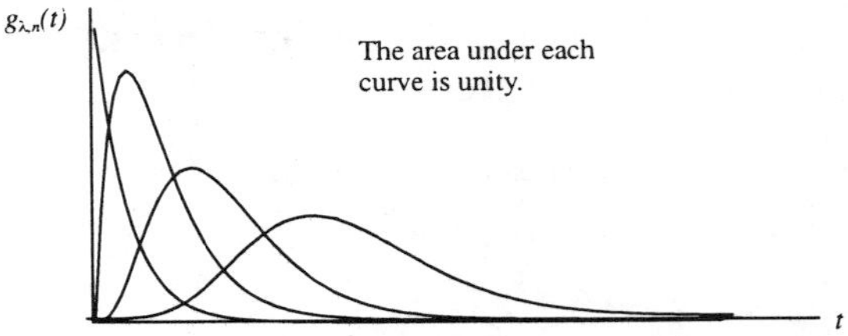

FIGURE 2. Examples of several members from the Gamma distribution family. Each family member corresponds to specific values of the distribution parameters λ and n.

From the moment generating function [11][20] of the Gamma distribution,

$$m_G(s) = \left(\frac{\lambda}{\lambda - s}\right)^n \quad (18)$$

we can express the moments in terms of the parameters n and λ:

$$m_1 = \frac{d}{ds} m_G(s)\Big|_{s=0} = \frac{n}{\lambda}$$
$$m_2 = \frac{d^2}{ds^2} m_G(s)\Big|_{s=0} = \frac{n(n+1)}{\lambda^2} \quad (19)$$
$$m_3 = \frac{d^3}{ds^3} m_G(s)\Big|_{s=0} = \frac{n(n+1)(n+2)}{\lambda^3}$$

Since Gamma has only two parameters, λ and n, matching two moments would completely characterize this model. Therefore, to match the third moment and capture the skewness of the distribution, we add a third variable, Δ, to include a degree of freedom in terms of time-shifting the function.

To begin, the shape information that is encapsulated in μ_2 and μ_3 can be used to fit the Gamma-pdf parameters λ and n. Moment fitting λ and n in this way corresponds to a system of two equations and two unknowns:

$$\mu_2 = \frac{n(n+1)}{\lambda^2} - \left(\frac{n}{\lambda}\right)^2 = \frac{n}{\lambda^2}$$
$$\mu_3 = \frac{n(n+1)(n+2)}{\lambda^3} - 3\left(\frac{n}{\lambda}\right)\frac{n(n+1)}{\lambda^2} + 2\left(\frac{n}{\lambda}\right)^3 = \frac{2n}{\lambda^3} \quad (20)$$

Rearranging (20) results in:

$$\lambda = \frac{2\mu_2}{\mu_3}; \qquad n = \frac{4(\mu_2)^3}{(\mu_3)^2} \quad (21)$$

The first moment is matched by translating the Gamma distribution by a time-shift Δ that is equal to $m_1 - \frac{n}{\lambda}$, as shown in Fig. 3. From (13) it is recognized that fitting μ_2, μ_3 and m_1 is equivalent to fitting the first three moments, m_1, m_2, m_3. This is equivalent to moment matching, in the AWE sense[15], to the moments of a time-shifted gamma function:

$$e^{-\Delta s}\left[1 - \frac{n}{\lambda}s + \frac{n(n+1)}{2!\cdot\lambda^2}s^2 - \frac{n(n+1)(n+2)}{3!\cdot\lambda^3}s^3 + \ldots\right] \quad (22)$$

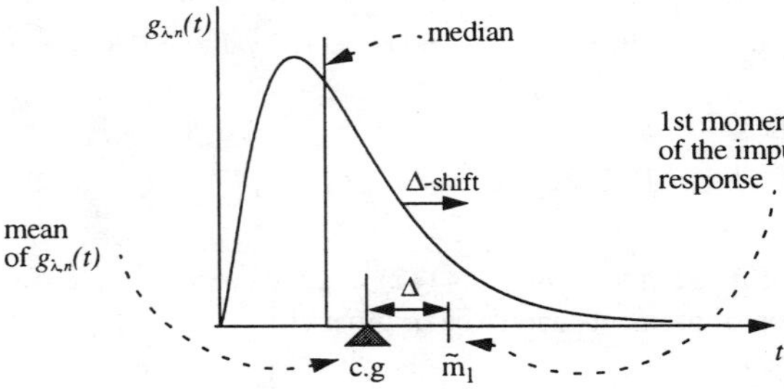

FIGURE 3. The variance and skewness, which are invariants of the origin location, are used to fit the shape of the curve. Then the curve is shifted such that its mean is aligned with the first moment of the impulse response.

Once the values of Δ, n and λ are obtained by matching the circuit response moments, the step response delay can be estimated from the probability density function $g_{\lambda,t}(t)$. For example, the approximate step response $\hat{y}(t)$ is given by

$$\hat{y}(t - \Delta) = \int_0^t g_{\lambda,n}(\tau)\, d\tau \quad (23)$$

Finding the delay point $t_{(\alpha)}$ for a particular waveform threshold voltage $\alpha \cdot V_{max}$ (*e.g.* if $\alpha = 0.5$ it is the 50% delay), is equivalent to finding a percentile α of a distribution,

$$\hat{y}(t_{(\alpha)}) = \alpha \quad (24)$$

Therefore, it is straightforward to use pre-compiled tables of Gamma distribution percentiles (with respect to the parameters n and λ) to solve (24) at the runtime cost of a *single* table-access. Moreover, we can normalize time by λ so that the percentile value can be obtained from a one-dimensional table in terms of n. An efficient computation scheme for the Gamma distribution function can be found in [16].

4: Properties and Implementation Issues

4.1: Stability

The Gamma probability density function in (17) is stable if $\lambda > 0$. From (21), $sgn(\lambda) = sgn(\mu_2) \cdot sgn(\mu_3)$. It was proven in [6] that both μ_2 and μ_3 are both positive for any RC tree. It was further shown in [5] that the point at which the skew changes from positive to negative for an RLC tree can be used as a measure of the response damping. We use these results to construct the following theorem:

Theorem of Stability: *The Gamma approximation is valid and stable for any RLC topology which has a response for which its derivative is a positively skewed bell shape distribution.*

4.2: Accuracy

Although it is provably stable, the shifted Gamma function may display significant error at times, particularly for the fitting of the early portion of the impulse response. As expected, fitting moments will tend to be most accurate for the tail portion of the response, since moments represent the series expansion about s=0. Fortunately, the response near t=0 is the least important portion of the response for delay calculation purposes. In addition, a large time-shift value, Δ, provides some indication of when the Gamma function is not fitting the impulse response in the expected manner.

4.3: Finite Rise Time Effects

Finite rise time effects can be captured by using two time-shifted gamma functions, separated by the risetime, to approximate the pulse response. The pre-compiled 2-D table is now a function of n and the risetime[12].

5: Experimental Results

As a first example, consider the RC tree in Fig. 4. We compare the

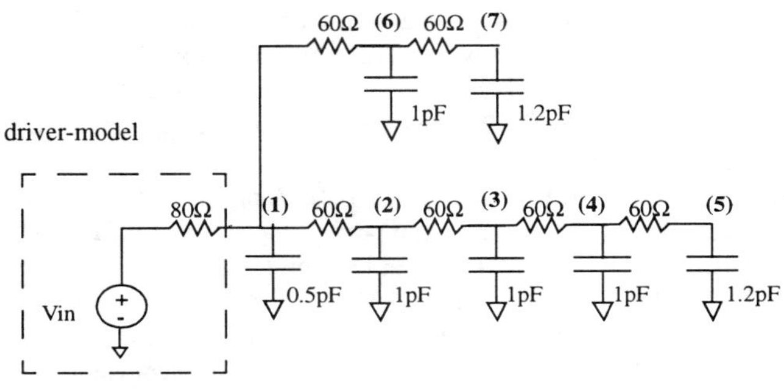

FIGURE 4. An RC tree example.

delays obtained from SPICE with those found using the Gamma approximation. The results for the 50% delay and 90% delay are summarized in Table 1.

Note that the differences between the PRIMO delays and the SPICE delays at the leaf nodes is about 1% or less. Excluding the driving point node, the largest error at any downstream node is about 4%. This worst case occurs for nodes that are closest to the root (node 2). These nodes correspond to the responses with the highest frequency content, hence one would expect the largest moment matching error there.

TABLE 1. Comparison of the 50% and 90% step response delays from SPICE and from PRIMO-Gamma (time in ns).

Node	SPICE (50%)	PRIMO (50%)	SPICE (90%)	PRIMO (90%)
2	0.477	0.497	2.02	2.01
3	0.700	0.699	2.27	2.27
4	0.845	0.836	2.42	2.43
5	0.919	0.909	2.50	2.50
6	0.375	0.376	1.75	1.74
7	0.452	0.450	1.82	1.82

The impulse response at the driving point node does not follow the bell shape, but rather starts out with a non-zero value and asymptotically approach zero in a multi-exponential decay form. Other distribution families may permit matching higher order moments, or more naturally capture these driving point response shapes. But as with all moment matching problems, the greatest challenge is to find a model that is provably stable and realizable.

The Beta function, for example, can be characterized in terms of its moments just like Gamma. But even with more shape fitting parameters, the Beta function is not always superior to the Gamma function for fitting downstream nodes. Some examples for driving point and near-end loads are shown in Table 2. Importantly, the utility of Beta is questionable, since the moment matching is sometimes non-realizable.

TABLE 2. The 50% step response delays obtained from Gamma and Beta approximations at the highest frequency nodes.

Node	SPICE	Gamma	Beta
1	0.196	0.243	0.193
2	0.477	0.498	0.489
6	0.375	0.378	0.355

An RLC Line

Since PRIMO requires a bell-shaped distribution for the impulse response, overdamped and critically damped RLC circuit responses can be approximated with this method. As a simple example, consider the 2000 micron distributed RLC line in Fig. 5, which was modeled with 20 lumped RLC sections. The results are compared with SPICE in Table 3.

TABLE 3. Comparison of SPICE and PRIMO for RLC line.

R_{driver}	C_{load}	Slope	SPICE 50%	SPICE 90%	Gamma 50%	Gamma 90%
25Ω	1pF	step	under-damped		under-damped	
35Ω	1pF	step	40.1ps	99.3ps	39.6ps	100ps

PRIMO matched the SPICE simulation very well for the overdamped responses. The underdamped cases can be recognized prior to analysis by a negative value for the third central moment[5]. Since delay is difficult to define for an underdamped response (since it may cross a threshold point multiple times), and because underdamped responses generally require design/circuit changes, simply detecting this condition is sufficient for many applications.

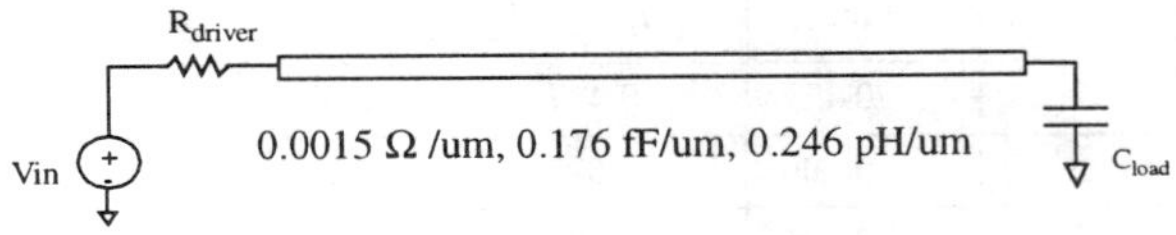

FIGURE 5. A distributed RLC line with a linear driver and capacitance load.

0.25 micron RC Tree Example

As a final example we considered a large, 10-fanout RC tree taken from a 0.25 micron commercial design. PRIMO is compared with SPICE and a 1-pole approximation in Table 4. Even for a step approximation the PRIMO results match SPICE in most cases, and is significantly better than a 1-pole approximation in almost all cases. The largest errors are observed at the near-end fanouts for both the PRIMO and the 1-pole models.

TABLE 4. Comparison of the 50% step response delays from SPICE, PRIMO (Gamma) and a 1-pole approximation (in ps).

Fan-out	SPICE	PRIMO	1-Pole	PRIMO % error	1-Pole % error
1	137.19	137.85	137.25	0.48	0.04
2	141.76	142.02	140.02	0.19	1.22
3	112.90	116.85	123.82	3.50	9.67
4	112.89	116.84	123.81	3.50	9.67
5	112.77	116.71	123.72	3.50	9.72
6	48.46	59.05	84.68	21.9	74.74
7	19.08	29.09	63.28	52.5	232.0
8	59.50	68.79	91.31	15.6	53.45
9	53.22	63.22	87.53	18.8	64.5
10	191.93	189.98	172.43	1.02	10.2

6: Conclusion

This paper describes a new delay metric based on a probability interpretation of moments of the circuit response. A time-shifted Gamma function is shown to provide good approximations of RC tree impulse responses so that one-dimensional probability tables can be used to evaluate the delays. Therefore, this new metric is able to capture higher order moment effects but with runtime efficiency that is comparable to that for the Elmore delay.

7: Acknowledgments

The authors would like to thank Tao Lin and Emrah Acar of CMU for their insightful contributions and for generating results for some of the examples.

8: Reference

[1]C. Chu and M. Horowitz, Charge-Sharing Models for Switch-Level Simulation, Trans. on CAD, June, 1987.

[2]C. C. Craig, "A new Exposition and Chart for the Pearson system of Frequency Curves", Ann. Math. Statistics, vol. 7, 1936.

[3]W. C. Elmore, "The Transient Response of Damped Linear Networks with Particular Regard to Wideband Amplifiers", J. Applied Physics, vol. 19, no. 1, Jan. 1948.

[4]P. Feldman and R. W. Freund, "Efficient Linear Circuit Analysis by Padé Approximation Via the Lanczos Process", IEEE Tran. Computer-Aided Design, vol. 14, no. 5, May 1995.

[5]R. Gupta, B.Krauter and L. Pileggi, "Transmission Line Synthesis via Constrained Multivariable Optimization", IEEE Trans. Computer-Aided Design, Vol. 16, No. 1, Jan. 1997.

[6]R. Gupta, B. Tutuianu and L. Pileggi, "The Elmore Delay as a Bound for RC Trees Generalized input Signals", IEEE Trans. Computer-Aided Design, vol. 16, no. 1, January 1997.

[7]M.A. Horowitz, *Timing Models for MOS Circuits*, Ph.D. thesis, Stanford University, January 1984.

[8]A. B. Kahng, K. Masuko and S. Muddu, "Analytical Delay Models for VLSI Interconnects Under Ramp Input", IEEE/ACM Inter. Conf. Computer-Aided Design, 1996.

[9]M. G. Kendall and A. Stuart, "The Advanced Theory of Statistics, vol. 1: Distribution Theory", New York: Hafner, 1969.

[10]K. J. Kerns, I. L. Wemple, A. T. Yang, "Stable and Efficient Reduction of Substrate Model Network using Congruence Transformation", ICCAD, Nov. 1995.

[11]H. J. Larson, "Introduction to Probability Theory and Statistical Inference", 3rd ed., John Wiley & Sons pub., 1982.

[12]T. Lin, E. Acar, and L. Pileggi, Carnegie Mellon University, CMU Technical Report in Preparation.

[13]H. L MacGillivray, "The Mean, Median, Mode Inequality and Skewness for a Class of Densities", Australian J. of Statistics, vol. 23 no. 2, 1981.

[14]A. Odabasioglu, M, Celik and L. Pileggi, "PRIMA: Passive Reduced-order Interconnect Macromodeling Algorithm, IEEE/ACM Inter. Conf. Computer-Aided Design, 1997.

[15]L. T. Pillage and R. A. Rohrer, "Asymptotic Waveform Evaluation for Timing Analysis", Tran. on CAD, Apr. 1990.

[16]W. H. Press, S. A. Teukolsky, W. T. Vetterling and B. P. Flannery, "Numerical Recipes in C", Cambridge University Press, 2nd ed., 1992.

[17]C. Ratzlaff and L. T. Pillage, "RICE: Rapid Interconnect Circuit Evaluator using Asymptotic Waveform Evaluation", IEEE Trans. Computer-Aided Design, Jun 1994.

[18]J. Rubenstein, P. Penfield and M. A. Horowitz, "Signal Delay in RC Tree Networks", Trans. on CAD, July 1983.

[19]L. M. Silveria, M. Kamon and J. White, "Efficient Reduced-Order Modeling of Frequency-Dependent Coupling Inductance Associated with 3-D Interconnect Structure", Proc. ACM/IEEE Design Automation Conf., Jun 1995.

[20]S. M. Stigler, "The History of Statistics: The Measurement of Uncertainty before 1900", Belknap Press of Harvard University Press, 1986.

[21]B. Tutuianu, F. Dartu and L. Pileggi, "An Explicit RC-Circuit Delay Approximation Based on the First Three Moments of the Impulse Response", Proc. ACM/IEEE Design Automation Conf., 1996.

ftd: An Exact Frequency to Time Domain Conversion for Reduced Order RLC Interconnect Models†

Ying Liu, Lawrence T. Pileggi, Andrzej J. Strojwas
Department of Electrical and Computer Engineering
Carnegie Mellon University
5000 Forbes Ave., Pittsburgh, PA 15213

Abstract

*Recursive convolution provides an exact solution for interfacing reduced-order frequency domain representations with discrete time domain models of piecewise linear voltage waveforms. The state-space method is more efficient, but not exact, and can sometimes produce large time domain errors. This paper presents a new algorithm, **ftd** (frequency to time domain), for incorporating linear frequency domain macromodels into time domain simulators. **ftd** provides accuracy equivalent to recursive convolution with efficiency that is superior to the state-space methods.*

1: Introduction

There are two popular methods for converting a frequency domain model into a discrete time domain model. The first is recursive convolution[1], where the input is assumed to be piece-wise linear and decomposed as different ramps at each time point. This permits the convolution to be carried out in a recursive way. The second approach is the state-space method[2], whereby a numerical integration scheme such as trapezoidal is applied to the state space realization of the frequency domain model to compute a discrete time domain model. Among the two, recursive convolution achieves the best possible accuracy[3], but can be inefficient due to the inherent complexity of the algorithm. In contrast, the state-space method requires fewer floating-point operations at each time point than recursive convolution, but has inferior accuracy due to the required numerical integration scheme. Moreover, both methods are theoretically complicated and, therefore, somewhat difficult to implement. This difficulty not only impacts their widespread use, but also makes it difficult to optimize the software routines, which indirectly impacts the runtime efficiency.

† This work was supported, in part, by the Semiconductor Research Corporation under Contract DC-068.067 and SGS-Thomson Microelectronics.

In this paper, we propose an efficient and accurate algorithm, called ***ftd***, for converting a frequency domain model to a discrete time domain model for circuit and timing simulation. ***ftd*** uses an efficient algorithm for direct computations of the non-zero state response at each time point. ***ftd*** provides accuracy equivalent to that of the recursive convolution but with the same complexity as the state-space method. Furthermore, due to the algorithmic simplicity of ***ftd***, empirical results demonstrate that its implementation is more efficient than the state-space method too.

2: Background

For timing simulation, a linear interconnect subcircuit can be pre-processed into a reduced-order **Y** matrix[4]. A two-port example is written as:

$$\begin{bmatrix} I_1(s) \\ I_2(s) \end{bmatrix} = \begin{bmatrix} y_{11}(s) & y_{12}(s) \\ y_{21}(s) & y_{22}(s) \end{bmatrix} \cdot \begin{bmatrix} V_1(s) \\ V_2(s) \end{bmatrix} \tag{1}$$

The first equation in (1) can be expressed as:

$$I_1(s) = y_{11}(s) \cdot V_1(s) + y_{12}(s) \cdot V_2(s) \tag{2}$$

where the general form of any $y_{ij}(s)$ term is:

$$y(s) = \frac{b_{q-1}s^{q-1} + \ldots + b_1 s + b_0}{s^q + a_{q-1}s^{q-1} + \ldots a_1 s + a_0} \tag{3}$$

Using recursive convolution[1], state space method[2], or the approach proposed in this paper, the discrete time domain model of (2) at time point t_k can be written as:

$$i_1(t_k) = g_{11}(t_k)v_1(t_k) + g_{12}(t_k)v_2(t_k) + I_{1eq}(t_k) \tag{4}$$

once a time step is chosen. It follows from (1) that:

$$\begin{bmatrix} i_1(t_k) \\ i_2(t_k) \end{bmatrix} = \begin{bmatrix} g_{11}(t_k) & g_{12}(t_k) \\ g_{21}(t_k) & g_{22}(t_k) \end{bmatrix} \cdot \begin{bmatrix} v_1(t_k) \\ v_2(t_k) \end{bmatrix} + \begin{bmatrix} I_{1eq}(t_k) \\ I_{2eq}(t_k) \end{bmatrix} \tag{5}$$

The equivalent circuit realization of (5) is shown in Fig.1, where each port includes an equivalent conductance, current source, and VCCS. During the simulation process, at each time point, these discrete time domain models are constructed based on a reduced-order frequency domain approximation. This N-port model is combined with the models connected to its ports via Modified Nodal Analysis(MNA). The solution of the MNA equations along with

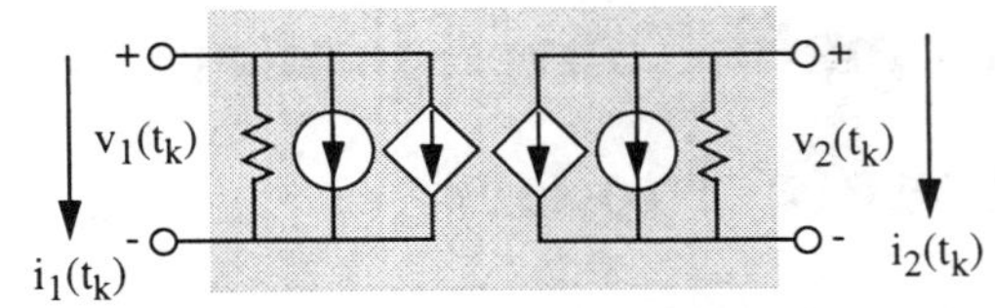

FIGURE 1: Discrete time domain model of a two-port.

the non-linear elements will yield the overall circuit solution.

It can be shown that for all three methods the values of conductances and VCCS's depend only on the time step. Therefore, when the time step is fixed for consecutive time points, these parameter values are constant. The equivalent current sources are always updated at each time point, thereby dominating the runtime for each method. For this reason we will focus on the updating procedure of the equivalent current source models when comparing the algorithmic complexity of the three methods.

In recursive convolution, it is assumed that the input voltage is piecewise linear(PWL) as multiple incremental ramps occur at each time point with different slopes. The overall response can be calculated as the sum of multiple zero-state ramp responses up to the present time point. By careful bookkeeping of these variables, the calculation can be carried out in a recursive and efficient way. A complete derivation of recursive convolution can be found in [1][7]. Recursive convolution is exact for PWL voltages, however this accuracy is achieved at the expense of efficiency. For an N-port with a q-th order approximation, the number of internal variables that are updated at each time point is $3N^2q$. The algorithm complexity is estimated as $O(7N^2q)$ at each time point. When the time step is changed, updating of the parameters requires $O(2N^2q)$ exponential evaluations and $O(4N^2q)$ floating-point multiplications/divisions.

In the state-space method, the admittances in the form of (3) are re-formatted into a controllable canonical state-space form[5]. Numerical integration such as the trapezoidal method is used to calculate the response. A complete description of the state-space method can be found in [2][7]. For the state-space method, the number of variables that must be updated at each time point is N^2q. The complexity of the algorithm is $O(3N^2q)$ at each time point. When the time step changes, updating of the parameters requires $O(q^2N^2)$ of floating-point multiplications/divisions. Because a numerical integration scheme is used, no exponential evaluation is needed. Since it requires fewer floating-point operations at each time point, the state-space method is faster than recursive convolution for a fixed time step. However, since a numerical integration scheme is used, the state-space method is not as accurate as recursive convolution.

3: Description of *ftd*

In *ftd*, each admittance of the form given by (3) is first decomposed into a sum of partial fractions. This pre-processing step represents some computational overhead for *ftd* and recursive convolution. However, most model order reduction schemes rely on calculation of the poles and residues to verify accuracy and stability.

The decomposition can be formulated in matrix notation as:

$$\begin{cases} \dot{\mathbf{x}}(t) = \mathbf{A}\mathbf{x}(t) + \mathbf{B}v(t) \\ i(t) = \mathbf{C}\mathbf{x}(t) \end{cases} \tag{6}$$

with

$$\mathbf{x} = \begin{bmatrix} x_1 \\ x_2 \\ \vdots \\ x_q \end{bmatrix},\ \mathbf{A} = \begin{bmatrix} p_1 & 0 & \dots & 0 \\ 0 & p_2 & \dots & 0 \\ \vdots & \vdots & \ddots & \vdots \\ 0 & 0 & \dots & p_q \end{bmatrix},\ \mathbf{B} = \begin{bmatrix} 1 \\ 1 \\ \vdots \\ 1 \end{bmatrix},\ \mathbf{C} = \begin{bmatrix} k_1 & \dots & k_q \end{bmatrix}$$

Unlike recursive convolution, *ftd* uses directly the **exact** solution of the differential equation (6), which can be written as[5]:

$$\mathbf{x}(t) = e^{\mathbf{A}t} \cdot \mathbf{x}(0) + e^{\mathbf{A}t}\int_0^t e^{-\mathbf{A}\tau}\mathbf{B}v(\tau)d\tau$$

Assuming that we know the solution at the time point t_{k-1}, then the exact solution at time point t_k can be expressed as a zero-input response term plus a zero-state response term:

$$\mathbf{x}(t_k) = e^{\mathbf{A}\Delta t_k} \cdot \mathbf{x}(t_{k-1}) + \int_0^{\Delta t_k} e^{\mathbf{A}\tau}\mathbf{B}v(t_k - \tau)d\tau \tag{7}$$

Since at the time point t_{k-1} we don't know the voltage response shape to the time point t_k, the integral in (7) cannot be computed explicitly. However, as with the recursive convolution, and as generally assumed in circuit and timing simulation, for a reasonably small timestep we can assume that the waveshape is piecewise linear. Thus, the voltage between the time points t_{k-1} and t_k is expressed as follows:

$$v(t_k - \tau) = v(t_k) - \left(\tau \cdot \frac{v(t_k) - v(t_{k-1})}{\Delta t_k}\right) \quad for\ 0 \le \tau \le \Delta t_k \tag{8}$$

Using the piecewise linear assumption in (8), the integral in (7) can be evaluated explicitly

$$\mathbf{x}(t_k) = e^{\mathbf{A}\Delta t_k}\mathbf{x}(t_{k-1}) + \left(\mathbf{A}^{-1}e^{\mathbf{A}\Delta t_k}\mathbf{B} - \frac{\mathbf{A}^{-2}}{\Delta t_k}(e^{\mathbf{A}\Delta t_k} - \mathbf{I})\mathbf{B}\right) \cdot$$

$$v(t_{k-1}) + \left(\frac{\mathbf{A}^{-2}}{\Delta t_k}(e^{\mathbf{A}\Delta t_k} - \mathbf{I})\mathbf{B} - \mathbf{A}^{-1}\mathbf{B}\right)v(t_k)$$

This formula appears to be formidable. For a general realization of {**A**, **B**, **C**}, it is quite difficult to carry out the above matrix operations. But since we expand the admittance terms into a sum of partial fractions, the **A** matrix is diagonal and all of the entries in **B** are equal to one. Therefore, the matrix operations in the above equation are actually scalar operations. After some simplification, the equation can be re-written as:

$$x_i(t_k) = \zeta_i(\Delta t_k) \cdot x_i(t_{k-1}) + \eta_i(\Delta t_k) \cdot v(t_{k-1}) + \theta_i(\Delta t_k) \cdot v(t_k)$$
$$for \quad i = 1, 2, \ldots, q$$

with the parameters ζ, η and θ defined as:

$$\zeta_i(\Delta t_k) = e^{p_i \Delta t_k}$$
$$\eta_i(\Delta t_k) = \frac{e^{p_i \Delta t_k}(p_i \Delta t_k - 1) + 1}{\Delta t_k \cdot p_i^2} \tag{9}$$
$$\theta_i(\Delta t_k) = \frac{1}{\Delta t_k}\left(-\frac{1}{p_i^2} - \frac{1}{p_i}\Delta t_k + \frac{1}{p_i^2} e^{p_i \Delta t_k}\right)$$

These parameters depend only on the poles and the time step. They remain constant when the time step is fixed for consecutive time points.

The current at the time point t_k can be expressed as:

$$i(t_k) = I_{eq}(t_k) + g(t_k) \cdot v(t_k)$$

where the equivalent current source and the equivalent conductance are defined as:

$$g(t_k) = \sum_{i=1}^{q} k_i \cdot \theta_i$$
$$I_{eq}(t_k) = \left(\sum_{i=1}^{q} k_i \cdot \eta_i\right) v(t_{k-1}) + \sum_{i=1}^{q} k_i \zeta_i x_i(t_{k-1}) \tag{10}$$

Once the voltage at time point t_k is known, the states should be updated to prepare for the next time point calculation.

When there are complex poles in the admittance, ***ftd*** involves the evaluation of complex numbers. Since the complex numbers cost twice as much in storage, they should be avoided whenever possible. The discussion of the complex pole algorithm can be found in [7].

4: Implementation for Multiports

ftd is very simple to implement. Following the 2-port example shown in Section 2, the goal is to calculate the discrete time-domain model of a two-port shown in (5) at each time point. We use superscripts to distinguish between different ports and subscripts to distinguish between different poles and residues for the four admittance terms. Upon choosing the appropriate time step, the parameters ζ, η, θ for each real pole are calculated based on the formulae in (9).

The number of states to update in ***ftd*** is 4q, which can be stored in a single vector **x**. At each time point, if the time step is unchanged, only the equivalent current sources have to be updated. The equivalent current source updating algorithm can be carried out as shown in Fig.2. Note that all the floating-point operations involved are SAXPY ($y \leftarrow a \cdot x + y$) type of operation. For a modern computer architecture, this can be carried out with the utmost efficiency.

At each time point:

◆Calculate intermediate variables, store them in vector **x**:

$$x_i^{(mn)} \leftarrow \zeta_i^{(mn)} \cdot x_i^{(mn)} + \eta_i^{(mn)} \cdot v_n(t_{k-1})$$

◆Calculate the equivalent current sources:

$$\begin{bmatrix} I_{1eq} \\ I_{2eq} \end{bmatrix} \leftarrow \begin{matrix} \sum_{i=1}^{q} k_i^{(11)} \cdot x_i^{(11)} + \sum_{i=1}^{q} k_i^{(12)} \cdot x_i^{(12)} \\ \sum_{i=1}^{q} k_i^{(21)} \cdot x_i^{(21)} + \sum_{i=1}^{q} k_i^{(22)} \cdot x_i^{(22)} \end{matrix}$$

◆Stamp equivalent conductances and current sources into MNA matrix and solve MNA matrix for the voltages.

◆Update the state vector for the next time point.

$$x_i^{(mn)} \leftarrow x_i^{(mn)} + \theta_i^{(mn)} \cdot v_n(t_k)$$

FIGURE 2: Flowchart of equivalent current source updating scheme.

5: LTE and Complexity of the Algorithms

Following the common assumption of LTE(Local Truncation Error) estimation, we assume that the exact value of the state at the time point t_{k-1} is known. Furthermore, we assume that the voltages are perfectly linear between time points t_k and t_{k-1}. A complete discussion of the LTE of the three methods can be found in [7]. The results are tabulated in Table 1.

We compare the complexity for a fixed time step in terms of the floating-point operations. The second complexity measure considers the number of floating-point operations when the time step changes. Details can be found in [7], but the results are summarized in Table 1.

	Recursive Convolution	State-Space Method	*ftd* Method
# of variables	$3N^2q$	N^2q	N^2q
Complexity[1]	$O(7N^2q)$	$O(3N^2q)$	$O(3N^2q)$
Complexity[2]	exp: $O(2N^2q)$	exp: none	exp: $O(N^2q)$
	mul: $O(5N^2q)$	mul: $O(N^2q^2)$	mul: $O(5N^2q)$
LTE	0	$p\Delta t_k^2(v_k - v_{k-1})$	0

TABLE 1: LTE and estimation of complexity for all three methods.

It is apparent that the recursive convolution requires the largest amount of floating-point computation for a fixed time step analysis. At first glance it would appear that the state-space method and ***ftd*** have identical complexity when the time step is fixed, however, for the state-space method the state $x_2(t_k)$ cannot be calculated until $x_1(t_k)$ is available. Similarly, $x_3(t_k)$ cannot be calculated until $x_2(t_k)$ is available, and so on. ***ftd*** does not suffer from this dependency, therefore with modern computers, and pipeline processing, the ***ftd*** updates will run much more efficiently.

When the time step changes, the state-space method requires the largest amount of multiplications/divisions, but avoids exponential evaluations. Moreover, when there are complex poles, both recursive convolution and ***ftd*** also require evaluation of triangular functions. In practice, the exponential and triangular functions can be stored in a look-up table. As a result, the recursive convolution and ***ftd*** may spend less time in re-calculating the parameters than the state-space method when the time step is changed.

6: Results

The first example is a clock tree with 64 leaf nodes. Each interconnect segment is modeled as a lumped RC segment providing a 65-port circuit. AWE[4] is used to calculate a 4th order model for all admittance terms. The resulting **Y**(s) macromodel is a 65×65 matrix. We compare recursive convolution, the state space method, and ***ftd*** in terms of a Matlab implementation on an IBM RS/6000 43P-140.

We plot the absolute errors as a function of time for one port node in Fig.2 by comparing to a SPICE simulation with a 0.1 picosecond timestep. The error plots are shown for two different timesteps. When the time step is 0.2 picosecond, the three methods have comparable error. However, when the time step grows to 1 picosecond, the state-space error is considerably larger than the other two methods.

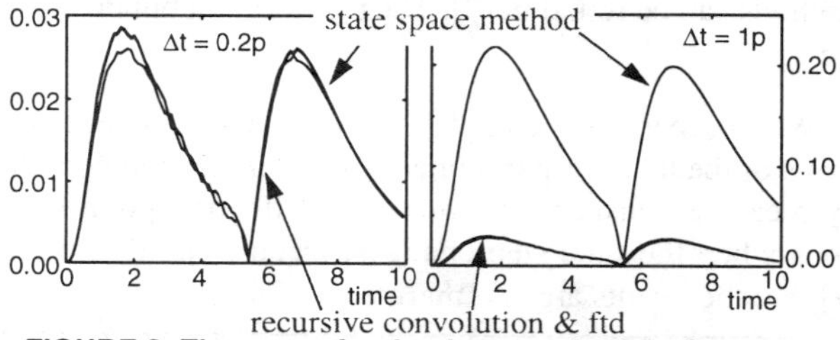

FIGURE 2: The errors for the three methods using two different time steps. Note the different plot scales.

Table 2 shows the runtimes for the three methods(without the use of table models for exponential function calls). As expected, ***ftd*** is the most efficient, but these results must be carefully interpreted. First of all, Matlab is an interpretive language, thus the running time calculation is inaccurate. Secondly, Matlab is optimized to execute vector/matrix operations, but the optimization relies on skillful programming. For the recursive convolution implementation, the complicated updating algorithm was by far the most difficult to optimize. The state-space method was more straightforward to implement, however, the serial dependency discussed in Section 5 does not allow for certain optimizations and slows down the simulation. We must emphasize, however, that the authors spent a significant amount of time in implementing and optimizing all three methods as much as possible. But as indicated above, the simplicity of the ***ftd*** algorithm can lead to better runtime efficiency because it is more easily optimized for a wide variety of computer platforms.

Time Step	Recursive Convolution	State Space Method	Our Method
Δt = 0.2p	291.90 sec	274.99 sec	141.49 sec
Δt = 1p	60.98 sec	54.52 sec	30.36 sec

TABLE 2: Simulation time of the three methods.

In the second example, ***ftd*** was used with TETA(Transistor-level Engine for Timing Analysis)[6] to simulate a simple coupled RLC line problem with SPICE-like nonlinear drivers. The loads at the far ends of the lumped transmission model are linear capacitors, as shown in Fig.3. The interconnect is a six-port with six identical nonlinear drivers. The reduced-order model of the two-port is calculated using AWE to obtain a 3rd order approximation. The simulation result is shown in Fig.3 for a timestep of 1 picosecond. The SPICE and ***ftd*** (using TETA for the nonlinear elements) plots are indistinguishable.

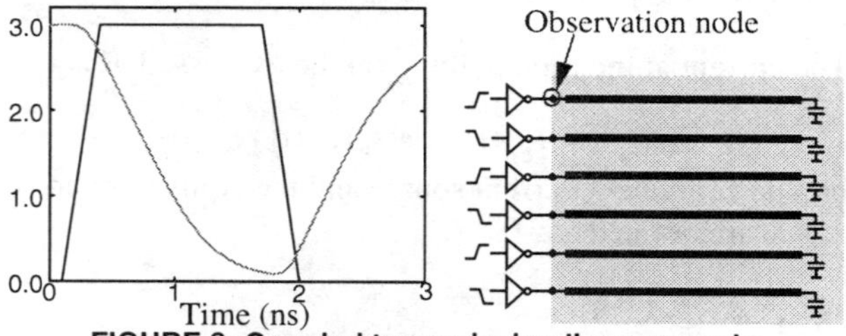

FIGURE 3: Coupled transmission lines example.

7: Conclusion

This paper presents a new algorithm for the converting frequency domain macromodels into the discrete time-domain models for circuit and timing simulation. Both the theoretical analysis and numerical results indicate that ***ftd*** achieves the highest possible accuracy while requiring the least computational resources. An important feature of ***ftd*** is that it is very straight-forward and can be easily implemented with increased efficiency on most modern hardware platforms.

References

[1]J. E. Bracken, V. Raghavan and R. A. Rohrer, "Interconnect Simulation with Asymptotic Waveform Evaluation", IEEE Trans. on Circuits and Systems, vol. 39, No. 11, Nov. 1992.

[2]S.-Y. Kim, N. Gopal and L. T. Pillage, "Time-Domain Macromodels for VLSI Interconnect Analysis", IEEE Trans. on CAD, vol. 13, No. 10, Oct. 1994.

[3]T. V. Nguyen, "Efficient Simulation of Lossy and Dispersive Transmission Lines", IEEE/ACM Proc. DAC, 1994.

[4]L. T. Pillage and R. A. Rohrer, "Asymptotic Waveform Evaluation for Timing Analysis", IEEE Trans. on CAD, vol. 9, no. 4, Apr. 1990.

[5]C.-T. Chen, "Linear System Theory and Design", Holt, Rinehart and Winston, Inc., 1984.

[6]F. Dartu and L.T. Pileggi, "TETA: Transistor-Level Engine for Timing Analysis", IEEE/ACM Proc. DAC, Jun. 1998.

[7]Y. Liu, L.T. Pileggi and A.J. Strojwas, "ftd: A Frequency to Time Domain Conversion Algorithm for Interconnect Simulation", submitted to IEEE Tran. CAD.

Extending Moment Computation to 2-Port Circuit Representations

Fang-Jou Liu and Chung-Kuan Cheng
Department of Computer Science and Engineering
University of California, San Diego
La Jolla, CA 92093
{fjliu, kuan}@cs.ucsd.edu

Abstract

In this paper, we present an extension of moment computation to 2-port circuits. Our formulas are applicable to both transfer function moments and driving-point admittance moments. Given the input admittances, output admittances, and transfer functions of two 2-ports, our formulas compute the input admittance, output admittance, and transfer function when these 2-ports are combined either in parallel or in series. A nice conclusion of our work is the discovery our formulas form an elegant framework integrating the results from two classical papers, Rubinstein et al. & O'Brien and Savarino, for computing the Elmore delay and driving-point admittance moments in RC trees.

1 Introduction and Motivations

In recent years, timing analysis methods based on moments have become increasingly popular [1] [4]. These methods are typically faster than SPICE by two orders of magnitude for digital and/or interconnect circuits, while retain a high degree of accuracy. The efficient computation of moments is crucial for these timing analysis methods, since this step can be very time-consuming, particularly for large circuits.

A majority of large circuits are composed of subcircuits, and it would be beneficial if moment computation can take advantage of the hierarchies formed by these subcircuits to shorten computation time. For example, if the moments of subcircuits are already computed, it would be desirable to reuse these moments in computing the moments of the overall circuit, instead of having to calculate the moments of the overall circuit from the ground up. What is needed is a way to calculate moments *hierarchically*.

This work is supported by NSF grant MIP-9529077 and grant 96-067 from the California MICRO program.

DAC 98, San Francisco, California
©1998 ACM 0-89791-964-5/98/06..$5.00

In this paper, we present an extension of moment computation to 2-port circuits. Our formulas are applicable to both transfer function moments and driving-point admittance moments. A nice conclusion of our work is the discovery our formulas form an elegant framework integrating the results from two classical papers, Rubinstein et al. & O'Brien and Savarino, for computing the Elmore delay and driving-point admittance moments in RC trees.

2 Applications of Our Formulas

Figure 1 illustrates the applications of our formulas. In Figure 1, 2-ports P_1 and P_2 are in parallel, with P_3 in series with both of them. *Driver* and *Output* are the driver and output stage, respectively. **Please note there is no loop in this circuit: signals can only propagate "down" the circuit, not "up".**

Given the transfer function, input admittance, and output admittance of every 2-port in this circuit, our formulas compute the input and output admittances of the entire circuit, as well as the transfer function between arbitrary nodes on either *Path*1 or *Path*2, and when *Path*1 and *Path*2 are combined. Thus with our formulas, we can compute the transfer function, input admittance, and output admittance of the overall circuit *hierarchically* from those of the individual constituent 2-ports.

3 Notations

In this section we introduce the notations used in the rest of this paper. For an illustration of these notations, please refer to Figure 3.

- $\bar{H}_{a,b}$, $\bar{H}_{c,d}$: transfer functions from a to b and from c to d of 2-ports P_1 and P_2, respectively. $\bar{H}_{a,b} = \left.\frac{V_b}{V_a}\right|_{I_b=0} = \sum_{i=0}^{\infty}\bar{h}_{a,b}^{(i)}s^i$; $\bar{H}_{c,d} = \left.\frac{V_d}{V_c}\right|_{I_d=0} = \sum_{i=0}^{\infty}\bar{h}_{c,d}^{(i)}s^i$. $\bar{h}_{a,b}^{(i)}$ and $\bar{h}_{c,d}^{(i)}$ are the ith moments of $\bar{H}_{a,b}$ and $\bar{H}_{c,d}$, respectively.

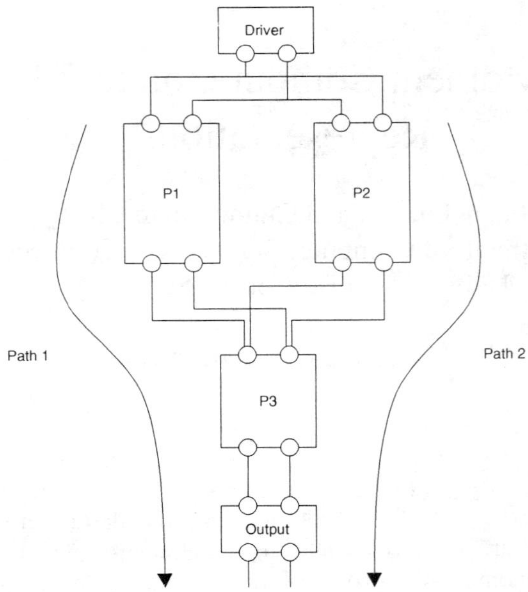

Figure 1: Illustration of the applications of our formulas.

- $H_{a,b}$: transfer function from a to b of 2-ports P_1 and P_2 *in series*. $H_{a,b} = \sum_{i=0}^{\infty} h_{a,b}^{(i)} s^i$, where $h_{a,b}^{(i)}$ is the ith moment of $H_{a,b}$.
- $\bar{Y}_a$, $\bar{Y}_c$: input admittances of 2-ports P_1 and P_2, respectively. $\bar{Y}_a = \frac{I_a}{V_a}\Big|_{I_b=0} = \sum_{i=0}^{\infty} \bar{y}_a^{(i)} s^i$; $\bar{Y}_c = \frac{I_c}{V_c}\Big|_{I_d=0} = \sum_{i=0}^{\infty} \bar{y}_c^{(i)} s^i$. $\bar{y}_a^{(i)}$ and $\bar{y}_c^{(i)}$ are the ith moments of $\bar{Y}_a$ and $\bar{Y}_c$, respectively.
- $\bar{Y}_b$, $\bar{Y}_d$: output admittances of 2-ports P_1 and P_2, respectively. $\bar{Y}_b = \frac{I_b}{V_b}\Big|_{V_a=0} = \sum_{i=0}^{\infty} \bar{y}_b^{(i)} s^i$; $\bar{Y}_d = \frac{I_d}{V_d}\Big|_{V_c=0} = \sum_{i=0}^{\infty} \bar{y}_d^{(i)} s^i$. $\bar{y}_b^{(i)}$ and $\bar{y}_d^{(i)}$ are the ith moments of $\bar{Y}_b$ and $\bar{Y}_d$, respectively
- Y_a, Y_d: input and output admittances, respectively, of 2-ports P_1 and P_2 *in series*. $Y_a = \sum_{i=0}^{\infty} y_a^{(i)} s^i$; $Y_d = \sum_{i=0}^{\infty} y_d^{(i)} s^i$. $y_a^{(i)}$ and $y_d^{(i)}$ are the ith moments of Y_a and Y_d, respectively.

4 Formula Derivations

4.1 Parallel Case

For 2-ports P_1 and P_2 in parallel as shown in Figure 2, the input admittance, Y, is the sum of the input admittances of P_1 and P_2, and we have $Y = \bar{Y}_a + \bar{Y}_c$. Similarly, the output admittance Y' is the sum of the output admittances of P_1 and P_2, with $Y' = \bar{Y}_b + \bar{Y}_d$.

Theorem 4.1 gives the formula for the new transfer function $H_{ac,bd}$ when P_1 and P_2 are in parallel:

Theorem 4.1 *When two 2-ports P_1 and P_2 are in parallel, the new transfer function $H_{ac,bd}$ is*

$$H_{ac,bd} = \frac{V_b}{V_a} = \frac{V_d}{V_c} = \frac{\bar{Y}_b \bar{H}_{a,b} + \bar{Y}_d \bar{H}_{c,d}}{\bar{Y}_b + \bar{Y}_d} \tag{1}$$

Expanding into moments, we have $(q \geq 1)$

$$\begin{aligned} h_{ac,bd}^{(q)} &= \frac{1}{\bar{y}_b^{(1)} + \bar{y}_d^{(1)}} \sum_{l=1}^{q} \left(\bar{h}_{a,b}^{(l)} \bar{y}_b^{(q+1-l)} + \bar{h}_{c,d}^{(l)} \bar{y}_d^{(q+1-l)} \right) - \\ & \quad \frac{1}{\bar{y}_b^{(1)} + \bar{y}_d^{(1)}} \sum_{l=1}^{q-1} h_{ac,bd}^{(l)} \left(\bar{y}_b^{(q+1-l)} + \bar{y}_d^{(q+1-l)} \right) \end{aligned} \tag{2}$$

where $h_{ac,bd}^{(q)}$ $(q \geq 1)$ is the qth moment of $H_{ac,bd}$, with $h_{ac,bd}^{(0)} = 1$. Since all quantities on the right-hand side of (1) are known, $H_{ac,bd}$ can be computed. Please refer to [2] for the proof of Theorem 4.1.

4.2 Series Case

Theorem 4.2 below computes $H_{a,b}$, Y_a, and Y_d, when P_1 and P_2 are in series:

Theorem 4.2 *For 2-ports P_1 and P_2 in series as shown in Figure 3, we have*

$$H_{a,b} = \frac{V_b}{V_a} = \frac{\bar{H}_{a,b}}{1 + \bar{Y}_c / \bar{Y}_b} \tag{3}$$

$$Y_a = \frac{I_a}{V_a} = \bar{Y}_a + \frac{\bar{Y}_b \bar{Y}_c \bar{H}_{a,b}^2}{\bar{Y}_b + \bar{Y}_c} \tag{4}$$

$$Y_d = \frac{I_d}{V_d} = \frac{\bar{Y}_d (\bar{Y}_b + \bar{Y}_c)}{\bar{Y}_d \bar{H}_{c,d}^2 + \bar{Y}_b + \bar{Y}_c} \tag{5}$$

Expanding into moments, we have $(q \geq 0)$

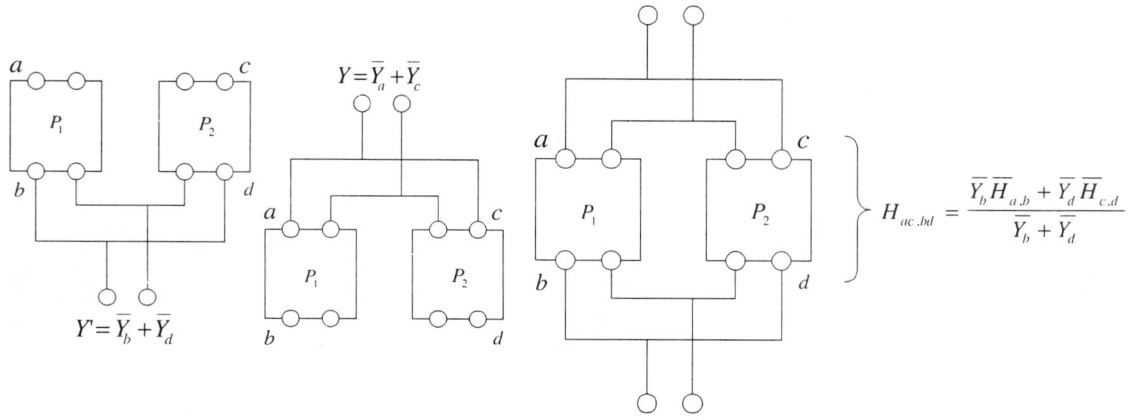

Figure 2: P_1 in parallel with P_2, illustrating the overall input (Y) and output (Y') admittances, as well as the overall transfer function ($H_{ac,bd}$).

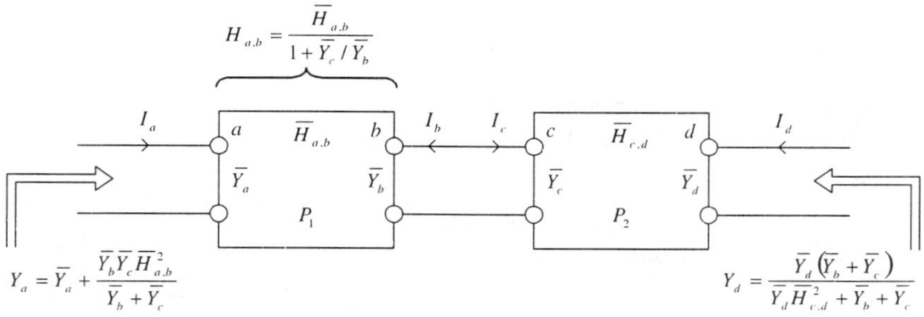

Figure 3: P_1 in series with P_2, illustrating the input ($\bar{Y}_a$, $\bar{Y}_c$) and output ($\bar{Y}_b$, $\bar{Y}_d$) admittances, as well as the transfer functions ($\bar{H}_{a,b}$, $\bar{H}_{c,d}$) of P_1 and P_2.

$$
\begin{aligned}
h_{a,b}^{(q)} &= \frac{1}{\bar{y}_b^{(0)} + \bar{y}_c^{(0)}} \left(\sum_{l=0}^{q} \bar{y}_b^{(l)} \bar{h}_{a,b}^{(q-l)} - \sum_{l=1}^{q} \left(\bar{y}_b^{(l)} + \bar{y}_c^{(l)} \right) h_{a,b}^{(q-l)} \right) \\
y_a^{(q)} &= \frac{1}{\bar{y}_b^{(0)} + \bar{y}_c^{(0)}} \left(\sum_{l=0}^{q} \left(\bar{y}_b^{(l)} + \bar{y}_c^{(l)} \right) \bar{y}_a^{(q-l)} - \sum_{l=1}^{q} \left(\bar{y}_b^{(l)} + \bar{y}_c^{(l)} \right) y_a^{(q-l)} + c_{q+1} \right) \\
y_d^{(q)} &= \frac{1}{d_{1,1}} \left(\sum_{l=0}^{q} \left(\bar{y}_b^{(l)} + \bar{y}_c^{(l)} \right) \bar{y}_d^{(q-l)} - \sum_{l=2}^{q+1} d_{l,1} y_d^{(q+1-l)} \right)
\end{aligned}
$$

where

$$
\begin{aligned}
c_{q+1} &= \sum_{m+n+p+r=q} \bar{y}_b^{(m)} \bar{y}_c^{(n)} \bar{h}_{a,b}^{(p)} \bar{h}_{a,b}^{(r)} \\
d_{1,1} &= \bar{y}_d^{(0)} \left(\bar{h}_{c,d}^{(0)} \right)^2 + \bar{y}_b^{(0)} + \bar{y}_c^{(0)} \\
d_{l,1} &= \sum_{n+p+r=l-1} \bar{y}_d^{(n)} \bar{h}_{c,d}^{(p)} \bar{h}_{c,d}^{(r)} + \bar{y}_b^{(l-1)} + \bar{y}_c^{(l-1)}
\end{aligned}
$$

Please refer to [2] for the proof of Theorem 4.2. Note all quantities on the right-hand side of (3), (4), and (5) are known. Thus, $H_{a,b}$, Y_a. and Y_d can be computed.

Expanding (3), (4), and (5) into moments and re-arranging terms, we get the polynomial expressions for $h_{a,b}^{(q)}$, $y_a^{(q)}$, and $y_d^{(q)}$ as stated in Theorem 4.2.

5 Equivalence with Rubinstein et al. & O'Brien and Savarino

In this section, we demonstrate that our formulas are a natural extension of Rubinstein et al. [5] and O'Brien and Savarino [3].

Theorem 5.1 below states the equivalence of our formulas with [5] and [3]. Theorem 5.1 can be obtained by properly simplifying the formulas in Theorem 4.2 when the circuit is a RC tree.

Theorem 5.1 *For the RC tree in Figure 4, we have ($i \geq 1$)*

$$h_{a,b}^{(i)} = -Ry_a^{(i)} \tag{6}$$

$$y_a^{(i)} = \sum_{l=0}^{i-1} h_{a,b}^{(l)} \hat{y}_c^{(i-l)} \tag{7}$$

$$\hat{y}_c^{(i)} = \bar{y}_c^{(i)} \quad (i \neq 1) \tag{8}$$

$$\hat{y}_c^{(i)} = \bar{y}_c^{(i)} + C \quad (i = 1) \tag{9}$$

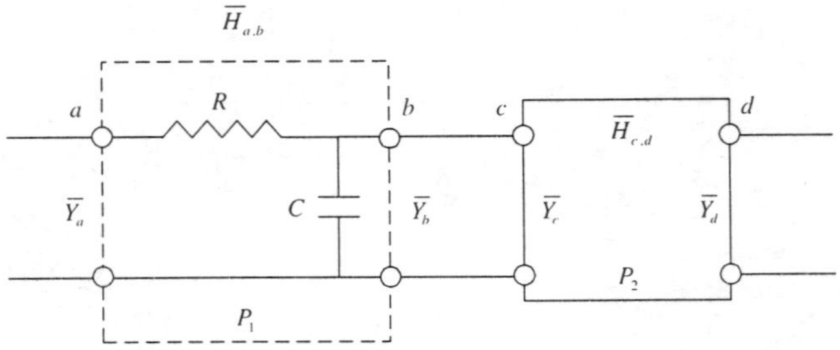

Figure 4: RC tree demonstrating the equivalence of our formulas with Rubinstein et al. [5] & O'Brien and Savarino [3]. P_1 is a single RC segment, while P_2 is an arbitrary RC tree.

- **Equivalence with Rubinstein et al. [5]:** (6) relates $h_{a,b}^{(i)}$, the ith ($i \geq 1$) moment of $H_{a,b}$, to $y_a^{(i)}$, the ith ($i \geq 1$) moment of Y_a. When $i = 1$, (6) becomes

$$h_{a,b}^{(1)} = -Ry_a^{(1)} \quad (10)$$

Since $y_a^{(1)}$ is the rooted capacitance at node a (see Figure 4), the absolute value of (10) is the Elmore delay for the segment from a to b.

To compute the Elmore delay of a path consisting of more than one segment, Theorem 5.2 below must be used in conjunction with (10). But first, we state our formulas' equivalence with O'Brien and Savarino [3].

- **Equivalence with O'Brien and Savarino [3]:** (6) and (7) form a recursive relation that can compute the input admittance moment of an arbitrary order, while [3] gives the formulas for only the first three input admittance moments. $\hat{y}_c^{(i)}$ is defined to simplify notation. Due to space limitations, we do not list the case-by-case equivalence with [3]. For details, please refer to [2].

Theorem 5.2 below is used with (6) to compute the Elmore delay of a multiple-segment path:

Theorem 5.2 *Suppose we have, for the RC tree in Figure 4:*

$$H_{a,b} = 1 + h_{a,b}^{(1)}s + h_{a,b}^{(2)}s^2 + \cdots = \sum_{l=0}^{\infty} h_{a,b}^{(l)}s^l$$

$$\bar{H}_{c,d} = 1 + \bar{h}_{c,d}^{(1)}s + \bar{h}_{c,d}^{(2)}s^2 + \cdots = \sum_{l=0}^{\infty} \bar{h}_{c,d}^{(l)}s^l$$

$$H_{a,d} = 1 + h_{a,d}^{(1)}s + h_{a,d}^{(2)}s^2 + \cdots = \sum_{l=0}^{\infty} h_{a,d}^{(l)}s^l$$

where $H_{a,b}$, $\bar{H}_{c,d}$, and $H_{a,d}$ are the transfer functions from a to b (with P_2 loaded on P_1), from c to d, and from a to d, respectively; $h_{a,b}^{(l)}$, $\bar{h}_{c,d}^{(l)}$, and $h_{a,d}^{(l)}$ are the lth moments of $H_{a,b}$, $\bar{H}_{c,d}$, and $H_{a,d}$, respectively.

Then we have ($l \geq 1$)

$$h_{a,d}^{(l)} = \sum_{m=0}^{l} h_{a,b}^{(m)}\bar{h}_{c,d}^{(l-m)} \quad (11)$$

When $l = 1$ in (11), we have

$$h_{a,d}^{(1)} = h_{a,b}^{(1)} + \bar{h}_{c,d}^{(1)} \quad (12)$$

with which the Elmore delay from a to d, $h_{a,d}^{(1)}$, can be calculated in terms of $h_{a,b}^{(1)}$, the Elmore delay from a to b, and $\bar{h}_{c,d}^{(1)}$, the Elmore delay from c to d.

Please refer to [2] for the proof of Theorem 5.2.

6 Conclusions

In this paper, we have presented a method to hierarchically compute moments for 2-port circuits. In addition, we demonstrated that our formulas form an elegant framework integrating results from two classical papers: Rubinstein et al. & O'Brien and Savarino.

References

[1] A. B. Kahng and S. Muddu, "Efficient Analyses and Models of VLSI and MCM Interconnects," *UCLA CS Dept. TR-960016.*

[2] F. J. Liu and C. K. Cheng, "Extending Moment Computation to 2-Port Circuit Representations," *Technical Report*, Department of Computer Science and Engineering, University of California, San Diego, March 1998. http://www.cs.ucsd.edu/users/fjliu/tr.ps.

[3] P. R. O'Brien and T. L. Savarino, "Modeling the driving-point characteristic of resistive interconnect for accurate delay estimation," *Proc. IEEE Intl. Conf. on Computer-Aided Design*, Nov. 1989, pp 512-515.

[4] L. T. Pillage and R. A. Rohrer, "Asymptotic Waveform Evaluation for Timing Analysis," *IEEE Trans. on CAD*, April 1990, vol. 9, pp 352-66.

[5] J. Rubinstein, P. Penfield Jr., and M.A. Horowitz, "Signal delay in RC tree networks," *IEEE Trans. on CAD*, July 1983, dol. 2, pp 202-11.

Adjoint Transient Sensitivity Computation in Piecewise Linear Simulation

Tuyen V. Nguyen*, Anirudh Devgan*, and Ognen J. Nastov**
*IBM Austin Research Laboratory, Austin, TX 78758
**Dept. of EE & CS, MIT, Cambridge, MA 02139

Abstract

This paper presents a general method for computing transient sensitivities using the adjoint method in event driven simulation algorithms that employ piecewise linear device models. Sensitivity information provides first order assessment of circuit variability with respect to design variables and parasitics. This information is particularly useful for noise stability analysis, timing rule generation, and circuit optimization. Techniques for incorporating adjoint transient sensitivity into ACES, a general piecewise linear simulator, are presented. Sensitivity computation includes algorithms to handle instantaneous charge redistribution due to the discontinuous conductance models of the piecewise linear elements, and the loss of simulation accuracy due to the non-monotonous responses in autonomous adjoint circuits with non-zero initial conditions. Results demonstrate the efficiency and accuracy of the proposed techniques.

1.Introduction

Sensitivity information is extremely useful in assessing the performance and robustness of VLSI circuits, especially as design trends move continuously towards higher performance at lower costs and faster turn around times. These high performance designs require better exploitation of process technology capabilities and more dependence on transistor level analysis and verification. Information on how the design responds to changes in design variables and parasitics is particularly useful for verification and optimization. Transient sensitivity information provides a first order approximation of the design variability. This information can be utilized for noise analysis, timing rule generation, and traditional use in circuit optimization. Noise margin of the circuit can be computed by knowing the sensitivity of the output voltage with respect to the input voltage. The use of transient sensitivity provides information on the more accurate and useful AC noise margin than the traditional DC noise margin. Transient sensitivity information can also be used to reduce the number of required simulations for the generation of timing rules. These timing rules are then used in a static timing analysis environment. Transient sensitivity has been particularly useful in circuit optimization and tuning. Transient sensitivity provides gradient information which reduces the number of simulations required to perform transistor and interconnect sizing, which has implications for timing, power, and manufacturability improvements. Gradients can also be approximated by finite differences instead of transient sensitivity computation. However, in a direct or finite difference scheme, the gradient computation of a given circuit performance with respect to each circuit parameter requires an extra simulation of the perturbed circuit in addition to the simulation of the nominal circuit. The cost of gradient computation, therefore, will quickly escalate as the number of circuit or design parameters increases. Efficiency of the transient sensitivity computation is critical for its application to noise analysis, timing rule generation and circuit optimization. Furthermore, this sensitivity computation should be provided within an efficient circuit/timing simulation environment that can handle large and complex VLSI circuits. In order to address these needs, this paper proposes a method for computing transient sensitivities in Adaptively Controlled Explicit Simulation (ACES)[1], an efficient event driven simulator that employs piecewise linear (PWL) device models.

In general, there are two well known methods for computing sensitivities: the direct[3] and the adjoint[2] methods. The advantages and disadvantages of these methods for sensitivity computation have been discussed in detail [3][4]. In general, the direct method is advantageous when the sensitivities of a large number of circuit responses with respect to a few circuit parameters are desired. On the other hand, the adjoint method is more advantageous when the sensitivities of a few performance functions with respect to a large number of circuit parameters are required. The latter is usually the case in practice, and the adjoint method is the preferred method. However, there are inherent limitations that make the computation of adjoint transient sensitivities expensive in traditional circuit simulation[3].

In such a simulation environment, the original or nominal circuit is simulated forward in time over the time interval of interest. Then, the adjoint circuit, which is constructed from the original circuit, is simulated backward in time. At each time point of this backward simulation, the transpose of the Jacobian matrix, that specifies the state or the appropriate linearized approximation of the original circuit, is used to analyze the resultant adjoint circuit. The adjoint method, therefore, incurs the cost of either storing the Jacobian at each time point of the forward simulation or reconstructing these Jacobians during the backward simulation. Another problem is that, for good error control, the simulation of the adjoint circuit will require a different time grid as compared with the time grid for the original simulation. This time grid mismatch entails the interpolation of the previously stored Jacobians. This interpolation may impact the accuracy of the adjoint simulation. On the other hand, recomputing the Jacobians during the backward simulation will undermine computational efficiency. These problems are the main reasons why the adjoint method has not been widely used for computing transient sensitivities in circuit simulation even though it is the method of choice for computing DC and frequency sensitivities. It is usually more convenient and possibly more efficient to incorporate the direct method for computing transient sensitivities in traditional or relaxation based circuit simulators[3][5]. As a result, the efficiency of the transient sensitivity computation is compromised in these circuit simulation environments.

The key to resolve the above problems for the adjoint method is to take advantage of the simplified device models and the event

DAC 98, San Francisco, California

driven nature of the more efficient circuit/timing simulators[4][6]. The event driven nature of these simulators allows one to store only a few circuit elements at each event time to reconstruct the Jacobians during the adjoint simulation. Moreover, the adjoint simulation can traverse the same event queue, which comes from the original simulation, except backward in time. The simplified models eliminate the need for interpolating the Jacobians. These advantages hold for ACES. This paper will also address two specific problems in incorporating adjoint transient sensitivity computation into ACES: the discontinuity of the device models and the non-monotonic responses in the adjoint circuit. The piecewise linear models are continuous in the original ACES simulation. However, the device models in the adjoint circuit, as derived from these PWL models are discontinuous. This discontinuity gives rise to the impulse conductance in the derivative space of the adjoint circuit. An algorithm has been developed to handle this type of impulse conductance to ensure the accuracy of the adjoint simulation. The next problem is the non-monotonic responses of the autonomous adjoint circuit with non-zero initial conditions. This problem is similar to the problem of underdamped RLC circuit responses, which can be handled by a special procedure that uses macromodeling of the linear circuits [1]. In this paper, we will develop a more general method that essentially makes the numerical integration in ACES less damping and allows it to capture these non-monotonic responses.

The paper is organized as follows: a brief overview of the adjoint method in ACES is presented in Section 2 together with the description of typical circuit elements in the adjoint circuit. Section 3 discusses the algorithms to handle problems specific to adjoint computation in ACES. Section 4 presents some results to demonstrate the effectiveness of the new algorithms. Section 5 concludes the paper.

2.Overview of Adjoint Sensitivity Computation in ACES

An overview of the procedure for computing transient sensitivities in ACES using the adjoint method is shown in Fig. 1. First, the orig-

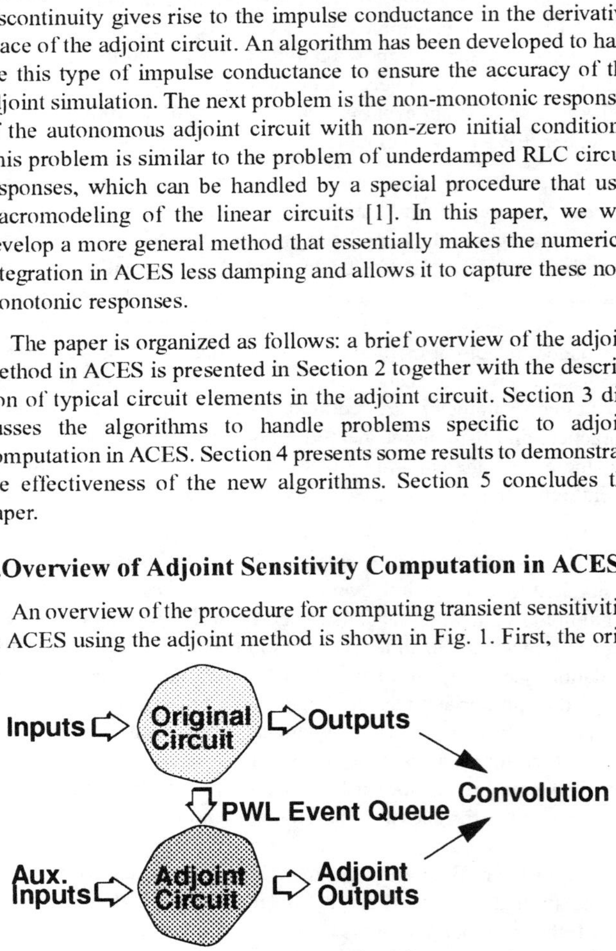

Fig. 1: *An overview of the adjoint method for computing transient sensitivities in ACES*

inal circuit is simulated forward in time. The events due to the PWL elements are stored in an event queue during the original simulation. At each PWL event, the index of the parent partition and all the necessary information about the PWL element causing the event are stored. Then the adjoint circuit is constructed from the original circuit and simulated backward in time. The inputs of the adjoint circuit are derived from the specified circuit performance function. These auxiliary excitations constitute the inputs for the adjoint simulation. The same partitions of the original circuit can be used for the adjoint simulation. During the adjoint simulation, the PWL event queue is traversed backward in time. At each event time, the information about the PWL element and its parent partition are retrieved for the analysis of the adjoint circuit at that particular time point. Finally, the results of the adjoint simulation are convolved with the results from the original simulation to provide the required sensitivities. A detailed example is presented in Section 4 to illustrate the adjoint sensitivity computation in ACES.

In the following subsections, the derived models in the adjoint circuit of some typical circuit elements used in ACES and the selection of the excitations for the adjoint circuit will be described. The following notations will be used throughout the paper. The voltages and currents in the in original circuit are denoted by v and i while the voltages and currents in the adjoint circuit are denoted by $\hat{v}$ and $\hat{i}$. The temporal variable in the original circuit is denoted by t, and in the adjoint circuit by $\tau = t_0 + t_f - t$, where t_0 and t_f are the initial and final times of the simulation.

2.1. Piecewise Linear Adjoint Circuit Elements

In this subsection, we will derive the characteristic equations of the elements in the adjoint circuit corresponding to the PWL devices, which model all the resistive elements in ACES. The adjoint characteristic equations for other circuit elements can be found in [2][4]. In order to simplify the discussion, consider a two terminal device, the PWL characteristic of which is shown in Fig. 2. Formally, the i-v

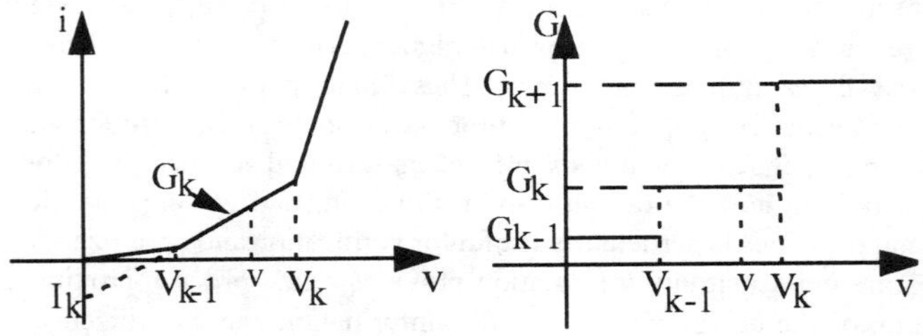

Fig. 2: *The i-v characteristic and the conductance of a two terminal PWL element*

characteristic of a PWL two terminal resistive element can be written as

$$i = \sum_{k=-\infty}^{\infty} (G_k v + I_k)[u(v - V_k) - u(v - V_{k+1})] \quad (1)$$

where V_k and V_{k+1} are the breakpoints in the PWL model, $(G_k v + I_k)$ is the equation representing the kth linear segment of the model, and $u(v)$ is the unit step function. Note that the index is extended to infinity to simplify the notation. Changing the index in Eq.(1) yields

$$\begin{aligned} i &= \sum_{k=-\infty}^{\infty} (G_k v + I_k) u(v - V_k) - \sum_{k=-\infty}^{\infty} (G_{k-1} v + I_{k-1}) u(v - V_k) \\ &= \sum_{k=-\infty}^{\infty} (\Delta G_k v + \Delta I_k) u(v - V_k) \end{aligned} \quad (2)$$

where $\Delta G_k = G_k - G_{k-1}$ and $\Delta I_k = I_k - I_{k-1}$. Hence, the conductance of the PWL model can be written as

$$G = \frac{\partial i}{\partial v} = \sum_{k=-\infty}^{\infty} (\Delta G_k) u(v - V_k) + \sum_{k=-\infty}^{\infty} (\Delta G_k v + \Delta I_k) \delta(v - V_k)$$
$$= \sum_{k=-\infty}^{\infty} (\Delta G_k) u(v - V_k) \tag{3}$$

where $\delta(v)$ is the Dirac delta function and $(\Delta G_k v + \Delta I_k)\delta(v - V_k) = 0$ due to the continuity of the current at the breakpoints of the PWL model.

It can be shown [4] that for a general two terminal device, $i(t) = i(v(t))$, the corresponding element in the adjoint circuit is described by

$$\hat{i}(\tau) = \left.\frac{\partial i}{\partial v}\right|_{\hat{v}(\tau)} \hat{v}(\tau) = G(\hat{v}(\tau))\hat{v}(\tau) \tag{4}$$

In ACES, it is more convenient and efficient to operate in the derivative space [1]. The corresponding element of the two terminal device described by Eq.(4) in the derivative space is given by

$$\frac{d\hat{i}}{d\tau} = G\frac{d\hat{v}}{d\tau} + \frac{\partial G}{\partial \tau}\hat{v} \tag{5}$$

Therefore, in the adjoint circuit, the impulse conductance of a PWL two terminal device in the derivative space can be written as

$$\frac{dG}{d\tau} = \frac{\partial G}{\partial \hat{v}}\frac{d\hat{v}}{d\tau} = \left\{ \sum_{k=-\infty}^{\infty} (\Delta G_k)\delta(\hat{v} - V_k) \right\} \dot{\hat{v}}(\tau)$$
$$= \left\{ \sum_{k=-\infty}^{\infty} (\Delta G_k)\delta[\dot{\hat{v}}(\tau)(\tau - \tau_k)] \right\} \dot{\hat{v}}(\tau) = \sum_{k=-\infty}^{\infty} (\Delta G_k)\delta(\tau - \tau_k) \tag{6}$$

where $\hat{v} - V_k = \dot{\hat{v}}(\tau)(\tau - \tau_k)$ with τ_k being the event time when the breakpoint V_k is reached during the simulation, and $\delta[\dot{\hat{v}}(\tau)(\tau - \tau_k)] = [\delta(\tau - \tau_k)]/\dot{\hat{v}}(\tau)$ by the scaling property of the impulse function. In other words, $(dG)/(d\tau)$ is a train of impulses, which is zero everywhere except at the event times τ_k's corresponding to the element reaching one of its breakpoints. This derivation can be extended in a straightforward manner to multi-terminal PWL elements.

2.2. Adjoint Circuit Excitations

The excitations in the adjoint circuit are derived from a specified circuit performance function. In general, the circuit performance will be a function of circuit variables such as currents and voltages. The performance function can be assumed to be of the form

$$\Phi(\mathbf{v}_I, \mathbf{i}_V) = \int_{t_0}^{t_f} f(\mathbf{v}_I, \mathbf{i}_V)dt \tag{7}$$

where $\mathbf{v}_I$ denotes the vector of all voltage responses and $\mathbf{i}_V$ denotes the vector of all current responses. This formulation is sufficiently general to express all the interesting performance functions in practice. The performance function variation can then be expanded as

$$\delta\Phi = \int_{t_0}^{t_f} \left[\sum_V \frac{\partial f}{\partial v_I}\delta v_I + \sum_I \frac{\partial f}{\partial i_V}\delta i_V \right] dt \tag{8}$$

If we assume that $\delta v_V = 0$ and $\delta i_I = 0$, and the adjoint circuit excitations are chosen such that $\hat{i}_I = \frac{\partial f}{\partial v_I}$ and $\hat{v}_V = -\frac{\partial f}{\partial i_V}$, then the left hand side of (8) is exactly the performance function variation $\delta\Phi$.

For example, suppose that the sensitivity, $\delta v_n(t_k)$, of a given node voltage at a particular time point, $v_n(t_k)$, is desired. First attach a zero valued current source I_n to node n to measure the node voltage in the original circuit, i.e. $v_{In}(t) = v_n(t)$. In other words,

$$\Phi(\mathbf{v}_I, \mathbf{i}_V) = v_{In}(t_k) = \int_{t_0}^{t_f} f(\mathbf{v}_I, \mathbf{i}_V)dt = \int_{t_0}^{t_f} v_{In}(t)\delta(t - t_k)dt$$

Then, $\hat{v}_V = -\frac{\partial f}{\partial i_V} = 0$ and $\hat{i}_{In} = \frac{\partial f}{\partial v_{In}} = \delta(t - t_k)$. For other circuit performance functions such as delay or power, the excitations in the adjoint circuit can be derived in a similar manner.

3.Adjoint Sensitivity Algorithms

This section discusses the additional algorithms that have been developed to address the specific problems of incorporating adjoint sensitivity computation into ACES: charge redistribution due to impulse conductances and non-monotonic responses of the adjoint circuit. A brief overview of the problems is presented before a detailed description in the next two subsections. The problem of impulse conductances arises from the nature of the PWL models used in ACES. As shown in the previous section, the BCR of a multi-terminal resistive element in the adjoint circuit is the transpose of the conductance matrix, $\frac{\partial i}{\partial v}$, of the corresponding element in the original circuit. Since the resistive elements in the original circuit are modeled by PWL i-v characteristics, the corresponding element in the adjoint circuit will be modeled by piecewise constant conductance. One more level of differentiation to operate in the derivative space yield the impulse conductances described by Eq.(6). These impulse conductances will cause instantaneous charge redistribution in the derivative space. An algorithm is presented in subsection 3.1. to compute accurately the charge redistribution.

The problem of non-monotonic responses of autonomous adjoint circuits arises when a voltage sensitivity at a specified time point is desired. As discussed in subsection 2.2., the excitation corresponding to this performance function in the adjoint circuit is an impulse current source. The effect of such an impulse excitation is to deposit non-zero initial conditions on node charges. The adjoint circuit is then autonomous or not driven by excitations. The responses of such an autonomous circuit with non-zero initial conditions are usually non-monotonic, and this type of non-monotonic response is not captured by the current quiescence algorithm in ACES [1]. A modified quiescence algorithm will be proposed in subsection 3.2. to handle this case.

3.1. Instantaneous Charge Redistribution in the Adjoint Circuit

In general, using the charge formulation, the circuit equation of the original circuit can be written as

$$\dot{q}_n(v_n) = i_{pwl}(v_n) + I_s \tag{9}$$

where q_n denotes the vector of node charges, i_{pwl} is the vector of currents due to the PWL elements, and I_s is the vector of current sources. For ACES, it is more convenient to perform the simulation in the derivative space [1]. Differentiating Eq. (9) with respect to time yields

$$\ddot{q}_n(v_n) = G_{pwl}\dot{v}_n(t) + \dot{I}_s \tag{10}$$

where $G_{pwl} = \dfrac{\partial i_{pwl}}{\partial v_{pwl}}$ is the piecewise constant conductances of the PWL elements. It can be shown that the state equation for the adjoint circuit can then be written as

$$\frac{d}{d\tau}\hat{q}(\hat{v}_n) = G^T_{pwl}(\hat{v}_n)\hat{v}_n(\tau) + \hat{I}_s \tag{11}$$

In the derivative space, Eq. (11) becomes

$$\frac{d^2}{d\tau^2}\hat{q}(\hat{v}_n) = G^T_{pwl}(\hat{v}_n)\frac{d\hat{v}_n}{d\tau} + \left(\frac{dG_{pwl}}{d\tau}\right)\hat{v}_n(\tau) + \frac{d\hat{I}_s}{d\tau} \tag{12}$$

where

$$\frac{dG_{pwl}}{d\tau} = \left(\frac{\partial G_{pwl}}{\partial v_{pwl}}\right)\left(\frac{d\hat{v}_{pwl}}{d\tau}\right) \tag{13}$$

represents a train of impulses because G_{pwl} are piecewise constant functions of voltages (and time). In other words, $(dG_{pwl})/(d\tau)$ is a train of impulses, which is zero everywhere except at the event times τ_k's corresponding to the element reaching one of its breakpoints. Therefore, between two successive PWL events τ_{e-1} and τ_e, Eq. (12) simplifies to

$$\frac{d^2}{d\tau^2}\hat{q}(\tau) = G_{pwl}(\tau)\frac{d\hat{v}_n}{d\tau} + \frac{d\hat{I}_s}{d\tau} \qquad \text{for} \quad \tau \in (\tau^+_{e-1}, \tau^-_e) \tag{14}$$

The numerical solution of this equation can be computed using standard ACES algorithm.

Next we need to compute the charge redistribution due to the impulse conductance at each PWL event τ_e. There is a more direct way to derive Eq. (6) for each PWL event, which is also more convenient for implementation. Note that at each event time τ_e corresponding to a PWL element changing its region, the conductance of the element can be written as

$$\begin{aligned} G(\tau) &= G(\tau^-_e) + [G(\tau^+_e) - G(\tau^-_e)]u(\tau - \tau_e) \\ &= G(\tau^-_e) + \Delta G(\tau_e)u(\tau - \tau_e) \qquad \text{for} \quad \tau \in (\tau^-_e, \tau^+_e) \end{aligned} \tag{15}$$

where $\tau^-_e = \lim_{\varepsilon \to 0}(\tau_e - \varepsilon)$ and $\tau^+_e = \lim_{\varepsilon \to 0}(\tau_e + \varepsilon)$ with $\varepsilon > 0$, and $\Delta G(\tau_e) = G(\tau^+_e) - G(\tau^-_e)$. Hence,

$$\frac{\partial G}{\partial \tau} = \Delta G(\tau_e)\delta(\tau - \tau_e) \qquad \text{for} \quad \tau \in (\tau^-_e, \tau^+_e) \tag{16}$$

Substituting Eq.(16) into Eq.(12) and integrating the resulting equation from τ^-_e to τ^+_e yields the following equation for updating $\dot{q}$ at each PWL event,

$$\dot{q}_n(\tau^+_e) = \dot{q}_n(\tau^-_e) + \Delta G(\tau_e)\hat{v}_n(\tau_e) \tag{17}$$

3.2. Non-monotonic Responses in the Autonomous Adjoint Circuit

The problem of non-monotonic responses in the adjoint circuit will be illustrated in this subsection by considering a simple 2RC example as shown in Fig. 3. The associated adjoint circuit is constructed to compute the sensitivity of v_2 at a given time t_f. The

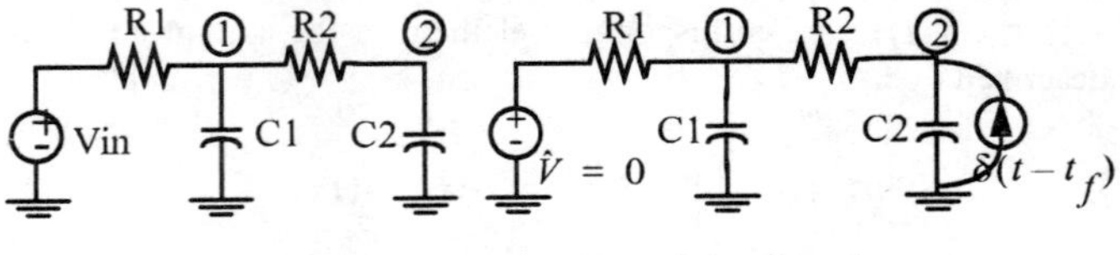

(a) Original Circuit (b) Adjoint Circuit

Fig. 3: *A simple 2RC example: (a) the original circuit and (b) the adjoint circuit to compute the sensitivity of the voltage at node 2.*

effect of the impulse source is to deposit a charge of 1 on C2. In other words, the above circuit is equivalent to the original 2RC circuit without any excitations, but with non-zero initial conditions, $v_1 = 0$ and $v_2 = 1$. The sensitivity with respect to various circuit parameters can then be computed by convolving appropriate simulation results of the original circuit shown in Fig. 3(a) with simulation results of the adjoint circuit shown in Fig. 3(b). Fig. 4 shows the cur-

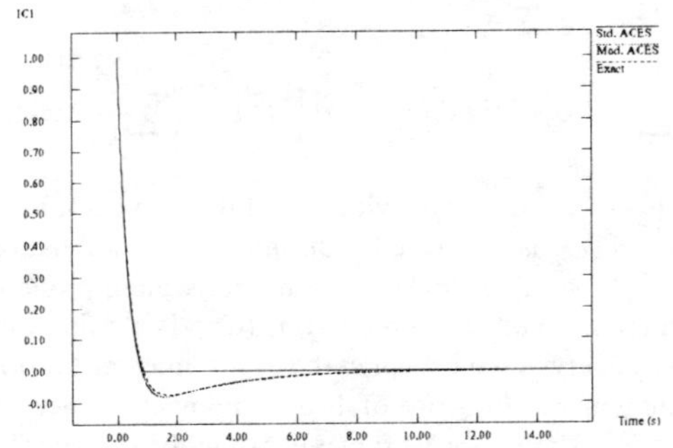

(a) Current waveforms (IC1)

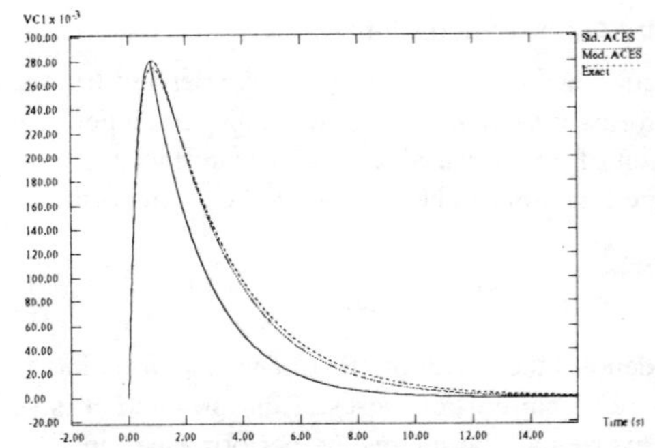

(b) Voltage Waveforms (VC1)

Fig. 4: *The current and voltage waveforms at node 1 obtained by (a) Exact solution, (b) ACES with standard quiescence algorithm, and (c) ACES with modified quiescence algorithm* rent and voltage waveforms at node 1 of the adjoint circuit obtained by exact solution, ACES with standard quiescence algorithm, and ACES with modified quiescence algorithm. The ACES simulations were performed with an error tolerance of 0.001. As shown in Fig. 4, both the current and voltage at node 1 approach steady state non-monotonically. In the standard ACES algorithm, when the current IC1 becomes zero at $t = 0.8s$, the (capacitance) state V1 is put into quiescence. This corresponds to the peak of the voltage waveform when the voltage slope becomes zero. As a result, the standard ACES algorithm missed the negative currents for $t > 0.8s$ resulting in a corresponding loss of accuracy for the voltage waveform.

This problem illustrates the overdamping nature of the standard quiescence algorithm in ACES. One way to resolve this problem is to make the quiescence algorithm less damping. Note that the goal of the quiescence algorithm in ACES is to ensure the stability of the underlying explicit simulation algorithm by not allowing the state variables to cross their steady state values. However, one needs to distinguish between stable and unstable equilibrium (or steady) state. The peak of the voltage waveform in Fig. 4(b) represents an unstable equilibrium point, where a small perturbation in either direction will move it out of equilibrium. On the other hand, the "true" steady state at $t \to \infty$ is a stable equilibrium. Hence, the solution to the problem of unstable equilibrium is not to put a given state into quiescence when it first reaches quiescence (or equilibrium point). We simply mark the state to be in potential quiescence (PQS). Note we do not ignore the computation of the quiescence (QS) time step. It is still used as a time step limiting mechanism to ensure that all quiescence (or equilibrium) points are captured. However, the state is not put into QS the first time it reaches an equilibrium point. Recall that in ACES once a state is in quiescence it is in quiescence at all successive time points until an input source changes its PWL segment or a PWL element changes its PWL region. In the modified quiescence algorithm, a state will stay in PQS unless the circuit has an event due an input source, a PWL element, or a local truncation error (LTE) step. These events indicate that the equilibrium point for the given state may not be stable. For input source or PWL events, all states should be kicked out of PQS or QS. For LTE events, an additional check is performed to make sure that the time derivative (slope) of a given PQS state has changed sign before it is kicked out of PQS. Otherwise, the state will be kept in PQS. Note that a state is still in PQS if other states enters PQS or QS. These conditions indicate that the circuit is approaching steady state; hence, the given state should be kept in PQS. The state will enter QS when it has a second QS time step. The modified quiescence algorithm is summarized below

```
if (event due to the QS time step of a state S) {
    if (state S is non-quiescent)
        mark S in PQS
    else /* state S already in PQS */
        mark S in QS
}
else if (event due to an LTE time step)
    mark PQS states non-quiescent if slopes changed signs
else /* event due to an input source or a PWL element */
    mark all PQS and QS states non-quiescent
```

The results in Fig. 4 shows that the modified quiescence algorithm indeed captures all the equilibrium points correctly and helps ensure the accuracy of the simulation.

4. Results

The first example is a simple CMOS circuit shown in Fig. 5. This

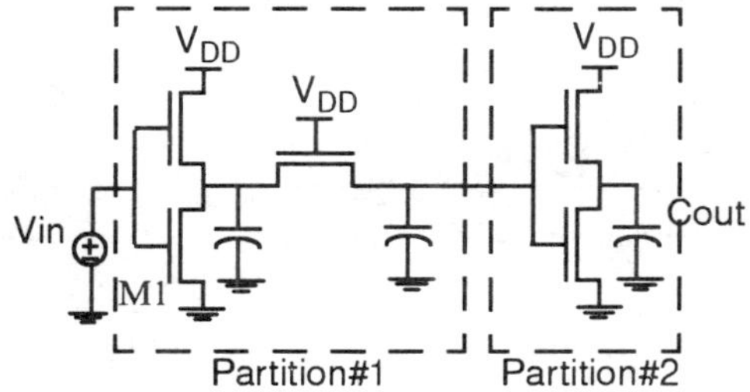

Fig. 5: *A simple two inverter circuit to demonstrate adjoint sensitivity computation with circuit partitioning.*

example is to demonstrate adjoint sensitivity computation with circuit partitioning. In the original circuit, the first inverter and the pass transistor is grouped into one partition, and the second inverter is grouped into a second partition. Suppose that the sensitivities of the output node with respect to various circuit parameters are desired. Then the information flow in the adjoint circuit is from the output to the input in opposition to the information flow from the input to the output for the original circuit. Since the controlling flow of the equivalent models of the MOSFETs in the adjoint circuit is also reversed, it can be argued that the adjoint circuit should be repartitioned to group the second inverter and the pass transistor into one partition and the first inverter into another partition. However, it is advantageous to retain the same partitioning to avoid the cost of partitioning and setting up the partitions. Moreover, the MOSFET model in the adjoint circuit does not have the nice property of the gate current being always zero [4]. Hence, there is no DC decoupling in the adjoint circuit as compared to the original circuit. Therefore, there are no clear advantages in repartitioning the circuit. In addition, the partitioning should affect only the efficiency and not the accuracy of the simulation. The sensitivity waveforms of the output voltage and the current through Vdd with respect to the length of the NMOS transistor (M1) of the first inverter as obtained by direct (finite) difference and ACES adjoint computation are shown in Fig. 6. Note that the adjoint waveforms are obtained by multiple adjoint simulations with the voltage or current at a set of timepoints as the performance function. These simulations are performed with an error tolerance of 0.01.

A set of realistic test circuits are then used to assess the performance of the adjoint analysis in ACES. One adjoint analysis was performed for each circuit. Table 1 summarizes the run times of the ACES simulation and the associated adjoint simulation.

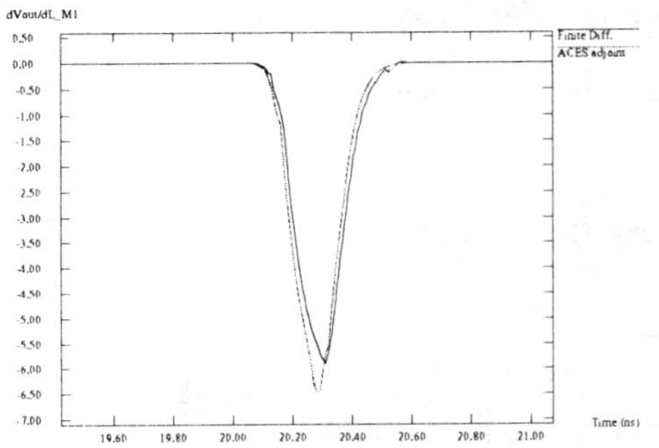

(a) Voltage sensitivity waveforms

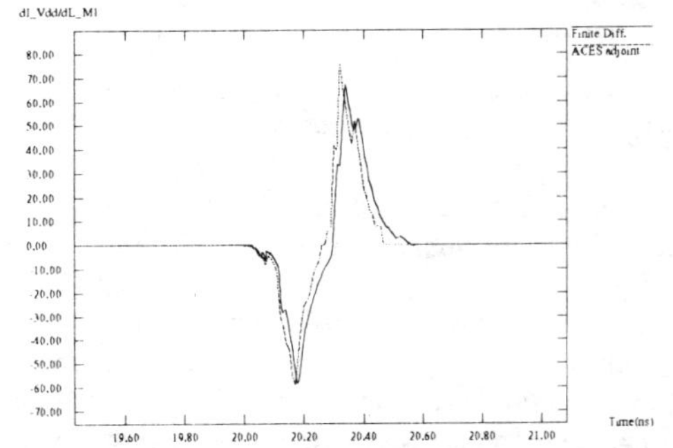

(b) Current sensitivity waveforms

Fig. 6: *The sensitivity waveforms of the output voltage and the current through Vdd with respect to the length of the NMOS transistor M1 of the circuit shown in Fig. 5.*

Table 1: Simulation run times for a number of test circuits

Circuit	No. of MOSFETs	Original simulation	Adjoint simulation
2bit error control circuit	104	3.81s	0.34s
Counter circuit	220	20.5s	2.2s
4bit ALU	282	6.64s	0.58s

Note that only the event processing of the original ACES simulation and the adjoint simulation are shown. The costs of setting up the circuit and computing the convolutions are not included. The cost of the convolution varies depends upon the number of timepoints in the waveforms. On the average, the cost of the adjoint simulation is about 10% of the cost of the original simulation with 0.01-0.02% overhead for each convolution. Note that these results are obtained with an error tolerance of 0.01 on an IBM RS600 390 workstation with 256M of RAM. A set of typical sensitivity waveforms obtained by finite difference and ACES adjoint computation is shown in Fig. 7.

5.Conclusions

A general method for computing adjoint transient sensitivities in event driven simulation algorithms employing piecewise linear models such as ACES has been presented. Accuracy and efficiency of the

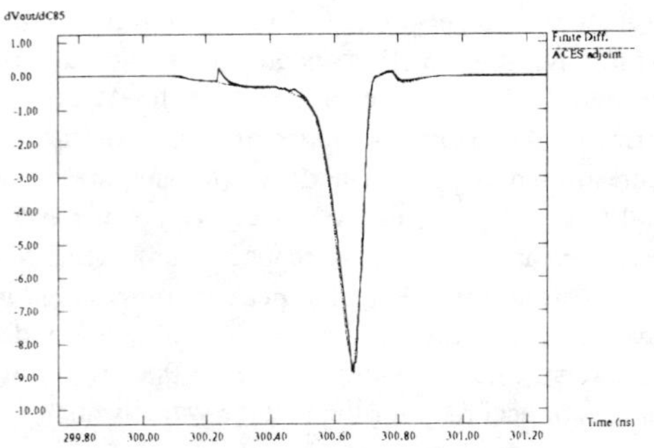

Fig. 7: *A set of sensitivity waveforms as obtained by finite (direct) difference and ACES adjoint computation.*

algorithms has been demonstrated on various examples. The cost of the adjoint simulation is only about 10% of the original nominal simulation. The sensitivities of both voltage and current responses with respect to circuit parameters can be computed. Sensitivities of general circuit performances which are functions of these circuit responses can be obtained by applying the chain rule. Two additional algorithms have been developed to address the charge redistribution from the impulse conductances due to the discontinuity in the PWL device models, and the non-monotonic responses of the autonomous adjoint circuit.

Acknowledgments

The authors would like to thank Dr. David LaPotin for his support during the course of this work, Dr. David Winston for his help with the interface code, and Dr. Ali Sadigh for his helpful discussion with the quiescence algorithm.

References

[1] A. Devgan and R. A. Rohrer, "Adaptively Controlled Explicit Simulation," *IEEE Trans. Computer-Aided Design*, Val. 13, pp.746-762, June 1994.

[2] S. W. Director and R. A. Rohrer, "The Generalized Adjoint Network and Network Sensitivities," *IEEE Tran. Circuit Theory*, vol.16, pp. 318-323, August 1969.

[3] D. A. Hocevar, P. Yang, T. N. Trick, and B. D. Epler, "Transient Sensitivity Computation for MOSFET Circuits," *IEEE Trans. Computer-Aided Design*, vol. 4, pp. 609-620, Oct. 1985.

[4] P. Feldmann, T. V. Nguyen, S. W. Director, and R. A. Rohrer, "Sensitivity Computation in Piecewise Approximate Circuit Simulation," *IEEE Trans. Computer-Aided Design*, vol. 10, pp. 171-183, Feb. 1991.

[5] C-J Chen and W-S Feng, "Relaxation-Based Transient Sensitivity Computation for MOSFET Circuits," *IEEE Trans. Computer-Aided Design*, vol. 14, pp. 173-185, Feb. 1995.

[6] T. V. Nguyen, P. Feldmann, S. W. Director, and R. A. Rohrer, "SPECS Simulation Validation with Efficient Transient Sensitivity Computation," *Proc. ICCAD*, 1989, pp. 252-255.

Design Methodology of Ultra Low-power MPEG4 Codec Core Exploiting Voltage Scaling Techniques

Kimiyoshi Usami, Mutsunori Igarashi, Takashi Ishikawa, Masahiro Kanazawa, Masafumi Takahashi, Mototsugu Hamada, Hideho Arakida, Toshihiro Terazawa and Tadahiro Kuroda

Toshiba Corporation
580-1, Horikawa-cho, Saiwai-ku, Kawasaki 210, JAPAN
Phone: +81-44-548-2346

kimiyoshi.usami@toshiba.co.jp

1. ABSTRACT

This paper describes a fully automated low-power design methodology in which three different voltage-scaling techniques are combined together. Supply voltage is scaled globally, selectively, and adaptively while keeping the performance. This methodology enabled us to design an MPEG4 codec core with 58% less power than the original in three week turn-around-time.

1.1 Keywords

Low power, voltage scaling, design automation, synthesis, placement, flip-flops, level converters, MPEG4, codec

2. INTRODUCTION

Mobile computing devices have been enjoying considerable success in the consumer electronics market. Market requirements are pushing the performance of those devices even higher, and the battery life even longer. In the near future, a wireless multi-media terminal on which we can enjoy moving pictures and stereo sound will emerge. MPEG4 is a standard on video and audio coding/decoding for mobile computing environment. Clearly, major requirements for MPEG4 codec LSI's are low power for longer battery life and high performance.

In CMOS LSI's, scaling supply voltage V_{DD} is very effective because power is reduced quadratically with V_{DD}. However, scaling the voltage poses an problem of degrading the performance. The following techniques have been reported which reduce power by voltage scaling without degrading the entire performance: The first is a technique in which supply voltage is globally scaled with scaling threshold voltage Vth, such as VTCMOS [2]. The second is a Dual-V_{DD} approach in which the reduced V_{DD} is selectively applied to non-critical paths [5][6]. The third is a variable supply voltage (VS) scheme in which V_{DD} is controlled adaptively using an on-chip DC-DC converter [3]. However, there have been few papers which present a methodology of making these techniques work together and show the level of power reduction we could reach.

In this paper, we propose a novel design methodology in which Dual-V_{DD}, VTCMOS and VS techniques are incorporated into a fully automated design flow. We discuss not only the total effectiveness in power reduction but also the contribution of each technique through an actual design of an MPEG4 codec core. Section 3 presents a background of the three techniques. Section 4 describes a design methodology we propose. Section 5 summarizes the feature of an MPEG4 codec core. Section 6 presents results from applying our design methodology to the core.

3. BACKGROUND

In this section, we present features of Dual-V_{DD}, VTCMOS and VS techniques in detail.

3.1 Dual-V_{DD} Approach

In the Dual-V_{DD} approach, a couple of different supply voltages are used. The reduced voltage (VDDL) is applied to the circuit on non-critical paths, while the original voltage (VDDH) is applied to the circuit on critical paths. So far, a Clustered Voltage Scaling (CVS) technique [5] and an Extended Clustered Voltage Scaling (ECVS) technique [6] have been reported as Dual-V_{DD} approaches. These techniques do not change the critical path delay, resulting in keeping the entire circuit performance. The Dual-V_{DD}

circuit structure resulting from this technique is shown in Fig.1.

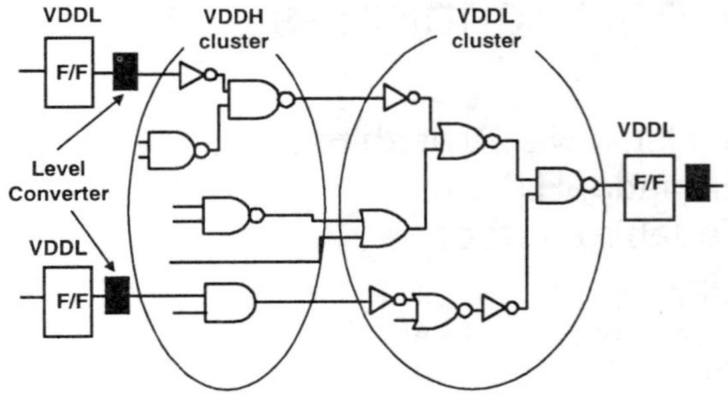

Figure 1 Dual-V_{DD} Structure

Level converters are inserted at the interface between the VDDL circuit and the VDDH circuit. The CVS technique can be further improved by optimizing the insertion points of level converters, leading to an ECVS technique. The ECVS technique allows us to increase the share of VDDL circuits, resulting in saving more power. An algorithm to quickly synthesize the clustered structure with dual supply voltages is presented in [6]. Layout architecture in the Dual-V_{DD} approach is depicted in Fig.2. VDDH cells and VDDL cells are placed in different rows in this architecture [6] .

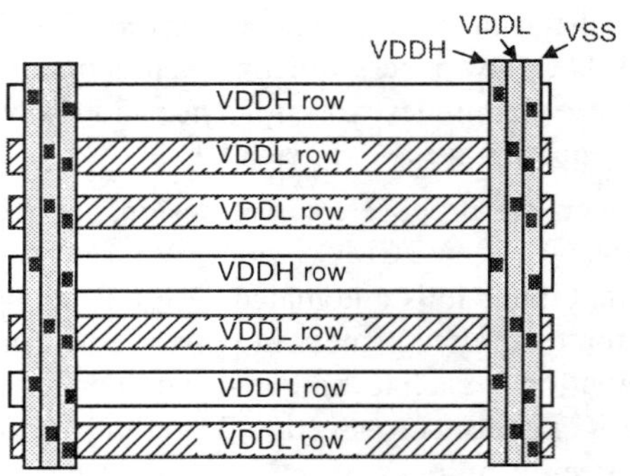

Figure 2 Row-by-Row Dual-V_{DD} Layout Architecture

The Dual-V_{DD} approach reduces power by utilizing excessive slack in a circuit. In actual designs, critical paths are only small portions of the circuit and excessive slack remains in the rest of the circuit. Designers have to struggle with critical paths so they meet the timing constraints. This work usually takes up their design time, resulting in excessive slack left untouched even if it remains. In the Dual-V_{DD} approach such excessive slack is automatically detected and utilized for power reduction.

3.2 VTCMOS

The VTCMOS technique [2] dynamically controls effective threshold voltage by applying substrate bias to MOS transistors. In the active mode, transistors are made to operate at low V_{DD} with low Vth. Meanwhile, in the standby mode, the effective threshold voltage is made to be larger by applying substrate bias to block the leakage current. Transistor performance in the active mode is kept as that in the conventional design by utilizing low V_{DD} and low Vth.

3.3 VS Scheme

In the VS scheme, the minimum supply voltage at which the speed requirements are just satisfied is generated on a chip. Major components in this scheme are an on-chip DC-DC converter and a replica circuit of the critical path [3]. Delay of the replica circuit is dynamically monitored and fed back to the DC-DC converter. The DC-DC converter is adaptively controlled such that it generate the supply voltage at which the delay of the replica circuit matches the required clock period.

Power savings in the VS scheme come from the following: An LSI chip is usually designed with a margin taking into account the voltage fluctuation of the external supply. In other words, a chip is designed so that it can operate even when the supply voltage goes down from V_{DD} to V_{DD} -V'. From the viewpoint of power consumption, however, the power is wasted when the chip operates at higher voltage than V_{DD} -V'. In the VS scheme, the chip is made to operate at the minimum supply voltage satisfying the speed requirements, resulting in saving power.

3.4 Potential Advantages by Combining Three Techniques

All of the three techniques described above reduce power by scaling the supply voltage. However, the targets they aim at are different. The VTCMOS and VS techniques aim at scaling supply voltage of critical paths, while they do not care excessive slack remaining in non-critical paths. On the other hand, the Dual-V_{DD} approach targets at reducing the voltage of non-critical paths, while it does not touch the critical paths. From these observations, it is anticipated that by combining the three techniques, power is reduced both at the critical paths and at the non-critical paths.

The combination of the three techniques has another big advantage. An advantage of implementing a DC-DC converter on a chip can be shared with the three techniques. In VTCMOS and the VS scheme, the minimum supply voltage satisfying the speed requirements is needed. In the Dual-V_{DD} approach, the reduced supply voltage VDDL is required. Those supply voltages are generated with DC-DC converters on a chip. This fact is a big advantage for the system design because a single external voltage source is just enough for the chip.

4. DESIGN METHODOLOGY

We present a new design flow in which Dual-V_{DD}, VTCMOS and VS techniques are combined together. Potential abilities of the three techniques are fully drawn out in the flow. We also present approaches for shortening turn-around-time (TAT) and those for reducing area overhead in the flow. Those approaches include novel circuit technology and CAD algorithm.

4.1 Design Flow

Figure 3 shows the design flow of our voltage scaling approaches. To avoid increasing design complexity for generating Dual-V_{DD} netlist, we separate the design flow into two major steps: One is an ordinary single-voltage design step and the other is a Dual-V_{DD} design step. At the first step, we perform logic synthesis using RT level description and timing constraints assuming a single supply voltage. Generated mapped netlist is sent to the second step. At the second step, we perform Dual-V_{DD} structure synthesis and layout using the same timing constraints as that of the first step.

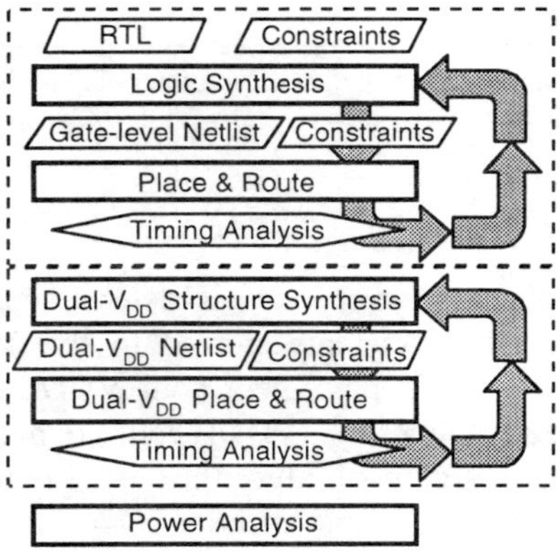

Figure 3 Design Flow

The VTCMOS technique allows the logic circuit in the core to operate at reduced supply voltage with the same performance as the original voltage. In our application of this paper, we determined the supply voltage of the circuit in the core as 2.5V while the original voltage as 3.3V. This is because the combination of 0.2V threshold voltage and the 2.5V supply voltage provides the same performance as that of the 0.55V threshold voltage and 3.3V supply voltage.

In the Dual-V_{DD} structure synthesis, we chose VDDL as 1.75V at which the power gets the minimum when VDDH is 2.5V. Layout results from the Dual-V_{DD} place and route and the layout data of the control circuits for the VS scheme are merged into the final layout data.

4.2 Approaches for Shortening TAT

In logic synthesis to generate single-voltage mapped netlist, we used a cell library which includes a flip-flop with level conversion function. We present why this flip-flop contributes to shortening the design TAT. Detailed features of the flip-flop will be described in the next section. Register elements in RT level description are mapped into those flip-flops even in single-voltage netlist in logic synthesis. The use of flip-flops with level conversion function is very helpful for minimizing the change between the netlist with single voltage and that with Dual-V_{DD}. This is because there is no need to insert level converters at the outputs of those flip-flops, resulting in minimizing the change in gate count and path delay. This allows us to utilize the placement result of the single V_{DD} layout as initial floorplanning information at Dual-V_{DD} layout, leading to shortening TAT.

4.3 Approaches for Reducing Area Overhead

Area penalty in the Dual-V_{DD} approach is reported as 15% [6], while penalties in VTCMOS and a VS scheme are reported as less than one percent [2][3]. This motivated us to study and develop new techniques for reducing area overhead in the Dual-V_{DD} approach. Major components of the area overhead are as follows [1]: The first is an increase in cell area resulting from inserted level converters. The second is an overhead caused by layout constraints such that an N-well of VDDH cells should be isolated from that of VDDL cells. The third is an overhead resulting from additional power lines for VDDL. It is also reported that the first two components are dominant: the first component causes area penalty by 10% while the second does by 5%. We present an approach for reducing the area penalty of inserted level converters by using a novel flip-flop circuit. For reducing the overhead caused by layout constraints, we propose a new cell-placement algorithm.

4.3.1 Development of Flip-Flop with Level Converter

The CVS and ECVS structure described in Section 3 minimizes the number of level converters. The number is at most the number of flip-flops because a level converter is basically inserted at the output of a flip-flop. However, in the case of a circuit with a lot of flip-flops, inserted level converters cause a significant area penalty. We propose an approach which reduces the penalty by using a flip-flop circuit with a level-conversion function (FFLC). The circuit is depicted at the right-most column in Table 1. A master latch of the FFLC is the same as a conventional flip-flop, while a slave latch has a level-conversion function. The master latch operates at VDDL, so that it can receive voltage swing of VDDL at the data-input D and the clock-input CLK. The flip-flop circuit works as follows: When CLK is low, data is loaded into the master latch. When CLK goes high, the data in the master latch is transmitted to the slave latch through transmission gates. Level conversion is performed in the slave latch and the data is output to Q. Since the transmission gates are cut off between both latches when CLK is low, the gate of an NMOS transistor connected to the ground in the slave latch becomes floating. To avoid this problem, NMOS transmission gates controlled by CKB are provided. These transmission gates turn on when CLK is low, so that data is statically held in the slave latch.

We compared area penalties among three cases as shown in Table 1. A conventional flip-flop is chosen as the base of comparison. As shown in the second column, area increases by 64% in the case of inserting a level-converter cell at the output of a flip-flop. Meanwhile, area penalty of the FFLC is only 21%. Therefore, by using the FFLC we can reduce the area overhead in the CVS or ECVS structure.

	Conventional flip-flop	Insertion of level-converter cell at output of flip-flop	Flip-flop with level-conversion function (FFLC)
Circuit structure	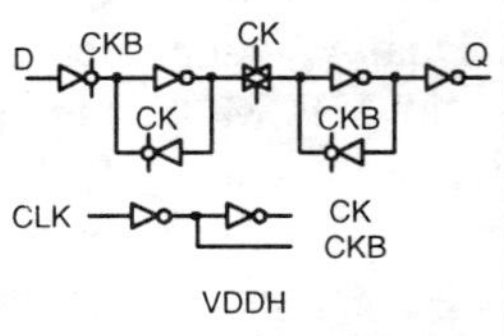	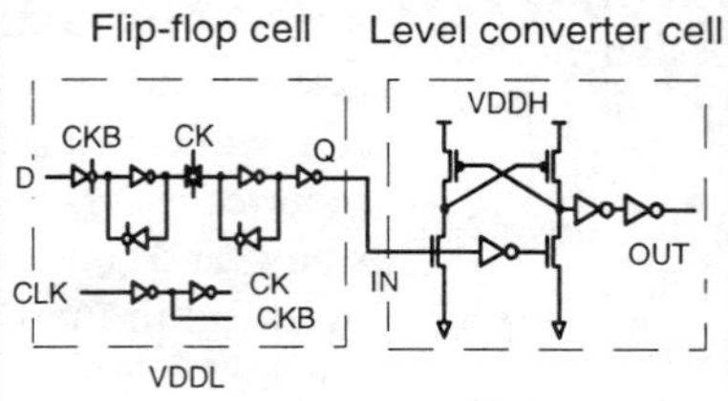	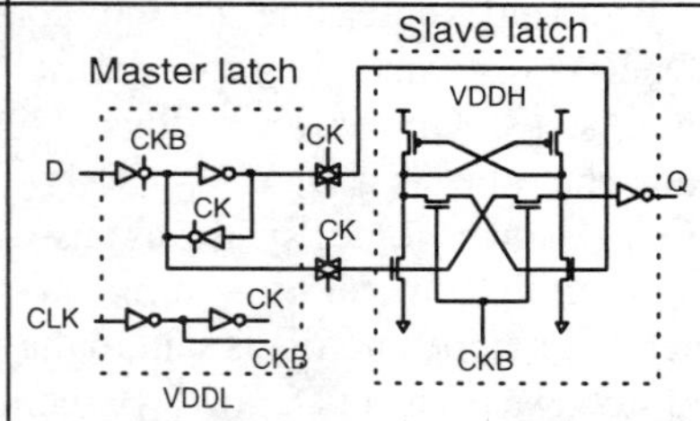
Area *	1	1.64	1.21
Power *	1	0.74	0.63
CLK-to-Q delay *	0.24 ns	0.59 ns **	0.42 ns

* Results on area and power are normalized by those on the conventional flip-flop. Power and delay were simulated at VDDH=2.5V, VDDL=1.75V, |Vth|=0.2V.
** CLK-to-OUT delay

Table 1 Comparisons on flip-flops

The FFLC has significant advantages also in power and performance. Data shown in Table 1 were obtained using a SPICE simulation at VDDH of 2.5V, VDDL of 1.75V, and threshold voltage of 0.2V. It is found that the FFLC reduces power to 63% compared to the conventional flip-flop. This results from the fact that the master latch in the FFLC operates at VDDL while the entire circuit of the conventional flip-flop operates at VDDH. CLK-to-Q delay of the FFLC is larger only by 0.18ns than that of the conventional flip-flop. Compared to the case of inserting a level-converter at the output of a flip-flop, the FFLC is superior also in power and performance. Thus, we have chosen to use the FFLC in the design.

4.3.2 Improvement in Cell Placement Algorithm

In the Dual-V_{DD} layout, we place cells considering timing constraints and routability as we do in an ordinary single-voltage design. In addition, the placement constraint is added to each cell corresponding to its supply voltage. In the layout architecture shown in Fig.2, voltage assignment to rows significantly affects the layout quality.

In this paper, we propose an improved voltage assignment technique. Basic framework is identical to that in [1]. First, cells are placed without considering the positional constraints on the supply voltage. VDDH cells and VDDL cells are mixed in any rows. Average sum of cell area within a row, A_{AVE} is computed at this stage. Next, the sum of the area of VDDL cells within the top row is computed. In [1], the authors determine that the top row should be a VDDL row if the sum reaches A_{AVE}. If not, they move to the second row and accumulate the area of VDDL cells. They keep scanning rows and accumulating the area of VDDL cells until the accumulated result reaches A_{AVE}. If it reaches A_{AVE} at the m-th row, they determine that the m-th row should be a VDDL row and the rows up to (m-1)th be VDDH ones. The same procedure is repeated again from the (m+1)th row. Thus the voltage of every row is determined. After this procedure, VDDL cells are moved to the VDDL rows because in the initial placement both VDDL and VDDH cells are mixed in any rows. VDDH cells are moved as well. This voltage assignment algorithm is simple and fast. However, area penalty was an issue. Our analysis showed the moving distances of the cells were fairly large. This indicates the final positions of the cells deviated from the optimum, resulting in area increase.

In an algorithm we propose, we assign the voltage to rows such that the moving distances of cells be minimized. The procedure of scanning rows and accumulating the area of VDDL cells is the same as that described above. The difference is that we compute the moving distances of the cells when the accumulated result of the area reaches A_{AVE} at the m-th row. In other words, we compute the sum of moving distances of cells assuming the i-th row is VDDL while changing i from 1 to m. Then we compare those sums and choose the voltage assignment with the minimum sum of moving distances.

5. OUTLINE OF MPEG4 CODEC CORE

We summarize features of an MPEG4 codec core to which we applied the design methodology described in Section 4. This MPEG4 video codec is the first design [4] which not only conforms to H.263 but implements essential functions in the MPEG4 Verification Model Version 7. The core consists of nine function blocks: a 16-bit RISC processor, DCT and IDCT blocks, a couple of motion estimation blocks, a motion compensation block, variable length encoding and decoding blocks, and a DMA controller. Most of them have local memories inside. Operation frequency is 30MHz. The core is fabricated in 0.3μm CMOS double-well triple-metal technology. Internal supply voltages of 2.5V and 1.75V are generated in a couple of on-chip DC-DC converters from an external 3.3V power supply.

6. RESULTS

We present results from applying our methodology to the core. Effectiveness is discussed from the viewpoint of power reduction, area overhead and design TAT.

6.1 Power Reduction by Dual-V_{DD} and VTCMOS Techniques

Power reductions in three designs are compared: 1) a conventional design, 2) a design to which only VTCMOS is applied, and 3) a design to which VTCMOS and Dual-V_{DD} techniques are both applied. Power dissipation was analyzed using a PowerMill simulation with real test vectors of the MPEG4 codec. Results are shown in Fig.4. VTCMOS using V_{DD}=2.5V and Vth=0.2V reduced power by 42% compared to the conventional design using V_{DD}=3.3V and Vth=0.55V. As shown in the figure, power is reduced in all of the components in the core such as logic gates, flip-flops, the clock tree and memories. This is because the supply voltage has been reduced from 3.3V to 2.5V while scaling Vth at every transistor in the core.

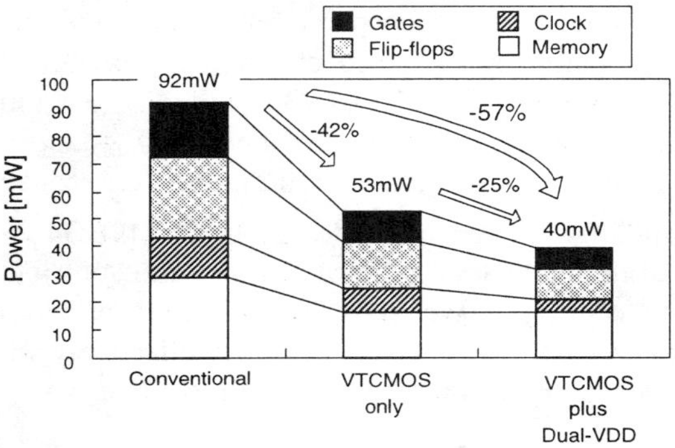

Figure 4 Power reduction resulting from VTCMOS and Dual-V_{DD} approach

Applying the Dual-V_{DD} approach in addition to VTCMOS reduces power further by 25%. The Dual-V_{DD} approach reduced power at logic gates by 30%, at flip-flops by 37%, and at the clock tree by 49%, while did not reduce power at memories. This is because we were obliged to use ASIC memory macros and those operating at VDDL were not supported. Therefore, we applied the Dual-V_{DD} approach to the portions except memories. At the portions to which we applied the Dual-V_{DD} approach, power was reduced by 35% even after applying VTCMOS.

Thus the power in the conventional design reduced by 57% while keeping the spec frequency 30MHz by combining both techniques together.

6.2 Effectiveness of Dual-V_{DD} Approach

We discuss effectiveness shown by the Dual-V_{DD} approach after applying VTCMOS in more detail. First we present power reduction at logic gates. Gates in non-critical paths are made to operate at VDDL in the Dual-V_{DD} approach, leading to reducing power. By applying the Dual-V_{DD} structure synthesis, 69% of the cells in the entire core have been made to operate at VDDL while meeting the timing constraints. In order to examine the share of non-critical paths in the total paths, we investigated 50 million paths in the entire core using a static path analyzer. Path-delay distribution is shown in Fig.5. The X-axis shows the path delay normalized by the cycle time, while the Y-axis shows the path count. In the design with only VTCMOS, the center of the distribution exists at the normalized path-delay of 0.5. By applying the Dual-V_{DD} technique to the design, the center of the distribution actually shifted toward the right, as shown in the figure. Thus, a lot of non-critical paths remain even in the design to which VTCMOS was applied. Excessive slack in those paths are effectively spent for power reduction in the Dual-V_{DD} approach.

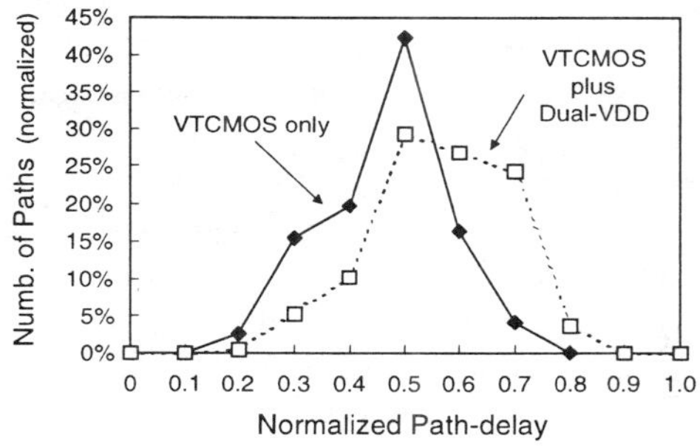

Figure 5 Path-delay distribution

Second, we present power reduction in flip-flops. In the design to which only VTCMOS is applied, conventional flip-flops are used. Meanwhile, in the design to which both the Dual-V_{DD} approach and VTCMOS are applied the FFLC's shown in Table 1 are used. The FFLC dissipates much less power than a conventional flip-flop, resulting in power reduction at flip-flops in the Dual-V_{DD} approach.

Third, we describe power reduction at the clock tree. The clock tree is made to operate at VDDH in the design with only VTCMOS, while the tree is made to operate at VDDL in the design with both the Dual-V_{DD} approach and VTCMOS. This leads to reducing power further at the clock tree in the design with the Dual-V_{DD} plus VTCMOS, compared to the design with only VTCMOS.

6.3 Power Overhead of DC-DC Converter

The VS scheme contains an on-chip DC-DC converter as a major component. The power consumption of a DC-DC converter was about 7.5mW. Two converters are needed to generate 2.5V and 1.75V from an external voltage of 3.3V, resulting in a power overhead of 15mW. The voltage scaling techniques presented in this paper can be applied to Intellectual Property (IP) modules such as an MPEG4 codec core, a PCI controller, and so on. When we apply the voltage scaling techniques to each IP module on a chip, the DC-DC converter is not needed for each module. Instead, the converter provided on a chip can be shared by all the IP modules. Compared to the total power savings resulting from the voltage scaling techniques with the DC-DC converter, the power overhead described above is considered to be small enough.

6.4 Area Overhead

Figure 6 shows the photograph of the entire chip. Circuits for VTCMOS and VS schemes including DC-DC converters are placed at the corners of the chip. Since the corner of a chip is usually a dead space, area overhead of these circuits is almost negligible. The Dual-V_{DD} approach was applied to the random logic circuit located at the center of the chip. The area overhead was around 5%. This shows that our methodology has effectively reduced the area overhead in the Dual-V_{DD} design.

6.5 Design TAT (Turn-Around-Time)

The total design time from logic synthesis to the completion of the Dual-V_{DD} layout was only three weeks. Major factors that enabled us to shorten design time are as follows. One is to divide the design step into two sub-steps: the single-voltage design step and the Dual-V_{DD} design step. This approach was very effective to reduce the design complexity of logic synthesis in generating Dual-V_{DD} mapped netlist. Second is a flip-flop containing functionality of level conversion. The use of this flip-flop enabled us to save an additional floorplanning effort in the Dual-V_{DD} layout design, because the increase of the gate count or gate area became very small. Third is the improvement of the placement algorithm shortening the wire length in the Dual-V_{DD} layout. This resulted in eliminating redundant design loops between the layout step and Dual-V_{DD} structure synthesis step. Fourth is the fact that we extensively utilized not only standard cells but also memory macros in an ASIC cell-library.

6.6 Evaluation of Test Chip

We measured power dissipation of fabricated chips using a test vector with a function of MPEG4 video encoding. Results have shown that the core dissipates 45mW excluding the overhead of level converters. The power was 58% less than that of the conventional design.

7. FURURE WORK

By applying the methodology we have presented in this paper, power has been significantly reduced at logic gates, flip-flops and the clock tree. Consequently the power at memories has become to occupy large share. Power reduction techniques of memories will be required. Circuit technology of memories will be needed in which the performance is not degraded even at lower supply voltage. A structure synthesis technique such that memory macros in non-critical paths are made to operate at VDDL will be also required.

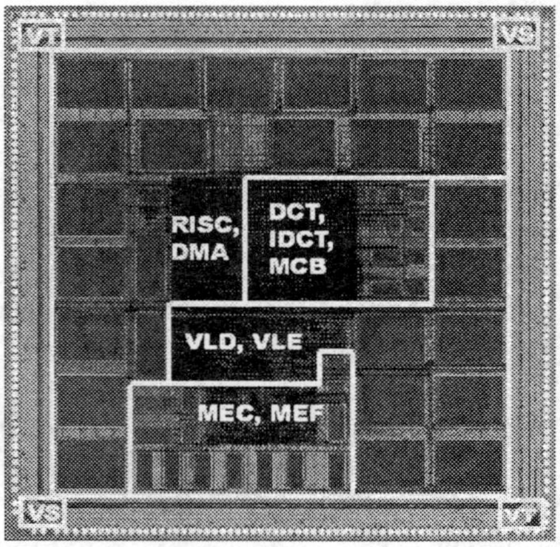

Figure 6 Chip micrograph of MPEG4 codec core

8. CONCLUSIONS

We have shown the potential advantages of our design methodology in which VTCMOS, a Dual-V_{DD} approach, and a VS scheme are combined together. These voltage scaling techniques work cooperatively at different design steps, and summation of each effectiveness becomes the total effectiveness. The VTCMOS and VS techniques reduce power at the entire circuit. The Dual-V_{DD} approach changes excessive slack in the circuit into power reduction.

We applied our design methodology to an MPEG4 codec core to examine the total power reduction and contribution of each technique. We have proved that our design methodology enables us to reduce power effectively with a small area overhead in a short design time.

9. REFERENCES

[1] Igarashi, M., et al. "A Low-power Design Method Using Multiple Supply Voltages", *ACM/IEEE ISLPED-97*, pp.36-41, Aug. 1997.

[2] Kuroda, T., et al., "A High-Speed Low-Power 0.3um CMOS Gate-Array with Variable Threshold Voltage (VT) Scheme", *IEEE CICC-96*, pp.53-56, May 1996.

[3] Suzuki, K., et al., "A 300MIPS/W RISC Core Processor with Variable Supply-Voltage Scheme in Variable Threshold-Voltage CMOS", *IEEE CICC-97*, pp.587-590, May 1997.

[4] Takahashi, M., et al, "A 60mW MPEG4 Video Codec Using Clustered Voltage Scaling with Variable Supply-Voltage Scheme", *ISSCC Digest of Technical Papers*, pp.36-37, Feb. 1998.

[5] Usami, K., et al., "Clustered Voltage Scaling Technique for Low-Power Design", *ACM/IEEE ISLPD-95*, pp.3-8, April 1995.

[6] Usami, K., et al., "Automated Low-power Technique Exploiting Multiple Supply Voltages Applied to a Media Processor", *IEEE CICC-97*, pp.131-134, May 1997.

Design and Optimization of Low Voltage High Performance Dual Threshold CMOS Circuits *

Liqiong Wei, Zhanping Chen, Mark Johnson and Kaushik Roy
School of Electrical and Computer Engineering
Purdue University, W. Lafayette, IN 47907-1285

Vivek De
Microcomputer Research Labs, Intel Corp., Hillsboro, OR 97124-6461

Abstract

Reduction in leakage power has become an important concern in low voltage, low power and high performance applications. In this paper, we use dual threshold technique to reduce leakage power by assigning high threshold voltage to some transistors in non-critical paths, and using low-threshold transistors in critical paths. In order to achieve the best leakage power saving under target performance constraints, an algorithm is presented for selecting and assigning an optimal high threshold voltage. A general standby leakage current model which has been verified by HSPICE is used to estimate standby leakage power. Results show that dual threshold technique is good for power reduction during both standby and active modes. The standby leakage power savings for some ISCAS benchmarks can be more than 50%.

1 Introduction

With the growing use of portable and wireless electronic systems, reduction in power consumption has become more and more important in today's VLSI circuit and system designs [1], [2], [3].

In CMOS digital circuits, power dissipation consists of dynamic and static components. Since dynamic power is approximately proportional to the square of supply voltage V_{dd} and static power is proportional to V_{dd}, lowering supply voltage is the most effective way to reduce power consumption as long as dynamic power is dominant. With the lowering of supply voltage, transistor threshold voltage should also be scaled in order to satisfy the performance requirements. Unfortunately, such scaling can lead to a dramatic increase in leakage current, which becomes an important concern in low voltage high performance circuit designs.

Multiple thresholds can be used to deal with the leakage problem. This technique has commonly been used in DRAM chips by raising threshold voltages of the array devices with a fixed body bias [5]. For LSI circuits, Multithreshold-Voltage CMOS (MTCMOS) circuit technology was proposed to reduce the standby leakage current by inserting high threshold devices in series to normal circuitry [8]. However, the large inserted MOSFETs will increase the area and delay.

For a logic circuit, a higher threshold voltage can be assigned to some transistors on non-critical paths so as to reduce leakage current, while the performance is maintained due to the low threshold transistors in the critical path(s). Therefore, no additional transistors are required, and both high performance and low power can be achieved simultaneously. Recently, a dual-Vth MOSFET process was developed [6], which makes the implementation of dual-Vth logic circuits more feasible.

However, due to the complexity of a circuit, not all the transistors in non-critical paths can be assigned a high threshold voltage. In order to achieve the best leakage power saving under performance constraints, we present a heuristic algorithm for selecting and assigning an optimal high threshold voltage. A standby leakage model which has been verified by HSPICE is used to estimate the standby leakage power of a circuit. The power dissipations of single-Vth and dual-Vth circuits in active mode are also compared using HSPICE simulations.

*Acknowledgment: This research is supported in part by DARPA (F33615-95-C-1625), NSF CAREER award (9501869-MIP) & Intel.

DAC 98, San Francisco, California
©1998 ACM 0-89791-964-5/98/06..$5.00

2 Delay Model

2.1 Definitions

A combinational circuit can be represented as a directed acyclic graph $G(V, E)$. Each node (except for primary inputs and outputs) in the graph maps to a logic gate in the circuit while each edge maps to a path.

The propagation delay through node x, denoted as $t_p(x)$, defines how quickly the output responds to a change in the input. The propagation delay of a path π_j, denoted as $Pd(\pi_j)$, is the sum of the propagation delays $t_p(i)$ of each node i along this path. It can be expressed as

$$Pd(\pi_j) = \sum t_p(i) \tag{1}$$

The arrival time ($T_a(x)$) is the propagation delay of each fan-in path of node x. Among all the fan-in paths, there exists a path (or paths) which has a maximum propagation delay value $T_{max}(x)$, where

$$T_{max}(x) = \max_{i \in all\ fanins} \{T_a(x)[i]\} \tag{2}$$

The departure time ($T_l(x)$) of node x is defined as

$$T_l(x) = T_{max}(x) + t_p(x) \tag{3}$$

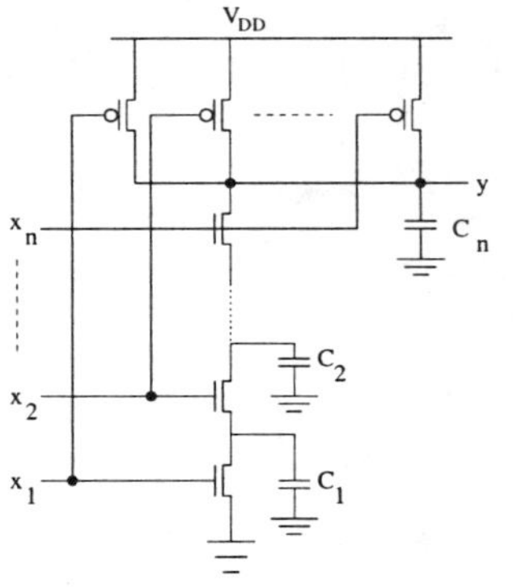

Figure 1: n-input NAND gate

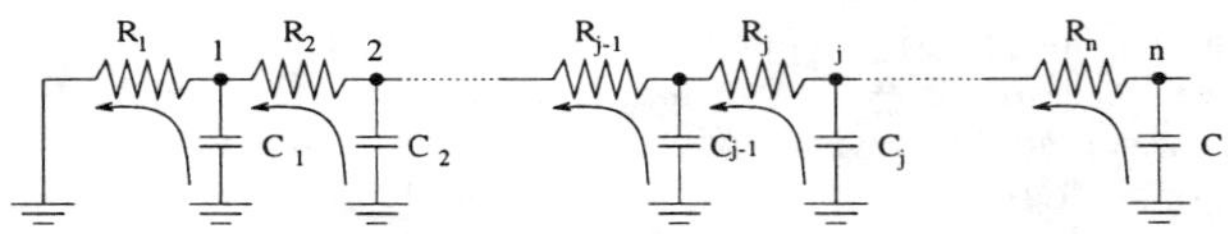

Figure 2: Equivalent pull-down network of the NAND gate

The path which determines the maximum speed of the circuit is called the critical path. There may be more than one critical path. Critical delay ($T_{critical}$) is the delay along the critical path.

2.2 Elmore delay model

Consider an n-input NAND gate (Figure 1). It can be analyzed using an equivalent RC circuit. Figure 2 shows the equivalent RC circuit of the pull-down network(PDN).

The worst case occurs when all C_j's are discharged simultaneously. Based on the Elmore delay model, the worst case delay (t_{PHL}) of the PDN is given by

$$t_{PHL} = 0.69 \sum_{j=1}^{n} (C_j \sum_{k=1}^{j} R_k) \tag{4}$$

The capacitance of each internal node j ($j = 1 \ldots n-1$) in the n-input NAND gate is given as follows,

$$C_j = 2\, C_{dN} \tag{5}$$

where C_{dN} is the diffusion capacitance of an NMOSFET. The capacitance of the gate at output is given by

$$C_n = F_O\,(C_{gP} + C_{gN}) + F_I\, C_{dP} + C_{dN} + F_O\, C_{int} \tag{6}$$

where C_{dP} is the diffusion capacitance of a PMOSFET. C_{gP} and C_{gN} are the gate capacitances of PMOS and NMOS transistors, respectively. C_{int} represents the interconnect capacitance per fan-out. F_O is the number of fan-outs of the gate, while F_I represents the number of fan-ins. For an n-input NAND gate, we have $F_I = n$.

Assuming that each NMOSFET has the same on-resistance, the worst-case delay of the PDN can be simplified as follows

$$t_{PHL} = 0.69[R_N\, C_{dN}\, F_I\,(F_I - 1) + F_I\, R_N\, C_n] \tag{7}$$

Although the on-resistance depends on the operation point and varies during the switching transient, we still can make a reasonable approximation by using a fixed value. This value is the average of the resistances at the end points of the transitions [7]. The on-resistance of an NMOSFET is given by

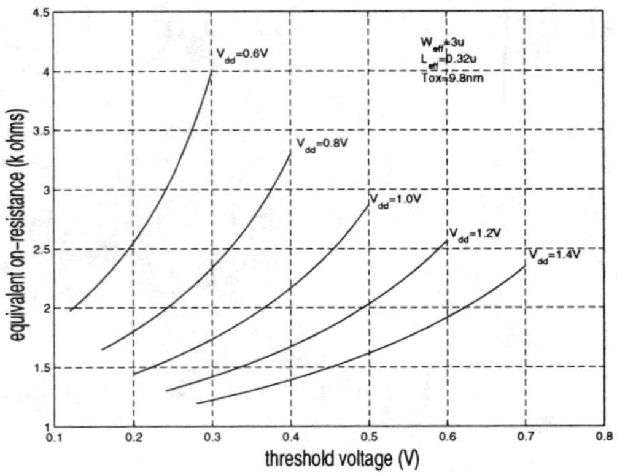

Figure 3: Relationship between R_N and Vth

$$\begin{aligned} R_N &= \frac{R_{NMOS}\,|_{V_{out}=V_{dd}} + R_{NMOS}\,|_{V_{out}=V_{dd}/2}}{2} \\ &= \frac{1}{2}(\frac{V_{DS}}{I_D}\,|_{V_{out}=V_{dd}} + \frac{V_{DS}}{I_D}\,|_{V_{out}=V_{dd}/2}) \\ &= \frac{V_{dd}}{k_N(V_{dd}-V_{TN})^{\alpha}} + \frac{V_{dd}}{k_N[2(V_{dd}-V_{TN})V_{dd} - \frac{V_{dd}^2}{2}]} \end{aligned} \tag{8}$$

where V_{TN} is the threshold voltage of an NMOSFET and k_N is the gain factor. The constant α is 2 and 1.3 for long channel and short channel MOSFETs, respectively. The relationship between R_N and V_{TN} at different supply voltages is shown in Figure 3. For a PMOSFET, the on-resistance(R_P) can be evaluated similarly. For simplicity, we assume that $|V_{TN}| = |V_{TP}| = |V_{th}|$ and $R_N = R_P$.

For the pull-up network (PUN), the worst case occurs when only one PMOS transistor is "on". The worst case delay (t_{PLH}) can be expressed by

$$t_{PLH} = 0.69 R_P\, C_n \tag{9}$$

The worst-case propagation delay of a CMOS gate is given by

$$t_p = (t_{PHL} + t_{PLH})/2 \tag{10}$$

Following similar procedure, we can get the worst-case propagation delay of the other gates.

3 Standby Leakage Power Estimation

In standby mode, the power dissipation is produced by the standby leakage current through each transistor. The leakage current has two sources: reversed-biased diode junction leakage current and subthreshold leakage current. Diode junction leakage is small and can be ignored [7]. Subthreshold leakage exponentially increases with the reduction of threshold voltage, making it critical for low voltage circuits [4]. Therefore, in our simulation, we focus on subthreshold leakage power estimation.

In order to estimate standby leakage power accurately, a general transistor model [11], which considers sub-zero gate-to source voltage (V_{GS}) for NMOS and super-zero V_{GS} for PMOS (occurs when multiple series connected transistors are turned off), body effect and drain induced barrier lowering (DIBL), is used. The following analysis is done for NMOSFETs, but is equally applicable to PMOSFETs.

From BSIM2 MOS transistor model [12], the subthreshold current of a MOSFET can be modeled as

$$I_{sub} = A\, e^{\frac{q}{n'kT}(V_G - V_S - V_{TH_0} - \gamma' V_S + \eta V_{DS})}\,(1 - e^{\frac{-qV_{DS}}{kT}}) \quad (11)$$

where $A = \mu_0 C_{ox} \frac{W_{eff}}{L_{eff}} (\frac{kT}{q})^2 e^{1.8}$. C_{ox} is the gate oxide capacitance per unit area. μ_0 is the zero bias mobility. n' is the subthreshold swing coefficient of the transistor. V_{TH_0} is the zero bias threshold voltage. The body effect for small values of V_S is very nearly linear. It is represented by the term $\gamma' V_S$, where γ' is the linearized body effect coefficient. η is the DIBL coefficient.

The standby leakage power of a logic circuit can be expressed as follows [7],

$$P_{stdby} = (\sum_i I_{stdby_i})\, V_{dd} \quad (12)$$

where I_{stdby_i} is the standby leakage current through each node i. It depends on the gate topology as well as input signal levels.

Let's consider an NAND gate. Assume the transistors which are turned on are short circuits. If all the inputs are "1", the PDN is shorted and the standby leakage current is determined by the PUN. We can get the source and drain voltages of each transistor in the PUN easily. Using equation (11), the leakage contribution of each transistor in the PUN can be calculated separately and added together. If at least one input is "0", the PUN is shorted and standby leakage current is the leakage though the PDN. Suppose there are n transistors which are turned off in the PDN, the quiescent subthreshold leakage current through each of them must be identical. By equating the leakage current of the transistors in the stack, V_S and V_{DS} of each transistor can be obtained. If $n = 1$, $V_{DS1} = V_{dd}$ and $V_{S1} = 0$. Otherwise, the following equations can be used

$$V_{DS_2} = \frac{n'kT}{q(1 + 2\eta + \gamma')} ln(\frac{A_1}{A_2} e^{\frac{q\eta V_{dd}}{n'kT}} + 1) \quad (13)$$

$$V_{DS_i} = \frac{n'kT}{q(1+\gamma')} ln(1 + \frac{A_{i-1}}{A_i}(1 - e^{-\frac{q}{kT}V_{DS_{i-1}}})) \quad (2 < i \leq n) \quad (14)$$

$$V_{S_i} = \sum_{j=i+1}^{n} V_{DS_j} \quad (1 \leq i \leq n) \quad (15)$$

$$V_{DS_1} = V_{dd} - V_{S_1} \quad (16)$$

where $i = 1$ represents the top transistor and $i = n$ represents the bottom transistor in the stack. A more detailed derivation of the above equations can be found in [11]. Now equation (11) can be used to calculate the quiescent leakage for any transistor in the stack, which is the leakage current of the PDN.

Considering the fact that standby leakage current depends on input signal levels, the average leakage power of a circuit can be evaluated with random patterns applied to primary inputs.

4 Algorithm

Due to the exponential relationship between threshold voltage and drain current in the weak inversion region, a higher threshold voltage will significantly reduce leakage current, thereby reducing leakage power. However, Figure 3 indicates that a higher threshold voltage will increase the equivalent on-resistance of each transistor, which results in a higher propagation delay. Normally, threshold voltage is empirically defined to be around 20% of supply voltage to maintain the performance of a circuit [9]. For low supply voltage circuits, the threshold voltage could be very small, leading to a large leakage current.

This problem can be circumvented by using dual threshold voltages. A low threshold voltage is assigned to the transistors in critical path(s) in order to achieve high performance, while a high threshold may be assigned to some transistors in non-critical paths to reduce leakage power. The lower bound of low threshold voltage is determined by noise margin. The possible high threshold value should be in the range from low threshold to $0.5V_{dd}$. However, not all the transistors in non-critical paths can be assigned the high threshold voltage. Otherwise, some non-critical paths may become critical. Whether a node can be modified depends on the value of the high threshold. If it is too small, there is little difference of propagation delay between low-Vth and high-Vth transistors. Hence, more nodes can be assigned high-Vth without influencing the critical delay, but the leakage current improvement for each high-Vth transistor would be small. On the other hand, if the high threshold voltage is too large, the leakage current can be reduced by a large amount for each such transistor. However, fewer nodes can be modified. Hence, among the allowable values for high threshold voltage, there exist an optimal one. We developed a breadth-first search(BFS) algorithm to search for the optimal high-Vth.

The first step in our algorithm is to initialize a circuit with a single low threshold. After initialization, all necessary parameters associated with each node ($t_p(x)$, $T_{max}(x)$, and $T_l(x)$) are computed. By checking all the primary outputs and then backtracing, the critical delay and critical path(s) can be identified using a first-in-first-out (FIFO) queue Q. The pseudo-code for the initialization procedure is shown below. Note that, primary output (PO) does not map to a gate in a circuit, and each PO has only one fan-in gate ($fanin(PO)$).

```
Initialization () {
  Assign a level number to each node
  Calculate the propagation delay t_p(x) of each node x
  Calculate T_max(x) and T_l(x) of each node x level by level
  Identify T_critical by checking the maximum T_l(fanin(PO))
  For each primary output PO {
    If (T_l(fanin(PO)) = T_critical)
      Mark fanin(PO) as a node in critical path
      Add node fanin(PO) into a FIFO queue Q
  }
  While (Q not empty) {
    Remove node x from Q
    For each fan-in y of node x {
      If ((T_l(y) = T_max(x)) && (y is not a primary input))
        Mark x as a node in critical path
        Add node y into queue Q
    }
  }
}
```

During the initialization procedure, in order to obtain $T_{max}(x)$ and $T_l(x)$, the circuit has to be levelized. Essentially, levelization assigns a number to each node to indicate the depth of the node in the graph. The level of each primary input is defined to be 0. The level of any node x, denoted as $l(x)$, can be calculated as follows,

$$l(x) = 1 + \max_{j \in all\ fanins} \{l(j)\} \quad (17)$$

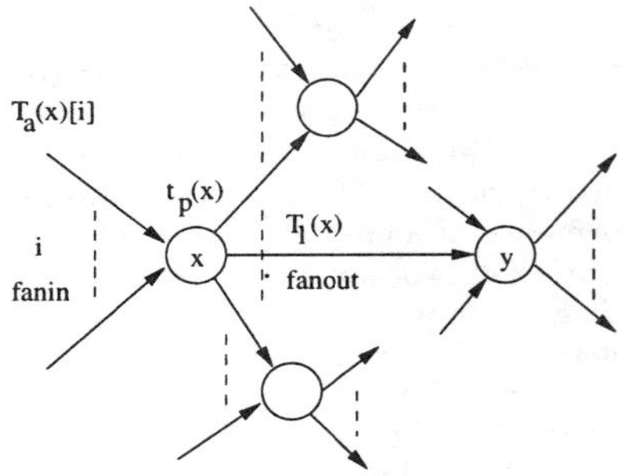

Figure 4: Diagram of a part of a logic circuit

where j varies for all fan-in nodes of node x.

For each primary input x, $t_p(x) = 0$, $T_a(x) = 0$, $T_l(x) = 0$ and $T_{max}(x) = 0$. For each node x in level 1, $T_a(x) = 0$, $T_{max}(x) = 0$ and $T_l = t_p(x)$. Therefore, the parameters ($t_p(x)$, $T_a(x)$, $T_l(x)$ and $T_{max}(x)$) associated with each node x can be computed by equations (2) and (3) level by level during the initialization procedure.

The next step is to assign a high threshold to some transistors on non-critical paths under performance constraints. This is performed by checking the slack of each node using a BFS-based backtracing algorithm. Slack of a node ($T_\delta(x)$) denotes the amount by which the gate can be slowed down. For the nodes in critical path(s), slack is 0. For a PO,

$$T_\delta(PO) = T_{critical} - T_l(f_{anin}(PO)) \qquad (18)$$

For any other node x (suppose x is traversed from node y during back-tracing (Figure 4)), $T_\delta(x)$ can be expressed as

$$T_\delta(x) = \min\{(T_\delta(y)+T_{max}(y)-T_l(x)), \min(T_{max}(z)-T_l(x))\}$$
$$\forall z = fanout(x) \neq y \qquad (19)$$

where $f_{anin}(x)$ and $f_{anout}(x)$ are the fan-in nodes and fan-out nodes of node x, respectively. Consider Figure 4. The first term in equation (19) ensures that the propagation delay of the path(s): $\ldots \rightarrow fanin(x) \rightarrow x \rightarrow y \rightarrow \ldots$ is no greater than the critical delay. The second term guarantees that the modification of the propagation delay of node x cannot affect the propagation delay of all the other fanout paths of node x. To make sure that the critical delay is not affected, $T_\delta(x)$ is taken as the minimum value of the two terms as shown in equation (19).

The procedure for choosing the nodes with a high threshold voltage works as follows. From each PO, BFS is used to explore every node on the breadth-first tree of $G(V, E)$. If a node has not been visited yet, by checking its slack, we can decide whether its threshold voltage should be changed. Once the node is visited, it is marked to avoid repeating assignment. By definition, for each node in a single threshold circuit, its slack(T_δ) is no less than 0. Increasing the threshold voltage of a node can result in a higher propagation delay and departure time of this node. Therefore, the slack will decrease. Whether a node should be assigned to a high threshold depends on whether its slack is still positive if its threshold is changed to high threshold. If slack is still positive, this node will be assigned to the high threshold and the number of high threshold NMOS-PMOS pairs (N_c) is incremented. Since the slack of each node in critical path is 0, the threshold voltage of these transistors will not be changed, and hence, the performance is maintained. The pseudo-code for the above procedure is shown below:

```
High-Vth-Assignment(Vth2){
  For each primary output PO {
    Explore each node x using breadth-first search
      If x has not been visited
        Calculate tp(x), Tl(x) and Tδ(x) for high threshold Vth2
        if Tδ(x) ≥ 0
          Assign Vth2 to x
          Assign tp(x), Tl(x) and Tδ(x) for Vth2 to x
          Nc++
        else
          Keep tp(x), Tl(x) and Tδ(x) for initial low Vth for x
        Mark x visited
  }
}
```

Finally, we will search the optimal high threshold voltage (opt_Vth_2) corresponding to the best saving of standby leakage power. The high thresholds are sampled according to the different on-resistances (R). $R(0)$ is the on-resistance for original low Vth. Step size ΔR of $0.1R(0)$ is chosen for the simulation. The relationship between on-resistance and threshold voltage is given by equation (8). Standby leakage power can be evaluated using the method described in section 3. After updating the network for opt_Vth_2, the circuit can be transfered into SPICE netlist and simulated using HSPICE to verify the results. The procedure is outlined below:

```
Optimal-High-Vth (){
  i=1 and R(i) = R(0) + ΔR
  Calculate Vth_i corresponding to R(i)
  While (Vth_i < 0.5V_dd){
    Initialization
    High-Vth-Assignment(Vth_i)
    Estimate standby leakage power P_stdby
    If standby leakage power is the least power so far
        P_stdby_min = P_stdby
        opt_Vth2 = Vth_i
    ++i and R_i = R_0 + i × ΔR
    Calculate Vth_i corresponding to R_i
  }
  Update network with opt_Vth2
  Transfer the network into SPICE netlist
}
```

The above algorithm can be easily extended to solve the other problems, such as multiple supply voltage design & optimization [13].

5 Implementation and Results

The method to reduce leakage power using dual-threshold-voltage transistors has been implemented in C under the Berkeley SIS environment. In order to simplify the analysis, technology-mapping was used to map the circuits to a library which contains NAND gates, NOR gates and Inverters. All the simulation results were based on a $0.5\mu m$ MOSIS process. The effective channel length was $0.32\mu m$ and the gate oxide thickness was 9.8nm. The effective channel widths for PMOSFETs and NMOSFETs were assumed to be $10.5\mu m$ and $3\mu m$, respectively.

Figure 5 gives an example circuit to show how our algorithm works. Figure 5(a) is the original single-Vth circuit, where the supply voltage is 1V and the threshold voltage is 0.2V. Figure 5(b), (c),and (d) show the dual-Vth circuits with the low Vth of $0.2V$ and the high-Vth of 0.25V, 0.396V,and 0.46V, respectively. Note that the critical paths and critical delay are maintained after the assignment.

Figure 6 shows the standby leakage power of the above example circuit with different high thresholds(Vth_2). The supply voltage is 1V, the low threshold voltage is 0.2V and the circuit temperature is 25°C. Vth_2 varies from 0.2V to

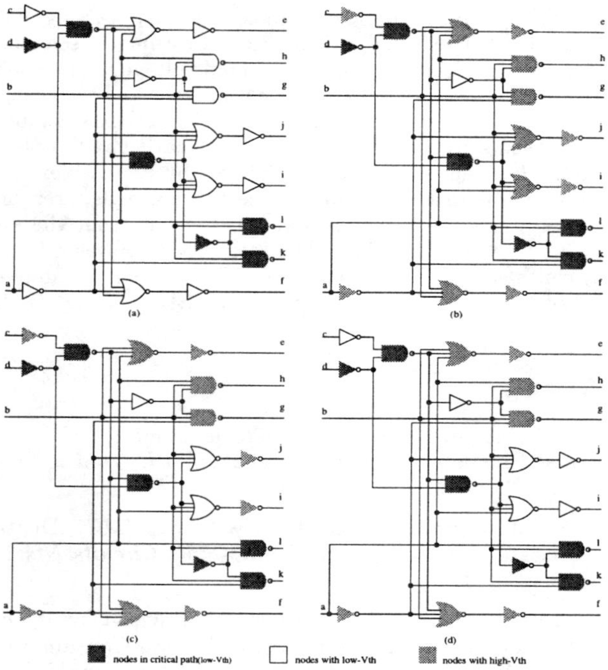

Figure 5: An example circuit (a) 1-Vth $V_{dd} = 1V$, $Vth = 0.2V$ (b)-(d) 2-Vth $V_{dd} = 1V$, $V_{th1} = 0.2V$ (b) $Vth_2 = 0.25V$ (c) $Vth_2 = 0.396V$ (d) $Vth_2 = 0.46V$

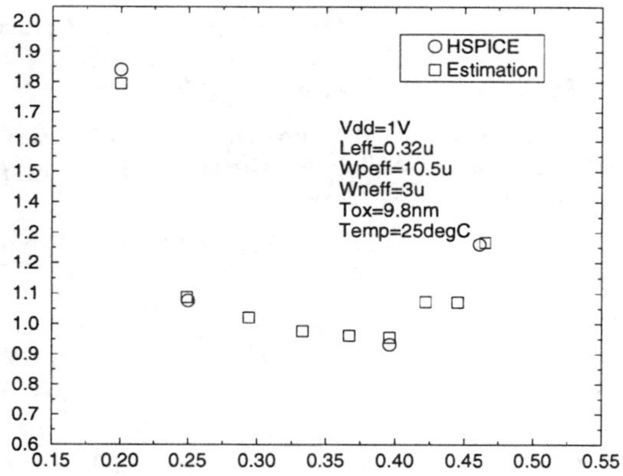

Figure 6: Standby Leakage power with different Vth_2

0.5V ($Vth_2 = 0.2V$ represents the single low threshold circuit). The squares represent the leakage power obtained by our estimation technique while the circles denote the leakage power obtained by HSPICE simulations. Clearly, the estimation results fit well with HSPICE simulation results. The convex point of the curve indicate that there exits an optimal high threshold voltage($0.396V$) which leads to a 50.67% savings in standby leakage power.

Table 1 and Figure 7 show the optimal high threshold and standby leakage power savings for ISCAS benchmark circuits. The percentages of high-Vth transistors and gates over total transistors and gates for different dual-Vth benchmark circuits are illustrated in Figure 8. In this experiment, supply voltage was 1V and the circuit temperature was 25^oC. The low threshold voltage was assumed to be

Table 1: Optimal High-Vth and Standby Leakage Power Saving (V_{dd}=1V, Temp=25°C, Single-Vth:$V_{th} = 0.2V_{dd}$, Dual-Vth: $Vth_1 = 0.2Vdd$, Vth_2=opt-Vth_2)

Circuit Chosen	Gate #	$P_{stdby}(\mu W)$ (1-Vth)	Vth_2 (mV)	$P_{stdby}(\mu W)$ (2-Vth)	Red. (%)
C432	206	27.97	333	11.60	58.51
C499	532	67.17	367	45.04	32.95
C880	353	46.29	333	17.64	61.90
C1355	517	64.86	333	37.14	42.73
C1908	615	77.68	333	37.58	51.63
C2670	807	106.96	333	54.98	48.60
C3540	1131	151.18	367	60.09	60.25
C5315	1778	228.58	367	102.69	55.08
C6288	2400	355.09	367	277.14	21.95
C7552	2803	378.95	367	201.01	46.96

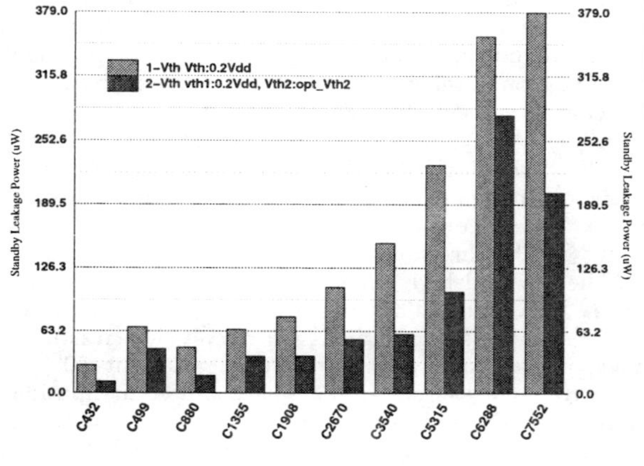

Figure 7: Standby leakage power savings for ISCAS benchmarks ($V_{dd} = 1V$)

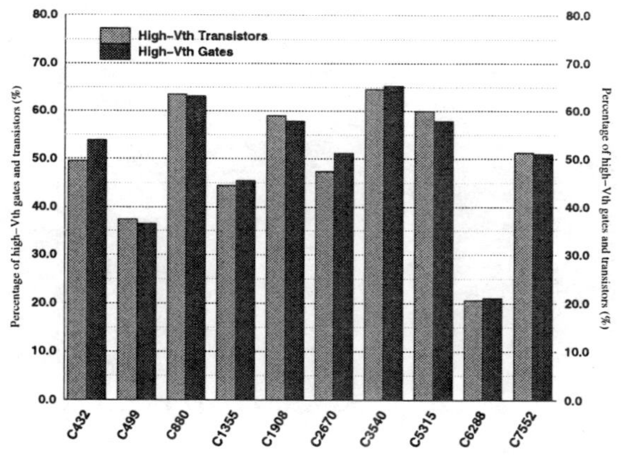

Figure 8: Percentage of high threshold gates and transistors for dual-Vth ISCAS Benchmarks($V_{dd} = 1V$)

0.2V and high threshold voltage was the optimal value obtained from our heuristic algorithm given in section 4. Results indicate that the percentage of high threshold voltage transistors can be more than 60% and standby leakage power can be reduced by around 50% for most of the circuits. Even though the optimal high threshold voltage varies for different circuits, for most of the circuits, it was between $0.3V_{dd}$ and $0.4V_{dd}$.

For a CMOS digital circuit, total power dissipation includes dynamic and static components in active mode. Ignoring power dissipation due to the short circuit current, total active power dissipation can be expressed as follows [10],

$$\begin{aligned} P_T &= P_{dyn} + P_{static} \\ &= \sum_i \alpha_i\, C_i\, V_i\, V_{dd}\, f_{clk} + I_{static} V_{dd} \end{aligned} \quad (20)$$

where α_i is the switching activity (the probability of switching), C_i is the load and parasitic capacitances, f_{clk} is the operating frequency and V_i is the voltage swing which equals to V_{dd} at the output node and $V_{dd} - V_{th}$ at internal nodes. The summation is taken over all nodes of the circuit. I_{static} is the leakage current through the circuit.

Consider the example circuit(Figure 5). Figure 9 shows the HSPICE simulation results of the total active power dissipations of single-Vth and dual-Vth circuits at different frequencies. The circuits were simulated at 1V supply voltage and 110°C. The threshold voltage of single-Vth circuit was 0.2V. The low and high threshold voltages of dual-Vth circuit were 0.2V and 0.396V, respectively. At low frequency, the active power saving of dual-Vth circuit, which is mainly because of the static power reduction, is about 50%. As for high frequency circuits, the active power dissipation is dominated by dynamic consumption. In addition to leakage power saving, the dynamic power is reduced due to the reduction of internal node voltage swing for high threshold gates. In our example, the total power saving can be around 13% at 100MHz frequency.

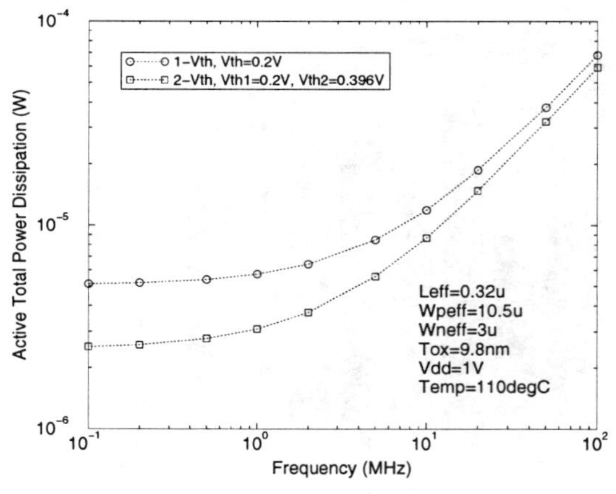

Figure 9: Active power dissipation at different frequencies

6 Conclusions

In this paper we present a method to design and optimize low voltage dual-Vth CMOS circuits. In order to reduce leakage power under performance constraints, starting with a single low Vth circuit, a heuristic algorithm for selecting and assigning an optimal high threshold voltage is proposed. For accurate leakage power estimation, a standby leakage current model which has been verified by HSPICE simulation is used. Results for ISCAS benchmark circuits show that the leakage power can be reduced by more than 50% under performance constraints. The optimal high threshold voltages are between $0.3V_{dd}$ and $0.4V_{dd}$, given that the low threshold voltage is $0.2V_{dd}$. The total active power dissipation can also be reduced using dual-Vth design technique. The total power saving can be about 13% for some circuit at 100MHz frequency.

References

[1] J. D. Meindl, "Low power Microelectronics: Retrospect and Prospect", *Proceedings of the IEEE*, Vol.83, No.4, pp.619, 1995.

[2] A. P.Chandrakasan, et al.,"Low-Power CMOS Digital Design", *IEEE Journal of Solid-State Circuits*, Vol.27, No.4, pp.473, 1992.

[3] P. Pant, V. De and A. Chstterjee, "Device-Circuit Optimization for Minimal Energy and Power Consumption in CMOS Random Logic Network", *DAC*, pp. 25.1.1-25.1.6, 1997.

[4] C. Hu, "Device and technology impact on low power electronics", in *Low Power Design Methodologies*, J.M.Rabaey and M.Pedram, Eds. Norwell, MA:Kluwer,pp21-36, 1996.

[5] B. Davari, et al., "CMOS Scaling for High Performance and Low Power-The Next Ten Years", *Proceedings of the IEEE*, Vol.83, No.4, pp595, 1995.

[6] Z. Chen, et al., "0.18um Dual Vt MOSFET Process and Energy-Delay Measurement", *IEDM Digest*, pp851, 1996.

[7] Jan M. Rabaey, "Digital Integrated Circuits", New Jersey: Prentice-Hall, 1996.

[8] S. Mutoh, et al., "1-V Power Supply High-Speed Digital Circuit Technology with Multithreshold-Voltage CMOS", *IEEE Journal of Solid-State Circuits*, Vol.30, No.8, pp. 847-853, 1993.

[9] H. Oyamatsu, et al., "Design Methodology of Deep Submicron CMOS Devices for 1V Operation",*IEICE Trans. Electron.*, VOL.E79-C, No.12, pp1720-1724, 1996

[10] A. Bellaouar and M. I. Elmasry, "Low-Power Digital VLSI Design", Kluwer Academic Publishers, 1995

[11] M.C. Johnson, K. Roy, and D. Somasekhar "A model for leakage control by transistor stacking", Technical Report TR-ECE 97-12, Purdue University, Dept. of ECE.

[12] J. Sheu, et. al., "BSIM: Berkeley Short-Channel IGFET Model for MOS Transistors", *IEEE J. Solid-State Circuits*, SC-22, 1987.

[13] M.C. Johnson and K. Roy, "Datapath scheduling with multiple supply voltages and level converters", ACM Transactions on Design Automation of Electronic Systems, July 1997.

MTCMOS Hierarchical Sizing Based on Mutual Exclusive Discharge Patterns

James Kao, Siva Narendra, Anantha Chandrakasan
Department of Electrical Engineering and Computer Science
Massachusetts Institute of Technology
{jkao, naren, anantha}@mtl.mit.edu

ABSTRACT

Multi-threshold CMOS is a popular circuit style that will provide high performance and low power operation. Optimally sizing the gating sleep transistor to provide adequate performance is difficult because the overall delay characteristics are strongly dependent on the discharge patterns of internal gates. This paper proposes a methodology for sizing the sleep transistor for a large module based on mutual exclusive discharge patterns of internal blocks. This algorithm can be applied at all levels of a circuit hierarchy, where the internal blocks can represent transistors, cells within an array, or entire modules. This methodology will give an upper bound for the sleep transistor size required to meet any performance constraint.

1. BACKGROUND

Multi-threshold CMOS is an emerging technology that provides high performance and low power operation by utilizing both high and low V_t transistors[1][2][3]. By using low V_t transistors in the signal path, the supply voltage can be lowered (while still maintaining performance) to reduce switching power dissipation. By reducing V_{dd}, the switching power can be reduced quadratically, but as V_t decreases to maintain performance, the subthreshold leakage current will increase exponentially. For ambitious scaling, the increased leakage power can actually dominate the switching power[4]. In many event driven applications, like a processor running an X-server, circuits spend most of their time in an idle state where no computation is being performed. During these "sleep" times, it is very wasteful to have large subthreshold leakage currents. Static power dissipation can be reduced in the sleep mode by using high V_t transistors with very low leakage currents to gate the power supply lines for the entire module.

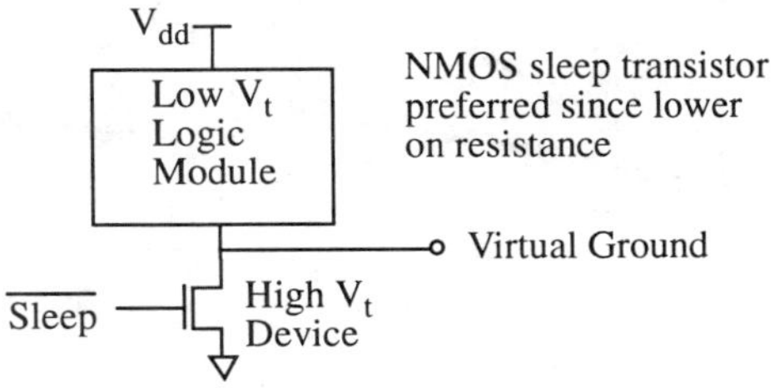

Figure 1. MTCMOS circuit structure.

DAC 98, San Francisco, California

Although it is easy to reduce leakage by using a high V_t gating device, it is difficult to size the sleep transistor large enough so that performance is maintained. Some initial work on MTCMOS circuits was presented in [5], and it was shown that the sleep transistor can be approximated very closely by a linear resistor that creates a finite voltage drop across the virtual ground node as gates are discharging. This virtual ground bounce causes the internal logic to slow down for two reasons: first, the gate drive is reduced and second, the internal transistor threshold voltages will increase due to the body effect. The worst case delay in an MTCMOS circuit is strongly dependent on the discharge patterns of internal gates, which will cause the virtual ground line to fluctuate depending on discharge patterns through this sleep transistor. The worst case input vector is difficult to predict and can even be different than a vector which exercises a critical path in an ordinary CMOS implementation. As a result, optimal sizing of the sleep transistor for an arbitrary circuit to meet a performance constraint can be difficult. A switch level simulator had been proposed to provide fast MTCMOS simulations to help narrow down this search space[5].

In this paper, we will explore another methodology for sizing the sleep transistor to meet a performance constraint. Rather than search for the worst case input vector to exercise the worst case discharge patterns in the MTCMOS circuit, we instead work from the bottom up, and synthesize a sleep transistor size based on mutual exclusive discharge patterns. Application of this sizing methodology will guarantee that the performance of a complex MTCMOS circuit will be within a chosen percentage of the original CMOS version for all possible inputs.

2. APPROACHES TO TRANSISTOR SIZING

The most straightforward (but difficult) way to correctly size the sleep transistor of an MTCMOS circuit is to exhaustively test for the worst case input vector and to ensure that the worst case delay meets a fixed performance constraint. However, individual gates within this critical path, and other paths within the circuit can degrade in percentage more or less than this fixed criteria. Figure 2b shows how individual gate degradations can vary (assuming both polarities of sleep transistor are used) while overall performance is maintained.

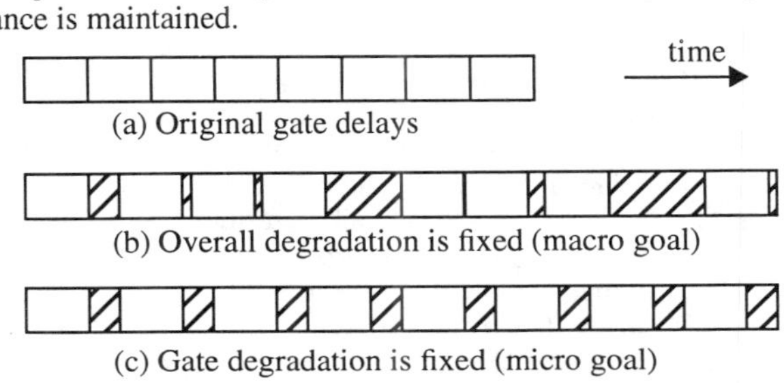

Figure 2. MTCMOS gate degradation scenarios to meet fixed specification.

A different way to satisfy a macro performance criteria is to ensure that every individual gate meets a local performance constraint (Figure 2c, which assumes both the high to low and low to high transitions are degraded). This will ensure that any combination of gates in a path will also meet the performance requirements. Forcing every single gate to meet a nominal performance measure is a much more demanding constraint than simply achieving an overall performance goal. However, in the context of MTCMOS circuits, it is much easier to implement this sizing strategy because one does not need to determine the worst case input vector pattern for the whole circuit. Instead, each individual gate can be assigned it's own high V_t sleep transistor, whose size will be locally determined through exhaustive SPICE simulations.

Once an MTCMOS circuit is sized with individual sleep transistors then one can systematically merge the sleep transistors together, because they can be shared among mutually exclusive gates, where no two gates can be discharging current at the same time. Finally, these sets of sleep transistors can then be combined to make a single sleep transistor for the whole circuit that guarantees that for any input vector, the MTCMOS circuit performance will be within the specified range of the corresponding CMOS circuit.

3. SLEEP TRANSISTOR SIZING AND MERGING TECHNIQUE

A good way to describe the sleep transistor sizing and merging technique is through an example. Figure 3 to the right shows how an MTCMOS circuit can initially be sized using individual sleep transistors that can be merged together at later steps. The circuit consists of three chains of five low V_t inverters, and measurements are made for the input to output delay, the delay for inverter I_5, and the virtual ground bounce transients. Table 1 below summarizes some simulation results.

R ohms	I_5 (a) %	Total (a) %	I_5 (c) %	Total (c) %	I_5 (e) %	Total (e) %
0	0	0	0	0	0	0
100	1.9	1.2	2.0	1.1	4.9	2.9
200	3.0	2.3	3.5	1.9	9.3	5.7
300	4.2	3.4	4.9	2.9	13.7	8.5
400	5.8	4.6	6.1	3.7	17.9	11.0
500	7.5	5.7	7.5	4.6	22.1	13.4

Table 1. Percent degradation for gate I5 and total delay (for cases a, b, c) as function of sleep resistance.

3.1 Individual sleep transistor sizing

Figure 3a shows the first step in the transistor sizing procedure, where an identical sleep resistor (which models a sleep transistor in the on state) is placed in series with each gate. As can be seen in columns 2 and 3 of Table 1, the overall performance of the inverter chain will be satisfied if the internal gates meet the required speed (i.e. the % delay in column 3 is always less than or equal to that of column 2). The overall delay degradation is less than the individual gate degradation of gate I5 namely because the low to high transitions of inverters I2 and I4 are not degraded by an NMOS sleep transistor. Figure 3b shows how the virtual ground lines (V_1, V_3, and V_5) for this circuit will fluctuate as a result of a rising step function applied to the input.

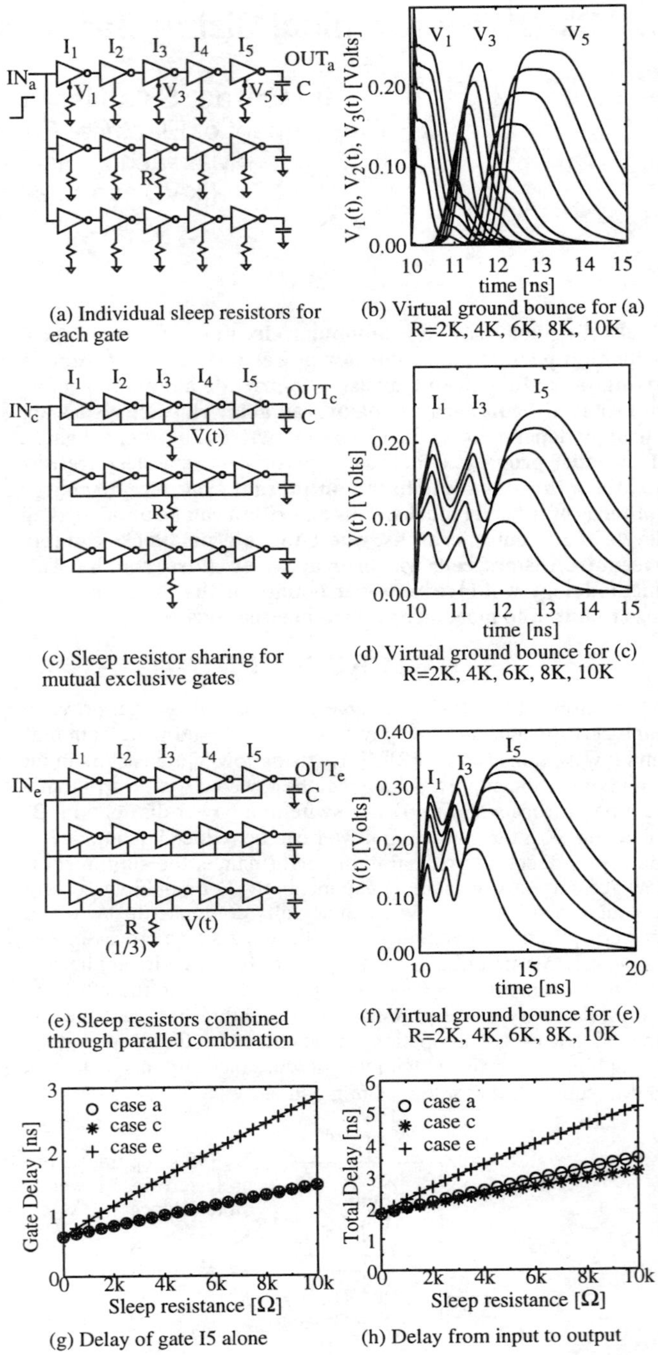

Figure 3. Inverter chain example showing the 3 steps for merging sleep resistors. Circuits use V_{dd}=1.0v, V_t=0.2v, C=50fF, l_{min}=0.7µm.

3.2 Sleep transistor merging based on mutual exclusive discharge patterns

Although it is relatively simple to develop an MTCMOS sizing strategy by individually adding high V_t transistors to each gate in a circuit, this can result in large overestimates in sleep transistor area and large overheads in wiring area. However, since not all gates in the circuit will switch at the same times, it is possible to merge sleep transistors together from mutual exclusive gates and thereby reduce circuit complexity. For a set of n such gates with equivalent sleep resistances r_1, r_2,... r_n, the sleep resistors can be combined and replaced by a single $r_{eff} = \min(r_1, r_2, \ldots r_n)$. These mutually exclusive gates will discharge currents through the sleep transistor at different times so that the virtual ground bounce that each transitioning gate experiences will still be the same or smaller than before. As a result, the delay of each gate sharing the common sleep transistor should also be the same or smaller than in the original circuit. An added benefit of replacing n sleep resistors with a single one is that the subthreshold leakage current will decrease by a factor of n, and also the increased parasitic capacitance on the virtual ground line can improve performance.

Figure 3c shows how the Figure 2a inverter tree's sleep resistors can be replaced by only 3 resistors by utilizing the same high V_t switch for mutual exclusive gates. Inverters I_1, I_2, I_3, I_4, and I_5 for example will never transition from high to low at the same times, and as a result can share a common sleep transistor. Figure 3g shows how the delay of inverter I_5 remains the same for both cases, which is to be expected. Upon closer inspection, one can see that the overall performance of the inverter chain actually improves when the transistors are merged together as in Figure 3h. This is because there is a larger parasitic capacitance on the virtual ground line for the merged case, which will tend to low pass filter the virtual ground bounce. As a result, inverter I_1 will be faster because the virtual ground bounce rises more slowly. As the parasitic capacitances charges up though, later gates will not see these beneficial effects since the capacitance does not have time to discharge again, as can be seen in Figure 3d.

3.3 Merging through parallel combination

Having separate sleep resistors for different groups of mutually exclusive gates can be cumbersome for the circuit layout. In many cases, it is possible to lump these sleep transistors together as a parallel combination, and performance will still be maintained. Although total transistor area will be the same, wiring and layout area can be reduced. To quantify this point, consider the circuit in Figure 4.

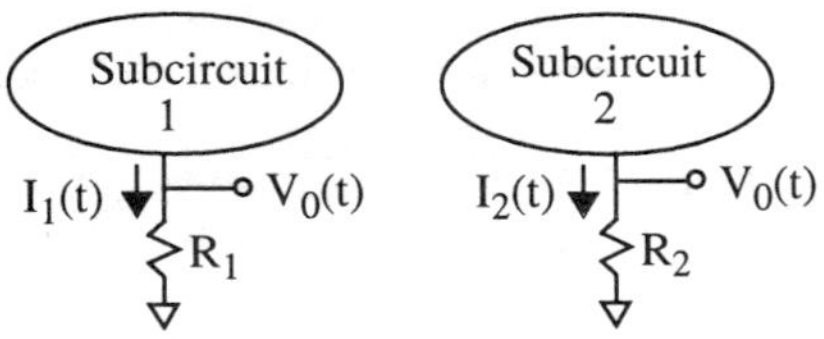

Figure 4. Circuit showing how sleep resistors can be combined in parallel.

If the virtual ground voltages for two different subcircuits is similar, then they can be modeled as two current sources, $i_1(t)$ and $i_2(t)$, connected to resistors r_1 and r_2 to give a voltage waveform $v_0(t)$ for both cases. However, if $i_1(t)$ and $i_2(t)$ are summed together and r_1 and r_2 are placed in parallel, then the new voltage over the resistor is:

$$v(t) = (i_1(t) + i_2(t)) * (r_1 // r_2)$$
$$= (i_1(t) * r_1 * r_2 + i_2(t) * r_1 * r_2) * (1 / r_1 + r_2)$$
$$= (v_0(t) * r_2 + v_0(t) * r_1) * (1 / r_1 + r_2)$$
$$= v_0(t)$$

which is the same as before. Thus, for two subcircuits with very similar virtual ground transient behaviors, combining the two systems together will result in unchanged virtual ground characteristics, so the overall performance should be unchanged. In general, if voltages $v_1(t)$ and $v_2(t)$ are very different, then the resistors should be combined such that v(t) will not exceed the minimum of $v_1(t)$ or $v_2(t)$. In this case, $r_{eq} = \min(v_1(t), v_2(t)) / (v_1(t)/r_1 + v_2(t)/ r_2)$.

In Figure 3e, the three separate sleep resistors from Figure 3c can be replaced by a single resistor with three times the conductance that now gates the entire circuit. Figures 3g and 3h show comparisons of the delay vs. sleep resistor size for these two cases, and that the resistance must be lowered by one third in order to achieve the same performance. Another way to appreciate this relationship is to examine the virtual ground transient response shown in Figure 3d and 3f. By scaling the resistance by 1/3 for the case with a single global sleep transistor, the virtual ground bounce shown in Figure 3f can be matched to the that of Figure 3g, which would give the same delay behavior.

In general, combining separate sleep transistors into a single common one will be beneficial. The increased parasitic capacitances will tend to speed up the circuit during the capacitor charging stage. More important, the worst case scenario where the subcircuits will all discharge simultaneously is not common. Because the larger resistances used in the original subcircuits are replaced by a smaller resistance applied to the combined circuit, in many cases individual gates will be faster than before. In some degenerate examples using pure parallel combination, it may be possible that two subcircuits with separate sleep transistors might have very different virtual ground transient responses. In such a case, combining sleep transistors by a simple parallel combination will speed up one case, but could possibly slow down the other (the one with a much smaller virtual ground bounce). However, this is most likely not going to affect the overall performance of the circuit as a whole.

3.4 Comparison with optimal sleep transistor size

As a concrete example, we simulated the MTCMOS inverter network where the sleep transistor was designed to provide only a 5% degradation in performance over a conventional CMOS implementation. By simulating a single inverter with a sleep resistor in SPICE, we discovered that a sleep transistor with an equivalent resistance of less than 340Ω was required for less than 5% individual degradation. When applied to the inverter chain network and merged together, the sleep transistor equivalent resistance was 113Ω, with a 3.13% degradation in delay. The predicted sleep transistor required was actually an overestimate, because direct simulation shows that one only needs a resistance of 180Ω in order to achieve a 5% degradation in performance. By using this transistor sizing methodology, the transistor width was overestimated by 60%. One major cause for this discrepancy is that in MTCMOS circuits with NMOS sleep transistors, typically only half the gates, those switching from high to low, are actually degraded. Thus even if the high to low transition degrades by 5%, the overall chain will degrade on average by only 2.5% if pulldown and pull up transitions are balanced. Although this inverter chain circuit is easy enough to size through brute simulation, the resistor synthesis approach can be applied to more complicated circuits where exhaustive simulation is not possible.

4. SLEEP TRANSISTOR ALGORITHM

The previous example demonstrated how MTCMOS sleep transistors can be sized individually for each gate and then shared among mutually exclusive gates, where no two gates can be discharging current at the same time. The primary value of this technique is in the sleep transistor reduction step, because area of the sleep transistor is of primary concern in MTCMOS circuits. One approach to develop a mutual exclusive set of gates in a circuit, is to use a criteria based on the structural interconnections in the network graph. Assuming a unit delay model for each gate, then one can tabulate all the possible times that any particular gate can switch. Mutually exclusive gates can then be grouped together whenever there is no intersection between the corresponding sets of times. In order to minimize total sleep transistor sizes, the number of these groupings of mutually exclusive gates should be minimized, and the sleep transistors chosen to be the largest transistors in each respective group.

Figure 5 shows a random logic circuit with arbitrary gate interconnections, where it is assumed that each gate has a corresponding sleep transistor (modeled as a resistor). Each gate is annotated using a unit delay model with all possible time slots that a transition can occur. Gates that do not have a time period in common will thus be mutually exclusive, and can be grouped together with a common sleep transistor. In cases where a gate can switch at multiple times, we further annotate the set of transition times by a subscript indicating the reference gate, because these two gates are also mutually exclusive even though they share a time slot. For example, gates g_7 and g_9 both show possible transitions at time 3, but this will never happen simultaneously because g_9 is always one time unit behind g_7. Ideally, the groupings should be selected to minimize the overall sleep transistor widths such that gates with very large sleep transistors should be lumped together.

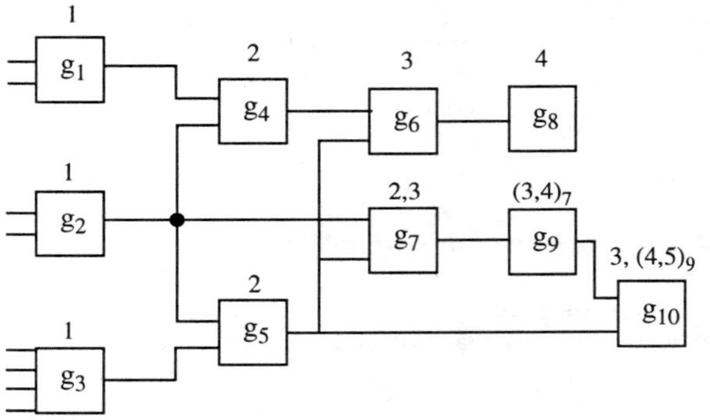

Grouping #1 = $\{g_1, g_4, g_6, g_8\}$ $R_a = \min(r_1, r_4, r_6, r_8)$

Grouping #2 = $\{g_2, g_7, g_9\}$ $R_b = \min(r_2, r_7, r_9)$

Grouping #3 = $\{g_3, g_5, g_{10}\}$ $R_c = \min(r_3, r_5, r_{10})$

$$R_{equivalent} = R_a \,//\, R_b \,//\, R_c$$

Figure 5. Logic gates annotated with all possible transition times, so that sleep resistors can be merged.

This merging technique based on mutual exclusive gate discharge patterns is most effective for balanced circuits with minimal glitching. Fortunately, a large class of circuits fall into this category, especially since less glitching is attractive from a low power point of view [6]. For circuits with more complicated interconnections and glitching, the merging technique can still be used, although the compression ratio would probably be lower. To further improve the sleep transistor reduction, we can also use more rigorous criteria to determine mutual exclusivity that is based on logic rather than the structural connections in a circuit. We are currently working on such an approach that utilizes boolean manipulation[7].

4.1 Hierarchical Transistor Sizing Methodology

Although the MTCMOS transistor sizing algorithm has been presented at the gate level, in fact it can be applied at many hierarchical levels of a circuit. The algorithm simply operates on generic circuit blocks that are elements within a larger module, and each block is assumed to have a local high V_t sleep transistor that is used for gating the power supply rails. The algorithm is applied to the network by combining the sleep transistors for mutual exclusive blocks. Thus, the blocks that the algorithm operates on can represent individual gates, cells within an array (like an adder cell in a multiplier), or even a module within a chip (like an ALU). In all these cases, a gating sleep transistor can be shared among several different blocks if those blocks have activity patterns that do not overlap in time.

In order to achieve the best results, one should initially use a detailed simulator like SPICE to simulate as large a block as possible and to exhaustively determine the optimal sleep transistor size. Next, the hierarchical merging technique can then be applied to these existing blocks to synthesize an overall sleep transistor for a larger module, where determining a worst case input vector would have been exceedingly difficult. Applying this hierarchical methodology too early can result in unnecessary overestimates for sleep transistor sizes however.

Using the hierarchical sizing methodology again, it is also possible to further apply this transistor merging technique on these existing modules into a larger system. However, by applying this nested algorithm at several levels of abstraction, we will tend to overestimate the minimum sleep transistor size required again, mainly because the granularity of our interactions between blocks will be much larger. For example, applying the algorithm at the cell level within an array might give a larger estimate for the sleep transistor size than if the algorithm had been pushed down in the hierarchy and applied to the gates directly. However, utilizing a hierarchical approach to sizing the sleep transistors is very attractive because detailed circuit complexity can be abstracted away at the expense of accuracy, a tradeoff which is very often desirable.

5. PARITY CHECKER EXAMPLE

As a practical example, the hierarchical sizing methodology was applied to a 32 bit MTCMOS parity checker circuit. The circuit consists of 31 XOR gates which are connected as a tree with 5 levels. Figure 7 shows a smaller 8 bit version of this circuit.

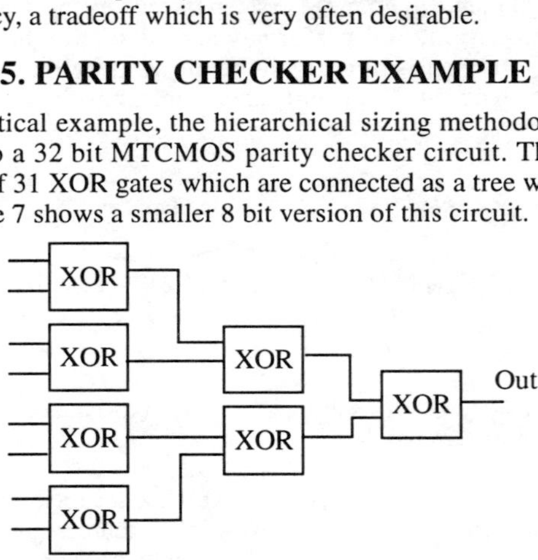

Figure 6. 8 bit parity checker.

The XOR gate was simulated by itself to determine the local sleep resistance needed for a single gate to meet performance requirements. For 20% degradation, the sleep transistor needs a resistance less than 4800Ω, and for 10% degradation the resistance must be less than 2400Ω. With an application of the merging algorithm based on mutual exclusive discharging gates, the total number of sleep transistors required could be reduced from 31 (one for each gate) to only 16. The resulting sleep transistor for the entire 32 bit parity checker was then calculated to be less than 300Ω for 20% degradation and less than 150Ω for 10% degradation.

Since there are too many vector pairs (2^{64}) to test exhaustively, Table 2 below shows simulation results for a subset of 5 input vectors. Each of these vectors was chosen to exercise a critical path through the top row of the parity checker. Furthermore the critical 2-input XOR gates each transition with the worst case inputs (x=0->1,y=0->0).

Input Vector	CMOS [ns]	R =150Ω [ns], %degr	R = 300Ω [ns], %degr
1	9.08ns	9.14ns, 0.7%	9.21ns, 1.4%
2	9.07ns	9.34ns, 3.0%	9.60ns, 5.5%
3	9.07ns	9.46ns, 4.3%	9.87ns, 8.8%
4	9.08ns	9.44ns, 4.0%	9.81ns, 8.0%
5	9.08ns	9.34ns, 2.9%	9.60ns, 5.7%

Table 2. Parity generator performance as function of sleep transistor width for different input vectors.

The SPICE simulation shows how the sleep transistor sizes (150Ω and 300Ω) ensure performance within 10% (9.99ns) and 20% (10.90ns) of the CMOS critical delay of 9.08ns. Vector #1 does not cause large currents to flow in adjacent gates, so its degradation in performance is not large (0.7% and 1.4%). However, vector #3 creates significant currents through adjacent gates, and as a result is more susceptible to degradation (4.3% and 8.8%). In all cases however, the delays are significantly faster than predicted. Although there are other vector combinations that will result in larger delays, typically the sleep transistor sizing from the algorithm will still be a conservative overestimate of the required sleep transistor size. This is due mainly to three factors. First, only one half of the gates, those switching from high to low, are actually degraded as described in section 3.4. As a result, ensuring all NMOS transistors degrade by no more than 20%, will likely cause only a 10% degradation in overall performance. Second, our gate partitioning can be further improved by using more sophisticated algorithms to determine mutual exclusivity, as only a structural logic independent grouping algorithm was used. Finally the requirement that each gate's high to low transition degrade by no more than a fixed amount is overly stringent, and also contributes to a conservative estimate of sleep transistor size.

6. WALLACE TREE MULTIPLIER EXAMPLE

As another example, the hierarchical sizing methodology was applied to a 6x6 bit Wallace tree multiplier circuit shown in Figure 7. This is a type of circuit is well suited for this algorithm because there are many mutually exclusive gates that cannot transition at the same time[8].

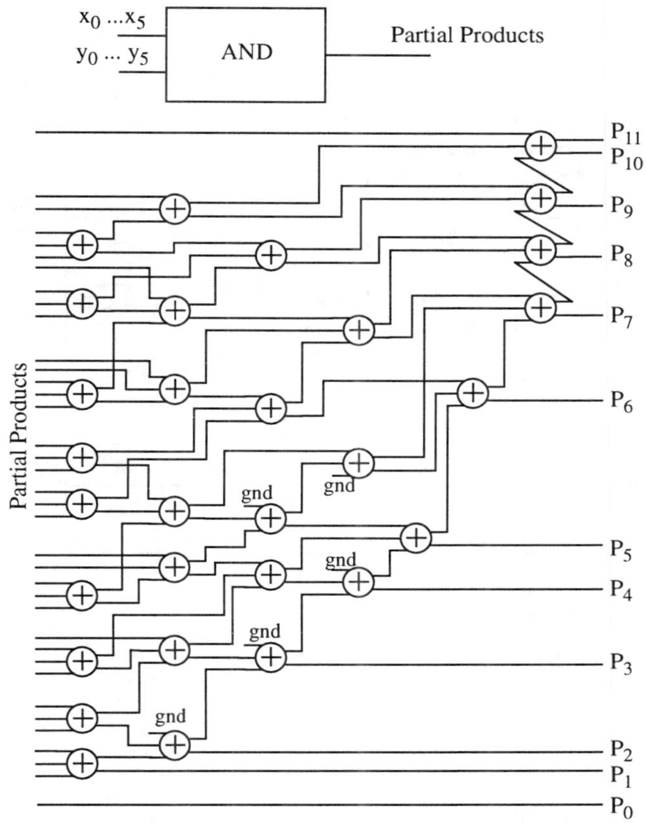

Figure 7. 6x6 Wallace Multiplier.

Initially, the AND gates and the carry save adder units (with representative loadings) were simulated in SPICE to determine optimal high V_t sleep transistor sizes (actually equivalent resistances) for each unit to give rise to a fixed degradation in performance. To achieve a degradation of 20% and 10%, the CSA required sleep transistors with equivalent resistances of 1600Ω and 800Ω, respectively. Likewise 20% and 10% degradation of the AND gates requires equivalent resistances of 2700Ω and 1350Ω, respectively.

Next, the sleep transistor reduction and merging steps were performed to give rise to an equivalent resistor that could gate the entire multiplier. By tabulating all possible time periods that each cell can transition as described in Figure 5, we were able to reduce the 36 AND cell and 30 adder cell sleep resistors into 21 AND cell and 15 adder cell sleep resistors. The total equivalent resistance for the multiplier could then be written as $(R_{add}/15) \,//\, (R_{and}/21)$, corresponding to 30Ω and 60Ω for 10% and 20% maximum degradation. The merged resistance is a factor of two greater than the case where no merging takes place, which would correspond to a factor of two decrease in sleep transistor sizing. The branches of this wallace tree structure were not completely balanced because adder cells at inner levels of the tree could actually receive inputs from two levels before. As a result, this implementation has fewer mutual exclusive gates because a fair amount of glitching can occur. Another implementation that balances the paths more carefully would result in larger compression results from the merging algorithm.

For a 6x6 bit multiplier, there are 2^{24} possible input vector pairs, so again it was not possible to exhaustively verify the circuit. However 6 representative vectors were simulated for output nodes P6 and P1 as shown in Table 3.

Input Vector	CMOS [ns]	R =30Ω [ns], %degr	R = 60Ω [ns], %degr
1	8.79ns	9.01ns, 2.6%	9.26ns, 5.4%
2	8.46ns	8.87ns, 4.9%	9.16ns, 8.3%
3	8.72ns	8.92ns, 2.2%	9.111ns, 4.4%
4	11.17ns	11.28ns, 1.0%	11.40ns, 2.1%
5	12.31ns	12.43ns, 1.0%	12.76ns, 3.7%
6	6.55ns	6.79ns, 3.6%	7.07ns, 8.0%

(a) P_6 Delay

Input Vector	CMOS [ns]	R =30Ω [ns], %degr	R = 60Ω [ns], %degr
1	3.19ns	3.24ns, 1.9%	3.40ns, 6.8%
2	3.19ns	3.25ns, 2.1%	3.41ns, 6.9%
3	2.98ns	3.09ns, 3.7%	3.17ns, 6.8%
4	2.98ns	3.05ns, 2.3%	3.10ns, 3.8%
5	2.98ns	3.12ns, 4.8%	3.21ns, 7.8%
6	3.16ns	3.31ns, 5.1%	3.46ns, 10.1%

(b) P_1 Delay

Table 3. Degradation of delays (P_6 and P_1) in multiplier circuit for various sleep resistances and vectors.

By using the hierarchical sizing algorithm, the degradation in any path within the multiplier should degrade by no more than the nominal amount (10% and 20%). As shown in Table 3, two very different paths, ending at P6 (which can include 6 cells), and P2 (which always includes 2 cells), are ensured to meet these performance requirements. Again, since only an NMOS sleep transistor was used, typically only 1/2 the transitions will actually be affected and total degradations will be limited to near 5% and 10%. Also, as described earlier, the restriction that all paths meet the same performance constraint will yield an overly conservative estimate of sleep transistor sizing. For example, an inherently slow path like P2 should be allowed to degrade more since it will unlikely be the worst case delay. By relaxing the degradation requirements for non critical gates, then the sleep transistor can be reduced in size. Nonetheless, the hierarchical sizing strategy provides at least an upper bound on the size of the sleep transistor needed to ensure performance.

7. CONCLUSION AND FUTURE WORK

A sleep transistor sizing methodology for MTCMOS circuits based on mutual exclusive gate discharge patterns was presented in this paper. This methodology provides an upper bound on the sleep transistor size to guarantee a performance level in an MTCMOS circuit by placing delay constraints on individual blocks. Although this sizing algorithm will give an overestimate compared to the optimal sleep transistor size, it is straightforward and can be applied at many hierarchical levels within a circuit. This algorithm is most useful when applied to a large module in conjunction with a detailed simulator that can provide more accurate sleep transistor sizing information for the module's sub-blocks.

The sleep transistor merging algorithm currently relies on a unit delay model and purely structural analysis to determine mutual exclusivity. By using a more complicated delay model and utilizing logic dependencies, the merging algorithm can be improved and more accurate sleep transistor size requirements can be computed. Work is also being done to reduce the constraint that all individual gates meet a fixed degradation requirement. By allowing more flexibility where the speed degradation of individual gates within the circuit can be greater or less than the nominal, one can more efficiently size the sleep transistor to maintain performance in MTCMOS circuits.

8. ACKNOWLEDGEMENTS

This work was funded by DARPA contract #DABT63-95-C-0088.

9. REFERENCES

[1] S. Mutoh, T. Douseki, Y. Matsuya, T. Aoki, S. Shigematsu, J. Yamada, "1-V Power Supply High-Speed Digital Circuit Technology with Multithreshold-Voltage CMOS," IEEE JSSC, vol. 30, no. 8, pp. 847-854, August 1995.

[2] S. Mutoh, S. Shigematsu, Y. Matsuya, H. Fukada, J. Yamada, "1V Multi-Threshold CMOS DSP with an Efficient Power Management Technique for Mobile Phone Application", IEEE ISSCC, pp. 168-169, 1996.

[3] W. Lee, et al., "A 1V DSP for Wireless Communications," ISSCC, pp. 92-93, Feb., 1997.

[4] A. Chandrakasan, I. Yang, C. Vieri, D. Antoniadis, "Design Considerations and Tools for Low-voltage Digital System Design," 33rd Design Automation Conference, pp. 113-118, June 1996.

[5] J. Kao, A. Chandrakasan, D. Antoniadis, "Transistor Sizing Issues and Tool For Multi-Threshold CMOS Technology," 34th Design Automation Conference, pp. 409-414, June 1997.

[6] T. Sakuta, W. Lee, P. Balsara, "Delay Balanced Multipliers for Low Power/ Low Voltage DSP Core," IEEE Symposium on Low Power Electronics, pp. 36-37, 1995.

[7] S. Devadas, K. Keutzer, J. White, "Estimation of Power Dissipation in CMOS Combinational Circuits Using Boolean Function Manipulation," IEEE JSSC, vol. 11, no. 3, pp. 373-383, March 1992.

[8] N. Weste, K. Eshraghian, "Principles of CMOS VLSI Design," Addison-Wesley, Reading MA., pp. 554-557, 1993.

PANEL: TECHNICAL CHALLENGES OF IP AND SYSTEM-ON-CHIP: THE ASIC VENDOR PERSPECTIVE

Chair: A. Richard Newton – University of California, Berkeley, CA

Organizers: Andrew Graham – Silicon Integration Initiative, Austin, TX
Andrew B. Kahng – University of California, Los Angeles, CA

Panelists: Bruce Beers – IBM, Essex Junction, VT
Jeffery Hilbert – LSI Logic Corporation, Milpitas, CA
Anand Naidu – Sand Microelectronics, Santa Clara, CA
Bob Payne – VLSI Technology, San Jose, CA
L.J. Reed – Motorola SPS, Austin, TX
Mark Stibitz – Lucent Technologies, Allentown, PA
Hitoshi Yoshizawa – NEC Corp., Kawasaki, Japan

SUMMARY

The vision of easily accessible IP that can be quickly integrated on silicon as "virtual components" is a compelling one, with deep implications for reuse methodology and EDA technology. Activities of the VSI Alliance, starting nearly two years ago, have fueled interest in IP and raised market expectations for value and reusability to very high levels. Indeed, the sheer number of new IP companies, in combination with third-party ASIC libraries and EDA tools offerings, suggests that the era of plug and play IP has arrived. Today, the claimed independence of IP from the underlying silicon has led to bold market claims for foundry manufactured system chips using internet sourced IP with Lego block, snap-together integration simplicity. Traditional roles of foundries, ASIC suppliers, and EDA vendors are blurring. To the end customer, the situation presents both opportunities and risks. Are we on the cusp of a new era, or a rude awakening?

From the perspective of leading ASIC vendors, the evolution of IP-based systems on silicon, along with the means and methods of producing them, are not new. Full use of rapidly-increasing raw ASIC gate counts – which is synonymous with systems on silicon – has been a driving force in the ASIC industry for many years. As developers of high-value IP and systems architectures, the leading ASIC companies have an established record of experience with design and system chip level integration. This experience – e.g., with system-level analysis, integration of cores designed to common bus-structures, etc. – can be leveraged toward practical realization of an IP vision that offers the silicon consumer more choice without unacceptable implementation risks. On the other hand, from the perspective of a pure IP integrator, IP is poised to change the way customers and ASIC vendors themselves produce systems-on-silicon, just as vendor-specific tools evolved to commercial CAE starting a decade ago. Early IP success stories are pointing the way to a new, more specialized business model where IP, tools, services, and silicon foundry are brought together based on the specific needs of a given project.

This panel will address technical challenges presented by the systems-chip opportunity, as well as practical expectations and key missing pieces of industry infrastructure. Such pieces include: customer expectations, EDA technology, standards, legal barriers and associated risks facing ASIC suppliers and EDA vendors, challenges of incorporating 3rd-party IP, and practical reuse methodologies. In this forum, a noteworthy challenge – which has remained constant throughout the evolution to deep-submicron – is the ability for customers to develop their systems when faced with a widening gap between process technology and commercial EDA tools. As silicon suppliers continue to focus on providing the industry with advanced technologies and products, EDA suppliers must also increase their focus on providing tools and infrastructure. Another example challenge, faced by ASIC vendors, lies in supplying the high-level core models required for emerging cycle-based simulators, and supporting the use of cores in emulators and accelerators, when no standards for model creation or protection exist. Standard model interfaces such as those that exist for event-based simulators must be developed to support cycle simulators, emulators, accelerators as well as hardware-software co-verification tools. The panel will also examine essential factors to consider in the use of IP, based on their experience with the latest silicon process technologies. Finally, the ASIC vendor participants will showcase their joint efforts to build an effective IP infrastructure for the industry.

Software Synthesis of Process-Based Concurrent Programs

Bill Lin
Electrical and Computer Engineering Department, University of California, San Diego, La Jolla, California, 92093-0407

Abstract— **We present a Petri net theoretic approach to the software synthesis problem that can synthesize ordinary C programs from process-based concurrent specifications without the need for a run-time multi-threading environment. The synthesized C programs can be readily retargeted to different processors using available optimizing C compilers. Our compiler can also generate sequential Java programs as output, which can also be readily mapped to a target processor without the need for a multi-threading environment. Initial results demonstrate significant potentials for improvements over current run-time solutions.**

I. Introduction

SOFTWARE is playing an increasingly important role in embedded systems. While high-level language compilers exist for implementing sequential programs on embedded processors [17]. e.g. starting from C [10], many embedded software applications are more naturally expressed as concurrent programs, specified in terms of communicating processes. This is because typically actual system applications are composed of multiple tasks.

Currently, the most widely deployed solution is to use an embedded operating system to manage the run-time scheduling of processes and to handle the inter-process communication. However, this solution tends to add significant overhead in program size, run-time memory requirements, and execution time.

Several alternative high-level approaches have been proposed. Static data-flow solutions [2], successfully used to design DSP-oriented systems, achieve compile-time scheduling at the expense of disallowing conditional and non-deterministic execution. Other researchers have considered hybrid approaches [7], [18] that generate application-specific run-time schedulers to handle the multi-tasking of conditional and non-deterministic computations. Reactive approaches, e.g. Esterel [1], rely on a strong synchrony hypothesis that makes two fundamental assumptions: the existence of a global clock abstraction to discretize computation over instances, and computation takes no time within each instance. This hypothesis is difficult to satisfy for distributed implementations and may not match naturally to many applications from a specification standpoint.

In contrast, our work is based on a model of *asynchrony* where the concurrent parts can evolve independently and only synchronize where specified. Recently, we introduced a new Petri net theoretic software synthesis method based on a new Petri net theoretic technique that can synthesize efficient embedded software implementations from asynchronous process-based specifications without the need for a run-time scheduler [11]. This approach has been implemented in a system under development called *Picasso*. In this paper, we further develop on our approach. We also briefly describe a new Java-based implementation, in addition to our earlier C-based implementation.

The remainder of this paper is organized as follows: Our specification model is introduced in Section II. The Petri net intermediate representation is introduced in Section III. The software synthesis method is detailed in Section IV. Initial results from an encryption example are presented in Section V.

II. Specification

Our programs are hierarchically composed of processes that communicate through synchronizing channels. The semantics is based on the CSP formalism [9], but the syntax is similar to C. Consider the following simple example composed of two processes called `ping` and `pong`.

```
1 ping (input chan(int) a, output chan(int) b) {
2   int x;
3   for (;;) {
4       x = <-a; /* receive */
5       if(x < 100) x = 10 - x;
6       else x = 10 + x;
7       b <-= x; /* send */
8 } }

9 pong (input chan(int) c, output chan(int) d) {
10  int y, z = 0;
11  for (;;) {
12      d <-= 10;/* send */
13      y = <-c; /* receive */
14      z = (z + y) % 345; /* send */
15 }}

16 system ( ) {
17   chan(int) c1, c2;
18   par {
19       ping (c2, c1);
20       pong (c1, c2);
21 } }
```

Channels are declared using the `chan` statement, as exemplified in `Line 1`. The unary receive operator, `<-`, receives data on the channel specified as its right operand. The received value may then be manipulated by other operators, e.g. it can be assigned to a variable, as exemplified in `Line 4`. The send operator, `<-=`, transmits the result of the expression provided as its right operand on the channel specified as its left operand, as exemplified in `Line 7`. Basic control-flow constructs, like `if-then-else`, `for-loops`, and `while-loops`, and basic arithmetic and relational operators, like `+`, `-`, `*`, `%`, and `>`, `>=`, `==`, `!=`, are the same as in C. There is also an `alt` construct [9], not used here, that provides a mechanism for non-deterministic execution. Finally, processes can be hierarchically composed to form larger systems, as exemplified by the process `system`. The `par` statement executes the statements in its body in parallel and joins the threads of execution at the end by waiting for all processes to terminate before proceeding.

III. Intermediate Representation

A. Petri nets

Let $G = \langle P, T, F, m_0 \rangle$ be a Petri net [14], where P is a set of places, T is a set of transitions, $F \subseteq (P \times T) \cup (T \times P)$ is the flow relation, and $m_0 : P \rightarrow N$ is the initial marking, where N is the set of natural numbers. The symbols $\bullet t$ and $t\bullet$ define, respectively, the set of input places and the set of output places of transition t. Similarly, $\bullet p$ and $p\bullet$ define, respectively, the set of input transitions and the set of output transitions of place p. A place p is called a *conflict place* if it has more than one output transition. Two transitions, t_i and t_j are said to be in *conflict* if and only if $\bullet t_i \cap \bullet t_j \neq \emptyset$. A state, or marking, $m : P \rightarrow N$, is an assignment of a non-negative number to each place. $m(p)$ denotes the number of tokens in the place p. A transition t can fire at marking m_1 if all its input places contain at least one token. The firing of t removes one token from each of its input places and adds a new token to each of its output places, leading to a new marking m_2. This firing is denoted by $m_1 \xrightarrow{t} m_2$. Given a Petri net G, the reachability set of G is the set of all markings reachable in G from the initial marking m_0 via the reflexive transitive closure of the above firing relation. The corresponding graphical representation is called a reachability graph. A Petri net G is said to be *live* if $\forall t \in T$, $\exists m$ reachable from the initial marking m_0 such that t is enabled. It is said to be *safe* if in every reachable marking, there is at most one token in any place. In this case, we can simply represent each marking $m : P \rightarrow \{0,1\}$ as a binary assignment.

B. Intermediate construction

In [4], [19], a process algebra was developed for constructing a Petri net model from a program of communication processes. Consider again the example presented in Section II. The derived Petri net models for processes `ping` and `pong` are shown in Fig. 1(a) and Fig. 1(b), respectively, along with their initial markings.

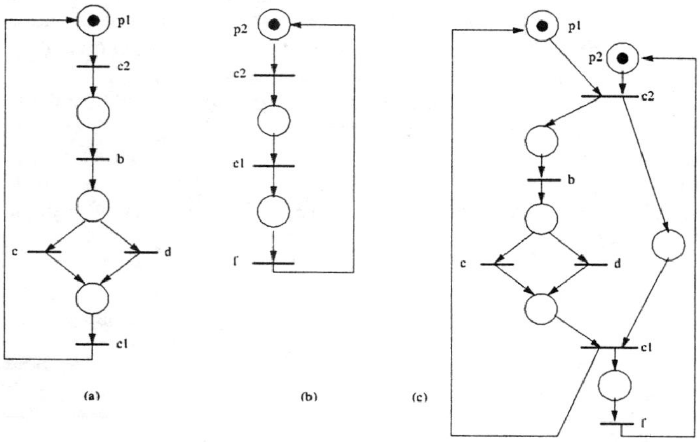

	(a)		(b)		(c)
c2:	x=<-a	c2:	d<-=10	c2:	x=10
b:	(x<0)	c1:	y=<-c	b:	(x<0)
c:	x=10-x	f:	z=(z+y)%345	c:	x=10-x
d:	x=10+x			d:	x=10+x
c1:	b<-=x			c1:	y=x;
				f:	z=(z+y)%345

Fig. 1. (a) `ping` (b) `pong` (c) `system = ping || pong`

Concurrent processes can be composed via *parallel composition*. In parallel composition, communication actions in fact form *synchronization points* and are joined together at their common transitions. This is illustrated in Fig. 1(c).

IV. Software Synthesis Method

A. Classes of Petri nets

A *Marked Graph (MG)* is a net $G = \langle P, T, F, m_0 \rangle$ such that $\forall p \in P : |\bullet p| = 1 = |p \bullet|$. MGs cannot model conficts.

A *State Machine (SM)* is a net $G = \langle P, T, F, m_0 \rangle$ such that $\forall t \in T : |\bullet t| = 1 = |t \bullet|$. SMs cannot model concurrency.

A *Free-Choice Net (FC-net)* is a net $G = \langle P, T, F, m_0 \rangle$ such that $\forall t_1 t_2 \in T, t_1 \neq t_2 : \bullet t_1 \cap \bullet t_2 \neq \emptyset \Rightarrow |\bullet t_1| = 1 = |\bullet t_2|$, or $\forall p_1 p_2 \in P, p_1 \neq p_2 : p_1 \bullet \cap p_2 \bullet \neq \emptyset \Rightarrow |p_1 \bullet| = 1 = |p_2 \bullet|$. Every MG and SM is a FC-net. For FC-nets, all conflicts can be decided *locally*.

Let G' be a subset of a net G generated by a *non-empty* set $X \subseteq P \cup T$. G' is a *MG-Component* of G if $\bullet t \cup t \bullet \subseteq X$ for every $t \in X$, and G' is a strongly connected MG.

Let G' be a subset of a net G generated by a *non-empty* set $X \subseteq P \cup T$. G' is a *SM-Component* of G if $\bullet p \cup p \bullet \subseteq X$ for every $p \in X$, and G' is a strongly connected SM.

G is said to be *covered* by a set of MG-Components if every transition of the net belongs to some MG-Component. G is said to be *covered* by a set of SM-Components if every place of the net belongs to some SM-Component. Hack [8] proved that a live safe FC-net can always be *covered* by a set of MG-Components or a set of SM-Components.

B. Expansions

Definition IV.1 (Expansion) An *expansion* is an acyclic Petri net with the following properties:

- There is one or more places without input transitions.
- There is one or more places without output transitions.
- There are no transitions without at least one input place or one output place.

The places without input transitions are called *initial places*. The places without output transitions are called *cut-off places*.

Definition IV.2 (Maximal expansion) Let G be a Petri net and let m be a marking of G. The *maximal expansion* of G with respect to m, E, is an acyclic Petri net with the following properties:

- The initial places correspond to m.
- The cut-off places correspond to the set of places encountered when a cycle has been reached.
- E is transitively closed: for each $t \in E$ or $p \in E$, all preceding places and transitions reachable from m are also in E.

m is referred to as the *initial marking*.

Intuitively, the maximal expansion of G with respect to a marking m corresponds to the largest *unrolling* of G from m before a cycle has been encountered. Consider the example shown in Fig. 2(a). The corresponding maximal expansion with $m = \langle p1, p2 \rangle$ is shown in Fig. 2(b).

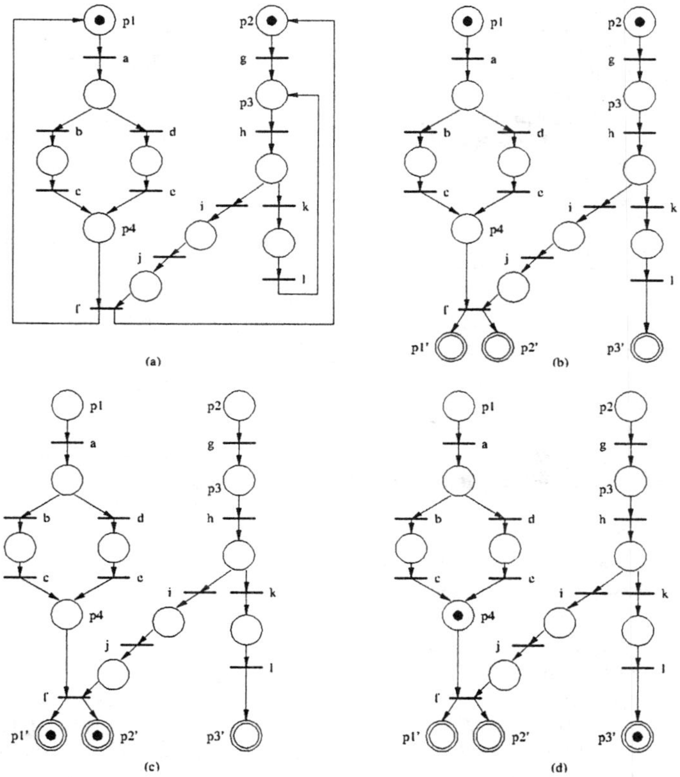

Fig. 2. (a) Petri net example. (b) Its maximal expansion. (c) A cut-off marking. (d) Another cut-off marking.

Definition IV.3 (Cut-off markings) Let G be a Petri net, and let E be a maximal expansion of G with respect to the initial marking m. A marking m_c is said to be a *cut-off marking* if it is reachable from m and no transitions are enabled to fire. The set of cut-off markings is denoted by $CM(E)$.

For the example shown in Fig. 2, there are two possible cut-off markings $m_{c_1} = \langle p1', p2' \rangle$ and $m_{c_2} = \langle p3', p4 \rangle$, shown respectively in Fig. 2(c) and Fig. 2(d).

Our synthesis procedure works by generating code from a maximal expansion segment E obtained by using the initial marking m_0 as the initial marking for the expansion. Then from each cut-off marking

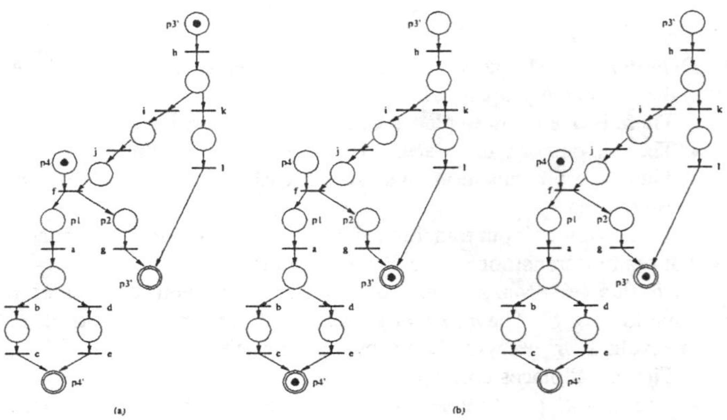

Fig. 3. (a) Maximal expansion. (b) Cut-off marking.

$m_{c_i} \in CM(E)$, a new maximal expansion segment E_i is generated using m_{c_i} as the initial marking. This iteration terminates when all cut-off markings have already been visited. The pseudo-code for the overall algorithm is shown below.

```
soft-synt (G, m0) {
   EM = {m0};
   push (m0);
   while ((m = pop()) ≠ ∅) {
      E = maximal-expansion (G, m);
      static-scheduling (E, m);
      foreach mc ∈ CM(E) {
         if mc ∉ EM {
            EM = EM ∪ mc;
            push (mc);
} } } }
```

The `static-scheduling` step is applied to each expansion segment to produce the actual code.

In the example shown in Fig. 2, only two expansion segments are needed. From the initial marking $m = \langle p1, p2\rangle$, the only cut-off markings reachable are $m_c = \langle p1, p2\rangle$ and $m_c = \langle p3, p4\rangle$. However, from $m = \langle p3, p4\rangle$, the only cut-off marking reachable is $m_c = \langle p3, p4\rangle$ itself, as shown in Fig. 3.

However, in the example shown in in Fig. 1, only *one* expansion segment is needed since the only cut-off marking reachable from the initial marking is the initial marking itself (i.e. $m = \langle p1, p2\rangle$)[1].

C. Properties

The expansion procedure described in Section IV-B is guaranteed to converge since the number of possible markings in a Petri net is finite. Hence, the number of expansions or iterations is also finite. Typically, very few expansions are required.

For certain classes of Petri nets, the convergence property is even stronger. In the case of a strongly connected live safe MG, the number of expansions is exactly one. This is because in the case of a strongly connected live safe MG, the initial marking m_0 forms a minimal feedback arc set. The number of tokens along any directed cycle in the MG in the initial marking is exactly one. Thus, according to Definition IV.2, the maximal expansion of a MG G with respect to its initial marking m_0 is exactly defined as the acyclic Petri net E where both the initial places and the cut-off places correspond exactly to the places marked by m_0. Thus, the set of cut-off markings for E contains only the initial marking m_0.

In the case of a strongly connected live safe FC-net G that can covered by a set of strongly connected live safe MG components $G_1 \ldots G_n$ such that the initial marking m_0 of G restricted to G_i is also a live safe initial marking for the MG component G_i, the number of expansions is also exactly one. The argument follows a similar line as the argument for the MG case. That is, the initial marking m_0 corresponds to both the initial places and cut-off places if we maximally expand G with respect to m_0. Thus, convergence is guaranteed after one expansion since the set of cut-off markings contains only m_0.

D. Static scheduling

Give an expansion segment E, represented as an acyclic Petri net fragment, our software synthesis method performs a *static scheduling* of the operations[2] in that segment. During scheduling, a *step* assigned to every operation in E. More formally, static scheduling is defined as follows:

Definition IV.4 (Static scheduling) Let E be an expansion segment. t_i is said to *precede* t_j in E, denoted as $t_i \prec t_j$, if there is a directed path from t_i to t_j. Let $\pi : T \rightarrow N$, be a *schedule function* that assigns a non-negative integer $\pi(t) \in N$ to every $t \in E$. A schedule is said to be *valid* iff it satisfies the following condition: $\forall t_i, t_j \in E$, if $t_i \prec t_j$, then $\pi(t_i) < \pi(t_j)$.

To illustrate this process, consider the expansion segment shown in Fig. 4(a) (corresponding to the example depicted in Fig. 4). Two valid schedules are shown in Fig. 4(b) and Fig. 4(c). It is not the intention of this paper to discuss in details the different possible scheduling heuristics. The interested reader can refer to [3], [5] for a survey of example techniques.

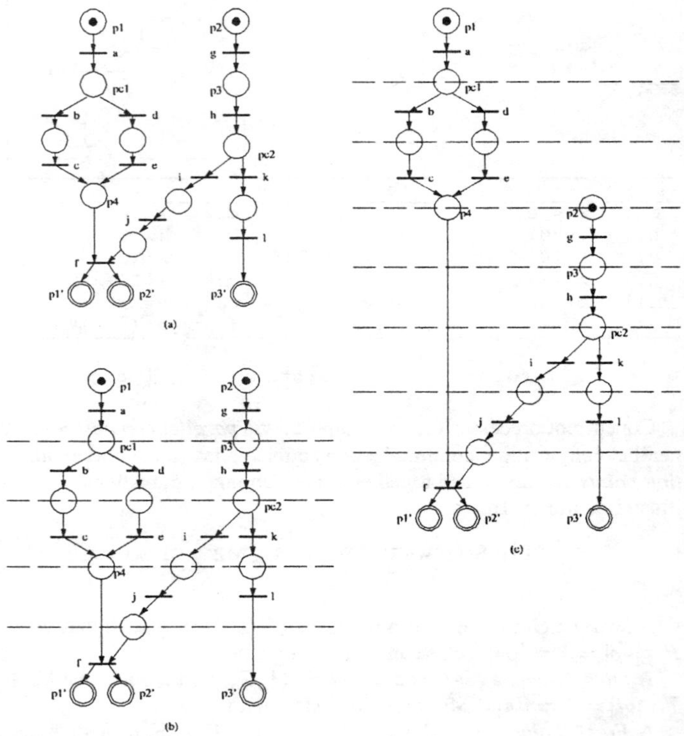

Fig. 4. (a) An expansion segment. (b) A valid static schedule. (c) Another valid static schedule.

Given a schedule π, a control-flow-graph fragment is constructed. In contrast to the traditional scheduling problem, where typically only *data-flow blocks* are considered, the control-flow-graph mapping step is much less straightforward. This is because we can have complex concurrent conditionals where the *firing* of a *transition* is dependent on the concurrent conflow flow and must obey Petri net firing rules. Essentially, the control-flow-graph generation step is based on a traveral of E, but we modify the Petri net firing rules so that we proceed in accordance to the levels defined by π. For example, the schedule shown in Fig. 4(b) will result in the control-flow-graph fragment depicted in

[1]Here, we do not distinguish between p_i and p_i' because they simply denote different instances of the same place.

[2]Previously called pre-ordering [11].

Fig. 5(a). Similarly, Fig. 5(b) shows the resulting control-flow-graph for the schedule shown in Fig. 4(c).

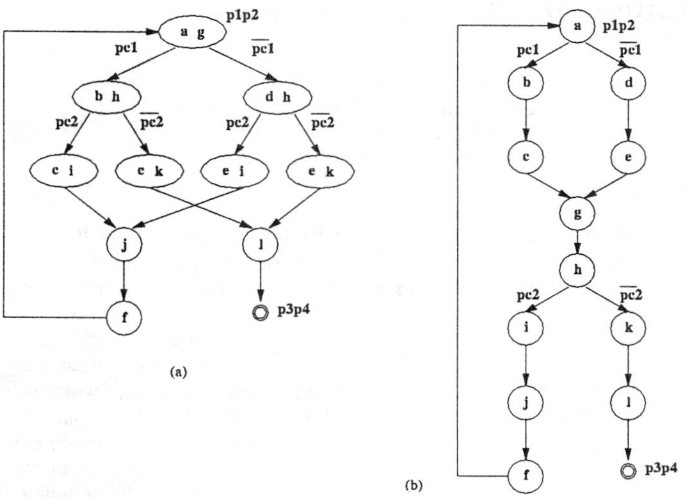

Fig. 5. (a) Control-flow-graph fragment. (b) Another control-flow-graph fragment.

E. Code generation

Once the overall control-flow-graph has been generated, it can be syntactically translated into plain C or sequential Java code for implementation. This last code generation step can leverage upon well-studied standard code optimization techniques [17].

V. IMPLEMENTATION AND RESULTS

A. Implementation

The synthesis method presented in this paper has been implemented in a system called *Picasso*. The compiler is implemented as a pre-processor that generates either plain C [10] or Java [6]. Both solutions are highly portable.

In the case of the C output, any available optimizing C compiler can be used to produce the target machine code. For comparisons, we have also implemented a multi-threading approach that uses the Solaris thread library to implement the processes. In principle, any real-time multi-threading packages may be used.

In the case of the Java output, the Java produced by our synthesis method is *sequential* in that it does not make use of any multi-threading feature in Java. Thus, a much lighter weight Java Virtual Machine without multi-threading support may be used. Also, any Just-In-Time compiler or Java-To-C translator (e.g. [12], [15]) can be used to produce native executables, again without the need for multi-threading support. For comparisons, we have also implemented a multi-threading approach in Java by mapping processes to Java threads.

B. Results

We have applied our C implementation to an example derived from the RC5 encryption algorithm. RC5 is widely used by RSA Data Security in a range of Internet security products [16]. A novel feature of RC5 is the heavy use of data-dependent rotations. The top-level view of the example is shown in Fig. 6. It consists of two encryption/decryption chains that are merged together by a monitor process.

We chose this example because it contains both data-dependent loops as well as non-deterministic choices. Table I compares the results of our method with a multi-threading library approach. The table compares the execution times of both approaches on different size input streams. The first row corresponds to a 40K bytes input file, the second row corresponds to a 400K byte input file, and the third row corresponds to a 4M byte input file. The CPU-times are reported in seconds

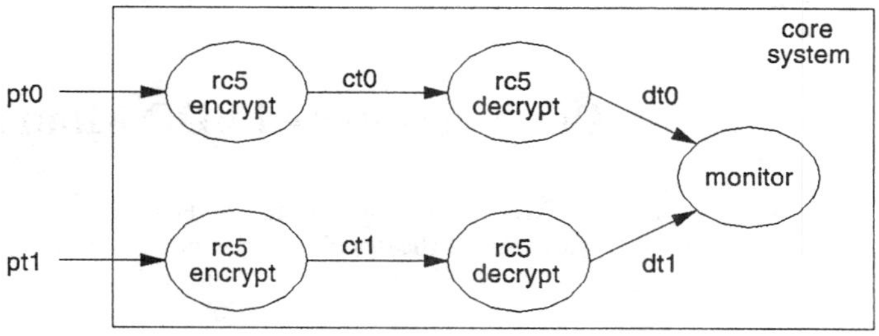

Fig. 6. RC5 encryption chain example.

size	threads	synthesis
40K	2.6	0.04
400K	26.0	0.33
4M	256.6	3.30
rate	15.4KB/s	1.21MB/s

TABLE I

COMPARATIVE RESULTS ON A SUN/SOLARIS ULTRA-1.

on a Sun Ultra-1 workstation running Solaris. The row labeled "rate" summarizes the execution of the two solutions in terms of bytes per second. Comparing CPU-times, the Solaris thread based implementation is significantly slower than our software code synthesis approach, due to the significant overhead introduced by Solaris threads.

REFERENCES

[1] G. Berry et al. "The synchronous approach to reactive and real-time systems", *IEEE Proceedings*, 1991.

[2] J. T. Buck et al. "Ptolemy: A framework for simulating and prototyping heterogeneous systems", *International Journal on Computer Simulation*, January 1994.

[3] R. Camposano and W. Wolf (editors), *Trends in High-Level Synthesis*, Kluwer Academic Publishers, 1993.

[4] G. de Jong, B. Lin. "A communicating Petri net model for the design of concurrent asynchronous modules", *ACM/IEEE Design Automation Conference*, 1994.

[5] H. De Man, F. Catthoor, G. Goossens, J. Vanhoof, J. Van Meerbergen, S. Note, J.A. Huisken, "Architecture-driven synthesis techniques for VLSI implementation of DSP algorithms", *Proceedings of IEEE*, vol.72, no.2, pp.319-335, February 1990.

[6] J. Gosling, B. Joy, and G. Steele. *The Java Language Specification*, Addison-Wesley, 1996.

[7] R. K. Gupta. "Hardware-software cosynthesis of microcontrollers", *Proc. Codes/CASHE*, 1996.

[8] M. Hack. *Analysis of production schemata by Petri nets*. M.S. Thesis, MIT, February 1972.

[9] C. A. R. Hoare. *Communicating Sequential Processes*. Prentice-Hall, 1985.

[10] B. W. Kernighan, D. M. Ritchie. *The C Programming Language*, Prentice-Hall, Englewood Cliffs, New Jersey, 1978.

[11] B. Lin. "Efficient compilation of process-based concurrent programs without run-time scheduling", *Proc. of DATE'98*, February 1998.

[12] B. Morgan. *Visual J++ Unleashed*. Sams.Net, 1996.

[13] J. Morse, S. Hargrave. "The increasing importance of software", *Electronic Design*, vol. 44, no. 1, Jan. 1996.

[14] J.L. Peterson. *Petri net Theory and Modeling of Systems*, Prentice Hall, 1981.

[15] T. A. Proebsing, G. Townsend, P. Bridges, J. H. Hartman, T. Newshan, S. A. Watterson. "Toba: Java for applications, a way ahead of time (WAT) compiler", *ftp://ftp.cs.arizona.edu/sumatra/report/toba.pdf*.

[16] R. L. Rivest. "The RC5 Encryption Algorithm", *Proceedings of the 1994 Leuven Workshop on Algorithms*, Springer-Verslag, 1994.

[17] R. M. Stallman, *Using and porting GNU CC*, Free Software Foundation, June 1993.

[18] F. Thoen et al. "Real-time multi-tasking in software synthesis for information processing systems", *Proc. of ISSS'95*, 1995.

[19] S. Vercauteren et al. "Derivation of formal representations from process-based specification and implementation models", *Proc. of ISSS'97*, September 1997.

Don't care-based BDD Minimization for Embedded Software

Youpyo Hong Peter A. Beerel
Department of Electrical Engineering - Systems
University of Southern California
Los Angeles, CA 90089

Luciano Lavagno Ellen M. Sentovich
Cadence Berkeley Laboratories
2001 Addison Street 3rd floor
Berkeley, CA 94704-1103

Abstract

This paper explores the use of don't cares in software synthesis for embedded systems. Embedded systems have extremely tight real-time and code/data size constraints, that make expensive optimizations desirable. We propose applying BDD minimization techniques in the presence of a don't care set to synthesize code for extended Finite State Machines from a BDD-based representation of the FSM transition function. The don't care set can be derived from local analysis (such as unused state codes or don't care inputs) as well as from external information (such as impossible input patterns). We show experimental results, discuss their implications, the interaction between BDD-based minimization and dynamic variable reordering, and propose directions for future work.

1 Introduction

Embedded systems are informally defined as a collection of programmable parts surrounded by ASICs and other standard components, that interact continuously with an environment through sensors and actuators. The programmable parts include micro-controllers and Digital Signal Processors (DSPs).

Real-time and cost constraints imply that manual optimization (often at the assembly code level) of software executed on the programmable parts has been considered imperative for this class of systems so far. However, the growing complexity of such systems and the ever-shrinking time-to-market requirements now mandate a higher level of abstraction for specification and design. These constraints and trends explain the resurgence of interest in *domain-specific software synthesis techniques*. Such techniques exploit knowledge of the target domain, as well as the limitations of embedded systems (e.g., absence of virtual memory and tight memory space severely limit programming techniques such as liberal use of recursion, "wild" pointers and so on) to achieve a quality that is comparable with expert designers, at a fraction of the programming time cost.

Software synthesis is especially popular in the Digital Signal Processing domain, where "code stitching" of highly optimized library function kernels allows the designer to quickly simulate, prototype and implement in hardware and software sophisticated data processing applications from Data Flow network specifications (see, e.g., [15]). Code synthesis for control-dominated applications, in which rapid reaction to external events and complex decision mechanisms dominate over pure data computations, require software or hardware to be efficiently synthesized from Finite State machine-like specifications [3, 13, 9, 8].

The aim of this paper is to exploit *don't care* (*DC*) information (coming, e.g., from impossible or irrelevant conditions -- limited controllability) in software code synthesis. The use of DCs, to the best of our knowledge, has been limited so far only to hardware synthesis [11].

Note that some compiler optimizations, such as variable lifetime analysis, constant value propagation, and so on [1] can be considered a form of DC exploitation. For example, avoiding assigning a variable that is not read before being assigned again, is exploiting a form of "observability DC", just as the elimination of an **if** statement with a constant condition is exploiting a form of "controllability DC". However, this paper uses DCs in a much more global and systematic sense than standard compilers that often take the *control flow* as given, and perform only global *data flow* analysis (if any).

The software synthesis technique that we use in order to exploit DCs is based on the use of Binary Decision Diagrams (BDDs) [5] to optimally synthesize software (in particular C code) from a specification in the form of Finite State Machines extended with integer arithmetic capabilities [7, 2]. That technique uses a direct mapping between BDD nodes and low-level C statements in order to derive a highly optimized implementation of the FSM transition relation. We use this synthesis path because it is publicly available as part of an embedded system design tool called POLIS [2].

The classes of DCs that we consider in this paper are as follows:

1. Internal DCs, arising from the FSM itself. In particular, we consider DCs that arise from setting FSM inputs that are irrelevant in a given state to a fixed and known value (all other combinations automatically become DCs).
2. External DCs, arising from the FSM environment. Such DCs generally belong to two categories:
   (a) input (also called "controllability") DCs, due to impossible input combinations. In particular, we consider input combinations that are specified as impossible by the designer, or input combinations that are forbidden by the specification language semantics.
   (b) output (also called "observability") DCs, due to output combinations that are ignored by the following FSMs.

In this paper we apply several different BDD minimization algorithms from the literature to DCs arising from various sources among those listed above, for the purpose of *minimizing code size* while preserving the execution speed.

The paper is organized as follows: relevant background material is presented in Section 2, our application of BDD minimization in Section 3, results in Section 4, and conclusions and future work in Section 5.

2 Background

This section summarizes some previous results that are used in the rest of the paper. The reader is referred to [5, 16, 2] for a complete explanation of notions and algorithms. We assume that the reader is familiar with Reduced Ordered Binary Decision Diagrams (simply called BDDs in the following).

2.1 Multioutput Functions

Multioutput functions (or, equivalently, sets of single-output functions on the same domain) can be represented by their *characteris-*

DAC 98, San Francisco, California

tic functions. A single-output binary-valued function χ_f: $(X \times Y) \rightarrow \{0,1\}$, where $X = X_1 \times \ldots X_m$ and $Y = Y_1 \times \ldots Y_l$, represents the multioutput multivalued (MV) function $f: X \rightarrow Y$ if $\chi_f(x, y) \Leftrightarrow (y = f(x))$.

2.2 Codesign Finite State Machines

Co-design Finite State Machines (CFSMs) are Extended Finite State Machines interconnected asynchronously by means of one-place buffers. In this paper we consider only the function of a *single CFSM* although more powerful DCs could be extracted from a *network* of CFSMs (we leave this extension to future work).

A CFSM is a reactive Extended Finite State Machine with a set of input and output events. States of the CFSM are events that are fed back. Events are entities that may occur at determinate instants of time and may or may not carry a value.

Given a BDD representation of the characteristic function of the set of transitions of a CFSM, the technique of [7] derives a fragment of C code, as follows:

- A BDD node associated with an input variable becomes an **if-then-else** statement, branching to the statements corresponding to the BDD children nodes.

- A BDD node associated with an output variable becomes an assignment of the value corresponding to the *single child* with a path to the BDD leaf labeled with 1, followed by a branch to the statement corresponding to that child.

This technique produces *extremely fast* code (each input variable is tested at most once in an execution of a transition of a CFSM) with a reasonable size (comparable with good quality hand-written code [2]). For example, it is faster and more compact than code generated in a more standard way from a CFSM, by producing two levels of **switch** statements (one based on the state, one on the inputs) followed by a sequence of assignments.

Another very important characteristic of this technique is that, since BDD nodes and code fragments are almost in 1-1 correspondence, the BDD size becomes an excellent *predictor* of the code size.

2.3 BDD Minimization Using Don't Cares

An incompletely specified function can be represented by a BDD pair BDD_f and BDD_c, denoting the on-set and the care-set of the function, respectively. A function g is a *cover* of $[f, c]$ if f and g have the same function values for all care-points, i.e., $f \cap c = g \cap c$.

Given BDD_f and BDD_c, finding the smallest BDD cover of $[f, c]$ is the problem of BDD minimization using DCs. The problem is known to be NP-complete [18], so heuristics for DC-based BDD minimization are typically used [10, 6, 12, 21, 14].

3 BDD Minimization in POLIS

3.1 Internal Don't Cares

The first class of DCs that we consider comes from the structure of the transition function of the CFSM itself.

Before giving a more precise definition of this transformation, we offer an intuitive explanation. The transition function can be described as a table of transitions, each with an input and an output part. A '-' in the input part is shorthand for specifying several input minterms for the same transition; a '-' in the output part is a DC. Our first minimization procedure effectively transfers a '-' in the input part to a '-' in the output part. This works as follows. Given a transition with a '-' for a particular input, for example *ab-d*, if we can guarantee that the value for that input is always fixed for this cube, for example *abcd*, we know that none of the other values ever occur, in this case $ab\bar{c}d$, and this is a DC. In order for this optimization to be correct, all input minterms contained by this cube must be mapped to this distinct identifier *before* calling the CFSM.

In our case, in order to preserve the specified CFSM behavior, this optimization strategy assumes that we make a simple modification to the RTOS. That is, we force the RTOS under certain conditions to feed only some input combinations to the CFSM; the remaining input combinations become DC conditions.

Specifically, we assume the RTOS uses a state-based mask which sets certain MV-input signals to a user-definable fixed value in a given state. Once a mask is defined, the BDD representation of the characteristic function of a CFSM can be minimized by taking into account that for that state, other values of the MV-input signal are not possible. Thus, the behavior of the CFSM for such an impossible input assignment becomes a DC.

Consider an MV-input variable i with possible values whose range is $I_i = \{i_1, i_2, \ldots i_l\}$. The transition relation f is *independent* of i in state s if $(f \cdot s)_{i_k} = f \cdot s$ for all $1 \le k \le l$ (where $(f \cdot s)_{i_k}$ denotes $f \cdot s$ evaluated for i_k). Thus, in this state the behavior of f does not depend on the value of i. We can utilize this independence by making the RTOS fix the value of i in this state. Then, the other state-value combinations become DCs with which f can be minimized.

Formally, the mask M is a function from states and inputs variables to the union of MV-values and a special *do not change value symbol* '-', as follows: *states* $\times\, i \;\rightarrow I_i \cup$ '-'. For example, for a given state s and input variable i, $M[s, i]$ can be set to a user-specifiable constant $a_{s,i}$, where $a_{s,i}$ is a possible value of i, $a_{s,i} \in I_i$. For those combinations of input variables and states where the MV-input signal should not be altered by the RTOS, the mask evaluates to '-'.

Our algorithm to compute the mask and the corresponding DC set is shown in Figure 1.

Algorithm Find_DC_and_Mask (T: characteristic function)
/* D: resulting DC function, d: a DC cube */
$D = 0$
foreach state s
  foreach input signal i
    $d = 0$
    if (T is independent of i in s) **then**
      $M[s,i] = a_{s,i}$ /* $a_{s,i}$: the user-defined value to assign to i in s */
      d = complement($a_{s,i}$)
      $D = D + d$

Figure 1: Computing the DC set and the Mask

The cost of this optimization is the overhead in the RTOS required to implement the mask M. Fortunately, in POLIS the variables are bit-packed. Thus, we can typically create the mask with an array of integers, one integer per state, and perform the input presence test by using that mask at the beginning of every CFSM transition, at an almost insignificant cost.

3.2 External Don't Cares

Another class of DCs that we explored is external *input* DCs. These DCs indicate that some input event combinations are impossible, either due to environmental constraints, or due to the functionality of other CFSMs. While it would be possible in principle to automatically extract such DCs due to other CFSMs from the CFSM network specification, in this work we consider only designer-specified input DCs embedded in the specification language. In particular, we focus on two input languages: Esterel [4] and SDL [17].

Esterel is a textual specification language for reactive systems, whose semantics is defined in terms of Extended Finite State Machines, and that is used as an input language for POLIS. An Esterel specification may contain relations that indicate limited controllability on the inputs. For example, the Esterel statement

relation a # b # c

indicates that none of inputs a, b, or c occur simultaneously. For any function of these variables, a valid DC set is $DC = ab + ac + bc$. The use of the '#' operator for specifying limited controllability does not allow the specification of *any* conceivable DC set; however, it is convenient for the reactive control-type designs that are designed using Esterel.

In the current Esterel compiler, which is used as a front-end by POLIS and hence by our BDD minimization procedure, these relations are taken into account during simulation (an error is signaled to the user if one of these conditions is given as an input), and during causality analysis. However, the compilation path does not by default use these relations to optimize the implementation.[1] Furthermore, in the Esterel compiler there is no exploration of and optimization using both sifting and relation-based minimization.

SDL is a graphical and textual language for telecommunication systems, especially suited for specification of network protocols. SDL also has an EFSM semantics (but it cannot currently be used as an input to POLIS). The semantics, however, requires that input signals are delivered to each EFSM *one at a time*, through a priority queue. This means that for a CFSM hypothetically specified in SDL, any condition with no input events or more than one input event present is DC (the former due to the reactivity of CFSMs, that are not activated unless they have at least one event, the latter due to the SDL single-event constraint). Since POLIS currently does not allow specification of CFSMs in SDL, we have analyzed the effect of BDD minimization on this kind of specification by using an Esterel relation that states that all CFSM inputs are mutually exclusive.

4 Experimental Results

In this section we report the results from the application of various minimization algorithms to the classes of DCs discussed in the previous section. We report minimization results as a single column (with the best result among all the algorithms), since there was no significant difference in performance or quality of results (it was at most 2%). We also report only examples for which we could find a significant effect of minimization. Such examples are generally characterized as having a *high degree of concurrency and symmetry*, so that the external DCs discussed in Section 3.2 can be especially effective.

In the following sections we describe the results of exploiting the external DCs from Esterel described in Section 3.2. In all cases we specified that at most one input event could be delivered to a CFSM at any given time. This, of course, caused a big reduction in size of the 12-element Distributed Mutual Access element modeled in **arbiter12**.

The main point of this paper, however, is to show that BDD minimization using *external DCs* is orthogonal to other minimization opportunities, e.g., by BDD variable reordering. Hence, they should be used whenever an appropriate DC set is available from the designer or from the language semantics.

On the other hand, the class of *internal DCs* described in Section 3.1 interacted heavily with minimization by dynamic variable ordering, specifically *sifting* [16] in our experiments. This is the reason why we do not report any results about that class of internal DCs in this paper. In particular, DC minimization *before* sifting (i.e., with the initial variable ordering in which all inputs appear before all outputs) reduces the BDD size by 20% on average and the code size by 10% on average. However, it also prevents sifting from obtaining the 50%-200% improvement it can achieve alone. This may be due to the effect of sifting on a CFSM in which some states are independent of some inputs. In that case, a good ordering will most likely put the present state variable at the top of the characteristic function BDD, and then the BDD will have one distinct branch for each state. Of course, DC variables will not occur in those branches, and the minimization effect due to them will be achieved.

It is not clear, though, if the absence of positive effect in the case of internal DCs is only due to the particular type of specifications that we considered in our benchmark set.

4.1 Optimization within Esterel

The next set of experiments was run to see how much improvement in the size of the C code could be obtained using external DCs with the current Esterel technology alone. Here, we can compare the percentage gains with the percentages we obtain using a combination of reordering and BDD minimization.

For each example under test, three different C files were produced: one for the original Esterel example, one for the original after optimization, and one for the original with a relation added and with subsequent optimization. To use optimization in Esterel, one must first extract the control portion using the Esterel compiler, run a logic optimization tool on it, and then reinsert this optimized control part and continue the Esterel compilation to C code. The optimization that is available through the Esterel compiler is a well-tested set of scripts that uses the SIS sequential logic synthesis program [19] in conjunction with the **rem_latch** program [20] which contains more sophisticated algorithms for latch removal. The relations that were added were always of the nature of severely restricting the inputs to arrive one at a time -- that is, all inputs are assumed to be mutually exclusive.

The results of these experiments are shown in Table 1. The C code sizes are given in bytes. Modest reductions in the size of the C code when optimization and optimization with a relation present are observed. In the case of the arbiter, clearly, the reduction is considerable.

Name	*Original*	*Opt*	*Rel & Opt*
abcd	16673	15903	14767
abcdef	24450	23329	22593
arbiter12	55439	18438	18442

Table 1: Effects of DCs on Esterel C-code Generation

4.2 Optimization within POLIS

The next set of experiments shows the results of applying DC-based BDD minimization *after* performing minimization by sifting. In all cases, we report the number of basic C statements (**if-then-else** and assignments), as it tracks the final code size very well especially in cases, such as the examples we used, in which there is no data computation (note: *this table does not use the same units as Table* [1]).

The meaning of the columns is as follows:

• **No Esterel Opt** does not include the effect of Esterel optimization, as described in the previous section.

• **Esterel Opt** includes the effect of Esterel optimization, as a first step *before* BDD construction. Note that, even though Esterel optimization reduces the code size produced by the Esterel compiler, this may not change the function at all, but only modify the Boolean network structure. Hence the BDD size may not change at all.

• **No DC Opt** shows the result of POLIS software synthesis without exploiting DCs.

• **DC Opt** shows the result of POLIS software synthesis after the

[1] One can use these relations in optimizing the control portion of the design after separating it from the data portion, but this optimization is not yet fully automatic.

best DC-based BDD minimization.

Examples whose name ends in **_rel** had external DCs already specified in Esterel. A simple conclusion that can be drawn from this table is that in general it is advisable *not* to use the DCs in the Esterel optimization stage, but only in the "standard" Esterel compilation procedure, since the former may reduce further possibilities of optimization. This is probably due to the same reasons as in the case of DC minimization *before* sifting, as discussed above.

	No Esterel Opt		*Esterel Opt*	
Name	*No DC Opt*	*DC Opt*	*No DC Opt*	*DC Opt*
abcd	496	469	496	473
abcd_rel	93	63	94	85
abcdef	4023	3959	4023	3959
abcdef_rel	171	105	162	149
arbiter12	406	172	406	172
arbiter12_rel	406	172	197	195

Table 2: Effect of DCs on POLIS C-code Generation

5 Conclusion

There are several very interesting directions to explore further. Here, we describe two that are particularly intriguing.

The RTOS mask-based minimization, while not producing significant results for this application, hints at a more general optimization method that could be useful in a slightly different application domain, or useful under different conditions. This optimization technique effectively transforms care-set cubes (cubes with '-'s in the input part) to a (unique, identifying) care-set minterm (no '-'s) plus a set of DC-set cubes (cubes with '-'s in both the input and output parts). The RTOS guarantees that each minterm in the original care-set cube is mapped to the unique identifying minterm before the CFSM is called to react. We have chosen a subset of the possible optimizations of this flavor that is efficient to implement: input cubes that are identified by a particular state. The more general optimization technique would transform the '-'s in the input part to '-'s in the output part based on more general conditions. This is an avenue for further exploration.

The tradeoff between variable reordering and BDD-based minimization is an area completely open for research. To the best of our knowledge, this subject has not been tackled thus far. It is clear that there is an important relationship between the two. Minimization is potentially more powerful from the standpoint that it explores a *set* of possible functions; reordering changes only the structure of the representation of a single function. Our experiments reveal that while minimization is potentially more powerful, variable ordering has such a strong influence on BDD size that fixed-order minimization must be considered a secondary optimization technique. We expect the observed results to vary widely with the application domain and the nature of the derived DCs; this is an analogy to the observation that BDD size varies widely with the type of BDD employed and the type of function represented.

References

[1] A. V. Aho and R. Sethi and J. D. Ullman, Compilers: Principles, Techniques and Tools. Addison-Wesley, 1988,

[2] F. Balarin et al. Hardware-Software Co-design of Embedded Systems - The POLIS approach. Kluwer Academic Publishers, 1997.

[3] G. Berry. The Constructive Semantics of Pure Esterel. To Appear, available now at ftp: //www.inria.fr/meije/esterel/papers/constructiveness.ps.gz, 1996.

[4] G. Berry and P. Couronn and G. Gonthier. The Synchronous Approach to Reactive and Real-time Systems. IEEE Proceedings, 79, September, 1991.

[5] R. Bryant. Graph-Based Algorithms for Boolean Function Manipulation. *IEEE Transactions on Computers*, C-35(8), pp. 677-691, August, 1986.

[6] S. Chang and D.I. Cheng and M. Marek-Sadowska. Minimizing ROBDD Size of Incompletely Specified Multiple Output Functions. In *Proc. European Design and Test Conference*, pp. 620-624, 1994.

[7] M. Chiodo et al. Synthesis of Software Programs from CFSM Specifications. In *Proc. Design Automation Conference*, June 1995.

[8] P. Chou and G. Borriello. Software Scheduling in the Co-synthesis of Reactive Real-time Systems. In *Proc. Design Automation Conference*, June 1994.

[9] C.N. Coelho and G. De Micheli. Analysis and Synthesis of Concurrent Digital Circuits using Control-flow Expressions. *IEEE Transactions on Computer-Aided Design*, 15(8), pp. 854-876, August 1996.

[10] O. Coudert, C. Berthet, and J.C. Madre. Verification of Synchronous Sequential Machines Based on Symbolic Execution. In *Automatic Verification Methods for Finite State Systems*, pp. 365-373. Springer-Verlag, 1989.

[11] S. Devadas, A. Ghosh, and K. Keutzer. *Logic Synthesis*. McGraw-Hill, 1994.

[12] R. Drechsler and N. Gockel. Minimization of BDDs by Evolutionaly Algorithms. In *Proc. Int. Workshop on Logic Synthesis*, 1997.

[13] R. K. Gupta, C. N. Coelho Jr., and G. De Micheli. Program Implementation Schemes for Hardware-software Systems. *IEEE Computer*, pp. 48-55, January 1994.-

[14] Y. Hong, P.A. Beerel, J.R. Burch, and K.L. McMillan. Safe BDD Minimization Using Don't Cares. In *Proc. Design Automation Conference*, pp. 208-213, June 1997.

[15] S. Ritz et al. Code Generation Techniques in the Block Diagram Oriented Design Tool COSSAP/DESCARTES. In *Proc. International Conference on Signal Processing Applications and Technology*, pp. 709-714, 1994.-

[16] R. Rudell. Dynamic Variable Ordering for Ordered Binary Decision Diagrams. In *Proc. Int. Conference on Computer-Aided Design*, pp. 42-47, Nov. 1993.

[17] R. Saracco, J.R.W. Smith, and R. Reed. *Telecommunications Systems Engineering Using SDL*. North-Holland - Elsevier, 1989.

[18] M. Sauerhoff and I. Wegener. On the Complexity of Minimizing the OBDD Size for Incompletely Specified Functions. *IEEE Transactions on Computer-Aided Design*, 15, pp. 1435-1437, November 1996.

[19] E.M. Sentovich et al. Sequential Circuit Design Using Synthesis and Optimization. In *Proc. Int. Conf. Computer Design*, pp. 328-333, October 1992.

[20] E.M. Sentovich, H. Toma, and G. Berry. Efficient Latch Optimization Using Exclusive Sets. In *Proc. Design Automation Conference*, pp. 8-11, June 1997.

[21] T. Shiple, R. Hojati, A. Sangiovanni-Vincentelli, and R.K. Brayton. Heuristic Minimization of BDDs Using Don't Cares. In *Proc. Design Automation Conference*, pp. 25-231, June 1994.

Instruction Selection, Resource Allocation, and Scheduling in the AVIV Retargetable Code Generator

Silvina Hanono
Department of EECS, MIT
silvina@lcs.mit.edu

Srinivas Devadas
Department of EECS, MIT
devadas@mit.edu

Abstract—
The AVIV retargetable code generator produces optimized machine code for target processors with different instruction set architectures. AVIV optimizes for minimum code size.

Retargetable code generation requires the development of heuristic algorithms for instruction selection, resource allocation, and scheduling. AVIV addresses these code generation subproblems concurrently, whereas most current code generation systems address them sequentially. It accomplishes this by converting the input application to a graphical (Split-Node DAG) representation that specifies all possible ways of implementing the application on the target processor. The information embedded in this representation is then used to set up a heuristic branch-and-bound step that performs functional unit assignment, operation grouping, register bank allocation, and scheduling concurrently. While detailed register allocation is carried out as a second step, estimates of register requirements are generated during the first step to ensure high quality of the final assembly code.

We show that near-optimal code can be generated for basic blocks for different architectures within reasonable amounts of CPU time. Our framework thus allows us to accurately evaluate the performance of different architectures on application code.

I. INTRODUCTION

A. *Hardware–Software Co-Design for Embedded Systems*

For a variety of reasons, manufacturers profit from integrating an *entire system* on a single integrated circuit. However, designing an entire complex system as an Application Specific Integrated Circuit (ASIC) is neither economical nor practical. Furthermore, as time-to-market requirements place greater burdens on designers for fast design cycles, an increasing amount of system functionality is implemented in *software* relative to hardware.

In a *hardware–software co-design* [1] methodology, designers first partition the system functionality into hardware and software. A target processor, called an *embedded processor*, is then selected from existing processor designs, or an ASIP (Application Specific Instruction-Set Processor) is designed to execute the software. The hardware, software, and ASIP are designed and the resulting system is evaluated using a hardware–software co-simulator. The partitioning and processor design are repeated until an acceptable system is developed.

As the complexity of embedded systems grows, programming in assembly language and optimization by hand are no longer practical. Further, hand coding virtually eliminates the possibility of changing the processor architecture. We argue that in order to be able to explore the processor design space, an automatically retargetable compilation strategy is required.

B. *Overview of the* AVIV *Retargetable Code Generator*

We are developing a retargetable code generator, AVIV, whose inputs are the application program and a machine description of the target processor. The machine description, written in ISDL (Instruction Set Description Language) [2], includes an instruction set specification and some architectural information. AVIV produces code, optimized for minimal size, that can run on the target processor. Code size is our optimization cost function because we are focusing on embedded processors where the size of the on-chip ROM is a critical issue.

DAC 98, San Francisco, California

By varying the machine description and evaluating the resulting object code, the design space of both hardware and software components can be effectively explored. AVIV focuses on architectures exhibiting instruction-level parallelism (ILP), including Very Long Instruction Word (VLIW) architectures.

Retargetable code optimization requires the development of heuristic algorithms for instruction selection, resource allocation, and scheduling. AVIV addresses these code generation subproblems concurrently, whereas most current code generation systems address them sequentially. The main reason why current code generators address these problems sequentially is to simplify decision-making in code generation. However, decisions made in one phase have a profound effect on the other phases.

AVIV converts the application code to a graphical (Split-Node DAG) representation that specifies all possible ways of implementing the application's basic blocks on the target processor. The information embedded in this representation is then used to set up a heuristic branch-and-bound step that performs functional unit assignment, operation grouping, register bank allocation, and scheduling concurrently. While detailed register allocation is carried out as a second step, required loads and spills due to limits on available registers are generated and scheduled during the first step. This ensures that a valid detailed register allocation can always be found without undoing the chosen operation groupings or functional unit assignments.

C. *Organization of this paper*

Section II describes the compiler framework of the AVIV code generator. Section III presents the Split-Node DAG formulation for retargetable code generation. Section IV formulates the code generation problem as a covering problem on the Split-Node DAG. Section V presents previous work on retargetability and code generation, as well as comparisons to our own work. Results using our code generator and ongoing work on our compiler project are presented in Section VI.

II. COMPILER FRAMEWORK

Figure 1 illustrates the compiler framework used by the AVIV code generator. The compiler front end receives a source program written in C or C++. It performs machine independent optimizations and generates an intermediate format description in SUIF[1] [3]. The ISDL [2] machine description of the target processor is generated either by hand or by a high-level CAD tool. The compiler back end takes the SUIF code and the ISDL description as inputs and produces assembly code specific to, and optimized for, the target processor. The ISDL description is also used to create an assembler. The automatically generated assembler transforms the code produced by the compiler to a binary file that is used as input to an instruction-level simulator. This paper focuses on the back-end of the AVIV code generator.

AVIV uses the SUIF compiler [3] in conjunction with the SPAM[2] [4] compiler as its front-end. The front-end converts the source program to an intermediate form consisting of expression DAGs and control flow information. It also performs machine independent optimizations such as loop unrolling and other transformations that extract machine independent parallelism. Thus, the starting point of the AVIV compiler is a number of basic block DAGs connected through control flow information. A sample basic block DAG is shown in Figure 2. AVIV then focuses on extracting machine

[1] Stanford University Intermediate Format
[2] Synopsys, Princeton, Aachen, MIT

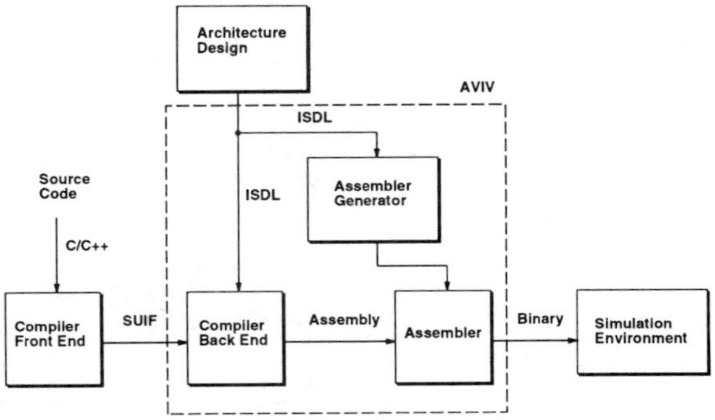

Fig. 1. **Retargetable Code Generation Framework using Aviv**

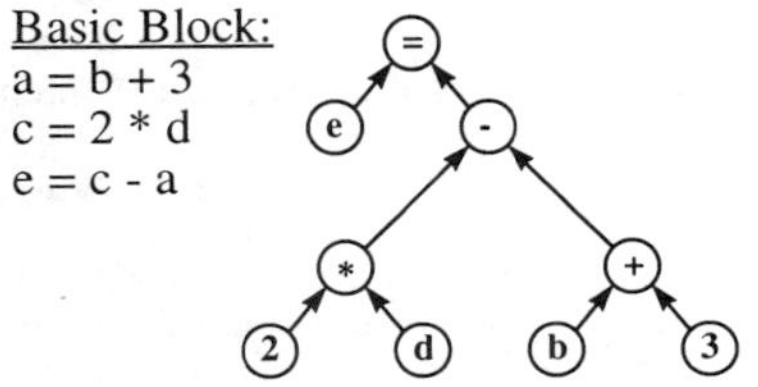

Fig. 2. **Basic Block DAG**

dependent parallelism using the ISDL description of the target processor. This is accomplished by converting the basic block DAGs, or expression trees, into Split-Node DAGs.

The instruction set information contained in the ISDL machine description is used to create several databases which are later used to create the Split-Node DAG, as described in Section III. One of these databases stores a correlation between the target processor operations and the SUIF basic operations such as ADD and SUB. This information comes from the RTL description of each instruction. A second database stores all possible data transfers explicitly stated in the target machine description, and is subsequently expanded to include multiple-step data transfers as well.

The machine description also describes constraints on the instructions that exist in the target processor. This information is used to remove the constrained operations from the databases ensuring that when creating the Split-Node DAG, it will only contain legal mappings from SUIF operations to target machine operations.

III. The Split-Node DAG

A. *The Split-Node DAG Definition*

The Split-Node DAG representation contains all the necessary information to generate code that will perform the operations of the original basic block DAG on the target processor. In addition, it allows for the exploration of parallelism, that is achievable on the target architecture, within the basic block. The Split-Node DAG provides information about the multiple ways that any operation can be performed on the target processor. For example, an ADD SUIF node will be split into several nodes representing the different functional units that can perform an ADD operation. The Split-Node DAG also includes data transfer information. It is important to include data transfer nodes because the cost of covering the Split-Node DAG should include the cost of transferring data between the units. Also, by including the transfer nodes, the transfers are automatically scheduled along with the operations.

Syntactically, a Split-Node DAG is similar to a DAG representing the operations performed in a block of code. A Split-Node DAG has two additional types of nodes: *split nodes* and *data transfer nodes*. A *split node* corresponds to an operation node in the original DAG. The immediate (non data transfer node) descendants of a split node correspond to all possible ways the operation may be performed on the target processor. The *data transfer nodes* are inserted on the path between a split node and its immediate operation descendants and correspond to data transfers required along that path.

The Split-Node DAG explicitly represents all possible *implementations* for a block of code on the target processor. Operation

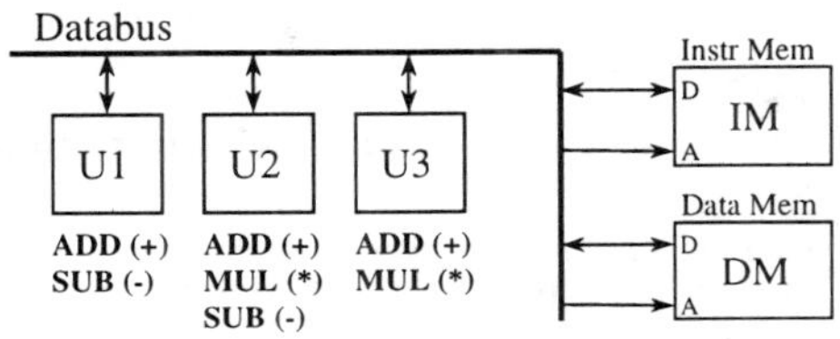

Fig. 3. **Example Target Architecture**

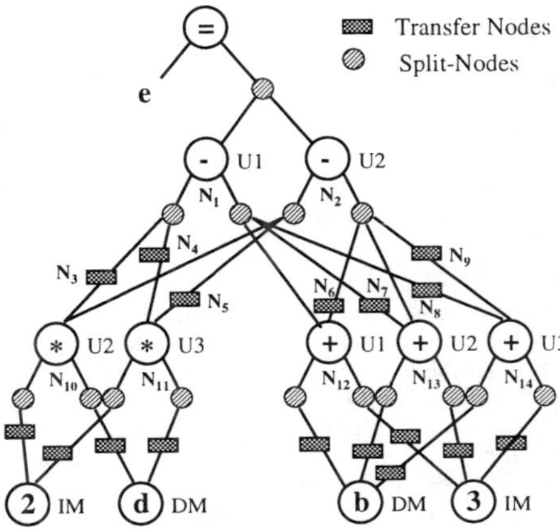

Fig. 4. **Split-Node DAG**

grouping, functional unit assignment[3], register bank allocation, and scheduling are performed simultaneously from the Split-Node DAG. An *implementation* of a block of code corresponds to a set of instructions in which all operations have been assigned an execution unit, and all data transfers[4] between these units are accounted for. At this stage, detailed register allocation has not been performed, however, we estimate the number of registers required from each register bank, and insert loads and spills if the available resources are exceeded. This guarantees that we do not have to modify the instruction selection during detailed register allocation.

B. *Creating the Split-Node DAG*

To clarify the structure of the Split-Node DAG, let us create one for the basic block DAG shown in Figure 2. We will use the example VLIW processor presented in Figure 3 as the target processor. In this processor, `Unit U1` can perform addition (ADD), and subtraction (SUB). `Unit U2` can perform addition (ADD), subtraction (SUB), and multiplication (MUL). `Unit U3` can perform addition (ADD), and multiplication (MUL). Each unit contains its own register file and can perform only one operation at a time. The architecture also includes an instruction memory (IM), a data memory (DM), and a databus that connects all units and memories.

All of the information required to generate the Split-Node DAG is extracted from the ISDL target machine description, as described in Section II. Figure 4 illustrates the result of converting the basic block DAG into a Split-Node DAG. Each Split-Node has children corresponding to the multiple units that can perform the ADD, MUL, or SUB operations on the target architecture. Note that paths from several split nodes can reconverge to one set of operation nodes. Any edge connecting two operations, A and B, in the original basic block DAG, is now split into multiple edges representing all possible paths from operation A to operation B. If these edges result in a transfer from one unit to another, then a data transfer node is added along that edge. This includes multi-level paths if a direct transfer path is not available on the target architecture.

The Split-Node DAG structure can easily incorporate complex instructions, as well as basic operations by utilizing an initial pattern matching phase that detects which nodes in the original expression

[3] Operation grouping and functional unit assignment are the main tasks in instruction selection for VLIW processors.

[4] These data transfers are inserted between pairs of operations executed on different functional units.

Explore possible split-node functional unit assignments
- Estimate cost of assignment
- Select several lowest cost assignments to explore in further detail

Foreach selected assignment
- Insert required data transfers
- Generate all maximal groupings of nodes that could be executed in parallel
- Select a minimal-cost set of maximal groupings that covers all nodes

Final solution is the lowest-cost solution found above

Fig. 5. **Overall Algorithm for Covering the Split-Node DAG**

DAG can be covered by a complex instruction supported by the target processor. When generating the Split-Node DAG, additional nodes and edges corresponding to the matched complex instructions are added in addition to the basic operation matches.

C. *Control Flow*

AVIV receives a collection of basic blocks connected by control flow information as an input. Code is generated for the basic blocks using the methods presented in Section IV. Code corresponding to control-flow instructions (e.g., jumps) needs to be generated. Conventional tree-covering algorithms are used for this step. Since AVIV currently targets minimum code size, control flow optimizations such as speculative execution which improve performance at the expense of code size are not incorporated.

IV. CODE GENERATION USING THE SPLIT-NODE DAG

The goal of the code generator is to cover the Split-Node DAG with a minimal-cost set of target processor instructions. Our methodology addresses the instruction selection, resource allocation, and scheduling phases of code generation concurrently because these phases are mutually dependent, thus performing them sequentially generally results in non-optimal code. Covering the Split-Node DAG, using the AVIV code generator, performs functional unit assignment, operation and data transfer grouping into instructions, register bank allocation, and scheduling. Detailed register allocation and peephole optimizations are performed as a second step.

Figure 5 describes our overall algorithm for covering the Split-Node DAG using a minimal-cost set of instructions. We introduce multiple heuristics in order to reduce the runtime of our algorithm without sacrificing the quality of the results. The algorithm begins by exploring the possible split-node functional unit assignments and selecting several of the lowest cost assignments to explore in further detail. The selection is made based on a heuristic cost function described in Section IV-A. Each of the selected assignments is then explored in detail. First, the data transfers required for the given functional unit assignment are added (Section IV-B). Next, the nodes in the current assignment, including the transfer nodes, are merged into *maximal groupings* of nodes that can be executed in parallel (Section IV-C). The maximal groupings correspond to VLIW instructions. A heuristic selection process (Section IV-D) then covers the nodes in the current assignment with a minimal-cost set of maximal groupings. An accurate cost reflecting the number of instructions required to cover all the nodes is maintained. Finally, the assignment resulting in the lowest cost is selected as the final solution.

A. *Exploring the Split-Node Functional Unit Assignments*

The first step in covering the Split-Node DAG is to select the target processor operation (i.e., functional unit) that should cover each split-node. For the Split-Node DAG of Figure 4, node N_1 or N_2 would be selected to cover the subtract split-node, node N_{10} or N_{11} would be selected to cover the multiply split-node, and so on. The number of possible split-node covering assignments can be quickly calculated by multiplying the number of possible target processor operations covering each split-node (i.e., for this example $2 \times 2 \times 3$). This is a very small basic block that results in few possible assignments. However, for typical basic blocks, the multiplicative growth in the number of possible split-node functional unit assignments quickly makes it prohibitively CPU-intensive to explore all possible cases.

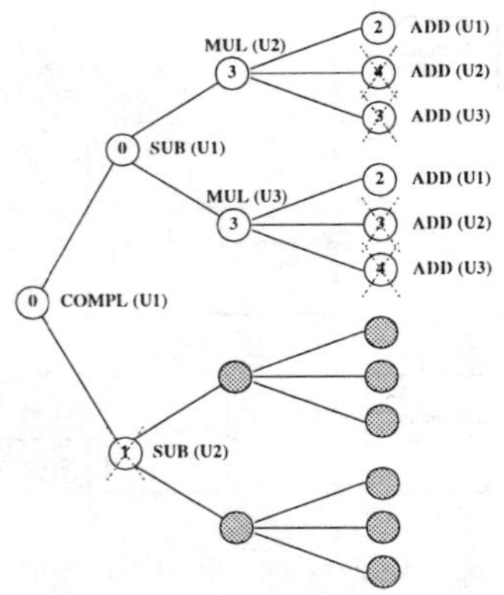

Fig. 6. **Pruning the Search Space of Split-Node Assignments**

Thus, the first step of our algorithm is to prune the search space by selecting only a few of the split-node functional unit assignments to explore in depth. This selection is made based on a cost function that takes into account the two main factors contributing to higher cost (measured in number of instructions) of covering the Split-Node DAG. These two factors are the amount of parallelism that is foregone due to the split-node covering assignments, and the number of data transfers required for the given assignment. It is important to include both factors in the cost function because this allows us to optimize the instruction selection and scheduling phases of code generation concurrently.

Calculating this cost function for all possible functional unit assignments can in itself take too long. Thus, we prune the search space of possible assignments by calculating an incremental cost for each split-node encountered and continue the search only for split-node assignments with minimum incremental cost. The split-nodes are tested in order of increasing level from the top of the Split-Node DAG. For the purpose of illustrating this cost function, let us assume that the `SUB` nodes in the Split-Node DAG of Figure 4 feed into a sink node that is a complement (`COMPL`) function that can only be executed on unit U1. Also let us assign a cost of 1 for each required transfer and for each node that cannot be executed in parallel due to the current functional unit assignment. Figure 6 shows the incremental cost at each node as well as the locations where the search space was pruned (marked by an X).

The incremental cost of the `SUB` node executed on unit U1 is 0 because it does not require a transfer to the `COMPL` node which is also executed on U1, and it does not preclude any possible parallel execution. The `SUB` on U2, however, has a cost of 1 because it requires a data transfer to unit U1 for the `COMPL` function. Thus, the search space is pruned at the `SUB` on U2 node. The incremental cost of implementing the `MUL` operation on unit U2 or U3 are the same, thus both paths are explored. Let us now examine the paths that include the `MUL` operation on unit U2. Performing the `ADD` on unit U1 introduces an incremental cost of 2 due to the two transfers required to load its operands. It does not require a transfer to the `SUB` operation because both are executed on unit U1. Also, it does not preclude the possible merging of the `ADD` with the `MUL` because the `MUL` operation is executed on unit U2. The `ADD` on unit U2, on the other hand, incurs an incremental cost of 4 (2 for its two operands, 1 because a transfer to the `SUB` operation is required, and an additional 1 because it precludes the merging of the `ADD` and `MUL` operations since both are executed on the same unit). Thus, the result, for this example, would be to select the two assignments where both the `SUB` and `ADD` operations are performed on unit U1.

B. *Adding Required Transfers*

For each selected split-node covering assignment, the required data transfer nodes are then added. In the case of a single data transfer path, this step is straightforward because for each pair of nodes that requires a data transfer between them, there is only one possible selection of data transfer node. However, in architectures containing multiple transfer paths, there may be more than one possibility

	N_2	N_9	N_{10}	N_{14}
N_2	0	1	1	1
N_9	1	0	0	1
N_{10}	1	0	0	0
N_{14}	1	1	0	0

Fig. 7. **Matrix for Finding Maximal Cliques**

```
gen_max_clique( clique, index ) {
    for (i = 0; i < number_of_nodes; i++) {
        if (i can be executed in parallel with all the nodes in the
            current clique)
            if (adding i will not preclude adding any other node)
                if (i < index)          // Pruning condition
                    return;
                else
                    add i to clique;
    }

    for (i = 0; i < number_of_nodes; i++) {
        if (i can be executed in parallel with all the nodes in the
            current clique)
            gen_max_clique(clique with i added, MAX(i, index));
    }
}
```

Fig. 8. **Pseudo-Code for Generating Maximal Cliques**

for any given data transfer. Thus, the problem resembles our initial problem of selecting the best split-node covering assignments where the number of options grows multiplicatively. Once again we apply our heuristic in order to select among the possible transfer paths, except that now the cost function is based solely on parallelism.

C. *Maximal Groupings*

Once we have selected a set of possible split-node covering assignments and their corresponding data transfer nodes, we want to explore these assignments in depth to find the one that will result in the minimum-cost set of target processor instructions to cover the nodes in the assignment.

The rest of this section will use the term *assignment* to refer to the collection of functional unit assignments made to cover all the split-nodes, along with their associated transfer nodes.

The goal is to examine the nodes in a given assignment and merge them into groups of nodes that can be executed in parallel on the target processor. Each grouping corresponds to a VLIW instruction. We then want to select the minimal-cost set of VLIW instructions that can cover all the nodes in the assignment.

In order to reduce the total cost of the VLIW instructions required to cover the Split-Node DAG, it is preferable to select instructions that make good use of the parallelism provided in the target processor. Thus, the groupings of nodes that we examine are all the maximal groupings, or **maximal cliques** of parallel nodes. In other words, subsets of a larger clique are not considered as a possible grouping. Note that every node in the assignment being explored is covered by at least one clique. It is possible for a clique to contain only one node. Also, the maximal cliques group operation nodes together with data transfer nodes.

C.1 Generating the Maximal Cliques

We create an $N \times N$ matrix representing pairwise parallelism, where N is the number of nodes in the current assignment of the Split-Node DAG. This matrix contains a 0 for element $[i, j]$ if N_i can be executed in parallel with N_j, and 1's elsewhere. Any two operations using different functional units and having no directed path between them can be performed in parallel. Figure 7 shows the matrix for a proposed assignment to the Split-Node DAG of Figure 4 consisting of nodes N_2, N_9, N_{10}, and N_{14}. Row N_{14}, for example, specifies that an ADD on unit U3 (N_{14}) can be performed in parallel with a MUL on unit U2 (N_{10}).

The maximal cliques generated for this example would be: $(C_1 : N_2)$, $(C_2 : N_{10}, N_9)$, $(C_3 : N_{10}, N_{14})$.

The pseudo-code shown in Figure 8 describes our algorithm for generating all the maximal cliques using the pairwise parallelism matrix. The routine **gen_max_clique** is called from a parent routine within a for-loop which iterates through all the nodes as the starting clique node. The value of the argument index is set to the starting clique node number.

Within the **gen_max_clique** routine, the first loop adds all the nodes that do not preclude the addition of any other nodes to the current clique. When no more nodes can be added that meet this criterion, the second loop spawns off several recursive calls to **gen_max_clique** where each call has one new node added to the clique. The new node can be executed in parallel with all the nodes in the current clique; however, it does preclude the addition of some other nodes. This process is repeated with the new clique until no new nodes can be added, at which point a maximal clique has been found.

The pruning condition follows from the fact that if i < index, then we will have already generated all the cliques that will be generated in this (recursive) call. Thus, we can terminate this branch. For example, suppose that we have a clique $c = \{j\}$ and we are adding a node i to the clique that does not preclude the selection of any other node. If $i < j$, then we would have already generated maximal cliques from clique $\{i, j\}$, when we started with clique $\{i\}$ as the seed.

C.2 Reducing the Number of Maximal Cliques Generated

Generating all of the maximal cliques is the most time consuming portion of our algorithm. In order to improve runtime, we have implemented a heuristic that only allows the merging of nodes whose level from the bottom and level from the top of the solution DAG are close to each other. It is generally not a good idea to merge nodes whose levels from top or bottom are very different because chances are that such a merging would preclude the merging of other nodes that would allow greater use of parallelism. We found that incorporating this heuristic improves runtime because fewer maximal cliques are generated, and maintains the quality of our results.

C.3 Eliminating Illegal Instructions

The maximal clique generation phase merges all nodes that have no data dependence between them and are executed on different functional units into single instructions (i.e., cliques). Merging based solely on this criterion is insufficient to guarantee legality of the instructions on the target processor. Illegal groupings are described in the Constraints section of the ISDL description. Thus, each proposed instruction must be compared with the constraints of the target processor. If the constraints are not met, then the illegal instruction, or maximal clique, is split into instructions with smaller cliques until all the constraints are met.

D. *Selecting a Minimum-Cost Set of Maximal Cliques*

Once the maximal cliques have been generated, the next step is to find the minimum-cost set of cliques that cover all the nodes in the assignment.

Our covering algorithm begins with an empty solution set. It then selects a maximal clique that covers the largest number of remaining uncovered nodes whose children have all been covered (i.e., nodes at the bottom of the Split-Node DAG will be scheduled before nodes that depend on them, thus creating a schedule as the cliques are selected) and whose register requirements do not exceed the available resources. The available resources are determined by performing a liveness analysis on the selected nodes and maintaining a running upper bound on the number of required registers for each register bank. After selecting the clique, the remaining cliques are shrunk so that they no longer include any of the covered nodes. This process is repeated until all the nodes have been covered.

In the case of a tie, where several cliques cover an equal number of nodes, the algorithm differentiates among them by using a lookahead cost function that estimates the number of cliques required to cover the rest of the nodes, if that particular clique was added to the solution set.

In the event that all the remaining cliques contain nodes that would require a spill to memory in order to satisfy the register resource constraints, the algorithm determines which covered node should be spilled. The decision is made based on the most needed resource, and the number of parent nodes that would later require the spilled value to be reloaded from memory. Once the node to be spilled is selected, the required load and spill nodes are added

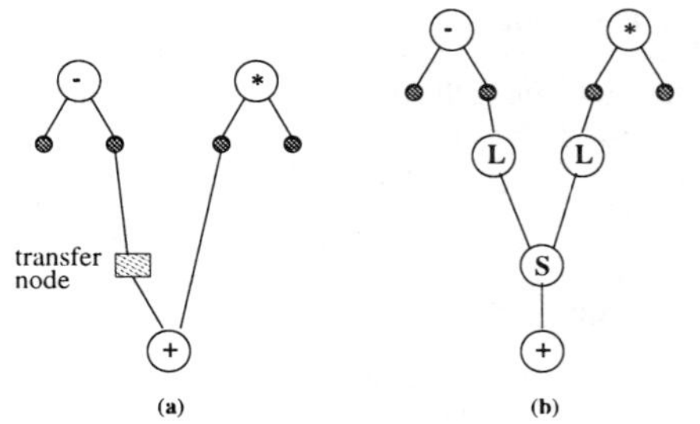

Fig. 9. **Inserting Loads and Spills into the Split-Node DAG (a)** the original Split-Node DAG, **(b)** the Split-Node DAG augmented with load (L) and spill (S) nodes

to the Split-Node DAG, and any transfer nodes that are no longer required are removed, as shown in Figure 9. Here the + node was selected as the node to spill, and the transfer node between the + and - nodes was removed. New maximal cliques are then generated for all the remaining uncovered nodes including the new load and spill nodes. The covering algorithm then continues with the new maximal cliques and the remaining uncovered nodes.

E. *The Covering Solution*

The assignment that required the minimum-cost set of cliques to cover all its nodes is selected as the final assignment. The order in which cliques were selected determines a schedule of the cliques (i.e., instructions). The final covering solution is thus a minimal-cost set of shrunk maximal cliques that cover the Split-Node DAG. It implies that unit assignment has been made, operation and data transfer nodes have been merged into VLIW instructions, register bank allocation has been performed including the addition of load and spill nodes when necessary, and a schedule has been determined. The remaining task is that of detailed register allocation.

F. *Detailed Register Allocation*

We perform detailed register allocation using conventional graph coloring algorithms [5]. We are guaranteed to be able to color each register bank graph using the given number of registers because we have analyzed the variable lifetimes in the instruction selection and scheduling step. However, it is possible that not all the inserted load and spill operations are required because the lifetime analysis used for inserting these operations is pessimistic.

G. *Peephole Optimization*

If, after performing detailed register allocation, it is determined that a particular load or spill is not needed, peephole optimization will be performed in an attempt to improve the schedule. It will remove the unnecessary loads and spills and try to compact the schedule by moving other operations into the empty slots if the dependency constraints allow it. This may, or may not, reduce the final number of required instructions.

V. Related Work

This section presents a brief survey of compiler literature that pertains to retargetable compilers for embedded systems, along with a comparison to our own work. We briefly review several representative research projects in this area: MIMOLA [6], FLEXWARE [7], CHESS [8], and the ILP-Based approach of Wilson et al. [9].

A. *Mimola*

MIMOLA's distinguishing feature is a microcode compiler that infers rules for code generation directly from a *structural* description (e.g., a net-list) of the target architecture instead of a *behavioral* description (e.g., the instruction set). The advantage of this approach is that the same machine description is used for both the synthesis of the target architecture and the generation of machine code. However, it is more difficult for the compiler to find code optimizations without having explicit behavioral information.

B. *FlexWare*

FLEXWARE includes a retargetable code generator, CODESYN, that takes an algorithm expressed in a high-level language and maps it onto a user defined instruction set to produce optimized machine code for a target ASIP or a commercial processor core. It uses a mixed structural and behavioral level model. This mixed model includes an enumeration of all possible partial instructions (i.e., microinstructions), an abstract netlist describing the data path topology, and a definition of the register classes.

CODESYN first converts the high-level source program into a hierarchy of Control-Data Flow Graphs (CDFGs). A distinguishing feature of CODESYN's pattern matching phase is its use of pre-sorted tree-like patterns that allow for pruning of the search tree of possible target processor instruction matches [10]. It determines the best implementation of the CDFG on the target processor, including the support of complex instruction recognition and utilization. Global scheduling and register allocation are performed. Finally, the compaction, assembly, and linkage phases produce the machine code.

The general framework of CodeSyn is similar to our own. However, AVIV is more focused on determining a globally optimum solution which considers instruction selection, resource allocation, and scheduling concurrently, rather than performing the various code generation steps sequentially.

C. *Chess*

CHESS is a retargetable code generation system targeting fixed-point digital signal processors and ASIPs. It generates machine code for the target processor, described in the nML language, and provides feedback as to how well suited the target processor is for the given application.

The nML target processor description is translated into an Instruction-Set Graph (ISG), a mixed structural and behavioral representation of the processor. The ISG models connectivity and encoding restrictions, as well as structural hazards.

The code generation process first translates the input algorithm into a CDFG. Code selection covers the CDFG with patterns that correspond to partial instructions supported by the instruction set, called bundles. Rather than making an exhaustive list of all possible bundles, they bundle instructions on the fly by searching for valid paths in the ISG. In CHESS each phase of code generation is performed separately; however, in order to ensure phase coupling, an intermediate scheduling view is constructed after each phase which takes into account constraints imposed by any previous phase.

CHESS is driven by nML which differs from ISDL, particularly in the way constraints are handled. Constraints in ISDL allow the code generator to treat all operations as completely orthogonal. Illegal operation combinations are then removed by comparing the maximal cliques to the constraints. nML, on the other hand, is an attributed grammar where production rules define the instruction set [11]. Therefore, all legal groupings of operations must be explicitly listed. Thus, ISDL descriptions can be more concise, and, as a result, easier for the code generator to handle. Further, as stated above, each phase of code generation is performed separately, though not independently. Again, AVIV is focused on solving the various code generation problems concurrently.

D. *Wilson et al.*

Wilson et al.'s Integer Linear Programming (ILP) based approach to code generation is based on a behavioral model of the target processor. Similar to FLEXWARE and CHESS, code generation begins by translating a high-level source language into a Data-Flow Graph (DFG). The designer can supply optimization hints. Pattern matching is used to recognize complex instructions. Next, a possible schedule that attempts to minimize overhead costs such as spills to memory and address calculations is identified. Register assignment, including necessary spills to memory, is then performed.

The cornerstone of all the optimizations is an ILP solver that can simultaneously do scheduling, instruction selection, register assignment, and compaction, as well as select from alternative spill and addressing candidates. Since all the constraints are considered simultaneously in the ILP formulation, trade offs can be made between the various optimizations leading to a globally optimal solution.

The problem with using ILP solvers for code generation is that finding the optimal solution is far too CPU intensive. The ILP solver does not have sufficient information about the structure of the problem in order to make intelligent decisions about how to prune

Basic Block	Original DAG #Nodes	Split-Node DAG #Nodes	#Registers per RegFile	#Spills Inserted	#Instr In Solution By Hand	#Instr In Solution Aviv	CPU Time (secs)
Ex1	8	30	4	0	7	7 (7)	0.1 (0.2)
Ex2	13	56	4	0	10	10 (10)	0.4 (37.2)
Ex3	11	55	4	0	13	13 (13)	0.9 (122.3)
Ex4	15	81	4	0	16	16 (16)	8.2 (41,466)
Ex5	16	106	4	0	14	16 (14)	10.7 (89,337)
Ex6	15	81	2	2	18	22 (18)	6.9 (29,072)
Ex7	16	106	2	1	15	18 (15)	9.9 (64,748)

TABLE I

Code Generation Experiments for the Example Target Architecture

Basic Block	Original DAG #Nodes	Split-Node DAG #Nodes	#Registers per RegFile	#Spills Inserted	#Instr In Solution By Hand	#Instr In Solution Aviv	CPU Time (secs)
Ex1	8	17	4	0	8	8	0.1
Ex2	13	28	4	0	11	12	0.2
Ex3	11	23	4	0	13	13	0.7
Ex4	15	29	4	0	16	17	3.0
Ex5	16	51	4	0	15	15	11.4

TABLE II

Code Generation Experiments for the Target Architecture II

the search space. Thus, in general, user supplied hints are required to produce good code within a reasonable amount of CPU time.

VI. Experiments and Ongoing Work

We have implemented a preliminary version of the retargetable code generator and run some experiments on several basic blocks. These examples are generic basic blocks that occur in DSP application code. Examples 1-2 are simple basic blocks that are found as part of a conditional statement or loop. Examples 3-5 are simple basic blocks of loops that have been unrolled twice.

Aviv generated code for these basic blocks targeting minimum code size. The results for the example target architecture of Figure 3 are summarized in Table I. The table summarizes the number of nodes in the original DAGs of the basic blocks examined, as well as, the number of nodes in the equivalent Split-Node DAGs for the example architecture. It compares the number of instructions found by hand-coding, to the solution found by Aviv. The hand-coded results are all optimal. The number of instructions equals the number of clock cycles required to execute the basic block on our target architecture. Examples 1-5 were run with four registers per register file. These examples did not require any spills to memory. Examples 4 and 5 were then rerun with two registers per register file to show what happens when the required number of registers exceeds the available resources. Both of these examples resulted in spills to memory, and their results are presented in examples 6 and 7 respectively. Note, however, that the optimal solutions for examples 6 and 7 did not require spills. These solutions were not found by Aviv because the initial functional unit assignment cost function did not detect that the functional unit assignments it made would result in spills to memory.

We see that Aviv's results are quite close to optimal. The CPU times shown for finding the solutions were measured on a Sun Microsystems Ultra-30/300. Aviv incorporates multiple heuristics that can be turned off if desired. For example, rather than selecting only a few of the possible split-node assignments, it can generate all possible assignments. Also, heuristics such as using the level from top and bottom to reduce the number of maximal cliques generated, can be turned off. In parentheses, we present the results of running these examples with the heuristics turned off. Note that turning off the heuristics does not result in an exact algorithm for code generation since we do not explore all possible schedules. It is clear that our pruning heuristics work very well, and generate the same quality results within a fraction of the CPU time required to find the optimum solution.

Our purpose in developing the Split-Node DAG structure, and heuristic algorithms for covering it, was to enable retargetable code generation. With this in mind, we changed the target architecture of Figure 3 by removing the `SUB` operation from functional unit `U1`, and completely removing functional unit `U3`. The results using this architecture are summarized in Table II. As can be seen, for several of these basic blocks, removing a functional unit does not degrade performance. The flexibility of the Aviv retargetable code generation system allows for the exploration of different architectures until the best one is found.

We are currently working on modifying the initial functional unit assignment cost function to incorporate register resource limits so that it can detect assignments that are likely to require spills to memory. In addition, we are working on peephole optimization methods that will be applied after detailed register allocation.

Acknowledgments

This research is supported by NSF Contract MIP-9612632.

References

[1] R. K. Gupta and G. De Micheli. Hardware–Software Cosynthesis for Digital Systems. *IEEE Design & Test of Computers*, pages 29–41, September 1993.

[2] G. Hadjiyiannis, S. Hanono, and S. Devadas. ISDL: An Instruction Set Description Language for Retargetability. In *Proceedings of the* 34th *Design Automation Conference*, pages 299–302, June 1997.

[3] Stanford Compiler Group. *The SUIF Library*, 1.0 edition, 1994.

[4] SPAM Research Group. *SPAM Compiler User's Manual*, 1.0 edition, 1997.

[5] G. Chaitin, M. Auslander, A. Chandra, J. Cocke, M. Hopkins, and P. Markstein. Register Allocation via Coloring. *Computer Languages*, 6:47–57, 1981.

[6] P. Marwedel. The MIMOLA Design System: Tools for the Design of Digital Processors. In *Proceedings of the 21th Design Automation Conference*, pages 587–593, 1984.

[7] P. Paulin et al. FlexWare: A Flexible Firmware Development Environment for Embedded Systems. In *Code Generation for Embedded Processors*, pages 67–84. Kluwer Academic Publishers, 1995.

[8] D. Lanneer et al. CHESS: Retargetable Code Generation for Embedded DSP Processors. In *Code Generation for Embedded Processors*, pages 85–102. Kluwer Academic Publishers, 1995.

[9] T. Wilson et al. An ILP-Based Approach to Code Generation. In *Code Generation for Embedded Processors*, pages 103–118. Kluwer Academic Publishers, 1995.

[10] C. Liem, T. May, and P. Paulin. Instruction-Set Matching and Selection for DSP and ASIP Code Generation. In *Proceedings of the European Design and Test Conference*, pages 31–37, Feb 1994.

[11] M. Freericks. The nML Machine Description Formalism. Technical report, TU Berlin, 1993.

Code Compression for Embedded Systems

Haris Lekatsas and Wayne Wolf
Department of Electrical Engineering
Princeton University
{lekatsas,wolf}@ee.princeton.edu

Abstract

Memory is one of the most restricted resources in many modern embedded systems. Code compression can provide substantial savings in terms of size. In a compressed code CPU, a cache miss triggers the decompression of a main memory block, before it gets transferred to the cache. Because the code must be decompressible starting from any point (or at least at cache block boundaries), most file-oriented compression techniques cannot be used. We propose two algorithms to compress code in a space-efficient and simple to decompress way, one which is independent of the instruction set and another which depends on the instruction set. We perform experiments on two instruction sets, a typical RISC (MIPS) and a typical CISC (x86) and compare our results to existing file-oriented compression algorithms.

1 Introduction

Many embedded computing systems are space and cost sensitive. As a result, available memory is limited, posing serious constraints on program size. We are studying ways of reducing the size of stored programs in such systems by using code compression techniques. The reduced executable size can result in significant savings in terms of cost, size, weight and power consumption.

RISC instructions were designed in part to reduce instruction decode overhead. However, in deep-submicron pipelined processors, we believe that most instruction decode time can be hidden (as in the Intel architectures) particularly in the case of compressed code. We decompress our code before inserting it in the cache, following the design first proposed by Wolfe and Chanin [13]. We therefore assume that the processor executes normal uncompressed code and that decompression of a cache line occurs only when there is a cache miss. The loss in performance should therefore depend on the instruction cache hit ratio.

Such a system poses some constraints on the compression algorithms that can be used. Since programs can have jumps, we must decompress individual cache blocks and not the whole program sequentially. Compression techniques which require sequential decompression starting from the beginning of the compressed file cannot be used here. Furthermore the decompression hardware should be fast enough not to severely reduce CPU performance and should be kept as small as possible.

There is an abundance of algorithms for data compression in the literature. Due to the constraints stated above most of the existing state of art algorithms cannot be used directly. In terms of compression ratios the the algorithms that use finite context modelling such as PPM (Prediction by Partial Match) [1], DMC [3], and WORD [8] seem to achieve the best performance. However they require large amounts of memory both for compression and decompression, making them unsuitable for program compression. The Ziv-Lempel family of algorithms [14] use pointers to previous occurences of strings which makes an individual block decompression scheme impossible.

An instruction-based compression approach which overcomes the above problems is described by Kozuch and Wolfe [5], where byte-based Huffman codes [4] are used, yielding a compression ratio around 0.73 (compressed size/original size). Such a scheme is simple to implement, and certainly allows decompression to take place from any place in the code. A shortcoming of this method is the the fact that 8-bit symbols have been used instead of 32-bit symbols corresponding to the size of RISC instructions. This means that all 4 bytes within the same 32-bit word are encoded using the same table. Since instructions have different fields which have different statistical characteristics such a choice increases the entropy of the source significantly. However, if we use 32-bit symbols we need a Huffman table with 2^{32} entries making the size of the decompressor prohibitively large. Futhermore, this method does not take into account dependencies between instructions, limiting the overall compression performance. Another dictionary-based approach is described by Liao et al. [6], where a dictionary is generated for TMS320C25 code. Their approach requires little or no extra hardware, but gives only modest compression ratios.

In this paper, we present two different approaches which produce significantly better compression. We concetrate on the algorithm side to understand the limits of program compressibility as a CAD problem. The related architecture question is the cost in time (and gates) of these techniques on CPUs. We present here some initial analyses which suggest our schemes can be reasonably implemented in hardware, but architectural details remain future work. In section 2, we describe design issues for embedded code compression and we introduce our techniques. Section 3 describes the details of our statistical coding scheme, which is independent of the instruction set and only assumes fixed-size instructions. Section 4 discusses an alternative approach which uses dictionary coding and is instruction set dependent. In section 5 we present our experimental results, and finally in section 6 we conclude and discuss possible future directions.

2 Design issues

An approach to reduce code size in embedded systems would be to design a new RISC or CISC architecture with denser encoding. Although such an approach may be useful, we chose to follow an alternative method where the memory sys-

DAC 98, San Francisco, California

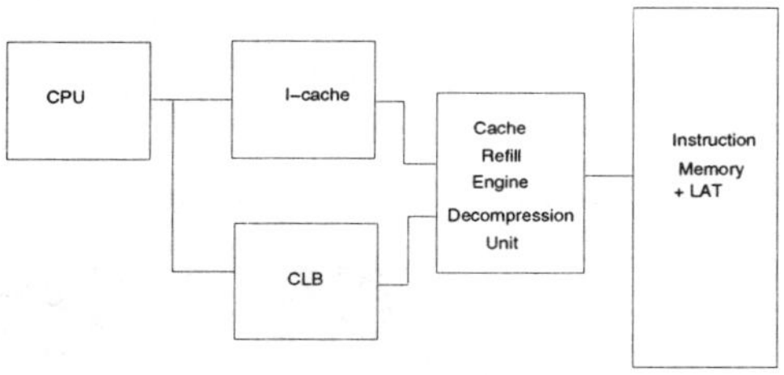

Figure 1: Memory System Organization

tem is redesigned while the processor architecture remains intact. Such an approach has several practical advantages: We can use well-characterized existing processor pipelines, take advantage of existing programming development tools, use existing workstations as development platforms, etc.

Wolfe and Chanin [13] designed a system with run time decompression. Figure 1 shows the memory organization for such a system. The instruction cache is used as a decompression buffer storing recently used decompressed cache blocks. The decompression is being done by the cache refill engine whenever a cache miss occurs. This implies that the compression algorithm must compress on a block-by-block basis. Since the engine must know where to find a compressed cache line in memory (different cache lines will generally have different compressed sizes) a table called LAT (line address table) has been stored in main memory along with the instruction code. This table maps program instruction block addresses into compressed code instruction block addresses. Since accessing the LAT will increase the cache refill time a CLB (Cache line Address Lookaside Buffer) can be used which is essentially identical to a TLB.

Note that we only compress the part of the executable which contains instructions, not any data, tables etc. Compressing data which may also be writable would complicate the design as we would have to provide a fast compressor as well when writing to main memory.

3 SAMC: An ISA-independent method using statistical coding

Our first method uses a binary arithmetic coder [12] driven by a Markov model. In the following we will refer to this method as *SAMC* for Semiadaptive Markov Compression. Since we are compressing cache blocks, an adaptive method cannot be used effectively as the coder will not be able to gather enough statistical information from just one block. A static model to feed the arithmetic coder is therefore required. We will use the term *stream* to refer to a group of bits within instructions. Figure 2 shows an example of stream subdivision for a MIPS instruction set. Although the bits in the figure are adjacent, this is not required by our algorithm. We performed several experiments on this stream subdivision idea, and found that a scheme that works particularly well is to divide instructions into 4 streams of 8 bits each and build Markov models for each one of them. This gives reasonable compression without requiring excessive storage. Here is the two-step compression scheme (n is the number of bits per instruction):

1. **Statistics-gathering phase.**
   (a) Instructions are divided into k streams of bits, each of the streams containing k_i bits, i=0,1,..,k-1, i.e. $n = \sum_{i=0}^{k-1} k_i$

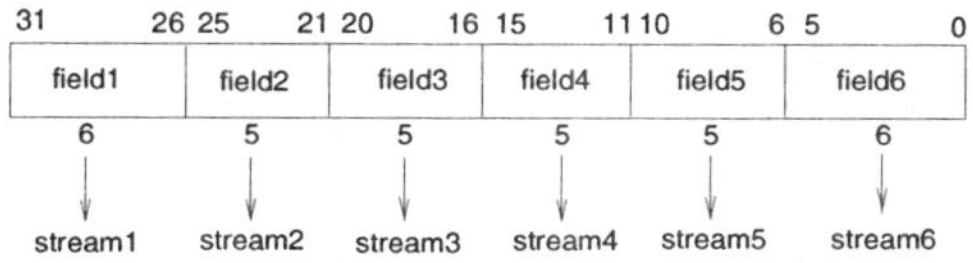

Figure 2: Example stream division for MIPS

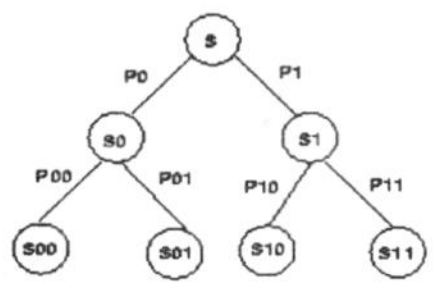

Figure 3: Binary Markov tree

   (b) For each stream we generate a binary Markov tree with $2^{k_i+1} - 1$ states. The first state is the initial state corresponding to no input bits. Its left and right child correspond to the states "0 input" and "1 input" respectively. Figure 3 shows such a tree for a stream with 2 bits. State S_0 corresponds to the state "0 input", state S_{01} corresponds to the state "0 input then 1 input" , etc. Each transition has a probability which is generated by going through the whole subject program. Probability $P_{b_0 b_1 b_2 \ldots b_l}$ corresponds to the probability that the bit sequence $b_0 b_1 b_2 \ldots b_l$ has arrived. We only need to store the probabilities on the left branches of the tree since the probabilities on the right branches are the complements of the probabilities on the left branches. Hence for the tree in figure 3 we need to store 3 probabilities: P_0, P_{00}, P_{10}. In general for a stream of k-bits we need to store $(2^{k+1} - 2)/2$ probabilities.

2. **Compression phase.** The probabilities calculated above are done as a preprocessing step. The compressor uses them as predictions on the coming bit to perform the binary arithmetic coding. Hence the compressor goes through the subject program a second time and encodes the message. The final storage requirements are the encoded message and the Markov trees for the all the streams.

Each stream is coded independently and any correlation between streams is not taken into account. Compression performance can be improved by connecting the Markov trees of adjacent streams. This provides some limited memory between streams to the model. The connection between trees is illustrated in figure 4, where for simplicity we have assumed that instructions are 4 bits wide and that we have divided them into 2 streams, 2 bits each.

There are two reasons for dividing instructions into streams. First, it would be impossible to generate one Markov tree for the whole instruction as the number of probabilities we would need to store would be $(2^{32+1} - 2)/2 = 2^{32} - 1$. The second reason is that we can chooose the streams appropriately so that they have low entropy and thus high compressibility. In order to do the stream division we calculated the correlation factor ρ_{ij} between all bits i and j, i,j = 0,1,...,31, and we tried to group bits that are strongly correlated. Strongly correlated bits will give either high (close to 1) or low (close to 0) probabilities for the Markov tree

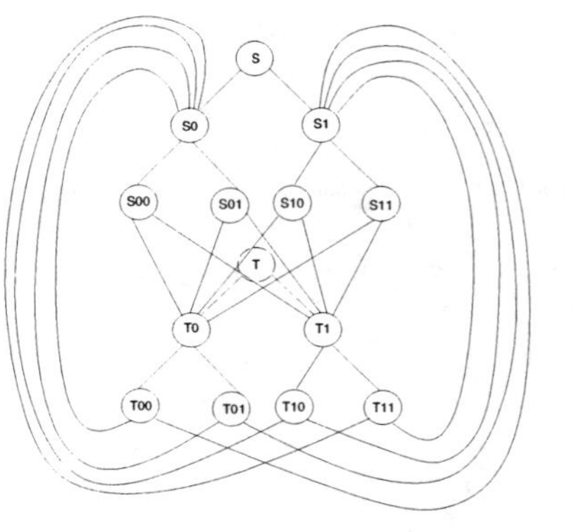

Figure 4: Connected Markov trees

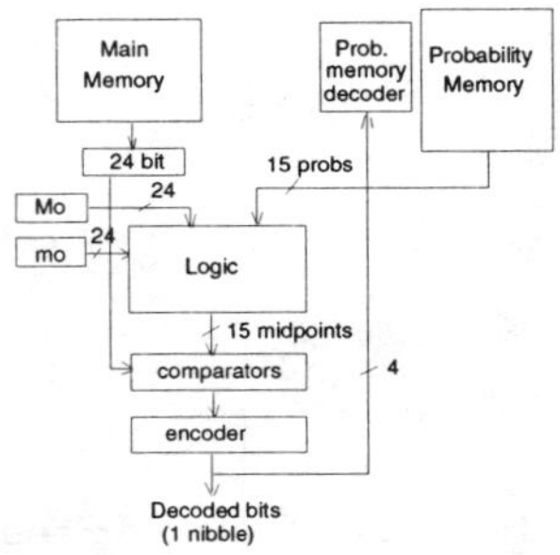

Figure 5: Decoder Block Diagram

improving the compression performance of the arithmetic coder. Our program combines bits with high correlation to streams and calculates their entropies. It then attempts to exchange some bits between streams randomly and recalculates the entropies. If the new average entropy is lower it accepts this step, otherwise it tries a different combination. In practice, we found that dividing 32-bit instructions into 4 8-bit streams (a stream does not necessarily have adjacent bits) produces results close to optimal.

Once the Markov trees are calculated for all streams the compressor and decompressor use the same steps to perform the arithmetic coding of one cache line. The following lines of pseudo-code illustrate the steps the decompressor must take to decompress one cache line. Note that the interval [min,max] used by our approach has to be reset upon compression and decompression of each block and we must also restart from the initial node of the Markov model. This ensures that we can decompress starting from any block boundary. For the same reason we reset the Markov model to its starting node for each block. We have assumed that the decompression engine works with 24 bits of the compressed code, and that 24 bits is the accuracy of our interval.

```
1   max = 0x1000000; min = 0  // use 24 bits
2   p = top_of_Markov_tree;   // make p point to the start
3   bytes_written = 0;        // of the model
4   val = get_24bits_of_compressed_code();
5   do {
6     c = 0;
7     if (val == (max-1)) break;
8     for (i=0; i<8;i++) {
9       mid = min + (max-min-1)*p->prediction;
10      if (mid == min) mid++;       // fix mid value if
11      if (mid == (max-1)) mid--; // p->prediction=0,1
12      if (val >= mid) {
13        bit = 1;                   // decode 1 bit
14        min = mid;                 // update interval
15      }
16      else {
17        bit = 0;                   // decode 1 bit
18        max = mid;                 // update interval
19      }
20      if (!bit) then p = p->left // update position
21      else p = p->right          // in Markov tree
22      c = c<<1 | bit;            // store decoded bit
23      while ((max-min) < 0xff) {
24        if(bit)max--;
25        bytes_written++;
26        val = (val << 8) | get_byte(); // get 8 more
27        min = min << 8;                // compressed
28        max = max << 8;                // bits
29        if (min >= max) max = 0x1000000;
30      }
31    }
32    write_byte_to_cache(c);
33  } while (bytes_written<block_size);
```

The above code performs binary arithmetic decoding. Starting with the interval [min,max] the algorithm uses the probability from the Markov tree (line 9) to calculate a midpoint value. The input value (compressed code) in val is compared with the midpoint and a 0 or a 1 is written to the output (decompressed code). The main drawback of the procedure above is that it works on a bit-by-bit basis. In order to speed it up we can perform midpoint calculations in parallel. However the new midpoint always depends on the previous one, as the new value of max or min are equal to mid (lines 9,14,18). In order to overcome this we can calculate all midpoints for both the cases when $val >= mid$ and $val < mid$. The procedure below gives all the midpoints, where the probabilities $p_{x_0 x_1 \ldots}$ are given by the Markov tree:

$$
\begin{array}{ll}
mid_0 = min_0 + (max_0 - min_0 - 1) \cdot p_0 & \\
mid_1 = mid_0 + (max_0 - mid_0 - 1) \cdot p_{00} & , mid_0 < val \\
mid_1 = min_0 + (mid_0 - min_0 - 1) \cdot p_{01} & , mid_0 > val \\
mid_2 = mid_1 + (max_0 - mid_1 - 1) \cdot p_{000} & , mid_0 < val \\
mid_2 = mid_0 + (max_0 - mid_0 - 1) \cdot p_{010} & , mid_0 < val \\
mid_2 = mid_1 + (mid_0 - min_0 - 1) \cdot p_{100} & , mid_0 > val \\
mid_2 = mid_0 + (mid_0 - min_0 - 1) \cdot p_{110} & , mid_0 > val
\end{array}
$$

...

We can see that all μ_i's can be expressed as functions of M_0, m_0 and p_{ij}'s. Clearly the number of μ_i's and p_{ij}'s grow exponentially. A reasonable solution is to decode 4-bit values which means we need 15 μ_i's and 15 p_{ij}'s. After calculating all μ_i's and p_{ij}'s, we choose the right ones by comparing them with val. Figure 6 shows a block diagram of the decompression engine using the algorithm described above. The "Logic" and "Comparators" calculate midpoints. Note that the output bits are used to determine which probabilities to fetch from the probability memory. To avoid the multiplication in the midpoint calculation unit we can constrain the probability of the less probable symbol to the nearest integral power of $\frac{1}{2}$, thus requiring only shifts. Witten et al [12] showed that the worst-case efficiency is about 95% when we pose this constraint. In our case, in order to save space, we only store the probability of a 0 input. If it is the less probable symbol then only a shift is required, otherwise a shift and a subtraction from the max_i point is required. This greatly simplifies the design of the midpoint calculation unit.

4 SADC: An ISA-dependent method using a dictionary

Another valid choice is to build a dictionary with common sequences of instructions. Dictionaries can be static, semi-adaptive, or dynamic [1]. Dynamic dictionaries are adaptive and change while the compressor or decompressor goes through the program. This technique cannot be used here because it does not allow for random access. Static dictionaries are built once and used for all programs, while semiadaptive are built for each subject program. Clearly a semiadaptive dictionary will achieve better compression for a given program as it is specifically designed for that program. The method of this section uses a semiadaptive dictionary to compress opcodes, opcode-register combinations and opcode-immediate combinations. The stream subdivision idea also applies here; instructions are divided into streams in a predetermined way which depends on the instruction set. For example, we divided MIPS instructions into 4 different streams: opcode, register, immediate,long immediate stream (streams have different bit widths in this case). In the following we will use the term *SADC* (Semiadaptive Dictionary Compression) for our compression scheme.

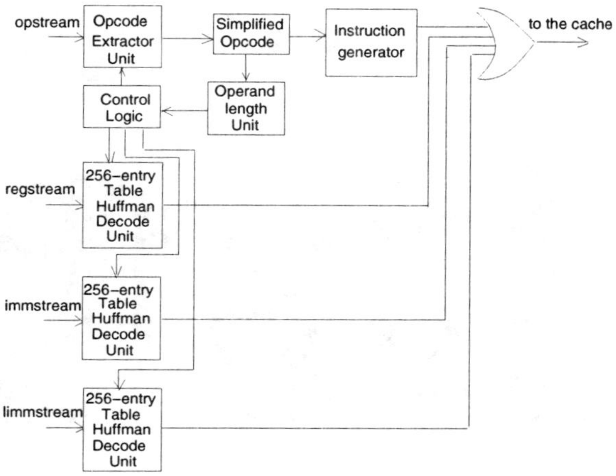

Figure 6: Decompressor Block Diagram for MIPS architecture

4.1 Compressing the streams

Opcodes in an instruction sequence, tend to exhibit some dependence between them since compilers typically generate code which uses the same sequences of instructions over and over. Having thrown away immediates and registers we can fully exploit such dependencies by generating new augmented opcodes which combine several opcodes together. We noticed that all our benchmark programs tend to use no more than 50 instructions. Since we cannot allow such opcode augmentation to become very large and in order to keep the opcode value in one byte, we restricted the maximum number of opcodes to 256. This means that we can augment the instruction set by about 200 new opcodes. We are effectively building for each subject program a semiadaptive dictionary which maps indices to opcodes or opcode combinations.

Finding the best dictionary for a given program is shown by Storer [10] to be NP-hard. We therefore do not attempt to generate an optimal dictionary. Once the dictionary is generated we are also faced with the problem of how to parse the subject program and replace occurences of opcode combinations with dictionary indices. Several heuristics for parsing have been proposed with *greedy* parsing being the most popular due to its simplicity and speed. Our approach does dictionary generation and parsing simultaneously. It inserts one candidate for dictionary insertion at a time, and then it modifies the subject file by inserting the dictionary index in the place of the candidate. It then works on the new modified file by finding new candidates and replacing them by their dictionary indices.

In order to decide which opcodes or opcode combinations should be inserted in the dictionary we test the following candidates:

- Adjacent opcode pairs. This attempts to take advantage the correlation between adjacent instructions.
- Generation of new opcodes for instructions which appear frequently with some specific registers or immediates. This means that if, for example, the register R31 in instruction jr R31 for MIPS code appears much more frequently than any other register, we can reduce the register stream size by introducing a new special opcode for jr R31.

Our dictionary generator works as follows:

1. The generator scans the subject program and creates a tree with all the opcodes and their frequencies, all the groups of 2 consecutive opcodes with their frequencies and finally all the groups of 3 consecutive opcodes with their frequencies. It does not attempt to search any bigger groups as that would result in large execution times and memory storage requirements.
2. Going through the tree generated in step 1, the generator inserts all single opcodes in the dictionary. It subsequently follows one of two paths: it either encodes the group of adjacent opcodes which has the largest gain in savings of the encoded opcode stream; or it encodes the opcode with a specific register or immediate which will gives the greatest reduction in register stream or immediate stream size. We explain how to calculate these gains in reduction below.
3. Using the dictionary generated above, the generator encodes the file by storing the dictionary indices for each opcode or group parsed.
4. All dictionary entries and the tree are erased. The generator repeats from step 1 until the dictionary generated has entries equal to the maximum allowed, or the new encoded file isn't smaller than the one of the previous cycle.

Our maximum dictionary size is 256 entries which means we need to store just one byte for the indices. The gain stated in step 2 of the algorithm is calculated as follows:

- Opcode groups. Suppose that the group to be inserted has length n. (Note that this length can be much larger than 2 or 3, since after several cycles of the algorithm the groups generated will contain longer sequences). Also suppose it occurs with frequency f, where f means non-overlapping occurences of the group. Since if it is inserted in the dictionary it will consume n bytes of space, the gain by inserting it in the dictionary and encoding the file will be $g = f \cdot (n-1) - n$.
- Register and immediate selection. The gain here is given by $g_{reg} = f \cdot n_{regs}$ and $g_{imm} = f$, where f is the frequency of an opcode, and n_{regs} is the number of registers selected for integration in the opcode.

The compressor calculates the above gains for all opcodes and picks the group of instructions with the largest gain to include in the dictionary. The final step of our compression is to encode all resulting compressed streams by using Huffman encoding. This tends to further improve overall

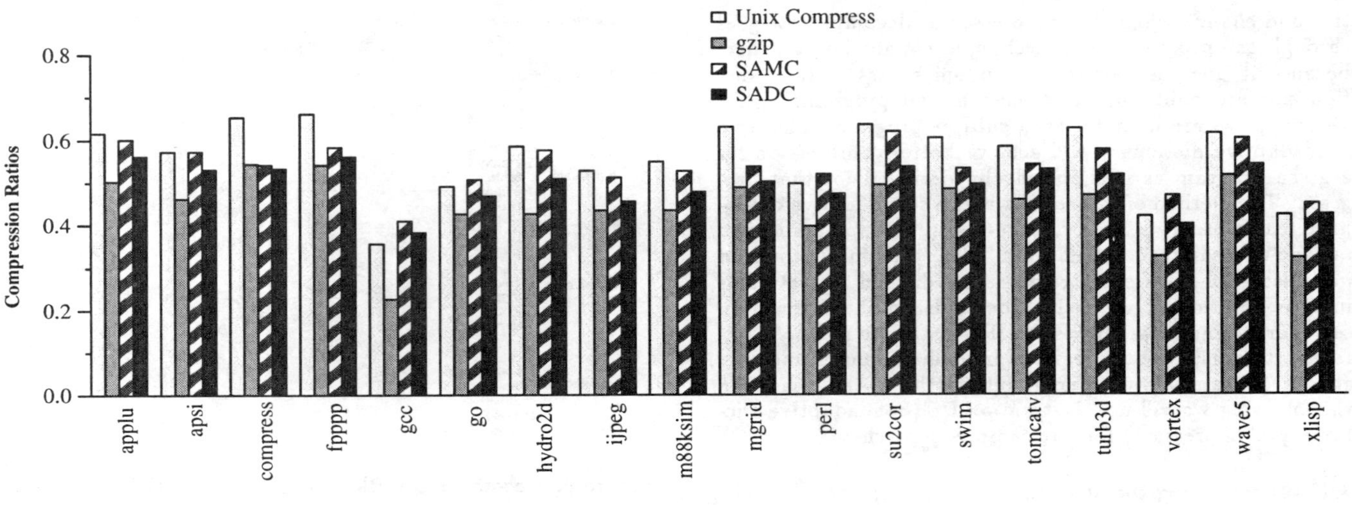

Figure 7: Compression results for MIPS

compression. Figure 6 shows a block diagrams for a MIPS decompression hardware:

- Opcode extractor unit: This takes as input the dictionary indices and produces the 8-bit opcode values contained in the dictionary. For MIPS an instruction generator unit is required since instructions are specified by one or two different groups of bits. The 8-bit opcode values are mapped to the corresponding bits in a real 32-bit instruction.
- Instruction generator. This maps 8-bit (simplified) opcodes to the format required by MIPS. The resulting 32-bit word coming as an output of this unit is ORed with the decompressed register and immediate streams and the final word is sent to the cache. This unit is needed because on MIPS the streams created are not composed from consecutive bits, hence we must put the decompressed bits to their right place.
- Operand length unit: This gives for each opcode how many registers and immediates (if any) must be combined with it to form an instruction.
- Control Logic unit: This sends the appropriate signals to the decompressing units, so that the right number of registers and immediates are combined with each opcode.

5 Experimental results

We performed experiments on two architectures, MIPS and x86 (Pentium Pro). For MIPS we divided instructions into 4 streams (opcode, register, 16-bit immediates, 26-bit immediates), while for Pentium we formed 3 streams (opcode, modR/M and SIB, immediate and displacement stream). The Pentium streams are 8 consecutive bits wide. This means the design of a Pentium decompressor would not need an instruction generator.

Figures 7 and 8 show our experimental results on MIPS and Pentium-Pro architectures. To measure compression performance we used the ratio of compressed program size over original program size. Thus a short bar means good compression performance. Since embedded code is hardly portable among architectures we used the SPEC95 benchmarks to measure the compressibility of our algorithms. On both machines we compiled the benchmarks using the default flags provided by the SPEC package. All of our experiments are done assuming a cache block size of 32 bytes. Different cache block sizes have a minimal impact on the results presented.

SAMC on MIPS tends to exhibit compression performance similar to UNIX compress. Gzip generally outperforms SAMC except for the compress benchmark. This is probably due to the small size of compress, which does not allow gzip to take advantage of long repeating sequences. SAMC however, sustains similar performance for both small and large files as it cannot see beyond cache blocks . SADC performs about 4-6% better than SAMC, and for some benchmarks on MIPS it comes close to gzip.

File compression algorithms tend to perform better on the CISC architecture. For SAMC this is expected as we cannot do any stream subdivision (SAMC was designed for fixed-sized RISC instructions). SADC performs better but still fails to come close to gzip. This may be attributed to the rather crude stream subdivision (only 3 byte-based streams). A different stream subdivision working with individual fields and not with whole bytes might improve compression, but one the other hand it would complicate the decompressor's logic.

Figure 9 compares our algorithms with byte-based Huffman coding as presented by Kozuch and Wolfe [5]. The compression ratios presented in figure 9 are the average compression ratios of all SPEC95 benchmarks. For MIPS SAMC and SADC perform substantially better than Huffman, while on Pentium the difference is not as big. SADC on Pentium does not do any stream subdivision and compresses bytes, hence performance similar to Huffman is expected. The results are slightly better than Huffman since the connected Markov trees exploit some of the dependencies between consecutive bytes, while Huffman does not take into account such dependencies. SADC performs much better than SAMC, but results might have been even better with more careful stream subdivision.

6 Conclusions and Future Work

This paper presents two methods to reduce code size for embedded systems. The compression performance results

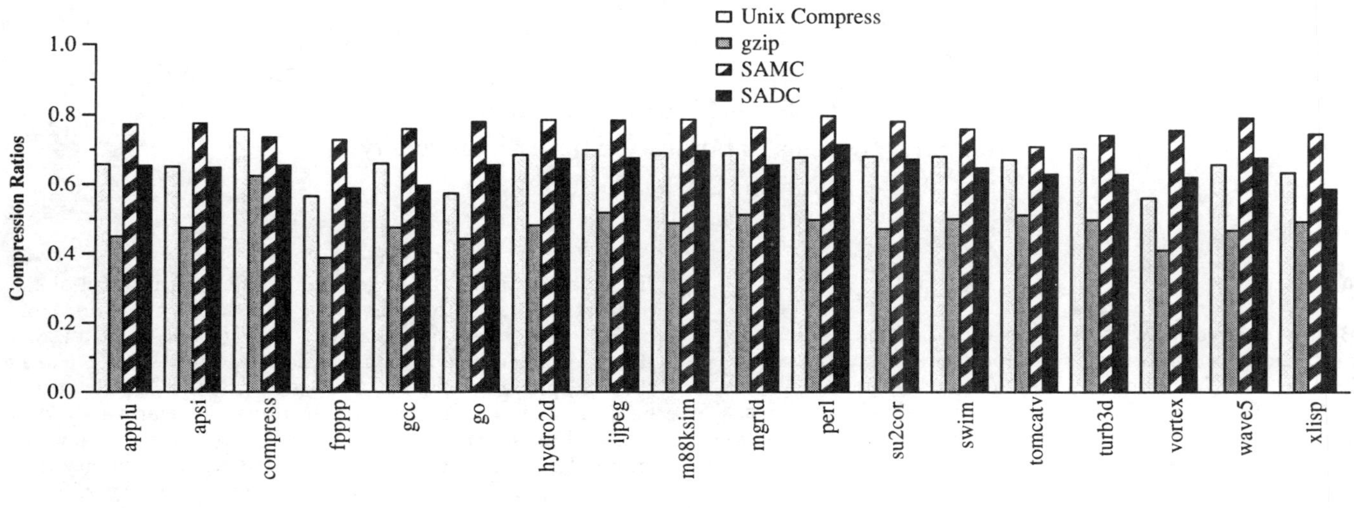

Figure 8: Compression results for Pentium Pro

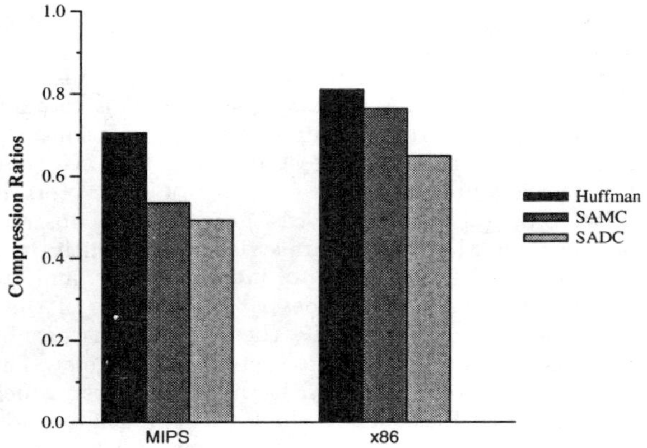

Figure 9: Instruction Compression Algorithms

show significant improvement over previous attempts. Our first method is targeted for RISC instruction sets with fixed-sized instructions and can work for any such architecture. Compression ratios are comparable to UNIX compress. Our second method works on a specific program and instruction set. As a result, it can achieve significantly better compression. Furthermore, as it is a dictionary method, it allows for fast hardware implementations. Its performance is much better than previous instruction compression schemes and is competitive with some file-oriented compression algorithms.

There are several areas that need further work. Some research can be done on how to generate the best Markov model given a subject program for compression. Finally, further research could be made into different and faster decompressor implementations.

Acknowledgments

This work has been supported by the DARPA Small Memory program.

References

[1] T. Bell, J. Cleary, and I. Witten. *Text Compression*. Prentice Hall, 1990.

[2] J. Bentley, D. Sleator, R. Tarjan, and V. Wei. A locally adaptive data compression scheme. *Communications of the ACM*, Vol. 29(No 4):pp 320–330, 1986.

[3] G. Cormack and R. Horspool. Data compression using dynamic markov modelling. *The Computer Journal*, Vol. 30(No 6), 1987.

[4] D. Huffman. A method for the construction of minimum-redundancy codes. In *Proceedings of the IRE*, volume Vol. 4D, pages pp 1098–1101, September 1952.

[5] M. Kozuch and A. Wolfe. Compression of embedded system programs. *1994 IEEE International Conference on Computer Design: VLSI in Computers & Processors*, 1994.

[6] S. Liao, S. Devadas, and K. Keutzer. Code density optimization for embedded dsp processors using data compression techniques. In *Proceedings of the 1995 Chapel Hill Conference on Advanced Research in VLSI*, pages pp 393–399, 1995.

[7] A. Mayne and E. James. Information compression by factoring common strings. *Computer Journal*, Vol. 18(No 2):pp 157–160, 1975.

[8] A. Moffat. Word based text compression. Technical report, Department of Computer Science, University of Melbourne,Parkville,Victoria,Australia, 1987.

[9] J. Rissanen and G. Landon. Universal modeling and coding. *IEEE Transactions on Information Theory*, Vol. 27:pp 12–23, 1981.

[10] J. Storer. NP-completeness results concerning data compression. Technical Report No 234, Department of Electrical Engineering and Computer Science, Princeton University, Princeton, N.J., 1977.

[11] T. Welch. A technique for high-performance data compression. *IEEE Computer*, pages pp 8–19, June 1984.

[12] I. Witten, R. Neal, and J. Cleary. Arithmetic coding for data compression. *Communications of the Association for Computing Machinery*, Vol. 30(No. 6):pp 520–540, June 1987.

[13] A. Wolfe and A. Chanin. Executing compressed programs on an embedded RISC architecture. *Proc. 25th Ann. International Symposium on Microarchitecture*, pages pp 81–91, December 1992.

[14] J. Ziv and A. Lempel. A universal algorithm for sequential data compression. *IEEE Transactions on Information Theory*, Vol. 23(No 3):pp 337–343, May 1977.

A Decision Procedure for Bit-Vector Arithmetic

Clark W. Barrett, David L. Dill, and Jeremy R. Levitt*
Computer Systems Laboratory, Stanford University
Stanford, CA 94305, USA

Abstract

Bit-vector theories with concatenation and extraction have been shown to be useful and important for hardware verification. We have implemented an extended theory which includes arithmetic. Although deciding equality in such a theory is NP-hard, our implementation is efficient for many practical examples. We believe this to be the first such implementation which is efficient, automatic, and complete.

1 Introduction

As designs grow in complexity, design verification becomes increasingly important and challenging. New and better verification techniques are critical to ensure correctness, maintain design cycle times, and protect designers from economic losses due to undiscovered bugs. Formal methods for verification are especially attractive because they have the potential to cover most or all of the behaviors in a design without having to exhaustively simulate it.

The Stanford Validity Checker (SVC) [2, 9] is an automatic verification tool which has been in development for several years at Stanford University. The input to SVC is a Boolean formula in a quantifier-free subset of first-order logic. It may also contain Boolean operators, uninterpreted functions, and various interpreted functions such as operations on infinite arrays and arithmetic. We have found these constructs to be useful for modeling hardware designs. Using a combination of case-splitting and cooperating decision procedures, SVC determines whether a formula is valid (i.e. equivalent to true in every possible interpretation). If the formula is not valid, SVC returns a counterexample. SVC is used as the final step in the automatic hardware verification paradigm of Burch and Dill [4]. In their approach a specification and an implementation are each symbolically simulated and the resulting states are then compared to see if they are equivalent. This method has been shown to be successful for verification of actual designs and is currently being applied to the TORCH microprocessor, an aggressive superscalar microprocessor developed for educational and research purposes at Stanford University [11]. The powerful and efficient decision procedures in SVC are critical for the success of this effort.

Other formal methods such as theorem proving and model checking have been used extensively, but theorem provers suffer from a lack of automation and model checking from the inability to handle large designs due to state explosion. SVC attempts to take the best (and avoid the worst) of both worlds. First of all, SVC is able to reason at an abstract level and has built-in decision procedures much like a theorem prover. However, the logic of SVC is restricted to be decidable, which enables it to prove or disprove all formulas automatically. Secondly, SVC uses a directed acyclic graph (DAG) structure much like that used for binary decision diagrams (BDDs) in model checkers. However, SVC does not require the DAG to be canonical. As a result, SVC is more robust and can handle formulas that would blow up if represented with BDDs. Such formulas may still take a very long time to verify, but various heuristics can be applied to speed up the verification without fear of a sudden failure due to BDD explosion.

As mentioned, a nice feature of theorem provers is their support for abstraction. A primary goal of SVC is to make abstraction easier by providing uninterpreted functions and various interpreted theories. Most recently, we have completed a decision procedure for a theory of bit-vectors in SVC. Bit-vectors (also called "words") are a critical abstraction for reasoning about hardware structures. Intuitively, a bit-vector is a fixed-length string of individual bits, and operations on bit-vectors can be described in terms of their effect on each bit. Alternatively, these operations can be viewed as transformations on bit-vectors as a whole. The advantage of the latter approach is that a property which may be complex and difficult at the bit-level (such as addition) can be expressed easily as an operation on bit-vectors. It is especially desirable to be able to reason about concatenation, extraction, bit-wise Boolean operations, and arithmetic, since these correspond to hardware structures.

As we describe in the next section, deciding equality of arbitrary combinations of these operations is NP-hard. In spite of this fact, we have developed an automatic algorithm for reasoning about fixed-size bit-vector addition, negation, concatenation, and extraction which avoids exponential blow-up on many practical examples.[1]

As mentioned above, the approach taken in SVC differs from that of both theorem provers and model checkers. However, there is closely related work in both areas. Recently, successful methods for reasoning about bit-vector operations in a model-checking paradigm have used Binary Moment Diagrams (*BMDs) [1, 3]. These have been able to automatically verify large arithmetic circuits. The set of problems solvable using *BMDs is comparable to those solvable by SVC's bit-vector canonizer and a comparison of the two is presented in Section 5 below.

Bit-vector libraries have also been developed for many theorem provers including Boyer-Moore [8], SDVS [10], HOL [13] and PVS [6]. All of these libraries implement the basic operations of concatenation and extraction, but none of

*Now at 0-in Design Automation.

[1] Although SVC is capable of representing other Boolean bit-vector operations, we will not discuss them in this paper.

DAC 98, San Francisco, California

$x_{[n]}$	A bit-vector of size n. We will sometimes omit the subscript if it is obvious from the context.
$\texttt{val}_{[n]}$	A constant bit-vector which is the binary representation of the decimal value `val`. If n is larger than is required to represent `val`, then the upper bits are assumed to be 0. If n is omitted, it is assumed to be the smallest value required to represent `val`. We only use values which are positive, except in the case of $(-1)_{[n]}$ which we use to represent the vector of size n containing all 1's (i.e. having value $2^n - 1$).
$x_{[n]}[i : j]$	The extraction of bits i through j of x. We require $0 \leq j \leq i < n$. We write $x[i]$ as an abbreviation when $i = j$.
$x_{[m]} \circ y_{[n]}$	The concatenation of x and y to yield a new bit-vector of size $m + n$.
NOT $x_{[n]}$	The bit-vector whose bits are the negation of the bits of x.
$x_{[m]} +_{[k]} y_{[n]}$	Addition of $x_{[m]}$ and $y_{[n]}$ modulo 2^k. If k is omitted, we assume it is equal to the larger of m and n. Also, if m is less than k, then x is implicitly zero-extended to size k, whereas if $k < m$, the intended meaning is that only the lowest k bits of x are to be used (and similarly for n and y). Because modular addition is associative, it is unnecessary to use parentheses when referring to more than two operands.
$x_{[m]} = y_{[n]}$	True if and only if $m = n$ and corresponding bits of x and y are equal.

Table 1: Bit-vector Theory and Definitions

them provide a complete and automatic implementation of bit-vector arithmetic. Probably the most closely related work is that of Cyrluk et al. in PVS. A comparison of of their work with SVC is included in Section 4 below.

The rest of the paper is organized as follows. Section 2 describes some notation and complexity results. Section 3 explains in some detail the theory behind the SVC implementation and contains the main contributions of the paper. Section 4, as mentioned, contains a comparison of our method with that presented in [6], and Section 5 gives experimental results obtained using SVC on microprocessor verification examples. Finally, Section 6 gives some conclusions and directions for future work.

2 Complexity of Bit-Vector Logics

In contrast to those approaches which convert bit-vectors into natural numbers, our approach is to remain in the bit-vector domain; all operations, therefore, take bit-vectors as arguments and return bit-vectors as results. Table 1 lists the elements that make up our theory of bit-vectors. Note that we refer to the bits in a bit-vector of size n by index, with the least significant (right-most) bit being indexed with 0 and the most significant bit being indexed with $n - 1$.

In order to evaluate our decision procedure it is necessary to answer the question of whether there exists a simple algorithm for deciding expressions in this bit-vector theory. Some complexity results are given in [6]. It is shown that equality of terms under the core theory of fixed-size bit-vectors with concatenation and extraction is decidable in polynomial time. A subsequent extension to include bit-vector Boolean operations such as **AND**, **OR** and **NOT**, however, can easily be shown to produce a theory in which deciding equality is NP-hard as follows. Consider an arbitrary instance of the Boolean satisfiability problem which is a well-known NP-complete problem. A general Boolean proposition can be encoded using 1-bit bit-vectors and the Boolean operators. Call this encoding P. Then the satisfiability problem can be solved by checking the validity of $P = FALSE$. If valid, the formula is unsatisfiable, otherwise it is satisfiable.

Alternatively, consider extending the core theory by including arithmetic operations. Unfortunately, even the most trivial extension can quickly be seen to be NP-hard. Allowing only the additional operation of adding one to any bit-vector immediately gives us the ability to express arbitrary propositional logic statements since

$$x_{[1]} + 1 = \textbf{NOT } x_{[1]} \quad \textbf{and}$$
$$(x_{[1]} \circ y_{[1]} +_{[3]} 1)[2] = x_{[1]} \textbf{ AND } y_{[1]}.$$

The same reduction as above shows that deciding equality in this simple extended theory is NP-hard. This gives some insight into the difficulty of handling bit-vector arithmetic automatically.

Finally, one additional complication when dealing with bit-vectors is that there are some bit-vector formulas which are valid only because each bit-vector variable has a finite number of possible values. For example,

$$a_{[1]} = b_{[1]} \vee b_{[1]} = c_{[1]} \vee a_{[1]} = c_{[1]}.$$

In order to correctly handle these formulas, we must manually force SVC to consider all possible values for each variable. Fortunately, the examples we have encountered in practice do not have this property.

3 Our Approach

As described in [2], SVC uses a framework for cooperating decision procedures very much like that developed by Shostak [5] and used in PVS. One of the requirements of this framework is that semantically equivalent terms should have a unique representation, which we refer to as a *canonical* form. We call the process of transforming terms into their canonical form *canonizing* and we call the algorithm which does it a *canonizer*. In SVC, not all terms need to be canonized. Only terms which do not contain Boolean subexpressions, which we call *atomic*, must be maintained in canonical form. This policy is acceptable because non-atomic expressions contain at least one term which SVC can use to perform a case-split. Since SVC exhausts all possible case-splits before reporting a counterexample, it is impossible for a false negative to result. As we will see below, the flexibility of not having to canonize non-atomic expressions can be exploited to delay canonization of complicated expressions until absolutely necessary. Our framework further requires that atomic equations be written in a specific form in which the left-hand side contains a single variable and the right-hand side containes the rest of the terms. We call the algorithm to do this a *solver*. Every time a new theory is added to SVC, a canonizer and solver for that theory must be provided. The canonizer and solver for bit-vectors in

Rule	Expression		Result	Condition
(1)	$\alpha_{[m]} \circ \beta_{[n]}$	$\rightarrow$	$2^n \cdot \alpha_{[m]} +_{[m+n]} \beta_{[n]}$	
(2)	$\mathbf{NOT}\ \alpha_{[m]}$	$\rightarrow$	$(-1)_{[m]} \cdot \alpha_{[m]} +_{[m]} (-1)_{[m]}$	
(3)	$w_{[m]}[i:j]$	$\rightarrow$	$y_{[i-j+1]}$ Additionally, the following equation is added to the current knowledge database : $w_{[m]} = 2^{(i+1)} \cdot x_{[m-i-1]} +_{[m]} 2^j \cdot y_{[i-j+1]} +_{[m]} z_{[j]}$ where x, y, and z are new variables. If $i = m-1$ or $j = 0$ then the appropriate terms are omitted.	
(4)	$(x_0 +_{[m]} \cdots x_s)[i:0]$	$\rightarrow$	$x_0 +_{[i+1]} \cdots x_s$	
(5)	$(x_0 +_{[m]} \cdots x_s)[i:j]$	$\rightarrow$	$x_0[i_0:j] +_{[i-j+1]} \cdots x_s[i_s:j] +_{[i-j+1]}$ $\mathbf{OVF}_{[i:j]}(x_0[j-1:0] \ldots x_k[j-1:0])$	if $j > 0$
(6)	$\alpha +_{[i]} (x_0 +_{[j]} \cdots x_s)$	$\rightarrow$	$\alpha +_{[i]} x_0 +_{[i]} \cdots x_s$	if $j \geq i$
(7)	$\alpha +_{[i]} (x_0 +_{[j]} \cdots x_s)$	$\rightarrow$	$\alpha +_{[i]} x_0 +_{[i]} \cdots x_s +_{[i]} 2^j \cdot (-1)_{[i-j]} \cdot \mathbf{OVF}_{[i-1:j]}(x_0 \ldots x_s)$	if $j < i$
(8)	$\mathbf{OVF}_{[i:j]}(\alpha_0 \ldots \alpha_k)$	$\rightarrow$	$\mathbf{OVF}_{[i:j]}(\alpha_0, (\alpha_1 +_{[j]} \cdots \alpha_k)) +_{[i-j+1]} \mathbf{OVF}_{[i:j]}(\alpha_1 \ldots \alpha_k)$	if $k > 1$
(9)	$\mathbf{OVF}_{[i:j]}(\alpha_{[m]}, \beta_{[n]})$	$\rightarrow$	$\alpha[m-1:j] +_{[i-j+1]} \beta[n-1:j] +_{[i-j+1]}$ $\mathbf{OVF}_{[j:j]}(\alpha[j-1:0], \beta[j-1:0])$	if $i > j$
(10)	$\mathbf{OVF}_{[n:n]}(\alpha_{[n]}, \beta_{[n]})$	$\rightarrow$	$\mathbf{OVF}_{[n-1:n-1]}(\mathbf{ite}(\alpha[n-1] = \beta[n-1]\ ,\ \alpha[n-1] \circ 0_{[n-2]}\ ,\ \alpha[n-2:0]),$ $\mathbf{ite}(\alpha[n-1] = \beta[n-1]\ ,\ \beta[n-1] \circ 0_{[n-2]}\ ,\ \beta[n-2:0]))$	if $n > 1$
	$\mathbf{OVF}_{[n:n]}(\alpha_{[n]}, \beta_{[n]})$	$\rightarrow$	$\mathbf{ite}(\alpha = \beta\ ,\ \alpha\ ,\ 0_{[1]})$	if $n = 1$

Table 2: Rules for eliminating concatenation, negation, and extraction, flattening addition, and converting **OVF** terms to non-atomic expressions. Note that α and β are arbitrary expressions and $i_k = min(i, n_k)$ where n_k is the size of x_k.

SVC are based on properties of hardware arithmetic. They constitute the major contribution of this paper and are described in the next two subsections.

3.1 Canonizer

Coming up with a canonical form for bit-vector expressions is complicated by the inclusion of bit-vector arithmetic. This is because the same expression may be represented in non-trivially different ways. For example, $(x_{[n]} +_{[n+1]} x_{[n]})$ is equivalent to $(x_{[n]} \circ 0_{[1]})$. Similarly, $(x_{[1]} +_{[1]} 1_{[1]})$ is equivalent to $(\mathbf{NOT}\ x_{[1]})$.

To avoid such redundancy, we translate all bit-vector expressions into a specific kind of arithmetic expression: the addition (modulo 2^n for some fixed bit-width n) of bit-vector variables with constant coefficients. We call these *bitplus* expressions. In order to ensure a unique representation, variables are ordered with duplicates eliminated, and each coefficient is reduced modulo 2^n. A set of transformations for converting bit-vector operations into bitplus expressions is shown in Table 2. Some of these rules make use of the **OVF** operator, which we define and explain below. The first two rules are simple transformations for dealing with concatenation and negation. Rule (3) shows how to eliminate extraction at the cost of introducing new variables. We refer to this process as *slicing*, and it is desirable to avoid it whenever possible. Intuitively, repeated slicing moves from the bit-vector abstraction to the bit-level, and in the worse case, each bit must be considered. Section 4 provides one illustration of how slicing can be avoided. Rules (4) and (5) show how to eliminate extraction when applied to a bitplus expression, and (6) and (7) show how to "flatten" bitplus expressions to ensure that other bitplus expressions do not appear as subexpressions.

However, a canonical form cannot always be obtained by simple transformations (as we would expect given the fact that the general problem is NP-hard). The difficulty comes from the interaction of extraction and addition. Consider the following two expressions.

$$(8 \cdot x_{[n]} +_{[4]} 7)[2]$$

$$(x_{[3]} +_{[4]} y_{[3]})[3]$$

In the first case, it is desirable to push the extraction inside the bitplus expression which will result in $1_{[1]}$. However, in the second case, there is no way to represent the result of pushing the extraction inside the bitplus, because the result depends on whether adding x and y overflows into the fourth (most significant) bit. To deal with such cases, we introduce a new *overflow* operator, $\mathbf{OVF}_{[i:j]}$ which represents bits i through j of the sum of its operands. Using this operator, we can rewrite the second expression above as

$$(x_{[3]} +_{[4]} y_{[3]})[3] = \mathbf{OVF}_{[3:3]}(x, y).$$

We define overflow for the general case as follows:

$$\mathbf{OVF}_{[i:j]}(x_{0[n_0]} \ldots x_{k[n_k]}) = (x_{0[n_0]} +_{[i+1]} \cdots x_{k[n_k]})[i:j].$$

Whenever the overflow operator is applied, the expression is first checked using a simple algorithm to see if it is equivalent to a concatenation of variables or their negations. If it is, then the appropriate bits are extracted and then converted back into a bitplus expression, for example,

$$\mathbf{OVF}_{[3:1]}(2 \cdot y_{[2]}, 9) = y_{[2]} + 4.$$

If the overflow expression cannot be written in a simple form as above, we break it down using the last three rules shown in Table 2.

To understand rule (8), notice that $\mathbf{OVF}_{[i:j]}(x, y, z)$ and $\mathbf{OVF}_{[i:j]}(x, (y +_{[j]} z))$ differ by exactly $\mathbf{OVF}_{[i:j]}(y, z)$. Rule (8) is the generalization of this property which we use to split an overflow expression with k arguments into two overflow expressions that have 2 and $k-1$ arguments respectively. This rule is applied repeatedly until all overflow expressions have only two arguments. Then rule (9) is applied to convert overflow expressions in which $i > j$ to overflow expressions in which $i = j$ (we show the case where $m > j$ and $n > j$, but the other cases are similar). Finally, rule (10) takes overflow expressions resulting from the application of rule (9) and turns them into non-atomic expressions. It does this by making use of the **ite** operator which is used in SVC to represent all Boolean operations. For arbitrary expressions, α, β, and γ,

$$\textbf{ite}(\alpha, \beta, \gamma) = \textbf{if } \alpha \textbf{ then } \beta \textbf{ else } \gamma.$$

The intuition behind rule (10) is that we are simply computing the carry bit of an n-bit adder. If the most significant bits are equal, then they determine and are equal to the carry bit. If they are different, then the carry bit is propagated from other $n-1$ bits. The use of the **ite** operator is of key importance. As we mentioned earlier, SVC does not require a canonical form for non-atomic expressions. A canonical form would require expressing the full logic of a ripple-carry adder and would require looking at all the bits of the arguments. But the decomposition we have given is incremental, suspended until SVC does a case-split on the equality of the most significant bits. If they are unequal, SVC will slice off the next most significant bits. In the worst case, we will have to look at every bit. But the incremental approach will avoid this unless absolutely necessary.

3.2 Solver

The other main contribution of this paper is a solver for equations involving bit-vector operations. The requirements for the solver are very similar to those of the canonizer. In fact, the solver can be viewed as a canonizer for equations. In SVC, canonical equations are required to have the most complex variable or uninterpreted function isolated on the left-hand side, with the rest of the terms on the right-hand side. Complexity is defined by a total ordering on expressions (see [2]). In the case of bit-vectors, we arrange for longer bit-vectors to be more complex than shorter ones, so that we solve for the longest bit-vector in the equation. This avoids slicing bit-vectors unnecessarily. In general, we must be able to solve arbitrary equations of the form

$$a_0 \cdot x_0 \ +_{[n]} \ \cdots \ a_p \cdot x_p = b_0 \cdot y_0 \ +_{[n]} \ \cdots \ b_q \cdot y_q.$$

Using arithmetic modulo 2^n, we can easily isolate the most complex variable, say $z_{[m]}$, with coefficient c on the left-hand side. The resulting equation has the following form:

$$c \cdot z_{[m]} = d_0 \cdot w_{0[m_0]} \ +_{[n]} \ \cdots \ d_j \cdot w_{j[m_j]}.$$

Now, we must eliminate c in order to isolate z. If c is odd, we can do this by finding its multiplicative inverse, which is c^{2^k-1} for some $k \le n-2$. We explain briefly why this is true. It is well-known [7] that the set of all positive integers less than and relatively prime to some positive integer p forms a group under multiplication modulo p, denoted $U(p)$. In particular, when $p = 2^n$, every odd positive integer less than p is in $U(p)$. Furthermore, $U(2^n)$ is isomorphic to the (external) direct product of the cyclic groups of order 2 and 2^{n-2}, which means that each element of $U(2^n)$ has order 2^k for some $k \le 2^{n-2}$. Thus, if c is odd, there exists $k \le n-2$ such that $c^{2^k} = 1$ modulo 2^n, and it follows that the inverse is c^{2^k-1}. To find the inverse i of c, we use the simple algorithm shown in Figure 1.

Thus, finding the inverse of c requires at most $2(n-2)$ n-bit multiplies, and assuming uniform distribution, the expected number of multiplies is about $2(n-3)$. In the examples we have done, however, the coefficient is almost always either 1 or $2^n - 1$, so the actual average number of multiplications is actually much less (between 0 and 2). After calculating the inverse, we simply multiply all terms

```
i := c;
while (c ≠ 1) do begin
  c := (c × c) mod 2^n;
  i := (i × c) mod 2^n;
end
```

Figure 1: Algorithm to find the multiplicitive inverse of c.

in the equation by the inverse and the resulting equation will have $z_{[m]}$ alone on the left-hand side.

Suppose on the other hand that c is even. Then we can write $c = 2^k \cdot b$ for some $k \ge 1$ and b odd. We can then split the equation into two equations as follows:

$$\begin{aligned} b \cdot z &= (d_0 \cdot w_0 \ +_{[n]} \ \cdots \ d_j \cdot w_j)[n-1:k] \quad \textbf{and} \\ 0_{[k]} &= d_0 \cdot w_0 \ +_{[k]} \ \cdots \ d_j \cdot w_j. \end{aligned}$$

The first equation can now be solved by calculating the inverse of b. Furthermore, though we will have to repeat the canonization process on the second equation, we have eliminated z from it without adding any variables, ensuring that the process will terminate.

Once we have an equation of the form

$$z_{[m]} = c_0 \cdot w_0 \ +_{[n]} \ \cdots \ c_j \cdot w_j,$$

there is one final step if $m < n$ (which is possible since, as mentioned in Table 1, we do not put any restrictions on the bit-widths of variables appearing in bitplus expressions). In this case, we know that the most significant bits of the right-hand side of the equation must be zero. So as above, we split it into two equations:

$$\begin{aligned} z_{[m]} &= c_0 \cdot w_0 \ +_{[m]} \ \cdots \ c_j \cdot w_j \quad \textbf{and} \\ 0_{[n-m]} &= (c_0 \cdot w_0 \ +_{[n]} \ \cdots \ c_j \cdot w_j)[n-1:m]. \end{aligned}$$

Again, we may have to canonize the second equation. Eventually, though, our initial equation will be transformed into a conjunction of equations, each solved for a different variable. We have found that it is very desirable for efficiency to transform these equations (via substitution) so that variables appearing on the left-hand side do not appear in the right-hand sides of any of the other equations.

4 Avoiding Bit-Slicing

As previously mentioned, whenever part of a bit-vector variable is extracted, that bit-vector is sliced into several parts, which reduces the level of abstraction and is thus to be avoided if possible. Our initial implementation of bit-vectors was based on the work done by Cyrluk et al. in [6]. However, our decision to use bitplus expressions as our internal representation was a significant departure from their method. As a result, we were able to increase the range of arithmetic examples which can be verified automatically.

However, it turns out that for some examples, even the original core theory of concatenation and extraction benefits from this change in the internal representation. This is because bit-slicing can be avoided in many cases. Suppose w is a bit-vector of size n and consider the following example:

$$w[n-1:1] = w[n-2:0] \ \Rightarrow \ w[n-1] = w[0]$$

In order to canonize this formula, the decision procedure of Cyrluk et al. generates a new variable for every bit of w,

$(x[0] = 0_{[1]})$	$\Rightarrow$	$((x[n-1:1]+1) \circ 0_{[1]})$	$=$	$x+2)$	(1)
$(x[0] = 1)$	$\Rightarrow$	$((x+(-1)_{[n]})[n-1:1]$	$=$	$x[n-1:1])$	(2)
		NOT $((x+y+(-1)_{[n]})[0])$	$=$	$(x+y)[0]$	(3)
		$((2^n-2)\cdot x_{[n-1]} +_{[n]} (-1)_{[n]})[n-1:1]$	$=$	**NOT** $x_{[n-1]}$	(4)
		$(y_{[1]} \circ x_{[n-2]} +_{[n]} 1)[n-1]$	$=$	$(x_{[n-2]} +_{[n]} y_{[1]})[n-2]$	(5)

Figure 2: Bit-Vector Arithmetic Verification Examples. x and y are bit-vectors of size n unless otherwise specified.

resulting in a conjunction of n equations:

$$\begin{array}{rcl} (w_{n-1} & = & w_{n-2}) \wedge \\ (w_{n-2} & = & w_{n-3}) \wedge \\ & \vdots & \\ (w_1 & = & w_0) \quad \wedge \\ (w & = & w_{n-1} \circ w_{n-2} \cdots w_0) \end{array}$$

where each w_i is a new bit-vector of length one. Using our canonizer, only the most and least significant bits are sliced, resulting in the following two equations:

$$\begin{array}{rcl} (2^{n-2} \cdot x_{[1]} +_{[n-1]} y_{[n-2]} & = & 2 \cdot y_{[n-2]} +_{[n-1]} z_{[1]}) \wedge \\ (w & = & x_{[1]} \circ y_{[n-2]} \circ z_{[1]}) \end{array}$$

We then invoke the solver and end up with:

$$\begin{array}{rcl} (y_{[n-2]} & = & (-1)_{[n-2]} \cdot z_{[1]}) \wedge \\ (x_{[1]} & = & z_{[1]}) \qquad\quad \wedge \\ (w & = & x_{[1]} \circ y_{[n-2]} \circ z_{[1]}) \end{array}$$

Thus, instead of producing n equations, we produce only three. The information from the other equations is stored in the coefficients. Most of w remains as an abstract bit-vector.

5 Results

The examples shown in Figure 2 demonstrate the kinds of formulas which can be verified using the methods described above. Formulas (1) through (3) are small pieces of much larger formulas from processor verification, and (4) and (5) are simply test benchmarks which we feel are representative of the complexity of the general problem. In general, SVC must solve many such small problems, as well as similar problems in other theories, as part of proving a larger formula.

Examples three through five are easily verified using only the SVC canonizer. These examples can also be verified using *BMDs (in contrast, the first two examples cannot be directly verified using only *BMDs because they include Boolean connectives). In Table 3, the column labeled "SVC" shows the time required to verify these examples running SVC on a 200 MHz Pentium Pro. The second column shows the time required to verify the same property on the same machine using the *BMD package from Bryant and Chen [3]. The third column shows the time required on a 300 MHz UltraSparc-30 using Laurent Arditi's *BMD implementation which has special support for bit-vector and Boolean operations [1]. For the two *BMD packages, the examples were run with four typical variable orderings and the best time for each example is reported. In order to better compare the two different *BMD implementations, a simple n-bit adder example was also done and the result is listed as example A.

Example	n	SVC	*BMD 1	*BMD 2
1	32	2	N/A	40
2	32	2	N/A	1100
3	8	2	440	30
3	16	2	265000	70
3	32	2	> 500000	180
4	8	2	112	80
4	16	2	26400	720
4	32	2	> 500000	8790
5	8	30	95	60
5	16	111	22700	390
5	32	520	> 500000	3780
A	32	0.2	32	80

Table 3: Results on Bit-Vector Arithmetic Verification Examples. Times are in milliseconds. Examples 1 through 5 are from Table 2. Example A is the expression $x_{[n]} + y_{[n]}$.

Several important observations can be made from these data. First, it is clear that *BMDs benefit greatly from the special-purpose algorithms in Arditi's implementation. From example A, which does not use the special-purpose algorithms, it also seems clear that Arditi's package could benefit further from implementation in a lower-level language (Bryant and Chen's package is in C, whereas Arditi's package is in Scheme). However, even given these possible improvements, SVC would still outperform *BMDs on all examples with the possible exception of example 5 (which we will come back to in a moment). More importantly, the performance of SVC is the same despite increasing bit width on two out of three of the examples. The reason for this is that SVC is able to maintain its bit-vector abstraction in these examples and thus does not need to consider each bit individually.

The reason SVC does poorly in example 5 is that it ends up slicing x and thus the execution time depends on the number of bits in x. However, if SVC were to split on y instead, it could avoid slicing x. This is a problem similar to variable ordering in *BMDs. We were able to write a slightly modified version of example 5 which forced SVC to split on y first. The result was an execution time of less than 10 ms independent of bit-width. One of the areas of ongoing research in SVC is how to automatically choose the best variables for case-splitting.

The primary application for SVC is microprocessor verification. As mentioned above, we are currently applying it to the TORCH microprocessor [11]. Table 4 shows the times required to verify several formulas from this effort. These formulas are large and require cooperating decision procedures from several theories including the bit-vector theory

Example	Size (KB)	Case Splits	Time (s)
PCUnitDataPath	29	1254	1.67
TakenBranch	31	38	0.15
IFetchControl	65	29676	24.1
IFetchPC	42	69374	317

Table 4: TORCH Microprocessor Verification Examples. Time is in seconds

(thus they cannot be verified using *BMDs alone and a direct comparison cannot be made). The first example, PCUnitDataPath, verifies that the program counter is calculated correctly. The second example, TakenBranch, verifies that the hardware correctly identifies when a conditional branch should be taken. The last two examples verify various properties of the instruction fetch unit. For each example, we show the size of the formula, the total number of case splits, and the time required to verify the formula. We would be unable to automatically verify these formulas without the bit-vector decision procedure.

6 Conclusions

Our method for dealing with bit-vector arithmetic has many advantages. First, it is complete and automatic, and although the complexity of the problem dictates that there are examples for which it will be exponentially slow, it is efficient on the examples we have encountered so far. Second, we reason at the word-level and avoid slicing bit-vectors whenever possible. This avoids some of the blow-ups in time and space experienced by other methods. Finally, our method is easy to implement and integrate into an environment of cooperating decision procedures such as that found in SVC.

As we have seen, one limitation of SVC is that the choice of which variable to split on can greatly influence its efficiency. Additionally, there is no way to represent non-linear multiplication of bit-vectors directly as there is with *BMDs. There is also currently no way to reason about bit-vectors of unknown size. We are working on addressing some of these limitations. For example, we have implemented a couple of automatic learning strategies for choosing splitters which dramatically increase the efficiency of many examples.

Some areas of future research include finding an efficient abstraction for Boolean bit-vector operations and finding more efficient ways of dealing with the overflow operator. We would also like to explore extensions to bit-vectors of unknown size and variable extraction indices, as well as non-linear arithmetic. We intend to continue to use SVC in our verification of TORCH and expect that it will continue to develop towards an important and powerful tool for automatic hardware verification.

Acknowledgments

We would like to thank Yirng-An Chen for his help in understanding and using *BMDs. We would like to thank other members of the TORCH verification effort for their work in specifying and translating examples for this paper: Jeffrey Su, Jens Skakkebæk, and especially Laurent Arditi who also provided insight and analysis for comparison with *BMDs. Also, we would like to thank the group at SRI International for valuable discussions, especially David Cyrluk whose work on bit-vectors provided a valuable starting point. This work was sponsored by DARPA contract number DABT63-96-C-0097 and by a National Defense Science and Engineering Graduate Fellowship. The content of this paper does not necessarily reflect the position of the policy of the government and no official endorsement should be inferred.

References

[1] Laurent Arditi. *BMDs Can Delay the Use of Theorem Proving for Verifying Arithmetic Assembly Instructions. In Srivas [12], pages 34–48.

[2] C. W. Barrett, D. L. Dill, and J. R. Levitt. Validity Checking for Combinations of Theories with Equality. In Srivas [12], pages 187–201.

[3] Randal E. Bryant and Yirng-An Chen. Verification of Arithmetic Circuits with Binary Moment Diagrams. In 32nd *ACM/IEEE Design Automation Conference*, pages 535–541, San Francisco, CA (USA), 1995.

[4] J. R. Burch and D. L. Dill. Automatic Verification of Microprocessor Control. In Dill, editor, *Computer-Aided Verification*, volume 818 of *Lecture Notes in Computer Science*, pages 68–79, Stanford, CA (USA), June 1994.

[5] D. Cyrluk, P. Lincoln, and N. Shankar. On Shostak's Decision Procedure for Combinations of Theories. In 13th *International Conference on Automated Deduction*, pages 463–477, Rutgers University, NJ (USA), July 1996.

[6] D. Cyrluk, O. Möller, and H. Rueß. An Efficient Decision Procedure for the Theory of Fixed-Sized Bitvectors. In 9th *International Conference on Computer-Aided Verification*, 1997.

[7] Joseph A. Gallian. *Contemporary Abstract Algebra*. D. C. Heath and Company, second edition, 1990.

[8] Warren A. Hunt Jr. Microprocessor Design Verification. *Journal of Automated Reasoning*, 5(4), December 1989.

[9] R. B. Jones, D. L. Dill, and J. R. Burch. Efficient Validity Checking for Processor Verification. In *IEEE Internationl Conference on Computer-Aided Design*, pages 2–6, San Jose, CA (USA), November 1995. IEEE Computer Society Press.

[10] Leo G. Marcus. *SDVS 13 Users' Manual*. The Aerospace Corporation, El Segundo, CA 90245, October 1994. Aerospace Report ATR-94(4778)-5.

[11] M. Smith, M. Lam, and M. Horowitz. Boosting Beyond Static Scheduling in a Superscalar Processor. In *International Symposium on Computer Architecture*, pages 344–354, Seattle, WA, May 1990. IEEE/ACM.

[12] Srivas, editor. *International Conference on Formal Methods in Computer-Aided Design*, volume 818 of *Lecture Notes in Computer Science*, Palo Alto, CA (USA), November 1996. Springer-Verlag.

[13] Wai Wong. Modelling Bit Vectors in HOL: the `word` Library. In Joyce and Seger, editors, *Higher Order Logic Theorem Proving and Its Applications*, volume 780 of *Lecture Notes in Computer Science*, pages 371–384, Vancouver, Canada, August 1993.

Functional Vector Generation for HDL models Using Linear Programming and 3-Satisfiability

Farzan Fallah, Srinivas Devadas
Department of EECS
MIT, Cambridge

Kurt Keutzer
Department of EECS
University of California, Berkeley

Abstract—
Our strategy for automatic generation of functional vectors is based on exercising selected paths in the given hardware description language (HDL) model. The HDL model describes interconnections of arithmetic, logic and memory modules. Given a path in the HDL model, the search for input stimuli that exercise the path can be converted into a standard satisfiability checking problem by expanding the arithmetic modules into logic-gates. However, this approach is not very efficient.

We present a new HDL-satisfiability checking algorithm that works directly on the HDL model. The primary feature of our algorithm is a seamless integration of linear-programming techniques for feasibility checking of arithmetic equations that govern the behavior of datapath modules, and 3-SAT checking for logic equations that govern the behavior of control modules. This feature is critically important to efficiency, since it avoids module expansion and allows us to work with logic and arithmetic equations whose cardinality tracks the size of the HDL model.

We describe the details of the HDL-satisfiability checking algorithm in this paper. Experimental results which show significant speedups over state-of-the-art gate-level satisfiability checking methods are included.

I. Introduction

The most common approach to checking correctness of a design is to verify that a description of the design in some hardware description language has the proper behavior as elicited by a series of simulation vectors. There are several drawbacks of this approach. Exhaustive simulation is required to guarantee correctness and is possible only for small circuits. If a manageable number of input stimuli are to be simulated, the problem of selecting or generating these input stimuli arises. Generating random or pseudo-random input stimuli is relatively easy; however, random input stimuli, typically, do not thoroughly exercise the design and therefore miss a lot of design errors.

One can base a functional vector generation strategy on coverage metrics. Commonly used coverage metrics correspond to instruction counts, i.e., how often, if at all, an instruction or statement in the hardware description language (HDL) code is exercised. Full statement coverage corresponds to the input stimuli exercising every statement in the given HDL code at least once. Full path coverage is a more comprehensive criterion, where every path in the HDL code is exercised by the input stimuli. Since the number of paths in the HDL model can grow exponentially with the size (number of lines) of the model, full path coverage is generally not attainable. Design-specific strategies for computing coverage have been proposed. Very often, designers manually generate functional vectors with full statement coverage, and which exercise all "interesting" paths in the model. The generation of input stimuli that satisfy coverage requirements is a tedious process. However, the automatic generation of functional vectors is a difficult computational problem.

We assume that we are given an HDL model as input which describes interconnections of arithmetic, logic and memory modules. We have chosen selective path coverage as a metric for functional vector generation. Given a set of paths in the HDL model, for each path, we solve a satisfiability problem to obtain the input stimuli that exercise the chosen path. Path sensitization in HDL models can be converted into a standard satisfiability checking problem by expanding the arithmetic modules into logic-gates. However, this approach is not very efficient.

We present a satisfiability algorithm that works directly on the HDL model. The primary feature of our HDL-satisfiability algorithm is a seamless integration of linear-programming techniques for arithmetic modules and 3-satisfiability (3-SAT) checking for logic modules. This integration is critically important to efficiency, since it avoids module expansion and allows us to work with logic equations and arithmetic equations that track the size of the HDL model. We use 3-SAT checking to solve the logic equations and linear-programming to check the feasibility of the arithmetic equations. Interactions between Boolean variables in the 3-SAT equations and integer variables in arithmetic equations are handled by increasing the number of integer variables while reducing their range.

We have chosen path coverage as our metric because exercising computation paths in a circuit is a tried-and-trusted way of design debugging. Since the number of paths in the circuit may be too large to enumerate, we select a subset of paths. Other coverage metrics, such as statement coverage, can also be used in our framework. This will only impact the specifics of the generated satisfiability problem.

Representative approaches for gate-level satisfiability include [8] and [13]. An excellent survey of satisfiability checking methods at the gate level that includes empirical comparisons is presented by Stephan *et al.* in [14]. Test generation for stuck-at faults given hierarchical or register-transfer level descriptions has been a subject of considerable attention. We review relevant previous work in hierarchical test generation in Section II.

We describe our overall strategy toward functional vector generation targeting selective path coverage in Section III.

The primary focus of this paper is the HDL-satisfiability checking algorithm. We first describe the basic 3-SAT formulation due to Larrabee [8] in Section IV. The HDL-satisfiability algorithm is described in detail in Section IV.

The public-domain 3-satisfiability (3-SAT)-based test pattern generator from [14] incorporates many different heuristics for fast 3-SAT checking. We have augmented this 3-SAT framework, and used a commercial linear-programming (LP) solver to implement our algorithm. Experimental results which show significant speedups over gate-level satisfiability checking are included in Section V.

II. Previous Work

A. Hierarchical Test Generation

Roy and Abraham [11] [12] use data flow descriptions to perform hierarchical test generation. Data flow graphs are recursively selected for propagating or justifying signals. The assumption of precomputed test sets for modules is adopted. However, conflicts due to reconvergence are not handled for fault effect propagation.

The use of nonlinear equation solving has been proposed in [7] and [9]. Both approaches work significantly better than gate-level approaches in some cases, but suffer from some drawbacks. Finding the solution to the assembled system of equations is not guaranteed, even if a solution exists. Further, the approaches have not been shown to work for circuits with complicated control and large datapaths.

B. Functional Test Generation for HDL Models

The extended finite state machine model is used to automatically generate functional tests in [6]. The extended finite state machine of [6] allows update transformations that include increment, and

DAC 98, San Francisco, California

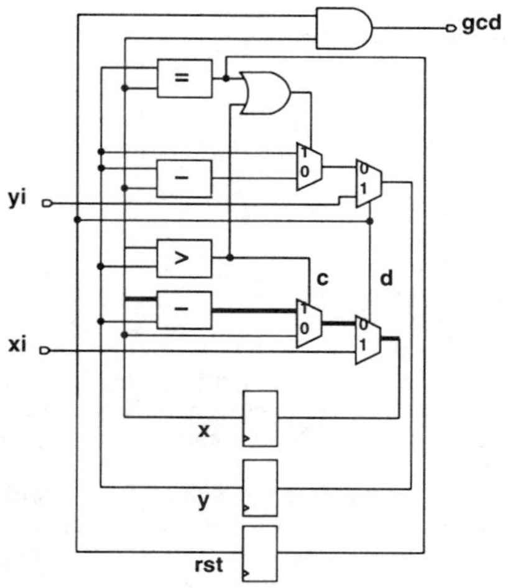

Fig. 1. Module-level circuit for a greatest common divisor example

comparison with a constant. It does not allow addition of several variables, or comparison between several variables.

Constraint solving for test generation is a well-researched problem in the software test area (e.g., [1], [10]). These systems typically attempt to solve nonlinear constraint systems heuristically. In many such systems the logic corresponding to the constraint system is undecidable. The high-level design validation system of [5] uses augmented linear-program solving to solve certain types of constraints. Nonlinear operators in constraints require exhaustive enumeration, random guesses, or the use of heuristics. Bultan *et al.* [4] present a method that uses Presburger formulas as a symbolic representation to perform model checking of infinite state systems. A Presburger solver is used to manipulate these formulas.

III. Functional Vector Generation Strategy

A. Introduction

In this section, we will describe our overall strategy toward functional vector generation given an HDL model. Our strategy involves choosing a coverage metric, setting up search problems given the coverage metric, and solving the search problems to obtain input stimuli. For the purposes of this paper, we have chosen selective path coverage as our metric.

B. HDL Models

We view an HDL model as a structural interconnection of modules. The modules can be comprised of combinational logic, and registers. The combinational logic can correspond to Boolean operators (e.g., AND, OR), or arithmetic operators (e.g., $+$, $>$).

We give an example of a module-level sequential circuit corresponding to a greatest common divisor in Figure 1.

C. Path Testing

We will first focus on combinational circuits. Our strategy of generating functional vectors is to **sensitize** paths from the circuit inputs to circuit outputs. The sensitization of a path can mean many things, and we will define it in the next section. Roughly speaking, sensitizing a path implies that the value at the input to the path should affect the value at the output of the path. An important point to note is that we do not expand busses or word-level variables into bit-level variables. Therefore, a path can be a collection of electrically-distinct signals.

Since there may be exponentially many paths in an HDL model, we will choose a subset. This subset is chosen such that at least one path in the subset passes through each module input to each module output. That is, if we have a module with two inputs A and B, and two outputs C and D, we will choose paths that pass through $A \rightarrow C$, $A \rightarrow D$, $B \rightarrow C$, and $B \rightarrow D$.

There are other criteria in choosing paths, and the quality of the generated vector set is dependent on these criteria. We will target full statement coverage in the HDL model, which can be achieved by sensitizing a set of paths that cover each module-input module-output subpath.

D. Path Sensitization

A path is a set of alternating modules and signals, call it $P = \{s_1, m_1, s_2, m_2, \cdots, s_n\}$. The first signal, s_1, is a circuit input, and the last, s_n, is a circuit output. For each i, s_i is an input to module m_i and s_{i+1} is an output of module m_i. Sensitizing a path simply corresponds to sensitizing each module m_i.

We now need to define the sensitization of a subpath through a module m_i, namely s_i, m_i, s_{i+1}. Sensitization of a subpath through a module will require values at the module side-inputs, i.e., the inputs other than s_i to m_i.

For logic-gates, we can say the following:

- INVERT: no values required.
- AND: Require side-inputs to be at 1.
- OR: Require side-inputs to be at 0.

The above rules can be used to determine sensitization for arbitrary collections of logic-gates, as well as logic macros such as multiplexors, and decoders.

For word-level operators, sensitization needs to be defined on the basis of the logic function of the operator. Arithmetic operators are easy to sensitize, we merely need to check for overflow or underflow. In the sequel, assume we wish to sensitize the path from A to the module output. Sensitization should reflect the condition that a (large enough) change in the value of A should result in a change in the module output.

- Adder $A + B$: If we are not concerned with overflow, there is no requirement. If we are, then we require $A + B < MAX$, where MAX is the largest represented number.
- Increment $A + k$: To avoid overflow $A < MAX - k$.
- Subtractor $A - B$: Similar to an adder, except we are concerned with underflow.
- Comparator $A > B$: We require $B < MAX$. Similarly, for other types of comparators.
- Scalar multiplication $A * k$: $A < \frac{MAX}{k}$ to avoid overflow.

While the above sensitization requirements for logic-gates and word-level arithmetic operators may appear simple, the reader should note that we are asking for requirements on intermediate signals in the circuit. Therefore, we may ask for a Boolean signal c to be a 1, which is the output of a comparator $A > B$, where $A = C + D$. (C and D may be circuit inputs or intermediate signals themselves.)

Our method of finding input values that sensitize paths is to:

1. Write sensitization requirements on (intermediate) signal values as described above.
2. Write module-input module-output relationships for **every** module in the circuit (cf. Section IV), and
3. Solve a satisfiability problem that corresponds to 1 and 2 above (cf. Section IV). This produces an input assignment that satisfies the sensitization requirement and is consistent with the behavior of the logic circuit.

We could easily choose different sensitization criteria without changing the overall framework described immediately above.

E. Sequential Circuits

We have considered only combinational HDL models thus far. In general, we will have sequential models such as the one in Figure 1, where registers are intermixed with Boolean and arithmetic operators. Our approach to handling sequential circuits is to use the conventional time-frame expansion strategy of sequential test generation [2].

A path may be selected from circuit inputs/register outputs to circuit outputs/register inputs. Once the satisfiability problem for path sensitization is set up and solved, we will require the circuit inputs to be at particular values, and we may require particular $\{0, 1\}$ values at the register outputs as well. (It may be that certain circuit inputs and register outputs can be left undefined.)

We now need to **justify** the required values at the register outputs. This is done using a time-frame expansion strategy. We check to see what the initial values of the registers are. If the initial values are consistent with the required values, there is no need for justification. Otherwise, we attempt to determine circuit input values that produce

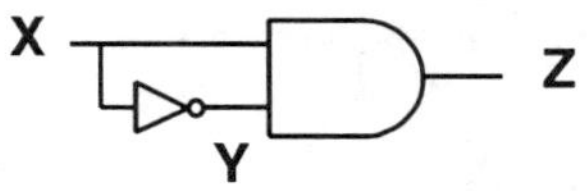

Fig. 2. Simple Circuit

the required values at the register *inputs*, so, in the next cycle, the register outputs have the right values. This is done by solving a new satisfiability problem. If no satisfying assignment can be found, we time-frame expand the circuit over 2 cycles, and attempt to find a sequence of input values of length 2 that produces the required values at the register inputs, solving a (larger) satisfiability problem. This process continues until a user-defined limit on the number of time-frames is reached.[1]

The vector generation method we currently use does not attempt to propagate values at register inputs to circuit outputs. Given an HDL model the registers are all assumed to be observable during functional simulation. If the user wishes to only observe circuit outputs, then additional steps of satisfiability are required to accomplish propagation.

F. Satisfiability Problems

As was described above, to generate vectors that sensitize a single path, we have to solve several satisfiability problems for a sequential HDL model. Further, the satisfiability problems encountered in the justification (and propagation) steps may result from a time-frame expanded circuit, which is larger than the original, since combinational logic is replicated. Thus, it is very important that the satisfiability algorithm be as powerful as possible. We describe our algorithm in the next section.

IV. A Hybrid Algorithm for Satisfiability Checking

A. Introduction

We will describe our hybrid algorithm for satisfiability checking in this section. This algorithm seamlessly integrates linear-programming feasibility and 3-satisfiability checking. Integration would be easy if there was no correlation between word-level variables and Boolean variables. But, in general, a word-level variable A may appear in an arithmetic operation (e.g., $C = A + B$), and a bit-masked version may appear in a Boolean operation (e.g., $d = a_{n-1} \vee b_0$, where $\vee$ denotes logical OR).

B. The 3-Satisfiability Approach

The 3-Satisfiability (3-SAT) approach to stuck-at fault test generation was pioneered by Larrabee [8]. Here, we are not interested in stuck-at fault testing, but rather in justifying values on a collection of signals. That is, given a logic circuit, we wish to find an input assignment that produces appropriate signal values at circuit outputs, or intermediate signals. The 3-SAT approach converts this problem into one of satisfying a conjunctive normal form (CNF) expression. Each of the clauses in the CNF have at most 3 variables in them, hence the name 3-SAT.

Consider the circuit of Figure 2. Assume we wish to find an input assignment that sets the output to a 1. We will write clauses for each gate in the circuit. These clauses model the input-output relationship for each gate, and are shown below.

$$(\overline{X} \vee \overline{Y}) \cdot (X \vee Y) \cdot (\overline{Z} \vee X) \cdot (\overline{Z} \vee Y) \cdot (\overline{X} \vee \overline{Y} \vee Z). \quad (1)$$

In order to produce a 1 at the output of the circuit we need to satisfy the CNF above with Z set to 1. This simplifies to

$$(\overline{X} \vee \overline{Y}) \cdot (X \vee Y) \cdot X \cdot Y, \quad (2)$$

which is clearly not satisfiable.

[1]The theoretical upper bound on the number of time-frames that may be required grows exponentially with the number of register bits in the circuit.

C. New Algorithm for Satisfiability

C.1 Problem

We assume that we are given an arbitrary interconnection of combinational modules which include Boolean and word-level operators:

- AND, OR and NOT gates
- Comparison: $>, \geq, <, \leq$
- Addition/subtraction: $+/-$
- Increment/decrement: $+/-\ k$
- Scalar multiplication: $*k$
- Left and right shift: $<<\ k$, $>>\ k$

Any other word-level or Boolean operator that is not one of those listed above has to be converted into a collection of the listed word-level and Boolean operators. Note that the word-level operators are all linear.

We are given a required set of values for an arbitrary subset of module outputs, and we may be given additional constraints as described in Section III-D. Our task is to find an input assignment that produces the required output values, or prove that none exists. We call this the **HSAT** problem.

C.2 Variables

There are two kinds of variables, Boolean variables, and word-level variables or integers. The word-level operators operate on word-level variables, whereas the Boolean operators work on Boolean (single-bit) variables.

It is possible that Boolean operators are applied to word-level variables. For example, a word-level variable A may operated upon as $A + B$, and also as A & C, where & corresponds to a bitwise Boolean AND. To represent the bitwise AND operator we will write $O(n)$ **Boolean clauses** over the bits in A and C, namely, $a_{n-1}, a_{n-2}, \cdots, a_0$, and $c_{n-1}, c_{n-2}, \cdots, c_0$. Here a_{n-1} is the most significant bit in A, and a_0 the least significant bit.

We will use the word A in a **linear arithmetic constraint** (LAC) to model the word-level operations on A. We will "remember" the correlation/equivalence between A and the a_i, but we will represent the two separately throughout the algorithm.

The **range** of a variable is the set of values that the variable can take. For example, an integer variable A may take on the values -15 to $+15$. A Boolean variable a_i can only take on the values $\{0, 1\}$.

Given the set of combinational modules listed in Section IV-C.1, for each module we will write a set of Boolean clauses defining the relationship between the inputs and outputs of that module. In the next two sections, we describe the clauses for Boolean operators, and linear constraints for word-level operators, respectively.

C.3 Boolean Clauses for Boolean Operators

We write 2-SAT and 3-SAT clauses for each logical gate as described in Section IV-B. For example,

$$z = AND(x, y): \quad (\overline{z} \vee x) \cdot (\overline{z} \vee y) \cdot (\overline{x} \vee \overline{y} \vee z) \quad (3)$$

D. Linear Constraints for Word-Level Operators

We write linear constraints for each of the word-level operators.

$$C = A + B: \quad A + B - C \leq 0 \ and \ A + B - C \geq 0 \quad (4)$$
$$C = A + k: \quad C - A \leq k \ and \ C - A \geq k \quad (5)$$

Similarly for $-$, and $-k$. For scalar multiplication, we have:

$$C = A * k: \quad C - kA \leq 0 \ and \ C - kA \geq 0 \quad (6)$$

The shift operators $<<$ and $>>$ can be viewed as scalar multiplication. In the case of shifting in 1's we may need to add a constant to the above equations. We can also model cyclic shifts, by adding the most significant bit a_{n-1} to the equations.

Now consider $c = A > B$. Note that c is a Boolean variable. In the sequel, U refers to 2^n where n is the maximum number of bits in A or B. We will write

$$A - B + U(1 - c) \geq 1. \quad (7)$$

In the above equation, if c is 1, then the constraint becomes $A - B \geq 1$, i.e., $A > B$. If c is 0, then we have $A - B \geq 1 - U$, which is always true, given that U is large enough. We also write,

$$A - B - Uc \leq 0. \quad (8)$$

If c is 0, then the constraint becomes $A - B \leq 0$. If c is 1, then the constraint is $A - B \leq U$, which is always true.

Thus, the two constraints above will express the input-output relationships of a $>$ comparator. We can write similar pairs of constraints for $\geq$, $\leq$, and $<$ comparators.

Nonlinear operators such as integer multiplication can be decomposed into linear operators. We can write $4n + 2$ linear constraints for an n-bit multiplier, $Z = X * Y$, by expanding the bits of one of the operands as shown below.

$$Z \geq \sum_{i=0}^{n-1} 2^i p_i, \qquad Z \leq \sum_{i=0}^{n-1} 2^i p_i$$

$$p_i - Ux_i \leq 0, \qquad 0 \leq i < n$$

$$p_i + U(1 - x_i) - Y \geq 0, \qquad 0 \leq i < n$$

$$0 \leq p_i \leq Y, \qquad 0 \leq i < n \quad (9)$$

Consider the modulus operator $Z = X \ mod \ m$, where m is a constant. We will model it using the following equations and variables.

$$X - Z - mY \leq 0$$

$$X - Z - mY \geq 0$$

$$0 \leq Z \leq m - 1$$

$$0 \leq Y \leq \lfloor \frac{maximum \ value \ of \ X}{m} \rfloor \quad (10)$$

We can handle integer division in a similar manner.

E. Transformation of LAC

Given an arbitrary set of modules, we have three types of constraints:

- 2-SAT clauses
- 3-SAT clauses and
- Linear arithmetic constraints (LAC).

The 2-SAT and 3-SAT clauses are over the Boolean variables. The LAC are over the word-level variables, and some Boolean variables.

HSAT = 2-SAT + 3-SAT + LAC.

In this section, we describe a transformation that allows the integration of linear arithmetic constraints and Boolean clauses in our satisfiability algorithm.

Note that there is correspondence or correlation between the word-level variables and the Boolean variables. For example, we may have A appear in an LAC, and a_i, a bit in A, appearing in a 2-SAT clause. We have to make sure that values of word-level variables in the LACs are **consistent** with the values of constituent bits in the Boolean clauses.

Initially, we have all the variables **unbound** or set to unknown values. During our search (cf. Section IV-E.1), we may set Boolean variables or word-level variables. However, if a word-level variable W is correlated to a Boolean variable w_i, we will only set the Boolean variable to $\{0, 1\}$, and transform the LACs that contain W (cf. Section IV-E.1). We describe the transformation below.

Assume that $0 \leq W \leq 2^n - 1$, i.e., W is a positive integer n bits wide. Suppose w_0 is set to 1 in W. Then, we introduce a new variable, call it, V, and we write $2V + 1$, instead of W. The range of V is smaller than that of W, and it is $2^{n-1} - 1$. In general, given a word-level variable W, and an arbitrary number of Boolean variables (bits) in W set to 1 or 0, we can always write a **linear expression** for the arithmetic value for W using

$$W = 2^{n-1} w_{n-1} + 2^{n-2} w_{n-2} + \cdots + 2w_1 + w_0$$

If W is of n bits, and we set r bits in W, then we will need in the worst case, $r + 1$ new word-level variables, assuming none of the r bits are adjacent. As an example, consider a four-bit W, with w_2 set. We may transform it to $8X + 4 + Y$, where $X \leq 1$, $Y \leq 3$. X is effectively a Boolean variable, and Y is a two-bit integer. We no longer need W, so the increase in the number of variables is at most r.

For sign/magnitude or two's complement integers, we merely use the appropriate equation that corresponds to the arithmetic value of the integer based on the individual bits. For example, the two's complement equation for an $n + 1$-bit number W, whose most significant bit is the sign bit, would be:

$$W = -2^n w_n + 2^{n-1} w_{n-1} + 2^{n-2} w_{n-2} + \cdots + 2w_1 + w_0$$

Thus, given an arbitrary setting of Boolean variables that correlate to word-level variables in an LAC, we can modify the LAC into another LAC using the method outlined above. The left-hand-side of the LAC is modified and the right hand side remains the same. Simple algebraic manipulation is applied to normalize the LAC, i.e., to convert it to a standard form $\sum_i a_i w_i \leq C$ or $\sum_i a_i w_i \geq C$. The modified LAC will typically have more variables than the original LAC, however, the ranges of the variables will be smaller. Further, there is no exponential increase in the number of variables or constraints; the total number of constraints and the total number of variables in the LAC will grow no more than the number of Boolean variables that are set to 0 or 1.

The above transformation is key to the seamless integration of Boolean clauses and LAC in our hybrid satisfiability algorithm. We next describe the satisfiability search which uses the above transformation, so word-level operators are only expanded when needed, resulting in greater efficiency.

E.1 Satisfiability Search

If the LAC is empty, we can check for various special cases in the 3-SAT clauses.

Essential variable : If a clause contains a single variable in true form (complemented form), clearly the variable has to be set to 1 (0) for the clause to be satisfied.

Dominating Clause : If two clauses $(a \vee b)$ and $(a \vee b \vee c)$ exist, we can delete the latter, since the former implies the latter.

Unate Variable : If a variable a appears only in true form, (complemented form) in all the clauses, we can immediately set the variable to 1 (0).

There are many other optimizations that can be applied for the 3-SAT problem, see [14].

In the general case, the LAC will be non-empty. We can still check for essential variables, and delete dominating clauses. However, we cannot set variables that are unate with respect to the Boolean clauses without checking for correlation in the LAC. For example, if we have

$$(w_0 \vee a \vee b)$$

$$W + X \leq 5$$

with w_0 only appearing in the one clause shown, we still cannot assume that $w_0 = 1$, because w_0 is correlated with W. In order to satisfy the clauses and LAC, we might need $w_0 = 0$, $a = 1$.

We will perform the standard branching search to find a satisfying assignment. We will only set variables corresponding to inputs to the circuit, and not intermediate signals. (The intermediate signals may be set by implication on the clauses.) We first perform special case pruning, such as the checks outlined above. There is a lot of structural information that can be exploited to simplify the 3-SAT clauses. The reader is referred to [8] and [14] for details.

Regardless of correlation among bit-level and word-level variables, we can assert the following:

- Polynomial-time checks using cycle detection can be performed to see if 2-SAT + 3-SAT is infeasible [8]. If 2-SAT + 3-SAT is infeasible, so is HSAT. Note that for 3-SAT even if cycle detection fails, the problem may still be infeasible.
- We can run full-fledged CNF satisfiability algorithms with a limit on the number of backtracks in search, to possibly prove infeasibility of 3-SAT.

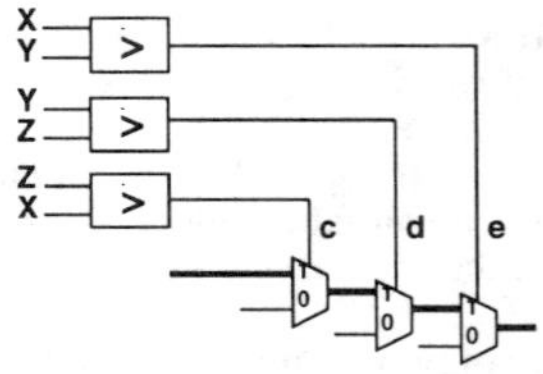

Fig. 3. A False Path

- The linear program relaxation (LPR) of the LACs corresponds to relaxing the assumption that the word-level variables are integral. (That is, they can be real.) If the LPR corresponding to HSAT = 2-SAT + 3-SAT + LAC is infeasible, then the HSAT problem is infeasible.

If the problem is not infeasible, we will heuristically select a Boolean variable, or a word-level variable. If we select a Boolean variable, we set it to 1, modify the clauses and LACs appropriately, and recursively call the satisfiability procedure. If we fail in finding a satisfying assignment, we will set the Boolean variable to 0, and recur. The selection of the variable is dependent on how many clauses and LACs it appears in. In the case of selecting a word-level variable, we check to see that no unknown Boolean variable correlated with that word-level variable exists. (If so, we will select another word-level variable, or the Boolean variable.) We split the range of the word-level variable and recur on the two smaller ranges. The split is based on the value of the variable returned by the LPR. That is, if the word-level variable's range is $[-7, 7]$, and the LPR returns a value of 4.5 for the variable, we will recur on the $[-7, 4]$ and $[5, 7]$ ranges. The only modification we need to perform, in this case, is to the LAC.

It is possible that LPR may return a feasible solution, but no integer solution may exist. If the LPR is feasible, the solution is integral, and is consistent across the word-level and bit-level variables, then we have a feasible solution to HSAT, and we can stop. When LPR returns a feasible, non-integral solution, we perform a simple perturbation of the the solution returned by LPR to see if we can make it integral, while satisfying the consistency requirements. If so, we have found a feasible solution to HSAT.

The CPU time allowed for search is limited by limiting the number of backtracks. Given a certain number of backtracks, three things may happen:

1. We find an input assignment that produces the required output values.
2. We determine that no input assignment that satisfies the required values exists.
3. We may run out of backtracks before 1 or 2 occurs above. In this case we have failed.

F. Examples

Consider the bold path in Figure 1. Assume we wish to sensitize this path. There are three requirements:

- Assuming unsigned numbers, to avoid underflow in the subtractor, we will require that `x >= y`.
- We will require `c = 1`. This will translate to `x > y` when the module-input module-output relationships are included.
- We will require `d = 0`. This directly translates to `rst = 0`.

Another example is depicted in Figure 3, where LPR quickly detects infeasibility, but gate-level satisfiability typically requires a large number of backtracks to prove infeasibility. The bold path in the figure is **false**, i.e., there is no assignment to the word-inputs X, Y, and Z that sensitizes it. One call to LPR will determine this. If we use gate-level satisfiability, the number of backtracks will depend on the order in which the bits of X, Y, and Z are assigned to 0/1 values. If the words are n bits wide, and we begin with the least significant bits, we will require $O(2^{3n})$ backtracks to establish infeasibility. If the most significant bits of X, Y, and Z are set first, gate-level satisfiability will still require $O(2^n)$ backtracks, because infeasibility will not detected for the $x_i = y_i = z_i = 1$, or $x_i = y_i = z_i = 0$ cases. It is possible that bits of word X or Y or Z appear in Boolean clauses due to additional circuitry. However, since we take correlation into account during the search, appropriately modifying the LAC will allow us to use LPR and outperform gate-level satisfiability.

If we use ordered Binary Decision Diagrams [3], under the right ordering, the BDDs for the outputs of the comparators will grow linearly with bit-width. To determine infeasibility we have to compute the intersection of the three BDDs, and the time required for this will grow super-linearly with bit-width. Further, for more complex sets of (linear) arithmetic constraints, it is difficult to find an ordering (assuming one exists) that results in BDDs of reasonable sizes.

Satisfiability problems generated during the justification step are similar to those generated in the path sensitization step. We have the module-input, module-output relationships, and additionally a requirement of particular values on register inputs.

V. Experiments and Comparisons

We present two sets of results on some example HDL models. The HDL models correspond to a parallel port for an embedded processor (`pport`), various finite state machines with some data operations, notably comparisons (`schsm`, `ctla`, `ctlbc`), and two protocols (`protrisc` and `protmpp`). The HDL models are in both VERILOG and VHDL and their size varies between 100-500 lines of code. We also present an example that corresponds to checking if the output of a 16-bit multiplier can be a prime number (`mult16`), restricting the operands to be different from 1.

A. Satisfiability

The first set of results compares our new HSAT algorithm to gate-level 3-SAT(GSAT). Our HSAT algorithm is implemented using the 3-SAT package from [14] and a commercial LP solver CPLEX. We compare the performance of HSAT to vanilla 3-satisfiability using the software from [14], adapted to read CNF expressions. During the process of vector generation, many satisfiability problems are solved. For the different HDL examples, we give data on the **hardest** satisfiability problem that was successfully [2] solved by HSAT.

In Table I we give the quantitative comparison between HSAT and GSAT. For HSAT, we report the number of clauses (2-SAT or 3-SAT), the number of LACs, the number of backtracks, and the amount of CPU time required for vector generation. For GSAT, we report the number of clauses, number of backtracks, the CPU time required to expand the LACs into Boolean clauses, and the CPU time required to solve the satisfiability problem. To compare overall performance the reader should compare the sum of the GSAT times against the HSAT time, since in order to run GSAT we have to perform the expansion of the LACs into clauses.

Just comparing the time for satisfiability checking, HSAT performs significantly better than GSAT for all the satisfiability problems. The improvement is due to the significant reduction in the number of backtracks, which directly translates to faster execution time. The reduction in backtracks is due to the LP feasibility check which prunes unsatisfiable branches of the tree early in the recursion. The CPU time required **per backtrack** is significantly higher for HSAT as compared to GSAT because we are performing an LP solve at each step. Note that Table I corresponds to the hardest satisfiability problems we encountered in vector generation. Many of the satisfiability problems encountered in vector generation are simple, and both HSAT and GSAT find satisfying assignments without backtracking. We did not encounter any satisfiability problems where GSAT was successful, and HSAT failed.

Other satisfiability approaches such as GRASP [13] may outperform GSAT in some cases; however, these approaches still have to deal with a substantially larger problem than HSAT, and do not use LPR feasibility checking. We were unable to create ordered Binary Decision Diagram [3] representations for the circuits corresponding to the large examples. Finally, one can envision using commercial ILP solvers once the problem has been set up, but this is also inefficient because any integer variable that appears in a Boolean clause has to be expanded in the initial problem specification; there is no notion of remembering correlation between different representations of a variable in ILP formulations.

[2] By successfully we mean that HSAT either found a satisfying assignment, or proved that none existed within the allotted number of backtracks.

Model	HSAT				GSAT			
	#clauses	#LACs	#backtracks	Sat. time	#clauses	#backtracks	Exp. time	Sat. time
pport	17	18	0	< 1 s	685	2,604	1.2s	4.7s
schsm	141	24	0	< 1 s	841	35,102	3.1s	261s
ctla	1	24	1	< 1 s	1355	> 100,000	12.0s	> 1000s
ctlbc	1	60	1	< 1 s	1341	17,477	4.9s	142s
mult16	0	104	4158	16 s	-	-	-	-

TABLE I
COMPARISON BETWEEN SATISFIABILITY APPROACHES

Model	Satisfiability Test			Random Test	
	#Vec	Cov %	Time	#Vec	Cov %
pport	23	90.9	34s	2000	90.9
schsm	24	88.1	56s	2000	85.6
ctla	66	100.0	146s	2000	91.2
ctlbc	32	100.0	82s	1157	100.0
protrisc	46	94.7	246s	2000	78.2
protmpp	76	100.0	140s	2000	64.0

TABLE II
COMPARISON BETWEEN TEST GENERATION APPROACHES

B. *Directed Test Generation*

We next report results on comparing the vector generation strategy against random pattern generation. The data is reported in Table II. For each HDL example, we give the number of vectors in the vector sequence generated by the HSAT algorithm, the amount of CPU time required, and the percentage of HDL statements exercised by the vectors. For random vector generation, we give the **best** coverage obtained by simulating 10 sets [3] of 2000 random vectors. CPU time for random vector generation and simulation grows proportionately with the number of vectors, and is not reported.

The directed vector generation method produces as good, or higher coverage, using a much smaller number of vectors in all cases. The circuits were initially assumed to be in a known state, since this makes it easier to generate random patterns. (If the circuit were in an unknown state, random pattern generation gives significantly worse results. However, using HSAT we can put the circuit into a known state and then perform vector generation.)

The CPU times in both tables correspond to seconds on an UltraSparc.

VI. ONGOING WORK

The two primary reasons why the hybrid satisfiability algorithm may fail for a particular problem are nonlinear operations and large sequential depth. In our methodology, nonlinear operators have to be decomposed into linear arithmetic or Boolean operators. We have choices in the decomposition process. For example, we may choose to decompose nonlinear multiplication $A * B$ into a large set of Boolean clauses. However, it is usually more efficient to transform $A * B$ into linear constraints as described in Section IV-D. Being able to perform the "optimal decomposition" is desirable.

Time-frame expansion works well for circuits with small sequential depth such as datapaths, and small finite state machines. One case where time-frame expansion may become prohibitively expensive is when there are timers or counters in the circuit. For example, we may have a 16-bit counter whose output is compared to 65535 to generate a Boolean signal. It is possible that we have to unroll the counter 65535 times in order to find the appropriate input stimuli. Unrolling the next state logic k times can be done efficiently by turning a $+1$ increment operator into a $+k$ increment operator. However, any reset logic in the counter may have to be explicitly unrolled k times to find the input conditions that force reset in the appropriate clock cycle. Methods to alleviate this computational expense are a subject of ongoing work.

[3]These sets were generated using different seeds.

ACKNOWLEDGEMENTS

F. Fallah and S. Devadas were supported in part by a grant from Schlumberger Foundation.

REFERENCES

[1] G. S. Avarunin, L. K. Dillon, and J. C. Wileden. Experiments with Automated Constrained Expression Analysis of Concurrent Software Systems. In *Proceedings of ACM SIGSOFT 89, 3rd Symposium on Software Testing, Analysis, and Verification*, pages 124–130, December 1989.

[2] M. A. Breuer and A. D. Friedman. *Diagnosis and Reliable Design of Digital Systems*. Computer Science Press, 1976.

[3] R. Bryant. Graph-Based Algorithms for Boolean Function Manipulation. *IEEE Transactions on Computers*, C-35(8):677–691, August 1986.

[4] T. Bultan, R. Gerber, and W. Pugh. Symbolic Model Checking of Infinite State Systems Using Presburger Arithmetic. In *Proceedings of the Workshop on Computer-Aided Verification*, pages 400–411, 1997.

[5] A. K. Chandra and V. Iyengar. Constraint Solving for Test Generation: A technique for high-level design verification. In *Proceedings of the Int'l Conference on Computer Design*, pages 245–248, October 1992.

[6] K-T. Cheng and A. S. Krishnakumar. Automatic Functional Test Generation Using the Extended Finite State Machine Model. In *Proceedings of the 30th Design Automation Conference*, pages 86–91, June 1993.

[7] A. Ghosh, S. Devadas, and A. R. Newton. *Sequential Logic Testing and Verification*. Kluwer Academic Publishers, Boston, Massachusetts, 1992.

[8] T. Larrabee. Test Pattern Generation Using Boolean Satisfiability. *IEEE Transactions on Computer-Aided Design*, 11:4–15, January 1992.

[9] J. Lee. *Architectural Level Test Generation and Fault Simulation*. PhD thesis, University of Illinois at Urbana-Champaign, 1992.

[10] K. M. Olender and L. J. Osterweil. Cesar: A Static Sequencing Constraint Analyzer. In *Proceedings of ACM SIGSOFT 89, 3rd Symposium on Software Testing, Analysis, and Verification*, pages 66–74, December 1989.

[11] K. Roy. The Use of RTL Descriptions in Accurate Timing Verification and Test Generation. *IEEE Journal of Solid-State Circuits*, 26(9):1230–1239, September 1991.

[12] K. Roy and J. Abraham. High Level Test Generation Using Data Flow Descriptions. In *Proceedings of the European Design Automation Conference*, pages 480–484, March 1990.

[13] J. Silva and K. Sakallah. GRASP: A New Search Algorithm for Satisfiability. In *Proceedings of the International Conference on Computer-Aided Design*, pages 220–227, November 1996.

[14] P. Stephan, R. K. Brayton, and A. Sangiovanni-Vincentelli. Combinational Test Pattern Generation Using Satisfiability. *IEEE Transactions on Computer-Aided Design*, 15(9):1167–1176, September 1996.

Automatic Generation of Assertions for Formal Verification of PowerPC™ * Microprocessor Arrays Using Symbolic Trajectory Evaluation

Li-C. Wang, Magdy S. Abadir, and Nari Krishnamurthy
Somerset PowerPC Design Center
Motorola, Inc., Austin, Texas

Abstract

For verifying complex sequential blocks such as microprocessor embedded arrays, the formal method of symbolic trajectory evaluation (STE) has achieved great success in the past [[3], [5], [6]]. Past STE methodology for arrays requires manual creation of "assertions" to which both RTL view and the actual design should be equivalent. In this paper, we describe a novel method to automate the assertion creation process which improves the efficiency and the quality of array verification. Encouraging results on recent PowerPC arrays will be presented.

1 Introduction

Arrays (such as cache, cache tag, tables, register files, etc.) are complex sequential logic blocks which represent one of the major challenges in modern microprocessor design environment. The challenge comes from two folds: the size of these arrays and the complexity of their designs. In today's RISC microprocessors, arrays may contain up to 80% of the total transistors in use. Verifying their correctness is obviously very important. Usually, the RTL view of an array design is first simulated as part of the full-chip verification effort. Then, the transistor level schematic view is verified to be equivalent to the corresponding RTL view during logic verification. In the past, logic verification relied on vector simulation and boolean equivalence checking. While exhaustive vector simulation is not feasible for arrays due to the exponentially large number of patterns required, boolean equivalence checking is not suitable either for it fails to capture the complex timing and sequential control [4].

In recent years, at the PowerPC Design Center, Somerset, an additional formal verification methodology has been deployed for arrays. The methodology utilizes the formal verification technique of Symbolic Trajectory Evaluation (STE) [7] to verify the equivalence between the RTL view and the schematic view [[3], [5], [6]].

*IBM and PowerPC are trademarks of IBM Corporation in the United States, or other countries, or both

DAC 98, San Francisco, California

In STE verification, at first a set of so-called *assertions* are **manaully** created according to the array specification, usually in the RTL view. Then, both the RTL view and the transistor level schematic view are proved to satisfy the same set of assertions under STE. Creating assertions usually requires a deep understanding of the overall functionality of the array so that each array operation can be expressed properly into an assertion. Such a process can be very tedious and potentially incomplete. In this paper, we will describe a novel framework for automatic generation of assertions directly from the RTL view for arrays and hence, provide a more efficient and robust approach for logic verification of arrays. Encouraging results on five selected array designs in recent PowerPC microprocessors will be reported.

2 Background

Arrays comprise storage elements, together with complex timing and control logic that determine how they are used and updated. An array design can be represented by different views, each for a particular set of tools and applications. For example, the RTL view is the most abstract, which serves as the high-level specification of a design and is mainly used for functional simulation. The actual implementation is represented by the transistor schematic view.

2.1 Logic Verification with STE

The main purpose of logic verification is to verify the equivalence between the RTL view and the schematic view. At Somerset, this is achieved by vector simulation, boolean equivalence checking and STE verification. For arrays, STE verification is used.

STE [7] is a formal verification technique which is a descendant of symbolic simulation [1]. At the core of symbolic simulation, Ordered Binary Decision Diagram [2] is used to efficiently manipulate boolean functions. In STE, a circuit is specified in terms of a set of properties in a restricted temporal logic form. Properties are expressed by so-called **assertions** which are of the form "**Antecedent** $\Longrightarrow$ **Consequent**" where both "Antecedent" and "Consequent" consist of a number of formulae. Formulae can be simple predicates such as "line A is 'a"' (line A holds the symbolic value 'a' at current time), or conjunctions of these simple predicates. Formulae can be applied with the next time operator in order to state the facts such as "line A remains 'a' in the next time step." It is also possible to apply *domain* operator so that "line A remains 'a' **when** D is true" can be expressed. This simple logic in STE is is true" can be expressed. This simple logic in STE is sufficient for array verification purposes. In STE, conditions specified in the antecedents are symbolically simulated, and the consequents are asserted. (Please refer to the previous papers [[3] [5] [6]] for details)

2.2 Assertion Creation

For complex array designs, assertion creation can be a complicated, tedious, and unreliable process. First, a person needs to fully understand the functionality of a unit before creating the assertions. Assertions themselves can be long and tedious in order to specify all aspects of a design in details. Very often, a verification engineer would create assertions to verify only what he/she understood as the normal operations. Hence, assertions can be incomplete. To avoid all these problems, a way to automatically generate assertions is required.

3 Automatic Assertion Generation for Arrays

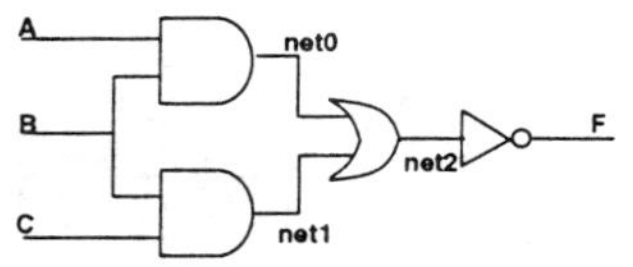

Figure 1: A simple combinational circuit

Automatic assertion generation for sequential designs in general is a difficult problem. Such a tool must be able to parse the statements in the RTL view, automatically extract the high level properties of the design, and then express them in the antecedent/consequent form. On the other hand, generating assertion for a combinational design is rather straightforward. Consider the combinational logic shown in Figure 1. The assertion for this circuit can be stated as the following.

Antecedent:	$\Longrightarrow$	**Consequent:**
$A = a$, $B = b$,	(lead to)	let $net0 = A\&B$
$C = c$		let $net1 = B\&C$
		let $net2 = net0 \mid net1$
		$F = \neg net2$

where $net0, net1, net2$ are temporary variables which hold the temporary symbolic expressions for the final consequent, and a, b, c are all symbolic values. Based on the simple observation above, the idea of our automatic assertion generation approach for arrays is to *abstract out the timing and sequential elements so that they can be treated as a combinational circuit at a given time.* At Somerset, a tool called AGA (automatic generation of assertion) is being developed for arrays.

3.1 Define Timing

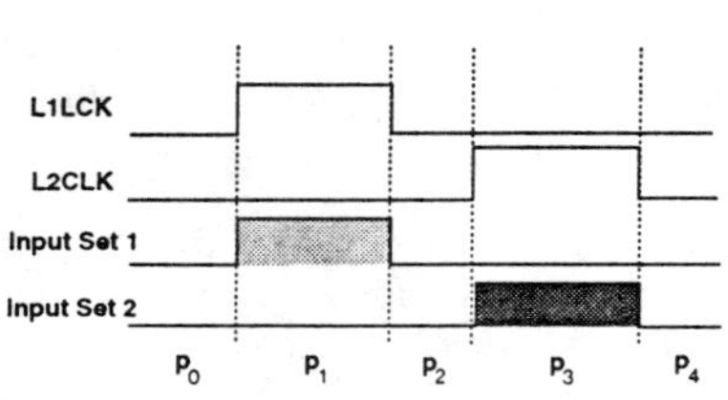

Figure 2: An example of clocks

The only necessary user input to AGA is a clock definition where the resulting assertion should be based on. For example, a typical array may operate based on two non-overlapping clocks, L1CLK and L2CLK. A clock definition is illustrated in Figure 2. The two clocks divide the cycle into five different phases of interest. Since arrays are clock-based, they behave like a combinational circuit within each phase. Without additional information, AGA will automatically assign different symbolic values to all primary inputs at all different phases, and then try to derive consequents for each phase *independently* based on the RTL view. Most of the time, this is unnecessary and quite inefficient. Therefore, as the Figure shows, we allow user to specify optional information regarding when to supply "meaningful" symbols to what inputs. Note that any unspecified inputs at a given phase will be assumed to have the unknown value X. An X at an input really means "don't care." Note that a different clock definition can be defined in the Figure by interchanging L1CLK and L2CLK.

3.2 Use Temporary Variables for Each Clock Phase

Basically, AGA behaves like a translator that translates each RTL construct into a sequence of intermediate symbolic expression assignments (as those in the combinational logic example). The translation process follows the topological order of a design from primary inputs to outputs. Final consequents will be formed at the primary outputs and at the memory elements as well.

For each variable name used in the RTL view, which can represent an input, an output, an internal line, or a memory block, a list of temporary variables will be created in AGA. For a variable A, these temporary variables will be involved in a list of symbolic expression assignments and will all be of the form "$A_{[p_i,cond]} = exp$" which means that A at the end of phase p_i will hold a symbolic value given by the boolean expression exp **when** $cond$ is true. As a simple example, supposed primary input Din is given a symbolic value $Data$ during $p2$. This results in five temporary variables inside AGA (one for each clock phase): $Din_{[p_0,1]} = Din_{[p_1,1]} = Din_{[p_3,1]} = Din_{[p_4,1]} = X$ and $Din_{[p_2,1]} = Data$.

3.3 Combinational Logic

Except for latch and memory element, other constructs in the RTL view can be treated as part of the combinational logic. These include assignment statement, if-then-else statement, case statement, tri-state buffer, etc. Note that Somerset's RTL design language does allow the for-loop statement. However, for-loop is considered a statement for macro expansion, which is commonly used to describe a piece of replicated logic. In the following, we will consider a few examples to illustrate the fundamental ideas. Note that the translation process handles combinational statements for every clock phase independently.

Expression involving X

Assume that $B_{[p_i,1]} = X$. The following rules tell how to deal with unknown values X during the translation process.

RTL statements		AGA results at p_i
$A = exp\&B$	$\longrightarrow$	$A_{[p_i,\neg exp]} = 0$
$A = exp \mid B$	$\longrightarrow$	$A_{[p_i,exp]} = 1$
$A = \neg B$	$\longrightarrow$	$A_{[p_i,1]} = X$
$A = B$	$\longrightarrow$	$A_{[p_i,1]} = X$

When conditions

Suppose that $A_{[p_i,C1]} = exp1$ and $B_{[p_i,C2]} = exp2$. The following rules tell how to merge the "when" conditions through logical expressions.

RTL		AGA results at p_i
$C = A\&B$	$\rightarrow$	$C_{[p_i,(\neg exp1\&C1\&\neg C2)\mid(\neg exp2\&C2\&\neg C1)]} = 0$
		$C_{[p_i,C1\&C2]} = exp1\&exp2$
$C = A \mid B$	$\rightarrow$	$C_{[p_i,(exp1\&C1\&\neg C2)\mid(exp2\&C2\&\neg C1)]} = 1$
		$C_{[p_i,C1\&C2]} = exp1 \mid exp2$
$C = \neg A$	$\rightarrow$	$C_{[p_i,C1]} = \neg exp1$
$C = A$	$\rightarrow$	$C_{[p_i,C1]} = exp1$

Note that all variables use the same phase index p_i. Since any boolean expression assignment can break down into a sequence of simple assignments involving binary AND and OR, and unary NOT, the above table is sufficient for more general cases.

If-Then-Else

Suppose that exp below has been evaluated first and resulted in a variable $E_{[p_i,C_0]} = exp_0$, and $A1_{[p_i,C1]} = exp_1$, $A2_{[p_i,C2]} = exp_2$. The following presents the idea of how to handle the if-then-else statement.

	RTL		AGA results at p_i
if	(exp) then	$\rightarrow$	$A_{[p_i,C_0\&C_1\&\neg C2\&exp_0]} = exp_1$
	$A = A1$		$A_{[p_i,C_0\&C_2\&\neg C1\&\neg exp_0]} = exp_2$
else	$A = A2$		$A_{[p_i,C_0\&C_1\&C_2]} =$
			$(exp_0\&exp_1) \mid (\neg exp_0\&exp_2)$

Tri-state buffer and bus

Let "$(exp) \leftarrow (select)$" denote a tri-state buffer with data input exp and switch line $select$. Also, let $A1_{[p_i,C1]} = exp1$, $A2_{[p_i,C2]} = exp2$. To simplify our presentation, we assume two sources with select lines $S1_{[p_i,1]} = s1, S2_{[p_i,1]} = s2$, respectively. In this example, we assume also that the bus is a wired-OR device.

RTL statements		AGA results at phase p_i
$Bus = (A1) \leftarrow S1$	$\rightarrow$	$Bus_{[p_i,s1\&\neg s2\&C1\&\neg C2]} = exp1$
		$Bus_{[p_i,s2\&\neg s1\&C2\&\neg C1]} = exp2$
$Bus = (A2) \leftarrow S2$		$Bus_{[p_i,s1\&s2\&C1\&\neg C2\&exp1]} = 1$
		$Bus_{[p_i,s1\&s2\&C2\&\neg C1\&exp2]} = 1$
		$Bus_{[p_i,(s1\mid s2)\&C1\&C2]} =$
		$s1\&exp1 \mid s2\&exp2$

Summary

In summary, processing a statement of combinational logic is straightforward if no "when" condition is involved. For a particular line (or variable) in the RTL view, a "when" condition defines an unique temporary variable in AGA at a given phase. In general, when a single statement involves multiple variables, each with a list of temporary variables, all possible combinations of their "when" conditions should be exhausted. In this case, all new "when" conditions produced should be mutually exclusive (see the Bus example).

3.4 Latch

The translation for latch is similar to that for the tri-state buffer. Normally, latches are controlled by the input clocks. Take the clock setting in Figure 2. A latch can be easily processed as the example shown below (Assume $A_{[p_i,C_i]} = exp$ has been obtained for A at phase p_i).

RTL statements		AGA results
$D = latch(A, L2CLK)$	$\rightarrow$	$D_{[p_0,1]} = D_{[p_1,1]}$
		$= D_{[p_2,1]} = X$
		$D_{[p_3,C_3]} = exp$
		$D_{[p_4,C_3]} = D_{[p_3,C_3]} = exp$

In other words, a latch will hold its state unless its input clock becomes active. Initially, its state is the unknown X. This example also illustrates how the temporary variables can be used to carry unchanged latch states across the clock phases ($D_{[p_4,C_3]} = D_{[p_3,C_3]}$).

3.5 Memory Element

The most difficult part in the translation process is to handle memory elements. However, this problem can be simplified by imposing some useful design rules on arrays. At Somerset, arrays in the RTL view are usually designed with predefined RAM primitives. In this way, all memory elements can be grouped into different RAM blocks, each implemented by a RAM primitive where its functions are well defined and understood. In AGA, a set of translation rules are associated with a RAM primitive. To illustrate the idea, we show the rules for a simple RAM primitive below. Figure 3 shows the RAM model, called as P-RAM (primitive RAM). Due to space limitations, other more complex primitives are omitted.

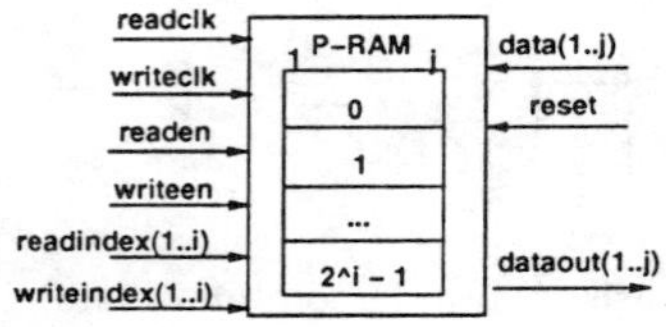

Figure 3: RAM primitives

The assumption for P-RAM is that read and write cannot happen simultaneously. However, this does not exclude us from verifying both read and write operations in one pass of the symbolic simulation. Usually, *readclk* and *writeclk* are derived from the input clocks and in a normal design, both can only be active during a particular clock phase. Note that the user can define how the primitive behaves when it is not doing a read. To simplify our presentation, we assume that *dataout* is precharged to 1 while the P-RAM is not doing a read.

Again, the goal of AGA translation is to find a sequence of symbolic expression assignments to model the behavior of P-RAM. To do so, we need to assume that the P-RAM contains some initial symbolic values. In STE, this can be achieved with symbolic index by declaring that

$$\text{P-RAM}[Addr(1..i)] = D(1..j)$$

Usually, this initialization is assumed to be done at the beginning of phase p_0.

Similarly, the translation process handles the P-RAM independently for each clock phase. For simplicity, assume that $\text{writeclk}_{[p_3,1]} = wexp$, $\text{readclk}_{[p_1,1]} = rexp$ (possible write in p_3 and read in p_1), $\text{readen}_{[p_1,1]} = ren$, $\text{writeen}_{[p_3,1]} = wen$, $\text{writeindex}_{[p_3,1]} = windex$, $\text{readindex}_{[p_1,1]} = rindex$, $\text{data}_{[p_3,1]} = da$, and $\text{reset}_{[p_i,1]} = re_i$ for $i = 0, 1, 2, 3, 4$. Note that it is possible to include a "when" condition in each of the above variables. Then, all possible combinations of these conditions will be exhausted and more temporary variables will be produced.

In order to properly specify the memory content, the temporary variables created for P-RAM will include a third entry in its footnote index, which is the index *Addr* used for initialization. Therefore, $\text{P-RAM}_{[p_i,C,Addr]} = D$ can be read as "P-RAM[$Addr$] $= D$ during p_i when C is true."

$$R_i = re_0 \mid \ldots \mid re_i \text{ for all } i$$

$$\begin{aligned} dataout_{[p_1,(rindex=Addr)\&rexp\&ren\&\neg R_1]} &= D \\ dataout_{[p_1,R_1]} &= 0 \\ dataout_{[p_j,1]}(j=0,2,3,4) &= 1 \end{aligned}$$

$$\begin{aligned} \text{P-RAM}_{[p_j,\neg R_j,Addr]} &= D\ (j=0,1,2) \\ \text{P-RAM}_{[p_3,wexp\&wen\&\neg R_3,Addr]} &= da(windex = Addr) \\ &\quad \mid D(windex \neq Addr) \end{aligned}$$

$$\begin{aligned} \text{P-RAM}_{[p_4,\neg R_4,Addr]} &= \text{P-RAM}_{[p_3,wexp\&wen\&\neg R_3,Addr]} \\ \text{P-RAM}_{[p_j,R_j,Addr]} &= 0\ (j=0,1,2,3,4) \end{aligned}$$

As discussed earlier, an array may contain multiple instances of the RAM primitives. One problem is how to initialize these RAM instances properly. AGA decides how to initialize multiple RAM instances by checking how their address inputs and data inputs are related. There are three possible scenarios: 1) Two RAMs share the same data inputs. Usually, this case represents a bad usage of RAM primitive where each RAM instance implements part of the same memory block, and it should be implemented with just one RAM. 2) Two RAMs share the same address inputs. Both RAMs will be initialized using the same symbolic address, but with different symbolic data. 3) No sharing occurs. In this case, each RAM is independent and hence, is initialized with an unique symbolic address and data.

3.6 Forming the Final Consequents

After the AGA translation process is done for all primary outputs, a list of temporary variables have been created for every line (variable) and every memory block (RAM primitive) used in the RTL view. Forming the final consequents is straightforward. For arrays, consequents are collected at the RAM instances and at all primary outputs. For each RAM instance, a temporary variable will be converted into a consequent statement in the STE. For example, "$\text{P-RAM}_{[p_4,R_4,Addr]} = 0$" will be translated into a consequent statement "P-RAM[$Addr$] are all 0 **when** R_4 is true before the end of phase p_4." Note that no consequent will be given to a variable which is assigned with the value unknown X. Moreover, each consequent only checks the corresponding result close to the end of each clock phase so that it is guaranteed all results due to input changes during that phase has stablized before reaching the phase end.

4 Summary and Results

Five selected arrays, a data segment register (DSR), an instruction cache tag (ITAG), a branch target instruction cache (BTIC), a translation lookaside buffer (TLB), and a block address translation (BAT) unit, are used as our benchmarks for the initial AGA exercise.

Table 1 presents the comparisons of the overall run time (on the schematic view) between manual assertions and automatic assertion (on an IBM RS6000/590 machine with 512M memory). A significant amount of time and efforts were spent to manually create these assertions while the costs of the automatic approach was minimal. Also note that a fixed OBDD variable ordering (which does not favor either automatic or manual assertions) is enforced for each array to make sure a fair comparison in our experiments.

For all cases, automatic assertions outperform manual assertions in the actual verification run time as well. The reason for these results is that manual assertions are usually not mutually exclusive with respect to which part of the circuit they try to verify.

Arrays	Manual	Automatic
DSR	84 secs	72 secs
BAT	7455 secs	6751 secs
TLB	1426 secs	575 secs
BTIC	1546 secs	768 secs
ITAG	5346 secs	3189 secs

Table 1: Run time comparison between manual assertions and automatic assertion for the five arrays

It is interesting to note that for each case, we have observed that some consequents generated for the automatic assertion are not considered by any of the manual assertions, but not vice versa. Hence, in every comparison result in the two Tables, the automatic assertion actually verifies more properties than the manual assertions.

5 Conclusion

STE has been shown to be an effective technique for logic verification of microprocessor embedded arrays. The past STE methodology for array verification requires manual creation of assertions. Since manual creation of assertions can be tedious and incomplete, in this paper we have presented a framework to automatically generate assertions for arrays from the RTL view. Our experience indicates that the new approach is very effective and can provide more complete verification results.

References

[1] Randal E. Bryant, *Symbolic simulation — techniques and applications*, 27th Design Automation Conference, 1990.

[2] Randal E. Bryant, *Symbolic Boolean Manipulation with Ordered Binary Decision Diagrams*, ACM Computing Surveys, Vol. 24, No. 3, Sep. 1992.

[3] Neeta Ganguly, Magdy S. Abadir, and Manish Pandey, *PowerPC Array Verification Methodology Using Formal Verification Techniques*, International Test Conference, Washington DC., 1996. pp. 857-864.

[4] Charles H. Malley and M. Dieudonne, *Logic Verification Methodology for PowerPC Microprocessors*, 32nd Design Automation Conference, 1995, pp. 234-240.

[5] Manish Pandey, Richard Raimi, Derek L. Beatty, and Randal E. Bryant, *Formal verification of PowerPC™ arrays using symbolic trajectory evaluation*, Proc. 33rd Design Automation Conference, Las Vegas, NV., 1996.

[6] Manish Pandey, Richard Raimi, Randal E. Bryant, and Magdy S. Abadir *Formal verification of Content Addressable Memories Using Symbolic Trajectory Evaluation*, Proc. 34rd Design Automation Conference, 1997.

[7] C,J.H. Seger and Randal E. Bryant, *Formal verification by symbolic evaluation of partially-ordered trajectories*, Formal Methods in System Design 6, 1995, pp. 147-189.

Combining Theorem Proving and Trajectory Evaluation in an Industrial Environment

Mark D. Aagaard, Robert B. Jones, Carl-Johan H. Seger
Strategic CAD Labs, Intel Corporation, Hillsboro, OR 97124, USA

Abstract

We describe the verification of the IM*: a large, complex (12,000 gates and 1100 latches) circuit that detects and marks the boundaries between Intel architecture (IA-32) instructions. We verified a gate-level model of the* IM *against an implementation-independent specification of IA-32 instruction lengths. We used theorem proving to to derive 56 model-checking runs and to verify that the model-checking runs imply that the* IM *meets the specification for all possible sequences of IA-32 instructions. Our verification discovered eight previously unknown bugs.*

1 Introduction

The Intel architecture (IA-32) instruction set has several hundred opcodes. The opcode length is variable, as are the lengths of operand and address displacement data. The architecture also includes the notion of *prefix bytes*, which change the semantics of the subsequent instruction. Two of the prefixes (`h66`, `h67`) can affect the length of the instruction. A single instruction may have multiple prefix bytes, but overall instruction length (including the prefix bytes) is limited to 15 bytes in recent IA-32 implementations.

Because of the instruction set complexity, decoding IA-32 instructions in a high-performance processor is a challenge. Recent Intel implementations have divided the instruction decoding process into multiple phases, including a phase that detects instruction boundaries.

This paper describes the verification of an instruction-length marker (the IM) against an implementation-independent specification of IA-32 instruction lengths. Because of the vast space of possible prefix and opcode combinations, exhaustive simulation of the IM would have been difficult, if not impossible. Formal methods seemed to be an attractive approach for verifying the IM. However, there were a number of challenges, including:

- The size and irregularity of the IA-32 instruction set made writing a specification an intimidating proposition.
- The IM was designed to meet timing, area, and power constraints. These factors led to optimizations in the design that sometimes made it difficult for us to relate our abstract notion of the IM's intended behavior to the actual hardware.
- The datapath is well beyond the capacity limits of current or even near-term model-checking technology.
- The gap between the specification of marking instruction boundaries over arbitrary-length streams and the gate-level design of a pipelined machine was so large that finding a feasible verification strategy was difficult.

We attribute our successful verification of the IM to the combination of symbolic trajectory evaluation (STE) model checking and lightweight theorem proving. This leveraged the benefits of model-checking automation and the reasoning power of theorem proving.

We employed theorem proving in three ways: 1) to guide the decomposition of the verification into tasks within the capacity limits of model checking, 2) to ensure that this decomposition covered all possible legal inputs, and 3) to check the validity of transformations used to optimize the proof obligations for improved model-checking efficiency.

We incorporated a variety of theorem-proving techniques that, while well understood within the theorem-proving community, have not been previously used in combination with model checking. We mechanized several case-splitting approaches and used them to control model-checking complexity. We used induction over time to relate individual model-checking runs to a top-level correctness property over arbitrary-length streams. In addition, we used automated tautology checking to check several side conditions that enabled us to optimize formulas to improve model-checking efficiency.

The purpose of this paper is to describe the verification techniques used and demonstrate their application on a large and complex industrial example. We present an overview of the IA-32 instruction set and the IM. We then describe our correctness statement and the proof itself. We conclude with a summary of the bugs we discovered.

2 Background

Trajectory evaluation [8] is based on traditional notions of digital circuit simulation and excels at datapath verification. The two keys to the efficiency of trajectory evaluation are the restricted language of the temporal formulas (there is no negation and the only temporal operator is "next") and the extended simulation model in which circuits operate over a lattice of logic values $\{\top, (1|0), X\}$.

Trajectory evaluation correctness statements are of the form $ckt \models A \Longrightarrow\!\!\!\!\!> C$ ($\Longrightarrow\!\!\!\!\!>$ is read "leads to"). The antecedent A gives the initial state and input stimuli to the circuit ckt, while the consequent C specifies the desired response of the circuit. Trajectory evaluation supports powerful abstraction mechanisms which are guaranteed to be safe and reduce the computational complexity.

Even with the continuing advances in model-checking technology, industrial scale verification efforts immediately encounter model-checking capacity limits. Using theorem proving to compose verification results offers the tantalizing possibility of mitigating model-checking capacity limitations. Several groups have experimented with combining theorem proving and model-checking technologies; most have found it to be a surprisingly problematic undertaking.

DAC 98, San Francisco, California

Joyce and Seger experimented with using trajectory evaluation as a decision procedure for the HOL proof system [6]. They concluded that for hardware verification, most of the user's interaction is with the model checker, not the proof system. Consequently, using a model checker as decision procedure in a proof system does not result in an effective hardware verification environment. Lessons learned from HOL-Voss resulted in VossProver, an experiment in implementing a lightweight proof tool on top of trajectory evaluation. The first demonstration of the effectiveness of this approach was the verification of a 64-bit integer multiplier by Hazelhurst and Seger [3]. Subsequently, Aagaard and Seger used VossProver to verify a pipelined, IEEE-compliant floating-point multiplier [1].

Our work builds on these efforts by using inference rules to combine verification results together. We extend the previous techniques in three ways. First, we use induction over time to elevate our results from showing that a single execution of the IM is correct to showing that the IM works correctly over arbitrary-length streams. Second, we use case-splits in the input space and state space to decompose the verification task into pieces that are feasible for model checking. Third, we use tautology checking to validate the correctness of transformations that optimize formulas for model-checking efficiency.

Other related work includes the PVS proof system. PVS includes support for BDDs and mu-calculus model checking. The AAMP-5 case study used the mu-calculus model checker in PVS for microcode verification [7]. Our work focuses on performing verification on a model that is as close to the actual hardware as feasible — for us, that means a gate-level model synthesized directly from the RTL code.

Camilleri [2] has used a theorem prover in conjunction with a model checker to verify properties of a cache-coherency algorithm. He used the separate tools to verify different properties but did not combine the results together mechanically. Jang *et al* [5] used CTL model checking to verify a collection of 76 properties about an embedded microcontroller. They point out that they relied on unmechanized arguments to bridge the gap between the 76 properties and the high-level specification.

3 Specification

The Intel architecture (IA-32) instruction-set has several hundred opcodes [4]. Opcodes are specified with one or two bytes. Following the opcode byte(s) is an optional `MODRM` byte which specifies register and memory modes. In some cases, the `MODRM` byte is followed by an `SIB` byte which specifies a scale, index, and displacement size for memory address computations. Instructions can have operand (immediate) data of 1, 2, or 4 bytes, and address (displacement) data of 1, 2, or 4 bytes.

The IA-32 architecture also includes the notion of *prefix bytes*, individual bytes which change the semantics of the subsequent instruction. The architecture has two primary modes: *real mode* (primarily 16-bit operations) and *protected mode* (primarily 32-bit operations). Memory can be addressed using either 16 or 32-bit addresses. Consequently, each instruction that uses memory addresses also has an associated address-size attribute. Similarly, instructions have an operand-size attribute for immediate data. The attribute values are determined by a combination of defaults, instruction prefixes, and size-specification bits in segment descriptors. Two of the prefix bytes (`h66`, `h67`) affect the *length* of the following instruction by toggling the default operand or address sizes. Instructions may have multiple prefix bytes, but overall instruction length is limited to 15 bytes. All instruction encodings are subsets of the general instruction format shown in Figure 1.

Our goal was to verify that the IM correctly marks the boundaries between IA-32 instructions. The IM is implemented as a pipelined

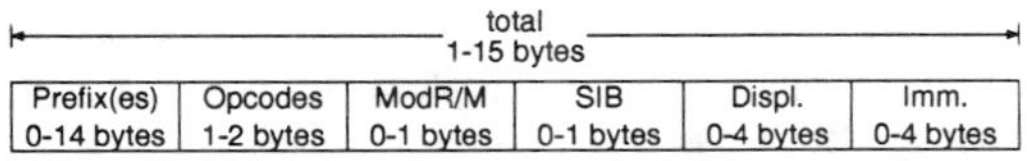

Figure 1: IA-32 Instruction Format.

datapath with internal state. It contains over 12,000 gates and 1100 latches. For obvious reasons, we do not discuss implementation details of the IM in this paper.

We developed a formal length specification of all opcodes in the IA-32 instruction set. We then used this specification to create a functional specification that operates over a byte stream of a given length. This stream specification returns a BDD describing all streams of that length that contain legal instructions. It also returns a BDD describing the boundaries between the instructions. Intuitively, the BDD can be viewed as a big disjunction of all the legal instructions and their corresponding boundaries.

The IM processes a fixed number of bytes simultaneously; we refer to these bytes as a *packet*. In Figure 2, the ADD and MOV instructions are completely contained in the first packet, but the AND instruction runs over into the next packet. Marking the boundaries between instructions in the second packet requires length information about the last instruction in the first packet. The IM records state information about the first packet and uses this state while processing the second packet (Figure 3). This recorded state serves as an *index* into the next instruction packet.

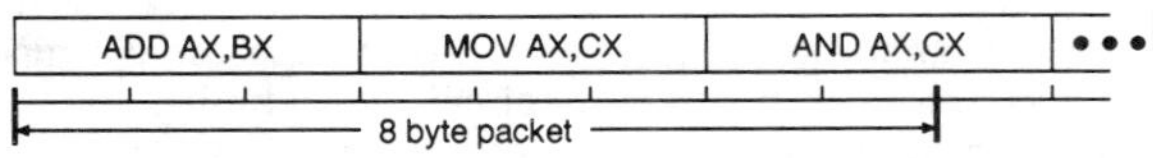

Figure 2: Instructions overlapping fixed packet boundary.

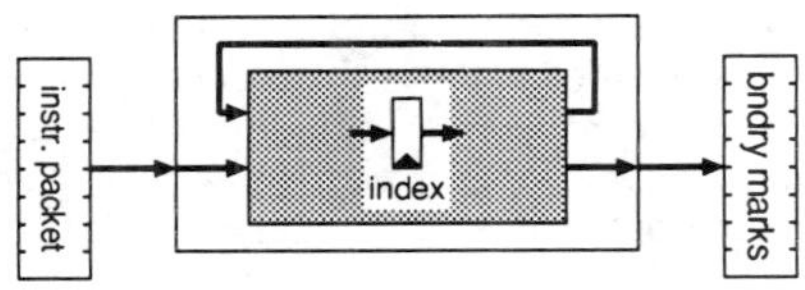

Figure 3: IM Functionality.

Our specification is essentially a functional program describing the length of each legal IA-32 instruction. It is based upon the information available in the Pentium(R) Processor User's Manual [4]. The specification is about 1500 lines. The verification compares the behavior of the specification directly against the behavior of the gate-level circuit.

4 Correctness Statement

Our top-level correctness statement describes the behavior of the IM over arbitrary-length streams. Trajectory-evaluation runs describe the execution of the IM on single packets. The theorem-proving part of the verification bridges the gap between the processing of individual packets and the correctness statement about arbitrary-length streams. This naturally requires induction over such streams. The formulation of the correctness statement affects what we can induct over and how the induction is carried out.

The general pattern of correctness statements in trajectory evaluation is shown in Equation 1.

$$\forall s, i.\ \mathit{Valid}\ s\ i \implies \mathit{ckt} \models (\mathit{Ant}\ s\ i) \Longrightarrow\!\!\!> (\mathit{Cons}\ s\ i) \qquad (1)$$

This statement says that for all states s and inputs i, if s and i are valid and if we stimulate the circuit *ckt* with the antecedent *Ant s i*, then the consequent *Cons s i* will hold.

In the final correctness statement (2), we strengthen Equation 1 to verify the behavior over any *sequence* of inputs (S and I are functions that map times to values) by quantifying over time t and adding separate quantifiers for t' for the validity condition and the trajectory formula. Equation 2 says that if the streams are valid for all time t' prior to time t, and if the circuit is stimulated with I and S for all time prior to t, then the consequent will hold for all time prior to t.

$$\begin{aligned} &\forall S, I.\ \forall t. \\ &\quad (\forall t'.\ t' \leq t \implies \mathit{Valid}\,(S\,t')\,(I\,t')) \implies \\ &\qquad \forall t'.\ t' \leq t \implies \\ &\qquad\quad \mathit{ckt} \models \mathit{Ant}\,(S\,t')\,(I\,t') \ggg \mathit{Cons}\,(S\,t')\,(I\,t') \end{aligned} \tag{2}$$

The predicate *Valid* says that the input stream contains only legal IA-32 instructions and that the control signals do not contain any resets. Our verification is complete with respect to our definition of *Valid*. This means that an error in *Valid* could lead to false positives or false negatives. Given the complexity of this predicate, we created a suite of tests to ensure that *Valid* covers all possible IA-32 instructions. This proved to be very useful and detected a number of specification bugs that would have otherwise weakened our results.

The antecedent drives the circuit with the index value, the input instructions and the control signals. The consequent checks that the boundaries and index match the specification. The functions that connect circuit nodes to specification values use a temporal mapping to convert abstract time to circuit signal timing.

Finding the next-state function for the index was an initial hurdle in our verification. The complexity of the datapath prevented us from using automated reachability analysis. Furthermore, the datapath and latches had been convoluted by several optimizations made for timing and area purposes. An additional challenge was that the value of the index affects the validity of the input stream. This is because the legality of a stream of bytes is dependent upon where the first instruction in the stream begins. Linking the validity conditions for the internal state and input stream while managing state space complexity was one of the most difficult tasks in the IM verification.

5 Verification

Our general strategy is to rely upon model checking as much as possible while limiting theorem proving to managing the decomposition of a proof into model-checking obligations. We do this for two reasons. First, model checking is usually less labor intensive than theorem proving. Second, a model-checking result is usually more robust to changes in the specification and circuit than a proof is. By robust, we mean that if the circuit satisfies the specification after a modification, then a model-checking run will still pass. In comparison, a proof may be dependent upon the portions of the specification or circuit that changed. As a result, the proof may break when run with the modified circuit and specification.

5.1 Trajectory Evaluation

Following the strategy of relying on model checking as our primary verification tool, we experimented with trajectory-evaluation runs to find good BDD variable orderings and to determine the maximum-size verifications that the model checker could handle. While performing these experiments we kept the top-level correctness statement in mind, because the proof that we eventually constructed needed to bridge the gap between our trajectory-evaluation runs and the correctness statement.

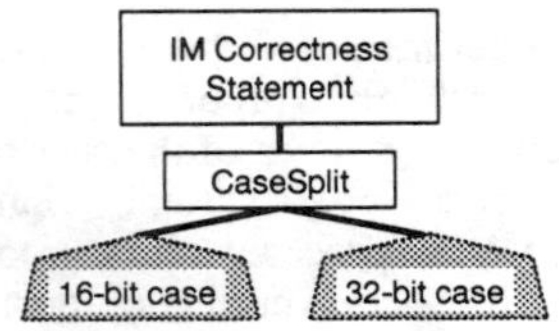

Figure 4: IM Proof skeleton: Top-level

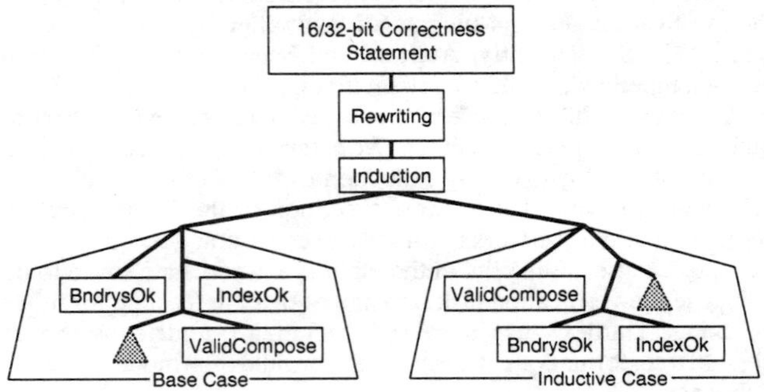

Figure 5: IM Proof skeleton: 16 and 32 bit cases

We discovered that we could not run trajectory evaluation if we left both the input instruction bytes and internal index state symbolic. After analyzing the circuit and specification, we determined that we could run the trajectory evaluations with non-symbolic mode and index inputs and stay within the capacity limits of our model checker. This results in a total of 56 trajectory-evaluation runs: 28 for each of 16- and 32-bit modes; which consist of 14 each for the boundary and index lemmas.

With this information, we proposed a theorem-proving strategy of decomposing the proof by case splitting on each of the different combinations for the mode and index. This approach is appealing from a theorem-proving point of view, because performing a series of case splits over the value of the index is simple and easily mechanized.

5.2 Theorem Proving

In theorem proving, we begin with the goal to be proved and then decompose it into a set of sub-goals. This process creates proof trees, as illustrated in figures 4–6. The leaves of the proof trees are solveable by decision procedures, trajectory-evaluation runs, lemmas and axioms. A proof is complete when all of its leaves have been solved.

Our top-level proof begins with a case split on whether the IM is in 16 or 32-bit mode. Each of these cases is a theorem, and the same proof script is used for each (Figure 4). The proof script begins with some preliminary manipulation (rewriting) of the correctness statement to get it into the right form for induction (Figure 5). The proofs center on an induction over time. In the base case we show that at time $t = 0$, the IM marks the boundaries for one packet of bytes and updates its index correctly. In the inductive case we show that if the IM has processed all of the previous bytes in the input stream correctly, then it will process the next packet correctly.

The base case and inductive case have different proofs, but both rely on the same three lemmas (*IndexOk*, *BndrysOk*, and *ValidCompose*). The two principal lemmas (*IndexOk* and *BndrysOk*) are where the "rubber meets the road", i.e. where the proof is connected with the IM circuit. They state that the index and boundaries are calculated correctly for each possible value of the index and inputs. The third lemma (*ValidCompose*) is an axiom about composing

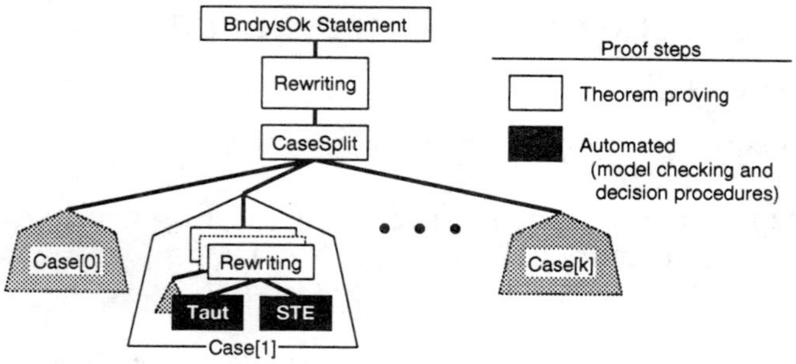

Figure 6: IM Proof skeleton: BndrysOk

Table 1: Proof Results

Theorem	Memory (MB)	CPU Time (min)
16-bit mode	203	113
32-bit mode	347	160

valid sequences of IA-32 instructions.

Trajectory evaluation is the connection between the circuit and the proof. Theorem proving uses the trajectory-evaluation results as lemmas. Because of this, when the circuit changes, only the trajectory-evaluation runs must be performed again.

The proof of *BndrysOk*, which is simpler than *IndexOk*, is sketched in Figure 6. Both proofs rely on equational rewriting and case splitting. We use rewriting to massage the correctness statement until it is ready for case splitting and then case split on the values for the index.

The resulting trajectory evaluation in each case split is still too complex for model checking. We use several rewrite equations to transform the formula with the aim of producing a formula that is more efficient to model check.

The validity of these transformations often relies on side conditions. This illustrates one of the significant advantages of using mechanized theorem proving in conjunction with model checking. When mechanizing these transformations, we are forced to consider exactly when these transformations are valid. In our verification, this process clarified the semantics of several transformations that we had previously applied in an ad-hoc manner.

Proving that the side conditions are satisfied is easy to do in practice, because they can usually be dispatched with automated techniques such as tautology checking. When the side conditions don't hold, we are able to detect errors that would otherwise compromise the soundness of our results. After completing the transformations and the associated side conditions, the trajectory evaluations match the model-checking results described in Section 5.1.

The execution times for the complete verification (theorem proving plus model checking) of the two theorems (one for 16-bit mode and one for 32-bit mode) are shown in Table 1. One proof was run for each theorem. As shown in the table, the 32-bit proof requires more time and memory than the 16-bit proof. The effective complexity of the instruction set is greater in 32-bit mode because of the richer addressing mechanisms provided.

6 Conclusion

A total of eight previously unknown IM bugs were discovered during our verification. Furthermore, four of the eight bugs were considered by the chip design team to be "high-quality"—meaning that they would have been difficult to find and diagnose with traditional validation techniques.

Two of the bugs involve relatively simple cases that would have been detected eventually with RTL simulation and testing. Two bugs involve boundary cases that were common enough that they also would likely have been found during normal testing. Four bugs involve an interesting mix of very narrow boundary cases that included both prefix operands (`h66`,`h67`), two-byte opcodes, and packet boundaries. These cases would have been particularly difficult to detect with traditional validation techniques because of implementation details that do not allow these bugs to be exercised except in the narrowest of cases.

The broad range of bugs detected by our efforts is a compelling example of the thorough design analysis facilitated by formal methods. Indeed, the space of possible behaviors analyzed by our studies would be impossible to analyze exhaustively with a scalar RTL simulator. The surprisingly reasonable run times and memory usage for our verification were stated previously in Table 1.

We also found several bugs in our specification. However, given the complexity of the IA-32 instruction set and the resultant specification complexity, the bugs were not unexpected. We developed some ad-hoc techniques to exercise the specification—essentially by creating a "specification of the specification". These properties were then used for regression testing as the specification evolved.

Our experience with one specification bug is particularly enlightening. When the initial specification was written, one of the "don't cares" in the IA-32 opcode map was minimized with an adjacent opcode. During the verification, a false bug was encountered that was the result of the hardware designer minimizing the same don't care in a different fashion. After this episode, we were careful to constrain the inputs to only include legal instruction sequences. Of course, in some situations, it would be more important to completely specify the behavior of the design—even for illegal inputs.

Acknowledgements

We would like to thank Clark Barrett for his help on an early version of the tool and proof, and Tom Melham for many insightful comments.

References

[1] M. D. Aagaard and C.-J. H. Seger. The formal verification of a pipelined double-precision IEEE floating-point multiplier. In *ICCAD*, pages 7–10. IEEE Comp. Soc. Press, Washington D.C. Nov. 1995.

[2] A. Camelleri. A hybrid approach to verifying liveness in a symmetric multi-processor. In E. L. Gunter and A. Felty, editors, *Theorem Proving in Higher Order Logics*, pages 49–68. Springer Verlag; New York, Sept. 1997.

[3] S. Hazelhurst and C.-J. H. Seger. A simple theorem prover based on symbolic trajectory evaluation and BDDs. *IEEE Trans. on CAD*, Apr. 1995.

[4] *Pentium Processor User's Manual, Volume 3: Architecture and Programming Manual.* Intel Corporation, 1993.

[5] J.-Y. Jang, S. Qadeer, M. Kaufmann, and C. Pixley. Formal verification of FIRE: A case study. In *DAC*, pages 173–177, June 1997.

[6] J. Joyce and C.-J. Seger. Linking BDD based symbolic evaluation to interactive theorem proving. In *DAC*, pages 469–474, June 1993.

[7] S. P. Miller and M. Srivas. Formal verification of the AAMP5 microprocessor: A case study in the industrial use of formal methods. In *Workshop on Industrial-Strength Formal Specification Techniques*, Apr. 1995.

[8] C.-J. H. Seger and R. E. Bryant. Formal verification by symbolic evaluation of partially-ordered trajectories. *Formal Methods in System Design*, 6(2):147, Mar. 1995.

A Fast and Low Cost Testing Technique for Core-based System-on-Chip

Indradeep Ghosh† Sujit Dey‡ Niraj K. Jha†

†Department of Electrical Engineering
Princeton University
Princeton, NJ 08544

‡Department of Electrical & Computer Engineering
University of California, San Diego
La Jolla, CA 92093-0407

Abstract

This paper proposes a new methodology for testing a core-based system-on-chip (SOC), targeting the simultaneous reduction of test area overhead and test application time. Testing of embedded cores is achieved using the transparency properties of surrounding cores. At the core level, testability and transparency can be achieved by reusing existing logic inside the core, and providing different versions of the core having different area overheads and transparency latencies. At the chip level, the technique analyzes the topology of the SOC to select the core versions that best meet the user's desired test area overhead and test application time objectives. Application of the method to example SOCs demonstrates the ability to design highly testable SOCs with minimized test area overhead, minimized test application time, or a desired trade-off between the two. Significant reduction in area overhead and test application time compared to an existing SOC testing technique is also demonstrated.

1 Introduction

System-level integration is evolving as a new paradigm in system design, allowing an entire system to be built on a single chip, using pre-designed functional blocks called cores. While SOC is proving to be very useful in meeting aggressive time-to-market, performance, and cost requirements of today's electronic products, testing such core-based SOCs poses a two-fold challenge. Core-level testing involves making each core testable - inserting the necessary design for testability (DFT) structures and generating test sequences. Typically, for hard and firm cores, testability is ensured, and precomputed test sets provided by the core provider. For soft cores, testability can be addressed and test sets generated by the user. When the cores are integrated into an SOC, chip-level testing needs to be addressed by the SOC designer. The main difficulty in chip-level testing is the problem of justifying precomputed test sequences of a core embedded deep in the design from the chip inputs, and propagating the test responses from the core outputs to the chip outputs.

Core-level testing can be done through automatic test pattern generation (ATPG) at the logic level and by employing a variety of DFT techniques like full or partial scan, and built-in self-test (BIST). At the chip level, a number of DFT techniques have been proposed [1]. In one existing DFT method, referred to as FSCAN-BSCAN, each core is made testable by full scan while chip-level testability is obtained by isolating each core using boundary scan [2]. This scheme may have large area and delay overheads, and prohibitively large test application times. Recently, some work has been done to address this problem [3]. Another existing DFT method, utilizes a combination of full scan and test bus. In this method, an added test bus runs from the PIs of the system to its POs and uses a series of multiplexers to isolate each embedded core, which is full scanned, during testing to provide system-level testability. In this case as well, the area and delay overheads can potentially be quite large. In addition, the test bus architecture is unable to test the interconnect that exists between cores.

In [4], a DFT method is described to test macro blocks inside a circuit with a heavy reliance on full/partial scan and boundary scan whose disadvantages have been stated above. Though functional information of modules is sometimes used to reduce test overhead by utilizing the concept of module transparency, the techniques for introducing transparency are *ad hoc*. In [5], a new technique is introduced to make each core transparent. In the test mode, a transparent core can propagate test data from its inputs to outputs without information loss. During testing of the SOC, the transparency property of cores is used to propagate precomputed core test sequences from the chip inputs to the core inputs, and test responses from the core outputs to the chip outputs. While successful in lowering the test area and delay overheads compared to existing techniques, the technique in [5] suffers from the drawback of relatively large test application times due to the potentially large transparency latency of each core (number of cycles needed to propagate test data from inputs to outputs of the core). Also, functional description of the cores are required to make them transparent which may not be available for many cores, including legacy cores.

In this work, we propose a new DFT technique for core-based SOC testing that simultaneously targets reduction of area overhead and test application time. It has two parts. The first part consists of core-level DFT and test generation to make each core testable and transparent, and generate a precomputed test set for the core. For the core-level DFT, we use the low-cost high-level scan (HSCAN) technique [6]. Transparency of a core is achieved by reusing existing paths in the core, including HSCAN paths. The core-level DFT and transparency method uses only structural information of the core, in contrast to functional information used in [5], which may be harder to extract for some cores. This task needs to be performed by the core provider in case of hard and firm cores, and the user in case of soft cores. During the core testability phase, various versions of the same core are synthesized. For each version, the following are specified: the appropriate DFT structures, testability and transparency area overheads, transparency latency for different input/output pairs, and the test sequence for testing each core. Note that this is a one-time cost incurred by the core provider or the user. The second part consists of chip-level DFT and test generation which is to be performed by the user. For this, we propose a technique that analyzes the given interconnection of various cores in the SOC, the test set size of each core, and the area overhead/transparency latency characteristics of the available core versions, and identifies the most suitable version of each core such that the desired test area overhead/test application time trade-off is achieved for the SOC. Testing of each embedded core can be performed by justifying (propagating) the test sequences (responses) from (to) the chip inputs (outputs) using the transparency mechanism of the cores in the SOC.

The proposed technique can be used to design an SOC with high fault coverage such that: (i) the test area overhead is minimized, (ii) the test application time for testing the SOC is minimized, or (iii) a desired trade-off is obtained between the test area overhead and test application time. This technique is suitable for testing the SOC in a hierarchical fashion. Even if soft cores are used in the SOC, sequential test pattern generation subsequent to integrating

DAC 98, San Francisco, California

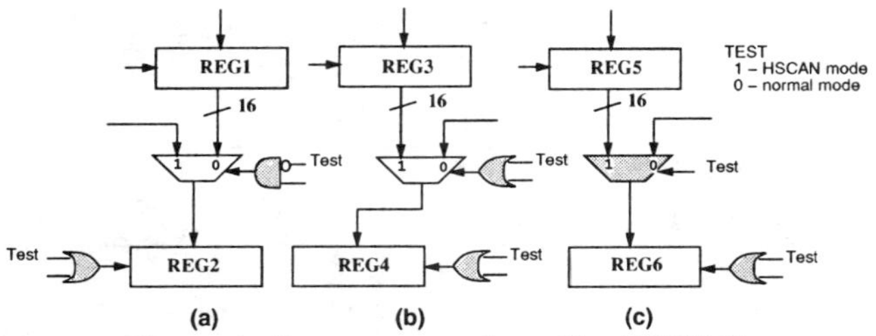

Figure 1: Structures employed by HSCAN

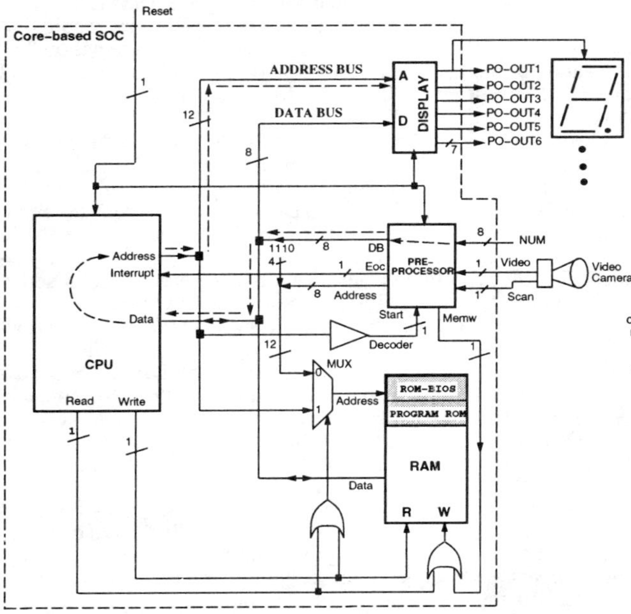

Figure 2: An example core-based embedded system

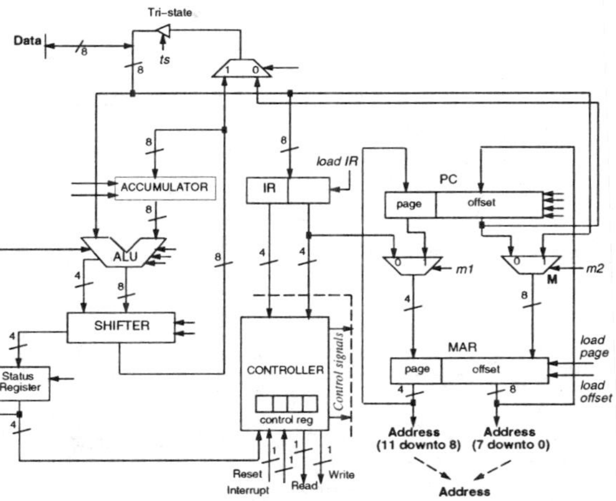

Figure 3: The CPU core (*Version 1*)

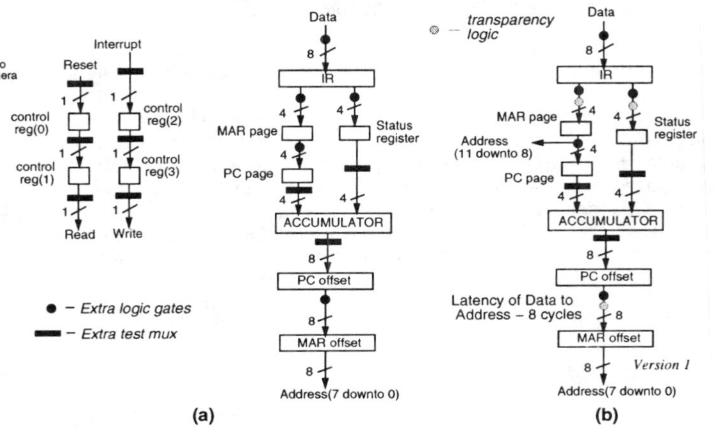

Figure 4: HSCAN and transparency chains in the CPU core

all cores into a single design can be computationally prohibitive. This problem does not arise in our method. Significant reduction in the area overhead and test application time compared to an existing SOC testing technique is demonstrated for two example SOCs.

2 Preliminaries

In this section, we briefly describe the register-transfer level (RTL) HSCAN methodology and illustrate its use as our underlying core-level testing technique [6]. It utilizes existing paths between registers, through multiplexers (mux paths), to connect registers in parallel scan chains from the circuit inputs to the circuit outputs. Consider the example RTL circuit shown in Figure 1, which shows 16-bit multiplexers and registers. Since a multiplexer path already exists between *REG1* and *REG2*, in the HSCAN mode these registers can be connected in 16 parallel scan chains by using just two extra logic gates, as shown in Figure 1(a). If the select-0 path of an existing multiplexer needs to be chosen during testing then a configuration like Figure 1(b) can be used. If a direct connection exists between two registers, only an OR gate is required at the load signal of the destination register. If no path exists between two registers, or if there is a conflict with already created HSCAN paths, then a scan path is created by adding a test multiplexer, as shown in Figure 1(c). This multiplexer can be integrated with the destination flip-flops to create a set of scan flip-flops to reduce test overheads further. Since it is a full scan technique, only combinational test pattern generation is required which makes it suitable for tackling large designs like cores.

To achieve transparency in a core, the existing parallel paths between registers as well as the ones introduced by HSCAN can be utilized to transfer data from the inputs of the core to its outputs making the HSCAN scheme suitable for obtaining low-cost transparency for each core. This is explained next.

3 Outline of the SOC Test Method

In this section, we describe our core-based testing method using an example core-based embedded system that implements a barcode scanning system. This embedded system consists of the following cores: CPU, PREPROCESSOR, DISPLAY, RAM and ROM, as shown in Figure 2. The PREPROCESSOR receives signals from a video scanner (signal *Video*), processes the barcode scanned, and if no errors are detected, writes the width of the black and white bars in consecutive locations in the RAM. Subsequently, the CPU uses an embedded program in the ROM to convert the barcode in the RAM to a cost. The DISPLAY then converts the binary coded decimal output of the CPU to a series of six seven-segment display codes. The memory space is 4KB long, organized as 16 pages of 256 byte each, and all ports are memory-mapped.

The internal structure of the CPU core is shown in Figure 3 [7]. In the figure, *PC* represents the program counter, *MAR* represents the memory address register, and *IR* represents the instruction register. To make the CPU testable, we use HSCAN to obtain the scan chain configuration shown in Figure 4(a). Each flip flop/register in the circuit now belongs to one scan chain which can shift in the tests and shift out the responses.

Next, we need to make the CPU transparent. This means that each output of the core can be justified from at least one input or a combination of inputs of the core, and each input of the core can be propagated to one output or a combination of outputs of the core, in a fixed number of cycles. First, we try to utilize the HSCAN chains for transparency. Some extra transparency logic may be required, so that we can freeze values along the chain whenever required. If HSCAN is not the underlying test methodology for the core, we can use other techniques given in Section 4 to achieve transparency. The transparency mechanism for the CPU is shown in Figure 4(b). Assume that the core output *Address(7 downto 0)* needs a symbolic value α. If α is applied at the core-input *Data*,

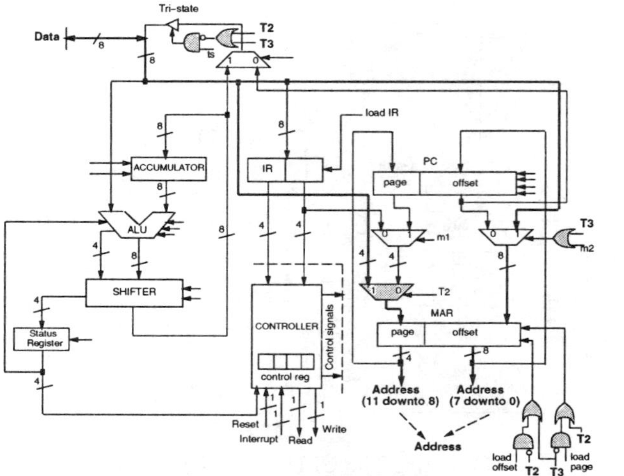

Figure 5: *Version 3* of the CPU core

then the core output *Address(7 downto 0)* will get α after six clock cycles. While α is being shifted from the core input to output, the data in the *Status register* needs to be frozen for a cycle so that the latency through the two parallel chains are balanced. This requires some extra transparency logic. Subsequently, the value required at the other core output *Address(11 downto 8)* can be applied at the core input *Data*. The required value appears in register *MAR page*, which is connected to *Address(11 downto 8)*), after two clock cycles. Also, the HSCAN chains can be used to transfer the value at input *Reset* to output *Read* in two cycles, and input *Interrupt* to output *Write* in two cycles. However, these single-bit chains are not explicitly shown in Figure 4(b). If all the bits of the core output *Address* need to be justified simultaneously from input *Data*, then some extra logic is required to freeze the values at output registers *MAR page* and *MAR offset*.

Next, we briefly describe our methodology of testing each embedded core, using transparency of the other cores in the SOC. We demonstrate the importance of the transparency latency of a core in determining the test application time of a core under test, and that of the entire SOC. We show how the transparency latency can be traded off for area overhead to reduce the test application time. Consider the DISPLAY in the SOC of Figure 2, which needs a test sequence of 525 HSCAN vectors for achieving 100% test efficiency (105 full-scan vectors are required and the sequential depth of the longest HSCAN chain is 4). To test the DISPLAY, its inputs *A* and *D* need to be controlled and its outputs *PO-PORT1* through *PO-PORT6* need to be observed. To control input *A* from the chip inputs, the test vector needs to be propagated through the PREPROCESSOR from its input *NUM* to its output *DB* and then through the CPU from its input *Data* to its output *Address*. This path is shown with the help of a dashed line in Figure 2. Input *D* of the DISPLAY can be controlled by propagating the desired vector through the PREPROCESSOR from its input *NUM* to its output *DB*. The outputs of the DISPLAY can be observed directly as they are the chip outputs.

Suppose the PREPROCESSOR can transfer data from input *NUM* to output *DB* in one cycle. The CPU takes a total of eight cycles to transfer a test vector from its input *Data* to output *Address* (six cycles to transfer a vector from *Data* to *Address(7 downto 0)*, and two cycles to transfer a vector from *Data* to *Address(11 downto 8)*). Here, we have assumed that test data cannot be pipelined through a core. Thus, if two transparency paths in a core share some logic between them, then data through one path can be propagated only after data has been completely propagated through the other path. Hence, a test vector can be propagated from the chip input *NUM* to the core-input *A* in nine cycles, where we have assumed that test data can flow through different cores simultaneously. In between, the test vector at input *D* can also be made ready. A subsequent test vector for the DISPLAY can be made ready at its inputs in another nine cycles. In the intermediate cycles, when the data at the inputs of the DISPLAY are not valid, the scan clock of the DISPLAY needs to be frozen. During this period, the test

CPU	Latency (cycles)			Overhead (cells)
	D -> A(7-0)	D -> A(11-8)	D -> A(11-0)	
Version 1	6	2	8	3
Version 2	1	2	3	10
Version 3	1	1	2	30

A– Address

Figure 6: Transparency latency *vs* overhead trade-off

response can be scanned out through the primary outputs. Thus, the test application time required for testing the DISPLAY alone is $525 \times 9 + 3 = 4,728$ cycles. The last three cycles are required to scan out the last test response. Since the DISPLAY core has 66 flip-flops and 20 internal inputs, an FSCAN-BSCAN approach would have needed $(66+20) \times 105 + (66+20) - 1 = 9,115$ cycles to do the same.

The transparency latency through the CPU can be reduced at the expense of area overhead, by exploiting existing alternative connections in the CPU and some extra transparency logic. In this case, output *Address(7 downto 0)* can be justified from the input *Data*, with some extra logic at the select line of multiplexer **M** in Figure 3, in one cycle. Also, *Address(11 downto 8)* can be justified from *Data* in two cycles. This transparency logic is added in addition to the HSCAN logic. Now, the CPU takes a total of three cycles to transfer a test vector from its input *Data* to output *Address*. If this version (named *Version 2*) of the CPU is used in the chip, the test application time for testing the DISPLAY is now $525 \times 4 + 3 = 2,103$ cycles only. The latency of the CPU core can be further reduced by using one extra transparency multiplexer to form *Version 3* of the CPU as shown in Figure 5. In this configuration, both outputs *Address(7 downto 0)* and *Address(11 downto 8)* can be justified from input *Data* in one cycle. The test application time for testing the DISPLAY is now reduced to $525 \times 3 + 3 = 1,578$ cycles. The results are summarized in the table in Figure 6. In the table, the area overheads are for the extra transparency logic only. The numbers are obtained after technology mapping with a .8μm cell library using an in-house synthesis tool.

In general, more expensive versions can be used for critical cores whose transparency latency affects the test application time of many other cores in the SOC.

4 Making Cores Transparent

In this section, we explain the method we have developed to make cores transparent with some given transparency latency. At first, if HSCAN is the underlying DFT methodology, we try to exploit the HSCAN chains to achieve transparency. If HSCAN is not the underlying test methodology or if transparency is not possible through the HSCAN chains or if any of the transparency latencies for an input or output is not acceptable, we introduce extra logic into the circuit to create transparency paths.

First, we extract a register connectivity graph (RCG) which consists of input nodes, output nodes, and register nodes of the core. An edge is present between two nodes if a direct or multiplexer path exists between them. Figure 7 shows the initial RCG of the CPU core. In the figure, the edges corresponding to the HSCAN paths are darkened. In many RTL circuits, bit-slicing is common, where a subset of bits (bit-slice) of a register is involved in an operation (as source or destination register), instead of the complete register. In such cases, the corresponding register node in the RCG is marked as a split node. When different bit-slices of a register must receive data from different sources, the register node is termed a C-split node. To control a desired value in a C-split node, the different bit-slices need to be controlled through different paths. In Figure 7, *ACCUMULATOR* is a C-split node. Similarly, if the fanout of a node is split into different bit-slices going to different destinations, the node is marked as an O-split node.

From the RCG, we try to find transparency paths from inputs to outputs. To do this, we do a breadth-first search (BFS) from each input node until we reach an output node. During this search, anytime we reach an O-split node, the BFS is done from each of the nodes at the fanout edges of the O-split node, as all the fanout edges need to be used for propagating the data at the O-split node. At first, we only use the HSCAN edges during this search. If we fail

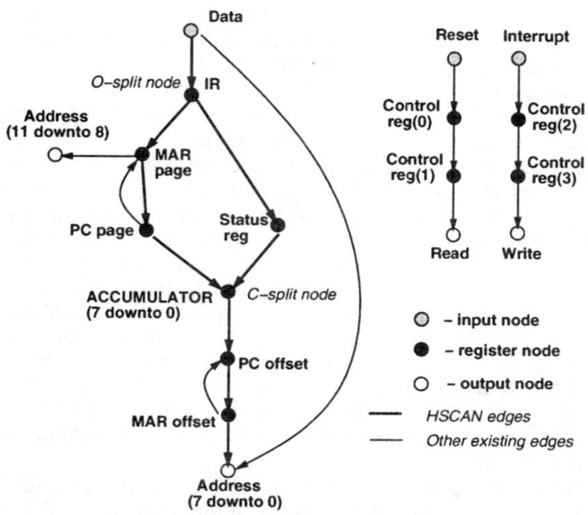

Figure 7: RCG of the CPU core

to get a transparency path, we try to use other existing edges in the RCG. For the CPU core, the BFS from input *Data* reaches node *IR*. Then, different BFSs start along the two fanout edges. The search from the left edge terminates at output node *Address(11 downto 8)*. The BFS from the right edge continues until we reach output node *Address(7 downto 0)*. Thus, we have found a transparency path for propagating the input to the output with a latency of six cycles. We delete this path and do a BFS for any remaining input whose transparency path has not yet been found. If no paths are found, the deleted paths are put back in the RCG and the BFS is repeated with the aim of finding a transparency path by reusing edges from prior transparency paths. If transparency paths are still not found, then a transparency path is created by adding extra logic. Any register reachable from the input in one cycle (found from the BFS) is connected to an output (or outputs if bit-widths mismatch) with a test multiplexer. To keep the transparency paths disjoint, preference is given to an output(s) which has not yet been used for providing transparency to other core inputs. If any edge is used in more than one transparency path, then it may not be possible to propagate data through the two paths in parallel. Finally, a solution is obtained where each input can be propagated to some output(s) in a fixed number of cycles (transparency latency) in the test mode.

Next, we try to find transparency paths so that each output of the core may be justified from some input of the core during the transparency mode. For this, we reverse all the edges in the RCG, and again do a BFS from each output node of the RCG. We select an output and continue the BFS until we reach an input node. For example, for the CPU, we first select *Address(7 downto 0)*, and do a BFS using only the HSCAN edges. When we reach a C-split node, the BFS is done from each of the nodes at the fanout edges (fanin in the original RCG) of the C-split node since all the fanout edges need to be used for justifying the data at the C-split node. When two BFSs originating at the fanout nodes of a C-split node converge at an O-split node, then they again constitute a single search. Thus, the search branches into two different searches at the C-split node *ACCUMULATOR* and reconverges to a single search at the O-split node *IR*, eventually reaching the input node *Data*. If the transparency path thus formed has split nodes, and if the parallel subpaths are not balanced at each fanin of a C-split node in the original RCG, for each fanin which does not fall on the longest subpath we add extra logic to freeze the data there. This balances the parallel sub-paths, so that the data, which have arrived early at the fanin of a C-split node, wait until the data at all the fanin edges are ready.

Next, we repeat the procedure for the other output nodes. For the CPU, we delete the transparency path and do BFS for *Address(11 downto 8)*. Since we cannot find a path, we put back the deleted path and try again. This time we succeed. Thus, the transparency paths of the two outputs are not disjoint, but rather reuse common edges (*MAR*, *IR*) and (*IR*, *Data*). When the same edge(s) in the RCG has to be reused in multiple transparency paths, we transfer data through them in a sequential manner.

If transparency paths are not found using HSCAN edges, then

BARCODE PREPRO-CESSOR	LATENCY (cycles)		Ovhd. (cells)
	NUM->DB	NUM->A	
Ver. 1	5	2	2
Ver. 2	1	2	19
Ver. 3	1	1	37

A – Address

(a)

DISPLAY	LATENCY (cycles)		Ovhd. (cells)
	D->OUT	A->OUT	
Ver. 1	2	3	5
Ver. 2	2	1	20
Ver. 3	1	1	55

OUT -> combination of output ports

(b)

Figure 8: Transparency latency-area overhead trade-off in the PREPROCESSOR and DISPLAY cores

other existing paths in the RCG are used to obtain transparency. In the case of the CPU, transparency paths for all inputs and outputs are achievable using HSCAN paths only. This gives rise to *Version 1* of the CPU core (see Figure 6).

The transparency latency of a core can be reduced by adding extra logic and using existing non-HSCAN paths or by adding transparency multiplexers. Consider Figure 7. Let the non-HSCAN edges existing in the RCG be included in the BFS this time. Since an edge exists from *Data* to *Address(7 downto 0)*, a transparency path can be found for output *Address(7 downto 0)* to input *Data* with a latency of just one cycle. A transparency path exists from *Address(11 downto 8)* to input *Data* with a latency of two cycles as before. Consequently, we have *Version 2* of the CPU core. When non-HSCAN edges are used for transparency, the type of transparency logic required is determined by analyzing the multiplexer tree or bus interconnect that created the edge in the RCG and adding the appropriate extra logic according to the data flow required. This is a straightforward mapping, as explained in Section 2.

Further reduction in transparency latency can be obtained by introducing transparency multiplexers. This is done one at a time for each input/output pair with a transparency latency greater than one. Figure 5 shows a transparency multiplexer (shaded) added to reduce the latency of *Data* $\rightarrow$ *Address(11 downto 8)* to one, leading to *Version 3* of the CPU. The trade-offs for different core versions for the other two cores in Figure 2 are shown in Figure 8.

Note that input control signals to a core can be treated as data inputs and if direct paths from control inputs to control registers are not present, the random logic in between can be simply bypassed with single-bit multiplexers during testing. The control registers can similarly connect to control outputs as shown in the *Read* and *Write* chain in Figure 4. Thus, expensive test generation techniques are avoided during the evaluation of transparency for such signals.

5 Trading Off Area and Test Application Time

We next present a method to select the best core versions to produce a testable SOC for the following objectives:
(i) Given a test area overhead constraint of x cells, the global test application time of the SOC is minimized.
(ii) Given a test application time constraint of y cycles, the overall test area overhead of the SOC is minimized.

We propose an iterative improvement technique, which can obtain solutions for either of the design objectives by varying the parameters of a cost function. At first, we create a core connectivity graph (CCG) from the given SOC. In this graph, there are primary input (PI) nodes, primary output (PO) nodes, core input nodes and core output nodes. Edges are added between input-output pairs of a core according to the transparency paths in the versions of the core currently in use. The cost associated with an edge represents the transparency latency between the corresponding input-output pair. An input node is split into separate nodes if its fanin is split, *i.e.*, if different bit-slices of the node receive data from different sources exclusively. Similar splitting is done for output nodes also by considering the fanout bits. The CCG of the barcode system in Figure 2 is shown in Figure 9 where the minimum area version (*Version 1*) of each core is used. We do not consider the memory cores in this discussion, as most memory cores use BIST [8].

5.1 Identification of test paths for a core

We next discuss how to identify the justification and propagation paths for each core such that the test application time needed for the

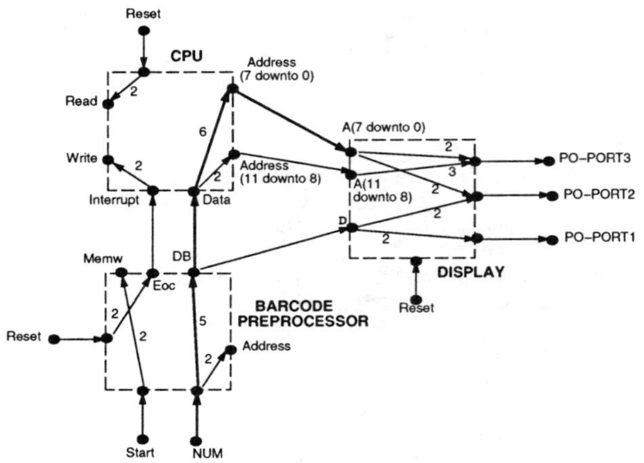

Figure 9: CCG of the barcode system

core is minimized. Let us take the DISPLAY core first. We have to control its inputs *A(7 downto 0)*, *A(11 downto 8)* and *D*. We find out all possible shortest paths from the chip inputs to *A(7 downto 0)* and choose the one with lowest cost (which takes the least number of clock cycles). Any shortest path algorithm like Dijkstra's algorithm can be used for this purpose. Here, only one path is possible (*NUM* $\rightarrow$ *DB* $\rightarrow$ *Data* $\rightarrow$ *Address(7 downto 0)* $\rightarrow$ *A(7 downto 0)*) which is highlighted in Figure 9. We mark this path and reserve the edges for the cycles in which they will be used. Hence, edge (*NUM*, *DB*) in the PREPROCESSOR is reserved from cycle 0 to 5, edge (*Data*, *Address(7 downto 0)*) is reserved from cycle 5 to 11, and so on. If there is no path possible, we add a system-level test multiplexer to connect the input of the core directly to a PI. Next, we try to find the shortest path from any chip PI to *A(11 downto 8)*. The shortest path algorithm is modified so that if any marked edge is reused, then the cost is automatically modified so that the edge is not reused in the reserved cycles. Here, part of the previous path needs to be reused. Therefore, the edge (*NUM*, *DB*) can only be utilized from cycle 6 onwards, the edge (*Data*, *Address(7 downto 0)*) can only be used from cycle 12 onwards, and so on. The minimum-cost path is again chosen among all possible paths and test multiplexers are added if necessary. Similarly, the path for controlling input *D* can be found.

Next, we try to find the paths for observing the core outputs at the POs of the chip. This process is very similar to the method of finding paths to inputs which is described above. For the DISPLAY, all paths are trivially found as they connect to POs of the chip. Again, test multiplexers are added to increase observability if necessary. This process is repeated for every core in the system. In Figure 9, the output *Address* of the PREPROCESSOR is connected to a PO with a system-level test multiplexer since there is no way of observing it by existing paths through the cores.

Once a test solution is found, the global test application time and area overheads for the system are calculated. If it does not meet the constraints that the user has set, then a different design point which provides the most improvement needs to be obtained.

5.2 Iterative improvement for core selection

Next, we outline the procedure to select the most suitable core to be replaced by the next higher overhead version. First, all core versions are ordered in terms of the area overhead from lowest upwards. Then a test application time improvement number is calculated for each core. To do this, the test solution is examined and the number of times each edge in a core is used is counted. Then this number is multiplied by the latency of the edge and is summed over all edges of the core. For example, the edge (*NUM*, *DB*) in the PREPROCESSOR is used twice to test the DISPLAY and once to test the CPU. Its latency is five. The edge (*NUM*, *Address*) is never used. The edge (*Reset*, *Eoc*) in the PREPROCESSOR is used once to test the CPU. Its latency is two. Hence, the initial latency number for the PREPROCESSOR is $3 \times 5 + 0 \times 2 + 1 \times 2 = 17$. Now if we replace the PREPROCESSOR with the next area-expensive version, the latency number changes to $3 \times 1 + 0 \times 2 + 1 \times 2 = 5$. Thus, the latency improvement, ΔTAT, in the PREPROCESSOR is $17 - 5 = 12$. ΔTAT gives a notion of the improvement in the global test application time if a core is replaced with another version. The area overhead increase, ΔA, for this replacement is 17 cells. For each core, the cost of replacing it with another version is calculated as $C = w_1 \times \Delta TAT + w_2 \times \Delta A$, where w_1 and w_2 are weights that can be adjusted according to the problem solved.

Next, suppose we are trying to meet objective (i) as specified before. We set w_1 to 1 and w_2 to 0 and improve the existing solution by replacing the core that has the highest C value with its next expensive version. We repeat this until the test overhead reaches the preset maximum or the test overhead becomes higher than the overhead of a system-level test multiplexer. In the latter case, we choose the core which contributes the most to the global test application time, examine the time it takes to get an input vector to each of its inputs from the PIs (observe an output vector at each of its outputs at the POs) and place the system-level multiplexer on the input or output that is the most critical. This process is stopped once the test overhead reaches the user-defined maximum allowable test overhead.

If we are trying to meet objective (ii) stated before, we set w_1 to 0 and w_2 to 1 and choose the core with the minimum C that has a non-zero ΔTAT value. This means that we replace the core which is the least expensive to replace but ensure that it will have some effect in reducing the global test application time. The global test application time is again calculated and the process repeated until the preset test application time limit is met. Once again, if the test overhead becomes higher than the overhead of a system-level test multiplexer, we place an extra test multiplexer as before. Thus, in the worst case, the solution will degenerate into a test bus like system in which core inputs are controllable from the PIs and core outputs observable at the POs. Note that this system will have the minimum possible test application time.

The proposed methodology requires that each core can be clocked independently of the other cores as sometimes test data needs to be frozen at certain points while it keeps on flowing at other points. This is done by freezing the clock of the core when necessary with the help of clock gating circuitry. During the test mode, these control signals as well as the transparency mode control signals like *T2* and *T3* in Figure 6 are provided by a test controller which is added to the chip. This usually consists of a small finite-state machine.

6 Experimental Results

We applied the proposed SOC testing methodology to two example core-based systems. The first example, System 1, is the barcode embedded system shown in Figure 2. System 2 consists of a graphics processor core [9], a GCD core [10], and a X25 protocol core [11]. Each core used in the systems has HSCAN DFT, and hence can be treated as a full-scan circuit and tested using combinational ATPG tools. The area numbers represent number of cells, and are obtained after synthesizing and technology mapping different versions of the cores and systems with an in-house synthesis tool. The fault coverage numbers are obtained by fault simulating the logic-level implementation of the system with the system-level test set using a commercial combinational ATPG tool.

The graph shown in Figure 10 demonstrates the chip-level trade-off between the area overhead and the chip test application time for 18 design points of System 1 which are obtained by different combinations of the core versions. Design point 1 (18) corresponds to the SOC consisting of a minimum area (minimum latency) version of each core. Details of some of the design points (1, 17, 18) are given in Table 1, with Columns 2, 3, 4, and 5 showing the area overhead, test application time, fault coverage, and test efficiency, respectively. The first two rows, representing design points 1 and 18, show that about 4.5 times reduction in test application time can be achieved with about two times increase in the test area overhead. The table also shows that selecting

Table 1: Design space exploration for System 1

Circuit description	*A. Ov.* (cells)	*TApp.* (cycles)	*FCov.* (%)	*TEff.* (%)
Each core has min. Area (1)	156	17,387	98.4	99.8
Each core has min. latency (18)	325	3,818	98.4	99.8
Min. chip TApp. (17)	307	3,806	98.4	99.8

Table 2: Area overheads

Circuit	Orig.	Core-level DFT		Chip-level DFT			Core + Chip-level DFT	
		FSCAN	HSCAN	BSCAN	SOCET		FSCAN-BSCAN	SOCET
	Area (cells)	Ovhd. (%)	Ovhd. (%)	Ovhd. (%)	Circuit type	Ovhd. (%)	Ovhd. (%)	Ovhd. (%)
System 1	8014	18.8	10.1	5.2	Min. Area	2.0	24.0	12.1
					Min. TApp.	3.8	24.0	13.9
System 2	5540	15.6	10.3	9.9	Min. Area	1.2	25.5	11.5
					Min. TApp.	4.7	25.5	15.0

Table 3: Testability results

Circuit	Orig.		HSCAN		FSCAN-BSCAN			SOCET			
	FC (%)	TEff. (%)	FC (%)	TEff. (%)	FC (%)	TEff. (%)	TApp. (cycles)	Circuit type	FC (%)	TEff. (%)	TApp. (cycles)
System 1	10.6	10.8	14.6	14.9	98.4	99.8	36,152	Min. Area	98.4	99.8	17,387
								Min. TApp.	98.4	99.8	3,806
System 2	11.2	11.3	13.8	13.8	98.2	99.9	46,394	Min. Area	98.2	99.9	16,435
								Min. TApp.	98.2	99.9	3,998

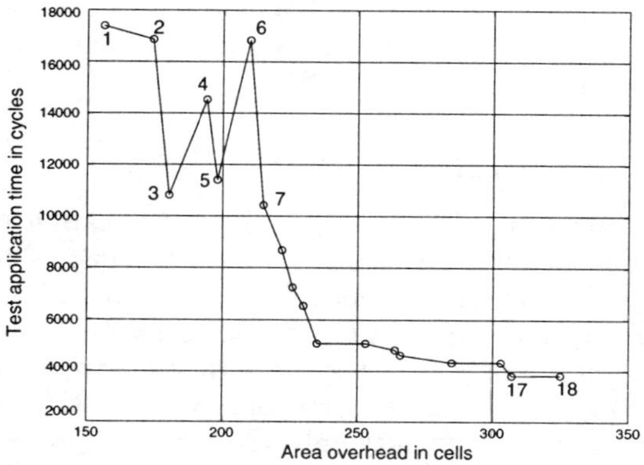

Figure 10: Test application time *vs* area ovhd. for System 1

the minimum latency cores does not necessarily lead to an SOC with minimum test application time. Design point 17 represents the SOC having minimum test application time though it uses a higher latency version of the PREPROCESSOR core than the one used in design point 18.

Table 2 compares the area overheads obtained by our method with the existing core-based SOC testing method FSCAN-BSCAN. We use two extreme design points obtained by our method, termed SOCET, for the comparison - the chip having least area overhead, and the chip having least test application time. In Column 2 of the table, the original area of the chip is given. Columns 3 and 4 give the area overhead needed to make the cores testable using FSCAN and HSCAN, respectively. The next two columns give the area overhead for the chip-level DFT needed to make the SOC testable: Column 5 shows the boundary scan area overhead, while Columns 6 and 7 list the overheads for the two extreme design points using SOCET. The total area overheads including the core-level and chip-level DFT for the two different methods are given in Columns 8 and 9. The area overhead required for SOCET is significantly lower than the area overhead required for FSCAN-BSCAN.

The testability results are reported in Table 3. Column 2 gives the fault coverage of the chip before the addition of any test hardware. This is obtained by running an in-house sequential test generation tool on the original circuit. As expected, the fault coverage is very poor. The test efficiency is given in Column 3. In Columns 4 and 5, the corresponding results are given for the chip when individual cores are made testable using HSCAN, but no chip-level DFT is added. This case shows that the overall fault coverage of the chip may be quite poor even if individual cores are testable. The fault coverage, test efficiency, and test application time for the chip made testable by FSCAN-BSCAN are given in Columns 6, 7, and 8, respectively. The corresponding results for SOCET are given in Columns 10, 11, and 12. While the fault coverage obtained by FSCAN-BSCAN and SOCET are equally high, the test application time required by SOCET is significantly lower than FSCAN-BSCAN.

7 Conclusions

In this paper, we have presented a novel technique for efficient testing of core-based SOCs. Individual cores are first made testable and then transparent using the existing logic inside the core and adding test hardware only if needed. Since the transparency latency of cores is a critical factor in the overall test application time of the core-based system, it is shown how different versions of the same core can be created with different transparency latencies and test overhead values. In the system, each core is tested with its precomputed test set which is propagated through other cores using their transparency modes. A technique is presented here that helps the user to explore the design space and choose a particular version of each core so that the final test overhead and chip test application time goals can be met. Application of the technique to example SOCs demonstrates the feasibility of our approach as well as significant reduction in test overhead and test application times compared to existing SOC test techniques.

Acknowledgments: We would like to thank A. Raghunathan for his help with the example embedded systems and NEC CCRL for supporting I. Ghosh with a summer internship.

References

[1] Y. Zorian, "Test requirements for embedded core-based systems and IEEE P1500," in *Proc. Int. Test Conf.*, Nov. 1997.

[2] R. Chandramouli and S. Pateras, "Testing systems on a chip," *IEEE Spectrum*, pp. 42-47, Nov. 1996.

[3] N.A. Touba and B. Pouya, "Testing embedded cores using partial isolation rings," in *Proc. VLSI Test Symp.* pp. 10-15, Apr. 1997.

[4] F. Bouwman *et al.*, "Macro testability; The results of production device applications," in *Proc. Int. Test Conf.*, pp. 232-241, Nov. 1992.

[5] I. Ghosh, N.K. Jha, and S. Dey, "A low overhead design for testability and test generation technique for core-based systems," in *Proc. Int. Test Conf.*, pp. 50-59, Nov. 1997.

[6] S. Bhattacharya and S. Dey "H-SCAN: A high level alternative to full-scan testing with reduced area and test application overheads," in *Proc. VLSI Test Symp.*, pp. 74-80, Apr. 1996.

[7] Z. Navabi, *VHDL Analysis and Modeling of Digital Systems*, McGraw-Hill, New York, 1993.

[8] Y. Zorian, "A distributed BIST control scheme for complex VLSI devices," in *Proc. VLSI Test Symp.*, pp. 6-11, Apr. 1993.

[9] A. Raghunathan, S. Dey, and N.K. Jha, "Power management techniques for control-flow intensive designs," in *Proc. Design Automation Conf.*, pp. 429-434, June 1997.

[10] P.R. Panda and N.D. Dutt, "1995 high-level synthesis design repository," in *Proc. Int. Symp. System Level Synthesis*, pp. 170-174, Sept. 1995.

[11] S. Bhattacharya, S. Dey, and F. Brglez, "Performance analysis and optimization of schedules for conditional and loop-intensive specifications," in *Proc. Design Automation Conf.*, pp. 491-496, June 1994.

Introducing Redundant Computations in a Behavior for Reducing BIST Resources*

Ishwar Parulkar†
Sun Microsystems
Sunnyvale, CA 94086
ishwar.parulkar@eng.sun.com

Sandeep K. Gupta and Melvin A. Breuer
University of Southern California
Los Angeles, CA 90089-2562
{sandeep, mb}@poisson.usc.edu

Abstract

The degree of freedom that can be exploited during scheduling and assignment to minimize BIST resources is often limited by the data dependencies of a behavior. We propose transformation of a behavior by introducing redundant computations such that the resulting data path requires few BIST resources. The transformation makes use of *spare capacity* of modules to add redundancy that enables test paths to be shared among the modules. A technique is presented for introducing redundant computations that reduce the BIST resource requirements of a data path without compromising the latency and functional resource constraints.

1 Introduction and Motivation

A typical data path contains registers and functional modules, such as adders and multipliers, that are selected from a pre-designed library. One cost-effective way of testing such data paths employs built-in self-test(BIST). An important consideration in the application of a BIST technique is the area overhead incurred in modifying functional registers to BIST resources, i.e. registers that generate test patterns and compress test responses. A data path synthesized without any consideration for BIST resources can have high BIST area overhead. The scheduling and assignment stages of high-level synthesis can incorporate test requirements thereby reducing BIST area overhead [1],[2],[3],[4]. However the degree of freedom available during scheduling and assignment for minimizing BIST resources is often limited by the data dependencies of a behavior. In such cases, alternate behavioral descriptions need to be explored. At the logic level, it has been shown that introduction of redundancy can be beneficial for certain testability objectives [5]. Transformations have been applied at the behavioral level for optimization of area, performance, fault tolerance and partial scan overhead [6],[7],[8],[9]. In this paper, we propose a transformation aimed at optimizing the cost of BIST resources required to make a synthesized data path self-testable.

1.1 Minimal Intrusion BIST

Minimal intrusion BIST involves the modification of only a subset of the functional data path registers to perform test functions. Depending on the test function, four different types of test registers (called BIST resources) are possible: 1) test pattern generation capability only (TPG), 2) test response compression or signature analysis capability only (SA), 3) test pattern generation and response compression capability at different times (BILBO), and 4) simultaneous test pattern generation and response compression capability (CBILBO). Different mappings of test register type to functional registers exist so that all functional modules in the data path are tested. Selection of registers for testing a module largely depends on how these registers are connected to the rest of the modules. For example, from the registers available for test response compression of a module, it is beneficial to choose one that can also compress responses from some other module, or maybe generate test patterns for some other module. For minimizing BIST area overhead, the design is analyzed globally to determine the BIST resources for each module such that all functional modules are tested with a minimum BIST area overhead.

1.2 Spare Capacity of Modules

Modules in most data paths perform useful computation in some clock cycles and are idle during other cycles. The percentage of time a module is idle is called the *spare capacity* of the module. For example, consider the DFG shown in Fig. 1(a). Suppose that a data path is synthesized from this DFG that executes in 3 clock cycles using one adder module A, one multiplier mod-

*This work was supported by the Advanced Research Projects Agency and monitored by the Department of the Army, Ft.Huachuca, under Contract No. DABT63-95-C-0042. The information reported here does not necessarily reflect the position or the policy of the Government and no official endorsement should be inferred.

†This work was done when the author was with the University of Southern California, Los Angeles, CA 90089.

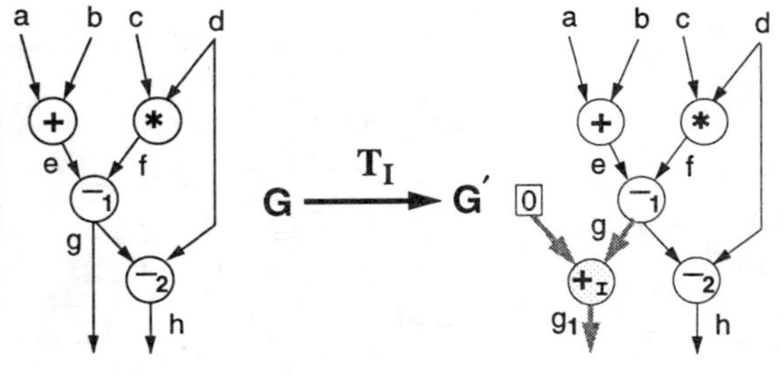

(a) (b)

Figure 1: I-transformation

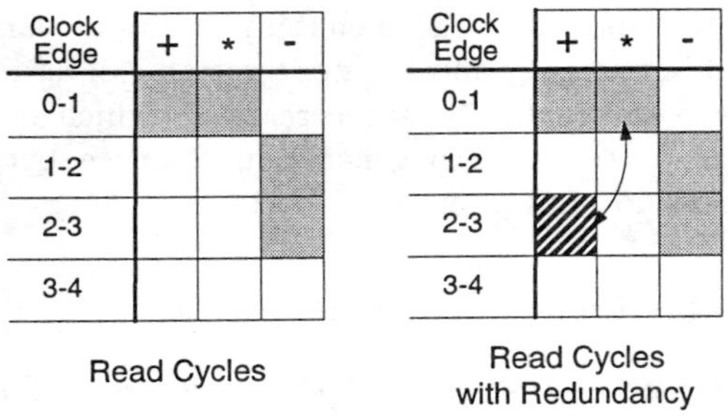

(a) (b)

Figure 2: Read Cycles of Operations

ule M and one subtractor module S. Modules A and M have a spare capacity of 66% since they perform computations only in clock cycle 1 and are idle during clock cycles 2 and 3. Module S has a spare capacity of 33%. By performing *redundant* computations in the modules during the clock cycles they are not performing *useful* computations, sharing of paths that carry test data can be increased.

Fig. 2(a) shows the state (busy or idle) of the read ports of operations in DFG of Fig. 1(a), assuming a minimum latency schedule. The shaded boxes represent when the input ports (read ports) are busy. Note that the read ports of the addition and multiplication operations are busy at the *same* time. A redundant read can be introduced for the addition operation during idle time (shown hashed in Fig. 2(b)). Since the redundant read of the addition occurs at a different time than the read of the multiplication, the two read actions can share hardware for data transfer which could be used to transfer test data for both the operations. A similar scenario exists for write cycles of operations.

2 I-transformation and BIST Resources

Many operations used in a behavioral specification have an identity value associated with them. For example, a multiplication operation has an identity value of 1 and addition operation has an identity value 0. If one of the operands of these operations is equal to the identity value then the output of the operation is the same as the other operand.

Definition 1 *An identity mode DFG node* (**I-node**) *is a node whose operation type has an identity value and one of its operands is set to a constant corresponding to the identity value.*

Definition 2 (**I-transformation**) *Replace edge e of a DFG with an I-node such that the source of e is the source of an input of the I-node and the destination of e is the destination of the output of the I-node. The other input of the I-node corresponds to the identity value of the I-node operation.*

We denote the transformation as $G \overset{T_I}{\to} G'$. Fig. 1(a) shows a DFG G which is transformed into the DFG G' in Fig. 1(b) by the addition of an I-node (shown shaded in the figure). The variable input of the I-node, g is the same as the output variable of $-_1$ and a new variable g_1 is created that is the output of the new addition I-node. The value of the variable g_1 in G' is the same as variable g in G and hence G' and G are functionally equivalent. The concept of exploiting identity modes of functional modules in *synthesized* data paths for transporting test data was suggested in [10]. An I-node is an analogous concept at the behavioral level with the distinction that the identity mode is used to move functional data without changing it.

We propose the use of I-transformations to modify DFGs which, when followed by scheduling and hardware assignment, will lead to cost-effective BIST data paths. The basic idea is to introduce redundant paths that can be used to 1) transport test data, or 2) enable sharing of non-redundant paths in transporting test data. Corresponding to 1) and 2), we have two types of I-transformations, Type 1 and Type 2, respectively.

2.1 Type 1 I-transformation

The DFG shown in Fig. 1(a) illustrates the Type 1 I-transformation. The minimum latency achievable for this DFG is 3. An adder, a multiplier and a subtractor are required to implement the scheduled behavior. A data path synthesized with the objective of minimizing BIST area overhead using assignment techniques in [3] is shown in Fig. 3(a). It can be seen that the adder and multiplier require distinct registers for test pattern generation. Also two distinct registers are required to compress their test responses. The scheduling and assignment stages of synthesis cannot improve the sharing of BIST resources between the modules.

Fig. 1(b) shows the same DFG with an I-node inserted. A data path synthesized using this modified DFG is shown in Fig. 3(b) where the functionally redundant path is shown highlighted. The redundant path from 1) variable g to the I-node, and 2) from the I-node to variable g_1, make it possible for the adder and multiplier to share BIST resources. Variables f (output variable of the multiplication operation) and g_1 (output variable of the addition operation created

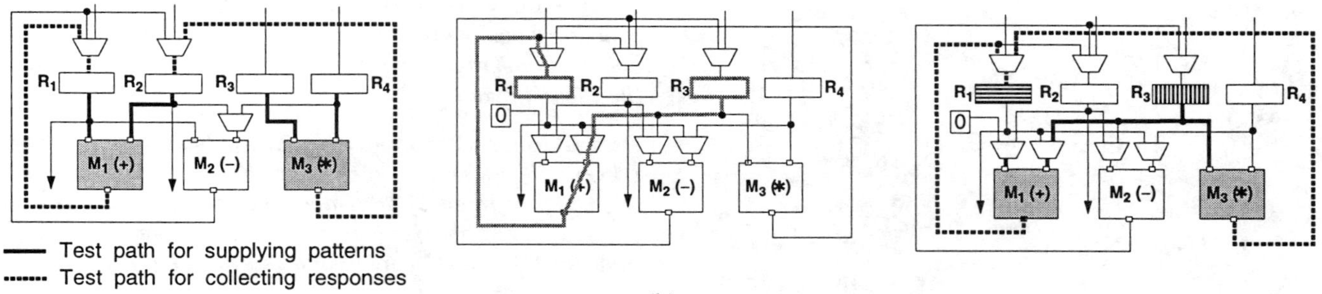

(a) (b) (c)

Figure 3: Data paths synthesized from DFG1 (Type 1 I-transformation)

by redundancy) can be assigned to the same register, R_1, which can compress responses for both the modules. Similarly, input variables of multiplication and addition, c and g, respectively, can be assigned to the same register, R_2, that can generate test patterns for both modules. As shown in Fig. 3(c), the redundant path through the adder can be shared with a functional path through the multiplier, resulting in common BIST resources.

2.2 Type 2 I-transformation

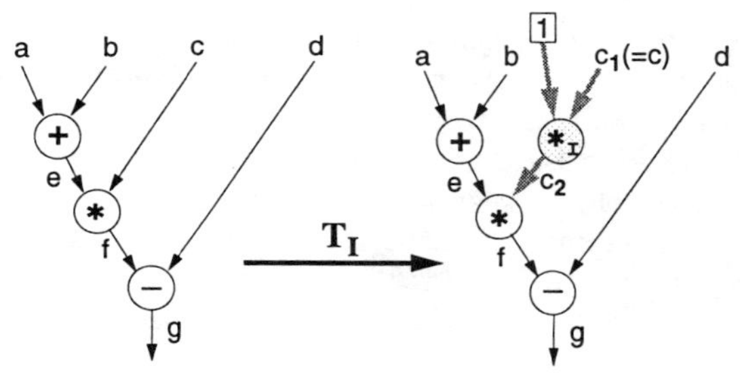

(a) (b)

Figure 4: Type 2 I-transformation on DFG2

In contrast to Type 1 transformation, Type 2 transformation creates redundant paths that *enable* sharing of other non-redundant functional paths as test paths. In this case the redundant path does not transport any test data. This is illustrated using the DFG in Fig. 4(a). The minimum latency is 3 and the module requirement is one adder, one multiplier and one subtractor. Fig. 5(a) shows a data path synthesized from this DFG. It can be seen that it is possible to assign registers such that the test response compression of the adder and multiplier can be done by the same register, namely, R_1. However, a distinct register is required for test pattern generation for the right input port of the adder and multiplier, R_2 and R_3, respectively. Fig. 4(b) shows the DFG modified using a multiplication I-node. The corresponding synthesized data path is shown in Fig. 5(b) with the redundant path highlighted. The addition of redundancy creates a data transfer to the right input port of the multiplier. This enables the input variable of multiplication, c_2, and an input variable of addition, b, to be assigned to the same register, which can be shared as a test pattern generation resource between the adder and multiplier (Fig. 5(c)).

3 Properties of I-transformation

The I-transformation changes the structure and some properties of a DFG. To apply the transformation in an efficient and beneficial manner, we have characterized the effect of the I-transformation on a DFG. A *critical path* in a DFG is the longest path in the DFG and the length of the critical path is a lower bound on the achievable latency of a DFG.

Property 1 *If an I-node is introduced in a non-critical path of a DFG, then the minimum latency of the DFG remains unchanged.*

This property enables the introduction of I-nodes such that they do not effect the minimum latency of the DFG. For optimizing BIST area overhead we need to maximize the sharing of registers as BIST resources between different modules. This depends on the flexibility available in assigning variables to registers.

Property 2 *An I-transformation increases the number of storage variables in the DFG by 1, but the minimum number of registers required to store all the variables remains unchanged.*

Property 2 demonstrates that introduction of I-nodes is not harmful in terms of storage resource requirement of a data path. The minimum number of registers required to implement a data path before and after any I-node transformation is the same. Thus we see that I-nodes can be introduced such that the number of modules, number of registers and the latency remains the same. However the transformation can have an adverse effect on the interconnect complexity.

Property 3 *The number of multiplexer inputs at the input ports of a module to which n_I I-nodes have been assigned increases by at most $n_I + 1$.*

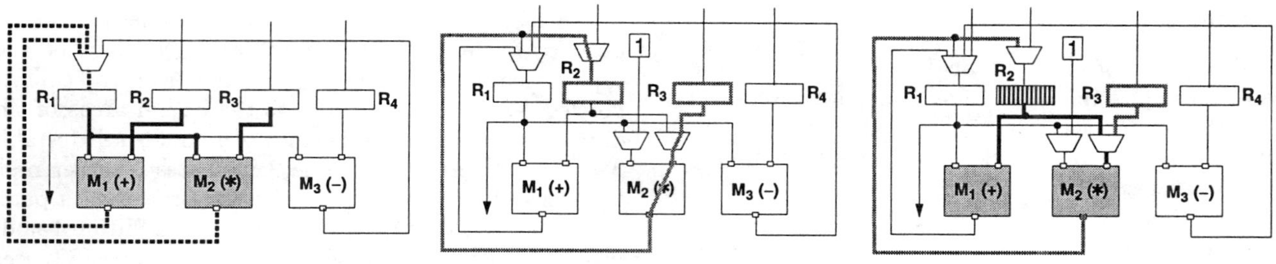

Figure 5: Data paths synthesized from DFG2 (Type 2 I-transformation)

The increase in the number of multiplexers is due to the fact that one additional operation (corresponding to the I-node) is assigned to a module and additional operands need to be supplied to the module. However, one of the operands of the I-node is a constant corresponding to the identity of the I-node operation. Taking advantage of this fact, multiplexer area can be optimized. This can be achieved in two ways.

1. Storing identity value in primary input register: The identity value of an I-node is required only during the clock cycle in which the redundant computation is performed. Hence a *primary input register*, such as R_1 in Fig. 6(a), that does not contain a valid value during that clock cycle can be used to supply the identity value during that clock cycle and the extra multiplexer input can be eliminated.

2. Logic optimization of multiplexers: Some times it is not possible to find a primary input register that does not hold a valid value during the clock cycle in which the redundant computation is performed. In such cases multiplexer overhead can be reduced by logic optimization as shown in Fig. 6(b).

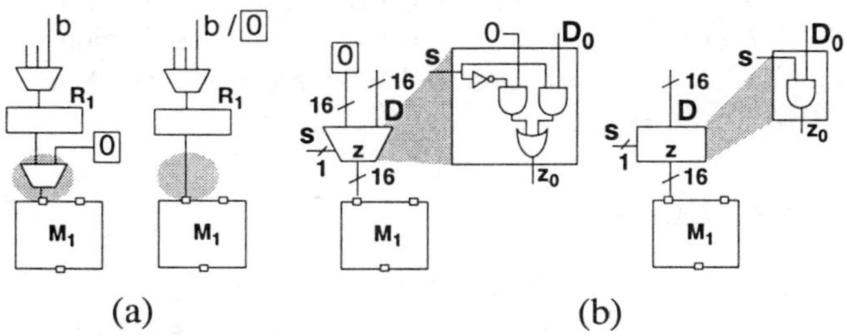

Figure 6: Optimization of multiplexers

4 Strategy for Introduction of I-nodes

The strategy for introduction of I-nodes involves two tasks: 1) identify pairs of operations that will have a potential problem with sharing of BIST resources in subsequent stages of synthesis; and 2) resolve the problem identified in Task 1 by introducing an I-node.

4.1 Identifying BIST Resource Sharing Problems (Task 1)

A BIST resource sharing problem between a pair of modules arises out of the inability to share registers by the input and output variables of pairs of operations assigned to those modules. Hence, the task of identifying BIST resource sharing problems in a DFG involves identification of pairs of operations that have 1) a high probability of being assigned to different modules, and 2) a high probability of having disjoint sets of BIST resources that cannot be shared.

First, we identify pairs of operations that get assigned to different modules with *certainty*. If the operations are of different types then they will definitely get assigned to different modules. If they are of the same type, then they will get assigned to different modules with certainty only if they are concurrent in all possible schedules. We target only such pairs of operations for the following two reasons. Firstly, since many parameters of the data path are not fixed at this point, it is efficient to use information that has a *high degree of certainty*. Secondly, from our experiments we find that introduction of I-nodes follows the *law of diminishing returns*. As more and more I-nodes are added, the resulting savings in BIST area overhead become marginal. Hence we target those pairs of operations that have the problem to the *highest degree* and with the *highest certainty*.

In [11], a procedure for estimating the number of *registers* for an unscheduled DFG is presented. Based on that we have derived a probabilistic estimate of the number of *BIST resources* required for a pair of operations [12]. A higher estimate of BIST resources corresponds to a higher degree of difficulty in sharing these resources between a pair of operations. From the pairs of operations that are assigned to different modules with certainty, a pair with the highest degree of difficulty of sharing BIST resources is selected as the bottleneck to be resolved in Task 2.

4.2 Choosing an I-node (Task 2)

The second task of choosing an I-node to resolve the problem identified in Task 1 has two parameters: 1) the operation type of the I-node , e.g. addition or multiplication, and 2) the position of the I-node in the DFG. As shown in section 3, the introduction of an I-node anywhere in the DFG preserves its functionality but changes some of the properties of the original DFG and affects the synthesized data path in different ways. To ensure that I-nodes are introduced in a manner that is beneficial for BIST but not harmful in terms of functional constraints, the introduction of an I-node should 1) not violate timing constraints, 2) not violate resource constraints, 3) not limit scheduling mobility, 4) maximize re-usage of interconnect to keep interconnect complexity low, and 5) optimize BIST area overhead. Criteria 1 and 2 are strict in the sense that I-nodes that violate them are not allowed. Criteria 3 and 4 are desirable and I-nodes with least impact on scheduling mobility and interconnect complexity are preferred. From the I-nodes that qualify for the first four criteria, an I-node with the highest impact on the BIST resource sharing problem is selected. The impact of an I-node on BIST resources is quantified using the same probabilistic BIST resource estimate that was used in Task 1 to identify a pair of operations that had difficulty in sharing BIST resources. For every candidate I-node the BIST resources are estimated for the pair of operations under consideration and an I-node that decreases the degree of difficulty of sharing BIST resources the most is selected for insertion into the DFG.

Introduction of I-nodes is an iterative process. The scheduling mobility of a DFG is used to determine whether more I-nodes should be added. I-nodes are introduced only while the scheduling mobility remains over a certain predetermined threshold. If the scheduling mobility drops below the threshold, no more transformations are performed.

5 Experimental Results

The I-node transformation was performed on some example DFGs. *ex* is the example from Fig. 1 and *diffeq*, *Tseng* and *AR_filter* are standard high-level synthesis benchmarks. All DFGs were analyzed for BIST resource sharing problems and the DFGs were transformed by introducing I-nodes. TOPS [13], a synthesis-for-test tool was used to synthesize data paths from the original and transformed DFGs which were then made self-testable using minimal intrusion BIST. Components designed using a macro-cell library supplied by LSI Logic Corp. were used for synthesis and the area numbers in the tables are given in cell units [14].

Table 1 shows the components of all synthesized data paths and the BIST resources required to make them self-testable. The name extension of '0' for a DFG refers to the original untransformed version while the rest are after performing I-transformations. The number and type of redundant operations are shown in the column labeled *I-nodes*. Table 2 shows the actual areas of the data paths before and after they were made self-testable The second column in Table 2 indicates whether multiplexer optimization was performed on the data path. It can be seen that introduction of one addition I-node in *ex*-1 results in a significant reduction in BIST area overhead and also total area. However, adding one more redundant addition in *ex*-2 does not give any significant improvement over *ex*-1. In *ex*-3, more redundant operations are added which result in a very large reduction in BIST area overhead. The multiplexer complexity goes up and the savings in total area are not as significant. The experiments on *ex* indicate that introduction of I-nodes follows the *law of diminishing returns*.

Version *diffeq*-1 has one redundant addition and one multiplication as compared to *diffeq*-0. The savings of about 6% in area are achieved at the expense of about 22% reduction in the scheduling mobility of the DFG. For the *AR_filter*, after adding 6 redundant computations, the BIST area overhead is reduced significantly (from 16% to 8.5%) sacrificing 60% of the scheduling mobility. However about only 2% saving in total area is achieved after multiplexer optimization. The *AR_filter* has a high degree of freedom in the scheduling and assignment stages and hence there is not much scope for BIST resource optimization by transforming the behavior. The last four columns in Table 1 show the number of different *types* of BIST resources in the minimal intrusion BIST solutions of all the synthesized data paths. As stated in section 1.1, depending on the test function, four different types of BIST resources with varying costs are possible. A significant reduction in BIST area overhead is achieved even if the number of cheaper BIST resources (such as TPG and SA) increases because the number of more expensive BIST resources (such as CBILBO) decreases.

6 Conclusions

We have demonstrated how the spare capacity of modules available in data paths can be exploited at the behavioral level for reducing BIST resources. A transformation that utilizes the spare resource capacity in a behavior and introduces redundant computations has been described. A technique based on identifying potential testability problems and resolving them by adding I-nodes has been presented. Experimental results demonstrate that very few well-placed I-nodes can significantly reduce the BIST resource requirements of a data path. The transformation is especially useful in cases where there is not enough freedom in the subsequent scheduling and assignment stages to optimize for BIST resources.

Table 1: Characteristics of synthesized data paths

DFG Version	Lat L	Reg	Modules					Muxes					I-nodes	BIST Resources			
			*	+	-	AND	OR	2:1	3:1	4:1	5:1	6:1		#C	#B	#T	#S
ex-0	3	4	1	1	1	-	-	6	0	0	-	-	-	1	1	2	0
ex-1	3	4	1	1	1	-	-	5	2	0	-	-	1+	0	1	2	1
ex-2	3	4	1	1	1	-	-	5	3	0	-	-	2+	0	0	2	2
ex-3	4	4	1	1	1	-	-	4	4	1	-	-	2+, 2*	0	0	2	1
diffeq-0	4	6	3	1	1	-	-	5	2	2	1	-	-	1	1	2	1
diffeq-1	4	6	3	1	1	-	-	10	2	1	-	-	1+,1*	0	1	3	2
Tseng-0	4	4	1	2	1	1	1	1	2	1	-	-	-	1	3	0	0
Tseng-1	4	4	1	2	1	1	1	2	2	1	-	-	1+	0	4	0	0
AR_filter-0	8	16	8	4	-	-	-	11	-	4	-	2	-	0	7	8	1
AR_filter-1	8	16	8	4	-	-	-	25	4	2	-	2	2+,4*	0	3	5	2

Table 2: Area comparisons of synthesized data paths

DFG Version	Mux opt.?	Area before BIST A	BIST overhead B	% BIST overhead $(B/A \cdot 100)$	Total area $C = A + B$	% decrease in total area $(C^0 - C)/C^0)^\dagger \cdot 100$
ex-0	-	6812	2272	25.01	9084	-
ex-1	No	7260	1088	13.03	8348	8.10
ex-2	No	7580	768	9.20	8348	8.10
ex-3	No	8188	576	6.57	8764	3.53
ex-1	Yes	**7228**	1088	**13.08**	**8316**	**8.46**
ex-2	Yes	**7324**	768	**9.49**	**8092**	**10.92**
ex-3	Yes	**7900**	576	**6.80**	**8476**	**6.70**
diffeq-0	-	6033	1072	15.09	7105	-
diffeq-1	No	5916	736	11.06	6652	6.38
diffeq-1	Yes	**5900**	736	**11.09**	**6636**	**6.61**
Tseng-0	-	4078	1296	24.12	5374	-
Tseng-1	No	4094	1024	20.01	5118	4.76
Tseng-1	Yes	**3998**	1024	**25.62**	**5022**	**6.56**
AR_filter-0	-	14000	2656	15.95	16656	-
AR_filter-1	No	15504	1440	8.50	16944	-1.73
AR_filter-1	Yes	**14928**	1440	**9.65**	**16368**	**1.73**

$^\dagger C^0$ is the total area of Version 0 of each DFG.

Acknowledgments

The authors would like to acknowledge Prof. Edward J. McCluskey and Dr. LaNae J. Avra of the Center for Reliable Computing at Stanford University for providing the TOPS synthesis system.

References

[1] H. Harmanani and C. Papachristou. An Improved Method for RTL Synthesis with Testability Tradeoffs. In *Proc. Intn'l Conf. on Computer-Aided Design*, Nov. 1993.

[2] L. Avra. Allocation and Assignment in High-level Synthesis for Self-testable Data Paths. In *Intn'l. Symp. on Circuits and Systems*, Aug. 1991.

[3] I. Parulkar, S.K. Gupta, and M.A. Breuer. Data Path Allocation for Synthesizing RTL Designs with Low BIST Area Overhead. In *Proc. 32nd Design Automation Conf.*, pages 395–401, June 1995.

[4] I. Parulkar, S.K. Gupta, and M.A. Breuer. Scheduling and Module Assignment for Reducing BIST Resources. In *Proc. Design Automation and Test in Europe*, Feb. 1998.

[5] A. Karasniewski. Can Redundancy Enhance Testability? In *Proc. Intn'l Test Conf.*, Oct. 1991.

[6] A.P. Chandrakasan, M. Potkonjak, J. Rabaey R. Mehra, and R. Brodersen. Optimizing Power Using Transformations. *IEEE Trans. on Computer-Aided Design*, Jan. 1995.

[7] L. Guerra, M. Potkonjak, and J. Rabaey. High-level Synthesis for Reconfigurable Datapath Structures. In *Proc. Intn'l Conf. on Computer-Aided Design*, pages 26–29, Nov. 1993.

[8] R. Karri and A. Oraiglu. Transformation-based High-level Synthesis of Fault-tolerant ASICs. In *Proc. 29th Design Automation Conf.*, June 1992.

[9] S. Dey and M. Potkonjak. Transforming Behavioral Secifications to Facilitate Synthesis of Testable Designs. In *Proc. Intn'l Test Conf.*, Oct. 1994.

[10] M.S. Abadir and M.A. Breuer. A Knowledge-Based System for Designing Testable VLSI Chips. *IEEE Design & Test of Computers*, Aug. 1985.

[11] R. Moreno, R. Hermida, and M. Fernandez. Short Note: Register Estimation in Unscheduled Dataflow Graphs. *ACM Trans. on Design Automation of Electronic Systems*, July 1996.

[12] I. Parulkar, S.K. Gupta, and M.A. Breuer. *Using Redundany to Minimize BIST Resources in Data Paths.* CEng Tech. Report 97-15, Univ. of Southern California, Oct. 1997.

[13] L.J. Avra, L. Gerbaux, J-C. Giomi, F. Martinolle, and E.J. McCluskey. *A Synthesis-for-Test Design System.* Tech. Report CSL TR 94-622, Stanford University, May 1994.

[14] Data Book. *G10-p Cell-Based ASIC Products.* LSI Logic Corp., May 1996.

A BIST Scheme for RTL Controller-Data Paths Based on Symbolic Testability Analysis *

Indradeep Ghosh† Niraj K. Jha† Sudipta Bhawmik‡

† Department of Electrical Engineering
Princeton University, Princeton, NJ 08544

‡ Engineering Research Center
Lucent Technologies, Princeton, NJ 08542

Abstract

This paper introduces a novel scheme for testing register-transfer level controller/data paths using built-in self-test (BIST). The scheme uses the controller netlist and the data path of a circuit to extract a test control/data flow (TCDF) which consists of operations mapped to modules in the circuit and variables mapped to registers. This TCDF is used to derive a set of symbolic justification and propagation paths (known as test environment) to test some of the operations and variables present in it. If it becomes difficult to generate such test environments with the derived TCDFs, a few test multiplexers are added at suitable points in the circuit to increase its controllability and observability. This test environment can then be used to exercise a module or register in the circuit with pseudorandom pattern generators which are placed only at the primary inputs of the circuit. The test responses can be analyzed with signature analyzers which are only placed at the primary outputs of the circuit. Every module in the module library is made random-pattern testable, whenever possible, using gate-level testability insertion techniques. Finally, a BIST controller is synthesized to provide the necessary control signals to form the different test environments during testing, and a BIST architecture is superimposed on the circuit. Experimental results on a number of industrial and university benchmarks show that high fault coverage (>99%) can be obtained with our scheme in a small number of test cycles at an average area (delay) overhead of only 6.4% (2.5%).

1 Introduction

Due to the increasing complexity of integrated circuits, BIST is emerging as a popular testing technique for large and complex designs. BIST has several advantages over other test approaches [1], [2], *e.g.*, it gets rid of the need to do test generation and enables at-speed testability. Previously, BIST techniques at the logic level have targeted better test pattern generators [3], and reduction in test overheads [4]. In this paper, the BIST problem is targeted at the register-transfer level (RTL). Owing to the drastic reduction in the number of circuit elements one needs to tackle at the RTL, the problem becomes more tractable. At the RTL, there have been some efforts to reduce test overheads by using a testability analysis scheme based on randomness and transparency of modules [5]. However, the complexity of this method becomes prohibitive as the bit-width increases. A BIST scheme for testing data paths of data-flow intensive RTL circuits was presented in [6]. A scheme for testing the controller was presented in [7]. However, in this work the class of circuits was restricted with the requirement that at most one multiplexer/bus exists along any register-to-register transfer path. Also, the scheme only dealt with register loads and multiplexer select lines coming out of the controller, and did not consider many other types of control lines like ALU select lines *etc*. Another BIST scheme targeting data paths of data-flow intensive RTL circuits was presented in [8]. It reduced the complexity of testability analysis by concentrating on subspace state coverage. However, since it relies on arithmetic units for test generation and compaction, its applicability to control-flow intensive circuits is limited.

In this paper, we present a BIST scheme for testing a controller-data path RTL circuit without imposing any major design restrictions. The only assumptions we make are that the circuit have a separate data path and controller, and a reset state be present in the controller. We can handle both data-flow intensive and control-flow intensive circuits. The scheme works by replacing RTL modules in the circuit which are random-pattern resistant with random-pattern testable versions. If the circuit is originally designed with a module library composed of random-pattern testable modules, then this step is unnecessary. Next, the control and data flow of the circuit are extracted form the controller/data path using the technique presented in [9]. This is termed as test control/data flow (TCDF). It consists of operations mapped to modules in the circuit and variables mapped to registers. The TCDF is used along with the functional information of modules to obtain a set of justification and propagation paths called *test environments* for some operations/variables in them [10]. The test environment of an operation (variable) guarantees the existence of a test path from the primary input ports of the circuit to the inputs of the module (register) to which the operation (variable) is mapped and a path from the output of the module (register) to a primary output port of the circuit. If it becomes difficult to generate such test environments with the existing TCDFs, a few test multiplexers are added at suitable points in the circuit to increase its controllability and observability [11]. Note that the methods presented in [9]-[11] are targeted towards deterministic test generation, not BIST. Since the search for a test environment is done symbolically, it is very fast and needs to be done only once for each module or register in the circuit. The test environment can then be used to exercise a module or register in the circuit with pseudorandom pattern generators (PRPGs) which are placed only at the primary inputs of the circuit. The test responses can be analyzed with multiple-input signature registers (MISRs), which are only placed at the primary outputs of the circuit. The test environments are also analyzed to see if primary input registers can themselves be converted to PRPGs and primary output registers converted to MISRs to reduce overheads further.

In Section 2, the procedure is explained with examples. The BIST architecture is discussed in Section 3. The method is formalized in Section 4. The experimental results and conclusions are in Sections 5 and 6, respectively.

2 Motivational Examples

In this section, we illustrate the complete RTL BIST scheme with a few examples.

2.1 Making RTL modules random-pattern testable

If elements present in the RTL design library are not individually random-pattern testable, then they can degrade the fault coverage of an RTL circuit. In practice, most RTL elements are random-pattern testable. For example, adders, subtracters, shifters, multiplexers, registers, and buses are all random-pattern testable. Even 32-bit versions of these elements are fully tested with less than 50 pseudorandom vectors. A 32-bit multiplier is completely tested

*Acknowledgments: This work was partially supported by NSF under Grant No. MIP-9319269.

DAC 98, San Francisco, California
©1998 ACM 0-89791-964-5/98/06..$5.00

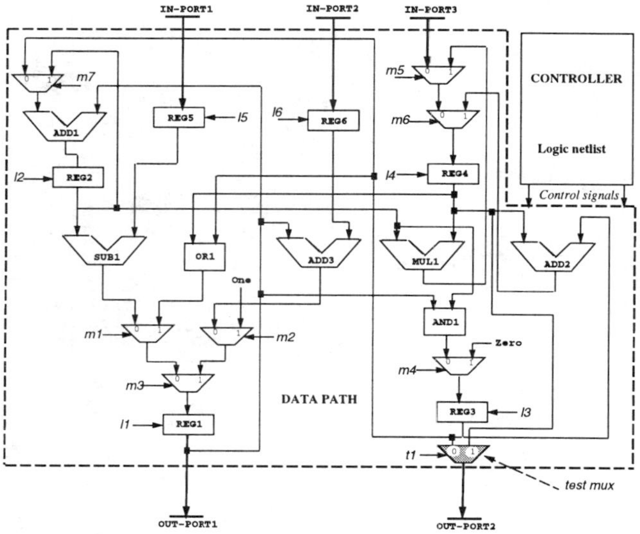

Figure 1: The RTL circuit for *Tseng*

INPUT	STATE		OUTPUTS												
reset	PS	NS	l1	l2	l3	l4	l5	l6	m1	m2	m3	m4	m5	m6	m7
1	ANY	S0	1	0	1	1	1	1	0	1	1	1	0	0	0
0	S0	S1	0	1	0	0	0	0	0	0	0	0	0	0	0
0	S1	S2	1	0	0	1	0	0	0	0	0	0	1	0	0
0	S2	S3	1	1	0	1	0	0	0	0	1	0	0	1	1
0	S3	S4	1	0	1	0	0	0	1	0	0	0	0	0	0
0	S4	S0	0	0	0	1	1	1	0	0	0	0	0	0	0

Figure 2: Controller state table of *Tseng*

with approximately 500 pseudorandom vectors. The real bottleneck lies with the comparators which are highly random-pattern resistant. Hence, in order to boost the random pattern testability of these modules, we resort to some logic-level techniques. This can be done with the help of some test points, *i.e.*, extra control/observe points [2]. For the scheme we have devised, a less-than (equal-to) comparator can be made fully random-pattern testable with only one (*n/8*) extra control and one extra observe point [12]. These extra control and observe points are connected to PRPGs and MISRs, respectively. Thus, they do not require any extra test pins when the comparator is part of an RTL circuit.

2.2 Extracting control/data flow information

Figure 1 shows the RTL circuit of benchmark *Tseng* taken from [13]. This particular RTL implementation was synthesized by the *Genesis* high-level synthesis system [10]. The shaded test multiplexer is not a part of the original circuit and will be discussed later.

The first step in the BIST process is to extract the TCDF using the RTL data path and controller netlist. For this, the procedure presented in [9] is used. First, the controller state table is extracted from the controller netlist. This can be done by a state machine extraction program, *e.g.*, SIS [14]. It starts from the controller reset state and extracts the input/output and state transition information of all reachable states. The extracted controller state table of *Tseng* is shown in Figure 2.

The basic idea behind extracting the TCDF from the RTL circuit is to extract operations executed in each cycle and keep track of variables present in each register or latch. We identify all registers that load in the first cycle by analyzing the load signals of all the registers in the data path from the outputs in the state table. If a load signal is 1 in any cycle, then the register loads in that cycle. In our example, in cycle 1, all registers except *REG2* load. If there are any latches, they load by default in each cycle. We next analyze the multiplexer tree that feeds each of the registers or latches and check if any input port is connected to the register/latch input in the first cycle. The multiplexer tree configuration in any cycle is obtained by looking at the values of the select signals of those multiplexers in the state table. We find that *IN-PORT1*, *IN-PORT2* and *IN-PORT3* are connected to *REG5*, *REG6*, and *REG4*, respectively, in the first cycle. Hence, three variables are born in these three registers. We call them *i1*, *i2* and *i3*, respectively. A variable is live

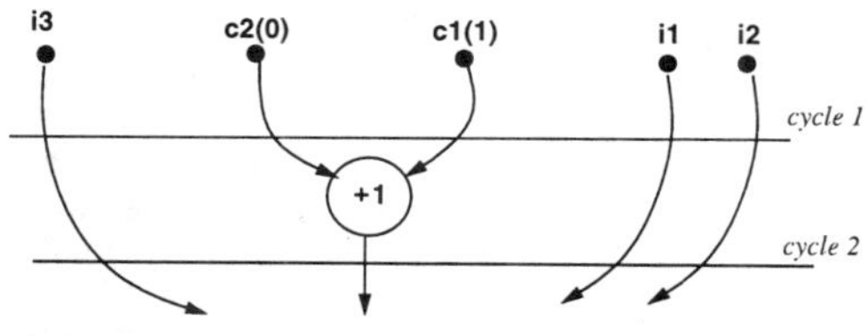

Figure 3: Partial TCDF after analyzing cycle 2 of *Tseng*

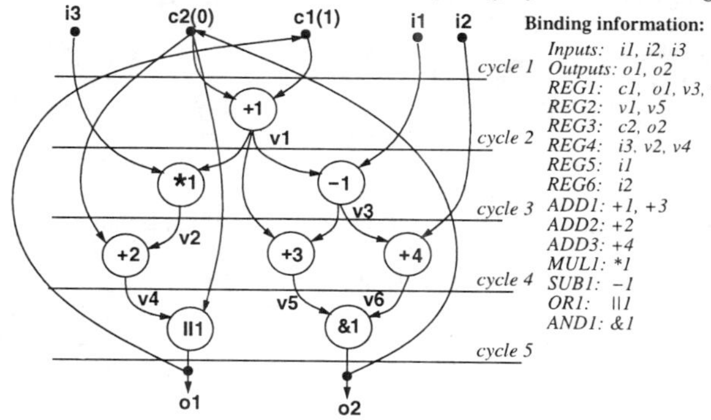

Figure 4: The TCDF for *Tseng*

until its register loads again. Thus, the variable-to-register binding information is developed in each cycle and the variables bound to a particular register noted. On the other hand, we find that in the first cycle, *REG1* and *REG3* are connected to hardwired constants. We name these constants *c1* and *c2* , respectively.

In cycle 2, we identify the operations that take place. For each module in the circuit, we find out the operand selected at each of its input ports by analyzing the multiplexer trees at its inputs. For example, the left input of the adder, *ADD1*, is connected to *REG3* and the right input to *REG1*. We check if both these registers have some live items in them. In this case, they do and they are *c2* and *c1*, respectively. In addition, we check if any of the registers (there may be many due to fanout) at the output of the module load at the end of the second cycle, and if the multiplexer configuration at the input of these registers is such that the output of the module is selected. In case of *ADD1*, *REG2* indeed loads at the end of the second cycle. Hence, we create a new variable *v1*, in *REG2*, and an operation in the TCDF, *c1* + *c2* = *v1*. We also label this operation as *+1*. Similarly, we analyze all other modules in order to obtain the set of operations that execute in the second cycle, as shown in Figure 3 (here the initial values of the constants are also shown in brackets). By repeating this procedure for five cycles, we generate the TCDF shown in Figure 4. In this figure, constants *c1* and *c2* are the initial values of the loop outputs. This method is general enough to handle operation chaining, multi-cycling, and structural pipelining.

2.3 Test environment generation

We use the obtained TCDF to generate *test environments* for all the RTL elements in the circuit. Before we can explain the test environment generation procedure, we need to define a few terms. The general controllability C_g of a TCDF variable is the ability to control it to any arbitrary value from the system primary inputs. The observability O_v of a TCDF variable is the ability to observe any arbitrary value at it at a system primary output. Similarly, we can define C_q (controllability to some arbitrary constant *q*). C_1 (controllability to 1), C_0 (controllability to 0), and C_{a1} (controllability to the all-1s vector) are special cases of C_q. The verifiability *V* of a TCDF variable is the ability to verify its value by either controlling or observing it. Controllability, observability and verifiability are all Boolean parameters and only take the values of 1 or 0, depending on whether the variable has the corresponding ability or not [10]. Next, we add another field to these parameters, which designates the cycle when the particular property of a variable is desired. Hence, $C_g(2)$ of a TCDF variable means that we need to control that variable to an arbitrary value in cycle 2.

Let us take the example of operation *+4*. The test environment of this operation will be used to test adder *ADD3* in the RTL circuit and is obtained as follows. For operation *+4*, we need $C_g(\alpha)$ of *v3*, $C_g(\alpha)$ of *i2* and $O_v(\alpha+1)$ of *v6*. $O_v(\alpha+1)$ of *v6* is trivially achieved since *v6* is mapped to *REG1* which is connected to primary output *OUT-PORT1*. $C_g(\alpha)$ of *i2* is transformed to $C_g(\alpha-2)$ of *i2* which is trivially satisfied as *i2* is a primary input variable. $C_g(\alpha)$ of *v3* is transformed through operation *–1* to $C_g(\alpha-1)$ of *v1* and $C_q(\alpha-1)$ of *i1*. $C_g(\alpha-1)$ of *v1* is transformed through operation *+1* to either $C_g(\alpha-2)$ of *c1* and $C_q(\alpha-2)$ of *c2*, or $C_g(\alpha-2)$ of *c2* and $C_q(\alpha-2)$ of *c1*. Since either objective is unachievable as both *c1* and *c2* are constants, we backtrack and transform $C_g(\alpha)$ of *v3* to $C_q(\alpha-1)$ of *v1* and $C_g(\alpha-1)$ of *i1* through operation *–1*. $C_g(\alpha-1)$ of *i1* is transformed to $C_g(\alpha-2)$ of *i1* which is trivially satisfied as *i1* is a primary input variable. $C_q(\alpha-1)$ of *v1* is transformed to $C_q(\alpha-2)$ of *c1* and $C_q(\alpha-2)$ of *c2*. Both of these are now satisfied and, hence, all objectives for testing *+4* are fulfilled. Next, we examine the test environment and assign a value of 1 to the lowest symbolic cycle value, which is $(\alpha-2)$, and calculate all the cycle values accordingly. Thus, if we control *i1* and *i2* in cycle 1 and observe *v6* in cycle 4, we have tested *+4* with a test vector applied at the primary inputs. From the BIST point of view, this means that if we place a pair of pseudorandom vectors at *PI-port1* and *PI-port2* in cycle 1 derived from PRPGs, and analyze the output at *OUT-PORT1* in cycle 4 with an MISR, we have tested adder *ADD3* with a pseudorandom pattern.

2.4 Test multiplexer insertion

Sometimes, it may not be possible to achieve a test environment for an RTL element using the existing behavior of the circuit. In the example *Tseng*, *ADD2* is symbolically untestable since operation *+2* mapped to *ADD2* does not have a test environment. This is because of the reconvergent fanout of constant *c2*. (Note that *c2* is a constant in the first iteration of the CDFG only, but it can be controlled to an arbitrary value at the beginning of the second iteration). *c2* needs to be non-zero for providing general controllability to the left input of +2, but also needs to be zero for propagating its output *v4* through $||1$. In such cases, we solve the problem by adding a test multiplexer, as shown in Figure 1. Thus, variable *v4*, which is mapped to *REG4*, is observable at the primary output *OUT-PORT2*. The procedure for adding test multiplexers is given in [11]. It utilizes a badness count for TCDF variables to identify testability bottlenecks in the TCDF, and thus the RTL circuit to which the TCDF is mapped. It then inserts appropriate test multiplexers at those points. This procedure tries to avoid the critical path, whenever possible, while adding the test multiplexers. Since a large number of solutions typically exist for overcoming a testability bottleneck, avoiding the critical path is usually possible. The select signals of these test multiplexers come from a BIST controller which is described later.

Registers may be similarly targeted for test environment generation by generating a test environment for any one variable mapped to it. Test environments can also be generated for multiplexers. However, test multiplexers are added to overcome test environment bottlenecks for modules only, since they become an overkill for the other RTL elements. The fault coverage is not compromised because of this, as will be evident from the experimental results. From the above discussion, it is clear that the test environment of an RTL element guarantees that a test path exists from the primary input(s) to the input(s) of the element and from the output of the element to a primary output. We will be utilizing this fact for the BIST scheme.

2.5 Modifications for control-flow intensive circuits

We explain the modifications for control-flow intensive designs with the help the *Barcode* example shown in Figure 5. This circuit is used as a preprocessor to read barcodes printed on objects. The main problem in extending the approach outlined before lies in the status signals fed by the data path to the controller. Now the TCDF for an operation is usually input vector dependent as the data and control flow of the data path change according to the input vectors applied. Hence, in order to generate a fixed TCDF for an operation during the test mode, the status inputs of the controller

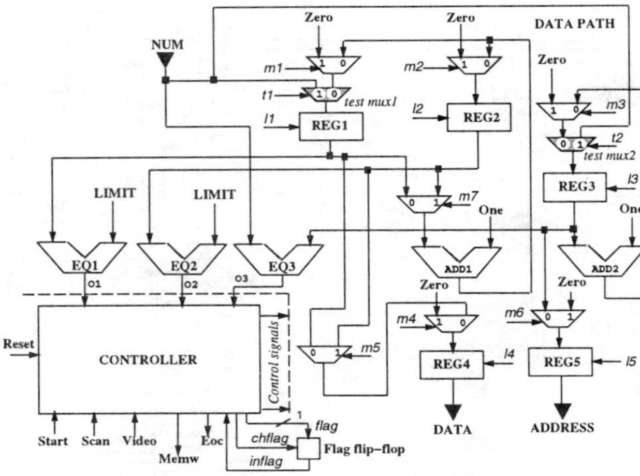

Figure 5: RTL circuit of *Barcode*

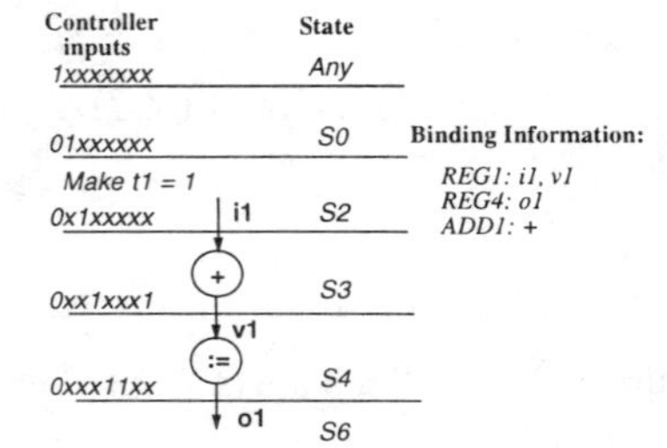

Figure 6: TCDF for testing ADD1 in *Barcode*

need to be controlled. The TCDF generated for the incrementer, *ADD1*, is shown in Figure 6. In the figure, the (:=) operation denotes a register-to-register transfer. Here, since there is no direct input possible from the primary input to the module input, the extraction procedure does a shortest path analysis on the controller state transition graph, and creates a test path by placing the test multiplexer *test mux1* as shown in Figure 5 [9]. Then, the TCDF is generated along with the controller input signals and the test multiplexer select signals required to generate such a TCDF. During the BIST mode, these signals are fed to the controller from a BIST controller. Thus, a TCDF for *ADD1* is guaranteed. After this, test environment generation can proceed as explained in Section 2.3. A single TCDF is usually not enough to test all data path elements in a control-flow intensive RTL circuit.

2.6 Synthesizing a BIST controller

The BIST controller is instrumental in guaranteeing the existence of test environments for all the RTL elements in the test mode. Let us take the example of *Barcode*. The external inputs of the controller, the status signals and the test multiplexer select signals required to test module *ADD1* are shown in Figure 7(a). The signals required to test module *ADD2* are shown in Figure 7(b). The BIST controller can be implemented as a finite-state machine (FSM) that just generates all these input signal sequences one after another. Hence, a trivial way of synthesizing the BIST controller would be to just append the input signal sequences required for the test environments of all the RTL elements, and derive an FSM for those many cycles, as shown in Figure 7(c). In the test mode, each RTL element is tested with pseudorandom vectors from the PRPGs placed at the primary inputs, and the MISRs at the primary outputs, in a sequential fashion. When the BIST controller steps through all its states, it resets to the first state and each RTL element continues getting its pseudorandom vectors, and so on. However, this trivial approach can lead to large overheads. To reduce the BIST controller overheads, we try to merge the input signal sequences for various test environments. For example, the input sequences of *ADD1* and *ADD2* can be merged to test both the modules simultaneously with the same pseudorandom vector at the primary inputs,

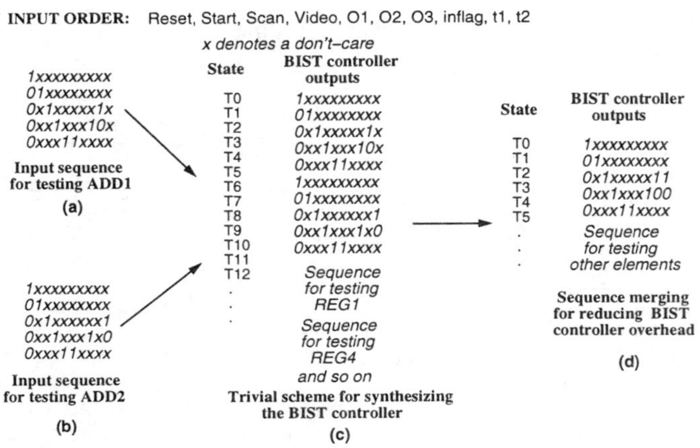

Figure 7: BIST controller synthesis for *Barcode*

as shown in Figure 7(d). This reduces the number of states in the BIST controller, and consequently the overhead.

The compaction technique that we use here is similar to the ones used in test vector compaction for automatic test pattern generation. First, a test environment and the corresponding controller and test multiplexer input sequence for an untested RTL element are generated. Such a sequence typically has many don't-care values for unspecified signals, as shown in Figure 7. Using this initial sequence as a constraint, static test compaction techniques are used to merge other input sequences with it to come up with compact sequences. During compaction, we use existing test set compaction techniques [15], [16].

During test environment generation for a module, we also check if other RTL elements like registers and multiplexers get tested in the process. For example, while testing module *ADD1* in *Barcode*, we see that registers *REG1* and *REG4* also get tested with the same test environment. This is because in the test environment, there exists at least one variable mapped to each of these registers which is controlled from a primary input and observed at a primary output. In such cases, separate test environments are not generated for testing these registers. This is useful in bringing down the BIST controller overhead further.

3 Imposing a BIST Architecture

Once the BIST controller is synthesized, the test hardware is integrated with the original RTL circuit to facilitate its testing during BIST mode by imposing a BIST architecture on it. A low-overhead solution to this problem is shown in Figure 8. In this figure, the added test hardware is shaded grey. (We assume one extra pin to provide the *Test* mode signal.) The controller outputs are multiplexed with a data path primary output port to facilitate testing of the controller [10]. This multiplexer does not result in a delay overhead, as the delay of a multiplexer or even a multiplexer tree at the output of a primary output register is usually much less than the clock period.

The status signals are made directly observable by multiplexing the status signal lines from the data path with an output port, and made controllable during BIST mode by feeding the status inputs of the controller from the BIST controller. The BIST controller is activated during test mode by the *Test* pin. Its *reset* signal is connected to the controller *reset*. The BIST controller feeds: (i) the *select* signals of the test multiplexers that are added to the circuit, (ii) two bits S_0 and S_1 that control the select signals of multiplexer **M** and the output multiplexer, respectively. In the normal mode, the test-multiplexer select lines, the *Test* signal, and signals S_0 and S_1 should be 0.

The data path and the controller are tested separately. In the test mode, the *Test* pin is set. While testing the data path, the BIST controller gives a suitable control flow which is achieved by controlling the controller input signals to a desired sequence. Signals S_1 and S_0 are both set to 0 for testing the data path elements other than the ones that produce the status signals for the controller. At

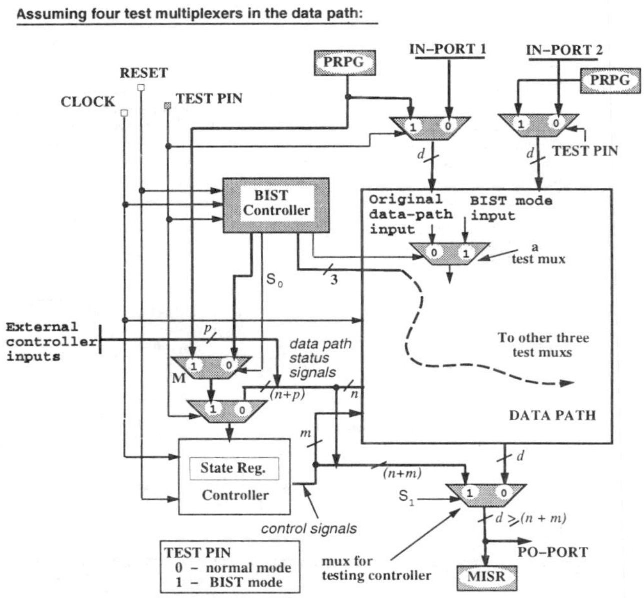

Figure 8: The BIST architecture

specified cycles, we need to reconfigure the data path using the test multiplexers. The BIST controller also provides these select signals. While testing the data path elements that produce the status signals, such as comparators, we need to observe the status signals coming out of the data path to get full observability of the element outputs. In this case, the BIST controller makes signals S_1 and S_0, 1 and 0 respectively. At each cycle, pseudorandom vectors are fed into the primary input ports from PRPGs and the outputs captured at the primary output ports with MISRs. While testing the controller, we can directly control the controller inputs using the PRPG at the primary input port and directly analyze controller outputs by the MISR at the primary output port. The BIST makes S_1=S_0=1 to obtain this particular configuration. In some rare cases, it might so happen that the sum of the number of controller output bits and the status output bits $(n+m)$ exceeds the number of output bits of the data path (d). In that case, the MISR width at the output port has to be increased to capture these extra bits.

It is frequently possible to convert existing input registers (registers connected to primary input ports) to PRPGs and output registers (registers connected to primary output ports) to MISRs, instead of adding extra PRPGs and MISRs. This reduces the overhead even further. However, this can only be done if the occurrence of a circular path can be avoided in the BIST mode. In other words, the input and output registers should not be involved in any test environment in a circular path [17]. For example, in the RTL circuit of *Tseng*, input registers *REG5* and *REG6* can be converted to PRPGs, whereas in the RTL circuit of *Barcode*, output registers *REG4* and *REG5* can be converted to MISRs.

4 The RTL BIST Scheme

In this section, we formalize our RTL BIST scheme. The complete scheme is summarized in Figure 9. First, we extract a controller state table from the controller logic netlist. This can be done by a state machine extraction program starting from the reset state of the controller. Then we select an untested RTL data path element. We extract a TCDF that can be used to test this element along with the controller input sequence required to generate it. This is done with the help of the procedure given in [9] using the RTL data path and the controller state table. Next, we perform a symbolic testability analysis (STA) of the TCDF and try to obtain test environments of as many untested data path elements as possible. This is done using the symbolic justification and propagation techniques presented in [10] and [11]. This procedure may add additional test multiplexers to the data path to increase the controllability and observability of the circuit at certain points where there are testability bottlenecks.

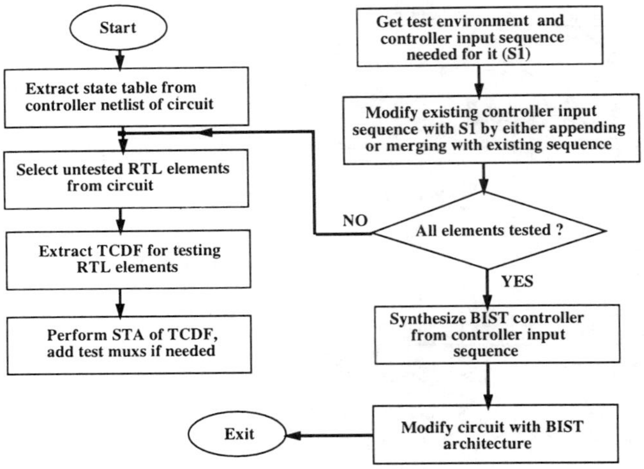

Figure 9: The RTL BIST Scheme

Table 1: Circuit size and DFT hardware statistics

Circuit	Bit-width	# lits.	# FFs	# test muxs	CPU time (sec.)
Tseng	16	5246	100	1	27.5
Paulin	16	6826	116	1	3.5
GCD	16	1329	50	0	0.0
Barcode	16	1518	84	2	1.3
Amvabs	12	1456	40	1	0.8
Amvdual	12	2056	46	1	0.9

The input sequence at the select signals of these test multiplexers needed for the particular test environment is noted. The controller input sequence and the test multiplexer select signal sequence are stored. If possible, they are merged with existing sequences generated for testing other data path elements. The above process is repeated until all data path elements are tested. Then a BIST controller is synthesized from the stored input sequences. Test environments are analyzed to detect the occurrence of circular paths. Input and output registers, for which circular paths are not a problem, are converted to PRPGs and MISRs, respectively. Otherwise, extra PRPGs and MISRs are added along with test multiplexers at the primary input ports and primary output ports of the circuit only. Finally, the circuit is integrated with the BIST components using the BIST architecture described in the previous section.

5 Experimental Results

In this section, we present experimental results obtained by applying our BIST scheme to six example circuits. Among these, *Tseng* and *Paulin* are data-flow intensive circuits literature [13]. *GCD* and *Barcode* are control-flow intensive circuits taken from [18]. *Amvabs* and *Amvdual* are two subcircuits of an MPEG decoder that was designed in the industry. The area and delay results were obtained after technology mapping the gate-level implementation of the RTL circuits using the *stdcell2_2.genlib* cell library in the SIS [14] logic synthesis system.

Table 1 shows the characteristics and specifications of these circuits. The BIST overheads are reported in Table 2 and the testability results are shown in Figure 10. In Table 1, Column 2 shows the bit-width of the circuit data paths. In Columns 3 and 4, the literal-counts of the original technology-mapped circuit and the number of flip-flops are given, respectively. The number of test multiplexers added to the data path by the DFT procedure is given in Column 5. Note that this number just shows the number of extra multiplexers in the data path and does not include the multiplexers in the BIST architecture which are added by default to all circuits, and which have been taken into account while calculating the overheads in Table 2. In Column 6, the CPU time required for symbolic testability analysis is given in seconds. This is on a SparcStation 20 with 256 MB DRAM. Since the analysis is symbolic, it is independent of the bit-width of the data path. Hence, the testability analysis time would remain the same even for 32-bit wide data paths. This is a major advantage over many previous testability analysis schemes [5] where the CPU time for analysis explodes with an increase in the data path bit-width.

Table 2: DFT hardware placement overheads

Circuit	*Area*			*Delay*		
	Orig.	*Mod.*	*Ovhd.* (%)	*Orig.* (ns)	*Mod.* (ns)	*Ovhd.* (%)
Tseng	58654	60531	3.2	70.2	71.8	2.3
Paulin	72345	74443	2.9	74.3	75.7	1.9
GCD	16734	18893	12.9	55.3	56.9	2.9
Barcode	20516	23573	14.9	34.2	35.9	5.0
Amvabs	19564	21168	8.2	29.7	30.2	1.7
Amvdual	28765	31728	10.3	32.3	32.9	1.9

In Column 2 of Table 2, the original area of the circuits after technology mapping is given. This is a relative figure obtained from the layouts of the standard cells used, and hence has no units. Column 3 shows the area after the circuits have been modified by the BIST architecture and test hardware. The area overhead is mainly due to the extra PRPGs and MISRs, the BIST controller, the *Test* pin, the extra logic for making the comparators random-pattern testable, and test multiplexers in the data path and BIST architecture. The percentage overheads are given in Column 4. In Columns 5, 6 and 7, the corresponding figures for delay are provided. The delay represents the clock period in *nanoseconds*. The average area and delay overheads are 6.4% and 2.5%, respectively (the average is calculated based on total area/delay of all the examples for the original and modified cases). The area and delay overheads are much smaller than conventional BIST techniques due to the utilization of functional information and existing data and control flows of the circuit in the test mode.

In Figure 10, the testability results for the original circuits (without any testability hardware) and the circuits augmented by our BIST scheme are shown. The fault coverage numbers are obtained by fault simulating the pseudorandom patterns obtained from the PRPGs (linear feedback shift registers were used as PRPGs) on the circuits using PROOFS [19]. The fault coverage numbers obtained by our method are above 99% for all the examples (with a minimum of 99.1% at 4,500 cycles for *Amvdual* and a maximum of 99.7% for *Paulin*), and the number of test cycles required to get over 99% fault coverage is never more than 4,000.

To check if most of the untested faults in the original circuits were random-pattern resistant, we increased the number of clock cycles to 40,000. By doing so, the increase in the fault coverage (compared to the fault coverage at 4,500 clock cycles) for the original circuits of *Tseng*, *Paulin*, *GCD*, *Barcode*, *Amvabs* and *Amvdual* was only 2.1%, 0.3%, 3.2%, 4.0%, 3.2%, and 5.5%, respectively. This points to the need of employing a BIST scheme, such as the one presented in this paper.

In order to see how different parameters vary with a change in bit-width, we evaluated the BIST architecture for 8, 16 and 32 bit wide data paths of the example *Tseng*. Figure 11 shows how area overhead (AO), delay overhead (DO), and number of cycles required to attain 99% fault coverage (#cycles_for_99%_fc) vary with bit-width. Similar results were obtained for other examples as well. The AO and DO curves are predictable since the testability hardware does not scale as fast as the data path size. The almost linear increase in #cycles_for_99%_fc is also very encouraging.

6 Conclusions

In this paper, we presented a new BIST scheme for testing RTL controller/data path circuits. The scheme uses functional information of modules and symbolic testability analysis techniques to obtain test environments of all the RTL elements in the circuit. The test environment of an element guarantees that the element can be fed by pseudorandom vectors provided at the primary inputs and its test response observed at a primary output. If it is not possible to generate the test environment of a module, then test multiplexers are added to the circuit at suitable points to increase the controllability and observability of the circuit. Then a BIST controller is

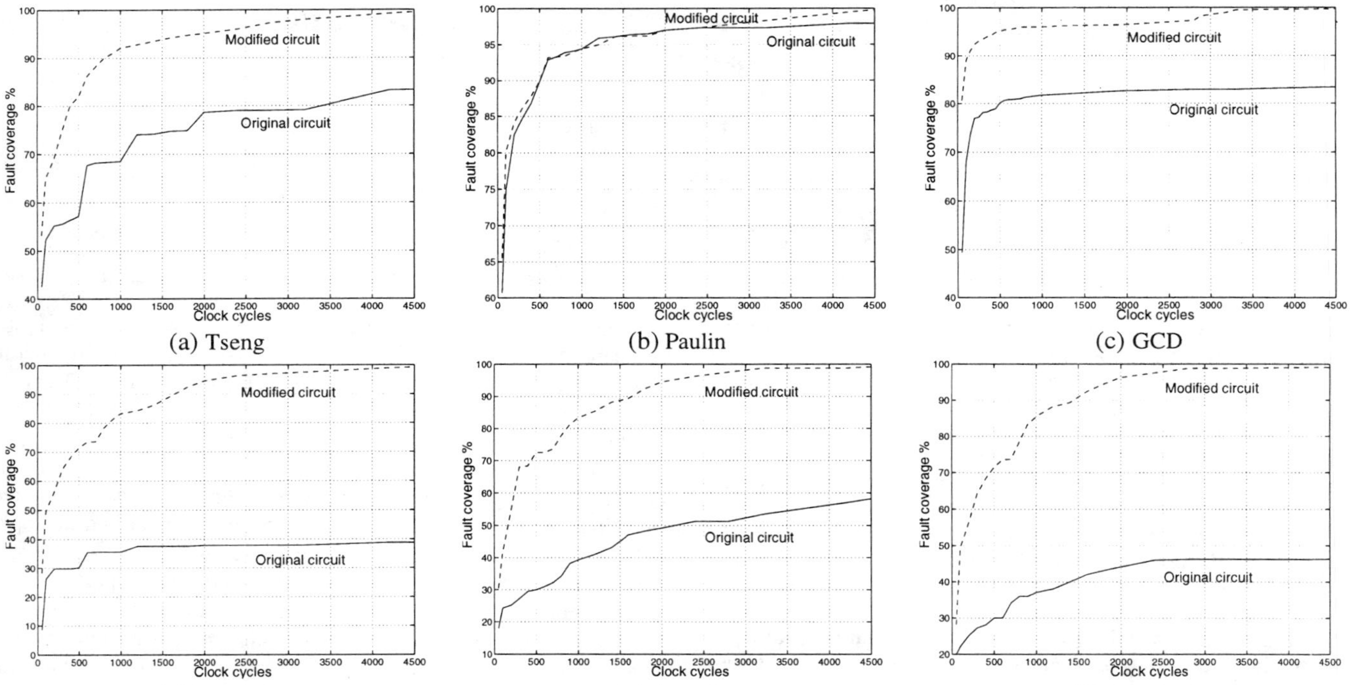

(a) Tseng (b) Paulin (c) GCD

(d) Barcode (e) Amvabs (f) Amvdual

Figure 10: Fault coverage curves for the six example circuits

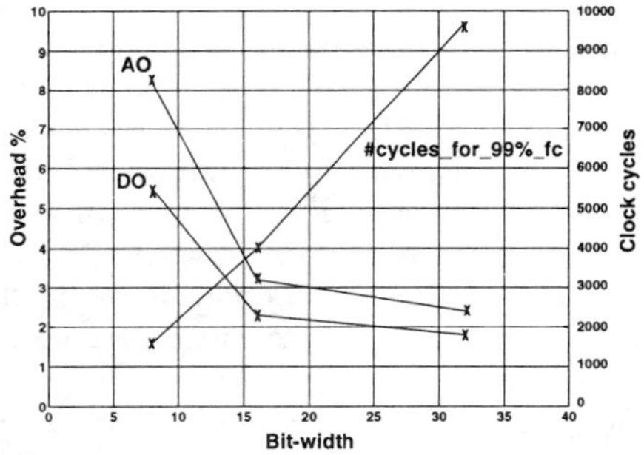

Figure 11: Effect of bit-width on the BIST parameters

synthesized to produce the necessary values of the status inputs of the controller in the test mode and the test multiplexer select signal values. Finally, PRPGs and MISRs are added at the primary inputs and primary outputs of the circuit respectively, and a BIST architecture imposed on the whole RTL circuit. The advantages of the scheme are: (i) comparatively low area and delay overheads, (ii) very high fault coverage, and (iii) low test application times.

References

[1] M. Abadir and M. Breuer, "Constructing optimal test schedules for VLSI circuits having built-in test hardware," in *Proc. Int. Symp. Fault Tolerant Computing*, pp. 165-170, June 1985.

[2] M. Abramovici, M.A. Breuer, and A.D. Friedman, *Digital Systems Testing and Testable Design*, IEEE Press, New York, 1990.

[3] H. Schnurmann, E. Lindblom, and R.G. Carpenter, "The weighted random test-pattern generator," *IEEE Trans. Comput.*, vol. C-24, pp. 695-700, July 1975.

[4] C.J. Lin, Y. Zorian, and S. Bhawmik, "PSBIST: A partial-scan based BIST scheme," in *Proc. Int. Test Conf.*, pp. 507-516, Oct. 1993.

[5] S.S.K. Chiu and C. Papachristou, "A built-in self-testing approach for minimizing hardware overhead," in *Proc. Int. Conf. Computer Design*, pp. 282-285, Oct. 1991.

[6] C. Papachristou and J. Carletta, "Test synthesis in the behavioral domain," in *Proc. Int. Test Conf.*, pp. 693-702, Oct. 1995.

[7] M. Nourani, J. Carletta, and C. Papachristou, "A scheme for integrated controller-data path fault testing," in *Proc. Design Automation Conf.*, pp. 546-551, June 1997.

[8] N. Mukherjee, M. Kassab, J. Rajski, and J. Tyszer, "Arithmetic built-in self test for high-level synthesis," in *Proc. VLSI Test Symp.*, pp. 132-139, May 1995.

[9] I. Ghosh, A. Raghunathan, and N.K. Jha, "A design for testability technique for RTL circuits using control/data flow extraction," in *Proc. Int. Conf. Computer-Aided Design*, pp. 329-336, Nov. 1996.

[10] S. Bhatia and N.K. Jha, "Behavioral synthesis for hierarchical testability of controller/data path circuits with conditional branches," in *Proc. Int. Conf. Computer Design*, pp. 91-96, Oct. 1994.

[11] I. Ghosh, A. Raghunathan, and N.K. Jha, "Design for hierarchical testability of RTL circuits obtained by behavioral synthesis," *IEEE Trans. Computer-aided Design*, vol. 16, pp. 1000-1001, Sept. 1997.

[12] I. Ghosh, N.K. Jha, and S. Bhawmik, "A BIST scheme for RTL controller-data paths based on symbolic testability analysis," Tech. Rep. CE-J98-005, EE Dept., Princeton Univ.

[13] L. Avra, "Allocation and assignment in high-level synthesis for self-testable data paths," in *Proc. Int. Test Conf.*, pp. 463-471, June 1991.

[14] E. Sentovich, K. Singh, C. Moon, H. Savoj, R. Brayton, and A. Sangiovanni-Vincentelli, "Sequential circuit design using synthesis and optimization," in *Proc. Int. Conf. Computer Design*, pp. 328-333, Oct. 1992.

[15] P. Goel and B.C. Rosales, "PODEM-X: An automatic test generation system for VLSI logic structures," in *Proc. Design Automation Conf.*, pp. 260-268, June 1981.

[16] I. Pomeranz, L.N. Reddy, and S.M. Reddy, "COMPACTEST: A method to generate compact test set for combinational circuits," *IEEE Trans. Computer-Aided Design*, vol. 12, pp. 1040-1049, July 1993.

[17] L.J. Avra and E.J. McCluskey, "Synthesis for scan dependence in built-in self-testable designs," in *Proc. Int. Test Conf.*, pp. 734-742, Oct. 1993.

[18] P.R. Panda and N.D. Dutt, "1995 high-level synthesis design repository," in *Proc. Int. Symp. System Level Synthesis*, pp. 170-174, Sept. 1995.

[19] T.M. Niermann, W.T. Cheng, and J.H. Patel, "PROOFS: A fast, memory efficient sequential circuit fault simulator," in *Proc. Design Automation Conf.*, pp. 535-540, June 1990.

Figures of Merit to Characterize the Importance of On-Chip Inductance

Yehea I. Ismail, Eby G. Friedman, and Jose L. Neves[1]

Department of Electrical Engineering
University of Rochester
Rochester, New York 14627

[1]IBM Microelectronics
1580 Route 52
East Fishkill, New York 12533

Abstract **- A closed form solution for the output signal of a CMOS inverter driving an *RLC* transmission line is presented. This solution is based on the alpha power law for deep submicrometer technologies. Two figures of merit are presented that are useful for determining if a section of interconnect should be modeled as either an *RLC* or an *RC* impedance. The damping factor of a lumped *RLC* circuit is shown to be a useful figure of merit. The second useful figure of merit considered in this paper is the ratio of the rise time of the input signal at the driver of an interconnect line to the time of flight of the signals across the line. AS/X circuit simulations of an *RLC* transmission line and a five section *RC* Π circuit based on a 0.25 μm IBM CMOS technology are used to quantify and determine the relative accuracy of an *RC* model. One primary result of this study is evidence demonstrating that a range for the length of the interconnect exists for which inductance effects are prominent. Furthermore, it is shown that under certain conditions, inductance effects are negligible despite the length of the section of interconnect.**

I. Introduction

It has become well accepted that interconnect delay dominates gate delay in current deep submicrometer VLSI circuits [1]-[4]. With the continuous scaling of technology and increased die area, this situation is expected to become worse. In order to properly design complex circuits, more accurate interconnect models and signal propagation characterization are required. Historically, interconnect has been modeled as a single lumped capacitance in the analysis of the performance of on-chip interconnects. With the scaling of technology and increased chip sizes, the cross-sectional area of wires has been scaled down while interconnect length has increased. The resistance of the interconnect has therefore become significant, requiring the use of more accurate *RC* delay models. At first interconnect was modeled as a lumped *RC* circuit. To further improve accuracy, the interconnect has been modeled as a distributed *RC* circuit (multiple T or Π sections) for those nets requiring more accurate delay models. A well known method used to determine which nets require more accurate delay models is to compare the driver resistance R_{tr} and the load capacitance C_L to the total resistance and capacitance of the interconnect line, R_t and C_t [5], [6]. Typically, those nets that require more accurate *RC* models are longer, more highly resistive nets.

Currently, inductance is becoming more important with faster on-chip rise times and longer wire lengths. Wide wires are frequently encountered in clock distribution networks and in upper metal layers. These wires are low resistive lines that can exhibit significant inductive effects. Furthermore, performance requirements are pushing the introduction of new materials for low resistance interconnect [7]. In the limiting case, high temperature superconductors may become commercially available [8]. With these trends it is becoming crucial to be able to determine which nets within a high speed VLSI circuit exhibit prominent inductive effects.

The focus of this paper is the introduction of simple figures of merit that can be used as criteria to determine which nets require more accurate transmission line models. The equations describing the signal behavior of an *RLC* transmission line are provided in section II, along with a closed form solution for the output signal of a CMOS inverter driving an *RLC* transmission line based on the alpha power law [9] for deep submicrometer (DSM) technologies. The damping factor of a lumped *RLC* circuit and the rise time of the input signal at the driver of the interconnect are used to derive two figures of merit that describe the relative significance of inductance of a local interconnect line. These figures of merit are presented in section II. In section III, the two figures of merit described in section II are combined to define a range of the length of interconnect at which inductance become important. AS/X circuit simulations [10] are also compared in this section to the analytical results presented in the previous section. Finally, some conclusions are offered in section IV.

II. Theoretical Analysis of Inductance Effects in *RLC* Interconnect

The behavior of waves traveling across an *RLC* transmission line is explained in this section. The attenuation that a wave exhibits as it travels along a line is compared to the damping factor of a lumped *RLC* circuit representation of the same line. It is shown in subsection A that the damping factor of a lumped *RLC* circuit representation of a line can be a useful figure of merit to determine the relative importance of inductance. A closed form solution for the output voltage of a CMOS gate driving an *RLC* transmission line is presented in subsection B. A closed form solution for a CMOS gate driving a single *RC* T section representation of the line is also presented. Both solutions are compared for different values of attenuation to further investigate the damping factor as a useful figure of merit. The two solutions are also compared with different input transition times which leads to the second figure of merit in this paper. It is shown that the ratio

This research was supported in part by the National Science Foundation under Grant No. MIP-9423886 and Grant No. MIP-9610108, the Army Research Office under Grant No. DAAH04-93-G-0323, a grant from the New York State Science and Technology Foundation to the Center for Advanced Technology - Electronic Imaging Systems, and by grants from the Xerox Corporation, IBM Corporation, and Intel Corporation.

DAC 98, San Francisco, California

of the transition time of the signal at the input of a CMOS gate driving an interconnect line to the time of flight of a wave across the interconnect is a second useful figure of merit to determine the relative importance of inductance to a specific line.

A. Damping Factor

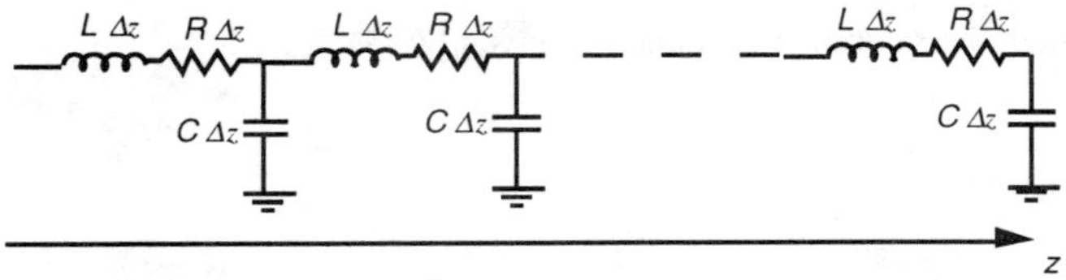

Fig. 1. *RLC* transmission line model of an interconnect line.

A single interconnect line can be modeled as an *RLC* transmission line as shown in Fig. 1, where *R*, *L*, and *C* are the resistance, inductance, and capacitance per unit length, respectively, and Δz is an incremental length segment of the line. For an *RLC* transmission line driven by a sinusoidal input $Re\{e^{j\omega t}\}$, the voltage across the transmission line [11] is

$$V(z,t) = \mathrm{Re}\{V_1 e^{(j\omega t - \gamma z)} + V_2 e^{(j\omega t + \gamma z)}\}. \quad (1)$$

The solution of $V(z,t)$ is the sum of two traveling waves, one in the positive z direction and the other in the negative z direction. V_1 is the summation of the original voltage wave and all the reflected voltage waves in the positive z direction. V_2 is the summation of all the reflected voltage waves traveling in the negative z direction. The propagation constant of the transmission line, γ, describes the characteristics of the wave propagation across the line. For an *RLC* transmission line, the propagation constant is complex [11] and is

$$\gamma = \alpha + j\beta, \quad (2)$$

where the real part α is the attenuation constant of the waves as the waves propagate across the line, and the imaginary part β is the phase constant which determines the speed of propagation of the waves across the line. Substituting (2) into (1), the real part of the voltage is given by

$$V(z,t) = V_1 e^{-\alpha z}\cos(\omega t - \beta z) + V_2 e^{\alpha z}\cos(\omega t + \beta z). \quad (3)$$

The attenuation of a traveling wave is exponentially dependent on the distance traveled by the wave, and both the attenuation and speed of the wave are frequency dependent.

The speed of propagation of the wave across the line [11] is

$$v = \frac{\omega}{\beta}. \quad (4)$$

The sequence of events that constitutes a transient response for an input wave begins with a portion of the wave launched into the line from the source end. This wave propagates across the line towards the load with a speed determined by (4). The wave attenuates as it travels across the lossy line. If a mismatch exists between the characteristic impedance of the transmission line and the load impedance, a reflected wave is generated and propagates towards the source to compensate for the mismatch. This reflected wave is further attenuated as it moves towards the source. The reflection process is repeated infinitely, but practically, the signal can be considered to be at steady state when the reflections become negligible. As the rate of attenuation increases, the reflections become negligible faster. This behavior can be explained by noting that the waves are multiplied by a factor of $e^{-2\alpha l}$ for a round trip across the line, where l is the length of the line. This aspect means that as the line becomes longer, the effect of the reflections becomes less and the line behaves as an *RC* line. This same behavior occurs if the resistance of the line increases, increasing the attenuation constant. The attenuation constant α of an *RLC* transmission line can be derived from the basic equations and is

$$\alpha = \omega\sqrt{LC}\sqrt{\frac{1}{2}\left(\sqrt{\left(1 + \left(\frac{R}{\omega L}\right)^2\right)} - 1\right)}. \quad (5)$$

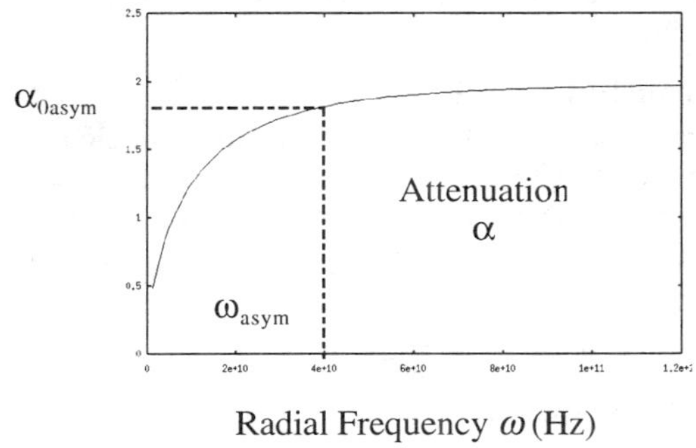

Fig. 2. The attenuation constant α versus the radial frequency ω.

The attenuation constant as a function of frequency is plotted in Fig. 2 with $L=10^{-8}$ *H/cm*, $R=400$ *Ω/cm*, *and* $C=10^{-12}$ *F/cm* [12]. The attenuation constant is shown to saturate with increasing frequency to an asymptotic value given by

$$\alpha_{asym} = \frac{R}{2}\sqrt{\frac{C}{L}}, \quad (6)$$

and the radial frequency at which this saturation begins is given by

$$\omega_{asym} \approx \frac{R}{L}. \quad (7)$$

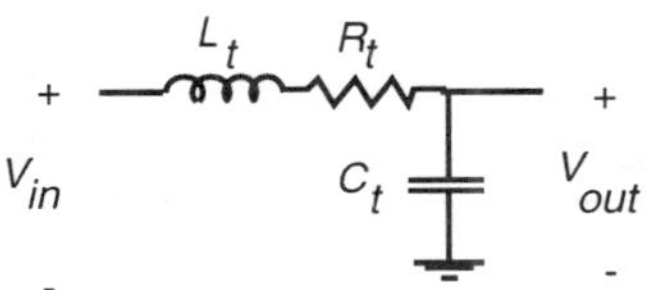

Fig. 3. Simple lumped *RLC* circuit model of an interconnect line.

This analysis of an *RLC* transmission line is compared to the analysis of a lumped *RLC* circuit (see Fig. 3 for a lumped *RLC* circuit). The interconnect is modeled as a single section *RLC* circuit with $R_t=Rl$, $L_t=Ll$, and $C_t=Cl$. The poles of this circuit are

$$p_{1,2} = \omega_0[-\xi \pm \sqrt{(\xi^2 - 1)}], \quad (8)$$

and the damping factor ξ is

$$\xi = \frac{Rl}{2}\sqrt{\frac{C}{L}} = l\alpha_{asym}. \quad (9)$$

As (8) implies, if ξ is greater than one, the poles are real and the effect of the inductance on the circuit is small. The greater the value of ξ, the more accurate the *RC* model becomes. On the other hand, as ξ becomes less than one, the poles become complex and oscillations occur. In that case, the inductance cannot be neglected. The strong analogy between the lumped *RLC* circuit and the *RLC* transmission line is illustrated by (9). This relationship is physically intuitive, since ξ represents the degree of attenuation the wave suffers as it propagates a distance equal to the length of the line. As this attenuation increases, the effects of the reflections decrease and the *RC* model becomes more accurate. Therefore ξ is a useful figure of merit that anticipates the importance of considering inductance in a particular interconnect line. This figure of merit is the same result as described in [13]-[15] but is derived in a different way. Note that if ξ in (9) is squared, this figure of merit becomes a comparison between the time constant *L/R* and the time constant *RC,* which is the same result as described in [1], [3], and [16].

B. Input Transition Time

The characteristic impedance of an *RLC* transmission line is complex with a portion that is negative and imaginary. Therefore, the characteristic impedance looks like a resistance in series with a capacitance. Thus, the characteristic impedance can be expressed as

$$Z_0 = R_0 - j\frac{1}{\omega C_0}, \tag{10}$$

where R_0 and C_0 are given by

$$R_0 = \sqrt{\frac{L}{C}}\sqrt{\frac{1}{2}\left(\sqrt{\left(1+\left(\frac{R}{\omega L}\right)^2\right)}+1\right)}, \tag{11}$$

$$C_0 = \frac{1}{\omega\sqrt{\frac{L}{C}}\sqrt{\frac{1}{2}\left(\sqrt{\left(1+\left(\frac{R}{\omega L}\right)^2\right)}-1\right)}}. \tag{12}$$

Plots of R_0 and C_0 versus frequency are shown in Figs. 4 and 5, respectively, with $L=10^{-8}$ *H/cm*, $R=400$ *Ω/cm, and* $C=10^{-12}$ *F/cm.*

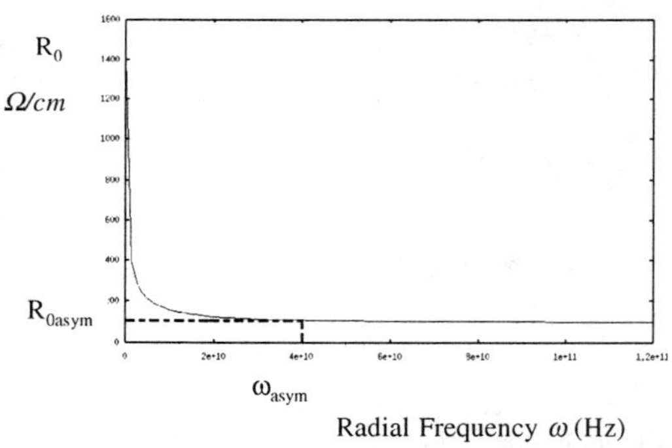

Fig. 4. Real part of the characteristic impedance of an *RLC* transmission line.

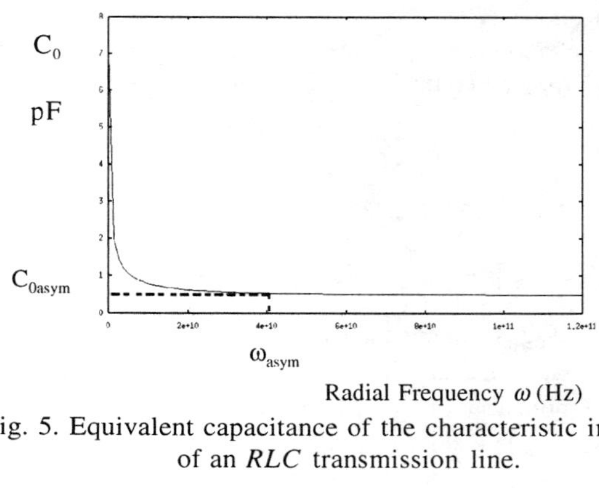

Fig. 5. Equivalent capacitance of the characteristic impedance of an *RLC* transmission line.

Both R_0 and C_0 saturate to the asymptotic values given by

$$R_{0asym} = \sqrt{\frac{L}{C}}, \tag{13}$$

$$C_{0asym} = \frac{2\sqrt{LC}}{R}, \tag{14}$$

where the saturation frequency is given by (7).

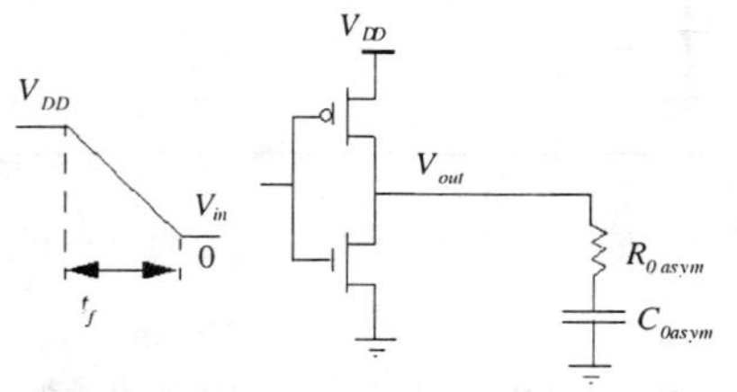

Fig. 6. A CMOS inverter driving the equivalent characteristic impedance of an *RLC* transmission line.

An *RLC* transmission line driven by a CMOS inverter can be approximated as shown in Fig. 6, for the period of time, $0<t<2T_0$, where T_0 is the time required for the waves to travel a distance equal to the length of the transmission line. This term is frequently described as the time of flight of the transmission line. The input is assumed to be a ramp with a fall time t_f. Asymptotic values for the characteristic impedance and the attenuation are assumed in the following analysis. The technology used in this analysis is an IBM 0.25 µm CMOS technology with a 2.5 volt power supply. The alpha power law is used to characterize the devices [9]. A pulse is generated at V_{out} for the period of time $0<t<2T_0$, where the time reference is chosen when the input signal reaches $V_{DD}+V_{Tp}$ where V_{Tp} is the threshold voltage of the P-channel devices and for an enhancement mode device is negative. Under the aforementioned conditions and the assumption that the PMOS transistor is saturated and neglecting the effect of the NMOS transistor, V_{out} is

$$V_{out}(t) = P_{Cp}\frac{W_p}{L_p}V'_{SGp}(t)^{\alpha p}\left[\sqrt{\frac{L}{C}}+\frac{Rt}{2\sqrt{LC}}\right]u(t) = P(t), \tag{15}$$

for $0<t<2T_0$, where $u(t)$ is the unit step function and $V'_{SGp}(t) = V_{DD} + V_{Tp} - V_{in}(t)$. P_C is a constant that characterizes the drive current of the transistor in saturation, *W* and *L* are the geometric width and length, respectively, of the transistor, and α is a constant between one (strong velocity saturation) and two (weak velocity saturation) [9]. *p* indicates the PMOS transistor.

The pulse propagates across the transmission line. At the load, the signal is completely reflected assuming an open circuit (or a small load capacitor) at the end of the line. This reflected signal propagates back towards the driver and reaches the driver at a time $t = 2T_0$. After this round trip is completed, the pulse that reaches the source is attenuated by a factor of $e^{-2\alpha l}$ and can be described mathematically by $P(t-2T_0)\, e^{-2\alpha l}$. As long as the transistor is in saturation, the transistor maintains a relatively constant current. Thus, the current reflection coefficient is -1 and consequently the voltage reflection coefficient is 1. Therefore the pulse is multiplied by two. This cycle repeats as long as the transistor is saturated. The complete solution for the period when the transistor remains saturated is

$$V_{out}(t) = P(t) + \sum_{i=1}^{n}\left[2P(t-2nT_0)e^{-2\alpha nl}\right], \tag{16}$$

for $2nT_0<t<2(n+1)T_0$. This solution is compared to an *RC* representation of the line, as shown in Fig. 7.

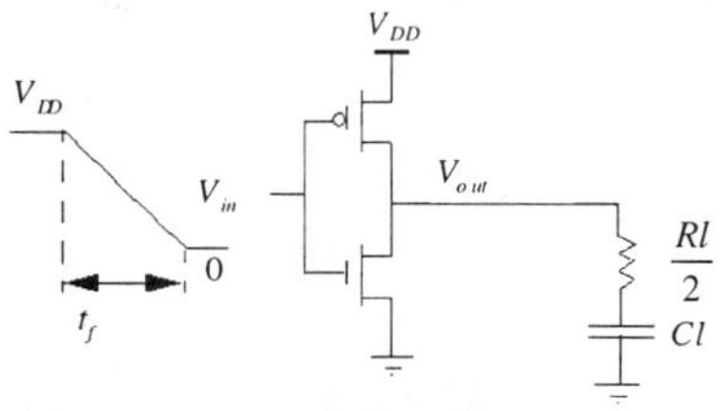

Fig. 7. A CMOS inverter driving an *RC* approximation of an interconnect line.

The solution for V_{out} based on this model when the PMOS transistor is in saturation is

$$V_{out} = P_{Cp}\frac{W_p}{L_p}V'_{SGp}(t)^{\alpha p}\left[\frac{Rl}{2}+\frac{t}{Cl}\right]. \tag{17}$$

The analytical solution in (16) is compared with AS/X simulation [10] results for the *RLC* transmission line characterized in Figs. 8 and 9, with $L=10^{-7}$ *H/cm* and $C=10^{-12}$ *F/cm*. The analytical solution agrees with the simulations of an *RLC* transmission line for a wide variety of interconnect resistances and input fall times. As implied by the analytical solution, the output signal follows the changing input signal. The period when the input signal is falling represents the fast rising parts of the response that depend on the transition time of the input signal. Once the input signal is settled, the current provided by the transistor is constant and the output signal changes slowly due to the charging of the equivalent capacitor of the transmission line. This period of time represents the slow rising part of the response that depends upon the value of the equivalent capacitance of the transmission line. Note in Fig. 8 that as *R* increases, the slope of those portions of the response increases since the value of the equivalent capacitor decreases, as given by (14). It can also be seen that as the resistance of the line increases, the attenuation of the reflections increases as given by (9), which makes the *RC* response approach the *RLC* transmission line response.

The output response of the *RLC* transmission line tracks the output response of the *RC* circuit. The points of intersection with the *RC* response can be calculated by equating (16) and (17), and are given by

$$t_n = \frac{\left[T_0(2K_n-1)+4T_0\xi\cdot\frac{dK_n}{d(2\alpha l)}-\frac{RCl^2}{2}\right]}{\left[1-\xi(2K_n-1)\right]}, \tag{18}$$

where $K_0=1$ and

$$K_n = \frac{e^{-2\alpha l(n+1)}-1}{e^{-2\alpha l}-1} \qquad n=1,2... \tag{19}$$

The interesting point to note is that those times at which the *RC* response intersects the *RLC* transmission line response are not dependent on the transition time of the input signal. This characteristic can be noticed in Fig. 9. Thus, as the transition time of the input signal increases, the slope of the fast changing portions of the response decreases, which reduces the width of the slowly varying parts of the response. This behavior makes the response of the *RLC* transmission line appear more continuous. Since the times at which the *RLC* response intersect with the *RC* response are constant, the *RLC* transmission line response approaches the *RC* circuit response.

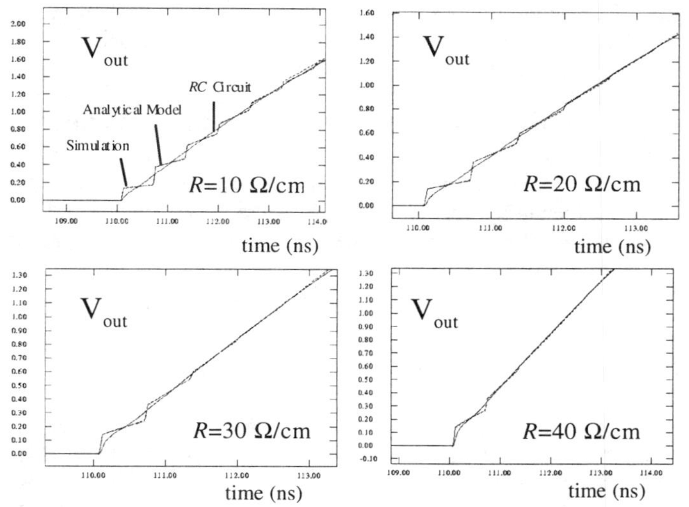

Fig. 8. Analytical solution in (16) compared to AS/X simulations and a five section *RC* Π circuit. The fall time of the input signal is held constant at 60 ps, while *R* is varied.

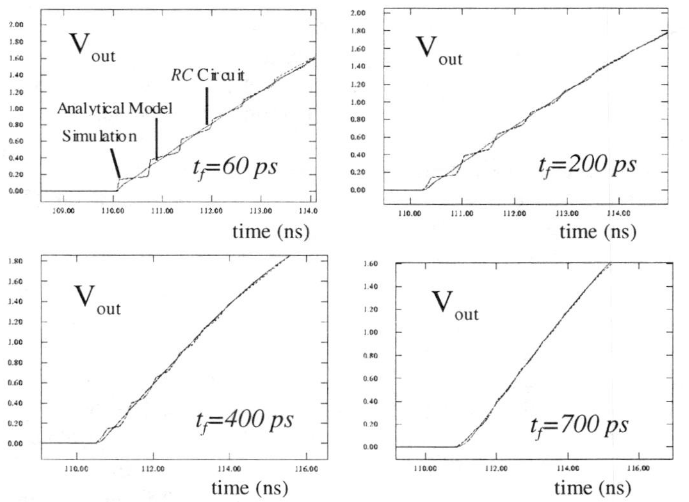

Fig. 9. Analytical solution in (16) compared to AS/X simulations and a five section *RC* Π circuit. *R* is held constant at *10* Ω/cm, while the fall time of the input signal is varied.

As the transition time of the input signal becomes equal to or greater than $2T_0$, the slowly varying portions of the *RLC* transmission line response disappear, and the response coincides with the *RC* approximation. This behavior is evident from Fig. 9 which leads to the second figure of merit given by

$$t_r > 2l\sqrt{LC}, \tag{20}$$

where the asymptotic value of T_0 is $l\sqrt{LC}$. When this inequality is satisfied, inductance becomes unimportant. Note that the figure of merit in (20) is accurate only if the line is matched (the width of the transistor driving the line is adjusted to match the transistor output impedance with the load impedance of the line to avoid reflections) or underdriven (the width of the transistor driving the line is less than is necessary to match the transistor impedance to the load impedance). However, this condition does not affect the validity of the results since in most practical cases it is undesirable to overdrive the line (using a transistor wider than the matched size). If the line is overdriven, overshoots occur which degrade performance. Also, to overdrive the line, wider transistors are needed which places a larger capacitive load on the previous stage.

III. Range of Interconnect for Significant Inductance Effects

The two figures of merit in (9) and (20) can be combined into a two sided inequality that determines the range of the length of interconnect in which inductance effects are significant. This condition is given by

$$\frac{t_r}{2\sqrt{LC}} < l < \frac{2}{R}\sqrt{\frac{L}{C}}. \qquad (21)$$

This range depends upon the parasitic impedances of the interconnect per unit length as well as on the rise time of the signal at the input of the CMOS circuit driving the interconnect. In certain cases, this range can be non-existent if the following condition is satisfied,

$$t_r > 4\frac{L}{R}. \qquad (22)$$

In this case, inductance is not important for any length of interconnect. For short lines, the time of flight across the line is too small compared to the transition time of the input signal. As the line becomes longer, the attenuation becomes large enough to make the inductance effects negligible. If the effect of the attenuation comes into play before the effect of the rise time vanishes, the inductance is not important for any length of interconnect

To demonstrate this behavior, AS/X simulations are shown in Fig. 10 for $L=10^{-7}$ *H/cm*, $R=400$ *Ω/cm*, $C=10^{-12}$ *F/cm*, and $t_r=0.25$ *ns*. With these values, (21) reduces to 0.3259 *cm* $< l < 1.58$ *cm*. This region defines the range of l for which an *RC* model is no longer accurate and the interconnect impedance model must include inductance. The response of a 5-section *RC* Π circuit compared to the response of an *RLC* transmission line is shown in Fig. 10. The *RC* model is inaccurate in the range indicated by (21). The limits are not sharp and the further l is out of the range defined by (21), the more accurate the *RC* model becomes. The case, $L=10^{-8}$ *H/cm*, $R=400$ *Ω/cm*, $C=10^{-12}$ *F/cm*, and $t_r=0.25$ *ns*, is shown in Fig. 11. In this case, (21) reduces to 1.25 *cm* $< l < 0.5$ *cm*, which demonstrates that no possible value of l exists for which the inductance effects are significant. The results depicted in Fig. 11 show that the response of an *RC* circuit model is accurate for all l for this set of device and interconnect parameters.

The region where inductance becomes important in terms of the transition time and the length of interconnect is depicted in Fig. 12. Note that as the inductance L increases, the upper bounding line shifts up and the slope of the lower bounding line decreases, thereby increasing the region where inductance is important. The effect of increasing the resistance is to shift the upper bounding line down, thereby decreasing the region where inductance is important. Increasing the capacitance shifts the upper bounding line down and decreases the slope of the lower bounding line. The transition time at which the two lines intersect remains constant at

$$t_r = 4\frac{L}{R}. \qquad (23)$$

Thus, as the capacitance increases, the area where inductance is important is reduced. Also note that this area may be non-existent if (22) is satisfied.

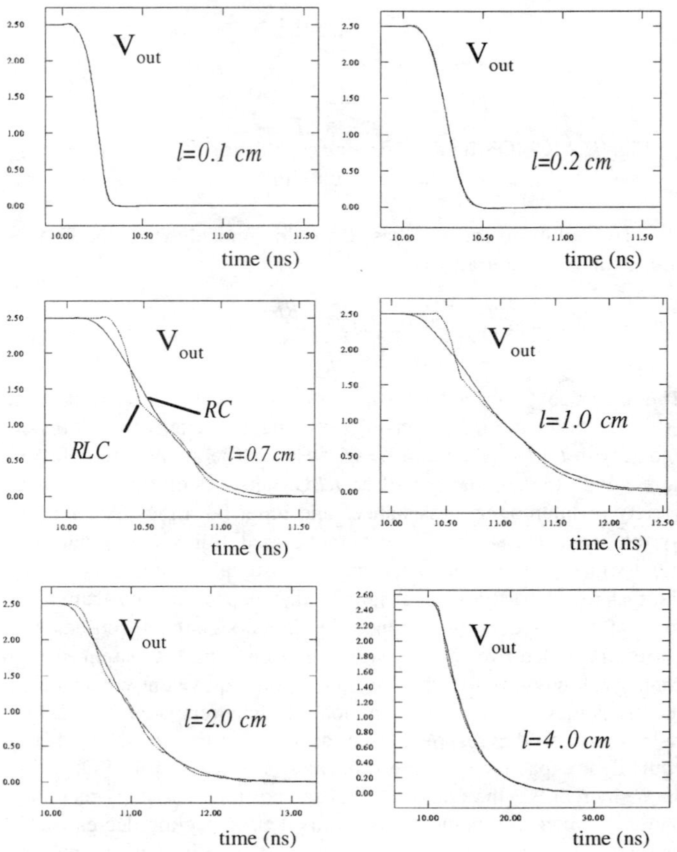

Fig. 10. AS/X simulations of the response of a 5-section *RC* Π model compared to the response of an *RLC* transmission line for different values of l. $L=10^{-7}$ *H/cm*, $R=400$ *Ω/cm*, $C=10^{-12}$ *F/cm*, and $t_r=0.25$ *ns*. The results of the circuit simulation demonstrate that inductance has a significant effect on the response of a signal propagating across an interconnect line for the range of length defined by (21). Note that the *RC* circuit model becomes more accurate for small l or large l.

IV. Conclusions

A closed form solution of the output response of a CMOS inverter driving an *RLC* transmission line is presented using the alpha power law for deep submicrometer technologies. Simple to use figures of merit have been developed that determine the relative accuracy of an *RC* impedance to model on-chip interconnect. The range of length of interconnect where a more accurate transmission line model becomes necessary is shown to be based on the parasitic impedances of the line (*R, L, and C*) and the rise time of the input signal at the gate driving the line. AS/X simulations with a 0.25 μm IBM CMOS technology exhibit good agreement with these figures of merit. These figures of merit can be used in CAD tools to determine which nets needs to be modeled more accurately by including the effects of on-chip inductance. These figures of merit can also be used to properly size the interconnect and buffers along a line during the initial design phase of a high frequency circuit.

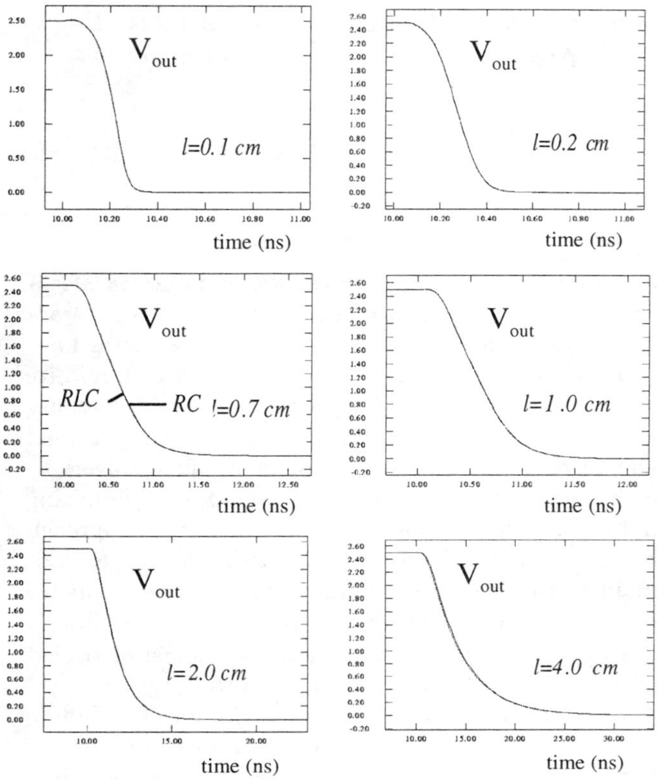

Fig. 11. AS/X simulations of the response of a 5-section *RC* Π model compared to the response of an *RLC* transmission line for different values of *l*. $L=10^{-8}$ *H/cm*, $R=400$ *Ω/cm*, $C=10^{-12}$ *F/cm*, and $t_r=0.25$ *ns*. The results of the circuit simulation demonstrate that inductance has a minimal effect on the response of a signal propagating across an interconnect line despite the length of interconnect as given by (21) for the *R*, *L*, *C*, and t_r values cited above.

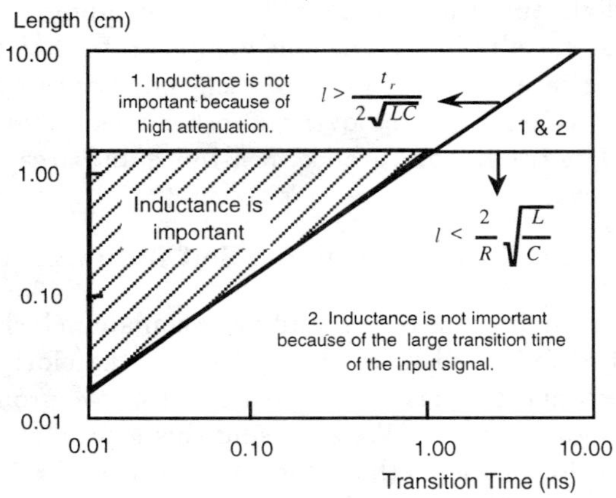

Fig. 12. Transition time (t_r) versus the length of the interconnect line (*l*). The crosshatched area denotes the region where inductance is important. $L=10^{-8}$ *H/cm*, $R=400$ *Ω/cm*, *and* $C=10^{-12}$ *F/cm*.

References

[1] J. M. Rabaey, *Digital Integrated Circuits, A Design Perspective,* Prentice Hall, Inc., New Jersey, 1996.

[2] D. A. Priore, "Inductance on Silicon for Sub-Micron CMOS VLSI," *Proceedings of the IEEE Symposium on VLSI Circuits*, pp. 17-18, May 1993.

[3] D. B. Jarvis, "The Effects of Interconnections on High-Speed Logic Circuits," *IEEE Transactions on Electronic Computers*, Vol. EC-10, No. 4, pp. 476 - 487, October 1963.

[4] M. P. May, A. Taflove, and J. Baron, "FD-TD Modeling of Digital Signal Propagation in 3-D Circuits with Passive and Active Loads," *IEEE Transactions on Microwave Theory and Techniques*, Vol. MTT-42, No. 8, pp. 1514 - 1523, August 1994.

[5] T. Sakurai, "Approximation of Wiring Delay in MOSFET LSI," *IEEE Journal of Solid-State Circuits,* Vol. SC-18, No. 4, pp. 418 - 426, August 1983.

[6] G. Y. Yacoub, H. Pham, M. Ma, and E. G. Friedman, "A System for Critical Path Analysis Based on Back Annotation and Distributed Interconnect Impedance Models," *Microelectronic Journal*, Vol. 18, No. 3, pp. 21 - 30, June 1988.

[7] J. Torres, "Advanced Copper Interconnections for Silicon CMOS Technologies," *Applied Surface Science,* Vol. 91, No. 1, pp. 112 - 123, October 1995.

[8] K. K. Likharev and V. K. Semenov, "RSFQ Logic/Memory Family: A New Josephson-Junction Technology for Sub-Terahertz-Clock Frequency Digital System," *IEEE Transactions on Applied Superconductivity*, Vol. AS-1, No. 1, pp. 3 - 28, March 1991.

[9] T. Sakurai and A. R. Newton "Alpha-Power Law MOSFET Model and its Applications to CMOS Inverter Delay and Other Formulas," *IEEE Journal of Solid-State Circuits*, Vol. SC-25, No. 2, pp. 584 - 593, April 1990.

[10] *AS/X User's Guide,* IBM Corporation, New York, 1996.

[11] L. N. Dworsky, *Modern Transmission Line Theory and Applications,* John Wiley & Sons, Inc., New York, 1979.

[12] Y. Eo and W. R. Eisenstadt, "High-Speed VLSI Interconnect Modeling Based on S-Parameter Measurement," *IEEE Transactions on Components, Hybrids, and Manufacturing Technology*, Vol. CHMT-16, No. 5, pp. 555 - 562, August 1993.

[13] A. Deutsch, *et al.*, "High-Speed Signal Propagation on lossy transmission lines," *IBM Journal of Research and Development*, Vol. 34, No. 4, pp. 601 - 615, July 1990.

[14] A. Deutsch, *et al.,* "Modeling and Characterization of Long Interconnections for High-Performance Microprocessors," *IBM Journal of Research and Development*, Vol. 39, No. 5, pp. 547 - 667, September 1995.

[15] A. Deutsch, *et al.,* "When are transmission-line effects important for on-chip interconnections?," *IEEE Transactions on Microwave Theory and Techniques,* Vol. 45, No. 10, pp. 1836 - 1846, October 1997.

[16] M. Shoji, *High-Speed Digital Circuits,* Addison Wesley, Massachusetts, 1996.

Layout Techniques for Minimizing On-Chip Interconnect Self Inductance

Yehia Massoud
Mass. Institute of Tech.
Dept. of Elect. Eng.
Cambridge, MA

Steve Majors *
Motorola Inc.
Somerset Design Center
Austin, TX

Tareq Bustami
Motorola Inc.
Somerset Design Center
Austin, TX

Jacob White
Mass. Institute of Tech.
Dept of Elect. Eng.
Cambridge, MA

Abstract

Because magnetic effects have a much longer spatial range than electrostatic effects, an interconnect line with large inductance will be sensitive to distant variations in interconnect topology. This long range sensitivity makes it difficult to balance delays in nets like clock trees, so for such nets inductance must be minimized. In this paper we use two- and three-dimensional electromagnetic field solvers to compare dedicated ground planes to a less area-consuming approach, interdigitating the signal line with ground lines. The surprising conclusion is that with very little area penalty, interdigitated ground lines are more effective at minimizing self-inductance than ground planes.

1 Introduction

The potential for inductance to become a factor in the calculation of interconnect delay and signal integrity on deep submicron IC's has been forecast due to well known technology trends [1]. In high performance microprocessor design inductive effects have already begun to influence the design process [2]. With the rate of advancement in process technology continuing and market pressures for functional integration leading to larger chips, inductance may be expected to not only become a factor but a standard second order effect that designers must trade-off with other performance variables in the DSM space.

Today, most extraction and delay analysis tools are limited to RC networks leaving an inherent unpredictability in the design process where inductive effects are suspected. At the same time, typical design practices, based on RC analysis, of widening metal lines and increasing separation on critical nets to minimize delay, cross-talk, and skew are ineffective for reducing inductance. In this paper we use two-dimensional and three-dimensional electromagnetic field solvers to demonstrate a highly effective method for minimizing inductance without increasing die area. This approach can be used to help constrain a design to the RC domain to maintain predictability at some performance cost, or it can be used as a basis for alternative design rules where inductance and capacitance must be traded off to optimize for specific performance targets.

We start, in section two, by examining the inductance of a signal line sandwiched between ground return lines, and show that line spacing can have only a limited impact on self-inductance. In section three we use three-dimensional magnetoquasistatic analysis to show that for integrated circuit interconnect operating at below twenty-five gigahertz, it is the low frequency inductance that predicts performance[3]. In section four we compare the performance of the sandwiched structure, using two dedicated ground planes, and interdigitating thinned signal lines with thinned ground lines. Results in section four demonstrate that the interdigitated approach reduces self-inductance by more than a factor of four over the other techniques, for a modest rise in capacitance, resistance and area.

*Now with Rockwell Semiconductor Systems, Austin, TX.

2 Two-Dimensional Self Inductance

The main focus of our study was coplanar clock trees, and as indicated in Figure (1), where the clock signal interconnect line is sandwiched between two ground return guard traces. We are using this typical structure to explore some methods for minimizing the self inductance, and therefore reduce the clock skew.

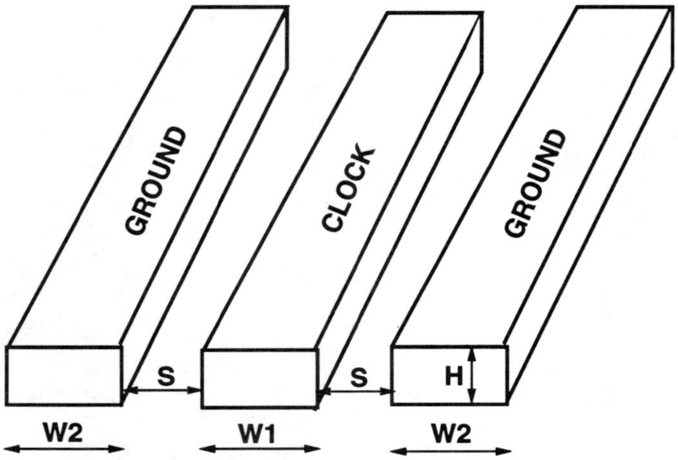

Figure 1: Coplanar Interconnect Clock Example

We used a two-dimensional field solver for the inductance and fixed the width of the clock signal line, W1, and the width of the ground return lines, W2, to 1um. As expected, we found that two-dimensional inductance is proportional to the separation distance between the clock signal line and the ground lines, S, as shown in Figure (2). Thus, in order to minimize the inductance, the separation distance between the clock signal line and the ground return lines, S, should be as small as possible. The two-dimensional inductance is calculated assuming high frequency meaning that fields exist only outside the conductors.

3 Three-Dimensional Self Inductance

It is often presumed, as was in the previous section, that near gigahertz clock rates imply that on-chip inductive effects can be analyzed by determining high frequency limit current distributions. In order to verify this assumption and get the specific frequency at which the conductors behave as perfect conductors, we did a frequency sweep on the inductance of the clock structure using the 3-D field solver FastHenry [4]. FastHenry employs multipole-accelerated Method-of-Moments techniques [5,6]. In Figure (3), we show the frequency dependence of the self inductance of the structure shown in Figure (1), where $W1 = W2 = S = 1\mu$. Note that the 2-D self inductance computed in the previous section is, as expected, the high frequency limit of the 3-D self inductance computed by FastHenry. Figure (3) also shows that for frequencies less than twenty-five gigahertz, it is the low-frequency self inductance not the high frequency self inductance that determines inductive effects. The corner frequency for self-inductance is determined by the

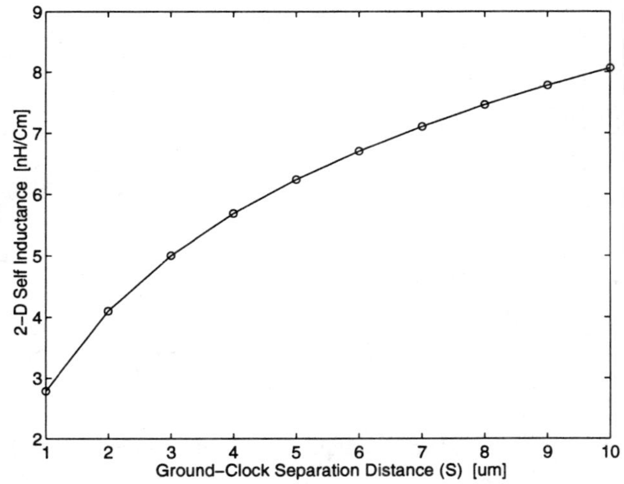

Figure 2: Variation of two-dimensional self inductance with the separation distance between the clock and the ground lines.

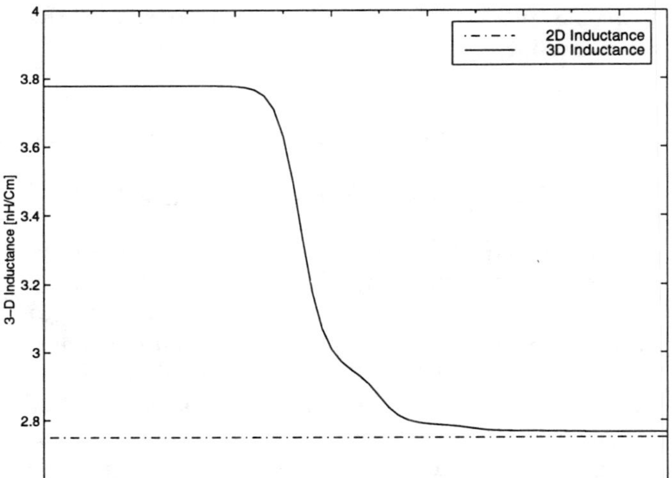

Figure 3: Three-dimensional self inductance frequency dependance for the clock structure in Figure (1). $W1 = W2 = S = 1\mu$.

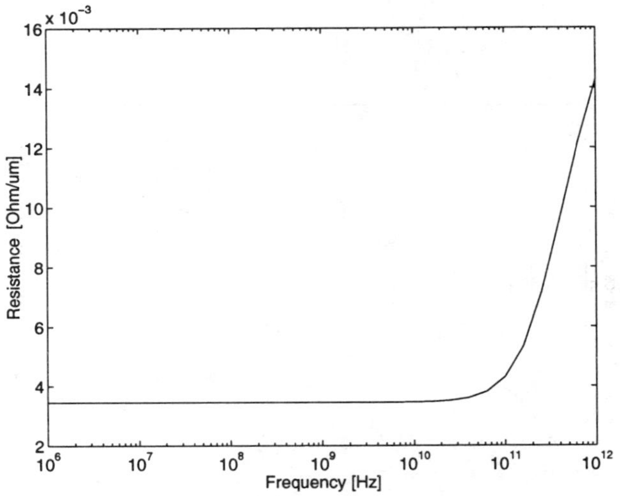

Figure 4: Resistance frequency dependance for the clock structure in Figure (1). $W1 = W2 = S = 1\mu$.

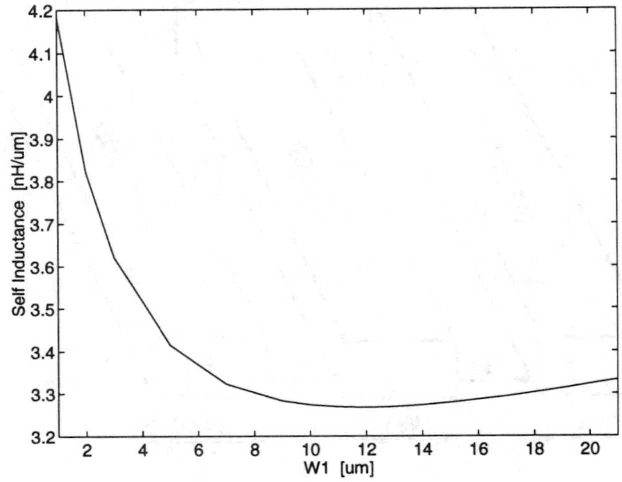

Figure 5: Variation of self inductance with W1 for the clock structure in Figure (1). $W2 = 3\mu, S = 1\mu$.

skin effect. The thinner the conductor is, the higher the corner frequency. Note that, in new technologies, conductor widths are getting to be much smaller than 1μ, and therefore the corner frequency is getting higher. This guarantees that the self inductance of interest is the low-frequency one. Figure (4) shows the frequency dependance of the resistance for the same structure. It shows that structure resistance has the DC resistance value for frequencies below the corner frequency and as frequency increases beyond the corner frequency, the resistance increases due to the skin effect.

4 Reducing Self Inductance

4.1 Optimizing Dimensions of Same Structure

In order to determine the optimum structure that minimizes the self inductance of the clock, we started by keeping the same structure in Figure (1) and trying to optimize its dimensions to reach minimum inductance. As shown in Figure (5), the inductance decreases as the width of the clock signal line, W1, increases, till it reaches a minimum value at $W1 = 12\mu$. After this minimum the inductance increases as W1 increases. The resistance of the clock is always decreasing as W1 is increasing as shown in Figure (6). This curve is not a linear function of W1 due to the constant resistance of the ground return lines, W2. The component of the total resistance from W1 continues to fall off linearly, but the total value saturates at the ground return value. The 3-D capacitance solver FastCap [7] was used to measure the structure capacitance. A floating plane was present to represent the density of a more realistic interconnect environment. Note that without including the floating plane, changes in capacitance between the conductors would be grossly over-estimated. Figure (7) shows that the capacitance is increasing linearly with W1.

Consider the two following structures, the first has $W1 = 3\mu$, $W2 = 3\mu$, and $S = 1\mu$, and the second has $W1 = 12\mu$, $W2 = 3\mu$, and $S = 1\mu$. The second structure has been optimized for minimum inductance given fixed $W2 = 3\mu$, its inductance is 10% less than the first structure. This 10% reduction in the inductance was achieved by using 2.3 times the original space, and 120% increase in the capacitance as shown in Figure (7).

Therefore, techniques for widening the clock to lower resistance, has little impact on the inductance. On the contrary, it might increase the capacitance significantly.

4.2 Using Dedicated Ground Plane Techniques

We also investigated using dedicated ground planes as return paths for the clock signal, as shown in Figure(8). Figure(9) shows the low frequency self inductance as a function of the ground plane width, Wg. Figure (9) also shows that, at around $Wg = 4\mu$, the inductance has a minimum value. After that minimum value, the low frequency self inductance increases monotonically as Wg increases. The current, at low frequency, is uniformly distributed on the ground plane, therefore, big current loops are formed when Wg is large. This increases the inductance. Figure(10) compares the self inductance frequency response of the dedicated ground planes case, with $W1 = H = Sg = 1\mu, Wg = 100\mu$,

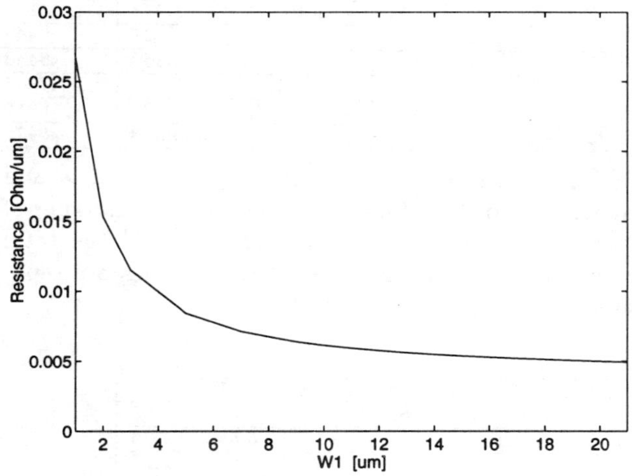

Figure 6: Variation of resistance with W1 for the clock structure in Figure (1). $W2 = 3\mu, S = 1\mu$.

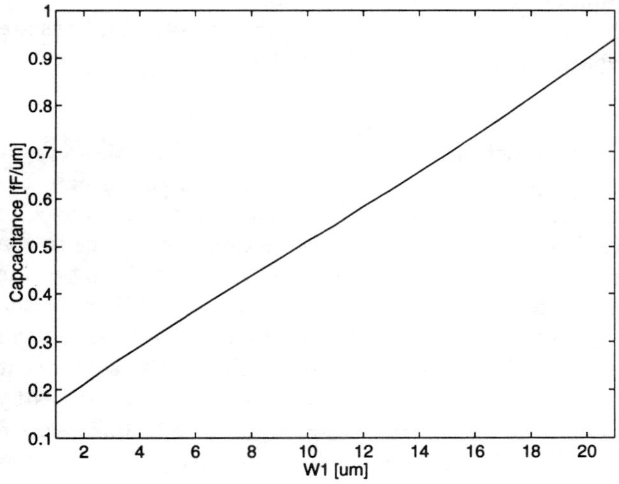

Figure 7: Variation of Capacitance with W1 for the clock structure in Figure (1). $W2 = 3\mu, S = 1\mu$.

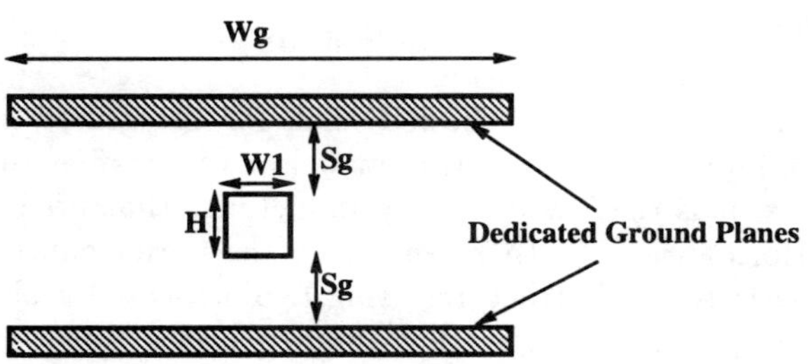

Figure 8: Using dedicated ground planes as return paths for the clock signal.

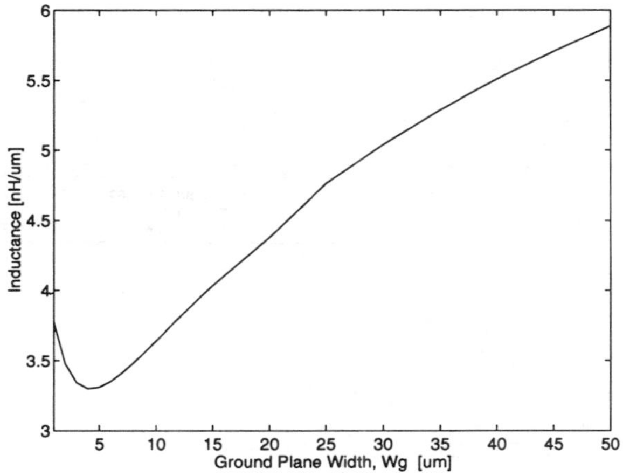

Figure 9: Self inductance for the structure shown in Figure(8). $W1 = H = Sg = 1\mu$.

and the two ground traces case, same as in Figure(3). Figure (10) also shows the frequency response when having both the ground traces and the ground planes as return paths. As shown in Figure (10), using only guard traces has the smallest inductance unless the frequencies of interest exceed several GHz. Above that frequency, dedicated ground planes have somewhat lower inductance, but since most of the energy in the signal is below several GHz, the use of dedicated ground planes is ineffective.

4.3 Using Interdigitated Techniques

As space is always a limiting factor for chip designs, it would be best if the inductance can be significantly reduced with only a limited increase of the total space allocated for the clock structure. In order to achieve that, one might think of distributing the clock signal on many lines, and doing the same for the ground return lines. As shown in Figure (11),

the 10μ signal line has been divided into two 5μ lines. Similarly, the two 3μ ground returns have been exchanged by three 2μ ground returns. This design resulted in no change in resistance, a 27% increase in capacitance, a 43% decrease in inductance, and only an 11% increase in area. The inductance is not reduced by 50%, as might be expected, due to the non-opposing mutual inductances.

We tried different interdigitated structures keeping the total structure width increase to less than 20%. Table (1) shows that a significant reduction of the self inductance of the clock can be achieved increasing the number interconnect lines in the clock structure. Table (2) shows the relative change in the RLC performance, for all structures.

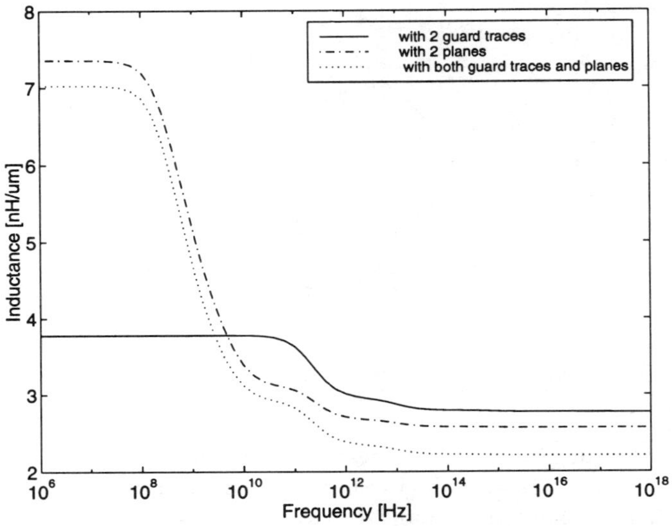

Figure 10: Self inductance Frequency response for the dedicated ground plane case shown in Figure(8), $W1 = H = Sg = 1\mu, Wg = 100\mu$, the guard traces case, as in Figure (3) , and both ground traces and ground plane case.

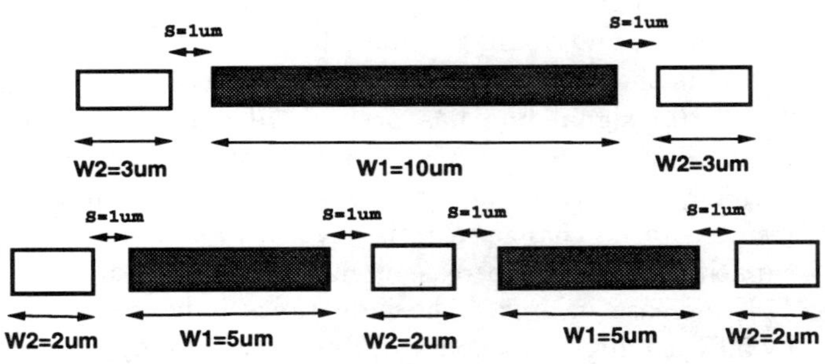

Figure 11: Interdigitated clock structure, using 5 lines instead of 3 lines. Total structure width has been increased from 18μ to 20μ.

N	W1	W2	Wt	R[$m\Omega/\mu$]	L[nH/Cm]	C[fF/μ]
3	10μ	3μ	18μ	R1=6.1	L1=3.273	C1=.5124
5	5μ	1μ	17μ	9.9	1.927	.6449
5	5μ	2μ	20μ	6.1	1.869	.6483
7	3μ	1μ	19μ	8.3	1.336	.7345
9	2μ	1μ	21μ	7.5	1.027	.8215
11	1μ	1μ	21μ	8.4	0.850	.8478

Table 1: Variation of the resistance, inductance and capacitance of the clock structure with the number of interconnect lines in the structure, N, and the total width of the clock structure, Wt. All lines are separated from adjacent lines by 1μ.

N	W1	W2	Wt	R/R1	L/L1	C/C1
3	10μ	3μ	18μ	1.00	1.00	1.00
5	5μ	1μ	17μ	1.62	0.59	1.26
5	5μ	2μ	20μ	1.00	0.57	1.27
7	3μ	1μ	19μ	1.36	0.41	1.43
9	2μ	1μ	21μ	1.23	0.31	1.60
11	1μ	1μ	21μ	1.38	0.26	1.66

Table 2: Relative Variation of the resistance, inductance and capacitance of the clock structure with the number of interconnect lines in the structure, N, and the total width of the clock structure, Wt. All lines are separated from adjacent lines by 1μ.

The resistance or the maximum space allowed for the clock, whichever is more critical in the design, determines the exact width for each line. As shown in Table (1), about the same reduction percentage in the inductance, can be achieved with two different 5 line structures. However, the five line structure with the 17μ wide clock structure has 62% more resistance than the 20μ wide clock structure. Table (2) also shows that the inductance can be reduced as low as 3.9 times by having the clock structure composed of 11 lines, where ground and signal lines are alternatively placed. This significant reduction in the inductance can be achieved with an insignificant increase in the total clock structure width, and clock resistance, and a modest increase in the capacitance.

5 Conclusions

In this paper we showed that for integrated circuit interconnect operating at below twenty-five gigahertz, it is the low frequency inductance that predicts performance. We then compared the performance of the sandwiched structure, using two dedicated ground planes, and interdigitating thinned signal lines with thinned ground lines. Our results demonstrate that the interdigitated approach reduces self-inductance by more than a factor of four over the other techniques, for a modest rise in capacitance, resistance and area.

6 Acknowledgments

The authors would like to thank Steve Weitzel with IBM from Somerset Design Center, Austin, TX. for the valuable discussions and his fruitful suggestions. The authors would also like to thank Matt Kamon and Keith Nabors, the authors of FastHenry and Fastcap, for making their programs publicly available. This work was supported under the DARPA packaging program and by the semiconductor research corporation.

References

[1] L. Pileggi, "Coping with RC(L) Interconnect Design Headaches," Proceedings of the International Conference on Computer-Aided Design, October, 1995.

[2] D.A. Priore, "Inductance on Silicon for Sub-Micron CMOS VLSI," IEEE Symposium on VLSI Circuits, 1993.

[3] Y. Massoud and J. White, "Simulation and Modeling of the Effect of Substrate Conductivity on Coupling Inductance," Proceedings of the International Electron Devices Meeting, Washington, DC, pp. 491-494, December 1995.

[4] M. Kamon, M. Tsuk, and J. White "FASTHENRY: A Multipole-Accelerated 3-D Inductance Extraction Program," IEEE Trans. on MTT, vol. 42, No. 9, pp. 1750-1758, September 1994.

[5] R. Harrington, Field Computations by Moment Methods. New York: MacMillan, 1968.

[6] L. Greengard, The Rapid Evaluation of Potential Fields in Particle Systems. Cambridge, Massachusetts: M.I.T. Press, 1988.

[7] K. Nabors and J. White, "FastCap: A Multipole-Accelerated 3-D Capacitance Extraction Program," IEEE Transactions on Computer-Aided Design, 1447-1459, November 1991.

A Practical Approach to Static Signal Electromigration Analysis

Nagaraj NS[1], Frank Cano[2], Haldun Haznedar[2], Duane Young[2]
{nsnr,f-cano,h-haznedar,d-young}@ti.com

[1] Texas Instruments, Inc.
8505 Forest Ln, MS 8635
Dallas, TX 75251

[2] Texas Instruments, Inc.
P.O. Box 1443, MS 714
Houston, TX 77251

ABSTRACT

Some literature suggests that sweep back effects may make electromigration (EM) a non-issue in signal lines. However this is only the case when the shape of the positive and negative current pulses are closely matched. Moreover, as performance pressures increase, the peak current values are exceeding the range for which electromigration models are valid. Thus, during the design of TI's TMS320c6201 DSP chip, it was determined that limits needed to be placed on the current densities in signal-line segments, and that every net in the design should be checked. Dynamic current density analysis on all nets of a large design is computationally very expensive. In this paper, we describe a practical CAD methodology for a static, signal electromigration analysis for large cell-based designs. We present results and some observations from application of this methodology on the TMS320c6201.

1. INTRODUCTION

For some time it has been known that signal-line electromigration (EM) effects can be significant even in the presence of reverse-recovery effects [10]. Furthermore, as chip clock periods decrease relative to the average signal transition time, the Joule heating of signal lines increases and can drastically reduce the life time. In addition, with higher performance goals designers are creating circuits that exceed the peak current limits for which existing EM models are valid [2,3]. This all leads to the need to validate the EM reliability of signal lines.

With market trends indicating higher levels of system level integration and the associated increasing complexity of designs [8,9], a static approach to resolving the signal-EM analysis problem is needed. Dynamic analysis of current densities on millions of nets in a large design is computationally very expensive and impractical.

This paper presents a design methodology to validate the current density in signal-line segments and vias for all signals in a VLSI design. Moreover, a floor planner based correct by construction methodology is proposed.

This paper is organized as follows: Section 2 briefly discusses EM and its impact on signal lines. Section 3 covers the proposed methodology followed by a description of the flow in section 4. Results from application of the design methodology on the design of the TMS320c6201 DSP core are presented in section 5 followed by future enhancements in section 6 and conclusions in section 7.

2. BACKGROUND

Electromigration is caused by the transport of metal ions through a conductor resulting from the passage of an electric current. Reliability of metal interconnect is commonly described by a lifetime experiment on a set of lines to obtain the *median time to failure* (MTF), whose general expression is given by the Black's formula [1] as follows:

$$\mathrm{MTF} = AJ^{-n}\exp(E_a / kT) \tag{1}$$

where A is a material constant based on the structure and geometric properties of the conductor ($A \approx wt$, where w is the width of the line and t is its thickness), J is the current density, E_a is the activation energy, k is the Boltzmann constant, and T is the temperature in Kelvin. A has dependence on geometry. It has been shown experimentally that the width dependence of MTF is a function of the grain size d of the film and the width of the conductor w. As w/d decreases, the MTF will increase due to the *bamboo effect.* The dependence of MTF on the line-width has significant implications from a design point of view and is usually characterized during the development phase of any new metallization process.

For signal lines, the current in the interconnect alternates as the output of the driver switches. A positive current flows into the lead, whose peak value and duration is determined by the drive strength of the pull-up transistor and impedance characteristics of the load. Similarly, for negative excursions of the output. If the current pulse through the interconnect is bi-directional (both positive and negative pulses), an annealing effect referred to as *sweep back* will take place. Sweep back can be modeled by a recovery term r in the equation for the average current [10].

$$I_{avg} = \max(I^{+}_{avg}, |I^{-}_{avg}|) - r \cdot \min(I^{+}_{avg}, |I^{-}_{avg}|) \qquad (2)$$

The value of r must be determined from experimental analysis of a given process and is typically between 0 and 1. I^{+}_{avg} can differ in magnitude from I^{-}_{avg}, yielding I_{avg} that is positive.

Short-time failure of metal interconnect caused by current pulses [2,3] necessitates peak current density measurements for EM analysis. In addition, the exponential dependency of MTF on temperature necessitates limiting the maximum Joule heating that may occur due to the current flowing in a signal line. The temperature rise is proportional to the RMS current, and hence by limiting the RMS current one can limit the temperature rise.

Thus, the EM guidelines implemented and used in the initial version of this flow were written in terms of limits for average, RMS and peak currents. For bi-directional pulses, J_{rms}, J_{avg} and J_{peak} for a particular metal system must be less than their maximum permitted values $J_{rms,max}$, $J_{avg,max}$ and $J_{peak,max}$ respectively.

3. METHODOLOGY DEVELOPMENT

This section describes the overall methodology of the proposed signal-EM analysis flow and discusses some of the observations made during its development. Electromigration of signal lines through dynamic simulation of the full design is computationally expensive since it is difficult to activate all the signals for worst-case transitions. A static approach with good heuristics is more promising to ensure compliance of all the signals in a reasonable amount of time.

3.1 Current Density Verification

The first step in the development process was to create a current density verification program that implemented the EM guidelines for the fabrication process. The inputs to the verifier were the calculated current density limits of the specific technology for each layer and via type, the maximum device cycle time, the simulated current waveform from SPICE and the simulation cycle time. The output was the required width and via count for each layer and via type in the technology. Subsequently, the verifier was upgraded to accept the actual route segment widths corresponding to each current waveform and identify cases where the width was not sufficient.

The current density checks implemented were based on the effects mentioned in the introduction: 1) average current limit with recovery; 2) RMS current limit; and 3) peak current limit.

3.2 Driver Modeling: Analytical vs. Simulation

The first attempt to model the driver interconnect was to extend the work described in [4]. This approach was based on modeling each CMOS cell by a linear resistance and applying a Driving Point Admittance (DPA) model of the interconnect at its output. Though this approach was performing reasonably well on average and RMS current densities, it failed to accurately model the peak current. Hence, a transistor level simulation approach was adopted. This was easily implemented by extending the existing SPICE-based cell characterization environment. The resulting system excites all possible input vectors and runs SPICE to obtain current waveforms and post-process these waveforms using the verifier described in Section 3.1.

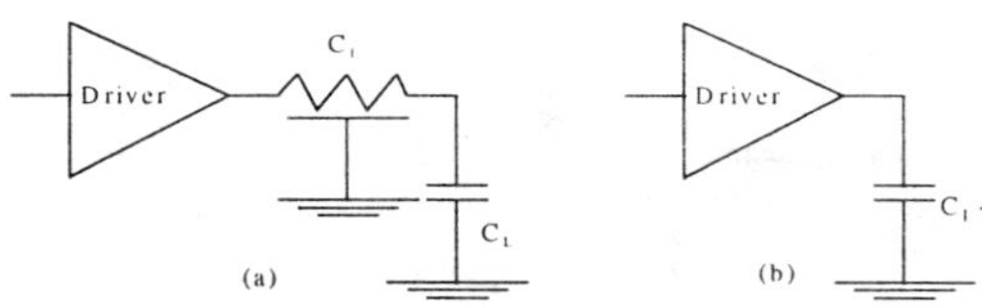

Figure 1. Distributed-RC and Lumped C models

3.3 Load Model: Lumped Capacitance vs. Distributed Capacitance

The next step in flow development was to choose a proper model for the interconnect load. A simulation was setup to determine how well a lumped capacitance would model the three current properties (average, RMS and peak) versus a point-to-point distributed RC network. The circuits in Figure 1 were used for this comparison. In both circuits the current supplied by the driver was monitored. Reasonable transistor parameters and parasitic parameters were selected consistent with a 1um metal pitch/.25um gate length CMOS process.

Figure 2 shows the difference in the various current properties at the driver output when the load is modeled as a single lumped capacitor versus modeled as a distributed

RC network. (All parameters in the graphs have been scaled to protect proprietary data). Figure 2(a), as expected, shows that the average current at each simulated length for the two circuits was identical. Thus, lumped capacitance is a very accurate model for average current. Figure 2(b) shows that significant error can be introduced in the RMS current calculation if the resistance of the line is not modeled. Likewise, Figure 2(c) shows an even larger error occurs in peak current when the resistance is not modeled. From this data, it was concluded that the full RC effects of the interconnect must be modeled to get accurate RMS and peak currents.

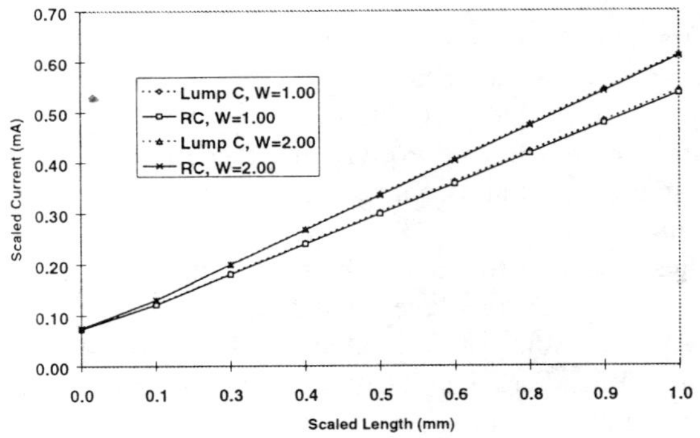

Fig 2(a)

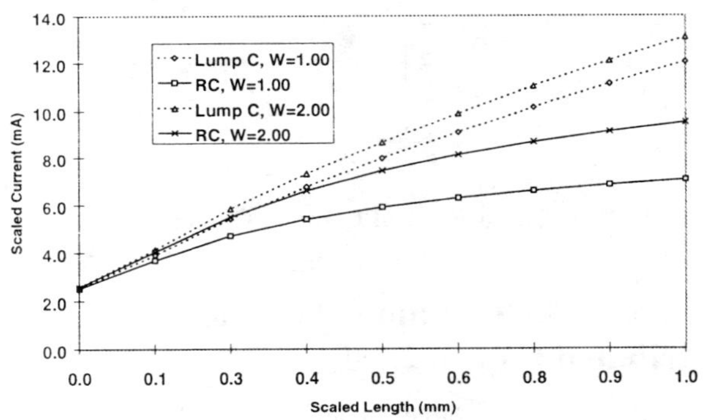

Fig 2(b)

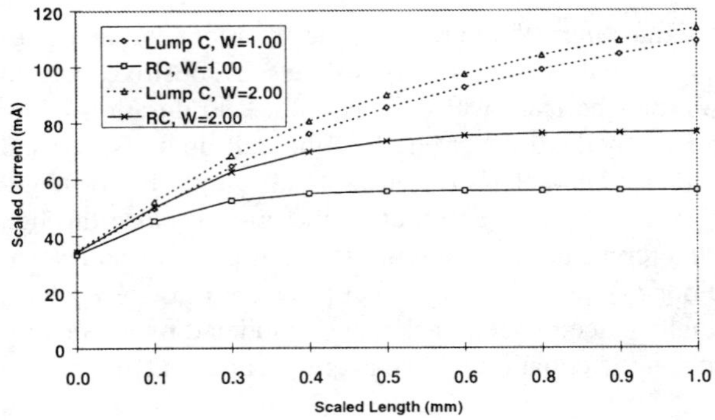

Fig 2(c)

Figure 2. Average (a), RMS (b) and peak (c) currents at the driver as a function of interconnect length for a point-to-point connection. Comparison of lump capacitance model to distributed interconnect model. Two typical route widths are shown.

Figures 3 and 4 are included here for additional insight into the behavior of RMS and peak current as a function of the end point load and route length. These plots reemphasize the effect of resistance in improving the EM performance of a design. Of particular note is that the peak and RMS currents will decrease with increase in interconnect length. This is due to the resistive shielding that the increase in length provides between the driver and the end point load.

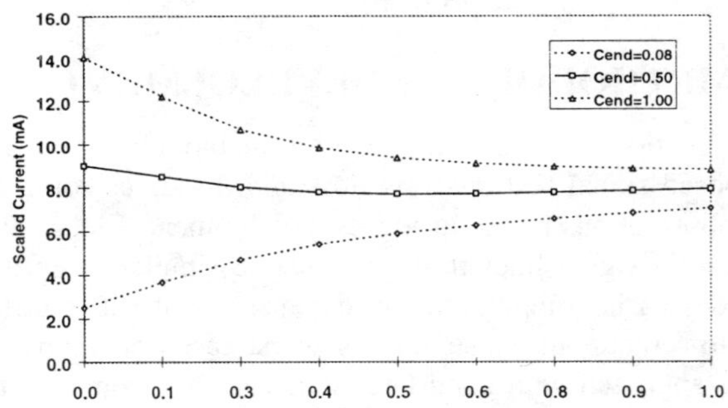

Figure 3. End point load effects on RMS current at the driver as a function of point-to-point interconnect length.

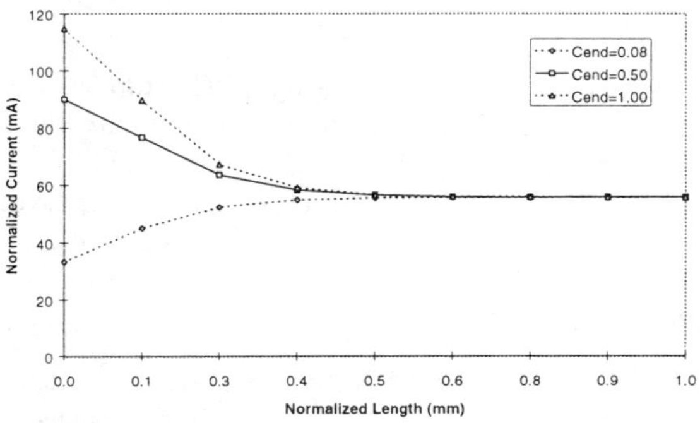

Figure 4. End point load effects on peak current at the driver as a function of point-to-point interconnect length.

3.4 Full RC network or DPA Model

The Driving Point Admittance(DPA) model [5] portrayed in Figure 5 approximates the driving point characteristics of interconnect RCs as a pi-model. Simulations have shown that the DPA model is sufficiently accurate to model average, RMS, and peak current densities for the current process technology and reasonable interconnect lengths.

A capability in RICE [6] can be used to compute the DPA parameters at the driver output. This is done at no extra computational cost since RICE computes the DPA during effective-capacitance calculation phase in interconnect delay analysis. DPA-based analysis helps in accurately finding current densities at the driver output. The DPA model allows for screening of driver circuits to some minimum criteria.

However, use of the DPA model does not fully validate the layout since the model does not retain any layer or width information. In order to properly validate the layout, the full RC network of the signal of interest must be simulated. Later enhancements to the extraction post-processing tools allowed upgrading the flow so that each segment of the signal line that had varying layer or width was checked for compliance with the guidelines.

3.5 Pruning

The above detailed analysis is very time consuming if tens or hundreds of thousands of nets are to be validated. Some means of reducing the set of nets for detailed analysis was needed.

As can be seen in Figures 2 (b) and (c), the lumped capacitance model for the interconnect yields an upper bound on the three current characteristics. Since for this design, the methodology was cell based, it was straight forward to extend the cell characterization environment to generate a maximum load for each driver given a specified minimum route width and via count. The resulting table of maximum safe loads was used along with extracted total capacitance information to determine which instances needed to be fully simulated.

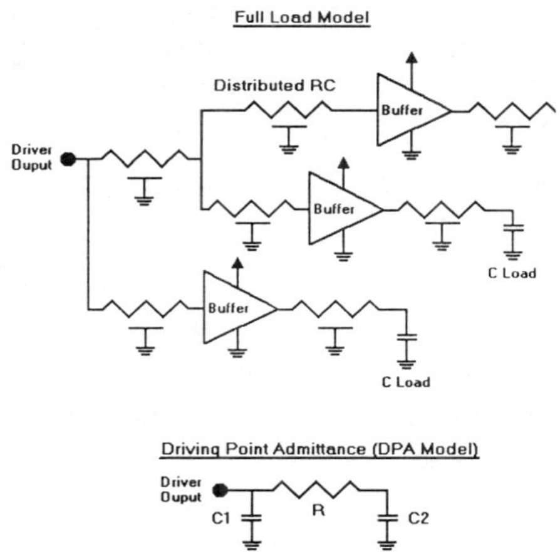

Figure 5. Illustration of DPA modeling at the driver output.

For peak current pruning, the effective capacitance computed during interconnect delay analysis[6] was also investigated for use as a first pass filter. It was found to be an optimistic model for peak current.

3.6 Reducing Pessimism

Even though accounting for resistance of the interconnect helps in eliminating false failures, it still may not be sufficient. It was observed that input transition times can also influence the current characteristics of the output. Assuming worst case input transition times (fastest rise/fall times) can be overly pessimistic. Hence, the actual input rise and fall times as computed during normal interconnect delay analysis can be used for more accurate results. This helped in reducing the number of false violations.

4. DESCRIPTION OF THE CAD FLOW

The CAD flow used to perform the static signal-EM analysis is shown in Figure 6. Key features of this flow are:

- All possible input combinations causing output transitions are exercised to determine the worst-case failures.
- The complete RC network is analyzed at the driver output. Peak, average and RMS current densities are validated for each interconnect layer, via and contact inside the driver.

- Bamboo and non-bamboo lines are identified based on interconnect width and separate current density thresholds are used in determining peak violations.
- For pre-layout analysis, the interconnect can be collapsed to its DPA model. DPA-based analysis could also be used for same-width routes.
- First-pass filter based on lumped capacitance is used to reduce the number of cases that need to be analyzed. Actual input-edge rates are used to reduce the pessimism in current density computation.
- Toggle frequency can be set on a signal-by-signal basis.
- To expedite the execution, simulation and analysis is distributed across the network.
- Incremental analysis can be performed when current density limits for a given technology change, without performing time-intensive simulations.

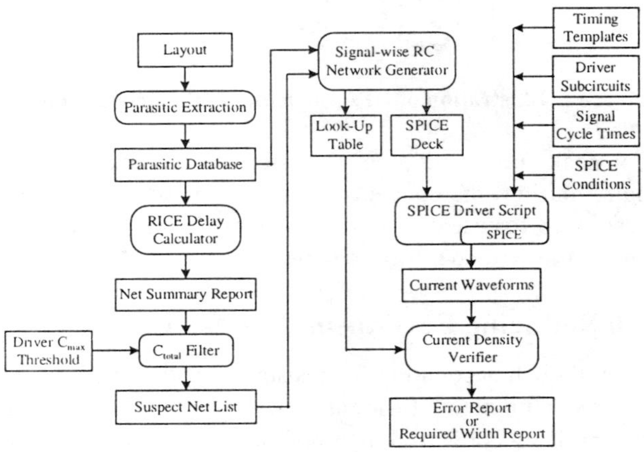

Figure 6. Design flow for signal-EM checker

4.1 Salient Aspects of the Flow

In this section, salient aspects of the CAD flow are discussed. The parasitic extraction includes extraction width and layer information for each segment in the RC network. Total capacitance on the driver output and also input transition times for each instance input pin are obtained during the RC-delay calculation. The total capacitance thresholds are determined for each library cell used in the design. This is a design-independent procedure and is performed only once for a given cell library.

Vectors designed to exhaust all valid combinations of driver input-output transitions are derived from the cell characterization timing templates. The simulation frequency for any signal could be specified in the signal frequency file. Design corners such as transistor models, temperature, supply voltage and metal temperature coefficients are specified by the SPICE conditions file. Tightening of the tolerance values in SPICE resulted in a 10%-15% increase in the peak current values.

During pre-layout stages of the design, RC estimates from the floor planner could be used to obtain minimum widths and via count constraints. These constraints could then be forward annotated to the detail routers to minimize layout iterations.

5. RESULTS

The design flow in its final form was used to validate a high performance DSP design. The design consisted of static CMOS in two layout styles. Half of the layout area was dedicated to standard cell place and route. The other half of the layout was composed of semi-custom cell based datapath blocks. The design was targeted at a 1.0um metal pitch, 0.25um gate length CMOS process.

The flow was setup to be run twice at two interconnect process corners. One corner was with thick-wide metal and the other was with thin-narrow metal. The metal interconnect sheet resistances were set consistent with the metal thickness. As well, the thickness/width assumptions were also taken into account in the calculation of the current density limits for the verification flow. The other conditions of voltage, temperature and transistor strength were chosen to be worst case for EM lifetime.

	Thick-Wide Metal		
Block Type	% Signals Fail Pass1	% Signals w/ Violations	% Pass1 w/ Violtions
Standard Cell	24%	8%	31%
Datapath	8%	3%	65%
Inter-block	35%	1%	4%

	Thin-Narrow Metal		
Block Type	% Signals Fail Pass1	% Signals w/ Violations	% Pass1 w/ Violations
Standard Cell	24%	1%	5%
Datapath	8%	5%	40%
Inter-block	33%	1%	3%

Table 1 Result statistics showing effectiveness of capacitance filter

Table 1 summarizes the results from running the flow on the standard cell, datapath and top level interconnect. The first column indicates the percentage of all signals that failed the total capacitance filter. The second column is the percentage of all signals that had failures on them. The third column is the percentage of signals that failed the total capacitance filter and also had violations on them.

One intriguing observation is that more violations were reported for thick-wide metal than for thin-narrow metal. This was analyzed and determined to be due to the increase in RMS and peak currents when the resistance of the interconnect decreased and the capacitance increased. Similar effects can be seen in Figures 3 and 4 in that for some cases the RMS and peak currents decrease with increasing length (resistance).

For the standard cell portion of the design, the detailed analysis with SPICE took an average of 4.5 min/signal. For the top level route, the average was much longer, at 24 min/signal.

For the current limits selected from the target process, the design that was analyzed contained primarily peak and RMS current violations. In only one case did a circuit need to be modified for an average current violation.

6. FUTURE ENHANCEMENTS

6.1 Additional Filtering

Currently all signals that fail the lumped capacitance threshold are scheduled for full RC simulation. For nets with a large number of elements, the run times of the signal may approach hours. In these cases, the DPA model could be run first, and the width and layer data could be scanned to see if any nets could possibly fail. Only then, if there was an element with a maximum current limit less than the current at the driver output, would the signal be submitted for detailed analysis.

6.2 Analytical Solution

Now that a comprehensive SPICE based benchmark is available, analytical and/or numerical solutions will be explored. Accurate modeling of the driver is needed to address the high degree of non-linearity in peak current characteristics.

7. CONCLUSION

A methodology has been presented for static EM analysis on signal lines in deep submicron designs. Results from application of this methodology on TI's TMS320c6201 DSP chip have been presented. The proposed methodology offers a practical solution to the EM analysis on signal-line segments.

Presently SPICE simulations to obtain current waveforms is distributed across the network to improve throughput. We are looking into accurate driver modeling from a current characteristics viewpoint and faster analytical and/or numerical solutions to obtain better accuracy vs. speed tradeoff with respect to SPICE. In order to get a first-pass clean design, it will be necessary for constraints to be placed on synthesis and physical design tools in future.

8. ACKNOWLEDGEMENTS

We would like to thank Prof. Larry Pileggi for providing us with the RICE software, which has been incorporated in the TI internal parasitics analysis environment known as Parasitic Workbench (PWb). We also would like to thank Tony Leigh, Jim Gallia, Joe McPherson and Larry Ting for providing expert advice and support during this work. Special thanks also go to the key user, Steve Weigand, for detailed feedback.

9. REFERENCES

[1] Black, J. R. "Electromigration Failure Modes in Aluminum Metallization for Semiconductor Devices." Proceedings of IEEE, 57, p. 1587, 1969.

[2] Maiz, J. A. "Characterization of Electromigration under Bidirectional (BX) and Pulsed Unidirectional (PDC) Currents," in Proceedings of 27th Annual International Reliability Physics Symposium, IEEE, 1989, pp, 220-228.

[3] Murguia, J. E., and Brenstien, J. B., "Short-Time Failure of Metal Interconnect Caused by Current Pulses," IEEE Electron Device Letters, p. 481-483 ,1993

[4] Nagaraj NS, "Approximate computation of signal characteristics of on-chip interconnect." Proc. of ISCAS 1994

[5] O'Brien, P., "Modeling of Driving Point Characteristics of Resistive Interconnect for accurate delay estimation.," Proc. of ICCAD 1991, pp 512-515

[6] Ratzlaff, C. L., and Pillage, L. "RICE: Rapid Interconnect Circuit Evaluation Using AWE," IEEE Transactions on Computer Aided Design of Integrated Circuits and Systems. Vol. 13. No. 6. 1994.

[7] Scarpulla, et al. "Reliability of Metal Interconnect After a High Current Pulse," Electron Device Letters, 1996

[8] The National Technology Roadmap for Semiconductors, Semiconductor Industry Association (SIA), 1997

[9] "Design Needs for the 21st Century: White Paper," Semiconductor Research Center (SRC), September, 1994

[10] Ting, L, May, J.S., Hunter, W.R., and McPherson, J.W., "AC Electromigration Characterization and Modeling of Multilayered Interconnections," Proceedings of IRPS, pp. 311-316 1993.

PANEL: DESIGN PRODUCTIVITY: HOW TO MEASURE IT, HOW TO IMPROVE IT

Chair: Carlos Dangelo – LSI Logic Corporation/SRC/MARCO, San Jose, CA
Organizers: Ronald E. Collett – Collett International, Santa Clara, CA
Andrew B. Kahng – University of California, Los Angeles, CA

"Design and Test" inputs and outputs can be viewed as generic **interfaces** between product creators (system designers) and IC manufacturing. The interface is actually a set of engineering **processes** which transform the design inputs (specifications and requirements) into a data set that drives IC manufacturing. There is a gap between advances in manufacturing and advances in design. Hence, the emphasis on design productivity. Design productivity can be viewed as a measure of the efficiency of the resources (people, skills, methodologies, computers, tools, libraries, etc.) when applied to these processes to meet **design input requirements**. Some resources and processes are "hard" and can be readily quantified, e.g., numbers of computers, tools, and people. Others are more elusive, e.g., organization skill set, "quality" of specification code, methodologies, etc. How to measure and improve design productivity is the topic addressed by the panel.

Andy Bechtolsheim
Cisco Systems, San Jose, CA

What is the future direction of deep-submicron system-on-chip design? Will the physical back-end integrate with logic design and synthesis such that system designers can go to a COT business model, or will the new generation of back-end tools benefit primarily the existing ASIC and custom design houses or lead to new business models?

The key objective, of course, is to improve time-to-market and designer productivity. It appears that a closer coupling between logical and physical design is required to take advantage of DSM technology. Many questions remain to be answered.

Ronald E. Collett
Collett International, Inc., Santa Clara, CA

The widely published "design productivity gap" highlights the growing disparity between advancement in IC manufacturing and in design capability. The ability to fully exploit our tremendous silicon fabrication technologies is potentially limited by lack of improvement in design productivity. However, the gap presumes a certain rate of advancement in both chip manufacturing and design capability. We can measure IC manufacturing improvements fairly easily, but the same cannot be said for design productivity. Productivity is simply output divided by labor input. Engineering labor input is relatively straightforward to measure, but what about output? What should be measured - gates, transistors, lines of RTL? The key to accurately measuring design capability is to take into consideration the differences in complexity between different designs. Without an accurate measurement system it is impossible to determine today's true design capability levels and whether we are making adequate improvements. Once we have such a system, we can rely on the measurements and can use them to determine what improvements are necessary.

Jeff Hilbert
LSI Logic, Milpitas, CA

With each new generation of deep-submicron silicon, the design productivity gap continues to widen. Innovation in process technology continues to outpace innovation in EDA and methodology, resulting in silicon capacities that far exceed our ability to fully utilize them under today's time-to-market constraints. Traditional approaches to measuring, and hence improving, design productivity have been hampered by the need to determine the minimal set of key design characterization parameters and by the inherent sensitive nature of the data. Most efforts to date have focused on gate-level analogs to the software development process, i.e., transistors per day, development cycle times, etc. Furthermore, measurement efforts have concentrated on the implementation portion of the product development process, which has also been the traditional focus of EDA technology.

To improve productivity in today's deep-submicron environment, we must shift the focus of our engineering efforts and the measurement thereof in three directions: (1) design for reuse and design reuse; (2) utilization of silicon capacity; and (3) product development methodology. In the area of design for reuse and design reuse, possible metrics are percent of total effort expended to craft reuseable entities and percentage of total gates which are reused. (One would need to accommodate the inevitable customization which seems to occur with reuse.) In the area of silicon capacity, measures could be taken against the maximum performance, density, etc. possible with the technology, i.e., how much silicon does this design leave on the table? And finally, in the area of methodology, a focus will be required on the end product. How "well" did we achieve the end product goals? What tradeoffs were made in the development process, and why? How "well" is the product specification implemented? These new directions will require a new set of tools and methodologies that allow designers to focus on design (the "what", not the "how"), that support the entirety of the product development process, and that deal with granularity well above the gate level. In short, a top-down, end-market product focus is required in an era where the silicon defines the system.

Chris Malachowsky
NVidia Corp., Sunnyvale, CA

Factors that influence design productivity:

Productivity can be: (1) fostered, (2) encouraged as well as discouraged, to a degree, (3) captured in a methodology, (4) staffed for, (5) rewarded/valued, (6) always improved and (7) compared and contrasted.

Productivity is the result of: hard work, attitude/desire, tools and methodologies, results oriented culture, laziness and impatience with the mundane.

Some Keys to Productive Design teams:

People - good skills (the broader the better), good attitude, give them a broad exposure to both low- and high-level goals/problem areas.

Goals - the better understood, the better the drive to accomplish them. Also, a longer goal horizon (i.e., build now for the future) is an encouragement to make productivity investments now rather than later.

Empowerment - everyone is responsible for their deliverables. Dependencies and critical decisions are their responsibility to drive, work around or make on their own.

Attitude - expect the unexpected (tools break - deal with it!); always look for improvement (nothing is perfect); hate dealing with the same problem twice; management has to recognize the value of and encourage automation efforts, methodology criticisms, CAD improvements. Place a high value on results and methodologies to get them. Be willing to invest behind it - reward results not effort.

Productivity measures:

(1) Allow for progress tracking, provide for general inter-project comparisons; (2) highlight areas of needed improvement; (3) direct a team to focus on certain deliverables - pick your measures carefully! (you get what you ask for...); (4) generally apply to only a portion of the design/verification effort at a time - need a whole set of measures to capture overall project productivity other simple duration measures.

Leif Rosqvist
Cadence Spectrum Design, San Jose, CA

Design Productivity encompasses downstream process requirements driven by the semiconductor industry's multi-billion dollar investments in the "Roadmap". Such requirements determine whether your design organization will be a viable player in an ever-increasingly competitive market where the escalating pressure for time-to-market reduction is a consequence of process technology (.35μm → .18μm), cost of fab ($1.5B → 3-4B), design cycle reduction (18mo → 8mo), silicon complexity (250K → 4-6M gates) including IP content, and the convergence of application (cellular, set top box, Internet applications).

Independent design services providers like Cadence have traditionally been operating under "time and materials" contracts that shelter them from this performance pressure. However, these contracts are rapidly being replaced by a "fixed fee for services" type of contract, which forces design services organizations to face the "design productivity" issue head-on in order to stay competitive.

This leads to significant investments in the development of world-class design processes and flows to be used in executing customer projects, and in meeting the ever-increasing demand for faster product development. A very good and well-published success story in this context is the "Motorola 7-Day CSIC Design Process", which led to a 7X turn-around improvement. This cannot be allowed to be a one-time event, but rather should be an established and continued process of refining the design processes and flows measured against a well-defined set of operating metrics.

To stay on top of the requirements for continued improvements against such operating metrics, we must constantly measure and update the organization's ability to deliver against expectations. Following is an example of a "Design Parameters Table" which illustrates what could be measured in a well-running Digital IC Design operation:

Description	Definition	Digital IC Design	Unit	Value
Design parameters	Baseline design	Design characteristics	gates, area, MHz, microns, etc.	-
Quality	Productivity	Design productivity	gate/person-month	-
		Verification productivity	cycles/day	-
Time	Cycle time	Total cycle time	calendar weeks	-
	Iteration time for all tools	Tool step time	hours	-

The actual set of values in this "Design Parameters Table" example is not necessarily the most important information; rather, the key is that a well thought-out set of operating metrics are developed, put to use, and constantly measured and updated to indicate whether the organization is making progress or not.

This needs to be combined with significant upgrades to appropriate design skills based on (1) the latest design processes, flows and design tools, (2) a targeted set of end product applications, and (3) the world-class compute and design management environment that is required to allow execution at the highest and most competitive levels.

Jim Thomas
Motorola M-Core Technology Center, Austin, TX

How productive designers can be is always an important question. Productivity for large designs drives design cycle time, the competitive weapon in today's time-to-market driven world. As the silicon capability has grown, the expectation and implementation of higher levels of integration are leading to huge, complex designs that must be completed in the most rapid and productive way to be competitive in the marketplace.

But is there a crisis in productivity? Can the transistors-per-man-day keep up with integration capabilities? Should a new business model emerge, requiring re-deployment of industry resource? Is the sky falling? Claims have always been made that the "next" design will be "too big" and take "too long". At Motorola, the most complex designs – 68000, 68020, 68040, and 60x PPC chips – have always involved teams of about 60-100 engineers and have been completed with increasing levels of complexity and integration. The keys to effectively managing this complexity and integration have always been design reuse, design tools and design hierarchy. The continuous wavefront of more and more productive design methods for complex ICs has enabled us to routinely design and build multi-million transistor chips.

So, then, what is meant by productivity? Measuring productivity by transistors per designer-day has become obfuscated by ASIC design style, re-use factors, and new types of design which make this measure relatively meaningless. Perhaps the real productivity measures are revenue return per designer-day and design cycle time for new and derivative products.

In any case, as systems become more and more complex, the real issue is not design productivity, but rather verification productivity. In the largest complex systems, verification of the design is the largest single task – a task that does not necessarily scale with hierarchy or design style. The challenge of design verification is to confirm both before and after silicon implementation that a design properly meets the required functional, electrical and reliability system requirements. We can put millions and millions of transistors together on a chip – but does it work in the system? This is a daunting task.

Hierarchical Functional Timing Analysis*

Yuji Kukimoto Robert K. Brayton
Department of Electrical Engineering and Computer Sciences
University of California, Berkeley, CA 94720
{kukimoto,brayton}@eecs.berkeley.edu

Abstract

We propose a hierarchical timing analysis technique for combinational circuits under the tightest known sensitization criterion, the XBD0 delay model. Given a hierarchical combinational circuit, a generalized delay model of each leaf module is characterized first. Since this timing characterization step takes into account false paths in each module, the delay model is more accurate than the one obtained by topological analysis. Then topological delay analysis is performed on the circuit composed of generalized gates replacing the leaf modules, where the "gate" delay model is the derived one. As far as the authors know, this is the first result that shows that hierarchical analysis is possible under state-of-the-art tight sensitization criteria. We demonstrate by experimental results that loss of accuracy in using the hierarchical approach is very minimal in practice. The theory developed in this paper also provides a foundation for incremental timing analysis under accurate sensitization criteria.

1 Introduction

A common assumption in functional timing analysis techniques in the literature is that a circuit under analysis must be a flat gate-level circuit without any hierarchy inside. Due to this restriction even if an original circuit has a hierarchy meaningful to designers, it must be flattened completely before analysis. This flattening step increases the size of the circuit passed to an analysis tool significantly. Existing flat analysis techniques are applicable only up to circuits of thousands of gates and are not scalable. Another drawback of flat analysis is that even a small change in a single module forces us to repeat an entire analysis from scratch. Thus no incremental analysis technique is known for tight sensitization criteria.

In topological timing analysis, hierarchical approaches have been used extensively in practice to manage the complexity of industrial circuits. The reason why hierarchical analysis is prevalent is that the delay of a module is completely independent of a surrounding environment since all paths are assumed to propagate a signal event. This assumption makes hierarchical analysis trivial. However, false paths are completely ignored in this approach, thereby making more accurate analysis difficult. Recently Belkhale and Suess [1] proposed topological timing analysis under known false subgraphs. They assume that designers give the correct information on false subgraphs, which is then used to perform analysis more accurately. The falsity of a subgraph, however, is in many cases relative to arrival times at primary inputs. Characterizing *manually* the correct condition of arrival times at the primary inputs that the falsity of a subgraph is guaranteed is error-prone. Our approach can be thought of as a way of automating this process.

Yalcin and Hayes recently proposed in [9] a hierarchical timing analysis technique using conditional delays. Even if their algorithm for constructing a delay model of a combinational module, called *tagged-mode analysis*, is correct for sensitization criteria *independent* of arrival times (e.g. static co-sensitization [3]), it is not applicable to tighter criteria like floating mode [2] or viability [5]. The main problem is that they construct a delay model assuming that all the inputs of a module arrive simultaneously and use it in different arrival time conditions[1]. Since this particular arrival time scenario may not exhibit the worst-case delay of the module in tight sensitization criteria, the resulting timing model may underapproximate true delays if it is used under different arrival times [2]. It is, however, difficult to see what is the worst-case arrival behavior before analysis. These points were not mentioned in [9]. Their experimental results on hierarchical analysis in fact are based on static sensitization, which is known to underapproximate delays also.

In this paper we propose a hierarchical timing analysis technique for combinational circuits under the XBD0 delay model [6][3]. This delay model is the underlying model for state-of-the-art sensitization criteria like floating mode [2] and viability [5]. We show that a conservative yet relatively accurate delay characterization of combinational modules is possible by using our recent result on "exact" required time analysis [4]. Delay models for leaf modules computed using this method can capture false path effects (as characterized by the tight XBD0 sensitization) inside modules, thus are guaranteed to be more accurate than those constructed by topological analysis. Using these derived delay models, we can then perform topological delay analysis on the graph (DAG) where each node is a leaf module. We prove that the result is a conservative approximation to the delay obtained by flat analysis.

The paper is organized as follows. Section 2 summarizes exact required time analysis [4], which forms the basis of delay characterization of combinational modules. Section 3 discusses how analysis can be done in two steps: 1) delay model construction of leaf modules and 2) top-level delay analysis. A simple example of hierarchical analysis is shown in Section 4. Based on the theory developed so far Section 5 introduces a more efficient timing analysis algorithm where delay model construction is performed in a demand-driven fashion. Experimental results of this improved algorithm are reported in Section 6. Section 7 concludes the paper.

2 Preliminaries

We overview exact required time analysis for combinational circuits proposed in [4]. This technique will be used in characterizing delays of combinational modules.

The problem studied in [4] is: given a gate-level combinational circuit and required times at primary outputs, when does each primary input need to be stabilized? The simplest conservative solution to this problem is to perform delay analysis backwards from primary outputs to primary inputs by propagating required times *topologically* without taking into account the functionality of the circuit. Every path in the circuit is assumed to propagate a signal event here. Since this analysis gives conservative required times in runtime linear in the circuit size, it is widely used in timing opti-

*This work is supported by SRC-98-DC-324.

DAC 98, San Francisco, California

[1]Given a different arrival time condition, one can choose the latest arrival time among the inputs and conservatively assume that all the other inputs arrive at that time. This way the delay model computed in [9] can be safely used without any underapproximation. However this is not the way they used the delay model. Even if this approach is taken, it is difficult to maintain accuracy since aligning all arrival times as in the above can make the analysis too conservative.

[2]A small example which exhibits this problem is available from the authors.

[3]Although stated for combinational circuits, the methods clearly apply to sequential circuits with edge triggered latches.

mization to identify critical paths. Although topological analysis suffices in many cases, more accurate analysis is often required in the design of high-performance circuits. [4] shows that such elaborate analysis is possible by considering false paths in a circuit. This new required time analysis leads to looser timing requirements at primary inputs, which can then relax the timing constraint of the circuit that drives the inputs. In the following, assume for simplicity that a combinational circuit under analysis has a single output.

The basic idea of [4] is that given a combinational circuit and a required time at a primary output, the stability of the output at the required time can be characterized by a function of the stability of primary inputs at potential required times. To make the output available by the required time, the stability of the output must be guaranteed for all possible primary input vectors. This in turn constrains the stability of primary inputs. By extracting the loosest condition at primary inputs that meets the above condition, the exact required times at primary inputs can be computed. The exact analysis gives required times of the primary inputs for each input vector. Notice that the notion of required times is extended from a single value to a relation between input vectors and required times at primary inputs.

Instead of illustrating the theory in detail, we will only discuss important results from [4] to clarify the basic concept since the analysis technique itself is irrelevant to this paper. Interested readers should refer to [4].

Consider a single-output combinational circuit. Let $X = \{x_1, \ldots, x_n\}$ be the set of primary inputs of the circuit and z be the primary output. Assume that a required time $t = 0$ is asserted at z. If topological analysis is performed, the required time at x_i is $-l_i$, where l_i is the longest topological path delay from x_i to z. Note that the required time of each input is completely independent of input values.

The technique developed in [4] computes a more sophisticated input-vector-dependent constraint on required times, which is guaranteed to be looser than the one obtained by topological analysis. The required times at the primary inputs are represented in terms of a relation $T^{exact} \subseteq B^n \times R^n$, where the first component corresponds to values of the primary inputs and the second one to the required times at the primary inputs. R denotes the set of real values. The interpretation of the relation is as follows. For each input vector $x \in B^n$, the relation specifies a set of n-tuples of real numbers. Each tuple is an ordered list of required times, where the i-th element corresponds to the required time of x_i; i.e., each tuple represents a valid timing constraint at the primary inputs under the vector x. There may exist more than one tuple specified for a single vector. The interpretation of this is that any one of those tuples meets the constraint. This situation happens when timing requirements are incomparable. Two timing tuples $s, t \in R^n$ are said to be comparable if either $\forall i, 1 \leq i \leq n : s_i \leq t_i$ or $\forall i, 1 \leq i \leq n : s_i \geq t_i$ holds. In the first case since t is a looser condition than s, we can safely drop the tuple s without losing accuracy. A symmetric argument holds for the second case. The relation only keeps the loosest conditions for each vector by removing all dominated constraints. Two timing tuples are called *incomparable* if they are not comparable. Consider a circuit with two inputs x_1 and x_2, whose functionality $x_1 x_2$ is implemented as a single AND gate. Assume the unit delay model. If $(x_1, x_2) = (0, 0)$, it is enough to have either of the inputs by time $t = -1$. This can be captured by two tuples $(-1, \infty)$, $(\infty, -1)$, which are incomparable. ∞ in a tuple means that the stability of the corresponding input is not even required.

Since the exact characterization is only practical for small circuits due to its inherent complexity, two approximate approaches were proposed in [4]. The main idea is to partially ignore interactions between inputs to simplify the analysis. The required times at the primary inputs computed in this approach are characterized as a set of timing tuples $T^{approx} \subseteq R^n$. The interpretation of T^{approx} is exactly the same as that of sets of timing tuples in the exact analysis. The only difference between the two methods is that T^{approx} is independent of input vectors. Conceptually T^{approx} can be obtained from the relation T^{exact} by computing a set of maximum required times valid for all vectors although this is not how it is computed. Since for each vector there may be several choices, the maximum over all vectors is not unique in general. [4] shows that an approximate algorithm based on repeated functional timing analysis can handle large circuits and that required times looser than topological analysis are obtained in many examples.

In this paper we use the approximate technique to simplify the illustration. How to adapt the entire methodology to exact required time analysis is omitted due to page limitation.

3 Hierarchical Timing Analysis

We consider the case where a given combinational circuit is described as connections of subcircuits, each of which has no hierarchy inside. In other words, the hierarchy depth of the circuit is 1[4]. We call such a subcircuit without hierarchy a *leaf module*. No glue logic is in the top level for the sake of simplicity. We also assume that a topological order of subcircuits exists, i.e. there is no path from an output of a subcircuit to an input of the same subcircuit[5]. Hierarchical timing analysis is performed in two steps. The first step constructs the delay model of each leaf module by a direct application of the required time analysis algorithm described in Section 2. Arrival times at subcircuit boundaries are then determined in a topological order from primary inputs to primary outputs using the delay models.

3.1 Timing Characterization of Leaf Modules

Given a hierarchical circuit, we first analyze each leaf module to characterize its timing property. If there is more than one instance of the same leaf module used in the circuit, it is analyzed once[6]. Let $\mathcal{M}$ be a leaf module. Let $X = (x_1, \ldots, x_n)$ and $Z = (z_1, \ldots, z_m)$ be the primary inputs and the primary outputs of $\mathcal{M}$ respectively. Consider a delay from the primary inputs X to a primary output, say $z_i \in Z$. The standard way to define the delay of a module is by computing the earliest stable time of each output given arrival times at all the primary inputs. The difference between the output stable time and the arrival time of each input gives the delay from the input to the output. This approach, however, is not applicable to our setting since we do not know when the primary inputs of a module in a hierarchy arrive. Our goal is to capture the timing characteristics of a given module valid under *any* surrounding environment. To achieve this the delay of a module is defined in a different way. We first set a required time, say $t = 0$, to the output and analyze the given circuit to see when the primary inputs are required so that the output becomes stable by the required time. The delay from an input to the output is then defined as the difference between the required time at the output ($t = 0$) and the derived required time of the input. This is exactly the same problem as the one discussed in Section 2.

Assume that the approximate required time analysis algorithm is used for the sake of simplicity. As seen in Section 2, the approximate analysis gives a set of timing tuples $T_{z_i}^{approx} \subseteq R^n$. Each n-tuple $t = (t_1, \ldots, t_n) \in T_{z_i}^{approx}$ is a set of valid required times at the inputs. The interpretation of a tuple t is that the output is guaranteed

[4] The analysis described here can be extended to circuits with multi-level hierarchies.

[5] Even if there does not exist a topological order among subcircuits, subcircuit outputs can be always ordered topologically unless combinational cycles exist. Therefore the general case can be handled in a similar way.

[6] If a load-dependent delay model is used, timing characterization must be done for each load at an output of the module. Even under a load-independent delay model, timing characterization can be done for each instance so that the SDC/ODC at the inputs of the instance is taken care of. This yields a more accurate customized timing model.

to be stable at $t = 0$ if the primary inputs $X = (x_1, \ldots, x_n)$ arrive at or before $t = (t_1, \ldots, t_n)$ respectively[7]. Remember that $T_{z_i}^{approx}$ may contain more than one timing tuple, in which case each of the timing tuples captures a different permissible arriving behavior at the primary inputs. This flexibility will be exploited fully later. The delay model[8] T_{z_i} of $\mathcal{M}$ is then defined as

$$T_{z_i} = \{(-t_1, \ldots, -t_n) \mid t = (t_1, \ldots, t_n) \in T^{approx}\}.$$

So far we have considered the delay from the primary inputs to a single output in $\mathcal{M}$. To compute the timing model of $\mathcal{M}$, one can apply the same analysis above for each output independently. Each output of a leaf module has the corresponding delay model at the end of the timing model construction.

3.2 Hierarchical Delay Computation

Assume that arrival times are given at the primary inputs of the top-level circuit under analysis. The goal of the second step is to compute the earliest signal stable time for each primary output of the top-level circuit.

Let $\mathcal{C}_1 < \mathcal{C}_2 < \ldots < \mathcal{C}_p$ be a topological order of the subcircuits. Delay analysis is performed by visiting subcircuits in this order and determining signal stable times at subcircuit outputs. This guarantees that when subcircuit $\mathcal{C}_i$ is visited, arrival times at the subcircuit inputs are known. We then combine these arrival times and the timing model of the corresponding leaf module to compute stable times at the subcircuit outputs.

The core of this computation is in how arrival times at subcircuit inputs are propagated through a subcircuit. Assume that $\mathcal{C}_i$ is under analysis. Let $X = (x_1, \ldots, x_n)$ and $Z = (z_1, \ldots, z_m)$ be the inputs and the outputs of $\mathcal{C}_i$ respectively. Let $A = (a_1, \ldots, a_n)$ be the arrival times at X. Let T be the timing model for the output z_k. The signal stable time at z_k is computed as:

$$\min_{t_i = (t_{i,1}, \ldots, t_{i,n}) \in T} \max_j (a_j + t_{i,j}).$$

The max operation corresponds to standard topological analysis under a timing tuple t_i in T. The min operation then examines all the timing tuples in T and chooses the *earliest* stable time at the output. The important differences between our analysis and topological analysis are that our timing models are more accurate than topological delay models and can maintain multiple timing tuples for an output to preserve accuracy. This min-max computation is linear in $n|T|$, where n and $|T|$ denote the number of inputs of the subcircuit and the number of timing tuples in T respectively. $|T|$ is typically a small constant like 1 or 2.

Theorem 1 *The above analysis gives a conservative approximation to the true delay of the entire circuit under the XBD0 model.*

3.3 Incremental Timing Analysis

Incremental timing analysis can be easily incorporated into our formulation. Once the delay model of a leaf module is obtained, it remains valid no matter what changes are made in other modules. Therefore a modification of a module only leads to 1) delay characterization of the modified module and 2) top-level analysis. Even without any change in a given circuit it is likely that one is interested in performing timing analysis of the same hierarchical circuit under different arrival time conditions. If flat analysis is employed, each arrival time condition requires a separate analysis while in hierarchical analysis the delay characterization stage can be shared by all the analyses. Thus once the delay models of all modules are computed, it is enough to perform top-level analysis for each case.

4 Example

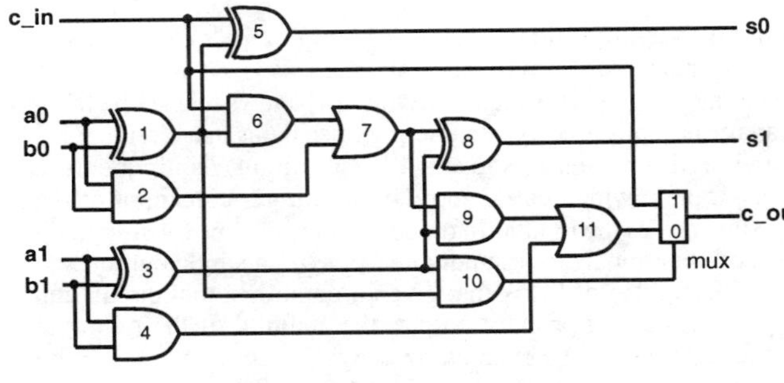

Figure 1: 2-bit Carry-Skip Adder

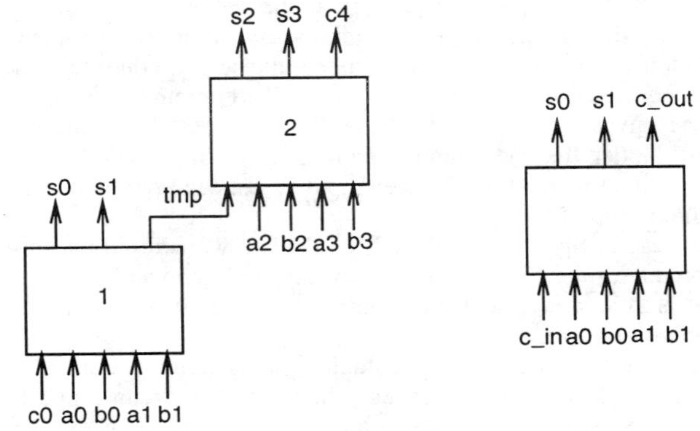

Figure 2: 4-bit Carry-Skip Adder Composed of Two 2-bit Adders

To illustrate the technique proposed in Section 3 we take a simple example.

Consider a 2-bit carry-skip adder [7](page 24) shown in Figure 1. Cascading this adder n times yields a carry-skip adder of $2n$ bits. We show how the performance of this cascade adder can be estimated accurately by hierarchical analysis. Figure 2 shows a 4-bit adder composed of two 2-bit carry-skip adders, where c_{out} of the first adder is fed into c_{in} of the second adder.

Assume that a gate delay of 1 for the AND gate and the OR gate and gate delays of 2 for the XOR gate and the MUX gate.

The approximate required time analysis of the 2-bit carry-skip adder gives the timing models T_{s_0}, T_{s_1} and $T_{c_{out}}$ as follows.

$$\begin{aligned} T_{s_0} &= \{(2, 4, 4, -\infty, -\infty)\} && \text{(topological delay)} \\ T_{s_1} &= \{(4, 6, 6, 4, 4)\} && \text{(topological delay)} \\ T_{c_{out}} &= \{(2, 8, 8, 6, 6)\} \end{aligned}$$

where primary inputs are ordered as $c_{in} < a_0 < b_0 < a_1 < b_1$. In this particular circuit each timing model has a single timing tuple. The timing models for s_0 and s_1 are exactly the same as those under topological analysis. $T_{c_{out}}$ is more accurate than its topological delay model since the delay from c_{in} to c_{out} is 2 in $T_{c_{out}}$ while the longest topological path is of length 6[9].

Let us proceed to the second step of the analysis. Assume that all the primary inputs of the cascade circuit in Figure 2 arrive at $t = 0$. We focus on computing the arrival time of c_4 since it is the most interesting output in terms of analysis. The arrival times for all the other primary outputs are the same as their topological delays. The arrival time at the intermediate signal tmp is first computed. Since all the inputs of the first adder arrive simultaneously at $t = 0$, the delay at tmp is determined as $t = 8$, where a_0 and b_0 are critical.

The arrival time of the carry output c_4 can now be computed by analyzing the second adder. The arrival time at tmp is $t = 8$ while the other inputs arrive at $t = 0$. Combining the arrival times

[7] Since the required time analysis is done in the XBD0 model, required times computed follow the monotone speedup property, i.e. if a signal arrives earlier than the required time specified, it never worsens the stability of the output.

[8] If T^{exact} is used instead of T^{approx}, one can construct the correct conditional delay [9] of the module under the XBD0 model. In general, each output has more than one conditional delay unlike the formulation in [9].

[9] $c_{in} \rightarrow g_6 \rightarrow g_7 \rightarrow g_9 \rightarrow g_{11} \rightarrow mux \rightarrow c_{out}$

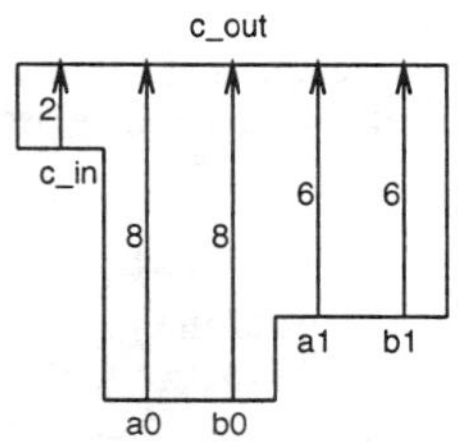

Figure 3: Timing Model for the 2-bit Adder

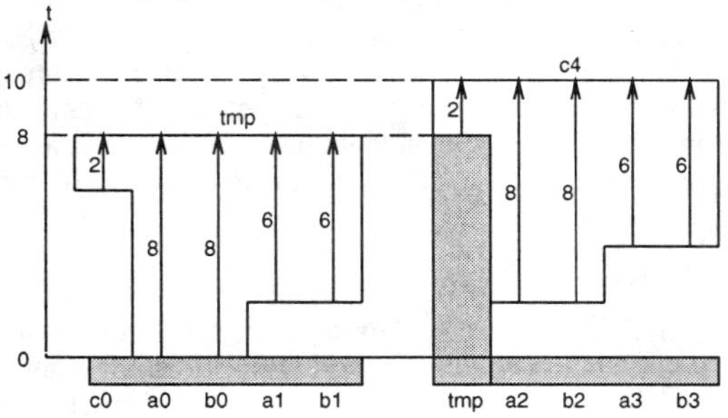

Figure 4: Hierarchical Analysis of the 4-bit Carry-Skip Adder Composed of Two 2-bit Adders

and the timing model $T_{c_{out}}$, we see that a path from tmp is critical. This gives the arrival time at c_4 $t = 8 + 2 = 10$, which matches the result of flat analysis.

It is more intuitive to see the analysis pictorially. Figure 3 illustrates $T_{c_{out}}$ as a polygon. The polygon describes the fact that to have the signal c_{out} stabilized, inputs c_{in}, a_0, b_0, a_1 and b_1 must arrive 2, 8, 8, 6 and 6 time units before respectively[10]. To determine the stable time of tmp, the polygon is pushed down from the top as much as possible, as in Figure 4, so as not to intersect a given arrival time constraint. In this example, all the primary inputs arrive simultaneously at $t = 0$. Therefore the bottom edges for a_0 and b_0 first touch the constraint, which gives the stable time of $t = 8$ at tmp. Shadowed regions represent arrival time constraints. Next, another polygon of the same shape for the second adder is stacked on the first polygon. This time the bottom edge of c_0 first hits the arrival time constraint of tmp. (Note that c_{in} in the module is connected to tmp and all the other inputs of the modules are connected to primary inputs, whose arrival times are $t = 0$.) Therefore, the arrival time of c_4 is 10.

It is easy to see from the analysis above that the delay of the last carry output of the circuit composed of n adders is $t = 8 + 2(n-1) = 2n + 6$ if all the primary inputs arrive at $t = 0$. We have verified that this delay estimation matches the results of the flat analysis at least up to $n = 8$. Parametric analysis like this is not possible with flat analysis.

In [7] the circuit in Figure 1 is analyzed under $arr(c_{in}) = 5, arr(a_0) = arr(b_0) = arr(a_1) = arr(b_1) = 0$. It is easy to see that a_0 and b_0 form a critical path in this case from Figure 5. The delay of c_{out} is $t = 0 + 8 = 8$, which is again the same as the result of flat analysis under the arrival times. From Figure 5 we can even claim that delaying c_{in} by one time unit does not change the signal arrival time at c_{out}, i.e. the slack of c_{in} is 1. This can be thought of as a "real" slack available at c_{in}. Notice that if the slack of this input is computed topologically, it is -3 indicating that the signal needs to be sped up 3 time units to meet the required time $t = 8$ at c_{out}, which is completely opposite to the above. This is because

[10] If there is more than one timing tuple in a timing model, we have multiple polygons. Whenever arrival times are propagated through a subcircuit, all the polygons are tried and the best one that gives the earliest arrival time is chosen.

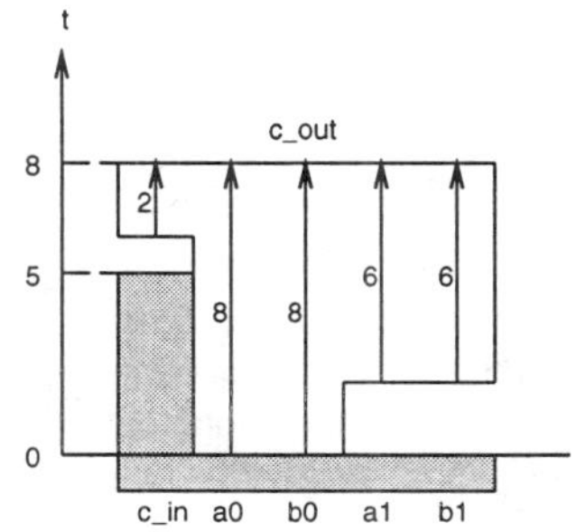

Figure 5: Timing Analysis of the 2-bit Adder under $arr(c_{in}) = 5, arr(\text{others}) = 0$

false paths are completely ignored in the latter case.

5 Improved Algorithm for Hierarchical Timing Analysis

So far we have seen a two-step hierarchical timing analysis algorithm where delay model construction is followed by arrival time computation. Although this separation makes the understanding of the entire flow easy, timing analysis can be performed more efficiently by interleaving the two steps in a smart way.

The main computational disadvantage of the two step approach is that the effective delay of each input-output pair of every leaf module is computed even if the input-output paths are never critical in any of the instances of the module. Therefore part of the additional accuracy achieved in the resulting timing models may not contribute to the accuracy of the estimated delay of the entire circuit. In this sense some of the CPU time spent in timing model construction is wasted. Since delay characterization of leaf modules is not always a cheap operation, this loss in CPU time is not negligible in general.

One way to alleviate this is to start from crude timing models based on topological delay analysis and to refine them gradually as we estimate the delay of a given hierarchical circuit. This way timing models are refined only if additional accuracy is needed to get a better delay estimation of the circuit under analysis. Hence timing models are never more accurate than necessary.

Assume that a circuit of hierarchical depth 1 is given, i.e. the top-level circuit consists of several leaf modules. Consider a directed acyclic graph $G = (V, E)$, called *timing graph*, constructed in the following way. Each input or output of the leaf modules forms a vertex in G. If an output of a module feeds an input of another module, these two nodes share the same vertex. A directed edge is added from a vertex to another vertex if there is a topological path within a single leaf module between the corresponding input-output pair in the circuit. The edge is initially labeled by the delay of the longest topological path between the nodes. The timing graph is simply an abstract timing representation of the hierarchical circuit.

We start by assigning arrival times to each vertex corresponding to a primary input of the top-level circuit. The arrival times are then propagated forward topologically based on the current edge weights. Once the latest stabilizing primary output is determined, its arrival time is asserted at each primary output vertex as a required time. These required times are then propagated backward topologically through the timing graph again using the edge weights. Once an arrival time and a required time are computed at each vertex, the slack at the vertex is defined as the difference between the required time and the arrival time. Any vertex whose slack is zero is on a critical path.

Once slacks are computed at all the vertices, critical edges of the timing graph are identified. A critical edge is an edge in the timing graph which connects two vertices whose slacks are zero. To get a better estimation of the delay of the entire circuit, one needs to have a better delay estimation of such critical edges. Therefore each critical edge is examined one by one to see if the corresponding input-output delay can be improved further by considering false paths. More specifically, the transitive fanin cone of the corresponding leaf module from the output is examined to see if the effective delay from the input to the output is smaller than the topological delay between the two nodes. This can be checked easily by performing false path analysis of the cone as follows.

Let $x_1, \ldots, x_n$ be the inputs of the cone and z be the output of the cone. Let l_i be the longest topological path length from x_i to z. Assume that x_k is the critical input and that the second longest topological path length from x_k to z is $l'_k < l_k$. Now consider the case where input $x_i (i \neq k)$ arrives at $t = -l_i$ and critical input x_k arrives at $t = -l'_k$. If z is still stabilized by $t = 0$ under these arrival times, the effective input-output delay between x_k and z is l'_k. We can then update the weight of the edge from l_k to l'_k and repeat timing analysis of the timing graph to see which edges are critical under this refined delay model[11]. Otherwise the current delay estimation of l_k is accurate and cannot be improved. To indicate this the edge is marked. Notice that the stabilization check at z can be realized by performing false path analysis of the cone under the arrival time condition above.

As we refine weights of critical edges this way, a more accurate delay estimation of the entire circuit is obtained. This iterative process stops once all critical edges are marked.

6 Experimental Results

We have implemented the improved timing analysis algorithm described in Section 5 on top of SIS. CPU time reported in this section was measured on DEC AlphaServer 8400 5/300 and is reported in seconds. In all experiments the unit delay model was used.

Table 1 shows the results of hierarchical timing analysis of various types of carry-skip adders. Each circuit csa$n.m$ is an n-bit adder structured as a cascade connection of n/m m-bit carry-skip adders. The same circuits were also analyzed by flat timing analysis [6] after expanding out all the existing hierarchy to get equivalent flat circuits. In all the circuits, primary inputs were assumed to arrive at $t = 0$. Accuracy of estimated delay was fully preserved in all cases. CPU time saving of hierarchical analysis was significant.

This example of carry-skip adders is suited very well for hierarchical analysis since the circuit structure is regular, i.e. the same leaf module is used repeatedly. Therefore once the accurate delay model of a component carry-skip adder is obtained, it can be used many times to improve the accuracy of the estimated delay of the entire circuit.

It is interesting to see how the algorithm performs on hierarchical circuits with irregular structures. Although ideally this experiment should be done on realistic hierarchical circuits, there is no hierarchical benchmark circuit available. Therefore we created artificial hierarchical circuits from ISCAS combinational circuits in the following way. A benchmark circuit was partitioned into two circuits in a cascade structure so that one circuit drives the other. We then assumed that each component is a leaf module. This way we constructed a simple hierarchical circuit from a benchmark circuit.

Each hierarchical circuit was then analyzed by the algorithm in Section 5. Original benchmark circuits were also analyzed by flat timing analysis to compare accuracy of estimated delays and CPU time. Table 2 summarizes the results of this experiment. We could confirm that accuracy was preserved well in hierarchical analysis although small overestimation occurred on some circuits. In our current approach only local false paths completely contained inside a single leaf module are detected. Therefore global false paths that are false due to the interaction of various leaf modules are overlooked. The fact that accuracy is maintained reasonably well in this experiment indicates that many false paths in real circuits are in fact somewhat local.

Since these circuits are not large enough, flat analysis can finish timing analysis very quickly. Therefore we could not see any speedup in terms of CPU time for the case where estimated delay of hierarchical analysis matches that of flat analysis. In fact, since false path analysis is performed repeatedly on smaller circuits in hierarchical analysis, it takes more time to complete the analysis than flat analysis in many cases. However this result should not be taken negatively. As we have seen, the underlying algorithm for both hierarchical and flat analyses is false path analysis of a combinational block. In hierarchical analysis, false path analysis is performed only on a single leaf module while in flat analysis it is done on an entire circuit. Given that false path analysis can only be applied up to circuits of a certain size, it is clear that hierarchical analysis is more scalable[12].

7 Conclusions

We have proposed a hierarchical timing analysis technique for combinational circuits under state-of-the-art sensitization criteria. The analysis consists of two steps: 1) delay model construction for leaf modules and 2) arrival time propagation using delay models. We have experimentally shown that this approach maintains enough accuracy even with approximate timing models.

The timing characterization step can be used in constructing the timing abstraction of black box modules, e.g. intellectual property blocks [8]. The delay models can be accurate without giving the internal details of black boxes.

References

[1] K. P. Belkhale and A. J. Suess. Timing analysis with known false sub graphs. In *Proceedings of IEEE/ACM International Conference on Computer-Aided Design*, pages 736–740, November 1995.

[2] H.-C. Chen and D. H.-C. Du. Path sensitization in critical path problem. *IEEE Transactions on Computer-Aided Design*, 12(2):196–207, February 1993.

[3] S. Devadas, K. Keutzer, and S. Malik. Computation of floating mode delay in combinational circuits: Theory and algorithms. *IEEE Transactions on Computer-Aided Design*, 12(12):1913–1923, December 1993.

[4] Y. Kukimoto and R. K. Brayton. Exact required time analysis via false path detection. In *Proceedings of 34th ACM/IEEE Design Automation Conference*, pages 220–225, June 1997.

[5] P. C. McGeer and R. K. Brayton. *Integrating Functional and Temporal Domains in Logic Design*. Kluwer Academic Publishers, 1991.

[6] P. C. McGeer, A. Saldanha, R. K. Brayton, and A. Sangiovanni-Vincentelli. Delay models and exact timing analysis. In T. Sasao, editor, *Logic Synthesis and Optimization*, pages 167–189. Kluwer Academic Publishers, 1993.

[7] A. Saldanha. Performance and testability interactions in logic synthesis. Technical Report UCB/ERL M91/100, University of California, Berkeley, October 1991.

[11] If there is more than one instance of the same module in the given circuit, edge weights are updated in all instances.

[12] Another alternative is to perform flat analysis of subcircuits in a topological order. The accuracy loss of this method is the same as that of hierarchical analysis. However, each instance of the same module must be analyzed separately given different arrival times at its inputs. Furthermore incremental analysis capability is very limited since a modification of a subcircuit invalidates all the analyses for its transitive fanout cone whereas in hierarchical timing analysis the delay models of the modules in the transitive fanout cone are still valid.

circuit	topological delay	hierarchical analysis		flat analysis	
		estimated delay	CPU time	estimated delay	CPU time
csa32.4	82	26	0.3	26	7.1
csa64.4	162	42	0.3	42	33.4
csa128.4	322	74	0.3	74	173.1
csa32.8	74	26	1.1	26	6.0
csa64.8	146	34	1.1	34	30.5
csa128.8	290	50	1.1	50	151.5
csa32.16	70	38	5.5	38	3.9
csa64.16	138	42	5.5	42	23.4
csa128.16	274	50	5.5	50	116.2

Table 1: Timing Analysis of Carry-Skip Adders – Hierarchical vs. Flat

circuit	topological delay	hierarchical analysis		flat analysis	
		estimated delay	CPU time	estimated delay	CPU time
C1908	40	38	62.7	37	5.7
C2670	32	31	5.5	30	19.8
C3540	47	46	24.2	46	8.1
C5315	49	47	7.5	47	2.2
C6288	124	124	51.0	123	273.9
C7552	43	42	4.0	42	1.2

Table 2: Timing Analysis of ISCAS Circuits – Hierarchical vs. Flat

[8] S. V. Venkatesh, R. Palermo, M. Mortazavi, and K. A. Sakallah. Timing abstraction of intellectual property blocks. In *Proceedings of Custom Integrated Circuit Conference*, pages 99–102, May 1997.

[9] H. Yalcin and J. P. Hayes. Hierarchical timing analysis using conditional delays. In *Proceedings of IEEE/ACM International Conference on Computer-Aided Design*, pages 371–377, November 1995.

Making Complex Timing Relationships Readable: Presburger Formula Simplification Using Don't Cares

http://www.cs.swt.edu/papers/dac98/

Tod Amon[†], Gaetano Borriello[‡], Jiwen Liu[†]

[†]Department of Computer Science
Southwest Texas State University
San Marcos, TX 78666
tod@cs.swt.edu

[‡]Department of Computer Science and Engineering
University of Washington
Seattle, WA 98195
gaetano@cs.washington.edu

Abstract

Solutions to timing relationship analysis problems are often reported using symbolic variables and inequalities which specify linear relationships between the variables. Complex relationships can be expressed using Presburger formulas which allow Boolean relations to be specified between the inequalities. This paper develops and applies a highly effective simplification approach for Presburger formulas based on logic minimization techniques.

1 Introduction

Many problems in computer-aided design (ranging from physical placement to scheduling in behavioral synthesis) can be formulated using systems of linear inequalities. This is especially the case for problems that deal with temporal information. In the simplest cases, the inequalities must all hold and are derived from the delay ranges in the specification and the constraints (imposed by the user or environment) that must be satisfied for the system to function correctly. Graph algorithms have been effectively applied to determine the satisfiability (the ranges of allowable values for each variable) of these systems of inequalities [7, 13].

In more complex cases, the temporal relationships may involve minimums and maximums which often lead to disjoint solution spaces and disjunctions between the linear constraints (e.g., {`[X,Y] : X <= 30 or Y <= 30 or X + Y <= 70`}) [1, 14]. Presburger formulas – which consist of affine constraints over integer variables, the logical connectives $\neg$, $\wedge$, $\vee$, and the quantifiers $\forall$ and $\exists$, can be used to specify these types of relationships.

Although Presburger formulas have not been widely used due to complexity concerns, they have been effectively used for problems such as dependence analysis for advanced compiler optimizations [9], model checking of infinite state programs [3], and mechanical verification in theorem provers [4]. Some of the more recent uses have arisen because new software that can efficiently manipulate Presburger formulas has been developed (i.e., the Omega libraries [5, 6, 10, 11]).

We have recently been using this software to perform symbolic timing verification of timing diagrams [2]. The promise of symbolic timing verification is that it can be used to derive constraints to be imposed on synthesis tools and/or enable the abstraction of complex timing relationships. Working with symbolic variables can also help humans understand and validate formal specifications of timing behavior.

One problem is that the formulas reported by the Omega tools (and we suspect other libraries which support Presburger formulas) are in some cases not easy to read. This is a problem both for a human designer that may be trying to interpret the result and make design decisions as well as for synthesis tools that may work inefficiently with a large Presburger formula that relates many variables (i.e., delays in the circuit) with each other. In many cases, the results reported by Omega can be simplified. Omega makes use of some important simplification strategies but these are often ineffective even on formulas that a human can simplify in a few minutes. Many formulas require several hours of human manipulation before reaching their simplest form.

In this paper, we report on a technique we have developed to automatically simplify quantifier free Presburger formulas obtained using the Omega libraries (our symbolic timing verification work results in these types of formulas) using techniques from logic minimization that exploit don't care information. Our results to date have been quite promising and we believe argue strongly that future implementations of Omega should support this simplification methodology. Because the formulas can be simplified efficiently, our results further support our belief that Presburger formulas can be effectively used to express and verify timing constraints for a variety of applications.

Section 2 briefly describes how we use Presburger formulas to perform symbolic timing verification and presents examples which help motivate our work. Section 3 describes our methodology for simplifying Presburger formulas and how we derive the crucial don't care information needed for minimization. Section 4 presents our results including examples of the simplifications our tools can accomplish and the execution times required. Section 5 presents our conclusions.

2 Symbolic Verification of Timing Diagrams

We have been using Presburger formulas to perform symbolic verification of timing diagrams, e.g., see Figure 1. Because we include support for symbolic variables, our verification problem consists of determining the requirements that the symbolic variables must meet in order for the timing constraints to be satisfied assuming that the other temporal relationships (e.g., specified propagation delays) are maintained.

The problem description takes the form of a set of integer n-tuples where n is the number of symbolic variables. Within the formula we quantify variables which represent times at which signal transitions take place and use inequalities to express all of the temporal relationships (see [2] for details). We use the Omega libraries to remove the quantifications (using Fourier-Motzkin variable elimination) to obtain a description of the requirements that the symbolic variables must meet in order for the constraints to always be satisfied. Some of these requirements are quite trivial while others represent important discoveries that arise from transforming the specification by removing the quantifications using Omega. Numeric bounds for the symbolic variables may result, or important relationships between the symbolic variables may be exposed. We present two examples.

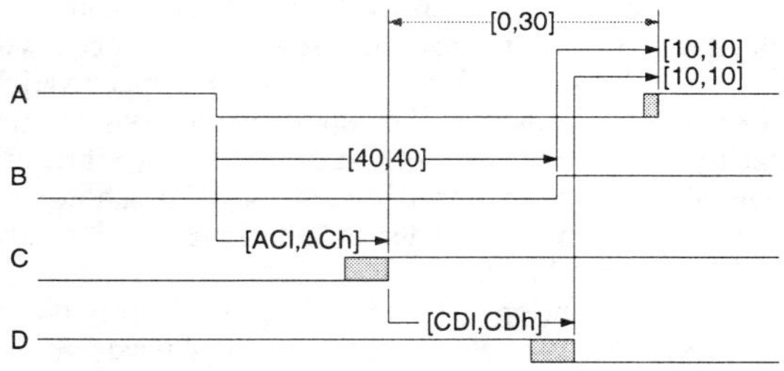

Figure 1

A timing diagram with a single constraint requiring that the second edge on signal A occurs no more than 30 nanoseconds after the first edge on signal C. All other times (in the range of [low,high]) are propagation delays some of which are "unknowns" expressed using symbolic variables. Presburger formulas are used to formally capture the semantics of the timing diagram. The Omega libraries are used to help perform verification.

2.1 A Simple Example

Figure 1 contains a simple example taken from [1] in which CLP(R) and symbolic linear programming are used to perform the verification. Using our methodology the timing diagram is analyzed and the following integer tuple set is obtained:

```
{ [ACl,ACh,CDl,CDh]:
Exists (A_1,A_2,B_1,C_1,D_1:
 0 <= ACl <= ACh and 0 <= CDl <= CDh and A_2 >= A_1 and
    ACl <= (C_1-A_1) <= ACh and
     40 <= (B_1-A_1) <= 40 and
    CDl <= (D_1-C_1) <= CDh and
  max(B_1+10,D_1+10) <= A_2 <= max(B_1+10,D_1+10)) and
Forall (A_1,A_2,B_1,C_1,D_1: not (
 0 <= ACl <= ACh and 0 <= CDl <= CDh and A_2 >= A_1 and
    ACl <= (C_1-A_1) <= ACh and
     40 <= (B_1-A_1) <= 40 and
    CDl <= (D_1-C_1) <= CDh and
 max(B_1+10,D_1+10) <= A_2 <= max(B_1+10,D_1+10)
) or (0 <= (A_2-C_1) <= 30  ))
};
```

Our verification tools transform the above formula into a Presburger formula (by removing the maximizations) and then use the Omega libraries to remove all of the quantified variables (i.e., the events in the timing diagram). The result is a greatly simplified Presburger formula which is obtained using .2 seconds of CPU time and specifies the requirements that the symbolic variables must meet in order to satisfy the constraint (i.e., $0 \leq A_2 - C_1 \leq 30$):

```
{[ACl,ACh,CDl,CDh]: 20 <= ACl <= ACh <= -CDh+39 and
                     0 <= CDl <= CDh} union
{[ACl,ACh,CDl,CDh]:  0 <= CDl <= CDh <= 20 and
                    20 <= ACl <= ACh and
                    40 <= ACh+CDh}
```

We can, however, use Omega to formally prove that this solution is equivalent to a solution obtained by manual inspection:

```
{[ACl,ACh,CDl,CDh]:  0 <= CDl <= CDh <= 20 and
                    20 <= ACl <= ACh}
```

Interestingly enough, we have on occasion found that the Omega libraries can produce this simplified answer. There are some simplification algorithms built into Omega but they are not as general as the one we describe in this paper. Omega always presents solutions as unions of sets containing no disjuncts (i.e., in sum of products form).

2.2 Interface Verification Example

This second example is taken from [15] which contains two specifications for a memory controller and a bus. Here, we consider only the read protocol and the combined specification as shown in Figure 2. Using symbolic variables we can easily perform experiments which might, for example, help us to select another memory controller. We have added a constraint (i.e., ResponseTime) to the specification to ask the question: "Given these values (some of which are symbolic) what will the overall length of the read cycle be?"

Omega reports (using .6 seconds of CPU time to analyze a fairly complex Presburger formula) an answer that is difficult to read. One extension we have added to our tools is the ability to *map* Presburger formulas by introducing *mapped variables* in place of inequalities. Our tool can also replace unions by disjunctions. Using these simple tools, we can report the result in a more readable form as:

```
{[ RRl , RRh , PWl , PWh , RTl , RTh ]: (
( a and  b and c and d and  e and f and g and h ) or
( i and  b and c and h and !f and d and g ) or
( b and !a and c and e and  f and g ) or
( b and  c and j and d and  k and e and  f and g ) or
( i and  b and c and j and  d and e and !f and g ))}
map
a : RTl <= RRl + 160
b : RRl >= 30
c : RRh >= RRl
d : PWl >= 0
e : PWh >= PWl
f : RTl <= 10 + RRl + PWl
g : RTh >= 30 + RRh + PWh
h : PWh >= 150
i : RTl <= RRl + 150
j : RTh >= 180 + RRh
k : PWl <= 150
```

By inspection one can note that many of the inequalities appear multiple times in the result. One of our original intentions was to use a logic minimization tool to support factoring and algebraic manipulation of the resulting formulas. This can help present answers in a more readable form and also help one to manually simplify the formula. Over the course of several days spent working with this resulting formula, we obtained a much simpler formula which Omega verified was identical to the result reported above (the same mapping applies):

```
{[ RRl , RRh , PWl , PWh , RTl , RTh ]:
b and c and d and e and g and j and (f  or ( i and k)) }
```

Automatically obtaining simplified answers such as this one is the purpose of the work reported in this paper. Using simple answers one can more easily and efficiently uncover important and interesting relationships than would be possible with the more complex results reported by Omega.

3 Simplification Methodology

Our simplification methodology is based upon mapping Presburger formulas and then extracting don't care information so that we can better use logic minimization tools such as SIS (Espresso) to perform the actual simplification. Using this technique we can obtain

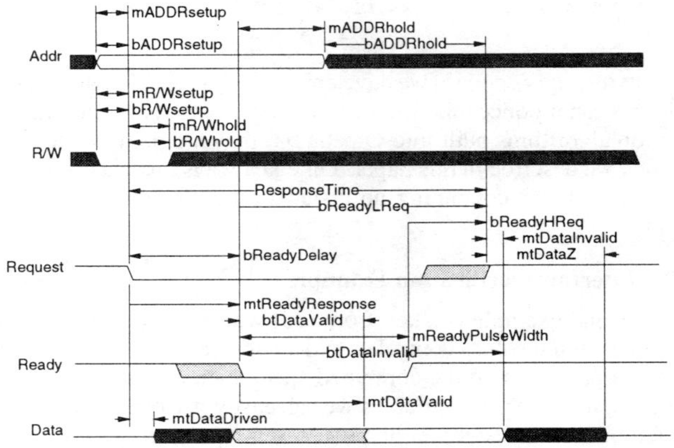

Type	Name	Bounds
constraint	mADDRsetup	[25,]
constraint	mR/Wsetup	[25,]
constraint	mADDRhold	[15,]
delay	mtReadyResponse	[RRl,RRh]
delay	mtDataValid	[40,90]
guarantee	mtDataDriven	[5,]
guarantee	mReadyPulseWidth	[PWl, PWh]
guarantee	mtDataInvalid	[0,]
guarantee	mtDataZ	[0,5]
guarantee	bADDRsetup	[30,]
guarantee	bR/Wsetup	[30,]
guarantee	bADDRhold	[20,]
guarantee	bR/Whold	[20,]
constraint	btDataValid	[,100]
constraint	btDataInvalid	[140,]
constraint	bReadyDelay	[30,]
delay	bReadyLReq	[150,180]
delay	bReadyHReq	[10,30]
constraint	ResponseTime	[RTl,RTh]

Figure 2

Combined timing diagram and tables for the bus and memory controller. The prefixes "b" and "m" indicate whether the relationship came from the bus specification or the memory specification. The response time constraint was added after the diagrams were composed.

not only mimized sum-of-products form but also factored forms that tend to be more readable because the number of literals is minimized.

As a simple example, consider the Presburger formula: { `[X] : X > 20 and X > 30`} which should obviously be simplified and reported as { `[X]: X > 30`}. Omega contains algorithms for eliminating redundant constraints and would in fact report this solution. We map this formula to produce:

```
{ [X]: (a and b) }
map
a : X > 20
b : X > 30
```

and then discover the don't care: $\mathbf{a}'\ \mathbf{b}$ (i.e., **not a and b**) which when included for the minimization of $\mathbf{a}\ \mathbf{b}$ (i.e., $\mathbf{a}\ \mathbf{b}\ +\ \mathbf{a}'\ \mathbf{b}$) produces $\mathbf{b}$ which corresponds to { `[X]: X > 30`}.

The don't cares we discover are reported as a mapped Presburger formula in sum-of-products form (each product is a don't care). Discovering this information is potentially quite time consuming. Our initial efforts have been based upon the rather obvious and rather inefficient method of considering every possible *combination* of mapped variables. If there are M mapped variables and one wishes to consider don't cares containing n mapped variables there are: $\binom{M}{n}$ combinations. Each combination represents 2^n potential don't cares because each mapped variable might appear in a don't care in either negated or non-negated form. Each potential don't care is analyzed using Omega which will report *false* if there is no way to satisfy the conjuncted inequalities (e.g., { `[X]: not (X > 20) and X > 30` }.

We have identified two optimizations we believe are quite important and useful for making this computationally feasible. We also believe other optimizations and improvements to our algorithm are quite likely to be discovered (we intend to investigate a more constructive approach to finding don't cares which would avoid the need to even generate all of the combinations).

One optimization uses previously discovered don't cares to avoid generating don't cares which are of no interest (e.g., if $\mathbf{a}\ \mathbf{b}'$ is a don't care then obviously so is $\mathbf{a}\ \mathbf{b}'\ \mathbf{x}$ where $\mathbf{x}$ is any mapped variable). For the examples we looked at, this optimization does not greatly reduce the time required to extract don't cares because it requires a fair amount of computation time to implement. It does, however, reduce the amount of time needed during logic minimization because fewer don't cares are identified.

Our most important optimization relies on an array that indicates which symbolic variables appear in each mapped variable. If there are M mapped variables and S symbolic variables we construct IN, an $M \times S$ array where:

$$IN[m][s] \quad = \quad \begin{cases} 1 & \text{if mapped variable } m \text{ includes} \\ & \text{symbolic variable } s \text{ in its inequality} \\ 0 & \text{otherwise} \end{cases}$$

This array is consulted for each combination of mapped variables to determine whether or not we need to use this combination to generate potential don't cares. Although we do not implement our algorithm this way, we essentially compute the sum of each row corresponding to a mapped variable in the combination being considered. The resulting vector indicates how many times each symbolic variable appears in the combination.

We do not construct potential don't cares for combinations that have a 1 somewhere in their summation vector. This optimization is based on the observation that if a symbolic variable appears only once then there is only one inequality constraining that variable and thus it is essentially a free variable. Given the presence of a free variable, Omega will not report *false*, unless irrespective of the value of the variable, the result is false (and thus at least one mapped variable is not needed in order to cover this don't care).

Our simplification methodology allows users to specify a range of *don't care levels*, to specify the minimum and maximum number of mapped variables appearing in potential don't cares. Users can adopt a simplification methodology based upon iteratively increasing the maximum don't care level until a reasonable solution is found or compute times grow to be too large.

We have two tools we use to move between mapped Presburger formulas and SIS, a logic optimization tool developed at UC Berkeley [12]. Our tool *eqnextract* takes a mapped Presburger formula as input and produces a `.eqn` file for use by SIS. Our tool *eqninsert* takes a `.eqn` file and a mapped Presburger formula and replaces the formula with that specified by the `.eqn` file.

Using SIS we can read an `.eqn` file and write the result back in the `.blif` format. We have a tool, *blifmerge*, which merges two `.blif` files assuming the second file specifies the onset of the don't cares for the first (i.e., we use the `.exdc` directive). We can then use SIS (Espresso) to read the combined blif file and minimize the function using the don't care information (i.e., using `full_simplify`). If a factored form is desired it can be created and the minimized logic can be written back to an `.eqn` file and then incorporated back into a Presburger formula. Although the underlying minimization metric used by SIS is geared towards reducing the size of the resulting logic we believe it performs a reasonable job of simplifying expressions for human readability.

4 Results

In this section we describe the result of applying our techniques to the examples described in Section 2. Our execution times are quite reasonable especially given that the code we have written is not very efficient and that we are running on a fairly old DEC 3000 workstation with 32 MB of memory. Even more importantly, our algorithms sit on top of Omega. We maintain an internal data structure similar to that used by Omega and a great deal of time is spenting copying between our data structure and the Omega data structure. If our methodology were to be implemented within the Omega libraries vast improvements in execution time should result. Mapping and logic minimization are quite efficient and took less than one second of CPU time for all of the examples reported below.

4.1 Interface Timing Verification

For the mapped formula in Section 2.2 we have:

$$IN = \begin{pmatrix} 1 & 0 & 0 & 0 & 1 & 0 \\ 1 & 0 & 0 & 0 & 0 & 0 \\ 1 & 1 & 0 & 0 & 0 & 0 \\ 0 & 0 & 1 & 0 & 0 & 0 \\ 0 & 0 & 1 & 1 & 0 & 0 \\ 1 & 0 & 1 & 0 & 1 & 0 \\ 0 & 1 & 0 & 1 & 0 & 1 \\ 0 & 0 & 0 & 1 & 0 & 0 \\ 1 & 0 & 0 & 0 & 1 & 0 \\ 0 & 1 & 0 & 0 & 0 & 1 \\ 0 & 0 & 1 & 0 & 0 & 0 \end{pmatrix}$$

The combination **a f k** would result in the summation of rows 1, 6, and 11 and produce:

$$\begin{pmatrix} 2 & 0 & 2 & 0 & 2 & 0 \end{pmatrix}$$

which indicates that symbolic variables `RRl`, `PWl`, and `RTl` each appear twice in the mapped inequalities of **a f k** and that the symbolic variables `RRh`, `PWh`, and `RTh` do not appear at all. Because no symbolic variable appears only once the combination **a f k** would be used to generate potential don't cares. Figure 3 summarizes the results of applying our don't care extraction methodology at various don't care levels. Using the results from don't care level 2, SIS simplified the formula but it still consisted of 5 product terms. Much to our surprise, using the additional results from don't care level 3, SIS reports an even better result than the one we obtained by hand:

```
{[ RRl , RRh , PWl , PWh , RTl , RTh ]:
b and c and d and e and g and j and (f  or i) }
```

4.2 Our Simple Example with More Symbolic Variables

As a test of our methodology we complicated our first example (see Figure 1) by introducing more symbolic variables. Specifically, all numeric values were replaced with symbolic variables including the upper bound (i.e., 30) of our constraint (we kept the lower bound of 0 because we are assuming $A_2 \geq C_1$). Using the Omega libraries to analyze this revised problem (which now contains 11 symbolic variables) we obtained the following solution (after mapping):

```
{[ ACl, ACh, ABl, ABh, CDl, CDh, DAl, DAh, BAl, BAh, CAh ]:
((a and b and c and d and e and f and g
    and h and i and j and k and l) or
  (a and b and c and d and e and f and g
    and h and i and j and m and n) or
  (a and b and c and d and e and f and g
    and h and i and j and !m and l and !k and n))}
map
```

level	analyzed	Omega	time	don't cares
2	2 / 55	8 / 220	0.04	$a'\,i + d'\,k'$
3	8 / 165	62 / 1,320	0.2	$a\,f'\,k' + a'\,f\,k + d'\,f\,i' + e\,h'\,k' + e'\,h\,k + g\,h\,j' + g'\,h'\,j$
4	14 / 330	174 / 5,280	.72	none
5	23 / 462	495 / 14,784	5.12	none
6	35 / 462	1496 / 29,568	17.24	none

Figure 3

Results for the example in Section 2.2 reported for each don't care level. We report the actual number of combinations that needed to be analyzed (based on our optimization) compared to the total number of combinations. We also report the number of Omega calls that were made compared to the number that would have been made if all of the combinations were considered. The CPU time and the don't cares discovered at each level are also reported.

```
a : ACl >= 0
b : ACh >= ACl
c : ABl >= 0
d : ABh >= ABl
e : CDl >= 0
f : CDh >= CDl
g : DAl >= 0
h : DAh >= DAl
i : BAl >= 0
j : BAh >= BAl
k : ACh + CDh < ABl + BAh - DAh
l : ACl >= ABh + BAh - CAh
m : ACl + CDl > ABh + BAh - DAh
n : CDh <= CAh - DAh
```

Don't care analysis at levels 2 and 3 revealed nothing. Don't care analysis at level 4 yielded two don't cares and SIS reduced the answer to two product terms. Don't care analysis at level 5 yielded twelve additional don't cares and using SIS we obtained the following simplified answer:

```
{[ ACl, ACh, ABl, ABh, CDl, CDh, DAl, DAh, BAl, BAh, CAh ]:
(a and b and c and d and e and f and g and
 h and i and j and l and n)}
```

Note that some of the mapped inequalities (i.e., **k** and **m**) disappear as a result of the simplification. The result is much easier to understand, because most of the inequalities are quite trivial (i.e., **a** to **j** simply state that the lower bounds on the delay ranges are greater than zero and that the upper bounds are greater than the lower bounds). The two non-trivial inequalities reflect the essence of the solution: in order for our constraint from C_1 to A_2 to be met we need an upper bound on the delay from C_1 to D_1 and D_1 to A_2 (i.e., inequality **n**) and we need a lower bound on the delay from A_1 to C_1 (i.e., inequality **l**) to ensure that C_1 does not occur too soon with respect to B_1. Readers may want to compare this solution to that reported in Section 2.1 (i.e., where `CAh=30`, `ABh=40`, `DAh=10`, and `BAh=10`) and note that this symbolic analysis tells us which numeric values were important in our original solution.

Our optimization worked particularly well for this example. At don't care level 4 we had to analyze one combination out of a possible 1,001 and at don't care level 5 we had to analyze four combinations out of a possible 2,002. The required execution times for this analysis were thus quite minimal. We believe this example demonstrates the benefits of using symbolic variables to obtain symbolic solutions to timing verification problems. Symbolic solutions can assist designers in evaluating tradeoffs and make it easier to validate and understand their formal specifications.

5 Conclusions

The best known upper bound on the performance of an algorithm for verifying Presburger formulas is $O(2^{2^{2^n}})$ [8] and Omega may have an even larger upper bound. For this reason, Presburger formulas have not been used very much for timing analysis. We believe that this will change because the Omega libraries have demonstrated that manipulations can often be efficiently performed and because the type of analysis that can be provided is quite useful. Our simplification algorithms also have terrible worst case complexity yet are quite capable (given simple optimizations) of greatly improving the results reported by Omega using reasonable amounts of computation.

We intend to make immediate use of our results to help us continue exploring symbolic timing verification of timing diagrams and other timing abstractions. Our simplification methodology will quite likely be of great benefit in other application areas as well, both for users of Omega (i.e., as of February 1994 over 400 researchers had obtained copies of the Omega software libraries [10]) and for other users of Presburger formulas (many theorem provers contain algorithms for analyzing Presburger formulas).

Our tools need to be further refined in addition to being made more efficient. The refinements are needed not only to more automatically translate between forms but also to give users a suite of manipulation routines so that results can be presented in the form most appropriate for the domain of application and intended use. For example, in interface circuits, it might be especially valuable to identify "trivial" inequalities (such as those that require lower bounds to be smaller than upper bounds or that require delays to be greater than zero) and provide users the option of omitting them from displayed solutions. Efficiency improvements revolve around identifying further optimizations in don't care extraction and more tight integration with the Omega libraries.

6 Acknowledgements

This work was supported by an NSF RIA Award (MIP-9410279). We wish to thank Wilbon Davis for several interesting discussions and Ellen Sentovich for help with SIS.

References

[1] T. Amon and G. Borriello. An approach to symbolic timing verification. In *Proc. ACM/IEEE Design Automation Conference (DAC)*, June 1992.

[2] T. Amon, G. Borriello, D. Hu, and J. Liu. Symbolic timing verification of timing diagrams using presburger formulas. In *Proc. ACM/IEEE Design Automation Conference (DAC)*, June 1997.

[3] Tevfik Bultan, Richard Gerber, and William Pugh. Symbolic model checking of infinite state programs using presburger arithmetic. Technical Report UMIACS-TR-96-66, University of Maryland, September 1996.

[4] D. C. Cooper. Programs for mechanical program verification. In *Machine Intelligence 6*, 1971.

[5] Wayne Kelly and William Pugh. Determining schedules based on performance estimation. In *Parallel Processing Letters*, September 1994.

[6] Wayne Kelly and William Pugh. Finding legal reordering transformations using mappings. In *Seventh International Workshop on Languages and Compilers for Parallel Computing*, August 1994.

[7] K. L. McMillan and D. L. Dill. Algorithms for interface timing verification. In *Proc. International Conf. Computer Design (ICCD)*, October 1992.

[8] Derek Oppen. A $2^{2^{2^{pn}}}$ upper bound on the complexity of presburger arithmetic. In *Journal of Computer and System Sciences*, July 1978.

[9] William Pugh. A practical algorithm for exact array dependence analysis. *Communications of the ACM*, 35(8), August 1992.

[10] William Pugh et al. The omega project. In *URL: http://www.cs.umd.edu/projects/omega*.

[11] William Pugh and David Wonnacott. An exact method for the analysis of value-based array data dependences. In *Proc. 6th Workshop on Programming Languages and Compilers for Parallel Computing*, August 1993.

[12] E. M. Sentovich et al. SIS: A system for sequential circuit synthesis. Technical Report UCB/ERL M92/41, UC Berkeley, May 1992. see http://www-cad.eecs.berkeley.edu/Software/.

[13] P. Vanbekbergen, G. Goossens, and H. De Man. Specification and analysis of timing constraints in signal transition graphs. In *European Design Automation Conference*, March 1992.

[14] E. Walkup and G. Borriello. Interface timing verification with application to synthesis. In *Proc. ACM/IEEE Design Automation Conference (DAC)*, June 1994.

[15] Elizabeth A. Walkup. *Optimization of Linear Max-Plus Systems with Application to Timing Analysis*. Ph.D. thesis, University of Washington, 1995.

Delay Estimation of VLSI Circuits from a High-Level View†

Mahadevamurty Nemani and Farid N. Najm

ECE Dept. and Coordinated Science Lab.
University of Illinois at Urbana-Champaign
Urbana, Illinois 61801

Abstract

Estimation of the delay of a Boolean function from its functional description is an important step towards design exploration at the register transfer level (RTL). This paper addresses the problem of estimating the delay of certain *optimal* multi-level implementations of combinational circuits, given only their functional description. The proposed delay model uses a new complexity measure called the *delay measure* to estimate the delay. It has an advantage that it can be used to predict both, the minimum delay (associated with an optimum delay implementation) and the maximum delay (associated with an optimum area implementation) of a Boolean function without actually resorting to logic synthesis. The model is empirical and results demonstrating its feasibility and utility are presented.

1. Introduction

Rapid increase in the design complexity and reduction in design time have resulted in a need for CAD tools that can help make important design decisions *early* in the design process. To do so, these tools must operate with a design description at a high level of abstraction. Performance being an important design criterion, there is an increasing need for high-level delay estimation.

In this paper we will focus on the estimation of the delay associated with a circuit from its high-level description, i.e., given only a *functional* view of the circuit, such as when a circuit is described only with Boolean equations. We will estimate the minimum, maximum and other feasible delay points, from a high level description of the circuit. The advantages of estimating the delay of a circuit from a high-level view are in the areas of high-level power estimation [1], high-level timing analysis and logic synthesis.

Our delay estimation technique is based on the novel concept of *delay measure* of a Boolean function, to be defined later in the paper. Based on this complexity measure we will provide a delay model which makes predictions of the minimum and maximum delays attainable by a circuit with reasonable accuracy.

† This work was supported in part by Intel, by Rockwell, and by the NSF under CAREER Award MIP-9623237.

DAC 98, San Francisco, California

This will be demonstrated with experimental results on a large set of benchmarks. An overall high-level delay estimation flow is given in Fig. 1. In our work, the model characterization is done using SIS, which uses a simple static timing analysis to estimate the delay. This, however, is not a limitation of the proposed model, which can be used in conjunction with more accurate timing models.

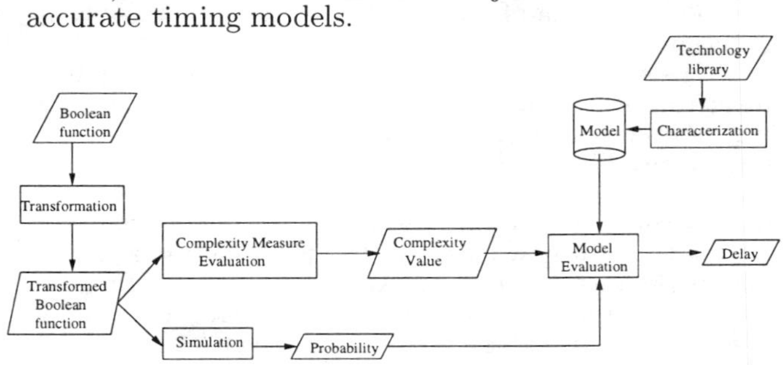

Figure 1. High-level delay estimation flow.

The proposed delay model, however, is inherently limited to circuits which do not include large exclusive-or arrays. One way around the problem of exclusive-or arrays is to require that the Boolean function specification explicitly list exclusive-or gates. In that case, these can be identified up-front and excluded from the analysis, so that the proposed method is applied only to the remaining circuitry. In any case, in this paper we will not consider circuits composed of large exclusive-or arrays.

Before leaving this section, we should mention some previous work on high-level delay estimation. In [2], a delay model is proposed for randomly generated Boolean functions (RGBF) (functions in which each point in the Boolean space is placed in the on- or off-set in a random fashion) based on empirical data collected on these functions ($n = 8, 9, 10$. Here n is the number of primary inputs). Their model relates the delay of a RGBF function to its entropy and primary-input count and exhibits an *exponential* in n behavior, leading to unrealistically large predictions in delay on realistic circuits. In [3], the authors propose a hierarchical timing estimation method. Their method relies on the concept of conditional delay matrix, which along with the delay information also contains conditions under which the delay information is true. However, in order to build a hierarchical model, each module is assumed to be accurately characterizable, which is unreasonable for random logic modules at higher levels of abstractions (as they are specified as Boolean equations). In [4], an approach for early layout-driven delay prediction is proposed, which includes the effect of wiring on the delay of a design. For an RT level design, this approach assumes that the pin-to-pin delays of the black-box combinational circuits are already known. In [5], the authors perform technology-independent synthesis operations on the RTL design to map it to a technology-independent library. Curve fitting techniques, using representative previous designs, are then used to relate the delays and areas of the technology-independent generating functions, and hence the overall function, to that of their actual implementations. Several technology-independent optimization steps may have to be performed to obtain the area-delay trade-off curve. Our approach does not resort to any optimization and is a one-shot scheme to predict the area-delay curve.

2. The Delay Complexity Measure

The problem of estimating the delay of a Boolean function pertains to estimation of the delay associated with the slowest path ($\mathcal{D}$), also called the critical path, in an implementation of the function, given only its high level description (Boolean equations) and a target technology library. It must be noted here that by "implementation" we mean a multi-level implementation of the Boolean function, optimal in a certain sense.

Let us, for now, assume that the function at hand is a single-output function. Also, let us assume for now that we are interested in predicting the minimum delay associated with a Boolean function, given its high-level description. In the results section (section 4), we will show that the proposed complexity measure along with the delay model (section 3) can be used to predict the delay at any point on the area-delay trade-off curve.

We will define a *delay measure* of a Boolean function, in terms of delay complexity measures associated with the on and off-sets of the function. For a single-output Boolean function f, let C^{on} be a minimal cover [6] of the on-set of f. Also let $c_i \in C^{on}$ denote a cube of C^{on}, i.e., a prime implicant, for $i = 1, 2, \ldots, |C^{on}|$, where $|C^{on}|$ is the size of C^{on} (number of cubes). Finally, let $|c_i|$ denote the size of cube c_i, defined as the number of literals in c_i. With this, the delay complexity measure of the on-set of f, denoted $\mathcal{M}_1(f)$, is defined as:

$$\mathcal{M}_1(f) = \log_2 \left(\max_{c_i \in C^{on}} |c_i| \right) + \log_2 (|C^{on}|) \qquad (1)$$

The intuitive reason for the above definition is the following. The size of a prime implicant of a function directly influences the delay associated with the function, as the larger the number of literals in the implicant, the longer the time it would take for a function to be evaluated. Also, the number of prime implicants that need to be evaluated influences the delay, as the larger the number, the longer it would take to combine the results of the individual prime implicants. Likewise, we can define the delay complexity measure of the off-set of f, denoted $\mathcal{M}_0(f)$. Using the delay complexity measures of the on and off-sets of f, we define the *delay measure* of f as follows:

$$\mathcal{M}(f) = \min \{\mathcal{M}_1(f), \mathcal{M}_0(f)\} \qquad (2)$$

This measure requires the computation of a minimal cover of a (single output) Boolean function, which was obtained using the heuristic minimizer ESPRESSO [6].

2.1. Extension to Multi-Output Functions

We now propose a method by which the above delay complexity measure (defined for single-output functions) can be extended to measure the delay of multi-output functions. We transform an n-input, m-output Boolean function, f, by the addition of an m to 1 multiplexor, into an equivalent $(n + \lceil \log_2 m \rceil)$ input, single-output function $\hat{f}$, where the $\lceil \log_2(m) \rceil$ inputs correspond to the control inputs of the multiplexor required to control the m outputs of the Boolean function f.

Since $\hat{f}$ is a single-output function, its delay complexity measure can be computed. However, in order to compute the delay of f from the predicted delay of $\hat{f}$, we must be able to compute the influence of the multiplexor on the delay of $\hat{f}$. This problem of recovery of delay of f from that of $\hat{f}$ will be discussed in the next section.

3. The Delay Model

In this section we present the delay model to compute the delay, $\mathcal{D}(f)$, of Boolean functions. The delay model uses the delay measure of the Boolean function along with its output entropy to estimate the delay of the function at hand. We begin by presenting the delay model corresponding to the prediction of delay associated with the minimum delay implementation.

For a given n, consider the set of all Boolean functions on n inputs *and* whose output entropy is $\mathcal{H}(f) = 1$, based on all inputs being independent and with 0.5 probability. For a number of RGBFs from this set, we computed their delay measure, $\mathcal{M}(f)$, using our algorithm and obtained an estimate of the delay $\mathcal{D}(f)$ from a minimum delay implementation of the function using SIS. It was observed that the set of RGBFs for each n is clustered around specific points. Moreover, the variation of delay $\mathcal{D}(f)$ of a minimum delay implementation with delay measure, $\mathcal{M}(f)$, was linear. We have found that not only do RGBFs fall on this linear curve, but also typical VLSI functions fall on it or close to it, but spread all over the curve. Thus these linear curves of $\mathcal{D}(f)$ (of minimum delay implementation) versus $\mathcal{M}(f)$ are very important and are in fact the essence of our minimum delay prediction model.

We generate a family of such curves capturing the variation of $\mathcal{D}$ with $\mathcal{M}$ for various values of output entropy. Hence given the output entropy of a Boolean function and $\mathcal{M}(f)$, we use the curve corresponding to that value of output entropy and predict $\mathcal{D}(f)$. Note that these curves need to be generated only once, which is an up-front once-only cost, and they can be used to predict the minimum delay of various functions. An important consideration is what the largest n should be for which these curves need to be generated. Obviously, the curves are going to be more difficult to generate for larger n because of the cost of running synthesis to obtain the $\mathcal{D}(f)$ values. Luckily, there are two reasons why this is not a problem so that considering $n \leq 12$ is sufficient. Firstly, we have found that for typical VLSI functions, the value of $\mathcal{M}(f)$ turns out to be much smaller than n in most cases. Indeed, all the test cases that we will present (for which n ranges from 4 to 70) had $\mathcal{M}(f) \leq 13$. Secondly, the curves corresponding to minimum delay seem to exhibit a linear variation of minimum delay with delay measure and given by

$$\mathcal{D}(f) = k_1\mathcal{M}(f) + k_2 \tag{3}$$

where k_1 and k_2 are the slope and intercept of the minimum least squares fit.

So far, we have discussed the delay model for predicting the minimum delay. A similar reasoning can be applied to derive the model for predicting the maximum delay or any other delay. For example, we found that the maximum delay versus delay measure curves exhibit a slightly nonlinear variation, for which the best fit was an exponential:

$$\mathcal{D}(f) = k_1\beta^{\mathcal{M}(f)} \tag{4}$$

where k_1 and β are constants ($\beta \approx 1.3$), which can be obtained by performing a least squares fit.

For a multi-output function, we apply the delay model on $\hat{f}$. We recover the delay of f from that of $\hat{f}$ by using the following empirical model:

$$\mathcal{D}(f) = \mathcal{D}(\hat{f}) - \alpha\mathcal{D}_{mux} \tag{5}$$

where $\mathcal{D}_{mux}$ is the delay of an m to 1 multiplexor, and $0 \leq \alpha \leq 1$ is a coefficient that represents the contribution of the multiplexor to the delay of $\hat{f}$.

Let A_i denote the number of (multiplexor) control inputs in the ith prime implicant, and define A_{on} to be the average number of control inputs in a prime implicant belonging to the on-set of $\hat{f}$, so that:

$$A_{on} = \frac{\sum_{i=1}^{K_{on}} A_i}{K_{on}}$$

where K_{on} is the number of prime implicants in the on-set of $\hat{f}$. Similarly, one can define A_{off}. It can be shown [1] that the delay measure of an m to 1 multiplexor is given by:

$$\mathcal{M}_{mux} = \log_2(\lceil \log_2 m \rceil + 1) + \log_2 m$$

and that the delay measure of a reduced multiplexor (in an optimized version of $\hat{f}$) is given by [1]:

$$\mathcal{M}_{mux}^{red} = \log_2\left\{\min\{A_{on}, A_{off}\}+1\right\}+\min\{A_{on}, A_{off}\}$$

Then α, at the minimum-delay point, can be computed as [1]:

$$\alpha = \frac{k_1\mathcal{M}_{mux}^{red} + k_2}{k_1\mathcal{M}_{mux} + k_2} \tag{6}$$

The same approach can be used to compute the delay contribution of the reduced multiplexor at the maximum or any delay point.

4. Implementation and Results

Empirical results on delay estimation will be reported for three delay points on the area-delay tradeoff curve, namely, the minimum, maximum and 50%-delay (delay point midway between minimum and maximum delays) points. The proposed delay models corresponding to these three points were tested on 31 ISCAS-89 [8] and MCNC [9] benchmark circuits, which are listed in Table 1. For each circuit, the table also includes the total execution time (in CPU seconds) required to predict the delay for that circuit using our approach, on a SUN Sparc-5 machine with 32MB RAM. Also, it includes the time taken by SIS to synthesize the circuits at the minimum area and minimum delay points. It must be noted that while a new run of SIS is required to synthesize a circuit to a new point on the area-delay curve, only one run of our tool would suffice to make predictions regarding the delay and area [1] at any point on the area-delay curve.

TABLE 1.
Benchmarks and run-times (sec).

Circuit Name	Circuit Function	#I	#O	CPU time	SIS time min-area	SIS time min-delay
s298	Controller	17	20	4.4	18.1	44.9
s386	Controller	13	13	4.2	22.1	58.3
s400	Controller	24	27	8.5	20.7	59.3
s444	Controller	24	27	8.5	20.6	60.3
s510	Controller	25	13	6.9	37.1	118
s526	Controller	24	27	10.4	26.3	72.6
s526n	Controller	24	27	10	26.0	71.3
s641	Controller	59	43	41.4	29.1	83.5
s713	Controller	58	42	42.3	29.1	88.4
s820	Controller	37	24	16.3	57.3	168.8
s832	Controller	37	24	16.5	57.3	160.4
s953	Controller	39	52	38.8	64.5	223.5
s1196	Logic	28	32	163	284	670.9
s1238	Logic	28	32	141	323	782.1
s1494	Controller	27	25	26.8	155.4	601.2
s1488	Controller	27	25	29.3	136.8	432.1
b9	Logic	41	21	5.7	10.6	29.8
c8	Logic	28	18	4.9	9.9	32.1
frg2	Logic	143	139	268	201.6	501.0
vda	Logic	17	39	39.3	278	497.2
i7	Logic	199	67	23.1	54.8	221.1
i6	Logic	138	67	17.5	40.2	151.5
example2	Logic	85	66	28	23.6	97.8
cht	Logic	47	36	6.5	10.8	35.4
x1	Logic	51	35	12.8	27.0	89.4
ttt2	Logic	24	21	6.25	20.4	65.9
x3	Logic	135	99	53	76	278.7
x4	Logic	94	71	28.6	50.0	108.1
apex6	Logic	135	99	45.3	62.0	236.9
apex7	Logic	49	37	20.3	17.8	61.0
k2	Logic	45	45	170.1	591.0	1352.6

In practice we have found that there is a tradeoff between run-time and the value of m (number of outputs to which the transformation is applied). A heuristic is to partition the outputs of a function into groups of 16 outputs each (of course, one group may be left with less than 16 outputs), and to apply the delay prediction to the resulting sub-functions, one at a time, before recombining the results to get the delay of the whole function. If we denote by g_i the sub-function corresponding to the ith group of $m = 16$ outputs, we propose the following model for recombining the results:

$$\mathcal{D}(f) = \max_i\{\mathcal{D}(g_i)\} \tag{7}$$

In order to generate the minimum delays, the circuits were optimized in SIS using *script.rugged* followed by *script.delay* [6, 7] for delay optimization, and mapped using the library *lib2.genlib*. A comparison of the actual delay with the predicted delay obtained using the model (7), for the benchmark circuits, is given in Fig. 2. It can be seen that the overall delay can be predicted with reasonable accuracy. In order to generate the maximum delay, the circuits were optimized in SIS using *script.rugged* for area optimization, and mapped using the library *lib2.genlib*. For the 50%-delay point, this was followed by speeding up the circuit using the command *speed_up* [7] in the SIS environment. A comparison of the actual maximum and 50%-delays with the predicted delays obtained using the model (7), for the benchmarks, is given in Fig. 3 and Fig. 4, respectively. The overall delay prediction is reasonably accurate for all except two circuits (*i6* and *i7*). This is due to the fact that these circuits have a small amount of control logic which controls all the outputs. Thus this method may not be suitable for circuits which have both control *and* data in them. Also note that the above is not a problem for the minimum delay point because the best way to reduce the delay is to ensure that the same control circuit is not overloaded by feeding multiple outputs.

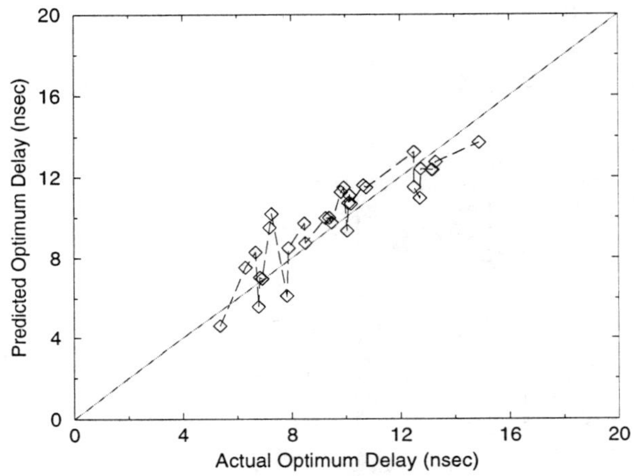

Figure 2. Actual versus predicted optimum delays.

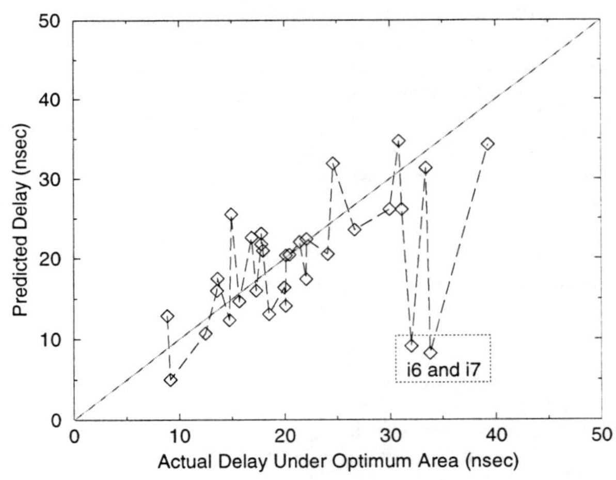

Figure 3. Actual versus predicted maximum delays.

5. Conclusions

In this paper, we have presented a new delay model to predict the delay of combinational logic, given only its functional specification and no structural information. This was achieved through a new delay complexity measure called *delay measure*. Empirical evidence showing the utility of the proposed approach in estimating any feasible delay from the given high-level description has been presented. While this approach is being further developed, certain potential limitations of this method were pointed out for circuits with control and data, with the same control logic controlling all outputs.

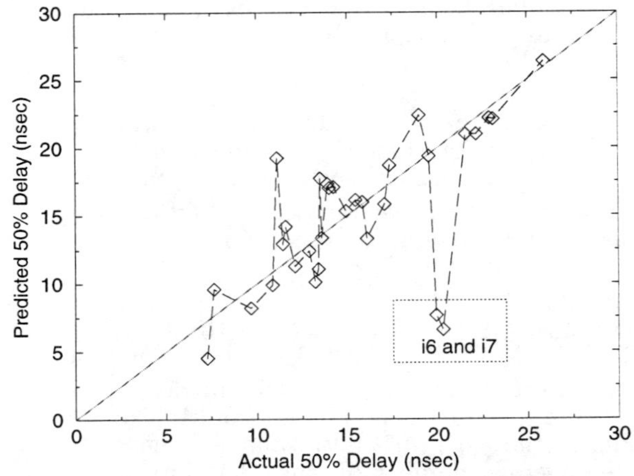

Figure 4. Actual versus predicted 50% delays.

References

[1] M. Nemani, "High-Level Power Estimation," Ph.D. dissertation, University of Illinois at Urbana-Champaign, 1998.

[2] K-T Cheng and V. Agrawal, "An entropy measure for the complexity of multi-output Boolean functions," *27th ACM/IEEE Design Automation Conference*, pp. 302–305, 1990.

[3] H. Yalcin, J. Hayes and K. Sakallah, "An approximate Timing Analysis Method for Datapath Circuits," *IEEE International Conference on Computer Aided Design*, pp. 114-118, 1996.

[4] C. Ramachandran and F. Kurdahi, "Combined Topological and Functionality-Based Delay Estimation Using a Layout-Driven Approach for High-Level Applications," *IEEE Transactions on Computer Aided Design of Integrated Circuits and System*, vol. 13, no. 12, pp. 1450-1460, Dec. 1994.

[5] A. Srinivasan, G. D. Huber and D. P. LaPotin, "Accurate Area and Delay Estimation from RTL Descriptions," *IEEE Transactions on VLSI systems*, vol. 6, no. 1, pp. 168-172, March 1998.

[6] G. De Micheli, *Synthesis and Optimization of Digital Circuits*, New York, NY: McGraw-Hill Inc., 1994.

[7] *SIS-1.2, Reference Manual*, University of California, Berkeley, 1992.

[8] F. Brglez, D. Bryan and K. Koźmiński, "Combinational profiles of sequential benchmark circuits," *IEEE International Symposium On Circuits and Systems*, pp. 1929-1934, 1989.

[9] S. Yang, "Logic Synthesis and Optimization Benchmarks User Guide Version 3.0," *Rep. Microelectronics Center of North Carolina*, 1991.

TETA: Transistor-Level Engine for Timing Analysis

Florentin Dartu*
Strategic CAD Labs
Intel Corporation
fdartu@ichips.intel.com

Lawrence T. Pileggi
Department of ECE
Carnegie Mellon University
pileggi@ece.cmu.edu

Abstract

TETA is an interconnect-centric waveform calculator that was optimized to achieve the utmost efficiency for analyzing logic stages comprised of transistors and large coupled RC(L) interconnect models. TETA applies a novel compaction for the transistor clusters and employs successive chord iterations to solve the resulting nonlinear equations. These algorithms permit the use of simple SIMO (single input multi-output) N-port interconnect models since macromodel passivity is not required. The successive chord analysis also enables TETA to avoid the N-port matrix factorization during nonlinear iterations and allows the use of simple table look-up models for MOS devices. Complex gates and nonlinear capacitors can be handled without loss of generality.

1: Introduction

For future CMOS technologies, a key feature of a transistor level timing engine will be the efficient handling of RLC interconnect. Coupling between signal paths transforms interconnect-tree problems into *N*-port problems forcing timing analyzers to handle RLC *N*-port macromodels. Model order reduction techniques are imperative for managing the overwhelming size and complexity of extracted RLC interconnect circuits, [1,2], and of primary importance is their simplicity and passivity. Recently it was shown that multi-input multi-output (MIMO) reduced order N-port methods can guarantee passivity [3], however, the total number of state variables for each *N*-port matrix entry is Nxq, where *N* is the number of ports, and *q* is the order of the approximation. Reduced-order multi-port models based on a single-input multi-output (SIMO) approaches can generate simpler models based on *q* state variables, but passivity cannot be guaranteed [3].

There are numerous transistor-level timing simulation algorithms which are used for the timing engine of a static timing analysis methodology [4]. Unfortunately, timing simulators demonstrate only a modest gain in performance as compared with SPICE for analyzing a simple logic stage where spatial latency does not matter. Therefore, for timing analysis, a fair efficiency comparison is made between SPICE and the timing engine only for both analyzing the partitioned logic stages. Clearly, the most challenging aspect of a timing analysis engine is the efficient handling of the coupled interconnect that appears in the logic stage.

In this paper we propose TETA, based on an interconnect-centric view of the waveform calculation problem. Instead of using traditional transistor models and forcing interconnect compatibility, TETA employs efficient interconnect models and analyses and derives new transistor models that can work with them. TETA is based on successive chord iterations for the nonlinear devices and a novel compaction scheme for the transistor clusters. Since the successive chord method is represented by a fixed conductance model for nonlinear iterations, stability is guaranteed with a SIMO interconnect macromodel by combining the chord conductance with the interconnect circuit prior to model order reduction. This provides the further benefit that the N-port model is factored only when the time step is changed (and for timing analysis applications fixed timestep engine works very well).

TETA *is not* a timing simulator, as its purpose is to simulate only relatively small circuits that result after the partitioning into strongly connected components. The efficiency for small problems is mainly attributable to the successive chord models which enable the use of table models for the transistors. The table models can be made simpler in TETA because the chord method requires only the function continuity [5]. The following section will briefly introduce the main components of TETA that are presented in more detail in [6]. Section 3 presents the structure of the simulation engine. The results are shown in Section 4.

2: Basic components

Circuit simulation accuracy is derived from the combination of detailed nonlinear device models, implicit numerical integration schemes, and powerful nonlinear solvers (some variations of Newton-Raphson) taken to convergence [7]. Assuming a partitioning of the full circuit into strongly connected components, a circuit simulator engine is competitive with most timing simulators in terms of efficiency and accuracy trade-offs.

A: Linear N-port macromodeling

Consider the situation shown in Fig.1, where the drivers at the R(L)C macromodel ports are represented by their time varying Norton equivalents. To simulate this circuit in time domain, every timestep will require stamping a new admittance matrix for the full linear *N*-port, factoring it and solving for the port voltages. The *N*-port is a reduced-order representation of the complete circuit, obtained using one of the many reduction schemes[3].

If the linear *N*-port contains significantly more energy storage elements than the number of ports (which is generally the case), using a reduced order *N*-port model yields a significant gain in efficiency [8]. But while the stability of the reduced order model is relatively easy to achieve, the passivity is not. Passivity of the

*Previously with Carnegie Mellon University, Department. of ECE.

This work was supported in part by the Semiconductor Research Corporation under contract 97-DC-068 and a grant from Intel Corporation.

DAC 98, San Francisco, California

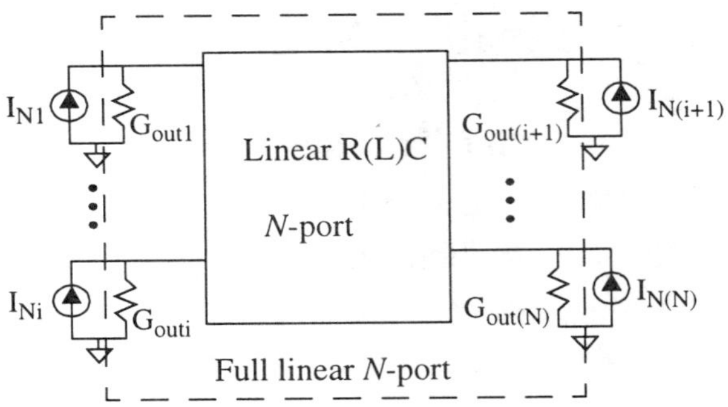

FIG. 1: Linear N-port driven by time-varying Norton equivalents.

reduced-order model is the only guarantee of stability under all impedances conditions at the ports. Passivity can be guaranteed with certain block reduction methods [2,3], however, these methods unfortunately generate $O(N^3)$ state variables for the reduced order model. This can increase the simulation time dramatically.

Therefore, from the standpoint of efficiency and simplicity of the N-port model, it is desirable for the port conductances (G_{out}'s in Fig.1) to be constant. The benefits are twofold. First, the admittance matrix of the full N-port can be factored only once, prior to the simulation. This is especially important in timing analysis applications which simulate the same logic stages many times. Secondly, passivity, and the complex block model-order reduction methods necessary to ensure it, are not required.

The z-matrix for this case has only to be *stable* (because it is driven by current sources) so we can use a *single input multiple output* (SIMO) order reduction technique [3] that will generate a much smaller number of state variables, $O(N)$.

B: Nonlinear relaxation

To describe the nonlinear iterations which maintain constant impedances at the ports, we begin by defining the following set of nonlinear equations:

$$\begin{cases} F_1(x_1, \dots, x_n) = 0 \\ \dots \\ F_n(x_1, \dots, x_n) = 0 \end{cases} \tag{1}$$

One method to solve this nonlinear system of equations is by nonlinear relaxation techniques. For example, in the standard Gauss-Seidel-Newton relaxation method [9], the equations in (1) are solved sequentially and independently using a unidimensional Newton-Raphson. This approach, although very successful for some applications, can require special attention regarding the equation ordering and can exhibit a slow rate of convergence.

We propose to use a variation of the Gauss-Seidel-Newton algorithm. The set of nodal equations representing our digital circuit will be split in two: one set containing the output nodes connected to the load (possibly an N-port) and another set formed by the rest of the nodes (internal nodes and inputs to the transistors). Note that we have partitioned the circuit at gate inputs, as it is done with dcccs. However, unlike dccs (dc coupled components), these interconnect models can include lines that are capacitively coupled.

The generalization to cells with n output nodes is straightforward. One example of which is a cell with complex pass transistor logic. The only major difference will be the use of a generalized Norton equivalent.

C: The successive chord method

Implicit integration schemes require the solution of a nonlinear system of equations for each timestep. SPICE-like simulators use various derivatives of the classical Newton-Raphson (N-R) method [10]. All these methods rely on linearizing the system around an initial guess and solving the resulting linear system to obtain a better approximation.

The major difference between a typical N-R algorithm and the successive chord (S-C) method [5] lies in the linearization step. In the standard N-R procedure the value of the linear conductance equals the derivative of the I-V characteristic at the operating point. In general, its value is a function of the voltage across the nonlinear resistor. For the S-C method, the linear conductance is the same for every operating point, and its value can be selected prior to the simulation.

Three observations are of particular importance. First, the standard N-R procedure requires the continuity of both I(V) and its derivative, while the S-C method needs only the continuity of I(V). For this reason it is much easier to employ table look-up I-V models in an S-C environment than when using standard N-R. Secondly, the convergence rate for the S-C method is linear, worse than the quadratic rate of the N-R. So it would appear that the S-C is less expensive per iteration, but may take more iterations to converge for the same level of accuracy. But importantly, based on Section 2:A and Section 2:B, the S-C method is more compatible with a simplified multi-port representation.

D: MOS device model

In SPICE [11], particularly for small circuits, model evaluation can take 80% or more of the simulation time. Piecewise (linear, constant or quadratic) models were introduced in timing simulators to provide speed-up over traditional nonlinear models. These models [4,11] enable trade-offs between accuracy and efficiency.

The simplest and most efficient representation of I_{DS} as a function of V_{GS} and V_{DS} is the table look-up model. Unfortunately, we cannot use the resulting piecewise constant model directly due to the continuity requirement for $I_{DS}(V_{GS}, V_{DS})$. In TETA we generate a continuous piecewise linear surface from a table look-up model using the method of triangulation [12]. The generalization of the triangularization procedure to 3D tables is straightforward.

It is very important to observe that the derivative of I_{DS} is not continuous, so these MOS models can be used in conjunction with the successive chord method, but not a Newton-Raphson procedure. At the same time, the accuracy of the table does not significantly affect the simulation time, as is the case for timing simulators which *compute* the time of the next event.

For implicit integration methods, the nonlinear MOS capacitors are replaced at every timestep by their companion model [10]. The nonlinear conductance from the companion model can depend on more than one MOS terminal voltage but it is always possible to use the chord method to linearize it.

3: The simulation engine

TETA is a simulation engine combining the basic components presented in the previous section. To better understand the way TETA works we will start by describing the case of a nonlinear gate driving a one-port interconnect circuit, as shown in Fig.2.

The outer loop of TETA resembles any other SPICE-like or timing simulation engine: using a numerical integration method (for TETA we use trapezoidal integration), discretize the simulation time and solve the corresponding nonlinear system at every time point. To solve the nonlinear system, TETA uses the compaction scheme introduced in Section 2:B in conjunction with the successive chord method presented in Section 2:C.

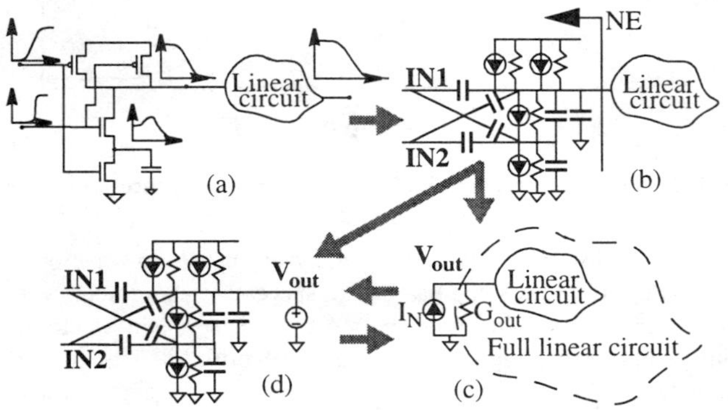

FIG. 2: The approach taken by TETA in solving nonlinear circuits.

For the circuit in Fig.2a, the first step will be to replace the capacitors by their companion models (not shown). Next, all transistors will be linearized using the successive chord method (described in Section 2:C), which yields the circuit shown in Fig.2b. Table look-up I-V characteristics are used to model the MOS transistors. The Norton equivalent for the linearized gate model is computed, and the circuit from Fig.2c is solved for V_{out}. The time domain *impedance* model for the full linear circuit in Fig.2c does not depend on time or nonlinear iteration, so it can be computed *a priori*. Using the newly found value for V_{out}, TETA iterates using a successive chord solver until convergence for the circuit shown in Fig.2d. A new set of internal voltages are found which specify a new value for I_N (the Norton equivalent current source from Fig.2c). The procedure is repeated until a convergence criterion is met.

To solve the circuit from Fig.2c in the time domain, TETA uses a classical state variable approach associating one state variable with every pole of $Z_{port}(s)$. Although a fixed timestep is not required by this procedure, it is generally more efficient to employ one since it avoids the need for any division operations.

Solving the circuit in Fig.2d, in what we call the *inner loop* of TETA, follows the SPICE approach with the exception of using an S-C solver instead of Newton-Raphson. A variable timestep can be implemented, but is genereally less efficient. In terms of computational cost, TETA gains efficiency over SPICE through:

- more efficient device models as shown in Section 2:D,
- not factorizing the MNA matrix at every nonlinear iteration.

The latter is an important feature for large N-port problems.

An apparently delicate problem related to the successive chord method is the choice for the equivalent conductance of the nonlinear transistors. Practically, for MOS transistors we only have to find two values, one for NMOS and one for PMOS, which scale with transistor dimensions. A reasonable value of equivalent conductance can be chosen for each transistor based on its average nonlinear conductance. But we found the PMOS and NMOS conductance values with an experimental procedure where we observed TETA's runtime for different gates [6].

A couple of observations are noteworthy. First, the G_{out} value used in Fig.2c is a by-product of the LU factorization of the circuit shown in Fig.2d, so its computation is basically free [6]. Secondly, I_N is obtained while solving the same circuit.

TETA was built to efficiently handle large N-port problems. For reasons outlined in Section 2:A, we want to have every dc problem expressed as the N-port problem in Fig.1, with each G_{outi} independent of time or nonlinear iteration. As for the one-port case presented above, the compaction scheme introduced in Section 2:B is used to obtain the Norton equivalents for each nonlinear driver. This approach also has the advantage of naturally decoupling the circuit at the dc solver level. Each nonlinear driver is solved separately. The MNA matrix for the internal nodes of each nonlinear driver is also independent of time or nonlinear iteration and can, therefore, be LU factored prior to the simulation.

4: Results

For all examples presented in this section we compare TETA with HSPICE using level 3 MOS models for a 0.5μ effective length MOSIS technology. Both programs were executed on an IBM Power Series 850 computer using a 133 MHz PowerPC microprocessor. For a fair comparison, we used the same capacitance models, node voltage tolerances and timestep for both simulators. The CPU times for HSPICE include only the "transient analysis" times as reported in its output. Moreover, we used the option FAST that optimizes HSPICE for further speed.

A prototype implementation of TETA consists of 2000 lines of C code. It is able to handle MOS transistors, resistors, capacitors and general N-port representations in z-parameter format. The MOS transistor I-V models were built as 3D tables in order to take the body effect into account.

We expect the smallest speed-up advantage for TETA as compared with HSPICE for simple gates without interconnect. Therefore, we consider as a first example a 4-input NAND gate driving a purely capacitive load. The output waveform generated by TETA is shown in Fig.3. It is worth observing that for this accuracy TETA is able to simulate approximately 1000 of these 4-input NAND gates per second.

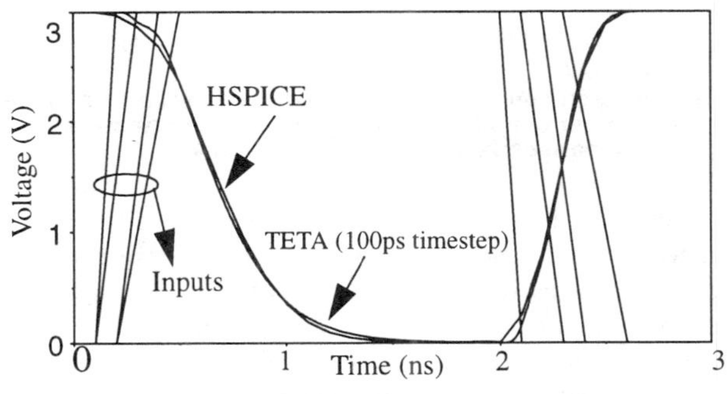

FIG. 3: Accuracy comparison for a NAND4 gate.

Table 1 shows a comparison between TETA and HSPICE for a set of CMOS gates. For this set of examples, the MOS models included linear grounded capacitors in both simulators. Also, both simulators used a fixed timestep of 10ps. The use of table models for the MOS I-V characteristics and the implementation of temporal latency (used to bypass MOS model evaluation) represent the main sources of run time gain for TETA in these cases. The results in Table 1 show a relative constant TETA speed-up over HSPICE with the circuit size.

TABLE 1 Runtime comparison between TETA and HSPICE (one pulse). Both simulators used a 10ps timestep for which TETA presents no discernible error.

Gate	HSPICE (ms)	TETA (ms)	speed-up
Inverter	95.9	1.4	68.5
NAND2	246.54	4.2	58.7
NAND3	354.6	6.1	58.13
NAND4	531.52	8.8	60.4
Complex CMOS gate (20 transistors)	1630.4	26.2	62.23

But most importantly, TETA yields its best efficiency when simulating large N-port problems generated by RLC circuits. To demonstrate this we analyzed a 16-bit bus line with every line capacitively coupled to adjacent lines. The complexity of the linear circuit is measured by the number of nodes of the linear circuit. Fig.4b shows the accuracy comparison between HSPICE and TETA using an aggressive timestep of 200ps (for 30ps there is no discoverable error in TETA results). The per transition run times for TETA are 1.44ms and 4.4 ms for timesteps of 200ps and 30ps, respectively. In Fig.4a we present the speed-up over HSPICE as a function of the number of nodes of the linear circuit.

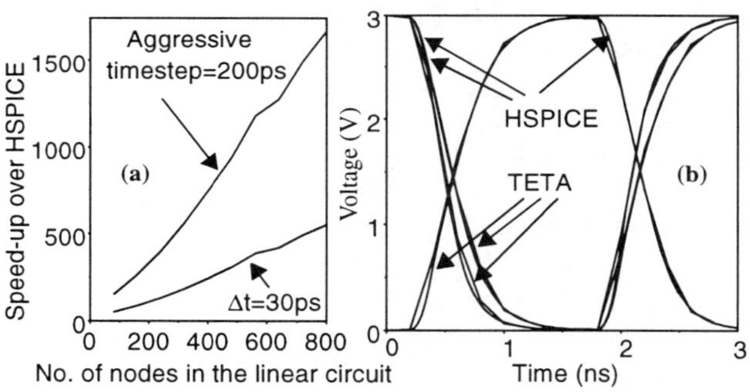

FIG. 4: A 16-bit bus example. a) Speed-up as a function of the number of nodes in the linear circuit; b) Accuracy comparison (200ps timestep for TETA).

Another interesting example is obtained by simulating an inverter driving a clock tree for which the inductive component plays a significant role. The waveforms for the driving point of this clock tree are presented in Fig.5. HSPICE simulated the full linear circuit containing 160 nodes. TETA simulated a 9th order macromodel having four pairs of complex poles and one real pole. Due to the small time constants presented by the linear circuit it was necessary to simulate the circuit with a time step of 1ps. TETA obtained the results shown in Fig.5 in 26.4ms, resulting in a speed-up over HSPICE of 327x.

5: Conclusion and Future work

TETA is an interconnect-centric transistor level engine

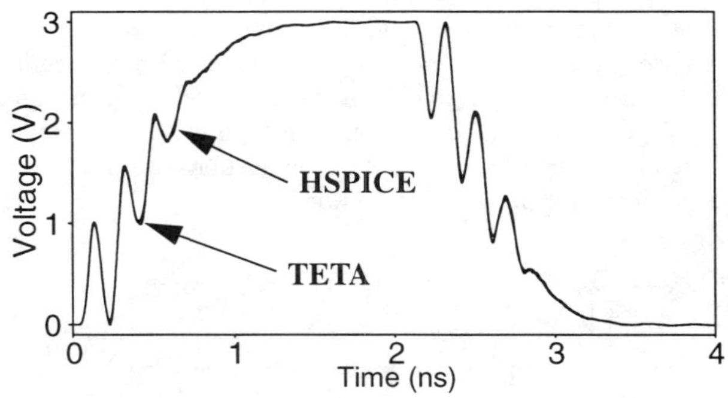

FIG. 5: A clock tree with significant inductive component.

intended for use inside of a timing analyzer to perform waveform calculation. It was created based on a successive chord nonlinear solver for compatibility with the most efficient *N*-port solution. Furthermore, since the successive chord algorithm is able to use table models for the nonlinear devices, TETA can be significantly faster than SPICE even for simple gates driving capacitor loads. Since the successive chord method is represented by a fixed conductance model for nonlinear iterations, *N*-port passivity is not an issue, and simple *N*-port interconnect models that are faster and easier to obtain and simulate can be employed.

We did expect better runtime results for larger transistor cells due to the fact that TETA doesn't do the LU factorization at every time step. But such runtime gains were not realized due to the slower convergence rate of the successive chord method. As future work we will investigate new compaction schemes for transistor clusters to avoid this problem.

BIBLIOGRAPHY

[1] C.L. Ratzlaff, L.T. Pillage, "RICE: Rapid interconnect circuit evaluation using AWE," *IEEE TCAD*, pp. 763-776, 1994.

[2] K.J. Kerns, "Accurate and stable reduction of RLC networks using split congruence transformations," Ph.D. thesis, U. of Washington, September 1996.

[3] A. Odabasioglu, "PRIMA: Passive Reduced-order Interconnect Macromodeling Algorithm," *Int'l Conference on Computer-Aided Design*, Nov. 1997.

[4] A. Devgan, R.A. Rohrer, "Adaptively controlled explicit integration," *IEEE Trans. on CAD*, pp. 746-762, June 1994.

[5] W.J. McCalla, "Fundamentals of computer-aided circuit simulation," Kluwer Academic Publishers, 1988.

[6] F. Dartu, "Gate and transistor level waveform calculation for timing analysis," Ph.D. thesis, Carnegie Mellon Univ., 1997.

[7] L.W. Nagel, "SPICE2, a computer program to simulate semiconductor circuits," Tech. Rep. Memo UCB/ERL M520. Univ. of California, Berkeley, May 1975.

[8] S.Y. Kim, N. Gopal, L.T. Pillage, "Time-domain macromodels for VLSI interconnect analysis," *IEEE Trans. on CAD*, pp. 1257-1270, October 1994.

[9] R.A. Saleh, A.R. Newton, "Mixed-mode simulation," Kluwer Academic Publishers, 1990.

[10] L.T. Pillage, R.A. Rohrer, C. Visweswariah, "Electronic circuit and system simulation methods," McGraw Hill, 1995.

[11] A. Devgan, "Dynamic timing simulation," DAC '96 tutorial.

[12] J.J. Rotman, "An introduction to algebraic topology," Springer-Verlag, 1988.

Validation with Guided Search of the State Space

C. Han Yang
Stanford University
Gates Building, Room 312
Stanford, CA 94305-9030
hyang@verify.stanford.edu

David L. Dill
Stanford University
Gates Building, Room 349
Stanford, CA 94305-9030
dill@cs.stanford.edu

ABSTRACT

In practice, model checkers are most useful when they find bugs, not when they prove a property. However, because large portions of the state space of the design actually satisfy the specification, model checkers devote much effort verifying correct portions of the design. In this paper, we enhance the bug-finding capability of a model checker by using heuristics to search the states that are most likely to lead to an error, first. Reductions of 1 to 3 orders of magnitude in the number of states needed to find bugs in industrial designs have been observed. Consequently, these heuristics can extend the capability of model checkers to find bugs in designs.

Keywords

Model checking, Guided search, Verification

1. Introduction

The complexity of modern chip designs has stretched the ability of the verification techniques and methodologies. Traditional verification techniques use simulators with handcrafted or random test vectors to validate the design. Unfortunately, generating handcrafted test vectors is very labor-intensive, and one is never certain which cases random testing have missed. Model checking techniques [2][7], like Murφ [3], SMV [8], are other means for checking the compliance between implementation and specification. In practice, the primary value of model checkers has been to search for design errors, not to prove correctness.

This paper proposes to optimize model checking for bug finding by using heuristics to search the part of the state space that is most likely to contain design flaws. The property being model checked is described in an assertion checker, which describes an anomalous condition. If the heuristics do find problems, they usually find the problem in substantially fewer states than conventional model checkers, which use breadth-first or depth-first search. When a design is too large for model checking to run to completion, our method is much more likely to find an error before the program exhausts available time or memory. Consequently, these guided search heuristics give the verification engineer another tool that can handle larger designs than traditional model checking.

Several heuristics are discussed in this paper. The first one is *Target Enlargement*, where the error states are enlarged so they can be found with less searching. The second technique is to use *Hamming distance* [4] as the search metric. The third technique, called *Tracks*, uses approximate preimages to help find the violations to assertions. Finally, the last technique uses explicit hints, called *Guideposts*, given by designer to help direct the search. Based on our experience with large industrial examples, we find that Target Enlargement combined with Tracks and Guideposts can consistently find errors much faster than breadth-first search.

The heuristics presented here are more sophisticated, more effective, and are applied to more designs than those found in [11] and [12]. Using Target Enlargement and Hamming distance for guided search were first proposed in [11]. Yuan et al. independently applied similar techniques to two simple designs. In this paper, Target Enlargement and Hamming distance are applied to a much larger set of design examples with wider range in performance.

In the following section, some terminology will be defined. Section 3 discusses the basic guided search algorithm. Section 4 discusses the designs to which these techniques were applied. In sections 5-8, each of the heuristics will be discussed. Section 9 discusses the overall experimental results. Finally, section 10 has some concluding remarks.

2. Background

A finite state machine (FSM) is a 6-tuple, $(S, I, O, \delta, \lambda, s_0)$, where S is the set of states, I is the set of input values, O is the set of output values. $\delta: S \times I \rightarrow S$ is the next state function, $\lambda : S \times I \rightarrow O$ is the set of output function, and $s_0 \in S$ is the initial state. In this paper, let $B = \{0, 1\}$. V is the set of Boolean state variables in the system. Each state is a bit vector, which we represent mathematically as a map from the set of state variables to Boolean values. Hence, S is the set of all such bit vectors, which we write as $[V \rightarrow B]$.

If A is a set of states in S, the preimage of A under δ is defined as follows.

Definition 1: Preimage

$$PreImage(\delta, A) = \{s \mid \exists i \in I . \exists s' \in A . s' = \delta(s, I)\}$$

In essence, *PreImage* is the set of states that can reach a state in A in one transition.

The projection of a set of state onto another set of states with respect to a set of variables V_i is defined as follows.

Definition 2: Projection

$$Proj_{V_i}(T) = \{t' \in [V \rightarrow B] \mid \exists t \in T . \forall v \in V_i . t'(v) = t(v)\}$$

DAC 98, San Francisco, California
©1998 ACM 0-89791-964-5/98/06..$5.00

In other words, the projection of T onto V_i is the set of bit vectors, which have the same values for variables in V_i as some bit vector in T (and have *any* value for variables not in V_i).

3. Guided Search Algorithm

The guided search algorithms are implemented in a verifier called Murφ++. Designs can be described directly in Murφ++ (which is similar to C++) or be translated from Verilog [5]. The algorithm for the verifier is shown in Figure 1. During the search, there are two types of states, *explored states* and *visited states*. Explored states are those that have their next states generated, and visited states are those that have been encountered in the search but whose next states have *not* been generated. Given an initial assignment to all the state variables that is in the PriorityQueue, Murφ++ retrieves a visited state and finds all next states. Because Murφ++ is able to represent all combinations of inputs symbolically, all next states are found at once. Then, the evaluation function is used to compute the score for each next state. If the state has not been seen before, then all values for the state variables are stored in a hash table, and the state is placed in the priority queue. For breadth-first search, the evaluation function is simply a constant. Otherwise, the search heuristic will provide a score for each next state, and the state is stored in the priority queue according to its score. Backtracking is used to find all reachable states. When no more states are found, the verification process is complete. Assertion checker, which describes what should not happen in the design, is also coded as part of the description. If an assertion is violated during the generation of the reachable states, an error trace is printed out.

```
1.  Put Start State in PriorityQueue
2.  While (PriorityQueue not empty)
3.        Dequeue State from Priority-Queue
4.        Find set of Next States
5.        Use Evaluation Function to find
          score for each Next State
6.        If Next States not in HashTable
7.            Put into PriorityQueue
8.        Check for Assertion violation for
              each Next State
```

Figure 1 Guided Search Algorithm

4. Benchmark Designs

Several realistic industrial designs are used to evaluate the effectiveness of the heuristics. They are briefly described below.

4.1 S3.mp Data Link Level Protocol

Sun Microsystems' S3.mp distributed shared memory computer uses workstations connected with a gigabit serial link to form a parallel computer [9]. The data link level protocol used in the serial link, which operates over a slotted ring [10], was first developed in the Murφ verification system [3]. The description of this protocol is about 1200 lines of Murφ++ code. For a ring with 4, 6, and 8 slots, there are 61, 85, and 107 state bits, respectively.

4.2 FLASH Examples

Many of the benchmarks are portions of the Stanford FLASH (FLexible Architecture for Shared memory) multiprocessor [6]. These designs were all converted from the original Verilog code that was annotated to extract the control logic [5]. Many of the assertions were also translated directly from the FLASH validation suite in Verilog. All of the bugs that we have found were actually encountered by the FLASH design team through their validation testing.

Table 1 lists some information about each design. There are two versions of MC2, big and small. The only difference between the two is the range of inputs that were considered. For the smaller designs, it is easier to discuss some of the behavior of the heuristics (in Section 7.1).

Design	State Bits	Min Cycles to Violate Assertion
Inbox with Cache Control (Inbox)	174	5
Memory Controller State Machine (MC1)	39	17
Memory Controller with SDRAM Control (MC2)	180	39
Load/Store Control (LSC)	112	49

Table 1 Flash Benchmarks

5. Target Enlargement

As the name suggests, Target Enlargement is an effort to make the set of states that will violate the assertion, called error states, bigger.[1] The preimage of these error states is the set of states that in one cycle can reach an error state. If it is possible to reach a state in the preimage from a start state, then it is also possible to reach an error state. Each successive preimage potentially describes an even larger set of states that can reach the error states. The larger target increases the opportunity for the guided search to find a path to the error states and consequently reduces the amount of searching that needs to be done. Because computing preimages can be a memory intensive operation, Target Enlargement computes successive preimages until a given computer memory limitation is reached.

5.1 Target Enlargement Experimental Results

Table 2 shows the number of cycles of target enlargement that was possible with each of the designs within 200 Mbytes of memory. While the size of the Binary Decision Diagrams(BDDs) is not very large, computing the next larger preimage for all the examples except Inbox, MC1, and MC2 (Small), exceeded the memory limit [1].

Design	Max Cycles of Target Enlargement	BDD Size of Largest Target Enlargement
CM (4 Slots)	4	55,301
CM (6 Slots)	4	33,374
CM (8 Slots)	4	25,235
Inbox	5	28
MC1	5	95
MC2 (Small)	4	1,457
MC2 (Big)	6	23,885
LSC	4	13,172

Table 2 Target Enlargement

Table 3 shows the number of visited states and explored states. The number of visited states includes those that eventually became explored states. The number of visited states is an indication of the size of the hash table that holds the reachable states. The number of explored states is an indication of the efficiency of the search heuristic. Fewer explored states would mean a better search heuristic. For some designs, like MC2 (Big),

---

[1] Yuan et al. refer to *target enlargement* as *retrograde analysis*.

it was not possible to find the set of reachable states within 160 Mbytes of memory.

Design	Visited States		Explored States	
	Breadth-first Search	Target Enlargement	Breadth-first Search	Target Enlargement
CM (4 Slots)	32,773	197	12,784	83
CM (6 Slots)	386,994	17,650	379,088	6,437
CM (8 Slots)	>2,362,324	>2,362,324	>432,177	>145,498
Inbox	77,313	38,913	35,842	12,802
MC1	117,889	57,089	90,320	47,874
MC2 (Small)	161,577	119,802	149,210	113,920
MC2 (Big)	>4,799,484	>4,799,484	>2,278,144	>2,278,144
LSC	246,553	224,767	237,508	223,744

Table 3 Target Enlargement Experimental Results

Table 3 shows that Target Enlargement can have significant reductions in the number of visited and explored states to find a violation of the assertions. While Yuan et al. also reported some reductions in the number of states needed to find bugs on two small examples, Table 3 illustrates the use of Target Enlargement on a larger and more realistic set of examples [12]. In addition, the reduction on this set of examples can be up to two orders of magnitude, which is much larger than shown by Yuan et al.. Unfortunately, the effectiveness of Target Enlargement varies dramatically from one design to another.

5.2 Analysis of Target Enlargement

Target Enlargement is most effective when it substantially increases the probability of visiting a target state, assuming a search that randomly generates inputs. Consider an example where Target Enlargement is not very effective: Suppose there is a sequence of n pipeline registers between the point where an erroneous signal is generated and the assertion that is eventually violated by the signal. In this case, the probability of hitting an error becomes 1 when the signal is generated n cycles before it is actually detected. Target Enlargement will not improve the search significantly until $n + 1$ cycles of preimages have been computed. One of our benchmarks, MC2, has this behavior. On the other hand, if one out of a thousand inputs reaches an error state in the next cycle, one cycle of target enlargement will increase the probability of visiting the error state by a thousand fold. MC2's first cycle of target enlargement exhibits this behavior. Because Target Enlargement's ability to increase the error probability strongly depends on the design, the location of the inputs and the location of the assertions, the performance of Target Enlargement varies widely across the benchmark designs. Despite the dependency on the design, Target Enlargement consistently reduces the number of states needed to find violation to the assertions; consequently, it is used in conjunction with all other heuristics described later in this paper.

6. Hamming Distance

The first search heuristic is Hamming distance, which is defined as the number of bits that are different between two bit vectors [4]. In generating the set of reachable states, Hamming distance can be used as a search metric [11][12]. Those states that have the lowest Hamming distance to the largest Target Enlargement are explored first. In essence, we hope states with very few bits differing from the enlarged target will require very few cycles to reach that target. The minimum Hamming distance between a single state and the enlarged target is computed using an algorithm that has a complexity which is linear with respect to the size of the BDD of the enlarged target.

6.1 Hamming Distance Experimental Results

Table 4 and Table 5 show the result for applying Hamming distance to the benchmark designs. At first, the Hamming distance is measured against the largest enlarged target. In the case of the communication module, this simple heuristic worked exceptionally well. In the case of CM with 6 and 8 slots, the reduction in the number of states found was about 1 to 2 orders of magnitude. In other designs, Hamming distance's performance was less stellar. For the FLASH examples, the reduction in the number of states for most designs ranged from 10-50%. For the case of LSC, the result was actually worse than Target Enlargement.

Design	Visited States		
	Target Enlargement	Hamming Distance (Large Target) + Target Enlargement	Hamming Distance (Small Target) + Target Enlargement
CM (4 Slots)	197	197	179
CM (6 Slots)	17,650	17,176	1,092
CM (8 Slots)	>2,362,324	417,671	5,947
Inbox	38,913	38,913	12,913
MC1	57,089	38,913	28,161
MC2 (Small)	119,802	119,802	94,618
MC2 (Big)	>4,799,484	>4,793,500	>4,793,500
LSC	224,767	343,128	225,344

Table 4 Hamming Distance Visited States

Design	Explored States		
	Target Enlargement	Hamming Distance (Large Target) + Target Enlargement	Hamming Distance (Small Target) + Target Enlargement
CM (4 Slots)	83	81	83
CM (6 Slots)	6,437	6,348	499
CM (8 Slots)	>145,498	901	2,735
Inbox	12,802	12,802	12,802
MC1	47,874	22,019	8,451
MC2 (Small)	113,920	113,594	88,442
MC2 (Big)	>2,278,144	>2,274,048	>2,260,480
LSC	223,744	245,817	222,357

Table 5 Hamming Distance Explored States

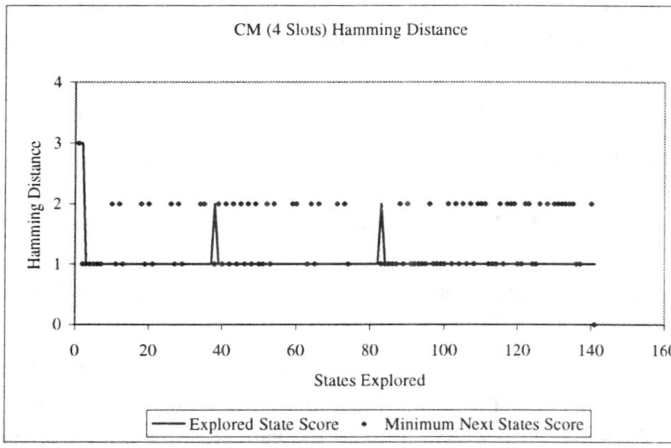

Figure 2 Hamming Distance Example

For example, Figure 2 shows the explored state's Hamming distance state and the minimum Hamming distance of its next states for a small example, CM (4 Slots). The start state has a relatively large Hamming distance. Soon, the search rapidly reduced the Hamming distance to 1. Unfortunately, these states did not lead to states that were even closer. Eventually, the search had to resort to states with Hamming distance of 2. After exhausting more states with Hamming distance of 1, the search again had to find a state with distance 2. Eventually, the search found the enlarged target. Throughout the search, all the next states had a Hamming distance of 1 or 2. This is typical of Hamming distance, which tends not to be very discriminating.

Our experiments show an interesting interaction between Hamming distance and Target Enlargement. While Hamming distance is "aiming" toward a target, Target Enlargement is really making the target larger and therefore less discriminating, which negates some of the effectiveness of Hamming distance. In fact, when Hamming distance is "aimed" toward the smallest target, which is the least fuzzy, the results were better. Using this technique, another order of magnitude of improvement can be achieved with the CM. In addition, all the FLASH benchmarks also showed improvement over Hamming distance with large target.

Based on these experimental results, it appears that Hamming distance can be a useful heuristic in finding the violations of assertions. On a few small examples, Yuan et al. reported modest reduction in number of steps needed to find the error [12]. This much larger set of design benchmarks suggests that while Hamming distance can reduce the number of states needed to find the error by up to several orders of magnitude, its performance is also very inconsistent across designs.

7. Tracks

As can be seen from the results in Section 5, for many of the designs, it is not possible to do Target Enlargement for many cycles. However, *approximate* preimages that are based on a subset of the state variables may be computed for more cycles than Target Enlargement. In practice, a subset of the state variables can control most of the behavior of the design. This subset may be the state variables from a few of the state machines within the design, or may be a few key state variables within a FSM that dictates its behavior. For example, in the control logic of FSMs, the Verilog description usually has a `switch` statement that depends on a state variable that determines much of the behavior. Track computes a series of approximate preimages based on a given set of variables that strongly control the behavior of the system. With multiple tracks implicitly conjoined, it may be possible to construct a sufficiently accurate preimage that aids the guided search.

Tracks are defined formally in Equation 1.

Equation 1 Track Computation

$$T_i^{k+1} = PreImage[\,\delta, Proj_{V_i}(T_i^k)\,]$$

where T is one layer of one track, i is the Track number, V_i is the variables in Track i, and k is the layer number, or number of cycles away from the largest enlarged target.

T_i^0 is the largest enlarged target. In essence, the next layer of a track is the preimage of the projection of the last layer onto the variables in the track. Because all the variables that are not in the track are projected out, the resulting preimage is always a superset of the true preimage. The projection also reduces the size of the BDD and thus enables computation of more preimages.

Our experience shows that the following guidelines are useful in choosing the variables in a track.

1. Identify the main FSMs in the design
2. Identify the main state variables in the FSMs
3. Rank other variables that the designer deem "important"
4. Start with the assertion and working backward toward primary inputs, include as many FSMs as possible. The size of the partition depends on the size of the BDDs for each of the preimages.

More than one track can be used to approximate the true preimage more accurately. In fact, tracks can have overlapping state variables. During the guided search, the state's score depends on to which layer the state belongs. The score is the *least* layer number in which *all* tracks contain the state. This score is the minimum cycle number that satisfies the implicit conjunction of all the tracks. Consequently, this evaluation function greedily chooses the layer number that is closest to the target. States that have the smallest score, or those that are closest to the enlarged target, are explored first.

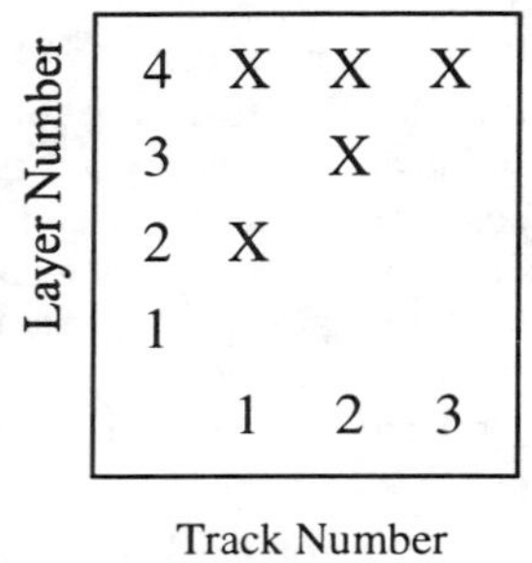

Figure 3 Tracks Evaluation Function Example

As an example of the evaluation function, Figure 3 shows a state that is in layers 2, 3, and 4 of three tracks. Because the tracks are approximate preimages, a state can actually belong to multiple layers of the same track. The evaluation function returns a value of 4 since all tracks agree that it is the minimum layer number.

7.1 Tracks Experimental Results

For most of the designs, Tracks worked quite well across all designs except LSC (Table 6).

Design	Visited States: Target Enlargement	Visited States: Tracks + Target Enlargement	Explored States: Target Enlargement	Explored States: Tracks + Target Enlargement
CM (4 Slots)	197	99	83	41
CM (6 Slots)	17,650	1,655	6,437	644
CM (8 Slots)	>2,362,324	2,313	>145,498	969
Inbox	38,913	4,162	12,802	5
MC1	57,089	11,394	47,874	76
MC2 (Small)	119,802	1,498	113,920	161
MC2 (Big)	>4,799,484	17,474	>2,278,144	4,283
LSC	224,767	210,946	223,744	208,228

Table 6 Tracks Experimental Results

Figure 4 shows the score for MC2 (Small) during the search. Because MC2 (Small) has relatively few nondeterministic inputs, it was possible to compute preimages from the assertion to the start state and plot the true distance of the explored states to the enlarged target. This example illustrates the effectiveness of the approximate preimages. As can be seen in Figure 4, the score produced by Tracks initially was routinely off by 10 cycles or more. If all states were uniformly inaccurate, then the score still would give the right guidance to the search. In fact, Figure 4 shows that the curve between the explored state score and the actual distance to have similar shape.

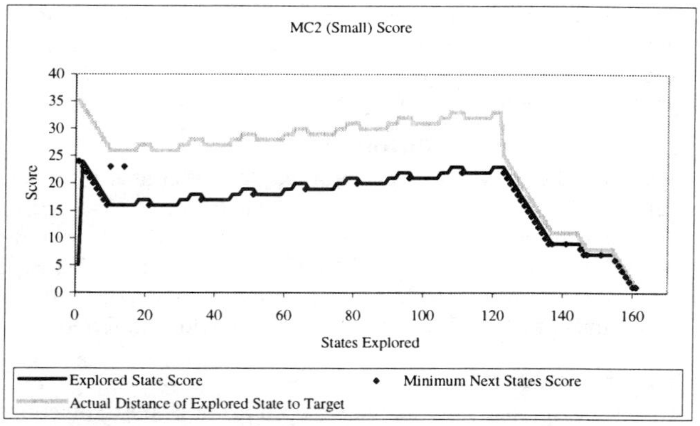

Figure 4 Tracks Score Compared with True Score

However, there was a local minimum introduced in the search due to varying amount of inaccuracy occurring at exploring states 10 and 14. Because of this local minimum, the search explored states that had good score, but were not able to lead to the enlarged target. At state 123, the search finally went back to the states that were found when exploring states 10 and 14. Eventually, exploring these resulted in finding states that eventually lead to the target zone.

Tracks allow the verification engineer to trade off between accuracy and size of BDDs. While a track with more state bits offers more accuracy, the size of the BDD usually grows faster. When the size of the BDD is too large, the verification engineer can remove some state bits and introduce several new overlapping tracks to compensate for the loss in accuracy. Consequently, Tracks enables the verification engineer to give high-level guidance to the search while working within the computer memory limitation.

Unfortunately, Tracks did not work very well with the LSC design. This particular design had many very small interacting FSMs. The other design usually had one large FSM that controlled the behavior of the system. Because of the large number of FSMs, it was very difficult to pick a series of tracks that were able to compute preimages with any accuracy. Consequently, Guidepost was added to help in the search for LSC.

8. Guidepost

In addition to more automated heuristics mentioned previously, designers can also provide *hints* to the guided search. The hints, called *Guideposts*, are a series of conditions that the designer believe to be interesting or even required preconditions for the assertion checker to be violated.

Guidepost encodes the number of hints that the search has gone through in the score itself. Consequently, the score for a child state is dependent on the score of the parent state *and* on the properties in the child state. Equation 2 shows the evaluation function for Guideposts. Just like the evaluation function for Tracks, a lower score is deemed to be a state that is closer to the target. In other words, the passage of more guideposts should indicate a lower score, which is simply the number of guideposts yet to be encountered or $(T_{Guideposts} - P_{Guideposts})$. The multiplication of this term with M_{Cycles} ensures that the number past guideposts dominates over the score from Tracks. In fact, the Track score only has an effect when the number of previous guideposts is the same. With this kind of evaluation function, the score is a summary of the history of the hints that the search has encountered.

Equation 2 Guidepost Evaluation Function

$$Score = (T_{Guideposts} - P_{Guideposts}) * M_{Cycles} + Score_{Tracks}$$

Where $T_{Guideposts}$ is the total number of guideposts

$P_{Guideposts}$ is the number of guideposts that the current state and its ancestors have past through

M_{Cycles} is the maximum number of cycles in all the tracks.

$Score_{Tracks}$ is the Score from Tracks evaluation function

This evaluation function biases the search towards going through the guideposts. As the search encounters a new hint, the score improves according to the number of hints that it has encountered in the past. Consequently, the score only improves when the hints are found in the prescribed total order.

8.1 Guidepost Experimental Results

Table 7 shows the results for Guidepost. Only one guidepost was used for each design benchmark. Many of the guideposts were only to help the search to get out of a local minimum that was introduced by Tracks.

In the LSC, many FSMs control the complex operations of handling normal loading and storing of data from the cache, managing the store buffers and exceptions such as loading before storing. In the LSC, the condition of the guidepost was the FSM in the state ready to start a cache refill operation and starting a counter that would indicate the completion of the cache refill. With this simple condition, the enlarged target for the LSC can be found in about 15% of the states needed for Target Enlargement.

	Visited States		Explored States	
Design	Target Enlargement	Guideposts + Target Enlargement	Target Enlargement	Guideposts + Target Enlargement
CM (4 Slots)	197	94	83	38
CM (6 Slots)	17,650	1,650	6,437	641
CM (8 Slots)	>2,362,324	2,296	>145,498	955
Inbox	38,913	4,162	12,802	5
MC1	57,089	11,394	47,874	12
MC2 (Small)	119,802	1,834	113,920	39
MC2 (Big)	>4,799,484	9,602	>2,278,144	289
LSC	224,767	37,070	223,744	30,503

Table 7 Guidepost Experimental Results

9. Comparison of Heuristics

Figure 5 shows the number of visited states as a percentage of the breadth-first states. Similarly, Figure 6 shows percentages for the explored states (Note that the vertical axes in both of these figures are in logarithmic scale). As can be seen, Target Enlargement always reduces the search space except CM (8 Slots) and MC2 (Big), which were simply too big to find the error using Target Enlargement alone. Hamming distance with the large target did not perform well consistently; however, Hamming distance with small target did show reductions in either explored or visited states for all designs. With Tracks and Guidepost, the error states was found consistently faster.

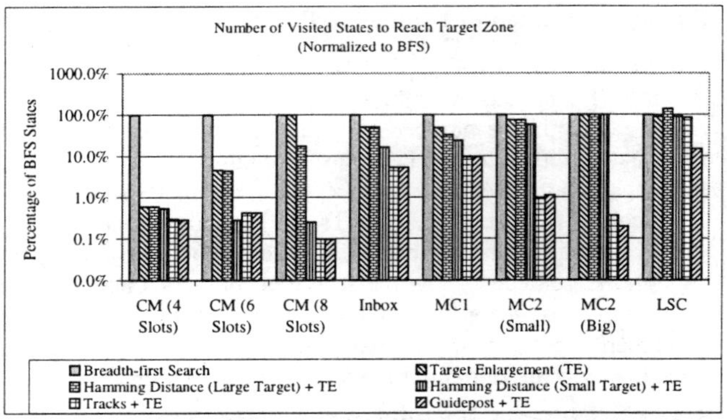

Figure 5 States Visited as a Percentage of Breadth-first Search

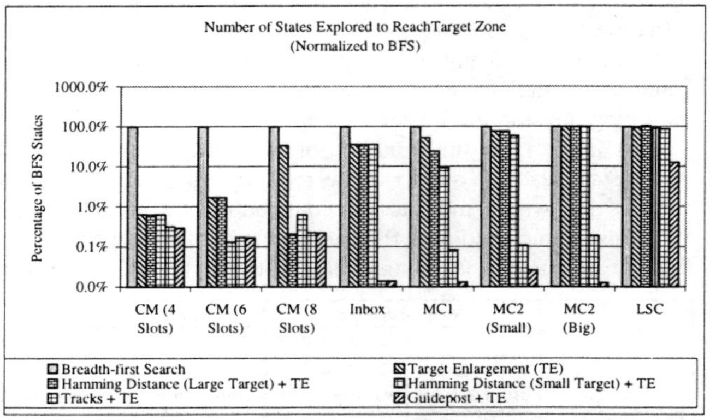

Figure 6 States Explored as a Percentage of Breadth-first Search

10. Conclusion and Future Work

Several heuristics have demonstrated to be very effective in guiding the search to find violations to assertions in realistic designs. These heuristics can find the errors in much fewer states than the breadth-first search that is used in model checking. These techniques extend the capability of model checking while providing a more powerful tool than conventional simulation.

Hamming distance is relatively easy to implement; however, its effectiveness is rather inconsistent. Techniques like Tracks and Guidepost offer more robustness. Unfortunately, they also require more designer input. This set of heuristics offers the verification engineer tradeoffs for manual labor and effectiveness in the guided search.

Guidepost can be extended to allow for creation of high level test vectors. The verification engineer only needs to specify the critical events that a test vector needs to go through. Then, the heuristic can go through the design and fill in a set of inputs that will lead the design through those critical events. Being able to specify such a high level test vectors should improve the productivity of verification engineers and merits further study.

11. Acknowledgements

Many thanks for David Nakahira and Jules Bergmann for their help in the memory controller and load/store control design examples, respectively. This work was sponsored under contract number DABT63-95-C-0049-P00005. The content of this paper does not necessarily reflect the position of the policy of the Government and no official endorsement should be inferred.

12. References

1. R. E. Bryant, Graph-based Algorithms for Boolean function Manipulation. IEEE Transactions on Computers, 6(C-35):677-691, August 1986
2. E. M. Clarke, E. A. Emerson, A. P. Sistla, *Automatic Verification of Finite-state Concurrent Systems using Temporal Logic Specifications*, ACM Transactions on Programming Languages and Systems, vol. 8, no. 2 pp. 244-63
3. D. L. Dill, A. J. Drexler, A. J. Hu, and C. H. Yang, *Protocol Verification as a Hardware Design Aid.* 1992 ICCD, pp.522-525
4. R. W. Hamming. *Error Detecting and Error Correcting Codes.* Bell System Tech Journal, 9:147-160, April 1950.
5. R. C. Ho, C. H. Yang, M. A. Horowitz, and D. L. Dill. *Architecture Validation for Processors.,* 1995 ISCA, pp. 403-413
6. J. Kuskin, D. Ofelt, M. Heinrich, J. Heinlein, R. Simoni, K. Gharachorloo, J. Chapin, D. Nakahira, J. Baxter, M. Horowitz, A. Gupta, M. Rosunblum, and J. Hennesy. *The Stanford FLASH Multiprocessor.* 1994 ISCA, pp. 302-313
7. F. Maruyama, M. Fujita, *Hardware Verification*, Computer, vol. 18, no. 2, pp 22-32, Feb 1985.
8. K. L. McMillan, *Symbolic Model Checking*. Ph.D. thesis, School of Computer Science, Carnegie Mellon University, May 1992.
9. A. G. Nowatzyk, M. C. Browne, E. J. Kelly, and M. Parkin. *S-connect: From Network of Workstations to Supercomputer Performance.* 1995 ISCA, pp. 71-82
10. A. S. Tanenbaum, *Computer Networks*, chapter 7, pp. 312-313. Prentice-Hall Inc., 1981
11. C. H. Yang, D. L. Dill, *SpotLight: Best-First Search of FSM State Space.* IEEE International High Level Design Validation and Test Workshop, presented on November 16, 1996, http://verify.stanford.edu/hyang/hldvtSlides.ps
12. J. Yuan, J. Shen, J. Abraham, A. Aziz, *On Combining Formal and Informal Verification*, 1997 CAV, pp. 376-387

Efficient State Classification of Finite State Markov Chains †

Aiguo Xie and Peter A. Beerel
Department of Electrical Engineering - Systems
University of Southern California
Los Angeles, CA 90089-2562

Abstract This paper presents an efficient method for state classification of finite state Markov chains using BDD-based symbolic techniques. The method exploits the fundamental properties of a Markov chain and classifies the state space by iteratively applying reachability analysis. We compare our method with the current state-of-the-art technique which requires the computation of the transitive closure of the transition relation of a Markov chain. Experiments in over a dozen synchronous and asynchronous systems demonstrate that our method dramatically reduces the CPU time needed, and solves much larger problems because of reduced memory requirements.

1 Introduction

Markov chains are an important class of stochastic processes that can effectively model randomly evolving systems. They are conceptually simple because they have *short memory*, i.e., their future evolution depends only on the current outcome and not on further history. Although they are simple, Markov chains have found vast applications in engineering including the analysis of system reliability [1, 2], control, e.g., [3], power consumption [4, 5] and performance [6, 7, 8, 9, 10]. The first two applications are usually concerned with the short-term (*transient*) behavior or *time to certain events* of the Markov chain whereas the latter two applications often involve knowledge of the long-term (*stationary*) behavior. An important and typically the first step in carrying out these analyses is the structural analysis of the Markov chain. It partitions the state space into two sets, namely, the set of *transient states* (those expected to be visited only a finite number of times) and the set of *recurrent states* (those expected to be re-visited infinitely often once they are visited). The set of recurrent states may be further classified into different classes which once entered trap the Markov chain inside the class indefinitely. Both transient and stationary analyses need these structural information in order to proceed or get simplified. The purpose of this paper is to present an efficient method that performs this state classification step.

The structure of the state space of a Markov chain may be represented by a directed graph where vertices correspond to the states and edges correspond to the transitions among the states. One method to perform structural analysis of the state space is to *explicitly* identify all the (maximal) *strongly connected components* (SCCs) of the graph. Components (those called terminal SCCs) that do not have edges leading to other components correspond to the recurrent state classes and all other components correspond to the set of transient states. Since Markov chains derived from real systems often possess huge state space [6, 11, 8], this method can be intolerably slow because it analyzes states *sequentially*. Another method for structural analysis is described in [11][1] which uses *implicit* (symbolic) techniques based on *binary decision diagrams* (BDDs) [12] so that all the SCCs of the corresponding graph are *concurrently* computed and the terminal SCCs are subsequently identified. A key step of this method uses the recursive paradigm in [13] to compute all the states that can be reached from individual states, or from a graph point of view, the *transitive closure* of the corresponding graph. Unfortunately, this step tends to be *memory-intensive* and *time-consuming*.

This paper also proposes a BDD-based symbolic technique but exploits the fundamental property of transience and recurrence of a Markov chain. Our technique *sequentially* explores the state space by repeatedly applying symbolic reachability analysis. Newly reached transient states and recurrence class are detected after each reachability analysis. Since most systems contain only a few recurrence classes and a large portion of their state spaces are recurrent, the number of times reachability analysis is performed is often very small. Finally, because reachability analysis is typically much faster and less memory-intensive than transitive closure computation, our method dramatically outperforms the one discussed in [11] on almost all the examples we have tested. In particular, we observe run time reduction of up to several orders of magnitude, and our method solves all the examples on which the previous method

†This work is funded in part by a NSF CAREER Award MIP-9502386 and a gift from the Intel Corporation.

DAC 98, San Francisco, California
©1998 ACM 0-89791-964-5/98/06..$5.00

[1]We note, however, that the main thrust of [11] is symbolic algorithms for stationary analysis with less emphasis on structural analysis.

fails due to insufficient memory.

The remaining of this paper is organized as follows. Section 2 reviews basics of Markov chains and the previous work on their structural analysis. Section 3 describes the principles, proofs, and the algorithm of our method. Experimental results are given in Section 4 and our conclusions are given in Section 5.

2 Preliminaries and Previous Work

2.1 Finite-state Markov chains

Consider a discrete state space $S = \{1, 2, \cdots\}$. A discrete-time stochastic process X on S is a sequence of random variables, $\{X_n \in S : n \geq 0\}$. We call the value assumed by random variable X_n the state of X at time instant n. X is a *Markov chain* (or *Markovian*) if its future evolution depends only on its current state. More formally, if X is Markovian and it is in state $i \in S$ at time n, then the probability that state $j \in S$ is visited at time $n+1$ is independent of its history before time n, i.e, $Pr(X_{n+1} = j \mid X_n = i, (X_{n-1}, \cdots, X_0) \in S) = Pr(X_{n+1} = j \mid X_n = i)$. X is *time-homogeneous* if the above conditional probability is independent of time n. In this case, the evolution of X is governed by a (1-step) *transition probability matrix* $P = (p_{ij}), i, j \in S$ where $p_{ij} = Pr(X_{n+1} = j \mid X_n = i) = Pr(X_1 = j \mid X_0 = i)$.

Real systems can often be modeled as Markov chains which are time-homogeneous. We assume the state space S is finite so that one may write $S = \{1, \cdots, N\}$ where N is the size of S. Let $M = (S, P)$ denote such a chain with probability transition matrix P. Systems with massive concurrency [6] or having correlations among system events [4] can all be modeled by Markov chains with properly defined state spaces.

To consider the relationship between states of S, one needs the notion of *communication*. A state $i \in S$ communicates with state $j \in S$, denoted by $i \rightarrow j$, if there is a positive probability that state j will eventually be visited once state i is visited. Or equivalently, $p_{ij}(n) = Pr(X_n = j \mid X_0 = i) > 0$ for some $n > 0$. Otherwise, state i does not communicate with state j, denoted by $i \not\rightarrow j$. States i and j *intercommunicate* if $i \rightarrow j$ and $j \rightarrow i$, written $i \leftrightarrow j$.

Definition 2.1 *State $i \in S$ is recurrent if $\forall j \in S$ such that $i \rightarrow j$, $j \rightarrow i$ as well. Otherwise, state i is transient, i.e, $\exists j \in S$ such that $i \rightarrow j$ but $j \not\rightarrow i$.*

It can be easily checked that a recurrent state is visited infinitely often once it is visited, while the expected number of the times M visits a transient state is bounded from above.

Many of the nice properties of Markov chain M can be derived from the famous *Chapman-Kolmogorov* theorem [14] which expresses the probability of M to go from one state to another in $n \geq 0$ time steps in terms of the 1-step transition matrix P. One such property concerning the communication relation is as follows.

Theorem 2.1 *If $i \rightarrow j$ and $j \rightarrow k$, then $i \rightarrow k$. Further, if $i \leftrightarrow j$, then $j \leftrightarrow i$.*

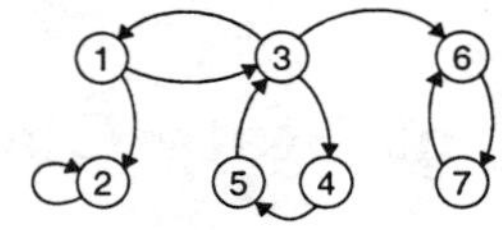

Figure 1: A small Markov chain example.

By Definition 2.1 and Theorem 2.1, it is easily verified that if states i and j intercommunicate, either both of them are recurrent or both of them are transient.

Theorem 2.1 ensures that $\leftrightarrow$ is an equivalence relation so that the state space can be partitioned into equivalence classes according to $\leftrightarrow$. That is, $S = T \cup R_1 \cup R_2 \cdots \cup R_K$ where T is the set of transient states, R_1 through R_K are K recurrence classes each containing recurrent states that do not communicate with states in other classes.

Example Figure 1 shows a Markov chain containing 7 states. State 1 is clearly transient because $1 \rightarrow 2$ but $2 \not\rightarrow 1$. States 3, 4 and 5 are also transient for similar reasons. The chain has two recurrence classes, namely, $\{2\}$ and $\{6, 7\}$. □

We have seen that the transience and recurrence property of states can be derived from the their communication relation. In particular, the structure of the state space can be represented by a directed graph $G = (V, E)$, where the vertices in V are in one-to-one correspondence with the states in S. There is an edge e_{uv} from vertex u to v iff $p_{ij} > 0$ where the vertices u and v correspond to the states i and j. Then, $i \rightarrow j, i, j \in S$ means there is a path from vertex u to v in G. Similarly, $i \leftrightarrow j$ means vertices u and v are strongly connected. We call graph G the boolean transition graph of M.

Many types of analysis may be done on the Markov chain M. These analyses include the computation of the probability distribution of M over its state space at a fixed time instant (*transient analysis*), the expected time and probability to enter a set of states (*first-passage-time and absorption probability analysis*) [15], and its asymptotic probability distribution after an infinite amount of time (*stationary analysis*). State classification plays plays an important role in these analyses. For instance, first-passage-time analysis is often performed against a particular recurrence class. By treating each recurrence class as a single state, absorption analysis may also be accelerated. Moreover, the stationary analysis is not possible without the identification of the recurrence classes (unless M is guaranteed to have only one recurrence class) because the system will be underdetermined if the chain has multiple recurrence classes. In fact, even when the number of the recurrence classes is expected to be one, it is often necessary to verify this is the case because modeling errors or design bugs may introduce extra recurrence classes.

2.2 Previous work

We focus on the recently proposed symbolic method [11] which finds the recurrent classes by *concurrently* identifying all terminal SCCs of the transition graph G

associated with the Markov chain.

The method symbolically computes *all* the SCCs in G by creating the transitive closure TC of G which describes the set of states that each state of S communicates with. One may think of TC as a boolean matrix of size $S \times S$, i.e, $TC_{ij} = \begin{cases} 1 & if \quad i \to j, \\ 0 & otherwise. \end{cases}$ Therefore, checking if $i \to j$ is equivalent to checking if the element $TC_{ij} = 1$. Suppose it is determined that $i \to j$ by the fact that $T_{ij} = 1$. To see if $j \to i$ so that i, j are strongly connected, one merely needs to check if $T_{ji} = 1$. Thus, the second step in the method is to *term-wise* multiply TC with its transpose, TC^t, yielding a symmetric matrix (let it be denoted by $TC * TC^t$). Clearly, i and j are in the same SCC *iff* $(TC * TC^t)_{ij} = (TC^t * TC)_{ji} = 1$. All above operations are performed symbolically.

Next, each SCC is represented by a particular state (called the representative state) of the SCC using the so called *datum priority function*[16] of the corresponding BDD. A new directed graph consisting of these representative states is then constructed so that it has an edge from one (representative) state i to another (representative) state j *iff* there is an edge from some state in the SCC represented by i to some state in the SCC represented by j. This new graph is *acyclic* and its leaf vertices represent the terminal SCCs.

The bottleneck of the method is the computation of the transitive closure TC of G. To this end, a recursive algorithm presented in [13] which also uses BDD-based symbolic operations is adopted. Nevertheless, it can still take intensive amount of CPU time and memory to compute TC when G is large.

3 State classification by iterative application of reachability analysis

The method we present in this section does not compute the transitive closure of the graph G. Rather, it relies on the definition of a transient/recurrent state. Once we determine a state to be transient/recurrent, the transience and recurrence properties of other states that communicate with this state are deduced and they are removed from further consideration.

3.1 The principles

We start by defining the *forward* and *backward* sets of a state.

Definition 3.1 *The forward set of state $i \in S$, denoted by $\mathcal{F}(i)$, is the set of states that i communicates with. That is, $\mathcal{F}(i) = \{j \in S \mid i \to j\}$. Similarly, the backward set of state i, denoted by $\mathcal{B}(i)$, is the set of states that communicate with i. That is, $\mathcal{B}(i) = \{j \in S \mid j \to i\}$.*

The following lemma shows a simple property of forward and backward sets.

Lemma 3.1 *Let $i, j \in S$. If $j \in \mathcal{F}(i)$, then $\mathcal{F}(j) \subseteq \mathcal{F}(i)$. Similarly, if $j \in \mathcal{B}(i)$, then $\mathcal{B}(j) \subseteq \mathcal{B}(i)$.*

Proof The proof is a direct consequence of the transitivity property of relation $\to$ given in Theorem 2.1. Suppose $j \in \mathcal{F}(i)$, i.e., $i \to j$, then, $\forall k \in \mathcal{F}(j)$, i.e., $j \to k$, we have $i \to k$, meaning $k \in \mathcal{F}(i)$ so that $\mathcal{F}(j) \subseteq \mathcal{F}(i)$. Similarly, if $j \in \mathcal{B}(i)$, i.e., $j \to i$, then, $\forall k \in \mathcal{B}(j)$, i.e., $k \to j$, we have $k \to i$, which implies $\mathcal{B}(j) \subseteq \mathcal{B}(i)$. □

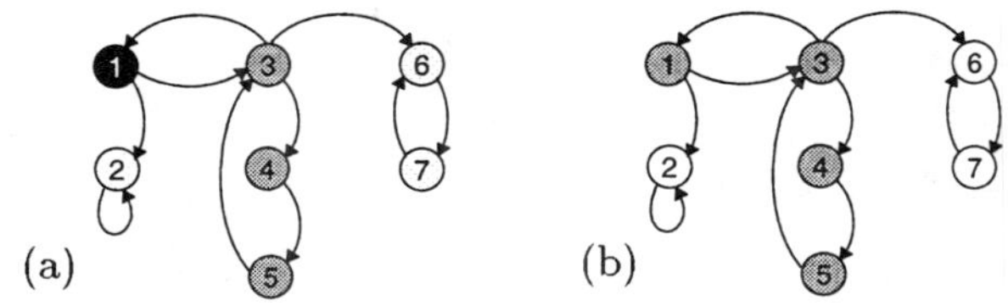

Figure 2: Determining if a state is recurrent: (a) computing $\mathcal{F}(1)$ and $\mathcal{B}(1)$, (b) $\{1\} \cup \mathcal{B}(1)$ all transient.

Theorem 3.1 *A state $i \in S$ is recurrent if and only if $\mathcal{F}(i) \subseteq \mathcal{B}(i)$. In other words, i is transient if and only if $\mathcal{F}(i) \not\subseteq \mathcal{B}(i)$.*

Proof Suppose $i \in S$ is recurrent, then $\forall j \in S$ such that $i \to j$, i.e., $j \in \mathcal{F}(i)$, we have $j \to i$ by Definition 2.1 of a recurrent state. That is to say, $j \in \mathcal{B}(i)$. Therefore, $\mathcal{F}(i) \subseteq \mathcal{B}(i)$. Conversely, if $\mathcal{F}(i) \subseteq \mathcal{B}(i)$, which means $\forall j \in S$ such that $i \to j$, we have $j \in \mathcal{F}(i) \subseteq \mathcal{B}(i)$, and hence $j \to i$. Then i is recurrent by Definition 2.1. The fact that a state is either recurrent or transient concludes the proof. □

Theorem 3.2 *If state $i \in S$ is transient, then states in $\mathcal{B}(i)$ are all transient. If state i is recurrent, on the other hand, states in $\mathcal{F}(i)$ are all recurrent. In the latter case, set $\mathcal{F}(i)$ is a recurrence class, and set $\mathcal{B}(i) - \mathcal{F}(i)$ (if not empty) contains only transient states.*

Proof Suppose state i is transient. By Theorem 3.1, $\mathcal{F}(i) \not\subseteq \mathcal{B}(i)$, i.e., $\exists k \in \mathcal{F}(i)$ such that $k \notin \mathcal{B}(i)$. Now, suppose state $j \in \mathcal{B}(i)$, then $j \to k$ so that $k \in \mathcal{F}(j)$. This is because $i \in \mathcal{F}(j)$ so that $\mathcal{F}(i) \subseteq \mathcal{F}(j)$ by Lemma 3.1. On the other hand, $\mathcal{B}(j) \subseteq \mathcal{B}(i)$ since $j \in \mathcal{B}(i)$. Therefore, we have state $k \in \mathcal{F}(j)$ but $k \notin \mathcal{B}(j)$ since $k \notin \mathcal{B}(i)$, which implies $\mathcal{F}(j) \not\subseteq \mathcal{B}(j)$ so that j is transient by Theorem 3.1.

If state i is recurrent, i.e., $\mathcal{F}(i) \subseteq \mathcal{B}(i)$ by Theorem 3.1, then, $\forall j \in \mathcal{F}(i) \implies i \leftrightarrow j$ by definition so that we have $\mathcal{F}(j) \subseteq \mathcal{F}(i)$ and $\mathcal{B}(i) \subseteq \mathcal{B}(j)$ by Lemma 3.1. Thus, $\mathcal{F}(j) \subseteq \mathcal{F}(i) \subseteq \mathcal{B}(i) \subseteq \mathcal{B}(j)$, which implies j is recurrent by Theorem 3.1. Finally, if i is recurrent and $\mathcal{B}(i) - \mathcal{F}(i)$ is not empty, let state $k \in \mathcal{B}(i) - \mathcal{F}(i)$, we merely need to show that $\mathcal{F}(k) \not\subseteq \mathcal{B}(k)$ so that k is transient. In fact, $k \in \mathcal{B}(i) \iff i \in \mathcal{F}(k)$, and $k \notin \mathcal{F}(i) \iff i \notin \mathcal{B}(k)$, which implies $\mathcal{F}(k) \not\subseteq \mathcal{B}(k)$. □.

Example Figure 2(a) shows the small example Markov chain in Section 2. To determine if state 1 is recurrent, we first compute $\mathcal{F}(1) = \{1, 2, 3, 4, 5, 6, 7\}$ and $\mathcal{B}(1) = \{1, 3, 4, 5\}$. Since $\mathcal{F}(1) \not\supseteq \mathcal{B}(1)$, not only is state 1 transient, all states in $\mathcal{B}(1)$ are transient as well. Thus, only states 2, 6 and 7 need to be further classified. □

Theorem 3.1 determines if a state i is recurrent by checking if its forward set is contained in its backward set. If it is, then a recurrence class has been found which equals the forward set. Moreover, according to Theorem

```
forward_set(s, Q) {
    RS^(0) ← ∅, FS^(0) ← {s};
    k ← 0;
    while(FS^(k) ≠ ∅) {
        FS^(k+1)(x) = ∃_y FS^(k)(y)Q(y,x) ∧ ¬RS^(k)(x);
        RS^(k+1)(x) = RS^(k)(x) ∨ FS^(k+1)(x);
        k ← k + 1; }
    return(F(s) ← RS^(k)); }

backward_set(s, Q) {
    RS^(0) ← ∅, FS^(0) ← {s};
    k ← 0;
    while(FS^(k) ≠ ∅) {
        FS^(k+1)(x) = ∃_y FS^(k)(y)Q(x,y) ∧ ¬RS^(k)(x);
        RS^(k+1)(x) = RS^(k)(x) ∨ FS^(k+1)(x);
        k ← k + 1; }
    return(B(s) ← RS^(k)); }
```

Figure 3: Computing the forward set and backward set.

3.2 if the backward set properly contains the forward set, those states in the backward set not belonging to the forward set are all found to be transient. In the case the forward set is not contained in the backward set, we have found a set of transient states equal to $\{i\} \cup \mathcal{B}(i)$.

The next subsection gives our symbolic algorithm for state classification based on the above theory.

3.2 The algorithm

We associate the transition graph $G = (S, E)$ with a boolean transition relation Q defined from $S \times S$ into $\{0, 1\}$ such that $Q(s, s') = 1$ iff $e_{s,s'} \in E$. We say s is the current state while s' is the next state. To represent the relation Q with BDDs, let X and Y be two sets of boolean variables encoding current states and next states, respectively. Details of BDDs and their operations can be found in [12].

3.2.1 Computing forward and backward sets

Computation of the forward set can be performed exactly as the symbolic reachability analysis for finite state machines using *fixed-point calculations* [17, 18].

The idea of fixed point calculation may be explained by the concept of the frontier set, FS. Initially, the reachable state set, RS, is set to $\emptyset$, and the frontier set is set to be the initial states. That is, $RS^{(0)} = \emptyset$, and $FS^{(0)} = I$, where I is the set of initial states. Next, the set of states, Z, that can be reached from the current frontier set $FS^{(0)}$ in one time step (or one state transition) is computed. Those states that are newly reached during this iteration are taken as the frontier for the next iteration. The process repeats until the frontier set is empty.

The first procedure given in Figure 3 computes the forward set of state s where $\exists$ and $\wedge$ denote BDD operations of existential quantification and AND, respectively. Applying the same idea, the backward set of s can be computed similarly.

3.2.2 Classifying states

Figure 4 shows our state classification algorithm. It *iteratively* finds all the transient states and the recurrence classes. During each iteration, a state from the remaining state space S' is taken as a *trial* state whose forward set and backward set are computed using the procedures given in Figure 3. Next, the trial state is determined to be either recurrent or transient by checking the containment of its forward and backward sets as discussed in the Section 3.1. States that communicate with this trial state are also determined for their recurrence/transience property. The iteration terminates when there is no state to be classified.

```
State_classification(S, Q) {
    S' ← S;
    while(S' ≠ ∅) {
        s ← random_take(S');
        F(s) ← forward_set(s, Q);
        B(s) ← backward_set(s, Q);
        if(F(s) ∧ ¬B(s) = ∅) {
            report  F(s)  a recurrence class;
            report  B(s) ∧ ¬F(s)  all transient; }
        else
            report  s ∨ B(s)  all transient;
        S' ← S' ∧ ¬(s ∨ B(s)); } }
```

Figure 4: The state classification algorithm.

The function *random_take*(S') works as follows. It randomly takes a state, say s_1 from S' as a start point, and searches for a state s_2 not yet considered to be a trial state but reachable from s_1 within a user definable number of steps. The intuition is that, state s_2 is more likely to be recurrent because $s_2 \in \mathcal{F}(s_1)$. Since this search can be done iteratively and after each iteration only one state is selected to be the new frontier set, the time spent in this function is negligible.

In general, our algorithm has much better complexity than the method in [11]. This is because the reachability analysis is performed only for part of the state space, i.e, for those trial states returned by function *random_take*() whereas the transitive closure of graph Q required by the latter is equivalent to the knowledge of the reachable sets for all the states in S.

The worst case of our algorithm may occur if each iteration determines the transience or recurrence property of only the corresponding trial state when its backward set contains at most itself. Then, exactly $|S|$ iterations would be required. This could happen for instance if (1) every state in S forms a single-state recurrence class or (2) all states except one are transient and each transient state reaches only the sole recurrent state. Since most real systems contains relatively few recurrence classes and a significant portion of the state space is recurrent, very few (typically only one) iterations are performed so that the worst case complexity is rarely observed. Another case where our algorithm may also be slow is when all states (nearly) form a loop. This case does not require a large number of iterations, but the time spent in the reachability analysis during each iteration could be significant. However, the time complexity of our algorithm still remains at the same order of that of the reachability analysis. The *counter* example in the next section is a case of this kind.

4 Experimental Results

We implemented both our method and the method in [11] based on the CUDD [19] and SMV [20] packages on a SPARC20 with 256 Megabytes of memory. The efficiencies of the two methods are compared on many circuits as well as a queueing network model. All of them are specified in SMV format.

4.1 Synchronous and asynchronous circuits

Our synchronous examples are the synchronous bus arbiter (*syncarb*) [20], the mutual exclusion element (*mutex*), the counter circuit (*counter*), and a 3-stage synchronized pipeline (*periodic*) which are all from the SMV package. Our asynchronous examples include the distributed mutual exclusion (*dme*) circuit [20], the pausible clock interface (*pci*) [21], the asynchronous differential equation solver [21] with its estimated version (*diffeq_est*) and its back-annotated version (*diffeq_bka*) both of which were analyzed symbolically for their average performance in [8] and finally the asynchronous FIFO circuit *fifo* also discussed in [8].

According to the results in Table 1, all the circuits contain a single recurrence class. The synchronous ones have no transient states. Among the asynchronous ones, only the *diffeq* and the *fifo* contain a significant fraction of transient states.

Our method finishes all examples while the previous method fails on half of them due to insufficient memory. Comparing the run time, except on the counter circuit, our method dramatically outperforms the previous method. With more experiments, we notice that for the *fifo* example, the CPU time required by our method increases polynomially while the state space grows exponentially. This is because the BDD size of its transition relation Q grows polynomially with each added stage. Similar performance is achieved in the *syncarb*, *pci* and *dme* examples. The exception of the counters is because their state spaces are complete graphs for which the hashed recursive TC computation is particularly efficient [13].

4.2 Closed queueing networks

Finally, we consider a behavioral model of a *closed queueing network* [22]. Each queue is modeled as a single server with finite storage capacity. For simplicity, we assume each server has no input/output buffers. We model the delay of the server with a bounded time interval with a discrete probability distribution. In practice, this delay model is more realistic than the typical models with exponential delays. Figure 5(a) shows the overall system structure and Figure 5(b) specifies the behavior of a single server as a finite state machine.

Notice that because of its closed nature of the system, once (randomly) initialized, the number of jobs in the system remains constant. Consequently, this example has multiple recurrence classes.

Initially, a server may or may not have a job to process, i.e, it is randomly initialized as either `idle` or

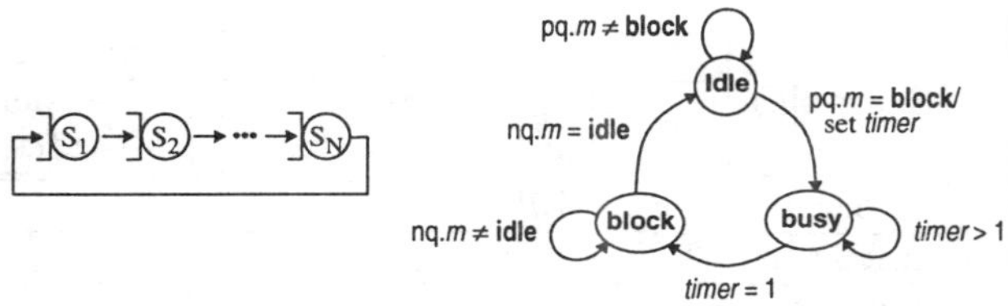

(a) system structure, (b) behavior of one server.

Figure 5: A closed queueing network.

`busy`. The server leaves `idle` mode if there is an incoming job (pq.m = `block`) and becomes `busy`. Meanwhile, a timer (modeled by a `timer` variable) is randomly set to a value from the specified service time interval. Variable `timer` decrements each time step, denoting the residual service time. One time step after the timer reaches 1, the server moves to `block` mode and stays there until the next queue (nq) is `idle`. Then, it returns to `idle` in the subsequent time step.

Suppose each server has a service time anywhere between 1 and 7 steps. The reachable state space grows roughly by a factor of 6.5 with each added queue. As shown in Table 1 (cqn's), the number of recurrence classes is $N + 1$ where N is the number of queues. In fact, under the given service time assumption, one can show that each recurrence class uniquely determines the long-run system behavior for a fixed number of jobs.

The results show that the run time of our method grows polynomially whereas that of the previous method increases exponentially, implying an exponential speed up. The previous method fails to handle systems with more than 6 queues due to insufficient memory while our method can handle systems with many more queues. For example, for a system with 16 queues, our method completes in less than 2 hours.

5 Conclusions

We have presented a BDD-based symbolic method for state classification of discrete-time finite state Markov chains along with the proofs of its correctness. The method iteratively identifies all the transient states and the recurrence classes by the standard reachability analysis. Typically, the number of times the reachability analysis required is very small. Compared with the previous proposed symbolic method, our method has been demonstrated to achieve dramatic improvements on the run time complexity as well as on the memory requirement on almost all examples tested.

References

[1] M. Malhotra and K. S. Trivedi. Power-hierarchy of dependability-model types. *IEEE Trans. on Reliability*, 43(3), Sept. 1994.

[2] O. Daniel Z. Simeu-Abazi and B. Descotes-Genon. Analytical method to evaluate the dependability of manufacturing systems. *Reliability Engineering & Systems Safety*, 55(2), Feb. 1997.

[3] A.M. Stankovic, G.C. Verghese, and D. J. Perreault. Randomized modulation of power converters via Markov chains.

Examples	State space					Run time (CPU sec)	
	Total states	Reachable states	Transient states	Recurrent states	Recurrent classes	Previous method	Our method
dme3cells	1.80×10^{16}	6,579	0	6,579	1	417	7.1
dme6cells	3.25×10^{32}	8.217×10^{6}	0	8.217×10^{6}	1	$8,140^{\dagger}$	123.7
syncarb5	32,768	5,120	0	5,120	1	0.63	0.1
syncarb5	1.07×10^{9}	1.049×10^{7}	0	1.049×10^{7}	1	2.97	0.28
mutex8	524, 288	2, 048	0	2, 048	1	0.48	0.18
periodic	6.87×10^{10}	3, 952	0	3, 952	1	$1,577^{\dagger}$	1.3
counter8	256	256	0	256	1	0.23	0.18
counter16	65,536	65,536	0	65,536	1	0.43	100.9
pci-2-1	4.29×10^{9}	14, 556	0	14, 556	1	$3,514^{\dagger}$	14.7
pci-4-1	2.88×10^{17}	1.76×10^{7}	0	1.76×10^{7}	1	$34,620^{\dagger}$	116.5
pci-8-1	1.30×10^{33}	5.24×10^{13}	0	5.24×10^{13}	1	$11h^{\ddagger}$	426.3
diffeq-est	4.29×10^{9}	7, 632	2,762	4,870	1	$3,180^{\dagger}$	4.0
diffeq-bka	2.75×10^{11}	11, 036	6,176	4,860	1	$1,059^{\dagger}$	6.5
fifo2	65, 536	98	2	96	1	8.0	0.23
fifo4	1.68×10^{7}	1,150	18	1,132	1	72.2	0.58
fifo8	1.10×10^{12}	171,518	730	170,788	1	$1,535^{\dagger}$	2.9
fifo16	4.72×10^{21}	4.258×10^{9}	2.48×10^{6}	4.256×10^{9}	1	$11h^{\ddagger}$	24.2
fifo32	5.44×10^{39}	7.736×10^{17}	2.88×10^{13}	7.736×10^{17}	1	$11h^{\ddagger}$	277.0
cqn2	1,024	21	3	18	3	0.45	0.1
cqn4	1.05×10^{6}	1,693	15	1,678	5	24.7	1.7
cqn6	1.07×10^{9}	80,874	63	80,811	7	1,097	14.3
cqn8	1.10×10^{12}	3.083×10^{6}	255	3.083×10^{6}	9	8, 138†	60.9
cqn10	1.26×10^{15}	1.328×10^{8}	1,043	1.328×10^{8}	11	3, 776†	230.8
cqn12	1.15×10^{18}	5.614×10^{9}	5,535	5.614×10^{9}	13	1, 808†	573.0

Table 1: Experimental results: † out of memory, ‡ incomplete.

IEEE Trans. on Control Systems Technology, 5(1), Jan. 1997.

[4] R. Marculescu, D. Marculescu, and M. Pedram. Efficient power estimation for highly correlated input streams. In *Proc. ACM/IEEE Design Automation Conf.*, 1995.

[5] P. A. Beerel and H. S. Wadekar. Energy estimation of speed-independent control circuits. *IEEE Trans. on Computer-Aided Design*, pp. 672–680, June 1996.

[6] M. Holiday and M. Vernon. A generalized timed Petri net model for performance analysis. *IEEE Trans. on Software Engineering*, 13:1297–1310, Dec. 1987.

[7] P. Kudva, G. Gopalakrishnan, E. Brunvand, and V. Akella. Performance analysis and optimization of asynchronous circuits. In *Proc. Int. Conf. Computer Design (ICCD)*, pp. 221–225, Oct. 1994.

[8] A. Xie and P. A. Beerel. Symbolic techniques for performance analysis of asynchronous systems based on average time separations of events. In *Proc. Int. Symp. on Adv. Res. in Asynchronous Circuits and Systems*, pp. 64–75, 1997.

[9] R. M. Izquierdo and D. S. Reeves. MPEG VBR slice layer model using linear predictive coding and generalized periodic Markov chains. In *IEEE Int. Performance, Computing and Communications Conf.*, 1997.

[10] T. Takine, T. Suda, and T. Hasegawa. Cell-loss and output process analysis of a finite-buff discrete-time atm queueing system with correlated arrival. *IEEE Trans. on Communications*, 43(2-4), Feb.-March 1995.

[11] G. D. Hachtel, E. Macii, A. Pardo, and F. Somenzi. Markovian analysis of large finite state machines. *IEEE Trans. on Computer-Aided Design*, pp. 1479–1493, Dec. 1996.

[12] R. E. Bryant. Graph-based algorithm for boolean function munipulation. *IEEE Trans. on Computers*, 35, Aug. 1986.

[13] Y. Matsunaga, P. C. McGeer, and R. K. Brayton. On computing the transitive closure of a state transition relation. In *Proc. ACM/IEEE Design Automation Conf.*, 1993.

[14] J. G. Kemeny and J. L. Snell. *Finite Markov Chains.* Springer, 1976.

[15] R. A. Howard. *Dynamic Probabilistic Systems, Volume I: Markov Models.* John Wiley & Sons, Inc, 1971.

[16] G. D. Hachtel and F. Somenzi. A symbolic algorithm for maximum flow in 0-1 networks. In *Proc. Int. Conf. Computer-Aided Design (ICCAD)*, 1993.

[17] J. R. Burch, E. M. Clarke, D. E. Long, K. L. McMillan, and D. L. Dill. Symbolic model checking for sequential circuit verification. *IEEE Trans. on Computer-Aided Design*, 13(4), Apr. 1994.

[18] O. Coudert, J. C. Madre, and C. Berthet. Verifying temporal properties of sequential machines without building their state diagrams. In E. M. Clarke and R. P. Kurshan, editors, *Computer-Aided Verification'90.* American Mathematical Society, June 1990.

[19] R. I. Bahar, E. A. Frohm, C. M. Gaona, and G. D. Hachtel. Algebraic decision diagrams and their applications. In *Proc. Int. Conf. Computer-Aided Design (ICCAD)*, 1993.

[20] K. L. McMillan. *Symbolic Model Checking.* Kluwer Academic Publishers, 1993.

[21] K. Y. Yun and R. P. Donohue. Pausible clocking: A first step toward heterogeneous systems. In *Proc. Int. Conf. Computer Design (ICCD)*, pp. 118–123. Oct. 1996.

[22] L. Kleinrock. *Queueing Systems, Volume II: Application.* Wiley, New York, 1975.

An Implicit Algorithm for Finding Steady States and its Application to FSM Verification

Gagan Hasteer† Ambit Design Systems
Anmol Mathur Silicon Graphics Inc.
Prithviraj Banerjee Northwestern University

Abstract

Finding the set of steady states of a machine has applications in formal verification, sequential synthesis and ATPG. Existing techniques assume the presence of a designated set of initial states which is impractical in a real design environment. The set of steady states of a design is defined by the terminally strongly connected components (tSCCs) of the underlying state transition graph (STG). We show that multiple tSCCs and non-terminal SCCs need to be handled in a real design environment especially for verification. We present a fully implicit algorithm to find the steady states of a machine without any knowledge of initial states. We demonstrate the utility of our algorithm by applying it to FSM equivalence checking.

1 Introduction

The steady states of an FSM capture its asymptotic behavior, after sufficiently long input sequences. In many applications, the behavior of the FSM outside the steady states is a don't care machine. Consequently, finding the set of steady states of a sequential machine has important application in sequential optimization, automatic test pattern generation (ATPG) and FSM verification[1, 2]. The set of transient (non-steady) states provides flexibility for performance optimizations such as removal of redundant latches [1], sequential redundancy removal and sequential resynthesis [3]. Backward state justification is the computationally dominant part of ATPG, especially for hard to test faults. The search space for backward state justification can be pruned with the knowledge of the steady states of the FSM. If the current state requirement consists only of transient states then it can never be justified starting from all possible states. Justification is stopped if this condition is true.

Due to the complexity of computing the steady states of an FSM, researchers have used the states reachable from a suitable set of designated initial states (DIS) as an approximation. The DIS assumption is restrictive in practice, since many industrial designs avoid having a reset line to all latches for area and performance. Previous approaches for computing the steady states of an FSM work under specific conditions. Under the DIS assumption, [2] can compute the steady states if the initial state set is contained in the steady states. In the absence of non-terminal SCCs in the STG of the FSM, an intersection based fixed point computation suffices starting from the set of all the states [3].

† This research was supported in part by the National Science Foundation under grant MIP-9320854 and the Semiconductor Research Corporation under contract SRC 96-DP-109.

A novel technique proposed in [1] becomes enumerative in the presence of multiple tSCCs in the STG. A tSCC is obtained by successively pruning the state space. A state is selected randomly from the currently unpruned space (initially equal to the entire state space). All the forward and backward reachable states are computed from this state. If the selected state belongs to some tSCC then the forward reachable set (which will be the tSCC) is contained in the backward reachable set. Otherwise the whole backward reachable set is pruned away, and another state is selected from the unpruned space, and the process repeated. This algorithm can be extended to iterate over the remaining state space to obtain all the tSCCs of a machine. In §3 we show that a design error can result in a machine having exponential tSCCs in its STG, bringing out the worst case of this algorithm.

The main contributions of this paper are :

1. We show that a product machine (a popular model for FSM verification) can have multiple tSCCs and non-terminal SCCs (ntSCCs), making a case for a general algorithm which can handle STGs with arbitrary topology.
2. We propose a fully implicit algorithm for computing the set of steady states of a machine without making the DIS assumption. The set of steady states is extracted implicitly by exploiting the graph structure of the underlying STG. Our algorithm can handle machines where STGs have any arbitrary topology.
3. We characterize Sequential Hardware Equivalence (SHE) using steady states of the product machine and give an efficient algorithm for checking SHE [4].

2 Preliminaries

An FSM M is represented by a quintuple, $(Q, I, O, \lambda, \delta)$, where Q is the set of states, I is the set of input values, O is the set of output values, λ is the output function, and δ is the next state function. For simplicity, we assume that M is fully specified i.e $\lambda : (Q \times I) \rightarrow O$ and $\delta : (Q \times I) \rightarrow Q$ are completely specified. We assume that the reader is familiar with BDD based manipulation of FSMs.

Definition 1 : A set of states $S \subset Q$ is a **strongly connected component**(SCC) if it is a maximal set of states such that for any two states $s, t \in S$, there exist input sequences π_0 and π_1 so that $\delta(s, \pi_0) = t$ and $\delta(t, \pi_1) = s$.
Definition 2 : A set of states $S \subset Q$ is a **terminal strongly connected component**(tSCC) if S is an SCC and for any state $s \in S$ and any input i : $\delta(s, i) \in S$. An ntSCC is an SCC which is not terminal.
Definition 3 : The set of tSCCs of a machine determine its steady state behavior. Thus the **steady states** of a machine M, $\mathbf{\Sigma(M)}$ is defined to be the union of the tSCCs of M. Henceforth, we use the term **steady states** and **tSCCs** of a machine interchangeably.

3 Arbitrary STG Topologies

In this section we show that there are many situations in synthesis and verification where steady states need to be computed for FSMs whose STGs have arbitrary topology. For embedded designs a component sub-machine may have multiple tSCCs even if the product machine has only one tSCC [5]. In the domain of FSM verification design errors in the implementation may cause multiple tSCCs even though the corresponding good machine specification has only one tSCC.

3.1 Case for Multiple tSCCs

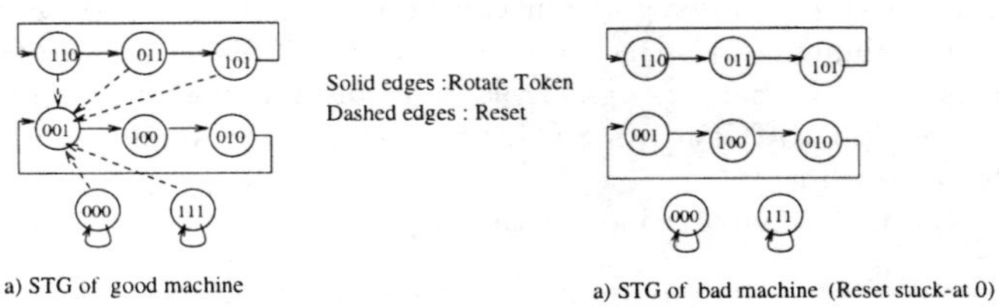

Figure 1: Exponential number of tSCCs in a machine due to design error

Consider an n latch design with one-hot encoding. Such sparsely encoded designs are popular in a high-performance design environment (e.g. to save the cost of decoding encoded state bits). These latches exchange a token in a round robin fashion. Figure 1 illustrates the STG for a 3 latch design. A reset signal brings the latches into a designated one-hot initial state. The design has one tSCC and $O(\frac{2^n}{n})$ ntSCCs (since each SCC in the STG has atmost n states). In the presence of a design error when the reset signal is stuck at 0 all the ntSCCs become tSCCs in the erroneous design. This example also shows why the algorithm of [1] is exponential. It can find atmost one tSCC for every randomly picked state. Hence it would have to iterate over the state space $O(\frac{2^n}{n})$ times to find all the steady states.

The example of Figure 2 illustrates two FSMs with one tSCC each. The machines enter steady state after the first clock cycle. Their product machine has two tSCCs which define its steady state behavior. The reason is the flexibility of entering the steady state in machine 2 from the transient state $b1$. This flexibility is afforded by the equivalence of states $b2$ and $b3$.

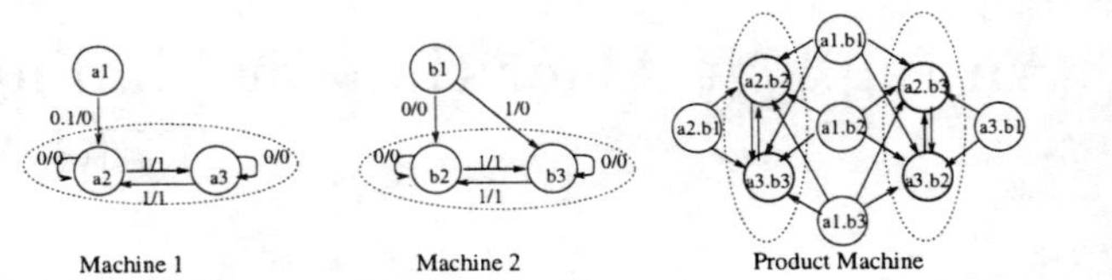

Figure 2: Multiple tSCCs in the product machine when the component machines have only one tSCC

3.2 Case for ntSCCs

Another issue which deserves attention is the existence of ntSCCs in STGs. As mentioned in §1 an intersection based fixed point computation can compute the steady states if there are no ntSCCs. There may be machines which do not have ntSCCs (e.g. counters). But even for such designs the product machine may have an ntSCC outside the steady state space. Figure 3 shows a 2-bit counter with only one SCC which is its tSCC. To verify two identical copies of this counter, we construct their product machine as shown in Figure 3c. The product machine has three ntSCCs and one tSCC. Each of the ntSCCs correspond to the states where the two counters power-up with distinct values. The tSCC corresponds to the states with the two counters having equal values. The reset signal drives the product machine into state (a_0, b_0) of the tSCC which represents their initialization to $(0, 0)$.

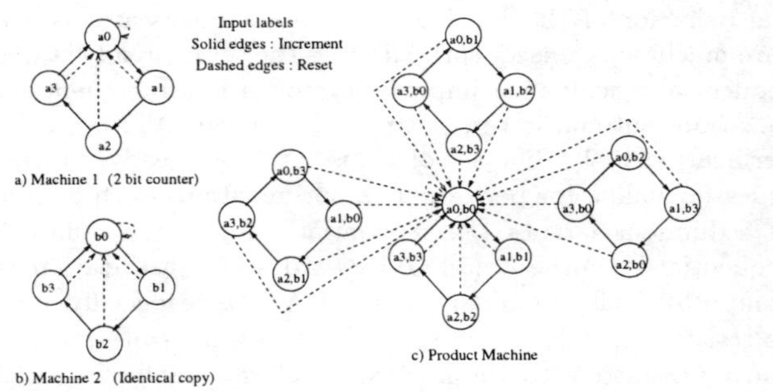

Figure 3: ntSCCs in the product machine when the component machines have no ntSCCs

The above examples demonstrate that assumptions of single tSCC and no ntSCCs are too restrictive when erroneous designs or designs generated as a composition of smaller sub-designs are considered. This is specially significant in the formal verification arena, where erroneous designs have to be handled and composition of FSMs through cross product is a common operation. This motivates the need for an efficient algorithm for finding steady states of a general design.

4 Computing Steady States

We describe *Implicit_SS*, an algorithm to compute the steady states of a machine which works on STGs having any topology. The intuition behind our algorithm is to induce a property unique to the set of steady states by applying a series of transformations based on the following lemma.

Lemma 1 : *The containment of forward reachable states in the backward reachable states is a property unique to the steady states of an FSM.*

Proof : Follows directly from the definition of a tSCC □

In the transitive closure, G^* of the STG G all the outgoing edges from a state terminate in the forward reachable set. If the edges of G^* are reversed to get a new graph G^*_R, then all the outgoing edges from a state in G^*_R terminate in its backward reachable set. Then by exploiting Lemma 1 if we remove all the edges in G^*_R from G^*, the steady states will have no outgoing transitions in $(G^* - G^*_R)$ whereas every non-steady state will have atleast one outgoing transition. The steady states can be subsequently extracted implicitly by exploiting the induced property. It is easier to explain these transformations using graph terminology on the STG of the machine. *However, none of our algorithms build or manipulate the STG explicitly.*

Let $G(V, E)$ be the STG for machine $M = (Q, I, O, \lambda, \delta)$. Then $V = Q$ and $(s,t) \in E$ if and only if $\delta(s,i) = t$ for some input $i \in I$. The main step of our algorithm are :

1. Compute the transitive closure of G, $G^* = (Q, \{(s,t)|\delta(s,\pi) = t \; for \; some \; \pi \in I^*\})$
2. Edge reversal, $G^*_R = (Q, \{(s,t)|(t,s) \in G^*\})$
3. Compute $G_\cap = (Q, \{(s,t)|(s,t) \in G^* and (s,t) \in G^*_R\})$
4. Compute the set difference $(G^* - G_\cap) = (Q, \{(s,t)|(s,t) \in G^* and (s,t) \notin G_\cap\})$
5. Compute the steady states of G, $S = \{s | \forall \; t \; (s,t) \notin (G^* - G_\cap)\}$

Theorem 1 : *A state $s \in \Sigma(M)$ if and only if s has outdegree zero in $(G^* - G_\cap)$.*

5 FSM Equivalence Checking

We choose the SHE paradigm of machine equivalence, which does not make the DIS assumption. We prove that two FSMs are SHE if their product machine always outputs zero in the set of its steady states. Subsequently, we give an efficient algorithm to compute the steady states of the product machine without constructing the transition relation for the product machine.

5.1 Sequential Hardware Equivalence

Definition 4 : Two states $s \in Q_{M_0}$ and $t \in Q_{M_1}$ are **equivalent** $(s \sim t)$ if for every sequence of inputs $\pi \in I^*, \lambda_{M_0}(s,\pi) = \lambda_{M_1}(t,\pi)$. If $s \sim t$, then for every $\pi \in I^*, \delta_{M_0}(s,\pi) \sim \delta_{M_1}(t,\pi)$.

When a machine powers up, its initial state cannot be predicted. The desired behavior is usually achieved by driving an initialization sequence through the machine. Based on this premise, Pixley proposed SHE in [4].

Definition 5 : Given two machines M_0 and M_1, a state pair $(s_0, s_1) \in Q_{M_0} \times Q_{M_1}$ is **alignable** if there is a sequence of inputs $\pi \in I^*$ such that $\delta_{M_0}(s_0,\pi) \sim \delta_{M_1}(s_1,\pi)$. The sequence π is called an **aligning sequence**.

Definition 6 : Machines M_0 and M_1 are **SHE** $(M_0 \approx M_1)$ if and only if all state pairs $(s_0, s_1) \in Q_{M_0} \times Q_{M_1}$ are alignable.

5.2 Modeling SHE

A popular approach to check FSM equivalence is to construct a product machine of the specification and the implementation. The two FSMs share the same set of primary inputs and their output pairs are XOR-ed to get the primary outputs of the product machine. Equivalence checking from DIS is performed by verifying that the outputs of the product machine are 0 in its set of reachable states. We extend the above paradigm to check for SHE. The following theorem characterizes SHE in terms of the steady states of the product machine.

Theorem 2 : *$M_1 \approx M_2$ if and only if the product machine $M = M_1 \times M_2 = (Q_M, I_M, O_M, \lambda_M, \delta_M)$ satisfies $\lambda_M(s,i) = 0$ for all states $s \in \Sigma(M)$, where $\Sigma(M) \subset Q_M$ is the steady state set of M (union of the tSCCs of M).*

A naive approach to prove $M_1 \approx M_2$ is to construct their product machine and find its tSCCs using the algorithm *Implicit_SS*. Computation of the transitive closure of a machine has been shown to be very expensive in terms of memory and CPU. The fact that the product machine has an order of magnitude more states than the individual machines may make this computation infeasible. We need a technique to prune the large state space of the product machine. The following theorem provides a characterization of the steady states of $M_1 \times M_2$ using the steady states of M_1 and M_2.

Theorem 3 : $\Sigma(M_1 \times M_2) = \Sigma(\Sigma(M_1) \times \Sigma(M_2))$.

Using Theorems 2 and 3, we find the tSCCs of the individual machines $\Sigma(M_1)$ and $\Sigma(M_2)$, and construct their product machine, $\Sigma(M_1) \times \Sigma(M_2)$. The tSCCs of $\Sigma(M_1) \times \Sigma(M_2)$ are the steady states of $M_1 \times M_2$. Thus the transitive closure is now performed only on a small subgraph of the STG of the product machine. This enables us to avoid building the product machine altogether resulting in significant savings. Subsequently the output functions of the two machines are XORed to give the output function of the product machine. If this output function evaluates to 0 when constrained to the steady state set, then the two machines are SHE.

6 Results

The algorithms described above have been implemented as a part of a prototype equivalence checking engine. Tables 2 and 3 show the performance of the algorithm on a set of industrial controllers on SGI Power Challenge machines with 200MHz clock and 2048MB of main memory. The Colorado BDD package was used for symbolic manipulation. The BDD variable are ordered by a depth first heuristic. Time is reported in seconds for the entire application. Memory is the maximum BDD memory size in MegaBytes(MB). Timeout for our experiments is set as 12 hours. *# Iter* is the number of iterations required by the algorithm to compute the transitive closure using iterative squaring.

Table 1 reports the statistics of our benchmark circuits. These circuits are two phase designs with one of the phases abstracted out by making the latches of that phase transparent. The validity of this transformation is proven in [6]. The first two columns give the number of latches of each phase in the input circuit. Heuristically the phase with lesser latches is retained during phase abstraction. Thus the size of the state space would be 2^9 and 2^{15} for *Ckt1* and *Ckt9* respectively. The third and fourth columns give the number of gates and the size of the steady state space respectively.

Table 1: Benchmark Statistics

Type of Controllers	Circuit	# Latches		# Gates	# SS
		Ph1	Ph2		
Cache	Ckt1	9	12	478	84
	Ckt2	16	22	823	296
Queue	Ckt3	11	14	340	34
	Ckt4	19	20	584	139364
Arithmetic	Ckt5	19	45	3104	288
	Ckt6	27	28	2645	324
	Ckt7	36	57	2915	23040
Memory	Ckt8	9	9	286	20
System Interface	Ckt9	20	15	560	432
	Ckt10	41	23	1152	512

Table 2: Steady State Computation Statistics

Circuit	Enum_SS[1]		Implicit_SS		
	Mem	Time	Mem	Time	# Iter
Ckt1	0.6	0.3	0.6	1.5	4
Ckt2	1.2	4.0	0.8	4.7	4
Ckt3	0.7	2.3	0.7	1.7	6
Ckt4	2.6	302.2	5.2	73.3	8
Ckt5	4.7	15685.5	1.0	14.09	1
Ckt6	1.2	149.1	1.0	6.7	1
Ckt7	1.7	51.6	1.7	126.2	2
Ckt8	0.7	1.0	0.9	1.5	5
Ckt9	0.9	5.5	2.0	41.0	6
Ckt10	*ovf*	-	1.0	14.0	3

Table 2 compares the performance of *Implicit_SS()* with the algorithm of [1](*Enum_SS()*). For arithmetic and queue controllers, *Implicit_SS()* does better than the enumerative technique. The biggest win is seen in a system interface controller (*Ckt10*) where the enumerative algorithm experiences a blowup, but our algorithm is able to finish in just 14 seconds. This is a typical example of a controller with a very sparse steady state space. In general, our algorithm outperforms *Enum_SS()* when the steady state space is small and the diameter of the underlying STG is small. This is evident from the results for *Ckt5*, *Ckt6* and *Ckt10*. Intuitively, small diameter results in faster transitive closure computation, which is the most compute intensive part of the algorithm. For all these cases the steady state space is a small fraction of the total state space. *Ckt6* for instance has 324 steady states out of 2^{27} states. This sparseness causes *Enum_SS()* to go through more pruning steps, each step incurring the cost of a forward and backward reachability computation. *Ckt4* is an exception with a dense steady state space and large diameter. It contains two embedded counters which make the state space amenable to the transitive closure computation. However, the counters make the reachability computations of *Enum_SS()* go though exponential image computation steps leading to larger runtimes. For the other examples, the time taken is marginally smaller for *Enum_SS*.

Table 3 compares the performance of Algorithm *Check_SHE* with the naive approach (of constructing the product machine and then computing the steady states). The steady state computation of the minimal machine is much faster than that of the product machine due to state space pruning. Cache controllers are a notable exception where the minimal machine loses some structure causing memory blowups during the transitive closure computation. For *Ckt8* and *Ckt9* the naive approach fails to complete. For *Ckt4* both the approaches timeout.

Table 3: Checking identical copies for SHE

Circuit	Naive Algorithm		Check_SHE	
	Mem	Time	Mem	Time
Ckt1	8.0	223.5	9.2	481.1
Ckt2	10.9	798.5	29.4	8456
Ckt3	2.3	44	2.0	26
Ckt4	-	timeout	-	timeout
Ckt5	1.2	25.0	1.2	23.6
Ckt6	1.1	15.8	1.3	17
Ckt7	29.4	13643	4.6	639
Ckt8	-	timeout	1.2	25.5
Ckt9	-	timeout	21.9	3016
Ckt10	2.1	127.5	1.0	35.5

7 Conclusion

Finding the set of steady states of an FSM has application in sequential optimization, ATPG, and FSM verification. We propose an implicit algorithm for computing the steady states of an FSM. We show that an erroneous machine can have multiple tSCCs making a case for a general algorithm which can extract all the tSCCs without using enumerative techniques. We prove that the tSCCs of the product machine are contained in the product of the tSCCs of the individual machines. This enables us to prune the state space and check for SHE without constructing the product machine. Future work involves applying the above algorithm to other domains.

References

[1] S. Qadeer, R. Brayton, V. Singhal, and C. Pixley, "Latch redundancy removal without global reset," *Proceedings of the International Conference on Computer Design*, Oct. 1996.

[2] O. Coudert and J. Madre, "A unified framework for formal verification of sequential circuits," *Proceedings of the International Conference on Computer Aided Design*, Nov. 1990.

[3] C. Pixley, V. Singhal, A. Aziz, and R. Brayton, "Multi-level synthesis for safe replacability," *Proceedings of the International Conference on Computer Aided Design*, Nov. 1994.

[4] C. Pixley, "A theory and implementation of sequential hardware equivalence," *IEEE Transactions on Computer-Aided Design*, pp. 1469–1494, Dec. 1992.

[5] V. Singhal, "Design replacements for safe replacability." PhD. thesis, University of California at Berkeley, Mar. 1996.

[6] G. Hasteer, A. Mathur, and P. Banerjee, "A framework for equivalence checking of multi-phase fsms," *Proceedings of High Level Design Validation and Test*, Nov. 1997.

Hybrid Verification Using Saturated Simulation*

Adnan Aziz
Dept. of Elec. and Comp. Engineering
Univ. of Texas at Austin

Jim Kukula Tom Shiple
Advanced Tech. Group
Synopsys, Inc.

Abstract — We develop a verification paradigm called *saturated simulation*, that is applicable to designs which can be decomposed into a set of interacting controllers. The core procedure is a symbolic algorithm that explores the space of controller interactions; heuristics for making this traversal efficient are described. Experiments demonstrate that our procedure explores substantially more of the controller interactions, and is more efficient than conventional symbolic reachability analysis.

1 Introduction

In this paper we are concerned with the problem of *design verification*; specifically, the problem of *invariant checking* over gate-level designs. Traditionally, designs have been verified by extensive simulation. A model is built (in software or hardware), and monitors may be added to check for bad behavior. Large numbers of test inputs are applied to this model. While offering the benefits of simplicity and scalability, simulation offers no guarantees of correctness; for large designs, the fraction of the design space which can be covered in this methodology is vanishingly small. Indeed, there are many examples of designs that passed extensive simulation, but were still found to contain bugs [5]. This has led to the proposal of "formal methods" for design verification.

The computational complexity of invariant checking on netlists is very high. In practice, many designs are well structured, and this can be exploited to devise heuristic procedures that perform well on specific classes of designs; in particular, BDDs have been used to successfully verify a large number of complex designs [7]. The primary limitation of BDD based approaches to FV is that is that for many designs, the resulting BDDs grow very large [7].

Practicing verifiers are less concerned with providing formal proofs of correctness for their designs than they are with finding bugs in them as early as possible; this is attested to by the complete lack of penetration of theorem-provers in the hardware verification community. Indeed, "falsification" is a more accurate description of the endeavor commonly called "verification".

Faced with the twin dilemmas of diminished coverage through simulation and the inability of symbolic methods to formally verify large designs, it is natural to ask how best to use symbolic methods to find bugs in designs. In this paper we provide an answer to this problem. It is to be stressed that the approach is not complete, i.e., not guaranteed to find a counterexample, if the design fails to satisfy its properties. However, our procedures are sound — all reported violations of the invariant are true bugs. Our goal in this work is to find more and different bugs more efficiently and cheaply than simulation or formal verification used in isolation.

1.1 Our Approach

Many designs consist of a set of interacting controllers. A common source of bugs in such designs is that the designer fails to take into account "corner-cases", which occur when two or more controllers are simultaneously in "exception-handling" states. An example of this is when a transmitter unit has a buffer overflow, and a receiver unit simultaneously times out. It is difficult for random simulation to find such corner conditions, since exceptions are rare. Similarly, it is very tedious for designers to craft tests that bring the design to such states. For these reasons, corner-cases are the single largest source of bugs in designs [8].

Our contribution is a procedure that performs a partial traversal of the state space while covering controller interactions. At each step, symbolic techniques are used to compute the full set of states reachable in one step from the current set. We then restrict our attention to a subset of the complete set of visited states; the subset is chosen heuristically to minimize the size of the symbolic representation while guaranteeing that for every pair of controllers, all states that the pair could be in remain in the subset. In this way we explore as many of the controller interactions as possible (which increases the likelihood of finding bugs), while controlling the BDD size explosion that takes place with a complete analysis.

Previous Work

We have been influenced by a number of related works. The basic notion of saturated simulation was put forth by Yuan et al. [15]. However, this work is not directly applicable to designs which consist of a number of interacting controllers; they decompose the entire design into a monolithic controller and a monolithic datapath. For individual processors, such an approach works well. However, when dealing with a number of distributed components, the entire control space is very large, and their techniques fail.

The work which is perhaps most closely related to ours is that of Ravi et al. [14]. Their approach attempts to pick "high-density" subsets of the state sets encountered during reachability analysis. These sets are chosen so as to minimize the BDD representation, while keeping as many states as possible. In contrast, in our approach a subset is chosen in which all distinct control-state pairs are preserved; minimization is performed subject to this requirement. We believe that ignoring the high-level control structure of the design when doing subsetting, and using simply the number of states as a measure of the approximation, is a significant limitation of their approach.

Literature distributed by 0-In [10] at DAC 1997 also influenced us. Specifically, they underlined the importance of exploring control-pair interactions. Of course, we can only speculate about their algorithms, since no description of the underlying procedures is given.

Other less closely related works include those of Ashar

DAC 98, San Francisco, California

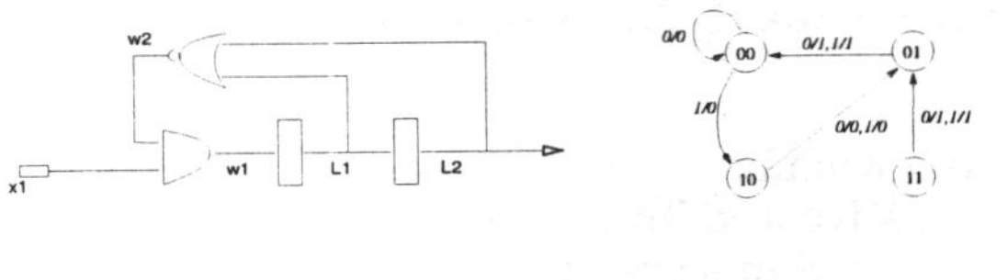

(a) Netlist (b) Finite State Machine

Figure 1: Examples: Netlist and FSM

and Malik [2], and McGeer et al. [12], who stress the importance of managing BDD size explosion, and the research of Ho et al. [8] and Hoskote et al. [9], who create test vectors using ideas from FV. Note that our approach is "liberal", i.e., does note guarantee that all bugs are found; "conservative" approaches include that of Cho et al. [6]. Details of these approaches are given in [1].

2 Background

In order to analytically reason about hardware, we first need to develop mathematical models for digital systems. We illustrate these models through the use of examples; refer to [7] for details.

Netlists and FSMs

Hardware can be modeled at the *structural* level using *netlists*, or at the *behavioral* level using *finite state machines* (FSMs). A netlist consists of an interconnected set of primary inputs, gates, and latches, as illustrated in Figure 1(a). Each gate has an associated Boolean function. A finite state machine can be represented by an edge-labeled directed graph, where the vertices correspond to *states*, and the labels are *input-output* pairs; an example is presented in Figure 1(b).

For a given netlist η, there is a natural way of deriving a finite state machine from it; states are evaluations to the set of latch variables, and the next-state/output functions are derived by composing the gate functions. This is illustrated in Figure 1.

Invariant Verification

A common verification problem for hardware designs is to determine if every state reachable from a designated set of initial states lies within a specified set of "good states" (referred to as the *invariant*). This problem is variously known as *invariant verification*, or *assertion checking*. Though conceptually simple, invariant checking can be used to verify all "safety" properties, making it a very general verification methodology[7].

Invariant checking can be performed by computing all states reachable from the initial states and determining that they all lie in the invariant. This can be achieved by traversing the STG of the design — successors of a state can be generated by iterating over all possible inputs. This procedure has a very high complexity; a design with n latches and m inputs may have as many as 2^n reachable states, and for each state, there are 2^m inputs to iterate over.

2.1 BDDs and Symbolic Invariant Checking

The Reduced Ordered Binary Decision Diagram [4] (BDD) is a graph based data structure that can compactly represent a large class of useful Boolean functions. Operations such as tautology checking, negation, conjunction, and existential quantification can be efficiently performed on BDDs.

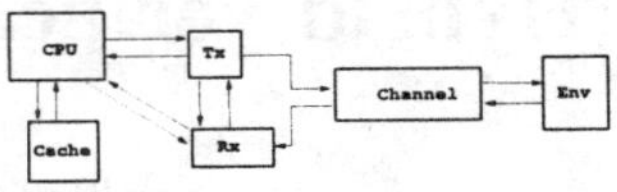

Figure 2: A design consisting of multiple functional units.

A set of states can be identified with a BDD, and the *Image* and *PreImage* operations can be performed directly on BDDs [13]. This is the basis for *symbolic model checking*; the primary limitation of this approach is the BDD size explosion problem.

3 Saturated Simulation

3.1 Introduction

Many designs can be separated into a set of interacting control units, as illustrated in Figure 2; furthermore, the designer is aware of this partition. As described in Section 1, a common source of bugs is when two controllers simultaneously enter exception handling states; the designer overlooks this possibility, and consequently the design may behave erratically. Since it is rare that controllers enter exception states, such bugs are hard to find with random simulation; indeed, a large effort is expended by designers trying manually to design test vectors that bring the design into precisely such states.

Note that for a design consisting of n controllers, each with m states, the total number of control-pair states is $O(m^2 \cdot n^2)$; this is a polynomial bound, as contrasted with the total number of states which is m^n. One way of visiting all reachable control-state pairs is by sequential ATPG. Of course this problem is as hard as computing the set of reachable states, and sequential ATPG is still in its infancy. In this section, we describe an approach we refer to as "control-pair saturated simulation", which heuristically attempts to symbolically explore as many pairs of control states as is computationally feasible.

In part, the feasibility of this approach follows from the fact that the construction of the BDD for the next-state logic can be restricted to the current set of states. This suggests that it may be possible to perform a "partial" reachability analysis, in which all distinct control pair states are preserved at each step.

3.2 Formalization

Let D be a design consisting of interacting control units $C_1, C_2, \ldots, C_n$. Let the set of state variables associated with controller i of the design be X_i. In this way the set of states corresponding to controller i is 2^{X_i}, and the global set of states is 2^X where $X = \bigcup_i X_i$.

Definition 1 Let A be a set of states. A subset $\hat{A}$ of A is *control-pair saturated* with respect to A if

$$(\forall i)(\forall j)\ (\forall \alpha_i \in 2^{X_i})(\forall \alpha_j \in 2^{X_j}) \\ \Big[(\alpha_i, \alpha_j) \in \exists Y_{ij}(A) \ \rightarrow \ (\alpha_i, \alpha_j) \in \exists Y_{ij}(\hat{A}) \Big]$$

where $Y_{ij} = X - (X_i \cup X_j)$.

Intuitively, $\hat{A}$ is a control-pair saturated subset of A if every pair of control states occurring in A occurs in $\hat{A}$. Thus control-pair saturated subsets of A preserve all the control-pair states present in A. Heuristically, a minimal control-pair saturated subset of A is a good representative set — it

```
/* A --- BDD for set over variables V.  */
/* V' ⊂ V --- variables being cprojected.     */
BDD_t function BDD_cproject(A, V') {

    v := topVar(A);
    if (v ∉ V') {
        return v · BDD_cproject(A_v) + v̄ · BDD_cproject(A_v̄);
    }

    T := ∃(V' − {v})A_v;
    if BDD_Equal(T, BDD_One) {
        return v · BDD_cproject(A_v, V');
    }
    else if BDD_Equal(T, BDD_Zero) {
        return v̄ · BDD_cproject(A_v̄, V');
    }
    else {
        /* In cproject_min T̄ is replaced by 1 */
        return v · BDD_cproject(A_v, V')
               +v̄ · T̄ · BDD_cproject(A_v̄, V');
    }
}
```

Figure 3: Pseudocode for `cproject` operator.

includes all the distinct control-pair configurations in A, and is as small as possible (in the sense of cardinality). An example illustrating this definition is given in [1].

We now address the problem of computing control-saturated subsets of A. Let f be a Boolean function on variables X. Lin et al. [11] presented an efficient algorithm (referred to as the `cproject` operator) that takes a BDD for f and a subset $X' \subset X$ of the variables, and returns a BDD for a function f^* which has the property that

1. for any assignment v to the variables in $X - X'$, so that $f(v) = 1$, there is exactly one valuation v' which agrees with the valuation v over the variables in $X - X'$ so that $f^*(v') = 1$, and furthermore

2. for all u, $f^*(u) = 1 \Rightarrow f(u) = 1$.

Pseudo-code for the `cproject` operator is given in Figure 3; an example of the application of `cproject` is given in [1]. Since sets can be represented by their characteristic functions, we will freely apply the `cproject` operator to sets. Observe that $\bigcup_{i,j}$ `cproject`$(A, (X_i \cup X_j))$ is a control-pair saturated subset of A (though not necessarily minimal, becaue of the union).

In Figure 4 we sketch a simple symbolic procedure for invariant verification. Reachable states are iteratively computed; at each step, a control-pair saturated subset ($satImg$) of the image of the current set of states set is computed using the *BDD_cproject* operator. The union of this set is taken with the current set of states (A) and is used as the next set of states. Much work has been done on improving the efficiency of the *Img* operator [7], and we can directly use it.

This procedure may reach a fixed point without visiting all reachable states because at each step only a subset of states is being explored. However, if a state not in the invariant is reached, this is guaranteed to indicate a bug.

```
/* A --- initialized to BDD for init states.  */
/* C --- list of state vars in each controller.     */
/* G --- BDD for invariant.     */
BDD_t function Cntrl_Pair_Sat_Sim(A, C, G) {

    if (BDD_Intersects(A, BDD_not(G))  /* Invariant fails!!  */
        assert FAIL;

    ImgA := BDD_Img(A);
    satImg := BDD_Zero();
    for_each(i, j) {
        satImg := BDD_Or(satImg, BDD_cproject( ImgA, ∪_{k≠i,j} C[k] ) );
    }

    R := BDD_Or( A, satImg );
    if (BDD_Equal(R, A))
        return R;

    return Cntrl_Pair_Sat_Sim(R, C, G);
}
```

Figure 4: Control-pair saturated simulation.

3.3 Extensions to saturated simulation

Biasing

The implementation of the cproject operator in Fig 3 is "biased towards 1", i.e., when presented with a choice for a projection variable, it sets it to 1. This biasing can result in dropping portions of the state space that may be significant (e.g., branch on zero). We overcome this by computing the union of two subsets; one biased towards 0, and one biased towards 1. While no longer minimal, the resulting set has cardinality no more than twice of the original set computed by `cproject`. More generally, we can define a `cproject_random` operator, wherein at each level of the recursion the bias for each variable is selected at random. The results of `cproject_random` can be added to the 0/1 biased subsets to obtain a rich, yet sparse subset.

BDD Minimization

Consider the expression for the final case of the *BDD_cproject* function listed in Figure 3:

$$v \cdot BDD\_cproject(A_v) \; + \; \bar{v} \cdot \overline{T} \cdot BDD\_cproject(A_{\bar{v}})$$

Replacing $\overline{T}$ by 1 results in a function which computes a superset of the result of *BDD_cproject*. Intuitively, since the expression is simplified, we reason that the BDD should also be simpler. We refer to the resulting operator as `BDD_cproject_min`; heuristically, it should give a larger set (in the sense of cardinality) with a smaller BDD.

Control-pair Edge Saturated Simulation

An extension to obtain enhanced coverage is to perform a partial reachability analysis and at each step pick a subset of the image that preserves all "controller transitions" to the image from the current set [8, 9]. Yuan et al [15] show how to compute minimal control-edge saturated sets by augmenting the design: for every control latch x_C, add a new latch x_S which "shadows" x_C, that is, the next state of x_S is the present state of x_C; hence, control-edge pair saturated simulation can be reduced to control-state pair saturated simulation.

4 Experimental Results – Saturated Simulation

We coded the routines described in the previous section as part of the VIS program [7]. We report experimental results on the viper microprocessor, which is distributed with the VIS system [3]. Viper is approximately 4000 gate-equivalents. It contains 252 latches; from these, we identified 6 control units, containing between 4 and 12 latches, for a total of 42 control latches. The variable ordering was derived by building the BDD for the decomposed transition relation and performing reordering; dynamic reordering was enable throughout.

In the previous section, we described several extensions of the basic saturated simulation procedure laid out in Figure 4. Some of these were independent (such as the use of control-pair edges, and the use of `cproject_min`); consequently, there are an exponential number of possible experiments. We performed a representative set, detailed results are given in [1]; we summarize them below. In particular, we experimented with the following —

Comp-Reach: Complete BDD based reachability analysis.

Sat-Reach: Saturated reachability analysis as in Figure 4.

0-Bias: *Sat-Reach*, with 0-biased *cproject* (§ 3.3).

Rnd-Bias: *Sat-Reach*, with random *cproject* (§ 3.3).

Min-Sat: *0-Bias*, with *BDD_cproject_min* instead of *BDD_cproject* (§ 3.3).

Min-Rnd-Sat: *Sat-Reach*, with randomly biased *BDD_cproject_min*.

All experiments were performed on a DEC AlphaServer 2100A 5/250 with 1Gb of memory.

Summary of results

1. **Comp-Reach** spaced-out after four iterations of reachability, after having visited 224,326 control pair states (CSPs).

2. **Sat-Reach** reached a fixed point in 64 iterations (144.6 seconds), having found 353,949 CSPs; peak BDD size was 5,512 nodes.

3. **0-Bias** reached a fixed point in 67 iterations (1269.5 seconds), having found 633,603 CSPs; peak BDD size was 38,033 nodes.

4. **Rnd-Bias** spaced-out in 15 iterations (2565.4 seconds), having found 1,152,080 CSPs; peak BDD size was 2,755,343 nodes.

5. **Min-Sat** reached a fixed point in 68 iterations ($4,951.8$ seconds), having found 1,063,720 CSPs; peak BDD size was 138,253 nodes.

6. **Min-Rnd-Bias** spaced-out in 15 iterations ($9,423.8$ seconds), having found 1,649,070 CSPs; peak BDD size was 3,308,410 nodes.

Comp-Reach exhibited unstable behavior — the first four iterations took 3.2 seconds, and the largest BDD encountered was $4,019$ nodes; the example spaced out on the fifth iteration.

All variants of saturated reachability were able to go through substantially more iterations. The 0-biased cproject routine found 80% more control pair states than the 1-biased version; we conjecture this is because there is more interesting behavior which can be reached from states where the datapath registers contain 0's.

Among the nonminimized routines, the random biased cproject discovered the most control state pairs, a total of 1.15×10^6, which is five times as many as complete analysis, and two/three times better than 0/1-biased analysis.

Somewhat to our surprise, the `BDD_cproject_min` operator usually gave larger BDDs than `BDD_cproject`. However, there was benefit in terms of exploring more control pair states: **Min-Rnd-Sat** found 1.65×10^6 control pair states.

For comparison we ran cycle simulation with random inputs for 100,000 seconds; we could not compute the number of control-state pairs, but the total number of states found was an order of magnitude less that the number of control-state pairs reached in any of the above experiments.

5 Conclusion

In summary, our primary contribution in this paper is a symbolic procedure for exploring the set of control pair states. The various heuristics developed in the course of this paper result in substantial improvement over complete BDD-based reachability and cycle simulation. The resulting procedure is robust with respect to time and memory, and requires only that the user specify the different controllers in the design.

We believe that it is fair to say that existing symbolic verification techniques, though complete, are limited. Coping with BDD blowup requires considerable expertise; we suggest more research should be performed on exploring how to best use computational resources.

In the future we plan to continue exploring ways in which to combine formal and informal verification; one possibility is adaptive simulation, in which random search is combined with backtrack. We are also looking at applications in software verification.

References

[1] www.ece.utexas.edu/~adnan/publications/AKS-DAC-98.ps.

[2] P. Ashar and S. Malik. Fast Functional Simulation Using Branching Programs. In *Proc. Intl. Conf. on Computer-Aided Design*, November 1995.

[3] UC Berkeley. www.cad.eecs.berkeley.edu/~vis.

[4] R. Bryant. Graph-based Algorithms for Boolean Function Manipulation. *IEEE Transactions on Computers*, C-35:677–691, August 1986.

[5] B. Chen, M. Yamazaki, and M. Fujita. Bug Identification of a Real Chip Design by Symbolic Model Checking. In *Proc. European Conf. on Design Automation*, pages 132–136, March 1994.

[6] H. Cho, G. D. Hachtel, E. Macii, M. Poncino, and F. Somenzi. A Structural Approach to State Space Decomposition for Approximate Reachability Analysis. In *Proc. Intl. Conf. on Computer Design*, October 1994.

[7] R. K. Brayton et al. VIS: A system for Verification and Synthesis. In *Proc. of the Computer Aided Verification Conf.*, July 1996.

[8] Richard C. Ho, C. Han Yang, Mark A. Horowitz, and David L. Dill. Architectural Validation for Processors. In *Proceedings of the International Symposium on Computer Architecture*, June 1995.

[9] Y. Hoskote, D. Moundanos, and J. Abraham. Automatic Extraction of the Control Flow Machine and Application to Evaluating Coverage of Verification Vectors. In *Proc. Intl. Conf. on Computer Design*, Austin, TX, October 1995.

[10] 0-in Inc. www.0in.com.

[11] B. Lin and R. Newton. Implicit Manipulation of Equivalence Classes Using Binary Decision Diagrams. In *Proc. Intl. Conf. on Computer Design*, Cambridge, MA, October 1991.

[12] P. McGeer, K. McMillan, A. Saldanha, A. Sangiovanni-Vincentelli, and P. Scaglia. Fast Discrete Function Evaluation. In *Proc. Intl. Conf. on Computer-Aided Design*, November 1995.

[13] Kenneth L. McMillan. *Symbolic Model Checking*. Kluwer Academic Publishers, 1993.

[14] K. Ravi and F. Somenzi. High Density Reachability Analysis. In *Proc. Intl. Conf. on Computer-Aided Design*, Santa Clara, CA, November 1995.

[15] J. Yuan, J. Shen, J. Abraham, and A. Aziz. On Combining Formal and Informal Verification. In *Proc. of the Computer Aided Verification Conf.*, July 1997.

Fast State Verification

Dechang Sun, Bapiraju Vinnakota, Wanli Jiang*
Dept. of Electrical and Computer Engineering,
University of Minnesota
Minneapolis, MN 55455, USA

Abstract

Unique input/output(UIO) sequences are used for state verification and functional test in finite state machines. A UIO sequence for a state s distinguishes it from other states in the FSM. Current algorithms to compute UIO sequences are limited in their applicability to FSMs with binary input symbols such as those found in control applications. Execution times of traditional approaches are exponential in the number of FSM inputs. We develop a new heuristic algorithm to generate UIO sequences for FSMs with binary inputs. Execution time is reduced significantly by reducing the size of the search space. When a UIO sequence cannot be generated, our algorithm generates a small number of functional faults for state verification.

1 Introduction

Efficient functional test techniques for finite state machines are of great importance in many applications. Unique Input/Output (UIO) sequences for states in FSMs are one solution to problems such as state verification for functional test and protocol conformance test[1, 3, 5]. A Unique Input/Output (UIO) sequence is a sequence, specific to a state s, that distinguishes it from all other states in the FSM. UIO sequence construction has been shown to be a difficult problem [1]. No efficient algorithms are known for finding UIO sequences. The methods currently used are usually based on appropriate successor trees and are only suitable for FSMs with a few input symbols, such as the FSMs used for representing computer network protocols [5]. For FSMs with binary inputs, if the number of input bits is high, the successor trees are too large to handle.

We present a guided heuristic method to compute UIO sequences. Typically, in FSMs designed for control applications the range of input labels is far less than the number of input symbols(which is exponential). However, forming successor trees will require the formation of all input symbols. Our basic approach is to search as much as is possible on input labels. This reduces the size of trees that need to be formed in a breadth first search. In addition, we use a preprocessing step to compute information that is used when computing UIO sequences for all states. This accelerates the search since there is less to do in each step. It is possible that some states of an FSM do not have UIO sequences [1]. When a UIO sequence cannot be found, our algorithm generates a set of distinguishable groups of states. Each distinguishable group can be distinguished from s by a single sequence. Our algorithm attempts to minimize the number of distinguishable state groups.

This paper is organized as follows. Section 2 reviews essential preliminaries and relevant previous work. The new heuristic is discussed in Section 3. Experimental results are presented in Section 4.

*This research was supported by the NSF through grant MIP-9502240

2 Preliminaries

A finite state machine is formally defined as follows [1].

Definition 1 *A finite state machine(FSM) is a quintuple* $(I, O, S, \delta, \lambda)$ *where* I, O *and* S *are finite and nonempty sets of input labels, output labels, and states, respectively.*
δ: $S \times I \rightarrow S$ *is the state transition function and*
λ: $S \times I \rightarrow O$ *is the output function.*
When the machine is in a current state $s \in S$ *and receives an input* $i \in I$, *it moves to the next state specified by* $\delta(s, i)$ *and produces an output given by* $\lambda(s, i)$. □

Definition 2 *An input symbol is an M-bit binary vector* $(x_{M-1}, x_{M-2}, ..., x_0)$ *where each* $x_i \in \{0, 1\}$.

Definition 3 *An input label is an M-bit vector* $(y_{M-1}, y_{M-2}, ..., y_0)$ *where each* $y_i \in \{0, 1, -\}$.

Definition 4 *An input label* i_l *covers or includes an input symbol* i_s *if* i_s *can be obtained from* i_l *by setting don't care bits to binary values. This is denoted by* $i_s \in i_l$.
An input label i_1 *intersects with another label* i_2 *if there exists an input symbol that is covered by both* i_1 *and* i_2.

Each transition in an FSM is a 4-tuple (i_t, s_t, d_t, o_t) where i_t, s_t, d_t, o_t denote the input label, source state, destination state and output label on the transition, respectively. Let the number of states be N, the number of input pins be M and the number of transitions be E. The maximum number of distinct input symbols is 2^M. It is possible to define up to 2^M outgoing transitions from each state. But in practice, because on many transitions several input pins are set to don't care values, E is far less than $2^M \times N$.

2.1 UIO sequences

Definition 5 *A UIO sequence of a state s is an input sequence x that uniquely identifies s. The output sequence produced by the machine in response to x when starting from any state other than s is different from that produced when starting from s,* $\lambda(s_i, x) \neq \lambda(s, x)$ *for any* $s_i \neq s$. □ *[1]*

In this paper we will consider only irredundant FSMs in which no two states are equivalent. (Obviously, if s_1 and s_2 are equivalent, neither will possess a UIO sequence). Further, to limit search times, we will only search for UIO sequences of a maximum length L. The straightforward method to compute a UIO sequence is to enumerate all possible sequences of input symbols and simulate them to find out whether one or more of them can distinguish the target state from all other states. For each state, there are 2^M possible input symbols. It's obvious that the number of all possible input sequences with length L is 2^{ML}. This number is too large to handle for many practical FSMs.

However, the number of outgoing transitions(or number of labels) from each state is usually a number much smaller than 2^M. If instead of enumerating all possible sequences of input symbols one need only enumerate all possible sequences of input labels, the problem would be much easier. Unfortunately, outgoing transitions from different states usually do not have the same set of labels. Further more, a label on one outgoing transition from a state s_1 may intersect with the labels of multiple transitions from other states. This makes a search process difficult. Consider the benchmark circuit *ex*4 [10]. There are only 11 different input labels. Assume we wish to compute a UIO sequence for state s_1. An input label "1 − − − − − −" will transfer the FSM from state s_1 to state s_3 as shown in Figure 1(a). The same input label is not defined for any transition from s_8, as shown in Figure 1(b). The destination state on the labels "1 − 0 − − −" and "1 − 10 − −" is s_3 whereas the destination state on the transition label "1 − 11 − −" is s_4. In other words, the next state cannot be uniquely defined if we attempt to search on the input label "1 − − − − − −". To define a search in which destination states on branches are unique one would have to search on the input labels "1 − 0 − − −", 1 − 10 − −" and "1 − 11 − −". This problem becomes more acute if the number of states in the search is increased. In the worst case, one may have to search on all input symbols. All don't care bits on all labels may have to be expanded to all possible values. In summary, the intersection of input labels makes the problem of finding UIO sequences much more complicated.

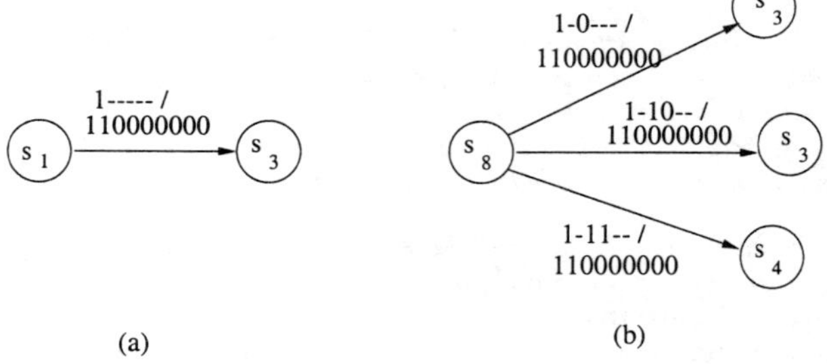

Figure 1: Input label intersection

Previous methods to find UIO sequences are based on the formation of successor trees or modified successor trees. A successor tree represents the behavior of the FSM from all possible initial states under all possible input sequences. Each node n in the successor tree is a partition P_n of the states. The root partition consists of a single block containing all states in the FSM. The number of outgoing branches from each node is equal to the number of input symbols. Each outgoing branch is labeled with a distinct input symbol. Consider a branch from node n_1 to node n_2 labeled with input symbol i_s. Two states s_1 and s_2 in the same block B in the partition P_{n1} are placed in different blocks in the partition P_{n2} if outgoing transitions, for an input i_s, from states s_1 and s_2 produce different outputs. That is, they are placed in different blocks if input i_s can be used to distinguish s_1 from s_2. Two states in different blocks in P_{n1} will continue to be in different blocks in P_{n2}. A UIO sequence can be obtained by searching the successor tree. A sequence of edges starting from the root that ends in a node n in which state s_i is a separate block is a UIO sequence for state s_i. There are two limitations to successor trees. Forming them is expensive since the number of branches at each node is 2^M. Searching the large tree to compute UIO sequences is also expensive.

2.2 Functional Test Techniques

The primary application of UIO sequences is in functional test for FSMs. The input to the functional test generator is a functional description of an FSM. The goal is to generate a test sequence that can be used to verify if an implementation of the FSM meets its specifications. If the response of the implementation differs from that of the specification, it indicates that the implementation does not conform to the specification [1]. A test sequence for an FSM is formed by composing test sequences for individual transitions. A test sequence for a transition (i_t, s_t, d_t, o_t) typically consists of three steps: (1) apply an input sequence to transfer the FSM to state s_t, (2) apply the input label i_t and verify that the output is o_t, and (3) apply an appropriate sequence to verify that the destination state of the transition is d_t. There may be 3 possible corruptions of a transition: (1) wrong output label, (2) wrong destination state, (3) both wrong output label and destination state. A type (2) fault dominates the other two kinds of faults[4].

For FSMs with binary inputs it may be difficult to generate UIO sequences. Further, UIO sequences may not exist. Cheng and Jou proposed a functional fault model[4] in which a fault is assumed to affect a transition in a specific manner. As mentioned above, only faults in the destination state need be considered. Thus, each transition is associated with $N-1$ transition faults. Each fault corresponds to a specific faulty destination state. A machine with E transitions will have $E(N-1)$ transition faults in its fault list. This number is very large for practical machines and may result in large test sets. Cheng and Jou suggested a fault collapsing technique based on finding *State Group Differentiating*(SGD) sequences. A SGD sequence, $T(s, SG)$, for a state s and a state group $SG = \{s_1, s_2, ..., s_i\}$ where $s \notin SG$, is an input sequence that can distinguish s from all the states in SG. A SGD sequence set for a state s is a collection of $K(s)$ SGD sequences, $T(s, SG_1)$, $T(s, SG_2)$, ..., $T(s, SG_{K(s)})$, such that s, SG_1, SG_2, ..., $SG_{K(s)}$ are disjoint subsets of all states S and their union is S.

In order to limit the size of functional test sets, some have proposed translating faults at the logic level to transition faults [7, 9]. Functional fault models which maximize the coverage of logic level faults have also been proposed. Researchers have also investigated functional fault models which reduce the total length of the test set [8]. These papers are concerned with using functional test techniques to achieve high logic-level fault coverage. Our focus is on the first part of the process.

3 A Heuristic Algorithm for Finding UIO Sequences

The problem of computing UIO sequences has recently been shown to be NP-complete[1]. A high performance deterministic algorithm may be difficult to find. Hence, we focus on a heuristic algorithm. For each transition, a preprocessing step generates a set of distinguishable state groups(DSG) and associated distinguishing vectors. UIO sequences are constructed by a breadth-first search of the transition space, starting at the state for which a UIO sequence is to be generated. The DSGs are used to identify which states can be distinguished from the target state at that stage of the search process. The search process is fast because we are searching only the space of transitions. However, the search is not complete because we do not search the space of input symbols. The algorithm is described informally as follows.

Consider a transition (i_t, s_t, d_t, o_t). This transition can be used to distinguish the source state s_t from part or all of the other $N-1$ states in the FSM. Recall that the input label i_t may actually cover more than one input symbol. The exact set of states distinguished by i_t will depend on the actual symbols covered by i_t. Usually, a set of symbols DVS will distinguish the same set of states in $S - \{s_t\}$ from s_t. Enumerating all vectors in a DVS in a search process would be redundant. The set of symbols covered by i_t can be partitioned into disjoint sets $\{DVS_i\}$ such that each DVS_i is a set of symbols covered by i_t and all symbols in each DVS_i distinguish the same set of states from s_t. Furthermore, Some DVS_i may not be useful because the set of states distinguished by them are a subset of those distinguished by another vector set DVS_j. Our heuristic method attempts to identify distinguishable state groups by listing only one vector from each DVS_i. We also eliminate DVS_i that are not useful. Thus each label is associated with a set of distinguishable state groups and distinguishing vectors. Most calculations are related to transitions and can be done before the breadth-first search in a short time. This method can be used to find long UIO sequences for large FSMs.

3.1 Distinguishable State Groups

The first step in our process is the generation of distinguishing state groups and the corresponding distinguishing vectors for each transition in the FSM. The information on each transition is used in the construction of UIO sequences of all states, not just for the source state of the transition.

Definition 6 *A* Distinguishable State Group *(DSG) of a transition (i_t, s_t, d_t, o_t) is a group of states SG such that there exists an input vector $v \in i_t$ which*

(1) can distinguish state s_t from all states in the group, i.e., for any $s_i \in SG$, $\lambda(s_i, v) \neq \lambda(s_t, v)$, and

(2) cannot distinguish s_t from any other states, i.e., for any $s_i \notin SG$, $\lambda(s_i, v) = \lambda(s_t, v)$.

v is called a Distinguishing Vector *(DV) of the DSG. All possible DVs of a DSG form a* Distinguishing Vector Set*(DVS).*□

A DSG of transition (i_t, s_t, d_t, o_t) may contain up to $N-1$ states, i.e., all except s_t. A transition may have more than one DSG, each associated with a different symbol. For each DSG there exists a corresponding *Un-distinguished States Group*(USG).

Definition 7 . *An* Un-distinguishable State Group*(USG) U of transition (i_t, s_t, d_t, o_t) corresponds to a Distinguishable State Group D of this transition where $U = S - \{ps\} - D$.* □

The algorithm shown in Figure 2 can be used to find all the DSGs of a transition.

Algorithm 1.
INPUT: transition table $\{T_i\}$, $i = 1, 2, ..., E$,
    a target transition (i_t, s_t, d_t, o_t)
OUTPUT: all DSGs of the target transition
FUNCTION: $T(s, i) = \{(i_p, s_p, d_p, o_p)$ with $s_p = s$ and
    $i_p \cap i \neq \emptyset\}$
$SU = S - \{s_p\}$, $ADSG = AUSG = \emptyset$, $k = 0$
/* step 1*/
for each state $s_u \in SU$ {
  if(all transitions in $T(s_u, i_t)$ have o/p
      labels different from o_t)
    $ADSG = ADSG + \{s\}$
  else if(all transitions in $T(s_u, i_t)$ have o/p label o_t)
    $AUSG = AUSG + \{s\}$
}
$SU = SU - ADSG - AUSG$, $SU_r = SU$
/* step 2*/
while(SU_r is not empty) do {
  $k = k + 1$, $DSG_k = ADSG$
  take out a state s_r from SU_r
  $SU_r = SU_r - \{s_r\}$, $DSG_k = DSG_k + \{s_r\}$
  $TP = \{T_r \in T(s_r, i_t)$ and o/p label of $T_r \neq o_t\}$
  $i_r = i_t \cap$ (union of input labels of TP), $DVS_k = i_r$
  for each state $s_u \in SU$ except s_r
    $TP = \{T_r \in T(s_u, i_t)$ and o/p label of $T_r \neq o_t\}$
    $temp = i_r \cap$(union of input labels of TP)
    if($temp \neq \emptyset$) {
      $DSG_k = DSG_k + \{s_u\}$, $DVS_k = temp$
      $SU_r = SU_r - \{s_u\}$
      $i_r = temp$
    }
  }
} □

Figure 2: Algorithm to compute DSGs

In Algorithm 1, a transition $t = (i_t, s_t, d_t, o_t)$ is processed in two steps as follows. This transition is compared with the outgoing transitions from every other state $s \neq s_t$. The function $T(s, i)$ returns a list of transitions with a source state s, and an input label i' such that $i' \cap i \neq \emptyset$. We will use the label $T(s, i)$ to refer to the set of transitions returned by $T(s, i)$. Consider the following special cases. If, for all $t' \in T(s', i_t)$, the output label on t' is not equal to o_t, then any symbol covered by i_t can distinguish s' from s_t. All states which satisfy this condition can be added to a state group named ADSG. The states in ADSG can be added to any DSG associated with the transition t. Similarly, if for all $t' \in T(s'', i_t)$, the output labels on t' are equal to o_t, no vector covered by i_t can distinguish s'' from s_t. All states which satisfy this condition are added to AUSG. These states need no longer be processed. This step reduces the complexity of step 2 considerably.

Distinct DSGs are formed iteratively in step 2 as follows. If a state s_u is in the set SU, it implies that for some $t \in$

$T(s_u, i_t)$ the output is equal to o_t, while it is not equal to o_t for another $t' \in T(s_u, i_t)$. The set SU_r is the set of states which have not yet been added to any DSG. (Every state not in AUSG has to belong to at least one DSG). Initially, SU_r and SU are identical. DSGs are formed as follows. A state s' is chosen from SU_r. Consider a transition $t' = (i', s', d', o' \neq o_t)$. The set of input symbols in i_t (the input label of the target transition) covered by i' and i_t is $i_r = i' \cap i_t$. This set consists of the vectors that can be associated with a DSG containing s'. Next, the algorithm verifies if other states in SU(not SU_r!) can be added to the DSG. When a state in SU_r is added to any DSG(in the nested loop) it is removed from SU_r. The algorithm terminates when SU_r is empty. That implies, every state in SU has been assigned to at least one DSG. Any vector in the set i_r is a distinguishing vector. The set i_r is pruned dynamically using boolean relationships.

Example 1: Consider the FSM shown in Table 1. Suppose we are trying to find the DSGs of transition $(1---, s_0, s_0, 00)$. In step 1 of algorithm 1, we compute ADSG and AUSG. This is done by first computing the function $T(s_i, 1---)$ for each state $s_i \neq s_0$. The result is shown in Table 2.

$T(s_1, 1---)$ contains two transitions. Neither of them has an output label of 00. All other states do not have this property. So $ADSG = \{s_1\}$.

$T(s_4, 1---)$ contains two transitions. Both of them have an output label of 00. All other states do not have this property. So $AUSG = \{s_4\}$.

So, the states needed to be considered in step 2 are $SU = SU_r = \{s_2, s_3\}$.

Step 2 of algorithm 1 is carried out as follows:

(a) $DSG_1 = ADSG = \{s_1\}$, $i_t = 1---$.

(b) Take out a state $s_r = s_2$ from SU_r,
$SU_r = SU_r - \{s_2\} = \{s_3\}$,
$DSG_1 = DSG_1 + \{s_2\} = \{s_1, s_2\}$
$TP = \{(11--, s_2, s_4, 10)\}$
$i_r = i_t \cap$ (union of input labels of TP) $= \{11--\}$, $DVS_1 = \{11--\}$

(c) performing the "for" loop. There is only one state s_3 in SU except $s_r = s_2$.
TP=$\{(100-, s_3, s_3, 01), (101-, s_3, s_4, 10)\}$
$temp = i_r \cap$(union of input labels of TP) $= \{11--\} \cap \{100-, 101-\} = \{\emptyset\}$
The "for" loop is over.

(d) Take out a state $s_r = s_3$ from SU_r,
$SU_r = SU_r - \{s_2\} = \{\emptyset\}$,
$DSG_2 = DSG_2 + \{s_3\} = \{s_1, s_3\}$
$TP = \{(100-, s_3, s_3, 01), (101-, s_3, s_4, 10)\}$
$i_r = i_t \cap$ (union of input labels of TP) $= \{100-, 101-\}$, $DVS_2 = \{100-, 101-\}$

(e) performing the "for" loop. There is only one state s_2 in SU except $s_r = s_3$.
TP=$\{(11--, s_2, s_4, 01)\}$
$temp = i_r \cap$(union of input labels of TP) $= \{100-, 101-\} \cap \{11--\} = \{\emptyset\}$
The "for" loop is over.

(f) Since SU_r is empty now, the process is finished. The result is: for transition $(1---, s_0, s_0, 00)$, there are two DSGs. They are $DSG_1 = \{s_1, s_2\}$ with a DVS of $DVS_1 = \{11--\}$ and $DSG_2 = \{s_1, s_3\}$ with a DVS of $DVS_2 = \{100-, 101-\}$.

The time needed to compute all the DSGs of a transition depends on the size of the set i_r in step 2 of algorithm 1. Usually, the maximum size of i_r is limited and the algorithm takes a short time. Under this condition, the time complexity of algorithm 1 is $O(M^2E^2)$.

Input	Source	Dest.	Output
$1---$	s_0	s_0	00
$0---$	s_0	s_1	11
$1--1$	s_1	s_0	01
$1--0$	s_1	s_2	10
$0---$	s_1	s_3	00
$-0--$	s_2	s_2	00
$-1--$	s_2	s_4	10
$-00-$	s_3	s_3	01
$-01-$	s_3	s_4	10
$-1--$	s_3	s_0	00
$--0-$	s_4	s_1	00
$--1-$	s_4	s_4	00

Table 1: Transition Table of an example FSM

$T(s_i, 1---)$	Input	Source	Dest.	Output
$T(s_1, 1---)$	$1--1$	s_1	s_0	01
	$1--0$	s_1	s_2	10
$T(s_2, 1---)$	$10--$	s_2	s_2	00
	$11--$	s_2	s_4	10
$T(s_3, 1---)$	$100-$	s_3	s_3	01
	$101-$	s_3	s_4	10
	$11--$	s_3	s_0	00
$T(s_4, 1---)$	$1-0-$	s_4	s_1	00
	$1-1-$	s_4	s_4	00

Table 2: $T(s_i, 1---)$ for the example FSM in Table 1

3.2 Computing UIO sequences

The purpose of a UIO sequence is to verify the destination state of a transition. An alternative view of the problem is as follows. Consider N copies of the FSM. Assume one is in an initial state s_0, the target state. Let the other $N-1$ copies of the FSM each have a distinct initial state $s' \in S - \{s_0\}$. We will refer to the copy of the FSM starting in state s_0 as the good FSM. We will refer to the remaining $N-1$ copies as faulty FSMs. A UIO sequence for state s_0 would cause the good FSM to produce an output sequence not produced by any of the faulty FSMs.

Assume the first vector i_0 in the UIO sequence has been chosen. Let TBD_k be the set of states to be distinguished from the target state after k vectors in the UIO sequence have been applied. Initially, $TBD_0 = S - \{s_0\}$. Some states in $S - \{s_0\}$ will be distinguished from s_0 by i_0. The good machine is transfered to a state $s_0' = \delta(s_0, i_0)$. The faulty machines are also transfered as well. Let $S_d{}^0$ be the set of states distinguished by i_0. The states which have not been distinguished will be $TBD_1 = \delta(S - S_d{}^0 - \{s_0\}, i_0)$. The rest of the sequence will have to distinguish the states in TBD_1 from s_0'. Thus, at each step k the states in TBD_k will have to be distinguished from the current state of the good FSM.

We construct a UIO sequence for a state s_0 as follows. A tree of transitions, whose root is state s_0, starting from state s_0 is constructed using a breadth-first search process. The transitions in the first level of the tree are the set of outgoing transitions from state s_0. The set of nodes in the first level is the set of destination states of the transitions from s_0. Each path in the tree is a sequence of transitions. The labels on the transitions may each cover several symbols. To generate a sequence of inputs, the labels have to be converted to symbols. The distinguishing state groups of each transition are used to choose the actual input vectors. Assume we are analyzing transition T_r with input label i_r in a sequence of r transitions $T_1, ..., T_{r-1}, T_r$ starting from the root of the tree. At each step the symbol associated with T_r is chosen using a greedy heuristic, that is, such that the number of states in TBD_r is minimized. Recall that each transition T_r is associated with a set of distinguishable state groups. Let the symbol associated with DSG_k be i_r^k. Then the input symbol $i_r^k \in i_r$ is chosen such that $DSG_k \cap TBD_r^k$ is maximized.

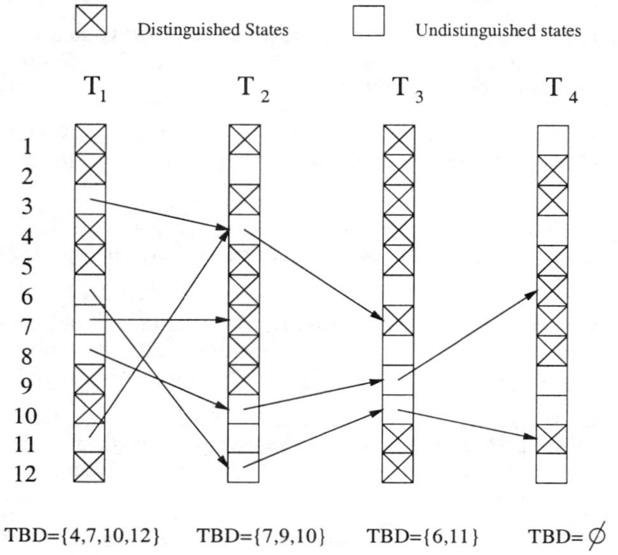

Figure 3: Finding TBD states

The algorithm for finding UIO sequences is omitted for brevity. An example showing reduction in the size of TBD is shown in Figure 3. Though the above process is not theoretically complete, experimental results will show that the heuristic is a good approach for most practical FSMs.

3.3 Valid Input Labels

When constructing a UIO sequence, if there are some states not distinguished from the target state, we need to extend the current sequence to a longer one. However, if the following situation occurs, we will never get a UIO sequence by extending the current sequence. Consider a vector sequence $T = v_1, v_2, ..., v_l$ applied to the FSM that transfers it from the target state s to a state d and produces an output sequence of $O = o_1, o_2, ..., o_l$. Assume T is also a transfer sequence from state s' to state d and produces the same output sequence. We will not be able to distinguish s from s' by extending T. This is the *state merging* problem.

Definition 8 *A* Merging State *of a transition* (i_t, p_t, n_t, o_t) *is a state* $s_m \neq s_t$ *and there exists an input vector* $v_m \in i_t$ *such that* $\delta(s_m, v_m) = d_t$ *and* $\lambda(s_m, v_m) = o_t$. □

The state merging problem makes the search for UIO sequences much more complicated.

To solve the state merging problem, we introduce the concept of *Valid Input Label.*

FSM	N	E	M	# UIO	L_{max}	CPU(s)
dk14	7	56	3	3	1	0.13
dk15	4	32	3	3	2	0.05
dk16	27	108	2	21	3	0.47
dk17	8	32	2	7	2	0.04
dk512	15	30	1	11	4	0.17
ex1	20	138	9	17	2	0.85
ex4	14	21	6	14	4	0.06
planet	48	115	7	47	4	0.76
opus	10	22	5	6	1	0.05
styr	30	166	9	28	4	7.26
bbara	10	60	4	6	7	3.83
cse	16	91	7	14	3	9.67
sand	32	184	11	32	18	17.97
scf	121	166	27	91	8	16.28

Table 3: Experimental Result of finding UIO sequences

Definition 9 *For a transition* (i_t, s_t, d_t, o_t)*, its* Valid Input Label i_v *is a set of input vectors such that*
(1) $i_v \in i_t$
(2) no vector $v \in i_v$ *will cause a state merging problem.*□

The valid input label of each transition in an FSM can be calculated in a preprocessing step. When we use valid input labels instead of the original labels on the transition, no state merging problems will occur. However, the size of DSGs will be reduced by doing so.

4 Application of UIO Sequence Generator to Functional Test

A program in C has been written to perform the heuristic search for UIO sequences. When applied to a group of benchmark FSMs, the result is quite satisfactory, as shown in Table 3. Most of the UIO sequences have been found, with some of them as long as 18 vectors in length. There are 7 columns in the table. The 5th column represents the number of states for which a UIO sequence has been found. The 6th column lists the maximum length of the UIO sequences found. Column 7 shows the CPU time needed to compute the UIO sequences on a Sun Station ULTRA 2 computer.

When a UIO sequence cannot be found, the algorithm returns a set of dominant state groups $\{DG_i\}$, such that all the states in a DG_i can be distinguished from the target state by a single sequence. UIO sequences and dominant state groups reduce the number of faults to be modeled for functional test. When a UIO sequence is found, all N-1 functional faults associated with a transition can be collapsed to a single fault. If a UIO sequence is not found, faults associated with all the states in one dominant group can be collapsed into one fault. Compared to [4], our algorithm results in a sharp reduction in the number of modeled faults. Table 4 shows some of the results. Column 4 is the number of transition faults before fault collapsing. Column 5 shows the result in [4] and column 6 the result of our program. It can be seen that the numbers of faults modeled in our algorithm is significantly smaller than those in [4].

4.1 Discussion

UIO sequences have been used extensively in functional test for protocols and sequential logic circuits. Unfortunately,

FSM	(N)	(E)	E(N-1)	Faults modeled in [4]	Faults modeled here
dk14	7	56	336	94	92
dk15	4	32	96	84	41
dk16	27	108	2808	292	120
dk17	8	32	224	70	41
dk512	15	30	420	48	33
ex1	20	138	2057	526	138
ex4	14	21	273	88	21
planet	48	115	5405	654	118
opus	10	22	232	72	26
styr	30	166	4814	1039	176
bbara	10	60	252	135	60
cse	16	91	1365	507	93
sand	32	184	5704	1227	184

Table 4: Experimental Result on number of modeled faults

almost no authors have reported similarly detailed UIO sequence generation results for FSMs with binary input pins. Most research on UIO sequences, and functional test, has concentrated on using UIO sequences and/or some form of group distinguishing sequences in functional test [11, 12]. Though UIO sequence generation times have been reported in [13], the exact FSMs used and the percentage of FSMs for which UIO sequences have been successfully generated was not specified clearly. Similarly, many papers on functional test report test set generation times. However, the procedures in [4, 7] focus on generating sequences which distinguish groups of states from the target state. They do not focus on generating UIO sequences. The procedures reported in [8] generate both UIO sequences, and structures similar to state group distinguishing sequences. Results are not presented separately for the two types of sequences.

To the best of our knowledge, our procedure is the first to be able to successfully generate UIO sequences for a large number of states in virtually all benchmark FSMs [10]. Readers may note that our procedure offers high performance for even the largest benchmarks in the benchmark set. We are able to generate very long UIO sequences, for a large fraction of the states, in the largest FSMs, very quickly.

5 Conclusion

Functional test generation algorithms are useful for FSM test and verification. Traditional algorithms for UIO sequence generation offer very poor performance when applied to FSMs with binary inputs. Consequently, functional test techniques for such FSMs are commonly based on a functional fault model. In FSMs designed for control applications, the input labels on most state transitions contain a large number of don't cares. Thus, the number of actual transitions is far less than the number of possible transitions. This fact is not exploited in current algorithms for UIO sequence generation. We have developed a new heuristic technique to generate UIO sequences for FSMs with binary inputs. The heuristic searches for UIO sequences on the space of transitions in the state transition graph. Thus, compared to current algorithms the size of the search space is significantly reduced. To the best of our knowledge, no previous algorithm could generate UIO sequences for such large FSMs in a reasonable period of time.

References

[1] D. Lee and M. Yannakakis, "Principles and Methods of Testing Finite State Machines— A Survey," *Proceedings of IEEE*, vol. 84, No.8, pp. 1090–1122, August, 1996.

[2] D. Lee and M. Yannakakis, "Testing finite state machines: state identification and verification," *IEEE Trans. Computers*, vol. 43, No.3, pp. 306–320, 1994.

[3] A. V. Aho, A. T. Dahbura, D. Lee and M. U. Uyar, "An Optimization technique for protocol conformance test generation based on UIO sequences and rural Chinese postman tours," *IEEE Trans. Communications*, vol. 39, No.3, pp. 1604–1615, 1991.

[4] K. T. Cheng and J. Y. Jou, "Functional Test Generation for Finite State Machines," *International Test Conference,* September 1990.

[5] K. Sabnani and A. Dahbura, "A Protocol Test Generation Procedure," *Computer Networks*, vol.15, pp285–297, 1988.

[6] E. Hsieh, "Checking Experiments for Sequential Machines," *IEEE Trans. Computers*, vol. C-20, pp.1152-1166, Oct. 1971.

[7] I. Pomeranz and S. M. Reddy, "On achieving complete fault coverage for sequential machines using the transition fault model," *Proceedings DAC*, pp. 341–346, 1991.

[8] G. Buonanno, F. Fummi, and D. Sciuto, F. Lombardi, "FSMTest: Functional test generation for sequential circuits," *INTEGRATION, the VLSI Journal 20*, pp.303–325, 1996.

[9] B. Vinnakota and J. Andrews, "Fast Fault Translation," *IEEE Trans. VLSI Systems*, pp.122-133, March 1998.

[10] R. Lisanke, "Finite-state machine benchmark set," *Preliminary benchmark collection*, 1987. Available at "http://www.cbl.ncsu.edu".

[11] A. Aho, A.T. Dahbura, D.Lee and M. UmitUyar, "An optimization technique for protocol conformance test generation based on UIO sequences and Rural Chinese Postman Tours," *Protocol Specification, Testing and Verification*, Vol. VIII, North-Holland, Amsterdam, pp.131-143, 1988.

[12] Y.-N. Shen, F. Lombardi and A.T. Dahbura, "Protocol conformance testing by multiple UIO sequence," *Protocol Specification, Testing and Verification*, Vol. VIX, North-Holland, Amsterdam, pp.131-143, 1990.

[13] D. Schin, Y.-N. Shen and F. Lombardi, "An approach for UIO generation for FSM verification and validation," *Proc. ISCAS*, vol. 4, pp.303-306, 1994.

A Fast Sequential Learning Technique for Real Circuits with Application to Enhancing ATPG Performance

Aiman El-Maleh, Mark Kassab, and Janusz Rajski

Mentor Graphics Corporation
8005 S.W. Boeckman Rd.
Wilsonville, OR 97070, USA

1. ABSTRACT

This paper presents an efficient and novel method for sequential learning of implications, invalid states, and tied gates. It can handle real industrial circuits, with multiple clock domains and partial set/reset. The application of this method to improve the efficiency of sequential ATPG is also demonstrated by achieving higher fault coverages and lower test generation times.

2. INTRODUCTION

It is well-known that the performance of a search process can be significantly enhanced by finding necessary assignments and identifying relations between signals. An effective way of identifying necessary assignments and signal relations is by learning. The use of learned information helps the search engine in recognizing value assignment conflicts sooner and pruning the search space, significantly reducing the number of backtracks. Learning relations in a circuit is typically performed by injecting both logic values on a gate and propagating them backward and forward. This can be done either statically in a pre-processing phase [1,2], or dynamically during the search process [3]. Dynamic learning can extract significantly more implications since the learning is done in the presence of assignments that have already been made, which allows the search space to be pruned further. However, this can be very computationally expensive since it is performed multiple times during the search process. Furthermore, the implications learned are only valid as long as backtracking has not occurred beyond the point where learning was performed. In most learning techniques, justification stops when a decision node is reached. Recursive learning [4] attempts to extract more relations by learning implications which would be valid regardless of the assignment made at the decision node. Furthermore, it can be applied statically or dynamically. However, unless the depth of the recursion is restricted, this technique can be even more expensive than the search process. Note that most learning techniques are performed in the combinational logic of the circuit. Only [5] attempts learning of relations by extending the analysis across sequential elements, but the learning is only performed across two time frames.

DAC 98, San Francisco, California
©1998 ACM 0-89791-964-5/98/06..$5.00

Learning is used in several areas of computer-aided design, most notably automatic test pattern generation (ATPG) [1-5], redundancy identification [2,6], logic verification [7], and logic optimization [8]. Although this paper focuses on the application of the learning method to ATPG, the proposed method is not restricted to test generation.

Sequential ATPG is a much more complex process than combinational ATPG due to signal dependencies across multiple time frames. Extracting such relations can significantly enhance the performance of sequential ATPG as backtracks across multiple time frames are reduced. Furthermore, it has recently been shown [9] that a key indicator of the complexity of sequential ATPG is the density of encoding of the circuit, which is the ratio of the number of valid states to the total number of states. With circuits that have a low density of encoding, the test generator is more likely to waste a lot of time trying to justify an invalid state. Therefore, extraction of invalid states can also significantly enhance the efficiency of sequential ATPG. Most methods which identify the valid or invalid states in a sequential circuit require the presence of a reset state (e.g., [10]), which is not usually available in real circuits. Some more recent techniques do not require a reset state [11,12]. However, since those methods usually attempt to enumerate all invalid states, they can only be used for small circuits.

In this paper, we present a fast, memory-efficient technique for learning across multiple time frames. We propose the sequential learning mechanism, and its application for enhancing the performance of sequential ATPG by identifying implications, invalid states, and tied values. Its efficiency enables it to be used on very large circuits. Furthermore, it can be used on real circuits including those with multiple clock domains, and partial or full set/reset.

3. SEQUENTIAL LEARNING TECHNIQUE

Sequential learning can identify relations between two nodes at different time frames or in the same time frame. Relations learned between nodes in the same time frame are of more interest since they represent the set of illegal assignments or don't care set for a circuit, which can be utilized by ATPG, logic optimization or verification. Many more relations can be learned between nodes at different time frames than those between nodes in the same time frame. However, such relations have limited application since they involve the time domain. For an ATPG to take advantage of such relations, it needs to work on a window equivalent to the number of time frames across which the relations hold. The proposed learning technique can learn relations of either type. However, in this paper we present results for relations learned in the same time frame. Relations are learned between pairs of sequential elements and between gates and sequential elements. Relations between pairs of gates follow directly from relations between gates and sequential elements, and are therefore not extracted.

The sequential learning technique proposed in this paper is mainly based on forward simulation across time frames. This is in contrast to combinational learning techniques, which inject both a 0 and 1 on each gate and perform backward and forward implications across non-decision nodes. Performing the implications both backward and forward can allow learning of relations that may not be learned using only forward or backward implications. This is because a decision node may cease to be a decision node due to other assignments made by implications, allowing further propagation. However, this approach requires the learning mechanism to be performed twice the number of gates in the circuit. Since sequential learning for each gate can traverse the circuit forward and backward across time frames several times, this could make such an approach less attractive for large circuits. Furthermore, the time frames across which learning is to be performed has to be decided and memory has to be allocated a priori.

3.1 Implication Learning

The proposed learning technique is comprised of several steps. First, fanout stems (i.e., nodes that have a fanout greater than one) are identified. Then, for each fanout stem, both a 0 and a 1 are injected and simulated forward across time frames. This step allows the extraction of relations between nodes that are implied by the different values on the same stem based on the contrapositive law. If $A{=}0 \rightarrow F_1{=}0$ and $A{=}1 \rightarrow F_2{=}0$, then by the contrapositive law, $F_1{=}1 \rightarrow A{=}1$ which results in the relation $F_1{=}1 \rightarrow F_2{=}0$. This technique is called ***single-node learning***. For example, consider the circuit in Figure 1. This circuit has five fanout stems, namely I_1, I_2, F_1, F_2, and F_3. The forward simulation results for each value on every stem and their corresponding implied gates are shown in Table 1. Simulation is allowed to continue forward for a specified number of time frames. However, the simulation is stopped if the same state is repeated over two consecutive time frames. For example, injecting a 1 on stem F_3 produces the same state after time frame 1 and hence simulation is stopped at time frame 2. Similarly for stem I_2, simulation stops at time frame 4. Relations between nodes at two different time frames can be easily extracted from the simulation results for each stem. For example, the relation $G_1{=}0$ at $T{=}i{+}1 \rightarrow I_2{=}0$ at $T{=}i$ is learned based on the stem I_2. Relations between nodes in the same time frame are extracted similarly. For example, the relation $F_6{=}1 \rightarrow F_4{=}0$ is extracted based on stem I_2 since $I_2{=}0 \rightarrow F_6{=}0$ in time frame 1 and $I_2{=}1 \rightarrow F_4{=}0$ in time frame 1 as well. By the contrapositive law, $F_6{=}1 \rightarrow I_2{=}1 \rightarrow F_4{=}0$. Relations extracted between sequential elements based on single-node learning are shown in Table 2. Relations between sequential elements are called *invalid-state relations* since they represent the presence of invalid states. The relation $F_6{=}1 \rightarrow F_4{=}0$ represents the set of invalid states $(F_1,F_2,F_3,F_4,F_5,F_6)=(X,X,X,1,X,1)$. Learning of invalid states by implication has the benefit that sets of invalid states are extracted concurrently and their representation is compact. However, invalid states that cannot be represented as implications cannot be learned using this approach.

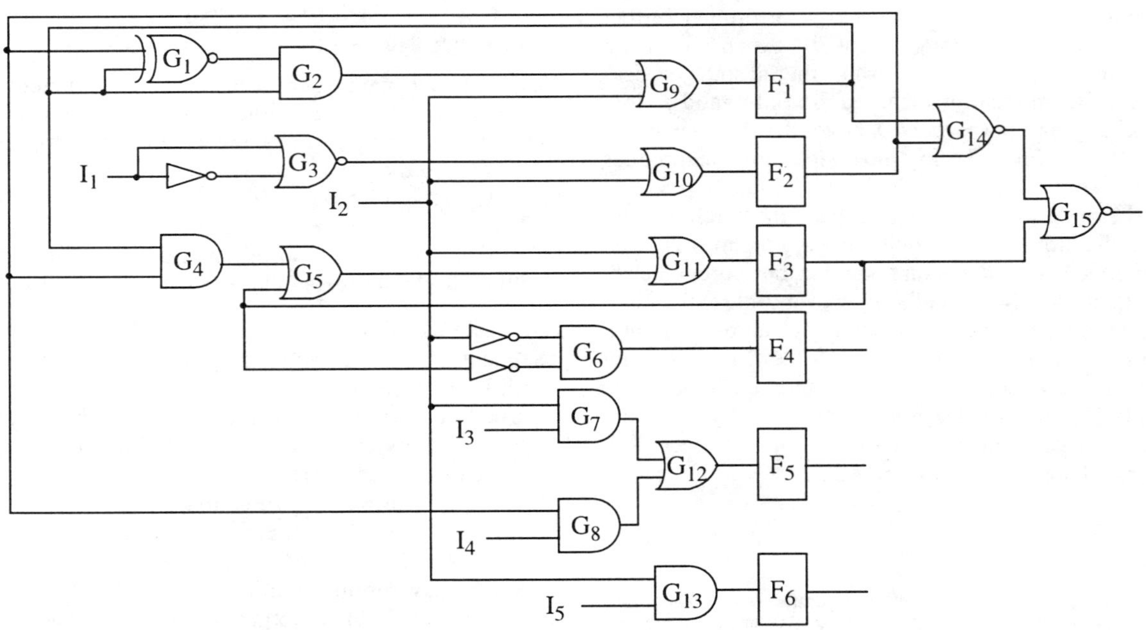

Figure 1: An example illustrating the sequential learning technique

Stem	T=0	T=1	T=2	T=3
$I_1=0$	$G_3=0$	{}	{}	{}
$I_1=1$	$G_3=0$	{}	{}	{}
$I_2=0$	$G_7=0$ $G_{13}=0$	F6=0	{}	{}
$I_2=1$	$G_6=0, G_9=1$ $G_{10}=1$ $G_{11}=1$	$F_1=1, F_2=1$ $F_3=1, F_4=0$ $G_1=1, G_2=1$ $G_4=1, G_5=1$ $G_6=0, G_9=1$ $G_{11}=1$ $G_{14}=0$ $G_{15}=0$	$F_1=1, F_3=1$ $F_4=0, G_5=1$ $G_6=0$ $G_{11}=1$ $G_{14}=0$ $G_{15}=0$	$F_3=1, F_4=0$ $G_5=1, G_6=0$ $G_{11}=1$ $G_{15}=0$
$F_1=0$	G2=0, G4=0	{}	{}	{}
$F_1=1$	G14=0	{}	{}	{}
$F_2=0$	G4=0, G8=0	{}	{}	{}
$F_2=1$	G14=0	{}	{}	{}
$F_3=0$	{}	{}	{}	{}
$F_3=1$	$G_5=1, G_6=0$ $G_{15}=0$ $G_{11}=1$	$F_3=1, F_4=0,$ $G_5=1, G_6=0$ $G_{11}=1$ $G_{15}=0$	{}	{}

Table 1: Simulation results for stems of the circuit in Figure 1.

Single-Node Relations	Additional Multiple-Node Relations	Additional Gate-Equivalence Relations
$F_6=1 \rightarrow F_4=0$	$F_1=0 \rightarrow F_2=0$	$F_3=0 \rightarrow F_1=0$
$F_6=1 \rightarrow F_3=1$	$F_1=0 \rightarrow F_5=0$	$F_4=1 \rightarrow F_1=0$
$F_6=1 \rightarrow F_2=1$	$F_3=0 \rightarrow F_2=0$	
$F_6=1 \rightarrow F_1=1$	$F_3=0 \rightarrow F_4=1$	
	$F_3=0 \rightarrow F_5=0$	
	$F_4=1 \rightarrow F_2=0$	
	$F_4=1 \rightarrow F_5=0$	
	$F_4=1 \rightarrow F_3=0$	

Table 2: Learned invalid state relations for the circuit in Figure 1.

Single-node learning can only learn relations due to a single stem. However, relations that can be learned due to multiple assignments are missed. This occurs when two or more stems result in a node having the same value. Thus, the contrapositive of the value on the node implies the contrapositive of the values on each of the stems. By injecting the contrapositive values of all the stems and performing forward logic simulation, relations can be identified between the node and other nodes set by the simulation. Injecting the values on multiple stems concurrently usually allows values to propagate further during simulation and hence allows more relations to be extracted. Thus, for each value on a gate, the set of stems that produce this value on the gate are stored. This analysis is called ***multiple-node learning***.

As an example, consider the value 1 on the sequential element F_3 in the circuit shown in Figure 1. This value is set on F_3 at T=1 by each of the stems $I_2=1$ and $F_3=1$, at T=2 by the stem $I_2=1$, and at T=3 by the stem $I_2=1$. Thus, the value $F_3=0$ at T=3 implies that $I_2=0$ at T=3-3=0, $I_2=0$ at T=3-2=1, and $I_2=0$ and $F_3=0$ at T=3-1=2. By injecting those assignments on the stems and performing forward simulation, this results in $F_2=0$ and $F_6=0$ at T=1, $F_2=0$, $F_5=0$, and $F_6=0$ at T=2, and $F_2=0$, $F_4=1$, $F_5=0$, and $F_6=0$ at T=3. Note that the fact that gate G_3 is tied to a 0 is taken advantage of during simulation. Extraction of tied gates will be discussed in the next subsection. This learning leads to the extraction of four invalid state relations, namely $F_3=0 \rightarrow F_2=0$, $F_3=0 \rightarrow F_4=1$, $F_3=0 \rightarrow F_5=0$, and $F_3=0 \rightarrow F_6=0$, the first three of which were not learned by single-node learning. Similarly, multiple-node learning for the other nodes can be performed. Additional invalid state relations learned by multiple-node learning are shown in Table 2.

Taking advantage of equivalent gates in combinational circuits can help values propagate forward during simulation. For example, gates G_2 and G_4 are equivalent combinationally. However, injecting a 0 on F_2 results in $G_4=0$ but does not set $G_2=0$, due to the limitation of three-valued simulation. Knowing that the two gates are equivalent allows the simulator to set G_2 to 0. This can help in identifying extra relations. For example, for the multiple-node learning of $F_3=0$, the value $I_2=0$ at T=0 results in $F_2=0$ and $F_6=0$ at T=1. Then, $F_2=0$ at T=1 results in $G_4=0$ and by equivalence $G_2=0$ at T=1. With $I_2=0$ injected at T=1, this leads to $F_1=0$, $F_2=0$, $F_5=0$, and $F_6=0$ at T=2. Injecting $I_2=0$ and $F_3=0$ at T=2 with the additional assignments on the sequential elements propagating from the previous time frame, this leads to the assignments $F_1=0$, $F_2=0$, $F_4=1$, $F_5=0$, and $F_6=0$ at T=3. Thus, a new relation is extracted between F_3 and F_1 by utilizing gate equivalence in combinational circuits, namely $F_3=0 \rightarrow F_1=0$. Equivalent combinational gates can be efficiently identified based on parallel pattern simulation techniques. Multiple-node learning for the other nodes can be performed similarly. Additional invalid state relations learned by multiple-node learning taking advantage of gate equivalence are shown in Table 2.

It is worth mentioning that the multiple-node learning technique can identify relations that are not learned by techniques based on backward/forward techniques. This is illustrated by the example shown in Figure 2. From the figure, it can be seen that each of the stems $I_2=0$ and $I_3=0$ at T=0 imply $G_9=1$ at T=1. Thus, $G_9=0$ at T=1 implies that $I_2=1$ and $I_3=1$ at T=0, which implies that $F_2=0$ at T=1. Thus, the relation $G_9=0 \rightarrow F_2=0$ is learned. Such a relation cannot be extracted by injecting a 0 or 1 on G_9 and performing backward/forward implications.

3.2 Tie Gate Learning

In addition to learning implications, this learning technique can effectively identify tie gates. A tie gate is one that can only assume a single known value. Tied gates can be tied combinationally or sequentially. A combinational gate tied to the value $v \in \{0, 1\}$ will always be set to the value v and cannot be set to the value $\bar{v}$. A sequential gate tied to the

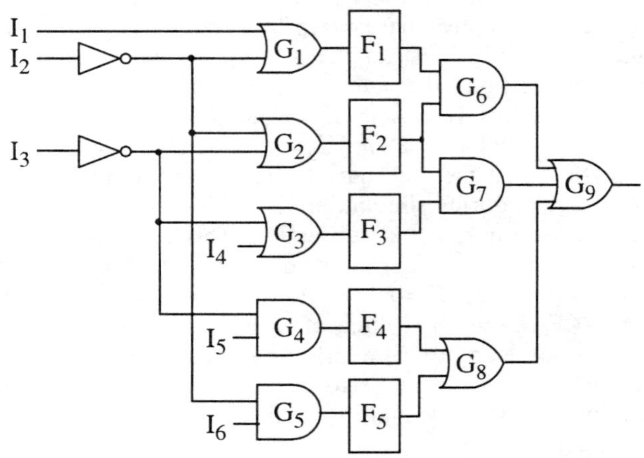

Figure 2: An example illustrating extraction of relations by multiple-node learning that are not extracted by backward/ forward approaches.

value $v \in \{0, 1\}$ will always be set to the value v for each of the reachable states. If the circuit is powered up in one of the unreachable states, then the sequentially tied gate can initially be set to $\bar{v}$ and after some number of cycles will be set to the value v. Under three-valued simulation, a sequentially tied gate will only get the value v once it is set to a known value. It is worth noting that sequentially tied gates are by the definition in [13] c-cycle redundant for some cycle c. Thus, identification of tied gates can help in identifying untestable faults in the circuit and also in optimizing the design by removing the redundant gates.

Our technique identifies tied gates based on both single- and multiple-node learning techniques. If both a value of 0 and a value of 1 on a stem produce the same value v on a node G at the same time frame, then this implies that node G is tied to the value v and hence the fault stuck-at-v is untestable. For example, consider the circuit in Figure 1. Both I_1=0 and I_1=1 imply that G_3=0. Thus, gate G_3 is tied to 0. This is considered a combinational tie since it is learned at T=0. Extraction of tied gates based on this criterion is performed after the single-node learning phase, so that the multiple-node learning phase can take advantage of learned tied gates.

Tied gates can also be learned during the multiple-node learning phase. This is illustrated by the example given in Figure 1. From Table 1, it can be seen that by the contrapositive law, G_{15}=1 implies I_2=0 at T=0, I_2=0 at T=1, I_2=0 and F_3=0 at T=2, and F_3=0 at T=3. Injecting those assignments in their respective time frames and simulating forward, this implies that F_2=0 and F_6=0 at T=1, F_1=0, F_2=0, F_5=0, and F_6=0 at T=2, and F_1=0, F_2=0, F_4=1, F_5=0, and F_6=0 at T=3. The assignments F_1=0 and F_2=0 at T=3 imply that G_{14}=1 which implies that G_{15}=0. Thus, we get a conflict which indicates that G_{15} cannot be assigned the value 1, and hence is learned to be sequentially tied to 0. It is also possible to learn that gate G_{15} is a tie gate in a different way during logic simulation. For example, if G_{14} is a stem, then we will have the implication $G_{15}=1 \rightarrow G_{14}=0$ at T=3. However, from the multiple-node learning we described above, simulation of the injected values results in G_{14}=1 at T=3, which is again a conflict indicating that G_{15} is a tie gate. Thus, conflicts occurring during logic simulation indicate that the target gate for which learning is performed is a tie gate. It should be observed that this gate would not have been learned to be a tie without taking advantage of the previously learned tie gate G_3 and the equivalence relation between gates G_2 and G_4.

3.3 Practical Issues

Industrial circuits are often featured with multiple clock domains, multiple-phase clocks, multiple-port latches, and partial set/reset. Such practical issues require special consideration during learning.

3.3.1 Multiple-port Latches

Multiple-port latches have several ports through which their values can be assigned. To guarantee that the extracted relations during learning are valid, values are not allowed to propagate across such latches. This is because the extracted relations would otherwise only be valid with respect to a certain port.

3.3.2 Multiple clock domains

To extract learned relations that are valid regardless of temporal information and the clock domain used, sequential elements are classified into classes where each class consists of those that are driven by the same or equivalent clock and at the same phase. A clock and a gated version of that clock are considered as separate clocks. Then, learning is performed for each class separately. Latches and flip-flops are also classified into different classes even if they are driven by the same clock and at the same phase, since their different capture times may invalidate the learned relations.

3.3.3 Set/Reset handling

Values are not allowed to propagate across sequential elements that have both unconstrained set and reset lines. This is because the learned information will be invalidated by the set/reset behavior since both a 0 and 1 can be obtained on the sequential element regardless of the value being propagated. However, if a sequential element contains either an unconstrained set or reset line, then a value is allowed to propagate across the sequential element if the value is the same as what could be produced by the set or reset line. This guarantees that the extracted relations cannot be invalidated by the set/reset behavior and that all the extracted relations are valid.

4. ENHANCING ATPG PERFORMANCE WITH LEARNED DATA

Implication learning can enhance ATPG performance by identifying conflicts early in the search process and by eliminating decision nodes whose values are implied. Implication relations can be used as known-value implications or forbidden-value implications. Forbidden-value implications are useful in identifying illegal assignments. However, they do not eliminate implied decision nodes as known-value implications do. We illustrate this by the example shown in Figure 2. To detect the stuck-at-1 fault on G_9, it is necessary to set G_9=0 to excite the fault. An implication relation exists between G_9 and F_2 namely, $G_9=0 \rightarrow F_2=0$. Without using the implication, G_6 and G_7 become decision nodes with two solutions each, F_1=0 or F_2=0 for justifying a 0 on G_6, and F_2=0 or F_3=0 for justifying a 0 on G_7. If a solution other than F_2=0 is selected, this results in additional unnecessary requirements to detect the fault. By taking advantage of the known-value implication, G_6=0 and G_7=0 are longer decision nodes and hence the search space is pruned. It should be observed that if the implication is used as forbidden-value implication i.e., $G_9=0 \rightarrow F_2=\bar{1}$, then no benefit is gained as G_6=0 and G_7=0 remain decision nodes.

While known-value implications are more useful than forbidden-value implications, they could result in unnecessary requirements. To illustrate this, consider the example in Figure 1. Suppose that we need to justify F_6=1. By known-value implication, this would require the ATPG to justify F_4=0, F_3=1, F_2=1, and F_1=1. If any of those requirements is targeted by the ATPG before the requirement F_6=1, this can complicate the ATPG further and result in unnecessary requirements. In addition to causing unnecessary requirements, known-value implications can cause the ATPG to declare testable faults as untestable, as demonstrated in [15].

To eliminate the unnecessary requirements caused by learned implications, we propose the following solution. Learned implications are used as forbidden-value implications. However, forbidden values are implied forward and backward dynamically during the ATPG process similar to known-value implications. Forbidden 0 ($\bar{0}$) is implied as 1, and forbidden 1 ($\bar{1}$) is implied as 0. Then, when a value on a decision node needs to be justified, the inputs of the node are examined and the input with the forbidden non-controlling value is selected. This provides the same benefit that known-value implications provide without justifying unnecessary values. However, while known-value implications can eliminate decisions forward, forbidden-value implications do not. To illustrate this, consider the example in Figure 2 again. Using the implication as forbidden-value implication, we have $G_9=0 \rightarrow F_2=\bar{1}$. To justify a 0 on G_6, the input with a $\bar{1}$ assignment is selected, which is F_2. Then, a 0 on F_2 is justified which is what a known-value implication would have required. However, if justifying a 0 on F_2 fails, then in the case of forbidden-value implications backtracking will occur and the other choices for G_6 and G_7 will be tried and will fail too. In this case, known-value implications will be more effective. Considering the example in Figure 1, justifying F_6=1 would not require justifying any other value since no assignments are required on F_4, F_3, F_2, and F_1.

Since the use of each of the known- and forbidden-value implications has relative advantages and disadvantages, neither method is expected to consistently perform better. However, it is expected that the use of forbidden-value implications can be enhanced to capture the full benefit of known-value implications without its associated problems.

5. Experimental Results

Experiments using the proposed learning method were conducted on several sequential circuits, including ISCAS 89 and 93 benchmark circuits, 4 retimed circuits, and 3 industrial circuits. The retimed circuits were chosen as it was shown that ATPG can be particularly inefficient on this class of circuits due to the large number of invalid states introduced by retiming [9]. The 4 retimed circuits chosen were those with the lowest density of encoding, and which consequently had the highest APTG complexity when run on HITEC [14]. The industrial circuits were used to demonstrate the method on real circuits with set/reset and multiple clock domains, and to show the high efficiency of the method when handling very large circuits. All experiments were run on a 167 MHz Sun Ultra 1.

5.1 Sequential Learning Results

Sequential learning is performed with simulation allowed to propagate a maximum number of 50 time frames. In the results shown in Table 3, the relations reported are those learned sequentially; the relations which can be learned in the combinational logic are excluded to isolate what can only be extracted by sequential learning. The learned relations between pairs of FFs and between gates and FFs are reported. The efficiency of the sequential learning technique is clearly demonstrated as for the largest circuit, which has over 680,000 gates, it requires less than 7 minutes.

To illustrate the effectiveness of our approach at identifying tie gates, we compare the number of untestable faults identified due to tie gates with those identified by FIRES [13]. FIRES uses sequential analysis to identify faults that cannot be activated or cannot be propagated due to conflicts on stems. So, it targets a more general class of untestable faults than those which cannot be activated or propagated due to tie gates. However, our method identifies untestable faults as a by-product of learning tie gates during the sequential learning process, and is not targeted for this purpose. Table 4 shows the number of untestable faults identified due to tie gates and those identified by FIRES. In three circuits, namely s5378, s3330, and s38417, more untestable faults are identified with tie gate information than FIRES was able to identify.

5.2 ATPG Results

Sequential ATPG experiments were used to demonstrate the benefit of the learned information in enhancing the efficiency of test generation. The experiments were run with and without the use of sequential learning. With sequential learning enabled, two scenarios were used: using the relations between signals as forbidden-value implications as proposed, or as known-value implications. Note that all the ATPG experiments performed make use of combinational learning. Therefore, the difference in the results reported is only due to the use of sequential learning.

For the smaller benchmark circuits in Table 3 (s382 through

Circuit	FFs	Gates	Relations		CPU (s)
			FF-FF	Gate-FF	
s382	21	158	9	37	0.06
s386	6	159	8	135	0.04
s400	21	164	12	47	0.07
s444	21	181	11	69	0.08
s641	19	377	36	197	0.04
s713	19	393	36	216	0.06
s953	29	424	145	1870	0.78
s967	29	395	126	1437	0.43
s1196	18	529	8	44	0.07
s1238	18	508	9	48	0.07
s1269	37	569	30	232	0.06
s1423	74	657	4	251	0.16
s3330	132	1789	367	1764	1.30
s3384	183	1685	31	48	0.19
s4863	104	2342	256	17398	4.15
s5378	179	2779	250	2233	6.42
s6669	239	3080	24	1603	0.39
s9234	228	5597	416	7321	4.38
s13207	638	7951	1566	35093	23.08
s15850	597	9772	1516	29378	42.04
s38417	1636	22179	1554	46981	30.24
s38584	1452	19253	2320	32372	41.93
s510jcsrre	26	243	127	891	0.10
s510josrre	28	243	50	484	0.07
s832jcsrre	27	195	125	743	0.11
scfjisdre	20	764	22	1980	0.56
indust1	460	8693	118	6774	2.74
indust2	7068	63156	2069	36397	24.31
indust3	15689	681595	8016	186930	403.30

Table 3: Sequential learning experiments

Circuit	Untestable faults	
	Tie Gates	FIRES
s5378	441	367
s3330	232	161
s9234	61	284
s13207	182	893
s15850	69	332
s38417	192	147
s38584	538	1437

Table 4: Comparison of untestable faults identified due to tie gates with those identified by FIRES[13].

s1269), no results were reported since the ATPG was originally efficient and there was no benefit from using the sequential learned data. Table 5 reports the test generation results on the larger benchmark circuits which the ATPG tool has difficulty with. The experiments were run in two stages, with different backtrack limits. This allows a better understanding of the trade-offs between ATPG results and CPU time. From the results summarized in the table, it can be seen that in most cases, sequential learning can increase the number of detected faults and those identified as untestable, as well as decrease the ATPG time. For example, based on forbidden-value implications, s5378 with a backtrack limit of 1000 achieves over 7% higher test coverage (fault coverage excluding untestable faults) with 42% less CPU time. Even using the 1000 abort limit and spending over 15 times more time, the ATPG without sequential learning fails to achieve the test coverage that ATPG with learning obtains using a 30 abort limit. Another example is s4863, in which the use of known-value implications gives approximately 4% higher test coverage in less than one-seventh the ATPG time.

For the retimed circuits, higher fault coverage and lower ATPG times were obtained using sequential learning. For example, based on forbidden-value implications, s832jcsrer with a backtrack limit of 1000 achieves over 31% higher test coverage in less than half the CPU time, compared to no learning. Note that for retimed circuits, a solution was proposed in [16] for enhancing ATPG performance in which the retimed circuit is transformed prior to ATPG to increase the density of encoding. Although that solution is very effective at enhancing ATPG results, the sequential learning solution has the advantage that it is not restricted to this class of circuits.

As predicted, experimental results indicate that neither known-value implications nor the current use of forbidden-value implications consistently performs better.

In a few cases, the use of sequential learning results in lower fault coverage. For example, for the circuit s510josrre, using forbidden-value implications resulted in fewer detected faults than without learning. This was analyzed to be caused by random effects. In ATPG, after a test sequence is generated for a fault, the sequence is fault-simulated and all detected faults are dropped. Therefore, it is possible to detect a fault which ATPG would fail to find a test for. For the cases where there were fewer faults detected with learning, the additional faults which the original ATPG detected were targeted individually. In all cases, the ATPG failed to find a test for them, proving they were detected by fault simulation of tests generated for other faults.

6. Conclusions

We have presented a novel and very efficient method for sequential learning, which can be used to identify implications, invalid states, and tied gates. While most learning techniques published in the literature perform learning in the combinational logic, the method presented also learns information which can only be extracted by performing the analysis across memory elements. The method can be used on large industrial circuits as it is extremely fast, and capable of handling real circuit issues such as multiple clock domains, and partial or full set/reset.

The learned information can be used in a variety of design automation tools, including test pattern generation, redundancy identification, logic verification, and logic optimization. The experiments performed focus on demonstrating the use of sequential learning to enhance sequential ATPG efficiency. Although the technique can also perform combinational learning, the focus of the experiments was to isolate the effect of sequential relations and demonstrate the benefit of doing the learning sequentially. It can be observed for most circuits that when ATPG uses the learned information, the number of detected faults and those identified as untestable is higher, and the

Circuit	Total faults	Backtrack limit	No learning			Forbidden values			Implications		
			Det	Untest	CPU(s)	Det	Untest	CPU(s)	Det	Untest	CPU(s)
s1423	1515	30	1329	19	685	1375	19	676	1385	19	716
		1000	1343	19	4029	1381	19	3711	1395	19	3674
s3330	2870	30	2123	464	368	2124	542	276	2124	359	701
		1000	2124	484	6801	2124	562	4848	2124	359	10572
s3384	3380	30	3141	1	500	3155	1	420	3155	1	414
		1000	3211	1	7235	3183	1	5807	3183	1	5784
s4863	4764	30	4495	0	250	4600	126	100	4615	126	103
		1000	4566	0	2069	4612	126	350	4623	126	271
s5378	4603	30	3411	289	1761	3540	630	1370	3519	642	1440
		1000	3454	432	21012	3564	646	12095	3561	646	15995
s6669	6684	30	6574	0	189	6624	0	156	6632	0	197
		1000	6596	0	2025	6654	0	969	6658	0	1168
s13207	9815	30	892	8887	14341	892	8896	16154	895	8903	15562
		1000	892	8908	15362	892	8916	16915	895	8913	16298
s510jcsrre	668	30	309	9	134	309	9	148	309	9	144
		1000	477	11	1775	563	11	1065	556	11	1134
s510josrre	656	30	179	4	142	179	4	150	179	4	157
		1000	556	4	1126	487	4	1745	565	4	1077
s832jcsrre	596	30	343	0	65	331	0	74	266	0	100
		1000	369	0	1496	556	0	659	525	0	850
scfjisdre	1920	30	507	6	680	503	6	811	503	6	764
		1000	1700	6	7281	1714	6	7929	1746	6	6645

Table 5: ATPG experiments with and without sequential learning

ATPG time is significantly reduced. The benefit of sequential learning can be particularly observed on retimed circuits, which have a large number of invalid state relations that can significantly increase ATPG complexity.

7. ACKNOWLEDGEMENTS

The authors would like to thank Wu-Tung Cheng, Rob Thompson, and Srinivas Patil of Mentor Graphics for their useful discussions.

8. REFERENCES

[1] M. H. Schulz, E. Trischler, and T. M. Sarfert, "SOCRATES: A Highly Efficient Automatic Test Pattern Generation System," *IEEE Transactions on Computer-Aided Design*, pp. 126-136, January 1988.

[2] J.-K. Zhao, E. M. Rudnick, and J. H. Patel, "Static Logic Implication with Application to Redundancy Identification," in *Proc. VLSI Test Symposium*, pp. 288-293, April 1997.

[3] W. Kunz and D. K. Pradhan, "Accelerated Dynamic Learning for Test Pattern Generation in Combinational Circuits," *IEEE Transactions on Computer-Aided Design*, pp. 684-694, May 1993.

[4] W. Kunz and D. K. Pradhan, "Recursive Learning: An Attractive Alternative to the Decision Tree for Test Generation in Digital Circuits," in *Proc. International Test Conference*, pp. 816-825, September 1992.

[5] E. Auth and M. H. Schulz, "A Test-Pattern-Generation Algorithm for Sequential Circuits," in *IEEE Design and Test of Computers*, pp. 72-85, June 1991.

[6] M. A. Iyer and M. Abramovici, "FIRE: A Fault-Independent Combinational Redundancy Identification Algorithm," *IEEE Transactions on VLSI Systems*, pp. 295-301, June 1996.

[7] W. Kunz, "HANNIBAL: An Efficient Tool for Logic Verification Based on Recursive Learning," in *Proc. International Conference on Computer-Aided Design*, pp. 538-543, November 1993.

[8] W. Kunz and P. R. Menon, "Multi-Level Logic Optimization by Implication Analysis," in *Proc. International Conference on Computer-Aided Design*, pp. 6-13, 1994.

[9] Thomas Marchok, Aiman El-Maleh, Wojciech Maly, and Janusz Rajski, "A Complexity Analysis of Sequential ATPG," *IEEE Transactions on Computer-Aided Design*, Vol. 15, pp. 1409-1423, Nov. 1996.

[10] H. Touati, H. Savoj, B. Lin, R. K. Brayton, and A. Sangiovanni-Vincentelli, "Implicit State Enumeration of Finite State Machines using BDD's," in *Proc. International Conference on Computer-Aided Design*, pp. 130-133, November 1990.

[11] D. E. Long, M. A. Iyer and M. Abramovici, "Identifying Sequentially Untestable Faults Using Illegal States," in *Proc. VLSI Test Symposium*, pp. 4-11, 1995.

[12] H.-C. Liang, C. L. Lee, and J. E. Chen, "Invalid State Identification for Sequential Circuit test Generation," in *Proc. Asian Test Symposium*, pp. 10-15, 1996.

[13] M. A. Iyer, D. E. Long, and M. Abramovici, "Identifying Sequential Redundancies Without Search," in *Proc. VLSI Test Symposium*, pp. 4-11, 1995.

[14] T. M. Niermann and J. H. Patel, "HITEC: A test generation package for sequential circuits," in *Proc. European Design Automation Conference*, pp. 214-218, 1991.

[15] Wu-Tung Cheng, Aiman El-Maleh, Rob Thompson, Don Ross, and Janusz Rajski, "The Pitfalls of Necessary Assignments", *Fourth International Test Synthesis Workshop*, May 1997, Santa Barbara, CA.

[16] Aiman El-Maleh, Thomas Marchok, Janusz Rajski, and Wojciech Maly, "Behavior and Testability Preservation Under the Retiming Transformation," *IEEE Transactions on Computer-Aided Design*, Vol. 16, pp. 528-543, May 1997.

Fault-Simulation Based Design Error Diagnosis for Sequential Circuits

Shi-Yu Huang
National Semiconductor Corp.
2900 Semiconductor Dr., D3-677
Santa Clara, CA 95052
e-mail: syhuang@galaxy.nsc.com

Kwang-Ting Cheng
ECE Department
Univ. of California, Santa Barbara
Santa Barbara, CA 93106
e-mail: timcheng@ece.ucsb.edu

Kuang-Chien Chen
Verplex Systems Inc.
Santa Clara, CA 950542
e-mail: kchen@best.com

Juin-Yeu Joseph Lu
National Semiconductor Corp.
2900 Semiconductor Dr., D3-677
Santa Clara, CA 95052
e-mail: jlu@galaxy.nsc.com

Abstract

This paper addresses the problem of locating design errors in a sequential circuit. For single-error circuits, we consider a signal *f* as a potential error source only if the circuit can be completely rectified by re-synthesizing *f* (i.e., changing the function of signal *f*). In order to handle larger circuits, we do not rely on Binary Decision Diagram. Instead, we search for potential error sources by a modified sequential fault simulation process. The main contributions of this paper are two-fold: (1) we derive the necessary and sufficient condition of whether an erroneous input sequence (i.e., an input sequence producing erroneous responses) can be corrected by changing the function of a particular internal signal; and (2) we propose a modified fault simulation procedure to check this condition. Our approach does not rely on any error model, and thus, is suitable for general types of errors. Furthermore, it can be easily extended to identify multiple errors. Experimental results on ISCAS89 benchmark circuits are presented to demonstrate its capability.

1. Introduction

During the VLSI design process, functional mismatches between a given specification and its implementation often occur. Once a functional mismatch is found by verification tools, the designer will face the daunting task of error diagnosis — a process that identifies or narrows down the error sources in the implementation, so as to assist the subsequent error correction process. Due to the difficulty of diagnosing sequential circuits, most previous approaches have focused on combinational circuits.

Most combinational error diagnosis approaches can be classified into two categories: (1) simulation-based approaches [17,18,10,9,14,15,21], and (2) symbolic approaches [13,11,5,12]. The simulation-based approaches first derive a number of input vectors that can differentiate the implementation and the specification. These binary or 3-valued input vectors are called *erroneous vectors* hereafter. By simulating each erroneous vector, the potential error region can then be trimmed down gradually. The conditions for eliminating those signals that cannot be an error source vary from one approach to another. On the other hand, the symbolic approaches do not enumerate the erroneous vectors. They primarily rely on Ordered Binary Decision Diagram (OBDD) [3] to characterize the necessary and sufficient condition of a potential error source as a boolean formula. Based on this formulation, every potential error source can be precisely identified. More recently, a method combining the above two approaches has also be proposed to reduce the run time of error diagnosis and correction [19]. In comparison, the symbolic approaches are accurate and extendible to multiple errors. However, constructing the required BDD representations may cause memory explosion when applied to large circuits. On the other hand, the simulation-based approaches, although scalable with the circuit size, are usually not accurate enough.

In [7], we proposed a fault simulation based approach for combinational error diagnosis. This approach considers the error diagnosis as a fault simulation process which takes a number of erroneous vectors as inputs. These erroneous vectors can be generated in advance by random simulation or modified ATPG tools. We defined a term called *correctable set* with respect to a signal. A particular signal's correctable set is the set of erroneous vectors that can be corrected by re-synthesizing the signal. Re-synthesizing a signal *f* is an operation that replaces the function of *f* with a new function as shown in Fig. 1.

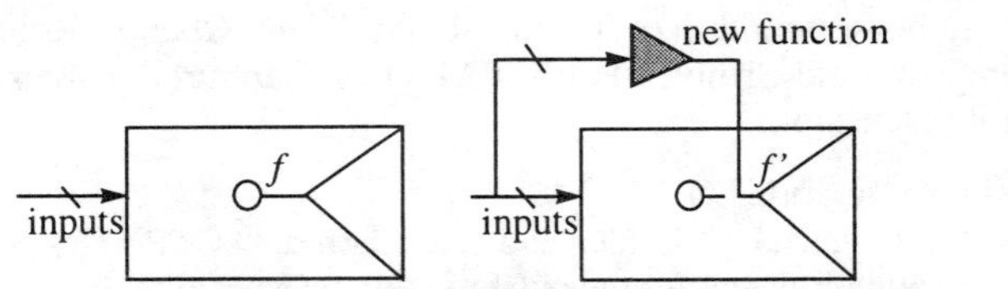

(a) original circuit. (b) circuit after *f* is re-synthesized.

Fig 1. Re-synthesizing an internal signal.

Let *v* be an erroneous vector and *f* be a signal of the erroneous implementation. We showed that whether *v* is correctable by *f* can be precisely determined by simulating *f* stuck-at faults with respect to the input vector *v*. Like other simulation-based approaches, our algorithm is a monotone filtering process. Initially, every signal is considered as a candidate of error source. We enumerate a number of erroneous vectors and simulate the circuits for stuck-at faults at each candidate signal. According to the fault simulation results, if a signal is unable to correct any simulated erroneous vector, then it is a false candidate. False candidates are removed from the candidate list immediately to speed up the process. This approach is highly accurate and applicable for large circuits. Also, it can deal with circuits with multiple errors. We have successfully handled the entire set of ISCAS85 benchmark circuits injected with one or two random errors. On the average, only 6.6 signals are identified as the potential error sources for an ISCAS85 benchmark circuit injected with a single error.

Consider the problem of diagnosing a customized design versus a specification described at a higher level of abstraction, such as

---

* The first author was with UC, Santa Barbara when part of this work was conducted under the support of National Science Foundation grant MIP-9503651, California MICRO, and Fujitsu Labs of America.

DAC 98, San Francisco, California
©1998 ACM 0-89791-964-5/98/06..$5.00

the register-transfer (RT) level. To deal with this problem, a fast-pass synthesis can be performed on the RT-level specification to derive its gate-level representation. Then, the gate-level to gate-level error diagnosis process is followed. It is likely that one-to-one flip-flop correspondence between the implementation and the gate-level specification may not exist, and thus, the combinational approaches cannot be applied. A sequential error diagnosis approach is needed for this kind of situation.

Due to the even higher difficulty and complexity, only a few papers in the literature have addressed the problem of diagnosing design errors in sequential circuits [6,15,20]. The method discussed in [15], modeling the error in the state transition table, only targets small controllers. The approach proposed in [6] is not very general in the sense that it only focuses on small feedback-free circuits, or finite state machines that have one-to-one state correspondence with their specifications. Another approach, extending a combinational backward error propagation heuristic [21] to the iterative array model, was proposed in [20]. In this approach, a restricted error hypothesis (containing three types of wrong-gate errors) is used. It has two major limitations. First, it relies on a restricted error hypothesis, and thus, it may fail when the design error is not modeled in the underlying hypothesis. Secondly, their approach cannot deal with multiple errors.

This paper extends our fault simulation-based idea to sequential circuits. Similar to the combinational case, our algorithm first generates a number of erroneous input sequences. Then, a modified fault simulation process is performed to eliminate those signals that cannot be a potential error source. There are two main contributions in this paper. First, we derive a necessary and sufficient condition of whether a signal can be held totally responsible for a particular *erroneous input sequence*. Secondly, we introduce a modified sequential fault simulation procedure to check this condition. Compared to the other approaches, our algorithm offers three advantages: (1) it does not require any error model, and thus, can be used to diagnose general errors; (2) it is highly accurate because we use a stringent condition for identifying potential error sources; and (3) it can be easily generalized for circuits with multiple errors.

The rest of this paper is organized as follows. Section 2 gives the basic definitions and assumptions reviews our fault simulation-based approach for diagnosing combinational circuits. Section 3 describes the extension to sequential circuits. Section 4 presents the experimental results. Section 5 concludes this paper.

2. Preliminaries

2.1 Definitions

In our algorithm, the specification can be regarded as a black box with completely specified input/output behavior. However, for the sake of simplicity without losing generality, we assume that both the specification and the erroneous implementation are given as gate-level circuits. As mentioned earlier, if the specification is given at a higher level of abstraction, a low-effort synthesis can be performed to generate its corresponding gate-level representation. The specification and the implementation share the same set of primary input signals, denoted as $\{x_1, x_2, ..., x_n\}$. In the sequel, we refer to the specification and the implementation as C_1 and C_2, respectively. The primary outputs of C_1 and C_2 are denoted as $\{S_1, S_2, ..., S_m\}$ and $\{I_1, I_2, ..., I_m\}$. The number of flip-flops of these two circuits could be different and no flip-flop correspondence is required in our algorithm.

Definition 1: (S_i, I_i) is called the i-th *primary output pair.*

Definition 2: A *joint network* is a network obtained by connecting the primary inputs of the specification and the implementation together as shown in Fig. 2.

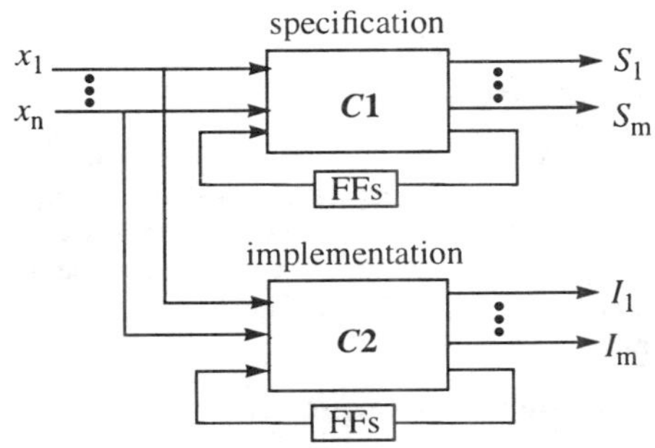

Fig 2. Joint network.

Definition 3: An *erroneous sequence* is a binary input sequence that can differentiate any primary output pair.

Definition 4: (*Correctable vector*): Suppose C_1 and C_2 are combinational circuits. An erroneous input vector v is called *correctable* by signal f in C_2 if there exists a new function for signal f in terms of the primary inputs such that v *is not* an erroneous vector for the resulting new circuit.

Definition 5: (*Single signal correctable*): If the implementation C_2 can be completely corrected by re-synthesizing a signal f, then C_2 is called single signal correctable and the signal f is called a *single-fix signal.*

2.2 Correctability check via fault simulation

In this sub-section we review the fault simulation-based techniques for diagnosing a combinational circuit in [7]. The algorithm is an iterative filtering process, which reduces the number of candidate fix signals gradually. We first run random simulation to collect a number of erroneous vectors. At the beginning, we assume every signal in C_2 is a single-fix candidate. Then this algorithm enters a two-level loop. The outer loop enumerates every erroneous vector. The inner loop iterates through every single-fix candidate signal. For each erroneous vector v and a target single-fix candidate f, we examine if v is correctable by f (as will be explained later). If signal f cannot correct vector v, then f cannot be a single-fix signal. Therefore, it is safe to remove f from the candidate list. After we have examine every erroneous vector to eliminate false single-fix candidates, our algorithm delivers a set of signals as the potential error sources.

Given an erroneous vector v and a signal f in C_2, *correctability check* is to decide whether v is correctable by f. This can be done precisely by simulating f stuck-at faults taking v as the input vector. In the following discussion, we use the notation $C_{2|f=0}$ ($C_{2|f=1}$) to denote the faulty implementation of f-stuck-at-0 (f-stuck-at-1).

Proposition 1: [7] Signal f can correct v if and only if stuck-at faulty circuit $C_{2|f=0}$ or $C_{2|f=1}$ has the same output responses as the specification C_1 under the application of input vector v.

Based on the above proposition, our algorithm becomes a fault simulation process that takes the generated erroneous vector set as inputs. The performance can be improved by exploiting the topological dominance relations between signals. Let f and *dom* be two signals, and *dom* is a topological dominator of f, which means that every path originated from f to any primary output will pass through signal *dom*. In [11], it was proven that if *dom* cannot correct an erroneous vector *v*, then *f* cannot correct *v*, either. Based on this property, we can further speed up the algorithm by dropping some false candidates early without running fault simulation on them.

3. Diagnosing Sequential Circuits

In this section we describe the idea of diagnosing a sequential circuit. First, we show the necessary and sufficient condition of whether an erroneous input sequence can be corrected by re-synthesizing a particular signal. Then, we discuss how to check this condition via fault simulation. The approach is then generalized for multiple errors.

3.1 Correctable input sequence

Definition 6: (*Correctable sequence*): An erroneous input sequence E is called *correctable* by signal f in C_2 if there exists a new function for signal f in terms of the primary inputs and the present state lines of C_2 such that E *is not* an erroneous sequence for the resulting new circuit (as illustrated in Fig. 3).

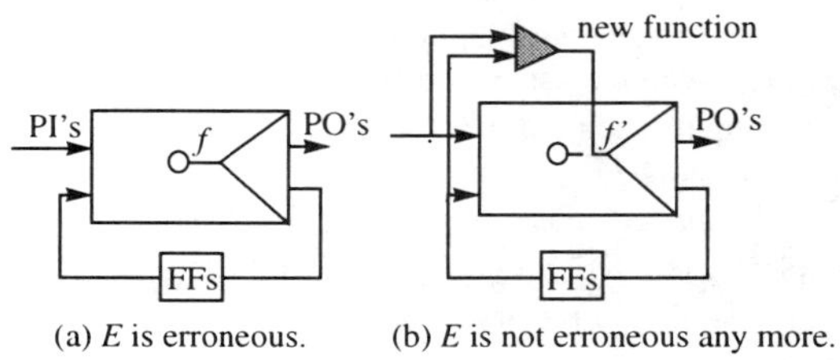

Fig 3. Correctable input sequence E by re-synthesizing a signal.

Based on this definition, we assume that an error source should be able to correct every erroneous input sequence. Once a signal is found unable to correct *any* erroneous sequence, it can be excluded from the candidate list of potential error sources without losing accuracy. It is worth mentioning that, for combinational circuits, if a signal can correct every erroneous input vector, then it is a single-fix signal. However, this statement is not true for sequential circuits. The reason is as follows. For a combinational circuit, every erroneous vector can be fixed *independently* [7], i.e., the requirement to fix every erroneous vector can be satisfied at the same time. But for sequential circuits, conflicts may occur as deriving the fix function that corrects every erroneous sequence, even though each of them can be fixed individually.

3.2 Necessary and sufficient condition

In the following discussions, we assume that both C_1 and C_2 have a known reset state, p_1 and q_1, respectively. The implementation C_2 is represented by the time-frame expansion model as shown in Fig. 4.

The number of combinational portions duplicated in the time-frame expansion model equals the length of the input sequence under consideration. Consider an erroneous input sequence with 3 input vectors, $E = \{v_1, v_2, v_3\}$. Suppose E brings C_2 through a sequence of states $\{q_1, q_2, q_3\}$, where q_1 is the initial state of C_2. Then we call $(v_1 \mid q_1)$ the *pseudo-input vector* for the first time-frame. Similarly, $(v_2 \mid q_2)$ and $(v_3 \mid q_3)$ are the pseudo-input vectors for the second and the third time-frames, respectively. Based on this model, it follows that if signal f can correct an erroneous input

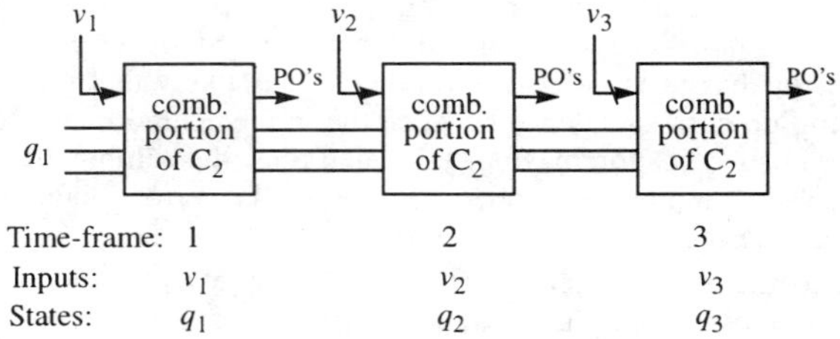

Fig 4. An example of the time-frame expansion model.

sequence E, then there exists a new function of f such that every primary output at every time-frame becomes error-free (i.e., every primary output has the same response as their corresponding primary output in C_1 with respect to the input sequence E). We define a term called *injection* before we derive the necessary and sufficient condition of correcting an erroneous input sequence.

Definition 7: (*Injection*): Given a signal f in C_2, a t *time-frame Injection* at f is a set of value assignments to the signal f for the first t time-frames. For example, $J = \{f^{(1)} = 0, f^{(2)} = 0, f^{(3)} = 0\}$ represents a 3 time-frame injection that injects value '0' at f for all 3 time-frames, where the superscript denotes the index of a time-frame.

An injection defines a new circuit. The output responses of the resulting new circuit are computed by treating the injected signal as an independent pseudo primary input line, taking the injected value as the input value. Since we can inject either '0' or '1' to a signal at each time-frame, there are total 2^t different t time-frame injection. The complexity grows exponentially with the number of time-frames. Based on the above definition, we have the following proposition.

Proposition 2: Let E be an erroneous input sequence with t input vectors, $E = \{v_1, v_2, ..., v_t\}$. A signal f in C_2 can correct E *only if* there exists a t time-frame injection at f such that E is not an erroneous input sequence for the resulting new circuit. (The injection is called a *cure injection* at f for E).

(Explanation) This proposition states that if there exists no cure injection at f, then there exists no new function for f that corrects the erroneous input sequence E, but not vice versa. In other words, to correct an erroneous input sequence, it is necessary to find a cure injection. However, a cure injection is not sufficient to assure that the input sequence is indeed correctable. This is due to the fact that, for every fix function at f, there always exists a cure injection. On the other hand, however, not every cure injection can be realized by a function.

Given a fix function at f that corrects an erroneous input sequence E, the corresponding cure injection can be derived as follows: Let the response of f with respect to E in the resulting new circuit be $\{\alpha_1, \alpha_2, ..., \alpha_t\}$, where α_i, $1 \le i \le t$, is a binary value. Then $J = \{f^{(1)} = \alpha_1, f^{(2)} = \alpha_2, ..., f^{(t)} = \alpha_t\}$ is a cure injection. On the other hand, there exists some injection that is not realizable. Fig. 5 shows an example. The pseudo-input vectors for the first and second time-frames are the same: (0 | 000). But the injected values at f for these two time frames are different ('0' and '1' respectively). A function realizing this injection needs to map the same pseudo-input vector, (0 | 000), to '0' and '1' at the same time. Therefore, a contradiction has occurred. It follows that there does not exist a new function for f in terms of the primary inputs and the present state lines to realize this injection. Whether an

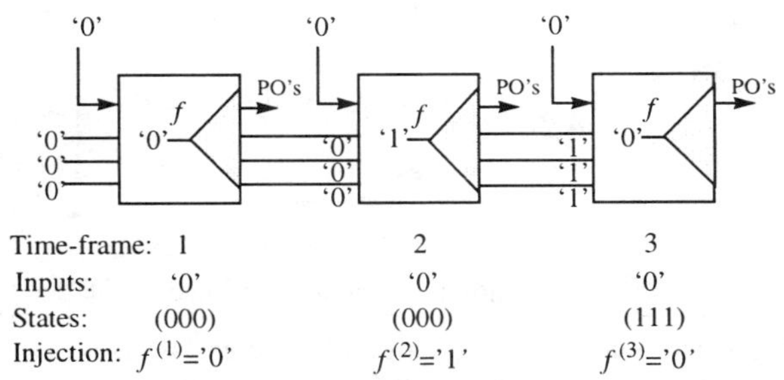

Fig 5. Unrealizable injection.

injection is realizable or not can be checked easily by simulating the input vector for the resulting circuit with the injection. After collecting the sequence of states encountered in the resulting circuit, the pseudo-input vector for each time-frame and the injected value can then be derived. If no conflict exists, then the injection is realizable.

Proposition 3: (Necessary and sufficient condition) Let E be an erroneous input sequence with t vectors, $E = \{v_1, v_2, ..., v_t\}$. A signal f in C_2 can correct E if and only if there exists a *realizable t time-frame cure injection.*

(Proof): As described above.

3.3 Correctability check via fault simulation

Based on the above proposition, we can determine if an erroneous input sequence is correctable by a signal by a two-step checking: (1) check if there exists a cure injection, and (2) check if it is realizable. Given an injection, determining whether it is a cure injection can be done via a modified sequential fault simulation process. Traditionally, sequential fault simulation assumes that the target signal is stuck at the same binary value for every time-frame. In our application, we need to modify the fault simulation algorithm, so that a signal is allowed to be stuck at different binary values at different time-frames to account for the fact that an injection may inject different values at different time-frames.

Theoretically, we need to enumerate every possible 2^t injections in the worst case to find a cure injection, or to conclude that one does not exist. However, like most branch-and-bound procedures, some criteria can be used to cut down the search space. In our application, the search space can be represented as a binary injection tree shown in Fig. 6.

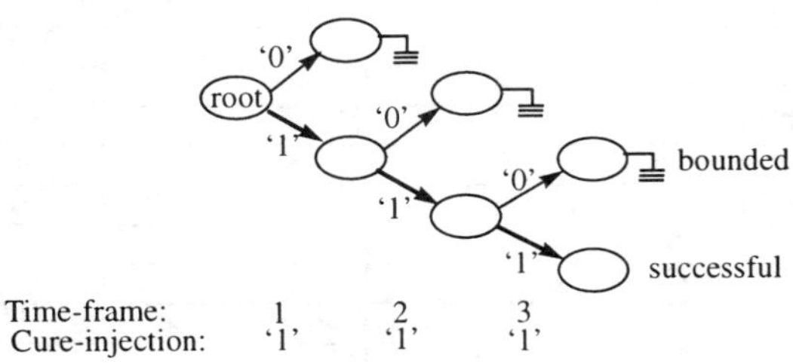

Fig 6. Injection tree.

The meaning of this tree is as follows. (1) Each node corresponds to a resulting circuit with a partial injection (i.e., t' time-frame injection where $t' < t$). The root node (level 0) corresponds to the original implementation C_2. (2) The level of each node corresponds to the index of the time-frame being considered for value injection. (3) The upper (lower) branch of each node represents injecting value '0' ('1') to the signal under consideration in the time-frame being considered. (4) Any path from the root node to a leaf node represents a complete t time-frame injection.

Based on this injection tree, it follows that if a node's corresponding partial t' time-frame injection cannot produce the correct responses for the first t' time-frames, then the sub-tree of this node need not be explored. This simple bounding criterion is useful to speed up the search for a cure injection. Once a cure injection is found, the pseudo-input vectors of the resulting circuit with respect to this injection can be derived. The realizability check can then be followed to determine if the erroneous input sequence is indeed correctable by the target signal.

3.4 Generalization for multiple errors

For diagnosing circuits with multiple errors, our algorithm searches for multiple signals that can jointly correct every generated erroneous input sequence. Fig. 7 shows an example of a 3 time-frame injection defined over a set of signals $\{f, g\}$. Similar

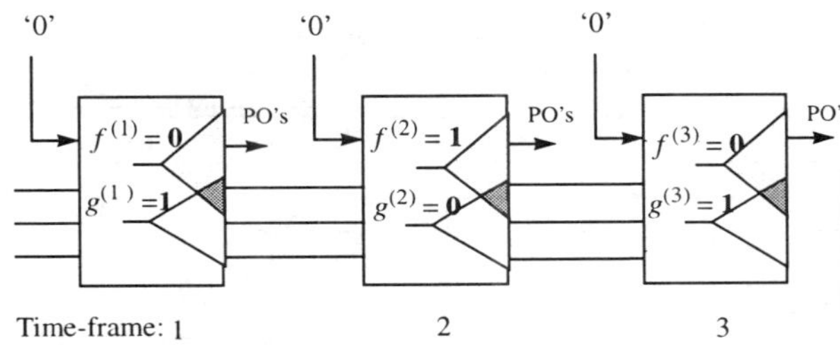

Fig 7. A 3 time-frame injection at signals $\{f, g\}$.

to the case of single-error diagnosis, if this injection is a realizable cure injection for the applied erroneous sequence $E = (0, 0, 0)$ then E is correctable by this set of signals. Again, this condition can be checked primarily via a modified fault simulation process. Given a set of k signals, σ, the worst-case complexity of deciding if an input sequence E with t vectors can be corrected by σ or not is proportional to the number of possible injections. There are 2^k possible value combinations for each time-frame, and thus, the total number of possible t time-frame injections defined over σ is $2^{(k \cdot t)}$.

In [7], we developed a two-stage multiple-error diagnosis algorithm for combinational circuits. This algorithm can take advantage of the dominance relation between two sets of signals to speed up the diagnosis process. Let $\sigma_1 = \{f_1, f_2, ..., f_k\}$ and $\sigma_2 = \{d_1, d_2, ..., d_k\}$. If f_i is *topologically dominated*[1] by d_i for $1 \le i \le k$, then σ_1 is called *dominated* by σ_2. It can be shown that if an erroneous sequence E cannot be corrected by re-synthesizing σ_2, then E cannot be corrected by re-synthesizing σ_1 either. The details of this two-stage algorithm can be found in [7].

1. A signal f is *topologically dominated* by another signal d if every path originated from f to any primary output or next state lines pass through the signal d.

4. Experimental Results

We have implemented the proposed extension in our error diagnosis program in the environment of SIS [16], called *Error-Tracer*. The experiments are performed on ISCAS-89 sequential benchmark circuits. For each circuit, we first optimize it using the optimization script in SIS, *script.rugged*, to obtain the implementation. Then we decompose it into AND/OR gates using SIS command *"tech_decomp -a 5 -o 5"*. For generating erroneous implementations, gate type errors are injected randomly using the program discussed in [9]. Our approach is based on the notion of re-synthesis, and thus, not restricted to the types of errors introduced in the implementation. Also, our algorithm does not require the knowledge about the number of flip-flops or the structures of the specification. Only the input/output functional behavior of the specification is needed.

Results of diagnosing single-error circuits

Table 1 shows the results of single-error diagnosis. In the pre-processing stage, we run random simulation to collect erroneous input sequences. We set 10 as the maximum limit on the length of the sequences. The random simulation terminates when 32 erroneous input sequences have been collected, or 32,000 sequences have been simulated.

TABLE 1. Results of diagnosing optimized ISCAS89 benchmark circuits injected with **one** error.

Circuit	C2 size	E-length	fix signals	bound	time (sec)
s298	65	10	3	2	49
s344	102	4	4	2	13
s349	101	3	2	1	11
s386	60	1	1	1	3
s510	146	5	7	4	13
s641	109	1	2	2	2
s713	112	1	2	2	6
s820	171	5	7	3	21
s832	174	2	2	2	9
s1196	327	1	6	5	14
s1238	348	1	1	1	9
s1488	387	7	18	2	245
s1494	375	2	11	2	29
s1423	422	10	42	2	1229
s5378	920	2	3	2	66
s9234	1075	6	6	2	1089
s13207	1325	2	4	2	216
Average	631	3.7	7.1	2.2	178

The meanings of some columns are described as follows. (1) E-length (min.): the minimal length of the erroneous input sequences found in our pre-processing step. (2) *fix signals*: is the number of potential single-fix signals produced by our program. On average, the number of potential single-fix signals produced by our program is 7.1 for those ISCAS-89 benchmark circuits listed in Table 1. The curves of the number of potential single-fix signals versus the number of simulated erroneous input sequences for s1196 and s5378 are shown in Fig. 8. It is very common in our experience that only a few erroneous input sequences are enough to drop most false single-fix candidates. Note that our algorithm will not overlook the real injected error signal in the sense that it is always included in the delivered set of potential single-fix signals. Our program fails on seven circuits: s208, s400, s444, s15850, s35932, s38417, and s38584. The reasons will be further discussed later. (3) *Lower bound*: is a *pessimistic* lower bound on the total

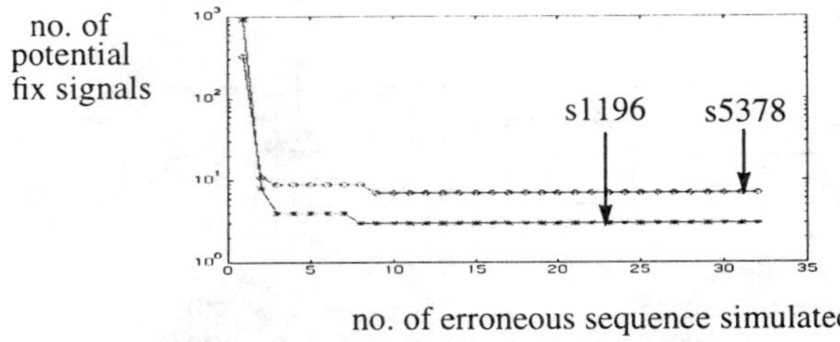

Fig 8. The number of potential single-fix signals versus the number of simulated erroneous sequences.

number of single-fix signals. This bound is obtained by counting the number of dominators of the injected error signal plus one. This is conservative because some single-fix signals may not dominate the injected error signal, and thus, are not included in the lower bound calculation. For ISCAS-89 benchmark circuits, the average is 2.2. (4) *Time*: is the CPU-time on a 150 MHz Sparc20 workstation. The total time consists of the random simulation time and the fault simulation time. The above results are obtained by using a modified differential fault simulation algorithm [4].

Results of diagnosing double-error circuits

Table 2 shows the results of diagnosing implementation injected with two random errors. Our program first searches for single-fix signals. If there exists no such signals, then we search for signal pairs that can jointly correct the implementation. Among the 15 circuits in Table 2, four of them are classified as single-signal correctable. Note that these circuits *may or may not be indeed* single-signal correctable. They could be mis-classified by our program because of inadequate accuracy. However, the other 11 are proven *not* single-signal correctable by our program (the number of potential single-fix signals is 0). For these circuits, we report the number of the potential double-fix pairs. The run time is longer than the case of diagnosing single error due to the rapid growth of the number of candidate signal pairs and the number of possible injections that need to be checked for correctability.

TABLE 2. Results of diagnosing optimized ISCAS89 benchmark circuits injected with **two** error.

circuit	C2 size	E-length	single-fix	double-fix	bound	time (sec)
s298	65	2	0	36	4	213
s344	102	3	0	16	6	612
s349	101	2	0	16	6	61
s386	60	1	3	-	-	44
s510	146	1	0	5	2	101
s641	109	1	0	15	6	61
s713	112	1	0	6	4	27
s820	171	2	1	-	-	11
s832	174	2	3	-	-	10
s1196	327	2	0	10	3	87
s1238	348	3	0	10	4	157
s1488	387	1	0	2	1	269
s1494	375	2	0	18	1	314
s1423	422	3	5	-	-	116
s5378	920	1	0	14	4	3883

Challenges

Based on this approach, there are several issues that need to be further addressed. (1) *Erroneous input sequence generation*. For combinational circuits, random simulation [8] or advanced Automatic Test Pattern Generation (ATPG) based techniques [1,21] have provided satisfactory solutions to generate erroneous vectors even for fairly large circuits. However, these techniques may not be adequate to generate erroneous input sequences for some sequential designs. Random simulation cannot find any erroneous input sequences for single-error circuits s15850, s35932, s38417, and s38584 in our experiments. For these large designs, if manually crafted functional sequences are available for simulation-based design validation, then most design errors are likely to be exposed and the erroneous sequences can be generated as a by-product. In that case, our approach is then applicable. Another possible solution to this problem is to explore the flip-flop correspondence or internal structural similarity between the specification and implementation. If a large number of corresponding flip-flops exists, then the sequentiality of the circuit can be reduced by treating the inputs (outputs) of some flip-flops as pseudo-primary outputs (inputs). In this way, the difficulty of generating erroneous input sequences can be reduced. (2) *High complexity for long erroneous input sequences*. If the errors occur in a highly sequential module (e.g., a counter) and cannot be detected by any input sequence with reasonable length (e.g., 20 vectors), then our approach may become too time-consuming. For s208, s400 and s444, our approach fails due to this reason. To address this problem, new techniques needs to be developed to deal with long erroneous input sequences for circuits that are manageable by BDD techniques. (3) *Difficulty for circuits with larger number of errors*. In practice, the complexity of diagnosing circuits with more than two errors is prohibitively high. Some heuristics are under investigation to estimate each signal's error probability and help the designer to locate the errors one at a time.

5. Conclusion

Traditionally, most error diagnosis approaches are only applicable to combinational circuits. In this paper, we present a fault simulation-based approach for diagnosing the general design errors in a sequential circuit. Our approach allows the implementation to have different number of flip-flops or a different state encoding from the specification. This capability is particularly useful when the specification is given at a higher level of abstraction, such as the register-transfer level. In our approach, we regard the signals that are totally responsible for the incorrectness of the implementation as the potential error sources. We prove that a modified fault simulation procedure can precisely determine if a signal can be held responsible for a particular erroneous input sequence. Based on this formulation, our algorithm excludes most signals that are impossible to be an error source by simulating a number of erroneous input sequences. Experimental results on diagnosing the ISCAS89 benchmark circuits with one and two random errors show that this could be an important step towards the sequential error diagnosis.

References

[1] H. Al-Asaad and J. P. Hayes, "Design Verification via Simulation and Automatic Test Pattern Generation," *Proc. of Int'l Conf. on Computer-Aided Design*, pp. 174-180, Nov. 1995.

[2] M. S. Abadir, J. Ferguson, and T. E. Kirkland, "Logic Design Verification via Test Generation," *IEEE Trans. on Computer,* pp. 138-148, Jan. 1988.

[3] R. E. Bryant, "Graph-based Algorithms for Boolean Function Manipulation," *IEEE Trans. on Computers,* vol. C-35, pp. 667-691, Aug. 1986.

[4] W.-T. Cheng and M.-L. Yu, "Differential Fault Simulation - A Fast Method Using Minimal Memory," *Proc. of Design Automation Conf.,* pp. 424-428, June 1989.

[5] P. Y. Chung, Y. M., Wang, and I. N., Hajj, "Logic design error diagnosis and correction", *IEEE Transactions on VLSI Systems,* vol. 2, no. 3, pp. 320-332, Sept. 1994.

[6] M. Fujita, "Methods for Automatic Design Error Correction in Sequential Circuits," *Proc. of European Conf. on Design Automation*, pp. 76-80, 1993.

[7] S.-Y. Huang, K.-T. Cheng, K.-C. Chen, and D.-I. Cheng, "ErrorTracer: A Fault Simulation Based Approach to Design Error Diagnosis," *Proc. of Int'l Test Conf.*, pp. 974-981, Nov. 1997.

[8] F. Krohm, A. Kuehlmann, and A. Mets, 'The Use of Random Simulation in Formal Verification," *Proc. of Int'l Conf. on Computer Design*, Oct. 1996.

[9] A. Kuehlmann, D. I. Cheng, A. Srinivasan, and D. P. LaPotin, "Error Diagnosis for Transistor-level verification," *Proc. of Design Automation Conf.,* pp. 218-224, June 1994.

[10] S. Y. Kuo, "Locating Logic Design Errors via Test Generation and Don't-Care Propagation," *Proc. of European Design Automation Conf.,* pp. 466-471, Sept. 1992.

[11] H.-T. Liaw, J.-H. Tsaih, and C.-S. Lin, "Efficient Automatic Diagnosis of Digital Circuits", *Proc. of Int'l Conf. on Computer Aided Design,* pp. 464-467, Nov. 1990.

[12] C.-C. Lin, K.-C. Chen, S.-C. Chang, M. Marek-Sadowska, and K.-T. Cheng, "Logic Synthesis for Engineering Change," *Proc. of Design Automation Conf,* pp. 647-652, June 1995.

[13] J. C. Madre, O. Coudert, J. P. Billon, "Automating the Diagnosis and the Rectification of the Design Errors with PRIAM," *Proc. of Int'l Conf. on Computer Aided Design,* pp. 30-33, Nov. 1989.

[14] I. Pomeranz and S. M. Reddy, "On Correction of Multiple Design Errors," *IEEE Trans. of Computer-Aided Design of Integrated Circuits and Systems*, pp. 255-264, vol. 14, no. 2, Feb. 1995.

[15] I. Pomeranz and S. M. Reddy, "A method for diagnosing implementation errors in synchronous sequential circuits and its implications on synthesis," *Proc. of European Design Automation Conference*, pp. 252-258, Sept. 1993.

[16] "SIS: A System for Sequential Circuit Synthesis," Report M92/41, University of California, Berkeley, 1992.

[17] K. A. Tamura, "Locating Functional Errors in Logic Circuits," *Proc. of Design Automation Conference,* pp. 185-191, June 1989.

[18] M. Tomita, H. H. Jiang, T. Tomamoto, and Y. Hayashi, "An Algorithm for Locating Logic Design Errors," *Proc. of Int'l Conf. on Computer Aided Design,* pp. 468-471, Nov. 1990.

[19] A. G. Veneris and I. N. Hajj, "A Fast Algorithm for Locating and Correcting Simple Design Errors in VLSI Digital Circuits," *Proc. of Great Lake Symposium on VLSI Design*, pp. 45-50, March 1997.

[20] A. M. Wahba and D. Borrione, "Design Error Diagnosis in Sequential Circuits," *Proc. of Correct Hardware Designs and Verification Methods, CHARME'95, Lecture Notes in Computer Science* No. 987, pp. 171-188, Springer Verlag, Oct. 1995.

[21] A. M. Wahba and D. Borrione, "A method for automatic design error location and correction in combinational logic circuits", *Journal of Electronic Testing: Theory and Applications*, vol.8, no.2, pp. 113-27, April 1996.

Functional Verification of a Multiple-issue, Out-of-Order, Superscalar Alpha Processor— The DEC Alpha 21264 Microprocessor

Scott Taylor, Michael Quinn,
Darren Brown, Nathan Dohm, Scot Hildebrandt, James Huggins, Carl Ramey
Digital Equipment Corporation

ABSTRACT

DIGITAL's Alpha 21264 processor is a highly out-of-order, superpipelined, superscalar implementation of the Alpha architecture, capable of a peak execution rate of six instructions per cycle and a sustainable rate of four per cycle. The 21264 also features a 500 MHz clock speed and a high-bandwidth system interface that channels up to 5.3 Gbytes/second of cache data and 2.6 Gbytes/second of main-memory data into the processor. Simulation-based functional verification was performed on the logic design using implementation-directed, pseudo-random exercisers, supplemented with implementation-specific, hand-generated tests. Extensive functional coverage analysis was performed to grade and direct the verification effort. The success of the verification effort was underscored by first prototype chips which were used to boot multiple operating systems across several different prototype systems.

Keywords

Alpha, microprocessor, verification, validation, architecture, pseudo-random, coverage analysis, 21264

1. INTRODUCTION

The Alpha 21264 microprocessor is a highly out-of-order, superscalar implementation of the Alpha architecture. The CPU required a rigorous verification effort to ensure that there were no logical bugs. World-class performance dictated the use of many advanced architectural features, including:

- four-wide instruction fetch
- instruction cache line prediction
- branch prediction (local and global history)
- out-of-order speculative-issue execution pipelines (4 integer pipelines, 2 floating-point pipelines)
- two out-of-order memory references per cycle
- up to sixteen in-flight off-chip memory references

DAC 98, San Francisco, California

In addition, on-chip caches include a 64-KByte virtually-indexed/virtually-tagged two-way set predicting instruction cache and a 64-KByte virtually-indexed/physically-tagged two-way set associative data cache. Each cache is paired with its own 128-entry fully associative instruction address translation buffer, and the data cache also features cache-hit prediction logic. The chip's pin interface features separate system and second-level (L2) cache interfaces, with support for various CPU-system clock ratios, system configurations, and L2-cache RAM types [1].

Functional verification of the Alpha 21264 design was performed by a team of verification engineers organized as a separate skill-based group. The primary responsibility of this team was to detect any logical errors in the chip design which would result in incorrect operation or negatively impact performance. The correction of these logical errors was the responsibility of the chip architects. The detection and correction of timing, electrical, and physical design errors were separate efforts conducted by the chip implementation team.

The chip verification team primarily utilized simulation-based techniques, which in turn relied heavily upon pseudo-random test generation to improve the quality and efficiency of the verification effort. These techniques have been in use at DIGITAL for more than nine years and are also used elsewhere in the industry and in academia [2-6]. The complexity of the Alpha 21264 design required that significant advancements be made in this area in order to guarantee meeting the key goal of working first-pass parts. This paper describes those advancements within the context of the overall verification effort.

1.1 Goals

Conflicting factors such as increased processor complexity and reduced time-to-market goals have resulted in first-pass silicon being relied upon to extensively verify and debug system-related hardware and software. It has become a typical goal that first-pass silicon be capable of booting one or more operating systems across several different prototype platforms. Achieving this first-pass goal provides a more effective mechanism for exercising the processor to its design limits and thereby exposing bugs that would not be readily detected using simulation-based verification techniques. However, as these new bugs are encountered, the original simulation environment can be used to isolate the bug source and to verify the solution. All of this acts to directly support the key goal of second-pass silicon, which is to be bug-free so that these chips can be shipped to customers for use in revenue-producing systems.

2. MODELING METHODOLOGY

The Alpha 21264 microprocessor was modeled in several levels of abstraction. These abstractions included a performance model, an RTL/behavioral model, a gate-level model, and a three-state simulation model.

The RTL model was written using a DIGITAL proprietary modeling language. This cycle-based simulator contained a command-line interface that enabled the user to access internal processor signals and call subroutines internal to the model. The signal interface also made it possible to dump binary traces of internal signal activity to a file for later processing.

The new simulation engine enabled the Alpha Standard Reference Model (also called the Instruction Set Processor or "ISP"), which was written in C++, to be directly incorporated in the Alpha 21264's RTL model. Verification of the RTL model was often accomplished by comparing its state with that of this reference model. Because these simulation checks could now be done at run-time, rather than as a post-processing task, the speed and extent of the checking was greatly increased.

The simulation engine was also used as a basis for a gate-level simulator. This simulation was also cycle-based, and could be run in lockstep with the RTL simulation. Various signal states could then be compared between the two models; any discrepancies between them would trigger an error condition.

2.1 Pseudo-system Models

The verification team created several modules that communicated with the RTL model under test. These routines typically modeled external logic such as main memory, caches, and I/O devices; external agents such as other processors or system demons; and basic pin interface functions such as reset and clocking. Most of this functionality was collectively referred to as the "Xbox." The Xbox implemented all possible system configurations that the Alpha 21264 microprocessor supported. It acted as the interface between the simulation engine, the RTL code, and the user. The Xbox represented the system and drove the pins of the RTL model, providing inputs and checking certain outputs.

Other pseudo-system models were created as well. One example was a "shadow" model of the Alpha 21264's branch prediction logic. Since branch prediction was inherently transparent to the user, it was difficult to verify that the logic was functionally correct. The shadow model was an abstracted version of the RTL logic, and used the same interface signals. These interface values were driven to both the RTL model and the shadow model on a cycle-by-cycle basis. Routines within the shadow model compared the state of the two models and flagged any errors.

3. STRATEGY

The verification strategy employed multiple techniques to achieve full functional verification of the Alpha 21264 chip. The primary technique used pseudo-random exercisers coupled with coverage analysis. These exercisers generated pseudo-random instruction sequences, executed the sequences on a model that included both the 21264 model and the ISP reference model, and compared the results during the execution. A second technique used focused, hand-generated tests to cover specific areas of logic. Other methods consisted of model code reviews, box-level testing, and executing Alpha architectural test suites and benchmarks. Figure 1 shows the general flow for a single simulation.

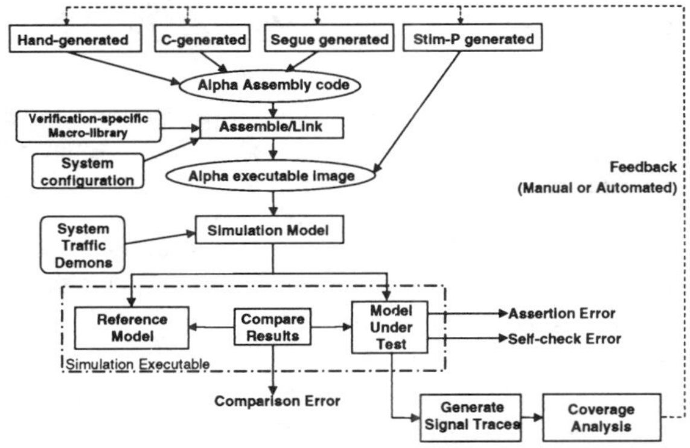

Figure 1. Verification Test Environment

This strategy was deployed in four parts: the pre-integration phase, the try-anything phase, the test-planning phase, and the structured completion phase. In the pre-integration phase, testing focused on individual functional units, or "boxes," of the design. These boxes were tested using transactors that drove all inputs and checked all outputs of the functional unit. This allowed for a smooth integration of the full-chip model. Once the full-chip model had been integrated, assembly-level code could be used to test the chip.

The main purpose of the try-anything phase was to exercise the majority of the design in a short period of time. Automated exercisers were the main source of stimulus during this part of the project. This allowed the verification team to spend most of its time debugging failures rather than writing tests. Existing Alpha architectural test suites and benchmarks were also executed to ensure base-level functionality. This debugging period produced an atmosphere of intensity and challenge. Everyone was required to interact constantly and identify problem sources, thus stabilizing the model quickly.

Once the design stabilized and the bug rate declined, the verification team took a step back and created a test plan. The purpose of the test plan was to ensure that the verification team understood what needed to be verified. The initial phases of the project allowed the verification team to gain insight into complex areas of the chip, leading to the creation of a better initial test plan. A set of joint test plan reviews were then held with the design teams for each section of the chip. These joint reviews ensured that the verification team did not miss important aspects of the design. Finally, the test plan provided a means for scheduling and prioritizing the rest of the verification effort.

The final verification step was the structured completion phase. During this time, each item from the test plan was analyzed and verified. The analysis consisted of deciding which mechanism was appropriate for covering that particular piece of the design. This might consist of a focused test, a pseudo-random exerciser with coverage analysis, or an assertion checker. As the verification of each section of the chip was completed, a review was held with the design and architecture teams to examine what was verified and how it was done. This review identified any verification coverage holes.

4. TEST STIMULUS GENERATION

4.1 Focused Testing

Verification of the 21264 relied heavily on pseudo-random test generation and, to a lesser extent, focused testing. About 300 focused tests were written during the verification effort. Additional post-silicon tests were added to increase manufacturing defect detection and isolate critical speed paths. Focused tests covered all major functional areas of the chip and provided repeatable, guaranteed coverage for regression testing.

A subset of the focused test suite was run as part of the model release procedure. For increased coverage the entire test suite was continuously run against the released model using random configurations. This test loop guaranteed that all model releases maintained consistent functionality, as well as providing additional verification coverage. The regression test loop was instrumental in maintaining gate-level and RTL model coherency for the implementation team.

Focused tests had advantages over pseudo-random tests not only in their directed areas of coverage, but in their self-checking abilities. Self-checking code aided in finding bugs in the reference model as well as in the RTL. Most of these tests were used as manufacturing patterns because of their ability to run at various system and L2 cache speed settings. Focused assembly-code tests were readable and directed at specific logic areas, so they were faster to debug.

4.2 Pseudo-random Testing

The complexity in the 21264 made pseudo-random testing critical. Defining the logical interactions that needed to be tested was extremely difficult in this chip. Creating these interactions by hand was even more difficult. The pseudo-random testing approach allowed the verification engineer to concentrate effort on defining areas of interest within the chip. Automated pseudo-random test programs called exercisers were created to test the complex interactions within those areas.

4.2.1 Pseudo-random Exercisers

Several dozen different pseudo-random exercisers were used in the verification of the 21264. Most of these focused on specific architectural areas within the chip; others were more generalized. Some areas targeted explicitly were: control-flow, out-of-order instruction processing, superscalar structures, cache and memory transactions, multiprocessing and synchronization code, floating point, traps, illegal instructions, and Privileged Architecture Library (PAL mode).

Fundamentally, each exerciser operated the same way. First, the exerciser created pseudo-random assembly-language code. Next it simultaneously ran the code on the model under test and on the ISP reference model. Any divergence of data or instruction stream between the two models signaled an error.

The out-of-order feature of the 21264 required new testing methods. Code had to be generated in such a way that the chip's full out-of-order potential was tested. This involved tight control of data dependencies in the code, which were matched to the issue depth and superscalar width of the 21264. New tools were developed to combine multiple independent code streams into one test case allowing maximum instruction throughput on the chip while maintaining consistent out-of-order instruction execution.

Many of the exercisers used the test generation engine known as Stim-P. With Stim-P, the user controlled the probability of certain events or instructions rather than specifying the actual code sequence. Stim-P generated pseudo-random Alpha assembly code based on the user's configuration and its own understanding of the Alpha architecture. This allowed each user to "tune" an exerciser to focus on a specific area of the chip. For example, a verification engineer trying to test integer unit traps would have turned up the frequency of integer instructions relative to floating-point instructions and also maximized the probability of generating overflows on a given operation. Similarly, a memory-focused exerciser may have used a reduced number of trapping instructions and an increased probability of re-using a previously generated memory address for loads and stores.

SEGUE, a text generation and expansion tool, was also used to create pseudo-random code. Similar to Stim-P, variables were passed in to determine the focus of a particular test case. SEGUE, however, relied solely on the use of user-defined templates to generate a code stream. Given a basic building block, SEGUE filled in the template with appropriate instructions. An important application of the SEGUE method was to test the 21264's PAL (Privileged Architecture Library) mode. Since PAL mode imposed many restrictions on the content and order of the instruction stream, SEGUE was a natural fit for generating pseudo-random code which followed the specification's coding rules.

Some exercisers combined the Stim-P and SEGUE tools to create specialized test cases. Illegal instructions, for example, were exhaustively generated by SEGUE and run within Stim-P tests. This helped to expose bugs that involved several random events occurring around an illegal instruction. Other hybrid exercisers used these tools to check PAL mode instructions within random user code.

4.2.2 Random Configuration Selection

Each test, whether pseudo-random or focused, made use of certain "configuration parameters" to describe both internal CPU state and the type of system that would be emulated for a given simulation. These parameters consisted of variables that could be weighted to make certain system and/or internal CPU events occur. Interesting "fixed configurations" could also be randomly chosen to better mimic a real system design.

Examples of parameters included single-bit error injection rate, external cache size and configuration, the ratio between the system clock and the CPU internal clock rate, external probe rates, floating point unit enable, branch prediction enable, page mapping method, and PALcode location. In all, there were over eight hundred parameters.

Once again, SEGUE scripts were utilized to create the command files that chose the various configuration parameter values. This was done during the compile stage of a test. This method insured that different configurations were chosen and tested for each simulation. This meant that any test, including focused tests, could be used as a regression test and still provide new verification coverage each time it was run.

Each test could override the randomization of any parameter, which allowed a test to consistently run in a particular mode if necessary. Configuration parameters were also loaded into the memory model to provide information accessible through assembly code regarding the configuration type.

4.2.3 Demons

Demons were pieces of model code that could query or modify internal or external states of the 21264 chip. These demons were all accessible via macrocode. The access method chosen allowed for quick and predictable demon operation but still allowed their use during test pattern generation time.

There were several varieties of demons. State-modifying demons were used to affect internal or external state of the processor—such as randomly injecting bit errors on fills, or asserting random interrupt requests. System traffic demons were used to simulate probe requests from other processors, such as Invalidate or Set-Dirty requests. There were also a large number of status/query demons, which performed such functions as printing user messages to a log file and counting the number of data cache hits/misses. Special demons were also created to reduce the size of exception flows. One example was a demon that directly looked up page mapping in a hash table structure rather than using macrocode to do the lookup in memory.

5. SIMULATION

Once the focused or pseudo-random test case had been generated, the test was compiled to produce one binary image. This image contained the test case along with the random configuration chosen and the PALcode. PALcode is a set of low-level firmware routines designed to handle events such as traps, interrupts, and page-table misses. This PALcode could be customized on a per-test basis.

At the beginning of the simulation, the binary image would be read by a simulation loader. This loader would then place the test throughout memory as directed, set up system configuration variables, and place PALcode and internal configuration state in memory.

After a test was loaded, the simulation would start. A PALcode reset handler was executed first. Information was read from the configuration and various internal registers were loaded with the specified values. The test was executed after the PALcode completed. During simulation, results were compared and any failures detected were reported to the user and stopped the simulation. Upon successful completion of the test, some post-processing steps were performed. Failing cases were saved to disk for debugging, while passing cases were generally discarded.

6. CORRECTNESS CHECKING

Several classes of mechanisms were used to ensure the RTL model was functioning correctly. Some focused tests used self-checking, hand-coded by the test author. This approach is time-consuming, prone to error, and not useful beyond a single focused test, so self-checking was kept to a minimum. More general techniques included assertion checkers, reference model comparisons, and memory coherency checkers.

6.1 Assertion Checkers

The new simulation engine utilized for Alpha 21264 verification provided many built-in assertion checkers for illegal signal states associated with its RTL modeling primitives, an example of which would be an invalid select combination on a multiplexer. Many other assertion checkers were added to the RTL code to check for illegal events such as an invalid transition in a state machine. These were added to the model by architects or verification engineers. RTL assertion checkers were easy to implement and prevented wasted simulation cycles by halting the simulation when an error occurred. Debugging information was also provided near the location of the error.

Assertion checkers were also included in the Xbox by verification engineers. These checkers tended to be more involved, and tracked the state of the machine. Examples include checkers for queue overruns and violation of the bus protocol.

Another class of assertion checkers was included in coverage analysis tests. The verification engineer would analyze a set of cross-products, and if any elements were illegal, an assertion checker was added. These checks were done as a post-processing step, but were able to provide useful debuging information, most importantly which cycle the violation occurred.

6.2 Reference Model Comparisons

A module was built to compare the ISP reference model with the RTL model in real time. Some comparisons were done at the end of a test, such as a full memory comparison. Most comparisons were done in real time; these include:

- Register values at instruction retire time
- Program counter at instruction retire or exception
- Memory ordering at the pins
- System lock flag at instruction retire
- Checks for success/failure of atomic operations (load_locked and store_conditional)

6.3 Memory Coherency Checkers

A mechanism was developed that could "snoop" through various levels of the cache hierarchy, including main memory, a second-level backside cache, the first-level data cache, buffers for data arriving on the chip, buffers for data being evicted from the chip, and the pipelines for fills and evicts. This enabled the verification team to check the flat memory state of the ISP against the complex cache hierarchy of the processor. For performance reasons, the entire memory space was not checked every cycle. Instead, selected areas of memory were checked at specific coherency points such as cache block evictions. These checks were invaluable and found many bugs in the design, but were also difficult to maintain.

7. COVERAGE ANALYSIS

The most difficult question within functional verification has always been: "Is the verification effort complete?" One could run test after test and still not have the confidence that all areas of the design were fully tested. One method of answering this question involves watching the bug rate. While this provides some interesting data, watching it fall may just mean that other problem areas are going untested. A more successful method of answering this question is the use of coverage analysis. Coverage analysis plays an important role by providing visibility into the completeness of the verification effort. On the 21264 microprocessor, extensive coverage analysis of both the focused tests and pseudo-random exercisers was done.

Signal traces were collected during a simulation. This data was then used by trace analysis tools to collect the coverage data which was accumulated across multiple simulation runs. That data was analyzed periodically, and areas that were lacking coverage were identified. This allowed the identification of trends in the coverage and provided an understanding as to how well the pseudo-random exercisers were exercising the chip. With this insight, pseudo-random exercisers were modified or new focused tests were created to improve the test coverage. Running pseudo-random exercisers with coverage analysis proved to be a very powerful technique in functional verification.

The coverage analysis done can be broken up into four major areas: state transitions, sequence, occurrence, and case.

7.1 State Transition Analysis

State transition analysis was used for more complex state machines-ones with dozens of states or arcs. A tool called CADET was used to describe the actual state diagram for the function. The tool then automatically generated the necessary coverage test to check all possible states and arcs. On the 21264 microprocessor, one such state machine checked this way was the instruction cache stream buffer control logic.

7.2 Sequence Analysis

CADET could also be used to check sequences of events. A set of possible events happening at time (t) can be "crossed" with a set of events happening at time (t+1). For example, it was verified that every type of system command heading out of the CPU was followed with every type of system command heading into the CPU.

These events were described within CADET graphically which reduced the time needed to create such coverage tests. These events were then organized in sets that described the cross-products. As the data was collected, it could be displayed in a matrix format. Figure 2 shows coverage data for cache blocks transitions from invalid to four possible valid states.

	V	~V	~V	~V	~V	Time (t-1)
Commands	~V	V	V/S	V/D	V/S/D	Time (t)
RdBlk	.	✔	7	✔	0	
RdBlkMod	.	.	.	✔	1	
Clean2Dirty	.	.	.	.	.	
Share2Dirty	.	.	.	.	.	
Inval2Dirty	.	.	.	✔	.	
Evict	✔	.	.	.	.	

V = Valid S = Shared D = Dirty

. = event cannot occur

✔ = more than 100 events of this type were seen

Figure 2. Sample Cadet Coverage Matrix

Illegal transitions (such as a data read causing a cache block to become invalid) can be tagged, such that any occurrence of the event would flag an error. Legal transitions increment the value at that matrix location; when the value reaches a pre-determined value (100 in the example), the value is considered "filled." In the example, a transition from invalid to valid-shared-dirty has not occurred. This was a legal, albeit rare, occurrence. The verification engineer could then take this information and tune an exerciser (or write a focused test) to fill in this coverage hole.

7.3 Occurrence Analysis

Coverage analysis doesn't always consist of time-related events. Many coverage issues were simple occurrences of "one-time" events. A CADET test might be written to check that a carry-out has been generated for every bit in an adder. Another would make sure each type of interrupt was taken.

7.4 Case Analysis

Case analysis consisted of collecting a set of statistics on each simulation run. Types of statistics included exerciser type, cycles simulated, issue rate, instruction types, cache hit/miss rate, specific system configuration and many more. This data was kept in database format using a tool called STATS. This same tool could then generate reports, for later analysis, for specific date ranges and/or exerciser types to gauge an exerciser's verification effectiveness.

8. BUG TRENDS

Throughout the 21264 verification effort many bugs were logged and tracked, most before first-pass parts were manufactured. Figure 3 shows the bugs found as a function of time for the duration of the project as given by a bug-tracking system. Tracking of bugs started after all the subsections of the RTL-level model had been integrated and a small subset of tests was run successfully. Since many areas of the model were ready early or tested at the box level before changes were integrated into the full model, the action tracking system does not represent all the issues raised. However, it is interesting to look at the trends presented by the data.

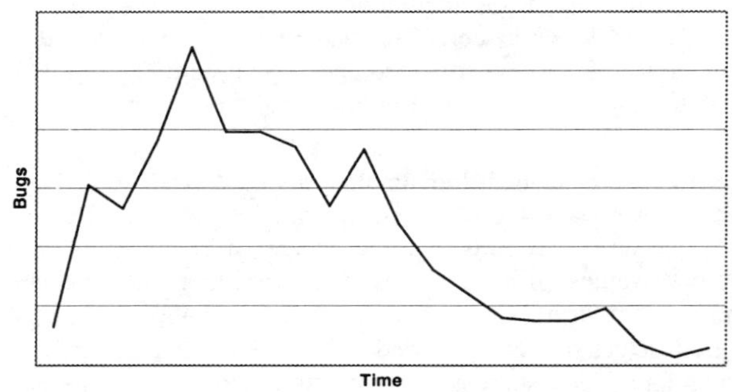

Figure 3. Bug Detection as a Function of Time

The first trend to consider is the effectiveness of the pseudo-random and focused efforts. As shown in Figure 4, more than three-quarters of the bugs were found using pseudo-random techniques. Errors found by the focused testing were typically in sections of the design that had sparse pseudo-random test coverage or where self-checking tests could be easily implemented.

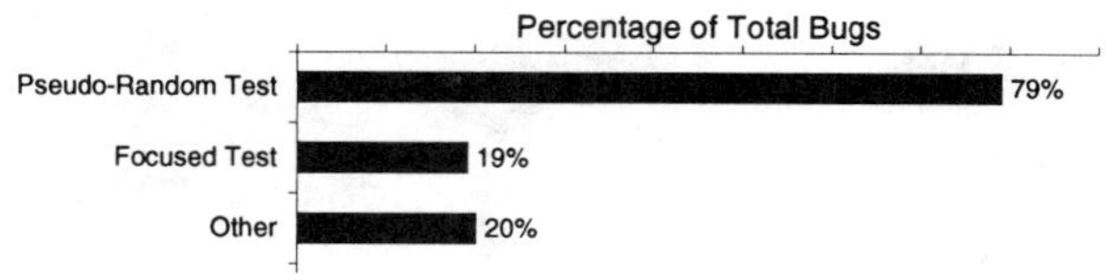

Figure 4. Effectiveness of Test Category

Bugs in the design had been introduced in a variety of ways. Figure 5 shows the breakdown of the causes of bugs, of which the majority occurred while implementing the architectural specifications for the project.

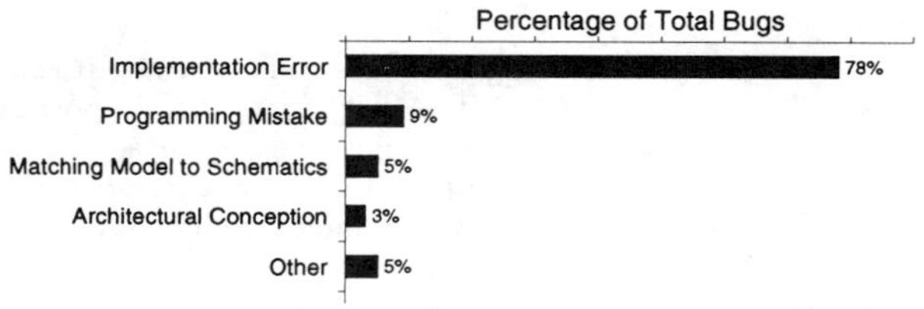

Figure 5. Introduction of Bugs

Figure 6 shows the various detection mechanisms that were used to find bugs. Assertion checkers placed in the design to quickly detect when something is not correct were the most successful. However, when viewed collectively, nearly 44% of all bugs were found by reference model comparison, which includes register, PC, and memory state comparisons.

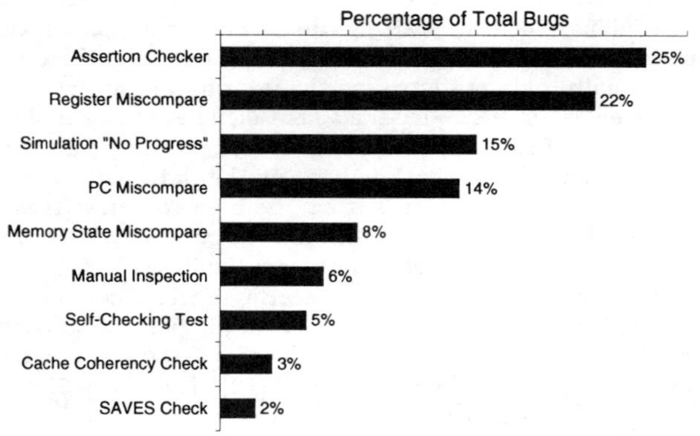

Figure 6. Bug Detection Mechanisms

9. RESULTS AND CONCLUSIONS

As of December, 1997, eight logical bugs were found in the first-pass prototype Alpha 21264 CPU design. None of these bugs prevented first-pass silicon from being used to boot multiple operating systems on several different prototype platforms. Two of the bugs were found while debugging test patterns on the tester; the others were all found during initial prototype system debugging. These bugs escaped detection via functional verification of the RTL model for the following reasons:

- Three of the bugs were due to the schematics not matching the RTL model, and comparisons between the RTL model and the structural model (generated from the schematics) did not detect the mismatch.
- One bug was due to electrical issues.
- One bug was due to insufficient randomization of the initial state of the RTL model during simulation.
- Two of the bugs were actual logical problems in the RTL model. In both cases checkers had been added to detect that exact problem, however the checkers were either incorrectly or insufficiently implemented.
- The last bug was also a logical problem in the RTL model. However, it involved an extremely complex combination of cross-products that were unlikely to be hit via pseudo-random testing, or even thought of in terms of a focused test. In a very real sense this was exactly the type of bug that first-pass silicon was intended to help find.

All of the above issues have been fixed in the version of the design that will be released to customers. They were all viewed as being relatively minor bugs, with the possible exception of the last one (which again was exactly the type of bug that was considered most likely to be found during first-pass silicon debugging). The success of the Alpha 21264 verification effort once again underscores the value and effectiveness of pseudo-random based testing.

10. FOR FURTHER INFORMATION

The authors can be contacted at the following address, c/o Digital Equipment Corporation:

334 South Street
Shrewsbury, MA 01545-4112

Email inquiries may be directed to Scott.Taylor@digital.com or Quinn@ad.enet.dec.com.

11. REFERENCES

[1] Linley Gwennap, "Digital 21264 Sets New Standard," *Microprocessor Report* (October 28, 1996): 11-16.

[2] Mike Kantrowitz and Lisa Noack, "Functional Verification of a Multiple-issue, Pipelined, Superscalar Alpha Processor — the Alpha 21164 CPU Chip," *Digital Technical Journal*, vol. 7, no. 1 (1995): 136-143.

[3] W. Anderson, "Logical Verification of the NVAX CPU Chip Design," *Digital Technical Journal*, vol. 4, no. 3 (Summer 1992): 38-46.

[4] A. Aharon, A. Bar-David, B. Dorfman, E. Gofman, M. Leibowitz, and V. Schwartzburd, "Verification of the IBM RISC System/6000 by a Dynamic Biased Pseudo-random Test Program Generator," *IBM Systems Journal*, vol. 30, no. 4 (1991): 527-538.

[5] A. Ahi, G. Burroughs, A. Gore, S. LaMar, C-Y. Lin, and A. Wiemann, "Design Verification of the HP 9000 Series 700 PA-RISC Workstations," *Hewlett-Packard Journal* (August 1992): 34-42.

[6] D. Wood, G. Gibson, and R. Katz, "Verifying a Multiprocessor Cache Controller Using Random Test Generation," *IEEE Design and Test of Computers* (August 1990): 13-25.

Design Reliability - Estimation through Statistical Analysis of Bug Discovery Data

Yossi Malka, Avi Ziv
IBM Research Lab in Haifa
MATAM
Haifa 31905, Israel
email: {yossi, aziv@vnet.ibm.com}

Abstract

Statistical analysis of bug discovery data is used in the software industry to check the quality of the testing process and estimate the reliability of the tested program. In this paper, we show that the same techniques are applicable to hardware design verification. We performed a study on two implementations of state-of-the-art PowerPC processors that shows that these techniques can provide quality information on the progress of verification and good predictions of the number of bugs left in the design and the future MTTF.

1 Introduction

Functional verification comprises a large portion of the effort in designing a processor [1]. The investment in expert time and computer resources is huge, and so is the cost of delivering faulty products [3]. In current industrial practice, random architectural test program generators and simulation methods are used to implement large portions of the verification plans [1, 2, 11]. Verification is done by comparing the behavior of the tested design to its expected behavior, as defined by the design's architecture and specification. Each time the behavior of a tested design diverges from the expected behavior, a failure occurs. When a failure occurs, the bug that caused the failure is tracked and, in most cases, the fault is removed or bypassed. Each time a bug is removed from the tested design, there are less elements that can cause failures of the design, and therefore, the mean time to the next failure (MTTF) increases or, in other words, the reliability of the design grows.

One of the major challenges and concerns that verification people are facing is how to measure the quality of the verification process and determine the reliability of the tested design. Verification people are often asked questions like: "When can a certain level of reliability be expected?" or "What are the effects of a new test generation tool on the verification process?"

Accurate predictions of the reliability of a design are extremely important when determining dates for tape-out or general release. The huge costs of a tape-out and the complexity of the hardware bring-up process that follows it leads to a small number of tape-outs. It is, therefore, important to tie tape-outs with reliability milestones, such as the expected time when the processor is able to boot its operating system. This requires estimation of the current reliability of the design. Moreover, since a tape-out needs to be planned months in advance, it is important to accurately predict when the desired reliability is going to be reached in order to avoid a premature tape-out.

During the verification process, a lot of information regarding the bugs in the tested design and the verification process is collected. This information includes data on when the bug was discovered, how it was discovered, and its severity. The information also includes the amount of simulation, measured in cycles, that was executed. This information is used to evaluate the reliability of the tested design, examine the verification process, and to make decisions regarding future plans, such as starting the next testing phase and tape-outs. Since the amount of data that is gathered is huge, usually only statistics on the information are used.

Currently, intuitive decisions are taken based on the shape of the curve of the number of bugs found versus the amount of simulation. Statistical analysis of the data can provide more concrete observations. It can be used to detect trends that occurred in the past, determine the current state of the tested design, and predict its future reliability.

The software reliability engineering (SRE) community considers statistical bug analysis a major part of software reliability. Statistical bug analysis of software is used in many organizations, such as AT&T [12], Tandem [15], and JPL [9]. The analysis is used to measure the reliability of the software in a wide variety of systems, such as operating systems, switching devices, and spacecraft controllers. Statistical bug analysis is used in these systems to measure the reliability of the software system, predict its future reliability, and optimal release date and to assist in design and testing plans.

The design, coding, and testing of software is closely related to design, implementation, and verification of hardware designs. In both hardware and software, the design process starts with high level design of the whole system, which is then broken into smaller units. Each unit is implemented and tested by itself, and then the whole system is constructed out of the units. Testing of software systems is also done in a similar way to hardware design verification. Input patterns are generated and the behavior of the module is compared to its expected behavior, which is defined by the specification and architecture. Finally, in both hardware and software, bugs are detected and removed from the

DAC 98, San Francisco, California
©1998 ACM 0-89791-964-5/98/06..$5.00

Design	Design Type	Duration of Testing [Month]	Number of bugs found	Simulation cycles [$\times 10^6$]
Processor A	Full processor	8.5	719	68052
Processor B	Execution units	8.5	236	2383

Table 1: Source bug discovery data summary

design during testing, causing reliability growth in the module. Therefore, the vast experience and knowledge gathered in statistical analysis of software bugs can be used (at least as the starting point) for statistical bug analysis during verification of hardware modules.

In this paper we describe the results of a study we did on the applicability of statistical bug analysis techniques used in software reliability engineering. We used bug discovery data obtained for two advanced PowerPC processors, and applied to it two types of statistical bug analysis that are commonly used is software reliability engineering. The techniques that we are using are *trend analysis* for reliability growth, and *modeling.* These techniques can provide answers to questions like:

- How does the introduction of a new test generator affect the reliability growth?
- What is the mean time to the next failure?
- When can a certain level of reliability be expected? That is, when will the probability of a failure in X time units be less than a given ϵ?

The answer to the first question can be provided using trend analysis of the reliability growth. For example, if after introduction of a new test generator, the reliability growth slowed down, we can then assume that the new generator is more efficient in detecting bugs, since it can produce more failures. Answering the second and third questions involve prediction of the future reliability of the tested design. The second question requires short term prediction about the time of the next failure, while the third question requires long term prediction.

The study reported in the paper was performed on two different designs. The first design, called Processor A in the paper, is a super-scalar PowerPC processor. The second design, called Processor B, includes all the execution units of a different super-scalar PowerPC processor. For both designs, we used bug discovery data that was collected over a period of more than 8 months. During this period, the designs were heavily tested (over 68 billion simulation cycles for Processor A and over 2 billion for Processor B), and several hundred bugs were found. The data used for the analysis, for both designs, was the number of passing simulation cycles executed and the number of bugs found every day. The data includes only bugs found using random simulation, and ignores bugs found using other verification techniques, such as formal verification and inspections. Table 1 summarizes the source bug discovery data that was used for the analysis.

The results of the analysis enable us to detect the affects of the introduction of a major function in one processor on its reliability growth, and to predict accurately the number of bugs discovered and the MTTF in the other processor three months in advance. Overall, our results show that these techniques can be very useful in monitoring the verification process and estimating its future progress.

Note that in this paper we concentrate on the analysis results, not on a description of the methods. A complete description of the methods used in this paper can be found in many text books, such as [9] and [13].

2 Trend Analysis

The task of the verification process is to increase the reliability of the tested design or, in other words, to decrease the bug intensity of the design. This is done by detecting and removing bugs that cause failures from the design. Therefore, the first indication we would like to get from the statistical analysis is whether there is a reliability growth in the tested design. This type of information can be provided by trend analysis.

Reliability growth means that the bug discovery intensity in the design is decreasing, and therefore the time between bug discoveries is increasing. This trend can be detected by looking at the cumulative number of bugs as a function of the testing time. Since the bug intensity is a derivative of the cumulative number of bugs, decreasing bug intensity means that the function of the cumulative number of bugs vs. simulation time should be concave.

Taking a look at the raw data provides only a rough indication of reliability growth for two main reasons:

- The raw data of time between bugs is only an instantiation of a random process. Therefore, the cumulative number of bugs is not a smooth concave function, but a function with many bumps and extreme points.
- The raw data information can very easily hide local changes in the trend of the reliability growth or strong local variations can hide the overall trend of reliability growth.

There are several analytical tests for reliability growth that are designed to overcome these problems. The idea behind these tests, known as trend tests, is to test a null hypothesis H_0 versus an alternative H_1. Usually, H_0 corresponds to the assumption that the reliability is not changing, while H_1 corresponds to the assumption that reliability undergoes a monotonic trend. One of the most widely used trend tests is the Laplace test [4]. The Laplace test assumes that the bug discovery intensity is equal to $\lambda(t) = e^{a+bt}$. The test compares the null hypothesis, constant bug intensity ($b = 0$), versus the hypothesis that the bug discovery rate is decreasing, that is $b < 0$. The test procedure is to calculate the Laplace factors $u(\cdot)$. Simply stated, the test procedure compares the midpoint of the tested interval with the median bug discovery time. Negative values of Laplace factors indicate that more bugs were discovered in the first half of the inspected interval, and therefore there is a reliability growth. Positive values of Laplace factors indicate that the bug intensity is increasing. If the values are in the range -2 to 2, then the reliability is stable or, more precisely, the null hypothesis cannot be rejected.

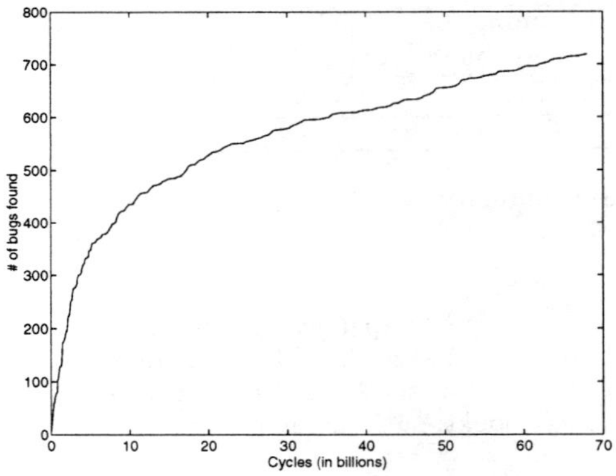

(a) Raw data

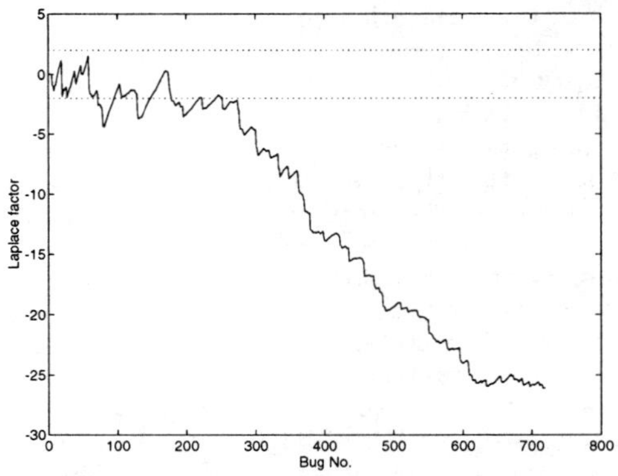

(b) Fault based trend analysis

Figure 1: Raw data and trend analysis for Processor A

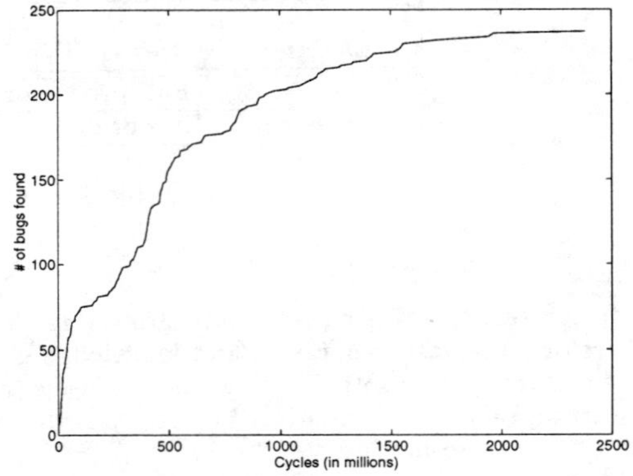

(a) Raw data

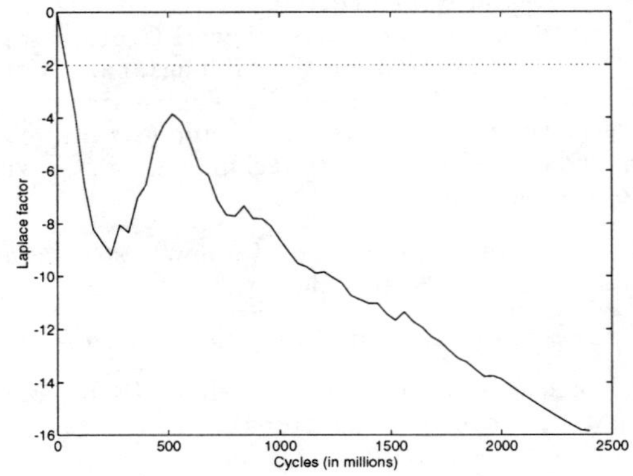

(b) Cycle based Trend analysis

Figure 2: Raw data and trend analysis for Processor B

Laplace factors can be calculated both as a function of the testing time or the number of bugs found. For example, Figure 1b shows the Laplace factors versus the number of bugs found for Processor A, while Figure 2b shows the Laplace factors versus the number of simulation cycles for Processor B. Often, using one form of the test helps to show things that are not clear in the other form. For example, The first region in Figure 1b, where there is almost no reliability growth, is hard to detect in a cycle based plot, since the region covers only 3% of the total amount of simulation cycles.

Figure 1 shows the raw bug discovery data and the result of the Laplace test for reliability growth for Processor A. The raw data shows that there is an overall reliability growth, since the plot of the total number of bugs vs. the total amount of simulation cycles is concave. This look at the raw data provides only a rough indication of the reliability growth. A better indication can be obtained using trend analysis. The trend analysis plot shows Laplace factors as a function of the bug number. From the figure, we can see that, overall, there is a continuous reliability growth, but there are three regions of growth. In the first region, which exists until about the 270th bug (which took about 3% of the simulation cycles), there is a small decrease in the Laplace factor which indicates a small reliability growth. The second region, between bug 270 and 610, shows a much faster drop in the value of the Laplace factor, which indicates a faster reliability growth. In the last region, the Laplace factors drop slowly, which indicates a slower growth in reliability.

The first region in the trend, with slow reliability growth, can be explained by difficulties in the start-up of the verification process, and therefore it might not be a good representative for the overall reliability growth of the processor and should be ignored. The slowdown in the reliability growth rate in the third region can be explained by the level of reliability that was achieved by that time. This type of behavior is observed in all existing models for reliability growth.

Figure 2 shows the raw bug discovery data and the Laplace factors for Processor B. The raw data plot shows that, overall, the reliability of the processor grows with time, with some anomaly in the beginning. A look at the Laplace factors clearly shows that, in the area between 200 and 500 million cycles, there is an increase in the Laplace factors, which indicates a decrease in the reliability of the tested design. After that period, the reliability starts to grow again and continues to grow continuously until the end of the examined period. When we tried to find the reason for this behavior, we found that this time corresponds to the time when testing of a major function of the processor started. This caused an increase in the bug rate due to the bugs in the new function, and thus a decrease in the reliability.

3 Modeling and Predictions

Trend analysis can provide information about trends that occurred during the verification process, and answer important questions, such as: does the reliability of the tested design grow, and how do specific events affect this growth? Still, trend analysis cannot answer questions like: how many bugs are left in the tested design, and when will a desired level of reliability be achieved? To answer such questions, predictions of the future behavior of the tested design and the verification process are needed. Usually, predictions of the future behavior of a process are obtained by matching the history information of the process to some predefined model (or models) and using the properties of the model to describe predicted future behavior [13].

The basic approach of modeling is to fit past data to a model that describes the expected behavior of the data, and to use the model to predict future behavior. It has been shown that, in order to provide good predictions, the model should accurately describe the observed process or at least be close to it. Predictions obtained using models that do not fit the observed data can be misleading. Therefore, selecting the proper model is probably the most important part of modeling.

Because of the similarities in the processes of hardware and software design and testing, bug models that have been tried and found applicable in software are good candidates to be applicable in hardware as well. In our study, we tested the bug discovery data of the processors against several commonly used models in software reliability engineering. A short description of three of the models is given below. A complete description of the models can be found in their references or in books like [9] and [13]. The models that we used are:

- Goel Okumoto non-homogeneous Poisson process model (GO) [8]. This model is a finite model that assumes that the number of bugs observed by time t follow a non-homogeneous Poisson process with a bug rate that drops exponentially.
- Duane's model [6]. This model is an infinite Poisson model that assumes that the cumulative number of bugs grows exponentially with time.
- Musa Okumoto logarithmic Poisson model (MO) [14]. This is an infinite Poisson model that assumes that the bug rate drops exponentially with the expected number of bugs experienced.

As stated earlier, modeling fits the observed data into a predefined model and uses the model to predict future behavior. Fitting the observed data to the model is done by estimating the parameters of the model based on the observed data. There are several estimation techniques, each finding the best estimator based on different criteria. The most commonly used estimators are the maximum likelihood estimator (MLE), the least square error (LSE) and methods of moments. After the parameters of the model are estimated, the model can be used to predict future behavior of the tested design.

The predictions provided by different models, based on the same bug discovery data, can differ considerably. For example, Table 2 compares the predictions made by the three models for the number of bugs found and the MTTF for Processor A after 120 billion cycles of simulation. The table shows that there is a huge difference in the predictions made by the three models. Therefore, in order to get quality predictions, we need to find the model that fits the bug discovery behavior of the tested design in the best way, and thus has the best chance to provide the most accurate predictions. Unfortunately, no one model is clearly better than

	Number of bugs found	MTTF [10^6 cycles]
Goel-Okumoto	810	750
Duane	886	355
Musa-Okumoto	756	4300

Table 2: Predictions for Processor A after 120 billions cycles of simulations

the other models, and can be used as the model for reliability predictions.

There are many possible ways to check how well a model fits the actual process. The first method is to compare the observed data with the expected behavior based on the model and the estimated parameters. Even when the observed data is used to estimate the parameters of the model, there could be a big differences between the behavior predicted by the model and the actual behavior when the actual behavior does not fit the model.

Figure 3a shows how the bug discovery data of Processor A is approximated by the three models described above. The figure shows that none of the models totally fit the bug discovery data, but some of them, specifically, the Musa Okumoto model, are fairly close to it.

While a plot that compares actual data with a given model (like Figure 3a) can provide an intuitive feeling of the fitness of the observed data, it cannot provide quantitative answers on the fitness level, or the level of certainty that the data fits the model. There are several statistical tests that can provide the answers to these questions, such as the *chi-square test* that compares sample data to a probability density function, and the *Kolmogorv-Smirnov test* that compares sample data to a cumulative distribution function [5]. The answers from the goodness-of-fit tests can be used to determine the quality of the fitness and the interval of confidence for estimated values.

Figure 3b shows the results of the Kolmogorv-Smirnov fitness test for the three models for Processor A. In this test, the closer the curve of the model to the diagonal, the better the fitness is. A model whose curve is inside the 10% significance lines is considered to have a good fit. The figure shows that only the curve for the Musa Okumoto model is inside that region, and thus not only is it the most fitting model, but it is also the only one with a good fit.

Table 3 compares the results of the Kolmogorv-Smirnov fitness test for the three models for Processor B in two cases. The first case is when all the bug discovery data is considered, and the second case is when only the data after 500 million cycles is considered. The table illustrates the importance of selecting the correct data for analysis. The complete bug discovery data for Processor B contains two regions, as Figure 2b shows, one before the major feature was activated, and one after it. If we consider both regions together, then none of the models has a good fit to the data but, if only the second region is considered, the fitness of all the models improves, and 2 out of the 3 models have a good fitness.

The Kolmogorv-Smirnov fitness test is not the only possible test of the fitness of models; there are many other tests that can be used to test the fitness. Since the goal of modeling is to provide predictions about future behavior, many such tests involve checking the quality of predictions made by the models. The most basic test is to compare the predicted median time to the next bug discovery, with the measured time. Other fitness tests include long-term prediction

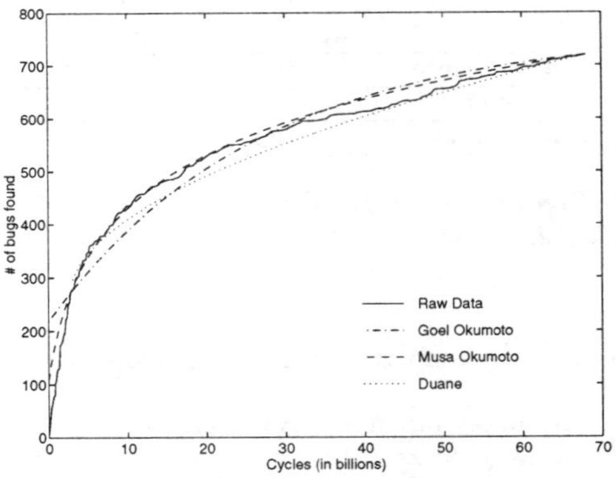

(a) Models Estimation

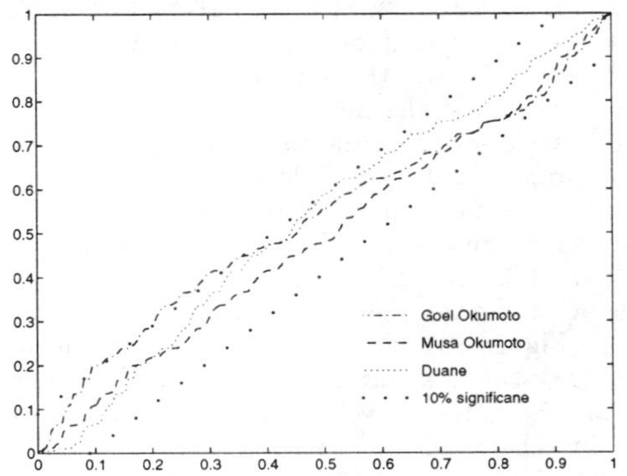

(b) Estimation fitness

Figure 3: Model estimation and estimation fitness for Processor A

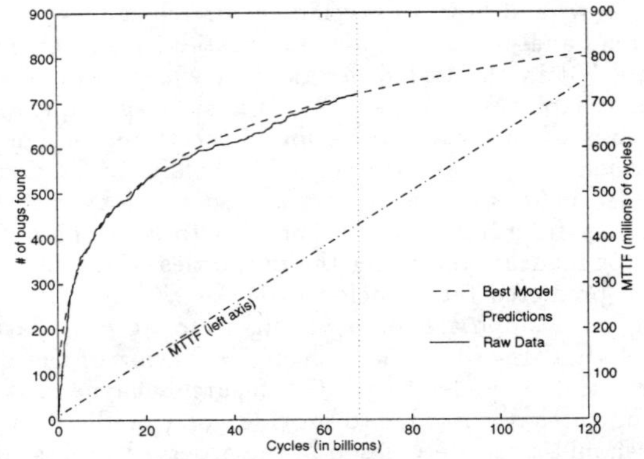

(a) Processor A

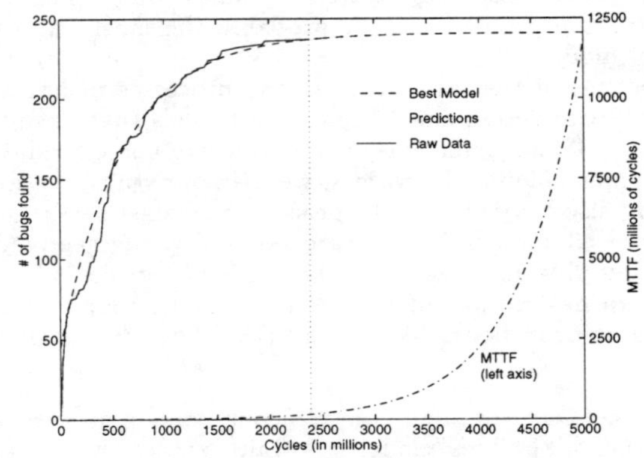

(b) Processor B

Figure 4: Long term predictions

fitness and stability of the parameters of the model over time [9].

After all the desired fitness tests are made, the models can be ranked according to their performance in the tests, and the model with the highest ranking should be the one used for future predictions. There are several possible criteria according to which the ranking can be done, each giving different weights to different fitness tests. After the best model is chosen, provided that its absolute fitness is good enough, it can be used to predict the future behavior of the tested design as regards the design's reliability.

Figure 4 shows the predictions for MTTF (mean time to failure) and the number of bugs discovered for both processors, using the best fitting models. The best fitting model for Processor A is the Musa Okumoto model, and the best fitting model for Processor B is the Goel Okumoto model. The figure shows that there is a big difference between the predictions of future behavior for the two processors. For Processor A, the best fitting model is an infinite model, which predicts a continuous increase in the number of bugs found and linear growth in the MTTF. The prediction for Processor B, on the other hand, is for a finite number of bugs and an exponential growth in the MTTF.

	All data	After 500M cycles
Goel-Okumoto	0.169	*0.104*
Duane	0.169	0.226
Musa-Okumoto	0.126	0.155
10% Significance	0.089	0.157

Table 3: Model estimation fitness for Processor B

We compared the total number of bugs and the MTTF for Processor A, as predicted by the Musa Okumoto model with the actual bug discovery data taken three months after the end of the study period. At the time of the comparison, a total of 105 billion cycles of simulation were executed and 794 bugs were found (that is 37 billion more cycles and 75 more bugs since the end of the study). The comparison results are summarized in Table 4. The table shows that there is a small difference of less than 7% in both the prediction of the MTTF and the number of bugs found since the prediction, despite the fact that the prediction was made 3 months in advance.

4 Conclusions

In this paper we described the results of statistical analysis of bug discovery data which we performed on two state-of-the-art PowerPC processors. The analysis we performed is similar to the analysis that is done in software reliability engineering. The results of the analysis enable us to detect the

	Musa-Okumoto Prediction	Actual	Difference [%]
Total Number of bugs found	789.6	794	0.4
Number of bugs since prediction	70.6	75	4.0
MTTF [$\times 10^6$ cycles]	646	620	6.7

Table 4: Comparison of Musa Okumoto long term predictions and actual data for Processor A

effects of the introduction of a major feature in one processor on its reliability growth, and to predict accurately the number of bugs and the MTTF in the other processor.

The conclusion we came to from our study is that statistical analysis of bug discovery data, in general, and techniques used in software reliability engineering, in particular, can and should be used to analyze bug discovery data during hardware verification, and that such an analysis can provide useful information on the state of the verification process and the reliability of the tested design.

This type of analysis is very useful in determining optimal release dates for processors, since it is capable of providing an estimation of the post-release bug discovery rate. This information, combined with data on the cost of such bugs and the cost of a delayed release, can be used to determine the optimal time to release the processor.

The cost of the statistical analysis is very low. Only accurate data on the times in which bugs were found and the amount of testing done are needed. The analysis itself, although based on sophisticated statistical concepts, can be done using one of several commercially available tools, such as CASRE [10], and SMERFS [7].

There are several open questions regarding the use of statistical analysis of bug discovery data in hardware design verification. First, we need to find out which models are applicable to the analysis. In our study we found out that some models used in SRE are also applicable to hardware. The question is whether all SRE models are applicable to hardware design and whether there are new models that are applicable to hardware design, but not to software.

A second interesting issue is the relationship between testing environment and real environment. In many cases, the reliability prediction is required in terms of real operation. Since statistical analysis is done in a testing environment, it can provide predictions only in this environment. Unlike software testing, which is done in conditions similar to real operation of the system, hardware testing is done in a much more stressful environment. Therefore, one cycle in testing is equal to many cycles in real operation. It is interesting to investigate the relationship between testing cycles and real cycles, and how the testing environment affects this relationship.

References

[1] A. Aharon, D. Goodman, M. Levinger, Y Lichtenstein, Y. Malka, C. Metzger, M. Molcho, and G. Shurek. Test program generation for functional verification of PowerPC processors in IBM. In *Proceedings of the 32nd Design Automation Conference*, pages 279–285, June 1995.

[2] A.M. Ahi, G.D. Burroughs, A.B. Gore, S.W. LaMar, C.R. Linand, and A.L. Wieman. Design verification of the HP9000 series 700 PA-RISC workstations. *Hewlett-Packard Journal*, 14(8), August 1992.

[3] B. Beizer. The Pentium bug, an industry watershed. *Testing Techniques Newsletter, On-Line Edition*, September 1995.

[4] D.R. Cox and P.A.W. Lewis. *The Statistical Analysis of a Series of Events*. Chapman and Hall, 1978.

[5] M.H. DeGroot. *Probability and Statistics*. Series in Statistics. Addison Wesley, 1986.

[6] J.T. Duane. Learning curve approach to reliability monitoring. *IEEE Transaction on Aerospace*, 2:563–566, 1964.

[7] W.H. Farr and O.D. Smith. Statistical modeling and estimation of reliability functions for software (SMERFS) user's guide. TR-84-373, revision 1, NSWC, 1988.

[8] A.L. Goel and K. Okumoto. Time-dependent error-detection rate model for software and other performance measure. *IEEE Transactions on Software Engineering*, 11(12):285–306, December 1985.

[9] M.R. Lyu. *The Handbook of Software Reliability Engineering*. McGraw Hill, 1996.

[10] M.R. Lyu and A.P. Nikora. CASRE: A computer-aided software reliability estimation tool. In *Proceedings of the 5th International Workshop on Computer-Aided Software Engineering*, pages 264–275, July 1992.

[11] J. Monaco, D. Holloway, and R. Raina. Functional verification methodology for the PowerPC 604(TM) microprocessor. In *Proceedings of the 33rd Design Automation Conference*, pages 319–324, June 1996.

[12] J.D. Musa. Software-reliability-engineered testing. *Computer*, 29(11):61–68, November 1996.

[13] J.D. Musa, A. Iannino, and K Okumoto. *Software Reliability: Measurement, Prediction, Application*. McGraw-Hill, 1987.

[14] J.D. Musa and K. Okumoto. A logarithmic Poisson execution time model for software reliability measurement. In *Proceedings of the Seventh International Conference on Software Engineering*, pages 230–238, March 1984.

[15] A. Wood. Predicting software reliability. *Computer*, 29(11):69–77, November 1996.

Functional Verification of Large ASICs

Adrian Evans, Allan Silburt, Gary Vrckovnik, Thane Brown,
Mario Dufresne, Geoffrey Hall, Tung Ho, Ying Liu
Nortel
P.O. Box 3511, Station C, Ottawa, Ontario, K1Y-4H7, Canada
1-613-{763-6328,763-7975,763-4365,763-5482,763-7898,763-9729,765-5007,763-6358}
{abe, silburt, garyvrck, thane, dufresne, hallg, tungho, yingliu}@nortel.ca

1. ABSTRACT

This paper describes the functional verification effort during a specific hardware development program that included three of the largest ASICs designed at Nortel. These devices marked a transition point in methodology as verification took front and centre on the critical path of the ASIC schedule. Both the simulation and emulation strategies are presented.

The simulation methodology introduced new techniques such as ASIC sub-system level behavioural modeling, large multi-chip simulations, and random pattern simulations. The emulation strategy was based on a plan that consisted of integrating parts of the real software on the emulated system. This paper describes how these technologies were deployed, analyzes the bugs that were found and highlights the bottlenecks in functional verification as systems become more complex.

1.1 Keywords

ASIC Verification, Simulation, Emulation

2. INTRODUCTION

This paper describes the ASIC functional verification techniques that were deployed on a recent large hardware development program. It details the simulation and emulation methodologies and provides an in-depth analysis of the bugs that were found with both technologies. Finally, some of the bottlenecks of each technology are presented with ideas on how to overcome them in the future.

The project consisted of three new circuit boards connected via a high speed backplane using a proprietary protocol. A full system consists of 26 boards. Various combinations of the three ASICs reside on each of the circuit boards. These ASICs interface with high speed memories and powerPC microprocessors. For the purposes of this paper, we will refer to the chips as Alpha, Beta and Gamma. They were fabricated in a 0.25 micron process and had gate counts of 482K gates, 824K gates and 635K gates, respectively.

3. Verification Methodology

The simulation methodology shown in figure 1, consisted of writing a behavioural model (BEH) for each of the three devices. A conformance test-plan (CTP) document was written for each ASIC, enumerating all the features and functions that required verification. Based on this document, a regressionable testbench suite was written for each ASIC. Initially, the testbenches were developed using the behavioural model and then as the RTL stabilized, both platforms were used for writing testbenches. [1]

Near the end of the ASIC testbench effort, system testbench development began in parallel using the behavioural models. The objective of these simulations was to prove the interfaces between ASICs and verify the protocol error conditions. Only functionality that required multiple ASICs to prove correct operation was simulated at this level.

As the directed ASIC and system simulation efforts completed, the TLSC (Top Level Simulation Complete) milestone was declared; however the bug find rates remained high. It was apparent that the feature driven test-plan had not sufficiently verified interactions when multiple features were enabled in combinations. To address these holes, an Extended Simulation Plan (ESP) was developed in order to create random and multi-feature tests and to target those lines of code that had not been simulated based on code-coverage reports. These tests proved extremely valuable in finding obscure bugs.

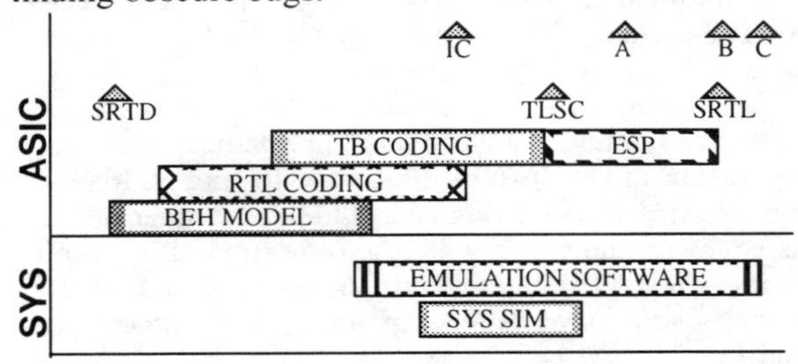

Legend :
IC=In Circuit (first functional hardware in emulation lab)
A,B,C=Emulation Software Milestones (section 3.2)
SRTD=Start of Design TLSC=Top Level Simulation Complete
SRTL=Start of Layout ESP=Extended Simulation

Figure 1. ASIC Development Gantt Chart

During the TLSC to SRTL (start of layout) interval, timing analysis, floorplanning, DFT and synthesis were completed.

3.1 Behavioural Model

The ASIC behavioural(BEH) models were written in Verilog, were pin compatible with the RTL model and implemented nearly all of the device functionality. They were written at a high level of abstraction, manipulating entire transactions rather than individual signals, and thus required less effort to write and simulated faster than the RTL. As the latency through the behavioural model was not the same as that of the RTL implementation, testbenches had to be tolerant of both sets of timing.

Early in the project, it was unclear how manageable full RTL level simulations of a single ASIC would be. In addition, considering the size of the design teams and the tight schedule, there was concern over the amount of time it might take to integrate the partially tested RTL blocks at the ASIC top level. Therefore, the behavioural models were written at the ASIC sub-system level. That is, the behavioural model was structurally decomposed into the same sub-systems as the RTL code for the ASIC. This approach enabled the possibility of mixed BEH / RTL chip level simulations. By modeling at the sub-system level, the verification team helped to formalize the internal interface specifications and maintain adherence to them. Finally, by modeling at this level, the verification team was able to deliver a ready-made test environment to each sub-system designer. These benefits enabled top level RTL integration times of between a day to a week from the first attempts at ASIC level simulations to successful testbench runs.

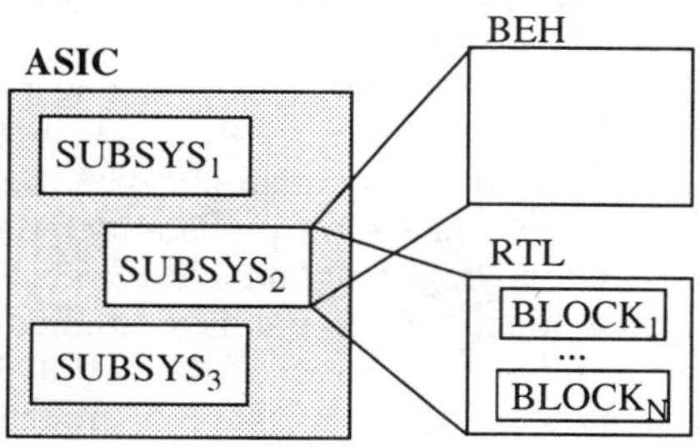

Figure 2. Sub-System Level Modeling

The choice to model at a lower level of hierarchy meant that the behavioural model simulation speed gain was limited by the additional simulation events from the internal interfaces. Furthermore, this approach necessitated additional work in order to track changes in the internal interfaces.

Other benefits of the behavioural model were that it found discrepancies between the RTL and the specification, as it provided an alternative view of the specification. Several times, a testbench initially developed on the RTL failed on the behavioural model due to an error in the RTL that was duplicated in the testbench; these would otherwise have gone undetected in simulation. The most important benefit of the model, however, was that it allowed the verification teams to become thoroughly familiar with the functionality of the devices very early in the project.

3.2 Emulation Methodology

The emulation system implemented 2.4 Million ASIC gates in a total system configuration of 5 circuit boards and multiple instances of each device. Each ASIC was reduced in size from the original by removing duplicated logic, non-mission mode logic, and in the case of the Beta, by synthesizing with area constraints and relaxed timing. The emulation elements were filled to capacities of between 68% and 97%, with most (60%) being filled to greater than 90% capacity. In some cases, this high utilization made it difficult to insert probes for debugging problems.

Production quality firmware was used on the circuit cards. Virtually all of this firmware is being reused in the production system. The firmware staff noted that the emulation environment was the most stable prototype platform available to them for developing their code.

Special high level software was written for the emulation platform in order to prove the hardware functionality. The software was rolled out in three phases (see figure 1) :

A. major features of the ASICs were tested and a software development environment was established.

B. every feature of each ASIC was tested in a stand alone fashion.

C. every feature of each ASIC was tested repeatedly and in random combinations over long durations.

The requirements specification of the software tests was driven primarily by the ASIC teams. Approximately 40% of the high level software that was developed is considered production code that will be reused in the real system. In addition, the software designers reported that several design years worth of experience on the system was gained and that in later development the most productive software designers were those that had worked with the emulation system.

Overall, the total number of netlist revisions that was emulated was 32 Alpha, 18 Beta and 16 Gamma. New netlists were compiled as functionality was added or bug fixes were addressed in the design. The average time to process each netlist (fully compile and go in-circuit) varied from 3 weeks with the first full netlists, to 12 hours with the final full netlists.

4. Project Retrospective

The graph in figure 3 shows a breakdown of the effort spent between the start of design (SRTD) and start of layout (SRTL) milestones, averaged across all three devices.

The lower sections represent tasks not associated with functional verification, the remainder is functional verification. The slice "RTL and block test" in figure 3 includes block and sub-system level design and testbench effort. At this level of hiearchy, it was not possible to separate these two tasks. In effect a portion of this slice could be grouped with verification. From figure 3, it is clear that functional verification represents well over 50% of the effort from SRTD to SRTL.

It is interesting to note that on the Alpha and Beta ASICs, functional verification was on the critical path. Timing clean and floor-planned netlists were available well in advance of their actual SRTL dates. In other words, the tape out was delayed in order to find and address functional bugs. As devices become more complex, functional verification is increasingly on the critical path[2,3].

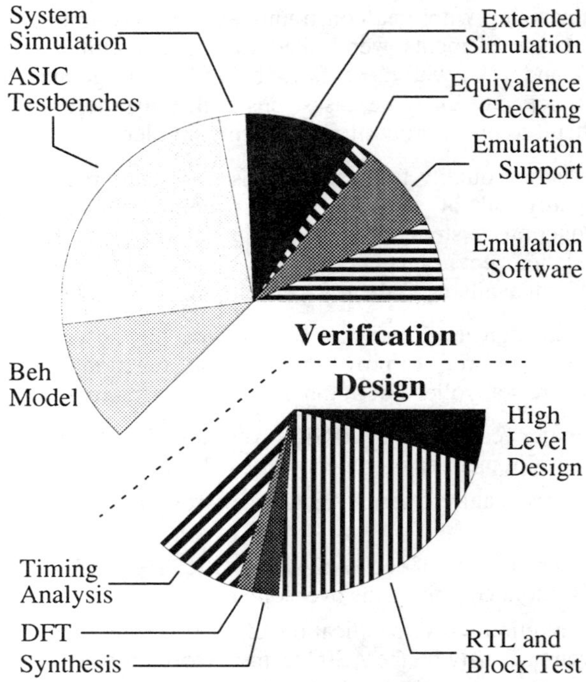

Figure 3. Breakdown of Effort

5. Bug Analysis

Accurate bug information was kept for emulation bugs from the beginning of the project and for simulation bugs from TLSC onwards. Prior to TLSC, it is difficult to separate bugs from code development and evolution. Each bug was classified by severity according to the effect if the bug had been left unfound in the actual silicon:

minor a simple software or board fix exists to workaround the bug

respin the chip could be used for lab work and initial software development with the bug present, but would have to be respun prior to shipping the product to the customer

DOA (dead on arrival) the device would be of limited use in the lab

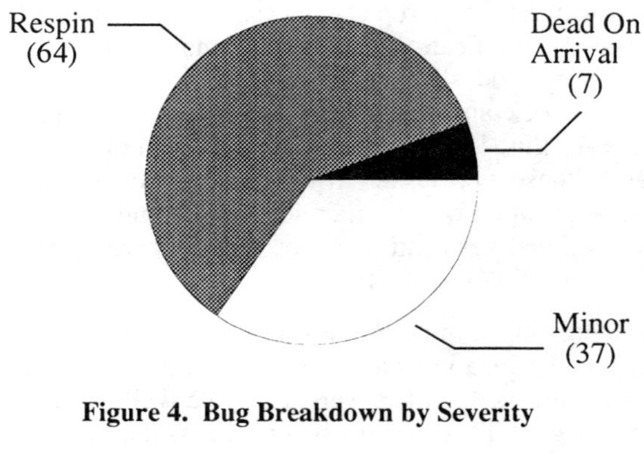

Figure 4. Bug Breakdown by Severity

Figure 4 shows a break-down of the bugs by severity. It is interesting to note that only a small number of bugs fit into the DOA category. The last DOA bug was found three months before the SRTL of that device

The designers were asked to rank the probability that each bug could have been found by an alternate technology on a scale of high, medium or low. The results are summarized in tables 1 and 2. Note that two bugs were found almost simultaneously by both simulation and emulation, and have thus been included under both categories.

5.1 Emulation Bugs

In table 1, we summarize the severity and probability of being found by simulation for each of the emulation bugs :

	Severity			Chance of Finding in Simulation		
	DOA	Respin	Minor	Hi	Med	Low
Alpha	0	18	6	8	13	3
Beta	3	6	4	5	4	4
Gamma	1	4	5	3	3	4

Table 1: Emulation Bug Classification

Overall, 47 bugs were found through emulation (24 Alpha, 13 Beta, 10 Gamma). The severity of the emulation bugs was classified as 4 DOA, 28 respin and 15 being minor problems. Of the emulation bugs, 34% had a high probability of being found in simulation, 43% had a medium probability and 23% had a low probability of being found by simulation. Of the bugs that had a low probability of being found in simulation, 55% would have resulted in a respin and 45% were minor. All three ASICs had at least one bug in the category: respin severity and low probability of finding it with simulation.

5.2 Simulation Bugs

In table 2, we summarize the severity and probability of being found by emulation for each of the simulation bugs.

	Severity			Chance of Finding in Emulation		
	DOA	Respin	Minor	Hi	Med	Low
Alpha	0	4	2	0	3	3
Beta	2	12	7	6	6	9
Gamma	1	22	13	2	12	22

Table 2: Simulation Bug Classification

Overall, 63 bugs were found by post-TLSC simulation (6 Alpha, 21 Beta, 36 Gamma). The severity of these simulation bugs was classified as 3 DOA, 38 respin and 22 minor. Of the simulation bugs, 13% had a high probability of being found in emulation, 33% had a medium probability and 54% had a low probability of being found by emulation.

Of the bugs that had a low probability of being found in emulation, 53% were respin severity and 3% were DOA severity. All three ASICs had at least one bug in the category : respin or DOA severity and low probability of finding it with emulation.

5.3 Coverage Overlap

In comparing the three devices, it is interesting to note the predominance of bugs found by emulation on Alpha and by simulation on Gamma. This is due to the fact that Alpha completed functional design several months ahead of the other two devices. A conscious decision was made to rely on early emulation, rather than extend the simulation effort on Alpha. Gamma, on the other hand, had the most simulation effort. As seen in tables 1 and 2, there is a strong correlation between where the effort was invested and where the bugs were found.

Simulation and emulation both have certain characteristics that make them better suited to finding certain types of bugs. These characteristics are summarized in table 3 :

Emulation	Simulation
Greater throughput	Precise event timing control
Gate level representation	Greater observability
Software Stimulus	Greater control of the environment
Board netlists and board environment	Runs at true clock frequency

Table 3: Characteristics of Emulation and Simulation

From the low probability slices, in figure 5, it is clear that there are bugs that have a low chance of being found with only one verification technology.

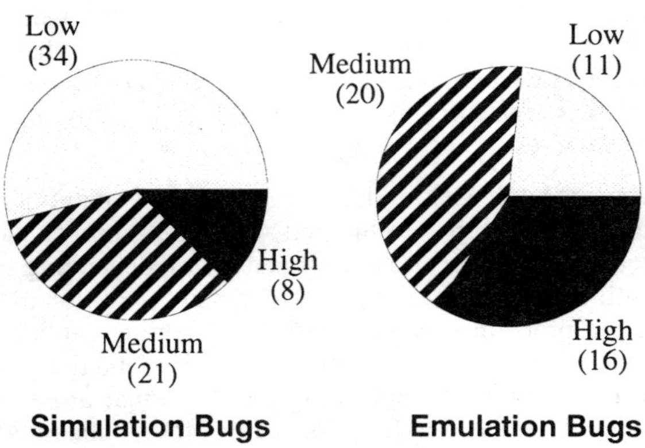

Figure 5. Alternate Bug Find Probability

Some of the bugs that are difficult to find in emulation are related to external memory timing violations under error cases. Such violations would not be detected in emulation unless they actually corrupt memory contents, which is unlikely with the reduced clock speed. Others are related to the fact that only a partial system is being emulated; thus certain traffic scenarios are not possible. Still others relate to scenarios that are difficult to setup from high level software.

The throughput afforded by emulation can find bugs that would require extremely long simulations. Since the emulation system used actual circuit boards, system level interconnect problems were found. Additionally, software usage of the hardware exercised the chips in manners the hardware verification team had not considered.

In table 4, we have classified the bugs found by emulation both by that attribute of emulation which enabled the bug to be found as well as by the probability that it could have been found by simulation. Those bugs in the right hand column are especially significant as they were very unlikely to have been found by simulation. There was one noteworthy bug in the "Board Environment" row, in the low column, which was not found earlier in simulation because the electrical characteristics of a certain bus were not correctly modeled in the simulation environment. The bug in the "Gate Level" row with low probability of being found in simulation relates to a problem in the re-synchronization of signals across an asynchronous boundary. This bug manifested itself in emulation, only during long traffic soaks. Interestingly, there are a large number of bugs in the "Richness" row related to richness of stimulus afforded by real software running on the emulation platform.

	Total	**Hi**	**Med**	**Low**
Throughput	6	0	5	1
Richness	16	7	9	0
Gate Level	7	4	2	1
Board Environment	7	3	1	3
Real S/W	9	2	3	4
Other	2	0	0	2

Table 4: Classification of Emulation Bugs

6. Bottlenecks

In order to compress the ASIC verification interval, it is important to understand where the true bottlenecks lie.

6.1 Simulation Time

There are two very different areas where simulation time impacts productivity : testbench development and running the regressions. Table 5 captures some simulation speed data for Gamma, which is also typical for the other devices. All simulations were done in Verilog and the data is averaged for Sparc-20 and Sparc Ultra-1 class machines :

	RTL	BEH	BEH (virtual i/o)
Average ASIC TB Compile/Link/Run Time	2 hr	52 min	12 min
Total Regression Suite Run Time[a]	8 days	14 hr	

a.Behavioural regression uses a mixture of normal and virtual interfaces.

Table 5: Simulation Speed Data

A behavioral modeling technique that was successfully applied was one which we call "virtual interfacing". When modeling a device that receives frames serially, it is possible to bypass the physical interface by transferring entire frames directly into the model with a single simulator event. Using this technique on the high speed interfaces, eliminated the simulation events associated with the high speed clocks and resulted in significant simulation speedups (right hand column in table 5). For certain tests involving the physical aspects of the interface, this technique was disabled. The behavioural regression used a mixture of real and virtual interfaces.

From the data in table 5, several observations can be made. The compile/run time for a RTL simulation is on the order of a few hours and with a behavioural model with virtual interfaces this is ten times faster. The typical ASIC testbench consisted of between two and three independent tests that could generally be developed in isolation. Thus with the behavioural model with virtual interfaces, the average simulation iteration time was approximately 5.5 minutes. With RTL this interval was 50 minutes. Therefore, simulation times are manageable even for extremely large devices, especially with behavioural models.

The second aspect of simulation consists of regularly running the regression suite during the later phases of the design cycle. The time required to run the regression is limited by two factors : the number of workstations on which to distribute the jobs and the duration of the longest testbench. Given this, a complete regression for any of the devices could essentially be run over a weekend, if spread across about 10 workstations. Here again, there isn't a real bottleneck : given sufficient workstations, simulator licenses and judicious sizing of the testbenches, the regression interval can be managed.

6.2 Testbench Development

The real bottleneck in functional verification was the development and debugging of the testbenches. From figure 3, we see that the ASIC, system and extended simulation testbenches alone represented over a third of the total effort. Testbench coding can not begin until a platform for writing and debugging testbenches is in place. In this case the platform was initially the behavioural model. In order to compress the duration of the testbench coding interval, a large team size was applied to the task. Figure 6 shows the testbench team sizes for Beta and Gamma over the course of the project :

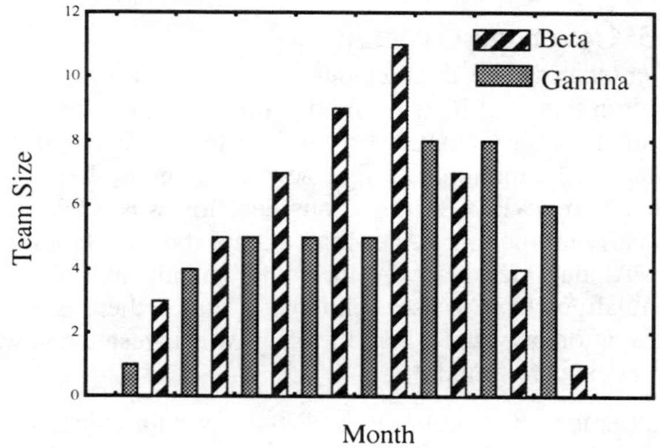

Figure 6. Testbench Team Size

Ramping-up the necessary people was a challenge. The testbench coding team consisted of behavioural modelers, dedicated testbench coders, RTL designers and contractors. For people joining the project as testbench coders, reading the ASIC specifications and learning about the system operation required from several weeks to over a month. New people often required help when debugging obscure problems, whether they lay in the testbench, behavioural or RTL code. We found that a dedicated verification team that started by writing behavioral models was effective in enabling a large, productive team to be built.

6.3 Emulation Bottlenecks

The emulation process can be broken up into three distinct phases : going in-circuit, debugging and recompiling the netlists.

Going in-circuit for the first time requires a target board to be manufactured and the first in-circuit netlist needs to be compiled. Each target board required about two person months of effort to produce. Each netlist took about three weeks to compile the first time. Once the netlist was compiled, an interconnection from the emulator to the target board was created. This typically took about one week to design and test.

The major bottleneck in debugging was the amount of time it took to add probes to the netlists. Adding probes would typically take between an hour and a day depending upon the utilization of the emulation hardware. To completely debug a problem several iterations of adding probes was often required. Another bottleneck in the debugging cycle was the difficulty of replicating problems that arose. This was due to the need to have identical software loads, ASIC netlists and starting conditions.

After the first compile of a netlist the emulation team could start writing scripts to automate the compile process. With each compile of the ASIC, the scripts were improved until compiles were being completed in under 12 hours. Since the compiles were done over the network, LAN traffic had a major effect on the time to compile an ASIC. With no network traffic the compiles could be completed in under 6 hours.

Another major bottleneck in emulation is the number of platforms available for development and debugging. In this project, software development, firmware development, bug tracing and lab issues all competed for the same resource. Had multiple platforms been available, then these tasks could have been parallelized and overall efficiency would have improved. However, the high cost of emulation equipment makes replicating the platform impractical. This remains one of the most serious impediments to productivity improvement using emulation.

7. Looking Forward

7.1 Simulation

We observe that the dominant activities in testbench development are the intellectual process of understanding the device behaviour, the feature to be tested and interpreting the results of simulation runs. As a result, the opportunity for productivity improvement through faster simulation or coding productivity tools is limited. Faster simulators are certainly welcome and will provide an incremental improvement in productivity, but they will not drastically reduce the overall verification effort.

Verification productivity tools largely address shortcomings in HDLs and facilitate the development of random tests. These tools may provide incremental improvements in testbench coding productivity, however, they do not address the issue of ramping-up a large verification team who understand the system.

In order to address the large testbench team problem, two things can be done. First, these people should be involved in the project as early as possible in order to learn about the system, either as behavioural modelers, RTL designers, firmware developers, etc. Second, good teaching material must be created in order to help new people understand the system being designed. This can be achieved through explanatory documentation, tutorials, walk-through documents or debugging guides.

Model Checking formal verification tools may provide a productivity gain for block level verification where the tools can handle the state space. It is unlikely that model checking will be applicable at the ASIC or system level on a project of this size, because of the state explosion problem. From isolated tests we have found this technology has the potential to find obscure bugs in communicating state machines early in the design cycle.

7.2 Emulation

The increased capacities of modules in newer emulation systems should decrease the time to compile a netlist for the first time, as partitioning will be much easier. Newer emulation systems are also evolving towards 100% visibility of all netlist nodes. This should greatly improve the debug cycle because probing will no longer require recompilation Finally, compile effort with the newer systems is also reduced, and faster workstations should allow for faster recompiles.

Given these improvements, software development and hardware debug, will be limited by the number of emulation platforms available prior to tape-out. Without a significant reduction in cost, the bottleneck associated with sharing a single platform between multiple users will remain an issue in using this technology.

8. Conclusions

Several key observations emerged during the course of the design of the ASICs. First, was that functional verification represents the largest single task in the design to layout interval. Second, based on the high bug find rates after the TLSC milestone, that a targeted, feature driven test plan is not sufficient when complex feature interactions are present. The directed tests must be supplemented with extensive random and heavy traffic tests.

We observed that there is value in emulating as early as possible, but the system should not be handed over to software staff until high quality netlists are available and all known lab issues have been resolved. With increased simulation coverage, the number of bugs that will be found in emulation is reduced. However, with devices of the complexity of those on this project, it appears that regardless of how much simulation is done, there is a significant probability that some respin severity bugs will remain, some of which can be found through emulation. Real software interacting with the ASICs generates a richness of stimulus that is difficult to achieve in simulation.

Finally, there is no revolutionary new verification technology on the horizon. Simulation will remain the workhorse for functional verification for the foreseeable future. Faster simulators will enable simulation to track ASIC growth, but will not significantly reduce the overall effort devoted to functional verification.

9. ACKNOWLEDGEMENTS

Our thanks go out to the entire project team, especially Joël Bolduc, David Wheelock and Randy Mullin, for the information they shared to make this paper possible.

10. REFERENCES

[1] Silburt, A., Perryman, I., Bergeron, J., Nichols, S., Dufresne, M., Ward, G., "Accelerating Concurrent Hardware Design with Behavioural Modelling and System Simulation", Design Automation Conference 1995.

[2] Miklosz, John, "Meeting the Crisis in Confidence That Complexity Causes", Integrated Systems Design, August 1997.

[3] Silburt, A., Evans, A., Vrckovnik, G., Dufresne, M., Brown, T., "Functional Verification of ASICs in Silicon-Intensive Systems", DesignCon98, January 1998, Santa Clara, California.

Panel: The EDA Start-up Experience: The First Product

Chair: Erach Desai - SoundView Financial Group, Inc., Stamford, Ct.

Organizer: Mike Murray - Acuson Corporation, Mountain View, Ca.

Panelists:

Rick Carlson - Synplicity, Inc., Sunnyvale, Ca.

Lorne Cooper - Sente, Inc., Acton, Ma.

Dean Drako - Design Acceleration, Inc., San Jose, Ca.

Rajeev Madhavan - Magma Design Automation, Inc., Palo Alto, Ca.

John Sanguinetti - Modellogic, Inc., Los Altos, Ca.

Curt Widdoes - 0-In Design Automation, Inc., San Jose, Ca.

Abstract

How does a novel EDA idea get transformed into a commercially successful product? Six veteran EDA entrepreneurs will discuss their experiences in bringing their companies' first products to market.Where did their ideas come from? How did they know their ideas would meet real customer needs? And how many customers would there be? How are evolutionary and revolutionary products developed and marketed differently? When did the entrepreneurs stop developing and start shipping? Who were their competitors and their partners? Did they adopt industry standards or create new ones? How did they use advertising, DAC, and the WWW to promote their products? What are the relative merits of direct, VAR, and OEM selling?

Position Statements

Rick Carlson, Synplicity

I have been involved in several EDA start-ups since 1982 (Mentor Graphics, Daisy Systems, EDA Systems, Escalade, and Synplicity) as a sales manager or sales executive, and I have found that the companies that were successful had a number of key ingredients that made their first products winners. These ingredients were:

1) Compelling technology that can be built into multiple products
2) Dynamic market
3) Strong, well-rounded management team that can work together
4) Well thought out sales and marketing strategy
5) Unrelenting desire to listen hard, tell the truth, and be flexible
6) An empowering Corporate Philosophy

In various personal experiences when too many of the above items are not in place the degree of success is variably mitigated.

Dean Drako, Design Acceleration

There are several ingredients to launching a successful EDA company:

1) solve a real customer problem that no one else is solving
2) stay focussed on solving the customer's problem
3) assemble the best team you can.

Customers have problems. To successfully sell a product you need to solve a REAL and specific problem for them and one that does not have another solution. Too many company ideas attempt to solve problems that are either only a nuisance to a customer, are non-existent in the customer's eyes, or already have a reasonable solution. It is hard to get customers to pay money for items like this. At DAI we have targeted solving the Waveform Viewing and RTL Debugging problem. When we entered the market we were the only supplier addressing this customer need. It was and remains a real customer need.

Once you understand your customer and the problem you are trying to solve, you need to stay focussed on solving that problem for that customer. You will encounter many other ideas and many other types of customers, but you need to resist the temptation to serve them. Specialization and expertise will eventually win in the market place. At DAI we have stayed focussed on providing the most powerful and complete Waveform Viewing and Source Code Debugging system. We continue to stay focussed on this. We continue to strive to keep our customers happy in this area. You cannot falter in your belief that solving your customer's problem well will make the company successful.

The team is extremely important. You need to assemble the strongest engineering, sales, marketing, and management team that you can. The road to success is extremely hard and only the best teams make it. Any weak link in the team will make the challenge even higher. At DAI we established a tremendous team early in the company.

Rajeev Madhavan, Magma Design Automation

EDA start-ups are successful when a product is created to address a compelling customer need. The company's potential financial success is directly related to the size of the customer need.

The company's focus on this customer need must be singular, directed, and complete. Start-ups need to focus on only one initial product in their core competency. A common start-up mistake is to attempt to financially bootstrap the company with consulting or smaller products. These efforts may provide needed cash but can seriously defocus the company away from its main product.

A start-ups real assets are its people: employees and potential customers. A crucial factor is recruiting an experienced capable team where each member has a clear purpose. Many start-ups fail to spend enough time understanding all of the detailed customer requirements. Successful EDA ventures use innovative approaches to address customer needs.

John Sanguinetti, Modellogic

The key ingredient of launching a successful EDA product is customers.

Having a particular type of customer in mind, and a particular customer if possible, and knowing what their needs are is the key. In my case, the original customer prototype was myself, since I had been a design verification engineer and used Verilog for regression testing. Very early on, we identified a particular customer, Sun, to be our target customer. We figured that if we made Sun happy, we would make other people happy, too. This turned out to be true.

We also identified the problem we were solving - simulation speed. We focused almost entirely on that, from company slogan (The Fast Verilog Company), to advertising, to customer benchmarks. The acceptance criterion for our product in competitive benchmarks was always "how much faster is it than the competition". This focus was as useful internally in making design decisions as it was externally in positioning the company and product against the competition.

If there is anything that can be generalized from Chronologic's experience it is the value of a single focus on a real customer problem.

Curt Widdoes, 0-In Design Automation

I have founded three successful EDA start-ups: Valid Logic, Logic Modeling, and 0-In, so I have learned quite a bit about what it takes to make one succeed.

To be a success in EDA today, a start-up needs a franchise product. A franchise product is one that has high value, is highly differentiated, and is difficult for competitors to duplicate. A franchise product can be priced to reflect its value and can support the growth of a new company. Dracula, Verilog, and Design Compiler were franchise products.

It is important for a new company to maintain its initial franchise long enough to build a strong channel. In this sense, EDA is a friendly environment, because the limited size of the market makes it relatively easy to see potential competition and to plan from the beginning to steer clear of it. These days in EDA, we no longer have multiple piles of venture capital all chasing the same idea, the large EDA companies have their plates full taking care of legacy product lines, and non-EDA companies with lots of capital find our market too difficult and too small. Even better, our customers always have tremendous needs for new products and technologies; our industry has continuously been behind the needs of its customers. As a result, there are many opportunities for wonderful new franchise products.

In EDA today, the first product must complement existing design methodology rather than attempt to replace it. There have been a number of "vision" start-ups in EDA, which attempted to convince customers to adopt major methodology shifts. These start-ups ran out of capital. Once a start-up's initial complementary product is successfully deployed, then its customers may be willing to try some incremental changes in methodology in order to increase the benefit derived from the new product. Over time, an EDA company can cause a sizable methodology shift in this way, and in doing so can provide substantial value to its customer base.

In the twenty years that I have been starting EDA companies, I have learned a couple of rules of thumb which probably transcend our industry. First, make sure that your customers want to buy the initial product. The most effective way to do this is to ask them.In EDA, at least, it is so difficult for the customers to get what they need that they are eager to provide guidance. Second, make absolutely no compromises in the quality of the team. This is, without any question, the most important rule. Every young company will encounter some problems; a good team will eventually figure out how to solve them.

Digital System Simulation: Methodologies and Examples

Kunle Olukotun, Mark Heinrich and David Ofelt

Computer Systems Laboratory
Stanford University
Stanford, CA 94305-4070

1 Introduction

Two major trends in the digital design industry are the increase in system complexity and the increasing importance of short design times. The rise in design complexity is motivated by consumer demand for higher performance products as well as increases in integration density which allow more functionality to be placed on a single chip. A consequence of this rise in complexity is a significant increase in the amount of simulation required to design digital systems. Simulation time typically scales as the square of the increase in system complexity [4]. Short design times are important because once a design has been conceived there is a limited time window in which to bring the system to market while its performance is competitive.

Simulation serves many purposes during the design cycle of a digital system. In the early stages of design, high-level simulation is used for performance prediction and analysis. In the middle of the design cycle, simulation is used to develop the software algorithms and refine the hardware. In the later stages of design, simulation is used make sure performance targets are reached and to verify the correctness of the hardware and software. The different simulation objectives require varying levels of modeling detail. To keep design time to a minimum, it is critical to structure the simulation environment to make it possible to trade-off simulation performance for model detail in a flexible manner that allows concurrent hardware and software development.

In this paper we describe the different simulation methodologies for developing complex digital systems, and give examples of one such simulation environment. The rest of this paper is organized as follows. In Section 2 we describe and classify the various simulation methodologies that are used in digital system design and describe how they are used in the various stages of the design cycle. In Section 3 we provide examples of the methodologies. We describe a sophisticated simulation environment used to develop a large ASIC for the Stanford FLASH multiprocessor.

2 Simulation Methodologies

2.1 Design Level

During the design cycle of an ASIC or processor that is a part of a complex digital system a variety of simulators are developed. These simulators model the design at different levels of abstraction and typically become slower as the design progresses to lower levels of abstraction.

DAC 98, San Francisco, California
©1998 ACM 0-89791-964-5/98/06..$5.00

Design level	Description language	Primitives	Simulation slowdown
Algorithm	HLL	Instructions	10–100
Architecture	HLL	Functional blocks	1K–10K
Register transfer	HLL, HDL	RTL primitives	1M–10M
Logic	HDL, Netlist	Logic gates	10M–100M

Table 1. Digital system design levels. The simulation slowdown assumes the processor is simulating a model of itself.

Table 1 shows the design levels typically used in the process of creating a digital system [9]. The highest level of system design is the specification of the algorithm or protocol that the system will perform. Many digital systems are programmable to make them flexible and to make the system easier to debug. The programmable element might be a general purpose processor or an application-specific instruction-set processor (ASIP) [5]. If the system contains an ASIP, the design of the instruction set will have a large impact on the performance and flexibility of the system. Instruction set design is typically performed with the aid of an instruction level simulator. Simulating the system at the instruction level can be done with instruction interpretation [19] or binary translation [3]. Due to the similarity between instructions of an ASIP and the instructions of general purpose processors, instruction level simulation only requires tens or hundreds of instructions for each simulated instruction. This results in high simulation performance of 1 million to 10 million instructions per second on a 100 MIPS workstation.

At the architecture level designers are concerned with defining and estimating the performance of the main functional blocks in the system. An example of this is the design of the interconnection network and communication protocols used to interconnect processors in a heterogeneous multiprocessor system. Another example is the design of a processor's execution pipeline together with its memory hierarchy and I/O system [11]. At the architectural level, designers are interested in estimating overall system performance and collecting performance data to guide architectural design trade-offs. The speed of simulation at this level depends on the level of detail that is modeled, but usually results in simulation speeds that vary between 10 and 100 kHz on a 100 MIPS workstation.

At the register transfer level the goal is to model all the components in the system in enough detail so that performance and correctness can be modeled to the nearest clock cycle. At this level of simulation the simulator accurately models all the concurrency and resource contention in the system. The level of detail required to do this for the complexity of current chips makes RTL simulation slow even for the fastest simulators [22]. Despite their slow speed, RTL simulators are heavily used to check the correctness of the system. The performance of RTL simulation varies between 10 and 100 Hz on a 100 MIPS workstation. The high demand for simulation cycles for both performance and correctness verification makes simulation speed at this level a critical factor in determining the quality of the system design.

At the logic level all the nets and gates in the system are defined. At this point the design could be used to drive the surrounding system. However, the number of primitives and the characteristics of gate level models constrain their simulation performance to between 1 and 10 Hz on a 100 MIPS workstation. This performance is typically too low to allow the simulation of significant system applications at the logic level.

2.2 Simulation Input

An important consideration in developing a simulation model is the manner in which simulation input is provided to the model. There are three basic techniques: statistics-driven simulation, trace-driven simulation and execution-driven simulation. These techniques differ in the completeness of the simulation model. In statistics-driven simulation the input is a set of statistics gathered from an existing machine or a more complete simulator. These statistics are combined with static analysis of the behavior of the instruction set or architecture to provide early performance estimates. For example, instruction-mix data can be combined with predicted instruction execution latencies to estimate the performance of a pipelined processor [11]. Statistics-driven simulation models are very simple because they leave much of the system unspecified.

In trace-driven simulation the input data is a sequence of events that is unaffected by the behavior of the model. Trace-driven simulation is used extensively in instruction set design where the input events are a sequence of instructions. However, it is possible to develop a trace-driven processor model at any of the design levels discussed above. Trace-driven memory-system simulators are also quite common. Here the trace is a sequence of memory address references. Trace-driven simulation is also used in the design of network routers and packet switches. Although trace-driven simulators are simpler and faster than execution-driven simulators because they do not model data transformations, trace-driven simulators must be fed from a trace that comes from an execution driven-simulator or an existing digital system that has been instrumented for trace collection [18]. Trace-driven simulation is primarily used for detailed architecture performance prediction.

In execution-driven simulation the exact sequence of input events is determined by the behavior of the simulation model as it executes. Execution-driven simulators are developed later in the design cycle because they tend to have more design detail. They require more design detail to model the data transformations that must occur to accurately simulate the behavior of the system. Execution-driven simulation is essential during the later stages of the design cycle when the design is being verified for correctness. For example, while it is possible to accurately estimate the performance of a branch prediction algorithm using trace-driven simulation, execution-driven simulation is required to verify that the branch miss-prediction recovery mechanisms work correctly.

2.3 Simulating Concurrency

Concurrency is a component of all digital systems. In fact, built-in support for concurrency is the main characteristic that distinguishes hardware description languages like Verilog and VHDL from high-level languages like C. The way concurrency is simulated has a large impact on both simulation performance and modeling flexibility. The classic way of simulating concurrency is to make use of a dynamic scheduler which orders the execution of user defined threads based on events. This is known as discrete-event simulation and is the most flexible method of simulating concurrency. This is the model of concurrency that is embodied in Verilog and VHDL semantics, but it can be added with library support to C. The drawback to discrete-event simulation is low simulation performance due to the runtime overhead of dynamic event-scheduling. A number of techniques have been used to improve the performance of discrete-event simulation by reducing the event scheduler overhead [14, 20]. These techniques sacrifice some of the generality of event-driven simulation to achieve higher performance.

The alternative to dynamic event-scheduling is cycle-based simulation. In cycle-based simulation all components are evaluated once each clock cycle in a statically-scheduled order that ensures all input events to a component have their final value by the time the component is executed. Cycle based logic simulators which are also called levelized compiled code logic simulators rely on the synchronous nature of digital logic to eliminate all considerations of time that are smaller than a clock cycle. Cycle based simulators have the potential to provide much higher simulation performance than discrete-event simulators because they eliminate much of the run time processing overhead associated with ordering and propagating events [1, 6, 22]. Some cycle-based logic simulators dispense with the accurate representation of the logic between the state elements altogether and transform this logic to speed up simulation even further [15]. The main disadvantage of cycle-based simulation techniques is that they are not general. These techniques will not work on asynchronous models. In practice, even though most digital systems are mostly synchronous, asynchronous chip interfaces are common.

2.4 Managing Simulation

Complex digital systems design requires a sophisticated simulation environment for performance prediction and correctness validation. Most systems are composed of hardware and software and it is desirable to develop the hardware and the software concurrently. In the later stages of the design once the components of the system have been described at the RTL level, it is often necessary to simulate the whole system with realistic input data; however, the slow simulation speed of RTL-level models makes this computationaly expensive and time consuming. Two simulation management techniques that have proved to be useful in providing solutions to these problems are the use of multi-level simulation in which models at different design levels are simulated together [2] and dynamic switching between levels of detail [17].

Multi-level simulation can be used to support concurrent software and hardware development of an ASIP with specialized functional units in the following way. Software development can use a fast instruction level simulator of the ASIP while hardware development of the functional units takes place at the RTL level. Multi-level simulation incorporates both types of simulator so that the RTL-level models can seamlessly replace the high-level descriptions of the specialized instructions.

Dynamic switching between levels of detail takes multi-level simulation one step further. It allows the models of the specialized functional units to be dynamically switched at simulation time. This ability makes it possible to run realistic workloads on a detailed model while keeping simulation time reasonable. Simulation typically begins using the high-level fast simulator, and then when an interesting region of the simulation is reached, the high-level model is switched to the detailed model. The detailed model can be switched back to the less detailed model to quickly simulate unimportant portions of the workload. This switching back and forth can take place repeatedly and may involve multiple model levels with different speed-detail characteristics. This technique is very effective at simulating realistic workloads in their entirety because, as Table 1 shows, the simulation speed difference between the algorithmic level and the logic levels can span five orders of magnitude.

3 FLASH Case Study

This section gives a brief description of the FLASH multiprocessor, concentrating on the design and verification of MAGIC, the large ASIP node controller for the FLASH machine. The simulation methodology used in the FLASH design is discussed in detail as a real-life example of the environment required to develop a complex digital system.

3.1 FLASH Overview

The Stanford FLASH multiprocessor [12] is a general-purpose multiprocessor designed to scale up to 4096 nodes. It supports applications which communicate implicitly via a shared address space (*shared memory* applications), as well as applications that communicate explicitly via the exchange of messages (*message passing* applications). Hardware support for both types of communication is implemented in the MAGIC chip, the communication controller of the FLASH node. MAGIC contains an embedded, programmable *protocol processor* that runs software code sequences to implement communication protocols. Since these protocols can be quite complex, this approach greatly simplifies the hardware design process, and reduces the risk of fabricating a chip that does not work. A programmable protocol processor also allows great flexibility in the types of communication protocols the machine can run, and permits debugging and tuning protocols or developing completely new protocols even after the machine is built.

The structure of a FLASH node, and the central location of the MAGIC chip is shown in Figure 1. The MAGIC chip integrates interfaces to the main processor, memory system, I/O system, and network with the programmable protocol processor. As the heart of a FLASH node, MAGIC must concurrently process requests from each of its external interfaces and perform its principal function—that of a data transfer crossbar. However, MAGIC also needs to perform bookkeeping to implement the communication protocol. It is the protocol processor which handles these control decisions, and properly directs data from one interface to another.

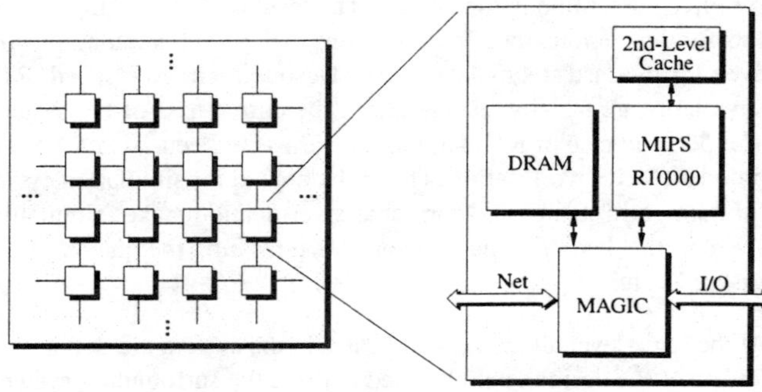

Figure 1. The Stanford FLASH multiprocessor.

MAGIC must control data movement in a protocol-dependent fashion without adversely affecting performance. To achieve this goal, the chip separates the data-transfer logic from the control logic. The data transfer logic is implemented in hardware to achieve low-latency, high-bandwidth data transfers. When data enters the MAGIC chip it is placed in a dedicated on-chip data buffer and remains there until it leaves the chip, avoiding unnecessary data copying. To manage their complexity, the shared memory and messaging protocols are implemented in software (called *handlers*) that runs on the embedded protocol processor. The protocol processor is a simple two-way issue 64-bit RISC processor with no hardware interlocks, no floating point capability, no TLB or virtual memory management, and no interrupts or restartable exceptions.

The MAGIC chip was fabricated as a standard-cell ASIC with the cooperation of LSI Logic. The die is 256mm^2 and contains approximately 250,000 gates and 220,000 bits of memory, making it large by ASIC standards. One of our 18 on-chip memories is a custom, 6-port memory that implements 16 128-byte data buffers, the key to our hardware data-transfer mechanism.

The combination of multiple interfaces, aggressive pipelining, and the necessity of handling multiple outstanding requests from the main processor makes MAGIC inherently multithreaded, posing complex verification and performance challenges.

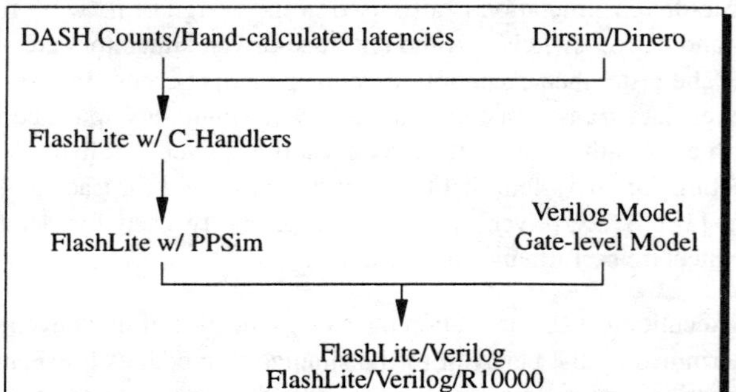

Figure 2. Simulation hierarchy used in the FLASH design.

As shown in Figure 2, the simulation strategy for FLASH went through several phases, moving from low-detail exploratory studies early in the design to very detailed gate-level simulations. The following sections describe each of these phases in more detail.

3.2 Initial Design Exploration

Early on in the FLASH design, we decided to use an ASIP to implement the communication controller and to use a data cache to reduce the amount of SRAM needed for protocol storage. The first studies we performed were to decide if the ASIP would be efficient enough to implement protocols in software at "hardware-like" speeds, and if a cache for protocol data could achieve high hit-rates.

3.2.1 ASIP Performance

The first design problem was to determine if an ASIP would be fast enough to act as a protocol engine for the FLASH machine. At this point, the architecture of the machine was very vague and we had no simulation environment, but we needed to generate performance numbers to guide the design decisions.

We used a previous research prototype built in our group, the Stanford DASH machine [13], to generate event counts from the SPLASH parallel application suite [21]. We wrote the important protocol handlers by hand in MIPS assembly to get an estimate for how long each of the protocol operations would take. Overall machine performance was obtained by multiplying the event counts by the handler estimates and summing over all handlers.

This approach gave us a good idea on what the average performance of the machine would be, and demonstrated that a general-purpose CPU core could come within a factor of two of the performance needed for MAGIC. Once we had these base performance estimates, we changed the handler cycle counts to account for our estimates of various hardware features that would help accelerate protocol handling, to see how the performance would change.

Unfortunately, since the input data is based on statistical averages, it is not possible to generate information about contention. In machines like FLASH, contention is a major performance issue, so it is important to model contention accurately in the simulation environment.

3.2.2 Cache Performance

The other major design issue we needed to confirm was the idea of caching the protocol data in the protocol processor's data cache. To study this we wrote a very simple discrete event model that implemented the general cache coherence protocol. This model was again driven by programs from the SPLASH application suite. The output from the model was a trace of addresses generated by the loads and stores executed by the protocol processor.

We fed this address trace into the Dinero [7] cache simulator to study cache miss rates over a wide range of standard cache parameters. Since we were interested in obtaining average cache statistics rather than absolute performance from this model trace-driven simulation was an acceptable choice. The results of this study demonstrated that a protocol data cache would work well.

3.3 FlashLite

Our initial simulation studies confirmed that implementing the communication controller as an ASIP was a good idea. We now needed to understand the relationship of MAGIC's performance to overall system performance. To discover this, we constructed FlashLite, a simulator for the entire FLASH machine. FlashLite uses an execution-driven processor model [8, 17] to run real applications on a simulated FLASH machine. FlashLite is written in a combination of C and C++ and uses a fine-grained threads package [16] that allows easy modeling of the functional units and independent state machines that comprise MAGIC.

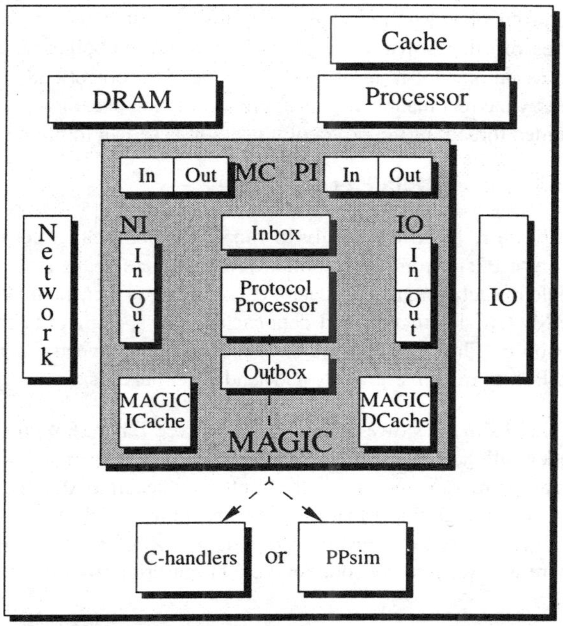

Figure 3. The different FlashLite threads that model the entire FLASH system.

Figure 3 shows the major FlashLite threads that make up a single FLASH node. The protocol processor thread can itself be modeled at two levels of detail. The next two sections describe why this is an important feature.

3.3.1 FlashLite - "C-handlers"

Initially the FlashLite threads comprising the communication controller were an abstract model of the MAGIC chip, parameterized to have the same basic interfaces. At that point, we were able to vary high-level parameters to determine what we could do well in the software running on the protocol processor, and what needed to be done with support from the surrounding hardware.

As the MAGIC architecture evolved, so did FlashLite, until eventually FlashLite had accurate models of all the on-chip hardware interfaces except for the protocol processor itself. This was natural, since the instruction set architecture of the protocol processor was not yet defined. However, FlashLite was still able to simulate the entire MAGIC chip by having a high-level model of the protocol processor that ran the communication protocol in C (the "C-handlers" and provided approximate delays to the basic protocol actions like sending messages and manipulating protocol state. The C-handlers played a crucial role in FLASH development, as they allowed the protocols to be written and debugged, and high-level design decisions to be made while the MAGIC hardware and the protocol processor instruction set were still being designed.

3.3.2 FlashLite - PPSim

The next major step in the project was the design of the instruction set architecture of the protocol processor. Once we had the basic instruction set designed (we built our instruction set on top of MIPS ISA) one of the first simulation tools we created to look at the trade-offs in instruction-set design was an instruction level simulator for the protocol processor. Along with this simulator, we developed a series of other software tools, including a C-compiler, scheduler, and assembler for translating the handlers into object code executable by the real machine. The simulator, PPsim, became FlashLite's detailed thread for the protocol processor as shown in Figure 3.

3.3.3 Speed-Detail Trade-offs

FlashLite retained the ability to run C-handlers, but it could now also use PPsim run the compiled protocol code sequences as they would execute on the real machine, complete with cache models for the MAGIC instruction and data caches. For speed and debugging flexibility, FlashLite had the ability to switch between C-handlers and PPSim on a fine-grain level (handler by handler).

While PPsim ran slower than C-handlers (see Table 2), it provided a model with accurate timing, and provided new information such as actual instruction counts, cycle counts, and accurate MAGIC cache behavior. Integrating PPsim into FlashLite also had the advantage of allowing us to debug the compiler, scheduler, and assembler before we ever ran any code on the RTL description of the MAGIC chip.

3.4 Verilog

Once the architecture started to solidify, work on the RTL design began in earnest. This work was done completely in parallel with the work on software tools and the fine-tuning of MAGIC's parameters with FlashLite and PPsim.

3.4.1 Verilog - RTL

The MAGIC chip design is implemented in well over 110,000 lines of Verilog. Most of the verification effort focused on this level of detail. We generated a large set of directed and random diagnostics that were run on the Verilog model. To make sure that no illegal conditions arose, the design was liberally annotated with sanity checking code that would not be part of the physical chip, but helped to indicate when a problem arose during verification.

3.4.2 Verilog - Gate-level

Using logic synthesis and libraries supplied by LSI Logic, we converted the RTL MAGIC description into a gate-level netlist description of the chip. This is the lowest-level description of MAGIC in our design methodology, so all diagnostics had to pass on this version before we sent MAGIC out for fabrication. As Table 2 shows though, the gate-level simulation is by far the slowest since it is the most detailed. For this reason, most diagnostic development and verification effort was spent at the behavioral RTL-level, and only when we had developed a large suite of random and direct diagnostics did we simulate at this lowest level to make sure the chip was really ready for fabrication.

Simulation Level	Speed (Hz)
FlashLite—C handlers	90,000
FlashLite—PPsim	80,000
Verilog—RTL	13
Verilog—Gate-Level	3

Table 2. Simulation Speed at Different Levels in the Hierarchy.

3.5 Putting it All Together

To test the RTL-level MAGIC design with realistic workloads we combined some of our simulation models in interesting ways. Two of the issues that lead us to do multi-level simulations for the hardware design were the correctness of the chip when running a realistic workload and the correctness of the chip interfaces when communicating with the other chips on the board.

3.5.1 FlashLite-Verilog

Ultimately we are interested in the overall performance of FLASH as a multiprocessor. To get this information we needed to see how a realistic workload would run on the hardware. We were particularly interested in how efficiently the protocol processor supported the multiprocessor cache coherency protocols.

Unfortunately, the performance of the RTL model is so poor, that it would be impractical to replicate the model N times to build a N-processor simulation. Instead, we used the FlashLite simulator to model N-1 of the nodes, and replaced a single node with the Verilog model of the MAGIC chip. Since FlashLite is much faster than the Verilog, the resulting simulation ran only slightly slower than the Verilog model by itself and allowed us to verify MAGIC with a realistic workload.

3.5.2 FlashLite-Verilog-R10000

The other major issue that lead us to use multi-level simulation was the task of confirming that MAGIC could communicate with the other chips on the node board. One of these chips was the compute processor for FLASH, the MIPS R10000.

In our simulation infrastructure, we replaced the high-level model of the processor we used in Section 3.5.1 with the real RTL model of the MIPS R10000. By running the realistic workloads within this simulation environment we had high confidence that the processor interface of MAGIC would work.

3.6 FLASH Conclusions

The evolution of the FLASH and MAGIC designs show that the simulation environment is intimately linked with the design task at hand. In the early stages of design low-detail statistics-driven simulation was used to quickly verify high-level architectural design decisions. Once the basic architecture was defined—an ASIP control path surrounded by dedicated data-movement hardware—more detailed execution-driven simulation was used to provide guidance for detailed architectural trade-offs. The result of these architectural trade-offs was an RTL model that could be used to drive ASIC design tools.

Two themes that emerge from the FLASH design are the need for simulation to support concurrent software and hardware development and the need to drive simulations with realistic workloads. The simulation environment was structured to allow the software development to commence using PPsim while the RTL development proceeded in parallel. Once the RTL-level design was complete, accurate performance evaluation was possible by using realistic workloads. Multi-level simulation is the key methodology that made this feasible.

Acknowledgments

This work is supported by DARPA contract number DABT63-94-C-0054. The authors also thank the entire FLASH development team.

References

[1] Z. Barzilai, J. L. Carter, B. K. Rosen, and J. D. Rutledge, "HSS– A high speed simulator," *IEEE Transactions on Computer-Aided Design*, vol. CAD-6, pp. 601–617, 1987.

[2] J. Buck, S. Ha, E. lee, and D. Messerchmitt, "Ptolemy: a framework for simulating and prototyping heterogeneous systems," *International Journal of Computer Simulation*, January, 1990.

[3] R. F. Cmelik and D. Keppel, "Shade: A Fast Instruction-Set Simulator for Execution Profiling," University of Washington, Technical Report UWCSE 93-06-06, June 1993.

[4] R. Collett, "Panel: Complex System Verification: The Challenge Ahead," *Proceedings of the 31st IEEE/ACM Design Automation Conference*, p. 320, San Diego, CA, June, 1994.

[5] G. DeMicheli, "Computer-Aided Hardware-Software Co-Design," *IEEE Micro*, vol. 14, August, 1994.

[6] R. S. French, M. S. Lam, J. R. Levitt, and K. Olukotun, "A general method for compiling event-driven simulations," *Proceedings of 32nd ACM/IEEE Design Automation Conference*, pp. 151–156, 1995.

[7] J. D. Gee et al. Cache Performance of the SPEC Benchmark Suite. *IEEE Micro*, vol. 3, no. 2, August 1993.

[8] S. Goldschmidt. Simulation of Multiprocessors: Accuracy and Performance. Ph.D. Thesis, Stanford University, June 1993.

[9] J. P. Hayes, *Computer Architecture and Organization 3rd Edition.* New York, NY: McGraw-Hill, 1998.

[10] M. Heinrich et al. "The Performance Impact of Flexibility in the Stanford FLASH Multiprocessor," In *Proceedings of the 6th International Conference on Architectural Support for Programming Languages and Operating Systems*, pp. 274-285, San Jose, CA, October 1994.

[11] J. L. Hennessy and D. A. Patterson, *Computer Architecture A Quantitative Approach 2nd Edition.* San Francisco, California: Morgan Kaufman Publishers, Inc., 1996.

[12] J. Kuskin et al. "The Stanford FLASH Multiprocessor," Proceedings of 21st Annual Int. Symp. Computer Architecture, pp. 302–313, April-, 1994.

[13] D. Lenoski et al. "The Stanford DASH Multiprocessor," *IEEE Computer*, 25(3):63-79, March 1992.

[14] D. M. Lewis, "A hierarchical compiled code even-driven logic simulator," *IEEE Transactions on Computer-Aided Design*, vol. 10, pp. 726–737, 1991.

[15] P. McGeer et al. "Fast discrete functional evaluation using decision diagrams," *Proceedings of IEEE/ACM International Conf. Computer-Aided Design*, pp. 402–407, San Jose, CA, November 1995.

[16] D.P. Reed and R K. Kanodia, "Synchronization with Eventcounts and Sequencers, "*Communication of the ACM*, vol. 22, no. 2, February 1979

[17] M. Rosenblum, S. Herrod, E. Witchel, and A. Gupta, "The SimOS approach," *IEEE Parallel and Distributed Technology*, vol. 4, pp. 34–43, 1995.

[18] M. D. Smith, "Tracing with Pixie," Stanford University, Computer Systems Laboratory, Technical CSL-TR-91-497, November 1991.

[19] J. Veenstra, "MINT a front end for efficient simulation of shared-memory multiprocessors," *Proceedings of International Workshop on Modeling, Analysis and Simulation of Computer and Telecommunication Systems*, pp. 201–207, January 1994.

[20] Z. Wang and P. M. Maurer, "LECSIM: A levelized Event driven compiled logic simulator," Proceedings of 27th ACM/IEEE Design Automation Conference, pp.491–496 Orlando. Florida, 1990.

[21] S. Woo et al. "The SPLASH-2 Programs: Characterization and Methodological Considerations." *Proceedings of the 22nd International Symposium on Computer Architecture*, June 1995

[22] J. Yim, et al. "A C-based RTL Design verification methodology for complex microprocessor," *Proceedings of 324th ACM/IEEE Design Automation Conference*, pp. 83–88, 1997

Hybrid Techniques for Fast Functional Simulation

Yufeng Luo
Design Tools Group
Synopsys, Inc.
Mountain View, CA
luo@synopsys.com

Tjahjadi Wongsonegoro Adnan Aziz
Electrical and Computer Engineering
University of Texas-Austin
Austin, TX
{tjahjadi, adnan}@ece.utexas.edu

Abstract

We implement and experiment with techniques for the functional simulation of very large digital systems. We consider techniques that are a hybrid of classical compiled code simulation and recent branching program based simulation in order to resolve memory performance problems inherent to BDD based cycle simulation. Specifically, predefined functional units ("macros") are extracted from the circuit and evaluated directly instead of building BDDs for them. The functionality of those macros, such as multipliers, filters, etc., can in turn be verified by simulation of their gate-level implementations respectively or by formal verification techniques. Our results demonstrate that this approach leads to considerably faster simulation.

1 Introduction

Design verification is a crucial step in the process of designing a digital system. The complexity of design has increased dramatically over the past few years, and this trend continues unabated in the deep sub-micron era. Consequently, a very large effort is concentrated on design verification and validation.

Formal verification techniques have been shown to be effective in proving the correctness of designs with respect to specified properties [9]. However, much more research is still required to make formal verification techniques capable of handling designs containing more than a few hundred flipflops [11]. The real-world design verification process still largely depends on semi-exhaustive functional simulation; thus new functional simulation techniques that are much faster than those available today are strongly needed.

In this paper, we present a new technique for fast functional simulation. We define functional simulation as delay-independent, cycle based simulation for synchronous digital circuits. Our approach is a hybridisation of a BDD-based cycle simulator with fast direct evaluation of high-level functional macros drawn from the macro library defined and provided by us. Such macros are common in designs with control and data-path (e.g., DSP applications).

DAC 98, San Francisco, California
©1998 ACM 0-89791-964-5/98/06..$5.00

We have implemented this idea as a proof-of-concept functional simulator in the Berkeley VIS package [4]; this is invoked by the `hsim` (hybrid simulation) command. The VIS package provides a framework and functions to decompose logic circuits and represent designs using BDDs [2].

The performance of our simulator is compared with a pure BDD-based cycle simulator, which is implemented based on the work of Ashar et al [1] and McGeer et al [10]. This simulator is available as the new `sim` command in the latest VIS release. Experimental results show the superiority of our simulator; both input-vector evaluation time and BDD model building time are greatly reduced.

2 BDD-based Functional Simulation

Traditional functional simulators are usually implemented using levelized compiled code simulation or event-driven simulation. Circuit activities are simulated by means of evaluating every gate in the circuit, therefore the time complexity of each evaluation is $O(N)$, where N is the number of gates in the circuit.

The use of branching programs for switch-level functional simulation was proposed by Cerny et al in [7]. Recent research shows that functional simulation with branching programs or decision diagrams on partitioned circuits can speedup the logic level functional simulation by an order of magnitude [1, 10].

BDD-based functional simulation of single-output circuits is straightforward. BDD representation for the function of the primary output is built in terms of all primary inputs in its support. Evaluation of the function is done by evaluating the BDD [5], i.e., branching at each node according to the value of the input vector, which has a time complexity of $O(I)$, where I representing the number of primary inputs.

BDD-based characteristic function representation of the circuit could be used to evaluate multiple outputs in one pass with time complexity of $O(I+O)$, where I is the number of primary inputs and O the number of primary outputs [1]. Compared with traditional techniques whose time complexity is proportional to the number of gates in the circuit, BDD-based functional simulation offers potentially much faster simulation.

In practice, BDD-based simulation has some inherent disadvantages. BDD representations of very large circuits usually use too much memory, and BDD evaluation can cause memory "thrashing". Memory performance could be improved by partitioning the circuit so a set of BDDs could

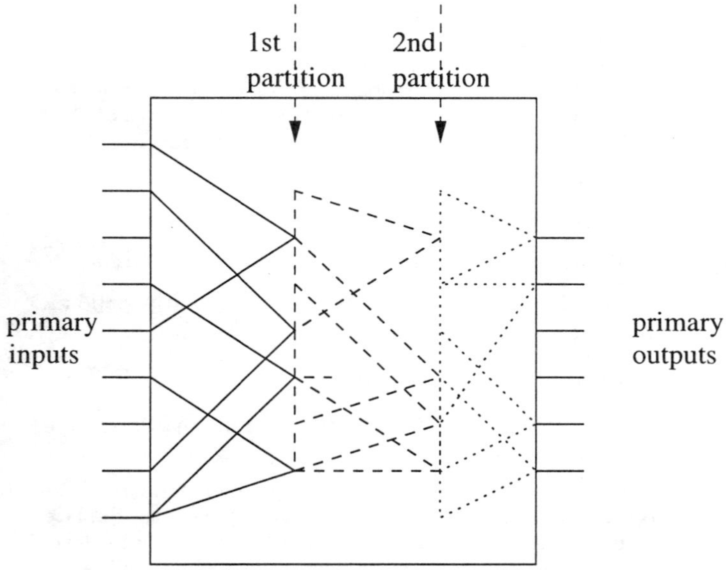

Figure 1: Circuit partition with intermediate BDDs

be built for intermediate points with a threshold for the size of a BDD, and by dynamic variable ordering [12].

However, controlling BDD blow-up by reducing the threshold size for partitions results in many partitions, and hence increases the number of functions to be evaluated in order to evaluate each vector, thus reduce the benefit of using BDDs. Moreover, any pure BDD-based simulation technique would inevitably suffer from BDD blow-up, which could simply be triggered by small circuits – like multipliers.

3 Hybrid Simulation Technique Using Macros

Large designs are hierarchical, i.e., built out of simpler subsystems [8]. Often, they consist of interacting control and data units. The control logic tends to be very irregular. In contrast, the data-path is extremely regular. It is natural to ask if the regular structure of the data-path can be exploited in simulation.

To exploit the regularity of the data-path, we have taken the following approach: we provide a library of predefined functional macros, such as multipliers, adders, ALU, etc. We purposely included adders as a macro module because even though adders have small BDD sizes, in the content of a larger design, this may not be true. The designer specifies his design in Verilog using these macros. When building the simulation model, the program identifies the functional macros. BDDs are then built for the control logic.

During simulation, macros spotted in the circuit are extracted and evaluated directly, which might be much faster than the evaluation of the same functional units done at the gate-level using BDDs. In addition, BDD representation of the circuit are much simpler and intermediate points for the control logic much less. This decreases the overall simulation time considerably.

It is obvious that the correctness of our hybrid approach is assured as long as the correctness of the macro functions can be verified separately, in which case a macro is no different with a "complex gate" and our hybrid simulation is no different than normal simulation. The correctness of the macros can be verified by BMDs [6], theorem provers [3]

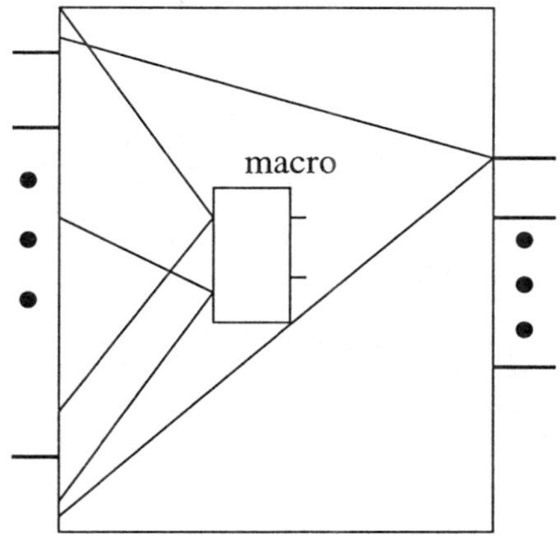

Figure 2: Hybrid simulation with macros

(which are especially well suited to hierarchical design verification) or other formal verification techniques.

In summary, our simulation proceeds as follows:

- Identify and extract macros from the circuit.
- Build BDDs for primary outputs as well as macro inputs, in terms of primary inputs and macro outputs. Introduce intermediate points if needed to control BDD size explosion.
- Macro inputs are evaluated from its BDD representation using the given primary input vector and possibly values of intermediate points and outputs of other macros on which it depends, which should be already evaluated (this is ensured by doing a topological sort of all the points which need to be evaluated – primary outputs, macro inputs, macro outputs and intermediate points – and evaluate them according to this order).
- Primary outputs are evaluated using values of primary inputs, macro outputs and values of intermediate points in its support.

4 Simulator Implementation

4.1 Macro Library

We provide a library of predefined functional units, which are specified as Verilog modules. We have to use a special naming convention which allows the simulator to extract macro inputs and outputs. Special care also has to be taken to resolve the naming conflicts resulting from the same net appearing in different level of hierarchies. In our implementation of macro modules, we assign to each macro input or output a buffer with the special naming format which encodes the pin number and the code representing the functionality of this macro.

Because the macro is evaluated directly by the simulator according to its predefined function, there is no need for the macro module to really be a gate-level implementation of its function. For example, in our implementation, the functionality is identified by the name of the macro. Instead, we can just represent each output of the macro by a simple function. Since we can evaluate the macro output only after we have values of all the macro inputs, this function

```
module Adder2MACRO(A, B, SUM);
  input  [1:0] A, B;
  output [1:0] SUM;
  wire   [1:0] MACROINPUT0_TYPE0,
               MACROINPUT1_TYPE0,
             MACROOUTPUT0_TYPE0;

  assign MACROINPUT0_TYPE0 = A;
  assign MACROINPUT1_TYPE0 = B;

  assign MACROOUTPUT0_TYPE0[0] =
    !MACROINPUT0_TYPE0 | !MACROINPUT1_TYPE0;
  assign MACROOUTPUT0_TYPE0[1] =
    !MACROINPUT0_TYPE0 | !MACROINPUT1_TYPE0;

  assign SUM = MACROOUTPUT0_TYPE0;
endmodule
```

Figure 3: 2-bit adder macro module

must have all the macro inputs in its support to reflect this dependency. In contrast, in the "real" implementation of a functional unit, such as an adder, some outputs may not depend on all the inputs.

The sample implementation of a two-bit adder module is shown in Figure 3. It is obvious that the functionality represented here is not at all an adder. The function is only to represent the dependency so that the topological sort will get the correct result. The "real" functionality is encoded in its name. The simulator will parse the name and give it an internal representation of its functionality accordingly.

4.2 Model Building

We call the process of building the BDD representation of a circuit the "module building" phase. We first traverse the network to extract all the macro inputs and macro outputs based on the name of the nodes. BDDs are built for primary outputs, intermediate points and macro inputs in terms of primary inputs and macro outputs. A threshold based heuristic is used to avoid BDD size explosion. We notice that there is less BDD size explosion since the BDDs for the macro-defined units are not built.

Macro inputs and primary outputs are sorted in topological order and will be evaluated in exactly this order; hence a BDD or a macro is evaluated only after all variables which are possibly in its support are evaluated.

The module building phase is an overhead to the simulation process with a time complexity of $O(N)$, where N representing the number of nodes in the network.

4.3 Function Evaluation

In the function evaluation phase, we evaluate the BDDs for macro inputs and primary outputs in topological order. After each evaluation, the results are saved as inputs for future evaluation if it is an intermediate point or macro input; or as the final result if it is a primary output. Once all the inputs of a macro are available, outputs of this macro are evaluated directly according to its function and the results saved.

5 Experimental Results

We compared our simulator — `hsim`, with the new `sim` command in the Berkeley VIS package, which is implemented based on the work of Ashar et al [1] and McGeer et al [10].

The benchmarks used in the experiment are:

1. `adder32.mv`: 32-bit adder
2. `mult16.mv`: 16-bit multiplier
3. `pwl16.mv`: 16-bit piece-wise linear transformation
4. `matrix8.mv`: 3x3 8-bit integer matrix multiplication
5. `alu8.mv`: 8-bit ALU
6. `filter.mv`: adaptive filter
7. `predictor.mv`: branch predictor for 32-bit addressable microprocessor

Among these examples, the multiplier exploits the best potential of our simulator because it is very easy to simulate as a pure macro, but the building BDD for its gate level representation will cause BDD blow-up; whereas the branch predictor serves as the worst case scenario since it has only adder and comparator macros, which are well-behaved in BDD, and most of the circuit is control logic.

The total run time is divided into model building time and evaluation time. The model building time is the time for building the BDDs, which is a fixed overhead to the overall simulation time. The evaluation time is the time for simulating all the vectors.

`sim` suffers from occasional BDD blow-up. Also, sometimes it takes forever for `sim` to finish the simulation, presumably due to poor memory performance. These results are also marked as "BDD blow-up".

It can be easily seen from the comparison that the performance of `hsim` is considerably better than that of `sim`. We see that the evaluation is much faster, thanks to the direct evaluation of functional level macros and better memory performance. The module building is usually much faster too, because the network is simplified due to the use of macro modules.

6 Conclusion and Future Work

We found that replacing gate-level representation of some functional units with predefined functional macros can achieve very fast evaluation, improve memory performance and reduce the overhead of building the BDDs.

We have implemented a proof-of-concept hybrid functional simulator using this approach. The experimental results show the superiority of this simulator. The important aspect of our work is that we employ fast BDD based techniques for simulating random logic, and use native machine code for structured components. Intuitively, such an approach is superior to simulating the entire design at the gate-level. This is shown in our experimental results. This approach can take advantage of formal verification techniques by rendering some functional units to formal verification then simulate the simplified circuits using macros to achieve faster simulation.

Possible future work includes automatic extraction of data-path elements from flattened gate-level net-lists, possibly with the help of links to synthesis; and to extend as well as to improve our macro library to make it dynamically linkable. We are also exploring the possibility of combining software simulation with hardware emulation.

simulator	sim		hsim	
time(ms)	model building	evaluation	model building	evaluation
adder32.mv	2300	20	180	30
mult16.mv	29900	100	90	20
pwl16.mv	BDD blow-up		21620	190
matrix8.mv	3140800	900	22780	170
alu.mv	19100	100	470	20
filter.mv	40700	200	188050	80
predictor.mv	11400	30	1190	60

Table 1: hsim vs. sim: 100 vectors

simulator	sim		hsim	
time(ms)	model building	evaluation	model building	evaluation
adder32.mv	5100	1900	200	1190
mult16.mv	30100	6400	100	820
pwl16.mv	BDD blow-up		21780	10480
matrix8.mv	3141100	52400	22650	8900
alu.mv	19300	6500	490	1300
filter.mv	40900	11400	188390	3940
predictor.mv	11600	3400	1200	3060

Table 2: hsim vs. sim: 5000 vectors

References

[1] P. Ashar and S. Malik. Fast Functional Simulation Using Branching Programs. *Proc. Intl. Conf. on Computer-Aided Design*, pages 408–412, 1995.

[2] UC Berkeley. `www-cad.eecs.berkeley.edu/~vis`.

[3] R. S. Boyer and J. S. Moore. *A Computational Logic Handbook*. Academic Press, New York, 1988.

[4] R. K. Brayton, G. D. Hachtel, A. Sangiovanni-Vincentelli, F. Somenzi, A. Aziz, S.-T. Cheng, S. Edwards, S. Khatri, Y. Kukimoto, A. Pardo, S. Qadeer, R. K. Ranjan, S. Sarwary, T. R. Shiple, G. Swamy, and T. Villa. VIS: A system for Verification and Synthesis. In *Proc. of the Computer Aided Verification Conf.*, July 1996.

[5] R. Bryant. Graph-based Algorithms for Boolean Function Manipulation. *IEEE Transactions on Computers*, C-35:677–691, August 1986.

[6] R. Bryant and Y.A. Chen. Verification of Arithmetic Circuits with Binary Moment Diagrams. In *Proc. of the Design Automation Conf.*, pages 535–541, June 1995.

[7] A. Cerny and J. Gecsei. Simulation of MOS Circuits by Decision Diagrams. *IEEE Transactions on Computer-Aided Design*, C-4:685–693, 1985.

[8] Berkeley CAD group. Revisiting BLIF-MV, An Intermediate Format for Verification and Synthesis of Hierarchical Networks of FSMs, 1993.

[9] A. Gupta. Formal Hardware Verification Methods: A Survey. *Formal Methods in System Design*, 1:151–238, October 1993.

[10] P. McGeer, K. McMillan, A. Saldanha, A. Sangiovanni-Vincentelli, and P. Scaglia. Fast Discrete Function Evaluation. *Proc. Intl. Conf. on Computer-Aided Design*, pages 402–407, 1995.

[11] K. McMillan. *Symbolic Model Checking*. Kluwer Academic Publishers, 1993.

[12] C. Scholl, R. Drechsler, and B. Becker. Functional Simulation using Binary Decision Diagram. *Proc. Intl. Conf. on Computer-Aided Design*, pages 8–12, 1997.

A Reconfigurable Logic Machine for Fast Event–Driven Simulation

Jerry Bauer, Michael Bershteyn, Ian Kaplan, Paul Vyedin
Quickturn Design Systems, Inc.
55 West Trimble Rd.
San Jose, CA 95131-1013
(408) 914-6000
jerry@quickturn.com, bmike@quickturn.com, iank@bearcave.com, paulv@quickturn.com

1. ABSTRACT

As the density of VLSI circuits increases, software techniques cannot effectively simulate designs through the millions of simulation cycles needed for verification. Emulation can supply the necessary capacity and performance, but emulation is limited to designs that are structural or can be synthesized. This paper discusses a new system architecture that dramatically accelerates event-driven behavioral simulation and describes how it is merged with emulation.

1.1 Keywords

reconfigurable computing, event-driven simulation

2. Introduction

As ASIC design complexity has increased, logic simulation has evolved. Early simulators compiled designs, written in Verilog or VHDL hardware description languages (HDL), into a pseudo-instruction set for interpretation by a software "machine". To improve performance, many simulators now compile HDL designs directly into a microprocessor instruction set (e.g., SPARC, HP PA-RISC, powerPC) [12]. These compiled simulators can be more than ten times the speed of interpreted simulators. The performance improvement yielded by compiled simulation is limited by the overhead of event processing. Cycle-based simulators take advantage of native code performance, but avoid event processing overhead by only simulating synchronous designs in structural or synthesizable form.

Although the performance of software simulators has constantly improved, the rate of improvement falls far short of the rate at which design density is increasing. Hardware acceleration, usually based on a network of custom processors, has been used to increase the speed and capacity of simulation by up to 100 times. Hardware emulation, where the hardware design is emulated on a large network of programmable logic (FPGAs) [14], has been used to increase simulation speed and capacity by up to 10,000 times [11].

Emulation sacrifices generality for performance; it cannot be used to simulate behavioral circuit models that contain delays or other constructs that are either non-structural or cannot be synthesized [2] into gate-level circuitry. The high speed of emulation also introduces its own constraints. To supply data (test vectors or other system input) at a rate that matches emulation performance, existing emulation systems must be used "in-circuit" with the target system hardware.

Both hardware accelerators and emulators have been combined with software simulators, which can be used to simulate the behavioral component of a design. Communication between the software simulator and the emulated design usually takes place over Ethernet or across a system bus, like VME. This approach has drawbacks:

- The software simulation component limits the overall verification speed. Behavioral testbenches that provide models of hardware components or supply test vectors cannot be used without slowing hardware emulation.
- Partitioning between emulation and software simulation must be handled explicitly, and cannot be done at a fine-grained level.

Reconfigurable computing systems using field programmable gate arrays (FPGAs) have been proposed by a number of research groups [3, 9]. Several research groups have coupled reconfigurable logic with processors [4, 9, 13]. The HARP system [9] couples FPGAs with a 32-bit transputer microprocessor [7]. The Garp processor [13], proposed by the BRASS research group at UC Berkeley, is a custom VLSI design that couples a MIPS II core with an on-chip FPGA array. However, the application of these techniques to event-driven simulation has not been adequately explored.

This paper describes a reconfigurable computing system that has been applied to the event driven simulaton of large VLSI circuit designs. The foundation for this reconfigurable computing system is an FPGA-based emulator, which provides large blocks of reconfigurable logic. The system hardware is supported by a compiler that can compile a behavioral Verilog HDL description of the design. This system has a number of advantages:

- Logic (in the FPGA array) can be synthesized at compile time to support event processing. This removes the event processing overhead from behavioral simulation.
- Low overhead communication. In this system, communication between the behavioral code and the structural portion of the design is supported by synthesized logic.
- The design testbench may be written in behavioral Verilog with none of the timing model restrictions typically imposed by cycle-based simulation or emulation [1].
- The capacity and speed of emulation are available for the structural or synthesizable portions of the design.

3. System Architecture Overview

The reconfigurable computer system is pictured in *Figure 1*, below. The system consists of one or more emulation modules. Each emulation module consists of:

- A powerPC 403GCX processor
- Local RAM
- A local FPGA array with programmable interconnect

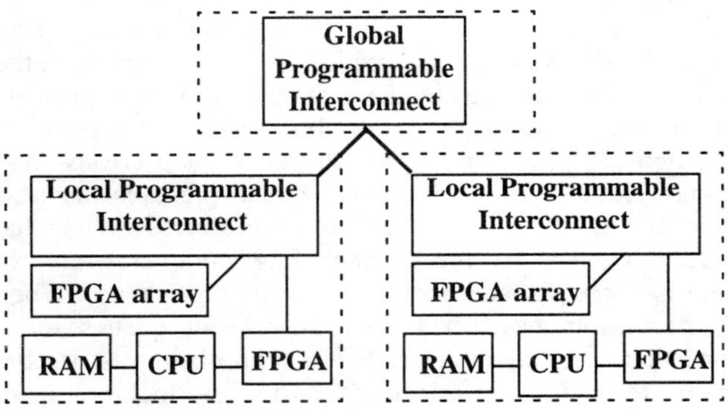

Figure 1.

The FPGA array in each emulation module is interconnected to all of the other emulation module FPGA arrays via a programmable interconnect. The programmable interconnect allows all FPGA arrays to be treated as if they were a single large FPGA array. Each emulation module runs a simple real time operating system to manage behavioral simulation and logic netlist emulation. A control processor (not shown) supports higher level operating system functions such as TCP/IP network access, disk, and the network file system (NFS).

To the user, the reconfigurable computer system appears to be a simulation server on the local area network. However, only one simulation may run on the system at any given time.

4. Compiler Overview

Compiling a behavioral Verilog design can be logically divided into two tasks: compiling behavioral Verilog into assembly language and synthesizing a netlist to program the FPGA array.

The Verilog [6] compiler serves as the front end for this process. The Verilog compiler scans and parses the input and produces an abstract syntax tree intermediate in memory. The module hierarchy is expanded; parameters, "defparams", and hierarchical names are resolved. Semantic analysis is then performed and functions are inlined. A transformation pass does tree-to-tree translation to produce an intermediate form that can be more easily broken up into schedulable processes. Later passes compile the processes into control flow graphs and into powerPC [5] assembly instructions.

Each schedulable process created by the Verilog compiler has a set of data inputs and outputs. The partitioning and netlist creation software treats the processes as design units (e.g., module instances) and enters them into a design unit database. The purely structural parts of a Verilog design skip many of these steps.

5. Data Partitioning

Event detection and scheduling logic is synthesized from the behavioral design. Verilog variables that are arguments to module instances or triggers for events are synthesized into latches in the FPGA array. Other variables are allocated in processor memory. Each variable allocated in hardware is addressed with a unique hardware tag.

The partitioning of variables between FPGA hardware and processor memory is illustrated by `module sum`, (see *Figure 2*), below. In this module, the variables `reset` and `clk` are event triggers and the variable `sum` is a module output. Since these variables are involved either with events or module interconnection, they are allocated in the FPGA hardware. The local variable `val` is not a trigger or a module argument, so it is allocated in the processor memory.

```
module sum( reset, clk, sum );
input reset, clk;
output [7:0] sum;
reg [3:0] val;
reg [7:0] sum;

  always @(posedge clk)
  begin
    sum = sum + val;
    val = val + 1;
  end

  always @(posedge reset)
  begin
    sum = 0;
    val = 1
  end
endmodule
```

Figure 2.

6. Netlist Generation

Each process created by the Verilog compiler consists of a block of code and a set of triggering conditions: for example `@(a or b)` or `#4`. While translating the Verilog design, the Verilog compiler passes information to the hardware mapping phase. Structural module definitions are ultimately mapped directly into gate equivalents in the FPGA arrays. For behavioral processes, the hardware mapping phase takes the Verilog compiler's definition for a process and creates a data base element for it. This consists of three parts: an interface specification (data inputs and outputs), the triggering conditions, and a pointer to the code for the process.

The process interface specification is compiled into a set of addressable registers in the FPGA array. Each triggering condition is synthesized into its hardware equivalent. When a process is instantiated, its synthesized triggering condition circuitry is instantiated along with it. The triggering condition is a boolean expression with additional operators:

- anyedge(x)
- posedge(x)
- negedge(x)
- delay(value, x)

Edge operators are implemented as hardware edge detectors. Delay is implemented as a down counter preset to the delay value at the positive edge of its input signal.

A simulation cycle is divided into four subcycles: active, inactive, non-blocking, and monitor. Each triggering condition is assigned to one of these subcycles. This information is retained for use in event queue generation.

A process may refer to variables allocated as registers in the FPGA array or in processor memory (RAM). A two-step process effects the transfer of variable information between the processor and the FPGA array:

- The hardware tag of the variable is written to the tag decoder.
- The hardware signals that represent the variable are read or written.

When all structural and process module instantiations are complete, a partitioning algorithm is run. This divides the design into sections, each of which will fit onto an emulation module (see *Figure 1*). The partitioning algorithm also defines the connections between the emulation modules. After the process components and structural components are placed into the emulation modules, a levelization algorithm provides a static prioritization which is used to generate the event queue. The synthesized logic for each triggering condition is connected to the associated event circuitry. Each separate event circuit in the event queue is given a unique identifier or tag. A software table, created by the code generation phase of the Verilog compiler, maps each process tag to a pointer to the start of the associated process. Each emulation module executable (consisting of one or more processes) is linked with a "main" section that contains a dispatch loop that waits for a hardware event tag. When the software dispatch loop reads the process tag from the tag register, it fetches the process address and branches to the start of the process.

Figure 3 diagrams the event queue mechanism. The block

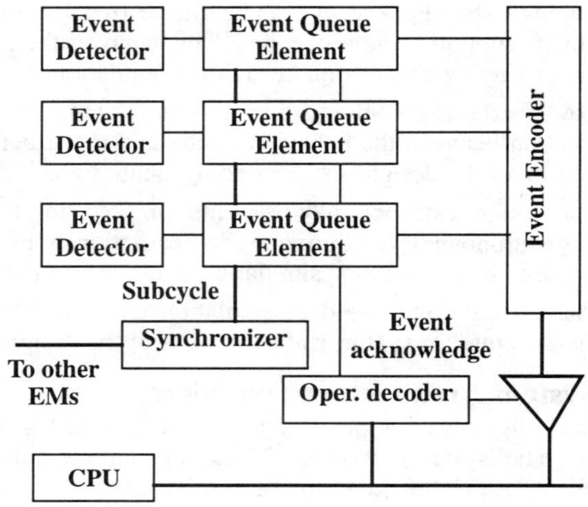

Figure 3.

labeled "Event Encoder" is synthesized as part of hardware compilation. The event encoder generates the process tags that identify processes to be executed in response to hardware events. When a triggering condition is met, the event detector associated with the synthesized triggering logic "fires". Subject to inhibition by the synchronizer, the asserted event captured by the event queue element presents its associated software tag to be read by the software dispatch loop. After the dispatch loop reads and acknowledges the tag, the event queue prepares the next process tag, if any other events are asserted. When the process invoked by the dispatch loop completes, control returns to the dispatch loop and another tag is read. When all processors have exhausted their event queues for a particular simulation cycle, the synchronizer re-starts the hardware simulation clock. Hardware clocking again proceeds until an event is triggered.

The synchronizer in *Figure 3* serves two related purposes:

- It maintains the subcycle timing for the hardware events
- It enforces inter-processor lock-step synchronization.

When an event is active on any processor, the hardware clock is halted until the event is serviced. This maintains simulation-clock and subcycle synchrony across all emulation modules — in essence, simulation time proceeds only when no event is active. This means that co-processing is limited to simultaneous execution of events that occur during the same simulation cycle time. During the time that a processor is executing a software process, the event queues remain available. A process can, by writing a hardware-mapped variable, trigger another event, either on the same processor or on another processor. Processes are never interrupted; they always run to completion.

7. Limitations

Hardware support for event triggers and process scheduling increases the performance of the behavioral code, but places limits on the Verilog implementation. Any variable may be allocated in hardware, so the Verilog computation model is limited to what can be efficiently supported in emulation. Emulation supports binary logic, although there is some support for tri-stated devices. The Verilog logic state **x** is not supported (all **x** values default to zero), and the logic state **z** is supported only in limited contexts: a **z** value can be written but will not be read as **z**. Verilog strength and switch modeling are not supported.

8. Experimental Results

The performance of event-driven simulation on our reconfigurable computing system was examined with designs that mixed gate-level netlists and circuit blocks implemented as behavioral models. The netlist portion of the design is emulated in the FPGA array. The behavioral component is compiled into powerPC object code, which runs closely coupled with the netlist emulation.

The designs, shown in *Table 1*, were compiled with the behavioral Verilog compiler, resulting in netlists that support event detection and generation and object code that implements all other behavioral statements. The size of these designs, in gates, was estimated by replacing the behavioral components with synthesizable components.

The tests were run on a reconfigurable computer with two emulation modules (see *Figure 1*).

Testcase	Size	Speed
test35	35KG	200K
test150	150KG	175.2K
test210	210KG	170.2K

Table 1: Simulation Speed Measurements

The speed is in simulation cycles per second. In each case, the fastest design clock had a period of two simulation cycles. Therefore, the speed in design clock cycles is half of the speed given in the table.

9. Conclusion

ASIC design complexity is increasing far more rapidly than the performance delivered by software simulators. Hardware acceleration and emulation deliver higher performance but require synchronous, synthesizable designs. This makes the design testbench and models of off-chip components difficult to develop. Our work shows that a reconfigurable computing system can be used to simulate behavioral level VLSI designs. This approach couples the flexibility of event driven behavioral simulation with the capacity and speed of emulation. Although most of IEEE standard Verilog can be supported, some limitations are imposed by emulation.

10. Acknowledgments

Viktor Salitrennik implemented the register allocation phase of the compiler and the runtime support for wide operations. Peter Donovan did early work on the Verilog compiler. Youngmin Hur helped us with testing. Arun Ramachandran did early work on Verilog system functions and PLI support which was continued by Jay Deleanu. David Morrison implemented networking support. Without their efforts, the results reported here would not have been possible.

11. References

[1] Casaubieilh, F., McIsaac, A., Benjamin, M., Bartley, M., Pogodalla, F., Rocheteau, F., Belhadj, M., Eggleton, J., Mas, G., Barrett, G., Berhet, C., *Functional Verification Methodology of Cameleon Processor*, Proc. 33ed ACM/IEEE Design Automation Conference, 1996, pp 421-426.

[2] De Micheli, G. *Synthesis and Optimization of Digital Circuits*, McGraw-Hill, 1994

[3] DeHon, A. *DPGA Utilization and Application*, Proceedings of the 1996 International Symposium on Field Programmable Gate Arrays, available from URL:http://www.ai.mit/projects/transit/tn129/tn129.html

[4] DeHon, A. DPGA-Coupled Microprocessors: Commodity ICs for the Early 21st Century, MIT Transit Project technical note 100, Jan. 1994, available from URL:http://www.ai.mit/projects/transit/tn100/tn100.htm.

[5] IBM Corporation, *PPC403GA Embedded Controller User's Manual*, Second Edition, March 1995

[6] IEEE, *Standard Verilog HDL Language Reference Manual*, IEEE Standard 1364, 1995

[7] Inmos, *The Transputer Databook*, Inmos document 72 TRN 203 02. Order code DBTRANST/3.

[8] McGeer,P. C., McMillan,K.L., Saldanha, A., Sangiovanni-Vincentelli, A.L., Scaglia, P. *Fast Discrete Function Evaluation using Decision Diagrams*, IEEE/ACM International Conference on Computer Aided Design, 1995, pp. 402-407

[9] Page, I. *The HARP Reconfigurable Computing System*, Oxford University Hardware Compilatoin Group, 1994, available from URL:http://www.comlab.ox.ac.uk/oucl/users/ian.page/papers.html

[10] Page, I. *Reconfigurable Processors*, Invited Keynote Address for Heathrow PLD Conference, April 1995, available from URL:http://www.comlab.ox.ac.uk/oucl/users/ian.page/papers.html

[11] Rowson, J. A. *Hardware/Software Co-Simulation*, Proc. 31st ACM/IEEE Design Automation Conference, 1996, pp 439-440

[12] Sanguinetti, J. *Language Considerations and Experimental Results Using A Verilog Compiler*, 2nd Annual International Verilog HDL Conference, 1993.

[13] Hauser, J. R and Wawrzynek, J., *Garp: A MIPS Processor with a Reconfigurable Coprocessor*, Proceedings of the IEEE Symposium on Field Programmable Custom Computing Machines, April 1997, pgs 24-33.

[14] Varghese J., Butts M., and Batcheller J., *An Efficient Logic Emulation System*, IEEE Trans. on VLSI Systems, vol. 1, No. 2, June 1993, pgs 171-174.

Parallel Algorithms for Power Estimation

Victor Kim and Prithviraj Banerjee *
Center for Parallel and Distributed Computing
Dept. of Electrical and Computer Engineering
Northwestern University, Evanston, Illinois
{vicjkim,banerjee}@ece.nwu.edu

Abstract

Several techniques currently exist for estimating the power dissipation of combinational and sequential circuits using exhaustive simulation, Monte Carlo sampling, and probabilistic estimation. Exhaustive simulation and Monte Carlo sampling techniques can be highly reliable but often require long runtimes. This paper presents a comprehensive study of *pattern-partitioning* and *circuit-partitioning* parallelization schemes for those two methodologies in the context of distributed-memory multiprocessing systems. Issues in *pipelined* event-driven simulation and dynamic load balancing are addressed. Experimental results are presented for an IBM SP-2 system and a network of HP-9000 workstations. For instance, runtimes have been reduced from over 3 hours to under 20 minutes in one case.

1 Introduction

Power consumption in VLSI circuits has become a major issue of concern in modern IC design. Power minimization is fast becoming a major optimization objective in VLSI design, along with area and speed. But these multiple objectives often conflict with one another, so engineers need fast and reliable power estimation tools to make the right compromises during different stages of the design cycle. The importance of power estimation and their integration into the design cycle have already been recognized and implemented to a certain extent in some commercial CAD products [11].

The focus of this study is on presenting a comprehensive demonstration of the design issues and performance attainable with parallelized execution of well-performing serial-run power estimation algorithms for combinational and sequential circuits. In particular, we address the parallel design issues for estimating the **average** power dissipation (as opposed to the **worst-case** power) using the following two methodologies:

- **Exhaustive simulation** [13]: A circuit simulator is driven by user-given input and is used to collect statistics about the circuit activity during a very lengthy period of operation. Although the technique makes no guarantees regarding the input-independence of the resulting power value, it is still often used to provide the baseline for an "accurate and reliable" power estimate in many studies.
- **Monte Carlo sampling**: A *sampling* procedure is used to refine a power estimate until certain statistical bounds are satisfied. Two recently published techniques, one for combinational [14] and another for sequential circuits [10], were selected for parallel implementation in this study because of their reliability, as reported by their respective publications. The reader is referred to [14] and [10] for a comprehensive treatment of each methodology. Essentially, Monte Carlo sampling employs a 3-phase framework:
  1. **Input generation phase**, in which input vectors are randomly generated.
  2. **Circuit simulation phase**.
  3. **Statistical analysis phase**, in which the *power sample* (power dissipation during the previous simulation phase) is computed and combined with previous power samples to determine whether the entire process needs to be repeated in order to satisfy certain bounds. (The "termination" criteria are computed from user-given constaints on statistical "error" and "confidence.")

  One major difference to note between the combinational and sequential methods chosen for this study is that, in the sequential Monte Carlo method, multiple ($2N = 100$) instances of the circuit are used to keep track of the circuit's state transitions according to multiple random input streams [10].

Numerous **probabilistic estimation** approaches have also been proposed in the literature, but they have also been reported to pose some reliability inconsistencies, especially for sequential circuits. Therefore, probabilistic methods were not considered for this study. The reader is referred to [8] for a general survey of the above methodologies.

In their respective studies, both exhaustive simulation and Monte Carlo sampling methods employ a simulation core to solve for power-related variables. The simulation core may be in the form of a circuit-level simulator such as SPICE [7] or PowerMill [3], but for CMOS circuits, the

*This research was supported in part by the National Science Foundation under grant MIP 9696164 and the Advanced Research Projects Agency under contract DAA-H04-94-G0273 administered by the Army Research Office.

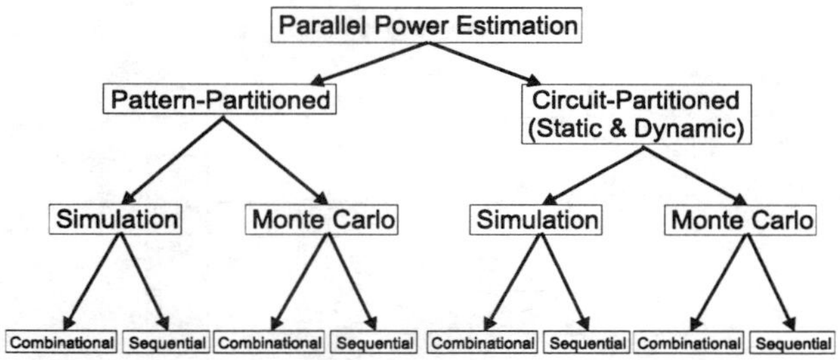

Figure 1: Overview of parallel power estimation approaches described in the paper.

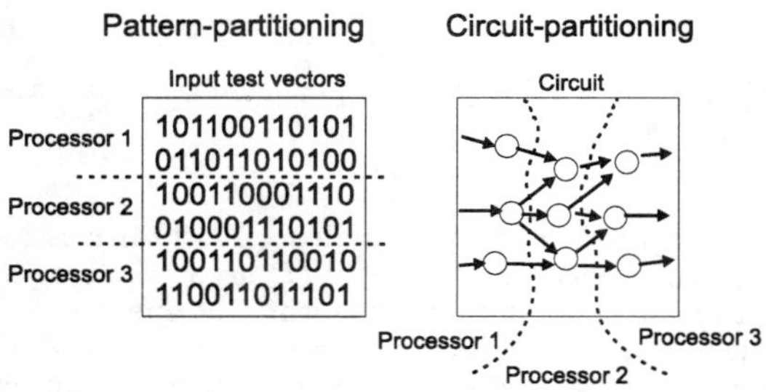

Figure 2: Illustration of pattern-partitioning and circuit-partitioning schemes.

static power component (which is highly technology dependent) is generally negligible in comparison to the *dynamic power* component (which is due to the capacitive switching activity) [13]. Hence, event-driven logic simulators have been used more extensively in related studies.

Logic simulators often run an order of magnitude faster than circuit-level simulators due to their simplistic modeling of power. Still, both exhaustive simulation and Monte Carlo sampling procedures using logic simulation cores can often require lengthy runtimes in order to ensure their reliability. For instance, transistor-level logic simulation of a relatively small sequential circuit (i.e. s35932) with a relatively small number of input vectors (i.e. 20,000) on a high-end workstation (HP-9000) can require more than 3 hours. Their runtimes are proportional to the runtime of the simulation core and the execution of the simulation framework.

The problem, then, of parallel power estimation in the context of this study is that of achieving speed-up and scalability for the logic simulation core while maintaining the overall execution framework in which the simulation core resides. This problem of parallel simulation is similar to that of parallel logic and fault simulation, for which numerous approaches have been reported [1, 6]. The approach to parallel power estimation in this study are based on schemes similar to those used in parallel logic/fault simulation. However, the following factors distinguish parallel logic simulation in the context of this study from that in the context of other parallel logic/fault simulation:

- The sought-after result is an *average* power value. Since we seek *average* power, *exact* tracking of circuit activity throughout each and every simulation step for a single stream of input vectors is not necessary for estimating the average power value, thereby loosening some of the restrictions commonly encountered in, for instance, parallel simulation of sequential circuits.

- The content and number of input vectors are either specified up-front or generated independently from one vector to another, thereby allowing for potential of speed-up without *necessitating* **concurrent** simulation over particular input vector(s).

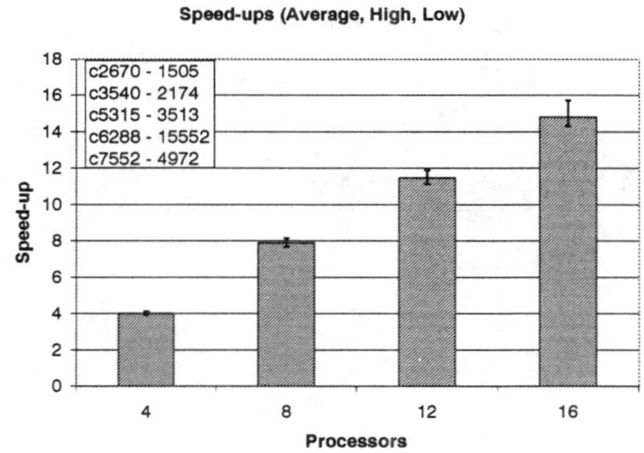

Figure 3: Speed-ups for pattern-partitioned combinational exhaustive simulation (corner box indicates the serial runtime in seconds for each benchmark)

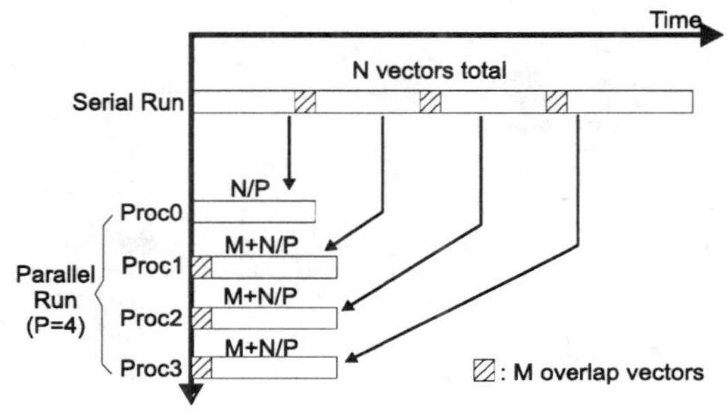

Figure 4: Pattern-partitioning for sequential exhaustive simulation.

To the best of the authors' knowledge, this is the first published report describing a thorough investigation of parallel logic simulation in the context of parallel power estimation. Two types of parallelization schemes were considered for this study (Figure 1). In the *pattern-partitioned* approach, the given circuit is replicated in each processor to perform concurrent simulation and power computation (Figure 2). For very large circuits, the *circuit-partitioned* approach distributes the circuit to each processor and performs pipelined computation. Each partitioning scheme is subdivided into handling exhaustive simulation and Monte Carlo sampling frameworks, each of which is further subdivided into combinational and sequential cases. A comprehensive treatment of this work appears in [4].

2 Experimental setup

Each of the following sections include experimental results obtained on benchmark circuits, which are comprised of the five largest circuits from each of the ISCAS'85 suite (c2670, c3540, c5315, c6288, c7552) and the ISCAS'89 suite (s13207.1, s15850.1, s35932, s38417, s38584.1). Average, high, and low speed-ups are displayed as bar charts, along with a corner box listing the serial runtimes of each benchmark in each methodology. The parallel implementations

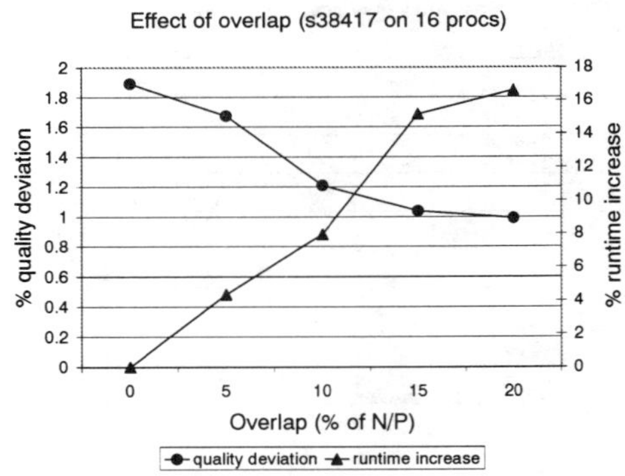

Figure 5: Effect of vector overlap on quality and runtime in sequential exhaustive simulation.

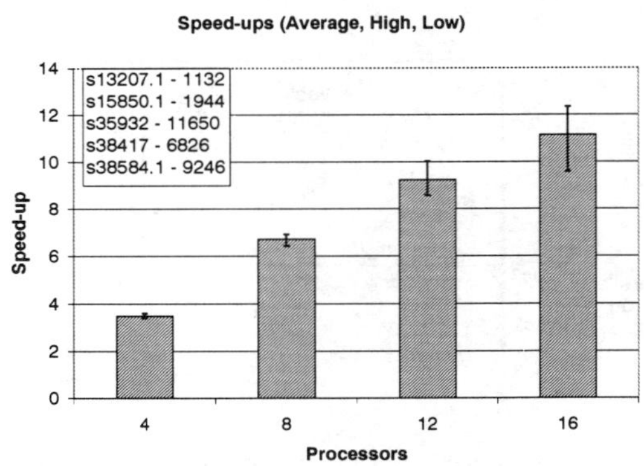

Figure 6: Speed-ups for pattern-partitioned sequential exhaustive simulation.

were developed using the Message Passing Interface (MPI) [2] on a 16-processor IBM SP-2 distributed-memory system. Portability of the implementation is demonstrated on a network of 8 HP-9000 workstations.

As for the simulation core, two different implementations are presented in this paper. For the pattern-partitioning sections (Section 3), the simulation core was adapted from IRSIM [12], a popular switch-level, variable-delay, event-driven logic simulator. For the circuit-partitioning sections (Section 4), a gate-level, variable-delay, event-driven logic simulation code from MED [14] was used instead, because our particular choice of circuit-partitioning scheme (described in Section 4) required that all signal flow be unidirectional.

For exhaustive simulation power estimation, a fixed sequence of 20,000 randomly-generated input vectors was used. For Monte Carlo sampling, baseline runtimes were established by modifying each framework to collect a fixed number of samples, which roughly averaged out to be equivalent to setting 99%-confidence/1%-error constraints.

3 Parallel power estimation via pattern-partitioning

3.1 Pattern-partitioning for exhaustive simulation

3.1.1 Combinational circuits

The entire circuit is replicated for each of P processors, and the total of N input vectors is divided equally into N/P

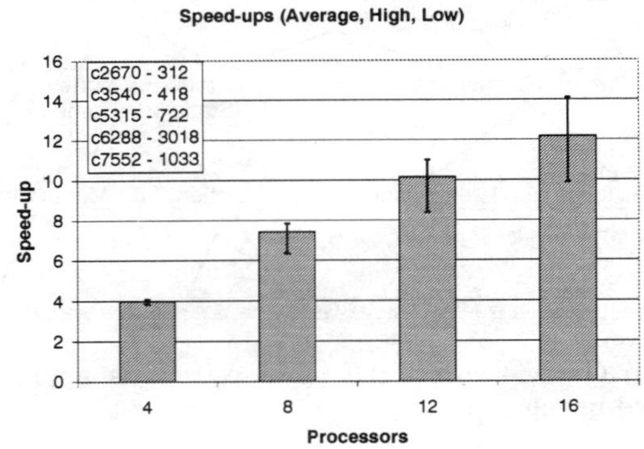

Figure 7: Speed-ups for pattern-partitioned combinational Monte Carlo sampling.

portions. Each processor then independently drives its local circuit copy with its share of N/P input vectors. Since communication is minimal in this setup, the speed-ups (Figure 3) were near-linear in most cases, as expected. Quality (in terms of absolute percentage error from serial-run power value) was also maintained highly accurately in all cases (less than 0.1% error).

3.1.2 Sequential circuits

With sequential circuits, one cannot simply partition the input sequences among processors and perform parallel simulation since the state dependencies over time will be violated, and may potentially lead to error in the power estimate. One recent study [5] addressed a similar problem in parallel fault simulation and proposed partitioning the input vectors along with some *overlap* between the partitions. A similar approach was adopted for parallel power estimation (Figure 4). Processors 1 through P-1 replicates M overlap vectors from the N/P portion owned by the "preceeding" processor. Figure 5 shows the effect of the overlap length on quality deviation (percentage error from serial power) and increase in runtime (percentage increase compared to no-overlap) for the largest sequential benchmark in our suite. It has been noted that, in general, the larger the circuit the more input vectors it requires to achieve a properly "warmed-up" state. Therefore, it is expected that, with larger circuits and more processors, the need for proper "warm-up" in the form of some extra input vectors will become evident. In the case shown in Figure 5, adding overlap does appear to decrease quality deviation without severely impacting the runtime. Since the benchmarks were too small to provide a clear indication of an optimal overlap size, 10% overlap was used to demonstrate our concept for the other benchmarks (Figure 6). Quality was maintained to within 1% in all but one benchmark, as noted above.

3.2 Pattern-partitioning for Monte Carlo sampling

3.2.1 Combinational circuits

In parallel Monte Carlo sampling, each processor performs *sampling* (the input-generation and simulation phases) independently with its own unique stream of random inputs. In order to minimize communication and synchronization overhead, processor 0 is designated to gather the power samples from the other processors and to determine whether

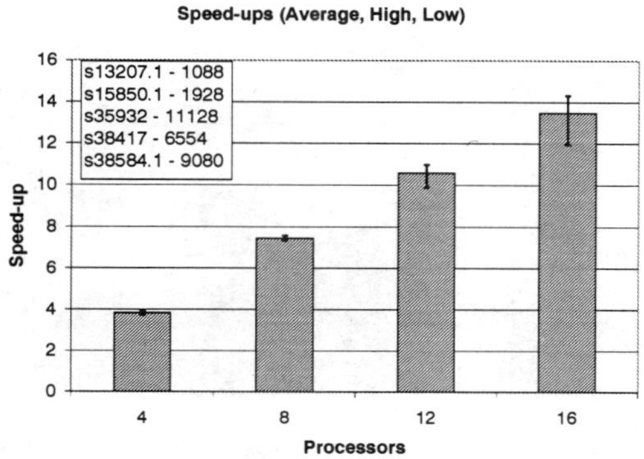

Figure 8: Speed-ups for pattern-partitioned sequential Monte Carlo sampling.

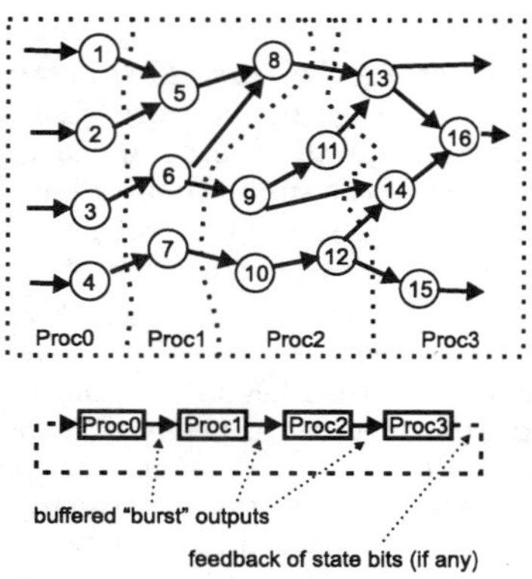

Figure 9: Partitioning a circuit in pipelined fashion.

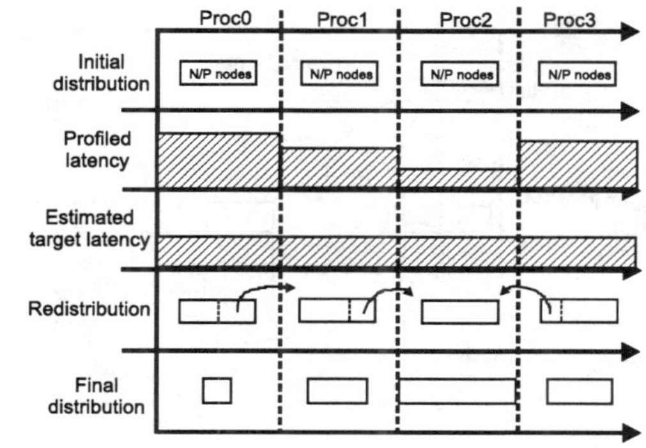

Figure 10: Profile-based dynamic repartitioning during run-time in the circuit-partitioning approach.

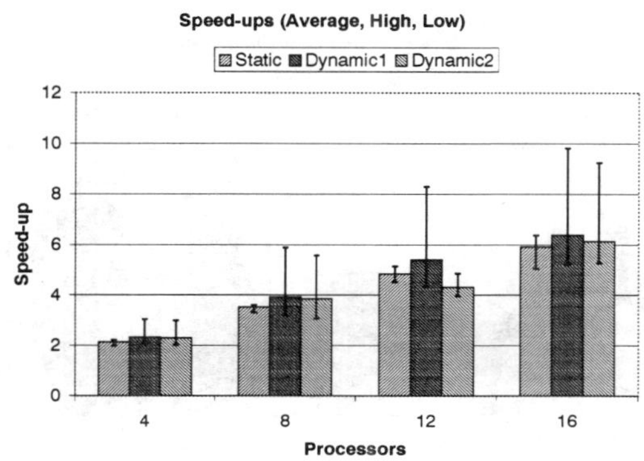

Figure 11: Speed-ups for circuit-partitioned combinational exhaustive simulation.

the stopping criteria have been satisfied. The submission of power samples proceed in an asynchronous "push" manner. The resulting speed-ups (Figure 7) were increasingly sub-linear as more processors were added due to the increased communication load on processor 0. However, quality was maintained to within 1% in all cases.

3.2.2 Sequential circuits

Parallel Monte Carlo sampling for sequential circuits is similar to that for combinational circuits, except for the provision that each instance of the sequential circuit (with its own state history) needs to be maintained in the same processor throughout the entire procedure. Hence, each processor is now responsible for $2N/P$[1] instances of the circuit (Figure 8). Each processor maintains the power samples for its share of the circuit instances and sends all the updates to processor 0, which evaluates the stopping criteria. Quality was maintained to within 1% in all except one case.

4 Parallel power estimation via circuit-partitioning

In the circuit-partitioning approach, a circuit is subdivided and distributed across the processors so as to accommodate for large circuits which may not fit in the memory space of a single processor. The circuit is partitioned in a *pipelined* fashion (Figure 9), wherein each processor acts as a "pipeline stage," which simulates its portion of the circuit in one phase, and then propagates the resulting circuit activity to the next stage during another phase. The partitioning is performed on a levelized graph of the circuit, so as to maintain the signal delay hierarchy and to streamline communication, which is restricted to one-way send-receive between neighboring stages only. A similar partitioning scheme has been reported in the context of parallel fault simulation with a zero-delay model in [9], but in this study, pipeline execution of variable-delay event-driven simulation is performed in an *asynchronous store-and-forward model* in order to avoid synchronization overhead and to minimize communication start-up overhead. Any message that needs to propagate beyond the following stage is bundled up with the rest of the downstream message packet, and each stage is responsible for sorting out the destination of the packet's content. We can then identify two major factors that would influence the actual speed-up:

- **Load imbalance** may be caused by one or more "bottleneck" stages. In an event-driven simulation environment, the *load* is a dynamic function of the fanout of each node, their switching activity, and the communication overhead associated with the partition-boundary nodes.
- **Communication overhead** adds extra latency according to the size and frequency of the message packets. The message frequency can be minimized by using

[1] $2N$ was set to 96 such that the total number of active circuit instances is evenly divisible by powers of 2.

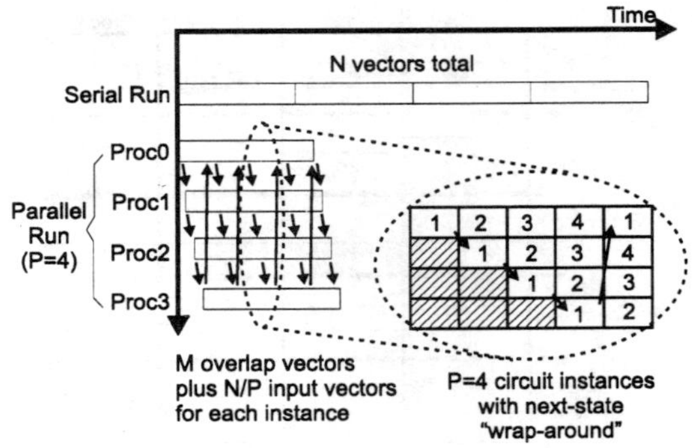

Figure 12: Circuit-partitioning for sequential simulation; the numbers indicate the instance of the circuit being simulated during that pipeline cycle.

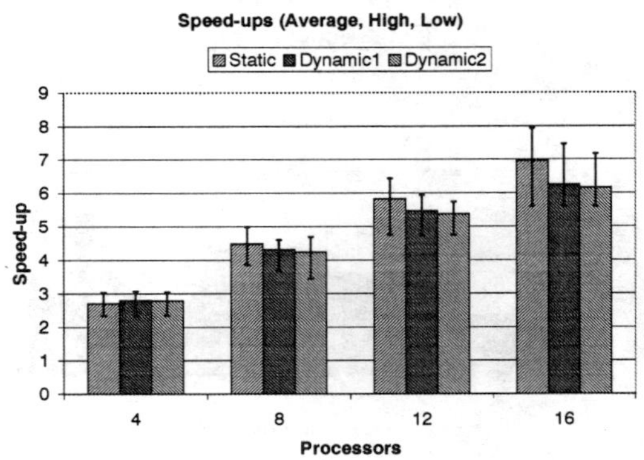

Figure 13: Speed-ups for circuit-partitioned sequential exhaustive simulation.

a store-and-forward approach, as described above. But the message size is determined by the number of edges that were "cut" during partitioning and the transition activity of those edges.

With event-driven simulation, these issues are compounded by the fact that the latency of any one stage can vary unpredictably depending on the inputs to that stage.

4.1 Circuit partitioning strategies

4.1.1 Static partitioning

A fast and simple approach, which has often been found in other parallel simulation work, is to distribute an equal number of nodes to each processor. Another approach may be to use a min-cut type of heuristic to minimize the partition "cut size." However, static partitioning of either kind does not take into account the actual switching activity which directly relates to the pipeline load. Since the computational cost of static partitioning may not be amortized when used inside a design optimization cycle (during which a complete re-partitioning may be necessary at every cycle), only a simple static partitioning strategy based on balanced node count was implemented for this study.

4.1.2 Dynamic partitioning

Since the load cannot be determined statically, we propose a dynamic profiling method of partitioning, which monitors a portion of its runtime (i.e. first K simulation cycles) for the average load (latency) in each processor, and then re-partitions and re-distributes the circuit before proceeding further with the rest of the simulations. Figure 10 illustrates the re-distribution process, in which each node is weighed as $\frac{profiled\ load\ on\ processor}{number\ of\ nodes\ in\ the\ processor}$ and then re-located to a new processor (if necessary) to meet a "target" average latency (as computed from the profile). The re-assignment and re-distribution procedure takes $O(n)$ time (where n is the number of nodes) so it adds relatively little overhead to the overall runtime. In order to determine whether the profiling load measure should include the accompanying communication latency, two heuristic measures were implemented: **Dynamic1**, which measured only the relevant simulation load; and **Dynamic2**, which measured both the simulation and communication latencies. Additional experiments on the profiling length did not reveal any conclusive clues about an *optimal* profiling length. Therefore, an empirical value of 30 pipeline cycles was demonstrated for the following sections.

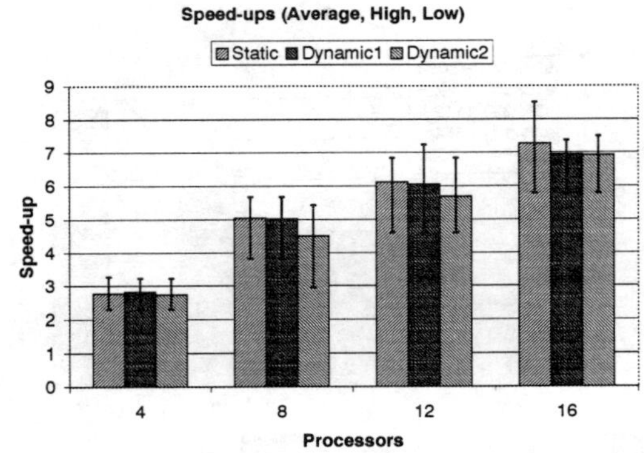

Figure 14: Speed-ups for circuit-partitioned combinational Monte Carlo sampling.

4.2 Circuit-partitioning for exhaustive simulation

4.2.1 Combinational circuits

The absence of feedback in combinational circuits allows for a streamlined pipeline operation, configured as shown in Figure 9. As illustrated by the high-low lined superimposed in the speed-up chart (Figure 11) there was a varied distribution of actual performance gain. It should be noted that the quality was identical to serial-run quality for all cases.

4.2.2 Sequential circuits

For sequential circuits, state feedback is allowed to wrap-around the pipeline (Figure 12,13). In order to overlap the workload while the state feedback is being propagated, we use the lesson learned from pattern-partitioned sequential exhaustive simulation by using multiple (P) instances of the circuit along with some overlap (same M as in section 3.1.2), yielding the same quality as that from pattern-partitioning.

4.3 Circuit-partitioning for Monte Carlo sampling

4.3.1 Combinational circuits

In pipelined combinational Monte Carlo sampling, the power samples are sent down the pipeline and eventually end up in

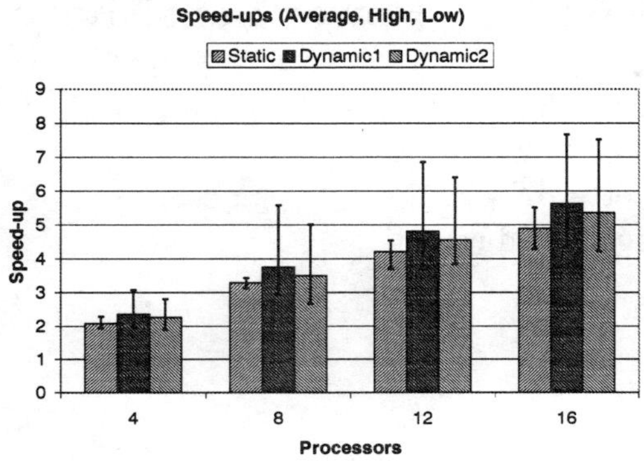

Figure 15: Speed-ups for circuit-partitioned sequential Monte Carlo sampling.

the last stage, which is responsible for evaluating the stopping criteria and broadcasting the termination signal. Resulting speed-ups are shown in Figure 14.

4.3.2 Sequential circuits

Pipelined sequential Monte Carlo sampling also requires state feedback, as shown in section 4.2.2. Each pipe stage cycles through the multiple instances of the circuit so as to overlap state propagation in one instance with simulation update in another. Resulting speed-ups are shown in Figure 15.

5 Portability of implementation

On an ethernet network of HP-9000 workstations, the same MPI code that ran on the IBM SP-2 achieved similar performance gains with pattern-partitioning, and up to factors of 3 on 4 processors and 3.5 on 8 processors with circuit-partitioning. Scalability of circuit-partitioning appears to level off after 4 processors due to increased communication costs.

6 Concluding remarks

In this paper, we have presented an overview of a comprehensive study on parallel power estimation using exhaustive simulation and Monte Carlo sampling techniques. The reader is referred to [4] for a detailed treatment of each case study. The pattern-partitioning strategies yielded good speed-ups for all cases on both a distributed-memory system and a network of workstations. With pipelined circuit-partitioning, consistent speed-ups of as much as 7 and individual speed-ups of as much as 10 on 16 processors were achieved. The performance gain was particularly susceptible to the nature of the circuit: a "deep/narrow" circuit (more levels and less "surface area" of the cuts) generally performed better with pipelined partitioning than a "shallow/wide" circuit. The impact of dynamic partitioning also varied. In cases where it did improve against static partitioning, it did so sizeably (as much as 65%), while in other cases it actually degraded the performance (by 15% or less).

Communication overhead is clearly one of the major factors that impacted the performance. The dynamic nature of an event-driven simulation core has shown itself to be a challenge to our load balancing efforts. A future study that integrates dynamic re-partitioning with cut minimization may prove to be more effective.

References

[1] P. Banerjee, *Parallel Algorithms for VLSI Computer-Aided Design*, Englewood Cliffs, NJ: PTR Prentice Hall, 1994.

[2] W. Gropp, E. Lusk, and A. Skjellum, *Using MPI: Portable Parallel Programming with the Message-Passing Interface*, MIT Press, 1994.

[3] C.X. Huang, B. Zhang, An-Chang Deng, and B. Swirski, "The design and implementation of PowerMill," *ACM/IEEE International Symposium on Low Power Design*, pp.105-109, 1995.

[4] V. Kim, "Parallel Algorithms for CMOS Power Estimation,"
Masters Thesis, Northwestern University, 1997, available at http://www.ece.nwu.edu/cpdc/TechReports/.

[5] D. Krishnaswamy, P. Banerjee, E. Rudnick and J. Patel, "Asynchronous parallel algorithms for test set partitioned parallel fault simulation," *Proc. Workshop on Parallel and Distributed Simulation (PADS-97)*, Jun. 1997.

[6] P. Luksch, "Evaluation of three approaches to parallel logic simulation on a distributed memory multiprocessor," *Proc. 26th Annual Simulation Symposium*, Arlington, pp.2-11, 1993.

[7] L.W. Nagel, "SPICE2: a computer program to simulate semiconductor circuits," *Memo*, ERL-M520, Dept. Electrical Engineering and Computer Science, University of California at Berkeley, May 9, 1975.

[8] F.N. Najm, "Power estimation techniques for integrated circuits," *IEEE Trans. VLSI*, vol.2, no.4, pp.492-499, Dec.1994.

[9] F. Ozguner, C. Aykanat, and O. Khalid, "Logic fault simulation on a vector hypercube multiprocessor", *Proc. 3rd Int. Conf. Hypercube Concurrent Computers Applications*, II:1108-1116, Jan. 1988.

[10] V. Saxena, "Statistical power estimation for sequential CMOS circuits," Masters Thesis, University of Illinois, 1996.

[11] Synopsys Corp. *http://www.synopsys.com/products/power/power.html*, 1997.

[12] C. Terman. "RSIM: A logic level timing simulator," *Proc. Int. Conf. Computer Design*, pages. 437-440, Oct. 1983.

[13] N.H.E. Weste and K. Eshraghian, *Principles of CMOS VLSI Design: A Systems Perspective*, Reading, MA: Addison-Wesley Publishing Company, 1993.

[14] M.G. Xakellis and F.N. Najm, "Statistical estimation of the switching activity in digital circuits," *31st ACM/IEEE Design Automation Conf.*, pp.728-733, 1995.

A Power Macromodeling Technique Based on Power Sensitivity

Zhanping Chen and Kaushik Roy
School of Electrical and Computer Engineering
Purdue University, West Lafayette, IN 47907
{zhanping,kaushik}@ecn.purdue.edu

Abstract

In this paper, we propose a novel power macromodeling technique for high level power estimation based on power sensitivity. Power sensitivity defines the change in average power due to changes in the input signal specification. The contribution of this work is that we can use only a few points to construct a complicated power surface in the specification-space. With such a power surface, we can easily obtain the power dissipation under any distribution of primary inputs. The advantages of our technique are two-fold. First, the required parameters corresponding to each representative point can be efficiently obtained by only one symbolic power estimation run or by only one Monte Carlo based statistical power estimation process. This stems from the fact that power sensitivity can be obtained as a by-product of probabilistic or statistical power estimation runs. Second, the memory requirements for the macromodel are reduced to $O(dn)$, where n is the number of primary inputs of a circuit and d is the number of representative points (d can be as small as 1 in some cases). Results on a number of benchmark circuits demonstrate the effectiveness of our technique.

1 Introduction

The increasing use of portable computing and communication systems makes power dissipation a critical parameter to be minimized during circuit and system design [1, 14]. Hence, there is a great need for tools to accurately estimate power dissipation at various levels of design abstraction.

Research on power estimation has started in earnest, however, most of the research concentrates at the logic level. In order to shorten design time and to reduce design iterations, we have to estimate the power dissipation at a high level of abstraction to ensure that the strict power requirement of a future design is satisfied.

High level power estimation can be categorized into: top-down methods and bottom-up methods. The top-down methods [8, 10] are not widely investigated because they require that a combinational circuit be specified as a Boolean function with no knowledge of the circuit structure or implementation. Bottom-up methods [7, 9, 11, 12, 13], on the other hand, have been exploited by several authors. However, one of the main concerns is how to develop a power macromodel for a module so that power dissipation can be easily obtained under any distribution of primary inputs.

Since power dissipation of a circuit strongly depends on the statistics of primary inputs, the relationship of power versus primary input probabilities (probability of a signal being logic ONE) and activities (probability of signal switching) is a complicated surface. Once such a surface set up, power dissipation under any distribution of primary inputs can be easily obtained. However, to construct such a power surface, a huge number of discrete points are required. If one chooses d representative values for the probability and activity of each primary input, the number of representative points in the specification-space can be d^{2n} (n is the number of primary inputs). This means that a symbolic or statistical power estimation process has to be repeated d^{2n} times. For large circuits with large number of inputs, such a process is obviously impractical due to the exponential growth of complexity. Moreover, the memory requirements are $O((2n+1)d^{2n}) = O(nd^{2n})$. This is also not affordable.

In this paper, we present a novel macromodeling technique based on *power sensitivity* [2, 3]. The basic idea of our technique is to use a limited number of representative points in the specification-space to approximate a complicated power surface. Power dissipation under any distribution of primary inputs is calculated by considering the representative points and power sensitivities. For each representative point, we need only four variables (probability, activity, power sensitivity to probability, and power sensitivity to activity) for each input. Thereby, memory requirements is reduced to $O(dn)$.

2 Preliminaries

2.1 Power Dissipation in CMOS Logic Circuits

Among the three sources of power dissipation – switching current, short-circuit current, and leakage current. The switching power is dominant in current technology. Thus the average power for a CMOS circuit can be approximated by $Power_{avg} = \frac{1}{2}V_{dd}^2 \sum_{j \in all\ nodes} C(j)\ A(j)$ where V_{dd} is the supply voltage, $C(j)$ is the node capacitance, $A(j)$ is the activity (probability of signal switching) at node j. If we assume that all primary inputs to the circuits under con-

*This research was supported in part by DARPA (F33615-95-C-1625), NSF CAREER award (9501869-MIP), IBM, AT&T, and Rockwell.

sideration switch only at the leading edge of the clock and that the circuits are delay-free, we can define *normalized activity*, denoted by $a(x_i)$, as $A(x_i)/f$, where f is the clock frequency. $C(j)$ is approximately proportional to the fanout at node j. Hence, we can define the *normalized power dissipation measure* Φ as: $\Phi = \sum_{j \in all\ nodes} f_{anout}(j)\ a(j)$ where $f_{anout}(j)$ is the fanout number at node j.

2.2 Power Sensitivity

To measure the effect of the variations of primary input specifications on power dissipation, we define *power sensitivity to primary input activity* $\zeta_{a(x_i)}$ and *power sensitivity to primary input probability* $\zeta_{P(x_i)}$ as follows:

$$\zeta_{a(x_i)} = \frac{\partial Power_{avg}}{\partial a(x_i)} = \sum_{j \in all\ nodes} f_{anout}(j) \frac{\partial a(j)}{\partial a(x_i)} \quad (1)$$

$$\zeta_{P(x_i)} = \frac{\partial Power_{avg}}{\partial P(x_i)} = \sum_{j \in all\ nodes} f_{anout}(j) \frac{\partial a(j)}{\partial P(x_i)} \quad (2)$$

where $a(x_i)$ and $P(x_i)$ are the activity and probability of primary input x_i, respectively, and $a(j)$ is the activity of node j. For simplicity, we can refer to $\partial a(j)/\partial a(x_i)$ and $\partial a(j)/\partial P(x_i)$ as activity sensitivities.

Let $\theta(x_i)$ be $P(x_i)$ or $a(x_i)$. We define second order power sensitivities to primary input x_i as

$$\frac{\partial^{(2)} Power_{avg}}{\partial \theta(x_k)\partial \theta(x_i)} = \sum_{j \in all\ nodes} f_{anout}(j) \frac{\partial^{(2)} a(j)}{\partial \theta(x_k)\partial \theta(x_i)} \quad (3)$$

The m-th order power sensitivities to primary input x_i can be defined as

$$\underbrace{\frac{\partial^{(m)} Power_{avg}}{\partial \theta(x_k)\ldots\partial \theta(x_i)}}_{m} = \sum_{j \in all\ nodes} f_{anout}(j) \underbrace{\frac{\partial^{(m)} a(j)}{\partial \theta(x_k)\ldots\partial \theta(x_i)}}_{m} \quad (4)$$

where x_k varies for all primary inputs except for x_i.

3 Power Macromodel Based on Power Sensitivity

3.1 Power Macromodel

If we can get the power sensitivities to each primary input, we can compute power under any distribution of primary inputs by the following expansion:

$$\begin{aligned} Power &= Power_{nom} + \sum_i \zeta_{\theta(x_i)} \Delta\theta(x_i) + \\ &\frac{1}{2!}\sum_{i,k} \frac{\partial^{(2)} Power_{avg}}{\partial \theta(x_k)\partial \theta(x_i)} \Delta\theta(x_k)\Delta\theta(x_i) + \\ &\frac{1}{3!}\sum_{i,l,k} \frac{\partial^{(3)} Power_{avg}}{\partial \theta(x_k)\theta(x_l)\partial \theta(x_i)} \Delta\theta(x_k)\Delta\theta(x_l)\Delta\theta(x_i) \\ &+ \ldots + \frac{1}{n!}(\ldots) + R_n \end{aligned} \quad (5)$$

where i, l, and k vary for all primary inputs (PI's). Equation (5) is the Taylor's expansion with Lagrange remainder R_n, the initial point being $(P(x_1)_{nom}, a(x_1)_{nom}, \ldots, P(x_n)_{nom}, a(x_n)_{nom})$ [5]. $Power_{nom}$ is the average power dissipation based on nominal probabilities $(P(x_i)_{nom})$ and activities $(a(x_i)_{nom})$.

For higher order power sensitivities, for any primary input x_i, we have $\partial^{(m)} Power_{avg}/\partial^{(k)}\theta(x_i) = 0$ for any $m \geq 2$ and $k \geq 2$. This means that only those higher order power sensitivities in which each $\theta(x_i)$ appears exactly once in the denominator will survive. This can be proved by induction.

For the first order approximation, we can ignore the second or higher order power sensitivities to obtain power by

$$Power = Power_{nom} + \sum_{i \in PI's} \left(\zeta_{a(x_i)}\Delta a(x_i) + \zeta_{P(x_i)}\Delta P(x_i)\right) \quad (6)$$

Thus, the power macromodel based on power sensitivities can be given by:

$$Power = f(P(x_i), a(x_i), \zeta_{a(x_i)}, \zeta_{P(x_i)}) \quad (\forall x_i \in PI's) \quad (7)$$

3.2 Construction of Power Surface

Power dissipation of a CMOS circuit is heavily dependent on the primary input probabilities and activities. If we plot the relationship between average power and input specifications in terms of signal probability and activity (in this paper we call such a plot power-specification plot), we will get a complicated surface. Since $2n$ (n is the number of primary inputs) dimensions are required to represent a certain specification of primary inputs, the power-specification plots are intrinsically $(2n+1)$-dimensional.

However, the process to obtain such a power-specification plot is nontrivial. A possible method is to use random number generators to generate a large number of distributions of primary inputs. Then use symbolic or statistical method to obtain the power dissipation corresponding to each set of such distributions, which is a representative point for the power surface in the specification-space. However, the effectiveness of this method strongly depends on the density of the chosen points. The more points one chooses, the more accurate result one can obtain. However, more points directly translate to longer CPU time. In practice, one must use a finite number of points to approximate a complex power surface. In order to obtain the power under a given specification of primary inputs, one has to approximate its power by the value of a simulated point. This means that for some inputs there exist differences between the actual values of the probabilities and activities and those corresponding to the representative points in the specification-space. In [2], it has been shown that for some circuit a small deviation in the statistics of some primary inputs may have a major effect on power dissipation. For those sensitive primary inputs, such approximations may make average power severely off the actual value. These errors can be effectively reduced by taking power sensitivities at different nominal values.

The basic idea of our method is to approximate a complex power surface with a number of small planes. Firstly, randomly choose d different statistics of primary inputs. Then determine average power and power sensitivities due to the d different statistics of primary inputs. Given a certain statistics of primary inputs, we use the nearest point in the specification-space as the nominal value of the needed power. Finally, the required power dissipation can be obtained by equation (6). Consequently, the vicinity of a representative point is approximately characterized by a plane.

To illustrate how our technique works, let us consider a 2-input AND gate $y = x_1 x_2$, where x_1 and x_2 are independent primary inputs. We have activity $a(y) = P(x_1)a(x_2) + P(x_2)a(x_1) - \frac{1}{2}a(x_1)a(x_2)$. Figure 1 gives a simplified activity surface for the 2-input AND with $P(x_1) = 0.3$ and

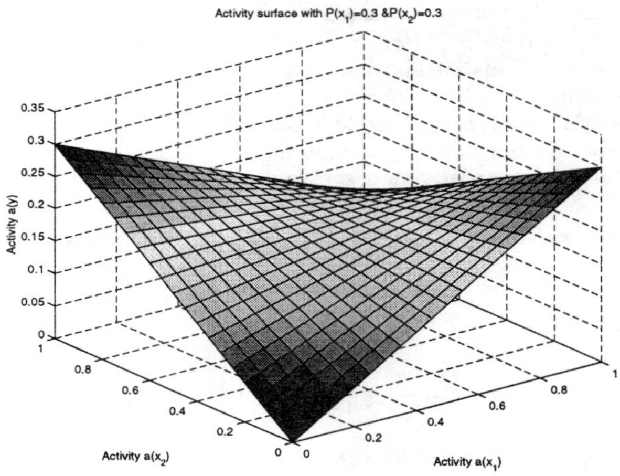

Figure 1: Activity surface for a 2-input AND

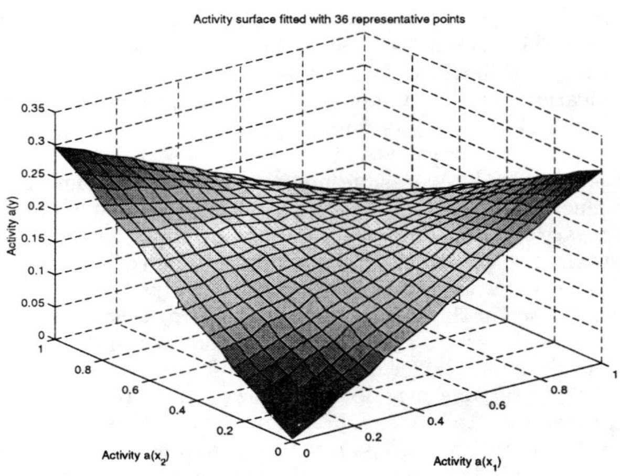

Figure 2: Approximated activity surface

$P(x_2) = 0.3$. Figure 2 gives the activity surface obtained by equation (6) with 36 representative points. It fits the actual surface quite well. It should be noted that for a circuit with n inputs, the $(2n+1)$-dimensional power surface is impossible to plot. Even for the 2-input AND, it is still impossible to plot the actual 5-dimensional activity surface when considering all possible probability values.

4 Efficient Techniques to Estimate Power Sensitivities

A naive approach to estimate power sensitivity would be to simulate a circuit to obtain the average power dissipation based on nominal values of primary input probabilities and activities. Then assign a small variation to only one primary input and re-simulate the circuit. After all the primary inputs have been exhausted, power sensitivity can be obtained using $\Delta Power_i/\Delta\theta(x_i)$. This naive method can be easily implemented. However, it involves $n+1$ times of power estimation processes. If the number of primary inputs is large, this method can be computationally expensive. Therefore, the naive method is impractical for large circuits with large number of primary inputs. In this section we present two efficient techniques to estimate power sensitivities.

4.1 Symbolic Technique to Obtain Power Sensitivities

The basic idea of the symbolic technique to obtain power sensitivities is to express the activity of each internal node or primary output in terms of the probabilities and activities of primary inputs. After we obtain the exact expression for signal activity, we can easily compute power sensitivities $\zeta_{a(x_i)}$ and $\zeta_{P(x_i)}$ by equations (1) and (2), and higher order power sensitivities by equation (4).

Note that accurate calculation of power sensitivity depends on whether we can accurately express the probability and activity of each internal node and primary output in terms of independent inputs. However, the size of symbolic probability expression and activity expression grow exponentially with respect to the number of independent inputs. Consequently, we resort to circuit partitioning to trade-off accuracy versus computation time. The circuit partitioning technique is similar to the one given in [4].

4.2 Statistical Technique to Estimate Power Sensitivities

The basic idea of the statistical technique is to obtain sensitivities as a by-product of statistical power estimation processes. A more detailed derivation can be found in [3].

The expected value of average power in a CMOS circuits can be expressed as follows:

$$E[Pwr] = \sum_{\mathbf{X^O},\mathbf{X^T}} Power(\mathbf{X^O}\ \mathbf{X^T})P(\mathbf{X^O}\ \mathbf{X^T}) \tag{8}$$

where $\mathbf{X^O} = (I_1^0, I_2^0, \cdots, I_n^0)$ and $\mathbf{X^T} = (I_1^T, I_2^T, \cdots, I_n^T)$ are primary input vectors, T is the clock cycle, and $I_i^0 = x_i^0$ or $\overline{x_i}^0$, and $I_i^T = x_i^T$ or $\overline{x_i}^T$. The power consumption in every clock cycle is a random variable and is denoted as Pwr. $Power(\mathbf{X^O}\ \mathbf{X^T})$ represents the power consumption due to the pair of input vectors $\mathbf{X^O}$ and $\mathbf{X^T}$. Power sensitivity can be expressed as follows:

$$\begin{aligned}\zeta_{\theta(x_i)} &= \sum_{\mathbf{X^O},\mathbf{X^T}} \left(Power(\mathbf{X^O X^T}) \frac{\partial P(\mathbf{X^O X^T})}{\partial \theta(x_i)} \right) && (9)\\ &= E\left[Pwr\ \frac{\partial ln(P(I_i^0\ I_i^T))}{\partial \theta(x_i)} \right] && (10)\end{aligned}$$

$P(I_i^0\ I_i^T)$ has four combinations: $P(x_i^0\ x_i^T)$, $P(\overline{x_i}^0\ \overline{x_i}^T)$, $P(x_i^0\ \overline{x_i}^T)$, and $P(\overline{x_i}^0\ x_i^T)$. $\frac{\partial ln(P(I_i^0\ I_i^T))}{\partial a(x_i)}$ and $\frac{\partial ln(P(I_i^0\ I_i^T))}{\partial P(x_i)}$ can be calculated as follows [3]:

$$\frac{\partial ln(P(x_i^0\ x_i^T))}{\partial a(x_i)} = \frac{-\frac{1}{2}}{P(x_i) - \frac{1}{2}a(x_i)} \tag{11}$$

$$\frac{\partial ln(P(\overline{x_i^0}\ \overline{x_i^T}))}{\partial a(x_i)} = \frac{-\frac{1}{2}}{1 - P(x_i) - \frac{1}{2}a(x_i)} \tag{12}$$

$$\frac{\partial ln(P(x_i^0\ \overline{x_i^T}))}{\partial a(x_i)} = \frac{\partial ln(P(\overline{x_i^0}\ x_i^T))}{\partial a(x_i)} = \frac{1}{a(x_i)} \tag{13}$$

$$\frac{\partial ln(P(x_i^0\ x_i^T))}{\partial P(x_i)} = \frac{1}{P(x_i) - \frac{1}{2}a(x_i)} \tag{14}$$

$$\frac{\partial ln(P(\overline{x_i^0}\ \overline{x_i^T}))}{\partial P(x_i)} = \frac{-1}{1 - P(x_i) - \frac{1}{2}a(x_i)} \tag{15}$$

$$\frac{\partial ln(P(x_i^0\ \overline{x_i^T}))}{\partial P(x_i)} = \frac{\partial ln(P(\overline{x_i^0}\ x_i^T))}{\partial P(x_i)} = 0 \tag{16}$$

To obtain higher order power sensitivities, we can substitute $\partial P(\mathbf{X^0X^T})/\partial\theta(x_i)$ in equation (9) by

$$\frac{\partial^{(2)}P(\mathbf{X^0X^T})}{\partial\theta(x_k)\theta(x_i)} = P(\mathbf{X^0X}^T)\frac{\partial ln(P(I_k^0 I_k^T))}{\partial\theta(x_k)}\frac{\partial ln(P(I_i^0 I_i^T))}{\partial\theta(x_i)} \tag{17}$$

$$\underbrace{\frac{\partial^{(m)}P(\mathbf{X^0X}^T)}{\partial\theta(x_k)...\partial\theta(x_i)}}_{m} = \underbrace{P(\mathbf{X^0X}^T)\frac{\partial ln(P(I_k^0 I_k^T))}{\partial\theta(x_k)}...\frac{\partial ln(P(I_i^0 I_i^T))}{\partial\theta(x_i)}}_{m} \tag{18}$$

where $\partial ln(P(I_k^0\ I_k^T))/\partial\theta(x_k)$ or $\partial ln(P(I_i^0\ I_i^T))/\partial\theta(x_i)$ has one of the eight combinations shown in equations (11)-(16).

5 Experimental Results

We have implemented the two techniques to estimate power sensitivities in C under the Berkeley SIS environment. Based on the power sensitivities, the power macromodel has been implemented to accurately estimate power under any signal statistics.

5.1 Results for Power Sensitivities

Results for power sensitivities indicate that for some circuits power is much more sensitive to some primary inputs than others. A small activity variation of such highly sensitive primary inputs will result in a dramatic change of the average power. Consider circuit *i*6. All the primary inputs are assumed to have probability and activity values (nominal) of 0.5 and 0.26, respectively. The power sensitivities to activities of the 1st, 2nd, 60th, and 124th primary input are 237, 276, 133, and 30, respectively while the power sensitivities of the other primary inputs are less than 4. If the activities of the 1st and 2nd primary input have a variation of ± 0.05, results indicate that power can change by 30%.

The naive method to compute power sensitivities proposed in the previous section acts as a figure of merit for *STOPS* (symbolic technique to obtain power sensitivity) and *STEPS* (statistical technique to estimate power sensitivity). The average percent difference between the results obtained by *STEPS* and those obtained by the naive method is about 4% [3]. For *STOPS*, circuit partitioning may introduce an error of the order of $5-15\%$ [2]. Both *STOPS* and *STEPS* can be several orders of magnitude faster than the naive method.

It should be noted that *STEPS* can handle different delay models for logic gates. Due to the delay of each logic gate, paths arriving at any internal gate may have different propagation delays. The delay mismatch of different paths can cause spurious transitions. This in turn increases power dissipation and power sensitivities. Another interesting result is that the relative values of power sensitivities to different primary inputs may change when considering non-zero delay. For circuit $C3540$, with zero-delay model, the average power is more sensitive to the third primary input than any others. However, with unit delay, the most sensitive primary input changes to the first one instead of the third.

5.2 Power under Any Distribution of Primary Inputs

After obtaining power sensitivities, we use equation (6) to compute the average power for each simulated circuit under certain specification of primary inputs. The power sensitivities used in this section are obtained by *STEPS*. For simplicity, we assume that each primary input has a fixed probability value of 0.5.

Table 1: Average power for $C432$ with $\Delta a = 0.05$

a	Power (Zero Delay)			Power (Unit Delay)		
	macro	*sim*	%	*macro*	*sim*	%
0.1	54.5	52.9	3.0	75.4	73.7	2.3
0.15	67.0	67.6	-0.9	95.4	98.9	-3.5
0.2	80.3	80.0	0.4	120.3	120.0	0.3
0.25	91.7	91.6	0.1	140.7	141.4	-0.5
0.3	102.5	101.3	1.2	161.5	159.3	1.4
0.35	111.5	112.4	-0.8	179.6	180.6	-0.6
0.4	121.7	121.1	0.5	199.0	197.6	0.7
0.45	130.0	130.3	-0.2	215.9	216.5	-0.3
0.5	138.7	138.6	0.0	233.9	233.4	0.2
0.55	147.2	146.4	0.5	250.3	250.6	-0.1
0.6	154.4	153.8	0.4	267.2	266.5	0.3
0.65	161.5	162.4	-0.6	282.9	281.4	0.5
0.7	169.8	169.2	0.4	297.8	295.9	0.6
0.75	176.6	176.2	0.2	312.2	313.3	-0.4
Avg	-	-	0.65	-	-	0.83

Table 1 shows the results for circuit $C432$ based on different nominal values for activity. Corresponding to each chosen nominal activity, all primary inputs are assumed to take this value. The column "macro" represents the power obtained by our technique with Δa equal to 0.05 for each primary input, while column "sim" represents the power obtained by a Monte Carlo based simulation method with the activity value of each primary input equal to $(a_{nom} + \Delta a)$. The values shown in the column " %" are obtained by $(P_{macro} - P_{sim})/P_{sim}$, which stands for the percent difference between the results obtained by the two methods. The last row stands for the average percent difference. Results indicate that with zero-delay model the percent difference never exceeds 3.0% and the average percent difference is only 0.65%. With unit delay model, the average percent difference is 0.83% with the maximum percent difference no greater than 3.5%.

Table 2: Average power with $a = 0.25$ and $\Delta a = 0.05$

Circuit	Power (Zero Delay)			Power (Unit Delay)		
Chosen	*macro*	*sim*	%	*macro*	*sim*	%
C432	91.7	91.6	0.1	140.7	141.4	-0.5
C499	152.0	151.5	0.3	236.3	236.4	0.0
C880	200.1	198.7	0.7	257.7	257.7	0
C1355	312.7	311.6	0.4	691.4	690.4	0.1
C1908	478.7	474.2	0.9	1096.2	1074.9	2.0
C2670	649.5	649.5	0	1099.5	1102.9	-0.3
C3540	812.2	808.3	0.5	1718.8	1710.6	0.5
C5315	1464.3	1446.4	1.2	3008.4	2952.5	1.9
C6288	1688.9	1688.0	0.1	46710.1	46874.8	-0.4
C7552	2084.9	2067.1	0.9	4479.6	4470.0	0.2
Avg	-	-	0.5	-	-	0.6

Table 2 gives the results for different benchmarks. In this case, each primary input of each simulated circuit has the same a_{nom} of 0.25 and the same Δa of 0.05. The values shown in columns "macro", "sim", and " %" are obtained in the same way as those shown in table 1. With zero delay model, compared with the Monte Carlo based simulation, the average percent difference is only 0.5% and the maxi-

Table 3: Average power with randomly a and $\Delta a = 0.05$

Circuit Chosen	*Power (Zero Delay)*			*Power (Unit Delay)*		
	macro	*sim*	%	*macro*	*sim*	%
C432	144.5	143.9	-0.4	246.7	246.4	-0.1
C499	201.2	201.9	0.3	287.5	289.2	0.6
C880	293.2	291.8	-0.5	416.2	415.7	-0.1
C1355	435.4	436.0	0.1	974.7	980.2	-0.6
C1908	678.1	680.0	0.3	1812.1	1818.6	-0.4
C2670	989.5	985.5	-0.4	2047.6	2048.7	0.0
C3540	1184.0	1182.6	-0.1	2878.8	2881.8	-0.1
C5315	2086.0	2083.5	-0.1	5070.4	5055.6	0.3
C6288	2042.0	2039.1	-0.1	57888.7	57845.0	0.1
C7552	2902.5	2895.6	-0.2	8578.0	8533.9	0.5
Avg	-	-	0.3	-	-	0.3

Table 4: Average power with randomly a and $\Delta a = 0.25$

Circuit Chosen	*Power (Zero Delay)*			*Power (Unit Delay)*		
	macro	*sim*	%	*macro*	*sim*	%
C432	158.1	151.4	4.4	281.8	260.1	8.3
C499	198.4	197.7	0.4	282.0	283.9	-0.7
C880	338.3	317.7	6.5	485.5	525.8	-7.7
C1355	435.1	425.0	2.4	1034.6	955.9	8.2
C1908	710.8	719.7	-1.2	2024.4	1983.0	2.1
C2670	1055.2	1000.2	5.5	2018.6	1986.6	1.6
C3540	1252.5	1216.1	3.0	3108.8	3024.4	2.8
C5315	2344.8	2238.4	4.8	5717.8	5590.6	2.3
C6288	2204.4	2142.3	2.9	63973.7	60721.9	5.4
C7552	3170.3	2871.3	10.4	7352.2	7117.2	3.3
Avg	-	-	4.2	-	-	4.2

mum value of the difference is under 1.5%. With unit delay model, the percent difference does not exceed 2.0% while the average value is about 0.6%.

For table 3, each primary input of a tested circuit has a randomly generated activity value between 0.05 and 0.95 while Δa is still equal to 0.05. The column "macro" still represents the power obtained by equation (6) based on randomly generated possible activities as the nominal values. The maximum percent difference is under 1.0%.

The difference between table 3 and table 4 is that we choose a relatively large value 0.25 for Δa for each primary input. Compared to the Monte Carlo based simulation method, the average percent difference is about 4%. This is encouraging. It demonstrates that even a relatively large activity change does not affect the accuracy of our technique. Hence, the number of nominal activity values used to construct a power surface can be really small. The fewer the number of simulated points, the less will be CPU time and memory requirements.

Table 5 gives the results when both a and Δa are randomly generated with values between 0 and 1. This means that each primary input may have a different activity and a different Δa. But for each primary input, there exists a constraint $a + \Delta a < 1.0$. For the zero delay case, results obtained by our method match extremely well with those obtained by the Monte Carlo method. The maximum percent difference is only 0.2%. For the unit delay case, the average percent difference is still under 6% while the maximum difference is less than 10%.

Table 5: Average power with randomly a and Δa

Circuit Chosen	*Power (Zero Delay)*			*Power (Unit Delay)*		
	macro	*sim*	%	*macro*	*sim*	%
C432	131.4	131.1	0.2	295.1	297.7	-0.9
C499	188.4	188.4	0	326.2	301.9	8.0
C880	269.6	269.2	0.1	504.6	493.4	2.3
C1355	396.1	396.0	0.0	1132.9	1034.7	9.5
C1908	612.4	612.2	0.0	2291.4	2105.3	8.8
C2670	897.5	897.6	0.0	2446.2	2323.5	5.3
C3540	1156.5	1153.7	0.2	3651.9	3446.8	6.0
C5315	1832.5	1832.7	0.0	6312.5	6133.4	2.9
C6288	1912.0	1914.4	-0.1	64609.1	61778.2	4.6
C7552	2751.1	2751.3	0.0	10521.8	9839.3	6.9
Avg	-	-	0.06	-	-	5.5

Finally, to show the efficiency and accuracy of our technique, we report in tables 6 and 7 the results based on the power surface approximated by **ten** randomly chosen points for each circuit, where tables 6 and 7 correspond to zero delay model and unit delay model, respectively. For each representative point, we randomly generate an activity value for each primary input, and then use *STEPS* to estimate the corresponding power and power sensitivities. Therefore, the power surface of a circuit is approximated by ten randomly generated points. In order to compare our technique with Monte Carlo based simulation, for each circuit, ten different distributions of primary inputs are randomly generated. For distribution Υ_i, we first find the nearest point which minimizes $\sqrt{\sum_j a(j) - a(j)_{nom(i)})^2}$, where j varies for all primary inputs and i varies for the ten representative points in the specification space. Then by equation (6) we use the corresponding power as the nominal value to compute the power corresponding to distribution Υ_i. The percent differences between the results obtained by the two methods are shown in columns 2 to 11 in tables 6 and 7. Column "Average" represents the average percent difference over the ten distributions. The results are promising. For each simulated circuit, our technique can yield 95% accuracy on an average with only ten randomly chosen representative points.

We conclude this section with a few comments. Firstly, when we show the above results we assume a fixed probability for each primary input. However, our technique is not limited to such an assumption. Secondly, the power dissipation shown in the "macro" columns in the above tables are obtained by only considering (first order) power sensitivities. In order to improve accuracy, higher order power sensitivities can be included without much overhead, which is due to the fact that power sensitivities can be obtained as by-products during power estimation process. Thirdly, we can further reduce memory requirements by storing only those parameters associated with very sensitive primary inputs. Finally, in some cases, the number of representative points used to construct the power surface can be reduced to **ONE**, which means only **one** symbolic run or only **one** statistical power estimation run is needed. For example, if the properties of each primary input falls into a relatively small range, for each primary input, we can use the middle point of its range as the nominal value to obtain all the necessary parameters: nominal average power, power sensitivities to activities, and power sensitivities to probabilities. Based on these parameters, power under any input distribution within

Table 6: Comparison of our technique and Monte Carlo simulation with zero delay model

Circuit Chosen	*Percent Difference %*										
	Υ_1	Υ_2	Υ_3	Υ_4	Υ_5	Υ_6	Υ_7	Υ_8	Υ_9	Υ_{10}	Average
C432	-1.9	-2.1	-1.0	-0.6	2.7	-1.3	-2.4	0.9	2.0	10.8	2.6
C499	3.9	-0.6	0.4	0.3	0.5	2.1	-0.8	3.6	3.0	-0.4	1.6
C880	3.2	1.0	4.4	-1.4	-3.0	-2.1	-0.5	1.4	-3.0	-3.4	2.3
C1355	5.4	1.8	1.4	2.1	0.9	2.4	0.0	-1.7	3.5	4.9	2.4
C1908	3.3	-3.3	1.3	-2.1	1.1	0.2	4.3	-2.9	2.5	-1.5	2.3
C2670	0.4	4.3	-5.1	-3.9	-3.1	0.3	-1.9	1.8	-1.4	2.6	2.5
C3540	2.0	3.7	-1.5	0.6	2.7	3.8	7.0	-1.5	3.2	2.4	2.8
C5315	2.8	-5.4	-3.1	3.0	3.8	3.1	3.2	-3.4	0.4	3.3	3.2
C6288	-7.8	0.6	0.1	-1.1	1.4	-0.6	-1.5	-2.7	-0.1	-1.4	1.7
C7552	1.8	-0.2	-3.1	1.9	6.3	10.9	-2.1	1.9	3.2	2.7	3.4

Table 7: Comparison of our technique and Monte Carlo simulation with unit delay model

Circuit Chosen	*Percent Difference %*										
	Υ_1	Υ_2	Υ_3	Υ_4	Υ_5	Υ_6	Υ_7	Υ_8	Υ_9	Υ_{10}	Average
C432	-4.8	-2.8	0.0	1.5	2.2	-0.8	2.0	-1.4	-5.8	-5.1	2.6
C499	1.3	-0.8	-1.8	2.0	2.2	1.8	-1.5	1.7	2.1	-5.9	2.1
C880	-3.5	-4.1	-1.0	1.1	3.6	0.4	-8.1	-2.9	2.5	-0.8	2.8
C1355	3.5	1.5	1.1	-0.4	6.4	-2.7	-1.7	-6.8	-2.4	2.8	3.0
C1908	-4.2	1.2	1.5	1.7	-1.3	3.5	-0.3	-1.0	-0.5	2.5	2.9
C2670	0.4	7.3	7.7	-5.0	-4.8	5.5	4.4	3.3	1.0	0.6	4.0
C3540	-3.7	1.4	3.3	-0.9	8.3	-1.7	-2.0	1.8	-5.8	-1.4	3.0
C5315	3.4	-4.0	2.6	3.5	4.4	-0.8	-10.4	-7.3	1.5	-3.0	4.2
C6288	-1.0	1.7	0.0	-1.7	-8.0	-1.1	-0.6	-1.5	0.4	-1.5	1.8
C7552	5.6	3.1	3.7	6.1	-3.4	8.2	-2.4	-2.3	-5.3	4.2	4.4

the given range can be easily computed.

6 Conclusions

In this paper we have presented a novel power macromodel technique for high level power estimation based on power sensitivities. A key advantage of this work is that we can use a relatively small number of representative points to construct a complicated power surface which can then be used to determine the average power under any statistics of primary inputs. This makes the implementation of our technique very efficient. Another feature of our technique is that the memory requirements for the power macromodel can be reduced to $O(n)$ (n is the number of primary inputs). The feasibility and effectiveness of our technique have been verified on a large number of benchmark examples.

References

[1] A.P.Chandrakasan, S.Sheng, and R.W.Brodersen, "Low-Power CMOS Digital Design," *Journal of Solid-State Circuits*, Vol.27, No.4, pp.473-483, April,1992.

[2] Z. Chen, K. Roy, and T.-L. Chou, "Sensitivity of Power Dissipation to Uncertainties in Primary Input Specification," *Custom Integrated Circuits Conference*, pp. 487-490, 1997.

[3] Z. Chen, K. Roy, and T.-L. Chou, "Power Sensitivity - A New Method to Estimate Power Considering Uncertain Specifications of Primary Inputs," *IEEE Intl. Conf. on Computer-Aided-Designed*, pp.40-44, 1997.

[4] T.-L. Chou, K. Roy, and S. Prasad, "Estimation of Circuit Activity Considering Signal Correlations and Simultaneous Switching," *IEEE Intl. Conf. on Computer-Aided-Designed*, pp. 300-303, 1994.

[5] Michael D. Greenberg, "Advanced Engineering Mathematics," New Jersey: Prentice-Hall, 1988.

[6] S. Gupta and F. Najm, "Power Macromodeling for High Level Power Estimation," *34rd ACM/IEEE Design Automation Conference*, 1997.

[7] P. E. Landman and J. M. Rabaey,"Architectural Power Analysis:The Dual Bit Type Method," *IEEE Transactions on VLSI*, pp. 173-187, 1995.

[8] D. Marculescu, R. Marculescu, and M. Pedram, "Information Theoretic Measures of Energy Consumption at Register Transfer Level," *ACM/IEEE Intl. Symposium on Low Power Design*, pp. 87-92, 1995.

[9] H. Mehta, R. M. Owens and M. J. Irwin, "Energy Characterization based on Clustering," *33rd ACM/IEEE Design Automation Conference*, pp. 702-707, 1996.

[10] M. Nemani and F. Najm, "Towards a High-Level Power Estimation Capability," *IEEE Trans. on CAD*, pp. 588-598, June 1996.

[11] S. R. Powell and P. M. Chau, "Estimating Power Dissipation of VLSI Signal Processing Chips: The PFA Technique," *VLSI Signal Processing IV*, pp. 250-259, 1990.

[12] Q. Qiu, Q. Wu, and M. Pedram, "Cycle-Accurate Macro-Models for RT-Level Power Analysis," *Proceedings of 1997 International Symposium on Low Power Electronics and Design*, pp. 125-130, August 1997.

[13] A. Raghunathan, S. Dey and N. K. Jha, "Register-Transfer Level Estimation Techniques for Switching Activity and Power Consumption," *IEEE Intl. Conf. on Computer-Aided-Design*, pp. 158-165, 1996.

[14] K. Roy and S. Prasad, "Circuit Activity Based Logic Synthesis for Low Power Reliable Operations," *IEEE Trans. on VLSI Systems*, pp. 503-513, Dec. 1993.

Maximum Power Estimation Using the Limiting Distributions of Extreme Order Statistics*

Qinru Qiu, Qing Wu and Massoud Pedram

Department of Electrical Engineering – Systems
University of Southern California
Los Angeles, CA 90089
EMAIL: {qinru, qingwu, massoud}@zugros.usc.edu

Abstract

In this paper we present a statistical method for estimating the maximum power consumption in VLSI circuits. The method is based on the theory of extreme order statistics applied to the probabilistic distribution of the cycle-based power consumption, maximum likelihood estimation, and Monte-Carlo simulation. The method can predict the maximum power in the constrained space of given input vector pairs as well as the complete space of all possible input vector pairs. The simulation-based nature of the proposed method allows one to avoid the limitations imposed by simple gate-level delay models and handle arbitrary circuit structures. The proposed method can produce maximum power estimates to satisfy user-specified error and confidence levels. Experimental results show that this method provides maximum power estimates within 5% of the actual value and with a 90% confidence level by simulating, on average, about 2500 vector pairs.

I. Introduction

There are many factors that may cause circuit failure, including excessive power dissipation over a short period of time. High current flow over a short time frame may lead to permanent circuit damage, voltage change on the power/ground distribution nets, circuit slow down, or double clocking hazards. To design a reliable circuit, designers have to depend on efficient and accurate estimation of maximum power dissipation in the circuit. The maximum power may be averaged over the clock cycle, resulting in the cycle-based maximum power value. This is the focus of our paper.

In much of the previous work, the maximum power estimation problem refers to the problem of estimating the maximum power (or current) that the circuit may consume within any clock cycle. The problem is thus equivalent to finding the maximum-power-consuming vector pair among all possible input vector pairs. Previous techniques, therefore, focus on finding lower bound and upper bounds on the maximum power dissipation in the circuit. In many applications, however, the space of input vectors that can be applied to a VLSI chip is constrained. This gives rise to the following classification:

I.1 The maximum power for all possible vector pairs applied to the inputs of the circuit. We refer to this as the *unconstrained maximum power* estimation problem.

I.2 The maximum power for given transition/joint-transition probability specification for the circuit inputs. We refer to this as the *constrained maximum power* estimation problem.

A number of techniques have been developed to solve the problems in Category I.1 [1]-[10]. Category I.2 has however not been explored. The method proposed in [1] propagates the signal uncertainty through the circuits to obtain a loose upper bound on the maximum power. The bound is then made tighter by doing analysis of the circuit structure. The bound tightening method tends to be time consuming when the number of the primary inputs is large.

Automatic test pattern generation (ATPG) based techniques [5]-[6] attempt to generate an input vector pair that produces the largest switched capacitance in the circuit. The power consumption by the vector pair is then used as a lower bound on the maximum power of the circuit. The ATPG based techniques are very efficient and generate a tighter lower bound than that generated by random vector generation. The limitations are, however, that the ATPG based techniques can only handle simple delay models such as the zero-delay and unit-delay models, and that the analysis is done at the gate-level. Consequently, the estimation accuracy is not high. A continuous optimization method was proposed in [7], which treats the input vector space as a continuous real-valued vector space and then performs a gradient search to find the maximum. Similar to the ATPG based techniques, the estimation accuracy is not high.

The authors of [8] proposed a technique for finding the maximum power-consuming vector using a genetic search algorithm. The limitation of this approach is that efficiency is not high. Statistical methods have also been studied for maximum power estimation. In [5] a statistical technique for maximum current estimation was briefly discussed. The method randomly generates high-activity vector pairs and maximum power is then estimated by simulation. This method also suffers from low efficiency. The theory of order statistics has been applied in [9][10] to estimate maximum power by estimating the high quantile point. The efficiency is however as low as the random vector generation technique.

In this paper, we present a simulation-based statistical method for maximum power estimation for combinational circuits. It is a method of estimating the maximum power using the theory of *Asymptotic Extreme Order Statistics*. Compared to previous work, this approach makes the following contributions:

* This work was supported in part by DARPA under contract No. F33615-95-C1627, SRC under contract No. 98-DJ-606, and NSF under contract No. MIP-9628999.

DAC 98, San Francisco, California
©1998 ACM 0-89791-964-5/98/06..$5.00

1. This approach can estimate the maximum power defined in both categories I.1 and I.2.
2. Because it is a simulation-based technique, the delay model and circuit structure do not limit the accuracy of our method.
3. The approach is the first to provide the confidence interval for the estimated maximum power for the user-specified confidence level.
4. This is the first approach which can provide maximum power estimation for any given error and confidence level.
5. By efficient statistical estimation of the extreme distributions, the estimation efficiency is significantly improved compared to existing statistical methods (including simple random sampling or quantile estimation).

On average, our method can perform maximum power estimation by simulating only about 2500 vector pairs to achieve a 5% error at a confidence level of 90%.

This paper is organized as follows. Section II introduces the theory of asymptotic extreme order statistics and maximum likelihood estimation. Section III describes our approach for maximum power estimation. Sections IV and V present our experimental results and concluding remarks.

II. Background

2.1 The asymptotic theory of extreme order statistics

The (cumulative) distribution function (in short d.f.) of a random variable (in short r.v.) x is defined as:

$$F(t) = P\{x \le t\} \tag{2.1}$$

The quantile function (in short q.f.) of a d.f. F is defined as:

$$F^{-1}(q) = \inf\{t : F(t) \ge q\}, \quad q \underset{random}{\in} [0,1] \tag{2.2}$$

where $inf(S)$ calculates the lower bound of set S. Notice that the q.f. F^{-1} is a real-valued function and $F^{-1}(q)$ is the smallest q quantile of F, that is, if Z is a r.v. with d.f. F, then $F^{-1}(q)$ is the smallest value t such that $P\{Z < t\} \le q \le P\{Z \le t\}$. We remark that $F(x)=\sup\{q \in [0,1]: F^{-1}(q) \le x\}$. Let $z_1, z_2, ..., z_n$ be n random units drawn from a common distribution. If they are drawn in at random, they are called independent identically distributed (in short i.i.d.) r.v.'s. If there is no interest in the order in which $z_1, z_2, ..., z_n$ are drawn, but interest in the order of the magnitude of their values, the ordered sample values can be examined:

$$X_{1:n} \le X_{2:n} \le \cdots \le X_{n:n}$$

which are the *order statistics* of a sample of size n.

$X_{r:n}$ is called the rth order statistic and the random vector $(X_{1:n}, X_{2:n}, ..., X_{n:n})$ is called the order statistic. Note that $X_{1:n}$ is the sample minimum and $X_{n:n}$ is the sample maximum. $X_{1:n}$ is called the *minima order statistic* and $X_{n:n}$ is called the *maxima order statistic*, or in general, they are called the *extreme order statistics* of a sample of size n.

The distribution function of the sample maxima $X_{n:n}$, is given by:

$$P\{X_{n:n} \le t) = F(t)^n \tag{2.3}$$

Three distribution functions are given for studying the limiting d.f. of sample maxima (in other words: extreme value d.f.'s):

$$G_{1,\alpha}(x) = \begin{cases} 0 & \text{if } x \le 0 \\ \exp(-x^{-\alpha}) & \quad\; x > 0 \end{cases}, \quad \text{“Fréchet”} \tag{2.4}$$

$$G_{2,\alpha}(x) = \begin{cases} \exp(-(-x)^{-\alpha}) & \text{if } x \le 0 \\ 0 & \quad\; x > 0 \end{cases}, \quad \text{“Weibull”} \tag{2.5}$$

$$G_3(x) = \exp(-e^{-x}) \quad \text{for every } x, \quad \text{“Gumbel”} \tag{2.6}$$

Definition 1 [12] F is said to belong to the weak domain of attraction of limiting d.f. G, if there exist series of constants $a_n > 0$ and reals b_n such that:

$$F^n(b_n + xa_n) \to G(x), \qquad n \to \infty \tag{2.7}$$

for every continuity point of G.

Let us define the *right endpoint* of d.f. F as:

$$\omega(F) = \sup\{x : F(x) < 1\} = F^{-1}(1) \tag{2.8}$$

Theorem 1 [12] A d.f. F belongs to the weak domain of attraction of an extreme value d.f. $G_{i,\alpha}$ if, and only if, one of the following conditions holds:

$$(1,\alpha): \omega(F) = \infty \text{ and } \lim_{t\to\infty}\left[\frac{1-F(tx)}{1-F(t)}\right] = x^{-\alpha}, x > 0 \tag{2.9}$$

$$(2,\alpha): \omega(F) < \infty \text{ and } \lim_{t\downarrow 0}\left[\frac{1-F(\omega(F)+xt)}{1-F(\omega(F)-t)}\right] = (-x)^{\alpha}, x < 0 \tag{2.10}$$

$$(3): \lim_{t\uparrow\omega(F)}\left[\frac{1-F(t+xg(t))}{1-F(t)}\right] = e^{-x}, \quad -\infty < x < \infty \tag{2.11}$$

and $\downarrow$ represents approaching decreasingly, $\uparrow$ represents approaching increasingly.

Moreover, constants a_n and b_n can be chosen in the following way:

$$(1,\alpha): \quad b_n^* = 0, \quad a_n^* = F^{-1}(1-1/n) \tag{2.12}$$

$$(2,\alpha): \quad b_n^* = \omega(F), \quad a_n^* = \omega(F) - F^{-1}(1-1/n) \tag{2.13}$$

$$(3): \quad b_n^* = F^{-1}(1-1/n), \quad a_n^* = g(b_n^*) \tag{2.14}$$

If a distribution F satisfies one of the conditions in Theorem 1, we simply call the corresponding $G_{i,\alpha}$ the asymptotic distribution of the sample maxima of distribution F. Theorem 1 not only gives conditions under which the extreme value d.f. G will converge, but also guidelines for selecting the correct asymptotic extreme distribution for different application.

Theorem 2 [12] The weak convergence to the limiting d.f. G holds for other choices of constants a_n and b_n if, and only if,

$$a_n / a_n^* \to 1 \quad \text{and} \quad (b_n - b_n^*)/a_n \to 0 \text{ as } n \to \infty \tag{2.15}$$

Theorem 2 gives other possibilities of choosing a_n and b_n in theorem 1. In special cases when $F(x)$ has a finite right endpoint, by Theorem 2, the choice of b_n $(n\to\infty)$ in Eqn.(2.13) is unique.

2.2 Maximum-likelihood estimation for parameters of the Weibull distribution when $\alpha > 2$

For reasons that will be made clear in a later section, we are interested in developing a maximum-likelihood estimator for parameters of a generalized Weibull distribution defined as:

$$G(x;\alpha,\beta,\mu) = \begin{cases} \exp(-\beta(\mu - x)^{-\alpha}) & \text{if } x \le \mu \\ 0 & \quad\; x > \mu \end{cases} \tag{2.16}$$

where μ is the location parameter which determines the right endpoint (i.e. maximum) of the distribution, $\beta > 0$ is the scale parameter, and α is the shape parameter.

The maximum-likelihood estimation problem is defined as follows: Given m independent random samples $x_1, x_2, \ldots, x_m$ of $G(x; \alpha, \beta, \mu)$, find the values of α, β, μ which maximize the likelihood function [11]:

$$L_m(\alpha,\beta,\mu) = \frac{1}{m}\sum_{i=1}^{m}\log[G(x_i;\alpha,\beta,\mu)] \tag{2.17}$$

This maximum likelihood estimator, when it exists, will be denoted by the vector $(\hat{\alpha}_m, \hat{\beta}_m, \hat{\mu}_m)$ and satisfies:

$$\frac{\partial L_m}{\partial\{\alpha,\beta,\mu\}}(\hat{\alpha}_m,\hat{\beta}_m,\hat{\mu}_m)=0 \tag{2.18}$$

Let α_0, β_0, μ_0 denote the actual values of parameters of the distribution, G. It was proved in [11] that, when $\alpha > 2$, $\sqrt{m}(\hat{\alpha}_m-\alpha_0), \sqrt{m}(\hat{\beta}_m-\beta_0), \sqrt{m}(\hat{\mu}_m-\mu_0)$ will all converge to normal distributions with mean 0, when $m\to\infty$.

III. The Estimation Approach

The problem of maximum power estimation can be stated as follows: Given a set V (called *population*) of input vector pairs, estimate the maximum power dissipation that the circuit may exhibit for any vector pair in the population. A vector pair in V is called a *unit* of the population. In this paper, the population may include all possible input vector pairs applied to a circuit, or all possible vector pairs under some transition probability constraints. Although there may be a finite number of distinct vector pairs in the population, the size of V, represented by $|V|$, is assumed to be infinite because repeating the vector pairs is allowed.

3.1 The asymptotic distribution of the sample maximum power

If we regard power consumption of a vector pair as a random variable p, a distribution of p is then formed by the power consumption values of vector pairs in set V. The average power is the mean value of the distribution. The maximum power is then the right endpoint of the distribution. We assume that the d.f. of power consumption in large circuits is continuous distribution.

Given population V, the ith sample for maximum power estimation is formed by the power values of n randomly selected units:

$$p_{i,1}, p_{i,2}, \cdots, p_{i,n} \qquad i=1,2,\cdots,m$$

where n is called the sample size and m the number of samples. The maximum power in each sample is defined as:

$$p_{i,MAX}=\max\{p_{i,1},p_{i,2},\cdots,p_{i,n}\} \quad i=1,2,\cdots,m \tag{3.1}$$

According to Eqn.(2.3), the d.f. of $p_{i,MAX}$ can be written as: $H(p_{i,MAX}) = F(p)^n$. As mentioned in Theorem 1, $H(b_n+ p_{i,MAX}\cdot a_n)$ asymptotically converges to one of the three distributions defined in Eqn.'s(2.4), (2.5) and (2.6).

In the remainder of this paper, we will use $\omega(F)$ to denote the actual maximum power of the population.

We know that power consumption in an LSI circuit is a finite value, i.e., $\omega(F) < \infty$. Therefore the condition in (2.9) is not met and $H(b_n+ p_{i,MAX}\cdot a_n)$ cannot converge to $G_{1,\alpha}$. Also, because the upper bound of the supporting domain for G_3 is infinite whereas that of $G_{2,\alpha}$ is finite, the condition in (2.10) is more likely to hold than that in (2.11). Therefore $H(b_n+ p_{i,MAX}\cdot a_n)$ is more likely to converge to $G_{2,\alpha}$ rather than G_3.

It is pointed out in [12] that, most frequently used continuous distributions with finite right endpoint ($\omega(F)<\infty$) satisfy the condition in Eqn.(2.10). Therefore, in many engineering applications of maxima estimation, it is assumed that the distribution under study belongs to the weak convergence domain of $G_{2,\alpha}$. This statement has also been empirically proved to be true by our experiments (cf. later this section). Therefore, we state that the distribution of $p_{i,MAX}$ asymptotically follows the Weibull distribution $G_{2,\alpha}$. This means that there exist a_n and b_n such that:

$$F(b_n+a_n\cdot p_{i,MAX})=F^n(b_n+a_n p)\to G_{2,\alpha}(p_{i,MAX}), n\to\infty \tag{3.2}$$

or,
$$F(p_{i,MAX})\to G_{2,\alpha}(\frac{p_{i,MAX}-b_n}{a_n}), n\to\infty, i=1,\cdots,m \tag{3.3}$$

From Eqn.'s (2.13) and (2.15), we get $b_n=\omega(F)$ where $\omega(F)$ is the maximum power consumption of the population. If we substitute the generalized Weibull distribution defined in Eqn.(2.16) into Eqn.(3.3), we get

$$F(p_{i,MAX})\to G(p_{i,MAX};\alpha,\beta,\mu),\ n\to\infty,\ i=1,2,\cdots,m$$

where $\beta=(1/a_n)^\alpha$ and $\mu=b_n$.

Experiments have been designed to verify the asymptotic distribution of sample maxima. The distributions of sample maxima for different sample size (n = 2, 20, 30, 50) was formed by 1,000 random samples from the population. Their closest Weibull distributions are obtained by using least mean squared error fit. Figure 1 shows the results for circuit C3540.

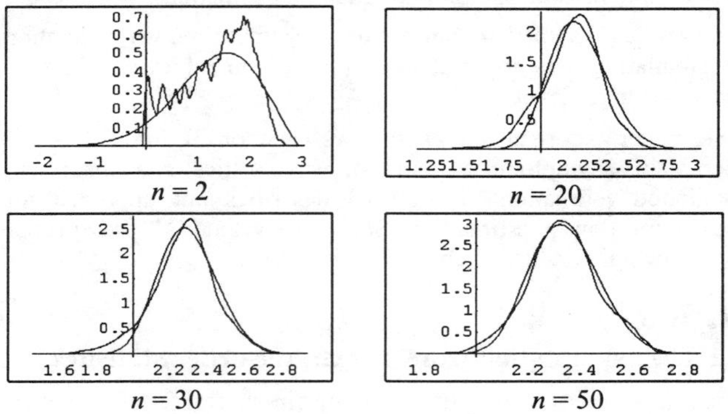

Figure 1 Comparison between distribution of sample maxima and Weibull distribution

Experiments are done for other circuits and populations and similar results are obtained. From these results we concluded that the difference between distributions of $p_{i,MAX}$ and Weibull distribution in the region near the maximum power is negligible when $n \geq 30$. Since we are only interested in estimating the maximum power, we fix the sample size n to 30 and assume that the distribution of $p_{i,MAX}$ follows Weibull distribution when $n \geq 30$. Consequently, $p_{i,MAX}$ (i=1,2,…,m) (n = 30) become the samples of the generalized Weibull distribution in (2.16). More importantly, if previous assumptions hold, we have: $\omega(F)=\mu$.

The problem of maximum power estimation is thus equivalent to that of estimating the location parameter μ of a generalized Weibull distribution from random samples. The simplest way of doing this is to curve-fit the samples to Eqn.(2.16) to get values of α, β, and μ. However, our study shows that the curve fitting approach is unstable since the problem becomes difficult when we try to construct the distribution from a small number of samples. We therefore choose another estimation method that is more robust and has a solid theoretical support.

3.2 A maximum-likelihood estimator of maximum power dissipation

The maximum-likelihood estimators for parameters of generalized Weibull distribution for $\alpha > 2$ have been introduced in Section II. In fact, α is always larger than 2 if the sample size n is much smaller than the population size $|V|$. Consequently, let $\hat{\alpha}_m, \hat{\beta}_m, \hat{\mu}_m$ be the estimators that satisfy Eqn. (2.18), we can prove the following result.

Theorem 3 $\hat{\alpha}_m, \hat{\beta}_m, \hat{\mu}_m$ ($m\to\infty$) are the unbiased estimators of α, β, μ of the Weibull distribution, which means that $\hat{\alpha}_m, \hat{\beta}_m, \hat{\mu}_m$ ($m\to\infty$) follow normal distributions with mean values of α_0, β_0, μ_0 and covariance matrix **VAR**. The matrix **VAR** is defined as:

$$\mathbf{VAR} = \frac{1}{m}\begin{bmatrix} \sigma_\alpha^2 & \sigma_{\alpha,\beta} & \sigma_{\alpha,\mu} \\ \sigma_{\beta,\alpha} & \sigma_\beta^2 & \sigma_{\beta,\mu} \\ \sigma_{\mu,\alpha} & \sigma_{\mu,\beta} & \sigma_\mu^2 \end{bmatrix} \tag{3.4}$$

From Theorem 3 it is shown that the maximum power estimator $\hat{\mu}_m$ converges to a normal distribution with mean of μ_0 (which is the actual maximum power $\omega(F)$) and variance of σ_μ^2/m.

Theorem 4 $\hat{\mu}_m$ is an unbiased estimator for maximum power $\omega(F)$. Given confidence level l ($l\in(0,1)$), the *confidence interval* of the estimated maximum power $\hat{\mu}_m$ ($m\to\infty$) is given by:

$$[\omega(F) - u_l\cdot\sqrt{\sigma_\mu^2/m},\quad \omega(F) + u_l\cdot\sqrt{\sigma_\mu^2/m}] \tag{3.5}$$

where $\omega(F)$ is the actual maximum power, m is the number of samples, σ_μ^2 is defined in Eqn.(3.4), and u_l is defined as:

$$\int_{-u_l}^{u_l}\frac{1}{\sqrt{2\pi}}e^{-\frac{x^2}{2}}dx = l \tag{3.6}$$

Theorem 4 states the probability that the estimated maximum power falls into the interval defined in Eqn.(3.5) is l. For a given l, smaller confidence interval means higher estimation accuracy. Therefore, the relative estimation error is inversely proportional to the square root of the variance of the estimator.

In practice, the theoretical confidence interval cannot be calculated directly because σ_μ^2 is unknown. Therefore, we do not know *a priori* how many samples are needed to achieve certain confidence interval at given confidence level. An iterative (Monte-Carlo) method has been designed to solve this problem.

3.3 The iterative estimation procedure

Experiments have been designed to study the distribution of the maximum likelihood estimator for maximum power in cases when the number of samples m is finite (we know from Theorem 3 that when $m\to\infty$ this maximum likelihood estimator for $\omega(F)$ follows a normal distribution). The sample size is fixed at n=30 and a different number of samples are used (m=10,50). During each single experiment, m samples with sample size n are randomly selected from the population. Maximum power is then estimated by using the maximum likelihood estimator $\hat{\mu}_m$. For each distinct m, the sampling-estimation procedure is repeated 100 times to form the distribution of estimated value. The distributions of estimated maximum power for different values of m are then formed and their nearest normal distributions are obtained by least-square curve fitting. The results for circuit C3540 are shown in Figure 2.

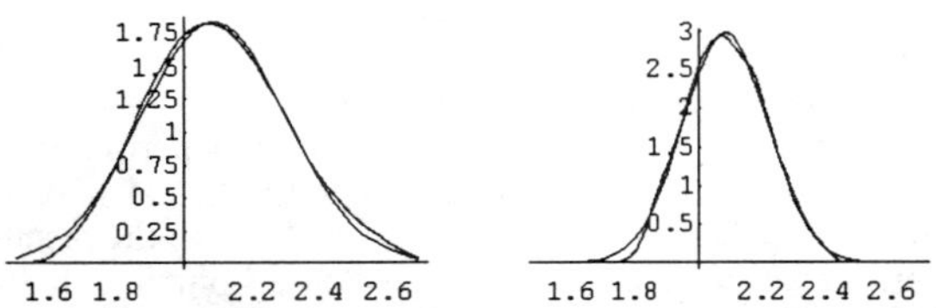

Figure 2 The distributions of estimated maximum power compared with the nearest normal distribution

Similar results are obtained for other circuits. From the experimental results, we can conclude that the estimator for maximum power is approximately normally distributed when the number of samples is large enough ($m\geq10$). Therefore, we assume normal distribution of estimator for maximum power when $m\geq10$.

Before we introduce our practical maximum power estimation procedure, we summarize our discussions in the beginning of this section as shown in Figure 3.

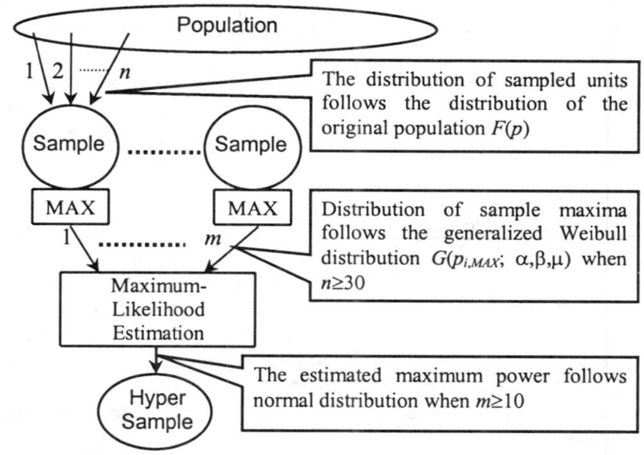

Figure 3 Synopsis of maximum power estimation method

In Figure 3, a hyper-sample is defined as the result of one run of maximum power estimation for m samples with size n. We fix the value of n to 30 and value of m to 10, so the number of units which is needed to form a hyper-sample is 300.

Theorem 5 Let $\hat{P}_{i,MAX}$ (i=1,2,...,k) denote the ith hyper-sample, for n=30 and m=10, $\hat{P}_{i,MAX}$ follows the normal distribution with mean value of $\omega(F)$ and variance of $\sigma_\mu^2/10$, where σ_μ^2 is defined in Eqn.(3.4). Let us define:

$$\overline{P}_{MAX} = \frac{1}{k}\sum_{i=1}^{k}\hat{P}_{i,MAX} \text{ and } s^2 = \frac{1}{k-1}\sum_{i=1}^{k}(\hat{P}_{i,MAX} - \overline{P}_{MAX})^2 \tag{3.7}$$

Theorem 6 $\overline{P}_{MAX}$ and s^2 are unbiased estimators of the actual maximum power $\omega(F)$ and $\sigma_\mu^2/10$, respectively. Given confidence level l, the confidence interval for the actual maximum power is given by:

$$[\overline{P}_{MAX} - \frac{t_{l,k-1}\cdot s}{\sqrt{k}}, \overline{P}_{MAX} + \frac{t_{l,k-1}\cdot s}{\sqrt{k}}] \tag{3.8}$$

where $t_{l,k-1}$ is the $l\times100\%$ percentile point of the t distribution with k-1 degrees of freedom.

Theorem 6 gives a guideline for designing an iterative procedure for maximum power estimation subject to the required accuracy (relative error less than or equal to ε) at given confidence level l. The basic workflow is shown in Figure 4.

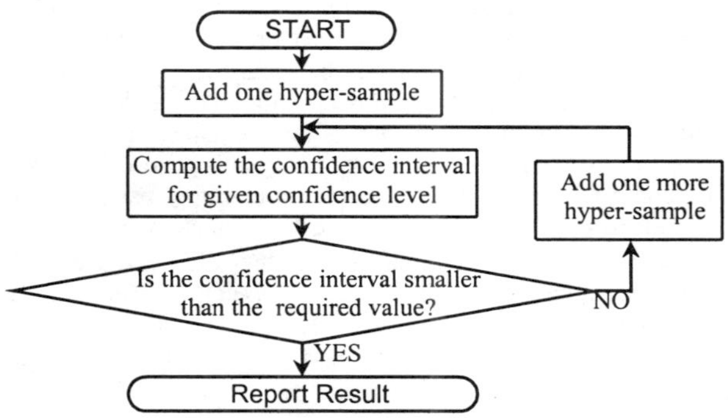

Figure 4 Iterative flow of maximum power estimation

In Figure 4, the generation of a hyper-sample follows the procedure shown in Figure 3. Confidence interval is calculated using Eqn.(3.8). The maximum relative error is calculated using the confidence interval as $\frac{t_{l,k-1} \cdot s}{\sqrt{k}} \Big/ \overline{P}_{MAX}$. If this quantity is larger than the required ε, then the estimated value has not converged and one more hyper-sample is added; otherwise, the estimation has converged and the estimation result is reported.

3.4 Practical issue: finite population versus infinite population

The approach discussed earlier in this section is designed for estimating the maximum power of an infinite population. However, we must deal with a finite population in real applications. As an example, our experimental setup in the next section uses finite populations. Experimental results shows that, if we use the same approach for finite population as for the infinite population, there will be a bias in the maximum likelihood estimation in the sense that the mean of the estimated value will always be larger than the actual maximum power of the population. This happens because μ_m estimates the maximum power of an infinite population that should have (with some probability) a tail even after the actual maximum power value for the population. Obviously, this tail does not exist in the case of a finite population.

To solve this problem, we can regard the finite population V as a sample of size $|V|$ selected randomly from the assumed continuous distribution for the infinite population. Assume there is only one unit in the finite population which consumes the maximum power, then the maximum power of the finite population becomes the estimated (1-1/$|V|$) quantile point of the assumed continuous distribution. According to the tail-equivalence property between a distribution and the limiting distribution of its sample maxima [12], estimating the (1-1/$|V|$) quantile point of the original population is equivalent to estimating the (1-1/$|V|$) quantile point of the generalized Weibull distribution of the sample maxima. Therefore, when estimating the maximum power of a finite population, instead of using the theoretical $\hat{\mu}_m$ (which is the 100% quantile point of the estimated generalized Weibull distribution), we use the (1-1/$|V|$) quantile point of the Weibull distribution (whose parameters are calculated by using the maximum likelihood estimator) as the estimator for the maximum power. We call this the "finite population estimator". Experimental results show that the modified estimator gives an unbiased estimator for finite populations.

IV. Experimental Results

Category I.1. Estimating the unconstrained maximum power.

In this category, the goal is to estimate the maximum power of the circuit for all possible input vector pairs. Consequently, the simple random sampling procedure can be realized by randomly generating vector pairs, that is, the two methods of random vector generation and simple random sampling are equivalent in this case. Except for the fact that the sampling technique is replaced by the random vector generation, the remaining parts of the estimation flow (cf. Figure 3 and Figure 4) are unchanged.

Let us give a theoretical analysis of the efficiency of the estimation method of random vector generation (or simple random sampling). Assume we want to estimate the maximum power of a population with an error less than 5% at a confidence level of 90%. Let the size of the population be $|V|$. Define the "qualified units" as those units whose values are within 5% of the actual maximum. Assume the number of the "qualified units" is Z. The portion of the "qualifies units" in the whole population is then $Y=Z/|V|$. If we sample x units from the population, the probability that there is at least one "qualified unit" in these x units is given by: $P = 1-(1-Y)^x$. For P to be larger than or equal to 90%, we need on average x = log(0.1)/log(1-Y) sampled units. From our experiments, we have observed that Y is very small (e.g., <0.0001). This leads to very large x (e.g., >23,000). Thus we conclude that maximum power estimation using simple random sampling is not efficient.

The experimental setup is as follows. The population contains 160,000 randomly generated high activity (average switching activity larger that 0.3) vector pairs. Random vector generation is equivalent to random sampling of vector pairs from the population. The whole population is simulated using Powermill [13] to get the power consumption value for each unit and in the process the actual maximum power. Our approach (n=30, m=10) and simple random sampling (SRS) have been applied to do maximum power estimation for relative error < 5% at confidence level 90%. We perform maximum power estimation using our approach one hundred times for each circuit. Experimental results are shown in Table 1 and Table 2.

Table 1 shows the efficiency comparison between our approach and simple random sampling (SRS). The portion of the "qualified units" in the whole population is given in the 2nd column. The maximum, minimum, and average (over 100 runs) number of units needed for our approach to converge are reported in the 3rd, 4th, and 5th columns, respectively. The 6th column gives the theoretically calculated (according to the discussion in the second paragraph of this section) number of units needed by simple random sampling to achieve the same error (5%) and confidence (90%) level. The 7th and 8th columns give the absolute value of the maximum and minimum estimation error of our approach. The relative error for SRS is not given because the SRS technique cannot predict the maximum power subject to given error and confidence levels.

Circuit	Portion of the "qualified units"	# of units needed				Relative error	
		Our approach			SRS	Our approach	
		MAX	MIN	AVE	AVE	MAX	MIN
C1355	0.0001	2700	900	1924	23024	6.0%	0.3%
C1908	0.00015	3600	1500	2410	15349	5.3%	2.4%
C2670	0.000288	1500	600	924	7993	6.2%	0.6%
C3540	0.000094	5100	600	2553	24494	5.2%	1.2%
C432	0.000038	5400	2100	3544	60593	7.7%	1.7%
C5315	0.000194	2700	600	1653	11868	5.8%	0.8%
C6288	0.000163	900	600	676	14125	6.2%	0.05%
C7552	0.00005	4500	3300	3825	46050	8.2%	0.6%
C880	0.000063	3000	2700	2859	36547	5.4%	2.9%

Table 1 Efficiency comparison for unconstrained input sequences

Table 2 shows the comparison for the estimation quality. SRS techniques using 2500, 10K, and 20K units are performed 100 times, respectively. The 2nd column gives the actual maximum power of the population. Columns 3, 4, 5 and 6 give the results of largest-error estimates for different techniques. Column 7, 8, 9 and 10 give the results of the percentage of the time when the estimated value exceeds the error level.

The experimental results show that our approach is much more efficient than the SRS technique (about 12X speedup on average). More importantly, however, simple random sampling or similar techniques are not reliable because they cannot provide confidence interval and confidence level for maximum power estimation. Also the estimation quality of our approach is obviously better than simple random sampling. From the results of Table 2, if we compare our approach with simple random sampling with 20K units, the average largest error is 5.3% for our approach, and 10.4% for SRS. As for the percentage of estimated values with error larger than 5%, it is 4.3% for our approach and 23% for SRS. It can be seen that the advantage of our approach over SRS increases for infinite population.

Circuits	Actual max. power (mW)	Largest estimation error				% of estimates with error > 5%			
		Our appr.	SRS			Our appr.	SRS		
			2500	10K	20K		2500	10K	20K
C1355	2.145	-6.0%	-13%	-8.5%	-6.3%	6%	80%	52%	15%
C1908	2.745	-5.3%	-14%	7.5%	-6.3%	3%	73%	28%	8%
C2670	6.529	-6.2%	-8.6%	-5.4%	-2.5%	1%	38%	2%	0%
C3540	10.732	5.2%	-14%	-10%	-8.9%	5%	80%	52%	33%
C432	1.818	-7.7%	-22%	-13%	-14%	8%	89%	73%	57%
C5315	14.372	5.8%	-9.7%	-7.7%	-6.2%	2%	73%	27%	3%
C6288	126.62	6.2%	-21%	-21%	-21%	3%	76%	26%	5%
C7552	31.237	8.2%	-14%	-10%	-7.3%	7%	92%	69%	54%
C880	4.312	5.4%	-20%	-15%	-11%	4%	88%	42%	29%

Table 2 Estimation quality comparison for unconstrained input sequences

Category I.2. Estimating the constrained maximum power.

The estimation flow is similar to Category I.1, except that vector pairs are generated under given constraints.

We generate two populations (each of size 80,000) subject to the constraint that the average switching activity per input line is 0.7 and 0.3, respectively. A detailed comparison with simple random sampling has also been performed.

Circuit	Portion of the "qualified units"	# of units needed				Relative error	
		Our approach			SRS	Our approach	
		MAX	MIN	AVE	AVE	MAX	MIN
C1355	0.000241	3900	600	2112	9553	5.4%	1.8%
C1908	0.000378	3000	600	2403	6090	7.3%	2.0%
C2670	0.000778	900	600	675	2958	4.1%	0.5%
C3540	0.000196	1200	900	1054	11747	6.7%	4.0%
C432	0.000071	3300	1200	2259	32430	7.7%	2.2%
C5315	0.000488	1200	900	975	4717	7.1%	4.1%
C6288	0.000427	1200	600	1052	5391	4.5%	1.7%
C7552	0.000308	3900	900	2252	7475	8.0%	0.9%
C880	0.000135	2700	600	1703	17055	12%	2.1%

Table 3 Results for constrained input sequences (high activity)

Circuit	Portion of the "qualified units"	# of units needed				Relative error	
		Our approach			SRS	Our approach	
		MAX	MIN	AVE	AVE	MAX	MIN
C1355	0.000119	4800	1500	3348	19384	3.6%	2.2%
C1908	0.000246	2700	900	2001	9359	6.6%	3.5%
C2670	0.000313	3600	1500	2584	7355	5.3%	1.7%
C3540	0.000053	5100	600	3587	43444	7.4%	2.9%
C432	0.000179	3000	1500	2389	12862	6.8%	2.4%
C5315	0.000231	3600	1200	2623	9967	13%	3.4%
C6288	0.000079	6000	2700	5424	29145	5.1%	0.6%
C7552	0.000194	2400	1200	1976	16446	7.1%	3.3%
C880	0.000018	2700	900	1897	127920	5.0%	1.9%

Table 4 Results for constrained input sequences (low activity)

We provide only results for comparing efficiency and accuracy due to space limitation. The experimental results for populations of average switching activity 0.7 and 0.3 are shown in Table 3 and Table 4, respectively. The meaning of entries in different columns is the same as those in Table 1. The estimation quality comparison can be seen from the value of the portion of the "qualified units" in the 2nd columns of both tables. As expected when the number of qualified units in the population decreases, the number of units that are required to estimate the maximum power dissipation in the circuit increases.

V. Conclusion

A statistical approach for peak power estimation based on the asymptotic theory of extreme order statistics has been proposed. This is also the first approach which can do maximum power estimation for any user-specified error and confidence levels. The proposed approach can predict the maximum power in the space of constrained input vector pairs, as well as the complete space of all possible input vector pairs. It is an efficient simulation-based approach with high accuracy. The generality of this approach makes it applicable to other fields of VLSI design automation; for example, longest path delay estimation.

Reference

[1] S. Devadas, K. Keutzer, and J. White. "Estimation of power dissipation in CMOS combinational circuits using boolean function manipulation", *IEEE Transactions on Computer-Aided Design*, 11(3):373-383, March 1992.

[2] H. Kriplani, F. Najm, and I. Hajj, "Maximum Current Estimation in CMOS Circuits", *Proceedings of Design Automation Conference*, pp. 2-7, 1992.

[3] C.C. Teng, A. M. Hill, and S. M. Kang, "Estimation of Maximum Transition Counts at Internal Nodes in CMOS VLSI Circuits", *Proceedings of International Conference on Computer Aided Design*, pp. 366-370, 1995.

[4] S.Manne, A.Pardo, R.Bahar, G.Hachtel, F.Somenzi, E.Macii and M.Poncino, "Computing the Maximum Power Cycles of a Sequential Circuit", *Proceedings of Design Automation Conference*, pp. 23-28, 1995.

[5] C.-Y. Wang and K. Roy, "Maximum Current Estimation in CMOS Circuits Using Deterministic and Statistical Techniques," *IEEE Transactions on VLSI Systems*, pp. 134-140, Mar. 1998.

[6] C.-Y. Wang, T.-L. Chou, and K. Roy, "Maximum Power Estimation Under Arbitrary Delay Model," *Proceedings of IEEE International Symposium on Circuits and Systems*, 1996.

[7] C.-Y. Wang and K. Roy, "COSMOS: A Continuous Optimization Approach for Maximum Power Estimation of CMOS Circuits," *Proceedings of ACM/IEEE International Conference on Computer-Aided Design*, pp. 52-55, 1997.

[8] M.S. Hsiao, E.M. Rudnick, and J.H. Patel, "K2: An Estimator for Peak Sustainable Power of VLSI Circuits", *Proceedings of International Symposiunm on Low Power Electronics and Design*, pp.178-183, Aug., 1997.

[9] A. Hill, C.-C. Teng, and S.M. Kang, "Simulation-based maximum power estimation", *Proceedings of IEEE International Symposium on Circuits and Systems*, 1996.

[10] C.S. Ding, Q. Wu, C.T. Hsieh, and M. Pedram, "Statistical Estimation of the Cumulative Distribution Function for power dissipation in VLSI Circuits", *Proceedings of Design Automation Conference*, pp.371-376, Jun., 1997.

[11] R.L. Smith, "Maximum likelihood estimation in a class of nonregular cases", *Biometrika*, pp. 67-90, 1972.

[12] J. Galambos, *The Asymptotic Theory of Extreme Order Statistics*, Robert E. Krieger Publishing Company, 1987

[13] Analog Circuit Engine Release Notes, Synopsys, 1997.

An Optimization-Based Error Calculation for Statistical Power Estimation of CMOS Logic Circuits

Byunggyu Kwak

Samsung Data Systems
Seoul, 135-080 Korea

Eun Sei Park

School of Electrical and Computer Engineering
Hanyang University, Ansan, 425-791 Korea

Abstract

In this paper, we present a statistical power estimation method where estimation time and accuracy can be balanced by assigning smaller errors to the nodes with higher power dissipation and higher errors to the nodes with lower power dissipation. To calculate the error rates for individual nodes, a quadratic programming based problem will be formulated which incorporates the distribution data of all individual node switching activities. Also, an iterative statistical power estimation system will be presented. Finally, we will demonstrate experimental results which show drastic reduction in the number of simulation patterns compared to previous methods.

1. Introduction

With increasing complexity and operational frequency of VLSI circuits, power reduction has become one of most important design goals. Especially in the design of portable electronic equipment and mobile electronics, the failure to meet the specification of power consumption may result in repetitive redesign steps of the system because high power consumption can shorten the battery life and, even worse, may lower the reliability of the system.

Gate-level power estimation techniques are divided into two categories [1]; probabilistic and statistical techniques. In probabilistic techniques [2, 3], the average switching rates (called transition densities) of the primary inputs are propagated upstream to the primary outputs of a circuit using some gate-level transfer functions. Probabilistic techniques are usually fast in computing the power dissipation, but their accuracy may not be reliable enough to be used in the design process. It is mainly because the circuit's topological correlation due to the signal reconvergence can hardly be modeled using simple signal probabilities and probabilistic transfer functions of logic elements. On the other hand, in sampling-based simulation methods, either the average switching activities of individual circuit nodes [4] or the total switching activities of a circuit [5-8] is estimated. The advantages of the simulation-based techniques are the accuracy and the simplicity of implementation using existing logic simulator with realistic delay models, where glitches and signal correlations can be automatically taken care of. To determine the number of simulation patterns (sample size), a stopping criterion is derived under some statistical assumptions. For example, the normality on the distribution of the average transition densities of individual nodes [4] or the normality on the distribution of the total transition density [5] are often assumed. Then the average transition density is estimated with a user-specified error and a confidence level. Extensions of [4] and [5] approaches have been proposed subsequently [6-8].

It has been experimentally shown that the nodes with higher transition densities often tend to converge more rapidly to a given error bound than those with lower transition densities [4]. Thus, if a constant estimation error is specified for every individual node as in [4], the actual errors for the nodes with high switching activities will be much lower than the given target errors. And, because of slow convergence rate for the nodes with low transition densities, the number of simulation patterns could be unacceptably large. Fig.1 shows the above statement more clearly. For c6288 and s38417 benchmark circuits, the statistical technique of [4] has been used with 5% error and 99% confidence. In Fig.1, the entire nodes were partitioned into 20 groups according to their transition densities, and the resultant error in each group has been averaged. The results in Fig.1 indicate that a fixed estimation error for all nodes may not be adequate to meet reasonable simulation time budget.

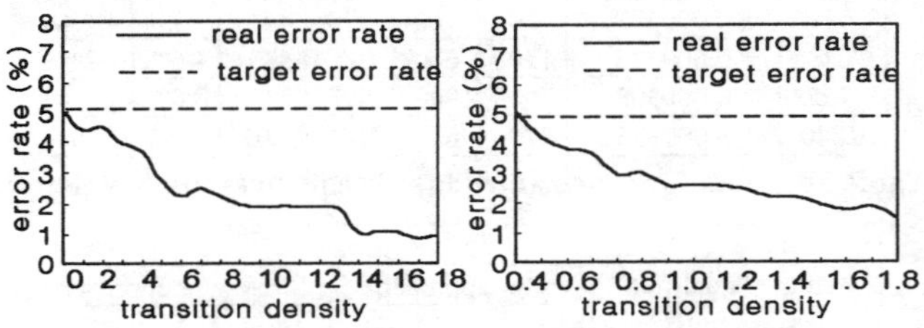

(a) c6288 (b) s38417

Fig. 1. Distribution of actual errors determined by MED

In [4], to alleviate the slow convergence of the noes with low transition densities, the circuit nodes are classified into regular- and low-density nodes. For the low-density nodes whose average switching densities are less than some value η_{min}, an absolute error bound $\eta_{min}\varepsilon$ is applied with the same confidence level, where ε is the user-specified percentage error for regular-density nodes. Although this approach can speed convergence of statistical power estimation, there is no clear way to determine η_{min} to separate low-density and regular-density nodes.

For the practical applications such as electromigration

DAC 98, San Francisco, California

analysis and power optimization, it is evident that the nodes with higher power dissipation should be identified with smaller errors even at the sacrifice of estimation errors for low power nodes. Therefore, for a fast sampling-based statistical power estimation, errors for individual nodes may need to be graded according to their relative contributions to the total power dissipation. That is, individual nodes are assigned gradually decreasing errors according to their contribution to the total power consumption. In fact, this type of error assignment for all individual nodes naturally matches with the characteristics of simulation-based power estimation as shown in Fig.1.

In this paper, we present a method to determine the errors which are dependent on individual node's relative contribution to the total power dissipation. Since it can be impractical to assign different error to every individual node in a circuit, we partition the entire nodes in a circuit into several groups according to their contribution to the total power dissipation. And gradually decreasing errors are assigned to the groups in such a way that the nodes of high power dissipation have a smaller error. For this, a quadratic programming based error calculation problem will be formulated. Also, an iterative statistical power estimation method will be proposed.

2. A Statistical Power Estimation Problem

2.1. General Ideas

The number of simulation patterns in a sampling-based statistical power estimation is totally dependent on the stopping criterion. The stopping criterion of MED approach [4] is,

$$N \geq \left(\frac{Z_{\alpha/2} S}{\varepsilon \overline{n}} \right)^2 \tag{1}$$

where N, ε, $\overline{n}$, and s are the number of simulation patterns, user-specified error, sample mean of the number of transition at a node, and the standard deviation, respectively. And $Z_{\alpha/2}$ is obtained from the normal distribution with confidence level α. From Eq.(1) with the constant values of ε and α, we can observe that the nodes with the smallest value of $(s / \overline{n})$ determines the largest N, which is the total number of simulation patterns.

Unlike the MED approach with a fixed error for all nodes, we use variable errors for individual nodes to balance the simulation time and the estimation accuracy. The variable errors for individual nodes will be determined such that the nodes with higher transition densities can be estimated more accurately by tightening the estimation error bound. Although the error rates are relaxed for low power nodes, we put a constraint on the relaxed errors so that the sum of estimated power dissipation for all nodes should be within a certain error ε_T of the total power dissipation. This additional constraint will bound the excessive errors for low density nodes.

2.2. An Optimization Problem Formulation

To reduce the complexity of the analysis, the entire nodes are partitioned into M groups according to their contributions to the total power dissipation. Let G_i and G_j be the *i-th* and *j-th* partitions. We further assume that ε_i and ε_j be the errors assigned to the nodes in G_i and G_j, respectively. It is also assumed that, if $i > j$, the nodes in G_i consume less power than the nodes in G_j. Thus, G_1 is the group of nodes with the highest power consumption. Let the estimated power of the nodes in G_i be P_i and the sum of the estimated power dissipation for all nodes be P_T. We further assume that ε_i and ε_j be the errors assigned to the nodes in G_i and G_j, respectively. And, ε_T is the user-specified estimation error of the total power P_T. Then, the following equations hold.

$$(1 - \varepsilon_T)P_T \leq \sum_{i=1}^{M} (1 - \varepsilon_i)P_i \tag{2}$$

$$(1 + \varepsilon_T)P_T \geq \sum_{i=1}^{M} (1 + \varepsilon_i)P_i \tag{3}$$

Since $P_T = \sum_{i=1}^{M} P_i$, Eqs. (2) and (3) are combined into one inequality equation as in Eq.(4)

$$\varepsilon_T \geq \frac{P_1}{\sum_{i=1}^{M} P_i}\varepsilon_1 + \frac{P_2}{\sum_{i=1}^{M} P_i}\varepsilon_2 + \ldots + \frac{P_M}{\sum_{i=1}^{M} P_i}\varepsilon_M \tag{4}$$

Obviously, if $\varepsilon_i = \varepsilon_T$, the Eq. (4) holds and it is the same case as in MED. Since our goal in this work is to determine gradually decreasing error rates so that the nodes with higher contribution to the total power dissipation can be more accurately estimated, we need a constraint as in Eq. (5).

$$\varepsilon_i < \varepsilon_{i+1} + \beta, \quad \text{for } i = 1, 2, \ldots, M, \tag{5}$$

where β is some arbitrarily small positive number. From the stopping criterion of Eq.(1), the sample size (the number of simulation pattern) N is proportional to $(s / \varepsilon \overline{n})^2$. The sample size N_i for the block G_i is also proportional to $(s_i / \varepsilon_i \overline{n}_i)^2$. Notice that the sample size of this problem is the maximum of the individual sample size for each group. Therefore, to achieve the maximum speedup for convergence, we need to have equal number of simulation patterns for the nodes in different partitions G_1 through G_M,

$$N_1 = N_2 = \cdots = N_M \tag{6}$$

Since $N_i \propto (s_i / \varepsilon_i \overline{n}_i)^2$, this leads us to a quadratic cost function Q to be minimized.

$$Q = \sum_{i=2}^{M} \left(\frac{\varepsilon_i \overline{n}_i}{s_i} - \frac{\varepsilon_1 \overline{n}_1}{s_1} \right)^2 \tag{7}$$

In Eq.(7), the index 1 represents the group of the nodes with highest power consumption and ε_1 can be specified by the user. The purpose of Eq. (7) is to equalize the sample sizes by preventing a certain term in the quadratic cost function from being excessively large, which determines the total sample size in our problem. Now, we can formulate an optimization problem as below.

<An Optimization-Based Error Calculation Problem>
Find the error rates $\varepsilon_1, \varepsilon_2, \ldots, \varepsilon_M$ for the partitions $G_1, G_2, \ldots, G_M$ which minimize the quadratic objective function of Eq.(7) subject to the linear constraint equations in Eq. (4) and (5).

3. Circuit Node Partitioning

The efficiency of our method is dependent on the way that the entire circuit nodes are partitioned. In this work, we try three different ways of partitioning after simulation with typically over 30 simulation samples from Central Limit Theorem.

(1) Type 1 : Interval based partitioning

From the simulation data of individual power consumption, the minimum and the maximum value of the power dissipation of all nodes can be determined. With the data, we divide the interval between the minimum value (V_{min}) and the maximum value (V_{max}) into fixed M groups. And let Δ be (V_{min} - V_{max})/M. Then, the nodes in G_i have power dissipation between $V_{min}+\Delta*(i-1)$ and $V_{min}+\Delta*i$. This partitioning method is simple, but it does not reflect the distribution of power dissipation.

(2) Type 2 : Distribution based partitioning with equal weights.

First, we sort the individual node power dissipation and construct its cumulative distribution. If the individual nodes are to be partitioned into M groups, the sum of power dissipation in each partition has *100/M* contribution in percentage value. By taking this approach, the size of each partition is dependent on the shape of cumulative distribution.

(3) Type 3: Distribution based partitioning with increasing weights

This partitioning approach is a modified version of Type 2. Unlike the equal weight as in Type 2, each partition in this approach has different contribution to the total power dissipation such that the sum of the power dissipation in G_1 is larger than that of G_2, etc. We give the weight (M+1-i) to G_i . So, the overall weight will become $M(M+1)/2$. For example, if M = 10, the aggregate power contribution percentage of the nodes in G_1 to the total power is about 18.2% and the aggregate power contribution of the nodes in G_{10} is about 1.8%. This partition can lead the quadratic programming tool to obtain higher errors on the nodes with lower power dissipation and lower errors on the nodes with higher power dissipation.

We have performed experiments with these three partitioning methods for c6288 and s38417 ISCAS benchmark circuits [9, 10], where 30 random patterns are simulated and M=10. The percentages of the nodes in each partition are shown in Fig. 2.

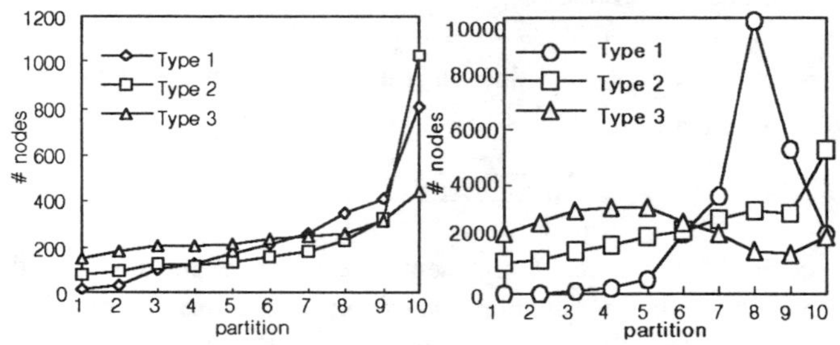

(a) c6288 (b) s38417

Fig. 2. Number of nodes per block partitioned by Type 1,2,3

4. Iterative Statistical Power Estimation

Since the initial partitioning is based upon small amount of simulation results, there may exist incorrect prediction that some nodes are placed in a wrong partition and thereby incorrect error rates are assigned to them. Therefore, we need to iterate the statistical power estimation until all nodes are verified to be correctly partitioned. The overall scheme of our statistical power estimation is shown in Fig. 3.

First, we simulate the circuit with 30 or more simulation patterns. Then, we partition the nodes in a circuit and determine their error rates by solving the quadratic programming based problem. Next, we simulate the circuit until the power estimation of all nodes can be done within their calculated error rates and confidence level. To make sure that all individual nodes are estimated with correct error rates, we have to repeat the circuit partitioning and calculate their error rates again. If the power dissipation of every node is already correctly estimated within the new calculated error, this procedure stops.

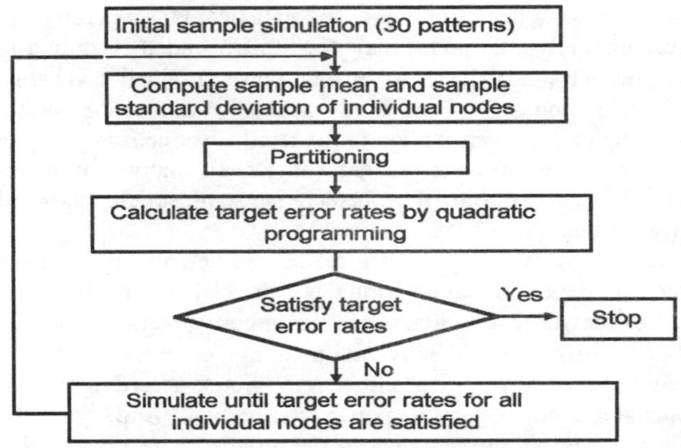

Fig. 3. Flow chart of suggested approach.

5. Experimental Results

We have implemented the proposed statistical power estimation method in C language and have performed several experiments on ISCAS benchmark circuits on a SUN Sparc-20 workstation. We use a gate-level logic simulator with unit-delay model. Although the unit-delay model may not fully represent the switching activity of a circuit, our method can be used in conjunction with any arbitrary logic simulators. Our main objective in the following experiments is to compare the relative efficiency between our method and the conventional statistical estimation methods which use a constant error for all individual node power estimation.

Table 1 shows the comparison results between the MED approach and ours for the ISCAS85 benchmark circuits. In the experiment of the MED approach, 99.9% confidence and 5% user-specified error have been specified for all nodes and no η_{min} (pre-defined lower bound absolute error) has been assumed. On the other hand, in our proposed method, overall confidence, total power estimation error, and the error for the partition with the highest power dissipation were specified 99.9%, 5%, and 2%, respectively. And all nodes in each circuit has been partitioned into 25 groups.

From the results in Table 1, our method requires much less simulation patterns in all three partitioning methods. The results obtained from type 2 and 3 partitioning methods outperform the type 1 result, which imply that the partitioning methods based upon the distribution data of individual node power are well suited for statistical power estimation. Especially, for the circuit c2670 in the type 2, the number of samples required is as low as 1.1% of the samples of the MED approach. On average, the type 2 approach requires only 3.2% samples compared to the sample size of MED. The CPU time spent for solving the quadratic programming is negligible compared to the overall simulation time. And the number of iterations are less than two times in all experiments. Obviously, the speedup ratio achieved from our method over the MED approach are close to the sample size reduction ratio. While the results in Table 1 may seem to be quite obvious, it can be claimed that the efficient control of error rates can lead to an accurate estimation of high power dissipation nodes and the total power consumption in significantly shorter CPU time than conventional statistical power estimation methods.

The number of partitions can affect the overall power

estimation efficiency. In general, more partitions produce higher (lower) errors for the groups of low (high) power nodes in solving the quadratic programming. However, if the errors for the high power nodes are too low, satisfying the errors for high power nodes can become a bottleneck rather than satisfying the errors for low power nodes. From our experience, 20 to 30 partitions yield reasonable results and the time spent on solving the quadratic programming is less than 30 seconds up to 50 partitions. In Fig. 4, the target error rates and the real errors rates in c6288 and c7552 circuits are compared against the error rates of the MED approach. In our method, the real error rates are roughly approaching the target error rates. This phenomena indicates that our method can avoid excessive simulation efforts and thus can accelerate the power estimation process.

Table 1. ISCAS85 benchmark result

ckt.	MED sample size	Type 1 (%)		Type 2 (%)		Type 3 (%)	
		sample size	type1/ MED	sample size	type2/ MED	sample size	type3/ MED
c1355	7397	5740	77.6	201	2.7	257	3.5
c1908	41603	1237	2.9	961	2.3	961	2.3
c2670	3524	1292	36.6	386	1.1	1045	29.6
c3540	11567	2446	21.1	265	2.3	422	3.6
c5315	2493	1473	59.1	301	12.1	433	17.3
c6288	906	595	65.6	73	8.1	101	11.1
c7552	7487	1411	18.8	206	2.7	402	5.4
avg.	10711	2028	18.9	342	3.2	517	4.8

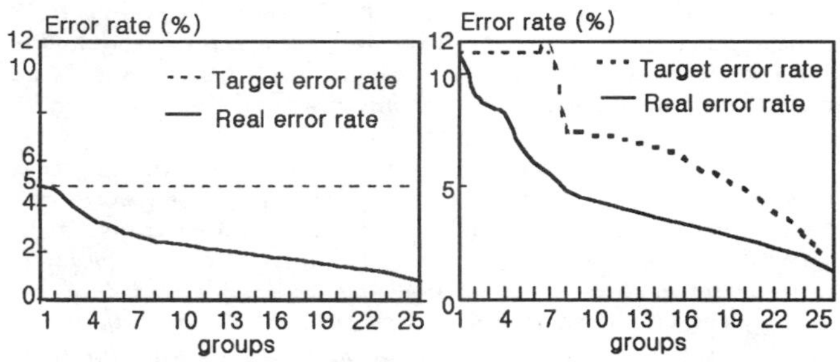

(a) Error rate given by MED (b) Error rate given by ours

Fig. 4. Variable error rate when using our approach

In Table 2, some benchmark results of the ISCAS89 circuits are shown for the MED approach and ours. For larger circuits, MED sets a threshold value to reduce the execution time to a practical level. In the MED approach, 20% of the circuit nodes with the lowest transition densities has been assumed as low-density node and η_{min} was chosen to include the low 20% nodes. In this experiment, we have followed the same procedure as in MED and 20% of the circuit nodes with the lowest transition densities has been assumed as low-density node. The stopping conditions are 99.9% of confidence, 5% user-specified error for the total power, and 2% error for the group of nodes with the highest power dissipation.

Table 2. Benchmark result on ISCAS89

	MED sample size	Type2 sample size	Type2/ MED (%)
s13207	1700	503	29.5
s38417	1524	1133	74.3
s38584	2187	1282	58.6

As shown in Table 2, our method needs about 30-70% simulation patterns of the MED approach. Because an absolute error has been applied to the 20% of low-density nodes, the saving of sample size in our method has been reduced. However, it is still possible to apply our method to any operating environment and to speed up the statistical power estimation process.

6. Conclusion

In this paper, we have proposed a new statistical estimation method for the individual node power as well as the total power in CMOS logic circuits. Unlike the conventional statistical power estimation methods, the error rate for each node is determined according to the relative contribution of each node to the total power dissipation. Hence, the nodes with higher power dissipation can have smaller error rates at the sacrifice of higher estimation errors for the nodes with lower power dissipation. For this, a quadratic programming based problem was formulated and an iterative statistical power estimation system has been developed. Compared to the previous methods which employ a constant error for all nodes, our proposed method will balance between the time and the estimation accuracy in simulation-based statistical power estimation. By using our method, the power dissipation of all individual nodes can be estimated with 3.2% of samples size of the MED technique.

Acknowledgements

The authors would like to acknowledge the support by MIC, the government of Korea, under the University Basic Research Support Program.

References

[1] F. N. Najm, "A Survey of Power Estimation Techniques in VLSI Circuits," *IEEE Trans. VLSI Syst.*, pp.446-455, 1994.

[2] M. A. Cirit, "Estimating dynamic power consumption of CMOS circuits.", *IEEE ICCAD*, pp.534-537, Nov. 9-12, 1987.

[3] S. Ramprasad, N.R. Shanbhag, and I.N. Hajj, "Analytical Estimation of Transition Activity from Word-Level Signature Statistics," *ACM/IEEE Proc. Design Automation Conf.*, pp.582-587, 1997.

[4] M. Xakellis and F. Najm, "Statistical estimation of the switching activity in digital circuits." *ACM/IEEE Proc. Design Automation Conf.*, pp. 728-733, 1994.

[5] R. Burch, F. Najm P. Yang and T. Trick, "A Monte Carlo approach for power estimation" *IEEE Trans. VLSI Syst.*, vol. 1, no. 1, pp. 63-71, Mar.1993.

[6] C.-S. Ding, Q. Wu, C.-T. Hsieh, and M. Pedram, "Statistical Estimation of the Cumulative Distribution Function for Power Dissipation in VLSI Circuit," *ACM/IEEE Proc. Design Automation Conf.*, pp.371-376, 1997.

[7] S. Y .Huang, K. T. Cheng, K. C. Chen and M. T-C. Lee, "A Novel Methodology for Transistor-Level Power Estimation" *ISLPED*, Monterey CA, pp.67-72, August 12-14, 1996.

[8] A. M. Hill and Sung-Mo Kang, "Determining Accuracy Bounds for Simulation-Based Switching Activity Estimation", *IEEE Trans. CAD*, vol. 15, no. 6, pp. 611-618, June 1996.

[9] F. Brglez and H. Fujiwara, "A Neutral Netlist of 10 Combinational Benchmark Circuits and a Target Translator in Fortran," *IEEE ISCAS*, pp.695-698, 1985.

[10] Brglez, D. Bryan, and K. Kozminski, "Combinational Profiles of Sequential Benchmark Circuits," *IEEE ISCAS*, pp.1929-1934, 1989.

Using Complementation And Resequencing To Minimize Transitions

Rajeev Murgai, Masahiro Fujita
Fujitsu Laboratories of America, Inc.

Arlindo Oliveira
Cadence European Labs./IST-INESC

Abstract

Recently, in [3], the following problem was addressed: *Given a set of data words or messages to be transmitted over a bus such that the sequence (order) in which they are transmitted is irrelevant, determine the optimum sequence that minimizes the total number of transitions on the bus.* In 1994, Stan and Burleson [5] presented the *bus-invert method* as a means of encoding words for reducing I/O power, in which a word may be inverted and then transmitted if doing so reduces the number of transitions. In this paper, we combine the two paradigms into one – that of sequencing words under the bus-invert scheme for the minimum transitions, i.e., words can be complemented, reordered and then transmitted. We prove that this problem DOPI – Data Ordering Problem with Inversion – is NP-complete. We present a polynomial-time approximation algorithm to solve DOPI that comes within a factor of 1.5 from the optimum. Experimental results show that, on average, the solutions generated by our algorithm were within 4.4% of the optimum, and that resequencing along with complementation leads to 34.4% reduction in switching activity.

1 Motivation

In several applications such as microprocessors and DSPs, switching activity on high-capacitance buses account for almost 30-40% of the power consumption [1]. Thus for reducing power dissipation, the problem of minimizing switching activity (i.e., the number of transitions) on a bus assumes great importance. In this context, consider the following problem. Given a set of n data words or messages, each of length k, to be transmitted over a bus (k-bit wide) such that the sequence in which they are transmitted is irrelevant, determine the optimum sequence that minimizes the total number of transitions on the bus. We call this the **Data Ordering Problem (DOP)**.

Definition 1.1 *If word w_r is transmitted, immediately followed by w_s, the total number of transitions is given by the number of bits that change. This is $d(w_r, w_s) = \sum_{j=1}^{k} w_{rj} \oplus w_{sj}$, also known as the* Hamming distance *between w_r and w_s. Here, w_{rj} denotes the jth bit of w_r, and $\oplus$ denotes the EX-OR operation. For instance,* $d(1001, 1110) = 3$.

Word reordering can change the number of transitions significantly. For instance, let $n = 3$, $w_1 = 00011, w_2 = 10110$, and $w_3 = 01011$. So $k = 5$. If words are transmitted in the order w_1, w_2, w_3, the total number of transitions on the bus is $d(w_1, w_2) + d(w_2, w_3) = 3 + 4 = 7$. If transmitted in the order w_1, w_3, w_2, the number of bus transitions would be $d(w_1, w_3) + d(w_3, w_2) = 1 + 4 = 5$, 28% fewer.

The data ordering problem was addressed in [3]. It was shown that this problem arises in a wide variety of design tasks when power dissipation is the main concern: instruction scheduling, die testing, sequencing of test patterns in built-in self test (BIST) systems, cache write-back, and scheduling in high-level synthesis. It was shown in [3] that DOP is NP-complete. Three approximation schemes were proposed: *Double Spanning Tree (DST)* – which provably comes within a factor of two of the optimum solution, *Minimum Spanning Tree Maximum Matching (ST-MM)* – which comes within 1.5 (plus a small additive term) of the optimum, and *greedy* – which was empirically found to be the best.

In [5], Stan and Burleson presented the *bus-invert method* to reduce the number of transitions. It works as follows. An extra bus line, called *invert*, is used. The method looks at two consecutive data words on the bus. If the Hamming distance between the next word and the current transmitted word is at most $k/2$, the next word is sent as it is, with *invert* set to 0. Otherwise, each bit of the next word is complemented (inverted) and *invert* is set to 1, indicating that the word has been complemented. This scheme reduces the number of transitions on the bus. Let $\overline{w_i}$ denote the complement of w_i. E.g., $\overline{10110} = 01001$. In the last example, if we transmit w_1, w_3, and then $\overline{w_2}$, it results in $d(w_1, w_3) + d(w_3, \overline{w_2}) = 1 + 1 = 2$ transitions, 3 less than the uncomplemented case. If we count the extra transition on the *invert* signal, the total number of transitions is 3, still 2 less than 5.

In this paper, we combine the two paradigms into one: that of sequencing words under the bus-invert scheme, i.e., words can be complemented (if it helps), reordered and then transmitted. *Given n data words $w_1, w_2, \ldots, w_n$, the problem then is to determine i) for each word w_i, if it should be complemented or left uncomplemented (i.e., its* **phase assignment***), and ii) the order in which words should be transmitted, so as to minimize the transitions on the bus.* We call this problem **DOPI – Data Ordering Problem with Inversion**. The next section discusses the relevant background graph-theoretic material. We prove in Section 3 that DOPI is NP-complete. In Section 4, we devise approximation schemes for DOPI that guarantee bounds from the optimum solution. Experimental results are presented in Section 5.

2 Preliminaries

A **spanning tree** of a graph $G = (V, E)$ is a subgraph of G that is a tree and spans all the vertices of G. If the edges have weights, it makes sense to talk of the **minimum-weight spanning tree (MST)** of G, where the weight of a tree is the sum of the weights of the edges in the tree. It turns out that a simple, greedy algorithm that starts from an empty tree and repeatedly selects the minimum-weight edge, adding it to the partially generated tree if the tree still remains acyclic generates an MST. For the weighted graph of Figure 1 (A), the minimum-weight spanning tree is shown in Figure 1 (B). A **Hamiltonian tour** of a graph $G = (V, E)$ is a walk (simple path) with the same beginning and end points that visits each vertex of V exactly once. A **Hamiltonian path** of G is a walk with different beginning and end points that visits each vertex of V exactly once. For instance, 1, 2, 3, 4, 6, 5, 7, 1 is a Hamiltonian tour of the graph of Figure 1 (A), whereas 1, 2, 3, 4, 6, 5, 7 is a Hamiltonian path. If the edges have weights, we

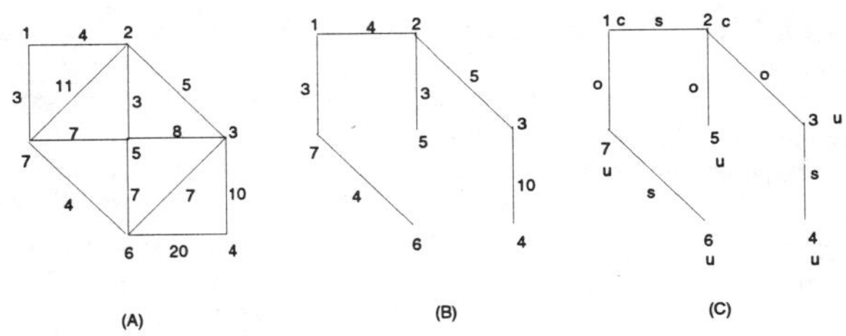

Figure 1: An example graph and its spanning tree

have the problems of finding a minimum-weight Hamiltonian tour (better known as the **traveling salesman problem** tour or TSP-tour) and minimum-weight Hamiltonian path (TSP-path). Both of these are known to be NP-complete, even for complete graphs.

A graph $G = (V, E)$ in which there is more than one edge joining a pair of nodes is called a **multigraph**. An **Eulerean cycle** of a multigraph is a walk with the same beginning and end points that contains each edge of the graph exactly once. E.g., in Figure 1(A), the tour 1, 2, 3, 4, 6, 3, 5, 6, 7, 5, 2, 7, 1 is an Eulerean cycle. It is well-known that *a multigraph G contains an Eulerean cycle if and only if each node of G is of even degree*. The following simple procedure finds an Eulerean cycle in such a multigraph G.

Procedure 2.1 (Finding an Eulerean cycle) *Start at any vertex v of G. Traverse any edge (v, w) whose removal does not disconnect the remaining (untraversed) graph. Delete (v, w). Repeat the procedure from w. Terminate when no more edges remain in G. This can happen only when at v. The order in which the edges are traversed yields the Eulerean cycle.*

Given a complete graph G and a spanning tree $T = (V, E')$ of G, the following procedure finds a Hamiltonian tour on G.

Procedure 2.2 (Constructing a Hamiltonian tour) *Construct the multigraph $\widehat{G}$ from T by duplicating each edge $e \in E'$. Since each node of $\widehat{G}$ is of even degree, $\widehat{G}$ contains an Eulerean cycle U. Construct U using Procedure 2.1. Delete all the node repetitions from U except for the final return to the first node. The resulting node sequence HT is a Hamiltonian tour on G.*

If we are interested in a minimum-weight Hamiltonian tour, we will start from an MST T and use Procedure 2.2.

Let each edge (v_i, v_j) of the graph have a non-negative weight $d(v_i, v_j)$. The weight $d(S)$ of a tour (path, tree) S is the sum of the weights of the edges in S. If the edge weights d satisfy **triangle inequality** (i.e., $d(v_1, v_2) + d(v_2, v_3) \geq d(v_1, v_3)$ for all v_1, v_2, v_3), it is easy to see that $d(HT) \leq d(U)$. Since $d(U) = 2d(T)$, we get

$$d(HT) \leq 2d(T) \tag{1}$$

A **matching** in a graph $G = (V, E)$ is a subset $E' \subseteq E$ such that no two edges of E' are incident to each other. A **perfect matching** is a matching that is incident to each vertex in V. Clearly, $|V|$ must be even for a perfect matching to exist.

3 DOPI is NP-complete

The Data Ordering Problem with Inversion (DOPI) stated as a decision problem is:

INSTANCE: A set of n k-bit data words, $w_1, w_2, \ldots, w_n$, each $w_i \in \{0, 1\}^k$ – k a positive integer, and a positive integer M.

QUESTION: Are there phase assignment ρ and permutation δ of $w_1, w_2, \ldots, w_n$, such that

$$\sum_{i=1}^{n-1} d(\rho(w_{\delta(i)}), \rho(w_{\delta(i+1)})) \leq M? \tag{2}$$

where $\rho(w_i) = w_i$ or $\rho(w_i) = \overline{w_i}$.

Since we know that DOP is hard – it was shown to be NP-complete in [4], it seems intuitive that DOPI, which has an extra degree of freedom of word-inversion, should be at least as hard. The following theorem shows that that is indeed the case.

Theorem 3.1 *DOPI is NP-complete.*

Sketch of Proof DOPI is easily seen to be in NP. To show NP-hardness of DOPI, we transform DOP to DOPI. DOP stated as a decision problem is:

INSTANCE: A set of N K-bit data words, $W_1, W_2, \ldots, W_N$, each $W_i \in \{0, 1\}^K$ – K a positive integer, and a positive integer L.

QUESTION: Is there a permutation σ of $W_1, W_2, \ldots, W_N$, i.e. a 1-1 function $\sigma : \{1, \ldots, N\} \rightarrow \{1, \ldots, N\}$ such that

$$\sum_{i=1}^{N-1} d(W_{\sigma(i)}, W_{\sigma(i+1)}) \leq L? \tag{3}$$

Given an instance of DOP, we generate the following instance of DOPI. Let $n = N$ and $M = L$. For each i, form w_i by concatenating KN zeros after W_i, i.e., $w_i = W_i000\ldots0$ (KN zeros). So, $k = K + KN$. It is easy to see that the transformation is polynomial-time. It can be shown that there exists a permutation σ of W_is such that (3) holds if and only if there exist ρ and δ of w_is such that (2) holds. Details can be found in [4]. ■

4 Solving DOPI

First, let us take care of the technicality of *invert* signal. It makes a transition when two successive words are in opposite phase. We should count also these transitions. This is easily handled in our formulation by appending a zero bit at the end of each word w_i. If the word is sent uncomplemented, the bit remains 0. Otherwise, this bit is set to 1. So, from now on, we will work on w_is augmented with the invert bit, but would still call them w_is.

There are 2^n possibilities, corresponding to two possible phase assignments ρ for each of the n words – for each word w_i, $\rho(w_i) = w_i$ or $\overline{w_i}$. Each phase assignment generates different words, which results in different edge weights and hence a different instance of DOP. One way to solve DOPI then is to generate these 2^n instances of DOP, solve them one by one (with DST, ST-MM, or greedy algorithm), and pick the best solution. Although feasible for small n, this scheme will clearly run out of steam for large n.

It turns out that it is not necessary to explicitly generate and solve 2^n instances of DOP. Due to special properties of the Hamming distance, *it is possible to come up with polynomial-time approximation algorithms (DST, ST-MM) for DOPI that guarantee the same performance bounds from optimality as the corresponding DOP algorithms and also have the same time complexity.*

4.1 DST Algorithm for DOPI

In the double spanning tree algorithm for DOP, the key idea for guaranteeing the upper bound was generating a minimum-weight spanning tree [3]. It turns out that the corresponding key concept in solving DOPI is *generating a minimum-weight spanning tree over all phase assignments*. Given n data words, create a multigraph G on n vertices – one vertex for each word. There are exactly two edges between each pair of vertices v_i and v_j, one labeled SAME (s) and the other OPPOSITE (o). The edge (v_i, v_j) labeled SAME corresponds to the case when the corresponding words w_i and w_j are either both complemented or both uncomplemented. The label OPPOSITE corresponds to when exactly one of w_i and w_j is complemented. These edges are assigned weights d_s and d_o respectively, as follows:

$$d_s(v_i, v_j) = d_s(w_i, w_j) = d(w_i, w_j) = d(\overline{w_i}, \overline{w_j})$$
$$d_o(v_i, v_j) = d_o(w_i, w_j) = d(\overline{w_i}, w_j) = d(w_i, \overline{w_j})$$

Consider $w_1 = 1000$ and $w_2 = 1111$. Then, $d_s(w_1, w_2) = d(w_1, w_2) = 3$, whereas $d_o(w_1, w_2) = d(\overline{w_1}, w_2) = 1$.

Construct a minimum spanning tree T of G. Some edges of T will be labeled s and the rest, o. To determine the phase in which each data word should be transmitted, we assign a label ℓ to each vertex. The **label $\ell(v_i)$ of a vertex v_i** is either COMPLEMENTED (c) or UNCOMPLEMENTED (u), implying that the corresponding data word w_i should be complemented and then sent or just sent uncomplemented respectively.

Definition 4.1 *Given a graph with all edges labeled s or o (edge labeling) and all vertices labeled c or u (vertex labeling ℓ). The vertex labeling ℓ is* **consistent** *with the edge labeling if for each edge (v_i, v_j) labeled s, $\ell(v_i) = \ell(v_j)$ (either both c or both u), and for each edge (v_i, v_j) labeled o, exactly one of v_i & v_j is labeled u.*

Proposition 4.1 *Given a spanning tree T with an edge labeling, there exists a vertex labeling consistent with the edge labeling.*

Proof The following procedure assigns labels to all the vertices of G (or T). Hang the tree T on any vertex, say v. So T becomes a rooted tree with root v. Let $\ell(v) = u$ (arbitrarily).[1] Assign vertex labels from the root to the leaves of T. For an edge $(w, x) \in T$, where w is x's parent, we can assume that w has already been labeled. If (w, x) is labeled s, assign $\ell(x) = \ell(w)$. Otherwise, (w, x) is labeled o, and assign $\ell(x) \neq \ell(w)$, i.e., if $\ell(w) = c$, assign $\ell(x) = u$, and vice-versa. It is easy to see that this procedure labels vertices consistent with the edge labeling. ■

Consider the tree T of Figure 1 (B). Assume edge-labels as in Figure 1 (C). To see that a vertex labeling consistent with this edge labeling exists, hang T on (arbitrarily chosen) vertex 6. $\ell(6) = u$. Since edge (6, 7) is labeled s, $\ell(7) = \ell(6) = u$. This implies $\ell(1) = c$, which implies $\ell(2) = c$. Both 5 and 3 are assigned phases opposite to that of 2. So $\ell(5) = \ell(3) = u$. Finally, $\ell(4) = u$.

Thus, from T and its edge labeling, a consistent vertex labeling is determined, from which the codes ρ that will be transmitted are computed. If $\ell(v_i) = c$, $\rho(w_i) = \overline{w_i}$. Otherwise, $\rho(w_i) = w_i$. The complete DST algorithm is as follows:

1. Form the multigraph G as explained above, with edges labeled s and o appropriately.
2. Find an MST T of G. Using the edge labels in T, determine vertex labels ℓ (c or u) – as in Proposition 4.1. From these labels, determine the phase assignment ρ.
3. Apply Procedure 2.2 to find a Hamiltonian tour HT. This procedure first duplicates each edge of T, finds an Eulerean cycle U using Procedure 2.1, and then constructs HT from U. Care must be taken while constructing HT from U. For instance, if (v_4, v_1) and (v_1, v_6) are being replaced by (v_4, v_6), appropriate edge (v_4, v_6) must be added, since there are two edges between v_4 and v_6. The edge consistent with the labels $\ell(v_4)$ and $\ell(v_6)$ should be added. For instance, if $\ell(v_4) = \ell(v_6) = c$, add the edge labeled SAME.
4. Delete the longest edge e in HT to get a Hamiltonian path HP. The sequence of vertices in this path corresponds to the desired permutation.

Theorem 4.2 *DST algorithm for DOPI yields a Hamiltonian path HP and phase assignment ρ such that $d(HP_\rho) \leq \frac{2(n-1)}{n} d(HP^*_{\rho^*})$, where $HP^*_{\rho^*}$ is the optimum solution to DOPI.*

[1] For $\ell(v) = c$, another labeling will be obtained.

Sketch of Proof The key observation is that T is in fact the MST over all possible phase assignments, i.e., if we generate 2^n graphs each corresponding to a different phase assignment (and hence different edge weights), then determine the MST for each assignment, and pick from all these trees the one with the minimum weight, it will have the same weight as T.

Consider the (complete) graph G^* corresponding to the phase assignment ρ^*. Let a minimum-weight spanning tree in G^* be $T^*_{\rho^*}$. Then, $d(HP^*_{\rho^*}) \geq d(T^*_{\rho^*}) \geq d(T) \geq 0.5 d(HT_\rho)$. The last inequality is from (1). The factor $\frac{n}{n-1}$ comes from deleting the longest edge in HT while obtaining HP. ■

Modified DST It turns out that to find T, it is not necessary to form a multigraph G. It suffices to have a simple graph $\widetilde{G}$ that has only one edge between any vertex pair. This edge is the one that has lower weight among the two edges labeled s and o in the multigraph G. This is because the minimum spanning tree algorithm will always select the lower-weight edge. It can be shown that the edge weights of $\widetilde{G}$ satisfy triangle inequality [4]. Then DST can be modified as follows. *Find MST T in $\widetilde{G}$. Use T to find a Hamiltonian tour HT. Delete the longest edge e in HT to get a Hamiltonian path HP. HP has labels s and o on the edges. Obtain phase assignment for all the vertices from Proposition 4.1, since HP is a spanning tree.* This algorithm also comes within $2\frac{(n-1)}{n}$ of the optimum [4].

4.2 ST-MM Algorithm

ST-MM is a theoretical improvement over DST, since it guarantees a bound within a factor of 1.5 from the optimum solution (plus an additive term). The algorithm is as follows. Find MST T in $\widetilde{G}$. Identify the set O of vertices in the tree T with odd degrees ($|O|$ must be even). Find a minimum-weight perfect matching M on $G(O) = (O, E(O))$, the subgraph of G induced on O.[2] Consider the graph $T \cup M$. Since each edge of M connects two different odd-degree vertices of T, $T \cup M$ is a connected graph with all vertices having even degrees. Then, there exists an Eulerean cycle U in $T \cup M$ – construct U using Procedure 2.1. From U, construct a Hamiltonian tour HT by Procedure 2.2. Delete the longest edge e in this tour to get a Hamiltonian path HP. The phase assignment ρ is determined exactly as in the modified DST.

Theorem 4.3 *Let $HP^*_{\rho^*}$ be the optimum solution for DOPI. Then, $d(HP_\rho) \leq \frac{n-1}{n}[\frac{3}{2} d(HP^*_{\rho^*}) + \frac{k}{2}]$, where k is the word size and HP_ρ is the Hamiltonian path produced by the ST-MM algorithm.*

Although T and M may correspond to different phase assignments, it does not matter since the final phase assignment is decided at the very end – on HP.

4.3 Greedy Heuristic

It works on $\widetilde{G}$ and selects the minimum-weight edge, say (i, j), of $\widetilde{G}$. The path P after this step is (i, j). Next, the minimum-weight edge incident on an end-point of P (i.e., on i or j) is selected. Let it be (i, k). Thus P becomes (k, i, j). Thus, at each step, it greedily selects the minimum-weight edge incident on an end-point of P that is not incident on an internal vertex of P. The heuristic terminates when all the vertices of $\widetilde{G}$ are in P, in which case a Hamiltonian path has been constructed. The phase assignment ρ is determined as in modified DST.

[2] $G(O)$ is a subgraph of G on the vertices O and has exactly those edges of G that have both their end-points in O.

5 Experimental Results

We generated a set of uniformly distributed n k-bit data words, for different values of n and k. Then we applied the following schemes: **random**, **greedy**, DST, and ST-MM. A subset of results is shown in Table 1 – for the complete set, please see [4]. In DOP, only resequencing is allowed. In DOPI, both complementation and resequencing are done. Each entry in the table denotes the total number of transitions for transmitting the n words. In **random** (**rand** in the table), words are transmitted in the same order as they were generated (which was random). This scheme models no resequencing. In DOPI **random**, a word is complemented if it reduces the number of transitions with respect to the last word sent. Other schemes were discussed in Section 4. We also provide a lower bound on the optimum solution (column **ℓ-bnd**) – which is the weight of the MST.

First we compare columns DOPI **greedy** – **grdy** in the table (which gives the best results) with DOPI **ℓ-bnd** to check how close our algorithms come to the optimum. On average, the **greedy** heuristic is within 4.4% of the **ℓ-bnd**, and hence the optimum. The standard deviation of the difference between **greedy** and the **ℓ-bnd** is merely 4.6%. Thus, most of the time, the greedy solution is very close to the optimum.

How much do we benefit by resequencing and complementation? For that, we compare columns DOP **random** and DOPI **greedy**. DOP **random** represents the case when the data words are transmitted uncomplemented and without resequencing. The average percentage reduction in transitions is 34.4%. The maximum reduction was as high as 80% (for $n = 40, k = 5$).

How useful is complementation alone? We make two comparisons. *In the absence of resequencing*: comparing columns **random** under DOP and DOPI (**random** models no resequencing), we see that complementation yields, on average, an 11.3% reduction in transitions, a significant number. *With resequencing*: on comparing resequencing with resequencing and complementation (columns **greedy** under DOP and DOPI), we find that complementation and resequencing yield, on average, an improvement of 3.9% over resequencing. The best improvement is 33%. Also, as k or n increase, the effectiveness of *bus-invert* decreases. This indicates that although complementation by itself is good, when combined with resequencing its impact is significant only for small values of n and k. This is still fine, since in most applications such as instruction sequencing, high-level scheduling, cache write-back, etc., n is not so large and k is at most 64 (machine's word-length).

Finally, we see that the **greedy** scheme is the best. DST results are omitted for lack of space; they were almost always inferior to **greedy** and ST-MM. DST and ST-MM are important nevertheless, since they guarantee bounds with respect to the optimum.

Cache Write-back Application In Table 1, we also present results for a real cache write-back application. Consider a computer system with a main memory and a cache. On a context switch, often the cache is "flushed," i.e., the dirty data lines in the cache are written back to main memory. The order in which lines are written back is often irrelevant. So we can use our data ordering and/or complementation methods to reduce transitions on the data and address buses. For these experiments, the data on cache flushes was obtained from code that performs LU-decomposition of a 30x30 matrix. This code was compiled using `gcc` and its execution was simulated in a system with the following characteristics: Harvard architecture with separate address and instruction buses, cache block size of 4 bytes, and write policy of write-back and write-allocate. For simulation, we used a modified version of the **DineroIII** cache simulator [2] and **dlxsim**, a simulator for the DLX processor. Three experiments were conducted, with unified caches of sizes 1024, 4096, and 65536 bytes, respectively. We assumed that a process switch took place every 2.5×10^5 instructions, causing a cache flush of all dirty blocks. Table 1 shows that resequencing and complementation reduce the number of transitions by 42.7%. We also note that using both complementation and resequencing does not buy us much over using resequencing alone.

In the future, we would like to study resequencing in conjunction with encoding schemes other than complementation. It would also be interesting to study the interaction among resequencing, encoding, and compression. This has interesting applications in network and satellite communication.

n	k	DOP		DOPI			
		rand	**grdy**	**ℓ-bnd**	**rand**	**grdy**	**ST-MM**
5	5	12	8	7	10	7	7
5	16	30	27	27	28	27	27
5	32	64	58	55	60	55	55
20	3	29	7	7	23	7	8
20	8	70	34	28	61	32	35
20	16	146	97	83	126	92	93
40	5	99	21	19	83	20	24
40	40	784	567	523	690	543	555
100	10	496	174	141	423	172	189
100	80	4019	3049	2896	3621	2987	3101
200	20	1998	957	848	1702	922	1011
200	150	15001	12407	11522	14115	11771	12182
400	100	20116	15419	13943	18517	14377	15090
400	200	39846	33392	31378	37811	32014	32999

Results for cache write-back: $k = 32$

n	cache size	DOP		DOPI			
		rand	**grdy**	**ℓ-bnd**	**rand**	**grdy**	**ST-MM**
4	1024	1	1	1	1	1	1
18	1024	191	148	137	184	143	153
10	1024	113	89	80	110	89	85
10	1024	66	27	25	61	25	25
617	4096	7851	3582	3260	7745	3582	3915
348	4096	4205	2076	1887	4168	2076	2254
373	4096	4699	2370	2163	4617	2358	2553
138	4096	1712	991	914	1679	979	1043
965	65536	11854	5270	4749	11676	5273	5679
756	65536	9329	4218	3827	9232	4218	4610
560	65536	7066	3394	3062	6960	3384	3678
181	65536	2222	1248	1136	2194	1243	1305

Table 1: Effect of Data Resequencing and Inversion on Number of transitions

References

[1] Experienced Motorola Designer. Personal comm., Apr. 95.

[2] J. L. Hennessy and D. A Patterson. *Computer Architecture: A Quantitative Approach.* Morgan Kaufmann, 96.

[3] R. Murgai, M. Fujita, and S. C. Krishnan. Data Sequencing For Minimum-transition Transmission. In *VLSI'97*, Brazil.

[4] R. Murgai, M. Fujita, and A. Oliveira. Using Complementation And Resequencing To Minimize Transitions. In *Internal Report, Fujitsu Labs of America, Inc.*, Nov. 97.

[5] M. R. Stan and W. P. Burleson. Limited-weight Codes for Low-power I/O. In *Int. Work. on Low Power Design*, Apr. 94.

Technology Mapping for Large Complex PLDs

Jason Helge Anderson and Stephen Dean Brown
Department of Electrical and Computer Engineering
University of Toronto
10 King's College Road
Toronto, Ontario, Canada M5S 3G4
{janders|brown}@eecg.toronto.edu

1. ABSTRACT

In this paper we present a new technology mapping algorithm for use with complex PLDs (CPLDs), which consist of a large number of PLA-style logic blocks. Although the traditional synthesis approach for such devices uses two-level minimization, the complexity of recently-produced CPLDs has resulted in a trend toward multi-level synthesis. We describe an approach that allows existing multi-level synthesis techniques [13] to be adapted to produce circuits that are well-suited for implementation in CPLDs. Our algorithm produces circuits that require up to 90% fewer logic blocks than the circuits produced by a recently-published algorithm.

1.1 Keywords

PLA-style logic blocks, programmable logic devices, technology mapping.

2. INTRODUCTION

Field-programmable gate arrays (FPGAs) and complex programmable logic devices (CPLDs) have become widely used for the implementation of digital circuits, with each type of chip representing about half of a two billion dollar industry. FPGAs and CPLDs are user-programmable chips that consist of an array of logic blocks and a configurable interconnection network. Most FPGAs have logic blocks based on look-up-tables (LUTs) [14][2], and some have multiplexer-based logic blocks [1]. CPLDs have PLA-style logic blocks [3][2].

In recent years, synthesis for look-up-table-based FPGAs has been an active area of research and development. Many LUT-based technology mapping algorithms have been proposed. Some of these algorithms focus on minimizing either area [8][9] or depth [4][10], while others allow a user to explore the area/depth trade-off [5]. By contrast, very little has been published on synthesis for CPLDs. Several CAD tool vendors offer products that can perform technology mapping for CPLDs, but the algorithms are proprietary and are not publicized. This paper describes a new CAD tool, called *TEMPLA*, that performs technology mapping for devices with PLA-style logic blocks.

The goal of our technology mapping algorithm is to minimize the number of logic blocks required to implement the resulting circuits, where each logic block has the structure shown in Figure 1. The block is characterized by the tuple (I, P, O) where the parameters I, P, and O represent the number of block inputs, product terms, and outputs, respectively. Each output of the block has an associated flip-flop which can either be used or by-passed. Each ✖ in the figure represents a programmable switch. The block in the figure can be adapted to represent the logic blocks in most commercially-available CPLDs. As an example, in Section 4.4 we discuss using our approach to technology mapping for the CPLDs available from Vantis (AMD) [3].

This paper is organized as follows: Background and related work is discussed in Section 3. We present our technology mapping algorithm in Section 4. The results of a comparative experimental study are given in Section 5. Final remarks and suggestions for future work are provided in Section 6.

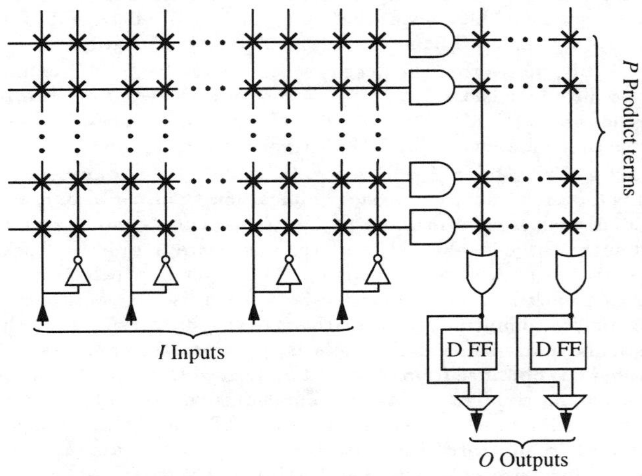

Figure 1. PLA-Style Logic Block.

3. BACKGROUND AND RELATED WORK

The combinational part of a digital circuit can be represented by a directed acyclic graph (DAG). Each node in the DAG represents a single logic function in the circuit; the edges in the DAG represent dependencies between logic functions.

The logic function associated with a node in a DAG can be represented in sum-of-products (SOP) form.

PLA-style blocks are referred to as coarse-grained logic blocks because they typically have a large number of inputs, and hence can realize a large number of different logic functions. Since a PLA-style block contains a programmable AND-array followed by an OR-array, the traditional approach for synthesis of circuits for implementation in such blocks uses two-level minimization. This means that the logic functions in a circuit are collapsed into sum-of-products form. Modern CPLDs contain a large number of PLA-style blocks, with currently-available devices containing up to about 100 blocks. Such large chips allow the implementation of complex circuits for which two-level minimization is not efficient. Hence, the recent trend for synthesis when targeting CPLDs is to make use of multi-level synthesis.

Our approach to synthesis of circuits for realization in PLA-style blocks represents a combination of techniques. We use multi-level logic optimization and traditional technology mapping as a first step. Then, since it is not feasible to directly use traditional technology mapping for PLA-based architectures, as discussed below, we *partially-collapse* the resulting circuit to make efficient use of PLA-style blocks. We describe our algorithm in detail in the next section.

In traditional technology mapping, a library is used to specify all of the logic functions that can be used in the resulting circuit. This approach is not feasible for PLA-based architectures, because the library would be too large to be repetitively searched during technology mapping. One of the few publicized approaches for mapping circuits into PLA-style blocks is found in [11]. It adapts a LUT-based technology mapper and sets the number of LUT inputs to the number of PLA-style block inputs. Then, any node containing more product terms than allowable in the PLA-style block is decomposed into smaller nodes. Finally, the nodes are packed into the multi-output PLA-style blocks.

4. TEMPLA: A TECHNOLOGY MAPPING ALGORITHM FOR LARGE COMPLEX PLDS

In TEMPLA, the technology mapping problem is broken into three phases: performing an optimal tree mapping, heuristic partial collapsing, and bin packing. This algorithmic flow is similar to that of the Chortle-crf technology mapper [9] for LUT-based architectures.

4.1 Phase I: Optimal Tree Mapping

Technology mapping begins by partitioning a circuit's DAG into a forest of fanout-free trees. This is accomplished by identifying the nodes within the DAG that have an out-degree greater than one, and using these nodes as 'breaking points', as illustrated in Figure 2. The reason for this step is to divide the technology mapping problem into smaller and simpler sub-problems. Technology mapping for fanout-free trees is simpler because no node in a fanout-free tree has an out-degree greater than one and, therefore, it is not necessary to consider replication of logic at this juncture of the algorithm.

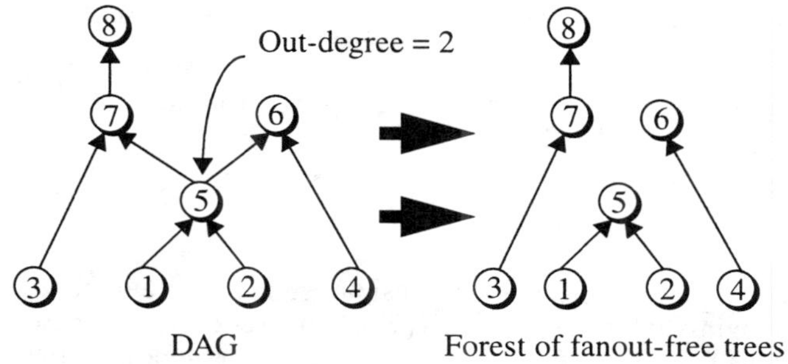

Figure 2. Partitioning a DAG into a forest of fanout-free trees.

Primary input nodes are added to each of the fanout-free trees by modifying them in the following way: for each leaf vertex, n, in a fanout-free tree, $T = (V, E)$, a new primary input node, p, is added to the vertex set V. An edge, $e = (p, n)$, is created and added to the edge set E. The primary input node p is a dummy node and implements no logic function. Before explaining the algorithm further, we must define several terms:

feasible node - a node implementing a single combinational logic function with the property that when simplified, it has less than or equal to I inputs and less than or equal to P product terms. We use the two-level logic minimization provided by Espresso [7] to simplify combinational nodes.

feasible subtree - a subtree of a fanout-free tree with the property that it can be collapsed into a single feasible node. Feasible subtrees are not allowed to possess any of the dummy primary input nodes.

cone at n - a subtree of a fanout-free tree consisting of a node, n, and all of n's predecessors.

size of node n - the size of a node with p product terms and i inputs is equal to $p \times i$.

After the partitioning of the DAG, dynamic programming [6] is used to map each fanout-free tree into a new tree possessing the minimum number of feasible nodes. The trees in the forest can be mapped in any order.

To map a fanout-free tree, $T = (V, E)$, it is traversed in a bottom-up (leaves to root) manner. As each node, n, is visited in turn, the algorithm proceeds to find the set, $S(n)$, of all feasible subtrees of T rooted at n. A cost is computed for each feasible subtree and the feasible subtree of minimum cost is selected and stored at node n. $Cost(n)$ is an integer that refers to this minimum cost.

Primary input nodes implement no logic function and are assigned a cost of zero. All other nodes in V initially have no cost assigned. Three steps are performed repetitively until all of the nodes in V have been assigned a cost.

Step 1: Select a node, n, from V that has not yet been assigned a cost but whose fan-in nodes have been assigned a cost (this implies a bottom-up tree traversal).

Step 2: Determine $S(n)$ - the set of all feasible subtrees rooted at n.

Step 3: Assign a cost to node n using the expression:

$$Cost(n) = min_{T' \in S(n)}\left(1 + \sum_{u \in FI(T')} Cost(u)\right) \quad (1)$$

where $T' = (V', E')$ is a feasible subtree rooted at n belonging to the set $S(n)$; $FI(T')$ is the set of nodes in the fanout-free tree, $T = (V, E)$, that are not nodes in the feasible subtree T' but that are inputs to nodes in T'. More formally:

$$FI(T') = \{v | v \in V, v \notin V', (v, w) \in E, w \in V'\} \quad (2)$$

In Equation (1), $Cost(n)$ represents the minimum number of feasible nodes needed to implement the cone at n. Each subtree, T', in $S(n)$ can be collapsed into one feasible node in the mapping solution; this is the reason for the 1 inside the brackets of Equation (1). The summation term tallys the costs of nodes in T that are inputs to nodes in the subtree T'. The min function selects the feasible subtree rooted at n that results in the minimum-cost mapping of the cone at n. Figure 3 shows a node, n, along with three feasible subtrees rooted at n. The cost of each of n's predecessors is shown internal to each node. The last node to be assigned a cost is the root of the fanout-free tree being mapped.

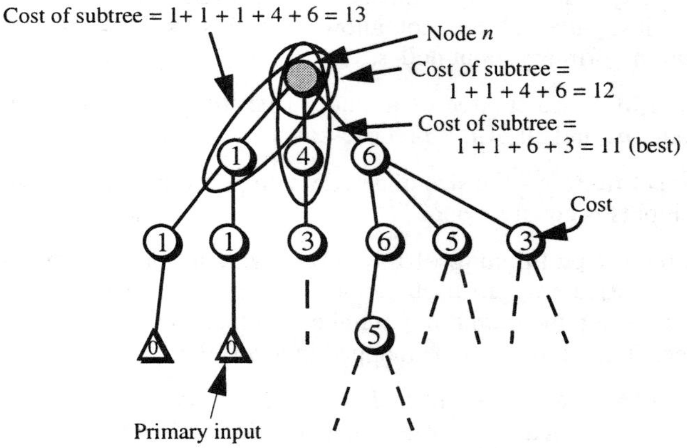

Figure 3. Computation of feasible subtree cost.

Steps 1 to 3 are repeated until all of the nodes have been assigned a cost, at which point the final mapping solution for the tree is specified according to the minimum-cost feasible subtree stored at each node. The final solution is extracted by first considering the root, r, of the original fanout-free tree. The minimum-cost feasible subtree stored at the root is implemented as a new feasible node in the mapping solution. Nodes in T that are inputs to this new feasible node are then added to a node set, M. Mapping proceeds by removing a node, m, from M, implementing the subtree stored at m as a new node in the mapping solution, and lastly, identifying nodes in T that are inputs to the newly created node and adding them to the set M. This process continues until M contains only primary input nodes at which time a network with the minimum number of feasible nodes has been created to implement the function of the original tree. An example of the mapping solution produced by phase I of TEMPLA for a tree in a real circuit is shown in Figure 4.

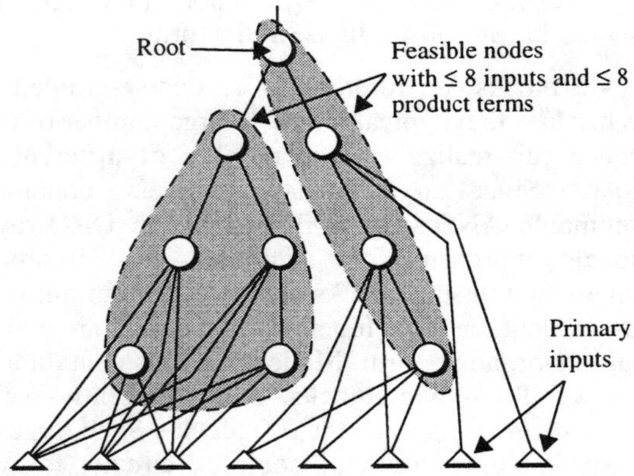

Figure 4. Mapping solution for a tree in MCNC circuit s5378 for I=8, P=8.

One aspect of Step 2 of the algorithm has to be mentioned. A recursive algorithm is used to find $S(n)$. The constraint that a feasible subtree must have a limited number of product terms adds substantial complexity to the problem. For example, consider the case of finding the set of feasible subtrees for a node, n, with two fanin nodes, A and B. Assume that the subtree consisting of n and A is a feasible subtree but that the subtree consisting of n and B is not feasible because it has more than P product terms. Complexity is introduced because the infeasibility of the n and B subtree does not imply the infeasibility of the n and A and B subtree. Specifically, the n and A and B subtree may be feasible because the subtree consisting of n and A may simplify into a feasible node containing fewer product terms than were originally in node n. If we compare this problem to technology mapping for look-up-tables, for which only the number of inputs to the block needs to be considered, it is clear that technology mapping for PLA-style blocks has added complexity.

After performing phase I on all of the fanout-free trees within a circuit's DAG, the mapping solutions for each tree are put back together into a complete circuit. The algorithm then considers replacing each node in the circuit with its complemented form. This can be beneficial if the complemented form of a node contains fewer product terms than its uncomplemented form, and therefore, this optimization is considered after each of our algorithm's three phases. Phase II attempts to reduce the number of nodes in the circuit by collapsing nodes across tree boundaries.

4.2 Phase II: Heuristic Partial Collapsing

Any node that can be collapsed into all of its fanout nodes

can be eliminated, provided that all nodes remain feasible after the collapsing. This introduces another optimization problem since collapsing some nodes into their fanout nodes may preclude the possibility of collapsing other nodes into their fanout nodes. Several criteria were studied empirically using 30 benchmark circuits to determine how best to choose nodes to collapse. A total of 19 of the 30 circuits used in this study are large MCNC benchmarks [15], 10 circuits are HDL specifications developed by the authors, and one circuit is a processor benchmark from the PREP synthesis suite [12]. The criteria listed below refer to nodes *to be collapsed* into their fanout nodes, and not the new node(s) that would exist after collapsing is done. The criteria considered are:

1. Inputs - prefer to collapse nodes with fewer inputs.

2. Product terms - prefer to collapse nodes with fewer product terms.

3. Node size - prefer to collapse small nodes.

4. Fanout - prefer to collapse nodes with low fanout.

To evaluate the criteria, each was applied individually as the selection criteria for partial collapsing. The number of circuit nodes before and after collapsing was determined and a percentage reduction was computed for each benchmark circuit. These percentages were then averaged; hence, each circuit was treated equally in the comparison. The results of this experiment showed that each of the criteria performed similarly, on average. We decided to use node size as the primary criteria for selecting nodes to collapse, with fanout used as a secondary criteria.

Since phase II uses a heuristic approach when choosing nodes to collapse, we attempted to evaluate the quality of our solution. To do this, we implemented an exhaustive algorithm based on the branch-and-bound method to perform the equivalent of phases I and II. Although the exhaustive approach could be used only for small circuits, due to its time-complexity, it revealed an interesting property of the problem being solved. Recall that the PLA-style blocks may have multiple outputs. For such blocks, comparing our heuristic solution to the exhaustive solution revealed that it is sometimes best to limit the size of nodes created during phase II to better facilitate the packing of nodes into multi-output blocks. TEMPLA was modified to deal with this issue as discussed below.

When a node is collapsed into its fanout nodes, the sum of the sizes of the resultant nodes after collapsing may be larger than the sum of the sizes of nodes before collapsing. The TEMPLA algorithm allows a user to limit this size increase by varying the parameter β in the following relation:

$$\sum_{t \in T} size(t) \le \beta \cdot \left(size(v) + \sum_{s \in S} size(s) \right) \qquad (3)$$

where v is the node to be collapsed into its fanout nodes; S is the set of v's fanout nodes before any collapsing; and T is the set of v's fanout nodes after v has been collapsed into them. The algorithm will not collapse a node into its fanout nodes if Relation (3) evaluates false.

When the logic blocks in the target architecture have only one output, β should be set to a large number, which places no restrictions on the size of nodes after collapsing. However, for multi-output logic blocks, it may be advantageous to set β to a smaller value. This was investigated experimentally in the context of the third phase of our algorithm, and the results are shown in the next section.

4.3 Phase III: Bin Packing

The final phase of TEMPLA packs circuit nodes into the multi-output blocks available in the target architecture. This is accomplished using a first-fit-decreasing bin packing algorithm that attempts to maximize the number of shared inputs between nodes that are packed into the same logic block. The bin packing algorithm used is described using pseudo code in Figure 5. Single input nodes that implement inverters are ignored during bin packing because signals can be inverted for 'free' in PLA-style logic blocks, since each input to the block is available in both true and complemented form.

To investigate what value of β in Relation (3) is appropriate for multi-output blocks, β was varied while 30 benchmark circuits were mapped into logic blocks with the parameters (10, 12, 4). The number of blocks needed to implement each circuit was compared to that attained when β was set to a large value (unrestricted collapsing) and a percentage decrease in the number of logic blocks was computed. The average percentage decrease over all 30 circuits is shown in Table 1.

The results in Table 1 suggest that it is not beneficial to pack as much logic as possible into each feasible node before packing the nodes into multi-output logic blocks. Setting β equal to 1.5 is the best choice when targeting blocks with four outputs.

Table 1: Effect of restricted partial collapsing.

β	Average % decrease in # of logic blocks
1.0	-1.4
1.25	3.7
1.5	4.3
1.75	3.6
2.0	2.9
2.25	3.0

4.4 Adapting TEMPLA to Target the Vantis (AMD) Mach 4 [3]

As an example of how TEMPLA can be applied to a commercial CPLD product, this section describes its use for the Vantis (AMD) Mach 4 CPLDs [3]. Each logic block in the Mach 4 has 33 inputs, 90 product terms, and 16 outputs. Eighty of the 90 product terms are grouped into 16 clusters of five and these product terms are available to implement combinational logic. Each of the 16 'OR gates' in the logic block is allocated a cluster of five product terms; however, clusters may be redirected from an OR gate to another adjacent OR gate (leaving one OR gate unused). A maximum of 20 product terms are allowed to feed a single OR gate.

Phases I and II of TEMPLA can be used to create nodes that pack efficiently into Mach 4 logic blocks by setting I equal to 33 and P equal to 20. These first two phases could be augmented with cost functions to reflect the notion that a node with between one and five product terms costs less to implement than a node with between six and ten product terms, since a product term cluster would need to be redirected in the latter case. The cost of a node would be proportional to the number of product term clusters that need to be redirected to implement it. Nodes could be packed into the Mach 4 logic blocks using a slightly modified version of phase III of TEMPLA. Hence, we believe that relatively few changes to our algorithm would be required for it to efficiently target the Mach 4 architecture.

5. EXPERIMENTAL RESULTS

Our algorithm has been implemented in the C language within the SIS [13] framework, allowing it to access the I/O routines and two-level logic minimization algorithms within SIS. To assess the quality of mapping solutions produced by our tool, it was compared with the approach used in [11], called DDMAP. The first step of DDMAP is to apply a look-up-table technology mapper; we used Level-Map [8], which minimizes the number of look-up-tables needed, to perform this initial mapping.

Table 2 shows the results when the technology mappers are used to map circuits into PLA-style blocks with the parameters (10, 12, 4). These parameters were chosen based on previous research in [11], which showed that logic blocks with 8-10 inputs, 12-13 product terms, and 3-4 outputs are the most area-efficient PLA-style logic blocks. The results for 15 of the 30 circuits used in this study are shown in the table. Before applying TEMPLA or DDMAP, each of the circuits was synthesized using the Synopsys Design Compiler. This resulted in an optimized multi-level netlist of gates from an intermediate library. The intermediate library consists of elements suitable for mapping circuits into PLA-based architectures. The first column in the table lists the name of each benchmark circuit. The second lists the number of logic blocks needed to implement each circuit when TEMPLA is used. The third column shows the running time in seconds for TEMPLA on a 300 MHz SPARCstation. The fourth column gives the results for DDMAP; a percentage is given in brackets which represents the amount of additional logic blocks needed to implement each circuit in comparison with TEMPLA. On average, when the 15 circuits shown in Table 2 are mapped using DDMAP, they require 90% more blocks than when TEMPLA is used. Over all 30 circuits considered, the DDMAP solutions contained 93.8% more blocks on average.

Notice that TEMPLA performs poorly for the benchmark 'ex5p'. For this circuit, DDMAP produces a solution with nearly 80% fewer blocks than TEMPLA. The ex5p circuit is a combinational circuit possessing 8 primary inputs, and 63 primary outputs. Since the number of inputs to the circuit is less than the number of inputs to the logic blocks in the (10, 12, 4) architecture, Level-Map produces a mapping containing 63 nodes: one node for each primary output. Level-Map produces such a mapping because it is able to deal effectively with reconvergent paths within circuits. Furthermore, for this circuit, most of the nodes in the Level-Map solution happen to be feasible nodes. Many of the nodes have common inputs, allowing several nodes to be packed into each 4-output logic block. To verify that the exploitation of reconvergent paths was the reason for the superior mapping, the circuit was mapped with the LUT-based technology mapper, Chortle-crf [9], which deals with reconvergence in only a limited way. Chortle-crf produced a mapping containing 363 nodes which is significantly greater than the 63 nodes in the Level-Map solution. Since TEMPLA breaks up a circuit into fanout-free trees, it is not able to exploit reconvergent paths effectively, and hence,

```
nodeSet ← Set of all nodes in network (minus single input inverters)
while (nodeSet is not empty) {
        plaBlock ← empty block /* allocate a new PLA-style logic block */
        nodeSel ← largest node in nodeSet (node size = number of inputs x number of product terms)
        Add nodeSel to plaBlock
        Remove nodeSel from nodeSet
        while (nodeSet is not empty and there are nodes in nodeSet that can fit into plaBlock) {
                nodeSel ← node from nodeSet that has the largest number of inputs in common with the nodes
                        already in plaBlock; the node must be able to fit into plaBlock; use node size to break ties
                Add nodeSel to plaBlock
                Remove nodeSel from nodeSet
                }
        }
```

Figure 5. Maximum shared input bin packing algorithm.

produces an inferior solution for this circuit.

6. FINAL REMARKS AND FUTURE WORK

In this paper we presented a new technology mapping CAD tool for CPLDs with a large number of PLA-style logic blocks. Our algorithm breaks the technology mapping problem into three phases: optimal tree mapping, heuristic partial collapsing, and bin packing. The experimental study described in the previous section shows that our tool produces mapping solutions containing fewer logic blocks than solutions produced by a recently-published technique.

One direction for future work is to modify our algorithm to target a specific commercially available CPLD, as discussed in Section 4.4, and compare the mapping results to those produced by existing commercial CAD tools. Since development of CPLDs that contain sufficient numbers of PLA-style blocks to benefit from multi-level synthesis has only recently come about, we believe that synthesis issues for such chips is an important area for future research.

Table 2: Experimental results for (10, 12, 4) architecture.

Circuit	TEMPLA (# blocks)	TEMPLA running time (seconds)	DDMAP [11] (# blocks (% more))
alu4	155	29.3	199 (28.4)
apex4	193	30.2	193 (0.0)
clma	957	815.4	1458 (52.4)
cps	120	18.3	159 (32.5)
dalu	64	9.0	102 (59.4)
ex5p	132	18.8	27 (-79.5)
misex3	154	27.8	214 (39.0)
pdc	618	281.6	1221 (97.6)
s38417	603	495.2	1208 (100.3)
seq	229	52.8	337 (47.2)
fir	249	123.3	1424 (471.9)
fsm8_8_13	49	5.5	58 (18.4)
pmac	237	126.6	911 (284.4)
psdes	151	37.0	301 (99.3)
sort	138	29.8	275 (99.3)
			Average over 15 circuits above: 90%
			Average over all 30 circuits: 93.8%

7. ACKNOWLEDGMENTS

The authors gratefully acknowledge the financial support of Chip Express Corporation and the Government of Ontario. Thanks to Dr. Jack Kouloheris for providing code for the DDMAP [11] algorithm.

8. REFERENCES

[1] *ACT 1 Series FPGAs Data Sheet*, Actel Corporation, 1996.

[2] *The Altera Data Book*, Altera Corporation, 1996.

[3] *The MACH 4 Family Data Sheet,* Advanced Micro Devices, 1996.

[4] J. Cong and Y. Ding, "FlowMap: An Optimal Technology Mapping Algorithm for Delay Optimization in Lookup-Table Based FPGA Designs", *IEEE Transactions on Computer-Aided Design of Integrated Circuits and Systems*, Vol. 13, No. 1, January 1994, pp. 1-11.

[5] J. Cong and Y. Hwang, "Simultaneous Depth and Area Minimization in LUT-based FPGA Mapping", *UCLA Department of Computer Science Technical Report*, CSD TR-9500001.

[6] T. H. Cormen, C. E. Leiserson and R. L. Rivest, *Introduction to Algorithms*, McGraw-Hill Book Company, Toronto, 1994.

[7] Giovanni De Micheli, *Synthesis and Optimization of Digital Circuits*, McGraw-Hill Inc., Toronto, 1994.

[8] A. H. Farrahi and M. Sarrafzadeh, "Complexity of the Lookup-Table Minimization Problem for FPGA Technology Mapping", *IEEE Transactions on Computer-Aided Design of Integrated Circuits and Systems*, Vol. 13, No. 11, November 1994, pp. 1319-1332.

[9] R.J Francis, J. Rose and Z. Vranesic, "Chortle-crf: Fast Technology Mapping for Lookup Table-Based FPGAs", *28th ACM/IEEE Design Automation Conference,* June 1991, pp. 227-233.

[10] R. J. Francis, J. Rose and Z. Vranesic, "Technology Mapping of Lookup Table-Based FPGAs for Performance", *1991 IEEE Conference on Computer-Aided Design,* pp. 568-571.

[11] J. L. Kouloheris, "Empirical Study of the Effect of Cell Granularity on FPGA Density and Performance", *Ph.D. Thesis,* Department of Electrical Engineering, Stanford University, 1993.

[12] *Programmable Electronics Performance Corporation Test Benches,* http://www.prep.org, 1996.

[13] E. M. Sentovice et al., "SIS: A System for Sequential Circuit Synthesis", *Technical Report UCB/ERL M92/41,* Electronics Research Laboratory, Department of Electrical Engineering and Computer Science, University of California, Berkeley, 1992.

[14] *The Programmable Logic Data Book,* Xilinx Corporation, 1994.

[15] S. Yang, "Logic Synthesis and Optimization Benchmarks", *Technical Report,* Microelectronics Center of North Carolina, 1991.

Delay-Optimal Technology Mapping for FPGAs with Heterogeneous LUTs

Jason Cong and Songjie Xu
Department of Computer Science
University of California, Los Angeles, CA 90095

Abstract

Recent generation of FPGAs take advantage of speed and density benefits resulted from *heterogeneous* FPGAs, which provide either an array of homogeneous programmable logic blocks (PLBs), each configured to implement circuits with lookup tables (LUTs) of different sizes, or an array of physically heterogeneous LUTs. LUTs with different sizes usually have different delays. This paper presents the *first* polynomial-time optimal technology mapping algorithm, named HeteroMap, for delay minimization in heterogeneous FPGA designs. For a heterogeneous FPGA consisting of K_1-LUTs, K_2-LUTs, ..., and K_c-LUTs, HeteroMap computes the minimum delay mapping solution in $O(\sum_{i=1}^{c} K_i \cdot n \cdot m \cdot \log n)$ time for a circuit netlist with n gates and m edges. The HeteroMap algorithm generates favorable results for Xilinx XC4000 series FPGAs and Lucent ORCA2C series FPGAs. Furthermore, the optimality of the HeteroMap algorithm enables us to quantitatively evaluate various heterogeneous architectures without the bias of mapping heuristics.

1 Introduction

In order to maximize performance and device utilization, recent generation of FPGAs take advantage of speed and density benefits resulted from *heterogeneous* FPGAs, which provide either an array of homogeneous PLBs, each configured to implement circuits with LUTs of different sizes, or an array of physically heterogeneous LUTs. For example, each PLB in XC4000 series FPGAs [10] can be configured to implement either two seperate 4-LUTs or a single 5-LUT, while each PLB in ORCA2C series FPGAs can be configured to implement either two seperate 5-LUTs or a single 6-LUT. Note that both XC4000 and ORCA2C devices provide heterogeneous LUTs through reconfiguration instead of presenting a physical array of LUTs of different sizes directly.

In a heterogeneous FPGA chip, larger LUTs can certainly cover more gates, but usually have longer delay. Therefore, an important problem is how to utilize the heterogeneous LUTs to minimize the overall circuit delay and/or area. For example, the circuit depicted in Figure 1(a) can be mapped into five homogeneous 4-LUTs, each of delay 1, with the total mapping delay of 3 (see Figure 1(b)). In Figure 1(c), the same circuit can be mapped into two 4-LUTs assuming each of delay 1 and one 5-LUT assuming of delay 1.5, with a total mapping delay of 2.5. In general, given a heterogeneous FPGA with LUTs of different sizes, we want to find an optimal mapping solution with the minimum delay or area.

DAC 98, San Francisco, California
©1998 ACM 0-89791-964-5/98/06..$5.00

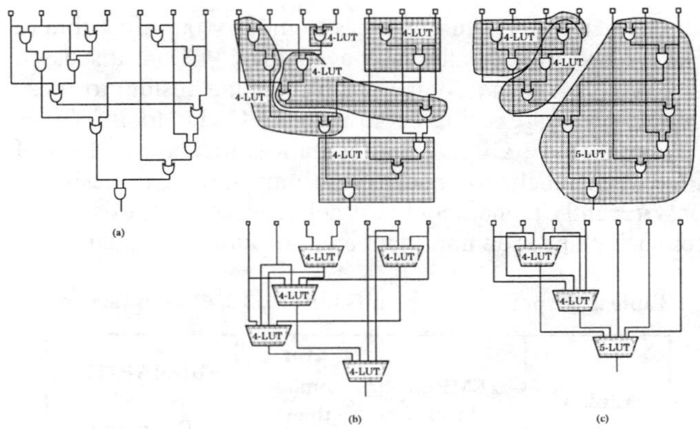

Figure 1: (a) The orginal circuit; (b) Mapping to 4-LUTs by Flowmap; (c) Mapping to 4-LUTs and 5-LUTs by HeteroMap.

Reference [4] provides a comprehensive survey of the existing technology mapping algorithms for homogeneous LUT-based FPGAs. However, none of these algorithms are able to deal with the delay optimization problem for heterogeneous FPGAs, as they assume the identical capacity and delay for every LUT. In [7], an approach for technology mapping into heterogeneous LUT-based FPGAs was presented for area minimization, but their architecture assumes a mixture of only two types of LUTs with a fixed ratio in one FPGA chip, and their algorithm can not be easily extended for the general heterogeneous FPGAs employing three or more types of logic blocks.

In this paper, we present a general technology mapping problem formulation for heterogeneous FPGAs and a polynomial-time optimal technology mapping algorithm, named HeteroMap, for delay minimization in heterogeneous FPGA designs. The remainder of this paper is organized as follows. Section 2 presents the problem formulation and preliminaries. Section 3 describes the HeteroMap algorithm. Experimental results and comparative study are presented in Section 4. Section 5 concludes the paper. Due to the page limitation, the detailed proofs of theorems are left out, which are available in [6].

2 Problem Formulation and Preliminaries

A Boolean network N can be represented as a directed acyclic graph (DAG) where each node represents a logic gate, and a directed edge (i, j) exists if the output of gate i is an input of gate j. Primary input (PI) nodes have no incoming edge and primary output (PO) nodes have no outgoing edge. We use $input(v)$ to denote the set of fanins of gate v. A boolean network is K-*bounded* if $|input(v)| \leq K$ for each node v in the network.

A general LUT-based heterogeneous FPGA consists of c types of LUTs, K_1-LUT, K_2-LUT, ..., and K_c-LUT ($K_1 < K_2 < \ldots < K_c$) with the delay of $d_1, d_2, \ldots,$ and d_c ($d_1 < d_2 < \ldots d_c$). For a circuit mapped into a heterogeneous FPGA, we assume different access delays for heterogeneous LUTs but a constant delay for the interconnection[1], which we call *heterogeneous LUT-delay model.* The unit-delay model used in [2] is a special case of heterogeneous LUT-delay model in homogeneous FPGAs. Given these definitions, the technology mapping problem for heterogeneous FPGAs can be formulated as follows: *given a K_1-bounded Boolean network N and the heterogeneous FPGA, transform N to an equivalent LUT network N' which consists of K_1-LUTs, K_2-LUTs, ..., and K_c-LUTs such that the circuit delay and/or area are minimized.* In this paper, our primary objective is to minimize the circuit mapping delay under the heterogeneous LUT-delay model through technology mapping. Therefore, a mapping solution is said to be *optimal* if the mapping delay is minimized.

Given a network $N = (V(N), E(N))$ with a source s and a sink t, a *cut*$(X, \bar{X})$ is a partition of the nodes in $V(N)$ such that $s \in X$ and $t \in \bar{X}$. The *node cut set of* of $(X, \bar{X})$, denoted $C(X, \bar{X})$, is the set of nodes in X that are adjacent to some node in $\bar{X}$, *i.e.*, $C(X, \bar{X}) = \{x : (x, y) \in E(N), x \in X \text{ and } y \in \bar{X}\}$. The node cut-size of $(X, \bar{X})$, denoted $n(X, \bar{X})$, is the number of nodes in $C(X, \bar{X})$. A cut $(X, \bar{X})$ is *K-feasible* if its node cut-size is no more than K, *i.e.*, $n(X, \bar{X}) \leq K$. Moreover, assuming that there is a given label $l(v)$ associated with each node v ($l(v)$ may not be an interger), the *height* of a cut $(X, \bar{X})$, denoted $h(X, \bar{X})$, is defined to be the maximum label of the nodes in $C(X, \bar{X})$. The *volume* of a cut $(X, \bar{X})$, denoted $vol(X, \bar{X})$, is the number of nodes in $\bar{X}$, *i.e.*, $vol(X, \bar{X}) = |\bar{X}|$.

3 Delay Optimal Mapping for Heterogeneous FPGAs

In this section, we present a delay optimal technology mapping algorithm for heterogeneous FPGAs under the heterogeneous LUT-delay model. Applicable to any *K-bounded* ($K \leq K_1$) Boolean network, HeteroMap runs in two phases. In the first phase, according to the topological order from PI to PO, HeteroMap uses the dynamic programing technique to compute the label for each node, which is the delay of the node if implemented by an LUT in a delay-optimal mapping solution. In the second phase, HeteroMap generates the heterogeneous LUT mapping solution based on the node labels and cuts computed in the first phase. Our algorithm also minimizes the circuit area by maximizing the volume of each cut and by the post-mapping heterogeneous packing operations. The details are discussed in the following subsections.

3.1 Labeling Phase

Given a K-bounded Boolean network N, let N_t denote the subnetwork consisting of node t and all the predecessors of t. The *label* of t, denoted $l(t)$, is the delay of t in the *optimal* heteroheneous LUT mapping solution of N_t. The first phase of our algorithm computes the labels for all the nodes in N, according to the topological order starting from the PIs. The topological order guarantees that every node is processed after all of its predecessors have been processed. For each PI node u, we assign $l(u) = 0$. Suppose that t is the current node being processed. Then, for each node $v \neq t$ in N_t, the label $l(v)$ must have been computed. By including in N_t an auxiliary node s and connecting s to all the PI nodes in N_t, we obtain a network with s as the source and t as the sink. For simplicity, we still denote it as N_t. Figure 2(a) shows a Boolean network in which gate t is to be labeled. Figure 2(b) shows the construction of the network N_t. Assuming that $LUT(t)$ is the K-LUT that implements node t in an optimal mapping solution of N_t. Since K may be any of K_1, K_2, ..., or K_c, we compute $l_i(t)$ for each $i = 1, 2, \ldots,$ and c, which is the minimum delay of node t in the optimal mapping solution of N_t if t is implemented by a K_i-LUT. We have

$$l(t) = \min_{1 \leq i \leq c} l_i(t) \quad (1)$$

In order to compute $l_i(t)$, suppose $LUT_i(t)$ be the K_i-LUT that implements node t. Let $\bar{X}_i$ denote the set of nodes in $LUT_i(t)$ and X_i denote the remaining nodes in N_t. Then, $(X_i, \bar{X}_i)$ forms a K_i-feasible cut between s and t in N_t because the number of inputs of $LUT_i(t)$ is no more than K_i. Moreover, let u_i be the node with the maximum label in $C(X_i, \bar{X}_i)$, then, $l_i(t)$, the delay of $LUT_i(t)$, is $l(u_i) + d_i$. According to the definition of the height of a cut in Section 2, $l(u_i) = h(X_i, \bar{X}_i)$. Therefore, in order to minimize $l_i(t)$, we want to find a minimum height K_i-feasible cut $(X_i, \bar{X}_i)$ in N_t. Since for each i ($i = 1, 2, \ldots, c$), we compute a minimum height K_i-feasible cut $(X_i, \bar{X}_i)$ in N_t and $l_i(t) = h(X_i, \bar{X}_i) + d_i$, Eq. 1 can be transformed to

$$l(t) = \min_{1 \leq i \leq c} \; \min_{(X_i, \bar{X}_i) \text{ is } K_i\text{-feasible}} (h(X_i, \bar{X}_i) + d_i) \quad (2)$$

Based on the above discussion, we have

Lemma 1 *The label $l(t)$ computed by Eq. 2 gives the minimum delay of any mapping solution of N_t under the heterogeneous LUT-delay model.*

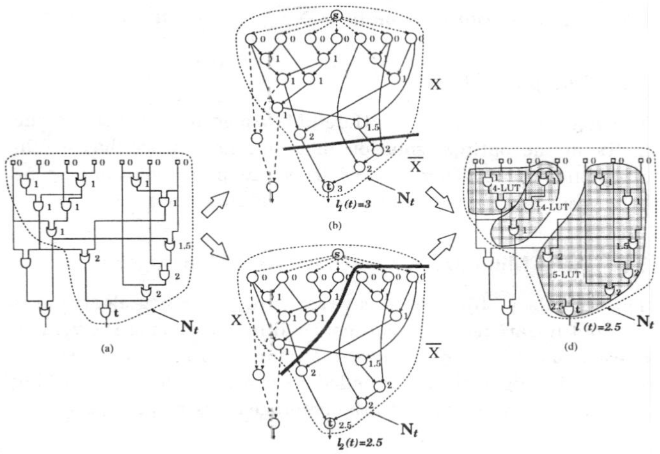

Figure 2: (a) Gate t to be labeled; (b) Construct N_t and compute the minimum height 4-feasible cut in N_t such that $l_1(t) = 3$; (c) Compute the minimum height 5-feasible cut in N_t such that $l_2(t) = 2.5$; (d) Determine $l(t) = min\{l_1(t), l_2(t)\} = 2.5$.

In Figure 2, we assume that there are two types of LUTs, 4-LUTs and 5-LUTs, with delays of 1 and 1.5, in a heterogeneous FPGA device. Through the minimum height 4-feasible cut computation, Figure 2(b) computes $l_1(t)$ to be 3, assuming that t is implemented by a 4-LUT. Similarly, Figure 2(c) computes $l_2(t)$ to be 2.5, assuming that t is implemented by a 5-LUT. In Figure 2(d), the label $l(t)$ is then set to be 2.5. The LUT that implements t in the optimal mapping solution is determined to be a 5-LUT which includes all the nodes in $\bar{X}$, where $(X, \bar{X})$ is the minimum height 5-feasible cut in N_t computed in Figure 2(c).

[1]The constant interconnection delay can be counted into the LUT delays so that the interconnection delay can be set to zero.

Lemma 2 (Monotone Property) *Let $l(t)$ be the label of node t and $l(t')$ be the label of a predecessor t' of t, then $l(t) \geq l(t')$.*

According to Eq. 2, the key problem in computing $l(t)$ is to compute the minimum height K_i-feasible cut $(X_i, \bar{X}_i)$ in N_t to get $l_i(t)$ (i = 1, 2, ..., and c). The computation of a minimum height K-feasible cut in the case of heterogeneous LUT-delay model is more complicated than that under the unit-delay model, although $l(t)$ is monotone. First of all, we need to determine the range of the minimum height of a K_i-feasible cut in N_t, denoted $H_t(i)$.

Lemma 3 *Let t be a non-PI node in network N, and $I_t(i) = \max\{l(u) : u \in input(t)\}$, then, $I_t(i) - d_i \leq H_t(i) \leq I_t(i)$.*

Based on Lemma 3, we get a height array, denoted $H_{t_{list}}(i)$, which includes all the distinct labels $l(u)$ ($u \in N_t$) in the specified range. Obviously, the maximum size of the height array is $|N_t|$. We can then perform a binary search over $H_{t_{list}}(i)$ to determin $H_t(i)$ and hence $l_i(t) = H_t(i) + d_i$. To compute a K-feasible cut of height H in N_t, we apply the approach proposed in [2].

Lemma 4 *A minimum height K_i-feasible cut in N_t under heterogeneous LUT-delay model can be found in $O(K_i \cdot m \cdot \log |N_t|)$ time where m is the number of edges in N_t.*

By computing $l_i(t)$ (i = 1, 2, ..., c), $l(t)$ can be determined in $O(\sum_{i=1}^{c} K_i \cdot m \cdot \log |N_t|)$ time, where m is the number of edges in N_t. Applying the label computation for each node in N according to the topological order, we have

Theorem 1 *Given a network N with n nodes and m edges, the labeling process can be done in $O(\sum_{i=1}^{c} K_i \cdot n \cdot m \cdot \log n)$ time.*

3.2 Mapping Phase

The mapping phase of the HeteroMap algorithm generates the LUTs in the optimal solution in a similar way as that in [2], exept that the LUTs generated in our mapping solution can be heterogeneous. The entire second phase takes linear time.

3.3 Area Minimization in the HeteroMap Algorithm

The secondary objective of our technology mapping algorithm is area optimization, which is considered by maxmizing the volume of each cut [2] and by the post-mapping heterogeneous predecessor packing operations, extended from the predecessor packing technique proposed in [1] to the heterogeneous mapping solution.

4 Experimental Results and Comparative Study

4.1 Mapping Comparison on Heterogeneous FPGAs

We first compare FlowMap [2] and HeteroMap on XC4000 series FPGAs, which can implement circuits with 4-LUTs and 5-LUTs of delay 1 and 1.5[2]. Match4K [3] is an intelligent post-mapping multi-step matching heuristic for XC4000 devices to reduce the PLB number. We compare FlowMap with HeteroMap both followed by Match4K on MCNC benchmarks, and perform layout using the same XC4000 part (4013MQ208-5) on XACT5.2 FPGA development system. For FlowMap, we set $K = 5$. Table 1 shows that HeteroMap reduces 19% of the mapping delays (M-Dly) and 7% of the post-layout delays (PL-D) with only 2% increase on the PLB numbers over FlowMap.

[2]This delay ratio is derived from [10].

Circuits	FlowMap + Match4K			HeteroMap + Match4K		
	M-Dly	PLB	PL-D (ns)	M-Dly	PLB	PL-D (ns)
5xp1	4.50	16	41.00	3.50	14	36.70
9sym	7.00	35	54.10	5.50	39	53.10
9symml	7.00	37	60.60	5.50	38	53.00
C499	6.50	83	80.70	5.50	88	76.80
C880	12.00	132	140.20	10.00	132	106.80
alu2	12.00	94	110.10	10.50	94	107.50
alu4	15.00	149	128.40	11.50	163	160.20
apex2	7.00	87	84.00	6.50	91	70.70
apex7	6.00	46	55.00	4.00	49	52.70
count	4.50	44	52.90	4.00	39	46.90
duke2	6.00	99	75.90	4.50	103	69.90
e64	4.50	83	68.00	3.50	91	63.10
misex1	3.00	10	30.30	2.50	11	31.00
rd84	6.00	29	53.70	4.50	26	49.80
vg2	5.00	23	55.00	4.00	23	44.30
z4ml	3.50	7	34.70	3.50	8	34.50
TOTAL	109.50	974	1124.60	89.00	1009	1057
A_Mean	6.84	60.88	70.29	5.56	63.06	66.06
A_Ratio	1	1	1	-19%	+3%	-6%
G_Mean	6.22	44.23	64.24	5.05	45.32	59.77
G_Ratio	1	1	1	-19%	+2%	-7%

Table 1: Comparison between FlowMap and HeteroMap on XC4000 series FPGAs

In order to compare HeteroMap with FlowMap on ORCA2C series FPGAs, HeteroMap takes the ORCA2C device as a heterogeneous FPGA which consists of 5-LUTs and 6-LUTs with the delays of 1 and 1.2[3]. For FlowMap, we set $K = 6$ such that after technology mapping each 6-input node is implemented by one 6-LUT which occupies one PLB, and all other nodes can be implemented in 5-LUTs every two of which are packed into one PLB. The comparison results are summarized in Table 2. Compared with FlowMap, HeteroMap reduces 9% of the mapping delays (M-Dly) and 8% of the area in terms of the PLB number.

4.2 Heterogeneous *v.s.* Homogeneous Architectures

The optimality of the HeteroMap algorithm enables us to quantitatively evaluate various heterogeneous architectures using MCNC benchmarks without the bias of mapping heuristics. As an example, we use HeteroMap to map the circuits into a heterogeneous FPGA with four types of logic blocks, 3-LUTs, 4-LUTs, 5-LUTs, and 6-LUTs with delay of 1, 1.33, 1.67, and 2. FlowMap maps circuits into the homogeneous K-LUT FPGAs with K = 3, 4, 5, and, 6 respectively. The comparision results on both area and mapping delay are summaried in Table 3. Let $r = \frac{\text{area of } (K+1)-\text{LUT}}{\text{area of } K-\text{LUT}}$. In order to compare the area in the mapping solution, we set $r = 2$ in Table 3, which assumes that the area of one 3-LUT, 4-LUT, 5-LUT and 6-LUT is 1, 2, 4 and 8, respectively. With HeteroMap, heterogeneous FPGAs always achieve much better mapping delay than that of homogeneous FPGAs. As for the area, the heterogeneous FPGA outperforms the homogeneous FPGAs with K = 4, 5, and 6. When $K = 3$, the homogeneous FPGA gets 34% better area than the heterogeneous FPGA but has 25% longer delays on average. To better understand the impact of area ratio r and the mapping results on both area and delay, Figure 3 is plotted with $1 \leq r \leq 3$ as the X-axis and $area \times delay^2$ as the Y-axis [9]. From Figure 3 we can see that the FPGA with heterogenous LUTs, 3-LUTs or 4-LUTs is more efficient than the one with 5-LUTs or 6-LUTs.

[3]This delay ratio is derived from [8].

Circuits	FlowMap M-Dly	FlowMap PLB	HeteroMap M-Dly	HeteroMap PLB
5xp1	2.20	12	2.20	12
9sym	4.60	30	4.20	30
9symml	4.60	28	4.20	30
C499	5.40	104	5.00	85
C880	8.20	141	8.00	138
alu2	8.40	95	8.00	97
alu4	9.40	191	8.60	168
apex2	5.60	98	5.00	97
apex4	6.00	506	5.20	406
apex6	4.80	143	4.00	132
apex7	3.40	53	3.20	49
count	3.60	42	3.00	36
des	3.60	457	3.40	458
duke2	4.40	110	4.00	106
e64	3.60	111	3.00	94
misex1	2.40	8	2.00	7
rd84	4.40	24	4.00	23
rot	6.00	165	5.60	153
vg2	3.40	24	3.20	25
z4ml	2.20	4	2.20	4
TOTAL	96.20	2336	88.00	2150
A_Mean	4.81	116.80	4.40	107.50
A_Ratio	1	1	-9%	-8%
G_Mean	4.44	61.89	4.04	58.40
G_Ratio	1	1	-9%	-6%

Table 2: Comparison between FlowMap and HeteroMap on ORCA2C series FPGAs

More detailed results are available in [6]. Table 4 shows how the heterogeneous LUTs with different capacities and delays are used in the delay optimal mapping solution of HeteroMap with different delay ratios, which could be very useful for the architecture design of future heterogeneous FPGAs. When the delays between different LUTs become closer, HeteroMap will make use of more larger LUTs in the delay optimal mapping solution.

Comparison	FlowMap on homogeneous FPGAs of K-LUTs			
	$K = 3$		$K = 4$	
	M-Dly	Area	M-Dly	Area
A_Mean	8.10	352.70	8.06	537.20
A_Ratio	+25%	-34%	+25%	+0.2%

FlowMap on homogeneous FPGAs of K-LUTs				HeteroMap on Hetero-FPGAs	
$K = 5$		$K = 6$			
M-Dly	Area	M-Dly	Area	M-Dly	Area
8.08	870.40	8.30	1273.60	6.45	536.15
+25%	+62%	+28%	+137%	1	1

Table 3: Homogeneous FPGAs *v.s.* heterogeneous FPGAs

Delay ratio = 1 : 1.33 : 1.67 : 2				Delay Ratio = 1 : 1.2 : 1.4 : 1.6			
3-lut	4-lut	5-lut	6-lut	3-lut	4-lut	5-lut	6-lut
1	1	1	1	-36%	-29%	+90%	+49%

Table 4: HeteroMap results with different delay ratios

5 Conclusions and Future Work

In this paper, we presented a delay optimal technology mapping algorithm, named HeteroMap, for heterogeneous FPGA designs.

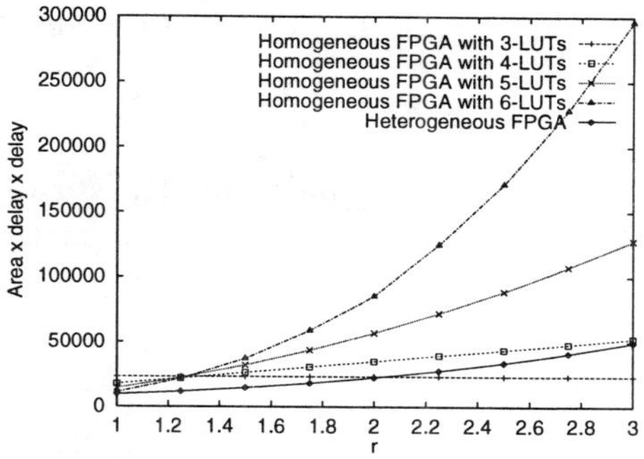

Figure 3: $Area \times delay^2$ Comparison between homogeneous and heterogeneous FPGAs with respect to the area ratio of r.

Taking different delays of heterogeneous LUTs into consideration, the HeteroMap algorithm computes the minimum mapping delay of a circuit in polynomial time based on a series of minimum height K-feasible cut computations on each node in the circuit. Some heterogeneous FPGAs may have resource limitations on the LUTs of specific sizes, which we call *heterogeneous FPGAs with bounded resources.* The delay-optimal technology mapping for this type of heterogeneous FPGAs is still an open problem, and we are in the process of developing optimal or near-optimal algorithms for this problem.

6 Acknowledgments

This work is partially supported by Lucent Technologies Inc. and Xilinx Inc. under the California MICRO Program and National Science Foundation Young Investigator Award MIP9357582.

References

[1] K. C. Chen, J. Cong, Y. Ding, A. B. Kahng, P. Trajmar, *"DAG-Map: Graph-based FPGA Technology Mapping for Delay Optimization"*, IEEE Design and Test of Computers, Sep. 1992, pp. 7-20.

[2] J. Cong, Y. Ding, *"FlowMap: An Optimal Technology Mapping Algorithm for Delay Optimization in Lookup-Table Based FPGA Designs"*, IEEE Transactions on Computer-Aided Design, Feb. 1994, Vol. 13, No. 1, pp. 1-12.

[3] J. Cong, J. Peck, Y. Ding, *"RASP: A General Logic Synthesis System for SRAM-based FPGAs"*, Proc. ACM International Symposium on FPGA, Feb. 1996, pp. 137-143.

[4] J. Cong, Y. Ding, *"Tutorial and Survey Paper — Combinational Logic Synthesis for LUT Based Field Programmable Gate Arrays"*, ACM Transactions on Design Automation of Electronic Systems, Vol. 1, No. 2, April 1996, pp. 145-204.

[5] J. Cong, S. Xu, *"Technology Mapping for FPGAs with Embedded Memory Blocks"*, Proc. ACM International Symposium on FPGA, Monterey, CA., Feb. 1998, pp. 179-188.

[6] J. Cong, S. Xu, *"Delay-Optimal Technology Mapping for FPGAs with Heterogeneous LUTs"*, UCLA Computer Science Department Technical Report CSD-980015.

[7] J. He, J. Rose, *"Technology Mapping for Heterogeneous FPGAs"*, Proc. ACM/SIGDA International Symposium on FPGA, Feb. 1994.

[8] Lucent Technologies, *"ORCA OR2C-A/OR2T-A Series FPGAs Data Sheet"*, Lucent Technologies, Inc., Allentown, PA, 1996.

[9] J. D. Ullman, *"Computational Aspects of VLSI"*, Computer Science Press, Inc., Rockville, MD, 1983.

[10] Xilinx, *"The Programmable Logic Data Book"*, Xilinx Inc., San Jose, CA, 1997.

Exact Tree-based FPGA Technology Mapping for Logic Blocks with Independent LUTs *

Madhukar R. Korupolu K. K. Lee D. F. Wong
Department of Computer Sciences
The University of Texas at Austin

Abstract

The logic blocks (CLBs) of a lookup table (LUT) based FPGA consist of one or more LUTs, possibly of different sizes. In this paper, we focus on technology mapping for CLBs with several *independent* LUTs of two different sizes (called ICLBs). The Actel *ES6500* family is an example of a class of commercially available ICLBs. Given a tree network with n nodes, the only previously known approach for minimum area tree-based mapping to ICLBs was a heuristic with running time $\Theta(n^{d+1})$, where d is the maximum indegree of any node. We give an $O(n^3)$ time exact algorithm for mapping a given tree network, an improvement over this heuristic in terms of run time and the solution quality. For general networks, an effective strategy is to break it into trees and combine them. We also give an $O(n^3)$ exact algorithm for combining the optimal solutions to these trees, under the condition that LUTs do not go across trees. The method can be extended to solve mapping onto CLBs that can be configured into different ICLBs, (e.g. Xilinx' *XC4000E*).

1 Introduction

FPGAs are becoming increasingly popular due to the short design cycle and the low cost they offer. A popular FPGA architecture is the LUT-based logic blocks (also called CLBs). Fig. 1 shows two different commercially available CLBs. A k-input LUT (k-LUT) can implement any function of up to k inputs. The CLB is a combination of one or more LUTs, usually of different sizes. When the LUTs are independent (i.e. not *logically* connected within the CLB), we call these *CLB with independent LUTs* (denoted ICLB). The CLB of Actel's *ES6500* family (Fig. 1(a)) is an example of a commercially available ICLB, where each ICLB has one 2-LUT and two 3-LUTs. Some other types of CLBs can also be configured into ICLBs. In the Xilinx *XC4000E* series (Fig. 1(b) & (c)), each CLB can be configured to (i) one 3-LUT and two 4-LUTs, (ii) one 5-LUT, or a few other non-ICLB configurations [9].

The minimum area LUT mapping problem has been studied extensively. The survey paper of [2] gives a good overview of the techniques used and the results obtained. This problem was shown to be NP-complete for $k \geq 3$ [3, 10] and polynomial time solvable for $k = 2$ [8]. A polynomial time algorithm (for any k) was given for tree networks in [3]. However, minimizing the number of LUTs does not necessarily minimize the number of ICLBs. In the tree network and ICLB

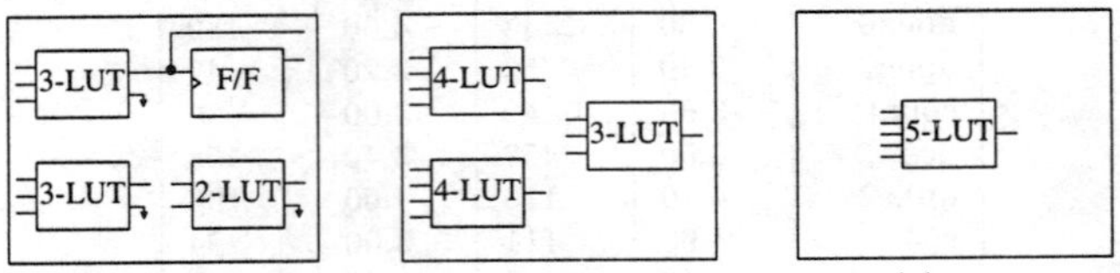

(a) (b) (c)

Figure 1: *(a) A commercial ICLB: Actel's ES6500, (b) & (c) Two possible ICLB configurations for Xilinx' XC4000E*

of Fig. 2, each ICLB has two 3-LUTs and one 4-LUT. Any mapping using only 3-LUTs requires four LUTs (and hence two CLBs), while any mapping using only 4-LUTs requires two LUTs (and hence two CLBs). On the other hand, the network can be mapped using just one logic block.

The ICLB mapping problem was also previously studied in [5]. Though their algorithm for combining tree solutions is optimal, each tree is mapped using a bin packing based heuristic running in time $\Theta(n^{d+1})$, where n is the number of nodes and d the maximum indegree of any node in the tree.

In this paper, we give an $O(n^3)$ time exact algorithm for mapping a tree network using the minimum number of ICLBs, an improvement over the heuristic of [5], in both solution quality and run time. To the best of our knowledge, this is the first polynomial time algorithm for the logic block mapping problem, on any nontrivial class of networks. Since the simpler LUT mapping problem is NP-Complete, it is very unlikely that a polynomial time algorithm would exist for the ICLB mapping problem on arbitrary networks.

For general networks, a common and effective strategy to obtaining a good mapping solution is to partition the network into trees, map each tree separately and then combine the mapping solutions of the trees. Such an approach was used in [3, 6]. Using a similar approach and our tree algorithm, we then give an $O(n^3)$ time algorithm for finding the optimal tree-based mapping of a general network, under the condition that LUTs do not go across trees.

This paper is organized as follows: Section 2 introduces some notation and gives a precise formulation of the problem. Section 3 gives the algorithm for trees and section 4 describes the optimal tree partitioning based mapping algorithm for general networks under the condition that LUTs do not go across trees. We present some experimental results in section 5 and finally discuss extensions to CLBs that can be configured into different ICLBs in section 6.

2 Notation and Problem Formulation

For integers i and j, we use $[i \ldots j]$ to denote the set of integers between i and j (inclusive).

A combinational network is a DAG (directed acyclic graph) $\mathcal{N} = (\mathcal{V}(\mathcal{N}), \mathcal{E}(\mathcal{N}))$ where each node is a gate implementing a boolean function. A directed edge (i, j) exists iff node i is an input to node j. Nodes that have no incoming edges are

*This work was partially supported by the Texas Advanced Research Program under Grant No. 003658288.

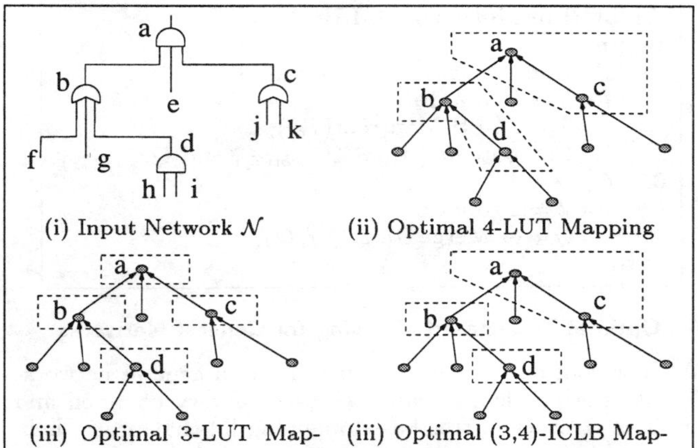

Figure 2: *Minimizing number of LUTs does not minimize the number of ICLBs*

called *primary input* (PI) nodes, and the set of such nodes is denoted by $\mathcal{PI}(\mathcal{N})$. Similarly, nodes that have no outgoing edges are called *primary output* (PO) nodes and the set of such nodes is denoted by $\mathcal{PO}(\mathcal{N})$. $\mathcal{N}$ is *k-bounded* if every node has at most k inputs. For a tree network T and a node v in T, T_v denotes the subtree of T rooted at v. The set of nodes that supply inputs to v is denoted by $input(v)$. Similarly, the set of nodes that supply inputs to a subnet H is denoted by $input(H)$. If there is a path from node u to node v, then u is called a *predecessor* of v. Given a node v, a *cone* of v (denoted C_v), is a subnetwork consisting of v and *some* of its non-PI predecessors s.t. for any node $w \in C_v$, there is a path from w to v that lies entirely within C_v. In Fig. 2, some cones are shown in dotted boxes. v is called the *root* of C_v. A cone C_v is k-feasible if $|input(C_v)| \leq k$. Since a k-LUT can implement any function of up to k inputs, it can match any k-feasible cone. Throughout this paper, we use the terms *cone* and LUT interchangeably.

Definition 1 (k-LUT Mapping Solution): Given a network $\mathcal{N}$, a set $S \subseteq \mathcal{V}(\mathcal{N})$ is a feasible k-LUT mapping solution for $\mathcal{N}$ iff (i) $\mathcal{PO}(\mathcal{N}) \subseteq S$, (ii) for every $v \in S$, $\exists$ a k-feasible cone C_v s.t. $input(C_v) \subseteq S \cup \mathcal{PI}(\mathcal{N})$, and (iii) every non-PI $u \in \mathcal{V}(\mathcal{N})$ is contained in C_v, for some $v \in S$.

The k-LUT mapping solution should be more precisely defined as $\{C_v : v \in S\}$. However, once S is known, the cones can be determined. Hence, we call S the mapping solution. The area needed is the number of k-LUTs used in S, i.e. $|S|$.

Problem 1 (k-LUT Mapping Problem): Given a k-bounded network $\mathcal{N}$, find a feasible k-LUT mapping solution S s.t. $|S|$ is minimized.

Suppose that each target ICLB consists of a_1 k_1-LUTs and a_2 k_2-LUTs which are independent. WLOG, assume $k_1 < k_2$.

Definition 2 ((k_1, k_2)-ICLB Mapping Solution): Given a k_2-bounded network $\mathcal{N}$, a feasible (k_1, k_2)-ICLB mapping solution of $\mathcal{N}$ is a pair (S_1, S_2) of subsets of $\mathcal{V}(\mathcal{N})$, s.t. (i) $\mathcal{PO}(\mathcal{N}) \subseteq S_1 \cup S_2$, (ii) for every $v \in S_1$, $\exists$ a k_1-feasible cone C_v s.t. $input(C_v) \subseteq S_1 \cup S_2 \cup \mathcal{PI}(\mathcal{N})$, (iii) for every $v \in S_2$, $\exists$ a k_2-feasible cone C_v s.t. $input(C_v) \subseteq S_1 \cup S_2 \cup \mathcal{PI}(\mathcal{N})$, and (iv) every non-PI $u \in \mathcal{V}(\mathcal{N})$ is contained in C_v, for some $v \in S_1 \cup S_2$.

Problem 2 ((k_1, k_2)-ICLB Mapping Problem): Given a k_2-bounded network $\mathcal{N}$, and a target ICLB architecture with a_1 k_1-LUTs and a_2 k_2-LUTs each, compute a feasible (k_1, k_2)-ICLB mapping solution (S_1, S_2) of $\mathcal{N}$, s.t. the number of ICLBs used is minimized.

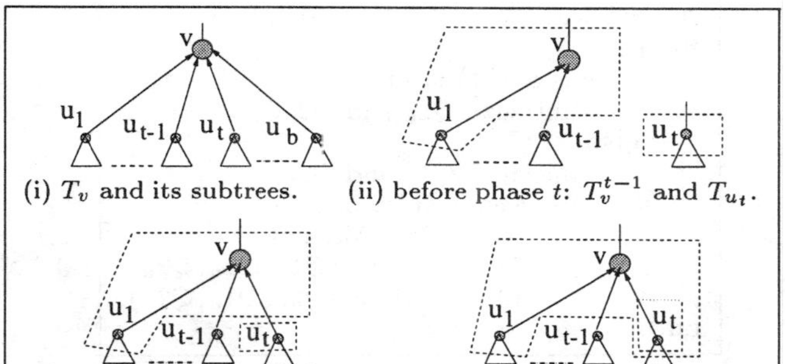

Figure 3: *Computation of $Q1_v$ and $Q2_v$. During phase t, we can either merge the LUT at u_t into the LUT at v or not.*

In the 4-LUT mapping solution in Fig. 2(ii), $S = \{a, b\}$, and has area two. In the $(3, 4)$-ICLB mapping solution in Fig. 2(iv), $S_1 = \{b, d\}$ and $S_2 = \{a\}$.

3 Mapping Trees Using ICLBs

In this section, we consider the following more general ICLB mapping problem on trees. Given a k_2-bounded tree network T with root r, our aim is to compute a (tradeoff) array Q_T with the following property: for each integer $i \in [0 \dots n]$, $Q_T[i]$ gives the minimum $|S_1|$ s.t. $\exists$ a feasible (k_1, k_2)-ICLB mapping solution (S_1, S_2) of T satisfying $|S_2| \leq i$. Informally, $Q_T[i]$ is the minimum number of k_1-LUTs needed to cover T using, additionally, at most i k_2-LUTs. Once Q_T is known, the minimum number of ICLBs needed to implement T can be computed easily. We give an $O(n^3)$ time algorithm for computing Q_T, visiting each node in postorder. At each node v, we compute two two-dimensional arrays, $Q1_v$ and $Q2_v$ of size $n \times (k_1 + 1)$ and $n \times (k_2 + 1)$ respectively from the $Q1$ and $Q2$ arrays of its children. The proofs of the lemmas and theorems following can be found in [7].

3.1 Semantics of Q1 and Q2

Definition For a node v in T and for $i \in [0 \dots n]$ and $j \in [0 \dots k_1]$, the entry $Q1_v[i, j]$ is the minimum $|S_1|$ s.t. $\exists$ a feasible (k_1, k_2)-ICLB mapping solution (S_1, S_2), of T_v, satisfying the following conditions: (i) $v \in S_1$ (i.e. v is covered by a k_1-LUT), (ii) $|S_2| \leq i$, and (iii) $|input(C_v)| \leq j$.

Definition Similarly, for a node v in T and for $i \in [0 \dots n]$ and $j \in [0 \dots k_2]$, the entry $Q2_v[i, j]$ is the minimum $|S_1|$ s.t. $\exists$ a feasible (k_1, k_2)-ICLB mapping solution (S_1, S_2), of T_v, satisfying the following conditions: (i) $v \in S_2$ (i.e. v is covered by a k_2-LUT), (ii) $|S_2| \leq i$, and (iii) $|input(C_v)| \leq j$.

Lemma 1 $Q_r[i] = \min\{Q1_r[i, k_1], Q2_r[i, k_2]\}$.

3.2 Computing Q1 and Q2

To compute $Q1_v$ and $Q2_v$ at every node v, it suffices to show how these are initialized at the leaves, and how at each internal node v these are computed using the arrays of its children.

Lemma 2 *For a leaf v, the arrays $Q1_v$ and $Q2_v$ can be initialized as follows:*

- *Set $Q1_v[i, j] = 1$ for all $i \in [0 \dots n]$ and $j \in [0 \dots k_1]$.*
- *Set $Q2_v[0, j] = \infty$ for all $j \in [0 \dots k_2]$, and*
- *Set $Q2_v[i, j] = 0$ for all $i \in [1 \dots n]$ and $j \in [0 \dots k_2]$.*

Computation is done in postorder. At each node v, the arrays for its children are available when we start the computation at v. Let $u_1, \dots, u_b$ be the children of v. We compute the

```
Algorithm Tree_ICLB_Map(v)
begin
    if (v is a leaf) then
        Initialize Q1_v and Q2_v;
    else
        Initialize Q1_v and Q2_v;
        for each child u of v do
            Tree_ICLB_Map(u);
            Q1_v ← Merge(Q1_v, Q1_u, Q2_u);
            Q2_v ← Merge(Q2_v, Q1_u, Q2_u);
end.
```

arrays $Q1_v$ and $Q2_v$ in b phases, one for each child of v. For $t \in [0 \ldots b]$, we use $Q1_v^t$ and $Q2_v^t$ to denote the arrays computed during phase t. For convenience, we define $T_v^0 = \{v\}$ and $T_v^t = \{v\} \cup T_{u_1} \cup \ldots \cup T_{u_t}$. Note that $T_v^b = T_v$. At the end of the tth phase, we maintain the *invariant* that $Q1_v^t[i,j]$ (and similarly $Q2_v^t[i,j]$) is the minimum $|S_1|$ s.t. $\exists$ a feasible (k_1, k_2)-ICLB mapping solution (S_1, S_2), of T_v^t, that satisfies the three conditions in the definition of $Q1_v[i,j]$ (and correspondingly $Q2_v[i,j]$) under the assumption that the only children of v are $u_1, \ldots, u_t$. Hence, at the end of phase b, $Q1_v^b$ and $Q2_v^b$ are the required arrays $Q1_v$ and $Q2_v$ respectively.

The arrays $Q1_v^0$ and $Q2_v^0$ will be initialized as in Lemma 2. As $T_v^0 = \{v\}$ is just a leaf (it is taken to have no children), it follows that $Q1_v^0$ and $Q2_v^0$ satisfy the base case of the above mentioned invariant. The following two lemmas show how to compute $Q1_v^t$ and $Q2_v^t$, during phase t, while maintaining the invariant. Fig. 3 gives an overview of a general phase t.

Lemma 3 *(Computing $Q1_v^t$) Let $i \in [0 \ldots n]$ and $j \in [1 \ldots k_1]$ be two integers and $t > 0$ and set $Q1_v^t[i, 0] = \infty$. $Q1_v^t$ satisfies the above mentioned invariant if we set $Q1_v^t[i,j]$ to the minimum, over all $i_1 \in [0 \ldots i]$ and all $j_1 \in [0 \ldots j]$, of:*

- $Q1_v^{t-1}[i - i_1, j - 1] + Q1_{u_t}[i_1, k_1]$,
- $Q1_v^{t-1}[i - i_1, j - 1] + Q2_{u_t}[i_1, k_2]$, *and*
- $Q1_v^{t-1}[i - i_1, j - j_1] + Q1_{u_t}[i_1, j_1] - 1$.

Lemma 4 *(Computing $Q2_v^t$) Let $i \in [0 \ldots n]$ and $j \in [1 \ldots k_2]$ be two integers, $t > 0$, and set $Q2_v^t[i, 0] = \infty$. $Q2_v^t$ satisfies the above mentioned invariant if we set $Q2_v^t[i,j]$ to the minimum, over all $i_1 \in [0 \ldots i - 1]$ and all $j_1 \in [0 \ldots j]$, of:*

- $Q2_v^{t-1}[i - i_1, j - 1] + Q1_{u_t}[i_1, k_1]$,
- $Q2_v^{t-1}[i - i_1, j - 1] + Q2_{u_t}[i_1, k_2]$, *and*
- $Q2_v^{t-1}[i - i_1, j - j_1] + Q2_{u_t}[i_1 + 1, j_1]$.

The overall recursive mapping algorithm for trees is outlined in Algorithm **Tree_ICLB_Map**. After computing $Q1_r$ and $Q2_r$, we can then apply Lemma 1 to get Q_r.

Note that the above algorithm only computes the number of k_1-LUTs and k_2-LUTs and not the actual mapping. However, the actual mapping can be easily computed by remembering the combination (of i_1, j_1 and term) that gives the minimum value in each application of Lemmas 3 and 4, and a downward pass suffices to determine the mapping.

Time Complexity: The $[i,j]$th entry of $Q1_v^t$ and $Q2_v^t$ can be computed in $O(n \cdot k_2)$ time, as $k_2 \geq k_1$. Hence $Q1_v^t$ and $Q2_v^t$ can be computed in $O(n^2 \cdot k_2^2)$ time. Therefore, the time taken for the b phases is $O(n^2 \cdot k_2^2 \cdot b)$, where b is the number of children of v. Hence the overall time taken to compute the arrays at the root of the tree is $O(n^2 \cdot k_2^2 \cdot \sum_v |input(v)|)$, which is $O(n^3 \cdot k_2^2)$. As k_2 is a constant, this is $O(n^3)$.

Theorem 1 *Given a k_2-bounded tree network, T, with n nodes and a target ICLB with LUTs of two different sizes k_1 and k_2 (say $k_1 < k_2$), then a mapping of T using the minimum number of ICLBs can be computed in $O(n^3)$ time.*

```
Algorithm Network_ICLB_Map(N)
begin
1.  Partition N into disjoint trees, T_1, ..., T_p;
2.  for t = 1 to p do
        Tree_ICLB_Map(root(T_t));
        Compute Q_{T_t} from Q1_{root(T_t)} and Q2_{root(T_t)}
3.  Q_N ← Q_{T_1};
    for t = 2 to p do
        Q_N ← MergeTrees(Q_N, Q_{T_t});
end.
```

4 Optimal Tree-Based Mapping for General Networks

We now consider technology mapping of a general network, $\mathcal{N}$. As the problem is hard, we use the very common and effective strategy of tree based mapping [6] that proceeds in three stages: $\mathcal{N}$ is first partitioned into disjoint trees: each PO node and each node with multiple fanout defines a tree in the partition [2]. Each tree is then mapped independently. In the final stage, the mapping solutions of all the trees are combined to give a good mapping solution for $\mathcal{N}$. A similar approach was studied in [5]. However, their method for mapping a tree is a heuristic, with no guarantees on the solution quality. As we now have an optimal algorithm for trees, this should give better solutions for the entire network. The run time also improves from $\Theta(n^{d+1})$ to $O(n^3)$.

Algorithm **Network_ICLB_Map** is based on this strategy. Let $T_1, \ldots, T_p$ be the trees obtained in stage 1. Algorithm **Tree_ICLB_Map** is used to compute each Q_{T_t} optimally. The aim of the third stage is to compute a (tradeoff) array $Q_\mathcal{N}$, for the entire network, where $Q_\mathcal{N}[i]$ is the minimum number of k_1-LUTs needed to cover $\mathcal{N}$ using, additionally, at most i k_2-LUTs. Under the restriction that LUTs do not go across different trees, the best value for $Q_\mathcal{N}[i]$ is the minimum of $\sum_{t=1}^{p} Q_{T_t}[i_t]$ over all $i_1, \ldots, i_p$ s.t. $i = i_1 + \ldots + i_p$. The straightforward approach to compute $Q_\mathcal{N}$ would therefore take $\Theta(n^p)$ time. To get an efficient implementation, we compute $Q_\mathcal{N}$ in p phases by combining the trees two at a time, similar to [5]. Let $Q_\mathcal{N}^t$ denote the array computed during phase t. We maintain the invariant that $Q_\mathcal{N}^t$ has the correct values under the assumption that the only trees in the graph are $T_1, \ldots, T_t$. Initially we set $Q_\mathcal{N}^1 = Q_{T_1}$. Lemma 5 below shows how to compute $Q_\mathcal{N}^t$ from $Q_\mathcal{N}^{t-1}$ and Q_{T_t}. Finally, $Q_\mathcal{N}^p$ is our required $Q_\mathcal{N}$.

Lemma 5 *Let $i \in [0 \ldots n]$ be an integer. Set $Q_\mathcal{N}^t[i]$ to the minimum, over all $i_1 \in [0 \ldots i]$, of $Q_\mathcal{N}^{t-1}[i - i_1] + Q_{T_t}[i_1]$. Then $Q_\mathcal{N}^t$ has the correct values under the assumption that the only trees in the network are $T_1, \ldots, T_t$.*

This implies that algorithm Network_ICLB_Map is optimal under the assumption of tree based mapping (i.e., under the constraint that the LUTs do not go across different trees).

Time Complexity: Each phase in stage 3 in the algorithm takes $O(n^2)$ time. Since $p = O(n)$, the stage 3 takes $O(n^3)$ time. Stage 1 can be done implicitly in $O(n)$ time. The time taken by stage 2 is $O(\sum_{t=1}^{p} |T_t|^3)$ which is $O(n^3)$, since $\sum_{t=1}^{p} |T_t|^3 \leq (\sum_{t=1}^{p} |T_t|)^3 = n^3$. Thus the overall time taken by this algorithm is $O(n^3)$.

Theorem 2 *If LUTs do not go across different trees, the algorithm* **Network_ICLB_Map** *gives the optimal mapping solution for general networks in $O(n^3)$ time.*

Remarks: Note that the tree based mapping strategy is strictly tree-based and does not exploit the reconvergence of the network, nor sharing of LUTs across trees. Employing some techniques for these could further improve the mapping solutions.

Circuit	n_A	n_B	%	CPU (s)
majority	2	2	0	0.01
parity	4	3	25	0.02
mux	11	8	27	0.15
decod	12	9	25	0.02
my_adder	27	22	18	0.10
b9	26	21	19	0.17
C432	43	33	23	0.98
9symml	44	33	23	2.93
apex7	49	37	24	0.24
C1355	130	97	25	0.87
rot	122	93	24	0.98
apex6	138	103	25	1.19
frg2	265	210	21	4.39
dalu	401	301	25	8.48
k2	474	430	9	15.28
i8	687	518	25	24.98
t481	905	684	25	1193.82

Table 1: *Results for MCNC circuits when $k_2 = k_1 + 1$ on a 180MHz Pentium Pro PC. The target ICLB has two 3-LUTs and one 4-LUT. Note that this is strictly tree-based and no preprocessing or postprocessing was performed.*

5 Implementation Issues and Experimental Results

We implemented the **Network_ICLB_Map** algorithm in C and tested it on several MCNC benchmark circuits. The circuits were first decomposed to 2-bounded networks with SIS (`xl_imp -n 2`). Though the algorithm Tree_ICLB_Map as described above requires $Q1_v$ and $Q2_v$ to be of size n x $(k_1 + 1)$ and n x $(k_2 + 1)$ respectively, the implementation uses much smaller arrays. Let s_v be the size of the subtree rooted at node v. Note that for $i > s_v$, $Q1_v[i, j] = Q1_v[s_v, j]$ (similarly, $Q2_v[i, j] = Q2_v[s_v, j]$). Similarly, when combining the arrays from the individual trees, many unncessary updates can be avoided. By carefully making only necessary table updates, the running time was sped up tremendously.

There remains an intriguing question: How does our algorithm compare with the LUT mapping algorithm, when k_1 and k_2 are very close to each other, e.g. $k_2 = k_1 + 1$ like the Actel *ES6500*? One reasonable way of getting an ICLB mapping solution is to map the network using k_1-LUTs only and then place $a_1 + a_2$ of these k_1-LUTs into each ICLB. Let n_A denote the number of ICLBs used by such a mapping solution, and n_B the number of ICLBs used by our solution. If k_2 is quite a bit bigger than k_1, n_B is expected to be much better than n_A. However if k_2 is close to k_1, it is not clear as to how significant is the difference between n_B and n_A. Inspired by this question, we tested our algorithm using a target ICLB with $k_1 = 3$ and $k_2 = 4$. Each ICLB has two 3-LUTs and one 4-LUT. Table 1 summarizes the results of this experiment. We observed that the improvement is about 25% for most of the circuits, which is quite significant.

6 Extension to CLB with multiple ICLB configurations

We now extend the algorithm to handle CLBs with several ICLB configurations (e.g. each CLB in Xilinx *XC4000E* has two possible ICLB configurations (Fig. 1).) If each CLB can be configured into one of the k ICLBs $A_1, A_2, \ldots, A_k$ where each ICLB configuration A_i has a_{i1} k_1-LUTs and a_{i2} k_2-LUTs. Our problem now is to determine the minimum number of CLBs to be used as well as the individual ICLB configuration for each CLB. Let p_i be the number of CLBs whose configurations are A_i, where $1 \leq i \leq k$, in a mapping solution. Given the tradeoff array Q_T computed in the previous section, the problem at hand can be formulated as :

$$\begin{aligned} \text{for each } j, \text{minimize} \quad & \textstyle\sum_i^k p_i \\ \text{subject to} \quad & \textstyle\sum_i^k p_i \cdot a_{i1} \geq Q_T[j] \\ \text{and} \quad & \textstyle\sum_i^k p_i \cdot a_{i2} \geq j, \end{aligned}$$

When $k = 2$, the straightforward way of trying all possible p_1 and p_2 uses only $O(n^2)$ time for each j. Hence the total time required is $O(n^3)$.

Another extension is when there are more than two different kinds of independent LUTs in each CLB. For example, we might have the following plausible configurations: (i) two 3-LUTs and one 4-LUT, (ii) two 4-LUTs or (iii) one 5-LUT. The extension of our algorithm to handle three different kinds of LUTs is direct: instead of computing a tradeoff array, we compute a two-dimensional tradeoff matrix. Thus $Q_T[i, j]$ gives us the minimum number of k_1-LUTs to be used in the mapping under the constraint that at most i additional k_2-LUTs and at most j k_3-LUTs are used.

7 Conclusions

We have presented an $O(n^3)$ time exact algorithm for mapping a tree network using the minimum number of ICLBs. This is the first polynomial time algorithm for this problem. The algorithm is applicable on both physically ICLBs with two different types of LUTs or CLBs that can be configured to have only at most two different types of independent LUTs. The algorithm can be extended to configurations with more than two different kinds of independent LUTs. Using this we presented an optimal tree based mapping algorithm for general networks. This can then be extended to incorporate some post-processing steps using techniques that exploit reconvergence (as in [4]). This should yield improved solutions for networks with a lot of reconvergence.

The problem of mapping a tree network using CLBs with *dependent* LUTs needs further study. The Xilinx *XC4000* CLB is one such example. See [2] for some pointers.

References

[1] Actel. Actel's Reprogrammable SPGAs. *Preliminary Advance Information*, Oct. 1996.

[2] Jason Cong and Yuzheng Ding. Combinational Logic Synthesis for LUT Based Field Programmable Gate Arrays. *ACM Trans. on Design Automation of Electronic Systems*, 1(2):145–204, Apr. 1996. Tutorial and Survey Paper.

[3] A. Farrahi and M. Sarrafzadeh. Complexity of the lookup-table minimization problem for FPGA technology mapping. *IEEE Trans. CAD/ICAS*, 13(11):1319–1332, 1994.

[4] R.J. Francis, J. Rose, and Z.G. Vranesic. Chortle-crf: Fast technology mapping for lookup table-based FPGAs. In *Proc. ACM/IEEE Design Automation Conference*, pp. 227–233, Jun. 1991.

[5] J. He and J. Rose. Technology mapping for heterogeneous FPGAs. In *Proc. of ACM/SIGDA Intl. Workshop on FPGAs*, 1994.

[6] K. Keutzer. DAGON: Technology binding and local optimization by DAG matching. In *Proc. ACM/IEEE Design Automation Conference*, pp. 341–347, 1987.

[7] Madhukar R. Korupolu, K. K. Lee and D. F. Wong, FPGA Technological Mapping Using Logic Blocks with Independent LUTs. In *Technical Reference TR-98-08*, Dept. of Computer Sciences, University of Texas at Austin, 1998.

[8] Shashidhar Thakur and D. F. Wong. Simultaneous area and delay minimum K-LUT mapping for K-exact networks. *Integration, the VLSI journal*, 20:287–302, 1996.

[9] Xilinx. XC4000E and XC4000X Series (EX/XL) Field Programmable Gate Arrays. *data sheet*, Jun. 1997.

[10] S. Zhang, Michael D. Miller, and Jon C. Muzio. Notes on Complexity of the Lookup-Table Minimization Problem for FPGA Technology Mapping. *IEEE Trans. CAD/ICAS*, 15(12):1588–1590, Dec. 1996.

Compatible Class Encoding in Hyper-Function Decomposition for FPGA Synthesis

Jie-Hong R. Jiang, Jing-Yang Jou, Juinn-Dar Huang

Department of Electronics Engineering
National Chiao Tung University
Hsinchu 300, Taiwan, R.O.C.

jiang@athena.ee.nctu.edu.tw, jyjou@bestmap.ee.nctu.edu.tw, jdhuang@ee.nctu.edu.tw

ABSTRACT

Recently, functional decomposition has been adopted for LUT based FPGA technology mapping with good results. In this paper, we propose a novel method for functional multiple-output decomposition. We first address a compatible class encoding method to minimize the compatible classes in the image function. After the encoding algorithm is applied, the decomposability will be improved in the subsequent decomposition of the image function. The above encoding algorithm is then extended to encode multiple-output functions through the construction of a hyper-function. Common sub-expressions among these multiple-output functions can be extracted during the decomposition of the hyper-function. Therefore, we can handle the multiple-output decomposition in the same manner as the single-output decomposition. Experimental results show that our algorithms are very promising.

1. INTRODUCTION

Field Programmable Gate Arrays (FPGA's) can provide programmability for users to implement their own logic in a short turnaround time. FPGA's become increasingly popular in rapid system prototyping. Look-up table (LUT) based architecture is a prevalent one among many FPGA architectures. LUT-based FPGA's consist of an array of LUT's, each of which can implement any Boolean function with up to k (typically 4 or 5) inputs. A Boolean network can be directly realized by a one-to-one mapping between nodes and LUT's if every node in the network has up to k supports.

Functional decomposition [13,14] is a pivotal decomposition technique for LUT-based FPGA logic synthesis. Three interesting problems in functional decomposition should be noticed:

1. How to select the bound set variables?
2. How to encode the compatible classes?
3. How to extract adequate sub-expressions among multiple-output functions?

Algorithms proposed in [1,2] provide solutions to choose good bound set variables. On the other hand, approaches suggested in [3-7] deal with the second problem. These encoding algorithms can be classified into three classes according to their objectives. The first kind of encoding algorithm such as [3] simplifies the **image function** (g-function). The second kind such as [4,5] makes some **decomposition functions** (α-functions) be able to be shared among multiple-output functions. The third kind such as [6,7] minimizes the supports of the decomposition functions. As Problem 3 is considered, approaches proposed in [4,5,8] extract common sub-expressions among multiple-output functions. Extracting common decomposition functions via compatible class encoding was suggested in [4,5], while the authors of [8] tried to resubstitute decomposition functions into other functions to reduce their supports.

In this paper, we adopt the variable partitioning algorithm proposed in [2], which takes advantage of Binary Decision Diagrams (BDD's) [4,10] to conduct functional decomposition, to solve Problem 1. We will thus focus on solving Problems 2 and 3 in this paper. A new encoding algorithm is proposed to simplify the image function. Instead of reducing the number of cubes or literals in the image function as suggested in [3], our encoding algorithm aims at reducing the compatible class count. To deal with Problem 3, we transform multiple-output functions into a single-output function by introducing the hyper-function approach. Consequently, multiple-output decomposition can be reduced to an equivalent single-output decomposition. It unifies the solutions of single-output and multiple-output decomposition. Besides, the extracted common sub-expression can be large.

The rest of this paper is organized as follows. Section 2 introduces the preliminaries. In Section 3, the compatible class encoding algorithm is proposed. Hyper-function decomposition is then discussed in Section 4. After experimental results are shown in Section 5, concluding remarks will be given in Section 6.

2. PRELIMINARIES

Let $B = \{0,1\}$. A single-output function f with n input variables $b_0,\ldots,b_{n-1}$ is denoted as $f : B^n \rightarrow B$. A function $f(b_0,\ldots,b_{n-1})$, is **decomposable** if it can be represented by another function $g(\vec{\alpha}(b_0,\ldots,b_{i-1}),b_j,\ldots,b_{n-1}) = g(\alpha_0(b_0,\ldots,b_{i-1}),\ldots,\alpha_{t-1}(b_0,\ldots,b_{i-1}), b_j,\ldots,b_{n-1})$, where $0 < t < j \le i$. The decomposition is disjoint if $j = i$. When $j = i$, $\{b_0,\ldots,b_{i-1}\}$ is called the **bound (λ) set** and $\{b_j,\ldots,b_{n-1}\}$ is called the **free (μ) set**. In this paper, only the disjoint decomposition is considered.

Definition 2.1 Let X and Y be two sets of binary variables, $X \cap Y = \varnothing$. Given a completely specified function f: $B^{|X|} \times B^{|Y|} \rightarrow B$, with X being the λ set and Y being the μ set. We say that $x_1, x_2 \in B^{|X|}$ are **compatible** with respect to f, denoted as $x_1 \sim x_2$, if $\forall\ y \in B^{|Y|}$, (x_1,y) and $(x_2,y) \in B^{|X|} \times B^{|Y|}$ such that $f(x_1,y) = f(x_2,y)$.❑

All mutually compatible elements form a **compatible class**.

DAC 98, San Francisco, California
©1998 ACM 0-89791-964-5/98/06..$5.00

Theorem 2.1 $\forall (x,y) \in B^{|X|} \times B^{|Y|}$, f: $B^{|X|} \times B^{|Y|} \rightarrow B$, $\vec{\alpha}$: $B^{|X|} \rightarrow W$, g: $W \times B^{|Y|} \rightarrow B$,

$$f(x,y) = g(\vec{\alpha}(x),y) \quad (1)$$

holds if and only if

$$\forall\ x_1, x_2 \in B^{|X|},\ \vec{\alpha}(x_1) = \vec{\alpha}(x_2) \Rightarrow x_1 \sim x_2 \quad (2)$$ ❑

$\vec{\alpha}$ is a function with binary inputs and a symbolic output. The number of the admissible values in W, $|W|$, must be not less than the number of compatible classes. To implement $\vec{\alpha}$ by binary logic, at least $t = \lceil \log_2 |W| \rceil$ binary functions, $\alpha_0,..., \alpha_{t-1}$, are required to encode $\vec{\alpha}$. When $t = \lceil \log_2 |W| \rceil$, we say that this encoding is **rigid**. Otherwise when $t > \lceil \log_2 |W| \rceil$, the encoding is **pliable**. Eq. (1) can be rewritten as $f(x,y) = g(\alpha_0(x),...,\alpha_{t-1}(x),y)$. For a single-output function, if each compatible class is assigned just one code, then this encoding is **strict** and Eq. (2) can be redefined as

$$\forall\ x_1, x_2 \in B^{|X|},\ \vec{\alpha}(x_1) = \vec{\alpha}(x_2) \Leftrightarrow x_1 \sim x_2.$$

In contrast, if there exists any compatible class encoded with more than one code, then the encoding is **non-strict**.

3. COMPATIBLE CLASS MINIMIZATION

Two important factors affect the decomposition quality: one is the variable partitioning and the other is the compatible class encoding. We solve the variable partitioning problem by using the algorithm proposed in [2]. Moreover, we intend to encode compatible classes to reduce the number of compatible classes in the next decomposition of the image function. Before our discussing the encoding technique in Subsection 3.2, don't care assignment that is used in the encoding is introduced first in Subsection 3.1.

3.1 Don't Care Assignment

The authors of [8] used the don't care assignment to minimize the supports of an incompletely specified function. However, we formulate the don't care assignment as the clique partitioning problem in order to reduce the number of compatible classes instead of minimizing the number of supports.

We record the compatible relationship among λ set vertices by using the compatibility graph. Each λ set vertex corresponds to a vertex in the graph. A pair of vertices are connected by an edge if and only if these two vertices cab be compatible under certain don't care assignments. After constructing the graph, we want to find the least number of cliques such that each vertex is covered by exactly one clique; thus the number of cliques equals to the number of compatible classes. Because the clique partitioning problem is NP-complete, we adopt the heuristics appeared in [9] to have a polynomial time solution.

3.2 Compatible Class Encoding

The authors of [3] assumed that the fewer cubes or literals in the image function, the better decomposition quality could be obtained. Hence, the compatible class encoding problem was modeled as the symbolic-input encoding problem to minimize the number of cubes or literals of the image function. However, those counts may not be a good cost function for LUT-based FPGA synthesis. In this paper, we formulate the encoding problem as minimizing the number of compatible classes generated at the subsequent decomposition of the image function. The new cost function has better meaning for LUT architecture.

After λ set selection and don't care assignment, compatible classes are fixed. The next step is to encode these compatible classes. In order to exploit more don't care set, we take the strict encoding policy. Example 3.1 illustrates why an encoding is relevant to the number of compatible classes in the decomposition of the image function.

Example 3.1 Assume that the targeted LUT can implement any 4-input functions. We want to decompose the function f in Figure 1(a) with $\{a,b,c\}$ as the λ set selection.

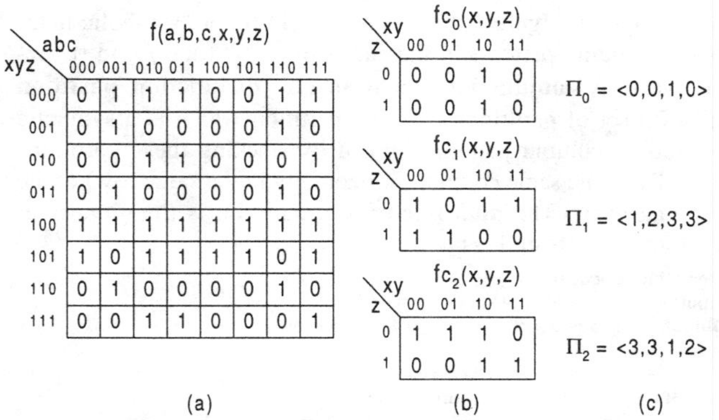

Figure 1: (a)Decomposition Chart of f (b)Compatible Class Functions (c)Symbolic Notations of Column Patterns

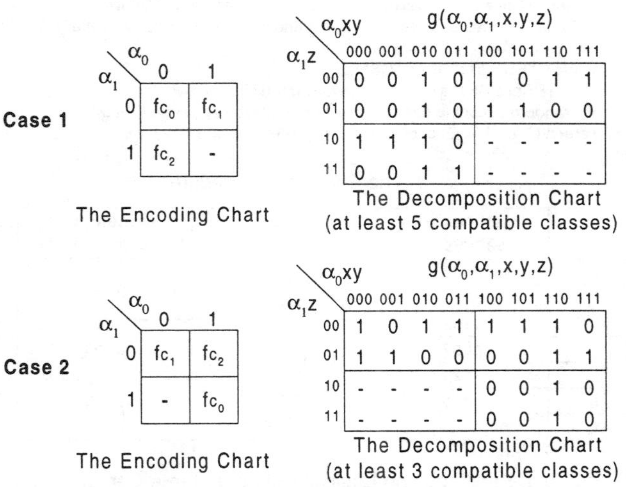

Figure 2: Encoding Chart and Decomposition Chart

In the decomposition of f if a, b and c are selected as the λ set variables, there are three compatible classes with functions as shown in Figure 1(b). Two α-functions, $\alpha_0(a,b,c)$ and $\alpha_1(a,b,c)$, are needed to encode three compatible classes. Suppose we choose α_0, x, and y as the λ set variables in the decomposition of $g(\alpha_0,\alpha_1,x,y,z)$. Examining the two encoding cases in Figure 2, we can see that the encoding may affect the number of compatible classes in the decomposition of the g-function. Note that "-" represents don't care.

Theorem 3.1 The encoding does not affect the number of compatible classes in the subsequent decomposition of the image function if all of the α-functions are selected together in the λ set or in the μ set of the subsequent decomposition of the image function. ❑

Theorem 3.2 After the λ set variables used in the decomposition of an image function have been selected, to reduce the number of compatible classes of the image function, we only have to determine which compatible class functions should belong to the same column or the same row in the encoding chart. The exact codes of these columns and rows do not influence the number of compatible classes of the image function. ❑

In the rest of this paper, we use symbolic notations (decimal numbers) to represent column patterns.

Definition 3.1 A **partition Π**, $<s_0,..,s_{n-1}>$, is a symbolic notation of n column patterns. Elements s_i equals to s_j if and only if i^{th} column pattern equals to j^{th} column pattern. ❑

For example in Figure 1, Π_0, Π_1 and Π_2 in (c) are symbolic notations of column patterns in the charts of fc_0, fc_1 and fc_2 in (b) respectively. A **conjunction partition Πc (disjunction partition Πd)** of a set of partitions is a new partition which is a symbolic notation of column patterns formed by stacking these partitions vertically in the same column (horizontally in the same row) of the encoding chart. The **multiplicity** of a partition is the number of different symbols in this partition.

```
Algorithm: Encoding
Input:     Compatible Class Functions fc's
Output:    Image Function with its λ Set Variables
begin
1   g' ← Encode compatible classes at random;
2   if ( g' is κ-feasible )  return {g',∅};
3   {λ',#compatible_classes} ← Variable_Partitioning (g');
    /*From g',λ' and fc's, we can derive the # of rows, #R, the # of
    columns, #C, in the encoding chart and the partitions of fc's, Π_fc's.*/
4   if ( #R==1 or #C==1 ) return {g',λ'};    /*According to Theorem 3.1*/
    /*Each Π_fc occupies a distinct row set and a distinct column set initially.*/
5   CSet's ← CombineColumnSets (Π_fc's);
6   while ( |RSet's| > #R or |CSet's| > #C )
7        {RSet's,CSet's} ← CombineRowSets (RSet's,CSet's);
8   if ( random encoding has less compatible classes )  return {g',λ'};
9   return ( {g,λ'} ← Encoding according to RSet's and CSet's );
end
```

Figure 3: The Encoding Procedure

Partition	Positions with the same content
Π_2	p0p3
Π_3	p1p3
Π_4	p1p3
Π_5	p0p2
Π_6	p1p2p3
Π_7	p0p1p3
Π_8	p0p2, p1p3

(a)

Partition	Positions with the same content
Πc of $\{\Pi_2,\Pi_7\}$	p0p3
Πc of $\{\Pi_3,\Pi_4,\Pi_6,\Pi_7,\Pi_8\}$	p1p3
Πc of $\{\Pi_5,\Pi_8\}$	p0p2

(b)

Figure 4: Partitions and Positions with the Same Contents

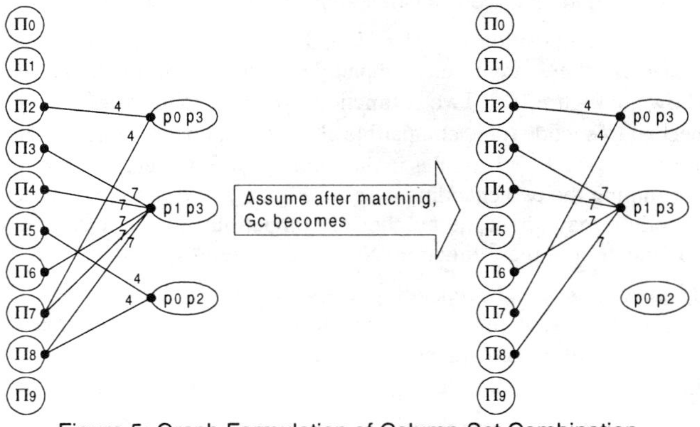

Figure 5: Graph Formulation of Column-Set Combination

The encoding draft is shown in Figure 3. We explain Steps 5 and 7 in detail with Example 3.2.

Example 3.2 Assume we have ten compatible class functions, $fc_0,\dots,fc_9$, with their partitions $\Pi_0,\dots,\Pi_9$ respectively as follows.

Π_0=<0,1,2,3> Π_1=<0,2,1,3> Π_2=<3,0,1,3> Π_3=<2,1,0,1> Π_4=<0,1,3,1> Π_5=<0,1,0,2> Π_6=<1,0,0,0> Π_7=<1,1,2,1> Π_8=<1,2,1,2> Π_9=<3,2,1,0>

Suppose we want to place these partitions (or these compatible class functions) in the encoding chart with #R = 4 and #C = 4.

Step 5 in Figure 3: *Evaluate which compatible classes should be bounded in the same column of the encoding chart.*

We represent position i in a partition as p_i for convenience. Since the contents of p_1 and p_3 in Π_4 are the same, we say that positions with the same content of Π_4 is p_1p_3. For the above ten partitions, Figure 4(a) records the information. Positions with the same content such as p_0p_3, we denote it as $\boldsymbol{Psc_{03}}$. $|Psc_{03}| = 2$ because there are two positions, p_0 and p_3, in Psc_{03}. As there are two partitions Π_2 and Π_7 having Psc_{03}, ***#Partitions***$(Psc_{03}) = 2$ and ***Partitions***$(Psc_{03}) = \{\Pi_2,\Pi_7\}$. Therefore, the conjunction partition of Π_2 and Π_7 has the same content in p_0 and p_3. For all *Psc*'s in Figure 4(a) (Psc_{03}, Psc_{13}, Psc_{02}, Psc_{123}, or Psc_{013}) with *#Partitions*(*Psc*) ≥ 2, we list them in Figure 4(b). We then build a **column-graph** *Gc*(*Vc*,*Uc*,*Ec*) as depicted in Figure 5, which is a bipartite graph. For each partition, there is a corresponding vertex in *Vc*; for each *Psc* in Figure 4(b), there are $\lceil$(*#Partitions*(*Psc*)–1)/#R$\rceil$ corresponding vertices, u_{Psc}'s, in *Uc*. (It is because that *Partitions*(*Psc*) may be collected in more than one column set.) A vertex in *Uc* corresponding to a *Psc* has *#Partitions*(*Psc*) edges connecting this vertex with vertices in *Vc* which are corresponding to *Partitions*(*Psc*). The weight of an edge connecting to u_{Psc} equals to (|*Psc*|+the number of edges connecting to u_{Psc}). We then find a *b-matching* [12], *Mc*, of maximum weight for *Gc*. For this *b*-matching, each vertex in *Vc* is connected by at most one edge; each vertex in *Uc* is connected by at most #R edges. After matching, the corresponding partitions of vertices ∈ *Vc* connected with the same u_{Psc} are grouped in the same column set. According to the matching result in Figure 5, we have six column sets, $\{\Pi_3,\Pi_4,\Pi_6,\Pi_8\}$, $\{\Pi_2,\Pi_7\}$, $\{\Pi_0\}$, $\{\Pi_1\}$, $\{\Pi_5\}$ and $\{\Pi_9\}$. After applying the above algorithm, we can reduce the multiplicity of the Πc of partitions in the same column set.

Step 7 in Figure 3: *Evaluate which compatible classes should be bounded in the same row and in adjacent columns of the encoding chart.*

Assume each partition initially occupies a separate row set. We calculate the benefits of merging pairs of row sets. For Π_i and Π_j in different column sets, we calculate the benefit of merging them as $\sigma\times Br_{ij}+\tau\times Bc_{ij}$; otherwise their merging benefit is calculated as $\sigma\times Br_{ij}+\tau\times Bc_{ij}$–the weight of the edge connecting the vertex corresponding to Π_i in *Vc* of *Gc*. Because when Π_i and Π_j are in the same column set, we don't want to tear them into different columns. The detailed calculations of σ, Br_{ij}, τ, and Bc_{ij} are as follows.

$Br_{ij} = n-(n_{ij}-n_i)-(n_{ij}-n_j)$

$Bc_{ij} = \sum_{\text{every symbol S in all }\Pi\text{'s}} ((\text{the \# of S in }\Pi_i\text{ and }\Pi_j) - k)^2$

σ = (# of row sets so far) – #R (if $\sigma < 0 \Rightarrow \sigma = 0$)
τ = (# of column sets so far) – #C (if $\tau < 0 \Rightarrow \tau = 0$)
$k = m/n$

m: There are *m* positions in the Πd of Π_i and Π_j.
n: There are totally *n* kinds of symbols in all partitions.
n_{ij}: There are totally n_{ij} kinds of symbols in the Πd of Π_i and Π_j.
n_i, n_j: There are n_i and n_j kinds of symbols in Π_i and Π_j respectively.

According to the calculated benefits, we construct a **row-graph** *Gr*(*Vr*,*Er*). Each partition Π_i has a corresponding vertex $v_i \in Vr$; each pair of vertices (v_i,v_j) is connected by an undirected edge whose weight is the benefit of merging Π_i and Π_j in the same row set. We then find the *maximum-cardinality matching* [12], *Mr*, of *Gr*. For each edge ∈ *Mr*, the corresponding partitions of its two end vertices are hopefully to be combined together in a row set. We combine these pairs of partitions iteratively with benefits from high to low until the number of current row sets is not greater than #R or all edges $\in$ *Mr* have been selected. In this example, $\{\Pi_7,\Pi_8\}$, $\{\Pi_5,\Pi_6\}$, $\{\Pi_2,\Pi_4\}$, $\{\Pi_0,\Pi_9\}$ and $\{\Pi_1,\Pi_3\}$ are therefore selected in succession. According to column sets derived in Step 5, we stack these pairs of partitions properly. If there are some conflicts between Step 5 and Step 7, we assume that the decisions of Step 7 have higher priority than those of Step 5. So far we have 5 row sets $\{\Pi_7,\Pi_8\}$, $\{\Pi_5,\Pi_6\}$, $\{\Pi_2,\Pi_4\}$, $\{\Pi_0,\Pi_9\}$, $\{\Pi_1,\Pi_3\}$ and 4 column sets $\{\Pi_3,\Pi_4,\Pi_6,\Pi_8\}$, $\{\Pi_1,\Pi_2,\Pi_5,\Pi_7\}$, $\{\Pi_0\}$, $\{\Pi_9\}$. We have row-column relation as illustrated in Figure 6(a).

(a)

Π_7 <1,1,2,1>	Π_8 <1,2,1,2>	?	?
Π_5 <0,1,0,2>	Π_6 <1,0,0,0>	?	?
Π_2 <3,0,1,3>	Π_4 <0,1,3,1>	?	?
Π_1 <0,2,1,3>	Π_3 <2,1,0,1>	?	?

Π_0 <0,1,2,3>	Π_9 <3,2,1,0>

(b)

Πd_{78} <1,1,2,1,1,2,1,2>	?
Πd_{56} <0,1,0,2,1,0,0,0>	?
Πd_{24} <3,0,1,3,0,1,3,1>	?
Πd_{13} <0,2,1,3,2,1,0,1>	?

Πd_{09} <0,1,2,3,3,2,1,0>

Figure 6: Row-Column Relation

(a)

Π_7 <1,1,2,1>	Π_8 <1,2,1,2>	–	–
Π_5 <0,1,0,2>	Π_6 <1,0,0,0>	–	–
Π_2 <3,0,1,3>	Π_4 <0,1,3,1>	–	–
Π_1 <0,2,1,3>	Π_3 <2,1,0,1>	Π_0 <0,1,2,3>	Π_9 <3,2,1,0>

(b)

$\alpha_2\alpha_3$ \ $\alpha_0\alpha_1$	00	01	10	11
00	fc_7	fc_8	-	-
01	fc_5	fc_6	-	-
10	fc_2	fc_4	-	-
11	fc_1	fc_3	fc_0	fc_9

Figure 7: (a)Final Row-Column Relation (b)Final Encoding

Step 7 must be repeated until the number of row sets $\leq$ #R and the number of column sets $\leq$ #C. In this example so far, since there are 5 row sets, 5 > #R, we thus iterate Step 7 to calculate the benefits of combining pairs of row sets. For each row set, we represent it by the disjunction partition of the partitions in this row set as shown in Figure 6(b). After constructing the new *Gr'*(*Vr'*,*Er'*) and finding the *Mr'* for it, we combine the pair of row sets with maximum benefits ($\{\Pi_1,\Pi_3\}$,$\{\Pi_0,\Pi_9\}$) into a row set $\{\Pi_1,\Pi_3,\Pi_0,\Pi_9\}$. Since the number of row sets becomes 4 (< #R), we don't have to combine row sets further. Finally as shown in Figure 7(a), we have 4 row sets and 4 column sets. According to Theorem 3.2, we know that the codes of columns and rows do not affect the number of compatible classes in the decomposition of the image function. So we can encode these compatible classes as that indicated in Figure 7(b). After encoding, we can derive the real image function. In the decomposition of this image function, we will select α_0, α_1 and some original bound set variables in the previous decomposition as the bound set variables according to Step 3 in Figure 3. Consequently, we will have 4 compatible classes in the decomposition of the image function.

4. HYPER-FUNCTION DECOMPOSITION

To solve multiple-output decomposition in the same manner as single-output decomposition, we propose a solution, which is called hyper-function decomposition.

Definition 4.1 A set of distinct Boolean functions $\{f_0,\ldots,f_{n-1}\}$, called **ingredients**, can be combined together to form a single-output **hyper-function** by using additional $i = \lceil \log_2 n \rceil$ binary bits $\eta_0,\ldots,\eta_{i-1}$, called **pseudo primary inputs**, to encode these ingredients. The supports of a hyper-function include pseudo primary inputs and the union of supports of its ingredients. ❑

Hyper-function transforms multiple-output functions into a single-output function. It can perform the function of any individual ingredient by assigning the corresponding code to the pseudo primary inputs. In Subsection 4.1, we discuss how to encode the ingredients to form a hyper-function with better decomposability.

4.1 Encoding of Ingredients

Actually, a hyper-function can be considered as an image function and its ingredients can be viewed as compatible class functions. Consequently, Theorem 3.1 and Theorem 3.2 can be extended as follows.

Theorem 4.1 The encoding does not affect the number of compatible classes in the subsequent decomposition of the hyper-function if all of the pseudo primary inputs are selected together in the λ set or in the μ set of the subsequent decomposition of the hyper-function. ❑

Theorem 4.2 After the λ set variables used in the decomposition of a hyper-function have been selected, to reduce the number of compatible classes of the hyper-function, we only have to determine which ingredients should belong to the same column or the same row in the encoding chart. The exact codes of these columns and rows do not influence the number of compatible classes of the hyper-function. ❑

We thus use the same encoding strategy to encode these ingredients as that used in the compatible class encoding.

4.2 The Decomposition of Hyper-Function

Single-output decomposition can be easily applied to the decomposition of a hyper-function. Via hyper-function decomposition, any algorithm that is proposed for single-output decomposition can be applied to multiple-output decomposition.

Definition 4.2 The **transitive fanout** of a node *j*, denoted as ***TFOj***, is defined as $TFOj = \{\text{node } i \mid i = j \text{ or } \exists \text{ path from } j \text{ to } i\}$.❑

Definition 4.3 The **duplication source** (***DS***) after a hyper-function decomposition is the set of nodes which have at least one pseudo primary input as their direct fanin. ❑

Note that after we have decomposed a hyper-function, every node $\notin DS$ must be *k*-feasible. However, every node $\in DS$ with *t* pseudo primary inputs as its direct fanins must be (*t*+*k*)-feasible.

Definition 4.4 The **duplication cone** (***DC***) after a hyper-function decomposition is defined as $DC = \{\cup_j TFOj \mid j \in DS\}$. ❑

Definition 4.5 The **m^{th} layer duplication set** (***$DSet_m$***) after a hyper-function decomposition is defined as $DSet_m = \{\text{node } j \mid j \text{ is in } TFO\text{'s of } m \text{ pseudo primary inputs}\}$. ❑

After we have decomposed a hyper-function, the *DC* should be duplicated to implement the ingredients. Assume that the hyper-function has *n* pseudo primary inputs and *i* ingredients. A node $\in DSet_m$ $(m < n)$ must be duplicated $(2^m - 1)$ additional copies; a node $\in DSet_n$ must be duplicated to have additional $(i - 1)$ copies. To implement each ingredient, we then assign its corresponding code to the pseudo primary inputs. These pseudo primary inputs, assigned with constant values, can be collapsed into their fanout nodes. Nodes $\in DS$ are thus reduced by eliminating the extra pseudo primary inputs. After a hyper-function decomposition, all new generated nodes $\notin DC$ can be shared by these ingredients.

Example 4.1 Assume that four distinct Boolean functions, $f_0(i_0,i_1,i_2,i_3,i_4,i_5,i_7,i_8)$, $f_1(i_0,i_1,i_2,i_3,i_4,i_5,i_6)$, $f_2(i_0,i_1,i_2,i_3,i_4,i_5)$ and $f_3(i_0,i_1,i_2,i_3,i_4,i_5)$, form a hyper-function $F: B^{11} \rightarrow B$. Suppose the coding of each ingredient is derived by applying our compatible classes encoding algorithm and is shown in Figure 8(a). To implement F with 5-input LUT's, assume that F is decomposed as what is depicted in Figure 8(b), in which nodes $\in$ duplication cone are filled with gray. After duplicating the duplication cone, as demonstrated in Figure 9(a), we assign (0,0) to (η_0,η_1) to recover f_0, (1,0) to recover f_1, (0,1) to recover f_2 and (1,1) to recover f_3.

$$F(\eta_0,\eta_1,i_0,i_1,i_2,i_3,i_4,i_5,i_6,i_7,i_8)$$
$$=\bar{\eta}_0\bar{\eta}_1 f_0+\eta_0\bar{\eta}_1 f_1+\bar{\eta}_0\eta_1 f_2+\eta_0\eta_1 f_3$$
$$f_0=F(0,0,i_0,i_1,i_2,i_3,i_4,i_5,i_6,i_7,i_8)$$
$$f_1=F(1,0,i_0,i_1,i_2,i_3,i_4,i_5,i_6,i_7,i_8)$$
$$f_2=F(0,1,i_0,i_1,i_2,i_3,i_4,i_5,i_6,i_7,i_8)$$
$$f_3=F(1,1,i_0,i_1,i_2,i_3,i_4,i_5,i_6,i_7,i_8)$$

(a)

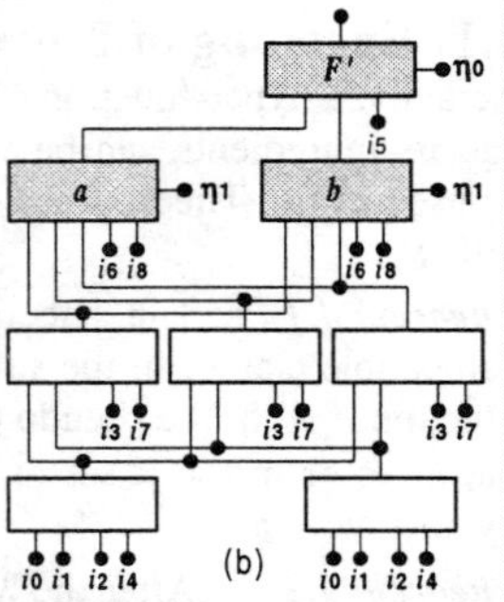

Figure 8: (a)Hyper-Function and Ingredients (b)Decomposition of F

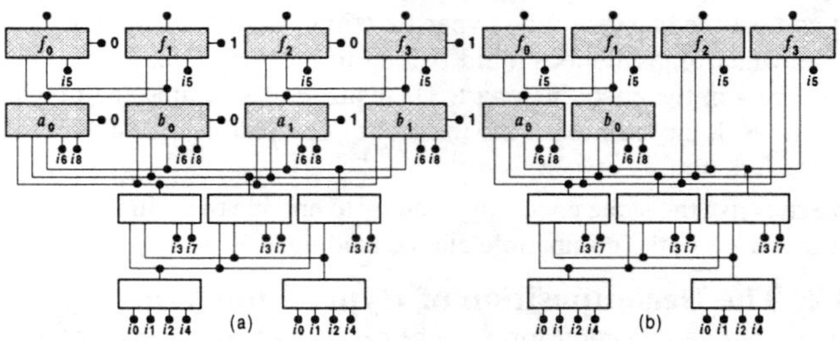

Figure 9: (a)Duplication for Ingredients (b)Further Reductions

After our collapsing these constant input signals in Figure 9(a) into their fanout nodes, the resultant network as is indicated in Figure 9(b) will be independent of these pseudo primary inputs. Since f_2 and f_3 are independent of i_6, i_7 and i_8 according to their original function expressions, we can further simplify the network by collapsing nodes a_1 and b_1 into f_2 and f_3 as shown in Figure 9(b). Nodes $\notin$ duplication cone can be shared by the ingredients of F.

4.3 Properties of Hyper-Function Decomposition

Because nodes in the duplication cone must be duplicated, fewer nodes in the cone are preferred. Consequently, we should keep the pseudo primary inputs as close to the output as possible during the decomposition process. In other words, pseudo primary inputs are preferred to be kept in the μ set during decomposition. In the extreme case, if we always select the pseudo primary inputs in the μ set, hyper-function decomposition can be considered as the *column encoding method* in [4]. Hence, the column encoding approach in [4] is a special case of our hyper-function decomposition. Hyper-function decomposition provides a more generic and flexible means to extract common sub-logic than column encoding. Since multiple-output functions can be decomposed as easily as a single-output function, the λ set size doesn't have to be restricted to a small value. Moreover, the shared logic can cross many levels; the extracted common sub-expressions can be large.

Although a hyper-function is decomposed by applying single-output decomposition, two differences exist between hyper-function decomposition and single-output decomposition. First, strict encoding for a single-output decomposition may become non-strict for a hyper-function decomposition. Because for each ingredient of the hyper-function, a compatible class can be encoded with more than one code. (When a hyper-function is constructed, conjunction partitions may be performed on the partitions of ingredients. Hence a pattern in a partition may be broken into several patterns in a conjunction partition.) Secondly, an encoding that is rigid for a single-output decomposition may become pliable for a hyper-function decomposition. Because the compatible classes of an ingredient may be encoded by more bits than necessary. IMODEC [5] provides a good approach to guarantee that each function is rigidly encoded. However, pliable encoding can save more areas than rigid one in the cases discussed below.

Definition 4.6 A partition **A** is **contain**ed by another partition **B** if the multiplicity of **B** equals to the multiplicity of the conjunction partition of {**A**,**B**}. ❑

Theorem 4.3 Given two partitions **A** of function f_a and **B** of function f_b with respect to the same λ set selection and both of $\lceil \log_2|\text{multiplicity of } \mathbf{A}| \rceil$ and $\lceil \log_2|\text{multiplicity of } \mathbf{B}| \rceil$ are less than the λ set size. **A** is contained by **B** if and only if the decomposition functions of f_b (which identify the column patterns in **B** by strict encoding) can be used as the decomposition functions of f_a. ❑

Theorem 4.4 Given two partitions **A** of function f_a and **B** of function f_b with respect to the same λ set selection and both of $\lceil \log_2|\text{multiplicity of } \mathbf{A}| \rceil$ and $\lceil \log_2|\text{multiplicity of } \mathbf{B}| \rceil$ are less than the λ set size. If **A** is contained by **B**, then the decomposition functions of f_b can be used as the decomposition functions of f_a. ❑

Example 4.2 Given three functions $f_0(x_0,x_1,x_2,x_3,y_0,y_1)$, $f_1(x_0,x_1,x_2,x_3,y_2,y_3)$ and $f_2(x_0,x_1,x_2,x_3,y_3,y_4)$ with λ set selection as $\{x_0,x_1,x_2,x_3\}$, assume therefore we have three partitions:
Π_0 = <0,0,1,0,1,2,2,0,3,2,0,0,0,0,0,2> of f_0,
Π_1 = <0,1,2,0,2,3,3,2,4,3,0,2,1,5,1,3> of f_1, and
Π_2 = <0,1,1,0,1,2,2,3,3,2,0,3,1,4,5,2> of f_2.
If f_1 and f_2 are combined to construct a hyper-function h_{12} with λ set $\{x_0,x_1,x_2,x_3\}$, then the hyper-function has partition Πc of $\{\Pi_1,\Pi_2\}$, Πc_{12}. Because Πc of $\{\Pi_0,\Pi_1,\Pi_2\}$, Πc_{012}, has the same multiplicity as Πc_{12}, Π_0 is contained by Πc_{12} by Definition 4.6. According to Theorem 4.3 or Theorem 4.4, the decomposition functions of h_{12} can be used as the decomposition functions of f_0. Therefore, if f_0, f_1 and f_2 are combined to form a hyper-function with λ set $\{x_0,x_1,x_2,x_3\}$, there are three decomposition functions (because of the multiplicity of $\Pi c_{012} = 8$) shared by the three functions as shown in Figure 10(a). Because f_0 uses three decomposition functions instead of two decomposition functions to encode four compatible classes, the encoding becomes pliable. On the other hand, if the encoding is restricted to being rigid, such as [5], it may derive the result as shown in Figure 10(b). In this case, two more LUT's are consumed.

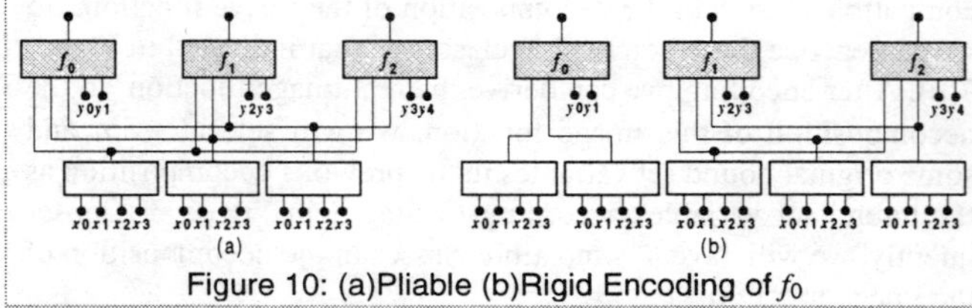

Figure 10: (a)Pliable (b)Rigid Encoding of f_0

5. EXPERIMENTAL RESULTS

Our algorithm, HYDE, has been implemented in SIS environment [11]. Experiments are conducted over a set of benchmark circuits. To prepare the initial circuits for the following technology mapping, small circuits are collapsed, while large circuits are optimized by applying SIS algebraic script. (The benchmark circuit **des** is, in addition, partially collapsed such that several nodes can share the same supports.) After the initial circuits are derived, the technology mapping script used for 2-level circuits is: our decomposition, xl_partition -tm, xl_cover and the script for multi-level circuits is: (full_)simplify, our decomposition, xl_partition -tm, xl_cover. For multi-level circuits, the script are applied several times to improve the results by taking advantage of extracting the local don't care set. The experiments are run under SUN SPARC 20 workstation. We compare these results with other state-of-the-art FPGA synthesis techniques [4,5,8] in Table 1 and Table 2.

Table 1: Experimental Results for XC3000 Device

Circuit	IMODEC[5] CLB	FGSyn[4] CLB	HYDE CLB	CPU Time sec.
5xp1	9	9	10	1.3
9sym	7	7	6	22.8
alu2	46	55	43	554.4
alu4	168	56	140	911.7
apex6	129	181	135	108.7
apex7	41	43	39	9.6
clip	12	18	11	407.2
count	26	23	24	1.6
des	489	-	408	236.6
duke2	122	85	75	28.0
e64	55	44	48	0.0
f51m	8	8	8	10.4
misex1	9	8	9	11.8
misex2	21	22	22	3.3
rd73	5	5	5	3.0
rd84	8	8	7	16.0
rot	127	136	125	132.7
sao2	17	25	17	117.5
vg2	19	17	18	3.6
z4ml	4	4	4	2.7
C499	50	54	50	2.9
C880	81	87	68	69.8
Total	**1453**		**1272**	
Subtotal	**964**	**895**	**864**	

Table 2: Experimental Results for 5-input 1-output LUT's

Circuit	without resub. [8] LUT	with resub. [8] LUT	resub. PO[8] LUT	HYDE LUT
5xp1	15	11	10	13
9sym	7	7	7	6
alu2	48	48	48	50
alu4	172	90	56	206
apex4	374	374	374	354
apex6	192	161	155	186
apex7	120	61	54	54
b9	53	39	37	36
clip	18	11	14	14
count	52	31	31	31
des	-	-	-	561
duke2	175	155	150	116
e64	-	-	-	80
f51m	12	10	8	12
misex1	12	10	10	13
misex2	40	36	36	29
misex3	195	213	120	131
rd73	8	6	6	6
rd84	12	7	8	9
rot	-	-	-	185
sao2	23	21	21	22
vg2	44	21	17	18
z4ml	6	5	4	5
C499	-	-	-	70
C880	-	-	-	81
Total	**1578**	**1317**	**1166**	**1311**
Subtotal(-alu4)	**1406**	**1227**	**1110**	**1105**

In Table 1, the target architecture is the Xilinx XC3000 FPGA. Our algorithm is compared with IMODEC [5] and FGSyn [4]. In Table 2, the resultant circuits are constructed by 5-input 1-output LUT. Without much difference in the consumed CPU time, it is not shown in Table 2. In column 2, 3 and 4 of Table 2, we repeat the results reported in [8]. Although not mapping **alu4** well, HYDE still outperforms IMODEC and FGSyn. Excluding **alu4**, our algorithm produces slightly better results than those in column 4 of Table 2. Due to the disability of handling large circuits such as **C880** in [8], our algorithm is considered more practical to handle large circuits.

6. CONCLUSIONS

Compatible class encoding and hyper-function decomposition techniques have been proposed. The former improves the decomposability of the image function and the hyper-function, while the latter extracts common sub-expressions among multiple-output functions. By transforming these multiple-output functions into a single-output hyper-function, the problem of multiple-output decomposition can thus be reduced to that of single-output decomposition. As a result, previous algorithms intended for single-output decomposition can be easily extended to solve multiple-output decomposition. Experimental results show that our approach is practical and promising. We believe that hyper-function decomposition can be not only used for FPGA synthesis but also exploited to identify common sub-logic in the technology-independent optimization phase of logic synthesis. Another possibility of application is the time-multiplexed reconfigurable computing. For time-multiplexed functions, we can combine them together as a hyper-function. After decomposition, we don't have to duplicate the duplication cone at all. Instead, we can use the pseudo primary inputs to recover the time-multiplexed functions. More applications and theoretical works need to be explored in the future.

REFERENCES

[1] Wen-Zen Shen, Juinn-Dar Huang and Shih-Min Chao, "Lambda Set Selection in Roth-Karp Decomposition for LUT-Based FPGA Technology Mapping," *Proc. 32nd DAC*, pp.65-69, June 1995.

[2] Jie-Hong Jiang, Jing-Yang Jou, Juinn-Dar Huang and Jung-Shang Wei, "BDD Based Lambda Set Selection in Roth-Karp Decomposition for LUT Architecture," *Proc. ASP-DAC*, pp.259-264, January 1997.

[3] Rajeev Murgai, Robert K. Brayton and Alberto Sangiovanni-Vincentelli, "Optimum Functional Decomposition Using Encoding," *Proc. 31st DAC*, pp.408-414, June 1994.

[4] Yung-Te Lai, Kuo-Rueih Ricky Pan and Massoud Pedram, "OBDD-Based Function Decomposition: Algorithms and Implementation," *IEEE Trans. CAD*, vol. 15, pp.977-990, August 1996.

[5] Bernd Wurth, Klaus Eckl and Kurt Antreich, "Functional Multiple-Output Decomposition: Theory and an Implicit Algorithm," *Proc. 32nd DAC*, pp.54-59, June 1995.

[6] Juinn-Dar Huang, Jing-Yang Jou and Wen-Zen Shen, "Compatible Class Encoding in Roth-Karp Decomposition for Two-Output LUT Architecture," *Proc. Intl. Conf. on CAD*, pp.359-363, Nov. 1995.

[7] Christian Legl, Bernd Wurth and Klaus Eckl, "An Implicit Algorithm for Support Minimization during Functional Decomposition," *Proc. European Design and Test Conf.*, 1996.

[8] Hiroshi Sawada, Takayuki Suyama and Akira Nagoya, "Logic Synthesis for Look-Up Table based FPGAs using Functional Decomposition and Support Minimization," *Proc. Intl. Conf. on CAD*, pp.353-358, Nov. 1995.

[9] Daniel D. Gajski, Nikil D. Dutt, Allen C-H Wu and Steve Y-L Lin, *High-Level Synthesis*, Kluwer Ac. Pub., 1992.

[10] Randal E. Bryant, "Graph-Based Algorithms for Boolean Function Manipulation," *IEEE Trans. on Computers*, pp.677-691, August 1986.

[11] R. K. Brayton, R. Rudell, A. Sangiovanni-Vincentelli and A. R. Wang, "MIS: A Multiple-Level Logic Optimization System," *IEEE Trans. on CAD*, pp.1062-1081, Nov. 1987.

[12] G. L. Nemhauser and L. A. Wolsey, *Integer and Combinatorial Optimization*, John Wiley, New York, 1988.

[13] R. L. Ashenhurt, "The Decomposition of Switching Functions," *Ann. Computation Lab. of Harvard Univ.*, vol.29, pp.74-116, 1959.

[14] J. P. Roth and R. M. Karp, "Minimization Over Boolean Graphs," *IBM Journal*, pp.227-238, 1962.

In-Place Power Optimization for LUT-Based FPGAs

Balakrishna Kumthekar # Luca Benini * Enrico Macii ‡ Fabio Somenzi #

University of Colorado
Boulder, CO 80309

* Universitá di Bologna
Bologna, ITALY 40122

‡ Politecnico di Torino
Torino, ITALY 10129

Abstract

This paper presents a new technique to perform power-oriented re-configuration of a system implemented using LUT FPGAs. The main features of our approach are: Accurate exploitation of degrees of freedom, concurrent optimization of multiple LUTs based on Boolean relations, and in-place re-programming without re-routing. Our tool optimizes the combinational component of the CLBs after layout, and does not require any re-wiring. Hence, delay and CLB usage are left unchanged, while power is minimized. As the algorithm operates locally on the various LUT clusters, it best performs on large examples as demonstrated by our experimental results: An average power reduction of 20.6% *has been obtained on standard benchmarks.*

1 Introduction

Synthesis and optimization of FPGA circuits is a challenging problem. The freedom provided by the programmable architectures reduce the effectiveness of most of the design automation tools adopted for standard cell devices. Therefore, ad-hoc techniques need to be adopted to handle area, delay, and power optimization of FPGA-based designs.

In this paper, we focus our attention on power minimization. The method we propose, based on the concept of *neighborhood* of a *cluster of functions*, targets power optimization through *in-place re-programming* of one or more configurable logic blocks (CLBs) used in the circuit implementation. This distinctive feature of our approach makes it particularly suitable to deal with post-layout descriptions, i.e., to designs for which the internal routing. In fact, even though for some architectures (e.g., the Xilinx LUT-based FPGAs) the internal routing can be modified by re-configuring the switch and connection blocks, the first-time choice of the interconnections is usually made so that tight timing and resource (i.e., number of used CLBs) constraints are met. Successive power-oriented modifications that affect the routing may thus negatively impact the timing behavior or the feasibility of the resulting solution.

A further advantage of the post-layout re-configuration procedure we propose is that, since full information on wiring capacitance and load is available even for designs that span multiple FPGAs, accurate power estimation is possible; thus, the iterative optimization procedure can exploit such information to re-program the most power-critical CLBs.

Concerning previous work, several approaches to FPGA power minimization have been proposed in the literature [1, 2, 3, 4]. Our technique is not mutually exclusive with any of these methods, but it is applied at a later stage in the design flow.

DAC 98, San Francisco, California

Relevant to us is also the method of [6], where a re-programming technique for solving the *engineering change* problem, when the FPGA implementation of the design is already in place, is presented. Differently from the engineering change problem, our purpose is to obtain a circuit whose input-output behavior is equivalent to the original one, but whose power dissipation is reduced. Similar to the work of [6], however, we enforce the constraint that the connectivity of the design remains unchanged, and only the CLB functionality can be altered.

Finally, another work which is related to ours is the *generalized matching* approach to gate re-mapping proposed in [7]. The idea here is to replace groups of logic gates with functionally equivalent (but less power consuming) groups; as in our case, this objective is achieved by exploiting the concurrent replacement of more than one gate to achieve a more aggressive global power optimization. However, our target technology and the core optimization algorithm differ from those in [7].

2 Background

In this section, we review some basic concepts and terminology used throughout the discussion. We assume that the reader is familiar with Boolean algebras and BDD-based Boolean function manipulation. We denote vectors and matrices in bold, i.e., $\mathbf{x} = [x_1, x_2, \ldots, x_n]^T$. We use the symbols:

$$\forall_x f = f|_x \cdot f|_{x'} \quad \text{and} \quad \exists_x f = f|_x + f|_{x'}$$

to designate the *consensus* (or *universal quantification*) and the *smoothing* (or *existential quantification*) of Boolean function f with respect to variable x, respectively.

We consider Boolean functions that model a portion (or *cluster*) of a combinational circuit, and we refer to them as *cluster functions*. We denote by $\mathbf{f} = [f_1, f_2, \ldots, f_n]^T$ a generic multi-output cluster function.

The concept of Boolean relation is at the basis of the optimization algorithms detailed in Section 3. For the sake of brevity, we provide here only some basic definitions.

Definition 1 *A Boolean relation $\mathcal{F}$ is a subset of the Cartesian product of two Boolean spaces B^r and B^n, where $B = \{0, 1\}$: $\mathcal{F} \subseteq B^r \times B^n$.*

Boolean relations can be used to represent sets of Boolean functions. In this case, the Boolean relation is said to be *well defined*. The formal definition of this property is the following:

Definition 2 *A Boolean relation $\mathcal{F}$ is* well defined *if, $\forall x \in B^r$, $\exists y \in B^n$ such that $(x, y) \in \mathcal{F}$.*

Furthermore, we define the property of a function contained in a Boolean relation as follows:

Definition 3 *For a given Boolean relation $\mathcal{F}$, a Boolean function $f : B^r \mapsto B^n$ is* compatible *with $\mathcal{F}$ if, for every minterm $x \in B^r, (x, f(x)) \in \mathcal{F}$. Otherwise, f is* incompatible *with $\mathcal{F}$.*

3 Power Optimization Algorithm

The starting point for our optimization algorithm is a netlist of K-input LUTs. The netlist is completely specified, and it has been back-annotated with post-layout information concerning loads, parasitics, and wiring capacitances.
Given a generic LUT, we denote with $\mathbf{i}$ the vector of its input variables, with o the output variable, and with $f(\mathbf{i})$ the function it implements. In addition, we denote with C the output load, whose value is known with high accuracy, since it comes from post-layout analysis.

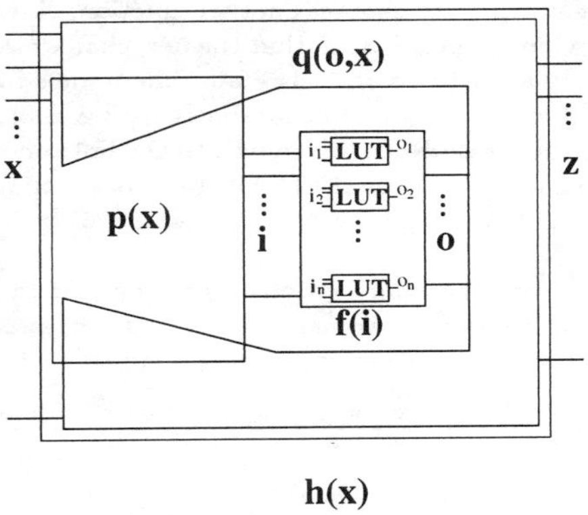

Figure 1: A Multi-Output Cluster and Its Neighborhood.

The optimization algorithm iteratively selects clusters of two or more LUTs, and finds low-power alternative personalizations (i.e., functions that can be implemented by the LUTs) exploiting the degrees of freedom produced by the environment around the target cluster, as shown in Figure 1. For the sake of clarity, in the following we consider a simple two-output cluster (that is, two LUTs). However, the algorithms and equations we present generalize to clusters with any number of outputs.
The degrees of freedom on the functions in the cluster are represented by a Boolean relation whose characteristic function can be computed with the following symbolic equation [8]:

$$\mathcal{F}(\mathbf{i},\mathbf{o}) = \forall_{\mathbf{x},\mathbf{z}}[(P(\mathbf{x},\mathbf{i}) \cdot Q(\mathbf{o},\mathbf{x},\mathbf{z})) \Rightarrow H(\mathbf{z},\mathbf{x})] \tag{1}$$

where P, Q and H are the characteristic functions corresponding to the Boolean functions $\mathbf{p}$, $\mathbf{q}$ and $\mathbf{h}$ describing the *neighborhood* of the cluster (see Figure 1).
Boolean relation $\mathcal{F}$ represents a set of compatible functions that is richer than the set represented by *don't cares*. The LUTs are optimized for low-power by finding the minimum-power pair of functions f_1^{opt} and f_2^{opt} that are compatible with $\mathcal{F}$ and that *have the same support* $\mathbf{i}_1$, $\mathbf{i}_2$ as the original functions $f_1(\mathbf{i}_1)$ and $f_2(\mathbf{i}_2)$, respectively. Notice that the constraint on the support of f_1^{opt} and f_2^{opt} is enforced because we want to perform *in-place* optimization of CLBs without changing the connectivity of the FPGA implementation we start with.
Several points must be addressed to achieve an effective implementation of in-place power optimization. First, we need to specify how the candidate clusters of the LUTs and their neighborhoods are selected. Second, we need to formulate a strategy for estimating the power improvements and to decide when to stop the optimization loop. Third, we need to be able to find f_1^{opt} and f_2^{opt} given $\mathcal{F}$ and to enforce the constant support constraint. The first two issues are discussed in Section 3.1, where the main optimization loop is described. The third problem is described in Section 3.2.

3.1 Main Loop

The block diagram of the main loop of the optimization algorithm is shown in Figure 2. The first step consists of simulating the FPGA network to estimate the power dissipation. The user can provide typical long input pattern streams, possibly coming from behavioral/RT-level simulation. Alternatively, the input probability distributions can be supplied. Within the loop, the network is re-simulated every N_{int} iterations in order to update the switching statistics and to ascertain that there has been a decrease in the power consumption after every N_{int} optimization steps. In order to speed-up the procedure, only a few patterns (m_{int}% of the total) are used for the in-loop simulations. Both N_{int} and m_{int} are user-definable parameters.

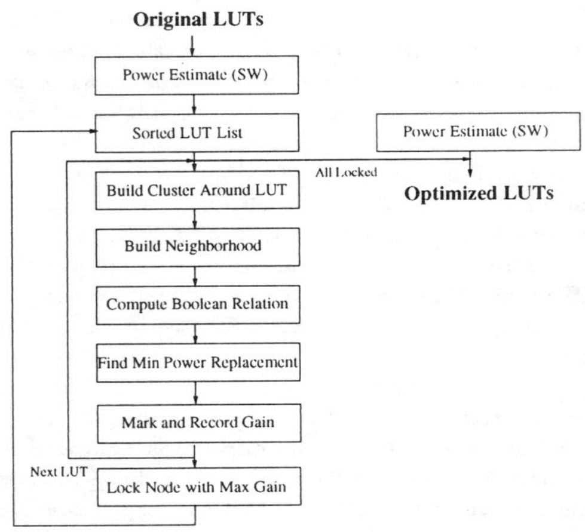

Figure 2: Main Power Optimization Loop.

The nodes in the network (each LUT is a network node, and to each node we associate the function f stored in the LUT) are first sorted according to the product of their switching activity times output load ($SW \times C$). Let these sorted nodes constitute the set S. The LUTs are then picked according to this order. For each selected LUT, λ, the companion cluster members and the neighborhoods are computed. Only the members of S which have not been *locked* due to re-programming in earlier iterations of the algorithm are chosen. (The procedures for cluster selection and neighborhood construction are not described here for space reasons; however, the interested reader can find details on such procedures in [9].)
With the selection of the cluster and its neighborhood, functions P, Q and H (Equation 1) can be computed. The stage is then ready for the construction of Boolean relation $\mathcal{F}$ and the computation of the minimum-power compatible functions f^{opt} to be implemented by the LUTs in the cluster. These are the key steps of the optimization algorithm, and will be described in the following subsection.
Cluster and neighborhood selection, and minimum-power compatible function computation are repeated for each of the nodes in the set S. After all the nodes in S have been processed, an LUT is selected which has the highest gain. The LUT is re-programmed with the minimum-power compatible function and then it is *locked*. This means that the LUT cannot be re-programmed in future iterations. The algorithm iterates (Figure 2) until the improvement is below a minimum user-defined threshold d, or all nodes have been locked. Finally, full-delay simulation with the complete set of input patterns is run again on the optimized network to confirm the reduction in power.

3.2 Optimization Core

Once the neighborhood and the cluster members are identified, the Boolean relation $\mathcal{F}$ is computed through Equation 1. The key optimization performed by our procedure consists of finding minimum-power compatible functions for re-programming the LUTs in the cluster. In the following, let $f_j(\mathbf{i}_j)$, $f_j^{opt}(\mathbf{i}_j)$, $j = 1, \cdots,$ `maxCluster` represent the functions implemented by the LUTs in the cluster before and after re-programming, respectively. Notice that the support variables $\mathbf{i}$ of the multi-output cluster function $\mathbf{f}(\mathbf{i})$ are the union of the $\mathbf{i}_j$.

To enforce the constraint that the network connectivity be left unchanged, the Boolean relation $\mathcal{F}$ must be restricted according to constant support constraint to yield $\mathcal{R} \subseteq \mathcal{F}$. Relation $\mathcal{R}$ is a restriction of Boolean relation $\mathcal{F}$ with the property that if functions f_j^{opt}, $j = 1, \cdots,$ `maxCluster` compatible with $\mathcal{F}$ and with the same support as the original f_j exist, then they are compatible with $\mathcal{R}$ as well. The usefulness of $\mathcal{R}$ is that it eliminates many compatible functions of $\mathcal{F}$ that do not satisfy support constraints, without excluding any valid solution. We have followed the algorithm proposed by Kukimoto and Fujita [6] for the computation of $\mathcal{R}$. Unfortunately, although $\mathcal{R}$ does not contain functions which are compatible with $\mathcal{F}$ and do not meet the support constraint, it is not guaranteed to contain only valid compatible functions either. Hence, the correctness of the solutions extracted from Boolean relation $\mathcal{R}$ must be checked against the support constraints.

After $\mathcal{R}$ has been computed, the LUTs in the cluster are ordered for decreasing $SW \times C$. Starting from the LUT with highest switched capacitance, the min-power compatible functions are computed. Assume that LUT j has been selected. We determine the ***lower bound*** $l_j(\mathbf{i}_j)$ and the ***upper bound*** $u_j(\mathbf{i}_j)$ for it. Function l_j has the same support as f_j and it is the function with minimum ON-set compatible with $\mathcal{R}$, while u_j is the function with maximum ON-set (and same support as f_j) compatible with $\mathcal{R}$. In symbols: $l_j(\mathbf{i}_j) \leq h(\mathbf{i}_j) \leq u_j(\mathbf{i}_j)$, $\forall h(\mathbf{i}_j)$ compatible with $\mathcal{R}$.

To compute l_j and u_j we first extract from $\mathcal{R}$ the Boolean relation $\mathcal{R}_j \subseteq \mathcal{R}$ which contains all and only the functions with support $\mathbf{i}_j$:

$$\mathcal{R}_j(\mathbf{i}_j, o_j) = \forall_{i \in (\mathbf{i} - \mathbf{i}_j)} \exists_{o \in (\mathbf{o} - o_j)} \mathcal{R}(\mathbf{i}, \mathbf{o}) \tag{2}$$

where o_j is the output variable corresponding to LUT j.

If $\mathcal{R}_j$ is well-defined, the two bounds are easily computed as follows:

$$\begin{aligned} l_j(\mathbf{i}_j) &= \mathcal{R}_j|_{o_j}(\mathbf{i}_j, o_j) \cdot \overline{\mathcal{R}_j|_{\overline{o_j}}(\mathbf{i}_j, o_j)} && (3) \\ u_j(\mathbf{i}_j) &= \mathcal{R}_j|_{o_j}(\mathbf{i}_j, o_j) && (4) \end{aligned}$$

The minimum-power function f_j^{opt} for re-programming LUT j is then selected to be either u_j or l_j. The rationale for this choice is the following. Assuming no temporal correlation, the switching activity of output j can be computed as $2p_j(1 - p_j)$, where p_j is the probability of the output j to be one. Notice that this is a single-maximum function. Since l_j is smaller than any function with support $\mathbf{i}_j$ compatible with $\mathcal{R}$, its probability is guaranteed to be minimum. On the same lines, since u_j is larger than any function with support $\mathbf{i}_j$ compatible with $\mathcal{R}$, its probability is guaranteed to be maximum. Consequently, the minimum switching activity is attained either in u_j or in l_j or both. This conclusion is valid in absence of temporal correlation, and may not be accurate in some cases. However, it provides an efficient way of getting function f_j^{opt}, as shown by the results of Section 4.

Once the choice between l_j and u_j is made, node j is marked as ***potential candidate for re-programming*** and its gain is recorded. It is also marked as ***processed***, so that it is not considered again in the same iteration. The relation $\mathcal{R}$ is restricted by taking into account the choice of f_j^{opt} made. Then, another cluster output is considered.

It should be noted that our choice of f_j^{opt} is greedy. We try to get the best possible re-programming for LUT j disregarding the ones that come after j. As a consequence, the choice for j may result in the Boolean relation being ill-defined when making a choice for, say $j + 1$. If this is the case, the re-programming algorithm backtracks and chooses another function, different from the one previously chosen, such that the new choice also reduces the switching activity. If this also fails, the node's function is not changed and the algorithm continues with a choice for the next cluster member. Node j is moved to the bottom of the list of nodes yet to be processed. It will be processed again, but it will not be moved again and it will not be allowed to cause backtrack.

Having described the main optimization loop and the core re-programming algorithm, we can now provide the pseudo-code of the in-place optimization procedure (Figure 3).

```
stop = iter = 0;
frequency = N_int;
do {
   if(iter % frequency == 0) {
      if(iter == 0)
         PerformCompleteSimulation();
      else
         PerformPartialSimulation();
      if(Power is increased) {
         Undo the last N_int changes;
         Leave nodes locked;
      }
   }
   sortedNodes = SortUnlockedNodes(network);
   foreach(node ∈ sortedNodes) {
      if(node is not processed) {
         cluster = SelectCluster(node,maxCluster);
         neighbor = SelectNeighborhood(cluster,maxZNodes);
         F(i, o) = ComputeRelation(cluster,neighbor);
         R(i, o) = RestrictAccordingToSupporConstraints(F(i, o));
         ChooseCompatibleFunctions(F_r(i, o),potNodeTable);
      }
   }
   if (size of potNodeTable > 0)
      ReProgramAndLockBestNode(potNodeTable);
   else
      stop = 1;
   iter++;
} while(stop == 0);
PerformCompleteSimulation();
```

Figure 3: In-Place LUT Re-Programming Algorithm.

The first part of the loop dispatches simulations to assess the quality of the optimization. Simulation of the complete input stream is performed at the first iteration, and simulations of shorter streams are performed every N_{int} iterations.

After simulation, the nodes are sorted based on their output power consumption. The sorted list `sortedNodes` is then processed, one node at a time. For each node, if the node has not been locked in previous iterations, cluster and neighborhood are constructed. Boolean relations $\mathcal{F}$ and $\mathcal{R}$ are then computed, and the min-power compatible functions are obtained. The variable `potNodeTable` returns a set of nodes for potential re-programming. The node in the cluster for which power savings are maximum is selected, re-programmed and locked. The iteration stops when no more savings or unlocked nodes can be found. The procedure is guaranteed to terminate and has a number of iterations which is linear in the number of LUTs.

4 Implementation and Results

The flow we have used for benchmarking the capabilities of our in-place LUT re-programming algorithm is illustrated in Figure 4. The original `.blif` description is fed to `SIS` for logic optimization and mapping onto 5-input LUTs. The script we used for this task is the one suggested in the SIS manual. The mapped circuit (`.fpga`) is supplied to a tool (`PREX`) that generates a file (`.cap`) containing the post-layout capacitances for the nodes in the circuit. The `.fpga` description and the corresponding capacitance file are input to `VIS`, the framework in which the optimization algorithm has been developed. `VIS` is interfaced with a power simulator (`PSIM`), whose task is that of providing power estimates at all stages of execution of the re-programming algorithm using a set of user-supplied patterns (file `.pat`), or user-specified primary input statistics (file `.i_sat`). The output of the flow is the optimized LUT netlist (file `.opt`).

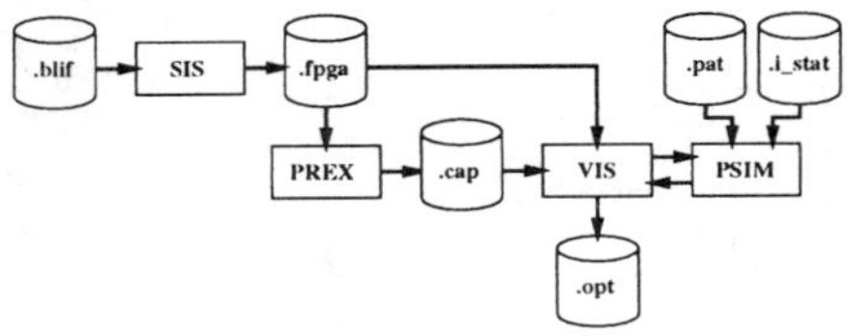

Figure 4: Experimental Flow.

Program `PREX` mimics the job of the tools for placement, routing, and capacitance extraction usually distributed by the FPGA vendors together with their parts; in fact, it produces a capacitance file to be used for accurate timing and power simulation. For our experiments, `PREX` generates capacitances based on the fanout of the LUTs and some additive random noise (that models the uncertainties in the routing process). This is for the purpose of running experiments which are not biased by the results of different layout tools. Notice, however, that any capacitance distribution would be acceptable, as far as it is not modified before and after the optimization, because the interconnections of the various LUTs never change.

Several parameters can be specified to control the optimizer: The depth of the neighborhood, the number of nodes in the cluster, and the number of nodes in the neighborhood. Our goal was to be able to show a practical realization of the in-place power optimization algorithm. Therefore, we implemented the algorithm targeting robustness and conservatively set the parameters to avoid failure even in corner cases and produce results in a reasonable amount of time. Needless to say, with this choice we gave up the opportunity to explore the entire space spanned by the various degrees of freedom. A case in point being the limit set on the number of nodes in the cluster to 2 and the depth of the neighborhood limited to 1.

We have assumed the propagation delay through the CLBs to be constant. The pin-to-pin block delay of the CLB is then assumed to be constant. The total delay of CLB j, including the delay due to the output capacitance, can thus be approximated as:

$$D(j) = \texttt{BlockDelay} + C_j \times \texttt{fanoutCount}(j) \qquad (5)$$

where C_j is the capacitance of node j taken from the `.cap` file. Our algorithm disregards the internal power of CLBs. A recent work [5] has shown that, for the same function, the internal power dissipation of a CLB can be reduced by reordering the inputs to the CLB. As we do not allow any change in the interconnect, the above procedure cannot be used, and the constant power model seems reasonable. In addition, it should be noted that the output switching power of the CLB usually dominates the internal power.

Table 1 reports the results we have obtained on a sample of the large Mcnc'91 combinational multi-level circuits [10]. Columns *CLB*, *PI* and *PO* report the number of CLBs, primary inputs and primary outputs of the circuit. Columns *Init.* and *Fin.* give the power dissipation of the circuit before and after optimization. Column *R* gives the number of CLBs that have been re-programmed, column *Sav.* the percentage of power saving, and column *Time* the CPU time, measured in seconds on a 200MHz Pentium Pro Linux machine with 128MB of RAM, required by the algorithm to complete.

Circ.	CLB	PI	PO	Init.	Fin.	R	Sav.	Time
c499	70	41	32	155	118	22	23.8	343
c1355	94	41	32	199	101	77	48.9	765
alu2	113	10	6	156	117	52	25.3	221
c880	115	60	26	167	151	26	9.4	881
c1908	150	33	25	254	200	52	21.4	1570
alu4	214	14	8	327	246	91	24.8	1696
c2670	307	233	140	330	313	59	5.1	1110
pair	440	173	137	572	562	58	1.7	1582
c3540	470	50	22	652	482	206	26.2	17222
c5315	515	178	123	907	723	207	20.2	12443
t481	628	16	1	652	627	39	3.8	1558
i8	774	133	81	1170	787	102	32.7	1475
c7552	801	207	108	1393	1188	319	14.7	24920
i10	830	257	224	919	782	163	14.9	6948
des	1406	256	245	1887	1341	461	28.9	28675

Table 1: Power Optimization Results.

On all the benchmarks, power was estimated through real-delay simulation of user specified input patterns. 20,000 to 50,000 vectors were used for the initial and final simulations, while only an average of 50% of such vectors was simulated during the optimization phase. The internal simulations were performed every 5 to 10 iterations of the algorithm. The peak power reduction we have obtained is around 48.9%, and the average power reduction is around 20.6%.

5 Conclusions

We have presented a technique to perform power-oriented reconfiguration of a system implemented using LUT-based FPGAs. Our approach has the distinctive property of being applicable to designs for which the layout has already been generated. Our method operates locally on the various LUT clusters of the original network and performs best on large examples, as demonstrated by the experimental results we have reported.

References

[1] A. Farrahi, M. Sarrafzadeh, "FPGA Technology Mapping for Power Minimization," IWFPLA, pp. 66-77, Sep. 1994.

[2] K-R. Pan, M. Pedram, "FPGA Synthesis for Minimum Area, Delay and Power Consumption," EDTC-96, page 603, Mar. 1996.

[3] C-C. Wang, C-P. Kwan, "Low Power Technology Mapping by Hiding High-Transition Paths in Invisible Edges for LUT-Based FPGAs," ISCAS-97, pp. 1536-1540, May 1997.

[4] C-S. Chen, T. Hwang, C. L. Liu, "Low Power FPGA Design – A Re-Engineering Approach," DAC-34, pp. 656-661, Jun. 1997.

[5] M. Alexander, "Power Optimization for FPGA Look-Up Tables," ISPD-97, pp. 156-162, Apr. 1997.

[6] Y. Kukimoto, M. Fujita, "Rectification Method for Lookup-Table Type FPGAs," ICCAD-92, pp. 54-61, Nov. 1992.

[7] P. Vuillod, L. Benini, G. De Micheli, "Re-Mapping for Low Power Under Tight Timing Constraints," ISLPED-97, pp. 287-292, Aug. 1997.

[8] Y. Watanabe, L. Guerra, R. K. Brayton, "Permissible Functions for Multi-Output Components in Combinational Logic Optimization," IEEE TCAD, Vol. 15, No. 7, pp. 734-744, Jul. 1996.

[9] B. Kumthekar, L. Benini, E. Macii, F. Somenzi, In-Place Power Optimization for LUT-Based FPGAs Tech. Rep., Dept. of ECE, Univ. of Colorado, Oct. 1997.

[10] S. Yang, Logic Synthesis and Optimization Benchmarks User Guide Version 3.0, Tech. Rep., MCNC, Jan. 1991.

A Re-engineering Approach to Low Power FPGA Design Using SPFD

Jan-Min Hwang, Feng-Yi Chiang, and TingTing Hwang
Department of Computer Science, Tsing Hua University
Hsinchu, Taiwan, 300, Republic of China

Abstract

In this paper, we present a method to re-synthesize Look-Up Table (LUT) based Field Programmable Gate Arrays (FPGAs) for low power design after technology mapping, placement and routing are performed. We use *Set of Pairs of Functions to be Distinguished (SPFD)* to express functional permissibility of each signal. Using different propagations of SPFD to fan-in signals, we change the functionality of a PLB (Programmable Logic Block) which drives large loading into one with low transition density. Experimental results show that our method can reduce on average 12% power consumption compared to the original circuits without affecting placement and routing.

1 Introduction

Because of its short design cycle and low manufacturing cost, Field Programmable Gate Array technology offers the best solution to time-to-market pressure. LUT-based FPGA is the most popular one[1]. Meanwhile, power consumption becomes one of the crucial design considerations in circuit designs recently [2, 3, 4].

In this paper, we propose a re-engineering approach to low power FPGA designs after technology mapping, placement, and routing are performed. We take advantage of the characteristic of LUT which can implement an arbitrary k-variable boolean function. A newly developed don't care representation, *Set of Pairs of Functions to be Distinguished (SPFD)*[5], is used to represent functional permissibility of PLBs in an FPGA. The use of SPFDs facilitates the alternations of internal logic function of a PLB.

After placement and routing of an FPGA circuit are performed, we first generate an SPFD for each PLB from outputs to inputs. Then, with the objective of power reduction, we select the function which is a permissible function specified by the SPFD with the lowest transition density. The main advantage of our approach is that the re-synthesis is performed after optimization steps of technology mapping, placement and routing are applied. While keeping the placement and routing of a circuit intact, we can reduce the power consumption of the design.

This work was supported by the National Science Council of R.O.C. under Grant NSC 87-2215-E-007-016.

DAC 98, San Francisco, California

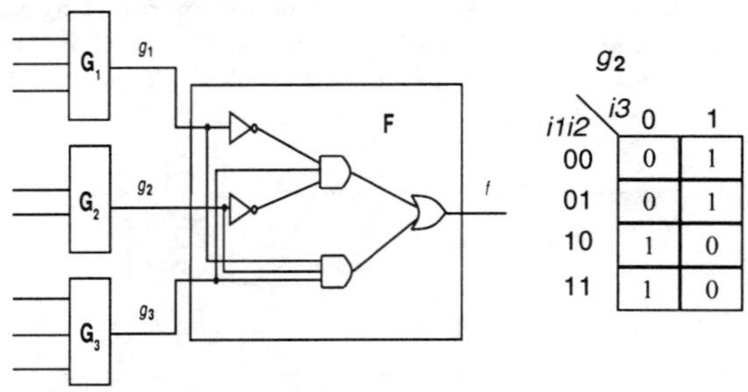

Figure 1: Part of a Mapped Network

2 Expressing Functional Permissibility by SPFD

In this section, we will review how *Set of Pairs of Functions to be Distinguished (SPFD)*[5] is used to express functional permissibility.

SPFD is a novel way to represent functional permissibility. At each signal line s, permissible functions are represented by sets of on-off pairs. As long as a synthesized function satisfies the pairs of constraints, the function is a permissible function for signal s. For example, Figure 1 depicts a LUT-based FPGA circuit example, where $i1$, $i2$ and $i3$ denote primary input variables and g_2 and f are two realized functions. Let SPFD of g_2 be computed as shown in Figure 2(a). In this case, there are two pairs of on-off pairs, $(a_{010},\ a_{001})$ and $(a_{101},\ a_{111})$. Consider the pair $(a_{010},\ a_{001})$ first. The function realized at g_2 must distinguish the pair specified by on-sets of a_{010} and a_{001}. That is, as long as the function values of g_2 represented by A and $\bar{A}$ shown in Figure 2(b) are selected complementally to each other, the constraint is satisfied. Similarly, the constraint imposed by the other function pair $(a_{101},\ a_{111})$ is represented by B and $\bar{B}$. Hence, functions satisfying the constraints in Figure 2(b) are permissible functions at signal g_2.

In [5], an algorithm to compute SPFDs of a signal from *primary outputs (POs)* to *primary inputs (PIs)* was proposed. We outline the basic steps as follows. For a pair (f_0, f_1) at signal s:

1. Calculate 2^n functions of logical products for n fanins of s represented as $\{b_{000}, b_{001}, ..., b_m\}$ where m is $(2^n)_2$.
2. For all b_i, calculate $a_i = b_i \cdot (f_0 + f_1)$.
3. Discard zero functions of a_i and separate the remaining a_i into two groups $F0$ and $F1$ by checking if its on-set is included in f_0 or f_1. Then calculate the Cartesian product $F_c = F0 \times F1$.

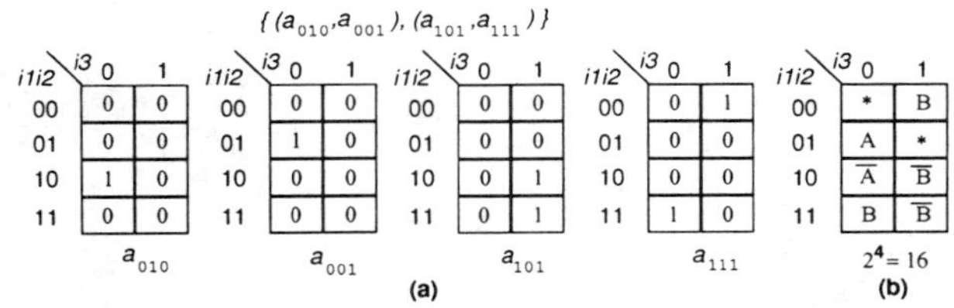

Figure 2: Functions Satisfying the SPFD: (a) SPFD(g_2), (b) Permissible Functions Specified by SPFD(g_2)

4. For each element $(a_i,\ a_j)$ in F_c, add this pair into the SPFD of the k-th fanin if the k-th bit of i and j are different.
5. Do steps 1-4 for all PLBs from POs to PIs. If a PLB has more than two fanouts, the SPFDs of the PLB are calculated by the union of the fanouts' SPFDs.

3 Power Reduction by Functional Transformation for PLBs

The commonly used power dissipation model shown below will be used as our cost function for power minimization,

$$P_{avg} = 0.5 \times C_{load} \times \frac{V_{dd}^2}{T_{cyc}} \times E(switchings) \qquad (1)$$

where P_{avg} is the average power dissipation, and C_{load} the load capacitance, V_{dd} the supply voltage, T_{cyc} the global clock period, and E(switchings) the expected value of the output switching count per clock cycle.

As mentioned in the previous section, SPFDs are a powerful method to express functional permissibility. Moreover, propagating SPFDs of an output to its input wires is flexible. As shown in *step 4* of the procedure presented in Section 2, if there are more than one different bits of i and j, we may propagate a pair (a_i, a_j) to any fan-in signal as long as it corresponds to a different bit. This flexibility of selecting fan-in lines to propagate SPFDs allows us to take power minimization into consideration. To reduce power consumption, the most effective way is to select a permissible function with the lowest switching activities.

3.1 Procedure Overview

We propose a functional transformation method to reduce the power consumption of an FPGA design after performing technology mapping, and placement-and-routing. Let's start with a circuit $G(V, E)$ (V is the set of PLBs $\{P_1, P_2, ..., P_n\}$, E is the set of interconnections) which has been mapped into PLBs of an FPGA. The procedure is performed in two phases. In the first phase, PLBs are processed from POs to PIs. First, at the output signal of a PLB, we propagate its SPFD to its fan-in signals. A heuristic is used to determine which fan-in signal to be selected for propagation. The objective is to synthesize a function with the lowest transition density. At each signal, we will keep just one pair of function to be distinguished. Therefore, the next step is to merge pairs of functions in an SPFD to a single one if a fanin signal has more than one pairs. In this merging step, power reduction will also be taken into consideration. Finally, we apply a simulated annealing based method to further improve the selection of fan-in signal for propagating SPFD.

In the second phase, we generate new functions for PLBs based on the merged SPFD computed in the first phase from PIs to POs.

3.2 Heuristics for Propagating SPFD to Fanin Signals

Pairs of functions to be distinguished in SPFDs of a PLB must be propagated to fan-in signals in such a way that all pairs of output values in $F0$ and $F1$ can be distinguished. In many cases, there are many propagation ways. In the algorithm presented in [5], a fanin signal is arbitrarily selected. However, a heuristic can be used to guide the selection of fanin signal for power reduction.

It is preferable that a signal with high capacitance loading has low switching activity. Therefore, when selecting fanin signal we should avoid selecting those signals that have high capacitance loading. The reason is that pairs of SPFD represent constraints. The more pairs of functions in SPFDs imposed on a signal, the more constraints imposed on function that can be synthesized to implement the signal. That is, the less *don't-care* condition this signal has. To reduce power consumption, we would like to have more flexibility in synthesizing signals that have large loading. Therefore, the first consideration to select a fan-in signal is the capacitance loading of signals.

The second consideration to select a fanin signal is the level of fanin signals. The reason why we take the level of a signal into consideration is that the value of level represents the complexity of the fan-in cone of a signal. The signal with a large level means that many SPFDs may be generated for its fanin cone before we reach the primary inputs. The more pairs of functions in an SPFD of a signal, the less flexible a signal can be synthesized into one with a low transition density. Therefore, when selecting fanin signals, we should avoid selecting those signals that have a large value of level.

Based on the above two considerations, our cost function to select a fanin signal s is defined as follows:

$$cost(s) = (1 - w) \times LC(s) + w \times Level(s) \qquad (2)$$

where LC is the load capacitance of the signal s, $Level$ is the level and w is a constant value between 0 to 1 which represents the weight of loading capacitance and level in the cost function. In our experimental results, we found that for those circuits with large values of level, better solutions can be produced if w is set to a larger value.

The *cost* being computed is used as a priority in selecting assignment of input signals. The higher the cost of a signal, the lower priority the signal will be selected.

3.3 Merging Pairs of Functions in an SPFD

As described in the previous section, many pairs of functions may be derived for a signal. The SPFD of a signal is the union of these pairs. Since merging of pairs resulting in the function with the lowest power consumption in the local space can be computed, we need to keep only one pair in an SPFD set for each signal.

However, the merging of pairs of functions in an SPFD is not arbitrary. Certain merging constraints must be satisfied. First the merged pair must satisfy the property of transitive relations. That is, if $(A,\ B)$ and $(A,\ C)$ are two pairs of relations of a signal, then $(A,\ B + C)$ is the resultant merged pair of the relations. For example, consider the two pairs of functions in an SPFD for g_2, (a_{000}, a_{111}) and (a_{101}, a_{111}), shown in Figure 3(a).

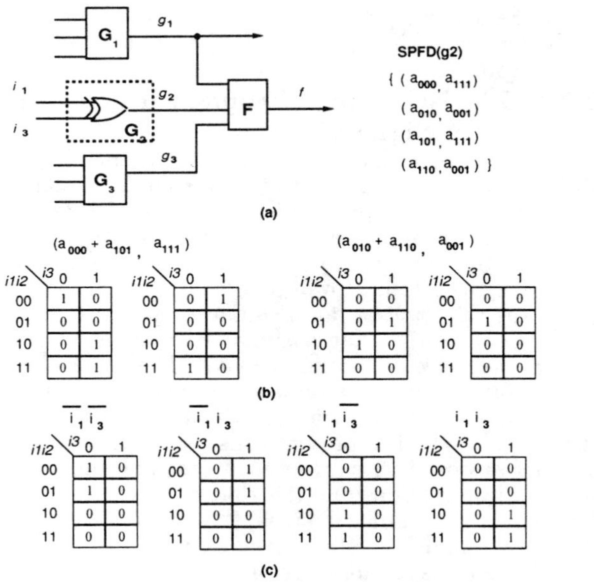

Figure 3: Merging Pairs of Functions in the SPFD for g_2

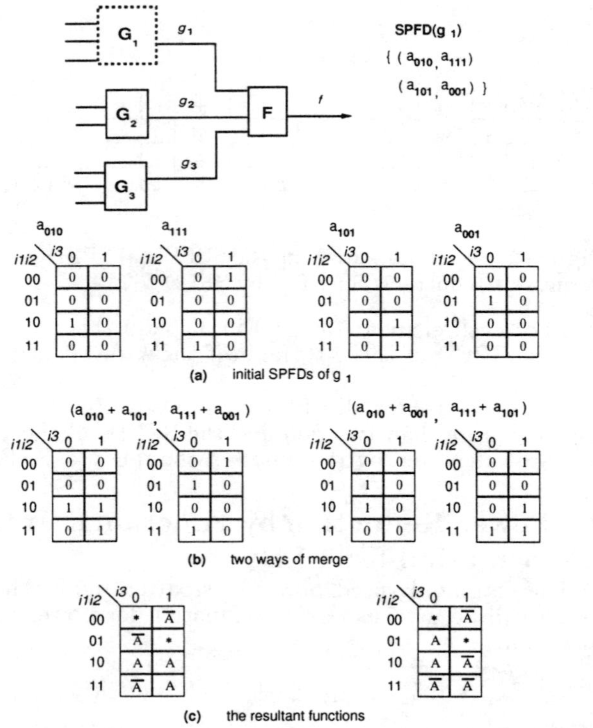

Figure 4: Merging Pairs of Functions in the SPFD for g_1

To satisfy the property of transitive relations, the two pairs of functions are merged to pair $(a_{000}+a_{101},\ a_{111})$. Similarly, the two pairs of functions, $(a_{010},\ a_{001})$ and $(a_{110},\ a_{001})$, are merged to $(a_{010}+a_{110}, a_{001})$. The resultant function after performing the two merges is shown in Figure 3(b).

The second merging constraint is that the pairs must be merged so that the function computed at a PLB is synthesizable by the local inputs to the PLB in the second phase. That is, we generate functions computed by minterms of the local input signals of a PLB and merge pairs according to the functions computed by the minterms. Let $(A,\ B)$ and $(C,\ D)$ be two pairs of relations for the output signal of a PLB and E be a function computed by one minterm of local input signals to the PLB. If $A \cap E \neq \emptyset$ and $C \cap E \neq \emptyset$, then the pairs must be merged as $(A+C,\ B+D)$. For example, let g_2 has two pairs of function pairs $(a_{000}+a_{101},\ a_{111})$ and $(a_{010}+a_{110},\ a_{001})$ shown in Figure 3(b) and the local input signals to G_2 be i_1 and i_3. First, we compute all functions computed by the minterms of the signals i_1 and i_3 as shown in Figure 3(c). Since function $\bar{i_1}\bar{i_3}$ intersects function $a_{000}+a_{101}$, and function $\bar{i_1}i_3$ intersects function a_{001}, pairs of functions are merged as $((a_{000}+a_{101})+a_{001},\ a_{111}+(a_{010}+a_{110}))$. The reason behind this merge rule is that a function represented by a minterm of local input signals is indistinguishable. Hence, the function specified by the minterm must be in one set. Note that the number of inputs to a PLB is a constant, k, and the maximum number of minterm for a PLB is 2^k. In the case of $k=5$, the number of minterms needs to be considered is at most $2^5=32$.

In other cases, the merging process will be performed by taking the transition density into consideration. We will merge function pairs so that the function synthesized from the resultant pair has a minimal transition density.

Let's explain the merging process by taking the signal g_1 shown in Figure 4 as an example. Two pairs of functions (a_{010}, a_{111}) and (a_{101}, a_{001}) in an SPFD for g_1 can be merged arbitrarily because the two merging constraints do not impose on them. There are two ways to merge these two pairs. The first way is to merge a_{010} with a_{101} and a_{111} with a_{001}. The second way is to merge a_{010} with a_{001} and a_{111} with a_{101}. The results of merging are shown in Figure 4(b) and their permissible functions are shown in Figure 4(c). Assume the probabilities of input signals to be 0 and 1 are equal. Then, the function with the lowest transition density for the first merge is synthesized by assigning A to be 0 (1) and $\bar{A}$ and $*$ (*don't care*) to be 1 (0), which results in transition density of 0.469 at the signal g_1. Similarly, the function with the lowest transition density for the second merge will have transition density of 0.375 at the signal g_1. From the above example, we conclude that the merge should be selected to produce a resultant pair in an SPFD with uneven probability for A and $\bar{A}$. Thus, when merging two pairs, the two minterms with most-frequent occurrences and the two with least-frequent occurrences should be merged separately.

The above merging process is applied to input signals of a PLB to which many pairs of functions to be distinguished are derived from the output of a PLB. The same merging process is also applied at the output of a signal when the output signal is not fanout-free. In this case, one pair of SPFDs is derived for each fanout, the SPFD of the output signal is the union of SPFDs of all its fanouts.

3.4 Applying Simulated Annealing to Improve the Initial Solution

As described in Section 3.2, the selection of input signals to propagate pairs of output function in F_c is based on a heuristic algorithm. Since the selection of input

Table 1: Experimental Results

Circuit	#CLBs	FPGA	Init Power	Cost 1	Cost 2	SA	Ratio
9symml	66	3030PC84	48.98	48.14	47.42	42.25	86%
C1908	134	3090PC84	193.28	191.35	184.25	167.88	87%
C432	59	3090PC84	103.54	103.76	97.62	92.14	89%
C880	91	3064PP132	96.36	93.42	91.24	88.75	92%
alu2	119	3090PC84	116.60	116.72	112.64	95.93	82%
alu4	250	3195PP175	307.44	301.01	290.35	270.10	88%
apex7	61	3042PP132	30.45	30.00	29.81	28.57	94%
b9	36	3020CB100	15.50	15.19	15.19	15.19	98%
comp	28	3020PC68	31.17	30.95	30.95	30.95	99%
dalu	372	3720PC68	505.68	507.32	471.24	436.25	86%
example2	110	3195PQ208	46.78	45.24	44.18	44.18	94%
frg1	32	3020PC68	25.10	24.96	24.96	24.96	99%
i9	308	3195PQ208	203.90	203.93	201.90	192.08	94%
t481	155	3090PC84	107.48	106.19	103.24	94.83	88%
term1	48	3030PC68	41.56	35.20	33.79	33.79	81%
too_large	107	3090PC84	123.34	122.92	110.56	106.72	87%
ttt2	47	3030PC68	23.03	22.76	22.76	22.76	99%
vda	194	3195PC84	165.02	165.35	151.03	138.61	84%
x1	99	3090PQ160	77.15	76.12	76.12	73.12	95%
x4	127	3195PQ208	53.11	52.98	49.80	47.60	90%
TOTAL	2457		2315.47			2046.66	88.4%

signal is not optimal, we want to further improve the selection by using a simulated annealing algorithm.

The selection described in Section 3.2 is used to derive an initial assignment. Since pairs of functions in an SPFD are merged to a single one in the following step, we can compute the power consumption exactly by summing the product of load-capacitance and switching-density at each fanin signal. The sum will be used as the cost function. A move in the simulated annealing process is to try other assignment for the propagation of an SPFD which is generated by randomly selecting a different SPFD assignment. Then, the merge process is applied to the assignment, and the cost is evaluated.

3.5 Generating New Functions

The last step of our procedure is to generate the new function realized at each PLB from PIs to POs. The steps are shown below:

Given a merged SPFD (f_0, f_1) at the output of a PLB F:

1. Select either function f_0 or f_1 as f_i.
2. Generate 2^n minterms in terms of n local inputs of F.
3. Select a minterm p and check if its on-set intersects f_i. If so, update f_i to $f_i \cdot \bar{p}$ and add p into the sum of product of the new function.
4. If f_i is zero function, end the process. Otherwise, goto step 3.

4 Experimental Results

We used **BDD** to represent each function pair in SPFDs. The AND and OR operations of BDDs are applied to check the inclusion (exclusion) between on-set of functions and merging of SPFDs.

First, we optimized the benchmark circuits by *SIS* with the script of *script.alg* and mapped them into different FPGAs of *Xilinx* (here we selected the type of FPGA which has the most suitable IO pins and number of PLBs for each circuit). Then we used *Xilinx*'s APR tools to realize the placement of FPGA circuits. The routing distance of each net is estimated by using the *half-perimeter* rule.

Table 1 summarizes the experimental results. The cost of power consumption of each circuit is evaluated by the sum of product of loading capacitance and switching density for each net. The column "cost 1" is the result of just using LC term in the cost function to determine the priority of propagating SPFD to fanins. The column "cost 2" is the result of using both LC and $Level$ terms in the cost function. The column "SA" lists the results of applying simulated annealing to the initial solutions obtained by using "cost 2" as the cost function. The last column shows the ratio of the power consumption by applying a simulated annealing procedure to those of initial circuits. From the table, we can see that a fixed ordering (priority) of propagating SPFD does not produce good results. However, dynamically adjusting the assignment (simulated annealing) can improve the results. We also found that for deep circuits, better solutions can be obtained by cost function 2. In general, the larger the circuit, the more power reduction can be achieved because propagating SPFDs is more flexible.

References

[1] S.D.Brown, R.J.Francis, J.Rose and Z.G.Vranesic. Field-Programmable Gate Arrays. *Kluwer Academic Publishers, 1992.*

[2] V.Tiwari, P.Ashar and S.Malik. Technology Mapping for Low Power. *Proc. of Design Automation Conference, 1993, pp.74-79.*

[3] M.Alidina, J.Monteiro, S.Devadas, A.Ghosh and M.Papaefthymiou. Precomputation-Based Sequential Logic Optimization for Low Power. *IEEE Transactions on VLSI Systems, Vol.2, NO.4, December 1994.*

[4] Chao-Shen Chen, Ting-Ting Hwang. Low Power FPGA Design - A Re-engineering Approach. *Proc. of Design Automation Conference, 1997.*

[5] Shigeru Yamashita, Hiroshi Sawada and Akira Nagoya. A New Method to Express Functional Permissibilities for LUT based FPGAs and Its Applications. *Proc. of International Conference on Computer Aided Design, 1996.*

Power Considerations in the Design of the Alpha 21264 Microprocessor

Michael K. Gowan, Larry L. Biro, Daniel B. Jackson
Digital Equipment Corporation
Hudson, Massachusetts

1. ABSTRACT

Power dissipation is rapidly becoming a limiting factor in high performance microprocessor design due to ever increasing device counts and clock rates. The 21264 is a third generation Alpha microprocessor implementation, containing 15.2 million transistors and operating at 600 MHz. This paper describes some of the techniques the Alpha design team utilized to help manage power dissipation. In addition, the electrical design of the power, ground, and clock networks is presented.

2. INTRODUCTION

Digital introduced the Alpha 21064 [1] microprocessor in 1992, thus delivering the industry's highest performance at that time. Manufacturing process technology advancements, architectural innovations, and full-custom circuit design techniques have been significant contributors to Digital's delivery of two additional generations of performance leadership Alpha microprocessors [2,3]. Between the first generation Alpha 21064 and the third generation Alpha 21264 designs, device counts have grown by a factor of 9, from 1.68 million to 15.2 million, while clock frequencies have increased from 200 MHz to 600 MHz, or a factor of 3. These rapid increases in available chip real estate and clock rates, combined with Digital's goal of delivering the highest performance possible, have caused significant increases in power dissipation and supply currents, which have had far-reaching effects on the clock, power, and ground networks, and on reliability. Consequently, power dissipation has joined clock frequency and die size as a first order design constraint. As die sizes and clock rates continue to increase, power dissipation may eventually limit the amount of hardware included on a microprocessor.

DAC 98, San Francisco, California
©1998 ACM 0-89791-964-5/98/06..$5.00

3. SOURCES OF POWER DISSIPATION

Power is dissipated in CMOS circuits during both static and dynamic operating conditions. Ideally, a non-transitioning complementary CMOS gate dissipates no power because the PMOS and NMOS devices are never simultaneously conducting. However, in reality, a small leakage current flows through the reverse biased diodes formed by the source and drain diffusions and the substrate. More significant static leakage currents result from subthreshold conduction of the PMOS and NMOS devices. High performance CMOS processes specifically tuned for speed critical applications often feature very low threshold voltage devices. These low threshold devices have significantly increased subthreshold currents. Circuits designed with static loads may also dissipate power under DC operating frequencies.

Dynamic power is dissipated when the device output capacitance is charged to VDD through the PMOS device and discharged to VSS through the NMOS device. This power is given by the well-known equation $p = cv^2 f$. Note that while dynamic power dissipation is directly proportional to switching capacitance and frequency, there is a quadratic dependence on supply voltage. An additional source of dynamic power dissipation results from finite input signal edge rates. As an input signal transitions, there is a region of input voltages where the PMOS and NMOS devices are conducting simultaneously, thereby creating a low-resistance path from VDD to VSS. Careful control of the input and output edge rates of CMOS circuits can keep this crossover current component to less than 20% of the total dynamic power dissipation. Dynamic power also results from the glitching hazards inherent in cascaded complementary logic chains. Finite propagation delays in the logic stages can cause spurious glitches at the outputs of the complementary gates. In some cases, this glitching activity can be a significant source of power dissipation. Dynamic power is also consumed when subthreshold leakage currents experience exponential increases due to power supply noise.

4. ALPHA STATISTICS AND POWER AND CURRENT TRENDS

Table 1 lists some of the key features of the three generations of Alpha microprocessors.

	21064	21164	21264
Introduction	1992	1995	1998
SpecInt / FP95	3.6 / 4.0	7.5 / 12.0	40 / 60 (est.)
Transistor Count (millions)	1.68	9.3	15.2
Die Size (cm^2)	2.33	2.99	3.14
Process Technology (μm)	0.75	0.50	0.35
Power Supply (Volts)	3.3	3.3	2.2
Avg. Power Dissipation (Watts)	30	50	72
Avg. Supply Current (Amps)	9.1	15.2	32.7
Target Design Frequency (MHz)	200	300	600
Typical Gate Delays / Cycle	16	14	12
On-chip Cache	8 KB L1-I 8 KB L1-D	8 KB L1-I 8 KB L1-D 96 KB L2-U	64 KB L1-I 64 KB L1-D
Instruction Issue / Cycle	2	4	6
Execution Flow	In-Order	In-Order	Out-of-Order

Table 1: Alpha Statistics and Trends

Process scaling has enabled a significant growth in the number of devices on a die. For example, a reduction in feature size from .50μm to .35μm results in an increase of approximately 60% more devices on a similarly sized die. Also note the nearly linear increase in average power dissipation of the microprocessors over time, despite the use of advanced processing technologies and scaled power supply voltages. Power supply currents have grown at an even faster rate. The table shows that supply current is nearly doubling with each new Alpha microprocessor generation. This trend of increased supply currents can be explained by a number of factors.

Each succeeding generation of microprocessors has been designed with an increasing number of execution units. The 21264, for example, integrates 6 execution units that can operate simultaneously. In addition, architectural features such as out-of-order execution, speculative execution algorithms, improved branch prediction, and compiler enhancements insure that these units are more effectively utilized than in previous generations. Therefore, relative to previous generation Alpha microprocessors, the 21264 has a higher switching activity rate in addition to the increased number of devices.

Another cause for the increase in currents is that each succeeding Alpha microprocessor generation has relied on advancements in circuit design techniques to increase the operating frequency by more than would be predicted by scaling alone. This effect can be seen by the reduction in typical gate delays per cycle from 16 for the 21064 to 12 for the 21264. This reduction in the number of gate delays per cycle was accomplished without reducing the effective work per cycle. The extensive use of large dynamic arrays, dual-rail dynamic logic, static ratioed logic, and other advanced circuit techniques has significantly contributed to this increased performance at the expense of increased power consumption. In addition, the 21264 has a larger software dependent variation of supply currents than in previous generations. These large increases in power and currents, combined with the increased variation of the magnitude of the currents, require careful design of the clock, power, and ground networks.

5. ALPHA 21264 POWER RESULTS AND ANALYSIS

A plot of average power versus operating frequency for the 21264 at a junction temperature of 100°C when running typical programs is shown in Figure 1.

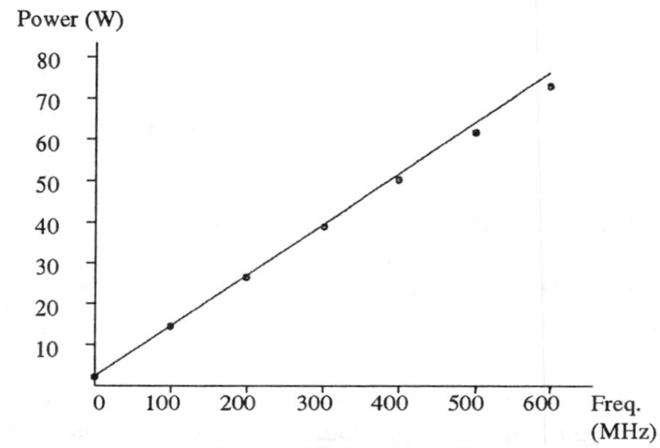

Figure 1: Alpha 21264 Power vs. Frequency

Note that power dissipated by the 21264 is approximately 2 Watts at static operating conditions. DC power dissipation of the 21264 is nearly fully attributed to subthreshold conduction. A limited number of circuits on the 21264, such as portions of the analog Phase Locked Loop (PLL) and pad input receivers, dissipate power continuously, independent of clock frequency. With the aid of CAD tools, the 21264 designers insured that no unwanted conduction paths exist between power and ground at DC operating conditions. In order to maintain functionality and limit power at DC frequencies, keeper devices were added to all dynamic nodes. These keeper devices supply the current necessary to keep the voltage of the dynamic nodes from varying significantly from the power supply rails. Power supply current at low frequencies would be expected to be much higher without these keeper devices because the degraded dynamic nodes driving other structures would otherwise create large crossover currents.

As expected, the graph shows power increasing nearly linearly with increasing operating frequency, with the power reaching approximately 50 Watts at 400 MHz. Power then starts to rise slightly less rapidly with increasing frequency until it reaches 72 Watts at 600 MHz. This slight fall-off in the rate of increasing power may be explained by the extensive use of low-swing differential busses in the design. As the chip frequency approaches its functionality limit, these busses experience reduced voltage swings.

Table 2 estimates the components of total power in the 21264 design at the maximum operating frequency.

Component	Power
Global Clock Network	32%
Instruction Issue Units	18%
Caches	15%
Floating Execution Units	10%
Integer Execution Units	10%
Memory Management Unit	8%
I/O	5%
Miscellaneous Logic	2%

Table 2: Alpha 21264 Power Components

Note that the clocking network accounts for about one third of the chip's power dissipation. For previous generation Alpha microprocessors, nearly 40% of the chips power was attributed to the clocks. As will be described later, the reduction of clock power was the primary power savings focus for the 21264 designers. The table also shows that the out-of-order instruction issue logic dissipates nearly 20% of the total power, or about as much as the floating and integer execution units combined.

It is difficult to precisely determine the maximum power dissipation of a large microprocessor because the effective operating frequency of each circuit is dependent on its switching activity. A running program that maximally utilizes all of the circuits in a microprocessor will result in substantially higher power dissipation than a program that leaves much of the processor idle. In order to maximize switching activity within the 21264 to create a worst case for power testing, a special executable was created. This executable attempts to maximize switching activity of all internal components of the microprocessor by fully and continuously utilizing the instruction issue logic, all of the execution units, and major busses. It should be noted that this program performs no useful task and it generates no output. Because the development of this program required an intimate understanding of details of the internals of the microprocessor, the level of switching activity accomplished with this program is far higher than would be generated while running normal applications. The maximum power dissipated in the current 0.35μ process under these absolute worst case conditions was found to be about 95 Watts, which approaches the limit at which conventional air-cooling techniques can be used. The 21264 when running normal programs dissipates a far lower average power, as previously described.

In the next generation process, a VDD reduction significantly lowers chip power.

6. DESIGN TECHNIQUES TO REDUCE POWER ON THE ALPHA 21264

The 21264 designers used a number of techniques to help limit power dissipation. Lowering VDD has the largest impact on reducing power, but clock frequency is negatively impacted. As previously noted, the reduction of clock power is important to the management of overall chip power. The use of a hierarchical clocking scheme with conditional clocks was the primary design choice used to lower clock and logic power. Full-custom design methodologies also help enable power reduction. When circuits are designed at the transistor level, it is possible to use more innovative power reduction techniques

Previous Alpha designs utilized a single large clock grid that generated a timing reference for nearly all latches in the design. The Alpha 21164 did include some conditional clocking, however its use was limited to the large on-chip secondary cache. While straightforward and robust, a single wire clocking scheme causes many nodes to switch every cycle even though no useful work is being performed. The main benefit of conditional clocking is the lower average power that results from turning off sections of the chip that are not needed on a cycle-by-cycle basis. Moreover, peak power is also lowered because of logical exclusivities inherent in the design. Conditional clocking significantly complicates timing analysis of the design [4], however its use was justified because of the average and peak power savings.

The 21264 Floating Point Unit (Fbox) provides an example of the use of conditional clocking in datapaths. The basic clocking scheme for each operational unit (multiply, add, etc.) is indicated conceptually in Figure 2.

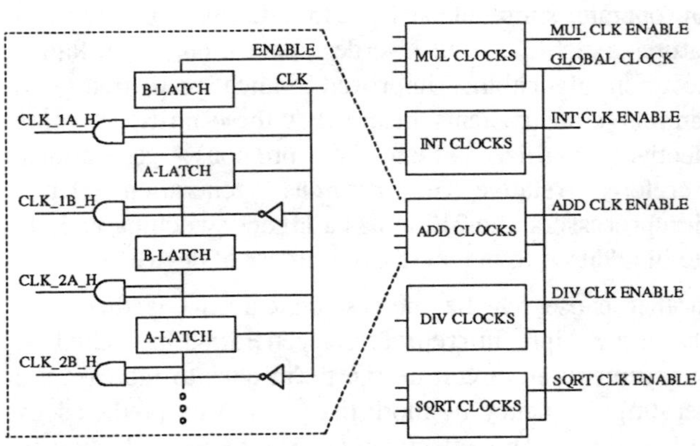

Figure 2: Floating Point Datapath Clocking

If, for example, the chip were to execute a single floating point ADD, the control logic would assert the adder clock-enable line for one cycle. The pulse would then propagate through a chain of alternating active-low (B-latch) and active-high (A-latch) level-sensitive latches. Each latch output is used to enable the latch bank clock driver, indicated conceptually by the AND buffer, for the corresponding datapath pipeline stage. The data flow down the datapath with the clock enable signal preceding them so the latch banks at successive stages are clocked as needed. As pipeline stages of the datapath are no longer required, the de-assertion edge of the enable input signal propagates down the latch chain, successively turning off the latch bank clocks and saving power that would otherwise be wasted in performing unneeded work. If the Fbox were performing only a floating point ADD, clocks for the other operational units would remain disabled. The Fbox control logic is unconditionally clocked so it can respond to incoming floating point instructions with minimal delay.

Due to difficulties in managing circuit race issues, the entire divider datapath is operated on one conditional clock, rather than by pipeline stage as in the other operational units. The net power saved by these techniques depends on the floating point instructions being executed. For the case where no floating point instructions are being executed, Fbox power dissipation is estimated to be about 25% of what it would have been without conditional clocking. As floating point instructions are added to the instruction stream, Fbox power dissipation increases.

Low-swing busses were used extensively on the 21264 in order to help meet performance goals. An additional benefit of these busses is their reduced power consumption. Low-swing busses save power by reducing the voltage excursions on large capacitances to hundreds of millivolts instead of the full power supply range. The Execution Unit (Ebox), as shown in Figure 3, provides an example of the usage of low-swing busses in the 21264 design.

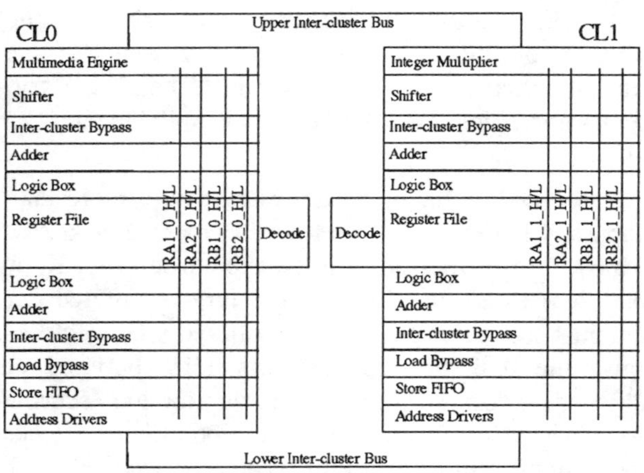

Figure 3: Execution Unit Low-Swing Busses

The Ebox is divided into two clusters, CL0 and CL1. Each cluster contains an 80-entry register file plus two independent execution pipelines containing arithmetic, shift, and multimedia functional units. Functional unit results, register file contents, load data, and inter-cluster results drive 64-bit limited-swing, differential operand busses (depicted as RA1_0, RA2_0, RB1_0, RB2_0, RA1_1, RA2_1, RB1_1, RB2_1 in the diagram).

Had these busses been implemented with standard full-swing techniques, it is estimated that their worst case dynamic power dissipation at the 600 MHz target operating frequency would have been about 1.5 Watts. The resulting power estimate with the use of limited swing busses under these same conditions is less than 15mW, or a reduction of two orders of magnitude. The bus receivers were designed to operate at 200mV of differential swing, thus requiring careful circuit and layout design to minimize all differential noise effects.

Performance requirements dictated the extensive use of high performance dynamic structures. However, dynamic structures have several properties that can help reduce power. Dynamic circuits have reduced fanin and fanout capacitances because there is no need for the complementary PMOS network. In addition, dynamic circuits have smaller layout areas relative to equivalent static complementary structures, resulting in lower wiring capacitances. Dynamic circuits also generate monotonic signal transitions, so power is not dissipated by glitching hazards. On the 21264 design, many dynamic nodes are conditionally precharged only a cycle or phase before the circuit is to evaluate. This optimization lowers the effective switching capacitance of the clock network by keeping the clock node from needlessly driving the gates of the precharge devices. In addition, extensive device-level simulations were performed to insure that the overlap between dynamic node precharge and evaluation is minimized, thereby reducing crossover currents.

Static complementary logic was used in many large control sections of the chip. Spurious input transitions to large logic sections were limited in order to keep unused signal transitions from propagating through the networks. Additionally, significant power savings was also achieved by reducing the I/O voltage from 3.3 Volts to 2.2 Volts.

7. IMPLEMENTATION TECHNIQUES FOR HIGH PERFORMANCE DESIGN

7.1 Power and Ground Distribution

As previously described, supply current is roughly doubling with each new generation. This dramatic increase in current has required the implementation of increasingly complex on-chip power distribution schemes. In addition, the scaling of power supply voltages to limit power dissipation has also reduced the amount of absolute supply noise that can be tolerated. A high performance power distribution scheme must allow for all circuits on the die to receive the same voltage reference. A variation of supply reference across the die can lead to problems such as reduced noise margins, reduced device switching speed, increased subthreshold conduction, or possibly even latchup. An additional factor which stresses the 21264 power distribution scheme is the extensive use of conditional clocks, which exaggerates cycle-to-cycle current variations. The 21264 power supply network was designed to handle a maximum variation in power supply current of up to 25 amps between adjacent cycles.

With these severe constraints, it was demonstrated that a standard distribution scheme of a two dimensional grid of course pitch interconnect for routing power and ground would be insufficient. The grid would not adequately prevent a power supply drop in the chip's center due to the chip's large average current. In addition, the program dependent variable peak currents further exacerbate the problem. For this reason two thick, low resistance aluminum reference planes were added to the .35µm 21264 process. Reference plane 1 was added between metal 2 and metal 3 and was connected to VSS. Reference plane 2, which was connected to VDD, was added above metal 4. The use of power planes had several beneficial effects. First, a stable power reference is available to circuits situated anywhere on the die. Secondly, the lower reference plane inductively and capacitively isolates the metal 2 and metal 3 signal lines, effectively reducing crosstalk.

The additional power plane layers significantly reduce the DC IR voltage drop to the center of the die, but the high clock frequencies and conditional clock buffers cause large cycle-to-cycle fluctuations in supply current. These variations of current flowing though the package and bond wires can cause significant power supply ringing on the chip. Therefore, substantial amounts of on-chip decoupling capacitance are used to smooth out the supply current fluctuations though the bond wires. These decoupling capacitors, which are constructed using the gate oxide of an NMOS transistor, are used extensively around large clock drivers, under large signal busses, and in other unoccupied areas of the die. A total of 0.32µF of decoupling capacitance was added the 21264 die, and it occupied roughly 15-20% of the die area. Extensive simulations showed that the on-chip decoupling capacitance alone was not sufficient to control the power and ground fluctuations. Therefore an additional 1µF, $2cm^2$ Wire-bond Attached Chip Capacitor, or WACC, was added to the chip and package network. It should be noted that advanced packaging techniques such as flip chip with solder bumps can be used to provide a more robust power distribution scheme than can be achieved with stitch wire bonding techniques.

7.2 Clocking Network

The high frequencies of the Alpha microprocessors have required the generation and distribution of very high quality clock signals. As more aggressive circuit techniques and complex micro-architectural features were implemented on the 21264, power consumption became a major concern in designing the clocking system. The 21264 uses a single wire global clock, called GCLK, as the chip's main timing reference [3,5]. Performance goals required that the overall GCLK skew be limited to less than 75ps, with steep, uniform edge rates, and a duty cycle approaching 50%. These goals were accomplished with a modified H-tree GCLK distribution scheme shown in Figure 4.

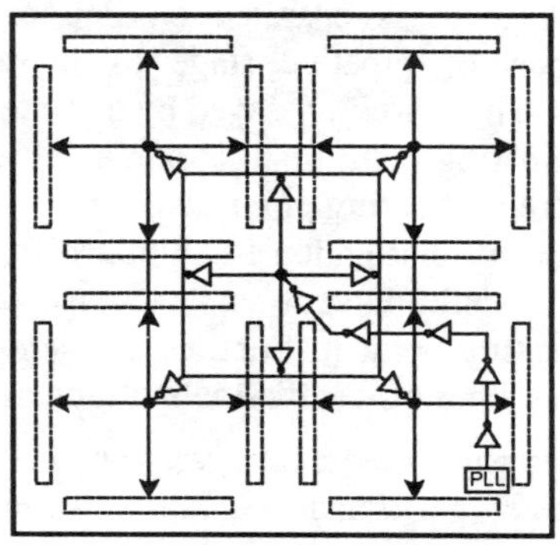

Figure 4: Alpha 21264 GCLK Network

The PLL output is repeated, driven to the approximate die center, and then routed vertically and horizontally to buffers midway between the die center and die edges. From these locations, the clock is driven to the centers of each quadrant of the chip where a second set of buffers is located. Here, the clock is inverted and driven to the peripheries of the quadrants where sixteen 2-stage distributed buffers drive the GCLK grid. The GCLK grid consumes approximately 3% of the total available metals 3 and 4.

GCLK was used as the basis of a hierarchy of more than ten thousand buffered and conditioned clocks used across the chip. Most clock load is isolated from GCLK by double buffering the load onto large buffered or conditioned section clocks. These buffered section clocks are also the basis for additional, smaller buffered or conditioned clocks. This clock distribution scheme allowed designers freedom to use conditioned clocks wherever significant power savings could be accomplished. In addition, the improved locality of the clock drivers with respect to their loads allowed for significant reductions in skews and metal usage. This clock generation and distribution scheme results in significantly lower metal 3 and 4 clock usage, saving an estimated 10 Watts over previously used gridded clocking schemes. Note that the choice of skew targets greatly affects overall clock network loading. A less aggressive skew target can be achieved with a sparser grid and smaller drivers.

An additional benefit of this distributed clock generation scheme is the reduced thermal gradient over the centralized clocking approach. The 21264 achieves a thermal resistance from the junction to the heatsink of 0.3°C/W, whereas the previous Alpha design using the same packaging technology achieved a θ of .4°C/W. The distributed clock drivers also relieve stress on the power and ground distribution schemes by spreading out the large drivers across the die. As previously described, the major disadvantages of this hierarchical clocking scheme is the potential for large current gradients on a cycle by cycle basis as the conditional clocks are successively enabled and disabled, and the increased complexity of the timing analysis.

7.3 Electromigration Reliability

The increase in supply currents in each succeeding generation of microprocessor design has necessitated more sophisticated electromigration analysis, especially in the power, ground, and clock networks. Traditionally, current densities through all interconnect segments of a signal were required to be below defined threshold limits that depended on the interconnect type. These thresholds were based on current densities that had been historically determined to be safe. If all interconnect segments of a design met this electromigration specification, the entire design was assumed to be safe against electromigration risk. Increased currents have caused this approach to be increasingly difficult to implement. Additionally, this approach does not take into account the increased risk of premature electromigration failure if an excessive number of interconnect segments of a design are at or near the maximum current limit.

The 21264 microprocessor design team used a technique called Statistical Electromigration Budgeting (SEB) [6]. SEB takes into account the statistical nature of electromigration failures. With SEB, an electromigration failure probability, which depends on the current density through the interconnect segment, is calculated for each interconnect segment. The probability of an electromigration failure of the entire design is then a function of the failure probabilities of all the individual interconnect segments. SEB allows a limited number of interconnect segments to have larger currents. Although any individual interconnect segment with a larger current density has a higher probability of failure, a small number of these segments will have little affect on the failure probability of the whole design. SEB does impose an absolute maximum current density limitation on each interconnect segment, which if exceeded, may cause a catastrophic failure due to Joule heating.

8. CONCLUSIONS

As demonstrated in the three generations of high performance Alpha microprocessors, increased clock frequencies and dramatically higher device counts have caused rapid increases in power and current consumption. In order for microprocessor performance to keep increasing at historical rates, careful consideration must be given to power consumption, power and ground routing, and the clocking networks. One of the main challenges in future microprocessor design will center around devising new circuit techniques that help lower power without impacting overall circuit performance.

9. ACKNOWLEDGMENTS

This paper represents the work of the many talented Alpha microprocessor design engineers at Digital's Alpha design center. The authors would like to thank William Bowhill, Randy Allmon, and Dan Bailey for their constructive feedback.

10. REFERENCES

[1] Dobberpuhl, D., et al., "A 200 MHz 64b Dual-Issue CMOS Microprocessor," IEEE Journal of Solid State Circuits, vol. 27, no. 11, Nov., 1992.

[2] Benschneider, B., et al., "A 300-MHz 64b Quad-Issue CMOS RISC Microprocessor," IEEE Journal of Solid State Circuits, vol. 30, no. 11, Nov., 1995.

[3] Gieseke, B., et al., "A 600 MHz Superscaler RISC Microprocessor With Out-of-Order Execution," ISSCC Digest of Technical Papers, pp. 222-223, Feb., 1996.

[4] Grundmann, W., et al., "Designing High Performance CMOS Microprocessors Using Full Custom Techniques," Proceedings of the 34th Design Automation Conference, pp. 722-727, June, 1997.

[5] Fair, H. and Bailey, D., "Clocking Design and Analysis for a 600 MHz Alpha Microprocessor," ISSCC Digest of Technical Papers, pp. 398-399, Feb., 1998.

[6] Kitchin, J., "Statistical Electromigration Risk Budgeting for Reliable Design and Verification in a 300MHz Microprocessor," Digest of Technical Papers, VLSI Circuits Symposium, 1995.

Reducing Power in High-performance Microprocessors

Vivek Tiwari, Deo Singh, Suresh Rajgopal, Gaurav Mehta, Rakesh Patel, Franklin Baez
Intel Corporation
2200 Mission College Blvd.
Santa Clara, CA 95052
408-765-6395
Email contact: vtiwari@scdt.intel.com

ABSTRACT

Power consumption has become one of the biggest challenges in high-performance microprocessor design. The rapid increase in the complexity and speed of each new CPU generation is outstripping the benefits of voltage reduction and feature size scaling. Designers are thus continuously challenged to come up with innovative ways to reduce power, while trying to meet all the other constraints imposed on the design. This paper presents an overview of the issues related to power consumption in the context of Intel CPUs. The main trends that are driving the increased focus on design for low power are described. System and benchmarking issues, and sources of power consumption in a high-performance CPU are briefly described. Techniques that have been tried on real designs in the past are described. The role of CAD tools and their limitations in this domain will also be discussed. In addition, areas that need increased research focus in the future are also pointed out.

1. INTRODUCTION

The drive towards increasing levels of performance has pushed frequencies higher and has increased the complexity of microprocessors. This has come at the cost of higher power consumption. The costs associated with packaging, cooling and power delivery have thus jumped to the forefront in the microprocessor industry. There is even concern that power consumption may set the limit to how much can be integrated on a chip, and how fast it can be clocked [5].

The challenges for power reduction in high-performance general-purpose CPUs are unique. First, the instruction-set and system architecture are designed for a wide market and for a wide range of applications. This restricts the search space for low-power solutions. Second, it is necessary that proposed solutions remain robust and scale gracefully across multiple technology generations. And finally, while significant power savings are desired, they must come at *little or no* performance impact.

DAC 98, San Francisco, California

The aim of this paper is to highlight the key issues associated with design for low power in this domain. The paper is organized as follows. Section 2 motivates the need for low power and shows the historical trends with respect to power consumption. It also explains some metrics associated with CPU power that are of interest from the system's perspective. In Section 3, the power consumption of a recent CPU is broken down by circuit type. This power breakdown provides a quantitative basis for judging the effectiveness of different power reduction techniques. Section 4 presents an overview of the techniques that have been used in the past. Among these, voltage scaling has been the primary weapon in the battle against power consumption. But it is shown that the limitations of this approach get amplified as technology pushes towards smaller geometries. Thus, it is also important to incorporate power as a design metric at all levels of the design process, and examples from the domains of circuit, logic, CAD tools and software are presented. Section 5 concludes with a discussion of directions that need increased research focus in the future in order to deal with the power challenge.

2. THE CPU POWER PROBLEM

Figure 1 shows the power consumption for Intel CPUs. The X-axis shows the technology generation and the Y-axis the maximum power consumption. As indicated by the dashed line in the main part of the curve, power consumption has been increasing for each new CPU generation. The points to the side of the main curve indicate newer versions of each processor family. These are implemented in newer semiconductor processes with smaller geometries than the lead processor in that family.

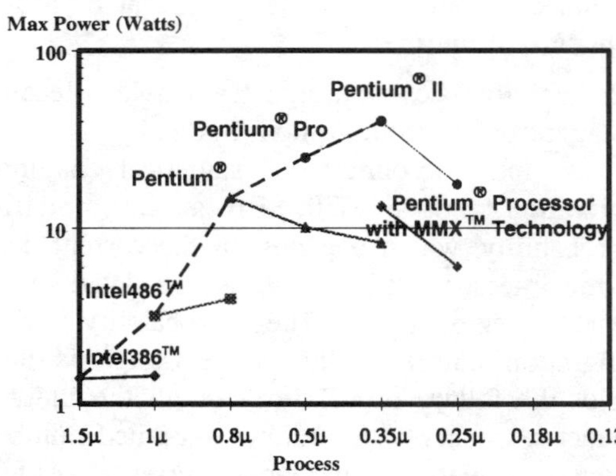

Figure 1: Power consumption for Intel CPUs.

Smaller feature sizes in conjunction with lower supply voltages lead to lower power consumption in the newer versions. However, moving to a new CPU generation in the same process is associated with an increase in the power consumption.

Why is this cause for concern? The reason is that increased power consumption directly impacts CPU and system cost. This cost has two components. The first is *thermal cost,* which is associated with keeping the devices below the specified operating temperature limits. Maintaining the integrity of packaging at higher temperatures also requires expensive solutions. The second component of the cost of power consumption is the cost of power delivery, i.e., the on-chip, on-package, and on-board decoupling capacitances and interconnect associated with the power distribution network. Increased power consumption at lower voltages increases the magnitude of the current drawn by the CPU. In addition, today's design trends are such that the variability in the amount of current drawn from cycle to cycle is also increasing. These factors combine to make resistive and inductive noise a big problem. Dealing with this is becoming increasingly costly.

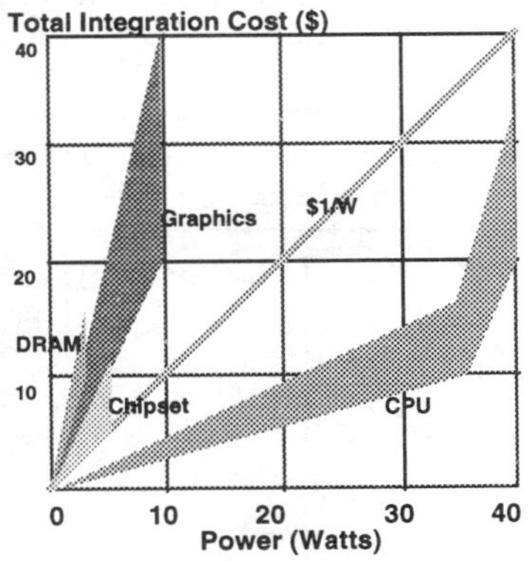

Figure 2: Thermal and power delivery cost in a PC

Figure 2 gives an idea of the range of dollar amounts associated with the above costs for different system components. As can be seen, when the CPU power is in the 35-40W range, the cost of each additional Watt goes above $1/W per chip. An interesting observation is that the power cost of the other system components (DRAM, chipsets, graphics) is on a steeper curve than the CPU. This is because the spatial layout of today's system chassis designs is such that these components are harder to cool. This is likely to change with new designs, further increasing the relative importance of the CPU power cost.

CPU activity is a new design attribute that is needed for power estimation as well as for the choice of power savings techniques. What kind of tests must be used to characterize this activity? They have to come from traces of real applications. But power dissipation inside a CPU is a complex scenario. Figure 3 shows the need for defining multiple power specifications in this regard, using illustrative power traces. The first trace, labeled *Max Power*, is associated with the power consumption while running an artificial piece of code specifically written to generate maximum CPU activity. Modern CPUs with increasingly sophisticated implementations have multiple parallel units that are not generally fully active simultaneously. Hence there is an increasing gap between artificial worst-case power and the peak demands required by *practical* applications, as shown by the second trace, labeled *Worst-Case App Trace*.

There are other power components that are of interest. The *Thermal Power* is the running average of the worst-case application power over several seconds, consistent with the component/system thermal time constant. It is used to determine cooling options and system integration and thermal design costs. The *Standby Power* (not shown here) is power consumption under a predefined idle condition. For battery life, the criterion is *Energy,* which is the product of the *Average Power* and the running time for a given task. In the most general case, *Average Power* is often taken as the long-term average power of typical applications, measured over intervals of hours. In addition, multiple on-chip units and power down techniques have increased the variability in power consumption. This is quantified by the *Transient Power,* whose time constant is one or few CPU clock cycles. This is an important parameter in designing the power delivery system.

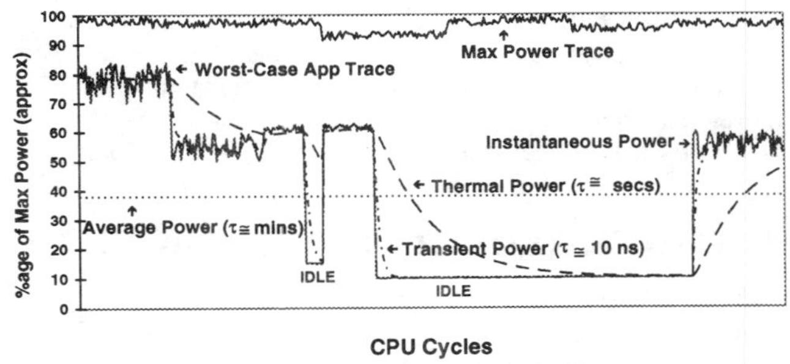

Figure 3: CPU power terminology

For each of the power components discussed above, a different trace may be needed in order to capture the desired nature of the CPU activity. Generating these traces for today's complex CPUs is itself a challenge. Another challenge is to translate these traces into test vectors for RTL and schematic power estimation. Since real traces are usually millions of clock cycles long, selection and compression are required. The selection/compression must ensure that the characteristics of an application trace that track a given power component (*viz.* worst-case, average, etc.) are maintained.

3. WHERE DOES THE POWER GO

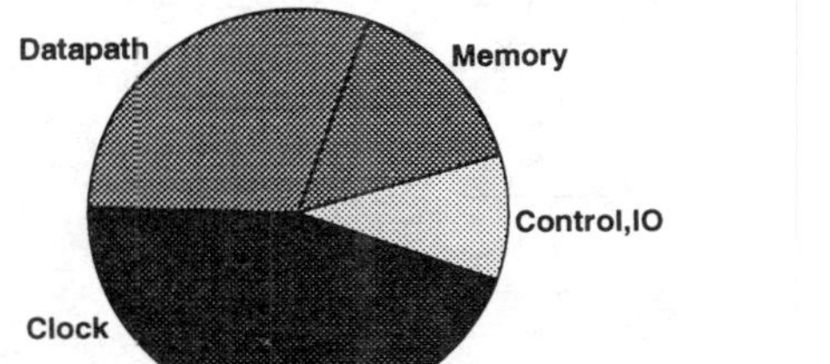

Figure 4: Power breakdown in a high-performance CPU

Figure 4 shows the power breakdown for a recent high-performance CPU, as obtained by detailed switch-level power simulation. As can be seen, the clock is the largest power consuming component. This includes the clock generator, the clock drivers, the clock distribution tree, the latches, and the clock loading due to all the clocked elements. The clock loading is actually the largest component of clock power. As shown in Figure 5, even a simple latch presents a certain amount of capacitive load to the clock network (gate capacitance of 4 clocked devices in this example). This capacitance switches on every clock tick, causing significant power consumption even when the data inputs have low activity factor (AF), or are even totally stable (AF = 0).

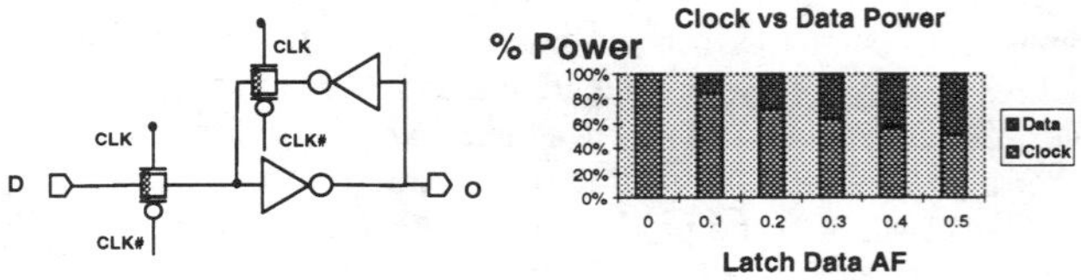

Figure 5: Power consumption in latches

After the clock, the datapath is the highest power consumer. This includes the large execution units, buses and the register files on the die. The datapath circuits in high-performance microprocessors typically have tight timing margins and are thus custom-designed. Fast and power-hungry dynamic logic is often used here. The power contribution of on-chip memories (caches, TLBs, branch buffers) is determined by the size as well as the circuit and physical structure of the memory arrays. The controller and I/O is usually a small contributor to the full-chip power. The logic blocks that constitute the control functionality are mostly non-critical from a timing standpoint. The logic and physical layout for these is thus mostly synthesized.

4. POWER REDUCTION TECHNIQUES

4.1 Voltage Scaling

Power is proportional to the square of the supply voltage (Vcc). This makes Vcc reduction as the most effective way for reducing power, and the industry has thus steadily moved to lower Vcc. This trend should and will continue. However, the drive for higher performance is outstripping the benefits of voltage scaling, as illustrated in Figure 6. The figure shows the power, feature size, voltage, frequency, and relative die size of some recent CPUs. Starting from a 5V part, there was an initial decrease in power when moving to a smaller technology at 3.3V. However, the power came back to original levels when the frequency was increased. The Pentium® Pro with its aggressive microarchitecture, saw an additional increase in power, even at 3.3V and a smaller technology.

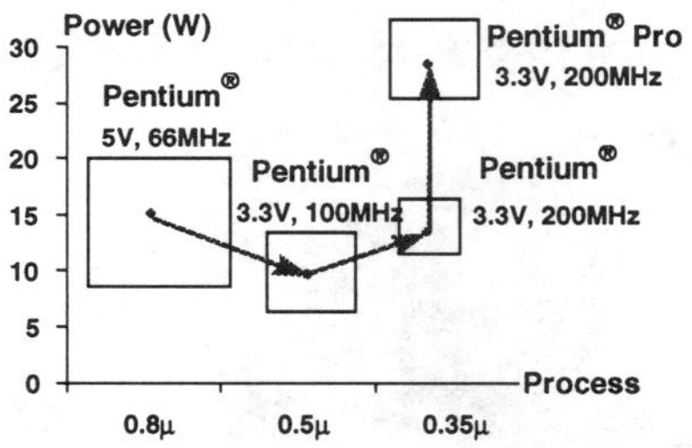

Figure 6: Power and Vcc for some recent CPUs

Another issue with voltage scaling is that to maintain performance, threshold voltage (Vt) also needs to be scaled. At low Vt, leakage power starts becoming a bigger factor [3]. A first order analysis, using constant electric field scaling of the process parameters, illustrates this as shown in Figure 7 [2]. At sub-0.1μm feature sizes, the leakage power starts eating into the benefits for lower Vcc. In addition, design of dynamic circuits, caches, sense-amps, PLAs, etc. becomes difficult at higher leakage currents. Lower Vcc also exacerbates noise and reliability concerns. Therefore, voltage scaling alone cannot be relied upon to solve the power problem. Design for low power is also necessary.

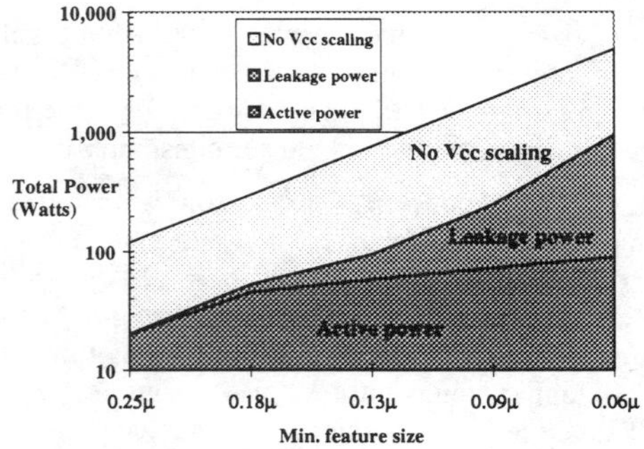

Figure 7: First order analysis for leakage power trends

4.2 Clock Gating

As discussed in Section 3, the clock is the largest contributor to the CPU power. Reducing the switched capacitance on the clocks will thus have the most impact on total power. A practical and effective way to do this is to partition the clock network and allow only those portions to toggle that are needed on each cycle. This is achieved through *clock gating* . It is implemented by qualifying the different clock partitions by special "enable" signals. It is well suited for CPUs since it can often be easily integrated into existing clock networks as illustrated in Figure 8. A regular clock buffer can be changed into a qualifying gate at low area and performance overhead.

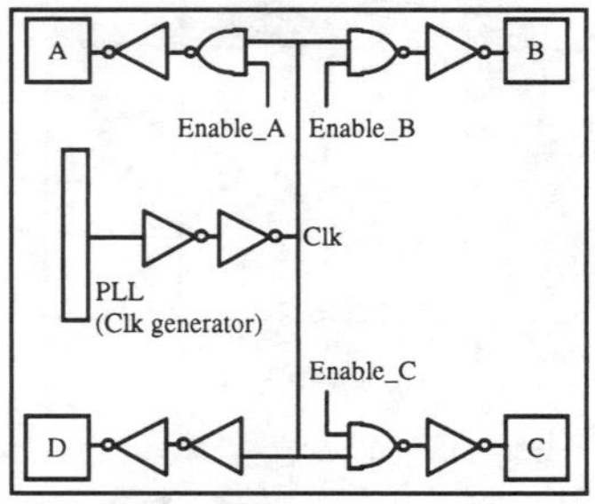

Figure 8: Clock gating and clock networks

However, clock gating cannot be used indiscriminately since there are some issues that need to be considered. An important concern is that the disabled block may not power up in time, or that modified clocks may generate glitches. This imposes strict timing constraints on the enable signals and calls for careful timing verification. Functional validation is becoming a greater challenge with each new CPU generation, and gated clocks make the problem even more difficult. In addition to this, at clock frequencies of 100MHz and above, clock skew becomes critical and every extra gate used to qualify the clock can potentially introduce timing critical skews. Thus, the granularity at which clock gating can be applied becomes a tradeoff against overall clock network design time and complexity.

Another concern with clock gating is the impact on current variations when large blocks of logic are switched on and off. A CPU may be at peak current levels for some cycles, when few blocks can be clock gated. But it may rapidly transition to low values of current if something like a stall of pipeline flush causes a number of units to be powered off. This increases the variation in transient power (*cf.* Figure 3). Switching between a normal operation mode and a standby mode, in which most of the internal clocks are turned off, also causes the same problem. This is illustrated in Figure 9 for the case of the Pentium® Processor [7]. Notice how the differences between peak and idle current are larger when power saving techniques are enabled. Thus, the power delivery system may have to be designed for higher inductive noise specifications, increasing the CPU and system cost. This power supply noise problem will get worse with higher frequencies and lower voltages.

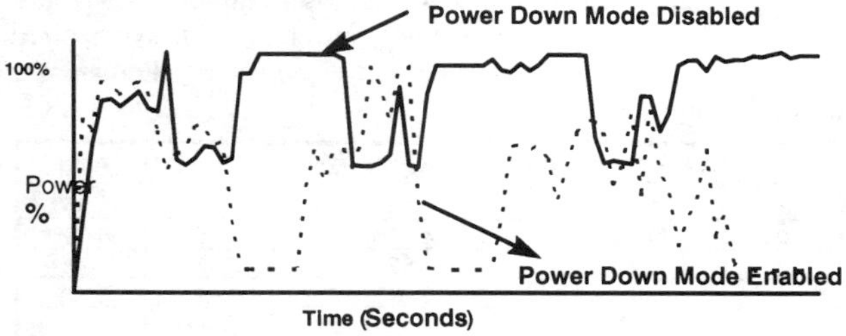

Figure 9: Power vs. time running a real application

A wide variety of techniques exist for reducing power at the logic and circuit level in a high-performance CPU. The most important of these target areas which have a large impact on the overall power of the chip (*cf* Figure 4). Redesign of latches and master slave flip flops, for example, has a large impact on full chip power. Other techniques like transistor sizing, low power logic synthesis, and specific circuit techniques also produce power savings. These are discussed below.

4.3 Libraries

Power savings from the redesign of cell libraries can come from two sources: device sizing and restructuring of the logic and physical layout of cell. Device sizing for optimizing switching energy vs. delay, ensures that the libraries are designed with power "in mind". Figure 10 shows a power vs. delay curve for two different standard cell libraries. The transistor sizes in Library 1 have been optimized for minimum delay as a target. If one were to change the optimization criteria of transistor sizes to that of energy *and* delay, one would get a lower power library (Library 2). For Library 2, the delay penalty is small but the power reduction is greater, since the original library was in the steep non-optimal part of the curve. It has been seen in practice, that this extra delay can often be absorbed during circuit design, especially for non-critical paths. One can, therefore, get the same performance with Library2 and still have overall lower power.

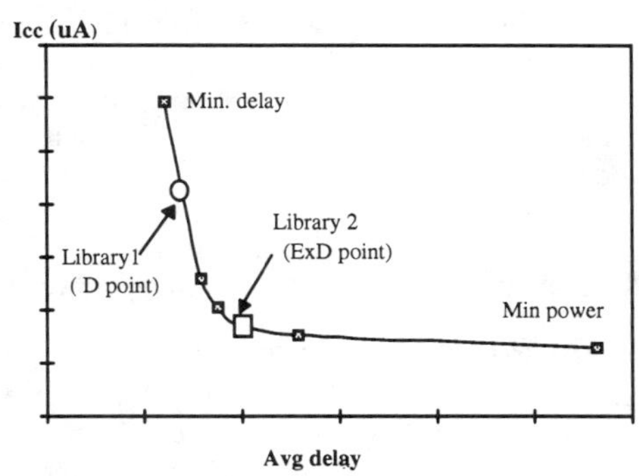

Figure 10: Power-delay curve for differently sized libraries

Figure 11 below shows some of the nodes in a latch cell that may change due to the different optimization criteria.

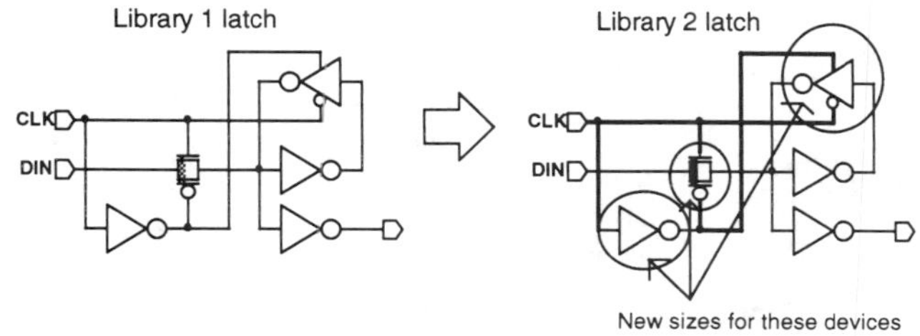

Figure 11: Illustration of energy vs. delay size optimization

The second way to optimize a cell library is to change the schematics of the most commonly used and most power hungry cells in the design. These typically consist of latches and master slave flip flops since these have clock nodes which switch on every clock edge. Figure 12 shows an example of latch redesign that gets rid of clock nodes while still maintaining functionality and performance. Table 1 shows results for some circuits where some sequential cells were replaced with more efficient ones. As can be seen, a small amount of redesign effort on some selected cells can have a significant power impact.

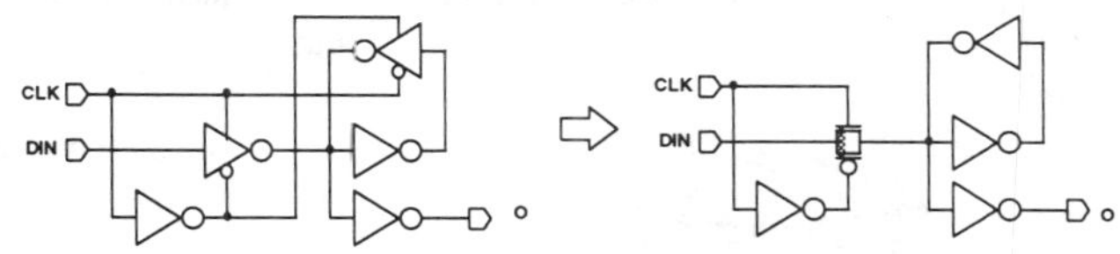

Figure12: Latch redesign example

	Ckt1	Ckt2	Ckt3	Ckt4
Num seq. cells replaced	148	13	18	15
Power change	-13.30%	-9.50%	-12.67%	-3.56%
Area change	0%	8%	1.30%	-1.50%

Table 1: Power Savings from sequential cell redesign

4.4 Power-Delay Curves and Automated Transistor Sizing

As discussed in Section 3, a large part of high-performance CPUs is typically custom designed. These designs typically involve manual tweaking of transistors to upsize drivers in critical paths. If too many transistors are upsized unnecessarily, certain designs can lie on the steep part of a circuit's power-delay curve. In addition, the choice of logic family used, e.g. static vs. dynamic logic, can also greatly influence the circuit's power consumption. Figure 13 illustrates these scenarios for a 32-bit adder. Suppose a designer has the data shown in the figure at his/her disposal. Knowledge of where in the power-delay curve the circuit lies can tell the designer whether he/she can trade a little performance for a larger power savings. In this example, 69% total savings can be gained by transistor sizing and going from domino to static. There is a 23% delay penalty. This extra delay penalty may be overcome by upsizing adjacent blocks at a much less power penalty and still end up with an overall power benefit. Experiments with this methodology have yielded 10% power savings with no delay increase in several circuits. CAD tools that enable this kind of logic and circuit design exploration for custom circuits can thus have a significant impact at the full-chip level.

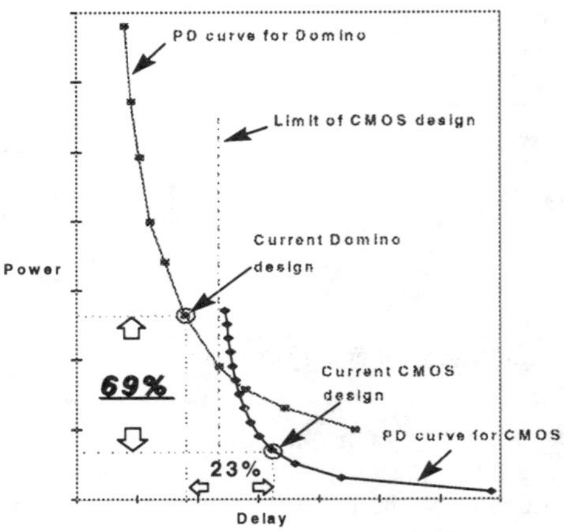

Figure 13: Power-Delay curves for two different 32 bit adders

Figure 14 shows an example of another kind of intelligent tradeoff for power/performance. It shows the results for a logic block whose 7 sub-blocks can either be implemented as PLAs or as synthesized random logic. Increasing the number of synthesized blocks leads to some increase in delay but for much larger power savings.

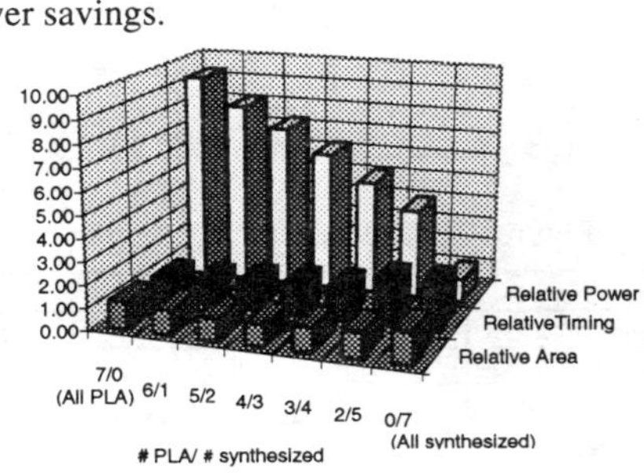

Figure 14: Power/Area/Delay tradeoffs for PLA vs. synth. logic

The traditional emphasis on performance often leads to over-design that is wasteful for power. An emphasis on power, however, motivates identification of such sources of power wastage. An example of this is paths that are designed faster than they ultimately need to be. For synthesized blocks, such paths can be automatically downsized by the synthesis tool. For manually designed blocks, on the other hand, downsizing may not always get done. Automated downsizing tools can thus have a big impact. Transistor width savings (with no delay increase) from the use of one such tool are shown in Table 2. The benefit of such tools is power savings, as well as productivity enhancement for manual designs.

	Ckt1	Ckt2	Ckt3	Ckt4
No of elements	4853	1953	18300	19756
Width Savings	40.00%	42.00%	17.00%	3.00%

Table 2: Transistor width savings with a sizing tool

4.5 Low Power Logic Synthesis

A lot of the CAD research in low power has been in the area of low power logic synthesis. Technology dependent phases of synthesis [8,10] are particularly suited for practical applications, since they have access to low-level circuit information. It is possible to save power with these methods, even when they are constrained to not increase the delay of the circuit. Table 3 shows the results of applying these techniques on a couple of sample circuits from a high-performance CPU. Column 2 shows the power savings when only the combinational part of the circuit is allowed to be changed. Column 3 shows the results when the sequential elements are also allowed to be changed. The area impact is low (Column 5) and there is no increase in delay. The results show ~10% power savings for synthesized blocks, but this translates into only 1% full-chip power savings. This is because only 10% of total power is from synthesized logic for CPUs such as the one whose power breakdown was shown in Figure 4.

Circuit	After LP Synth	After seq. dnsizing	Total	Area
1	1.14%	7.39%	8.52%	-1.85%
2	0.76%	4.89%	5.65%	-0.69%
3	8.40%	1.53%	9.92%	1.99%
4	6.13%	3.31%	9.44%	7.12%
5	11.03%	1.07%	12.09%	1.09%
Total	**5.16%**	**3.81%**	**8.97%**	**0.53%**

Table 3: Low power logic synthesis results

It is pertinent to note that the system power consumption problem also encompasses chipsets, i.e., devices such as the memory, I/O, and graphics controllers. These operate at a fraction of the CPU clock frequency, and large portions of these are well-suited to be implemented as ASICs. Low power synthesis thus has a much larger impact in this domain.

4.6 Specific Circuit Level Techniques

In addition to the techniques described above, specific optimizations on individual circuits can be done to reduce power. Besides the power, area, and delay impact, the use of such ideas depends on factors such as design complexity, schedule impact, noise and reliability issues, etc. A detailed discussion of this class of ideas is beyond the scope of this paper. However, an illustrative example for a domino multiplexer is shown in Figure 14 and is discussed below.

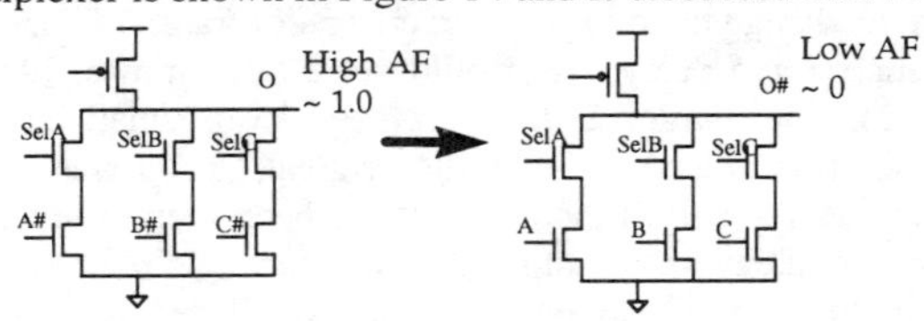

Figure 14: Domino AF reduction in muxes

In this circuit, at least one of *SelA*, *SelB*, *SelC* is high every cycle. The data inputs A#, B#, C# are mostly high (A,B,C have very low signal probability). In this situation, all the capacitance associated with the data nodes toggles almost every cycle. This total capacitance is almost 3 times that associated with the clocked nodes. Consequently, when the polarity of the data inputs is changed, as shown on the right, large power savings are seen. The use of a related idea on a large logic block in a recent CPU resulted in an estimated *3%* full-chip power savings.

4.7 System Power Management

The interaction of the CPU with the rest of the system also provides avenues for reducing average power. Often the CPU is waiting for inputs from peripherals and its power is being wasted. To reduce this waste, CPUs are now provided with a hierarchy of power states. Each state defines a certain level of activity on the CPUs and a certain time penalty for it to get back into a fully active state. Memory and I/O devices often also have similar power states. It is the *system power management* mechanism that monitors the system activity and enforces the movement of the system components between different power states [4].

System power management has its roots in mobile systems. However, EPA requirements under the Energy Star program motivated the migration of these techniques to desktop systems. A recent development in this area is a cross-company initiative called ACPI (Advanced Configuration & Power Interface) [1]. The recognition of the need to eliminate wasted power ensures that system power management will continue to be an area of high interest and active development.

4.8 Software Based Power Reduction

Traditionally the focus on low power design has been purely hardware based. This tends to ignore the fact that it is the software that executes on a CPU that determines its power consumption. The software perspective on power consumption has been the subject of recent work [9]. Here a detailed instruction-level power model of the Intel486DX2 was built. The impact of software on the CPU's power and energy consumption, and software optimizations to reduce these were studied. An important conclusion from this work was that in complex CPUs like the 486DX2, software energy and performance track each other, i.e., for a given task, a faster program implementation will also have lower energy. This is because the CPU power consumption is dominated by a large cost factor (clocks, caches, etc.) that for the most part, does not vary much from one cycle to the other.

There are some issues when this work is extended to recent CPUs. First, multiple-issue and out-of-order execution mechanisms make it hard to model power on a "per instruction" basis, and more complex power models are required. Also, increased use of clock gating implies that there is greater variation in power consumption from cycle to cycle. However, it is expected that the relationship between software energy and power that was observed before will continue to hold. In any case, it is important to realize that software directly impacts energy/power consumption, and thus it should be designed to be efficient with respect to these metrics.

A classic example of inefficient software is "busy wait loops". Consider an application such as a spreadsheet that requires frequent user input. During the times when the spreadsheet is recalculating values, high CPU activity is desired in order to complete the recalculation in a short time. In contrast, when the application is waiting for the user to type in values, the CPU should be inactive and in a low-power state. However, a busy wait loop will prevent this from happening, and will keep the CPU in a high-power state. The power wastage is significant. For example, a 166MHz Pentium® Processor with MMX™ technology draws over 7 Watts in normal operation but only 1 Watt when halted. The Intel Power Monitor (IPM) is a publicly available [6] software analysis tool that monitors system activity to provide information about software that may be wasting power in this and other cases.

5. CONCLUSIONS

High-performance CPU design presents unique challenges for research in power related issues. Certain directions here need increased research and development focus in the future.

The highest priority is to continue pushing the voltage scaling treadmill. However, the technological and design hurdles in the path of using sub-1V supply voltages in large, high-performance CPUs have to be removed. Circuit styles and methodologies suited for low voltage are also needed.

Microarchitectures in today's high-performance CPUs are aimed at exploiting ever-increasing amounts of instruction-level parallelism. Organizational choices and tradeoffs are not made with power in mind. This needs to change and power consumption has to become a primary consideration here, since higher levels of design have the greatest leverage on the overall power consumption. Investigation of the hardware-software interface in CPUs will yield additional avenues for power reduction.

As frequencies increase, the number of pipeline stages will most likely increase. Combined with the trend for wider machines, this means that the number of clocked elements will keep increasing. Thus more efficient clock distribution schemes and latches are needed. It also implies that a more aggressive use of clock gating would be required. CAD tools that help in identifying and implementing clock gating opportunities will be useful here. Clocks, high-performance datapaths, and memories are not well-served by existing CAD tools. To have a significant impact on power, CAD tools have to target these domains.

Lower voltages, higher power consumption, more devices, and more clock gating - all these imply that the inductive noise problem will get worse. Increased innovation in packaging and power supplies would be needed to make sure that power delivery does not become the limiting factor for high-performance CPUs.

ACKNOWLEDGMENTS

The authors would like to thank Shekhar Borkar, Ricardo Suarez, Guy Therenin, Michael Walz, Steve Gunther, Shishpal Rawat and Jeff Parkhurst for assistance in writing this paper.

REFERENCES

[1] ACPI home page. *http://www.teleport.com/~acpi.*

[2] S. Borkar, Intel Corp. Personal communication.

[3] A. Chandrakasan, R. Brodersen. Minimizing power consumption in digital CMOS circuits. *Proceedings of the IEEE* 83(4), April 1995.

[4] S. Ellis. Power management in notebook computers. *Proc. Silicon Valley Personal Computing Design Conference,* July 1991.

[5] L. Gwennap. Power issues may limit future CPUs. *Microprocessor Report*, 10(10), August 1996.

[6] Intel Power Monitor home page. *http://developer.intel.com/ial/ipm.*

[7] J. Schutz. A 3.3V 0.6um BiCMOS SuperScalar Microprocessor. *ISSCC Digest of Tech. Papers*, Feb 1994.

[8] V. Tiwari, P. Ashar, S. Malik. Technology mapping for low power. *Proc. Design Automation Conference*, June 1993.

[9] V. Tiwari, S. Malik, A. Wolfe, T.C. Lee. Instruction level power analysis and optimization of software. *Journal of VLSI Signal Processing*, 13(2), August 1996.

[10] C.Y. Tsui, M. Pedram, A. Despain. Technology decomposition and mapping targeting low power dissipation *Proc. Design Automation Conference*, June 1993.

Design and Analysis of Power Distribution Networks in PowerPC™ Microprocessors

Abhijit Dharchoudhury, Rajendran Panda, David Blaauw, Ravi Vaidyanathan
Advanced Tools Group, Advanced System Technologies Lab, Motorola, Austin, TX 78730.
email: abhijit_dharchoudhury@email.sps.mot.com

Bogdan Tutuianu and David Bearden
Somerset Design Center, Austin, TX 78730

Abstract

We present a methodology for the design and analysis of power grids in the PowerPC™ microprocessors. The methodology covers the need for power grid analysis across all stages of the design process. A case study showing the application of this methodology to the PowerPC™ 750 microprocessor is presented.

Keywords

power distribution network, PowerPC™, IR-drop, reliability

1. Overview

A robust power distribution network is vital in meeting performance guarantees and ensuring reliable operation of high performance microprocessors. Higher device densities and faster switching frequencies cause large switching currents to flow in the power and ground networks which degrade performance and reliability. Excessive voltage drops in the power grid reduce switching speeds and noise margins of circuits, and inject noise which might lead to functional failures. High average current densities lead to undesirable wearing out of metal wires due to electromigration[1]. Therefore, the challenge in the design of a power distribution network is in achieving excellent voltage regulation at the consumption points notwithstanding the wide fluctuations in power demand across the chip, and to build such a network using minimum area of the metal layers. These issues are prominent in high performance PowerPC™ processors as large amounts of power have to be distributed through a hierarchy of five or six metal layers. For example, in the PowerPC™ 750 processor, the average power dissipation for a nominal V_{dd} of 2.5V is 5W.

The crux of the problem in designing a power grid is that there are many unknowns until the very end of the design cycle. Nevertheless, decisions about the structure, size and layout of the power grid have to be made at very early stages when a large part of the chip design has not even begun. Unfortunately, most commercial tools focus on post-layout verification of the power grid when the entire chip design is complete and detailed information about the parasitics of the power and ground lines and the currents drawn by the transistors are known. Power grid problems revealed at this stage are usually very difficult or expensive to fix. The methodology described in this paper is centered around an initial power grid which is refined progressively as the chip design progresses. Paramount to such a methodology is an analysis tool that has a very wide choice of models for specifying the grid and the power consumption behavior of the chip. The models must have a range of accuracies that make them suitable for use at any stage of the overall chip design. Such an approach is necessary to identify and remedy any problems with the power grid early on so that final verification results in only minor fixes.

A critical issue in the analysis of power grids is the large size of the network (typically millions of nodes in a state-of-the-art microprocessor). Simulating all the non-linear devices in the chip together with the non-ideal power grid is computationally infeasible. To make the size manageable, the simulation is done in two steps. First, the non-linear devices are simulated assuming perfect supply voltages and the currents drawn by the devices are measured. Next, these devices are modeled as independent time-varying current sources for simulating the power grid and the voltage drops at the transistors are measured. Since voltage drops are typically less than 10% of the power supply voltage, the error incurred by ignoring the interaction between the device currents and the supply voltage is small. By doing these two steps, the power grid analysis problem reduces to solving a linear network which is still quite large. To further reduce the network size, we exploit the hierarchy in the chip design.

Note that the circuit currents are not independent due to signal correlations between blocks. This is addressed by deriving the inputs for individual blocks of the chip from the results of logic simulation using a common set of chip-wide input patterns. An important issue in power grid analysis is to determine what these input patterns should be. For IR-drop analysis, patterns that produce maximum instantaneous currents are required, whereas for electromigration purposes, patterns producing large sustained (average) currents are of interest. Determining either set of patterns is a difficult problem and has been addressed by many, including [2][3][4]. This issue will not be addressed in this paper, and we will assume that suitable input patterns are available for simulation. Moreover, in this paper, we will not discuss electromigration issues, but concentrate on the issue of computing the worst voltage drops due to IR-drop in the power network.

This paper is organized as follows. The next section describes the various analysis modes implemented in our methodology. Section 3 discusses linear solution techniques in the context of power distribution networks. A case study of IR-drop analysis on the PowerPC™ 750 processor is shown in Section 4, and the paper concludes with a discussion of open issues.

2. IR-Drop Analysis Modes

To apply the IR-drop analysis methodology described in this paper across all stages of the design of a complex microprocessor, we define several modes of operation of the tool. These modes are distinguished by different models of the power distribution network and of the currents being drawn by the functional blocks. In this paper, we describe three different *modes* of operation: *early* or *pre-floorplan* mode, *post-floorplan* mode and *post-layout* mode. As the design proceeds, IR-drop analysis is run in these different modes

DAC 98, San Francisco, California
©1998 ACM 0-89791-964-5/98/06..$5.00

using more and more accurate models of the power grid and the block currents.

2.1 Early Mode Analysis

At the very early stages of the design of a microprocessor, there are a number of issues related to the power distribution network that have to be addressed. These include locations of the clean VDD/ GND pads, nominal pitches and widths of metal layers, via styles (point or bar vias), and parameters of the chip package. Since at this early stage of the design, the power network has not yet been synthesized and the location and logic content of the blocks are not known, IR-drop analysis is performed using very simplistic models of the grid topology and the block currents.

1. A mock power grid down to the lowest metal layer is constructed using a simple uniform grid topology, where the metal lines in each layer have a user-specified pitch (separation) and width. At the areas where the metal lines of adjacent layers cross over, vias are placed according to user-specified via geometries and via styles. Other topologies such as rings can also be modeled. The clean VDD/GND pads can be placed at the periphery of the chip or on the surface of the chip using C4 pads (for flip-chip packages).
2. To model the currents drawn by the devices, a simple area-based DC estimate of the current is used. This is obtained by taking the current estimate of a previous chip and scaling it by the power supply voltage, operating frequency, complexity, size and technology variables. This estimate is inflated 3-7 times to account for differences between the average and maximum instantaneous currents and to obtain a robust grid. The current sinks are placed on the lines of the lowest metal layer at points midway between adjacent vias that connect the lowest layer to the upper layer. The value of the current is obtained by multiplying the per-unit-area current by the product of the pitches of the two metal layers.

Using simple length-based resistance formulae, a resistive electrical network is constructed from the mock grid topology. DC analysis of this network yields the IR-drops at various locations of the chip. This analysis is very fast and allows the designer to evaluate a large number of different topologies and to trade-off robustness and metal utilization in the power grid. This analysis is used to design the locations of the C4 pads and nominal pitches and widths of the metal layers. Moreover, if the processing technology allows different width and thickness combinations for some of the top metal layers, the user can determine the best values of these in terms of IR-drop. Even though the real power grid will not be as regular as the mock grid, and all the devices will not be drawing the estimated current simultaneously, important design decisions are made from the results of this simple analysis and an early picture of the robustness of the grid is obtained. An example of this analysis is given in Section 4.

2.2 Post-Floorplan Analysis

In this mode, the global power distribution network has been designed and the blocks have been placed. The locations and geometries of the power lines and the blocks are read from the design database. Even though the blocks are placed, the power grids within them have not yet been wired. The *power service terminals* (PST's) of a block are the wires in the topmost metal layer within the block that connect the global and intra-block power networks. In this mode, the PST's for a block may or may not be known - if they are not known, mock PST's are constructed. Next, the block *ports* are determined by the intersection or overlays between the global lines and the block PST's. In our hierarchy, *blocks* are custom datapath components, synthesized random logic macros (RLM), and off-the-shelf (OTS) components (custom components that can be reused). Custom components are small but RLM's can be large. OTS components can range from small blocks (nands, nors, muxes, etc.) to large blocks (adders, comparators, etc.). A *functional block* (e.g., floating point unit, memory management unit) consists of several instances of custom, synthesized and off-the-shelf components.

Each block current can be independently described in one of the following ways, thus allowing a mixture of them (see Fig. 1).

1. If the logic content of a block is not yet defined, the current model is a DC estimate based on the block area. The total block current is divided equally among all the ports. Since the area-based numbers are calculated such that they reflect peak expected currents, the analysis where every block has area-based currents is likely to be pessimistic since it assumes that each block draws this current simultaneously.
2. The next more accurate block current model is derived from a full-chip gate-level power estimation tool. Given a set of chip-wide input vectors, this tool computes the average power consumed by each block over a cycle. From the average power consumed by a block, an average block current is computed and distributed equally among all of its ports. Hence in this mode, a multi-cycle DC current signature is used for a block. Since chip-wide vectors are used for the simulation, correlation among the blocks is preserved.
3. The most accurate current model comes from a detailed transistor-level simulation of a *functional block* using PowerMill [5]. The input vectors for the functional block are derived from the chip-wide vectors through logic simulation. This ensures that correlation across functional blocks is maintained. The transistor level netlist of the block is available and capacitances are extracted for the signal nets. However, since the power grid within the block has not been designed, it is considered to be ideal. The transient current waveform drawn by each custom, RLM or OTS block within the functional block is obtained from the PowerMill simulation and is divided equally among all of their individual ports. Since the blocks are not very large and the power grid within them have not yet been wired, this block current model is quite accurate for this stage of design.

If all block current models are derived from methods 1 and 2 above, then a resistance-only electrical network is extracted from the geometrical information using length-based resistance formulae. DC analysis is then performed to yield the IR-drop values at each of the block ports (multiple DC analyses if multi-cycle DC current signatures are used). If transient current signatures are used

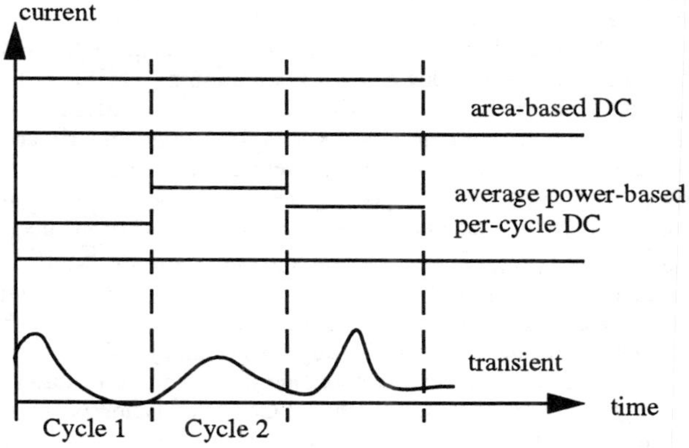

Fig. 1. Block current signature models

for some of the blocks, then an RC network is extracted from the global grid using length-based resistance extraction and statistical rule-based capacitance extraction (since global routing is not done, these statistical rules account for coupling between power and signal lines). An example of IR-drop analysis in this mode is given in Section 4.

2.3 Post-Layout Mode

This mode is used when the global and block-level grids have been completely designed. We employ hierarchical analysis, where each block is analyzed differently based on its size.

1. **Custom blocks**: We run Railmill [6] on the custom blocks using vectors that are obtained from the common set of chip-wide vectors and clean VDD and GND locations as supplied from the design database. The RailMill analysis a) verifies that the drop in the local grid satisfies the bounds for the block and b) supplies the currents at the block ports that are then promoted to the chip level for global grid analysis.
2. **Random Logic Macros**: RLM's are analyzed using PowerMill but unlike in Section 2.2, each RLM is broken up into its constituent standard cells. In other words, the standard cells are elevated to the chip-level in the hierarchy. The current drawn by each standard cell is measured separately and these currents are inserted into the ports of the standard cells for global grid analysis. Since the standard cells are small, this mode gives us visibility to the gate level, and the power grid within the RLM is also verified.
3. **Off-the-shelf components**: OTS's are modeled using PowerMill in the same manner as described in the previous section. Since the OTS's are small and their PST's are at the lowest layer of metal, we get good visibility into the OTS's as well.

In this mode, RC extraction of the global grid is performed using a commercial extraction tool.

3. Linear System Solution Techniques

Several direct and iterative approaches are available [7] to solve linear systems. In this section, we analyze the relative merits and limitations of these methods as applied to solving large power networks. The PowerPC™ processors use 6 layers of metal and the power grid is a very tight mesh. This implies that "crunching" of the global grid does not yield appreciable reduction in the size of the network and DC analysis must resort to conventional matrix methods. The size and structure of the conductance matrix of the power grid is important in determining the type of linear solution technique that should be used. Typically, the power grid contains millions of nodes, but the conductance matrix is sparse (typically, less than 5 entries per row/column). This matrix is also symmetric positive definite, but for a purely resistive network, it may be ill-conditioned. Sparsity favors the use of iterative methods, but convergence is slowed down by ill-conditioning and can be mitigated to some extent by preconditioning. Iterative methods do not suffer from size limitations so long as the (sparse) matrix and some iteration vectors can fit into the memory. The single-biggest problem with direct methods is the need for large amounts of memory to store the factors of the matrix. The number of fill-ins is of the order of $O(N^2)$, where N is the number of rows/columns in the matrix. However, if fixed time steps are used for transient analysis, then the initial factorization can be reused with subsequent current vectors, thus amortizing the large decomposition time. Iterative methods do not have this feature of reusability. Among the direct methods, Cholesky factorization is best suited since the conductance matrix is symmetric and positive definite. However, for a machine with 1 GB of memory, we could only simulate a network with 300K nodes with this technique. Because of this severe size limitation, we are currently using the conjugate gradient iterative scheme with incomplete Cholesky preconditioning to solve our matrices.

As mentioned earlier, a flat analysis of the entire power distribution network where each transistor that connects to the power lines is modeled as a current source would be computationally infeasible due to the large size. This size limitation can be mitigated by hierarchical analysis, in which each block has a macromodel which is used for the analysis of the global grid. A block macromodel consists of current sources at the block ports and an admittance matrix relating the currents and the voltages at the ports. For an exact equivalence with flat analysis, the admittance between every pair of ports must be modeled resulting in a dense admittance matrix for each block. However, this adversely affects the speed of an iterative method, reducing the efficiency gained by a hierarchical approach. To preserve the efficiency of the global analysis, we ignore the admittance between the ports of a block and model the blocks purely as current sources. For chip-level analysis, the error due to this assumption can be kept within bounds if the blocks are small. In our methodology, the blocks used for global power grid analysis are OTS's, standard cells and custom elements, all of which are small. An important mechanism implicit in this hierarchical analysis is that each block has an IR-drop budget, i.e., a maximum allowed voltage drop at its ports. If the voltage at the port of a block violates its budget, then the problem is fixed by making the power grid that supplies this block more robust.

4. Case Study

In this section, we provide examples of the application of the various stages of the IR-drop analysis methodology described above. As mentioned in Sec 2.1, the mock global grid in the early mode analysis consists of uniformly spaced power lines and clean VDD sources (C4 pads). Since the analysis will result in a symmetric voltage distribution, we confine the analysis to a representative area of the chip. The mock global grid is shown in Fig. 2(a): the uniformly spaced lines and C4 pads are clearly seen. The current sources are inserted at the lowest metal layer midway between the vias connecting it to the upper layer. We assume that the total current drawn by the chip is 6A, which makes the power of the chip 15W (three times the average power of 5W). By dividing the total current by the estimated area of the chip, we get the current per unit area. The voltage map using this current value is shown in Fig. 2(b). As expected, the worst voltage drop occurs at the points with the largest distances from the C4 pads; for this analysis, the worst voltage drop was measured to be 70mV. Now, if the pitch and width of the second layer of metal is halved (keeping the same amount of power grid metal utilization), the voltage drop reduces to 45mV. However, the trade-off with the reduced voltage drop is that the congestion for routing signal lines will be significantly more. This example illustrates the type of trade-offs that can be explored using the early analysis mode.

Fig 3(a) shows the placement of the blocks and the power grid in the PPC750 processor - there are approximately 15000 blocks. As mentioned earlier, these blocks consist of customs, RLM's and OTS's. The floating point unit (FPU) functional block is located at the bottom right corner of the chip, and is shown highlighted. The first run of the post-floorplan analysis mode was done with area-based DC current estimates for each of the blocks in the chip. The per-unit area current estimate was modified to yield a less pessimistic value of 4.5A for the total chip current. The voltage map for this run is shown in Fig. 3(b) - the worst drop in this case was measured to be 170 mV and was occurring in the IO pads near the left edge of the chip. This value is worse than the worst drop predicted by the early analysis even though the average pitches and widths of the metal layers are similar in both analyses. This is because the real grid is not as uniform as the mock grid and the "distance" from the nearest C4 pad is more for the IO pads. However, within the FPU, the worst drop was measured to be 110mV near the bottom right corner of the chip. A close-up of the voltage map within the FPU is shown in Fig. 3(c). The sum of the area-based DC currents for all of the blocks within the FPU functional unit was computed to be 0.7A.

For reasons of brevity, we will not show the results of IR-drop analysis when the area-based current estimate of the blocks is replaced by current estimates based on full-chip multi-cycle power simulation. Instead, we will show the results with transient current signatures obtained from PowerMill and also concentrate on the FPU area of the chip. As discussed earlier, PowerMill simulation is done on the transistor level netlist of all the blocks within the FPU with extracted parasitics. The input vectors for the FPU are obtained from the results of a full-chip logic simulation, using "high stress" chip-wide vectors. A subset of the total current waveform of the FPU is shown in Fig. 3(d). The first current spike peaks at approximately 2.1A, whereas the second spike peaks at approximately 0.7A. This shows that the area-based estimates agree well with the typical peak values (0.7A); however, there are input patterns which cause the peak values to be significantly higher as in the case of the first spike (2.1A). Since the sum of the FPU currents using the area-based estimate was also 0.7A, we expect that the worst voltage drop within the FPU for this time point will be comparable to the area-based analysis. In fact, the worst drop within the FPU in this case was measured to be 130mV. However, as shown in Fig. 3(e), the voltage map is different from the area-based analysis. In fact, the worst drop region is much more localized in this analysis. This is due to the non-uniform spatial distribution of the block currents, i.e., at this time point, the blocks near the worst drop region are drawing much larger currents compared to the other blocks. Fig. 3(f) shows the voltage map for the analysis at the time point when the total FPU current is 2.1A. In this case, the worst drop within the FPU is 300mV and the worst drop region has shifted upward. This is due to the fact that at this time instant, the locations of the high current blocks are different from the previous case.

This analysis shows that because of the safety margins built into the early and post-floorplan modes, the typical voltage drops in the final verification stage are well within the block budgets. However, as seen in this example, transient analysis may reveal cases when the worst drop is worse than expected. In these cases, the voltage drops can be improved by making local alterations to the power grid.

5. Conclusions

We presented a coherent design and analysis flow for designing the power grids of large, high performance processors. The usefulness of the multi-mode analysis capability in progressively refining the design of a power grid design was demonstrated through case studies. Several issues related to power grid analysis were discussed and a case study of the PPC750 processor was presented.

Areas of future work involve investigating fast solution techniques for power and ground networks, calibrating and reducing the error due to hierarchical analysis, incorporating inductance effects into the IR-drop analysis, and determining optimal locations of decoupling capacitors.

References

[1] Black, J.R. Electromigration failure modes in aluminum metallization for semiconductor devices. *Proc. IEEE*, pp. 1587-1594, Sept 1969.

[2] S. Chowdhry et al. Estimation of maximum currents in MOS IC logic circuits. *IEEE Trans. CAD*, pp. 642-654, June 1990.

[3] H. Kriplani et al. Pattern independent maximum current estimation in power and ground buses of CMOS VLSI circuits: algorithms, signal correlations, and their resolution. *IEEE Trans. CAD*, pp. 998-1012, Aug 1995.

[4] A. Krstic et al. Vector generation for maximum instantaneous current through supply lines for CMOS circuits. *Proc. 34th DAC*, pp. 383-388, June 1997.

[5] *PowerMill User Guide*, Synopsys Inc., 1997.

[6] *RailMill User Guide*, Synopsys Inc., 1997.

[7] Golub, G and Van Loan, C. *Matrix Computations*. The Johns Hopkins Univ. Press, 1989.

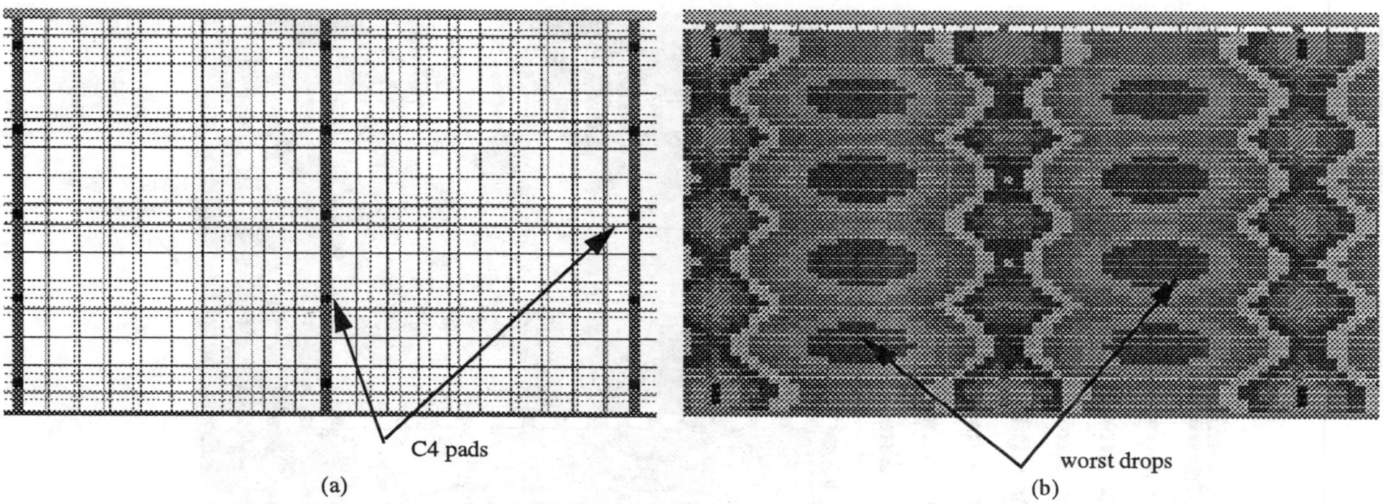

Fig. 2: Early mode analysis of PPC750: (a) mock grid, (b) corresponding voltage map.

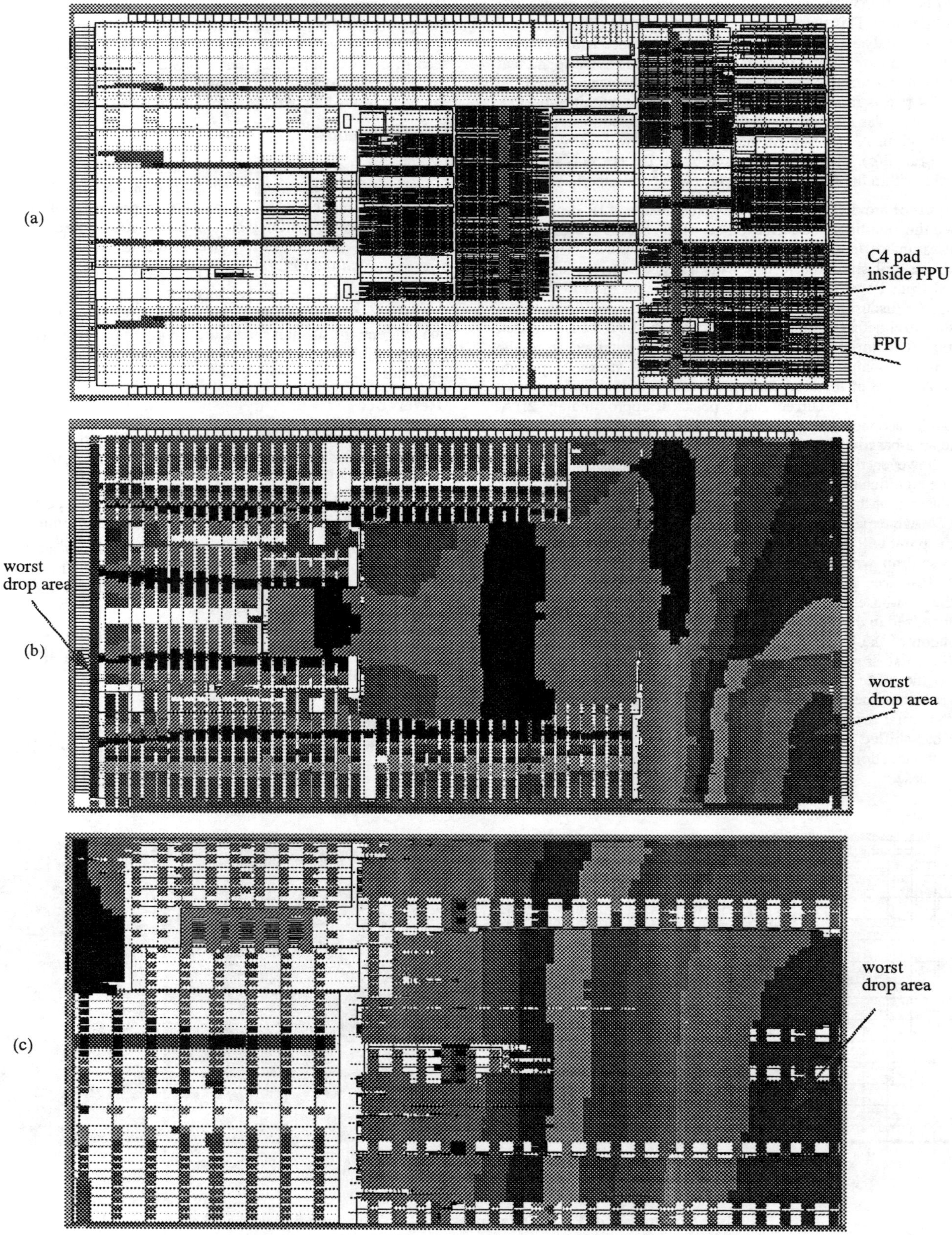
(a)
C4 pad
inside FPU
FPU
worst
drop area
(b)
worst
drop area
(c)
worst
drop area

(d)

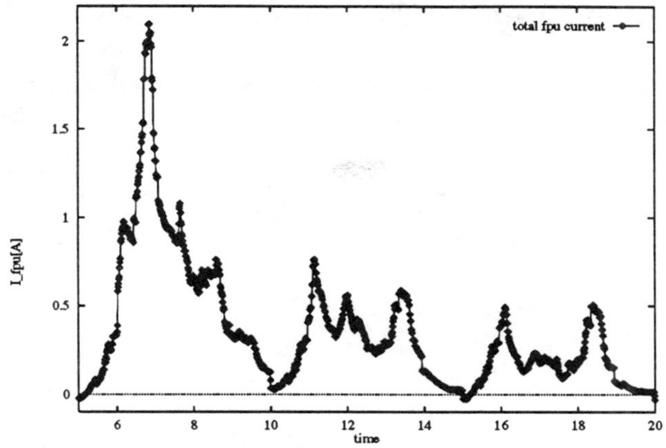

(e)

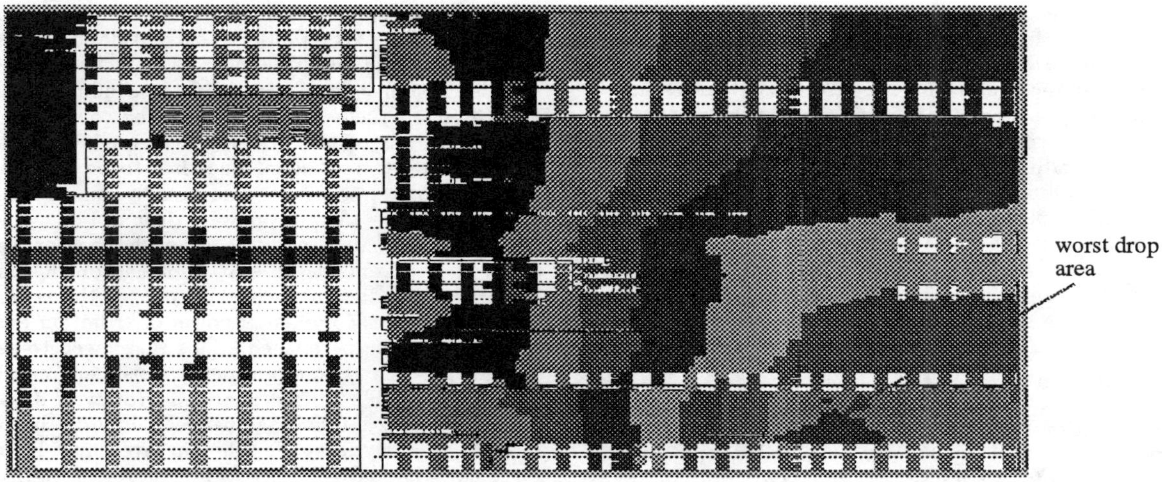

(f)

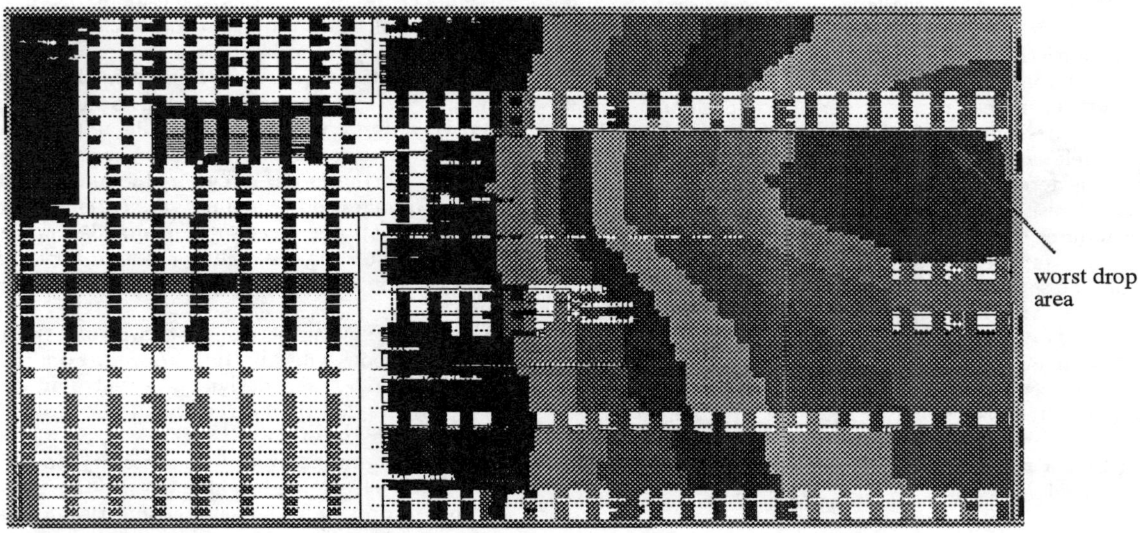

Fig. 3. PPC™ 750 case study: (a) power grid and block layout, (b) voltage map for area-based block currents, (c) voltage map inside the FPU for area-based block currents, (d) total transient currents in the FPU, (e) voltage map inside the FPU for total current of 0.7A, (f) voltage map inside the FPU for total current of 2.1A.

Full-Chip Verification Methods for DSM Power Distribution Systems

Gregory Steele, David Overhauser, Steffen Rochel, Syed Zakir Hussain
Simplex Solutions, Inc.

Abstract

Power distribution verification is rapidly becoming a necessary step in deep submicron (DSM) design of high performance integrated circuits. With the increased load and reduced tolerances of DSM circuits, more failures are being seen due to poorly designed power distribution systems. This paper describes an efficient approach for the verification of power distribution at the full-chip transistor level based on a combination of hierarchical static and dynamic techniques. Application of the methodology on practical design examples will be provided. We will also demonstrate the necessity of an analysis at the full-chip transistor level to verify the complex interactions between different design blocks based on static and dynamic effects.

1 Introduction

The comparison of the power density for an average microprocessor (ca. 20W / 400 mm^2 = 50kW / m^2) and the power density of a large city (ca. 500MW / 100 km^2 = 5W / m^2) reveals some of the challenges related to the full-chip power distribution verification. Not only is the power density high, but it is also increasing with shrinking transistor geometries and increasing device count. The reduced power supply voltage, increased power density, and thinner wires used in DSM designs are causing an increase in the number of failures related to the power distribution network. Failures are considered excessive voltage drops at the gates due to the sheet resistance of the power supply lines, so called IR drop failures, as well as the growth of voids and shorts at the power supply lines due to electromigration. With the narrow noise margins of a DSM design, severe IR drop due to missing vias or an inadequate power grid can cause functional failures. Less severe IR drop may cause timing problems. A five percent drop in the supply voltage of a gate can cause a fifteen percent increase in the gate delay. Proper gate supply voltage is critical to maintain functionality and performance. Power grid verification is a necessary step in ensuring that a new design will work at its intended speed once fabricated.

DAC 98, San Francisco, California

In DSM, parasitic resistance and capacitance of the power and signal lines have a considerable effect on the actual chip performance, therefore increasing the accuracy requirement for parasitic extraction. For current design technologies, the extraction of linear resistors and capacitors is in general considered sufficient but, in the future, inductive effects as well as frequency effects may have to be considered. The effective loading of the power grid caused by parasitic device capacitances and signal lines has to be taken into account during power analysis. Additionally, signal timing and signal slope are affecting power consumption, which increases the accuracy demands for power distribution verification.

It is reasonable to start looking for IR drop problems early in the design cycle. As soon as the layout of a functional block is completed, power distribution analysis of this block may be performed. This is referred to as block level analysis. Many common design mistakes may be found by performing block level analysis. However, due to the tight and complex interactions between different blocks of the design, an analysis at the full-chip level must be performed to ensure that no IR drop related problems will occur in the assembled design. At this stage all parasitic elements can be extracted based on layout and technology description.

The number of resistors in a power grid can be estimated roughly by the number of transistors times the number of metal layers, but the relationship is actually super-linear. Typically, for a design with 5 million transistors in a five metal layer process, the number of resistors is about 30-40 million. Furthermore, circuit reduction cannot be applied for electromigration reliability verification of the power grid. This demonstrates clearly the demands on the capacity of the verification tools as well as the need for an efficient post-processing mechanism to quickly pinpoint critical spots in the design.

In the following sections, we will use the term "power grid" for a power distribution network even if it is not necessarily a grid based design. We will omit the discussion of techniques for the extraction of the circuit description, the power distribution network itself, as well as the signal interconnect lines. But it is necessary to point out that to perform electromigration analysis, the extractor has to be able to provide layout and geometry information for the resistors comprising the power grid and must provide the location of all circuit elements.

Section 2 discusses block-based and full-chip power grid analysis methodology. In Section 3, we will discuss in detail static and dynamic techniques to determine the current loading at the power grid. Section 4 describes different methods for analyzing results of the power grid analysis.

Electromigration verification will be discussed in Section 5 and the application of the described methodology is demonstrated in Section 6.

2 Verification Methodology

Power grid analysis for either blocks or full-chip can be divided into two basic phases. Each phase can be varied in complexity and accuracy depending on the stage in the design process where the power distribution analysis is applied. Nevertheless, accurate verification requires consideration at the transistor level.

First, the circuit description (behavioral, gate or transistor level) as well as the power grid including parasitic elements have to be extracted from the design database. Secondly, the current distribution in the power grid has to be calculated. This can be done in two different ways. One possibility is to calculate the power grid and transistor circuit together. This takes into account the coupled nature of the problem and provides an accurate picture, but also limits the size of the problem that can be handled. The complexity of the problem and run times will limit the application of this method to a block level. Therefore, this technique cannot be applied for final verification of complex designs such as microprocessors and ASSP's and will not be discussed further.

Especially for digital circuits, a more efficient two-step technique can be applied. The first step is to calculate the current consumption for all circuit elements assuming ideal power supply values using static or dynamic current data, later on referred to as static and dynamic approaches. Different techniques are described in the literature and will be discussed below. The second step is to use computed loading currents to calculate the current distribution in the power grid. The resulting IR drop can be fed back into the first step to improve the accuracy of the current data drawn from the power grid. Usually, one iteration loop is sufficient to get realistic IR drop and current data for the power grid. The separation into two steps allows the exploitation of specialized algorithms during each step which in return improves speed and capacity.

Block-based analysis can be applied at different stages of the design process. Two possible scenarios can be described, the verification of the local voltage distribution inside a block or the verification of the global power grid based on a block loading abstraction. Both methods can be based on static and dynamic current data and allow a global verification to check for basic violations. This allows easy location of major layout mistakes like omitting a row of vias or contacts.

Static simulations are useful for fast analysis of the power distribution trends in the design, whereas dynamic simulations can pinpoint problems due to simultaneously switching gates within a block. Additionally, vector-based dynamic analysis provides a much better coverage and through-put on a block level versus full-chip simulation. This allows verification of the dynamic behavior in much greater detail.

In order to perform a block-based analysis, assumptions about the supply voltage have to be made where the block connects to either the global power grid or other blocks. Worst case assumptions should be used in any case for a block-based analysis. Later on, these assumptions have to be verified in a full-chip context.

The block can be simulated using current data generated under the assumption that each transistor was operating at the full supply voltage. This approach will overestimate the current data and the results will be slightly pessimistic. Again, the accuracy of the analysis can be improved by feeding back the calculated IR drop at the power grid and repeat the current calculation. Another way is to simulate the power grid and block together as long as the block size is small enough.

Block-based analysis should be performed using both static and dynamic current data as early in the design cycle as is practical. Static analysis should be performed, and changes to the grid structure made, until no further power grid problems appear during analysis. Then a dynamic analysis can be run to ensure that no additional problems occur due to simultaneously switching gates.

Even though block-based analysis provides valuable information about power distribution trends and local IR drop data, it is dangerous to base sign-off only on a block-based analysis. The assumptions which enabled block-based analysis may be violated in the full-chip context. The connection of neighboring blocks in the full chip design will frequently influence the IR drop within the block as current flows through the block to its neighbors. Additionally, it is hard to estimate the potential effects of electromigration in a block-based context, where current may be supplied to neighboring blocks. A full-chip analysis must therefore be performed to validate the assumptions used during block-based analysis, and for reliability verification after the final layout of the entire chip has been completed.

Since the power in a design is supplied to the blocks through metal lines which are often designed separately from the blocks, the effects of these metal lines on the block supply voltages must be examined. If the designer is only interested in verifying that the block supply voltage is within the specification used for block-based analysis, then it suffices to simulate only the upper layers of the grid, down to the level of the blocks. One method is to use the current data at the connection point of each block to its neighbors as well as the global power grid and analyze only the upper metal layers of the grid.

Full-chip IR drop analysis should be performed in both static and dynamic modes. Dynamic current data gathered during detailed block based dynamic verification may be reused during full-chip verification [1].

Full-chip analysis often reveals problems with the power grid which were not evident from the block-level analysis. The power supply for blocks near the center of the chip may be routed through neighboring blocks which are not prepared to handle the additional current load. These blocks which appeared fine during block level analysis could show excess voltage drop due to current drain of their neighbors. One of the worst cases is, for example, the supply of a high activity data path through a neighboring RAM for which the power supply was designed with the assumption of a low current drain.

3 Current Calculation

3.1 Static Approach

A static approach to calculate the currents is useful as a first step in power grid analysis. Practical experience has shown that most major power grid design problems will show up during analysis based on static current data. Additionally, performing static analysis is much faster than a dynamic one, allowing a designer to make several iterations quickly while making changes to a portion of the power grid.

Very simple and fast current estimates can be generated

by calculating the Id-sat of the devices connected to the power grid. Usually this current data must be calibrated (i.e. scaled) so that the average power consumption is equivalent to that of the actual circuit. The availability of parasitic capacitance data is not required, which allows a fast qualitative analysis on a power grid design. An amazingly large number of weak spots, design flaws, and mistakes in the power grid design may be found very quickly. The most appealing advantage of this approach is simply the achievable turn-around time.

However, the results have to be carefully interpreted. The current calculation based on Id-sat frequently indicates potential problems in portions of the design with low switching probabilities, such as cache RAMs. These errors are not real, since in reality only a few of the transistors in these parts of the design will be active at any given time. Therefore these IR drop problems must be ignored during Id-sat based static analysis.

A much more accurate static calculation can be performed by considering the amount of charge drawn from the power grid during each clock cycle based on the switching activity at the gate level. The accuracy of this approach depends heavily on accurate device and interconnect capacitance data as well as the accuracy of the actual switching probability. Switching data can be derived from probabilistic analysis as well as vector-based simulation at the behavioral, gate or transistor level [2], [3], [4], [5]. The average power dissipation for each gate can be expressed by the relatively accurate approximation [6]:

$$P_{avg} = 1/2 * C_L * V_{dd} * P_s * f \quad (1)$$

whereby C_L is the effective loading capacitance seen by the gate, V_{dd} the supply voltage, P_s the switching probability and f the clock frequency. Static current values estimated in this way are generally very close to actual average current values, and are therefore very useful for electromigration analysis. This also avoids some of the problems encountered using only Id-sat based current calculation in areas with low switching probabilities.

3.2 Dynamic Approach

Dynamic analysis has to be performed to verify instantaneous voltage drops at the power grid due to simultaneously switching gates. Just after a clock transition, many of the gates in a design begin to switch simultaneously, causing peak current consumption and peak IR drop on a power grid. This effect will not be seen using a static current calculation, which is the reason a dynamic verification must be performed for sign off. The dynamic peak IR drop will cause additional delays until the signals settle and may cause timing conflicts in aggressive designs, especially if cycle stealing techniques are used. A representative set of vectors has to be used for dynamic IR drop analysis. Ideally a large set of vectors would be simulated to ensure that no IR drop problems occur under operating conditions. However this is impractical due to the simulation time that would be required to run a large set of vectors. Different techniques have been proposed to either reduce the number of vectors to be considered for a dynamic analysis [2] or to find a subset from all possible input stimuli representing a similar power distribution [7], [8]. In practice, only a few tens of vectors are used to identify IR drop problems.

Dynamic power grid analysis presents some unique challenges due to the size of the problem as well as the time required to perform dynamic simulation at the transistor level. To reduce the time required for performing dynamic simulation at transistor level, macro modeling techniques based on gate-level simulation have been proposed [9], [10] and are utilized during synthesis and optimization. Because of the accuracy required for final verification, transistor level analysis that takes into account the interdependency of power consumption, signal delay and slope degradation is still required.

Due to the time required to simulate the entire power grid once, it is undesirable to simulate the power grid over many clock cycles. A compaction technique has been developed which reduces current data, calculated from a simulation lasting several clock cycles, to produce a current profile for a typical cycle [11]. The compaction method averages the current data, so the calculated IR drop is slightly pessimistic. This technique not only decreases the necessary time interval for the power grid simulation itself, but it also provides average data usable for electromigration analysis.

4 Analyzing Power Grid Data

Due to the immense quantity of data generated during power grid analysis, data pruning has to be applied before reporting results. Application-dependent data filtering ensures that only critical sections of the design need to be investigated in detail. A visual plot is extremely useful for tracking down the elements of a design that are contributing to excessive IR-drop or electromigration problems.

During static IR-drop analysis with Id-sat data, the absolute value of the voltage on each node is not particularly meaningful. The important aspect to observe is the voltage distribution trends in the design. Frequently one or two areas of a design are found to be operating at much lower voltages than the rest of the chip. This indicates that the portion of the power grid supplying these areas may be insufficient and that more accurate analysis, such as activity-based static analysis, has to be performed on those portions of the design. Examining the current distribution in the power grid shows the source of the supply current for a block. Especially for blocks with IR problems, the expected source from visual inspection of the power grid and the real source differ significantly.

Activity-based static analysis and full dynamic analysis produce meaningful data values that may be compared to design specifications for sign-off purposes. Any portions of the design that exhibit excessive voltage drop with activity-based current data should be carefully inspected to locate the cause. Changes should be made until this condition is corrected.

5 Electromigration Analysis

Full-chip reliability analysis has become more critical because advances in technology are yielding narrower interconnect structures and higher-frequency designs. This combination increases the risk of electromigration and joule heating failures in designs. Performing power grid reliability analysis on designs with tens of millions of resistors in the grid requires new techniques.

Traditionally, designers are given simple layout design rules based on wire current density limits to which they must adhere. These limits, set to provide reliability over a broad range of circuit configurations, can make high-speed designs excessively large or impossible to design. This indicates that a methodology for reliability budgeting is needed

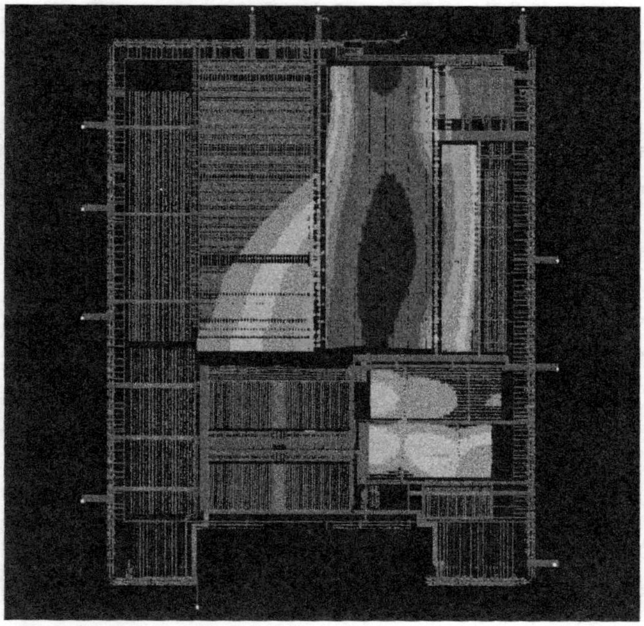

Figure 1: Static IR drop with removed connection between blocks

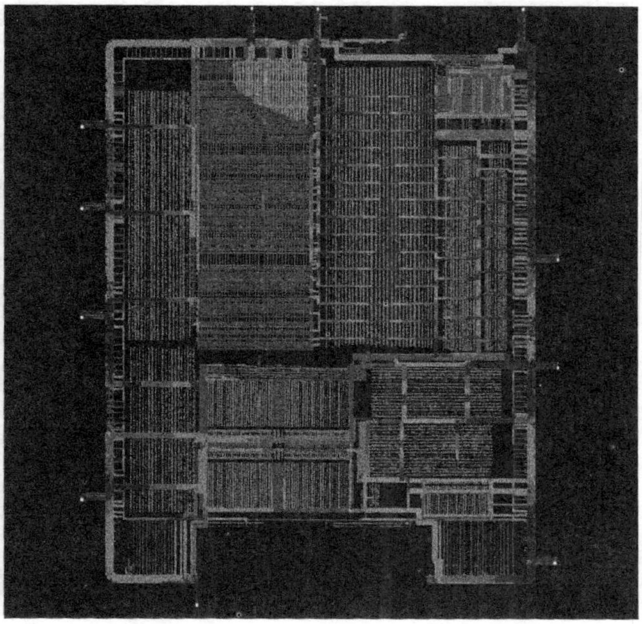

Figure 2: Static Full chip current distribution for the initial version of the design

to permit engineering tradeoffs between performance, design size and lifetime. This methodology must analyze the circuit to obtain realistic estimates of actual currents flowing in the circuit; apply advanced electromigration models to wire segments, usually based on Black's equation; and perform statistical analysis over the wires in the design to estimate the probability of the chip operating properly over its lifetime.

With growing chip complexities and power grid redundancies designers no longer understand exactly how their chip operates. Due to complex power grid and distributed blocks in a design, current flow from a chip pin to the gates cannot be determined without full-chip analysis. This is one of the reasons that Full-chip electromigration analysis finds design problems. The current flowing in the chip may be taking a completely unexpected route, through failure prone portions of the power grid.

A methodology has been developed to permit analysis of the complete power grid. This methodology includes extraction of chip interconnect data without the application of interconnect net reduction or contact clustering, static and dynamic full-chip analysis to determine current loading characteristics at the various device contacts to power grid, and modeling mechanisms to report either wire segments likely to fail or overall chip lifetime statistics. Full-chip reliability analysis is part of the power distribution verification process and can be carried out in parallel with IR drop analysis. The ability to apply reliability analysis at the full-chip level makes it possible to bring product reliability and reliability budgeting into designers hands.

Power grid electromigration analyses require the creation of models for a chip. Model data is provided for each metal and via layer in the chip. Each metal layer model provides the layer thickness; current density limits for peak, average, and RMS currents through wire segments (different foundries provide different rules) for simple threshold checks; and a set of model parameters for the mean time to failure (MTF) model. Different model parameters may be applied for narrow wires and wide wires, where an additional model parameter defines the boundary between narrow and wide wires. Each via and contact model provides the current limits for peak, average, and RMS currents through each via/contact for simple threshold checks, and a set of model parameters for the MTF model. An MTF model can be based on Black's equation or some other model. Since the actual MTF value for an individual component can be misleading, and because the MTF values of relatively reliable components can vary over a very wide range, it is more useful to examine the inverse of the MTF. The inverse of the MTF is referred to as EM risk - the risk that the design will fail during the time of operation. During reliability analysis, the components with larger EM risk values are the ones most likely to fail.

The power grid of a chip is operated primarily in a pulsed DC sense with respect to electromigration analysis. Therefore average current data throughout the circuit is used to perform electromigration analysis on the grid. Because the frequency is above 100kHz, the full-chip transistor analysis tool provides the average current drawn by each transistor connected to the power grid. One at a time, each power grid is modeled by voltage sources at the pins, providing power to the chip and the transistor tap currents at the device connection points. The large linear system is then solved to determine the precise current flowing through every wire segment and via in the chip. Once each wire segment current density has been determined, the simple checks are applied to identify those wires in the design which exceed the threshold. A more detailed analysis of the reliability is made by calculating the theoretical time to failure and EM-risk value for each element and using the proper failure statistics to obtain a failure probability as a function of time for the entire chip. The results are highly dependent on the choice of the statistical model used. Extreme value lognormal, also known as multilognormal statistics, is used, assuming the

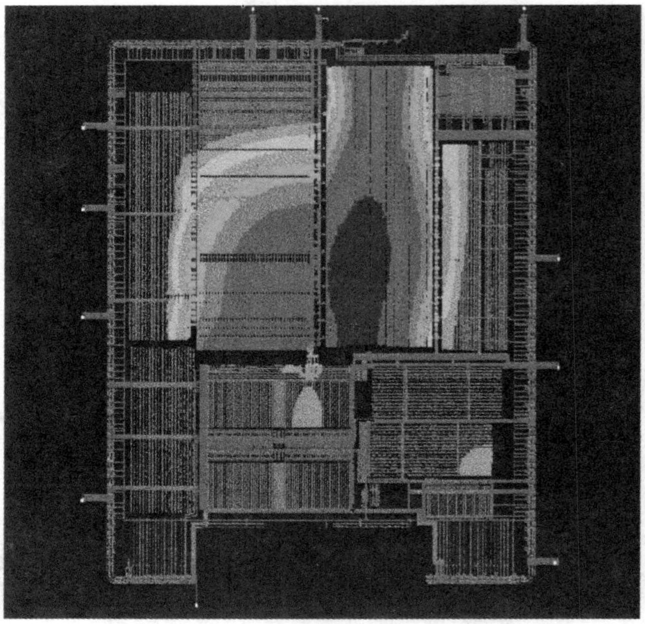

Figure 3: Static IR drop analysis results with improved power grid

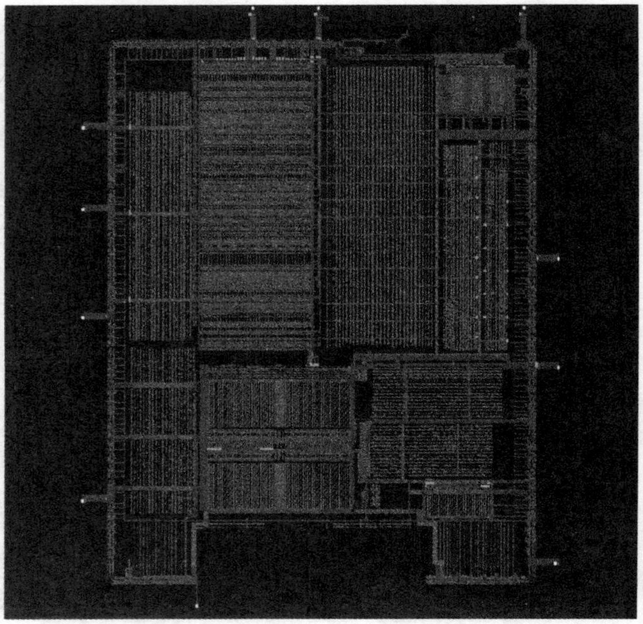

Figure 4: Electromigration analysis results on an improved power grid

failure probability to be that of a chain that fails at the weakest link. The links are assumed to be lognormally distributed. Since the wire segments with highest EM risk are identified, these can be provided to the designer for an engineering change order (ECO) if the overall chip probability is below specification. Improving the reliability of the least reliable elements in the design will drastically increase the overall MTF of the design.

6 Example

A modified design of a multi-media chip with 1.6M transistors has been used to demonstrate the capabilities of the proposed methodology. Figure 1 shows the result of an static IR drop analysis. Loading currents have been calculated as described above and are scaled according to the switching activity in each block. The highest voltage drop is represented in the darkest (red) color, giving an immediate feedback to the designer, which blocks are the most critical one as well the location of the critical spots inside the blocks.

It is clearly visible that the large block in the middle draws current from its neighbors, causing remarkable voltage drops in the surrounding blocks. It is interesting to see, that the missing power connection from the lower part of the chip to the upper right block causes an excessive IR drop in the lower left corner of that block. The result of a current distribution analysis is shown in Figure 2. It has to be pointed out, that this level of detail about the operation of the circuit can only be extracted by a full chip analysis. The current distribution analysis on the chip enables designer to gain knowledge of the current paths between the blocks as well as inside the blocks and in this way allowing them to correct the problem.

The results of a static IR drop analysis on an improved design version is shown in Figure 3. The upper and lower blocks are now connected. Current distribution changed in such a way, that the IR drop in the lower left logic block is reduced, but now an IR drop in the memory block is introduced.

Another interesting effect is revealed by performing electromigration analysis on the design as shown in Figure 4. First the analysis shows clearly, that the connection is not sufficient to carry the introduced high current density over a longer period of time and will eventually fail during the lifetime of the chip. Second electromigration problems have been detected at points far from the location of the modification. This problems are also marked with dark (red) colors to be able to easily spot the problems. Even if every block is properly designed by considering IR drop and electromigration, problems caused by the connection between the blocks cannot be detected by an analysis on the block level emphasizing the importance of final verification at full chip.

Figure 5 shows the IR drop on a further improved design version. In this case the attempt was made to avoid electromigration problems and to minimize the IR drop across the chip. The analysis results on the final design version show, that the attempt was successful and the excessive IR drop in the large block at the upper right side of the chip has been avoided. Additionally, the coupling to the neighboring blocks has been reduced and a more uniform IR drop distribution across the chip achieved.

A snapshot of the IR drop during dynamic analysis of a clock cycle is shown in Figure 6. This example was captured immediately after the clock switched. The comparision between static and dynamic analysis shows that the global IR drop is approximated very well. Besides the inherent differences of the static and dynamic approach, local differences are caused by inaccuracies of the underlying switching probabilities and the test vectors used for dynamic analysis. The results validate the experience of the authors, that most of the design problems related to power distribution can be detected using static techniques.

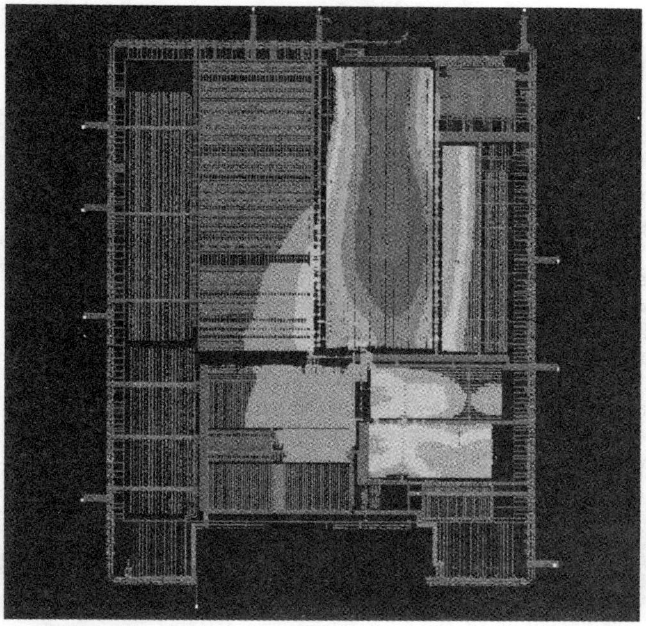

Figure 5: Static IR drop analysis results with proper power grid

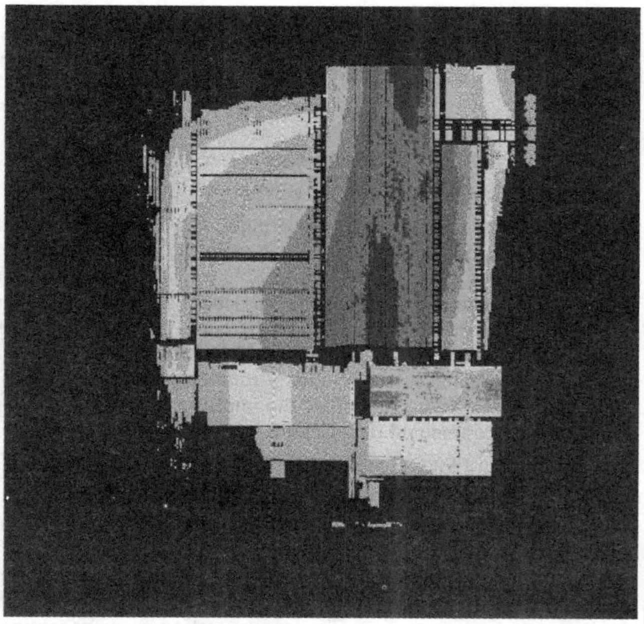

Figure 6: Snapshot of the dynamic IR drop immediatly after clock switching (power grid itself is not drawn)

7 Conclusion

Due to the advancement in fabrication and design technology, power grid analysis has become a necessary step in the verification of new designs. The quality of the power grid has a significant effect on the timing and reliability of a design. In the future, new materials such as copper, with lower sheet resistances and higher resistance to electromigration, will be used for power distribution and interconnect lines. This will increase performance and improve reliability, but this advantage will quickly be diminished by increasing device density and higher operating frequencies. Power grid analysis will continue to be an important part of the design verification process. A methodology has been described, which allows the application of full-chip power grid analysis, permitting the validation of power supply voltages at every gate in the design as well as the reliability of the power grid. The combination of different techniques makes it possible for designers to determine critical voltage drops and reliability weak spots at the full-chip level, taking into account the interdependency of adjacent blocks in terms of power flow.

The authors would like to thank R. Saleh and M. Benoit for valuable discussions and support.

References

[1] *Lightning V1.2 Power-Distribution Verification Manual.* Simplex Solutions, Inc., 1997.

[2] Chi-ying Tsui, Radu Marculescu, Diana Marculescu, and Massoud Pedram. Improving the efficiency of power simulators by input vector compaction. In *Design Automation Conf.*, pages 165–168, June 1997.

[3] Joseph N. Kozhaya and Farid N. Najm. Accurate Power Estimation for Large Sequential Circuits. In *Int'l Conf. on Computer Aided Design*, pages 488–493, November 1997.

[4] Yong Je Lim, Kyung-Im Son, Heung-Joon Park, and Mani Soma. A Statistical Approach to the Estimation of Delay-Dependent Switching Activities in CMOS Combinational Circuits. In *Design Automation Conf.*, pages 445–450, June 1996.

[5] Li-Pen Yuan, Chin-Chi Teng, and Sung-Mo Kang. Statistical Estimation of Average Power Dissipation in Sequential Circuits. In *Design Automation Conf.*, pages 377–382, June 1997.

[6] H.J.M. Veendrick. Short-circuit dissipation of static CMOS circuitry and its impact on the design of buffer circuits. *IEEE J. Solid-State Circuits*, SC-19:486–473, August 1984.

[7] Radu Marculescu, Diana Marculescu, and Massoud Pedram. Hierarchical Sequence Compaction for Power Estimation. In *Design Automation Conf.*, pages 570–575, June 1997.

[8] Angela Krstić and Kwang-Ting Cheng. Vector Generation for Maximum Instantaneous Current Through Supply Lines for CMOS Circuits. In *Design Automation Conf.*, pages 383–388, June 1997.

[9] Subodh Gupta and Farid N. Najm. Power Macromodeling for High Level Power Estimation". In *Design Automation Conf.*, pages 365–370, June 1997.

[10] Alessandro Bogliolo, Luca Benini, Giovanni De Micheli, and Bruno Riccò. Gate-Level Power and Current Simulation of CMOS Integrated Circuits. *IEEE Trans. VLSI Syst.*, 5(4):473–487, December 1997.

[11] *Thunder V1.2 Transistor-Level Verification Manual.* Simplex Solutions, Inc., 1997.

PANEL: SYSTEM CHIP TEST CHALLENGES, ARE THERE SOLUTIONS TODAY?

Chair: Prab Varma - Duet Technologies Inc.

Organizers: Prab Varma - Duet Technologies Inc., Takahide Inoue - Sony USISTC.

To meet the challenges involved in designing systems on silicon, IC designers are increasingly adopting a design re-use methodology, in which pre-designed logic modules, often called virtual components (VC) or intellectual property (IP) cores, are integrated together to construct system chips. Testing such increasingly complex systems comprising components that may be "black-boxes" is a major problem. The panel will discuss the challenges associated with testing system chips containing a diverse range of pre-designed VCs and will address the question of whether there are viable solutions today. It will discuss the impact of design re-use on test methodologies and will discuss whether the time to market gains offered by design re-use will be realized without test re-use. It will examine the role of scan, BIST, at-speed testing etc. as virtual component test techniques, and will discuss the issues involved in providing test access to facilitate test re-use. It will debate the impact of the SIA road-map on test requirements and will discuss the limitations in ATE that will have to be overcome and the issues in performance testing that will have to be addressed to test tomorrow's system chips. In addition, it will look at the role of standards (such as those being proposed by VSIA and IEEE) and discuss whether standardization is possible.

Erik Jan Marinissen, Philips Research Laboratories

Designers increasingly make use of large reusable modules, the so-called "cores", in order to efficiently use design resources and improve on time-to-market. To prevent testing from becoming the bottleneck in the entire development trajectory, the test community has set its hopes on the reuse paradigm as well. The potential benefits of design and test reuse w.r.t. time-to-market are obvious. However, many practical cases have demonstrated that design and test reuse might even increase development time if no special measures are taken to facilitate reuse. Additional features should enable "plug & play" in both design and test, and the costs of those features in terms of silicon area, performance degradation, power dissipation, etc. should be balanced against the benefits of reduced time-to-market.

Core-based design and test divide the IC development community into two groups: (1) core providers and (2) core users. To allow "mix & match" of cores from different sources standardization is needed. A core test standard should enable the integration of cores with different, possibly incompatible, core-internal tests onto one IC. Large semiconductor companies have internal core providers and core users; here, company-internal standardization is a solution. As an example, Philips has both design-for-test guidelines to ensure access to embedded cores, as well as CAD tools for test protocol expansion and scheduling in place. However, an industry-wide take off of core-based design requires an industry-wide standard w.r.t. core test, as pursued by IEEE P1500.

Industry-wide standardization is by definition slow and tedious. Fortunately, the P1500 Working Group has made good progress. Core providers, core users, and CAD companies will all benefit from one industry-wide standard as opposed to many different proprietary solutions. Therefore: let's make P1500 a success, it will be a key enabling factor for systems-on-chips!

Bruce Mathewson, Advanced RISC Machines Ltd.

As systems on a chip become increasingly complex it becomes more difficult to use a single test methodology for an entire system design. A number of different test methodologies exist and each has its own advantages and disadvantages. The following trade-offs must be balanced to determine the most suitable test approach to adopt:

o Fast / automated test vector generation

o High fault coverage

o Low silicon overhead

o Quick test time

o Ease of tracing faults from vector failures

With the increased size of silicon area available it becomes necessary to import IP blocks (or Virtual Components) from a number of different sources and it is possible that the test methodologies of the various components may not be compatible.

Additional factors are also introduced into the choice of test methodology, such as:

o How the virtual component can be tested when it is not easily accessible from the device pins.

o How the virtual component can be integrated into the test process without revealing the internal IP.

In future the industry will require a standard approach to interface the test facilities of different virtual components, as well as a method of bringing the internal test interfaces to the external pins of the device.

Rudy Garcia, Schlumberger Technologies

In order to completely test manufactured system chips, we need to address all of the following requirements:

VC Test - Each Virtual Component needs to tested in situ, to the desired quality level.

This requirement has further test implications:

Test Access- We need to access individual VCs from IC primary I/Os.

Test Isolation- That is, the need to be able to isolate one or more VCs from the rest of the chip for test purposes.

Interconnect Test- The ability to test the interconnect between VCs

Shadow Logic Test- The system on a chip integrator generally adds user defined logic to the collection of VCs. The ability to test this logic which may not accessible from the primary I/Os is also required.

Test Vector Integration- How to allow for the integration of VC and User Defined Logic (UDL) into a single system-chip test program.

Test Logic Integration- What logic circuits may be necessary to manage the test logic in an efficient manner.

Failure Identification- Because the VCs may in general come from different providers, we need to have the ability of isolating failures to individual VCs.

Finally, although most of the above techniques suggest a structural test kind of methodology, recent experiments conducted by Sematech and published last year conclude that a purely structural test regimen is not sufficient to ensure the quality of outgoing devices. There continues to be a need for at-speed functional testing to ensure the quality of the product.

Yervant Zorian, LogicVision Inc.

Issues in testing core-based system-chips differ considerably from the conventional ASICs and standard ICs on a board. A core-based system-chip has only a single manufacturing and test instance for the whole system-chip. The individual cores need to have predetermined and autonomous test solutions, especially with today's trends of:

- Mixing distinct technologies on a single chip, such as DRAM, analog and logic technologies, hence the need for a single ATE based test solution.

- Mixing diverse levels of hardware descriptions on a single chip, such as layout, transistor, gate, RTL, etc., hence the need for self-contained test solution.

- Increasing the dominance of non-mergeable cores in today's system chips, by using providing hardened soft cores, hence the need for autonomous test.

- Increasing the internal speed of system-chips versus the external (I/O speed), hence the need for at-speed test based on-chip clocking.

- Increasing need for plug-n-play of cores from diverse sources, hence the need for self-contained test solutions.

- Requiring IP protection, hence the need of a test solution with limited structural accessibility.

The above trends force us to use embedded test resources in IP cores. A good example of such resources would be cores with full BIST capabilities or at least BIST-ready feature. The system-chip test strategy has to leverage such resources and build the overall test solution accordingly.

Sujit Dey, Univ. of California, San Diego

The key features that make system-on-chips using IP cores an attractive system design methodology, namely design re-use, heterogeneity, configurability, and customizability, are also the ones that make testing and debugging system-on-chips a complex challenge. Design re-use of IPs from various sources, and integration of heterogeneous components - from processor cores to complex multimedia cores, digital and analog components, synchronous and asynchronous interfaces, and electrical and mechanical components, pose serious challenges to testing a system chip, with diverse test strategies employed by the heterogeneous components. While development of test standards by the IEEE P1500 WG and the VSIA are necessary steps to ease the tasks of chip-level test access and test integration, new test methodologies need to be developed to address the following issues:

- Testability characterization of soft IPs,

- Test analysis tools to determine optimal chip-level test access structures, and test area and test time trade-offs,

- Test planning and scheduling tools at the chip-level,

- Debugging and diagnosing ICs which have significant software content,

- Testing mixed-signal ICs, and future microelectromechanical systems (MEMS) on silicon.

In the system-on-chip design paradigm, the processor core, the associated peripheral units, and the underlying bus, memory and communication architectures, can all be configured and customized to best match the embedded system application, thereby giving tremendous cost and performance advantages over traditional system-on-board approaches. How are we going to address the issue of standardization of test and access mechanisms, when the cores themselves can be customized, and the system chip configured, by the system designer? We need to rethink our "standard test interface" requirements, and evolve a flexible, and customizable, test re-use strategy. We definitely need standards for today's needs, but we also need new test methodologies and tools to cope with the multi-million gate, heterogeneous, and customizable system chips of tomorrow.

Rob Roy, Intel Corp.

No advance position statement.

System-Chip Test Strategies

Yervant Zorian

LogicVision, Inc.
San Jose, California, USA
zorian@logicvision.com

Abstract

A major challenge in realizing core-based system-chips is the adoption of adequate test and diagnosis strategies. this paper focuses on the current industrial practices in test strategies for system-chips. It discusses the challenges in testing embedded cores, the testing requirements for individual cores, and their test access mechanisms. It also covers the integrated test strategies for system-chips based on reusable cores. In addition to the state-of-the-art practices in testability schemes, this paper covers the current standardization efforts for embedded core test interface mechanisms.

Keywords: Testing Embedded Core, Intellectual Property Test, System-on-Chip Test.

1: Introduction

Today's embedded cores incorporated into system-chips cover a very wide range of functions, while utilizing an unprecedented range of technologies, from logic to DRAM to analog. They often come in hierarchical compositions. For instance, a complex core may embed one or more simple cores. An example system-chip design is shown in Figure (1), where a cores of various functions are demonstrated. These cores typically come in a wide range of hardware description levels. They spread from fully optimized layouts in GDSII format to widely flexible RTL codes. Embedded cores are categorized into three major types based on their hardware description level: soft, firm and hard [15]. Each type of core has different modeling and test requirements. The three types of cores offer trade-off. Soft cores leave much of the implementation to the designer, but are flexible and process-independent. Hard cores have been optimized for predictable performance, but lack flexibility. Firm cores offer a compromise between the two.

The emerging process of plug-and-play with embedded cores from diverse sources faces numerous challenges in the areas of system-on-chip design, integration, and test. This paper analyzes the test and diagnosis challenges and discusses the current solutions to create testable core-based system-chips. The paper mainly covers the common practices in the industry and shed light on the ongoing efforts to contain today's challenges.

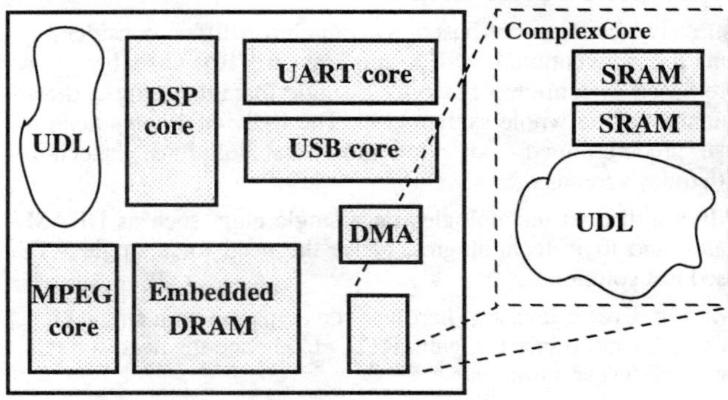

Figure (1) System-Chip consisting of hierarchical cores and User Defined Logic (UDL)

In section 2, the paper discusses the testing challenges in core-based system-chips. Section 3 describes the existing strategies to address core internal test. Section 4 covers embedded core peripheral access mechanisms. Section 5 discusses the overall system-chip test and diagnosis solutions. Finally, Section 5 concludes the paper.

2. System-Chip Test Challenges:

In this section, the main test challenges in system-chips are analyzed and compared to the conventional system-on-board (SOB) ones. In addition to the major differences here, we have to note that system-chips do also share the typical testing challenges of the traditional deep-submicron chips, such as defect/fault coverage, overall test cost and time-to-market.

Even though, the design processes in both paradigms, i.e. in system-chip (SC) and SOB, are conceptually analogous, but the test processes are quite different. For conventional SOB, the manufacturing and test of a chip takes place first, prior to the PCB assembly and test. A chip in this case could be a standard IC or a user defined ASIC; whereas in a core-based system-chip case, there is only a single manufacturing and test instance that takes place for the whole system-chip, see Figure (2). In this case, the individual cores and their surrounding User Defined Logic (UDL) are designed

DAC 98, San Francisco, California

individually first. Next, they are all integrated into an SC. And only then, the composite manufacturing and test steps do take place. See the comparison of the two processes in Figure (2).

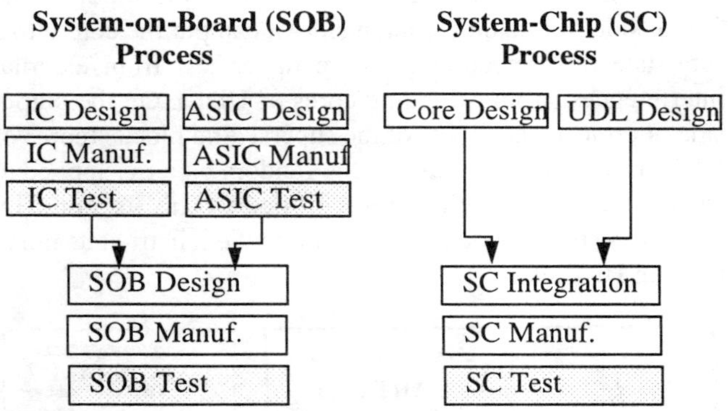

Figure (2) Multi-stage testing for conventional System-on-Board versus single stage System-Chip Test

2.1 Core Internal Test Challenges:

Various test techniques are used for core internal testing, such as functional test, scan, BIST, etc. They typically require two elements:

a. A set of test patterns to be applied to or captured from the core peripheries (core input/outputs). These test patterns are composed of data and control patterns and are generated and evaluated by a test resource (on-chip or off-chip).

b. Internal design-for-testability structures which correspond to the test technique used in the core (e.g. scan chains, test points), if any.

The data patterns are the actual stimulus and response signals; whereas the control patterns specify the test flow, i.e. how to apply and capture the data patterns. Selecting certain internal test mode is considered static control; whereas a shift and apply sequence in a scan protocol is considered dynamic control. Static and dynamic test control patterns, in addition to the data patterns, are applied to the core peripheries to execute a given internal test mode.

A core is typically the hardware description of today's standard ICs, e.g. DSP, RISC processor, or DRAM core. Even though, a given core is not tested individually as in standard ICs, and instead is tested as a part of the overall system-chip, see Figure (2), but still, the preparation of a core internal test is typically done by the core creator. Because, the core integrator in most cases, except for soft cores, has very limited knowledge about the structural content of the adopted core, and hence deals with it as a black box. Therefore he/she can not prepare the necessary test for it. This is especially true if a core is a hard one or is an encrypted Intellectual Property block. This necessitates that the core creator prepares the internal testability, i.e. the DFT or BIST structures and the corresponding test patterns, and delivers it with the core.

Another core creator function is to determine the internal test requirements of the core without knowing the target process and application. For instance, which test method needs to be adopted (e.g. BIST, scan, Iddq, functional test), what type of fault(s) (e.g. static, dynamic, parametric) to target and what level of fault coverage is desired. In SOB, a chip's ppm figures are known prior to the PCB assembly. Hence, the test method and desired fault coverages are predetermined accordingly. But in system-chips, a core creator might not know the target process and the desired quality level. Hence, the provided quality level might or might not be adequate. If the coverage is too low, the quality level of the system-chip is put at risk, and if it is too high the test cost may become prohibitive (e.g. test time, performance, area, power). Furthermore, different processes have different defect type distributions and levels of defect.

Furthermore, in system testing, the failure mechanisms differ from those in production and the test time and power requirements (e.g. for BIST) may be different. Hence, a core should be well characterized with respect to the test and fault coverage. Ideally, the internal test should be a composite one containing modular sections with different characteristics. These modular sections can be switched on or off according to the target process and application. Different internal core test approaches are discussed in Section 3.

Finally, the core internal test prepared by a core creator need to be adequately described, ported and ready for plug and play, i.e. for interoperability, with the system-chip test. For an internal test to accompany its corresponding core and be interoperable, it needs to be described in an commonly accepted, i.e. standard, format. Such a standard format is currently being developed by the IEEE P1500 and referred to as standardization of a core test description language (CTDL).

2.2 Core Test Access Challenges:

Another key difference between SOB and system-chip is the accessibility of component peripheries, i.e. accessing primary input/outputs of chips and cores, respectively. With SOB, direct physical access to chip peripheries, i.e. pins, is typically available to use probing during manufacturing test; whereas for cores, which are often deeply embedded in a system-chip, direct physical access to its peripheries is not available by default, hence, an electronic access mechanism is needed. This access mechanism requires additional logic and wiring to connect core peripheries to the test resources defined in Section 2.1. The logic block performs switching between normal mode and the test mode(s) and the wiring is meant to connect this logic block which surrounds the core to the test resource.

1. Peripheral Access Block: is the switching logic that surrounds the core, as shown in Figure (3). This is sometimes also called a test collar or wrapper. This block contains a Test Access Mechanism (TAM) dedicated to applying and

capturing test data patterns, and a Test Control Mechanism (TCM) for dynamic and static test control patterns. The content of a Peripheral Access Block is typically compatible with the core internal test technique and is optimized for area and performance constraints, are summarized in Section 4. Test control, as shown in Figure(3), is necessary for a given peripheral access block to select (i.e. enable and disable) the paths according to the desired mode.

2. Test Access Path: connects the input/output or bidirectional cells of the TAM and static and dynamic signals of the TCM to the test resource (off-chip or on-chip). The test resource as indicated earlier contains the necessary test data and control patterns. Figure (3) demonstrates three possible access paths connecting core inputs P1, P2, and P3 to three possible test resources T1, T2, T3, while passing via the TAM of the Peripheral Access Block. The first test resource is based off-chip, e.g. functional patterns applied from an Automatic Test Equipment. T2 is connected to an on-chip test response. This may be a BIST Test Pattern Generator [1], a functional resource [2], or an on-chip bus [14], etc. Finally, T3 is connected to a test resource (test access mechanism) internal to the Peripheral Access block, such as a scan flipflop from a neighboring cell supplying serialized (functional or structural) patterns. N1, N2, and N3 are normal mode inputs to the embedded core.

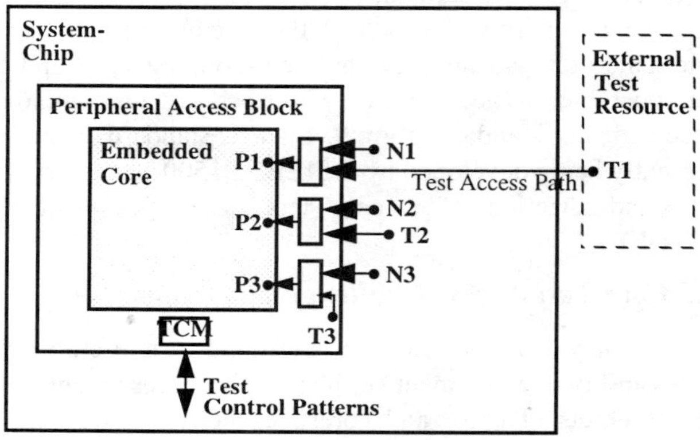

Figure (3) Embedded Core Accessibility

In addition to the normal mode and core internal test modes, a System-chip often requires several other test related modes. The TCM receives the static test control patterns and provides necessary modes, i.e. configuration, at the Peripheral Access Block. One such mode is the core external test. The role of an external test is to test the interconnect faults between the cores in a System-chip (e.g. interconnect faults between the DSP core and the Embedded DRAM in Figure(4)). Also, core external test is needed to participate in the UDL test, by providing controllability and observability to the UDL peripheries through the Peripheral Access Block of a neighboring core, such as the DSP core in Figure (4).

In addition to applying and capturing test patterns during internal and external test modes, the Peripheral Access Block can also be utilized for core isolation. Typically, a core needs to be isolated from its surroundings in certain test modes. Core isolation is often required on either the input or the output side. If it is on the input side, it can put the core into a safe-state by protecting the core under test from external interferences (from preceding cores or UDL). On the output side, it protects the inputs of the superseding blocks (cores or UDL) from undesired values (e.g. random patterns applied to tri-state buffers creating bus conflicts). This is done by controlling the core outputs and putting them in tri-state mode (core in Hi-Z state).

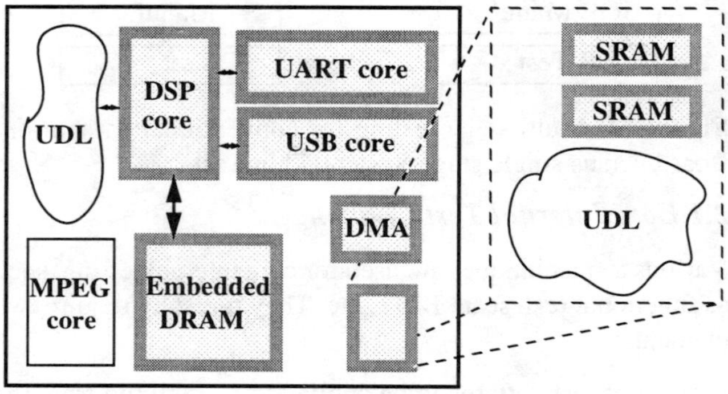

Figure (4) Embedded Cores with Peripheral Access

Sometimes, other core isolation modes are also required to put a core in a stand-by mode, or low power dissipation mode, or ByPass mode [5][11], or finally to control the current supply of a core, if Iddq testing is used in the System-on-chip.

The Peripheral Access Block being an interface block between a core and its surrounding logic (UDL or other cores) provides the necessary support for the above access and isolation modes. It allows these modes to be selected through the mode control protocols which control the switching capabilities of the Peripheral Access Block. In Section 4, this paper summarizes popular testability strategies to provide peripheral access for embedded cores. Because cores are getting imported from diverse sources, a common set of TAMs and TCMs are required. A scalable architecture with such capabilities is pursued by the IEEE P1500 standardization effort [12].

2.3 System-Chip Test and Diagnosis Challenges:

One of the major challenges in the SC realization process is the integration and coordination of the on-chip test and diagnosis capabilities. If compared to SOB, the system-chip test requirements are far more complex than the PCB assembly test, which consists of interconnect and pin toggling tests. The SC test, as in Figure (2), is a single composite test. This test is comprised of the individual core tests of each core, the UDL test, and the test of their interconnects. As discussed earlier, each individual core or

UDL test may involve surrounding components. Certain peripheral constraints (e.g. safe-mode, low power mode, bypass mode) are often required. This necessitates corresponding access and isolation modes.

In addition to the test integration and interdependence issues, the SC composite test requires adequate test scheduling. This is needed to meet a number of chip level requirements, such as total test time, power dissipation, area overhead, etc. Also, test scheduling is necessary to run intra-core and inter-core tests in certain order not to impact the initialization and final contents of individual cores. With the above scheduling constraints the schedule of the composite SOC test is created.

The static test control for individual Peripheral Access Blocks contribute to the execution of the composite test. A chip level static test control network is required to run the composite test and apply/capture the necessary on-chip and off-chip test patterns (i.e. data patterns and dynamic control).

3. Embedded Core Test Strategies:

The test preparation function for cores crosses the boundaries, from core creation phase, to SC integration, to silicon fabrication phase. In each phase a portion of the overall test is completed. Typically, the core internal test is prepared during core creation. The extent of the core preparation in this phase depends on the core type, i.e. soft, firm, or hard. In some cases, full DFT needs to be incorporated in the core creation phase, and in others only test scripts and benches are generated in this phase. The test information transfer from one phase to the other, i.e. from one player to the other, requires standardization in order to guarantee interoperability and ease of plug-and-play.

Selecting an adequate test strategy for an embedded core, such as BIST, ATPG/scan, or functional patterns, is not very different then selecting a test method for a conventional ASIC. Here, the issue is that the function of preparing a test for an embedded core is distributed between the core creator and user. The extent of their involvements depends on the hardware description level of the core. If the core is mergeable with the User Defined Logic (UDL), than the core integrator has more flexibility in identifying and implementing a core test method; whereas if the core is non-mergeable then the core creator has a much larger role in determining and incorporating the core test method(s).

3.1 Mergeable Cores:

A core is defined as mergeable if it is a soft or firm (RTL or gate level) and can be combined with its surrounding random logic via standard flow without the need to create an isolation colar (i.e. Peripheral Access Block) between itself and the UDL around it. In this case, the internal test method is not pre-defined during the core creation stage. The core integrator merges the core with the surrounding logic and then applies the same test method (e.g. scan, ATPG, BIST) on the merged logic, i.e. the core and the UDL [7]. Figure(5) shows the RTL level MPEG core of Figure (4) merged with the neighboring UDL to create an extended UDL for test purposes. In the creation phase, the extent of test preparation for mergeable cores is limited to passing audits, generating DFT scripts, running fault simulation with the selected test strategy to estimate the fault coverage and the vector counts. Describing the above test related information by the core creator and porting it to the integrator is required.

3.2 Non-Mergeable Cores:

A core is considered non-mergeable if it can not be absorbed by its surrounding logic. Hence, it remains as a distinct entity. This happens either because the core hardware description level is different from the one used in the surrounding logic. For example, a layout level DSP core shown in Figure (5) can not be merged with the RT level surrounding UDL; or the core uses a different technology, such as DRAM, Flash, RF or analog; or the core needs to remain isolated for IP protection. Hence, it is encrypted.

In the case of non-mergeable cores, typically the internal core test is incorporated mainly during the core creation phase and it is based on the target technology. In addition, the history of the core may have an impact on the internal test strategy. For instance, almost all memories today tend to use BIST. Hence, providers of embedded memory generators typically incorporate BIST wrappers in the memory core design [3].

The basic core test strategies used for mergeable and non-mergeable cores are typically the same as the ones used in conventional ASICs today. They either depend on external resources or are self-contained with minimal need for external test patterns. The tests that require external patterns are based on functional test patterns or structural (ATPG generated) patterns. The functional patterns are created for design verification purposes and typically do not provide high fault coverage, especially for large cores. Various cores today do provide functional vectors with the delivery of a core. Examples of such cases are RISC cores, some memory cores, and analog cores.

ATPG generated patterns are based on structural testing, hence provide measurable fault coverage. But they do not typically achieve high fault coverage unless coupled with a DFT approach, such as scan. In such a case, the scan chain is pre-synthesized during the core creation stage of a non-mergeable core, and during the integration stage for mergeable cores. ATPG patterns are often generated by the core user for un-encrypted cores. Typically, ATPG patterns target stuck-at faults, and sometimes performance and Iddq faults are also targeted. Certain processor and peripheral cores are delivered today with scans and ATPG patterns.

The problem with external test pattern based strategies is that they typically require large test volumes and need complex dynamic test control protocols to get stimulus to cores, to receive responses from cores and to set up the necessary control to execute the test on a core. A promising strategy that minimizes such requirements is BIST. It generates and evaluates the test patterns on-chip, hence does not require porting of test vectors from the core creator to the integrator and then to the fabricator. BIST is an autonomous testing method and is considered ideal for core-based systems.

In addition to providing the on-chip test resource, BIST simplifies peripheral access complications, since it requires a simple interface and allows re-use of the core as a self-contained block. Today's BIST schemes are mature enough in terms of providing very high fault coverages. They also enhance the diagnosability, while allowing IP protection.

The core test scheme has to provide modular interface to allow integration of hierarchical test control scheme and allow potential sharing of test resources at higher levels.

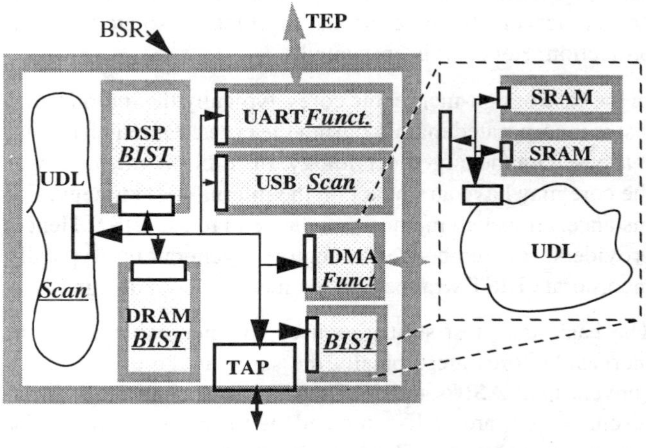

Figure (5) System-Chip with Core Internal Test, Peripheral Access Blocks and Core Test Interface

With a number of test strategies used for core internal test, there seems to be no possibility for standardizing an internal test method. However, the delivery mechanism of the core test from core providers to core integrators need to be standardized. This may happen through standardizing the interface mechanism between the core and the system-on-chip, as pursued by IEEE P1500 [12]. Such a common mechanism definitely helps create interoperability of cores from diverse sources.

4. Core Peripheral Access Mechanisms:

Typically, cores are deeply embedded in a system-chip. In order to apply or exercise the test of a given core, access to its peripheries is necessary. Such an access un-embeds the core, hence providing full controllability and observability to its peripheries. In addition to accessing, this provides effective isolation for IP protection and for putting the core and the surrounding logic in safe modes. A peripheral access around a core fully contributes to the test of the surrounding UDL, by providing observability to the core inputs and controllability to the core outputs.

Traditionally, there are three major peripheral access techniques.

4.1 Parallel Direct Access:

The technique is based on parallel access to the core I/Os by either using dedicated chip level terminals or by sharing chip terminals while multiplexing and demultiplexing them with other signals. This approach, which is termed as star approach, is popular when the number of cores is very limited and the number of chip pins is more than the core pins. But with the continuous increase in the ratio of core terminals to chip terminals, this peripheral access technique is reaching its limits. A number of proprietary techniques exist today to map core peripheries to chip primary input/outputs, even if the later is smaller in number. A major disadvantage in this access mechanism is its very high routing cost. Also, special attention need to be applied if at-speed testing is required due to problems with access path delays, skews, etc.

4.2. Serial Scan Access:

The second technique for peripheral access is based on serialization of test patterns. A scan chain around the core allows indirect but full access to all the I/Os of a core. This is called ring access mechanism. The rings can be internal or external to the core and can be implemented by the core creator or the integrator. In addition to accessing the core, such a ring also provides effective isolation from the surrounding host logic, which may be necessary to run different tests in the core and its surrounding logic. The complexity and hardware requirements for the ring approach is acceptable for today's system-on-chips and it remains independent of pin limitations. The use of Boundary-Scan IEEE 1149.1 as a serial access mechanism has been used by few core providers such as ARM and TI [16].

The routing cost of this serial approach is far less than the previous one. Also, the serial approach allows effective isolation and UDL test if the scan chain is used in its external test mode. The two negative aspects of this approach are the test time increase, due to serialization, and the arbitrary signal switching during scan (control and clock signals need to hold scan cell output stable).

Several DFT approaches were introduced to minimize the hardware cost of the serial scan ring around a core, by using existing functional flip-flops, mixing internal and external FFs, or using compaction and expansion [4] [13]. In realistic examples often a combination of serial and parallel access cells are used for a given core, where the data patterns are applied through the serial ring, whereas the clocks and control signals are asserted via parallel access.

4.3. Functional Access:

The third technique for peripheral access is based on using the surrounding logic to propagate and justify the necessary test patterns [2] This can be either based on existing normal mode logic put in transparency for core test purposes [2], or be based on using existing DFT mode, such as scan chains in the surrounding logic. Even though, the cost of extra hardware is very low, this approach becomes complicated if the surrounding logic has deep functional paths and requires sophisticated protocols for test pattern translation.

The timing impact of a peripheral access mechanism is critical. The path delay created by peripheral access logic and the skews introduced in the access path need to meet the requirements of core test timing constraints.

The ring based technique remains the most realistic option for peripheral access. The ring approach provides access to run internal BIST and internal scan for each core. It also helps testing the surrounding logic using the ring registers.

5. Test and Diagnosis for System-Chips:

At the system-chip level, several test related functions are performed. This includes the test for UDL, the assembly of the composite test, the creation of the test and diagnosis test data and control network. All the above capabilities are utilized during fabrication. Also, at this phase, the DFT circuits are used for silicon debug and diagnosis.

The system-chip test capabilities are integrated by connecting the test facilities of each core and the DFT of the surrounding logic under one chip level mechanism. This can be realized by either multiplexing the test access paths of individual cores with the primary input/outputs of the chip [8], or by connecting the test access paths to chip level test posting buses [14]. Typically, the chip level control circuitry is hooked to the Boundary-Scan port of the chip. A test session for the system-chip has to be composed of intra-core testing to guarantee that each core has been manufactured correctly, and an inter-core testing throughout the surrounding logic. The test execution schedule has to take into account several variables such as test time, power dissipation, noise during test, and area optimization [17].

In addition to the composite SC test, which is meant to be used as manufacturing test, other test, debug and diagnosis modes are often required. For instance, test modes are required for field testing and silicon debug modes for the SC development. Furthermore, diagnostic modes are often very critical to the production of SCs. They are required to identify failed cores in some cases to determine the internal failures to each core. Fault isolation requirements may impact the adopted internal core test strategy. In certain cases, the identification of a failed core may not be possible, due to the fact that a core passed its individual intra-core tests, but the composite test detects an inter-core fault, such as an interconnect fault or a delay fault.

6. Conclusion:

This paper presents a number of testing challenges facing core-based system-chips. It identifies three major aspects to testing core-based systems and describes the test strategies used for each case. These are creating adequate and portable core internal tests; providing core accessibility; and creating an integrated test and its control mechanism for the overall system-chip manufacturing.

References:

[1] V.D. Agrawal, C.J. Lin, P.W. Rutkowski, S. Wu, Y. Zorian, "Built-In Self-Test for Digital Integrated Circuits", AT&T Technical Journal, Vol. 73, No. 2, p. 30, March 1994.

[2] F. Beenker, B. Bennetts, L. Thijssen, "Testability Concepts for Digital ICs - Macro Test Approach" Volume 3 of Frontiers in Electronics Testing. Kluwer Academic Publishers, Boston, 1995.

[3] R. Chandramouli, S. Pateras, "Testing Systems on a Chip", IEEE Spectrum, pp. 42-47, Nov. 1996

[4] K. De, "Test Methodology for Embedded Cores which Protects Intellectual Property", VTS 97, pp. 2-9.

[5] I. Ghosh, N. Jha, S. Dey, "A Low Overhead Design for Testability and Test Generation Technique for Core-Based Systems", Proc. ITC, pp. 50-59, Nov. 1997.

[6] R.K. Gupta, Y. Zorian, "Introduction to Core-based System Design", IEEE Design & Test of Computers, Vol. 14, No. 4, Oct.-Dec. 1997.

[7] S.G. Hemmady, T.L. Anderson, Y. Zorian, "Verification and Testing of Embedded Cores", Proc. Design SuperCon, On-Chip Design, S122-1-19, Jan. 1997.

[8] V. Immaneni, S. Raman, "Direct Access Test Scheme- Design of Block and Core Cells for Embedded ASICs", Proc. IEEE ITC, pp. 488-492, Sept. 1990.

[9] M. Hunt, J.A. Rowson, "Blocking in a System on a Chip", IEEE Spectrum, pp. 35-41, Nov. 1996.

[10] B.K. Koenemann, K. Wagner, "Test Sockets: A Test Framework for System-On-Chip Designs", IEEE TTTC Embedded Core Test TAC Study Group, April 1997.

[11] E.J. Marinissen, M. Lousberg, "Macro Test: A Liberal Test Approach for Embedded Reusable Cores", IEEE TECS, Nov. 1997.

[12] Y. Zorian, "Test Requirements for Embedded Core-based Systems and IEEE P1500", Proc. IEEE ITC, pp. 191-199, Nov. 1997

[13] N. A. Touba, B. Pouya, "Testing Embedded Cores Using Partial Isolation Rings", Proc. of the 15th IEEE VLSI Test Symposium, pp. 10-15, April, 1997.

[14] Varma, P., Bhatia, S., "A Structured Test Re-Use Methodology for Systems on Silicon", IEEE TECS Workshop, Nov. 1997.

[15] "Virtual Socket Interface Architectural Document", VSI Alliance, Nov. 1996.

[16] L. Whetsel, "An IEEE 1149.1 based Test Access Architecture for ICs with Embedded IP Cores", Porc. Int'l Test Conference, Nov. 1997.

[17] Y. Zorian, "A Distributed BIST Control Scheme for Complex VLSI Devices", Proc. of the 11th IEEE VLSI Test Symposium, p. 6-11, April 1993.

Finite State Machine Decomposition For Low Power

José C. Monteiro
IST-INESC
Rua Alves Redol, 9
1000 Lisboa, Portugal
jcm@inesc.pt

Arlindo L. Oliveira
Cadence European Labs/IST-INESC
Rua Alves Redol, 9
1000 Lisboa, Portugal
aml@inesc.pt

Clock-gating techniques have been shown to be very effective in the reduction of the switching activity in sequential logic circuits. In this paper we describe a new clock-gating technique based on finite state machine (FSM) decomposition. We compute two sub-FSMs that together have the same functionality as the original FSM. For all the transitions within one sub-FSM, the clock for the other sub-FSM is disabled. To minimize the average switching activity, we search for a small cluster of states with high stationary state probability and use it to create the small sub-FSM. This way we will have a small amount of logic that is active most of the time, during which is disabling a much larger circuit, the other sub-FSM.

We provide a set of experimental results that show that power consumption can be substantially reduced, in some cases up to 80%.

I. Introduction

Power consumption has become a major design parameter in the project of integrated circuits. Two independent factors have contributed for this. On one hand, low power consumption is essential to achieve longer autonomy for portable devices. On the other hand, increasingly higher circuit density and higher clock frequencies are creating heat dissipation problems, which in turn raise reliability concerns and lead to more expensive packaging.

In static CMOS circuits, the probabilistic switching activity of nodes in the circuit is a good measure of the average power dissipation of the circuit. Methods that can efficiently compute the average switching activity, and thus power dissipation, in CMOS combinational [10] and sequential [13] circuits have been developed.

In this work, we are concerned with the problem of optimizing logic-level sequential circuits for low power. This problem has received some attention recently. Several techniques for state assignment have been presented which aim at reducing the average switching activity of the present state lines, and consequently of the internal nodes in the combinational logic block (see for example [12]). Retiming has also been tailored so that the distribution of the registers within the logic block minimizes the total amount of glitching in the sequential circuit [9].

Techniques based on disabling the input/state registers when some input conditions are met have been proposed and shown to be among the most effective in reducing the overall switching activity in sequential circuits [1], [2], [3], [5]. The disabling of the input/state registers is decided on a clock-cycle basis and can be done either by using a register load-enable signal or by gating the clock. A common feature of these methods is the addition of extra circuitry that is able to identify input conditions for which some or all of the input/state registers can be disabled. In this situation there will be zero switching activity in the logic driven by input signals coming from the disabled registers. This class of techniques is sometimes referred to as *dynamic power management*.

The method we propose in this paper falls into this class of techniques. We use finite state machine (FSM) decomposition to obtain the conditions for which a significant part of the registers in the circuit can be disabled. The original FSM is divided into two sub-FSMs, where one of them is significantly smaller than the other. Except for transitions that involve going from one state in one sub-machine to a state in the other, only one of the sub-machines needs to be clocked. By selecting for the small sub-FSM a cluster of states in which the original FSM has a high probability of being in, most of the time we will be disabling all the state registers in the larger sub-FSM.

The overhead associated with the FSM decomposition makes this method not very effective for typical FSMs with a small number of states. However, for large machines, impressive gains up to 80% power reduction are possible.

In Section II, we introduce some basic definitions used throughout the paper. We present related work on logic level power management in Section III. Section IV gives an overview of the problem of FSM decomposition. We describe our method in Section V. We first describe the architecture that allows for the disabling of the inputs/state registers and present an algorithm for the selection of the states for each of the sub-FSMs. In Section VI, we provide a set of experimental results and end with some conclusions and discussion of future work in section VII.

II. Basic Definitions

We use the standard definition of finite state machines:

Definition 1: A finite state machine is defined in the standard way as a tuple $M = (\Sigma, \Delta, Q, q_0, \delta, \lambda)$ where Σ is a finite set of input symbols, $\Delta \neq \emptyset$ is a finite set of output symbols, $Q \neq \emptyset$ is a finite set of states, $q_0 \in Q$ is the "reset" state, $\delta(q,a) : Q \times \Sigma \to Q$ is the transition function, and $\lambda(q,a) : Q \times \Sigma \to \Delta$ is the output function.

The specification of the finite state machine is equivalent to the specification of a state transition graph (STG), where each state corresponds to one node in the graph, and there exists an edge e_{ij} between two states q_i and q_j with label a/b iff $\delta(q_i, a) = q_j$ and $\lambda(q,a) = b$.

Under specific input line probabilities, it is possible to compute the stationary state and transition probabilities. The stationary state probability, $P(q_i)$, is the probability of the finite state machine being in state q_i in a given clock cycle. In this work, we will always use the

DAC 98, San Francisco, California

absolute transition probabilities, indicated as $P(e_{ij})$. $P(e_{ij})$ represents the probability that, at any specific time, transition e_{ij} is active. $P(e_{ij})$ can be computed using:

$$P(e_{ij}) = P(q_i) \times P(a) \tag{1}$$

where $P(a)$ represents the probability of the primary inputs combinations that trigger the transition e_{ij}.

For each state q_i we can write an equation:

$$P(q_i) = \sum_{k=1}^{|Q|} P(e_{ki}) \tag{2}$$

We obtain $|Q|$ equations out of which any one equation can be derived from the remaining $|Q| - 1$ equations. Since at any specific time the machine is in one and only one state, we have a final equation:

$$\sum_{k=1}^{|Q|} P(q_i) = 1 \tag{3}$$

This linear set of $|Q|$ equations can be solved to obtain the different $P(q_i)$'s.

This system of equations is known as the Chapman-Kolmogorov equations for a discrete-time discrete-transition Markov process, and, under some general conditions, has a unique solution [11].

III. Related Work

So called power management techniques that shutdown blocks of hardware for periods of time in which they are not producing useful data are effective methods to reduce the power consumption of a circuit. Shutdown can be accomplished by either turning off the power supply or by disabling the clock signal. A system-level approach is to identify idle periods for entire modules and turn off the clock lines for these modules for the duration of the idle periods ([4], Chapter 10).

Power management techniques have also been proposed at the logic level. The main difference is that the shutdown of hardware is decided on every clock cycle. The technique we propose in this paper falls under this category. In this section we describe previously proposed methods.

A. Precomputation

One of the first logic-level shutdown methods is *precomputation* [1]. In this method a simple combinational circuit (the precomputation logic) is added to the original circuit. Under certain input conditions, the precomputation logic disables the loading of all or a subset of the input registers. Under these input conditions, no power is dissipated in the portions of the original circuit with only disabled registers as inputs.

The choice of the number of inputs to use for the pre-computation logic is critical. The more inputs used the highest the probability the precomputation logic will be active, thus disabling logic in the original logic block. However, this also increases the size of the precomputation logic, a circuitry overhead that is active all the time, thus offsetting the gains obtained by disabling A a larger fraction of the time.

Once the number of inputs to the precomputation logic is fixed, the input selection is based on the probability that the outputs can be computed without the knowledge of a specific input, i.e., the size of the observability don't-care set. Inputs with lowest probabilities are selected to be in the precomputation logic.

B. Gated-Clock Finite State Machines

The *gated-clock finite state machines* approach [2] is based on identifying self-loops in a Moore FSM. If the FSM enters a state with a self loop, the clock is turned off. In this situation, the inputs to the combinational logic block do not switch, and thus we have virtually zero power dissipation in that block. When the input values cause the FSM to make a state transition, the clock signal is again enabled and the circuit resumes normal operation.

The fact that this procedure is only applicable to Moore FSMs can be very limiting. Techniques to locally transform a Mealy FSM into a Moore FSM are given in [2] so that the opportunity for gating the clock is increased.

The method for FSM decomposition that we describe in this paper can be seen as an extension of the gated-clock FSM approach. In FSM decomposition we can consider the cluster of states that we select for the small sub-FSM as a "super-state" and then transitions between states in this cluster are no more than self-loops in this "super-state". The decision as to what states make up the "super-block" basically gives the opportunity to maximize the number of self-loops.

C. Selectively-Clocked Systems

The implementation of a FSM in terms of sub-FSMs with the objective of achieving low power consumption by only enabling the clock signal of the active sub-FSM has been proposed recently [3]. However, the approach followed is completely different from the one we are proposing in this paper.

The work presented in [3] is concerned with the construction of controllers obtained during the process of behavioral synthesis. In this process, mutually exclusive sections of computation operations are detected. Each of these sections is then implemented as a separate FSM. The overall controller is thus made up of a set of interacting FSMs. Since by construction no two FSMs can be active simultaneously, only the clock signal for one of the FSMs needs to be working, therefore achieving a significant power reduction as compared to a monolithic implementation of the controller.

The technique we describe in the present paper is more flexible in the sense that it is not tied to the structure of some behavioral-level circuit description. However, the overhead in our approach may be higher. In any case, during the synthesis process we can take advantage of both techniques as the FSM decomposition approach can be applied to each individual FSM generated with the approach of [3].

D. Decomposition By Choice Of Disjoint Encodings

An approach that is very closely related to our work is based on the selection of encodings that are mutually orthogonal. This method can also be viewed as a state transition graph decomposition approach [5], although the resulting hardware does not follow strictly the standard structure for FSM decomposition, as described in the next section.

In this method, the STG is partitioned in two (or more) sets of states. All the states in one of the sets are assigned encodings with the first bit at 0, while the states in the other set are assigned encodings with that bit at 1. The combinational logic can now be broken into two separate blocks: one that is active when the first state bit is 0, and one that is active when the first state bit is 1. Since only one of these blocks can be active at any time, the other block can be disabled, leading to potentially significant power savings, if the state partition is chosen carefully.

The authors propose two methods to chose the state partition. The first one is based on the computation of the co-factors of the sequential circuit with respect to one of the state bits, and arbitrarily splits

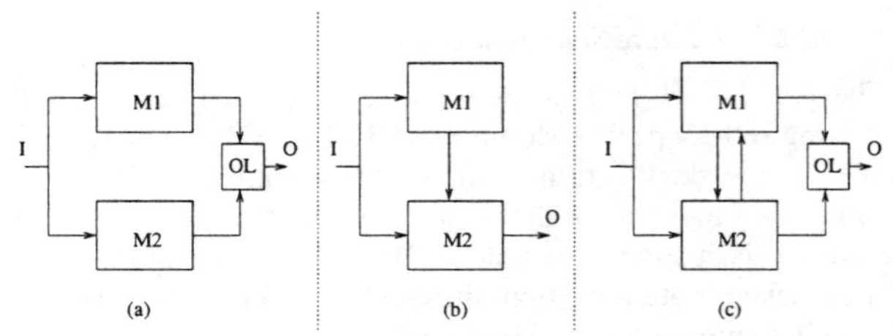

Fig. 1. Parallel, cascade and general approaches to finite state machine decomposition.

the set of two states into two disjoint sets. The second method is related to our own, in that it uses the stationary state probabilities to achieve a partition that leads to a final circuit with low power dissipation. However, the methodology used to implement the final circuit is very different, since only one set of registers is used, and therefore all the registers have to be clocked at every time instant.

IV. Decomposition Of Finite State Machines

The decomposition of finite state machines has been addressed by a number of authors, the first work dating back to 1960 by Hartmanis [7]. Several decomposition strategies have been proposed and are described in detail elsewhere [6]. Parallel and cascade decomposition, shown in Figure 1 (a) and (b), respectively, are simple, but of limited use with finite state machines that do not have a very regular structure.

On the other hand, the more general form of finite state machine decomposition, shown in Figure 1 (c), is applicable to any finite state machine. The structure of the resulting finite state machine is shown in Figure 2. Each of the sub-machines receives, as inputs, not only the

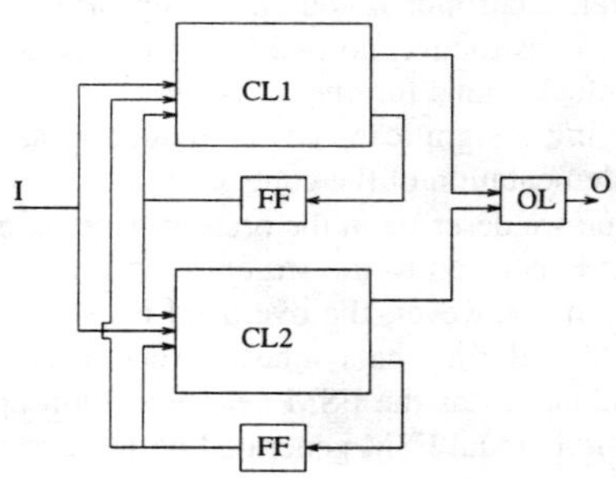

Fig. 2. Structure of decomposed finite state machine.

primary inputs and its own state variables, but also the state variables of the other sub-machine. Although there are several ways in which the STGs of each machine can be built, the following approach is simple and intuitive and will be used in this work.

1. Select a subset of the states in the original STG to belong to the first sub-machine, and let the remaining states belong to the second machine.
2. Generate two STGs, one for each sub-machine. Create a new state, the RESET state, in each of the two new STGs.
3. All transitions entirely inside each of these STGs are copied unmodified from the original STG.
4. Transitions between a state in the first sub-STG and a state in the second sub-STG are replaced by two transitions: one to the RESET state in the first sub-machine, and one from the RESET state in the second sub-machine. The reverse is true for the symmetric case.

We illustrate the procedure with the following simple example depicted in Figure 3. Consider the following state transition graph for a simple sequence detector and assume that a decomposition of the corresponding FSM is desired, with states A and B to belong to machine M2, and all the other states to belong to machine M1.

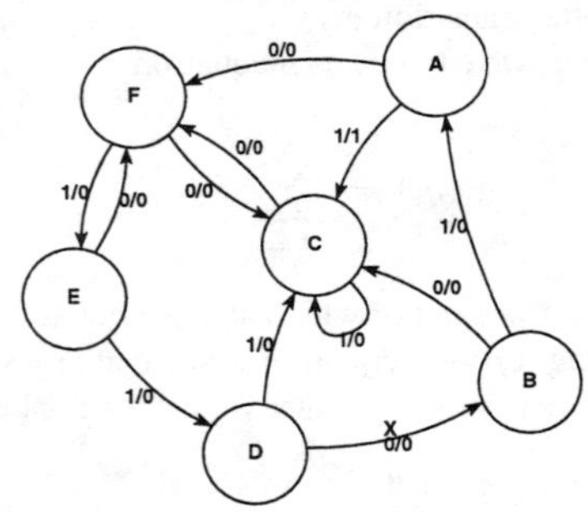

Fig. 3. State transition graph of the sequence detector.

The original STG is transformed into two smaller STGs. Transitions between the states in M1, as well as transitions between the states in M2, are copied without transformation. As shown in Figure 2, the number of inputs to each sub-FSM is enlarged with the value of the state lines of the other sub-FSM, each FSM also providing as outputs the value of the state lines. Transition between a states in M1

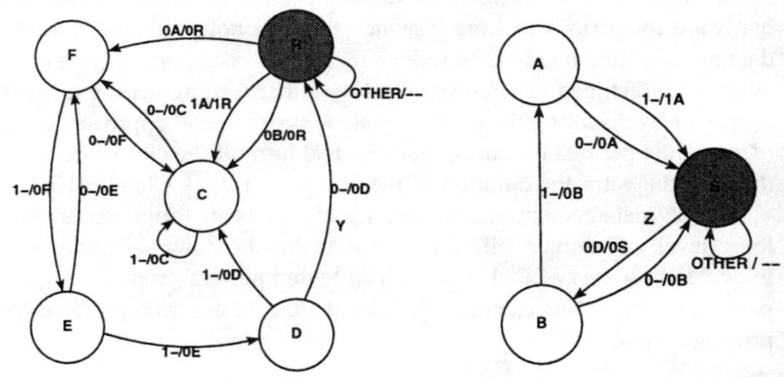

Fig. 4. STGs for the sequence detector after decomposition.

and a state in M2 require a slightly more complex treatment. Consider, for example, the transition marked with **X**, in Figure 3, between states D and B. This transition corresponds to two new transitions, one in each of the STG of the sub-FSMs. In M1, it corresponds to the transition marked **Y**, between state D and state R (the newly created RESET state in M1). This transition is labeled with the original value of the input (0) and has, as output, the value of the source state in M1, D. When machine M1 performs this state transition, it becomes inactive. In M2, it corresponds to the transition marked **Z**, between state S (again, the newly created RESET state in M2) and state B. This transition is labeled with 0 D (the original value of the input plus the encoding of state D), and signals that machine M2 should become active. All input combinations not explicitly shown in the edges leaving the reset state of each machine are self-loops, and they signal the fact that the given machine is to stay inactive until it receives the right input combination.

In practice, the procedure works in a slightly different way from the one illustrated here, because the codification of the sub-FSMs is not known at the time the STG is being partitioned. In fact, using the same state codification for the sub-FSMS and for the original FSM would be sub-optimal, specially in the case where one of the sub-FSMs is much smaller than the other one and requires a much smaller number of registers. In the decomposition of large machines, this may represent a significant advantage over an approach that uses the same set of registers to implement both sub-machines [5]. The exact way by which this information is exchanged between the sub-machines is detailed in section V-C.1.

V. Finite State Machine Decomposition For Low Power

The main objective of this work is to use finite state machine decomposition techniques to achieve low power dissipation. The basic idea, described in this section, is to decompose the original machine into two machines, one of them with a reduced number of states and low power dissipation that performs needed computations for a large fraction of the time.

A. *Targeting The Partition Strategy For Low Power*

Consider the decomposition shown in Figure 2 in Section IV and the state transition diagrams shown in Figures 3 and 4. Notice that, for any transition that takes place between two states other than the RESET state in the same factor machine, only this machine needs to be active. This means that, for these transitions, we can disable entirely the other machine, avoiding all the power dissipation it incurs.

On the other hand, transitions that involve the RESET state and another state are active simultaneously in both machines, because when one machine is leaving the RESET state, the other one is entering it. These transitions tend to generate the largest power dissipation, because both machines are active at once.

The potential for the gains in power dissipation obtainable from the decomposition technique described in Section IV are larger if one selects a partition that exhibits the following characteristics:

1. One of the machines (the *small* machine) has a state transition diagram with a small number of states, and is therefore simple and dissipates a small amount of power.
2. The sum of the transition probabilities between two states in the *small* machine other than the RESET state is as large as possible.
3. The sum of the transition probabilities involving the RESET state in the *small* machine is as small as possible.

The motivation for each of these factors is relatively straightforward. The first and second requirements specify that one of the machines, the *small* machine, will be operating a large fraction of the time with a relatively small amount of power dissipation. The third requirement minimizes the fraction of time both machines are operating, a situation that is undesirable because it corresponds to maximum power dissipation.

Clearly, a partition that satisfies these requirements is not guaranteed to perform well, because the power dissipation of the final system depends on a variety of factors that cannot be considered at this abstraction level, like state encoding, combinational circuit synthesis and technology mapping. However, the three requirements listed above can be considered as useful heuristics in the selection of a partition that will achieve the desired reduction in power dissipation.

B. *Selecting A State Partition For Low Power*

The problem of selecting a graph partition that maximizes a given cost function has many applications and has been addressed in a variety of ways. One of the best known algorithms for this problem is the Kernighan-Lin algorithm [8]. This algorithm has been applied with success in a variety of applications, and was the algorithm used in this work.

The algorithm can be applied in the selection of both balanced or unbalanced partitions, with either a fixed or variable number of states in each partition. Assume, for the moment, that the number of states in each partition is fixed beforehand and that the STG has $N = |Q|$ states. We wish to obtain a partition in two sets with N_1 states on one set and N_2 states on the other set, with $N_1 <= N_2$.

The algorithm starts with an arbitrary partition that has two sets with, respectively, N_1 and N_2 states. It then selects a pair of states to swap, in a greedy way. Specifically, it selects the pair of states to be swapped that lead to higher increase (or smaller decrease) in the objective function. The states are then swapped, and the value of the resulting partition is recorded. These states are also recorded as already swapped, and are therefore fixed in their new partitions. After N_1 states have been swapped, the algorithm terminates, and outputs the best partition seen in the process.

B.1 Selection Of The Objective Function

The application of the Kernighan-Lin algorithm with a cost function that reflects the objectives stated above generates the desired finite state machine decomposition. The result can then be used to generate a sequential network that has a functionality equivalent to the original one, but dissipates, in general, less power.

Given the general objectives described above, we use the Kernighan-Lin algorithm to select a partition of the original set of states Q into two subsets Q_1 and Q_2, that maximize the following objective function:

$$F = \sum_{q_i \in Q_1, q_j \in Q_1} P(e_{ij}) - \beta\Big(\sum_{q_i \in Q_1, q_j \in Q_2} P(e_{ij}) + \sum_{q_j \in Q_1, q_i \in Q_2} P(e_{ij})\Big) \quad (4)$$

The first summation represents the total probability of the transitions occurring entirely inside the STG of the *small* machine, while the second and third summations represent the total probability of the transitions occurring between the two machines.

Empirically, we found that values between 0.5 and 1.0 for the β coefficient worked best, and we selected a value of 0.7 for all experiments.

B.2 Complexity of the Kernighan-Lin procedure

The partition algorithm uses the standard data structures for the representation of state transition graphs. With this simple data structures, a total of $N_1 \times N_2$ pairings of states need to be considered for each step[1] of the Kernighan-Lin algorithm. The evaluation of the objective function can be done in time that is, under some general assumptions, linear on the number of states in the small partition, N_1. Finally, a total of N_1 steps need to be executed to finish the algorithm.

This means the procedure has complexity $O(N_1^3 N_2)$. However, N_1 is always a small number, because it is the number of states in the *small* machine. Typically, N_1 is never superior to 15, and N_1^3 is therefore bounded from above by a constant, leading to a total complexity that is linear in the total number of states in the original STG.

Empirically, we found that the procedure terminates in less than 30 seconds for all the circuits reported, taking less than a second for the majority of them. In any case, the execution of this procedure was never a bottleneck given that, in the majority of the circuits that we were unable to test, the STG extraction step was the limiting factor.

[1]Each step of this algorithm performs a state swap between the two partitions.

C. Generating The Final Network From The State Partition

The FSM decomposition method described in Section IV requires small changes to be effective in this application. These changes are motivated by the following facts:

- Because the *small* machine has a reduced number of states, it is possible to minimize the amount of information sent from the large machine to the small machine.
- Maximum power savings will be achieved if the machines are disabled when they are in a self-loop condition in the RESET state.

C.1 Minimizing Information Exchange Between The Two Machines

In the partition scheme described in Section IV and shown in Figure 1, the communication between the two sub-machines is completely symmetrical.

However, if one of the machines is much smaller than the other one (as measured by the number of states in the STG), this communication mechanism can be changed to our advantage in the following way:

- When the *small* machine makes a transition to the RESET state, the extra outputs of this machine encode information describing which state the machine was before reaching the RESET state.
- When the *large* machine makes a transition to the RESET state, the extra outputs of this machine encode information describing to which state the *small* machine should go from its RESET state.

Because the small machine has a smaller number of states, this change reduces the number of extra outputs and inputs required, thereby reducing the total power dissipation.

C.2 Generation Of The Disabling Network

In this application, it is important to have a sub-FSM disabled when it is in a loop in its reset state. Given the partition methodology outlined in Section IV, it is easy to generate an extra output for each sub-FSM that is used to enable the other machine.

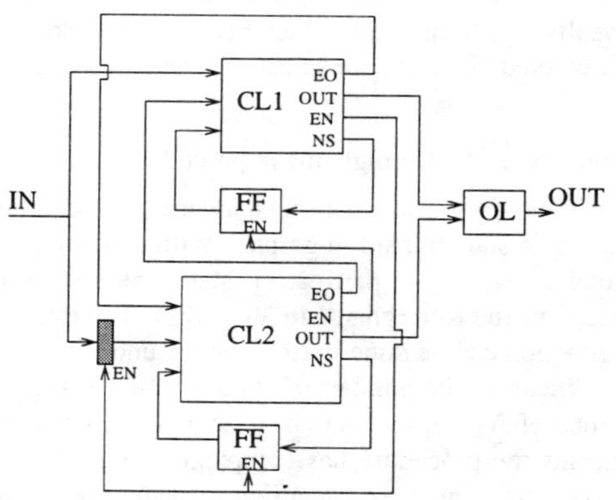

Fig. 5. Structure of decomposed finite state machine with enable circuits and blocking devices.

The structure of the decomposed machine with the enable circuits in place is the one shown in Figure 5. In this scheme, each machine generates a set of extra output variables (*EO*) that encode the information needed to transmit to the other machine which state it must go to, as described in Section V-C.1. Each sub-FSM also generates an enable signal (*EN*) that is used to disable the state registers in the other machine.

TABLE I
FINITE STATE MACHINE STATISTICS.

Circuit	# PI	# PO	# states	# literals	Power (μW)
opus	5	6	10	122	416
mark1	5	14	13	129	386
s386	7	7	13	190	422
sse	7	7	13	174	458
ex4	5	9	14	122	482
cse	7	7	16	255	407
kirkman	12	6	16	297	670
ex2	2	2	19	209	723
keyb	7	2	19	306	550
ex1	8	19	20	317	570
donfile	2	5	24	261	685
s820	18	19	25	433	808
s832	18	19	25	454	836
dk16	2	3	27	382	1087
styr	9	10	30	647	909
sand	11	9	32	718	1146
tbk	6	3	32	1019	1726
s510	19	7	47	381	831
planet	7	19	48	729	1832
s1488	8	19	48	942	1680
s1494	8	19	48	951	1675
scf	27	56	115	1058	1117

For one machine (M2, the *large* one), the primary inputs also go through a set of latches or AND gates, thereby stopping the propagation of transitions from these inputs before they reach the combinational logic. This is represented by the shaded block in Figure 5. We decided not to have the primary inputs of the *small* machine disabled in the same way because this machine is, in general, small, and because the extra number of latches added to this small machine may increase instead of decreasing the total power dissipation.

VI. EXPERIMENTAL RESULTS

We have applied our FSM decomposition technique on circuits from the MCNC91 and ISCAS89 benchmark set. We present a set of preliminary results on the power savings we have obtained. The results were obtained with SIS running on a 170MHZ Ultra I Sun Workstation, using the `power_estimate` command [13] to compute the dissipated power. All circuits were first optimized using `script.algebraic` and mapped using the msu library.

In Table I, we show statistics of the circuits we present results for. The name, number of inputs, outputs and states for each of the circuits used is given in the first four columns. We give the number of literals in column five, which is a good measure of circuit area. The power dissipation of the original circuit is shown in the last column, assuming a supply voltage of 5V, a clock frequency of 20MHz, a zero-delay model and uniform input probabilities.

Table II presents the results we obtained for each FSM. Since the algorithm for state selection described in Section V-B.1 is very efficient, we run the algorithm for different values of n, the number of states in the small sub-FSM. The cost function given in Eqn. 4 was used with $\beta = 0.7$, which we have found empirically to give best results. For each FSM in the benchmark set, we present results for the value of n with which the best power savings were obtained.

TABLE II
RESULTS OBTAINED AFTER DECOMPOSITION.

Circuit name	n	Blocking latches			Blocking ANDs		
		% A	Pwr	% P	% A	Pwr	% P
opus	4	47.4	277	33.5	42.2	281	32.6
mark1	6	55.3	274	29.1	49.1	304	21.4
s386	6	18.5	330	21.9	13.4	342	18.9
sse	5	40.3	231	49.6	34.5	252	44.9
ex4	5	30.5	268	44.2	24.0	234	51.5
cse	5	57.8	231	43.3	53.7	238	41.6
kirkman	3	17.6	461	31.3	10.3	420	37.3
ex2	8	52.6	437	39.6	51.0	462	36.0
keyb	8	27.2	343	37.7	23.7	271	50.7
ex1	6	38.1	459	19.6	34.7	467	18.2
donfile	5	54.8	485	29.2	53.5	574	16.2
s820	4	21.8	344	57.5	14.2	321	60.2
s832	4	19.6	442	47.1	12.3	429	48.7
styr	6	34.9	587	35.5	32.9	611	32.8
sand	15	42.5	1030	10.2	39.8	972	15.2
tbk	10	-22.1	928	46.2	-23.2	975	43.5
s510	7	45.7	372	55.3	37.8	224	73.1
planet	9	31.7	896	51.1	29.9	888	51.6
s1488	4	-1.1	345	79.5	-2.7	438	73.9
s1494	6	-3.4	392	76.6	-5.0	435	74.0
scf	14	47.1	582	47.9	42.8	734	34.3

We have experimented using latches and AND gates as blocking devices for the primary inputs (see Figure 5). In Table II, we give the percentage area increase, power dissipation of the decomposed FSM and the corresponding percentage power reduction using latches and and gates. As we can observe from the table, impressive power savings can be obtained, up to 80%. As it should be expected, the higher savings correspond to larger FSMs. In fact, for smaller FSMs there is no gain. This is due to the fact that the decomposed FSM has some extra logic circuitry and two extra states, the RESET state in each sub-FSM. This can represent a large overhead for small machines, but is less important for larger FSMs.

As should be expected, using latches leads to larger increase in area, with (sometimes significantly) higher power savings. This is not true for all circuits as latches have larger input and internal capacitances than AND gates, thus offsetting the reduced switching activity at the outputs.

VII. CONCLUSIONS AND FUTURE WORK

We presented a methodology for the decomposition of finite state machines targeted towards low power dissipation. The methodology uses well known decomposition techniques to obtain a state machine that, for the majority of the large examples tested, exhibits a much smaller power dissipation than the original. The results show that power savings of up to 80% are possible in some of the examples.

Despite the good quality of the results obtained, there are several directions for future work that may improve these results and extend the range of applicability of the technique.

The single most important direction is probably the extension of this work to the case where the state transition graph cannot be explicitly described. In our experiments, we verified that the bottleneck resides in the extraction of the STG from the sequential circuit description. Because the decomposition algorithm works with an explicit description of the STG, it is not possible to apply the method to machines where the STG cannot even be extracted.

However, it may be possible to apply the decomposition idea even in this case, if a set of transitions with high enough probability exist to justify the method. The basic idea is that, in many cases, this set of transitions can be identified using Monte Carlo methods even without doing a complete traversal of the STG. With this set of transitions available, it is possible to perform a partition of the STG, considering only the states involved in these transitions and using only an implicit representation of the STG. Once the STG is partitioned, the rest of the method is directly applicable, making it feasible to use it in cases where the machine is too large to permit the extraction of the STG.

Another interesting direction for future research is on the automatic selection of the size of the *small* machine. Although, in some cases, significant gains can be obtained with a variety of sizes for this machine, in other cases the result depends strongly on the adequate selection of the value of this parameter. It may be possible to modify the cost function described above to automatically include a term that depends on the number of states, thereby removing the need for the user to specify the value of this parameter or to perform a search for its right value.

Finally, it is clear that the results obtained in this paper can be improved if a more efficient implementation of the decomposition strategy is selected. In particular, it should be possible to reduce the overhead incurred by the addition of the extra outputs to each sub-FSM, by encoding these outputs in a different way that take into account the encoding of each of the sub-machines.

REFERENCES

[1] M. Alidina, J. Monteiro, S. Devadas, A. Ghosh, and M. Papaefthymiou. Precomputation-Based Sequential Logic Optimization for Low Power. *IEEE Transactions on VLSI Systems*, 2(4):426–436, December 1994.

[2] L. Benini, P. Siegel, and G. De Micheli. Automatic Synthesis of Low-Power Gated-Clock Finite-State Machines. *IEEE Transactions on Computer-Aided Design*, 15(6):630–643, June 1996.

[3] L. Benini, P. Vuillod, C. Coelho, and G. De Micheli. Synthesis of Low-Power Partially-Clocked Systems from High-Level Specifications. In *9th International Symposium on System Synthesis*, November 1996.

[4] A. Chandrakasan and R. Brodersen. *Low Power Digital CMOS Design*. Kluwer Academic Publishers, 1995.

[5] S-H. Chow, Y-C. Ho, and T. Hwang. Low Power Realization of Finite State Machines – A Decomposition Approach. *ACM Transactions on Design Automation of Electronic Systems*, 1(3):315–340, July 1996.

[6] S. Devadas and A. Newton. Decomposition and Factorization of Sequential Finite State Machines. *IEEE Transactions on Computer-Aided Design*, 8(11):1206–1217, November 1989.

[7] J. Hartmanis. Symbolic Analysis of a Decomposition of Information Processing. *Information Control*, 3:154–178, June 1960.

[8] B. W. Kernighan and S. Lin. An Efficient Heuristic Procedure for Partitioning Graphs. *The Bell System Technical Journal*, pages 291–307, February 1970.

[9] J. Monteiro, S. Devadas, and A. Ghosh. Retiming Sequential Circuits for Low Power. In *Proceedings of the International Conference on Computer-Aided Design*, pages 398–402, November 1993.

[10] F. Najm. A Survey of Power Estimation Techniques in VLSI Circuits (*Invited Paper*). *IEEE Transactions on VLSI Systems*, 2(4):446–455, December 1994.

[11] A. Papoulis. *Probability, Random Variables and Stochastic Processes*. McGraw-Hill, 2nd edition, 1984.

[12] K. Roy and S. Prasad. Circuit Activity Based Logic Synthesis for Low Power Reliable Operations. *IEEE Transactions on VLSI Systems*, 1(4):503–513, December 1993.

[13] C-Y. Tsui, J. Monteiro, M. Pedram, S. Devadas, A. Despain, and B. Lin. Power Estimation Methods for Sequential Logic Circuits. *IEEE Transactions on VLSI Systems*, 3(3):404–416, September 1995.

Computational Kernels and their Application to Sequential Power Optimization

L. Benini # G. De Micheli # A. Lioy ‡ E. Macii ‡ G. Odasso ‡ M. Poncino ‡

Stanford University
Computer Systems Laboratory
Stanford, CA 94305

‡ Politecnico di Torino
Dip. di Automatica e Informatica
Torino, ITALY 10129

Abstract

We introduce a new sequential optimization paradigm based on the extraction of computational kernels, i.e., logic blocks whose behavior mimics the steady-state behavior of the original circuit. We present a procedure for the automatic extraction of such kernels directly from the gate-level description of the design. The advantage of this solution with respect to extraction algorithms based on STG analysis is that it can be applied to large circuits, since it does not require to manipulate the STG specification.
We exploit computational kernels for optimization purposes; in particular, we describe an architectural decomposition paradigm whose template is reminiscent of the mux-based scheme adopted in parallel implementations of logic-level descriptions.
We show the usefulness of the new optimization style by applying it to the problem of reducing the power dissipated by a sequential circuit. Experimental results, obtained on standard benchmarks, demonstrate the merit of the proposed approach.

1 Introduction

In spite of the large number of states that are potentially reachable by a sequential design, it is a well established fact that, during normal operation, the circuit tends to run through only a few of such states. An informal, though convincing proof of this statement is given by the results of the probabilistic analysis of the finite state machines associated to large networks: Only a few states have sizable steady-state occupation probability [1]. A similar situation occurs at the network primary outputs. While the circuit walks through the most probable states, a limited number of output patterns is generated.
We call *computational kernel* of a sequential circuit a logic block whose behavior mimics the steady-state behavior of the original network. Usually, such block is much smaller, faster, and less power consuming than the circuit it is extracted from. Nevertheless, it can replace the original network for a large fraction of the operation time.
For circuits with a few registers, computational kernels can be calculated in an exact fashion through symbolic procedures for FSM reachability analysis. However, when the network size increases over a few tens of memory elements, the above approaches are no longer applicable for both memory and time reasons. Resorting to approximate, simulation-based techniques that rely on structural circuit analysis, is then mandatory.
In this paper, we discuss heuristics for the automatic extraction of computational kernels, we propose an optimization paradigm based on the exploitation of such kernels, and we illustrate how the technique can be applied to sequential power minimization.

DAC 98, San Francisco, California

The extraction procedure constructs the computational kernel of a given sequential network incrementally by iteratively reducing the size of the logic block implementing the next-state and the output functions through implication analysis and redundancy removal. The extraction algorithm is driven by a cost function, whose objective is that of monitoring the quality of the kernel with respect to a given optimization target (e.g., delay or power) and thus providing a criterion for stopping the iterations.
Both the original circuit and its kernel are fed by the same primary inputs, and the state and output values are computed by the low cost logic (i.e., the kernel) anytime this is possible. By definition of computational kernel, there are high chances for this situation to happen. Therefore, the average computational cost decreases with respect to the original design. Moreover, once the computational kernel is available as a logic network, it can be exploited for optimizing the original circuit as well, because the input conditions when the kernel is active are *controllability don't cares* for the original circuit.
We show the usefulness of kernel-based optimization by applying it to the problem of reducing the power dissipated by a sequential circuit. In this case, the cost function which drives the kernel extraction procedure must take into account both the switching activity and the capacitive load of the reduced network. Experimental results, obtained on a set of standard benchmarks, demonstrate the validity of the proposed approach.

2 Definition of Computational Kernel

For the sake of simplicity, we provide the definition of computational kernel of a sequential design by referring to the FSM which models the behavior of the circuit. This definition is then extended to the more interesting case of circuits described at the structural-level, that is, to designs whose manipulation is not constrained by the size of the available computer memory.
A ***finite state machine*** (FSM), M, is defined as the 5-tuple:

$$M = (X, Z, S, S^0, R)$$

where X and Z are the input and output alphabets, S is the finite set of states, S^0 is the unique reset state, and $R \subseteq X \times S \times S \times Z \rightarrow \{0,1\}$ is the *global relation*. $R(x,s,t,z) = 1$ if and only if, under input $x \in X$, M moves from present state $s \in S$ to next state $t \in S$ with output $z \in Z$.
Mealy-type FSMs produce the output value z when the transition from state $s \in S$ to state $t \in S$ labeled x is taken, while Moore-type FSMs produce the output value z when a given state s is reached. Therefore, in Moore-type machines, states, rather than edges, are labeled with output symbols. M can be represented by a ***state transition graph*** (STG); for Mealy-type FSMs, the vertices of the STG are elements of $s \in S$, and the edges are labeled with pairs $(x, z) \in X \times Z$. On the other hand, the vertices of Moore-type FSMs are labeled with $z \in Z$, while the edges are labeled with $x \in X$.

Given a Moore-type finite state machine, $M = (X, Z, S, S^0, R)$, modeling the behavior of a sequential circuit, and given a probability threshold, p, we define the p-order ***computational kernel*** of M, denoted as M_p, as the following finite state machine:

$$M_p = (X_p, Z, S_p, S_p^0, R_p)$$

$S_p = S_M \cup idle$, where $S_M = \{s \in S : P(s) \geq p\}$ is the subset of the states of M whose steady-state occupation probabilities are larger than p, and ***idle*** is an additional state which corresponds to the remaining states of M. S_p^0 equals S^0 in case $S^0 \in S_M$, otherwise $S_p^0 = idle$. X_p is the set of input signals which includes X. Additional input signals are required to specify the destination state in transitions from ***idle*** to other states in S_p. Finally, $R_p(x, s, t, z) = R(x, s, t, z) - \{(x, s, t, z) : s \notin S_M \wedge t \notin S_M\} + \{(x, s, t, z) : s = idle \wedge t \in S_M\} + \{(x, s, t, z) : s \in S_M \wedge t = idle\} + (x, idle, idle, z)$.
As an example, consider the simple finite state machine shown in Figure 1-a, in which the input and output values are omitted for the sake of simplicity, and the states are annotated with the steady-state occupation probabilities calculated through Markovian analysis of the STG [1]. If we specify a probability threshold $p = 0.25$, the computational kernel of the machine is depicted in Figure 1-b.

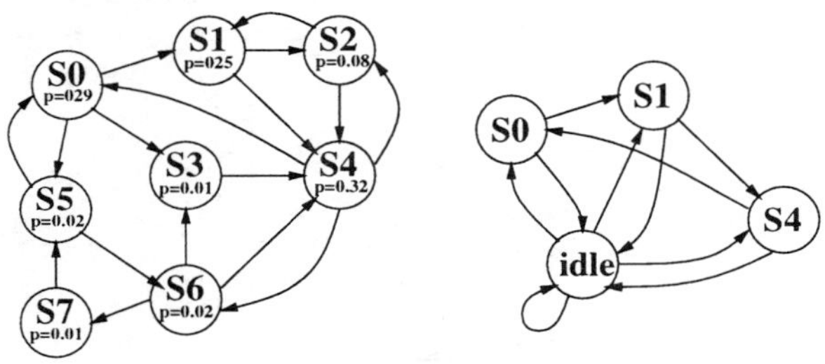

(a) (b)

Figure 1: Moore FSM (a); 0.25-Order Computational Kernel (b).

When the FSM modeling the design is of Mealy-type, the definition of computational kernel seen above becomes more complex, since the output and the transition relations of M_p must be determined separately. The reason for this is that the output values generated by the FSM are associated to the edges, rather than to the STG vertices. Then, the way to proceed for determining the kernel consists of extracting the two relations above from the global relation through universal quantification, applying the definition given for Moore-type FSMs to the STGs of such relations, and forcing the two kernels to run in lock-step. From the discussion above, it is apparent that calculating the computational kernel of a FSM can be done easily through symbolic manipulation, by handling simultaneously sets of states represented as characteristics functions [2]. Well-established technology developed in the context of exact [3, 4, 5, 6] and approximate [7, 8] FSM reachability analysis can be easily applied to solve the kernel calculation problem in the case the STG of the FSM is available or it can be extracted from the circuit description.
If the computational kernel has been determined following a top-down path starting from the STG (implicitly or explicitly represented), state assignment must be performed, and existing logic synthesis and optimization techniques [9] can be applied to generate a sequential network functioning exactly as M_p.

A rather different situation must be faced when the STG construction is prevented by the memory required to (implicitly) store the STG itself. First, exact probabilistic analysis can not be performed in absence of the FSM specification at the STG-level. Approximate algorithms, such as those discussed in [10], must then be employed; consequently, some states may be incorrectly considered as belonging to M_p. Second, state encoding for S_p is not performed, because it may be too time consuming. In the next section, we propose a heuristic approach for simultaneously determining the output and the next-state kernels of a sequential circuit, and to directly obtain a logic-level sequential network for such kernels. The method does not require explicit nor implicit STG manipulation; therefore, it is not memory intensive, and can deal with large designs.

3 Heuristics for Kernel Extraction

In this section, we present an iterative procedure for determining the computational kernels of a sequential circuit that works on the structural description of the network. The algorithm is heuristic by nature, and it exploits concepts such as ***logic implication*** and ***redundancy removal*** which have been in use for a long time in applications like test generation and logic optimization. More specifically, it takes the initial sequential circuit, A, and it iteratively computes the kernel, K, by removing gates and connections from the combinational logic CL of A.
The elementary step of the iterative transformation is shown in Figure 2. A connection w is selected inside network A. Signal w is replaced with either the constant value 0 (shown in the figure) or 1, and the circuit is simplified by propagating the constant value in the fanout cone of w and by removing all the fanout-free logic gates in the transitive fanin of w. Notice that w can be a primary input or a primary output of the logic network.

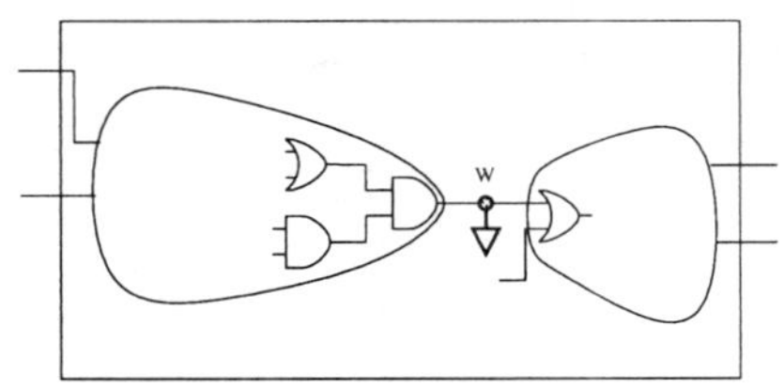

Figure 2: Basic Transformation for Kernel Construction.

The computational kernel K is obtained from the original circuit A by applying a sequence of these elementary transformations, until a stopping criterion is satisfied. We denote with K^i the computational kernel K after i applications of the above transformation. Initially, $K^0 \equiv A$; then, the i-th transformation yields a new circuit K^i from the previous one K^{i-1}. The pseudo-code of the iterative extraction procedure is shown in Figure 3.

```
procedure Compute_Kernel(A,F,U_w) {
    i = 0; K^i =A;
    do {
        w = SelectNode(A,U_w);
        v = ChooseBestValue(A,w);
        K^{i+1} = PropagateValues(K^i,A,w,v);
        i + +;
    } while (!StopTest(K^i,F))
    return(K^i);
}
```

Figure 3: The Compute_Kernel Algorithm.

The procedure receives, as inputs, the original circuit, A, the cost function, F, which controls the stopping criterion, and the utility function, U_w, which drives the node selection process, and it returns the computational kernel, K^i, obtained after i iterations of the **do while** loop.
It is easy to realize that the selection of the nodes in the network to which the transformation is applied (procedure `SelectNode` in the pseudo-code) heavily impacts the result of the kernel extraction. The node selection strategy can thus be customized for different cost functions to be optimized. As an example, in Section 5 we present a heuristics for reducing the power dissipated by the kernel under the constraint that K "covers" most of the behaviors of the original circuit A.
The computational kernel can be seen as a "dense" implementation of the circuit it has been extracted from. In other words, K implements the core functions of the original circuit, and because of its reduced complexity, it usually implements such functions in a faster and more efficient way. In the next section, we propose an innovative optimization paradigm that takes advantage of this fact.

4 Kernel-Based Optimization

The computational kernels of a sequential circuit, extracted as discussed in Section 3, provide us with an extremely powerful device to be used for various types of logic optimization.
Given a sequential circuit with the well-known topology of Figure 4-a, the paradigm we propose for improving its quality with respect to a given cost function (e.g, power dissipation, timing) is based on the architecture shown in Figure 4-b.

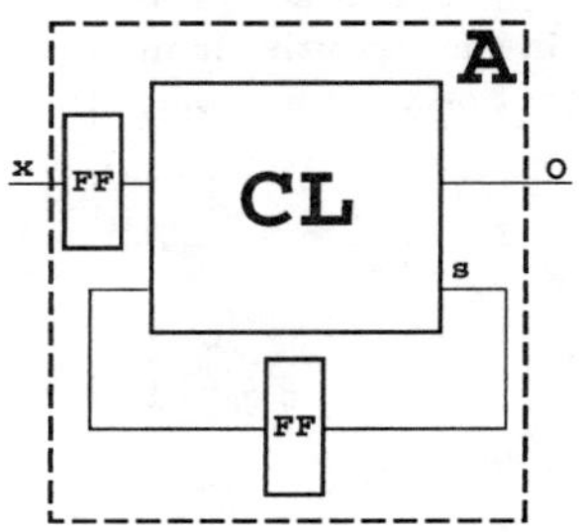

(a)

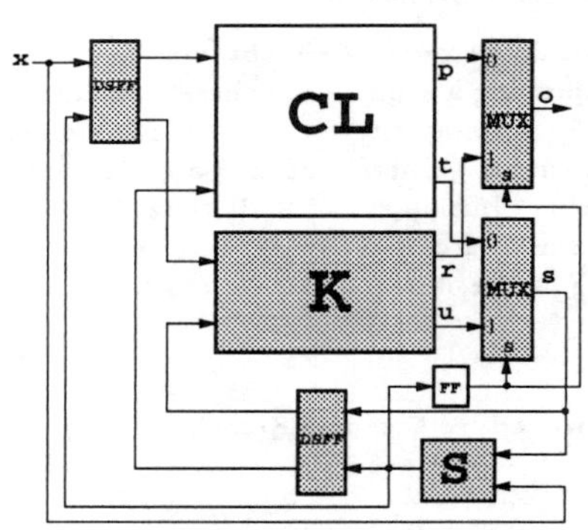

(b)

Figure 4: Sequential Circuit (a); Kernel-Based Architecture (b).

The essential elements of the architecture are the following: The combinational portion of the original circuit (block CL), the kernel (block K), the selector function (block S), the double-state flip-flops ($DSFF$), and the output multiplexors (MUX).

The purpose of the selector function S is that of deciding what logic block, between CL and K, will provide the output value and the next-state in the following clock cycle. To take a decision, S examines the values of the primary and present-state inputs that will be fed to blocks CL and K in cycle $n+1$. If the output values in clock cycle $n+1$ can be computed by the kernel K, then S takes on the value 1. Otherwise, it takes on the value 0. The value of S is fed to a flip-flop, whose output is connected to the multiplexors that select which block produces the output and the next-state. The optimized implementation is functionally equivalent to the original one.
The sequential elements indicated as $DSFF$s are called *dual-state* flip-flops, and they replace the ordinary flip-flops present in the original design. A dual-state flip-flop is functionally equivalent to the schematic of Figure 5, and it operates as follows: When $S=1$, flip-flop F_2 is loaded with a new value coming from the external data input, while flip-flop F_1 holds its state. The opposite happens when $S=0$. In this way, the state of either the kernel or the original network can be kept unchanged (i.e., frozen in the dual-state flip-flops) while the other logic block is being used to produce the output and the next state.
Clearly, Figure 5 only describes the functional behavior of a dual-state flip-flop, whose actual implementation can be properly optimized to reduce the overhead with respect to the standard flip-flops used in the original sequential network.

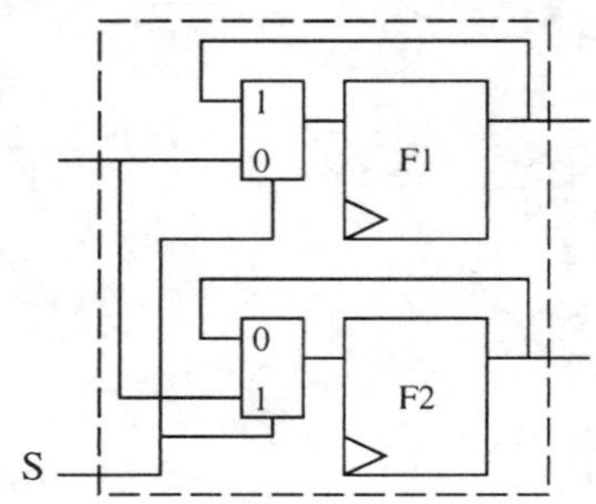

Figure 5: Functional Model of a Dual-State Flip-Flop.

It can be observed in first place that, besides the automatic generation of the kernel (this task is accomplished by procedure `Compute_Kernel` of Section 3), also the logic for the selector function S needs to be synthesized. Fortunately, the logic for S can be obtained as a by-product of the kernel extraction process. In other words, procedure `Compute_Kernel` can be modified to incorporate the automatic synthesis of S.
This is because the ON-set of the selection function, at the i-th iteration of the kernel extraction loop, is defined as the set of primary input and present-state conditions for which each primary and next-state output of K^i is equal to the corresponding primary and next-state output of CL (that is, K^i can be used in place of CL to compute the output and next-state values). Thus, S_i can be simply computed in an exact fashion by the following equation:

$$S_i(x,s) = \prod_{j=1}^{N} (\rho_j^i \equiv \omega_j) \quad (1)$$

where the product stands for logic conjunction, ρ_j^i is the j-th output of network K^i (that is, an element of set (r,u)), ω_j is the j-th output of network CL (that is, an element of set (p,t)), x are the primary inputs, and N is the number of primary and next-state outputs.

Since binary decision diagrams (BDDs) [11] are used to represent both K and S, S_i can be computed as long as it is possible to construct its BDD. There may exist circuits for which the BDD of S can not be constructed. In this case, we have to resort to approximate methods. For performance reasons, such methods must be customarily designed depending on the specific target optimization. In Section 5, we describe a solution which is particularly suited to power minimization.

The revisited pseudo-code is shown in Figure 6. It should be noticed that an extra input parameter is passed to the procedure: The timing constraint T to be used in the synthesis of S. This is because the speed of S must always be kept under consideration, since its delay adds up to the critical delay of the original sequential circuit.

```
procedure Compute_Kernel_and_Select(A,F,U_w,T) {
    i = 0; K^i = A;
    do {
        w = SelectNode(A,U_w);
        v = ChooseBestValue(A,w);
        K^{i+1} = PropagateValues(K^i,A,w,v);
        S_i = ∏_{j=1}^{N} (ρ_j^{i+1} ≡ ω_j);
        BuildSelectionLogic(S_i,T);
        i++;
    } while (!StopTest(K^i,S_i,F))
    return(K^i,S_i);
}
```

Figure 6: The Compute_Kernel_and_Select Algorithm.

5 Application to Power Optimization

Research on logic-level power minimization techniques has always been very active [12]. Most of the optimization approaches at the logic-level target combinational circuits, and the literature on the subject is vast. Concerning sequential networks, on the other hand, only a few solutions have appeared to be successful, namely, state re-encoding, re-timing, clock-gating, and guarded-evaluation. For a detailed discussion of these techniques the interested reader may refer to [13, 14].

In this section, we propose the application of the computational kernel based paradigm discussed in Section 4 to solve the sequential power optimization problem. In other words, we discuss how the basic operations of algorithm Compute_Kernel_and_Select can be implemented when the target of the optimization is a low-power realization of the original circuit.

A few techniques similar to ours have been proposed in the recent literature. *Precomputation* [15, 16] is an approach to power minimization that relies on the idea of computing some of the output values of a circuit through a simplified (and low power consuming) logic running in parallel with the original circuit. The technique works well for circuits with pipelined structure; on the other hand, the extension to the case of sequential circuits with feed-back proposed in [15] is not applicable in practice, as shown by the data presented in [17, 18].

FSM decomposition for low power [19, 20, 21] can be seen as a top-down kernel extraction procedure that starts from explicit STG specifications (e.g., state tables). Although these techniques reported sizable power reductions, they can be applied only to small circuits for which the explicit STG description can be manipulated in reasonable time and memory space. The advantage of our approach to kernel extraction is that it can be applied to circuits for which not even the implicit representation of the STG can be constructed. Kernel extraction for low power is a specialization of the general algorithm of Figure 6.

5.1 Node and Value Selection

In order to perform the selection of the candidate node, w, procedure SelectNode within the algorithm of Figure 6 requires a utility function U_w that estimates the savings that can be achieved if w is replaced by either 0 or 1.

The function U_w we adopt in our method is an approximate one, since we are interested in fast, yet reasonably accurate estimates of the power savings. It is a combination of the expected power savings on w (χ_w) and the expected power savings on the fanin and fanout (ξ_w):

$$U_w = \chi_w + \alpha\xi_w \qquad (2)$$

where $0 \leq \alpha \leq 1$ is a scaling coefficient (by default, $\alpha = 1$).

ξ_w can be efficiently computed, given the transition probability $Tp[w]$ and the load capacitance C_w^{load} of node w, as $\chi_w = Tp[w] \cdot C_w^{load}$. On the other hand, ξ_w can be determined by finding how many gates will be eliminated from A by the transformation, and summing the total power:

$$Pow_{save} = \sum_{\forall w_i \text{ eliminated}} Tp[i] \cdot C_i^{load} \qquad (3)$$

The number of gates eliminated can be easily determined by evaluating how many gates in the fanin and fanout cones have their outputs fixed to some constant value.

Even though the approximate U_w discussed above usually provides accurate enough estimates, for critical cases (e.g., portions of logic which are always under stress, controllers), it may be required to resort to exact calculations, at the price of an increased run-time of the kernel extraction procedure. The exact solution consists of evaluating the actual power savings resulting by the substitution in the network, by running some power estimation tool, such as symbolic simulation [22], or any available commercial tool. This is clearly much more time and memory demanding, and it is applicable for small circuits, since it would be invoked for every new substitution. In the current implementation, it is possible to trigger symbolic simulation to evaluate the exact power savings.

Once a candidate node w has been selected, there are two options for the selection of which value is the best to force on w (procedure ChooseBestValue in Figure 6). For small to medium-sized circuits, it may be feasible to force first 0 and then 1 on the network A, and then evaluate U_w using Equation 2 for each of the two choices. The selected value v will be the one with the highest value of U_w.

For larger circuits, in which the propagation of a value can require a certain amount of time, it is better to *estimate* the best value to assign to w. This can be done as follows: The node w will usually have a transition probability $Tp[w]$ that is close to 0.5; this is because, under the assumption of temporally uncorrelated circuit inputs, the transition probability can be obtained as:

$$Tp[w] = p_w(1 - p_w) + (1 - p_w)p_w = 2p_w(1 - p_w) \qquad (4)$$

where p_w is the signal probability of node w. The first term in Equation 4 indicates the probability of a $1 \rightarrow 0$ transition occurring on node w, while the second term represents the probability of the other transition ($0 \rightarrow 1$). This function has a maximum at $p_w = 0.5$; therefore, values of the signal probability close to 0.5 imply a high transition probability. This means that, once w has been selected, the value v to be assigned should be obtained by the logic expression $p_w > 0.5$. In other terms, w should be assigned to 0 if $p_w < 0.5$, to 1 otherwise, that is, to the value which is more likely to occur.

5.2 Stopping Criterion

The iterative construction of K and S will tend to yield increasingly small S's. Intuitively, this corresponds to lowering the probability of using K instead of A to compute the outputs and the next states. After some iterations within the **do while** loop, this probability might become too small, reducing the fraction of time in which K is selected. The process should then be stopped.

An estimate of the power, Pow_i, dissipated by the i-th version of the architecture can be computed as:

$$Pow_i = Pow(K^i) \cdot P(S_i) + Pow(A) \cdot (1 - P(S_i)) + Pow(S_i) \quad (5)$$

Equation 5 can be interpreted as follows: When $S_i = 1$, the kernel K is operating; thus, it dissipates $Pow(K^i)$ for a fraction of time equal to $P(S_i)$. Conversely, when $S_i = 0$, the original circuit A is operating; thus, it dissipates $Pow(A)$ for a fraction of time equal to $1 - P(S_i)$. Then, the condition to be checked for terminating the iterations occurs when Δ_i stops decreasing from one iteration to the next one.

In order to prevent an early termination, in the implementation of procedure `StopTest` we have added to the right-hand side of Equation 5 a term, O_i, representing an overhead value, that may be adjusted iteration by iteration. In the first few iterations, it should be negative, to force the first transformations. Then, as i increases, it should tend to the value $Pow_{DSFF} + Pow_{MUX}$, where Pow_{DFSS} is the incremental (with respect to regular flip-flops) power dissipation of the dual-state flip-flops, and Pow_{MUX} is the power dissipation of the output multiplexors. Alternatively, the stopping condition may not be tested for the first few iterations, to ensure that the K^i is sufficiently simpler than A, and it is then thereafter to make sure than $P(S_i)$ does not become too small.

5.3 Synthesis of the Selection Logic

The selection logic S should be synthesized for power under timing constraints. In most cases, especially for large circuits, the function S obtained by Equation 1 is too large to be synthesized as is. In these cases, we use a ***sub-setting*** algorithm [7] that provides an implementation K^r of the selection logic whose ON-set is smaller than that of K (and whose implementation is fast (or small) enough), but with maximum probability. With subsetting, we sacrifice probability for performance and/or area.

Finally, one additional optimization can be obtained by observing that the circuit A can be optimized using S_i as controllability ***don't care*** set, to reduce the area overhead and to save additional power. This is because, when $S = 1$, the functionality implemented by A is already computed by K.

6 Experimental Results

We have implemented the procedure for the extraction of the computational kernels of a sequential circuit and to build from it the architecture of Figure 4 as an extension of SIS [23] using CUDD [24] as the underlying BDD package. Experiments have been run on a DEC AXP 1000/400 with 256 MB of memory.

The benchmarks we have used to check the effectiveness of the kernel-based power optimization approach are ALL the large Iscas'89 sequential circuits [25], including the *addendum* (that is, a total of 15 examples). The circuits were initially optimized for area using `script.rugged` (whenever possible), and mapped for area with `map -m0 -AFG` onto a library containing two to four-input NAND and NOR gates, inverters and buffers with three different drive strengths, and a flip-flop.

Table 1 reports the data for the examples for which some power savings have been obtained (10 cases). In particular, columns *In*, *Out*, *FF*, *Gates*, and *Delay* report the characteristics of the reference circuits. Column *Power* shows the power, in μW, of the original circuit A (column *Reference*), and that of the kernel-based architecture (column *Optimized*), as well as the obtained savings (column *Savings*). Column *P(S)* tells the probability of the selector function S to be 1, that is, the probability of kernel K to be active. Column *Area Overhead* shows the area cost of the modified architecture, expressed as a percentage of the initial gate count. Similarly, column *Delay Overhead* shows the performance penalty introduced by the use of the kernel-based architecture. Finally, column *CPU Time* indicates the execution time, in seconds, required by the optimization procedure to complete.

Power estimates within the kernel extraction procedure have been computed using symbolic simulation, while those for the initial and final circuits have been determined using the Irsim switch-level simulator [26].

The results we have obtained are promising. An average power savings of approximately 13.6% has been achieved, with a peak of 29.3% on circuit s1512.

Five of the benchmarks in the suite did not produce acceptable results. More specifically, the application of our technique to examples s3330 and s9234 did not produce any noticeable power improvement. For circuits s35932, s38417, and s38584, on the other hand, the execution did not complete, due to the size of the original network to be handled.

Concerning the timing of the optimized circuits, in spite of the fact that the delay of the selector function S adds up to the largest delay between CL and K, the penalty is limited (3.71% on average). This is because CL is optimized using S as *don't care set*, thus its delay usually reduces with respect to the original circuit.

As expected, the area penalty is relevant (58.9% on average). This is because the kernel-based approach we have proposed suffers, in principle, from the same overhead (logic duplication) that affects any type of parallel implementation.

7 Conclusions and Future Work

The computational kernels of a sequential circuit are blocks of logic whose behavior mimics the steady-state behavior of the original circuit. The computation of the kernel of a design specified through the state graph of the corresponding finite state machine is feasible only when the state space is of limited size, even though BDD-based, symbolic representation techniques are adopted.

In this paper, we have presented a heuristic method for the automatic extraction of approximate computational kernels directly from the gate-level description of a sequential design. The solution is simulation based, and it is thus applicable to large networks for which the STG can not be extracted and/or manipulated.

We have then proposed the exploitation of computational kernel as a device for general-purpose logic optimization. The method requires the modification of the original circuit according to an architectural decomposition paradigm whose template is reminiscent of the mux-based scheme adopted in most parallel implementations of logic-level descriptions.

We have proved the effectiveness of this optimization paradigm by using it for reducing the power dissipated by a sequential circuit; experimental results, obtained on standard benchmarks, are very promising.

Circuit	In	Out	FF	Gates	Delay	Power			P(S)	Area Overhead	Delay Overhead	CPU Time
						Reference	Optimized	Savings				
s1269	18	10	37	468	50.42	1412.96	1339.02	5.23 %	0.45	69%	3.34%	10
s1423	17	5	74	602	73.58	2037.88	1473.64	27.68 %	0.25	38%	0.87%	43
s1512	29	21	57	475	42.71	804.51	568.10	29.38 %	0.93	36%	1.69%	85
s3271	26	14	116	1045	35.21	3436.60	3209.33	6.61 %	0.44	79%	5.43%	132
s3384	43	26	183	1393	92.47	4651.68	4383.40	5.77 %	0.25	85%	6.65%	127
s4863	49	16	104	2022	86.67	5926.34	5618.15	5.20 %	0.99	81%	7.58%	321
s5378	35	49	164	1132	22.30	1891.85	1706.55	9.79 %	0.90	44%	1.61%	452
s6669	83	55	239	2703	163.59	8083.82	7304.13	9.64 %	0.98	63%	4.53%	398
s13207	31	121	669	2462	44.59	3918.45	3230.56	17.55 %	0.97	43%	2.13%	494
s15850	14	87	597	3417	72.73	4971.02	4022.97	19.07 %	0.99	51%	3.28%	795
Avg.								13.59 %	0.72	58.9%	3.71 %	285.7

Table 1: Experimental Results.

We believe that the concept of computational kernel can be exploited in several directions. Performance optimization is an interesting future application of this technology. At this regard, computational kernels can be seen as an extension of the ideas presented in [27]. Moreover, we are considering the possibility of applying computational kernel extraction at higher levels of abstraction.

References

[1] G. D. Hachtel, E. Macii, A. Pardo, F. Somenzi, "Markovian Analysis of Large Finite State Machines," *IEEE Transactions on CAD*, Vol. CAD-15, No. 12, pp. 1479-1493, December 1996.

[2] G. D. Hachtel, F. Somenzi, *Algorithms for Logic Synthesis and Verification*, Kluwer Academic Publishers, 1996.

[3] J. R. Burch, E. M. Clarke, K. L. McMillan, D. L. Dill, "Sequential Circuit Verification Using Symbolic Model Checking," *DAC-27*, pp. 46-51, Orlando, FL, June 1990.

[4] O. Coudert, J. C. Madre, "A Unified Framework for the Formal Verification of Sequential Circuits," *ICCAD-90*, pp. 126-129, Santa Clara, CA, November 1990.

[5] H. Touati, H. Savoj, B. Lin, R. K. Brayton, A. Sangiovanni-Vincentelli, "Implicit Enumeration of Finite State Machines Using BDDs," *ICCAD-90*, pp. 130-133, Santa Clara, CA, November 1990.

[6] H. Cho, G. D. Hachtel, S. W. Jeong, B. Plessier, E. Schwarz, F. Somenzi, "ATPG Aspects of FSM Verification," *ICCAD-90*, pp. 134-137, Santa Clara, CA, November 1990.

[7] K. Ravi, F. Somenzi, "High-Density Reachability Analysis," *ICCAD-95*, pp. 154-158, San Jose, CA, November 1995.

[8] H. Cho, G. D. Hachtel, E. Macii, B. Plessier, F. Somenzi, "Algorithms for Approximate FSM Traversal Based on State Space Decomposition,", *IEEE Transactions on CAD*, Vol. CAD-15, No. 12, pp. 1465-1478, December 1996.

[9] G. De Micheli, *Synthesis and Optimization of Digital Circuits*, McGraw-Hill, 1994.

[10] C-Y. Tsui, J. Monteiro, M. Pedram, S. Devadas, A. M. Despain, B. Lin, "Power Estimation Methods for Sequential Logic Circuits," *IEEE Transactions on VLSI*, Vol. VLSI-3, No. 3, pp. 404-416, September 1995.

[11] R. Bryant, "Graph-Based Algorithms for Boolean Function Manipulation," *IEEE Transactions on Computers*, Vol. C-35, No. 8, pp. 79-85, August 1986.

[12] M. Pedram, "Power Minimization in IC Design: Principles and Applications," *ACM Transactions on Design Automation of Electronic Systems*, Vol. 1, No. 1, pp. 3-56, 1996.

[13] W. Nebel and J. Mermet Editors, *Low-Power Design in Deep Sub-Micron Electronics*, Kluwer Academic Publishers, 1997.

[14] L. Benini, G. De Micheli, *Dynamic Power Management: Design Techniques and CAD Tools*. Kluwer Academic Publishers, 1998.

[15] M. Alidina, J. Monteiro, S. Devadas, A. Ghosh, M. Papaefthymiou, "Precomputation-Based Sequential Logic Optimization for Low Power," *IEEE Transactions on VLSI Systems*, Vol. VLSI-2, No. 4, pp. 426-436, 1994.

[16] J. Monteiro, J. Rinderknecht, S. Devadas, A. Ghosh, "Optimization of Combinational and Sequential Circuits for Low Power Using Precomputation," *1995 Chapel Hill Conference on Advanced Research in VLSI*, pp. 430-444, Chapel Hill, NC, March 1995.

[17] L. Benini, G. De Micheli, E. Macii, M. Poncino, R. Scarsi, "Symbolic Synthesis of Clock-Gating Logic for Power Optimization of Control-Oriented Synchronous Networks," *EDTC-97*, pp. 514-520, Paris, France, March 1997.

[18] L. Benini, G. De Micheli, E. Macii, M. Poncino, R. Scarsi, "Integrating Logic-Level Power Management Techniques," *SASIMI-97*, pp. 59-65, Osaka, Japan, December 1997.

[19] S. H. Chow, Y. C. Ho, T. Hwang, C. L. Liu, "Lower Power Realization of Finite State Machines - A Decomposition Approach," *ACM Transactions on Design Automation of Electronic Systems*, Vol. 1, No. 3, pp. 315-340, July 1996.

[20] L. Benini, P. Vuillod, C. Coelho, G. De Micheli, "Synthesis of Low-Power Selectively-Clocked Systems from High-Level Specification," *ISSS-96*, pp. 57-62, La Jolla, CA, October 1996.

[21] L. Benini, F. Vermeulen, G. De Micheli, "Finite-State Machine Partitioning for Low Power," *ISCAS-98*, Monterey, CA, May 1998, To Appear.

[22] J. Monteiro, A. Ghosh, S. Devadas, K. Keutzer, J. White, "Estimation of Average Switching Activity in Combinational Logic Circuits Using Symbolic Simulation," *IEEE Transactions on CAD*, Vol. CAD-16, No. 1, pp. 121-127, January 1997.

[23] E. M. Sentovich, K. J. Singh, C. W. Moon, H. Savoj, R. K. Brayton, A. Sangiovanni-Vincentelli, "Sequential Circuits Design Using Synthesis and Optimization," *ICCD-92*, pp. 328-333, Cambridge, MA, October 1992.

[24] F. Somenzi, *CUDD: University of Colorado Decision Diagram Package*, Release 2.1.2, Technical Report, Dept. of ECE, University of Colorado, Boulder, CO, April 1997.

[25] F. Brglez, D. Bryan, K. Koźmiński, "Combinational Profiles of Sequential Benchmark Circuits," *ISCAS-89*, pp. 1929-1934, Portland, OR, May 1989.

[26] A. Salz, M. Horowitz, "IRSIM: An Incremental MOS Switch-Level Simulator," *DAC-26*, pp. 173-178, Las Vegas, NV, June 1989.

[27] L. Benini, E. Macii, M. Poncino, "Telescopic Units: Increasing the Average Throughput of Pipelined Designs by Adaptive Latency Control", *DAC-34*, pp. 22-27, Anaheim, CA, June 1997.

Partitioning and Optimizing Controllers Synthesized from Hierarchical High-Level Descriptions

Andrew Seawright and Wolfgang Meyer
Synopsys, Inc. 700 E. Middlefield Road Mountain View, CA 94043

Abstract - *This paper describes methods for partitioning and optimizing controllers described by hierarchical high-level descriptions. The methods utilize the structure of the high-level description, provide flexible exploration of the trade-off between combinational logic and registers to reduce implementation cost, and allow the designer to control the synthesis process. Results are presented using industrial examples.*

1. Introduction

To tackle the exponential growth in the complexity of digital circuits, designers are moving to higher levels of abstraction in the design process. In control dominated applications, several abstractions are popular for managing the complexity of the design of sequential control logic. Examples include: hierarchical state machine description styles such as State Charts [Har87] and the many variants, the reactive programming language ESTEREL [Ber92], and other compositional description styles such as used in the PBS system [Sea94] and used in the Protocol Compiler design environment (DALI) [Sea96] [Mey97].

In these abstractions, the high-level controller description is typically described in a hierarchical fashion as depicted by the tree in Figure 1. The nodes of the tree represent the compositional operators of the control abstraction. For example, a particular node might represent the sequencing of the sub-behaviors or the concurrent execution of the sub-behaviors, etc.

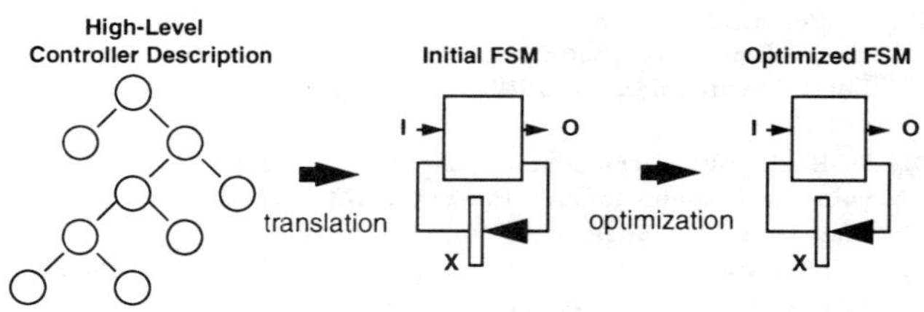

Figure 1. FSM synthesized from a high-level description

Synthesis from these abstractions involves the translation of the high-level controller description into an FSM. Typically, the synthesis is performed as an initial translation step followed by optimization steps. This is the case in the ESTEREL compilers and in Protocol Compiler. The translation ensures correct implementation of the high-level semantics into an FSM, and the subsequent optimizations aim to reduce the cost of the implementation while preserving the observable sequential behavior.

DAC 98, San Francisco, California
©1998 ACM 0-89791-964-5/98/06..$5.00

Separation of the translation and optimization phases can however lead to the loss of information about the high-level description that is useful for optimization.

This work presents methods for partitioning and optimizing controllers synthesized from high-level descriptions which 1) utilize the structure of the high-level description, 2) provide flexible exploration of the trade-off between sequential and combinational logic to reduce implementation cost, 3) and enable the designer to control the synthesis process.

This paper is organized as follows: Section 2 summarizes related work. Section 3 describes the overall compilation and partitioning process. Automatic partitioning is discussed in Section 4. Section 5 presents experimental results and conclusions are drawn in Section 6.

2. Relation to Prior Work

Most of the work involving the optimization of controllers operates on the circuit (or state graph) after translation from the high level description, and without any additional guidance from the structure of the high level description. The classical FSM optimization techniques -- state minimization, state assignment, state encoding -- and sequential circuit level optimizations such as retiming fall into this category. In the work presented here, the structure of the high-level description is central in guiding the partitioning and optimization of the controller.

Touati and Berry [Tou93] studied the optimization of ESTEREL generated controller circuits using specialized techniques for removing redundant registers using reachability don't cares. Specific optimizations such as replacing registers by combinational functions of other registers, detecting and exploiting net equivalences, and performing local retimings at the circuit level are described. Continuing in this area, Sentovich, Toma, and Berry [Sen96] [Sen97] presented additional methods for optimizing controller circuits generated from high-level descriptions such as ESTEREL. Single register, multiple register, and related re-encoding optimizations are described to improve the register / combinational logic ratio to reduce the cost of implementation or formal verification of the designs. These methods are simply suitable for the characteristics of the circuits generated by the high-level description translation process. The high-level representation is only used to calculate an over approximate reachable state set and is not central in the optimization process.

Crews and Brewer [Cre96] presented various techniques for optimizing protocol controllers generated from hierarchical regular expression descriptions. The input regular expression DAG (directed acyclic graph) is minimized before circuit translation resulting in a reduction of registers after circuit generation. Other techniques for removing registers during circuit translation are described. These techniques do in fact utilize the structure of the input description. However, since the information about the global reachability and observability is unknown, the optimizations are generally local in scope.

The above techniques do not consider partitioning the controller, based on the high level description, into multiple sub-machines and do not allow designer control of the optimization process. However, these techniques can and are combined with the approach presented here.

3. Controller Compilation

Optimizing the controller by extracting a single state transition graph for the entire controller is generally only feasible for small controllers (less than a few hundred states). Optimizing the controller using techniques such as redundant register removal and local re-timing tends to result in optimizations which are local in scope (each transformation only re-encodes a small part of the logic). To avoid these limitations and allow architectural exploration of large controllers, it is advantageous to allow the optimization and re-encoding of "chunks" or *partitions* of the controller logic as illustrated in Figure 2.

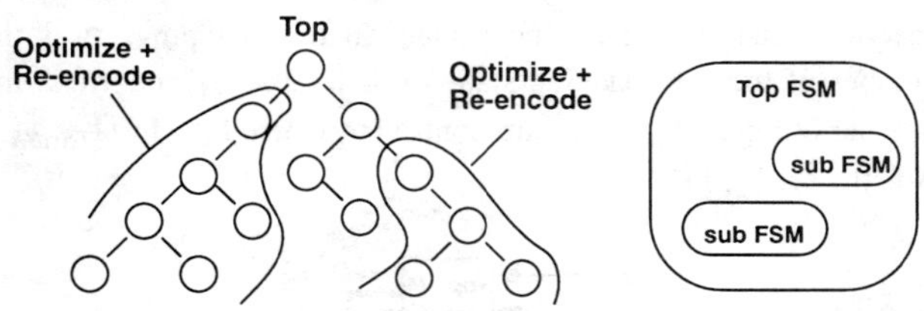

Figure 2. Optimizing and re-encoding portions of the controller related to portions of the high level description

To select the portions of the circuit to optimize, two methods are used: automatic and manual. In automatic partitioning, reachability information is used *in conjunction with* the high level description to automatically identify portions of the circuit to optimize. Manual partitioning allows the designer to select portions of the circuit to optimize by placing attributes on the nodes of the high level description.

In the compilation process, the relationship between the high-level description and the initial translated FSM is maintained. This allows the partitioning and optimization to utilize the high-level structure. Data structures are created to identify the logic and registers translated for each node of the high level description (Figure 3).

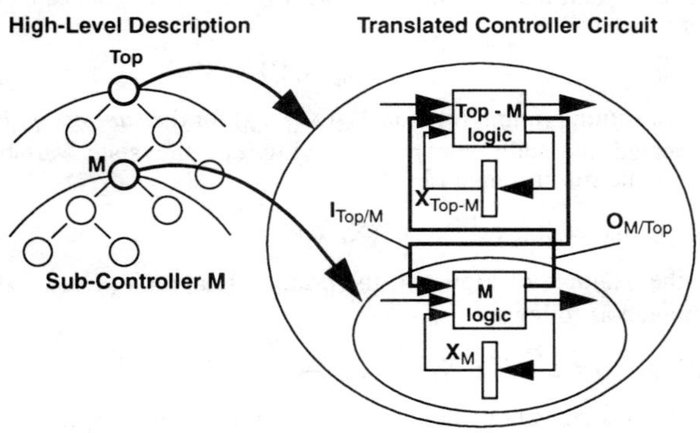

Figure 3. Relationship between the high-level description and the translated controller logic

Each partition sub-machine is optimized and re-encoded utilizing sequential don't care information. This optimization is performed by extracting and re-encoding the state transition graph of the sub-machine in the context of the overall controller. This is performed using a *transition pruning relation* to prune impossible state transitions of the sub-machine in the context of the overall controller.

Advantages of this transition pruning relation approach are that 1) the transition pruning relation can be constructed from either the implicitly computed reachability information *or* from certain designer specified properties of the nodes of the high level description (if the exact reachable states can't be computed) and 2) the avoidance of creating an explicit Cartesian product state transition graph of the sub-machine and the rest of the controller.

3.1 Overall Compilation Process

The overall compile process is illustrated in Figure 4. After the initial translation of the high-level description (`top_node`) into an initial circuit (`top`), a reachability analysis is performed to calculate the set of reachable states R(X) of the entire circuit using an implicit BDD-based approach [Cou89] [Tou90]. If the reachable states computation is too expensive, this phase is skipped. In this case manually specified partitions and designer specified property attributes are required.

```
PartitionedCompile(description top_node,
                   circuit top) {
   R(X) = AnalyzeController(top)
   ChoosePartitions(top_node, top, R(X))
   foreach partition p in top {
     c = ExtractSubCircuit(top, p)
     g = CreateStateGraph(c, R(X))
     g'= MinimizeStateGraph(g)
     EncodeStateGraph(g')
     c'= StateGraphToCircuit(g')
     MergeBackSubCircuit(top, p, c')
   }
}
```

Figure 4. Overall Compilation Process

Controller partitions are selected by the routine `ChoosePartitions()`. Automatic partition selection is described in detail later in Section 4. For manual partitioning, `ChoosePartitions()` picks the nodes designated by the designer.

Each controller partition is optimized by extracting the logic for the designated node as a sub-circuit. This sub-circuit is then converted into a state transition graph which is minimized and re-encoded. During this conversion, sequential don't care information is used to prune the state transition graph as described in the next section. The minimized and re-encoded state transition graph is converted back into a circuit and merged with the rest of the controller.

3.2 Sub-Controller State Graph Pruning

In order to effectively optimize a sub-controller, the communication between the sub-circuit and the rest of the controller must be carefully considered in the state graph generation process. In other words, the sequential don't care information should be utilized. Consider the diagram in Figure 3 which shows the logic for a sub-controller **M** in relation to the rest of the controller **Top**. The sub-circuit extracted for **M** comprises all the combinational logic (**M logic**) and registers ($\mathbf{X_M}$) for the sub-tree **M** in the high-level description. A portion of the inputs to **M** are internal wires from other parts of the controller and are not primary inputs. These inputs are designated $\mathbf{I_{Top/M}}$. In a similar way, a portion of the outputs of **M** are internal wires to the rest of the controller and are not primary outputs. These outputs are designated $\mathbf{O_{M/Top}}$.

Optimization of **M** must consider the sequential behavior of the inputs $I_{Top/M}$ as these inputs are internally generated control signals and their signal values are typically strongly correlated with the sequential behavior of **M** itself. The sequential don't care information about $I_{Top/M}$ should be considered in the optimization of **M**. If the don't care information is not considered, then the extracted state graph for **M** can be exponentially larger than necessary.

The state graph for a sub-controller is generated in a breadth first search process. Starting with the initial state of the sub-circuit, the next states and next state transitions are enumerated by forward image calculations of the present states through the next state logic [Cou89] [Tou90]. Each new state transition traversed is tested with respect to a *transition pruning relation* to determine if the state transition of the sub-circuit is consistent with the overall controller. States and transitions traversed consistent with the overall controller are allocated as the nodes and edges of the state transition graph. States and transitions inconsistent with the global controller are *pruned*. The pruning process is how the sequential don't cares of $I_{Top/M}$ are incorporated in the optimization in an implicit way.

State graph pruning is illustrated by the example in Figure 4. The sub-controller **M** has one input **start** from the rest of the controller which signals when the sub-controller should start, and one output **done** to the rest of the controller which signals when **M** is done ($I_{Top/M}$ = {**start**}, $O_{M/Top}$ = {**done**}).

Assume **M** waits in its initial state until **start** is asserted and that **start** can only be true when **M** is in the initial state and can't be true again until after **done** has been asserted and M returns back to the initial state. The execution of **M** can't overlap with itself, it is *not pipelined*. A sub-machine that *can* be activated while it is in the process of execution is referred to as being *pipelined*. When the state graph for **M** is generated, the state X_{M2} is unreachable and is pruned because the transition t_3 is taken from X_{M1} when **start** is true, which can't occur.

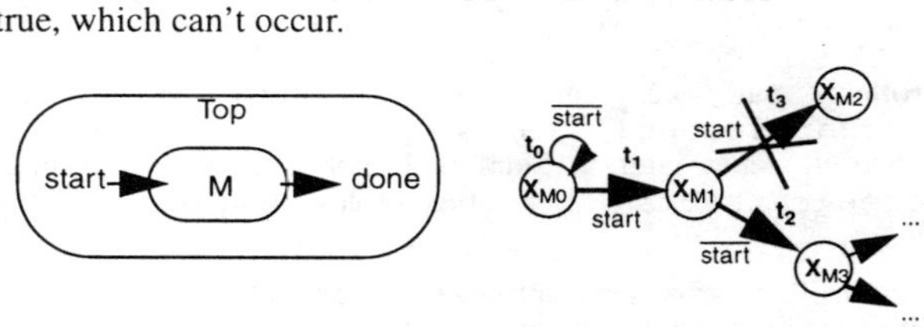

Optimization of sub-machine M — State Graph Generated for M

Figure 4. State Graph Pruning Example

Under the same assumptions about the behavior of the **start** signal, suppose the circuit extracted for sub-machine **M** is the simple shift register delay line illustrated in Figure 5. Without pruning, the number of states of the generated state graph would 2^n, where n is the number of registers in X_M (the size of the delay line). The state graph generation algorithm will assume that the shift register can fill with any binary value without using the sequential behavior of the **start** signal to prune the graph. However, the number of states in the pruned state graph is just n, as only one register in X_M can be set at a time based on the sequential behavior of the input **start**.

The above assumption about the behavior of the **start** signal is just a simple example to illustrate the pruning process. Under different assumptions about the sequential nature of **start**, the pruning can be more complicated.

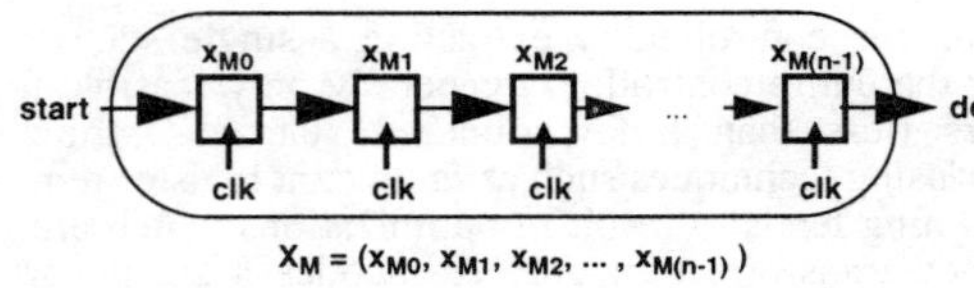

Figure 5. An Example Sub-Machine **M**

The transition pruning relation is a Boolean function $\mathbf{T} : \mathbf{B}^{n+m} \rightarrow \mathbf{B}$. It returns true if and only if the current state and input condition specified leads to a valid next state. In this mapping, **n** is the number of the sub-machine state variables in X_M and **m** is the number of inputs I_M to the sub-controller, where $I_M = I \cup I_{Top/M}$. **B** is the set {0, 1}.

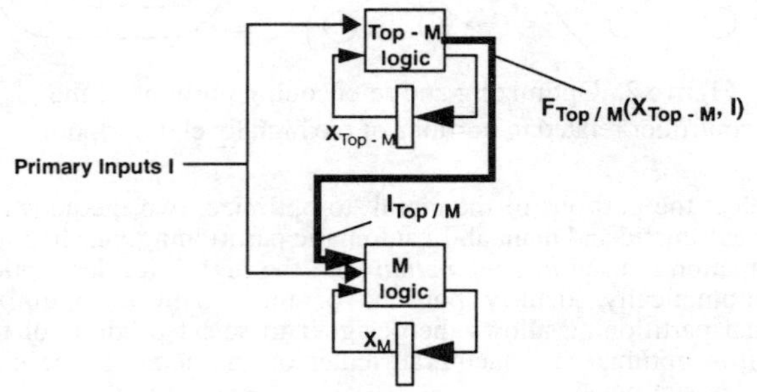

Figure 6. Functions $F_{Top/M}(X_{Top/M}, I)$ drive sub-controller inputs $I_{Top/M}$

The transition pruning relation $T(X, I_M)$ of the overall controller is computed by setting the sub-machine internal inputs $I_{Top/M}$ equivalent to the functions driving the inputs $F_{Top/M}(X_{Top-M}, I)$ (Figure 6) and *anding* this result with the reachable states of the entire circuit R(X):

$$T(X, I_M) = (I_{Top/M} \Leftrightarrow F_{Top/M}(X_{Top-M}, I))\, R(X)$$

The transition pruning relation $T_M(X_M, I_M)$ for the *sub-machine* **M** is created by additionally quantifying out the state variables outside the sub-machine **M**:

$$T_M(X_M, I_M) = \exists(x_i\text{'s} \notin X_M)\; T(X, I_M)$$

For the example of Figure 4, the pruning relation $T_M(X_M, I_M)$ is computed as follows:

$$T_M(X_M, I_M) = \exists(x_i\text{'s} \notin X_M)\; (start \Leftrightarrow f_{start}(X_{Top-M}, I))\, R(X)$$

Table 1 describes how the transition pruning relation $T_M(X, I_M)$ is used to determine if the transitions of the example in Figure 4 are valid or invalid.

Table 1. Transitions of Pruning Example

Transition	Description	Reason
t_0	Valid	$T_M(X_M, I_M) X_{M0} \overline{start} \neq 0$
t_1	Valid	$T_M(X_M, I_M) X_{M1} start \neq 0$
t_2	Valid	$T_M(X_M, I_M) X_{M0} \overline{start} \neq 0$
t_3	Invalid	$T_M(X_M, I_M) X_{M1} start = 0$

If the exact reachable states information is not known, the transition pruning relations can be created by using 1) an over approximate R(X) calculated in a method similar to the approach in [Sen97] (beyond the scope of this paper), or by 2) using certain properties of the design set by the designer on the high level description. For an example of the latter case that is used in Protocol Compiler, consider the example of Figure 4. Suppose the designer -- based on knowledge of the design -- attaches a special property attribute on the node **M** of the high level description of the controller specifying that the sub-controller **M** is not pipelined. This property is then converted into a transition pruning relation $T_M(X_M, I_M)$ where **start** must be false in all states other than $\mathbf{X_{M0}}$. A suitable transition pruning relation can be directly constructed as follows:

$$T_M(X_M, I_M) = X_{M0} + \overline{X}_{M0}\overline{start}$$

3.3 Sub-Controller Optimization

After the state transition graphs are extracted for each sub-controller, the sub-controllers are optimized by minimizing and re-encoding their state graphs. During the state minimization process, the sub-controller outputs $\mathbf{O_{M/Top}}$ are treated as primary outputs so the state merging is done correctly as these outputs are observable by the other parts of the controller. After state minimization, state encoding methods are applied. For manual partitioning, the designer can select a particular style of encoding for each partition. For automatic partitioning, a minimum width binary code state encoding algorithm is used.

4. Automatic Partitioning

Section 3 described the overall compilation flow and how individual sub-controllers are extracted and optimized. This section describes an effective heuristic algorithm for automatically identifying the sub-controllers to extract and optimize.

The automatic partitioning approach uses the structure of the high-level description and information about the reachability of the controller.

The idea is to partition the controller, based on the high-level description, into the fewest sub-controllers each having the largest number of states below some *maximum* state threshold. This method is based on the observation that the re-encoding of state graphs is effective for state graphs below a certain size. Above this threshold, the combinational logic complexity for encoding and decoding the state generally outweighs the savings in register cost. The threshold should be set such that state graphs of this size or less -- when optimized and re-encoded -- will be implemented in efficiently.

The pseudocode for the automatic partitioning algorithm `AutoPart()` is shown in Figure 7. The first step of the algorithm `CollectData()`, collects data about the number of states contributed by each node of the high-level specification. This routine recursively traverses each node of the high-level description and calculates the number of states $\texttt{num\_states}_n$ of the sub-controller `Mn` corresponding to the node `n` of the specification. The number of states for a node is computed by quantifying out the state variables of the parent node's reachable state set $R_p(X)$ passed down which are not state variables of node `n` (line 9).

The number of states for a node `n` is computed from the node's reachable state set by computing the number of minterms in $R_{Mn}(X_{Mn})$ (line 10). The algorithm recurses for each child node and passes down the node's reachable state set $R_{Mn}(X_{Mn})$ (line 12).

```
1.AutoPart(description top_node,
          circuit top, R(X))
2. {
3.    CollectData(top_node, R(X))
4.    PickPartitions(top_node)
5. }
6.
7. CollectData(node n, R_p(X))
8. {
9.    R_Mn(X_Mn) = ∃ (x_i's ∉ X_Mn) R_P(X)
10.   num_states_n = num_minterms(R_Mn(X_Mn))
11.   foreach child node c of node n {
12.      CollectData(c, R_Mn(X_Mn))
13.   }
14.}
15.
16.PickPartitions(node n)
17.{
18.   if (num_states > MaxStates) {
19.      foreach child node c of node n {
20.      PickPartitions(c)
21.   } else {
22.      MarkPartition(c);
23.      return;
24.   }
25.}
```

Figure 7. Maximum State Threshold Algorithm

Concurrent behaviors which are independent or loosely interacting are generally selected as independent partitions as their concurrent execution leads to the Cartesian product of the state space of the individual concurrent behaviors. Thus, the partitioning algorithm tends to break the Cartesian products and each concurrent behavior is extracted and optimized separately.

5. Results

These techniques for partitioning and optimizing sub-controllers have been incorporated into Protocol Compiler [Sea96] [Mey97]. Controller partitioning was applied in the synthesis of two ATM controllers and 11 modules of a 70,000 gate commercial telecommunications ASIC (Tables 2-4).

Architecture exploration of the first ATM controller is shown in Table 1. The design is a receiver controller in an ATM switch. The controller extracts cell header information and performs routing lookup functions. The hierarchical high-level specification for this design comprised about 70 compositional constructs. Without changing the high-level controller description, seven different architectures were created using a variety of synthesis techniques for comparison. The architectures varied in the number of control state registers, from 135 registers in the initial translated circuit (first row), to 9 registers in the design implemented by optimizing it as a single large state transition graph (last row). In Protocol Compiler, the initial translated circuit is characterized by a large number of registers and with many small next state functions. This encoding style is referred to as a "distributed" encoding and is usually a fast implementation.

Table 2. ATM1 Architecture Exploration

architecture	time sec.	total state vars	# parts	final states	area	delay ns
initial translation (distributed)	2.2	135	--	751	1874	12.2
distributed + replace	14.3	120	--	751	1834	11.94
distributed + remove	82.1	117	--	751	2684	59.82
auto part @ 50	21.9	68	3	(31,49,1)	1964	15.28
manual part	29.4	21	4	(6,21,96,1)	1577	18.22
auto part @ 200	35.9	15	2	(31,96)	1719	18.40
single state graph	60	9	--	495	3795	30.97

The middle rows in the table compare architectures optimized by applying different implicit redundant register removal techniques and architectures optimized by different partitioning approaches. The implicit redundant register removal techniques (rows 2 and 3) are similar to the techniques described in [Tou93] and [Sen96] for removing redundant registers. The "distributed + replace" method removes redundant registers in the initial distributed encoding which can be replaced by wires to other registers. The "distributed + remove" method removes registers which can be replaced by combinational functions of other registers. The partitioned controller architectures were created by applying manual and automatic partitioning with different maximum state thresholds (50 and 200 states).

In the table, the times are for Protocol Compiler and are measured in CPU seconds on a Sun SparcStation 20 machine. The number of sub-controller partitions is shown (# part) along with the final number of states and transitions for each partition after state minimization for the partitioning examples. The final columns in the table show the gate level area and delay for the synthesized architectures after logic synthesis. The designs were mapped to the LSI 10k library.

The best design in terms of area was achieved by manually partitioning the design into four sub-controllers (row 5). In terms of delay, the best architecture is the "distributed + replace" design, where 15 registers were replaced by wires. It is important to point out that the single state graph design is the worst in area and the second worst in terms of delay. Optimizing this design as a single state graph minimized to 495 states and re-encoding resulted in very complex next state logic functions. The "distributed + remove" design architecture suffered, because the logic functions introduced to remove the additional 3 registers significantly impacted the area and critical path.

Results for architecture exploration of the second ATM controller design (ATM2) is shown in Table 3. ATM2 is a larger version of ATM1 which also performs operations and maintenance (OAM) and available bit-rate (ABR) control functions. This design is larger, its high-level description comprises over 270 compositional constructs. The number of states of the overall controller is 11,542, compiling the controller as a single state graph is not appropriate. In addition, manual partition of the design is tedious as the description is large. However, the auto partitioning approach can easily partition this controller and produced a far smaller design than the applicable non-partitioned approaches.

Table 4 shows the results of using Protocol Compiler to design 11 modules of a commercial telecommunications ASIC. The table compares the initial translated controller designs, optimized architectures selected manually by the designers, and optimized architectures created by applying automatic partitioning.

Table 3. ATM2 Architecture Exploration

architecture	time sec.	total state vars	# parts	final states	area	delay ns
initial translation (distributed)	24	414	--	11,542	3939	49.92
distributed + replace	647	402	--	11,542	3838	48.11
auto part @ 100	1213	55	9	(31,96,1,46, 46,46,46,50,49)	2813	22.76
auto part @ 200	1402	34	6	(31,96,1, 103,50,49)	3292	27.73

Partitioning was critical in achieving high quality of results in these designs. Initially, the automatic partitioning algorithm was not available to the designers at the time the ASIC implementations were synthesized. Thus, for each module, the designers studied a variety of architectures to achieve the best quality of results. The designers compared the final architectures selected for the ASIC implementation to similar or equivalent RTL descriptions to check the overall quality of results and they were satisfied that the Protocol Compiler implementations were comparable to what could be achieved by the conventional RTL methods.

For seven of the 11 modules, partitioning was superior to the other methods. In three other cases (d4, d7, and d10), optimizing the designs as a single state graph achieved the best results. The module (d11) was optimized by using the "distributed + replace" method. In the manual partitioning process, a full reachability analysis was not performed. Instead, the designers placed design property attributes on the high-level description to indicate how the sub-controllers interacted. From this information, transition pruning relations were created and used in the sub-controller optimizations.

After the automatic partitioning algorithm was implemented in Protocol Compiler, it was applied to all of these modules without changing the high-level controller description. The maximum state threshold was set to 200 states and a full reachablility analysis was performed during the automatic partitioning for use in optimizing the sub-controllers.

The automatic partitioning algorithm, in almost all cases, selected the same or very similar partitions to the designer's manual partitions. For the cases where the designers did not use manual partitioning, the automatic partitioning results were close. The difference in the CPU times between manual partitioning and automatic partitioning are due to the full reachability analysis performed only during the automatic partitioning.

6. Conclusions

This paper presented methods for partitioning and optimizing hierarchically described controllers. These methods exploited the hierarchical high-level description. Partitioning of the controller can be performed either manually or automatically. The results showed that the designer can explore a wide variety of architectures and demonstrated an effective algorithm for automatically partitioning these controllers.

References

[Ber92] G. Berry and G. Gonthier, "The ESTEREL synchronous programming language: design, semantics, implementation", in Science of Computer Programming, vol. 19 (no. 2), pp. 87-152, November 1992.

[Cou89] O. Coudert, C. Berthet, and J. C. Madre. "Verification of Synchronous Sequential Machines Based on Symbolic Execution", in *Automatic Verification Methods for Finite State Systems, International Workshop,* Grenoble France, vol. 407, Lecture Notes in Computer Science, Springer-Verlag, June 1989.

[Cre96] A. Crews and F. Brewer. "Controller Optimization for Protocol Intensive Applications", proceedings of Euro-DAC 96, Geneva, Switzerland, pp. 140-145, September 1996.

[Har87] D. Harel, "Statecharts: A Visual Approach to Complex Systems", in Science of Computer Programming, vol. 8 (no.3), pp. 231-275, August 1987.

[Mey97] W. Meyer, A. Seawright, and F. Tada, "Design and Synthesis of Array Structured Telecommunication Processing Applications", proceedings of the 34th Design Automation Conference, pp. 486-491, June 1997.

[Sea94] A. Seawright and F. Brewer. "Clairyoyant: A System for Production-Based Specifications", in the proceedings of IEEE Transactions on VLSI Systems, vol. 2., (no. 2), pp. 172-185, June 1994.

[Sea96] A. Seawright, U. Holtmann, W. Meyer, B. Pangrle, R. Verbrugghe, and J. Buck. "A System for Compiling and Debugging Structured Data Processing Controllers", proceedings of Euro-DAC 96, Geneva, Switzerland, pp. 86-91. September 1996.

[Sen96] E. Sentovich, H. Toma, G. Berry, "Latch Optimization in Circuits Generated from High-level Descriptions", in the proceedings of the International Conference on Computer-Aided Design ICCAD96, pp. 428-435, November 1996.

[Sen97] E. Sentovich, H. Toma, G. Berry, "Efficient Latch Optimization Using Exclusive Sets", proceedings of 34th Design Automation Conference, pp. 8-11, June 1997.

[Tou90] H. Touati et al. "Implicit State Enumeration of Finite State Machines using BDD's" in the proceedings of International Conference on Computer-Aided Design ICCAD90, pp. 130-133, November 1990.

[Tou93] H. Touati and G. Berry, "Optimized Controller Synthesis Using Esterel", proceedings of the International Workshop on Logic Synthesis IWLS93, Lake Tahoe, 1993.

Table 4. Telecommunications ASIC Results

design	initial translation				manually selected architecture				auto partition @ 200 (full reachablility)			
	time sec.	total state vars	states	area	time sec.	total state vars	final states	area	time sec.	total state vars	final states / trans	area
d1	3	62	414	1,722 (668+1054)	8	15	(18,11,29)	1,572 (833+739)	22	15	(18,11,29)	1,549 (824+725)
d2	31	453	7,430	4,560 (851+3709)	395	13	(59,59)	1,156 (523+633)	3,157	13	(59,59)	1,175 (542+633)
d3	16	344	3,482	4,343 (665+3678)	113	13	(59,59)	1,836 (633+1203)	1,884	13	(59,59)	1,812 (611+1201)
d4	8	217	125	3,186 (877+2309)	99	7	min_encoded 114	1,353 (514+839)	305	8	(115)	1,416 (570+846)
d5	8	240	11,316,504	2,832 (312+2520)	39	28	(8,58,58,58,58)	1,524 (484+1040)	233	28	(8,58,58,58,58)	1,539 (496+1043)
d6	2	44	4,906	1,694 (609+1085)	6	15	(7,17,6,5)	1,585 (703+882)	15	15	(7,17,6,5)	1,582 (700+882)
d7	11	261	124	2,321 (282+2039)	187	6	min_encoded 54	601 (319+282)	694	7	(54)	620 (331+289)
d8	40	567	13,456	6,549 (1422+5127)	491	13	(59,61)	2,032 (756+1276)	3,437	13	(59,61)	2,033 (763+1270)
d9	5	153	3,970	2,260 (609+1651)	22	13	(62,62)	1,159 (568+591)	111	13	(62,62)	1,150 (559+591)
d10	8	218	125	3,656 (1070+2586)	111	7	min_encoded 117	1,966 (857+1109)	292	8	(118)	1,933 (817+1116)
d11	3	63	52,225	1,330 (478+852)	244	62	distributed +replace 52,225	1,206 (443+763)	107	9	(2,2,17)	1,125 (661+464)
totals	135			34,453	1,715			15,990	10,257			15,934

Watermarking Techniques for Intellectual Property Protection*

A. B. Kahng, J. Lach[†], W. H. Mangione-Smith[†], S. Mantik, I. L. Markov,
M. Potkonjak, P. Tucker[‡], H. Wang and G. Wolfe

UCLA Computer Science Dept., Los Angeles, CA 90095-1596
(†) UCLA Electrical Engineering Dept., Los Angeles, CA 90095-1594
(‡) UCSD Computer Science & Engineering Dept., La Jolla, CA 92093-0114

Abstract

Digital system designs are the product of valuable effort and know-how. Their embodiments, from software and HDL program down to device-level netlist and mask data, represent carefully guarded intellectual property (IP). Hence, design methodologies based on IP reuse require new mechanisms to protect the rights of IP producers and owners. This paper establishes principles of *watermarking-based* IP protection, where a *watermark* is a mechanism for identification that is (i) nearly invisible to human and machine inspection, (ii) difficult to remove, and (iii) permanently embedded as an integral part of the design. We survey related work in cryptography and design methodology, then develop desiderata, metrics and example approaches – centering on *constraint-based* techniques – for watermarking at various stages of the VLSI design process.

1 Introduction

The advance of processing technology has led to a rapid increase in IC design complexity. The economic drivers are compelling: only by putting more integration and more function on a single die, and by achieving more revenue per wafer, can multi-billion dollar foundries be amortized over their useful lifespan. At the same time, market forces have led to more design starts, shorter design cycle times and greater time-to-market pressures. Industry organizations have documented a compounding "design productivity shortfall" [26], which demands ever-larger design teams with each successive process generation just to maintain a given level of design competitiveness.[1] Finally, system design costs are increasingly impacted by software, which accounts for up to 70% of total development cost in recent design projects.

In response to these trends, *reuse-based* design methodologies for both hardware and software have been embraced as a means of achieving design productivity on par with the underlying silicon technology. The reuse-based vision is predicated on easily accessible, easily integrable "virtual components". Pure IP companies, third-party ASIC libraries, tools for IP integration, and industry organizations such as the VSI Alliance [33] have created high expectations for the value and reusability of design IP. Nonetheless, a recognized obstacle to reuse-based methodologies is the lack of mechanisms to protect the rights of IP creators and owners.[2]

From both the research and implementation points of view, *intellectual property protection* (IPP) poses a unique set of new requirements that must be addressed by mathematically sound, yet practical, techniques. In this work, we establish principles for development of new *watermarking-based* IPP procedures. These principles, which are centered around the use of *constraints* to "sign" the output of a given design synthesis or optimization, are compatible with current IP development tools infrastructure. Furthermore, our principles apply to the protection of both hardware and software (e.g., Verilog or C++ code) IP.

A Motivating Example: 3SAT

We illustrate key ideas behind watermarking-based IPP using the *satisfiability* (SAT) problem [10]:

SAT (U,C)

Instance: A finite set of variables U and a collection $C = \{c_1, c_2, \ldots, c_m\}$ of clauses over U.

Question: Is there a truth assignment for U that satisfies all the clauses in C?

For example, $U = \{u_1, u_2\}$ and $C = \{\{u_1, u_2\}, \{\bar{u}_1\}, \{\bar{u}_1, \bar{u}_2\}\}$ is a SAT instance for which the answer is positive (a satisfying truth assignment is $t(u_1) = F$ and $t(u_2) = T$). On the other hand, if we have collection $C' = \{\{\bar{u}_1, u_2\}, \{\bar{u}_1, \bar{u}_2\}, \{u_1\}\}$, the answer is negative. SAT is well-known as the first problem shown to be NP-complete, and the starting point for establishing the known body of NP-completeness results [10]. Problems from many application domains have been modeled as SAT instances. In VLSI CAD, SAT formulations have been used in testing [4, 8, 11, 20], logic synthesis and physical design [8].

We now illustrate the *constraint-based watermarking* of a SAT solution. For convenience, we assume the 3SAT restriction of the problem where each clause has exactly three variables. Consider the following 3SAT instance:

$$U = \{u_1, u_2, \ldots, u_{14}\}$$

$$\begin{aligned} C = \{&\{\bar{u}_1\bar{u}_2u_9\}, \{\bar{u}_1\bar{u}_3\bar{u}_4\}, \{\bar{u}_1u_2\bar{u}_5\}, \{u_1\bar{u}_2u_{10}\}, \{\bar{u}_1\bar{u}_3u_8\}, \{\bar{u}_1\bar{u}_3u_7\}, \\ &\{u_1\bar{u}_5u_7\}, \{\bar{u}_1\bar{u}_6\bar{u}_{12}\}, \{\bar{u}_1u_{10}u_{12}\}, \{\bar{u}_1u_6u_9\}, \{\bar{u}_2\bar{u}_3\bar{u}_{10}\}, \{u_2\bar{u}_5\bar{u}_{14}\}, \\ &\{\bar{u}_2u_7u_8\}, \{u_2\bar{u}_8u_9\}, \{u_3u_4u_8\}, \{u_3u_5\bar{u}_7\}, \{\bar{u}_3u_8u_{13}\}, \{u_3\bar{u}_9\bar{u}_{11}\}, \\ &\{u_3u_{10}\bar{u}_{12}\}, \{\bar{u}_4\bar{u}_7\bar{u}_8\}, \{\bar{u}_5\bar{u}_8\bar{u}_{12}\}, \{u_4\bar{u}_7u_{13}\}, \{\bar{u}_5\bar{u}_9\bar{u}_{11}\}, \{\bar{u}_5u_7u_9\}, \\ &\{u_6u_{10}u_{11}\}, \{u_6\bar{u}_8\bar{u}_{12}\}, \{u_7u_9\bar{u}_{12}\}, \{u_7u_9\bar{u}_{13}\}, \{u_9u_{11}\bar{u}_{14}\}, \{u_{10}u_{11}\bar{u}_{12}\}\} \end{aligned}$$

Our goal is to alter the given 3SAT instance such that (i) any satisfying assignment ("solution") to the modified instance is a solution to the original instance, and (ii) both the modified instance and the solution contain information (i.e., a "signature") that uniquely identifies the author of the solution.

* Work by M. Potkonjak and G. Wolfe supported in part by DARPA under grant N66001-97-2-8901. Work by W. H. Mangione-Smith and J. Lach supported by the Defense Advanced Research Projects Agency of the United States of America, under contract DAB763-95-C-0102 and subcontract QS5200 from Sanders, a Lockheed Martin company. Work by A. B. Kahng, S. Mantik, I. L. Markov, P. Tucker and H. Wang supported by a grant from Cadence Design Systems, Inc.

[1] According to [26], the available transistor density has increased by 58%/year over the last 20 years; designer efficiency (measured in transistors designed per staff-month) has increased by only 21%/year over the same period.

[2] For example, the Virtual Socket Interface (VSI) Alliance has identified six key technologies that must be in place to enable industrial-strength virtual component-based synthesis. In addition to system verification, mixed-signal design integration, on-chip bus, manufacturing-related test, and system-level design, intellectual property protection is considered to be a crucial enabling technology [29].

Enumeration of the solution space indicates that the given 3SAT instance has 556 different satisfying assignments. We impose additional *constraints* in the form of extra 3-literal clauses, using the simple (case-insensitive) encoding $A - u_1, B - \bar{u}_1, C - u_2, D - \bar{u}_2, \ldots, Y - u_{13}, Z - \bar{u}_{13}, space - u_{14}$ to encode a signature.[3] We choose the signature "Watermarking Techniques for Intellectual Property Protection University of California at Los Angeles VLSI CAD LAB", which adds 38 new clauses to the instance. After adding these constraints to the initial instance, the number of satisfying assignments decreases to 2. We claim that any satisfying assignment for this augmented 3SAT instance contains our signature, and that the likelihood of someone else generating such a solution by chance is 2 in 556, or 0.00496. In this example, the addition of a watermark incurs no overhead; it simply guides which solution is selected. The watermarking strategy is also *non-intrusive* (more specifically, it is based on *pre-processing* of the input instance), in that any existing solution strategy remains applicable to the augmented (watermarked) 3SAT instance.[4] In our experience, many commonly encountered NP-complete formulations can also be watermarked using similar constraints.

1.1 Organization of the Paper

The remainder of this paper is organized as follows. Related concepts in artifact watermarking, cryptography and IP-based synthesis are surveyed in Section 2. Principles and desiderata (e.g., protection requirements) of non-intrusive, constraint-based IP watermarking are discussed in Section 3. A review of cryptography background and supporting tools (e.g., one-way functions, pseudo-random encrypted streams, digital signatures, and strength analyses) is given in Section 4. Section 5, as well as the above discussion in Section 1, suggests that non-instrusive (pre- or post-processing based) IP watermarking with constraints is often easy to implement with no significant added complexity or loss of solution quality. While an analysis of typical attacks in Section 4 shows that not all possible protection levels can be achieved with known algorithms, we nonetheless conclude in Section 6 that constraint-based watermarking has significant potential to protect IP and support design reuse.

2 Related Work

We now survey related work in watermarking-based IPP, cryptography, and IP-based synthesis.

2.1 Artifact vs. IP Protection Watermarking

A *watermark* is a mark that is (i) embedded into an artifact (text, image, video, audio) or piece of intellectual property (hardware, software, algorithm, data organization), (ii) designed to identify the author, the source, the used tools and techniques and/or recipient of the artifact or the intellectual property, and (iii) difficult to detect and remove. It is important to distinguish traditional requirements for *artifact watermarking* from those governing the *IP protection* applications that we address. Artifact watermarking simply adds a signature into a given artifact *without regard to maintaining correctness or function.* "Transparency" of the signature stems from imperfections in human auditory and visual systems: the artifact (e.g., a digitized photograph) is actually changed, but the human eye cannot perceive the change.[5]

In contrast, watermarking for IP protection imposes much stronger constraints because the watermarked IP must remain *functionally correct.* For example, one cannot arbitrarily introduce extra lines of code into a Verilog program, or extra devices and interconnects into a transistor-level layout. Our discussion is centered around the following key idea: watermarking for IPP is most practically accomplished by imposing a set of additional *constraints* during the design and implementation of IP, so as to uniquely encode the signature of the author. Since 1996, the effectiveness of this generic scheme for watermarking-based IPP has been demonstrated at the level of algorithms [15], behavior [14], logic synthesis and physical design [16], as well as in FPGA designs [18, 19].

2.2 Cryptography

Modern age cryptography grew from the seminal work of Diffie and Hellman [9], who introduced public-key cryptography based on the computational intractability of certain mathematical tasks. Since 1976, cryptographic algorithms and techniques have evolved through vigorous innovation and public scrutiny, resulting in a variety of digital signature mechanisms, as well as protocols for secret splitting, timestamping, proxy signatures, group signatures, key escrow, oblivious transfer, oblivious signatures, digital cash, etc. [27, 22]. The link between cryptography and watermarking is fundamental: cryptography provides the theoretical foundations as well as the algorithmic and protocol infrastructure that support watermarking-based IPP and provide a wide spectrum of authorship protection services.

2.3 IP-Based Synthesis

As noted above, short design times, increased device counts and design starts, and foundry amortization have together forced a change in design methodology. The new semiconductor business regime is based on IP reuse. No other regime is compatible with rapid turnaround and high device counts; no other regime enables ASIC vendors to keep their foundries full of high-value product.[6]

Less than two years ago, the VSI Alliance and CFI Component Information Library Project were first announced. Today, at least three major industry organizations – RAPID (IP providers) [32], SI^2 [34] (ASIC vendors), and VSIA [33] (a large organization of EDA vendors, ASIC vendors, system houses and IP providers) – are actively building the industry infrastructure for IP-based design.[7] Several missing infrastructure pieces are technical, with deep implications for the associated EDA technology and design methodologies.[8] Other missing pieces include the standards for representing design IP. However, arguably the most pressing infrastructure issues are legal: what are the risks faced by ASIC suppliers and EDA tools vendors as they incorporate third-party IP? Who holds accountability for design success? How will the rights of IP creators and owners be protected? It is notable that despite their varying perspectives, each of the three major industry organizations has a working group for legal issues.

3 Precepts and an Approach for Constraint-Based IPP

In this section we develop basic precepts, and a general constraint-based approach, for watermarking IP protection. Our discussion will abstract the design process as a form of optimization, and we will focus on opportunities for non-intrusive watermarking (i.e.,

[3] For example, the signature "cat dog fox" would be encoded using the extra clauses $\{\{u_2, u_1, \bar{u}_{10}\}, \{u_{14}, \bar{u}_2, u_8\}, \{u_4, u_{14}, \bar{u}_3\}, \{u_8, \bar{u}_{12}, u_{14}\}\}$ (we pad the end of the message with an extra space to maintain three literals per clause).

[4] This observation holds for the three major classes of SAT heuristics: (i) random search [25, 7], (ii) nonlinear programming relaxation and rounding [12], and (iii) a variety of BDD-based techniques [3].

[5] Artifact watermarking has been used for thousands of years. Only with the proliferation of digital media has it attracted wide research and economic interest, e.g., for protection of audio [1, 15], text [21, 2], image [5], and video.

[6] Interestingly, the vision of pervasive IP reuse can be viewed as simply a refinement of earlier visions which saw the inability of the structured-custom methodology to scale with design complexity as a main driver for "methodology convergence".

[7] The early CFI effort spawned the Pinnacles Component Information Standard, and CFI subsequently became SI^2 (Silicon Integration Initiative).

[8] For example, how reusable IP will be bundled with standardized test and simulation "envelopes", or the form of reusable IP and the manner in which it will "mix and match", remains unclear. Current visions encompass varying degrees of "hardness" of the IP, e.g., soft (HDL program), medium (HDL program + floorplan), hard (GDSII stream file), etc. Harder forms of IP might have greater value since they would embody greater amounts of design effort. At the same time, hard IP is less reusable due to its well-defined shape and inherent timing/noise/thermal context; it also allows less flexibility in floorplanning and routing due to constraints on over-the-block routing (e.g., timing and signal integrity margins). It remains to be seen how "parameterizable" an IP block can be in terms of area-time tradeoffs, migration to alternate processes, routing resource utilization, etc.

methods that can be transparently integrated within existing design flows via pre- or post-processing).

3.1 Context for Watermarking

The following ingredients form the *context* for a non-intrusive watermarking procedure:

- An *optimization problem* with known difficult complexity, corresponding to some design synthesis task. By difficult, we mean that either achieving an acceptable solution, or enumerating enough acceptable solutions, is prohibitively costly. The solution space of the optimization problem should be large enough to accommodate a digital watermark.
- A well-defined *interpretation* of the solutions of the optimization problem as intellectual property.
- Existing *algorithms* and/or *off-the-shelf software* that solve the *optimization problem*, likely without any kind of watermarking involved. Typically, the "black box" software model is appropriate, and is moreover compatible with defining the watermarking procedure by composition with pre- and post-processing stages.[9]
- *Protection requirements* that are largely similar to well-understood protection requirements for currency watermarking. Examples of such requirements, which are discussed in Section 4 below, include: (i) removing or forging a watermark must be as hard as re-creating the design; and (ii) tampering with a watermark must be provable in court.

A non-intrusive watermarking procedure then applies to any given instance of the optimization problem, and can be attached to any specific algorithms and/or software solving it. Such a procedure can be described by the following components:

- A *use model* or *protocols* for the watermarking procedure. This is not the same as algorithm descriptions; it is less formal, and can be helpful in revealing possible attacks beyond the generic types noted above. For example, algorithms assume a cell numbering, while renumbering cells can defeat a watermarking procedure (something that can be seen only at the protocol level). In general, each watermarking scheme must be aware of attacks based on design symmetries, renaming, reordering, small perturbations (which may set requirements for the structure of the solution space), etc.
- Algorithmic descriptions of the *pre-* and *post-processing* steps of the watermarking procedure.
- *Strength and feasibility analyses* showing that the procedure satisfies given protection requirements on a given instance. Strength analysis requires metrics, as well as structural understanding of the solution space (e.g., "barriers" (with respect to local search) between acceptable solutions). Feasibility analysis requires measures of solution quality, whether a watermarked solution remains well-formed, etc.
- *General robustness analyses*, including discussion of susceptibility to typical attacks, discussion of possible new attacks, performance guarantees (including complexity analysis) and implementation feasibility.

Before describing a general strategy for embedding digital watermarks, we observe that optimization problems with known watermarking procedures share several common features: (i) having multiple acceptable solutions (we typically accept suboptimal solutions for NP-hard problems); (ii) solved by optimization heuristics; and (iii) discrete in nature.[10]

[9] Watermarking the results of non-deterministic and/or unknown algorithms – or even "hand-made" results – is possible as well. IP protection can even be achieved, to some extent, with black-box off-the-shelf software that is viewed as a one-way function mapping inputs to design solutions. In this discussion, we focus only on the simple model involving known deterministic algorithms.

3.2 General Strategy for Constraint-Based IPP

Our general strategy is to map an author's signature into a set of constraints ("desired relations") which can independently hold for a particular solution (or independence can be assumed, via some manipulations). If disproportionately many of these constraints are satisfied the presence of the signature is indicated, and vice versa. Choosing the type of constraints, and the tactic (e.g., pre- or post-processing) by which we make it likely for more of them to be satisfied than would otherwise be expected, is what instantiates a particular watermarking algorithm from the general strategy. These choices can dramatically affect the strength of the watermark and the degradation of solution quality caused by watermarking. To facilitate later discussion, we now describe generic watermarking and signature verification procedures using "Alice (and Bob)" scenarios, where Alice uses watermarking to protect some IP (below, Bob will attempt to subvert such protection).

Generic Watermarking Procedure. Alice wishes to protect some IP that involves many stages of processing. She chooses to watermark one or more of these stages. The results of these stages now carry a watermark which will propagate down to the output of further stages all the way down to the final result. Clearly the amount of watermarking she imposes on a particular stage trades off with the degree of degradation of quality of the final result. Alice watermarks each stage by selecting a set of "constraints", then using preprocessing of the stage's input and postprocessing of the stage's output to encourage a disproportionate number of these constraints to be satisfied. Note that Alice need not tell anyone which constraints correspond to her signature.

Generic Signature Verification Procedure. To demonstrate that a particular stage was watermarked Alice must show that its solution (which may have been passed on undisturbed to other stages and perhaps all the way to the final result) satisfies a disproportionate number of her watermarking constraints. By identifying the watermarking constraints, determining how many of them are satisfied, and calculating P_c – the probability of so many (or more) of the constraints being satisfied by coincidence – Alice can verify that her signature is present. A strong proof of authorship corresponds to a low value for P_c. Note that to show this to other people, Alice must reveal her signature and, hence, the chosen constraints.

4 Analysis of Constraint-Based Watermarking

We now analyze the constraint-based watermarking strategy and present the cryptographic background and support tools that are necessary for the analysis. We first describe how to map a signature into a set of constraints and the method by which we determine the strength of the watermark in a watermarked solution. We then discuss typical forms of attack on our scheme and the obstacles that prevent these attacks from succeeding.

4.1 Selection of Constraints

Given a pseudorandom number generator and a particular type of constraint it is a simple matter to select a set of X constraints where each one is determined independently. It is only slightly more work to select a set of X constraints with no constraint repeated. Thus, the task of mapping an author's signature into a set of constraints can be reduced to the task of seeding a pseudorandom number generator with the signature. This is also easy. Suppose that the author's signature is a particular text message. We can convert this message into a cryptographically sound pseudorandom bit stream by simply hashing the message (using a one-way hash function such as MD5 [24]) and using the hash as a seed for a stream cipher such as RC4 [35].

[10] We believe "continuous watermarking" is possible as well, e.g., by mapping into "discrete watermarking" by Fourier transform. However, watermarking has traditionally been of more relevance to discrete problems.

4.2 Proof of Authorship

A watermark's *proof of authorship* is expressed as a single value, P_c, which is the probability of so many (or more) of the selected constraints being satisfied. Essentially, P_c is the probability of a non-watermarked solution carrying our watermark by coincidence. We wish this probability to be convincingly low so as to have a strong proof of authorship. When we cannot compute P_c exactly it is acceptable to overestimate it so that we actually report an upper bound on P_c. Computing such an upper bound on P_c is typically straightforward. Let p be the probability of satisfying a single random constraint by coincidence. This value, or a fairly tight upper bound on it, is usually obvious from the definition of a constraint. Here we assume that p is independent of whether the other constraints were satisfied. Let C be the number of imposed constraints. Let b be the number of these constraints that were *not* satisfied. Let X be a random variable that represents how many of the C constraints were not satisfied. Now P_c can be computed as a sum of binomials, i.e., the probability that coincidentally only b or less of C constraints were not satisfied is given by $P_c \equiv P(X \leq b) = \sum_{i=0}^{b}(\frac{C!}{(C-i)!\cdot i!}\cdot(p)^{C-i}\cdot(1-p)^i)$. This analysis assumes that p is independent of whether other constraints are satisfied, an assumption that is often untrue. However, when the number of imposed constraints (C) is sufficiently small, we have a very good approximation.

4.3 Typical Attacks

There are several general ways of attacking our watermarking scheme. Here we discuss the more prominent ones: finding "ghost signatures", tampering, and forging. We analyze these attacks using "Alice and Bob" scenarios.

Attack: Finding Ghosts. Bob wishes to steal IP from Alice and claim it as his own. He knows that Alice has protected her IP (i.e., the solution to a particular stage of the design process) with a watermark, but will claim that the IP *also* contains his own watermark. Bob thus attempts to find a *ghost signature*, namely, a signature that corresponds to a set of constraints that yields a favorable P_c, but which was discovered after fact instead of being actually watermarked into the solution. To be convincing, Bob must find a ghost signature that yields a sufficiently convincing value P_c.

Bob has only two approaches. He may choose a set of constraints (presumably ones that yield a good proof of authorship P_c) and then attempt to find a signature that corresponds to this set. This requires reversing the cryptographically secure one-way functions that convert a signature into a set of constraints, which is hard. Alternatively, Bob may try a brute-force approach to find a signature that corresponds to a set of constraints that yields a convincing proof of authorship P_c. However, this brute-force attack becomes computationally infeasible if the threshold for proof of authorship is set sufficiently low (e.g., $P_c \leq 2^{-56}$).

Attack: Tampering. If Bob cannot find a convincing ghost signature, he may decide to *tamper* with Alice's solution. Ideally, such tampering would completely remove Alice's signature and add Bob's own signature. Bob can do this by simply re-solving the problem from scratch with his own watermark encoded, then continuing through subsequent processing stages based upon the output he obtains. Nothing can be done to stop this directly. However, we believe that in realistic scenarios, Bob cannot afford to redo all of the subsequent phases of the design process, particularly if the watermarking occurred relatively early in the process.

There are realistic means by which Bob can tamper with a solution without having to re-solve every subsequent stage of the process. Generally, these amount to transforming the solution output by the last phase of the design process, where the transformation has a similar effect on the output of the watermarked phase of the design process. Specific changes that Bob makes to the final solution will likely correspond to (i) local perturbations of the solution to the watermarked phase, or to (ii) global-scale transformations such as those which exploit a symmetry of the design representation. Given that Bob is limited to these kinds of tampering attacks, it is critical that Alice's watermarking technique be resistant to such transformations.[11]

Attack: Forging. Finally, Bob may attempt to subvert Alice's watermark by inappropriately watermarking other solutions with Alice's watermark. In other words, Bob wishes to *forge* Alice's signature. To do this, Bob needs a signature that he can convince others belongs to Alice. If a signature corresponds simply to a text message (as it has so far in this discussion) then Bob's task is easy: he simply chooses a text message resembling one that Alice would use. However, such attacks can be easily prevented by using a public key encryption system [23]. Any message actually signed by Alice would be encrypted with her private key, yet verifiable with her public key. Notice that the private key is not compromised even if messages that encoded with it are compromised, so Alice may still demonstrate the presence of her watermark to anyone who knows her public key, without compromising her private key. Thus, Bob is able to forge a message from Alice only if he knows her private key.

5 IP Watermarking Synthesis Examples

In this section, we sketch three examples of IP watermarking approaches, in three very distinct domains. Our intent is to illustrate the wide-ranging applicability of the principles developed above.

5.1 Preprocessing in System-Level Design

At the system level, instruction and data caches consume a significant portion of the overall area, and often have crucial impact on system timing and power consumption [17]. Much effort has been devoted to allocating minimal cache structures and optimizing code for effective cache utilization [30]. A particularly successful technique is *cache line coloring* [13].

Given a code segment and input data benchmarks, cache line coloring code optimization seeks a permutation of basic block code segments such that the mapping of code to cache entries minimizes the cache miss ratio over the given benchmarks. The problem can be modeled as follows. The program is profiled with respect to the benchmark data, and spatial (frequent sequences of sequentially executed code) and temporal (frequent control sequences) correlations noted among basic blocks of code. The program is modeled using a control data flow graph, where a graph node corresponds to a set of instructions that are encompassed in a single basic block and fit exactly one cache line. Weighted edges between nodes correspond to spatial or temporal correlations that exceed given threshold values (modeling accuracy thus depends on the thresholds for edge inclusion). The problem of minimizing cache misses is equivalent to finding a solution to graph coloring using a given fixed number of colors (corresponding to available cache lines).[12]

To watermark such designs, the initial design constraints may be augmented with additional constraints corresponding to the digital signature of the designer. For example, following the technique for watermarking of graph coloring solutions proposed by Hong and Potkonjak [14], one may add additional edges to the graph according to some encrypted signature of the author. Therefore, the signature will be embedded in the activation path which transfers data between two levels of hierarchy.

[11] Note that since the attacker does not know which constraints correspond to the author's signature, tampering attacks might not be able to ruin the proof of authorship before they significantly degrade the quality of the final solution (at which point the tampered solution ceases to be useful); see [16] for experience in the physical design realm. However, it seems possible for an attacker to use tampering methods to remove a signature that is known to him, or to add an entirely new signature.

[12] Kirovski et al. [17] have experimentally shown that this optimization results in significant performance increase. In general, such optimizations can play an important role in the design of modern multimedia, communications, or low-power systems-on-silicon.

5.2 Postprocessing in Physical-Level FPGA Design

One method of watermarking an FPGA at the physical level involves manipulating unused portions of the configuration bitstream. Informed parties can then extract the mark from the bitstream. There is no effect on the function of the design during insertion or extraction because only unused portions of the design are altered. This approach can be implemented through pre-processing, iterative, or post- processing techniques. The advantage of post-processing is that it does not impact other CAD design tools, and has zero impact on design peformance, area or power consumption. The disadvantage of this approach is that the watermark is not embedded in the functional part of the design; given enough information, the watermark can be removed without affecting design functionality. An example of an iterative approach can be found in the work by Lach et al. [18, 19].

An example of a purely post-processing approach involves inserting the watermark into the control bits for unused outputs from configurable logic blocks (CLBs). Certain bits in the configuration bitstream that control multiplexers for the CLB outputs can be replaced by watermark bits if the CLB outputs are not used. For example, the Xilinx 4000 family of FPGAs contain CLBs with four outputs [31]. Two outputs (X and Y) are combinational, while the others (XQ and YQ) can be used in sequential designs. The two combinatorial outputs are each controlled by a 2-to-1 multiplexer, and the two sequential outputs are each controlled by three 2-to-1 multiplexers and one 4-to-1 multiplexer. Figure 1 shows the control layout of the 4000 family's CLB outputs.

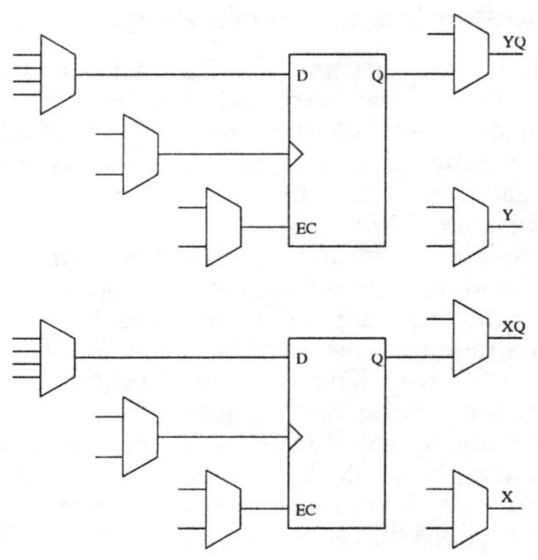

Figure 1: Control directly attributed to CLB outputs.

The number of configuration bits associated with a multiplexer is equal to (or greater than) the number of required control bits. Therefore, one and two watermark bits can be inserted at each unused 2-to-1 and 4-to-1 multiplexer respectively. Thus, each unused combinatorial output can store one watermark bit and each unused sequential output can store five watermark bits. The total number of watermark bits that can be inserted in an entirely unused CLB is twelve. Table 1 shows the number of watermark bits that can be inserted into various devices within the 4000 family given certain percentages of unused CLB outputs. The numbers calculated here are for an even number of unused combinatorial and sequential outputs.

The process of watermark insertion in this approach is an entirely post-processing step and requires very little added design effort. The tool methodically scans the bitstream searching for unused outputs by finding CLB output pinwires that do not attach to any external CLB interconnect. Upon the detection of unused outputs, the next bits of the watermark are inserted in place of the corresponding multiplexer configuration bits. The size of the watermark is limited by the number of bits made available by this approach. Extracting the watermark is an almost identical process.

% Outputs Unused	Part/# CLBs				
	4006 /256	4010 /400	4013 /576	4020 /784	4025 /1024
1	30	48	69	94	122
5	153	240	345	470	614
10	307	480	691	940	1228
20	614	960	1382	1881	2457

Table 1: Number of bits available for watermarking.

The tool finds unused CLB outputs the same way as was done in insertion and pieces the watermark back together by examining the corresponding multiplexer configuration bits.

This FPGA watermarking approach requires little extra design effort, can store fairly large watermarks, allows for easy mark extraction, and has no overhead in terms of design area or performance. However, because a mark is nonfunctional, it may be removed by reverse engineering a design to a stage in the flow before the mark has been applied. Fortunately, most FPGA vendors will not reveal the specification of their configuration streams, specifically to complicate the task of reverse engineering and thus protect the investment of their customers. For example, the Xilinx XC4000 devices follow a form of Pareto's rule: the first 80% of the configuration information can be determined relatively easily by inspection, the next 18% is much more difficult, etc. The complexity is enhanced by an irregular pattern that is not consistent between rows or columns, as a result of the hierarchical interconnect network. Xilinx does not take any specific actions to make their configurations difficult to reverse engineer. However, they do believe that it is difficult to do in general, and they promise their customers that they will keep the bitstream specification confidential in order to raise the bar for reverse engineering [28].

5.3 Preprocessing in Physical Design

Finally, in the context of physical design, we present a new preprocessing based approach for design watermarking. Our approach exploits the flexibility with which *path-based timing constraints* can be satisfied.

Consider the typical elements of an input instance for timing-driven placement and routing.

- physical floorplan, library of physical cell masters, and cell-level netlist
- cell-level performance macromodels for each cell master (e.g., non-linear table models (Synopsys .lib, Cadence .ctlf, OVI ALF, etc.)) for timing and power dissipation analysis
- technology file (models of interconnect RC characteristics, design rules, etc.)
- constraints, which are chiefly (i) "direct" placement and routing constraints (e.g., region-based location constraints arising from the floorplanner, and transmitted in PDEF format), and (ii) performance constraints (e.g., SDF latch-to-latch path timing upper and lower bounds, with false path and multicycle constraints specially annotated)

We watermark a design by selecting path timing constraints and replacing them with "subpath" timing constraints. Suppose that we have the path timing constraint $t(C_1 - C_2 - C_3 - \ldots - C_{10}) \leq 50ns$ ($C_i \equiv$ cells). We can allocate the timing bound between two subpaths and replace this constraint by two constraints $t(C_1 - \ldots - C_5) \leq 20ns$ and $t(C_5 - \ldots - C_{10}) \leq 30ns$. All else being equal, the chance that satisfying the original constraint happens to satisfy both of these subpath constraints is at most one-half.[13] Constraining on

[13]Note that the allocation would be done with respect to available slack on the path, e.g., path delay upper bound minus sum of "intrinsic" cell delays. Also note that constraint satisfaction will likely be determined in the context of final layout.

the order of hundreds of timing paths (from the several millions one finds in typical verbose SDF specifications) is transparent to timing-driven design tools, yet affords strong proofs of authorship. Similar techniques can be applied in the regime of compact SDF timing constraints, or at the budgeting stages of timing-driven design.[14]

6 Conclusions

Motivations and antecedents for *watermarking-based* protection of hardware and software design IP arise in reuse-centric system design, artifact watermarking, and cryptography. In this paper, we have described fundamental precepts, a canonical technique, and example applications for watermarking-based IPP. Several key ideas are as follows.

- Stages of the (hardware, software) design process can typically be viewed as (difficult) *optimization instances* whose solutions constitute intellectual property to be protected.
- IP watermarking can typically be achieved by adding *constraints* (e.g., interpreted from a cryptographically secure encoding of the IP owner's signature) to any given design optimization instance.
- The addition of constraints can typically be achieved using *pre-* or *post-processing* of the inputs and outputs, respectively, for a given design optimization. In this way, the watermarking is often transparent to existing algorithms and tools, i.e., it is *non-intrusive*.

We have also noted other aspects of the watermarking context, e.g., protection requirements against typical forms of attack, and cryptography background (one-way functions, cipher streams, and digital signatures). Problem formulations from several domains (high-level design, FPGA design, physical design, as well as SAT) illustrate the general applicability of our techniques, and suggest that non-instrusive IP watermarking with constraints can typically be implemented with no significant added complexity or loss of solution quality. Thus, constraint-based watermarking appears to have significant potential to protect IP and support design reuse.

Our ongoing work develops watermarking-based IPP techniques for many other domains, with particular attention to robustness under various attacks. We also address a number of variant requirements, including fingerprinting, copy detection, and proportionate watermarking (e.g., of hierarchical designs).

References

[1] W. Bender, D Gruhl, N. Morimoto and A. Lu, "Techniques for Data Hiding", *IBM Systems Journal*, 35(3-4), 1996, pp. 313-336.

[2] J.T. Brassil, S. Low, N.F. Maxemchuk and L. O'Gorman, "Electronic Marking and Identification Techniques to Discourage Document Copying", *IEEE Journal on Selected Areas in Communications*, 13(4) (1995), pp. 1495-1504.

[3] R.E. Bryant, "Binary Decision Diagrams and Beyond: Enabling Technologies for Formal Verification", *Proc. of the International Conference on Computer-Aided Design*, 1995, pp. 236-243.

[4] S.T. Chakradhar, V.D. Agrawal and S.G. Rothweiler, "A Transitive Closure Algorithm for Test Generation", *IEEE Transactions on Computer-Aided Design of Integrated Circuits and Systems*, 12(7) (1993), pp.1015-28.

[5] I.J. Cox and M.L. Miller, "A Review of Watermarking and the Importance of Perceptual Modeling", *SPIE Conf. on Human Vision and Electronic Imaging II*, 3016(1997), pp.92-99.

[14] In general, the physical design context presents a rich environment for constraint-based watermarking. For example, the physical library information and/or design rules allow variant pin access models for a cell, which will constrain how interconnects attach to pins; extra blockage geometries in cell instances or masters can also be used to constrain the routing; and via and stub rules can again encode a signature within the output of a constraint-driven router [16]. Simple parity-based schemes abound, e.g., based on mirroring of cells, parity of row indices to which cells are assigned [16], routing of wires to the left or right of shield wires, etc. Even performance macromodels (nonlinear table models for timing and power) can be perturbed (thus constraining the performance-driven layout) to influence the layout tool's output.

[6] I.J.Cox, J. Kilian, F.T. Leighton and T. Shamoon, "Secure Spread Spectrum Watermarking for Multimedia", *Transactions on Image Processing*, 6(12) (1997), pp.1673-87.

[7] M. Davis and H. Putnam, "A Computing Procedure for Quantification Theory", *Journal of the ACM*, 7(3) (1960), pp. 201-215.

[8] S. Devadas, "Optimal Layout Via Boolean Satisfiability", *IEEE International Conference on Computer-Aided Design*, 1989, pp. 294-7.

[9] W.Diffie and M. Hellman, "New Directions in Cryptography", *IEEE Transactions on Information Theory*, IT-22(6), 1976, pp. 644-654.

[10] M. E. Garey and D. S. Johnson, *Computers and Intractability: a Guide to the Theory of NP-Completeness*, Freeman, 1979.

[11] J. Giraldi and M.L. Bushnell, "Search State Equivalence for Redundancy Identification and Test Generation", *Proc. Intl. Test Conf.*, 1991, pp. 184-193.

[12] M. X. Goemans and D. P. Williamson, "Improved Approximation Algorithms for Maximum Cut and Satisfiability Problems Using Semidefinite Programming", *Journal of the ACM*, 42(6) (1995), pp.1115-45.

[13] A.H. Hashemi, D.R. Kaeli and B. Calder, "Efficient Procedure Mapping Using Cache Line Coloring", SIGPLAN Notices, 32(5) (1997), pp.171-182.

[14] I. Hong and M. Potkonjak, "Behavioral Synthesis Techniques for Intellectual Property Protection", unpublished manuscript, 1997.

[15] L. Honey, A.H. Tewfik and K.N. Hamdy, "Digital Watermarks for Audio Signals", *Proc. of the International Conference on Multimedia Computing and Systems*, 1998, pp. 473-480.

[16] A. B. Kahng, S. Mantik, I. L. Markov, M. Potkonjak, P. Tucker, H. Wang and G. Wolfe, "Robust IP Watermarking Methodologies for Physical Design", *Proc. ACM/IEEE Design Automation Conf.*, 1998.

[17] D. Kirovski and M. Potkonjak, "System-Level Synthesis of Low-Power Hard Real-Time Systems", *Proc. ACM/IEEE Design Automation Conference*, 1997, pp. 697-702.

[18] J. Lach, W. H. Mangione-Smith and M. Potkonjak, "Fingerprinting Digital Circuits on Programmable Hardware", *Proc. Workshop on Information Hiding*, 1998.

[19] J. Lach, W. H. Mangione-Smith and M. Potkonjak, "FPGA Fingerprinting Techniques for Protecting Intellectual Property", *Proc. CICC*, 1998.

[20] T. Larrabee, "Test Pattern Generation Using Boolean Satisfiability", *Proc. International Test Conf.*, 11(1) (1992), pp.4-15.

[21] S.H. Low, N.F. Maxemchuk, J.T. Brassil and L. O'Gorman, "Document Marking and Identification Using both Line and Word Shifting", *Proc. INFOCOM*, 1995, pp.853-60.

[22] A.J. Menezes, P.C. van Oorschot and S.A. Vanstone, *Handbook of Applied Cryptography*, Boca Raton, CRC Press, 1997.

[23] Pretty Good (tm) Privacy, Phil Zimmerman, Users Guide: Vol. 1, Vol. 2, and Internal Data Structures from Phil's Pretty Good Software, copy: `http://www-atp.llnl.gov/atp/papers/HRM/references/pgp-refs.html`

[24] R.L.Rivest, "RFC 1321: the MD5 Message-Digest Algorithm", *Internet Activities Board*, April 1992.

[25] B. Selman, "Stochastic Search and Phase Transitions: AI Meets Physics", *IJCAI*, 1 (1995), pp.998-1002.

[26] Semiconductor Industry Association, *National Technology Roadmap for Semiconductors*, revised November 1997.

[27] D.R. Stinson, *Cryptography: Theory and Practice*, Boca Raton, CRC Press, 1995.

[28] S. Trimberger, *personal communication*, 1997.

[29] VSI Alliance, "Fall Worldwide Member Meeting: a Year of Achievement", Santa Clara, CA, October 1997.

[30] H. Wang, T. Sun, Q. Yang, "Minimizing Area Cost of on-Chip Cache Memories by Caching Address Tags", *IEEE Transactions on Computers*, 46(11) (1997), pp. 1187-1201.

[31] Xilinx, *The Programmable Logic Data Book*, Xilinx Corporation, San Jose, 1996.

[32] `http://www.rapid.org`

[33] `http://www.vsi.org`

[34] `http://www.si2.org`

[35] `http://dcs.ex.ac.uk/~aba/rsa/rc4.html`

Robust IP Watermarking Methodologies for Physical Design*

Andrew B. Kahng, Stefanus Mantik, Igor L. Markov, Miodrag Potkonjak,
Paul Tucker†, Huijuan Wang and Gregory Wolfe

UCLA Computer Science Dept., Los Angeles, CA 90095-1596
† UCSD Computer Science & Engineering Dept., La Jolla, CA 92093-0114

Abstract

Increasingly popular reuse-based design paradigms create a pressing need for authorship enforcement techniques that protect the intellectual property rights of designers. We develop the first intellectual property protection protocols for embedding design *watermarks* at the physical design level. We demonstrate that these protocols are transparent with respect to existing industrial tools and design flows, and that they can embed watermarks into real-world industrial designs with very low implementation overhead (as measured by such standard metrics as wirelength, layout area, number of vias, routing congestion and CPU time). On several industrial test cases, we obtain extremely strong, tamper-resistant proofs of authorship for placement and routing solutions.

1 Introduction

Due to rapidly growing device counts, shortened design cycle times and a compounding "design productivity shortfall" [9], core-based design and software reuse strategies are widely believed to be the only viable implementation alternatives for the next level of integrated circuits and their system integration. As a result, development of intellectual property protection (IPP) techniques and tools has emerged as a very prominent open research topic. In this work, we develop the first protocols for IPP at the *physical design* level, using the concept of constraint-based watermarking.

As defined in [5], a *design watermark* is an invisible (i.e., imperceptible to human or machine analysis) identification code that is permanently embedded as an integral part within a design. Various criteria for a given watermarking-based IPP technique, as determined by a leading industry organization, include [12]: (i) maintenance of functional correctness, (ii) transparency to existing design flows; (iii) minimal overhead cost; (iv) enforceability; (v) flexibility in providing a spectrom of protection levels; (vi) persistency; (vii) invisibility; and (viii) proportional component protection. In [5], we list other watermarking desiderata, and describe a canonical approach to watermarking-based IPP where additional *constraints* encoding the IP author's signature are added into a given design optimization instance. The solution of the optimization instance, in satisfying these constraints (which would be unlikely in a random solution to the original instance), implicitly contains a proof of authorship. The approach described in [5] is transparent to existing design flows in that it relies on *preprocessing* (of inputs) or *postprocessing* (of solutions) with respect to any given design optimization.

We center on the physical design phase for several reasons.

- Physical design is traditionally viewed as a "difficult" domain, where even a small percentage variation in solution quality can make or break a design, and where high-quality solutions are known to have strong structural resemblance to each other [4]. Devising a watermarking technique that can make a solution "unique", without compromising solution quality, is quite challenging in such a domain.
- With deep-submicron technology, many performance constraints (e.g., budgeted edge delays consistent with path timing bounds) cannot be considered satisfied until they are satisfied in the physical design. Thus, for example, it is disingenuous to "watermark" a design by constraining timing budgets, without verifying that such constraints are satisfied after physical design.
- Other trends – IP reuse methodologies, higher perceived valuation of "hard IP", increasing availability of multiple foundry sources, difficulty of performance validation before physical design, changing handoff models, etc. – all point to physical design as an appropriate juncture in the design cycle for watermarking.

Contributions of Our Work

To our knowledge, this work gives the first solution for IPP at the physical design level. For placement, we propose a *postprocessing* flow that encodes a signature as specified parity (i.e., odd- or even-index) of the cell row within which particular standard cells must be placed. For routing, we propose a *preprocessing* flow that encodes a signature as upper bounds on the wrong-way wiring used to route particular signal nets.[1] Using real industrial design examples and commercial layout tools, we demonstrate the effectiveness of both the preprocessing- and the postprocessing-based watermarking. In particular, strong signatures are achieved without compromising any of the standard metrics for solution quality (routability, wirelength, number of vias, CPU time, etc.). We also demonstrate that these signatures are tamper-resistant. We conclude that addressing IP protection at a lower level of abstraction has an advantage: designs inherently have orders of magnitude more components, allowing significantly stronger proofs of authorship as well as lower overhead. We also conclude that the postprocessing approach is not only feasible, but indeed quite attractive for several reasons: (i) it enables watermarking of already existing designs; (ii) it enables direct calculationg of the hardware overhead incurred by IPP;[2] and (iii) is may be likelier to find acceptance among designers and managers, since the complete design process is not altered in any way.

* Work by M. Potkonjak and G. Wolfe supported in part by DARPA under grant N66001-97-2-8901. Work by A. B. Kahng, S. Mantik, I. L. Markov, P. Tucker and H. Wang supported by a grant from Cadence Design Systems, Inc.

DAC 98, San Francisco, California

[1] The particular constraints used to encode the watermark affect the strength of the authorship proof as well as metrics of the layout solution. Devising constraint types such that strong signatures can be achieved transparently using existing flows and tools is, in our experience, a very nontrivial task.

[2] When preprocessing is used, two designs (one without additional signature constraints, and one with) must be realized in order to determine the hardware overhead incurred by IPP.

2 Related Work

Related work in artifact watermarking and cryptography is reviewed in [5].[3] We therefore focus our survey of related concepts within the physical design realm.

No previous work in the literature deals with watermarking of physical design solutions. However, constraint specification and management now receive close attention through all phases of chip implementation, including physical design. Standard chip implementation flows begin with such high-level constraints as clock cycle times and offsets, I/O boundary timing, power dissipation bounds, and choice of packaging and implementation technology. Derived constraints[4] then arise throughout the RTL floorplanning, block placement, and routing phases. Within physical design, the following constraint types are most common.

- Timing constraints. Path delay constraints are often expressed in some form of Standard Delay Format (SDF), with heuristic "path cover" techniques used to reduce data volume and improve convergence of timing-driven layout tools. A static timing analysis engine may operate directly from the clock cycle times/offsets and I/O boundary timing to evaluate timing correctness, without explicit enumeration of timing path constraints. For purposes of layout design, path delay constraints are typically budgeted into individual constraints on source-sink edges [11].
- Physical (floorplanning) constraints. To improve timing convergence of the design process, assumptions made during RTL floorplanning or block floorplanning must be propagated to downstream flow stages (e.g., placement and global routing). This is often accomplished via region constraints: a given cell must be located in a given region of the layout, a set of cells must be co-located as a "group", etc. Such constraints may be captured using PDEF or equivalent formats which allow specification of assumed routing topology, layer usage, etc. at the level of global routing.

The mechanisms by which physical design tools enforce such constraints vary widely. However, classic paradigms such as top-down min-cut placement synthesis or ripup-and-reroute interconnect synthesis generally do not support constraints well. One reason is that iterative search mechanisms used today were originally adopted for regimes with "smooth" cost surfaces, while adding constraints induces more 0-1 "discontinuities" in the cost surface. Another reason is that many performance constraints are "global" (e.g., path delay, power dissipation, EMC, signal integrity) while current layout approaches rely on local optimizations. Most importantly, "good" solutions to hard combinatorial problems are often quite similar.[5] The implications for watermarking in physical design are that (i) current tools do not easily support too many "extra" watermarking constraints, and (ii) introduction of too many watermarking constraints will likely degrade solution quality. These issues complicate the choice of watermarking technique.

[3]Our present techniques use well-established ingredients – namely, the cryptographic hash function MD5, the public-key cryptosystem RSA, and the stream cipher RC4 [8, 7] – on which many state-of-the-art commercial cryptographic programs are based.

[4]Derived constraints are of two basic types. *Inferred* constraints can often be viewed as "transformed", e.g., when a signal net's wirelength upper bound is inferred from a signal propagation delay upper bound. *Refined* constraints can often be viewed as created by a "budgeting" or "allocation" process, e.g., when a global path delay constraint is broken up into separate edge delay constraints.

[5]This has been generally characterized as a "big valley" [3] or "massif central" [6]; the phenomenon has also been specifically documented for standard-cell placements under the minimum wirelength objective [4].

3 Watermarking Standard-Cell Place and Route

We have considered a variety of mechanisms by which standard-cell physical design can be constrained. Our goal has been to develop a watermarking protocol that, beyond satisfying criteria listed above, is (i) consistent with existing design practices and tools, (ii) relatively easy to implement, and (iii) acceptable in terms of its impact on real-world layout metrics.

3.1 Row-Based Placement

For row-based placement, traditional physical embedding constraint types (i.e., region and grouping constraints) are straightfoward to realize. Region constraints are transparent to top-down placers, since iterative partitioners accomodate "fixed" preassignments (see [2] for a review). Annealing placers (see [10] for a review) also support such constraints by restricting move generation, and analytic placers support region constraints via inequalities or center-of-gravity constraints (see [1] for a review). Grouping constraints are typically enforced by inducing contracted netlists over clustered representations of the design. However, region and grouping constraints are not well-suited to placement-based watermarking: when made without any structural knowledge of the netlist or of good placement solutions, they can lead to substantial deterioration of solution quality.[6]

Our approach bases the watermarking constraints on the underlying fine-grain placement substrate, which is well-defined prior to the placement phase of design. By "placement substrate", we mean the row structure of legal site locations in the physical floorplan. In particular, we constrain individual cells to be placed with specified *cell row parity*. For example, cell INV4_10 might be constrained to be placed in a cell row that has EVEN index; cell RKPPX1Y might be constrained to an ODD-index row. Our approach has the following advantages:

- Very few constraints are needed to make a strong signature. E.g., if the signature constrains 50 cells with specific row parities, and if the placer realizes all 50 constraints, the chances are 2^{-50} that this could have occurred by accident. Typical placement instances have tens of thousands of standard cells.
- It is compatible with region and group constraint types, and can be applied as soon as a gate-level netlist exists (no specific row/site plan is needed). A priori, the only time failure is guaranteed is when a "watermark cell" is constrained to be in, e.g., an EVEN row, and is simultaneously constrained to be in a region that contains only a single ODD row.
- It is not easy to tamper with the signature via local perturbations: the small signature size implies that many cells must be perturbed before the signature becomes unrecoverable. Furthermore, local perturbations that shift cells between cell rows are difficult to make without worsening the solution quality.
- It is not affected by downstream stages of the design flow. Many current design methodologies do not significantly change the row assignments (let alone the locations) of existing cells during routing; hence, our proposed watermarking scheme will remain intact.
- It allows the watermark to be realized completely during the placement phase. (For schemes such as the watermarking of budgeted timing constraints, the realization remains incomplete until after routing [5].)

[6]The impact on solution quality is even worse when the design is performance-constrained. Imagine the effect of arbitrarily constraining a cell in a timing-critical path to be in the "wrong" region of the layout.

3.2 Routing

For standard-cell routing, applicable constraints usually involve performance (e.g., crosstalk and delay bounds) or reliability (antenna rules, electromigration and self-heat limits, hot-electron rules). How these constraints are represented, how they are enforced, and what degrees of freedom (e.g., shielding, tapering, spacing, repeater insertion, driver sizing, topology design, etc.) are exploited depends on the routing tool.

We considered constraint types involving segment widths, spacings, and choice of topology. These not only are difficult to enforce within current routing approaches, but also have potentially harmful interactions with performance constraints (e.g., a watermarking constraint might require a net to be routed at minimum separation from its closest neighbors; a crosstalk constraint might dictate otherwise). We also considered various "parity" watermarking schemes based on, e.g., the orientation of the "L" for two-pin connections, the parity of the number of segments, the parity of path lengths in the routing, etc. These were dismissed as highly unnatural (e.g., pin access clearly dictates which "L" the router will choose), difficult to enforce using known routing methodologies (e.g., parity of total tree length), or vulnerable to simple tampering (e.g., tampering by compaction would ruin length-parity schemes).

Our approach bases the watermarking constraints on the (per-net) *costing* of the underlying routing resource. Specifically, for each watermark net we impose unusual costs on "wrong-way" and/or via resources, and hope that the watermark nets are provably unusual in their utilization of such resources. Many commercial routers already accept such control of the routing cost structure on a per-net basis. Our approach has the following advantages:

- Very few constraints are needed to make a strong signature, assuming that the resource costing is reflected in the routing result for each watermark net.
- It is compatible with many existing routing constraints, e.g., those that are based on wire width, spacing or shielding.[7]
- It is not easy to disturb the signature with local perturbations: the small signature size implies that many nets will need to be rerouted before the signature is likely to be unrecoverable. Furthermore, as designs are increasingly limited in terms of the interconnect resource, the routing of watermark nets is likely to be "locked in" by the routing of the remaining non-watermark nets. Hence, destroying the watermark requires rerouting of the design.

3.3 Commercial Tools

To demonstrate practicality and allow evaluation with respect to real-world design metrics, we test watermarking protocols that are transparent to existing layout tools. Our implementation of the proposed placement and routing watermarking methodologies uses commercial tools from Cadence Design Systems: placement watermarking is built around QPlace v4.0.27 and WarpRoute v1.0.10, and routing watermarking is built around the IC Craftsman v2.1.3 router using a standard constraint type in this tool ("limit way" rule). We now give details of the experimental protocol.

4 Experimental Protocol

4.1 Placement

Our experimental methodology is designed to show how easily an existing tool can be modified to offer watermarking capability. The basic comparison is shown in Figure 1. A traditional non-watermarked placement flow reads library and design information

[7]Some potential conflicts exist with respect to wrong-way routing, but these have not been an issue in our experience, particularly since we impose *upper* bounds on the use of wrong-way routing.

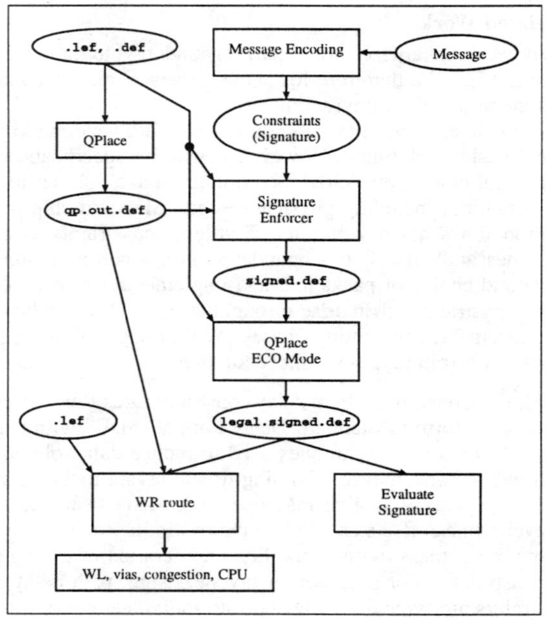

Figure 1: A postprocessing-based watermarking protocol for standard-cell placement, using the Cadence Design Systems QPlace v4.0.27 and WarpRoute v1.0.10 tools.

via LEF/DEF formats, executes QPlace, then executes WarpRoute to evaluate the placement quality. This is shown on the left side of Figure 1. Our *postprocessing-based* watermarking flow, shown on the right of the figure, consists of the following steps:

1. We read the default QPlace placement result (a DEF file with location data) and the LEF file into our internal design database.
2. We ask the user for a message (e.g., "Placed by QP on 10-10-97") which we then transform into row-parity constraints for some subset of the core (non-pad, standard) cells of the design.
3. We enforce all the row-parity constraints by local changes to the placement (e.g., pair-swap operations), generating a "signed DEF" file.
4. We ensure that the resulting placement is ready for routing, by re-running in "ECO mode"; this makes only minimal changes to the placement, and only if necessary (typically, to avoid illegal overlaps with fixed obstacles or other cells). The output of this step is a "legal signed DEF" file.
5. We execute WarpRoute, and evaluate the placement quality.

We make the following observations. (1) Our postprocessing approach is absolutely equivalent to what might be implemented in a modification of the actual commercial tool. Alternatively, our watermarking flow is trivially implemented by scripting and standard capabilities of the commercial placer (LEF/DEF manipulation, ECO placement, etc.). (2) We begin with a high-quality solution and retrospectively impose constraints. Not only is this a good approach to maintaining solution quality, but from the outside one cannot tell whether the watermarked placement is created from scratch or by postprocessing of a nonwatermarked placement. (3) The "final list of core cells" is a well-defined concept in all existing design flows including those that invoke "in-place optimization" or

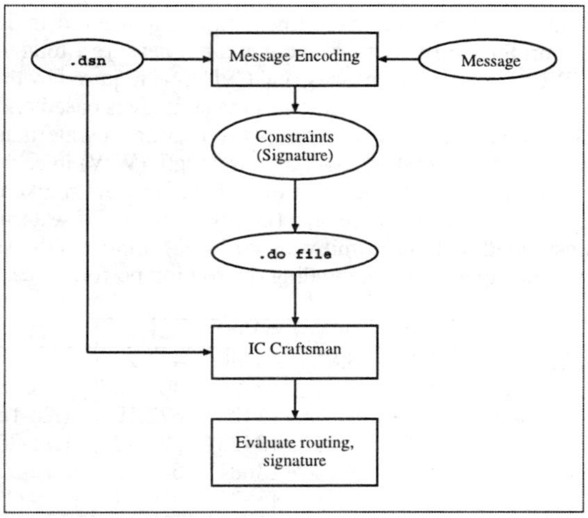

Figure 2: A preprocessing-based watermarking protocol for (standard-cell) gridless area routing, using the Cadence Design Systems IC Craftsman v2.1.3 tool. The tool is used in its standard context, controlled by a ".do" file using standard rules syntax.

"placement-based synthesis". Thus, generating a watermark based on core cell indices and row parities is also well-defined.[8]

4.2 Routing

A traditional non-watermarked routing flow using the IC Craftsman router reads library and placed design information via the .dsn file format, then executes the router under the control of a ".do file". This is shown on the left side of Figure 2. Our *preprocessing-based* watermarking flow, shown on the right side of the Figure, consists of the following steps:

1. We identify all unique signal net names in the .dsn file.
2. We ask the user for a message (e.g., "Routed by ICC on 10-10-97") which we then transform into a list of "watermark nets" (some subset of the net names in the design).
3. We then constrain the watermark nets using IC Craftsman rules in the "do file". Specifically, our methodology applies a "limit way = 1" rule to each of the watermark nets.
4. We execute the ICC router.

4.3 Evaluation of Signature Strength

Each constraint involves some "random" choice, e.g., choosing a random cell or signal net. (Such choices are not actually random, but use a cryptographically strong pseudorandom number generator that is seeded with a binary signature file.) The choices may occur either with or without replacement. If there is replacement, then constraints will be independent of each other. Even if there is no replacement, the constraints may very nearly act as if they were independent, especially if the pool of constraints to choose from is large relative to the number of constraints actually chosen. As long as the constraints are either independent or nearly so, the probability P_c of a solution carrying an author's watermark purely by coincidence can be computed by a simple binomial. We use P_c to measure the strength of the authorship proof.

Let X be the number of constraints imposed, let x be the number of these that are *not* satisfied, and let p be the probability of a constraint being satisfied purely by coincidence.[9] The probability that x or fewer out of X constraints are satisfied by coincidence is given by $P_c = \sum_{i=0}^{x}(C(X,i) \cdot (p)^{X-i} \cdot (1-p)^i)$.

- For our placement watermarks, the signature consists of a certain subset of cells, each constrained to be in a cell row with specified-parity index. We use $p = 0.5$ as the chance that a given cell will satisfy its constraint by coincidence.
- For our routing watermarks, the signature consists of a certain subset of signal nets, each with an "unusually low" limit on the amount of wrong-way wiring that can be used to route the net. We use the following methodology to establish a binary indicator of whether a given net has been "successfully watermarked". Given a routed design, we evaluate the total wirelength (WL_{tot}) and the wrong-way wirelength (WL_{way}) for each signal net. We then rank all nets in order of increasing value of the ratio WL_{way}/WL_{tot}. The watermark nets are expected to occur earlier in this ranking, while non-watermark nets are expected to occur later in this ranking. We then establish a cutoff rank below which a watermark net is considered "successfully watermarked", and above which a watermark net is considered "not successfully watermarked". In the routing experimental results reported below, we always set the threshold rank at the 40th percentile, i.e., $p = 0.4$. (Stronger results can be obtained by more carefully choosing the value of p; this is noted below.)

4.4 Resistance to Tampering Attacks

Another way to evaluate the strength of a given watermark is to assess its resistance to attacks (see [5] for a survey of prototypical attacks). Thus, in addition to reporting P_c values, we also report the resistance of our watermarking schemes to the following *tampering* attacks.[10] In these scenarios, the *attacker* is trying to erase the watermark by small layout perturbations.

- **Placement**
  - Assumptions: (i) the attacker has access only to an incremental ("legalizing") placement tool such as QPlace ECO mode;[11] (ii) the watermarking scheme is unknown to the attacker; and (iii) original design constraints are retained.
  - Attack: (i) select N random pairs of cells and swap the locations of each cell pair; and (ii) run the legalizing placer to legalize the design (continue with routing, etc.).
- **Routing**
  - Assumptions: (i) the attacker has access only to incremental (single-net) auto-routing; (ii) the watermarking scheme is unknown to the attacker; and (iii) original design constraints are retained.
  - Attack: select N random nets, then reroute these nets with only the original design constraints (if any).

[8]If one desires a netlist-dependent, *floorplan-independent* watermark, then one must assume a canonical row indexing (e.g., top-down for horizontal rows and left-right for vertical rows). One can also simply define the row parity constraints in terms of two equivalence classes of constrained cells. We also note that the implicit assumption of fixed cell names is also reasonable; any methodology allowing arbitrary renaming of cells would likely have some overhead for verification.

[9]If constraints are not sufficiently independent of each other, p is not well-defined. Overestimating the value of p always makes P_c larger, which we interpret as weaker strength of the authorship proof. Since overestimating p can never improve the supposed strength of our watermark, we can typically obtain useful estimates for p even when its exact value is not known.

[10]A tampering attack attempts to remove the rightful IP owner's signature, and possibly introduce the attacker's own signature into the IP.

[11]Recall that the attacker could always re-solve the placement problem from scratch to remove the IP owner's signature, but we assume this is too expensive.

5 Experimental Results

We applied our proposed physical design watermarking protocols to six industry test cases, four in placement and two in routing. Aspects of the test cases are given in Table 1. The routing test case *sc2* has a relatively small number of nets relative to cells because many signals are pre-routed, and hence not included in the netlist.

	Placement				Routing	
	test1	test2	test3	test4	sc1	sc2
#cells	9011	12133	12857	20577	1653	4250
#nets	11962	11828	10880	25634	1802	1597

Table 1: Numbers of cells and nets in the six industry test cases.

5.1 Watermark Strength P_c

Results for the placement experiments are summarized in Table 2. We report five post-routing layout quality measures for each test case. These measures are: total wirelength, total number of vias, percentage of overcongested "global routing cells" (as reported by the placer), and CPU time in (mm:ss) required by the router (all CPU times are for a 140MHz Sun Ultra-1). Together, these measures provide a fairly complete picture of the utility of each placement. In the Table, the subscript *orig* indicates the default unwatermarked solution; the subscript $wm-x,y$ indicates a watermarked solution with x cells in the signature, of which y ended up being successfully watermarked. There is essentially no solution quality overhead to introducing the placement watermark. Our placement watermarking protocol also shows graceful degradation of solution quality if extremely strong signatures are required.

Test Case	WL	# Vias	Cong	CPU	P_c
$t1_{orig}$	6.38	86072	1.52%	11:38	
$t1_{wm-56,52}$	6.40	86595	1.52%	12:03	5.5e-12
$t1_{wm-112,96}$	6.40	86449	1.52%	12:13	2.2e-15
$t1_{wm-224,189}$	6.42	86712	1.54%	12:10	4.8e-27
$t1_{wm-448,389}$	6.44	87143	1.53%	12:08	5.7e-61
$t1_{wm-896,786}$	6.51	87716	1.52%	13:02	7.3e-127
$t1_{wm-1792,1526}$	6.62	88955	1.55%	13:25	8.3e-215
$t2_{orig}$	3.32	95601	0.86%	9:30	
$t2_{wm-56,53}$	3.33	95811	0.78%	9:20	4.1e-13
$t2_{wm-112,110}$	3.33	95978	0.80%	9:06	1.2e-30
$t2_{wm-224,208}$	3.34	95913	0.77%	9:11	4.5e-44
$t2_{wm-448,409}$	3.35	96554	0.91%	9:15	3.4e-79
$t2_{wm-896,837}$	3.38	97902	0.94%	9:28	3.2e-177
$t2_{wm-1792,1678}$	3.42	99467	1.13%	9:56	2.9e-357
$t3_{orig}$	3.13	52401	4.47%	13:58	
$t3_{wm-56,55}$	3.15	52433	4.57%	11:32	7.9e-16
$t3_{wm-112,110}$	3.14	52636	4.60%	11:45	1.2e-30
$t3_{wm-224,219}$	3.16	52529	4.65%	11:53	1.7e-58
$t3_{wm-448,443}$	3.17	53003	4.57%	11:44	2.1e-124
$t3_{wm-896,879}$	3.21	53559	5.00%	11:45	7.2e-235
$t3_{wm-1792,1740}$	3.25	54279	5.14%	11:58	3.2e-439
$t4_{orig}$	8.13	179526	0.02%	17:07	
$t4_{wm-56,50}$	8.14	179680	0.02%	17:35	5.1e-10
$t4_{wm-112,102}$	8.14	179678	0.02%	14:42	1.2e-20
$t4_{wm-224,201}$	8.16	180052	0.02%	15:45	5.7e-37
$t4_{wm-448,407}$	8.18	180590	0.02%	23:49	3.5e-77
$t4_{wm-896,797}$	8.25	182224	0.02%	19:59	1.6e-136
$t4_{wm-1792,1597}$	8.32	183783	0.02%	17:44	6.9e-274

Table 2: Watermarking results for placement. Wirelengths are scaled to 10^8 user database units.

Results for the routing experiments are summarized in Table 3. We report three post-routing layout quality measures: total wirelength (WL), total number of vias, and CPU time required by the IC Craftsman router. Since our watermarking strategy is based on limiting the length of acceptable wrong-way routing in watermarked nets, we also report total wrong-way wirelength (WW) in each solution. Finally, we report the value of P_c for each watermarked design. Increasing the signature size (i.e., the number of watermark nets constrained with the "limit way = 1" rule) improves the value of P_c without significantly degrading the routing performance.

Test Case	WL	WW	# Vias	CPU	P_c
$sc1_{orig}$	5.48	9.67	13440	422:27	
$sc1_{wm-20,20}$	5.46	9.56	13320	402:18	1.1e-8
$sc1_{wm-40,40}$	5.49	9.55	13426	672:19	1.2e-16
$sc1_{wm-80,78}$	5.48	9.23	13433	503:48	1.1e-28
$sc1_{wm-160,159}$	5.46	8.32	13681	641:32	5.2e-62
$sc1_{wm-320,315}$	5.47	7.51	13921	450:34	9.5e-117
$sc2_{orig}$	2.29	3.54	5590	23:52	
$sc2_{wm-20,20}$	2.29	3.60	5547	21:29	1.1e-8
$sc2_{wm-40,40}$	2.30	3.77	5716	18:58	1.2e-16
$sc2_{wm-80,79}$	2.30	3.47	5713	16:50	1.8e-30
$sc2_{wm-160,151}$	2.29	3.23	5650	20:43	1.3e-48
$sc2_{wm-320,294}$	2.30	2.93	5706	20:43	2.2e-85

Table 3: Watermarking results for routing. Total wirelengths (WL) are scaled to 10^7 user database units, and wrong-way wirelengths (WW) are scaled to 10^6 units.

As a side note, recall from above that the value of p (a consequence of the threshold rank) may be chosen to optimize the signature strength measure P_c. Table 4 shows how calculated P_c values can vary as p varies from 0.2 to 0.4. In the Table, the second column gives the size of the signature (number of watermark nets), and each entry $x(y)$ represents the P_c value (x) and the number of unsuccessfully watermarked nets (y). We observe that fine-tuning of p (e.g., choosing $p = 0.35$) could potentially improve our results.

5.2 Resistance to Tampering

Tables 5 and 6 present results of experiments where we attempt to tamper with placement and routing watermarks, respectively. respectively. In each Table, the second column indicates the original signature size, and the third column (Init) gives the original watermarked solution quality (total WL). Subsequent columns indicate the number of cell pair-swap (net ripup and reroute) operations performed in the placement (routing) tampering, expressed as a percentage of the total number of cells (nets) in the design. We report P_c values for the 10% column to show that the watermarks remain strong even after tampering. For placement (Table 5), the solution quality degrades much faster than the signature strength, even though we restricted all random pair swaps to occur over Manhattan

test case	# of nets	p values				
		0.2	0.25	0.3	0.35	0.4
sc1	20	1.1e-14(0)	9.1e-13(0)	3.5e-11(0)	7.6e-10(0)	1.1e-8(0)
	40	1.4e-24(2)	1.0e-22(1)	1.2e-21(0)	5.8e-19(0)	1.2e-16(0)
	80	3.0e-46(5)	8.9e-41(4)	2.6e-38(2)	3.7e-33(2)	1.1e-28(2)
	160	3.5e-91(10)	4.9e-82(7)	1.7e-75(4)	3.4e-71(1)	5.2e-62(1)
	320	6.2e-91(88)	2.2e-115(46)	3.2e-129(20)	8.9e-124(11)	9.5e-117(5)
sc2	20	1.5e-5(7)	1.6e-9(2)	3.8e-8(2)	7.6e-10(0)	1.1e-8(0)
	40	7.9e-10(14)	2.5e-15(6)	5.3e-18(2)	4.4e-17(1)	1.2e-16(0)
	80	1.9e-13(34)	7.0e-32(10)	2.5e-33(5)	3.7e-33(2)	1.8e-30(1)
	160	2.6e-17(80)	5.8e-32(49)	3.3e-47(24)	2.3e-45(18)	1.3e-48(9)
	320	2.5e-8(214)	2.4e-34(137)	7.7e-61(81)	1.5e-83(41)	2.2e-85(26)

Table 4: P_c values corresponding to different values of p. Each entry $x(y)$ represents the P_c value (x) and the number of unsuccessfully watermarked nets (y).

Test	# cells	Init	1 %	2 %	5 %	10 %	P_c
t3	56	3.15	3.24	3.34	3.61	4.00	4.1e-13
	112	3.14	3.25	3.33	3.60	4.03	4.9e-25
	224	3.16	3.26	3.34	3.61	4.02	4.5e-44
	448	3.17	3.27	3.36	3.62	4.04	9.1e-95
	896	3.21	3.30	3.39	3.65	4.08	1.1e-180
	1792	3.25	3.35	3.44	3.69	4.12	1.1e-334
t4	56	8.14	8.28	8.68	9.83	11.57	1.3e-7
	112	8.14	8.31	8.75	9.80	11.52	1.3e-14
	224	8.16	8.34	8.71	9.86	11.64	2.7e-29
	448	8.18	8.36	8.76	9.80	11.62	3.4e-48
	896	8.25	8.42	8.88	9.96	11.63	6.2e-94
	1792	8.32	8.48	8.88	9.94	11.64	6.1e-196

Table 5: Result from tampering with the placement watermark. Solution quality degrades much faster than signature strength; hence, tampering does not appear to be a viable form of attack.

Test	# nets	Init	1%	2%	5%	10%	P_c
sc1	20	5.46	5.47	5.47	5.49	5.53	1.1e-5
	40	5.49	5.50	5.50	5.51	5.55	1.3e-11
	80	5.48	5.48	5.48	5.49	5.54	4.2e-21
	160	5.46	5.46	5.47	5.48	5.54	3.1e-38
	320	5.47	5.48	5.48	5.49	5.54	1.4e-61
	CPU	534:06	28:30	51:58	118:26	127:25	
sc2	20	2.29	2.29	2.29	2.29	2.30	7.7e-8
	40	2.30	2.30	2.30	2.31	2.31	8.6e-13
	80	2.30	2.29	2.30	2.30	2.30	1.1e-23
	160	2.29	2.29	2.29	2.30	2.30	2.3e-36
	320	2.30	2.30	2.30	2.31	2.31	1.7e-67
	CPU	19:45	1:24	2:18	3:57	8:39	

Table 6: Result from tampering with the routing watermark. CPU times for ripup and reroute approach the time required to re-solve the problem from scratch; hence, tampering does not appear to be a viable form of attack.

distances less than twice the cell row height. (ECO placement CPU times were consistent and small, and we do not report them.) For routing (Table 6), the solution quality appears relatively immune to tampering (other measures such as number of vias also remained constant). However, the CPU time required to tamper with a large number of nets approaches the cost of redoing the entire solution from scratch (at which point tampering is not needed). We conclude that our watermarking schemes are quite robust with respect to random tampering.

Finally, Figure 3 shows the watermarked layout of test case *sc*1 (56 watermark nets), and Figure 4 shows the unwatermarked layout of the same design.

6 Conclusions

In conclusion, we have developed the first IPP protocols for embedding design watermarks at the physical design level. We have implemented these protocols transparently to existing design flows, using leading industrial tools. On real designs, we show strong proofs of authorship with very acceptable cost overhead for the watermarking. We also show the robustness of our watermarking scheme with respect to random tampering attacks. Our current research is aimed at extending these ideas to alternate watermarking strategies that are appropriate for various application scenarios in physical design, and that will remain robust under various specific forms of attack.

Figure 3: Routing solution (watermarked) for the sc1 test case.

Figure 4: Routing solution (unwatermarked) for the sc1 test case.

References

[1] C. J. Alpert, T. Chan, D. J. Huang, A. B. Kahng, I. Markov, P. Mulet and K. Yan, "Faster Minimization of Linear Wirelength for Global Placement", in *Proc. ACM/IEEE Intl. Symp. on Physical Design*, Napa, April 1997, pp. 4-11.

[2] C. J. Alpert and A. B. Kahng, "Recent Directions in Netlist Partitioning: A Survey", *Integration: The VLSI Journal* 19 (1995), pp. 1-81.

[3] K. D. Boese, A. B. Kahng and S. Muddu, "New Adaptive Multistart Techniques for Combinatorial Global Optimizations", *Operations Research Letters* 16(2) (1994), pp. 101-113.

[4] D. C.-M. Chi, *Improving Upon Local Search Heuristics for VLSI Standard Cell Placement*, M.S. Thesis, UCLA Computer Science Department, 1995.

[5] A. B. Kahng, J. Lach, W. H. Mangione-Smith, S. Mantik, I. L. Markov, M. Potkonjak, P. Tucker, H. Wang and G. Wolfe, "Watermarking Techniques for Intellectual Property Protection", *Proc. ACM/IEEE Design Automation Conf.*, 1998.

[6] S. A. Kauffman, *At Home in the Universe: The Search for Laws of Self-Organization and Complexity*, New York, Oxford University Press, 1995.

[7] A. J. Menezes, P. C. van Oorschot and S.A. Vanstone, *Handbook of Applied Cryptography*, Boca Raton, CRC Press, 1997.

[8] B. Schneier, *Applied Cryptography : Protocols, Algorithms, and Source Code in C*, 2nd ed., New York, Wiley, 1996.

[9] Semiconductor Industry Association, *The National Technology Roadmap for Semiconductors: Technology Needs*, December 1997.

[10] W. Swartz and C. Sechen, "Timing Driven Placement for Large Standard Cell Circuits", *Proc. ACM/IEEE Design Automation Conf.*, June 1995, pp. 211-215.

[11] G. E. Tellez, D. A. Knol and M. Sarrafzadeh, "A Performance-Driven Placement Technique Based on a New Budgeting Criterion", *Proc. IEEE Intl. Symp. on Circuits and Systems*, May 1996, vol. 4, pp. 504-507.

[12] VSI Alliance, *Fall Worldwide Member Meeting: A Year Of Achievement* (guidelines proposed by VSIA development working group on intellectual property protection), Santa Clara, Oct. 1997.

Data Security for Web-based CAD

Scott Hauck, Stephen Knol
Department of Electrical and Computer Engineering
Northwestern University
Evanston, IL 60208-3118 USA
{hauck, sknol}@ece.nwu.edu

ABSTRACT

Internet-based computing has significant potential for improving most high-performance computing, including VLSI CAD. In this paper we consider the ramifications of the Internet on electronics design, and develop two models for Web-based CAD. We also investigate the security of these systems, and propose methods for protection against threats both from unrelated users, as well as from the CAD tools and tool developers themselves. These techniques provide methods for hiding unnecessary information. Such techniques will be key to the development of future Internet-based CAD applications, since serious CAD users will be unwilling to use any CAD methodology that risks exposing their designs to outsiders. By enabling Web-based CAD, these techniques can improve CAD performance, enable collaboratory design, and create a usage-based pricing methodology.

Keywords

Encryption, Web-based CAD, Internet, Data Security

1. Introduction

With the development of fabrication processes that can put tens of millions of transistors on a single chip, the design process has gotten so complex that human designers cannot create circuits on their own. VLSI CAD research efforts and software vendors have successfully created tools to handle many of these mapping tasks, making software an integral part of all complex circuit design efforts. However, while such tools can handle much of the design process, making VLSI designers more productive, these tools do not come without a price. Due to the complexity of the tasks undertaken by these VLSI CAD tools, the software can take a significant amount of time to compute. Runtimes of hours or days are common for many tasks, and may grow to weeks with future 100 Million transistor designs, slowing down the development of these circuits. For example, tools for using logic emulators to speed up the mapping process can take a day or more to operate [6], meaning that before simulation acceleration hardware can be used the designer must wait a significant amount of time. For tasks such as design space exploration, where multiple candidate circuit structures are synthesized in order to determine the best approach to adopt, multi-day runtimes are not surprising. These delays increase time-to-market, and reduce circuit quality by restricting the number of alternative implementations that can be considered.

It is possible to speed up CAD algorithms via high-performance computing systems. Parallel Supercomputers and Networks-of-Workstations can be brought to bear on most CAD problems, speeding up the algorithms. Custom hardware such as logic emulators and simulation accelerators can speed up some algorithms, particularly circuit simulation. However, these hardware resources can be extremely expensive, making them unavailable to a large segment of the CAD user community.

A similar dynamic has occurred in general-purpose computing, leading to the creation of Supercomputing centers. Such centers provide supercomputer access to a broad user community by hooking these machines to a network and allowing remote access. We believe a solution to the performance problem for VLSI CAD is similar: develop a Web-based CAD Compute Center, providing both hardware and software for high-performance computing, and make it accessible over the Internet via a World-Wide-Web interface. This will allow the cost of the high-performance computing resources (supercomputers, networks-of-workstations, and simulation accelerators/logic emulators) to be amortized over a large user base, and give even small VLSI design houses access to the computation resources that are becoming a requirement for VLSI design.

Beyond just providing affordable access to high-performance computing resources, a Web-based CAD Compute Center can have several benefits for VLSI design. By providing an Internet interface to a design project's CAD resources, it will enable user groups distributed geographically to use a single set of tools, and collaborate on a complex design project. Collaborative design is becoming much more important because the integration of complete systems onto a single chip requires a large variety of skills, skills that may not all be found at a single site.

The final benefit of a centralized CAD compute center is the ability to implement a per-use pricing policy for CAD software. Current CAD software is sold on a per-license (per-seat) basis, meaning that an organization that ever wishes to use N copies of a CAD tool simultaneously must pay for continuous use of those N tools. Thus, under the current pricing scheme there is no differentiation between a casual user that uses a given capability rarely, and thus is unwilling to pay a significant amount for a tool, and power users that constantly use that tool. Both software companies and users are promoting a usage-based pricing structure, where users are billed based upon the actual value (measured in time executing the algorithm) the software delivers to the user. By creating a centralized CAD compute center such a per-use pricing model can be implemented (and will be required in order to fairly allocate computer time to the center's users), something that is much more difficult to do in today's model of software distributed to each end-user's machine.

While such a CAD compute center has tremendous potential for supporting current and future VLSI design, there are several issues that must be addressed to make such a system a reality. In this paper we consider one of these problems, the development of security protocols for maintaining the privacy of user designs within an Internet-based computing paradigm.

2. Web-based CAD

There have been several research projects that have harnessed the World Wide Web for providing access to VLSI CAD tools [2, 3, 7 - 11]. In general, there are two models for such systems: the Applet model, and the Compute Center model. Under the Applet model the CAD algorithm is sent to the user's machine as Java Applets (or any other similar technology), and the algorithm executes on the user's machine. Thus, in this model the user must provide the compute resources for performing the CAD task, meaning that there is no way to amortize the cost of high-performance computing resources. Under the Compute Center model the design files are transferred to a centralized, shared compute center, where the CAD algorithms are applied to the design. Thus, the tools operate on the resources of the center.

As with anything that operates over the Internet, data security is an important concern for both Web-based CAD models. The design files for a VLSI chip represent a significant intellectual property resource, and most users will be unwilling to use any computing model that does not protect the secrecy of these files. Note that even within a company Intranet, where CAD resources are provided within a protected network accessible only to that company's computers, data security is still a significant concern since access to design files is usually restricted to only those employees working directly on that project.

The security challenges for a Web-based CAD system can be broken into three categories: *Client Security*, *Transmission Security*, and *Server Security*. *Client Security* encompasses all of the concerns of protecting data on the user's machine from attacks unrelated to the Web-based CAD system. *Transmission Security* deals with the need for ensuring that data transmitted between the user and a Compute Server is not intercepted and decrypted, and includes both information sent to and received from the server. Note that the Applet model does not include these concerns since the data never leaves the user's machine. Finally, *Server Security* represents the concern that the Web-based CAD tools and developers themselves may not be trusted, and thus steps must be taken to minimize the amount of information that can be obtained by the CAD tools.

Client and Transmission Security are faced by most Internet systems, and thus we can use standard solutions to these problems. In order to protect machines connected to the Internet techniques such as firewalls and other security protocols have been developed. Securing transmissions between the compute center and the user can be handled by public-key encryption, such as PGP. All communications from the user to the compute center will be encrypted by the center's public key (ensuring that only the compute center can decrypt the information), and communication back uses the user's public key. Thus, we can handle the Client Security and Transmission Security issues with standard solutions from the Internet community, and therefore they do not pose special problems for Web-based CAD.

Server Security is a problem somewhat unique to Web-based CAD (and, as we will see, the solutions are very application-specific). Most security methods for Web-based systems concentrate on keeping Java Applets and other Web-based tools from gaining access to data they are not intended to view. Thus, restrictions are placed on file read and write operations. However, the problem with Web-based CAD is that the user must reveal parts of their design to the applets or compute center, since the CAD tools need this information to perform their optimizations. However, the users do not want this information divulged to others, since the circuit designs are often very valuable intellectual property, and are kept as trade secrets. Thus, even though the user may give information to the CAD tools in order to allow them to optimize the design, there must be some guarantees that the entire design is not then revealed to the CAD tool owners, who may not be trusted. For example, the owners of the computer server or CAD applets may be university researchers, and commercial VLSI design projects may be unwilling to use such a system without strong guarantees that the design data will be kept confidential (often requiring guarantees stronger than even nondisclosure or confidentiality agreements).

In the rest of this paper we present new methods for Server Security for Web-based CAD. As we will see, these methods primarily revolve around the idea that only the information necessary to perform a given optimization should be supplied to the CAD tool, while all other information should be hidden or removed. Note that similar techniques can be used for benchmarking, releasing only those features of a circuit necessary to test a given optimization, as well as the sale of cores and other circuit components, where designs sold to customers must be made as resistant to reverse-engineering as possible.

3. Data Security Techniques

The main challenge for Sever Security for Web-based CAD is the fact that the CAD user must divulge information to the CAD tool for it to perform its optimization, yet the user needs assurances that this does not let others learn the details of their design. Although intellectual property agreements may help, relying on these agreements is too risky for many potential users. The solution to this dilemma is to make sure that only the information necessary for the tool to perform its optimization is given to the center. For example, comments in an file do nothing to help the CAD algorithms operate, and thus stripping the comments from all files before submitting them to the Web-based CAD algorithms helps limit the amount of information divulged.

While comment stripping is a universal information-hiding transformation for circuits, since this removes information unnecessary for any optimization algorithm, many other transformations are not universal. As we will see, some of the methods for Server Security eliminate information that is unused in some optimization steps, yet is required for others. Thus, as we consider different methods for Server Security, we must realize that not all techniques can be used before each CAD algorithm, and some methods will be only useful in specific circumstances. Also, some CAD optimizations, such as logic synthesis and logic simulation, require a great deal of information about the input circuit, and thus there are only limited possibilities for information hiding.

To apply the information-hiding transformations proposed in this paper it is necessary to develop programs to perform these operations before the information is sent to the Web-based CAD algorithm, as well as reintegration operations to replace the hidden information into the results from the Web-based CAD tools. Note that this program cannot itself be part of the Web-based CAD toolsuite, since it will have access to the raw design files, and can itself be viewed as a security vulnerability. The answer to this problem is to provide simple information hiding algorithms in source code, allowing users to verify that the algorithm does not contain any security vulnerabilities.

In the sections that follow, we will discuss numerous methods for providing Server Security for Web-based CAD.

3.1 Current Techniques

There are techniques that have already been developed for protecting simulatable files, such as processor models and other cores. For example, Cadence provides an encryption method that allows encrypted files to be distributed to end-users with some hope that the design information cannot be recovered. However, this methodology protects the source files from direct observation (similar to Transmission Security), but does not protect the information from the Cadence CAD tools themselves. In fact, methods have already been demonstrated for coaxing the design information out of Cadence's Verilog-XL simulator [4].

A more promising approach is provided by LEDA's KRYPTON code scrambler. The system hides information by renaming all variable names in a circuit to a hard to read series of "I", "1", "0", and "O" characters, removing formatting and comments, applying constant propagation, and other optimizations. Note that not only do these optimizations make it harder for a human to understand the source code, by eliminating the original variable names from the code this eliminates information from the circuit, since in (well-written) code the variable names carry information about the values they contain. This alteration to variable names should also be a universal transformation, since most or all CAD algorithms should not need this information.

3.2 Netlist Flattening

Hierarchical circuit descriptions, such as those found in hardware description languages, group together common subcircuits to condense the circuit description, and highlight commonalties in the circuit. By replacing calls to subcircuits by the logic itself, and removing hierarchy levels, this information can be removed from the circuit without changing the circuit functionality.

3.3 Functionality Removal

CAD optimizations such as partitioning, placement, and routing do not care about the exact functions computed in a circuit, but instead view the circuit netlist as a graph consisting of only connectivity information, and perhaps node size. Thus, while the CAD tool may take in circuit descriptions containing the exact logic functionality, it is immediately thrown away. Thus, when using such a tool we can eliminate the logic functions, or replace them with a default (i.e. all nodes become AND gates), before submitting the design. This removes most information from the data provided to the CAD tool, while still allowing the tool to perform its optimizations.

3.4 Net Direction Alteration

Even after the functionality in a circuit is removed, the direction of signal flow can give hints to circuit behavior, such as bus widths and other structures. In some circuit representations it is possible to simply remove the direction information, turning the circuit into an undirected graph. Other formats may require signal directionality, making removing this information difficult. For example, the XNF format (input to the Xilinx FPGA tools [12]) requires each gate to have a single output, and generally requires each net to have exactly one source. Although we cannot remove net directionality from such a circuit, we can destroy the value of this information by reversing the direction of some nets. Simply reversing the direction of a single net will not work, since the original source will no longer generate any signals, and the original sink will generate two signals. The solution is to find a path in the circuit such that reversing all of the signals on the path maintains the format requirements. One possibility is to find a directed path from a circuit input to any output, and then reverse these signals. Since along this path every node will have been the source of one signal, and the sink of another, when the signals are reversed this node will still be the source of exactly one signal. Another method is to perform the same process for state bits instead of circuit I/Os, except that a directed path must be found starting from the state bit's output, and leading to an input of that stateholding element (a directed cycle containing the stateholding function).

It is important to realize that removing or changing the net direction can only be done before certain optimizations. The main problem is that by removing net direction, or changing signal directions, critical path information can be lost. Thus, this technique can be applied before area-optimization algorithms for partitioning, placement, and routing, but may not be valid for delay optimizing tools. Slack assignment algorithms, where critical path information is added to the circuit before the main algorithm is run, may be tolerant of this data hiding technique as long as the slacks are computed before nets are reversed.

3.5 Net Removal

Perhaps the most easily identifiable parts of a circuit are the high-fanout signals, particularly clock and reset signals. Once these signals are found, the registers are easily determined, giving an idea of the overall circuit structure. However, for some CAD algorithms these high-fanout nets are unnecessary, and in fact their removal may improve the speed of the CAD tool. For example, partitioning and placement algorithms can normally ignore very high fanout nets, since it is impossible to minimize their length. Routing algorithms for FPGAs can also ignore these signals, since they will be carried on special resources, and need not be considered during normal routing. Thus, high fanout nets can be removed before sending a design to such CAD tools, helping to hide the stateholding functions in the circuit.

3.6 Circuit Partitioning

One method for speeding up CAD tools is to break down a large design problem into smaller pieces. For example, the placement problem for a large VLSI chip can be broken down to a series of subproblems, where each run of the placement tool places only a fraction of the overall chip area. This technique can also be effective for hiding the details of the design being optimized.

Instead of giving a CAD tool the entire circuit to optimize, the circuit is broken down into subproblems and separately optimized. As long as there is a way to break the problem into subproblems without introducing significant inefficiencies this methodology can be applied. Note that it is critical to ensure that the information needed to reconnect the pieces is hidden, so reintegration information is retained at the user site.

Not only can splitting up a problem hide information by making it hard to reassemble the circuit, it is possible to take this further by adding bogus data to the processing, or optimizing unrelated circuits at the same time. Specifically, even though the CAD tools may only see one part of the circuit at the time, they could collect all input files received within a given time period and try to merge them into a complete circuit. Optimizing two or more unrelated circuits at the same time will make it difficult to decide which elements belong to which circuit. However, details of the circuit domain may leave characteristic features on the circuit, making separation of unrelated subcircuits possible. Bogus data can instead be added to the data stream. Although this will slow down processing because the CAD tools must also optimize the bogus data, this overhead can be kept small. For example, if a circuit is split into 100 subproblems, and 10 bogus subproblems are also added, this means only a 10% overhead, yet significantly complicates reverse-engineering. A similar technique is used in chip design, where bogus circuitry is added to otherwise unused portions of the chip, complicating reverse-engineering attempts. In order to generate this bogus data, modified portions of the current design, or portions from other designs, can be used.

3.7 Component Simplification

High fanin nodes and high fanout signals can provide convenient points of reference, aiding reverse-engineering. For example, if a repeated component is one of the few circuit structures using a 7-input logic gate, finding all 7-input gates in the circuit helps find all instances of that component. This can also be a landmark for connecting data from different steps in the mapping process. For example, a partitioning algorithm may be given the entire circuit, but with all information except connectivity removed. After partitioning a technology mapper would receive complete information on the logic in each of the subcomponents. With information from both steps, a user could look for characteristic node fanins and net fanouts to determine which inputs to the technology mapping algorithm correspond to which partitions created by the partitioner, allowing for reintegration of the subcomponents. Finally, the grouping of logic into high-fanin gates is the result of design decisions, containing information similar to the hierarchy information described earlier.

The answer is to break all components down to smaller elements, hiding this information. For logic functions this is often easy, since multi-input gates can be converted into trees of 2-input functions. Breaking up high fanout signals can be more difficult. However, there are several approaches. First, buffers can be inserted, splitting the fanout among multiple drivers. Also, the source logic for this net can be duplicated, splitting the fanout amongst the copies. Finally, some algorithms require that all multi-terminal nets be replaced with 2-terminal nets. For example, Spectral methods for logic partitioning often convert multi-terminal nets to two terminal nets via clique expansion or other methods [1]. This can be taken advantage of for Server Security for Web-based CAD by performing the multi-terminal to two terminal net conversion on the user's machine with trusted software, eliminating this information from the input netlists.

3.8 I/O Encryption & Splitting

For the specific case of logic simulation/emulation there are modifications that can be done to the circuit to improve Server Security. For simulation the circuit does not have to have exactly the same behavior as the real circuit, but the user must be able to know from the simulation how the real circuit will behave. For example, we can add an inverter to one of the outputs of the simulation, knowing that whatever the simulation does, the real circuit will produce the opposite value on that output. We can extend this transformation further, radically changing the interface behavior. One simple method is to take two output signals and replace one of them with the XOR of the two values. Thus, if the circuit used to compute A and B, and now computes A and A XOR B, we can recover the value of B by computing the XOR of the two signals. Similar optimizations can be applied to the circuit inputs, and can be applied multiple times to further encrypt the I/O behavior. Even more complex I/O modifications can be applied, as long as there is a relatively simple method for recovering the circuit I/O behavior given the simulation behavior.

If we just add this extra logic as a wrapper around the original circuit, someone aware of this transformation should be able to recover the original circuit. However, instead of XOR gates we can use the equivalent AND-OR form, and then migrate these gates into the circuit itself. Specifically, the new gates can be moved into the circuit via algebraic transformations to the circuit. Distributive and associative laws of Boolean Algebra can be used, integrating this logic into the complete simulation. Applying these rules a random number of times (and thus pushing the logic into the circuit a random distance) should make it quite difficult to remove this I/O encryption.

One optimization to the I/O behavior that should be extremely hard for an attacker to undo is to split the circuit inputs into several separate signals. For example, if an input to the circuit fans out to three destinations, it can be replaced with three separate inputs, each going to only one destination. Thus, after this transformation the CAD tool no longer knows that those three destinations should always receive the same value, yet the user of the circuit can easily ensure that this always happens.

For all of these I/O transformations, the effects can be hidden from the user quite simply. The user specifies the inputs to the original circuit, and a simple software algorithm can generate the corresponding inputs to the transformed circuit. Once the CAD tools create output for the simulation, the results are fed through a similar transformation algorithm on the user's machine, which takes the simulation output and recreates the corresponding output from the original circuit. In this way, I/O encryption and splitting can be used to help improve Server Security, while the user will see no changes in the simulation behavior.

4. Data Security Technique Interactions

As discussed in the previous section, there are numerous ways to provide Server Security for Web-based CAD. These techniques alter the circuit information sent to the CAD tool, minimizing the amount of information that can be compromised if the CAD tools or Compute Server proves to be untrustworthy or malicious.

Source:

```
SYM, <8112>, OUTFF
PIN, C, I, Clk_internal
PIN, D, I, cntrl/$MemWriter349
PIN, Q, O, MemWrite
END
SYM, $Instructionr11<31>_DFF_EN, DFF
PIN, C, I, Clk_internal
PIN, D, I, MemData<31>_internal
PIN, CE, I, cntrl/$stater370<0>
PIN, Q, O, $Instructionr11<31>
END
SYM, $ReadAddr27<63>, OR
PIN, 1, I, <5133>
PIN, 2, I, <5134>
PIN, O, O, $ReadAddr27<63>
END
SYM, $cntrl/$MemReadr136, OR
PIN, 1, I, $Instructionr11<29>
PIN, 2, I, $Instructionr11<26>, , INV
PIN, 3, I, <1500>
PIN, 4, I, <1662>
PIN, O, O, $cntrl/$MemReadr136
END
```

Encrypted:

```
SYM, I000000, AND
PIN, 1, I, I000000
PIN, 2, I, IO00000
PIN, O, O, I100000
END
SYM, II00000, AND
PIN, 1, I, I0O0000
PIN, 2, I, II00000
PIN, O, O, IOO0000
END
SYM, I1O0000, AND
PIN, 1, I, IIO0000
PIN, 2, I, I010000
PIN, O, O, I1O0000
END
```

Dictionary:

```
I11O00O ReadData2<0> A -1
I1101001 ReadData2<3> A -1
I01O011 ReadData2<4> OR 3696 { 1 <3338> 2 <3339> O ReadData2<4> }
IO1OI001 ReadData2<1> OR 3705 { 1 <3344> 2 <3345> O ReadData2<1> }
I011O00 ReadData2<2> A -1
```

Figure 1. Encryption example, containing excerpts from the source, encrypted, and dictionary files. Note that with the file order randomization feature, the portions of the files shown do not correspond to the same regions of the circuit. The source file is altered to create the encrypted version, which in this case retains only the connectivity information. A dictionary file is also created which allows the source to be recreated from the encrypted version. The dictionary would be retained by the user, while the encrypted version is sent to the CAD center.

Although these techniques cannot protect all data about a design, since the CAD tools require information about the design being optimized in order to operate, these techniques make it more difficult to reconstruct the complete circuit description, and minimize the vulnerability introduced by Web-based CAD.

As pointed out in the description of some of the security techniques, not all optimizations can be used before some CAD steps. For example, although comment removal and variable renaming can be applied before submitting the design to any CAD tool, since these tools should not depend on these features of the input netlist, I/O encryption and splitting can only be applied in a limited set of circumstances since it alters the behavior of the circuit being manipulated.

While some security techniques are simply inappropriate in some circumstances, use of others may imply a tradeoff between mapping quality or performance and data security. For example, net direction alteration changes the directionality of signals in a circuit. Such a transformation can be applied before most mincut based partitioners, since these algorithms do not use information on net directionality. However, such a partitioner might use logic replication, the process of duplicating logic nodes to reduce the communication between partitions. Such an optimization does require circuit directionality, since when replicating a logic node the output of that node is always removed from the cutset because the signal is computed in each partition simultaneously, while inputs to the logic node are forced into the cutset unless their sources are also replicated. This means that if a circuit undergoes net direction alteration before partitioning, any replication features in the partitioner must be turned off, adversely affecting the resulting quality.

Some security techniques may impact CAD runtimes instead of quality. For example, techniques such as I/O encryption alter the topology, reducing locality by connecting unrelated portions of the circuit. When this circuit is mapped to a logic emulator, the loss of locality may increase the required communications in the system, slowing down the emulator's performance. Thus, when deciding which security transformations should be applied to a circuit, the resulting impact on the quality and runtimes of the CAD algorithms must also be considered.

5. Web-based CAD Data Security Testbed

In order to test the feasibility of our data security techniques, we have developed a prototype encryption system. Our program begins with a circuit in XNF format, the format used by the Xilinx FPGA toolsuite. The algorithm transforms the circuit into an encrypted form. It also creates a dictionary file, which contains the information needed to reverse the encryption steps (Figure 1). Via command line options various encryption techniques can be applied. Our system includes variable renaming, node order alteration, functionality removal, and net direction alteration. The resulting file is a valid XNF file, which can be submitted to CAD algorithms that accept XNF.

Results for running our encryption algorithm on some standard benchmark circuits is shown in Table 1. The results include the runtimes of the algorithm, as well as the percent of the nets reversed. Note that while we choose to randomly reverse half of the circuit outputs and flip flops, this tends to result in an overall reversal of about one third of the nets in the circuit. However, this is more than sufficient to make the direction of the nets in the circuit useless for reverse engineering.

Circuit	Nodes	Time (sec)	% Reversed
S38584	22,451	956.94	27%
s38417	25,589	769.86	27%
s35932	19,880	378.53	45%
s15850	11,071	107.41	29%
s9234	6,098	48.39	23%
s5378	3,225	22.09	28%
s1423	831	4.86	28%
s208	129	1.34	30%

Table 1. Results from protecting benchmarks with prototype data security algorithm. Circuits had their signal names encrypted, their order in the file randomized, the functions removed, and half of all outputs and flip-flops were used as starting points for net reversal. The number of nodes, runtimes, and the portion of nets reversed is listed. The runtimes include the time to create the dictionary files to allow for undoing the encryption.

6. Conclusions

The Internet is revolutionizing most communication and computation systems. It also has the potential to radically change the way future electronic systems are designed by creating a new methodology for VLSI CAD. However, just like every other domain, security concerns with an Internet based computing paradigm may overwhelm the potential benefits.

In this paper we presented two models of Web-based CAD: Applets and Compute Centers. We also demonstrated how the security concerns of these systems can be broken down to Client, Transmission, and Server security concerns. Client security entails the securing of the user's machine against external intrusion, and can be handled by firewall techniques currently in place. Transmission security encompasses concerns over transporting information to a Compute Center, and can be supported by public key encryption techniques. The remaining problem is server security, the concern that the CAD tools themselves (or the creators of these tools) are not trustworthy, and thus may divulge information given to these tools.

In this paper we proposed numerous new techniques for server security for Web-based CAD, as well as demonstrated how existing techniques can be harnessed for this new domain. Techniques such as variable renaming and formatting and comment removal eliminate the self-documenting code features of hardware description languages, while netlist flattening and component simplification removes grouping information. Further information hiding can be accomplished by understanding what information in a circuit is not needed by a given CAD tool. These techniques include functionality removal, net direction alteration, and net removal. Circuit partitioning can also serve to hide information by only providing subportions of the circuit, and potentially providing bogus data. Finally, I/O encryption and splitting can alter the interface of circuits under simulation, complicating reverse-engineering attempts.

By combining these techniques together, a complete security methodology for Web-based CAD can be provided. This will help create a Web-based CAD methodology that can provide significant benefits for future VLSI design efforts. Applet-based CAD can enable collaborative design between geographically distant designers and create a usage-based pricing methodology for CAD software. Compute Centers can provide these same benefits, as well as providing high-performance computation resources in an affordable manner by amortizing their costs across a large user base.

7. Acknowledgements

This research was supported in part by a grant from the National Science Foundation, and software donations from Xilinx, Inc. We also wish to thank Mohammed Khalid from the University of Toronto for providing example XNF files for use in this research.

8. References

[1] C. J. Alpert, A. B. Kahng, "Recent Directions in Netlist Partitioning: A Survey", *Integration: the VLSI Journal*, Vol. 19, No. 1-2, pp. 1-81, 1995.

[2] C. Barrett, D. Dill, J. Levitt, *SVC*, http://agamemnon.stanford.edu/levitt/vc.html, 1997.

[3] D. Chakrabarti, P. Joisha, P. Banerjee, *The WADE Project*, http://cpdcser.ece.nwu.edu:8080/WADE/wade.html, 1997.

[4] J. Cooley, "SPY vs. SPY: the VMC Story", *Integrated System Design*, Vol. 8, No. 90, pp. 56-64, December, 1996.

[5] M. Enos, S. Hauck, M. Sarrafzadeh, "Replication for Logic Bipartitioning", *ICCAD*, pp. 342-349, 1997.

[6] J. Gateley, "Logic Emulation Aids Design Process", *ASIC & EDA*, July, 1994.

[7] D. Lidsky, J. M. Rabaey "Early Power Exploration--a World Wide Web Application", *Design Automation Conference*, June 1996.

[8] David Lidsky, *PowePlay*, http://infopad.eecs.berkeley.edu/PowerPlay, 1997.

[9] P. Losleben, D. Boning, A. Bhandarkar, A. Gower, H. Huang, E. Jhong, J. Jovanovic, P.-F. Lam, M. McIlrath, W. Moyne, S. Narayan, L. Wang, S. Wang, T. Wu, E. Zhou, *National Research Enterprise*, http://www-snf.stanford.edu/ComputationalPrototyping/comproto/nre.html, 1997.

[10] A. R. Newton, F. Chan, N. Ghazal, M. Horton, H. Hse, A. Hsia, S. Leung, K. Nguyen, M. Shilman, M. Spiller, S. Szollar, J. Young, J. Zhang, *Weld*, http://www-cad.eecs.berkeley.edu/Respep/Research/weld, 1997.

[11] K. Saraswat, J. P. McVittie, W. Abdel-Ati, S. Abdollahi-Alibeik, D. Bang, P. Canupp, J. Han, Z.-K. Hsiau, M. Joshi, P. Kapur, S. Ma, B. Shieh, *Speedie*, http://speedie.stanford.edu, 1997.

[12] *The Programmable Logic Data Book*, San Jose, CA: Xilinx, Inc., 1994

Design of a SPDIF Receiver using Protocol Compiler

Ulrich Holtmann
Synopsys, Inc.

700 E. Middlefield Rd., 2.31
Mountain View, CA 94043-4033
+1 (650) 528-4732
ulrich@synopsys.com

Peter Blinzer
Technical University of Braunschweig
Dept. of Integrated Circuit Design (E.I.S.)
P.O. Box 33 29
D-38023 Braunschweig
+49 (531) 391-2380
blinzer@eis.cs.tu-bs.de

1. ABSTRACT

This paper describes the design of a receiver for the digital audio signal SPDIF used by CD-ROM players. The design was done with Protocol Compiler, a high-level synthesis tool for the design of structured data stream processing controllers.

Compared to traditional RTL design, Protocol Compiler makes entry, debugging, and re-use easier. Design time was cut by factor 2 while the results in terms of area and delay are competitive.

1.1 Keywords

High-level synthesis, telecommunication.

2. INTRODUCTION

CD-ROM, DCC or DAT players transmit digital audio signals in the SPDIF protocol [Son83, IEC958]. At the Technical University of Braunschweig, Department for Integrated Circuit Design (E.I.S.), a SPDIF decoder mapped onto a Xilinx FPGA was designed. Design was done at the RTL level in Verilog. Experience showed that the hardest part was to get the controllers of the various modules right. Design of the decoder requires separation of the three SPDIF layers (bit level, block, and frame) and writing the individual controllers for each. Implementing and verifying these controllers as FSMs at the RTL level is an error-prone and time-consuming task. While the design requirements at the protocol level ("*what*" to do) are well structured, the state transition graphs at the implementation level ("*how*" to do it) lack structures. Furthermore, the individual controllers must communicate with each other, so an FSM often contains, besides the functionality required by the protocol, additional states and transitions implementing hand-shakes, stalling, etc.

In the light of these design issues, a joint research project was undertaken to apply the high-level synthesis tool "Protocol Compiler" to the decoder design and determine how the issues are addressed. This paper describes the FPGA design of the SPDIF receiver using Protocol Compiler.

The rest of the paper is structured as follows: Section 3 gives a short description of Protocol Compiler and related languages. The SPDIF protocol and design specification is given in Section 4. Then the design implementation is described in Section 5 and results are presented in Section 6. Finally, we give a summary and conclusions.

3. RELATED WORK

Protocol Compiler is a design environment for the processing of structured data streams such as the ATM, SDH/SONET or MPEG protocol. It is based on previous work by Seawright and Brewer, and later Crews, on logic synthesis from grammatical productions [Sea94, Cre96]. Protocol Compiler was first described in [Sea96] and an application for SDH/SONET is shown in [Mey97]. The design environment consists of entry, debugging support (by means of annotating simulation results back onto the source), protocol synthesis, and HDL generation (Both VHDL and Verilog are supported). The Protocol Compiler language is conceptually higher than RTL and, together with protocol synthesis and intuitive debugging capabilities, is the key element resulting in a productivity gain for design and verification of controllers. Due to limited space please see [Sea96, Mey97] for a detailed description.

Protocol Compiler is designed to model controllers for structured data streams where data processing is done on a continuous basis. Modeling structured data over time is the key aspect. Reactive systems on the other hand focus on external or internal events with different urgency. Each event requires certain actions to be performed and the key aspect is to cope with the complexity of many simultaneous events. The most prominent high-level synthesis systems are Esterel [Ber92] and Statecharts [Har87].

4. DESIGN REQUIREMENTS

Figure 1 gives an overview of the decoder and environment. The decoder receives the SPDIF signal, transmits audio values to the Digital Analog Converter (DAC) over a serial interface, and also drives a few status LEDs which are not shown in the figure.

The target device is a Xilinx XC3042 FPGA clocked at 16 MHz which imposes tough constraints to area (only 144 CLBs available) and timing.

The SPDIF protocol consists of the three layers Bit, Block, and Frame. Figure 2 shows the format of the lower layers Bit and Block. The Block layer carries the audio values and some control bits. Each block consist of a header telling whether the audio value is for the left or right channel, four reserved bits, the 20-bit value with the LSB first, two control bits, a status bit used in the upper frame layer, and a parity bit.

The Bit layer uses a Bi-Phase coding to transport bits. Each data bit is coded with two code bits so that the polarity of the SPDIF signal changes with every data bit (see Figure 2). In case a '1' is coded, then the signal also changes between the code bits, in the case of a

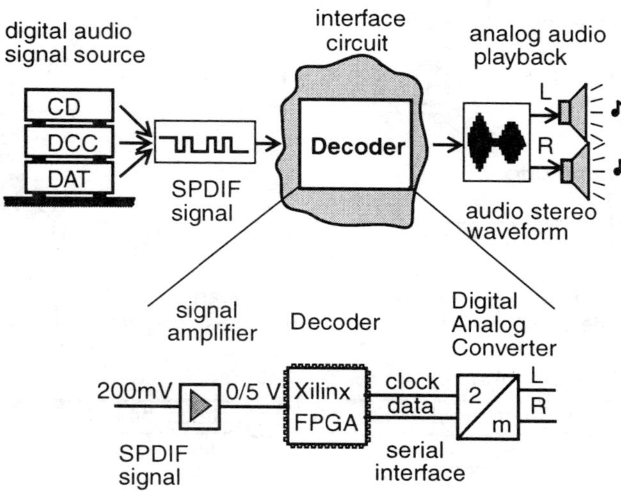

Figure 1: System Overview

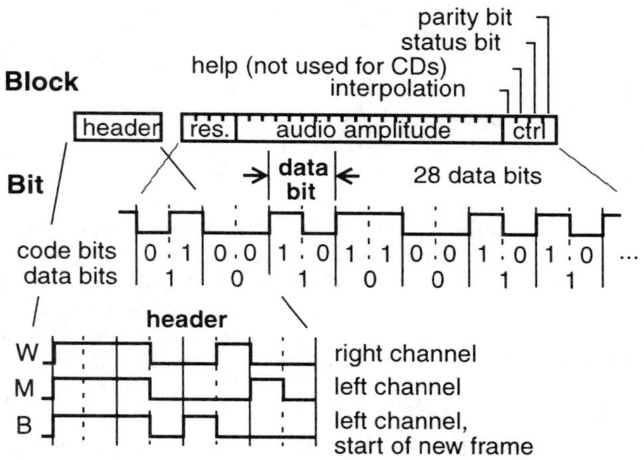

Figure 2: SPDIF Protocol, Bit + Block Layer

'0' it doesn't change. Since the polarity is constantly changed, a '0' may be coded with two high or two low code bits. A '1' is either coded with code bits high+low or opposite polarity. Headers W, M, B are coded differently to allow simple detection of the next block (for sake of simplicity, Figure 2 shows them only in positive polarity). W tells that the value is going to the right channel, M and B indicate the left channel. Header B furthermore indicates the beginning of a new frame. The data bit rate is roughly 2.8 MHz allowing transmissions of audio samples with frequencies up to 44 KHz. The clock speed of the code bits is consequently about 5.6 MHz.

The third, Frame, layer provides status information such as whether it is allowed to copy the CD or not. A frame consists of 192 pairs of left and right blocks, starting with a left block indicated by an B header. Due to limited paper size, we will not explain more details of the Frame layer, however, the receiver must decode it in order to drive three status LEDs (copy permission, copy status, and data is valid).

The Digital Analog Converter (DAC) has two separate converters and a common serial interface shown in Figure 3. The decoder must generate the CLK clock and SDI data signal as well as the two control signals. SDI is sampled on the rising edge of CLK and once all 16 bits (4 control + 12 amplitude value) are shifted in, it needs a falling edge on nLD to apply the values to the converters.

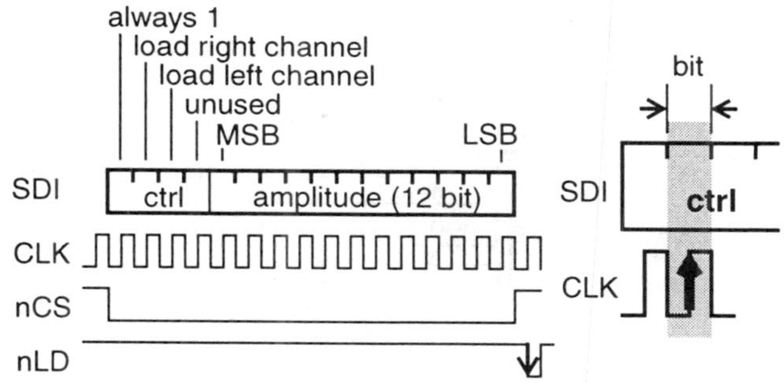

Figure 3: Interface to the Digital Analog Converter

The SPDIF decoder in the FPGA environment described above may not be a common design. However, we found that certain design requirements of the SPDIF decoder are typical for ASIC designs in the networking area such as ATM, SDH/SONET, MPEG, 1394, etc.:

- Recognition of header patterns and synchronization (Bit layer)
- Parsing a structured data stream (Block and Frame layer)
- Generating a waveform (DAC interface)
- Interface issues between modules such as synchronization or stall.

We therefore think that the solutions and conclusions described below can be applied there as well.

5. DESIGN IMPLEMENTATION

The design is broken into four modules, one for each layer of the SPDIF format and one for the DAC interface. Figure 4 shows the block diagram including all important signals.

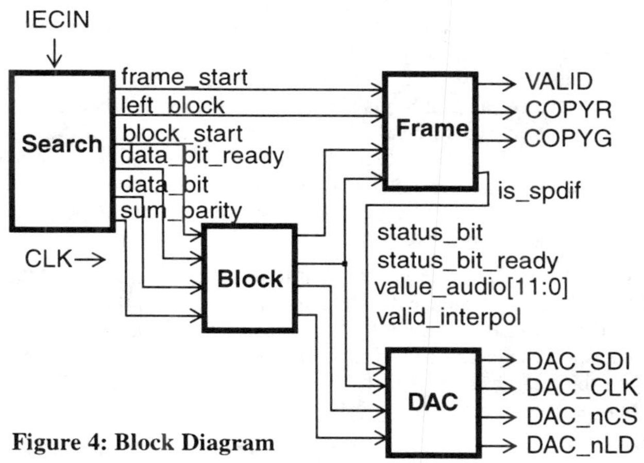

Figure 4: Block Diagram

Signals frame_start, block_start, and data_bit_ready are flags, which means they are set high for only one clock cycle to indicate the corresponding event. For example, when a data bit '1' has been detected, then data_bit is set high and signal data_bit_ready is set high for one clock cycle. Please note that the decoder runs with a clock speed of 16 MHz, several times faster than the data bit rate of the SPDIF source, so during most cycles no header or data bit will be detected.

A snapshot of the specification in Protocol Compiler is given in figure 5. The figure shows the specification in exact the same way (besides color) that one sees it in Protocol Compiler's graphical frame editor. Each of the four boxes is a *reference frame*, abstracting a complete module of the design modeled in the Protocol Compiler language. (Please see [Sea96] for a detailed description of individual language elements, the so called "frames"). By double-clicking, one "jumps" into such a reference frame.

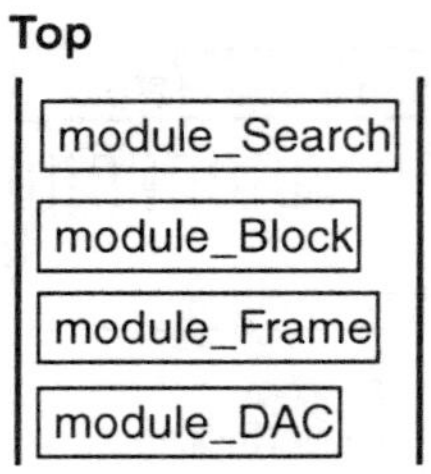

Figure 5: Specification in Protocol Compiler (Top Level)

5.1 Block Module

Figure 6 shows the definition of the module_Block reference frame implementing the Block layer. Please compare Figure 2 how the sequence {reserved..parity} directly reflects the structure of the block layer. The close similarity between specification and implementation is a strong feature of Protocol Compiler's high-level language. Understanding *what* is implemented or incorporating a specification change is very easy thus making the designer more productive.

The fields reserved, amplitude, ... are abstracted into individual frame definitions shown in Figure 7. All fields expand to *terminal "1" frames* describing a delay of one clock cycle each. A *repeat frame* $[\,]^N$ means to repeat the inner frame N times. Some terminal frames have *actions* (the rounded boxes) attached to them which are executed in the corresponding clock cycle. Most actions like valid_interpol=data_bit just store the received data bit into internal registers. The concat action "{ }" shifts the audio value into val_audio[11:0] for later use by the DAC-Interface module. The audio value is transmitted with 20-bit resolution but the DAC has only 12 bits, so the first 8 bits shifted in (the least significant ones) are overwritten, which means simply ignored. The last of the four control bits is the parity which should always be even. In case the parity check is successful, the conditional action "!sum_parity: set(status_bit_ready)" executes, telling the Frame module that the next status bit is available.

The two *template frames* RunIdle and Proc_StartAfter surrounding the sequence describe internal interface issues between the Search and Block module. Since the decoder runs at a higher clock speed than the SPDIF source, data bits are only available when the Search layer detects them and indicates so with signals block_start and data_bit_ready.

Template frames are, like reference frames, defined by sub-frames but can be customized. The RunIdle template basically stalls its inner frame when the condition data_bit_ready || block_start is false, so time proceeds only during cycles when one of the flags is set. Template Proc_StartAfter is an endless loop, waiting for condition block_start to be true, then executing its body (the sequence) and then waiting again. Modification of sequence {reserved..parity} with the two template frames allows us to substitute "1 cycle = 1 data bit" with respect to the sequence.

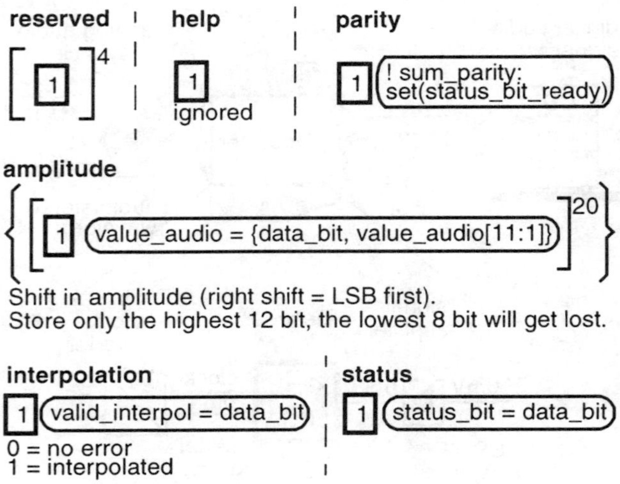

Figure 7: Submodules of the Block Module

Please note that the Protocol Compiler description clearly distinguishes between the internal interface issues and the pure structure of the Block layer as defined by the SPDIF protocol. This not only increases the readability but also allows easy re-use of frames in similar designs. Furthermore, the orthogonal description of inter-block interface and protocol structure isolates any changes between these two parts, thus speeding up design changes. An RTL or schematic implementation doesn't allow a clear separation and would make later re-use of the design far more difficult.

5.2 Frame Module

The Frame module is very similar to the Block module and due to limited paper space not explained. Its task is to extract certain status information and set the ports VALID, COPYR, and COPYG which drive status LEDs.

5.3 DAC-Interface Module

The task of the DAC interface is to sequentially send the audio value plus a few control bits to the DAC. The interface (Figure 3) requires that the data signal SDI must be stable during the rising edge of CLK. A simple way to guarantee this is to spend two clock cycles per bit as done here (see the two terminal frames within the $[\,]^{16}$ frame in Figure 8). During the first cycle, SDI is set and CLK is cleared. In the next cycle, SDI stays but CLK is set, so SDI is stable during the rising edge of CLK.

Besides the 12-bit audio value, four control bits must be transmitted

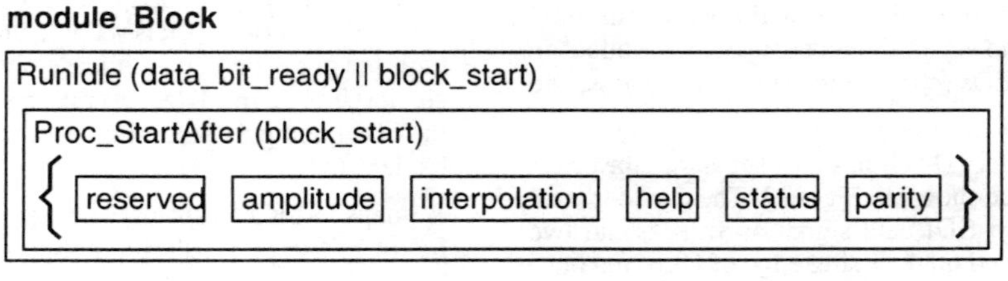

Figure 6: Implementation of the Block Module

as well. In this implementation, the shift register is extended to 16 bits and the four control bits are loaded into it before the shifting starts. This way the underlying controller stays very simple.

Port DAC_nLD is set every clock cycle by a *default action*. When all 16 bits have been sent and the block is valid, then DAC_nLD is cleared, overriding the default set and thus generating the negative edge required to load the audio value into the DAC internal converter.

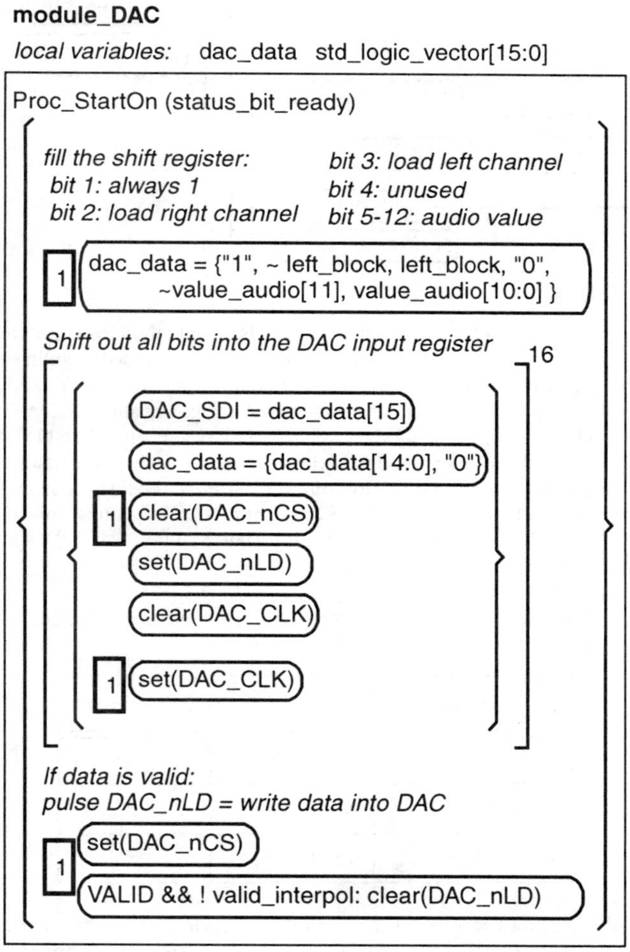

Figure 8: Implementation of the DAC Module

5.4 The Search Module

The Search module processes the Bit layer and indicates received headers and data bits, and calculates the parity bit. The Search module directly receives the IECIN input and must detect headers and data bits of the Bit layer. Its main task is to search the input stream for certain patterns which are derived from the SPDIF protocol. Figure 9 shows how to compute the search pattern for the M header. Since the receiver runs on a higher clock speed than the SPDIF source and both clocks are not synchronized, one has to allow for variances in the pattern at the receiver side. For example, a header lasts exactly 8 cycles at the source but may need either 22 *or* 23 cycles at the receiver. Determining good search patterns is a s complicated task and not explained in this paper. It has to be done by the designer and we merely describe here the chosen patterns.

A "1" ("0") in the pattern means that the input stream must be high (low) during the corresponding cycle, while a don't care "-" means that the input stream is not checked.

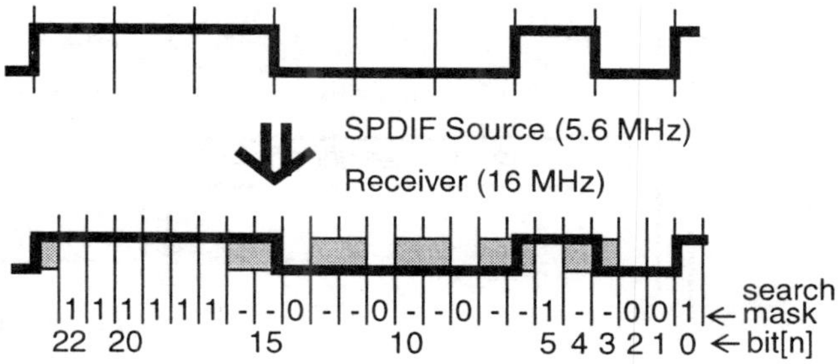

Figure 9: Search Pattern for the M Header

There are several ways to implement the Search module. The chosen architecture uses a 22-bit wide shift register and compares the register contents in two consecutive cycles. This way only 6 bits at a time must be compared and the underlying FSM is very simple. Figure 10 shows the continuous pattern search modeled with a [1]* frame and Figure 11 shows the pattern broken into two consecutive comparisons. The action pattern={pattern[20:0],IECIN} realizes the 22-bit shift register. Other implementations with shorter shift registers and more compare cycles were tried as well but due to limited space are not discussed in this paper.

Specifying the pattern search in Protocol Compiler is a straightforward task due to its ability to understand *pipelined control flow*. For example, when a M header is received, then the terminal frame pattern[21:13]=="111111--0" will match first, and terminal pattern[11:0]=="0--0--1--001" matches in the next cycle. The complete comparison consist of a two-cycle sequence. Because it is not known when the first match will occur, the sequence must be tried every clock cycle thus requiring to overlap, or pipeline, two instances of the sequence. In Protocol Compiler, pipelined control is specified with a leading [1]* frame followed by terminals containing search patterns as seen in Figure 10.

Manually implementing pipelined control as an FSM or connection of stage registers is a tedious and error prone task. Fortunately, Protocol Compiler does this task automatically and furthermore can guarantee with its distributed encoding style [Sea96] a very efficient implementation.

After a header has been detected, 28 data bits will follow and must be detected as well. Like the header detection, the designer must determine patterns recognizing the bi-phase codes. Due to limited space these bi_phase_pattern frames are not shown in the paper.

Figure 10: Implementation of the Search Module

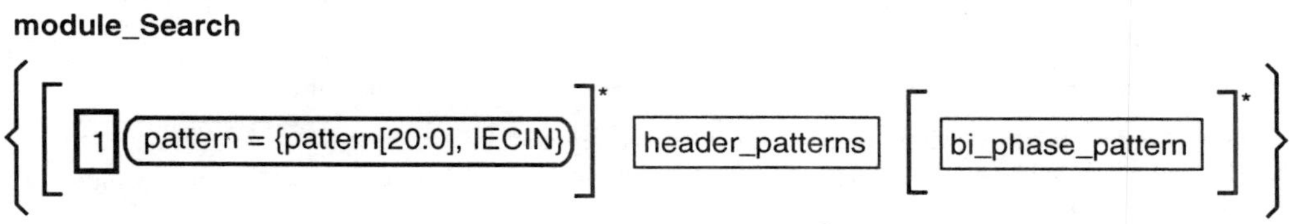

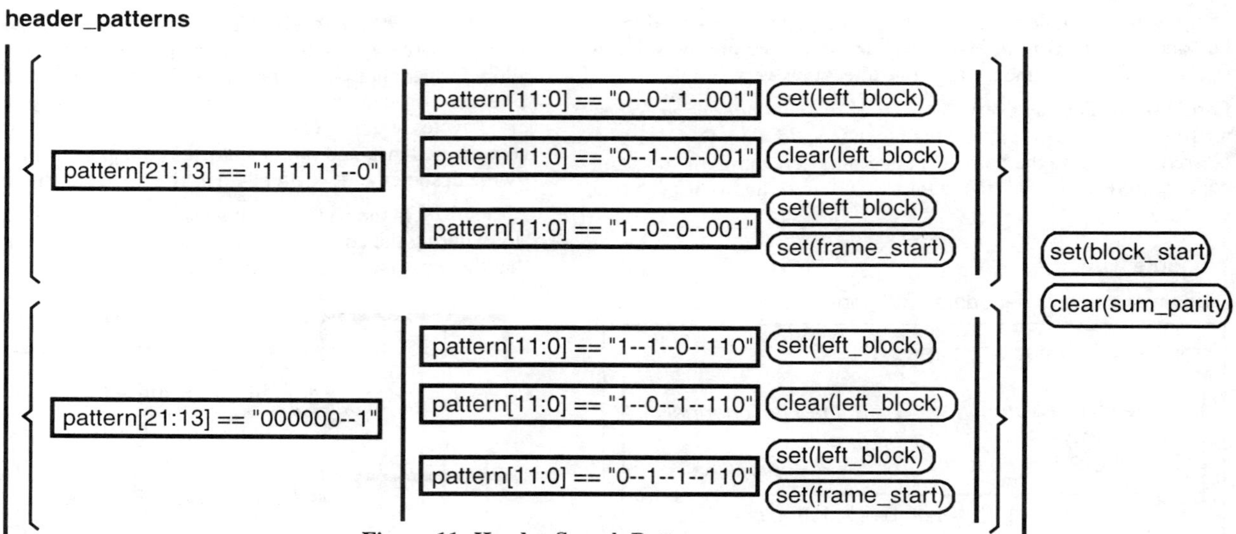

Figure 11: Header Search Patterns

6. RESULTS

The design flow of the decoder involved Protocol Compiler (Synopsys), Verilog-XL (Cadence), Design Compiler v3.5a (Synopsys) and XACT 5.2.1 (Xilinx). Design entry was done with Protocol Compiler. Simulation used Verilog-XL + Protocol Compiler for annotating the simulation results back onto the source. Compilation into an FPGA required protocol synthesis with Protocol Compiler first, then logic synthesis with Design Compiler and finally P&R by XACT.

Protocol Compiler compiles the frame description into a set of FSMs during protocol synthesis. By defining the set of FSMs and their individual coding style, the user can easily explore different architectures without the need to change the input description. For example, repeat $[\ ]^N$ frames can either be unrolled or implemented by an LFSR counter resulting in different architectures.

The architectural exploration of different coding styles and counter usage is demonstrated by the results in Table 1 for the DAC-Interface module. The repeat $[\ ]^{16}$ frame was either unrolled or implemented by an LFSR counter. All results refer to mapping onto an XC30xx target with 64 CLBs.

Table 1: Different Architectures of DAC Module

Counter	Coding Style	Area CLBs	Critical Path
LFSR	distributed	48	74.2ns
LFSR	min-encoded	39	46.1ns
LFSR	1-hot	46	64.4ns
unrolled	distributed	59	49.4ns
unrolled	min-encoded	35	49.3ns
unrolled	1-hot	57	75.7ns

Table 2 shows the results of the whole receiver for different target devices. The receiver is split into 4 modules according to the block structure, all repeat frames are unrolled, modules Block and DAC are min-encoded, the Frame module is 1-hot encoded, and the Search module is distributed encoded. These protocol synthesis settings were found by trying out different settings for each module and using that one with the best results, e.g. unrolled and min-encoding for the DAC module as Table 1 shows. Protocol and logic synthesis are done for each module individually, followed by a final logic synthesis run with "-ungroup all". Area and delay values for FPGAs were obtained after P&R with XACT; values for the LCA10K library were obtained after logic synthesis with Design Compiler.

Table 2: Protocol Compiler Results for Different Target Architectures

Target	Area	Critical Path
XC3042A-7	103 CLBs	53.4ns
XC4003-6	81 CLBs	57.6ns
LCA10K	1350 gates	34.8ns

Table 3 shows results for different design styles. The values in the table refer to the whole receiver and also refer to experienced designers. Schematics are derived from the Verilog RTL solution,

Table 3: Results for Different Design Styles

Design Style	Area	Critical Path	Design Time
Protocol Compiler	103 CLBs	53.4ns	2-3 days
Verilog-RTL	90 CLBs	53.2ns	1 week
Verilog-RTL +Schematic	114 CLBs	33.6ns	4-5 weeks

therefore the 4-5 weeks include 1 week for RTL. About 2/3 of the

Protocol Compiler frames describing the design are shown in this paper. The Verilog RTL solution consists of 912 lines, 570 of them comments or blank lines. The results show that the design time with Protocol Compiler compared to RTL methodology was improved by 1.5x -2x and compared to RTL+schematic by 10x.

7. SUMMARY AND CONCLUSIONS

This paper described the design of a digital audio receiver for the SPDIF format used by CD-ROM or DAT devices. The design was done with Protocol Compiler, a high-level synthesis tool for the design of controllers. By comparing Protocol Compiler with a traditional design style of writing a Verilog RTL file and then either automatically (logic synthesis) or manually (schematic entry) mapping it onto the FPGA the following conclusions were obtained:

- The Protocol Compiler description is far more readable than Verilog RTL, thus allowing faster entry, debugging, and changes. While the Protocol Compiler design is suitable for later re-use in a similar project, the RTL and schematic solutions are not.
- Protocol Compiler cut the design time compared to RTL design style by 1.5x - 2x.
- Protocol Compiler designs can always be synthesized and their quality (area, timing) is acceptable. Furthermore, design exploration is much simpler.

Altogether, applying Protocol Compiler to this design showed that it is a productive high-level synthesis tool for controllers. We think that our conclusions apply to other ASIC designs in the networking area such as (ATM, SDH/SONET, MPEG, 1394) as well because the design requirements are similar.

8. REFERENCES

[Ber92] G. Berry and G. Gainsayer, "The ESTEREL synchronous programming language: design, semantics, implementation", in *Science of Computer Programming*, Nov. 1992, vol. 19 (no. 2): pp. 87-152.

[Cre96] A. Crews and F. Brewer, "Controller Optimization for Protocol Intensive Applications", in Proceedings of the European Design Automation Conference 1996, Geneva, Switzerland, September 1996, pp. 140-145

[Har87] D. Harel, "Statecharts: A Visual Approach to Complex Systems," in *Science of Computer Programming*, Aug. 1987, vol. 8 (no. 3), pp. 231-275.

[IEC958] "Digital Audio Interface", International Standard IEC958, First edition 1989-03.

[Mey97] Wolfgang Meyer, Andrew Seawright, Fumiya Tada: "Design and Synthesis of Array Structured Telecommunication Processing Applications." in Proceedings of the Design Automation Conference 1997, Anaheim, June1997, pp. 486-491.

[Sea94] A. Seawright and F. Brewer, "Clairvoyant: A Synthesis System For Production-Based Specification," in *IEEE Trans. on VLSI Systems*, June 1994, pp. 172-85.

[Sea96] A. Seawright, U. Holtmann, W. Meyer, B. Pangrle, R. Verbrugghe, and J. Buck, "A System for Compiling and Debugging Structured Data Processing Controllers", in Proceedings of the European Design Automation Conference 1996, Geneva, Switzerland, September 1996, pp. 86-91

[Son83] Sony/Philips: "Digital Audio Interface for Domestic Use", September 1983.

[Tou93] H. Touati and G. Berry, "Optimized Controller Synthesis Using Esterel", in *Proc. International Workshop on Logic Synthesis IWLS'93*, Lake Tahoe, 1993.

MetaCore: An Application Specific DSP Development System

Jin-Hyuk Yang, Byoung-Woon Kim, Sang-Jun Nam, Jang-Ho Cho, Sung-Won Seo, Chang-Ho Ryu, Young-Su Kwon, Dae-Hyun Lee, Jong-Yeol Lee, Jong-Sun Kim, Hyun-Dhong Yoon, Jae-Yeol Kim, Kun-Moo Lee, Chan-Soo Hwang, In-Hyung Kim, Jun-Sung Kim, Kwang-Il Park, Kyu-Ho Park, Yong-Hoon Lee, Seung-Ho Hwang, In-Cheol Park, and Chong-Min Kyung

Department of Electrical Engineering
KAIST, Taejon, 305-701, Korea

Abstract

This paper describes the MetaCore system which is an ASIP(Application-Specific Instruction set Processor) development system targeted for DSP applications. The goal of MetaCore system is to offer an efficient design methodology meeting specifications given as a combination of performance, cost and design turnaround time.

MetaCore system consists of two major design stages: design exploration and design generation. In the design exploration stage, MetaCore system accepts a set of benchmark programs and a formal specification of ISA(Instruction Set Architecture), and estimates the hardware cost and performance for each hardware configuration being explored. Once a hardware configuration is chosen, the system helps generate a VLSI processor design in the form of HDL along with the application program development tools such as C compiler, assembler and instruction set simulator.

1 Introduction

Each DSP application requires different arithmetic functions and bit-rates according to the algorithm to be executed.[1] The characteristics of DSP algorithms must be fully reflected in designing an application-specific processor architecture.[2]

Instruction set of ASIP(Application-Specific Instruction set Processor) is designed to maximize the performance on specific applications, while that of general-purpose microprocessor is designed to provide an acceptable performance on a wide variety of applications.

The most important issue in the design of ASIP is the determination of micro-architecture and instruction set, which are closely affected by each other. In other words, the design of instruction set cannot be viewed as a design process independent of the design of the micro-architecture. Generally, instruction sets designed without the knowledge of sufficient details of the underlying micro-architecture are not efficient in terms of execution time. The matching between micro-architecture and the instruction set can only be achieved through a diverse exploration of the design space and the understanding of the interaction between micro-architecture(hardware) and algorithm of the application(software).

The design turnaround time of an ASIP depends on how efficiently the system specification is transferred to lower-level design to be implemented as a chip layout and system software implementation. The traditional design approach of describing design entity using HDL language or schematic and verifying this using simulation needs a significant number of iterations between higher-level specification and lower-level implementation and usually results in an increase of design turnaround time.

In this paper, we present a systematic approach called MetaCore to develop an ASIP for DSP applications. MetaCore system consists of two major design stages: design exploration and design generation. MetaCore system accepts a set of benchmark programs and a formal specification of ISA(Instruction-Set-Architecture) as its inputs, and produces information on the design such as gate count, use record of each instruction, and so on. Then MetaCore system is used to generate an ASIP design in the form of HDL along with a set of application program development tools such as C compiler, assembler and instruction set simulator.

2 Design Philosophy and Framework

Major issues in the design of MetaCore system are:

1. how to design programmable yet efficient micro-architecture for DSP applications.
2. how to extract features(such as instruction set and special functional blocks) from statistical data(such as instruction usage and block usage).
3. how to generate automatically the chip equipped with the extracted features.

In order to solve these problems, we have developed a reconfigurable micro-architecture and a formal representation language. The micro-architecture is designed to meet the general characteristics of various DSP applications. The reconfigurability of micro-architecture offers high degree of performance optimization as required for each specific application. Recently, various reconfigurable micro-architectures were introduced such as EPICS[3] and PEAS-I[4]. Those systems retain their reconfigurability by changing some system parameters such as bit-width of hardware functional blocks, and selecting instructions from predefined superset. However, such instruction customization as selecting a subset of instructions from a superset can not fully satisfy the demands of various algorithms in DSP applications.

The formal approach, adopted in MetaCore system, enables the designer to construct original instruction set for a given application. It is started with a specification of ISA and generates performance/cost parameters to support the designer in the choice among alternatives. The performance/cost parameter is generated by analyzing the relationship between the behavior of benchmark programs and hardware cost. The new design knowledge, the user-defined instructions and macro-blocks supporting the instructions, if

DAC 98, San Francisco, California

necessary, is simplely incorporated in automatic design environment only by specifying the ISA and updating the system library.

2.1 Micro-architecture

DSP applications are generally characterized as computationally intensive with large data set, loop-dominant control flow behavior, and accumulation-based operation. To support these characteristics, it is necessary that the micro-architecture supports effective data communication between memory system and execution units, low overhead loop control, and accumulator based instruction set architecture.

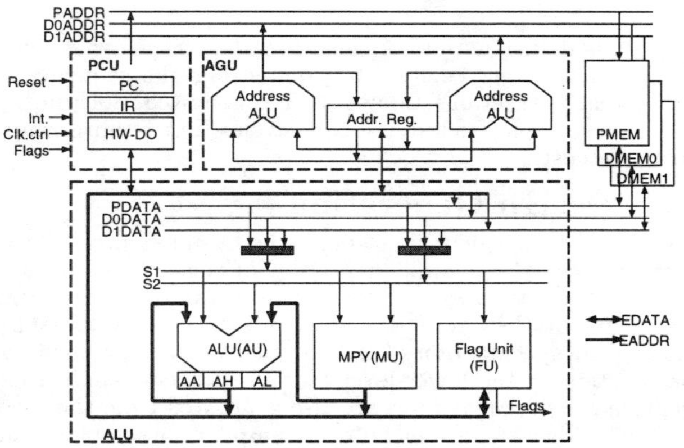

Figure 1: The micro-architecture of MetaCore.

The heart of the proposed micro-architecture shown in Figure 1 consists of three execution units operating in parallel, i.e., ALU(Arithmetic and Logical Unit), AGU(Address Generation Unit), and PCU (Program Control Unit). The ALU, which consists of MAC(Multiply-ACcumulate) unit and various basic hardware units, has been designed to optimize the time-critical inner-loop functions of DSP algorithms. The AGU is responsible for effective indirect addressing of data operands in memory. The AGU contains dedicated interconnections among address registers(ARs) and increment/decrement capabilities for efficient traversal of the special memory structures. The PCU is a pipelined instruction decoder with branching(direct/indirect jump, call/return), interrupt handling(goto/return-from interrupt) and zero-overhead hardware do-loop capability.

The design style supported by the MetaCore system is pipelined and parameterized micro-architecture. The pipeline is controlled in a data stationary fashion. It consists of stages for *instruction fetch, instruction decode, operand read, execute,* and *result write.* The first two stages are identical for all instructions. The *operand read, execute* and *result write* stages are dependent on the semantics of the instructions. The target micro-architecture can be fully described by specifying the ISA and a set of parameters such as register file size, width of bus, size and address space of memory, and bit-width of functional unit.

MetaCore system has a predefined instruction set, which consists of instructions generally used for the implementation of DSP algorithms. Table 1 shows a summary of the predefined instruction set, where instructions are classified into two classes, i.e., *primitive* class, and *optional* class. The instructions of the *primitive* class are essential, and are not omissible in any ASIP designed under the MetaCore environment, while the instructions of the *optional* class can be eliminated to satisfy the given design constraint.

class	category	instructions
primitive	arithmetic	add adc and andc ash ldi lhi lsh nop not or orc sub subi subr xor
	data move	ld st pop push
	control	d.cnd bd.cnd call callr for rep reti rets trap
optional	arithmetic	bf0 bf1 bts bcr bst clip ext idiv mac mpy mrg

Table 1: The predefined instruction set of the MetaCore system.

2.2 Formal Specifications

In the MetaCore system, the target ASIP is specified using formal representation language, from which the instruction set of the target ASIP can be synthesized for the purpose of simulation and lower-level implementation. The descriptive power of the formal representation language corresponds to that of ISPS[5].

Figure 2 shows an example of machine specification. The *hardware* clause describes the hardware configuration of EM1. The EM1 includes a 36 bit-wide accumulator, four 16 bit-wide address pointer register, 2 Kword sized pmem(program memory), and two 1 Kword sized dmem(data memory).

During each instruction cycle, *instruction fetch* and *instruction decode* stages are identical for all instructions. Thus, the behaviors of these two stages are omitted from the specification of ISA. The expected machine behaviors in the *operand read* stage, *execute* stage and *result write* stage of each instruction are specified in the *def_inst* clause and *operand* clause as follows: The *operand* clause defines the kind of operand. The *def_inst* clause represents the instruction behavior in detail: arithmetic function, flag fields which are affected by execution of the instruction, and the number of execution stages in the pipeline to execute the instruction.

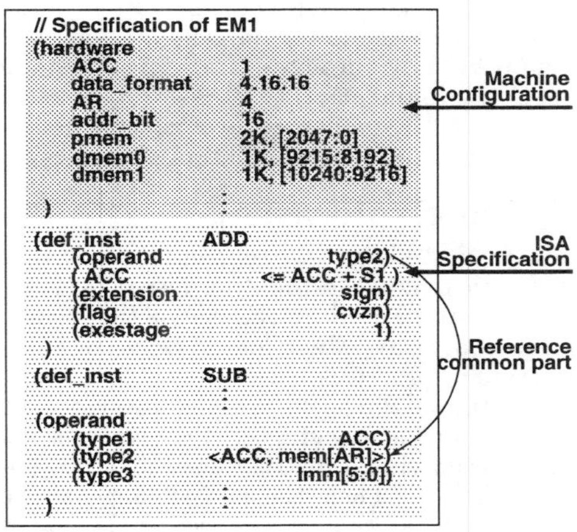

Figure 2: Formal specification of EM1.

2.3 Benchmarks

MetaCore is a benchmark-driven ASIP design system. The benchmark programs, which are programmed in C language, are compiled into machine code and simulated on MISS(MetaCore Instruction Set Simulator). The information extracted from profiles of benchmark programs is used for decision making of target ASIP design. Performance evaluation from benchmarks is more realistic than analytic estimation models.

3 Design Flow

Figure 3 shows the ASIP design flow under the MetaCore environment. The design exploration at the initial design stage directly affects the quality of the final design. The design philosophy of formal specification-based approach, adopted in the MetaCore system, is guided by two principles: flexibility and simplicity. These can be achieved through the use of design abstraction and leads to an extensive design exploration. During the design exploration, suggestions obtained from the MetaCore system are used to motivate the designer to improve the design quality.

While the design exploration can be done at very abstract levels, the definition of lower-level details and constraints is described in the system library. This library consists of descriptions of micro-architecture and macro-blocks in the form of HDL, formal specification, and physical details(cost, speed, and so on).

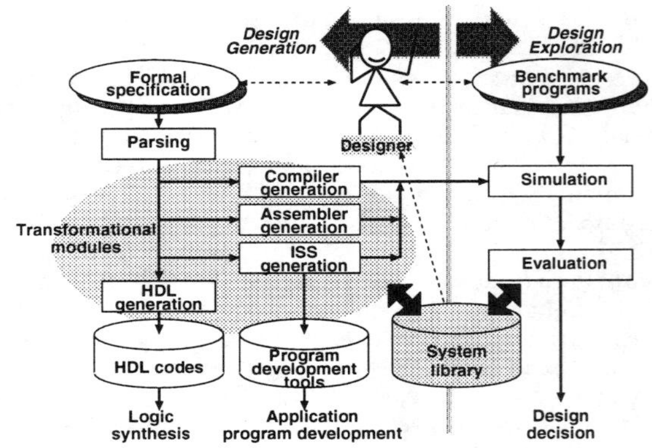

Figure 3: The ASIP design flow under MetaCore environment.

After successful design exploration, the design generation corresponds to the design transformation from the upper(abstract) level to the lower(implementation) level. This process is supported by automatic transformational modules in the design environment. The transformational modules are HDL generator, compiler generator, assembler generator, and ISS(Instruction Set Simulator) generator.

3.1 Design Exploration

The main task of design exploration stage is to evaluate the target design through estimating lower-level implementation characteristics based on the instruction-level analysis of benchmark programs. MetaCore system, as an ASIP development system, produces useful information which helps the designer optimize the hardware and develop application-specific modules and instructions. The analyzer calculates the frequency of individual instructions and sequence of contiguous instructions. Then the performance and cost are calculated for each instruction set variation resulting from instruction substitution. Based upon our experiment(see section 4), information on the frequency of instructions is very useful to develop application-specific instructions. Application-specific instructions can be developed by condensing frequent instruction sequence into a user-defined instruction or a combination of user-defined instructions.

Figure 4 shows an example of the instruction substitution and condensation of frequent sequence of contiguous instructions. The instruction substitution is performed by the analyzer automatically. In this example, the original instruction sequence, (a), is transformed into a new sequence, (b), by substitution of *abs* instruction. The new sequence takes more clock cycles while requiring less hardware cost.

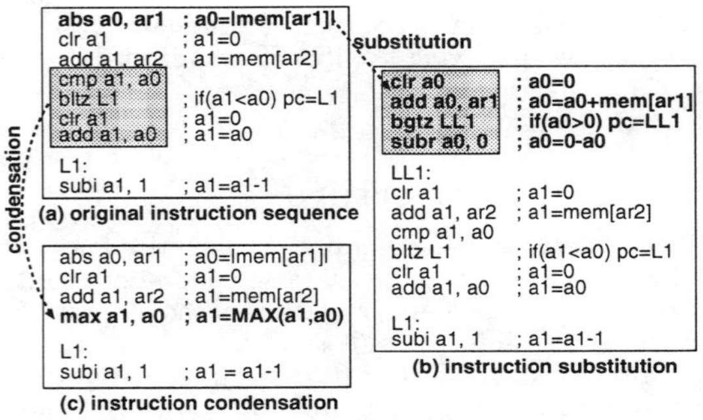

Figure 4: An example of instruction substitution and instruction condensation.

The original instruction sequence is also transformed into another new sequence, (c), by condensing the sequence of contiguous instructions, shown as the shadowed block in (a). This new sequence takes less clock cycles, but requires more hardware cost.

3.2 Design Generation

In the design generation stage, the designer proceeds to derive a hierarchical implementation of the processor starting from the previously described specification. The design generation in the MetaCore system corresponds to transforming the specification of ISA into a structural/behavioral specification in the lower-level. In this process, application program development tools are also generated from the specification. These tools include C-compiler, assembler, and ISS. The application program development tools are critical element of ASIP system.

The HDL generation performed by HDL generator of the MetaCore system is divided into three steps, which are macro-block generation, control-path synthesis, and connectivity synthesis. In the macro-block generation step, the HDL generator generates each functional block which supports instructions described in the given specification. MetaCore system accesses necessary descriptions of macro-blocks from the MetaCore system library and instantiates the parameters such as the size of register/memory, bit-width of multipliers, bit-width of data-path, and so on. In the control-path synthesis a data stationary control model is used for the pipeline design. In this model the opcode is transmitted through each stage of pipeline in synchronization with the data. The opcodes are forwarded to next stages synchronously. At each stage, the opcode, together with possible status bits from the data path is decoded to generate the control signals necessary to drive the data path. The decode logics are derived from the relationship between instructions and functional blocks. After all the control signals are defined, connectivity synthesis is performed.

The C-compiler generation, performed by the compiler generator of the MetaCore system, is based on compiler-compiler method introduced by *gcc(GNU C compiler)*. This task can be performed by generating *Target Description Macros* and *Machine Description File*. The *Target Description Macros* define architectural information such as bus width, register configuration, stack area, and so on. The *Machine Description File* defines the mapping rule between intermediate representation of source program, which is output of *gcc*, and the instructions of target ASIP.

The assembler of MetaCore(MASM) translates the source code, output of C-compiler, into an object code in COFF(Common Object File Format). The translation rule is described in TRF(Translation Rule File), which contains mnemonics, encodings of instructions and memory configuration, and so on. By generating TRF, target assembler can be obtained.

The ISS can be obtained by generating IDF(Instruction Definition File), which contains instruction behavior and encoding. The ISS accepts object code of benchmarks, then generates profiles of the benchmarks.

4 Experiment

In this section, a design of an ASIP targeting for mobile communication application, the GSM(Global System for Mobile communications), is presented to show the effectiveness of MetaCore system and proposed design methodology. Each algorithm constructing GSM was programmed in C language and used as benchmark to drive design direction. The initial design was started with the specification of predefined instruction set(see section 2). Table 2 lists some instructions from predefined instruction set of MetaCore system. The *cost* in Table 2 represents the hardware cost of each instructions in equivalent gate count. The *freq.(frequency)* denotes the number of occurrence of each instructions in profiles of benchmarks. The *CF(condensing factor)* represents substitution cost: the *bcr* can be substituted by the sequence of *immediate load* and *and* instructions.

The design decision based on the performance and cost can be made in various ways. For example, the MIPS project formulated this into their 1% rule[6]: *any instruction added for performance reasons has to provide a verifiable 1% performance gain over a range of application or else the instruction was rejected.* The acceptance of added instruction according to the 1% rule can be formulated as follows:

$$100 * \Delta C/C + 1 < 0 \qquad \text{eq.(1)}$$

The acceptance of each instruction used in this experiment is formulated as follows:

$$100 * \Delta C/C + \Delta G/G_m < 0 \qquad \text{eq.(2)}$$

Eq.(2) is an upgraded form of the 1% rule, where C and G mean cycle count and gate count, respectively. The G_m means average gate count necessary to implement each instruction. The G_m in the experiment is about $18064/40 \doteq 452$, where 18064 is the total gate counts necessary to implement the initial design specification(see Table 4), and 40 is the number of instructions in the initial design specification. The eq.(2) reflects the hardware cost directly in it. The acceptance of each instruction according to the eq.(2) is listed in the last column in Table 2.

inst.	cost	freq.	CF	performance effect(%)	acceptance
bcr	64	1390	2	0.018	accept
bf0	808	1926	10	0.128	reject
ext	140	119	8	0.006	reject
idiv	114	29330	12	2.335	accept
mrg	340	0	7	0.000	reject

Table 2: Acceptance of the predefined instructions according to their cost and performance effect.

From profiles of benchmarks, the frequent sequences of contiguous instructions was extracted. Further design refinement was performed based on this information. We have developed application-specific instructions which can condense the sequence of contiguous instructions. The results are listed in Table 3. For example, The SDIS1 and SDIS2 calculate square distance of two complex numbers, which is used for Viterbi equalizer to remove the ISI(Inter-Symbol Interference) of the channel. The SDIS1 and SDIS2 condense 10-cycle instruction sequence into 2 cycles. Thus the condensing factor is 8. The multiplication of *freq.* and *CF* represents the benefit of the instruction.

inst.	description	cost	freq.	CF
CMPY1(2)	complex MAC	2096	220464	4
SDIS1(2)	square distance	1808	271238	8
HDIS	hamming distance	87	99876	17
CVENC	convolutional encoding	306	48526	23
MIN,MAX	min, max	101	100034	3

Table 3: The user-defined instructions for GSM application.

Table 4 shows the summary of this experiment. The EM0 is the initial design containing all predefined instructions of MetaCore. The EM1 is the selected design by instruction substitution method under performance/cost constraint, given by eq.(2). The EM2 is the refined design containing application-specific instructions. The EM1 has about 20% advantage over EM0 in term of performance/cost matrix. The EM2 has 79% and 48% advantage over EM0 and EM1, respectively.

machine spec.	cycle count	cost
EM0	15069959	18064
EM1	15078241	14993
EM2	7819155	19391

Table 4: Design summary of EM0, EM1, and EM2.

We selected the EM2 as the final design for the given GSM application. From the ISA specification and user-defined macro-blocks in the form of HDL, MetaCore generated HDL codes for EM2 and a set of application development tools. The lower-level implementation, logic synthesis and placement and routing, was performed from the HDL codes and verified with benchmark programs. The EM2 was fabricated using $0.6\mu m$ TLM CMOS technology and has 55 MIPS peak performance.

5 Conclusions

In this paper, we have presented a full range of ASIP design automation system called MetaCore. It is demonstrated that by giving benchmark programs, MetaCore system can optimize its predefined instruction set and provide valuable information to guide the development of application-specific instructions. The formal specification-based design methodology enables MetaCore system to generate automatically the design equipped with the user-defined application-specific instructions. Future research on MetaCore system includes automatic synthesis of application-specific instructions, power optimization and supporting VLIW(Very-Long-Instruction-Word) architecture.

References

[1] E. A. Lee, “Programmable DSPs: a Brief Overview”, *IEEE Micro*, Oct. 1990, Vol. 10, No. 5, pp. 14-16.

[2] G. Intrater, D. Biran, “Application Specific Microprocessors”, *Proc. IEEE Int'l Conference on Computer Design: VLSI in Computers & Processors*, 1990.

[3] R. Woudsma, R. A. M. Beltman, et al.,“EPICS, a Flexible Approach to Embedded DSP Cores”, *Proc. Int'l Conference on Signal Processing Applications & Technology (ICSPAT)*, Oct. 1994, pp. 506-511.

[4] Jun Sato, et al., "PEAS-I: A Hardware/Software Codesign System for ASIP Development", *IEICE Trans. Fundamentals*, Mar. 1994, pp. 483-491.

[5] Mario R. Barbacci, "Instruction Set Processor Specifications(ISPS): The Notation and Its Applications", *IEEE Trans. CAD*, Jan. 1981, pp. 24-40.

[6] G. Kane. MIPS RISC architecture, Prentice-Hall, 1989.

A Case Study in Embedded System Design: an Engine Control Unit

Tullio Cuatto, Politecnico di Torino, Italy
Claudio Passerone, Politecnico di Torino – Cadence Europeans Labs, Italy
Luciano Lavagno, Politecnico di Torino – Cadence Europeans Labs, Italy
Attila Jurecska, Magneti Marelli, Venaria Reale, Italy
Antonino Damiano, Magneti Marelli, Venaria Reale, Italy
Claudio Sansoè, Politecnico di Torino, Italy
Alberto Sangiovanni-Vincentelli, Dept. of EECS, University of California at Berkeley, USA

Abstract

A number of techniques and software tools for embedded system design have been recently proposed. However, the current practice in the designer community is heavily based on manual techniques and on past experience rather than on a rigorous approach to design. To advance the state of the art it is important to address a number of relevant design problems and solve them to demonstrate the power of the new approaches.

We chose an industrial example in automotive electronics to validate our design methodology: an existing commercially available Engine Control Unit. We discuss in detail the specification, the implementation philosophy, and the architectural trade-off analysis. We analyze the results obtained with our approach and compare them with the existing design underlining the advantages offered by a systematic approach to embedded system design in terms of performance and design time.

1 Introduction

Hardware/software co-design and embedded system design techniques, such as those presented in [6], advocate a formal design procedure for embedded systems, based on unbiased specification, simulation-based validation, various forms of automated partitioning, and module and interface synthesis.

However, the design methodology followed in practice is far from being at the sophistication level implied by the approaches listed above. The common resistance offered by designers to innovation is made even stronger in the case of embedded system design because of the safety concerns and of the strict constraints on implementation costs. To demonstrate the applicability of the design methodology and tools that our group has proposed ([1]) we tackled an industrially relevant design: an Engine Control Unit for a commercial vehicle. The case study is relevant because it represents an actual product and because of its complexity. In particular, we were able to compare the results obtained with our design methodology with the present implementation showing its advantages both in terms of performance and design time.

DAC 98, San Francisco, California

The paper is organized as follows. In Section 2 we describe the specifications of our case study. In Section 3 we show how architectural decisions, such as partitioning and processor choice, are made for that example. In Section 4 we discuss some lessons learned in the course of this project, about specification styles and trade-offs. In Section 5 we draw some conclusions.

2 Specification of the ECU

In this paper we present an application of the *POLIS* [1] environment: the formal specification and hardware/software partitioning analysis of a functional subset of an Engine Control Unit (ECU), a typical automotive embedded system. We used as a reference an already existing device from Magneti Marelli, a worldwide supplier of automotive electronic components. Both a functional Structured Analysis specification [4], and a target architecture were available for this device, that was originally designed as a prototype to experiment with On-Board Diagnosis control strategies.

An electronic *Engine Control Unit* (ECU) consists of a set of sensors which periodically measure the engine status, an electronic unit which processes data coming from the sensors and drives the actuators, and the actuators themselves which execute the commands received from the control unit. A control strategy is implemented in the electronic unit to optimize the fuel injection and ignition; in particular it should minimize fuel consumption, minimize emissions of polluting substances and maximize torque and power, when possible. These requirements are usually competing, so the algorithm must find the best compromise for each situation.

The two main tasks of an ECU are the control of injection and ignition. The control specifications for these two tasks are as follows:

Injection : in order to burn completely and correctly the fuel, the ratio between the air and the fuel which go into each piston should be kept constant and close to the value 14.7 (for a gasoline engine). This is achieved by controlling the opening time of each injector.

Ignition : in order to give the fuel enough time to burn completely, the spark should be fired in advance with respect to the instant when the piston is at its highest point. This parameter also affects consumption and emissions, and it is basically computed from the engine RPM.

Both injection and ignition can be adapted dynamically with a very high precision, by processing the inputs signals coming from the sensors.

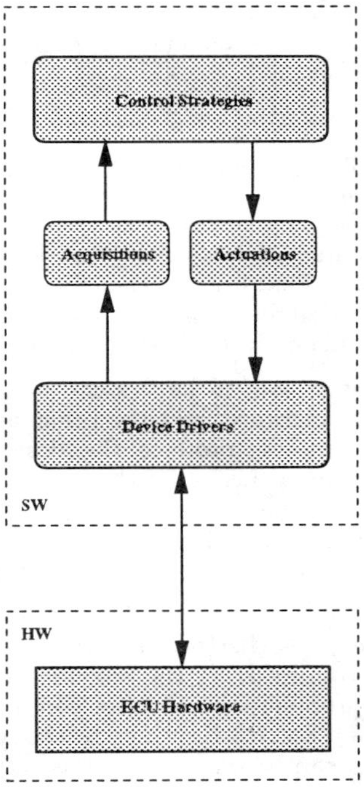

Figure 1: Functional architecture of an ECU

The functional architecture of the ECU is represented in Figure 1. The control layer implements the overall control strategy, aimed at satisfying the normative requirements in an optimal way. The signal acquisition and actuation layer transforms the raw data coming from the sensors (both binary-valued and derived from an A/D conversion) into filtered variables, scaled into the appropriate engineering units (like RPM) used by the control layer, while handling possible sensor errors and failures. The same layer also converts commands for the actuators, measured in engineering units (like seconds) into the appropriate signals (e.g. pulse streams) required by the actuation coils, lights and so on. The device driver layer further separates the signal acquisition and actuation layer, to provide some level of processor independence. The RTOS layer coordinates the overall functionality, by managing time-triggered (e.g., fixed-rate sensor sampling) and event-triggered (e.g., injection, triggered by the piston position) tasks.

The architecture of the embedded controller as implemented by Magneti Marelli contains a Motorola 68332 microprocessor (16 MHz), an ASIC implementing the PWM functions, 33 KBytes of used RAM and 194 KBytes of used ROM. The software written for the microcontroller consists of 35,000 lines written in C. The design effort was 8 man/year.

3 Design of the ECU

3.1 High level functional specification

Starting from the functional description of the ECU mentioned in the previous section, we chose a subset of relevant functionalities to implement using the *POLIS* environment. The ECU we implemented can manage the basic control operation of a 4 cylinder Multi-point Injection and Ignition System. We eliminated secondary tasks such as the interactions with the Catalytic Gas Sensor.

The control algorithm itself was taken from the original specifications untouched. This corresponds to the present organization of Magneti Marelli where two separate divisions handle the control algorithms and the actual implementation in terms of software and hardware modules. This separation favors some degree of architecture independence and the sharing of control algorithms between systems designed for different cars.

The specification has been partitioned into a hierarchical network of functional blocks, each described using the ESTEREL language. At the top level of the specification there are four main blocks:

1. acquisition of analog signals,
2. acquisition of frequency signals,
3. engine control strategies for the injection and ignition subsystems
4. control of the actuators.

The acquisition blocks can handle several types of sensors, including detection and recovery strategies in case of faulty sensors. The most important ones are the battery voltmeter, the intake manifold air pressure sensor, the throttle position sensor, the air and water temperature sensors, the position sensor of the phonic wheel and the camshaft position sensor.

After measuring and filtering the above external signals, the acquisition block calculates the values of system variables, such as RPM, air flow, derivatives of input signals, and so on. These variables are used by the control strategy layer, whose algorithm was taken from the original specifications almost untouched.

3.2 Functional simulation

Once the functionality of our system has been captured as a set of interacting Extended Finite State Machines, we can use *POLIS* to verify the behavior of the overall algorithm. In this step, timing is not considered since the implementation architecture has not been chosen, but instead a maximally parallel solution is analyzed.

The top level of the simulation model contains the testbench, using special functional blocks that model (in a simplified way, for reasons of performance) the engine and the sensors, thus providing inputs and monitoring outputs of the ECU.

The output of the simulation, as well as the state of several internal variables for debugging purposes, was monitored and compared with the expected data in order to determine the functional correctness of the specification.

3.3 Architecture Selection and Mapping

One of the most important component of our methodology is the architecture selection step. Part of the architecture selection step is the decision on how to partition the functional behavior into hardware and software. Another part is the organization of several software modules into tasks, to be executed together by the RTOS, based on expected input event frequency and priority. One last aspect is the selection of the scheduling algorithm (cyclic or priority-based, pre-emptive or non-pre-emptive, ...).

The performance of the simulation, in terms of simulated clock cycles per second on an unloaded ULTRASparc1, is described in Table 1. The column labeled "no graphics" was obtained by eliminating the graphical displays. The

engine speed (RPM)	cycles/second	
	graphics	no graphics
1000	366K	433K
6666	116K	216K

Table 1: Simulation performance

function	code size	
	manual	synthesized
acquisition	18K	13K
actuation	12K	12K

Table 2: Code size comparison

relatively small difference between the cases with an engine speed of 1000 RPM and 6666 RPM is due to the fact that only some acquisition tasks are triggered by the engine phase, while some are triggered by periodic timer interrupts.

While our system can perform an analysis of different architectural choices involving a number of microcontrollers, in this case, we selected the Motorola 68332 microcontroller, running at 16 MHz, with an on-chip Time Processing Unit (TPU [7]), to be able to compare our results with the already available ECU. Part of the TPU function was specified and simulated at the behavioral level, as described in [5]. The more complex functions implemented using the microprogramming capabilities of the TPU, like recognizing the engine phase from the phonic wheel sensor sample, were, on the other hand, modeled in ESTEREL.

Partitioning was performed by starting from the functional simulation model, in which every block has a delay of 1 clock cycle, and thus corresponds to a maximally parallel hardware implementation. We then moved selected groups of modules to software, based on the timing constraints and feedback from the simulation.

POLIS models functional components as Finite State Machines (extended with integer arithmetic capabilities) communicating via one-place buffers. Any time one of those buffers is overwritten, this is often a sign that a deadline has been violated, because the receiving FSM was too slow. Missed deadlines (in the form of lost events) are logged to a file, and can be used as a guidance for system performance analysis.

Since the system is composed of three main functional blocks, which in turn consist of several modules, we decided to start the partitioning process by changing the implementation of each unit as a whole; only after several experiments we had to refine our strategy by looking more closely at the lower levels of the hierarchy, to identify the critical path.

1. The first experiment was a complete hardware implementation. This is equivalent to the functional verification through simulation described in paragraph 3.1, and yielded the expected results. The system at 16 MHz flawlessly performs its task, but the implementation is very expensive due to the presence of many multipliers and dividers.

2. As a second step, the Analog Acquisition block was entirely implemented as software. The simulation showed that even with a slower CPU (i.e. 4 MHz), the results were correct.

3. The driver blocks were therefore also implemented as software, both for injection and ignition; however, the Frequency Acquisitions unit was left in hardware, and the low level counting task were assigned to the TPU. Even with the engine running at the highest speed (6600 RPM) no event was lost, and the algorithm always managed to correctly control the engine.

4. Except for the 68332 TPU, a full software implementation was then simulated. This time, even with the engine at a moderate speed, a lot of events coming from the crankshaft position sensor were lost. We identified two critical modules, those that filter the signals coming from the engine, since they are scheduled at the highest frequency in the entire system.

5. The next step was to implement one of the critical modules as hardware, and the other as software, but this again did not produce the desired result. By closely examining a plot of the task schedule (in the same form as shown in Figure 2), and comparing it with the missed deadline log, we discovered that the problem was the high latency introduced by the Round Robin scheduler we were using.

6. The solution was therefore to drive the acquisition of the signal coming from the engine through an interrupt, thus stopping the ongoing computation to start the Interrupt Service Routine. In this way, both critical modules can be implemented as software routines, without losing any functionality.

A plot of the scheduling charts with the engine running at full speed for the final partition is shown in Figure 2: the bars which reach the value 1 are the two interrupt routines, and the rest are all the other tasks.

The CPU load, due almost totally to the acquisition and actuation layer, is about 40% in the final partition. This number is close to the estimated load due to that layer in the manual design, thus indicating that the (estimated) performance of the synthesized code is very close to that of hand-written code.

The total code size in bytes (for the 68332 microcontroller) in the final partition is compared with the manual design in Table 2.

The entire partitioning process was performed within two days. The tool lets the designer change the implementation of each single block by just updating a parameter and then running a new simulation. No further compilation is required. Also the scheduling charts and a log of the missed deadlines are automatically generated, and were very helpful to drive the partitioning process.

4 Lessons learned and design guidelines

During this project, we had to choose a method to translate the existing formal functional specification, done with Structured Analysis ([4, 3]) into an executable specification using CFSMs. This required to solve a number of trade-offs about the *granularity* of the CFSMs.

One main aspect of the compilation mechanism used by ESTEREL (and hence by the *POLIS* front-end) is that it *fully abstracts* the user-given specification. This means that the control structure in the ESTEREL specification (e.g., `loop` and `await` statements) is translated into a Finite State Machine. The transition function of each CFSM is then represented as a Binary Decision Diagram [2], and the optimal sequence of `if`, `goto` and assignment statements is computed [1]. This is a very expensive procedure, that yields automatically results that are comparable with hand-optimized code, but is sensitive to module size.

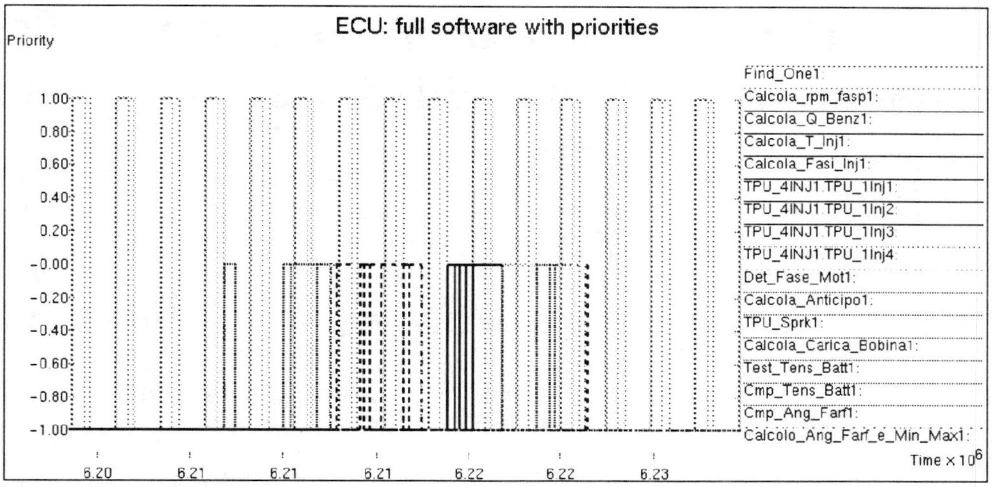

Figure 2: Task scheduling chart for the final implementation

Small modules are easier to understand, and guaranteed not to blow up at compilation time. However, the resulting code, with one procedure per module, can impose an excessive overhead in terms of code size and execution time on the target processor, due to the procedure call and return sequences of short procedures.

On the other hand, large modules offer better opportunities for optimization as described above, but may require too much memory or CPU time during compilation.

POLIS, however, provides the designer with several techniques for exploring these trade-offs in an assisted manner. In particular, it provides methods for

- chaining the execution of several modules, without returning control to the Real Time Operating System,
- merging several CFSMs, by computing their synchronous product, into a single CFSM.

The first technique is relatively cheap, does not cause any potential blowup in code size, and is especially suitable for data-intensive modules. The second technique is especially useful for control-dominated modules with a high degree of interaction, because it abstracts away all the internal communication between the collapsed CFSMs, thus yielding a dramatically simpler implementation in several cases.

For example, we applied the first technique to the submodule Test_Tens_Batt, which consists of two cascaded CFSMs. The first one checks the value of the battery voltage and implements a recovery strategy, while the second one computes the value of an internal variable. In this way we speeded up the code by 11% while keeping the same code size. We estimated that this kind of optimization can be applied to about 20% of the modules of the design.

In another case, we resorted to the opposite transformation, that is splitting a single CFSM, that was updating several variables concurrently, into four cascaded CFSMs, reducing the total code size from 11 Kbytes to 1.5 Kbytes.

5 Conclusions

New methodologies are required in the design of digital electronic systems, due to the increased complexity and reduced time-to-market. Moreover, a product should be flexible to adapt to changes during its lifetime, which is best obtained by using software, and must also meet tight timing constraints, which is most suitable for hardware components. Partitioning between hardware and software is therefore a critical step in the design flow, and CAD tools should help the designer in making the right choices.

In this paper, we have shown how to perform partitioning by using fast co-simulation and software estimation. We have also shown how to optimize the system with respect to code size and running time, by selectively collapsing or dividing modules, and by moving the threshold between control and data-flow in conditional statements. The results obtained are very promising, and were achieved on a real design and in a relative short period of time.

References

[1] F. Balarin, E. Sentovich, M. Chiodo, P. Giusto, H. Hsieh, B. Tabbara, A. Jurecska, L. Lavagno, C. Passerone, K. Suzuki, and A. Sangiovanni-Vincentelli. *Hardware-Software Co-design of Embedded Systems – The POLIS experience.* Kluwer Academic Publishers, 1997.

[2] R. Bryant. Graph-based algorithms for boolean function manipulation. *IEEE Transactions on Computers*, C-35(8):677–691, August 1986.

[3] A. Damiano and P. Mortara. Problematiche software nei sistemi elettronici per applicazioni automotive. *Alta Frequenza – Rivista di Elettronica*, 7(3):10–16, May-June 1993.

[4] T. DeMarco. *Structured Analysis and System Specification.* Yourdon Press, USA, 1988.

[5] H. Hsieh, L. Lavagno, C. Passerone, C. Sansoè, and A. Sangiovanni-Vincentelli. Modeling micro-controller peripherals for high-level co-simulation and synthesis. In *Proceedings of the International Workshop on Hardware-Software Codesign*, March 1997.

[6] G. De Micheli and M. G. Sami, editors. *Nato Advanced Study Insitute on Hardware/Software Codesign.* Kluwer Academic, 1996.

[7] Motorola Inc. *M68300 Family. TPU Time Processor Unit. Reference Manual*, 1990.

HW/SW CoVerification Performance Estimation & Benchmark for a 24 Embedded RISC Core Design

Thomas W. Albrecht
Siemens, Austria
Erdberger Laende 26
A-1030 Vienna, Austria
+43 1 1707 - 35850
thomas.albrecht@siemens.at

Johann Notbauer
Siemens, Austria
Erdberger Laende 26
A-1030 Vienna, Austria
+43 1 1707 - 36087
johann.notbauer@siemens.at

Stefan Rohringer
Siemens, Austria
Erdberger Laende 26
A-1030 Vienna, Austria
+43 1 1707 - 37630
stefan.rohringer@siemens.at

1 ABSTRACT

This paper describes the benchmarking of a HW/SW-coverification design strategy. The benchmark results were the base for making a principal verification decision for an already ongoing project at Siemens AG, Public Communication Network Group. The intention for this benchmark was to verify whether commercial available coverification tools can handle the design complexity of an embedded system containing 24 embedded RISC cores and provides the necessary performance in terms of simulation speed and throughput.

2 INTRODUCTION

2.1 Why Aiming HW/SW-CoVerification

The design to be coverified consists of

- 3 different types of ASICs each containing an embedded RISC core, SDRAM arbiter, protocol specific controller and interface to serve the board external interface, system internal controller and interface to serve board-to-board serial links. The gate count of a single ASIC is about 250kGates each.
- 3 different types of printed-circuit boards each containing a number of one type ASIC and off-the-shelf-components.
- Each type of those boards has one of the following external interfaces:
  200 MHz ATM serial interface, 8x HDLC serial interfaces, proprietary system interface
- 2 boards of each type in the minimum configuration (equals 24 embedded core ASICs) and a sum of 26 boards in the maximum configuration (equals 76 embedded core ASICs) will be delivered to our customer.
- Each board is connected by a serial link with each other of the boards.

At the time when the benchmark was carried out the design described above was in specification phase.

The reasons for doing the benchmark and to ensure high performance coverification capabilities for the design are several:

- No visibility and controllability of the RISC core's signals since they are all inside ASICs
- Limited access via a special type of In-Circuit-Emulator
- Early Integration of HW and SW on a virtual base
- Higher SW quality available for HW prototype tests
- „Breaking“ the wall between SW- and HW-engineers from the system specification phase on
- Very tight project schedule

The goal of our coverification is to simulate the whole SW startup routine, operating system, HW-device driver and protocol stack with reasonable performance in the minimum configuration of the system to be delivered to our customer - this means to coverify a 24-core embedded RISC core design to be the largest coverification configuration. This design configuration equals to 6,000 kGates.

2.2 Performance Factors in HW/SW CoVerification

The traditional verification methodology [7] is based on simulation of the design mainly described in a HDL. Processors are either modeled in a HDL or they are integrated into simulation via a hardware model. Both of them have the disadvantages that the simulation performance is mainly given by the processor and that there is no possibility to connect a debugger to the processor.

Figure 1 shows a principal block diagram of a HW/SW-coverification tool [1].

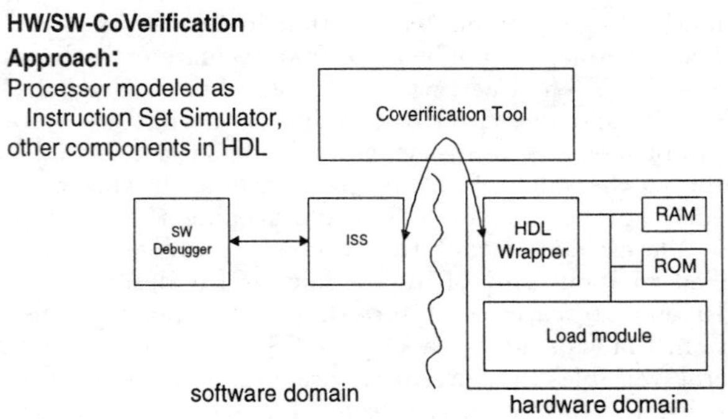

Figure 1 Blockdiagram of a HW/SW-coverification tool

The HDL model of the RISC core is replaced by an InstructionSetSimulator (ISS) of the core. The ISS is connected to the HW simulator by the coverification tool and instanciated

DAC 98, San Francisco, California

by a HDL wrapper. This HDL wrapper has to behave like a bus-functional model controlled by the ISS in order to provide the HW-design with cycles. The remaining HW is still modeled in the HDL simulator. Beside the execution unit the ISS is also able to model local memory areas. The key of simulation speedup is to reduce the number of events in the HDL-simulator. The percentage of memory cycles running across the HW-simulator to the HDL-type of memory model is called „core-I/O-activity". 100% core-I/O-activity means every single memory access is running across the HW-simulator, no portion of memory is modeled inside the ISS. The portion of memory area to be modeled either in the ISS or to be accessed via the HW-simulator to the HDL-type of memory must be defined in a mapping file accordingly to the design's memory layout.

The ISS has also an interface to a commercial SW-debugger. This is beside simulation throughput essential to ensure acceptance by SW engineers.

3 ESTIMATION OF REQUIRED HW/SW COVERIFICATION PERFORMANCE

A set of coverification testcases has been defined for different HW-configurations (see Table 1). For each testcase a core-I/O-activity and the number of assembler commands has been assumed.

Testcases	# embedded cores	assumed core-I/O activity	# Assembler instructions per testcase	# memory accesses	# memory cycles running accross the HW-simulator
Driver	1	10	50.000	5.000	30.000
Startup	1	27	14.000	3.780	22.680
Diagnostics	1	10	200.000	20.000	120.000
RTOS	1	1	17.000	170	1.020
ASIC integration, 1 core	1	10	65.000	6.500	39.000
system integration, 2 cores	2	10	65.000	13.000	78.000
system integration, 8 cores	8	10	65.000	52.000	312.000
system integration, 24 cores	24	10	65.000	156.000	936.000

Table 1: Testcase Scenarios

In order to stress the protocol stack SW a certain amount of data has to be transferred via the serial interfaces (ATM or HDLC) of the boards into the design. For 1-,2-, 8- and 24-core integration testcases it has been considered to play 50 HDLC frames, each with a message length of 70 byte, and to play 50 ATM cells to the design. It has also been considered that the HDLC interface will be accelerated for simulation purposes by 10x. For this acceleration the original VHDL HDLC interface block will be replaced in the VHDL database by a speeded-up VHDL HDLC interface block that allows to run the HDLC-protocol 10 times faster in simulation than in reality.

It is important to consider the impact of the serial interfaces with respect to the real design because protocol payload will be stored in a common memory area. Therefore this memory area must be mapped to the HDL-type of memory modeled in the HW-simulator rather than in the local ISS memory.

Table 2 gives the estimated performance numbers in 1000 Instructions-per-second and simulation run time for a scenario of simulated clock-cycles-per second and for a given ISS performance.

System Clock (MHz)		50						
ISS performance (1.000 instr/se		100						
speed up of seriell HDLC links		10	max speed up with min add. effort					
Scenarios	Clock cycles simulated/sec, 1-core configuration							
	10	20	30	40	50	100	150	200
Results								
CoVerification Performance (IPS)								
Driver	17	33	50	67	83	166	249	332
Start up	6	12	19	25	31	62	93	123
Diagnostics	17	33	50	67	83	166	249	332
RTOS	166	332	498	662	826	1.639	2.439	3.226
ASIC integration	6	13	19	25	31	63	94	126
System integr. 2 cores	3	6	9	13	16	31	47	63
System integr. 8 cores	1	2	2	3	4	8	12	16
System integr. 24 cores	0	1	1	1	1	3	4	5
CoVerification Run Time (min)								
Driver	50	25	17	13	10	5	3	3
Start up	38	19	13	9	8	4	3	2
Diagnostics	200	100	67	50	40	20	13	10
RTOS	2	1	1	0	0	0	0	0
Asic integration	172	86	57	43	34	17	11	9
System integr. 2 cores	345	172	115	86	69	34	23	17
System integr. 8 cores	1.378	689	459	345	276	138	92	69
System integr. 24 cores	4.135	2.037	1.378	1.034	827	413	276	207

Table 2 Performance Estimation Sheet

Most of the testcases will be carried out of the ASIC-integration testcase group. A simulation run time of 10 to 20 minutes seems to be reasonable for this 1-core configuration. Using the performance estimation sheet it became obvious that the HW simulator must provide a performance in the range of 150 to 200 simulated clock cycles per second.
It turns out that the HW-simulator is the bottleneck in overall coverification performance. Therefore we have considered a cycle-based HDL simulator in our benchmark as well.

A variation of the ISS performance shows that at a given HW-simulator performance the ISS has not a substantial impact on the overall coverification performance. Only for testcases having a very low core-I/O-activity (around 1%) the ISS performance has some impact on the overall performance when the ISS is low performant (<5k Instructions per second).

4 THE BENCHMARK

4.1 The Benchmark Design

Due to the lack of having the real design available as described above, we had the need to build up a benchmark design.
The benchmark design is modeled in a way allowing easy scaling of the hardware's complexity. We decided to build up a basic module consisting of all the components necessary for a successful simulation run. Essentially these components are the RISC core, a clock generation unit, some memories and a load

module. The scaling of the hardware's complexity is done by simply multiple instanciations of the above mentioned basic module. That means, our benchmark design consists of 24 slots, each of them capable to embed a basic module. The number of basic modules in a specific configuration of the benchmark design is determined by a specific HDL configuration.

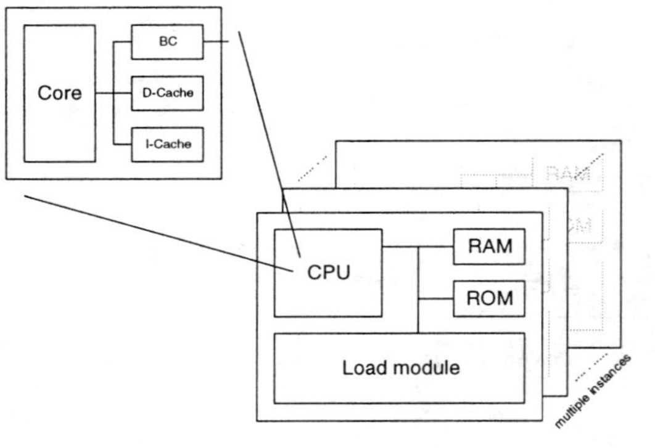

Figure 2 The Benchmark Design Configurations

When replacing the HDL model of the CPU by an ISS the load for the hardware simulator is only given by the memory models. In order to get a more realistic modeling of the target design we added a so-called load module to the main components of the basic module. It's aim is to represent the load of the hardware simulator generated by the specific interfaces (ATM, HDLC, ...) in the target design. The gate complexity of the load module (approx. 250 kGates) was chosen to be approximately equivalent to the estimated gate count numbers of the target design. The benchmark simulations were done with configurations consisting of 1, 2, 8 and 24 cores (basic modules).

4.2 Benchmark Testcases

The Instruction Set Simulator of the RISC core gives the possibility to map specific memory ranges either in the workstations memory (internal memory) or in the memory simulated by the hardware simulator (external memory). Accesses to the internal memory do not require any bus cycles and can therefore be done very fast. Typically code fragments are stored in this internal memory. The mapping of specific memory ranges to the external memory is necessary for data exchange between the software and the hardware.

The Benchmark testcases are designed to allow an easy modification of the access rate to the external memory. There are two loops which are performing read and write cycles to the both memories. By changing the maximum loop counts the density of the access cycles to the external memory can be adapted. The benchmark simulation runs were made for the following core-I/O-activities: 0%, 1%, 5%, 10% and 100 %.

The benchmark testcases have been defined in a way that the verification runs provide answers to the following questions:

- What is the simulation performance that can be achieved in HW/SW-CoVerification? Is it acceptable for SW engineers?
- What's the difference of the simulation performance when using either a HDL model for the processor or an ISS?
- How large is the performance gain when using the cycle based simulator Cyclone instead of the event driven QuickHDL?
- Is there a linear dependence between number of cores and simulated clock cycles per second?
- Is there a linear growth of the simulation runtime with the core-I/O-activity?
- Is there an impact from the number of CPUs in the workstation used for the simulation?

A total of 120 simulation runs have been done in order to present the performance data given in this paper.

4.3 Tools that Have Been Used

The following commercial available tools have been used:
EagleI CoVerification Tool by Viewlogic [1]
Cycle-Based Simulator CYCLONE by Synopsys [2] [3]
Event-Driven Simulator QuickHDL by MentorGraphics
ISS MiniSim by LSI Logic [4]
SW-Debugger CrossView by Tasking [6]

5 BENCHMARK RESULTS

Benchmark simulations were done in various configurations (Table 3) to answer the questions from section 4.2.

The carried out benchmark simulations gave clear performance figures and answers to the above mentioned questions:

- The required HW-simulator performance of 150 to 200 cycles-per-second and therefore the overall coverification performance has been achieved when using Cyclone as the HW-simulator (Figure 3 and Figure 4)
- The simulation performance when using the ISS of the processor instead of the VHDL model is twice as high (Figure 3 and Figure 4).
- There is a linear reciprocal dependency between the number of cores in the design and the number of simulated clock cycles per second, double number of cores half number of simulated clock cycles (Figure 4).
- The verification run-time increases linear to the core-I/O-activity.
- The HW-simulator performance in cycles-per-second is independent of the core-I/O activity. The core-I/O activity has its impact on the overall verification runtime. The higher the core-I/O activity the more memory cycles must run across the HW-simulator.
- There is a very little impact of the number of CPUs in the simulation workstation on the simulation drun-time. The main part of the simulation effort is consumed by the hardware simulation process which can not be splitted to run on different CPUs.(Figure 3 and Figure 4)
- The performance gain when using the Synopsys' Cyclone (cycle-bases simulator) instead of Mentor's QuickHDL (event-driven simulator) is approximately 10x when simulating in Cyclone 4 state debug mode, and approximately 20x when simulating in 4 state performance mode (Figure 3 and Figure 4).

Case 1	QuickHDL, pure VHDL
Case 2	QuickHDL, ISS, 1 CPU WS
Case 3	QuickHDL, ISS, 2 CPU WS
Case 4	Cyclone 4state, debug, ISS, 2 CPU WS
Case 5	Cyclone 4state, perf, ISS, 1 CPU WS

Table 3: Legend for the cases in Figure 3 and Figure 4

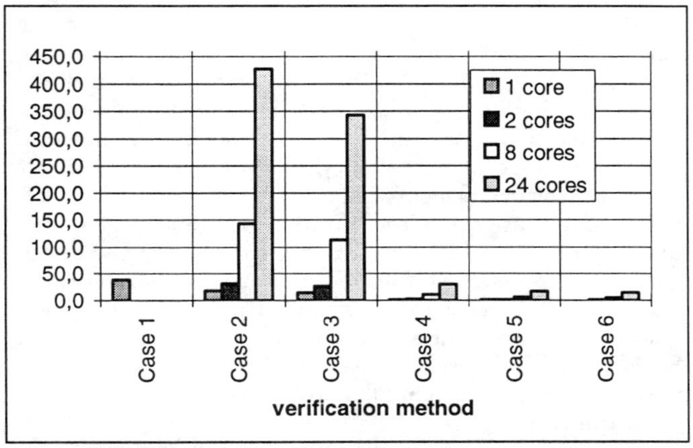

Figure 3 Verification RunTime in minutes depending on # of cores and verfication toolset at a given core I/O-activity of 10%

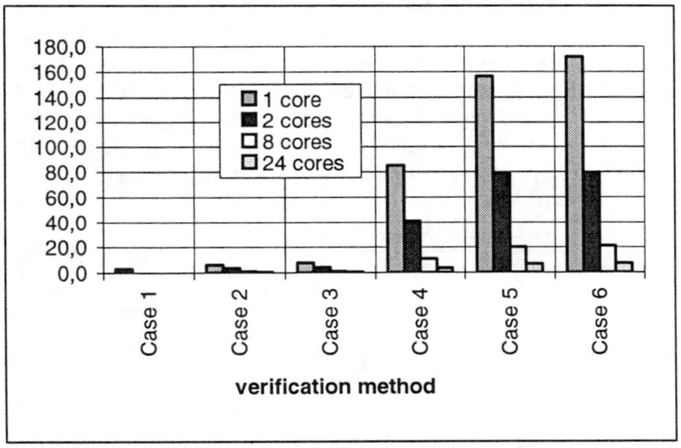

Figure 4 HW-simulator performance in simulated clock-cycles per second depending on # of cores and verification toolset

6 CONCLUSIONS

It turned out that HW/SW-coverification is a very powerful verification strategy. It has been decided to use HW/SW-coverfication in the ongoing project. It enables SW-engineers to start HW/SW integration already on a virtual base with no need to wait for physical HW samples. Also HW engineers make use of the cosimulation environment to debug their testcases written in Assembler or C-code to stimulate the design from the processor core site.

Despite the speed in terms of throughput performance that can be achieved with emulation, it has been decided not to use emulation for this project.

The reasons for this decision were:

- not capable to emulate a 24 core embedded RISC core design
- Emulation database and HW environment cannot be replicated very easily
- Emulation maintenance effort is high compared to a virtual based solution on workstations
- Man power costs for emulation-type of printed-circuit boards
- Emulation not flexible, e.g. in case of late ASIC-pinout changes (respin of boards required, ...)
- Emulation cannot start that early in the design process than HW/SW coverification can do, even for this type of system design.

The conclusion is that for this type of complex system designs the cost parameter for emulation is significantly higher than for HW/SW-CoVerification.

At the time this paper has been submitted are already first results of our real-world HW/SW-coverification application available. The coverification performance achieved using Eaglei/Cyclone is slightly better than the benmark results.

7 REFERENCES

[1] Viewlogic Systems Inc, „Hardware/Software Co-development with Eagle Tools", http://www.viewlogic.com/products/eagletools.html

[2] Synopsys Inc, „Cyclone VHDL Coding Style Guide V1.1b", 1997

[3] Synopsys Inc. „Cyclone VHDL Reference V1.1b", 1997

[4] LSI Logic, „MiniSIM Architectural Simulator Family User's Guide", November 1995

[5] Gerry Kane, Joe Heinrich, „MIPS RISC Architecture", Prentice Hall 1992 Mips Technology

[6] Tasking Inc. „R3000 v3.2 CrossView Debugger User's Guide", 1996

[7] Thomas Albrecht, „Concurrent Design Methodology and Configuration Management of the Siemens EWSD-CCS7E Processor System Simulation", Proceedings of the 32nd ACM/IEEE Design Automation Conference, pages 222-227

[8] Matthias Bauer, Wolfgang Ecker, „Hardware/Software Co-Simulation in a VHDL-Based Test Bench Approach", Proceedings of the 34th ACM/IEEE Design Automation Conference, Pages 774-779

System-Level Exploration with SpecSyn

Daniel D. Gajski
Inform. and Computer Science
University of California
Irvine, CA 92717
gajski@ics.uci.edu

Frank Vahid
Computer Science and Eng.
University of California
Riverside, CA 92521
vahid@cs.ucr.edu

Sanjiv Narayan
Ambit Design Systems
2500 Augustine Drive
Santa Clara, CA 95054
sanjiv@ambit.com

Jie Gong
Qualcomm Inc
6455 Lusk Boulevard
San Diego, CA 92121
jiej@qualcomm.com

Abstract

We present the SpecSyn system-level design environment supporting the specify-explore-refine (SER) design paradigm. This three-step approach includes precise specification of system functionality, rapid exploration of numerous system-level design options, and refinement of the specification into one reflecting the chosen option. A system-level design option consists of an allocation of system components like standard and custom processors, and a partitioning of functionality among those components. Focusing on SpecSyn's exploration techniques, we emphasize its two-phase estimation approach and highlight experiments using SpecSyn.

1 Introduction

The focus of design effort on higher abstraction levels, driven by increasing system complexity, shorter design times, and migration of entire systems onto a single chip, demands a system-level design methodology and supporting tools. We can isolate three tasks in such a methodology. First, we must specify the system's functionality and constraints. Second, we must explore various system-level design alternatives, each an interconnection of system components and an assignment of functionality to them. System components include standard software processors, custom hardware processors, memories, and buses. Third, we must refine the original specification into a new system-level description with components. Subsequently, we implement each component, where software components require compilation and possibly thread scheduling, while hardware components require behavioral and register-transfer synthesis.

In current practice, these three tasks are carried out in an informal manner. The *specify-explore-refine* paradigm aims to provide a more precise approach to these tasks, enabling automated assistance. We have developed the SpecSyn environment, illustrated in Figure 1, to support the specify-explore-refine paradigm. In this paper, we focus on SpecSyn's design exploration.

2 Related work

Several system-level design environments have evolved recently. TOSCA [1] focuses on control-dominated systems. It parses a hierarchical finite-state machine into a process algebra internal format, which it partitions among system components manually or by clustering, supporting some formal transformations. It outputs software in a virtual instruction set to achieve processor independence. COSYMA [2] focuses on microcontroller-based systems. It converts an extended C input to a syntax graph, dynamically modifying the graph's granularity [3]. Fast indirect design metrics guide a simulated annealing partitioning among a software and hardware component with shared memory, with direct design metrics from a subsequent implementation guiding further iterations. Vulcan II [4] uses a similar architecture and applies a greedy partitioning heuristic with fast indirect metrics. Ptolemy [5] focuses on cosimulating inputs from different computation models, but also partitions a task-level dataflow graph among a similar architecture using a custom heuristic. Approaches in [6, 7] simultaneously allocate processors from a diverse library while partitioning a task graph among them. Summaries of these and other approaches can be found in [8]. Recent ideas on the system exploration problem were provided in [9], and to a refinement-based approach in [10].

SpecSyn possesses several unique features. First, SpecSyn uses a two-level estimation method to obtain fast yet accurate estimations, using direct design metrics, like total hardware size or process execution time, rather than indirect metrics. Second, SpecSyn supports a variety of system architectures, heuristics, estimation models, and cost functions; no one version of any of these is advocated for all systems. For example, a suite of heuristics is provided, and new ones can be added. Third, SpecSyn outputs a system-level description that ideally can be simulated, edited, and synthesized, thus supporting the SER methodology.

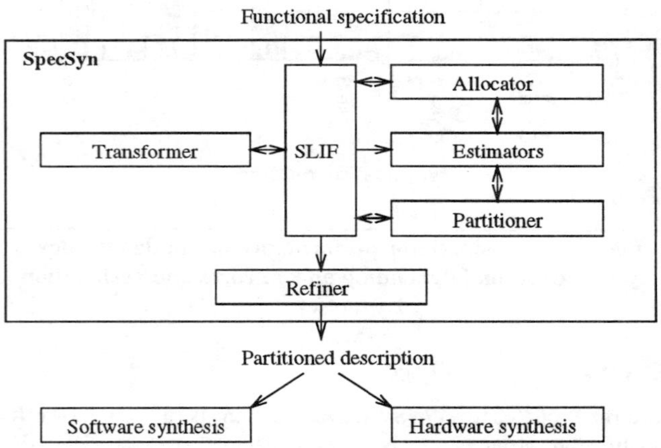

Fig. 1: The SpecSyn system-design environment

3 Specification

3.1 Computation model

We have focused on the Program-State Machine (PSM) computation model [11]. PSM can be thought of as Statecharts [12] with arbitrarily complex sequential programs in the leaf states. Thus, PSM represents a powerful computation model for both hardware and software, able to describe

DAC 98, San Francisco, California

hierarchical/concurrent finite-state machines, a sequential program, or even communicating sequential processes.

The designer can specify constraints on behavior execution times and channel bitrates. Other design constraints, such as a component size and I/O limitations, are derived from each component's library entry.

3.2 Internal representation

The specification must be converted into an internal representation on which subsequent tools can operate. Because system specifications tend to be highly procedural, we use a representation similar to a call-graph used in software compilation, introduced by an example. Figure 2 shows a partial VHDL specification of a fuzzy-logic controller. Inputs *in1* and *in2* must be converted to output *out1* using fuzzy logic. The main process *FuzzyMain* repeatedly reads inputs into variables *in1val* and *in2val*, twice calls procedure *EvalRule* to fill arrays (*tmr1* or *tmr2*) based on the input and on another predefined array (*mr1* or *mr2*), convolves the *tmr* arrays, and computes and outputs a centroid value.

```
entity FuzzyControllerE is
  port ( in1, in2 : in integer;  out1:  out integer );
end;
...
FuzzyMain: process
    variable in1val, in2val : integer;
    type mr_array is array (1 to 384) of integer;
    variable mr1, mr2: mr_array; -- membership rules
    type tmr_array is array (1 to 128) of integer;
    variable tmr1, tmr2: tmr_array; -- truncated memb. rules
    function Min ...
    ...
begin
    in1val := in1;  in2val := in2;
    EvaluateRule(1);
    EvaluateRule(2);
    Convolve;
    out1 <= ComputeCentroid;
    wait until ...
end process;

procedure EvaluateRule(num : in integer) is
  variable trunc : integer;  -- truncated value
begin
  if (num = 1) then
    trunc := Min(mr1(in1val), mr1(128+in1val));
  elsif (num = 2) then
    trunc := Min(mr2(in2val), mr2(128+in2val));
  end if;

  for i in 1 to 128 loop
    if (num = 1) then
      tmr1(i) := Min(trunc, mr1(256+i));
    elsif (num = 2) then
      tmr2(i) := Min(trunc, mr2(256+i));
    end if;
  end loop;
end;
```

Fig. 2: Fuzzy-logic controller example.

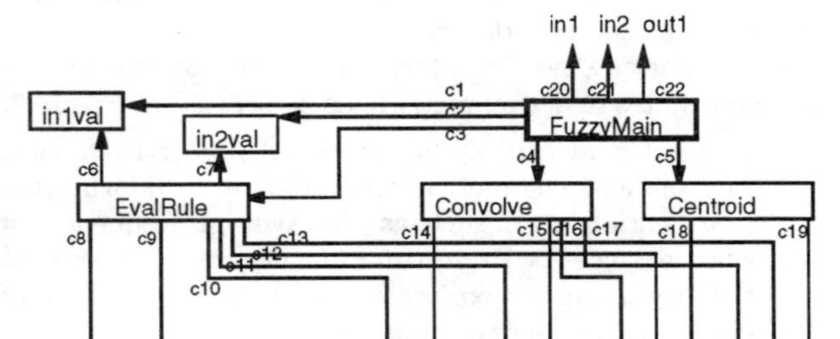

Fig. 3: Access graph for the example.

Figure 3 shows the internal representation. Each graph node represents a **behavior** or **variable**, where a behavior is a process or procedure, though for finer granularity we can exline [13] statement blocks like loops into new procedures. Each graph directed-edge represents a communication **channel**, which is a procedure call, a variable or port read or write, or a message pass specified using send/receive constructs. For example, process *FuzzyMain*, procedure *EvalRule* and variable *in1val* are each represented by a node. The write of *in1val* in *FuzzyMain* translates to a single edge, while the two calls of *EvalRule* by *FuzzyMain* translate to another single edge. We call this the **Specification-level intermediate format (SLIF)**, since

Object	ict.8051	ict.xc4020	size.8051	size.xc4020
FuzzyMain	6	8	80	500
in1val	0	0	2	80
in2val	0	0	2	80
EvalRule	778	522	500	1600
Convolve	800	600	900	2000
...				

Object	accreq	bits	src	dst
c1	1	6	FuzzyMain	in1val
c2	1	6	FuzzyMain	in2val
c3	2	9	FuzzyMain	EvalRule
c4	1	0	FuzzyMain	Convolve
c5	1	0	FuzzyMain	Centroid
c6	1	16	EvalRule	in1val
...				

Fig. 4: Slif behavior/variable and channel annotations

its granularity is that of objects explicit in the specification. The SLIF part shown above is an **access graph**, or **AG**, since it represents the accesses among objects.

The SLIF is annotated with numerous values, as shown in Figure 4. A behavior and variable object has a list of size weights, one weight for each component type to which the object may be assigned. For example, a variable object has the number of memory words for each library memory component, while a behavior has square microns, gates, combinational-logic blocks, and bytes for each custom chip, ASIC, FPGA, and standard processor, respectively, in the component library. Behaviors and variables are annotated with computation times on each component. Each edge has access frequency weights obtained through profiling, and a bit weight representing the number of bits per transfer. Annotations are computed during pre-estimation, and are combined into quality metric estimates during online estimation, as discussed in Section 4.3.

4 Exploration

Exploration is the task of finding a set of potential architectures that satisfy constrained metrics and optimize other metrics. It consists of allocation, partitioning, transformation and estimation. These problems can be solved in various orders, and we usually iterate several times.

4.1 Allocation

Allocation is the task of adding components to the design. The SpecSyn allocator permits allocation of any number of standard processors, custom processors, memories, and buses. Each component is characterized in a library by its constraints, and by a technology file. For example, a custom processor might be characterized by the maximum I/O pins and gates, and by a technology file describing an RT-component library. A standard processor is characterized by a maximum program memory size, a bus size, a maximum bus bitrate, and a technology file describing how to map a generic instruction set to the processor's instruction set [11]. A memory is characterized by the number of ports, number of words, word width, and access time. A bus is characterized by the number of wires, protocol, and maximum bitrates. Allocation is currently manual, though simple scripts can be used to automatically sequence through numerous possible allocations and apply partitioning.

Figure 5 demonstrates an example allocation. *StandardProc1* is an Intel 8051 with 4 kilobytes of on-chip memory, and *CustomProc1* is a Xilinx XC4010 FPGA with 160 I/O pins and 10,000 gates. Two 1 kbyte memories are also allocated.

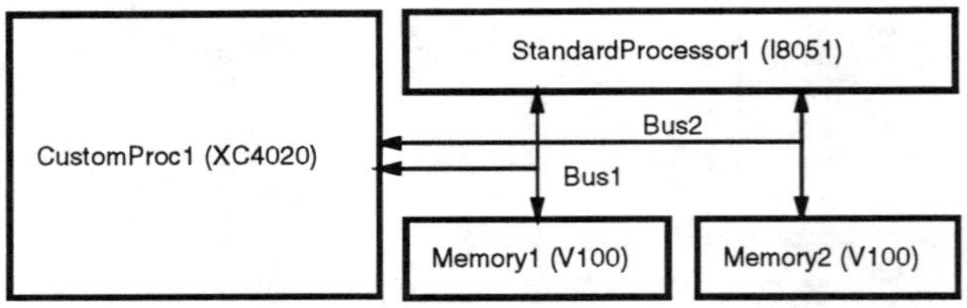

Fig. 5: An example allocation of components

4.2 Partitioning

Given a functional specification and an allocation of system components, we need to partition the specification and assign each part to one of the allocated components. In fact, we can distinguish three types of *functional objects* that must be partitioned. *Variables* store data values and are assigned to memory components. *Behaviors* transform data values and are assigned to processors. *Channels* transfer data from one behavior to another and are assigned to buses.

SpecSyn uses a partitioning engine that takes a data structure, cost function, and data structure update procedure as inputs, and applies any of numerous built-in heuristics. In this way, heuristics, data structures and cost functions stay distinct, and thus can easily be added or improved.

4.2.1 Manual partitioning and hints

SpecSyn supports designer interaction by providing the ability to manually relocate objects, allowing user control of the relative weights of various metrics in the cost function, and automatically providing hints of what changes might yield improvements to the current partition. SpecSyn currently supports two types of hints.

Closeness hints are based on a weighted function of various closeness metrics. Seven behavior closeness metrics have been defined. For example, *connectivity* is based on the number of wires shared between the sets of behaviors. Grouping behaviors that share wires should result in fewer pins. Three variable/channel closeness metrics have also been defined, such as *sequentiality of access* of the variables/channels by behaviors, which aims to group items to avoid resource conflicts and performance degradation.

The second type of hints are *lookahead* hints. Here, we generate all possible n modifications of the current partition, where an n modification is a sequence of n moves of any objects from one group to another (n is user-defined). We again provide a list of such modifications, sorted by the partition improvement gained by each as measured by a cost function.

4.2.2 Cost functions

Partitioning heuristics are guided by cost functions. A variety of cost functions can be supported. The following supported cost function focuses on satisfying constraints:

$$\begin{aligned} Costfct \;=\;& k_1 \cdot F(c1.size, c1.size\_constr) \\ +\;& k_2 \cdot F(c2.size, c2.size\_constr) \\ +\;& k_3 \cdot F(c1.IO, c1.IO\_constr) \\ +\;& k_4 \cdot F(b1.exectime, b1.exectime\_constr) \\ & \ldots \end{aligned} \qquad (1)$$

where the k's are user-provided constants indicating the relative importance of each metric, F is a function indicating the desirability of a metric's value, $c1, c2$ are components, and $b1$ is a behavior. A common form of F returns the degree of constraint violation, normalized such that 0 indicates no violation, and 1 indicates very large violation.

The above cost function is very general, permitting us to satisfy constraints as well as to optimize certain metrics (through heavier weights), without requiring specific knowledge in a heuristic of the constraints or optimization metrics.

As an example of the results of partitioning, Figure 6 shows a partition of several of the previous example's nodes among two memories, a custom processor, a standard processor and a bus. Note that four communication channels have been partitioned onto *bus1*.

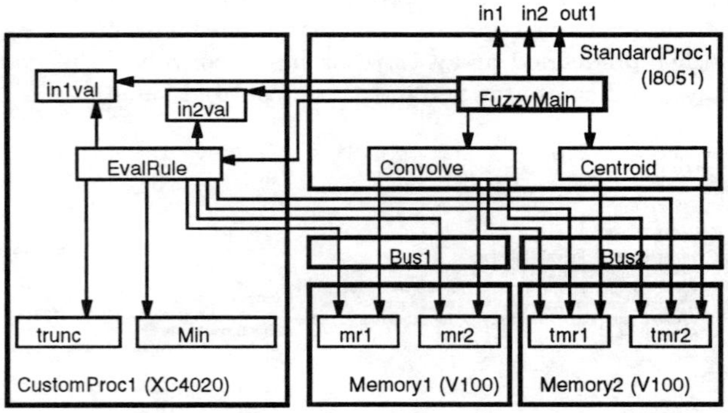

Fig. 6: Partitioning AG nodes among components

4.3 Estimation

Estimation of design quality metrics is required to determine if a particular system-level design (a partition of functions among allocated components) satisfies constraints, and to compare alternative designs.

SpecSyn uses a two-level technique to obtain fast yet accurate estimates of design metrics, as illustrated in Figure 7:

1. **Pre-estimation**: Each functional object (behavior, variable and channel) is annotated with information (see Section 3.2), such as the average frequency of channel access. Pre-estimation occurs only once at the beginning of exploration and is independent of any particular partition and allocation.
2. **Online-estimation**: Pre-estimated annotations are combined in complex expressions to rapidly obtain metric values for a particular partition and allocation (usually in constant time [14]). Online-estimation occurs hundreds or thousands of times during manual or automated exploration.

In most other approaches, exploration consists of only one level of estimation, with another level coming only after RT-level design. We now discuss SpecSyn estimation models for three metric types: performance, hardware size, and software size.

4.3.1 Performance

In SpecSyn's performance model, a behavior's execution time is calculated as the sum of the behavior's *internal computation time (ict)* and *communication time*. The *ict* is the

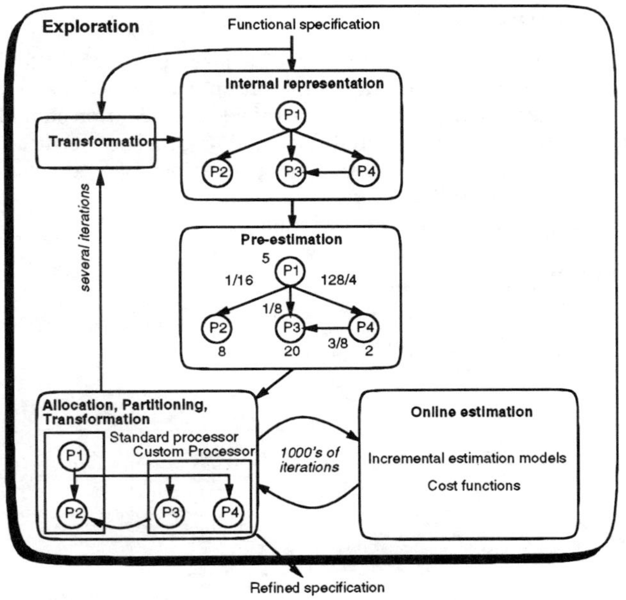

Fig. 7: Subtasks during exploration

execution time on a particular component, assuming all accesses to other behaviors and variables take zero time. The communication time includes time to transfer data to/from accessed behaviors and variables, as well as the time for such accessed behaviors to execute (e.g., the time for a called procedure to execute and return).

More precisely, execution time is computed as follows:

$$b.exectime = b.ict + b.commtime \qquad (2)$$

$$b.commtime = \sum_{c_k \in b.outchannels} c_k.accfreq \times (c_k.ttime_{bus} + (c_k.dst).exectime)$$

$$c_k.ttime_{bus} = \lceil bus.time \times (c_k.bits \div bus.width) \rceil$$

In other words, a behavior b's execution time equals its *ict* on the current component (*b.ict*), plus its communication time (*b.commtime*). The communication time equals the transfer time over a channel for each accessed object ($c_k.ttime_{bus}$), plus the execution time of each accessed object ($(c_k.dst).exectime$), times the number of such accesses ($c_k.accfreq$). The transfer time over a channel is determined from the bus data transfer time (*bus.time*) and the width of that bus (*bus.width*). If the data bits exceeds the bus width, then multiple transfers are used (as computed by the division). The *bus_time* is usually less when the communication is within one component.

For example, the execution-time equation for *FuzzyMain* of Figure 3 would be:

$$\begin{aligned} FuzzyMain.et &= FuzzyMain.ict \\ &+ c1.accfreq * (c1.tt + in1val.et) \\ &+ c2.accfreq * (c2.tt + in2val.et) \\ &+ c3.accfreq * (c3.tt + EvalRule.et) \\ &+ c4.accfreq * (c4.tt + Convolve.et) \\ &+ c5.accfreq * (c5.tt + Centroid.et) \\ in1val.et &= in1val.ict + 0 \\ in2val.et &= in2val.ict + 0 \\ EvalRule.et &= EvalRule.ict \\ &+ c8.accfreq * (c8.tt + trunc.et) \\ &+ \ldots \end{aligned} \qquad (3)$$

Pre-estimation – A behavior's internal computation time can be computed during pre-estimation through profiling and scheduling [15]. Profiling determines the execution count of each basic block. A schedule for each basic block is then estimated for each possible processor component, using compilation for standard processors and synthesis for custom processors. The summation over all blocks of each block's execution count times steps yields the total steps for the behavior. Multiplying by the step time, i.e., the clock period, yields an *ict* value. Channel access frequencies are also determined through profiling. Bus times and widths are already associated with each bus.

Online estimation – Given a partition of every functional object to a component, the actual *ict*, bus widths, and bus times become known. Thus, a behavior's execution time equation can be evaluated. When a partitioning heuristic moves an object, the object's *ict* value will change, and bus times may also change since objects previously on the same component may now be on different components. We only need to change those values and re-evaluate the equation. In addition, any other equations that include the object's execution time must also be updated.

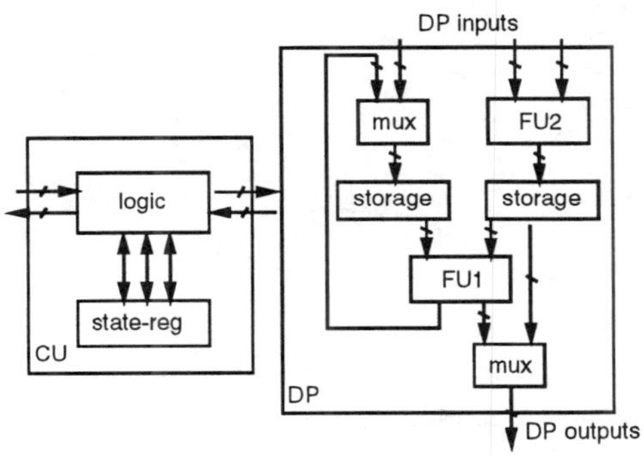

Fig. 8: CU/DP area model

		area factor	is a function of
CU/DP	CU	state_reg	# states
		+ logic	# states, # ctrl_lines, # states each ctrl_line is active
	+ DP	storage	# bits and # words of each storage
		+ func_units	# bits and type of each FU
		+ muxes	# sources of each storage or FU input, or DP output port
		+ wires	# DP connections, # DP components

Fig. 9: Equation and terms for computing CU/DP area

4.3.2 Hardware size

SpecSyn uses a hardware design model similar to those in [11, 16, 17], consisting of a control-unit/datapath (CU/DP) as shown in Figure 8. We present our estimation technique [14] briefly here. The CU/DP area can be computed as the sum of the following terms: *Functional-unit (FU) size*; *Storage-unit size* including registers, register files and memories; *Multiplexer size*; *State-register size*; *Control-logic size*; and *Wiring-size*. As shown in Figure 9, each term is a function of basic parameters, such as the number of possible states and control lines.

Pre-estimation – The parameters relevant to each functional object are computed (through approximate synthesis) and annotated to each object. Given an initial partition of functional objects among custom processors, we can obtain a rough design of each processor by intelligently combining its objects' parameter annotations. For example, we can determine the number of FU's by taking the union of the objects' FU's (since sequential behaviors can share FU's).

Online-estimation – When a partitioning heuristic removes an object from a processor, we update that processor's terms. Some terms can be updated simply by examining the object's annotations. For example, the number of possible processor states is reduced by the object's number, and the state register size recomputed using the log function. On the other hand, other terms require further examination. For example, an object might require a particular FU, but removing that object only removes that FU if no other object uses the FU; thus, we keep track of which objects use each FU; similar analysis can be performed for muxes.

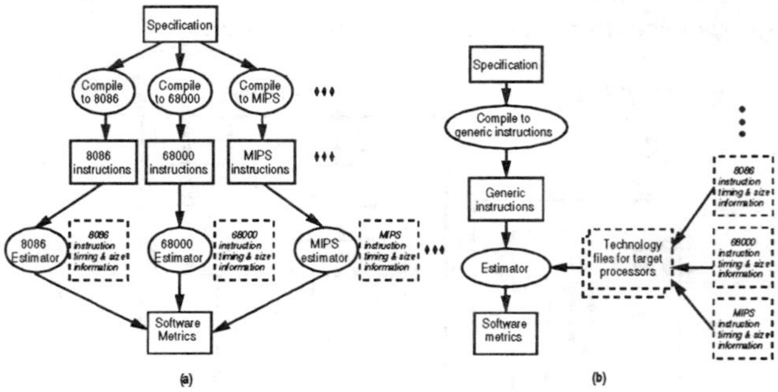

Fig. 10: Software size estimation: (a) processor-specific model, (b) generic model.

4.3.3 Software size

Software size refers to the number of bytes of program memory required when a behavior is compiled into the instruction set of a specific processor.

Pre-estimation – While compilation can be used to obtain program size for a behavior on a given processor, SpecSyn uses a generic processor model as shown in Figure 10(b). A functional object's size is first compiled into generic three-address instructions. Using available processor-specific technology files listing the number of bytes that each generic instruction would require in each processor, the estimator computes the software size. A target processor's technology file can be developed based on the size information of the processor's instruction set. Details on derivation of technology files for specific processors are given in [15]. Some experiments comparing the generic model with the processor-specific model yielded inaccuracy of roughly 7% [15].

Note that the same generic processor approach would be applied for software performance estimation. Specifically, the technology file of the target processor would include not only the bytes but also the number of steps for each generic instruction.

Online estimation – Online software size estimation consists simply of increasing or decreasing the processor size by the size of the added or removed functional object.

5 Refinement

Refinement is the generation of a new specification for each system component after exploration has yielded a suitable allocation and partition. The refined specification should be both readable and simulatable, enabling further verification and synthesis. Tasks include generations of interfaces [11], memories, and arbiters. The refined output can serve as input to simulation and cosimulation tools.

```
entity FuzzyControllerE is
  port ( in1, in2 : in integer;  out1:  out integer );
end;
...

    component ASIC1E is
      port ( in1, in2 : in integer; startEvalRule : in bit;
             doneEvalRule : out bit;  num_chan : int_chan;
             mr_chan, tmr_chan : addr_int_chan; ...);
    end;

    component Memory1E is
      port (mr_chan : addr_int_chan);
    end;

    component Memory2E is ...
    component Processor1E is ...

    < port maps > ...
...

entity ASIC1E is
  port ( in1, in2 : in integer; startEvalRule : in bit;
         doneEvalRule : out bit;  num_chan : int_chan;
         mr_chan, tmr_chan : addr_int_chan; ...);
end;
...

    process
      variable in1val, in2val : integer;
      function Min ...
      variable num : integer;
      ...
    begin
      wait until startEvalRule='1';
      num := ReadNum(num_chan);
      EvaluateRule(num);
      doneEvalRule='1';
      ....

procedure EvaluateRule(num : in integer) is
  variable trunc : integer;  -- truncated value
  variable mr_val1, mr_val2 : integer;
  variable tmr_val : integer;
begin
  if (num = 1) then
    mr_val1 := ReadMemory1(in1val + MR1OFFSET);
    mr_val2 := ReadMemory1(128 + in1val + MR1OFFSET);
    trunc := Min(mr1_val1, mr1_val2));
  elsif (num = 2) then
    mr_val1 := ReadMemory1(in1val + MR2OFFSET);
    mr_val2 := ReadMemory1(128 + in1val + MR2OFFSET);
    trunc := Min(mr_val1, mr_val2);
  end if;

  for i in 1 to 128 loop
    if (num = 1) then
      mr_val1 := ReadMemory1(256+i + MR1OFFSET);
      tmr_val := Min(trunc, mr_val1);
      WriteMemory2(i + TMR1OFFSET, tmr_val);
    elsif (num = 2) then
      mr_val1 := ReadMemory1(256+i + MR2OFFSET);
      tmr_val := Min(trunc, mr_val1);
      WriteMemory2(i+TMR2OFFSET, tmr_val);
    end if;
  end loop;
end;
```

Fig. 11: Refined fuzzy-logic controller VHDL description.

In Figure 11, we show part of a refined specification for the system design shown in Figure 6. The interface of the fuzzy controller remains unchanged. However, its contents now consist of more details than in the original specification of Figure 2, such as new subroutines that read data from *Memory1* and write data to *Memory2* using detailed communication protocols for memory accesses. Note the large amount of detail that must be added to the specification. Also, note that the designer has access to that detail, since a refined specification was generated, which can be viewed and edited. Due to space limitations, we refer the reader to [11, 18] for details on refinement.

6 Experiments

SpecSyn, under development since 1989, consists of over 150,000 lines of C code. Its main interface is a spreadsheet-like display showing each component and functional object along with annotations, constraints and metric values for each. Menu options permit designers to perform various design tasks, with results reflected in the displayed values, and violated constraints highlighted. SpecSyn has been released to several universities and companies.

We conducted experiments exploring design alternatives for several industrial examples. We present results for a fuzzy-logic controller example [19]. Four library components were available: a standard processor (Intel 8051) and three custom processors with 50k, 100k, and 150k gates, each with a cost. We automatically generated all possible allocations of these components below a certain cost. For each allocation, we partitioned using simulated annealing and a cost function that sought to meet size and pin constraints and minimize execution time.

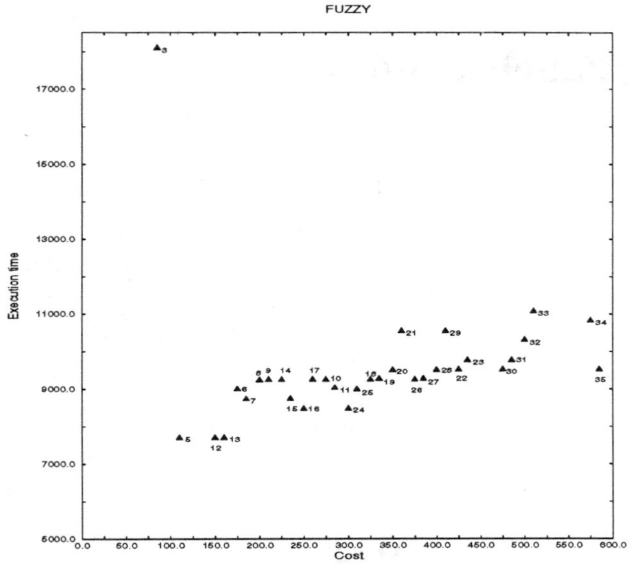

Fig. 12: Exploration for the fuzzy-logic controller

Figure 12 shows results for 35 allocations. Constraint-violating allocations are omitted. Allocation 3 was one 8051 standard and one 50k custom processor, resulting in a fuzzy-controller execution time of 18,115 microseconds. Allocation 5 was one 8051 and one 100k processor, yielding 7721 microseconds. Higher-cost allocations yielded no better execution time. SpecSyn thus aids the designer to understand the design space, enabling a focus on promising points. The above data was generated in 1 hour on a Sparc 2.

SpecSyn was used by an industry engineer to design the fuzzy-logic controller [19]. The partitioning results matched favorably with those obtained by another engineer who did a manual partition. The entire implementation was obtained in roughly 100 man-hours with the aid of SpecSyn and high-level synthesis, which is nearly a 10 times reduction from the 6-months required for the manual design.

SpecSyn was used in the design of a recent commercial product: a 1 ms scan, 10k step programmable controller [20]. Additional SpecSyn information can be found in [21].

7 Conclusions

We described the SpecSyn environment supporting the specify-explore-refine system-design paradigm. Our exploration approach uses pre-estimation and online-estimation to achieve fast and accurate estimates, supports various partitioning heuristics, and is intended to be continually extended, enabling a designer to quickly examine many alternative designs. This paradigm and tool may eventually result in a 100-hour design cycle, and our experiments demonstrate the feasibility of such a design-time reduction. Future work may involve estimation models for pipelining and caching, new transformations, support of fixed cores, and incorporation of post-synthesis metrics into design iterations.

References

[1] A. Balboni, W. Fornaciari, and D. Sciuto, "Partitioning and exploration strategies in the tosca co-design flow," in *Int. Workshop on Hardware-Software Co-Design*, pp. 62–69, 1996.

[2] R. Ernst, J. Henkel, and T. Benner, "Hardware-software cosynthesis for microcontrollers," in *IEEE Design & Test of Computers*, pp. 64–75, December 1994.

[3] J. Henkel and R. Ernst, "A hardware/software partitioner using a dynamically determined granularity," in *DAC*, 1997.

[4] R. Gupta and G. DeMicheli, "Hardware-software cosynthesis for digital systems," in *IEEE Design & Test of Computers*, pp. 29–41, October 1993.

[5] A. Kalavade and E. Lee, "A hardware/software codesign methodology for DSP applications," in *IEEE Design & Test of Computers*, 1993.

[6] J. Hou and W. Wolf, "Process partitioning for distributed systems," in *Int. Workshop on Hardware-Software Co-Design*, pp. 70–75, 1996.

[7] S.Prakash and A. Parker, "Synthesis of application-specific multiprocessor architectures," in *DAC*, pp. 8–13, 1991.

[8] W. Wolf, "Hardware-software co-design of embedded systems," *Proceedings of the IEEE*, vol. 82, no. 7, pp. 967–989, 1994.

[9] O. Bentz, J. Rabaey, and D. Lidsky, "A dynamic design estimation and exploration environment," in *DAC*, 1997.

[10] J. Rowson and A. Sangiovanni-Vincentelli, "Interface-based design," in *DAC*, 1997.

[11] D. Gajski, F. Vahid, S. Narayan, and J. Gong, *Specification and design of embedded systems*. New Jersey: Prentice Hall, 1994.

[12] D. Harel, "Statecharts: A visual formalism for complex systems," *Science of Computer Programming 8*, pp. 231–274, 1987.

[13] F. Vahid, "Procedure exlining: A transformation for improved system and behavioral synthesis," in *Int. Symposium on System Synthesis*, pp. 84–89, 1995.

[14] F. Vahid and D. Gajski, "Incremental hardware estimation during hardware/software functional partitioning," *IEEE Transactions on VLSI Systems*, vol. 3, no. 3, pp. 459–464, 1995.

[15] J. Gong, D. Gajski, and S. Narayan, "Software estimation using a generic processor model," in *Proceedings of the European Design and Test Conference (EDTC)*, pp. 498–502, 1995.

[16] M. McFarland and T. Kowalski, "Incorporating bottom-up design into hardware synthesis," *IEEE Transactions on CAD*, pp. 938–950, September 1990.

[17] E. Lagnese and D. Thomas, "Architectural partitioning for system level synthesis of integrated circuits," *IEEE Transactions on CAD*, vol. 10, pp. 847–860, July 1991.

[18] J. Gong, D. Gajski, and S. Bakshi, "Model refinement for hardware-software codesign," in *Proceedings of the European Design and Test Conference (EDTC)*, 1996.

[19] L. Ramachandran, D. Gajski, S. Narayan, F. Vahid, and P. Fung, "Towards achieving a 100-hour design cycle: A test case," in *EuroDAC*, pp. 144–149, 1994.

[20] L. Matsushita Electric Works, "Fp10sh programmable controller." Datasheet, 1996.

[21] D. Gajski, F. Vahid, S. Narayan, and J. Gong, "Specsyn: An environment supporting the specify-explore-refine paradigm for hardware/software system design," *IEEE Transactions on VLSI Systems*, vol. 6, no. 1, pp. 84–100, 1998.

Conference Author/Panelist Index

I

J

K

L

M

N

O

P

Q

R

S

T

U

V

W

X

Y

Z